P9-DBR-861

Basic Differentiation Rules

1. $\dfrac{d}{dx}[cu] = cu'$

2. $\dfrac{d}{dx}[u \pm v] = u' \pm v'$

3. $\dfrac{d}{dx}[uv] = uv' + vu'$

4. $\dfrac{d}{dx}\left[\dfrac{u}{v}\right] = \dfrac{vu' - uv'}{v^2}$

5. $\dfrac{d}{dx}[c] = 0$

6. $\dfrac{d}{dx}[u^n] = nu^{n-1}u'$

7. $\dfrac{d}{dx}[x] = 1$

8. $\dfrac{d}{dx}[|u|] = \dfrac{u}{|u|}(u'), \quad u \neq 0$

9. $\dfrac{d}{dx}[\ln u] = \dfrac{u'}{u}$

10. $\dfrac{d}{dx}[e^u] = e^u u'$

11. $\dfrac{d}{dx}[\log_a u] = \dfrac{u'}{(\ln a)u}$

12. $\dfrac{d}{dx}[a^u] = (\ln a)a^u u'$

13. $\dfrac{d}{dx}[\sin u] = (\cos u)u'$

14. $\dfrac{d}{dx}[\cos u] = -(\sin u)u'$

15. $\dfrac{d}{dx}[\tan u] = (\sec^2 u)u'$

16. $\dfrac{d}{dx}[\cot u] = -(\csc^2 u)u'$

17. $\dfrac{d}{dx}[\sec u] = (\sec u \tan u)u'$

18. $\dfrac{d}{dx}[\csc u] = -(\csc u \cot u)u'$

19. $\dfrac{d}{dx}[\arcsin u] = \dfrac{u'}{\sqrt{1 - u^2}}$

20. $\dfrac{d}{dx}[\arccos u] = \dfrac{-u'}{\sqrt{1 - u^2}}$

21. $\dfrac{d}{dx}[\arctan u] = \dfrac{u'}{1 + u^2}$

22. $\dfrac{d}{dx}[\text{arccot } u] = \dfrac{-u'}{1 + u^2}$

23. $\dfrac{d}{dx}[\text{arcsec } u] = \dfrac{u'}{|u|\sqrt{u^2 - 1}}$

24. $\dfrac{d}{dx}[\text{arccsc } u] = \dfrac{-u'}{|u|\sqrt{u^2 - 1}}$

25. $\dfrac{d}{dx}[\sinh u] = (\cosh u)u'$

26. $\dfrac{d}{dx}[\cosh u] = (\sinh u)u'$

27. $\dfrac{d}{dx}[\tanh u] = (\text{sech}^2 u)u'$

28. $\dfrac{d}{dx}[\coth u] = -(\text{csch}^2 u)u'$

29. $\dfrac{d}{dx}[\text{sech } u] = -(\text{sech } u \tanh u)u'$

30. $\dfrac{d}{dx}[\text{csch } u] = -(\text{csch } u \coth u)u'$

31. $\dfrac{d}{dx}[\sinh^{-1} u] = \dfrac{u'}{\sqrt{u^2 + 1}}$

32. $\dfrac{d}{dx}[\cosh^{-1} u] = \dfrac{u'}{\sqrt{u^2 - 1}}$

33. $\dfrac{d}{dx}[\tanh^{-1} u] = \dfrac{u'}{1 - u^2}$

34. $\dfrac{d}{dx}[\coth^{-1} u] = \dfrac{u'}{1 - u^2}$

35. $\dfrac{d}{dx}[\text{sech}^{-1} u] = \dfrac{-u'}{u\sqrt{1 - u^2}}$

36. $\dfrac{d}{dx}[\text{csch}^{-1} u] = \dfrac{-u'}{|u|\sqrt{1 + u^2}}$

Basic Integration Formulas

1. $\displaystyle\int kf(u)\, du = k\int f(u)\, du$

2. $\displaystyle\int [f(u) \pm g(u)]\, du = \int f(u)\, du \pm \int g(u)\, du$

3. $\displaystyle\int du = u + C$

4. $\displaystyle\int u^n\, du = \dfrac{u^{n+1}}{n+1} + C, \quad n \neq -1$

5. $\displaystyle\int \dfrac{du}{u} = \ln|u| + C$

6. $\displaystyle\int e^u\, du = e^u + C$

7. $\displaystyle\int a^u\, du = \left(\dfrac{1}{\ln a}\right)a^u + C$

8. $\displaystyle\int \sin u\, du = -\cos u + C$

9. $\displaystyle\int \cos u\, du = \sin u + C$

10. $\displaystyle\int \tan u\, du = -\ln|\cos u| + C$

11. $\displaystyle\int \cot u\, du = \ln|\sin u| + C$

12. $\displaystyle\int \sec u\, du = \ln|\sec u + \tan u| + C$

13. $\displaystyle\int \csc u\, du = -\ln|\csc u + \cot u| + C$

14. $\displaystyle\int \sec^2 u\, du = \tan u + C$

15. $\displaystyle\int \csc^2 u\, du = -\cot u + C$

16. $\displaystyle\int \sec u \tan u\, du = \sec u + C$

17. $\displaystyle\int \csc u \cot u\, du = -\csc u + C$

18. $\displaystyle\int \dfrac{du}{\sqrt{a^2 - u^2}} = \arcsin \dfrac{u}{a} + C$

19. $\displaystyle\int \dfrac{du}{a^2 + u^2} = \dfrac{1}{a}\arctan \dfrac{u}{a} + C$

20. $\displaystyle\int \dfrac{du}{u\sqrt{u^2 - a^2}} = \dfrac{1}{a}\text{arcsec} \dfrac{|u|}{a} + C$

TRIGONOMETRY

Definition of the Six Trigonometric Functions

Right triangle definitions, where $0 < \theta < \pi/2$.

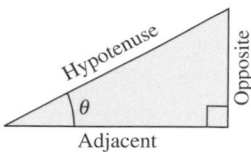

$$\sin \theta = \frac{\text{opp}}{\text{hyp}} \qquad \csc \theta = \frac{\text{hyp}}{\text{opp}}$$

$$\cos \theta = \frac{\text{adj}}{\text{hyp}} \qquad \sec \theta = \frac{\text{hyp}}{\text{adj}}$$

$$\tan \theta = \frac{\text{opp}}{\text{adj}} \qquad \cot \theta = \frac{\text{adj}}{\text{opp}}$$

Circular function definitions, where θ is any angle.

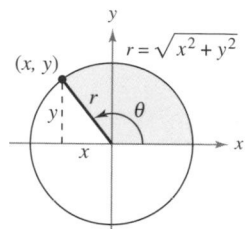

$$\sin \theta = \frac{y}{r} \qquad \csc \theta = \frac{r}{y}$$

$$\cos \theta = \frac{x}{r} \qquad \sec \theta = \frac{r}{x}$$

$$\tan \theta = \frac{y}{x} \qquad \cot \theta = \frac{x}{y}$$

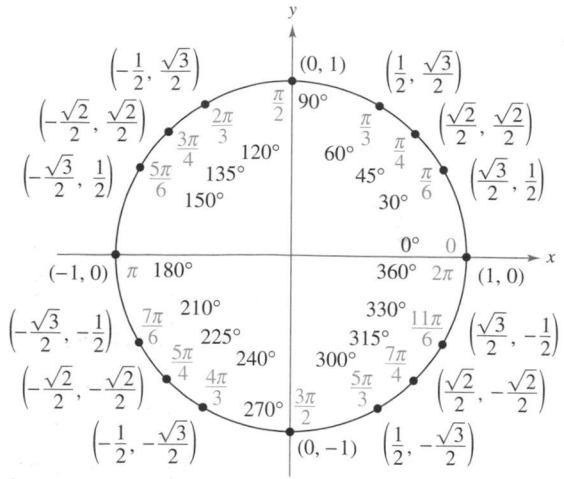

Reciprocal Identities

$$\sin x = \frac{1}{\csc x} \quad \sec x = \frac{1}{\cos x} \quad \tan x = \frac{1}{\cot x}$$

$$\csc x = \frac{1}{\sin x} \quad \cos x = \frac{1}{\sec x} \quad \cot x = \frac{1}{\tan x}$$

Quotient Identities

$$\tan x = \frac{\sin x}{\cos x} \quad \cot x = \frac{\cos x}{\sin x}$$

Pythagorean Identities

$$\sin^2 x + \cos^2 x = 1$$

$$1 + \tan^2 x = \sec^2 x \qquad 1 + \cot^2 x = \csc^2 x$$

Cofunction Identities

$$\sin\left(\frac{\pi}{2} - x\right) = \cos x \qquad \cos\left(\frac{\pi}{2} - x\right) = \sin x$$

$$\csc\left(\frac{\pi}{2} - x\right) = \sec x \qquad \tan\left(\frac{\pi}{2} - x\right) = \cot x$$

$$\sec\left(\frac{\pi}{2} - x\right) = \csc x \qquad \cot\left(\frac{\pi}{2} - x\right) = \tan x$$

Even/Odd Identities

$$\sin(-x) = -\sin x \qquad \cos(-x) = \cos x$$

$$\csc(-x) = -\csc x \qquad \tan(-x) = -\tan x$$

$$\sec(-x) = \sec x \qquad \cot(-x) = -\cot x$$

Sum and Difference Formulas

$$\sin(u \pm v) = \sin u \cos v \pm \cos u \sin v$$

$$\cos(u \pm v) = \cos u \cos v \mp \sin u \sin v$$

$$\tan(u \pm v) = \frac{\tan u \pm \tan v}{1 \mp \tan u \tan v}$$

Double-Angle Formulas

$$\sin 2u = 2 \sin u \cos u$$

$$\cos 2u = \cos^2 u - \sin^2 u = 2 \cos^2 u - 1 = 1 - 2 \sin^2 u$$

$$\tan 2u = \frac{2 \tan u}{1 - \tan^2 u}$$

Power-Reducing Formulas

$$\sin^2 u = \frac{1 - \cos 2u}{2}$$

$$\cos^2 u = \frac{1 + \cos 2u}{2}$$

$$\tan^2 u = \frac{1 - \cos 2u}{1 + \cos 2u}$$

Sum-to-Product Formulas

$$\sin u + \sin v = 2 \sin\left(\frac{u + v}{2}\right) \cos\left(\frac{u - v}{2}\right)$$

$$\sin u - \sin v = 2 \cos\left(\frac{u + v}{2}\right) \sin\left(\frac{u - v}{2}\right)$$

$$\cos u + \cos v = 2 \cos\left(\frac{u + v}{2}\right) \cos\left(\frac{u - v}{2}\right)$$

$$\cos u - \cos v = -2 \sin\left(\frac{u + v}{2}\right) \sin\left(\frac{u - v}{2}\right)$$

Product-to-Sum Formulas

$$\sin u \sin v = \frac{1}{2}[\cos(u - v) - \cos(u + v)]$$

$$\cos u \cos v = \frac{1}{2}[\cos(u - v) + \cos(u + v)]$$

$$\sin u \cos v = \frac{1}{2}[\sin(u + v) + \sin(u - v)]$$

$$\cos u \sin v = \frac{1}{2}[\sin(u + v) - \sin(u - v)]$$

CALCULUS
EARLY TRANSCENDENTAL FUNCTIONS
with **CalcChat**® and **CalcView**®

7e

Ron Larson

The Pennsylvania State University
The Behrend College

Bruce Edwards

University of Florida

CENGAGE

Australia • Brazil • Mexico • Singapore • United Kingdom • United States

Calculus: Early Transcendental Functions, **Seventh Edition**
Ron Larson, Bruce Edwards

Product Director: Mark Santee

Product Manager: Gary Whalen

Content Developer: Powell Vacha

Product Assistant: Abby DeVeuve

Marketing Manager: Ryan Ahern

Content Project Manager: Jennifer Risden

Production Service: Larson Texts, Inc.

Photo Researcher: Lumina Datamatics

Text Researcher: Lumina Datamatics

Illustrator: Larson Texts, Inc.

Text Designer: Larson Texts, Inc.

Compositor: Larson Texts, Inc.

Cover Designer: Larson Texts, Inc.

Cover Photograph: Caryn B. Davis | carynbdavis.com

Cover Background Image: iStockphoto.com/briddy_

Umbilic Torus by Helaman Ferguson, donated to Stony Brook University

The cover image is the Umbilic Torus statue created in 2012 by the famed sculptor and mathematician Dr. Helaman Ferguson. This statue weighs 10 tons and has a height of 24 feet. It is located at Stony Brook University in Stony Brook, New York.

For product information and technology assistance, contact us at
Cengage Customer & Sales Support, 1-800-354-9706.

For permission to use material from this text or product, submit all requests online at **www.cengage.com/permissions.**
Further permissions questions can be emailed to
permissionrequest@cengage.com.

Library of Congress Control Number: 2017951789

Student Edition:
ISBN: 978-1-337-55251-6

Loose-leaf Edition:
ISBN: 978-1-337-55303-2

Cengage
20 Channel Center Street
Boston, MA 02210
USA

Cengage is a leading provider of customized learning solutions with employees residing in nearly 40 different countries and sales in more than 125 countries around the world. Find your local representative at **www.cengage.com**.

Cengage products are represented in Canada by Nelson Education, Ltd.

To learn more about Cengage platforms and services, visit **www.cengage.com**.
To register or access your online learning solution or purchase materials for your course, visit **www.cengagebrain.com**.

QR Code is a registered trademark of Denso Wave Incorporated

Printed in the United States of America
Print Number: 03   Print Year: 2019

Contents

*Available at the text-specific website *www.cengagebrain.com*

Preface

Welcome to *Calculus: Early Transcendental Functions*, Seventh Edition. We are excited to offer you a new edition with even more resources that will help you understand and master calculus. This textbook includes features and resources that continue to make *Calculus: Early Transcendental Functions*, Seventh Edition, a valuable learning tool for students and a trustworthy teaching tool for instructors.

Calculus: Early Transcendental Functions, Seventh Edition, provides the clear instruction, precise mathematics, and thorough coverage that you expect for your course. Additionally, this new edition provides you with **free** access to three companion websites:

- **CalcView.com**—video solutions to selected exercises
- **CalcChat.com**—worked-out solutions to odd-numbered exercises and access to online tutors
- **LarsonCalculus.com**—companion website with resources to supplement your learning

These websites will help enhance and reinforce your understanding of the material presented in this text and prepare you for future mathematics courses. CalcView® and CalcChat® are also available as free mobile apps.

Features

NEW CalcView®

The website *CalcView.com* contains video solutions of selected exercises. Watch instructors progress step-by-step through solutions, providing guidance to help you solve the exercises. The CalcView mobile app is available for free at the Apple® App Store® or Google Play™ store. The app features an embedded QR Code® reader that can be used to scan the on-page codes and go directly to the videos. You can also access the videos at CalcView.com.

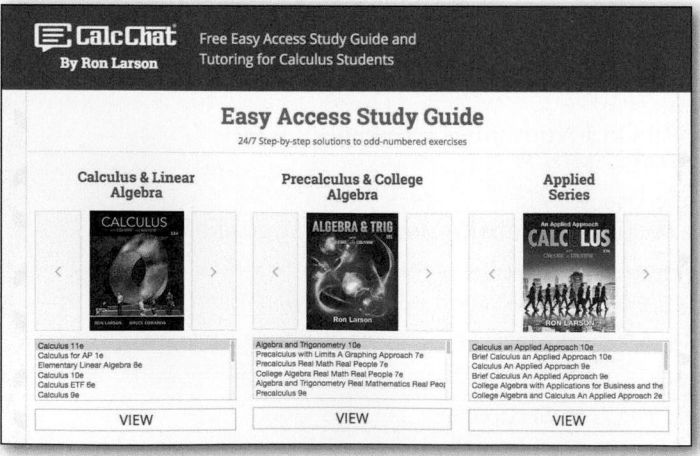

UPDATED CalcChat®

In each exercise set, be sure to notice the reference to *CalcChat.com*. This website provides free step-by-step solutions to all odd-numbered exercises in many of our textbooks. Additionally, you can chat with a tutor, at no charge, during the hours posted at the site. For over 15 years, millions of students have visited this site for help. The CalcChat mobile app is also available as a free download at the Apple® App Store® or Google Play™ store and features an embedded QR Code® reader.

REVISED LarsonCalculus.com

All companion website features have been updated based on this revision. Watch videos explaining concepts or proofs from the book, explore examples, view three-dimensional graphs, download articles from math journals, and much more.

NEW Conceptual Exercises

The *Concept Check* exercises and *Exploring Concepts* exercises appear in each section. These exercises will help you develop a deeper and clearer knowledge of calculus. Work through these exercises to build and strengthen your understanding of the calculus concepts and to prepare you for the rest of the section exercises.

REVISED Exercise Sets

The exercise sets have been carefully and extensively examined to ensure they are rigorous and relevant and to include topics our users have suggested. The exercises are organized and titled so you can better see the connections between examples and exercises. Multi-step, real-life exercises reinforce problem-solving skills and mastery of concepts by giving you the opportunity to apply the concepts in real-life situations.

REVISED Section Projects

Projects appear in selected sections and encourage you to explore applications related to the topics you are studying. We have added new projects, revised others, and kept some of our favorites. All of these projects provide an interesting and engaging way for you and other students to work and investigate ideas collaboratively.

Table of Contents Changes

Based on market research and feedback from users, we have made several changes to the table of contents.

- We added a review of trigonometric functions (Section 1.4) to Chapter 1.

- To cut back on the length of the text, we moved previous Section 1.4 *Fitting Models to Data* (now Appendix G in the Seventh Edition) to the text-specific website at *CengageBrain.com*.

- To provide more flexibility to the order of coverage of calculus topics, Section 4.5 *Limits at Infinity* was revised so that it can be covered after Section 2.5 *Infinite Limits*. As a result of this revision, some exercises moved from Section 4.5 to Section 4.6 *A Summary of Curve Sketching*.

- We moved Section 5.6 *Numerical Integration* to Section 8.6.

- We moved Section 8.7 *Indeterminate Forms and L'Hôpital's Rule* to Section 5.6.

Chapter Opener

Each Chapter Opener highlights real-life applications used in the examples and exercises.

Section Objectives

A bulleted list of learning objectives provides you with the opportunity to preview what will be presented in the upcoming section.

Theorems

Theorems provide the conceptual framework for calculus. Theorems are clearly stated and separated from the rest of the text by boxes for quick visual reference. Key proofs often follow the theorem and can be found at *LarsonCalculus.com.*

Definitions

As with theorems, definitions are clearly stated using precise, formal wording and are separated from the text by boxes for quick visual reference.

Explorations

Explorations provide unique challenges to study concepts that have not yet been formally covered in the text. They allow you to learn by discovery and introduce topics related to ones presently being studied. Exploring topics in this way encourages you to think outside the box.

Remarks

These hints and tips reinforce or expand upon concepts, help you learn how to study mathematics, caution you about common errors, address special cases, or show alternative or additional steps to a solution of an example.

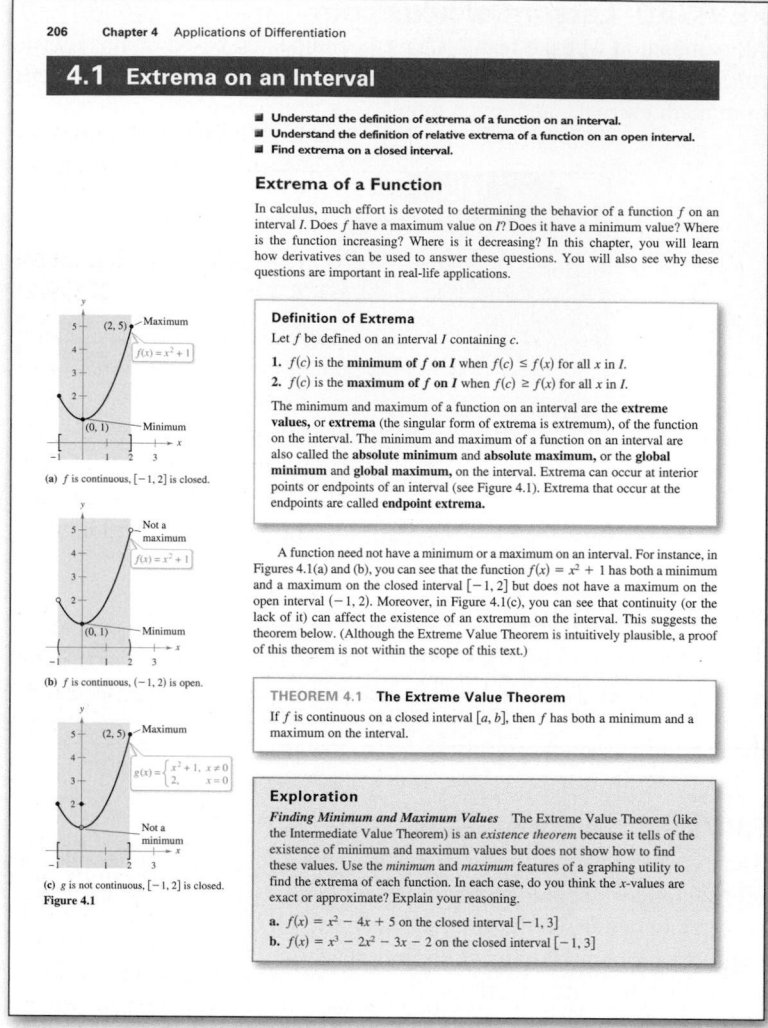

How Do You See It? Exercise

The How Do You See It? exercise in each section presents a problem that you will solve by visual inspection using the concepts learned in the lesson. This exercise is excellent for classroom discussion or test preparation.

Applications

Carefully chosen applied exercises and examples are included throughout to address the question, "When will I use this?" These applications are pulled from diverse sources, such as current events, world data, industry trends, and more, and relate to a wide range of interests. Understanding where calculus is (or can be) used promotes fuller understanding of the material.

Historical Notes and Biographies

Historical Notes provide you with background information on the foundations of calculus. The Biographies introduce you to the people who created and contributed to calculus.

Technology

Throughout the book, technology boxes show you how to use technology to solve problems and explore concepts of calculus. These tips also point out some pitfalls of using technology.

Putnam Exam Challenges

Putnam Exam questions appear in selected sections. These actual Putnam Exam questions will challenge you and push the limits of your understanding of calculus.

Student Solutions Manual for Calculus of a Single Variable: Early Transcendental Functions
ISBN-13: 978-1-337-55256-1

Student Solutions Manual for Multivariable Calculus
ISBN-13: 978-1-337-27539-2

Need a leg up on your homework or help to prepare for an exam? The *Student Solutions Manuals* contain worked-out solutions for all odd-numbered exercises in *Calculus of a Single Variable: Early Transcendental Functions* (Chapters 1–10 of *Calculus: Early Transcendental Functions*, Seventh Edition) and *Multivariable Calculus* (Chapters 11–16 of *Calculus*, Eleventh Edition and *Calculus: Early Transcendental Functions*, Seventh Edition). These manuals are great resources to help you understand how to solve those tough problems.

CengageBrain.com
To access additional course materials, please visit *www.cengagebrain.com*. At the *CengageBrain.com* home page, search for the ISBN of your title (from the back cover of your book) using the search box at the top of the page. This will take you to the product page where these resources can be found.

WEBASSIGN From Cengage **www.webassign.com**

Prepare for class with confidence using WebAssign from Cengage *Calculus: Early Transcendental Functions*, Seventh Edition. This online learning platform fuels practice, so you truly absorb what you learn—and are better prepared come test time. Videos and tutorials walk you through concepts and deliver instant feedback and grading, so you always know where you stand in class. Focus your study time and get extra practice where you need it most. Study smarter with WebAssign!

Ask your instructor today how you can get access to WebAssign, or learn about self-study options at *www.webassign.com*.

Complete Solutions Manual for Calculus of a Single Variable: Early Transcendental Functions, Vol. 1
ISBN-13: 978-1-337-55264-6

Complete Solutions Manual for Calculus of a Single Variable: Early Transcendental Functions, Vol. 2
ISBN-13: 978-1-337-55265-3

Complete Solutions Manual for Multivariable Calculus
ISBN-13: 978-1-337-27542-2

The *Complete Solutions Manuals* contain worked-out solutions to all exercises in the text. They are posted on the instructor companion website.

Instructor's Resource Guide **(on instructor companion site)**
This robust manual contains an abundance of instructor resources keyed to the textbook at the section and chapter level, including section objectives, teaching tips, and chapter projects.

Cengage Learning Testing Powered by Cognero
CLT is a flexible online system that allows you to author, edit, and manage test bank content; create multiple test versions in an instant; and deliver tests from your LMS, your classroom, or wherever you want. This is available online via *www.cengage.com/login.*

Instructor Companion Site
Everything you need for your course in one place! This collection of book-specific lecture and class tools is available online via *www.cengage.com/login.* Access and download PowerPoint® presentations, images, instructor's manual, and more.

Test Bank **(on instructor companion site)**
The Test Bank contains text-specific multiple-choice and free-response test forms.

⁂ WEBASSIGN From Cengage   **www.webassign.com**

WebAssign from Cengage *Calculus: Early Transcendental Functions*, Seventh Edition, is a fully customizable online solution for STEM disciplines that empowers you to help your students learn, not just do homework. Insightful tools save you time and highlight exactly where your students are struggling. Decide when and what type of help students can access while working on assignments—and incentivize independent work so help features are not abused. Meanwhile, your students get an engaging experience, instant feedback, and better outcomes. A total win-win!

To try a sample assignment, learn about LMS integration, or connect with our digital course support, visit *www.webassign.com/cengage.*

Acknowledgments

We would like to thank the many people who have helped us at various stages of *Calculus: Early Transcendental Functions*, Seventh Edition, over the years. Their encouragement, criticisms, and suggestions have been invaluable.

Reviewers of the Seventh Edition

Roberto Cabezas, *Miami Dade College–Kendall;* Justina Castellanos, *Miami Dade College–Kendall;* Christine Cole, *Moorpark College;* Scott Demsky, *Broward College–Central*

Reviewers of Previous Editions

Stan Adamski, *Owens Community College;* Alexander Arhangelskii, *Ohio University;* Seth G. Armstrong, *Southern Utah University;* Jim Ball, *Indiana State University;* Denis Bell, *University of Northern Florida;* Marcelle Bessman, *Jacksonville University;* Abraham Biggs, *Broward Community College;* Jesse Blosser, *Eastern Mennonite School;* Linda A. Bolte, *Eastern Washington University;* James Braselton, *Georgia Southern University;* Harvey Braverman, *Middlesex County College;* Mark Brittenham, *University of Nebraska;* Tim Chappell, *Penn Valley Community College;* Mingxiang Chen, *North Carolina A&T State University;* Oiyin Pauline Chow, *Harrisburg Area Community College;* Julie M. Clark, *Hollins University;* P.S. Crooke, *Vanderbilt University;* Jim Dotzler, *Nassau Community College;* Murray Eisenberg, *University of Massachusetts at Amherst;* Donna Flint, *South Dakota State University;* Michael Frantz, *University of La Verne;* Sudhir Goel, *Valdosta State University;* Arek Goetz, *San Francisco State University;* Donna J. Gorton, *Butler County Community College;* John Gosselin, *University of Georgia;* Shahryar Heydari, *Piedmont College;* Guy Hogan, *Norfolk State University;* Marcia Kleinz, *Atlantic Cape Community College;* Ashok Kumar, *Valdosta State University;* Kevin J. Leith, *Albuquerque Community College;* Maxine Lifshitz, *Friends Academy;* Douglas B. Meade, *University of South Carolina;* Bill Meisel, *Florida State College at Jacksonville;* Teri Murphy, *University of Oklahoma;* Darren Narayan, *Rochester Institute of Technology;* Susan A. Natale, *The Ursuline School, NY;* Martha Nega, *Georgia Perimeter College;* Terence H. Perciante, *Wheaton College;* James Pommersheim, *Reed College;* Laura Ritter, *Southern Polytechnic State University;* Leland E. Rogers, *Pepperdine University;* Paul Seeburger, *Monroe Community College;* Edith A. Silver, *Mercer County Community College;* Howard Speier, *Chandler-Gilbert Community College;* Desmond Stephens, *Florida A&M University;* Jianzhong Su, *University of Texas at Arlington;* Patrick Ward, *Illinois Central College;* Chia-Lin Wu, *Richard Stockton College of New Jersey;* Diane M. Zych, *Erie Community College*

Many thanks to Robert Hostetler, The Behrend College, The Pennsylvania State University, and David Heyd, The Behrend College, The Pennsylvania State University, for their significant contributions to previous editions of this text.

We would also like to thank the staff at Larson Texts, Inc., who assisted in preparing the manuscript, rendering the art package, typesetting, and proofreading the pages and supplements.

On a personal level, we are grateful to our wives, Deanna Gilbert Larson and Consuelo Edwards, for their love, patience, and support. Also, a special note of thanks goes out to R. Scott O'Neil.

If you have suggestions for improving this text, please feel free to write to us. Over the years we have received many useful comments from both instructors and students, and we value these very much.

Ron Larson

Bruce Edwards

1 Preparation for Calculus

Automobile Aerodynamics *(Exercise 95, p. 30)*

Ferris Wheel
(Exercise 74, p. 40)

Conveyor Design *(Exercise 26, p. 16)*

Cell Phone Subscribers
(Exercise 68, p. 9)

Modeling Carbon Dioxide Concentration *(Example 6, p. 7)*

1.1 Graphs and Models

- **Sketch the graph of an equation.**
- **Find the intercepts of a graph.**
- **Test a graph for symmetry with respect to an axis and the origin.**
- **Find the points of intersection of two graphs.**
- **Interpret mathematical models for real-life data.**

RENÉ DESCARTES (1596–1650)

Descartes made many contributions to philosophy, science, and mathematics. The idea of representing points in the plane by pairs of real numbers and representing curves in the plane by equations was described by Descartes in his book *La Géométrie*, published in 1637.

See LarsonCalculus.com to read more of this biography.

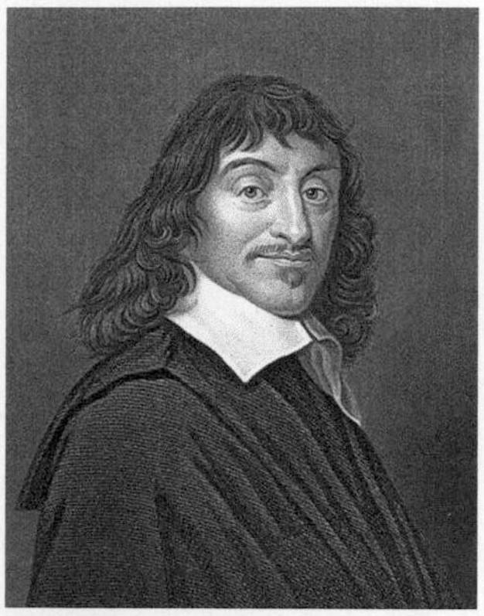

The Graph of an Equation

In 1637, the French mathematician René Descartes revolutionized the study of mathematics by combining its two major fields—algebra and geometry. With Descartes's coordinate plane, geometric concepts could be formulated analytically and algebraic concepts could be viewed graphically. The power of this approach was such that within a century of its introduction, much of calculus had been developed.

The same approach can be followed in your study of calculus. That is, by viewing calculus from multiple perspectives—*graphically*, *analytically*, and *numerically*—you will increase your understanding of core concepts.

Consider the equation $3x + y = 7$. The point $(2, 1)$ is a **solution point** of the equation because the equation is satisfied (is true) when 2 is substituted for x and 1 is substituted for y. This equation has many other solutions, such as $(1, 4)$ and $(0, 7)$. To find other solutions systematically, solve the original equation for y.

$$y = 7 - 3x \qquad \text{Analytic approach}$$

Then construct a **table of values** by substituting several values of x.

x	0	1	2	3	4
y	7	4	1	-2	-5

Numerical approach

From the table, you can see that $(0, 7)$, $(1, 4)$, $(2, 1)$, $(3, -2)$, and $(4, -5)$ are solutions of the original equation $3x + y = 7$. Like many equations, this equation has an infinite number of solutions. The set of all solution points is the **graph** of the equation, as shown in Figure 1.1. Note that the sketch shown in Figure 1.1 is referred to as the graph of $3x + y = 7$, even though it really represents only a *portion* of the graph. The entire graph would extend beyond the page.

In this course, you will study many sketching techniques. The simplest is point plotting—that is, you plot points until the basic shape of the graph seems apparent.

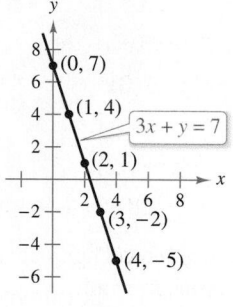

Graphical approach: $3x + y = 7$
Figure 1.1

EXAMPLE 1 **Sketching a Graph by Point Plotting**

To sketch the graph of $y = x^2 - 2$, first construct a table of values. Next, plot the points shown in the table. Then connect the points with a smooth curve, as shown in Figure 1.2. This graph is a **parabola.** It is one of the conics you will study in Chapter 10.

x	-2	-1	0	1	2	3
y	2	-1	-2	-1	2	7

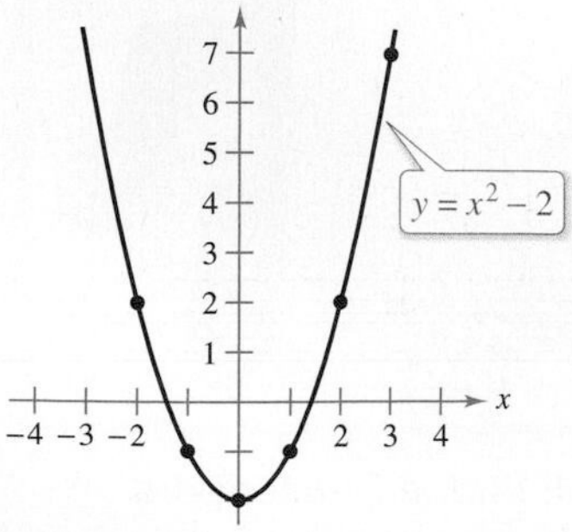

The parabola $y = x^2 - 2$
Figure 1.2

One disadvantage of point plotting is that to get a good idea about the shape of a graph, you may need to plot many points. With only a few points, you could badly misrepresent the graph. For instance, to sketch the graph of

$$y = \frac{1}{30}x(39 - 10x^2 + x^4)$$

you plot five points:

$$(-3, -3), \quad (-1, -1), \quad (0, 0), \quad (1, 1), \quad \text{and} \quad (3, 3)$$

as shown in Figure 1.3(a). From these five points, you might conclude that the graph is a line. This, however, is not correct. By plotting several more points, you can see that the graph is more complicated, as shown in Figure 1.3(b).

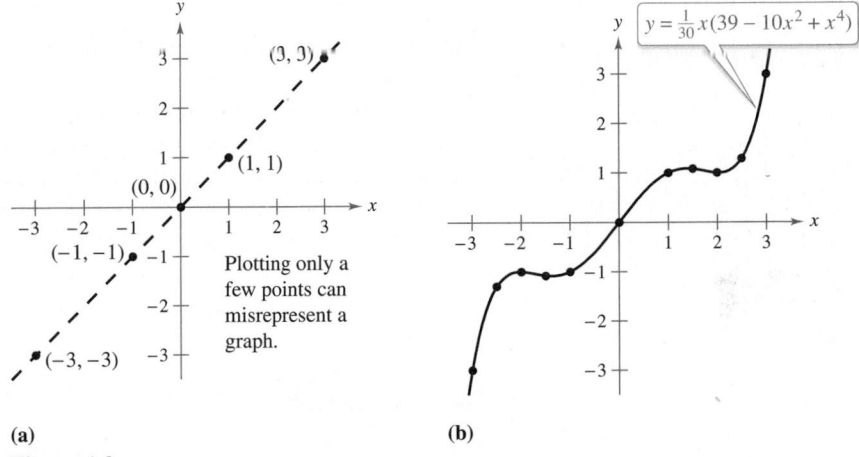

(a)

(b)

Figure 1.3

▷ **TECHNOLOGY**   Graphing an equation has been made easier by technology. Even with technology, however, it is possible to misrepresent a graph badly. For instance, each of the graphing utility* screens in Figure 1.4 shows a portion of the graph of

$$y = x^3 - x^2 - 25.$$

From the screen on the left, you might assume that the graph is a line. From the screen on the right, however, you can see that the graph is not a line. So, whether you are sketching a graph by hand or using a graphing utility, you must realize that different "viewing windows" can produce very different views of a graph. In choosing a viewing window, your goal is to show a view of the graph that fits well in the context of the problem.

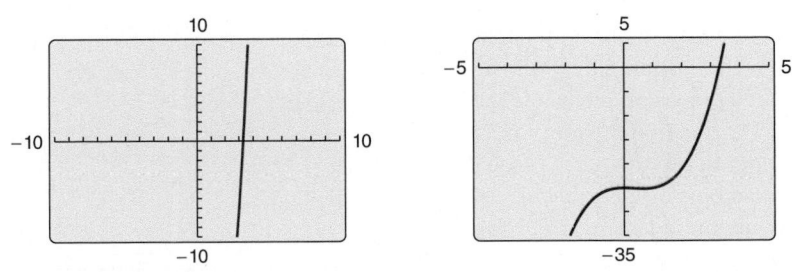

Graphing utility screens of $y = x^3 - x^2 - 25$

Figure 1.4

---

*In this text, the term *graphing utility* means either a graphing calculator, such as the *TI-Nspire*, or computer graphing software, such as *Maple* or *Mathematica*.

Exploration

Comparing Graphical and Analytic Approaches

Use a graphing utility to graph each equation. In each case, find a viewing window that shows the important characteristics of the graph.

a. $y = x^3 - 3x^2 + 2x + 5$

b. $y = x^3 - 3x^2 + 2x + 25$

c. $y = -x^3 - 3x^2 + 20x + 5$

d. $y = 3x^3 - 40x^2 + 50x - 45$

e. $y = -(x + 12)^3$

f. $y = (x - 2)(x - 4)(x - 6)$

A purely graphical approach to this problem would involve a simple "guess, check, and revise" strategy. What types of things do you think an analytic approach might involve? For instance, does the graph have symmetry? Does the graph have turns? If so, where are they? As you proceed through Chapters 2, 3, and 4 of this text, you will study many new analytic tools that will help you analyze graphs of equations such as these.

Intercepts of a Graph

REMARK Some texts denote the *x*-intercept as the *x*-coordinate of the point $(a, 0)$ rather than the point itself. Unless it is necessary to make a distinction, when the term *intercept* is used in this text, it will mean either the point or the coordinate.

Two types of solution points that are especially useful in graphing an equation are those having zero as their *x*- or *y*-coordinate. Such points are called **intercepts** because they are the points at which the graph intersects the *x*- or *y*-axis. The point $(a, 0)$ is an **x-intercept** of the graph of an equation when it is a solution point of the equation. To find the *x*-intercepts of a graph, let *y* be zero and solve the equation for *x*. The point $(0, b)$ is a **y-intercept** of the graph of an equation when it is a solution point of the equation. To find the *y*-intercepts of a graph, let *x* be zero and solve the equation for *y*.

It is possible for a graph to have no intercepts, or it might have several. For instance, consider the four graphs shown in Figure 1.5.

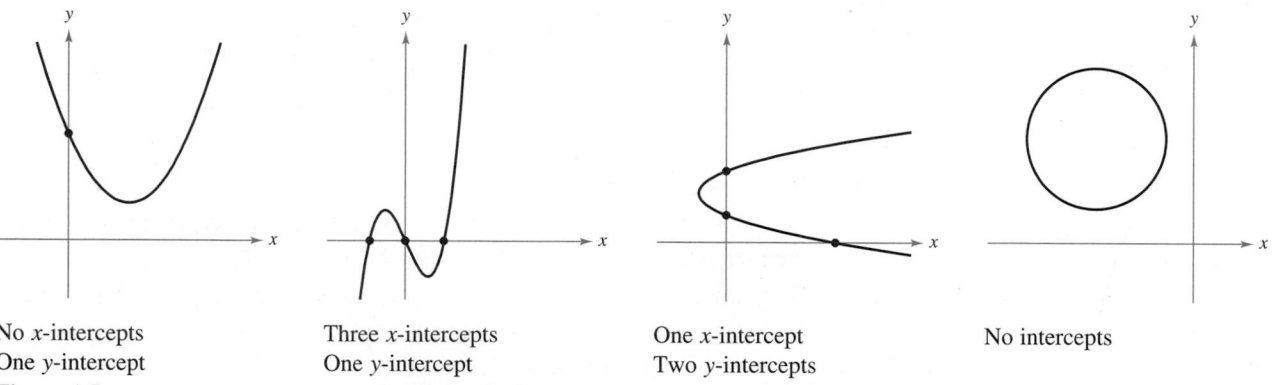

No *x*-intercepts
One *y*-intercept
Figure 1.5

Three *x*-intercepts
One *y*-intercept

One *x*-intercept
Two *y*-intercepts

No intercepts

EXAMPLE 2 Finding *x*- and *y*-Intercepts

Find the *x*- and *y*-intercepts of the graph of $y = x^3 - 4x$.

Solution To find the *x*-intercepts, let *y* be zero and solve for *x*.

$$x^3 - 4x = 0 \qquad \text{Let } y \text{ be zero.}$$
$$x(x - 2)(x + 2) = 0 \qquad \text{Factor.}$$
$$x = 0, 2, \text{ or } -2 \qquad \text{Solve for } x.$$

Because this equation has three solutions, you can conclude that the graph has three *x*-intercepts:

$$(0, 0), \quad (2, 0), \quad \text{and} \quad (-2, 0). \qquad \text{*x*-intercepts}$$

▷ **TECHNOLOGY** Example 2 uses an analytic approach to finding intercepts. When an analytic approach is not possible, you can use a graphical approach by finding the points at which the graph intersects the axes. Use the *trace* feature of a graphing utility to approximate the intercepts of the graph of the equation in Example 2. Note that your utility may have a built-in program that can find the *x*-intercepts of a graph. (Your utility may call this the *root* or *zero* feature.) If so, use the program to find the *x*-intercepts of the graph of the equation in Example 2.

To find the *y*-intercepts, let *x* be zero. Doing this produces $y = 0$. So, the *y*-intercept is

$$(0, 0). \qquad \text{*y*-intercept}$$

(See Figure 1.6.)

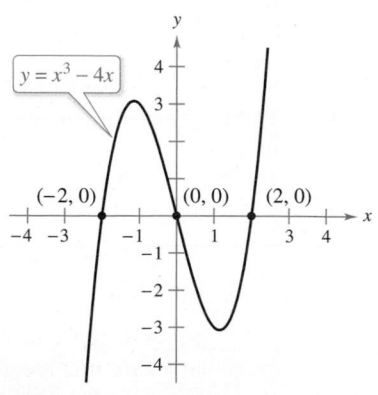

Intercepts of a graph
Figure 1.6

Symmetry of a Graph

Knowing the symmetry of a graph before attempting to sketch it is useful because you need only half as many points to sketch the graph. The three types of symmetry listed below can be used to help sketch the graphs of equations (see Figure 1.7).

1. A graph is **symmetric with respect to the *y*-axis** if, whenever (x, y) is a point on the graph, then $(-x, y)$ is also a point on the graph. This means that the portion of the graph to the left of the *y*-axis is a mirror image of the portion to the right of the *y*-axis.

2. A graph is **symmetric with respect to the *x*-axis** if, whenever (x, y) is a point on the graph, then $(x, -y)$ is also a point on the graph. This means that the portion of the graph below the *x*-axis is a mirror image of the portion above the *x*-axis.

3. A graph is **symmetric with respect to the origin** if, whenever (x, y) is a point on the graph, then $(-x, -y)$ is also a point on the graph. This means that the graph is unchanged by a rotation of $180°$ about the origin.

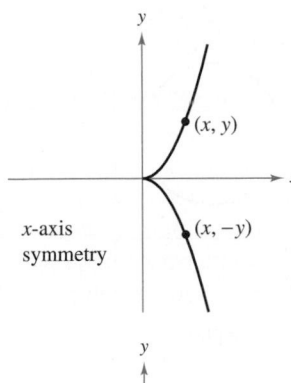

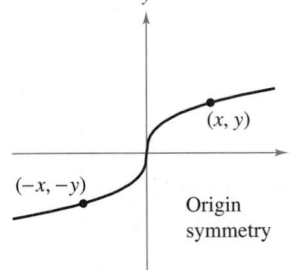

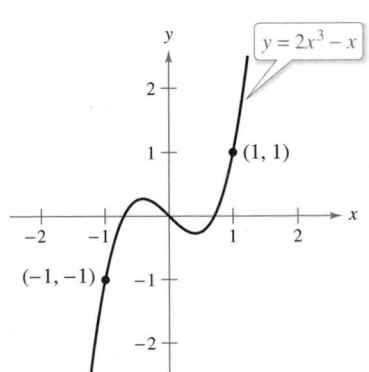

Figure 1.7

Tests for Symmetry

1. The graph of an equation in *x* and *y* is symmetric with respect to the *y*-axis when replacing *x* by $-x$ yields an equivalent equation.

2. The graph of an equation in *x* and *y* is symmetric with respect to the *x*-axis when replacing *y* by $-y$ yields an equivalent equation.

3. The graph of an equation in *x* and *y* is symmetric with respect to the origin when replacing *x* by $-x$ and *y* by $-y$ yields an equivalent equation.

The graph of a polynomial has symmetry with respect to the *y*-axis when each term has an even exponent (or is a constant). For instance, the graph of

$$y = 2x^4 - x^2 + 2$$

has symmetry with respect to the *y*-axis. Similarly, the graph of a polynomial has symmetry with respect to the origin when each term has an odd exponent, as illustrated in Example 3.

EXAMPLE 3   **Testing for Symmetry**

Test the graph of $y = 2x^3 - x$ for symmetry with respect to (a) the *y*-axis and (b) the origin.

Solution

a. $y = 2x^3 - x$ Write original equation.

$\quad y = 2(-x)^3 - (-x)$ Replace *x* by $-x$.

$\quad y = -2x^3 + x$ Simplify. The result is *not* an equivalent equation.

Because replacing *x* by $-x$ does *not* yield an equivalent equation, you can conclude that the graph of $y = 2x^3 - x$ is *not* symmetric with respect to the *y*-axis.

b. $\quad y = 2x^3 - x$ Write original equation.

$\quad -y = 2(-x)^3 - (-x)$ Replace *x* by $-x$ and *y* by $-y$.

$\quad -y = -2x^3 + x$ Simplify.

$\quad\quad y = 2x^3 - x$ Equivalent equation

Because replacing *x* by $-x$ and *y* by $-y$ yields an equivalent equation, you can conclude that the graph of $y = 2x^3 - x$ is symmetric with respect to the origin, as shown in Figure 1.8.

Origin symmetry
Figure 1.8

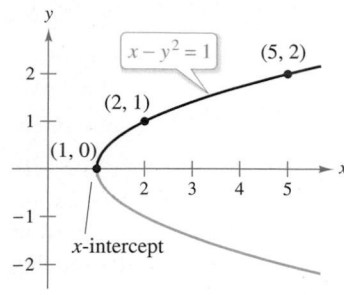

Figure 1.9

EXAMPLE 4 **Using Intercepts and Symmetry to Sketch a Graph**

$\cdots\cdots\triangleright$ *See LarsonCalculus.com for an interactive version of this type of example.*

Sketch the graph of $x - y^2 = 1$.

Solution The graph is symmetric with respect to the x-axis because replacing y by $-y$ yields an equivalent equation.

$$x - y^2 = 1 \qquad \text{Write original equation.}$$
$$x - (-y)^2 = 1 \qquad \text{Replace } y \text{ by } -y.$$
$$x - y^2 = 1 \qquad \text{Equivalent equation}$$

This means that the portion of the graph below the x-axis is a mirror image of the portion above the x-axis. To sketch the graph, first plot the x-intercept and the points above the x-axis. Then reflect in the x-axis to obtain the entire graph, as shown in Figure 1.9. ∎

$\triangleright$ **TECHNOLOGY** Graphing utilities are designed so that they most easily graph equations in which y is a function of x (see Section 1.3 for a definition of *function*). To graph other types of equations, you need to split the graph into two or more parts *or* you need to use a different graphing mode. For instance, to graph the equation in Example 4, you can split it into two parts.

$$y_1 = \sqrt{x - 1} \qquad \text{Top portion of graph}$$
$$y_2 = -\sqrt{x - 1} \qquad \text{Bottom portion of graph}$$

Points of Intersection

A **point of intersection** of the graphs of two equations is a point that satisfies both equations. You can find the point(s) of intersection of two graphs by solving their equations simultaneously.

EXAMPLE 5 **Finding Points of Intersection**

Find all points of intersection of the graphs of

$$x^2 - y = 3 \quad \text{and} \quad x - y = 1.$$

Solution Begin by sketching the graphs of both equations in the *same* rectangular coordinate system, as shown in Figure 1.10. From the figure, it appears that the graphs have two points of intersection. You can find these two points as follows.

$$y = x^2 - 3 \qquad \text{Solve first equation for } y.$$
$$y = x - 1 \qquad \text{Solve second equation for } y.$$
$$x^2 - 3 = x - 1 \qquad \text{Equate } y\text{-values.}$$
$$x^2 - x - 2 = 0 \qquad \text{Write in general form.}$$
$$(x - 2)(x + 1) = 0 \qquad \text{Factor.}$$
$$x = 2 \text{ or } -1 \qquad \text{Solve for } x.$$

The corresponding values of y are obtained by substituting $x = 2$ and $x = -1$ into either of the original equations. Doing this produces two points of intersection:

$$(2, 1) \quad \text{and} \quad (-1, -2). \qquad \text{Points of intersection}$$
∎

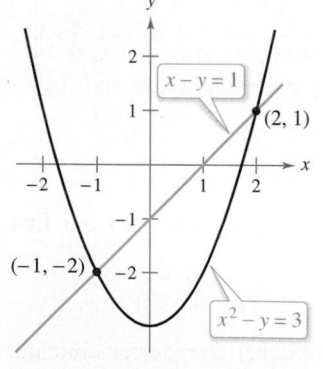

Two points of intersection
Figure 1.10

You can check the points of intersection in Example 5 by substituting into *both* of the original equations or by using the *intersect* feature of a graphing utility.

Mathematical Models

Real-life applications of mathematics often use equations as **mathematical models.** In developing a mathematical model to represent actual data, you should strive for two (often conflicting) goals—accuracy and simplicity. That is, you want the model to be simple enough to be workable, yet accurate enough to produce meaningful results. Appendix G explores these goals more completely.

EXAMPLE 6   Comparing Two Mathematical Models

The Mauna Loa Observatory in Hawaii records the carbon dioxide concentration y (in parts per million) in Earth's atmosphere. The January readings for various years are shown in Figure 1.11. In the July 1990 issue of *Scientific American*, these data were used to predict the carbon dioxide level in Earth's atmosphere in the year 2035, using the quadratic model

$$y = 0.018t^2 + 0.70t + 316.2 \qquad \text{Quadratic model for 1960–1990 data}$$

where $t = 0$ represents 1960, as shown in Figure 1.11(a). The data shown in Figure 1.11(b) represent the years 1980 through 2014 and can be modeled by

$$y = 0.014t^2 + 0.66t + 320.3 \qquad \text{Quadratic model for 1980–2014 data}$$

where $t = 0$ represents 1960. What was the prediction given in the *Scientific American* article in 1990? Given the second model for 1980 through 2014, does this prediction for the year 2035 seem accurate?

The Mauna Loa Observatory in Hawaii has been measuring the increasing concentration of carbon dioxide in Earth's atmosphere since 1958.

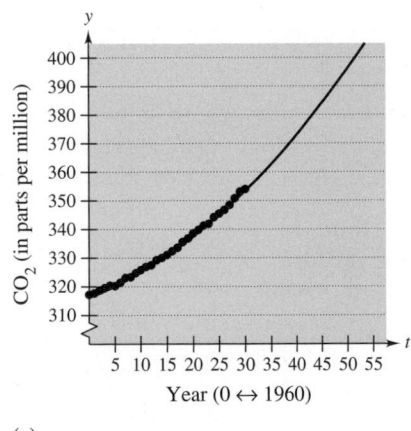

(a)

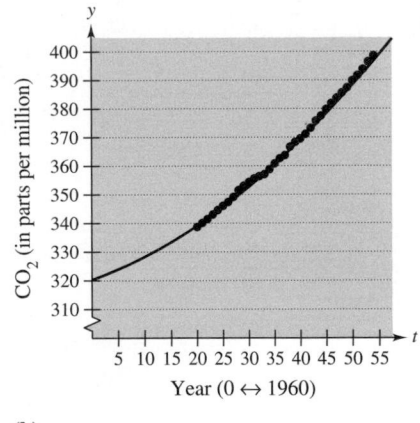

(b)

Figure 1.11

Solution   To answer the first question, substitute $t = 75$ (for 2035) into the first model.

$$y = 0.018(75)^2 + 0.70(75) + 316.2 = 469.95 \qquad \text{Model for 1960–1990 data}$$

So, the prediction in the *Scientific American* article was that the carbon dioxide concentration in Earth's atmosphere would reach about 470 parts per million in the year 2035. Using the model for the 1980–2014 data, the prediction for the year 2035 is

$$y = 0.014(75)^2 + 0.66(75) + 320.3 = 448.55. \qquad \text{Model for 1980–2014 data}$$

So, based on the model for 1980–2014, it appears that the 1990 prediction was too high. ∎

The models in Example 6 were developed using a procedure called *least squares regression* (see Section 13.9). The older model has a correlation of $r^2 \approx 0.997$, and for the newer model it is $r^2 \approx 0.999$. The closer r^2 is to 1, the "better" the model.

1.1 Exercises

See **CalcChat.com** for tutorial help and worked-out solutions to odd-numbered exercises.

CONCEPT CHECK

1. **Finding Intercepts** Describe how to find the x- and y-intercepts of the graph of an equation.

2. **Verifying Points of Intersection** How can you check that an ordered pair is a point of intersection of two graphs?

Matching In Exercises 3–6, match the equation with its graph. [The graphs are labeled (a), (b), (c), and (d).]

(a)
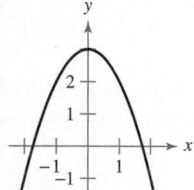

(b)
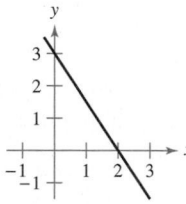

(c)
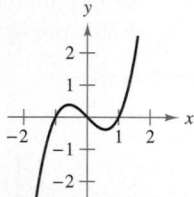

(d)
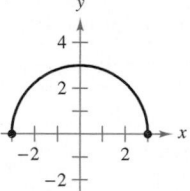

3. $y = -\frac{3}{2}x + 3$

4. $y = \sqrt{9 - x^2}$

5. $y = 3 - x^2$

6. $y = x^3 - x$

 **Sketching a Graph by Point Plotting** In Exercises 7–16, sketch the graph of the equation by point plotting.

7. $y = \frac{1}{2}x + 2$

8. $y = 5 - 2x$

9. $y = 4 - x^2$

10. $y = (x - 3)^2$

11. $y = |x + 1|$

12. $y = |x| - 1$

13. $y = \sqrt{x} - 6$

14. $y = \sqrt{x + 2}$

15. $y = \frac{3}{x}$

16. $y = \frac{1}{x + 2}$

Approximating Solution Points Using Technology In Exercises 17 and 18, use a graphing utility to graph the equation. Move the cursor along the curve to approximate the unknown coordinate of each solution point accurate to two decimal places.

17. $y = \sqrt{5 - x}$

   (a) $(2, y)$

   (b) $(x, 3)$

18. $y = x^5 - 5x$

   (a) $(-0.5, y)$

   (b) $(x, -4)$

 **Finding Intercepts** In Exercises 19–28, find any intercepts.

19. $y = 2x - 5$

20. $y = 4x^2 + 3$

21. $y = x^2 + x - 2$

22. $y^2 = x^3 - 4x$

23. $y = x\sqrt{16 - x^2}$

24. $y = (x - 1)\sqrt{x^2 + 1}$

25. $y = \frac{2 - \sqrt{x}}{5x + 1}$

26. $y = \frac{x^2 + 3x}{(3x + 1)^2}$

27. $x^2y - x^2 + 4y = 0$

28. $y = 2x - \sqrt{x^2 + 1}$

 **Testing for Symmetry** In Exercises 29–40, test for symmetry with respect to each axis and to the origin.

29. $y = x^2 - 6$

30. $y = 9x - x^2$

31. $y^2 = x^3 - 8x$

32. $y = x^3 + x$

33. $xy = 4$

34. $xy^2 = -10$

35. $y = 4 - \sqrt{x + 3}$

36. $xy - \sqrt{4 - x^2} = 0$

37. $y = \frac{x}{x^2 + 1}$

38. $y = \frac{x^5}{4 - x^2}$

39. $y = |x^3 + x|$

40. $|y| - x = 3$

 **Using Intercepts and Symmetry to Sketch a Graph** In Exercises 41–56, find any intercepts and test for symmetry. Then sketch the graph of the equation.

41. $y = 2 - 3x$

42. $y = \frac{2}{3}x + 1$

43. $y = 9 - x^2$

44. $y = 2x^2 + x$

45. $y = x^3 + 2$

46. $y = x^3 - 4x$

47. $y = x\sqrt{x + 5}$

48. $y = \sqrt{25 - x^2}$

49. $x = y^3$

50. $x = y^4 - 16$

51. $y = \frac{8}{x}$

52. $y = \frac{10}{x^2 + 1}$

53. $y = 6 - |x|$

54. $y = |6 - x|$

55. $3y^2 - x = 9$

56. $x^2 + 4y^2 = 4$

 **Finding Points of Intersection** In Exercises 57–62, find the points of intersection of the graphs of the equations.

57. $x + y = 8$
    $4x - y = 7$

58. $3x - 2y = -4$
    $4x + 2y = -10$

59. $x^2 + y = 15$
    $-3x + y = 11$

60. $x = 3 - y^2$
    $y = x - 1$

The symbol  and a red exercise number indicate that a video solution can be seen at *CalcView.com*.

The symbol ⨍⨏ indicates an exercise in which you are instructed to use graphing technology or a symbolic computer algebra system. The solutions of other exercises may also be facilitated by the use of appropriate technology.

61. $x^2 + y^2 = 5$

$x - y = 1$

62. $x^2 + y^2 = 16$

$x + 2y = 4$

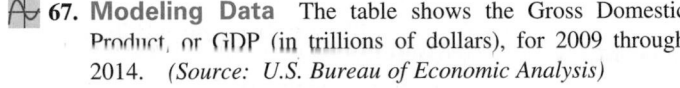 **Finding Points of Intersection Using Technology** In Exercises 63–66, use a graphing utility to find the points of intersection of the graphs of the equations. Check your results analytically.

63. $y = x^3 - 2x^2 + x - 1$

$y = -x^2 + 3x - 1$

64. $y = x^4 - 2x^2 + 1$

$y = 1 - x^2$

65. $y = \sqrt{x + 6}$

$y = \sqrt{-x^2 - 4x}$

66. $y = -|2x - 3| + 6$

$y = 6 - x$

67. Modeling Data The table shows the Gross Domestic Product, or GDP (in trillions of dollars), for 2009 through 2014. *(Source: U.S. Bureau of Economic Analysis)*

Year	2009	2010	2011	2012	2013	2014
GDP	14.4	15.0	15.5	16.2	16.7	17.3

(a) Use the regression capabilities of a graphing utility to find a mathematical model of the form $y = at + b$ for the data. In the model, y represents the GDP (in trillions of dollars) and t represents the year, with $t = 9$ corresponding to 2009.

(b) Use a graphing utility to plot the data and graph the model. Compare the data with the model.

(c) Use the model to predict the GDP in the year 2024.

68. Modeling Data

The table shows the numbers of cell phone subscribers (in millions) in the United States for selected years. *(Source: CTIA-The Wireless Association)*

Year	2000	2002	2004	2006
Number	109	141	182	233

Year	2008	2010	2012	2014
Number	270	296	326	355

(a) Use the regression capabilities of a graphing utility to find a mathematical model of the form $y = at^2 + bt + c$ for the data. In the model, y represents the number of subscribers (in millions) and t represents the year, with $t = 0$ corresponding to 2000.

(b) Use a graphing utility to plot the data and graph the model. Compare the data with the model.

(c) Use the model to predict the number of cell phone subscribers in the United States in the year 2024.

69. Break-Even Point Find the sales necessary to break even $(R = C)$ when the cost C of producing x units is $C = 2.04x + 5600$ and the revenue R from selling x units is $R = 3.29x$.

70. Using Solution Points For what values of k does the graph of $y^2 = 4kx$ pass through the point?

(a) $(1, 1)$ (b) $(2, 4)$

(c) $(0, 0)$ (d) $(3, 3)$

EXPLORING CONCEPTS

71. Using Intercepts Write an equation whose graph has intercepts at $x = -\frac{3}{2}$, $x = 4$, and $x = \frac{5}{2}$. (There is more than one correct answer.)

72. Symmetry A graph is symmetric with respect to the x-axis and to the y-axis. Is the graph also symmetric with respect to the origin? Explain.

73. Symmetry A graph is symmetric with respect to one axis and to the origin. Is the graph also symmetric with respect to the other axis? Explain.

74. HOW DO YOU SEE IT? Use the graphs of the two equations to answer the questions below.

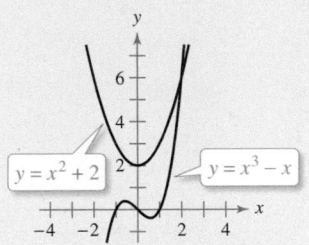

(a) What are the intercepts for each equation?

(b) Determine the symmetry for each equation.

(c) Determine the point of intersection of the two equations.

True or False? In Exercises 75–78, determine whether the statement is true or false. If it is false, explain why or give an example that shows it is false.

75. If $(-4, -5)$ is a point on a graph that is symmetric with respect to the x-axis, then $(4, -5)$ is also a point on the graph.

76. If $(-4, -5)$ is a point on a graph that is symmetric with respect to the y-axis, then $(4, -5)$ is also a point on the graph.

77. If $b^2 - 4ac > 0$ and $a \neq 0$, then the graph of

$$y = ax^2 + bx + c$$

has two x-intercepts.

78. If $b^2 - 4ac = 0$ and $a \neq 0$, then the graph of

$$y = ax^2 + bx + c$$

has only one x-intercept.

1.2 Linear Models and Rates of Change

■ Find the slope of a line passing through two points.
■ Write the equation of a line with a given point and slope.
■ Interpret slope as a ratio or as a rate in a real-life application.
■ Sketch the graph of a linear equation in slope-intercept form.
■ Write equations of lines that are parallel or perpendicular to a given line.

The Slope of a Line

The **slope** of a nonvertical line is a measure of the number of units the line rises (or falls) vertically for each unit of horizontal change from left to right. Consider the two points (x_1, y_1) and (x_2, y_2) on the line in Figure 1.12. As you move from left to right along this line, a vertical change of

$$\Delta y = y_2 - y_1 \qquad \text{Change in } y$$

units corresponds to a horizontal change of

$$\Delta x = x_2 - x_1 \qquad \text{Change in } x$$

units. (The symbol Δ is the uppercase Greek letter delta, and the symbols Δy and Δx are read "delta y" and "delta x.")

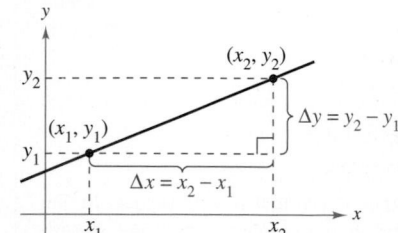

$\Delta y = y_2 - y_1 = $ change in y
$\Delta x = x_2 - x_1 = $ change in x
Figure 1.12

Definition of the Slope of a Line

The **slope** m of the nonvertical line passing through (x_1, y_1) and (x_2, y_2) is

$$m = \frac{\Delta y}{\Delta x} = \frac{y_2 - y_1}{x_2 - x_1}, \quad x_1 \neq x_2.$$

Slope is not defined for vertical lines.

When using the formula for slope, note that

$$\frac{y_2 - y_1}{x_2 - x_1} = \frac{-(y_1 - y_2)}{-(x_1 - x_2)} = \frac{y_1 - y_2}{x_1 - x_2}.$$

So, it does not matter in which order you subtract *as long as* you are consistent and both "subtracted coordinates" come from the same point.

Figure 1.13 shows four lines: one has a positive slope, one has a slope of zero, one has a negative slope, and one has an "undefined" slope. In general, the greater the absolute value of the slope of a line, the steeper the line. For instance, in Figure 1.13, the line with a slope of -5 is steeper than the line with a slope of $\frac{1}{5}$.

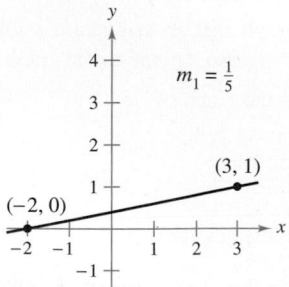

If m is positive, then the line rises from left to right.
Figure 1.13

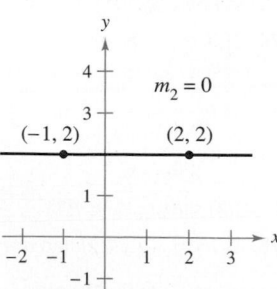

If m is zero, then the line is horizontal.

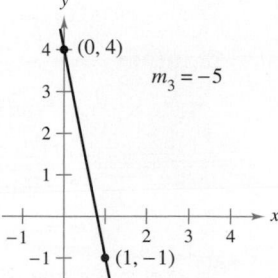

If m is negative, then the line falls from left to right.

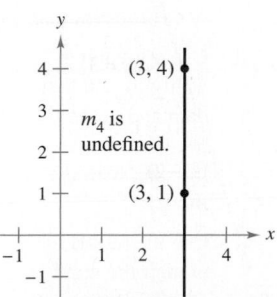

If m is undefined, then the line is vertical.

Equations of Lines

Any two points on a nonvertical line can be used to calculate its slope. This can be verified from the similar triangles shown in Figure 1.14. (Recall that the ratios of corresponding sides of similar triangles are equal.)

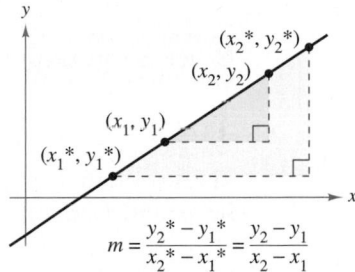

$$m = \frac{y_2{}^* - y_1{}^*}{x_2{}^* - x_1{}^*} = \frac{y_2 - y_1}{x_2 - x_1}$$

Any two points on a nonvertical line can be used to determine its slope.

Figure 1.14

If (x_1, y_1) is a point on a nonvertical line that has a slope of m and (x, y) is *any other* point on the line, then

$$\frac{y - y_1}{x - x_1} = m.$$

This equation in the variables x and y can be rewritten in the form

$$y - y_1 = m(x - x_1)$$

which is the **point-slope form** of the equation of a line.

Point-Slope Form of the Equation of a Line

The **point-slope form** of the equation of the line that passes through the point (x_1, y_1) and has a slope of m is

$$y - y_1 = m(x - x_1).$$

REMARK   Remember that only nonvertical lines have a slope. Consequently, vertical lines cannot be written in point-slope form. For instance, the equation of the vertical line passing through the point $(1, -2)$ is $x = 1$.

EXAMPLE 1   **Finding an Equation of a Line**

Find an equation of the line that has a slope of 3 and passes through the point $(1, -2)$. Then sketch the line.

Solution

$$
\begin{aligned}
y - y_1 &= m(x - x_1) && \text{Point-slope form} \\
y - (-2) &= 3(x - 1) && \text{Substitute } -2 \text{ for } y_1, 1 \text{ for } x_1, \text{ and } 3 \text{ for } m. \\
y + 2 &= 3x - 3 && \text{Simplify.} \\
y &= 3x - 5 && \text{Solve for } y.
\end{aligned}
$$

To sketch the line, first plot the point $(1, -2)$. Then, because the slope is $m = 3$, you can locate a second point on the line by moving one unit to the right and three units upward, as shown in Figure 1.15.

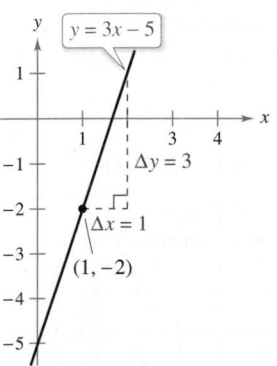

The line with a slope of 3 passing through the point $(1, -2)$

Figure 1.15

Ratios and Rates of Change

The slope of a line can be interpreted as either a *ratio* or a *rate*. If the x- and y-axes have the same unit of measure, then the slope has no units and is a **ratio.** If the x- and y-axes have different units of measure, then the slope is a rate or **rate of change.** In your study of calculus, you will encounter applications involving both interpretations of slope.

EXAMPLE 2   Using Slope as a Ratio

The maximum recommended slope of a wheelchair ramp is $\frac{1}{12}$. A business installs a wheelchair ramp that rises to a height of 22 inches over a length of 24 feet, as shown in Figure 1.16. Is the ramp steeper than recommended?   *(Source: ADA Standards for Accessible Design)*

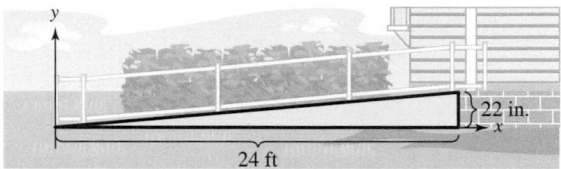

Figure 1.16

Solution   The length of the ramp is 24 feet or $12(24) = 288$ inches. The slope of the ramp is the ratio of its height (the rise) to its length (the run).

$$\text{Slope of ramp} = \frac{\text{rise}}{\text{run}}$$
$$= \frac{22 \text{ in.}}{288 \text{ in.}}$$
$$\approx 0.076$$

Because the slope of the ramp is less than $\frac{1}{12} \approx 0.083$, the ramp is not steeper than recommended. Note that the slope is a ratio and has no units.

EXAMPLE 3   Using Slope as a Rate of Change

The population of Oregon was about 3,831,000 in 2010 and about 3,970,000 in 2014. Find the average rate of change of the population over this four-year period. What will the population of Oregon be in 2024?   *(Source: U.S. Census Bureau)*

Solution   Over this four-year period, the average rate of change of the population of Oregon was

$$\text{Rate of change} = \frac{\text{change in population}}{\text{change in years}}$$
$$= \frac{3,970,000 - 3,831,000}{2014 - 2010}$$
$$= 34,750 \text{ people per year.}$$

Assuming that Oregon's population continues to increase at this same rate for the next 10 years, it will have a 2024 population of about 4,318,000. (See Figure 1.17.)   ◾

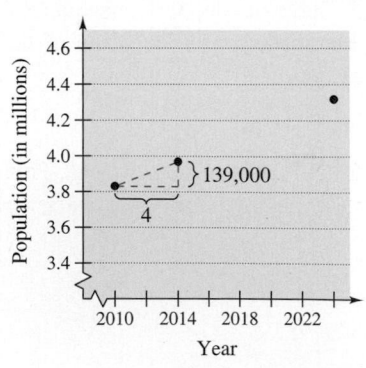

Population of Oregon
Figure 1.17

The rate of change found in Example 3 is an **average rate of change.** An average rate of change is always calculated over an interval. In this case, the interval is [2010, 2014]. In Chapter 3, you will study another type of rate of change called an *instantaneous rate of change.*

Graphing Linear Models

Many problems in coordinate geometry can be classified into two basic categories.

1. Given a graph (or parts of it), find its equation.
2. Given an equation, sketch its graph.

For lines, problems in the first category can be solved by using the point-slope form. The point-slope form, however, is not especially useful for solving problems in the second category. The form that is better suited to sketching the graph of a line is the **slope-intercept** form of the equation of a line.

The Slope-Intercept Form of the Equation of a Line

The graph of the linear equation

$$y = mx + b \qquad \text{Slope-intercept form}$$

is a line whose slope is m and whose y-intercept is $(0, b)$.

EXAMPLE 4   **Sketching Lines in the Plane**

Sketch the graph of each equation.

a. $y = 2x + 1$

b. $y = 2$

c. $3y + x - 6 = 0$

Solution

a. Because $b = 1$, the y-intercept is $(0, 1)$. Because the slope is $m = 2$, you know that the line rises two units for each unit it moves to the right, as shown in Figure 1.18(a).

b. By writing the equation $y = 2$ in slope-intercept form

$$y = (0)x + 2$$

you can see that the slope is $m = 0$ and the y-intercept is $(0, 2)$. Because the slope is zero, you know that the line is horizontal, as shown in Figure 1.18(b).

c. Begin by writing the equation in slope-intercept form.

$$3y + x - 6 = 0 \qquad \text{Write original equation.}$$
$$3y = -x + 6 \qquad \text{Isolate } y\text{-term on the left.}$$
$$y = -\tfrac{1}{3}x + 2 \qquad \text{Slope-intercept form}$$

In this form, you can see that the y-intercept is $(0, 2)$ and the slope is $m = -\tfrac{1}{3}$. This means that the line falls one unit for every three units it moves to the right, as shown in Figure 1.18(c).

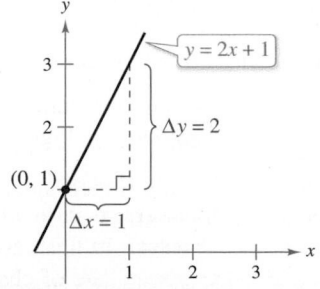

(a) $m = 2$; line rises

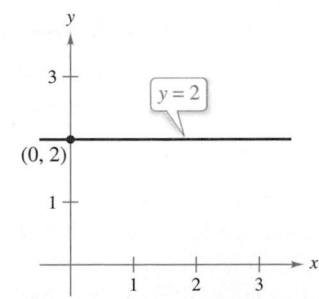

(b) $m = 0$; line is horizontal

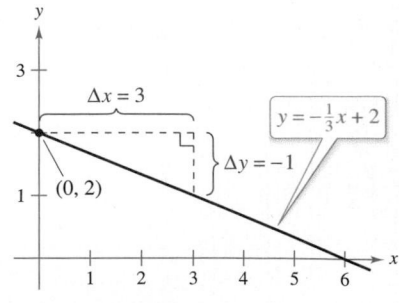

(c) $m = -\tfrac{1}{3}$; line falls

Figure 1.18

Because the slope of a vertical line is not defined, its equation cannot be written in slope-intercept form. However, the equation of any line can be written in the **general form**

$$Ax + By + C = 0$$ General form of the equation of a line

where A and B are not *both* zero. For instance, the vertical line

$$x = a$$ Vertical line

can be represented by the general form

$$x - a = 0.$$ General form

SUMMARY OF EQUATIONS OF LINES

1. General form: $Ax + By + C = 0$
2. Vertical line: $x = a$
3. Horizontal line: $y = b$
4. Slope-intercept form: $y = mx + b$
5. Point-slope form: $y - y_1 = m(x - x_1)$

Parallel and Perpendicular Lines

The slope of a line is a convenient tool for determining whether two lines are parallel or perpendicular, as shown in Figure 1.19. Specifically, nonvertical lines with the same slope are parallel, and nonvertical lines whose slopes are negative reciprocals are perpendicular.

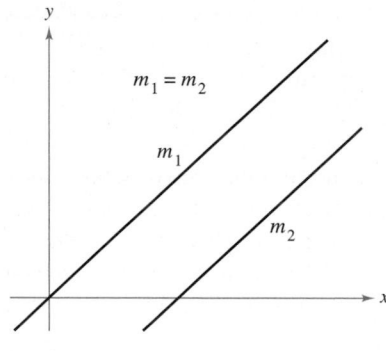

Parallel lines
Figure 1.19

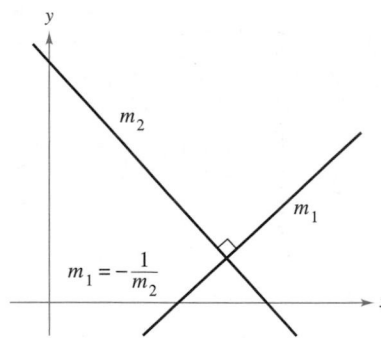

Perpendicular lines

> **·· REMARK** In mathematics, the phrase "if and only if" is a way of stating two implications in one statement. For instance, the first statement at the right could be rewritten as the following two implications.
>
> **a.** If two distinct nonvertical lines are parallel, then their slopes are equal.
>
> **b.** If two distinct nonvertical lines have equal slopes, then they are parallel.

Parallel and Perpendicular Lines

1. Two distinct nonvertical lines are **parallel** if and only if their slopes are equal—that is, if and only if

    $$m_1 = m_2.$$ Parallel ⟺ Slopes are equal.

2. Two nonvertical lines are **perpendicular** if and only if their slopes are negative reciprocals of each other—that is, if and only if

    $$m_1 = -\frac{1}{m_2}.$$ Perpendicular ⟺ Slopes are negative reciprocals.

EXAMPLE 5   Finding Parallel and Perpendicular Lines

⋮ ⋯ ▷ *See LarsonCalculus.com for an interactive version of this type of example.*

Find the general forms of the equations of the lines that pass through the point $(2, -1)$ and are (a) parallel to and (b) perpendicular to the line $2x - 3y = 5$.

Solution   Begin by writing the linear equation $2x - 3y = 5$ in slope-intercept form.

$$2x - 3y = 5 \qquad \text{Write original equation.}$$
$$y = \tfrac{2}{3}x - \tfrac{5}{3} \qquad \text{Slope-intercept form}$$

So, the given line has a slope of $m = \tfrac{2}{3}$. (See Figure 1.20.)

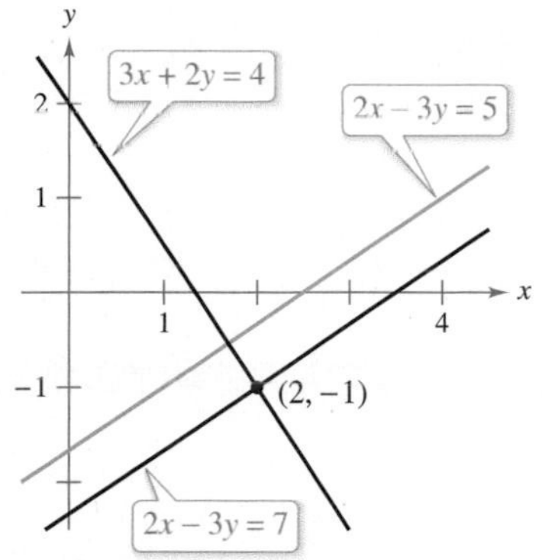

a. The line through $(2, -1)$ that is parallel to the given line also has a slope of $\tfrac{2}{3}$.

$$y - y_1 = m(x - x_1) \qquad \text{Point-slope form}$$
$$y - (-1) = \tfrac{2}{3}(x - 2) \qquad \text{Substitute.}$$
$$3(y + 1) = 2(x - 2) \qquad \text{Simplify.}$$
$$3y + 3 = 2x - 4 \qquad \text{Distributive Property}$$
$$2x - 3y - 7 = 0 \qquad \text{General form}$$

Note the similarity to the equation of the given line, $2x - 3y = 5$.

b. Using the negative reciprocal of the slope of the given line, you can determine that the slope of a line perpendicular to the given line is $-\tfrac{3}{2}$.

$$y - y_1 = m(x - x_1) \qquad \text{Point-slope form}$$
$$y - (-1) = -\tfrac{3}{2}(x - 2) \qquad \text{Substitute.}$$
$$2(y + 1) = -3(x - 2) \qquad \text{Simplify.}$$
$$2y + 2 = -3x + 6 \qquad \text{Distributive Property}$$
$$3x + 2y - 4 = 0 \qquad \text{General form}$$

▷ **TECHNOLOGY PITFALL**   The slope of a line will appear distorted if you use different tick-mark spacing on the *x*- and *y*-axes. For instance, the graphing utility screens in Figures 1.21(a) and 1.21(b) both show the lines

$$y = 2x \quad \text{and} \quad y = -\tfrac{1}{2}x + 3.$$

Because these lines have slopes that are negative reciprocals, they must be perpendicular. In Figure 1.21(a), however, the lines do not appear to be perpendicular because the tick-mark spacing on the *x*-axis is not the same as that on the *y*-axis. In Figure 1.21(b), the lines appear perpendicular because the tick-mark spacing on the *x*-axis is the same as on the *y*-axis. This type of viewing window is said to have a *square setting.*

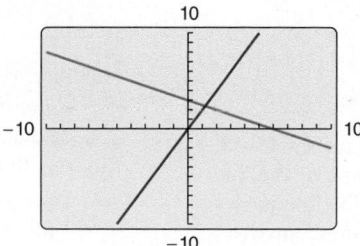

(a) Tick-mark spacing on the *x*-axis is not the same as tick-mark spacing on the *y*-axis.

(b) Tick-mark spacing on the *x*-axis is the same as tick-mark spacing on the *y*-axis.

Figure 1.21

Left margin:

Lines parallel and perpendicular to $2x - 3y = 5$

Figure 1.20

1.2 Exercises

See **CalcChat.com** for tutorial help and worked-out solutions to odd-numbered exercises.

CONCEPT CHECK

1. Slope-Intercept Form In the form $y = mx + b$, what does m represent? What does b represent?

2. Perpendicular Lines Is it possible for two lines with positive slopes to be perpendicular? Why or why not?

Estimating Slope In Exercises 3–6, estimate the slope of the line from its graph. To print an enlarged copy of the graph, go to *MathGraphs.com*.

3.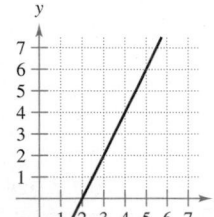

4.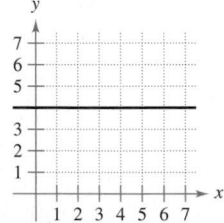

5.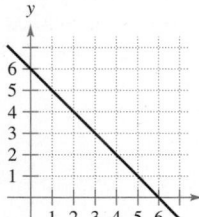

6.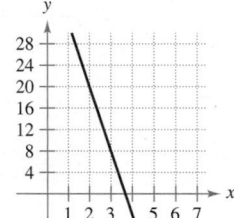

 **Finding the Slope of a Line** In Exercises 7–12, plot the pair of points and find the slope of the line passing through them.

7. $(3, -4), (5, 2)$ **8.** $(0, 0), (-2, 3)$

9. $(4, 6), (4, 1)$ **10.** $(3, -5), (5, -5)$

11. $\left(-\frac{1}{2}, \frac{2}{3}\right), \left(-\frac{3}{4}, \frac{1}{6}\right)$ **12.** $\left(\frac{7}{8}, \frac{3}{4}\right), \left(\frac{5}{4}, -\frac{1}{4}\right)$

 **Sketching Lines** In Exercises 13 and 14, sketch the lines through the point with the indicated slopes. Make the sketches on the same set of coordinate axes.

Point	Slopes
13. $(3, 4)$	(a) 1 (b) -2 (c) $-\frac{3}{2}$ (d) Undefined
14. $(-2, 5)$	(a) 3 (b) -3 (c) $\frac{1}{3}$ (d) 0

 **Finding Points on a Line** In Exercises 15–18, use the point on the line and the slope of the line to find three additional points that the line passes through. (There is more than one correct answer.)

Point	Slope	Point	Slope
15. $(6, 2)$	$m = 0$	**16.** $(-4, 3)$	m is undefined.
17. $(1, 7)$	$m = -3$	**18.** $(-2, -2)$	$m = 2$

 **Finding an Equation of a Line** In Exercises 19–24, find an equation of the line that passes through the point and has the indicated slope. Then sketch the line.

Point	Slope
19. $(0, 3)$	$m = \frac{3}{4}$
20. $(-5, -2)$	$m = \frac{6}{5}$
21. $(1, 2)$	m is undefined.
22. $(0, 4)$	$m = 0$
23. $(3, -2)$	$m = 3$
24. $(-2, 4)$	$m = -\frac{3}{5}$

25. Road Grade You are driving on a road that has a 6% uphill grade. This means that the slope of the road is $\frac{6}{100}$. Approximate the amount of vertical change in your position when you drive 200 feet.

26. Conveyor Design

A moving conveyor is built to rise 1 meter for each 3 meters of horizontal change.

(a) Find the slope of the conveyor.

(b) Suppose the conveyor runs between two floors in a factory. Find the length of the conveyor when the vertical distance between floors is 10 feet.

27. Modeling Data The table shows the populations y (in millions) of the United States for 2009 through 2014. The variable t represents the time in years, with $t = 9$ corresponding to 2009. *(Source: U.S. Census Bureau)*

t	9	10	11	12	13	14
y	307.0	309.3	311.7	314.1	316.5	318.9

(a) Plot the data by hand and connect adjacent points with a line segment. Use the slope of each line segment to determine the year when the population increased least rapidly.

(b) Find the average rate of change of the population of the United States from 2009 through 2014.

(c) Use the average rate of change of the population to predict the population of the United States in 2025.

28. Biodiesel Production The table shows the biodiesel productions y (in thousands of barrels per day) for the United States for 2007 through 2012. The variable t represents the time in years, with $t = 7$ corresponding to 2007. *(Source: U.S. Energy Information Administration)*

t	7	8	9	10	11	12
y	32	44	34	22	63	64

(a) Plot the data by hand and connect adjacent points with a line segment. Use the slope of each line segment to determine the year when biodiesel production increased most rapidly.

(b) Find the average rate of change of biodiesel production for the United States from 2007 through 2012.

(c) Should the average rate of change be used to predict future biodiesel production? Explain.

 **Finding the Slope and y-Intercept** In Exercises 29–34, find the slope and the y-intercept (if possible) of the line.

29. $y = 4x - 3$

30. $-x + y = 1$

31. $5x + y = 20$

32. $6x - 5y = 15$

33. $x = 4$

34. $y = -1$

 Sketching a Line in the Plane In Exercises 35–42, sketch the graph of the equation.

35. $y = -3$

36. $x = 4$

37. $y = -2x + 1$

38. $y = \frac{1}{3}x - 1$

39. $y - 2 = \frac{3}{2}(x - 1)$

40. $y - 1 = 3(x + 4)$

41. $3x - 3y + 1 = 0$

42. $x + 2y + 6 = 0$

 **Finding an Equation of a Line** In Exercises 43–50, find an equation of the line that passes through the points. Then sketch the line.

43. $(4, 3), (0, -5)$

44. $(-2, -2), (1, 7)$

45. $(2, 8), (5, 0)$

46. $(-3, 6), (1, 2)$

47. $(6, 3), (6, 8)$

48. $(1, -2), (3, -2)$

49. $(3, 1), (5, 1)$

50. $(2, 5), (2, 7)$

51. Writing an Equation Write an equation for the line that passes through the points $(0, b)$ and $(3, 1)$.

52. Using Intercepts Show that the line with intercepts $(a, 0)$ and $(0, b)$ has the following equation.

$$\frac{x}{a} + \frac{y}{b} = 1, \quad a \neq 0, b \neq 0$$

Writing an Equation in General Form In Exercises 53–56, use the result of Exercise 52 to write an equation of the line with the given characteristics in general form.

53. x-intercept: $(2, 0)$

y-intercept: $(0, 3)$

54. x-intercept: $\left(-\frac{2}{3}, 0\right)$

y-intercept: $(0, -2)$

55. Point on line: $(9, -2)$

x-intercept: $(2a, 0)$

y-intercept: $(0, a)$

$(a \neq 0)$

56. Point on line: $\left(-\frac{2}{3}, -2\right)$

x-intercept: $(a, 0)$

y-intercept: $(0, -a)$

$(a \neq 0)$

 Finding Parallel and Perpendicular Lines In Exercises 57–62, write the general forms of the equations of the lines that pass through the point and are (a) parallel to the given line and (b) perpendicular to the given line.

Point	Line
57. $(-7, -2)$	$x = 1$
58. $(-1, 0)$	$y = -3$
59. $(-3, 2)$	$x + y = 7$
60. $(2, 5)$	$x - y = -2$
61. $\left(\frac{3}{4}, \frac{7}{8}\right)$	$5x - 3y = 0$
62. $\left(\frac{5}{6}, -\frac{1}{2}\right)$	$7x + 4y = 8$

Rate of Change In Exercises 63 and 64, you are given the dollar value of a product in 2016 *and* the rate at which the value of the product is expected to change during the next 5 years. Write a linear equation that gives the dollar value V of the product in terms of the year t. (Let $t = 0$ represent 2010.)

2016 Value	Rate
63. $1850	$250 increase per year
64. $17,200	$1600 decrease per year

Collinear Points In Exercises 65 and 66, determine whether the points are collinear. (Three points are *collinear* if they lie on the same line.)

65. $(-2, 1), (-1, 0), (2, -2)$

66. $(0, 4), (7, -6), (-5, 11)$

EXPLORING CONCEPTS

67. Square Show that the points $(-1, 0), (3, 0), (1, 2),$ and $(1, -2)$ are vertices of a square.

68. Analyzing a Line A line is represented by the equation $ax + by = 4$.

(a) When is the line parallel to the x-axis?

(b) When is the line parallel to the y-axis?

(c) Give values for a and b such that the line has a slope of $\frac{5}{8}$.

(d) Give values for a and b such that the line is perpendicular to $y = \frac{2}{5}x + 3$.

(e) Give values for a and b such that the line coincides with the graph of $5x + 6y = 8$.

69. Tangent Line Find an equation of the line tangent to the circle $x^2 + y^2 = 169$ at the point $(5, 12)$.

70. Tangent Line Find an equation of the line tangent to the circle $(x - 1)^2 + (y - 1)^2 = 25$ at the point $(4, -3)$.

71. Finding Points of Intersection Find the coordinates of the point of intersection of the given segments. Explain your reasoning.

(a) Perpendicular bisectors   (b) Medians

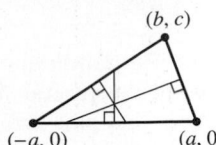   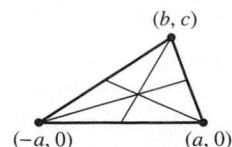

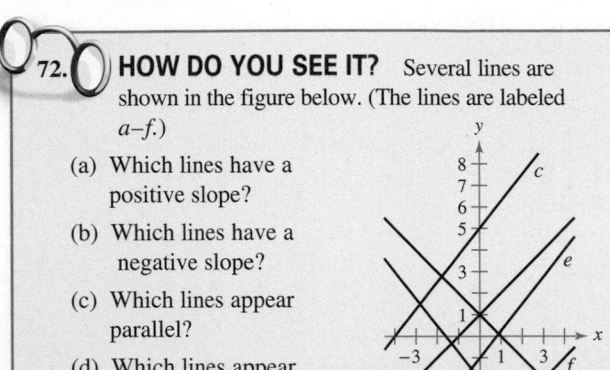

72. HOW DO YOU SEE IT? Several lines are shown in the figure below. (The lines are labeled *a–f.*)

(a) Which lines have a positive slope?

(b) Which lines have a negative slope?

(c) Which lines appear parallel?

(d) Which lines appear perpendicular?

73. Temperature Conversion Find a linear equation that expresses the relationship between the temperature in degrees Celsius C and degrees Fahrenheit F. Use the fact that water freezes at 0°C (32°F) and boils at 100°C (212°F). Use the equation to convert 72°F to degrees Celsius.

74. Choosing a Job As a salesperson, you receive a monthly salary of $2000, plus a commission of 7% of sales. You are offered a new job at $2300 per month, plus a commission of 5% of sales.

(a) Write linear equations for your monthly wage W in terms of your monthly sales s for your current job and your job offer.

(b) Use a graphing utility to graph each equation and find the point of intersection. What does it signify?

(c) You think you can sell $20,000 worth of a product per month. Should you change jobs? Explain.

75. Apartment Rental A real estate office manages an apartment complex with 50 units. When the rent is $780 per month, all 50 units are occupied. However, when the rent is $825, the average number of occupied units drops to 47. Assume that the relationship between the monthly rent p and the demand x is linear. (*Note:* The term *demand* refers to the number of occupied units.)

(a) Write a linear equation giving the demand x in terms of the rent p.

(b) *Linear extrapolation* Use a graphing utility to graph the demand equation and use the *trace* feature to predict the number of units occupied when the rent is raised to $855.

(c) *Linear interpolation* Predict the number of units occupied when the rent is lowered to $795. Verify graphically.

76. Modeling Data An instructor gives regular 20-point quizzes and 100-point exams in a mathematics course. Average scores for six students, given as ordered pairs (x, y), where x is the average quiz score and y is the average exam score, are (18, 87), (10, 55), (19, 96), (16, 79), (13, 76), and (15, 82).

(a) Use the regression capabilities of a graphing utility to find the least squares regression line for the data.

(b) Use a graphing utility to plot the points and graph the regression line in the same viewing window.

(c) Use the regression line to predict the average exam score for a student with an average quiz score of 17.

(d) Interpret the meaning of the slope of the regression line.

(e) The instructor adds 4 points to the average exam score of everyone in the class. Describe the changes in the positions of the plotted points and the change in the equation of the line.

77. Distance Show that the distance between the point (x_1, y_1) and the line $Ax + By + C = 0$ is

$$\text{Distance} = \frac{|Ax_1 + By_1 + C|}{\sqrt{A^2 + B^2}}.$$

78. Distance Write the distance d between the point $(3, 1)$ and the line $y = mx + 4$ in terms of m. Use a graphing utility to graph the equation. When is the distance 0? Explain the result geometrically.

Distance In Exercises 79 and 80, use the result of Exercise 77 to find the distance between the point and the line.

79. Point: $(-2, 1)$
Line: $x - y - 2 = 0$

80. Point: $(2, 3)$
Line: $4x + 3y = 10$

81. Proof Prove that the diagonals of a rhombus intersect at right angles. (A rhombus is a quadrilateral with sides of equal lengths.)

82. Proof Prove that the figure formed by connecting consecutive midpoints of the sides of any quadrilateral is a parallelogram.

83. Proof Prove that if the points (x_1, y_1) and (x_2, y_2) lie on the same line as (x_1^*, y_1^*) and (x_2^*, y_2^*), then

$$\frac{y_2^* - y_1^*}{x_2^* - x_1^*} = \frac{y_2 - y_1}{x_2 - x_1}.$$

Assume $x_1 \neq x_2$ and $x_1^* \neq x_2^*$.

84. Proof Prove that if the slopes of two nonvertical lines are negative reciprocals of each other, then the lines are perpendicular.

True or False? In Exercises 85 and 86, determine whether the statement is true or false. If it is false, explain why or give an example that shows it is false.

85. The lines represented by

$$ax + by = c_1 \quad \text{and} \quad bx - ay = c_2$$

are perpendicular. Assume $a \neq 0$ and $b \neq 0$.

86. If a line contains points in both the first and third quadrants, then its slope must be positive.

1.3 Functions and Their Graphs

- Use function notation to represent and evaluate a function.
- Find the domain and range of a function.
- Sketch the graph of a function.
- Identify different types of transformations of functions.
- Classify functions and recognize combinations of functions.

Functions and Function Notation

A **relation** between two sets X and Y is a set of ordered pairs, each of the form (x, y), where x is a member of X and y is a member of Y. A **function** from X to Y is a relation between X and Y that has the property that any two ordered pairs with the same x-value also have the same y value. The variable x is the **independent variable,** and the variable y is the **dependent variable.**

Many real-life situations can be modeled by functions. For instance, the area A of a circle is a function of the circle's radius r.

$$A = \pi r^2 \qquad \text{\small A is a function of r.}$$

In this case, r is the independent variable and A is the dependent variable.

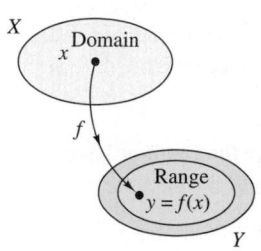

A real-valued function f of a real variable

Figure 1.22

Definition of a Real-Valued Function of a Real Variable

Let X and Y be sets of real numbers. A **real-valued function f of a real variable x** from X to Y is a correspondence that assigns to each number x in X exactly one number y in Y.

The **domain** of f is the set X. The number y is the **image** of x under f and is denoted by $f(x)$, which is called the **value of f at x.** The **range** of f is a subset of Y and consists of all images of numbers in X. (See Figure 1.22.)

Functions can be specified in a variety of ways. In this text, however, you will concentrate primarily on functions that are given by equations involving the dependent and independent variables. For instance, the equation

$$x^2 + 2y = 1 \qquad \text{\small Equation in implicit form}$$

defines y, the dependent variable, as a function of x, the independent variable. To **evaluate** this function (that is, to find the y-value that corresponds to a given x-value), it is convenient to isolate y on the left side of the equation.

$$y = \tfrac{1}{2}(1 - x^2) \qquad \text{\small Equation in explicit form}$$

Using f as the name of the function, you can write this equation as

$$f(x) = \tfrac{1}{2}(1 - x^2). \qquad \text{\small Function notation}$$

The original equation

$$x^2 + 2y = 1$$

implicitly defines y as a function of x. When you solve the equation for y, you are writing the equation in **explicit** form.

Function notation has the advantage of clearly identifying the dependent variable as $f(x)$ while at the same time telling you that x is the independent variable and that the function itself is "f." The symbol $f(x)$ is read "f of x." Function notation allows you to be less wordy. Instead of asking "What is the value of y that corresponds to $x = 3$?" you can ask "What is $f(3)$?"

LEONHARD EULER (1707–1783)

In addition to making major contributions to almost every branch of mathematics, Euler was one of the first to apply calculus to real-life problems in physics. His extensive published writings include such topics as shipbuilding, acoustics, optics, astronomy, mechanics, and magnetism.
See LarsonCalculus.com to read more of this biography.

In an equation that defines a function of x, the role of the variable x is simply that of a placeholder. For instance, the function

$$f(x) = 2x^2 - 4x + 1$$

can be described by the form

$$f(\boxed{}) = 2(\boxed{})^2 - 4(\boxed{}) + 1$$

where rectangles are used instead of x. To evaluate $f(-2)$, replace each rectangle with -2.

$$\begin{aligned} f(-2) &= 2(-2)^2 - 4(-2) + 1 && \text{Substitute } -2 \text{ for } x. \\ &= 2(4) + 8 + 1 && \text{Simplify.} \\ &= 17 && \text{Simplify.} \end{aligned}$$

Although f is often used as a convenient function name with x as the independent variable, you can use other symbols. For instance, these three equations all define the same function.

$$\begin{aligned} f(x) &= x^2 - 4x + 7 && \text{Function name is } f, \text{ independent variable is } x. \\ f(t) &= t^2 - 4t + 7 && \text{Function name is } f, \text{ independent variable is } t. \\ g(s) &= s^2 - 4s + 7 && \text{Function name is } g, \text{ independent variable is } s. \end{aligned}$$

EXAMPLE 1   Evaluating a Function

For the function f defined by $f(x) = x^2 + 7$, evaluate each expression.

a. $f(3a)$ **b.** $f(b - 1)$ **c.** $\dfrac{f(x + \Delta x) - f(x)}{\Delta x}$

Solution

a. $\begin{aligned}[t] f(3a) &= (3a)^2 + 7 && \text{Substitute } 3a \text{ for } x. \\ &= 9a^2 + 7 && \text{Simplify.} \end{aligned}$

b. $\begin{aligned}[t] f(b - 1) &= (b - 1)^2 + 7 && \text{Substitute } b - 1 \text{ for } x. \\ &= b^2 - 2b + 1 + 7 && \text{Expand binomial.} \\ &= b^2 - 2b + 8 && \text{Simplify.} \end{aligned}$

> **REMARK** The expression in Example 1(c) is called a *difference quotient* and has a special significance in calculus. You will learn more about this in Chapter 3.

c. $\begin{aligned}[t] \frac{f(x + \Delta x) - f(x)}{\Delta x} &= \frac{[(x + \Delta x)^2 + 7] - (x^2 + 7)}{\Delta x} \\ &= \frac{x^2 + 2x\Delta x + (\Delta x)^2 + 7 - x^2 - 7}{\Delta x} \\ &= \frac{2x\Delta x + (\Delta x)^2}{\Delta x} \\ &= \frac{\cancel{\Delta x}(2x + \Delta x)}{\cancel{\Delta x}} \\ &= 2x + \Delta x, \quad \Delta x \neq 0 \end{aligned}$

In calculus, it is important to specify the domain of a function or expression clearly. For instance, in Example 1(c), the two expressions

$$\frac{f(x + \Delta x) - f(x)}{\Delta x} \quad \text{and} \quad 2x + \Delta x, \quad \Delta x \neq 0$$

are equivalent because $\Delta x = 0$ is excluded from the domain of each expression. Without a stated domain restriction, the two expressions would not be equivalent.

The Domain and Range of a Function

The domain of a function can be described explicitly, or it may be described *implicitly* by an equation used to define the function. The **implied domain** is the set of all real numbers for which the equation is defined, whereas an explicitly defined domain is one that is given along with the function. For example, the function

$$f(x) = \frac{1}{x^2 - 4}, \quad 4 \le x \le 5$$

has an explicitly defined domain given by $\{x: 4 \le x \le 5\}$. On the other hand, the function

$$g(x) = \frac{1}{x^2 - 4}$$

has an implied domain that is the set $\{x: x \ne \pm 2\}$.

EXAMPLE 2   Finding the Domain and Range of a Function

Find the domain and range of each function.

a. $f(x) = \sqrt{x - 1}$ **b.** $g(x) = \sqrt{4 - x^2}$

Solution

a. The domain of the function

$$f(x) = \sqrt{x - 1}$$

is the set of all x-values for which $x - 1 \ge 0$, which is the interval $[1, \infty)$. To find the range, observe that $f(x) = \sqrt{x - 1}$ is never negative. So, the range is the interval $[0, \infty)$, as shown in Figure 1.23(a).

b. The domain of the function

$$g(x) = \sqrt{4 - x^2}$$

is the set of all values for which $4 - x^2 \ge 0$, or $x^2 \le 4$. So, the domain of g is the interval $[-2, 2]$. To find the range, observe that $g(x) = \sqrt{4 - x^2}$ is never negative and is at most 2. So, the range is the interval $[0, 2]$, as shown in Figure 1.23(b). Note that the graph of g is a *semicircle* of radius 2.

EXAMPLE 3   A Function Defined by More than One Equation

For the piecewise-defined function

$$f(x) = \begin{cases} 1 - x, & x < 1 \\ \sqrt{x - 1}, & x \ge 1 \end{cases}$$

f is defined for $x < 1$ and $x \ge 1$. So, the domain is the set of all real numbers. On the portion of the domain for which $x \ge 1$, the function behaves as in Example 2(a). For $x < 1$, the values of $1 - x$ are positive. So, the range of the function is the interval $[0, \infty)$. (See Figure 1.24.)

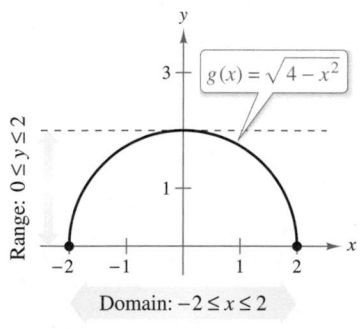

Range: $y \ge 0$

$f(x) = \sqrt{x - 1}$

Domain: $x \ge 1$

(a) The domain of f is $[1, \infty)$, and the range is $[0, \infty)$.

Range: $0 \le y \le 2$

$g(x) = \sqrt{4 - x^2}$

Domain: $-2 \le x \le 2$

(b) The domain of g is $[-2, 2]$, and the range is $[0, 2]$.

Figure 1.23

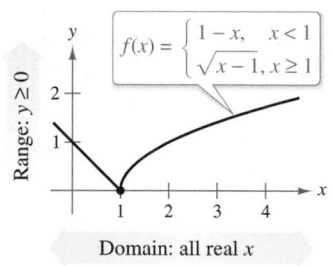

Range: $y \ge 0$

$f(x) = \begin{cases} 1 - x, & x < 1 \\ \sqrt{x - 1}, x \ge 1 \end{cases}$

Domain: all real x

The domain of f is $(-\infty, \infty)$, and the range is $[0, \infty)$.

Figure 1.24

THE SQUARE ROOT SYMBOL

The first use of a symbol to denote the square root can be traced to the sixteenth century. Mathematicians first used the symbol $\sqrt{\ }$, which had only two strokes. The symbol was chosen because it resembled a lowercase *r*, to stand for the Latin word *radix*, meaning root.

A function from X to Y is **one-to-one** when to each y-value in the range there corresponds exactly one x-value in the domain. For instance, the function in Example 2(a) is one-to-one, whereas the functions in Examples 2(b) and 3 are not one-to-one. A function from X to Y is **onto** when its range consists of all of Y.

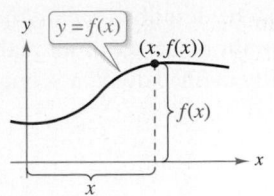

The graph of a function
Figure 1.25

The Graph of a Function

The graph of the function $y = f(x)$ consists of all points $(x, f(x))$, where x is in the domain of f. In Figure 1.25, note that

$$x = \text{the directed distance from the } y\text{-axis}$$

and

$$f(x) = \text{the directed distance from the } x\text{-axis}.$$

A vertical line can intersect the graph of a function of x at most *once*. This observation provides a convenient visual test, called the **Vertical Line Test,** for functions of x. That is, a graph in the coordinate plane is the graph of a function of x if and only if no vertical line intersects the graph at more than one point. For example, in Figure 1.26(a), you can see that the graph does not define y as a function of x because a vertical line intersects the graph twice, whereas in Figures 1.26(b) and (c), the graphs do define y as a function of x.

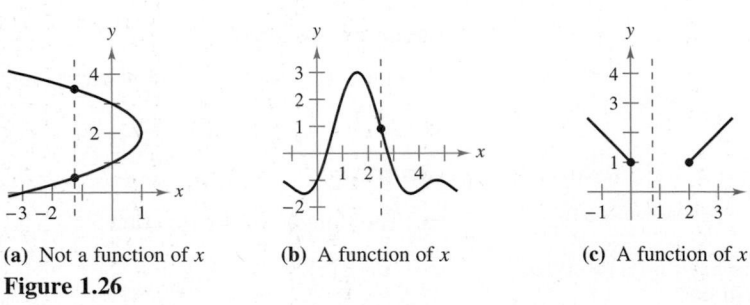

(a) Not a function of x (b) A function of x (c) A function of x

Figure 1.26

Figure 1.27 shows the graphs of six basic functions. You should be able to recognize these graphs. (The graphs of the six basic trigonometric functions are shown in Section 1.4.)

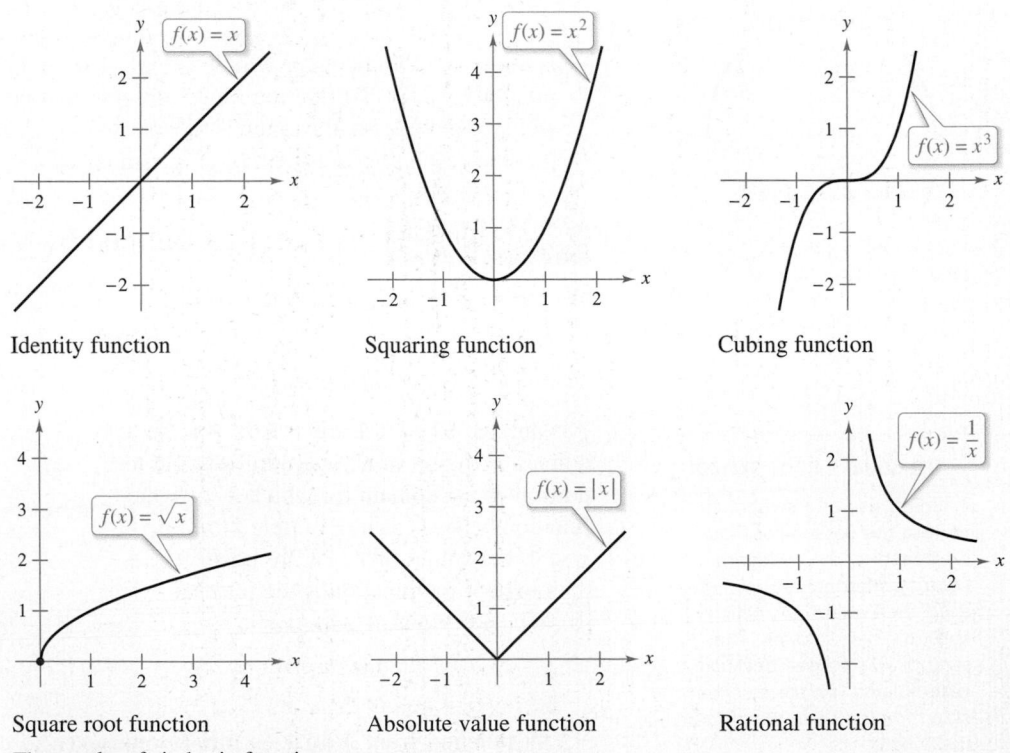

Identity function   Squaring function   Cubing function

Square root function   Absolute value function   Rational function

The graphs of six basic functions
Figure 1.27

Transformations of Functions

Some families of graphs have the same basic shape. For example, compare the graph of $y = x^2$ with the graphs of the four other quadratic functions shown in Figure 1.28.

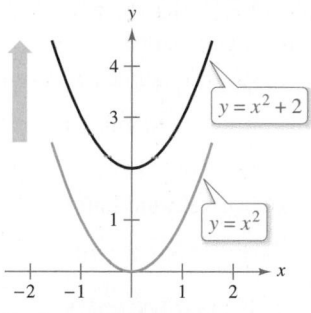

(a) Vertical shift upward

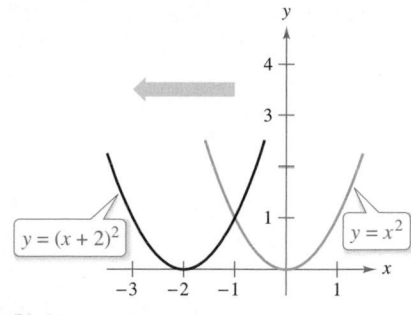

(b) Horizontal shift to the left

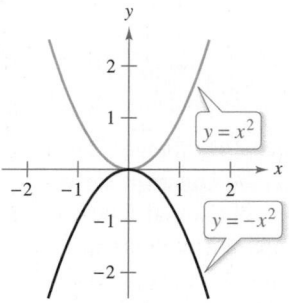

(c) Reflection

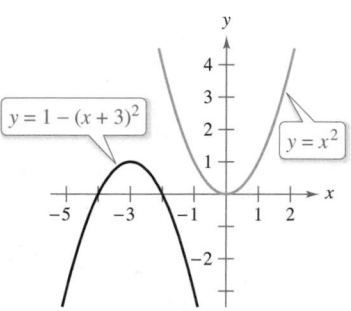

(d) Shift left, reflect, and shift upward

Figure 1.28

Each of the graphs in Figure 1.28 is a **transformation** of the graph of $y = x^2$. The three basic types of transformations illustrated by these graphs are vertical shifts, horizontal shifts, and reflections. Function notation lends itself well to describing transformations of graphs in the plane. For instance, using

$$f(x) = x^2 \qquad \text{Original function}$$

as the original function, the transformations shown in Figure 1.28 can be represented by these equations.

a. $y = f(x) + 2$ Vertical shift up two units

b. $y = f(x + 2)$ Horizontal shift to the left two units

c. $y = -f(x)$ Reflection about the x-axis

d. $y = -f(x + 3) + 1$ Shift left three units, reflect about the x-axis, and shift up one unit

---

Basic Types of Transformations $(c > 0)$

Original graph:	$y = f(x)$
Horizontal shift c units to the **right:**	$y = f(x - c)$
Horizontal shift c units to the **left:**	$y = f(x + c)$
Vertical shift c units **downward:**	$y = f(x) - c$
Vertical shift c units **upward:**	$y = f(x) + c$
Reflection (about the x-axis):	$y = -f(x)$
Reflection (about the y-axis):	$y = f(-x)$
Reflection (about the origin):	$y = -f(-x)$

Classifications and Combinations of Functions

■ **FOR FURTHER INFORMATION**
For more on the history of the concept of a function, see the article "Evolution of the Function Concept: A Brief Survey" by Israel Kleiner in *The College Mathematics Journal.* To view this article, go to *MathArticles.com.*

The modern notion of a function is derived from the efforts of many seventeenth- and eighteenth-century mathematicians. Of particular note was Leonhard Euler, who introduced the function notation $y = f(x)$. By the end of the eighteenth century, mathematicians and scientists had concluded that many real-world phenomena could be represented by mathematical models taken from a collection of functions called **elementary functions.** Elementary functions fall into three categories.

1. Algebraic functions (polynomial, radical, rational)
2. Trigonometric functions (sine, cosine, tangent, and so on)
3. Exponential and logarithmic functions

You will review the trigonometric functions in the next section. The other nonalgebraic functions, such as the inverse trigonometric functions and the exponential and logarithmic functions, are introduced in Sections 1.5 and 1.6.

The most common type of algebraic function is a **polynomial function**

$$f(x) = a_n x^n + a_{n-1} x^{n-1} + \cdots + a_2 x^2 + a_1 x + a_0$$

where n is a nonnegative integer. The numbers a_i are **coefficients,** with a_n the **leading coefficient** and a_0 the **constant term** of the polynomial function. If $a_n \neq 0$, then n is the **degree** of the polynomial function. The zero polynomial $f(x) = 0$ is not assigned a degree. It is common practice to use subscript notation for coefficients of general polynomial functions, but for polynomial functions of low degree, these simpler forms are often used. (Note that $a \neq 0$.)

Zeroth degree:	$f(x) = a$	Constant function
First degree:	$f(x) = ax + b$	Linear function
Second degree:	$f(x) = ax^2 + bx + c$	Quadratic function
Third degree:	$f(x) = ax^3 + bx^2 + cx + d$	Cubic function

Although the graph of a nonconstant polynomial function can have several turns, eventually the graph will rise or fall without bound as x moves to the right or left. Whether the graph of

$$f(x) = a_n x^n + a_{n-1} x^{n-1} + \cdots + a_2 x^2 + a_1 x + a_0$$

eventually rises or falls can be determined by the function's degree (odd or even) and by the leading coefficient a_n, as indicated in Figure 1.29. Note that the dashed portions of the graphs indicate that the **Leading Coefficient Test** determines *only* the right and left behavior of the graph.

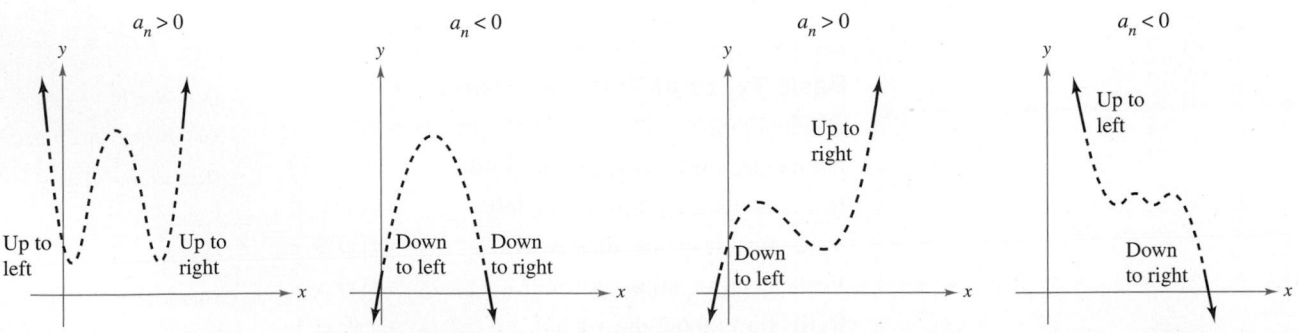

Graphs of polynomial functions of even degree   Graphs of polynomial functions of odd degree

The Leading Coefficient Test for polynomial functions
Figure 1.29

Just as a rational number can be written as the quotient of two integers, a **rational function** can be written as the quotient of two polynomials. Specifically, a function f is rational when it has the form

$$f(x) = \frac{p(x)}{q(x)}, \quad q(x) \neq 0$$

where $p(x)$ and $q(x)$ are polynomials.

Polynomial functions and rational functions are examples of **algebraic functions.** An algebraic function of x is one that can be expressed as a finite number of sums, differences, multiples, quotients, and radicals involving x^n. For example, $f(x) = \sqrt{x + 1}$ is algebraic. Functions that are not algebraic are **transcendental.** For instance, the trigonometric functions (see Section 1.4) are transcendental.

Two functions can be combined in various ways to create new functions. For example, given $f(x) = 2x - 3$ and $g(x) = x^2 + 1$, you can form the functions shown.

$$(f + g)(x) = f(x) + g(x) = (2x - 3) + (x^2 + 1) \qquad \text{Sum}$$
$$(f - g)(x) = f(x) - g(x) = (2x - 3) - (x^2 + 1) \qquad \text{Difference}$$
$$(fg)(x) = f(x)g(x) = (2x - 3)(x^2 + 1) \qquad \text{Product}$$
$$(f/g)(x) = \frac{f(x)}{g(x)} = \frac{2x - 3}{x^2 + 1} \qquad \text{Quotient}$$

You can combine two functions in yet another way, called **composition.** The resulting function is called a **composite function.**

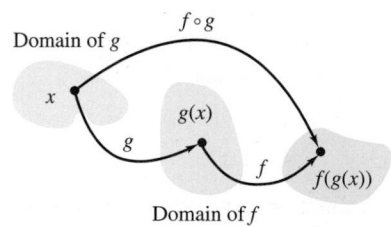

Domain of g

$f \circ g$

x

$g(x)$

g

f

$f(g(x))$

Domain of f

The domain of the composite function $f \circ g$

Figure 1.30

Definition of Composite Function

Let f and g be functions. The function $(f \circ g)(x) = f(g(x))$ is the **composite** of f with g. The domain of $f \circ g$ is the set of all x in the domain of g such that $g(x)$ is in the domain of f (see Figure 1.30).

The composite of f with g is generally not the same as the composite of g with f. This is shown in the next example.

EXAMPLE 4   Finding Composite Functions

⋯▷ *See LarsonCalculus.com for an interactive version of this type of example.*

For $f(x) = 2x - 3$ and $g(x) = x^2 + 1$, find each composite function.

a. $f \circ g$ **b.** $g \circ f$

Solution

a. $(f \circ g)(x) = f(g(x))$ Definition of $f \circ g$

$\qquad\qquad = f(x^2 + 1)$ Substitute $x^2 + 1$ for $g(x)$.

$\qquad\qquad = 2(x^2 + 1) - 3$ Definition of $f(x)$

$\qquad\qquad = 2x^2 - 1$ Simplify.

b. $(g \circ f)x = g(f(x))$ Definition of $g \circ f$

$\qquad\qquad = g(2x - 3)$ Substitute $2x - 3$ for $f(x)$.

$\qquad\qquad = (2x - 3)^2 + 1$ Definition of $g(x)$

$\qquad\qquad = 4x^2 - 12x + 10$ Simplify.

Note that $(f \circ g)(x) \neq (g \circ f)(x)$.

In Section 1.1, an x-intercept of a graph was defined to be a point $(a, 0)$ at which the graph crosses the x-axis. If the graph represents a function f, then the number a is a **zero** of f. In other words, *the zeros of a function f are the solutions of the equation* $f(x) = 0$. For example, the function

$$f(x) = x - 4$$

has a zero at $x = 4$ because $f(4) = 0$.

In Section 1.1, you also studied different types of symmetry. In the terminology of functions, a function $y = f(x)$ is **even** when its graph is symmetric with respect to the y-axis, and is **odd** when its graph is symmetric with respect to the origin. The symmetry tests in Section 1.1 yield the following test for even and odd functions.

Test for Even and Odd Functions

The function $y = f(x)$ is **even** when

$$f(-x) = f(x).$$

The function $y = f(x)$ is **odd** when

$$f(-x) = -f(x).$$

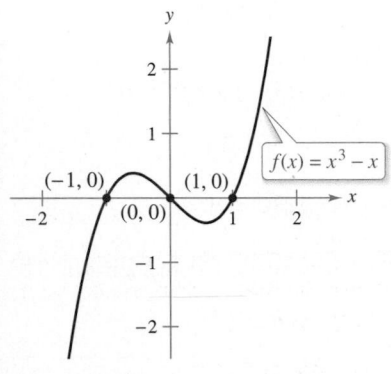

(a) Odd function

EXAMPLE 5 **Even and Odd Functions and Zeros of Functions**

Determine whether each function is even, odd, or neither. Then find the zeros of the function.

a. $f(x) = x^3 - x$ **b.** $g(x) = \dfrac{1}{x^2}$ **c.** $h(x) = -x^2 - x - 1$

Solution

a. This function is odd because

$$f(-x) = (-x)^3 - (-x) = -x^3 + x = -(x^3 - x) = -f(x).$$

The zeros of f are

$$
\begin{aligned}
x^3 - x &= 0 && \text{Let } f(x) = 0.\\
x(x^2 - 1) &= 0 && \text{Factor.}\\
x(x - 1)(x + 1) &= 0 && \text{Factor.}\\
x &= 0, 1, -1. && \text{Zeros of } f
\end{aligned}
$$

See Figure 1.31(a).

b. This function is even because

$$g(-x) = \frac{1}{(-x)^2} = \frac{1}{x^2} = g(x).$$

This function does not have zeros because $1/x^2$ is positive for all x in the domain, as shown in Figure 1.31(b).

c. Substituting $-x$ for x produces

$$h(-x) = -(-x)^2 - (-x) - 1 = -x^2 + x - 1.$$

Because $h(x) = -x^2 - x - 1$ and $-h(x) = x^2 + x + 1$, you can conclude that

$$h(-x) \neq h(x). \qquad \text{Function is } not \text{ even.}$$

and

$$h(-x) \neq -h(x). \qquad \text{Function is } not \text{ odd.}$$

So, the function is neither even nor odd. This function does not have zeros because $-x^2 - x - 1$ is negative for all x, as shown in Figure 1.31(c).

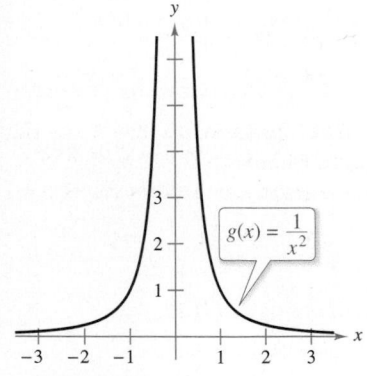

(b) Even function

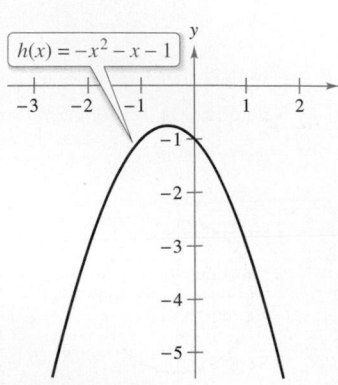

(c) Neither even nor odd
Figure 1.31

1.3 Exercises

See CalcChat.com for tutorial help and worked-out solutions to odd-numbered exercises.

CONCEPT CHECK

1. **Writing** Describe how a relation and a function are different.

2. **Domain and Range** In your own words, explain the meanings of *domain* and *range*.

3. **Transformations** What are the three basic types of function transformations?

4. **Right and Left Behavior** Describe the four cases of the Leading Coefficient Test.

Evaluating a Function In Exercises 5–12, evaluate the function at the given value(s) of the independent variable. Simplify the results.

5. $f(x) = 3x - 2$

   (a) $f(0)$ (b) $f(5)$ (c) $f(b)$ (d) $f(x - 1)$

6. $f(x) = 7x - 4$

   (a) $f(0)$ (b) $f(-3)$ (c) $f(b)$ (d) $f(x + 2)$

7. $f(x) = \sqrt{x^2 + 4}$

   (a) $f(-2)$ (b) $f(3)$ (c) $f(2)$ (d) $f(x + bx)$

8. $f(x) = \sqrt{x + 5}$

   (a) $f(-4)$ (b) $f(11)$ (c) $f(4)$ (d) $f(x + \Delta x)$

9. $g(x) = 5 - x^2$

   (a) $g(0)$ (b) $g(\sqrt{5})$ (c) $g(-2)$ (d) $g(t - 1)$

10. $g(x) = x^2(x - 4)$

   (a) $g(4)$ (b) $g(\frac{3}{2})$ (c) $g(c)$ (d) $g(t + 4)$

11. $f(x) = x^3$

$$\frac{f(x + \Delta x) - f(x)}{\Delta x}$$

12. $f(x) = 3x - 1$

$$\frac{f(x) - f(1)}{x - 1}$$

Finding the Domain and Range of a Function In Exercises 13–22, find the domain and range of the function.

13. $f(x) = 4x^2$ 14. $g(x) = x^2 - 5$

15. $f(x) = x^3$ 16. $h(x) = 4 - x^2$

17. $g(x) = \sqrt{6x}$ 18. $h(x) = -\sqrt{x + 3}$

19. $f(x) = \sqrt{16 - x^2}$ 20. $f(x) = |x - 3|$

21. $f(x) = \dfrac{3}{x}$ 22. $f(x) = \dfrac{x - 2}{x + 4}$

Finding the Domain of a Function In Exercises 23–26, find the domain of the function.

23. $f(x) = \sqrt{x} + \sqrt{1 - x}$ 24. $f(x) = \sqrt{x^2 - 3x + 2}$

25. $f(x) = \dfrac{1}{|x + 3|}$ 26. $g(x) = \dfrac{1}{|x^2 - 4|}$

Finding the Domain and Range of a Piecewise Function In Exercises 27–30, evaluate the function at the given value(s) of the independent variable. Then find the domain and range.

27. $f(x) = \begin{cases} 2x + 1, & x < 0 \\ 2x + 2, & x \geq 0 \end{cases}$

   (a) $f(-1)$ (b) $f(0)$ (c) $f(2)$ (d) $f(t^2 + 1)$

28. $f(x) = \begin{cases} x^2 + 2, & x \leq 1 \\ 2x^2 + 2, & x > 1 \end{cases}$

   (a) $f(-2)$ (b) $f(0)$ (c) $f(1)$ (d) $f(s^2 + 2)$

29. $f(x) = \begin{cases} |x| + 1, & x < 1 \\ -x + 1, & x \geq 1 \end{cases}$

   (a) $f(-3)$ (b) $f(1)$ (c) $f(3)$ (d) $f(b^2 + 1)$

30. $f(x) = \begin{cases} \sqrt{x + 4}, & x \leq 5 \\ (x - 5)^2, & x > 5 \end{cases}$

   (a) $f(-3)$ (b) $f(0)$ (c) $f(5)$ (d) $f(10)$

Sketching a Graph of a Function In Exercises 31–38, sketch a graph of the function and find its domain and range. Use a graphing utility to verify your graph.

31. $f(x) = 4 - x$ 32. $f(x) = x^2 + 5$

33. $g(x) = \dfrac{1}{x^2 + 2}$ 34. $f(t) = \dfrac{2}{7 + t}$

35. $h(x) = \sqrt{x - 6}$ 36. $f(x) = \frac{1}{4}x^3 + 3$

37. $f(x) = \sqrt{9 - x^2}$ 38. $f(x) = x + \sqrt{4 - x^2}$

Using the Vertical Line Test In Exercises 39–42, use the Vertical Line Test to determine whether y is a function of x. To print an enlarged copy of the graph, go to *MathGraphs.com*.

39. $x - y^2 = 0$ 40. $\sqrt{x^2 - 4} - y = 0$

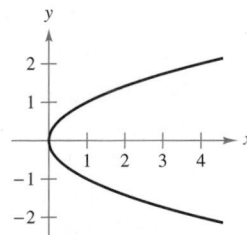   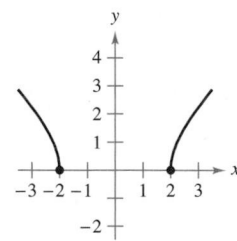

41. $y = \begin{cases} x + 1, & x \leq 0 \\ -x + 2, & x > 0 \end{cases}$ 42. $x^2 + y^2 = 4$

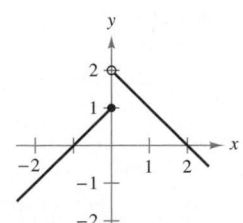   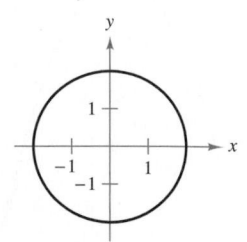

Deciding Whether an Equation Is a Function In Exercises 43–46, determine whether y is a function of x.

43. $x^2 + y^2 = 16$ **44.** $x^2 + y = 16$

45. $y^2 = x^2 - 1$ **46.** $x^2y - x^2 + 4y = 0$

Transformation of a Function In Exercises 47–50, the graph shows one of the six basic functions on page 22 and a transformation of the function. Describe the transformation. Then use your description to write an equation for the transformation.

47.

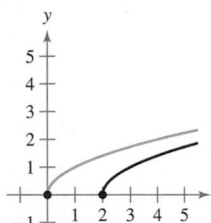

48.

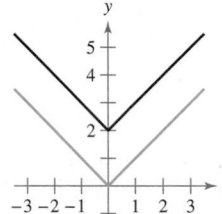

49.

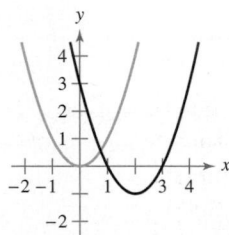

50.

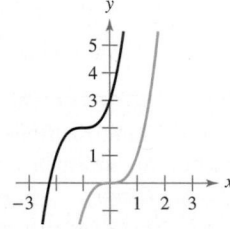

Matching In Exercises 51–56, use the graph of $y = f(x)$ to match the function with its graph.

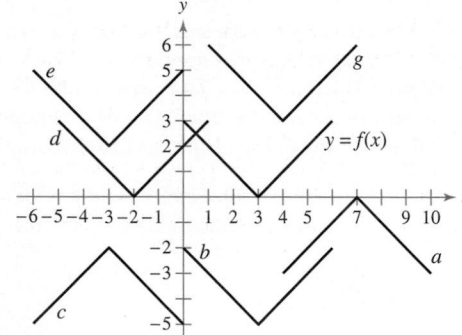

51. $y = f(x + 5)$ **52.** $y = f(x) - 5$

53. $y = -f(-x) - 2$ **54.** $y = -f(x - 4)$

55. $y = f(x + 6) + 2$ **56.** $y = f(x - 1) + 3$

57. Sketching Transformations Use the graph of f shown in the figure to sketch the graph of each function. To print an enlarged copy of the graph, go to *MathGraphs.com*.

(a) $f(x + 3)$ (b) $f(x - 1)$

(c) $f(x) + 2$ (d) $f(x) - 4$

(e) $3f(x)$ (f) $\frac{1}{4}f(x)$

(g) $-f(x)$ (h) $-f(-x)$

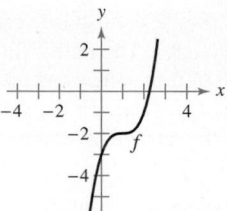

Figure for 57

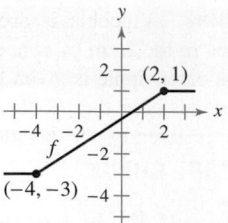

Figure for 58

58. Sketching Transformations Use the graph of f shown in the figure to sketch the graph of each function. To print an enlarged copy of the graph, go to *MathGraphs.com*.

(a) $f(x - 4)$ (b) $f(x + 2)$

(c) $f(x) + 4$ (d) $f(x) - 1$

(e) $2f(x)$ (f) $\frac{1}{2}f(x)$

(g) $f(-x)$ (h) $-f(x)$

Combinations of Functions In Exercises 59 and 60, find (a) $f(x) + g(x)$, (b) $f(x) - g(x)$, (c) $f(x) \cdot g(x)$, and (d) $f(x)/g(x)$.

59. $f(x) = 2x - 5$ **60.** $f(x) = x^2 + 5x + 4$

$g(x) = 4 - 3x$ $g(x) = x + 1$

61. Evaluating Composite Functions Given $f(x) = \sqrt{x}$ and $g(x) = x^2 - 1$, evaluate each expression.

(a) $f(g(1))$ (b) $g(f(1))$ (c) $g(f(0))$

(d) $f(g(-4))$ (e) $f(g(x))$ (f) $g(f(x))$

62. Evaluating Composite Functions Given $f(x) = 2x^3$ and $g(x) = 4x + 3$, evaluate each expression.

(a) $f(g(0))$ (b) $f\left(g\left(\frac{1}{2}\right)\right)$ (c) $g(f(0))$

(d) $g\left(f\left(-\frac{1}{4}\right)\right)$ (e) $f(g(x))$ (f) $g(f(x))$

Finding Composite Functions In Exercises 63–66, find the composite functions $f \circ g$ and $g \circ f$. Find the domain of each composite function. Are the two composite functions equal?

63. $f(x) = x^2$ **64.** $f(x) = x^2 - 1$

$g(x) = \sqrt{x}$ $g(x) = -x$

65. $f(x) = \dfrac{3}{x}$ **66.** $f(x) = \dfrac{1}{x}$

$g(x) = x^2 - 1$ $g(x) = \sqrt{x + 2}$

67. Evaluating Composite Functions Use the graphs of f and g to evaluate each expression. If the result is undefined, explain why.

(a) $(f \circ g)(3)$

(b) $g(f(2))$

(c) $g(f(5))$

(d) $(f \circ g)(-3)$

(e) $(g \circ f)(-1)$

(f) $f(g(-1))$

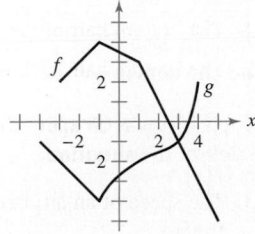

68. Ripples   A pebble is dropped into a calm pond, causing ripples in the form of concentric circles. The radius (in feet) of the outer ripple is given by $r(t) = 0.6t$, where t is the time in seconds after the pebble strikes the water. The area of the circle is given by the function $A(r) = \pi r^2$. Find and interpret $(A \circ r)(t)$.

Think About It   In Exercises 69 and 70, $F(x) = f \circ g \circ h$. Identify functions for f, g, and h. (There are many correct answers.)

69. $F(x) = \sqrt{2x - 2}$ **70.** $F(x) = \dfrac{1}{4x^6}$

Think About It   In Exercises 71 and 72, find the coordinates of a second point on the graph of a function f when the given point is on the graph and the function is (a) even and (b) odd.

71. $\left(-\frac{3}{2}, 4\right)$ **72.** $(4, 9)$

73. Even and Odd Functions   The graphs of f, g, and h are shown in the figure. Decide whether each function is even, odd, or neither.

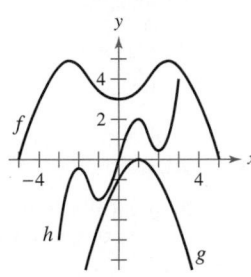

Figure for 73   Figure for 74

74. Even and Odd Functions   The domain of the function f shown in the figure is $-6 \le x \le 6$.

(a) Complete the graph of f given that f is even.

(b) Complete the graph of f given that f is odd.

 **Even and Odd Functions and Zeros of Functions**   In Exercises 75–78, determine whether the function is even, odd, or neither. Then find the zeros of the function. Use a graphing utility to verify your result.

75. $f(x) = x^2(4 - x^2)$ **76.** $f(x) = \sqrt[3]{x}$

77. $f(x) = 2\sqrt[6]{x}$ **78.** $f(x) = 4x^4 - 3x^2$

Writing Functions   In Exercises 79–82, write an equation for a function that has the given graph.

79. Line segment connecting $(-2, 4)$ and $(0, -6)$

80. Line segment connecting $(3, 1)$ and $(5, 8)$

81. The bottom half of the parabola $x + y^2 = 0$

82. The bottom half of the circle $x^2 + y^2 = 36$

Sketching a Graph   In Exercises 83–86, sketch a possible graph of the situation.

83. The speed of an airplane as a function of time during a 5-hour flight

84. The height of a baseball as a function of horizontal distance during a home run

85. A student commutes 15 miles to attend college. After driving for a few minutes, she remembers that a term paper that is due has been forgotten. Driving faster than usual, she returns home, picks up the paper, and once again starts toward school. Consider the student's distance from home as a function of time.

86. A person buys a new car and keeps it for 6 years. During year 4, he buys several expensive upgrades. Consider the value of the car as a function of time.

87. Domain   Find the value of c such that the domain of

$$f(x) = \sqrt{c - x^2}$$

is $[-5, 5]$.

88. Domain   Find all values of c such that the domain of

$$f(x) = \frac{x + 3}{x^2 + 3cx + 6}$$

is the set of all real numbers.

EXPLORING CONCEPTS

89. One-to-One Functions   Can the graph of a one-to-one function intersect a horizontal line more than once? Explain.

90. Composite Functions   Give an example of functions f and g such that $f \circ g = g \circ f$ and $f(x) \ne g(x)$.

91. Polynomial Functions   Does the degree of a polynomial function determine whether the function is even or odd? Explain.

92. Think About It   Determine whether the function $f(x) = 0$ is even, odd, both, or neither. Explain.

---

93. Graphical Reasoning   An electronically controlled thermostat is programmed to lower the temperature during the night automatically (see figure). The temperature T in degrees Celsius is given in terms of t, the time in hours on a 24-hour clock.

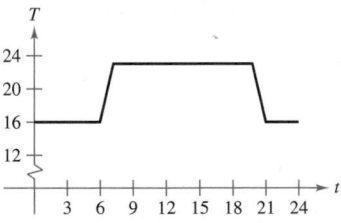

(a) Approximate $T(4)$ and $T(15)$.

(b) The thermostat is reprogrammed to produce a temperature $H(t) = T(t - 1)$. How does this change the temperature? Explain.

(c) The thermostat is reprogrammed to produce a temperature $H(t) = T(t) - 1$. How does this change the temperature? Explain.

94. **HOW DO YOU SEE IT?** Water runs into a vase of height 30 centimeters at a constant rate. The vase is full after 5 seconds. Use this information and the shape of the vase shown to answer the questions when d is the depth of the water in centimeters and t is the time in seconds (see figure).

30 cm

d

(a) Explain why d is a function of t.

(b) Determine the domain and range of the function.

(c) Sketch a possible graph of the function.

(d) Use the graph in part (c) to approximate $d(4)$. What does this represent?

95. Automobile Aerodynamics

The horsepower H required to overcome wind drag on a certain automobile is

$$H(x) = 0.00004636x^3$$

where x is the speed of the car in miles per hour.

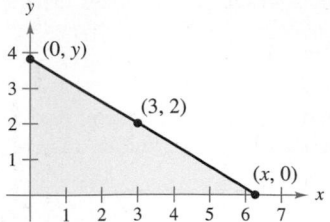

(a) Use a graphing utility to graph H.

(b) Rewrite H so that x represents the speed in kilometers per hour. [*Hint:* Find $H(x/1.6)$.]

96. Writing Use a graphing utility to graph the polynomial functions

$$p_1(x) = x^3 - x + 1 \quad \text{and} \quad p_2(x) = x^3 - x.$$

How many zeros does each function have? Is there a cubic polynomial that has no zeros? Explain.

97. Proof Prove that the function is odd.

$$f(x) = a_{2n+1}x^{2n+1} + \cdots + a_3x^3 + a_1x$$

98. Proof Prove that the function is even.

$$f(x) = a_{2n}x^{2n} + a_{2n-2}x^{2n-2} + \cdots + a_2x^2 + a_0$$

99. Proof Prove that the product of two even (or two odd) functions is even.

100. Proof Prove that the product of an odd function and an even function is odd.

101. Length A right triangle is formed in the first quadrant by the x- and y-axes and a line through the point $(3, 2)$ (see figure). Write the length L of the hypotenuse as a function of x.

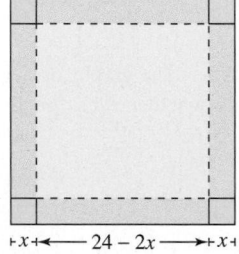

102. Volume An open box of maximum volume is to be made from a square piece of material 24 centimeters on a side by cutting equal squares from the corners and turning up the sides (see figure).

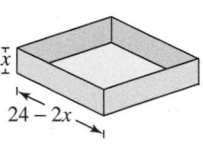

$24 - 2x$

x

x $24 - 2x$ x

(a) Write the volume V as a function of x, the length of the corner squares. What is the domain of the function?

(b) Use a graphing utility to graph the volume function and approximate the dimensions of the box that yield a maximum volume.

True or False? In Exercises 103–108, determine whether the statement is true or false. If it is false, explain why or give an example that shows it is false.

103. If $f(a) = f(b)$, then $a = b$.

104. A vertical line can intersect the graph of a function at most once.

105. If $f(x) = f(-x)$ for all x in the domain of f, then the graph of f is symmetric with respect to the y-axis.

106. If f is a function, then $f(ax) = af(x)$.

107. The graph of a function of x cannot have symmetry with respect to the x-axis.

108. If the domain of a function consists of a single number, then its range must also consist of only one number.

PUTNAM EXAM CHALLENGE

109. Let R be the region consisting of the points (x, y) of the Cartesian plane satisfying both $|x| - |y| \le 1$ and $|y| \le 1$. Sketch the region R and find its area.

110. Consider a polynomial $f(x)$ with real coefficients having the property $f(g(x)) = g(f(x))$ for every polynomial $g(x)$ with real coefficients. Determine and prove the nature of $f(x)$.

These problems were composed by the Committee on the Putnam Prize Competition. © The Mathematical Association of America. All rights reserved.

1.4 Review of Trigonometric Functions

■ Describe angles and use degree measure.
■ Use radian measure.
■ Understand the definitions of the six trigonometric functions.
■ Evaluate trigonometric functions.
■ Solve trigonometric equations.
■ Graph trigonometric functions.

Angles and Degree Measure

An **angle** has three parts: an **initial ray** (or side), a **terminal ray,** and a **vertex** (the point of intersection of the two rays), as shown in Figure 1.32(a). An angle is in **standard position** when its initial ray coincides with the positive *x*-axis and its vertex is at the origin, as shown in Figure 1.32(b).

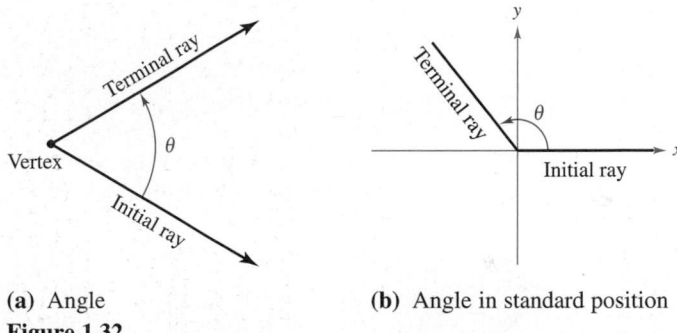

(a) Angle   **(b)** Angle in standard position

Figure 1.32

It is assumed that you are familiar with the degree measure of an angle.* It is common practice to use θ (the lowercase Greek letter theta) to represent both an angle and its measure. Angles between $0°$ and $90°$ are **acute,** and angles between $90°$ and $180°$ are **obtuse.**

Positive angles are measured *counterclockwise*, and negative angles are measured *clockwise*. For instance, Figure 1.33 shows an angle whose measure is $-45°$. You cannot assign a measure to an angle by simply knowing where its initial and terminal rays are located. To measure an angle, you must also know how the terminal ray was revolved. For example, Figure 1.33 shows that the angle measuring $-45°$ has the same terminal ray as the angle measuring $315°$. Such angles are **coterminal.** In general, if θ is any angle, then $\theta + n(360)$, n is a nonzero integer, is coterminal with θ.

An angle that is larger than $360°$ is one whose terminal ray has been revolved more than one full revolution counterclockwise, as shown in Figure 1.34(a). You can form an angle whose measure is less than $-360°$ by revolving a terminal ray more than one full revolution clockwise, as shown in Figure 1.34(b).

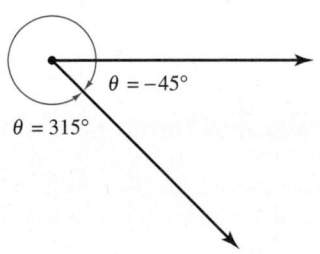

Coterminal angles
Figure 1.33

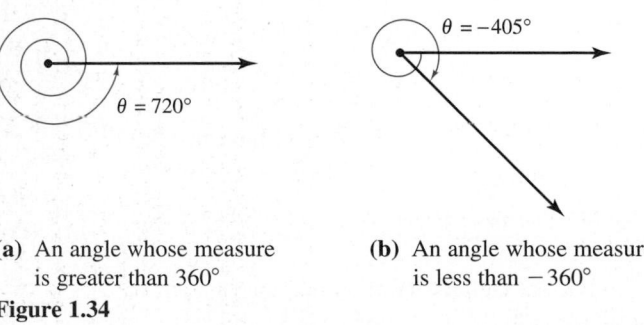

(a) An angle whose measure is greater than $360°$ **(b)** An angle whose measure is less than $-360°$

Figure 1.34

---

*For a more complete review of trigonometry, see *Precalculus,* 10th edition, or *Precalculus: Real Mathematics, Real People,* 7th edition, both by Ron Larson (Boston, Massachusetts: Brooks/Cole, Cengage Learning, 2018 and 2016, respectively).

Radian Measure

To assign a radian measure to an angle θ, consider θ to be a central angle of a circle of radius 1, as shown in Figure 1.35. The **radian measure** of θ is then defined to be the length of the arc of the sector. Because the circumference of a circle is $2\pi r$, the circumference of a **unit circle** (of radius 1) is 2π. This implies that the radian measure of an angle measuring 360° is 2π. In other words, 360° = 2π radians.

Using radian measure for θ, the length s of a circular arc of radius r is $s = r\theta$, as shown in Figure 1.36.

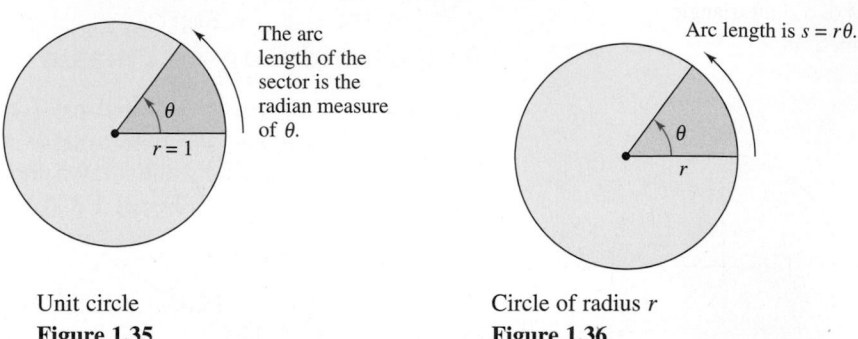

The arc length of the sector is the radian measure of θ.

Arc length is $s = r\theta$.

Unit circle
Figure 1.35

Circle of radius r
Figure 1.36

You should know the conversions of the common angles shown in Figure 1.37. For other angles, use the fact that 180° is equal to π radians.

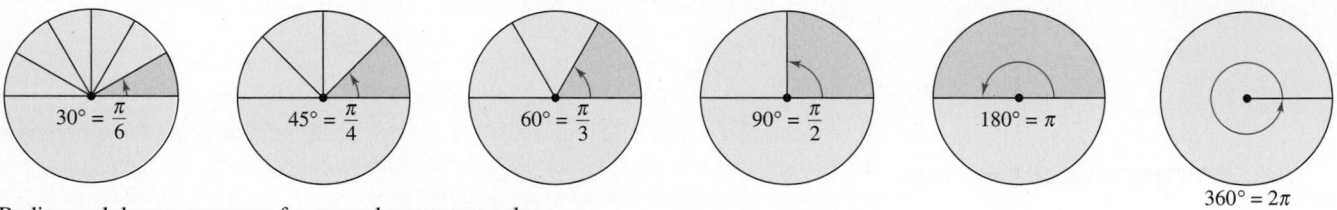

$30° = \dfrac{\pi}{6}$ $45° = \dfrac{\pi}{4}$ $60° = \dfrac{\pi}{3}$ $90° = \dfrac{\pi}{2}$ $180° = \pi$ $360° = 2\pi$

Radian and degree measures for several common angles
Figure 1.37

EXAMPLE 1   **Conversions Between Degrees and Radians**

a. $40° = (40 \text{ deg})\left(\dfrac{\pi \text{ rad}}{180 \text{ deg}}\right) = \dfrac{2\pi}{9}$ radian

b. $540° = (540 \text{ deg})\left(\dfrac{\pi \text{ rad}}{180 \text{ deg}}\right) = 3\pi$ radians

c. $-270° = (-270 \text{ deg})\left(\dfrac{\pi \text{ rad}}{180 \text{ deg}}\right) = -\dfrac{3\pi}{2}$ radians

d. $-\dfrac{\pi}{2}$ radians $= \left(-\dfrac{\pi}{2} \text{ rad}\right)\left(\dfrac{180 \text{ deg}}{\pi \text{ rad}}\right) = -90°$

e. 2 radians $= (2 \text{ rad})\left(\dfrac{180 \text{ deg}}{\pi \text{ rad}}\right) = \left(\dfrac{360}{\pi}\right)° \approx 114.59°$

f. $\dfrac{9\pi}{2}$ radians $= \left(\dfrac{9\pi}{2} \text{ rad}\right)\left(\dfrac{180 \text{ deg}}{\pi \text{ rad}}\right) = 810°$

▷ **TECHNOLOGY**   Most graphing utilities have both *degree* and *radian* modes. You should learn how to use your graphing utility to convert from degrees to radians, and vice versa. Use a graphing utility to verify the results of Example 1.

The Trigonometric Functions

There are two common approaches to the study of trigonometry. In one, the trigonometric functions are defined as ratios of two sides of a right triangle. In the other, these functions are defined in terms of a point on the terminal ray of an angle in standard position. The six trigonometric functions, **sine, cosine, tangent, cotangent, secant,** and **cosecant** (abbreviated as sin, cos, tan, cot, sec, and csc, respectively), are defined below from both viewpoints.

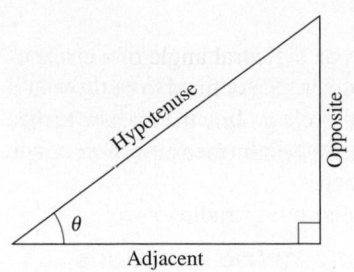

Sides of a right triangle
Figure 1.38

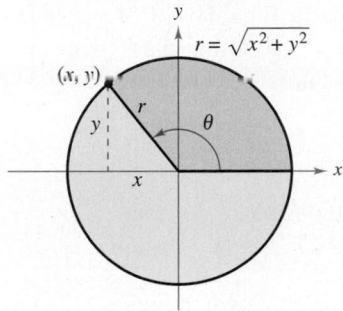

An angle in standard position
Figure 1.39

Definition of the Six Trigonometric Functions

Right triangle definitions, where $0 < \theta < \dfrac{\pi}{2}$ (see Figure 1.38)

$$\sin \theta = \frac{\text{opposite}}{\text{hypotenuse}} \qquad \cos \theta = \frac{\text{adjacent}}{\text{hypotenuse}} \qquad \tan \theta = \frac{\text{opposite}}{\text{adjacent}}$$

$$\csc \theta = \frac{\text{hypotenuse}}{\text{opposite}} \qquad \sec \theta = \frac{\text{hypotenuse}}{\text{adjacent}} \qquad \cot \theta = \frac{\text{adjacent}}{\text{opposite}}$$

Circular function definitions, where θ *is any angle* (see Figure 1.39)

$$\sin \theta = \frac{y}{r} \qquad \cos \theta = \frac{x}{r} \qquad \tan \theta = \frac{y}{x}, \; x \neq 0$$

$$\csc \theta = \frac{r}{y}, \; y \neq 0 \qquad \sec \theta = \frac{r}{x}, \; x \neq 0 \qquad \cot \theta = \frac{x}{y}, \; y \neq 0$$

The trigonometric identities listed below are direct consequences of the definitions. [Note that ϕ is the lowercase Greek letter phi and $\sin^2 \theta$ is used to represent $(\sin \theta)^2$.]

TRIGONOMETRIC IDENTITIES

Pythagorean Identities

$\sin^2 \theta + \cos^2 \theta = 1$

$1 + \tan^2 \theta = \sec^2 \theta$

$1 + \cot^2 \theta = \csc^2 \theta$

Even/Odd Identities

$\sin(-\theta) = -\sin \theta$

$\cos(-\theta) = \cos \theta$

$\tan(-\theta) = -\tan \theta$

$\csc(-\theta) = -\csc \theta$

$\sec(-\theta) = \sec \theta$

$\cot(-\theta) = -\cot \theta$

Sum and Difference Formulas

$\sin(\theta \pm \phi) = \sin \theta \cos \phi \pm \cos \theta \sin \phi$

$\cos(\theta \pm \phi) = \cos \theta \cos \phi \mp \sin \theta \sin \phi$

$\tan(\theta \pm \phi) = \dfrac{\tan \theta \pm \tan \phi}{1 \mp \tan \theta \tan \phi}$

Power-Reducing Formulas

$\sin^2 \theta = \dfrac{1 - \cos 2\theta}{2}$

$\cos^2 \theta = \dfrac{1 + \cos 2\theta}{2}$

$\tan^2 \theta = \dfrac{1 - \cos 2\theta}{1 + \cos 2\theta}$

Double-Angle Formulas

$\sin 2\theta = 2 \sin \theta \cos \theta$

$\begin{aligned} \cos 2\theta &= 2 \cos^2 \theta - 1 \\ &= 1 - 2 \sin^2 \theta \\ &= \cos^2 \theta - \sin^2 \theta \end{aligned}$

$\tan 2\theta = \dfrac{2 \tan \theta}{1 - \tan^2 \theta}$

Law of Cosines

$a^2 = b^2 + c^2 - 2bc \cos A$

Reciprocal Identities

$\csc \theta = \dfrac{1}{\sin \theta}$

$\sec \theta = \dfrac{1}{\cos \theta}$

$\cot \theta = \dfrac{1}{\tan \theta}$

Quotient Identities

$\tan \theta = \dfrac{\sin \theta}{\cos \theta}$

$\cot \theta = \dfrac{\cos \theta}{\sin \theta}$

Evaluating Trigonometric Functions

There are two ways to evaluate trigonometric functions: (1) decimal approximations with a graphing utility and (2) exact evaluations using trigonometric identities and formulas from geometry. When using a graphing utility to evaluate a trigonometric function, remember to set the graphing utility to the appropriate mode—*degree* mode or *radian* mode.

EXAMPLE 2 **Exact Evaluation of Trigonometric Functions**

Evaluate the sine, cosine, and tangent of $\pi/3$.

Solution Because $60° = \pi/3$ radians, you can draw an equilateral triangle with sides of length 1 and θ as one of its angles, as shown in Figure 1.40. Because the altitude of this triangle bisects its base, you know that $x = \frac{1}{2}$. Using the Pythagorean Theorem, you obtain

$$y = \sqrt{r^2 - x^2} = \sqrt{1 - \left(\frac{1}{2}\right)^2} = \sqrt{\frac{3}{4}} = \frac{\sqrt{3}}{2}.$$

Now, knowing the values of x, y, and r, you can write the following.

$$\sin\frac{\pi}{3} = \frac{y}{r} = \frac{\sqrt{3}/2}{1} = \frac{\sqrt{3}}{2}$$

$$\cos\frac{\pi}{3} = \frac{x}{r} = \frac{1/2}{1} = \frac{1}{2}$$

$$\tan\frac{\pi}{3} = \frac{y}{x} = \frac{\sqrt{3}/2}{1/2} = \sqrt{3}$$

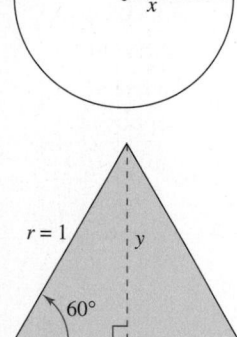

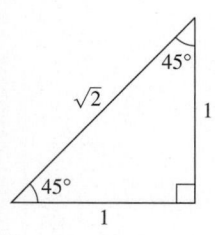

Figure 1.40

Note that all angles in this text are measured in radians unless stated otherwise. For example, when $\sin 3$ is written, the sine of 3 radians is meant, and when $\sin 3°$ is written, the sine of 3 degrees is meant.

The degree and radian measures of several common angles are shown in the table below, along with the corresponding values of the sine, cosine, and tangent (see Figure 1.41).

Trigonometric Values of Common Angles

θ (degrees)	0°	30°	45°	60°	90°	180°	270°
θ (radians)	0	$\frac{\pi}{6}$	$\frac{\pi}{4}$	$\frac{\pi}{3}$	$\frac{\pi}{2}$	π	$\frac{3\pi}{2}$
$\sin\theta$	0	$\frac{1}{2}$	$\frac{\sqrt{2}}{2}$	$\frac{\sqrt{3}}{2}$	1	0	-1
$\cos\theta$	1	$\frac{\sqrt{3}}{2}$	$\frac{\sqrt{2}}{2}$	$\frac{1}{2}$	0	-1	0
$\tan\theta$	0	$\frac{\sqrt{3}}{3}$	1	$\sqrt{3}$	Undefined	0	Undefined

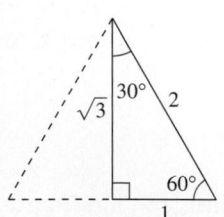

Common angles
Figure 1.41

EXAMPLE 3 **Using Trigonometric Identities**

a. $\sin\left(-\frac{\pi}{3}\right) = -\sin\frac{\pi}{3} = -\frac{\sqrt{3}}{2}$ $\quad\quad \sin(-\theta) = -\sin\theta$

b. $\sec 60° = \frac{1}{\cos 60°} = \frac{1}{1/2} = 2$ $\quad\quad \sec\theta = \frac{1}{\cos\theta}$

The quadrant signs for the sine, cosine, and tangent functions are shown in Figure 1.42. To extend the use of the table on the preceding page to angles in quadrants other than the first quadrant, you can use the concept of a **reference angle** (see Figure 1.43), with the appropriate quadrant sign. For instance, the reference angle for $3\pi/4$ is $\pi/4$, and because the sine is positive in Quadrant II, you can write

$$\sin \frac{3\pi}{4} = \sin \frac{\pi}{4} = \frac{\sqrt{2}}{2}.$$

Similarly, because the reference angle for $330°$ is $30°$, and the tangent is negative in Quadrant IV, you can write

$$\tan 330° = -\tan 30° = -\frac{\sqrt{3}}{3}.$$

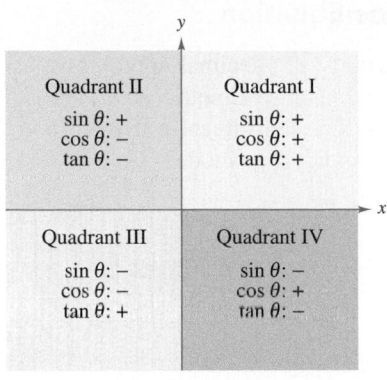

Quadrant signs for trigonometric functions

Figure 1.42

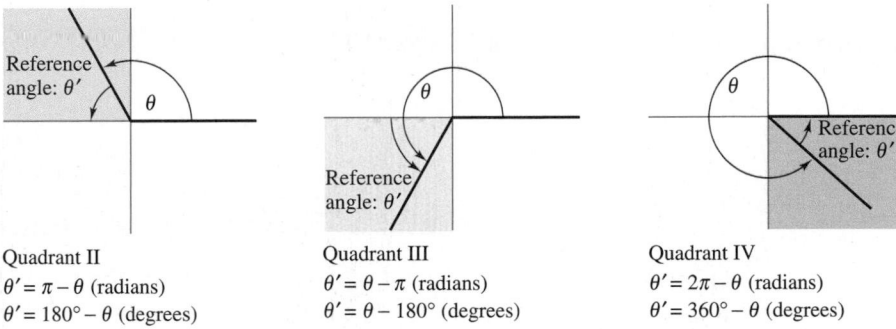

Quadrant II
$\theta' = \pi - \theta$ (radians)
$\theta' = 180° - \theta$ (degrees)

Quadrant III
$\theta' = \theta - \pi$ (radians)
$\theta' = \theta - 180°$ (degrees)

Quadrant IV
$\theta' = 2\pi - \theta$ (radians)
$\theta' = 360° - \theta$ (degrees)

Figure 1.43

Solving Trigonometric Equations

How would you solve the equation $\sin \theta = 0$? You know that $\theta = 0$ is one solution, but this is not the only solution. Any one of the following values of θ is also a solution.

$$\ldots, -3\pi, -2\pi, -\pi, 0, \pi, 2\pi, 3\pi, \ldots$$

You can write this infinite solution set as $\{n\pi : n \text{ is an integer}\}$.

EXAMPLE 4   Solving a Trigonometric Equation

Solve the equation $\sin \theta = -\dfrac{\sqrt{3}}{2}$.

Solution To solve the equation, you should consider that the sine function is negative in Quadrants III and IV and that

$$\sin \frac{\pi}{3} = \frac{\sqrt{3}}{2}.$$

So, you are seeking values of θ in the third and fourth quadrants that have a reference angle of $\pi/3$. In the interval $[0, 2\pi]$, the two angles fitting these criteria are

$$\theta = \pi + \frac{\pi}{3} = \frac{4\pi}{3} \quad \text{and} \quad \theta = 2\pi - \frac{\pi}{3} = \frac{5\pi}{3}.$$

By adding integer multiples of 2π to each of these solutions, you obtain the following general solution.

$$\theta = \frac{4\pi}{3} + 2n\pi \quad \text{or} \quad \theta = \frac{5\pi}{3} + 2n\pi, \quad \text{where } n \text{ is an integer.}$$

See Figure 1.44.

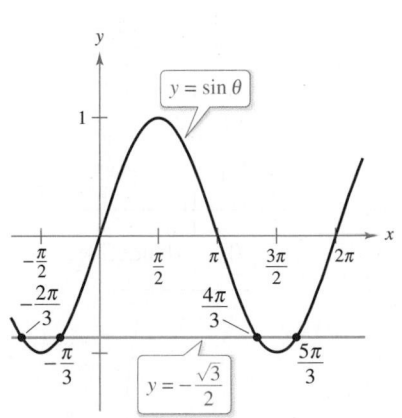

Solution points of $\sin \theta = -\dfrac{\sqrt{3}}{2}$

Figure 1.44

| EXAMPLE 5 | **Solving a Trigonometric Equation** |

Solve

$$\cos 2\theta = 2 - 3 \sin \theta$$

where $0 \le \theta \le 2\pi$.

Solution Using the double-angle formula $\cos 2\theta = 1 - 2 \sin^2 \theta$, you can rewrite the equation as follows.

$\cos 2\theta = 2 - 3 \sin \theta$	Write original equation.
$1 - 2 \sin^2 \theta = 2 - 3 \sin \theta$	Double-angle formula
$0 = 2 \sin^2 \theta - 3 \sin \theta + 1$	Quadratic form
$0 = (2 \sin \theta - 1)(\sin \theta - 1)$	Factor.

If $2 \sin \theta - 1 = 0$, then $\sin \theta = 1/2$ and $\theta = \pi/6$ or $\theta = 5\pi/6$. If $\sin \theta - 1 = 0$, then $\sin \theta = 1$ and $\theta = \pi/2$. So, for $0 \le \theta \le 2\pi$, the solutions are

$$\theta = \frac{\pi}{6}, \quad \frac{5\pi}{6}, \quad \text{or} \quad \frac{\pi}{2}.$$

Graphs of Trigonometric Functions

A function f is **periodic** when there exists a positive real number p such that $f(x + p) = f(x)$ for all x in the domain of f. The least such positive value of p is the **period** of f. The sine, cosine, secant, and cosecant functions each have a period of 2π, and the other two trigonometric functions, tangent and cotangent, have a period of π, as shown in Figure 1.45.

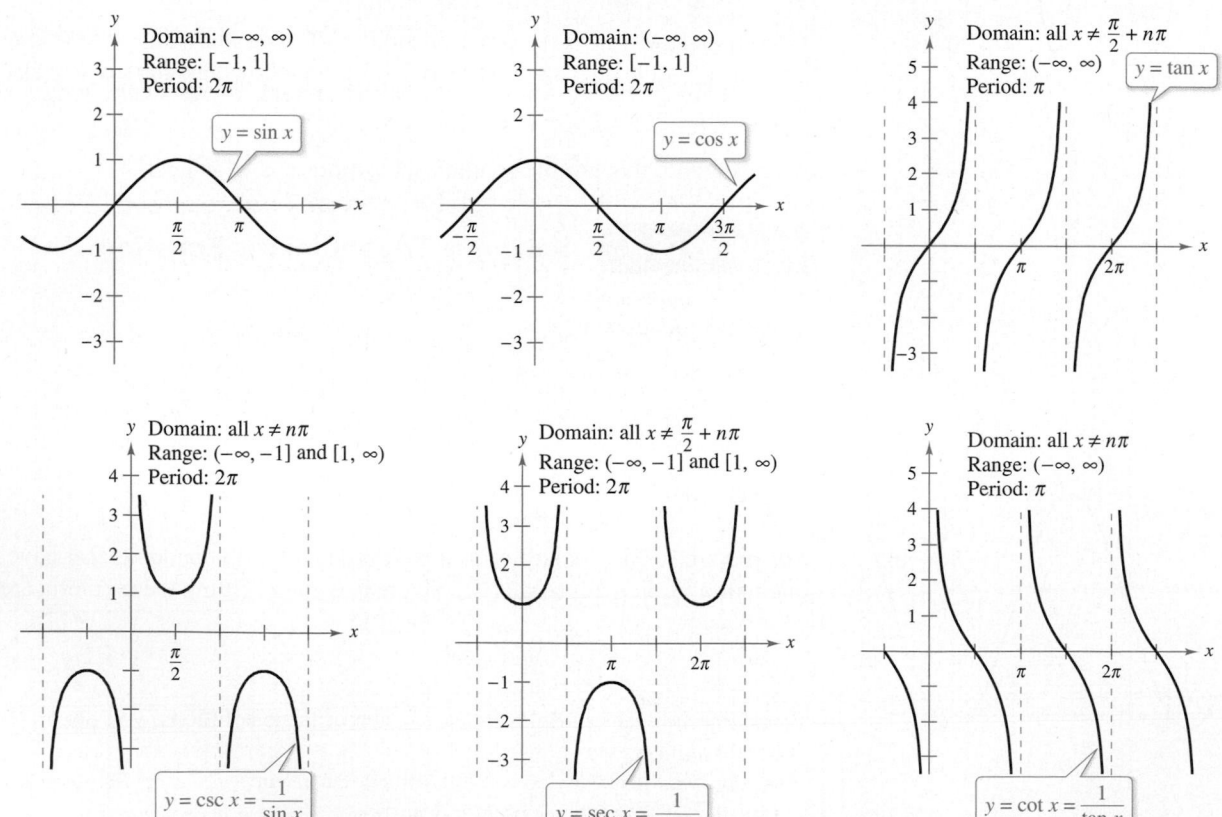

The graphs of the six trigonometric functions
Figure 1.45

▷ **TECHNOLOGY** To produce the graphs shown in Figure 1.45 with a graphing utility, make sure you set the graphing utility to *radian* mode.

Note in Figure 1.45 that the maximum value of $\sin x$ and $\cos x$ is 1 and the minimum value is -1. The graphs of the functions $y = a \sin bx$ and $y = a \cos bx$ oscillate between $-a$ and a, and so have an **amplitude** of $|a|$. Furthermore, because $bx = 0$ when $x = 0$ and $bx = 2\pi$ when $x = 2\pi/b$, it follows that the functions $y = a \sin bx$ and $y = a \cos bx$ each have a period of $2\pi/|b|$. The table below summarizes the amplitudes and periods of some types of trigonometric functions.

Function	Period	Amplitude
$y = a \sin bx$ or $y = a \cos bx$	$\dfrac{2\pi}{|b|}$	$|a|$
$y = a \tan bx$ or $y = a \cot bx$	$\dfrac{\pi}{|b|}$	Not applicable
$y = a \sec bx$ or $y = a \csc bx$	$\dfrac{2\pi}{|b|}$	Not applicable

EXAMPLE 6   **Sketching the Graph of a Trigonometric Function**

Sketch the graph of $f(x) = 3 \cos 2x$.

Solution   The graph of $f(x) = 3 \cos 2x$ has an amplitude of 3 and a period of $2\pi/2 = \pi$. Using the basic shape of the graph of the cosine function, sketch one period of the function on the interval $[0, \pi]$, using the following pattern.

Maximum: $(0, 3)$

Minimum: $\left(\dfrac{\pi}{2}, -3\right)$

Maximum: $(\pi, 3)$

By continuing this pattern, you can sketch several cycles of the graph, as shown in Figure 1.46.

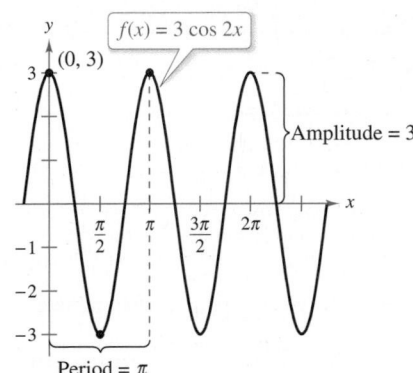

Figure 1.46

EXAMPLE 7   **Shifts of Graphs of Trigonometric Functions**

a. To sketch the graph of $f(x) = \sin(x + \pi/2)$, shift the graph of $y = \sin x$ to the left $\pi/2$ units, as shown in Figure 1.47(a).

b. To sketch the graph of $f(x) = 2 + \sin x$, shift the graph of $y = \sin x$ upward two units, as shown in Figure 1.47(b).

c. To sketch the graph of $f(x) = 2 + \sin(x - \pi/4)$, shift the graph of $y = \sin x$ upward two units and to the right $\pi/4$ units, as shown in Figure 1.47(c).

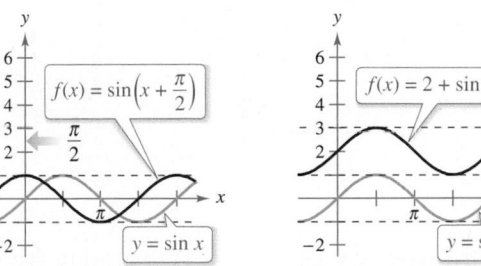

(a) Horizontal shift to the left   (b) Vertical shift upward   (c) Horizontal and vertical shifts

Transformations of the graph of $y = \sin x$

Figure 1.47

1.4 Exercises
See **CalcChat.com** for tutorial help and worked-out solutions to odd-numbered exercises.

CONCEPT CHECK

1. **Coterminal Angles** Explain how to find coterminal angles in degrees.

2. **Degrees to Radians** Explain how to convert from degrees to radians.

3. **Trigonometric Functions** Find $\sin\theta$, $\cos\theta$, and $\tan\theta$.

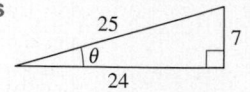

4. **Characteristics of a Graph** In your own words, describe the meaning of *amplitude* and *period*.

 **Coterminal Angles in Degrees** In Exercises 5 and 6, determine two coterminal angles in degree measure (one positive and one negative) for each angle.

5. (a) 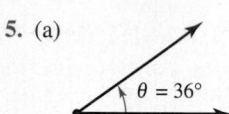 (b)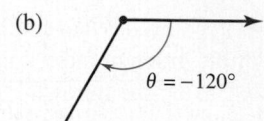

6. (a) 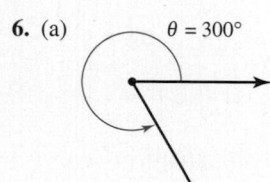 (b)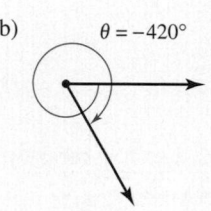

 **Coterminal Angles in Radians** In Exercises 7 and 8, determine two coterminal angles in radian measure (one positive and one negative) for each angle.

7. (a) 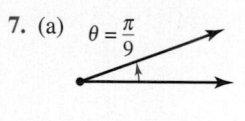 (b)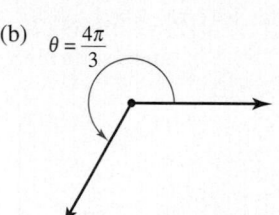

8. (a) 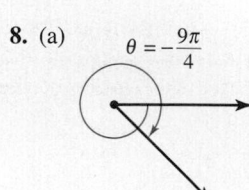 (b)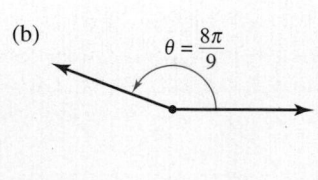

 **Degrees to Radians** In Exercises 9 and 10, convert the degree measure to radian measure as a multiple of π and as a decimal accurate to three decimal places.

9. (a) 30°   (b) 150°   (c) 315°   (d) 120°

10. (a) −20°   (b) −240°   (c) −270°   (d) 144°

 **Radians to Degrees** In Exercises 11 and 12, convert the radian measure to degree measure.

11. (a) $\dfrac{3\pi}{2}$ (b) $\dfrac{7\pi}{6}$ (c) $-\dfrac{7\pi}{12}$ (d) −2.367

12. (a) $\dfrac{7\pi}{3}$ (b) $-\dfrac{11\pi}{30}$ (c) $\dfrac{11\pi}{6}$ (d) 0.438

13. **Completing a Table** Let r represent the radius of a circle, θ the central angle (measured in radians), and s the length of the arc subtended by the angle. Use the relationship $s = r\theta$ to complete the table.

r	8 ft	15 in.	85 cm		
s	12 ft			96 in.	8642 mi
θ		1.6	$\dfrac{3\pi}{4}$	4	$\dfrac{2\pi}{3}$

14. **Angular Speed** A car is moving at the rate of 50 miles per hour, and the diameter of its wheels is 2.5 feet.

   (a) Find the number of revolutions per minute that the wheels are rotating.

   (b) Find the angular speed of the wheels in radians per minute.

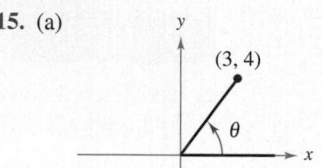 **Evaluating Trigonometric Functions** In Exercises 15 and 16, evaluate the six trigonometric functions of the angle θ.

15. (a) 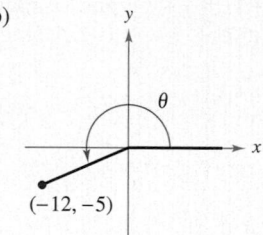 (b)

16. (a) 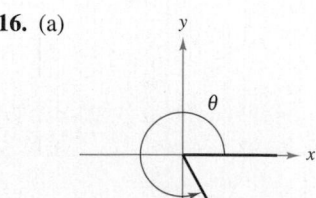 (b)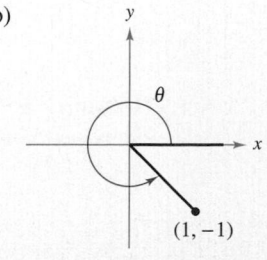

Evaluating Trigonometric Functions In Exercises 17–20, sketch a right triangle corresponding to the trigonometric function of the acute angle θ. Then evaluate the other five trigonometric functions of θ.

17. $\sin\theta = \dfrac{1}{2}$ 18. $\sin\theta = \dfrac{1}{3}$

19. $\cos\theta = \dfrac{4}{5}$ 20. $\sec\theta = \dfrac{13}{5}$

Evaluating Trigonometric Functions In Exercises 21–24, evaluate the sine, cosine, and tangent of each angle. Do not use a calculator.

21. (a) $60°$ (b) $120°$ (c) $\dfrac{\pi}{4}$ (d) $\dfrac{5\pi}{4}$

22. (a) $-30°$ (b) $150°$ (c) $-\dfrac{\pi}{6}$ (d) $\dfrac{\pi}{2}$

23. (a) $225°$ (b) $-225°$ (c) $\dfrac{5\pi}{3}$ (d) $\dfrac{11\pi}{6}$

24. (a) $750°$ (b) $510°$ (c) $\dfrac{10\pi}{3}$ (d) $\dfrac{17\pi}{3}$

Evaluating Trigonometric Functions Using Technology In Exercises 25–28, use a calculator to evaluate each trigonometric function. Round your answers to four decimal places.

25. (a) $\sin 10°$
    (b) $\csc 10°$

26. (a) $\sec 225°$
    (b) $\sec 135°$

27. (a) $\tan \dfrac{\pi}{9}$
    (b) $\tan \dfrac{10\pi}{9}$

28. (a) $\cot(1.35)$
    (b) $\tan(1.35)$

Determining a Quadrant In Exercises 29 and 30, determine the quadrant in which θ lies.

29. (a) $\sin\theta < 0$ and $\cos\theta < 0$
    (b) $\sec\theta > 0$ and $\cot\theta < 0$

30. (a) $\sin\theta > 0$ and $\cos\theta < 0$
    (b) $\csc\theta < 0$ and $\tan\theta > 0$

Solving a Trigonometric Equation In Exercises 31–34, find two solutions of each equation. Give your answers in radians $(0 \le \theta \le 2\pi)$. Do not use a calculator.

31. (a) $\cos\theta = \dfrac{\sqrt{2}}{2}$
    (b) $\cos\theta = -\dfrac{\sqrt{2}}{2}$

32. (a) $\sec\theta = 2$
    (b) $\sec\theta = -2$

33. (a) $\tan\theta = 1$
    (b) $\cot\theta = -\sqrt{3}$

34. (a) $\sin\theta = \dfrac{\sqrt{3}}{2}$
    (b) $\sin\theta = -\dfrac{\sqrt{3}}{2}$

Solving a Trigonometric Equation In Exercises 35–42, solve the equation for θ, where $0 \le \theta \le 2\pi$.

35. $2\sin^2\theta = 1$

36. $\tan^2\theta = 3$

37. $\tan^2\theta - \tan\theta = 0$

38. $2\cos^2\theta - \cos\theta = 1$

39. $\sec\theta\csc\theta = 2\csc\theta$

40. $\sin\theta = \cos\theta$

41. $\cos^2\theta + \sin\theta = 1$

42. $\cos\dfrac{\theta}{2} - \cos\theta = 1$

43. Airplane Ascent An airplane leaves the runway climbing at an angle of $18°$ with a speed of 275 feet per second (see figure). Find the altitude a of the plane after 1 minute.

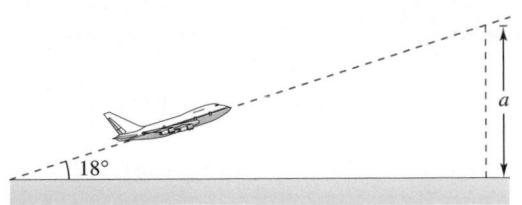

44. Height of a Mountain While traveling across flat land, you notice a mountain directly in front of you. Its angle of elevation (to the peak) is $3.5°$. After you drive 13 miles closer to the mountain, the angle of elevation is $9°$. Approximate the height of the mountain.

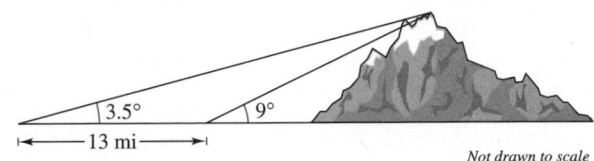

Not drawn to scale

Period and Amplitude In Exercises 45–48, determine the period and amplitude of each function.

45. $y = 2\sin 2x$

46. $y = \dfrac{3}{2}\cos\dfrac{x}{2}$

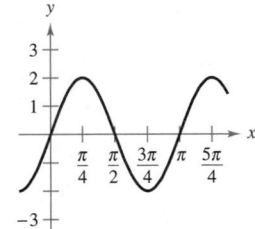
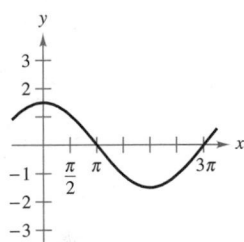

47. $y = -3\sin 4\pi x$

48. $y = \dfrac{2}{3}\cos\dfrac{\pi x}{10}$

Period In Exercises 49–52, find the period of the function.

49. $y = 5\tan 2x$

50. $y = 7\tan 2\pi x$

51. $y = \sec 5x$

52. $y = \csc 4x$

Writing In Exercises 53 and 54, use a graphing utility to graph each function f in the same viewing window for $c = -2$, $c = -1$, $c = 1$, and $c = 2$. Give a written description of the change in the graph caused by changing c.

53. (a) $f(x) = c\sin x$
    (b) $f(x) = \cos(cx)$
    (c) $f(x) = \cos(\pi x - c)$

54. (a) $f(x) = \sin x + c$
    (b) $f(x) = -\sin(2\pi x - c)$
    (c) $f(x) = c\cos x$

Sketching the Graph of a Trigonometric Function In Exercises 55–66, sketch the graph of the function.

55. $y = \sin \dfrac{x}{2}$

56. $y = 2 \cos 2x$

57. $y = -\sin \dfrac{2\pi x}{3}$

58. $y = 2 \tan x$

59. $y = \csc \dfrac{x}{2}$

60. $y = \tan 2x$

61. $y = 2 \sec 2x$

62. $y = \csc 2\pi x$

63. $y = \sin(x + \pi)$

64. $y = \cos\left(x - \dfrac{\pi}{3}\right)$

65. $y = 1 + \cos\left(x - \dfrac{\pi}{2}\right)$

66. $y = 1 + \sin\left(x + \dfrac{\pi}{2}\right)$

Graphical Reasoning In Exercises 67 and 68, find a, b, and c such that the graph of the function matches the graph in the figure.

67. $y = a \cos(bx - c)$

68. $y = a \sin(bx - c)$

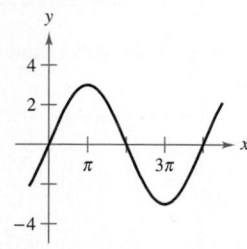

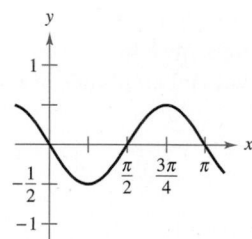

EXPLORING CONCEPTS

69. Think About It You are given the value of $\tan \theta$. Is it possible to find the value of $\sec \theta$ without finding the measure of θ? Explain.

70. Restricted Domain Explain how to restrict the domain of the sine function so that it becomes a one-to-one function.

71. Think About It How do the ranges of the cosine function and the secant function compare?

72. HOW DO YOU SEE IT? Consider an angle in standard position with $r = 12$ centimeters, as shown in the figure. Describe the changes in the values of x, y, $\sin \theta$, $\cos \theta$, and $\tan \theta$ as θ increases continually from $0°$ to $90°$.

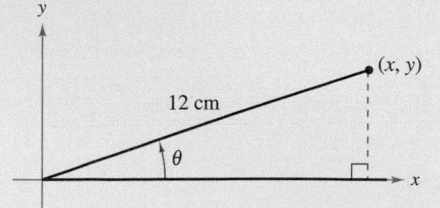

73. Think About It Sketch the graphs of

$$f(x) = \sin x, \quad g(x) = |\sin x|, \quad \text{and} \quad h(x) = \sin(|x|).$$

In general, how are the graphs of $|f(x)|$ and $f(|x|)$ related to the graph of f?

74. Ferris Wheel

The model for the height h of a Ferris wheel car is

$$h = 51 + 50 \sin 8\pi t$$

where t is measured in minutes. (The Ferris wheel has a radius of 50 feet.) This model yields a height of 51 feet when $t = 0$. Alter the model so that the height of the car is 1 foot when $t = 0$.

75. Sales The monthly sales S (in thousands of units) of a seasonal product are modeled by

$$S = 58.3 + 32.5 \cos \dfrac{\pi t}{6}$$

where t is the time (in months), with $t = 1$ corresponding to January. Use a graphing utility to graph the model for S and determine the months when sales exceed 75,000 units.

76. Pattern Recognition Use a graphing utility to compare the graph of

$$f(x) = \dfrac{4}{\pi}\left(\sin \pi x + \dfrac{1}{3} \sin 3\pi x\right)$$

with the given graph. Try to improve the approximation by adding a term to $f(x)$. Use a graphing utility to verify that your new approximation is better than the original. Can you find other terms to add to make the approximation even better? What is the pattern? (*Hint:* Use sine terms.)

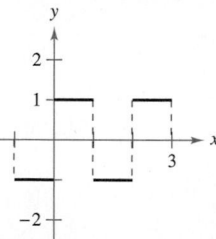

True or False? In Exercises 77–80, determine whether the statement is true or false. If it is false, explain why or give an example that shows it is false.

77. A measurement of 4 radians corresponds to two complete revolutions from the initial side to the terminal side of an angle.

78. Amplitude is always positive.

79. The function $y = \frac{1}{2} \sin 2x$ has an amplitude that is twice that of the function $y = \sin x$.

80. The function $y = 3 \cos(x/3)$ has a period that is three times that of the function $y = \cos x$.

1.5 Inverse Functions

■ Verify that one function is the inverse function of another function.
■ Determine whether a function has an inverse function.
■ Develop properties of the six inverse trigonometric functions.

Inverse Functions

Recall from Section 1.3 that a function can be represented by a set of ordered pairs. For instance, the function $f(x) = x + 3$ from $A = \{1, 2, 3, 4\}$ to $B = \{4, 5, 6, 7\}$ can be written as

$$f: \{(1, 4), (2, 5), (3, 6), (4, 7)\}.$$

By interchanging the first and second coordinates of each ordered pair, you can form the **inverse function** of f. This function is denoted by f^{-1}, which is read as "f inverse." It is a function from B to A, and can be written as $f^{-1}: \{(4, 1), (5, 2), (6, 3), (7, 4)\}$. Note that the domain of f is equal to the range of f^{-1}, and vice versa, as shown in Figure 1.48. The functions f and f^{-1} have the effect of "undoing" each other. That is, when you form the composition of f with f^{-1} or the composition of f^{-1} with f, you obtain the identity function.

$$f(f^{-1}(x)) = x \quad \text{and} \quad f^{-1}(f(x)) = x$$

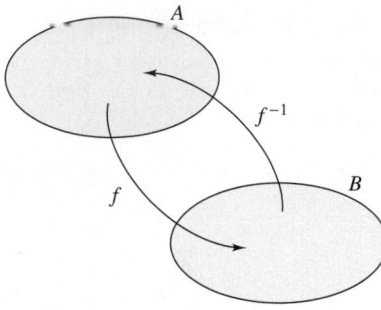

Domain of f = range of f^{-1}
Domain of f^{-1} = range of f
Figure 1.48

> **•• REMARK** Although the notation used to denote an inverse function resembles *exponential notation*, it is a different use of -1 as a superscript. That is, in general,
>
> $$f^{-1}(x) \neq \frac{1}{f(x)}.$$

Exploration

Finding Inverse Functions
Explain how to "undo" each of the functions below. Then use your explanation to write the inverse function of f.

a. $f(x) = x - 5$

b. $f(x) = 6x$

c. $f(x) = \dfrac{x}{2}$

d. $f(x) = 3x + 2$

e. $f(x) = x^3$

f. $f(x) = 4(x - 2)$

Use a graphing utility to graph each function and its inverse function in the same "square" viewing window. What observation can you make about each pair of graphs?

Definition of Inverse Function

A function g is the **inverse function** of the function f when

$$f(g(x)) = x \text{ for each } x \text{ in the domain of } g$$

and

$$g(f(x)) = x \text{ for each } x \text{ in the domain of } f.$$

The function g is denoted by f^{-1}.

Here are some important observations about inverse functions.

1. If g is the inverse function of f, then f is the inverse function of g.

2. The domain of f^{-1} is equal to the range of f, and the range of f^{-1} is equal to the domain of f.

3. A function need not have an inverse function, but when it does, the inverse function is unique (see Exercise 148).

You can think of f^{-1} as undoing what has been done by f. For example, subtraction can be used to undo addition, and division can be used to undo multiplication. So,

$$f(x) = x + c \quad \text{and} \quad f^{-1}(x) = x - c \qquad \text{Subtraction can be used to undo addition.}$$

are inverse functions of each other and

$$f(x) = cx \quad \text{and} \quad f^{-1}(x) = \frac{x}{c}, \quad c \neq 0 \qquad \text{Division can be used to undo multiplication.}$$

are inverse functions of each other.

EXAMPLE 1   Verifying Inverse Functions

Show that the functions are inverse functions of each other.

$$f(x) = 2x^3 - 1 \quad \text{and} \quad g(x) = \sqrt[3]{\frac{x+1}{2}}$$

Solution   Because the domains and ranges of both f and g consist of all real numbers, you can conclude that both composite functions exist for all x. The composition of f with g is

$$f(g(x)) = 2\left(\sqrt[3]{\frac{x+1}{2}}\right)^3 - 1$$

$$= 2\left(\frac{x+1}{2}\right) - 1$$

$$= x + 1 - 1$$

$$= x.$$

The composition of g with f is

$$g(f(x)) = \sqrt[3]{\frac{(2x^3-1)+1}{2}} = \sqrt[3]{\frac{2x^3}{2}} = \sqrt[3]{x^3} = x.$$

Because $f(g(x)) = x$ and $g(f(x)) = x$, you can conclude that f and g are inverse functions of each other (see Figure 1.49).

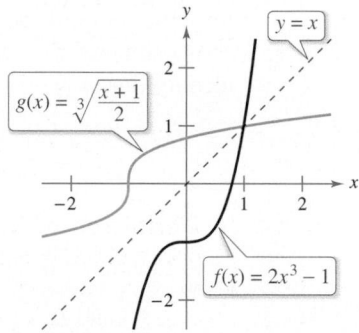

f and *g* are inverse functions of each other.
Figure 1.49

> **REMARK** In Example 1, try comparing the functions f and g verbally.
>
> For f: First cube x, then multiply by 2, then subtract 1.
>
> For g: First add 1, then divide by 2, then take the cube root.
>
> Do you see the "undoing pattern?"

In Figure 1.49, the graphs of f and $g = f^{-1}$ appear to be mirror images of each other with respect to the line $y = x$. The graph of f^{-1} is a **reflection** of the graph of f in the line $y = x$. This idea is generalized in the next definition.

Reflective Property of Inverse Functions

The graph of f contains the point (a, b) if and only if the graph of f^{-1} contains the point (b, a).

To see the validity of the Reflective Property of Inverse Functions, consider the point (a, b) on the graph of f. This implies $f(a) = b$ and you can write

$$f^{-1}(b) = f^{-1}(f(a)) = a.$$

So, (b, a) is on the graph of f^{-1}, as shown in Figure 1.50. A similar argument will verify this result in the other direction.

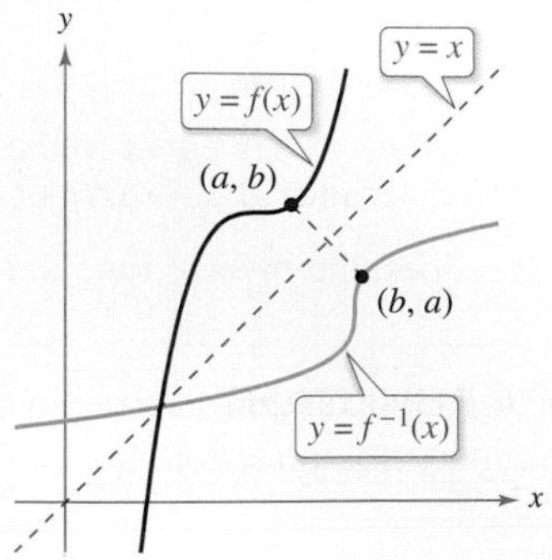

The graph of f^{-1} is a reflection of the graph of f in the line $y = x$.
Figure 1.50

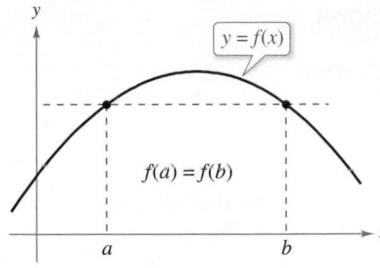

If a horizontal line intersects the graph of f twice, then f is not one-to-one.
Figure 1.51

Existence of an Inverse Function

Not every function has an inverse function, and the Reflective Property of Inverse Functions suggests a graphical test for those that do—the **Horizontal Line Test** for an inverse function. This test states that a function f has an inverse function if and only if every horizontal line intersects the graph of f at most once (see Figure 1.51). The next definition formally states why the Horizontal Line Test is valid.

The Existence of an Inverse Function

A function has an inverse function if and only if it is one-to-one.

EXAMPLE 2 The Existence of an Inverse Function

Which of the functions has an inverse function?

a. $f(x) = x^3 - 1$

b. $f(x) = x^3 - x + 1$

Solution

a. From the graph of f shown in Figure 1.52(a), it appears that f is one-to-one over its entire domain. To verify this, suppose that there exist x_1 and x_2 such that $f(x_1) = f(x_2)$. By showing that $x_1 = x_2$, it follows that f is one-to-one.

$$f(x_1) = f(x_2)$$
$$x_1^3 - 1 = x_2^3 - 1$$
$$x_1^3 = x_2^3$$
$$\sqrt[3]{x_1^3} = \sqrt[3]{x_2^3}$$
$$x_1 = x_2$$

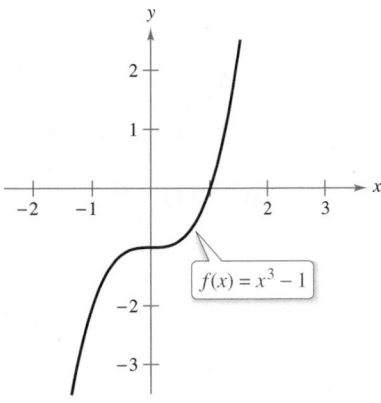

(a) Because f is one-to-one over its entire domain, it has an inverse function.

Because f is one-to-one, you can conclude that f must have an inverse function.

b. From the graph of f shown in Figure 1.52(b), you can see that the function does not pass the Horizontal Line Test. In other words, it is not one-to-one. For instance, f has the same value when $x = -1, 0$, and 1.

$$f(-1) = f(1) = f(0) = 1 \qquad \text{Not one-to-one}$$

Therefore, f does not have an inverse function. ■

Often it is easier to prove that a function has an inverse function than to find the inverse function. For instance, by sketching the graph of

$$f(x) = x^3 + x - 1$$

you can see that it is one-to-one. Yet it would be difficult to determine the inverse of this function algebraically.

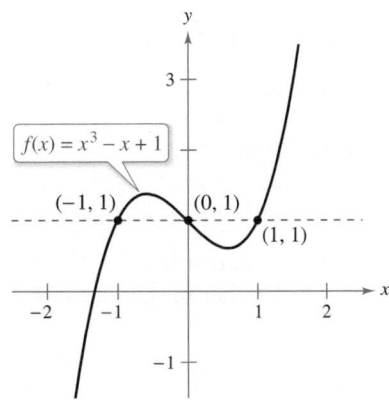

(b) Because f is not one-to-one, it does not have an inverse function.
Figure 1.52

GUIDELINES FOR FINDING AN INVERSE OF A FUNCTION

1. Determine whether the function given by $y = f(x)$ has an inverse function.
2. Solve for x as a function of y: $x = g(y) = f^{-1}(y)$.
3. Interchange x and y. The resulting equation is $y = f^{-1}(x)$.
4. Define the domain of f^{-1} as the range of f.
5. Verify that $f(f^{-1}(x)) = x$ and $f^{-1}(f(x)) = x$.

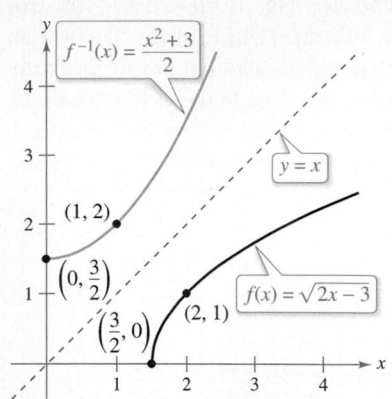

The domain of f^{-1}, $[0, \infty)$, is the range of f.
Figure 1.53

| EXAMPLE 3 | **Finding an Inverse Function** |

Find the inverse function of

$$f(x) = \sqrt{2x - 3}.$$

Solution   The function has an inverse function because it is one-to-one on its entire domain, $\left[\frac{3}{2}, \infty\right)$, as shown in Figure 1.53. To find an equation for the inverse function, let $y = f(x)$ and solve for x in terms of y.

$$\sqrt{2x - 3} = y \qquad \text{Let } y = f(x).$$
$$2x - 3 = y^2 \qquad \text{Square each side.}$$
$$x = \frac{y^2 + 3}{2} \qquad \text{Solve for } x.$$
$$y = \frac{x^2 + 3}{2} \qquad \text{Interchange } x \text{ and } y.$$
$$f^{-1}(x) = \frac{x^2 + 3}{2} \qquad \text{Replace } y \text{ by } f^{-1}(x).$$

The domain of f^{-1} is the range of f, which is $[0, \infty)$. You can verify this result by showing that $f(f^{-1}(x)) = x$ and $f^{-1}(f(x)) = x$.

$$f(f^{-1}(x)) = \sqrt{2\left(\frac{x^2 + 3}{2}\right) - 3} = \sqrt{x^2} = x, \quad x \geq 0$$
$$f^{-1}(f(x)) = \frac{\left(\sqrt{2x - 3}\right)^2 + 3}{2} = \frac{2x - 3 + 3}{2} = x, \quad x \geq \frac{3}{2}$$

Consider a function that is *not* one-to-one on its entire domain. By restricting the domain to an interval on which the function is one-to-one, you can conclude that the new function has an inverse function on the restricted domain.

| EXAMPLE 4 | **Testing Whether a Function Is One-to-One** |

⋮ • • • ▷ *See LarsonCalculus.com for an interactive version of this type of example.*

Show that the sine function $f(x) = \sin x$ is not one-to-one on the entire real number line. Then show that f is one-to-one on the closed interval $[-\pi/2, \pi/2]$.

Solution   It is clear that f is not one-to-one, because many different x-values yield the same y-value. For instance,

$$\sin 0 = 0 = \sin \pi.$$

Moreover, from the graph of $f(x) = \sin x$ in Figure 1.54, you can see that when f is restricted to the interval $[-\pi/2, \pi/2]$, the restricted function is one-to-one.

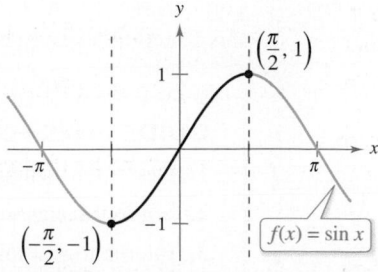

f is one-to-one on the interval $[-\pi/2, \pi/2]$.
Figure 1.54

Inverse Trigonometric Functions

From the graphs of the six basic trigonometric functions (see Section 1.4), you can see that they do not have inverse functions. The functions that are called "inverse trigonometric functions" are actually inverses of trigonometric functions whose domains have been restricted.

For instance, in Example 4, you saw that the sine function is one-to-one on the interval $[-\pi/2, \pi/2]$, as shown in Figure 1.55. On this interval, you can define the inverse of the *restricted* sine function as

$$y = \arcsin x \qquad \text{if and only if} \qquad \sin y = x$$

where

$$-1 \leq x \leq 1 \qquad \text{and} \qquad -\frac{\pi}{2} \leq \arcsin x \leq \frac{\pi}{2}.$$

From Figures 1.55 (a) and (b), you can see that you can obtain the graph of $y = \arcsin x$ by reflecting the graph of $y = \sin x$ in the line $y = x$ on the interval $[-\pi/2, \pi/2]$.

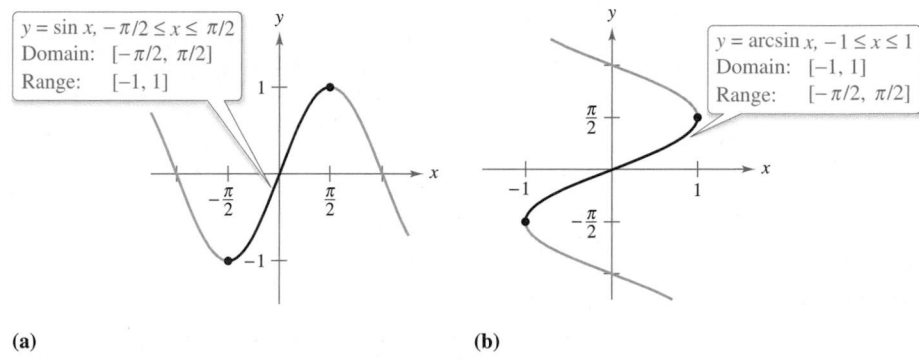

(a)   **(b)**

Figure 1.55

Under suitable restrictions, each of the six trigonometric functions is one-to-one and so has an inverse function, as indicated in the next definition. (The term "iff" is used to represent the phrase "if and only if.")

Definitions of Inverse Trigonometric Functions

Function	Domain	Range
$y = \arcsin x$ iff $\sin y = x$	$-1 \leq x \leq 1$	$-\dfrac{\pi}{2} \leq y \leq \dfrac{\pi}{2}$
$y = \arccos x$ iff $\cos y = x$	$-1 \leq x \leq 1$	$0 \leq y \leq \pi$
$y = \arctan x$ iff $\tan y = x$	$-\infty < x < \infty$	$-\dfrac{\pi}{2} < y < \dfrac{\pi}{2}$
$y = \operatorname{arccot} x$ iff $\cot y = x$	$-\infty < x < \infty$	$0 < y < \pi$
$y = \operatorname{arcsec} x$ iff $\sec y = x$	$\lvert x \rvert \geq 1$	$0 \leq y \leq \pi, \quad y \neq \dfrac{\pi}{2}$
$y = \operatorname{arccsc} x$ iff $\csc y = x$	$\lvert x \rvert \geq 1$	$-\dfrac{\pi}{2} \leq y \leq \dfrac{\pi}{2}, \quad y \neq 0$

The term $\arcsin x$ is read as "the arcsine of x" or sometimes "the angle whose sine is x." An alternative notation for the inverse sine function is $\sin^{-1} x$, which is consistent with the inverse function notation $f^{-1}(x)$.

Exploration

Inverse Secant Function

In the definition at the right, the inverse secant function is defined by restricting the domain of the secant function to the intervals

$$\left[0, \frac{\pi}{2}\right) \cup \left(\frac{\pi}{2}, \pi\right].$$

Most other texts and reference books agree with this, but some disagree. What other domains might make sense? Explain your reasoning graphically. Most calculators do not have a key for the inverse secant function. How can you use a calculator to evaluate the inverse secant function?

The graphs of the six inverse trigonometric functions are shown in Figure 1.56.

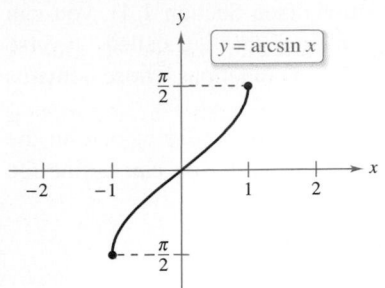

Domain: $[-1, 1]$
Range: $[-\pi/2, \pi/2]$

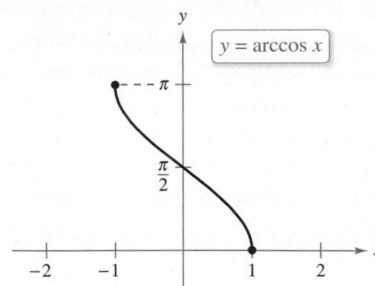

Domain: $[-1, 1]$
Range: $[0, \pi]$

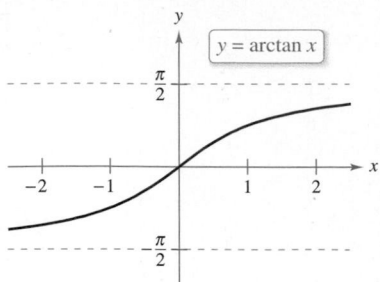

Domain: $(-\infty, \infty)$
Range: $(-\pi/2, \pi/2)$

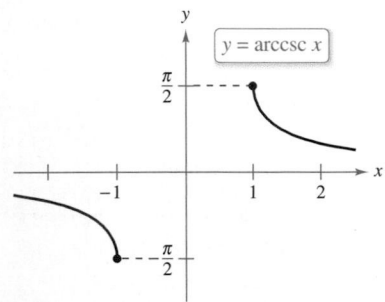

Domain: $(-\infty, -1] \cup [1, \infty)$
Range: $[-\pi/2, 0) \cup (0, \pi/2]$

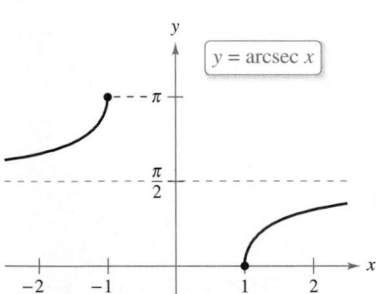

Domain: $(-\infty, -1] \cup [1, \infty)$
Range: $[0, \pi/2) \cup (\pi/2, \pi]$

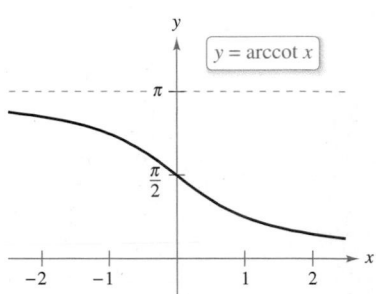

Domain: $(-\infty, \infty)$
Range: $(0, \pi)$

Figure 1.56

When evaluating inverse trigonometric functions, remember that they denote angles in *radian measure*.

EXAMPLE 5   **Evaluating Inverse Trigonometric Functions**

Evaluate each expression.

a. $\arcsin\left(-\dfrac{1}{2}\right)$ **b.** $\arccos 0$ **c.** $\arctan \sqrt{3}$ **d.** $\arcsin(0.3)$

Solution

a. By definition, $y = \arcsin\left(-\frac{1}{2}\right)$ implies that $\sin y = -\frac{1}{2}$. In the interval $[-\pi/2, \pi/2]$, the correct value of y is $-\pi/6$.

$$\arcsin\left(-\frac{1}{2}\right) = -\frac{\pi}{6}$$

b. By definition, $y = \arccos 0$ implies that $\cos y = 0$. In the interval $[0, \pi]$, you have $y = \pi/2$.

$$\arccos 0 = \frac{\pi}{2}$$

c. By definition, $y = \arctan \sqrt{3}$ implies that $\tan y = \sqrt{3}$. In the interval $(-\pi/2, \pi/2)$, you have $y = \pi/3$.

$$\arctan \sqrt{3} = \frac{\pi}{3}$$

d. Using a calculator set in *radian* mode produces

$$\arcsin(0.3) \approx 0.3047.$$

Inverse functions have the properties

$$f(f^{-1}(x)) = x \quad \text{and} \quad f^{-1}(f(x)) = x.$$

When applying these properties to inverse trigonometric functions, remember that the trigonometric functions have inverse functions only in restricted domains. For x-values outside these domains, these two properties do not hold. For example, $\arcsin(\sin \pi)$ is equal to 0, not π.

Properties of Inverse Trigonometric Functions

1. If $-1 \le x \le 1$ and $-\pi/2 \le y \le \pi/2$, then

$$\sin(\arcsin x) = x \quad \text{and} \quad \arcsin(\sin y) = y.$$

2. If $-\pi/2 < y < \pi/2$, then

$$\tan(\arctan x) = x \quad \text{and} \quad \arctan(\tan y) = y.$$

3. If $|x| \ge 1$ and $0 \le y < \pi/2$ or $\pi/2 < y \le \pi$, then

$$\sec(\text{arcsec } x) = x \quad \text{and} \quad \text{arcsec}(\sec y) = y.$$

Similar properties hold for the other inverse trigonometric functions.

EXAMPLE 6   Solving an Equation

Solve $\arctan(2x - 3) = \dfrac{\pi}{4}$ for x.

Solution

$$\arctan(2x - 3) = \frac{\pi}{4} \qquad \text{Write original equation.}$$

$$\tan[\arctan(2x - 3)] = \tan \frac{\pi}{4} \qquad \text{Take tangent of each side.}$$

$$2x - 3 = 1 \qquad \text{tan(arctan } x) = x$$

$$x = 2 \qquad \text{Solve for } x.$$

Some problems in calculus require that you evaluate expressions such as $\cos(\arcsin x)$, as shown in Example 7.

EXAMPLE 7   Using Right Triangles

a. Given $y = \arcsin x$, where $0 < y < \pi/2$, find $\cos y$.

b. Given $y = \text{arcsec}\left(\sqrt{5}/2\right)$, find $\tan y$.

Solution

a. Because $y = \arcsin x$, you know that $\sin y = x$. This relationship between x and y can be represented by a right triangle, as shown in Figure 1.57(a).

$$\cos y = \cos(\arcsin x) = \frac{\text{adj.}}{\text{hyp.}} = \sqrt{1 - x^2}$$

(This result is also valid for $-\pi/2 < y < 0$.)

b. Use the right triangle shown in Figure 1.57(b).

$$\tan y = \tan\left(\text{arcsec } \frac{\sqrt{5}}{2}\right) = \frac{\text{opp.}}{\text{adj.}} = \frac{1}{2}$$

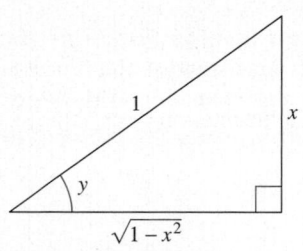

(a) $y = \arcsin x$

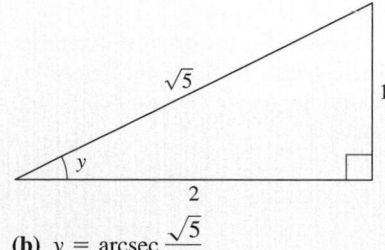

(b) $y = \text{arcsec } \dfrac{\sqrt{5}}{2}$

Figure 1.57

1.5 Exercises

See **CalcChat.com** for tutorial help and worked-out solutions to odd-numbered exercises.

CONCEPT CHECK

1. **Reflective Property of Inverse Functions**
Describe the relationship between the graph of a function and the graph of its inverse function.

2. **Domain of an Inverse Function** The function f has an inverse function f^{-1}. Is the domain of f the same as the domain of f^{-1}? Explain.

3. **Inverse Trigonometric Function** Describe the meaning of arccos x in your own words.

4. **Restricted Domain** What is a restricted domain? Why are restricted domains necessary to define inverse trigonometric functions?

Matching In Exercises 5–8, match the graph of the function with the graph of its inverse function. [The graphs of the inverse functions are labeled (a), (b), (c), and (d).]

(a)
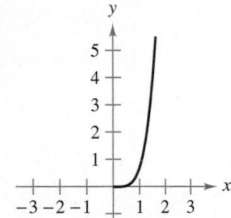

(b)
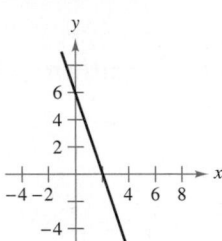

(c)
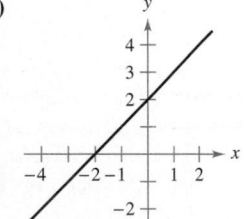

(d)
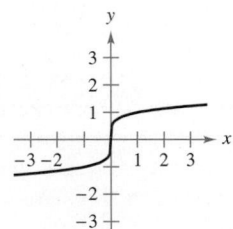

5.
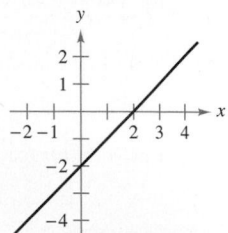

6.
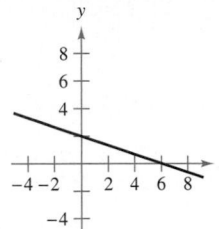

7.
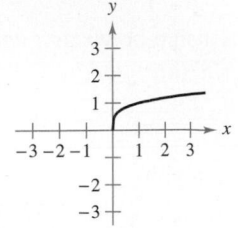

8.
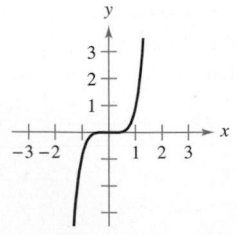

Verifying Inverse Functions In Exercises 9–16, show that f and g are inverse functions (a) analytically and (b) graphically.

9. $f(x) = 5x + 1$,　　　　$g(x) = \dfrac{x-1}{5}$

10. $f(x) = 3 - 4x$,　　　　$g(x) = \dfrac{3-x}{4}$

11. $f(x) = x^3$,　　　　　　$g(x) = \sqrt[3]{x}$

12. $f(x) = \sqrt[3]{x-3}$,　　$g(x) = 3 + x^3$

13. $f(x) = \sqrt{x-4}$,　　　$g(x) = x^2 + 4$,　$x \ge 0$

14. $f(x) = 16 - x^2$,　$x \ge 0$,　$g(x) = \sqrt{16-x}$

15. $f(x) = \dfrac{1}{x}$,　　　　　$g(x) = \dfrac{1}{x}$

16. $f(x) = \dfrac{1}{1+x}$,　$x \ge 0$,　$g(x) = \dfrac{1-x}{x}$,　$0 < x \le 1$

Using the Horizontal Line Test In Exercises 17 and 18, use the Horizontal Line Test to determine whether the function is one-to-one on its entire domain and therefore has an inverse function. To print an enlarged copy of the graph, go to *MathGraphs.com*.

17. $f(\theta) = \sin \theta$

18. $f(x) = 5x - 3$

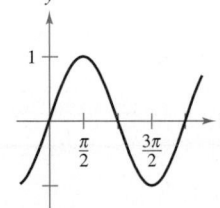

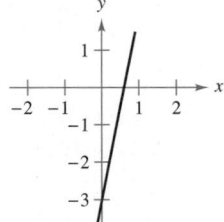

The Existence of an Inverse Function In Exercises 19–24, determine whether the function is one-to-one on its entire domain and therefore has an inverse function.

19. $f(x) = 2 - x - x^3$

20. $f(x) = \dfrac{x^4}{4} - 2x^2$

21. $f(x) = \dfrac{1}{3x+1}$

22. $f(x) = \sqrt[3]{x+1}$

23. $f(x) = \tan 2\pi x$

24. $f(x) = \sin \dfrac{3x}{2}$

The Existence of an Inverse Function In Exercises 25–30, use a graphing utility to graph the function. Determine whether the function is one-to-one on its entire domain and therefore has an inverse function.

25. $h(s) = \dfrac{1}{s-2} - 3$

26. $f(x) = \dfrac{6x}{x^2+4}$

27. $g(t) = \dfrac{1}{\sqrt{t^2+1}}$

28. $f(x) = 5x\sqrt{x-1}$

29. $g(x) = (x+5)^3$

30. $h(x) = |x+4| - |x-4|$

 **Finding an Inverse Function** In Exercises 31–38, (a) find the inverse function of f, (b) graph f and f^{-1} on the same set of coordinate axes, (c) describe the relationship between the graphs, and (d) state the domains and ranges of f and f^{-1}.

31. $f(x) = 2x - 3$ **32.** $f(x) = 9 - 5x$

33. $f(x) = x^5$ **34.** $f(x) = x^3 - 1$

35. $f(x) = \sqrt{x}$ **36.** $f(x) = x^4, \quad x \geq 0$

37. $f(x) = \sqrt{4 - x^2}, \quad 0 \leq x \leq 2$

38. $f(x) = \sqrt{x^2 - 4}, \quad x \geq 2$

 Finding an Inverse Function In Exercises 39–44, (a) find the inverse function of f, (b) use a graphing utility to graph f and f^{-1} in the same viewing window, (c) describe the relationship between the graphs, and (d) state the domains and ranges of f and f^{-1}.

39. $f(x) = \sqrt[3]{x - 1}$ **40.** $f(x) = 3\sqrt[5]{2x - 1}$

41. $f(x) = x^{2/3}, \quad x \geq 0$ **42.** $f(x) = x^{3/5}$

43. $f(x) = \dfrac{x}{\sqrt{x^2 + 7}}$ **44.** $f(x) = \dfrac{x + 2}{x}$

Finding an Inverse Function In Exercises 45 and 46, use the graph of the function f to make a table of values for the given points. Then make a second table that can be used to find f^{-1}, and sketch the graph of f^{-1}. To print an enlarged copy of the graph, go to *MathGraphs.com.*

45. 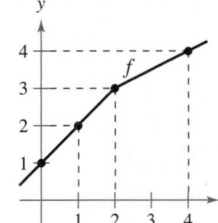 **46.**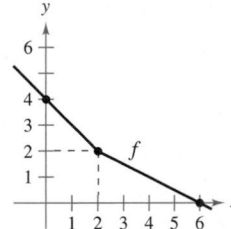

47. Cost You need a total of 50 pounds of two commodities costing $1.25 and $2.75 per pound.

(a) Verify that the total cost is $y = 1.25x + 2.75(50 - x)$, where x is the number of pounds of the less expensive commodity.

(b) Find the inverse function of the cost function. What does each variable represent in the inverse function?

(c) What is the domain of the inverse function? Validate or explain your answer using the context of the problem.

(d) Determine the number of pounds of the less expensive commodity purchased when the total cost is $73.

48. Temperature The formula $C = \frac{5}{9}(F - 32)$, where $F \geq -459.6$, represents the Celsius temperature C as a function of the Fahrenheit temperature F.

(a) Find the inverse function of C.

(b) What does the inverse function represent?

(c) What is the domain of the inverse function? Validate or explain your answer using the context of the problem.

(d) The temperature is 22°C. What is the corresponding temperature in degrees Fahrenheit?

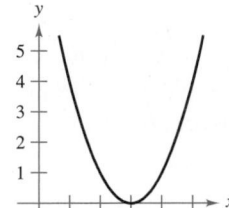 **Testing Whether a Function Is One-to-One** In Exercises 49–54, determine whether the function is one-to-one. If it is, find its inverse function.

49. $f(x) = \sqrt{x - 2}$ **50.** $f(x) = \sqrt{9 - x^2}$

51. $f(x) = -3$ **52.** $f(x) = |x - 2|, \quad x \leq 2$

53. $f(x) = ax + b, \quad a \neq 0$ **54.** $f(x) = (x + a)^3 + b$

Showing a Function Is One-to-One In Exercises 55–60, show that f is one-to-one on the given interval and therefore has an inverse function on that interval.

55. $f(x) = (x - 4)^2, \quad [4, \infty)$ **56.** $f(x) = |x + 2|, \quad [-2, \infty)$

57. $f(x) = \dfrac{4}{x^2}, \quad (0, \infty)$ **58.** $f(x) = \cot x, \quad (0, \pi)$

59. $f(x) = \cos x, \quad [0, \pi]$ **60.** $f(x) = \sec x, \quad \left[0, \dfrac{\pi}{2}\right)$

Making a Function One-to-One In Exercises 61 and 62, the function is not one-to-one. Delete part of the domain so that the function that remains is one-to-one. Find the inverse function of the remaining function and give the domain of the inverse function. (*Note:* There is more than one correct answer.)

61. $f(x) = (x - 3)^2$ **62.** $f(x) = |x - 3|$

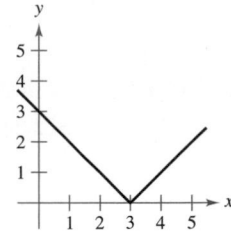

Finding an Inverse Function In Exercises 63–68, (a) sketch a graph of the function f, (b) determine an interval on which f is one-to-one, (c) find the inverse function of f on the interval found in part (b), and (d) give the domain of the inverse function. (*Note:* There is more than one correct answer.)

63. $f(x) = (x + 5)^2$ **64.** $f(x) = (7 - x)^2$

65. $f(x) = \sqrt{x^2 - 4x}$ **66.** $f(x) = -\sqrt{25 - x^2}$

67. $f(x) = 3 \cos x$ **68.** $f(x) = \sin 3x$

Finding Values In Exercises 69–74, find $f^{-1}(a)$ for the function f and real number a.

Function	Real Number
69. $f(x) = x^3 + 2x - 1$	$a = 2$
70. $f(x) = 2x^5 + x^3 + 1$	$a = -2$
71. $f(x) = 5 \sin x, \quad -\dfrac{\pi}{2} \leq x \leq \dfrac{\pi}{2}$	$a = -\dfrac{5}{2}$
72. $f(x) = \cos 2x, \quad 0 \leq x \leq \dfrac{\pi}{2}$	$a = 1$
73. $f(x) = x^3 - \dfrac{4}{x}, \quad x > 0$	$a = 6$
74. $f(x) = \sqrt{x + 4}$	$a = 3$

Using Composite and Inverse Functions In Exercises 75–78, use the functions $f(x) = \frac{1}{8}x - 3$ and $g(x) = x^3$ to find the indicated value.

75. $(f^{-1} \circ g^{-1})(1)$ **76.** $(g^{-1} \circ f^{-1})(-3)$

77. $(f^{-1} \circ f^{-1})(-2)$ **78.** $(g^{-1} \circ g^{-1})(8)$

Using Composite and Inverse Functions In Exercises 79–82, use the functions $f(x) = x + 4$ and $g(x) = 2 - x^3$ to find the indicated function.

79. $g^{-1} \circ f^{-1}$ **80.** $f^{-1} \circ g^{-1}$

81. $(f \circ g)^{-1}$ **82.** $(g \circ f)^{-1}$

Graphical Reasoning In Exercises 83 and 84, (a) use the graph of the function f to determine whether f is one-to-one, (b) state the domain of f^{-1}, and (c) estimate the value of $f^{-1}(2)$.

83. 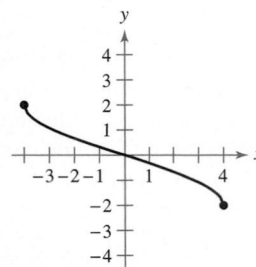   **84.**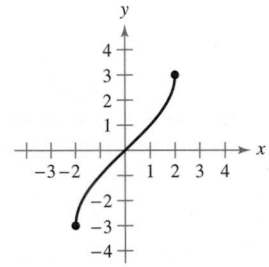

Graphical Reasoning In Exercises 85 and 86, use the graph of the function f to sketch the graph of f^{-1}. To print an enlarged copy of the graph, go to *MathGraphs.com*.

85. 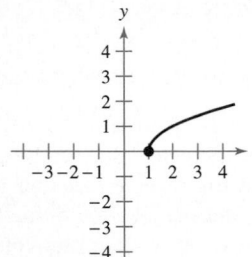   **86.**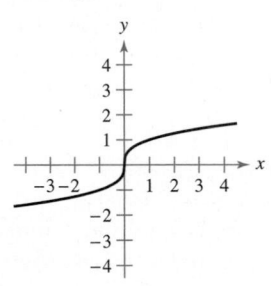

Numerical and Graphical Analysis In Exercises 87 and 88, (a) use a graphing utility to complete the table, (b) plot the points in the table and graph the function by hand, (c) use a graphing utility to graph the function and compare the result with your hand-drawn graph in part (b), and (d) determine any intercepts and symmetry of the graph.

x	-1	-0.8	-0.6	-0.4	-0.2	0	0.2	0.4	0.6	0.8	1
y											

87. $y = \arcsin x$ **88.** $y = \arccos x$

89. Missing Coordinates Determine the missing coordinates of the points on the graph of the function.

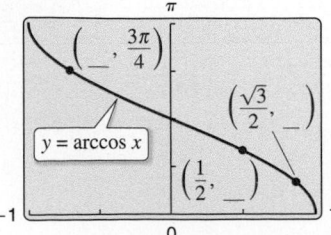

90. 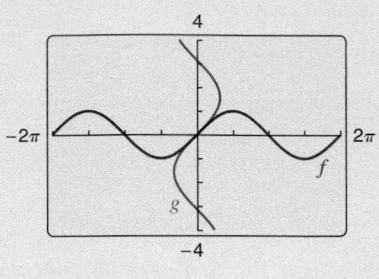 **HOW DO YOU SEE IT?** You use a graphing utility to graph $f(x) = \sin x$ and then use the *draw inverse* feature to graph g (see figure). Is g the inverse function of f? Why or why not?

Sketching a Graph In Exercises 91–94, sketch the graph of the function. Use a graphing utility to verify your graph.

91. $f(x) = \arcsin(x - 1)$ **92.** $f(x) = \text{arcsec } 2x$

93. $f(x) = \arctan x + \dfrac{\pi}{2}$ **94.** $f(x) = \arccos \dfrac{x}{4}$

 **Evaluating Inverse Trigonometric Functions** In Exercises 95–102, evaluate the expression without using a calculator.

95. $\arcsin \frac{1}{2}$ **96.** $\arcsin 0$

97. $\arccos \frac{1}{2}$ **98.** $\arccos(-1)$

99. $\arctan \dfrac{\sqrt{3}}{3}$ **100.** $\text{arccot}\left(-\sqrt{3}\right)$

101. $\text{arccsc}\left(-\sqrt{2}\right)$ **102.** $\text{arcsec } 2$

Approximating Inverse Trigonometric Functions In Exercises 103–106, use a calculator to approximate the value. Round your answer to two decimal places.

103. $\arccos(0.051)$ **104.** $\arcsin(-0.39)$

105. $\text{arcsec } 1.269$ **106.** $\text{arccsc}(-4.487)$

Using Properties In Exercises 107 and 108, use the properties of inverse trigonometric functions to evaluate the expression.

107. $\cos[\arccos(-0.1)]$ **108.** $\arcsin\left(\sin \dfrac{\pi}{3}\right)$

EXPLORING CONCEPTS

109. Think About It Does adding a constant term to a function affect the existence of an inverse function? Explain.

110. Inverse Trigonometric Functions Determine whether
$$\frac{\arcsin x}{\arccos x} = \arctan x.$$

111. Inverse Trigonometric Functions Explain why $\tan \pi = 0$ does not imply that $\arctan 0 = \pi$.

112. Think About It Use a graphing utility to graph $f(x) = \sin x$ and $g(x) = \arcsin(\sin x)$. Is the graph of g the line $y = x$? Explain.

Solving an Equation In Exercises 113–116, solve the equation for x.

113. $\arcsin(3x - \pi) = \frac{1}{2}$ **114.** $\arctan(2x - 5) = -1$

115. $\arcsin \sqrt{2x} = \arccos \sqrt{x}$

116. $\arccos x = \text{arcsec } x$

Using a Right Triangle In Exercises 117–122, use the figure to write the expression in algebraic form given $y = \arccos x$, where $0 < y < \pi/2$.

117. $\cos y$

118. $\sin y$

119. $\tan y$

120. $\cot y$

121. $\sec y$

122. $\csc y$

Evaluating an Expression In Exercises 123–126, evaluate each expression without using a calculator. [*Hint:* Sketch a right triangle, as demonstrated in Example 7.]

123. (a) $\sin\left(\arctan \frac{3}{4}\right)$ **124.** (a) $\tan\left(\arccos \frac{\sqrt{2}}{2}\right)$

(b) $\sec\left(\arcsin \frac{4}{5}\right)$ (b) $\cos\left(\arcsin \frac{5}{13}\right)$

125. (a) $\cot\left[\arcsin\left(-\frac{1}{2}\right)\right]$ **126.** (a) $\sec\left[\arctan\left(-\frac{3}{5}\right)\right]$

(b) $\csc\left[\arctan\left(-\frac{5}{12}\right)\right]$ (b) $\tan\left[\arcsin\left(-\frac{5}{6}\right)\right]$

Simplifying an Expression In Exercises 127–132, write the expression in algebraic form. [*Hint:* Sketch a right triangle, as demonstrated in Example 7.]

127. $\cos(\arcsin 2x)$ **128.** $\sec(\arctan 6x)$

129. $\sin(\text{arcsec } x)$ **130.** $\sec[\arcsin(x - 1)]$

131. $\tan\left(\text{arcsec } \frac{x}{3}\right)$ **132.** $\csc\left(\arctan \frac{x}{\sqrt{2}}\right)$

Fill in the Blank In Exercises 133 and 134, fill in the blank.

133. $\arctan \frac{9}{x} = \arcsin(\,\blacksquare\,)$, $x > 0$

134. $\arcsin \frac{\sqrt{36 - x^2}}{6} = \arccos(\,\blacksquare\,)$

Verifying an Identity In Exercises 135 and 136, verify each identity.

135. (a) $\text{arccsc } x = \arcsin \frac{1}{x}$, $|x| \geq 1$

(b) $\arctan x + \arctan \frac{1}{x} = \frac{\pi}{2}$, $x > 0$

136. (a) $\arcsin(-x) = -\arcsin x$, $|x| \leq 1$

(b) $\arccos(-x) = \pi - \arccos x$, $|x| \leq 1$

137. Verifying an Identity Verify each identity.

(a) $\text{arccot } x = \begin{cases} \pi + \arctan(1/x), & x < 0 \\ \pi/2, & x = 0 \\ \arctan(1/x), & x > 0 \end{cases}$

(b) $\text{arcsec } x = \arccos(1/x)$, $|x| \geq 1$

(c) $\text{arccsc } x = \arcsin(1/x)$, $|x| \geq 1$

138. Using an Identity Use the results of Exercise 137 and a graphing utility to evaluate each expression.

(a) $\text{arccot } 0.5$ (b) $\text{arcsec } 2.7$

(c) $\text{arccsc}(-3.9)$ (d) $\text{arccot}(-1.4)$

True or False? In Exercises 139–144, determine whether the statement is true or false. If it is false, explain why or give an example that shows it is false.

139. If f is an even function, then f^{-1} exists.

140. If the inverse function of f exists, then the y-intercept of f is an x-intercept of f^{-1}.

141. $\arcsin^2 x + \arccos^2 x = 1$

142. The range of $y = \arcsin x$ is $[0, \pi]$.

143. If $f(x) = x^n$ where n is odd, then f^{-1} exists.

144. There exists no function f such that $f = f^{-1}$.

145. Proof Prove that if f and g are one-to-one functions, then $(f \circ g)^{-1}(x) = (g^{-1} \circ f^{-1})(x)$.

146. Proof Prove that if f has an inverse function, then $(f^{-1})^{-1} = f$.

147. Proof Prove that $\cos(\sin^{-1} x) = \sqrt{1 - x^2}$.

148. Proof Prove that if a function has an inverse function, then the inverse function is unique.

149. Proof Prove that

$$\arctan x + \arctan y = \arctan \frac{x + y}{1 - xy}, \quad xy \neq 1.$$

Use this formula to show that

$$\arctan \frac{1}{2} + \arctan \frac{1}{3} = \frac{\pi}{4}.$$

150. Think About It The function $f(x) = k(2 - x - x^3)$ is one-to-one and $f^{-1}(3) = -2$. Find k.

151. Existence of an Inverse Determine the values of k such that the function $f(x) = kx + \sin x$ has an inverse function.

152. Determining Conditions Determine conditions on the constants a, b, and c such that the graph of $f(x) = \dfrac{ax + b}{cx - a}$ is symmetric about the line $y = x$.

153. Determining Conditions Determine conditions on the constants a, b, c, and d such that $f(x) = \dfrac{ax + b}{cx + d}$ has an inverse function. Then find f^{-1}.

154. Determining Conditions Let $f(x) = ax^2 + bx + c$, where $a > 0$ and the domain is all real numbers such that $x \leq -\dfrac{b}{2a}$. Find f^{-1}.

1.6 Exponential and Logarithmic Functions

■ Develop and use properties of exponential functions.
■ Understand the definition of the number e.
■ Understand the definition of the natural logarithmic function, and develop and use properties of the natural logarithmic function.

Exponential Functions

An **exponential function** involves a constant raised to a power, such as $f(x) = 2^x$. You already know how to evaluate 2^x for *rational* values of x. For instance,

$$2^0 = 1, \quad 2^2 = 4, \quad 2^{-1} = \frac{1}{2}, \quad \text{and} \quad 2^{1/2} = \sqrt{2} \approx 1.4142136.$$

For *irrational* values of x, you can define 2^x by considering a sequence of rational numbers that approach x. A full discussion of this process would not be appropriate now, but here is the general idea. To define the number $2^{\sqrt{2}}$, note that

$$\sqrt{2} = 1.414213\ldots$$

and consider the numbers below (which are of the form 2^r, where r is rational).

$$2^1 = 2 < 2^{\sqrt{2}} < 4 = 2^2$$
$$2^{1.4} = 2.639015\ldots < 2^{\sqrt{2}} < 2.828427\ldots = 2^{1.5}$$
$$2^{1.41} = 2.657371\ldots < 2^{\sqrt{2}} < 2.675855\ldots = 2^{1.42}$$
$$2^{1.414} = 2.664749\ldots < 2^{\sqrt{2}} < 2.666597\ldots = 2^{1.415}$$
$$2^{1.4142} = 2.665119\ldots < 2^{\sqrt{2}} < 2.665303\ldots = 2^{1.4143}$$
$$2^{1.41421} = 2.665137\ldots < 2^{\sqrt{2}} < 2.665156\ldots = 2^{1.41422}$$
$$2^{1.414213} = 2.665143\ldots < 2^{\sqrt{2}} < 2.665144\ldots = 2^{1.414214}$$

From these calculations, it seems reasonable to conclude that

$$2^{\sqrt{2}} \approx 2.66514.$$

In practice, you can use a calculator to approximate numbers such as $2^{\sqrt{2}}$.

In general, you can use any positive base a, $a \neq 1$, to define an exponential function. So, the exponential function with base a is written as $f(x) = a^x$. Exponential functions, even those with irrational values of x, obey the familiar properties of exponents.

• •REMARK Note that the base $a = 1$ is excluded because it yields $f(x) = 1^x = 1$. This is a *constant* function, not an *exponential* function.

Properties of Exponents

Let a and b be positive real numbers, and let x and y be any real numbers.

1. $a^0 = 1$ **2.** $a^x a^y = a^{x+y}$ **3.** $(a^x)^y = a^{xy}$ **4.** $(ab)^x = a^x b^x$

5. $\dfrac{a^x}{a^y} = a^{x-y}$ **6.** $\left(\dfrac{a}{b}\right)^x = \dfrac{a^x}{b^x}$ **7.** $a^{-x} = \dfrac{1}{a^x}$

EXAMPLE 1   Using Properties of Exponents

a. $(2^2)(2^3) = 2^{2+3} = 2^5$ **b.** $\dfrac{2^2}{2^3} = 2^{2-3} = 2^{-1} = \dfrac{1}{2}$

c. $(3^x)^3 = 3^{3x}$ **d.** $\left(\dfrac{1}{3}\right)^{-x} = (3^{-1})^{-x} = 3^x$

EXAMPLE 2 Sketching Graphs of Exponential Functions

⋮ ▷ *See LarsonCalculus.com for an interactive version of this type of example.*

Sketch the graphs of the functions

$$f(x) = 2^x, \quad g(x) = \left(\frac{1}{2}\right)^x = 2^{-x}, \quad \text{and} \quad h(x) = 3^x.$$

Solution To sketch the graphs of these functions by hand, you can complete a table of values, plot the corresponding points, and connect the points with smooth curves.

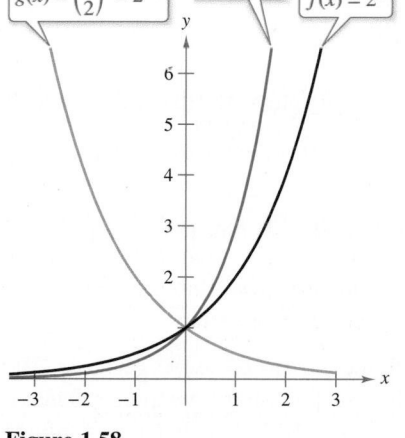

$g(x) = \left(\frac{1}{2}\right)^x = 2^{-x}$ $h(x) = 3^x$ $f(x) = 2^x$

Figure 1.58

x	-3	-2	-1	0	1	2	3	4
2^x	$\frac{1}{8}$	$\frac{1}{4}$	$\frac{1}{2}$	1	2	4	8	16
2^{-x}	8	4	2	1	$\frac{1}{2}$	$\frac{1}{4}$	$\frac{1}{8}$	$\frac{1}{16}$
3^x	$\frac{1}{27}$	$\frac{1}{9}$	$\frac{1}{3}$	1	3	9	27	81

Another way to graph these functions is to use a graphing utility. In either case, you should obtain graphs similar to those shown in Figure 1.58. Note that the graphs of f and h are increasing, and the graph of g is decreasing. Also, the graph of h is increasing more rapidly than the graph of f. ∎

The shapes of the graphs in Figure 1.58 are typical of the exponential functions $f(x) = a^x$ and $g(x) = a^{-x}$ where $a > 1$, as shown in Figure 1.59.

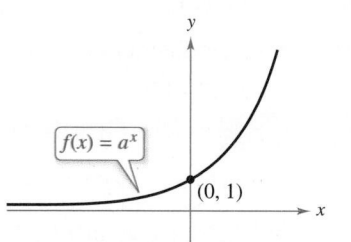

$f(x) = a^x$ $(0, 1)$

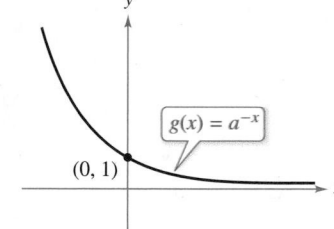

$g(x) = a^{-x}$ $(0, 1)$

Figure 1.59

Properties of Exponential Functions

Let a be a real number that is greater than 1.

1. The domain of $f(x) = a^x$ and $g(x) = a^{-x}$ is $(-\infty, \infty)$.
2. The range of $f(x) = a^x$ and $g(x) = a^{-x}$ is $(0, \infty)$.
3. The y-intercept of $f(x) = a^x$ and $g(x) = a^{-x}$ is $(0, 1)$.
4. The functions $f(x) = a^x$ and $g(x) = a^{-x}$ are one-to-one.

▷ **TECHNOLOGY** Functions of the form $h(x) = b^{cx}$ have the same types of properties and graphs as functions of the form $f(x) = a^x$ and $g(x) = a^{-x}$. To see why this is true, notice that

$$b^{cx} = (b^c)^x.$$

For instance, $f(x) = 2^{3x}$ can be written as

$$f(x) = (2^3)^x \quad \text{or} \quad f(x) = 8^x.$$

Try confirming this by graphing $f(x) = 2^{3x}$ and $g(x) = 8^x$ in the same viewing window.

The Number e

In calculus, the natural (or convenient) choice for a base of an exponential number is the irrational number e, whose decimal approximation is

$$e \approx 2.71828182846.$$

This choice may seem anything but natural. The convenience of this particular base, however, will become apparent as you continue in this course.

EXAMPLE 3 **Investigating the Number e**

Describe the behavior of the function $f(x) = (1 + x)^{1/x}$ at values of x that are close to 0.

Solution One way to examine the values of $f(x)$ near 0 is to construct a table.

x	-0.01	-0.001	-0.0001	0.0001	0.001	0.01
$(1 + x)^{1/x}$	2.7320	2.7196	2.7184	2.7181	2.7169	2.7048

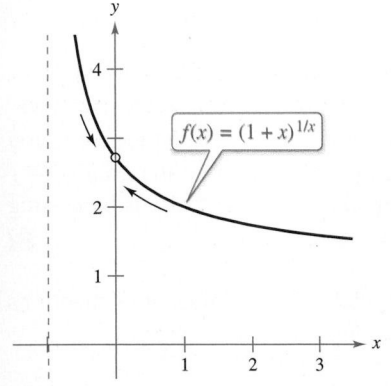

Figure 1.60

From the table, it appears that the closer x gets to 0, the closer $(1 + x)^{1/x}$ gets to e. The graph of f shown in Figure 1.60 supports this conclusion. Try using a graphing calculator to obtain this graph. Then zoom in closer and closer to $x = 0$. Although f is not defined when $x = 0$, it is defined for x-values that are arbitrarily close to zero. By zooming in, you can see that the value of $f(x)$ gets closer and closer to $e \approx 2.71828182846$ as x gets closer and closer to 0. Later, when you study limits, you will learn that this result can be written as

$$\lim_{x \to 0} (1 + x)^{1/x} = e$$

which is read as "the limit of $(1 + x)^{1/x}$ as x approaches 0 is e."

EXAMPLE 4 **The Graph of the Natural Exponential Function**

Sketch the graph of $f(x) = e^x$.

Solution To sketch the graph of f by hand, you can complete a table of values, plot the corresponding points, and connect the points with a smooth curve (see Figure 1.61).

x	-2	-1	0	1	2
e^x	$\dfrac{1}{e^2} \approx 0.135$	$\dfrac{1}{e} \approx 0.368$	1	$e \approx 2.718$	$e^2 \approx 7.389$

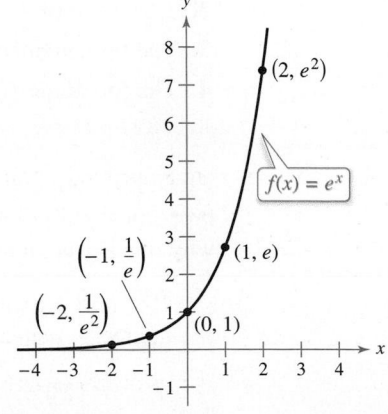

Figure 1.61

The Natural Logarithmic Function

Because the natural exponential function

$$f(x) = e^x$$

is one-to-one, it must have an inverse function. This inverse function is called the **natural logarithmic function.** The domain of the natural logarithmic function is the set of positive real numbers.

•• **REMARK** The notation ln x is read as "el en of x" or "the natural log of x."

> ### Definition of the Natural Logarithmic Function
>
> Let x be a positive real number. The **natural logarithmic function,** denoted by ln x, is defined as
>
> $$\ln x = b \quad \text{if and only if} \quad e^b = x.$$

The definition of the natural logarithmic function tells you that a logarithmic equation can be written in an equivalent exponential form, and vice versa. Here are some examples.

Logarithmic Form	Exponential Form
$\ln 1 = 0$	$e^0 = 1$
$\ln e = 1$	$e^1 = e$
$\ln e^{-1} = -1$	$e^{-1} = \dfrac{1}{e}$

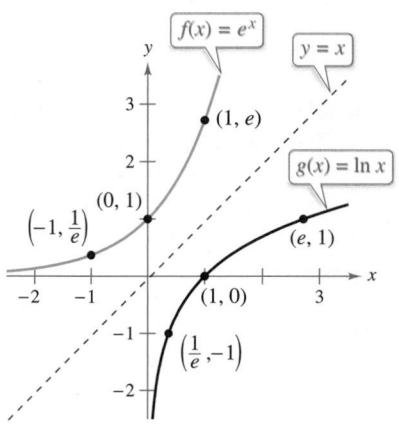

Figure 1.62

Because the function $g(x) = \ln x$ is defined to be the inverse function of $f(x) = e^x$, it follows that the graph of the natural logarithmic function is a reflection of the graph of the natural exponential function in the line $y = x$, as shown in Figure 1.62. Several other properties of the natural logarithmic function also follow directly from its definition as the inverse of the natural exponential function.

> ### Properties of the Natural Logarithmic Function
>
> **1.** The domain of $g(x) = \ln x$ is $(0, \infty)$.
> **2.** The range of $g(x) = \ln x$ is $(-\infty, \infty)$.
> **3.** The x-intercept of $g(x) = \ln x$ is $(1, 0)$.
> **4.** The function $g(x) = \ln x$ is one-to-one.

Because $f(x) = e^x$ and $g(x) = \ln x$ are inverse functions of each other, you can conclude that

$$\ln e^x = x \quad \text{and} \quad e^{\ln x} = x.$$

One of the properties of exponents states that when you multiply two exponential functions (having the same base), you add their exponents. For instance,

$$e^x e^y = e^{x+y}.$$

The logarithmic version of this property states that the natural logarithm of the product of two numbers is equal to the sum of the natural logs of the numbers. That is,

$$\ln xy = \ln x + \ln y.$$

This property and the properties dealing with the natural log of a quotient and the natural log of a power are listed on the next page.

Exploration

A graphing utility is used to graph $f(x) = \ln x^2$ and $g(x) = 2 \ln x$. Which of the graphs below is the graph of f? Which is the graph of g?

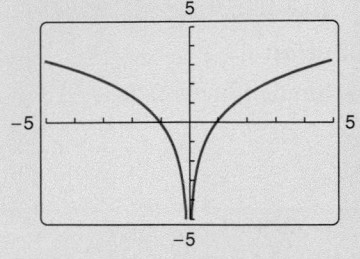

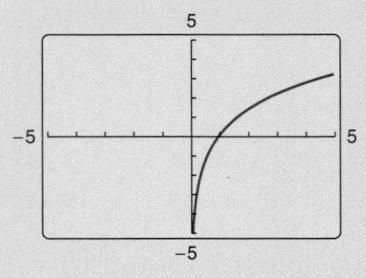

Properties of Logarithms

Let x, y, and z be real numbers such that $x > 0$ and $y > 0$.

1. $\ln xy = \ln x + \ln y$

2. $\ln \dfrac{x}{y} = \ln x - \ln y$

3. $\ln x^z = z \ln x$

EXAMPLE 5 Expanding Logarithmic Expressions

a. $\ln \dfrac{10}{9} = \ln 10 - \ln 9$ ⟶ Property 2

b. $\ln \sqrt{3x + 2} = \ln(3x + 2)^{1/2}$ ⟶ Rewrite with rational exponent.

$\qquad\qquad = \dfrac{1}{2} \ln(3x + 2)$ ⟶ Property 3

c. $\ln \dfrac{6x}{5} = \ln(6x) - \ln 5$ ⟶ Property 2

$\qquad\quad = \ln 6 + \ln x - \ln 5$ ⟶ Property 1

d. $\ln \dfrac{(x^2 + 3)^2}{x\sqrt[3]{x^2 + 1}} = \ln(x^2 + 3)^2 - \ln\left(x\sqrt[3]{x^2 + 1}\right)$

$\qquad\qquad = 2\ln(x^2 + 3) - \left[\ln x + \ln(x^2 + 1)^{1/3}\right]$

$\qquad\qquad = 2\ln(x^2 + 3) - \ln x - \ln(x^2 + 1)^{1/3}$

$\qquad\qquad = 2\ln(x^2 + 3) - \ln x - \dfrac{1}{3}\ln(x^2 + 1)$

When using the properties of logarithms to rewrite logarithmic functions, you must check to see whether the domain of the rewritten function is the same as the domain of the original function. For instance, the domain of $f(x) = \ln x^2$ is all real numbers except $x = 0$, and the domain of $g(x) = 2 \ln x$ is all positive real numbers.

EXAMPLE 6 Solving Exponential and Logarithmic Equations

Solve for x.

a. $7 = e^{x+1}$ **b.** $\ln(2x - 3) = 5$

Solution

a. $\qquad 7 = e^{x+1}$ ⟶ Write original equation.

$\qquad \ln 7 = \ln(e^{x+1})$ ⟶ Take natural log of each side.

$\qquad \ln 7 = x + 1$ ⟶ Apply inverse property.

$\quad -1 + \ln 7 = x$ ⟶ Solve for x.

$\qquad 0.946 \approx x$ ⟶ Use a calculator.

b. $\ln(2x - 3) = 5$ ⟶ Write original equation.

$\quad e^{\ln(2x-3)} = e^5$ ⟶ Exponentiate each side.

$\qquad 2x - 3 = e^5$ ⟶ Apply inverse property.

$\qquad\qquad x = \dfrac{1}{2}(e^5 + 3)$ ⟶ Solve for x.

$\qquad\qquad x \approx 75.707$ ⟶ Use a calculator.

1.6 Exercises

See **CalcChat.com** for tutorial help and worked-out solutions to odd-numbered exercises.

CONCEPT CHECK

1. **Stating Properties** In your own words, state the properties of the natural exponential function.

2. **Stating Properties** In your own words, state the properties of the natural logarithmic function.

3. **Think About It** Explain why $\ln e^x = x$.

4. **Logarithmic Properties** What is the value of n?

$$\ln 4 + \ln(n^{-1}) = \ln 4 - \ln 7$$

Evaluating an Expression In Exercises 5 and 6, evaluate the expressions.

5. (a) $25^{3/2}$ (b) $81^{1/2}$ (c) 3^{-2} (d) $27^{-1/3}$

6. (a) $64^{1/3}$ (b) 5^{-4} (c) $\left(\frac{1}{8}\right)^{1/3}$ (d) $\left(\frac{1}{4}\right)^3$

 Using Properties of Exponents In Exercises 7–10, use the properties of exponents to simplify the expressions.

7. (a) $(5^2)(5^3)$ (b) $(5^2)(5^{-3})$ (c) $\frac{5^3}{25^2}$ (d) $\left(\frac{1}{4}\right)^2 2^6$

8. (a) $(2^2)^3$ (b) $(5^4)^{1/2}$
   (c) $[(27^{-1})(27^{2/3})]^3$ (d) $(25^{3/2})(3^2)$

9. (a) $e^2(e^4)$ (b) $(e^3)^4$ (c) $(e^3)^{-2}$ (d) $\left(\frac{e^{-6}}{e^{-2}}\right)^2$

10. (a) $\left(\frac{1}{e}\right)^{-2}$ (b) $\left(\frac{e^5}{e^2}\right)^{-1}$ (c) e^0 (d) $\frac{1}{e^{-3}}$

 Solving an Equation In Exercises 11–26, solve for x.

11. $3^x = 81$

12. $4^x = 64$

13. $6^{x-2} = 36$

14. $5^{x+1} = 125$

15. $\left(\frac{1}{2}\right)^x = 32$

16. $\left(\frac{1}{4}\right)^x = 16$

17. $\left(\frac{1}{3}\right)^{x-1} = 27$

18. $\left(\frac{1}{5}\right)^{2x+1} = 25$

19. $4^3 = (x + 2)^3$

20. $343 = (5x - 7)^3$

21. $x^{3/4} = 8$

22. $(x + 3)^{4/3} = 16$

23. $e^x = e^{2x+1}$

24. $e^x = 1$

25. $e^{-2x} = e^5$

26. $e^{3x+1} = e^7$

 **Sketching the Graph of an Exponential Function** In Exercises 27–40, sketch the graph of the function.

27. $y = 4^x$

28. $y = 3^{x-1}$

29. $y = \left(\frac{1}{3}\right)^x$

30. $y = 2^{-x^2}$

31. $f(x) = 3^{-x^2}$

32. $f(x) = 3^{|x|}$

33. $y = e^{-x}$

34. $y = \frac{1}{2}e^x$

35. $y = e^x + 2$

36. $y = e^{x-1}$

37. $h(x) = e^{x-2}$

38. $g(x) = -e^{x/2}$

39. $y = e^{-x^2}$

40. $y = e^{-x/4}$

 **Finding the Domain** In Exercises 41–46, find the domain of the function.

41. $f(x) = \frac{1}{3 + e^x}$

42. $f(x) = \frac{1}{2 - e^x}$

43. $f(x) = \sqrt{1 - 4^x}$

44. $f(x) = \sqrt{1 + 3^{-x}}$

45. $f(x) = \sin e^{-x}$

46. $f(x) = \cos(2^{1-x})$

Matching In Exercises 47–50, match the equation with the correct graph. Assume that a and C are positive real numbers. [The graphs are labeled (a), (b), (c), and (d).]

(a)   (b)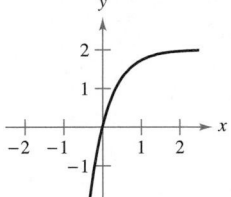

(c) 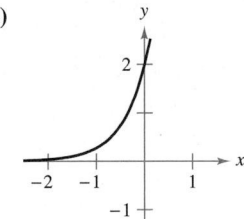 (d)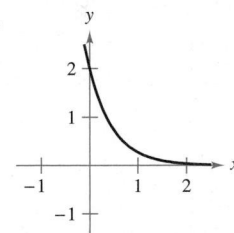

47. $y = Ce^{ax}$

48. $y = Ce^{-ax}$

49. $y = C(1 - e^{-ax})$

50. $y = \frac{C}{1 + e^{-ax}}$

Finding an Exponential Function In Exercises 51 and 52, find the exponential function $y = Ca^x$ that fits the graph.

51. 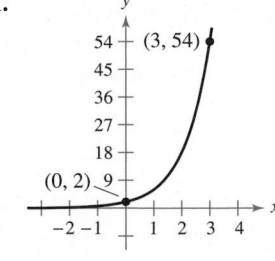 52.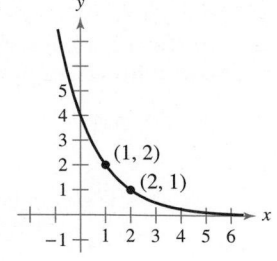

Writing Exponential or Logarithmic Equations In Exercises 53–56, write the exponential equation as a logarithmic equation, or vice versa.

53. $e^0 = 1$

54. $e^{-2} = 0.1353 \ldots$

55. $\ln 4.15 = 1.4231 \ldots$

56. $\ln 0.5 = -0.6931 \ldots$

Matching In Exercises 57–60, match the function with its graph. [The graphs are labeled (a), (b), (c), and (d).]

(a)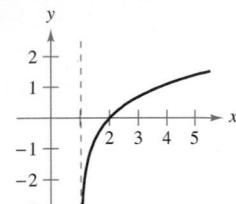

(b)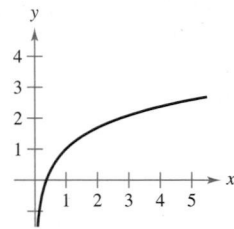

(c)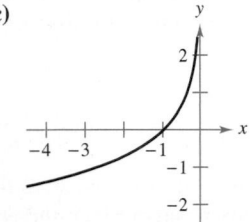

(d)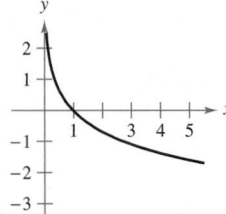

57. $f(x) = \ln x + 1$

58. $f(x) = -\ln x$

59. $f(x) = \ln(x - 1)$

60. $f(x) = -\ln(-x)$

 Sketching the Graph of a Logarithmic Function In Exercises 61–68, sketch the graph of the function and state its domain.

61. $f(x) = 3 \ln x$

62. $f(x) = -2 \ln x$

63. $f(x) = \ln 2x$

64. $f(x) = \ln|x|$

65. $f(x) = \ln(x - 3)$

66. $f(x) = \ln x - 4$

67. $f(x) = -\ln(x + 2)$

68. $f(x) = 4 \ln(x - 2) + 1$

Writing an Equation In Exercises 69–72, write an equation for the function having the given characteristics.

69. The shape of $f(x) = e^x$, but shifted eight units upward and reflected in the x-axis

70. The shape of $f(x) = e^x$, but shifted two units to the left and six units downward

71. The shape of $f(x) = \ln x$, but shifted five units to the right and one unit downward

72. The shape of $f(x) = \ln x$, but shifted three units upward and reflected in the y-axis

 Inverse Functions In Exercises 73–76, illustrate that the functions f and g are inverse functions of each other by using a graphing utility to graph them in the same viewing window.

73. $f(x) = e^{2x}, \ g(x) = \ln\sqrt{x}$

74. $f(x) = e^{x/3}, \ g(x) = \ln x^3$

75. $f(x) = e^x - 1, \ g(x) = \ln(x + 1)$

76. $f(x) = e^{x-1}, \ g(x) = 1 + \ln x$

 **Finding Inverse Functions** In Exercises 77–80, (a) find the inverse of the function, (b) use a graphing utility to graph f and f^{-1} in the same viewing window, and (c) verify that $f^{-1}(f(x)) = x$ and $f(f^{-1}(x)) = x$.

77. $f(x) = e^{4x-1}$

78. $f(x) = 3e^{-x}$

79. $f(x) = 2 \ln(x - 1)$

80. $f(x) = 3 + \ln 2x$

Applying Inverse Properties In Exercises 81–86, apply the inverse properties of $\ln x$ and e^x to simplify the given expression.

81. $\ln e^{x^2}$

82. $-4 \ln e^{2x-1}$

83. $e^{\ln(5x+2)}$

84. $e^{\ln\sqrt{x}}$

85. $-1 + \ln e^{2x}$

86. $-8 + e^{\ln x^3}$

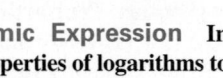 **Using Properties of Logarithms** In Exercises 87 and 88, use the properties of logarithms to approximate the indicated logarithms, given that $\ln 2 \approx 0.6931$ and $\ln 3 \approx 1.0986$.

87. (a) $\ln 6$ (b) $\ln \frac{2}{3}$ (c) $\ln 81$ (d) $\ln \sqrt{3}$

88. (a) $\ln 0.25$ (b) $\ln 24$ (c) $\ln \sqrt[3]{12}$ (d) $\ln \frac{1}{72}$

Expanding a Logarithmic Expression In Exercises 89–98, use the properties of logarithms to expand the logarithmic expression.

89. $\ln \dfrac{x}{4}$

90. $\ln \sqrt{x^5}$

91. $\ln \dfrac{xy}{z}$

92. $\ln(xyz)$

93. $\ln\left(x\sqrt{x^2 + 5}\right)$

94. $\ln \sqrt[3]{z + 1}$

95. $\ln \sqrt{\dfrac{x - 1}{x}}$

96. $\ln z(z - 1)^2$

97. $\ln(3e^2)$

98. $\ln \dfrac{1}{e}$

 **Condensing a Logarithmic Expression** In Exercises 99–106, write the expression as the logarithm of a single quantity.

99. $\ln x + \ln 7$

100. $\ln y + \ln x^2$

101. $\ln(x - 2) - \ln(x + 2)$

102. $3 \ln x + 2 \ln y - 4 \ln z$

103. $\frac{1}{3}[2 \ln(x + 3) + \ln x - \ln(x^2 - 1)]$

104. $2[\ln x - \ln(x + 1) - \ln(x - 1)]$

105. $2 \ln 3 - \frac{1}{2} \ln(x^2 + 1)$

106. $\ln(x^2 - 4) - \ln(x + 2) + \ln 4$

Solving an Exponential or Logarithmic Equation In Exercises 107–118, solve for x accurate to three decimal places.

107. $e^x = 12$

108. $5e^x = 36$

109. $9 - 2e^x = 7$

110. $8e^x - 12 = 7$

111. $50e^{-x} = 30$

112. $100e^{-2x} = 35$

113. $\ln x = 2$

114. $\ln x^2 = -8$

115. $\ln(x - 3) = 2$

116. $\ln 4x = 1$

117. $\ln\sqrt{x + 2} = 1$

118. $\ln(x - 2)^2 = 12$

Solving an Inequality In Exercises 119–122, solve the inequality for x.

119. $e^{2x+1} > 3$ **120.** $e^{1-x} < 6$

121. $-2 < \ln x < 0$ **122.** $1 < \ln(x^2 + 1) < 100$

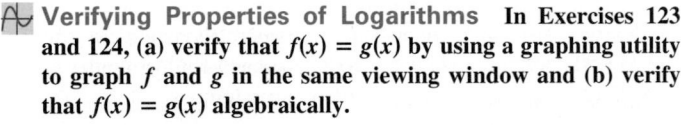 **Verifying Properties of Logarithms** In Exercises 123 and 124, (a) verify that $f(x) = g(x)$ by using a graphing utility to graph f and g in the same viewing window and (b) verify that $f(x) = g(x)$ algebraically.

123. $f(x) = \ln\dfrac{x^2}{4}, \; x > 0$

    $g(x) = 2\ln x - \ln 4$

124. $f(x) = \ln\sqrt{x(x^2 + 1)}$

    $g(x) = \frac{1}{2}[\ln x + \ln(x^2 + 1)]$

EXPLORING CONCEPTS

125. Think About It Is it always true that $\ln(a^b) = b \ln a$? Explain.

126. Think About It Is $\ln xy = \ln x \ln y$ a valid property of logarithms, where $x > 0$ and $y > 0$? Explain.

127. Analyze a Statement The table of values below was obtained by evaluating a function. Determine which of the statements may be true and which must be false. Explain your reasoning.

(a) y is an exponential function of x.

(b) x is an exponential function of y.

(c) y is a logarithmic function of x.

(d) y is a linear function of x.

x	1	2	8
y	0	1	3

128.  **HOW DO YOU SEE IT?** The figure below shows the graph of $y_1 = \ln e^x$ or $y_2 = e^{\ln x}$. Which graph is it? What are the domains of y_1 and y_2? Does $\ln e^x = e^{\ln x}$ for all real values of x? Explain.

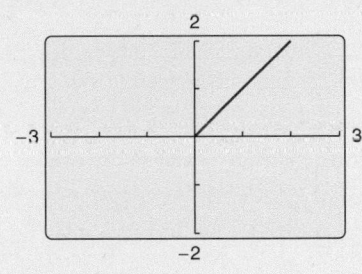

Sound Intensity In Exercises 129 and 130, use the following information. The relationship between the number of decibels β and the intensity I of a sound in watts per square centimeter is

$$\beta = \frac{10}{\ln 10} \ln\left(\frac{I}{10^{-16}}\right).$$

129. Use the properties of logarithms to write the formula in simpler form.

130. Determine the number of decibels of a sound with an intensity of 10^{-5} watt per square centimeter.

 131. Comparing Functions Use a graphing utility to graph the functions $f(x) = 6^x$ and $g(x) = x^6$ in the same viewing window. Where do these graphs intersect? As x increases, which function grows more rapidly?

 132. Comparing Functions Use a graphing utility to graph the functions $f(x) = \ln x$ and $g(x) = x^{1/4}$ in the same viewing window. Where do these graphs intersect? As x increases, which function grows more rapidly?

133. Analyzing a Function Let $f(x) = \ln(x + \sqrt{x^2 + 1})$.

 (a) Use a graphing utility to graph f and determine its domain.

(b) Show that f is an odd function.

(c) Find the inverse function of f.

134. Think About It Consider $f(x) = \ln x$. Show that $f(e^{n+1}) - f(e^n) = 1$ for any value of n.

Stirling's Formula For large values of n,

$$n! = 1 \cdot 2 \cdot 3 \cdot 4 \cdots (n - 1) \cdot n$$

can be approximated by Stirling's Formula,

$$n! \approx \left(\frac{n}{e}\right)^n \sqrt{2\pi n}.$$

In Exercises 135 and 136, find the exact value of $n!$, and then approximate $n!$ using Stirling's Formula.

135. $n = 12$ **136.** $n = 15$

137. Proof Prove that $\ln(x/y) = \ln x - \ln y, \quad x > 0, y > 0$.

138. Proof Prove that $\ln x^y = y \ln x$.

Review Exercises   See **CalcChat.com** for tutorial help and worked-out solutions to odd-numbered exercises.

Finding Intercepts   In Exercises 1–4, find any intercepts.

1. $y = 5x - 8$

2. $y = x^2 - 8x + 12$

3. $y = \dfrac{x - 3}{x - 4}$

4. $y = (x - 3)\sqrt{x + 4}$

Testing for Symmetry   In Exercises 5–8, test for symmetry with respect to each axis and to the origin.

5. $y = x^2 + 4x$

6. $y = x^4 - x^2 + 3$

7. $y^2 = x^2 - 5$

8. $xy = -2$

Using Intercepts and Symmetry to Sketch a Graph   In Exercises 9–14, find any intercepts and test for symmetry. Then sketch the graph of the equation.

9. $y = -\frac{1}{2}x + 3$

10. $y = -x^2 + 4$

11. $y = 9x - x^3$

12. $y^2 = 9 - x$

13. $y = 2\sqrt{4 - x}$

14. $y = |x - 4| - 4$

Finding Points of Intersection   In Exercises 15–18, find the points of intersection of the graphs of the equations.

15. $5x + 3y = -1$
     $x - y = -5$

16. $2x + 4y = 9$
     $6x - 4y = 7$

17. $x - y = -5$
     $x^2 - y = 1$

18. $x^2 + y^2 = 1$
     $-x + y = 1$

Finding the Slope of a Line   In Exercises 19 and 20, plot the pair of points and find the slope of the line passing through them.

19. $\left(\frac{3}{2}, 1\right), \left(5, \frac{5}{2}\right)$

20. $(-7, 8), (-1, 8)$

Finding an Equation of a Line   In Exercises 21–24, find an equation of the line that passes through the point and has the indicated slope. Then sketch the line.

Point	Slope	Point	Slope
21. $(3, -5)$	$m = \frac{7}{4}$	**22.** $(-8, 1)$	m is undefined.
23. $(-3, 0)$	$m = -\frac{2}{3}$	**24.** $(5, 4)$	$m = 0$

Finding the Slope and y-Intercept   In Exercises 25 and 26, find the slope and the y-intercept (if possible) of the line.

25. $y - 3x = 5$

26. $9 - y = x$

Sketching a Line in the Plane   In Exercises 27–30, sketch the graph of the equation.

27. $y = 6$

28. $x = -3$

29. $y = 4x - 2$

30. $3x + 2y = 12$

Finding an Equation of a Line   In Exercises 31 and 32, find an equation of the line that passes through the points. Then sketch the line.

31. $(0, 0), (8, 2)$

32. $(-5, 5), (10, -1)$

33. Finding Equations of Lines   Find equations of the lines passing through $(-3, 5)$ and having the following characteristics.

(a) Slope of $\frac{7}{16}$

(b) Parallel to the line $5x - 3y = 3$

(c) Perpendicular to the line $3x + 4y = 8$

(d) Parallel to the y-axis

34. Finding Equations of Lines   Find equations of the lines passing through $(2, 4)$ and having the following characteristics.

(a) Slope of $-\frac{2}{3}$

(b) Perpendicular to the line $x + y = 0$

(c) Parallel to the line $3x - y = 0$

(d) Parallel to the x-axis

35. Rate of Change   The purchase price of a new machine is $12,500, and its value will decrease by $850 per year. Use this information to write a linear equation that gives the value V of the machine t years after it is purchased. Find its value at the end of 3 years.

36. Break-Even Analysis   A contractor purchases a piece of equipment for $36,500 that costs an average of $9.25 per hour for fuel and maintenance. The equipment operator is paid $13.50 per hour, and customers are charged $30 per hour.

(a) Write a linear equation for the cost C of operating this equipment for t hours.

(b) Write a linear equation for the revenue R derived from t hours of use.

(c) Find the break-even point for this equipment by finding the time at which $R = C$.

Evaluating a Function   In Exercises 37–40, evaluate the function at the given value(s) of the independent variable. Simplify the results.

37. $f(x) = 5x + 4$

(a) $f(0)$ (b) $f(5)$ (c) $f(-3)$ (d) $f(t + 1)$

38. $f(x) = x^3 - 2x$

(a) $f(-3)$ (b) $f(2)$ (c) $f(-1)$ (d) $f(c - 1)$

39. $f(x) = 4x^2$
     $\dfrac{f(x + \Delta x) - f(x)}{\Delta x}$

40. $f(x) = 2x - 6$
     $\dfrac{f(x) - f(1)}{x - 1}$

Finding the Domain and Range of a Function   In Exercises 41 and 42, find the domain and range of the function.

41. $f(x) = x^2 + 3$

42. $f(x) = -|x + 1|$

Sketching a Graph of a Function   In Exercises 43 and 44, sketch a graph of the function and find its domain and range. Use a graphing utility to verify your graph.

43. $f(x) = \dfrac{4}{2x - 1}$

44. $g(x) = \sqrt{x + 1}$

Using the Vertical Line Test In Exercises 45 and 46, use the Vertical Line Test to determine whether y is a function of x. To print an enlarged copy of the graph, go to *MathGraphs.com*.

45. $x + y^2 = 2$ **46.** $x^2 - y = 0$

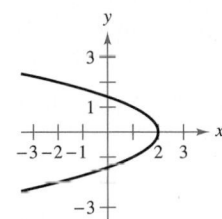 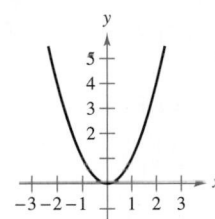

Deciding Whether an Equation Is a Function In Exercises 47 and 48, determine whether y is a function of x.

47. $xy + x^3 - 2y = 0$ **48.** $x = 9 - y^2$

 49. Transformations of Functions Use a graphing utility to graph $f(x) = x^3 - 3x^2$. Use the graph to write a formula for the function g shown in the figure.

(a)   (b)

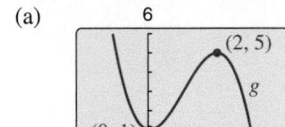 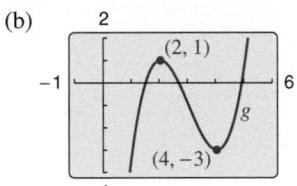

50. Think About It What is the minimum degree of the polynomial function whose graph approximates the given graph? What sign must the leading coefficient have?

(a)   (b)

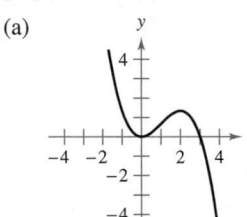 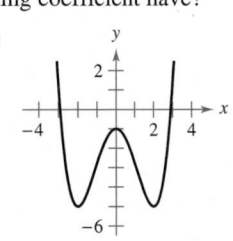

(c)   (d)

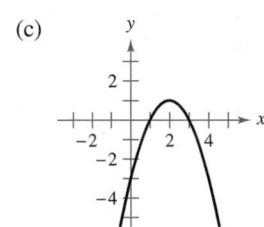 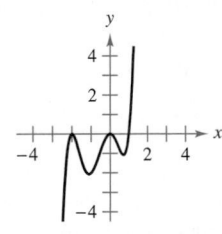

Finding Composite Functions In Exercises 51 and 52, find the composite functions $f \circ g$ and $g \circ f$. Find the domain of each composite function. Are the two composite functions equal?

51. $f(x) = 3x + 1$ **52.** $f(x) = \sqrt{x - 2}$

    $g(x) = -x$ $g(x) = x^2$

Even and Odd Functions and Zeros of Functions In Exercises 53 and 54, determine whether the function is even, odd, or neither. Then find the zeros of the function. Use a graphing utility to verify your result.

53. $f(x) = x^4 - x^2$ **54.** $f(x) = \sqrt{x^3 + 1}$

Degrees to Radians In Exercises 55–58, convert the degree measure to radian measure as a multiple of π and as a decimal accurate to three decimal places.

55. $340°$ **56.** $300°$

57. $-480°$ **58.** $-900°$

Radians to Degrees In Exercises 59–62, convert the radian measure to degree measure.

59. $\dfrac{\pi}{6}$ **60.** $\dfrac{11\pi}{4}$

61. $-\dfrac{2\pi}{3}$ **62.** $-\dfrac{13\pi}{6}$

Evaluating Trigonometric Functions In Exercises 63–68, evaluate the sine, cosine, and tangent of the angle. Do not use a calculator.

63. $-45°$ **64.** $240°$

65. $\dfrac{13\pi}{6}$ **66.** $-\dfrac{4\pi}{3}$

67. $405°$ **68.** $180°$

Evaluating Trigonometric Functions Using Technology In Exercises 69–74, use a calculator to evaluate the trigonometric function. Round your answers to four decimal places.

69. $\tan 33°$ **70.** $\cot 401°$

71. $\sec \dfrac{12\pi}{5}$ **72.** $\csc \dfrac{2\pi}{9}$

73. $\sin\left(-\dfrac{\pi}{9}\right)$ **74.** $\cos\left(-\dfrac{3\pi}{7}\right)$

Solving a Trigonometric Equation In Exercises 75–80, solve the equation for θ, where $0 \le \theta \le 2\pi$.

75. $2 \cos \theta + 1 = 0$ **76.** $2 \cos^2 \theta = 1$

77. $2 \sin^2 \theta + 3 \sin \theta + 1 = 0$

78. $\cos^3 \theta = \cos \theta$

79. $\sec^2 \theta - \sec \theta - 2 = 0$

80. $2 \sec^2 \theta + \tan^2 \theta - 5 = 0$

Sketching the Graph of a Trigonometric Function In Exercises 81–88, sketch the graph of the function.

81. $y = 9 \cos x$ **82.** $y = \sin \pi x$

83. $y = 3 \sin \dfrac{2x}{5}$ **84.** $y = 8 \cos \dfrac{x}{4}$

85. $y = \dfrac{1}{3} \tan x$ **86.** $y = \cot \dfrac{x}{2}$

87. $y = -\sec 2\pi x$ **88.** $y = -4 \csc 3x$

Verifying Inverse Functions In Exercises 89 and 90, show that f and g are inverse functions (a) analytically and (b) graphically.

89. $f(x) = 4x - 1$, $g(x) = \dfrac{x + 1}{4}$

90. $f(x) = \sqrt{x + 3}$, $g(x) = x^2 - 3$, $x \ge 0$

The Existence of an Inverse Function In Exercises 91–94, determine whether the function is one-to-one on its entire domain and therefore has an inverse function.

91. $f(x) = x^2 + 2x - 1$ **92.** $f(x) = 2 - x^3$

93. $f(x) = \csc 3\pi x$ **94.** $f(x) = 2x + \cos x$

Finding an Inverse Function In Exercises 95–100, (a) find the inverse function of f, (b) graph f and f^{-1} on the same set of coordinate axes, (c) verify that $f^{-1}(f(x)) = x$ and $f(f^{-1}(x)) = x$, and (d) state the domains and ranges of f and f^{-1}.

95. $f(x) = \frac{1}{2}x - 3$ **96.** $f(x) = 5x - 7$

97. $f(x) = \sqrt{x + 1}$ **98.** $f(x) = x^3 + 2$

99. $f(x) = \sqrt[3]{x + 1}$ **100.** $f(x) = x^2 - 5, \; x \geq 0$

Testing Whether a Function Is One-to-One In Exercises 101 and 102, determine whether the function is one-to-one. If it is, find its inverse function.

101. $f(x) = \sin 3\pi x$ **102.** $f(x) = |x + 3|, \; x \geq -3$

Showing a Function Is One-to-One In Exercises 103 and 104, show that f is one-to-one on the given interval and therefore has an inverse function on that interval.

103. $f(x) = x^4 + 2x^2 + 2, \; [0, \infty)$

104. $f(x) = \sin \dfrac{x}{2}, \; [-\pi, \pi]$

Sketching a Graph In Exercises 105 and 106, sketch the graph of the function. Use a graphing utility to verify your graph.

105. $f(x) = 2 \arctan(x + 3)$

106. $h(x) = -3 \arcsin 2x$

Evaluating Inverse Trigonometric Functions In Exercises 107 and 108, evaluate the expression without using a calculator.

107. $\arctan 1$ **108.** $\operatorname{arccsc}\left(-\dfrac{2\sqrt{3}}{3}\right)$

Solving an Equation In Exercises 109 and 110, solve the equation for x.

109. $\arccos(2x + 1) = 2$ **110.** $\operatorname{arcsec} 2x = \arctan x$

Evaluating an Expression In Exercises 111 and 112, evaluate each expression without using a calculator. (*Hint:* Make a sketch of a right triangle.)

111. (a) $\sin\left(\arcsin \frac{1}{2}\right)$ **112.** (a) $\tan(\operatorname{arccot} 2)$

(b) $\cos\left(\arcsin \frac{1}{2}\right)$ (b) $\cos(\operatorname{arcsec} \sqrt{5})$

Simplifying an Expression In Exercises 113 and 114, write the expression in algebraic form.

113. $\sin(\arctan 2x)$ **114.** $\cos\left(\operatorname{arccsc} \dfrac{x}{4}\right)$

Using Properties of Exponents In Exercises 115 and 116, use the properties of exponents to simplify the expressions.

115. (a) $(3^3)(3^{-1})$ (b) $(3^2)^4$ (c) $\dfrac{3^4}{3^{-2}}$ (d) $\left(\dfrac{1}{3}\right)^2 9^2$

116. (a) $(e^2)^{-2}$ (b) $(e^6)^{1/3}$ (c) $\dfrac{e^{-1}}{e^{-2}}$ (d) $\left(\dfrac{e}{4}\right)^{-2}$

Solving an Equation In Exercises 117 and 118, solve for x.

117. $3^{x/2} = 81$ **118.** $\left(\dfrac{1}{7}\right)^{x+1} = 49$

Sketching the Graph of an Exponential Function In Exercises 119 and 120, sketch the graph of the function.

119. $y = e^{-x/2}$ **120.** $y = 4e^{-x^2}$

Matching In Exercises 121–124, match the function with its graph. [The graphs are labeled (a), (b), (c), and (d).]

(a) 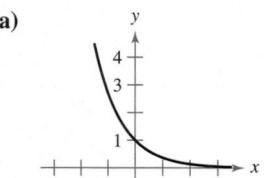   (b)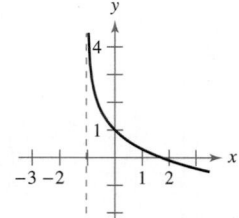

(c) 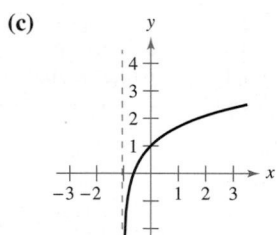   (d)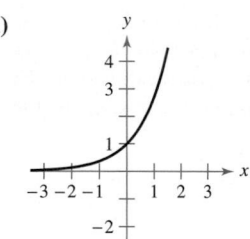

121. $f(x) = e^x$ **122.** $f(x) = e^{-x}$

123. $f(x) = \ln(x + 1) + 1$ **124.** $f(x) = -\ln(x + 1) + 1$

Sketching the Graph of a Logarithmic Function In Exercises 125 and 126, sketch the graph of the function and state its domain.

125. $f(x) = \ln x + 3$ **126.** $f(x) = \ln(x - 1)$

Expanding a Logarithmic Expression In Exercises 127 and 128, use the properties of logarithms to expand the logarithmic expression.

127. $\ln \sqrt[5]{\dfrac{4x^2 - 1}{4x^2 + 1}}$ **128.** $\ln[(x^2 + 1)(x - 1)]$

Condensing a Logarithmic Expression In Exercises 129 and 130, write the expression as the logarithm of a single quantity.

129. $\ln 3 + \frac{1}{3}\ln(4 - x^2) - \ln x$

130. $3[\ln x - 2\ln(x^2 + 1)] + 2\ln 5$

Solving an Exponential or Logarithmic Equation In Exercises 131 and 132, solve for x accurate to three decimal places.

131. $-4 + 3e^{-2x} = 6$ **132.** $\ln x + \ln(x - 3) = 0$

P.S. Problem Solving

See **CalcChat.com** for tutorial help and worked-out solutions to odd-numbered exercises.

1. Finding Tangent Lines Consider the circle

$$x^2 + y^2 - 6x - 8y = 0$$

as shown in the figure.

(a) Find the center and radius of the circle.

(b) Find an equation of the tangent line to the circle at the point $(0, 0)$.

(c) Find an equation of the tangent line to the circle at the point $(6, 0)$.

(d) Where do the two tangent lines intersect?

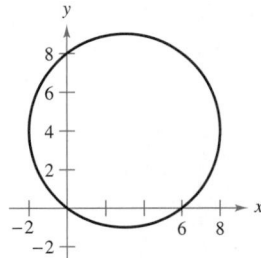

Figure for 1   Figure for 2

2. Finding Tangent Lines There are two tangent lines from the point $(0, 1)$ to the circle $x^2 + (y + 1)^2 = 1$ (see figure). Find equations of these two lines by using the fact that each tangent line intersects the circle at *exactly* one point.

3. Heaviside Function The Heaviside function $H(x)$ is widely used in engineering applications.

$$H(x) = \begin{cases} 1, & x \geq 0 \\ 0, & x < 0 \end{cases}$$

Sketch the graph of the Heaviside function and the graphs of the following functions by hand.

(a) $H(x) - 2$ (b) $H(x - 2)$ (c) $-H(x)$

(d) $H(-x)$ (e) $\frac{1}{2}H(x)$ (f) $-H(x - 2) + 2$

OLIVER HEAVISIDE (1850–1925)

Heaviside was a British mathematician and physicist who contributed to the field of applied mathematics, especially applications of mathematics to electrical engineering. The *Heaviside function* is a classic type of "on-off" function that has applications to electricity and computer science.

4. Sketching Transformations Consider the graph of the function f shown below. Use this graph to sketch the graphs of the following functions. To print an enlarged copy of the graph, go to *MathGraphs.com*.

(a) $f(x + 1)$ (b) $f(x) + 1$

(c) $2f(x)$ (d) $f(-x)$

(e) $-f(x)$ (f) $|f(x)|$

(g) $f(|x|)$

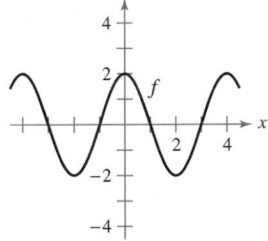

5. Maximum Area A rancher plans to fence a rectangular pasture adjacent to a river. The rancher has 100 meters of fencing, and no fencing is needed along the river (see figure).

(a) Write the area A of the pasture as a function of x, the length of the side parallel to the river. What is the domain of A?

(b) Graph the area function and estimate the dimensions that yield the maximum amount of area for the pasture.

(c) Find the dimensions that yield the maximum amount of area for the pasture by completing the square.

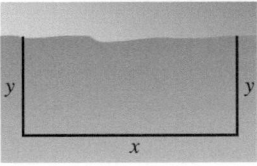   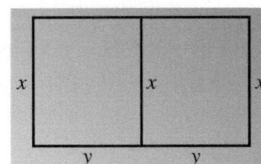

Figure for 5   Figure for 6

6. Maximum Area A rancher has 300 feet of fencing to enclose two adjacent pastures (see figure).

(a) Write the total area A of the two pastures as a function of x. What is the domain of A?

(b) Graph the area function and estimate the dimensions that yield the maximum amount of area for the pastures.

(c) Find the dimensions that yield the maximum amount of area for the pastures by completing the square.

7. Writing a Function You are in a boat 2 miles from the nearest point on the coast. You will travel to a point Q located 3 miles down the coast and 1 mile inland (see figure). You can row at 2 miles per hour and walk at 4 miles per hour. Write the total time T of the trip as a function of x.

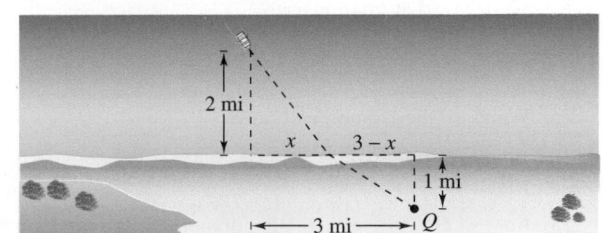

8. Analyzing a Function Graph the function

$$f(x) = e^x - e^{-x}.$$

From the graph, the function appears to be one-to-one. Assuming that the function has an inverse, find $f^{-1}(x)$.

9. Slope of a Tangent Line One of the fundamental themes of calculus is to find the slope of the tangent line to a curve at a point. To see how this can be done, consider the point $(2, 4)$ on the graph of $f(x) = x^2$ (see figure).

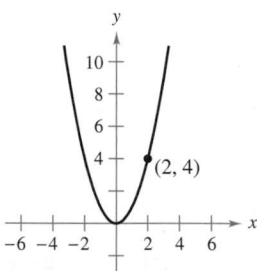

(a) Find the slope of the line joining $(2, 4)$ and $(3, 9)$. Is the slope of the tangent line at $(2, 4)$ greater than or less than this number?

(b) Find the slope of the line joining $(2, 4)$ and $(1, 1)$. Is the slope of the tangent line at $(2, 4)$ greater than or less than this number?

(c) Find the slope of the line joining $(2, 4)$ and $(2.1, 4.41)$. Is the slope of the tangent line at $(2, 4)$ greater than or less than this number?

(d) Find the slope of the line joining $(2, 4)$ and $(2 + h, f(2 + h))$ in terms of the nonzero number h. Verify that $h = 1, -1$, and 0.1 yield the solutions to parts (a)–(c) above.

(e) What is the slope of the tangent line at $(2, 4)$? Explain how you arrived at your answer.

10. Slope of a Tangent Line Sketch the graph of the function $f(x) = \sqrt{x}$ and label the point $(4, 2)$ on the graph.

(a) Find the slope of the line joining $(4, 2)$ and $(9, 3)$. Is the slope of the tangent line at $(4, 2)$ greater than or less than this number?

(b) Find the slope of the line joining $(4, 2)$ and $(1, 1)$. Is the slope of the tangent line at $(4, 2)$ greater than or less than this number?

(c) Find the slope of the line joining $(4, 2)$ and $(4.41, 2.1)$. Is the slope of the tangent line at $(4, 2)$ greater than or less than this number?

(d) Find the slope of the line joining $(4, 2)$ and $(4 + h, f(4 + h))$ in terms of the nonzero number h.

(e) What is the slope of the tangent line at $(4, 2)$? Explain how you arrived at your answer.

11. Composite Functions Let $f(x) = \dfrac{1}{1 - x}$.

(a) What are the domain and range of f?

(b) Find the composition $f(f(x))$. What is the domain of this function?

(c) Find $f(f(f(x)))$. What is the domain of this function?

(d) Graph $f(f(f(x)))$. Is the graph a line? Why or why not?

12. Graphing an Equation Explain how you would graph the equation

$$y + |y| = x + |x|.$$

Then sketch the graph.

13. Sound Intensity A large room contains two speakers that are 3 meters apart. The sound intensity I of one speaker is twice that of the other, as shown in the figure. (To print an enlarged copy of the graph, go to *MathGraphs.com*.) Suppose the listener is free to move about the room to find those positions that receive equal amounts of sound from both speakers. Such a location satisfies two conditions: (1) the sound intensity at the listener's position is directly proportional to the sound level of a source, and (2) the sound intensity is inversely proportional to the square of the distance from the source.

(a) Find the points on the x-axis that receive equal amounts of sound from both speakers.

(b) Find and graph the equation of all locations (x, y) where one could stand and receive equal amounts of sound from both speakers.

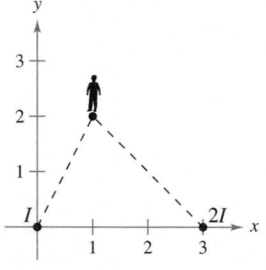

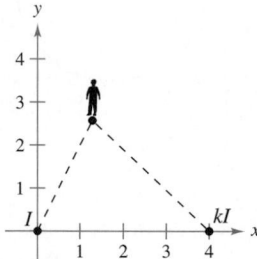

Figure for 13   Figure for 14

14. Sound Intensity Suppose the speakers in Exercise 13 are 4 meters apart and the sound intensity of one speaker is k times that of the other, as shown in the figure. To print an enlarged copy of the graph, go to *MathGraphs.com*.

(a) Find the equation of all locations (x, y) where one could stand and receive equal amounts of sound from both speakers.

(b) Graph the equation for the case $k = 3$.

(c) Describe the set of locations of equal sound as k becomes very large.

15. Lemniscate Let d_1 and d_2 be the distances from the point (x, y) to the points $(-1, 0)$ and $(1, 0)$, respectively, as shown in the figure. Show that the equation of the graph of all points (x, y) satisfying $d_1 d_2 = 1$ is

$$(x^2 + y^2)^2 = 2(x^2 - y^2).$$

This curve is called a **lemniscate**. Graph the lemniscate and identify three points on the graph.

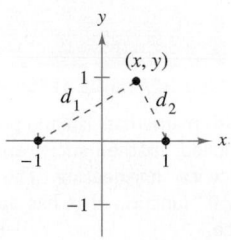

2 Limits and Their Properties

Average Speed *(Exercise 64, p. 113)*

Charles's Law and Absolute Zero *(Example 5, p. 98)*

Free-Falling Object *(Exercises 105 and 106, p. 93)*

Sports *(Exercise 70, p. 81)*

Bicyclist *(Exercise 5, p. 71)*

2.1 A Preview of Calculus

■ Understand what calculus is and how it compares with precalculus.
■ Understand that the tangent line problem is basic to calculus.
■ Understand that the area problem is also basic to calculus.

What Is Calculus?

Calculus is the mathematics of change. For instance, calculus is the mathematics of velocities, accelerations, tangent lines, slopes, areas, volumes, arc lengths, centroids, curvatures, and a variety of other concepts that have enabled scientists, engineers, and economists to model real-life situations.

Although precalculus mathematics also deals with velocities, accelerations, tangent lines, slopes, and so on, there is a fundamental difference between precalculus mathematics and calculus. Precalculus mathematics is more static, whereas calculus is more dynamic. Here are some examples.

• An object traveling at a constant velocity can be analyzed with precalculus mathematics. To analyze the velocity of an accelerating object, you need calculus.

• The slope of a line can be analyzed with precalculus mathematics. To analyze the slope of a curve, you need calculus.

• The curvature of a circle is constant and can be analyzed with precalculus mathematics. To analyze the variable curvature of a general curve, you need calculus.

• The area of a rectangle can be analyzed with precalculus mathematics. To analyze the area under a general curve, you need calculus.

Each of these situations involves the same general strategy—the reformulation of precalculus mathematics through the use of a limit process. So, one way to answer the question "What is calculus?" is to say that calculus is a "limit machine" that involves three stages. The first stage is precalculus mathematics, such as the slope of a line or the area of a rectangle. The second stage is the limit process, and the third stage is a new calculus formulation, such as a derivative or integral.

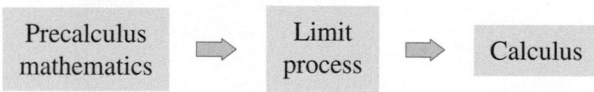

Some students try to learn calculus as if it were simply a collection of new formulas. This is unfortunate. If you reduce calculus to the memorization of differentiation and integration formulas, you will miss a great deal of understanding, self-confidence, and satisfaction.

On the next two pages are listed some familiar precalculus concepts coupled with their calculus counterparts. Throughout the text, your goal should be to learn how precalculus formulas and techniques are used as building blocks to produce the more general calculus formulas and techniques. Do not worry if you are unfamiliar with some of the "old formulas" listed on the next two pages—you will be reviewing all of them.

As you proceed through this text, come back to this discussion repeatedly. Try to keep track of where you are relative to the three stages involved in the study of calculus. For instance, note how these chapters relate to the three stages.

Chapter 1: Preparation for Calculus — Precalculus

Chapter 2: Limits and Their Properties — Limit process

Chapter 3: Differentiation — Calculus

This cycle is repeated many times on a smaller scale throughout the text.

Without Calculus	**With Differential Calculus**
Value of $f(x)$ when $x = c$	Limit of $f(x)$ as x approaches c
Slope of a line	Slope of a curve
Secant line to a curve	Tangent line to a curve
Average rate of change between $t = a$ and $t = b$	Instantaneous rate of change at $t = c$
Curvature of a circle	Curvature of a curve
Height of a curve when $x = c$	Maximum height of a curve on an interval
Tangent plane to a sphere	Tangent plane to a surface
Direction of motion along a line	Direction of motion along a curve

Without Calculus	**With Integral Calculus**
Area of a rectangle	Area under a curve
Work done by a constant force	Work done by a variable force
Center of a rectangle	Centroid of a region
Length of a line segment	Length of an arc
Surface area of a cylinder	Surface area of a solid of revolution
Mass of a solid of constant density	Mass of a solid of variable density
Volume of a rectangular solid	Volume of a region under a surface
Sum of a finite number of terms $\qquad a_1 + a_2 + \cdots + a_n = S$	Sum of an infinite number of terms $\qquad a_1 + a_2 + a_3 + \cdots = S$

The Tangent Line Problem

The notion of a limit is fundamental to the study of calculus. The following brief descriptions of two classic problems in calculus—*the tangent line problem* and *the area problem*—should give you some idea of the way limits are used in calculus.

In the tangent line problem, you are given a function f and a point P on its graph and are asked to find an equation of the tangent line to the graph at point P, as shown in Figure 2.1.

Except for cases involving a vertical tangent line, the problem of finding the **tangent line** at a point P is equivalent to finding the *slope* of the tangent line at P. You can approximate this slope by using a line through the point of tangency and a second point on the curve, as shown in Figure 2.2(a). Such a line is called a **secant line.** If $P(c, f(c))$ is the point of tangency and

$$Q(c + \Delta x, f(c + \Delta x))$$

is a second point on the graph of f, then the slope of the secant line through these two points can be found using precalculus and is

$$m_{sec} = \frac{f(c + \Delta x) - f(c)}{c + \Delta x - c} = \frac{f(c + \Delta x) - f(c)}{\Delta x}.$$

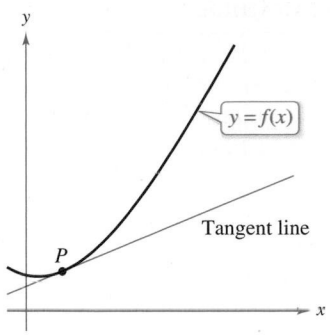

The tangent line to the graph of f at P
Figure 2.1

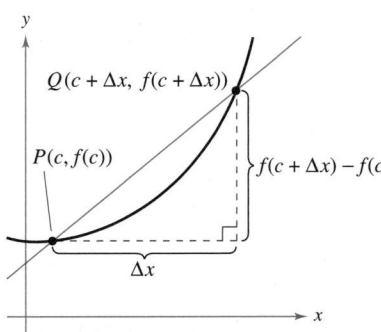
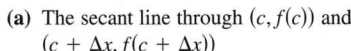

(a) The secant line through $(c, f(c))$ and $(c + \Delta x, f(c + \Delta x))$

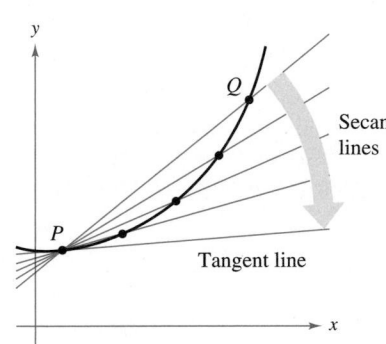

(b) As Q approaches P, the secant lines approach the tangent line.

Figure 2.2

As point Q approaches point P, the slopes of the secant lines approach the slope of the tangent line, as shown in Figure 2.2(b). When such a "limiting position" exists, the slope of the tangent line is said to be the **limit** of the slopes of the secant lines. (Much more will be said about this important calculus concept in Chapter 3.)

**GRACE CHISHOLM YOUNG
(1868–1944)**

Grace Chisholm Young received her degree in mathematics from Girton College in Cambridge, England. Her early work was published under the name of William Young, her husband. Between 1914 and 1916, Grace Young published work on the foundations of calculus that won her the Gamble Prize from Girton College.

Exploration

The following points lie on the graph of $f(x) = x^2$.

$$Q_1(1.5, f(1.5)), \quad Q_2(1.1, f(1.1)), \quad Q_3(1.01, f(1.01)),$$
$$Q_4(1.001, f(1.001)), \quad Q_5(1.0001, f(1.0001))$$

Each successive point gets closer to the point $P(1, 1)$. Find the slopes of the secant lines through Q_1 and P, Q_2 and P, and so on. Graph these secant lines on a graphing utility. Then use your results to estimate the slope of the tangent line to the graph of f at the point P.

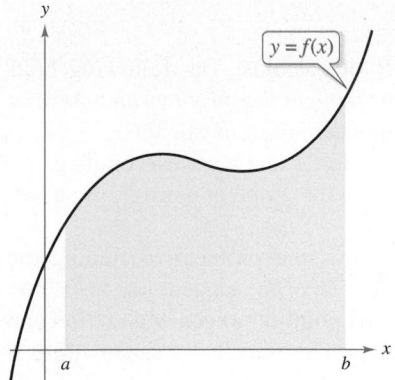

Area under a curve
Figure 2.3

The Area Problem

In the tangent line problem, you saw how the limit process can be applied to the slope of a line to find the slope of a general curve. A second classic problem in calculus is finding the area of a plane region that is bounded by the graphs of functions. This problem can also be solved with a limit process. In this case, the limit process is applied to the area of a rectangle to find the area of a general region.

As a simple example, consider the region bounded by the graph of the function $y = f(x)$, the x-axis, and the vertical lines $x = a$ and $x = b$, as shown in Figure 2.3. You can approximate the area of the region with several rectangular regions, as shown in Figure 2.4. As you increase the number of rectangles, the approximation tends to become better and better because the amount of area missed by the rectangles decreases. Your goal is to determine the limit of the sum of the areas of the rectangles as the number of rectangles increases without bound.

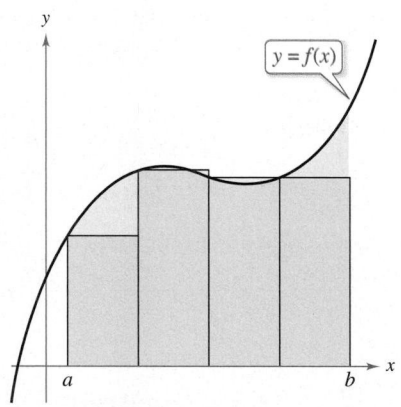

Approximation using four rectangles

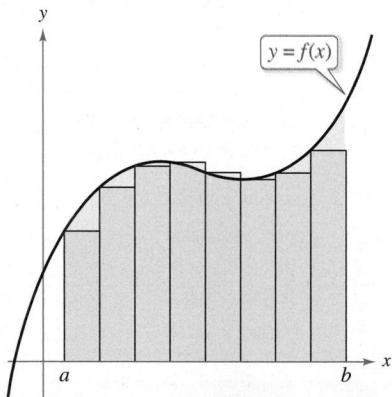

Approximation using eight rectangles

Figure 2.4

Exploration

Consider the region bounded by the graphs of

$$f(x) = x^2, \quad y = 0, \quad \text{and} \quad x = 1$$

as shown in part (a) of the figure. The area of the region can be approximated by two sets of rectangles—one set inscribed within the region and the other set circumscribed over the region, as shown in parts (b) and (c). Find the sum of the areas of each set of rectangles. Then use your results to approximate the area of the region.

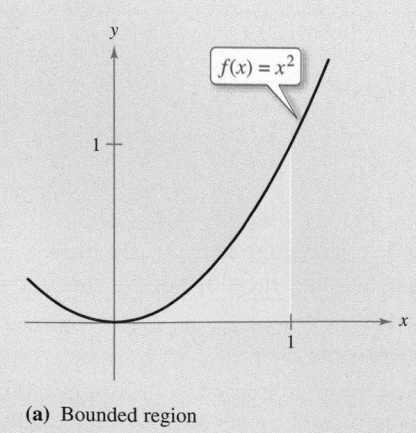

(a) Bounded region

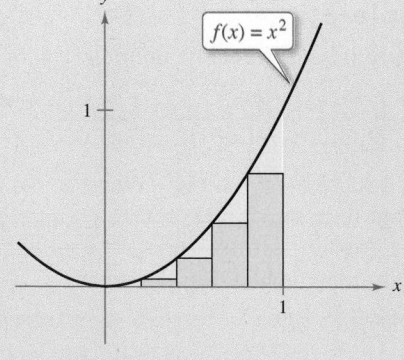
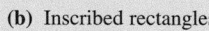

(b) Inscribed rectangles

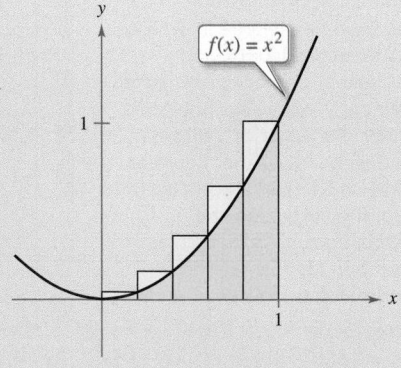

(c) Circumscribed rectangles

2.1 Exercises

See **CalcChat.com** for tutorial help and worked-out solutions to odd-numbered exercises.

CONCEPT CHECK

1. **Precalculus and Calculus** Describe the relationship between precalculus and calculus. List three precalculus concepts and their corresponding calculus counterparts.

2. **Secant and Tangent Lines** Discuss the relationship between secant lines through a fixed point and a corresponding tangent line at that fixed point.

 **Precalculus or Calculus** In Exercises 3–6, decide whether the problem can be solved using precalculus or whether calculus is required. If the problem can be solved using precalculus, solve it. If the problem seems to require calculus, explain your reasoning and use a graphical or numerical approach to estimate the solution.

3. Find the distance traveled in 15 seconds by an object traveling at a constant velocity of 20 feet per second.

4. Find the distance traveled in 15 seconds by an object moving with a velocity of $v(t) = 20 + 7 \cos t$ feet per second.

5. **Rate of Change**

A bicyclist is riding on a path modeled by the function $f(x) = 0.04(8x - x^2)$, where x and $f(x)$ are measured in miles (see figure). Find the rate of change of elevation at $x = 2$.

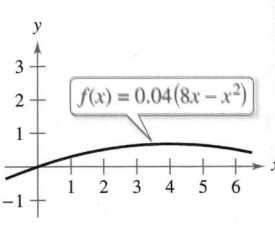

6. A bicyclist is riding on a path modeled by the function $f(x) = 0.08x$, where x and $f(x)$ are measured in miles (see figure). Find the rate of change of elevation at $x = 2$.

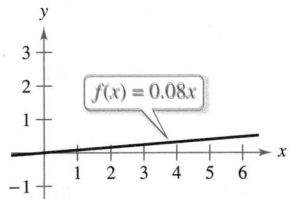

7. **Secant Lines** Consider the function $f(x) = \sqrt{x}$ and the point $P(4, 2)$ on the graph of f.

(a) Graph f and the secant lines passing through $P(4, 2)$ and $Q(x, f(x))$ for x-values of 1, 3, and 5.

(b) Find the slope of each secant line.

(c) Use the results of part (b) to estimate the slope of the tangent line to the graph of f at $P(4, 2)$. Describe how to improve your approximation of the slope.

8. **Secant Lines** Consider the function $f(x) = 6x - x^2$ and the point $P(2, 8)$ on the graph of f.

(a) Graph f and the secant lines passing through $P(2, 8)$ and $Q(x, f(x))$ for x-values of 3, 2.5, and 1.5.

(b) Find the slope of each secant line.

(c) Use the results of part (b) to estimate the slope of the tangent line to the graph of f at $P(2, 8)$. Describe how to improve your approximation of the slope.

9. **Approximating Area** Use the rectangles in each graph to approximate the area of the region bounded by $y = 5/x$, $y = 0$, $x = 1$, and $x = 5$. Describe how you could continue this process to obtain a more accurate approximation of the area.

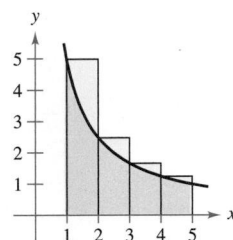 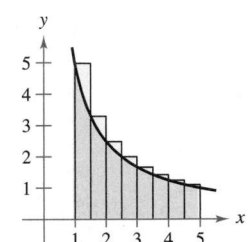

10. **HOW DO YOU SEE IT?** How would you describe the instantaneous rate of change of an automobile's position on a highway?

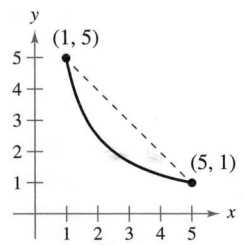

EXPLORING CONCEPTS

11. **Length of a Curve** Consider the length of the graph of $f(x) = 5/x$ from $(1, 5)$ to $(5, 1)$.

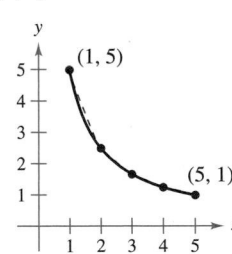

(a) Approximate the length of the curve by finding the distance between its two endpoints, as shown in the first figure.

(b) Approximate the length of the curve by finding the sum of the lengths of four line segments, as shown in the second figure.

(c) Describe how you could continue this process to obtain a more accurate approximation of the length of the curve.

2.2 Finding Limits Graphically and Numerically

■ Estimate a limit using a numerical or graphical approach.
■ Learn different ways that a limit can fail to exist.
■ Study and use a formal definition of limit.

An Introduction to Limits

To sketch the graph of the function

$$f(x) = \frac{x^3 - 1}{x - 1}$$

for values other than $x = 1$, you can use standard curve-sketching techniques. At $x = 1$, however, it is not clear what to expect. To get an idea of the behavior of the graph of f near $x = 1$, you can use two sets of x-values—one set that approaches 1 from the left and one set that approaches 1 from the right, as shown in the table.

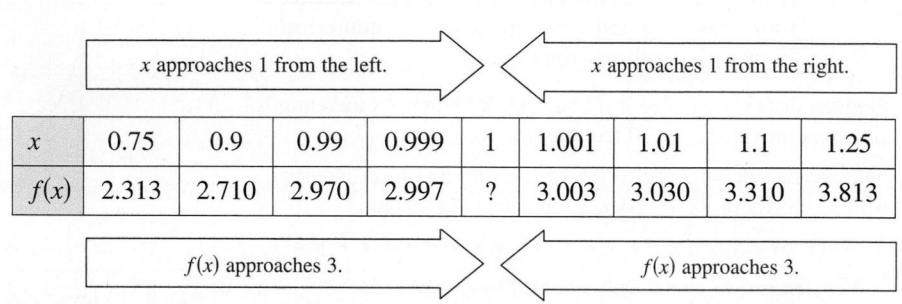

	x approaches 1 from the left.					x approaches 1 from the right.			
x	0.75	0.9	0.99	0.999	1	1.001	1.01	1.1	1.25
$f(x)$	2.313	2.710	2.970	2.997	?	3.003	3.030	3.310	3.813
	f(x) approaches 3.					f(x) approaches 3.			

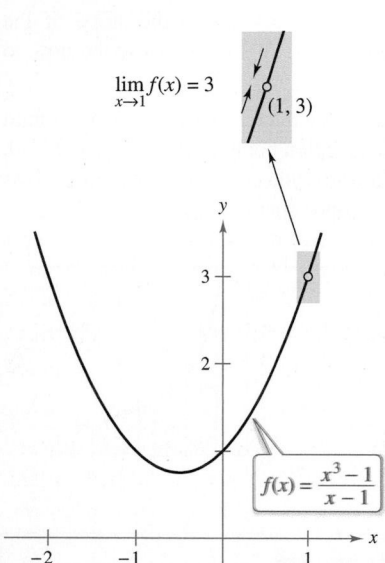

$\lim\limits_{x \to 1} f(x) = 3$

(1, 3)

$f(x) = \dfrac{x^3 - 1}{x - 1}$

The limit of $f(x)$ as x approaches 1 is 3.
Figure 2.5

The graph of f is a parabola that has a hole at the point $(1, 3)$, as shown in Figure 2.5. Although x cannot equal 1, you can move arbitrarily close to 1, and as a result $f(x)$ moves arbitrarily close to 3. Using limit notation, you can write

$$\lim_{x \to 1} f(x) = 3. \qquad \text{This is read as "the limit of } f(x) \text{ as } x \text{ approaches 1 is 3."}$$

This discussion leads to an informal definition of limit. If $f(x)$ becomes arbitrarily close to a single number L as x approaches c from either side, then the **limit** of $f(x)$ as x approaches c is L. This limit is written as

$$\lim_{x \to c} f(x) = L.$$

Exploration

The discussion above gives an example of how you can estimate a limit *numerically* by constructing a table and *graphically* by drawing a graph. Estimate the following limit numerically by completing the table.

$$\lim_{x \to 2} \frac{x^2 - 3x + 2}{x - 2}$$

x	1.75	1.9	1.99	1.999	2	2.001	2.01	2.1	2.25
$f(x)$	?	?	?	?	?	?	?	?	?

Then use a graphing utility to estimate the limit graphically.

EXAMPLE 1   **Estimating a Limit Numerically**

Evaluate the function $f(x) = x/(\sqrt{x+1} - 1)$ at several x-values near 0 and use the results to estimate the limit

$$\lim_{x \to 0} \frac{x}{\sqrt{x+1} - 1}.$$

Solution   The table lists the values of $f(x)$ for several x-values near 0.

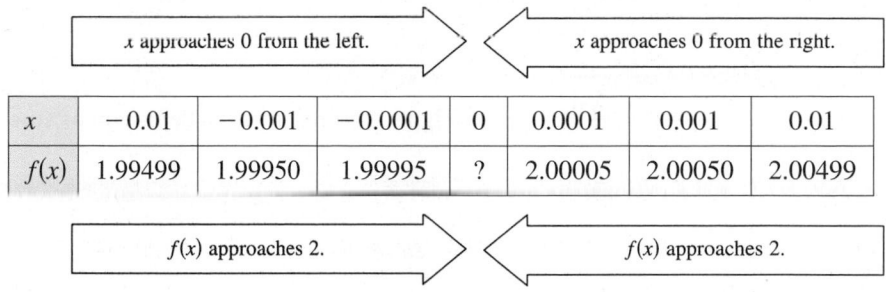

	x approaches 0 from the left.				x approaches 0 from the right.		
x	-0.01	-0.001	-0.0001	0	0.0001	0.001	0.01
$f(x)$	1.99499	1.99950	1.99995	?	2.00005	2.00050	2.00499

$f(x)$ approaches 2. $f(x)$ approaches 2.

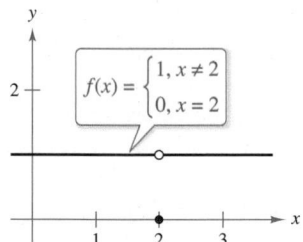

f is undefined at $x = 0$.

$$f(x) = \frac{x}{\sqrt{x+1} - 1}$$

The limit of $f(x)$ as x approaches 0 is 2.
Figure 2.6

From the results shown in the table, you can estimate the limit to be 2. This limit is reinforced by the graph of f shown in Figure 2.6. ∎

In Example 1, note that the function is undefined at $x = 0$, and yet $f(x)$ appears to be approaching a limit as x approaches 0. This often happens, and it is important to realize that *the existence or nonexistence of $f(x)$ at $x = c$ has no bearing on the existence of the limit of $f(x)$ as x approaches c.*

EXAMPLE 2   **Finding a Limit**

Find the limit of $f(x)$ as x approaches 2, where

$$f(x) = \begin{cases} 1, & x \neq 2 \\ 0, & x = 2 \end{cases}.$$

Solution   Because $f(x) = 1$ for all x other than $x = 2$, you can estimate that the limit is 1, as shown in Figure 2.7. So, you can write

$$\lim_{x \to 2} f(x) = 1.$$

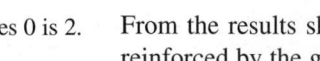

$$f(x) = \begin{cases} 1, & x \neq 2 \\ 0, & x = 2 \end{cases}$$

The limit of $f(x)$ as x approaches 2 is 1.
Figure 2.7

The fact that $f(2) = 0$ has no bearing on the existence or value of the limit as x approaches 2. For instance, as x approaches 2, the function

$$g(x) = \begin{cases} 1, & x \neq 2 \\ 2, & x = 2 \end{cases}$$

has the same limit as f. ∎

So far in this section, you have been estimating limits numerically and graphically. Each of these approaches produces an estimate of the limit. In Section 2.3, you will study analytic techniques for evaluating limits. Throughout the course, try to develop a habit of using this three-pronged approach to problem solving.

1. Numerical approach   Construct a table of values.
2. Graphical approach   Draw a graph by hand or using technology.
3. Analytic approach   Use algebra or calculus.

Limits That Fail to Exist

In the next three examples, you will examine some limits that fail to exist.

EXAMPLE 3   Different Right and Left Behavior

Show that the limit $\lim\limits_{x \to 0} \dfrac{|x|}{x}$ does not exist.

Solution   Consider the graph of the function

$$f(x) = \frac{|x|}{x}.$$

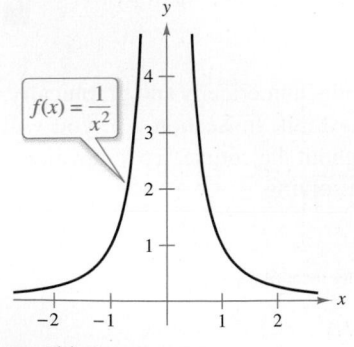

$\lim\limits_{x \to 0} f(x)$ does not exist.

Figure 2.8

In Figure 2.8 and from the definition of absolute value,

$$|x| = \begin{cases} x, & x \geq 0 \\ -x, & x < 0 \end{cases} \qquad \text{Definition of absolute value}$$

you can see that

$$\frac{|x|}{x} = \begin{cases} 1, & x > 0 \\ -1, & x < 0 \end{cases}.$$

So, no matter how close x gets to 0, there will be both positive and negative x-values that yield $f(x) = 1$ or $f(x) = -1$. Specifically, if δ (the lowercase Greek letter delta) is a positive number, then for x-values satisfying the inequality $0 < |x| < \delta$, you can classify the values of $|x|/x$ as -1 or 1 on the intervals

$$(-\delta, 0) \qquad \text{or} \qquad (0, \delta).$$

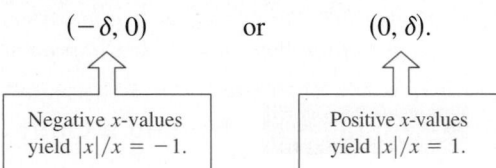

Negative x-values yield $|x|/x = -1$.

Positive x-values yield $|x|/x = 1$.

Because $|x|/x$ approaches a different number from the right side of 0 than it approaches from the left side, the limit $\lim\limits_{x \to 0} (|x|/x)$ does not exist.

EXAMPLE 4   Unbounded Behavior

Discuss the existence of the limit $\lim\limits_{x \to 0} \dfrac{1}{x^2}$.

Solution   Consider the graph of the function

$$f(x) = \frac{1}{x^2}.$$

In Figure 2.9, you can see that as x approaches 0 from either the right or the left, $f(x)$ increases without bound. This means that by choosing x close enough to 0, you can force $f(x)$ to be as large as you want. For instance, $f(x)$ will be greater than 100 when you choose x within $\frac{1}{10}$ of 0. That is,

$$0 < |x| < \frac{1}{10} \quad \Longrightarrow \quad f(x) = \frac{1}{x^2} > 100.$$

Similarly, you can force $f(x)$ to be greater than 1,000,000, as shown.

$$0 < |x| < \frac{1}{1000} \quad \Longrightarrow \quad f(x) = \frac{1}{x^2} > 1,000,000$$

$\lim\limits_{x \to 0} f(x)$ does not exist.

Figure 2.9

Because $f(x)$ does not become arbitrarily close to a single number L as x approaches 0, you can conclude that the limit does not exist.

EXAMPLE 5   **Oscillating Behavior**

⋯▷ *See LarsonCalculus.com for an interactive version of this type of example.*

Discuss the existence of the limit $\lim\limits_{x \to 0} \sin \dfrac{1}{x}$.

Solution   Let $f(x) = \sin(1/x)$. In Figure 2.10, you can see that as x approaches 0, $f(x)$ oscillates between -1 and 1. So, the limit does not exist because no matter how small you choose δ, it is possible to choose x_1 and x_2 within δ units of 0 such that $\sin(1/x_1) = 1$ and $\sin(1/x_2) = -1$, as shown in the table.

x	$\dfrac{2}{\pi}$	$\dfrac{2}{3\pi}$	$\dfrac{2}{5\pi}$	$\dfrac{2}{7\pi}$	$\dfrac{2}{9\pi}$	$\dfrac{2}{11\pi}$	$x \to 0$
$\sin \dfrac{1}{x}$	1	-1	1	-1	1	-1	Limit does not exist.

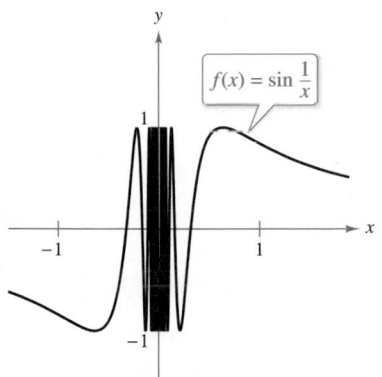

$\lim\limits_{x \to 0} f(x)$ does not exist.
Figure 2.10

Common Types of Behavior Associated with Nonexistence of a Limit

1. $f(x)$ approaches a different number from the right side of c than it approaches from the left side.
2. $f(x)$ increases or decreases without bound as x approaches c.
3. $f(x)$ oscillates between two fixed values as x approaches c.

In addition to $f(x) = \sin(1/x)$, there are many other interesting functions that have unusual limit behavior. An often cited one is the *Dirichlet function*

$$f(x) = \begin{cases} 0, & \text{if } x \text{ is rational} \\ 1, & \text{if } x \text{ is irrational} \end{cases}.$$

Because this function has *no limit* at any real number c, it is *not continuous* at any real number c. You will study continuity more closely in Section 2.4.

▷ **TECHNOLOGY PITFALL**   When you use a graphing utility to investigate the behavior of a function near the x-value at which you are trying to evaluate a limit, remember that you cannot always trust the graphs that graphing utilities draw. When you use a graphing utility to graph the function in Example 5 over an interval containing 0, you will most likely obtain an incorrect graph such as that shown in Figure 2.11. The reason that a graphing utility cannot show the correct graph is that the graph has infinitely many oscillations over any interval that contains 0.

PETER GUSTAV DIRICHLET (1805–1859)

In the early development of calculus, the definition of a function was much more restricted than it is today, and "functions" such as the Dirichlet function would not have been considered. The modern definition of function is attributed to the German mathematician Peter Gustav Dirichlet.
See LarsonCalculus.com to read more of this biography.

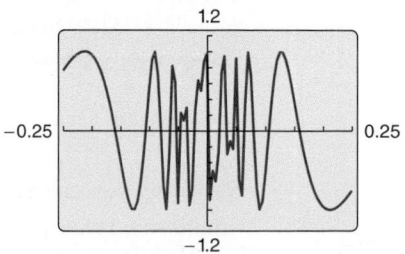

Incorrect graph of $f(x) = \sin(1/x)$
Figure 2.11

■ **FOR FURTHER INFORMATION**
For more on the introduction of
rigor to calculus, see "Who Gave
You the Epsilon? Cauchy and the
Origins of Rigorous Calculus"
by Judith V. Grabiner in *The
American Mathematical Monthly*.
To view this article, go to
MathArticles.com.

A Formal Definition of Limit

Consider again the informal definition of limit. If $f(x)$ becomes arbitrarily close to a single number L as x approaches c from either side, then the limit of $f(x)$ as x approaches c is L, written as

$$\lim_{x \to c} f(x) = L.$$

At first glance, this definition looks fairly technical. Even so, it is informal because exact meanings have not yet been given to the two phrases

"$f(x)$ becomes arbitrarily close to L"

and

"x approaches c."

The first person to assign mathematically rigorous meanings to these two phrases was Augustin-Louis Cauchy. His **ε-δ definition of limit** is the standard used today.

In Figure 2.12, let ε (the lowercase Greek letter epsilon) represent a (small) positive number. Then the phrase "$f(x)$ becomes arbitrarily close to L" means that $f(x)$ lies in the interval $(L - \varepsilon, L + \varepsilon)$. Using absolute value, you can write this as

$$|f(x) - L| < \varepsilon.$$

Similarly, the phrase "x approaches c" means that there exists a positive number δ such that x lies in either the interval $(c - \delta, c)$ or the interval $(c, c + \delta)$. This fact can be concisely expressed by the double inequality

$$0 < |x - c| < \delta.$$

The first inequality

$$0 < |x - c| \qquad \text{The distance between } x \text{ and } c \text{ is more than } 0.$$

expresses the fact that $x \neq c$. The second inequality

$$|x - c| < \delta \qquad x \text{ is within } \delta \text{ units of } c.$$

says that x is within a distance δ of c.

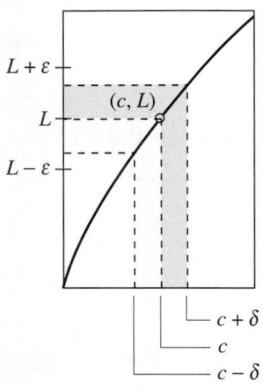

The ε-δ definition of the limit of $f(x)$ as x approaches c

Figure 2.12

Definition of Limit

Let f be a function defined on an open interval containing c (except possibly at c), and let L be a real number. The statement

$$\lim_{x \to c} f(x) = L$$

means that for each $\varepsilon > 0$ there exists a $\delta > 0$ such that if

$$0 < |x - c| < \delta$$

then

$$|f(x) - L| < \varepsilon.$$

REMARK Throughout this text, the expression

$$\lim_{x \to c} f(x) = L$$

implies two statements—the limit exists *and* the limit is L.

Some functions do not have limits as x approaches c, but those that do cannot have two different limits as x approaches c. That is, *if the limit of a function exists, then the limit is unique* (see Exercise 83).

The next three examples should help you develop a better understanding of the ε-δ definition of limit.

EXAMPLE 6   Finding a δ for a Given ε

Given the limit

$$\lim_{x \to 3} (2x - 5) = 1$$

find δ such that

$$|(2x - 5) - 1| < 0.01$$

whenever

$$0 < |x - 3| < \delta.$$

· · REMARK   In Example 6, note that 0.005 is the *largest* value of δ that will guarantee

$$|(2x - 5) - 1| < 0.01$$

whenever

$$0 < |x - 3| < \delta.$$

Any *smaller* positive value of δ would also work.

Solution   In this problem, you are working with a given value of ε—namely, $\varepsilon = 0.01$. To find an appropriate δ, try to establish a connection between the absolute values

$$|(2x - 5) - 1| \quad \text{and} \quad |x - 3|.$$

Notice that

$$|(2x - 5) - 1| = |2x - 6| = 2|x - 3|.$$

Because the inequality $|(2x - 5) - 1| < 0.01$ is equivalent to $2|x - 3| < 0.01$, you can choose

$$\delta = \tfrac{1}{2}(0.01) = 0.005.$$

This choice works because

$$0 < |x - 3| < 0.005$$

implies that

$$|(2x - 5) - 1| = 2|x - 3| < 2(0.005) = 0.01.$$

As you can see in Figure 2.13, for x-values within 0.005 of 3 ($x \neq 3$), the values of $f(x)$ are within 0.01 of 1.

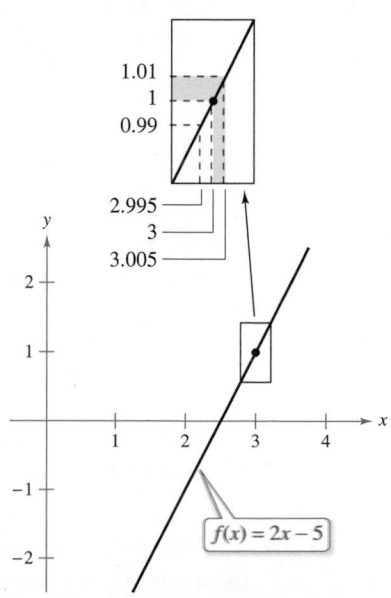

The limit of $f(x)$ as x approaches 3 is 1.
Figure 2.13

In Example 6, you found a δ-value for a *given* ε. This does not prove the existence of the limit. To do that, you must prove that you can find a δ for *any* ε, as shown in the next example.

EXAMPLE 7 Using the ε-δ Definition of Limit

Use the ε-δ definition of limit to prove that

$$\lim_{x \to 2} (3x - 2) = 4.$$

Solution You must show that for each $\varepsilon > 0$, there exists a $\delta > 0$ such that

$$|(3x - 2) - 4| < \varepsilon$$

whenever

$$0 < |x - 2| < \delta.$$

Because your choice of δ depends on ε, you need to establish a connection between the absolute values $|(3x - 2) - 4|$ and $|x - 2|$.

$$|(3x - 2) - 4| = |3x - 6| = 3|x - 2|$$

So, for a given $\varepsilon > 0$, you can choose $\delta = \varepsilon/3$. This choice works because

$$0 < |x - 2| < \delta = \frac{\varepsilon}{3}$$

implies that

$$|(3x - 2) - 4| = 3|x - 2| < 3\left(\frac{\varepsilon}{3}\right) = \varepsilon.$$

As you can see in Figure 2.14, for x-values within δ of 2 ($x \neq 2$), the values of $f(x)$ are within ε of 4.

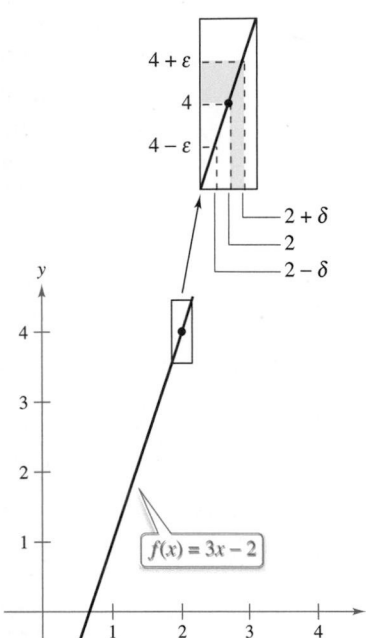

The limit of $f(x)$ as x approaches 2 is 4.
Figure 2.14

EXAMPLE 8 Using the ε-δ Definition of Limit

Use the ε-δ definition of limit to prove that $\lim_{x \to 2} x^2 = 4$.

Solution You must show that for each $\varepsilon > 0$, there exists a $\delta > 0$ such that

$$|x^2 - 4| < \varepsilon$$

whenever

$$0 < |x - 2| < \delta.$$

To find an appropriate δ, begin by writing $|x^2 - 4| = |x - 2||x + 2|$. You are interested in values of x close to 2, so choose x in the interval $(1, 3)$. To satisfy this restriction, let $\delta < 1$. Furthermore, for all x in the interval $(1, 3)$, $x + 2 < 5$ and thus $|x + 2| < 5$. So, letting δ be the minimum of $\varepsilon/5$ and 1, it follows that, whenever $0 < |x - 2| < \delta$, you have

$$|x^2 - 4| = |x - 2||x + 2| < \left(\frac{\varepsilon}{5}\right)(5) = \varepsilon.$$

As you can see in Figure 2.15, for x-values within δ of 2 ($x \neq 2$), the values of $f(x)$ are within ε of 4.

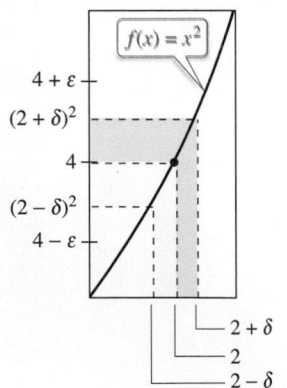

The limit of $f(x)$ as x approaches 2 is 4.
Figure 2.15

Throughout this chapter, you will use the ε-δ definition of limit primarily to prove theorems about limits and to establish the existence or nonexistence of particular types of limits. For *finding* limits, you will learn techniques that are easier to use than the ε-δ definition of limit.

2.2 Exercises

See **CalcChat.com** for tutorial help and worked-out solutions to odd-numbered exercises.

CONCEPT CHECK

1. Describing Notation Write a brief description of the meaning of the notation $\lim\limits_{x\to 8} f(x) = 25$.

2. Limits That Fail to Exist Identify three types of behavior associated with the nonexistence of a limit. Illustrate each type with a graph of a function.

3. Formal Definition of Limit Given the limit

$$\lim_{x\to 2} (2x + 1) = 5$$

use a sketch to show the meaning of the phrase "$0 < |x - 2| < 0.25$ implies $|(2x + 1) - 5| < 0.5$."

4. Functions and Limits Is the limit of $f(x)$ as x approaches c always equal to $f(c)$? Why or why not?

Estimating a Limit Numerically In Exercises 5–10, complete the table and use the result to estimate the limit. Use a graphing utility to graph the function to confirm your result.

5. $\lim\limits_{x\to 4} \dfrac{x - 4}{x^2 - 5x + 4}$

x	3.9	3.99	3.999	4	4.001	4.01	4.1
$f(x)$				?			

6. $\lim\limits_{x\to 0} \dfrac{\sqrt{x + 1} - 1}{x}$

x	−0.1	−0.01	−0.001	0	0.001	0.01	0.1
$f(x)$				?			

7. $\lim\limits_{x\to 0} \dfrac{\sin x}{x}$

x	−0.1	−0.01	−0.001	0	0.001	0.01	0.1
$f(x)$				?			

8. $\lim\limits_{x\to 0} \dfrac{\cos x - 1}{x}$

x	−0.1	−0.01	−0.001	0	0.001	0.01	0.1
$f(x)$				?			

9. $\lim\limits_{x\to 0} \dfrac{e^x - 1}{x}$

x	−0.1	−0.01	−0.001	0	0.001	0.01	0.1
$f(x)$				?			

10. $\lim\limits_{x\to 0} \dfrac{\ln(x + 1)}{x}$

x	−0.1	−0.01	−0.001	0	0.001	0.01	0.1
$f(x)$				?			

Estimating a Limit Numerically In Exercises 11–20, create a table of values for the function and use the result to estimate the limit. Use a graphing utility to graph the function to confirm your result.

11. $\lim\limits_{x\to 1} \dfrac{x - 2}{x^2 + x - 6}$

12. $\lim\limits_{x\to -4} \dfrac{x + 4}{x^2 + 9x + 20}$

13. $\lim\limits_{x\to 1} \dfrac{x^4 - 1}{x^6 - 1}$

14. $\lim\limits_{x\to -3} \dfrac{x^3 + 27}{x + 3}$

15. $\lim\limits_{x\to -6} \dfrac{\sqrt{10 - x} - 4}{x + 6}$

16. $\lim\limits_{x\to 2} \dfrac{[x/(x + 1)] - (2/3)}{x - 2}$

17. $\lim\limits_{x\to 0} \dfrac{\sin 2x}{x}$

18. $\lim\limits_{x\to 0} \dfrac{\tan x}{\tan 2x}$

19. $\lim\limits_{x\to 2} \dfrac{\ln x - \ln 2}{x - 2}$

20. $\lim\limits_{x\to 1} \dfrac{1 - x}{e - e^x}$

Limits That Fail to Exist In Exercises 21 and 22, create a table of values for the function and use the result to explain why the limit does not exist.

21. $\lim\limits_{x\to 0} \dfrac{4}{1 + e^{1/x}}$

22. $\lim\limits_{x\to 0} \dfrac{3|x|}{x^2}$

Finding a Limit Graphically In Exercises 23–30, use the graph to find the limit (if it exists). If the limit does not exist, explain why.

23. $\lim\limits_{x\to 3} (4 - x)$

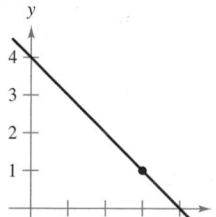

24. $\lim\limits_{x\to 0} \sec x$

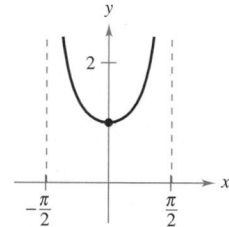

25. $\lim\limits_{x\to 2} f(x)$

$$f(x) = \begin{cases} 4 - x, & x \ne 2 \\ 0, & x = 2 \end{cases}$$

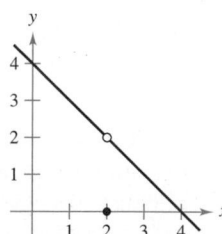

26. $\lim\limits_{x\to 1} f(x)$

$$f(x) = \begin{cases} x^2 + 3, & x \ne 1 \\ 2, & x = 1 \end{cases}$$

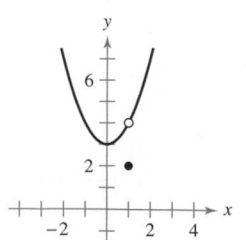

27. $\lim\limits_{x \to 2} \dfrac{|x-2|}{x-2}$

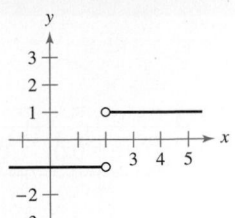

28. $\lim\limits_{x \to 0} \dfrac{4}{2 + e^{1/x}}$

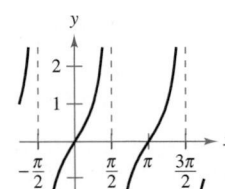

29. $\lim\limits_{x \to 0} \cos \dfrac{1}{x}$

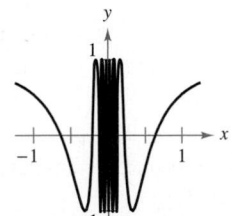

30. $\lim\limits_{x \to \pi/2} \tan x$

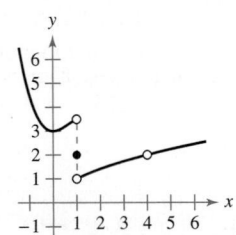

Graphical Reasoning In Exercises 31 and 32, use the graph of the function f to decide whether the value of the given quantity exists. If it does, find it. If not, explain why.

31. (a) $f(1)$

(b) $\lim\limits_{x \to 1} f(x)$

(c) $f(4)$

(d) $\lim\limits_{x \to 4} f(x)$

32. (a) $f(-2)$

(b) $\lim\limits_{x \to -2} f(x)$

(c) $f(0)$

(d) $\lim\limits_{x \to 0} f(x)$

(e) $f(2)$

(f) $\lim\limits_{x \to 2} f(x)$

(g) $f(4)$

(h) $\lim\limits_{x \to 4} f(x)$

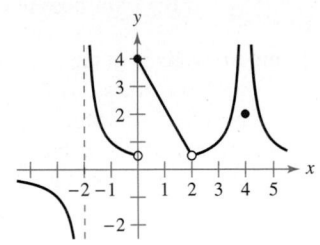

Limits of a Piecewise Function In Exercises 33 and 34, sketch the graph of f. Then identify the values of c for which $\lim\limits_{x \to c} f(x)$ exists.

33. $f(x) = \begin{cases} x^2, & x \le 2 \\ 8 - 2x, & 2 < x < 4 \\ 4, & x \ge 4 \end{cases}$

34. $f(x) = \begin{cases} \sin x, & x < 0 \\ 1 - \cos x, & 0 \le x \le \pi \\ \cos x, & x > \pi \end{cases}$

Sketching a Graph In Exercises 35 and 36, sketch a graph of a function f that satisfies the given values. (There are many correct answers.)

35. $f(0)$ is undefined.

$\lim\limits_{x \to 0} f(x) = 4$

$f(2) = 6$

$\lim\limits_{x \to 2} f(x) = 3$

36. $f(-2) = 0$

$f(2) = 0$

$\lim\limits_{x \to -2} f(x) = 0$

$\lim\limits_{x \to 2} f(x)$ does not exist.

37. Finding a δ for a Given ε The graph of $f(x) = x + 1$ is shown in the figure. Find δ such that if $0 < |x - 2| < \delta$, then $|f(x) - 3| < 0.4$.

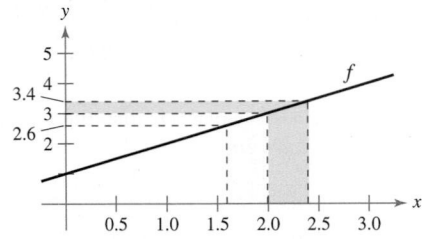

38. Finding a δ for a Given ε The graph of

$$f(x) = \frac{1}{x - 1}$$

is shown in the figure. Find δ such that if $0 < |x - 2| < \delta$, then $|f(x) - 1| < 0.01$.

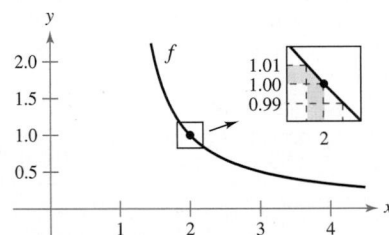

39. Finding a δ for a Given ε The graph of

$$f(x) = 2 - \frac{1}{x}$$

is shown in the figure. Find δ such that if $0 < |x - 1| < \delta$, then $|f(x) - 1| < 0.1$.

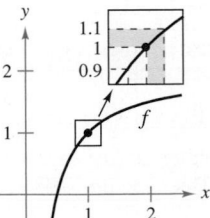

40. Finding a δ for a Given ε Repeat Exercise 39 for $\varepsilon = 0.05, 0.01,$ and 0.005. What happens to the value of δ as the value of ε gets smaller?

Finding a δ for a Given ε In Exercises 41–46, find the limit L. Then find δ such that $|f(x) - L| < \varepsilon$ whenever $0 < |x - c| < \delta$ for (a) $\varepsilon = 0.01$ and (b) $\varepsilon = 0.005$.

41. $\lim\limits_{x \to 2} (3x + 2)$

42. $\lim\limits_{x \to 6} \left(6 - \dfrac{x}{3}\right)$

43. $\lim\limits_{x \to 2} (x^2 - 3)$

44. $\lim\limits_{x \to 4} (x^2 + 6)$

45. $\lim\limits_{x \to 4} (x^2 - x)$

46. $\lim\limits_{x \to 3} x^2$

Using the ε-δ Definition of Limit In Exercises 47–58, find the limit L. Then use the ε-δ definition to prove that the limit is L.

47. $\lim\limits_{x \to 4} (x + 2)$

48. $\lim\limits_{x \to -2} (4x + 5)$

49. $\lim\limits_{x \to -4} \left(\frac{1}{2}x - 1\right)$

50. $\lim\limits_{x \to 3} \left(\frac{3}{4}x + 1\right)$

51. $\lim\limits_{x \to 6} 3$

52. $\lim\limits_{x \to 2} (-1)$

53. $\lim\limits_{x \to 0} \sqrt[3]{x}$

54. $\lim\limits_{x \to 4} \sqrt{x}$

55. $\lim\limits_{x \to -5} |x - 5|$

56. $\lim\limits_{x \to 3} |x - 3|$

57. $\lim\limits_{x \to 1} (x^2 + 1)$

58. $\lim\limits_{x \to -4} (x^2 + 4x)$

59. Finding a Limit What is the limit of $f(x) = 4$ as x approaches π?

60. Finding a Limit What is the limit of $g(x) = x$ as x approaches π?

Writing In Exercises 61 and 62, use a graphing utility to graph the function and estimate the limit (if it exists). What is the domain of the function? Can you detect a possible error in determining the domain of a function solely by analyzing the graph generated by a graphing utility? Write a short paragraph about the importance of examining a function analytically as well as graphically.

61. $f(x) = \dfrac{\sqrt{x + 5} - 3}{x - 4}$

$\lim\limits_{x \to 4} f(x)$

62. $f(x) = \dfrac{e^{x/2} - 1}{x}$

$\lim\limits_{x \to 0} f(x)$

63. Modeling Data For a long-distance phone call, a hotel charges $9.99 for the first minute and $0.79 for each additional minute or fraction thereof. A formula for the cost is given by

$$C(t) = 9.99 - 0.79[\![1 - t]\!], \quad t > 0$$

where t is the time in minutes.

(*Note:* $[\![x]\!]$ = greatest integer n such that $n \le x$. For example, $[\![3.2]\!] = 3$ and $[\![-1.6]\!] = -2$.)

(a) Evaluate $C(10.75)$. What does $C(10.75)$ represent?

(b) Use a graphing utility to graph the cost function for $0 < t \le 6$. Does the limit of $C(t)$ as t approaches 3 exist? Explain.

64. Modeling Data Repeat Exercise 63 for

$$C(t) = 5.79 - 0.99[\![1 - t]\!], \quad t > 0.$$

EXPLORING CONCEPTS

65. Finding δ When using the definition of limit to prove that L is the limit of $f(x)$ as x approaches c, you find the largest satisfactory value of δ. Why would any smaller positive value of δ also work?

66. Using the Definition of Limit The definition of limit on page 76 requires that f is a function defined on an open interval containing c, except possibly at c. Why is this requirement necessary?

67. Comparing Functions and Limits If $f(2) = 4$, can you conclude anything about the limit of $f(x)$ as x approaches 2? Explain your reasoning.

68. Comparing Functions and Limits If the limit of $f(x)$ as x approaches 2 is 4, can you conclude anything about $f(2)$? Explain your reasoning.

69. Jewelry A jeweler resizes a ring so that its inner circumference is 6 centimeters.

(a) What is the radius of the ring?

(b) The inner circumference of the ring varies between 5.5 centimeters and 6.5 centimeters. How does the radius vary?

(c) Use the ε-δ definition of limit to describe this situation. Identify ε and δ.

70. Sports

A sporting goods manufacturer designs a golf ball having a volume of 2.48 cubic inches.

(a) What is the radius of the golf ball?

(b) The volume of the golf ball varies between 2.45 cubic inches and 2.51 cubic inches. How does the radius vary?

(c) Use the ε-δ definition of limit to describe this situation. Identify ε and δ.

71. Estimating a Limit Consider the function

$$f(x) = (1 + x)^{1/x}.$$

Estimate

$$\lim\limits_{x \to 0} (1 + x)^{1/x}$$

by evaluating f at x-values near 0. Sketch the graph of f.

The symbol indicates an exercise in which you are instructed to use graphing technology or a symbolic computer algebra system. The solutions of other exercises may also be facilitated by the use of appropriate technology.

72. Estimating a Limit Consider the function

$$f(x) = \frac{|x+1| - |x-1|}{x}.$$

Estimate

$$\lim_{x \to 0} \frac{|x+1| - |x-1|}{x}$$

by evaluating f at x-values near 0. Sketch the graph of f.

 **73. Graphical Reasoning** The statement

$$\lim_{x \to 2} \frac{x^2 - 4}{x - 2} = 4$$

means that for each $\varepsilon > 0$ there corresponds a $\delta > 0$ such that if $0 < |x - 2| < \delta$, then

$$\left| \frac{x^2 - 4}{x - 2} - 4 \right| < \varepsilon.$$

If $\varepsilon = 0.001$, then

$$\left| \frac{x^2 - 4}{x - 2} - 4 \right| < 0.001.$$

Use a graphing utility to graph each side of this inequality. Use the *zoom* feature to find an interval $(2 - \delta, 2 + \delta)$ such that the inequality is true.

74. HOW DO YOU SEE IT? Use the graph of f to identify the values of c for which $\lim_{x \to c} f(x)$ exists.

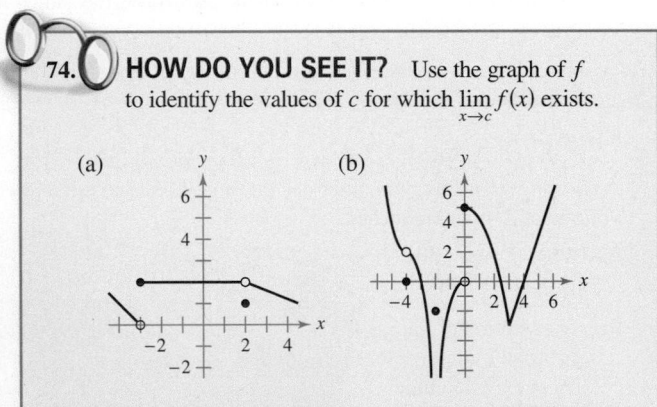

(a)

(b)

True or False? In Exercises 75–78, determine whether the statement is true or false. If it is false, explain why or give an example that shows it is false.

75. If f is undefined at $x = c$, then the limit of $f(x)$ as x approaches c does not exist.

76. If the limit of $f(x)$ as x approaches c is 0, then there must exist a number k such that $f(k) < 0.001$.

77. If $f(c) = L$, then $\lim_{x \to c} f(x) = L$.

78. If $\lim_{x \to c} f(x) = L$, then $f(c) = L$.

Determining a Limit In Exercises 79 and 80, consider the function $f(x) = \sqrt{x}$.

79. Is $\lim_{x \to 0.25} \sqrt{x} = 0.5$ a true statement? Explain.

80. Is $\lim_{x \to 0} \sqrt{x} = 0$ a true statement? Explain.

81. Evaluating Limits Use a graphing utility to evaluate

$$\lim_{x \to 0} \frac{\sin nx}{x}$$

for several values of n. What do you notice?

82. Evaluating Limits Use a graphing utility to evaluate

$$\lim_{x \to 0} \frac{\tan nx}{x}$$

for several values of n. What do you notice?

83. Proof Prove that if the limit of $f(x)$ as x approaches c exists, then the limit must be unique. [*Hint:* Let $\lim_{x \to c} f(x) = L_1$ and $\lim_{x \to c} f(x) = L_2$ and prove that $L_1 = L_2$.]

84. Proof Consider the line $f(x) = mx + b$, where $m \neq 0$. Use the ε-δ definition of limit to prove that $\lim_{x \to c} f(x) = mc + b$.

85. Proof Prove that

$$\lim_{x \to c} f(x) = L$$

is equivalent to

$$\lim_{x \to c} [f(x) - L] = 0.$$

86. Proof

(a) Given that

$$\lim_{x \to 0} (3x + 1)(3x - 1)x^2 + 0.01 = 0.01$$

prove that there exists an open interval (a, b) containing 0 such that $(3x + 1)(3x - 1)x^2 + 0.01 > 0$ for all $x \neq 0$ in (a, b).

(b) Given that $\lim_{x \to c} g(x) = L$, where $L > 0$, prove that there exists an open interval (a, b) containing c such that $g(x) > 0$ for all $x \neq c$ in (a, b).

PUTNAM EXAM CHALLENGE

87. Inscribe a rectangle of base b and height h in a circle of radius one, and inscribe an isosceles triangle in a region of the circle cut off by one base of the rectangle (with that side as the base of the triangle). For what value of h do the rectangle and triangle have the same area?

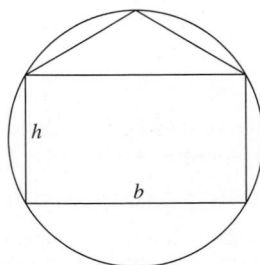

88. A right circular cone has base of radius 1 and height 3. A cube is inscribed in the cone so that one face of the cube is contained in the base of the cone. What is the side-length of the cube?

These problems were composed by the Committee on the Putnam Prize Competition. © The Mathematical Association of America. All rights reserved.

2.3 Evaluating Limits Analytically

■ Evaluate a limit using properties of limits.
■ Develop and use a strategy for finding limits.
■ Evaluate a limit using the dividing out technique.
■ Evaluate a limit using the rationalizing technique.
■ Evaluate a limit using the Squeeze Theorem.

Properties of Limits

In Section 2.2, you learned that the limit of $f(x)$ as x approaches c does not depend on the value of f at $x = c$. It may happen, however, that the limit is precisely $f(c)$. In such cases, you can evaluate the limit by **direct substitution.** That is,

$$\lim_{x \to c} f(x) = f(c). \qquad \text{Substitute } c \text{ for } x.$$

Such *well-behaved* functions are **continuous at c.** You will examine this concept more closely in Section 2.4.

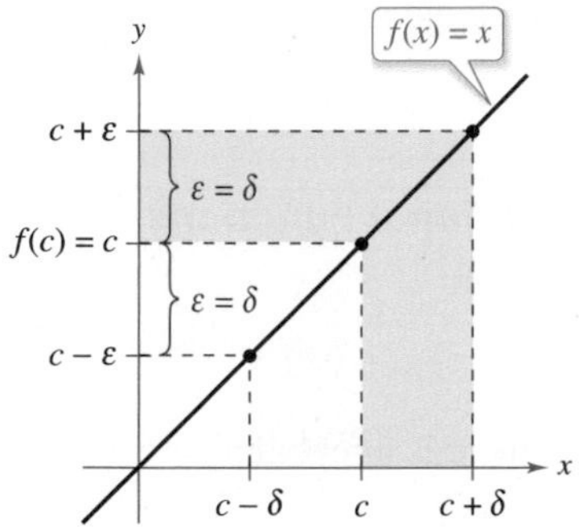

Figure 2.16

> **THEOREM 2.1 Some Basic Limits**
>
> Let b and c be real numbers, and let n be a positive integer.
>
> **1.** $\lim_{x \to c} b = b$ **2.** $\lim_{x \to c} x = c$ **3.** $\lim_{x \to c} x^n = c^n$

Proof The proofs of Properties 1 and 3 of Theorem 2.1 are left as exercises (see Exercises 111 and 112). To prove Property 2, you need to show that for each $\varepsilon > 0$ there exists a $\delta > 0$ such that $|x - c| < \varepsilon$ whenever $0 < |x - c| < \delta$. To do this, choose $\delta = \varepsilon$. The second inequality then implies the first, as shown in Figure 2.16.

EXAMPLE 1 Evaluating Basic Limits

a. $\lim_{x \to 2} 3 = 3$ **b.** $\lim_{x \to -4} x = -4$ **c.** $\lim_{x \to 2} x^2 = 2^2 = 4$

> **THEOREM 2.2 Properties of Limits**
>
> Let b and c be real numbers, let n be a positive integer, and let f and g be functions with the limits
>
> $$\lim_{x \to c} f(x) = L \quad \text{and} \quad \lim_{x \to c} g(x) = K.$$
>
> **1.** Scalar multiple: $\lim_{x \to c} [b f(x)] = bL$
>
> **2.** Sum or difference: $\lim_{x \to c} [f(x) \pm g(x)] = L \pm K$
>
> **3.** Product: $\lim_{x \to c} [f(x)g(x)] = LK$
>
> **4.** Quotient: $\lim_{x \to c} \dfrac{f(x)}{g(x)} = \dfrac{L}{K}, \quad K \neq 0$
>
> **5.** Power: $\lim_{x \to c} [f(x)]^n = L^n$
>
>
>
> The proof of Property 1 is left as an exercise (see Exercise 113). The proofs of the other properties are given in Appendix A.

The symbol 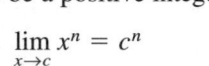 indicates that a video of this proof is available at *LarsonCalculus.com.*

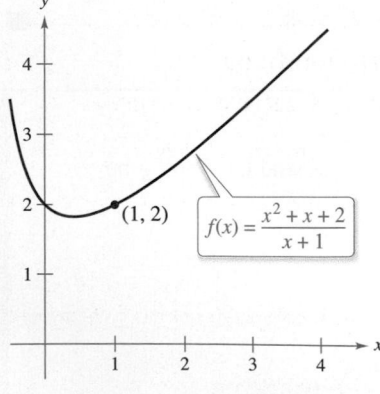

The limit of $f(x)$ as x approaches 2 is 19.
Figure 2.17

EXAMPLE 2 **The Limit of a Polynomial**

Find the limit: $\lim\limits_{x \to 2} (4x^2 + 3)$.

Solution

$$\lim_{x \to 2} (4x^2 + 3) = \lim_{x \to 2} 4x^2 + \lim_{x \to 2} 3 \qquad \text{Property 2, Theorem 2.2}$$

$$= 4\left(\lim_{x \to 2} x^2\right) + \lim_{x \to 2} 3 \qquad \text{Property 1, Theorem 2.2}$$

$$= 4(2^2) + 3 \qquad \text{Properties 1 and 3, Theorem 2.1}$$

$$= 19 \qquad \text{Simplify.}$$

This limit is reinforced by the graph of $f(x) = 4x^2 + 3$ shown in Figure 2.17. ■

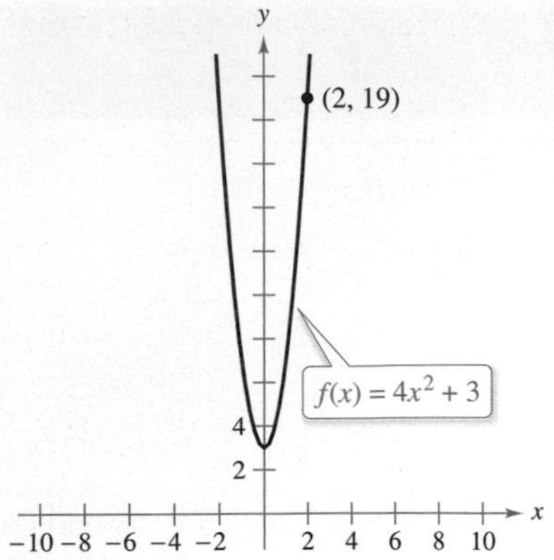

In Example 2, note that the limit (as x approaches 2) of the *polynomial function* $p(x) = 4x^2 + 3$ is simply the value of p at $x = 2$.

$$\lim_{x \to 2} p(x) = p(2) = 4(2^2) + 3 = 19$$

This *direct substitution* property is valid for all polynomial and rational functions with nonzero denominators.

THEOREM 2.3 **Limits of Polynomial and Rational Functions**

If p is a polynomial function and c is a real number, then

$$\lim_{x \to c} p(x) = p(c).$$

If r is a rational function given by $r(x) = p(x)/q(x)$ and c is a real number such that $q(c) \neq 0$, then

$$\lim_{x \to c} r(x) = r(c) = \frac{p(c)}{q(c)}.$$

EXAMPLE 3 **The Limit of a Rational Function**

Find the limit: $\lim\limits_{x \to 1} \dfrac{x^2 + x + 2}{x + 1}$.

Solution Because the denominator is not 0 when $x = 1$, you can apply Theorem 2.3 to obtain

$$\lim_{x \to 1} \frac{x^2 + x + 2}{x + 1} = \frac{1^2 + 1 + 2}{1 + 1} = \frac{4}{2} = 2. \qquad \text{See Figure 2.18.} ■$$

Polynomial functions and rational functions are two of the three basic types of algebraic functions. The next theorem deals with the limit of the third type of algebraic function—one that involves a radical.

The limit of $f(x)$ as x approaches 1 is 2.
Figure 2.18

THEOREM 2.4 **The Limit of a Function Involving a Radical**

Let n be a positive integer. The limit below is valid for all c when n is odd, and is valid for $c > 0$ when n is even.

$$\lim_{x \to c} \sqrt[n]{x} = \sqrt[n]{c}$$

A proof of this theorem is given in Appendix A.

Exploration

Your goal in this section is to become familiar with limits that can be evaluated by direct substitution. In the following list of elementary functions, what are the values of c for which

$$\lim_{x \to c} f(x) = f(c)?$$

Polynomial function:

$$f(x) = a_n x^n + \cdots + a_1 x + a_0$$

Rational function (p and q are polynomials):

$$f(x) = \frac{p(x)}{q(x)}$$

Trigonometric functions:

$$f(x) = \sin x, \quad f(x) = \cos x$$
$$f(x) = \tan x, \quad f(x) = \cot x$$
$$f(x) = \sec x, \quad f(x) = \csc x$$

Exponential functions:

$$f(x) = a^x, \quad f(x) = e^x$$

Natural logarithmic function:

$$f(x) = \ln x$$

The next theorem greatly expands your ability to evaluate limits because it shows how to analyze the limit of a composite function.

THEOREM 2.5   The Limit of a Composite Function

If f and g are functions such that $\lim_{x \to c} g(x) = L$ and $\lim_{x \to L} f(x) = f(L)$, then

$$\lim_{x \to c} f(g(x)) = f\left(\lim_{x \to c} g(x) \right) = f(L).$$

A proof of this theorem is given in Appendix A.

EXAMPLE 4   The Limit of a Composite Function

$\vdots \cdots \triangleright$ *See LarsonCalculus.com for an interactive version of this type of example.*

a. Because

$$\lim_{x \to 0} (x^2 + 4) = 0^2 + 4 = 4 \quad \text{and} \quad \lim_{x \to 4} \sqrt{x} = \sqrt{4} = 2$$

you can conclude that

$$\lim_{x \to 0} \sqrt{x^2 + 4} = \sqrt{4} = 2.$$

b. Because

$$\lim_{x \to 3} (2x^2 - 10) = 2(3^2) - 10 = 8 \quad \text{and} \quad \lim_{x \to 8} \sqrt[3]{x} = \sqrt[3]{8} = 2$$

you can conclude that

$$\lim_{x \to 3} \sqrt[3]{2x^2 - 10} = \sqrt[3]{8} = 2.$$

You have seen that the limits of many algebraic functions can be evaluated by direct substitution. The basic transcendental functions (trigonometric, exponential, and logarithmic) also possess this desirable quality, as shown in the next theorem (presented without proof).

THEOREM 2.6   Limits of Transcendental Functions

Let c be a real number in the domain of the given transcendental function.

1. $\displaystyle\lim_{x \to c} \sin x = \sin c$ **2.** $\displaystyle\lim_{x \to c} \cos x = \cos c$ **3.** $\displaystyle\lim_{x \to c} \tan x = \tan c$

4. $\displaystyle\lim_{x \to c} \cot x = \cot c$ **5.** $\displaystyle\lim_{x \to c} \sec x = \sec c$ **6.** $\displaystyle\lim_{x \to c} \csc x = \csc c$

7. $\displaystyle\lim_{x \to c} a^x = a^c, \ a > 0$ **8.** $\displaystyle\lim_{x \to c} \ln x = \ln c$

EXAMPLE 5   Limits of Transcendental Functions

a. $\displaystyle\lim_{x \to 0} \tan x = \tan(0) = 0$

b. $\displaystyle\lim_{x \to 0} \sin^2 x = \lim_{x \to 0} (\sin x)^2 = 0^2 = 0$

c. $\displaystyle\lim_{x \to -1} xe^x = \left(\lim_{x \to -1} x \right)\left(\lim_{x \to -1} e^x \right) = (-1)(e^{-1}) = -e^{-1}$

d. $\displaystyle\lim_{x \to e} \ln x^3 = \lim_{x \to e} 3 \ln x = 3 \ln(e) = 3(1) = 3$

A Strategy for Finding Limits

On the previous three pages, you studied several types of functions whose limits can be evaluated by direct substitution. This knowledge, together with the next theorem, can be used to develop a strategy for finding limits.

> **THEOREM 2.7 Functions That Agree at All but One Point**
>
> Let c be a real number, and let $f(x) = g(x)$ for all $x \neq c$ in an open interval containing c. If the limit of $g(x)$ as x approaches c exists, then the limit of $f(x)$ also exists and
>
> $$\lim_{x \to c} f(x) = \lim_{x \to c} g(x).$$
>
> A proof of this theorem is given in Appendix A.

EXAMPLE 6   **Finding the Limit of a Function**

Find the limit.

$$\lim_{x \to 1} \frac{x^3 - 1}{x - 1}$$

Solution Let $f(x) = (x^3 - 1)/(x - 1)$. By factoring and dividing out common factors, you can rewrite f as

$$f(x) = \frac{(x - 1)(x^2 + x + 1)}{(x - 1)} = x^2 + x + 1 = g(x), \quad x \neq 1.$$

So, for all x-values other than $x = 1$, the functions f and g agree, as shown in Figure 2.19. Because $\lim_{x \to 1} g(x)$ exists, you can apply Theorem 2.7 to conclude that f and g have the same limit at $x = 1$.

$$\lim_{x \to 1} \frac{x^3 - 1}{x - 1} = \lim_{x \to 1} \frac{(x - 1)(x^2 + x + 1)}{x - 1} \qquad \text{Factor.}$$

$$= \lim_{x \to 1} \frac{(x - 1)(x^2 + x + 1)}{(x - 1)} \qquad \text{Divide out common factor.}$$

$$= \lim_{x \to 1} (x^2 + x + 1) \qquad \text{Apply Theorem 2.7.}$$

$$= 1^2 + 1 + 1 \qquad \text{Use direct substitution.}$$

$$= 3 \qquad \text{Simplify.}$$

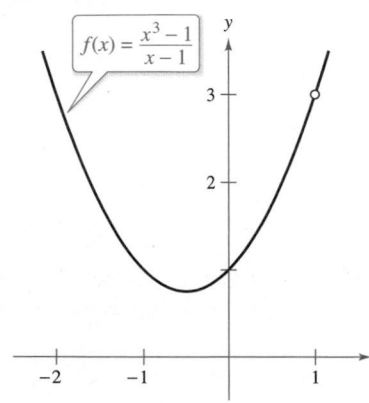

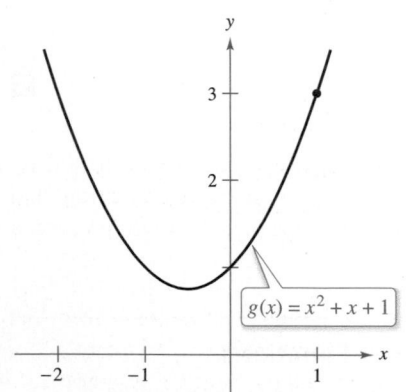

f and g agree at all but one point.
Figure 2.19

• • • • • • • • • • • • • ▷

•• **REMARK** When applying this strategy for finding a limit, remember that some functions do not have a limit (as x approaches c). For instance, the limit below does not exist.

$$\lim_{x \to 1} \frac{x^3 + 1}{x - 1}$$

A Strategy for Finding Limits

1. Learn to recognize which limits can be evaluated by direct substitution. (These limits are listed in Theorems 2.1 through 2.6.)

2. When the limit of $f(x)$ as x approaches c *cannot* be evaluated by direct substitution, try to find a function g that agrees with f for all x other than $x = c$. [Choose g such that the limit of $g(x)$ *can* be evaluated by direct substitution.] Then apply Theorem 2.7 to conclude *analytically* that

   $$\lim_{x \to c} f(x) = \lim_{x \to c} g(x) = g(c).$$

3. Use a *graph* or *table* to reinforce your conclusion.

Dividing Out Technique

Another procedure for finding a limit analytically is the **dividing out technique.** This technique involves dividing out common factors, as shown in Example 7.

EXAMPLE 7 Dividing Out Technique

⋮ • • ▷ *See LarsonCalculus.com for an interactive version of this type of example.*

Find the limit: $\displaystyle\lim_{x \to -3} \frac{x^2 + x - 6}{x + 3}$.

Solution Although you are taking the limit of a rational function, you *cannot* apply Theorem 2.3 because the limit of the denominator is 0.

$$\lim_{x \to -3} \frac{x^2 + x - 6}{x + 3} \quad\begin{array}{l} \nearrow \displaystyle\lim_{x \to -3}(x^2 + x - 6) = 0 \\[2ex] \searrow \displaystyle\lim_{x \to -3}(x + 3) = 0 \end{array}$$

Direct substitution fails.

Because the limit of the numerator is also 0, the numerator and denominator have a *common factor* of $(x + 3)$. So, for all $x \neq -3$, you can divide out this factor to obtain

$$f(x) = \frac{x^2 + x - 6}{x + 3} = \frac{\cancel{(x + 3)}(x - 2)}{\cancel{x + 3}} = x - 2 = g(x), \quad x \neq -3.$$

Using Theorem 2.7, it follows that

$$\lim_{x \to -3} \frac{x^2 + x - 6}{x + 3} = \lim_{x \to -3}(x - 2) \qquad \text{Apply Theorem 2.7.}$$

$$= -5. \qquad \text{Use direct substitution.}$$

This result is shown graphically in Figure 2.20. Note that the graph of the function f coincides with the graph of the function $g(x) = x - 2$, except that the graph of f has a hole at the point $(-3, -5)$. ◼

In Example 7, direct substitution produced the meaningless fractional form 0/0. An expression such as 0/0 is called an **indeterminate form** because you cannot (from the form alone) determine the limit. When you try to evaluate a limit and encounter this form, remember that you must rewrite the fraction so that the new denominator does not have 0 as its limit. One way to do this is to *divide out common factors.* Another way is to use the *rationalizing technique* shown on the next page.

▷ **TECHNOLOGY PITFALL** A graphing utility can give misleading information about the graph of a function. For instance, try graphing the function from Example 7

$$f(x) = \frac{x^2 + x - 6}{x + 3}$$

on a graphing utility. On some graphing utilities, the graph may appear to be defined at every real number, as shown in the figure at the right. However, because f is undefined when $x = -3$, you know that the graph of f has a hole at $x = -3$. You can verify this on a graphing utility using the *trace* or *table* feature.

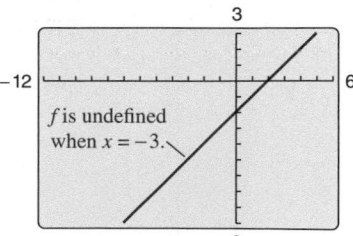
Misleading graph of f

⋮ • **REMARK** In the solution to Example 7, be sure you see the usefulness of the Factor Theorem of Algebra. This theorem states that if c is a zero of a polynomial function, then $(x - c)$ is a factor of the polynomial. So, when you apply direct substitution to a rational function and obtain

$$r(c) = \frac{p(c)}{q(c)} = \frac{0}{0}$$

you can conclude that $(x - c)$ must be a common factor of both $p(x)$ and $q(x)$.

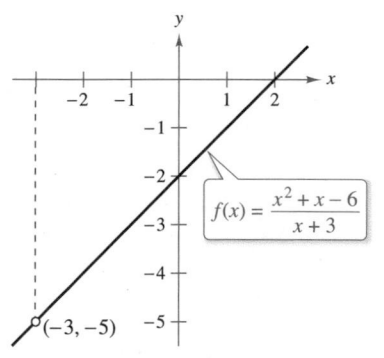

f is undefined when $x = -3$. The limit of $f(x)$ as x approaches -3 is -5.
Figure 2.20

Rationalizing Technique

Another way to find a limit analytically is the **rationalizing technique,** which involves rationalizing either the numerator or denominator of a fractional expression. Recall that rationalizing the numerator (denominator) means multiplying the numerator and denominator by the conjugate of the numerator (denominator). For instance, to rationalize the numerator of

$$\frac{\sqrt{x + 4}}{x}$$

multiply the numerator and denominator by the conjugate of $\sqrt{x} + 4$, which is

$$\sqrt{x} - 4.$$

EXAMPLE 8 Rationalizing Technique

Find the limit: $\displaystyle\lim_{x\to 0} \frac{\sqrt{x + 1} - 1}{x}$.

Solution By direct substitution, you obtain the indeterminate form $0/0$.

$$\lim_{x\to 0} \frac{\sqrt{x + 1} - 1}{x} \quad \begin{cases} \nearrow \lim_{x\to 0}\left(\sqrt{x + 1} - 1\right) = 0 \\[2mm] \searrow \lim_{x\to 0} x = 0 \end{cases}$$

Direct substitution fails.

In this case, you can rewrite the fraction by rationalizing the numerator.

$$\frac{\sqrt{x + 1} - 1}{x} = \left(\frac{\sqrt{x + 1} - 1}{x}\right)\left(\frac{\sqrt{x + 1} + 1}{\sqrt{x + 1} + 1}\right)$$

$$= \frac{(x + 1) - 1}{x\left(\sqrt{x + 1} + 1\right)}$$

$$= \frac{\cancel{x}}{\cancel{x}\left(\sqrt{x + 1} + 1\right)}$$

$$= \frac{1}{\sqrt{x + 1} + 1}, \quad x \neq 0$$

.................▷

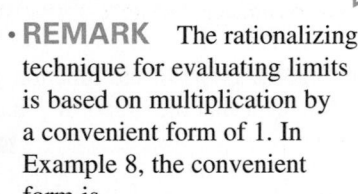

REMARK The rationalizing technique for evaluating limits is based on multiplication by a convenient form of 1. In Example 8, the convenient form is

$$1 = \frac{\sqrt{x + 1} + 1}{\sqrt{x + 1} + 1}.$$

Now, using Theorem 2.7, you can evaluate the limit as shown.

$$\lim_{x\to 0} \frac{\sqrt{x + 1} - 1}{x} = \lim_{x\to 0} \frac{1}{\sqrt{x + 1} + 1}$$

$$= \frac{1}{1 + 1}$$

$$= \frac{1}{2}$$

A table or a graph can reinforce your conclusion that the limit is $\frac{1}{2}$. (See Figure 2.21.)

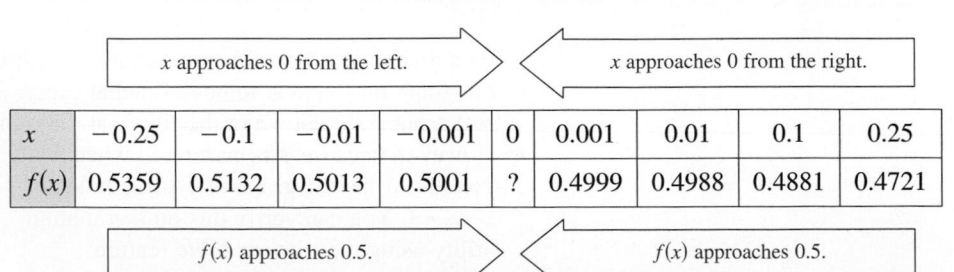

	x approaches 0 from the left.				x approaches 0 from the right.			
x	-0.25	-0.1	-0.01	-0.001	0	0.001	0.01	0.1	0.25
$f(x)$	0.5359	0.5132	0.5013	0.5001	?	0.4999	0.4988	0.4881	0.4721

$f(x)$ approaches 0.5. $f(x)$ approaches 0.5.

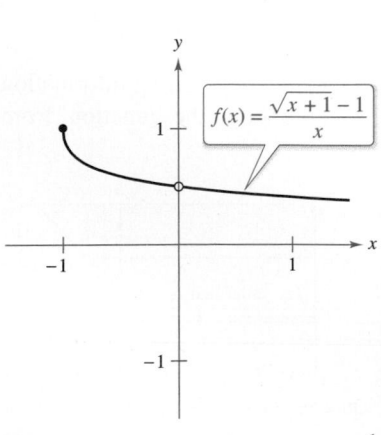

$$f(x) = \frac{\sqrt{x + 1} - 1}{x}$$

The limit of $f(x)$ as x approaches 0 is $\frac{1}{2}$.
Figure 2.21

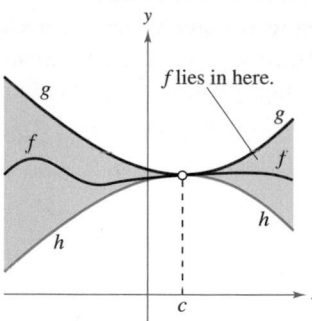

$h(x) \leq f(x) \leq g(x)$

f lies in here.

The Squeeze Theorem
Figure 2.22

•• REMARK   The third limit
of Theorem 2.9 will be used in
Section 3.2 in the development
of the formula for the derivative
of the exponential function
$f(x) = e^x$.

The Squeeze Theorem

The next theorem concerns the limit of a function that is squeezed between two other functions, each of which has the same limit at a given *x*-value, as shown in Figure 2.22.

> **THEOREM 2.8   The Squeeze Theorem**
>
> If $h(x) \leq f(x) \leq g(x)$ for all x in an open interval containing c, except possibly at c itself, and if
>
> $$\lim_{x \to c} h(x) = L - \lim_{x \to c} g(x)$$
>
> then $\lim_{x \to c} f(x)$ exists and is equal to L.
>
> A proof of this theorem is given in Appendix A.

You can see the usefulness of the Squeeze Theorem (also called the Sandwich Theorem or the Pinching Theorem) in the proof of Theorem 2.9.

> **THEOREM 2.9   Three Special Limits**
>
> **1.** $\lim_{x \to 0} \dfrac{\sin x}{x} = 1$ **2.** $\lim_{x \to 0} \dfrac{1 - \cos x}{x} = 0$ **3.** $\lim_{x \to 0} (1 + x)^{1/x} = e$

Proof   The proof of the second limit is left as an exercise (see Exercise 125). Recall from Section 1.6 that the third limit is actually the definition of the number *e*. To avoid the confusion of two different uses of *x*, the proof of the first limit is presented using the variable θ, where θ is an acute positive angle *measured in radians*. Figure 2.23 shows a circular sector that is squeezed between two triangles.

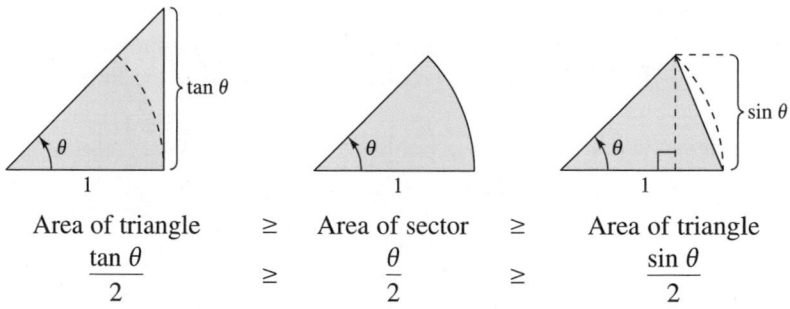

Area of triangle	$\geq$	Area of sector	$\geq$	Area of triangle
$\dfrac{\tan \theta}{2}$	$\geq$	$\dfrac{\theta}{2}$	$\geq$	$\dfrac{\sin \theta}{2}$

Multiplying each expression by $2/\sin \theta$ produces

$$\frac{1}{\cos \theta} \geq \frac{\theta}{\sin \theta} \geq 1$$

and taking reciprocals and reversing the inequalities yields

$$\cos \theta \leq \frac{\sin \theta}{\theta} \leq 1.$$

Because $\cos \theta = \cos(-\theta)$ and $(\sin \theta)/\theta = [\sin(-\theta)]/(-\theta)$, you can conclude that this inequality is valid for *all* nonzero θ in the open interval $(-\pi/2, \pi/2)$. Finally, because $\lim_{\theta \to 0} \cos \theta = 1$ and $\lim_{\theta \to 0} 1 = 1$, you can apply the Squeeze Theorem to conclude that

$$\lim_{\theta \to 0} \frac{\sin \theta}{\theta} = 1.$$

$(\cos \theta, \sin \theta)$

$(1, \tan \theta)$

$(1, 0)$

A circular sector is used to prove Theorem 2.9.
Figure 2.23

EXAMPLE 9 **A Limit Involving a Trigonometric Function**

Find the limit: $\lim\limits_{x \to 0} \dfrac{\tan x}{x}$.

Solution Direct substitution yields the indeterminate form 0/0. To solve this problem, you can write $\tan x$ as $(\sin x)/(\cos x)$ and obtain

$$\lim_{x \to 0} \frac{\tan x}{x} = \lim_{x \to 0} \left(\frac{\sin x}{x} \right) \left(\frac{1}{\cos x} \right).$$

Now, because

$$\lim_{x \to 0} \frac{\sin x}{x} = 1$$

and

$$\lim_{x \to 0} \frac{1}{\cos x} = 1$$

you can obtain

$$\lim_{x \to 0} \frac{\tan x}{x} = \left(\lim_{x \to 0} \frac{\sin x}{x} \right) \left(\lim_{x \to 0} \frac{1}{\cos x} \right)$$
$$= (1)(1)$$
$$= 1.$$

(See Figure 2.24.)

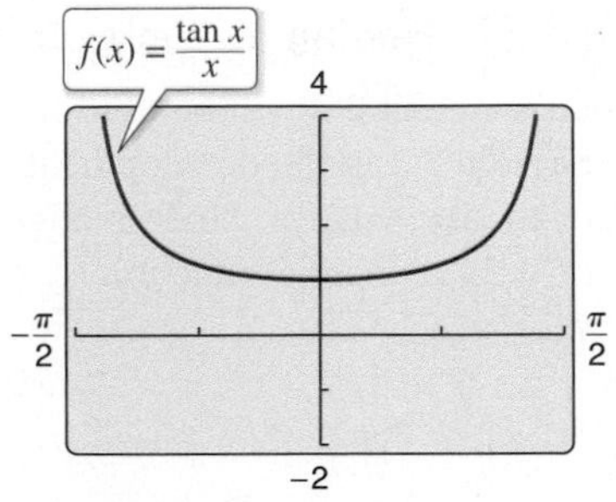

The limit of $f(x)$ as x approaches 0 is 1.
Figure 2.24

> •• **REMARK** Be sure you understand the mathematical conventions regarding parentheses and trigonometric functions. For instance, in Example 10, $\sin 4x$ means $\sin(4x)$.

EXAMPLE 10 **A Limit Involving a Trigonometric Function**

Find the limit: $\lim\limits_{x \to 0} \dfrac{\sin 4x}{x}$.

Solution Direct substitution yields the indeterminate form 0/0. To solve this problem, you can rewrite the limit as

$$\lim_{x \to 0} \frac{\sin 4x}{x} = 4 \left(\lim_{x \to 0} \frac{\sin 4x}{4x} \right). \qquad \text{Multiply and divide by 4.}$$

Now, by letting $y = 4x$ and observing that x approaches 0 if and only if y approaches 0, you can write

$$\lim_{x \to 0} \frac{\sin 4x}{x} = 4 \left(\lim_{x \to 0} \frac{\sin 4x}{4x} \right)$$
$$= 4 \left(\lim_{y \to 0} \frac{\sin y}{y} \right) \qquad \text{Let } y = 4x.$$
$$= 4(1) \qquad \text{Apply Theorem 2.9(1).}$$
$$= 4.$$

(See Figure 2.25.)

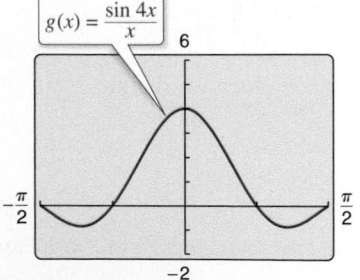

The limit of $g(x)$ as x approaches 0 is 4.
Figure 2.25

▷ **TECHNOLOGY** Use a graphing utility to confirm the limits in the examples and in the exercise set. For instance, Figures 2.24 and 2.25 show the graphs of

$$f(x) = \frac{\tan x}{x} \quad \text{and} \quad g(x) = \frac{\sin 4x}{x}.$$

Note that the first graph appears to contain the point $(0, 1)$ and the second graph appears to contain the point $(0, 4)$, which lends support to the conclusions obtained in Examples 9 and 10.

2.3 Exercises

See CalcChat.com for tutorial help and worked-out solutions to odd-numbered exercises.

CONCEPT CHECK

1. Polynomial Function Describe how to find the limit of a polynomial function $p(x)$ as x approaches c.

2. Indeterminate Form What is meant by an indeterminate form?

3. Squeeze Theorem In your own words, explain the Squeeze Theorem.

4. Special Limits List the three special limits.

Finding a Limit In Exercises 5–18, find the limit.

5. $\lim\limits_{x \to -3} (2x + 5)$

6. $\lim\limits_{x \to 9} (4x - 1)$

7. $\lim\limits_{x \to -3} (x^2 + 3x)$

8. $\lim\limits_{x \to 1} (2x^3 - 6x + 5)$

9. $\lim\limits_{x \to 3} \sqrt{x + 8}$

10. $\lim\limits_{x \to 2} \sqrt[3]{12x + 3}$

11. $\lim\limits_{x \to -4} (1 - x)^3$

12. $\lim\limits_{x \to 0} (3x - 2)^4$

13. $\lim\limits_{x \to 2} \dfrac{3}{2x + 1}$

14. $\lim\limits_{x \to -5} \dfrac{5}{x + 3}$

15. $\lim\limits_{x \to 1} \dfrac{x}{x^2 + 4}$

16. $\lim\limits_{x \to 1} \dfrac{3x + 5}{x + 1}$

17. $\lim\limits_{x \to 7} \dfrac{3x}{\sqrt{x + 2}}$

18. $\lim\limits_{x \to 3} \dfrac{\sqrt{x + 6}}{x + 2}$

Finding Limits In Exercises 19–22, find the limits.

19. $f(x) = 5 - x,\ g(x) = x^3$

  (a) $\lim\limits_{x \to 1} f(x)$ (b) $\lim\limits_{x \to 4} g(x)$ (c) $\lim\limits_{x \to 1} g(f(x))$

20. $f(x) = x + 7,\ g(x) = x^2$

  (a) $\lim\limits_{x \to -3} f(x)$ (b) $\lim\limits_{x \to 4} g(x)$ (c) $\lim\limits_{x \to -3} g(f(x))$

21. $f(x) = 4 - x^2,\ g(x) = \sqrt{x + 1}$

  (a) $\lim\limits_{x \to 1} f(x)$ (b) $\lim\limits_{x \to 3} g(x)$ (c) $\lim\limits_{x \to 1} g(f(x))$

22. $f(x) = 2x^2 - 3x + 1,\ g(x) = \sqrt[3]{x + 6}$

  (a) $\lim\limits_{x \to 4} f(x)$ (b) $\lim\limits_{x \to 21} g(x)$ (c) $\lim\limits_{x \to 4} g(f(x))$

Finding a Limit of a Transcendental Function In Exercises 23–36, find the limit of the transcendental function.

23. $\lim\limits_{x \to \pi/2} \sin x$

24. $\lim\limits_{x \to \pi} \tan x$

25. $\lim\limits_{x \to 1} \cos \dfrac{\pi x}{3}$

26. $\lim\limits_{x \to 2} \sin \dfrac{\pi x}{12}$

27. $\lim\limits_{x \to 0} \sec 2x$

28. $\lim\limits_{x \to \pi} \cos 3x$

29. $\lim\limits_{x \to 5\pi/6} \sin x$

30. $\lim\limits_{x \to 5\pi/3} \cos x$

31. $\lim\limits_{x \to 3} \tan \dfrac{\pi x}{4}$

32. $\lim\limits_{x \to 7} \sec \dfrac{\pi x}{6}$

33. $\lim\limits_{x \to 0} e^x \cos 2x$

34. $\lim\limits_{x \to 0} e^{-x} \sin \pi x$

35. $\lim\limits_{x \to 1} (\ln 3x + e^x)$

36. $\lim\limits_{x \to 1} \ln\left(\dfrac{x}{e^x}\right)$

Evaluating Limits In Exercises 37–40, use the information to evaluate the limits.

37. $\lim\limits_{x \to c} f(x) = \dfrac{2}{5}$

  $\lim\limits_{x \to c} g(x) = 2$

  (a) $\lim\limits_{x \to c} [5g(x)]$

  (b) $\lim\limits_{x \to c} [f(x) + g(x)]$

  (c) $\lim\limits_{x \to c} [f(x)g(x)]$

  (d) $\lim\limits_{x \to c} \dfrac{f(x)}{g(x)}$

38. $\lim\limits_{x \to c} f(x) = 2$

  $\lim\limits_{x \to c} g(x) = \dfrac{3}{4}$

  (a) $\lim\limits_{x \to c} [4f(x)]$

  (b) $\lim\limits_{x \to c} [f(x) + g(x)]$

  (c) $\lim\limits_{x \to c} [f(x)g(x)]$

  (d) $\lim\limits_{x \to c} \dfrac{f(x)}{g(x)}$

39. $\lim\limits_{x \to c} f(x) = 16$

  (a) $\lim\limits_{x \to c} [f(x)]^2$

  (b) $\lim\limits_{x \to c} \sqrt{f(x)}$

  (c) $\lim\limits_{x \to c} [3f(x)]$

  (d) $\lim\limits_{x \to c} [f(x)]^{3/2}$

40. $\lim\limits_{x \to c} f(x) = 27$

  (a) $\lim\limits_{x \to c} \sqrt[3]{f(x)}$

  (b) $\lim\limits_{x \to c} \dfrac{f(x)}{18}$

  (c) $\lim\limits_{x \to c} [f(x)]^2$

  (d) $\lim\limits_{x \to c} [f(x)]^{2/3}$

Finding a Limit In Exercises 41–46, write a simpler function that agrees with the given function at all but one point. Then find the limit of the function. Use a graphing utility to confirm your result.

41. $\lim\limits_{x \to -1} \dfrac{x^2 - 1}{x + 1}$

42. $\lim\limits_{x \to -2} \dfrac{3x^2 + 5x - 2}{x + 2}$

43. $\lim\limits_{x \to 2} \dfrac{x^3 - 8}{x - 2}$

44. $\lim\limits_{x \to -1} \dfrac{x^3 + 1}{x + 1}$

45. $\lim\limits_{x \to -4} \dfrac{(x + 4) \ln(x + 6)}{x^2 - 16}$

46. $\lim\limits_{x \to 0} \dfrac{e^{2x} - 1}{e^x - 1}$

Finding a Limit In Exercises 47–62, find the limit.

47. $\lim\limits_{x \to 0} \dfrac{x}{x^2 - x}$

48. $\lim\limits_{x \to 0} \dfrac{7x^3 - x^2}{x}$

49. $\lim\limits_{x \to 4} \dfrac{x - 4}{x^2 - 16}$

50. $\lim\limits_{x \to 5} \dfrac{5 - x}{x^2 - 25}$

51. $\lim\limits_{x \to -3} \dfrac{x^2 + x - 6}{x^2 - 9}$

52. $\lim\limits_{x \to 2} \dfrac{x^2 + 2x - 8}{x^2 - x - 2}$

53. $\lim\limits_{x \to 4} \dfrac{\sqrt{x + 5} - 3}{x - 4}$

54. $\lim\limits_{x \to 3} \dfrac{\sqrt{x + 1} - 2}{x - 3}$

55. $\lim\limits_{x \to 0} \dfrac{\sqrt{x + 5} - \sqrt{5}}{x}$

56. $\lim\limits_{x \to 0} \dfrac{\sqrt{2 + x} - \sqrt{2}}{x}$

57. $\lim\limits_{x \to 0} \dfrac{[1/(3 + x)] - (1/3)}{x}$ **58.** $\lim\limits_{x \to 0} \dfrac{[1/(x + 4)] - (1/4)}{x}$

59. $\lim\limits_{\Delta x \to 0} \dfrac{2(x + \Delta x) - 2x}{\Delta x}$ **60.** $\lim\limits_{\Delta x \to 0} \dfrac{(x + \Delta x)^2 - x^2}{\Delta x}$

61. $\lim\limits_{\Delta x \to 0} \dfrac{(x + \Delta x)^2 - 2(x + \Delta x) + 1 - (x^2 - 2x + 1)}{\Delta x}$

62. $\lim\limits_{\Delta x \to 0} \dfrac{(x + \Delta x)^3 - x^3}{\Delta x}$

 **Finding a Limit of a Transcendental Function** In Exercises 63–76, find the limit of the transcendental function.

63. $\lim\limits_{x \to 0} \dfrac{\sin x}{5x}$ **64.** $\lim\limits_{x \to 0} \dfrac{3(1 - \cos x)}{x}$

65. $\lim\limits_{x \to 0} \dfrac{(\sin x)(1 - \cos x)}{x^2}$ **66.** $\lim\limits_{\theta \to 0} \dfrac{\cos \theta \tan \theta}{\theta}$

67. $\lim\limits_{x \to 0} \dfrac{\sin^2 x}{x}$ **68.** $\lim\limits_{x \to 0} \dfrac{\tan^2 x}{x}$

69. $\lim\limits_{h \to 0} \dfrac{(1 - \cos h)^2}{h}$ **70.** $\lim\limits_{\phi \to \pi} \phi \sec \phi$

71. $\lim\limits_{x \to 0} \dfrac{6 - 6 \cos x}{3}$ **72.** $\lim\limits_{x \to 0} \dfrac{\cos x - \sin x - 1}{2x}$

73. $\lim\limits_{x \to 0} \dfrac{1 - e^{-x}}{e^x - 1}$ **74.** $\lim\limits_{x \to 0} \dfrac{4(e^{2x} - 1)}{e^x - 1}$

75. $\lim\limits_{t \to 0} \dfrac{\sin 3t}{2t}$

76. $\lim\limits_{x \to 0} \dfrac{\sin 2x}{\sin 3x}$ $\left[Hint\text{: Find } \lim\limits_{x \to 0} \left(\dfrac{2 \sin 2x}{2x} \right)\left(\dfrac{3x}{3 \sin 3x} \right). \right]$

 **Graphical, Numerical, and Analytic Analysis** In Exercises 77–86, use a graphing utility to graph the function and estimate the limit. Use a table to reinforce your conclusion. Then find the limit by analytic methods.

77. $\lim\limits_{x \to 0} \dfrac{\sqrt{x + 2} - \sqrt{2}}{x}$ **78.** $\lim\limits_{x \to 16} \dfrac{4 - \sqrt{x}}{x - 16}$

79. $\lim\limits_{x \to 0} \dfrac{[1/(2 + x)] - (1/2)}{x}$ **80.** $\lim\limits_{x \to 2} \dfrac{x^5 - 32}{x - 2}$

81. $\lim\limits_{t \to 0} \dfrac{\sin 3t}{t}$ **82.** $\lim\limits_{x \to 0} \dfrac{\cos x - 1}{2x^2}$

83. $\lim\limits_{x \to 0} \dfrac{\sin x^2}{x}$ **84.** $\lim\limits_{x \to 0} \dfrac{\sin x}{\sqrt[3]{x}}$

85. $\lim\limits_{x \to 1} \dfrac{\ln x}{x - 1}$ **86.** $\lim\limits_{x \to \ln 2} \dfrac{e^{3x} - 8}{e^{2x} - 4}$

 **Finding a Limit** In Exercises 87–94, find

$$\lim\limits_{\Delta x \to 0} \dfrac{f(x + \Delta x) - f(x)}{\Delta x}.$$

87. $f(x) = 3x - 2$ **88.** $f(x) = -6x + 3$

89. $f(x) = x^2 - 4x$ **90.** $f(x) = 3x^2 + 1$

91. $f(x) = 2\sqrt{x}$ **92.** $f(x) = \sqrt{x} - 5$

93. $f(x) = \dfrac{1}{x + 3}$ **94.** $f(x) = \dfrac{1}{x^2}$

Using the Squeeze Theorem In Exercises 95 and 96, use the Squeeze Theorem to find $\lim\limits_{x \to c} f(x)$.

95. $c = 0$

$$4 - x^2 \le f(x) \le 4 + x^2$$

96. $c = a$

$$b - |x - a| \le f(x) \le b + |x - a|$$

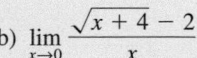 **Using the Squeeze Theorem** In Exercises 97–100, use a graphing utility to graph the given function and the equations $y = |x|$ and $y = -|x|$ in the same viewing window. Using the graphs to observe the Squeeze Theorem visually, find $\lim\limits_{x \to 0} f(x)$.

97. $f(x) = |x| \sin x$ **98.** $f(x) = |x| \cos x$

99. $f(x) = x \sin \dfrac{1}{x}$ **100.** $f(x) = x \cos \dfrac{1}{x}$

EXPLORING CONCEPTS

101. Functions That Agree at All but One Point

(a) In the context of finding limits, discuss what is meant by two functions that agree at all but one point.

(b) Give an example of two functions that agree at all but one point.

102. Writing Functions Write a function of each specified type that has a limit of 4 as x approaches 8.

(a) linear   (b) polynomial of degree 2

(c) rational   (d) radical

(e) cosine   (f) sine

(g) exponential   (h) natural logarithmic

103. Writing Use a graphing utility to graph

$$f(x) = x, \quad g(x) = \sin x, \quad \text{and} \quad h(x) = \dfrac{\sin x}{x}$$

in the same viewing window. Compare the magnitudes of $f(x)$ and $g(x)$ when x is close to 0. Use the comparison to write a short paragraph explaining why $\lim\limits_{x \to 0} h(x) = 1$.

104. HOW DO YOU SEE IT? Would you use the dividing out technique or the rationalizing technique to find the limit of the function? Explain your reasoning.

(a) $\lim\limits_{x \to -2} \dfrac{x^2 + x - 2}{x + 2}$ (b) $\lim\limits_{x \to 0} \dfrac{\sqrt{x + 4} - 2}{x}$

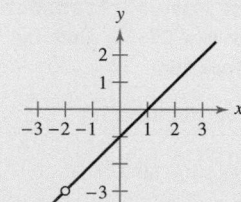

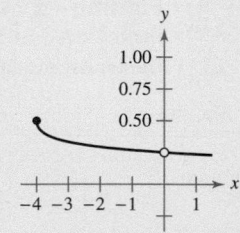

• • Free-Falling Object • • • • • • • • • • • • • • •

In Exercises 105 and 106, use the position function $s(t) = -16t^2 + 500$, which gives the height (in feet) of an object that has fallen for t seconds from a height of 500 feet. The velocity at time $t = a$ seconds is given by

$$\lim_{t \to a} \frac{s(a) - s(t)}{a - t}.$$

105. A construction worker drops a full paint can from a height of 500 feet. How fast will the paint can be falling after 2 seconds?

106. A construction worker drops a full paint can from a height of 500 feet. When will the paint can hit the ground? At what velocity will the paint can impact the ground?

Free-Falling Object In Exercises 107 and 108, use the position function $s(t) = -4.9t^2 + 200$, which gives the height (in meters) of an object that has fallen for t seconds from a height of 200 meters. The velocity at time $t = a$ seconds is given by

$$\lim_{t \to a} \frac{s(a) - s(t)}{a - t}.$$

107. Find the velocity of the object when $t = 3$.

108. At what velocity will the object impact the ground?

109. Finding Functions Find two functions f and g such that $\lim_{x \to 0} f(x)$ and $\lim_{x \to 0} g(x)$ do not exist, but

$$\lim_{x \to 0} [f(x) + g(x)]$$

does exist.

110. Proof Prove that if $\lim_{x \to c} f(x)$ exists and $\lim_{x \to c} [f(x) + g(x)]$ does not exist, then $\lim_{x \to c} g(x)$ does not exist.

111. Proof Prove Property 1 of Theorem 2.1.

112. Proof Prove Property 3 of Theorem 2.1. (You may use Property 3 of Theorem 2.2.)

113. Proof Prove Property 1 of Theorem 2.2.

114. Proof Prove that if $\lim_{x \to c} f(x) = 0$, then $\lim_{x \to c} |f(x)| = 0$.

115. Proof Prove that if $\lim_{x \to c} f(x) = 0$ and $|g(x)| \leq M$ for a fixed number M and all $x \neq c$, then $\lim_{x \to c} [f(x)g(x)] = 0$.

116. Proof

(a) Prove that if $\lim_{x \to c} |f(x)| = 0$, then $\lim_{x \to c} f(x) = 0$.

   (*Note:* This is the converse of Exercise 114.)

(b) Prove that if $\lim_{x \to c} f(x) = L$, then $\lim_{x \to c} |f(x)| = |L|$.

   [*Hint:* Use the inequality $\big| |f(x)| - |L| \big| \leq |f(x) - L|$.]

117. Think About It Find a function f to show that the converse of Exercise 116(b) is not true. [*Hint:* Find a function f such that $\lim_{x \to c} |f(x)| = |L|$ but $\lim_{x \to c} f(x)$ does not exist.]

118. Think About It When using a graphing utility to generate a table to approximate

$$\lim_{x \to 0} \frac{\sin x}{x}$$

a student concluded that the limit was 0.01745 rather than 1. Determine the probable cause of the error.

True or False? In Exercises 119–124, determine whether the statement is true or false. If it is false, explain why or give an example that shows it is false.

119. $\lim_{x \to 0} \dfrac{|x|}{x} = 1$ **120.** $\lim_{x \to \pi} \dfrac{\sin x}{x} = 1$

121. If $f(x) = g(x)$ for all real numbers other than $x = 0$ and $\lim_{x \to 0} f(x) = L$, then $\lim_{x \to 0} g(x) = L$.

122. If $\lim_{x \to c} f(x) = L$, then $f(c) = L$.

123. $\lim_{x \to 2} f(x) = 3$, where $f(x) = \begin{cases} 3, & x \leq 2 \\ 0, & x > 2 \end{cases}$

124. If $f(x) < g(x)$ for all $x \neq a$, then $\lim_{x \to a} f(x) < \lim_{x \to a} g(x)$.

125. Proof Prove the second part of Theorem 2.9.

$$\lim_{x \to 0} \frac{1 - \cos x}{x} = 0$$

126. Piecewise Functions Let

$$f(x) = \begin{cases} 0, & \text{if } x \text{ is rational} \\ 1, & \text{if } x \text{ is irrational} \end{cases}$$

and

$$g(x) = \begin{cases} 0, & \text{if } x \text{ is rational} \\ x, & \text{if } x \text{ is irrational} \end{cases}$$

Find (if possible) $\lim_{x \to 0} f(x)$ and $\lim_{x \to 0} g(x)$.

127. Graphical Reasoning Consider $f(x) = \dfrac{\sec x - 1}{x^2}$.

(a) Find the domain of f.

(b) Use a graphing utility to graph f. Is the domain of f obvious from the graph? If not, explain.

(c) Use the graph of f to approximate $\lim_{x \to 0} f(x)$.

(d) Confirm your answer to part (c) analytically.

128. Approximation

(a) Find $\lim_{x \to 0} \dfrac{1 - \cos x}{x^2}$.

(b) Use your answer to part (a) to derive the approximation $\cos x \approx 1 - \frac{1}{2}x^2$ for x near 0.

(c) Use your answer to part (b) to approximate $\cos(0.1)$.

(d) Use a calculator to approximate $\cos(0.1)$ to four decimal places. Compare the result with part (c).

2.4 Continuity and One-Sided Limits

■ Determine continuity at a point and continuity on an open interval.
■ Determine one-sided limits and continuity on a closed interval.
■ Use properties of continuity.
■ Understand and use the Intermediate Value Theorem.

Continuity at a Point and on an Open Interval

In mathematics, the term *continuous* has much the same meaning as it has in everyday usage. Informally, to say that a function f is continuous at $x = c$ means that there is no interruption in the graph of f at c. That is, its graph is unbroken at c, and there are no holes, jumps, or gaps. Figure 2.26 identifies three values of x at which the graph of f is *not* continuous. At all other points in the interval (a, b), the graph of f is uninterrupted and **continuous.**

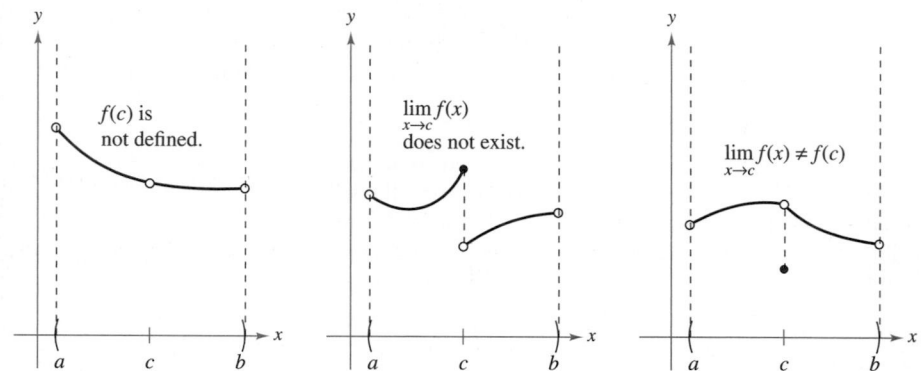

Three conditions exist for which the graph of f is not continuous at $x = c$.
Figure 2.26

In Figure 2.26, it appears that continuity at $x = c$ can be destroyed by any one of three conditions.

1. The function is not defined at $x = c$.
2. The limit of $f(x)$ does not exist at $x = c$.
3. The limit of $f(x)$ exists at $x = c$, but it is not equal to $f(c)$.

If *none* of the three conditions is true, then the function f is called **continuous at c,** as indicated in the important definition below.

Definition of Continuity

Continuity at a Point
A function f is **continuous at c** when these three conditions are met.

1. $f(c)$ is defined.
2. $\lim\limits_{x \to c} f(x)$ exists.
3. $\lim\limits_{x \to c} f(x) = f(c)$

Continuity on an Open Interval
A function is **continuous on an open interval (a, b)** when the function is continuous at each point in the interval. A function that is continuous on the entire real number line $(-\infty, \infty)$ is **everywhere continuous.**

Exploration

Informally, you might say that a function is *continuous* on an open interval when its graph can be drawn with a pencil without lifting the pencil from the paper. Use a graphing utility to graph each function on the given interval. From the graphs, which functions would you say are continuous on the interval? Do you think you can trust the results you obtained graphically? Explain your reasoning.

Function	Interval
a. $y = x^2 + 1$	$(-3, 3)$
b. $y = \dfrac{1}{x - 2}$	$(-3, 3)$
c. $y = \dfrac{\sin x}{x}$	$(-\pi, \pi)$
d. $y = \dfrac{x^2 - 4}{x + 2}$	$(-3, 3)$

■ **FOR FURTHER INFORMATION**
For more information on the concept of continuity, see the article "Leibniz and the Spell of the Continuous" by Hardy Grant in *The College Mathematics Journal*. To view this article, go to *MathArticles.com*.

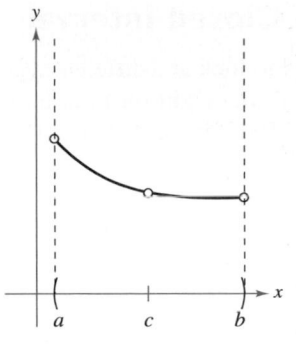

(a) Removable discontinuity

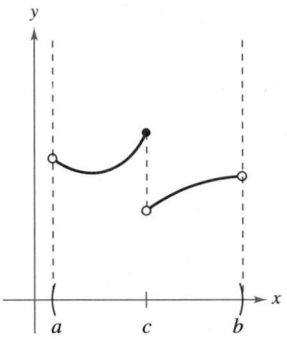

(b) Nonremovable discontinuity

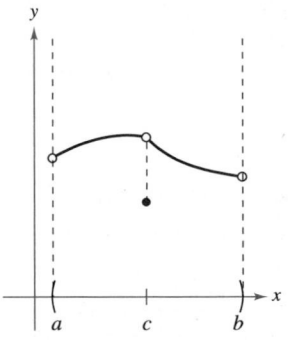

(c) Removable discontinuity

Figure 2.27

Consider an open interval I that contains a real number c. If a function f is defined on I (except possibly at c), and f is not continuous at c, then f is said to have a **discontinuity** at c. Discontinuities fall into two categories: **removable** and **nonremovable**. A discontinuity at c is called removable when f can be made continuous by appropriately defining (or redefining) $f(c)$. For instance, the functions shown in Figures 2.27(a) and (c) have removable discontinuities at c and the function shown in Figure 2.27(b) has a nonremovable discontinuity at c.

EXAMPLE 1   **Continuity of a Function**

Discuss the continuity of each function.

a. $f(x) = \dfrac{1}{x}$ **b.** $g(x) = \dfrac{x^2 - 1}{x - 1}$ **c.** $h(x) = \begin{cases} x + 1, & x \le 0 \\ e^x, & x > 0 \end{cases}$ **d.** $y = \sin x$

Solution

a. The domain of f is all nonzero real numbers. From Theorem 2.3, you can conclude that f is continuous at every x-value in its domain. At $x = 0$, f has a nonremovable discontinuity, as shown in Figure 2.28(a). In other words, there is no way to define $f(0)$ so as to make the function continuous at $x = 0$.

b. The domain of g is all real numbers except $x = 1$. From Theorem 2.3, you can conclude that g is continuous at every x-value in its domain. At $x = 1$, the function has a removable discontinuity, as shown in Figure 2.28(b). By defining $g(1)$ as 2, the "redefined" function is continuous for all real numbers.

c. The domain of h is all real numbers. The function h is continuous on $(-\infty, 0)$ and $(0, \infty)$, and because

$$\lim_{x \to 0} h(x) = 1$$

h is continuous on the entire real number line, as shown in Figure 2.28(c).

d. The domain of y is all real numbers. From Theorem 2.6, you can conclude that the function is continuous on its entire domain, $(-\infty, \infty)$, as shown in Figure 2.28(d).

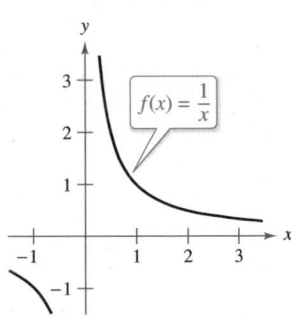

(a) Nonremovable discontinuity at $x = 0$

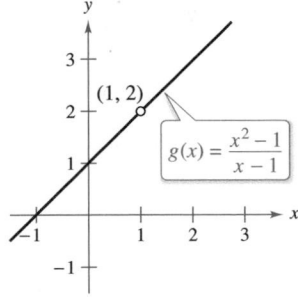

(b) Removable discontinuity at $x = 1$

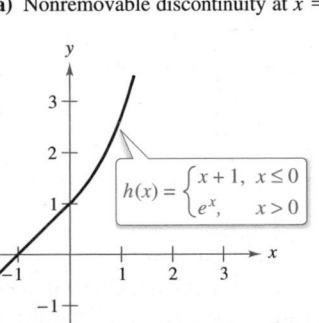

(c) Continuous on entire real number line

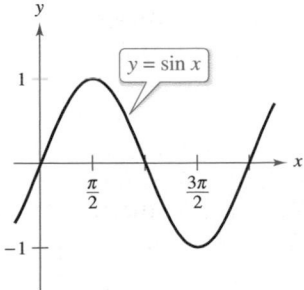

(d) Continuous on entire real number line

Figure 2.28

•••**REMARK** Some people may refer to the function in Example 1(a) as "discontinuous," but this terminology can be confusing. Rather than saying that the function is discontinuous, it is more precise to say that the function has a discontinuity at $x = 0$.

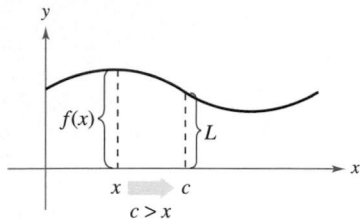

(a) Limit as x approaches c from the right.

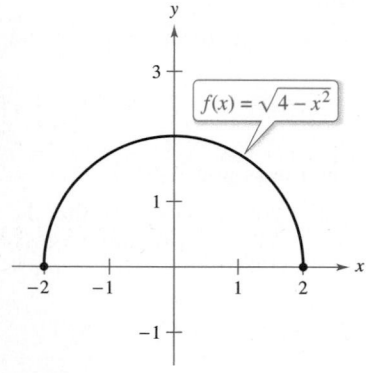

(b) Limit as x approaches c from the left.
Figure 2.29

One-Sided Limits and Continuity on a Closed Interval

To understand continuity on a closed interval, you first need to look at a different type of limit called a **one-sided limit.** For instance, the **limit from the right** (or right-hand limit) means that x approaches c from values greater than c [see Figure 2.29(a)]. This limit is denoted as

$$\lim_{x \to c^+} f(x) = L. \qquad \text{Limit from the right}$$

Similarly, the **limit from the left** (or left-hand limit) means that x approaches c from values less than c [see Figure 2.29(b)]. This limit is denoted as

$$\lim_{x \to c^-} f(x) = L. \qquad \text{Limit from the left}$$

One-sided limits are useful in taking limits of functions involving radicals. For instance, if n is an even integer, then

$$\lim_{x \to 0^+} \sqrt[n]{x} = 0.$$

EXAMPLE 2 **A One-Sided Limit**

Find the limit of $f(x) = \sqrt{4 - x^2}$ as x approaches -2 from the right.

Solution As shown in Figure 2.30, the limit as x approaches -2 from the right is

$$\lim_{x \to -2^+} \sqrt{4 - x^2} = 0. \qquad \blacksquare$$

One-sided limits can be used to investigate the behavior of **step functions.** One common type of step function is the **greatest integer function** $[\![x]\!]$, defined as

$$[\![x]\!] = \text{greatest integer } n \text{ such that } n \leq x. \qquad \text{Greatest integer function}$$

For instance, $[\![2.5]\!] = 2$ and $[\![-2.5]\!] = -3$.

The limit of $f(x)$ as x approaches -2
from the right is 0.
Figure 2.30

EXAMPLE 3 **The Greatest Integer Function**

Find the limit of the greatest integer function $f(x) = [\![x]\!]$ as x approaches 0 from the left and from the right.

Solution As shown in Figure 2.31, the limit as x approaches 0 *from the left* is

$$\lim_{x \to 0^-} [\![x]\!] = -1$$

and the limit as x approaches 0 *from the right* is

$$\lim_{x \to 0^+} [\![x]\!] = 0.$$

So, f has a discontinuity at zero because the left- and right-hand limits at zero are different. By similar reasoning, you can see that the greatest integer function has a discontinuity at any integer n.

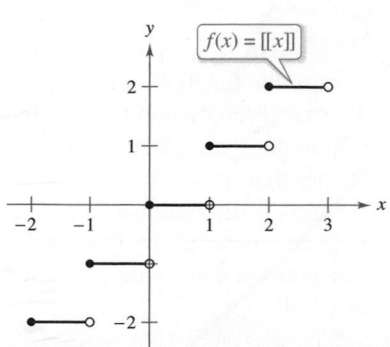

Greatest integer function
Figure 2.31

When the limit from the left is not equal to the limit from the right, the (two-sided) limit *does not exist*. The next theorem makes this more explicit. The proof of this theorem follows directly from the definition of a one-sided limit.

THEOREM 2.10 The Existence of a Limit

Let f be a function, and let c and L be real numbers. The limit of $f(x)$ as x approaches c is L if and only if

$$\lim_{x \to c^-} f(x) = L \quad \text{and} \quad \lim_{x \to c^+} f(x) = L.$$

The concept of a one-sided limit allows you to extend the definition of continuity to closed intervals. Basically, a function is continuous on a closed interval when it is continuous in the interior of the interval and exhibits one-sided continuity at the endpoints. This is stated formally in the next definition.

Definition of Continuity on a Closed Interval

A function f is **continuous on the closed interval** $[a, b]$ when f is continuous on the open interval (a, b) and

$$\lim_{x \to a^+} f(x) = f(a)$$

and

$$\lim_{x \to b^-} f(x) = f(b).$$

The function f is **continuous from the right** at a and **continuous from the left** at b (see Figure 2.32).

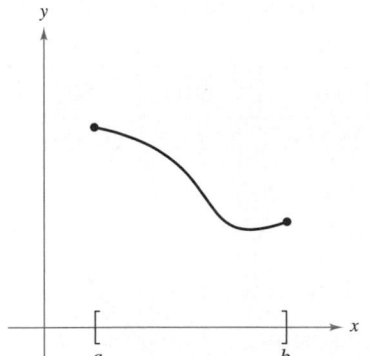

Continuous function on a closed interval
Figure 2.32

Similar definitions can be made to cover continuity on intervals of the form $(a, b]$ and $[a, b)$ that are neither open nor closed, or on infinite intervals. For example,

$$f(x) = \sqrt{x}$$

is continuous on the infinite interval $[0, \infty)$, and the function

$$g(x) = \sqrt{2 - x}$$

is continuous on the infinite interval $(-\infty, 2]$.

EXAMPLE 4   **Continuity on a Closed Interval**

Discuss the continuity of

$$f(x) = \sqrt{1 - x^2}.$$

Solution   The domain of f is the closed interval $[-1, 1]$. At all points in the open interval $(-1, 1)$, the continuity of f follows from Theorems 2.4 and 2.5. Moreover, because

$$\lim_{x \to -1^+} \sqrt{1 - x^2} = 0 = f(-1) \qquad \text{Continuous from the right}$$

and

$$\lim_{x \to 1^-} \sqrt{1 - x^2} = 0 = f(1) \qquad \text{Continuous from the left}$$

you can conclude that f is continuous on the closed interval $[-1, 1]$, as shown in Figure 2.33.

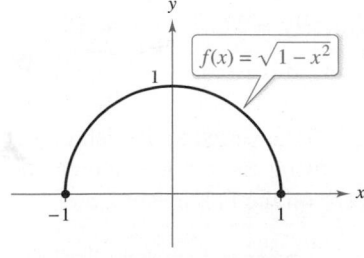

f is continuous on $[-1, 1]$.
Figure 2.33

The next example shows how a one-sided limit can be used to determine the value of absolute zero on the Kelvin scale.

- - - - - - - - - - - - - - - - ▷

REMARK Charles's Law for gases (assuming constant pressure) can be stated as

$$V = kT$$

where V is volume, k is a constant, and T is temperature.

EXAMPLE 5 **Charles's Law and Absolute Zero**

On the Kelvin scale, *absolute zero* is the temperature 0 K. Although temperatures very close to 0 K have been produced in laboratories, absolute zero has never been attained. In fact, evidence suggests that absolute zero *cannot* be attained. How did scientists determine that 0 K is the "lower limit" of the temperature of matter? What is absolute zero on the Celsius scale?

Solution The determination of absolute zero stems from the work of the French physicist Jacques Charles (1746–1823). Charles discovered that the volume of gas at a constant pressure increases linearly with the temperature of the gas. The table illustrates this relationship between volume and temperature. To generate the values in the table, one mole of hydrogen is held at a constant pressure of one atmosphere. The volume V is approximated and is measured in liters, and the temperature T is measured in degrees Celsius.

T	-40	-20	0	20	40	60	80
V	19.1482	20.7908	22.4334	24.0760	25.7186	27.3612	29.0038

The points represented by the table are shown in the figure at the right. Moreover, by using the points in the table, you can determine that T and V are related by the linear equation

$$V = 0.08213T + 22.4334.$$

Solving for T, you get an equation for the temperature of the gas.

$$T = \frac{V - 22.4334}{0.08213}$$

By reasoning that the volume of the gas can approach 0 (but can never equal or go below 0), you can determine that the "least possible temperature" is

$$\lim_{V \to 0^+} T = \lim_{V \to 0^+} \frac{V - 22.4334}{0.08213}$$

$$= \frac{0 - 22.4334}{0.08213} \qquad \text{Use direct substitution.}$$

$$\approx -273.15.$$

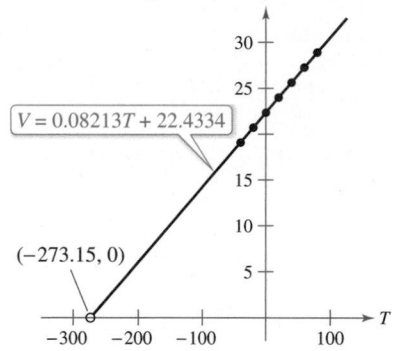

$V = 0.08213T + 22.4334$

$(-273.15, 0)$

The volume of hydrogen gas depends on its temperature.

So, absolute zero on the Kelvin scale (0 K) is approximately $-273.15°$ on the Celsius scale. ∎

Liquid helium is used to cool superconducting magnets, such as those used in magnetic resonance imaging (MRI) machines or in the Large Hadron Collider (see above). The magnets are made with materials that only superconduct at temperatures a few degrees above absolute zero. These temperatures are possible with liquid helium because helium becomes a liquid at $-269°C$, or 4.15 K.

The table below shows the temperatures in Example 5 converted to the Fahrenheit scale. Try repeating the solution shown in Example 5 using these temperatures and volumes. Use the result to find the value of absolute zero on the Fahrenheit scale.

T	-40	-4	32	68	104	140	176
V	19.1482	20.7908	22.4334	24.0760	25.7186	27.3612	29.0038

Properties of Continuity

In Section 2.3, you studied several properties of limits. Each of those properties yields a corresponding property pertaining to the continuity of a function. For instance, Theorem 2.11 follows directly from Theorem 2.2.

AUGUSTIN-LOUIS CAUCHY
(1789–1857)

The concept of a continuous function was first introduced by Augustin-Louis Cauchy in 1821. The definition given in his text *Cours d'Analyse* stated that indefinite small changes in *y* were the result of indefinite small changes in *x*. "... $f(x)$ will be called a *continuous* function if ... the numerical values of the difference $f(x + \alpha) - f(x)$ decrease indefinitely with those of α...."

See LarsonCalculus.com to read more of this biography.

> **THEOREM 2.11   Properties of Continuity**
>
> If b is a real number and f and g are continuous at $x = c$, then the functions listed below are also continuous at c.
>
> **1.** Scalar multiple: bf **2.** Sum or difference: $f \pm g$
>
> **3.** Product: fg **4.** Quotient: $\dfrac{f}{g}$, $g(c) \neq 0$
>
> A proof of this theorem is given in Appendix A.

It is important for you to be able to recognize functions that are continuous at every point in their domains. The list below summarizes the functions you have studied so far that are continuous at every point in their domains.

1. Polynomial: $p(x) = a_n x^n + a_{n-1} x^{n-1} + \cdots + a_1 x + a_0$

2. Rational: $r(x) = \dfrac{p(x)}{q(x)}, \quad q(x) \neq 0$

3. Radical: $f(x) = \sqrt[n]{x}$

4. Trigonometric: $\sin x, \cos x, \tan x, \cot x, \sec x, \csc x$

5. Exponential and logarithmic: $f(x) = a^x, f(x) = e^x, f(x) = \ln x$

By combining Theorem 2.11 with this list, you can conclude that a wide variety of elementary functions are continuous at every point in their domains.

EXAMPLE 6   **Applying Properties of Continuity**

⋅ ⋅ ⋅ ▷ *See LarsonCalculus.com for an interactive version of this type of example.*

By Theorem 2.11, it follows that each of the functions below is continuous at every point in its domain.

$$f(x) = x + e^x, \quad f(x) = 3 \tan x, \quad f(x) = \frac{x^2 + 1}{\cos x}$$ ■

The next theorem, which is a consequence of Theorem 2.5, allows you to determine the continuity of *composite* functions such as

$$f(x) = \ln 3x, \quad f(x) = \sqrt{x^2 + 1}, \quad \text{and} \quad f(x) = \tan \frac{1}{x}.$$

. . . . ▷

⋅⋅**REMARK**   One consequence of Theorem 2.12 is that when f and g satisfy the given conditions, you can determine the limit of $f(g(x))$ as x approaches c to be

$$\lim_{x \to c} f(g(x)) = f(g(c)).$$

> **THEOREM 2.12   Continuity of a Composite Function**
>
>
>
> If g is continuous at c and f is continuous at $g(c)$, then the composite function given by $(f \circ g)(x) = f(g(x))$ is continuous at c.

Proof   By the definition of continuity, $\lim\limits_{x \to c} g(x) = g(c)$ and $\lim\limits_{x \to g(c)} f(x) = f(g(c))$.

Apply Theorem 2.5 with $L = g(c)$ to obtain $\lim\limits_{x \to c} f(g(x)) = f\left(\lim\limits_{x \to c} g(x)\right) = f(g(c))$. So, $(f \circ g)(x) = f(g(x))$ is continuous at c.   ■

EXAMPLE 7 **Testing for Continuity**

Describe the interval(s) on which each function is continuous.

a. $f(x) = \tan x$ **b.** $g(x) = \begin{cases} \sin \dfrac{1}{x}, & x \neq 0 \\ 0, & x = 0 \end{cases}$ **c.** $h(x) = \begin{cases} x \sin \dfrac{1}{x}, & x \neq 0 \\ 0, & x = 0 \end{cases}$

Solution

a. The tangent function $f(x) = \tan x$ is undefined at

$$x = \frac{\pi}{2} + n\pi, \quad n \text{ is an integer.}$$

At all other points, f is continuous. So, $f(x) = \tan x$ is continuous on the open intervals

$$\ldots, \left(-\frac{3\pi}{2}, -\frac{\pi}{2}\right), \left(-\frac{\pi}{2}, \frac{\pi}{2}\right), \left(\frac{\pi}{2}, \frac{3\pi}{2}\right), \ldots$$

as shown in Figure 2.34(a).

b. Because $y = 1/x$ is continuous except at $x = 0$ and the sine function is continuous for all real values of x, it follows from Theorem 2.12 that

$$y = \sin \frac{1}{x}$$

is continuous at all real values except $x = 0$. At $x = 0$, the limit of $g(x)$ does not exist (see Example 5, Section 2.2). So, g is continuous on the intervals $(-\infty, 0)$ and $(0, \infty)$, as shown in Figure 2.34(b).

c. This function is similar to the function in part (b) except that the oscillations are damped by the factor x. Using the Squeeze Theorem, you obtain

$$-|x| \leq x \sin \frac{1}{x} \leq |x|, \quad x \neq 0$$

and you can conclude that

$$\lim_{x \to 0} h(x) = 0.$$

So, h is continuous on the entire real number line, as shown in Figure 2.34(c).

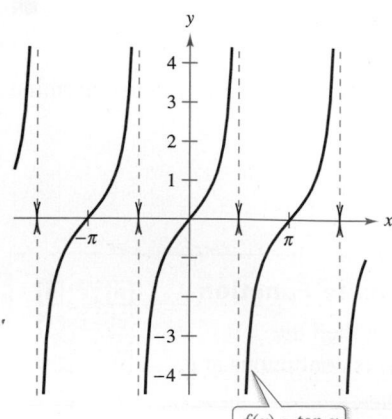

(a) f is continuous on each open interval in its domain.

Figure 2.34

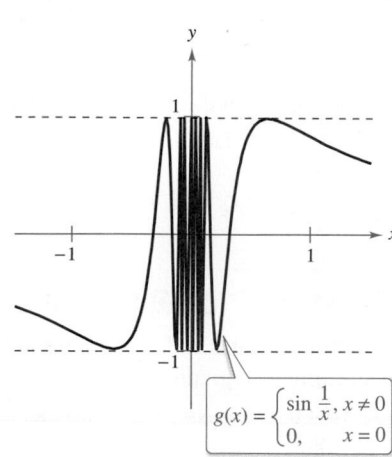

(b) g is continuous on $(-\infty, 0)$ and $(0, \infty)$.

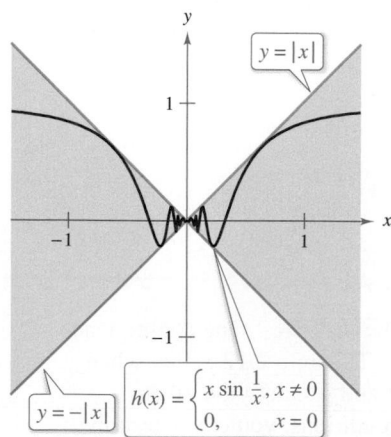

(c) h is continuous on the entire real number line.

The Intermediate Value Theorem

Theorem 2.13 is an important theorem concerning the behavior of functions that are continuous on a closed interval.

> **THEOREM 2.13 Intermediate Value Theorem**
>
> If f is continuous on the closed interval $[a, b]$, $f(a) \neq f(b)$, and k is any number between $f(a)$ and $f(b)$, then there is at least one number c in $[a, b]$ such that
>
> $$f(c) = k.$$

REMARK The Intermediate Value Theorem tells you that at least one number c exists, but it does not provide a method for finding c. Such theorems are called **existence theorems.** By referring to a text on advanced calculus, you will find that a proof of this theorem is based on a property of real numbers called *completeness.* The Intermediate Value Theorem states that for a continuous function f, if x takes on all values between a and b, then $f(x)$ must take on all values between $f(a)$ and $f(b)$.

As an example of the application of the Intermediate Value Theorem, consider a person's height. A girl is 5 feet tall on her thirteenth birthday and 5 feet 2 inches tall on her fourteenth birthday. Then, for any height h between 5 feet and 5 feet 2 inches, there must have been a time t when her height was exactly h. This seems reasonable because human growth is continuous and a person's height does not abruptly change from one value to another.

The Intermediate Value Theorem guarantees the existence of *at least one* number c in the closed interval $[a, b]$. There may, of course, be more than one number c such that

$$f(c) = k$$

as shown in Figure 2.35. A function that is not continuous does not necessarily exhibit the intermediate value property. For example, the graph of the function shown in Figure 2.36 jumps over the horizontal line

$$y = k$$

and for this function there is no value of c in $[a, b]$ such that $f(c) = k$.

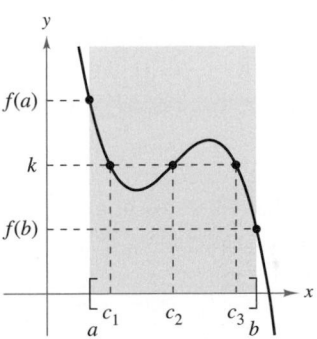

f is continuous on $[a, b]$.
[There exist three c's such that $f(c) = k$.]
Figure 2.35

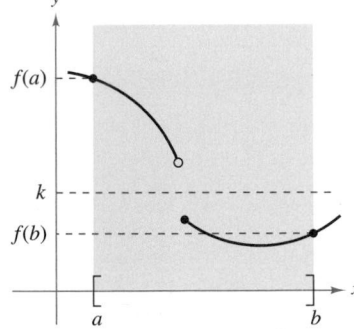

f is not continuous on $[a, b]$.
[There are no c's such that $f(c) = k$.]
Figure 2.36

The Intermediate Value Theorem often can be used to locate the zeros of a function that is continuous on a closed interval. Specifically, if f is continuous on $[a, b]$ and $f(a)$ and $f(b)$ differ in sign, then the Intermediate Value Theorem guarantees the existence of at least one zero of f in the closed interval $[a, b]$.

EXAMPLE 8 **An Application of the Intermediate Value Theorem**

Use the Intermediate Value Theorem to show that the polynomial function

$$f(x) = x^3 + 2x - 1$$

has a zero in the interval $[0, 1]$.

Solution Note that f is continuous on the closed interval $[0, 1]$. Because

$$f(0) = 0^3 + 2(0) - 1 = -1 \quad \text{and} \quad f(1) = 1^3 + 2(1) - 1 = 2$$

it follows that $f(0) < 0$ and $f(1) > 0$. You can therefore apply the Intermediate Value Theorem to conclude that there must be some c in $[0, 1]$ such that

$$f(c) = 0 \qquad \text{\small{f has a zero in the closed interval $[0, 1]$.}}$$

as shown in Figure 2.37.

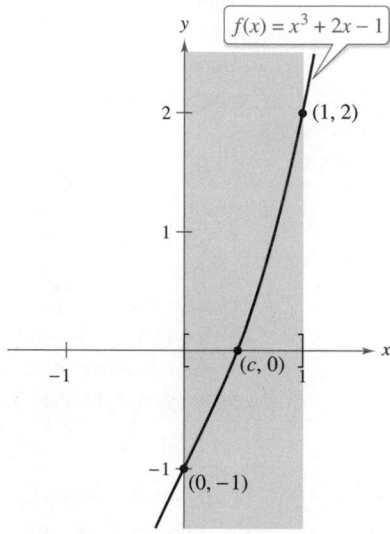

f is continuous on $[0, 1]$ with $f(0) < 0$ and $f(1) > 0$.
Figure 2.37

The **bisection method** for approximating the real zeros of a continuous function is similar to the method used in Example 8. If you know that a zero exists in the closed interval $[a, b]$, then the zero must lie in the interval $[a, (a + b)/2]$ or $[(a + b)/2, b]$. From the sign of $f([a + b]/2)$, you can determine which interval contains the zero. By repeatedly bisecting the interval, you can "close in" on the zero of the function.

▷ **TECHNOLOGY** You can use the *root* or *zero* feature of a graphing utility to approximate the real zeros of a continuous function. Using this feature, the zero of the function in Example 8, $f(x) = x^3 + 2x - 1$, is approximately 0.453, as shown in the figure.

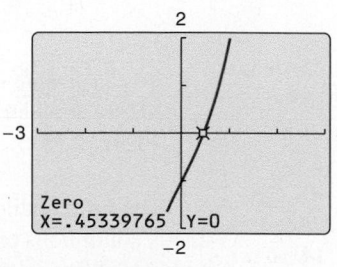

Zero of $f(x) = x^3 + 2x - 1$

2.4 Exercises

See **CalcChat.com** for tutorial help and worked-out solutions to odd-numbered exercises.

CONCEPT CHECK

1. **Continuity** In your own words, describe what it means for a function to be continuous at a point.

2. **One-Sided Limits** What is the value of c?

$$\lim_{x \to c^+} 2\sqrt{x + 1} = 0$$

3. **Existence of a Limit** Determine whether $\lim_{x \to 3^-} f(x)$ exists. Explain.

$$\lim_{x \to 3^-} f(x) = 1 \quad \text{and} \quad \lim_{x \to 3^+} f(x) = 1$$

4. **Intermediate Value Theorem** In your own words, explain the Intermediate Value Theorem.

 **Limits and Continuity** In Exercises 5–10, use the graph to determine each limit, and discuss the continuity of the function.

(a) $\lim\limits_{x \to c^+} f(x)$ (b) $\lim\limits_{x \to c^-} f(x)$ (c) $\lim\limits_{x \to c} f(x)$

5.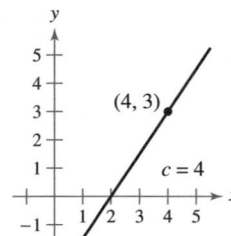

6.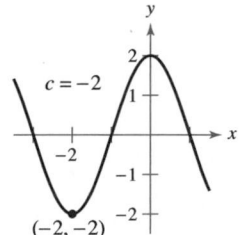

7.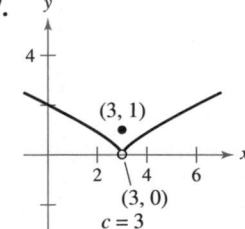

8.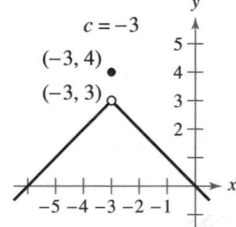

9.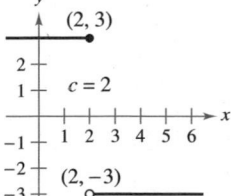

10.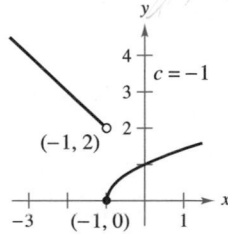

  Finding a Limit In Exercises 11–32, find the limit (if it exists). If it does not exist, explain why.

11. $\lim\limits_{x \to 8^+} \dfrac{1}{x + 8}$

12. $\lim\limits_{x \to 3^+} \dfrac{2}{x + 3}$

13. $\lim\limits_{x \to 5^+} \dfrac{x - 5}{x^2 - 25}$

14. $\lim\limits_{x \to 4^+} \dfrac{4 - x}{x^2 - 16}$

15. $\lim\limits_{x \to -3^-} \dfrac{x}{\sqrt{x^2 - 9}}$

16. $\lim\limits_{x \to 4^-} \dfrac{\sqrt{x} - 2}{x - 4}$

17. $\lim\limits_{x \to 0^-} \dfrac{|x|}{x}$

18. $\lim\limits_{x \to 10^+} \dfrac{|x - 10|}{x - 10}$

19. $\lim\limits_{\Delta x \to 0^-} \dfrac{\dfrac{1}{x + \Delta x} - \dfrac{1}{x}}{\Delta x}$

20. $\lim\limits_{\Delta x \to 0^+} \dfrac{(x + \Delta x)^2 + x + \Delta x - (x^2 + x)}{\Delta x}$

21. $\lim\limits_{x \to 3^-} f(x)$, where $f(x) = \begin{cases} \dfrac{x + 2}{2}, & x < 3 \\ \dfrac{12 - 2x}{3}, & x > 3 \end{cases}$

22. $\lim\limits_{x \to 3} f(x)$, where $f(x) = \begin{cases} x^2 - 4x + 6, & x < 3 \\ -x^2 + 4x - 2, & x \geq 3 \end{cases}$

23. $\lim\limits_{x \to \pi} \cot x$

24. $\lim\limits_{x \to \pi/2} \sec x$

25. $\lim\limits_{x \to 4^-} (5[\![x]\!] - 7)$

26. $\lim\limits_{x \to 2^+} (2x - [\![x]\!])$

27. $\lim\limits_{x \to -1} \left(\left[\!\left[\dfrac{x}{3}\right]\!\right] + 3 \right)$

28. $\lim\limits_{x \to 1} \left(1 - \left[\!\left[-\dfrac{x}{2} \right]\!\right] \right)$

29. $\lim\limits_{x \to 3^+} \ln(x - 3)$

30. $\lim\limits_{x \to 6^-} \ln(6 - x)$

31. $\lim\limits_{x \to 2^-} \ln[x^2(3 - x)]$

32. $\lim\limits_{x \to 5^+} \ln \dfrac{x}{\sqrt{x - 4}}$

 Continuity of a Function In Exercises 33–36, discuss the continuity of the function.

33. $f(x) = \dfrac{1}{x^2 - 4}$

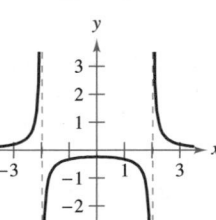

34. $f(x) = \dfrac{x^2 - 1}{x + 1}$

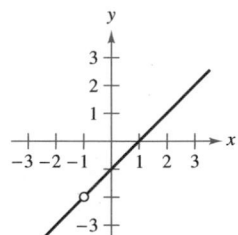

35. $f(x) = \frac{1}{2}[\![x]\!] + x$

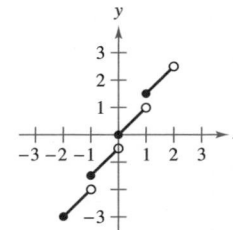

36. $f(x) = \begin{cases} x, & x < 1 \\ 2, & x = 1 \\ 2x - 1, & x > 1 \end{cases}$

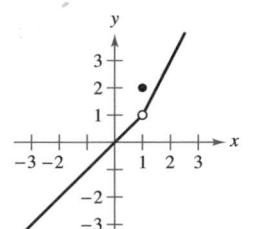

Continuity on a Closed Interval In Exercises 37–40, discuss the continuity of the function on the closed interval.

Function	Interval
37. $g(x) = \sqrt{49 - x^2}$	$[-7, 7]$
38. $f(t) = 3 - \sqrt{9 - t^2}$	$[-3, 3]$
39. $f(x) = \begin{cases} 3 - x, & x \le 0 \\ 3 + \frac{1}{2}x, & x > 0 \end{cases}$	$[-1, 4]$
40. $g(x) = \dfrac{1}{x^2 - 4}$	$[-1, 2]$

Removable and Nonremovable Discontinuities In Exercises 41–60, find the x-values (if any) at which f is not continuous. Which of the discontinuities are removable?

41. $f(x) = \dfrac{4}{x - 6}$

42. $f(x) = \dfrac{1}{x^2 + 1}$

43. $f(x) = 3x - \cos x$

44. $f(x) = \sin x - 8x$

45. $f(x) = \dfrac{x}{x^2 - x}$

46. $f(x) = \dfrac{x}{x^2 - 4}$

47. $f(x) = \dfrac{x + 2}{x^2 - 3x - 10}$

48. $f(x) = \dfrac{x + 2}{x^2 - x - 6}$

49. $f(x) = \dfrac{|x + 7|}{x + 7}$

50. $f(x) = \dfrac{2|x - 3|}{x - 3}$

51. $f(x) = \begin{cases} \frac{1}{2}x + 1, & x \le 2 \\ 3 - x, & x > 2 \end{cases}$

52. $f(x) = \begin{cases} -2x, & x \le 2 \\ x^2 - 4x + 1, & x > 2 \end{cases}$

53. $f(x) = \begin{cases} \tan \frac{\pi x}{4}, & |x| < 1 \\ x, & |x| \ge 1 \end{cases}$

54. $f(x) = \begin{cases} \csc \frac{\pi x}{6}, & |x - 3| \le 2 \\ 2, & |x - 3| > 2 \end{cases}$

55. $f(x) = \begin{cases} \ln(x + 1), & x \ge 0 \\ 1 - x^2, & x < 0 \end{cases}$

56. $f(x) = \begin{cases} 10 - 3e^{5-x}, & x > 5 \\ 10 - \frac{3}{5}x, & x \le 5 \end{cases}$

57. $f(x) = \csc 2x$

58. $f(x) = \tan \dfrac{\pi x}{2}$

59. $f(x) = [\![x - 8]\!]$

60. $f(x) = 5 - [\![x]\!]$

Making a Function Continuous In Exercises 61–66, find the constant a such that the function is continuous on the entire real number line.

61. $f(x) = \begin{cases} x^3, & x \le 2 \\ ax^2, & x > 2 \end{cases}$

62. $f(x) = \begin{cases} 3x^2, & x \ge 1 \\ ax - 4, & x < 1 \end{cases}$

63. $g(x) = \begin{cases} \dfrac{x^2 - a^2}{x - a}, & x \ne a \\ 8, & x = a \end{cases}$

64. $g(x) = \begin{cases} \dfrac{4 \sin x}{x}, & x < 0 \\ a - 2x, & x \ge 0 \end{cases}$

65. $f(x) = \begin{cases} ae^{x-1} + 3, & x < 1 \\ \arctan(x - 1) + 2, & x \ge 1 \end{cases}$

66. $f(x) = \begin{cases} 2e^{ax} - 2, & x \le 4 \\ \ln(x - 3) + x^2, & x > 4 \end{cases}$

Continuity of a Composite Function In Exercises 67–70, discuss the continuity of the composite function $h(x) = f(g(x))$.

67. $f(x) = \dfrac{1}{x - 6}$

$g(x) = x^2 + 5$

68. $f(x) = \dfrac{1}{\sqrt{x}}$

$g(x) = x - 1$

69. $f(x) = \tan x$

$g(x) = \dfrac{x}{2}$

70. $f(x) = \sin x$

$g(x) = x^2$

Finding Discontinuities Using Technology In Exercises 71–74, use a graphing utility to graph the function. Use the graph to determine any x-values at which the function is not continuous.

71. $f(x) = [\![x]\!] - x$

72. $h(x) = \dfrac{1}{x^2 + 2x - 15}$

73. $g(x) = \begin{cases} x^2 - 3x, & x > 4 \\ 2x - 5, & x \le 4 \end{cases}$

74. $f(x) = \begin{cases} \dfrac{\cos x - 1}{x}, & x < 0 \\ 5x, & x \ge 0 \end{cases}$

Testing for Continuity In Exercises 75–82, describe the interval(s) on which the function is continuous.

75. $f(x) = \dfrac{x}{x^2 + x + 2}$

76. $f(x) = \dfrac{x + 1}{\sqrt{x}}$

77. $f(x) = 3 - \sqrt{x}$

78. $f(x) = x\sqrt{x + 3}$

79. $f(x) = \sec \dfrac{\pi x}{4}$

80. $f(x) = \cos \dfrac{1}{x}$

81. $f(x) = \begin{cases} \dfrac{x^2 - 1}{x - 1}, & x \ne 1 \\ 2, & x = 1 \end{cases}$

82. $f(x) = \begin{cases} 2x - 4, & x \ne 3 \\ 1, & x = 3 \end{cases}$

Existence of a Zero In Exercises 83–88, explain why the function has at least one zero in the given interval.

Function	Interval
83. $f(x) = \frac{1}{12}x^4 - x^3 + 4$	$[1, 2]$
84. $f(x) = x^3 + 5x - 3$	$[0, 1]$
85. $f(x) = x^2 - 2 - \cos x$	$[0, \pi]$
86. $f(x) = -\dfrac{5}{x} + \tan \dfrac{\pi x}{10}$	$[1, 4]$
87. $h(x) = -2e^{-x/2} \cos 2x$	$\left[0, \dfrac{\pi}{2}\right]$
88. $g(t) = (t^3 + 2t - 2) \ln(t^2 + 4)$	$[0, 1]$

Existence of Multiple Zeros In Exercises 89 and 90, explain why the function has at least two zeros in the interval $[1, 5]$.

89. $f(x) = (x - 3)^2 - 2$

90. $f(x) = 2 \cos x$

Using the Intermediate Value Theorem In Exercises 91–98, use the Intermediate Value Theorem and a graphing utility to approximate the zero of the function in the interval $[0, 1]$. Repeatedly "zoom in" on the graph of the function to approximate the zero accurate to two decimal places. Use the *zero* or *root* feature of the graphing utility to approximate the zero accurate to four decimal places.

91. $f(x) = x^3 + x - 1$

92. $f(x) = x^4 - x^2 + 3x - 1$

93. $f(x) = \sqrt{x^2 + 17x + 19} - 6$

94. $f(x) = \sqrt{x^4 + 39x + 13} - 4$

95. $g(t) = 2 \cos t - 3t$ **96.** $h(\theta) = \tan \theta + 3\theta - 4$

97. $f(x) = x + e^x - 3$ **98.** $g(x) = 5 \ln(x + 1) - 2$

Using the Intermediate Value Theorem In Exercises 99–104, verify that the Intermediate Value Theorem applies to the indicated interval and find the value of c guaranteed by the theorem.

99. $f(x) = x^2 + x - 1$, $[0, 5]$, $f(c) = 11$

100. $f(x) = x^2 - 6x + 8$, $[0, 3]$, $f(c) = 0$

101. $f(x) = \sqrt{x + 7} - 2$, $[0, 5]$, $f(c) = 1$

102. $f(x) = \sqrt[3]{x} + 8$, $[-9, -6]$, $f(c) = 6$

103. $f(x) = \dfrac{x - x^3}{x - 4}$, $[1, 3]$, $f(c) = 3$

104. $f(x) = \dfrac{x^2 + x}{x - 1}$, $\left[\dfrac{5}{2}, 4\right]$, $f(c) = 6$

EXPLORING CONCEPTS

105. Writing a Function Write a function that is continuous on (a, b) but not continuous on $[a, b]$.

106. Sketching a Graph Sketch the graph of any function f such that

$$\lim_{x \to 3^+} f(x) = 1 \quad \text{and} \quad \lim_{x \to 3^-} f(x) = 0.$$

Is the function continuous at $x = 3$? Explain.

107. Continuity of Combinations of Functions If the functions f and g are continuous for all real x, is $f + g$ always continuous for all real x? Is f/g always continuous for all real x? If either is not continuous, give an example to verify your conclusion.

108. Removable and Nonremovable Discontinuities Describe the difference between a discontinuity that is removable and a discontinuity that is nonremovable. Then give an example of a function that satisfies each description.

(a) A function with a nonremovable discontinuity at $x = 4$

(b) A function with a removable discontinuity at $x = -4$

(c) A function that has both of the characteristics described in parts (a) and (b)

True or False? In Exercises 109–114, determine whether the statement is true or false. If it is false, explain why or give an example that shows it is false.

109. If $\lim_{x \to c} f(x) = L$ and $f(c) = L$, then f is continuous at c.

110. If $f(x) = g(x)$ for $x \neq c$ and $f(c) \neq g(c)$, then either f or g is not continuous at c.

111. The Intermediate Value Theorem guarantees that $f(a)$ and $f(b)$ differ in sign when a continuous function f has at least one zero on $[a, b]$.

112. The limit of the greatest integer function as x approaches 0 from the left is -1.

113. A rational function can have infinitely many x-values at which it is not continuous.

114. The function $f(x) = \dfrac{|x - 1|}{x - 1}$ is continuous on $(-\infty, \infty)$.

115. Think About It Describe how the functions

$$f(x) = 3 + [\![x]\!] \quad \text{and} \quad g(x) = 3 - [\![-x]\!]$$

differ.

116. HOW DO YOU SEE IT? Every day you dissolve 28 ounces of chlorine in a swimming pool. The graph shows the amount of chlorine $f(t)$ in the pool after t days. Estimate and interpret $\lim_{t \to 4^-} f(t)$ and $\lim_{t \to 4^+} f(t)$.

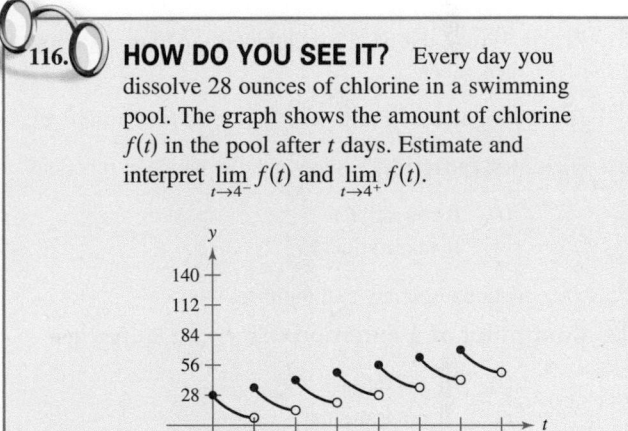

117. Data Plan A cell phone service charges $10 for the first gigabyte (GB) of data used per month and $7.50 for each additional gigabyte or fraction thereof. The cost of the data plan is given by

$$C(t) = 10 - 7.5[\![1 - t]\!], \quad t > 0$$

where t is the amount of data used (in GB). Sketch the graph of this function and discuss its continuity.

118. Inventory Management The number of units in inventory in a small company is given by

$$N(t) = 25\left(2\left[\!\left[\dfrac{t + 2}{2}\right]\!\right] - t\right)$$

where t is the time in months. Sketch the graph of this function and discuss its continuity. How often must this company replenish its inventory?

119. Déjà Vu At 8:00 A.M. on Saturday, a man begins running up the side of a mountain to his weekend campsite (see figure). On Sunday morning at 8:00 A.M., he runs back down the mountain. It takes him 20 minutes to run up but only 10 minutes to run down. At some point on the way down, he realizes that he passed the same place at exactly the same time on Saturday. Prove that he is correct. [*Hint:* Let $s(t)$ and $r(t)$ be the position functions for the runs up and down, and apply the Intermediate Value Theorem to the function $f(t) = s(t) - r(t)$.]

Saturday 8:00 A.M.   Sunday 8:00 A.M.

Not drawn to scale

120. Volume Use the Intermediate Value Theorem to show that for all spheres with radii in the interval $[5, 8]$, there is one with a volume of 1500 cubic centimeters.

121. Proof Prove that if f is continuous and has no zeros on $[a, b]$, then either

$$f(x) > 0 \text{ for all } x \text{ in } [a, b] \quad \text{or} \quad f(x) < 0 \text{ for all } x \text{ in } [a, b].$$

122. Dirichlet Function Show that the Dirichlet function

$$f(x) = \begin{cases} 0, & \text{if } x \text{ is rational} \\ 1, & \text{if } x \text{ is irrational} \end{cases}$$

is not continuous at any real number.

123. Continuity of a Function Show that the function

$$f(x) = \begin{cases} 0, & \text{if } x \text{ is rational} \\ kx, & \text{if } x \text{ is irrational} \end{cases}$$

is continuous only at $x = 0$. (Assume that k is any nonzero real number.)

124. Signum Function The **signum function** is defined by

$$\text{sgn}(x) = \begin{cases} -1, & x < 0 \\ 0, & x = 0 \\ 1, & x > 0 \end{cases}.$$

Sketch a graph of $\text{sgn}(x)$ and find the following (if possible).

(a) $\lim\limits_{x \to 0^-} \text{sgn}(x)$ (b) $\lim\limits_{x \to 0^+} \text{sgn}(x)$ (c) $\lim\limits_{x \to 0} \text{sgn}(x)$

125. Modeling Data The table lists the frequency F (in Hertz) of a musical note at various times t (in seconds).

t	0	1	2	3	4	5
F	436	444	434	446	433	444

(a) Plot the data and connect the points with a curve.

(b) Does there appear to be a limiting frequency of the note? Explain.

126. Creating Models A swimmer crosses a pool of width b by swimming in a straight line from $(0, 0)$ to $(2b, b)$. (See figure.)

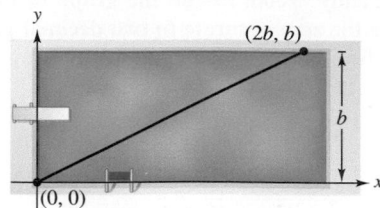

(a) Let f be a function defined as the y-coordinate of the point on the long side of the pool that is nearest the swimmer at any given time during the swimmer's crossing of the pool. Determine the function f and sketch its graph. Is f continuous? Explain.

(b) Let g be the minimum distance between the swimmer and the long sides of the pool. Determine the function g and sketch its graph. Is g continuous? Explain.

127. Making a Function Continuous Find all values of c such that f is continuous on $(-\infty, \infty)$.

$$f(x) = \begin{cases} 1 - x^2, & x \le c \\ x, & x > c \end{cases}$$

128. Proof Prove that for any real number y there exists x in $(-\pi/2, \pi/2)$ such that $\tan x = y$.

129. Making a Function Continuous Let

$$f(x) = \frac{\sqrt{x + c^2} - c}{x}, \quad c > 0.$$

What is the domain of f? How can you define f at $x = 0$ in order for f to be continuous there?

130. Proof Prove that if

$$\lim_{\Delta x \to 0} f(c + \Delta x) = f(c)$$

then f is continuous at c.

131. Continuity of a Function Discuss the continuity of the function $h(x) = x[\![x]\!]$.

132. Proof

(a) Let $f_1(x)$ and $f_2(x)$ be continuous on the closed interval $[a, b]$. If $f_1(a) < f_2(a)$ and $f_1(b) > f_2(b)$, prove that there exists c between a and b such that $f_1(c) = f_2(c)$.

(b) Show that there exists c in $\left[0, \frac{\pi}{2}\right]$ such that $\cos x = x$. Use a graphing utility to approximate c to three decimal places.

PUTNAM EXAM CHALLENGE

133. Prove or disprove: If x and y are real numbers with $y \ge 0$ and $y(y + 1) \le (x + 1)^2$, then $y(y - 1) \le x^2$.

134. Determine all polynomials $P(x)$ such that

$$P(x^2 + 1) = (P(x))^2 + 1 \text{ and } P(0) = 0.$$

These problems were composed by the Committee on the Putnam Prize Competition. © The Mathematical Association of America. All rights reserved.

2.5 Infinite Limits

■ Determine infinite limits from the left and from the right.
■ Find and sketch the vertical asymptotes of the graph of a function.

Infinite Limits

Consider the function $f(x) = 3/(x - 2)$. From Figure 2.38 and the table, you can see that $f(x)$ *decreases without bound* as x approaches 2 from the left, and $f(x)$ *increases without bound* as x approaches 2 from the right.

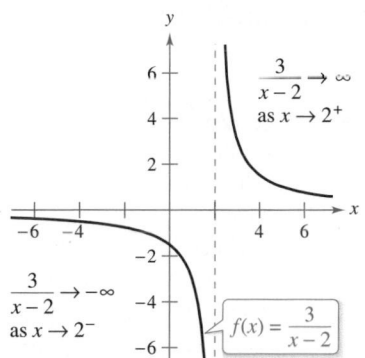

$f(x)$ increases and decreases without bound as x approaches 2.

Figure 2.38

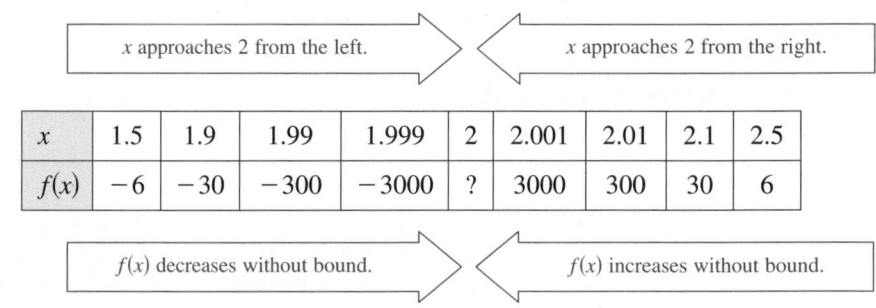

x	1.5	1.9	1.99	1.999	2	2.001	2.01	2.1	2.5
$f(x)$	-6	-30	-300	-3000	?	3000	300	30	6

This behavior is denoted as

$$\lim_{x \to 2^-} \frac{3}{x - 2} = -\infty \qquad f(x) \text{ decreases without bound as } x \text{ approaches 2 from the left.}$$

and

$$\lim_{x \to 2^+} \frac{3}{x - 2} = \infty. \qquad f(x) \text{ increases without bound as } x \text{ approaches 2 from the right.}$$

The symbols ∞ and $-\infty$ refer to positive infinity and negative infinity, respectively. These symbols do not represent real numbers. They are convenient symbols used to describe unbounded conditions more concisely. A limit in which $f(x)$ increases or decreases without bound as x approaches c is called an **infinite limit.**

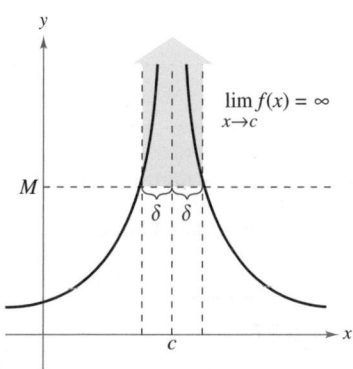

Infinite limits

Figure 2.39

Definition of Infinite Limits

Let f be a function that is defined at every real number in some open interval containing c (except possibly at c itself). The statement

$$\lim_{x \to c} f(x) = \infty$$

means that for each $M > 0$ there exists a $\delta > 0$ such that $f(x) > M$ whenever $0 < |x - c| < \delta$ (see Figure 2.39). Similarly, the statement

$$\lim_{x \to c} f(x) = -\infty$$

means that for each $N < 0$ there exists a $\delta > 0$ such that $f(x) < N$ whenever

$$0 < |x - c| < \delta.$$

To define the **infinite limit from the left,** replace $0 < |x - c| < \delta$ by $c - \delta < x < c$. To define the **infinite limit from the right,** replace $0 < |x - c| < \delta$ by $c < x < c + \delta$.

Be sure you see that the equal sign in the statement $\lim f(x) = \infty$ does not mean that the limit exists! On the contrary, it tells you how the limit **fails to exist** by denoting the unbounded behavior of $f(x)$ as x approaches c.

EXAMPLE 1 **Determining Infinite Limits from a Graph**

Determine the limit of each function shown in Figure 2.40 as x approaches 1 from the left and from the right.

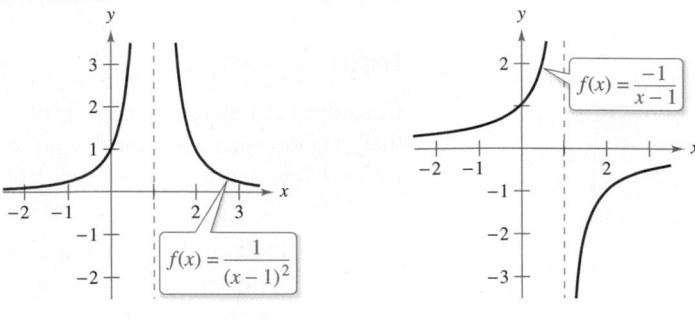

(a)   (b)

Each graph has an asymptote at $x = 1$.

Figure 2.40

Solution

a. When x approaches 1 from the left or the right, $(x-1)^2$ is a small positive number. Thus, the quotient $1/(x-1)^2$ is a large positive number, and $f(x)$ approaches infinity from each side of $x = 1$. So, you can conclude that

$$\lim_{x \to 1} \frac{1}{(x-1)^2} = \infty. \qquad \text{Limit from each side is infinity.}$$

Figure 2.40(a) confirms this analysis.

b. When x approaches 1 from the left, $x - 1$ is a small negative number. Thus, the quotient $-1/(x-1)$ is a large positive number, and $f(x)$ approaches infinity from the left of $x = 1$. So, you can conclude that

$$\lim_{x \to 1^-} \frac{-1}{x-1} = \infty. \qquad \text{Limit from the left side is infinity.}$$

When x approaches 1 from the right, $x - 1$ is a small positive number. Thus, the quotient $-1/(x-1)$ is a large negative number, and $f(x)$ approaches negative infinity from the right of $x = 1$. So, you can conclude that

$$\lim_{x \to 1^+} \frac{-1}{x-1} = -\infty. \qquad \text{Limit from the right side is negative infinity.}$$

Figure 2.40(b) confirms this analysis. ∎

▷ **TECHNOLOGY**   Remember that you can use a numerical approach to analyze a limit. For instance, you can use a graphing utility to create a table of values to analyze the limit in Example 1(a), as shown in the figure below.

Enter x-values using *ask* mode.

X	Y₁	
.9	100	
.99	10000	
.999	1E6	
1	ERROR	
1.001	1E6	
1.01	10000	
1.1	100	
X=1		

As x approaches 1 from the left, $f(x)$ increases without bound.

As x approaches 1 from the right, $f(x)$ increases without bound.

Use a graphing utility to make a table of values to analyze the limit in Example 1(b).

Vertical Asymptotes

If it were possible to extend the graphs in Figure 2.40 toward positive and negative infinity, you would see that each graph becomes arbitrarily close to the vertical line $x = 1$. This line is a **vertical asymptote** of the graph of f. (You will study other types of asymptotes in Sections 4.5 and 4.6.)

> **Definition of Vertical Asymptote**
>
> If $f(x)$ approaches infinity (or negative infinity) as x approaches c from the right or the left, then the line $x = c$ is a **vertical asymptote** of the graph of f.

In Example 1, note that each of the functions is a *quotient* and that the vertical asymptote occurs at a number at which the denominator is 0 (and the numerator is not 0). The next theorem generalizes this observation.

> **THEOREM 2.14   Vertical Asymptotes**
>
> Let f and g be continuous on an open interval containing c. If $f(c) \neq 0$, $g(c) = 0$, and there exists an open interval containing c such that $g(x) \neq 0$ for all $x \neq c$ in the interval, then the graph of the function
>
> $$h(x) = \frac{f(x)}{g(x)}$$
>
> has a vertical asymptote at $x = c$.
>
> A proof of this theorem is given in Appendix A.

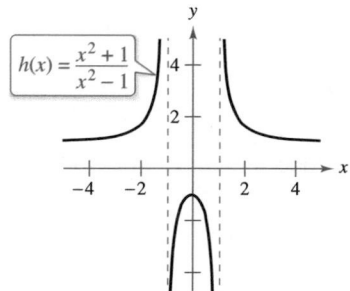

(a)

(b)

(c)

Functions with vertical asymptotes
Figure 2.41

EXAMPLE 2   Finding Vertical Asymptotes

•••••▷ *See LarsonCalculus.com for an interactive version of this type of example.*

a. When $x = -1$, the denominator of

$$h(x) = \frac{1}{2(x + 1)}$$

is 0 and the numerator is not 0. So, by Theorem 2.14, you can conclude that $x = -1$ is a vertical asymptote, as shown in Figure 2.41(a).

b. By factoring the denominator as

$$h(x) = \frac{x^2 + 1}{x^2 - 1} = \frac{x^2 + 1}{(x - 1)(x + 1)}$$

you can see that the denominator is 0 at $x = -1$ and $x = 1$. Also, because the numerator is not 0 at these two points, you can apply Theorem 2.14 to conclude that the graph of f has two vertical asymptotes, as shown in Figure 2.41(b).

c. By writing the cotangent function in the form

$$h(x) = \cot x = \frac{\cos x}{\sin x}$$

you can apply Theorem 2.14 to conclude that vertical asymptotes occur at all values of x such that $\sin x = 0$ and $\cos x \neq 0$, as shown in Figure 2.41(c). So, the graph of this function has infinitely many vertical asymptotes. These asymptotes occur at $x = n\pi$, where n is an integer.

Theorem 2.14 requires that the value of the numerator at $x = c$ be nonzero. When both the numerator and the denominator are 0 at $x = c$, you obtain the *indeterminate form* 0/0, and you cannot determine the limit behavior at $x = c$ without further investigation, as illustrated in Example 3.

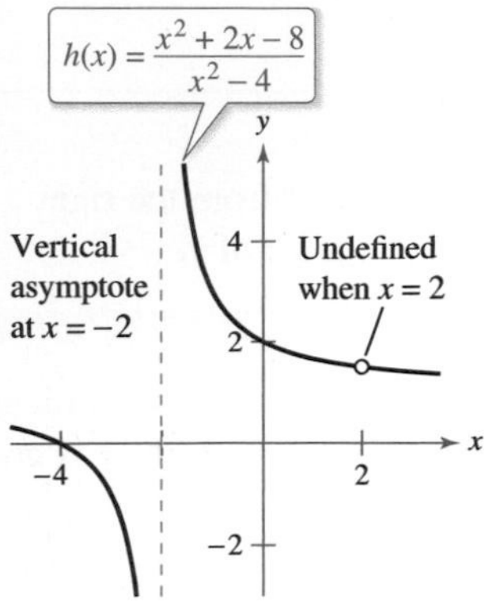

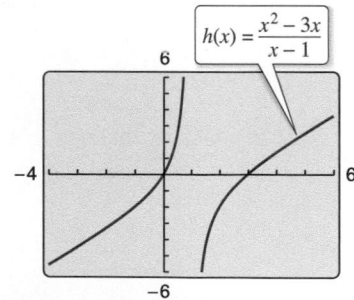

$h(x) = \dfrac{x^2 + 2x - 8}{x^2 - 4}$

Vertical asymptote at $x = -2$

Undefined when $x = 2$

$h(x)$ increases and decreases without bound as x approaches -2.

Figure 2.42

EXAMPLE 3 A Rational Function with Common Factors

Determine all vertical asymptotes of the graph of

$$h(x) = \frac{x^2 + 2x - 8}{x^2 - 4}.$$

Solution Begin by simplifying the expression, as shown.

$$
\begin{aligned}
h(x) &= \frac{x^2 + 2x - 8}{x^2 - 4} \\
&= \frac{(x + 4)(x - 2)}{(x + 2)(x - 2)} \\
&= \frac{x + 4}{x + 2}, \quad x \neq 2
\end{aligned}
$$

At all x-values other than $x = 2$, the graph of h coincides with the graph of $k(x) = (x + 4)/(x + 2)$. So, you can apply Theorem 2.14 to k to conclude that there is a vertical asymptote at $x = -2$, as shown in Figure 2.42. From the graph, you can see that

$$\lim_{x \to -2^-} \frac{x^2 + 2x - 8}{x^2 - 4} = -\infty \quad \text{and} \quad \lim_{x \to -2^+} \frac{x^2 + 2x - 8}{x^2 - 4} = \infty.$$

Note that $x = 2$ is *not* a vertical asymptote.

EXAMPLE 4 Determining Infinite Limits

Find each limit.

$$\lim_{x \to 1^-} \frac{x^2 - 3x}{x - 1} \quad \text{and} \quad \lim_{x \to 1^+} \frac{x^2 - 3x}{x - 1}$$

Solution Because the denominator is 0 when $x = 1$ (and the numerator is not 0), you know that the graph of

$$h(x) = \frac{x^2 - 3x}{x - 1}$$

$h(x) = \dfrac{x^2 - 3x}{x - 1}$

The graph of h has a vertical asymptote at $x = 1$.

Figure 2.43

has a vertical asymptote at $x = 1$. This means that each of the given limits is either ∞ or $-\infty$. You can determine the result by analyzing h at values of x close to 1 or by using a graphing utility. From the graph of h shown in Figure 2.43, you can see that the graph approaches ∞ from the left of $x = 1$ and approaches $-\infty$ from the right of $x = 1$. So, you can conclude that

$$\lim_{x \to 1^-} \frac{x^2 - 3x}{x - 1} = \infty \qquad \text{The limit from the left is infinity.}$$

and

$$\lim_{x \to 1^+} \frac{x^2 - 3x}{x - 1} = -\infty. \qquad \text{The limit from the right is negative infinity.}$$

▷ **TECHNOLOGY PITFALL** When using a graphing utility, be careful to interpret correctly the graph of a function with a vertical asymptote—some graphing utilities have difficulty drawing this type of graph.

THEOREM 2.15 Properties of Infinite Limits

Let c and L be real numbers, and let f and g be functions such that

$$\lim_{x \to c} f(x) = \infty \quad \text{and} \quad \lim_{x \to c} g(x) = L.$$

1. Sum or difference: $\displaystyle \lim_{x \to c} [f(x) \pm g(x)] = \infty$

2. Product: $\displaystyle \lim_{x \to c} [f(x)g(x)] = \infty, \quad L > 0$

   $\displaystyle \lim_{x \to c} [f(x)g(x)] = -\infty, \quad L < 0$

3. Quotient: $\displaystyle \lim_{x \to c} \frac{g(x)}{f(x)} = 0$

Similar properties hold for one-sided limits and for functions for which the limit of $f(x)$ as x approaches c is $-\infty$ [see Example 5(d)].

REMARK Be sure you understand that Property 2 of Theorem 2.15 is not valid when $\lim_{x \to c} g(x) = 0$.

Proof Here is a proof of the sum property. (The proofs of the remaining properties are left as an exercise [see Exercise 72].) To show that the limit of $f(x) + g(x)$ is infinite, choose $M > 0$. You then need to find $\delta > 0$ such that $[f(x) + g(x)] > M$ whenever $0 < |x - c| < \delta$. For simplicity's sake, you can assume L is positive. Let $M_1 = M + 1$. Because the limit of $f(x)$ is infinite, there exists δ_1 such that $f(x) > M_1$ whenever $0 < |x - c| < \delta_1$. Also, because the limit of $g(x)$ is L, there exists δ_2 such that $|g(x) - L| < 1$ whenever $0 < |x - c| < \delta_2$. By letting δ be the smaller of δ_1 and δ_2, you can conclude that $0 < |x - c| < \delta$ implies $f(x) > M + 1$ and $|g(x) - L| < 1$. The second of these two inequalities implies that $g(x) > L - 1$, and adding this to the first inequality, you can write

$$f(x) + g(x) > (M + 1) + (L - 1) = M + L > M.$$

So, you can conclude that

$$\lim_{x \to c} [f(x) + g(x)] = \infty.$$

EXAMPLE 5 Determining Limits

a. Because $\displaystyle \lim_{x \to 0} 1 = 1$ and $\displaystyle \lim_{x \to 0} \frac{1}{x^2} = \infty$, you can write

$$\lim_{x \to 0} \left(1 + \frac{1}{x^2} \right) = \infty. \qquad \text{Property 1, Theorem 2.15}$$

b. Because $\displaystyle \lim_{x \to 1^-} (x^2 + 1) = 2$ and $\displaystyle \lim_{x \to 1^-} (\cot \pi x) = -\infty$, you can write

$$\lim_{x \to 1^-} \frac{x^2 + 1}{\cot \pi x} = 0. \qquad \text{Property 3, Theorem 2.15}$$

c. Because $\displaystyle \lim_{x \to 0^+} 3 = 3$ and $\displaystyle \lim_{x \to 0^+} \ln x = -\infty$, you can write

$$\lim_{x \to 0^+} 3 \ln x = -\infty. \qquad \text{Property 2, Theorem 2.15 (See Figure 2.44.)}$$

d. Because $\displaystyle \lim_{x \to 0^-} x^2 = 0$ and $\displaystyle \lim_{x \to 0^-} \frac{1}{x} = -\infty$, you can write

$$\lim_{x \to 0^-} \left(x^2 + \frac{1}{x} \right) = -\infty. \qquad \text{Property 1, Theorem 2.15}$$

With a graphing utility, you can confirm that the natural logarithmic function has a vertical asymptote at $x = 0$. This implies that $\lim_{x \to 0^+} \ln x = -\infty$.

Figure 2.44

REMARK Note that the solution to Example 5(d) uses Property 1 from Theorem 2.15 for which the limit of $f(x)$ as x approaches c is $-\infty$.

2.5 Exercises

See **CalcChat.com** for tutorial help and worked-out solutions to odd-numbered exercises.

CONCEPT CHECK

1. **Infinite Limit** In your own words, describe the meaning of an infinite limit. What does ∞ represent?

2. **Vertical Asymptote** In your own words, describe what is meant by a vertical asymptote of a graph.

 **Determining Infinite Limits from a Graph** In Exercises 3–6, determine whether $f(x)$ approaches ∞ or $-\infty$ as x approaches -2 from the left and from the right.

3. $f(x) = 2\left|\dfrac{x}{x^2 - 4}\right|$

4. $f(x) = \dfrac{1}{x + 2}$

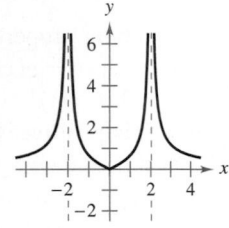 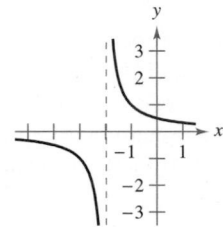

5. $f(x) = \tan\dfrac{\pi x}{4}$

6. $f(x) = \sec\dfrac{\pi x}{4}$

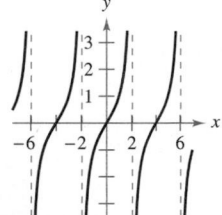 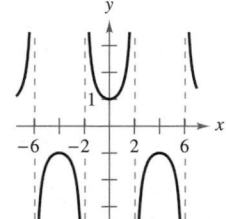

 **Determining Infinite Limits** In Exercises 7–10, determine whether $f(x)$ approaches ∞ or $-\infty$ as x approaches 4 from the left and from the right.

7. $f(x) = \dfrac{1}{x - 4}$

8. $f(x) = \dfrac{-1}{x - 4}$

9. $f(x) = \dfrac{1}{(x - 4)^2}$

10. $f(x) = \dfrac{-1}{(x - 4)^2}$

Numerical and Graphical Analysis In Exercises 11–16, create a table of values for the function and use the result to determine whether $f(x)$ approaches ∞ or $-\infty$ as x approaches -3 from the left and from the right. Use a graphing utility to graph the function to confirm your answer.

11. $f(x) = \dfrac{1}{x^2 - 9}$

12. $f(x) = \dfrac{x}{x^2 - 9}$

13. $f(x) = \dfrac{x^2}{x^2 - 9}$

14. $f(x) = -\dfrac{1}{3 + x}$

15. $f(x) = \cot\dfrac{\pi x}{3}$

16. $f(x) = \tan\dfrac{\pi x}{6}$

 **Finding Vertical Asymptotes** In Exercises 17–32, find the vertical asymptotes (if any) of the graph of the function.

17. $f(x) = \dfrac{x^2}{x^2 - 4}$

18. $g(t) = \dfrac{t - 1}{t^2 + 1}$

19. $f(x) = \dfrac{3}{x^2 + x - 2}$

20. $g(x) = \dfrac{x^2 - 5x + 25}{x^3 + 125}$

21. $f(x) = \dfrac{4x^2 + 4x - 24}{x^4 - 2x^3 - 9x^2 + 18x}$

22. $h(x) = \dfrac{x^2 - 9}{x^3 + 3x^2 - x - 3}$

23. $f(x) = \dfrac{e^{-2x}}{x - 1}$

24. $g(x) = xe^{-2x}$

25. $h(t) = \dfrac{\ln(t^2 + 1)}{t + 2}$

26. $f(z) = \ln(z^2 - 4)$

27. $f(x) = \dfrac{1}{e^x - 1}$

28. $f(x) = \ln(x + 3)$

29. $f(x) = \csc \pi x$

30. $f(x) = \tan \pi x$

31. $s(t) = \dfrac{t}{\sin t}$

32. $g(\theta) = \dfrac{\tan \theta}{\theta}$

 **Vertical Asymptote or Removable Discontinuity** In Exercises 33–36, determine whether the graph of the function has a vertical asymptote or a removable discontinuity at $x = -1$. Graph the function using a graphing utility to confirm your answer.

33. $f(x) = \dfrac{x^2 - 1}{x + 1}$

34. $f(x) = \dfrac{x^2 - 2x - 8}{x + 1}$

35. $f(x) = \dfrac{\cos(x^2 - 1)}{x + 1}$

36. $f(x) = \dfrac{\ln(x^2 + 1)}{x + 1}$

Finding a One-Sided Limit In Exercises 37–52, find the one-sided limit (if it exists).

37. $\displaystyle\lim_{x \to 2^+} \dfrac{x}{x - 2}$

38. $\displaystyle\lim_{x \to 2^-} \dfrac{x^2}{x^2 + 4}$

39. $\displaystyle\lim_{x \to -3^-} \dfrac{x + 3}{x^2 + x - 6}$

40. $\displaystyle\lim_{x \to (-1/2)^+} \dfrac{6x^2 + x - 1}{4x^2 - 4x - 3}$

41. $\displaystyle\lim_{x \to 0^-} \left(1 + \dfrac{1}{x}\right)$

42. $\displaystyle\lim_{x \to 0^+} \left(6 - \dfrac{1}{x^3}\right)$

43. $\displaystyle\lim_{x \to -4^-} \left(x^2 + \dfrac{2}{x + 4}\right)$

44. $\displaystyle\lim_{x \to 0^+} \left(x - \dfrac{1}{x} + 3\right)$

45. $\displaystyle\lim_{x \to 0^+} \left(\sin x + \dfrac{1}{x}\right)$

46. $\displaystyle\lim_{x \to (\pi/2)^+} \dfrac{-2}{\cos x}$

47. $\displaystyle\lim_{x \to 8^-} \dfrac{e^x}{(x - 8)^3}$

48. $\displaystyle\lim_{x \to 4^+} \ln(x^2 - 16)$

49. $\displaystyle\lim_{x \to (\pi/2)^-} \ln|\cos x|$

50. $\displaystyle\lim_{x \to 0^+} e^{-0.5x} \sin x$

51. $\displaystyle\lim_{x \to (1/2)^-} x \sec \pi x$

52. $\displaystyle\lim_{x \to (1/2)^+} x^2 \tan \pi x$

 **Finding a One-Sided Limit Using Technology** In Exercises 53 and 54, use a graphing utility to graph the function and determine the one-sided limit.

53. $\lim\limits_{x \to 1^+} \dfrac{x^2 + x + 1}{x^3 - 1}$

54. $\lim\limits_{x \to 1^-} \dfrac{x^3 - 1}{x^2 + x + 1}$

Determining Limits In Exercises 55 and 56, use the information to determine the limits.

55. $\lim\limits_{x \to c} f(x) = \infty$

$\lim\limits_{x \to c} g(x) = -2$

(a) $\lim\limits_{x \to c} [f(x) + g(x)]$

(b) $\lim\limits_{x \to c} [f(x)g(x)]$

(c) $\lim\limits_{x \to c} \dfrac{g(x)}{f(x)}$

56. $\lim\limits_{x \to c} f(x) = -\infty$

$\lim\limits_{x \to c} g(x) = 3$

(a) $\lim\limits_{x \to c} [f(x) + g(x)]$

(b) $\lim\limits_{x \to c} [f(x)g(x)]$

(c) $\lim\limits_{x \to c} \dfrac{g(x)}{f(x)}$

EXPLORING CONCEPTS

57. Writing a Rational Function Write a rational function with vertical asymptotes at $x = 6$ and $x = -2$, and with a zero at $x = 3$.

58. Rational Function Does the graph of every rational function have a vertical asymptote? Explain.

59. Sketching a Graph Use the graph of the function f (see figure) to sketch the graph of $g(x) = 1/f(x)$ on the interval $[-2, 3]$. To print an enlarged copy of the graph, go to *MathGraphs.com*.

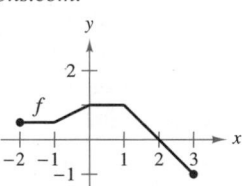

---

60. Relativity According to the theory of relativity, the mass m of a particle depends on its velocity v. That is,

$$m = \frac{m_0}{\sqrt{1 - (v^2/c^2)}},$$ where m_0 is the mass when the particle is at rest and c is the speed of light. Find the limit of the mass as v approaches c from the left.

 **61. Numerical and Graphical Reasoning** Use a graphing utility to complete the table for each function and graph each function to estimate the limit. What is the value of the limit when the power of x in the denominator is greater than 3?

x	1	0.5	0.2	0.1	0.01	0.001	0.0001
$f(x)$							

(a) $\lim\limits_{x \to 0^+} \dfrac{x - \sin x}{x}$

(b) $\lim\limits_{x \to 0^+} \dfrac{x - \sin x}{x^2}$

(c) $\lim\limits_{x \to 0^+} \dfrac{x - \sin x}{x^3}$

(d) $\lim\limits_{x \to 0^+} \dfrac{x - \sin x}{x^4}$

---

62. **HOW DO YOU SEE IT?** For a quantity of gas at a constant temperature, the pressure P is inversely proportional to the volume V. What is the limit of P as V approaches 0 from the right? Explain what this means in the context of the problem.

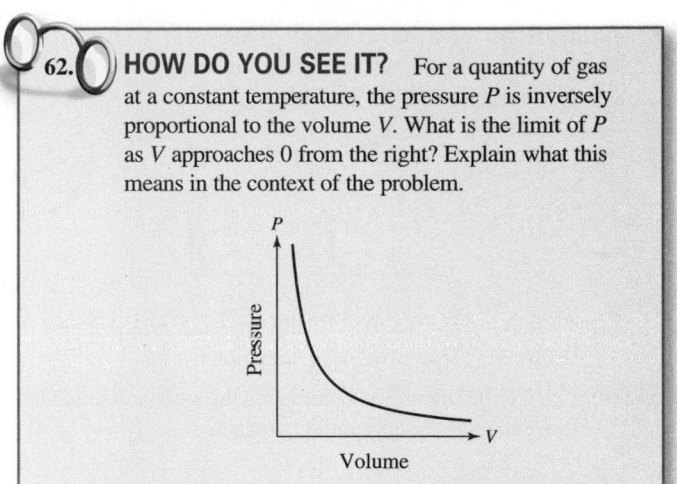

63. Rate of Change A 25-foot ladder is leaning against a house (see figure). If the base of the ladder is pulled away from the house at a rate of 2 feet per second, then the top will move down the wall at a rate of

$$r = \frac{2x}{\sqrt{625 - x^2}} \text{ ft/sec}$$

where x is the distance between the base of the ladder and the house, and r is the rate in feet per second.

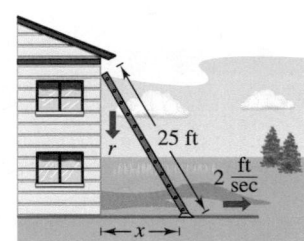

(a) Find the rate r when x is 7 feet.

(b) Find the rate r when x is 15 feet.

(c) Find the limit of r as x approaches 25 from the left.

64. Average Speed

On a trip of d miles to another city, a truck driver's average speed was x miles per hour. On the return trip, the average speed was y miles per hour. The average speed for the round trip was 50 miles per hour.

(a) Verify that

$$y = \frac{25x}{x - 25}.$$

What is the domain?

(b) Complete the table.

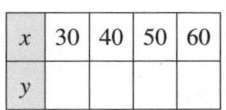

x	30	40	50	60
y				

Are the values of y different than you expected? Explain.

(c) Find the limit of y as x approaches 25 from the right and interpret its meaning.

65. Numerical and Graphical Analysis Consider the shaded region outside the sector of a circle of radius 10 meters and inside a right triangle (see figure).

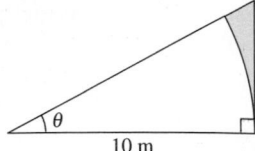

(a) Write the area $A = f(\theta)$ of the region as a function of θ. Determine the domain of the function.

(b) Use a graphing utility to complete the table and graph the function over the appropriate domain.

θ	0.3	0.6	0.9	1.2	1.5
$f(\theta)$					

(c) Find the limit of A as θ approaches $\pi/2$ from the left.

66. Numerical and Graphical Reasoning A crossed belt connects a 20-centimeter pulley (10-cm radius) on an electric motor with a 40-centimeter pulley (20-cm radius) on a saw arbor (see figure). The electric motor runs at 1700 revolutions per minute.

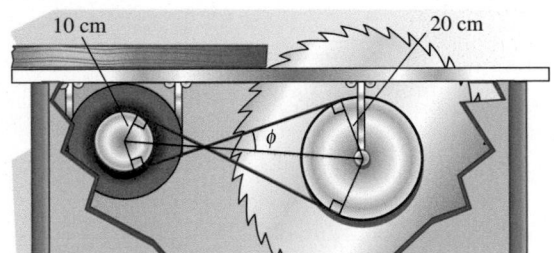

(a) Determine the number of revolutions per minute of the saw.

(b) How does crossing the belt affect the saw in relation to the motor?

(c) Let L be the total length of the belt. Write L as a function of ϕ, where ϕ is measured in radians. What is the domain of the function? (*Hint:* Add the lengths of the straight sections of the belt and the length of the belt around each pulley.)

(d) Use a graphing utility to complete the table.

ϕ	0.3	0.6	0.9	1.2	1.5
L					

(e) Use a graphing utility to graph the function over the appropriate domain.

(f) Find $\displaystyle\lim_{\phi \to (\pi/2)^-} L$.

(g) Use a geometric argument as the basis of a second method of finding the limit in part (f).

(h) Find $\displaystyle\lim_{\phi \to 0^+} L$.

True or False? In Exercises 67–70, determine whether the statement is true or false. If it is false, explain why or give an example that shows it is false.

67. The graph of a function cannot cross a vertical asymptote.

68. The graphs of polynomial functions have no vertical asymptotes.

69. The graphs of trigonometric functions have no vertical asymptotes.

70. If f has a vertical asymptote at $x = 0$, then f is undefined at $x = 0$.

71. Finding Functions Find functions f and g such that $\displaystyle\lim_{x \to c} f(x) = \infty$ and $\displaystyle\lim_{x \to c} g(x) = \infty$, but $\displaystyle\lim_{x \to c} [f(x) - g(x)] \neq 0$.

72. Proof Prove the difference, product, and quotient properties in Theorem 2.15.

73. Proof Prove that if $\displaystyle\lim_{x \to c} f(x) = \infty$, then $\displaystyle\lim_{x \to c} \frac{1}{f(x)} = 0$.

74. Proof Prove that if

$$\lim_{x \to c} \frac{1}{f(x)} = 0$$

then $\displaystyle\lim_{x \to c} f(x)$ does not exist.

Infinite Limits In Exercises 75–78, use the ε–δ definition of infinite limits to prove the statement.

75. $\displaystyle\lim_{x \to 3^+} \frac{1}{x - 3} = \infty$

76. $\displaystyle\lim_{x \to 5^-} \frac{1}{x - 5} = -\infty$

77. $\displaystyle\lim_{x \to 8^+} \frac{3}{8 - x} = -\infty$

78. $\displaystyle\lim_{x \to 9^-} \frac{6}{9 - x} = \infty$

SECTION PROJECT

Graphs and Limits of Trigonometric Functions

Recall from Theorem 2.9 that the limit of

$$f(x) = \frac{\sin x}{x}$$

as x approaches 0 is 1.

(a) Use a graphing utility to graph the function f on the interval $-\pi \leq x \leq \pi$. Explain how the graph helps confirm this theorem.

(b) Explain how you could use a table of values to confirm the value of this limit numerically.

(c) Graph $g(x) = \sin x$ by hand. Sketch a tangent line at the point $(0, 0)$ and visually estimate the slope of this tangent line.

(d) Let $(x, \sin x)$ be a point on the graph of g near $(0, 0)$, and write a formula for the slope of the secant line joining $(x, \sin x)$ and $(0, 0)$. Evaluate this formula at $x = 0.1$ and $x = 0.01$. Then find the exact slope of the tangent line to g at the point $(0, 0)$.

(e) Sketch the graph of the cosine function $h(x) = \cos x$. What is the slope of the tangent line at the point $(0, 1)$? Use limits to find this slope analytically.

(f) Find the slope of the tangent line to $k(x) = \tan x$ at $(0, 0)$.

Review Exercises   See CalcChat.com for tutorial help and worked-out solutions to odd-numbered exercises.

Precalculus or Calculus In Exercises 1 and 2, decide whether the problem can be solved using precalculus or whether calculus is required. If the problem can be solved using precalculus, solve it. If the problem seems to require calculus, explain your reasoning and use a graphical or numerical approach to estimate the solution.

1. Find the distance between the points $(1, 1)$ and $(3, 9)$ along the curve $y = x^2$.

2. Find the distance between the points $(1, 1)$ and $(3, 9)$ along the line $y = 4x - 3$.

Estimating a Limit Numerically In Exercises 3 and 4, complete the table and use the result to estimate the limit. Use a graphing utility to graph the function to confirm your result.

3. $\lim\limits_{x \to 3} \dfrac{x - 3}{x^2 - 7x + 12}$

x	2.9	2.99	2.999	3	3.001	3.01	3.1
$f(x)$				?			

4. $\lim\limits_{x \to 0} \dfrac{\sqrt{x + 4} - 2}{x}$

x	-0.1	-0.01	-0.001	0	0.001	0.01	0.1
$f(x)$				?			

Finding a Limit Graphically In Exercises 5 and 6, use the graph to find the limit (if it exists). If the limit does not exist, explain why.

5. $h(x) = \left[\!\!\left[-\dfrac{x}{2} \right]\!\!\right] + x^2$ 6. $f(t) = \dfrac{\ln(t + 2)}{t}$

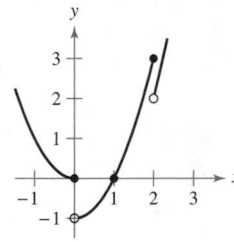   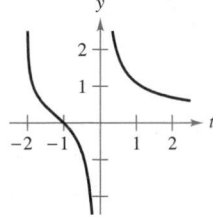

(a) $\lim\limits_{x \to 2} h(x)$ (b) $\lim\limits_{x \to 1} h(x)$ (a) $\lim\limits_{t \to 0} f(t)$ (b) $\lim\limits_{t \to -1} f(t)$

Using the ε-δ Definition of a Limit In Exercises 7–10, find the limit L. Then use the ε-δ definition to prove that the limit is L.

7. $\lim\limits_{x \to 1} (x + 4)$ 8. $\lim\limits_{x \to 9} \sqrt{x}$

9. $\lim\limits_{x \to 2} (1 - x^2)$ 10. $\lim\limits_{x \to 5} 9$

Finding a Limit In Exercises 11–28, find the limit.

11. $\lim\limits_{x \to -6} x^2$ 12. $\lim\limits_{x \to 0} (5x - 3)$

13. $\lim\limits_{x \to 27} \left(\sqrt[3]{x} - 1 \right)^4$ 14. $\lim\limits_{x \to 2} \sqrt{x^3 + 1}$

15. $\lim\limits_{x \to 4} \dfrac{4}{x - 1}$ 16. $\lim\limits_{x \to 2} \dfrac{x}{x^2 + 1}$

17. $\lim\limits_{x \to -3} \dfrac{2x^2 + 11x + 15}{x + 3}$ 18. $\lim\limits_{t \to 4} \dfrac{t^2 - 16}{t - 4}$

19. $\lim\limits_{x \to 4} \dfrac{\sqrt{x - 3} - 1}{x - 4}$ 20. $\lim\limits_{x \to 0} \dfrac{\sqrt{9 + x} - 3}{x}$

21. $\lim\limits_{x \to 0} \dfrac{[1/(x + 1)] - 1}{x}$ 22. $\lim\limits_{s \to 0} \dfrac{(1/\sqrt{1 + s}) - 1}{s}$

23. $\lim\limits_{x \to 0} \dfrac{1 - \cos x}{\sin x}$ 24. $\lim\limits_{x \to \pi/4} \dfrac{4x}{\tan x}$

25. $\lim\limits_{x \to 1} e^{x-1} \sin \dfrac{\pi x}{2}$ 26. $\lim\limits_{x \to 2} \dfrac{\ln(x - 1)^2}{\ln(x - 1)}$

27. $\lim\limits_{\Delta x \to 0} \dfrac{\sin[(\pi/6) + \Delta x] - (1/2)}{\Delta x}$

   [*Hint:* $\sin(\theta + \phi) = \sin \theta \cos \phi + \cos \theta \sin \phi$]

28. $\lim\limits_{\Delta x \to 0} \dfrac{\cos(\pi + \Delta x) + 1}{\Delta x}$

   [*Hint:* $\cos(\theta + \phi) = \cos \theta \cos \phi - \sin \theta \sin \phi$]

Evaluating a Limit In Exercises 29–32, evaluate the limit given $\lim\limits_{x \to c} f(x) = -6$ and $\lim\limits_{x \to c} g(x) = \frac{1}{2}$.

29. $\lim\limits_{x \to c} [f(x)g(x)]$ 30. $\lim\limits_{x \to c} \dfrac{f(x)}{g(x)}$

31. $\lim\limits_{x \to c} [f(x) + 2g(x)]$ 32. $\lim\limits_{x \to c} [f(x)]^2$

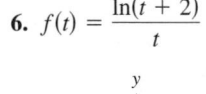 **Graphical, Numerical, and Analytic Analysis** In Exercises 33–36, use a graphing utility to graph the function and estimate the limit. Use a table to reinforce your conclusion. Then find the limit by analytic methods.

33. $\lim\limits_{x \to 0} \dfrac{\sqrt{2x + 9} - 3}{x}$ 34. $\lim\limits_{x \to 0} \dfrac{[1/(x + 4)] - (1/4)}{x}$

35. $\lim\limits_{x \to -9} \dfrac{x^3 + 729}{x + 9}$ 36. $\lim\limits_{x \to 0} \dfrac{\cos x - 1}{x}$

Free-Falling Object In Exercises 37 and 38, use the position function $s(t) = -4.9t^2 + 250$, which gives the height (in meters) of an object that has fallen for t seconds from a height of 250 meters. The velocity at time $t = a$ seconds is given by

$$\lim\limits_{t \to a} \dfrac{s(a) - s(t)}{a - t}.$$

37. Find the velocity of the object when $t = 4$.

38. When will the object hit the ground? At what velocity will the object impact the ground?

Finding a Limit In Exercises 39–50, find the limit (if it exists). If it does not exist, explain why.

39. $\lim\limits_{x \to 3^+} \dfrac{1}{x + 3}$ 40. $\lim\limits_{x \to 6^-} \dfrac{x - 6}{x^2 - 36}$

41. $\lim\limits_{x \to 25^+} \dfrac{\sqrt{x} - 5}{x - 25}$

42. $\lim\limits_{x \to 3^-} \dfrac{|x - 3|}{x - 3}$

43. $\lim\limits_{x \to 2^-} f(x)$, where $f(x) = \begin{cases} (x - 2)^2, & x \le 2 \\ 2 - x, & x > 2 \end{cases}$

44. $\lim\limits_{x \to 1^+} g(x)$, where $g(x) = \begin{cases} \sqrt{1 - x}, & x \le 1 \\ x + 1, & x > 1 \end{cases}$

45. $\lim\limits_{t \to 1} h(t)$, where $h(t) = \begin{cases} t^3 + 1, & t < 1 \\ \frac{1}{2}(t + 1), & t \ge 1 \end{cases}$

46. $\lim\limits_{s \to -2} f(s)$, where $f(s) = \begin{cases} -s^2 - 4s - 2, & s \le -2 \\ s^2 + 4s + 6, & s > -2 \end{cases}$

47. $\lim\limits_{x \to 2^-} (2[\![x]\!] + 1)$

48. $\lim\limits_{x \to 4} [\![x - 1]\!]$

49. $\lim\limits_{x \to 2^-} \dfrac{x^2 - 4}{|x - 2|}$

50. $\lim\limits_{x \to 1^+} \sqrt{x(x - 1)}$

Continuity on a Closed Interval In Exercises 51 and 52, discuss the continuity of the function on the closed interval.

51. $g(x) = \sqrt{8 - x^3}$, $[-2, 2]$

52. $h(x) = \dfrac{3}{5 - x}$, $[0, 5]$

Removable and Nonremovable Discontinuities In Exercises 53–58, find the x-values (if any) at which f is not continuous. Which of the discontinuities are removable?

53. $f(x) = x^4 - 81x$

54. $f(x) = x^2 - x + 20$

55. $f(x) = \dfrac{4}{x - 5}$

56. $f(x) = \dfrac{1}{x^2 - 9}$

57. $f(x) = \dfrac{x}{x^3 - x}$

58. $f(x) = \dfrac{x + 3}{x^2 - 3x - 18}$

59. Making a Function Continuous Find the value of c such that the function is continuous on the entire real number line.

$$f(x) = \begin{cases} x + 3, & x \le 2 \\ cx + 6, & x > 2 \end{cases}$$

60. Making a Function Continuous Find the value of c such that the function is continuous on the entire real number line.

$$f(x) = \begin{cases} cx^2 - 2x, & x < -1 \\ e^{x+1} + 3, & x \ge -1 \end{cases}$$

Testing for Continuity In Exercises 61–68, describe the intervals on which the function is continuous.

61. $f(x) = -3x^2 + 7$

62. $f(x) = \dfrac{4x^2 + 7x - 2}{x + 2}$

63. $f(x) = \sqrt{x} + \cos x$

64. $f(x) = [\![x + 3]\!]$

65. $g(x) = 2e^{[\![x]\!]/4}$

66. $h(x) = -2 \ln|5 - x|$

67. $f(x) = \begin{cases} \dfrac{3x^2 - x - 2}{x - 1}, & x \ne 1 \\ 0, & x = 1 \end{cases}$

68. $f(x) = \begin{cases} 5 - x, & x \le 2 \\ 2x - 3, & x > 2 \end{cases}$

69. Using the Intermediate Value Theorem Use the Intermediate Value Theorem to show that $f(x) = 2x^3 - 3$ has a zero in the interval $[1, 2]$.

70. Using the Intermediate Value Theorem Use the Intermediate Value Theorem to show that $f(x) = x^2 + x - 2$ has at least two zeros in the interval $[-3, 3]$.

Using the Intermediate Value Theorem In Exercises 71 and 72, verify that the Intermediate Value Theorem applies to the indicated interval and find the value of c guaranteed by the theorem.

71. $f(x) = x^2 + 5x - 4$, $[-1, 2]$, $f(c) = 2$

72. $f(x) = (x - 6)^3 + 4$, $[4, 7]$, $f(c) = 3$

Determining Infinite Limits In Exercises 73 and 74, determine whether $f(x)$ approaches ∞ or $-\infty$ as x approaches 6 from the left and from the right.

73. $f(x) = \dfrac{1}{x - 6}$

74. $f(x) = \dfrac{-1}{(x - 6)^2}$

Finding Vertical Asymptotes In Exercises 75–82, find the vertical asymptotes (if any) of the graph of the function.

75. $f(x) = \dfrac{3}{x}$

76. $f(x) = \dfrac{5}{(x - 2)^4}$

77. $f(x) = \dfrac{x^3}{x^2 - 9}$

78. $h(x) = \dfrac{6x}{36 - x^2}$

79. $f(x) = \sec \dfrac{\pi x}{2}$

80. $f(x) = \csc \pi x$

81. $g(x) = \ln(25 - x^2)$

82. $f(x) = 7e^{-3/x}$

Finding a One-Sided Limit In Exercises 83–94, find the one-sided limit (if it exists).

83. $\lim\limits_{x \to 1^-} \dfrac{x^2 + 2x + 1}{x - 1}$

84. $\lim\limits_{x \to (1/2)^+} \dfrac{x}{2x - 1}$

85. $\lim\limits_{x \to -1^+} \dfrac{x + 1}{x^3 + 1}$

86. $\lim\limits_{x \to -1^-} \dfrac{x + 1}{x^4 - 1}$

87. $\lim\limits_{x \to 0^+} \left(x - \dfrac{1}{x^3} \right)$

88. $\lim\limits_{x \to 2^-} \dfrac{1}{\sqrt[3]{x^2 - 4}}$

89. $\lim\limits_{x \to 0^+} \dfrac{\sin 4x}{5x}$

90. $\lim\limits_{x \to 0} \dfrac{\sec x^3}{2x}$

91. $\lim\limits_{x \to 0^+} \dfrac{\csc 2x}{x}$

92. $\lim\limits_{x \to 0^-} \dfrac{\cos^2 x}{x}$

93. $\lim\limits_{x \to 0^+} \ln(\sin x)$

94. $\lim\limits_{x \to 0^-} 12e^{-2/x}$

95. Environment A utility company burns coal to generate electricity. The cost C in dollars of removing $p\%$ of the air pollutants in the stack emissions is

$$C = \dfrac{80{,}000p}{100 - p}, \quad 0 \le p < 100.$$

(a) Find the cost of removing 50% of the pollutants.

(b) Find the cost of removing 90% of the pollutants.

(c) Find the limit of C as p approaches 100 from the left and interpret its meaning.

P.S. Problem Solving

1. Perimeter Let $P(x, y)$ be a point on the parabola $y = x^2$ in the first quadrant. Consider the triangle $\triangle PAO$ formed by P, $A(0, 1)$, and the origin $O(0, 0)$, and the triangle $\triangle PBO$ formed by P, $B(1, 0)$, and the origin (see figure).

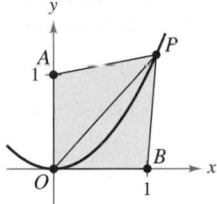

(a) Write the perimeter of each triangle in terms of x.

(b) Let $r(x)$ be the ratio of the perimeters of the two triangles,

$$r(x) = \frac{\text{Perimeter } \triangle PAO}{\text{Perimeter } \triangle PBO}.$$

Complete the table. Calculate $\lim\limits_{x \to 0^+} r(x)$.

x	4	2	1	0.1	0.01
Perimeter $\triangle PAO$					
Perimeter $\triangle PBO$					
$r(x)$					

2. Area Let $P(x, y)$ be a point on the parabola $y = x^2$ in the first quadrant. Consider the triangle $\triangle PAO$ formed by P, $A(0, 1)$, and the origin $O(0, 0)$, and the triangle $\triangle PBO$ formed by P, $B(1, 0)$, and the origin (see figure).

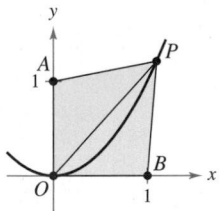

(a) Write the area of each triangle in terms of x.

(b) Let $a(x)$ be the ratio of the areas of the two triangles,

$$a(x) = \frac{\text{Area } \triangle PBO}{\text{Area } \triangle PAO}.$$

Complete the table. Calculate $\lim\limits_{x \to 0^+} a(x)$.

x	4	2	1	0.1	0.01
Area $\triangle PAO$					
Area $\triangle PBO$					
$a(x)$					

3. Area of a Circle

(a) Find the area of a regular hexagon inscribed in a circle of radius 1 (see figure). How close is this area to that of the circle?

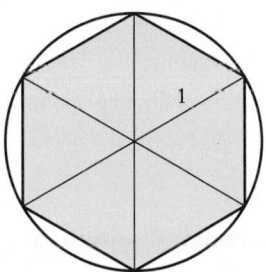

(b) Find the area A_n of an n-sided regular polygon inscribed in a circle of radius 1. Write your answer as a function of n.

(c) Complete the table. What number does A_n approach as n gets larger and larger?

n	6	12	24	48	96
A_n					

4. Tangent Line Let $P(3, 4)$ be a point on the circle $x^2 + y^2 = 25$ (see figure).

(a) What is the slope of the line joining P and $O(0, 0)$?

(b) Find an equation of the tangent line to the circle at P.

(c) Let $Q(x, y)$ be another point on the circle in the first quadrant. Find the slope m_x of the line joining P and Q in terms of x.

(d) Calculate $\lim\limits_{x \to 3} m_x$. How does this number relate to your answer in part (b)?

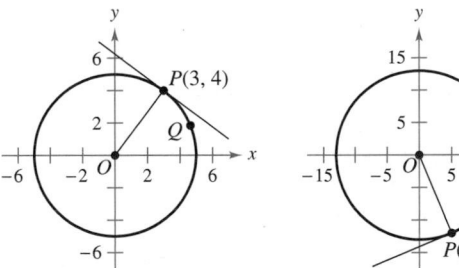

Figure for 4   Figure for 5

5. Tangent Line Let $P(5, -12)$ be a point on the circle $x^2 + y^2 = 169$ (see figure).

(a) What is the slope of the line joining P and $O(0, 0)$?

(b) Find an equation of the tangent line to the circle at P.

(c) Let $Q(x, y)$ be another point on the circle in the fourth quadrant. Find the slope m_x of the line joining P and Q in terms of x.

(d) Calculate $\lim\limits_{x \to 5} m_x$. How does this number relate to your answer in part (b)?

6. Finding Values Find the values of the constants a and b such that

$$\lim_{x \to 0} \frac{\sqrt{a + bx} - \sqrt{3}}{x} = \sqrt{3}.$$

7. Finding Limits Consider the function

$$f(x) = \frac{\sqrt{3 + x^{1/3}} - 2}{x - 1}.$$

(a) Find the domain of f.

(b) Use a graphing utility to graph the function.

(c) Find $\displaystyle \lim_{x \to -27^+} f(x)$.

(d) Find $\displaystyle \lim_{x \to 1} f(x)$.

8. Making a Function Continuous Find all values of the constant a such that f is continuous for all real numbers.

$$f(x) = \begin{cases} \dfrac{ax}{\tan x}, & x \ge 0 \\ a^2 - 2, & x < 0 \end{cases}$$

9. Choosing Graphs Consider the graphs of the four functions g_1, g_2, g_3, and g_4.

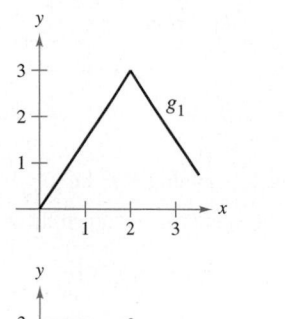

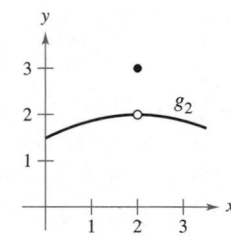

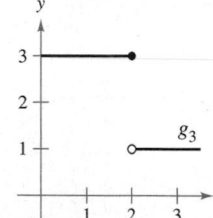

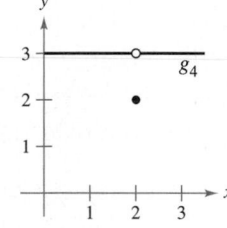

For each given condition of the function f, which of the graphs could be the graph of f?

(a) $\displaystyle \lim_{x \to 2} f(x) = 3$

(b) f is continuous at 2.

(c) $\displaystyle \lim_{x \to 2^-} f(x) = 3$

10. Limits and Continuity Sketch the graph of the function

$$f(x) = \left[\!\!\left[\frac{1}{x}\right]\!\!\right].$$

(a) Evaluate $f\left(\frac{1}{4}\right)$, $f(3)$, and $f(1)$.

(b) Evaluate the limits $\displaystyle \lim_{x \to 1^-} f(x)$, $\displaystyle \lim_{x \to 1^+} f(x)$, $\displaystyle \lim_{x \to 0^-} f(x)$, and $\displaystyle \lim_{x \to 0^+} f(x)$.

(c) Discuss the continuity of the function.

11. Limits and Continuity Sketch the graph of the function $f(x) = [\![x]\!] + [\![-x]\!]$.

(a) Evaluate $f(1)$, $f(0)$, $f\left(\frac{1}{2}\right)$, and $f(-2.7)$.

(b) Evaluate the limits $\displaystyle \lim_{x \to 1^-} f(x)$, $\displaystyle \lim_{x \to 1^+} f(x)$, and $\displaystyle \lim_{x \to 1/2} f(x)$.

(c) Discuss the continuity of the function.

12. Escape Velocity To escape Earth's gravitational field, a rocket must be launched with an initial velocity called the **escape velocity**. A rocket launched from the surface of Earth has velocity v (in miles per second) given by

$$v = \sqrt{\frac{2GM}{r} + v_0^2 - \frac{2GM}{R}} \approx \sqrt{\frac{192{,}000}{r} + v_0^2 - 48}$$

where v_0 is the initial velocity, r is the distance from the rocket to the center of Earth, G is the gravitational constant, M is the mass of Earth, and R is the radius of Earth (approximately 4000 miles).

(a) Find the value of v_0 for which you obtain an infinite limit for r as v approaches zero. This value of v_0 is the escape velocity for Earth.

(b) A rocket launched from the surface of the moon has velocity v (in miles per second) given by

$$v = \sqrt{\frac{1920}{r} + v_0^2 - 2.17}.$$

Find the escape velocity for the moon.

(c) A rocket launched from the surface of a planet has velocity v (in miles per second) given by

$$v = \sqrt{\frac{10{,}600}{r} + v_0^2 - 6.99}.$$

Find the escape velocity for this planet. Is the mass of this planet larger or smaller than that of Earth? (Assume that the mean density of this planet is the same as that of Earth.)

13. Pulse Function For positive numbers $a < b$, the **pulse function** is defined as

$$P_{a,b}(x) = H(x - a) - H(x - b) = \begin{cases} 0, & x < a \\ 1, & a \le x < b \\ 0, & x \ge b \end{cases}$$

where $H(x) = \begin{cases} 1, & x \ge 0 \\ 0, & x < 0 \end{cases}$ is the Heaviside function.

(a) Sketch the graph of the pulse function.

(b) Find the following limits:

  (i) $\displaystyle \lim_{x \to a^+} P_{a,b}(x)$ (ii) $\displaystyle \lim_{x \to a^-} P_{a,b}(x)$

  (iii) $\displaystyle \lim_{x \to b^+} P_{a,b}(x)$ (iv) $\displaystyle \lim_{x \to b^-} P_{a,b}(x)$

(c) Discuss the continuity of the pulse function.

(d) Why is $U(x) = \dfrac{1}{b - a} P_{a,b}(x)$ called the **unit pulse function?**

14. Proof Let a be a nonzero constant. Prove that if $\displaystyle \lim_{x \to 0} f(x) = L$, then $\displaystyle \lim_{x \to 0} f(ax) = L$. Show by means of an example that a must be nonzero.

3 Differentiation

Rate of Change
(Example 2, p. 186)

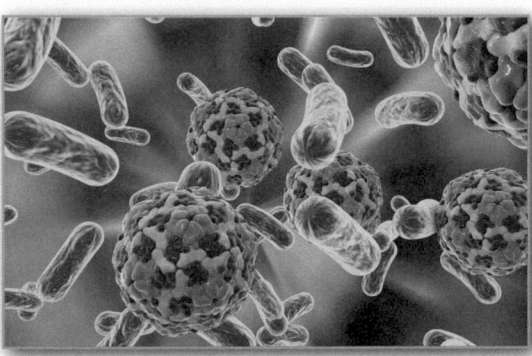

Bacteria *(Exercise 165, p. 168)*

Acceleration Due to Gravity *(Example 10, p. 149)*

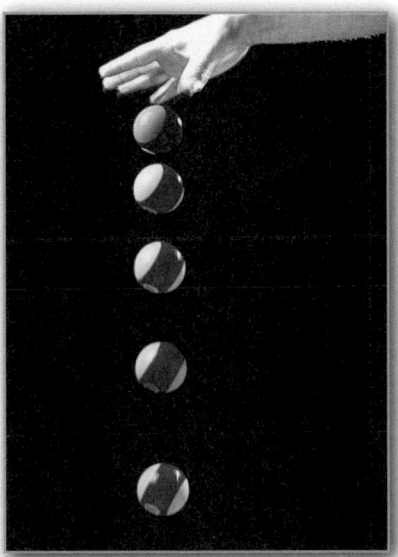

Velocity of a Falling Object
(Example 10, p. 137)

Stopping Distance *(Exercise 109, p.142)*

3.1 The Derivative and the Tangent Line Problem

■ Find the slope of the tangent line to a curve at a point.
■ Use the limit definition to find the derivative of a function.
■ Understand the relationship between differentiability and continuity.

The Tangent Line Problem

Calculus grew out of four major problems that European mathematicians were working on during the seventeenth century.

1. The tangent line problem (Section 2.1 and this section)
2. The velocity and acceleration problem (Sections 3.2 and 3.3)
3. The minimum and maximum problem (Section 4.1)
4. The area problem (Sections 2.1 and 5.2)

Each problem involves the notion of a limit, and calculus can be introduced with any of the four problems.

A brief introduction to the tangent line problem is given in Section 2.1. Although partial solutions to this problem were given by Pierre de Fermat (1601–1665), René Descartes (1596–1650), Christian Huygens (1629–1695), and Isaac Barrow (1630–1677), credit for the first general solution is usually given to Isaac Newton (1642–1727) and Gottfried Leibniz (1646–1716). Newton's work on this problem stemmed from his interest in optics and light refraction.

ISAAC NEWTON (1642–1727)

In addition to his work in calculus, Newton made revolutionary contributions to physics, including the Law of Universal Gravitation and his three laws of motion.
See LarsonCalculus.com to read more of this biography.

What does it mean to say that a line is tangent to a curve at a point? For a circle, the tangent line at a point P is the line that is perpendicular to the radial line at point P, as shown in Figure 3.1.

For a general curve, however, the problem is more difficult. For instance, how would you define the tangent lines shown in Figure 3.2? You might say that a line is tangent to a curve at a point P when it touches, but does not cross, the curve at point P. This definition would work for the first curve shown in Figure 3.2 but not for the second. *Or* you might say that a line is tangent to a curve when the line touches or intersects the curve at exactly one point. This definition would work for a circle but not for more general curves, as the third curve in Figure 3.2 shows.

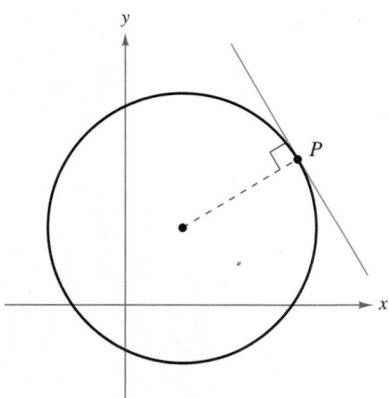

Tangent line to a circle
Figure 3.1

Exploration

Use a graphing utility to graph $f(x) = 2x^3 - 4x^2 + 3x - 5$. On the same screen, graph $y = x - 5$, $y = 2x - 5$, and $y = 3x - 5$. Which of these lines, if any, appears to be tangent to the graph of f at the point $(0, -5)$? Explain your reasoning.

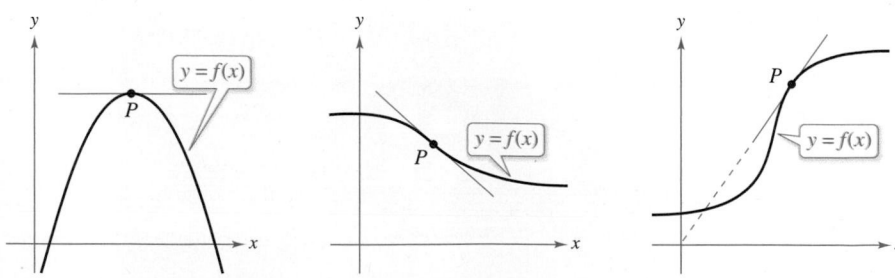

Tangent line to a curve at a point
Figure 3.2

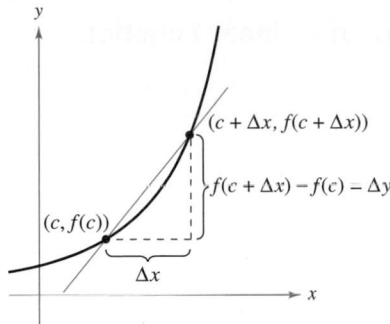

The secant line through $(c, f(c))$ and $(c + \Delta x, f(c + \Delta x))$

Figure 3.3

Essentially, the problem of finding the tangent line at a point P boils down to the problem of finding the *slope* of the tangent line at point P. You can approximate this slope using a **secant line*** through the point of tangency and a second point on the curve, as shown in Figure 3.3. If $(c, f(c))$ is the point of tangency and

$$(c + \Delta x, f(c + \Delta x))$$

is a second point on the graph of f, then the slope of the secant line through the two points is given by substitution into the slope formula

$$m = \frac{y_2 - y_1}{x_2 - x_1}$$

$$m_{\text{sec}} = \frac{f(c + \Delta x) - f(c)}{(c + \Delta x) - c} \qquad \begin{array}{l}\text{Change in } y\\ \text{Change in } x\end{array}$$

$$m_{\text{sec}} = \frac{f(c + \Delta x) - f(c)}{\Delta x}. \qquad \text{Slope of secant line}$$

The right-hand side of this equation is a **difference quotient.** The denominator Δx is the **change in x,** and the numerator

$$\Delta y = f(c + \Delta x) - f(c)$$

is the **change in y.**

The beauty of this procedure is that you can obtain more and more accurate approximations of the slope of the tangent line by choosing points closer and closer to the point of tangency, as shown in Figure 3.4.

THE TANGENT LINE PROBLEM

In 1637, mathematician René Descartes stated this about the tangent line problem:

"And I dare say that this is not only the most useful and general problem in geometry that I know, but even that I ever desire to know."

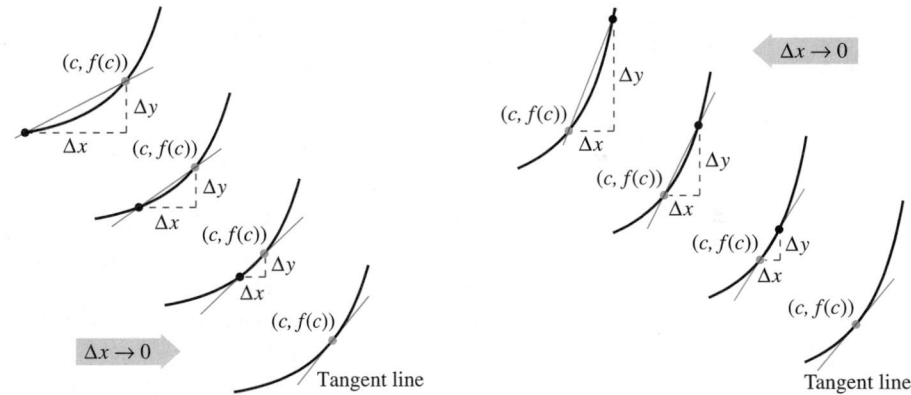

Tangent line approximations

Figure 3.4

Definition of Tangent Line with Slope m

If f is defined on an open interval containing c, and if the limit

$$\lim_{\Delta x \to 0} \frac{\Delta y}{\Delta x} = \lim_{\Delta x \to 0} \frac{f(c + \Delta x) - f(c)}{\Delta x} = m$$

exists, then the line passing through $(c, f(c))$ with slope m is the **tangent line** to the graph of f at the point $(c, f(c))$.

The slope of the tangent line to the graph of f at the point $(c, f(c))$ is also called the **slope of the graph of f at $x = c$.**

---

* This use of the word *secant* comes from the Latin *secare*, meaning to cut, and is not a reference to the trigonometric function of the same name.

EXAMPLE 1 The Slope of the Graph of a Linear Function

To find the slope of the graph of $f(x) = 2x - 3$ when $c = 2$, you can apply the definition of the slope of a tangent line, as shown.

$$\lim_{\Delta x \to 0} \frac{f(2 + \Delta x) - f(2)}{\Delta x} = \lim_{\Delta x \to 0} \frac{[2(2 + \Delta x) - 3] - [2(2) - 3]}{\Delta x}$$

$$= \lim_{\Delta x \to 0} \frac{4 + 2\Delta x - 3 - 4 + 3}{\Delta x}$$

$$= \lim_{\Delta x \to 0} \frac{2\Delta x}{\Delta x}$$

$$= \lim_{\Delta x \to 0} 2$$

$$= 2$$

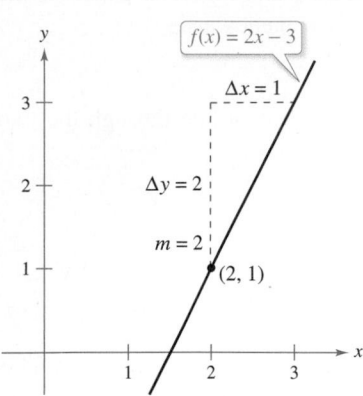

The slope of f at $(2, 1)$ is $m = 2$.
Figure 3.5

The slope of f at $(c, f(c)) = (2, 1)$ is $m = 2$, as shown in Figure 3.5. Notice that the limit definition of the slope of f agrees with the definition of the slope of a line as discussed in Section 1.2.

The graph of a linear function has the same slope at any point. This is not true of nonlinear functions, as shown in the next example.

EXAMPLE 2 Tangent Lines to the Graph of a Nonlinear Function

Find the slopes of the tangent lines to the graph of $f(x) = x^2 + 1$ at the points $(0, 1)$ and $(-1, 2)$, as shown in Figure 3.6.

Solution Let $(c, f(c))$ represent an arbitrary point on the graph of f. Then the slope of the tangent line at $(c, f(c))$ can be found as shown below. [Note in the limit process that c is held constant (as Δx approaches 0).]

$$\lim_{\Delta x \to 0} \frac{f(c + \Delta x) - f(c)}{\Delta x} = \lim_{\Delta x \to 0} \frac{[(c + \Delta x)^2 + 1] - (c^2 + 1)}{\Delta x}$$

$$= \lim_{\Delta x \to 0} \frac{c^2 + 2c(\Delta x) + (\Delta x)^2 + 1 - c^2 - 1}{\Delta x}$$

$$= \lim_{\Delta x \to 0} \frac{2c(\Delta x) + (\Delta x)^2}{\Delta x}$$

$$= \lim_{\Delta x \to 0} (2c + \Delta x)$$

$$= 2c$$

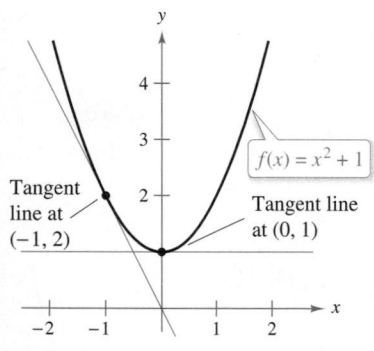

The slope of f at any point $(c, f(c))$ is $m = 2c$.
Figure 3.6

So, the slope at *any* point $(c, f(c))$ on the graph of f is $m = 2c$. At the point $(0, 1)$, the slope is $m = 2(0) = 0$, and at $(-1, 2)$, the slope is $m = 2(-1) = -2$.

The definition of a tangent line to a curve does not cover the possibility of a vertical tangent line. For vertical tangent lines, you can use the following definition. If f is continuous at c and

$$\lim_{\Delta x \to 0} \frac{f(c + \Delta x) - f(c)}{\Delta x} = \infty \quad \text{or} \quad \lim_{\Delta x \to 0} \frac{f(c + \Delta x) - f(c)}{\Delta x} = -\infty$$

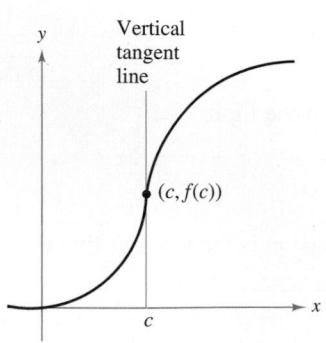

The graph of f has a vertical tangent line at $(c, f(c))$.
Figure 3.7

then the vertical line $x = c$ passing through $(c, f(c))$ is a **vertical tangent line** to the graph of f. For example, the function shown in Figure 3.7 has a vertical tangent line at $(c, f(c))$. When the domain of f is the closed interval $[a, b]$, you can extend the definition of a vertical tangent line to include the endpoints by considering continuity and limits from the right (for $x = a$) and from the left (for $x = b$).

The Derivative of a Function

You have now arrived at a crucial point in the study of calculus. The limit used to define the slope of a tangent line is also used to define one of the two fundamental operations of calculus—**differentiation.**

Definition of the Derivative of a Function

The **derivative** of f at x is

$$f'(x) = \lim_{\Delta x \to 0} \frac{f(x + \Delta x) - f(x)}{\Delta x}$$

provided the limit exists. For all x for which this limit exists, f' is a function of x.

REMARK The notation $f'(x)$ is read as "f prime of x."

Be sure you see that the derivative of a function of x is also a function of x. This "new" function gives the slope of the tangent line to the graph of f at the point $(x, f(x))$, provided that the graph has a tangent line at this point. The derivative can also be used to determine the **instantaneous rate of change** (or simply the **rate of change**) of one variable with respect to another.

The process of finding the derivative of a function is called **differentiation.** A function is **differentiable** at x when its derivative exists at x and is **differentiable on an open interval (a, b)** when it is differentiable at every point in the interval.

In addition to $f'(x)$, other notations are used to denote the derivative of $y = f(x)$. The most common are

$$f'(x), \quad \frac{dy}{dx}, \quad y', \quad \frac{d}{dx}[f(x)], \quad D_x[y]. \qquad \text{Notations for derivatives}$$

FOR FURTHER INFORMATION For more information on the crediting of mathematical discoveries to the first "discoverers," see the article "Mathematical Firsts— Who Done It?" by Richard H. Williams and Roy D. Mazzagatti in *Mathematics Teacher.* To view this article, go to *MathArticles.com.*

The notation dy/dx is read as "the derivative of y *with respect to* x" or simply "dy, dx." Using limit notation, you can write

$$\frac{dy}{dx} = \lim_{\Delta x \to 0} \frac{\Delta y}{\Delta x} = \lim_{\Delta x \to 0} \frac{f(x + \Delta x) - f(x)}{\Delta x} = f'(x).$$

EXAMPLE 3 Finding the Derivative by the Limit Process

▷ *See LarsonCalculus.com for an interactive version of this type of example.*

To find the derivative of $f(x) = x^3 + 2x$, use the definition of the derivative as shown.

$$\begin{aligned}
f'(x) &= \lim_{\Delta x \to 0} \frac{f(x + \Delta x) - f(x)}{\Delta x} \qquad \text{Definition of derivative}\\[2mm]
&= \lim_{\Delta x \to 0} \frac{(x + \Delta x)^3 + 2(x + \Delta x) - (x^3 + 2x)}{\Delta x}\\[2mm]
&= \lim_{\Delta x \to 0} \frac{x^3 + 3x^2\Delta x + 3x(\Delta x)^2 + (\Delta x)^3 + 2x + 2\Delta x - x^3 - 2x}{\Delta x}\\[2mm]
&= \lim_{\Delta x \to 0} \frac{3x^2\Delta x + 3x(\Delta x)^2 + (\Delta x)^3 + 2\Delta x}{\Delta x}\\[2mm]
&= \lim_{\Delta x \to 0} \frac{\Delta x[3x^2 + 3x\Delta x + (\Delta x)^2 + 2]}{\Delta x}\\[2mm]
&= \lim_{\Delta x \to 0} [3x^2 + 3x\Delta x + (\Delta x)^2 + 2]\\[2mm]
&= 3x^2 + 2
\end{aligned}$$

REMARK When using the definition to find a derivative of a function, the key is to rewrite the difference quotient so that Δx does not occur as a factor of the denominator.

· · · · · · · · · · · · · · ▷

· · REMARK Remember that the derivative of a function f is itself a function, which can be used to find the slope of the tangent line at the point $(x, f(x))$ on the graph of f.

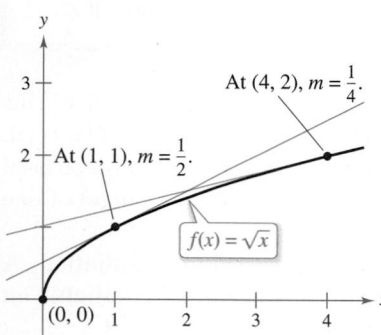

For $x > 0$, the slope of f at $(x, f(x))$ is $m = 1/(2\sqrt{x})$.
Figure 3.8

· · · · · · · · · · · · · · ▷

· · REMARK In many applications, it is convenient to use a variable other than x as the independent variable, as shown in Example 5.

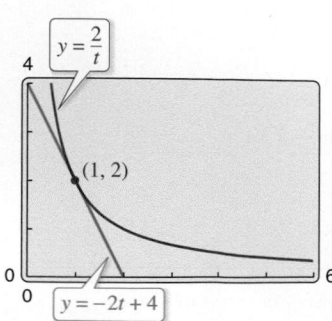

At the point $(1, 2)$, the line $y = -2t + 4$ is tangent to the graph of $y = 2/t$.
Figure 3.9

EXAMPLE 4 **Using the Derivative to Find the Slope at a Point**

Find $f'(x)$ for $f(x) = \sqrt{x}$. Then find the slopes of the graph of f at the points $(1, 1)$ and $(4, 2)$. Discuss the behavior of f at $(0, 0)$.

Solution Use the procedure for rationalizing numerators, as discussed in Section 2.3.

$$f'(x) = \lim_{\Delta x \to 0} \frac{f(x + \Delta x) - f(x)}{\Delta x} \qquad \text{Definition of derivative}$$

$$= \lim_{\Delta x \to 0} \frac{\sqrt{x + \Delta x} - \sqrt{x}}{\Delta x}$$

$$= \lim_{\Delta x \to 0} \left(\frac{\sqrt{x + \Delta x} - \sqrt{x}}{\Delta x} \right) \left(\frac{\sqrt{x + \Delta x} + \sqrt{x}}{\sqrt{x + \Delta x} + \sqrt{x}} \right)$$

$$= \lim_{\Delta x \to 0} \frac{(x + \Delta x) - x}{\Delta x \left(\sqrt{x + \Delta x} + \sqrt{x} \right)}$$

$$= \lim_{\Delta x \to 0} \frac{\Delta x}{\Delta x \left(\sqrt{x + \Delta x} + \sqrt{x} \right)}$$

$$= \lim_{\Delta x \to 0} \frac{1}{\sqrt{x + \Delta x} + \sqrt{x}}$$

$$= \frac{1}{2\sqrt{x}}$$

At the point $(1, 1)$, the slope is $f'(1) = \frac{1}{2}$. At the point $(4, 2)$, the slope is $f'(4) = \frac{1}{4}$. See Figure 3.8. The domain of f' is all $x > 0$, so the slope of f is undefined at $(0, 0)$. Moreover, the graph of f has a vertical tangent line at $(0, 0)$.

EXAMPLE 5 **Finding the Derivative of a Function**

· · · · ▷ *See LarsonCalculus.com for an interactive version of this type of example.*

Find the derivative with respect to t for the function $y = 2/t$.

Solution Considering $y = f(t)$, you obtain

$$\frac{dy}{dt} = \lim_{\Delta t \to 0} \frac{f(t + \Delta t) - f(t)}{\Delta t} \qquad \text{Definition of derivative}$$

$$= \lim_{\Delta t \to 0} \frac{\dfrac{2}{t + \Delta t} - \dfrac{2}{t}}{\Delta t} \qquad f(t + \Delta t) = \frac{2}{t + \Delta t} \text{ and } f(t) = \frac{2}{t}$$

$$= \lim_{\Delta t \to 0} \frac{\dfrac{2t - 2(t + \Delta t)}{t(t + \Delta t)}}{\Delta t} \qquad \text{Combine fractions in numerator.}$$

$$= \lim_{\Delta t \to 0} \frac{-2\Delta t}{\Delta t(t)(t + \Delta t)} \qquad \text{Divide out common factor of } \Delta t.$$

$$= \lim_{\Delta t \to 0} \frac{-2}{t(t + \Delta t)} \qquad \text{Simplify.}$$

$$= -\frac{2}{t^2}. \qquad \text{Evaluate limit as } \Delta t \to 0.$$

▷ **TECHNOLOGY** A graphing utility can be used to reinforce the result given in Example 5. For instance, using the formula $dy/dt = -2/t^2$, you know that the slope of the graph of $y = 2/t$ at the point $(1, 2)$ is $m = -2$. Using the point-slope form, you can find that the equation of the tangent line to the graph at $(1, 2)$ is

$$y - 2 = -2(t - 1) \quad \text{or} \quad y = -2t + 4. \qquad \text{See Figure 3.9.}$$

You can also verify the result using the *tangent* feature of the graphing utility.

Differentiability and Continuity

The alternative limit form of the derivative shown below is useful in investigating the relationship between differentiability and continuity. The derivative of f at c is

$$f'(c) = \lim_{x \to c} \frac{f(x) - f(c)}{x - c}$$ Alternative form of derivative

REMARK A proof of the equivalence of the alternative form of the derivative is given in Appendix A.

provided this limit exists (see Figure 3.10).

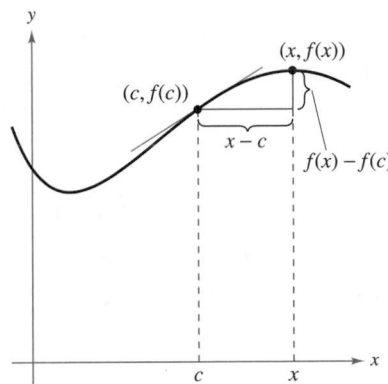

As x approaches c, the secant line approaches the tangent line.
Figure 3.10

Note that the existence of the limit in this alternative form requires that the one-sided limits

$$\lim_{x \to c^-} \frac{f(x) - f(c)}{x - c}$$

and

$$\lim_{x \to c^+} \frac{f(x) - f(c)}{x - c}$$

exist and are equal. These one-sided limits are called the **derivatives from the left and from the right,** respectively. It follows that f is **differentiable on the closed interval** $[a, b]$ when it is differentiable on (a, b) and when the derivative from the right at a and the derivative from the left at b both exist.

When a function is not continuous at $x = c$, it is also not differentiable at $x = c$. For instance, the greatest integer function

$$f(x) = [\![x]\!]$$

is not continuous at $x = 0$, and so it is not differentiable at $x = 0$ (see Figure 3.11 and Exercise 91). You can verify this by observing that

$$\lim_{x \to 0^-} \frac{f(x) - f(0)}{x - 0} = \lim_{x \to 0^-} \frac{[\![x]\!] - 0}{x} = \infty$$ Derivative from the left

and

$$\lim_{x \to 0^+} \frac{f(x) - f(0)}{x - 0} = \lim_{x \to 0^+} \frac{[\![x]\!] - 0}{x} = 0.$$ Derivative from the right

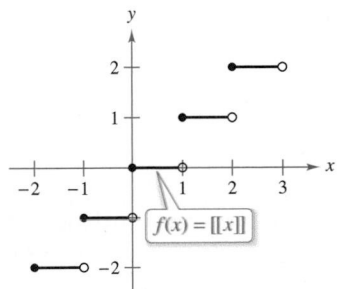

The greatest integer function is not differentiable at $x = 0$ because it is not continuous at $x = 0$.
Figure 3.11

Although it is true that differentiability implies continuity (as shown in Theorem 3.1 on the next page), the converse is not true. That is, it is possible for a function to be continuous at $x = c$ and *not* differentiable at $x = c$. Examples 6 and 7 illustrate this possibility.

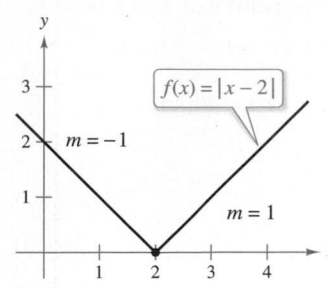

f is not differentiable at $x = 2$ because the derivatives from the left and from the right are not equal.

Figure 3.12

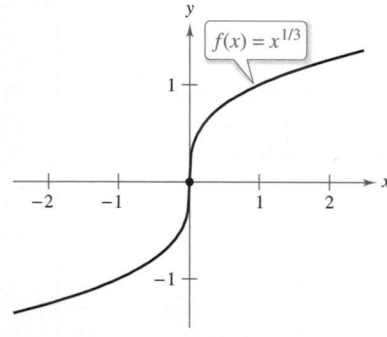

f is not differentiable at $x = 0$ because f has a vertical tangent line at $x = 0$.

Figure 3.13

EXAMPLE 6 **A Graph with a Sharp Turn**

$\cdots\cdots\triangleright$ *See LarsonCalculus.com for an interactive version of this type of example.*

The function $f(x) = |x - 2|$, shown in Figure 3.12, is continuous at $x = 2$. The one-sided limits, however,

$$\lim_{x \to 2^-} \frac{f(x) - f(2)}{x - 2} = \lim_{x \to 2^-} \frac{|x - 2| - 0}{x - 2} = -1 \qquad \text{Derivative from the left}$$

and

$$\lim_{x \to 2^+} \frac{f(x) - f(2)}{x - 2} = \lim_{x \to 2^+} \frac{|x - 2| - 0}{x - 2} = 1 \qquad \text{Derivative from the right}$$

are not equal. So, f is not differentiable at $x = 2$ and the graph of f does not have a tangent line at the point $(2, 0)$.

EXAMPLE 7 **A Graph with a Vertical Tangent Line**

The function $f(x) = x^{1/3}$ is continuous at $x = 0$, as shown in Figure 3.13. However, because the limit

$$\lim_{x \to 0} \frac{f(x) - f(0)}{x - 0} = \lim_{x \to 0} \frac{x^{1/3} - 0}{x} = \lim_{x \to 0} \frac{1}{x^{2/3}} = \infty$$

is infinite, you can conclude that the tangent line is vertical at $x = 0$. So, f is not differentiable at $x = 0$. ■

From Examples 6 and 7, you can see that a function is not differentiable at a point at which its graph has a sharp turn *or* a vertical tangent line.

THEOREM 3.1 **Differentiability Implies Continuity**

If f is differentiable at $x = c$, then f is continuous at $x = c$.

▷ **TECHNOLOGY** Some graphing utilities, such as *Maple*, *Mathematica*, and the *TI-Nspire*, perform symbolic differentiation. Some have a *derivative* feature that performs *numerical differentiation* by finding values of derivatives using the formula

$$f'(x) \approx \frac{f(x + \Delta x) - f(x - \Delta x)}{2\Delta x}$$

where Δx is a small number such as 0.001. Can you see any problems with this definition? For instance, using this definition, what is the value of the derivative of $f(x) = |x|$ when $x = 0$?

Proof You can prove that f is continuous at $x = c$ by showing that $f(x)$ approaches $f(c)$ as $x \to c$. To do this, use the differentiability of f at $x = c$ and consider the following limit.

$$\lim_{x \to c} [f(x) - f(c)] = \lim_{x \to c} \left[(x - c)\left(\frac{f(x) - f(c)}{x - c}\right) \right]$$

$$= \left[\lim_{x \to c} (x - c) \right]\left[\lim_{x \to c} \frac{f(x) - f(c)}{x - c} \right]$$

$$= (0)[f'(c)]$$

$$= 0$$

Because the difference $f(x) - f(c)$ approaches zero as $x \to c$, you can conclude that $\lim_{x \to c} f(x) = f(c)$. So, f is continuous at $x = c$. ■

The relationship between continuity and differentiability is summarized below.

1. If a function is differentiable at $x = c$, then it is continuous at $x = c$. So, differentiability implies continuity.

2. It is possible for a function to be continuous at $x = c$ and not be differentiable at $x = c$. So, continuity does not imply differentiability (see Examples 6 and 7).

3.1 Exercises

CONCEPT CHECK

1. **Tangent Line** Describe how to find the slope of the tangent line to the graph of a function at a point.

2. **Notation** List four notation alternatives to $f'(x)$.

3. **Derivative** Describe how to find the derivative of a function using the limit process.

4. **Continuity and Differentiability** Describe the relationship between continuity and differentiability.

Estimating Slope In Exercises 5 and 6, estimate the slope of the graph at the points (x_1, y_1) and (x_2, y_2).

5.

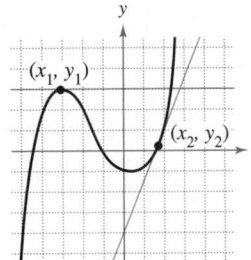

6.

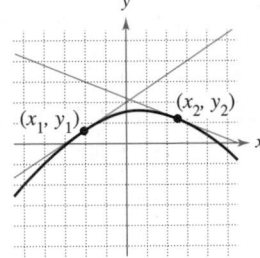

Slopes of Secant Lines In Exercises 7 and 8, use the graph shown in the figure. To print an enlarged copy of the graph, go to *MathGraphs.com*.

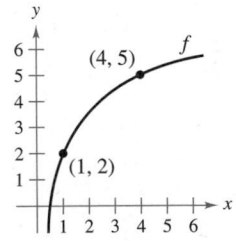

7. Identify or sketch each of the quantities on the figure.

(a) $f(1)$ and $f(4)$ (b) $f(4) - f(1)$

(c) $4 - 1$ (d) $y - 2 = \dfrac{f(4) - f(1)}{4 - 1}(x - 1)$

8. Insert the proper inequality symbol $(< \text{ or } >)$ between the given quantities.

(a) $\dfrac{f(4) - f(1)}{4 - 1}$ ▆ $\dfrac{f(4) - f(3)}{4 - 3}$

(b) $\dfrac{f(4) - f(1)}{4 - 1}$ ▆ $f'(1)$

 **Finding the Slope of a Tangent Line** In Exercises 9–14, find the slope of the tangent line to the graph of the function at the given point.

9. $f(x) = 3 - 5x$, $(-1, 8)$ 10. $g(x) = \frac{3}{2}x + 1$, $(-2, -2)$

11. $f(x) = 2x^2 - 3$, $(2, 5)$ 12. $f(x) = 5 - x^2$, $(3, -4)$

13. $f(t) = 3t - t^2$, $(0, 0)$ 14. $h(t) = t^2 + 4t$, $(1, 5)$

 **Finding the Derivative by the Limit Process** In Exercises 15–28, find the derivative of the function by the limit process.

15. $f(x) = 7$ 16. $g(x) = -3$

17. $f(x) = -5x$ 18. $f(x) = 7x - 3$

19. $h(s) - 3 + \frac{2}{3}s$ 20. $f(x) - 5 - \frac{2}{3}x$

21. $f(x) = x^2 + x - 3$ 22. $f(x) = x^2 - 5$

23. $f(x) = x^3 - 12x$ 24. $g(t) = t^3 + 4t$

25. $f(x) = \dfrac{1}{x - 1}$ 26. $f(x) = \dfrac{1}{x^2}$

27. $f(x) = \sqrt{x + 4}$ 28. $h(s) = -2\sqrt{s}$

Finding an Equation of a Tangent Line In Exercises 29–36, (a) find an equation of the tangent line to the graph of f at the given point, (b) use a graphing utility to graph the function and its tangent line at the point, and (c) use the *tangent* feature of a graphing utility to confirm your results.

29. $f(x) = x^2 + 3$, $(-1, 4)$ 30. $f(x) = x^2 + 2x - 1$, $(1, 2)$

31. $f(x) = x^3$, $(2, 8)$ 32. $f(x) = x^3 + 1$, $(-1, 0)$

33. $f(x) = \sqrt{x}$, $(1, 1)$ 34. $f(x) = \sqrt{x - 1}$, $(5, 2)$

35. $f(x) = x + \dfrac{4}{x}$, $(-4, -5)$ 36. $f(x) = x - \dfrac{1}{x}$, $(1, 0)$

Finding an Equation of a Tangent Line In Exercises 37–42, find an equation of the line that is tangent to the graph of f *and* parallel to the given line.

Function	Line
37. $f(x) = -\dfrac{1}{4}x^2$	$x + y = 0$
38. $f(x) = 2x^2$	$4x + y + 3 = 0$
39. $f(x) = x^3$	$3x - y + 1 = 0$
40. $f(x) = x^3 + 2$	$3x - y - 4 = 0$
41. $f(x) = \dfrac{1}{\sqrt{x}}$	$x + 2y - 6 = 0$
42. $f(x) = \dfrac{1}{\sqrt{x - 1}}$	$x + 2y + 7 = 0$

Sketching a Derivative In Exercises 43–48, sketch the graph of f'. Explain how you found your answer.

43.

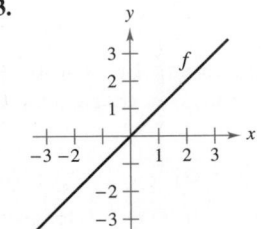

44.

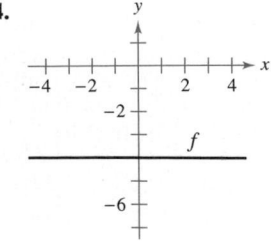

45.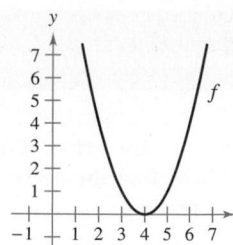

46.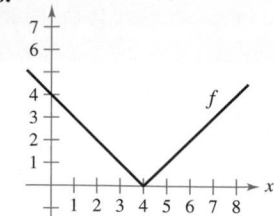

47.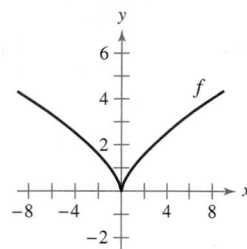

48.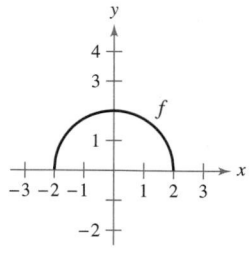

EXPLORING CONCEPTS

49. Sketching a Graph Sketch a graph of a function whose derivative is always negative. Explain how you found the answer.

50. Sketching a Graph Sketch a graph of a function whose derivative is zero at exactly two points. Explain how you found the answer.

51. Domain of the Derivative Do f and f' always have the same domain? Explain.

52. Symmetry of a Graph A function f is symmetric with respect to the origin. Is f' necessarily symmetric with respect to the origin? Explain.

---

53. Using a Tangent Line The tangent line to the graph of $y = g(x)$ at the point $(4, 5)$ passes through the point $(7, 0)$. Find $g(4)$ and $g'(4)$.

54. Using a Tangent Line The tangent line to the graph of $y = h(x)$ at the point $(-1, 4)$ passes through the point $(3, 6)$. Find $h(-1)$ and $h'(-1)$.

 **Working Backwards** In Exercises 55–58, the limit represents $f'(c)$ for a function f and a number c. Find f and c.

55. $\lim\limits_{\Delta x \to 0} \dfrac{[5 - 3(1 + \Delta x)] - 2}{\Delta x}$

56. $\lim\limits_{\Delta x \to 0} \dfrac{(-2 + \Delta x)^3 + 8}{\Delta x}$

57. $\lim\limits_{x \to 6} \dfrac{-x^2 + 36}{x - 6}$

58. $\lim\limits_{x \to 9} \dfrac{2\sqrt{x} - 6}{x - 9}$

Writing a Function Using Derivatives In Exercises 59 and 60, identify a function f that has the given characteristics. Then sketch the function.

59. $f(0) = 2;\ f'(x) = -3$ for $-\infty < x < \infty$

60. $f(0) = 4;\ f'(0) = 0;\ f'(x) < 0$ for $x < 0;\ f'(x) > 0$ for $x > 0$

Finding an Equation of a Tangent Line In Exercises 61 and 62, find equations of the two tangent lines to the graph of f that pass through the indicated point.

61. $f(x) = 4x - x^2$

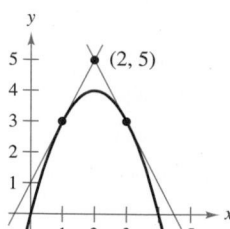

62. $f(x) = x^2$

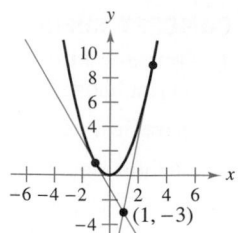

63. Graphical Reasoning Use a graphing utility to graph each function and its tangent lines at $x = -1$, $x = 0$, and $x = 1$. Based on the results, determine whether the slopes of tangent lines to the graph of a function at different values of x are always distinct.

(a) $f(x) = x^2$ (b) $g(x) = x^3$

64. **HOW DO YOU SEE IT?** The figure shows the graph of g'.

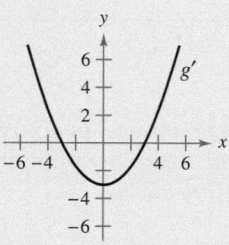

(a) $g'(0) =$ ▢   (b) $g'(3) =$ ▢

(c) What can you conclude about the graph of g knowing that $g'(1) = -\frac{8}{3}$?

(d) What can you conclude about the graph of g knowing that $g'(-4) = \frac{7}{3}$?

(e) Is $g(6) - g(4)$ positive or negative? Explain.

(f) Is it possible to find $g(2)$ from the graph? Explain.

65. Graphical Reasoning Consider the function $f(x) = \frac{1}{2}x^2$.

(a) Use a graphing utility to graph the function and estimate the values of $f'(0)$, $f'\left(\frac{1}{2}\right)$, $f'(1)$, and $f'(2)$.

(b) Use your results from part (a) to determine the values of $f'\left(-\frac{1}{2}\right)$, $f'(-1)$, and $f'(-2)$.

(c) Sketch a possible graph of f'.

(d) Use the definition of derivative to find $f'(x)$.

66. Graphical Reasoning Consider the function $f(x) = \frac{1}{3}x^3$.

(a) Use a graphing utility to graph the function and estimate the values of $f'(0)$, $f'\left(\frac{1}{2}\right)$, $f'(1)$, $f'(2)$, and $f'(3)$.

(b) Use your results from part (a) to determine the values of $f'\left(-\frac{1}{2}\right)$, $f'(-1)$, $f'(-2)$, and $f'(-3)$.

(c) Sketch a possible graph of f'.

(d) Use the definition of derivative to find $f'(x)$.

Approximating a Derivative In Exercises 67 and 68, evaluate $f(2)$ and $f(2.1)$ and use the results to approximate $f'(2)$.

67. $f(x) = x(4 - x)$ **68.** $f(x) = \frac{1}{4}x^3$

 **Using the Alternative Form of the Derivative** In Exercises 69–76, use the alternative form of the derivative to find the derivative at $x = c$, if it exists.

69. $f(x) = x^3 + 2x^2 + 1, c = -2$

70. $g(x) = x^2 - x, c = 1$

71. $g(x) = \sqrt{|x|}, c = 0$ **72.** $f(x) = 3/x, c = 4$

73. $f(x) = (x - 6)^{2/3}, c = 6$ **74.** $g(x) = (x + 3)^{1/3}, c = -3$

75. $h(x) = |x + 7|, c = -7$ **76.** $f(x) = |x - 6|, c = 6$

 **Determining Differentiability** In Exercises 77–80, describe the x-values at which f is differentiable.

77. $f(x) = (x + 4)^{2/3}$ **78.** $f(x) = \dfrac{x^2}{x^2 - 4}$

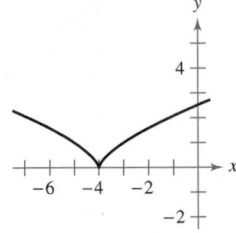 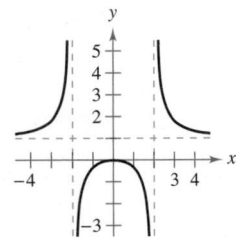

79. $f(x) = \sqrt{x + 1} + 1$ **80.** $f(x) = \begin{cases} x^2 - 4, & x \le 0 \\ 4 - x^2, & x > 0 \end{cases}$

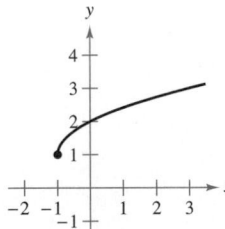 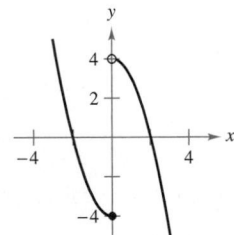

 **Graphical Analysis** In Exercises 81–84, use a graphing utility to graph the function and find the x-values at which f is differentiable.

81. $f(x) = |x - 5|$ **82.** $f(x) = \dfrac{4x}{x - 3}$

83. $f(x) = x^{2/5}$

84. $f(x) = \begin{cases} x^3 - 3x^2 + 3x, & x \le 1 \\ x^2 - 2x, & x > 1 \end{cases}$

Determining Differentiability In Exercises 85–88, find the derivatives from the left and from the right at $x = 1$ (if they exist). Is the function differentiable at $x = 1$?

85. $f(x) = |x - 1|$ **86.** $f(x) = \sqrt{1 - x^2}$

87. $f(x) = \begin{cases} (x - 1)^3, & x \le 1 \\ (x - 1)^2, & x > 1 \end{cases}$ **88.** $f(x) = (1 - x)^{2/3}$

Determining Differentiability In Exercises 89 and 90, determine whether the function is differentiable at $x = 2$.

89. $f(x) = \begin{cases} x^2 + 1, & x \le 2 \\ 4x - 3, & x > 2 \end{cases}$ **90.** $f(x) = \begin{cases} \frac{1}{2}x + 2, & x < 2 \\ \sqrt{2x}, & x \ge 2 \end{cases}$

91. Greatest Integer Function and Differentiability Use a graphing utility to graph $g(x) = [\![x]\!]/x$. Then let $f(x) = [\![x]\!]$ and show that

$$\lim_{x \to 0^-} \frac{f(x) - f(0)}{x - 0} = \infty \quad \text{and} \quad \lim_{x \to 0^+} \frac{f(x) - f(0)}{x - 0} = 0.$$

Is f differentiable? Explain.

92. Conjecture Consider the functions $f(x) = x^2$ and $g(x) = x^3$.

(a) Graph f and f' on the same set of axes.

(b) Graph g and g' on the same set of axes.

(c) Identify a pattern between f and g and their respective derivatives. Use the pattern to make a conjecture about $h'(x)$ if $h(x) = x^n$, where n is an integer and $n \ge 2$.

(d) Find $f'(x)$ if $f(x) = x^4$. Compare the result with the conjecture in part (c). Is this a proof of your conjecture? Explain.

True or False? In Exercises 93–96, determine whether the statement is true or false. If it is false, explain why or give an example that shows it is false.

93. The slope of the tangent line to the differentiable function f at the point $(2, f(2))$ is

$$\frac{f(2 + \Delta x) - f(2)}{\Delta x}.$$

94. If a function is continuous at a point, then it is differentiable at that point.

95. If a function has derivatives from both the right and the left at a point, then it is differentiable at that point.

96. If a function is differentiable at a point, then it is continuous at that point.

97. Differentiability and Continuity Let

$$f(x) = \begin{cases} x \sin \dfrac{1}{x}, & x \ne 0 \\ 0, & x = 0 \end{cases}$$

and

$$g(x) = \begin{cases} x^2 \sin \dfrac{1}{x}, & x \ne 0 \\ 0, & x = 0 \end{cases}.$$

Show that f is continuous, but not differentiable, at $x = 0$. Show that g is differentiable at 0 and find $g'(0)$.

98. Writing Use a graphing utility to graph the two functions $f(x) = x^2 + 1$ and $g(x) = |x| + 1$ in the same viewing window. Use the *zoom* and *trace* features to analyze the graphs near the point $(0, 1)$. What do you observe? Which function is differentiable at this point? Write a short paragraph describing the geometric significance of differentiability at a point.

3.2   Basic Differentiation Rules and Rates of Change

- ■ Find the derivative of a function using the Constant Rule.
- ■ Find the derivative of a function using the Power Rule.
- ■ Find the derivative of a function using the Constant Multiple Rule.
- ■ Find the derivative of a function using the Sum and Difference Rules.
- ■ Find the derivatives of the sine function and of the cosine function.
- ■ Find the derivatives of exponential functions.
- ■ Use derivatives to find rates of change.

The Constant Rule

In Section 3.1, you used the limit definition to find derivatives. In this and the next two sections, you will be introduced to several "differentiation rules" that allow you to find derivatives without the *direct* use of the limit definition.

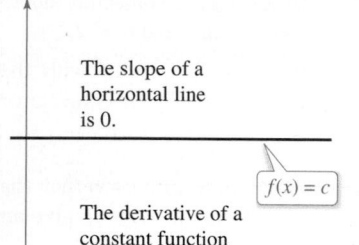

The slope of a horizontal line is 0.

$f(x) = c$

The derivative of a constant function is 0.

Notice that the Constant Rule is equivalent to saying that the slope of a horizontal line is 0. This demonstrates the relationship between slope and derivative.

Figure 3.14

> **THEOREM 3.2   The Constant Rule**
>
> The derivative of a constant function is 0. That is, if c is a real number, then
>
> $$\frac{d}{dx}[c] = 0. \qquad \text{See Figure 3.14.}$$

Proof   Let $f(x) = c$. Then, by the limit definition of the derivative,

$$\frac{d}{dx}[c] = f'(x)$$

$$= \lim_{\Delta x \to 0} \frac{f(x + \Delta x) - f(x)}{\Delta x}$$

$$= \lim_{\Delta x \to 0} \frac{c - c}{\Delta x}$$

$$= \lim_{\Delta x \to 0} 0$$

$$= 0.$$

EXAMPLE 1   **Using the Constant Rule**

Function	Derivative
a. $y = 7$	$dy/dx = 0$
b. $f(x) = 0$	$f'(x) = 0$
c. $s(t) = -3$	$s'(t) = 0$
d. $y = k\pi^2$, k is constant	$dy/dx = 0$

> **Exploration**
>
> *Writing a Conjecture*   Use the definition of the derivative given in Section 3.1 to find the derivative of each function. What patterns do you see? Use your results to write a conjecture about the derivative of $f(x) = x^n$.
>
> **a.** $f(x) = x^1$ **b.** $f(x) = x^2$ **c.** $f(x) = x^3$
>
> **d.** $f(x) = x^4$ **e.** $f(x) = x^{1/2}$ **f.** $f(x) = x^{-1}$

The Power Rule

Before proving the next rule, it is important to review the procedure for expanding a binomial.

$$(x + \Delta x)^2 = x^2 + 2x\Delta x + (\Delta x)^2$$
$$(x + \Delta x)^3 = x^3 + 3x^2\Delta x + 3x(\Delta x)^2 + (\Delta x)^3$$
$$(x + \Delta x)^4 = x^4 + 4x^3\Delta x + 6x^2(\Delta x)^2 + 4x(\Delta x)^3 + (\Delta x)^4$$
$$(x + \Delta x)^5 = x^5 + 5x^4\Delta x + 10x^3(\Delta x)^2 + 10x^2(\Delta x)^3 + 5x(\Delta x)^4 + (\Delta x)^5$$

The general binomial expansion for a positive integer n is

$$(x + \Delta x)^n = x^n + nx^{n-1}(\Delta x) + \underbrace{\frac{n(n-1)x^{n-2}}{2}(\Delta x)^2 + \cdots + (\Delta x)^n}.$$

$(\Delta x)^2$ is a factor of these terms.

This binomial expansion is used in proving a special case of the Power Rule.

THEOREM 3.3 The Power Rule

If n is a rational number, then the function $f(x) = x^n$ is differentiable and

$$\frac{d}{dx}[x^n] = nx^{n-1}.$$

For f to be differentiable at $x = 0$, n must be a number such that x^{n-1} is defined on an interval containing 0.

> ....▷
>
> •• **REMARK** From Example 7 in Section 3.1, you know that the function $f(x) = x^{1/3}$ is defined at $x = 0$ but is not differentiable at $x = 0$. This is because $x^{-2/3}$ is not defined on an interval containing 0.

Proof If n is a positive integer greater than 1, then the binomial expansion produces

$$\frac{d}{dx}[x^n] = \lim_{\Delta x \to 0} \frac{(x + \Delta x)^n - x^n}{\Delta x}$$

$$= \lim_{\Delta x \to 0} \frac{x^n + nx^{n-1}(\Delta x) + \frac{n(n-1)x^{n-2}}{2}(\Delta x)^2 + \cdots + (\Delta x)^n - x^n}{\Delta x}$$

$$= \lim_{\Delta x \to 0} \left[nx^{n-1} + \frac{n(n-1)x^{n-2}}{2}(\Delta x) + \cdots + (\Delta x)^{n-1} \right]$$

$$= nx^{n-1} + 0 + \cdots + 0$$

$$= nx^{n-1}.$$

This proves the case for which n is a positive integer greater than 1. It is left to you to prove the case for $n = 1$. Example 7 in Section 3.3 proves the case for which n is a negative integer. The cases for which n is rational and n is irrational are left as an exercise (see Section 3.5, Exercise 92). ∎

When using the Power Rule, the case for which $n = 1$ is best thought of as a separate differentiation rule. That is,

$$\frac{d}{dx}[x] = 1. \qquad \text{Power Rule when } n = 1$$

This rule is consistent with the fact that the slope of the line $y = x$ is 1, as shown in Figure 3.15.

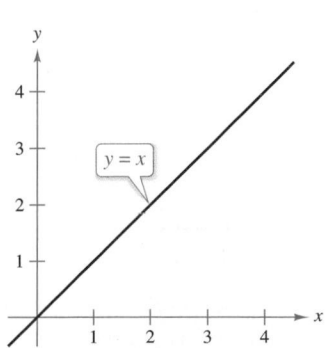

The slope of the line $y = x$ is 1.
Figure 3.15

EXAMPLE 2 **Using the Power Rule**

Function	Derivative
a. $f(x) = x^3$	$f'(x) = 3x^2$
b. $g(x) = \sqrt[3]{x}$	$g'(x) = \dfrac{d}{dx}[x^{1/3}] = \dfrac{1}{3}x^{-2/3} = \dfrac{1}{3x^{2/3}}$
c. $y = \dfrac{1}{x^2}$	$\dfrac{dy}{dx} = \dfrac{d}{dx}[x^{-2}] = (-2)x^{-3} = -\dfrac{2}{x^3}$

In Example 2(c), note that *before* differentiating, $1/x^2$ was rewritten as x^{-2}. Rewriting is the first step in *many* differentiation problems.

Given:		Rewrite:		Differentiate:		Simplify:
$y = \dfrac{1}{x^2}$	⇨	$y = x^{-2}$	⇨	$\dfrac{dy}{dx} = (-2)x^{-3}$	⇨	$\dfrac{dy}{dx} = -\dfrac{2}{x^3}$

EXAMPLE 3 **Finding the Slope of a Graph**

⋅⋅⋅⋅▷ *See LarsonCalculus.com for an interactive version of this type of example.*

Find the slope of the graph of $f(x) = x^4$ for each value of x.

a. $x = -1$ **b.** $x = 0$ **c.** $x = 1$

Solution The slope of a graph at a point is the value of the derivative at that point. The derivative of f is $f'(x) = 4x^3$.

a. When $x = -1$, the slope is $f'(-1) = 4(-1)^3 = -4$.   Slope is negative.

b. When $x = 0$, the slope is $f'(0) = 4(0)^3 = 0$.   Slope is zero.

c. When $x = 1$, the slope is $f'(1) = 4(1)^3 = 4$.   Slope is positive.

See Figure 3.16.

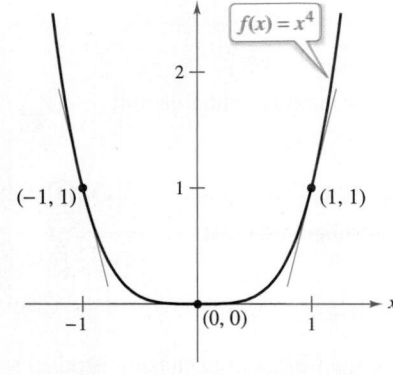

Note that the slope of the graph is negative at the point $(-1, 1)$, the slope is zero at the point $(0, 0)$, and the slope is positive at the point $(1, 1)$.
Figure 3.16

EXAMPLE 4 **Finding an Equation of a Tangent Line**

⋅⋅⋅⋅▷ *See LarsonCalculus.com for an interactive version of this type of example.*

Find an equation of the tangent line to the graph of $f(x) = x^2$ when $x = -2$.

Solution To find the *point* on the graph of f, evaluate the original function at $x = -2$.

$$(-2, f(-2)) = (-2, 4) \qquad \text{Point on graph}$$

To find the *slope* of the graph when $x = -2$, evaluate the derivative, $f'(x) = 2x$, at $x = -2$.

$$m = f'(-2) = -4 \qquad \text{Slope of graph at } (-2, 4)$$

Now, using the point-slope form of the equation of a line, you can write

$$y - y_1 = m(x - x_1) \qquad \text{Point-slope form}$$
$$y - 4 = -4[x - (-2)] \qquad \text{Substitute for } y_1, m, \text{ and } x_1.$$
$$y = -4x - 4. \qquad \text{Simplify.}$$

You can check this result using the *tangent* feature of a graphing utility, as shown in Figure 3.17.

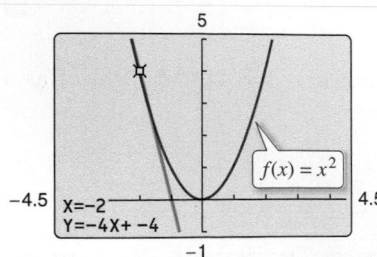

The line $y = -4x - 4$ is tangent to the graph of $f(x) = x^2$ at the point $(-2, 4)$.
Figure 3.17

The Constant Multiple Rule

> **THEOREM 3.4   The Constant Multiple Rule**
>
> If f is a differentiable function and c is a real number, then cf is also differentiable and
>
> $$\frac{d}{dx}[cf(x)] = cf'(x).$$

Proof

$$\frac{d}{dx}[cf(x)] = \lim_{\Delta x \to 0} \frac{cf(x + \Delta x) - cf(x)}{\Delta x} \qquad \text{Definition of derivative}$$

$$= \lim_{\Delta x \to 0} c\left[\frac{f(x + \Delta x) - f(x)}{\Delta x}\right]$$

$$= c\left[\lim_{\Delta x \to 0} \frac{f(x + \Delta x) - f(x)}{\Delta x}\right] \qquad \text{Apply Theorem 2.2.}$$

$$= cf'(x)$$

Informally, the Constant Multiple Rule states that constants can be factored out of the differentiation process, even when the constants appear in the denominator.

$$\frac{d}{dx}[cf(x)] = c\frac{d}{dx}[(\ )f(x)] = cf'(x)$$

$$\frac{d}{dx}\left[\frac{f(x)}{c}\right] = \frac{d}{dx}\left[\left(\frac{1}{c}\right)f(x)\right] = \left(\frac{1}{c}\right)\frac{d}{dx}[(\ )f(x)] = \left(\frac{1}{c}\right)f'(x)$$

EXAMPLE 5   Using the Constant Multiple Rule

Function	Derivative
a. $y = 5x^3$	$\dfrac{dy}{dx} = \dfrac{d}{dx}[5x^3] = 5\dfrac{d}{dx}[x^3] = 5(3)x^2 = 15x^2$
b. $y = \dfrac{2}{x}$	$\dfrac{dy}{dx} = \dfrac{d}{dx}[2x^{-1}] = 2\dfrac{d}{dx}[x^{-1}] = 2(-1)x^{-2} = -\dfrac{2}{x^2}$
c. $f(t) = \dfrac{4t^2}{5}$	$f'(t) = \dfrac{d}{dt}\left[\dfrac{4}{5}t^2\right] = \dfrac{4}{5}\dfrac{d}{dt}[t^2] = \dfrac{4}{5}(2t) = \dfrac{8}{5}t$
d. $y = 2\sqrt{x}$	$\dfrac{dy}{dx} = \dfrac{d}{dx}[2x^{1/2}] = 2\left(\dfrac{1}{2}x^{-1/2}\right) = x^{-1/2} = \dfrac{1}{\sqrt{x}}$
e. $y = \dfrac{1}{2\sqrt[3]{x^2}}$	$\dfrac{dy}{dx} = \dfrac{d}{dx}\left[\dfrac{1}{2}x^{-2/3}\right] = \dfrac{1}{2}\left(-\dfrac{2}{3}\right)x^{-5/3} = -\dfrac{1}{3x^{5/3}}$
f. $y = -\dfrac{3x}{2}$	$y' = \dfrac{d}{dx}\left[-\dfrac{3}{2}x\right] = -\dfrac{3}{2}(1) = -\dfrac{3}{2}$

• • REMARK Before differentiating functions involving radicals, rewrite the function with rational exponents.

The Constant Multiple Rule and the Power Rule can be combined into one rule. The combination rule is

$$\frac{d}{dx}[cx^n] = cnx^{n-1}.$$

EXAMPLE 6 **Using Parentheses When Differentiating**

	Original Function	Rewrite	Differentiate	Simplify
a.	$y = \dfrac{5}{2x^3}$	$y = \dfrac{5}{2}(x^{-3})$	$y' = \dfrac{5}{2}(-3x^{-4})$	$y' = -\dfrac{15}{2x^4}$
b.	$y = \dfrac{5}{(2x)^3}$	$y = \dfrac{5}{8}(x^{-3})$	$y' = \dfrac{5}{8}(-3x^{-4})$	$y' = -\dfrac{15}{8x^4}$
c.	$y = \dfrac{7}{3x^{-2}}$	$y = \dfrac{7}{3}(x^2)$	$y' = \dfrac{7}{3}(2x)$	$y' = \dfrac{14x}{3}$
d.	$y = \dfrac{7}{(3x)^{-2}}$	$y = 63(x^2)$	$y' = 63(2x)$	$y' = 126x$

The Sum and Difference Rules

> **THEOREM 3.5** **The Sum and Difference Rules**
>
> The sum (or difference) of two differentiable functions f and g is itself differentiable. Moreover, the derivative of $f + g$ (or $f - g$) is the sum (or difference) of the derivatives of f and g.
>
> $$\frac{d}{dx}[f(x) + g(x)] = f'(x) + g'(x) \qquad \text{Sum Rule}$$
>
> $$\frac{d}{dx}[f(x) - g(x)] = f'(x) - g'(x) \qquad \text{Difference Rule}$$

Proof A proof of the Sum Rule follows from Theorem 2.2. (The Difference Rule can be proved in a similar way.)

$$\frac{d}{dx}[f(x) + g(x)] = \lim_{\Delta x \to 0} \frac{[f(x + \Delta x) + g(x + \Delta x)] - [f(x) + g(x)]}{\Delta x}$$

$$= \lim_{\Delta x \to 0} \frac{f(x + \Delta x) + g(x + \Delta x) - f(x) - g(x)}{\Delta x}$$

$$= \lim_{\Delta x \to 0} \left[\frac{f(x + \Delta x) - f(x)}{\Delta x} + \frac{g(x + \Delta x) - g(x)}{\Delta x} \right]$$

$$= \lim_{\Delta x \to 0} \frac{f(x + \Delta x) - f(x)}{\Delta x} + \lim_{\Delta x \to 0} \frac{g(x + \Delta x) - g(x)}{\Delta x}$$

$$= f'(x) + g'(x)$$

The Sum and Difference Rules can be extended to any finite number of functions. For instance, if $F(x) = f(x) + g(x) - h(x)$, then $F'(x) = f'(x) + g'(x) - h'(x)$.

• • REMARK In Example 7(c), note that before differentiating,

$$\frac{3x^2 - x + 1}{x}$$

was rewritten as

$$3x - 1 + \frac{1}{x}.$$

▷

EXAMPLE 7 **Using the Sum and Difference Rules**

Function	Derivative
a. $f(x) = x^3 - 4x + 5$	$f'(x) = 3x^2 - 4$
b. $g(x) = -\dfrac{x^4}{2} + 3x^3 - 2x$	$g'(x) = -2x^3 + 9x^2 - 2$
c. $y = \dfrac{3x^2 - x + 1}{x} = 3x - 1 + \dfrac{1}{x}$	$y' = 3 - \dfrac{1}{x^2} = \dfrac{3x^2 - 1}{x^2}$

■ **FOR FURTHER INFORMATION**
For the outline of a geometric proof of the derivatives of the sine and cosine functions, see the article "The Spider's Spacewalk Derivation of sin′ and cos′" by Tim Hesterberg in *The College Mathematics Journal*. To view this article, go to *MathArticles.com*.

Derivatives of the Sine and Cosine Functions

In Section 2.3, you studied the limits

$$\lim_{\Delta x \to 0} \frac{\sin \Delta x}{\Delta x} = 1 \quad \text{and} \quad \lim_{\Delta x \to 0} \frac{1 - \cos \Delta x}{\Delta x} = 0.$$

These two limits can be used to prove differentiation rules for the sine and cosine functions. (The derivatives of the other four trigonometric functions are discussed in Section 3.3.)

THEOREM 3.6 Derivatives of Sine and Cosine Functions

$$\frac{d}{dx}[\sin x] = \cos x \qquad \frac{d}{dx}[\cos x] = -\sin x$$

Proof Here is a proof of the first rule. (The proof of the second rule is left as an exercise [see Exercise 120].) In the proof, note the use of the trigonometric identity $\sin(x + \Delta x) = \sin x \cos \Delta x + \cos x \sin \Delta x$.

$$\frac{d}{dx}[\sin x] = \lim_{\Delta x \to 0} \frac{\sin(x + \Delta x) - \sin x}{\Delta x} \qquad \text{Definition of derivative}$$

$$= \lim_{\Delta x \to 0} \frac{\sin x \cos \Delta x + \cos x \sin \Delta x - \sin x}{\Delta x}$$

$$= \lim_{\Delta x \to 0} \frac{\cos x \sin \Delta x - (\sin x)(1 - \cos \Delta x)}{\Delta x}$$

$$= \lim_{\Delta x \to 0} \left[(\cos x)\left(\frac{\sin \Delta x}{\Delta x}\right) - (\sin x)\left(\frac{1 - \cos \Delta x}{\Delta x}\right) \right]$$

$$= (\cos x)\left(\lim_{\Delta x \to 0} \frac{\sin \Delta x}{\Delta x}\right) - (\sin x)\left(\lim_{\Delta x \to 0} \frac{1 - \cos \Delta x}{\Delta x}\right)$$

$$= (\cos x)(1) - (\sin x)(0)$$

$$= \cos x$$

This differentiation rule is shown graphically in Figure 3.18. Note that for each x, the *slope* of the sine curve is equal to the value of the cosine. ■

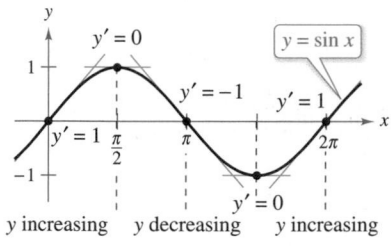

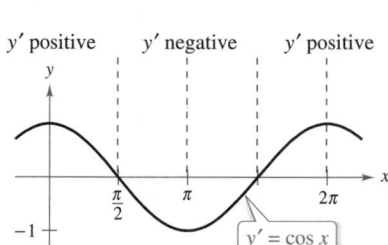

The derivative of the sine function is the cosine function.
Figure 3.18

EXAMPLE 8   Derivatives Involving Sines and Cosines

⋮ ····▷ *See LarsonCalculus.com for an interactive version of this type of example.*

Function	**Derivative**
a. $y = 2 \sin x$	$y' = 2 \cos x$
b. $y = \dfrac{\sin x}{2} = \dfrac{1}{2}\sin x$	$y' = \dfrac{1}{2}\cos x = \dfrac{\cos x}{2}$
c. $y = x + \cos x$	$y' = 1 - \sin x$
d. $y = \cos x - \dfrac{\pi}{3}\sin x$	$y' = -\sin x - \dfrac{\pi}{3}\cos x$

■

▷ **TECHNOLOGY** A graphing utility can provide insight into the interpretation of a derivative. For instance, Figure 3.19 shows the graphs of

$$y = a \sin x$$

for $a = \frac{1}{2}, 1, \frac{3}{2}$, and 2. Estimate the slope of each graph at the point $(0, 0)$. Then verify your estimates analytically by evaluating the derivative of each function when $x = 0$.

$$\frac{d}{dx}[a \sin x] = a \cos x$$

Figure 3.19

Derivatives of Exponential Functions

One of the most intriguing (and useful) characteristics of the natural exponential function is that *it is its own derivative*. Consider the following argument.

Let $f(x) = e^x$.

$$f'(x) = \lim_{\Delta x \to 0} \frac{f(x + \Delta x) - f(x)}{\Delta x}$$

$$= \lim_{\Delta x \to 0} \frac{e^{x + \Delta x} - e^x}{\Delta x}$$

$$= \lim_{\Delta x \to 0} \frac{e^x(e^{\Delta x} - 1)}{\Delta x}$$

The definition of e

$$\lim_{\Delta x \to 0} (1 + \Delta x)^{1/\Delta x} = e$$

tells you that for small values of Δx, you have $e \approx (1 + \Delta x)^{1/\Delta x}$, which implies that

$$e^{\Delta x} \approx 1 + \Delta x.$$

Replacing $e^{\Delta x}$ by this approximation produces the following.

$$f'(x) = \lim_{\Delta x \to 0} \frac{e^x[e^{\Delta x} - 1]}{\Delta x}$$

$$= \lim_{\Delta x \to 0} \frac{e^x[(1 + \Delta x) - 1]}{\Delta x}$$

$$= \lim_{\Delta x \to 0} \frac{e^x \, \Delta x}{\Delta x}$$

$$= e^x$$

This result is stated in the next theorem.

REMARK The key to the formula for the derivative of $f(x) = e^x$ is the limit

$$\lim_{x \to 0} (1 + x)^{1/x} = e.$$

This important limit was introduced on page 54 and formalized later on page 89. It is used to conclude that for $\Delta x \approx 0$,

$$(1 + \Delta x)^{1/\Delta x} \approx e$$

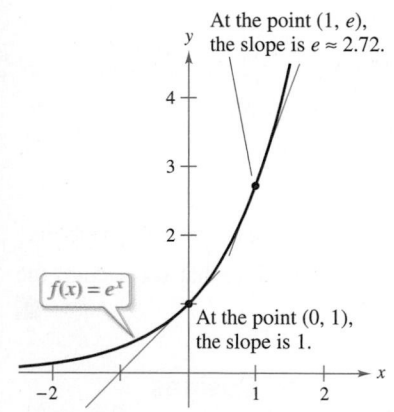

At the point $(1, e)$, the slope is $e \approx 2.72$.

At the point $(0, 1)$, the slope is 1.

$f(x) = e^x$

Figure 3.20

THEOREM 3.7   Derivative of the Natural Exponential Function

$$\frac{d}{dx}[e^x] = e^x$$

You can interpret Theorem 3.7 graphically by saying that the slope of the graph of $f(x) = e^x$ at any point (x, e^x) is equal to the y-coordinate of the point, as shown in Figure 3.20.

EXAMPLE 9   Derivatives Involving Exponential Functions

Find the derivative of each function.

a. $f(x) = 3e^x$ **b.** $f(x) = x^2 + e^x$ **c.** $f(x) = \sin x - e^x$

Solution

a. $f'(x) = 3\dfrac{d}{dx}[e^x] = 3e^x$

b. $f'(x) = \dfrac{d}{dx}[x^2] + \dfrac{d}{dx}[e^x] = 2x + e^x$

c. $f'(x) = \dfrac{d}{dx}[\sin x] - \dfrac{d}{dx}[e^x] = \cos x - e^x$

Rates of Change

You have seen how the derivative is used to determine slope. The derivative can also be used to determine the rate of change of one variable with respect to another. Applications involving rates of change, sometimes referred to as instantaneous rates of change, occur in a wide variety of fields. A few examples are population growth rates, production rates, water flow rates, velocity, and acceleration.

A common use for rate of change is to describe the motion of an object moving in a straight line. In such problems, it is customary to use either a horizontal or a vertical line with a designated origin to represent the line of motion. On such lines, movement to the right (or upward) is considered to be in the positive direction, and movement to the left (or downward) is considered to be in the negative direction.

The function s that gives the position (relative to the origin) of an object as a function of time t is called a **position function.** If, over a period of time Δt, the object changes its position by the amount

$$\Delta s = s(t + \Delta t) - s(t)$$

then, by the familiar formula

$$\text{Rate} = \frac{\text{distance}}{\text{time}}$$

the **average velocity** is

$$\frac{\text{Change in distance}}{\text{Change in time}} = \frac{\Delta s}{\Delta t}. \qquad \text{Average velocity}$$

EXAMPLE 10 **Finding Average Velocity of a Falling Object**

A billiard ball is dropped from a height of 100 feet. The ball's height s at time t is the position function

$$s = -16t^2 + 100 \qquad\qquad \text{Position function}$$

where s is measured in feet and t is measured in seconds. Find the average velocity over each time interval.

a. $[1, 2]$ **b.** $[1, 1.5]$ **c.** $[1, 1.1]$

Solution

a. For the interval $[1, 2]$, the object falls from a height of $s(1) = -16(1)^2 + 100 = 84$ feet to a height of $s(2) = -16(2)^2 + 100 = 36$ feet. The average velocity is

$$\frac{\Delta s}{\Delta t} = \frac{36 - 84}{2 - 1} = \frac{-48}{1} = -48 \text{ feet per second.}$$

b. For the interval $[1, 1.5]$, the object falls from a height of 84 feet to a height of $s(1.5) = -16(1.5)^2 + 100 = 64$ feet. The average velocity is

$$\frac{\Delta s}{\Delta t} = \frac{64 - 84}{1.5 - 1} = \frac{-20}{0.5} = -40 \text{ feet per second.}$$

c. For the interval $[1, 1.1]$, the object falls from a height of 84 feet to a height of $s(1.1) = -16(1.1)^2 + 100 = 80.64$ feet. The average velocity is

$$\frac{\Delta s}{\Delta t} = \frac{80.64 - 84}{1.1 - 1} = \frac{-3.36}{0.1} = -33.6 \text{ feet per second.}$$

Note that the average velocities are *negative*, indicating that the object is moving downward. ∎

Time-lapse photograph of a free-falling billiard ball

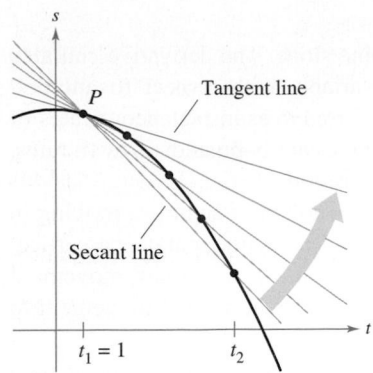

The average velocity between t_1 and t_2 is the slope of the secant line, and the instantaneous velocity at t_1 is the slope of the tangent line.

Figure 3.21

Suppose that in Example 10, you wanted to find the *instantaneous* velocity (or simply the velocity) of the object when $t = 1$. Just as you can approximate the slope of the tangent line by calculating the slope of the secant line, you can approximate the velocity at $t = 1$ by calculating the average velocity over a small interval $[1, 1 + \Delta t]$ (see Figure 3.21). By taking the limit as Δt approaches zero, you obtain the velocity when $t = 1$. Try doing this—you will find that the velocity when $t = 1$ is -32 feet per second.

In general, if $s = s(t)$ is the position function for an object moving along a straight line, then the **velocity** of the object at time t is

$$v(t) = \lim_{\Delta t \to 0} \frac{s(t + \Delta t) - s(t)}{\Delta t} = s'(t). \qquad \text{Velocity function}$$

In other words, the velocity function is the derivative of the position function. Velocity can be negative, zero, or positive. The **speed** of an object is the absolute value of its velocity. Speed cannot be negative.

The position of a free-falling object (neglecting air resistance) under the influence of gravity can be represented by the equation

$$s(t) = -\frac{1}{2} g t^2 + v_0 t + s_0 \qquad \text{Position function}$$

where s_0 is the initial height of the object, v_0 is the initial velocity of the object, and g is the acceleration due to gravity. On Earth, the value of g is approximately 32 feet per second per second or 9.8 meters per second per second.

EXAMPLE 11 **Using the Derivative to Find Velocity**

At time $t = 0$ seconds, a diver jumps from a platform diving board that is 32 feet above the water (see Figure 3.22). The initial velocity of the diver is 16 feet per second. When does the diver hit the water? What is the diver's velocity at impact?

Solution

Begin by writing an equation to represent the position of the diver. Using the position function given above with $g = 32$ feet per second per second, $v_0 = 16$ feet per second, and $s_0 = 32$ feet, you can write

$$s(t) = -\frac{1}{2}(32)t^2 + 16t + 32$$

$$= -16t^2 + 16t + 32. \qquad \text{Position function}$$

To find the time t when the diver hits the water, let $s = 0$ and solve for t.

$-16t^2 + 16t + 32 = 0$	Set position function equal to 0.
$-16(t + 1)(t - 2) = 0$	Factor.
$t = -1$ or 2	Solve for t.

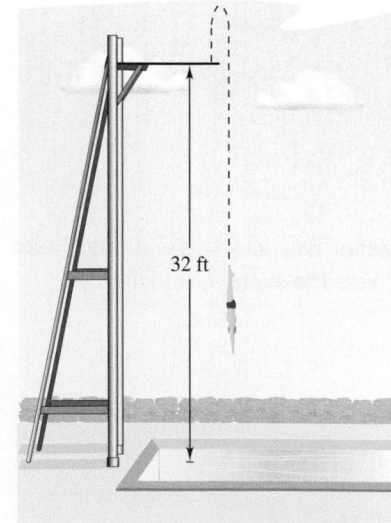

Velocity is positive when an object is rising and is negative when an object is falling. Notice that the diver moves upward for the first half-second because the velocity is positive for $0 < t < \frac{1}{2}$. When the velocity is 0, the diver has reached the maximum height of the dive.

Figure 3.22

Because $t \geq 0$, choose the positive value to conclude that the diver hits the water at $t = 2$ seconds. The velocity at time t is given by the derivative

$$s'(t) = -32t + 16. \qquad \text{Velocity function}$$

So, the velocity at time $t = 2$ is

$$s'(2) = -32(2) + 16 = -48 \text{ feet per second.}$$

Notice that the unit for $s'(t)$ is the unit for s (feet) divided by the unit for t (seconds). In general, the unit for $f'(x)$ is the unit for f divided by the unit for x.

CONCEPT CHECK

1. Finding a Derivative Explain how to find the derivative of the function $f(x) = cx^n$.

2. Derivatives of Trigonometric Functions What are the derivatives of the sine and cosine functions?

3. Functions Find an infinite number of functions which are equal to their derivatives.

4. Average Velocity and Velocity Describe the difference between average velocity and velocity.

 **Estimating Slope** In Exercises 5 and 6, use the graph to estimate the slope of the tangent line to $y = x^n$ at the point $(1, 1)$. Verify your answer analytically. To print an enlarged copy of the graph, go to *MathGraphs.com*.

5. (a) $y = x^{1/2}$ (b) $y = x^3$

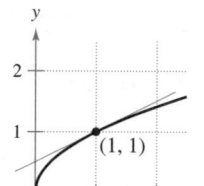

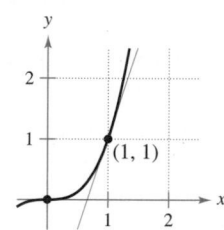

6. (a) $y = x^{-1/2}$ (b) $y = x^{-1}$

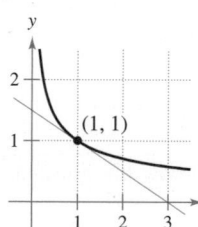

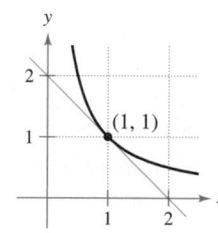

 **Finding a Derivative** In Exercises 7–26, use the rules of differentiation to find the derivative of the function.

7. $y = 12$

8. $f(x) = -9$

9. $y = x^7$

10. $y = x^{12}$

11. $y = \dfrac{1}{x^5}$

12. $y = \dfrac{3}{x^7}$

13. $f(x) = \sqrt[9]{x}$

14. $g(x) = \sqrt[4]{x}$

15. $f(x) = x + 11$

16. $g(x) = 6x + 3$

17. $f(t) = -3t^2 + 2t - 4$

18. $y = t^2 - 3t + 1$

19. $s(t) = t^3 + 5t^2 - 3t + 8$

20. $y = 2x^3 + 6x^2 - 1$

21. $y = \dfrac{\pi}{2}\sin\theta$

22. $g(t) = \pi\cos t$

23. $y = x^2 - \tfrac{1}{2}\cos x$

24. $y = 7x^4 + 2\sin x$

25. $y = \tfrac{1}{2}e^x - 3\sin x$

26. $y = \tfrac{3}{4}e^x + 2\cos x$

 Rewriting a Function Before Differentiating In Exercises 27–30, complete the table to find the derivative of the function.

Original Function	Rewrite	Differentiate	Simplify
27. $y = \dfrac{2}{7x^4}$			
28. $y = \dfrac{8}{5x^{-5}}$			
29. $y = \dfrac{6}{(5x)^3}$			
30. $y = \dfrac{3}{(2x)^{-2}}$			

 **Finding the Slope of a Graph** In Exercises 31–40, find the slope of the graph of the function at the given point. Use the *derivative* feature of a graphing utility to confirm your results.

Function	Point
31. $f(x) = \dfrac{8}{x^2}$	$(2, 2)$
32. $f(t) = 2 - \dfrac{4}{t}$	$(4, 1)$
33. $f(x) = -\tfrac{1}{2} + \tfrac{7}{5}x^3$	$\left(0, -\tfrac{1}{2}\right)$
34. $y = 2x^4 - 3$	$(1, -1)$
35. $y = (4x + 1)^2$	$(0, 1)$
36. $f(x) = 2(x - 4)^2$	$(2, 8)$
37. $f(\theta) = 4\sin\theta - \theta$	$(0, 0)$
38. $g(t) = -2\cos t + 5$	$(\pi, 7)$
39. $f(t) = \tfrac{3}{4}e^t$	$\left(0, \tfrac{3}{4}\right)$
40. $g(x) = -4e^x$	$(1, -4e)$

 Finding a Derivative In Exercises 41–58, find the derivative of the function.

41. $f(x) = x^2 + 5 - 3x^{-2}$

42. $f(x) = x^3 - 2x + 3x^{-3}$

43. $g(t) = t^2 - \dfrac{4}{t^3}$

44. $f(x) = 8x + \dfrac{3}{x^2}$

45. $f(x) = \dfrac{x^3 - 3x^2 + 4}{x^2}$

46. $h(x) = \dfrac{4x^3 + 2x + 5}{x}$

47. $g(t) = \dfrac{3t^2 + 4t - 8}{t^{3/2}}$

48. $h(s) = \dfrac{s^5 + 2s + 6}{s^{1/3}}$

49. $y = x(x^2 + 1)$

50. $y = x^2(2x^2 - 3x)$

51. $f(x) = \sqrt{x} - 6\sqrt[3]{x}$

52. $f(t) = t^{2/3} - t^{1/3} + 4$

53. $f(x) = 6\sqrt{x} + 5\cos x$

54. $f(x) = \dfrac{2}{\sqrt[3]{x}} + 3\cos x$

55. $y = \dfrac{1}{(3x)^{-2}} - 5\cos x$

56. $y = \dfrac{3}{(2x)^3} + 2\sin x$

57. $f(x) = x^{-2} - 2e^x$

58. $g(x) = \sqrt{x} - 3e^x$

Finding an Equation of a Tangent Line In Exercises 59–62, (a) find an equation of the tangent line to the graph of the function at the given point, (b) use a graphing utility to graph the function and its tangent line at the point, and (c) use the *tangent* feature of a graphing utility to confirm your results.

Function	Point
59. $f(x) = -2x^4 + 5x^2 - 3$	$(1, 0)$
60. $f(x) = \dfrac{2}{\sqrt[4]{x^3}}$	$(1, 2)$
61. $g(x) = x + e^x$	$(0, 1)$
62. $h(t) = \sin t + \frac{1}{2}e^t$	$\left(\pi, \frac{1}{2}e^\pi\right)$

Horizontal Tangent Line In Exercises 63–70, determine the point(s) (if any) at which the graph of the function has a horizontal tangent line.

63. $y = x^4 - 2x^2 + 3$ **64.** $y = x^3 + x$

65. $y = \dfrac{1}{x^2}$ **66.** $y = x^2 + 9$

67. $y = -4x + e^x$ **68.** $y = x + 4e^x$

69. $y = x + \sin x, \quad 0 \le x < 2\pi$

70. $y = \sqrt{3}x + 2\cos x, \quad 0 \le x < 2\pi$

Finding a Value In Exercises 71–74, find k such that the line is tangent to the graph of the function.

Function	Line
71. $f(x) = k - x^2$	$y = -6x + 1$
72. $f(x) = kx^2$	$y = -2x + 3$
73. $f(x) = \dfrac{k}{x}$	$y = -\dfrac{3}{4}x + 3$
74. $f(x) = k\sqrt{x}$	$y = x + 4$

EXPLORING CONCEPTS

Exploring a Relationship In Exercises 75 and 76, the relationship between f and g is given. Explain the relationship between f' and g'.

75. $g(x) = f(x) + 6$ **76.** $g(x) = 3f(x) - 1$

A Function and Its Derivative In Exercises 77 and 78, the graphs of a function f and its derivative f' are shown on the same set of coordinate axes. Label the graphs as f or f' and write a short paragraph stating the criteria you used in making your selection. To print an enlarged copy of the graph, go to *MathGraphs.com*.

77. 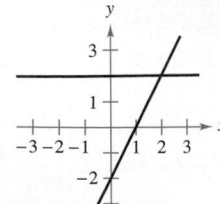   **78.**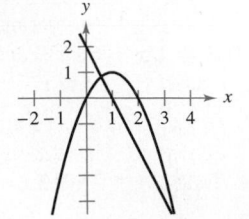

79. Sketching a Graph Sketch the graph of a function f such that $f' > 0$ for all x and the rate of change of the function is decreasing.

80. HOW DO YOU SEE IT? Use the graph of f to answer each question. To print an enlarged copy of the graph, go to *MathGraphs.com*.

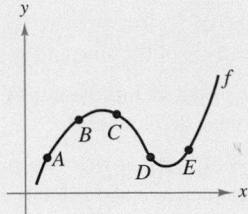

(a) Between which two consecutive points is the average rate of change of the function greatest?

(b) Is the average rate of change of the function between A and B greater than or less than the instantaneous rate of change at B?

(c) Sketch a tangent line to the graph between C and D such that the slope of the tangent line is the same as the average rate of change of the function between C and D.

81. Finding Equations of Tangent Lines Sketch the graphs of $y = x^2$ and $y = -x^2 + 6x - 5$, and sketch the two lines that are tangent to both graphs. Find equations of these lines.

82. Tangent Lines Show that the graphs of the two equations

$$y = x \quad \text{and} \quad y = \frac{1}{x}$$

have tangent lines that are perpendicular to each other at their point of intersection.

83. Horizontal Tangent Line Show that the graph of the function

$$f(x) = 3x + \sin x + 2$$

does not have a horizontal tangent line.

84. Tangent Line Show that the graph of the function

$$f(x) = x^5 + 3x^3 + 5x$$

does not have a tangent line with a slope of 3.

Finding an Equation of a Tangent Line In Exercises 85 and 86, find an equation of the tangent line to the graph of the function f through the point (x_0, y_0) not on the graph. To find the point of tangency (x, y) on the graph of f, solve the equation

$$f'(x) = \frac{y_0 - y}{x_0 - x}.$$

85. $f(x) = \sqrt{x}$ **86.** $f(x) = \dfrac{2}{x}$

    $(x_0, y_0) = (-4, 0)$ $(x_0, y_0) = (5, 0)$

 **87. Linear Approximation** Consider the function $f(x) = x^{3/2}$ with the solution point $(4, 8)$.

(a) Use a graphing utility to graph f. Use the *zoom* feature to obtain successive magnifications of the graph in the neighborhood of the point $(4, 8)$. After zooming in a few times, the graph should appear nearly linear. Use the *trace* feature to determine the coordinates of a point near $(4, 8)$. Find an equation of the secant line $S(x)$ through the two points.

(b) Find the equation of the line $T(x) = f'(4)(x - 4) + f(4)$ tangent to the graph of f passing through the given point. Why are the linear functions S and T nearly the same?

(c) Use a graphing utility to graph f and T on the same set of coordinate axes. Note that T is a good approximation of f when x is close to 4. What happens to the accuracy of the approximation as you move farther away from the point of tangency?

(d) Demonstrate the conclusion in part (c) by completing the table.

Δx	-3	-2	-1	-0.5	-0.1	0
$f(4 + \Delta x)$						
$T(4 + \Delta x)$						

Δx	0.1	0.5	1	2	3
$f(4 + \Delta x)$					
$T(4 + \Delta x)$					

 **88. Linear Approximation** Repeat Exercise 87 for the function $f(x) = x^3$, where $T(x)$ is the line tangent to the graph at the point $(1, 1)$. Explain why the accuracy of the linear approximation decreases more rapidly than in Exercise 87.

True or False? In Exercises 89–94, determine whether the statement is true or false. If it is false, explain why or give an example that shows it is false.

89. If $f'(x) = g'(x)$, then $f(x) = g(x)$.

90. If $y = x^{a+2} + bx$, then $dy/dx = (a + 2)x^{a+1} + b$.

91. If $y = \pi^2$, then $dy/dx = 2\pi$.

92. If $f(x) = -g(x) + b$, then $f'(x) = -g'(x)$.

93. If $f(x) = 0$, then $f'(x)$ is undefined.

94. If $f(x) = \dfrac{1}{x^n}$, then $f'(x) = \dfrac{1}{nx^{n-1}}$.

Finding Rates of Change In Exercises 95–100, find the average rate of change of the function over the given interval. Compare this average rate of change with the instantaneous rates of change at the endpoints of the interval.

95. $f(t) = 3t + 5$, $[1, 2]$ **96.** $f(t) = t^2 - 7$, $[3, 3.1]$

97. $f(x) = \dfrac{-1}{x}$, $[1, 2]$ **98.** $f(x) = \sin x$, $\left[0, \dfrac{\pi}{6}\right]$

99. $g(x) = x^2 + e^x$, $[0, 1]$ **100.** $h(x) = x^3 - \frac{1}{2}e^x$, $[0, 2]$

Vertical Motion In Exercises 101 and 102, use the position function $s(t) = -16t^2 + v_0 t + s_0$ for free-falling objects.

101. A silver dollar is dropped from the top of a building that is 1362 feet tall.

(a) Determine the position and velocity functions for the coin.

(b) Determine the average velocity on the interval $[1, 2]$.

(c) Find the instantaneous velocities when $t = 1$ and $t = 2$.

(d) Find the time required for the coin to reach ground level.

(e) Find the velocity of the coin at impact.

102. A ball is thrown straight down from the top of a 220-foot building with an initial velocity of -22 feet per second. What is its velocity after 3 seconds? What is its velocity after falling 108 feet?

Vertical Motion In Exercises 103 and 104, use the position function $s(t) = -4.9t^2 + v_0 t + s_0$ for free-falling objects.

103. A projectile is shot upward from the surface of Earth with an initial velocity of 120 meters per second. What is its velocity after 5 seconds? After 10 seconds?

104. A rock is dropped from the edge of a cliff that is 214 meters above water.

(a) Determine the position and velocity functions for the rock.

(b) Determine the average velocity on the interval $[2, 5]$.

(c) Find the instantaneous velocities when $t = 2$ and $t = 5$.

(d) Find the time required for the rock to reach the surface of the water.

(e) Find the velocity of the rock at impact.

105. Think About It The graph of the position function (see figure) represents the distance in miles that a person drives during a 10-minute trip to work. Make a sketch of the corresponding velocity function.

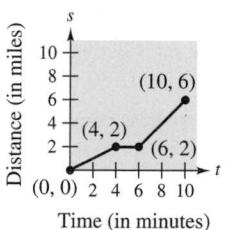
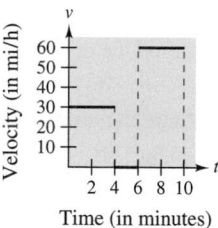

Figure for 105   Figure for 106

106. Think About It The graph of the velocity function (see figure) represents the velocity in miles per hour during a 10-minute trip to work. Make a sketch of the corresponding position function.

107. Volume The volume of a cube with sides of length s is given by $V = s^3$. Find the rate of change of the volume with respect to s when $s = 6$ centimeters.

108. Area The area of a square with sides of length s is given by $A = s^2$. Find the rate of change of the area with respect to s when $s = 6$ meters.

109. Modeling Data

The stopping distance of an automobile, on dry, level pavement, traveling at a speed v (in kilometers per hour) is the distance R (in meters) the car travels during the reaction time of the driver plus the distance B (in meters) the car travels after the brakes are applied (see figure). The table shows the results of an experiment.

Reaction time | Braking distance

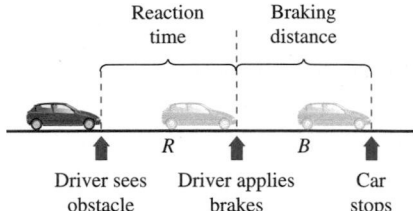

	Driver sees obstacle	Driver applies brakes	Car stops
	R	B	

Speed, v	20	40	60	80	100
Reaction Time Distance, R	8.3	16.7	25.0	33.3	41.7
Braking Time Distance, B	2.3	9.0	20.2	35.8	55.9

(a) Use the regression capabilities of a graphing utility to find a linear model for reaction time distance R.

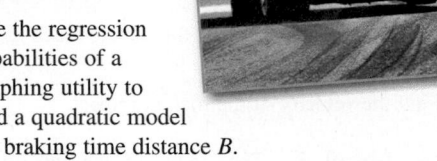

(b) Use the regression capabilities of a graphing utility to find a quadratic model for braking time distance B.

(c) Determine the polynomial giving the total stopping distance T.

(d) Use a graphing utility to graph the functions R, B, and T in the same viewing window.

(e) Find the derivative of T and the rates of change of the total stopping distance for $v = 40$, $v = 80$, and $v = 100$.

(f) Use the results of this exercise to draw conclusions about the total stopping distance as speed increases.

110. Fuel Cost   A car is driven 15,000 miles a year and gets x miles per gallon. Assume that the average fuel cost is \$3.48 per gallon. Find the annual cost of fuel C as a function of x and use this function to complete the table.

x	10	15	20	25	30	35	40
C							
dC/dx							

Who would benefit more from a one-mile-per-gallon increase in fuel efficiency—the driver of a car that gets 15 miles per gallon or the driver of a car that gets 35 miles per gallon? Explain.

Tumar/Shutterstock.com

111. Velocity   Verify that the average velocity over the time interval $[t_0 - \Delta t, t_0 + \Delta t]$ is the same as the instantaneous velocity at $t = t_0$ for the position function

$$s(t) = -\frac{1}{2}at^2 + c.$$

112. Inventory Management   The annual inventory cost C for a manufacturer is

$$C = \frac{1,008,000}{Q} + 6.3Q$$

where Q is the order size when the inventory is replenished. Find the change in annual cost when Q is increased from 350 to 351 and compare this with the instantaneous rate of change when $Q = 350$.

113. Finding an Equation of a Parabola   Find an equation of the parabola $y = ax^2 + bx + c$ that passes through $(0, 1)$ and is tangent to the line $y = x - 1$ at $(1, 0)$.

114. Proof   Let (a, b) be an arbitrary point on the graph of $y = 1/x$, $x > 0$. Prove that the area of the triangle formed by the tangent line through (a, b) and the coordinate axes is 2.

115. Tangent Line   Find the equation(s) of the tangent line(s) to the graph of the curve $y = x^3 - 9x$ through the point $(1, -9)$ not on the graph.

116. Tangent Line   Find the equation(s) of the tangent line(s) to the graph of the parabola $y = x^2$ through the given point not on the graph.

(a) $(0, a)$ (b) $(a, 0)$

Are there any restrictions on the constant a?

Making a Function Differentiable   **In Exercises 117 and 118, find a and b such that f is differentiable everywhere.**

117. $f(x) = \begin{cases} ax^3, & x \leq 2 \\ x^2 + b, & x > 2 \end{cases}$

118. $f(x) = \begin{cases} \cos x, & x < 0 \\ ax + b, & x \geq 0 \end{cases}$

119. Determining Differentiability   Where are the functions $f_1(x) = |\sin x|$ and $f_2(x) = \sin |x|$ differentiable?

120. Proof   Prove that $\dfrac{d}{dx}[\cos x] = -\sin x$.

■ **FOR FURTHER INFORMATION**   For a geometric interpretation of the derivatives of trigonometric functions, see the article "Sines and Cosines of the Times" by Victor J. Katz in *Math Horizons*. To view this article, go to *MathArticles.com*.

PUTNAM EXAM CHALLENGE

121. Find all differentiable functions $f : \mathbb{R} \to \mathbb{R}$ such that

$$f'(x) = \frac{f(x + n) - f(x)}{n}$$

for all real numbers x and all positive integers n.

3.3 Product and Quotient Rules and Higher-Order Derivatives

■ Find the derivative of a function using the Product Rule.
■ Find the derivative of a function using the Quotient Rule.
■ Find the derivative of a trigonometric function.
■ Find a higher-order derivative of a function.

The Product Rule

In Section 3.2, you learned that the derivative of the sum of two functions is simply the sum of their derivatives. The rules for the derivatives of the product and quotient of two functions are not as simple.

▷

REMARK A version of the Product Rule that some people prefer is

$$\frac{d}{dx}[f(x)g(x)] = f'(x)g(x) + f(x)g'(x).$$

The advantage of this form is that it generalizes easily to products of three or more factors.

> **THEOREM 3.8   The Product Rule**
>
> The product of two differentiable functions f and g is itself differentiable. Moreover, the derivative of fg is the first function times the derivative of the second, plus the second function times the derivative of the first.
>
> $$\frac{d}{dx}[f(x)g(x)] = f(x)g'(x) + g(x)f'(x)$$

Proof Some mathematical proofs, such as the proof of the Sum Rule, are straightforward. Others involve clever steps that may appear unmotivated to a reader. This proof involves such a step—subtracting and adding the same quantity—which is shown in color.

$$\frac{d}{dx}[f(x)g(x)] = \lim_{\Delta x \to 0} \frac{f(x + \Delta x)g(x + \Delta x) - f(x)g(x)}{\Delta x}$$

$$= \lim_{\Delta x \to 0} \frac{f(x + \Delta x)g(x + \Delta x) - f(x + \Delta x)g(x) + f(x + \Delta x)g(x) - f(x)g(x)}{\Delta x}$$

$$= \lim_{\Delta x \to 0} \left[f(x + \Delta x)\frac{g(x + \Delta x) - g(x)}{\Delta x} + g(x)\frac{f(x + \Delta x) - f(x)}{\Delta x} \right]$$

$$= \lim_{\Delta x \to 0} \left[f(x + \Delta x)\frac{g(x + \Delta x) - g(x)}{\Delta x} \right] + \lim_{\Delta x \to 0} \left[g(x)\frac{f(x + \Delta x) - f(x)}{\Delta x} \right]$$

$$= \lim_{\Delta x \to 0} f(x + \Delta x) \cdot \lim_{\Delta x \to 0} \frac{g(x + \Delta x) - g(x)}{\Delta x} + \lim_{\Delta x \to 0} g(x) \cdot \lim_{\Delta x \to 0} \frac{f(x + \Delta x) - f(x)}{\Delta x}$$

$$= f(x)g'(x) + g(x)f'(x)$$

Note that $\lim_{\Delta x \to 0} f(x + \Delta x) = f(x)$ because f is given to be differentiable and therefore is continuous. ■

▷

REMARK The proof of the Product Rule for products of more than two factors is left as an exercise (see Exercise 145).

The Product Rule can be extended to cover products involving more than two factors. For example, if f, g, and h are differentiable functions of x, then

$$\frac{d}{dx}[f(x)g(x)h(x)] = f'(x)g(x)h(x) + f(x)g'(x)h(x) + f(x)g(x)h'(x).$$

So, the derivative of $y = x^2 \sin x \cos x$ is

$$\frac{dy}{dx} = 2x \sin x \cos x + x^2 \cos x \cos x + x^2(\sin x)(-\sin x)$$

$$= 2x \sin x \cos x + x^2(\cos^2 x - \sin^2 x).$$

THE PRODUCT RULE

When Leibniz originally wrote a formula for the Product Rule, he was motivated by the expression

$(x + dx)(y + dy) - xy$

from which he subtracted $dx\, dy$ (as being negligible) and obtained the differential form $x\, dy + y\, dx$. This derivation resulted in the traditional form of the Product Rule. *(Source: The History of Mathematics by David M. Burton)*

The derivative of a product of two functions is not (in general) given by the product of the derivatives of the two functions. To see this, try comparing the product of the derivatives of

$$f(x) = 3x - 2x^2$$

and

$$g(x) = 5 + 4x$$

with the derivative in Example 1.

EXAMPLE 1   Using the Product Rule

Find the derivative of $h(x) = (3x - 2x^2)(5 + 4x)$.

Solution

$$h'(x) = \overbrace{(3x - 2x^2)}^{\text{First}} \overbrace{\frac{d}{dx}[5 + 4x]}^{\substack{\text{Derivative of} \\ \text{second}}} + \overbrace{(5 + 4x)}^{\text{Second}} \overbrace{\frac{d}{dx}[3x - 2x^2]}^{\substack{\text{Derivative} \\ \text{of first}}} \qquad \text{Apply Product Rule.}$$

$$= (3x - 2x^2)(4) + (5 + 4x)(3 - 4x)$$
$$= (12x - 8x^2) + (15 - 8x - 16x^2)$$
$$= -24x^2 + 4x + 15$$

In Example 1, you have the option of finding the derivative with or without the Product Rule. To find the derivative without the Product Rule, you can write

$$D_x[(3x - 2x^2)(5 + 4x)] = D_x[-8x^3 + 2x^2 + 15x]$$
$$= -24x^2 + 4x + 15.$$

In the next example, you must use the Product Rule.

EXAMPLE 2   Using the Product Rule

Find the derivative of $y = xe^x$.

Solution

$$\frac{d}{dx}[xe^x] = x\frac{d}{dx}[e^x] + e^x\frac{d}{dx}[x] \qquad \text{Apply Product Rule.}$$
$$= xe^x + e^x(1)$$
$$= e^x(x + 1)$$

· · REMARK In Example 3, notice that you use the Product Rule when both factors of the product are variable, and you use the Constant Multiple Rule when one of the factors is a constant. ▷

EXAMPLE 3   Using the Product Rule

Find the derivative of $y = 2x \cos x - 2 \sin x$.

Solution

$$\frac{dy}{dx} = \overbrace{(2x)\left(\frac{d}{dx}[\cos x]\right) + (\cos x)\left(\frac{d}{dx}[2x]\right)}^{\text{Product Rule}} - \overbrace{2\frac{d}{dx}[\sin x]}^{\text{Constant Multiple Rule}}$$
$$= (2x)(-\sin x) + (\cos x)(2) - 2(\cos x)$$
$$= -2x \sin x$$

The Quotient Rule

> **THEOREM 3.9 The Quotient Rule**
>
> The quotient f/g of two differentiable functions f and g is itself differentiable at all values of x for which $g(x) \neq 0$. Moreover, the derivative of f/g is given by the denominator times the derivative of the numerator minus the numerator times the derivative of the denominator, all divided by the square of the denominator.
>
> $$\frac{d}{dx}\left[\frac{f(x)}{g(x)}\right] = \frac{g(x)f'(x) - f(x)g'(x)}{[g(x)]^2}, \quad g(x) \neq 0$$

• • • • • • • • • • • • • • • • ▷

•**REMARK** From the Quotient Rule, you can see that the derivative of a quotient is not (in general) the quotient of the derivatives.

Proof As with the proof of Theorem 3.8, the key to this proof is subtracting and adding the same quantity—which is shown in color.

$$\frac{d}{dx}\left[\frac{f(x)}{g(x)}\right] = \lim_{\Delta x \to 0} \frac{\dfrac{f(x + \Delta x)}{g(x + \Delta x)} - \dfrac{f(x)}{g(x)}}{\Delta x} \qquad \text{Definition of derivative}$$

$$= \lim_{\Delta x \to 0} \frac{g(x)f(x + \Delta x) - f(x)g(x + \Delta x)}{\Delta x g(x)g(x + \Delta x)}$$

$$= \lim_{\Delta x \to 0} \frac{g(x)f(x + \Delta x) - f(x)g(x) + f(x)g(x) - f(x)g(x + \Delta x)}{\Delta x g(x)g(x + \Delta x)}$$

$$= \frac{\displaystyle\lim_{\Delta x \to 0} \frac{g(x)[f(x + \Delta x) - f(x)]}{\Delta x} - \lim_{\Delta x \to 0} \frac{f(x)[g(x + \Delta x) - g(x)]}{\Delta x}}{\displaystyle\lim_{\Delta x \to 0} [g(x)g(x + \Delta x)]}$$

$$= \frac{g(x)\left[\displaystyle\lim_{\Delta x \to 0} \frac{f(x + \Delta x) - f(x)}{\Delta x}\right] - f(x)\left[\displaystyle\lim_{\Delta x \to 0} \frac{g(x + \Delta x) - g(x)}{\Delta x}\right]}{\displaystyle\lim_{\Delta x \to 0} [g(x)g(x + \Delta x)]}$$

$$= \frac{g(x)f'(x) - f(x)g'(x)}{[g(x)]^2}$$

Note that $\displaystyle\lim_{\Delta x \to 0} g(x + \Delta x) = g(x)$ because g is given to be differentiable and therefore is continuous. ∎

▷ **TECHNOLOGY** A graphing utility can be used to compare the graph of a function with the graph of its derivative. For instance, in the figure below, the graph of the function in Example 4 appears to have two points that have horizontal tangent lines. What are the values of y' at these two points?

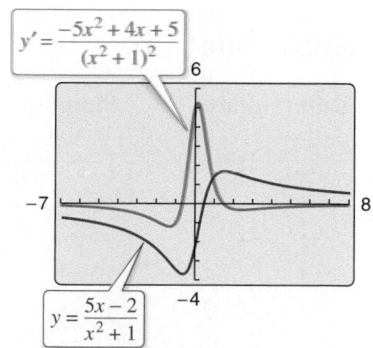

$$y' = \frac{-5x^2 + 4x + 5}{(x^2 + 1)^2}$$

$$y = \frac{5x - 2}{x^2 + 1}$$

Graphical comparison of a function and its derivative

EXAMPLE 4 Using the Quotient Rule

Find the derivative of $y = \dfrac{5x - 2}{x^2 + 1}$.

Solution

$$\frac{d}{dx}\left[\frac{5x - 2}{x^2 + 1}\right] = \frac{(x^2 + 1)\dfrac{d}{dx}[5x - 2] - (5x - 2)\dfrac{d}{dx}[x^2 + 1]}{(x^2 + 1)^2} \qquad \text{Apply Quotient Rule.}$$

$$= \frac{(x^2 + 1)(5) - (5x - 2)(2x)}{(x^2 + 1)^2}$$

$$= \frac{(5x^2 + 5) - (10x^2 - 4x)}{(x^2 + 1)^2}$$

$$= \frac{-5x^2 + 4x + 5}{(x^2 + 1)^2}$$

Note the use of parentheses in Example 4. A liberal use of parentheses is recommended for *all* types of differentiation problems. For instance, with the Quotient Rule, it is a good idea to enclose all factors and derivatives in parentheses and to pay special attention to the subtraction required in the numerator.

When differentiation rules were introduced in the preceding section, the need for rewriting *before* differentiating was emphasized. The next example illustrates this point with the Quotient Rule.

EXAMPLE 5   Rewriting Before Differentiating

Find an equation of the tangent line to the graph of $f(x) = \dfrac{3 - (1/x)}{x + 5}$ at $(-1, 1)$.

Solution Begin by rewriting the function.

$$f(x) = \frac{3 - (1/x)}{x + 5}$$ Write original function.

$$= \frac{x\left(3 - \dfrac{1}{x}\right)}{x(x + 5)}$$ Multiply numerator and denominator by x.

$$= \frac{3x - 1}{x^2 + 5x}$$ Rewrite.

Next, apply the Quotient Rule.

$$f'(x) = \frac{(x^2 + 5x)(3) - (3x - 1)(2x + 5)}{(x^2 + 5x)^2}$$ Quotient Rule

$$= \frac{(3x^2 + 15x) - (6x^2 + 13x - 5)}{(x^2 + 5x)^2}$$

$$= \frac{-3x^2 + 2x + 5}{(x^2 + 5x)^2}$$ Simplify.

To find the slope at $(-1, 1)$, evaluate $f'(-1)$.

$$f'(-1) = 0$$ Slope of graph at $(-1, 1)$

Then, using the point-slope form of the equation of a line, you can determine that the equation of the tangent line at $(-1, 1)$ is $y = 1$. See Figure 3.23.

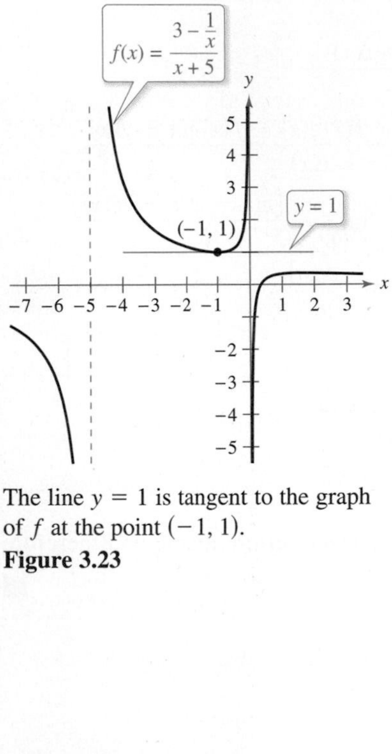

$f(x) = \dfrac{3 - \dfrac{1}{x}}{x + 5}$

$y = 1$

$(-1, 1)$

The line $y = 1$ is tangent to the graph of f at the point $(-1, 1)$.

Figure 3.23

Not every quotient needs to be differentiated by the Quotient Rule. For instance, each quotient in the next example can be considered as the product of a constant times a function of x. In such cases, it is more convenient to use the Constant Multiple Rule.

▷ **REMARK** To see the benefit of using the Constant Multiple Rule for some quotients, try using the Quotient Rule to differentiate the functions in Example 6. You should obtain the same results but with more work.

EXAMPLE 6   Using the Constant Multiple Rule

Original Function	Rewrite	Differentiate	Simplify
a. $y = \dfrac{x^2 + 3x}{6}$	$y = \dfrac{1}{6}(x^2 + 3x)$	$y' = \dfrac{1}{6}(2x + 3)$	$y' = \dfrac{2x + 3}{6}$
b. $y = \dfrac{5x^4}{8}$	$y = \dfrac{5}{8}x^4$	$y' = \dfrac{5}{8}(4x^3)$	$y' = \dfrac{5}{2}x^3$
c. $y = \dfrac{-3(3x - 2x^2)}{7x}$	$y = -\dfrac{3}{7}(3 - 2x)$	$y' = -\dfrac{3}{7}(-2)$	$y' = \dfrac{6}{7}$
d. $y = \dfrac{9}{5x^2}$	$y = \dfrac{9}{5}(x^{-2})$	$y' = \dfrac{9}{5}(-2x^{-3})$	$y' = -\dfrac{18}{5x^3}$

In Section 3.2, the Power Rule was proved only for the case in which the exponent n is a positive integer greater than 1. The next example extends the proof to include negative integer exponents.

EXAMPLE 7 **Power Rule: Negative Integer Exponents**

If n is a negative integer, then there exists a positive integer k such that $n = -k$. So, by the Quotient Rule, you can write

$$\frac{d}{dx}[x^n] = \frac{d}{dx}\left[\frac{1}{x^k}\right]$$

$$= \frac{x^k(0) - (1)(kx^{k-1})}{(x^k)^2} \qquad \text{Quotient Rule and Power Rule}$$

$$= \frac{0 - kx^{k-1}}{x^{2k}}$$

$$= -kx^{-k-1}$$

$$= nx^{n-1}. \qquad\qquad n = -k$$

So, the Power Rule

$$\frac{d}{dx}[x^n] = nx^{n-1} \qquad\qquad \text{Power Rule}$$

is valid for any integer n. The cases for which n is rational and n is irrational are left as an exercise (see Section 3.5, Exercise 92). ∎

Derivatives of Trigonometric Functions

Knowing the derivatives of the sine and cosine functions, you can use the Quotient Rule to find the derivatives of the four remaining trigonometric functions.

THEOREM 3.10 **Derivatives of Trigonometric Functions**

$$\frac{d}{dx}[\tan x] = \sec^2 x \qquad\qquad \frac{d}{dx}[\cot x] = -\csc^2 x$$

$$\frac{d}{dx}[\sec x] = \sec x \tan x \qquad\qquad \frac{d}{dx}[\csc x] = -\csc x \cot x$$

Proof Considering $\tan x = (\sin x)/(\cos x)$ and applying the Quotient Rule, you obtain

$$\frac{d}{dx}[\tan x] = \frac{d}{dx}\left[\frac{\sin x}{\cos x}\right]$$

$$= \frac{(\cos x)(\cos x) - (\sin x)(-\sin x)}{\cos^2 x} \qquad \text{Apply Quotient Rule.}$$

$$= \frac{\cos^2 x + \sin^2 x}{\cos^2 x}$$

$$= \frac{1}{\cos^2 x}$$

$$= \sec^2 x.$$

The proofs of the other three parts of the theorem are left as an exercise (see Exercise 93). ∎

EXAMPLE 8 **Differentiating Trigonometric Functions**

⋮ ⋯ ▷ *See LarsonCalculus.com for an interactive version of this type of example.*

| **Function** | **Derivative** |

a. $y = x - \tan x$ $\quad\quad\quad\quad\quad \dfrac{dy}{dx} = 1 - \sec^2 x$

b. $y = x \sec x$ $\quad\quad\quad\quad\quad\quad y' = x(\sec x \tan x) + (\sec x)(1)$
$$= (\sec x)(1 + x \tan x)$$

EXAMPLE 9 **Different Forms of a Derivative**

⋯⋯⋯⋯⋯⋯⋯▷

REMARK Because of trigonometric identities, the derivative of a trigonometric function can take many forms. This presents a challenge when you are trying to match your answers to those given in the back of the text.

Differentiate both forms of

$$y = \frac{1 - \cos x}{\sin x} = \csc x - \cot x.$$

Solution

First form: $\quad y = \dfrac{1 - \cos x}{\sin x}$

$$y' = \frac{(\sin x)(\sin x) - (1 - \cos x)(\cos x)}{\sin^2 x}$$

$$= \frac{\sin^2 x - \cos x + \cos^2 x}{\sin^2 x}$$

$$= \frac{1 - \cos x}{\sin^2 x} \quad\quad\quad\quad \sin^2 x + \cos^2 x = 1$$

Second form: $\quad y = \csc x - \cot x$

$$y' = -\csc x \cot x + \csc^2 x$$

To show that the two derivatives are equal, you can write

$$\frac{1 - \cos x}{\sin^2 x} = \frac{1}{\sin^2 x} - \frac{\cos x}{\sin^2 x}$$

$$= \frac{1}{\sin^2 x} - \left(\frac{1}{\sin x}\right)\left(\frac{\cos x}{\sin x}\right)$$

$$= \csc^2 x - \csc x \cot x. \quad\blacksquare$$

The summary below shows that much of the work in obtaining a simplified form of a derivative occurs *after* differentiating. Note that two characteristics of a simplified form are the absence of negative exponents and the combining of like terms.

	$f'(x)$ After Differentiating	$f'(x)$ After Simplifying
Example 1	$(3x - 2x^2)(4) + (5 + 4x)(3 - 4x)$	$-24x^2 + 4x + 15$
Example 3	$(2x)(-\sin x) + (\cos x)(2) - 2(\cos x)$	$-2x \sin x$
Example 4	$\dfrac{(x^2 + 1)(5) - (5x - 2)(2x)}{(x^2 + 1)^2}$	$\dfrac{-5x^2 + 4x + 5}{(x^2 + 1)^2}$
Example 5	$\dfrac{(x^2 + 5x)(3) - (3x - 1)(2x + 5)}{(x^2 + 5x)^2}$	$\dfrac{-3x^2 + 2x + 5}{(x^2 + 5x)^2}$
Example 9	$\dfrac{(\sin x)(\sin x) - (1 - \cos x)(\cos x)}{\sin^2 x}$	$\dfrac{1 - \cos x}{\sin^2 x}$

Exploration

For which of the functions

$$y = e^x, \qquad y = \frac{1}{e^x}$$

$$y = \sin x, \qquad y = \cos x$$

are the equations below true?

a. $y = y'$ **b.** $y = y''$

c. $y = y'''$ **d.** $y = y^{(4)}$

Without determining the actual derivative, is $y = y^{(8)}$ for $y = \sin x$ true? What conclusion can you draw from this?

Higher-Order Derivatives

Just as you can obtain a velocity function by differentiating a position function, you can obtain an **acceleration** function by differentiating a velocity function. Another way of looking at this is that you can obtain an acceleration function by differentiating a position function *twice*.

$s(t)$	Position function
$v(t) = s'(t)$	Velocity function
$a(t) = v'(t) = s''(t)$	Acceleration function

The function $a(t)$ is the **second derivative** of $s(t)$ and is denoted by $s''(t)$.

The second derivative is an example of a **higher-order derivative.** You can define derivatives of any positive integer order. For instance, the **third derivative** is the derivative of the second derivative. Higher-order derivatives are denoted as shown below.

First derivative:	y',	$f'(x)$,	$\dfrac{dy}{dx}$,	$\dfrac{d}{dx}[f(x)]$,	$D_x[y]$
Second derivative:	y'',	$f''(x)$,	$\dfrac{d^2y}{dx^2}$,	$\dfrac{d^2}{dx^2}[f(x)]$,	$D_x^2[y]$
Third derivative:	y''',	$f'''(x)$,	$\dfrac{d^3y}{dx^3}$,	$\dfrac{d^3}{dx^3}[f(x)]$,	$D_x^3[y]$
Fourth derivative:	$y^{(4)}$,	$f^{(4)}(x)$,	$\dfrac{d^4y}{dx^4}$,	$\dfrac{d^4}{dx^4}[f(x)]$,	$D_x^4[y]$
$\vdots$					
nth derivative:	$y^{(n)}$,	$f^{(n)}(x)$,	$\dfrac{d^ny}{dx^n}$,	$\dfrac{d^n}{dx^n}[f(x)]$,	$D_x^n[y]$

EXAMPLE 10   Finding the Acceleration Due to Gravity

Because the moon has no atmosphere, a falling object on the moon encounters no air resistance. In 1971, astronaut David Scott demonstrated that a feather and a hammer fall at the same rate on the moon. The position function for each of these falling objects is

$$s(t) = -0.81t^2 + 2$$

where $s(t)$ is the height in meters and t is the time in seconds, as shown in the figure at the right. What is the ratio of Earth's gravitational force to the moon's?

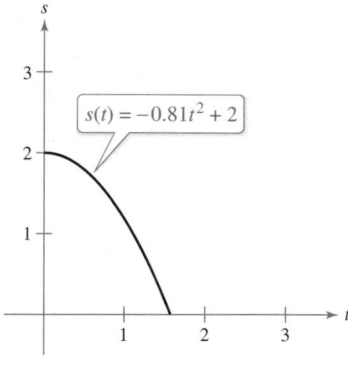

$$s(t) = -0.81t^2 + 2$$

The moon's mass is 7.349×10^{22} kilograms, and Earth's mass is 5.976×10^{24} kilograms. The moon's radius is 1737 kilometers, and Earth's radius is 6378 kilometers. Because the gravitational force on the surface of a planet is directly proportional to its mass and inversely proportional to the square of its radius, the ratio of the gravitational force on Earth to the gravitational force on the moon is

$$\frac{(5.976 \times 10^{24})/6378^2}{(7.349 \times 10^{22})/1737^2} \approx 6.0.$$

Solution To find the acceleration, differentiate the position function twice.

$s(t) = -0.81t^2 + 2$	Position function
$s'(t) = -1.62t$	Velocity function
$s''(t) = -1.62$	Acceleration function

Because $s''(t) = -g$, the acceleration due to gravity on the moon is $g = 1.62$ meters per second per second. The acceleration due to gravity on Earth is 9.8 meters per second per second, so the ratio of Earth's gravitational force to the moon's is

$$\frac{\text{Earth's gravitational force}}{\text{Moon's gravitational force}} = \frac{9.8}{1.62}$$

$$\approx 6.0.$$

NASA

3.3 Exercises

See **CalcChat.com** for tutorial help and worked-out solutions to odd-numbered exercises.

CONCEPT CHECK

1. **Product Rule** Describe the Product Rule in your own words.

2. **Quotient Rule** Describe the Quotient Rule in your own words.

3. **Trigonometric Functions** What are the derivatives of $\tan x$, $\cot x$, $\sec x$, and $\csc x$?

4. **Higher-Order Derivative** What is a higher-order derivative?

 **Using the Product Rule** In Exercises 5–10, use the Product Rule to find the derivative of the function.

5. $g(x) = (2x - 3)(1 - 5x)$ 6. $y = (3x - 4)(x^3 + 5)$

7. $h(t) = \sqrt{t}(1 - t^2)$ 8. $g(s) = \sqrt{s}(s^2 + 8)$

9. $f(x) = e^x \cos x$ 10. $g(x) = \sqrt{x} \sin x$

 **Using the Quotient Rule** In Exercises 11–16, use the Quotient Rule to find the derivative of the function.

11. $f(x) = \dfrac{x}{x - 5}$ 12. $g(t) = \dfrac{3t^2 - 1}{2t + 5}$

13. $h(x) = \dfrac{\sqrt{x}}{x^3 + 1}$ 14. $f(x) = \dfrac{x^2}{2\sqrt{x} + 1}$

15. $g(x) = \dfrac{\sin x}{e^x}$ 16. $f(t) = \dfrac{\cos t}{t^3}$

 **Finding and Evaluating a Derivative** In Exercises 17–24, find $f'(x)$ and $f'(c)$.

17. $f(x) = (x^3 + 4x)(3x^2 + 2x - 5), \quad c = 0$

18. $f(x) = (2x^2 - 3x)(9x + 4), \quad c = -1$

19. $f(x) = \dfrac{x^2 - 4}{x - 3}, \quad c = 1$ 20. $f(x) = \dfrac{x - 4}{x + 4}, \quad c = 3$

21. $f(x) = x \cos x, \quad c = \dfrac{\pi}{4}$ 22. $f(x) = \dfrac{\sin x}{x}, \quad c = \dfrac{\pi}{6}$

23. $f(x) = e^x \sin x, \quad c = 0$ 24. $f(x) = \dfrac{\cos x}{e^x}, \quad c = 0$

 **Using the Constant Multiple Rule** In Exercises 25–30, complete the table to find the derivative of the function without using the Quotient Rule.

Function	Rewrite	Differentiate	Simplify
25. $y = \dfrac{x^3 + 6x}{3}$			
26. $y = \dfrac{5x^2 - 3}{4}$			

Function	Rewrite	Differentiate	Simplify
27. $y = \dfrac{6}{7x^2}$			
28. $y = \dfrac{10}{3x^3}$			
29. $y = \dfrac{4x^{3/2}}{x}$			
30. $y = \dfrac{2x}{x^{1/3}}$			

Finding a Derivative In Exercises 31–42, find the derivative of the algebraic function.

31. $f(x) = \dfrac{4 - 3x - x^2}{x^2 - 1}$ 32. $f(x) = \dfrac{x^2 + 5x + 6}{x^2 - 4}$

33. $f(x) = x\left(1 - \dfrac{4}{x + 3}\right)$ 34. $f(x) = x^4\left(1 - \dfrac{2}{x + 1}\right)$

35. $f(x) = \dfrac{3x - 1}{\sqrt{x}}$ 36. $f(x) = \sqrt[3]{x}(\sqrt{x} + 3)$

37. $f(x) = \dfrac{2 - \dfrac{1}{x}}{x - 3}$ 38. $h(x) = \dfrac{\dfrac{1}{x^2} + 5x}{x + 1}$

39. $g(s) = s^3\left(5 - \dfrac{s}{s + 2}\right)$ 40. $g(x) = x^2\left(\dfrac{2}{x} - \dfrac{1}{x + 1}\right)$

41. $f(x) = (2x^3 + 5x)(x - 3)(x + 2)$

42. $f(x) = (x^3 - x)(x^2 + 2)(x^2 + x - 1)$

 **Finding a Derivative of a Transcendental Function** In Exercises 43–60, find the derivative of the transcendental function.

43. $f(t) = t^2 \sin t$ 44. $f(\theta) = (\theta + 1) \cos \theta$

45. $f(t) = \dfrac{\cos t}{t}$ 46. $f(x) = \dfrac{\sin x}{x^3}$

47. $f(x) = -e^x + \tan x$ 48. $y = e^x - \cot x$

49. $g(t) = \sqrt[4]{t} + 6 \csc t$ 50. $h(x) = \dfrac{1}{x} - 12 \sec x$

51. $y = \dfrac{3(1 - \sin x)}{2 \cos x}$ 52. $y = \dfrac{\sec x}{x}$

53. $y = -\csc x - \sin x$ 54. $y = x \sin x + \cos x$

55. $f(x) = x^2 \tan x$ 56. $f(x) = \sin x \cos x$

57. $y = 2x \sin x + x^2 e^x$ 58. $h(x) = 2e^x \cos x$

59. $y = \dfrac{e^x}{4\sqrt{x}}$ 60. $y = \dfrac{2e^x}{x^2 + 1}$

Finding a Derivative Using Technology In Exercises 61 and 62, use a computer algebra system to find the derivative of the function.

61. $g(x) = \left(\dfrac{x + 1}{x + 2}\right)(2x - 5)$ 62. $f(x) = \dfrac{\cos x}{1 - \sin x}$

Finding the Slope of a Graph In Exercises 63–66, find the slope of the graph of the function at the given point. Use the *derivative* feature of a graphing utility to confirm your results.

Function	Point
63. $y = \dfrac{1 + \csc x}{1 - \csc x}$	$\left(\dfrac{\pi}{6}, -3\right)$
64. $f(x) = \tan x \cot x$	$(1, 1)$
65. $h(t) = \dfrac{\sec t}{t}$	$\left(\pi, -\dfrac{1}{\pi}\right)$
66. $f(x) = (\sin x)(\sin x + \cos x)$	$\left(\dfrac{\pi}{4}, 1\right)$

Finding an Equation of a Tangent Line In Exercises 67–74, (a) find an equation of the tangent line to the graph of f at the given point, (b) use a graphing utility to graph the function and its tangent line at the point, and (c) use the *tangent* feature of a graphing utility to confirm your results.

67. $f(x) = (x^3 + 4x - 1)(x - 2)$, $(1, -4)$

68. $f(x) = (x - 2)(x^2 + 4)$, $(1, -5)$

69. $f(x) = \dfrac{x}{x + 4}$, $(-5, 5)$ 70. $f(x) = \dfrac{x + 3}{x - 3}$, $(4, 7)$

71. $f(x) = \tan x$, $\left(\dfrac{\pi}{4}, 1\right)$ 72. $f(x) = \sec x$, $\left(\dfrac{\pi}{3}, 2\right)$

73. $f(x) = (x - 1)e^x$, $(1, 0)$ 74. $f(x) = \dfrac{e^x}{x + 4}$, $\left(0, \dfrac{1}{4}\right)$

Famous Curves In Exercises 75–78, find an equation of the tangent line to the graph at the given point. (The graphs in Exercises 75 and 76 are called *Witches of Agnesi*. The graphs in Exercises 77 and 78 are called *serpentines*.)

75.
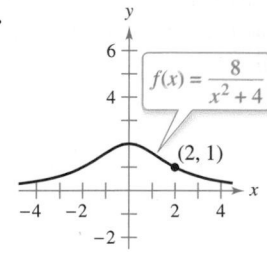
$f(x) = \dfrac{8}{x^2 + 4}$, $(2, 1)$

76.
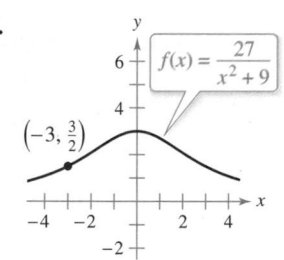
$f(x) = \dfrac{27}{x^2 + 9}$, $\left(-3, \dfrac{3}{2}\right)$

77.
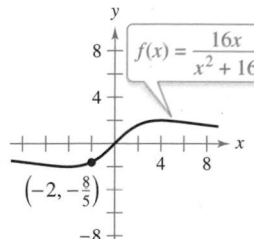
$f(x) = \dfrac{16x}{x^2 + 16}$, $\left(-2, -\dfrac{8}{5}\right)$

78.
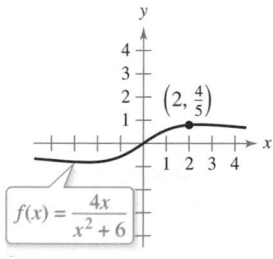
$f(x) = \dfrac{4x}{x^2 + 6}$, $\left(2, \dfrac{4}{5}\right)$

 Horizontal Tangent Line In Exercises 79–82, determine the point(s) at which the graph of the function has a horizontal tangent line.

79. $f(x) = \dfrac{x^2}{x - 1}$ 80. $f(x) = \dfrac{x - 4}{x^2 - 7}$

81. $g(x) = \dfrac{8(x - 2)}{e^x}$ 82. $f(x) = e^x \sin x$, $[0, \pi]$

83. **Tangent Lines** Find equations of the tangent lines to the graph of

$$f(x) = \frac{x + 1}{x - 1}$$

that are parallel to the line $2y + x = 6$. Then graph the function and the tangent lines.

84. **Tangent Lines** Find equations of the tangent lines to the graph of $f(x) = x/(x - 1)$ that pass through the point $(-1, 5)$. Then graph the function and the tangent lines.

Exploring a Relationship In Exercises 85 and 86, verify that $f'(x) = g'(x)$ and explain the relationship between f and g.

85. $f(x) = \dfrac{3x}{x + 2}$, $g(x) = \dfrac{5x + 4}{x + 2}$

86. $f(x) = \dfrac{\sin x - 3x}{x}$, $g(x) = \dfrac{\sin x + 2x}{x}$

Finding Derivatives In Exercises 87 and 88, use the graphs of f and g. Let $p(x) = f(x)g(x)$ and $q(x) = f(x)/g(x)$.

87. (a) Find $p'(1)$.   88. (a) Find $p'(4)$.

     (b) Find $q'(4)$.   (b) Find $q'(7)$.

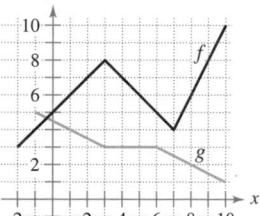 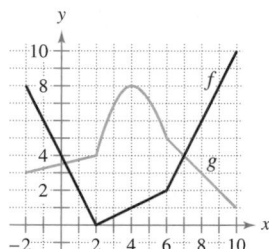

89. **Area** The length of a rectangle is given by $6t + 5$ and its height is $\sqrt{t}$, where t is time in seconds and the dimensions are in centimeters. Find the rate of change of the area with respect to time.

90. **Volume** The radius of a right circular cylinder is given by $\sqrt{t + 2}$ and its height is $\dfrac{1}{2}\sqrt{t}$, where t is time in seconds and the dimensions are in inches. Find the rate of change of the volume with respect to time.

91. **Inventory Replenishment** The ordering and transportation cost C for the components used in manufacturing a product is

$$C = 100\left(\frac{200}{x^2} + \frac{x}{x + 30}\right), \quad x \geq 1$$

where C is measured in thousands of dollars and x is the order size in hundreds. Find the rate of change of C with respect to x when (a) $x = 10$, (b) $x = 15$, and (c) $x = 20$. What do these rates of change imply about increasing order size?

92. **Population Growth** A population of 500 bacteria is introduced into a culture and grows in number according to the equation

$$P(t) = 500\left(1 + \frac{4t}{50 + t^2}\right)$$

where t is measured in hours. Find the rate at which the population is growing when $t = 2$.

93. Proof Prove each differentiation rule.

(a) $\dfrac{d}{dx}[\sec x] = \sec x \tan x$ (b) $\dfrac{d}{dx}[\csc x] = -\csc x \cot x$

(c) $\dfrac{d}{dx}[\cot x] = -\csc^2 x$

94. Rate of Change Determine whether there exist any values of x in the interval $[0, 2\pi)$ such that the rate of change of $f(x) = \sec x$ and the rate of change of $g(x) = \csc x$ are equal.

95. Modeling Data The table shows the national health care expenditures h (in billions of dollars) in the United States and the population p (in millions) of the United States for the years 2008 through 2013. The year is represented by t, with $t = 8$ corresponding to 2008. (*Source: U.S. Centers for Medicare & Medicaid Services and U.S. Census Bureau*)

Year, t	8	9	10	11	12	13
h	2414	2506	2604	2705	2817	2919
p	304	307	309	311	313	315

(a) Use a graphing utility to find linear models for the health care expenditures $h(t)$ and the population $p(t)$.

(b) Use a graphing utility to graph $h(t)$ and $p(t)$.

(c) Find $A = h(t)/p(t)$, then graph A using a graphing utility. What does this function represent?

(d) Find and interpret $A'(t)$ in the context of the problem.

96. Satellites When satellites observe Earth, they can scan only part of Earth's surface. Some satellites have sensors that can measure the angle θ shown in the figure. Let h represent the satellite's distance from Earth's surface, and let r represent Earth's radius.

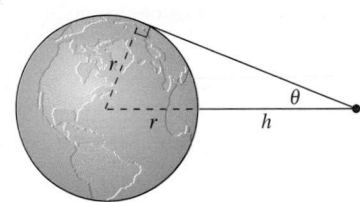

(a) Show that $h = r(\csc \theta - 1)$.

(b) Find the rate at which h is changing with respect to θ when $\theta = 30°$. (Assume $r = 4000$ miles.)

 **Finding a Second Derivative** In Exercises 97–108, find the second derivative of the function.

97. $f(x) = x^2 + 7x - 4$

98. $f(x) = 4x^5 - 2x^3 + 5x^2$

99. $f(x) = 4x^{3/2}$

100. $f(x) = x^2 + 3x^{-3}$

101. $f(x) = \dfrac{x}{x-1}$

102. $f(x) = \dfrac{x^2 + 3x}{x - 4}$

103. $f(x) = x \sin x$

104. $f(x) = x \cos x$

105. $f(x) = \csc x$

106. $f(x) = \sec x$

107. $g(x) = \dfrac{e^x}{x}$

108. $h(t) = e^t \sin t$

 **Finding a Higher-Order Derivative** In Exercises 109–112, find the given higher-order derivative.

109. $f'(x) = x^3 - x^{2/5}$, $\quad f^{(3)}(x)$

110. $f^{(3)}(x) = \sqrt[5]{x^4}$, $\quad f^{(4)}(x)$

111. $f''(x) = -\sin x$, $\quad f^{(8)}(x)$

112. $f^{(4)}(t) = t \cos t$, $\quad f^{(5)}(t)$

Using Relationships In Exercises 113–116, use the given information to find $f'(2)$.

$g(2) = 3$ and $g'(2) = -2$

$h(2) = -1$ and $h'(2) = 4$

113. $f(x) = 2g(x) + h(x)$

114. $f(x) = 4 - h(x)$

115. $f(x) = \dfrac{g(x)}{h(x)}$

116. $f(x) = g(x)h(x)$

EXPLORING CONCEPTS

117. Higher-Order Derivatives Polynomials of what degree satisfy $f^{(n)} = 0$? Explain your reasoning.

118. Differentiation of Piecewise Functions Describe how you would differentiate a piecewise function. Use your approach to find the first and second derivatives of $f(x) = x|x|$. Explain why $f''(0)$ does not exist.

Identifying Graphs In Exercises 119 and 120, the graphs of f, f', and f'' are shown on the same set of coordinate axes. Identify each graph. Explain your reasoning. To print an enlarged copy of the graph, go to *MathGraphs.com*.

119.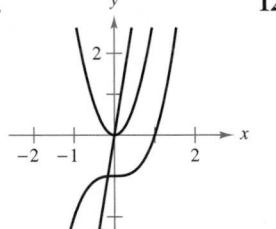

120.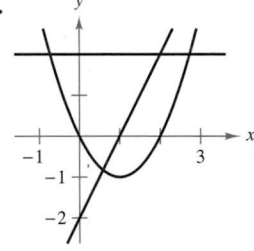

Sketching Graphs In Exercises 121 and 122, the graph of f is shown. Sketch the graphs of f' and f''. To print an enlarged copy of the graph, go to *MathGraphs.com*.

121.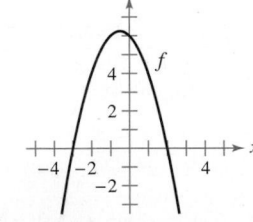

122.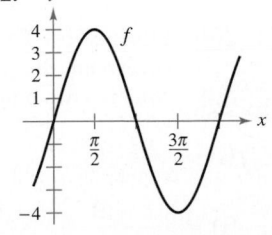

123. Sketching a Graph Sketch the graph of a differentiable function f such that $f(2) = 0$, $f' < 0$ for $-\infty < x < 2$, and $f' > 0$ for $2 < x < \infty$. Explain how you found your answer.

124. Sketching a Graph Sketch the graph of a differentiable function f such that $f > 0$ and $f' < 0$ for all real numbers x. Explain how you found your answer.

125. Acceleration The velocity of an object is

$$v(t) = 36 - t^2, \quad 0 \le t \le 6$$

where v is measured in meters per second and t is the time in seconds. Find the velocity and acceleration of the object when $t = 3$. What can be said about the speed of the object when the velocity and acceleration have opposite signs?

126. Acceleration The velocity of an automobile starting from rest is

$$v(t) = \frac{100t}{2t + 15}$$

where v is measured in feet per second and t is the time in seconds. Find the acceleration at (a) 5 seconds, (b) 10 seconds, and (c) 20 seconds.

127. Stopping Distance A car is traveling at a rate of 66 feet per second (45 miles per hour) when the brakes are applied. The position function for the car is

$$s(t) = -8.25t^2 + 66t$$

where s is measured in feet and t is measured in seconds. Use this function to complete the table and find the average velocity during each time interval.

t	0	1	2	3	4
$s(t)$					
$v(t)$					
$a(t)$					

128. HOW DO YOU SEE IT? The figure shows the graphs of the position, velocity, and acceleration functions of a particle.

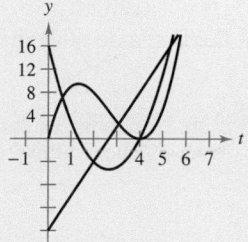

(a) Copy the graphs of the functions shown. Identify each graph. Explain your reasoning. To print an enlarged copy of the graph, go to *MathGraphs.com*.

(b) On your sketch, identify when the particle speeds up and when it slows down. Explain your reasoning.

Finding a Pattern In Exercises 129 and 130, develop a general rule for $f^{(n)}(x)$ given $f(x)$.

129. $f(x) = x^n$

130. $f(x) = \dfrac{1}{x}$

131. Finding a Pattern Consider the function $f(x) = g(x)h(x)$.

(a) Use the Product Rule to generate rules for finding $f''(x)$, $f'''(x)$, and $f^{(4)}(x)$.

(b) Use the results of part (a) to write a general rule for $f^{(n)}(x)$.

132. Finding a Pattern Develop a general rule for the nth derivative of $xf(x)$, where f is a differentiable function of x.

Finding a Pattern In Exercises 133 and 134, find the derivatives of the function f for $n = 1, 2, 3,$ and 4. Use the results to write a general rule for $f'(x)$ in terms of n.

133. $f(x) = x^n \sin x$

134. $f(x) = \dfrac{\cos x}{x^n}$

Differential Equations In Exercises 135–138, verify that the function satisfies the differential equation. (A *differential equation* in x and y is an equation that involves x, y, and derivatives of y.)

Function	Differential Equation
135. $y = \dfrac{1}{x}, x > 0$	$x^3y'' + 2x^2y' = 0$
136. $y = 2x^3 - 6x + 10$	$-y''' - xy'' - 2y' = -24x^2$
137. $y = 2 \sin x + 3$	$y'' + y = 3$
138. $y = 3 \cos x + \sin x$	$y'' + y = 0$

True or False? In Exercises 139–144, determine whether the statement is true or false. If it is false, explain why or give an example that shows it is false.

139. If $y = f(x)g(x)$, then $\dfrac{dy}{dx} = f'(x)g'(x)$.

140. If $y = (x + 1)(x + 2)(x + 3)(x + 4)$, then $\dfrac{d^5y}{dx^5} = 0$.

141. If $f'(c)$ and $g'(c)$ are zero and $h(x) = f(x)g(x)$, then $h'(c) = 0$.

142. If the position function of an object is linear, then its acceleration is zero.

143. The second derivative represents the rate of change of the first derivative.

144. The function $f(x) = \sin x + c$ satisfies $f^{(n)} = f^{(n+4)}$ for all integers $n \ge 1$.

145. Proof Use the Product Rule twice to prove that if f, g, and h are differentiable functions of x, then

$$\frac{d}{dx}[f(x)g(x)h(x)] = f'(x)g(x)h(x) + f(x)g'(x)h(x) + f(x)g(x)h'(x).$$

146. Think About It Let f and g be functions whose first and second derivatives exist on an interval I. Which of the following formulas is (are) true?

(a) $fg'' - f''g = (fg' - f'g)'$ 　　(b) $fg'' + f''g = (fg)''$

3.4 The Chain Rule

■ Find the derivative of a composite function using the Chain Rule.
■ Find the derivative of a function using the General Power Rule.
■ Simplify the derivative of a function using algebra.
■ Find the derivative of a transcendental function using the Chain Rule.
■ Find the derivative of a function involving the natural logarithmic function.
■ Define and differentiate exponential functions that have bases other than e.

The Chain Rule

This text has yet to discuss one of the most powerful differentiation rules—the **Chain Rule.** This rule deals with composite functions and adds a surprising versatility to the rules discussed in the two preceding sections. For example, compare the functions shown below. Those on the left can be differentiated without the Chain Rule, and those on the right are best differentiated with the Chain Rule.

Without the Chain Rule	**With the Chain Rule**
$y = x^2 + 1$	$y = \sqrt{x^2 + 1}$
$y = \sin x$	$y = \sin 6x$
$y = 3x + 2$	$y = (3x + 2)^5$
$y = e^x + \tan x$	$y = e^{5x} + \tan x^2$

Basically, the Chain Rule states that if y changes dy/du times as fast as u, and u changes du/dx times as fast as x, then y changes $(dy/du)(du/dx)$ times as fast as x.

EXAMPLE 1   **The Derivative of a Composite Function**

A set of gears is constructed so that the second and third gears are on the same axle (see Figure 3.24). As the first axle revolves, it drives the second axle, which in turn drives the third axle. Let y, u, and x represent the numbers of revolutions per minute of the first, second, and third axles, respectively. Find dy/du, du/dx, and dy/dx, and show that

$$\frac{dy}{dx} = \frac{dy}{du} \cdot \frac{du}{dx}.$$

Solution   Because the circumference of the second gear is three times that of the first, the first axle must make three revolutions to turn the second axle once. Similarly, the second axle must make two revolutions to turn the third axle once, and you can write

$$\frac{dy}{du} = 3 \quad \text{and} \quad \frac{du}{dx} = 2.$$

Combining these two results, you know that the first axle must make six revolutions to turn the third axle once. So, you can write

$$\frac{dy}{dx} = \boxed{\begin{array}{l}\text{Rate of change of first axle} \\ \text{with respect to second axle}\end{array}} \cdot \boxed{\begin{array}{l}\text{Rate of change of second axle} \\ \text{with respect to third axle}\end{array}}$$

$$= \frac{dy}{du} \cdot \frac{du}{dx}$$

$$= 3 \cdot 2$$

$$= 6$$

$$= \boxed{\begin{array}{l}\text{Rate of change of first axle} \\ \text{with respect to third axle}\end{array}}.$$

In other words, the rate of change of y with respect to x is the product of the rate of change of y with respect to u and the rate of change of u with respect to x. ■

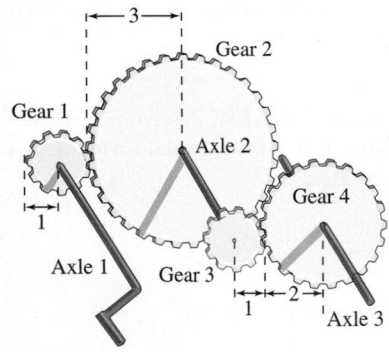

Axle 1: y revolutions per minute
Axle 2: u revolutions per minute
Axle 3: x revolutions per minute
Figure 3.24

Example 1 illustrates a simple case of the Chain Rule. The general rule is stated in the next theorem.

Exploration

Using the Chain Rule Each of the following functions can be differentiated using rules that you studied in Sections 3.2 and 3.3. For each function, find the derivative using those rules. Then find the derivative using the Chain Rule. Compare your results. Which method is simpler?

a. $y = \dfrac{2}{3x + 1}$

b. $y = (x + 2)^3$

c. $y = \sin 2x$

THEOREM 3.11   The Chain Rule

If $y = f(u)$ is a differentiable function of u and $u = g(x)$ is a differentiable function of x, then $y = f(g(x))$ is a differentiable function of x and

$$\frac{dy}{dx} = \frac{dy}{du} \cdot \frac{du}{dx}$$

or, equivalently,

$$\frac{d}{dx}[f(g(x))] = f'(g(x))g'(x).$$

Proof Let $h(x) = f(g(x))$. Then, using the alternative form of the derivative, you need to show that, for $x = c$,

$$h'(c) = f'(g(c))g'(c).$$

An important consideration in this proof is the behavior of g as x approaches c. A problem occurs when there are values of x, other than c, such that

$$g(x) = g(c).$$

Appendix A shows how to use the differentiability of f and g to overcome this problem. For now, assume that $g(x) \neq g(c)$ for values of x other than c. In the proofs of the Product Rule and the Quotient Rule, the same quantity was added and subtracted to obtain the desired form. This proof uses a similar technique—multiplying and dividing by the same (nonzero) quantity. Note that because g is differentiable, it is also continuous, and it follows that $g(x)$ approaches $g(c)$ as x approaches c.

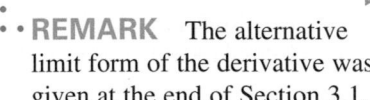

> **REMARK** The alternative limit form of the derivative was given at the end of Section 3.1.

$$
\begin{aligned}
h'(c) &= \lim_{x \to c} \frac{f(g(x)) - f(g(c))}{x - c} \qquad \text{Alternative form of derivative} \\[2mm]
&= \lim_{x \to c} \left[\frac{f(g(x)) - f(g(c))}{x - c} \cdot \frac{g(x) - g(c)}{g(x) - g(c)} \right], \ g(x) \neq g(c) \\[2mm]
&= \lim_{x \to c} \left[\frac{f(g(x)) - f(g(c))}{g(x) - g(c)} \cdot \frac{g(x) - g(c)}{x - c} \right] \\[2mm]
&= \left[\lim_{x \to c} \frac{f(g(x)) - f(g(c))}{g(x) - g(c)} \right] \left[\lim_{x \to c} \frac{g(x) - g(c)}{x - c} \right] \\[2mm]
&= f'(g(c))g'(c)
\end{aligned}
$$

When applying the Chain Rule, it is helpful to think of the composite function $f \circ g$ as having two parts—an inner part and an outer part.

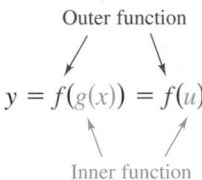

The derivative of $y = f(u)$ is the derivative of the outer function (at the inner function u) *times* the derivative of the inner function.

$$y' = f'(u) \cdot u'$$

EXAMPLE 2 **Decomposition of a Composite Function**

$y = f(g(x))$	$u = g(x)$	$y = f(u)$
a. $y = \dfrac{1}{x + 1}$	$u = x + 1$	$y = \dfrac{1}{u}$
b. $y = \sin 2x$	$u = 2x$	$y = \sin u$
c. $y = \sqrt{3x^2 - x + 1}$	$u = 3x^2 - x + 1$	$y = \sqrt{u}$
d. $y = \tan^2 x$	$u = \tan x$	$y = u^2$

EXAMPLE 3 **Using the Chain Rule**

Find dy/dx for

$$y = (x^2 + 1)^3.$$

Solution For this function, you can consider the inside function to be $u = x^2 + 1$ and the outer function to be $y = u^3$. By the Chain Rule, you obtain

$$\frac{dy}{dx} = \underbrace{3(x^2 + 1)^2}_{\frac{dy}{du}}\underbrace{(2x)}_{\frac{du}{dx}} = 6x(x^2 + 1)^2.$$

REMARK You could also solve the problem in Example 3 without using the Chain Rule by observing that

$$y = x^6 + 3x^4 + 3x^2 + 1$$

and

$$y' = 6x^5 + 12x^3 + 6x.$$

Verify that this is the same as the derivative in Example 3. Which method would you use to find

$$\frac{d}{dx}[(x^2 + 1)^{50}]?$$

The General Power Rule

The function in Example 3 is an example of one of the most common types of composite functions, $y = [u(x)]^n$. The rule for differentiating such functions is called the **General Power Rule,** and it is a special case of the Chain Rule.

THEOREM 3.12 **The General Power Rule**

If $y = [u(x)]^n$, where u is a differentiable function of x and n is a rational number, then

$$\frac{dy}{dx} = n[u(x)]^{n-1}\frac{du}{dx}$$

or, equivalently,

$$\frac{d}{dx}[u^n] = nu^{n-1}u'.$$

Proof Because $y = [u(x)]^n = u^n$, you apply the Chain Rule to obtain

$$\frac{dy}{dx} = \left(\frac{dy}{du}\right)\left(\frac{du}{dx}\right)$$

$$= \frac{d}{du}[u^n]\frac{du}{dx}.$$

By the (Simple) Power Rule in Section 3.2, you have $D_u[u^n] = nu^{n-1}$, and it follows that

$$\frac{dy}{dx} = nu^{n-1}\frac{du}{dx}.$$

EXAMPLE 4 **Applying the General Power Rule**

Find the derivative of $f(x) = (3x - 2x^2)^3$.

Solution Let $u = 3x - 2x^2$. Then

$$f(x) = (3x - 2x^2)^3 = u^3$$

and, by the General Power Rule, the derivative is

$$f'(x) = 3(3x - 2x^2)^2 \frac{d}{dx}[3x - 2x^2] \qquad \text{Apply General Power Rule.}$$

$$= 3(3x - 2x^2)^2(3 - 4x). \qquad \text{Differentiate } 3x - 2x^2.$$

EXAMPLE 5 **Differentiating Functions Involving Radicals**

Find all points on the graph of

$$f(x) = \sqrt[3]{(x^2 - 1)^2}$$

for which $f'(x) = 0$ and those for which $f'(x)$ does not exist.

Solution Begin by rewriting the function as

$$f(x) = (x^2 - 1)^{2/3}.$$

Then, applying the General Power Rule (with $u = x^2 - 1$) produces

$$f'(x) = \frac{2}{3}(x^2 - 1)^{-1/3}(2x) \qquad \text{Apply General Power Rule.}$$

$$= \frac{4x}{3\sqrt[3]{x^2 - 1}}. \qquad \text{Write in radical form.}$$

So, $f'(x) = 0$ when $x = 0$, and $f'(x)$ does not exist when $x = \pm1$, as shown in Figure 3.25.

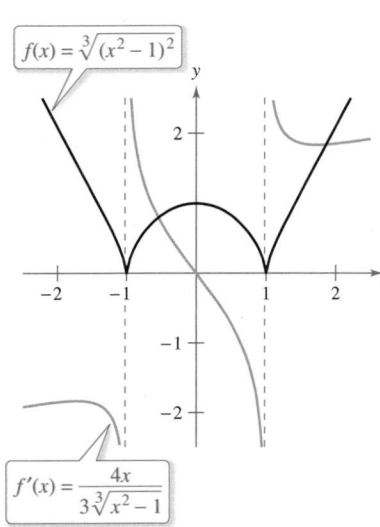

$f(x) = \sqrt[3]{(x^2 - 1)^2}$

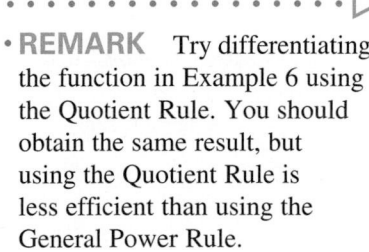

$f'(x) = \dfrac{4x}{3\sqrt[3]{x^2 - 1}}$

The derivative of f is 0 at $x = 0$ and is undefined at $x = \pm1$.
Figure 3.25

EXAMPLE 6 **Differentiating Quotients: Constant Numerators**

Differentiate the function

$$g(t) = \frac{-7}{(2t - 3)^2}.$$

Solution Begin by rewriting the function as

$$g(t) = -7(2t - 3)^{-2}.$$

Then, applying the General Power Rule (with $u = 2t - 3$) produces

$$g'(t) = (-7)(-2)(2t - 3)^{-3}(2) \qquad \text{Apply General Power Rule.}$$

$$\underbrace{}_{\substack{\text{Constant} \\ \text{Multiple Rule}}}$$

$$= 28(2t - 3)^{-3} \qquad \text{Simplify.}$$

$$= \frac{28}{(2t - 3)^3}. \qquad \text{Write with positive exponent.}$$

REMARK Try differentiating the function in Example 6 using the Quotient Rule. You should obtain the same result, but using the Quotient Rule is less efficient than using the General Power Rule.

Simplifying Derivatives

The next three examples demonstrate techniques for simplifying the "raw derivatives" of functions involving products, quotients, and composites.

EXAMPLE 7 **Simplifying by Factoring Out the Least Powers**

Find the derivative of $f(x) = x^2\sqrt{1 - x^2}$.

Solution

$$f(x) = x^2\sqrt{1 - x^2} \qquad \text{Write original function.}$$
$$= x^2(1 - x^2)^{1/2} \qquad \text{Rewrite.}$$
$$f'(x) = x^2 \frac{d}{dx}[(1 - x^2)^{1/2}] + (1 - x^2)^{1/2}\frac{d}{dx}[x^2] \qquad \text{Product Rule}$$
$$= x^2\left[\frac{1}{2}(1 - x^2)^{-1/2}(-2x)\right] + (1 - x^2)^{1/2}(2x) \qquad \text{General Power Rule}$$
$$= -x^3(1 - x^2)^{-1/2} + 2x(1 - x^2)^{1/2} \qquad \text{Simplify.}$$
$$= x(1 - x^2)^{-1/2}[-x^2(1) + 2(1 - x^2)] \qquad \text{Factor.}$$
$$= \frac{x(2 - 3x^2)}{\sqrt{1 - x^2}} \qquad \text{Simplify.}$$

EXAMPLE 8 **Simplifying the Derivative of a Quotient**

▷ **TECHNOLOGY** Symbolic
differentiation utilities are
capable of differentiating very
complicated functions. Often,
however, the result is given in
unsimplified form. If you have
access to such a utility, use it
to find the derivatives of the
functions given in Examples
7, 8, and 9. Then compare the
results with those given in
these examples.

$$f(x) = \frac{x}{\sqrt[3]{x^2 + 4}} \qquad \text{Original function}$$
$$= \frac{x}{(x^2 + 4)^{1/3}} \qquad \text{Rewrite.}$$
$$f'(x) = \frac{(x^2 + 4)^{1/3}(1) - x(1/3)(x^2 + 4)^{-2/3}(2x)}{(x^2 + 4)^{2/3}} \qquad \text{Quotient Rule}$$
$$= \frac{1}{3}(x^2 + 4)^{-2/3}\left[\frac{3(x^2 + 4) - (2x^2)(1)}{(x^2 + 4)^{2/3}}\right] \qquad \text{Factor.}$$
$$= \frac{x^2 + 12}{3(x^2 + 4)^{4/3}} \qquad \text{Simplify.}$$

EXAMPLE 9 **Simplifying the Derivative of a Power**

····▷ *See LarsonCalculus.com for an interactive version of this type of example.*

$$y = \left(\frac{3x - 1}{x^2 + 3}\right)^2 \qquad \text{Original function}$$

$$\overset{n}{\mid}\ \overbrace{\ }^{u^{n-1}}\ \overbrace{\ }^{u'}$$
$$y' = 2\left(\frac{3x - 1}{x^2 + 3}\right)\frac{d}{dx}\left[\frac{3x - 1}{x^2 + 3}\right] \qquad \text{General Power Rule}$$
$$= \left[\frac{2(3x - 1)}{x^2 + 3}\right]\left[\frac{(x^2 + 3)(3) - (3x - 1)(2x)}{(x^2 + 3)^2}\right] \qquad \text{Quotient Rule}$$
$$= \frac{2(3x - 1)(3x^2 + 9 - 6x^2 + 2x)}{(x^2 + 3)^3} \qquad \text{Multiply.}$$
$$= \frac{2(3x - 1)(-3x^2 + 2x + 9)}{(x^2 + 3)^3} \qquad \text{Simplify.}$$

Transcendental Functions and the Chain Rule

The "Chain Rule versions" of the derivatives of the six trigonometric functions and the natural exponential function are shown below.

$$\frac{d}{dx}[\sin u] = (\cos u)u' \qquad\qquad \frac{d}{dx}[\cos u] = -(\sin u)u'$$

$$\frac{d}{dx}[\tan u] = (\sec^2 u)u' \qquad\qquad \frac{d}{dx}[\cot u] = -(\csc^2 u)u'$$

$$\frac{d}{dx}[\sec u] = (\sec u \tan u)u' \qquad\qquad \frac{d}{dx}[\csc u] = -(\csc u \cot u)u'$$

$$\frac{d}{dx}[e^u] = e^u u'$$

• • **REMARK** Be sure you understand the mathematical conventions regarding parentheses and trigonometric functions. For instance, in Example 10(a), $\sin 2x$ is written to mean $\sin(2x)$.

$\triangleright$

EXAMPLE 10 The Chain Rule and Transcendental Functions

a. $y = \sin 2x$ $\qquad$ $\overbrace{}^{u}$ $\qquad$ $y' = \overbrace{\cos 2x}^{\cos u} \overbrace{\frac{d}{dx}[2x]}^{u'} = (\cos 2x)(2) = 2\cos 2x$

b. $y = \cos(x - 1)$ $\qquad$ $\overbrace{}^{u}$ $\qquad$ $y' = \overbrace{-\sin(x-1)}^{-(\sin u)} \overbrace{\frac{d}{dx}[x-1]}^{u'} = -\sin(x-1)$

c. $y = e^{3x}$ $\qquad$ $\overbrace{}^{u}$ $\qquad$ $y' = \overbrace{e^{3x}}^{e^u} \overbrace{\frac{d}{dx}[3x]}^{u'} = 3e^{3x}$

EXAMPLE 11 Parentheses and Trigonometric Functions

a. $y = \cos 3x^2 = \cos(3x^2)$ $\qquad$ $y' = (-\sin 3x^2)(6x) = -6x \sin 3x^2$

b. $y = (\cos 3)x^2$ $\qquad$ $y' = (\cos 3)(2x) = 2x \cos 3$

c. $y = \cos(3x)^2 = \cos(9x^2)$ $\qquad$ $y' = (-\sin 9x^2)(18x) = -18x \sin 9x^2$

d. $y = \cos^2 x = (\cos x)^2$ $\qquad$ $y' = 2(\cos x)(-\sin x) = -2\cos x \sin x$

e. $y = \sqrt{\cos x} = (\cos x)^{1/2}$ $\qquad$ $y' = \frac{1}{2}(\cos x)^{-1/2}(-\sin x) = -\frac{\sin x}{2\sqrt{\cos x}}$ ◼

To find the derivative of a function of the form $k(x) = f(g(h(x)))$, you need to apply the Chain Rule twice, as shown in Example 12.

EXAMPLE 12 Repeated Application of the Chain Rule

$$f(t) = \sin^3 4t \qquad\qquad \text{Original function}$$
$$\quad = (\sin 4t)^3 \qquad\qquad \text{Rewrite.}$$

$$f'(t) = 3(\sin 4t)^2 \frac{d}{dt}[\sin 4t] \qquad\qquad \text{Apply Chain Rule once.}$$

$$\quad = 3(\sin 4t)^2(\cos 4t)\frac{d}{dt}[4t] \qquad\qquad \text{Apply Chain Rule a second time.}$$

$$\quad = 3(\sin 4t)^2(\cos 4t)(4)$$

$$\quad = 12\sin^2 4t \cos 4t \qquad\qquad \text{Simplify.}$$

◼

The Derivative of the Natural Logarithmic Function

Up to this point in the text, derivatives of algebraic functions have been algebraic and derivatives of transcendental functions have been transcendental. The next theorem looks at an unusual situation in which the derivative of a transcendental function is algebraic. Specifically, the derivative of the natural logarithmic function is the algebraic function $1/x$.

THEOREM 3.13 **Derivative of the Natural Logarithmic Function**

Let u be a differentiable function of x.

1. $\dfrac{d}{dx}[\ln x] = \dfrac{1}{x}, \quad x > 0$

2. $\dfrac{d}{dx}[\ln u] = \dfrac{1}{u}\dfrac{du}{dx} = \dfrac{u'}{u}, \quad x > 0$

Proof To prove the first part, let $y = \ln x$, which implies that $e^y = x$. Differentiating both sides of this equation produces the following.

$$y = \ln x$$

$$e^y = x$$

$$\frac{d}{dx}[e^y] = \frac{d}{dx}[x]$$

$$e^y \frac{dy}{dx} = 1 \qquad\qquad \text{Chain Rule}$$

$$\frac{dy}{dx} = \frac{1}{e^y}$$

$$\frac{dy}{dx} = \frac{1}{x}$$

The second part of the theorem can be obtained by applying the Chain Rule to the first part. ∎

EXAMPLE 13   **Differentiation of Logarithmic Functions**

⋮····▷ *See LarsonCalculus.com for an interactive version of this type of example.*

a. $\dfrac{d}{dx}[\ln 2x] = \dfrac{u'}{u} = \dfrac{2}{2x} = \dfrac{1}{x}$ $\qquad\qquad u = 2x$

b. $\dfrac{d}{dx}[\ln(x^2 + 1)] = \dfrac{u'}{u} = \dfrac{2x}{x^2 + 1}$ $\qquad\qquad u = x^2 + 1$

c. $\dfrac{d}{dx}[x \ln x] = x\left(\dfrac{d}{dx}[\ln x]\right) + (\ln x)\left(\dfrac{d}{dx}[x]\right)$ $\qquad$ Product Rule

$\qquad\qquad\quad = x\left(\dfrac{1}{x}\right) + (\ln x)(1)$

$\qquad\qquad\quad = 1 + \ln x$

d. $\dfrac{d}{dx}[(\ln x)^3] = 3(\ln x)^2 \dfrac{d}{dx}[\ln x]$ $\qquad\qquad$ Chain Rule

$\qquad\qquad\quad = 3(\ln x)^2 \dfrac{1}{x}$

JOHN NAPIER (1550–1617)

Logarithms were invented by the Scottish mathematician John Napier. Although he did not introduce the *natural* logarithmic function, it is sometimes called the *Napierian* logarithm. *See LarsonCalculus.com to read more of this biography.*

John Napier used logarithmic properties to simplify *calculations* involving products, quotients, and powers. Of course, given the availability of calculators, there is now little need for this particular application of logarithms. However, there is great value in using logarithmic properties (see Section 1.6) to simplify *differentiation* involving products, quotients, and powers.

EXAMPLE 14   Logarithmic Properties as Aids to Differentiation

Differentiate $f(x) = \ln \sqrt{x + 1}$.

Solution   Because $\ln x^z = z \ln x$, you can rewrite the function as

$$f(x) = \ln \sqrt{x + 1} = \ln(x + 1)^{1/2} = \frac{1}{2} \ln(x + 1).$$ Rewrite before differentiating.

So, the first derivative is

$$f'(x) = \frac{1}{2}\left(\frac{1}{x + 1}\right) = \frac{1}{2(x + 1)}.$$ Differentiate.

EXAMPLE 15   Logarithmic Properties as Aids to Differentiation

Differentiate $f(x) = \ln \dfrac{x(x^2 + 1)^2}{\sqrt{2x^3 - 1}}$.

Solution   Notice how the logarithmic properties

$$\ln xy = \ln x + \ln y, \quad \ln \frac{x}{y} = \ln x - \ln y, \quad \text{and} \quad \ln x^z = z \ln x$$

are used to rewrite the function before differentiating.

$$f(x) = \ln \frac{x(x^2 + 1)^2}{\sqrt{2x^3 - 1}}$$ Write original function.

$$= \ln x + 2 \ln(x^2 + 1) - \frac{1}{2} \ln(2x^3 - 1)$$ Rewrite before differentiating.

$$f'(x) = \frac{1}{x} + 2\left(\frac{2x}{x^2 + 1}\right) - \frac{1}{2}\left(\frac{6x^2}{2x^3 - 1}\right)$$ Differentiate.

$$= \frac{1}{x} + \frac{4x}{x^2 + 1} - \frac{3x^2}{2x^3 - 1}$$ Simplify. ∎

········▷

··REMARK   In Examples 14 and 15, be sure that you see the benefit of applying logarithmic properties *before* differentiation. Consider, for instance, the difficulty of direct differentiation of the function given in Example 15.

Because the natural logarithm is undefined for negative numbers, you will often encounter expressions of the form $\ln|u|$. Theorem 3.14 states that you can differentiate functions of the form $y = \ln|u|$ as though the absolute value notation was not present.

THEOREM 3.14   Derivative Involving Absolute Value

If u is a differentiable function of x such that $u \neq 0$, then

$$\frac{d}{dx}[\ln|u|] = \frac{u'}{u}.$$

Proof   If $u > 0$, then $|u| = u$, and the result follows from Theorem 3.13. If $u < 0$, then $|u| = -u$, and you have

$$\frac{d}{dx}[\ln|u|] = \frac{d}{dx}[\ln(-u)] = \frac{-u'}{-u} = \frac{u'}{u}.$$ ∎

Bases Other than e

The **base** of the natural exponential function is e. This "natural" base can be used to assign a meaning to a general base a.

Definition of Exponential Function to Base *a*

If a is a positive real number ($a \neq 1$) and x is any real number, then the **exponential function to the base a** is denoted by a^x and is defined by

$$a^x = e^{(\ln a)x}.$$

If $a = 1$, then $y = 1^x = 1$ is a constant function.

Logarithmic functions to bases other than e can be defined in much the same way as exponential functions to other bases are defined.

Definition of Logarithmic Function to Base *a*

If a is a positive real number ($a \neq 1$) and x is any positive real number, then the **logarithmic function to the base a** is denoted by $\log_a x$ and is defined as

$$\log_a x = \frac{1}{\ln a} \ln x.$$

To differentiate exponential and logarithmic functions to other bases, you have two options: (1) use the definitions of a^x and $\log_a x$ and differentiate using the rules for the natural exponential and logarithmic functions, or (2) use the differentiation rules for bases other than e given in the next theorem.

▷

REMARK These differentiation rules are similar to those for the natural exponential function and the natural logarithmic function. In fact, they differ only by the constant factors $\ln a$ and $1/\ln a$. This points out one reason why, for calculus, e is the most convenient base.

THEOREM 3.15 Derivatives for Bases Other than *e*

Let a be a positive real number ($a \neq 1$) and let u be a differentiable function of x.

1. $\dfrac{d}{dx}[a^x] = (\ln a)a^x$ 　　2. $\dfrac{d}{dx}[a^u] = (\ln a)a^u \dfrac{du}{dx}$

3. $\dfrac{d}{dx}[\log_a x] = \dfrac{1}{(\ln a)x}$ 　　4. $\dfrac{d}{dx}[\log_a u] = \dfrac{1}{(\ln a)u}\dfrac{du}{dx}$

Proof By definition, $a^x = e^{(\ln a)x}$. Therefore, you can prove the first rule by letting

$$u = (\ln a)x$$

and differentiating with base e to obtain

$$\frac{d}{dx}[a^x] = \frac{d}{dx}[e^{(\ln a)x}] = e^u \frac{du}{dx} = e^{(\ln a)x}(\ln a) = (\ln a)a^x.$$

To prove the third rule, you can write

$$\frac{d}{dx}[\log_a x] = \frac{d}{dx}\left[\frac{1}{\ln a}\ln x\right] = \frac{1}{\ln a}\left(\frac{1}{x}\right) = \frac{1}{(\ln a)x}.$$

The second and fourth rules are simply the Chain Rule versions of the first and third rules. ∎

EXAMPLE 16 **Differentiating Functions to Other Bases**

a. $y' = \dfrac{d}{dx}[2^x] = (\ln 2)2^x$

b. $y' = \dfrac{d}{dx}[2^{3x}] = (\ln 2)2^{3x}(3) = (3 \ln 2)2^{3x}$

c. $y' = \dfrac{d}{dx}[\log_{10} \cos x] = \dfrac{-\sin x}{(\ln 10)\cos x} = -\dfrac{1}{\ln 10}\tan x$

REMARK Try writing 2^{3x} as 8^x and differentiating to see that you obtain the same result.

d. After rewriting the function below using logarithmic properties

$$y = \log_3 \frac{\sqrt{x}}{x + 5} = \frac{1}{2}\log_3 x - \log_3(x + 5)$$

you can apply Theorem 3.15 to find the derivative of the function.

$$y' = \frac{d}{dx}\left[\frac{1}{2}\log_3 x - \log_3(x + 5)\right] = \frac{1}{2(\ln 3)x} - \frac{1}{(\ln 3)(x + 5)} = \frac{5 - x}{2(\ln 3)x(x + 5)}$$

This section concludes with a summary of the differentiation rules studied so far. To become skilled at differentiation, you should memorize each rule in words, not symbols. As an aid to memorization, note that the cofunctions (cosine, cotangent, and cosecant) require a negative sign as part of their derivatives.

Summary of Differentiation Rules

General Differentiation Rules

Let c be a real number, let n be a rational number, let u and v be differentiable functions of x, let f be a differentiable function of u, and let a be a positive real number ($a \neq 1$).

Constant Rule:

$$\frac{d}{dx}[c] = 0$$

(Simple) Power Rule:

$$\frac{d}{dx}[x^n] = nx^{n-1}, \quad \frac{d}{dx}[x] = 1$$

Constant Multiple Rule:

$$\frac{d}{dx}[cu] = cu'$$

Sum or Difference Rule:

$$\frac{d}{dx}[u \pm v] = u' \pm v'$$

Product Rule:

$$\frac{d}{dx}[uv] = uv' + vu'$$

Quotient Rule:

$$\frac{d}{dx}\left[\frac{u}{v}\right] = \frac{vu' - uv'}{v^2}$$

Chain Rule:

$$\frac{d}{dx}[f(u)] = f'(u)u'$$

General Power Rule:

$$\frac{d}{dx}[u^n] = nu^{n-1}u'$$

Derivatives of Trigonometric Functions

$$\frac{d}{dx}[\sin x] = \cos x$$

$$\frac{d}{dx}[\cos x] = -\sin x$$

$$\frac{d}{dx}[\tan x] = \sec^2 x$$

$$\frac{d}{dx}[\cot x] = -\csc^2 x$$

$$\frac{d}{dx}[\sec x] = \sec x \tan x$$

$$\frac{d}{dx}[\csc x] = -\csc x \cot x$$

Derivatives of Exponential and Logarithmic Functions

$$\frac{d}{dx}[e^x] = e^x$$

$$\frac{d}{dx}[a^x] = (\ln a)a^x$$

$$\frac{d}{dx}[\ln x] = \frac{1}{x}$$

$$\frac{d}{dx}[\log_a x] = \frac{1}{(\ln a)x}$$

3.4 Exercises

See **CalcChat.com** for tutorial help and worked-out solutions to odd-numbered exercises.

CONCEPT CHECK

1. Chain Rule Describe the Chain Rule for the composition of two differentiable functions in your own words.

2. General Power Rule What is the difference between the (Simple) Power Rule and the General Power Rule?

3. Natural Logarithmic Functions Explain why the derivatives of $f(x) = \ln 2x$ and $g(x) = \ln 3x$ are the same.

4. Derivatives for Bases Other than e What are the values of a and b?

$$\frac{d}{dx}[6^{4x}] = a(\ln b)6^{4x}$$

 **Decomposition of a Composite Function** In Exercises 5–12, complete the table.

$y = f(g(x))$	$u = g(x)$	$y = f(u)$
5. $y = (6x - 5)^4$		
6. $y = \sqrt[3]{4x + 3}$		
7. $y = \dfrac{1}{3x + 5}$		
8. $y = \dfrac{2}{\sqrt{x^2 + 10}}$		
9. $y = \csc^3 x$		
10. $y = \sin\dfrac{5x}{2}$		
11. $y = e^{-2x}$		
12. $y = (\ln x)^3$		

 **Finding a Derivative** In Exercises 13–32, find the derivative of the function.

13. $y = (2x - 7)^3$

14. $y = 5(2 - x^3)^4$

15. $g(x) = 3(4 - 9x)^{5/6}$

16. $f(t) = (9t + 2)^{2/3}$

17. $h(s) = -2\sqrt{5s^2 + 3}$

18. $y = \sqrt[3]{6x^2 + 1}$

19. $y = \dfrac{1}{x - 2}$

20. $s(t) = \dfrac{1}{4 - 5t - t^2}$

21. $g(s) = \dfrac{6}{(s^3 - 2)^3}$

22. $y = -\dfrac{3}{(t - 2)^4}$

23. $y = \dfrac{1}{\sqrt{3x + 5}}$

24. $g(t) = \dfrac{1}{\sqrt{t^2 - 2}}$

25. $f(x) = x(2x - 5)^3$

26. $y = x^2\sqrt{16 - x^2}$

27. $y = \dfrac{x}{\sqrt{x^2 + 1}}$

28. $y = \dfrac{x}{\sqrt{x^4 + 4}}$

29. $g(x) = \left(\dfrac{x + 5}{x^2 + 2}\right)^2$

30. $g(x) = \left(\dfrac{3x^2 - 2}{2x + 3}\right)^{-2}$

31. $f(x) = [(x^2 + 3)^5 + x]^2$

32. $g(x) = [2 + (x^2 + 1)^4]^3$

 **Finding a Derivative of a Trigonometric Function** In Exercises 33–50, find the derivative of the trigonometric function.

33. $y = \cos 4x$

34. $y = \sin \pi x$

35. $g(x) = 5 \tan 3x$

36. $h(x) = \sec 6x$

37. $y = \sin(\pi x)^2$

38. $y = \csc(1 - 2x)^2$

39. $h(x) = \sin 2x \cos 2x$

40. $g(\theta) = \sec(\tfrac{1}{2}\theta) \tan(\tfrac{1}{2}\theta)$

41. $f(x) = \dfrac{\cot x}{\sin x}$

42. $g(v) = \dfrac{\cos v}{\csc v}$

43. $f(\theta) = \tfrac{1}{4}\sin^2 2\theta$

44. $h(t) = 2 \cot^2(\pi t + 2)$

45. $f(t) = 3 \sec(\pi t - 1)^2$

46. $y = 5 \cos(\pi x)^2$

47. $y = \sin(3x^2 + \cos x)$

48. $y = \cos(5x + \csc x)$

49. $y = \sin\sqrt{\cot 3\pi x}$

50. $y = \cos\sqrt{\sin(\tan \pi x)}$

Finding a Derivative Using Technology In Exercises 51–56, use a computer algebra system to find the derivative of the function. Then use the utility to graph the function and its derivative on the same set of coordinate axes. Describe the behavior of the function that corresponds to any zeros of the graph of the derivative.

51. $y = \dfrac{\sqrt{x} + 1}{x^2 + 1}$

52. $y = \sqrt{\dfrac{2x}{x + 1}}$

53. $y = \sqrt{\dfrac{x + 1}{x}}$

54. $g(x) = \sqrt{x - 1} + \sqrt{x + 1}$

55. $y = \dfrac{\cos \pi x + 1}{x}$

56. $y = x^2 \tan \dfrac{1}{x}$

 **Finding a Derivative of an Exponential Function** In Exercises 57–72, find the derivative of the exponential function.

57. $y = e^{5x}$

58. $y = e^{-x^2}$

59. $y = e^{\sqrt{x}}$

60. $y = x^2 e^{-x}$

61. $g(t) = (e^{-t} + e^t)^3$

62. $g(t) = e^{-3/t^2}$

63. $y = x^2 e^x - 2xe^x + 2e^x$

64. $y = xe^x - e^x$

65. $y = \dfrac{2}{e^x + e^{-x}}$

66. $y = \dfrac{e^x - e^{-x}}{2}$

67. $y = \dfrac{e^x + 1}{e^x - 1}$

68. $y = \dfrac{e^{2x}}{e^{2x} + 1}$

69. $y = e^x(\sin x + \cos x)$

70. $y = e^{2x} \tan 2x$

71. $g(x) = e^{\csc x}$

72. $f(x) = e^{x + \sec x}$

Finding a Derivative of a Logarithmic Function In Exercises 73–94, find the derivative of the logarithmic function.

73. $f(x) = \ln(x^2 + 3)$

74. $h(x) = \ln(2x^2 + 1)$

75. $y = (\ln x)^4$

76. $y = x^2 \ln x$

77. $y = \ln(t + 1)^2$

78. $y = \ln\sqrt{x^2 - 4}$

79. $y = \ln(x\sqrt{x^2 - 1})$

80. $y = \ln[t(t^2 + 3)^3]$

81. $f(x) = \ln\dfrac{x}{x^2 + 1}$

82. $f(x) = \ln\dfrac{2x}{x + 3}$

83. $g(t) = \dfrac{\ln t}{t^2}$

84. $h(t) = \dfrac{\ln t}{t^3 + 5}$

85. $y = \ln(\ln x^2)$

86. $y = \ln(\ln x)$

87. $y = \ln\sqrt{\dfrac{x + 1}{x - 1}}$

88. $y = \ln\sqrt[3]{\dfrac{x - 1}{x + 1}}$

89. $y - \ln|\sin x|$

90. $y = \ln|\csc x|$

91. $y = \ln\left|\dfrac{\cos x}{\cos x - 1}\right|$

92. $y = \ln|\sec x + \tan x|$

93. $f(x) = \ln(1 + e^{-3x})$

94. $y = \ln\left(\dfrac{1 + e^x}{1 - e^x}\right)$

Slope of a Tangent Line   In Exercises 95 and 96, find the slope of the tangent line to the sine function at the origin. Compare this value with the number of complete cycles in the interval $[0, 2\pi]$.

95.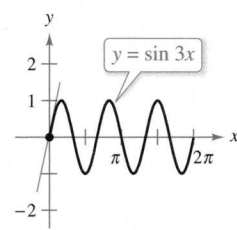
$y = \sin 3x$

96.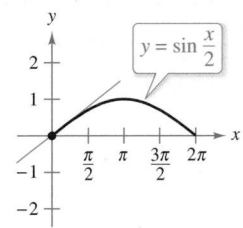
$y = \sin \dfrac{x}{2}$

Slope of a Tangent Line   In Exercises 97–100, find the slope of the tangent line to the graph of the function at the given point.

97. $y = e^{3x}$
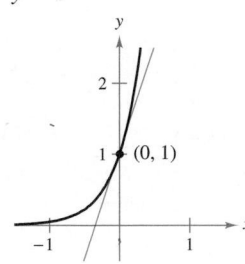
$(0, 1)$

98. $y = e^{-3x}$
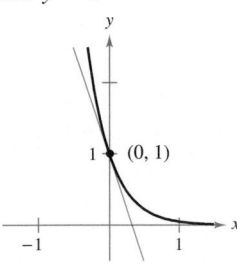
$(0, 1)$

99. $y = \ln x^3$
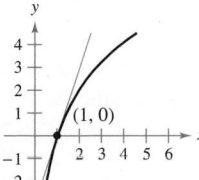
$(1, 0)$

100. $y = \ln x^{3/2}$
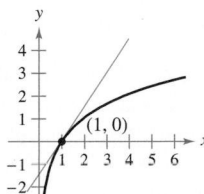
$(1, 0)$

Finding the Slope of a Graph   In Exercises 101–108, find the slope of the graph of the function at the given point. Use the *derivative* feature of a graphing utility to confirm your results.

101. $y = \sqrt{x^2 + 8x}, \quad (1, 3)$

102. $y = \sqrt[5]{3x^3 + 4x}, \quad (2, 2)$

103. $f(x) = \dfrac{5}{x^3 - 2}, \quad \left(-2, -\dfrac{1}{2}\right)$

104. $f(x) = \dfrac{1}{(x^2 - 3x)^2}, \quad \left(4, \dfrac{1}{16}\right)$

105. $y = \dfrac{4}{(x + 2)^2}, \quad (0, 1)$

106. $y = \dfrac{4}{(x^2 - 2x)^3}, \quad (1, -4)$

107. $y = 26 - \sec^3 4x, \quad (0, 25)$

108. $y = \dfrac{1}{x} + \sqrt{\cos x}, \quad \left(\dfrac{\pi}{2}, \dfrac{2}{\pi}\right)$

 **Finding an Equation of a Tangent Line   In Exercises 109–116, (a) find an equation of the tangent line to the graph of the function at the given point, (b) use a graphing utility to graph the function and its tangent line at the point, and (c) use the *tangent* feature of a graphing utility to confirm your results.**

109. $y = (4x^3 + 3)^2, \quad (-1, 1)$ **110.** $f(x) = (9 - x^2)^{2/3}, \quad (1, 4)$

111. $f(x) = \sin 8x, \quad (\pi, 0)$ **112.** $y = \cos 3x, \quad \left(\dfrac{\pi}{4}, -\dfrac{\sqrt{2}}{2}\right)$

113. $f(x) = \tan^2 x, \quad \left(\dfrac{\pi}{4}, 1\right)$ **114.** $y = 2\tan^3 x, \quad \left(\dfrac{\pi}{4}, 2\right)$

115. $y = 4 - x^2 - \ln(\tfrac{1}{2}x + 1), \quad (0, 4)$

116. $y = 2e^{1-x^2}, \quad (1, 2)$

Famous Curves   In Exercises 117 and 118, find an equation of the tangent line to the graph at the given point. Then use a graphing utility to graph the function and its tangent line at the point in the same viewing window.

117. Semicircle
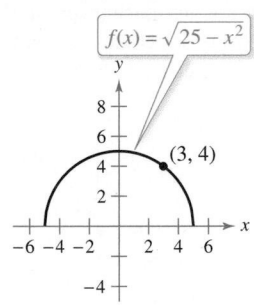
$f(x) = \sqrt{25 - x^2}$
$(3, 4)$

118. Bullet-nose curve
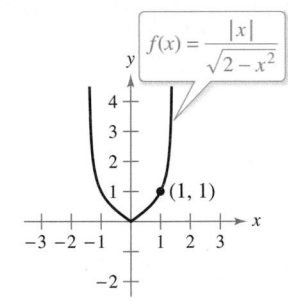
$f(x) = \dfrac{|x|}{\sqrt{2 - x^2}}$
$(1, 1)$

119. Horizontal Tangent Line   Determine the point(s) in the interval $(0, 2\pi)$ at which the graph of $f(x) = 2\cos x + \sin 2x$ has a horizontal tangent.

120. Horizontal Tangent Line   Determine the point(s) at which the graph of

$$f(x) = \dfrac{-4x}{\sqrt{2x - 1}}$$

has a horizontal tangent.

Finding a Second Derivative   In Exercises 121–128, find the second derivative of the function.

121. $f(x) = 5(2 - 7x)^4$ **122.** $f(x) = 6(x^3 + 4)^3$

123. $f(x) = \dfrac{1}{11x - 6}$ **124.** $f(x) = \dfrac{8}{(x - 2)^2}$

125. $f(x) = \sin x^2$ **126.** $f(x) = \sec^2 \pi x$

127. $f(x) = (3 + 2x)e^{-3x}$ **128.** $g(x) = \sqrt{x} + e^x \ln x$

Evaluating a Second Derivative In Exercises 129–132, evaluate the second derivative of the function at the given point. Use a computer algebra system to verify your result.

129. $h(x) = \frac{1}{9}(3x + 1)^3$, $\left(1, \frac{64}{9}\right)$ **130.** $f(x) = \frac{1}{\sqrt{x + 4}}$, $\left(0, \frac{1}{2}\right)$

131. $f(x) = \cos x^2$, $(0, 1)$ **132.** $g(t) = \tan 2t$, $\left(\frac{\pi}{6}, \sqrt{3}\right)$

 **Finding a Derivative** In Exercises 133–148, find the derivative of the function.

133. $f(x) = 4^x$

134. $g(x) = 5^{-x}$

135. $g(t) = t^2 2^t$

136. $y = x(6^{-2x})$

137. $f(t) = \frac{-2t^2}{8^t}$

138. $f(t) = \frac{3^{2t}}{t}$

139. $h(\theta) = 2^{-\theta} \cos \pi\theta$

140. $g(\alpha) = 5^{-\alpha/2} \sin 2\alpha$

141. $y = \log_3 x$

142. $y = \log_{10} 2x$

143. $y = \log_5 \sqrt{x^2 - 1}$

144. $f(x) = \log_2 \sqrt[3]{2x + 1}$

145. $f(x) = \log_2 \frac{x^2}{x - 1}$

146. $h(x) = \log_3 \frac{x\sqrt{x - 1}}{2}$

147. $g(t) = \frac{10 \log_4 t}{t}$

148. $f(t) = t^{3/2} \log_2 \sqrt{t + 1}$

EXPLORING CONCEPTS

Identifying Graphs In Exercises 149 and 150, the graphs of a function f and its derivative f' are shown. Label the graphs as f or f' and write a short paragraph stating the criteria you used in making your selection. To print an enlarged copy of the graph, go to *MathGraphs.com.*

149.
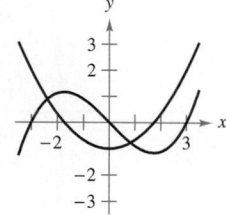

150.
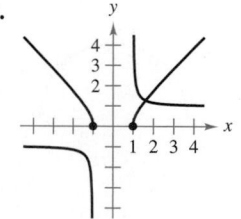

151. Describing Relationships Describe the relationship between f' and g' when (a) $g(x) = f(3x)$ and (b) $g(x) = f(x^2)$.

152. Comparing Methods Consider the function

$$r(x) = \frac{2x - 5}{(3e^x + 1)^2}.$$

(a) In general, how do you find the derivative of

$h(x) = \frac{f(x)}{g(x)}$ using the Product Rule, where g is a

composite function?

(b) Find $r'(x)$ using the Product Rule.

(c) Find $r'(x)$ using the Quotient Rule.

(d) Which method do you prefer? Explain.

---

153. Think About It The table shows some values of the derivative of an unknown function f. Complete the table by finding the derivative of each transformation of f, if possible.

(a) $g(x) = f(x) - 2$ (b) $h(x) = 2f(x)$

(c) $r(x) = f(-3x)$ (d) $s(x) = f(x + 2)$

x	-2	-1	0	1	2	3
$f'(x)$	4	$\frac{2}{3}$	$-\frac{1}{3}$	-1	-2	-4
$g'(x)$						
$h'(x)$						
$r'(x)$						
$s'(x)$						

154. Using Relationships Given that $g(5) = -3$, $g'(5) = 6$, $h(5) = 3$, and $h'(5) = -2$, find $f'(5)$ for each of the following, if possible. If it is not possible, state what additional information is required.

(a) $f(x) = g(x)h(x)$ (b) $f(x) = g(h(x))$

(c) $f(x) = \frac{g(x)}{h(x)}$ (d) $f(x) = [g(x)]^3$

Finding Derivatives In Exercises 155 and 156, the graphs of f and g are shown. Let $h(x) = f(g(x))$ and $s(x) = g(f(x))$. Find each derivative, if it exists. If the derivative does not exist, explain why.

155. (a) Find $h'(1)$.

(b) Find $s'(5)$.

156. (a) Find $h'(3)$.

(b) Find $s'(9)$.

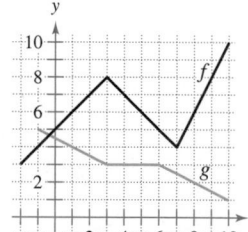

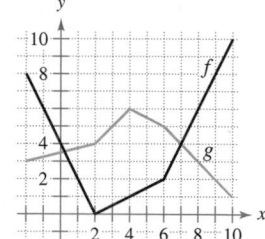

157. Pendulum A 15-centimeter pendulum moves according to the equation

$$\theta = 0.2 \cos 8t$$

where θ is the angular displacement from the vertical in radians and t is the time in seconds. Determine the maximum angular displacement and the rate of change of θ when $t = 3$ seconds.

158. Harmonic Motion The displacement from equilibrium of an object in harmonic motion on the end of a spring is

$$y = \frac{1}{3} \cos 12t - \frac{1}{4} \sin 12t$$

where y is measured in feet and t is the time in seconds. Determine the position and velocity of the object when $t = \pi/8$.

159. Doppler Effect The frequency F of a fire truck siren heard by a stationary observer is

$$F = \frac{132,400}{331 \pm v}$$

where $\pm v$ represents the velocity of the accelerating fire truck in meters per second (see figure). Find the rate of change of F with respect to v when

(a) the fire truck is approaching at a velocity of 30 meters per second (use $-v$).

(b) the fire truck is moving away at a velocity of 30 meters per second (use $+v$).

$$F = \frac{132,400}{331 + v} \qquad\qquad F = \frac{132,400}{331 - v}$$

160. **HOW DO YOU SEE IT?** The cost C (in dollars) of producing x units of a product is $C = 60x + 1350$. For one week, management determined that the number of units produced x at the end of t hours can be modeled by $x = -1.6t^3 + 19t^2 - 0.5t - 1$. The graph shows the cost C in terms of the time t.

Cost of Producing a Product

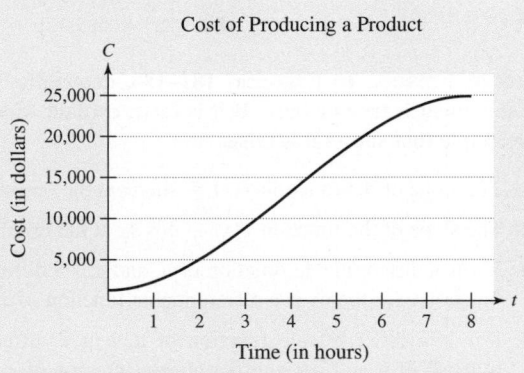

Time (in hours)

(a) Using the graph, which is greater, the rate of change of the cost after 1 hour or the rate of change of the cost after 4 hours?

(b) Explain why the cost function is not increasing at a constant rate during the eight-hour shift.

161. Inflation When the annual rate of inflation averages 5% over the next 10 years, the approximate cost C of goods or services during any year in that decade is $C(t) = P(1.05)^t$, where t is the time in years and P is the present cost.

(a) The price of an oil change for your car is presently $29.95. Estimate the price 10 years from now.

(b) Find the rates of change of C with respect to t when $t = 1$ and $t = 8$.

(c) Verify that the rate of change of C is proportional to C. What is the constant of proportionality?

162. Wave Motion A buoy oscillates in simple harmonic motion $y = A \cos \omega t$ as waves move past it. The buoy moves a total of 3.5 feet (vertically) from its low point to its high point. It returns to its high point every 10 seconds.

(a) Write an equation describing the motion of the buoy if it is at its high point at $t = 0$.

(b) Determine the velocity of the buoy as a function of t.

163. Modeling Data The table shows the temperatures T (in degrees Fahrenheit) at which water boils at selected pressures p (in pounds per square inch). (*Source: Standard Handbook of Mechanical Engineers*)

p	5	10	14.696 (1 atm)	20
T	162.24	193.21	212.00	227.96

p	30	40	60	80	100
T	250.33	267.25	292.71	312.03	327.81

A model that approximates the data is

$$T = 87.97 + 34.96 \ln p + 7.91 \sqrt{p}.$$

(a) Use a graphing utility to plot the data and graph the model.

(b) Find the rates of change of T with respect to p when $p = 10$ and $p = 70$.

(c) Use a graphing utility to graph T'. Find $\lim\limits_{p \to \infty} T'(p)$ and interpret the result in the context of the problem.

164. Modeling Data The normal daily maximum temperatures T (in degrees Fahrenheit) for Chicago, Illinois, are shown in the table. (*Source: National Oceanic and Atmospheric Administration*)

Month	Jan	Feb	Mar	Apr
Temperature	31.0	35.3	46.6	59.0

Month	May	Jun	Jul	Aug
Temperature	70.0	79.7	84.1	81.9

Month	Sep	Oct	Nov	Dec
Temperature	74.8	62.3	48.2	34.8

(a) Use a graphing utility to plot the data and find a model for the data of the form

$$T(t) = a + b \sin(ct - d)$$

where T is the temperature and t is the time in months, with $t = 1$ corresponding to January.

(b) Use a graphing utility to graph the model. How well does the model fit the data?

(c) Find T' and use a graphing utility to graph T'.

(d) Based on the graph of T', during what times does the temperature change most rapidly? Most slowly? Do your answers agree with your observations of the temperature changes? Explain.

165. Biology

The number N of bacteria in a culture after t days is modeled by

$$N = 400\left[1 - \frac{3}{(t^2 + 2)^2}\right].$$

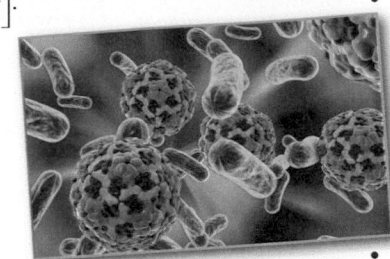

Find the rate of change of N with respect to t when (a) $t = 0$, (b) $t = 1$, (c) $t = 2$, (d) $t = 3$, and (e) $t = 4$. (f) What can you conclude?

166. Depreciation The value V of a machine t years after it is purchased is inversely proportional to the square root of $t + 1$. The initial value of the machine is \$10,000.

(a) Write V as a function of t.

(b) Find the rate of depreciation when $t = 1$.

(c) Find the rate of depreciation when $t = 3$.

167. Finding a Pattern Consider the function $f(x) = \sin \beta x$, where β is a constant.

(a) Find the first-, second-, third-, and fourth-order derivatives of the function.

(b) Verify that the function and its second derivative satisfy the equation $f''(x) + \beta^2 f(x) = 0$.

(c) Use the results of part (a) to write general rules for the even- and odd-order derivatives $f^{(2k)}(x)$ and $f^{(2k-1)}(x)$.

[*Hint:* $(-1)^k$ is positive if k is even and negative if k is odd.]

168. Conjecture Let f be a differentiable function of period p.

(a) Is the function f' periodic? Verify your answer.

(b) Consider the function $g(x) = f(2x)$. Is the function $g'(x)$ periodic? Verify your answer.

169. Think About It Let $r(x) = f(g(x))$ and $s(x) = g(f(x))$, where f and g are shown in the figure. Find (a) $r'(1)$ and (b) $s'(4)$.

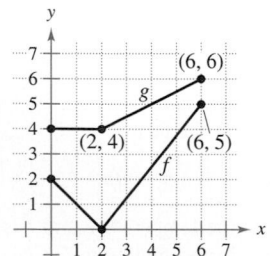

170. Using Trigonometric Functions

(a) Find the derivative of the function $g(x) = \sin^2 x + \cos^2 x$ in two ways.

(b) For $f(x) = \sec^2 x$ and $g(x) = \tan^2 x$, show that

$$f'(x) = g'(x).$$

171. Even and Odd Functions

(a) Show that the derivative of an odd function is even. That is, if $f(-x) = -f(x)$, then $f'(-x) = f'(x)$.

(b) Show that the derivative of an even function is odd. That is, if $f(-x) = f(x)$, then $f'(-x) = -f'(x)$.

172. Proof Let u be a differentiable function of x. Use the fact that $|u| = \sqrt{u^2}$ to prove that

$$\frac{d}{dx}[|u|] = u'\frac{u}{|u|}, \quad u \neq 0.$$

Using Absolute Value In Exercises 173–176, use the result of Exercise 172 to find the derivative of the function.

173. $g(x) = |3x - 5|$ **174.** $f(x) = |x^2 - 9|$

175. $h(x) = |x| \cos x$ **176.** $f(x) = |\sin x|$

Linear and Quadratic Approximations The linear and quadratic approximations of a function f at $x = a$ are

$$P_1(x) = f'(a)(x - a) + f(a) \quad \text{and}$$
$$P_2(x) = \tfrac{1}{2}f''(a)(x - a)^2 + f'(a)(x - a) + f(a).$$

In Exercises 177–180, (a) find the specified linear and quadratic approximations of f, (b) use a graphing utility to graph f and the approximations, (c) determine whether P_1 or P_2 is the better approximation, and (d) state how the accuracy changes as you move farther from $x = a$.

177. $f(x) = \tan x; \quad a = \dfrac{\pi}{4}$ **178.** $f(x) = \sec x; \quad a = \dfrac{\pi}{6}$

179. $f(x) = e^x; \quad a = 0$ **180.** $f(x) = \ln x; \quad a = 1$

True or False? In Exercises 181–184, determine whether the statement is true or false. If it is false, explain why or give an example that shows it is false.

181. The slope of the function $f(x) = \sin ax$ at the origin is a.

182. The slope of the function $f(x) = \cos bx$ at the origin is $-b$.

183. If y is a differentiable function of u, and u is a differentiable function of x, then y is a differentiable function of x.

184. If y is a differentiable function of u, u is a differentiable function of v, and v is a differentiable function of x, then

$$\frac{dy}{dx} = \frac{dy}{du}\frac{du}{dv}\frac{dv}{dx}.$$

PUTNAM EXAM CHALLENGE

185. Let $f(x) = a_1 \sin x + a_2 \sin 2x + \cdots + a_n \sin nx$, where $a_1, a_2, \ldots, a_n$ are real numbers and where n is a positive integer. Given that $|f(x)| \leq |\sin x|$ for all real x, prove that $|a_1 + 2a_2 + \cdots + na_n| \leq 1$.

186. Let k be a fixed positive integer. The nth derivative of $\dfrac{1}{x^k - 1}$ has the form $\dfrac{P_n(x)}{(x^k - 1)^{n+1}}$ where $P_n(x)$ is a polynomial. Find $P_n(1)$.

3.5   Implicit Differentiation

- Distinguish between functions written in implicit form and explicit form.
- Use implicit differentiation to find the derivative of a function.
- Find derivatives of functions using logarithmic differentiation.

Implicit and Explicit Functions

Up to this point in the text, most functions have been expressed in **explicit form.** For example, in the equation $y = 3x^2 - 5$, the variable y is explicitly written as a function of x. Some functions, however, are only implied by an equation. For instance, the function $y = 1/x$ is defined **implicitly** by the equation

$$xy = 1. \qquad \text{Implicit form}$$

To find dy/dx for this equation, you can write y explicitly as a function of x and then differentiate.

Implicit Form	**Explicit Form**	**Derivative**
$xy = 1$	$y = \dfrac{1}{x} = x^{-1}$	$\dfrac{dy}{dx} = -x^{-2} = -\dfrac{1}{x^2}$

This strategy works whenever you can solve for the function explicitly. You cannot, however, use this procedure when you are unable to solve for y as a function of x. For instance, how would you find dy/dx for the equation

$$x^2 - 2y^3 + 4y = 2?$$

For this equation, it is difficult to express y as a function of x explicitly. To find dy/dx, you can use **implicit differentiation.**

To understand how to find dy/dx implicitly, you must realize that the differentiation is taking place *with respect to x*. This means that when you differentiate terms involving x alone, you can differentiate as usual. However, when you differentiate terms involving y, you must apply the Chain Rule, because you are assuming that y is defined implicitly as a differentiable function of x.

EXAMPLE 1   Differentiating with Respect to x

a. $\dfrac{d}{dx}[x^3] = 3x^2$ Variables agree: use Simple Power Rule.

Variables agree

b. $\dfrac{d}{dx}[y^3] = 3y^2 \dfrac{dy}{dx}$ Variables disagree: use Chain Rule.

$\overbrace{\quad}^{u^n} \quad \overbrace{\quad}^{nu^{n-1}} \quad \overbrace{\quad}^{u'}$

Variables disagree

c. $\dfrac{d}{dx}[x + 3y] = 1 + 3\dfrac{dy}{dx}$ Chain Rule: $\dfrac{d}{dx}[3y] = 3y'$

d. $\dfrac{d}{dx}[xy^2] = x\dfrac{d}{dx}[y^2] + y^2\dfrac{d}{dx}[x]$ Product Rule

$\qquad\qquad = x\left(2y\dfrac{dy}{dx}\right) + y^2(1)$ Chain Rule

$\qquad\qquad = 2xy\dfrac{dy}{dx} + y^2$ Simplify.

Implicit Differentiation

GUIDELINES FOR IMPLICIT DIFFERENTIATION

1. Differentiate both sides of the equation *with respect to x*.
2. Collect all terms involving dy/dx on the left side of the equation and move all other terms to the right side of the equation.
3. Factor dy/dx out of the left side of the equation.
4. Solve for dy/dx.

In Example 2, note that implicit differentiation can produce an expression for dy/dx that contains both x and y.

EXAMPLE 2   Implicit Differentiation

Find dy/dx given that $y^3 + y^2 - 5y - x^2 = -4$.

Solution

1. Differentiate both sides of the equation with respect to x.

$$\frac{d}{dx}[y^3 + y^2 - 5y - x^2] = \frac{d}{dx}[-4]$$

$$\frac{d}{dx}[y^3] + \frac{d}{dx}[y^2] - \frac{d}{dx}[5y] - \frac{d}{dx}[x^2] = \frac{d}{dx}[-4]$$

$$3y^2\frac{dy}{dx} + 2y\frac{dy}{dx} - 5\frac{dy}{dx} - 2x = 0$$

2. Collect the dy/dx terms on the left side of the equation and move all other terms to the right side of the equation.

$$3y^2\frac{dy}{dx} + 2y\frac{dy}{dx} - 5\frac{dy}{dx} = 2x$$

3. Factor dy/dx out of the left side of the equation.

$$\frac{dy}{dx}(3y^2 + 2y - 5) = 2x$$

4. Solve for dy/dx by dividing by $(3y^2 + 2y - 5)$.

$$\frac{dy}{dx} = \frac{2x}{3y^2 + 2y - 5}$$

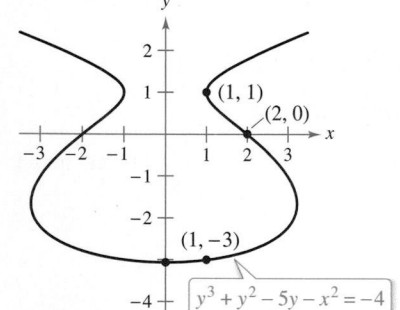

Point on Graph	Slope of Graph
$(2, 0)$	$-\frac{4}{5}$
$(1, -3)$	$\frac{1}{8}$
$x = 0$	0
$(1, 1)$	Undefined

The implicit equation

$$y^3 + y^2 - 5y - x^2 = -4$$

has the derivative

$$\frac{dy}{dx} = \frac{2x}{3y^2 + 2y - 5}.$$

Figure 3.26

To see how you can use an *implicit derivative*, consider the graph shown in Figure 3.26. From the graph, you can see that y is not a function of x. Even so, the derivative found in Example 2 gives a formula for the slope of the tangent line at a point on this graph. The slopes at several points on the graph are shown below the graph.

▷ **TECHNOLOGY** With most graphing utilities, it is easy to graph an equation that explicitly represents y as a function of x. Graphing other equations, however, can require some ingenuity. For instance, to graph the equation given in Example 2, use a graphing utility, set in *parametric* mode, to graph the parametric representations $x = \sqrt{t^3 + t^2 - 5t + 4}$, $y = t$, and $x = -\sqrt{t^3 + t^2 - 5t + 4}$, $y = t$, for $-5 \leq t \leq 5$. How does the result compare with the graph shown in Figure 3.26? (You will learn more about this type of representation when you study parametric equations in Section 10.2.)

It is meaningless to solve for dy/dx in an equation that has no solution points. (For example, $x^2 + y^2 = -4$ has no solution points.) If, however, a segment of a graph can be represented by a differentiable function, then dy/dx will have meaning as the slope at each point on the segment. Recall that a function is not differentiable at (a) points with vertical tangents and (b) points at which the function is not continuous.

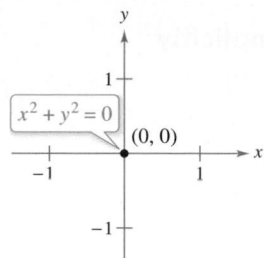

(a)

EXAMPLE 3   Graphs and Differentiable Functions

If possible, represent y as a differentiable function of x.

a. $x^2 + y^2 = 0$ **b.** $x^2 + y^2 = 1$ **c.** $x + y^2 = 1$

Solution

a. The graph of this equation is a single point. So, it does not define y as a differentiable function of x. See Figure 3.27(a).

b. The graph of this equation is the unit circle centered at $(0, 0)$. The upper semicircle is given by the differentiable function

$$y = \sqrt{1 - x^2}, \quad -1 < x < 1$$

and the lower semicircle is given by the differentiable function

$$y = -\sqrt{1 - x^2}, \quad -1 < x < 1.$$

At the points $(-1, 0)$ and $(1, 0)$, the slope of the graph is undefined. See Figure 3.27(b).

c. The upper half of this parabola is given by the differentiable function

$$y = \sqrt{1 - x}, \quad x < 1$$

and the lower half of this parabola is given by the differentiable function

$$y = -\sqrt{1 - x}, \quad x < 1.$$

At the point $(1, 0)$, the slope of the graph is undefined. See Figure 3.27(c).

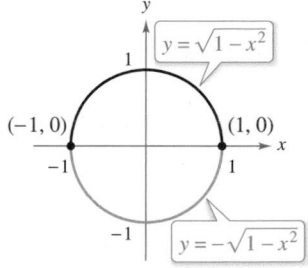

(b)

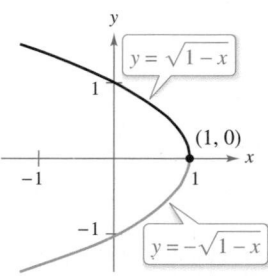

(c)

Some graph segments can be represented by differentiable functions.

Figure 3.27

EXAMPLE 4   Finding the Slope of a Graph Implicitly

$\cdots \triangleright$ *See LarsonCalculus.com for an interactive version of this type of example.*

Determine the slope of the tangent line to the graph of $x^2 + 4y^2 = 4$ at the point $\left(\sqrt{2}, -1/\sqrt{2}\right)$. See Figure 3.28.

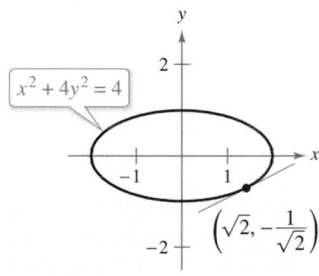

Figure 3.28

Solution

$$x^2 + 4y^2 = 4 \qquad \text{Write original equation.}$$

$$2x + 8y\frac{dy}{dx} = 0 \qquad \text{Differentiate with respect to } x.$$

$$\frac{dy}{dx} = \frac{-2x}{8y} \qquad \text{Solve for } \frac{dy}{dx}.$$

$$= \frac{-x}{4y} \qquad \text{Simplify.}$$

So, at $\left(\sqrt{2}, -1/\sqrt{2}\right)$, the slope is

$$\frac{dy}{dx} = \frac{-\sqrt{2}}{-4/\sqrt{2}} = \frac{1}{2}. \qquad \text{Evaluate } \frac{dy}{dx} \text{ when } x = \sqrt{2} \text{ and } y = -\frac{1}{\sqrt{2}}.$$

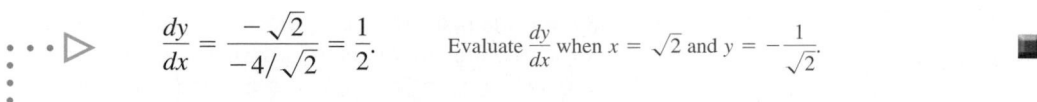

REMARK   To see the benefit of implicit differentiation, try doing Example 4 using the explicit function $y = -\frac{1}{2}\sqrt{4 - x^2}$.

Finding the Slope of a Graph Implicitly

Determine the slope of the graph of

$$3(x^2 + y^2)^2 = 100xy$$

at the point $(3, 1)$.

Solution

$$\frac{d}{dx}[3(x^2 + y^2)^2] = \frac{d}{dx}[100xy]$$

$$3(2)(x^2 + y^2)\left(2x + 2y\frac{dy}{dx}\right) = 100\left[x\frac{dy}{dx} + y(1)\right]$$

$$12y(x^2 + y^2)\frac{dy}{dx} - 100x\frac{dy}{dx} = 100y - 12x(x^2 + y^2)$$

$$[12y(x^2 + y^2) - 100x]\frac{dy}{dx} = 100y - 12x(x^2 + y^2)$$

$$\frac{dy}{dx} = \frac{100y - 12x(x^2 + y^2)}{-100x + 12y(x^2 + y^2)}$$

$$= \frac{25y - 3x(x^2 + y^2)}{-25x + 3y(x^2 + y^2)}$$

At the point $(3, 1)$, the slope of the graph is

$$\frac{dy}{dx} = \frac{25(1) - 3(3)(3^2 + 1^2)}{-25(3) + 3(1)(3^2 + 1^2)} = \frac{25 - 90}{-75 + 30} = \frac{-65}{-45} = \frac{13}{9}$$

as shown in Figure 3.29. This graph is called a **lemniscate**.

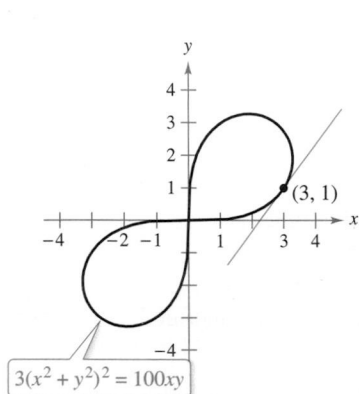

$3(x^2 + y^2)^2 = 100xy$

Lemniscate
Figure 3.29

Determining a Differentiable Function

Find dy/dx implicitly for the equation $\sin y = x$. Then find the largest interval of the form $-a < y < a$ on which y is a differentiable function of x (see Figure 3.30).

Solution

$$\frac{d}{dx}[\sin y] = \frac{d}{dx}[x]$$

$$\cos y \frac{dy}{dx} = 1$$

$$\frac{dy}{dx} = \frac{1}{\cos y}$$

The largest interval about the origin for which y is a differentiable function of x is $-\pi/2 < y < \pi/2$. To see this, note that $\cos y$ is positive for all y in this interval and is 0 at the endpoints. When you restrict y to the interval $-\pi/2 < y < \pi/2$, you should be able to write dy/dx explicitly as a function of x. To do this, you can use

$$\cos y = \sqrt{1 - \sin^2 y}$$

$$= \sqrt{1 - x^2}, \quad -\frac{\pi}{2} < y < \frac{\pi}{2}$$

and conclude that

$$\frac{dy}{dx} = \frac{1}{\sqrt{1 - x^2}}.$$

You will study this example further when derivatives of inverse trigonometric functions are defined in Section 3.6.

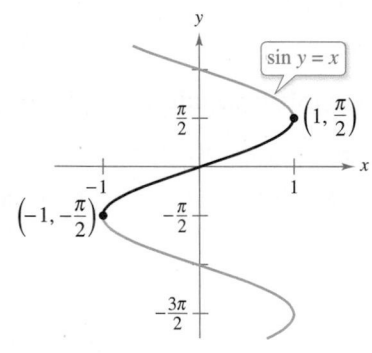

$\sin y = x$

$\left(1, \frac{\pi}{2}\right)$

$\left(-1, -\frac{\pi}{2}\right)$

The derivative is $\dfrac{dy}{dx} = \dfrac{1}{\sqrt{1 - x^2}}$.

Figure 3.30

With implicit differentiation, the form of the derivative often can be simplified (as in Example 6) by an appropriate use of the *original* equation. A similar technique can be used to find and simplify higher-order derivatives obtained implicitly.

ISAAC BARROW (1630–1677)

The graph in Figure 3.31 is called the **kappa curve** because it resembles the Greek letter kappa, κ. The general solution for the tangent line to this curve was discovered by the English mathematician Isaac Barrow. Newton was Barrow's student, and they corresponded frequently regarding their work in the early development of calculus.
See LarsonCalculus.com to read more of this biography.

EXAMPLE 7   **Finding the Second Derivative Implicitly**

Given $x^2 + y^2 = 25$, find $\dfrac{d^2y}{dx^2}$.

Solution   Differentiating each term with respect to x produces

$$2x + 2y\frac{dy}{dx} = 0$$

$$2y\frac{dy}{dx} = -2x$$

$$\frac{dy}{dx} = \frac{-2x}{2y}$$

$$= -\frac{x}{y}.$$

Differentiating a second time with respect to x yields

$$\frac{d^2y}{dx^2} = -\frac{(y)(1) - (x)(dy/dx)}{y^2} \qquad \text{Quotient Rule}$$

$$= -\frac{y - (x)(-x/y)}{y^2} \qquad \text{Substitute } -\frac{x}{y} \text{ for } \frac{dy}{dx}.$$

$$= -\frac{y^2 + x^2}{y^3} \qquad \text{Simplify.}$$

$$= -\frac{25}{y^3}. \qquad \text{Substitute 25 for } x^2 + y^2.$$

EXAMPLE 8   **Finding a Tangent Line to a Graph**

Find the tangent line to the graph of $x^2(x^2 + y^2) = y^2$ at the point $\left(\sqrt{2}/2,\ \sqrt{2}/2\right)$, as shown in Figure 3.31.

Solution   By rewriting and differentiating implicitly, you obtain

$$x^4 + x^2y^2 - y^2 = 0$$

$$4x^3 + x^2\left(2y\frac{dy}{dx}\right) + 2xy^2 - 2y\frac{dy}{dx} = 0$$

$$2y(x^2 - 1)\frac{dy}{dx} = -2x(2x^2 + y^2)$$

$$\frac{dy}{dx} = \frac{x(2x^2 + y^2)}{y(1 - x^2)}.$$

At the point $\left(\sqrt{2}/2,\ \sqrt{2}/2\right)$, the slope is

$$\frac{dy}{dx} = \frac{\left(\sqrt{2}/2\right)[2(1/2) + (1/2)]}{\left(\sqrt{2}/2\right)[1 - (1/2)]} = \frac{3/2}{1/2} = 3$$

and the equation of the tangent line at this point is

$$y - \frac{\sqrt{2}}{2} = 3\left(x - \frac{\sqrt{2}}{2}\right)$$

$$y = 3x - \sqrt{2}.$$

The kappa curve
Figure 3.31

$x^2(x^2 + y^2) = y^2$

$\left(\dfrac{\sqrt{2}}{2}, \dfrac{\sqrt{2}}{2}\right)$

Logarithmic Differentiation

On occasion, it is convenient to use logarithms as aids in differentiating nonlogarithmic functions. This procedure is called **logarithmic differentiation.**

EXAMPLE 9 **Logarithmic Differentiation**

Find the derivative of

$$y = \frac{(x - 2)^2}{\sqrt{x^2 + 1}}, \quad x \neq 2.$$

Solution Note that $y > 0$ for all $x \neq 2$. So, $\ln y$ is defined. Begin by taking the natural logarithm of each side of the equation. Then apply logarithmic properties and differentiate implicitly. Finally, solve for y'.

$$y = \frac{(x - 2)^2}{\sqrt{x^2 + 1}}, \quad x \neq 2 \qquad \text{Write original equation.}$$

$$\ln y = \ln \frac{(x - 2)^2}{\sqrt{x^2 + 1}} \qquad \text{Take natural log of each side.}$$

$$\ln y = 2 \ln(x - 2) - \frac{1}{2} \ln(x^2 + 1) \qquad \text{Logarithmic properties}$$

$$\frac{y'}{y} = 2\left(\frac{1}{x - 2}\right) - \frac{1}{2}\left(\frac{2x}{x^2 + 1}\right) \qquad \text{Differentiate.}$$

$$\frac{y'}{y} = \frac{x^2 + 2x + 2}{(x - 2)(x^2 + 1)} \qquad \text{Simplify.}$$

$$y' = y\left[\frac{x^2 + 2x + 2}{(x - 2)(x^2 + 1)}\right] \qquad \text{Solve for } y'.$$

$$y' = \frac{(x - 2)^2}{\sqrt{x^2 + 1}}\left[\frac{x^2 + 2x + 2}{(x - 2)(x^2 + 1)}\right] \qquad \text{Substitute for } y.$$

$$y' = \frac{(x - 2)(x^2 + 2x + 2)}{(x^2 + 1)^{3/2}} \qquad \text{Simplify.}$$

• • REMARK You could also solve the problem in Example 9 without using logarithmic differentiation by using the Power and Quotient Rules. Use these rules to find the derivative and show that the result is equivalent to the one in Example 9. Which method do you prefer?

EXAMPLE 10 **Logarithmic Differentiation**

Find the derivative of $y = x^{2x}$, $x > 0$.

Solution Note that $y > 0$ for all $x > 0$. So, $\ln y$ is defined.

$$y = x^{2x} \qquad \text{Write original equation.}$$

$$\ln y = \ln(x^{2x}) \qquad \text{Take natural log of each side.}$$

$$\ln y = (2x)(\ln x) \qquad \text{Logarithmic property}$$

$$\frac{y'}{y} = 2x\left(\frac{1}{x}\right) + 2 \ln x \qquad \text{Differentiate.}$$

$$\frac{y'}{y} = 2(1 + \ln x) \qquad \text{Simplify.}$$

$$y' = 2y(1 + \ln x) \qquad \text{Solve for } y'.$$

$$y' = 2x^{2x}(1 + \ln x) \qquad \text{Substitute for } y.$$

Here are some guidelines for using logarithmic differentiation. In general, use logarithmic differentiation when differentiating (1) a function involving many factors or (2) a function having both a variable base and a variable exponent.

3.5 Exercises

See **CalcChat.com** for tutorial help and worked-out solutions to odd-numbered exercises.

CONCEPT CHECK

1. **Explicit and Implicit Functions** Describe the difference between the explicit form of a function and an implicit equation. Give an example of each.

2. **Implicit Differentiation** Explain when you have to use implicit differentiation to find a derivative.

3. **Chain Rule** How is the Chain Rule applied when finding dy/dx implicitly?

4. **Logarithmic Differentiation** When is it beneficial to use logarithmic differentiation?

 **Finding a Derivative** In Exercises 5–26, find dy/dx by implicit differentiation.

5. $x^2 + y^2 = 9$ 　　　　**6.** $x^2 - y^2 = 25$

7. $x^5 + y^5 = 16$ 　　　　**8.** $2x^3 + 3y^3 = 64$

9. $x^3 - xy + y^2 = 7$ 　　**10.** $x^2y + y^2x = -2$

11. $x^3y^3 - y = x$ 　　　**12.** $\sqrt{xy} = x^2y + 1$

13. $x^3 - 3x^2y + 2xy^2 = 12$ 　**14.** $x^4y - 8xy + 3xy^2 = 9$

15. $xe^y - 10x + 3y = 0$ 　**16.** $e^{xy} + x^2 - y^2 = 10$

17. $\sin x + 2\cos 2y = 1$ 　**18.** $(\sin \pi x + \cos \pi y)^2 = 2$

19. $\csc x = x(1 + \tan y)$ 　**20.** $\cot y = x - y$

21. $y = \sin xy$ 　　　　**22.** $x = \sec \dfrac{1}{y}$

23. $x^2 - 3\ln y + y^2 = 10$ 　**24.** $\ln xy + 5x = 30$

25. $4x^3 + \ln y^2 + 2y = 2x$ 　**26.** $4xy + \ln x^2y = 7$

 **Finding Derivatives Implicitly and Explicitly** In Exercises 27–30, (a) find two explicit functions by solving the equation for y in terms of x, (b) sketch the graph of the equation and label the parts given by the corresponding explicit functions, (c) differentiate the explicit functions, and (d) find dy/dx implicitly and show that the result is equivalent to that of part (c).

27. $x^2 + y^2 = 64$ 　　**28.** $25x^2 + 36y^2 = 300$

29. $16y^2 - x^2 = 16$ 　**30.** $x^2 + y^2 - 4x + 6y + 9 = 0$

 **Finding the Slope of a Graph** In Exercises 31–38, find dy/dx by implicit differentiation. Then find the slope of the graph at the given point.

31. $xy = 6$, $(-6, -1)$ 　**32.** $3x^3y = 6$, $(1, 2)$

33. $y^2 = \dfrac{x^2 - 49}{x^2 + 49}$, $(7, 0)$ 　**34.** $4y^3 = \dfrac{x^2 - 36}{x^3 + 36}$, $(6, 0)$

35. $\tan(x + y) = x$, $(0, 0)$ 　**36.** $x\cos y = 1$, $\left(2, \dfrac{\pi}{3}\right)$

37. $3e^{xy} - x = 0$, $(3, 0)$ 　**38.** $y^2 = \ln x$, $(e, 1)$

 **Famous Curves** In Exercises 39–42, find the slope of the tangent line to the graph at the given point.

39. Witch of Agnesi: 　　**40.** Cissoid:

$(x^2 + 4)y = 8$ 　　　　$(4 - x)y^2 = x^3$

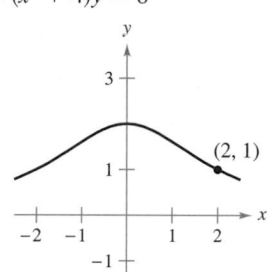 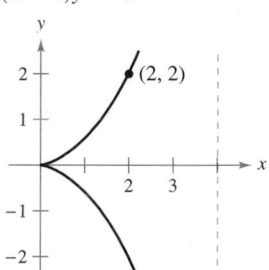

41. Bifolium: 　　　　**42.** Folium of Descartes:

$(x^2 + y^2)^2 = 4x^2y$ 　　$x^3 + y^3 - 6xy = 0$

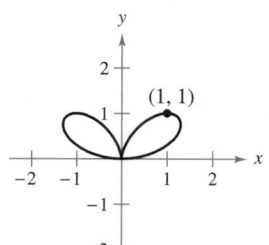 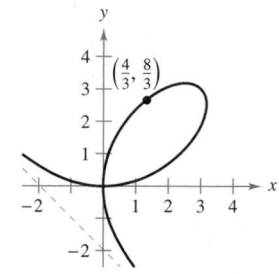

 **Famous Curves** In Exercises 43–48, find an equation of the tangent line to the graph at the given point. To print an enlarged copy of the graph, go to *MathGraphs.com*.

43. Parabola 　　　　**44.** Circle

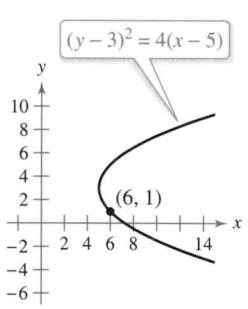 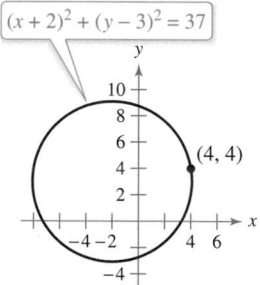

45. Cruciform 　　　　**46.** Astroid

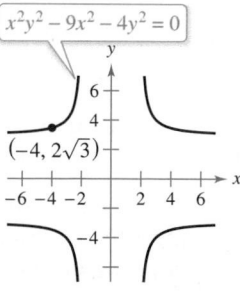 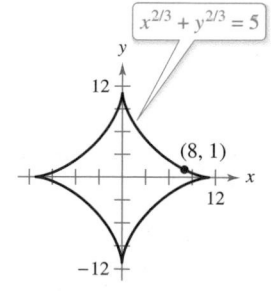

47. Lemniscate

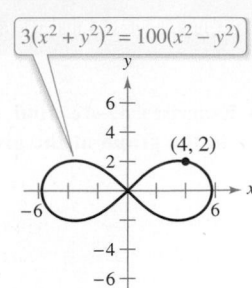

$3(x^2 + y^2)^2 = 100(x^2 - y^2)$

48. Kappa curve

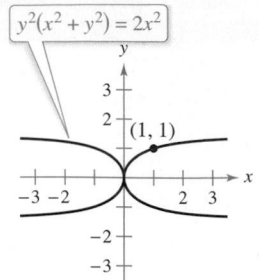

$y^2(x^2 + y^2) = 2x^2$

Finding an Equation of a Tangent Line In Exercises 49–52, use implicit differentiation to find an equation of the tangent line to the graph at the given point.

49. $4xy = 9$, $\left(1, \frac{9}{4}\right)$

50. $x^2 + xy + y^2 = 4$, $(2, 0)$

51. $x + y - 1 = \ln(x^2 + y^2)$, $(1, 0)$

52. $y^2 + \ln xy = 2$, $(e, 1)$

EXPLORING CONCEPTS

53. Implicit and Explicit Forms Write two different equations in implicit form that you can write in explicit form. Then write two different equations in implicit form that you cannot write in explicit form.

54. Think About It Explain why the derivative of $x^2 + y^2 + 2 = 1$ does not mean anything.

---

55. Ellipse

(a) Use implicit differentiation to find an equation of the tangent line to the ellipse $\frac{x^2}{2} + \frac{y^2}{8} = 1$ at $(1, 2)$.

(b) Show that the equation of the tangent line to the ellipse $\frac{x^2}{a^2} + \frac{y^2}{b^2} = 1$ at (x_0, y_0) is $\frac{x_0 x}{a^2} + \frac{y_0 y}{b^2} = 1$.

56. Hyperbola

(a) Use implicit differentiation to find an equation of the tangent line to the hyperbola $\frac{x^2}{6} - \frac{y^2}{8} = 1$ at $(3, -2)$.

(b) Show that the equation of the tangent line to the hyperbola $\frac{x^2}{a^2} - \frac{y^2}{b^2} = 1$ at (x_0, y_0) is $\frac{x_0 x}{a^2} - \frac{y_0 y}{b^2} = 1$.

 **Determining a Differentiable Function** In Exercises 57 and 58, find dy/dx implicitly and find the largest interval of the form $-a < y < a$ or $0 < y < a$ such that y is a differentiable function of x. Write dy/dx as a function of x.

57. $\tan y = x$

58. $\cos y = x$

 **Finding a Second Derivative** In Exercises 59–62, find d^2y/dx^2 implicitly in terms of x and y.

59. $x^2 y - 4x = 5$

60. $xy - 1 = 2x + y^2$

61. $7xy + \sin x = 2$

62. $3xy - 4\cos x = -6$

 **Tangent Lines and Normal Lines** In Exercises 63 and 64, find equations for the tangent line and normal line to the circle at each given point. (The *normal line* at a point is perpendicular to the tangent line at the point.) Use a graphing utility to graph the circle, the tangent lines, and the normal lines.

63. $x^2 + y^2 = 25$
    $(4, 3), (-3, 4)$

64. $x^2 + y^2 = 36$
    $(6, 0), (5, \sqrt{11})$

65. Normal Lines Show that the normal line at any point on the circle $x^2 + y^2 = r^2$ passes through the origin.

66. Circles Two circles of radius 4 are tangent to the graph of $y^2 = 4x$ at the point $(1, 2)$. Find equations of these two circles.

Vertical and Horizontal Tangent Lines In Exercises 67 and 68, find the points at which the graph of the equation has a vertical or horizontal tangent line.

67. $25x^2 + 16y^2 + 200x - 160y + 400 = 0$

68. $4x^2 + y^2 - 8x + 4y + 4 = 0$

Logarithmic Differentiation In Exercises 69–80, use logarithmic differentiation to find dy/dx.

69. $y = x\sqrt{x^2 + 1}$, $x > 0$

70. $y = \sqrt{x^2(x + 2)}$, $x > 0$

71. $y = \frac{x^2\sqrt{3x - 2}}{(x + 1)^2}$, $x > \frac{2}{3}$

72. $y = \sqrt{\frac{x^2 - 1}{x^2 + 1}}$, $x > 1$

73. $y = \frac{x(x - 1)^{3/2}}{\sqrt{x + 1}}$, $x > 1$

74. $y = \frac{(x + 1)(x - 2)}{(x - 1)(x + 2)}$, $x > 2$

75. $y = x^{2/x}$, $x > 0$

76. $y = x^{x-1}$, $x > 0$

77. $y = (x - 2)^{x+1}$, $x > 2$

78. $y = (1 + x)^{1/x}$, $x > 0$

79. $y = x^{\ln x}$, $x > 0$

80. $y = (\ln x)^{\ln x}$, $x > 1$

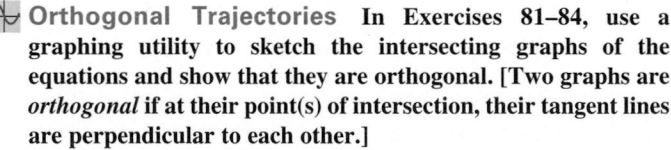 **Orthogonal Trajectories** In Exercises 81–84, use a graphing utility to sketch the intersecting graphs of the equations and show that they are orthogonal. [Two graphs are *orthogonal* if at their point(s) of intersection, their tangent lines are perpendicular to each other.]

81. $2x^2 + y^2 = 6$
    $y^2 = 4x$

82. $y^2 = x^3$
    $2x^2 + 3y^2 = 5$

83. $x + y = 0$
    $x = \sin y$

84. $x^3 = 3(y - 1)$
    $x(3y - 29) = 3$

Orthogonal Trajectories In Exercises 85 and 86, verify that the two families of curves are orthogonal, where C and K are real numbers. Use a graphing utility to graph the two families for two values of C and two values of K.

85. $xy = C$, $x^2 - y^2 = K$

86. $x^2 + y^2 = C^2$, $y = Kx$

87. True or False? Determine whether the statement is true. If it is false, explain why and correct it. For each statement, assume y is a function of x.

(a) $\frac{d}{dx}[\cos(x^2)] = -2x\sin(x^2)$ (b) $\frac{d}{dy}[\cos(y^2)] = 2y\sin(y^2)$

(c) $\frac{d}{dx}[\cos(y^2)] = -2y\sin(y^2)$

88. **HOW DO YOU SEE IT?** Use the graph to answer the questions.

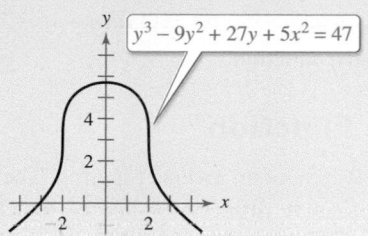

$$y^3 - 9y^2 + 27y + 5x^2 = 47$$

(a) Which is greater, the slope of the tangent line at $x = -3$ or the slope of the tangent line at $x = -1$?

(b) Estimate the point(s) where the graph has a vertical tangent line.

(c) Estimate the point(s) where the graph has a horizontal tangent line.

 89. Finding Equations of Tangent Lines Consider the equation $x^4 = 4(4x^2 - y^2)$.

(a) Use a graphing utility to graph the equation.

(b) Find and graph the four tangent lines to the curve for $y = 3$.

(c) Find the exact coordinates of the point of intersection of the two tangent lines in the first quadrant.

90. Tangent Lines and Intercepts Let L be any tangent line to the curve

$$\sqrt{x} + \sqrt{y} = \sqrt{c}.$$

Show that the sum of the x- and y-intercepts of L is c.

91. Slope Find all points on the circle $x^2 + y^2 = 100$ where the slope is $\frac{3}{4}$.

92. Proof

(a) Prove (Theorem 3.3) that $d/dx\,[x^n] = nx^{n-1}$ for the case in which n is a rational number. (*Hint:* Write $y = x^{p/q}$ in the form $y^q = x^p$ and differentiate implicitly. Assume that p and q are integers, where $q > 0$.)

(b) Prove part (a) for the case in which n is an irrational number. (*Hint:* Let $y = x^r$, where r is a real number, and use logarithmic differentiation.)

93. Tangent Lines Find equations of both tangent lines to the graph of the ellipse $\dfrac{x^2}{4} + \dfrac{y^2}{9} = 1$ that pass through the point $(4, 0)$ not on the graph.

94. Normals to a Parabola The graph shows the normal lines from the point $(2, 0)$ to the graph of the parabola $x = y^2$. How many normal lines are there from the point $(x_0, 0)$ to the graph of the parabola if (a) $x_0 = \frac{1}{4}$, (b) $x_0 = \frac{1}{2}$, and (c) $x_0 = 1$? (d) For what value of x_0 are two of the normal lines perpendicular to each other?

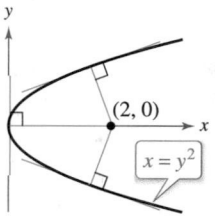

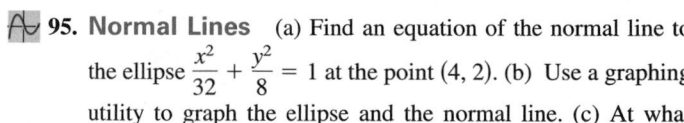

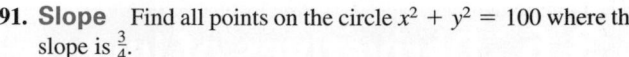 **95. Normal Lines** (a) Find an equation of the normal line to the ellipse $\dfrac{x^2}{32} + \dfrac{y^2}{8} = 1$ at the point $(4, 2)$. (b) Use a graphing utility to graph the ellipse and the normal line. (c) At what other point does the normal line intersect the ellipse?

SECTION PROJECT

Optical Illusions

In each graph below, an optical illusion is created by having lines intersect a family of curves. In each case, the lines appear to be curved. Find the value of dy/dx for the given values.

(a) Circles: $x^2 + y^2 = C^2$
$x = 3, y = 4, C = 5$

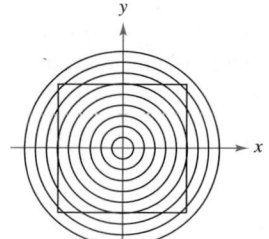

(b) Hyperbolas: $xy = C$
$x = 1, y = 4, C = 4$

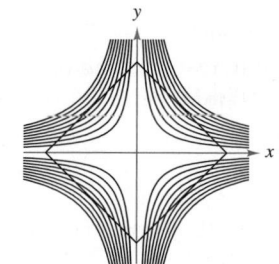

(c) Lines: $ax = by$
$x = \sqrt{3}, y = 3,$
$a = \sqrt{3}, b = 1$

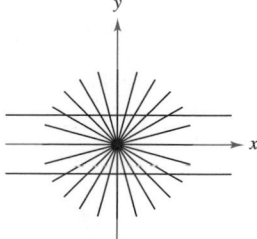

(d) Cosine curves: $y = C \cos x$
$x = \dfrac{\pi}{3}, y = \dfrac{1}{3}, C = \dfrac{2}{3}$

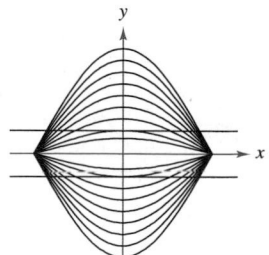

■ **FOR FURTHER INFORMATION** For more information on the mathematics of optical illusions, see the article "Descriptive Models for Perception of Optical Illusions" by David A. Smith in *The UMAP Journal.*

3.6 Derivatives of Inverse Functions

■ Find the derivative of an inverse function.
■ Differentiate an inverse trigonometric function.

Derivative of an Inverse Function

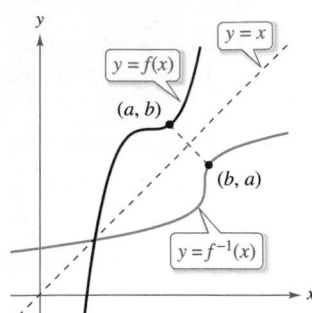

The graph of f^{-1} is a reflection of the graph of f in the line $y = x$.

Figure 3.32

The next two theorems discuss the derivative of an inverse function. The reasonableness of Theorem 3.16 follows from the reflective property of inverse functions, as shown in Figure 3.32.

THEOREM 3.16   Continuity and Differentiability of Inverse Functions

Let f be a function whose domain is an interval I. If f has an inverse function, then the following statements are true.

1. If f is continuous on its domain, then f^{-1} is continuous on its domain.
2. If f is differentiable on an interval containing c and $f'(c) \neq 0$, then f^{-1} is differentiable at $f(c)$.

A proof of this theorem is given in Appendix A.

THEOREM 3.17   The Derivative of an Inverse Function

Let f be a function that is differentiable on an interval I. If f has an inverse function g, then g is differentiable at any x for which $f'(g(x)) \neq 0$. Moreover,

$$g'(x) = \frac{1}{f'(g(x))}, \quad f'(g(x)) \neq 0.$$

A proof of this theorem is given in Appendix A.

EXAMPLE 1   Evaluating the Derivative of an Inverse Function

Let $f(x) = \frac{1}{4}x^3 + x - 1$.

a. What is the value of $f^{-1}(x)$ when $x = 3$?

b. What is the value of $(f^{-1})'(x)$ when $x = 3$?

Solution   Notice that f is one-to-one and therefore has an inverse function.

a. Because $f(2) = 3$, you know that $f^{-1}(3) = 2$.

b. Because the function f is differentiable and has an inverse function, you can apply Theorem 3.17 (with $g = f^{-1}$) to write

$$(f^{-1})'(3) = \frac{1}{f'(f^{-1}(3))} = \frac{1}{f'(2)}.$$

Moreover, using $f'(x) = \frac{3}{4}x^2 + 1$, you can conclude that

$$(f^{-1})'(3) = \frac{1}{f'(2)} = \frac{1}{\frac{3}{4}(2^2) + 1} = \frac{1}{4}. \qquad \text{See Figure 3.33.}$$

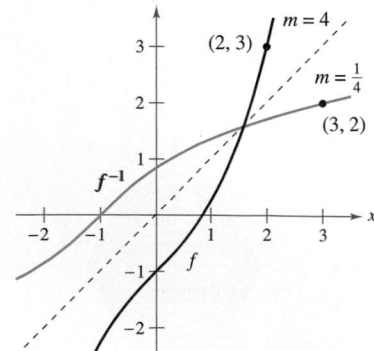

The graphs of the inverse functions f and f^{-1} have reciprocal slopes at points (a, b) and (b, a).

Figure 3.33

In Example 1, note that at the point $(2, 3)$, the slope of the graph of f is $m = 4$ and at the point $(3, 2)$, the slope of the graph of f^{-1} is $m = \frac{1}{4}$, as shown in Figure 3.33. This reciprocal relationship (which follows from Theorem 3.17) is sometimes written as

$$\frac{dy}{dx} = \frac{1}{dx/dy}.$$

EXAMPLE 2 **Graphs of Inverse Functions Have Reciprocal Slopes**

⋯▷ *See LarsonCalculus.com for an interactive version of this type of example.*

Let $f(x) = x^2$ (for $x \geq 0$) and let $f^{-1}(x) = \sqrt{x}$. Show that the slopes of the graphs of f and f^{-1} are reciprocals at each of the following points.

a. $(2, 4)$ and $(4, 2)$ **b.** $(3, 9)$ and $(9, 3)$

Solution The derivatives of f and f^{-1} are $f'(x) = 2x$ and $(f^{-1})'(x) = \dfrac{1}{2\sqrt{x}}$.

a. At $(2, 4)$, the slope of the graph of f is $f'(2) = 2(2) = 4$. At $(4, 2)$, the slope of the graph of f^{-1} is

$$(f^{-1})'(4) = \frac{1}{2\sqrt{4}} = \frac{1}{2(2)} = \frac{1}{4}.$$

b. At $(3, 9)$, the slope of the graph of f is $f'(3) = 2(3) = 6$. At $(9, 3)$, the slope of the graph of f^{-1} is

$$(f^{-1})'(9) = \frac{1}{2\sqrt{9}} = \frac{1}{2(3)} = \frac{1}{6}.$$

So, in both cases, the slopes are reciprocals, as shown in Figure 3.34. ■

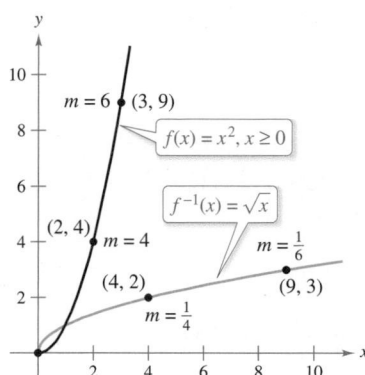

At $(0, 0)$, the derivative of f is 0 and the derivative of f^{-1} does not exist.

Figure 3.34

When determining the derivative of an inverse function, you have two options: (1) you can apply Theorem 3.17, or (2) you can use implicit differentiation. The first approach is illustrated in Example 3, and the second in the proof of Theorem 3.18.

■ **FOR FURTHER INFORMATION**
For more on the derivative of the arctangent function, see the article "Differentiating the Arctangent Directly" by Eric Key in *The College Mathematics Journal*. To view this article, go to *MathArticles.com*.

EXAMPLE 3 **Finding the Derivative of an Inverse Function**

Find the derivative of the inverse tangent function.

Solution Let $f(x) = \tan x$, $-\pi/2 < x < \pi/2$. Then let $g(x) = \arctan x$ be the inverse tangent function. To find the derivative of $g(x)$, use the fact that $f'(x) = \sec^2 x = \tan^2 x + 1$, and apply Theorem 3.17 as follows.

$$g'(x) = \frac{1}{f'(g(x))} = \frac{1}{f'(\arctan x)} = \frac{1}{[\tan(\arctan x)]^2 + 1} = \frac{1}{x^2 + 1}$$ ■

Derivatives of Inverse Trigonometric Functions

In Section 3.4, you saw that the derivative of the *transcendental* function $f(x) = \ln x$ is the *algebraic* function $f'(x) = 1/x$. You will now see that the derivatives of the inverse trigonometric functions also are algebraic (even though the inverse trigonometric functions are transcendental).

The next theorem lists the derivatives of the six inverse trigonometric functions. Note that the derivatives of arccos u, arccot u, and arccsc u are the *negatives* of the derivatives of arcsin u, arctan u, and arcsec u, respectively.

> **THEOREM 3.18** **Derivatives of Inverse Trigonometric Functions**
>
> Let u be a differentiable function of x.
>
> $$\frac{d}{dx}[\arcsin u] = \frac{u'}{\sqrt{1-u^2}} \qquad \frac{d}{dx}[\arccos u] = \frac{-u'}{\sqrt{1-u^2}}$$
>
> $$\frac{d}{dx}[\arctan u] = \frac{u'}{1+u^2} \qquad \frac{d}{dx}[\operatorname{arccot} u] = \frac{-u'}{1+u^2}$$
>
> $$\frac{d}{dx}[\operatorname{arcsec} u] = \frac{u'}{|u|\sqrt{u^2-1}} \qquad \frac{d}{dx}[\operatorname{arccsc} u] = \frac{-u'}{|u|\sqrt{u^2-1}}$$

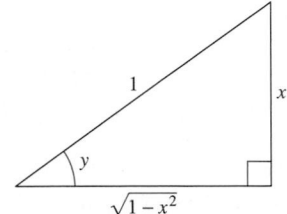

$y = \arcsin x$

Figure 3.35

Proof Let $y = \arcsin x$, $-\pi/2 \le y \le \pi/2$ (see Figure 3.35). So, $\sin y = x$, and you can use implicit differentiation as follows.

$$\sin y = x$$

$$(\cos y)\left(\frac{dy}{dx}\right) = 1$$

$$\frac{dy}{dx} = \frac{1}{\cos y}$$

$$\frac{dy}{dx} = \frac{1}{\sqrt{1-\sin^2 y}}$$

$$\frac{dy}{dx} = \frac{1}{\sqrt{1-x^2}}$$

Because u is a differentiable function of x, you can use the Chain Rule to write

$$\frac{d}{dx}[\arcsin u] = \frac{u'}{\sqrt{1-u^2}}, \quad \text{where} \quad u' = \frac{du}{dx}.$$

Proofs of the other differentiation rules are left as an exercise (see Exercise 83). ■

There is no common agreement on the definition of $\operatorname{arcsec} x$ (or $\operatorname{arccsc} x$) for negative values of x. For this text, the range of arcsecant was defined to preserve the reciprocal identity $\operatorname{arcsec} x = \arccos(1/x)$. For example, to evaluate $\operatorname{arcsec}(-2)$, you can write

$$\operatorname{arcsec}(-2) = \arccos(-0.5) \approx 2.09.$$

One of the consequences of the definition of the inverse secant function given in this text is that its graph has a positive slope at every x-value in its domain. This accounts for the absolute value sign in the formula for the derivative of $\operatorname{arcsec} x$.

EXAMPLE 4 Differentiating Inverse Trigonometric Functions

a. $\dfrac{d}{dx}[\arcsin(2x)] = \dfrac{2}{\sqrt{1-(2x)^2}} = \dfrac{2}{\sqrt{1-4x^2}}$

b. $\dfrac{d}{dx}[\arctan(3x)] = \dfrac{3}{1+(3x)^2} = \dfrac{3}{1+9x^2}$

c. $\dfrac{d}{dx}\left[\arcsin\sqrt{x}\right] = \dfrac{(1/2)x^{-1/2}}{\sqrt{1-x}} = \dfrac{1}{2\sqrt{x}\sqrt{1-x}} = \dfrac{1}{2\sqrt{x-x^2}}$

d. $\dfrac{d}{dx}[\operatorname{arcsec} e^{2x}] = \dfrac{2e^{2x}}{e^{2x}\sqrt{(e^{2x})^2-1}} = \dfrac{2}{\sqrt{e^{4x}-1}}$

The absolute value sign is not necessary because $e^{2x} > 0$. ■

EXAMPLE 5   **A Derivative That Can Be Simplified**

$$y = \arcsin x + x\sqrt{1 - x^2}$$

$$y' = \frac{1}{\sqrt{1 - x^2}} + x\left(\frac{1}{2}\right)(-2x)(1 - x^2)^{-1/2} + \sqrt{1 - x^2}$$

$$= \frac{1}{\sqrt{1 - x^2}} - \frac{x^2}{\sqrt{1 - x^2}} + \sqrt{1 - x^2}$$

$$= \sqrt{1 - x^2} + \sqrt{1 - x^2}$$

$$= 2\sqrt{1 - x^2}$$

GALILEO GALILEI (1564–1642)

Galileo's approach to science departed from the accepted Aristotelian view that nature had describable *qualities*, such as "fluidity" and "potentiality." He chose to describe the physical world in terms of measurable *quantities*, such as time, distance, force, and mass. *See LarsonCalculus.com to read more of this biography.*

In the 1600s, Europe was ushered into the scientific age by such great thinkers as Descartes, Galileo, Huygens, Newton, and Kepler. These men believed that nature is governed by basic laws—laws that can, for the most part, be written in terms of mathematical equations. One of the most influential publications of this period— *Dialogue on the Great World Systems*, by Galileo Galilei—has become a classic description of modern scientific thought.

As mathematics has developed during the past few hundred years, a small number of elementary functions has proven sufficient for modeling most* phenomena in physics, chemistry, biology, engineering, economics, and a variety of other fields. An **elementary function** is a function from the following list or one that can be formed as the sum, product, quotient, or composition of functions in the list.

Algebraic Functions	**Transcendental Functions**
Polynomial functions	Logarithmic functions
Rational functions	Exponential functions
Functions involving radicals	Trigonometric functions
	Inverse trigonometric functions

With the differentiation rules introduced so far in the text, you can differentiate *any* elementary function. For convenience, these differentiation rules are summarized below.

BASIC DIFFERENTIATION RULES FOR ELEMENTARY FUNCTIONS

1. $\dfrac{d}{dx}[cu] = cu'$

2. $\dfrac{d}{dx}[u \pm v] = u' \pm v'$

3. $\dfrac{d}{dx}[uv] = uv' + vu'$

4. $\dfrac{d}{dx}\left[\dfrac{u}{v}\right] = \dfrac{vu' - uv'}{v^2}$

5. $\dfrac{d}{dx}[c] = 0$

6. $\dfrac{d}{dx}[u^n] = nu^{n-1}u'$

7. $\dfrac{d}{dx}[x] = 1$

8. $\dfrac{d}{dx}[|u|] = \dfrac{u}{|u|}(u'), \quad u \neq 0$

9. $\dfrac{d}{dx}[\ln u] = \dfrac{u'}{u}$

10. $\dfrac{d}{dx}[e^u] = e^u u'$

11. $\dfrac{d}{dx}[\log_a u] = \dfrac{u'}{(\ln a)u}$

12. $\dfrac{d}{dx}[a^u] = (\ln a)a^u u'$

13. $\dfrac{d}{dx}[\sin u] = (\cos u)u'$

14. $\dfrac{d}{dx}[\cos u] = -(\sin u)u'$

15. $\dfrac{d}{dx}[\tan u] = (\sec^2 u)u'$

16. $\dfrac{d}{dx}[\cot u] = -(\csc^2 u)u'$

17. $\dfrac{d}{dx}[\sec u] = (\sec u \tan u)u'$

18. $\dfrac{d}{dx}[\csc u] = -(\csc u \cot u)u'$

19. $\dfrac{d}{dx}[\arcsin u] = \dfrac{u'}{\sqrt{1 - u^2}}$

20. $\dfrac{d}{dx}[\arccos u] = \dfrac{-u'}{\sqrt{1 - u^2}}$

21. $\dfrac{d}{dx}[\arctan u] = \dfrac{u'}{1 + u^2}$

22. $\dfrac{d}{dx}[\text{arccot } u] = \dfrac{-u'}{1 + u^2}$

23. $\dfrac{d}{dx}[\text{arcsec } u] = \dfrac{u'}{|u|\sqrt{u^2 - 1}}$

24. $\dfrac{d}{dx}[\text{arccsc } u] = \dfrac{-u'}{|u|\sqrt{u^2 - 1}}$

---

* Some important functions used in engineering and science (such as Bessel functions and gamma functions) are not elementary functions.

3.6 Exercises

See **CalcChat.com** for tutorial help and worked-out solutions to odd-numbered exercises.

CONCEPT CHECK

1. Evaluating the Derivative of an Inverse Function Given the point (x_1, y_1) on the graph of $y = f(x)$, explain how to find $(f^{-1})'(y_1)$. (Assume f^{-1} exists.)

2. Inverse Trigonometric Functions Are the derivatives of the inverse trigonometric functions algebraic or transcendental functions? Explain.

 **Evaluating the Derivative of an Inverse Function** In Exercises 3–12, verify that f has an inverse function. Then use the function f and the given real number a to find $(f^{-1})'(a)$. (*Hint:* See Example 1.)

3. $f(x) = 5 - 2x^3$, $a = 7$

4. $f(x) = x^3 + 3x - 1$, $a = -5$

5. $f(x) = x^3 - 1$, $a = 26$

6. $f(x) = \frac{1}{27}(x^5 + 2x^3)$, $a = -11$

7. $f(x) = \sin x$, $-\frac{\pi}{2} \le x \le \frac{\pi}{2}$, $a = \frac{1}{2}$

8. $f(x) = \cos 2x$, $0 \le x \le \frac{\pi}{2}$, $a = 1$

9. $f(x) = \frac{x + 6}{x - 2}$, $x > 2$, $a = 3$

10. $f(x) = \frac{x + 3}{x + 1}$, $x > -1$, $a = 2$

11. $f(x) = x^3 - \frac{4}{x}$, $x > 0$, $a = 6$

12. $f(x) = \sqrt{x - 4}$, $a = 2$

 **Graphs of Inverse Functions Have Reciprocal Slopes** In Exercises 13–16, show that the slopes of the graphs of f and f^{-1} are reciprocals at the given points.

Function	Point
13. $f(x) = x^3$	$\left(\frac{1}{2}, \frac{1}{8}\right)$
$f^{-1}(x) = \sqrt[3]{x}$	$\left(\frac{1}{8}, \frac{1}{2}\right)$
14. $f(x) = 3 - 4x$	$(1, -1)$
$f^{-1}(x) = \frac{3 - x}{4}$	$(-1, 1)$
15. $f(x) = \sqrt{x - 4}$	$(5, 1)$
$f^{-1}(x) = x^2 + 4$, $x \ge 0$	$(1, 5)$
16. $f(x) = \frac{4}{1 + x^2}$, $x \ge 0$	$(1, 2)$
$f^{-1}(x) = \sqrt{\frac{4 - x}{x}}$	$(2, 1)$

 **Finding a Derivative** In Exercises 17–42, find the derivative of the function.

17. $f(x) = \arcsin(x - 1)$

18. $f(t) = \text{arccsc}(-t^2)$

19. $g(x) = 3 \arccos \frac{x}{2}$

20. $f(x) = \text{arcsec } 2x$

21. $f(x) = \arctan e^x$

22. $f(x) = \text{arccot} \sqrt{x}$

23. $g(x) = \frac{\arcsin 3x}{x}$

24. $g(x) = \frac{\arccos x}{x + 1}$

25. $g(x) = e^{2x} \arcsin x$

26. $h(x) = x^2 \arctan 5x$

27. $h(x) = \text{arccot } 6x$

28. $f(x) = \text{arccsc } 3x$

29. $h(t) = \sin(\arccos t)$

30. $f(x) = \arcsin x + \arccos x$

31. $y = x \arccos x - 2\sqrt{x}$

32. $y = \ln t^2 - \arctan \frac{t}{2}$

33. $y = \ln \frac{x + 1}{x - 1} + \arctan x$

34. $y = x\sqrt{4 - x^2} + \arcsin \frac{x}{3}$

35. $g(t) = \tan(\arcsin t)$

36. $f(x) = \text{arcsec } x + \text{arccsc } x$

37. $y = x \arcsin x + \sqrt{1 - x^2}$

38. $y = x \arctan 2x - \frac{1}{4}\ln(1 + 4x^2)$

39. $y = 8 \arcsin \frac{x}{4} - \frac{x\sqrt{16 - x^2}}{2}$

40. $y = 25 \arcsin \frac{x}{5} - x\sqrt{25 - x^2}$

41. $y = \arctan x + \frac{x}{1 + x^2}$

42. $y = \arctan \frac{x}{2} - \frac{1}{2(x^2 + 4)}$

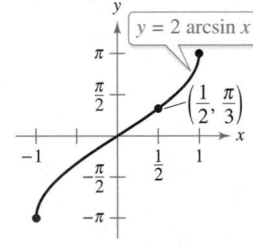 **Finding an Equation of a Tangent Line** In Exercises 43–48, find an equation of the tangent line to the graph of the function at the given point.

43. $y = 2 \arcsin x$

44. $y = \frac{1}{2} \arccos x$

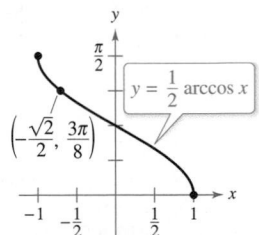
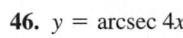

45. $y = \arctan \frac{x}{2}$

46. $y = \text{arcsec } 4x$

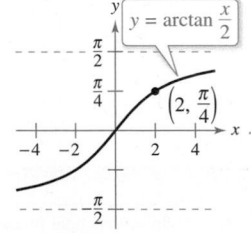
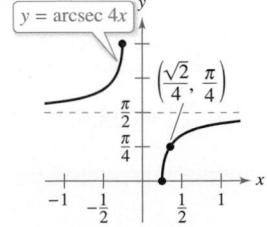

47. $y = 4x \arccos(x - 1)$ **48.** $y = 3x \arcsin x$

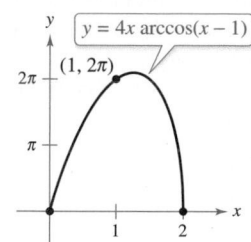

$y = 4x \arccos(x-1)$

$(1, 2\pi)$

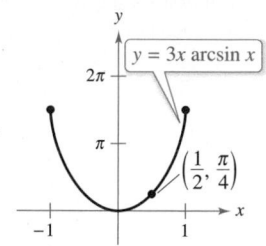

$y = 3x \arcsin x$

$\left(\frac{1}{2}, \frac{\pi}{4}\right)$

Finding an Equation of a Tangent Line In Exercises 49–52, (a) find an equation of the tangent line to the graph of the function at the given point, (b) use a graphing utility to graph the function and its tangent line at the point, and (c) use the *tangent* feature of a graphing utility to confirm your results.

49. $f(x) = \arccos x^2$, $\left(0, \frac{\pi}{2}\right)$ **50.** $f(x) = \arctan x$, $\left(-1, -\frac{\pi}{4}\right)$

51. $f(x) = \arcsin 3x$, $\left(\frac{\sqrt{2}}{6}, \frac{\pi}{4}\right)$ **52.** $f(x) = \arcsec x$, $\left(\sqrt{2}, \frac{\pi}{4}\right)$

53. Tangent Lines Find equations of all tangent lines to the graph of $f(x) = \arccos x$ that have slope -2.

54. Tangent Lines Find equations of all tangent lines to the graph of $f(x) = \arctan x$ that have slope $1/4$.

Linear and Quadratic Approximations The linear and quadratic approximations of a function f at $x = a$ are

$$P_1(x) = f'(a)(x - a) + f(a) \text{ and}$$

$$P_2(x) = \tfrac{1}{2}f''(a)(x - a)^2 + f'(a)(x - a) + f(a).$$

In Exercises 55–58, (a) find the specified linear and quadratic approximations of f, and (b) use a graphing utility to graph f and the approximations.

55. $f(x) = \arctan x$, $a = 0$ **56.** $f(x) = \arccos x$, $a = 0$

57. $f(x) = \arcsin x$, $a = \frac{1}{2}$ **58.** $f(x) = \arctan x$, $a = 1$

Finding the Slope of a Graph In Exercises 59–62, find dy/dx by implicit differentiation. Then find the slope of the graph at the given point.

59. $x = y^3 - 7y^2 + 2$, $(-4, 1)$

60. $x = 2\ln(y^2 - 3)$, $(0, 2)$

61. $x \arctan x = e^y$, $\left(1, \ln\frac{\pi}{4}\right)$

62. $\arcsin xy = \frac{2}{3}\arctan 2x$, $\left(\frac{1}{2}, 1\right)$

Finding the Equation of a Tangent Line In Exercises 63–66, use implicit differentiation to find an equation of the tangent line to the graph of the equation at the given point.

63. $x^2 + x \arctan y = y - 1$, $\left(-\frac{\pi}{4}, 1\right)$

64. $\arctan xy = \arcsin(x + y)$, $(0, 0)$

65. $\arcsin x + \arcsin y = \frac{\pi}{2}$, $\left(\frac{\sqrt{2}}{2}, \frac{\sqrt{2}}{2}\right)$

66. $\arctan(x + y) = y^2 + \frac{\pi}{4}$, $(1, 0)$

EXPLORING CONCEPTS

67. Inverse Secant Function Some calculus textbooks define the inverse secant function using the range $[0, \pi/2) \cup [\pi, 3\pi/2)$.

(a) Sketch the graph of $y = \arcsec x$ using this range.

(b) Show that $y' = \dfrac{1}{x\sqrt{x^2 - 1}}$.

68. Derivatives of Inverse Trigonometric Functions Determine whether the derivative of each inverse trigonometric function is odd or even.

69. Tangent Lines Consider the tangent line j to the graph of f at any point (a, b) and the tangent line k to the graph of f^{-1} at the point (b, a). Identify the line where the lines j and k intersect. Explain your reasoning.

70. HOW DO YOU SEE IT? Use the information in the graph of f below.

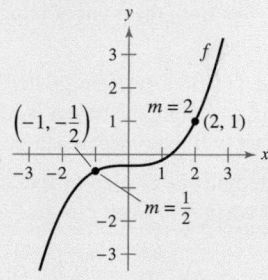

(a) What is the slope of the tangent line to the graph of f^{-1} at the point $\left(-\frac{1}{2}, -1\right)$? Explain.

(b) What is the slope of the tangent line to the graph of f^{-1} at the point $(1, 2)$? Explain.

71. Think About It The point $(1, 3)$ lies on the graph of f, and the slope of the tangent line through this point is $m = 2$. Assume f^{-1} exists. What is the slope of the tangent line to the graph of f^{-1} at the point $(3, 1)$?

72. Linear Approximation To find a linear approximation to the graph of the function in Example 5

$$y = \arcsin x + x\sqrt{1 - x^2}$$

you decide to use the tangent line at the origin, as shown below. Use a graphing utility to describe an interval about the origin where the tangent line is within 0.01 unit of the graph of the function. What might a person mean by saying that the original function is "locally linear"?

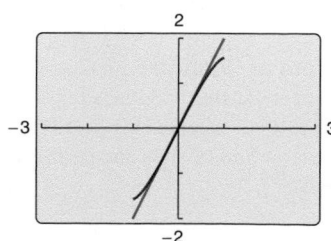

73. Angular Rate of Change An airplane flies at an altitude of 5 miles toward a point directly over an observer. Consider θ and x as shown in the figure.

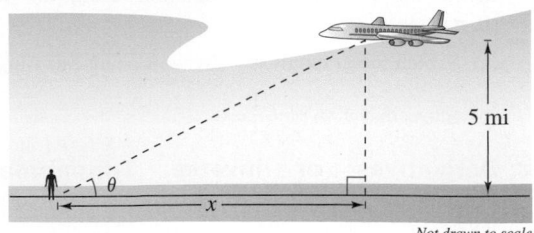

5 mi

x

Not drawn to scale

(a) Write θ as a function of x.

(b) The speed of the plane is 400 miles per hour. Find $d\theta/dt$ when $x = 10$ miles and $x = 3$ miles.

74. Angular Rate of Change Repeat Exercise 73 for an altitude of 3 miles and describe how the altitude affects the rate of change of θ.

75. Angular Rate of Change In a free-fall experiment, an object is dropped from a height of 256 feet. A camera on the ground 500 feet from the point of impact records the fall of the object (see figure).

(a) Find the position function giving the height of the object at time t, assuming the object is released at time $t = 0$. At what time will the object reach ground level?

(b) Find the rates of change of the angle of elevation of the camera when $t = 1$ and $t = 2$.

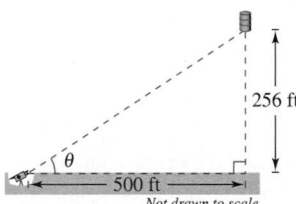

256 ft

θ

500 ft

Not drawn to scale

76. Angular Rate of Change A television camera at ground level is filming the lift-off of a rocket at a point 800 meters from the launch pad. Let θ be the angle of elevation of the rocket and let s be the distance between the camera and the rocket (see figure). Write θ as a function of s for the period of time when the rocket is moving vertically. Differentiate the result to find $d\theta/dt$ in terms of s and ds/dt.

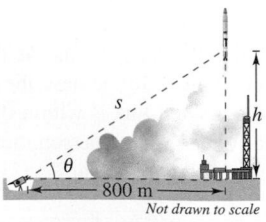

s

h

θ

800 m

Not drawn to scale

77. Angular Rate of Change An observer is standing 300 feet from the point at which a balloon is released. The balloon rises at a rate of 5 feet per second. How fast is the angle of elevation of the observer's line of sight increasing when the balloon is 100 feet high?

78. Angular Speed A patrol car is parked 50 feet from a long warehouse (see figure). The revolving light on top of the car turns at a rate of 30 revolutions per minute. Write θ as a function of x. How fast is the light beam moving along the wall when the beam makes an angle of $\theta = 45°$ with the line perpendicular from the light to the wall?

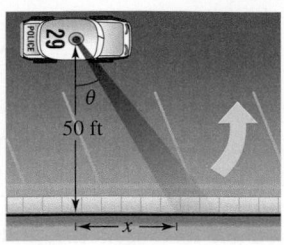

θ

50 ft

x

True or False? In Exercises 79–82, determine whether the statement is true or false. If it is false, explain why or give an example that shows it is false.

79. The derivative of arccsc x is the negative of the derivative of arcsec x.

80. The slope of the graph of the inverse tangent function is positive for all x.

81. $\dfrac{d}{dx}[\arctan(\tan x)] = 1$ for all x in the domain.

82. If $y = \arcsin x$, then

$$\frac{dy}{dx} = \frac{1}{dx/dy}$$

for all x in $[-1, 1]$.

83. Proof Prove each differentiation formula.

(a) $\dfrac{d}{dx}[\arccos u] = \dfrac{-u'}{\sqrt{1 - u^2}}$

(b) $\dfrac{d}{dx}[\arctan u] = \dfrac{u'}{1 + u^2}$

(c) $\dfrac{d}{dx}[\text{arcsec } u] = \dfrac{u'}{|u|\sqrt{u^2 - 1}}$

(d) $\dfrac{d}{dx}[\text{arccot } u] = \dfrac{-u'}{1 + u^2}$

(e) $\dfrac{d}{dx}[\text{arccsc } u] = \dfrac{-u'}{|u|\sqrt{u^2 - 1}}$

84. Proof Prove that

$$\arccos x = \frac{\pi}{2} - \arctan\left(\frac{x}{\sqrt{1 - x^2}}\right), \quad |x| < 1.$$

85. Proof Prove that

$$\arcsin x = \arctan\left(\frac{x}{\sqrt{1 - x^2}}\right), \quad |x| < 1.$$

86. Proof Show that the function

$$f(x) = \arcsin\frac{x - 2}{2} - 2\arcsin\frac{\sqrt{x}}{2}$$

is constant for $0 \le x \le 4$.

3.7   Related Rates

■ Find a related rate.
■ Use related rates to solve real-life problems.

Finding Related Rates

You have seen how the Chain Rule can be used to find dy/dx implicitly. Another important use of the Chain Rule is to find the rates of change of two or more related variables that are changing with respect to *time*.

For example, when water is drained out of a conical tank (see Figure 3.36), the volume V, the radius r, and the height h of the water level are all functions of time t. Knowing that these variables are related by the equation

$$V = \frac{\pi}{3} r^2 h \qquad \text{Original equation}$$

you can differentiate implicitly with respect to t to obtain the **related-rate** equation

$$\frac{d}{dt}[V] = \frac{d}{dt}\left[\frac{\pi}{3}r^2 h\right]$$

$$\frac{dV}{dt} = \frac{\pi}{3}\left[r^2 \frac{dh}{dt} + h\left(2r \frac{dr}{dt}\right)\right] \qquad \text{Differentiate with respect to } t.$$

$$= \frac{\pi}{3}\left(r^2 \frac{dh}{dt} + 2rh \frac{dr}{dt}\right).$$

From this equation, you can see that the rate of change of V is related to the rates of change of both h and r.

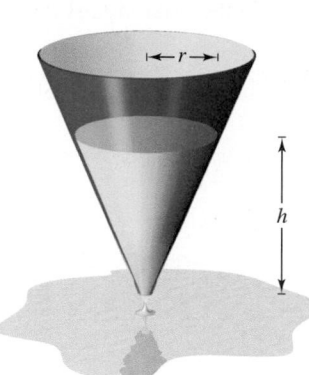

Exploration

Finding a Related Rate   In the conical tank shown in Figure 3.36, the height of the water level is changing at a rate of -0.2 foot per minute and the radius is changing at a rate of -0.1 foot per minute. What is the rate of change of the volume when the radius is $r = 1$ foot and the height is $h = 2$ feet? Does the rate of change of the volume depend on the values of r and h? Explain.

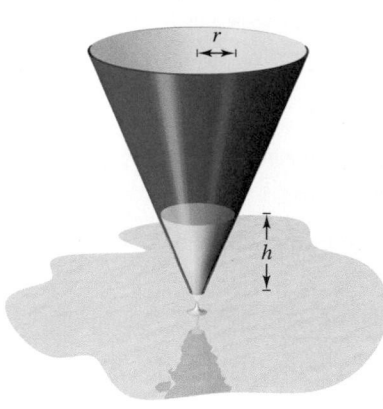

EXAMPLE 1   Two Rates That Are Related

The variables x and y are both differentiable functions of t and are related by the equation $y = x^2 + 3$. Find dy/dt when $x = 1$, given that $dx/dt = 2$ when $x = 1$.

Volume is related to radius and height.
Figure 3.36

Solution   Using the Chain Rule, you can differentiate both sides of the equation *with respect to t*.

$$y = x^2 + 3 \qquad \text{Write original equation.}$$

$$\frac{d}{dt}[y] = \frac{d}{dt}[x^2 + 3] \qquad \text{Differentiate with respect to } t.$$

$$\frac{dy}{dt} = 2x \frac{dx}{dt} \qquad \text{Chain Rule}$$

When $x = 1$ and $dx/dt = 2$, you have

$$\frac{dy}{dt} = 2(1)(2) = 4.$$

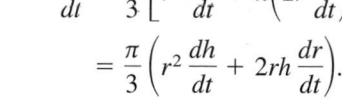

Problem Solving with Related Rates

In Example 1, you were *given* an equation that related the variables x and y and were asked to find the rate of change of y when $x = 1$.

Equation: $y = x^2 + 3$

Given rate: $\dfrac{dx}{dt} = 2$ when $x = 1$

Find: $\dfrac{dy}{dt}$ when $x = 1$

In each of the remaining examples in this section, you must *create* a mathematical model from a verbal description.

EXAMPLE 2   Ripples in a Pond

A pebble is dropped into a calm pond, causing ripples in the form of concentric circles, as shown in Figure 3.37. The radius r of the outer ripple is increasing at a constant rate of 1 foot per second. When the radius is 4 feet, at what rate is the total area A of the disturbed water changing?

Solution   The variables r and A are related by $A = \pi r^2$. The rate of change of the radius r is $dr/dt = 1$.

Equation: $A = \pi r^2$

Given rate: $\dfrac{dr}{dt} = 1$ foot per second

Find: $\dfrac{dA}{dt}$ when $r = 4$ feet

With this information, you can proceed as in Example 1.

$$\frac{d}{dt}[A] = \frac{d}{dt}[\pi r^2] \qquad \text{Differentiate with respect to } t.$$

$$\frac{dA}{dt} = 2\pi r \frac{dr}{dt} \qquad \text{Chain Rule}$$

$$= 2\pi(4)(1) \qquad \text{Substitute 4 for } r \text{ and 1 for } \frac{dr}{dt}.$$

$$= 8\pi \text{ square feet per second} \qquad \text{Simplify.}$$

When the radius is 4 feet, the area is changing at a rate of 8π square feet per second.

Total area increases as the outer radius increases.
Figure 3.37

REMARK   When using these guidelines, be sure you perform Step 3 before Step 4. Substituting the known values of the variables before differentiating will produce an inappropriate derivative.

GUIDELINES FOR SOLVING RELATED-RATE PROBLEMS

1. Identify all *given* quantities and quantities *to be determined*. Make a sketch and label the quantities.

2. Write an equation involving the variables whose rates of change either are given or are to be determined.

3. Using the Chain Rule, implicitly differentiate both sides of the equation *with respect to time t*.

4. *After* completing Step 3, substitute into the resulting equation all known values for the variables and their rates of change. Then solve for the required rate of change.

■ **FOR FURTHER INFORMATION**
To learn more about the history of related-rate problems, see the article "The Lengthening Shadow: The Story of Related Rates" by Bill Austin, Don Barry, and David Berman in *Mathematics Magazine*. To view this article, go to *MathArticles.com*.

The table below lists examples of mathematical models involving rates of change. For instance, the rate of change in the first example is the velocity of a car.

Verbal Statement	Mathematical Model
The velocity of a car after traveling for 1 hour is 50 miles per hour.	x = distance traveled $\dfrac{dx}{dt} = 50 \text{ mi/h}$ when $t = 1$
Water is being pumped into a swimming pool at a rate of 10 cubic meters per hour.	V = volume of water in pool $\dfrac{dV}{dt} = 10 \text{ m}^3/\text{h}$
A gear is revolving at a rate of 25 revolutions per minute (1 revolution = 2π rad).	θ = angle of revolution $\dfrac{d\theta}{dt} = 25(2\pi) \text{ rad/min}$
A population of bacteria is increasing at a rate of 2000 per hour.	x = number in population $\dfrac{dx}{dt} = 2000 \text{ bacteria per hour}$

EXAMPLE 3 An Inflating Balloon

Air is being pumped into a spherical balloon at a rate of 4.5 cubic feet per minute. Find the rate of change of the radius when the radius is 2 feet.

Solution Let V be the volume of the balloon, and let r be its radius. Because the volume is increasing at a rate of 4.5 cubic feet per minute, you know that at time t the rate of change of the volume is $dV/dt = \frac{9}{2}$. So, the problem can be stated as shown.

Given rate: $\dfrac{dV}{dt} = \dfrac{9}{2}$ cubic feet per minute (constant rate)

Find: $\dfrac{dr}{dt}$ when $r = 2$ feet

To find the rate of change of the radius, you must find an equation that relates the radius r to the volume V.

▷ ***Equation:*** $V = \dfrac{4}{3}\pi r^3$ Volume of a sphere

REMARK The formula for the volume of a sphere and other formulas from geometry are listed on the formula cards for this text.

Differentiating both sides of the equation with respect to t produces

$$\dfrac{dV}{dt} = 4\pi r^2 \dfrac{dr}{dt}$$ Differentiate with respect to t.

$$\dfrac{dr}{dt} = \dfrac{1}{4\pi r^2}\left(\dfrac{dV}{dt}\right).$$ Solve for $\frac{dr}{dt}$.

Finally, when $r = 2$, the rate of change of the radius is

$$\dfrac{dr}{dt} = \dfrac{1}{4\pi(2)^2}\left(\dfrac{9}{2}\right) \approx 0.09 \text{ foot per minute.}$$ ■

In Example 3, note that the volume is increasing at a *constant* rate, but the radius is increasing at a *variable* rate. Just because two rates are related does not mean that they are proportional. In this particular case, the radius is growing more and more slowly as t increases. Do you see why?

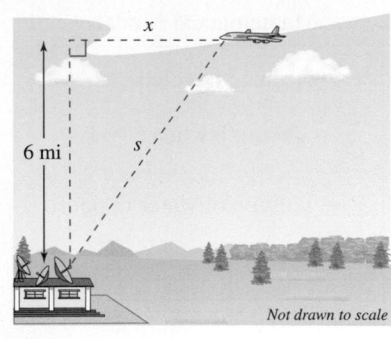

An airplane is flying at an altitude of 6 miles, s miles from the station.
Figure 3.38

EXAMPLE 4 **The Speed of an Airplane Tracked by Radar**

•••▷ *See LarsonCalculus.com for an interactive version of this type of example.*

An airplane is flying on a flight path that will take it directly over a radar tracking station, as shown in Figure 3.38. The distance s is decreasing at a rate of 400 miles per hour when $s = 10$ miles. What is the speed of the plane?

Solution Let x be the horizontal distance from the station, as shown in Figure 3.38. Notice that when $s = 10$, $x = \sqrt{10^2 - 6^2} = 8$.

Given rate: $ds/dt = -400$ miles per hour when $s = 10$ miles

Find: dx/dt when $s = 10$ miles and $x = 8$ miles

You can find the velocity of the plane as shown.

Equation: $\quad x^2 + 6^2 = s^2$	Pythagorean Theorem
$\qquad 2x\dfrac{dx}{dt} = 2s\dfrac{ds}{dt}$	Differentiate with respect to t.
$\qquad \dfrac{dx}{dt} = \dfrac{s}{x}\left(\dfrac{ds}{dt}\right)$	Solve for $\dfrac{dx}{dt}$.
$\qquad = \dfrac{10}{8}(-400)$	Substitute for s, x, and $\dfrac{ds}{dt}$.
$\qquad = -500$ miles per hour	Simplify.

•••▷ Because the velocity is -500 miles per hour, the *speed* is 500 miles per hour. ■

•••••••**REMARK** The velocity in Example 4 is negative because x represents a distance that is decreasing.

EXAMPLE 5 **A Changing Angle of Elevation**

Find the rate of change in the angle of elevation of the camera shown in Figure 3.39 at 10 seconds after lift-off.

Solution Let θ be the angle of elevation, as shown in Figure 3.39. When $t = 10$, the height s of the rocket is $s = 50t^2 = 50(10)^2 = 5000$ feet.

Given rate: $ds/dt = 100t =$ velocity of rocket (in feet per second)

Find: $d\theta/dt$ when $t = 10$ seconds and $s = 5000$ feet

Using Figure 3.39, you can relate s and θ by the equation $\tan\theta = s/2000$.

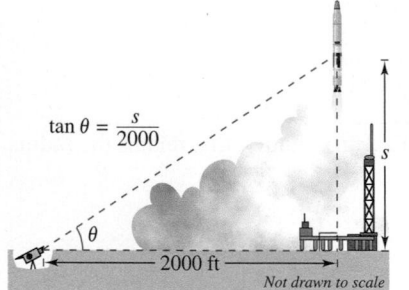

A television camera at ground level is filming the lift-off of a rocket that is rising vertically according to the position equation $s = 50t^2$, where s is measured in feet and t is measured in seconds. The camera is 2000 feet from the launch pad.
Figure 3.39

Equation: $\qquad \tan\theta = \dfrac{s}{2000}$	See Figure 3.39.
$\qquad (\sec^2\theta)\dfrac{d\theta}{dt} = \dfrac{1}{2000}\left(\dfrac{ds}{dt}\right)$	Differentiate with respect to t.
$\qquad \dfrac{d\theta}{dt} = \cos^2\theta\dfrac{100t}{2000}$	Substitute $100t$ for $\dfrac{ds}{dt}$.
$\qquad = \left(\dfrac{2000}{\sqrt{s^2 + 2000^2}}\right)^2 \dfrac{100t}{2000}$	$\cos\theta = \dfrac{2000}{\sqrt{s^2 + 2000^2}}$

When $t = 10$ and $s = 5000$, you have

$$\frac{d\theta}{dt} = \frac{2000(100)(10)}{5000^2 + 2000^2} = \frac{2}{29} \text{ radian per second.}$$

So, when $t = 10$, θ is changing at a rate of $\frac{2}{29}$ radian per second. ■

EXAMPLE 6 The Velocity of a Piston

In the engine shown in Figure 3.40, a 7-inch connecting rod is fastened to a crank of radius 3 inches. The crankshaft rotates counterclockwise at a constant rate of 200 revolutions per minute. Find the velocity of the piston when $\theta = \pi/3$.

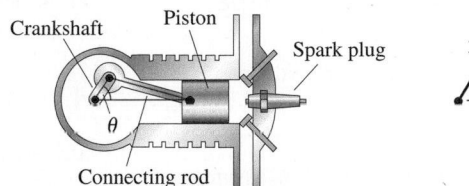
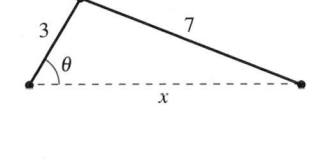

The velocity of a piston is related to the angle of the crankshaft.
Figure 3.40

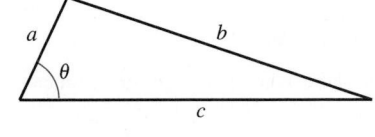

Law of Cosines:
$b^2 = a^2 + c^2 - 2ac \cos \theta$
Figure 3.41

Solution   Label the distances as shown in Figure 3.40. Because a complete revolution corresponds to 2π radians, it follows that $d\theta/dt = 200(2\pi) = 400\pi$ radians per minute.

Given rate: $\dfrac{d\theta}{dt} = 400\pi$ radians per minute   (constant rate)

Find: $\dfrac{dx}{dt}$ when $\theta = \dfrac{\pi}{3}$

You can use the Law of Cosines (see Figure 3.41) to find an equation that relates x and θ.

Equation:
$$7^2 = 3^2 + x^2 - 2(3)(x) \cos \theta$$
$$0 = 2x \frac{dx}{dt} - 6\left(-x \sin \theta \frac{d\theta}{dt} + \cos \theta \frac{dx}{dt}\right)$$
$$(6 \cos \theta - 2x)\frac{dx}{dt} = 6x \sin \theta \frac{d\theta}{dt}$$
$$\frac{dx}{dt} = \frac{6x \sin \theta}{6 \cos \theta - 2x}\left(\frac{d\theta}{dt}\right)$$

When $\theta = \pi/3$, you can solve for x as shown.

$$7^2 = 3^2 + x^2 - 2(3)(x) \cos \frac{\pi}{3}$$
$$49 = 9 + x^2 - 6x\left(\frac{1}{2}\right)$$
$$0 = x^2 - 3x - 40$$
$$0 = (x - 8)(x + 5)$$
$$x = 8 \text{ inches} \qquad\qquad \text{Choose positive solution.}$$

So, when $x = 8$ and $\theta = \pi/3$, the velocity of the piston is

$$\frac{dx}{dt} = \frac{6(8)\left(\sqrt{3}/2\right)}{6(1/2) - 16}(400\pi)$$
$$= \frac{9600\pi\sqrt{3}}{-13}$$
$$\approx -4018 \text{ inches per minute.}$$

REMARK   The velocity in Example 6 is negative because x represents a distance that is decreasing.

3.7 Exercises

See **CalcChat.com** for tutorial help and worked-out solutions to odd-numbered exercises.

CONCEPT CHECK

1. Related-Rate Equation What is a related-rate equation?

2. Related Rates In your own words, state the guidelines for solving related-rate problems.

 **Using Related Rates** In Exercises 3–6, assume that x and y are both differentiable functions of t and find the required values of dy/dt and dx/dt.

Equation	Find	Given
3. $y = \sqrt{x}$	(a) $\dfrac{dy}{dt}$ when $x = 4$	$\dfrac{dx}{dt} = 3$
	(b) $\dfrac{dx}{dt}$ when $x = 25$	$\dfrac{dy}{dt} = 2$
4. $y = 3x^2 - 5x$	(a) $\dfrac{dy}{dt}$ when $x = 3$	$\dfrac{dx}{dt} = 2$
	(b) $\dfrac{dx}{dt}$ when $x = 2$	$\dfrac{dy}{dt} = 4$
5. $xy = 4$	(a) $\dfrac{dy}{dt}$ when $x = 8$	$\dfrac{dx}{dt} = 10$
	(b) $\dfrac{dx}{dt}$ when $x = 1$	$\dfrac{dy}{dt} = -6$
6. $x^2 + y^2 = 25$	(a) $\dfrac{dy}{dt}$ when $x = 3, y = 4$	$\dfrac{dx}{dt} = 8$
	(b) $\dfrac{dx}{dt}$ when $x = 4, y = 3$	$\dfrac{dy}{dt} = -2$

 **Moving Point** In Exercises 7–10, a point is moving along the graph of the given function at the rate dx/dt. Find dy/dt for the given values of x.

7. $y = 2x^2 + 1;\ \dfrac{dx}{dt} = 2$ centimeters per second

   (a) $x = -1$ (b) $x = 0$ (c) $x = 1$

8. $y = \dfrac{1}{1 + x^2};\ \dfrac{dx}{dt} = 6$ inches per second

   (a) $x = -2$ (b) $x = 0$ (c) $x = 2$

9. $y = \tan x;\ \dfrac{dx}{dt} = 3$ feet per second

   (a) $x = -\dfrac{\pi}{3}$ (b) $x = -\dfrac{\pi}{4}$ (c) $x = 0$

10. $y = \cos x;\ \dfrac{dx}{dt} = 4$ centimeters per second

   (a) $x = \dfrac{\pi}{6}$ (b) $x = \dfrac{\pi}{4}$ (c) $x = \dfrac{\pi}{3}$

11. Area The radius r of a circle is increasing at a rate of 4 centimeters per minute. Find the rate of change of the area when $r = 37$ centimeters.

12. Area The length s of each side of an equilateral triangle is increasing at a rate of 13 feet per hour. Find the rate of change of the area when $s = 41$ feet. $\left(\textit{Hint: }\text{The formula for the area of an equilateral triangle is}\right.$

$$A = \frac{s^2\sqrt{3}}{4}.\Big)$$

13. Volume The radius r of a sphere is increasing at a rate of 3 inches per minute.

(a) Find the rates of change of the volume when $r = 9$ inches and $r = 36$ inches.

(b) Explain why the rate of change of the volume of the sphere is not constant even though dr/dt is constant.

14. Radius A spherical balloon is inflated with gas at the rate of 800 cubic centimeters per minute.

(a) Find the rates of change of the radius when $r = 30$ centimeters and $r = 85$ centimeters.

(b) Explain why the rate of change of the radius of the sphere is not constant even though dV/dt is constant.

15. Volume All edges of a cube are expanding at a rate of 6 centimeters per second. How fast is the volume changing when each edge is (a) 2 centimeters and (b) 10 centimeters?

16. Surface Area All edges of a cube are expanding at a rate of 6 centimeters per second. How fast is the surface area changing when each edge is (a) 2 centimeters and (b) 10 centimeters?

17. Height At a sand and gravel plant, sand is falling off a conveyor and onto a conical pile at a rate of 10 cubic feet per minute. The diameter of the base of the cone is approximately three times the altitude. At what rate is the height of the pile changing when the pile is 15 feet high? $\left(\textit{Hint: }\text{The formula for the volume of a cone is } V = \tfrac{1}{3}\pi r^2 h.\right)$

18. Height The volume of oil in a cylindrical container is increasing at a rate of 150 cubic inches per second. The height of the cylinder is approximately ten times the radius. At what rate is the height of the oil changing when the oil is 35 inches high? $\left(\textit{Hint: }\text{The formula for the volume of a cylinder is } V = \pi r^2 h.\right)$

19. Depth A swimming pool is 12 meters long, 6 meters wide, 1 meter deep at the shallow end, and 3 meters deep at the deep end (see figure). Water is being pumped into the pool at $\tfrac{1}{4}$ cubic meter per minute, and there is 1 meter of water at the deep end.

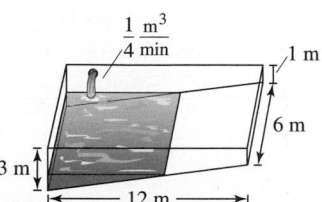

(a) What percent of the pool is filled?

(b) At what rate is the water level rising?

20. Depth A trough is 12 feet long and 3 feet across the top (see figure). Its ends are isosceles triangles with altitudes of 3 feet.

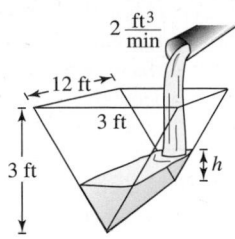

(a) Water is being pumped into the trough at 2 cubic feet per minute. How fast is the water level rising when the depth h is 1 foot?

(b) The water is rising at a rate of $\frac{3}{8}$ inch per minute when $h = 2$ feet. Determine the rate at which water is being pumped into the trough.

21. Moving Ladder A ladder 25 feet long is leaning against the wall of a house (see figure). The base of the ladder is pulled away from the wall at a rate of 2 feet per second.

(a) How fast is the top of the ladder moving down the wall when its base is 7 feet, 15 feet, and 24 feet from the wall?

(b) Consider the triangle formed by the side of the house, the ladder, and the ground. Find the rate at which the area of the triangle is changing when the base of the ladder is 7 feet from the wall.

(c) Find the rate at which the angle between the ladder and the wall of the house is changing when the base of the ladder is 7 feet from the wall.

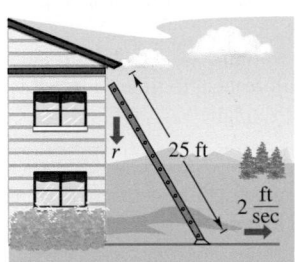

Figure for 21   Figure for 22

■ **FOR FURTHER INFORMATION** For more information on the mathematics of moving ladders, see the article "The Falling Ladder Paradox" by Paul Scholten and Andrew Simoson in *The College Mathematics Journal.* To view this article, go to *MathArticles.com.*

22. Construction A construction worker pulls a five-meter plank up the side of a building under construction by means of a rope tied to one end of the plank (see figure). Assume the opposite end of the plank follows a path perpendicular to the wall of the building and the worker pulls the rope at a rate of 0.15 meter per second. How fast is the end of the plank sliding along the ground when it is 2.5 meters from the wall of the building?

23. Construction A winch at the top of a 12-meter building pulls a pipe of the same length to a vertical position, as shown in the figure. The winch pulls in rope at a rate of -0.2 meter per second. Find the rate of vertical change and the rate of horizontal change at the end of the pipe when $y = 6$ meters.

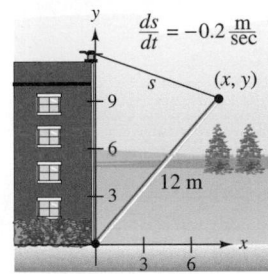   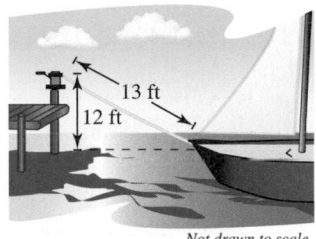

Figure for 23   Figure for 24

24. Boating A boat is pulled into a dock by means of a winch 12 feet above the deck of the boat (see figure).

(a) The winch pulls in rope at a rate of 4 feet per second. Determine the speed of the boat when there is 13 feet of rope out. What happens to the speed of the boat as it gets closer to the dock?

(b) Suppose the boat is moving at a constant rate of 4 feet per second. Determine the speed at which the winch pulls in rope when there is a total of 13 feet of rope out. What happens to the speed at which the winch pulls in rope as the boat gets closer to the dock?

25. Air Traffic Control An air traffic controller spots two planes at the same altitude converging on a point as they fly at right angles to each other (see figure). One plane is 225 miles from the point, moving at 450 miles per hour. The other plane is 300 miles from the point, moving at 600 miles per hour.

(a) At what rate is the distance s between the planes decreasing?

(b) How much time does the air traffic controller have to get one of the planes on a different flight path?

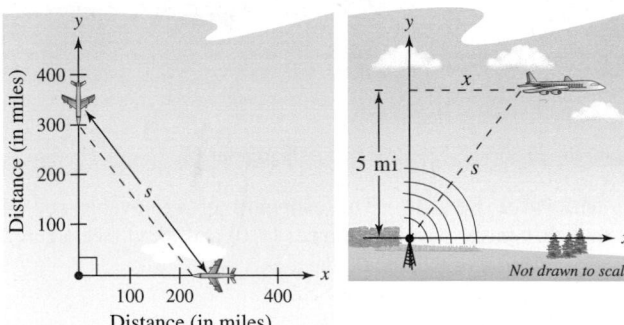

Figure for 25   Figure for 26

26. Air Traffic Control An airplane is flying at an altitude of 5 miles and passes directly over a radar antenna (see figure). When the plane is 10 miles away ($s = 10$), the radar detects that the distance s is changing at a rate of 240 miles per hour. What is the speed of the plane?

27. Sports A baseball diamond has the shape of a square with sides 90 feet long (see figure). A player running from second base to third base at a speed of 25 feet per second is 20 feet from third base. At what rate is the player's distance from home plate changing?

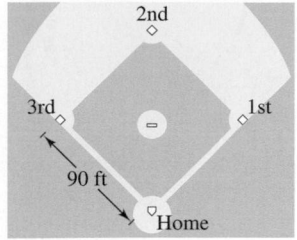

Figure for 27 and 28   Figure for 29

28. Sports For the baseball diamond in Exercise 27, suppose the player is running from first base to second base at a speed of 25 feet per second. Find the rate at which the distance from home plate is changing when the player is 20 feet from second base.

29. Shadow Length A man 6 feet tall walks at a rate of 5 feet per second away from a light that is 15 feet above the ground (see figure).

(a) When he is 10 feet from the base of the light, at what rate is the tip of his shadow moving?

(b) When he is 10 feet from the base of the light, at what rate is the length of his shadow changing?

30. Shadow Length Repeat Exercise 29 for a man 6 feet tall walking at a rate of 5 feet per second *toward* a light that is 20 feet above the ground (see figure).

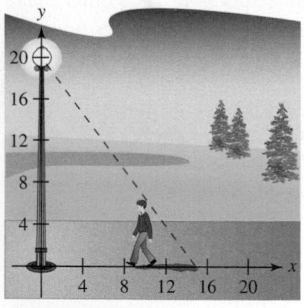 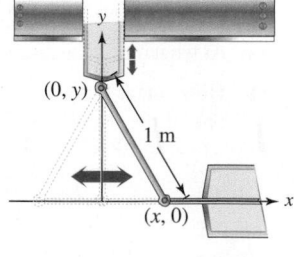

Figure for 30   Figure for 31

31. Machine Design The endpoints of a movable rod of length 1 meter have coordinates $(x, 0)$ and $(0, y)$ (see figure). The position of the end on the x-axis is

$$x(t) = \frac{1}{2} \sin \frac{\pi t}{6}$$

where t is the time in seconds.

(a) Find the time of one complete cycle of the rod.

(b) What is the lowest point reached by the end of the rod on the y-axis?

(c) Find the speed of the y-axis endpoint when the x-axis endpoint is $\left(\frac{1}{4}, 0\right)$.

32. Machine Design Repeat Exercise 31 for a position function of $x(t) = \frac{3}{5} \sin \pi t$. Use the point $\left(\frac{3}{10}, 0\right)$ for part (c).

33. Evaporation As a spherical raindrop falls, it reaches a layer of dry air and begins to evaporate at a rate that is proportional to its surface area $(S = 4\pi r^2)$. Show that the radius of the raindrop decreases at a constant rate.

34. HOW DO YOU SEE IT? Using the graph of f, (a) determine whether dy/dt is positive or negative given that dx/dt is negative, and (b) determine whether dx/dt is positive or negative given that dy/dt is positive. Explain.

(i)

(ii)

EXPLORING CONCEPTS

35. Think About It Describe the relationship between the rate of change of y and the rate of change of x in each expression. Assume all variables and derivatives are positive.

(a) $\dfrac{dy}{dt} = 3\dfrac{dx}{dt}$ (b) $\dfrac{dy}{dt} = x(L - x)\dfrac{dx}{dt}$, $\quad 0 \le x \le L$

36. Volume Let V be the volume of a cube of side length s that is changing with respect to time. If ds/dt is constant, is dV/dt constant? Explain.

37. Electricity The combined electrical resistance R of two resistors R_1 and R_2, connected in parallel, is given by

$$\frac{1}{R} = \frac{1}{R_1} + \frac{1}{R_2}$$

where R, R_1, and R_2 are measured in ohms. R_1 and R_2 are increasing at rates of 1 and 1.5 ohms per second, respectively. At what rate is R changing when $R_1 = 50$ ohms and $R_2 = 75$ ohms?

38. Electrical Circuit The voltage V in volts of an electrical circuit is $V = IR$, where R is the resistance in ohms and I is the current in amperes. R is increasing at a rate of 2 ohms per second, and V is increasing at a rate of 3 volts per second. At what rate is I changing when $V = 12$ volts and $R = 4$ ohms?

39. Flight Control An airplane is flying in still air with an airspeed of 275 miles per hour. The plane is climbing at an angle of $18°$. Find the rate at which the plane is gaining altitude.

40. Angle of Elevation A balloon rises at a rate of 4 meters per second from a point on the ground 50 meters from an observer. Find the rate of change of the angle of elevation of the balloon from the observer when the balloon is 50 meters above the ground.

41. Angle of Elevation A fish is reeled in at a rate of 1 foot per second from a point 10 feet above the water (see figure). At what rate is the angle θ between the line and the water changing when there is a total of 25 feet of line from the end of the rod to the water?

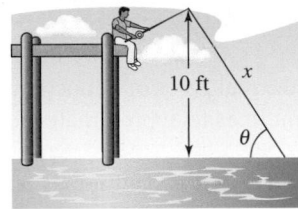

Figure for 41

Figure for 42

42. Angle of Elevation An airplane flies at an altitude of 5 miles toward a point directly over an observer (see figure). The speed of the plane is 600 miles per hour. Find the rates at which the angle of elevation θ is changing when the angle is (a) $\theta = 30°$, (b) $\theta = 60°$, and (c) $\theta = 75°$.

43. Linear vs. Angular Speed A patrol car is parked 50 feet from a long warehouse (see figure). The revolving light on top of the car turns at a rate of 30 revolutions per minute. How fast is the light beam moving along the wall when the beam makes angles of (a) $\theta = 30°$, (b) $\theta = 60°$, and (c) $\theta = 70°$ with the perpendicular line from the light to the wall?

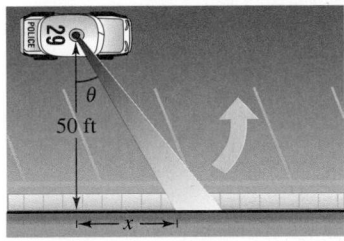

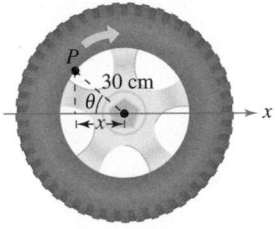

Figure for 43

Figure for 44

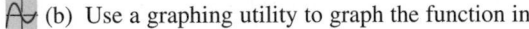

44. Linear vs. Angular Speed A wheel of radius 30 centimeters revolves at a rate of 10 revolutions per second. A dot is painted at a point P on the rim of the wheel (see figure).

(a) Find dx/dt as a function of θ.

(b) Use a graphing utility to graph the function in part (a).

(c) When is the absolute value of the rate of change of x greatest? When is it least?

(d) Find dx/dt when $\theta = 30°$ and $\theta = 60°$.

45. Area Consider the rectangle shown in the figure.

(a) Find the area of the rectangle as a function of x.

(b) Find the rate of the change of the area when $x = 4$ centimeters if $dx/dt = 4$ centimeters per minute.

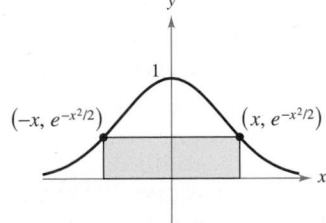

46. Area The included angle of the two sides of constant equal length s of an isosceles triangle is θ.

(a) Show that the area of the triangle is given by $A = \frac{1}{2}s^2 \sin \theta$.

(b) The angle θ is increasing at the rate of $\frac{1}{2}$ radian per minute. Find the rates of change of the area when $\theta = \pi/6$ and $\theta = \pi/3$.

47. Relative Humidity When the dewpoint is 65° Fahrenheit, the relative humidity H is

$$H = \frac{4347}{400,000,000}e^{369,444/(50t+19,793)}$$

where t is the temperature in degrees Fahrenheit.

(a) Determine the relative humidity when $t = 65°F$ and $t = 80°F$.

(b) At 10 A.M., the temperature is 75°F and increasing at the rate of 2°F per hour. Find the rate at which the relative humidity is changing.

48. Security Camera A security camera is centered 50 feet above a 100-foot hallway (see figure). It is easiest to design the camera with a constant angular rate of rotation, but this results in recording the images of the surveillance area at a variable rate. So, it is desirable to design a system with a variable rate of rotation and a constant rate of movement of the scanning beam along the hallway. Find a model for the variable rate of rotation when $|dx/dt| = 2$ feet per second.

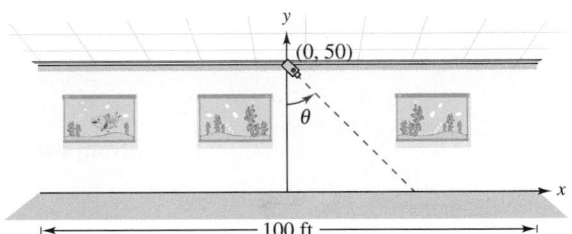

Acceleration In Exercises 49 and 50, find the acceleration of the specified object. (*Hint:* Recall that if a variable is changing at a constant rate, then its acceleration is zero.)

49. Find the acceleration of the top of the ladder described in Exercise 21 when the base of the ladder is 7 feet from the wall.

50. Find the acceleration of the boat in Exercise 24(a) when there is a total of 13 feet of rope out.

51. Moving Shadow A ball is dropped from a height of 20 meters, 12 meters away from the top of a 20-meter lamppost (see figure). The ball's shadow, caused by the light at the top of the lamppost, is moving along the level ground. How fast is the shadow moving 1 second after the ball is released? (*Submitted by Dennis Gittinger, St. Philips College, San Antonio, TX*)

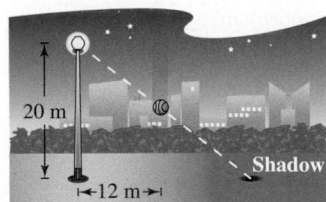

3.8 Newton's Method

■ Approximate a zero of a function using Newton's Method.

Newton's Method

In this section, you will study a technique for approximating the real zeros of a function. The technique is called **Newton's Method,** and it uses tangent lines to approximate the graph of the function near its x-intercepts.

To see how Newton's Method works, consider a function f that is continuous on the interval $[a, b]$ and differentiable on the interval (a, b). If $f(a)$ and $f(b)$ differ in sign, then, by the Intermediate Value Theorem, f must have at least one zero in the interval (a, b). To estimate this zero, you choose

$$x = x_1 \qquad \text{First estimate}$$

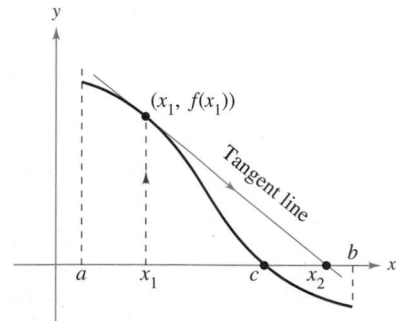

(a)

as shown in Figure 3.42(a). Newton's Method is based on the assumption that the graph of f and the tangent line at $(x_1, f(x_1))$ both cross the x-axis at *about* the same point. Because you can easily calculate the x-intercept for this tangent line, you can use it as a second (and, usually, better) estimate of the zero of f. The tangent line passes through the point $(x_1, f(x_1))$ with a slope of $f'(x_1)$. In point-slope form, the equation of the tangent line is

$$y - f(x_1) = f'(x_1)(x - x_1)$$
$$y = f'(x_1)(x - x_1) + f(x_1).$$

Letting $y = 0$ and solving for x produces

$$x = x_1 - \frac{f(x_1)}{f'(x_1)}.$$

So, from the initial estimate x_1, you obtain a new estimate

$$x_2 = x_1 - \frac{f(x_1)}{f'(x_1)}. \qquad \text{Second estimate [See Figure 3.42(b).]}$$

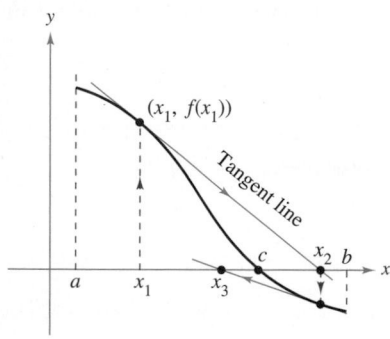

(b)
The x-intercept of the tangent line approximates the zero of f.
Figure 3.42

You can improve on x_2 and calculate yet a third estimate

$$x_3 = x_2 - \frac{f(x_2)}{f'(x_2)}. \qquad \text{Third estimate}$$

Repeated application of this process is called Newton's Method.

NEWTON'S METHOD

Isaac Newton first described the method for approximating the real zeros of a function in his text *Method of Fluxions.* Although the book was written in 1671, it was not published until 1736. Meanwhile, in 1690, Joseph Raphson (1648–1715) published a paper describing a method for approximating the real zeros of a function that was very similar to Newton's. For this reason, the method is often referred to as the Newton-Raphson Method.

Newton's Method for Approximating the Zeros of a Function

Let $f(c) = 0$, where f is differentiable on an open interval containing c. Then, to approximate c, use these steps.

1. Make an initial estimate x_1 that is close to c. (A graph is helpful.)
2. Determine a new approximation

$$x_{n+1} = x_n - \frac{f(x_n)}{f'(x_n)}.$$

3. When $|x_n - x_{n+1}|$ is within the desired accuracy, let x_{n+1} serve as the final approximation. Otherwise, return to Step 2 and calculate a new approximation.

Each successive application of this procedure is called an **iteration.**

· · · · · · · · · · · · · · · · · ▷

REMARK For many functions, just a few iterations of Newton's Method will produce approximations having very small errors, as shown in Example 1.

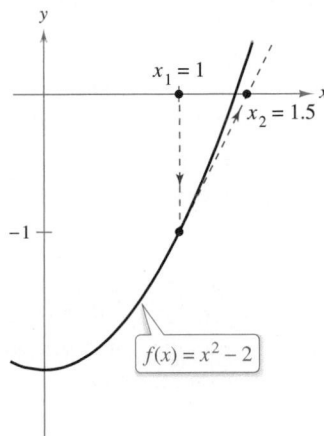

The first iteration of Newton's Method
Figure 3.43

EXAMPLE 1 Using Newton's Method

Calculate three iterations of Newton's Method to approximate a zero of $f(x) = x^2 - 2$. Use $x_1 = 1$ as the initial guess.

Solution Because $f(x) = x^2 - 2$, you have $f'(x) = 2x$, and the iterative formula is

$$x_{n+1} = x_n - \frac{f(x_n)}{f'(x_n)} = x_n - \frac{x_n^2 - 2}{2x_n}.$$

The calculations for three iterations are shown in the table.

n	x_n	$f(x_n)$	$f'(x_n)$	$\dfrac{f(x_n)}{f'(x_n)}$	$x_n - \dfrac{f(x_n)}{f'(x_n)}$
1	1.000000	−1.000000	2.000000	−0.500000	1.500000
2	1.500000	0.250000	3.000000	0.083333	1.416667
3	1.416667	0.006945	2.833334	0.002451	1.414216
4	1.414216				

Of course, in this case you know that the two zeros of the function are $\pm\sqrt{2}$. To six decimal places, $\sqrt{2} = 1.414214$. So, after only three iterations of Newton's Method, you have obtained an approximation that is within 0.000002 of an actual root. The first iteration of this process is shown in Figure 3.43.

EXAMPLE 2 Using Newton's Method

· · · · ▷ *See LarsonCalculus.com for an interactive version of this type of example.*

Use Newton's Method to approximate the zero(s) of

$$f(x) = e^x + x.$$

Continue the iterations until two successive approximations differ by less than 0.0001.

Solution Begin by sketching a graph of f, as shown in Figure 3.44. From the graph, you can observe that the function has only one zero, which occurs near $x = -0.6$. Next, differentiate f and form the iterative formula

$$x_{n+1} = x_n - \frac{f(x_n)}{f'(x_n)} = x_n - \frac{e^{x_n} + x_n}{e^{x_n} + 1}.$$

The calculations are shown in the table.

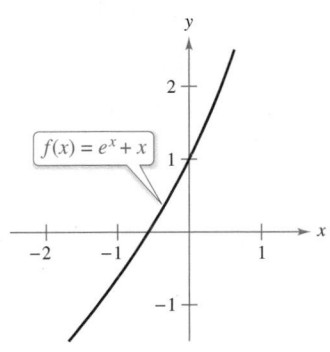

After three iterations of Newton's Method, the zero of f is approximated to the desired accuracy.
Figure 3.44

n	x_n	$f(x_n)$	$f'(x_n)$	$\dfrac{f(x_n)}{f'(x_n)}$	$x_n - \dfrac{f(x_n)}{f'(x_n)}$
1	−0.60000	−0.05119	1.54881	−0.03305	−0.56695
2	−0.56695	0.00030	1.56725	0.00019	−0.56714
3	−0.56714	0.00000	1.56714	0.00000	−0.56714
4	−0.56714				

Because two successive approximations differ by less than the required 0.0001, you can estimate the zero of f to be −0.56714. ∎

When, as in Examples 1 and 2, the approximations approach a limit, the sequence of approximations

$$x_1, \, x_2, \, x_3, \, \ldots, \, x_n, \, \ldots$$

is said to **converge.** Moreover, when the limit is c, it can be shown that c must be a zero of f.

Newton's Method does not always yield a convergent sequence. One way it can fail to do so is shown in the figure at the right. Because Newton's Method involves division by $f'(x_n)$, it is clear that the method will fail when the derivative is zero for any x_n in the sequence. When you encounter this problem, you can usually overcome it by choosing a different value for x_1. Another way Newton's Method can fail is shown in the next example.

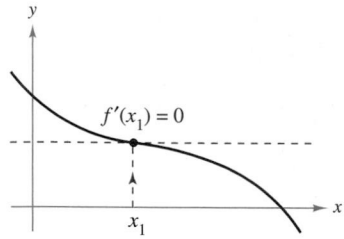

Newton's Method fails to converge when $f'(x_n) = 0$.

■ **FOR FURTHER INFORMATION**
For more on when Newton's Method fails, see the article "No Fooling! Newton's Method Can Be Fooled" by Peter Horton in *Mathematics Magazine*. To view this article, go to *MathArticles.com*.

EXAMPLE 3 **An Example in Which Newton's Method Fails**

The function $f(x) = x^{1/3}$ is not differentiable at $x = 0$. Show that Newton's Method fails to converge using $x_1 = 0.1$.

Solution Because $f'(x) = \frac{1}{3}x^{-2/3}$, the iterative formula is

$$x_{n+1} = x_n - \frac{f(x_n)}{f'(x_n)} = x_n - \frac{x_n^{1/3}}{\frac{1}{3}x_n^{-2/3}} = x_n - 3x_n = -2x_n.$$

The calculations are shown in the table. This table and Figure 3.45 indicate that x_n continues to increase in magnitude as $n \to \infty$, and so the limit of the sequence does not exist.

n	x_n	$f(x_n)$	$f'(x_n)$	$\dfrac{f(x_n)}{f'(x_n)}$	$x_n - \dfrac{f(x_n)}{f'(x_n)}$
1	0.10000	0.46416	1.54720	0.30000	-0.20000
2	-0.20000	-0.58480	0.97467	-0.60000	0.40000
3	0.40000	0.73681	0.61401	1.20000	-0.80000
4	-0.80000	-0.92832	0.38680	-2.40000	1.60000

• • • • • • • • • • • • • • • ▷
•
• • **REMARK** In Example 3, the initial estimate $x_1 = 0.1$ fails to produce a convergent sequence. Try showing that Newton's Method also fails for every other choice of x_1 (other than the actual zero).

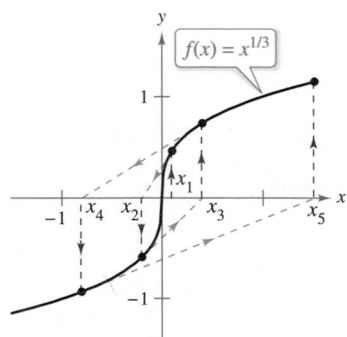

Newton's Method fails to converge for every x-value other than the actual zero of f.
Figure 3.45

It can be shown that a condition sufficient to produce convergence of Newton's Method to a zero of f is that

$$\left|\frac{f(x)f''(x)}{[f'(x)]^2}\right| < 1 \qquad \text{Condition for convergence}$$

on an open interval containing the zero. For instance, in Example 1, this test would yield

$$f(x) = x^2 - 2, \quad f'(x) = 2x, \quad f''(x) = 2,$$

and

$$\left|\frac{f(x)f''(x)}{[f'(x)]^2}\right| = \left|\frac{(x^2 - 2)(2)}{4x^2}\right| = \left|\frac{1}{2} - \frac{1}{x^2}\right|. \qquad \text{Example 1}$$

On the interval $(1, 3)$, this quantity is less than 1 and therefore the convergence of Newton's Method is guaranteed. On the other hand, in Example 3, you have

$$f(x) = x^{1/3}, \quad f'(x) = \frac{1}{3}x^{-2/3}, \quad f''(x) = -\frac{2}{9}x^{-5/3}$$

and

$$\left|\frac{f(x)f''(x)}{[f'(x)]^2}\right| = \left|\frac{x^{1/3}(-2/9)(x^{-5/3})}{(1/9)(x^{-4/3})}\right| = 2 \qquad \text{Example 3}$$

which is not less than 1 for any value of x, so you cannot conclude that Newton's Method will converge.

You have learned several techniques for finding the zeros of functions. The zeros of some functions, such as

$$f(x) = x^3 - 2x^2 - x + 2$$

can be found by simple algebraic techniques, such as factoring. The zeros of other functions, such as

$$f(x) = x^3 - x + 1$$

cannot be found by *elementary* algebraic methods. This particular function has only one real zero, and by using more advanced algebraic techniques, you can determine the zero to be

$$x = -\sqrt[3]{\frac{3 - \sqrt{23/3}}{6}} - \sqrt[3]{\frac{3 + \sqrt{23/3}}{6}}.$$

Because the *exact* solution is written in terms of square roots and cube roots, it is called a **solution by radicals.**

The determination of radical solutions of a polynomial equation is one of the fundamental problems of algebra. The earliest such result is the Quadratic Formula, which dates back at least to Babylonian times. The general formula for the zeros of a cubic function was developed much later. In the sixteenth century, an Italian mathematician, Jerome Cardan, published a method for finding radical solutions to cubic and quartic equations. Then, for 300 years, the problem of finding a general quintic formula remained open. Finally, in the nineteenth century, the problem was answered independently by two young mathematicians. Niels Henrik Abel, a Norwegian mathematician, and Evariste Galois, a French mathematician, proved that it is not possible to solve a *general* fifth- (or higher-) degree polynomial equation by radicals. Of course, you can solve particular fifth-degree equations, such as $x^5 - 1 = 0$, but Abel and Galois were able to show that no general *radical* solution exists.

NIELS HENRIK ABEL (1802–1829)

EVARISTE GALOIS (1811–1832)

Although the lives of both Abel and Galois were brief, their work in the fields of analysis and abstract algebra was far-reaching.

See LarsonCalculus.com to read a biography about each of these mathematicians.

3.8 Exercises

See **CalcChat.com** for tutorial help and worked-out solutions to odd-numbered exercises.

CONCEPT CHECK

1. **Newton's Method** In your own words and using a sketch, describe Newton's Method for approximating the zeros of a function.

2. **Failure of Newton's Method** Why does Newton's Method fail when $f'(x_n) = 0$? What does this mean graphically?

Using Newton's Method In Exercises 3–6, calculate two iterations of Newton's Method to approximate a zero of the function using the given initial guess.

3. $f(x) = x^2 - 5$, $x_1 = 2$

4. $f(x) = x^3 - 3$, $x_1 = 1.4$

5. $f(x) = \cos x$, $x_1 = 1.6$

6. $f(x) = \tan x$, $x_1 = 0.1$

Using Newton's Method In Exercises 7–18, use Newton's Method to approximate the zero(s) of the function. Continue the iterations until two successive approximations differ by less than 0.001. Then find the zero(s) using a graphing utility and compare the results.

7. $f(x) = x^3 + 4$

8. $f(x) = 2 - x^3$

9. $f(x) = x^3 + x - 1$

10. $f(x) = x^5 + x - 1$

11. $f(x) = 5\sqrt{x-1} - 2x$

12. $f(x) = x - 2\sqrt{x+1}$

13. $f(x) = x - e^{-x}$

14. $f(x) = x - 3 + \ln x$

15. $f(x) = x^3 - 3.9x^2 + 4.79x - 1.881$

16. $f(x) = -x^3 + 2.7x^2 + 3.55x - 2.422$

17. $f(x) = 1 - x + \sin x$

18. $f(x) = x^3 - \cos x$

Points of Intersection In Exercises 19–22, apply Newton's Method to approximate the x-value(s) of the indicated point(s) of intersection of the two graphs. Continue the iterations until two successive approximations differ by less than 0.001. [*Hint:* Let $h(x) = f(x) - g(x)$.]

19. $f(x) = 2x + 1$
    $g(x) = \sqrt{x+4}$

20. $f(x) = \arccos x$
    $g(x) = \arctan x$

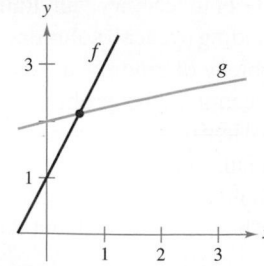

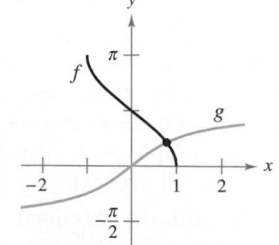

21. $f(x) = x$
    $g(x) = \tan x$

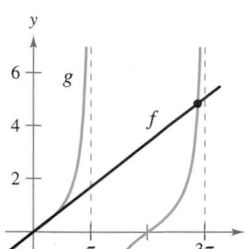

22. $f(x) = 2 - x^2$
    $g(x) = e^{x/2}$

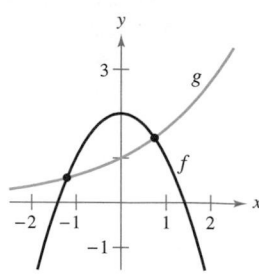

23. **Using Newton's Method** Consider the function $f(x) = x^3 - 3x^2 + 3$.

    (a) Use a graphing utility to graph f.

    (b) Use Newton's Method to approximate a zero with $x_1 = 1$ as the initial guess.

    (c) Repeat part (b) using $x_1 = \frac{1}{4}$ as the initial guess and observe that the result is different.

    (d) To understand why the results in parts (b) and (c) are different, sketch the tangent lines to the graph of f at the points $(1, f(1))$ and $\left(\frac{1}{4}, f\left(\frac{1}{4}\right)\right)$. Describe why it is important to select the initial guess carefully.

24. **Using Newton's Method** Repeat the steps in Exercise 23 for the function $f(x) = \sin x$ with initial guesses of $x_1 = 1.8$ and $x_1 = 3$.

Failure of Newton's Method In Exercises 25 and 26, apply Newton's Method using the given initial guess, and explain why the method fails.

25. $y = 2x^3 - 6x^2 + 6x - 1$, $x_1 = 1$

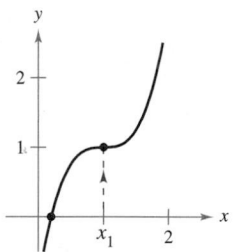

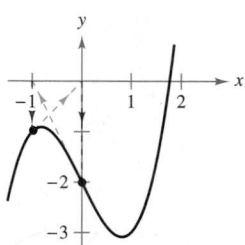

Figure for 25   Figure for 26

26. $y = x^3 - 2x - 2$, $x_1 = 0$

Fixed Point In Exercises 27–30, approximate the fixed point of the function to two decimal places. [A *fixed point* of a function f is a real number c such that $f(c) = c$.]

27. $f(x) = \cos x$

28. $f(x) = \cot x$, $0 < x < \pi$

29. $f(x) = e^{x/10}$

30. $f(x) = -\ln x$

31. Advertising Costs   A company that produces digital audio players estimates that the profit for selling a particular model is

$$P = -76x^3 + 4830x^2 - 320,000, \quad 0 \le x \le 60$$

where P is the profit in dollars and x is the advertising expense in 10,000s of dollars (see figure). Use Newton's Method to approximate the smaller of two advertising amounts that yield a profit of $2,500,000.

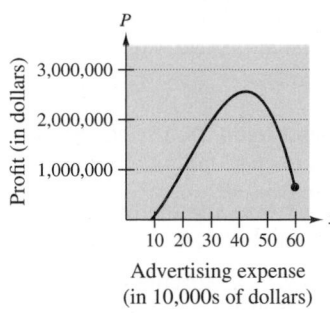

Figure for 31   Figure for 32

32. Engine Power   The torque produced by a compact automobile engine is approximated by the model

$$T = 0.808x^3 - 17.974x^2 + 71.248x + 110.843, \quad 1 \le x \le 5$$

where T is the torque in foot-pounds and x is the engine speed in thousands of revolutions per minute (see figure). Use Newton's Method to approximate the two engine speeds that yield a torque of 170 foot-pounds.

EXPLORING CONCEPTS

33. Newton's Method   What will be the values of future guesses for x if your initial guess is a zero of f? Explain.

34. Newton's Method   Does Newton's Method fail when the initial guess is a relative maximum of f? Explain.

35. Comparing Functions   Determine whether the zeros of $f(x) = p(x)/q(x)$ always coincide with the zeros of $p(x)$. Explain.

36. HOW DO YOU SEE IT?   For what value(s) will Newton's Method fail to converge for the function shown in the graph? Explain your reasoning.

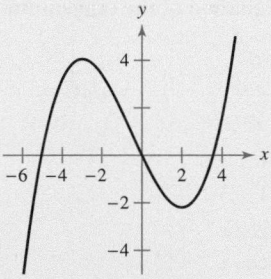

37. Mechanic's Rule   The Mechanic's Rule for approximating $\sqrt{a}$, $a > 0$, is

$$x_{n+1} = \frac{1}{2}\left(x_n + \frac{a}{x_n}\right), \quad n = 1, 2, 3, \ldots$$

where x_1 is an approximation of $\sqrt{a}$.

(a) Use Newton's Method and the function $f(x) = x^2 - a$ to derive the Mechanic's Rule.

(b) Use the Mechanic's Rule to approximate $\sqrt{5}$ and $\sqrt{7}$ to three decimal places.

38. Approximating Radicals

(a) Use Newton's Method and the function $f(x) = x^n - a$ to obtain a general rule for approximating $x = \sqrt[n]{a}$.

(b) Use the general rule found in part (a) to approximate $\sqrt[4]{6}$ and $\sqrt[3]{15}$ to three decimal places.

39. Approximating Reciprocals   Use Newton's Method to show that the equation $x_{n+1} = x_n(2 - ax_n)$ can be used to approximate $1/a$ when x_1 is an initial guess of the reciprocal of a. Note that this method of approximating reciprocals uses only the operations of multiplication and subtraction. (*Hint:* Consider

$$f(x) = \frac{1}{x} - a.\Big)$$

40. Approximating Reciprocals   Use the result of Exercise 39 to approximate (a) $\frac{1}{3}$ and (b) $\frac{1}{11}$ to three decimal places.

True or False?   **In Exercises 41–44, determine whether the statement is true or false. If it is false, explain why or give an example that shows it is false.**

41. The roots of $\sqrt{f(x)} = 0$ coincide with the roots of $f(x) = 0$.

42. If the coefficients of a polynomial function are all positive, then the polynomial has no positive zeros.

43. If $f(x)$ is a cubic polynomial such that $f'(x)$ is never zero, then any initial guess will force Newton's Method to converge to the zero of f.

44. Newton's Method fails when the initial guess x_1 corresponds to a horizontal tangent line for the graph of f at x_1.

45. Tangent Lines   The graph of $f(x) = -\sin x$ has infinitely many tangent lines that pass through the origin. Use Newton's Method to approximate to three decimal places the slope of the tangent line having the greatest slope.

46. Point of Tangency   The graph of $f(x) = \cos x$ and a tangent line to f through the origin are shown. Find the coordinates of the point of tangency to three decimal places.

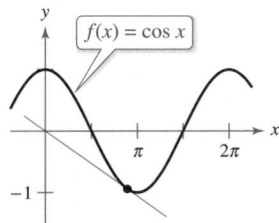

Review Exercises   See CalcChat.com for tutorial help and worked-out solutions to odd-numbered exercises.

Finding the Derivative by the Limit Process In Exercises 1–4, find the derivative of the function by the limit process.

1. $f(x) = 12$

2. $f(x) = 5x - 4$

3. $f(x) = x^3 - 2x + 1$

4. $f(x) = \dfrac{6}{x}$

Using the Alternative Form of the Derivative In Exercises 5 and 6, use the alternative form of the derivative to find the derivative at $x = c$, if it exists.

5. $g(x) = 2x^2 - 3x,\quad c = 2$

6. $f(x) = \dfrac{1}{x + 4},\quad c = 3$

Determining Differentiability In Exercises 7 and 8, describe the x-values at which f is differentiable.

7. $f(x) = (x - 3)^{2/5}$

8. $f(x) = \dfrac{3x}{x + 1}$

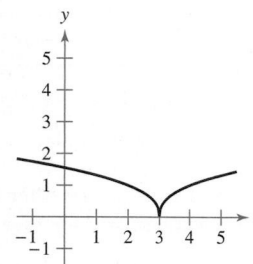

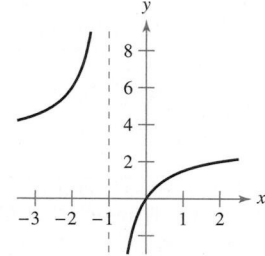

Finding a Derivative In Exercises 9–20, use the rules of differentiation to find the derivative of the function.

9. $y = 25$

10. $f(t) = \pi/6$

11. $f(x) = x^3 - 11x^2$

12. $g(s) = 3s^5 - 2s^4$

13. $h(x) = 6\sqrt{x} + 3\sqrt[3]{x}$

14. $f(x) = x^{1/2} - x^{-5/6}$

15. $g(t) = \dfrac{2}{3t^2}$

16. $h(x) = \dfrac{8}{5x^4}$

17. $f(\theta) = 4\theta - 5\sin\theta$

18. $g(\alpha) = 4\cos\alpha + 6$

19. $f(t) = 3\cos t - 4e^t$

20. $g(s) = \frac{5}{3}\sin s - 2e^s$

Finding the Slope of a Graph In Exercises 21–24, find the slope of the graph of the function at the given point.

21. $f(x) = \dfrac{27}{x^3},\quad (3, 1)$

22. $f(x) = 3x^2 - 4x,\quad (1, -1)$

23. $f(x) = 4x^5 + 3x - \sin x,\quad (0, 0)$

24. $f(x) = 5\cos x - 9x,\quad (0, 5)$

25. Vibrating String When a guitar string is plucked, it vibrates with a frequency of $F = 200\sqrt{T}$, where F is measured in vibrations per second and the tension T is measured in pounds. Find the rates of change of the frequency when (a) $T = 4$ pounds and (b) $T = 9$ pounds.

26. Surface Area The surface area of a cube with sides of length x is given by $S = 6x^2$. Find the rate of change of the surface area with respect to x when $x = 4$ inches.

Vertical Motion In Exercises 27 and 28, use the position function $s(t) = -16t^2 + v_0 t + s_0$ for free-falling objects.

27. A ball is thrown straight down from the top of a 600-foot building with an initial velocity of -30 feet per second.

   (a) Determine the position and velocity functions for the ball.

   (b) Determine the average velocity on the interval $[1, 3]$.

   (c) Find the instantaneous velocities when $t = 1$ and $t = 3$.

   (d) Find the time required for the ball to reach ground level.

   (e) Find the velocity of the ball at impact.

28. A block is dropped from the top of a 450-foot platform. What is its velocity after 2 seconds? After 5 seconds?

Finding a Derivative In Exercises 29–42, use the Product Rule or the Quotient Rule to find the derivative of the function.

29. $f(x) = (5x^2 + 8)(x^2 - 4x - 6)$

30. $g(x) = (2x^3 + 5x)(3x - 4)$

31. $f(x) = (9x - 1)\sin x$

32. $f(t) = 2t^5 \cos t$

33. $f(x) = \dfrac{x^2 + x - 1}{x^2 - 1}$

34. $f(x) = \dfrac{2x + 7}{x^2 + 4}$

35. $y = \dfrac{x^4}{\cos x}$

36. $y = \dfrac{\sin x}{x^4}$

37. $y = 3x^2 \sec x$

38. $y = -x^2 \tan x$

39. $y = x\cos x - \sin x$

40. $g(x) = x^4 \cot x + 3x\cos x$

41. $y = 4xe^x - \cot x$

42. $y = \dfrac{8x + 1}{e^x}$

Finding an Equation of a Tangent Line In Exercises 43–46, find an equation of the tangent line to the graph of f at the given point.

43. $f(x) = (x + 2)(x^2 + 5),\quad (-1, 6)$

44. $f(x) = (x - 4)(x^2 + 6x - 1),\quad (0, 4)$

45. $f(x) = \dfrac{x + 1}{x - 1},\quad \left(\dfrac{1}{2}, -3\right)$

46. $f(x) = \dfrac{1 + \cos x}{1 - \cos x},\quad \left(\dfrac{\pi}{2}, 1\right)$

Finding a Second Derivative In Exercises 47–54, find the second derivative of the function.

47. $g(t) = -8t^3 - 5t + 12$ **48.** $h(x) = 6x^{-2} + 7x^2$

49. $f(x) = 15x^{5/2}$ **50.** $f(x) = 20\sqrt[5]{x}$

51. $f(\theta) = 3 \tan \theta$ **52.** $h(t) = 10 \cos t - 15 \sin t$

53. $g(x) = 4 \cot x$

54. $h(t) = -12 \csc t$

55. Acceleration The velocity of an object is $v(t) = 20 - t^2$, $0 \le t \le 6$, where v is measured in meters per second and t is the time in seconds. Find the velocity and acceleration of the object when $t = 3$.

56. Acceleration The velocity of an automobile starting from rest is

$$v(t) = \frac{90t}{4t + 10}$$

where v is measured in feet per second and t is the time in seconds. Find the acceleration at (a) 1 second, (b) 5 seconds, and (c) 10 seconds.

Finding a Derivative In Exercises 57–82, find the derivative of the function.

57. $y = (7x + 3)^4$ **58.** $y = (x^2 - 6)^3$

59. $y = \dfrac{1}{(x^2 + 5)^3}$ **60.** $f(x) = \dfrac{1}{(5x + 1)^2}$

61. $y = 5 \cos(9x + 1)$ **62.** $y = -6 \sin 3x^4$

63. $y = \dfrac{x}{2} - \dfrac{\sin 2x}{4}$ **64.** $y = \dfrac{\sec^7 x}{7} - \dfrac{\sec^5 x}{5}$

65. $y = x(6x + 1)^5$ **66.** $f(s) = (s^2 - 1)^{5/2}(s^3 + 5)$

67. $f(x) = \left(\dfrac{x}{\sqrt{x+5}}\right)^3$ **68.** $h(x) = \left(\dfrac{x+5}{x^2+3}\right)^2$

69. $h(z) = e^{-z^2/2}$ **70.** $y = 3e^{-3/t}$

71. $g(t) = t^2 e^{1/4}$ **72.** $g(x) = \dfrac{x^2}{e^{9-x}}$

73. $y = \sqrt{e^{2x} + e^{-2x}}$ **74.** $f(\theta) = \dfrac{1}{2}e^{\sin 2\theta}$

75. $g(x) = \ln \sqrt{x}$ **76.** $f(x) = \ln(2 - x^2)$

77. $f(x) = x\sqrt{\ln x}$ **78.** $y = \dfrac{x+1}{\ln x}$

79. $g(x) = \ln\left(x^3\sqrt{4x + 1}\right)$ **80.** $f(x) = \ln[x(x^2 - 2)^{2/3}]$

81. $f(x) = \ln\dfrac{x+4}{x(x+5)}$ **82.** $h(x) = \ln\dfrac{x(x-1)}{x-2}$

Finding the Slope of a Graph In Exercises 83–88, find the slope of the graph of the function at the given point.

83. $f(x) = \sqrt{1 - x^3}$, $(-2, 3)$ **84.** $f(x) = \sqrt[3]{x^2 - 1}$, $(3, 2)$

85. $f(x) = \dfrac{x+8}{\sqrt{3x+1}}$, $(0, 8)$ **86.** $f(x) = \dfrac{3x+1}{(4x-3)^3}$, $(1, 4)$

87. $y = \dfrac{1}{2}\csc 2x$, $\left(\dfrac{\pi}{4}, \dfrac{1}{2}\right)$ **88.** $y = \csc 3x + \cot 3x$, $\left(\dfrac{\pi}{6}, 1\right)$

Finding a Second Derivative In Exercises 89–92, find the second derivative of the function.

89. $y = (8x + 5)^3$ **90.** $y = \dfrac{1}{5x + 1}$

91. $f(x) = \cot x$ **92.** $y = x \sin^2 x$

93. Refrigeration The temperature T (in degrees Fahrenheit) of food in a freezer is

$$T = \frac{700}{t^2 + 4t + 10}$$

where t is the time in hours. Find the rate of change of T with respect to t at each of the following times.

(a) $t = 1$ (b) $t = 3$ (c) $t = 5$ (d) $t = 10$

94. Harmonic Motion The displacement from equilibrium of an object in harmonic motion on the end of a spring is

$$y = \frac{1}{4}\cos 8t - \frac{1}{4}\sin 8t$$

where y is measured in feet and t is the time in seconds. Determine the position and velocity of the object when $t = \pi/4$.

95. Modeling Data The atmospheric pressure decreases with increasing altitude. At sea level, the average air pressure is one atmosphere (1.033227 kilograms per square centimeter). The table gives the pressures p (in atmospheres) at various altitudes h (in kilometers).

h	0	5	10	15	20	25
p	1	0.55	0.25	0.12	0.06	0.02

(a) Use a graphing utility to find a model of the form $p = a + b \ln h$ for the data. Explain why the result is an error message.

(b) Use a graphing utility to find the logarithmic model $h = a + b \ln p$ for the data.

(c) Use a graphing utility to plot the data and graph the model from part (b).

(d) Use the model to estimate the altitude at which the pressure is 0.75 atmosphere.

(e) Use the model to estimate the pressure at an altitude of 13 kilometers.

(f) Use the model to find the rates of change of pressure when $h = 5$ and $h = 20$. Interpret the results in the context of the problem.

96. Tractrix A person walking along a dock drags a boat by a 10-meter rope. The boat travels along a path known as a *tractrix*. The equation of this path is

$$y = 10 \ln\left(\frac{10 + \sqrt{100 - x^2}}{x}\right) - \sqrt{100 - x^2}.$$

(a) Use a graphing utility to graph the function.

(b) What are the slopes of this path when $x = 5$ and $x = 9$?

(c) What does the slope of the path approach as x approaches 10 from the left?

Finding a Derivative In Exercises 97–102, find dy/dx by implicit differentiation.

97. $x^2 + y^2 = 64$

98. $x^2 + 4xy - y^3 = 6$

99. $x^3y - xy^3 = 4$

100. $\sqrt{xy} = x - 4y$

101. $x \sin y = y \cos x$

102. $\cos(x + y) = x$

Tangent Lines and Normal Lines In Exercises 103–106, find equations for the tangent line and the normal line to the graph of the equation at the given point. (The *normal line* at a point is perpendicular to the tangent line at the point.) Use a graphing utility to graph the equation, the tangent line, and the normal line.

103. $x^2 + y^2 = 10$, $(3, 1)$

104. $x^2 - y^2 = 20$, $(6, 4)$

105. $y \ln x + y^2 = 0$, $(e, -1)$

106. $\ln(x + y) = x$, $(0, 1)$

Logarithmic Differentiation In Exercises 107 and 108, use logarithmic differentiation to find dy/dx.

107. $y = \dfrac{x\sqrt{x^2 + 1}}{x + 4}$

108. $y = \dfrac{(2x + 1)^3(x^2 - 1)^2}{x + 3}$

Evaluating the Derivative of an Inverse Function In Exercises 109–112, verify that f has an inverse function. Then use the function f and the given real number a to find $(f^{-1})'(a)$. (*Hint:* Use Theorem 3.17.)

109. $f(x) = x^3 + 2$, $a = -1$

110. $f(x) = x\sqrt{x - 3}$, $a = 4$

111. $f(x) = \tan x$, $-\dfrac{\pi}{4} \le x \le \dfrac{\pi}{4}$, $a = \dfrac{\sqrt{3}}{3}$

112. $f(x) = \cos x$, $0 \le x \le \pi$, $a = 0$

Finding a Derivative In Exercises 113–118, find the derivative of the function.

113. $y = \sin(\arctan 2x)$

114. $y = \arctan(2x^2 - 3)$

115. $y = x \operatorname{arcsec} x$

116. $y = \frac{1}{2} \arctan e^{2x}$

117. $y = x(\arcsin x)^2 - 2x + 2\sqrt{1 - x^2} \arcsin x$

118. $y = \sqrt{x^2 - 4} - 2 \operatorname{arcsec} \dfrac{x}{2}$, $2 < x < 4$

119. Rate of Change A point moves along the curve $y = \sqrt{x}$ in such a way that the y-component of the position of the point is increasing at a rate of 2 units per second. At what rate is the x-component changing for each of the following values?

(a) $x = \frac{1}{2}$ (b) $x = 1$ (c) $x = 4$

120. Surface Area All edges of a cube are expanding at a rate of 8 centimeters per second. How fast is the surface area changing when each edge is 6.5 centimeters?

121. Linear vs. Angular Speed A rotating beacon is located 1 kilometer off a straight shoreline (see figure). The beacon rotates at a rate of 3 revolutions per minute. How fast (in kilometers per hour) does the beam of light appear to be moving to a viewer who is $\frac{1}{2}$ kilometer down the shoreline?

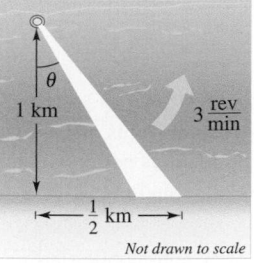

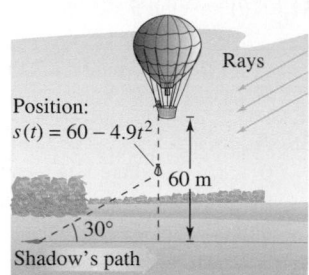

Figure for 121   Figure for 122

122. Moving Shadow A sandbag is dropped from a balloon at a height of 60 meters when the angle of elevation to the sun is $30°$ (see figure). The position of the sandbag is

$$s(t) = 60 - 4.9t^2.$$

Find the rate at which the shadow of the sandbag is traveling along the ground when the sandbag is at a height of 35 meters.

Using Newton's Method In Exercises 123–128, use Newton's Method to approximate the zero(s) of the function. Continue the iterations until two successive approximations differ by less than 0.001. Then find the zero(s) using a graphing utility and compare the results.

123. $f(x) = x^3 - 3x - 1$

124. $f(x) = x^3 + 2x + 1$

125. $g(x) = xe^x - 4$

126. $f(x) = 3 - x \ln x$

127. $f(x) = x^4 + x^3 - 3x^2 + 2$

128. $f(x) = 3\sqrt{x - 1} - x$

Points of Intersection In Exercises 129 and 130, apply Newton's Method to approximate the x-value(s) of the point(s) of intersection of the two graphs. Continue the iterations until two successive approximations differ by less than 0.001. [*Hint:* Let $h(x) = f(x) - g(x)$.]

129. $f(x) = -x$
$g(x) = \ln x$

130. $f(x) = x^4$
$g(x) = x + 3$

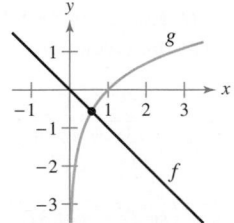

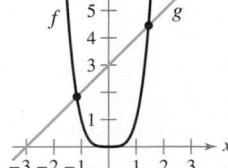

Figure for 129   Figure for 130

P.S. Problem Solving

See **CalcChat.com** for tutorial help and worked-out solutions to odd-numbered exercises.

1. Finding Equations of Circles Consider the graph of the parabola $y = x^2$.

(a) Find the radius r of the largest possible circle centered on the y-axis that is tangent to the parabola at the origin, as shown in the figure. This circle is called the **circle of curvature** (see Section 12.5). Find the equation of this circle. Use a graphing utility to graph the circle and parabola in the same viewing window to verify your answer.

(b) Find the center $(0, b)$ of the circle of radius 1 centered on the y-axis that is tangent to the parabola at two points, as shown in the figure. Find the equation of this circle. Use a graphing utility to graph the circle and parabola in the same viewing window to verify your answer.

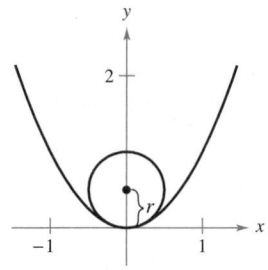

Figure for 1(a)   Figure for 1(b)

2. Finding Equations of Tangent Lines Graph the two parabolas

$$y = x^2 \quad \text{and} \quad y = -x^2 + 2x - 5$$

in the same coordinate plane. Find equations of the two lines that are simultaneously tangent to both parabolas.

3. Finding a Polynomial Find a third-degree polynomial $p(x)$ that is tangent to the line $y = 14x - 13$ at the point $(1, 1)$, and tangent to the line $y = -2x - 5$ at the point $(-1, -3)$.

4. Finding a Function Find a function of the form $f(x) = a + b \cos cx$ that is tangent to the line $y = 1$ at the point $(0, 1)$, and tangent to the line

$$y = x + \frac{3}{2} - \frac{\pi}{4}$$

at the point $\left(\dfrac{\pi}{4}, \dfrac{3}{2}\right)$.

5. Tangent Lines and Normal Lines

(a) Find an equation of the tangent line to the parabola $y = x^2$ at the point $(2, 4)$.

(b) Find an equation of the normal line to $y = x^2$ at the point $(2, 4)$. (The *normal line* at a point is perpendicular to the tangent line at the point.) Where does this line intersect the parabola a second time?

(c) Find equations of the tangent line and normal line to $y = x^2$ at the point $(0, 0)$.

(d) Prove that for any point $(a, b) \neq (0, 0)$ on the parabola $y = x^2$, the normal line intersects the graph a second time.

6. Finding Polynomials

(a) Find the polynomial $P_1(x) = a_0 + a_1 x$ whose value and slope agree with the value and slope of $f(x) = \cos x$ at the point $x = 0$.

(b) Find the polynomial $P_2(x) = a_0 + a_1 x + a_2 x^2$ whose value and first two derivatives agree with the value and first two derivatives of $f(x) = \cos x$ at the point $x = 0$. This polynomial is called the second-degree **Taylor polynomial** of $f(x) = \cos x$ at $x = 0$.

(c) Complete the table comparing the values of $f(x) = \cos x$ and $P_2(x)$. What do you observe?

x	-1.0	-0.1	-0.001	0	0.001	0.1	1.0
$\cos x$							
$P_2(x)$							

(d) Find the third-degree Taylor polynomial of $f(x) = \sin x$ at $x = 0$.

7. Famous Curve The graph of the **eight curve**

$$x^4 = a^2(x^2 - y^2), \quad a \neq 0$$

is shown below.

(a) Explain how you could use a graphing utility to graph this curve.

(b) Use a graphing utility to graph the curve for various values of the constant a. Describe how a affects the shape of the curve.

(c) Determine the points on the curve at which the tangent line is horizontal.

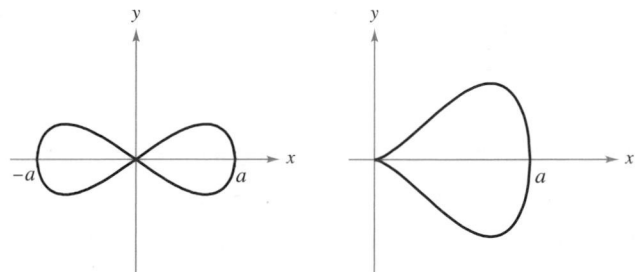

Figure for 7   Figure for 8

8. Famous Curve The graph of the **pear-shaped quartic**

$$b^2 y^2 = x^3(a - x), \quad a, b > 0$$

is shown above.

(a) Explain how you could use a graphing utility to graph this curve.

(b) Use a graphing utility to graph the curve for various values of the constants a and b. Describe how a and b affect the shape of the curve.

(c) Determine the points on the curve at which the tangent line is horizontal.

9. Shadow Length A man 6 feet tall walks at a rate of 5 feet per second toward a streetlight that is 30 feet high (see figure). The man's 3-foot-tall child follows at the same speed, but 10 feet behind the man. At times, the shadow behind the child is caused by the man, and at other times, by the child.

(a) Suppose the man is 90 feet from the streetlight. Show that the man's shadow extends beyond the child's shadow.

(b) Suppose the man is 60 feet from the streetlight. Show that the child's shadow extends beyond the man's shadow.

(c) Determine the distance d from the man to the streetlight at which the tips of the two shadows are exactly the same distance from the streetlight.

(d) Determine how fast the tip of the man's shadow is moving as a function of x, the distance between the man and the streetlight. Discuss the continuity of this shadow speed function.

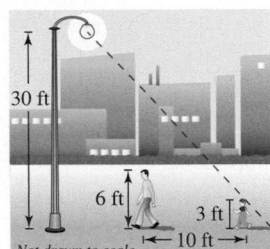

| Figure for 9 | Figure for 10 |

10. Moving Point A particle is moving along the graph of $y = \sqrt[3]{x}$ (see figure). When $x = 8$, the y-component of the position of the particle is increasing at the rate of 1 centimeter per second.

(a) How fast is the x-component changing at this moment?

(b) How fast is the distance from the origin changing at this moment?

(c) How fast is the angle of inclination θ changing at this moment?

11. Projectile Motion An astronaut standing on the moon throws a rock upward. The height of the rock is

$$s = -\frac{27}{10}t^2 + 27t + 6$$

where s is measured in feet and t is measured in seconds.

(a) Find expressions for the velocity and acceleration of the rock.

(b) Find the time when the rock is at its highest point by finding the time when the velocity is zero. What is the height of the rock at this time?

(c) How does the acceleration of the rock compare with the acceleration due to gravity on Earth?

12. Proof Let E be a function satisfying $E(0) = E'(0) = 1$. Prove that if

$$E(a + b) = E(a)E(b)$$

for all a and b, then E is differentiable and $E'(x) = E(x)$ for all x. Find an example of a function satisfying $E(a + b) = E(a)E(b)$.

13. Padé Approximation To approximate e^x, you can use a function of the form $f(x) = \dfrac{a + bx}{1 + cx}$. (This function is known as a **Padé approximation.**) The values of $f(0)$, $f'(0)$, and $f''(0)$ are equal to the corresponding values of e^x. Show that these values are equal to 1 and find the values of a, b, and c such that $f(0) = f'(0) = f''(0) = 1$. Then use a graphing utility to compare the graphs of f and e^x.

14. Radians and Degrees The fundamental limit

$$\lim_{x \to 0} \frac{\sin x}{x} = 1$$

assumes that x is measured in radians. Suppose you assume that x is measured in degrees instead of radians.

(a) Set your calculator to *degree* mode and complete the table.

z (in degrees)	0.1	0.01	0.0001
$\dfrac{\sin z}{z}$			

(b) Use the table to estimate

$$\lim_{z \to 0} \frac{\sin z}{z}$$

for z in degrees. What is the exact value of this limit? (*Hint:* $180° = \pi$ radians)

(c) Use the limit definition of the derivative to find $D_z[\sin z]$ for z in degrees.

(d) Define the new functions $S(z) = \sin cz$ and $C(z) = \cos cz$, where $c = \pi/180$. Find $S(90)$ and $C(180)$. Use the Chain Rule to calculate $D_z[S(z)]$.

(e) Explain why calculus is made easier by using radians instead of degrees.

15. Acceleration and Jerk If a is the acceleration of an object, then the *jerk* j is defined by $j = a'(t)$.

(a) Use this definition to give a physical interpretation of j.

(b) Find j for the slowing vehicle in Exercise 127 in Section 3.3 and interpret the result.

(c) The figure shows the graphs of the position, velocity, acceleration, and jerk functions of a vehicle. Identify each graph and explain your reasoning.

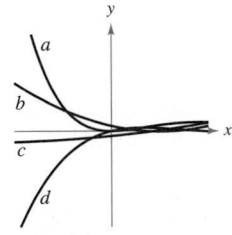

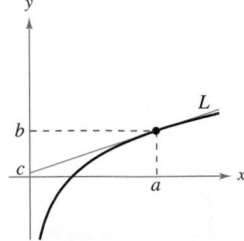

| Figure for 15 | Figure for 16 |

16. Distance Let L be the tangent line to the graph of the function $y = \ln x$ at the point (a, b), where c is the y-intercept of the tangent line, as shown in the figure. Show that the distance between b and c is always equal to 1.

4 Applications of Differentiation

Offshore Oil Well *(Exercise 39, p. 268)*

Estimation of Error *(Example 3, p. 273)*

Engine Efficiency *(Exercise 57, p. 247)*

Path of a Projectile *(Example 5, p. 226)*

Speed *(Exercise 69, p. 220)*

4.1 Extrema on an Interval

- Understand the definition of extrema of a function on an interval.
- Understand the definition of relative extrema of a function on an open interval.
- Find extrema on a closed interval.

Extrema of a Function

In calculus, much effort is devoted to determining the behavior of a function f on an interval I. Does f have a maximum value on I? Does it have a minimum value? Where is the function increasing? Where is it decreasing? In this chapter, you will learn how derivatives can be used to answer these questions. You will also see why these questions are important in real-life applications.

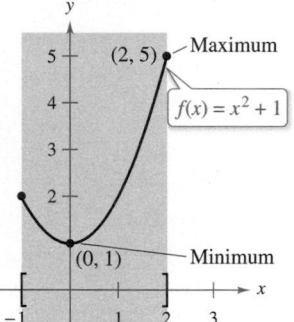

(a) f is continuous, $[-1, 2]$ is closed.

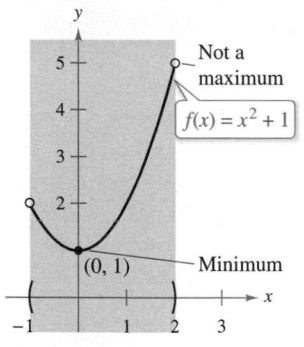

(b) f is continuous, $(-1, 2)$ is open.

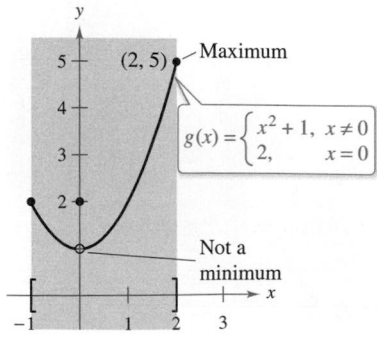

(c) g is not continuous, $[-1, 2]$ is closed.

Figure 4.1

> ### Definition of Extrema
>
> Let f be defined on an interval I containing c.
>
> 1. $f(c)$ is the **minimum of f on I** when $f(c) \le f(x)$ for all x in I.
> 2. $f(c)$ is the **maximum of f on I** when $f(c) \ge f(x)$ for all x in I.
>
> The minimum and maximum of a function on an interval are the **extreme values,** or **extrema** (the singular form of extrema is extremum), of the function on the interval. The minimum and maximum of a function on an interval are also called the **absolute minimum** and **absolute maximum,** or the **global minimum** and **global maximum,** on the interval. Extrema can occur at interior points or endpoints of an interval (see Figure 4.1). Extrema that occur at the endpoints are called **endpoint extrema.**

A function need not have a minimum or a maximum on an interval. For instance, in Figures 4.1(a) and (b), you can see that the function $f(x) = x^2 + 1$ has both a minimum and a maximum on the closed interval $[-1, 2]$ but does not have a maximum on the open interval $(-1, 2)$. Moreover, in Figure 4.1(c), you can see that continuity (or the lack of it) can affect the existence of an extremum on the interval. This suggests the theorem below. (Although the Extreme Value Theorem is intuitively plausible, a proof of this theorem is not within the scope of this text.)

> ### THEOREM 4.1 The Extreme Value Theorem
>
> If f is continuous on a closed interval $[a, b]$, then f has both a minimum and a maximum on the interval.

> ### Exploration
>
> ***Finding Minimum and Maximum Values*** The Extreme Value Theorem (like the Intermediate Value Theorem) is an *existence theorem* because it tells of the existence of minimum and maximum values but does not show how to find these values. Use the *minimum* and *maximum* features of a graphing utility to find the extrema of each function. In each case, do you think the x-values are exact or approximate? Explain your reasoning.
>
> **a.** $f(x) = x^2 - 4x + 5$ on the closed interval $[-1, 3]$
> **b.** $f(x) = x^3 - 2x^2 - 3x - 2$ on the closed interval $[-1, 3]$

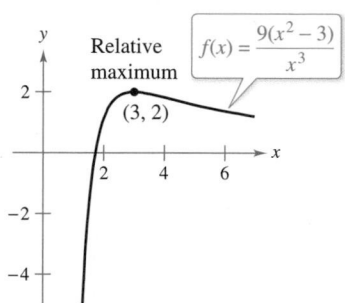

Hill
(0, 0)
$f(x) = x^3 - 3x^2$

Valley
(2, −4)

f has a relative maximum at (0, 0) and a relative minimum at (2, −4).

Figure 4.2

Relative Extrema and Critical Numbers

In Figure 4.2, the graph of $f(x) = x^3 - 3x^2$ has a **relative maximum** at the point (0, 0) and a **relative minimum** at the point (2, −4). Informally, for a continuous function, you can think of a relative maximum as occurring on a "hill" on the graph, and a relative minimum as occurring in a "valley" on the graph. Such a hill and valley can occur in two ways. When the hill (or valley) is smooth and rounded, the graph has a horizontal tangent line at the high point (or low point). When the hill (or valley) is sharp and peaked, the graph represents a function that is not differentiable at the high point (or low point).

Definition of Relative Extrema

1. If there is an open interval containing *c* on which $f(c)$ is a maximum, then $f(c)$ is called a **relative maximum** of *f*, or you can say that *f* has a **relative maximum at** $(c, f(c))$.

2. If there is an open interval containing *c* on which $f(c)$ is a minimum, then $f(c)$ is called a **relative minimum** of *f*, or you can say that *f* has a **relative minimum at** $(c, f(c))$.

The plural of relative maximum is relative maxima, and the plural of relative minimum is relative minima. Relative maximum and relative minimum are sometimes called **local maximum** and **local minimum,** respectively.

Example 1 examines the derivatives of functions at *given* relative extrema. (Much more is said about *finding* the relative extrema of a function in Section 4.3.)

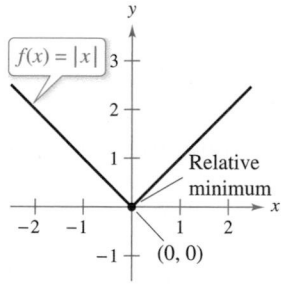

Relative maximum
$f(x) = \dfrac{9(x^2 - 3)}{x^3}$

(3, 2)

(a) $f'(3) = 0$

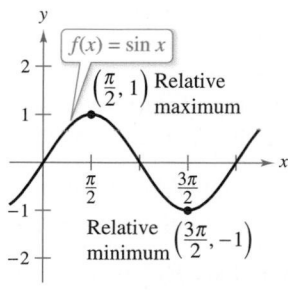

$f(x) = |x|$

Relative minimum
(0, 0)

(b) $f'(0)$ does not exist.

$f(x) = \sin x$

$\left(\dfrac{\pi}{2}, 1\right)$ Relative maximum

Relative minimum $\left(\dfrac{3\pi}{2}, -1\right)$

(c) $f'\left(\dfrac{\pi}{2}\right) = 0;\ f'\left(\dfrac{3\pi}{2}\right) = 0$

Figure 4.3

EXAMPLE 1   **The Value of the Derivative at Relative Extrema**

Find the value of the derivative at each relative extremum shown in Figure 4.3.

Solution

a. The derivative of $f(x) = \dfrac{9(x^2 - 3)}{x^3}$ is

$$f'(x) = \frac{x^3(18x) - (9)(x^2 - 3)(3x^2)}{(x^3)^2}$$ Differentiate using Quotient Rule.

$$= \frac{9(9 - x^2)}{x^4}.$$ Simplify.

At the point (3, 2), the value of the derivative is $f'(3) = 0$. [See Figure 4.3(a).]

b. At $x = 0$, the derivative of $f(x) = |x|$ *does not exist* because the following one-sided limits differ. [See Figure 4.3(b).]

$$\lim_{x \to 0^-} \frac{f(x) - f(0)}{x - 0} = \lim_{x \to 0^-} \frac{|x|}{x} = -1$$ Limit from the left

$$\lim_{x \to 0^+} \frac{f(x) - f(0)}{x - 0} = \lim_{x \to 0^+} \frac{|x|}{x} = 1$$ Limit from the right

c. The derivative of $f(x) = \sin x$ is

$$f'(x) = \cos x.$$

At the point $(\pi/2, 1)$, the value of the derivative is $f'(\pi/2) = \cos(\pi/2) = 0$. At the point $(3\pi/2, -1)$, the value of the derivative is $f'(3\pi/2) = \cos(3\pi/2) = 0$. [See Figure 4.3(c).]

Note in Example 1 that at each relative extremum, the derivative either is zero or does not exist. The x-values at these special points are called **critical numbers.** Figure 4.4 illustrates the two types of critical numbers. Notice in the definition that the critical number c has to be in the domain of f, but c does not have to be in the domain of f'.

Definition of a Critical Number

Let f be defined at c. If $f'(c) = 0$ or if f is not differentiable at c, then c is a **critical number** of f.

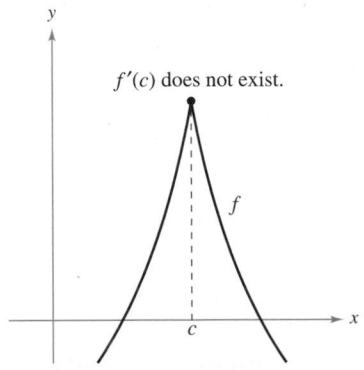
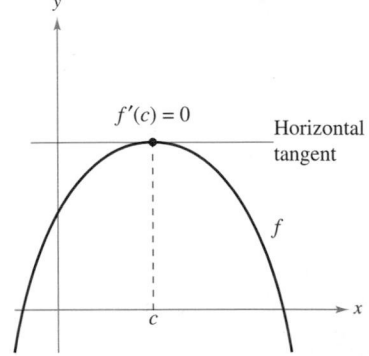

c is a critical number of f.
Figure 4.4

THEOREM 4.2   Relative Extrema Occur Only at Critical Numbers

If f has a relative minimum or relative maximum at $x = c$, then c is a critical number of f.

Proof

Case 1: If f is *not* differentiable at $x = c$, then, by definition, c is a critical number of f and the theorem is valid.

Case 2: If f is differentiable at $x = c$, then $f'(c)$ must be positive, negative, or 0. Suppose $f'(c)$ is positive. Then

$$f'(c) = \lim_{x \to c} \frac{f(x) - f(c)}{x - c} > 0$$

which implies that there exists an interval (a, b) containing c such that

$$\frac{f(x) - f(c)}{x - c} > 0, \text{ for all } x \neq c \text{ in } (a, b). \qquad \text{See Exercise 86(b), Section 2.2.}$$

Because this quotient is positive, the signs of the denominator and numerator must agree. This produces the following inequalities for x-values in the interval (a, b).

Left of c: $x < c$ and $f(x) < f(c)$ ⟹ $f(c)$ is not a relative minimum.

Right of c: $x > c$ and $f(x) > f(c)$ ⟹ $f(c)$ is not a relative maximum.

So, the assumption that $f'(c) > 0$ contradicts the hypothesis that $f(c)$ is a relative extremum. Assuming that $f'(c) < 0$ produces a similar contradiction, you are left with only one possibility—namely, $f'(c) = 0$. So, by definition, c is a critical number of f and the theorem is valid. ∎

PIERRE DE FERMAT (1601–1665)

For Fermat, who was trained as a lawyer, mathematics was more of a hobby than a profession. Nevertheless, Fermat made many contributions to analytic geometry, number theory, calculus, and probability. In letters to friends, he wrote of many of the fundamental ideas of calculus, long before Newton or Leibniz. For instance, Theorem 4.2 is sometimes attributed to Fermat.

See LarsonCalculus.com to read more of this biography.

Finding Extrema on a Closed Interval

Theorem 4.2 states that the relative extrema of a function can occur *only* at the critical numbers of the function. Knowing this, you can use these guidelines to find extrema on a closed interval.

GUIDELINES FOR FINDING EXTREMA ON A CLOSED INTERVAL

To find the extrema of a continuous function f on a closed interval $[a, b]$, use these steps.

1. Find the critical numbers of f in (a, b).
2. Evaluate f at each critical number in (a, b).
3. Evaluate f at each endpoint of $[a, b]$.
4. The least of these values is the minimum. The greatest is the maximum.

The next three examples show how to apply these guidelines. Be sure you see that finding the critical numbers of the function is only part of the procedure. Evaluating the function at the critical numbers *and* the endpoints is the other part.

EXAMPLE 2   **Finding Extrema on a Closed Interval**

Find the extrema of

$$f(x) = 3x^4 - 4x^3$$

on the interval $[-1, 2]$.

Solution   Begin by differentiating the function.

$$f(x) = 3x^4 - 4x^3 \qquad \text{Write original function.}$$
$$f'(x) = 12x^3 - 12x^2 \qquad \text{Differentiate.}$$

To find the critical numbers of f in the interval $(-1, 2)$, you must find all x-values for which $f'(x) = 0$ and all x-values for which $f'(x)$ does not exist.

$$12x^3 - 12x^2 = 0 \qquad \text{Set } f'(x) \text{ equal to 0.}$$
$$12x^2(x - 1) = 0 \qquad \text{Factor.}$$
$$x = 0, 1 \qquad \text{Critical numbers}$$

Because f' is defined for all x, you can conclude that these are the only critical numbers of f. By evaluating f at these two critical numbers and at the endpoints of $[-1, 2]$, you can determine that the maximum is $f(2) = 16$ and the minimum is $f(1) = -1$, as shown in the table. The graph of f is shown in Figure 4.5.

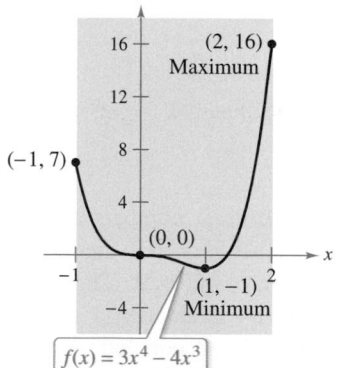

On the closed interval $[-1, 2]$, f has a minimum at $(1, -1)$ and a maximum at $(2, 16)$.

Figure 4.5

Left Endpoint	Critical Number	Critical Number	Right Endpoint
$f(-1) = 7$	$f(0) = 0$	$f(1) = -1$ Minimum	$f(2) = 16$ Maximum

In Figure 4.5, note that the critical number $x = 0$ does not yield a relative minimum or a relative maximum. This tells you that the converse of Theorem 4.2 is not true. In other words, *the critical numbers of a function need not produce relative extrema.*

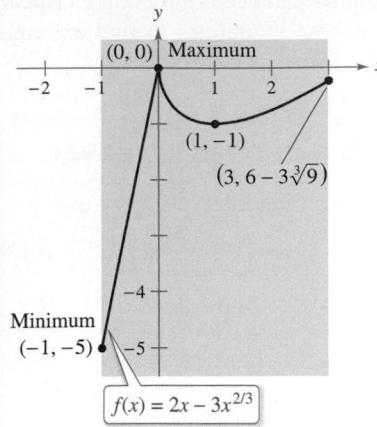

On the closed interval $[-1, 3]$, f has a minimum at $(-1, -5)$ and a maximum at $(0, 0)$.

Figure 4.6

EXAMPLE 3 **Finding Extrema on a Closed Interval**

Find the extrema of $f(x) = 2x - 3x^{2/3}$ on the interval $[-1, 3]$.

Solution Begin by differentiating the function.

$$f(x) = 2x - 3x^{2/3} \qquad \text{Write original function.}$$

$$f'(x) = 2 - \frac{2}{x^{1/3}} \qquad \text{Differentiate.}$$

$$= 2\left(\frac{x^{1/3} - 1}{x^{1/3}}\right) \qquad \text{Simplify.}$$

From this derivative, you can see that the function has two critical numbers in the interval $(-1, 3)$. The number 1 is a critical number because $f'(1) = 0$, and the number 0 is a critical number because $f'(0)$ does not exist. By evaluating f at these two numbers and at the endpoints of the interval, you can conclude that the minimum is $f(-1) = -5$ and the maximum is $f(0) = 0$, as shown in the table. The graph of f is shown in Figure 4.6.

Left Endpoint	Critical Number	Critical Number	Right Endpoint
$f(-1) = -5$ **Minimum**	$f(0) = 0$ **Maximum**	$f(1) = -1$	$f(3) = 6 - 3\sqrt[3]{9} \approx -0.24$

EXAMPLE 4 **Finding Extrema on a Closed Interval**

• • • • ▷ *See LarsonCalculus.com for an interactive version of this type of example.*

Find the extrema of

$$f(x) = 2 \sin x - \cos 2x$$

on the interval $[0, 2\pi]$.

Solution Begin by differentiating the function.

$$f(x) = 2 \sin x - \cos 2x \qquad \text{Write original function.}$$

$$f'(x) = 2 \cos x + 2 \sin 2x \qquad \text{Differentiate.}$$

$$= 2 \cos x + 4 \cos x \sin x \qquad \sin 2x = 2 \cos x \sin x$$

$$= 2(\cos x)(1 + 2 \sin x) \qquad \text{Factor.}$$

Because f is differentiable for all real x, you can find all critical numbers of f by finding the zeros of its derivative. Considering $2(\cos x)(1 + 2 \sin x) = 0$ in the interval $(0, 2\pi)$, the factor $\cos x$ is zero when $x = \pi/2$ and when $x = 3\pi/2$. The factor $(1 + 2 \sin x)$ is zero when $x = 7\pi/6$ and when $x = 11\pi/6$. By evaluating f at these four critical numbers and at the endpoints of the interval, you can conclude that the maximum is $f(\pi/2) = 3$ and the minimum occurs at *two* points, $f(7\pi/6) = -3/2$ and $f(11\pi/6) = -3/2$, as shown in the table. The graph is shown in Figure 4.7.

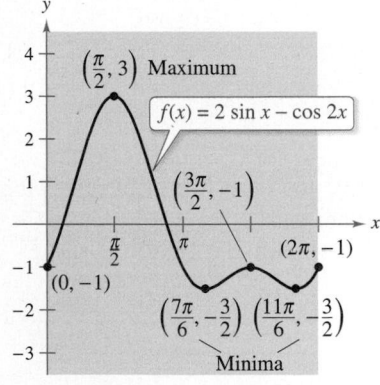

On the closed interval $[0, 2\pi]$, f has two minima at $(7\pi/6, -3/2)$ and $(11\pi/6, -3/2)$ and a maximum at $(\pi/2, 3)$.

Figure 4.7

Left Endpoint	Critical Number	Critical Number	Critical Number	Critical Number	Right Endpoint
$f(0) = -1$	$f\left(\dfrac{\pi}{2}\right) = 3$ **Maximum**	$f\left(\dfrac{7\pi}{6}\right) = -\dfrac{3}{2}$ **Minimum**	$f\left(\dfrac{3\pi}{2}\right) = -1$	$f\left(\dfrac{11\pi}{6}\right) = -\dfrac{3}{2}$ **Minimum**	$f(2\pi) = -1$

4.1 Exercises

See **CalcChat.com** for tutorial help and worked-out solutions to odd-numbered exercises.

CONCEPT CHECK

1. **Extreme Value Theorem** In your own words, describe the Extreme Value Theorem.

2. **Maximum** What is the difference between a relative maximum and an absolute maximum on an interval *I*?

3. **Critical Numbers** Use a graphing utility to graph the following four functions. Only one of the functions has critical numbers. Which is it? Explain.

    $f(x) = e^x$ $f(x) = \ln x$ $f(x) = \sin x$ $f(x) = \tan x$

4. **Extrema on a Closed Interval** Explain how to find the extrema of a continuous function on a closed interval $[a, b]$.

 **The Value of the Derivative at Relative Extrema** In Exercises 5–10, find the value of the derivative (if it exists) at each indicated extremum.

5. $f(x) = \dfrac{x^2}{x^2 + 4}$
6. $f(x) = \cos \dfrac{\pi x}{2}$

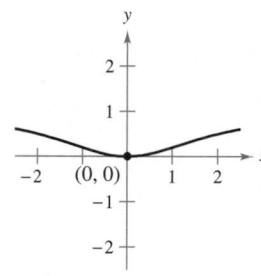
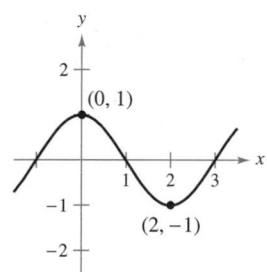

7. $g(x) = x + \dfrac{4}{x^2}$
8. $f(x) = -3x\sqrt{x + 1}$

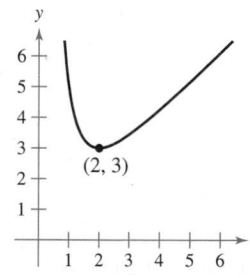
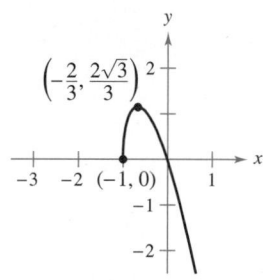

9. $f(x) = (x + 2)^{2/3}$
10. $f(x) = 4 - |x|$

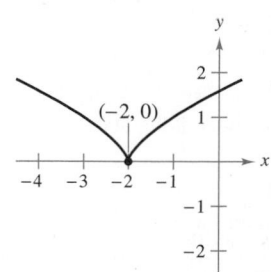
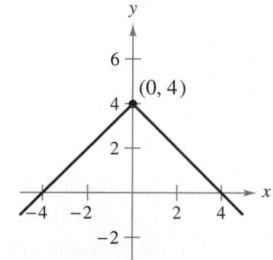

Approximating Critical Numbers In Exercises 11–14, approximate the critical numbers of the function shown in the graph. Determine whether the function has a relative maximum, a relative minimum, an absolute maximum, an absolute minimum, or none of these at each critical number on the interval shown.

11.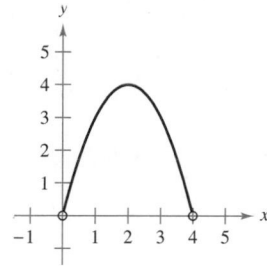
12.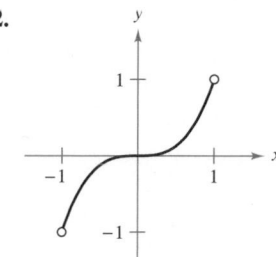

13.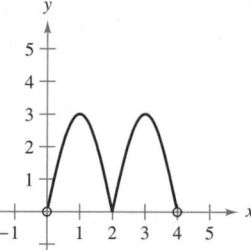
14.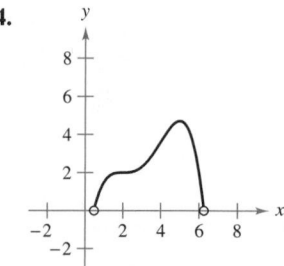

 **Finding Critical Numbers** In Exercises 15–24, find the critical numbers of the function.

15. $f(x) = 4x^2 - 6x$
16. $g(x) = x - \sqrt{x}$

17. $g(t) = t\sqrt{4 - t}, \ t < 3$
18. $f(x) = \dfrac{4x}{x^2 + 1}$

19. $h(x) = \sin^2 x + \cos x$
   $0 < x < 2\pi$
20. $f(\theta) = 2 \sec \theta + \tan \theta$
   $0 < \theta < 2\pi$

21. $f(t) = te^{-2t}$
22. $g(x) = 4x^2(3^x)$

23. $f(x) = x^2 \log_2(x^2 + 1)$
24. $g(t) = 2t \ln t$

Finding Extrema on a Closed Interval In Exercises 25–46, find the absolute extrema of the function on the closed interval.

25. $f(x) = 3 - x, \ [-1, 2]$
26. $f(x) = \dfrac{3}{4}x + 2, \ [0, 4]$

27. $h(x) = 5 - 2x^2, \ [-3, 1]$
28. $f(x) = 7x^2 + 1, \ [-1, 2]$

29. $f(x) = x^3 - \dfrac{3}{2}x^2, \ [-1, 2]$
30. $f(x) = 2x^3 - 6x, \ [0, 3]$

31. $y = 3x^{2/3} - 2x, \ [-1, 1]$
32. $g(x) = \sqrt[3]{x}, \ [-8, 8]$

33. $g(x) = \dfrac{6x^2}{x - 2}, \ [-2, 1]$
34. $h(t) = \dfrac{t}{t + 3}, \ [-1, 6]$

35. $y = 3 - |t - 3|, \ [-1, 5]$
36. $g(x) = |x + 4|, \ [-7, 1]$

37. $f(x) = [\![x]\!], \ [-2, 2]$
38. $h(x) = [\![2 - x]\!], \ [-2, 2]$

39. $y = 3 \cos x$, $[0, 2\pi]$

40. $y = \tan \dfrac{\pi x}{8}$, $[0, 2]$

41. $f(x) = \arctan x^2$, $[-2, 1]$

42. $g(x) = \dfrac{\ln x}{x}$, $[1, 4]$

43. $h(x) = 5e^x - e^{2x}$, $[-1, 2]$

44. $y = x^2 - 8 \ln x$, $[1, 5]$

45. $y = e^x \sin x$, $[0, \pi]$

46. $y = x \ln(x + 3)$, $[0, 3]$

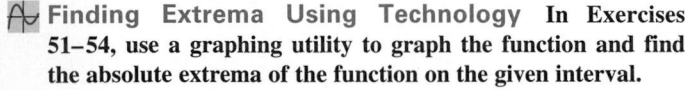 **Finding Extrema on an Interval** In Exercises 47–50, find the absolute extrema of the function (if any exist) on each interval.

47. $f(x) = 2x - 3$
  (a) $[0, 2]$ (b) $[0, 2)$
  (c) $(0, 2]$ (d) $(0, 2)$

48. $f(x) = 5 - x$
  (a) $[1, 4]$ (b) $[1, 4)$
  (c) $(1, 4]$ (d) $(1, 4)$

49. $f(x) = x^2 - 2x$
  (a) $[-1, 2]$ (b) $(1, 3]$
  (c) $(0, 2)$ (d) $[1, 4)$

50. $f(x) = \sqrt{4 - x^2}$
  (a) $[-2, 2]$ (b) $[-2, 0)$
  (c) $(-2, 2)$ (d) $[1, 2)$

 Finding Extrema Using Technology In Exercises 51–54, use a graphing utility to graph the function and find the absolute extrema of the function on the given interval.

51. $f(x) = \dfrac{3}{x - 1}$, $(1, 4]$

52. $f(x) = \dfrac{2}{2 - x}$, $[0, 2)$

53. $f(x) = \sqrt{x + 4}\, e^{x^2/10}$, $[-2, 2]$

54. $f(x) = -x + \cos 3\pi x$, $\left[0, \dfrac{\pi}{6}\right]$

 Finding Extrema Using Technology In Exercises 55–58, (a) use a computer algebra system to graph the function and approximate any absolute extrema on the given interval. (b) Use the utility to find any critical numbers, and use them to find any absolute extrema not located at the endpoints. Compare the results with those in part (a).

55. $f(x) = 3.2x^5 + 5x^3 - 3.5x$, $[0, 1]$

56. $f(x) = \dfrac{4}{3}x\sqrt{3 - x}$, $[0, 3]$

57. $f(x) = (x^2 - 2x) \ln(x + 3)$, $[0, 3]$

58. $f(x) = (x - 4) \arcsin \dfrac{x}{4}$, $[-2, 4]$

Finding Maximum Values Using Technology In Exercises 59–62, use a computer algebra system to find the maximum value of $|f''(x)|$ on the closed interval. (This value is used in the error estimate for the Trapezoidal Rule, as discussed in Section 8.6.)

59. $f(x) = \sqrt{1 + x^3}$, $[0, 2]$

60. $f(x) = \dfrac{1}{x^2 + 1}$, $\left[\dfrac{1}{2}, 3\right]$

61. $f(x) = e^{-x^2/2}$, $[0, 1]$

62. $f(x) = x \ln(x + 1)$, $[0, 2]$

Finding Maximum Values Using Technology In Exercises 63 and 64, use a computer algebra system to find the maximum value of $|f^{(4)}(x)|$ on the closed interval. (This value is used in the error estimate for Simpson's Rule, as discussed in Section 8.6.)

63. $f(x) = (x + 1)^{2/3}$, $[0, 2]$

64. $f(x) = \dfrac{1}{x^2 + 1}$, $[-1, 1]$

65. Think About It Explain why the function $f(x) = \tan x$ has a maximum on $[0, \pi/4]$ but not on $[0, \pi]$.

66.  **HOW DO YOU SEE IT?** Determine whether each labeled point is an absolute maximum or minimum, a relative maximum or minimum, or none of these.

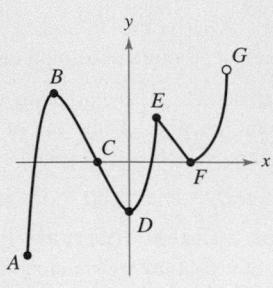

EXPLORING CONCEPTS

Using Graphs In Exercises 67 and 68, determine from the graph whether f has a minimum in the open interval (a, b). Explain your reasoning.

67. (a)   (b)

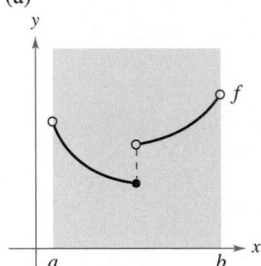 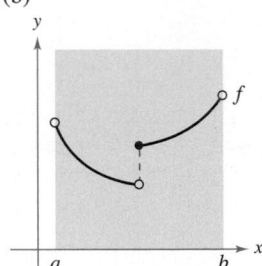

68. (a)   (b)

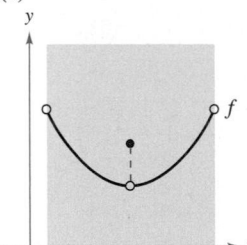 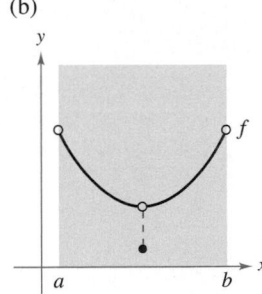

69. Critical Numbers Consider the function
$$f(x) = \dfrac{x - 4}{x + 2}.$$
Is $x = -2$ a critical number of f? Why or why not?

70. Creating the Graph of a Function Graph a function on the interval $[-2, 5]$ having the given characteristics.

Relative minimum at $x = -1$
Critical number (but no extremum) at $x = 0$
Absolute maximum at $x = 2$
Absolute minimum at $x = 5$

71. Power The formula for the power output P of a battery is

$$P = VI - RI^2$$

where V is the electromotive force in volts, R is the resistance in ohms, and I is the current in amperes. Find the current that corresponds to a maximum value of P in a battery for which $V = 12$ volts and $R = 0.5$ ohm. Assume that a 15-ampere fuse bounds the output in the interval $0 \le I \le 15$. Could the power output be increased by replacing the 15-ampere fuse with a 20-ampere fuse? Explain.

72. Lawn Sprinkler A lawn sprinkler is constructed in such a way that $d\theta/dt$ is constant, where θ ranges between 45° and 135° (see figure). The distance the water travels horizontally is

$$x = \frac{v^2 \sin 2\theta}{32}, \quad 45° \le \theta \le 135°$$

where v is the speed of the water. Find dx/dt and explain why this lawn sprinkler does not water evenly. What part of the lawn receives the most water?

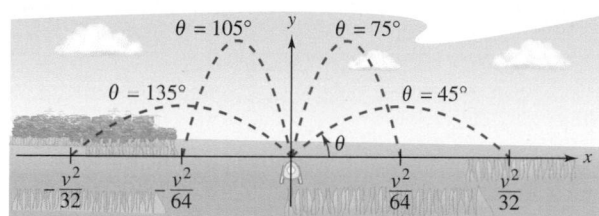

Water sprinkler: $45° \le \theta \le 135°$

■ **FOR FURTHER INFORMATION** For more information on the "calculus of lawn sprinklers," see the article "Design of an Oscillating Sprinkler" by Bart Braden in *Mathematics Magazine*. To view this article, go to *MathArticles.com*.

73. Honeycomb The surface area of a cell in a honeycomb is

$$S = 6hs + \frac{3s^2}{2}\left(\frac{\sqrt{3} - \cos\theta}{\sin\theta}\right)$$

where h and s are positive constants and θ is the angle at which the upper faces meet the altitude of the cell (see figure). Find the angle θ ($\pi/6 \le \theta \le \pi/2$) that minimizes the surface area S.

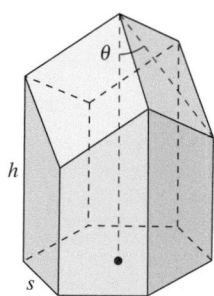

■ **FOR FURTHER INFORMATION** For more information on the geometric structure of a honeycomb cell, see the article "The Design of Honeycombs" by Anthony L. Peressini in UMAP Module 502, published by COMAP, Inc., Suite 210, 57 Bedford Street, Lexington, MA.

74. Highway Design In order to build a highway, it is necessary to fill a section of a valley where the grades (slopes) of the sides are 9% and 6% (see figure). The top of the filled region will have the shape of a parabolic arc that is tangent to the two slopes at the points A and B. The horizontal distances from A to the y-axis and from B to the y-axis are both 500 feet.

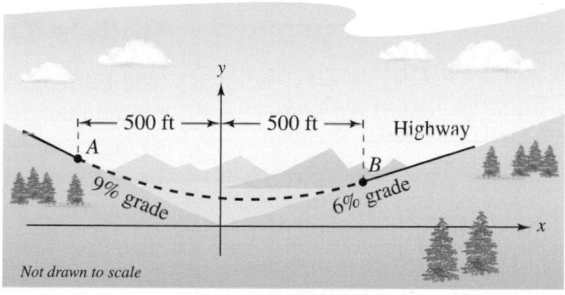

Not drawn to scale

(a) Find the coordinates of A and B.

(b) Find a quadratic function $y = ax^2 + bx + c$ for $-500 \le x \le 500$ that describes the top of the filled region.

(c) Construct a table giving the depths d of the fill for $x = -500, -400, -300, -200, -100, 0, 100, 200, 300, 400,$ and 500.

(d) What will be the lowest point on the completed highway? Will it be directly over the point where the two hillsides come together?

True or False? In Exercises 75–78, determine whether the statement is true or false. If it is false, explain why or give an example that shows it is false.

75. The maximum of $y = x^2$ on the open interval $(-3, 3)$ is 9.

76. If a function is continuous on a closed interval, then it must have a minimum on the interval.

77. If $x = c$ is a critical number of the function f, then it is also a critical number of the function $g(x) = f(x) + k$, where k is a constant.

78. If $x = c$ is a critical number of the function f, then it is also a critical number of the function $g(x) = f(x - k)$, where k is a constant.

79. Functions Let the function f be differentiable on an interval I containing c. If f has a maximum value at $x = c$, show that $-f$ has a minimum value at $x = c$.

80. Critical Numbers Consider the cubic function $f(x) = ax^3 + bx^2 + cx + d$, where $a \ne 0$. Show that f can have zero, one, or two critical numbers and give an example of each case.

PUTNAM EXAM CHALLENGE

81. Determine all real numbers $a > 0$ for which there exists a nonnegative continuous function $f(x)$ defined on $[0, a]$ with the property that the region $R = \{(x, y); 0 \le x \le a, 0 \le y \le f(x)\}$ has perimeter k units and area k square units for some real number k.

This problem was composed by the Committee on the Putnam Prize Competition.
© The Mathematical Association of America. All rights reserved.

4.2 Rolle's Theorem and the Mean Value Theorem

■ Understand and use Rolle's Theorem.
■ Understand and use the Mean Value Theorem.

Rolle's Theorem

The Extreme Value Theorem (see Section 4.1) states that a continuous function on a closed interval $[a, b]$ must have both a minimum and a maximum on the interval. Both of these values, however, can occur at the endpoints. **Rolle's Theorem,** named after the French mathematician Michel Rolle (1652–1719), gives conditions that guarantee the existence of an extreme value in the *interior* of a closed interval.

> **THEOREM 4.3 Rolle's Theorem**
>
> Let f be continuous on the closed interval $[a, b]$ and differentiable on the open interval (a, b). If $f(a) = f(b)$, then there is at least one number c in (a, b) such that $f'(c) = 0$.

Proof Let $f(a) = d = f(b)$.

Case 1: If $f(x) = d$ for all x in $[a, b]$, then f is constant on the interval and, by Theorem 3.2, $f'(x) = 0$ for all x in (a, b).

Case 2: Consider $f(x) > d$ for some x in (a, b). By the Extreme Value Theorem, you know that f has a maximum at some c in the interval. Moreover, because $f(c) > d$, this maximum does not occur at either endpoint. So, f has a maximum in the *open* interval (a, b). This implies that $f(c)$ is a *relative* maximum and, by Theorem 4.2, c is a critical number of f. Finally, because f is differentiable at c, you can conclude that $f'(c) = 0$.

Case 3: When $f(x) < d$ for some x in (a, b), you can use an argument similar to that in Case 2 but involving the minimum instead of the maximum. ■

From Rolle's Theorem, you can see that if a function f is continuous on $[a, b]$ and differentiable on (a, b), and if $f(a) = f(b)$, then there must be at least one x-value between a and b at which the graph of f has a horizontal tangent [see Figure 4.8(a)]. When the differentiability requirement is dropped from Rolle's Theorem, f will still have a critical number in (a, b), but it may not yield a horizontal tangent. Such a case is shown in Figure 4.8(b).

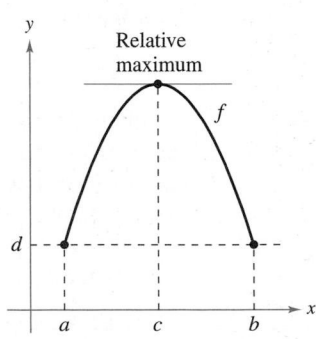

(a) f is continuous on $[a, b]$ and differentiable on (a, b).

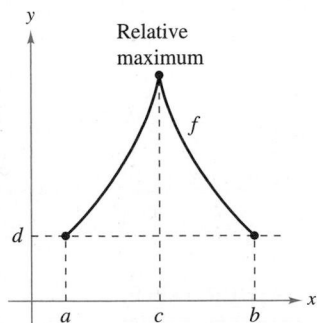

(b) f is continuous on $[a, b]$ but *not* differentiable on (a, b).

Figure 4.8

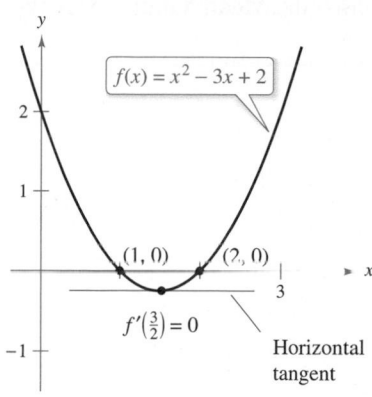

The x-value for which $f'(x) = 0$ is between the two x-intercepts.
Figure 4.9

EXAMPLE 1   Illustrating Rolle's Theorem

Find the two x-intercepts of

$$f(x) = x^2 - 3x + 2$$

and show that $f'(x) = 0$ at some point between the two x-intercepts.

Solution   Note that f is differentiable on the entire real number line. Setting $f(x)$ equal to 0 produces

$$x^2 - 3x + 2 = 0 \qquad \text{Set } f(x) \text{ equal to 0.}$$
$$(x - 1)(x - 2) = 0 \qquad \text{Factor.}$$
$$x = 1, 2. \qquad \text{Solve for } x.$$

So, $f(1) = f(2) = 0$, and from Rolle's Theorem you know that there *exists* at least one c in the interval $(1, 2)$ such that $f'(c) = 0$. To *find* such a c, differentiate f to obtain

$$f'(x) = 2x - 3 \qquad \text{Differentiate.}$$

and then determine that $f'(x) = 0$ when $x = \frac{3}{2}$. Note that this x-value lies in the open interval $(1, 2)$, as shown in Figure 4.9. ■

Rolle's Theorem states that when f satisfies the conditions of the theorem, there must be *at least* one point between a and b at which the derivative is 0. There may, of course, be more than one such point, as shown in the next example.

EXAMPLE 2   Illustrating Rolle's Theorem

Let $f(x) = x^4 - 2x^2$. Find all values of c in the interval $(-2, 2)$ such that $f'(c) = 0$.

Solution   To begin, note that the function satisfies the conditions of Rolle's Theorem. That is, f is continuous on the interval $[-2, 2]$ and differentiable on the interval $(-2, 2)$. Moreover, because $f(-2) = f(2) = 8$, you can conclude that there exists at least one c in $(-2, 2)$ such that $f'(c) = 0$. Because

$$f'(x) = 4x^3 - 4x \qquad \text{Differentiate.}$$

setting the derivative equal to 0 produces

$$4x^3 - 4x = 0 \qquad \text{Set } f'(x) \text{ equal to 0.}$$
$$4x(x - 1)(x + 1) = 0 \qquad \text{Factor.}$$
$$x = 0, 1, -1. \qquad x\text{-values for which } f'(x) = 0$$

So, in the interval $(-2, 2)$, the derivative is zero when $x = -1$, 0, and 1, as shown in Figure 4.10. ■

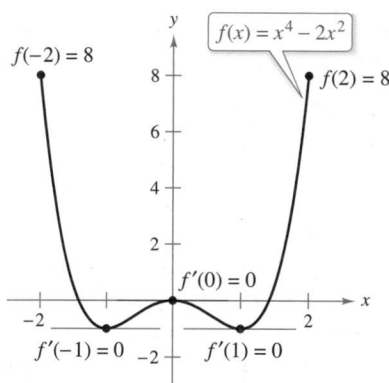

$f'(x) = 0$ for more than one x-value in the interval $(-2, 2)$.
Figure 4.10

▷ **TECHNOLOGY PITFALL**   A graphing utility can be used to indicate whether the points on the graphs in Examples 1 and 2 are relative minima or relative maxima of the functions. When using a graphing utility, however, you should keep in mind that it can give misleading pictures of graphs. For example, use a graphing utility to graph

$$f(x) = 1 - (x - 1)^2 - \frac{1}{1000(x - 1)^{1/7} + 1}.$$

With most viewing windows, it appears that the function has a maximum of 1 when $x = 1$, as shown in Figure 4.11. By evaluating the function at $x = 1$, however, you can see that $f(1) = 0$. To determine the behavior of this function near $x = 1$, you need to examine the graph analytically to get the complete picture.

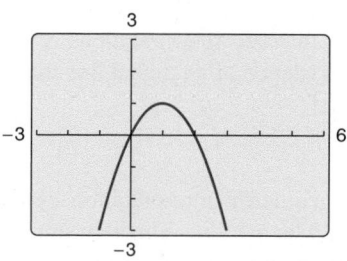

Figure 4.11

The Mean Value Theorem

Rolle's Theorem can be used to prove another theorem—the **Mean Value Theorem.**

> **THEOREM 4.4 The Mean Value Theorem**
>
> If f is continuous on the closed interval $[a, b]$ and differentiable on the open interval (a, b), then there exists a number c in (a, b) such that
>
> $$f'(c) = \frac{f(b) - f(a)}{b - a}.$$

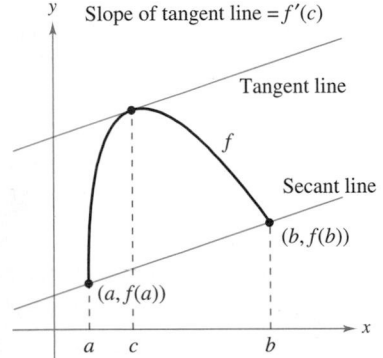

Slope of tangent line $= f'(c)$

Tangent line

f

Secant line

$(b, f(b))$

$(a, f(a))$

Figure 4.12

Proof Refer to Figure 4.12. The equation of the secant line that passes through the points $(a, f(a))$ and $(b, f(b))$ is

$$y = \left[\frac{f(b) - f(a)}{b - a}\right](x - a) + f(a).$$

Let $g(x)$ be the difference between $f(x)$ and y. Then

$$g(x) = f(x) - y$$
$$= f(x) - \left[\frac{f(b) - f(a)}{b - a}\right](x - a) - f(a).$$

By evaluating g at a and b, you can see that

$$g(a) = 0 = g(b).$$

Because f is continuous on $[a, b]$, it follows that g is also continuous on $[a, b]$. Furthermore, because f is differentiable, g is also differentiable, and you can apply Rolle's Theorem to the function g. So, there exists a number c in (a, b) such that $g'(c) = 0$, which implies that

$$g'(c) = 0$$
$$f'(c) - \frac{f(b) - f(a)}{b - a} = 0.$$

So, there exists a number c in (a, b) such that

$$f'(c) = \frac{f(b) - f(a)}{b - a}.$$ ∎

**JOSEPH-LOUIS LAGRANGE
(1736–1813)**

The Mean Value Theorem was first proved by the famous mathematician Joseph-Louis Lagrange. Born in Italy, Lagrange held a position in the court of Frederick the Great in Berlin for 20 years.
See LarsonCalculus.com to read more of this biography.

Although the Mean Value Theorem can be used directly in problem solving, it is used more often to prove other theorems. In fact, some people consider this to be the most important theorem in calculus—it is closely related to the Fundamental Theorem of Calculus discussed in Section 5.4. For now, you can get an idea of the versatility of the Mean Value Theorem by looking at the results stated in Exercises 87–95 in this section.

The Mean Value Theorem has implications for both basic interpretations of the derivative. Geometrically, the theorem guarantees the existence of a tangent line that is parallel to the secant line through the points

$$(a, f(a)) \quad \text{and} \quad (b, f(b))$$

as shown in Figure 4.12. Example 3 illustrates this geometric interpretation of the Mean Value Theorem. In terms of rates of change, the Mean Value Theorem implies that there must be a point in the open interval (a, b) at which the instantaneous rate of change is equal to the average rate of change over the interval $[a, b]$. This is illustrated in Example 4.

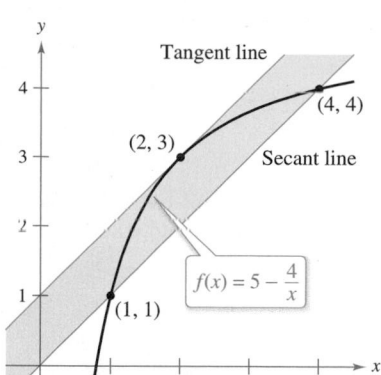

The tangent line at $(2, 3)$ is parallel to the secant line through $(1, 1)$ and $(4, 4)$.

Figure 4.13

EXAMPLE 3   Finding a Tangent Line

:····▷ *See LarsonCalculus.com for an interactive version of this type of example.*

For $f(x) = 5 - (4/x)$, find all values of c in the open interval $(1, 4)$ such that

$$f'(c) = \frac{f(4) - f(1)}{4 - 1}.$$

Solution   The slope of the secant line through $(1, f(1))$ and $(4, f(4))$ is

$$\frac{f(4) - f(1)}{4 - 1} = \frac{4 - 1}{4 - 1} = 1.$$ Slope of secant line

Note that the function satisfies the conditions of the Mean Value Theorem. That is, f is continuous on the interval $[1, 4]$ and differentiable on the interval $(1, 4)$. So, there exists at least one number c in $(1, 4)$ such that $f'(c) = 1$. Solving the equation $f'(x) = 1$ yields

$$\frac{4}{x^2} = 1$$ Set $f'(x)$ equal to 1.

which implies that

$$x = \pm 2.$$

So, in the interval $(1, 4)$, you can conclude that $c = 2$, as shown in Figure 4.13.

EXAMPLE 4   Finding an Instantaneous Rate of Change

Two stationary patrol cars equipped with radar are 5 miles apart on a highway, as shown in Figure 4.14. As a truck passes the first patrol car, its speed is clocked at 55 miles per hour. Four minutes later, when the truck passes the second patrol car, its speed is clocked at 50 miles per hour. Prove that the truck must have exceeded the speed limit (of 55 miles per hour) at some time during the 4 minutes.

Solution   Let $t = 0$ be the time (in hours) when the truck passes the first patrol car. The time when the truck passes the second patrol car is

$$t = \frac{4}{60} = \frac{1}{15} \text{ hour.}$$

By letting $s(t)$ represent the distance (in miles) traveled by the truck, you have $s(0) = 0$ and $s\left(\frac{1}{15}\right) = 5$. So, the average velocity of the truck over the five-mile stretch of highway is

$$\text{Average velocity} = \frac{s(1/15) - s(0)}{(1/15) - 0} = \frac{5}{1/15} = 75 \text{ miles per hour.}$$

Assuming that the position function is differentiable, you can apply the Mean Value Theorem to conclude that the truck must have been traveling at a rate of 75 miles per hour sometime during the 4 minutes. ∎

A useful alternative form of the Mean Value Theorem is: If f is continuous on $[a, b]$ and differentiable on (a, b), then there exists a number c in (a, b) such that

$$f(b) = f(a) + (b - a)f'(c).$$ Alternative form of Mean Value Theorem

When doing the exercises for this section, keep in mind that polynomial functions, rational functions, and transcendental functions are differentiable at all points in their domains.

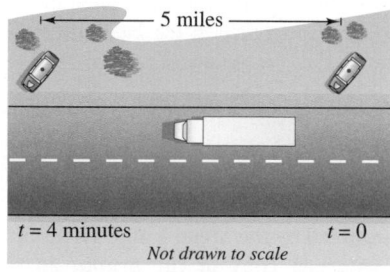

At some time t, the instantaneous velocity is equal to the average velocity over 4 minutes.

Figure 4.14

4.2 Exercises

See **CalcChat.com** for tutorial help and worked-out solutions to odd-numbered exercises.

CONCEPT CHECK

1. **Rolle's Theorem** In your own words, describe Rolle's Theorem.

2. **Mean Value Theorem** In your own words, describe the Mean Value Theorem.

Writing In Exercises 3–6, explain why Rolle's Theorem does not apply to the function even though there exist a and b such that $f(a) = f(b)$.

3. $f(x) = \left| \dfrac{1}{x} \right|$, $[-1, 1]$

4. $f(x) = \cot \dfrac{x}{2}$, $[\pi, 3\pi]$

5. $f(x) = 1 - |x - 1|$, $[0, 2]$

6. $f(x) = \sqrt{(2 - x^{2/3})^3}$, $[-1, 1]$

 **Using Rolle's Theorem** In Exercises 7–10, find the two x-intercepts of the function f and show that $f'(x) = 0$ at some point between the two x-intercepts.

7. $f(x) = x^2 - x - 2$

8. $f(x) = x^2 + 6x$

9. $f(x) = x\sqrt{x + 4}$

10. $f(x) = -3x\sqrt{x + 1}$

 **Using Rolle's Theorem** In Exercises 11–26, determine whether Rolle's Theorem can be applied to f on the closed interval $[a, b]$. If Rolle's Theorem can be applied, find all values of c in the open interval (a, b) such that $f'(c) = 0$. If Rolle's Theorem cannot be applied, explain why not.

11. $f(x) = -x^2 + 3x$, $[0, 3]$

12. $f(x) = x^2 - 8x + 5$, $[2, 6]$

13. $f(x) = (x - 1)(x - 2)(x - 3)$, $[1, 3]$

14. $f(x) = (x - 4)(x + 2)^2$, $[-2, 4]$

15. $f(x) = x^{2/3} - 1$, $[-8, 8]$

16. $f(x) = 3 - |x - 3|$, $[0, 6]$

17. $f(x) = \dfrac{x^2 - 2x - 3}{x + 2}$, $[-1, 3]$

18. $f(x) = \dfrac{x^2 - 4}{x - 1}$, $[-2, 2]$

19. $f(x) = \sin x$, $[0, 2\pi]$

20. $f(x) = \cos x$, $[\pi, 3\pi]$

21. $f(x) = \cos \pi x$, $[0, 2]$

22. $f(x) = \sin 3x$, $\left[\dfrac{\pi}{2}, \dfrac{7\pi}{6} \right]$

23. $f(x) = \tan x$, $[0, \pi]$

24. $f(x) = \sec x$, $[\pi, 2\pi]$

25. $f(x) = (x^2 - 2x)e^x$, $[0, 2]$

26. $f(x) = x - 2 \ln x$, $[1, 3]$

Using Rolle's Theorem In Exercises 27–32, use a graphing utility to graph the function on the closed interval $[a, b]$. Determine whether Rolle's Theorem can be applied to f on the interval and, if so, find all values of c in the open interval (a, b) such that $f'(c) = 0$.

27. $f(x) = |x| - 1$, $[-1, 1]$

28. $f(x) = x - x^{1/3}$, $[0, 1]$

29. $f(x) = \dfrac{x}{2} - \sin \dfrac{\pi x}{6}$, $[-1, 0]$

30. $f(x) = x - \tan \pi x$, $\left[-\dfrac{1}{4}, \dfrac{1}{4} \right]$

31. $f(x) = 2 + (x^2 - 4x)(2^{-x/4})$, $[0, 4]$

32. $f(x) = 2 + \arcsin(x^2 - 1)$, $[-1, 1]$

33. **Vertical Motion** The height of a ball t seconds after it is thrown upward from a height of 6 feet and with an initial velocity of 48 feet per second is $f(t) = -16t^2 + 48t + 6$.

(a) Verify that $f(1) = f(2)$.

(b) According to Rolle's Theorem, what must the velocity be at some time in the interval $(1, 2)$? Find that time.

34. **Reorder Costs** The ordering and transportation cost C for components used in a manufacturing process is approximated by

$$C(x) = 10\left(\dfrac{1}{x} + \dfrac{x}{x + 3} \right)$$

where C is measured in thousands of dollars and x is the order size in hundreds.

(a) Verify that $C(3) = C(6)$.

(b) According to Rolle's Theorem, the rate of change of the cost must be 0 for some order size in the interval $(3, 6)$. Find that order size.

 **Mean Value Theorem** In Exercises 35 and 36, copy the graph and sketch the secant line to the graph through the points $(a, f(a))$ and $(b, f(b))$. Then sketch any tangent lines to the graph for each value of c guaranteed by the Mean Value Theorem. To print an enlarged copy of the graph, go to *MathGraphs.com*.

35.
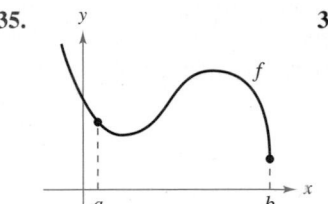

36.
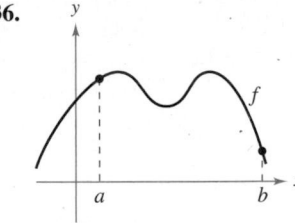

Writing In Exercises 37–40, explain why the Mean Value Theorem does not apply to the function f on the interval $[0, 6]$.

37.
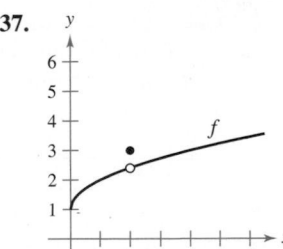

38.
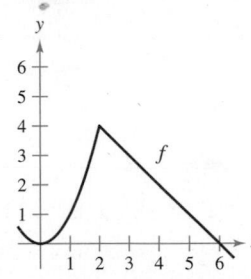

39. $f(x) = \dfrac{1}{x - 3}$

40. $f(x) = |x - 3|$

41. Mean Value Theorem   Consider the graph of the function $f(x) = -x^2 + 5$ (see figure).

(a) Find the equation of the secant line joining the points $(-1, 4)$ and $(2, 1)$.

(b) Use the Mean Value Theorem to determine a point c in the interval $(-1, 2)$ such that the tangent line at c is parallel to the secant line.

(c) Find the equation of the tangent line through c.

(d) Use a graphing utility to graph f, the secant line, and the tangent line.

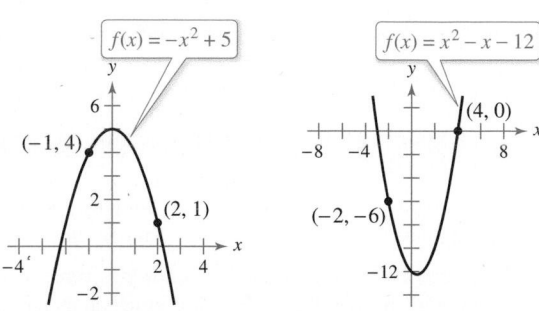

Figure for 41   Figure for 42

42. Mean Value Theorem   Consider the graph of the function $f(x) = x^2 - x - 12$ (see figure).

(a) Find the equation of the secant line joining the points $(-2, -6)$ and $(4, 0)$.

(b) Use the Mean Value Theorem to determine a point c in the interval $(-2, 4)$ such that the tangent line at c is parallel to the secant line.

(c) Find the equation of the tangent line through c.

(d) Use a graphing utility to graph f, the secant line, and the tangent line.

   **Using the Mean Value Theorem**   In Exercises 43–56, determine whether the Mean Value Theorem can be applied to f on the closed interval $[a, b]$. If the Mean Value Theorem can be applied, find all values of c in the open interval (a, b) such that

$$f'(c) = \frac{f(b) - f(a)}{b - a}.$$

If the Mean Value Theorem cannot be applied, explain why not.

43. $f(x) = 6x^3$, $[1, 2]$ **44.** $f(x) = x^6$, $[-1, 1]$

45. $f(x) = x^3 + 2x$, $[-1, 1]$ **46.** $f(x) = x^4 - 8x$, $[0, 2]$

47. $f(x) = \dfrac{x + 2}{x - 1}$, $[-3, 3]$ **48.** $f(x) = \dfrac{x}{x - 5}$, $[1, 4]$

49. $f(x) = |2x + 1|$, $[-1, 3]$ **50.** $f(x) = \sqrt{2 - x}$, $[-7, 2]$

51. $f(x) = \sin x$, $[0, \pi]$ **52.** $f(x) = e^{-3x}$, $[0, 2]$

53. $f(x) = \cos x + \tan x$, $[0, \pi]$

54. $f(x) = (x + 3) \ln(x + 3)$, $[-2, -1]$

55. $f(x) = x \log_2 x$, $[1, 2]$ **56.** $f(x) = \arctan(1 - x)$, $[0, 1]$

Using the Mean Value Theorem   In Exercises 57–62, use a graphing utility to (a) graph the function f on the given interval, (b) find and graph the secant line through points on the graph of f at the endpoints of the given interval, and (c) find and graph any tangent lines to the graph of f that are parallel to the secant line.

57. $f(x) = \dfrac{x}{x + 1}$, $\left[-\dfrac{1}{2}, 2\right]$ **58.** $f(x) = \sqrt{x}$, $[1, 9]$

59. $f(x) = x - 2 \sin x$, $[-\pi, \pi]$

60. $f(x) = x^4 - 2x^3 + x^2$, $[0, 6]$

61. $f(x) = 2e^{x/4} \cos \dfrac{\pi x}{4}$, $[0, 2]$ **62.** $f(x) = \ln|\sec \pi x|$, $\left[0, \dfrac{1}{4}\right]$

63. Vertical Motion   The height of an object t seconds after it is dropped from a height of 300 meters is $s(t) = -4.9t^2 + 300$.

(a) Find the average velocity of the object during the first 3 seconds.

(b) Use the Mean Value Theorem to verify that at some time during the first 3 seconds of fall, the instantaneous velocity equals the average velocity. Find that time.

64. Sales   A company introduces a new product for which the number of units sold S is

$$S(t) = 200\left(5 - \frac{9}{2 + t}\right)$$

where t is the time in months.

(a) Find the average rate of change of S during the first year.

(b) During what month of the first year does $S'(t)$ equal the average rate of change?

EXPLORING CONCEPTS

65. Converse of Rolle's Theorem   Let f be continuous on $[a, b]$ and differentiable on (a, b). If there exists c in (a, b) such that $f'(c) = 0$, does it follow that $f(a) = f(b)$? Explain.

66. Rolle's Theorem   Let f be continuous on $[a, b]$ and differentiable on (a, b). Also, suppose that $f(a) = f(b)$ and that c is a real number in the interval (a, b) such that $f'(c) = 0$. Find an interval for the function g over which Rolle's Theorem can be applied, and find the corresponding critical number of g, where k is a constant.

(a) $g(x) = f(x) + k$ (b) $g(x) = f(x - k)$

(c) $g(x) = f(kx)$

67. Rolle's Theorem   The function

$$f(x) = \begin{cases} 0, & x = 0 \\ 1 - x, & 0 < x \le 1 \end{cases}$$

is differentiable on $(0, 1)$ and satisfies $f(0) = f(1)$. However, its derivative is never zero on $(0, 1)$. Does this contradict Rolle's Theorem? Explain.

68. Mean Value Theorem   Can you find a function f such that $f(-2) = -2$, $f(2) = 6$, and $f'(x) < 1$ for all x? Why or why not?

69. Speed

A plane begins its takeoff at 2:00 P.M. on a 2500-mile flight. After 5.5 hours, the plane arrives at its destination. Explain why there are at least two times during the flight when the speed of the plane is 400 miles per hour.

70. Temperature When an object is removed from a furnace and placed in an environment with a constant temperature of 90°F, its core temperature is 1500°F. Five hours later, the core temperature is 390°F. Explain why there must exist a time in the interval $(0, 5)$ when the temperature is decreasing at a rate of 222°F per hour.

71. Velocity Two bicyclists begin a race at 8:00 A.M. They both finish the race 2 hours and 15 minutes later. Prove that at some time during the race, the bicyclists are traveling at the same velocity.

72. Acceleration At 9:13 A.M., a sports car is traveling 35 miles per hour. Two minutes later, the car is traveling 85 miles per hour. Prove that at some time during this two-minute interval, the car's acceleration is exactly 1500 miles per hour squared.

73. Think About It Sketch the graph of an arbitrary function f that satisfies the given condition but does not satisfy the conditions of the Mean Value Theorem on the interval $[-5, 5]$.

(a) f is continuous.   (b) f is not continuous.

74. HOW DO YOU SEE IT? The figure shows two parts of the graph of a continuous differentiable function f on $[-10, 4]$. The derivative f' is also continuous. To print an enlarged copy of the graph, go to *MathGraphs.com*.

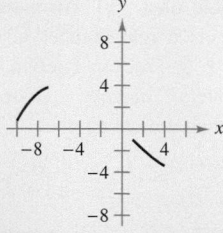

(a) Explain why f must have at least one zero in $[-10, 4]$.

(b) Explain why f' must also have at least one zero in the interval $[-10, 4]$. What are these zeros called?

(c) Make a possible sketch of the function, where f' has one zero on the interval $[-10, 4]$.

Finding a Solution In Exercises 75–78, use the Intermediate Value Theorem and Rolle's Theorem to prove that the equation has exactly one real solution.

75. $x^5 + x^3 + x + 1 = 0$ **76.** $2x^5 + 7x - 1 = 0$

77. $3x + 1 - \sin x = 0$ **78.** $2x - 2 - \cos x = 0$

Using a Derivative In Exercises 79–82, find a function f that has the derivative $f'(x)$ and whose graph passes through the given point. Explain your reasoning.

79. $f'(x) = 0$, $(2, 5)$ **80.** $f'(x) = 4$, $(0, 1)$

81. $f'(x) = 2x$, $(1, 0)$ **82.** $f'(x) = 6x - 1$, $(2, 7)$

True or False? In Exercises 83–86, determine whether the statement is true or false. If it is false, explain why or give an example that shows it is false.

83. The Mean Value Theorem can be applied to

$$f(x) = \frac{1}{x}$$

on the interval $[-1, 1]$.

84. If the graph of a function has three x-intercepts, then it must have at least two points at which its tangent line is horizontal.

85. If the graph of a polynomial function has three x-intercepts, then it must have at least two points at which its tangent line is horizontal.

86. The Mean Value Theorem can be applied to $f(x) = \tan x$ on the interval $[0, \pi/4]$.

87. Proof Prove that if $a > 0$ and n is any positive integer, then the polynomial function $p(x) = x^{2n+1} + ax + b$ cannot have two real roots.

88. Proof Prove that if $f'(x) = 0$ for all x in an interval (a, b), then f is constant on (a, b).

89. Proof Let $p(x) = Ax^2 + Bx + C$. Prove that for any interval $[a, b]$, the value c guaranteed by the Mean Value Theorem is the midpoint of the interval.

90. Using Rolle's Theorem

(a) Let $f(x) = x^2$ and $g(x) = -x^3 + x^2 + 3x + 2$. Then $f(-1) = g(-1)$ and $f(2) = g(2)$. Show that there is at least one value c in the interval $(-1, 2)$ where the tangent line to f at $(c, f(c))$ is parallel to the tangent line to g at $(c, g(c))$. Identify c.

(b) Let f and g be differentiable functions on $[a, b]$, where $f(a) = g(a)$ and $f(b) = g(b)$. Show that there is at least one value c in the interval (a, b) where the tangent line to f at $(c, f(c))$ is parallel to the tangent line to g at $(c, g(c))$.

91. Proof Prove that if f is differentiable on $(-\infty, \infty)$ and $f'(x) < 1$ for all real numbers, then f has at most one fixed point. [A *fixed point* of a function f is a real number c such that $f(c) = c$.]

92. Fixed Point Use the result of Exercise 91 to show that $f(x) = \frac{1}{2}\cos x$ has at most one fixed point.

93. Proof Prove that $|\cos a - \cos b| \le |a - b|$ for all a and b.

94. Proof Prove that $|\sin a - \sin b| \le |a - b|$ for all a and b.

95. Using the Mean Value Theorem Let $0 < a < b$. Use the Mean Value Theorem to show that

$$\sqrt{b} - \sqrt{a} < \frac{b - a}{2\sqrt{a}}.$$

4.3 Increasing and Decreasing Functions and the First Derivative Test

■ Determine intervals on which a function is increasing or decreasing.
■ Apply the First Derivative Test to find relative extrema of a function.

Increasing and Decreasing Functions

In this section, you will learn how derivatives can be used to *classify* relative extrema as either relative minima or relative maxima. First, it is important to define increasing and decreasing functions.

Definitions of Increasing and Decreasing Functions

A function f is **increasing** on an interval when, for any two numbers x_1 and x_2 in the interval, $x_1 < x_2$ implies $f(x_1) < f(x_2)$.

A function f is **decreasing** on an interval when, for any two numbers x_1 and x_2 in the interval, $x_1 < x_2$ implies $f(x_1) > f(x_2)$.

A function is increasing when, *as x moves to the right*, its graph moves up, and is decreasing when its graph moves down. For example, the function in Figure 4.15 is decreasing on the interval $(-\infty, a)$, is constant on the interval (a, b), and is increasing on the interval (b, ∞). As shown in Theorem 4.5 below, a positive derivative implies that the function is increasing, a negative derivative implies that the function is decreasing, and a zero derivative on an entire interval implies that the function is constant on that interval.

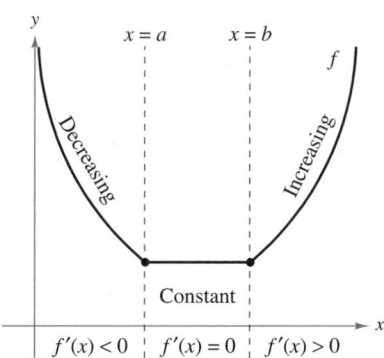

The derivative is related to the slope of a function.

Figure 4.15

THEOREM 4.5 Test for Increasing and Decreasing Functions

Let f be a function that is continuous on the closed interval $[a, b]$ and differentiable on the open interval (a, b).

1. If $f'(x) > 0$ for all x in (a, b), then f is increasing on $[a, b]$.
2. If $f'(x) < 0$ for all x in (a, b), then f is decreasing on $[a, b]$.
3. If $f'(x) = 0$ for all x in (a, b), then f is constant on $[a, b]$.

▷

· · REMARK The conclusions in the first two cases of Theorem 4.5 are valid even when $f'(x) = 0$ at a finite number of x-values in (a, b).

Proof To prove the first case, assume that $f'(x) > 0$ for all x in the interval (a, b) and let $x_1 < x_2$ be any two points in the interval. By the Mean Value Theorem, you know that there exists a number c such that $x_1 < c < x_2$, and

$$f'(c) = \frac{f(x_2) - f(x_1)}{x_2 - x_1}.$$

Because $f'(c) > 0$ and $x_2 - x_1 > 0$, you know that $f(x_2) - f(x_1) > 0$, which implies that $f(x_1) < f(x_2)$. So, f is increasing on the interval. The second case has a similar proof (see Exercise 113), and the third case is a consequence of Exercise 88 in Section 4.2. ■

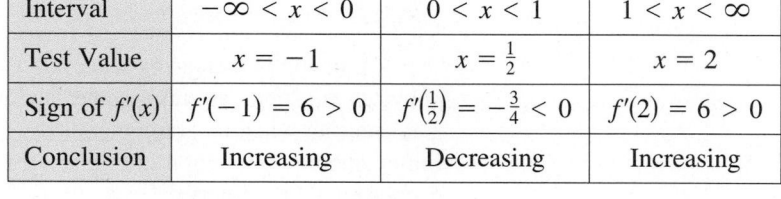

EXAMPLE 1 **Intervals on Which f Is Increasing or Decreasing**

Find the open intervals on which $f(x) = x^3 - \frac{3}{2}x^2$ is increasing or decreasing.

Solution Note that f is differentiable on the entire real number line and the derivative of f is

$$f(x) = x^3 - \frac{3}{2}x^2 \qquad \text{Write original function.}$$
$$f'(x) = 3x^2 - 3x. \qquad \text{Differentiate.}$$

To determine the critical numbers of f, set $f'(x)$ equal to zero.

$$3x^2 - 3x = 0 \qquad \text{Set } f'(x) \text{ equal to 0.}$$
$$3(x)(x - 1) = 0 \qquad \text{Factor.}$$
$$x = 0, 1 \qquad \text{Critical numbers}$$

Because there are no points for which f' does not exist, you can conclude that $x = 0$ and $x = 1$ are the only critical numbers. The table summarizes the testing of the three intervals determined by these two critical numbers.

Interval	$-\infty < x < 0$	$0 < x < 1$	$1 < x < \infty$
Test Value	$x = -1$	$x = \frac{1}{2}$	$x = 2$
Sign of $f'(x)$	$f'(-1) = 6 > 0$	$f'(\frac{1}{2}) = -\frac{3}{4} < 0$	$f'(2) = 6 > 0$
Conclusion	Increasing	Decreasing	Increasing

By Theorem 4.5, f is increasing on the intervals $(-\infty, 0)$ and $(1, \infty)$ and decreasing on the interval $(0, 1)$, as shown in Figure 4.16. ∎

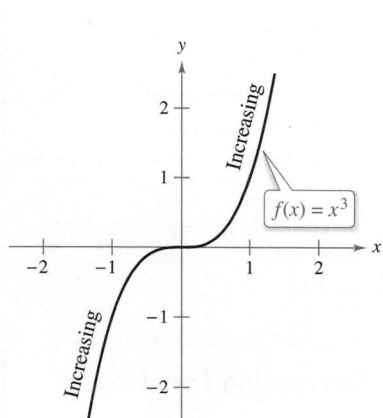

Figure 4.16

Example 1 gives you one instance of how to find intervals on which a function is increasing or decreasing. The guidelines below summarize the steps followed in that example.

GUIDELINES FOR FINDING INTERVALS ON WHICH A FUNCTION IS INCREASING OR DECREASING

Let f be continuous on the interval (a, b). To find the open intervals on which f is increasing or decreasing, use the following steps.

1. Locate the critical numbers of f in (a, b), and use these numbers to determine test intervals.
2. Determine the sign of $f'(x)$ at one test value in each of the intervals.
3. Use Theorem 4.5 to determine whether f is increasing or decreasing on each interval.

These guidelines are also valid when the interval (a, b) is replaced by an interval of the form $(-\infty, b)$, (a, ∞), or $(-\infty, \infty)$.

(a) Strictly monotonic function

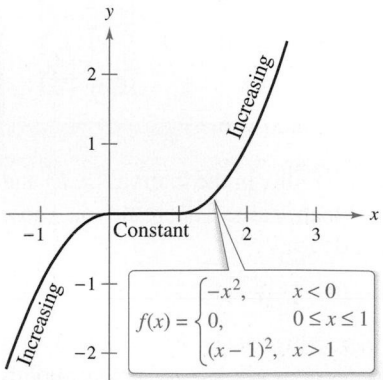

(b) Not strictly monotonic

Figure 4.17

A function is **strictly monotonic** on an interval when it is either increasing on the entire interval or decreasing on the entire interval. For instance, the function $f(x) = x^3$ is strictly monotonic on the entire real number line because it is increasing on the entire real number line, as shown in Figure 4.17(a). The function shown in Figure 4.17(b) is not strictly monotonic on the entire real number line because it is constant on the interval $[0, 1]$.

The First Derivative Test

After you have determined the intervals on which a function is increasing or decreasing, it is not difficult to locate the relative extrema of the function. For instance, in Figure 4.18 (from Example 1), the function

$$f(x) = x^3 - \frac{3}{2}x^2$$

has a relative maximum at the point $(0, 0)$ because f is increasing immediately to the left of $x = 0$ and decreasing immediately to the right of $x = 0$. Similarly, f has a relative minimum at the point $\left(1, -\frac{1}{2}\right)$ because f is decreasing immediately to the left of $x = 1$ and increasing immediately to the right of $x = 1$. The next theorem makes this more explicit.

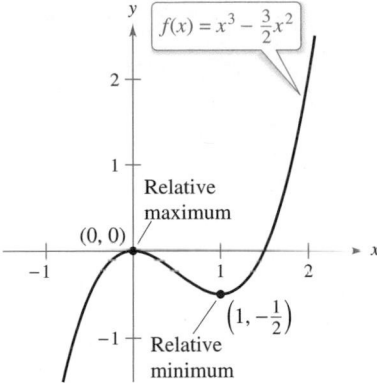

Relative extrema of f
Figure 4.18

THEOREM 4.6 The First Derivative Test

Let c be a critical number of a function f that is continuous on an open interval I containing c. If f is differentiable on the interval, except possibly at c, then $f(c)$ can be classified as follows.

1. If $f'(x)$ changes from negative to positive at c, then f has a *relative minimum* at $(c, f(c))$.

2. If $f'(x)$ changes from positive to negative at c, then f has a *relative maximum* at $(c, f(c))$.

3. If $f'(x)$ is positive on both sides of c or negative on both sides of c, then $f(c)$ is neither a relative minimum nor a relative maximum.

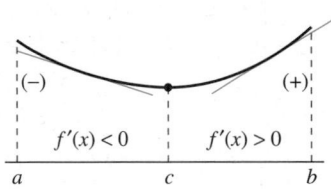

Relative minimum

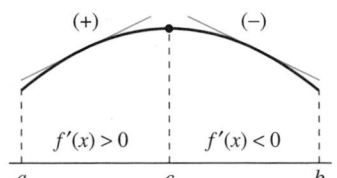

Relative maximum

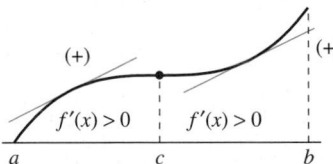

Neither relative minimum nor relative maximum

Proof Assume that $f'(x)$ changes from negative to positive at c. Then there exist a and b in I such that

$$f'(x) < 0 \text{ for all } x \text{ in } (a, c) \quad \text{and} \quad f'(x) > 0 \text{ for all } x \text{ in } (c, b).$$

By Theorem 4.5, f is decreasing on $[a, c]$ and increasing on $[c, b]$. So, $f(c)$ is a minimum of f on the open interval (a, b) and, consequently, a relative minimum of f. This proves the first case of the theorem. The second case can be proved in a similar way (see Exercise 114). ∎

EXAMPLE 2 **Applying the First Derivative Test**

Find the relative extrema of $f(x) = \frac{1}{2}x - \sin x$ in the interval $(0, 2\pi)$.

Solution Note that f is continuous on the interval $(0, 2\pi)$. The derivative of f is $f'(x) = \frac{1}{2} - \cos x$. To determine the critical numbers of f in this interval, set $f'(x)$ equal to 0.

$$\frac{1}{2} - \cos x = 0 \qquad \text{Set } f'(x) \text{ equal to 0.}$$

$$\cos x = \frac{1}{2}$$

$$x = \frac{\pi}{3}, \frac{5\pi}{3} \qquad \text{Critical numbers}$$

Because there are no points for which f' does not exist, you can conclude that $x = \pi/3$ and $x = 5\pi/3$ are the only critical numbers. The table summarizes the testing of the three intervals determined by these two critical numbers. By applying the First Derivative Test, you can conclude that f has a relative minimum at the point where $x = \pi/3$ and a relative maximum at the point where $x = 5\pi/3$, as shown in Figure 4.19.

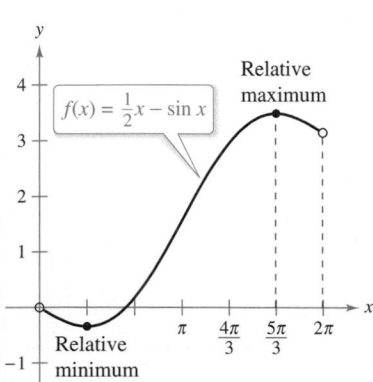

A relative minimum occurs where f changes from decreasing to increasing, and a relative maximum occurs where f changes from increasing to decreasing.
Figure 4.19

Interval	$0 < x < \dfrac{\pi}{3}$	$\dfrac{\pi}{3} < x < \dfrac{5\pi}{3}$	$\dfrac{5\pi}{3} < x < 2\pi$
Test Value	$x = \dfrac{\pi}{4}$	$x = \pi$	$x = \dfrac{7\pi}{4}$
Sign of $f'(x)$	$f'\left(\dfrac{\pi}{4}\right) < 0$	$f'(\pi) > 0$	$f'\left(\dfrac{7\pi}{4}\right) < 0$
Conclusion	Decreasing	Increasing	Decreasing

EXAMPLE 3 **Applying the First Derivative Test**

Find the relative extrema of $f(x) = (x^2 - 4)^{2/3}$.

Solution Begin by noting that f is continuous on the entire real number line. The derivative of f

$$f'(x) = \frac{2}{3}(x^2 - 4)^{-1/3}(2x) \qquad \text{General Power Rule}$$

$$= \frac{4x}{3(x^2 - 4)^{1/3}} \qquad \text{Simplify.}$$

is 0 when $x = 0$ and does not exist when $x = \pm 2$. So, the critical numbers are $x = -2$, $x = 0$, and $x = 2$. The table summarizes the testing of the four intervals determined by these three critical numbers. By applying the First Derivative Test, you can conclude that f has a relative minimum at the point $(-2, 0)$, a relative maximum at the point $\left(0, \sqrt[3]{16}\right)$, and another relative minimum at the point $(2, 0)$, as shown in Figure 4.20.

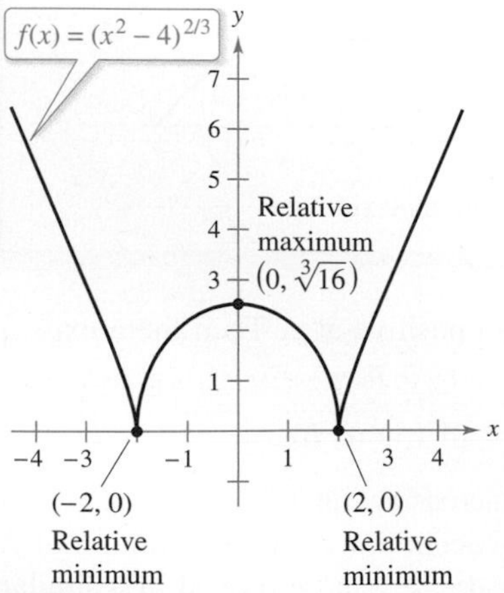

$f(x) = (x^2 - 4)^{2/3}$

Relative maximum $\left(0, \sqrt[3]{16}\right)$

$(-2, 0)$ Relative minimum

$(2, 0)$ Relative minimum

Figure 4.20

Interval	$-\infty < x < -2$	$-2 < x < 0$	$0 < x < 2$	$2 < x < \infty$
Test Value	$x = -3$	$x = -1$	$x = 1$	$x = 3$
Sign of $f'(x)$	$f'(-3) < 0$	$f'(-1) > 0$	$f'(1) < 0$	$f'(3) > 0$
Conclusion	Decreasing	Increasing	Decreasing	Increasing

Note that in Examples 1 and 2, the given functions are differentiable on the entire real number line. For such functions, the only critical numbers are those for which $f'(x) = 0$. Example 3 concerns a function that has two types of critical numbers—those for which $f'(x) = 0$ and those for which f is not differentiable.

When using the First Derivative Test, be sure to consider the domain of the function. For instance, in the next example, the function

$$f(x) = \frac{x^4 + 1}{x^2}$$

is not defined when $x = 0$. This x-value must be used with the critical numbers to determine the test intervals.

EXAMPLE 4 **Applying the First Derivative Test**

· · · · ▷ *See LarsonCalculus.com for an interactive version of this type of example.*

Find the relative extrema of $f(x) = \dfrac{x^4 + 1}{x^2}$.

Solution Note that f is not defined when $x = 0$.

$$f(x) = x^2 + x^{-2} \qquad\qquad \text{Rewrite original function.}$$

$$f'(x) = 2x - 2x^{-3} \qquad\qquad \text{Differentiate.}$$

$$= 2x - \frac{2}{x^3} \qquad\qquad \text{Rewrite with positive exponent.}$$

$$= \frac{2(x^4 - 1)}{x^3} \qquad\qquad \text{Simplify.}$$

$$= \frac{2(x^2 + 1)(x - 1)(x + 1)}{x^3} \qquad\qquad \text{Factor.}$$

So, $f'(x)$ is zero at $x = \pm 1$. Moreover, because $x = 0$ is not in the domain of f, you should use this x-value along with the critical numbers to determine the test intervals.

$$x = \pm 1 \qquad\qquad \text{Critical numbers, } f'(\pm 1) = 0$$

$$x = 0 \qquad\qquad \text{0 is not in the domain of } f.$$

The table summarizes the testing of the four intervals determined by these three x-values. By applying the First Derivative Test, you can conclude that f has one relative minimum at the point $(-1, 2)$ and another at the point $(1, 2)$, as shown in Figure 4.21.

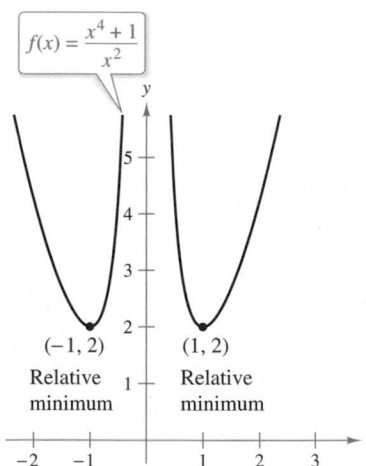

$f(x) = \dfrac{x^4 + 1}{x^2}$

$(-1, 2)$ Relative minimum

$(1, 2)$ Relative minimum

x-values that are not in the domain of *f*, as well as critical numbers, determine test intervals for *f'*.
Figure 4.21

Interval	$-\infty < x < -1$	$-1 < x < 0$	$0 < x < 1$	$1 < x < \infty$
Test Value	$x = -2$	$x = -\frac{1}{2}$	$x = \frac{1}{2}$	$x = 2$
Sign of $f'(x)$	$f'(-2) < 0$	$f'(-\frac{1}{2}) > 0$	$f'(\frac{1}{2}) < 0$	$f'(2) > 0$
Conclusion	Decreasing	Increasing	Decreasing	Increasing

▷ **TECHNOLOGY** The most difficult step in applying the First Derivative Test is finding the values for which the derivative is equal to 0. For instance, the values of x for which the derivative of

$$f(x) = \frac{x^4 + 1}{x^2 + 1}$$

is equal to zero are $x = 0$ and $x = \pm\sqrt{\sqrt{2} - 1}$. If you have access to technology that can perform symbolic differentiation and solve equations, use it to apply the First Derivative Test to this function.

When a projectile is propelled from ground level and air resistance is neglected, the object will travel farthest with an initial angle of 45°. When, however, the projectile is propelled from a point above ground level, the angle that yields a maximum horizontal distance is not 45° (see Example 5).

EXAMPLE 5 **The Path of a Projectile**

Neglecting air resistance, the path of a projectile that is propelled at an angle θ is

$$y = -\frac{g \sec^2 \theta}{2v_0^2}x^2 + (\tan \theta)x + h, \quad 0 \le \theta \le \frac{\pi}{2}$$

where y is the height, x is the horizontal distance, g is the acceleration due to gravity, v_0 is the initial velocity, and h is the initial height. (This equation is derived in Section 12.3.) Let $g = 32$ feet per second per second, $v_0 = 24$ feet per second, and $h = 9$ feet. What value of θ will produce a maximum horizontal distance?

Solution To find the distance the projectile travels, let $y = 0$, $g = 32$, $v_0 = 24$, and $h = 9$. Then substitute these values in the given equation as shown.

$$-\frac{g \sec^2 \theta}{2v_0^2}x^2 + (\tan \theta)x + h = y$$

$$-\frac{32 \sec^2 \theta}{2(24^2)}x^2 + (\tan \theta)x + 9 = 0$$

$$-\frac{\sec^2 \theta}{36}x^2 + (\tan \theta)x + 9 = 0$$

Next, solve for x using the Quadratic Formula with $a = (-\sec^2 \theta)/36$, $b = \tan \theta$, and $c = 9$.

$$x = \frac{-b \pm \sqrt{b^2 - 4ac}}{2a}$$

$$x = \frac{-\tan \theta \pm \sqrt{(\tan \theta)^2 - 4[(-\sec^2 \theta)/36](9)}}{2[(-\sec^2 \theta)/36]}$$

$$x = \frac{-\tan \theta \pm \sqrt{\tan^2 \theta + \sec^2 \theta}}{(-\sec^2 \theta)/18}$$

$$x = 18(\cos \theta)\left(\sin \theta + \sqrt{\sin^2 \theta + 1}\right), \quad x \ge 0$$

At this point, you need to find the value of θ that produces a maximum value of x. Applying the First Derivative Test by hand would be very tedious. Using technology to solve the equation $dx/d\theta = 0$, however, eliminates most of the messy computations. The result is that the maximum value of x occurs when

$$\theta \approx 0.61548 \text{ radian}, \quad \text{or} \quad 35.3°.$$

This conclusion is reinforced by sketching the path of the projectile for different values of θ, as shown in Figure 4.22. Of the three paths shown, note that the distance traveled is greatest for $\theta = 35°$.

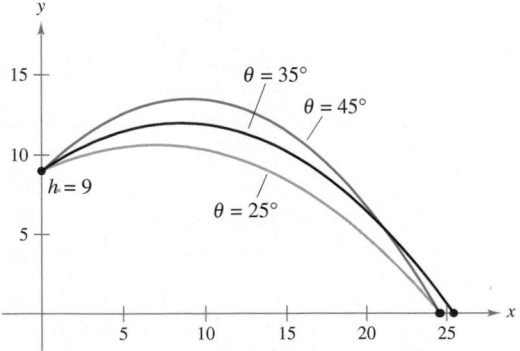

The path of a projectile with initial angle θ
Figure 4.22

4.3 Exercises

See **CalcChat.com** for tutorial help and worked-out solutions to odd-numbered exercises.

CONCEPT CHECK

1. **Increasing and Decreasing Functions** Describe the Test for Increasing and Decreasing Functions in your own words.

2. **First Derivative Test** Describe the First Derivative Test in your own words.

Using a Graph In Exercises 3 and 4, use the graph of f to find (a) the largest open interval on which f is increasing and (b) the largest open interval on which f is decreasing.

3.
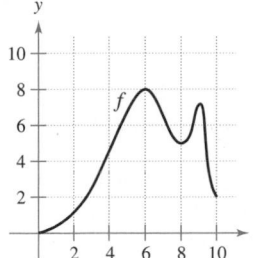

4.
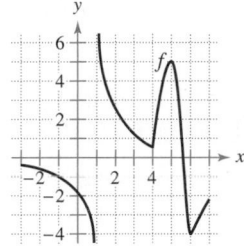

 **Using a Graph** In Exercises 5–10, use the graph to estimate the open intervals on which the function is increasing or decreasing. Then find the open intervals analytically.

5. $y = -(x + 1)^2$

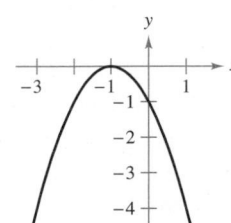

6. $f(x) = x^2 - 6x + 8$

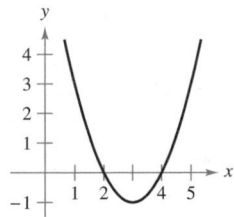

7. $y = \dfrac{x^3}{4} - 3x$

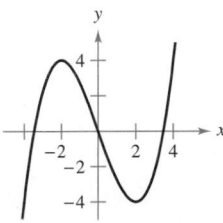

8. $f(x) = x^4 - 2x^2$

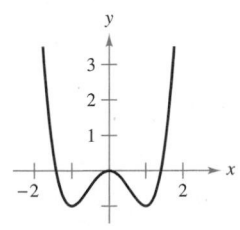

9. $f(x) = \dfrac{1}{(x + 1)^2}$

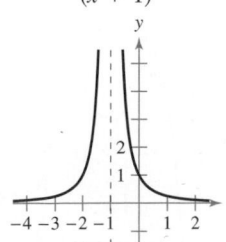

10. $y = \dfrac{x^2}{2x - 1}$

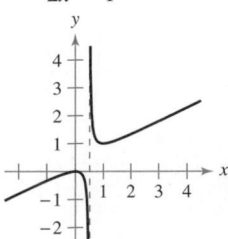

 **Intervals on Which a Function Is Increasing or Decreasing** In Exercises 11–22, find the open intervals on which the function is increasing or decreasing.

11. $g(x) = x^2 - 2x - 8$

12. $h(x) = 12x - x^3$

13. $y = x\sqrt{16 - x^2}$

14. $y = x + \dfrac{9}{x}$

15. $f(x) = \sin x - 1, \quad 0 < x < 2\pi$

16. $f(x) = \cos \dfrac{3x}{2}, \quad 0 < x < 2\pi$

17. $y = x - 2\cos x, \quad 0 < x < 2\pi$

18. $f(x) = \sin^2 x + \sin x, \quad 0 < x < 2\pi$

19. $g(x) = e^{-x} + e^{3x}$

20. $h(x) = \sqrt{x}e^{-x}$

21. $f(x) = x^2 \ln \dfrac{x}{2}$

22. $f(x) = \dfrac{\ln x}{\sqrt{x}}$

 **Applying the First Derivative Test** In Exercises 23–56, (a) find the critical numbers of f, if any, (b) find the open intervals on which the function is increasing or decreasing, (c) apply the First Derivative Test to identify all relative extrema, and (d) use a graphing utility to confirm your results.

23. $f(x) = x^2 - 8x$

24. $f(x) = x^2 + 6x + 10$

25. $f(x) = -2x^2 + 4x + 3$

26. $f(x) = -3x^2 - 4x - 2$

27. $f(x) = -7x^3 + 21x + 3$

28. $f(x) = x^3 - 6x^2 + 15$

29. $f(x) = (x - 1)^2(x + 3)$

30. $f(x) = (8 - x)(x + 1)^2$

31. $f(x) = \dfrac{x^5 - 5x}{5}$

32. $f(x) = \dfrac{-x^6 + 6x}{10}$

33. $f(x) = x^{1/3} + 1$

34. $f(x) = x^{2/3} - 4$

35. $f(x) = (x + 2)^{2/3}$

36. $f(x) = (x - 3)^{1/3}$

37. $f(x) = 5 - |x - 5|$

38. $f(x) = |x + 3| - 1$

39. $f(x) = 2x + \dfrac{1}{x}$

40. $f(x) = \dfrac{x}{x - 5}$

41. $f(x) = \dfrac{x^2}{x^2 - 9}$

42. $f(x) = \dfrac{x^2 - 2x + 1}{x + 1}$

43. $f(x) = \begin{cases} 4 - x^2, & x \le 0 \\ -2x, & x > 0 \end{cases}$

44. $f(x) = \begin{cases} 2x + 1, & x \le -1 \\ x^2 - 2, & x > -1 \end{cases}$

45. $f(x) = (3 - x)e^{x-3}$

46. $f(x) = (x - 1)e^x$

47. $f(x) = 4(x - \arcsin x)$

48. $f(x) = x \arctan x$

49. $f(x) = (x)3^{-x}$

50. $f(x) = 2^{x^2 - 3}$

51. $f(x) = x - \log_4 x$

52. $f(x) = \dfrac{x^3}{3} - \ln x$

53. $f(x) = \dfrac{e^{2x}}{e^{2x} + 1}$

54. $f(x) = \ln(2 - \ln x)$

55. $f(x) = e^{-1/(x-2)}$

56. $f(x) = e^{\arctan x}$

Applying the First Derivative Test In Exercises 57–62, consider the function on the interval $(0, 2\pi)$. (a) Find the open intervals on which the function is increasing or decreasing. (b) Apply the First Derivative Test to identify all relative extrema. (c) Use a graphing utility to confirm your results.

57. $f(x) = x - 2 \sin x$ **58.** $f(x) = \sin x \cos x + 5$

59. $f(x) = \sin x + \cos x$ **60.** $f(x) = \dfrac{x}{2} + \cos x$

61. $f(x) = \cos^2(2x)$ **62.** $f(x) = \sin x - \sqrt{3} \cos x$

Finding and Analyzing Derivatives Using Technology In Exercises 63–70, (a) use a computer algebra system to differentiate the function, (b) sketch the graphs of f and f' on the same set of coordinate axes over the given interval, (c) find the critical numbers of f in the open interval, and (d) find the interval(s) on which f' is positive and the interval(s) on which f' is negative. Compare the behavior of f and the sign of f'.

63. $f(x) = 2x\sqrt{9 - x^2}$, $[-3, 3]$
64. $f(x) = 10(5 - \sqrt{x^2 - 3x + 16})$, $[0, 5]$
65. $f(t) = t^2 \sin t$, $[0, 2\pi]$
66. $f(x) = \dfrac{x}{2} + \cos\dfrac{x}{2}$, $[0, 4\pi]$
67. $f(x) = -3 \sin\dfrac{x}{3}$, $[0, 6\pi]$
68. $f(x) = 2 \sin 3x + 4 \cos 3x$, $[0, \pi]$
69. $f(x) = \frac{1}{2}(x^2 - \ln x)$, $(0, 3]$
70. $f(x) = (4 - x^2)e^x$, $[0, 2]$

Comparing Functions In Exercises 71 and 72, use symmetry, extrema, and zeros to sketch the graph of f. How do the functions f and g differ?

71. $f(x) = \dfrac{x^5 - 4x^3 + 3x}{x^2 - 1}$ **72.** $f(t) = \cos^2 t - \sin^2 t$

   $g(x) = x(x^2 - 3)$ $g(t) = 1 - 2 \sin^2 t$

Think About It In Exercises 73–78, the graph of f is shown in the figure. Sketch a graph of the derivative of f. To print an enlarged copy of the graph, go to *MathGraphs.com*.

73.

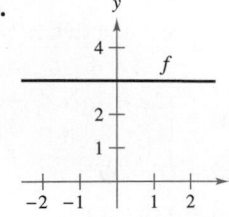

74.

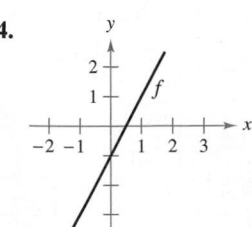

75.

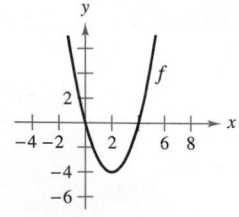

76.

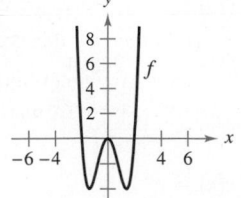

77.

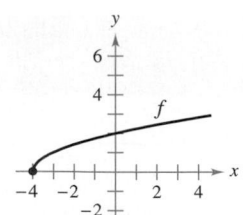

78.

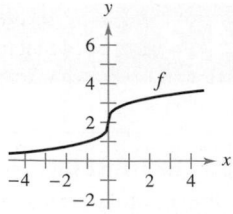

EXPLORING CONCEPTS

Transformations of Functions In Exercises 79–82, assume that f is differentiable for all x, where $f'(x) > 0$ on $(-\infty, -4)$, $f'(x) < 0$ on $(-4, 6)$, and $f'(x) > 0$ on $(6, \infty)$. Supply the appropriate inequality sign for the indicated value of c.

Function	Sign of $g'(c)$
79. $g(x) = f(x) + 5$	$g'(0)$ ▨ 0
80. $g(x) = 3f(x) - 3$	$g'(-5)$ ▨ 0
81. $g(x) = -f(x)$	$g'(-6)$ ▨ 0
82. $g(x) = f(x - 10)$	$g'(0)$ ▨ 0

83. Sketching a Graph Sketch the graph of the arbitrary function f such that $f'(x) > 0$ on $(-\infty, 4)$, $f'(x) < 0$ on $(4, \infty)$, and $f'(x)$ is undefined at $x = 4$.

84. Think About It Is it possible to find a differentiable function f where $f(x) > 0$ and $f'(x) < 0$? If so, give an example. If not, explain why not.

85. Increasing Functions Consider two increasing functions f and g.

(a) Is $f + g$ always increasing? Explain.

(b) Is fg always increasing? Explain.

86. 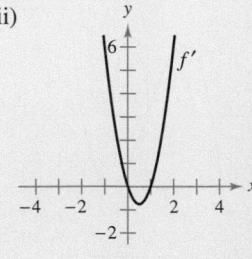 **HOW DO YOU SEE IT?** Use the graph of f' to (a) identify the critical numbers of f, (b) identify the open intervals on which f is increasing or decreasing, and (c) determine whether f has a relative maximum, a relative minimum, or neither at each critical number.

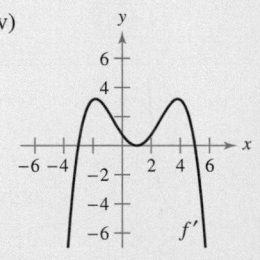

87. Analyzing a Critical Number A differentiable function f has one critical number at $x = 5$. Identify the relative extrema of f at the critical number when $f'(4) = -2.5$ and $f'(6) = 3$.

88. Analyzing a Critical Number A differentiable function f has one critical number at $x = 2$. Identify the relative extrema of f at the critical number when $f'(1) = 2$ and $f'(3) = 6$.

Think About It In Exercises 89 and 90, the function f is differentiable on the indicated interval. The table shows $f'(x)$ for selected values of x. (a) Sketch the graph of f, (b) approximate the critical numbers, and (c) identify the relative extrema.

89. f is differentiable on $[-1, 1]$.

x	-1	-0.75	-0.50	-0.25	0
$f'(x)$	-10	-3.2	-0.5	0.8	5.6

x	0.25	0.50	0.75	1
$f'(x)$	3.6	-0.2	-6.7	-20.1

90. f is differentiable on $[0, \pi]$.

x	0	$\pi/6$	$\pi/4$	$\pi/3$	$\pi/2$
$f'(x)$	3.14	-0.23	-2.45	-3.11	0.69

x	$2\pi/3$	$3\pi/4$	$5\pi/6$	π
$f'(x)$	3.00	1.37	-1.14	-2.84

91. Rolling a Ball Bearing A ball bearing is placed on an inclined plane and begins to roll. The angle of elevation of the plane is θ. The distance (in meters) the ball bearing rolls in t seconds is $s(t) = 4.9(\sin \theta)t^2$.

(a) Determine the speed of the ball bearing after t seconds.

(b) Complete the table and use it to determine the value of θ that produces the maximum speed at a particular time.

θ	0	$\pi/4$	$\pi/3$	$\pi/2$	$2\pi/3$	$3\pi/4$	π
$s'(t)$							

92. Modeling Data The end-of-year assets of the Medicare Hospital Insurance Trust Fund (in billions of dollars) for the years 2006 through 2014 are shown.

2006: 305.4   2007: 326.0   2008: 321.3
2009: 304.2   2010: 271.9   2011: 244.2
2012: 220.4   2013: 205.4   2014: 197.3

(Source: U.S. Centers for Medicare and Medicaid Services)

(a) Use the regression capabilities of a graphing utility to find a model of the form $M = at^3 + bt^2 + ct + d$ for the data. Let $t = 6$ represent 2006.

(b) Use a graphing utility to plot the data and graph the model.

(c) Find the maximum value of the model and compare the result with the actual data.

93. Numerical, Graphical, and Analytic Analysis The concentration C of a chemical in the bloodstream t hours after injection into muscle tissue is

$$C(t) = \frac{3t}{27 + t^3}, \quad t \geq 0.$$

(a) Complete the table and use it to approximate the time when the concentration is greatest.

t	0	0.5	1	1.5	2	2.5	3
$C(t)$							

(b) Use a graphing utility to graph the concentration function and use the graph to approximate the time when the concentration is greatest.

(c) Use calculus to determine analytically the time when the concentration is greatest.

94. Numerical, Graphical, and Analytic Analysis Consider the functions $f(x) = x$ and $g(x) = \sin x$ on the interval $(0, \pi)$.

(a) Complete the table and make a conjecture about which is the greater function on the interval $(0, \pi)$.

x	0.5	1	1.5	2	2.5	3
$f(x)$						
$g(x)$						

(b) Use a graphing utility to graph the functions and use the graphs to make a conjecture about which is the greater function on the interval $(0, \pi)$.

(c) Prove that $f(x) > g(x)$ on the interval $(0, \pi)$. [*Hint*: Show that $h'(x) > 0$, where $h = f - g$.]

95. Trachea Contraction Coughing forces the trachea (windpipe) to contract, which affects the velocity v of the air passing through the trachea. The velocity of the air during coughing is

$$v = k(R - r)r^2, \quad 0 \leq r < R$$

where k is a constant, R is the normal radius of the trachea, and r is the radius during coughing. What radius will produce the maximum air velocity?

96. Electrical Resistance The resistance R of a certain type of resistor is

$$R = \sqrt{0.001T^4 - 4T + 100}$$

where R is measured in ohms and the temperature T is measured in degrees Celsius.

(a) Use a computer algebra system to find dR/dT and the critical number of the function. Determine the minimum resistance for this type of resistor.

(b) Use a graphing utility to graph the function R and use the graph to approximate the minimum resistance for this type of resistor.

Motion Along a Line In Exercises 97–100, the function $s(t)$ describes the motion of a particle along a line. (a) Find the velocity function of the particle at any time $t \geq 0$. (b) Identify the time interval(s) on which the particle is moving in a positive direction. (c) Identify the time interval(s) on which the particle is moving in a negative direction. (d) Identify the time(s) at which the particle changes direction.

97. $s(t) = 6t - t^2$

98. $s(t) = t^2 - 10t + 29$

99. $s(t) = t^3 - 5t^2 + 4t$

100. $s(t) = t^3 - 20t^2 + 128t - 280$

Motion Along a Line In Exercises 101 and 102, the graph shows the position of a particle moving along a line. Describe how the position of the particle changes with respect to time.

101.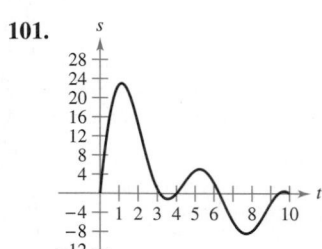

102.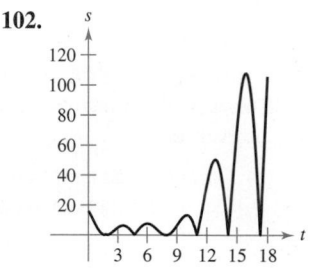

Creating Polynomial Functions In Exercises 103–106, find a polynomial function

$$f(x) = a_n x^n + a_{n-1} x^{n-1} + \cdots + a_2 x^2 + a_1 x + a_0$$

that has only the specified extrema. (a) Determine the minimum degree of the function and give the criteria you used in determining the degree. (b) Using the fact that the coordinates of the extrema are solution points of the function, and that the x-coordinates are critical numbers, determine a system of linear equations whose solution yields the coefficients of the required function. (c) Use a graphing utility to solve the system of equations and determine the function. (d) Use a graphing utility to confirm your result graphically.

103. Relative minimum: $(0, 0)$; Relative maximum: $(2, 2)$

104. Relative minimum: $(0, 0)$; Relative maximum: $(4, 1000)$

105. Relative minima: $(0, 0)$, $(4, 0)$; Relative maximum: $(2, 4)$

106. Relative minimum: $(1, 2)$; Relative maxima: $(-1, 4)$, $(3, 4)$

True or False? In Exercises 107–112, determine whether the statement is true or false. If it is false, explain why or give an example that shows it is false.

107. There is no function with an infinite number of critical points.

108. The function $f(x) = x$ has no extrema on any open interval.

109. Every nth-degree polynomial has $(n - 1)$ critical numbers.

110. An nth-degree polynomial has at most $(n - 1)$ critical numbers.

111. There is a relative extremum at each critical number.

112. The relative maxima of the function f are $f(1) = 4$ and $f(3) = 10$. Therefore, f has at least one minimum for some x in the interval $(1, 3)$.

113. Proof Prove the second case of Theorem 4.5.

114. Proof Prove the second case of Theorem 4.6.

115. Proof Let $x > 0$ and $n > 1$ be real numbers. Prove that $(1 + x)^n > 1 + nx$.

116. Proof Use the definitions of increasing and decreasing functions to prove that $f(x) = x^3$ is increasing on $(-\infty, \infty)$.

117. Proof Use the definitions of increasing and decreasing functions to prove that

$$f(x) = \frac{1}{x}$$

is decreasing on $(0, \infty)$.

118. Finding Values Consider $f(x) = axe^{bx^2}$. Find a and b such that the relative maximum of f is $f(4) = 2$.

PUTNAM EXAM CHALLENGE

119. Find the minimum value of

$$|\sin x + \cos x + \tan x + \cot x + \sec x + \csc x|$$

for real numbers x.

This problem was composed by the Committee on the Putnam Prize Competition.
© The Mathematical Association of America. All rights reserved.

SECTION PROJECT

Even Fourth-Degree Polynomials

(a) Graph each of the fourth-degree polynomials below. Then find the critical numbers, the open intervals on which the function is increasing or decreasing, and the relative extrema.

   (i) $f(x) = x^4 + 1$

   (ii) $f(x) = x^4 + 2x^2 + 1$

   (iii) $f(x) = x^4 - 2x^2 + 1$

(b) Consider the fourth-degree polynomial

   $$f(x) = x^4 + ax^2 + b.$$

   (i) Show that there is one critical number when $a = 0$. Then find the open intervals on which the function is increasing or decreasing.

   (ii) Show that there is one critical number when $a > 0$. Then find the open intervals on which the function is increasing or decreasing.

   (iii) Show that there are three critical numbers when $a < 0$. Then find the open intervals on which the function is increasing or decreasing.

   (iv) Show that there are no real zeros when

   $$a^2 < 4b.$$

   (v) Determine the possible number of zeros when

   $$a^2 \geq 4b.$$

   Explain your reasoning.

4.4 Concavity and the Second Derivative Test

■ **Determine intervals on which a function is concave upward or concave downward.**
■ **Find any points of inflection of the graph of a function.**
■ **Apply the Second Derivative Test to find relative extrema of a function.**

Concavity

You have already seen that locating the intervals on which a function f increases or decreases helps to describe its graph. In this section, you will see how locating the intervals on which f' increases or decreases can be used to determine where the graph of f is *curving upward* or *curving downward*.

> **Definition of Concavity**
>
> Let f be differentiable on an open interval I. The graph of f is **concave upward** on I when f' is increasing on the interval and **concave downward** on I when f' is decreasing on the interval.

The following graphical interpretation of concavity is useful. (See Appendix A for a proof of these results.)

1. Let f be differentiable on an open interval I. If the graph of f is concave *upward* on I, then the graph of f lies *above* all of its tangent lines on I.
   [See Figure 4.23(a).]

2. Let f be differentiable on an open interval I. If the graph of f is concave *downward* on I, then the graph of f lies *below* all of its tangent lines on I.
   [See Figure 4.23(b).]

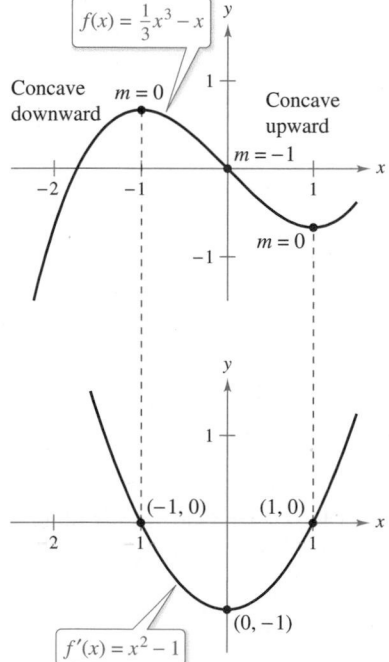

f' is decreasing. f' is increasing.

The concavity of f is related to the slope of the derivative.

Figure 4.24

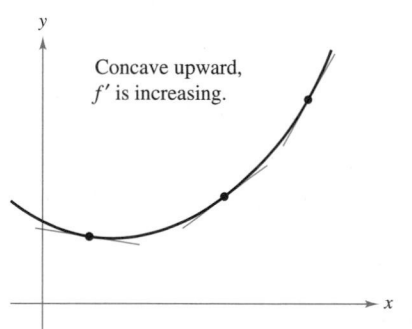

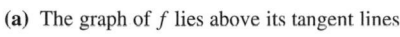

(a) The graph of f lies above its tangent lines.

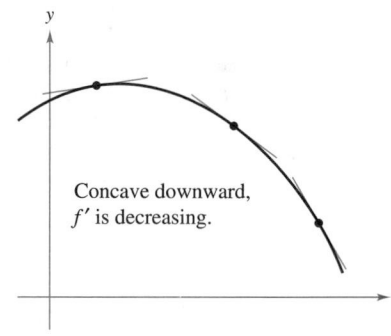

(b) The graph of f lies below its tangent lines.

Figure 4.23

To find the open intervals on which the graph of a function f is concave upward or concave downward, you need to find the intervals on which f' is increasing or decreasing. For instance, the graph of

$$f(x) = \frac{1}{3}x^3 - x$$

is concave downward on the open interval $(-\infty, 0)$ because

$$f'(x) = x^2 - 1$$

is decreasing there. (See Figure 4.24.) Similarly, the graph of f is concave upward on the interval $(0, \infty)$ because f' is increasing on $(0, \infty)$.

The next theorem shows how to use the *second* derivative of a function f to determine intervals on which the graph of f is concave upward or concave downward. A proof of this theorem follows directly from Theorem 4.5 and the definition of concavity.

> **THEOREM 4.7** **Test for Concavity**
>
> Let f be a function whose second derivative exists on an open interval I.
>
> 1. If $f''(x) > 0$ for all x in I, then the graph of f is concave upward on I.
> 2. If $f''(x) < 0$ for all x in I, then the graph of f is concave downward on I.
>
> A proof of this theorem is given in Appendix A.

• **REMARK** A third case of Theorem 4.7 could be that if $f''(x) = 0$ for all x in I, then f is linear. Note, however, that concavity is not defined for a line. In other words, a straight line is neither concave upward nor concave downward.

To apply Theorem 4.7, locate the x-values at which $f''(x) = 0$ or $f''(x)$ does not exist. Use these x-values to determine test intervals. Finally, test the sign of $f''(x)$ in each of the test intervals.

EXAMPLE 1 Determining Concavity

Determine the open intervals on which the graph of

$$f(x) = e^{-x^2/2}$$

is concave upward or concave downward.

Solution Begin by observing that f is continuous on the entire real number line. Next, find the second derivative of f.

$$f'(x) = -xe^{-x^2/2} \qquad \text{First derivative}$$
$$f''(x) = (-x)(-x)e^{-x^2/2} + e^{-x^2/2}(-1) \qquad \text{Differentiate.}$$
$$= e^{-x^2/2}(x^2 - 1) \qquad \text{Second derivative}$$

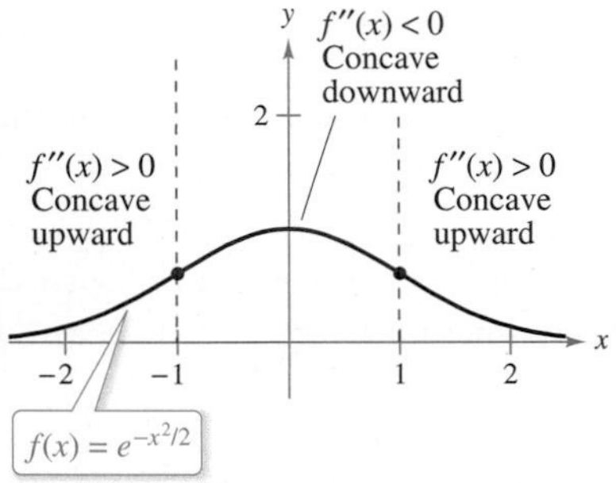

Because $f''(x) = 0$ when $x = \pm 1$ and f'' is defined on the entire real number line, you should test f'' in the intervals $(-\infty, -1)$, $(-1, 1)$, and $(1, \infty)$. The results are shown in the table and in Figure 4.25.

From the sign of $f''(x)$, you can determine the concavity of the graph of f.

Figure 4.25

Interval	$-\infty < x < -1$	$-1 < x < 1$	$1 < x < \infty$
Test Value	$x = -2$	$x = 0$	$x = 2$
Sign of $f''(x)$	$f''(-2) > 0$	$f''(0) < 0$	$f''(2) > 0$
Conclusion	Concave upward	Concave downward	Concave upward

Note that the function in Example 1 is similar to the general form of a **normal probability density function** (whose mean is 0),

$$f(x) = \frac{1}{\sigma\sqrt{2\pi}} e^{-x^2/(2\sigma^2)} \qquad \text{Normal probability density function}$$

where σ is the **standard deviation** (σ is the lowercase Greek letter sigma). This "bell-shaped" curve is concave downward on the interval $(-\sigma, \sigma)$.

The function given in Example 1 is continuous on the entire real number line. When there are x-values at which a function is not continuous, these values should be used, along with the points at which $f''(x) = 0$ or $f''(x)$ does not exist, to form the test intervals.

EXAMPLE 2 **Determining Concavity**

Determine the open intervals on which the graph of

$$f(x) = \frac{x^2 + 1}{x^2 - 4}$$

is concave upward or concave downward.

Solution Differentiating twice produces the following.

$$f(x) = \frac{x^2 + 1}{x^2 - 4}$$ Write original function.

$$f'(x) = \frac{(x^2 - 4)(2x) - (x^2 + 1)(2x)}{(x^2 - 4)^2}$$ Differentiate.

$$= \frac{-10x}{(x^2 - 4)^2}$$ First derivative

$$f''(x) = \frac{(x^2 - 4)^2(-10) - (-10x)(2)(x^2 - 4)(2x)}{(x^2 - 4)^4}$$ Differentiate.

$$= \frac{10(3x^2 + 4)}{(x^2 - 4)^3}$$ Second derivative

There are no points at which $f''(x) = 0$, but at $x = \pm 2$, the function f is not continuous. So, test for concavity in the intervals $(-\infty, -2)$, $(-2, 2)$, and $(2, \infty)$, as shown in the table. The graph of f is shown in Figure 4.26.

Interval	$-\infty < x < -2$	$-2 < x < 2$	$2 < x < \infty$
Test Value	$x = -3$	$x = 0$	$x = 3$
Sign of $f''(x)$	$f''(-3) > 0$	$f''(0) < 0$	$f''(3) > 0$
Conclusion	Concave upward	Concave downward	Concave upward

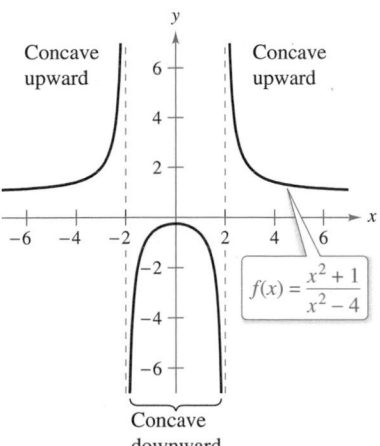

Concave upward   Concave upward

$$f(x) = \frac{x^2 + 1}{x^2 - 4}$$

Concave downward

Figure 4.26

Points of Inflection

The graph in Figure 4.25 has two points at which the concavity changes. If the tangent line to the graph exists at such a point, then that point is a **point of inflection.** Three types of points of inflection are shown in Figure 4.27.

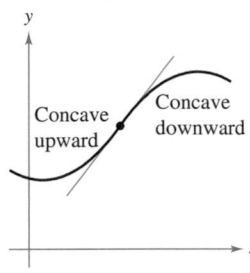

Concave upward   Concave downward

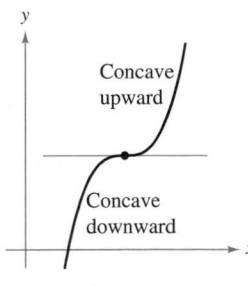

Concave upward

Concave downward

> **Definition of Point of Inflection**
>
> Let f be a function that is continuous on an open interval, and let c be a point in the interval. If the graph of f has a tangent line at the point $(c, f(c))$, then this point is a **point of inflection** of the graph of f when the concavity of f changes from upward to downward (or downward to upward) at the point.

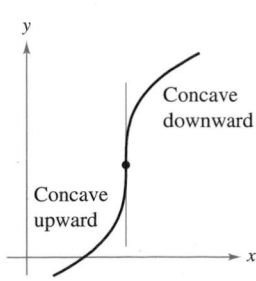

Concave downward

Concave upward

The concavity of f changes at a point of inflection. Note that the graph crosses its tangent line at a point of inflection.

Figure 4.27

The definition of *point of inflection* requires that the tangent line exists at the point of inflection. Some calculus texts do not require this. For instance, after applying the definition above to the function

$$f(x) = \begin{cases} x^3, & x < 0 \\ x^2 + 2x, & x \geq 0 \end{cases}$$

you would conclude that f does *not* have a point of inflection at the origin, even though the concavity of the graph changes from concave downward to concave upward.

To locate *possible* points of inflection, you can determine the values of x for which $f''(x) = 0$ or $f''(x)$ does not exist. This is similar to the procedure for locating relative extrema of f.

THEOREM 4.8 Points of Inflection

If $(c, f(c))$ is a point of inflection of the graph of f, then either $f''(c) = 0$ or $f''(c)$ does not exist.

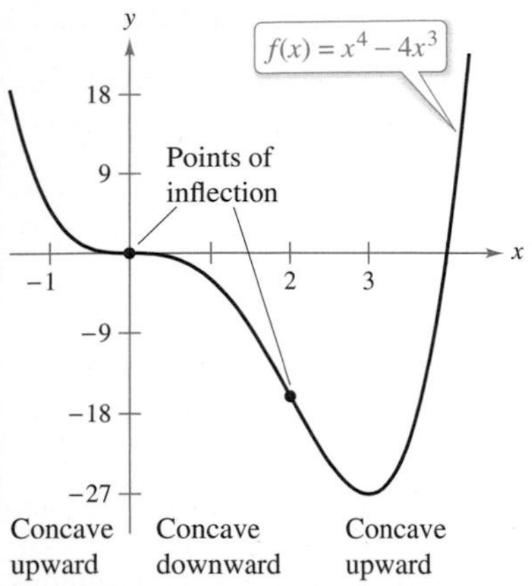

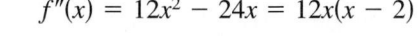

EXAMPLE 3 **Finding Points of Inflection**

Determine the points of inflection and discuss the concavity of the graph of

$$f(x) = x^4 - 4x^3.$$

Solution Differentiating twice produces the following.

$f(x) = x^4 - 4x^3$	Write original function.
$f'(x) = 4x^3 - 12x^2$	Find first derivative.
$f''(x) = 12x^2 - 24x = 12x(x - 2)$	Find second derivative.

Setting $f''(x) = 0$, you can determine that the possible points of inflection occur at $x = 0$ and $x = 2$. By testing the intervals determined by these x-values, you can conclude that they both yield points of inflection. A summary of this testing is shown in the table, and the graph of f is shown in Figure 4.28.

Points of inflection can occur where $f''(x) = 0$ or f'' does not exist.

Figure 4.28

Interval	$-\infty < x < 0$	$0 < x < 2$	$2 < x < \infty$
Test Value	$x = -1$	$x = 1$	$x = 3$
Sign of $f''(x)$	$f''(-1) > 0$	$f''(1) < 0$	$f''(3) > 0$
Conclusion	Concave upward	Concave downward	Concave upward

The converse of Theorem 4.8 is not generally true. That is, it is possible for the second derivative to be 0 at a point that is *not* a point of inflection. For instance, the graph of $f(x) = x^4$ is shown in Figure 4.29. The second derivative is 0 when $x = 0$, but the point $(0, 0)$ is not a point of inflection because the graph of f is concave upward on the intervals $-\infty < x < 0$ and $0 < x < \infty$.

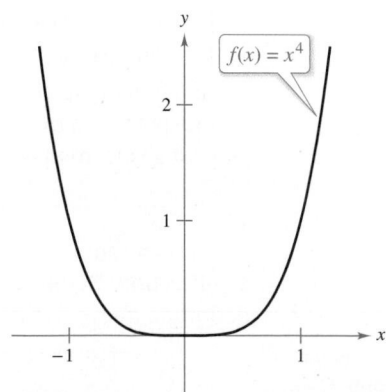

$f''(x) = 0$, but $(0, 0)$ is not a point of inflection.

Figure 4.29

Exploration

Consider a general cubic function of the form

$$f(x) = ax^3 + bx^2 + cx + d.$$

You know that the value of d has a bearing on the location of the graph but has no bearing on the value of the first derivative at given values of x. Graphically, this is true because changes in the value of d shift the graph up or down but do not change its basic shape. Use a graphing utility to graph several cubics with different values of c. Then give a graphical explanation of why changes in c do not affect the values of the second derivative.

The Second Derivative Test

In addition to testing for concavity, the second derivative can be used to perform a simple test for relative maxima and minima. The test is based on the fact that if the graph of a function f is concave upward on an open interval containing c, and $f'(c) = 0$, then $f(c)$ must be a relative minimum of f. Similarly, if the graph of a function f is concave downward on an open interval containing c, and $f'(c) = 0$, then $f(c)$ must be a relative maximum of f. (See Figure 4.30.)

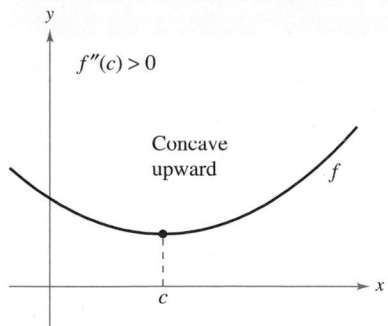

If $f'(c) = 0$ and $f''(c) > 0$, then $f(c)$ is a relative minimum.

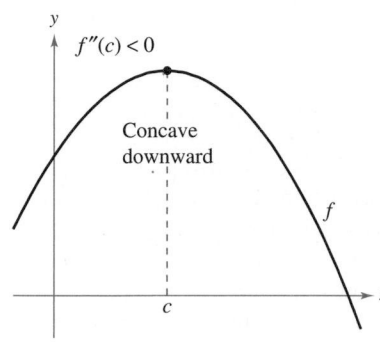

If $f'(c) = 0$ and $f''(c) < 0$, then $f(c)$ is a relative maximum.
Figure 4.30

THEOREM 4.9 Second Derivative Test

Let f be a function such that $f'(c) = 0$ and the second derivative of f exists on an open interval containing c.

1. If $f''(c) > 0$, then f has a relative minimum at $(c, f(c))$.
2. If $f''(c) < 0$, then f has a relative maximum at $(c, f(c))$.

If $f''(c) = 0$, then the test fails. That is, f may have a relative maximum, a relative minimum, or neither. In such cases, you can use the First Derivative Test.

Proof If $f'(c) = 0$ and $f''(c) > 0$, then there exists an open interval I containing c for which

$$\frac{f'(x) - f'(c)}{x - c} = \frac{f'(x)}{x - c} > 0$$

for all $x \neq c$ in I. If $x < c$, then $x - c < 0$ and $f'(x) < 0$. Also, if $x > c$, then $x - c > 0$ and $f'(x) > 0$. So, $f'(x)$ changes from negative to positive at c, and the First Derivative Test implies that $f(c)$ is a relative minimum. A proof of the second case is left to you. ∎

EXAMPLE 4 **Using the Second Derivative Test**

⋯▷ *See LarsonCalculus.com for an interactive version of this type of example.*

Find the relative extrema of

$$f(x) = -3x^5 + 5x^3.$$

Solution Begin by finding the first derivative of f.

$$f'(x) = -15x^4 + 15x^2 = 15x^2(1 - x^2)$$

From this derivative, you can see that $x = -1$, 0, and 1 are the only critical numbers of f. By finding the second derivative

$$f''(x) = -60x^3 + 30x = 30x(1 - 2x^2)$$

you can apply the Second Derivative Test as shown below.

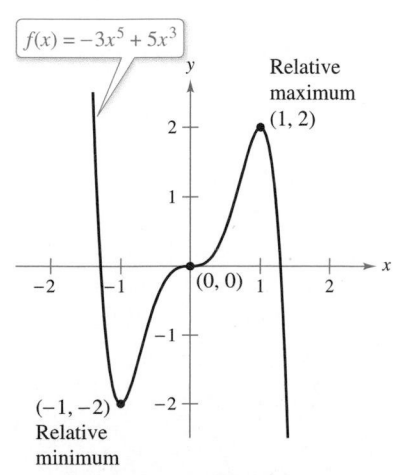

$(0, 0)$ is neither a relative minimum nor a relative maximum.
Figure 4.31

Point	$(-1, -2)$	$(0, 0)$	$(1, 2)$
Sign of $f''(x)$	$f''(-1) > 0$	$f''(0) = 0$	$f''(1) < 0$
Conclusion	Relative minimum	Test fails	Relative maximum

Because the Second Derivative Test fails at $(0, 0)$, you can use the First Derivative Test and observe that f increases to the left and right of $x = 0$. So, $(0, 0)$ is neither a relative minimum nor a relative maximum (even though the graph has a horizontal tangent line at this point). The graph of f is shown in Figure 4.31. ∎

4.4 Exercises   See **CalcChat.com** for tutorial help and worked-out solutions to odd-numbered exercises.

CONCEPT CHECK

1. **Test for Concavity**   Describe the Test for Concavity in your own words.

2. **Second Derivative Test**   Describe the Second Derivative Test in your own words.

Determining Concavity   In Exercises 3–14, determine the open intervals on which the graph of the function is concave upward or concave downward.

3. $f(x) = x^2 - 4x + 8$

4. $g(x) = 3x^2 - x^3$

5. $f(x) = x^4 - 3x^3$

6. $h(x) = x^5 - 5x + 2$

7. $f(x) = \dfrac{24}{x^2 + 12}$

8. $f(x) = \dfrac{2x^2}{3x^2 + 1}$

9. $f(x) = \dfrac{x - 2}{6x + 1}$

10. $f(x) = \dfrac{x + 8}{x - 7}$

11. $f(x) = \dfrac{x^2 + 1}{x^2 - 1}$

12. $h(x) = \dfrac{x^2 - 1}{2x - 1}$

13. $y = 2x - \tan x, \quad \left(-\dfrac{\pi}{2}, \dfrac{\pi}{2}\right)$

14. $y = x + \dfrac{2}{\sin x}, \quad (-\pi, \pi)$

Finding Points of Inflection   In Exercises 15–36, find the points of inflection and discuss the concavity of the graph of the function.

15. $f(x) = x^3 - 9x^2 + 24x - 18$

16. $f(x) = -x^3 + 6x^2 - 5$

17. $f(x) = 2 - 7x^4$

18. $f(x) = 4 - x - 3x^4$

19. $f(x) = x(x - 4)^3$

20. $f(x) = (x - 2)^3(x - 1)$

21. $f(x) = x\sqrt{x + 3}$

22. $f(x) = x\sqrt{9 - x}$

23. $f(x) = \dfrac{6 - x}{\sqrt{x}}$

24. $f(x) = \dfrac{x + 3}{\sqrt{x}}$

25. $f(x) = \sin \dfrac{x}{2}, \quad [0, 4\pi]$

26. $f(x) = 2 \csc \dfrac{3x}{2}, \quad (0, 2\pi)$

27. $f(x) = \sec\left(x - \dfrac{\pi}{2}\right), \quad (0, 4\pi)$

28. $f(x) = \sin x + \cos x, \quad [0, 2\pi]$

29. $f(x) = 2 \sin x + \sin 2x, \quad [0, 2\pi]$

30. $f(x) = x + 2 \cos x, \quad [0, 2\pi]$

31. $y = e^{-3/x}$

32. $y = \tfrac{1}{2}(e^x - e^{-x})$

33. $y = x - \ln x$

34. $y = \ln\sqrt{x^2 + 9}$

35. $f(x) = \arcsin x^{4/5}$

36. $f(x) = \arctan x^2$

Using the Second Derivative Test   In Exercises 37–58, find all relative extrema of the function. Use the Second Derivative Test where applicable.

37. $f(x) = 6x - x^2$

38. $f(x) = -x^3 + 7x^2 - 15x$

39. $f(x) = x^4 - 4x^3 + 2$

40. $f(x) = -x^4 + 2x^3 + 8x$

---

41. $f(x) = x^{2/3} - 3$

42. $f(x) = \sqrt{x^2 + 1}$

43. $f(x) = x + \dfrac{4}{x}$

44. $f(x) = \dfrac{9x - 1}{x + 5}$

45. $f(x) = \cos x - x, \quad [0, 4\pi]$

46. $f(x) = 2 \sin x + \cos 2x, \quad [0, 2\pi]$

47. $y = 8x^2 - \ln x$

48. $y = x \ln x$

49. $y = \dfrac{x}{\ln x}$

50. $y = x^2 \ln \dfrac{x}{4}$

51. $f(x) = \dfrac{e^x + e^{-x}}{2}$

52. $g(x) = \dfrac{1}{\sqrt{2\pi}} e^{-(x-3)^2/2}$

53. $f(x) = x^2 e^{-x}$

54. $f(x) = xe^{-x}$

55. $f(x) = 8x(4^{-x})$

56. $y = x^2 \log_3 x$

57. $f(x) = \text{arcsec } x - x$

58. $f(x) = \arcsin x - 2x$

Finding Extrema and Points of Inflection Using Technology   In Exercises 59–62, use a computer algebra system to analyze the function over the given interval. (a) Find the first and second derivatives of the function. (b) Find any relative extrema and points of inflection. (c) Graph f, f', and f'' on the same set of coordinate axes and state the relationship between the behavior of f and the signs of f' and f''.

59. $f(x) = 0.2x^2(x - 3)^3, \quad [-1, 4]$

60. $f(x) = x^2\sqrt{6 - x^2}, \quad [-\sqrt{6}, \sqrt{6}]$

61. $f(x) = \sin x - \tfrac{1}{3}\sin 3x + \tfrac{1}{5}\sin 5x, \quad [0, \pi]$

62. $f(x) = \sqrt{2x}\sin x, \quad [0, 2\pi]$

EXPLORING CONCEPTS

63. **Sketching a Graph**   Consider a function f such that f' is increasing. Sketch graphs of f for (a) $f' < 0$ and (b) $f' > 0$.

64. **Think About It** S represents weekly sales of a product. What can be said of S' and S'' for each of the following statements?

    (a) The rate of change of sales is increasing.

    (b) The rate of change of sales is constant.

    (c) Sales are steady.

    (d) Sales are declining but at a slower rate.

---

65. **Sketching Graphs**   In parts (a) and (b), the graph of f is shown. Graph f, f', and f'' on the same set of coordinate axes. To print an enlarged copy of the graph, go to *MathGraphs.com*.

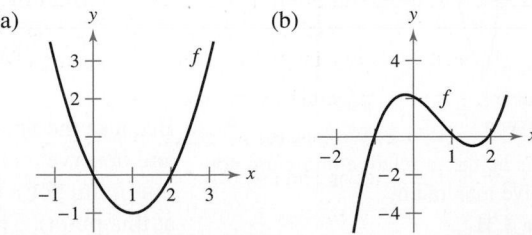

66. **HOW DO YOU SEE IT?** Using the graph of f, state the signs of f' and f'' on the interval $(0, 2)$.

(a)

(b)

Think About It In Exercises 67–70, sketch the graph of a function f having the given characteristics.

67. $f(0) = f(2) = 0$
$f'(x) > 0$ for $x < 1$
$f'(1) = 0$
$f'(x) < 0$ for $x > 1$
$f''(x) < 0$

68. $f(0) = f(2) = 0$
$f'(x) < 0$ for $x < 1$
$f'(1) = 0$
$f'(x) > 0$ for $x > 1$
$f''(x) > 0$

69. $f(2) = f(4) = 0$
$f'(x) < 0$ for $x < 3$
$f'(3)$ does not exist.
$f'(x) > 0$ for $x > 3$
$f''(x) < 0, x \neq 3$

70. $f(1) = f(3) = 0$
$f'(x) > 0$ for $x < 2$
$f'(2)$ does not exist.
$f'(x) < 0$ for $x > 2$
$f''(x) > 0, x \neq 2$

71. Think About It The figure shows the graph of f''. Sketch a graph of f. (The answer is not unique.) To print an enlarged copy of the graph, go to *MathGraphs.com*.

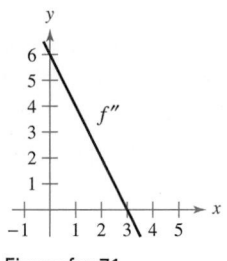

Figure for 71

Figure for 72

72. Think About It Water is running into the vase shown in the figure at a constant rate.

(a) Graph the depth d of water in the vase as a function of time.

(b) Does the function have any extrema? Explain.

(c) Interpret the inflection points of the graph of d.

73. Conjecture Consider the function $f(x) = (x - 2)^n$.

(a) Use a graphing utility to graph f for $n = 1, 2, 3$, and 4. Use the graphs to make a conjecture about the relationship between n and any inflection points of the graph of f.

(b) Verify your conjecture in part (a).

74. Inflection Point Consider the function $f(x) = \sqrt[3]{x}$.

(a) Graph the function and identify the inflection point.

(b) Does f'' exist at the inflection point? Explain.

Finding a Cubic Function In Exercises 75 and 76, find a, b, c, and d such that the cubic function

$$f(x) = ax^3 + bx^2 + cx + d$$

satisfies the given conditions.

75. Relative maximum: $(3, 3)$
Relative minimum: $(5, 1)$
Inflection point: $(4, 2)$

76. Relative maximum: $(2, 4)$
Relative minimum: $(4, 2)$
Inflection point: $(3, 3)$

77. Aircraft Glide Path A small aircraft starts its descent from an altitude of 1 mile, 4 miles west of the runway (see figure).

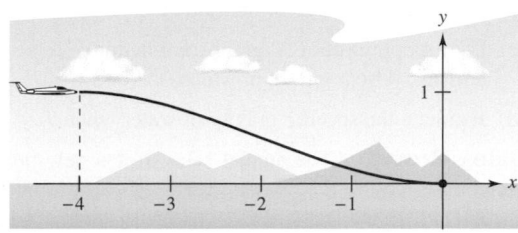

(a) Find the cubic function $f(x) = ax^3 + bx^2 + cx + d$ on the interval $[-4, 0]$ that describes a smooth glide path for the landing.

(b) The function in part (a) models the glide path of the plane. When would the plane be descending at the greatest rate?

■ **FOR FURTHER INFORMATION** For more information on this type of modeling, see the article "How Not to Land at Lake Tahoe!" by Richard Barshinger in *The American Mathematical Monthly*. To view this article, go to *MathArticles.com*.

78. Highway Design A section of highway connecting two hillsides with grades of 6% and 4% is to be built between two points that are separated by a horizontal distance of 2000 feet (see figure). At the point where the two hillsides come together, there is a 50-foot difference in elevation.

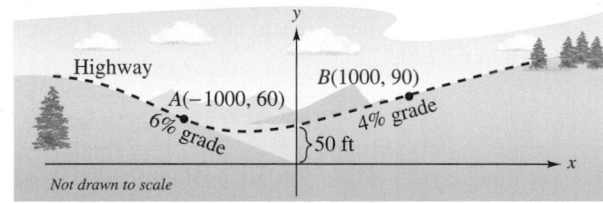

(a) Find the cubic function

$$f(x) = ax^3 + bx^2 + cx + d, \quad -1000 \leq x \leq 1000$$

that describes the section of highway connecting the hillsides. At points A and B, the slope of the model must match the grade of the hillside.

(b) Use a graphing utility to graph the model.

(c) Use a graphing utility to graph the derivative of the model.

(d) Determine the grade at the steepest part of the transitional section of the highway.

79. Average Cost A manufacturer has determined that the total cost C of operating a factory is

$$C = 0.5x^2 + 15x + 5000$$

where x is the number of units produced. At what level of production will the average cost per unit be minimized? (The average cost per unit is C/x.)

80. Specific Gravity A model for the specific gravity of water S is

$$S = \frac{5.755}{10^8}T^3 - \frac{8.521}{10^6}T^2 + \frac{6.540}{10^5}T + 0.99987, \quad 0 < T < 25$$

where T is the water temperature in degrees Celsius.

(a) Use the second derivative to determine the concavity of S.

(b) Use a computer algebra system to find the coordinates of the maximum value of the function.

(c) Use a graphing utility to graph the function over the specified domain. (Use a setting in which $0.996 \le S \le 1.001$.)

(d) Estimate the specific gravity of water when $T = 20°$.

81. Sales Growth The annual sales S of a new product are given by

$$S = \frac{5000t^2}{8 + t^2}, \quad 0 \le t \le 3$$

where t is time in years.

(a) Complete the table. Then use it to estimate when the annual sales are increasing at the greatest rate.

t	0.5	1	1.5	2	2.5	3
S						

(b) Use a graphing utility to graph the function S. Then use the graph to estimate when the annual sales are increasing at the greatest rate.

(c) Find the exact time when the annual sales are increasing at the greatest rate.

82. Modeling Data The average typing speeds S (in words per minute) of a typing student after t weeks of lessons are shown in the table.

t	5	10	15	20	25	30
S	28	56	79	90	93	94

A model for the data is

$$S = \frac{100t^2}{65 + t^2}, \quad t > 0.$$

(a) Use a graphing utility to plot the data and graph the model.

(b) Use the second derivative to determine the concavity of S. Compare the result with the graph in part (a).

(c) What is the sign of the first derivative for $t > 0$? By combining this information with the concavity of the model, what inferences can be made about the typing speed as t increases?

 Linear and Quadratic Approximations In Exercises 83–86, use a graphing utility to graph the function. Then graph the linear and quadratic approximations

$$P_1(x) = f(a) + f'(a)(x - a)$$

and

$$P_2(x) = f(a) + f'(a)(x - a) + \tfrac{1}{2}f''(a)(x - a)^2$$

in the same viewing window. Compare the values of f, P_1, and P_2 and their first derivatives at $x = a$. How do the approximations change as you move farther away from $x = a$?

Function	Value of a
83. $f(x) = 2(\sin x + \cos x)$	$a = \dfrac{\pi}{4}$
84. $f(x) = 2(\sin x + \cos x)$	$a = 0$
85. $f(x) = \arctan x$	$a = -1$
86. $f(x) = \dfrac{\sqrt{x}}{x - 1}$	$a = 2$

87. Determining Concavity Use a graphing utility to graph

$$y = x \sin \frac{1}{x}.$$

Show that the graph is concave downward to the right of

$$x = \frac{1}{\pi}.$$

88. Point of Inflection and Extrema Show that the point of inflection of

$$f(x) = x(x - 6)^2$$

lies midway between the relative extrema of f.

True or False? In Exercises 89–92, determine whether the statement is true or false. If it is false, explain why or give an example that shows it is false.

89. The graph of every cubic polynomial has precisely one point of inflection.

90. The graph of

$$f(x) = \frac{1}{x}$$

is concave downward for $x < 0$ and concave upward for $x > 0$, and thus it has a point of inflection at $x = 0$.

91. If $f'(c) > 0$, then f is concave upward at $x = c$.

92. If $f''(2) = 0$, then the graph of f must have a point of inflection at $x = 2$.

Proof In Exercises 93 and 94, let f and g represent differentiable functions such that $f'' \ne 0$ and $g'' \ne 0$.

93. Show that if f and g are concave upward on the interval (a, b), then $f + g$ is also concave upward on (a, b).

94. Prove that if f and g are positive, increasing, and concave upward on the interval (a, b), then fg is also concave upward on (a, b).

4.5   Limits at Infinity

- ■ Determine (finite) limits at infinity.
- ■ Determine the horizontal asymptotes, if any, of the graph of a function.
- ■ Determine infinite limits at infinity.

Limits at Infinity

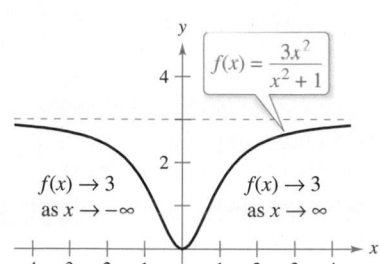

$f(x) \to 3$ as $x \to -\infty$ $f(x) \to 3$ as $x \to \infty$

The limit of $f(x)$ as x approaches $-\infty$ or ∞ is 3.

Figure 4.32

This section discusses the "end behavior" of a function on an *infinite* interval. Consider the graph of

$$f(x) = \frac{3x^2}{x^2 + 1}$$

as shown in Figure 4.32. Graphically, you can see that $f(x)$ appears to approach 3 as x increases without bound or decreases without bound. You can come to the same conclusions numerically, as shown in the table.

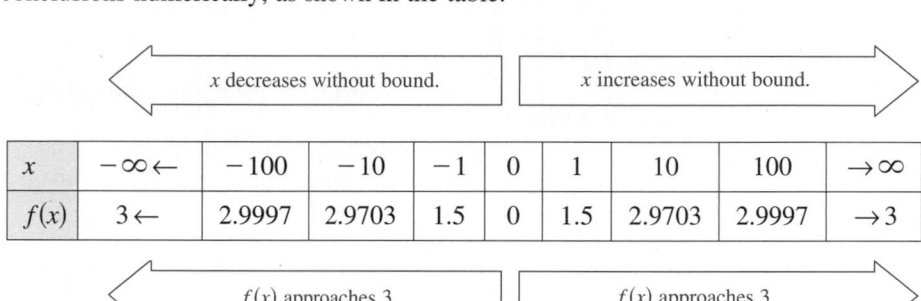

	x decreases without bound.					x increases without bound.			
x	$-\infty \leftarrow$	-100	-10	-1	0	1	10	100	$\to \infty$
$f(x)$	$3 \leftarrow$	2.9997	2.9703	1.5	0	1.5	2.9703	2.9997	$\to 3$
	$f(x)$ approaches 3.					$f(x)$ approaches 3.			

The table suggests that $f(x)$ approaches 3 as x increases without bound $(x \to \infty)$. Similarly, $f(x)$ approaches 3 as x decreases without bound $(x \to -\infty)$. These **limits at infinity** are denoted by

$$\lim_{x \to -\infty} f(x) = 3 \qquad \text{Limit at negative infinity}$$

and

$$\lim_{x \to \infty} f(x) = 3. \qquad \text{Limit at positive infinity}$$

> **REMARK**   The statement
> $$\lim_{x \to -\infty} f(x) = L \text{ or } \lim_{x \to \infty} f(x) = L$$
> means that the limit exists *and* the limit is equal to L.

To say that a statement is true as x increases *without bound* means that for some (large) real number M, the statement is true for *all* x in the interval $\{x: x > M\}$. The next definition uses this concept.

Definition of Limits at Infinity

Let L be a real number.

1. The statement $\lim\limits_{x \to \infty} f(x) = L$ means that for each $\varepsilon > 0$ there exists an $M > 0$ such that $|f(x) - L| < \varepsilon$ whenever $x > M$.

2. The statement $\lim\limits_{x \to -\infty} f(x) = L$ means that for each $\varepsilon > 0$ there exists an $N < 0$ such that $|f(x) - L| < \varepsilon$ whenever $x < N$.

$f(x)$ is within ε units of L as $x \to \infty$.

Figure 4.33

The definition of a limit at infinity is shown in Figure 4.33. In this figure, note that for a given positive number ε, there exists a positive number M such that, for $x > M$, the graph of f will lie between the horizontal lines

$$y = L + \varepsilon \quad \text{and} \quad y = L - \varepsilon.$$

Horizontal Asymptotes

In Figure 4.33, the graph of f approaches the line $y = L$ as x increases without bound. The line $y = L$ is called a **horizontal asymptote** of the graph of f.

Definition of a Horizontal Asymptote

The line $y = L$ is a **horizontal asymptote** of the graph of f when

$$\lim_{x \to -\infty} f(x) = L \quad \text{or} \quad \lim_{x \to \infty} f(x) = L.$$

Note that from this definition, it follows that the graph of a *function* of x can have at most two horizontal asymptotes—one to the right and one to the left.

Limits at infinity have many of the same properties of limits discussed in Section 2.3. For example, if $\lim\limits_{x \to \infty} f(x)$ and $\lim\limits_{x \to \infty} g(x)$ both exist, then

$$\lim_{x \to \infty} [f(x) + g(x)] = \lim_{x \to \infty} f(x) + \lim_{x \to \infty} g(x)$$

and

$$\lim_{x \to \infty} [f(x)g(x)] = \left[\lim_{x \to \infty} f(x) \right]\left[\lim_{x \to \infty} g(x) \right].$$

Similar properties hold for limits at $-\infty$.

When evaluating limits at infinity, the next theorem is helpful.

THEOREM 4.10   Limits at Infinity

1. If r is a positive rational number and c is any real number, then

$$\lim_{x \to \infty} \frac{c}{x^r} = 0 \quad \text{and} \quad \lim_{x \to -\infty} \frac{c}{x^r} = 0.$$

The second limit is valid only if x^r is defined when $x < 0$.

2. $\lim\limits_{x \to -\infty} e^x = 0 \quad \text{and} \quad \lim\limits_{x \to \infty} e^{-x} = 0$

A proof of the first part of this theorem is given in Appendix A.

EXAMPLE 1   **Finding Limits at Infinity**

Find each limit.

a. $\lim\limits_{x \to \infty} \left(5 - \dfrac{2}{x^2} \right)$ **b.** $\lim\limits_{x \to \infty} \dfrac{3}{e^x}$

Solution

a. $\lim\limits_{x \to \infty} \left(5 - \dfrac{2}{x^2} \right) = \lim\limits_{x \to \infty} 5 - \lim\limits_{x \to \infty} \dfrac{2}{x^2}$ Property of limits

$\qquad\qquad\qquad\quad = 5 - 0$ Apply Theorem 4.10.

$\qquad\qquad\qquad\quad = 5$

b. $\lim\limits_{x \to \infty} \dfrac{3}{e^x} = \lim\limits_{x \to \infty} 3e^{-x}$

$\qquad\quad = 3 \lim\limits_{x \to \infty} e^{-x}$ Property of limits

$\qquad\quad = 3(0)$ Apply Theorem 4.10.

$\qquad\quad = 0$

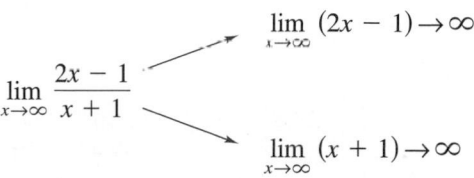

Finding a Limit at Infinity

Find the limit: $\displaystyle\lim_{x\to\infty} \frac{2x-1}{x+1}$.

Solution Note that both the numerator and the denominator approach infinity as x approaches infinity.

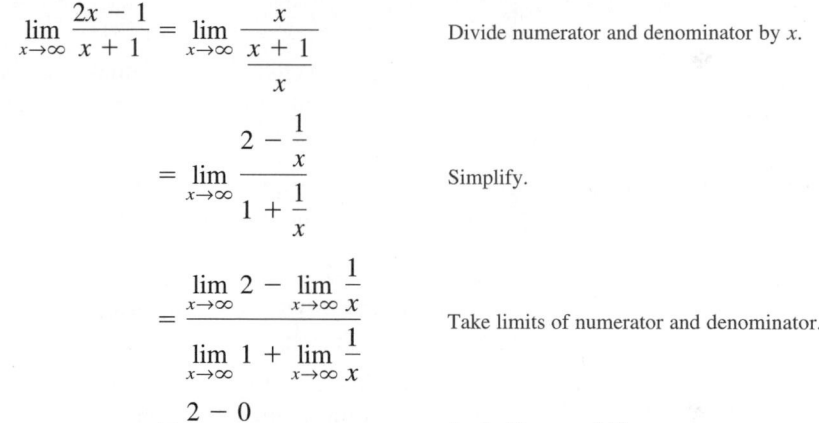

REMARK When you encounter an indeterminate form such as the one in Example 2, you should divide the numerator and denominator by the highest power of x in the *denominator*.

This results in ∞/∞, an **indeterminate form.** To resolve this problem, you can divide both the numerator and the denominator by x. After dividing, the limit may be evaluated as shown.

$$\lim_{x\to\infty} \frac{2x-1}{x+1} = \lim_{x\to\infty} \frac{\dfrac{2x-1}{x}}{\dfrac{x+1}{x}} \qquad \text{Divide numerator and denominator by } x.$$

$$= \lim_{x\to\infty} \frac{2 - \dfrac{1}{x}}{1 + \dfrac{1}{x}} \qquad \text{Simplify.}$$

$$= \frac{\displaystyle\lim_{x\to\infty} 2 - \lim_{x\to\infty}\dfrac{1}{x}}{\displaystyle\lim_{x\to\infty} 1 + \lim_{x\to\infty}\dfrac{1}{x}} \qquad \text{Take limits of numerator and denominator.}$$

$$= \frac{2 - 0}{1 + 0} \qquad \text{Apply Theorem 4.10.}$$

$$= 2$$

So, the line $y = 2$ is a horizontal asymptote to the right. By taking the limit as $x \to -\infty$, you can see that $y = 2$ is also a horizontal asymptote to the left. The graph of the function is shown in Figure 4.34.

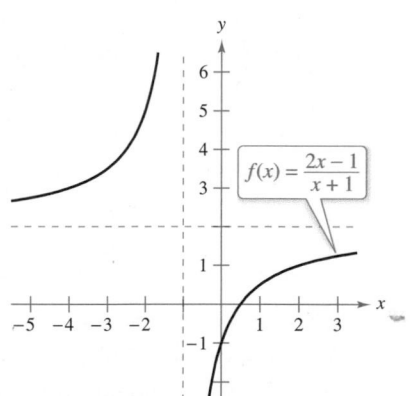

$y = 2$ is a horizontal asymptote.
Figure 4.34

TECHNOLOGY You can test the reasonableness of the limit found in Example 2 by evaluating $f(x)$ for a few large positive values of x. For instance,

$$f(100) \approx 1.9703, \quad f(1000) \approx 1.9970,$$
$$\text{and} \quad f(10{,}000) \approx 1.9997.$$

Another way to test the reasonableness of the limit is to use a graphing utility. For instance, in Figure 4.35, the graph of

$$f(x) = \frac{2x-1}{x+1}$$

is shown with the horizontal line $y = 2$. Note that as x increases, the graph of f moves closer and closer to its horizontal asymptote.

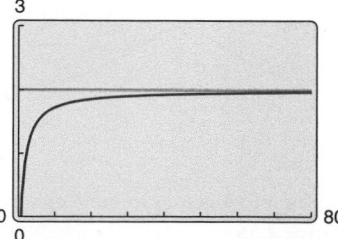

As x increases, the graph of f moves closer and closer to the line $y = 2$.
Figure 4.35

**MARIA GAETANA AGNESI
(1718–1799)**

Agnesi was one of a handful of women to receive credit for significant contributions to mathematics before the twentieth century. In her early twenties, she wrote the first text that included both differential and integral calculus. By age 30, she was an honorary member of the faculty at the University of Bologna.
See LarsonCalculus.com to read more of this biography.
For more information on the contributions of women to mathematics, see the article "Why Women Succeed in Mathematics" by Mona Fabricant, Sylvia Svitak, and Patricia Clark Kenschaft in *Mathematics Teacher*. To view this article, go to *MathArticles.com*.

EXAMPLE 3 **A Comparison of Three Rational Functions**

• • • • ▷ *See LarsonCalculus.com for an interactive version of this type of example.*

Find each limit.

a. $\lim\limits_{x \to \infty} \dfrac{2x + 5}{3x^2 + 1}$ **b.** $\lim\limits_{x \to \infty} \dfrac{2x^2 + 5}{3x^2 + 1}$ **c.** $\lim\limits_{x \to \infty} \dfrac{2x^3 + 5}{3x^2 + 1}$

Solution In each case, attempting to evaluate the limit produces the indeterminate form ∞ / ∞.

a. Divide both the numerator and the denominator by x^2.

$$\lim_{x \to \infty} \frac{2x + 5}{3x^2 + 1} = \lim_{x \to \infty} \frac{(2/x) + (5/x^2)}{3 + (1/x^2)} = \frac{0 + 0}{3 + 0} = \frac{0}{3} = 0$$

b. Divide both the numerator and the denominator by x^2.

$$\lim_{x \to \infty} \frac{2x^2 + 5}{3x^2 + 1} = \lim_{x \to \infty} \frac{2 + (5/x^2)}{3 + (1/x^2)} = \frac{2 + 0}{3 + 0} = \frac{2}{3}$$

c. Divide both the numerator and the denominator by x^2.

$$\lim_{x \to \infty} \frac{2x^3 + 5}{3x^2 + 1} = \lim_{x \to \infty} \frac{2x + (5/x^2)}{3 + (1/x^2)} = \frac{\infty}{3}$$

You can conclude that the limit *does not exist* because the numerator increases without bound while the denominator approaches 3. ∎

Example 3 suggests the guidelines below for finding limits at infinity of rational functions. Use these guidelines to check the results in Example 3.

GUIDELINES FOR FINDING LIMITS AT $\pm\infty$ OF RATIONAL FUNCTIONS

1. If the degree of the numerator is *less than* the degree of the denominator, then the limit of the rational function is 0.
2. If the degree of the numerator is *equal to* the degree of the denominator, then the limit of the rational function is the ratio of the leading coefficients.
3. If the degree of the numerator is *greater than* the degree of the denominator, then the limit of the rational function does not exist.

The guidelines for finding limits at infinity of rational functions seem reasonable when you consider that for large values of x, the highest-power term of the rational function is the most "influential" in determining the limit. For instance,

$$\lim_{x \to \infty} \frac{1}{x^2 + 1}$$

is 0 because the denominator overpowers the numerator as x increases or decreases without bound, as shown in Figure 4.36.

The function shown in Figure 4.36 is a special case of a type of curve studied by the Italian mathematician Maria Gaetana Agnesi. The general form of this function is

$$f(x) = \frac{8a^3}{x^2 + 4a^2} \qquad \text{Witch of Agnesi}$$

and, through a mistranslation of the Italian word *vertéré*, the curve has come to be known as the Witch of Agnesi. Agnesi's work with this curve first appeared in a comprehensive text on calculus that was published in 1748.

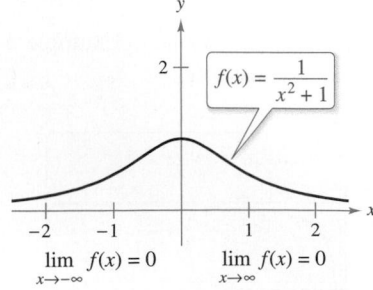

f has a horizontal asymptote at $y = 0$.
Figure 4.36

In Figure 4.36, you can see that the function

$$f(x) = \frac{1}{x^2 + 1}$$

approaches the same horizontal asymptote to the right and to the left. This is always true of rational functions. Functions that are not rational, however, may approach different horizontal asymptotes to the right and to the left. A common example of such a function is the **logistic function** shown in the next example. (You will learn more about the logistic function in Section 6.4.)

EXAMPLE 4   **A Function with Two Horizontal Asymptotes**

Show that the *logistic function*

$$f(x) = \frac{1}{1 + e^{-x}}$$

has different horizontal asymptotes to the left and to the right.

Solution   Begin by sketching a graph of the function. From Figure 4.37, it appears that

$$y = 0 \quad \text{and} \quad y = 1$$

are horizontal asymptotes to the left and to the right, respectively. The table shows the same results numerically.

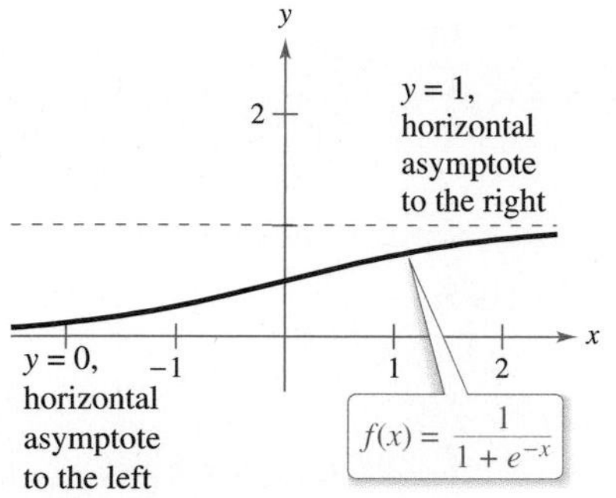

x	-10	-5	-2	-1	1	2	5	10
$f(x)$	0.000	0.007	0.119	0.269	0.731	0.881	0.993	1.000

You can obtain the same results analytically, as follows.

$$\lim_{x \to \infty} \frac{1}{1 + e^{-x}} = \frac{\lim\limits_{x \to \infty} 1}{\lim\limits_{x \to \infty} (1 + e^{-x})}$$

$$= \frac{1}{1 + 0}$$

$$= 1 \qquad\qquad y = 1 \text{ is a horizontal asymptote to the right.}$$

For the horizontal asymptote to the left, note that as $x \to -\infty$ the denominator of

$$\frac{1}{1 + e^{-x}}$$

approaches infinity. So, the quotient approaches 0 and thus the limit is 0. You can conclude that $y = 0$ is a horizontal asymptote to the left. ∎

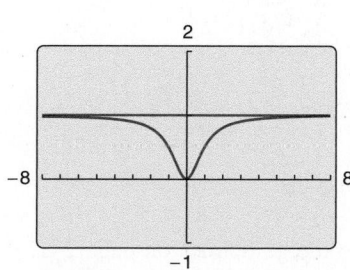

$y = 1$, horizontal asymptote to the right

$y = 0$, horizontal asymptote to the left

$f(x) = \dfrac{1}{1 + e^{-x}}$

Functions that are not rational may have different right and left horizontal asymptotes.

Figure 4.37

▷ **TECHNOLOGY PITFALL**   If you use a graphing utility to estimate a limit, be sure that you also confirm the estimate analytically—the graphs shown by a graphing utility can be misleading. For instance, Figure 4.38 shows one view of the graph of

$$y = \frac{2x^3 + 1000x^2 + x}{x^3 + 1000x^2 + x + 1000}.$$

From this view, one could be convinced that the graph has $y = 1$ as a horizontal asymptote. An analytical approach shows that the horizontal asymptote is actually $y = 2$. Confirm this by enlarging the viewing window on the graphing utility.

The horizontal asymptote appears to be the line $y = 1$, but it is actually the line $y = 2$.

Figure 4.38

In Section 2.4, Example 7(c), you used the Squeeze Theorem to evaluate a limit involving a trigonometric function. The Squeeze Theorem is also valid for limits at infinity.

EXAMPLE 5   Limits Involving Trigonometric Functions

Find each limit.

a. $\lim\limits_{x \to \infty} \sin x$ **b.** $\lim\limits_{x \to \infty} \dfrac{\sin x}{x}$

Solution

a. As x approaches infinity, the sine function oscillates between 1 and -1. So, this limit does not exist.

b. Because $-1 \le \sin x \le 1$, it follows that for $x > 0$,

$$-\frac{1}{x} \le \frac{\sin x}{x} \le \frac{1}{x}$$

where

$$\lim_{x \to \infty} \left(-\frac{1}{x}\right) = 0 \quad \text{and} \quad \lim_{x \to \infty} \frac{1}{x} = 0.$$

So, by the Squeeze Theorem, you obtain

$$\lim_{x \to \infty} \frac{\sin x}{x} = 0$$

as shown in Figure 4.39.

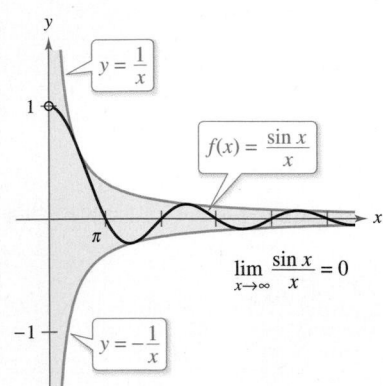

As x increases without bound, $f(x)$ approaches 0.
Figure 4.39

EXAMPLE 6   Oxygen Level in a Pond

Let $f(t)$ measure the level of oxygen in a pond, where $f(t) = 1$ is the normal (unpolluted) level and the time t is measured in weeks. When $t = 0$, organic waste is dumped into the pond, and as the waste material oxidizes, the level of oxygen in the pond is

$$f(t) = \frac{t^2 - t + 1}{t^2 + 1}.$$

What percent of the normal level of oxygen exists in the pond after 1 week? After 2 weeks? After 10 weeks? What is the limit as t approaches infinity?

Solution   When $t = 1, 2,$ and 10, the levels of oxygen are as shown.

$$f(1) = \frac{1^2 - 1 + 1}{1^2 + 1} = \frac{1}{2} = 50\% \qquad \text{1 week}$$

$$f(2) = \frac{2^2 - 2 + 1}{2^2 + 1} = \frac{3}{5} = 60\% \qquad \text{2 weeks}$$

$$f(10) = \frac{10^2 - 10 + 1}{10^2 + 1} = \frac{91}{101} \approx 90.1\% \qquad \text{10 weeks}$$

To find the limit as t approaches infinity, you can use the guidelines on page 242, or you can divide the numerator and the denominator by t^2 to obtain

$$\lim_{t \to \infty} \frac{t^2 - t + 1}{t^2 + 1} = \lim_{t \to \infty} \frac{1 - (1/t) + (1/t^2)}{1 + (1/t^2)} = \frac{1 - 0 + 0}{1 + 0} = 1 = 100\%.$$

See Figure 4.40.

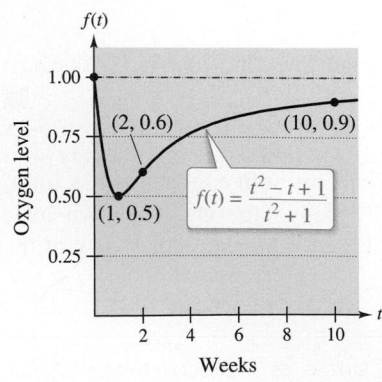

The level of oxygen in a pond approaches the normal level of 1 as t approaches ∞.
Figure 4.40

Infinite Limits at Infinity

Many functions do not approach a finite limit as x increases (or decreases) without bound. For instance, no polynomial function has a finite limit at infinity. The next definition is used to describe the behavior of polynomial and other functions at infinity.

REMARK Determining whether a function has an infinite limit at infinity is useful in analyzing the "end behavior" of its graph. You will see examples of this in Section 4.6 on curve sketching.

> **Definition of Infinite Limits at Infinity**
>
> Let f be a function defined on the interval (a, ∞).
>
> 1. The statement $\lim\limits_{x \to \infty} f(x) = \infty$ means that for each positive number M, there is a corresponding number $N > 0$ such that $f(x) > M$ whenever $x > N$.
> 2. The statement $\lim\limits_{x \to \infty} f(x) = -\infty$ means that for each negative number M, there is a corresponding number $N > 0$ such that $f(x) < M$ whenever $x > N$.
>
> Similar definitions can be given for the statements
>
> $$\lim_{x \to -\infty} f(x) = \infty \quad \text{and} \quad \lim_{x \to -\infty} f(x) = -\infty.$$

EXAMPLE 7 Finding Infinite Limits at Infinity

Find each limit.

a. $\lim\limits_{x \to \infty} x^3$ **b.** $\lim\limits_{x \to -\infty} x^3$

Solution

a. As x increases without bound, x^3 also increases without bound. So, you can write

$$\lim_{x \to \infty} x^3 = \infty.$$

b. As x decreases without bound, x^3 also decreases without bound. So, you can write

$$\lim_{x \to -\infty} x^3 = -\infty.$$

The graph of $f(x) = x^3$ in Figure 4.41 illustrates these two results. These results agree with the Leading Coefficient Test for polynomial functions as described in Section 1.3.

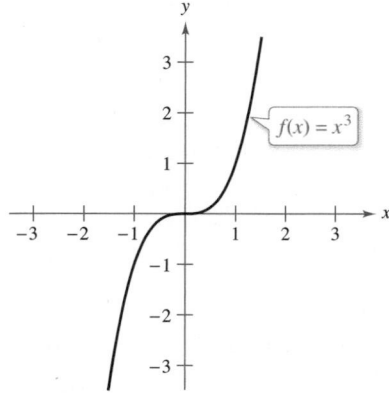

Figure 4.41

EXAMPLE 8 Finding Infinite Limits at Infinity

Find each limit.

a. $\lim\limits_{x \to \infty} \dfrac{2x^2 - 4x}{x + 1}$ **b.** $\lim\limits_{x \to -\infty} \dfrac{2x^2 - 4x}{x + 1}$

Solution One way to evaluate each of these limits is to use long division to rewrite the improper rational function as the sum of a polynomial and a rational function.

a. $\lim\limits_{x \to \infty} \dfrac{2x^2 - 4x}{x + 1} = \lim\limits_{x \to \infty} \left(2x - 6 + \dfrac{6}{x + 1} \right) = \infty$

b. $\lim\limits_{x \to -\infty} \dfrac{2x^2 - 4x}{x + 1} = \lim\limits_{x \to -\infty} \left(2x - 6 + \dfrac{6}{x + 1} \right) = -\infty$

The statements above can be interpreted as saying that as x approaches $\pm\infty$, the function $f(x) = (2x^2 - 4x)/(x + 1)$ behaves like the function $g(x) = 2x - 6$. In Section 4.6, you will see that this is graphically described by saying that the line $y = 2x - 6$ is a *slant asymptote* of the graph of f, as shown in Figure 4.42.

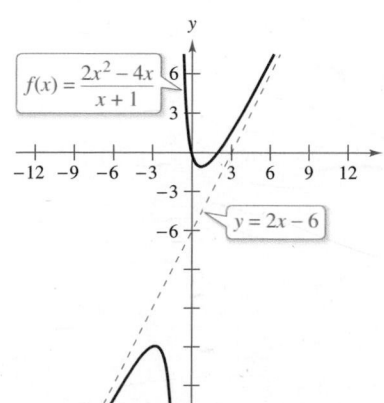

Figure 4.42

4.5 Exercises

See **CalcChat.com** for tutorial help and worked-out solutions to odd-numbered exercises.

CONCEPT CHECK

1. **Writing** Describe in your own words what each statement means.

   (a) $\lim\limits_{x \to \infty} f(x) = -5$

   (b) $\lim\limits_{x \to -\infty} f(x) = 3$

2. **Horizontal Asymptote** What does it mean for the graph of a function to have a horizontal asymptote?

3. **Horizontal Asymptote** A graph can have a maximum of how many horizontal asymptotes? Explain.

4. **Limits at Infinity** In your own words, summarize the guidelines for finding limits at infinity of rational functions.

Matching In Exercises 5–10, match the function with its graph using horizontal asymptotes as an aid. [The graphs are labeled (a), (b), (c), (d), (e), and (f).]

(a)
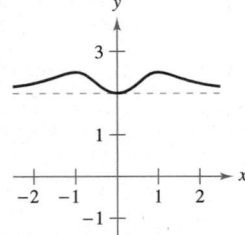

(b)
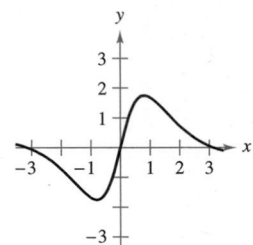

(c)
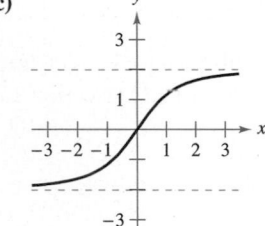

(d)
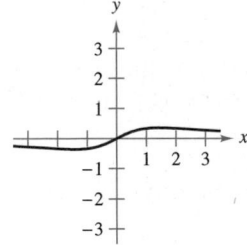

(e)
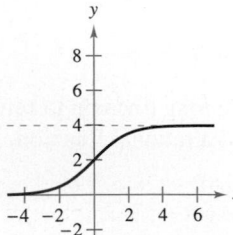

(f)
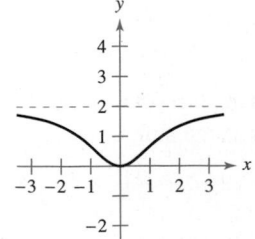

5. $f(x) = \dfrac{2x^2}{x^2 + 2}$

6. $f(x) = \dfrac{2x}{\sqrt{x^2 + 2}}$

7. $f(x) = \dfrac{x}{x^2 + 2}$

8. $f(x) = 2 + \dfrac{x^2}{x^4 + 1}$

9. $f(x) = \dfrac{4 \sin x}{x^2 + 1}$

10. $f(x) = \dfrac{4}{1 + e^{-x}}$

 **Finding Limits at Infinity** In Exercises 11 and 12, find $\lim\limits_{x \to \infty} h(x)$, if it exists.

11. $f(x) = 5x^3 - 3$

   (a) $h(x) = \dfrac{f(x)}{x^2}$

   (b) $h(x) = \dfrac{f(x)}{x^3}$

   (c) $h(x) = \dfrac{f(x)}{x^4}$

12. $f(x) = -4x^2 + 2x - 5$

   (a) $h(x) = \dfrac{f(x)}{x}$

   (b) $h(x) = \dfrac{f(x)}{x^2}$

   (c) $h(x) = \dfrac{f(x)}{x^3}$

 **Finding Limits at Infinity** In Exercises 13–16, find each limit, if it exists.

13. (a) $\lim\limits_{x \to \infty} \dfrac{x^2 + 2}{x^3 - 1}$

   (b) $\lim\limits_{x \to \infty} \dfrac{x^2 + 2}{x^2 - 1}$

   (c) $\lim\limits_{x \to \infty} \dfrac{x^2 + 2}{x - 1}$

14. (a) $\lim\limits_{x \to \infty} \dfrac{3 - 2x}{3x^3 - 1}$

   (b) $\lim\limits_{x \to \infty} \dfrac{3 - 2x}{3x - 1}$

   (c) $\lim\limits_{x \to \infty} \dfrac{3 - 2x^2}{3x - 1}$

15. (a) $\lim\limits_{x \to \infty} \dfrac{5 - 2x^{3/2}}{3x^2 - 4}$

   (b) $\lim\limits_{x \to \infty} \dfrac{5 - 2x^{3/2}}{3x^{3/2} - 4}$

   (c) $\lim\limits_{x \to \infty} \dfrac{5 - 2x^{3/2}}{3x - 4}$

16. (a) $\lim\limits_{x \to \infty} \dfrac{5x^{3/2}}{4x^2 + 1}$

   (b) $\lim\limits_{x \to \infty} \dfrac{5x^{3/2}}{4x^{3/2} + 1}$

   (c) $\lim\limits_{x \to \infty} \dfrac{5x^{3/2}}{4\sqrt{x} + 1}$

Finding a Limit In Exercises 17–42, find the limit, if it exists.

17. $\lim\limits_{x \to \infty} \left(4 + \dfrac{3}{x}\right)$

18. $\lim\limits_{x \to -\infty} \left(\dfrac{5}{x} - \dfrac{x}{3}\right)$

19. $\lim\limits_{x \to \infty} \dfrac{7x + 6}{9x - 4}$

20. $\lim\limits_{x \to -\infty} \dfrac{4x^2 + 5}{x^2 + 3}$

21. $\lim\limits_{x \to -\infty} \dfrac{2x^2 + x}{6x^3 + 2x^2 + x}$

22. $\lim\limits_{x \to \infty} \dfrac{5x^3 + 1}{10x^3 - 3x^2 + 7}$

23. $\lim\limits_{x \to -\infty} \dfrac{-4}{3 + 3e^{-2x}}$

24. $\lim\limits_{x \to \infty} \dfrac{6}{5 + 2e^{-4x}}$

25. $\lim\limits_{x \to -\infty} \dfrac{x}{\sqrt{x^2 - x}}$

26. $\lim\limits_{x \to -\infty} \dfrac{x}{\sqrt{x^2 + 1}}$

27. $\lim\limits_{x \to -\infty} \dfrac{2x + 1}{\sqrt{x^2 - x}}$

28. $\lim\limits_{x \to \infty} \dfrac{5x^2 + 2}{\sqrt{x^2 + 3}}$

29. $\lim\limits_{x \to \infty} \dfrac{\sqrt{x^2 - 1}}{2x - 1}$

30. $\lim\limits_{x \to -\infty} \dfrac{\sqrt{x^4 - 1}}{x^3 - 1}$

31. $\lim\limits_{x \to \infty} \dfrac{x + 1}{(x^2 + 1)^{1/3}}$

32. $\lim\limits_{x \to -\infty} \dfrac{2x}{(x^6 - 1)^{1/3}}$

33. $\lim\limits_{x \to \infty} \dfrac{1}{2x + \sin x}$

34. $\lim\limits_{x \to 0} \cos \dfrac{1}{x}$

35. $\lim\limits_{x \to \infty} \dfrac{\sin 2x}{x}$

36. $\lim\limits_{x \to \infty} \dfrac{x - \cos x}{x}$

37. $\lim\limits_{x\to\infty} (2 - 5e^{-x})$

38. $\lim\limits_{x\to\infty} \dfrac{8}{4 - 10^{-x/2}}$

39. $\lim\limits_{x\to\infty} \log_{10}(1 + 10^{-x})$

40. $\lim\limits_{x\to\infty} \left(\dfrac{5}{2} + \ln\dfrac{x^2 + 1}{x^2}\right)$

41. $\lim\limits_{t\to\infty} (8t^{-1} - \arctan t)$

42. $\lim\limits_{u\to\infty} \operatorname{arcsec}(u + 1)$

Finding Horizontal Asymptotes Using Technology
In Exercises 43–46, use a graphing utility to graph the function and identify any horizontal asymptotes.

43. $f(x) = \dfrac{|x|}{x + 1}$

44. $f(x) = \dfrac{5}{3 + 6e^{-4x}}$

45. $f(x) = \dfrac{3x}{\sqrt{x^2 + 2}}$

46. $f(x) = \dfrac{\sqrt{9x^2 - 2}}{2x + 1}$

Finding a Limit In Exercises 47 and 48, find the limit. (*Hint:* Let $x = 1/t$ and find the limit as $t \to 0^+$.)

47. $\lim\limits_{x\to\infty} x \sin\dfrac{1}{x}$

48. $\lim\limits_{x\to\infty} x \tan\dfrac{1}{x}$

Finding a Limit In Exercises 49–52, find the limit. Use a graphing utility to verify your result. (*Hint:* Treat the expression as a fraction whose denominator is 1, and rationalize the numerator.)

49. $\lim\limits_{x\to-\infty} \left(x + \sqrt{x^2 + 3}\right)$

50. $\lim\limits_{x\to\infty} \left(x - \sqrt{x^2 + x}\right)$

51. $\lim\limits_{x\to-\infty} \left(3x + \sqrt{9x^2 - x}\right)$

52. $\lim\limits_{x\to\infty} \left(4x - \sqrt{16x^2 - x}\right)$

Numerical, Graphical, and Analytic Analysis In Exercises 53–56, use a graphing utility to complete the table and estimate the limit as x approaches infinity. Then use a graphing utility to graph the function and estimate the limit. Finally, find the limit analytically and compare your results with the estimates.

x	10^0	10^1	10^2	10^3	10^4	10^5	10^6
$f(x)$							

53. $f(x) = x - \sqrt{x(x - 1)}$

54. $f(x) = x^2 - x\sqrt{x(x - 1)}$

55. $f(x) = x \sin\dfrac{1}{2x}$

56. $f(x) = \dfrac{x + 1}{x\sqrt{x}}$

57. Engine Efficiency

The efficiency (in percent) of an internal combustion engine is

$$\text{Efficiency} = 100\left[1 - \dfrac{1}{(v_1/v_2)^c}\right]$$

where v_1/v_2 is the ratio of the uncompressed gas to the compressed gas and c is a positive constant dependent on the engine design. Find the limit of the efficiency as the compression ratio approaches infinity.

58. Physics Newton's First Law of Motion and Einstein's Special Theory of Relativity differ concerning the behavior of a particle as its velocity approaches the speed of light c. In the graph, functions N and E represent the velocity v, with respect to time t, of a particle accelerated by a constant force as predicted by Newton and Einstein, respectively. Write limit statements that describe these two theories.

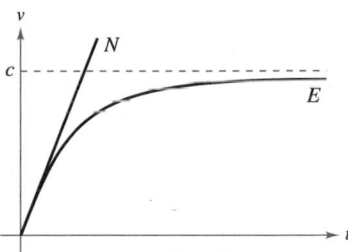

EXPLORING CONCEPTS

59. Limits Explain the differences between limits at infinity and infinite limits.

60. Horizontal Asymptote Can the graph of a function cross a horizontal asymptote? Explain.

61. Using Symmetry to Find Limits If f is a continuous function such that $\lim\limits_{x\to\infty} f(x) = 5$, find, if possible, $\lim\limits_{x\to-\infty} f(x)$ for each specified condition.

(a) The graph of f is symmetric with respect to the y-axis.

(b) The graph of f is symmetric with respect to the origin.

62. **HOW DO YOU SEE IT?** The graph shows the temperature T, in degrees Fahrenheit, of molten glass t seconds after it is removed from a kiln.

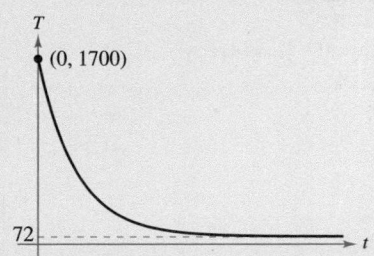

(a) Find $\lim\limits_{t\to 0^+} T$. What does this limit represent?

(b) Find $\lim\limits_{t\to\infty} T$. What does this limit represent?

63. Learning Theory In a group project in learning theory, a mathematical model for the proportion P of correct responses after n trials was found to be

$$P = \dfrac{0.83}{1 + e^{-0.2n}}.$$

(a) Use a graphing utility to graph P.

(b) Find the limiting proportion of correct responses as n approaches infinity.

64. Modeling Data A heat probe is attached to the heat exchanger of a heating system. The temperature T (in degrees Celsius) is recorded t seconds after the furnace is started. The results for the first 2 minutes are recorded in the table.

t	0	15	30	45	60
T	25.2	36.9	45.5	51.4	56.0

t	75	90	105	120
T	59.6	62.0	64.0	65.2

(a) Use the regression capabilities of a graphing utility to find a model of the form $T_1 = at^2 + bt + c$ for the data.

(b) Use a graphing utility to graph T_1.

(c) A rational model for the data is

$$T_2 = \frac{1451 + 86t}{58 + t}.$$

Use a graphing utility to graph T_2.

(d) Find $\lim_{t \to \infty} T_2$.

(e) Interpret the result in part (d) in the context of the problem. Is it possible to do this type of analysis using T_1? Explain.

65. Using the Definition of Limits at Infinity The graph of

$$f(x) = \frac{2x^2}{x^2 + 2}$$

is shown (see figure).

(a) Find $L = \lim_{x \to \infty} f(x)$.

(b) Determine x_1 and x_2 in terms of ε.

(c) Determine M, where $M > 0$, such that $|f(x) - L| < \varepsilon$ for $x > M$.

(d) Determine N, where $N < 0$, such that $|f(x) - L| < \varepsilon$ for $x < N$.

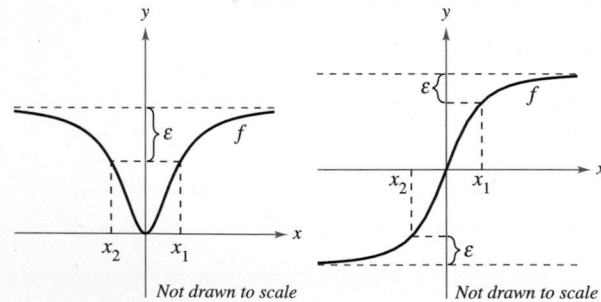

Figure for 65   Figure for 66

66. Using the Definition of Limits at Infinity The graph of

$$f(x) = \frac{6x}{\sqrt{x^2 + 2}}$$ is shown (see figure).

(a) Find $L = \lim_{x \to \infty} f(x)$ and $K = \lim_{x \to -\infty} f(x)$.

(b) Determine x_1 and x_2 in terms of ε.

(c) Determine M, where $M > 0$, such that $|f(x) - L| < \varepsilon$ for $x > M$.

(d) Determine N, where $N < 0$, such that $|f(x) - K| < \varepsilon$ for $x < N$.

67. Using the Definition of Limits at Infinity Consider

$$\lim_{x \to \infty} \frac{3x}{\sqrt{x^2 + 3}}.$$

(a) Use the definition of limits at infinity to find the value of M that corresponds to $\varepsilon = 0.5$.

(b) Use the definition of limits at infinity to find the value of M that corresponds to $\varepsilon = 0.1$.

68. Using the Definition of Limits at Infinity Consider

$$\lim_{x \to -\infty} \frac{3x}{\sqrt{x^2 + 3}}.$$

(a) Use the definition of limits at infinity to find the value of N that corresponds to $\varepsilon = 0.5$.

(b) Use the definition of limits at infinity to find the value of N that corresponds to $\varepsilon = 0.1$.

Proof In Exercises 69–72, use the definition of limits at infinity to prove the limit.

69. $\lim_{x \to \infty} \dfrac{1}{x^2} = 0$

70. $\lim_{x \to \infty} \dfrac{2}{\sqrt{x}} = 0$

71. $\lim_{x \to -\infty} \dfrac{1}{x^3} = 0$

72. $\lim_{x \to -\infty} \dfrac{1}{x - 2} = 0$

73. Distance A line with slope m passes through the point $(0, 4)$.

(a) Write the distance d between the line and the point $(3, 1)$ as a function of m. (*Hint:* See Section 1.2, Exercise 77.)

(b) Use a graphing utility to graph the equation in part (a).

(c) Find $\lim_{m \to \infty} d(m)$ and $\lim_{m \to -\infty} d(m)$. Interpret the results geometrically.

74. Distance A line with slope m passes through the point $(0, -2)$.

(a) Write the distance d between the line and the point $(4, 2)$ as a function of m. (*Hint:* See Section 1.2, Exercise 77.)

(b) Use a graphing utility to graph the equation in part (a).

(c) Find $\lim_{m \to \infty} d(m)$ and $\lim_{m \to -\infty} d(m)$. Interpret the results geometrically.

75. Proof Prove that if

$$p(x) = a_n x^n + \cdots + a_1 x + a_0$$

and

$$q(x) = b_m x^m + \cdots + b_1 x + b_0$$

where $a_n \neq 0$ and $b_m \neq 0$, then

$$\lim_{x \to \infty} \frac{p(x)}{q(x)} = \begin{cases} 0, & n < m \\ \dfrac{a_n}{b_m}, & n = m \\ \pm\infty, & n > m \end{cases}.$$

76. Proof Use the definition of infinite limits at infinity to prove that $\lim_{x \to \infty} x^3 = \infty$.

4.6 A Summary of Curve Sketching

■ Analyze and sketch the graph of a function.

Analyzing the Graph of a Function

It would be difficult to overstate the importance of using graphs in mathematics. Descartes's introduction of analytic geometry contributed significantly to the rapid advances in calculus that began during the mid-seventeenth century. In the words of Lagrange, "As long as algebra and geometry traveled separate paths their advance was slow and their applications limited. But when these two sciences joined company, they drew from each other fresh vitality and thenceforth marched on at a rapid pace toward perfection."

So far, you have studied several concepts that are useful in analyzing the graph of a function.

• x-intercepts and y-intercepts	(Section 1.1)
• Symmetry	(Section 1.1)
• Domain and range	(Section 1.3)
• Continuity	(Section 2.4)
• Vertical asymptotes	(Section 2.5)
• Differentiability	(Section 3.1)
• Relative extrema	(Section 4.1)
• Increasing and decreasing functions	(Section 4.3)
• Concavity	(Section 4.4)
• Points of inflection	(Section 4.4)
• Horizontal asymptotes	(Section 4.5)
• Infinite limits at infinity	(Section 4.5)

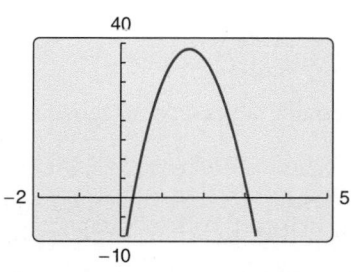

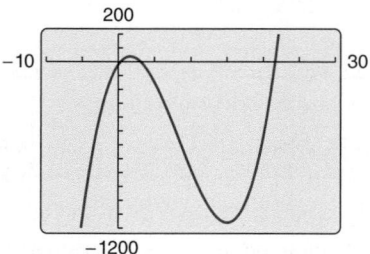

Different viewing windows for the graph of $f(x) = x^3 - 25x^2 + 74x - 20$
Figure 4.43

When you are sketching the graph of a function, either by hand or with a graphing utility, remember that normally you cannot show the *entire* graph. The decision as to which part of the graph you choose to show is often crucial. For instance, which of the viewing windows in Figure 4.43 better represents the graph of

$$f(x) = x^3 - 25x^2 + 74x - 20?$$

By seeing both views, it is clear that the second viewing window gives a more complete representation of the graph. But would a third viewing window reveal other interesting portions of the graph? To answer this, you need to use calculus to interpret the first and second derivatives. To determine a good viewing window for a function, use these guidelines to analyze its graph.

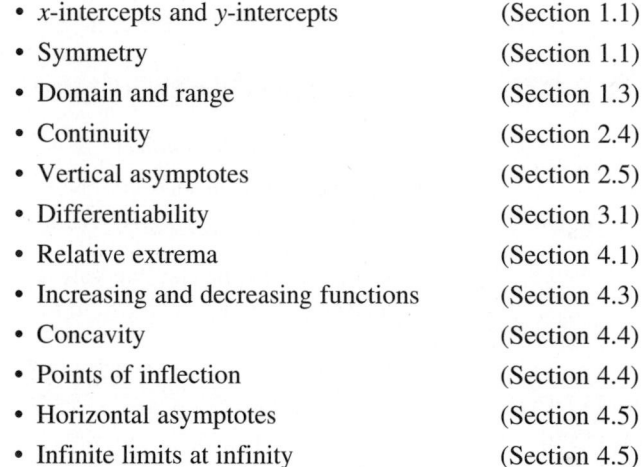

GUIDELINES FOR ANALYZING THE GRAPH OF A FUNCTION

1. Determine the domain and range of the function.
2. Determine the intercepts, asymptotes, and symmetry of the graph.
3. Locate the x-values for which $f'(x)$ and $f''(x)$ either are zero or do not exist. Use the results to determine relative extrema and points of inflection.

REMARK In these guidelines, note the importance of *algebra* (as well as calculus) for solving the equations $f(x) = 0$, $f'(x) = 0$, and $f''(x) = 0$.

EXAMPLE 1 **Sketching the Graph of a Rational Function**

Analyze and sketch the graph of

$$f(x) = \frac{2(x^2 - 9)}{x^2 - 4}.$$

Solution

Domain:	All real numbers except $x = \pm 2$
Range:	$(-\infty, 2) \cup \left[\frac{9}{2}, \infty\right)$
x-intercepts:	$(-3, 0), (3, 0)$
y-intercept:	$\left(0, \frac{9}{2}\right)$
Vertical asymptotes:	$x = -2, x = 2$
Horizontal asymptote:	$y = 2$
Symmetry:	With respect to y-axis
First derivative:	$f'(x) = \dfrac{20x}{(x^2 - 4)^2}$
Second derivative:	$f''(x) = \dfrac{-20(3x^2 + 4)}{(x^2 - 4)^3}$
Critical number:	$x = 0$
Possible points of inflection:	None
Test intervals:	$(-\infty, -2), (-2, 0), (0, 2), (2, \infty)$

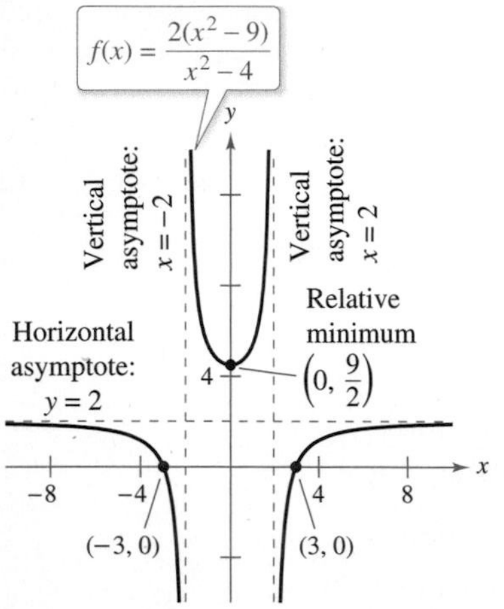

The table shows how the test intervals are used to determine several characteristics of the graph. The graph of f is shown in Figure 4.44.

	$f(x)$	$f'(x)$	$f''(x)$	Characteristic of Graph
$-\infty < x < -2$		$-$	$-$	Decreasing, concave downward
$x = -2$	Undef.	Undef.	Undef.	Vertical asymptote
$-2 < x < 0$		$-$	$+$	Decreasing, concave upward
$x = 0$	$\frac{9}{2}$	0	$+$	Relative minimum
$0 < x < 2$		$+$	$+$	Increasing, concave upward
$x = 2$	Undef.	Undef.	Undef.	Vertical asymptote
$2 < x < \infty$		$+$	$-$	Increasing, concave downward

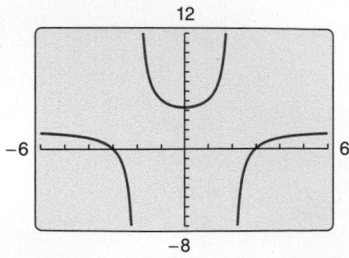

$$f(x) = \frac{2(x^2 - 9)}{x^2 - 4}$$

Vertical asymptote: $x = -2$
Vertical asymptote: $x = 2$
Horizontal asymptote: $y = 2$
Relative minimum $\left(0, \frac{9}{2}\right)$
$(-3, 0)$ $(3, 0)$

Using calculus, you can be certain that you have determined all characteristics of the graph of f.
Figure 4.44

■ FOR FURTHER INFORMATION For more information on the use of technology to graph rational functions, see the article "Graphs of Rational Functions for Computer Assisted Calculus" by Stan Byrd and Terry Walters in *The College Mathematics Journal*. To view this article, go to *MathArticles.com*.

Be sure you understand all of the implications of creating a table such as that shown in Example 1. By using calculus, you can be *sure* that the graph has no relative extrema or points of inflection other than those shown in Figure 4.44.

▷ **TECHNOLOGY PITFALL** Without using the type of analysis outlined in Example 1, it is easy to obtain an incomplete view of the basic characteristics of a graph. For instance, Figure 4.45 shows a view of the graph of

$$g(x) = \frac{2(x^2 - 9)(x - 20)}{(x^2 - 4)(x - 21)}.$$

From this view, it appears that the graph of g is about the same as the graph of f shown in Figure 4.44. The graphs of these two functions, however, differ significantly. Try enlarging the viewing window to see the differences.

By not using calculus, you may overlook important characteristics of the graph of g.
Figure 4.45

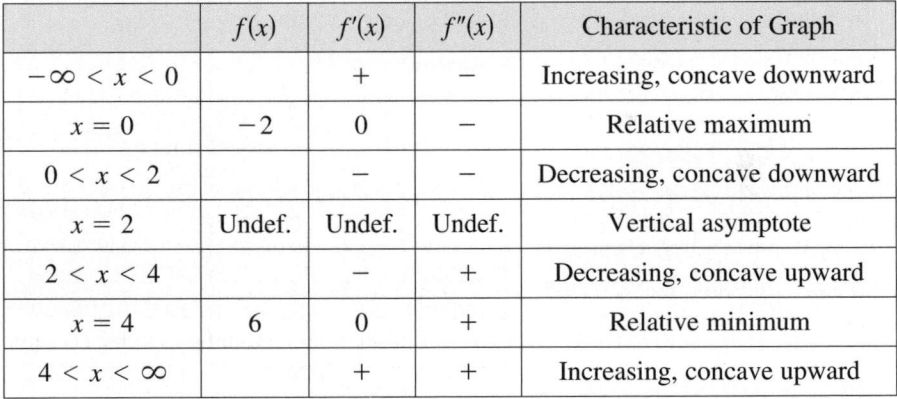

EXAMPLE 2   Sketching the Graph of a Rational Function

Analyze and sketch the graph of $f(x) = \dfrac{x^2 - 2x + 4}{x - 2}$.

Solution

Domain:	All real numbers except $x = 2$
Range:	$(-\infty, -2] \cup [6, \infty)$
x-intercepts:	None
y-intercept:	$(0, -2)$
Vertical asymptote:	$x = 2$
Horizontal asymptotes:	None
Symmetry:	None
End behavior:	$\displaystyle\lim_{x \to -\infty} f(x) = -\infty, \ \lim_{x \to \infty} f(x) = \infty$
First derivative:	$f'(x) = \dfrac{x(x - 4)}{(x - 2)^2}$
Second derivative:	$f''(x) = \dfrac{8}{(x - 2)^3}$
Critical numbers:	$x = 0, \ x = 4$
Possible points of inflection:	None
Test intervals:	$(-\infty, 0), \ (0, 2), \ (2, 4), \ (4, \infty)$

The analysis of the graph of f is shown in the table, and the graph is shown in Figure 4.46.

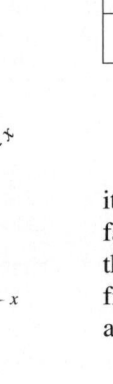

$f(x) = \dfrac{x^2 - 2x + 4}{x - 2}$

Figure 4.46

	$f(x)$	$f'(x)$	$f''(x)$	Characteristic of Graph
$-\infty < x < 0$		$+$	$-$	Increasing, concave downward
$x = 0$	-2	0	$-$	Relative maximum
$0 < x < 2$		$-$	$-$	Decreasing, concave downward
$x = 2$	Undef.	Undef.	Undef.	Vertical asymptote
$2 < x < 4$		$-$	$+$	Decreasing, concave upward
$x = 4$	6	0	$+$	Relative minimum
$4 < x < \infty$		$+$	$+$	Increasing, concave upward

Although the graph of the function in Example 2 has no horizontal asymptote, it does have a slant asymptote. The graph of a rational function (having no common factors and whose denominator is of degree 1 or greater) has a **slant asymptote** when the degree of the numerator exceeds the degree of the denominator by exactly 1. To find the slant asymptote, use long division to rewrite the rational function as the sum of a first-degree polynomial (the slant asymptote) and another rational function.

$$f(x) = \frac{x^2 - 2x + 4}{x - 2} \qquad \text{Write original equation.}$$

$$= x + \frac{4}{x - 2} \qquad \text{Rewrite using long division.}$$

In Figure 4.47, note that the graph of f approaches the slant asymptote $y = x$ as x approaches $-\infty$ or ∞.

A slant asymptote
Figure 4.47

| EXAMPLE 3 | **Sketching the Graph of a Logistic Function** |

Analyze and sketch the graph of the *logistic function* $f(x) = \dfrac{1}{1 + e^{-x}}$.

Solution

$$f'(x) = \frac{e^{-x}}{(1 + e^{-x})^2} \qquad \text{Find first derivative.}$$

$$f''(x) = \frac{e^{-x}(e^{-x} - 1)}{(1 + e^{-x})^3} \qquad \text{Find second derivative.}$$

The graph has only one intercept, $\left(0, \frac{1}{2}\right)$. It has no vertical asymptotes, but it has two horizontal asymptotes: $y = 1$ (to the right) and $y = 0$ (to the left). The function has no critical numbers and one possible point of inflection (at $x = 0$). The domain of the function is all real numbers. The analysis of the graph of f is shown in the table, and the graph is shown in Figure 4.48.

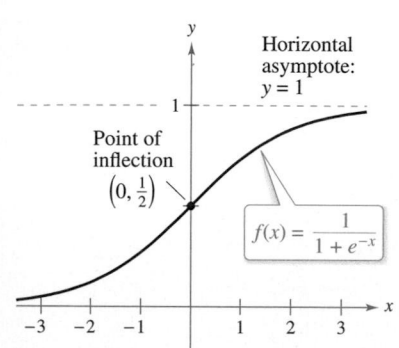

Horizontal asymptote: $y = 1$

Point of inflection $\left(0, \frac{1}{2}\right)$

$f(x) = \dfrac{1}{1 + e^{-x}}$

Figure 4.48

	$f(x)$	$f'(x)$	$f''(x)$	Characteristic of Graph
$-\infty < x < 0$		$+$	$+$	Increasing, concave upward
$x = 0$	$\dfrac{1}{2}$	$+$	0	Point of inflection
$0 < x < \infty$		$+$	$-$	Increasing, concave downward

| EXAMPLE 4 | **Sketching the Graph of a Radical Function** |

Analyze and sketch the graph of $f(x) = 2x^{5/3} - 5x^{4/3}$.

Solution

$$f'(x) = \frac{10}{3} x^{1/3}(x^{1/3} - 2) \qquad \text{Find first derivative.}$$

$$f''(x) = \frac{20(x^{1/3} - 1)}{9x^{2/3}} \qquad \text{Find second derivative.}$$

The function has two intercepts: $(0, 0)$ and $\left(\frac{125}{8}, 0\right)$. There are no horizontal or vertical asymptotes. The function has two critical numbers ($x = 0$ and $x = 8$) and two possible points of inflection ($x = 0$ and $x = 1$). The domain is all real numbers. The analysis of the graph of f is shown in the table, and the graph is shown in Figure 4.49.

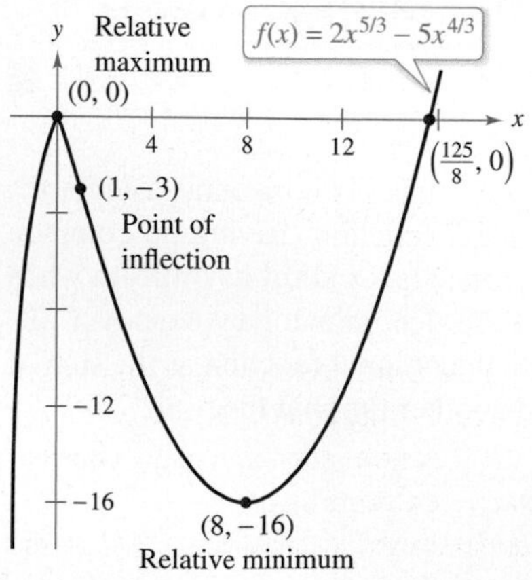

y Relative maximum $(0, 0)$

$f(x) = 2x^{5/3} - 5x^{4/3}$

$(1, -3)$ Point of inflection

$\left(\frac{125}{8}, 0\right)$

$(8, -16)$ Relative minimum

Figure 4.49

	$f(x)$	$f'(x)$	$f''(x)$	Characteristic of Graph
$-\infty < x < 0$		$+$	$-$	Increasing, concave downward
$x = 0$	0	0	Undef.	Relative maximum
$0 < x < 1$		$-$	$-$	Decreasing, concave downward
$x = 1$	-3	$-$	0	Point of inflection
$1 < x < 8$		$-$	$+$	Decreasing, concave upward
$x = 8$	-16	0	$+$	Relative minimum
$8 < x < \infty$		$+$	$+$	Increasing, concave upward

EXAMPLE 5 Sketching the Graph of a Polynomial Function

⋮ ···▷ *See LarsonCalculus.com for an interactive version of this type of example.*

Analyze and sketch the graph of

$$f(x) = x^4 - 12x^3 + 48x^2 - 64x.$$

Solution Begin by factoring to obtain

$$f(x) = x^4 - 12x^3 + 48x^2 - 64x$$
$$= x(x - 4)^3.$$

Then, using the factored form of $f(x)$, you can perform the following analysis.

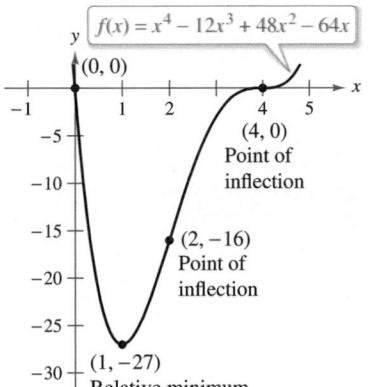

Domain:	All real numbers
Range:	$[-27, \infty)$
x-intercepts:	$(0, 0), (4, 0)$
y-intercept:	$(0, 0)$
Vertical asymptotes:	None
Horizontal asymptotes:	None
Symmetry:	None
End behavior:	$\lim\limits_{x \to -\infty} f(x) = \infty$, $\lim\limits_{x \to \infty} f(x) = \infty$
First derivative:	$f'(x) = 4(x - 1)(x - 4)^2$
Second derivative:	$f''(x) = 12(x - 4)(x - 2)$
Critical numbers:	$x = 1, x = 4$
Possible points of inflection:	$x = 2, x = 4$
Test intervals:	$(-\infty, 1), (1, 2), (2, 4), (4, \infty)$

(a)

The analysis of the graph of f is shown in the table, and the graph is shown in Figure 4.50(a). Using a computer algebra system such as *Maple* [see Figure 4.50(b)] can help you verify your analysis.

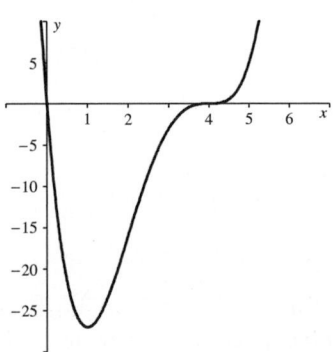

Generated by Maple

(b)
A polynomial function of even degree must have at least one relative extremum.
Figure 4.50

	$f(x)$	$f'(x)$	$f''(x)$	Characteristic of Graph
$-\infty < x < 1$		$-$	$+$	Decreasing, concave upward
$x = 1$	-27	0	$+$	Relative minimum
$1 < x < 2$		$+$	$+$	Increasing, concave upward
$x = 2$	-16	$+$	0	Point of inflection
$2 < x < 4$		$+$	$-$	Increasing, concave downward
$x = 4$	0	0	0	Point of inflection
$4 < x < \infty$		$+$	$+$	Increasing, concave upward

The fourth-degree polynomial function in Example 5 has one relative minimum and no relative maxima. In general, a polynomial function of degree n can have *at most* $n - 1$ relative extrema and *at most* $n - 2$ points of inflection. Moreover, polynomial functions of even degree must have *at least* one relative extremum.

Remember from the Leading Coefficient Test described in Section 1.3 that the "end behavior" of the graph of a polynomial function is determined by its leading coefficient and its degree. For instance, because the polynomial in Example 5 has a positive leading coefficient, the graph rises to the right. Moreover, because the degree is even, the graph also rises to the left.

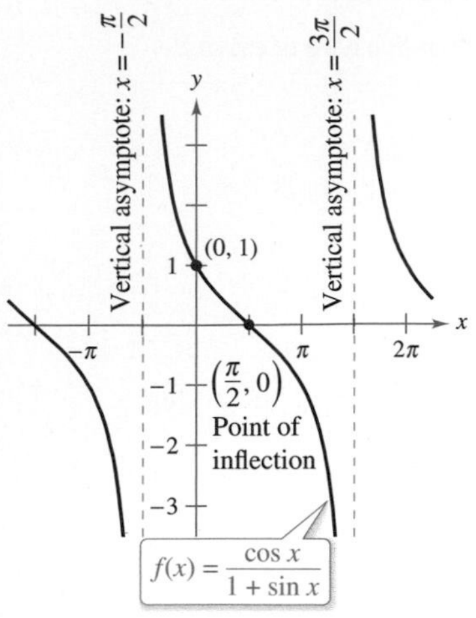

(a)

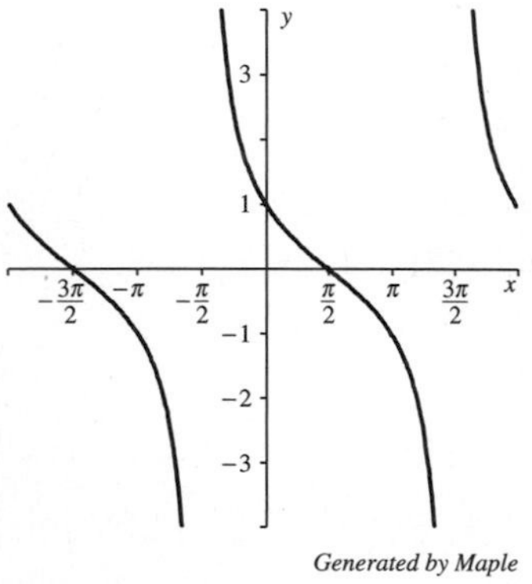

Generated by Maple

(b)

Figure 4.51

EXAMPLE 6 **Sketching the Graph of a Trigonometric Function**

Analyze and sketch the graph of $f(x) = (\cos x)/(1 + \sin x)$.

Solution Because the function has a period of 2π, you can restrict the analysis of the graph to any interval of length 2π. For convenience, choose $[-\pi/2, 3\pi/2]$.

Domain:	All real numbers except $x = \dfrac{3 + 4n}{2}\pi$
Range:	All real numbers
Period:	2π
x-intercept:	$\left(\dfrac{\pi}{2}, 0\right)$
y-intercept:	$(0, 1)$
Vertical asymptotes:	$x = -\dfrac{\pi}{2}, x = \dfrac{3\pi}{2}$ See Remark below.
Horizontal asymptotes:	None
Symmetry:	None
First derivative:	$f'(x) = -\dfrac{1}{1 + \sin x}$
Second derivative:	$f''(x) = \dfrac{\cos x}{(1 + \sin x)^2}$
Critical numbers:	None
Possible points of inflection:	$x = \dfrac{\pi}{2}$
Test intervals:	$\left(-\dfrac{\pi}{2}, \dfrac{\pi}{2}\right), \left(\dfrac{\pi}{2}, \dfrac{3\pi}{2}\right)$

The analysis of the graph of f on the interval $[-\pi/2, 3\pi/2]$ is shown in the table, and the graph is shown in Figure 4.51(a). Compare this with the graph generated by the computer algebra system *Maple* in Figure 4.51(b).

	$f(x)$	$f'(x)$	$f''(x)$	Characteristic of Graph
$x = -\dfrac{\pi}{2}$	Undef.	Undef.	Undef.	Vertical asymptote
$-\dfrac{\pi}{2} < x < \dfrac{\pi}{2}$		$-$	$+$	Decreasing, concave upward
$x = \dfrac{\pi}{2}$	0	$-$	0	Point of inflection
$\dfrac{\pi}{2} < x < \dfrac{3\pi}{2}$		$-$	$-$	Decreasing, concave downward
$x = \dfrac{3\pi}{2}$	Undef.	Undef.	Undef.	Vertical asymptote

REMARK By substituting $-\pi/2$ or $3\pi/2$ into the function, you obtain the indeterminate form $0/0$, which you will study in Section 5.6. To determine that the function has vertical asymptotes at these two values, rewrite f as

$$f(x) = \frac{\cos x}{1 + \sin x} = \frac{(\cos x)(1 - \sin x)}{(1 + \sin x)(1 - \sin x)} = \frac{(\cos x)(1 - \sin x)}{\cos^2 x} = \frac{1 - \sin x}{\cos x}.$$

In this form, it is clear that the graph of f has vertical asymptotes at $x = -\pi/2$ and $3\pi/2$.

EXAMPLE 7 **Analyzing an Inverse Trigonometric Graph**

Analyze the graph of $y = (\arctan x)^2$.

Solution From the derivative

$$y' = 2\,(\arctan x)\left(\frac{1}{1 + x^2}\right)$$

$$= \frac{2\arctan x}{1 + x^2}$$

you can see that the only critical number is $x = 0$. By the First Derivative Test, this value corresponds to a relative minimum at $(0, 0)$. You can use the first derivative to conclude that the graph is decreasing on the interval $(-\infty, 0)$ and increasing on $(0, \infty)$. From the second derivative

$$y'' = \frac{(1 + x^2)\left(\dfrac{2}{1 + x^2}\right) - (2\arctan x)(2x)}{(1 + x^2)^2}$$

$$= \frac{2(1 - 2x\arctan x)}{(1 + x^2)^2}$$

it follows that points of inflection occur when

$$2x\arctan x = 1.$$

Using Newton's Method, these points occur when $x \approx \pm 0.765$. Finally, because

$$\lim_{x \to \pm\infty} (\arctan x)^2 = \frac{\pi^2}{4}$$

it follows that the graph has a horizontal asymptote at

$$y = \frac{\pi^2}{4}.$$

The graph is shown in Figure 4.52.

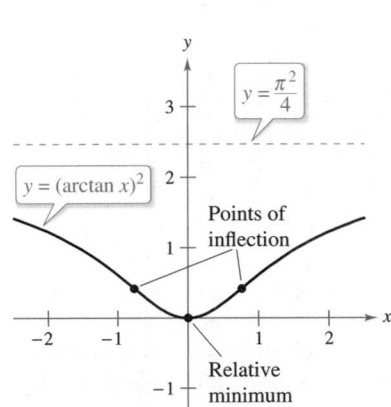

The graph of $y = (\arctan x)^2$ has a horizontal asymptote at $y = \pi^2/4$.

Figure 4.52

EXAMPLE 8 **Analyzing a Logarithmic Graph**

Analyze the graph of $f(x) = \ln(x^2 + 2x + 3)$.

Solution Note that the domain of f is all real numbers. The graph of f has no x-intercepts, but it does have a y-intercept at $(0, \ln 3)$. From the derivative

$$f'(x) = \frac{2x + 2}{x^2 + 2x + 3}$$

you can see that the only critical number is $x = -1$. By the First Derivative Test, this value corresponds to a relative minimum at $(-1, \ln 2)$. You can use the first derivative to conclude that the graph of f is decreasing on the interval $(-\infty, -1)$ and increasing on $(-1, \infty)$. From the second derivative

$$f''(x) = \frac{(x^2 + 2x + 3)(2) - (2x + 2)(2x + 2)}{(x^2 + 2x + 3)^2}$$

$$= \frac{-2(x^2 + 2x - 1)}{(x^2 + 2x + 3)^2}$$

it follows that points of inflection occur when $x^2 + 2x - 1 = 0$. Using the Quadratic Formula, these points occur when $x = -1 \pm \sqrt{2}$. Also, the graph of f is concave downward on the intervals $\left(-\infty, -1 - \sqrt{2}\right)$ and $\left(-1 + \sqrt{2}, \infty\right)$, and concave upward on $\left(-1 - \sqrt{2}, -1 + \sqrt{2}\right)$. The graph of f is shown in Figure 4.53.

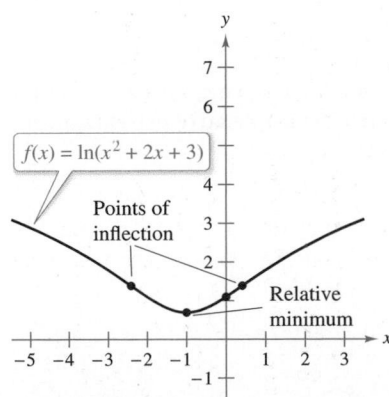

Figure 4.53

4.6 Exercises

See **CalcChat.com** for tutorial help and worked-out solutions to odd-numbered exercises.

CONCEPT CHECK

1. Analyzing the Graph of a Function Name several of the concepts you have learned that are useful for analyzing the graph of a function.

2. Analyzing a Graph Explain how to create a table to determine characteristics of a graph. What elements do you include?

3. Slant Asymptote Which type of function can have a slant asymptote? How do you determine the equation of a slant asymptote?

4. Polynomial What are the maximum numbers of relative extrema and points of inflection that a fifth-degree polynomial can have? Explain.

 **Analyzing the Graph of a Function** In Exercises 5–34, analyze and sketch a graph of the function. Label any intercepts, relative extrema, points of inflection, and asymptotes. Use a graphing utility to verify your results.

5. $y = \dfrac{1}{x-2} - 3$ 　　**6.** $y = \dfrac{x}{x^2+1}$

7. $y = \dfrac{x}{1-x}$ 　　**8.** $y = \dfrac{x-4}{x-3}$

9. $y = \dfrac{x+1}{x^2-4}$ 　　**10.** $y = \dfrac{2x}{9-x^2}$

11. $y = \dfrac{x^2}{x^2+3}$ 　　**12.** $y = \dfrac{x^2+1}{x^2-4}$

13. $y = 3 + \dfrac{2}{x}$ 　　**14.** $f(x) = \dfrac{x-3}{x}$

15. $f(x) = x + \dfrac{32}{x^2}$ 　　**16.** $y = \dfrac{4}{x^2} + 1$

17. $y = \dfrac{3x}{x^2-1}$ 　　**18.** $f(x) = \dfrac{x^3}{x^2-9}$

19. $y = \dfrac{x^2 - 6x + 12}{x-4}$ 　　**20.** $y = \dfrac{-x^2 - 4x - 7}{x+3}$

21. $y = \dfrac{x^3}{\sqrt{x^2-4}}$ 　　**22.** $y = \dfrac{x}{\sqrt{x^2-4}}$

23. $y = x\sqrt{4-x}$ 　　**24.** $g(x) = x\sqrt{9-x^2}$

25. $y = 3x^{2/3} - 2x$ 　　**26.** $y = (x+1)^2 - 3(x+1)^{2/3}$

27. $y = 2 - x - x^3$ 　　**28.** $y = -\frac{1}{3}(x^3 - 3x + 2)$

29. $y = 3x^4 + 4x^3$

30. $y = -2x^4 + 3x^2$

31. $xy^2 = 9$

32. $x^2y = 9$

33. $y = |2x - 3|$

34. $y = |x^2 - 6x + 5|$

 **Analyzing the Graph of a Trigonometric Function** In Exercises 35–42, analyze and sketch a graph of the function over the given interval. Label any intercepts, relative extrema, points of inflection, and asymptotes. Use a graphing utility to verify your results.

Function	Interval
35. $f(x) = 2x - 4\sin x$	$0 \le x \le 2\pi$
36. $f(x) = -x + 2\cos x$	$0 \le x \le 2\pi$
37. $y = \sin x - \frac{1}{18}\sin 3x$	$0 \le x \le 2\pi$
38. $y = 2(x - 2) + \cot x$	$0 < x < \pi$
39. $y = 2(\csc x + \sec x)$	$0 < x < \dfrac{\pi}{2}$
40. $y = \sec^2\dfrac{\pi x}{8} - 2\tan\dfrac{\pi x}{8} - 1$	$-3 < x < 3$
41. $g(x) = x\tan x$	$-\dfrac{3\pi}{2} < x < \dfrac{3\pi}{2}$
42. $g(x) = x\cot x$	$-2\pi < x < 2\pi$

Analyzing the Graph of a Transcendental Function In Exercises 43–54, analyze and sketch a graph of the function. Label any intercepts, relative extrema, points of inflection, and asymptotes. Use a graphing utility to verify your results.

43. $f(x) = e^{3x}(2 - x)$ 　　**44.** $f(x) = -2 + e^{3x}(4 - 2x)$

45. $g(t) = \dfrac{10}{1 + 4e^{-t}}$ 　　**46.** $h(x) = \dfrac{8}{2 + 3e^{-x/2}}$

47. $y = (x - 1)\ln(x - 1)$ 　　**48.** $y = \frac{1}{24}x^3 - \ln x$

49. $g(x) = 6\arcsin\left[\left(\dfrac{x-2}{2}\right)^2\right]$

50. $h(x) = 7\arctan(x + 1) - \ln(x^2 + 2x + 2)$

51. $f(x) = \dfrac{x}{3^{x-3}}$

52. $g(t) = (5 - t)5^t$

53. $g(x) = \log_4(x - x^2)$

54. $f(x) = \log_2|x^2 - 4x|$

Analyzing the Graph of a Function Using Technology In Exercises 55–62, use a computer algebra system to analyze and graph the function. Identify any relative extrema, points of inflection, and asymptotes.

55. $f(x) = \dfrac{20x}{x^2+1} - \dfrac{1}{x}$

56. $f(x) = \dfrac{-2x}{\sqrt{x^2+7}}$

57. $y = \cos x - \frac{1}{4}\cos 2x, \quad 0 \le x \le 2\pi$

58. $y = 2x - \tan x, \quad -\dfrac{\pi}{2} < x < \dfrac{\pi}{2}$

59. $y = \dfrac{x}{2} + \ln\dfrac{x}{x+3}$

60. $y = \dfrac{3x}{2}(1 + 4e^{-x/3})$

61. $f(x) = 2 + (x^2 - 3)e^{-x}$

62. $f(x) = \dfrac{10 \ln x}{x^2\sqrt{x}}$

Identifying Graphs In Exercises 63 and 64, the graphs of f, f', and f'' are shown on the same set of coordinate axes. Identify each graph. Explain your reasoning. To print an enlarged copy of the graph, go to *MathGraphs.com*.

63.

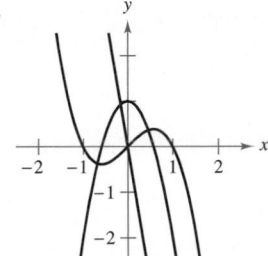

64.

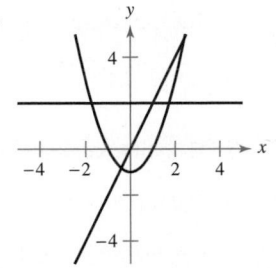

Graphical Reasoning In Exercises 65–68, use the graph of f' to sketch a graph of f and the graph of f''. To print an enlarged copy of the graph, go to *MathGraphs.com*.

65.

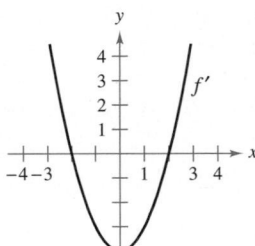

66.

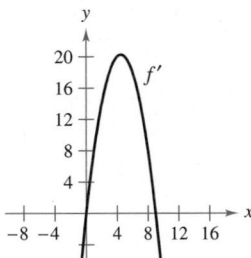

67.

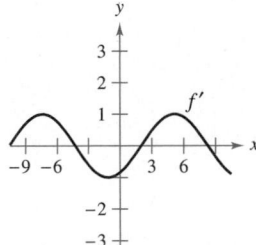

68.

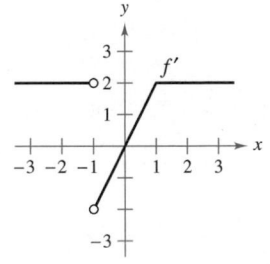

(Submitted by Bill Fox, Moberly Area Community College, Moberly, MO)

69. Graphical Reasoning Consider the function

$$f(x) = \dfrac{\cos^2 \pi x}{\sqrt{x^2 + 1}}, \quad 0 < x < 4.$$

(a) Use a computer algebra system to graph the function and use the graph to approximate the critical numbers visually.

(b) Use a computer algebra system to find f' and approximate the critical numbers. Are the results the same as the visual approximation in part (a)? Explain.

70. Graphical Reasoning Consider the function

$$f(x) = \tan(\sin \pi x).$$

(a) Use a graphing utility to graph the function.

(b) Identify any symmetry of the graph.

(c) Is the function periodic? If so, what is the period?

(d) Identify any extrema on $(-1, 1)$.

(e) Use a graphing utility to determine the concavity of the graph on $(0, 1)$.

71. Conjecture Use a graphing utility to graph f and g in the same viewing window and determine which is increasing at the faster rate for "large" values of x. What can you conclude about the rate of growth of the natural logarithmic function?

(a) $f(x) = \ln x, \ g(x) = \sqrt{x}$

(b) $f(x) = \ln x, \ g(x) = \sqrt[4]{x}$

72. Comparing Functions Let f be a function that is positive and differentiable on the entire real number line and let $g(x) = \ln f(x)$.

(a) When g is increasing, must f be increasing? Explain.

(b) When the graph of f is concave upward, must the graph of g be concave upward? Explain.

EXPLORING CONCEPTS

73. Sketching a Graph Sketch a graph of a differentiable function f such that $x = 2$ is the only critical number, $f'(x) < 0$ on $(-\infty, 2)$, $f'(x) > 0$ on $(2, \infty)$, $\lim\limits_{x \to -\infty} f(x) = 6$, and $\lim\limits_{x \to \infty} f(x) = 6$.

74. Points of Inflection Is it possible to sketch a graph of a function that satisfies the conditions of Exercise 73 and has *no* points of inflection? Explain.

75. Using a Derivative Let $f'(t) < 0$ for all t in the interval $(2, 8)$. Explain why $f(3) > f(5)$.

76. Using a Derivative Let $f(0) = 3$ and $2 \leq f'(x) \leq 4$ for all x in the interval $[-5, 5]$. Determine the greatest and least possible values of $f(2)$.

77. A Function and Its Derivative The graph of a function f is shown below. To print an enlarged copy of the graph, go to *MathGraphs.com*.

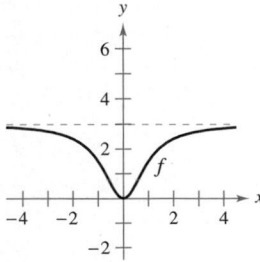

(a) Sketch f'.

(b) Use the graph to estimate $\lim\limits_{x \to \infty} f(x)$ and $\lim\limits_{x \to \infty} f'(x)$.

(c) Explain the answers you gave in part (b).

78. **HOW DO YOU SEE IT?** The graph of f is shown in the figure.

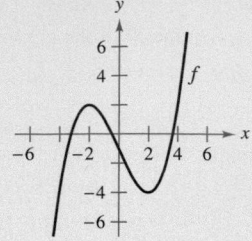

(a) For which values of x is $f'(x)$ zero? Positive? Negative? What do these values mean?

(b) For which values of x is $f''(x)$ zero? Positive? Negative? What do these values mean?

(c) On what open interval is f' an increasing function?

(d) For which value of x is $f'(x)$ minimum? For this value of x, how does the rate of change of f compare with the rates of change of f for other values of x? Explain.

Horizontal and Vertical Asymptotes In Exercises 79–82, use a graphing utility to graph the function. Use the graph to determine whether it is possible for the graph of a function to cross its horizontal asymptote. Do you think it is possible for the graph of a function to cross its vertical asymptote? Why or why not?

79. $f(x) = \dfrac{4(x-1)^2}{x^2 - 4x + 5}$

80. $g(x) = \dfrac{3x^4 - 5x + 3}{x^4 + 1}$

81. $h(x) = \dfrac{\sin 2x}{x}$

82. $f(x) = \dfrac{\cos 3x}{4x}$

Examining a Function In Exercises 83 and 84, use a graphing utility to graph the function. Explain why there is no vertical asymptote when a superficial examination of the function may indicate that there should be one.

83. $h(x) = \dfrac{6 - 2x}{3 - x}$

84. $g(x) = \dfrac{x^2 + x - 2}{x - 1}$

Slant Asymptote In Exercises 85–90, use a graphing utility to graph the function and determine the slant asymptote of the graph analytically. Zoom out repeatedly and describe how the graph on the display appears to change. Why does this occur?

85. $f(x) = -\dfrac{x^2 - 3x - 1}{x - 2}$

86. $g(x) = \dfrac{2x^2 - 8x - 15}{x - 5}$

87. $f(x) = \dfrac{2x^3}{x^2 + 1}$

88. $h(x) = \dfrac{-x^3 + x^2 + 4}{x^2}$

89. $f(x) = \dfrac{x^3 - 3x^2 + 2}{x(x - 3)}$

90. $f(x) = -\dfrac{x^3 - 2x^2 + 2}{2x^2}$

91. Investigation Let $P(x_0, y_0)$ be an arbitrary point on the graph of f that $f'(x_0) \neq 0$, as shown in the figure. Verify each statement.

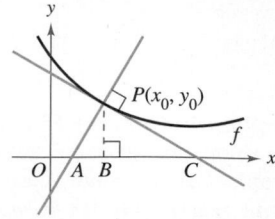

(a) The x-intercept of the tangent line is

$$\left(x_0 - \frac{f(x_0)}{f'(x_0)}, 0\right).$$

(b) The y-intercept of the tangent line is

$$(0, f(x_0) - x_0 f'(x_0)).$$

(c) The x-intercept of the normal line is

$$(x_0 + f(x_0)f'(x_0), 0).$$

(The *normal line* at a point is perpendicular to the tangent line at the point.)

(d) The y-intercept of the normal line is

$$\left(0, y_0 + \frac{x_0}{f'(x_0)}\right).$$

(e) $|BC| = \left|\dfrac{f(x_0)}{f'(x_0)}\right|$

(f) $|PC| = \left|\dfrac{f(x_0)\sqrt{1 + [f'(x_0)]^2}}{f'(x_0)}\right|$

(g) $|AB| = |f(x_0)f'(x_0)|$

(h) $|AP| = |f(x_0)|\sqrt{1 + [f'(x_0)]^2}$

92. Graphical Reasoning Identify the real numbers x_0, x_1, x_2, x_3, and x_4 in the figure such that each of the following is true.

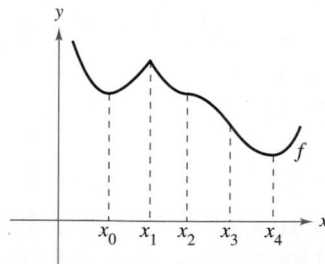

(a) $f'(x) = 0$

(b) $f''(x) = 0$

(c) $f'(x)$ does not exist.

(d) f has a relative maximum.

(e) f has a point of inflection.

Think About It   In Exercises 93–96, create a function whose graph has the given characteristics. (**There is more than one correct answer.**)

93. Vertical asymptote: $x = 3$

Horizontal asymptote: $y = 0$

94. Vertical asymptote: $x = -5$

Horizontal asymptote: None

95. Vertical asymptote: $x = 3$

Slant asymptote: $y = 3x + 2$

96. Vertical asymptote: $x = 2$

Slant asymptote: $y = -x$

True or False?   In Exercises 97–100, determine whether the statement is true or false. If it is false, explain why or give an example that shows it is false.

97. If $f'(x) > 0$ for all real numbers x, then f increases without bound.

98. If $f''(x) < 0$ for all real numbers x, then f decreases without bound.

99. Every rational function has a slant asymptote.

100. Every polynomial function has an absolute maximum and an absolute minimum on $(-\infty, \infty)$.

101. Graphical Reasoning   The graph of the first derivative of a function f on the interval $[-7, 5]$ is shown. Use the graph to answer each question.

(a) On what interval(s) is f decreasing?

(b) On what interval(s) is the graph of f concave downward?

(c) At what x-value(s) does f have relative extrema?

(d) At what x-value(s) does the graph of f have a point of inflection?

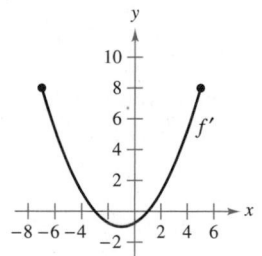

Figure for 101

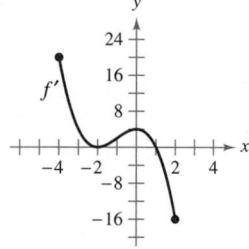

Figure for 102

102. Graphical Reasoning   The graph of the first derivative of a function f on the interval $[-4, 2]$ is shown. Use the graph to answer each question.

(a) On what interval(s) is f increasing?

(b) On what interval(s) is the graph of f concave upward?

(c) At what x-value(s) does f have relative extrema?

(d) At what x-value(s) does the graph of f have a point of inflection?

103. Graphical Reasoning   Consider the function

$$f(x) = \frac{ax}{(x - b)^2}.$$

Determine the effect on the graph of f as a and b are changed. Consider cases where a and b are both positive or both negative and cases where a and b have opposite signs.

104. Graphical Reasoning   Consider the function

$$f(x) = \frac{1}{2}(ax)^2 - ax, \quad a \neq 0.$$

(a) Determine the changes (if any) in the intercepts, extrema, and concavity of the graph of f when a is varied.

(b) In the same viewing window, use a graphing utility to graph the function for four different values of a.

Slant Asymptotes   In Exercises 105 and 106, the graph of the function has two slant asymptotes. Identify each slant asymptote. Then graph the function and its asymptotes.

105. $y = \sqrt{4 + 16x^2}$ **106.** $y = \sqrt{x^2 + 6x}$

107. Investigation   Consider the function

$$f(x) = \frac{2x^n}{x^4 + 1}$$

for nonnegative integer values of n.

(a) Discuss the relationship between the value of n and the symmetry of the graph.

(b) For which values of n will the x-axis be the horizontal asymptote?

(c) For which value of n will $y = 2$ be the horizontal asymptote?

(d) What is the asymptote of the graph when $n = 5$?

(e) Use a graphing utility to graph f for the indicated values of n in the table. Use the graph to determine the number of extrema M and the number of inflection points N of the graph.

n	0	1	2	3	4	5
M						
N						

PUTNAM EXAM CHALLENGE

108. Let $f(x)$ be defined for $a \leq x \leq b$. Assuming appropriate properties of continuity and derivability, prove for $a < x < b$ that

$$\frac{\dfrac{f(x) - f(a)}{x - a} - \dfrac{f(b) - f(a)}{b - a}}{x - b} = \frac{1}{2}f''(\varepsilon),$$

where ε is some number between a and b.

4.7 Optimization Problems

■ Solve applied minimum and maximum problems.

Applied Minimum and Maximum Problems

One of the most common applications of calculus involves the determination of minimum and maximum values. Consider how frequently you hear or read terms such as greatest profit, least cost, least time, greatest voltage, optimum size, least size, greatest strength, and greatest distance. Before outlining a general problem-solving strategy for such problems, consider the next example.

EXAMPLE 1   **Finding Maximum Volume**

A manufacturer wants to design an open box having a square base and a surface area of 108 square inches, as shown in Figure 4.54. What dimensions will produce a box with maximum volume?

Solution   Because the box has a square base, its volume is

$$V = x^2h. \qquad \text{Primary equation}$$

This equation is called the **primary equation** because it gives a formula for the quantity to be optimized. The surface area of the box is

$$S = (\text{area of base}) + (\text{area of four sides})$$
$$108 = x^2 + 4xh. \qquad \text{Secondary equation}$$

Because V is to be maximized, you want to write V as a function of just one variable. To do this, you can solve the equation $x^2 + 4xh = 108$ for h in terms of x to obtain $h = (108 - x^2)/(4x)$. Substituting into the primary equation produces

$$V = x^2h \qquad \text{Function of two variables}$$
$$= x^2\left(\frac{108 - x^2}{4x}\right) \qquad \text{Substitute for } h.$$
$$= 27x - \frac{x^3}{4}. \qquad \text{Function of one variable}$$

Before finding which x-value will yield a maximum value of V, you should determine the *feasible domain*. That is, what values of x make sense in this problem? You know that $V \geq 0$. You also know that x must be nonnegative and that the area of the base ($A = x^2$) is at most 108. So, the feasible domain is

$$0 \leq x \leq \sqrt{108}. \qquad \text{Feasible domain}$$

To maximize V, find its critical numbers on the interval $\left(0, \sqrt{108}\right)$.

$$\frac{dV}{dx} = 27 - \frac{3x^2}{4} \qquad \text{Differentiate with respect to } x.$$
$$27 - \frac{3x^2}{4} = 0 \qquad \text{Set derivative equal to 0.}$$
$$3x^2 = 108 \qquad \text{Simplify.}$$
$$x = \pm 6 \qquad \text{Critical numbers}$$

So, the critical numbers are $x = \pm 6$. You do not need to consider $x = -6$ because it is outside the domain. Evaluating V at the critical number $x = 6$ and at the endpoints of the domain produces $V(0) = 0$, $V(6) = 108$, and $V\left(\sqrt{108}\right) = 0$. So, V is maximum when $x = 6$, and the dimensions of the box are 6 inches by 6 inches by 3 inches. ■

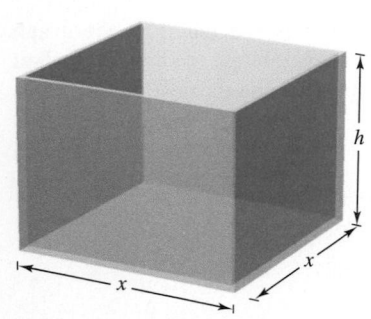

Open box with square base:
$S = x^2 + 4xh = 108$
Figure 4.54

▷ **TECHNOLOGY**   You can verify your answer in Example 1 by using a graphing utility to graph the volume function

$$V = 27x - \frac{x^3}{4}.$$

Use a viewing window in which $0 \leq x \leq \sqrt{108} \approx 10.4$ and $0 \leq y \leq 120$, and use the *maximum* or *trace* feature to determine the value of x that produces a maximum volume.

In Example 1, you should realize that there are infinitely many open boxes having 108 square inches of surface area. To begin solving the problem, you might ask yourself which basic shape would seem to yield a maximum volume. Should the box be tall, squat, or nearly cubical?

You might even try calculating a few volumes, as shown in Figure 4.55, to determine whether you can get a better feeling for what the optimum dimensions should be. Remember that you are not ready to begin solving a problem until you have clearly identified what the problem is.

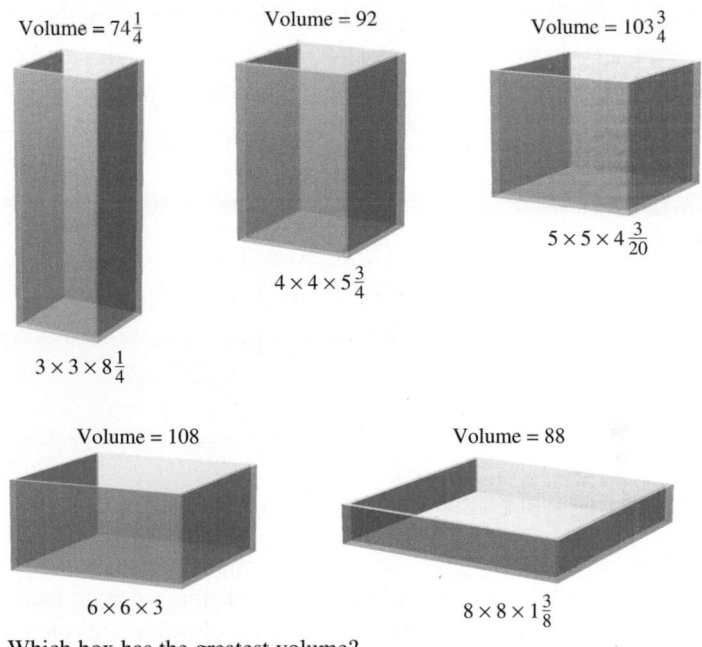

Volume $= 74\frac{1}{4}$

Volume $= 92$

Volume $= 103\frac{3}{4}$

$5 \times 5 \times 4\frac{3}{20}$

$4 \times 4 \times 5\frac{3}{4}$

$3 \times 3 \times 8\frac{1}{4}$

Volume $= 108$

Volume $= 88$

$6 \times 6 \times 3$

$8 \times 8 \times 1\frac{3}{8}$

Which box has the greatest volume?
Figure 4.55

Example 1 illustrates the following guidelines for solving applied minimum and maximum problems.

GUIDELINES FOR SOLVING APPLIED MINIMUM AND MAXIMUM PROBLEMS

1. Identify all *given* quantities and all quantities *to be determined*. If possible, make a sketch.

2. Write a **primary equation** for the quantity that is to be maximized or minimized. (A review of several useful formulas from geometry is presented on the formula card inside the back cover.)

3. Reduce the primary equation to one having a *single independent variable*. This may involve the use of **secondary equations** relating the independent variables of the primary equation.

4. Determine the feasible domain of the primary equation. That is, determine the values for which the stated problem makes sense.

5. Determine the desired maximum or minimum value by the calculus techniques discussed in Sections 4.1 through 4.4.

REMARK For Step 5, recall that to determine the maximum or minimum value of a continuous function f on a closed interval, you should compare the values of f at its critical numbers with the values of f at the endpoints of the interval.

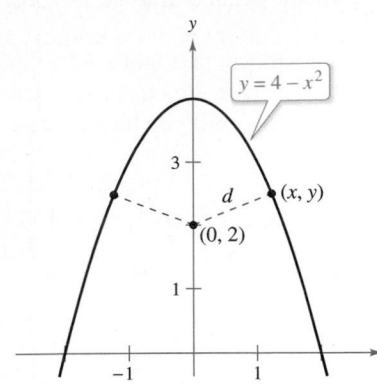

The quantity to be minimized is
distance: $d = \sqrt{(x-0)^2 + (y-2)^2}$.
Figure 4.56

EXAMPLE 2 **Finding Minimum Distance**

⋮⋯▷ *See LarsonCalculus.com for an interactive version of this type of example.*

Which points on the graph of $y = 4 - x^2$ are closest to the point $(0, 2)$?

Solution Figure 4.56 shows that there are two points at a minimum distance from the point $(0, 2)$. The distance between the point $(0, 2)$ and a point (x, y) on the graph of $y = 4 - x^2$ is

$$d = \sqrt{(x-0)^2 + (y-2)^2}.$$ Primary equation

Using the secondary equation $y = 4 - x^2$, you can rewrite the primary equation as

$$d = \sqrt{x^2 + (4 - x^2 - 2)^2}$$
$$= \sqrt{x^4 - 3x^2 + 4}.$$

Because d is smallest when the expression inside the radical is smallest, you need only find the critical numbers of $f(x) = x^4 - 3x^2 + 4$. Note that the domain of f is the entire real number line. So, there are no endpoints of the domain to consider. Moreover, the derivative of f

$$f'(x) = 4x^3 - 6x$$
$$= 2x(2x^2 - 3)$$

is zero when

$$x = 0, \quad \sqrt{\frac{3}{2}}, \quad -\sqrt{\frac{3}{2}}.$$

Testing these critical numbers using the First Derivative Test verifies that $x = 0$ yields a relative maximum, whereas both $x = \sqrt{3/2}$ and $x = -\sqrt{3/2}$ yield a minimum distance. So, the closest points are $\left(\sqrt{3/2}, 5/2\right)$ and $\left(-\sqrt{3/2}, 5/2\right)$.

EXAMPLE 3 **Finding Minimum Area**

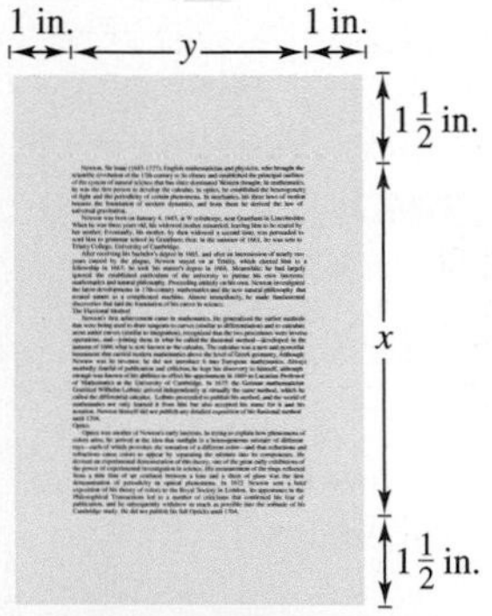

A rectangular page is to contain 24 square inches of print. The margins at the top and bottom of the page are to be $1\frac{1}{2}$ inches, and the margins on the left and right are to be 1 inch (see Figure 4.57). What should the dimensions of the page be so that the least amount of paper is used?

Solution Let A be the area to be minimized.

$$A = (x + 3)(y + 2)$$ Primary equation

The printed area inside the margins is

$$24 = xy.$$ Secondary equation

Solving this equation for y produces $y = 24/x$. Substituting into the primary equation produces

$$A = (x + 3)\left(\frac{24}{x} + 2\right) = 30 + 2x + \frac{72}{x}.$$ Function of one variable

Because x must be positive, you are interested only in values of A for $x > 0$. To find the critical numbers, differentiate with respect to x

$$\frac{dA}{dx} = 2 - \frac{72}{x^2}$$

and note that the derivative is zero when $x^2 = 36$, or $x = \pm 6$. So, the critical numbers are $x = \pm 6$. You do not have to consider $x = -6$ because it is outside the domain. The First Derivative Test confirms that A is a minimum when $x = 6$. So, $y = \frac{24}{6} = 4$ and the dimensions of the page should be $x + 3 = 9$ inches by $y + 2 = 6$ inches. ∎

The quantity to be minimized is area:
$A = (x + 3)(y + 2)$.
Figure 4.57

EXAMPLE 4 **Finding Minimum Length**

Two posts, one 12 feet high and the other 28 feet high, stand 30 feet apart. They are to be stayed by two wires, attached to a single stake, running from ground level to the top of each post. Where should the stake be placed to use the least amount of wire?

Solution Let W be the wire length to be minimized. Using the figure at the right, you can write

$$W = y + z. \qquad \text{Primary equation}$$

In this problem, rather than solving for y in terms of z (or vice versa), you can solve for both y and z in terms of a third variable x, as shown in the figure at the right. From the Pythagorean Theorem, you obtain

$$x^2 + 12^2 = y^2$$
$$(30 - x)^2 + 28^2 = z^2$$

which implies that

$$y = \sqrt{x^2 + 144}$$
$$z = \sqrt{x^2 - 60x + 1684}.$$

So, you can rewrite the primary equation as

$$W = y + z$$
$$= \sqrt{x^2 + 144} + \sqrt{x^2 - 60x + 1684}, \quad 0 \le x \le 30.$$

Differentiating W with respect to x yields

$$\frac{dW}{dx} = \frac{x}{\sqrt{x^2 + 144}} + \frac{x - 30}{\sqrt{x^2 - 60x + 1684}}.$$

By letting $dW/dx = 0$, you obtain

$$\frac{x}{\sqrt{x^2 + 144}} + \frac{x - 30}{\sqrt{x^2 - 60x + 1684}} = 0$$
$$\frac{x}{\sqrt{x^2 + 144}} = \frac{30 - x}{\sqrt{x^2 - 60x + 1684}}$$
$$x\sqrt{x^2 - 60x + 1684} = (30 - x)\sqrt{x^2 + 144}$$
$$x^2(x^2 - 60x + 1684) = (30 - x)^2(x^2 + 144)$$
$$x^4 - 60x^3 + 1684x^2 = x^4 - 60x^3 + 1044x^2 - 8640x + 129{,}600$$
$$640x^2 + 8640x - 129{,}600 = 0$$
$$320(x - 9)(2x + 45) = 0$$
$$x = 9, -22.5.$$

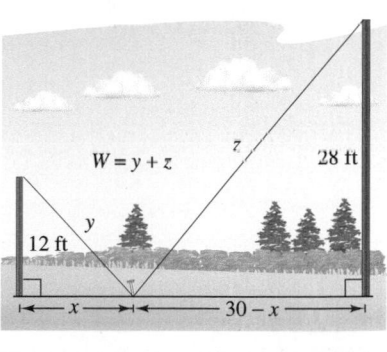

The quantity to be minimized is length. From the diagram, you can see that x varies between 0 and 30.

Because $x = -22.5$ is not in the domain and

$$W(0) \approx 53.04, \quad W(9) = 50, \quad \text{and} \quad W(30) \approx 60.31$$

you can conclude that the wires should be staked at 9 feet from the 12-foot pole. ■

▷ **TECHNOLOGY** From Example 4, you can see that applied optimization problems can involve a lot of algebra. If you have access to a graphing utility, you can confirm that $x = 9$ yields a minimum value of W by graphing

$$W = \sqrt{x^2 + 144} + \sqrt{x^2 - 60x + 1684}$$

as shown in Figure 4.58.

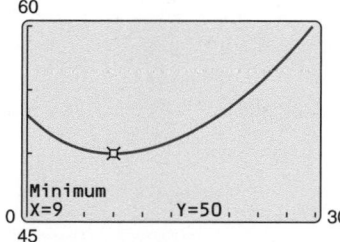

You can confirm the minimum value of W with a graphing utility.
Figure 4.58

In each of the first four examples, the extreme value occurred at a critical number. Although this happens often, remember that an extreme value can also occur at an endpoint of an interval, as shown in Example 5.

EXAMPLE 5   **An Endpoint Maximum**

Four feet of wire is to be used to form a square and a circle. How much of the wire should be used for the square and how much should be used for the circle to enclose the maximum total area?

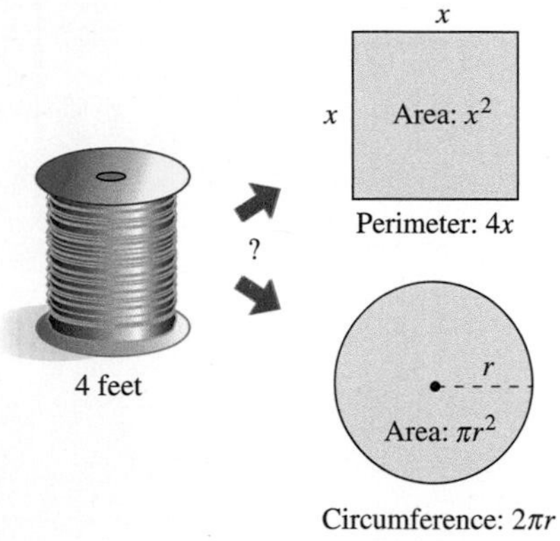

Solution   The total area (see Figure 4.59) is

$$A = \text{(area of square)} + \text{(area of circle)}$$
$$A = x^2 + \pi r^2. \qquad \text{Primary equation}$$

Because the total length of wire is 4 feet, you obtain

$$4 = \text{(perimeter of square)} + \text{(circumference of circle)}$$
$$4 = 4x + 2\pi r. \qquad \text{Secondary equation}$$

So, $r = 2(1 - x)/\pi$, and by substituting into the primary equation you have

$$A = x^2 + \pi \left[\frac{2(1 - x)}{\pi} \right]^2$$
$$= x^2 + \frac{4(1 - x)^2}{\pi}$$
$$= \frac{1}{\pi}(\pi x^2 + 4 - 8x + 4x^2)$$
$$= \frac{1}{\pi}[(\pi + 4)x^2 - 8x + 4].$$

The feasible domain is $0 \le x \le 1$, restricted by the square's perimeter. Because

$$\frac{dA}{dx} = \frac{2(\pi + 4)x - 8}{\pi}$$

the only critical number in $(0, 1)$ is $x = 4/(\pi + 4) \approx 0.56$. So, using

$$A(0) \approx 1.27, \quad A(0.56) \approx 0.56, \quad \text{and} \quad A(1) = 1$$

you can conclude that the maximum area occurs when $x = 0$. That is, *all* the wire is used for the circle. ∎

Before doing the section exercises, review the primary equations developed in Examples 1–5. As applications go, these five examples are fairly simple, and yet the resulting primary equations are quite complicated.

$$V = 27x - \frac{x^3}{4} \qquad \text{Example 1}$$
$$d = \sqrt{x^4 - 3x^2 + 4} \qquad \text{Example 2}$$
$$A = 30 + 2x + \frac{72}{x} \qquad \text{Example 3}$$
$$W = \sqrt{x^2 + 144} + \sqrt{x^2 - 60x + 1684} \qquad \text{Example 4}$$
$$A = \frac{1}{\pi}[(\pi + 4)x^2 - 8x + 4] \qquad \text{Example 5}$$

You must expect that real-life applications often involve equations that are *at least as complicated* as these five. Remember that one of the main goals of this course is to learn to use calculus to analyze equations that initially seem formidable.

---

x

x Area: x^2

Perimeter: $4x$

r

Area: πr^2

Circumference: $2\pi r$

4 feet

?

The quantity to be maximized is area:
$A = x^2 + \pi r^2$.

Figure 4.59

Exploration

What would the answer be if Example 5 asked for the dimensions needed to enclose the *minimum* total area?

EXAMPLE 6   Maximizing an Angle

•••▷ *See LarsonCalculus.com for an interactive version of this type of example.*

A photographer is taking a picture of a 4-foot painting hung in an art gallery. The camera lens is 1 foot below the lower edge of the painting, as shown in Figure 4.60. How far should the camera be from the painting to maximize the angle subtended by the camera lens?

Solution   In Figure 4.60, let β be the angle to be maximized.

$$\beta = \theta - \alpha \qquad \text{Primary equation}$$

From Figure 4.60, you can see that $\cot \theta = \dfrac{x}{5}$ and $\cot \alpha = \dfrac{x}{1}$. Therefore, $\theta = \operatorname{arccot} \dfrac{x}{5}$ and $\alpha = \operatorname{arccot} x$. So,

$$\beta = \operatorname{arccot} \frac{x}{5} - \operatorname{arccot} x.$$

Differentiating β with respect to x produces

$$\frac{d\beta}{dx} = \frac{-1/5}{1 + (x^2/25)} - \frac{-1}{1 + x^2}$$

$$= \frac{-5}{25 + x^2} + \frac{1}{1 + x^2}$$

$$= \frac{4(5 - x^2)}{(25 + x^2)(1 + x^2)}.$$

Because $d\beta/dx = 0$ when $x = \sqrt{5}$, you can conclude from the First Derivative Test that this distance yields a maximum value of β. So, the distance is $x \approx 2.236$ feet and the angle is $\beta \approx 0.7297$ radian $\approx 41.81°$.

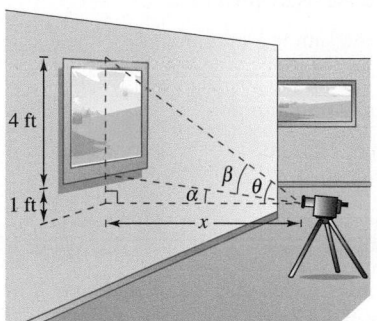

The camera should be 2.236 feet from the painting to maximize the angle β.
Figure 4.60

EXAMPLE 7   Finding a Maximum Revenue

The demand function for a product is modeled by

$$p = 56e^{-0.000012x} \qquad \text{Demand function}$$

where p is the price per unit (in dollars) and x is the number of units. What price will yield a maximum revenue?

Solution   The revenue function is given by

$$R = xp. \qquad \text{Revenue function}$$

Substituting for p (from the demand function) produces

$$R = 56xe^{-0.000012x}. \qquad \text{Primary equation}$$

The rate of change of revenue R with respect to the number of units sold x is called the *marginal revenue* and is given by

$$\frac{dR}{dx} = 56x(e^{-0.000012x})(-0.000012) + e^{-0.000012x}(56).$$

Setting the marginal revenue equal to zero,

$$56x(e^{-0.000012x})(-0.000012) + e^{-0.000012x}(56) = 0$$

and solving for x yields $x \approx 83{,}333$ units. From this, you can conclude that the maximum revenue occurs when the price is

$$p = 56e^{-0.000012(83,333)} = \$20.60.$$

So, a price of \$20.60 will yield a maximum revenue (see Figure 4.61).

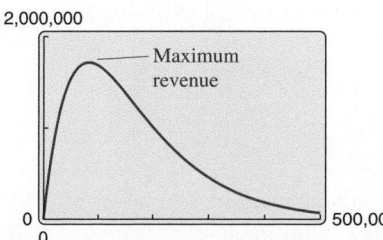

Figure 4.61

4.7 Exercises

CONCEPT CHECK

1. **Writing** In your own words, describe *primary equation*, *secondary equation*, and *feasible domain*.

2. **Optimization Problems** In your own words, describe the guidelines for solving applied minimum and maximum problems.

3. **Numerical, Graphical, and Analytic Analysis** Find two positive numbers whose sum is 110 and whose product is a maximum.

 (a) Analytically complete six rows of a table such as the one below. (The first two rows are shown.) Use the table to guess the maximum product.

First Number, x	Second Number	Product, P
10	$110 - 10$	$10(110 - 10) = 1000$
20	$110 - 20$	$20(110 - 20) = 1800$

 (b) Write the product P as a function of x.

 (c) Use calculus to find the critical number of the function in part (b). Then find the two numbers.

 (d) Use a graphing utility to graph the function in part (b) and verify the solution from the graph.

4. **Numerical, Graphical, and Analytic Analysis** An open box of maximum volume is to be made from a square piece of material, 24 inches on a side, by cutting equal squares from the corners and turning up the sides (see figure).

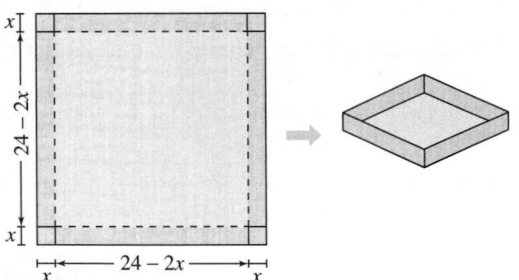

 (a) Analytically complete six rows of a table such as the one below. (The first two rows are shown.) Use the table to guess the maximum volume.

Height, x	Length and Width	Volume, V
1	$24 - 2(1)$	$1[24 - 2(1)]^2 = 484$
2	$24 - 2(2)$	$2[24 - 2(2)]^2 = 800$

 (b) Write the volume V as a function of x.

 (c) Use calculus to find the critical number of the function in part (b). Then find the maximum volume.

 (d) Use a graphing utility to graph the function in part (b) and verify the maximum volume from the graph.

 **Finding Numbers** In Exercises 5–10, find two positive numbers that satisfy the given requirements.

5. The sum is S and the product is a maximum.

6. The product is 185 and the sum is a minimum.

7. The product is 147 and the sum of the first number plus three times the second number is a minimum.

8. The sum of the first number squared and the second number is 54 and the product is a maximum.

9. The sum of the first number and twice the second number is 108 and the product is a maximum.

10. The sum of the first number cubed and the second number is 500 and the product is a maximum.

 **Maximum Area** In Exercises 11 and 12, find the length and width of a rectangle that has the given perimeter and a maximum area.

11. Perimeter: 80 meters   12. Perimeter: P units

 Minimum Perimeter In Exercises 13 and 14, find the length and width of a rectangle that has the given area and a minimum perimeter.

13. Area: 49 square feet   14. Area: A square centimeters

 **Minimum Distance** In Exercises 15 and 16, find the points on the graph of the function that are closest to the given point.

15. $y = x^2$, $(0, 3)$ 16. $y = x^2 - 2$, $(0, -1)$

17. **Minimum Area** A rectangular poster is to contain 648 square inches of print. The margins at the top and bottom of the poster are to be 2 inches, and the margins on the left and right are to be 1 inch. What should the dimensions of the poster be so that the least amount of poster is used?

18. **Minimum Area** A rectangular page is to contain 36 square inches of print. The margins on each side are to be $1\frac{1}{2}$ inches. Find the dimensions of the page such that the least amount of paper is used.

19. **Minimum Length** A farmer plans to fence a rectangular pasture adjacent to a river (see figure). The pasture must contain 405,000 square meters in order to provide enough grass for the herd. No fencing is needed along the river. What dimensions will require the least amount of fencing?

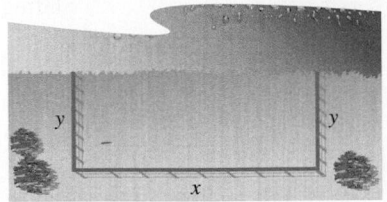

20. Maximum Volume A rectangular solid (with a square base) has a surface area of 337.5 square centimeters. Find the dimensions that will result in a solid with maximum volume.

21. Maximum Area A Norman window is constructed by adjoining a semicircle to the top of an ordinary rectangular window (see figure). Find the dimensions of a Norman window of maximum area when the total perimeter is 16 feet.

22. Maximum Area A rectangle is bounded by the x- and y-axes and the graph of $y = (6 - x)/2$ (see figure). What length and width should the rectangle have so that its area is a maximum?

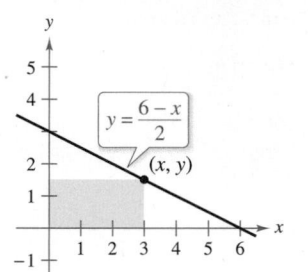

Figure for 22   Figure for 23

23. Minimum Length and Minimum Area A right triangle is formed in the first quadrant by the x- and y-axes and a line through the point $(1, 2)$ (see figure).

(a) Write the length L of the hypotenuse as a function of x.

(b) Use a graphing utility to approximate x graphically such that the length of the hypotenuse is a minimum.

(c) Find the vertices of the triangle such that its area is a minimum.

24. Maximum Area Find the area of the largest isosceles triangle that can be inscribed in a circle of radius 6 (see figure).

(a) Solve by writing the area as a function of h.

(b) Solve by writing the area as a function of α.

(c) Identify the type of triangle of maximum area.

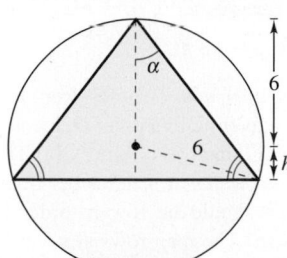

Figure for 24

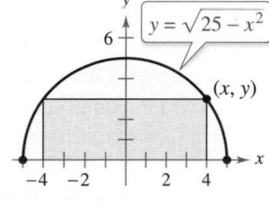

Figure for 25

25. Maximum Area A rectangle is bounded by the x-axis and the semicircle

$$y = \sqrt{25 - x^2}$$

(see figure). What length and width should the rectangle have so that its area is a maximum?

26. Maximum Area Find the dimensions of the largest rectangle that can be inscribed in a semicircle of radius r (see Exercise 25).

27. Numerical, Graphical, and Analytic Analysis An exercise room consists of a rectangle with a semicircle on each end. A 200-meter running track runs around the outside of the room.

(a) Draw a figure to represent the problem. Let x and y represent the length and width of the rectangle, respectively.

(b) Analytically complete six rows of a table such as the one below. (The first two rows are shown.) Use the table to guess the maximum area of the rectangular region.

Length, x	Width, y	Area, xy
10	$\frac{2}{\pi}(100 - 10)$	$(10)\frac{2}{\pi}(100 - 10) \approx 573$
20	$\frac{2}{\pi}(100 - 20)$	$(20)\frac{2}{\pi}(100 - 20) \approx 1019$

(c) Write the area A of the rectangular region as a function of x.

(d) Use calculus to find the critical number of the function in part (c). Then find the maximum area and the dimensions that yield the maximum area.

(e) Use a graphing utility to graph the function in part (c) and verify the maximum area from the graph.

28. Numerical, Graphical, and Analytic Analysis A right circular cylinder is designed to hold 22 cubic inches of a soft drink (approximately 12 fluid ounces).

(a) Analytically complete six rows of a table such as the one below. (The first two rows are shown.)

Radius, r	Height	Surface Area, S
0.2	$\dfrac{22}{\pi(0.2)^2}$	$2\pi(0.2)\left[0.2 + \dfrac{22}{\pi(0.2)^2}\right] \approx 220.3$
0.4	$\dfrac{22}{\pi(0.4)^2}$	$2\pi(0.4)\left[0.4 + \dfrac{22}{\pi(0.4)^2}\right] \approx 111.0$

(b) Use a graphing utility to generate additional rows of the table. Use the table to estimate the minimum surface area.

(c) Write the surface area S as a function of r.

(d) Use calculus to find the critical number of the function in part (c). Then find the minimum surface area and the dimensions that yield the minimum surface area.

(e) Use a graphing utility to graph the function in part (c) and verify the minimum surface area from the graph.

29. Maximum Volume A rectangular package to be sent by a postal service can have a maximum combined length and girth (perimeter of a cross section) of 108 inches (see figure). Find the dimensions of the package of maximum volume that can be sent. (Assume the cross section is square.)

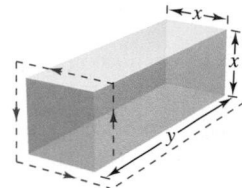

30. Maximum Volume Rework Exercise 29 for a cylindrical package. (The cross section is circular.)

EXPLORING CONCEPTS

31. Surface Area and Volume A shampoo bottle is a right circular cylinder. Because the surface area of the bottle does not change when it is squeezed, is it true that the volume remains the same? Explain.

32. Area and Perimeter The perimeter of a rectangle is 20 feet. Of all possible dimensions, the maximum area is 25 square feet when its length and width are both 5 feet. Are there dimensions that yield a minimum area? Explain.

---

33. Minimum Surface Area A solid is formed by adjoining two hemispheres to the ends of a right circular cylinder. The total volume of the solid is 14 cubic centimeters. Find the radius of the cylinder that produces the minimum surface area.

34. Minimum Cost An industrial tank of the shape described in Exercise 33 must have a volume of 4000 cubic feet. The hemispherical ends cost twice as much per square foot of surface area as the sides. Find the dimensions that will minimize cost.

35. Minimum Area The sum of the perimeters of an equilateral triangle and a square is 10. Find the dimensions of the triangle and the square that produce a minimum total area.

36. Maximum Area Twenty feet of wire is to be used to form two figures. In each of the following cases, how much wire should be used for each figure so that the total enclosed area is maximum?

(a) Equilateral triangle and square

(b) Square and regular pentagon

(c) Regular pentagon and regular hexagon

(d) Regular hexagon and circle

What can you conclude from this pattern? {*Hint:* The area of a regular polygon with n sides of length x is $A = (n/4)[\cot(\pi/n)]x^2$.}

37. Beam Strength A wooden beam has a rectangular cross section of height h and width w (see figure). The strength S of the beam is directly proportional to the width and the square of the height. What are the dimensions of the strongest beam that can be cut from a round log of diameter 20 inches? (*Hint:* $S = kh^2w$, where k is the proportionality constant.)

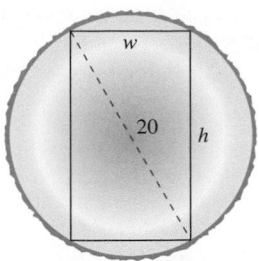

Figure for 37

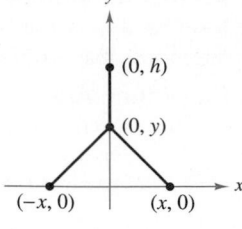

Figure for 38

38. Minimum Length Two factories are located at the coordinates $(-x, 0)$ and $(x, 0)$, and their power supply is at $(0, h)$, as shown in the figure. Find y such that the total length of power line from the power supply to the factories is a minimum.

39. Minimum Cost An offshore oil well is 2 kilometers off the coast. The refinery is 4 kilometers down the coast. Laying pipe in the ocean is twice as expensive as laying it on land. What path should the pipe follow in order to minimize the cost?

40. Illumination A light source is located over the center of a circular table of diameter 4 feet (see figure). Find the height h of the light source such that the illumination I at the perimeter of the table is maximum when

$$I = \frac{k \sin \alpha}{s^2}$$

where s is the slant height, α is the angle at which the light strikes the table, and k is a constant.

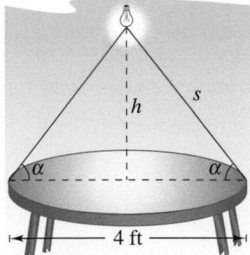

Figure for 40

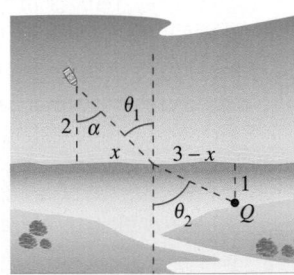

Figure for 41

41. Minimum Time A man is in a boat 2 miles from the nearest point on the coast. He is traveling to a point Q, located 3 miles down the coast and 1 mile inland (see figure). (a) The man rows at 2 miles per hour and walks at 4 miles per hour. Toward what point on the coast should he row in order to reach point Q in the least time? (b) The man rows at v_1 miles per hour and walks at v_2 miles per hour. Let θ_1 and θ_2 be the magnitudes of the angles. Show that the man will reach point Q in the least time when $(\sin \theta_1)/v_1 = (\sin \theta_2)/v_2$.

42. Population Growth Fifty elk are introduced into a game preserve. It is estimated that their population will increase according to the model $p(t) = 250/(1 + 4e^{-t/3})$, where t is measured in years. At what rate is the population increasing when $t = 2$? After how many years is the population increasing most rapidly?

43. Minimum Distance Sketch the graph of

$$f(x) = 2 - 2 \sin x$$

on the interval $[0, \pi/2]$.

(a) Find the distance from the origin to the y-intercept and the distance from the origin to the x-intercept.

(b) Write the distance d from the origin to a point on the graph of f as a function of x.

(c) Use calculus to find the value of x that minimizes the function d on the interval $[0, \pi/2]$. What is the minimum distance? Use a graphing utility to verify your results.

(Submitted by Tim Chapell, Penn Valley Community College, Kansas City, MO)

44. Minimum Time When light waves traveling in a transparent medium strike the surface of a second transparent medium, they change direction. This change of direction is called *refraction* and is defined by **Snell's Law of Refraction,**

$$\frac{\sin \theta_1}{v_1} = \frac{\sin \theta_2}{v_2}$$

where θ_1 and θ_2 are the magnitudes of the angles shown in the figure and v_1 and v_2 are the velocities of light in the two media. Show that this problem is equivalent to that in Exercise 41(b), and that light waves traveling from P to Q follow the path of minimum time.

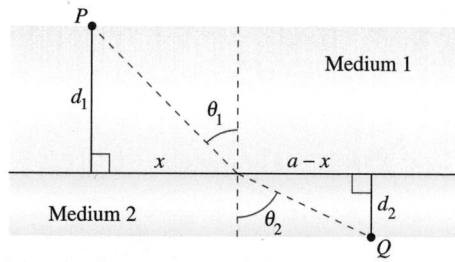

45. Maximum Volume A sector with central angle θ is cut from a circle of radius 12 inches (see figure), and the edges of the sector are brought together to form a cone. Find the magnitude of θ such that the volume of the cone is a maximum.

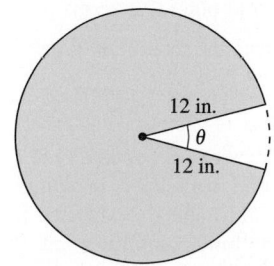

Figure for 45

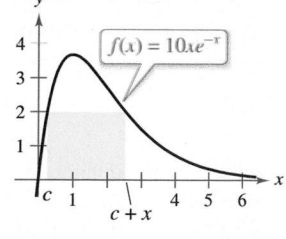

Figure for 46

46. Area Perform the following steps to find the maximum area of the rectangle shown in the figure.

(a) Solve for c in the equation $f(c) = f(c + x)$.

(b) Use the result in part (a) to write the area A as a function of x. [*Hint*: $A = xf(c)$]

(c) Use a graphing utility to graph the area function. Use the graph to approximate the dimensions of the rectangle of maximum area. Determine the required area.

(d) Use a graphing utility to graph the expression for c found in part (a). Use the graph to approximate

$$\lim_{x \to 0^+} c \quad \text{and} \quad \lim_{x \to \infty} c.$$

Use this result to describe the changes in the dimensions and position of the rectangle for $0 < x < \infty$.

47. Maximum Profit Assume that the amount of money deposited in a bank is proportional to the square of the interest rate the bank pays on this money. Furthermore, the bank can reinvest this money at 8%. Find the interest rate the bank should pay to maximize profit. (Use the simple interest formula.)

48. HOW DO YOU SEE IT? The graph shows the profit P (in thousands of dollars) of a company in terms of its advertising cost x (in thousands of dollars).

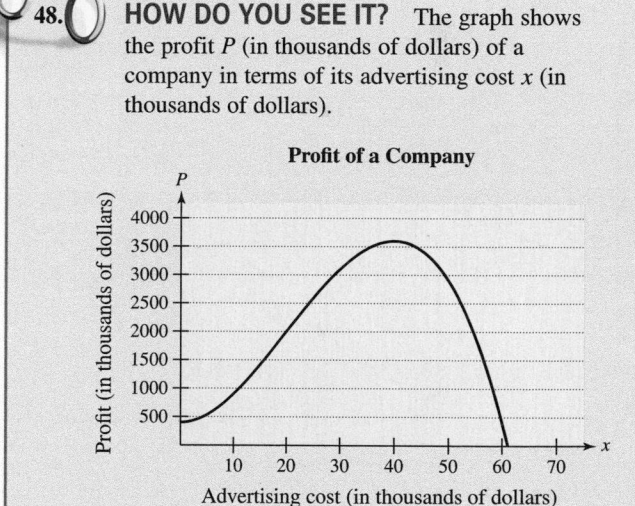

Profit of a Company

(a) Estimate the interval on which the profit is increasing.

(b) Estimate the interval on which the profit is decreasing.

(c) Estimate the amount of money the company should spend on advertising in order to yield a maximum profit.

(d) The *point of diminishing returns* is the point at which the rate of growth of the profit function begins to decline. Estimate the point of diminishing returns.

49. Maximum Rate Verify that the function

$$y = \frac{L}{1 + ae^{-x/b}}, \quad a > 0, \, b > 0, \, L > 0$$

increases at the maximum rate when $y = L/2$.

50. Area Find the area of the largest rectangle that can be inscribed under the curve $y = e^{-x^2}$ in the first and second quadrants.

Minimum Distance In Exercises 51–53, consider a fuel distribution center located at the origin of the rectangular coordinate system (units in miles; see figures). The center supplies three factories with coordinates $(4, 1)$, $(5, 6)$, and $(10, 3)$. A trunk line will run from the distribution center along the line $y = mx$, and feeder lines will run to the three factories. The objective is to find m such that the lengths of the feeder lines are minimized.

51. Minimize the sum of the squares of the lengths of the vertical feeder lines (see figure) given by

$$S_1 = (4m - 1)^2 + (5m - 6)^2 + (10m - 3)^2.$$

Find the equation of the trunk line by this method and then determine the sum of the lengths of the feeder lines.

 52. Minimize the sum of the absolute values of the lengths of the vertical feeder lines (see figure) given by

$$S_2 = |4m - 1| + |5m - 6| + |10m - 3|.$$

Find the equation of the trunk line by this method and then determine the sum of the lengths of the feeder lines. (*Hint:* Use a graphing utility to graph the function S_2 and approximate the required critical number.)

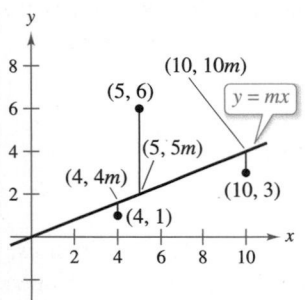

Figure for 51 and 52   Figure for 53

53. Minimize the sum of the lengths of the perpendicular feeder lines (see figure above and Exercise 77 in Section 1.2) from the trunk line to the factories given by

$$S_3 = \frac{|4m - 1|}{\sqrt{m^2 + 1}} + \frac{|5m - 6|}{\sqrt{m^2 + 1}} + \frac{|10m - 3|}{\sqrt{m^2 + 1}}.$$

Find the equation of the trunk line by this method and then determine the sum of the lengths of the feeder lines. (*Hint:* Use a graphing utility to graph the function S_3 and approximate the required critical number.)

54. **Maximum Area** Consider a symmetric cross inscribed in a circle of radius r (see figure).

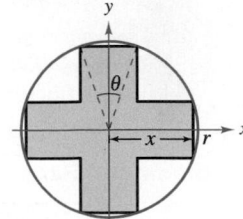

(a) Write the area A of the cross as a function of x and find the value of x that maximizes the area.

(b) Write the area A of the cross as a function of θ and find the value of θ that maximizes the area.

(c) Show that the critical numbers of parts (a) and (b) yield the same maximum area. What is that area?

55. **Minimum Distance** Find the point on the graph of the equation

$$16x = y^2$$

that is closest to the point $(6, 0)$.

56. **Minimum Distance** Find the point on the graph of the function

$$x = \sqrt{10y}$$

that is closest to the point $(0, 4)$. (*Hint:* Consider the domain of the function.)

PUTNAM EXAM CHALLENGE

57. Find, with explanation, the maximum value of $f(x) = x^3 - 3x$ on the set of all real numbers x satisfying $x^4 + 36 \le 13x^2$.

58. Find the minimum value of

$$\frac{(x + 1/x)^6 - (x^6 + 1/x^6) - 2}{(x + 1/x)^3 + (x^3 + 1/x^3)} \quad \text{for} \quad x > 0.$$

These problems were composed by the Committee on the Putnam Prize Competition. © The Mathematical Association of America. All rights reserved.

SECTION PROJECT

Minimum Time

A woman is at point A on the shore of a circular lake of radius 2 kilometers (see figure). She wants to walk around the lake to point B and then swim to point C in the least amount of time. Point C lies on the diameter through point A. Assume that she can walk at v_1 kilometers per hour and swim at v_2 kilometers per hour, and that $0 \le \theta \le \pi$.

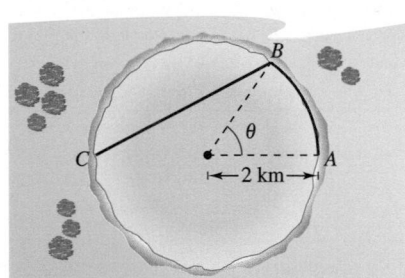

(a) Find the distance walked from point A to point B in terms of θ.

(b) Find the distance swam from point B to point C in terms of θ.

(c) Write the function $f(\theta)$ that represents the total time to move from point A to point C.

(d) Find $f'(\theta)$.

(e) If $v_1 = 5$ and $v_2 = 2$, approximate the critical number(s) of f. Does the critical number(s) correspond to a relative maximum or a relative minimum? Where should point B be located in order to minimize the time for the trip from point A to point C? Explain.

(f) Repeat part (e) for $v_1 = 3$ and $v_2 = 2$.

4.8 Differentials

■ Understand the concept of a tangent line approximation.
■ Compare the value of the differential, dy, with the actual change in y, Δy.
■ Estimate a propagated error using a differential.
■ Find the differential of a function using differentiation formulas.

Tangent Line Approximations

Newton's Method (see Section 3.8) is an example of the use of a tangent line to approximate the graph of a function. In this section, you will study other situations in which the graph of a function can be approximated by a straight line.

To begin, consider a function f that is differentiable at c. The equation for the tangent line at the point $(c, f(c))$ is

$$y - f(c) = f'(c)(x - c)$$

$$\boxed{y = f(c) + f'(c)(x - c)}$$

and is called the **tangent line approximation** (or **linear approximation**) **of** f **at** c. Because c is a constant, y is a linear function of x. Moreover, by restricting the values of x to those sufficiently close to c, the values of y can be used as approximations (to any desired degree of accuracy) of the values of the function f. In other words, as x approaches c, the limit of y is $f(c)$.

EXAMPLE 1 **Using a Tangent Line Approximation**

•••▷ *See LarsonCalculus.com for an interactive version of this type of example.*

Find the tangent line approximation of $f(x) = 1 + \sin x$ at the point $(0, 1)$. Then use a table to compare the y-values of the linear function with those of $f(x)$ on an open interval containing $x = 0$.

Solution The derivative of f is

$$f'(x) = \cos x. \qquad \text{First derivative}$$

So, the equation of the tangent line to the graph of f at the point $(0, 1)$ is

$$y = f(0) + f'(0)(x - 0)$$
$$y = 1 + (1)(x - 0)$$
$$y = 1 + x. \qquad \text{Tangent line approximation}$$

The table compares the values of y given by this linear approximation with the values of $f(x)$ near $x = 0$. Notice that the closer x is to 0, the better the approximation. This conclusion is reinforced by the graph shown in Figure 4.62.

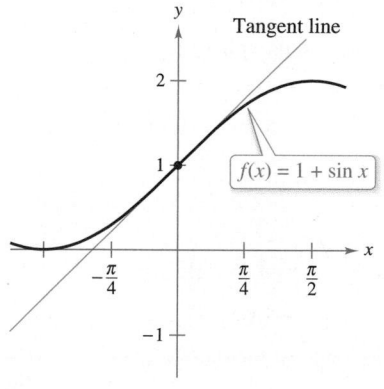

The tangent line approximation of f at the point $(0, 1)$

Figure 4.62

x	-0.5	-0.1	-0.01	0	0.01	0.1	0.5
$f(x) = 1 + \sin x$	0.521	0.9002	0.9900002	1	1.0099998	1.0998	1.479
$y = 1 + x$	0.5	0.9	0.99	1	1.01	1.1	1.5

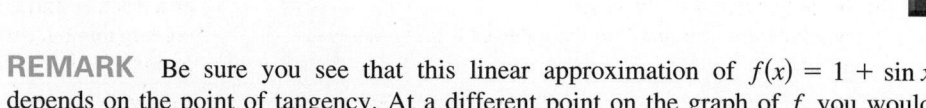

REMARK Be sure you see that this linear approximation of $f(x) = 1 + \sin x$ depends on the point of tangency. At a different point on the graph of f, you would obtain a different tangent line approximation.

Exploration

Tangent Line Approximation
Use a graphing utility to graph $f(x) = x^2$. In the same viewing window, graph the tangent line to the graph of f at the point $(1, 1)$. Zoom in twice on the point of tangency. Does your graphing utility distinguish between the two graphs? Use the *trace* feature to compare the two graphs. As the x-values get closer to 1, what can you say about the y-values?

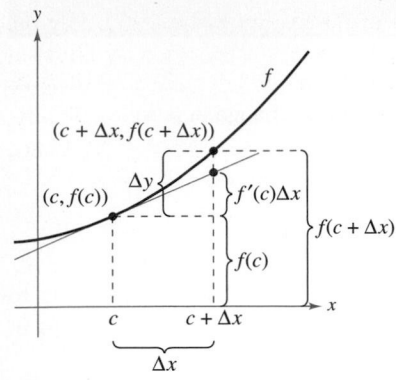

When Δx is small,
$\Delta y = f(c + \Delta x) - f(c)$ is
approximated by $f'(c)\Delta x$.

Figure 4.63

Differentials

When the tangent line to the graph of f at the point $(c, f(c))$

$$y = f(c) + f'(c)(x - c)$$ Tangent line at $(c, f(c))$

is used as an approximation of the graph of f, the quantity $x - c$ is called the change in x, and is denoted by Δx, as shown in Figure 4.63. When Δx is small, the change in y (denoted by Δy) can be approximated as shown.

$$\Delta y = f(c + \Delta x) - f(c)$$ Actual change in y

$$\approx f'(c)\Delta x$$ Approximate change in y

For such an approximation, the quantity Δx is traditionally denoted by dx and is called the **differential of x.** The expression $f'(x)\, dx$ is denoted by dy and is called the **differential of y.**

Definition of Differentials

Let $y = f(x)$ represent a function that is differentiable on an open interval containing x. The **differential of x** (denoted by dx) is any nonzero real number. The **differential of y** (denoted by dy) is

$$dy = f'(x)\, dx.$$

In many types of applications, the differential of y can be used as an approximation of the change in y. That is,

$$\Delta y \approx dy \quad \text{or} \quad \Delta y \approx f'(x)\, dx.$$

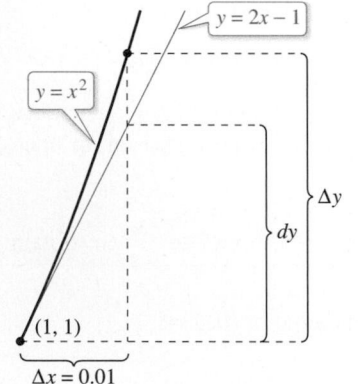

The change in y, Δy, is approximated by the differential of y, dy.

Figure 4.64

EXAMPLE 2   **Comparing Δy and dy**

Let $y = x^2$. Find dy when $x = 1$ and $dx = 0.01$. Compare this value with Δy for $x = 1$ and $\Delta x = 0.01$.

Solution   Because $y = f(x) = x^2$, you have $f'(x) = 2x$, and the differential dy is

$$dy = f'(x)\, dx = f'(1)(0.01) = 2(0.01) = 0.02.$$ Differential of y

Now, using $\Delta x = 0.01$, the change in y is

$$\Delta y = f(x + \Delta x) - f(x) = f(1.01) - f(1) = (1.01)^2 - 1^2 = 0.0201.$$

Figure 4.64 shows the geometric comparison of dy and Δy. Try comparing other values of dy and Δy. You will see that the values become closer to each other as dx (or Δx) approaches 0.

In Example 2, the tangent line to the graph of $f(x) = x^2$ at $x = 1$ is

$$y = 2x - 1.$$ Tangent line to the graph of f at $x = 1$.

For x-values near 1, this line is close to the graph of f, as shown in Figure 4.64 and in the table.

x	0.5	0.9	0.99	1	1.01	1.1	1.5
$f(x) = x^2$	0.25	0.81	0.9801	1	1.0201	1.21	2.25
$y = 2x - 1$	0	0.8	0.98	1	1.02	1.2	2

Error Propagation

Physicists and engineers tend to make liberal use of the approximation of Δy by dy. One way this occurs in practice is in the estimation of errors propagated by physical measuring devices. For example, if you let x represent the measured value of a variable and let $x + \Delta x$ represent the exact value, then Δx is the *error in measurement*. Finally, if the measured value x is used to compute another value $f(x)$, then the difference between $f(x + \Delta x)$ and $f(x)$ is the **propagated error.**

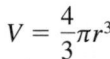

$$\underbrace{f(\overbrace{x + \Delta x}^{\text{Measurement error}})}_{\text{Exact value}} - \underbrace{f(x)}_{\text{Measured value}} = \overbrace{\Delta y}^{\text{Propagated error}}$$

EXAMPLE 3   **Estimation of Error**

The measured radius of a ball bearing is 0.7 inch, as shown in the figure. The measurement is correct to within 0.01 inch. Estimate the propagated error in the volume V of the ball bearing.

Solution   The formula for the volume of a sphere is

$$V = \frac{4}{3}\pi r^3$$

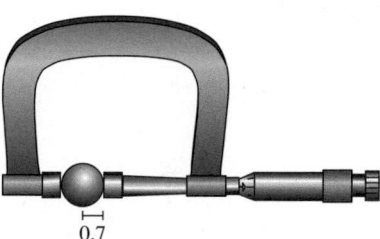

Ball bearing with measured radius that is correct to within 0.01 inch.

where r is the radius of the sphere. So, you can write

$$r = 0.7 \qquad \text{Measured radius}$$

and

$$-0.01 \le \Delta r \le 0.01. \qquad \text{Possible error}$$

To approximate the propagated error in the volume, differentiate V to obtain $dV/dr = 4\pi r^2$ and write

$$\begin{aligned} \Delta V &\approx dV &&\text{Approximate } \Delta V \text{ by } dV.\\ &= 4\pi r^2\, dr \\ &= 4\pi(0.7)^2(\pm0.01) &&\text{Substitute for } r \text{ and } dr.\\ &\approx \pm0.06158 \text{ cubic inch.} \end{aligned}$$

So, the volume has a propagated error of about 0.06 cubic inch. ■

Would you say that the propagated error in Example 3 is large or small? The answer is best given in *relative* terms by comparing dV with V. The ratio

$$\begin{aligned} \frac{dV}{V} &= \frac{4\pi r^2\, dr}{\frac{4}{3}\pi r^3} &&\text{Ratio of } dV \text{ to } V\\ &= \frac{3\, dr}{r} &&\text{Simplify.}\\ &\approx \frac{3(\pm0.01)}{0.7} &&\text{Substitute for } dr \text{ and } r.\\ &\approx \pm0.0429 \end{aligned}$$

is called the **relative error.** The corresponding **percent error** is approximately 4.29%.

Ball bearings are used to reduce friction between moving machine parts.

Calculating Differentials

Each of the differentiation rules that you studied in Chapter 3 can be written in **differential form.** For example, let u and v be differentiable functions of x. By the definition of differentials, you have

$$du = u' \, dx$$

and

$$dv = v' \, dx.$$

So, you can write the differential form of the Product Rule as shown below.

$$d[uv] = \frac{d}{dx}[uv]\, dx \qquad \text{Differential of } uv$$

$$= [uv' + vu']\, dx \qquad \text{Product Rule}$$

$$= uv'\, dx + vu'\, dx$$

$$= u\, dv + v\, du$$

Differential Formulas

Let u and v be differentiable functions of x.

Constant multiple: $\quad d[cu] = c\, du$

Sum or difference: $\quad d[u \pm v] = du \pm dv$

Product: $\quad d[uv] = u\, dv + v\, du$

Quotient: $\quad d\left[\dfrac{u}{v}\right] = \dfrac{v\, du - u\, dv}{v^2}$

EXAMPLE 4   **Finding Differentials**

Function	Derivative	Differential
a. $y = x^2$	$\dfrac{dy}{dx} = 2x$	$dy = 2x\, dx$
b. $y = \sqrt{x}$	$\dfrac{dy}{dx} = \dfrac{1}{2\sqrt{x}}$	$dy = \dfrac{dx}{2\sqrt{x}}$
c. $y = 2\sin x$	$\dfrac{dy}{dx} = 2\cos x$	$dy = 2\cos x\, dx$
d. $y = xe^x$	$\dfrac{dy}{dx} = e^x(x+1)$	$dy = e^x(x+1)\, dx$
e. $y = \dfrac{1}{x}$	$\dfrac{dy}{dx} = -\dfrac{1}{x^2}$	$dy = -\dfrac{dx}{x^2}$

GOTTFRIED WILHELM LEIBNIZ
(1646–1716)

Both Leibniz and Newton are credited with creating calculus. It was Leibniz, however, who tried to broaden calculus by developing rules and formal notation. He often spent days choosing an appropriate notation for a new concept.
See LarsonCalculus.com to read more of this biography.

The notation in Example 4 is called the **Leibniz notation** for derivatives and differentials, named after the German mathematician Gottfried Wilhelm Leibniz. The beauty of this notation is that it provides an easy way to remember several important calculus formulas by making it seem as though the formulas were derived from algebraic manipulations of differentials. For instance, in Leibniz notation, the *Chain Rule*

$$\frac{dy}{dx} = \frac{dy}{du}\frac{du}{dx}$$

would appear to be true because the du's divide out. Even though this reasoning is *incorrect*, the notation does help one remember the Chain Rule.

EXAMPLE 5 **Finding the Differential of a Composite Function**

$$y = f(x) = \sin 3x \qquad \text{Original function}$$
$$f'(x) = 3 \cos 3x \qquad \text{Apply Chain Rule.}$$
$$dy = f'(x)\, dx = 3 \cos 3x\, dx \qquad \text{Differential form}$$

EXAMPLE 6 **Finding the Differential of a Composite Function**

$$y = f(x) = (x^2 + 1)^{1/2} \qquad \text{Original function}$$
$$f'(x) = \frac{1}{2}(x^2 + 1)^{-1/2}(2x) = \frac{x}{\sqrt{x^2 + 1}} \qquad \text{Apply Chain Rule.}$$
$$dy = f'(x)\, dx = \frac{x}{\sqrt{x^2 + 1}}\, dx \qquad \text{Differential form}$$

Differentials can be used to approximate function values. To do this for the function given by $y = f(x)$, use the formula

$$f(x + \Delta x) \approx f(x) + dy = f(x) + f'(x)\, dx$$

which is derived from the approximation

$$\Delta y = f(x + \Delta x) - f(x) \approx dy.$$

The key to using this formula is to choose a value for x that makes the calculations easier, as shown in Example 7.

> **REMARK** This formula is equivalent to the tangent line approximation given earlier in this section.

EXAMPLE 7 **Approximating Function Values**

Use differentials to approximate $\sqrt{16.5}$.

Solution Using $f(x) = \sqrt{x}$, you can write

$$f(x + \Delta x) \approx f(x) + f'(x)\, dx = \sqrt{x} + \frac{1}{2\sqrt{x}}\, dx.$$

Now, choosing $x = 16$ and $dx = 0.5$, you obtain the following approximation.

$$f(x + \Delta x) = \sqrt{16.5} \approx \sqrt{16} + \frac{1}{2\sqrt{16}}(0.5) = 4 + \left(\frac{1}{8}\right)\left(\frac{1}{2}\right) = 4.0625$$

So, $\sqrt{16.5} \approx 4.0625$.

The tangent line approximation to $f(x) = \sqrt{x}$ at $x = 16$ is the line $g(x) = \frac{1}{8}x + 2$. For x-values near 16, the graphs of f and g are close together, as shown in Figure 4.65. For instance,

$$f(16.5) = \sqrt{16.5} \approx 4.0620$$

and

$$g(16.5) = \frac{1}{8}(16.5) + 2 = 4.0625.$$

In fact, if you use a graphing utility to zoom in near the point of tangency $(16, 4)$, you will see that the two graphs appear to coincide. Notice also that as you move farther away from the point of tangency, the linear approximation becomes less accurate.

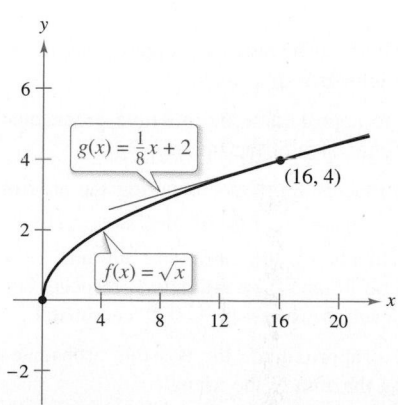

Figure 4.65

4.8 Exercises

See **CalcChat.com** for tutorial help and worked-out solutions to odd-numbered exercises.

CONCEPT CHECK

1. **Tangent Line Approximations** What is the equation of the tangent line approximation to the graph of a function f at the point $(c, f(c))$?

2. **Differentials** What do the differentials of x and y mean?

3. **Describing Terms** When using differentials, what is meant by the terms *propagated error, relative error,* and *percent error*?

4. **Finding Differentials** Explain how to find a differential of a function.

 **Using a Tangent Line Approximation** In Exercises 5–12, find the tangent line approximation T to the graph of f at the given point. Then complete the table.

x	1.9	1.99	2	2.01	2.1
$f(x)$					
$T(x)$					

5. $f(x) = x^2$, $(2, 4)$

6. $f(x) = \dfrac{6}{x^2}$, $\left(2, \dfrac{3}{2}\right)$

7. $f(x) = x^5$, $(2, 32)$

8. $f(x) = \sqrt{x}$, $(2, \sqrt{2})$

9. $f(x) = \sin x$, $(2, \sin 2)$

10. $f(x) = \csc x$, $(2, \csc 2)$

11. $f(x) = 3^x$, $(2, 9)$

12. $f(x) = \log_2 x$, $(2, 1)$

 **Verifying a Tangent Line Approximation** In Exercises 13 and 14, verify the tangent line approximation of the function at the given point. Then use a graphing utility to graph the function and its approximation in the same viewing window.

Function	Approximation	Point
13. $f(x) = \sqrt{x + 4}$	$y = 2 + \dfrac{x}{4}$	$(0, 2)$
14. $f(x) = \tan x$	$y = x$	$(0, 0)$

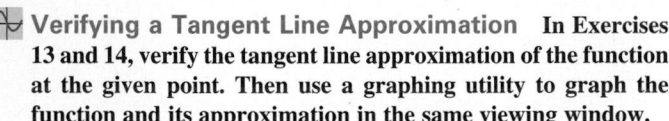 **Comparing Δy and dy** In Exercises 15–20, use the information to find and compare Δy and dy.

Function	x-Value	Differential of x
15. $y = 0.5x^3$	$x = 1$	$\Delta x = dx = 0.1$
16. $y = 6 - 2x^2$	$x = -2$	$\Delta x = dx = 0.1$
17. $y = x^4 + 1$	$x = -1$	$\Delta x = dx = 0.01$
18. $y = 2 - x^4$	$x = 2$	$\Delta x = dx = 0.01$
19. $y = x - 2x^3$	$x = 3$	$\Delta x = dx = 0.001$
20. $y = 7x^2 - 5x$	$x = -4$	$\Delta x = dx = 0.001$

 **Finding a Differential** In Exercises 21–32, find the differential dy of the given function.

21. $y = 3x^2 - 4$

22. $y = 3x^{2/3}$

23. $y = x \tan x$

24. $y = 3x - \sin^2 x$

25. $y = \dfrac{x + 1}{2x - 1}$

26. $y = \sqrt{x} + \dfrac{1}{\sqrt{x}}$

27. $y = \sqrt{9 - x^2}$

28. $y = x\sqrt{1 - x^2}$

29. $y = \ln\sqrt{4 - x^2}$

30. $y = e^{-0.5x} \cos 4x$

31. $y = x \arcsin x$

32. $y = \arctan(x - 2)$

Using Differentials In Exercises 33 and 34, use differentials and the graph of f to approximate (a) $f(1.9)$ and (b) $f(2.04)$. To print an enlarged copy of the graph, go to *MathGraphs.com*.

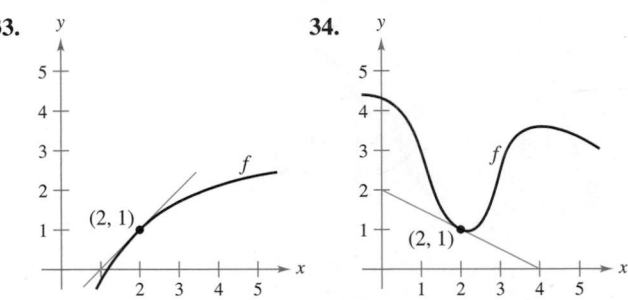

Using Differentials In Exercises 35 and 36, use differentials and the graph of g' to approximate (a) $g(2.93)$ and (b) $g(3.1)$ given that $g(3) = 8$.

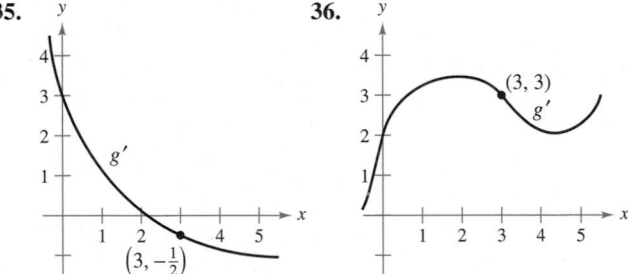

37. **Area** The measurement of the side of a square floor tile is 10 inches, with a possible error of $\frac{1}{32}$ inch.

(a) Use differentials to approximate the possible propagated error in computing the area of the square.

(b) Approximate the percent error in computing the area of the square.

38. **Area** The measurements of the base and altitude of a triangle are found to be 36 and 50 centimeters, respectively. The possible error in each measurement is 0.25 centimeter.

(a) Use differentials to approximate the possible propagated error in computing the area of the triangle.

(b) Approximate the percent error in computing the area of the triangle.

39. Volume and Surface Area The measurement of the edge of a cube is found to be 15 inches, with a possible error of 0.03 inch.

(a) Use differentials to approximate the possible propagated error in computing the volume of the cube.

(b) Use differentials to approximate the possible propagated error in computing the surface area of the cube.

(c) Approximate the percent errors in parts (a) and (b).

40. Volume and Surface Area The radius of a spherical balloon is measured as 8 inches, with a possible error of 0.02 inch.

(a) Use differentials to approximate the possible propagated error in computing the volume of the sphere.

(b) Use differentials to approximate the possible propagated error in computing the surface area of the sphere.

(c) Approximate the percent errors in parts (a) and (b).

41. Stopping Distance The total stopping distance T of a vehicle is

$$T = 2.5x + 0.5x^2$$

where T is in feet and x is the speed in miles per hour. Approximate the change and percent change in total stopping distance as speed changes from $x = 25$ to $x = 26$ miles per hour.

42. HOW DO YOU SEE IT? The graph shows the profit P (in dollars) from selling x units of an item. Use the graph to determine which is greater, the change in profit when the production level changes from 400 to 401 units or the change in profit when the production level changes from 900 to 901 units. Explain your reasoning.

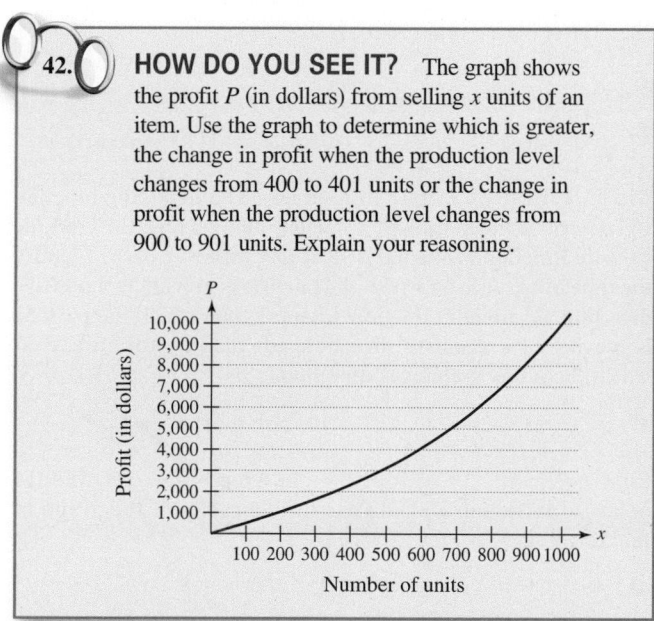

43. Pendulum The period of a pendulum is given by

$$T = 2\pi \sqrt{\frac{L}{g}}$$

where L is the length of the pendulum in feet, g is the acceleration due to gravity, and T is the time in seconds. The pendulum has been subjected to an increase in temperature such that the length has increased by $\frac{1}{2}\%$.

(a) Find the approximate percent change in the period.

(b) Using the result in part (a), find the approximate error in this pendulum clock in 1 day.

44. Ohm's Law A current of I amperes passes through a resistor of R ohms. **Ohm's Law** states that the voltage E applied to the resistor is

$$E = IR.$$

The voltage is constant. Show that the magnitude of the relative error in R caused by a change in I is equal in magnitude to the relative error in I.

45. Relative Humidity When the dewpoint is 65° Fahrenheit, the relative humidity H is modeled by

$$H = \frac{4347}{400,000,000} e^{369,444/(50t + 19,793)}$$

where t is the air temperature in degrees Fahrenheit. Use differentials to approximate the change in relative humidity when t is changed from 72°F to 73°F.

46. Surveying A surveyor standing 50 feet from the base of a large tree measures the angle of elevation to the top of the tree as 71.5°. How accurately must the angle be measured if the percent error in estimating the height of the tree is to be less than 6%?

 **Approximating Function Values In Exercises 47–50, use differentials to approximate the value of the expression. Compare your answer with that of a calculator.**

47. $\sqrt{99.4}$ **48.** $\sqrt[3]{26}$

49. $\sqrt[4]{624}$ **50.** $(2.99)^3$

EXPLORING CONCEPTS

51. Comparing Δy and dy Describe the change in accuracy of dy as an approximation for Δy when Δx approaches 0. Use a graph to support your answer.

52. Think About It For what value(s) of a would you use $y = x$ to approximate $f(x) = \sin x$ near $x = a$?

Using Differentials In Exercises 53 and 54, give a short explanation of why the approximation is valid.

53. $\sqrt{4.02} \approx 2 + \frac{1}{4}(0.02)$

54. $\tan 0.05 \approx 0 + 1(0.05)$

True or False? In Exercises 55–59, determine whether the statement is true or false. If it is false, explain why or give an example that shows it is false.

55. If $y = x + c$, then $dy = dx$.

56. If $y = ax + b$, then $\dfrac{\Delta y}{\Delta x} = \dfrac{dy}{dx}$.

57. If y is differentiable, then $\lim\limits_{\Delta x \to 0} (\Delta y - dy) = 0$.

58. If $y = f(x)$, f is increasing and differentiable, and $\Delta x > 0$, then $\Delta y \geq dy$.

59. The tangent line approximation at any point for any linear equation is the linear equation itself.

Review Exercises
See **CalcChat.com** for tutorial help and worked-out solutions to odd-numbered exercises.

Finding Extrema on a Closed Interval   In Exercises 1–8, find the absolute extrema of the function on the closed interval.

1. $f(x) = x^2 + 5x$, $[-4, 0]$ **2.** $f(x) = x^3 + 6x^2$, $[-6, 1]$

3. $f(x) = \sqrt{x} - 2$, $[0, 4]$ **4.** $h(x) = x - 3\sqrt{x}$, $[0, 9]$

5. $f(x) = \dfrac{4x}{x^2 + 9}$, $[-4, 4]$ **6.** $f(x) = \dfrac{x}{\sqrt{x^2 + 1}}$, $[0, 2]$

7. $g(x) = 2x + 5 \cos x$, $[0, 2\pi]$

8. $f(x) = \sin 2x$, $[0, 2\pi]$

Using Rolle's Theorem   In Exercises 9–12, determine whether Rolle's Theorem can be applied to f on the closed interval $[a, b]$. If Rolle's Theorem can be applied, find all values of c in the open interval (a, b) such that $f'(c) = 0$. If Rolle's Theorem cannot be applied, explain why not.

9. $f(x) = x^3 - 3x - 6$, $[-1, 2]$

10. $f(x) = (x - 2)(x + 3)^2$, $[-3, 2]$

11. $f(x) = \dfrac{x^2}{1 - x^2}$, $[-2, 2]$

12. $f(x) = \sin 2x$, $[-\pi, \pi]$

Using the Mean Value Theorem   In Exercises 13–18, determine whether the Mean Value Theorem can be applied to f on the closed interval $[a, b]$. If the Mean Value Theorem can be applied, find all values of c in the open interval (a, b) such that

$$f'(c) = \dfrac{f(b) - f(a)}{b - a}.$$

If the Mean Value Theorem cannot be applied, explain why not.

13. $f(x) = x^{2/3}$, $[1, 8]$

14. $f(x) = \dfrac{1}{x}$, $[1, 4]$

15. $f(x) = |5 - x|$, $[2, 6]$

16. $f(x) = 2x - 3\sqrt{x}$, $[-1, 1]$

17. $f(x) = x - \cos x$, $\left[-\dfrac{\pi}{2}, \dfrac{\pi}{2}\right]$

18. $f(x) = \ln(2x + 1)$, $[0, 3]$

19. Mean Value Theorem   Can the Mean Value Theorem be applied to the function

$$f(x) = \dfrac{1}{x^2}$$

on the interval $[-2, 1]$? Explain.

20. Using the Mean Value Theorem

(a) For the function $f(x) = Ax^2 + Bx + C$, determine the value of c guaranteed by the Mean Value Theorem on the interval $[x_1, x_2]$.

(b) Demonstrate the result of part (a) for $f(x) = 2x^2 - 3x + 1$ on the interval $[0, 4]$.

Intervals on Which a Function Is Increasing or Decreasing   In Exercises 21–28, find the open intervals on which the function is increasing or decreasing.

21. $f(x) = x^2 + 3x - 12$ **22.** $h(x) = (x + 2)^{1/3} + 8$

23. $f(x) = (x - 1)^2(2x - 5)$ **24.** $g(x) = (x + 1)^3$

25. $h(x) = \sqrt{x}(x - 3)$, $x > 0$

26. $f(x) = \sin x + \cos x$, $0 < x < 2\pi$

27. $f(t) = (2 - t)2^t$ **28.** $g(x) = 2x \ln x$

Applying the First Derivative Test   In Exercises 29–38, (a) find the critical numbers of f, if any, (b) find the open intervals on which the function is increasing or decreasing, (c) apply the First Derivative Test to identify all relative extrema, and (d) use a graphing utility to confirm your results.

29. $f(x) = x^2 - 6x + 5$ **30.** $f(x) = 4x^3 - 5x$

31. $f(t) = \dfrac{1}{4}t^4 - 8t$ **32.** $f(x) = \dfrac{x^3 - 8x}{4}$

33. $f(x) = \dfrac{x + 4}{x^2}$ **34.** $f(x) = \dfrac{x^2 - 3x - 4}{x - 2}$

35. $f(x) = \cos x - \sin x$, $(0, 2\pi)$

36. $f(x) = \dfrac{3}{2} \sin\left(\dfrac{\pi x}{2} - 1\right)$, $(0, 4)$

37. $f(x) = \ln x - x^2$ **38.** $f(x) = 2(x + \arccos x)$

Motion Along a Line   In Exercises 39 and 40, the function $s(t)$ describes the motion of a particle along a line. (a) Find the velocity function of the particle at any time $t \geq 0$. (b) Identify the time interval(s) on which the particle is moving in a positive direction. (c) Identify the time interval(s) on which the particle is moving in a negative direction. (d) Identify the time(s) at which the particle changes direction.

39. $s(t) = 3t - 2t^2$ **40.** $s(t) = 6t^3 - 8t + 3$

Finding Points of Inflection   In Exercises 41–48, find the points of inflection and discuss the concavity of the graph of the function.

41. $f(x) = x^3 - 9x^2$ **42.** $f(x) = 6x^4 - x^2$

43. $g(x) = x\sqrt{x + 5}$ **44.** $f(x) = 3x - 5x^3$

45. $f(x) = x + \cos x$, $[0, 2\pi]$

46. $f(x) = \tan \dfrac{x}{4}$, $(0, 2\pi)$

47. $f(x) = e^{4/x}$ **48.** $f(x) = 3x(3^x)$

Using the Second Derivative Test   In Exercises 49–56, find all relative extrema of the function. Use the Second Derivative Test where applicable.

49. $f(x) = (x + 9)^2$

50. $f(x) = x^4 - 2x^2 + 6$

51. $g(x) = 2x^2(1 - x^2)$ **52.** $h(t) = t - 4\sqrt{t + 1}$

53. $f(x) = 2x + \dfrac{18}{x}$

54. $h(x) = x - 2\cos x$, $[0, 4\pi]$

55. $g(x) = x \log_5 \dfrac{2}{x}$ **56.** $f(x) = \arctan\sqrt{5x + 2}$

Think About It   In Exercises 57 and 58, sketch the graph of a function f having the given characteristics.

57. $f(0) = f(6) = 0$

$f'(3) = f'(5) = 0$

$f'(x) > 0$ for $x < 3$

$f'(x) > 0$ for $3 < x < 5$

$f'(x) < 0$ for $x > 5$

$f''(x) < 0$ for $x < 3$ or $x > 4$

$f''(x) > 0$ for $3 < x < 4$

58. $f(0) = 4$, $f(6) = 0$

$f'(x) < 0$ for $x < 2$ or $x > 4$

$f'(2)$ does not exist.

$f'(4) = 0$

$f'(x) > 0$ for $2 < x < 4$

$f''(x) < 0$ for $x \neq 2$

59. Writing   A newspaper headline states that "The rate of growth of the national deficit is decreasing." What does this mean? What does it imply about the graph of the deficit as a function of time?

60. Inventory Cost   The cost of inventory C depends on the ordering and storage costs according to the inventory model

$$C = \left(\dfrac{Q}{x}\right)s + \left(\dfrac{x}{2}\right)r.$$

Determine the order size that will minimize the cost, assuming that sales occur at a constant rate, Q is the number of units sold per year, r is the cost of storing one unit for one year, s is the cost of placing an order, and x is the number of units per order.

 61. Modeling Data   Outlays for national defense D (in billions of dollars) for 2006 through 2014 are shown in the table, where t is the time in years, with $t = 6$ corresponding to 2006. *(Source: U.S. Office of Management and Budget)*

t	6	7	8	9	10
D	521.8	551.3	616.1	661.0	693.5

t	11	12	13	14
D	705.6	677.9	633.4	603.5

(a) Use the regression capabilities of a graphing utility to find a model of the form

$$D = at^4 + bt^3 + ct^2 + dt + e$$

for the data.

(b) Use a graphing utility to plot the data and graph the model.

(c) For the years shown in the table, when does the model indicate that the outlay for national defense was at a maximum? When was it at a minimum?

(d) For the years shown in the table, when does the model indicate that the outlay for national defense was increasing at the greatest rate?

 62. Modeling Data   The manager of a store recorded the annual sales S (in thousands of dollars) of a product over a period of 7 years, as shown in the table, where t is the time in years, with $t = 8$ corresponding to 2008.

t	8	9	10	11	12	13	14
S	8.1	7.3	7.8	9.2	11.3	12.8	12.9

(a) Use the regression capabilities of a graphing utility to find a model of the form

$$S = at^3 + bt^2 + ct + d$$

for the data.

(b) Use a graphing utility to plot the data and graph the model.

(c) Use calculus and the model to find the time t when sales were increasing at the greatest rate.

(d) Do you think the model would be accurate for predicting future sales? Explain.

Finding a Limit   In Exercises 63–74, find the limit, if it exists.

63. $\displaystyle\lim_{x \to \infty}\left(8 + \dfrac{1}{x}\right)$ **64.** $\displaystyle\lim_{x \to -\infty}\dfrac{1 - 4x}{x + 1}$

65. $\displaystyle\lim_{x \to \infty}\dfrac{x^2}{1 - 8x^2}$ **66.** $\displaystyle\lim_{x \to -\infty}\dfrac{9x^3 + 5}{7x^4}$

67. $\displaystyle\lim_{x \to -\infty}\dfrac{3x^2}{x + 5}$ **68.** $\displaystyle\lim_{x \to -\infty}\dfrac{\sqrt{x^2 + x}}{-2x}$

69. $\displaystyle\lim_{x \to \infty}\dfrac{5\cos x}{x}$ **70.** $\displaystyle\lim_{x \to \infty}\dfrac{x^3}{\sqrt{x^2 + 2}}$

71. $\displaystyle\lim_{x \to -\infty}\dfrac{6x}{x + \cos x}$ **72.** $\displaystyle\lim_{x \to -\infty}\dfrac{x}{2\sin x}$

73. $\displaystyle\lim_{x \to \infty}\left(e^{-x} + 4\ln\dfrac{x + 1}{x}\right)$ **74.** $\displaystyle\lim_{x \to \infty}\dfrac{-3}{1 + 6^{-6x}}$

Finding Horizontal Asymptotes Using Technology   In Exercises 75–78, use a graphing utility to graph the function and identify any horizontal asymptotes.

75. $f(x) = \dfrac{3}{x} + 4$ **76.** $f(x) = \dfrac{\sqrt{4x^2 - 1}}{8x + 1}$

77. $f(x) = \dfrac{5}{3 + 2e^{-x}}$ **78.** $h(x) = 10\ln\dfrac{x}{x + 1}$

79. Modeling Data   The average typing speeds S (in words per minute) of a typing student after t weeks of lessons are shown in the table.

t	5	10	15	20	25	30
S	28	56	79	90	93	94

A model for the data is $S = \dfrac{100t^2}{65 + t^2}$, $t > 0$.

(a) Use a graphing utility to plot the data and graph the model.

(b) Does there appear to be a limiting typing speed? Explain.

80. Using the Definition of Limits at Infinity Consider

$$\lim_{x \to \infty} \frac{2x}{\sqrt{x^2 + 5}}.$$

(a) Use the definition of limits at infinity to find the value of M that corresponds to $\varepsilon = 0.5$.

(b) Use the definition of limits at infinity to find the value of M that corresponds to $\varepsilon = 0.1$.

Analyzing the Graph of a Function In Exercises 81–92, analyze and sketch a graph of the function. Label any intercepts, relative extrema, points of inflection, and asymptotes. Use a graphing utility to verify your results.

81. $f(x) = 4x - x^2$

82. $f(x) = x^4 - 2x^2 + 6$

83. $f(x) = x\sqrt{16 - x^2}$

84. $f(x) = x^{1/3}(x + 3)^{2/3}$

85. $f(x) = \dfrac{5 - 3x}{x - 2}$

86. $f(x) = \dfrac{2x}{1 + x^2}$

87. $f(x) = x^3 + x + \dfrac{4}{x}$

88. $f(x) = x^2 + \dfrac{1}{x}$

89. $f(x) = \dfrac{1}{1 + \cos x}$, $(-\pi, \pi)$

90. $f(x) = \ln(x^2 - 3)$

91. $f(x) = e^{1 - x^3}$

92. $f(x) = \operatorname{arccot} x^2$

93. Finding Numbers Find two positive numbers such that the sum of twice the first number and three times the second number is 216 and the product is a maximum.

94. Minimum Distance Find the point on the graph of $f(x) = \sqrt{x}$ that is closest to the point $(6, 0)$.

95. Maximum Area A rancher has 400 feet of fencing with which to enclose two adjacent rectangular corrals (see figure). What dimensions should be used so that the enclosed area will be a maximum?

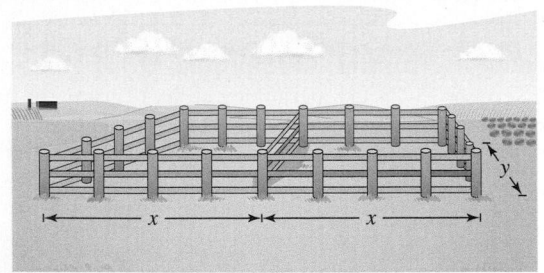

96. Maximum Area Find the dimensions of the rectangle of maximum area, with sides parallel to the coordinate axes, that can be inscribed in the ellipse given by

$$\frac{x^2}{144} + \frac{y^2}{16} = 1.$$

97. Minimum Length A right triangle in the first quadrant has the coordinate axes as sides, and the hypotenuse passes through the point $(1, 8)$. Find the vertices of the triangle such that the length of the hypotenuse is minimum.

98. Minimum Length The wall of a building is to be braced by a beam that must pass over a parallel fence 5 feet high and 4 feet from the building. Find the length of the shortest beam that can be used.

99. Maximum Length Find the length of the longest pipe that can be carried level around a right-angle corner at the intersection of two corridors of widths 4 feet and 6 feet.

100. Maximum Length A hallway of width 6 feet meets a hallway of width 9 feet at right angles. Find the length of the longest pipe that can be carried level around this corner. [*Hint:* If L is the length of the pipe, show that

$$L = 6 \csc \theta + 9 \csc\left(\frac{\pi}{2} - \theta\right)$$

where θ is the angle between the pipe and the wall of the narrower hallway.]

101. Maximum Volume Find the volume of the largest right circular cone that can be inscribed in a sphere of radius r.

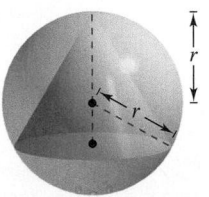

102. Maximum Volume Find the volume of the largest right circular cylinder that can be inscribed in a sphere of radius r.

Comparing Δy and dy In Exercises 103 and 104, use the information to find and compare Δy and dy.

	Function	x-Value	Differential of x
103.	$y = 4x^3$	$x = 2$	$\Delta x = dx = 0.1$
104.	$y = x^2 - 5x$	$x = -3$	$\Delta x = dx = 0.01$

Finding a Differential In Exercises 105 and 106, find the differential dy of the given function.

105. $y = x(1 - \cos x)$

106. $y = \sqrt{36 - x^2}$

107. Volume and Surface Area The radius of a sphere is measured as 9 centimeters, with a possible error of 0.025 centimeter.

(a) Use differentials to approximate the possible propagated error in computing the volume of the sphere.

(b) Use differentials to approximate the possible propagated error in computing the surface area of the sphere.

(c) Approximate the percent errors in parts (a) and (b).

108. Demand Function A company finds that the demand for its commodity is

$$p = 75 - \frac{1}{4}x$$

where p is the price in dollars and x is the number of units. Find and compare the values of Δp and dp as x changes from 7 to 8.

109. Profit The profit P for a company is $P = 100xe^{-x/400}$, where x is the number of units sold. Approximate the change and percent change in profit as sales increase from $x = 115$ to $x = 120$ units.

P.S. Problem Solving

See **CalcChat.com** for tutorial help and worked-out solutions to odd-numbered exercises.

1. Relative Extrema   Graph the fourth-degree polynomial

$$p(x) = x^4 + ax^2 + 1$$

for various values of the constant a.

(a) Determine the values of a for which p has exactly one relative minimum.

(b) Determine the values of a for which p has exactly one relative maximum.

(c) Determine the values of a for which p has exactly two relative minima.

(d) Show that the graph of p cannot have exactly two relative extrema.

2. Relative Extrema

(a) Graph the fourth-degree polynomial $p(x) = ax^4 - 6x^2$ for $a = -3, -2, -1, 0, 1, 2,$ and 3. For what values of the constant a does p have a relative minimum or relative maximum?

(b) Show that p has a relative maximum for all values of the constant a.

(c) Determine analytically the values of a for which p has a relative minimum.

(d) Let $(x, y) = (x, p(x))$ be a relative extremum of p. Show that (x, y) lies on the graph of $y = -3x^2$. Verify this result graphically by graphing $y = -3x^2$ together with the seven curves from part (a).

3. Relative Minimum   Let

$$f(x) = \frac{c}{x} + x^2.$$

Determine all values of the constant c such that f has a relative minimum, but no relative maximum.

4. Points of Inflection

(a) Let $f(x) = ax^2 + bx + c$, $a \neq 0$, be a quadratic polynomial. How many points of inflection does the graph of f have?

(b) Let $f(x) = ax^3 + bx^2 + cx + d$, $a \neq 0$, be a cubic polynomial. How many points of inflection does the graph of f have?

(c) Suppose the function $y = f(x)$ satisfies the equation

$$\frac{dy}{dx} = ky\left(1 - \frac{y}{L}\right)$$

where k and L are positive constants. Show that the graph of f has a point of inflection at the point where $y = L/2$. (This equation is called the **logistic differential equation.**)

5. Extended Mean Value Theorem   Prove the **Extended Mean Value Theorem:** If f and f' are continuous on the closed interval $[a, b]$, and if f'' exists in the open interval (a, b), then there exists a number c in (a, b) such that

$$f(b) = f(a) + f'(a)(b - a) + \frac{1}{2}f''(c)(b - a)^2.$$

6. Illumination   The amount of illumination of a surface is proportional to the intensity of the light source, inversely proportional to the square of the distance from the light source, and proportional to $\sin \theta$, where θ is the angle at which the light strikes the surface. A rectangular room measures 10 feet by 24 feet, with a 10-foot ceiling (see figure). Determine the height at which the light should be placed to allow the corners of the floor to receive as much light as possible.

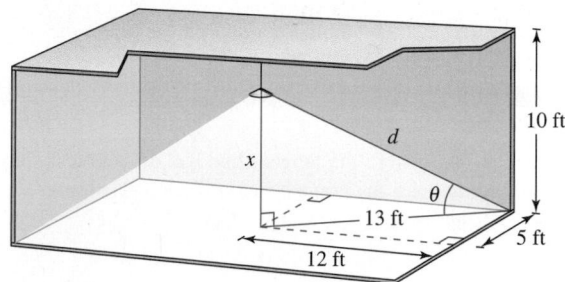

7. Minimum Distance   Consider a room in the shape of a cube, 4 meters on each side. A bug at point P wants to walk to point Q at the opposite corner, as shown in the figure. Use calculus to determine the shortest path. Explain how you can solve this problem without calculus. (*Hint:* Consider the two walls as one wall.)

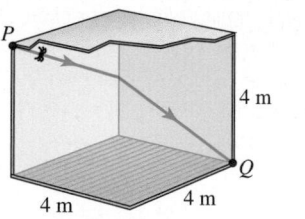

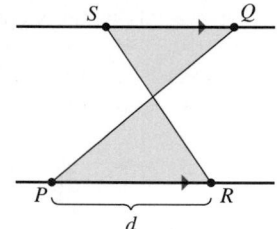

Figure for 7   Figure for 8

8. Areas of Triangles   The line joining P and Q crosses the two parallel lines, as shown in the figure. The point R is d units from P. How far from Q should the point S be positioned so that the sum of the areas of the two shaded triangles is a minimum? So that the sum is a maximum?

9. Mean Value Theorem   Determine the values a, b, and c such that the function f satisfies the hypotheses of the Mean Value Theorem on the interval $[0, 3]$.

$$f(x) = \begin{cases} 1, & x = 0 \\ ax + b, & 0 < x \leq 1 \\ x^2 + 4x + c, & 1 < x \leq 3 \end{cases}$$

10. Mean Value Theorem   Determine the values a, b, c, and d such that the function f satisfies the hypotheses of the Mean Value Theorem on the interval $[-1, 2]$.

$$f(x) = \begin{cases} a, & x = -1 \\ 2, & -1 < x \leq 0 \\ bx^2 + c, & 0 < x \leq 1 \\ dx + 4, & 1 < x \leq 2 \end{cases}$$

11. Proof Let f and g be functions that are continuous on $[a, b]$ and differentiable on (a, b). Prove that if $f(a) = g(a)$ and $g'(x) > f'(x)$ for all x in (a, b), then $g(b) > f(b)$.

12. Proof

(a) Prove that $\lim\limits_{x \to \infty} x^2 = \infty$.

(b) Prove that $\lim\limits_{x \to \infty} \dfrac{1}{x^2} = 0$.

(c) Let L be a real number. Prove that if $\lim\limits_{x \to \infty} f(x) = L$, then

$$\lim_{y \to 0^+} f\!\left(\frac{1}{y}\right) = L.$$

13. Tangent Lines Find the point on the graph of

$$y = \frac{1}{1 + x^2}$$

(see figure) where the tangent line has the greatest slope, and the point where the tangent line has the least slope.

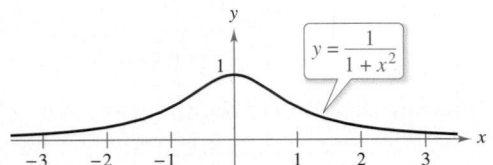

14. Stopping Distance The police department must determine the speed limit on a bridge such that the flow rate of cars is maximum per unit time. The greater the speed limit, the farther apart the cars must be in order to keep a safe stopping distance. Experimental data on the stopping distances d (in meters) for various speeds v (in kilometers per hour) are shown in the table.

v	20	40	60	80	100
d	5.1	13.7	27.2	44.2	66.4

(a) Convert the speeds v in the table to speeds s in meters per second. Use the regression capabilities of a graphing utility to find a model of the form $d(s) = as^2 + bs + c$ for the data.

(b) Consider two consecutive vehicles of average length 5.5 meters, traveling at a safe speed on the bridge. Let T be the difference between the times (in seconds) when the front bumpers of the vehicles pass a given point on the bridge. Verify that this difference in times is given by

$$T = \frac{d(s)}{s} + \frac{5.5}{s}.$$

(c) Use a graphing utility to graph the function T and estimate the speed s that minimizes the time between vehicles.

(d) Use calculus to determine the speed that minimizes T. What is the minimum value of T? Convert the required speed to kilometers per hour.

(e) Find the optimal distance between vehicles for the speed found in part (d).

15. Darboux's Theorem Prove **Darboux's Theorem:** Let f be differentiable on the closed interval $[a, b]$ such that $f'(a) = y_1$ and $f'(b) = y_2$. If d lies between y_1 and y_2, then there exists c in (a, b) such that $f'(c) = d$.

16. Maximum Area The figures show a rectangle, a circle, and a semicircle inscribed in a triangle bounded by the coordinate axes and the first-quadrant portion of the line with intercepts $(3, 0)$ and $(0, 4)$. Find the dimensions of each inscribed figure such that its area is maximum. State whether calculus was helpful in finding the required dimensions. Explain your reasoning.

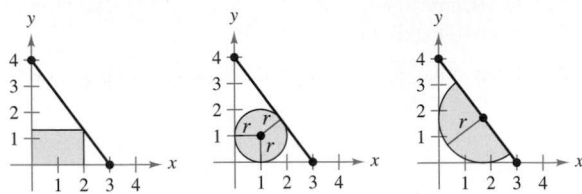

17. Point of Inflection Show that the cubic polynomial $p(x) = ax^3 + bx^2 + cx + d$ has exactly one point of inflection (x_0, y_0), where

$$x_0 = \frac{-b}{3a} \quad \text{and} \quad y_0 = \frac{2b^3}{27a^2} - \frac{bc}{3a} + d.$$

Use these formulas to find the point of inflection of $p(x) = x^3 - 3x^2 + 2$.

18. Minimum Length A legal-sized sheet of paper (8.5 inches by 14 inches) is folded so that corner P touches the opposite 14-inch edge at R (see figure). $\left(\textit{Note: } PQ = \sqrt{C^2 - x^2}.\right)$

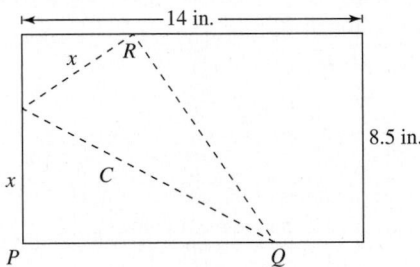

(a) Show that $C^2 = \dfrac{2x^3}{2x - 8.5}$.

(b) What is the domain of C?

(c) Determine the x-value that minimizes C.

(d) Determine the minimum length C.

19. Using a Function Let $f(x) = \sin(\ln x)$.

(a) Determine the domain of the function f.

(b) Find two values of x satisfying $f(x) = 1$.

(c) Find two values of x satisfying $f(x) = -1$.

(d) What is the range of the function f?

(e) Find $f'(x)$ and use calculus to find the maximum value of f on the interval $[1, 10]$.

(f) Use a graphing utility to graph f in the viewing window $0 \le x \le 5$, $-2 \le y \le 2$. Estimate $\lim\limits_{x \to 0^+} f(x)$, if it exists.

(g) Determine $\lim\limits_{x \to 0^+} f(x)$ analytically, if it exists.

5 Integration

Heat Transfer
(Exercise 93, p. 364)

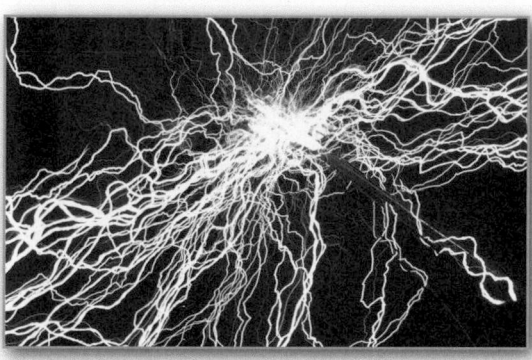

Electricity *(Exercise 106, p. 343)*

The Speed of Sound *(Example 5, p. 322)*

Amount of Chemical
Flowing into a Tank
(Example 9, p. 326)

Grand Canyon *(Exercise 62, p. 293)*

5.1 Antiderivatives and Indefinite Integration

■ Write the general solution of a differential equation and use indefinite integral notation for antiderivatives.
■ Use basic integration rules to find antiderivatives.
■ Find a particular solution of a differential equation.

Antiderivatives

To find a function F whose derivative is $f(x) = 3x^2$, you might use your knowledge of derivatives to conclude that

$$F(x) = x^3 \quad \text{because} \quad \frac{d}{dx}[x^3] = 3x^2.$$

The function F is an *antiderivative* of f.

> **Definition of Antiderivative**
>
> A function F is an **antiderivative** of f on an interval I when $F'(x) = f(x)$ for all x in I.

Note that F is called *an* antiderivative of f rather than *the* antiderivative of f. To see why, observe that

$$F_1(x) = x^3, \quad F_2(x) = x^3 - 5, \quad \text{and} \quad F_3(x) = x^3 + 97$$

are all antiderivatives of $f(x) = 3x^2$. In fact, for any constant C, the function $F(x) = x^3 + C$ is an antiderivative of f.

> **THEOREM 5.1 Representation of Antiderivatives**
>
> If F is an antiderivative of f on an interval I, then G is an antiderivative of f on the interval I if and only if G is of the form $G(x) = F(x) + C$ for all x in I, where C is a constant.

Proof The proof of Theorem 5.1 in one direction is straightforward. That is, if $G(x) = F(x) + C, F'(x) = f(x)$, and C is a constant, then

$$G'(x) = \frac{d}{dx}[F(x) + C] = F'(x) + 0 = f(x).$$

To prove this theorem in the other direction, assume that G is an antiderivative of f. Define a function H such that

$$H(x) = G(x) - F(x).$$

For any two points a and b $(a < b)$ in the interval, H is continuous on $[a, b]$ and differentiable on (a, b). By the Mean Value Theorem,

$$H'(c) = \frac{H(b) - H(a)}{b - a}$$

for some c in (a, b). However, $H'(c) = 0$, so $H(a) = H(b)$. Because a and b are arbitrary points in the interval, you know that H is a constant function C. So, $G(x) - F(x) = C$ and it follows that $G(x) = F(x) + C$. ∎

Using Theorem 5.1, you can represent the entire family of antiderivatives of a function by adding a constant to a *known* antiderivative. For example, knowing that

$$D_x[x^2] = 2x$$

you can represent the family of *all* antiderivatives of $f(x) = 2x$ by

$$G(x) = x^2 + C \qquad \text{Family of all antiderivatives of } f(x) = 2x$$

where C is a constant. The constant C is called the **constant of integration.** The family of functions represented by G is the **general antiderivative** of f, and $G(x) = x^2 + C$ is the **general solution** of the *differential equation*

$$G'(x) = 2x. \qquad \text{Differential equation}$$

A **differential equation** in x and y is an equation that involves x, y, and derivatives of y. For instance,

$$y' = 3x \quad \text{and} \quad y' = x^2 + 1$$

are examples of differential equations.

EXAMPLE 1   Solving a Differential Equation

Find the general solution of the differential equation $dy/dx = 2$.

Solution   To begin, you need to find a function whose derivative is 2. One such function is

$$y = 2x. \qquad 2x \text{ is } an \text{ antiderivative of } 2.$$

Now, you can use Theorem 5.1 to conclude that the general solution of the differential equation is

$$y = 2x + C. \qquad \text{General solution}$$

The graphs of several functions of the form $y = 2x + C$ are shown in Figure 5.1. ■

When solving a differential equation of the form

$$\frac{dy}{dx} = f(x)$$

it is convenient to write it in the equivalent differential form

$$dy = f(x)\, dx.$$

The operation of finding all solutions of this equation is called **antidifferentiation** (or **indefinite integration**) and is denoted by an integral sign $\int$. The general solution is denoted by

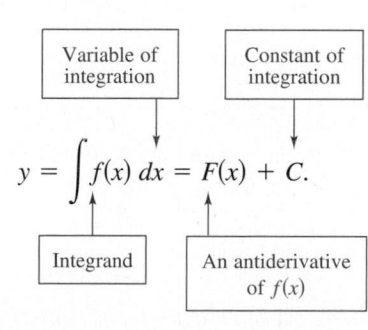

$$y = \int f(x)\, dx = F(x) + C.$$

The expression $\int f(x)\, dx$ is read as the *antiderivative of f with respect to x*. So, the differential dx serves to identify x as the variable of integration. The term **indefinite integral** is a synonym for antiderivative.

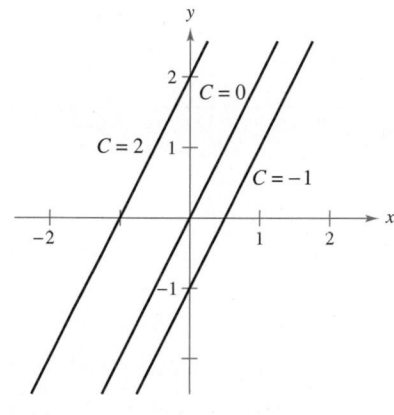

Functions of the form $y = 2x + C$
Figure 5.1

•• **REMARK**   In this text, the notation $\int f(x)\, dx = F(x) + C$ means that F is an antiderivative of f *on an interval.*

Basic Integration Rules

The inverse nature of integration and differentiation can be verified by substituting $F'(x)$ for $f(x)$ in the indefinite integration definition to obtain

$$\int F'(x)\, dx = F(x) + C.$$

Integration is the "inverse" of differentiation.

Moreover, if $\int f(x)\, dx = F(x) + C$, then

$$\frac{d}{dx}\left[\int f(x)\, dx\right] = f(x).$$

Differentiation is the "inverse" of integration.

REMARK The Power Rule for Integration has the restriction that $n \ne -1$. To evaluate $\int x^{-1}\, dx$, you must use the natural log rule. (See Exercise 77.)

These two equations allow you to obtain integration formulas directly from differentiation formulas, as shown in the following summary.

Basic Integration Rules

Differentiation Formula	Integration Formula
$\dfrac{d}{dx}[C] = 0$	$\displaystyle\int 0\, dx = C$
$\dfrac{d}{dx}[kx] = k$	$\displaystyle\int k\, dx = kx + C$
$\dfrac{d}{dx}[kf(x)] = kf'(x)$	$\displaystyle\int kf(x)\, dx = k\int f(x)\, dx$
$\dfrac{d}{dx}[f(x) \pm g(x)] = f'(x) \pm g'(x)$	$\displaystyle\int [f(x) \pm g(x)]\, dx = \int f(x)\, dx \pm \int g(x)\, dx$
$\dfrac{d}{dx}[x^n] = nx^{n-1}$	$\displaystyle\int x^n\, dx = \frac{x^{n+1}}{n+1} + C, \quad n \ne -1$ Power Rule
$\dfrac{d}{dx}[\sin x] = \cos x$	$\displaystyle\int \cos x\, dx = \sin x + C$
$\dfrac{d}{dx}[\cos x] = -\sin x$	$\displaystyle\int \sin x\, dx = -\cos x + C$
$\dfrac{d}{dx}[\tan x] = \sec^2 x$	$\displaystyle\int \sec^2 x\, dx = \tan x + C$
$\dfrac{d}{dx}[\sec x] = \sec x \tan x$	$\displaystyle\int \sec x \tan x\, dx = \sec x + C$
$\dfrac{d}{dx}[\cot x] = -\csc^2 x$	$\displaystyle\int \csc^2 x\, dx = -\cot x + C$
$\dfrac{d}{dx}[\csc x] = -\csc x \cot x$	$\displaystyle\int \csc x \cot x\, dx = -\csc x + C$
$\dfrac{d}{dx}[e^x] = e^x$	$\displaystyle\int e^x\, dx = e^x + C$
$\dfrac{d}{dx}[a^x] = (\ln a)a^x$	$\displaystyle\int a^x\, dx = \left(\frac{1}{\ln a}\right)a^x + C$
$\dfrac{d}{dx}[\ln x] = \dfrac{1}{x}, \quad x > 0$	$\displaystyle\int \frac{1}{x}\, dx = \ln|x| + C$

· · · · · · · · · · · · · · · · · ▷

REMARK In Example 2, note that the general pattern of integration is similar to that of differentiation.

Original integral

↓

Rewrite

↓

Integrate

↓

Simplify

EXAMPLE 2 **Describing Antiderivatives**

$$\int 3x \, dx = 3 \int x \, dx \qquad \text{Constant Multiple Rule}$$

$$= 3 \int x^1 \, dx \qquad \text{Rewrite } x \text{ as } x^1.$$

$$= 3\left(\frac{x^2}{2}\right) + C \qquad \text{Power Rule } (n = 1)$$

$$= \frac{3}{2}x^2 + C \qquad \text{Simplify.}$$

The antiderivatives of $3x$ are of the form $\frac{3}{2}x^2 + C$, where C is any constant. ■

When finding indefinite integrals, a strict application of the basic integration rules tends to produce complicated constants of integration. For instance, in Example 2, the solution could have been written as

$$\int 3x \, dx = 3 \int x \, dx = 3\left(\frac{x^2}{2} + C\right) = \frac{3}{2}x^2 + 3C.$$

Because C represents *any* constant, it is both cumbersome and unnecessary to write $3C$ as the constant of integration. So, $\frac{3}{2}x^2 + 3C$ is written in the simpler form $\frac{3}{2}x^2 + C$.

EXAMPLE 3 **Rewriting Before Integrating**

· · · · ▷ *See LarsonCalculus.com for an interactive version of this type of example.*

REMARK The properties of logarithms presented on page 56 can be used to rewrite antiderivatives in different forms. For instance, the antiderivative in Example 3(d) can be rewritten as

$$3 \ln|x| + C = \ln|x|^3 + C.$$

· · · · · · · · · · · · · · · · · ▷

Original Integral	Rewrite	Integrate	Simplify
a. $\int \dfrac{1}{x^3} \, dx$	$\int x^{-3} \, dx$	$\dfrac{x^{-2}}{-2} + C$	$-\dfrac{1}{2x^2} + C$
b. $\int \sqrt{x} \, dx$	$\int x^{1/2} \, dx$	$\dfrac{x^{3/2}}{3/2} + C$	$\dfrac{2}{3}x^{3/2} + C$
c. $\int 2 \sin x \, dx$	$2 \int \sin x \, dx$	$2(-\cos x) + C$	$-2 \cos x + C$
d. $\int \dfrac{3}{x} \, dx$	$3 \int \dfrac{1}{x} \, dx$	$3(\ln|x|) + C$	$3 \ln|x| + C$

REMARK The basic integration rules allow you to integrate any polynomial function.

EXAMPLE 4 **Integrating Polynomial Functions**

a. $\displaystyle\int dx = \int 1 \, dx$ Integrand is understood to be 1.

$$= x + C \qquad \text{Integrate.}$$

b. $\displaystyle\int (x + 2) \, dx = \int x \, dx + \int 2 \, dx$

$$= \frac{x^2}{2} + C_1 + 2x + C_2 \qquad \text{Integrate.}$$

$$= \frac{x^2}{2} + 2x + C \qquad C = C_1 + C_2$$

The second line in the solution is usually omitted.

c. $\displaystyle\int (3x^4 - 5x^2 + x) \, dx = 3\left(\frac{x^5}{5}\right) - 5\left(\frac{x^3}{3}\right) + \frac{x^2}{2} + C = \frac{3}{5}x^5 - \frac{5}{3}x^3 + \frac{1}{2}x^2 + C$ ■

▷

•• REMARK Before you begin
the exercise set, be sure you
realize that one of the most
important steps in integration
is *rewriting the integrand* in
a form that fits one of the
basic integration rules.

EXAMPLE 5 Rewriting Before Integrating

$$\int \frac{x+1}{\sqrt{x}}\, dx = \int \left(\frac{x}{\sqrt{x}} + \frac{1}{\sqrt{x}}\right) dx \qquad \text{Rewrite as two fractions.}$$

$$= \int (x^{1/2} + x^{-1/2})\, dx \qquad \text{Rewrite with fractional exponents.}$$

$$= \frac{x^{3/2}}{3/2} + \frac{x^{1/2}}{1/2} + C \qquad \text{Integrate.}$$

$$= \frac{2}{3}x^{3/2} + 2x^{1/2} + C \qquad \text{Simplify.}$$

$$= \frac{2}{3}\sqrt{x}(x+3) + C$$

When integrating quotients, do not integrate the numerator and denominator separately. This is no more valid in integration than it is in differentiation. For instance, in Example 5, be sure you understand that

$$\int \frac{x+1}{\sqrt{x}}\, dx = \frac{2}{3}\sqrt{x}(x+3) + C$$

is not the same as

$$\frac{\int (x+1)\, dx}{\int \sqrt{x}\, dx} = \frac{\frac{1}{2}x^2 + x + C_1}{\frac{2}{3}x\sqrt{x} + C_2}.$$

EXAMPLE 6 Rewriting Before Integrating

$$\int \frac{\sin x}{\cos^2 x}\, dx = \int \left(\frac{1}{\cos x}\right)\left(\frac{\sin x}{\cos x}\right) dx \qquad \text{Rewrite as a product.}$$

$$= \int \sec x \tan x\, dx \qquad \text{Rewrite using trigonometric identities.}$$

$$= \sec x + C \qquad \text{Integrate.}$$

▷ **TECHNOLOGY** Some
software programs, such as
Maple and *Mathematica*,
are capable of performing
integration symbolically. If
you have access to such a
symbolic integration utility,
try using it to evaluate the
indefinite integrals in Example 7.

EXAMPLE 7 Rewriting Before Integrating

Original Integral	Rewrite	Integrate	Simplify
a. $\displaystyle\int \frac{2}{\sqrt{x}}\, dx$	$2\int x^{-1/2}\, dx$	$2\left(\dfrac{x^{1/2}}{1/2}\right) + C$	$4x^{1/2} + C$
b. $\displaystyle\int (t^2 + 1)^2\, dt$	$\int (t^4 + 2t^2 + 1)\, dt$	$\dfrac{t^5}{5} + 2\left(\dfrac{t^3}{3}\right) + t + C$	$\dfrac{1}{5}t^5 + \dfrac{2}{3}t^3 + t + C$
c. $\displaystyle\int \frac{x^3 + 3}{x^2}\, dx$	$\int (x + 3x^{-2})\, dx$	$\dfrac{x^2}{2} + 3\left(\dfrac{x^{-1}}{-1}\right) + C$	$\dfrac{1}{2}x^2 - \dfrac{3}{x} + C$
d. $\displaystyle\int \sqrt[3]{x}(x - 4)\, dx$	$\int (x^{4/3} - 4x^{1/3})\, dx$	$\dfrac{x^{7/3}}{7/3} - 4\left(\dfrac{x^{4/3}}{4/3}\right) + C$	$\dfrac{3}{7}x^{7/3} - 3x^{4/3} + C$

As you do the exercises, note that you can check your answer to an antidifferentiation problem by differentiating. For instance, in Example 7(a), you can check that $4x^{1/2} + C$ is the correct antiderivative by differentiating the answer to obtain

$$D_x[4x^{1/2} + C] = 4\left(\frac{1}{2}\right)x^{-1/2} = \frac{2}{\sqrt{x}}. \qquad \text{Use differentiation to check antiderivative.}$$

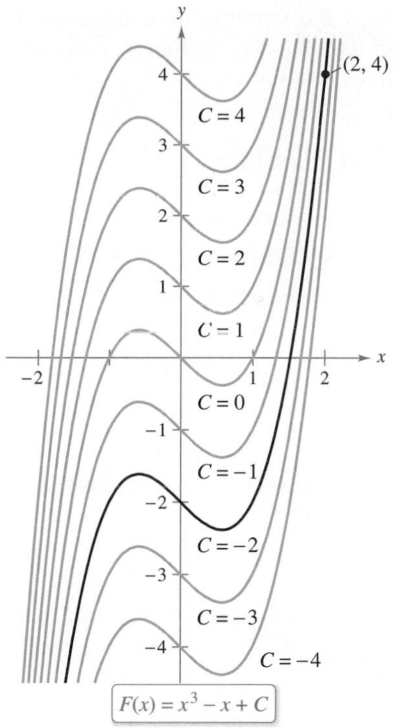

Initial Conditions and Particular Solutions

You have already seen that the equation $y = \int f(x)\,dx$ has many solutions (each differing from the others by a constant). This means that the graphs of any two antiderivatives of f are vertical translations of each other. For example, Figure 5.2 shows the graphs of several antiderivatives of the form

$$y = \int (3x^2 - 1)\,dx = x^3 - x + C \qquad \text{General solution}$$

for various integer values of C. Each of these antiderivatives is a solution of the differential equation

$$\frac{dy}{dx} = 3x^2 - 1.$$

In many applications of integration, you are given enough information to determine a **particular solution.** To do this, you need only know the value of $y = F(x)$ for one value of x. This information is called an **initial condition.** For example, in Figure 5.2, only one curve passes through the point $(2, 4)$. To find this curve, you can use the general solution

$$F(x) = x^3 - x + C \qquad \text{General solution}$$

and the initial condition

$$F(2) = 4. \qquad \text{Initial condition}$$

By using the initial condition in the general solution, you can determine that

$$F(2) = 8 - 2 + C = 4$$

which implies that $C = -2$. So, you obtain

$$F(x) = x^3 - x - 2. \qquad \text{Particular solution}$$

The particular solution that satisfies the initial condition $F(2) = 4$ is $F(x) = x^3 - x - 2$.

Figure 5.2

EXAMPLE 8   **Finding a Particular Solution**

Find the general solution of

$$F'(x) = e^x \qquad \text{Differential equation}$$

and find the particular solution that satisfies the initial condition

$$F(0) = 3. \qquad \text{Initial condition}$$

Solution   To find the general solution, integrate to obtain

$$F(x) = \int e^x\,dx$$

$$= e^x + C. \qquad \text{General solution}$$

Using the initial condition $F(0) = 3$, you can solve for C as follows.

$$F(0) = e^0 + C$$
$$3 = 1 + C$$
$$2 = C$$

So, the particular solution is

$$F(x) = e^x + 2 \qquad \text{Particular solution}$$

as shown in Figure 5.3. Note that Figure 5.3 also shows the solution curves that correspond to $C = -3, -2, -1, 0, 1,$ and 3.

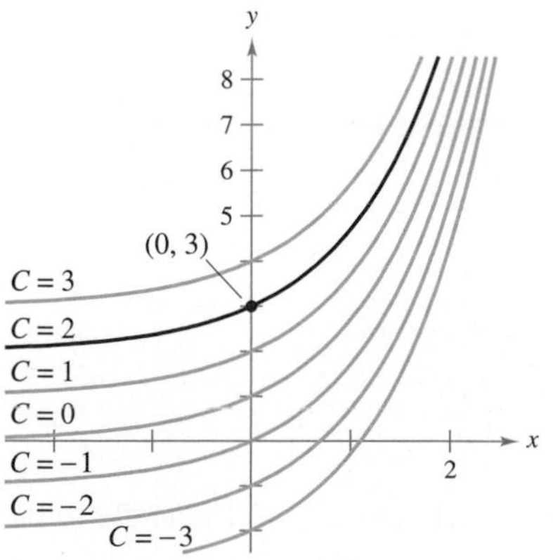

The particular solution that satisfies the initial condition $F(0) = 3$ is $F(x) = e^x + 2$.

Figure 5.3

So far in this section, you have been using x as the variable of integration. In applications, it is often convenient to use a different variable. For instance, in the next example, involving *time*, the variable of integration is t.

EXAMPLE 9 Solving a Vertical Motion Problem

A ball is thrown upward with an initial velocity of 64 feet per second from an initial height of 80 feet. [Assume the acceleration is $a(t) = -32$ feet per second per second.]

a. Find the position function giving the height s as a function of the time t.

b. When does the ball hit the ground?

Solution

a. Let $t = 0$ represent the initial time. The two given initial conditions can be written as follows.

$$s(0) = 80 \qquad \text{Initial height is 80 feet.}$$
$$s'(0) = 64 \qquad \text{Initial velocity is 64 feet per second.}$$

Recall that $a(t) = s''(t)$. So, you can write

$$s''(t) = -32$$
$$s'(t) = \int s''(t)\, dt = \int -32\, dt = -32t + C_1.$$

Using the initial velocity, you obtain $s'(0) = 64 = -32(0) + C_1$, which implies that $C_1 = 64$. Next, by integrating $s'(t)$, you obtain

$$s(t) = \int s'(t)\, dt = \int (-32t + 64)\, dt = -16t^2 + 64t + C_2.$$

Using the initial height, you obtain

$$s(0) = 80 = -16(0^2) + 64(0) + C_2$$

which implies that $C_2 = 80$. So, the position function is

$$s(t) = -16t^2 + 64t + 80. \qquad \text{See Figure 5.4.}$$

b. Using the position function found in part (a), you can find the time at which the ball hits the ground by solving the equation $s(t) = 0$.

$$-16t^2 + 64t + 80 = 0$$
$$-16(t + 1)(t - 5) = 0$$
$$t = -1, 5$$

Because t must be positive, you can conclude that the ball hits the ground 5 seconds after it was thrown.

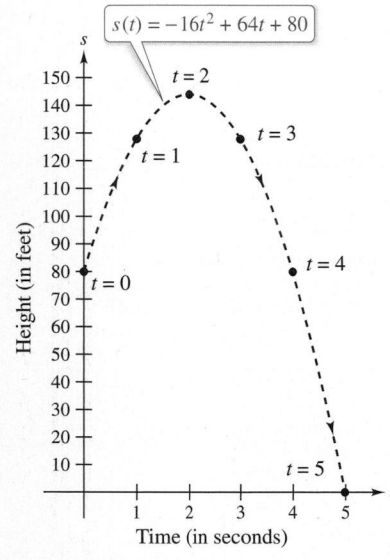

$s(t) = -16t^2 + 64t + 80$

Height (in feet)

Time (in seconds)

Height of a ball at time t
Figure 5.4

In Example 9, note that the position function has the form

$$s(t) = -\frac{1}{2}gt^2 + v_0 t + s_0$$

where g is the acceleration due to gravity, v_0 is the initial velocity, and s_0 is the initial height, as presented in Section 3.2.

Example 9 shows how to use calculus to analyze vertical motion problems in which the acceleration is determined by a gravitational force. You can use a similar strategy to analyze other linear motion problems (vertical or horizontal) in which the acceleration (or deceleration) is the result of some other force, as you will see in Exercises 65–72.

5.1 Exercises

See **CalcChat.com** for tutorial help and worked-out solutions to odd-numbered exercises.

CONCEPT CHECK

1. Antiderivative What does it mean for a function F to be an antiderivative of a function f on an interval I?

2. Antiderivatives Can two different functions both be antiderivatives of the same function? Explain.

3. Particular Solution What is a particular solution of a differential equation?

4. General and Particular Solutions Describe the difference between the general solution and a particular solution of a differential equation.

Integration and Differentiation In Exercises 5 and 6, verify the statement by showing that the derivative of the right side equals the integrand on the left side.

5. $\int \left(-\dfrac{6}{x^4} \right) dx = \dfrac{2}{x^3} + C$

6. $\int \left(8x^3 + \dfrac{1}{2x^2} \right) dx = 2x^4 - \dfrac{1}{2x} + C$

 **Solving a Differential Equation** In Exercises 7–10, find the general solution of the differential equation and check the result by differentiation.

7. $\dfrac{dy}{dt} = 9t^2$

8. $\dfrac{dy}{dt} = 5$

9. $\dfrac{dy}{dx} = x^{3/2}$

10. $\dfrac{dy}{dx} = 2x^{-3}$

 **Rewriting Before Integrating** In Exercises 11–14, complete the table to find the indefinite integral.

Original Integral	Rewrite	Integrate	Simplify
11. $\int \sqrt[3]{x}\, dx$			
12. $\int \dfrac{1}{4x^2}\, dx$			
13. $\int \dfrac{1}{x\sqrt{x}}\, dx$			
14. $\int \dfrac{1}{(3x)^2}\, dx$			

 **Finding an Indefinite Integral** In Exercises 15–36, find the indefinite integral and check the result by differentiation.

15. $\int (x + 7)\, dx$

16. $\int (x^5 + 1)\, dx$

17. $\int (x^{3/2} + 2x + 1)\, dx$

18. $\int \left(\sqrt{x} + \dfrac{1}{2\sqrt{x}} \right) dx$

19. $\int (x + 1)(3x - 2)\, dx$

20. $\int (4t^2 + 3)^2\, dt$

21. $\int \dfrac{1}{x^5}\, dx$

22. $\int \left(2 - \dfrac{3}{x^{10}} \right) dx$

23. $\int \dfrac{x + 6}{\sqrt{x}}\, dx$

24. $\int \dfrac{x^4 - 3x^2 + 5}{x^4}\, dx$

25. $\int (5 \cos x + 4 \sin x)\, dx$

26. $\int (\sin x - 6 \cos x)\, dx$

27. $\int (\sec^2 \theta - \sin \theta)\, d\theta$

28. $\int (\sec y)(\tan y - \sec y)\, dy$

29. $\int (\tan^2 y + 1)\, dy$

30. $\int (4x - \csc^2 x)\, dx$

31. $\int (2 \sin x - 5e^x)\, dx$

32. $\int (e^x - x)\, dx$

33. $\int (2x - 4^x)\, dx$

34. $\int (\cos x + 3^x)\, dx$

35. $\int \left(x - \dfrac{5}{x} \right) dx$

36. $\int \left(\dfrac{4}{x} + \sec^2 x \right) dx$

 **Finding a Particular Solution** In Exercises 37–44, find the particular solution of the differential equation that satisfies the initial condition(s).

37. $f'(x) = 6x,\ f(0) = 8$

38. $g'(x) = 4x^2,\ g(-1) = 3$

39. $f''(x) = 2,\ f'(2) = 5,\ f(2) = 10$

40. $f''(x) = 3x^2,\ f'(-1) = -2,\ f(2) = 3$

41. $f''(x) = x^{-3/2},\ f'(4) = 2,\ f(0) = 0$

42. $f''(x) = \sin x,\ f'(0) = 1,\ f(0) = 6$

43. $f''(x) = e^x,\ f'(0) = 2,\ f(0) = 5$

44. $f''(x) = \dfrac{2}{x^2},\ f'(1) = 4,\ f(1) = 3$

Slope Field In Exercises 45 and 46, a differential equation, a point, and a slope field are given. A *slope field* (or *direction field*) consists of line segments with slopes given by the differential equation. These line segments give a visual perspective of the slopes of the solutions of the differential equation. (a) Sketch two approximate solutions of the differential equation on the slope field, one of which passes through the indicated point. (To print an enlarged copy of the graph, go to *MathGraphs.com*.) (b) Use integration and the given point to find the particular solution of the differential equation and use a graphing utility to graph the solution. Compare the result with the sketch in part (a) that passes through the given point.

45. $dy/dx = x^2 - 1,\ (-1, 3)$

46. $dy/dx = -1/x^2,\ (1, 3)$

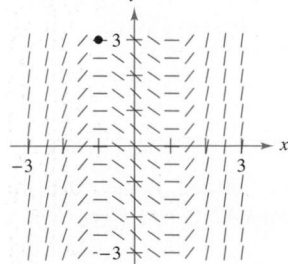

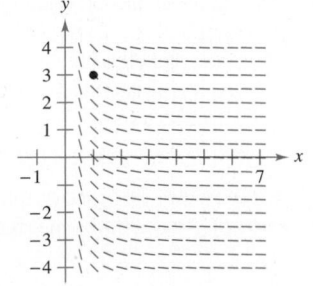

Slope Field In Exercises 47 and 48, (a) use a graphing utility to graph a slope field for the differential equation, (b) use integration and the given point to find the particular solution of the differential equation, and (c) graph the particular solution and the slope field in the same viewing window.

47. $\dfrac{dy}{dx} = 2x, (-2, -2)$

48. $\dfrac{dy}{dx} = 2\sqrt{x}, (4, 12)$

EXPLORING CONCEPTS

Sketching a Graph In Exercises 49 and 50, the graph of the derivative of a function is given. Sketch the graphs of *two* functions that have the given derivative. (There is more than one correct answer.) To print an enlarged copy of the graph, go to *MathGraphs.com.*

49.
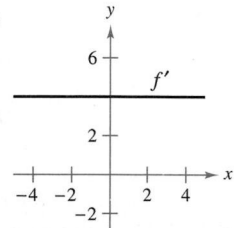

50.
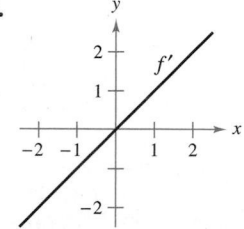

51. Comparing Functions Consider $f(x) = \tan^2 x$ and $g(x) = \sec^2 x$. What do you notice about the derivatives of f and g? What can you conclude about the relationship between f and g?

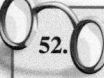 **52. HOW DO YOU SEE IT?** Use the graph of f' shown in the figure to answer the following.

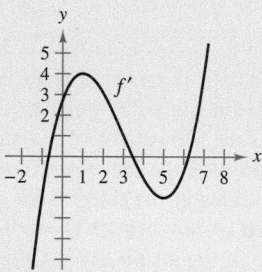

(a) Approximate the slope of f at $x = 4$. Explain.

(b) Is $f(5) - f(4) > 0$? Explain.

(c) Approximate the value of x where f is maximum. Explain.

(d) Approximate any open intervals on which the graph of f is concave upward and any open intervals on which it is concave downward. Approximate the x-coordinates of any points of inflection.

---

53. Horizontal Tangent Find a function f such that the graph of f has a horizontal tangent at $(2, 0)$ and $f''(x) = 2x$.

54. Sketching Graphs The graphs of f and f' each pass through the origin. Use the graph of f'' shown in the figure to sketch the graphs of f and f'. To print an enlarged copy of the graph, go to *MathGraphs.com.*

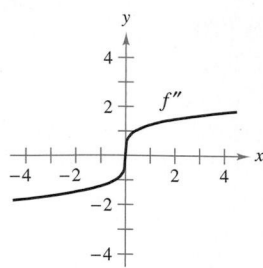

55. Tree Growth An evergreen nursery usually sells a certain type of shrub after 6 years of growth and shaping. The growth rate during those 6 years is approximated by $dh/dt = 1.5t + 5$, where t is the time in years and h is the height in centimeters. The seedlings are 12 centimeters tall when planted ($t = 0$).

(a) Find the height after t years.

(b) How tall are the shrubs when they are sold?

56. Population Growth The rate of growth dP/dt of a population of bacteria is proportional to the square root of t, where P is the population size and t is the time in days $(0 \le t \le 10)$. That is,

$$\frac{dP}{dt} = k\sqrt{t}.$$

The initial size of the population is 500. After 1 day, the population has grown to 600. Estimate the population after 7 days.

Vertical Motion In Exercises 57–59, assume the acceleration of the object is $a(t) = -32$ feet per second per second. (Neglect air resistance.)

57. A ball is thrown vertically upward from a height of 6 feet with an initial velocity of 60 feet per second. How high will the ball go?

58. With what initial velocity must an object be thrown upward (from ground level) to reach the top of the Washington Monument (approximately 550 feet)?

59. A balloon, rising vertically with a velocity of 16 feet per second, releases a sandbag at the instant it is 64 feet above the ground.

(a) How many seconds after its release will the bag strike the ground?

(b) At what velocity will the bag hit the ground?

Vertical Motion In Exercises 60–62, assume the acceleration of the object is $a(t) = -9.8$ meters per second per second. (Neglect air resistance.)

60. A baseball is thrown upward from a height of 2 meters with an initial velocity of 10 meters per second. Determine its maximum height.

61. With what initial velocity must an object be thrown upward (from a height of 2 meters) to reach a maximum height of 200 meters?

• •62. Grand Canyon • • • • • • • • • • • • • • • •

The Grand Canyon is 1800 meters deep at its deepest point. A rock is dropped from the rim above this point. How long will it take the rock to hit the canyon floor?

63. Lunar Gravity On the moon, the acceleration of a free-falling object is $a(t) = -1.6$ meters per second per second. A stone is dropped from a cliff on the moon and hits the surface of the moon 20 seconds later. How far did it fall? What was its velocity at impact?

64. Escape Velocity The minimum velocity required for an object to escape Earth's gravitational pull is obtained from the solution of the equation

$$\int v\, dv = -GM \int \frac{1}{y^2}\, dy$$

where v is the velocity of the object projected from Earth, y is the distance from the center of Earth, G is the gravitational constant, and M is the mass of Earth. Show that v and y are related by the equation

$$v^2 = v_0^2 + 2GM\left(\frac{1}{y} - \frac{1}{R}\right)$$

where v_0 is the initial velocity of the object and R is the radius of Earth.

Rectilinear Motion **In Exercises 65–68, consider a particle moving along the x-axis, where $x(t)$ is the position of the particle at time t, $x'(t)$ is its velocity, and $x''(t)$ is its acceleration.**

65. $x(t) = t^3 - 6t^2 + 9t - 2, \quad 0 \le t \le 5$

(a) Find the velocity and acceleration of the particle.

(b) Find the open t-intervals on which the particle is moving to the right.

(c) Find the velocity of the particle when the acceleration is 0.

66. Repeat Exercise 65 for the position function $x(t) = (t - 1)(t - 3)^2, 0 \le t \le 5$.

67. A particle moves along the x-axis at a velocity of $v(t) = 1/\sqrt{t}$, $t > 0$. At time $t = 1$, its position is $x = 4$. Find the acceleration and position functions for the particle.

68. A particle, initially at rest, moves along the x-axis such that its acceleration at time $t > 0$ is given by $a(t) = \cos t$. At time $t = 0$, its position is $x = 3$.

(a) Find the velocity and position functions for the particle.

(b) Find the values of t for which the particle is at rest.

69. Acceleration The maker of an automobile advertises that it takes 13 seconds to accelerate from 25 kilometers per hour to 80 kilometers per hour. Assume the acceleration is constant.

(a) Find the acceleration in meters per second per second.

(b) Find the distance the car travels during the 13 seconds.

70. Deceleration A car traveling at 45 miles per hour is brought to a stop, at constant deceleration, 132 feet from where the brakes are applied.

(a) How far has the car moved when its speed has been reduced to 30 miles per hour?

(b) How far has the car moved when its speed has been reduced to 15 miles per hour?

(c) Draw the real number line from 0 to 132. Plot the points found in parts (a) and (b). What can you conclude?

71. Acceleration At the instant the traffic light turns green, a car that has been waiting at an intersection starts with a constant acceleration of 6 feet per second per second. At the same instant, a truck traveling with a constant velocity of 30 feet per second passes the car.

(a) How far beyond its starting point will the car pass the truck?

(b) How fast will the car be traveling when it passes the truck?

72. Acceleration Assume that a fully loaded plane starting from rest has a constant acceleration while moving down a runway. The plane requires 0.7 mile of runway and a speed of 160 miles per hour in order to lift off. What is the plane's acceleration?

True or False? **In Exercises 73 and 74, determine whether the statement is true or false. If it is false, explain why or give an example that shows it is false.**

73. The antiderivative of $f(x)$ is unique.

74. Each antiderivative of an nth-degree polynomial function is an $(n + 1)$th-degree polynomial function.

75. Proof Let $s(x)$ and $c(x)$ be two functions satisfying $s'(x) = c(x)$ and $c'(x) = -s(x)$ for all x. If $s(0) = 0$ and $c(0) = 1$, prove that $[s(x)]^2 + [c(x)]^2 = 1$.

76. Think About It Find the general solution of

$$f'(x) = -2x \sin x^2.$$

77. Verification Verify the natural log rule $\int \frac{1}{x}\, dx = \ln|x| + C$ by showing that the derivative of $\ln|x| + C$ is $1/x$.

78. Verification Verify the natural log rule $\int \frac{1}{x}\, dx = \ln|Cx|$, $C \ne 0$, by showing that the derivative of $\ln|Cx|$ is $1/x$.

PUTNAM EXAM CHALLENGE

79. Suppose f and g are non-constant, differentiable, real-valued functions defined on $(-\infty, \infty)$. Furthermore, suppose that for each pair of real numbers x and y,

$$f(x + y) = f(x)f(y) - g(x)g(y) \quad \text{and}$$
$$g(x + y) = f(x)g(y) + g(x)f(y).$$

If $f'(0) = 0$, prove that $(f(x))^2 + (g(x))^2 = 1$ for all x.

This problem was composed by the Committee on the Putnam Prize Competition.
© The Mathematical Association of America. All rights reserved.

5.2 Area

■ Use sigma notation to write and evaluate a sum.
■ Understand the concept of area.
■ Approximate the area of a plane region.
■ Find the area of a plane region using limits.

Sigma Notation

In the preceding section, you studied antidifferentiation. In this section, you will look further into a problem introduced in Section 2.1—that of finding the area of a region in the plane. At first glance, these two ideas may seem unrelated, but you will discover in Section 5.4 that they are closely related by an extremely important theorem called the Fundamental Theorem of Calculus.

This section begins by introducing a concise notation for sums. This notation is called **sigma notation** because it uses the uppercase Greek letter sigma, written as Σ.

Sigma Notation

The sum of n terms $a_1, a_2, a_3, \ldots, a_n$ is written as

$$\sum_{i=1}^{n} a_i = a_1 + a_2 + a_3 + \cdots + a_n$$

where i is the **index of summation,** a_i is the **ith term** of the sum, and the **upper and lower bounds of summation** are n and 1.

REMARK The upper and lower bounds must be constant with respect to the index of summation. However, the lower bound does not have to be 1. Any integer less than or equal to the upper bound is legitimate.

EXAMPLE 1 **Examples of Sigma Notation**

a. $\displaystyle\sum_{i=1}^{6} i = 1 + 2 + 3 + 4 + 5 + 6$

b. $\displaystyle\sum_{i=0}^{5} (i + 1) = 1 + 2 + 3 + 4 + 5 + 6$

c. $\displaystyle\sum_{j=3}^{7} j^2 = 3^2 + 4^2 + 5^2 + 6^2 + 7^2$

d. $\displaystyle\sum_{j=1}^{5} \frac{1}{\sqrt{j}} = \frac{1}{\sqrt{1}} + \frac{1}{\sqrt{2}} + \frac{1}{\sqrt{3}} + \frac{1}{\sqrt{4}} + \frac{1}{\sqrt{5}}$

e. $\displaystyle\sum_{k=1}^{n} \frac{1}{n}(k^2 + 1) = \frac{1}{n}(1^2 + 1) + \frac{1}{n}(2^2 + 1) + \cdots + \frac{1}{n}(n^2 + 1)$

f. $\displaystyle\sum_{i=1}^{n} f(x_i)\,\Delta x = f(x_1)\,\Delta x + f(x_2)\,\Delta x + \cdots + f(x_n)\,\Delta x$

From parts (a) and (b), notice that the same sum can be represented in different ways using sigma notation. ■

Although any variable can be used as the index of summation, i, j, and k are often used. Notice in Example 1 that the index of summation does not appear in the terms of the expanded sum.

The properties of summation shown below can be derived using the Associative and Commutative Properties of Addition and the Distributive Property of Addition over Multiplication. (In the first property, k is a constant.)

1. $\displaystyle\sum_{i=1}^{n} ka_i = k \sum_{i=1}^{n} a_i$ **2.** $\displaystyle\sum_{i=1}^{n} (a_i \pm b_i) = \sum_{i=1}^{n} a_i \pm \sum_{i=1}^{n} b_i$

The next theorem lists some useful formulas for sums of powers.

THEOREM 5.2   Summation Formulas

1. $\displaystyle\sum_{i=1}^{n} c = cn,\ c$ is a constant   **2.** $\displaystyle\sum_{i=1}^{n} i = \frac{n(n+1)}{2}$

3. $\displaystyle\sum_{i=1}^{n} i^2 = \frac{n(n+1)(2n+1)}{6}$ **4.** $\displaystyle\sum_{i=1}^{n} i^3 = \frac{n^2(n+1)^2}{4}$

A proof of this theorem is given in Appendix A.

EXAMPLE 2   Evaluating a Sum

Evaluate $\displaystyle\sum_{i=1}^{n} \frac{i+1}{n^2}$ for $n = 10, 100, 1000,$ and $10,000$.

Solution

$$\sum_{i=1}^{n} \frac{i+1}{n^2} = \frac{1}{n^2} \sum_{i=1}^{n} (i+1) \qquad \text{Factor the constant } 1/n^2 \text{ out of sum.}$$

$$= \frac{1}{n^2} \left(\sum_{i=1}^{n} i + \sum_{i=1}^{n} 1 \right) \qquad \text{Write as two sums.}$$

$$= \frac{1}{n^2} \left[\frac{n(n+1)}{2} + n \right] \qquad \text{Apply Theorem 5.2.}$$

$$= \frac{1}{n^2} \left[\frac{n^2 + 3n}{2} \right] \qquad \text{Simplify.}$$

$$= \frac{n+3}{2n} \qquad \text{Simplify.}$$

Now you can evaluate the sum by substituting the appropriate values of n, as shown in the table below.

n	10	100	1000	10,000
$\displaystyle\sum_{i=1}^{n} \frac{i+1}{n^2} = \frac{n+3}{2n}$	0.65000	0.51500	0.50150	0.50015

FOR FURTHER INFORMATION
For a geometric interpretation of summation formulas, see the article "Looking at $\displaystyle\sum_{k=1}^{n} k$ and $\displaystyle\sum_{k=1}^{n} k^2$ Geometrically" by Eric Hegblom in *Mathematics Teacher*. To view this article, go to *MathArticles.com*.

In the table, note that the sum appears to approach a limit as n increases. Although the discussion of limits at infinity in Section 4.5 applies to a variable x, where x can be any real number, many of the same results hold true for limits involving the variable n, where n is restricted to positive integer values. So, to find the limit of $(n+3)/2n$ as n approaches infinity, you can write

$$\lim_{n \to \infty} \frac{n+3}{2n} = \lim_{n \to \infty} \left(\frac{n}{2n} + \frac{3}{2n} \right) = \lim_{n \to \infty} \left(\frac{1}{2} + \frac{3}{2n} \right) = \frac{1}{2} + 0 = \frac{1}{2}.$$

Area

In Euclidean geometry, the simplest type of plane region is a rectangle. Although people often say that the *formula* for the area of a rectangle is

$$A = bh$$

it is actually more proper to say that this is the *definition* of the **area of a rectangle.**

From this definition, you can develop formulas for the areas of many other plane regions. For example, to determine the area of a triangle, you can form a rectangle whose area is twice that of the triangle, as shown in Figure 5.5. Once you know how to find the area of a triangle, you can determine the area of any polygon by subdividing the polygon into triangular regions, as shown in Figure 5.6.

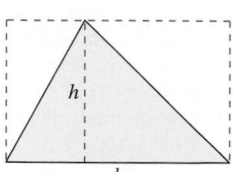

Triangle: $A = \frac{1}{2}bh$
Figure 5.5

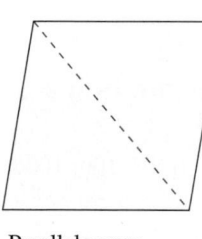
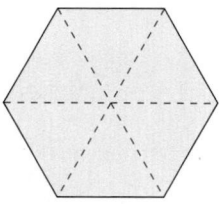
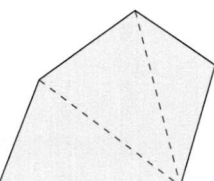

Parallelogram   Hexagon   Polygon
Figure 5.6

Finding the areas of regions other than polygons is more difficult. The ancient Greeks were able to determine formulas for the areas of some general regions (principally those bounded by conics) by the *exhaustion* method. The clearest description of this method was given by Archimedes. Essentially, the method is a limiting process in which the area is squeezed between two polygons—one inscribed in the region and one circumscribed about the region.

For instance, in Figure 5.7, the area of a circular region is approximated by an n-sided inscribed polygon and an n-sided circumscribed polygon. For each value of n, the area of the inscribed polygon is less than the area of the circle, and the area of the circumscribed polygon is greater than the area of the circle. Moreover, as n increases, the areas of both polygons become better and better approximations of the area of the circle.

ARCHIMEDES (287–212 B.C.)

Archimedes used the method of exhaustion to derive formulas for the areas of ellipses, parabolic segments, and sectors of a spiral. He is considered to have been the greatest applied mathematician of antiquity.
See LarsonCalculus.com to read more of this biography.

■ **FOR FURTHER INFORMATION**
For an alternative development of the formula for the area of a circle, see the article "Proof Without Words: Area of a Disk is πR^2" by Russell Jay Hendel in *Mathematics Magazine.* To view this article, go to *MathArticles.com.*

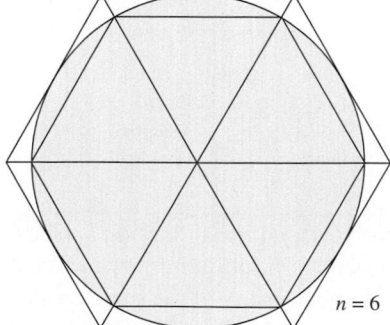
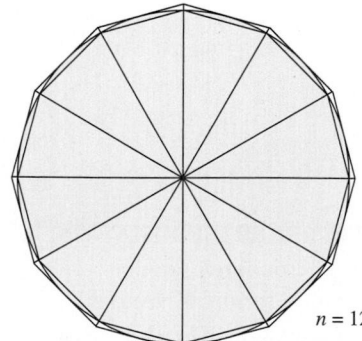

$n = 6$ $n = 12$

The exhaustion method for finding the area of a circular region
Figure 5.7

A process that is similar to that used by Archimedes to determine the area of a plane region is used in the remaining examples in this section.

The Area of a Plane Region

Recall from Section 2.1 that the origins of calculus are connected to two classic problems: the tangent line problem and the area problem. Example 3 begins the investigation of the area problem.

EXAMPLE 3   Approximating the Area of a Plane Region

Use the five rectangles in Figures 5.8(a) and (b) to find *two* approximations of the area of the region lying between the graph of

$$f(x) = -x^2 + 5$$

and the *x*-axis between $x = 0$ and $x = 2$.

Solution

a. The right endpoints of the five intervals are

$$\frac{2}{5}i \qquad \text{Right endpoints}$$

where $i = 1, 2, 3, 4, 5$. The width of each rectangle is $\frac{2}{5}$, and the height of each rectangle can be obtained by evaluating f at the right endpoint of each interval.

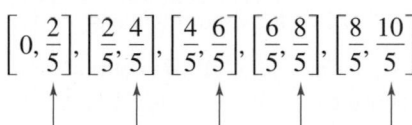

Evaluate f at the right endpoints of these intervals.

The sum of the areas of the five rectangles is

$$\sum_{i=1}^{5} f\!\left(\frac{2i}{5}\right)\!\left(\frac{2}{5}\right) = \sum_{i=1}^{5}\left[-\left(\frac{2i}{5}\right)^2 + 5\right]\!\left(\frac{2}{5}\right) = \frac{162}{25} = 6.48.$$

Because each of the five rectangles lies inside the parabolic region, you can conclude that the area of the parabolic region is greater than 6.48.

b. The left endpoints of the five intervals are

$$\frac{2}{5}(i - 1) \qquad \text{Left endpoints}$$

where $i = 1, 2, 3, 4, 5$. The width of each rectangle is $\frac{2}{5}$, and the height of each rectangle can be obtained by evaluating f at the left endpoint of each interval. So, the sum is

$$\sum_{i=1}^{5} f\!\left(\frac{2i - 2}{5}\right)\!\left(\frac{2}{5}\right) = \sum_{i=1}^{5}\left[-\left(\frac{2i - 2}{5}\right)^2 + 5\right]\!\left(\frac{2}{5}\right) = \frac{202}{25} = 8.08.$$

Because the parabolic region lies within the union of the five rectangular regions, you can conclude that the area of the parabolic region is less than 8.08.

By combining the results in parts (a) and (b), you can conclude that

$$6.48 < (\text{Area of region}) < 8.08.$$

By increasing the number of rectangles used in Example 3, you can obtain closer and closer approximations of the area of the region. For instance, using 25 rectangles of width $\frac{2}{25}$ each, you can conclude that

$$7.1712 < (\text{Area of region}) < 7.4912.$$

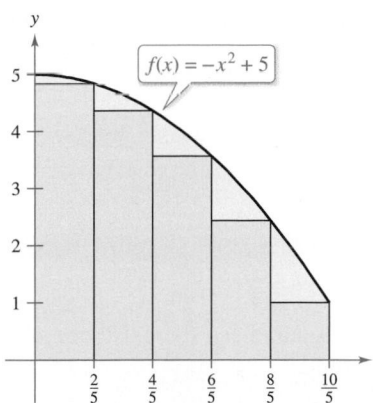

(a) The area of the parabolic region is greater than the area of the rectangles.

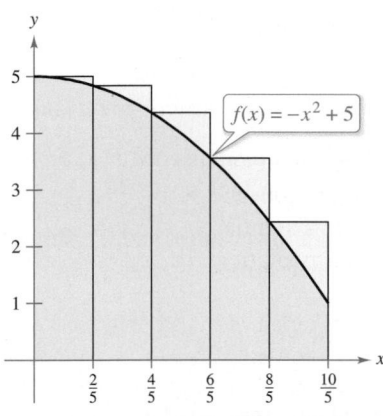

(b) The area of the parabolic region is less than the area of the rectangles.

Figure 5.8

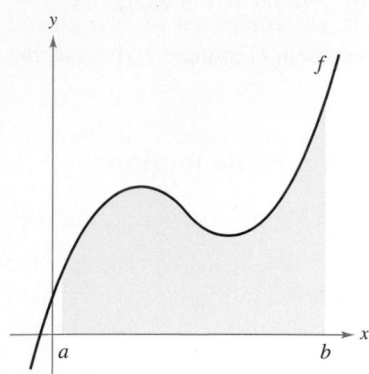

The region under a curve
Figure 5.9

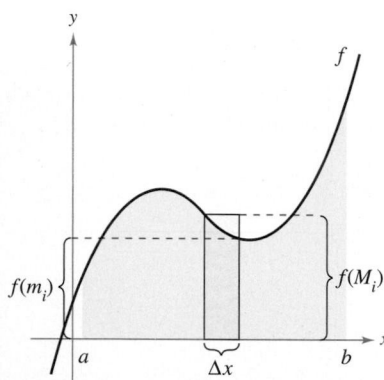

The interval $[a, b]$ is divided into n subintervals of width $\Delta x = \dfrac{b - a}{n}$.

Figure 5.10

Finding Area by the Limit Definition

The procedure used in Example 3 can be generalized as follows. Consider a plane region bounded above by the graph of a nonnegative, continuous function

$$y = f(x)$$

as shown in Figure 5.9. The region is bounded below by the x-axis, and the left and right boundaries of the region are the vertical lines $x = a$ and $x = b$.

To approximate the area of the region, begin by subdividing the interval $[a, b]$ into n subintervals, each of width

$$\Delta x = \frac{b - a}{n}$$

as shown in Figure 5.10. The endpoints of the intervals are

$$\underbrace{a = x_0}_{} \qquad \underbrace{x_1}_{} \qquad \underbrace{x_2}_{} \qquad \underbrace{x_n = b}_{}$$
$$a + 0(\Delta x) < a + 1(\Delta x) < a + 2(\Delta x) < \cdots < a + n(\Delta x).$$

Because f is continuous, the Extreme Value Theorem guarantees the existence of a minimum and a maximum value of $f(x)$ in *each* subinterval.

$$f(m_i) = \text{Minimum value of } f(x) \text{ in } i\text{th subinterval}$$
$$f(M_i) = \text{Maximum value of } f(x) \text{ in } i\text{th subinterval}$$

Next, define an **inscribed rectangle** lying *inside* the ith subregion and a **circumscribed rectangle** extending *outside* the ith subregion. The height of the ith inscribed rectangle is $f(m_i)$ and the height of the ith circumscribed rectangle is $f(M_i)$. For *each* i, the area of the inscribed rectangle is less than or equal to the area of the circumscribed rectangle.

$$\begin{pmatrix} \text{Area of inscribed} \\ \text{rectangle} \end{pmatrix} = f(m_i)\,\Delta x \le f(M_i)\,\Delta x = \begin{pmatrix} \text{Area of circumscribed} \\ \text{rectangle} \end{pmatrix}$$

The sum of the areas of the inscribed rectangles is called a **lower sum,** and the sum of the areas of the circumscribed rectangles is called an **upper sum.**

$$\text{Lower sum} = s(n) = \sum_{i=1}^{n} f(m_i)\,\Delta x \qquad \text{Area of inscribed rectangles}$$

$$\text{Upper sum} = S(n) = \sum_{i=1}^{n} f(M_i)\,\Delta x \qquad \text{Area of circumscribed rectangles}$$

From Figure 5.11, you can see that the lower sum $s(n)$ is less than or equal to the upper sum $S(n)$. Moreover, the actual area of the region lies between these two sums.

$$s(n) \le (\text{Area of region}) \le S(n)$$

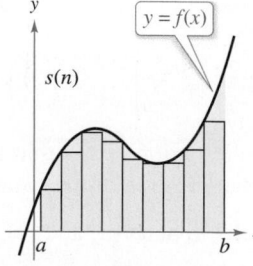

Area of inscribed rectangles is less than area of region.

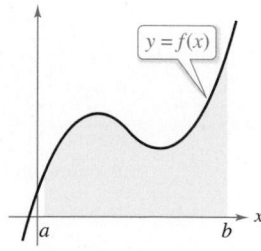

Area of region

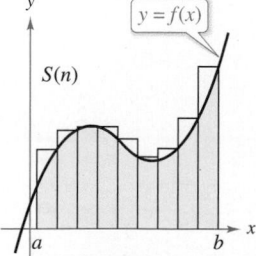

Area of circumscribed rectangles is greater than area of region.

Figure 5.11

EXAMPLE 4   **Finding Upper and Lower Sums for a Region**

Find the upper and lower sums for the region bounded by the graph of $f(x) = x^2$ and the x-axis between $x = 0$ and $x = 2$.

Solution   To begin, partition the interval $[0, 2]$ into n subintervals, each of width

$$\Delta x = \frac{b - a}{n} = \frac{2 - 0}{n} = \frac{2}{n}.$$

Figure 5.12 shows the endpoints of the subintervals and several inscribed and circumscribed rectangles. Because f is increasing on the interval $[0, 2]$, the minimum value on each subinterval occurs at the left endpoint, and the maximum value occurs at the right endpoint.

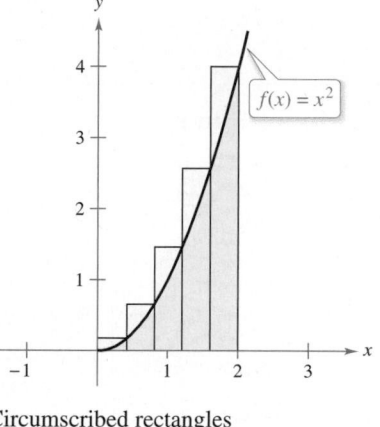

Inscribed rectangles

Circumscribed rectangles
Figure 5.12

Left Endpoints

$$m_i = 0 + (i - 1)\left(\frac{2}{n}\right) = \frac{2(i - 1)}{n}$$

Right Endpoints

$$M_i = 0 + i\left(\frac{2}{n}\right) = \frac{2i}{n}$$

Using the left endpoints, the lower sum is

$$
\begin{aligned}
s(n) &= \sum_{i=1}^{n} f(m_i)\, \Delta x \\
&= \sum_{i=1}^{n} f\left[\frac{2(i - 1)}{n}\right]\left(\frac{2}{n}\right) \\
&= \sum_{i=1}^{n} \left[\frac{2(i - 1)}{n}\right]^2\left(\frac{2}{n}\right) \\
&= \sum_{i=1}^{n} \left(\frac{8}{n^3}\right)(i^2 - 2i + 1) \\
&= \frac{8}{n^3}\left(\sum_{i=1}^{n} i^2 - 2\sum_{i=1}^{n} i + \sum_{i=1}^{n} 1\right) \\
&= \frac{8}{n^3}\left\{\frac{n(n + 1)(2n + 1)}{6} - 2\left[\frac{n(n + 1)}{2}\right] + n\right\} \\
&= \frac{4}{3n^3}(2n^3 - 3n^2 + n) \\
&= \frac{8}{3} - \frac{4}{n} + \frac{4}{3n^2}. \qquad \text{Lower sum}
\end{aligned}
$$

Using the right endpoints, the upper sum is

$$
\begin{aligned}
S(n) &= \sum_{i=1}^{n} f(M_i)\, \Delta x \\
&= \sum_{i=1}^{n} f\left(\frac{2i}{n}\right)\left(\frac{2}{n}\right) \\
&= \sum_{i=1}^{n} \left(\frac{2i}{n}\right)^2\left(\frac{2}{n}\right) \\
&= \sum_{i=1}^{n} \left(\frac{8}{n^3}\right)i^2 \\
&= \frac{8}{n^3}\left[\frac{n(n + 1)(2n + 1)}{6}\right] \\
&= \frac{4}{3n^3}(2n^3 + 3n^2 + n) \\
&= \frac{8}{3} + \frac{4}{n} + \frac{4}{3n^2}. \qquad \text{Upper sum}
\end{aligned}
$$

Example 4 illustrates some important things about lower and upper sums. First, notice that for any value of n, the lower sum is less than (or equal to) the upper sum.

$$s(n) = \frac{8}{3} - \frac{4}{n} + \frac{4}{3n^2} < \frac{8}{3} + \frac{4}{n} + \frac{4}{3n^2} = S(n)$$

Second, the difference between these two sums lessens as n increases. In fact, when you take the limits as $n \to \infty$, both the lower sum and the upper sum approach $\frac{8}{3}$.

$$\lim_{n \to \infty} s(n) = \lim_{n \to \infty} \left(\frac{8}{3} - \frac{4}{n} + \frac{4}{3n^2} \right) = \frac{8}{3} \qquad \text{Lower sum limit}$$

and

$$\lim_{n \to \infty} S(n) = \lim_{n \to \infty} \left(\frac{8}{3} + \frac{4}{n} + \frac{4}{3n^2} \right) = \frac{8}{3} \qquad \text{Upper sum limit}$$

The next theorem shows that the equivalence of the limits (as $n \to \infty$) of the upper and lower sums is not mere coincidence. It is true for all functions that are continuous and nonnegative on the closed interval $[a, b]$. The proof of this theorem is best left to a course in advanced calculus.

THEOREM 5.3   Limits of the Lower and Upper Sums

Let f be continuous and nonnegative on the interval $[a, b]$. The limits as $n \to \infty$ of both the lower and upper sums exist and are equal to each other. That is,

$$\lim_{n \to \infty} s(n) = \lim_{n \to \infty} \sum_{i=1}^{n} f(m_i)\, \Delta x$$

$$= \lim_{n \to \infty} \sum_{i=1}^{n} f(M_i)\, \Delta x$$

$$= \lim_{n \to \infty} S(n)$$

where $\Delta x = (b - a)/n$ and $f(m_i)$ and $f(M_i)$ are the minimum and maximum values of f on the ith subinterval.

In Theorem 5.3, the same limit is attained for both the minimum value $f(m_i)$ and the maximum value $f(M_i)$. So, it follows from the Squeeze Theorem (Theorem 2.8) that the choice of x in the ith subinterval does not affect the limit. This means that you are free to choose an *arbitrary* x-value in the ith subinterval, as shown in the *definition of the area of a region in the plane.*

Definition of the Area of a Region in the Plane

Let f be continuous and nonnegative on the interval $[a, b]$. (See Figure 5.13.) The area of the region bounded by the graph of f, the x-axis, and the vertical lines $x = a$ and $x = b$ is

$$\text{Area} = \lim_{n \to \infty} \sum_{i=1}^{n} f(c_i)\, \Delta x$$

where $x_{i-1} \le c_i \le x_i$ and

$$\Delta x = \frac{b - a}{n}.$$

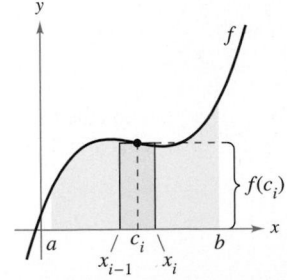

The width of the ith subinterval is $\Delta x = x_i - x_{i-1}$.

Figure 5.13

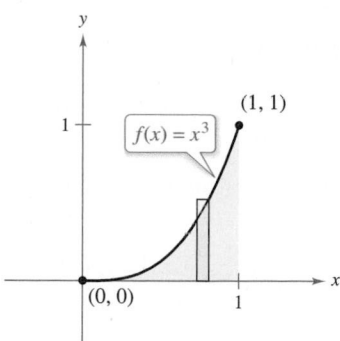

The area of the region bounded by the graph of f, the x-axis, $x = 0$, and $x = 1$ is $\frac{1}{4}$.

Figure 5.14

EXAMPLE 5 **Finding Area by the Limit Definition**

Find the area of the region bounded by the graph of $f(x) = x^3$, the x-axis, and the vertical lines $x = 0$ and $x = 1$, as shown in Figure 5.14.

Solution Begin by noting that f is continuous and nonnegative on the interval $[0, 1]$. Next, partition the interval $[0, 1]$ into n subintervals, each of width $\Delta x = 1/n$. According to the definition of area, you can choose any x-value in the ith subinterval. For this example, the right endpoints $c_i = i/n$ are convenient.

$$\text{Area} = \lim_{n \to \infty} \sum_{i=1}^{n} f(c_i) \, \Delta x$$

$$= \lim_{n \to \infty} \sum_{i=1}^{n} \left(\frac{i}{n}\right)^3 \left(\frac{1}{n}\right) \qquad \text{Right endpoints: } c_i = \frac{i}{n}$$

$$= \lim_{n \to \infty} \frac{1}{n^4} \sum_{i=1}^{n} i^3$$

$$= \lim_{n \to \infty} \frac{1}{n^4} \left[\frac{n^2(n + 1)^2}{4} \right]$$

$$= \lim_{n \to \infty} \left(\frac{1}{4} + \frac{1}{2n} + \frac{1}{4n^2} \right)$$

$$= \frac{1}{4}$$

The area of the region is $\frac{1}{4}$.

EXAMPLE 6 **Finding Area by the Limit Definition**

⋯▷ *See LarsonCalculus.com for an interactive version of this type of example.*

Find the area of the region bounded by the graph of $f(x) = 4 - x^2$, the x-axis, and the vertical lines $x = 1$ and $x = 2$, as shown in Figure 5.15.

Solution Note that the function f is continuous and nonnegative on the interval $[1, 2]$. So, begin by partitioning the interval into n subintervals, each of width $\Delta x = 1/n$. Choosing the right endpoint

$$c_i = a + i\Delta x = 1 + \frac{i}{n} \qquad \text{Right endpoints}$$

of each subinterval, you obtain

$$\text{Area} = \lim_{n \to \infty} \sum_{i=1}^{n} f(c_i) \, \Delta x$$

$$= \lim_{n \to \infty} \sum_{i=1}^{n} \left[4 - \left(1 + \frac{i}{n} \right)^2 \right] \left(\frac{1}{n} \right)$$

$$= \lim_{n \to \infty} \sum_{i=1}^{n} \left(3 - \frac{2i}{n} - \frac{i^2}{n^2} \right) \left(\frac{1}{n} \right)$$

$$= \lim_{n \to \infty} \left(\frac{1}{n} \sum_{i=1}^{n} 3 - \frac{2}{n^2} \sum_{i=1}^{n} i - \frac{1}{n^3} \sum_{i=1}^{n} i^2 \right)$$

$$= \lim_{n \to \infty} \left[3 - \left(1 + \frac{1}{n} \right) - \left(\frac{1}{3} + \frac{1}{2n} + \frac{1}{6n^2} \right) \right]$$

$$= 3 - 1 - \frac{1}{3}$$

$$= \frac{5}{3}.$$

The area of the region is $\frac{5}{3}$.

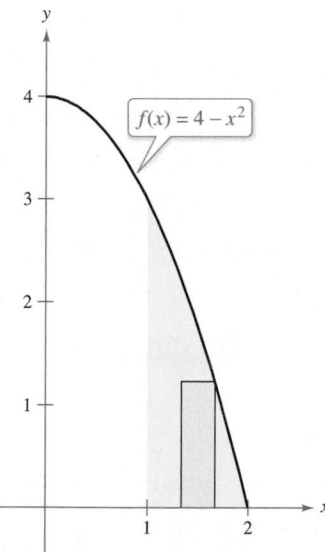

The area of the region bounded by the graph of f, the x-axis, $x = 1$, and $x = 2$ is $\frac{5}{3}$.

Figure 5.15

The next example looks at a region that is bounded by the y-axis (rather than by the x-axis).

EXAMPLE 7   A Region Bounded by the *y*-axis

Find the area of the region bounded by the graph of $f(y) = y^2$ and the y-axis for $0 \le y \le 1$, as shown in Figure 5.16.

Solution   When f is a continuous, nonnegative function of y, you can still use the same basic procedure shown in Examples 5 and 6. Begin by partitioning the interval $[0, 1]$ into n subintervals, each of width $\Delta y = 1/n$. Then, using the upper endpoints $c_i = i/n$, you obtain

$$\text{Area} = \lim_{n \to \infty} \sum_{i=1}^{n} f(c_i)\, \Delta y$$

$$= \lim_{n \to \infty} \sum_{i=1}^{n} \left(\frac{i}{n}\right)^2 \left(\frac{1}{n}\right) \qquad \text{Upper endpoints: } c_i = \frac{i}{n}$$

$$= \lim_{n \to \infty} \frac{1}{n^3} \sum_{i=1}^{n} i^2$$

$$= \lim_{n \to \infty} \frac{1}{n^3} \left[\frac{n(n+1)(2n+1)}{6}\right]$$

$$= \lim_{n \to \infty} \left(\frac{1}{3} + \frac{1}{2n} + \frac{1}{6n^2}\right)$$

$$= \frac{1}{3}.$$

The area of the region is $\frac{1}{3}$.

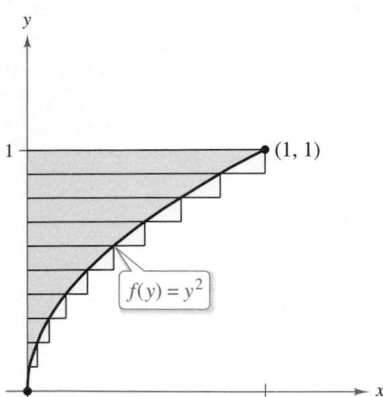

The area of the region bounded by the graph of f and the y-axis for $0 \le y \le 1$ is $\frac{1}{3}$.

Figure 5.16

In Examples 5, 6, and 7, c_i is chosen to be a value that is convenient for calculating the limit. Because each limit gives the exact area for *any* c_i, there is no need to find values that give good approximations when n is small. For an *approximation*, however, you should try to find a value of c_i that gives a good approximation of the area of the ith subregion. In general, a good value to choose is the midpoint of the interval, $c_i = (x_{i-1} + x_i)/2$, and apply the **Midpoint Rule**.

$$\text{Area} \approx \sum_{i=1}^{n} f\left(\frac{x_{i-1} + x_i}{2}\right) \Delta x. \qquad \text{Midpoint Rule}$$

·· REMARK   You will study other approximation methods in Section 8.6. One of the methods, the Trapezoidal Rule, is similar to the Midpoint Rule.

EXAMPLE 8   Approximating Area with the Midpoint Rule

Use the Midpoint Rule with $n = 4$ to approximate the area of the region bounded by the graph of $f(x) = \sin x$ and the x-axis for $0 \le x \le \pi$, as shown in Figure 5.17.

Solution   For $n = 4$, $\Delta x = \pi/4$. The midpoints of the subregions are shown below.

$$c_1 = \frac{0 + (\pi/4)}{2} = \frac{\pi}{8} \qquad\qquad c_2 = \frac{(\pi/4) + (\pi/2)}{2} = \frac{3\pi}{8}$$

$$c_3 = \frac{(\pi/2) + (3\pi/4)}{2} = \frac{5\pi}{8} \qquad\qquad c_4 = \frac{(3\pi/4) + \pi}{2} = \frac{7\pi}{8}$$

So, the area is approximated by

$$\text{Area} \approx \sum_{i=1}^{n} f(c_i)\, \Delta x = \sum_{i=1}^{4} (\sin c_i)\left(\frac{\pi}{4}\right) = \frac{\pi}{4}\left(\sin\frac{\pi}{8} + \sin\frac{3\pi}{8} + \sin\frac{5\pi}{8} + \sin\frac{7\pi}{8}\right)$$

which is about 2.052.

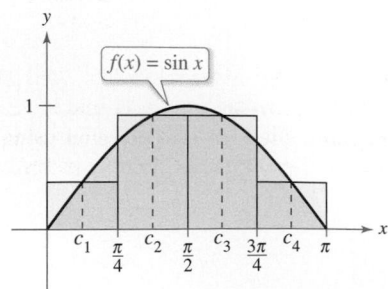

The area of the region bounded by the graph of $f(x) = \sin x$ and the x-axis for $0 \le x \le \pi$ is about 2.052.

Figure 5.17

5.2 Exercises

See **CalcChat.com** for tutorial help and worked-out solutions to odd-numbered exercises.

CONCEPT CHECK

1. Sigma Notation What are the index of summation, the upper bound of summation, and the lower bound of summation for $\sum_{i=3}^{8} (i - 4)$?

2. Sums What is the value of n?

(a) $\sum_{i=1}^{n} i = \frac{5(5 + 1)}{2}$ (b) $\sum_{i=1}^{n} i^2 = \frac{20(20 + 1)[2(20) + 1]}{6}$

3. Upper and Lower Sums In your own words and using appropriate figures, describe the methods of upper sums and lower sums in approximating the area of a region.

4. Finding Area by the Limit Definition Explain how to find the area of a plane region using limits.

 **Finding a Sum** In Exercises 5–10, find the sum by adding each term together. Use the summation capabilities of a graphing utility to verify your result.

5. $\sum_{i=1}^{6} (3i + 2)$

6. $\sum_{k=3}^{9} (k^2 + 1)$

7. $\sum_{k=0}^{4} \frac{1}{k^2 + 1}$

8. $\sum_{j=2}^{5} \frac{1}{2j}$

9. $\sum_{k=0}^{7} c$

10. $\sum_{i=1}^{4} [(i - 1)^2 + (i + 1)^3]$

Using Sigma Notation In Exercises 11–16, use sigma notation to write the sum.

11. $\frac{1}{5(1)} + \frac{1}{5(2)} + \frac{1}{5(3)} + \cdots + \frac{1}{5(11)}$

12. $\frac{6}{2 + 1} + \frac{6}{2 + 2} + \frac{6}{2 + 3} + \cdots + \frac{6}{2 + 11}$

13. $\left[7\left(\frac{1}{6}\right) + 5\right] + \left[7\left(\frac{2}{6}\right) + 5\right] + \cdots + \left[7\left(\frac{6}{6}\right) + 5\right]$

14. $\left[1 - \left(\frac{1}{4}\right)^2\right] + \left[1 - \left(\frac{2}{4}\right)^2\right] + \cdots + \left[1 - \left(\frac{4}{4}\right)^2\right]$

15. $\left[\left(\frac{2}{n}\right)^3 - \frac{2}{n}\right]\left(\frac{2}{n}\right) + \cdots + \left[\left(\frac{2n}{n}\right)^3 - \frac{2n}{n}\right]\left(\frac{2}{n}\right)$

16. $\left[2\left(1 + \frac{3}{n}\right)^2\right]\left(\frac{3}{n}\right) + \cdots + \left[2\left(1 + \frac{3n}{n}\right)^2\right]\left(\frac{3}{n}\right)$

 **Evaluating a Sum** In Exercises 17–24, use the properties of summation and Theorem 5.2 to evaluate the sum. Use the summation capabilities of a graphing utility to verify your result.

17. $\sum_{i=1}^{12} 7$

18. $\sum_{i=1}^{20} -8$

19. $\sum_{i=1}^{24} 4i$

20. $\sum_{i=1}^{16} (5i - 4)$

21. $\sum_{i=1}^{20} (i - 1)^2$

22. $\sum_{i=1}^{10} (i^2 - 1)$

23. $\sum_{i=1}^{7} i(i + 3)^2$

24. $\sum_{i=1}^{25} (i^3 - 2i)$

 **Evaluating a Sum** In Exercises 25–28, use the summation formulas to rewrite the expression without the summation notation. Use the result to find the sums for $n = 10, 100, 1000,$ and $10,000$.

25. $\sum_{i=1}^{n} \frac{2i + 1}{n^2}$

26. $\sum_{j=1}^{n} \frac{7j + 4}{n^2}$

27. $\sum_{k=1}^{n} \frac{6k(k - 1)}{n^3}$

28. $\sum_{i=1}^{n} \frac{2i^3 - 3i}{n^4}$

 **Approximating the Area of a Plane Region** In Exercises 29–34, use left and right endpoints and the given number of rectangles to find two approximations of the area of the region between the graph of the function and the x-axis over the given interval.

29. $f(x) = 2x + 5$, $[0, 2]$, 4 rectangles

30. $f(x) = 9 - x$, $[2, 4]$, 6 rectangles

31. $g(x) = 2x^2 - x - 1$, $[2, 5]$, 6 rectangles

32. $g(x) = x^2 + 1$, $[1, 3]$, 8 rectangles

33. $f(x) = \cos x$, $\left[0, \frac{\pi}{2}\right]$, 4 rectangles

34. $g(x) = \sin x$, $[0, \pi]$, 6 rectangles

Using Upper and Lower Sums In Exercises 35 and 36, bound the area of the shaded region by approximating the upper and lower sums. Use rectangles of width 1.

35.

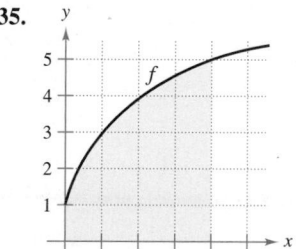

36.

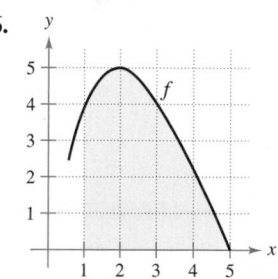

 **Finding Upper and Lower Sums for a Region** In Exercises 37–40, use upper and lower sums to approximate the area of the region using the given number of subintervals (of equal width).

37. $y = \sqrt{x}$

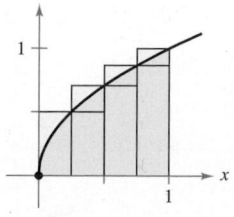

38. $y = 4e^{-x}$

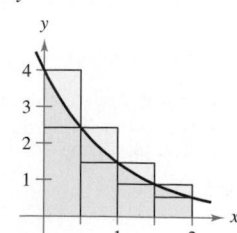

39. $y = \dfrac{1}{x}$

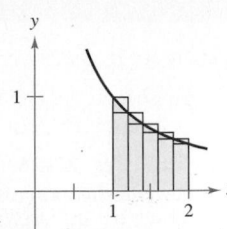

40. $y = \sqrt{1 - x^2}$

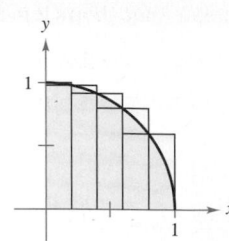

Finding Upper and Lower Sums for a Region In Exercises 41–44, find the upper and lower sums for the region bounded by the graph of the function and the x-axis on the given interval. Leave your answer in terms of n, the number of subintervals.

Function	Interval
41. $f(x) = 3x$	$[0, 4]$
42. $f(x) = 6 - 2x$	$[1, 2]$
43. $f(x) = 5x^2$	$[0, 1]$
44. $f(x) = 9 - x^2$	$[0, 2]$

45. Numerical Reasoning Consider a triangle of area 2 bounded by the graphs of $y = x$, $y = 0$, and $x = 2$.

(a) Sketch the region.

(b) Divide the interval $[0, 2]$ into n subintervals of equal width and show that the endpoints are

$$0 < 1\left(\dfrac{2}{n}\right) < \cdots < (n-1)\left(\dfrac{2}{n}\right) < n\left(\dfrac{2}{n}\right).$$

(c) Show that $s(n) = \displaystyle\sum_{i=1}^{n}\left[(i-1)\left(\dfrac{2}{n}\right)\right]\left(\dfrac{2}{n}\right).$

(d) Show that $S(n) = \displaystyle\sum_{i=1}^{n}\left[i\left(\dfrac{2}{n}\right)\right]\left(\dfrac{2}{n}\right).$

(e) Find $s(n)$ and $S(n)$ for $n = 5, 10, 50,$ and 100.

(f) Show that $\displaystyle\lim_{n\to\infty} s(n) = \lim_{n\to\infty} S(n) = 2.$

46. Numerical Reasoning Consider a trapezoid of area 4 bounded by the graphs of $y = x$, $y = 0$, $x = 1$, and $x = 3$.

(a) Sketch the region.

(b) Divide the interval $[1, 3]$ into n subintervals of equal width and show that the endpoints are

$$1 < 1 + 1\left(\dfrac{2}{n}\right) < \cdots < 1 + (n-1)\left(\dfrac{2}{n}\right) < 1 + n\left(\dfrac{2}{n}\right).$$

(c) Show that $s(n) = \displaystyle\sum_{i=1}^{n}\left[1 + (i-1)\left(\dfrac{2}{n}\right)\right]\left(\dfrac{2}{n}\right).$

(d) Show that $S(n) = \displaystyle\sum_{i=1}^{n}\left[1 + i\left(\dfrac{2}{n}\right)\right]\left(\dfrac{2}{n}\right).$

(e) Find $s(n)$ and $S(n)$ for $n = 5, 10, 50,$ and 100.

(f) Show that $\displaystyle\lim_{n\to\infty} s(n) = \lim_{n\to\infty} S(n) = 4.$

Finding Area by the Limit Definition In Exercises 47–56, use the limit process to find the area of the region bounded by the graph of the function and the x-axis over the given interval. Sketch the region.

47. $y = -4x + 5$, $\quad [0, 1]$ **48.** $y = 3x - 2$, $\quad [2, 5]$

49. $y = x^2 + 2$, $\quad [0, 1]$ **50.** $y = 5x^2 + 1$, $\quad [0, 2]$

51. $y = 25 - x^2$, $\quad [1, 4]$ **52.** $y = 4 - x^2$, $\quad [-2, 2]$

53. $y = 27 - x^3$, $\quad [1, 3]$ **54.** $y = 2x - x^3$, $\quad [0, 1]$

55. $y = x^2 - x^3$, $\quad [-1, 1]$ **56.** $y = 2x^3 - x^2$, $\quad [1, 2]$

Finding Area by the Limit Definition In Exercises 57–62, use the limit process to find the area of the region bounded by the graph of the function and the y-axis over the given y-interval. Sketch the region.

57. $f(y) = 4y$, $\quad 0 \le y \le 2$

58. $g(y) = \frac{1}{2}y$, $\quad 2 \le y \le 4$

59. $f(y) = y^2$, $\quad 0 \le y \le 5$

60. $y = 3y - y^2$, $\quad 2 \le y \le 3$

61. $g(y) = 4y^2 - y^3$, $\quad 1 \le y \le 3$

62. $h(y) = y^3 + 1$, $\quad 1 \le y \le 2$

Approximating Area with the Midpoint Rule In Exercises 63–68, use the Midpoint Rule with $n = 4$ to approximate the area of the region bounded by the graph of the function and the x-axis over the given interval.

63. $f(x) = x^2 + 3$, $\quad [0, 2]$ **64.** $f(x) = x^2 + 4x$, $\quad [0, 4]$

65. $f(x) = \tan x$, $\quad \left[0, \dfrac{\pi}{4}\right]$ **66.** $f(x) = \cos x$, $\quad \left[0, \dfrac{\pi}{2}\right]$

67. $f(x) = \ln x$, $\quad [1, 5]$ **68.** $f(x) = xe^x$, $\quad [0, 2]$

EXPLORING CONCEPTS

69. Approximation Determine which value best approximates the area of the region bounded by the graph of $f(x) = 4 - x^2$ and the x-axis over the interval $[0, 2]$. Make your selection on the basis of a sketch of the region, not by performing calculations.

(a) -2 (b) 6 (c) 10 (d) 3 (e) 8

70. Approximation A function is continuous, nonnegative, concave upward, and decreasing on the interval $[0, a]$. Does using the right endpoints of the subintervals produce an overestimate or an underestimate of the area of the region bounded by the function and the x-axis?

71. Midpoint Rule Explain why the Midpoint Rule almost always results in a better area approximation in comparison to the endpoint method.

72. Midpoint Rule Does the Midpoint Rule ever give the exact area between a function and the x-axis? Explain.

73. Graphical Reasoning Consider the region bounded by the graphs of $f(x) = 8x/(x + 1)$, $x = 0$, $x = 4$, and $y = 0$, as shown in the figure. To print an enlarged copy of the graph, go to *MathGraphs.com*.

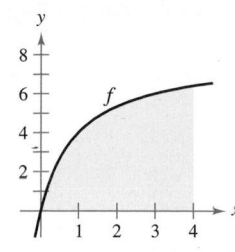

(a) Redraw the figure, and complete and shade the rectangles representing the lower sum when $n = 4$. Find this lower sum.

(b) Redraw the figure, and complete and shade the rectangles representing the upper sum when $n = 4$. Find this upper sum.

(c) Redraw the figure, and complete and shade the rectangles whose heights are determined by the function values at the midpoint of each subinterval when $n = 4$. Find this sum using the Midpoint Rule.

(d) Verify the following formulas for approximating the area of the region using n subintervals of equal width.

Lower sum: $s(n) = \sum_{i=1}^{n} f\left[(i - 1)\frac{4}{n}\right]\left(\frac{4}{n}\right)$

Upper sum: $S(n) = \sum_{i=1}^{n} f\left[(i)\frac{4}{n}\right]\left(\frac{4}{n}\right)$

Midpoint Rule: $M(n) = \sum_{i=1}^{n} f\left[\left(i - \frac{1}{2}\right)\frac{4}{n}\right]\left(\frac{4}{n}\right)$

 (e) Use a graphing utility to create a table of values of $s(n)$, $S(n)$, and $M(n)$ for $n = 4, 8, 20, 100$, and 200.

(f) Explain why $s(n)$ increases and $S(n)$ decreases for increasing values of n, as shown in the table in part (e).

74. HOW DO YOU SEE IT? The function shown in the graph below is increasing on the interval $[1, 4]$. The interval will be divided into 12 subintervals.

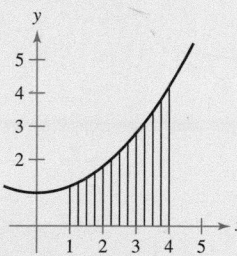

(a) What are the left endpoints of the first and last subintervals?

(b) What are the right endpoints of the first two subintervals?

(c) When using the right endpoints, do the rectangles lie above or below the graph of the function?

(d) What can you conclude about the heights of the rectangles when the function is constant on the given interval?

True or False? In Exercises 75 and 76, determine whether the statement is true or false. If it is false, explain why or give an example that shows it is false.

75. The sum of the first n positive integers is $n(n + 1)/2$.

76. If f is continuous and nonnegative on $[a, b]$, then the limits as $n \to \infty$ of its lower sum $s(n)$ and upper sum $S(n)$ both exist and are equal.

77. Writing Use the figure to write a short paragraph explaining why the formula

$$1 + 2 + \cdots + n = \frac{1}{2}n(n + 1)$$

is valid for all positive integers n.

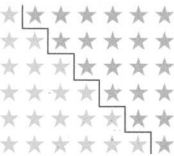

78. Graphical Reasoning Consider an n-sided regular polygon inscribed in a circle of radius r. Join the vertices of the polygon to the center of the circle, forming n congruent triangles (see figure).

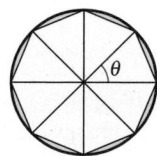

(a) Determine the central angle θ in terms of n.

(b) Show that the area of each triangle is $\frac{1}{2}r^2 \sin \theta$.

(c) Let A_n be the sum of the areas of the n triangles. Find $\lim_{n \to \infty} A_n$.

79. Seating Capacity A teacher places n seats to form the back row of a classroom layout. Each successive row contains two fewer seats than the preceding row. Find a formula for the number of seats used in the layout. (*Hint:* The number of seats in the layout depends on whether n is odd or even.)

80. Proof Prove each formula by mathematical induction. (You may need to review the method of proof by induction from a precalculus text.)

(a) $\sum_{i=1}^{n} 2i = n(n + 1)$ (b) $\sum_{i=1}^{n} i^3 = \frac{n^2(n + 1)^2}{4}$

PUTNAM EXAM CHALLENGE

81. A dart, thrown at random, hits a square target. Assuming that any two parts of the target of equal area are equally likely to be hit, find the probability that the point hit is nearer to the center than to any edge. Write your answer in the form $(a\sqrt{b} + c)/d$, where a, b, c, and d are integers.

5.3 Riemann Sums and Definite Integrals

■ Understand the definition of a Riemann sum.
■ Evaluate a definite integral using limits and geometric formulas.
■ Evaluate a definite integral using properties of definite integrals.

Riemann Sums

In the definition of area given in Section 5.2, the partitions have subintervals of *equal width*. This was done only for computational convenience. The next example shows that it is not necessary to have subintervals of equal width.

EXAMPLE 1 **A Partition with Subintervals of Unequal Widths**

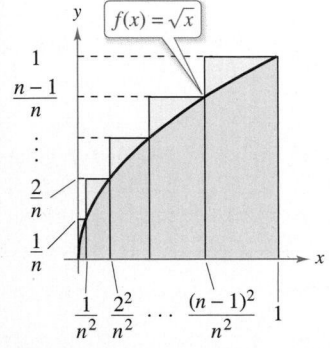

The subintervals do not have equal widths.

Figure 5.18

Consider the region bounded by the graph of $f(x) = \sqrt{x}$ and the x-axis for $0 \leq x \leq 1$, as shown in Figure 5.18. Evaluate the limit

$$\lim_{n \to \infty} \sum_{i=1}^{n} f(c_i)\, \Delta x_i$$

where c_i is the right endpoint of the partition given by $c_i = i^2/n^2$ and Δx_i is the width of the ith interval.

Solution The width of the ith interval is

$$\Delta x_i = \frac{i^2}{n^2} - \frac{(i-1)^2}{n^2}$$

$$= \frac{i^2 - i^2 + 2i - 1}{n^2}$$

$$= \frac{2i - 1}{n^2}.$$

So, the limit is

$$\lim_{n \to \infty} \sum_{i=1}^{n} f(c_i)\, \Delta x_i = \lim_{n \to \infty} \sum_{i=1}^{n} \sqrt{\frac{i^2}{n^2}} \left(\frac{2i - 1}{n^2} \right)$$

$$= \lim_{n \to \infty} \frac{1}{n^3} \sum_{i=1}^{n} (2i^2 - i)$$

$$= \lim_{n \to \infty} \frac{1}{n^3} \left[2 \left(\frac{n(n+1)(2n+1)}{6} \right) - \frac{n(n+1)}{2} \right]$$

$$= \lim_{n \to \infty} \frac{4n^3 + 3n^2 - n}{6n^3}$$

$$= \lim_{n \to \infty} \left(\frac{2}{3} + \frac{1}{2n} - \frac{1}{6n^2} \right)$$

$$= \frac{2}{3}.$$

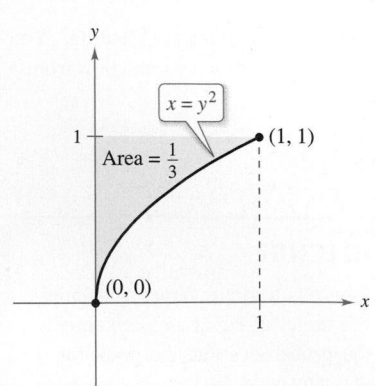

The area of the region bounded by the graph of $x = y^2$ and the y-axis for $0 \leq y \leq 1$ is $\frac{1}{3}$.

Figure 5.19

From Example 7 in Section 5.2, you know that the region shown in Figure 5.19 has an area of $\frac{1}{3}$. Because the square bounded by $0 \leq x \leq 1$ and $0 \leq y \leq 1$ has an area of 1, you can conclude that the area of the region shown in Figure 5.18 has an area of $\frac{2}{3}$. This agrees with the limit found in Example 1, even though that example used a partition having subintervals of unequal widths. The reason this particular partition gave the proper area is that as n increases, the *width of the largest subinterval approaches zero*. This is a key feature of the development of definite integrals.

In Section 5.2, the limit of a sum was used to define the area of a region in the plane. Finding area by this method is only one of *many* applications involving the limit of a sum. A similar approach can be used to determine quantities as diverse as arc lengths, average values, centroids, volumes, work, and surface areas. The next definition is named after Georg Friedrich Bernhard Riemann. Although the definite integral had been defined and used long before Riemann's time, he generalized the concept to cover a broader category of functions.

In the definition of a Riemann sum below, note that the function f has no restrictions other than being defined on the interval $[a, b]$. (In Section 5.2, the function f was assumed to be continuous and nonnegative because you were finding the area under a curve.)

Definition of Riemann Sum

Let f be defined on the closed interval $[a, b]$, and let Δ be a partition of $[a, b]$ given by

$$a = x_0 < x_1 < x_2 < \cdots < x_{n-1} < x_n = b$$

where Δx_i is the width of the ith subinterval

$$[x_{i-1}, x_i].$$ *ith subinterval*

If c_i is *any* point in the ith subinterval, then the sum

$$\sum_{i=1}^{n} f(c_i) \, \Delta x_i, \quad x_{i-1} \le c_i \le x_i$$

is called a **Riemann sum** of f for the partition Δ. (The sums in Section 5.2 are examples of Riemann sums, but there are more general Riemann sums than those covered there.)

The width of the largest subinterval of a partition Δ is the **norm** of the partition and is denoted by $\|\Delta\|$. If every subinterval is of equal width, then the partition is **regular** and the norm is denoted by

$$\|\Delta\| = \Delta x = \frac{b - a}{n}.$$ Regular partition

For a **general partition,** the norm is related to the number of subintervals of $[a, b]$ in the following way.

$$\frac{b - a}{\|\Delta\|} \le n$$ General partition

So, the number of subintervals in a partition approaches infinity as the norm of the partition approaches 0. That is, $\|\Delta\| \to 0$ implies that $n \to \infty$.

The converse of this statement is not true. For example, let Δ_n be the partition of the interval $[0, 1]$ given by

$$0 < \frac{1}{2^n} < \frac{1}{2^{n-1}} < \cdots < \frac{1}{8} < \frac{1}{4} < \frac{1}{2} < 1.$$

As shown in Figure 5.20, for any positive value of n, the norm of the partition Δ_n is $\frac{1}{2}$. So, letting n approach infinity does not force $\|\Delta\|$ to approach 0. In a regular partition, however, the statements

$$\|\Delta\| \to 0 \quad \text{and} \quad n \to \infty$$

are equivalent.

$\|\Delta\| = \frac{1}{2}$

$n \to \infty$ does not imply that $\|\Delta\| \to 0$.
Figure 5.20

Definite Integrals

■ **FOR FURTHER INFORMATION**
For insight into the history of the definite integral, see the article "The Evolution of Integration" by A. Shenitzer and J. Steprāns in *The American Mathematical Monthly.* To view this article, go to *MathArticles.com.*

To define the definite integral, consider the limit

$$\lim_{\|\Delta\| \to 0} \sum_{i=1}^{n} f(c_i)\, \Delta x_i = L.$$

To say that this limit exists means there exists a real number L such that for each $\varepsilon > 0$, there exists a $\delta > 0$ such that for every partition with $\|\Delta\| < \delta$, it follows that

$$\left| L - \sum_{i=1}^{n} f(c_i)\, \Delta x_i \right| < \varepsilon$$

regardless of the choice of c_i in the ith subinterval of each partition Δ.

- - - - - - - - - - - - - - - ▷

·· REMARK Later in this chapter, you will learn convenient methods for calculating $\int_a^b f(x)\, dx$ for continuous functions. For now, you must use the limit definition.

Definition of Definite Integral

If f is defined on the closed interval $[a, b]$ and the limit of Riemann sums over partitions Δ

$$\lim_{\|\Delta\| \to 0} \sum_{i=1}^{n} f(c_i)\, \Delta x_i$$

exists (as described above), then f is said to be **integrable** on $[a, b]$ and the limit is denoted by

$$\lim_{\|\Delta\| \to 0} \sum_{i=1}^{n} f(c_i)\, \Delta x_i = \int_a^b f(x)\, dx.$$

The limit is called the **definite integral** of f from a to b. The number a is the **lower limit** of integration, and the number b is the **upper limit** of integration.

It is not a coincidence that the notation for definite integrals is similar to that used for indefinite integrals. You will see why in the next section when the Fundamental Theorem of Calculus is introduced. For now, it is important to see that definite integrals and indefinite integrals are different concepts. A definite integral is a *number*, whereas an indefinite integral is a *family of functions.*

Though Riemann sums were defined for functions with very few restrictions, a sufficient condition for a function f to be integrable on $[a, b]$ is that it is continuous on $[a, b]$. A proof of this theorem is beyond the scope of this text.

THEOREM 5.4   Continuity Implies Integrability

If a function f is continuous on the closed interval $[a, b]$, then f is integrable on $[a, b]$. That is, $\int_a^b f(x)\, dx$ exists.

Exploration

The Converse of Theorem 5.4   Is the converse of Theorem 5.4 true? That is, when a function is integrable, does it have to be continuous? Explain your reasoning and give examples.

Describe the relationships among continuity, differentiability, and integrability. Which is the strongest condition? Which is the weakest? Which conditions imply other conditions?

Evaluating a Definite Integral as a Limit

Evaluate the definite integral $\displaystyle\int_{-2}^{1} 2x\,dx$.

Solution The function $f(x) = 2x$ is integrable on the interval $[-2, 1]$ because it is continuous on $[-2, 1]$. Moreover, the definition of integrability implies that any partition whose norm approaches 0 can be used to determine the limit. For computational convenience, define Δ by subdividing $[-2, 1]$ into n subintervals of equal width

$$\Delta x_i = \Delta x = \frac{b - a}{n} = \frac{3}{n}.$$

Choosing c_i as the right endpoint of each subinterval produces

$$c_i = a + i(\Delta x) = -2 + \frac{3i}{n}.$$

So, the definite integral is

$$\int_{-2}^{1} 2x\,dx = \lim_{\|\Delta\|\to 0} \sum_{i=1}^{n} f(c_i)\,\Delta x_i$$

$$= \lim_{n\to\infty} \sum_{i=1}^{n} f(c_i)\,\Delta x$$

$$= \lim_{n\to\infty} \sum_{i=1}^{n} 2\left(-2 + \frac{3i}{n}\right)\left(\frac{3}{n}\right)$$

$$= \lim_{n\to\infty} \frac{6}{n} \sum_{i=1}^{n}\left(-2 + \frac{3i}{n}\right)$$

$$= \lim_{n\to\infty} \frac{6}{n}\left(-2\sum_{i=1}^{n} 1 + \frac{3}{n}\sum_{i=1}^{n} i\right)$$

$$= \lim_{n\to\infty} \frac{6}{n}\left\{-2n + \frac{3}{n}\left[\frac{n(n+1)}{2}\right]\right\}$$

$$= \lim_{n\to\infty}\left(-12 + 9 + \frac{9}{n}\right)$$

$$= -3.$$

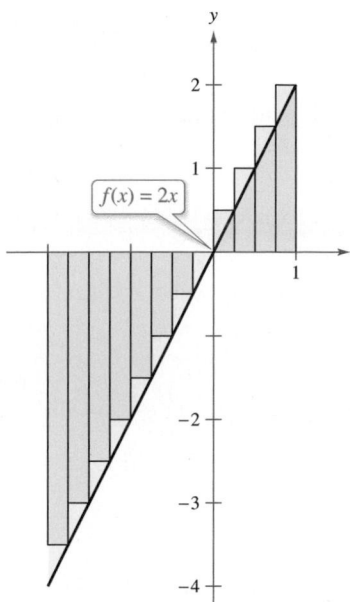

$f(x) = 2x$

Because the definite integral is negative, it does not represent the area of the region.
Figure 5.21

Because the definite integral in Example 2 is negative, it *does not* represent the area of the region shown in Figure 5.21. Definite integrals can be positive, negative, or zero. For a definite integral to be interpreted as an area (as defined in Section 5.2), the function f must be continuous and nonnegative on $[a, b]$, as stated in the next theorem. The proof of this theorem is straightforward—you simply use the definition of area given in Section 5.2, because it is a Riemann sum.

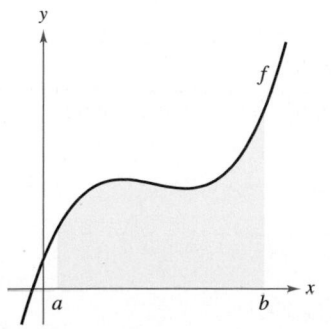

You can use a definite integral to find the area of the region bounded by the graph of f, the x-axis, $x = a$, and $x = b$.
Figure 5.22

THEOREM 5.5 **The Definite Integral as the Area of a Region**

If f is continuous and nonnegative on the closed interval $[a, b]$, then the area of the region bounded by the graph of f, the x-axis, and the vertical lines $x = a$ and $x = b$ is

$$\text{Area} = \int_{a}^{b} f(x)\,dx.$$

(See Figure 5.22.)

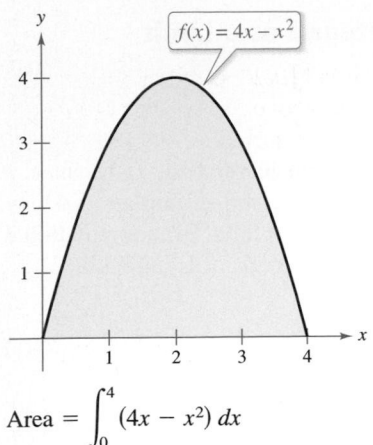

$$\text{Area} = \int_0^4 (4x - x^2)\, dx$$

Figure 5.23

As an example of Theorem 5.5, consider the region bounded by the graph of

$$f(x) = 4x - x^2$$

and the x-axis, as shown in Figure 5.23. Because f is continuous and nonnegative on the closed interval $[0, 4]$, the area of the region is

$$\text{Area} = \int_0^4 (4x - x^2)\, dx.$$

A straightforward technique for evaluating a definite integral such as this will be discussed in Section 5.4. For now, however, you can evaluate a definite integral in two ways—you can use the limit definition *or* you can check to see whether the definite integral represents the area of a common geometric region, such as a rectangle, triangle, or semicircle.

EXAMPLE 3   **Areas of Common Geometric Figures**

Sketch the region corresponding to each definite integral. Then evaluate each integral using a geometric formula.

a. $\displaystyle\int_1^3 4\, dx$ **b.** $\displaystyle\int_0^3 (x + 2)\, dx$ **c.** $\displaystyle\int_{-2}^2 \sqrt{4 - x^2}\, dx$

Solution   A sketch of each region is shown in Figure 5.24.

a. This region is a rectangle of height 4 and width 2.

$$\int_1^3 4\, dx = (\text{Area of rectangle}) = 4(2) = 8$$

b. This region is a trapezoid with an altitude of 3 and parallel bases of lengths 2 and 5. The formula for the area of a trapezoid is $\frac{1}{2}h(b_1 + b_2)$.

$$\int_0^3 (x + 2)\, dx = (\text{Area of trapezoid}) = \frac{1}{2}(3)(2 + 5) = \frac{21}{2}$$

c. This region is a semicircle of radius 2. The formula for the area of a semicircle is $\frac{1}{2}\pi r^2$.

$$\int_{-2}^2 \sqrt{4 - x^2}\, dx = (\text{Area of semicircle}) = \frac{1}{2}\pi(2^2) = 2\pi$$

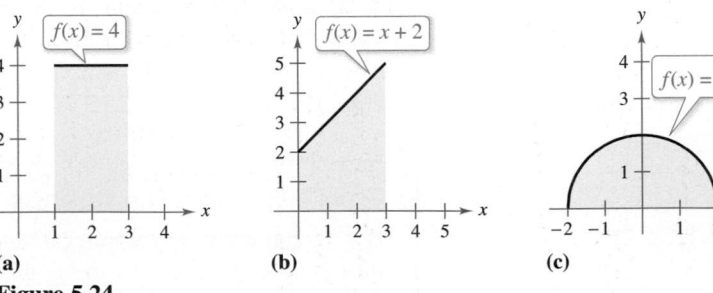

(a)   (b)   (c)
Figure 5.24

The variable of integration in a definite integral is sometimes called a *dummy variable* because it can be replaced by any other variable without changing the value of the integral. For instance, the definite integrals

$$\int_0^3 (x + 2)\, dx \quad \text{and} \quad \int_0^3 (t + 2)\, dt$$

have the same value.

Properties of Definite Integrals

The definition of the definite integral of f on the interval $[a, b]$ specifies that $a < b$. Now, however, it is convenient to extend the definition to cover cases in which $a = b$ or $a > b$. Geometrically, the next two definitions seem reasonable. For instance, it makes sense to define the area of a region of zero width and finite height to be 0.

Definitions of Two Special Definite Integrals

1. If f is defined at $x = a$, then $\displaystyle\int_a^a f(x)\, dx = 0$.

2. If f is integrable on $[a, b]$, then $\displaystyle\int_b^a f(x)\, dx = -\int_a^b f(x)\, dx$.

EXAMPLE 4 **Evaluating Definite Integrals**

⋮····▷ *See LarsonCalculus.com for an interactive version of this type of example.*

Evaluate each definite integral.

a. $\displaystyle\int_\pi^\pi \sin x\, dx$ **b.** $\displaystyle\int_3^0 (x + 2)\, dx$

Solution

a. Because the sine function is defined at $x = \pi$, and the upper and lower limits of integration are equal, you can write

$$\int_\pi^\pi \sin x\, dx = 0.$$

b. The integral $\int_3^0 (x + 2)\, dx$ is the same as that given in Example 3(b) except that the upper and lower limits are interchanged. Because the integral in Example 3(b) has a value of $\frac{21}{2}$, you can write

$$\int_3^0 (x + 2)\, dx = -\int_0^3 (x + 2)\, dx = -\frac{21}{2}.$$ ∎

In Figure 5.25, the larger region can be divided at $x = c$ into two subregions whose intersection is a line segment. Because the line segment has zero area, it follows that the area of the larger region is equal to the sum of the areas of the two smaller regions.

THEOREM 5.6 Additive Interval Property

If f is integrable on the three closed intervals determined by a, b, and c, then

$$\int_a^b f(x)\, dx = \int_a^c f(x)\, dx + \int_c^b f(x)\, dx. \qquad \text{See Figure 5.25.}$$

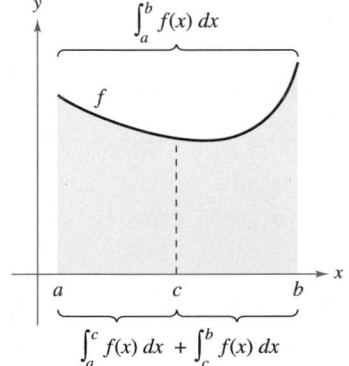

Figure 5.25

EXAMPLE 5 **Using the Additive Interval Property**

$$\int_{-1}^1 |x|\, dx = \int_{-1}^0 -x\, dx + \int_0^1 x\, dx \qquad \text{Theorem 5.6}$$

$$= \frac{1}{2} + \frac{1}{2} \qquad \text{Area of a triangle}$$

$$= 1$$ ∎

Because the definite integral is defined as the limit of a sum, it inherits the properties of summation given at the top of page 295.

THEOREM 5.7   Properties of Definite Integrals

If f and g are integrable on $[a, b]$ and k is a constant, then the functions kf and $f \pm g$ are integrable on $[a, b]$, and

1. $\displaystyle \int_a^b kf(x)\, dx = k \int_a^b f(x)\, dx$

2. $\displaystyle \int_a^b [f(x) \pm g(x)]\, dx = \int_a^b f(x)\, dx \pm \int_a^b g(x)\, dx.$

• • • • • • • • • • • • • • • ▷

· · REMARK   Property 2 of Theorem 5.7 can be extended to cover any finite number of functions (see Example 6).

EXAMPLE 6   **Evaluation of a Definite Integral**

Evaluate $\displaystyle \int_1^3 (-x^2 + 4x - 3)\, dx$ using each of the following values.

$$\int_1^3 x^2\, dx = \frac{26}{3}, \qquad \int_1^3 x\, dx = 4, \qquad \int_1^3 dx = 2$$

Solution

$$\int_1^3 (-x^2 + 4x - 3)\, dx = \int_1^3 (-x^2)\, dx + \int_1^3 4x\, dx + \int_1^3 (-3)\, dx$$

$$= -\int_1^3 x^2\, dx + 4\int_1^3 x\, dx - 3\int_1^3 dx$$

$$= -\left(\frac{26}{3}\right) + 4(4) - 3(2)$$

$$= \frac{4}{3}$$

If f and g are continuous on the closed interval $[a, b]$ and $0 \le f(x) \le g(x)$ for $a \le x \le b$, then the following properties are true. First, the area of the region bounded by the graph of f and the x-axis (between a and b) must be nonnegative. Second, this area must be less than or equal to the area of the region bounded by the graph of g and the x-axis (between a and b), as shown in Figure 5.26. These two properties are generalized in Theorem 5.8.

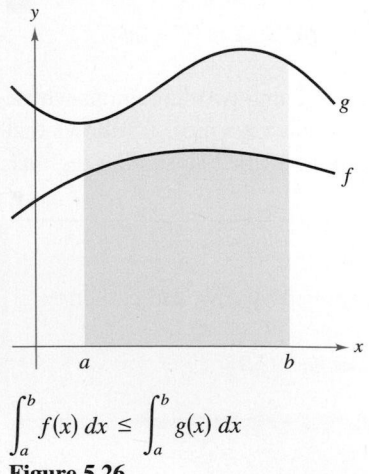

$\displaystyle \int_a^b f(x)\, dx \le \int_a^b g(x)\, dx$

Figure 5.26

THEOREM 5.8   Preservation of Inequality

1. If f is integrable and nonnegative on the closed interval $[a, b]$, then

$$0 \le \int_a^b f(x)\, dx.$$

2. If f and g are integrable on the closed interval $[a, b]$ and $f(x) \le g(x)$ for every x in $[a, b]$, then

$$\int_a^b f(x)\, dx \le \int_a^b g(x)\, dx.$$

A proof of this theorem is given in Appendix A.

5.3 Exercises

See **CalcChat.com** for tutorial help and worked-out solutions to odd-numbered exercises.

CONCEPT CHECK

1. Riemann Sum What does a Riemann sum represent?

2. Definite Integral Explain how to find the area of a region using a definite integral in your own words.

Evaluating a Limit In Exercises 3 and 4, use Example 1 as a model to evaluate the limit

$$\lim_{n \to \infty} \sum_{i=1}^{n} f(c_i) \, \Delta x_i$$

over the region bounded by the graphs of the equations.

3. $f(x) = \sqrt{x}, \quad y = 0, \quad x = 0, \quad x = 3 \quad \left(\text{Hint: Let } c_i = \dfrac{3i^2}{n^2}.\right)$

4. $f(x) = \sqrt[3]{x}, \quad y = 0, \quad x = 0, \quad x = 1 \quad \left(\text{Hint: Let } c_i = \dfrac{i^3}{n^3}.\right)$

Evaluating a Definite Integral as a Limit In Exercises 5–10, evaluate the definite integral by the limit definition.

5. $\displaystyle\int_{2}^{6} 8 \, dx$

6. $\displaystyle\int_{-2}^{3} x \, dx$

7. $\displaystyle\int_{-1}^{1} x^3 \, dx$

8. $\displaystyle\int_{1}^{4} 4x^2 \, dx$

9. $\displaystyle\int_{1}^{2} (x^2 + 1) \, dx$

10. $\displaystyle\int_{-2}^{1} (2x^2 + 3) \, dx$

Writing a Limit as a Definite Integral In Exercises 11–14, write the limit as a definite integral on the given interval, where c_i is any point in the ith subinterval.

11. $\displaystyle\lim_{\|\Delta\| \to 0} \sum_{i=1}^{n} (3c_i + 10) \, \Delta x_i, \quad [-1, 5]$

12. $\displaystyle\lim_{\|\Delta\| \to 0} \sum_{i=1}^{n} \sqrt{c_i^2 + 4} \, \Delta x_i, \quad [0, 3]$

13. $\displaystyle\lim_{\|\Delta\| \to 0} \sum_{i=1}^{n} \left(1 + \dfrac{3}{c_i}\right) \Delta x_i, \quad [1, 5]$

14. $\displaystyle\lim_{\|\Delta\| \to 0} \sum_{i=1}^{n} (2^{-c_i} \sin c_i) \, \Delta x_i, \quad [0, \pi]$

Writing a Definite Integral In Exercises 15–26, write a definite integral that represents the area of the region. (Do not evaluate the integral.)

15. $f(x) = 5$

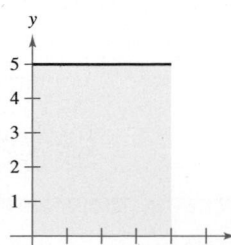

16. $f(x) = 6 - 3x$

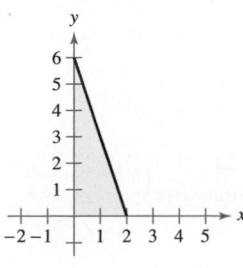

17. $f(x) = 4 - |x|$

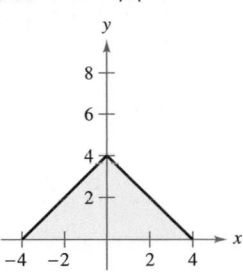

18. $f(x) = x^2$

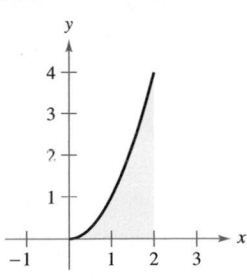

19. $f(x) = 25 - x^2$

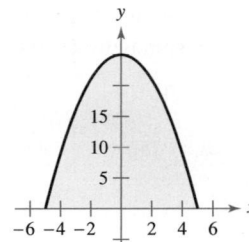

20. $f(x) = \dfrac{4}{x^2 + 2}$

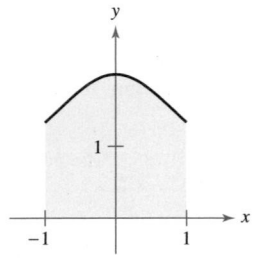

21. $f(x) = \cos x$

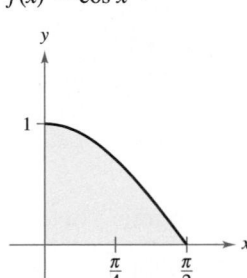

22. $f(x) = \tan x$

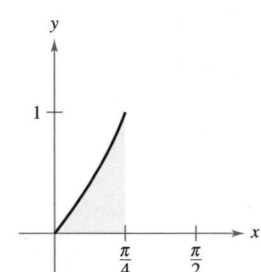

23. $g(y) = y^3$

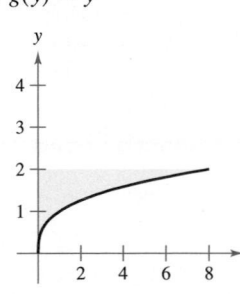

24. $f(y) = (y - 2)^2$

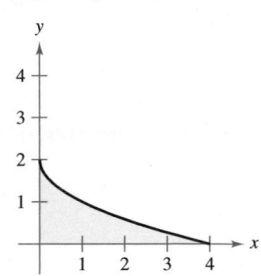

25. $f(x) = \dfrac{2}{x}$

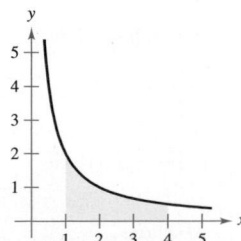

26. $f(x) = 2e^{-x}$

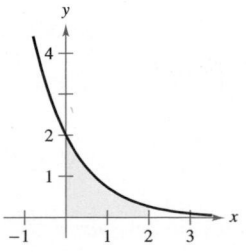

 **Evaluating a Definite Integral Using a Geometric Formula** In Exercises 27–36, sketch the region whose area is given by the definite integral. Then use a geometric formula to evaluate the integral $(a > 0, r > 0)$.

27. $\displaystyle\int_0^3 4 \, dx$

28. $\displaystyle\int_{-3}^4 9 \, dx$

29. $\displaystyle\int_0^4 x \, dx$

30. $\displaystyle\int_0^8 \frac{x}{4} \, dx$

31. $\displaystyle\int_0^2 (3x + 4) \, dx$

32. $\displaystyle\int_0^3 (8 - 2x) \, dx$

33. $\displaystyle\int_{-1}^1 (1 - |x|) \, dx$

34. $\displaystyle\int_{-a}^a (a - |x|) \, dx$

35. $\displaystyle\int_{-7}^7 \sqrt{49 - x^2} \, dx$

36. $\displaystyle\int_{-r}^r \sqrt{r^2 - x^2} \, dx$

 **Using Properties of Definite Integrals** In Exercises 37–44, evaluate the definite integral using the values below.

$$\int_2^6 x^3 \, dx = 320, \qquad \int_2^6 x \, dx = 16, \qquad \int_2^6 dx = 4$$

37. $\displaystyle\int_6^2 x^3 \, dx$

38. $\displaystyle\int_2^2 x \, dx$

39. $\displaystyle\int_2^6 \frac{1}{4}x^3 \, dx$

40. $\displaystyle\int_2^6 -3x \, dx$

41. $\displaystyle\int_2^6 (x - 14) \, dx$

42. $\displaystyle\int_2^6 \left(6x - \frac{1}{8}x^3\right) dx$

43. $\displaystyle\int_2^6 (2x^3 - 10x + 7) \, dx$

44. $\displaystyle\int_2^6 (21 - 5x - x^3) \, dx$

45. **Using Properties of Definite Integrals** Given $\displaystyle\int_0^5 f(x) \, dx = 10$ and $\displaystyle\int_5^7 f(x) \, dx = 3$, evaluate

(a) $\displaystyle\int_0^7 f(x) \, dx.$

(b) $\displaystyle\int_5^0 f(x) \, dx.$

(c) $\displaystyle\int_5^5 f(x) \, dx.$

(d) $\displaystyle\int_0^5 3f(x) \, dx.$

46. **Using Properties of Definite Integrals** Given $\displaystyle\int_0^3 f(x) \, dx = 4$ and $\displaystyle\int_3^6 f(x) \, dx = -1$, evaluate

(a) $\displaystyle\int_0^6 f(x) \, dx.$

(b) $\displaystyle\int_6^3 f(x) \, dx.$

(c) $\displaystyle\int_3^3 f(x) \, dx.$

(d) $\displaystyle\int_3^6 -5f(x) \, dx.$

47. **Using Properties of Definite Integrals** Given $\displaystyle\int_2^6 f(x) \, dx = 10$ and $\displaystyle\int_2^6 g(x) \, dx = -2$, evaluate

(a) $\displaystyle\int_2^6 [f(x) + g(x)] \, dx.$

(b) $\displaystyle\int_2^6 [g(x) - f(x)] \, dx.$

(c) $\displaystyle\int_2^6 2g(x) \, dx.$

(d) $\displaystyle\int_2^6 3f(x) \, dx.$

48. **Using Properties of Definite Integrals** Given $\displaystyle\int_{-1}^1 f(x) \, dx = 0$ and $\displaystyle\int_0^1 f(x) \, dx = 5$, evaluate

(a) $\displaystyle\int_{-1}^0 f(x) \, dx.$

(b) $\displaystyle\int_0^1 f(x) \, dx - \int_{-1}^0 f(x) \, dx.$

(c) $\displaystyle\int_{-1}^1 3f(x) \, dx.$

(d) $\displaystyle\int_0^1 3f(x) \, dx.$

49. **Estimating a Definite Integral** Use the table of values to find lower and upper estimates of

$$\int_0^{10} f(x) \, dx.$$

Assume that f is a decreasing function.

x	0	2	4	6	8	10
$f(x)$	32	24	12	-4	-20	-36

50. **Estimating a Definite Integral** Use the table of values to estimate

$$\int_0^6 f(x) \, dx.$$

Use three equal subintervals and the (a) left endpoints, (b) right endpoints, and (c) midpoints. When f is an increasing function, how does each estimate compare with the actual value? Explain your reasoning.

x	0	1	2	3	4	5	6
$f(x)$	-6	0	8	18	30	50	80

51. **Think About It** The graph of f consists of line segments and a semicircle, as shown in the figure. Evaluate each definite integral by using geometric formulas.

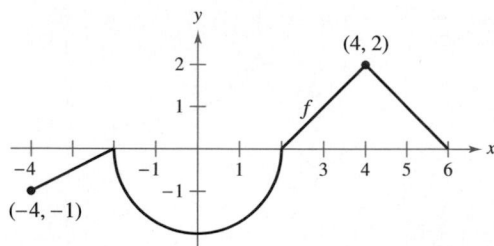

(a) $\displaystyle\int_0^2 f(x) \, dx$

(b) $\displaystyle\int_2^6 f(x) \, dx$

(c) $\displaystyle\int_{-4}^2 f(x) \, dx$

(d) $\displaystyle\int_{-4}^6 f(x) \, dx$

(e) $\displaystyle\int_{-4}^6 |f(x)| \, dx$

(f) $\displaystyle\int_{-4}^6 [f(x) + 2] \, dx$

52. Think About It The graph of f consists of line segments, as shown in the figure. Evaluate each definite integral by using geometric formulas.

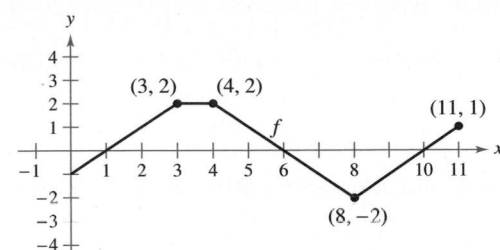

(a) $\int_0^1 -f(x)\,dx$ (b) $\int_3^4 3f(x)\,dx$

(c) $\int_0^7 f(x)\,dx$ (d) $\int_5^{11} f(x)\,dx$

(e) $\int_0^{11} f(x)\,dx$ (f) $\int_4^{10} f(x)\,dx$

53. Think About It Consider a function f that is continuous on the interval $[-5, 5]$ and for which

$$\int_0^5 f(x)\,dx = 4.$$

Evaluate each integral.

(a) $\int_0^5 [f(x) + 2]\,dx$ (b) $\int_{-2}^3 f(x + 2)\,dx$

(c) $\int_{-5}^5 f(x)\,dx,\ f$ is even (d) $\int_{-5}^5 f(x)\,dx,\ f$ is odd

54. HOW DO YOU SEE IT? Use the figure to fill in the blank with the symbol $<$, $>$, or $=$. Explain your reasoning.

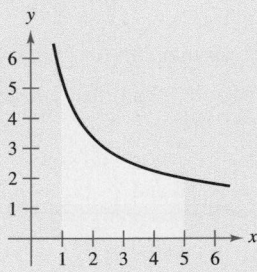

(a) The interval $[1, 5]$ is partitioned into n subintervals of equal width Δx, and x_i is the left endpoint of the ith subinterval.

$$\sum_{i=1}^n f(x_i)\,\Delta x \quad \rule{1cm}{0.4pt} \quad \int_1^5 f(x)\,dx$$

(b) The interval $[1, 5]$ is partitioned into n subintervals of equal width Δx, and x_i is the right endpoint of the ith subinterval.

$$\sum_{i=1}^n f(x_i)\,\Delta x \quad \rule{1cm}{0.4pt} \quad \int_1^5 f(x)\,dx$$

55. Think About It A function f is defined below. Use geometric formulas to find $\int_0^8 f(x)\,dx$.

$$f(x) = \begin{cases} 4, & x < 4 \\ x, & x \ge 4 \end{cases}$$

56. Think About It A function f is defined below. Use geometric formulas to find $\int_0^{12} f(x)\,dx$.

$$f(x) = \begin{cases} 6, & x > 6 \\ -\frac{1}{2}x + 9, & x \le 6 \end{cases}$$

EXPLORING CONCEPTS

Approximation In Exercises 57–60, determine which value best approximates the definite integral. Make your selection on the basis of a sketch.

57. $\int_0^4 \sqrt{x}\,dx$

(a) 5 (b) -3 (c) 10 (d) 2 (e) 8

58. $\int_0^{1/2} 4\cos \pi x\,dx$

(a) 4 (b) $\frac{4}{3}$ (c) 16 (d) 2π (e) -6

59. $\int_0^2 2e^{-x^2}\,dx$

(a) $\frac{1}{3}$ (b) 6 (c) 2 (d) 4

60. $\int_1^2 \ln x\,dx$

(a) $\frac{1}{3}$ (b) 1 (c) 4 (d) 3

61. Verifying a Rule Use a graph to explain why

$$\int_a^a f(x)\,dx = 0 \text{ if } f \text{ is defined at } x = a.$$

62. Verifying a Property Use a graph to explain why

$$\int_a^b kf(x)\,dx = k\int_a^b f(x)\,dx$$

if f is integrable on $[a, b]$ and k is a constant.

63. Using Different Methods Describe two ways to evaluate $\int_{-1}^3 (x + 2)\,dx$. Verify that each method gives the same result.

64. Finding a Function Give an example of a function that is integrable on the interval $[-1, 1]$ but not continuous on $[-1, 1]$.

Finding Values In Exercises 65–68, find possible values of a and b that make the statement true. If possible, use a graph to support your answer. (There may be more than one correct answer.)

65. $\int_{-2}^1 f(x)\,dx + \int_1^5 f(x)\,dx = \int_a^b f(x)\,dx$

66. $\int_{-3}^3 f(x)\,dx + \int_3^6 f(x)\,dx - \int_a^b f(x)\,dx = \int_{-1}^6 f(x)\,dx$

67. $\int_a^b \sin x\,dx < 0$ **68.** $\int_a^b \cos x\,dx = 0$

True or False? In Exercises 69–74, determine whether the statement is true or false. If it is false, explain why or give an example that shows it is false.

69. $\int_a^b [f(x) + g(x)] \, dx = \int_a^b f(x) \, dx + \int_a^b g(x) \, dx$

70. $\int_a^b f(x)g(x) \, dx = \left[\int_a^b f(x) \, dx \right]\left[\int_a^b g(x) \, dx \right]$

71. If the norm of a partition approaches zero, then the number of subintervals approaches infinity.

72. If f is increasing on $[a, b]$, then the minimum value of f on $[a, b]$ is $f(a)$.

73. The value of

$$\int_a^b f(x) \, dx$$

must be positive.

74. The value of

$$\int_2^2 \sin x^2 \, dx$$

is 0.

75. **Finding a Riemann Sum** Find the Riemann sum for $f(x) = x^2 + 3x$ over the interval $[0, 8]$, where

$$x_0 = 0, \quad x_1 = 1, \quad x_2 = 3, \quad x_3 = 7, \quad \text{and} \quad x_4 = 8$$

and where

$$c_1 = 1, \quad c_2 = 2, \quad c_3 = 5, \quad \text{and} \quad c_4 = 8.$$

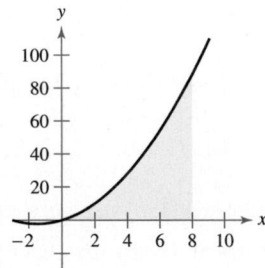

76. **Finding a Riemann Sum** Find the Riemann sum for $f(x) = \sin x$ over the interval $[0, 2\pi]$, where

$$x_0 = 0, \quad x_1 = \frac{\pi}{4}, \quad x_2 = \frac{\pi}{3}, \quad x_3 = \pi, \quad \text{and} \quad x_4 = 2\pi$$

and where

$$c_1 = \frac{\pi}{6}, \quad c_2 = \frac{\pi}{3}, \quad c_3 = \frac{2\pi}{3}, \quad \text{and} \quad c_4 = \frac{3\pi}{2}.$$

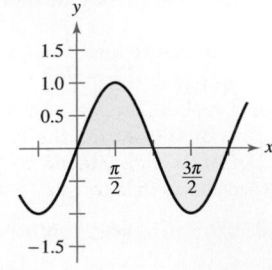

77. **Proof** Prove that $\int_a^b x \, dx = \frac{b^2 - a^2}{2}$.

78. **Proof** Prove that $\int_a^b x^2 \, dx = \frac{b^3 - a^3}{3}$.

79. **Think About It** Determine whether the Dirichlet function

$$f(x) = \begin{cases} 1, & x \text{ is rational} \\ 0, & x \text{ is irrational} \end{cases}$$

is integrable on the interval $[0, 1]$. Explain.

80. **Finding a Definite Integral** The function

$$f(x) = \begin{cases} 0, & x = 0 \\ \dfrac{1}{x}, & 0 < x \le 1 \end{cases}$$

is defined on $[0, 1]$, as shown in the figure. Show that

$$\int_0^1 f(x) \, dx$$

does not exist. Does this contradict Theorem 5.4? Why or why not?

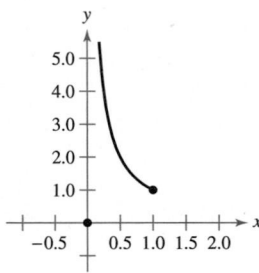

81. **Finding Values** Find the constants a and b that maximize the value of

$$\int_a^b (1 - x^2) \, dx.$$

Explain your reasoning.

82. **Finding Values** Find the constants a and b, where $a < 4 < b$, such that

$$\left| \int_a^b (x - 4) \, dx \right| = 16 \quad \text{and} \quad \int_a^b |x - 4| \, dx = 20.$$

83. **Think About It** When is

$$\int_a^b f(x) \, dx = \int_a^b |f(x)| \, dx?$$

Explain.

84. **Step Function** Evaluate, if possible, the integral

$$\int_0^2 [\![x]\!] \, dx.$$

85. **Using a Riemann Sum** Determine

$$\lim_{n \to \infty} \frac{1}{n^3}(1^2 + 2^2 + 3^2 + \cdots + n^2)$$

by using an appropriate Riemann sum.

5.4   The Fundamental Theorem of Calculus

- Evaluate a definite integral using the Fundamental Theorem of Calculus.
- Understand and use the Mean Value Theorem for Integrals.
- Find the average value of a function over a closed interval.
- Understand and use the Second Fundamental Theorem of Calculus.
- Understand and use the Net Change Theorem.

The Fundamental Theorem of Calculus

You have now been introduced to the two major branches of calculus: differential calculus (introduced with the tangent line problem) and integral calculus (introduced with the area problem). So far, these two problems might seem unrelated—but there is a very close connection. The connection was discovered independently by Isaac Newton and Gottfried Leibniz and is stated in the **Fundamental Theorem of Calculus.**

Informally, the theorem states that differentiation and (definite) integration are inverse operations, in the same sense that division and multiplication are inverse operations. To see how Newton and Leibniz might have anticipated this relationship, consider the approximations shown in Figure 5.27. The slope of the tangent line was defined using the *quotient* $\Delta y/\Delta x$ (the slope of the secant line). Similarly, the area of a region under a curve was defined using the *product* $\Delta y\Delta x$ (the area of a rectangle). So, at least in the primitive approximation stage, the operations of differentiation and definite integration appear to have an inverse relationship in the same sense that division and multiplication are inverse operations. The Fundamental Theorem of Calculus states that the limit processes (used to define the derivative and definite integral) preserve this inverse relationship.

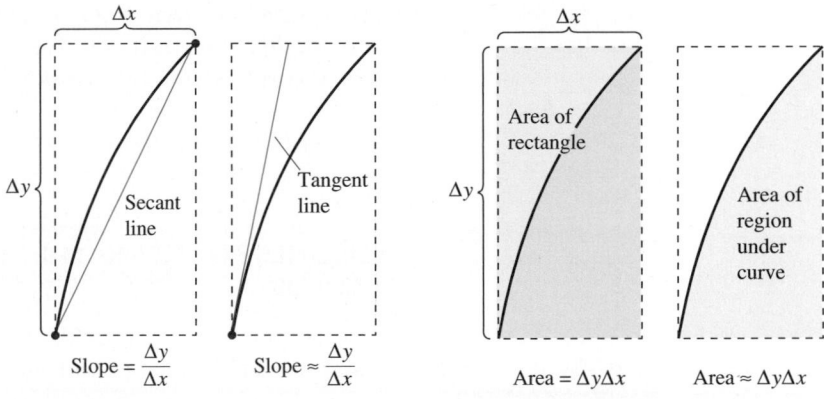

(a) Differentiation   (b) Definite integration

Differentiation and definite integration have an "inverse" relationship.

Figure 5.27

ANTIDIFFERENTIATION AND DEFINITE INTEGRATION

Throughout this chapter, you have been using the integral sign to denote an antiderivative (a family of functions) and a definite integral (a number).

$$\text{Antidifferentiation: } \int f(x)\, dx \qquad \text{Definite integration: } \int_a^b f(x)\, dx$$

The use of the same symbol for both operations makes it appear that they are related. In the early work with calculus, however, it was not known that the two operations were related. The symbol $\int$ was first applied to the definite integral by Leibniz and was derived from the letter S. (Leibniz calculated area as an infinite sum, thus, the letter S.)

THEOREM 5.9 The Fundamental Theorem of Calculus

If a function f is continuous on the closed interval $[a, b]$ and F is an antiderivative of f on the interval $[a, b]$, then

$$\int_a^b f(x)\, dx = F(b) - F(a).$$

Proof The key to the proof is writing the difference $F(b) - F(a)$ in a convenient form. Let Δ be any partition of $[a, b]$.

$$a = x_0 < x_1 < x_2 < \cdots < x_{n-1} < x_n = b$$

By pairwise subtraction and addition of like terms, you can write

$$F(b) - F(a) = F(x_n) - F(x_{n-1}) + F(x_{n-1}) - \cdots - F(x_1) + F(x_1) - F(x_0)$$

$$= \sum_{i=1}^{n} [F(x_i) - F(x_{i-1})].$$

By the Mean Value Theorem, you know that there exists a number c_i in the ith subinterval such that

$$F'(c_i) = \frac{F(x_i) - F(x_{i-1})}{x_i - x_{i-1}}.$$

Because $F'(c_i) = f(c_i)$, you can let $\Delta x_i = x_i - x_{i-1}$ and obtain

$$F(b) - F(a) = \sum_{i=1}^{n} f(c_i)\, \Delta x_i.$$

This important equation tells you that by repeatedly applying the Mean Value Theorem, you can always find a collection of c_i's such that the *constant* $F(b) - F(a)$ is a Riemann sum of f on $[a, b]$ for any partition. Theorem 5.4 guarantees that the limit of Riemann sums over the partition with $\|\Delta\| \to 0$ exists. So, taking the limit (as $\|\Delta\| \to 0$) produces

$$F(b) - F(a) = \int_a^b f(x)\, dx. \qquad \blacksquare$$

GUIDELINES FOR USING THE FUNDAMENTAL THEOREM OF CALCULUS

1. *Provided you can find* an antiderivative of f, you now have a way to evaluate a definite integral without having to use the limit of a sum.

2. When applying the Fundamental Theorem of Calculus, the notation shown below is convenient.

$$\int_a^b f(x)\, dx = F(x)\Big]_a^b = F(b) - F(a)$$

For instance, to evaluate $\int_1^3 x^3\, dx$, you can write

$$\int_1^3 x^3\, dx = \frac{x^4}{4}\Big]_1^3 = \frac{3^4}{4} - \frac{1^4}{4} = \frac{81}{4} - \frac{1}{4} = 20.$$

3. It is not necessary to include a constant of integration C in the antiderivative.

$$\int_a^b f(x)\, dx = \Big[F(x) + C\Big]_a^b = [F(b) + C] - [F(a) + C] = F(b) - F(a)$$

EXAMPLE 1   **Evaluating a Definite Integral**

:..⊳ *See LarsonCalculus.com for an interactive version of this type of example.*

Evaluate each definite integral.

a. $\int_1^2 (x^2 - 3)\, dx$ **b.** $\int_1^4 3\sqrt{x}\, dx$ **c.** $\int_0^{\pi/4} \sec^2 x\, dx$

Solution

a. $\int_1^2 (x^2 - 3)\, dx = \left[\dfrac{x^3}{3} - 3x\right]_1^2 = \left(\dfrac{8}{3} - 6\right) - \left(\dfrac{1}{3} - 3\right) = -\dfrac{2}{3}$

b. $\int_1^4 3\sqrt{x}\, dx = 3\int_1^4 x^{1/2}\, dx = 3\left[\dfrac{x^{3/2}}{3/2}\right]_1^4 = 2(4)^{3/2} - 2(1)^{3/2} = 14$

c. $\int_0^{\pi/4} \sec^2 x\, dx = \tan x\,\Big]_0^{\pi/4} = 1 - 0 = 1$

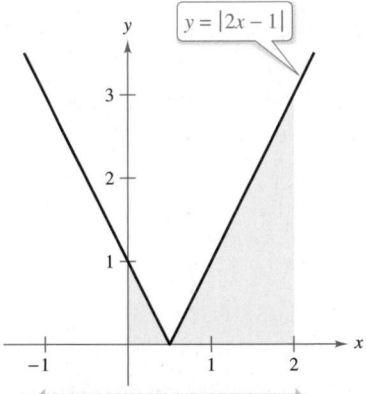

The definite integral of y on $[0, 2]$ is $\frac{5}{2}$.
Figure 5.28

EXAMPLE 2   **A Definite Integral Involving Absolute Value**

Evaluate $\int_0^2 |2x - 1|\, dx$.

Solution   Using Figure 5.28 and the definition of absolute value, you can rewrite the integrand as shown.

$$|2x - 1| = \begin{cases} -(2x - 1), & x < \frac{1}{2} \\ 2x - 1, & x \geq \frac{1}{2} \end{cases}$$

From this, you can rewrite the integral in two parts.

$$\int_0^2 |2x - 1|\, dx = \int_0^{1/2} -(2x - 1)\, dx + \int_{1/2}^2 (2x - 1)\, dx$$

$$= \left[-x^2 + x\right]_0^{1/2} + \left[x^2 - x\right]_{1/2}^2$$

$$= \left(-\dfrac{1}{4} + \dfrac{1}{2}\right) - (0 + 0) + (4 - 2) - \left(\dfrac{1}{4} - \dfrac{1}{2}\right)$$

$$= \dfrac{5}{2}$$

EXAMPLE 3   **Using the Fundamental Theorem to Find Area**

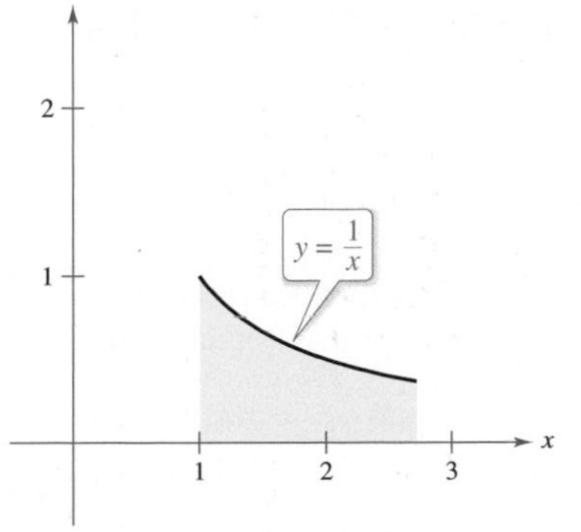

Find the area of the region bounded by the graph of

$$y = \dfrac{1}{x}$$

the x-axis, and the vertical lines $x = 1$ and $x = e$, as shown in Figure 5.29.

Solution   Note that $y > 0$ on the interval $[1, e]$.

$\text{Area} = \displaystyle\int_1^e \dfrac{1}{x}\, dx$ Integrate between $x = 1$ and $x = e$.

$= \Big[\ln x\Big]_1^e$ Find antiderivative.

$= \ln e - \ln 1$ Apply Fundamental Theorem of Calculus.

$= 1$ Simplify.

The area of the region bounded by the graph of $y = 1/x$, the x-axis, $x = 1$, and $x = e$ is 1.
Figure 5.29

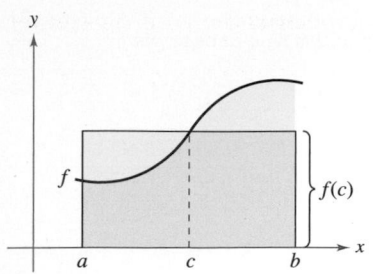

Mean value rectangle:

$$f(c)(b - a) = \int_a^b f(x)\,dx$$

Figure 5.30

The Mean Value Theorem for Integrals

In Section 5.2, you saw that the area of a region under a curve is greater than the area of an inscribed rectangle and less than the area of a circumscribed rectangle. The Mean Value Theorem for Integrals states that somewhere "between" the inscribed and circumscribed rectangles, there is a rectangle whose area is precisely equal to the area of the region under the curve, as shown in Figure 5.30.

THEOREM 5.10 **Mean Value Theorem for Integrals**

If f is continuous on the closed interval $[a, b]$, then there exists a number c in the closed interval $[a, b]$ such that

$$\int_a^b f(x)\,dx = f(c)(b - a).$$

Proof

Case 1: If f is constant on the interval $[a, b]$, then the theorem is clearly valid because c can be any point in $[a, b]$.

Case 2: If f is not constant on $[a, b]$, then, by the Extreme Value Theorem, you can choose $f(m)$ and $f(M)$ to be the minimum and maximum values of f on $[a, b]$. Because

$$f(m) \le f(x) \le f(M)$$

for all x in $[a, b]$, you can apply Theorem 5.8 to write the following.

$$\int_a^b f(m)\,dx \le \int_a^b f(x)\,dx \le \int_a^b f(M)\,dx \qquad \text{See Figure 5.31.}$$

$$f(m)(b - a) \le \int_a^b f(x)\,dx \le f(M)(b - a) \qquad \text{Apply Fundamental Theorem.}$$

$$f(m) \le \frac{1}{b - a}\int_a^b f(x)\,dx \le f(M) \qquad \text{Divide by } b - a.$$

From the third inequality, you can apply the Intermediate Value Theorem to conclude that there exists some c in $[a, b]$ such that

$$f(c) = \frac{1}{b - a}\int_a^b f(x)\,dx \quad \text{or} \quad f(c)(b - a) = \int_a^b f(x)\,dx.$$

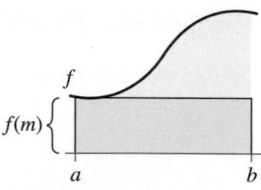

Inscribed rectangle
(less than actual area)

$$\int_a^b f(m)\,dx = f(m)(b - a)$$

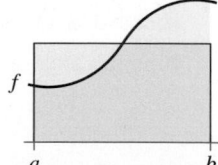

Mean value rectangle
(equal to actual area)

$$\int_a^b f(x)\,dx$$

Circumscribed rectangle
(greater than actual area)

$$\int_a^b f(M)\,dx = f(M)(b - a)$$

Figure 5.31

Notice that Theorem 5.10 does not specify how to determine c. It merely guarantees the existence of at least one number c in the interval.

Average Value of a Function

The value of $f(c)$ given in the Mean Value Theorem for Integrals is called the **average value** of f on the interval $[a, b]$.

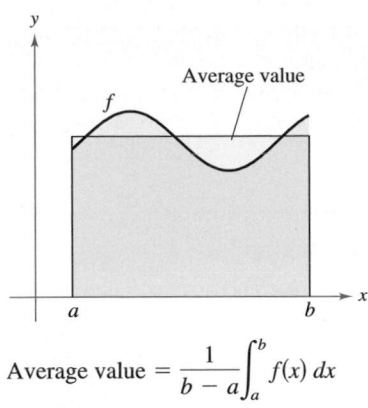

Average value $= \dfrac{1}{b-a}\displaystyle\int_a^b f(x)\,dx$

Figure 5.32

> ### Definition of the Average Value of a Function on an Interval
>
> If f is integrable on the closed interval $[a, b]$, then the **average value** of f on the interval is
> $$\frac{1}{b-a}\int_a^b f(x)\,dx.$$ See Figure 5.32.

To see why the average value of f is defined in this way, partition $[a, b]$ into n subintervals of equal width $\Delta x = (b-a)/n$. If c_i is any point in the ith subinterval, then the arithmetic average (or mean) of the function values at the c_i's is

$$a_n = \frac{1}{n}[f(c_1) + f(c_2) + \cdots + f(c_n)]. \qquad \text{Average of } f(c_1), \ldots, f(c_n)$$

By writing the sum using summation notation and then multiplying and dividing by $(b-a)$, you can write the average as

$$a_n = \frac{1}{n}\sum_{i=1}^n f(c_i) \qquad \text{Rewrite using summation notation.}$$

$$= \frac{1}{n}\sum_{i=1}^n f(c_i)\left(\frac{b-a}{b-a}\right) \qquad \text{Multiply and divide by } (b-a).$$

$$= \frac{1}{b-a}\sum_{i=1}^n f(c_i)\left(\frac{b-a}{n}\right) \qquad \text{Rewrite.}$$

$$= \frac{1}{b-a}\sum_{i=1}^n f(c_i)\,\Delta x. \qquad \Delta x = \frac{b-a}{n}$$

Finally, taking the limit as $n \to \infty$ produces the average value of f on the interval $[a, b]$, as given in the definition above. In Figure 5.32, notice that the area of the region under the graph of f is equal to the area of the rectangle whose height is the average value.

This development of the average value of a function on an interval is only one of many practical uses of definite integrals to represent summation processes. In Chapter 7, you will study other applications, such as volume, arc length, centers of mass, and work.

EXAMPLE 4 Finding the Average Value of a Function

Find the average value of $f(x) = 3x^2 - 2x$ on the interval $[1, 4]$.

Solution The average value is

$$\frac{1}{b-a}\int_a^b f(x)\,dx = \frac{1}{4-1}\int_1^4 (3x^2 - 2x)\,dx$$

$$= \frac{1}{3}\Big[x^3 - x^2\Big]_1^4$$

$$= \frac{1}{3}[64 - 16 - (1 - 1)]$$

$$= \frac{48}{3}$$

$$= 16. \qquad \text{See Figure 5.33.}$$

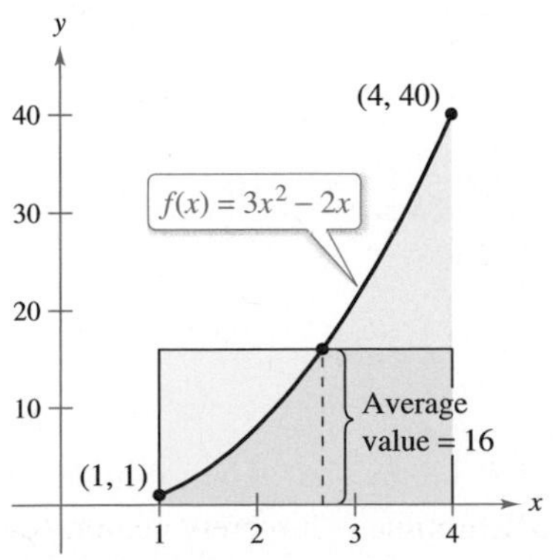

Figure 5.33

The first person to fly at a speed greater than the speed of sound was Charles Yeager. On October 14, 1947, Yeager was clocked at 295.9 meters per second at an altitude of 12.2 kilometers. If Yeager had been flying at an altitude below 11.275 kilometers, this speed would not have "broken the sound barrier." The photo shows an F/A-18F Super Hornet, a supersonic twin-engine strike fighter. A "green Hornet" using a 50/50 mixture of biofuel made from camelina oil became the first U.S. naval tactical aircraft to exceed 1 mach (the speed of sound).

EXAMPLE 5   The Speed of Sound

At different altitudes in Earth's atmosphere, sound travels at different speeds. The speed of sound $s(x)$, in meters per second, can be modeled by

$$s(x) = \begin{cases} -4x + 341, & 0 \le x < 11.5 \\ 295, & 11.5 \le x < 22 \\ \frac{3}{4}x + 278.5, & 22 \le x < 32 \\ \frac{3}{2}x + 254.5, & 32 \le x < 50 \\ -\frac{3}{2}x + 404.5, & 50 \le x \le 80 \end{cases}$$

where x is the altitude in kilometers (see Figure 5.34). What is the average speed of sound over the interval $[0, 80]$?

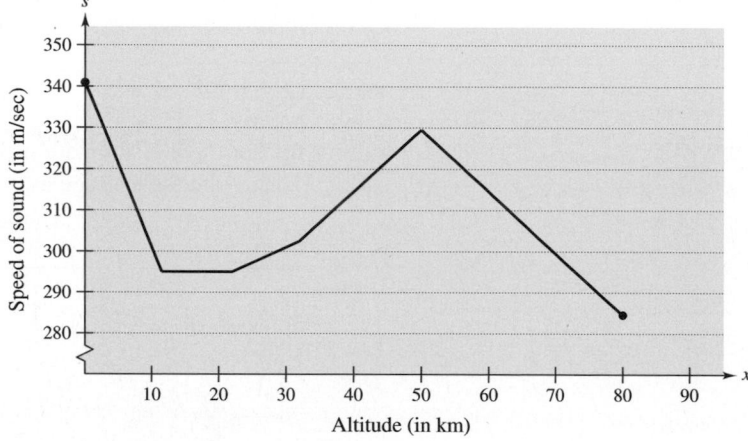

Speed of sound depends on altitude.
Figure 5.34

Solution Begin by integrating $s(x)$ over the interval $[0, 80]$. To do this, you can break the integral into five parts.

$$\int_0^{11.5} s(x)\, dx = \int_0^{11.5} (-4x + 341)\, dx = \left[-2x^2 + 341x\right]_0^{11.5} = 3657$$

$$\int_{11.5}^{22} s(x)\, dx = \int_{11.5}^{22} 295\, dx = \left[295x\right]_{11.5}^{22} = 3097.5$$

$$\int_{22}^{32} s(x)\, dx = \int_{22}^{32} \left(\tfrac{3}{4}x + 278.5\right) dx = \left[\tfrac{3}{8}x^2 + 278.5x\right]_{22}^{32} = 2987.5$$

$$\int_{32}^{50} s(x)\, dx = \int_{32}^{50} \left(\tfrac{3}{2}x + 254.5\right) dx = \left[\tfrac{3}{4}x^2 + 254.5x\right]_{32}^{50} = 5688$$

$$\int_{50}^{80} s(x)\, dx = \int_{50}^{80} \left(-\tfrac{3}{2}x + 404.5\right) dx = \left[-\tfrac{3}{4}x^2 + 404.5x\right]_{50}^{80} = 9210$$

By adding the values of the five integrals, you have

$$\int_0^{80} s(x)\, dx = 24{,}640.$$

So, the average speed of sound from an altitude of 0 kilometers to an altitude of 80 kilometers is

$$\text{Average speed} = \frac{1}{80} \int_0^{80} s(x)\, dx = \frac{24{,}640}{80} = 308 \text{ meters per second.}$$

The Second Fundamental Theorem of Calculus

Earlier you saw that the definite integral of f on the interval $[a, b]$ was defined using the constant b as the upper limit of integration and x as the variable of integration. However, a slightly different situation may arise in which the variable x is used in the upper limit of integration. To avoid the confusion of using x in two different ways, t is temporarily used as the variable of integration. (Remember that the definite integral is *not* a function of its variable of integration.)

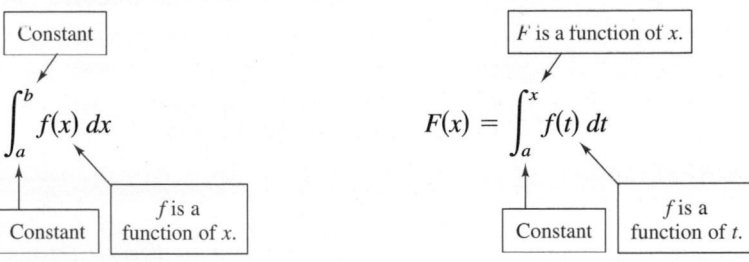

The Definite Integral as a Number

Constant

$$\int_a^b f(x)\,dx$$

Constant f is a function of x.

The Definite Integral as a Function of x

F is a function of x.

$$F(x) = \int_a^x f(t)\,dt$$

Constant f is a function of t.

Exploration

Use a graphing utility to graph the function

$$F(x) = \int_0^x \cos t\,dt$$

for $0 \le x \le 2\pi$. Do you recognize this graph? Explain.

EXAMPLE 6   The Definite Integral as a Function

Evaluate the function

$$F(x) = \int_0^x \cos t\,dt$$

at $x = 0, \dfrac{\pi}{6}, \dfrac{\pi}{4}, \dfrac{\pi}{3}$, and $\dfrac{\pi}{2}$.

Solution You could evaluate five different definite integrals, one for each of the given upper limits. However, it is much simpler to fix x (as a constant) temporarily to obtain

$$\int_0^x \cos t\,dt = \sin t\,\Big]_0^x$$
$$= \sin x - \sin 0$$
$$= \sin x.$$

Now, using $F(x) = \sin x$, you can obtain the results shown in Figure 5.35.

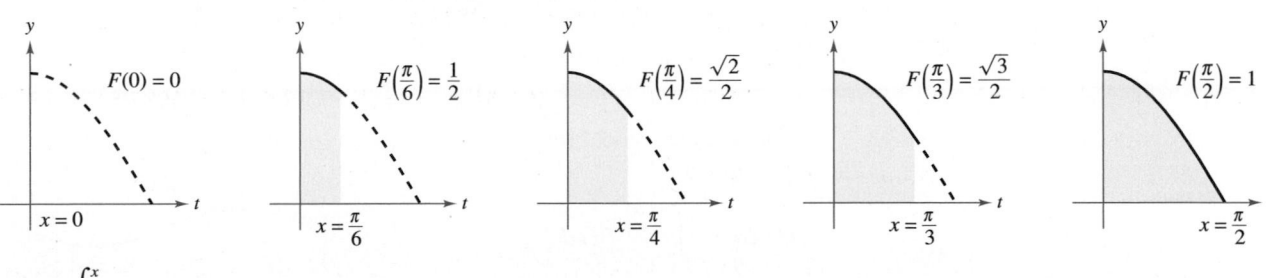

$F(x) = \displaystyle\int_0^x \cos t\,dt$ is the area under the curve $f(t) = \cos t$ from 0 to x.

Figure 5.35

You can think of the function $F(x)$ as *accumulating* the area under the curve $f(t) = \cos t$ from $t = 0$ to $t = x$. For $x = 0$, the area is 0 and $F(0) = 0$. For $x = \pi/2$, $F(\pi/2) = 1$ gives the accumulated area under the cosine curve on the entire interval $[0, \pi/2]$. This interpretation of an integral as an **accumulation function** is used often in applications of integration.

In Example 6, note that the derivative of F is the original integrand (with only the variable changed). That is,

$$\frac{d}{dx}[F(x)] = \frac{d}{dx}[\sin x] = \frac{d}{dx}\left[\int_0^x \cos t \, dt\right] = \cos x.$$

This result is generalized in the next theorem, called the **Second Fundamental Theorem of Calculus.**

THEOREM 5.11   The Second Fundamental Theorem of Calculus

If f is continuous on an open interval I containing a, then, for every x in the interval,

$$\frac{d}{dx}\left[\int_a^x f(t) \, dt\right] = f(x).$$

Proof   Begin by defining F as

$$F(x) = \int_a^x f(t) \, dt.$$

Then, by the definition of the derivative, you can write

$$F'(x) = \lim_{\Delta x \to 0} \frac{F(x + \Delta x) - F(x)}{\Delta x}$$

$$= \lim_{\Delta x \to 0} \frac{1}{\Delta x}\left[\int_a^{x+\Delta x} f(t) \, dt - \int_a^x f(t) \, dt\right]$$

$$= \lim_{\Delta x \to 0} \frac{1}{\Delta x}\left[\int_a^{x+\Delta x} f(t) \, dt + \int_x^a f(t) \, dt\right]$$

$$= \lim_{\Delta x \to 0} \frac{1}{\Delta x}\left[\int_x^{x+\Delta x} f(t) \, dt\right].$$

From the Mean Value Theorem for Integrals (assuming $\Delta x > 0$), you know there exists a number c in the interval $[x, x + \Delta x]$ such that the integral in the expression above is equal to $f(c) \, \Delta x$. Moreover, because $x \le c \le x + \Delta x$, it follows that $c \to x$ as $\Delta x \to 0$. So, you obtain

$$F'(x) = \lim_{\Delta x \to 0}\left[\frac{1}{\Delta x} f(c) \, \Delta x\right] = \lim_{\Delta x \to 0} f(c) = f(x).$$

A similar argument can be made for $\Delta x < 0$. ∎

Using the area model for definite integrals, the approximation

$$f(x) \, \Delta x \approx \int_x^{x+\Delta x} f(t) \, dt$$

can be viewed as saying that the area of the rectangle of height $f(x)$ and width Δx is approximately equal to the area of the region lying between the graph of f and the x-axis on the interval

$$[x, x + \Delta x]$$

as shown in the figure at the right.

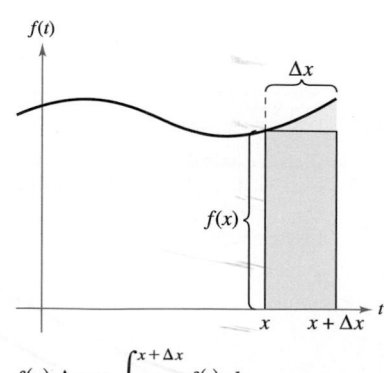

$$f(x) \, \Delta x \approx \int_x^{x+\Delta x} f(t) \, dt$$

Note that the Second Fundamental Theorem of Calculus tells you that when a function is continuous, you can be sure that it has an antiderivative. This antiderivative need not, however, be an elementary function. (Recall the discussion of elementary functions in Section 1.3.)

EXAMPLE 7 **The Second Fundamental Theorem of Calculus**

Evaluate $\dfrac{d}{dx}\left[\displaystyle\int_0^x \sqrt{t^2 + 1}\; dt\right]$.

Solution Note that $f(t) = \sqrt{t^2 + 1}$ is continuous on the entire real number line. So, using the Second Fundamental Theorem of Calculus, you can write

$$\frac{d}{dx}\left[\int_0^x \sqrt{t^2 + 1}\; dt\right] = \sqrt{x^2 + 1}. \qquad \blacksquare$$

The differentiation shown in Example 7 is a straightforward application of the Second Fundamental Theorem of Calculus. The next example shows how this theorem can be combined with the Chain Rule to find the derivative of a function.

EXAMPLE 8 **The Second Fundamental Theorem of Calculus**

Find the derivative of $F(x) = \displaystyle\int_{\pi/2}^{x^3} \cos t \; dt$.

Solution Using $u = x^3$, you can apply the Second Fundamental Theorem of Calculus with the Chain Rule as shown.

$$
\begin{aligned}
F'(x) &= \frac{dF}{du}\frac{du}{dx} && \text{Chain Rule}\\[4pt]
&= \frac{d}{du}[F(x)]\frac{du}{dx} && \text{Definition of } \frac{dF}{du}\\[4pt]
&= \frac{d}{du}\left[\int_{\pi/2}^{x^3}\cos t\; dt\right]\frac{du}{dx} && \text{Substitute } \int_{\pi/2}^{x^3}\cos t\; dt \text{ for } F(x).\\[4pt]
&= \frac{d}{du}\left[\int_{\pi/2}^{u}\cos t\; dt\right]\frac{du}{dx} && \text{Substitute } u \text{ for } x^3.\\[4pt]
&= (\cos u)(3x^2) && \text{Apply Second Fundamental Theorem of Calculus.}\\[4pt]
&= (\cos x^3)(3x^2) && \text{Rewrite as function of } x. \qquad \blacksquare
\end{aligned}
$$

Because the integrand in Example 8 is easily integrated, you can verify the derivative as follows.

$$
\begin{aligned}
F(x) &= \int_{\pi/2}^{x^3} \cos t\; dt\\[4pt]
&= \sin t \Big]_{\pi/2}^{x^3}\\[4pt]
&= \sin x^3 - \sin\frac{\pi}{2}\\[4pt]
&= \sin x^3 - 1
\end{aligned}
$$

In this form, you can apply the Chain Rule to verify that the derivative of F is the same as that obtained in Example 8.

$$\frac{d}{dx}[\sin x^3 - 1] = (\cos x^3)(3x^2) \qquad \text{Derivative of } F$$

Net Change Theorem

The Fundamental Theorem of Calculus (Theorem 5.9) states that if f is continuous on the closed interval $[a, b]$ and F is an antiderivative of f on $[a, b]$, then

$$\int_a^b f(x)\, dx = F(b) - F(a).$$

But because $F'(x) = f(x)$, this statement can be rewritten as

$$\int_a^b F'(x)\, dx = F(b) - F(a)$$

where the quantity $F(b) - F(a)$ represents the *net change of* $F(x)$ on the interval $[a, b]$.

THEOREM 5.12   The Net Change Theorem

If $F'(x)$ is the rate of change of a quantity $F(x)$, then the definite integral of $F'(x)$ from a to b gives the total change, or **net change**, of $F(x)$ on the interval $[a, b]$.

$$\int_a^b F'(x)\, dx = F(b) - F(a) \qquad \text{Net change of } F(x)$$

EXAMPLE 9   Using the Net Change Theorem

A chemical flows into a storage tank at a rate of $(180 + 3t)$ liters per minute, where t is the time in minutes and $0 \le t \le 60$. Find the amount of the chemical that flows into the tank during the first 20 minutes.

Solution   Let $c(t)$ be the amount of the chemical in the tank at time t. Then $c'(t)$ represents the rate at which the chemical flows into the tank at time t. During the first 20 minutes, the amount that flows into the tank is

$$\int_0^{20} c'(t)\, dt = \int_0^{20} (180 + 3t)\, dt$$
$$= \left[180t + \frac{3}{2}t^2\right]_0^{20}$$
$$= 3600 + 600$$
$$= 4200.$$

So, the amount of the chemical that flows into the tank during the first 20 minutes is 4200 liters.   ■

Another way to illustrate the Net Change Theorem is to examine the velocity of a particle moving along a straight line, where $s(t)$ is the position at time t. Then its velocity is $v(t) = s'(t)$ and

$$\int_a^b v(t)\, dt = s(b) - s(a).$$

This definite integral represents the net change in position, or **displacement**, of the particle.

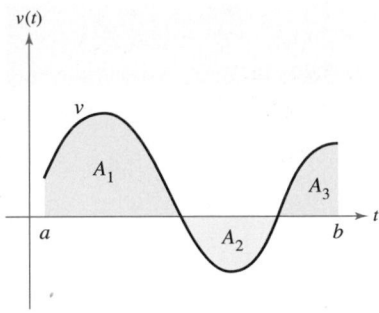

A_1, A_2, and A_3 are the areas of the shaded regions.

Figure 5.36

When calculating the *total* distance traveled by the particle, you must consider the intervals where $v(t) \leq 0$ and the intervals where $v(t) \geq 0$. When $v(t) \leq 0$, the particle moves to the left, and when $v(t) \geq 0$, the particle moves to the right. To calculate the total distance traveled, integrate the absolute value of velocity $|v(t)|$. So, the **displacement** of the particle on the interval $[a, b]$ is

$$\text{Displacement on } [a, b] = \int_a^b v(t) \, dt = A_1 - A_2 + A_3$$

and the **total distance traveled** by the particle on $[a, b]$ is

$$\text{Total distance traveled on } [a, b] = \int_a^b |v(t)| \, dt = A_1 + A_2 + A_3.$$

(See Figure 5.36.)

EXAMPLE 10   **Solving a Particle Motion Problem**

The velocity (in feet per second) of a particle moving along a line is

$$v(t) = t^3 - 10t^2 + 29t - 20$$

where t is the time in seconds.

a. What is the displacement of the particle on the time interval $1 \leq t \leq 5$?

b. What is the total distance traveled by the particle on the time interval $1 \leq t \leq 5$?

Solution

a. By definition, you know that the displacement is

$$\int_1^5 v(t) \, dt = \int_1^5 (t^3 - 10t^2 + 29t - 20) \, dt$$

$$= \left[\frac{t^4}{4} - \frac{10}{3}t^3 + \frac{29}{2}t^2 - 20t \right]_1^5$$

$$= \frac{25}{12} - \left(-\frac{103}{12} \right)$$

$$= \frac{128}{12}$$

$$= \frac{32}{3}.$$

So, the particle moves $\frac{32}{3}$ feet to the right.

b. To find the total distance traveled, calculate $\int_1^5 |v(t)| \, dt$. Using Figure 5.37 and the fact that $v(t)$ can be factored as $(t - 1)(t - 4)(t - 5)$, you can determine that $v(t) \geq 0$ on $[1, 4]$ and $v(t) \leq 0$ on $[4, 5]$. So, the total distance traveled is

$$\int_1^5 |v(t)| \, dt = \int_1^4 v(t) \, dt - \int_4^5 v(t) \, dt$$

$$= \int_1^4 (t^3 - 10t^2 + 29t - 20) \, dt - \int_4^5 (t^3 - 10t^2 + 29t - 20) \, dt$$

$$= \left[\frac{t^4}{4} - \frac{10}{3}t^3 + \frac{29}{2}t^2 - 20t \right]_1^4 - \left[\frac{t^4}{4} - \frac{10}{3}t^3 + \frac{29}{2}t^2 - 20t \right]_4^5$$

$$= \frac{45}{4} - \left(-\frac{7}{12} \right)$$

$$= \frac{71}{6} \text{ feet.}$$

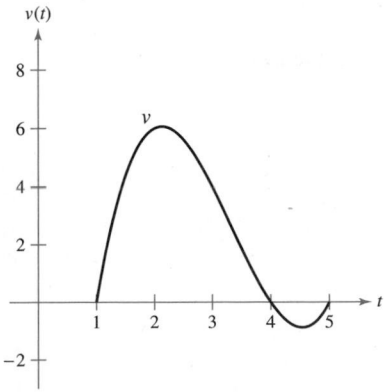

Figure 5.37

5.4 Exercises   See **CalcChat.com** for tutorial help and worked-out solutions to odd-numbered exercises.

CONCEPT CHECK

1. Fundamental Theorem of Calculus Explain how to evaluate a definite integral using the Fundamental Theorem of Calculus.

2. Mean Value Theorem Describe the Mean Value Theorem for Integrals in your own words.

3. Average Value of a Function Describe the average value of a function on an interval in your own words.

4. Accumulation Function Why is

$$F(x) = \int_0^x f(t)\, dt$$

considered an accumulation function?

 Graphical Reasoning In Exercises 5–8, use a graphing utility to graph the integrand. Use the graph to determine whether the definite integral is positive, negative, or zero.

5. $\int_0^\pi \dfrac{4}{x^2 + 1}\, dx$

6. $\int_0^\pi \cos x\, dx$

7. $\int_{-2}^2 x\sqrt{x^2 + 1}\, dx$

8. $\int_{-2}^2 x\sqrt{2 - x}\, dx$

 **Evaluating a Definite Integral** In Exercises 9–34, evaluate the definite integral. Use a graphing utility to verify your result.

9. $\int_{-1}^0 (2x - 1)\, dx$

10. $\int_1^2 (6x^2 - 3x)\, dx$

11. $\int_0^1 (2t - 1)^2\, dt$

12. $\int_1^4 (8x^3 - x)\, dx$

13. $\int_1^2 \left(\dfrac{3}{x^2} - 1\right) dx$

14. $\int_{-2}^{-1} \left(u - \dfrac{1}{u^2}\right) du$

15. $\int_1^4 \dfrac{u - 2}{\sqrt{u}}\, du$

16. $\int_{-8}^8 x^{1/3}\, dx$

17. $\int_{-1}^1 \left(\sqrt[3]{t} - 2\right) dt$

18. $\int_1^8 \sqrt{\dfrac{2}{x}}\, dx$

19. $\int_{-1}^0 (t^{1/3} - t^{2/3})\, dt$

20. $\int_{-8}^{-1} \dfrac{x - x^2}{2\sqrt[3]{x}}\, dx$

21. $\int_0^5 |2x - 5|\, dx$

22. $\int_0^4 |x^2 - 4x + 3|\, dx$

23. $\int_0^\pi (\sin x - 7)\, dx$

24. $\int_0^\pi (2 + \cos x)\, dx$

25. $\int_0^{\pi/4} \dfrac{1 - \sin^2\theta}{\cos^2\theta}\, d\theta$

26. $\int_0^{\pi/4} \dfrac{\sec^2\theta}{\tan^2\theta + 1}\, d\theta$

27. $\int_{-\pi/6}^{\pi/6} \sec^2 x\, dx$

28. $\int_{\pi/4}^{\pi/2} (2 - \csc^2 x)\, dx$

29. $\int_{-\pi/3}^{\pi/3} 4\sec\theta\tan\theta\, d\theta$

30. $\int_{-\pi/2}^{\pi/2} (2t + \cos t)\, dt$

31. $\int_0^2 (2^x + 6)\, dx$

32. $\int_0^3 (t - 5^t)\, dt$

33. $\int_{-1}^1 (e^\theta + \sin\theta)\, d\theta$

34. $\int_e^{2e} \left(\cos x - \dfrac{1}{x}\right) dx$

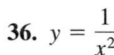 **Finding the Area of a Region** In Exercises 35–38, find the area of the given region.

35. $y = x - x^2$

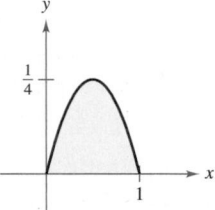

36. $y = \dfrac{1}{x^2}$

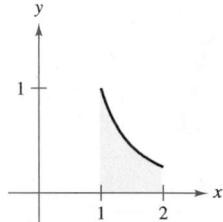

37. $y = \cos x$

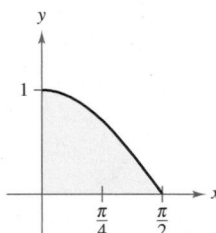

38. $y = x + \sin x$

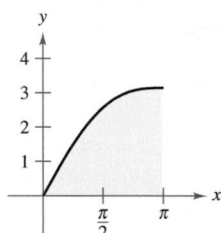

 **Finding the Area of a Region** In Exercises 39–44, find the area of the region bounded by the graphs of the equations.

39. $y = 5x^2 + 2, \quad x = 0, \quad x = 2, \quad y = 0$

40. $y = x^3 + 6x, \quad x = 2, \quad y = 0$

41. $y = 1 + \sqrt[3]{x}, \quad x = 0, \quad x = 8, \quad y = 0$

42. $y = -x^2 + 4x, \quad y = 0$

43. $y = \dfrac{4}{x}, \quad x = 1, \quad x = e, \quad y = 0$

44. $y = e^x, \quad x = 0, \quad x = 2, \quad y = 0$

 **Using the Mean Value Theorem for Integrals** In Exercises 45–50, find the value(s) of c guaranteed by the Mean Value Theorem for Integrals for the function over the given interval.

45. $f(x) = x^3, \quad [0, 3]$

46. $f(x) = \sqrt{x}, \quad [4, 9]$

47. $f(x) = 5 - \dfrac{1}{x}, \quad [1, 4]$

48. $f(x) = 10 - 2^x, \quad [0, 3]$

49. $f(x) = 2\sec^2 x, \quad \left[-\dfrac{\pi}{4}, \dfrac{\pi}{4}\right]$

50. $f(x) = \cos x, \quad \left[-\dfrac{\pi}{3}, \dfrac{\pi}{3}\right]$

 **Finding the Average Value of a Function** In Exercises 51–56, find the average value of the function over the given interval and all values of x in the interval for which the function equals its average value.

51. $f(x) = 4 - x^2$, $[-2, 2]$ **52.** $f(x) = \dfrac{4(x^2 + 1)}{x^2}$, $[1, 3]$

53. $f(x) = 2e^x$, $[-1, 1]$ **54.** $f(x) = \dfrac{1}{2x}$, $[1, 4]$

55. $f(x) = \sin x$, $[0, \pi]$ **56.** $f(x) = \cos x$, $[0, \pi]$

57. Force The force F (in newtons) of a hydraulic cylinder in a press is proportional to the square of sec x, where x is the distance (in meters) that the cylinder is extended in its cycle. The domain of F is $[0, \pi/3]$, and $F(0) = 500$.

(a) Find F as a function of x.

(b) Find the average force exerted by the press over the interval $[0, \pi/3]$.

58. Respiratory Cycle The volume V, in liters, of air in the lungs during a five-second respiratory cycle is approximated by the model $V = 0.1729t + 0.1522t^2 - 0.0374t^3$, where t is the time in seconds. Approximate the average volume of air in the lungs during one cycle.

59. Buffon's Needle Experiment A horizontal plane is ruled with parallel lines 2 inches apart. A two-inch needle is tossed randomly onto the plane. The probability that the needle will touch a line is

$$P = \frac{2}{\pi} \int_0^{\pi/2} \sin \theta \, d\theta$$

where θ is the acute angle between the needle and any one of the parallel lines. Find this probability.

60.  **HOW DO YOU SEE IT?** The graph of f is shown in the figure. The shaded region A has an area of 1.5, and $\int_0^6 f(x)\, dx = 3.5$. Use this information to fill in the blanks.

(a) $\displaystyle\int_0^2 f(x)\, dx = $ ▮

(b) $\displaystyle\int_2^6 f(x)\, dx = $ ▮

(c) $\displaystyle\int_0^6 |f(x)|\, dx = $ ▮

(d) $\displaystyle\int_0^2 -2f(x)\, dx = $ ▮

(e) $\displaystyle\int_0^6 [2 + f(x)]\, dx = $ ▮

(f) The average value of f over the interval $[0, 6]$ is ▮.

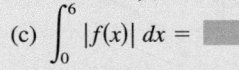
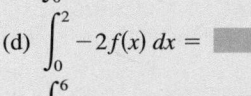

 **Evaluating a Definite Integral** In Exercises 61 and 62, find F as a function of x and evaluate it at $x = 2$, $x = 5$, and $x = 8$.

61. $F(x) = \displaystyle\int_1^x \dfrac{20}{v^2}\, dv$ **62.** $F(x) = \displaystyle\int_2^x (t^3 + 2t - 2)\, dt$

Evaluating a Definite Integral In Exercises 63 and 64, find F as a function of x and evaluate it at $x = 0$, $x = \pi/4$, and $x = \pi/2$.

63. $F(x) = \displaystyle\int_0^x \cos \theta \, d\theta$ **64.** $F(x) = \displaystyle\int_{-\pi}^x \sin \theta \, d\theta$

65. Analyzing a Function Let $g(x) = \displaystyle\int_0^x f(t)\, dt$, where f is the function whose graph is shown in the figure.

(a) Estimate $g(0)$, $g(2)$, $g(4)$, $g(6)$, and $g(8)$.

(b) Find the largest open interval on which g is increasing. Find the largest open interval on which g is decreasing.

(c) Identify any extrema of g.

(d) Sketch a rough graph of g.

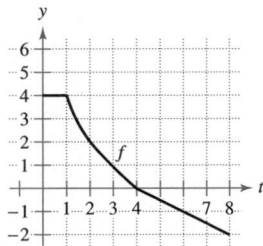 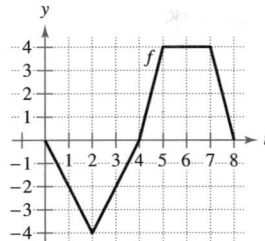

Figure for 65   Figure for 66

66. Analyzing a Function Let $g(x) = \displaystyle\int_0^x f(t)\, dt$, where f is the function whose graph is shown in the figure.

(a) Estimate $g(0)$, $g(2)$, $g(4)$, $g(6)$, and $g(8)$.

(b) Find the largest open interval on which g is increasing. Find the largest open interval on which g is decreasing.

(c) Identify any extrema of g.

(d) Sketch a rough graph of g.

 **Finding and Checking an Integral** In Exercises 67–74, (a) integrate to find F as a function of x, and (b) demonstrate the Second Fundamental Theorem of Calculus by differentiating the result in part (a).

67. $F(x) = \displaystyle\int_0^x (t + 2)\, dt$ **68.** $F(x) = \displaystyle\int_0^x t(t^2 + 1)\, dt$

69. $F(x) = \displaystyle\int_8^x \sqrt[3]{t}\, dt$ **70.** $F(x) = \displaystyle\int_4^x t^{3/2}\, dt$

71. $F(x) = \displaystyle\int_{\pi/4}^x \sec^2 t \, dt$ **72.** $F(x) = \displaystyle\int_{\pi/3}^x \sec t \tan t \, dt$

73. $F(x) = \displaystyle\int_{-1}^x e^t \, dt$ **74.** $F(x) = \displaystyle\int_1^x \dfrac{1}{t}\, dt$

Using the Second Fundamental Theorem of Calculus In Exercises 75–80, use the Second Fundamental Theorem of Calculus to find $F'(x)$.

75. $F(x) = \int_{-2}^{x} (t^2 - 2t)\, dt$ **76.** $F(x) = \int_{1}^{x} \frac{t^2}{t^2 + 1}\, dt$

77. $F(x) = \int_{-1}^{x} \sqrt{t^4 + 1}\, dt$ **78.** $F(x) = \int_{1}^{x} \sqrt[4]{t}\, dt$

79. $F(x) = \int_{1}^{x} \sqrt{t} \csc t\, dt$ **80.** $F(x) = \int_{0}^{x} \sec^3 t\, dt$

Finding a Derivative In Exercises 81–86, find $F'(x)$.

81. $F(x) = \int_{x}^{x+2} (4t + 1)\, dt$ **82.** $F(x) = \int_{-x}^{x} t^3\, dt$

83. $F(x) = \int_{0}^{\sin x} \sqrt{t}\, dt$ **84.** $F(x) = \int_{2}^{x^2} \frac{1}{t^3}\, dt$

85. $F(x) = \int_{0}^{x^3} \sin t^2\, dt$ **86.** $F(x) = \int_{0}^{2x} \cos t^4\, dt$

87. Graphical Analysis Sketch an approximate graph of g on the interval $0 \le x \le 4$, where

$$g(x) = \int_{0}^{x} f(t)\, dt.$$

Identify the x-coordinate of an extremum of g. To print an enlarged copy of the graph, go to *MathGraphs.com*.

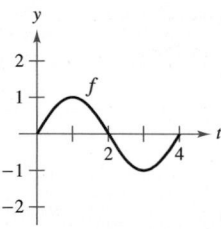

88. Area The area A between the graph of the function

$$g(t) = 4 - \frac{4}{t^2}$$

and the t-axis over the interval $[1, x]$ is

$$A(x) = \int_{1}^{x} \left(4 - \frac{4}{t^2}\right) dt.$$

(a) Find the horizontal asymptote of the graph of g.

(b) Integrate to find A as a function of x. Does the graph of A have a horizontal asymptote? Explain.

89. Water Flow Water flows from a storage tank at a rate of $(500 - 5t)$ liters per minute. Find the amount of water that flows out of the tank during the first 18 minutes.

90. Oil Leak At 1:00 P.M., oil begins leaking from a tank at a rate of $(4 + 0.75t)$ gallons per hour.

(a) How much oil is lost from 1:00 P.M. to 4:00 P.M.?

(b) How much oil is lost from 4:00 P.M. to 7:00 P.M.?

(c) Compare your answers to parts (a) and (b). What do you notice?

91. Velocity The graph shows the velocity, in feet per second, of a car accelerating from rest. Use the graph to estimate the distance the car travels in 8 seconds.

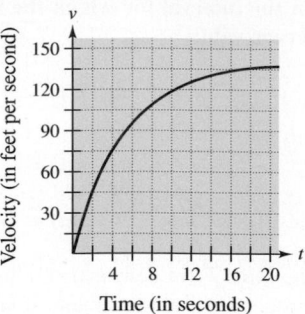

Time (in seconds)

92. Velocity The graph shows the velocity, in feet per second, of a decelerating car after the driver applies the brakes. Use the graph to estimate how far the car travels before it comes to a stop.

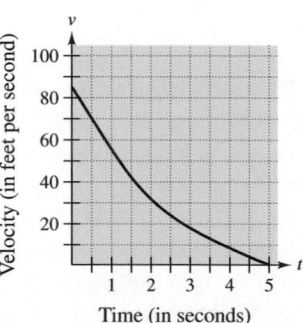

Time (in seconds)

Particle Motion In Exercises 93–98, the velocity function, in feet per second, is given for a particle moving along a straight line, where t is the time in seconds. Find (a) the displacement and (b) the total distance that the particle travels over the given interval.

93. $v(t) = 5t - 7, \quad 0 \le t \le 3$

94. $v(t) = t^2 - t - 12, \quad 1 \le t \le 5$

95. $v(t) = t^3 - 10t^2 + 27t - 18, \quad 1 \le t \le 7$

96. $v(t) = t^3 - 8t^2 + 15t, \quad 0 \le t \le 5$

97. $v(t) = \frac{1}{\sqrt{t}}, \quad 1 \le t \le 4$ **98.** $v(t) = \cos t, \quad 0 \le t \le 3\pi$

EXPLORING CONCEPTS

99. Particle Motion Describe a situation where the displacement and the total distance traveled for a particle are equal.

100. Rate of Growth Let $r'(t)$ represent the rate of growth of a dog, in pounds per year. What does $r(t)$ represent? What does $\int_{2}^{6} r'(t)\, dt$ represent about the dog?

101. Fundamental Theorem of Calculus Explain why the Fundamental Theorem of Calculus cannot be used to integrate

$$f(x) = \frac{1}{x - c}$$

on any interval containing c.

102. Modeling Data An experimental vehicle is tested on a straight track. It starts from rest, and its velocity v (in meters per second) is recorded every 10 seconds for 1 minute (see table).

t	0	10	20	30	40	50	60
v	0	5	21	40	62	78	83

(a) Use a graphing utility to find a model of the form $v = at^3 + bt^2 + ct + d$ for the data.

(b) Use a graphing utility to plot the data and graph the model.

(c) Approximate the distance traveled by the vehicle during the test.

103. Particle Motion A particle is moving along the x-axis. The position of the particle at time t is given by

$$x(t) = t^3 - 6t^2 + 9t - 2, \quad 0 \le t \le 5.$$

Find the total distance the particle travels in 5 units of time.

104. Particle Motion Repeat Exercise 103 for the position function given by

$$x(t) = (t - 1)(t - 3)^2, \quad 0 \le t \le 5.$$

Error Analysis In Exercises 105–108, describe why the statement is incorrect.

105. $\displaystyle\int_{-1}^{1} x^{-2}\, dx = \left[-x^{-1}\right]_{-1}^{1} = (-1) - 1 = -2$ ✗

106. $\displaystyle\int_{-2}^{1} \frac{2}{x^3}\, dx = \left[-\frac{1}{x^2}\right]_{-2}^{1} = -\frac{3}{4}$ ✗

107. $\displaystyle\int_{\pi/4}^{3\pi/4} \sec^2 x\, dx = \left[\tan x\right]_{\pi/4}^{3\pi/4} = -2$ ✗

108. $\displaystyle\int_{\pi/2}^{3\pi/2} \csc x \cot x\, dx = \left[-\csc x\right]_{\pi/2}^{3\pi/2} = 2$ ✗

True or False? In Exercises 109 and 110, determine whether the statement is true or false. If it is false, explain why or give an example that shows it is false.

109. If $F'(x) = G'(x)$ on the interval $[a, b]$, then

$$F(b) - F(a) = G(b) - G(a).$$

110. If $F(b) - F(a) = G(b) - G(a)$, then $F'(x) = G'(x)$ on the interval $[a, b]$.

111. Analyzing a Function Show that the function

$$f(x) = \int_0^{1/x} \frac{1}{t^2 + 1}\, dt + \int_0^{x} \frac{1}{t^2 + 1}\, dt$$

is constant for $x > 0$.

112. Finding a Function Find the function $f(x)$ and all values of c such that

$$\int_c^{x} f(t)\, dt = x^2 + x - 2.$$

113. Finding Values Let

$$G(x) = \int_0^{x}\left[s\int_0^{s} f(t)\, dt\right] ds$$

where f is continuous for all real t. Find (a) $G(0)$, (b) $G'(0)$, (c) $G''(x)$, and (d) $G''(0)$.

114. Proof Prove that

$$\frac{d}{dx}\left[\int_{u(x)}^{v(x)} f(t)\, dt\right] = f(v(x))v'(x) - f(u(x))u'(x).$$

PUTNAM EXAM CHALLENGE

115. For each continuous function $f: [0, 1] \to \mathbb{R}$, let

$$I(f) = \int_0^{1} x^2 f(x)\, dx$$

and

$$J(x) = \int_0^{1} x(f(x))^2\, dx.$$

Find the maximum value of $I(f) - J(f)$ over all such functions f.

SECTION PROJECT

Demonstrating the Fundamental Theorem

Use a graphing utility to graph the function

$$y_1 = \sin^2 t$$

on the interval $0 \le t \le \pi$. Let F be the following function of x.

$$F(x) = \int_0^{x} \sin^2 t\, dt$$

(a) Complete the table. Explain why the values of F are increasing.

x	0	$\dfrac{\pi}{6}$	$\dfrac{\pi}{3}$	$\dfrac{\pi}{2}$	$\dfrac{2\pi}{3}$	$\dfrac{5\pi}{6}$	π
$F(x)$							

(b) Use the integration capabilities of a graphing utility to graph F.

(c) Use the differentiation capabilities of a graphing utility to graph F'. How is this graph related to the graph in part (b)?

(d) Verify that the derivative of

$$y = \frac{1}{2}t - \frac{1}{4}\sin 2t$$

is $\sin^2 t$. Graph y and write a short paragraph about how this graph is related to those in parts (b) and (c).

5.5 Integration by Substitution

■ Use pattern recognition to find an indefinite integral.
■ Use a change of variables to find an indefinite integral.
■ Use the General Power Rule for Integration to find an indefinite integral.
■ Use a change of variables to evaluate a definite integral.
■ Evaluate a definite integral involving an even or odd function.

Pattern Recognition

In this section, you will study techniques for integrating composite functions. The discussion is split into two parts—*pattern recognition* and *change of variables*. Both techniques involve a **u-substitution.** With pattern recognition, you perform the substitution mentally, and with change of variables, you write the substitution steps.

The role of substitution in integration is comparable to the role of the Chain Rule in differentiation. Recall that for the differentiable functions

$$y = F(u) \quad \text{and} \quad u = g(x)$$

the Chain Rule states that

$$\frac{d}{dx}[F(g(x))] = F'(g(x))g'(x).$$

From the definition of an antiderivative, it follows that

$$\int F'(g(x))g'(x)\, dx = F(g(x)) + C.$$

These results are summarized in the next theorem.

REMARK The statement of Theorem 5.13 does not tell how to distinguish between $f(g(x))$ and $g'(x)$ in the integrand. As you become more experienced at integration, your skill in doing this will increase. Of course, part of the key is familiarity with derivatives.

THEOREM 5.13 Antidifferentiation of a Composite Function

Let g be a function whose range is an interval I, and let f be a function that is continuous on I. If g is differentiable on its domain and F is an antiderivative of f on I, then

$$\int f(g(x))g'(x)\, dx = F(g(x)) + C.$$

Letting $u = g(x)$ gives $du = g'(x)\, dx$ and

$$\int f(u)\, du = F(u) + C.$$

Examples 1 and 2 show how to apply Theorem 5.13 *directly*, by recognizing the presence of $f(g(x))$ and $g'(x)$. Note that the composite function in the integrand has an *outside function* f and an *inside function* g. Moreover, the derivative $g'(x)$ is present as a factor of the integrand.

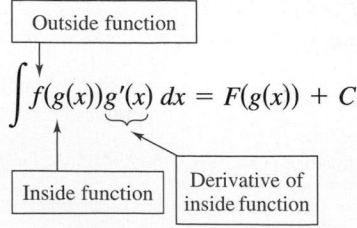

EXAMPLE 1 **Recognizing the $f(g(x))g'(x)$ Pattern**

Find $\int (x^2 + 1)^2(2x)\, dx$.

Solution Letting $g(x) = x^2 + 1$, you obtain

$$g'(x) = 2x$$

and

$$f(g(x)) - f(x^2 + 1) - (x^2 + 1)^2.$$

From this, you can recognize that the integrand follows the $f(g(x))g'(x)$ pattern. Using the Power Rule for Integration and Theorem 5.13, you can write

$$\int \overbrace{(x^2 + 1)^2}^{f(g(x))}\overbrace{(2x)}^{g'(x)}\, dx = \frac{1}{3}(x^2 + 1)^3 + C.$$

Try using the Chain Rule to check that the derivative of $\frac{1}{3}(x^2 + 1)^3 + C$ is the integrand of the original integral.

EXAMPLE 2 **Recognizing the $f(g(x))g'(x)$ Pattern**

Find $\int 5e^{5x}\, dx$.

Solution Letting $g(x) = 5x$, you obtain

$$g'(x) = 5$$

and

$$f(g(x)) = f(5x) = e^{5x}.$$

From this, you can recognize that the integrand follows the $f(g(x))g'(x)$ pattern. Using the Exponential Rule for Integration and Theorem 5.13, you can write

$$\int \overset{f(g(x))}{\overset{|}{e^{5x}}}\overset{g'(x)}{\overset{/}{(5)}}\, dx = e^{5x} + C.$$

You can check this by differentiating $e^{5x} + C$ to obtain the original integrand. ∎

▷ **TECHNOLOGY** Try using a computer algebra system, such as *Maple, Mathematica,* or the *TI-Nspire,* to find the integrals given in Examples 1 and 2. Do you obtain the same antiderivatives that are listed in the examples?

Exploration

Recognizing Patterns The integrand in each of the integrals labeled (a)–(c) fits the pattern $f(g(x))g'(x)$. Identify the pattern and use the result to find the integral.

a. $\int 2x(x^2 + 1)^4\, dx$ **b.** $\int 3x^2\sqrt{x^3 + 1}\, dx$ **c.** $\int (\sec^2 x)(\tan x + 3)\, dx$

The integrals labeled (d)–(f) are similar to (a)–(c). Show how you can multiply and divide by a constant to find these integrals.

d. $\int x(x^2 + 1)^4\, dx$ **e.** $\int x^2\sqrt{x^3 + 1}\, dx$ **f.** $\int (2\sec^2 x)(\tan x + 3)\, dx$

The integrands in Examples 1 and 2 fit the $f(g(x))g'(x)$ pattern exactly—you only had to recognize the pattern. You can extend this technique considerably with the Constant Multiple Rule

$$\int kf(x)\,dx = k \int f(x)\,dx.$$

Many integrands contain the essential part (the variable part) of $g'(x)$ but are missing a constant multiple. In such cases, you can multiply and divide by the necessary constant multiple, as shown in Example 3.

EXAMPLE 3   Multiplying and Dividing by a Constant

Find the indefinite integral.

$$\int x(x^2 + 1)^2\,dx$$

Solution   This is similar to the integral given in Example 1, except that the integrand is missing a factor of 2. Recognizing that $2x$ is the derivative of $x^2 + 1$, you can let

$$g(x) = x^2 + 1$$

and supply the $2x$ as shown.

$$\int x(x^2 + 1)^2\,dx = \int (x^2 + 1)^2 \left(\tfrac{1}{2}\right)(2x)\,dx \qquad \text{Multiply and divide by 2.}$$

$$= \frac{1}{2} \int \overbrace{(x^2 + 1)^2}^{f(g(x))}\,\overbrace{(2x)}^{g'(x)}\,dx \qquad \text{Constant Multiple Rule}$$

$$= \frac{1}{2}\left[\frac{(x^2 + 1)^3}{3}\right] + C \qquad \text{Integrate.}$$

$$= \frac{1}{6}(x^2 + 1)^3 + C \qquad \text{Simplify.} \qquad \blacksquare$$

In practice, most people would not write as many steps as are shown in Example 3. For instance, you could evaluate the integral by simply writing

$$\int x(x^2 + 1)^2\,dx = \frac{1}{2}\int (x^2 + 1)^2\,(2x)\,dx$$

$$= \frac{1}{2}\left[\frac{(x^2 + 1)^3}{3}\right] + C$$

$$= \frac{1}{6}(x^2 + 1)^3 + C.$$

Be sure you see that the *Constant* Multiple Rule applies only to *constants*. You cannot multiply and divide by a variable and then move the variable outside the integral sign. For instance,

$$\int (x^2 + 1)^2\,dx \neq \frac{1}{2x}\int (x^2 + 1)^2\,(2x)\,dx.$$

After all, if it were legitimate to move variable quantities outside the integral sign, you could move the entire integrand out and simplify the whole process. But the result would be incorrect.

Change of Variables for Indefinite Integrals

With a formal **change of variables,** you completely rewrite the integral in terms of u and du (or any other convenient variable). Although this procedure can involve more written steps than the pattern recognition illustrated in Examples 1 through 3, it is useful for complicated integrands. The change of variables technique uses the Leibniz notation for the differential. That is, if $u = g(x)$, then $du = g'(x)\, dx$, and the integral in Theorem 5.13 takes the form

$$\int f(g(x))g'(x)\, dx = \int f(u)\, du - F(u) + C.$$

EXAMPLE 4   **Change of Variables**

Find $\int \sqrt{2x - 1}\, dx$.

Solution   First, let u be the inner function, $u = 2x - 1$. Then calculate the differential du to be $du = 2\, dx$. Now, using $\sqrt{2x - 1} = \sqrt{u}$ and $dx = du/2$, substitute to obtain

$$\int \sqrt{2x - 1}\, dx = \int \sqrt{u} \left(\frac{du}{2} \right) \qquad \text{Integral in terms of } u$$

$$= \frac{1}{2} \int u^{1/2}\, du \qquad \text{Constant Multiple Rule}$$

$$= \frac{1}{2} \left(\frac{u^{3/2}}{3/2} \right) + C \qquad \text{Antiderivative in terms of } u$$

$$= \frac{1}{3} u^{3/2} + C \qquad \text{Simplify.}$$

$$= \frac{1}{3} (2x - 1)^{3/2} + C. \qquad \text{Antiderivative in terms of } x$$

•• **REMARK** Because integration is usually more difficult than differentiation, you should always check your answer to an integration problem by differentiating. For instance, in Example 4, you should differentiate $\frac{1}{3}(2x - 1)^{3/2} + C$ to verify that you obtain the original integrand.

EXAMPLE 5   **Change of Variables**

••••▷ *See LarsonCalculus.com for an interactive version of this type of example.*

Find $\int x\sqrt{2x - 1}\, dx$.

Solution   As in the previous example, let $u = 2x - 1$ and obtain $dx = du/2$. Because the integrand contains a factor of x, you must also solve for x in terms of u, as shown.

$$u = 2x - 1 \quad \Longrightarrow \quad x = \frac{u + 1}{2} \qquad \text{Solve for } x \text{ in terms of } u.$$

Now, using substitution, you obtain

$$\int x\sqrt{2x - 1}\, dx = \int \left(\frac{u + 1}{2} \right) u^{1/2} \left(\frac{du}{2} \right)$$

$$= \frac{1}{4} \int (u^{3/2} + u^{1/2})\, du$$

$$= \frac{1}{4} \left(\frac{u^{5/2}}{5/2} + \frac{u^{3/2}}{3/2} \right) + C$$

$$= \frac{1}{10}(2x - 1)^{5/2} + \frac{1}{6}(2x - 1)^{3/2} + C. \qquad ■$$

To complete the change of variables in Example 5, you solved for x in terms of u. Sometimes this is very difficult. Fortunately, it is not always necessary, as shown in the next example.

EXAMPLE 6 **Change of Variables**

Find $\displaystyle\int \sin^2 3x \cos 3x \, dx$.

Solution Because $\sin^2 3x = (\sin 3x)^2$, you can let $u = \sin 3x$. Then

$$du = (\cos 3x)(3) \, dx.$$

Now, because $\cos 3x \, dx$ is part of the original integral, you can write

$$\frac{du}{3} = \cos 3x \, dx.$$

Substituting u and $du/3$ in the original integral yields

$$\int \sin^2 3x \cos 3x \, dx = \int u^2 \frac{du}{3}$$

$$= \frac{1}{3}\int u^2 \, du$$

$$= \frac{1}{3}\left(\frac{u^3}{3}\right) + C$$

$$= \frac{1}{9} \sin^3 3x + C.$$

• • • • • • • • • • • • • • • ▷

REMARK When making a change of variables, be sure that your answer is written using the same variables as in the original integrand. For instance, in Example 6, you should not leave your answer as

$$\frac{1}{9}u^3 + C$$

but rather, you should replace u by $\sin 3x$.

You can check this by differentiating.

$$\frac{d}{dx}\left[\frac{1}{9}\sin^3 3x + C\right] = \left(\frac{1}{9}\right)(3)(\sin 3x)^2(\cos 3x)(3)$$

$$= \sin^2 3x \cos 3x$$

Because differentiation produces the original integrand, you know that you have obtained the correct antiderivative. ∎

The steps used for integration by substitution are summarized in the following guidelines.

GUIDELINES FOR MAKING A CHANGE OF VARIABLES

1. Choose a substitution $u = g(x)$. Usually, it is best to choose the *inner* part of a composite function, such as a quantity raised to a power.

2. Compute $du = g'(x) \, dx$.

3. Rewrite the integral in terms of the variable u.

4. Find the resulting integral in terms of u.

5. Replace u by $g(x)$ to obtain an antiderivative in terms of x.

6. Check your answer by differentiating.

So far, you have seen two techniques for applying substitution, and you will see more techniques in the remainder of this section. Each technique differs slightly from the others. You should remember, however, that the goal is the same with each technique—*you are trying to find an antiderivative of the integrand.*

The General Power Rule for Integration

One of the most common u-substitutions involves quantities in the integrand that are raised to a power. Because of the importance of this type of substitution, it is given a special name—the **General Power Rule for Integration.** A proof of this rule follows directly from the (simple) Power Rule for Integration, together with Theorem 5.13.

THEOREM 5.14   The General Power Rule for Integration

If g is a differentiable function of x, then

$$\int [g(x)]^n \, g'(x) \, dx = \frac{[g(x)]^{n+1}}{n+1} + C, \quad n \ne -1.$$

Equivalently, if $u = g(x)$, then

$$\int u^n \, du = \frac{u^{n+1}}{n+1} + C, \quad n \ne -1.$$

EXAMPLE 7   **Substitution and the General Power Rule**

a. $\displaystyle \int 3(3x - 1)^4 \, dx = \int \overbrace{(3x - 1)^4}^{u^4} \overbrace{(3) \, dx}^{du} = \overbrace{\frac{(3x - 1)^5}{5}}^{u^5/5} + C$

b. $\displaystyle \int (e^x + 1)(e^x + x) \, dx = \int \overbrace{(e^x + x)}^{u^1} \overbrace{(e^x + 1) \, dx}^{du} = \overbrace{\frac{(e^x + x)^2}{2}}^{u^2/2} + C$

c. $\displaystyle \int 3x^2 \sqrt{x^3 - 2} \, dx = \int \overbrace{(x^3 - 2)^{1/2}}^{u^{1/2}} \overbrace{(3x^2) \, dx}^{du} = \overbrace{\frac{(x^3 - 2)^{3/2}}{3/2}}^{u^{3/2}/(3/2)} + C = \frac{2}{3}(x^3 - 2)^{3/2} + C$

d. $\displaystyle \int \frac{-4x}{(1 - 2x^2)^2} \, dx = \int \overbrace{(1 - 2x^2)^{-2}}^{u^{-2}} \overbrace{(-4x) \, dx}^{du} = \overbrace{\frac{(1 - 2x^2)^{-1}}{-1}}^{u^{-1}/(-1)} + C = -\frac{1}{1 - 2x^2} + C$

e. $\displaystyle \int \cos^2 x \sin x \, dx = -\int \overbrace{(\cos x)^2}^{u^2} \overbrace{(-\sin x) \, dx}^{du} = -\overbrace{\frac{(\cos x)^3}{3}}^{u^3/3} + C$ ■

Some integrals whose integrands involve quantities raised to powers cannot be found by the General Power Rule. Consider the two integrals

$$\int x(x^2 + 1)^2 \, dx \quad \text{and} \quad \int (x^2 + 1)^2 \, dx.$$

The substitution

$$u = x^2 + 1$$

works in the first integral but not in the second. In the second, the substitution fails because the integrand lacks the factor x needed for du. Fortunately, *for this particular integral,* you can expand the integrand as

$$(x^2 + 1)^2 = x^4 + 2x^2 + 1$$

and use the (simple) Power Rule to integrate each term.

Change of Variables for Definite Integrals

When using u-substitution with a definite integral, it is often convenient to determine the limits of integration for the variable u rather than to convert the antiderivative back to the variable x and evaluate at the original limits. This change of variables is stated explicitly in the next theorem. The proof follows from Theorem 5.13 combined with the Fundamental Theorem of Calculus.

THEOREM 5.15   Change of Variables for Definite Integrals

If the function $u = g(x)$ has a continuous derivative on the closed interval $[a, b]$ and f is continuous on the range of g, then

$$\int_a^b f(g(x))g'(x)\, dx = \int_{g(a)}^{g(b)} f(u)\, du.$$

EXAMPLE 8   **Change of Variables**

Evaluate $\displaystyle \int_0^1 x(x^2 + 1)^3\, dx$.

Solution   To evaluate this integral, let $u = x^2 + 1$. Then, you obtain

$$du = 2x\, dx.$$

Before substituting, determine the new upper and lower limits of integration.

Lower Limit	**Upper Limit**
When $x = 0$, $u = 0^2 + 1 = 1$.	When $x = 1$, $u = 1^2 + 1 = 2$.

Now, you can substitute to obtain

$$\int_0^1 x(x^2 + 1)^3\, dx = \frac{1}{2}\int_0^1 (x^2 + 1)^3(2x)\, dx \qquad \text{Integration limits for } x$$

$$= \frac{1}{2}\int_1^2 u^3\, du \qquad \text{Integration limits for } u$$

$$= \frac{1}{2}\left[\frac{u^4}{4}\right]_1^2$$

$$= \frac{1}{2}\left(4 - \frac{1}{4}\right)$$

$$= \frac{15}{8}.$$

Notice that you obtain the same result when you rewrite the antiderivative $\frac{1}{2}(u^4/4)$ in terms of the variable x and evaluate the definite integral at the original limits of integration, as shown below.

$$\frac{1}{2}\left[\frac{u^4}{4}\right]_1^2 = \frac{1}{2}\left[\frac{(x^2 + 1)^4}{4}\right]_0^1$$

$$= \frac{1}{2}\left(4 - \frac{1}{4}\right)$$

$$= \frac{15}{8}$$

EXAMPLE 9 **Change of Variables**

Evaluate the definite integral.

$$\int_1^5 \frac{x}{\sqrt{2x-1}}\, dx$$

Solution To evaluate this integral, let $u = \sqrt{2x-1}$. Then, you obtain

$$u^2 = 2x - 1$$
$$u^2 + 1 = 2x$$
$$\frac{u^2 + 1}{2} = x$$
$$u\, du = dx. \qquad \text{Differentiate each side.}$$

Before substituting, determine the new upper and lower limits of integration.

Lower Limit

When $x = 1,\ u = \sqrt{2-1} = 1$.

Upper Limit

When $x = 5,\ u = \sqrt{10-1} = 3$.

Now, substitute to obtain

$$\int_1^5 \frac{x}{\sqrt{2x-1}}\, dx = \int_1^3 \frac{1}{u}\left(\frac{u^2+1}{2}\right) u\, du$$
$$= \frac{1}{2}\int_1^3 (u^2 + 1)\, du$$
$$= \frac{1}{2}\left[\frac{u^3}{3} + u\right]_1^3$$
$$= \frac{1}{2}\left(9 + 3 - \frac{1}{3} - 1\right)$$
$$= \frac{16}{3}.$$

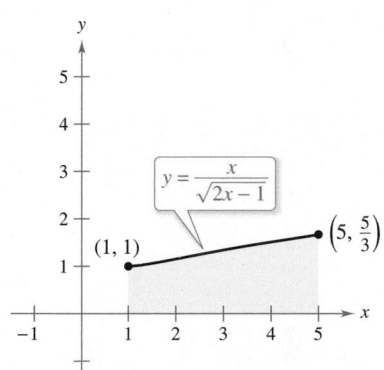

$y = \dfrac{x}{\sqrt{2x-1}}$

$(1, 1)$ $\left(5, \frac{5}{3}\right)$

The region before substitution has an area of $\frac{16}{3}$.

Figure 5.38

Geometrically, you can interpret the equation

$$\int_1^5 \frac{x}{\sqrt{2x-1}}\, dx = \int_1^3 \frac{u^2+1}{2}\, du$$

to mean that the two *different* regions shown in Figures 5.38 and 5.39 have the *same* area.

When evaluating definite integrals by substitution, it is possible for the upper limit of integration of the u-variable form to be smaller than the lower limit. When this happens, do not rearrange the limits. Simply evaluate as usual. For example, after substituting $u = \sqrt{1-x}$ in the integral

$$\int_0^1 x^2(1-x)^{1/2}\, dx$$

you obtain $u = \sqrt{1-0} = 1$ when $x = 0$, and $u = \sqrt{1-1} = 0$ when $x = 1$. So, the correct u-variable form of this integral is

$$-2\int_1^0 (1-u^2)^2 u^2\, du.$$

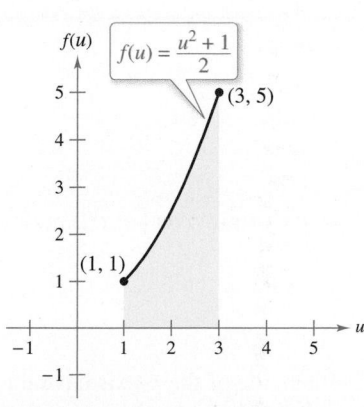

$f(u) = \dfrac{u^2+1}{2}$

$(3, 5)$

$(1, 1)$

The region after substitution has an area of $\frac{16}{3}$.

Figure 5.39

Expanding the integrand, you can evaluate this integral as shown.

$$-2\int_1^0 (u^2 - 2u^4 + u^6)\, du = -2\left[\frac{u^3}{3} - \frac{2u^5}{5} + \frac{u^7}{7}\right]_1^0 = -2\left(-\frac{1}{3} + \frac{2}{5} - \frac{1}{7}\right) = \frac{16}{105}$$

Integration of Even and Odd Functions

Even with a change of variables, integration can be difficult. Occasionally, you can simplify the evaluation of a definite integral over an interval that is symmetric about the y-axis or about the origin by recognizing the integrand to be an even or odd function (see Figure 5.40).

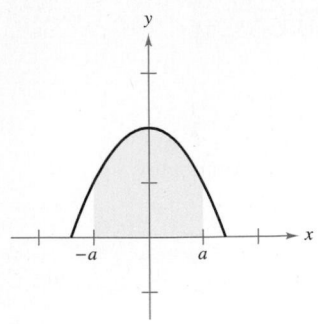

Even function

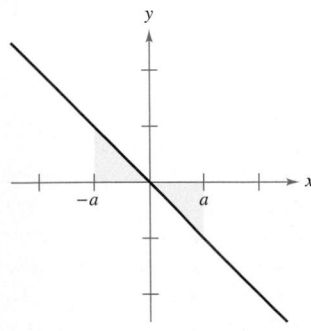

Odd function
Figure 5.40

> **THEOREM 5.16   Integration of Even and Odd Functions**
>
> Let f be integrable on the closed interval $[-a, a]$.
>
> 1. If f is an *even* function, then $\displaystyle\int_{-a}^{a} f(x)\, dx = 2\int_{0}^{a} f(x)\, dx$.
>
> 2. If f is an *odd* function, then $\displaystyle\int_{-a}^{a} f(x)\, dx = 0$.

Proof   Here is the proof of the first property. (The proof of the second property is left to you [see Exercise 121].) Because f is even, you know that

$$f(x) = f(-x).$$

Using Theorem 5.13 with the substitution $u = -x$ produces

$$\int_{-a}^{0} f(x)\, dx = \int_{a}^{0} f(-u)(-du) = -\int_{a}^{0} f(u)\, du = \int_{0}^{a} f(u)\, du = \int_{0}^{a} f(x)\, dx.$$

Finally, using Theorem 5.6, you obtain

$$\int_{-a}^{a} f(x)\, dx = \int_{-a}^{0} f(x)\, dx + \int_{0}^{a} f(x)\, dx$$
$$= \int_{0}^{a} f(x)\, dx + \int_{0}^{a} f(x)\, dx$$
$$= 2\int_{0}^{a} f(x)\, dx.$$

EXAMPLE 10   **Integration of an Odd Function**

Evaluate the definite integral.

$$\int_{-\pi/2}^{\pi/2} (\sin^3 x \cos x + \sin x \cos x)\, dx$$

Solution   Letting $f(x) = \sin^3 x \cos x + \sin x \cos x$ produces

$$f(-x) = \sin^3(-x)\cos(-x) + \sin(-x)\cos(-x)$$
$$= -\sin^3 x \cos x - \sin x \cos x$$
$$= -f(x).$$

So, f is an odd function, and because f is symmetric about the origin over $[-\pi/2, \pi/2]$, you can apply Theorem 5.16 to conclude that

$$\int_{-\pi/2}^{\pi/2} (\sin^3 x \cos x + \sin x \cos x)\, dx = 0.$$

From Figure 5.41, you can see that the two regions on either side of the y-axis have the same area. However, because one lies below the x-axis and one lies above it, integration produces a cancellation effect. (More will be said about areas below the x-axis in Section 7.1.)

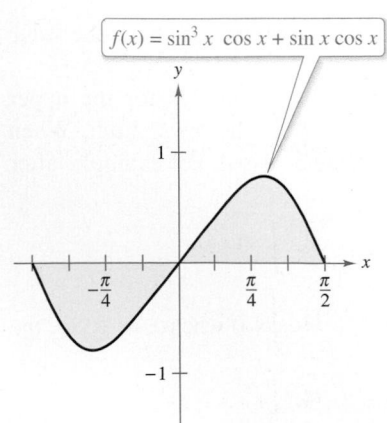

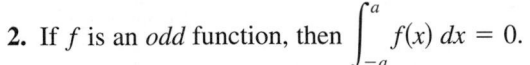
$f(x) = \sin^3 x \cos x + \sin x \cos x$

Because f is an odd function,
$$\int_{-\pi/2}^{\pi/2} f(x)\, dx = 0.$$

Figure 5.41

5.5 Exercises

See CalcChat.com for tutorial help and worked-out solutions to odd-numbered exercises.

CONCEPT CHECK

1. Constant Multiple Rule Explain how to use the Constant Multiple Rule when finding an indefinite integral.

2. Change of Variables In your own words, summarize the guidelines for making a change of variables when finding an indefinite integral.

3. The General Power Rule for Integration Describe the General Power Rule for Integration in your own words.

4. Analyzing the Integrand Without integrating, explain why

$$\int_{-2}^{2} x(x^2 + 1)^2 \, dx = 0.$$

Recognizing Patterns In Exercises 5–8, complete the table by identifying u and du for the integral.

$$\int f(g(x))g'(x) \, dx \qquad u = g(x) \qquad du = g'(x) \, dx$$

5. $\int (5x^2 + 1)^2 (10x) \, dx$

6. $\int x^2 \sqrt{x^3 + 1} \, dx$

7. $\int \tan^2 x \sec^2 x \, dx$

8. $\int \dfrac{\cos x}{\sin^2 x} \, dx$

Finding an Indefinite Integral In Exercises 9–28, find the indefinite integral and check the result by differentiation.

9. $\int (1 + 6x)^4 (6) \, dx$

10. $\int (x^2 - 9)^3 (2x) \, dx$

11. $\int \sqrt{25 - x^2} \, (-2x) \, dx$

12. $\int \sqrt[3]{3 - 4x^2} (-8x) \, dx$

13. $\int x^3 (x^4 + 3)^2 \, dx$

14. $\int x^2 (6 - x^3)^5 \, dx$

15. $\int x^2 (2x^3 - 1)^4 \, dx$

16. $\int x(5x^2 + 4)^3 \, dx$

17. $\int t \sqrt{t^2 + 2} \, dt$

18. $\int 5x \sqrt[3]{1 - x^2} \, dx$

19. $\int \dfrac{7x}{(1 - x^2)^3} \, dx$

20. $\int \dfrac{x^3}{(1 + x^4)^2} \, dx$

21. $\int \dfrac{x^2}{(1 + x^3)^2} \, dx$

22. $\int \dfrac{6x^2}{(4x^3 - 9)^3} \, dx$

23. $\int \dfrac{x}{\sqrt{1 - x^2}} \, dx$

24. $\int \dfrac{x^3}{\sqrt{1 + x^4}} \, dx$

25. $\int \left(1 + \dfrac{1}{t}\right)^3 \left(\dfrac{1}{t^2}\right) dt$

26. $\int \left(8 - \dfrac{1}{t^4}\right)^2 \left(\dfrac{1}{t^5}\right) dt$

27. $\int \dfrac{1}{\sqrt{2x}} \, dx$

28. $\int \dfrac{x}{\sqrt[3]{5x^2}} \, dx$

Differential Equation In Exercises 29–32, find the general solution of the differential equation.

29. $\dfrac{dy}{dx} = 4x + \dfrac{4x}{\sqrt{16 - x^2}}$

30. $\dfrac{dy}{dx} = \dfrac{10x^2}{\sqrt{1 + x^3}}$

31. $\dfrac{dy}{dx} = \dfrac{x + 1}{(x^2 + 2x - 3)^2}$

32. $\dfrac{dy}{dx} = \dfrac{18 - 6x^2}{\sqrt{x^3 - 9x + 7}}$

Slope Field In Exercises 33 and 34, a differential equation, a point, and a slope field are given. A *slope field* (or *direction field*) consists of line segments with slopes given by the differential equation. These line segments give a visual perspective of the slopes of the solutions of the differential equation. (a) Sketch two approximate solutions of the differential equation on the slope field, one of which passes through the given point. (To print an enlarged copy of the graph, go to *MathGraphs.com*.) (b) Use integration and the given point to find the particular solution of the differential equation and use a graphing utility to graph the solution. Compare the result with the sketch in part (a) that passes through the given point.

33. $\dfrac{dy}{dx} = x\sqrt{4 - x^2}$, $(2, 2)$ **34.** $\dfrac{dy}{dx} = e^{\sin x} \cos x$, $(\pi, 2)$

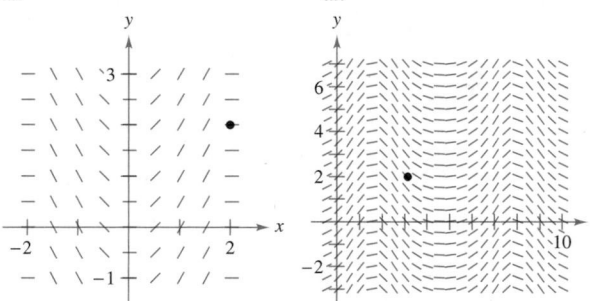

Finding an Indefinite Integral In Exercises 35–56, find the indefinite integral.

35. $\int \pi \sin \pi x \, dx$

36. $\int \sin 4x \, dx$

37. $\int \cos 6x \, dx$

38. $\int \csc^2 \left(\dfrac{x}{2}\right) dx$

39. $\int \dfrac{1}{\theta^2} \cos \dfrac{1}{\theta} \, d\theta$

40. $\int x \sin x^2 \, dx$

41. $\int \sin 2x \cos 2x \, dx$

42. $\int \sqrt[3]{\tan x} \sec^2 x \, dx$

43. $\int \dfrac{\csc^2 x}{\cot^3 x} \, dx$

44. $\int \dfrac{\sin x}{\cos^3 x} \, dx$

45. $\displaystyle\int e^{7x}(7)\,dx$

46. $\displaystyle\int (x+1)e^{x^2+2x}\,dx$

47. $\displaystyle\int e^x(e^x+1)^2\,dx$

48. $\displaystyle\int \frac{2e^x - 2e^{-x}}{(e^x + e^{-x})^2}\,dx$

49. $\displaystyle\int \frac{5 - e^x}{e^{2x}}\,dx$

50. $\displaystyle\int \frac{e^{2x} + 2e^x + 1}{e^x}\,dx$

51. $\displaystyle\int e^{\sin \pi x}\cos \pi x\,dx$

52. $\displaystyle\int e^{\tan 2x}\sec^2 2x\,dx$

53. $\displaystyle\int e^{-x}\sec^2(e^{-x})\,dx$

54. $\displaystyle\int \ln(e^{2x-1})\,dx$

55. $\displaystyle\int 3^{x/2}\,dx$

56. $\displaystyle\int (3-x)7^{(3-x)^2}\,dx$

 **Finding an Equation** In Exercises 57–64, find an equation for the function f that has the given derivative and whose graph passes through the given point.

57. $f'(x) = 2x(4x^2 - 10)^2$, $(2, 10)$

58. $f'(x) = -2x\sqrt{8 - x^2}$, $(2, 7)$

59. $f'(x) = -\sin\dfrac{x}{2}$, $(0, 6)$

60. $f'(x) = \sec^2 2x$, $\left(\dfrac{\pi}{2}, 2\right)$

61. $f'(x) = 2e^{-x/4}$, $(0, 1)$ **62.** $f'(x) = x^2 e^{-0.2x^3}$, $\left(0, \dfrac{3}{2}\right)$

63. $f'(x) = 0.4^{x/3}$, $\left(0, \dfrac{1}{2}\right)$ **64.** $f'(x) = 1.25^{2x}$, $\left(1, \dfrac{25}{16}\right)$

 **Change of Variables** In Exercises 65–72, find the indefinite integral by making a change of variables.

65. $\displaystyle\int x\sqrt{x+6}\,dx$

66. $\displaystyle\int x\sqrt{3x-4}\,dx$

67. $\displaystyle\int x^2\sqrt{1-x}\,dx$

68. $\displaystyle\int (x+1)\sqrt{2-x}\,dx$

69. $\displaystyle\int \frac{x^2 - 1}{\sqrt{2x-1}}\,dx$

70. $\displaystyle\int \frac{2x+1}{\sqrt{x+4}}\,dx$

71. $\displaystyle\int \cos^3 2x \sin 2x\,dx$

72. $\displaystyle\int \sec^5 7x \tan 7x\,dx$

 **Evaluating a Definite Integral** In Exercises 73–84, evaluate the definite integral. Use a graphing utility to verify your result.

73. $\displaystyle\int_{-1}^{1} x(x^2+1)^3\,dx$

74. $\displaystyle\int_{0}^{1} x^3(2x^4+1)^2\,dx$

75. $\displaystyle\int_{1}^{2} 2x^2\sqrt{x^3+1}\,dx$

76. $\displaystyle\int_{-1}^{0} x\sqrt{1-x^2}\,dx$

77. $\displaystyle\int_{0}^{4} \frac{1}{\sqrt{2x+1}}\,dx$

78. $\displaystyle\int_{0}^{2} \frac{x}{\sqrt{1+2x^2}}\,dx$

79. $\displaystyle\int_{1}^{9} \frac{1}{\sqrt{x}\left(1+\sqrt{x}\right)^2}\,dx$

80. $\displaystyle\int_{4}^{5} \frac{x}{\sqrt{2x-6}}\,dx$

81. $\displaystyle\int_{3}^{4} 4xe^{x^2}\,dx$

82. $\displaystyle\int_{1}^{2} e^{1-x}\,dx$

83. $\displaystyle\int_{1}^{3} \frac{e^{3/x}}{x^2}\,dx$

84. $\displaystyle\int_{0}^{\sqrt{2}} xe^{-(x^2/2)}\,dx$

 **Finding the Area of a Region** In Exercises 85–88, find the area of the region. Use a graphing utility to verify your result.

85. $\displaystyle\int_{0}^{7} x\sqrt[3]{x+1}\,dx$

86. $\displaystyle\int_{-2}^{6} x^2\sqrt[3]{x+2}\,dx$

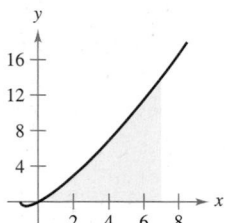   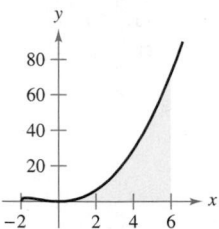

87. $\displaystyle\int_{\pi/2}^{2\pi/3} \sec^2\left(\frac{x}{2}\right)\,dx$

88. $\displaystyle\int_{\pi/12}^{\pi/4} \csc 2x \cot 2x\,dx$

Area In Exercises 89–92, find the area of the region bounded by the graphs of the equations. Use a graphing utility to graph the region and verify your result.

89. $y = e^x$, $y = 0$, $x = 0$, $x = 5$

90. $y = e^{-x}$, $y = 0$, $x = a$, $x = b$

91. $y = xe^{-x^2/4}$, $y = 0$, $x = 0$, $x = \sqrt{6}$

92. $y = e^{-2x} + 2$, $y = 0$, $x = 0$, $x = 2$

 **Even and Odd Functions** In Exercises 93–96, evaluate the integral using the properties of even and odd functions as an aid.

93. $\displaystyle\int_{-2}^{2} x^2(x^2+1)\,dx$

94. $\displaystyle\int_{-2}^{2} x(x^2+1)^3\,dx$

95. $\displaystyle\int_{-\pi/2}^{\pi/2} \sin x \cos x\,dx$

96. $\displaystyle\int_{-\pi/2}^{\pi/2} \sin^2 x \cos x\,dx$

97. Using an Even Function Use $\int_{0}^{6} x^2\,dx = 72$ to evaluate each definite integral without using the Fundamental Theorem of Calculus.

(a) $\displaystyle\int_{-6}^{6} x^2\,dx$

(b) $\displaystyle\int_{-6}^{0} x^2\,dx$

(c) $\displaystyle\int_{0}^{6} -2x^2\,dx$

(d) $\displaystyle\int_{-6}^{6} 3x^2\,dx$

98. Using Symmetry Use the symmetry of the graphs of the sine and cosine functions as an aid in evaluating each definite integral.

(a) $\displaystyle\int_{-\pi/4}^{\pi/4} \sin x\,dx$

(b) $\displaystyle\int_{-\pi/4}^{\pi/4} \cos x\,dx$

(c) $\displaystyle\int_{-\pi/2}^{\pi/2} \cos x\,dx$

(d) $\displaystyle\int_{-\pi/2}^{\pi/2} \sin x \cos x\,dx$

Even and Odd Functions In Exercises 99 and 100, write the integral as the sum of the integral of an odd function and the integral of an even function. Use this simplification to evaluate the integral.

99. $\displaystyle\int_{-3}^{3} (x^3 + 4x^2 - 3x - 6)\, dx$

100. $\displaystyle\int_{-\pi/2}^{\pi/2} (\sin 4x + \cos 4x)\, dx$

EXPLORING CONCEPTS

101. Choosing an Integral You are asked to find one of the integrals. Which one would you choose? Explain.

(a) $\displaystyle\int \sqrt{x^3 + 1}\, dx$ or $\displaystyle\int x^2 \sqrt{x^3 + 1}\, dx$

(b) $\displaystyle\int \cot 2x\, dx$ or $\displaystyle\int \cot^3 2x \csc^2 2x\, dx$

(c) $\displaystyle\int e^{4x-3}\, dx$ or $\displaystyle\int xe^{x+4}\, dx$

102. Comparing Methods Find the indefinite integral in two ways. Explain any difference in the forms of the answers.

(a) $\displaystyle\int (2x - 1)^2\, dx$ (b) $\displaystyle\int \sin x \cos x\, dx$

---

103. Depreciation The rate of depreciation dV/dt of a machine is inversely proportional to the square of $(t + 1)$, where V is the value of the machine t years after it was purchased. The initial value of the machine was $500,000, and its value decreased $100,000 in the first year. Estimate its value after 4 years.

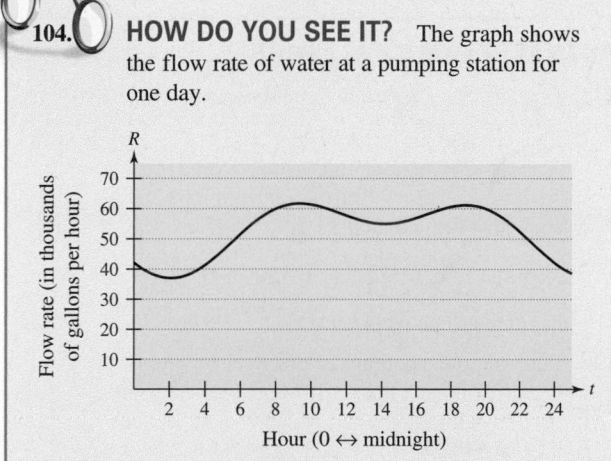

104. HOW DO YOU SEE IT? The graph shows the flow rate of water at a pumping station for one day.

(a) Approximate the maximum flow rate at the pumping station. At what time does this occur?

(b) Explain how you can find the amount of water used during the day.

(c) Approximate the two-hour period when the least amount of water is used. Explain your reasoning.

---

105. Sales The sales S (in thousands of units) of a seasonal product are given by the model

$$S = 74.50 + 43.75 \sin \frac{\pi t}{6}$$

where t is the time in months, with $t = 1$ corresponding to January. Find the average sales for each time period.

(a) The first quarter $(0 \le t \le 3)$

(b) The second quarter $(3 \le t \le 6)$

(c) The entire year $(0 \le t \le 12)$

106. Electricity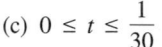

The oscillating current in an electrical circuit is

$$I = 2 \sin(60\pi t) + \cos(120\pi t)$$

where I is measured in amperes and t is measured in seconds. Find the average current for each time interval.

(a) $0 \le t \le \dfrac{1}{60}$

(b) $0 \le t \le \dfrac{1}{240}$

(c) $0 \le t \le \dfrac{1}{30}$

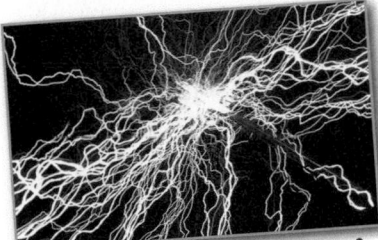

107. Graphical Analysis Consider the functions f and g, where

$$f(x) = 6 \sin x \cos^2 x \quad \text{and} \quad g(t) = \int_0^t f(x)\, dx.$$

(a) Use a graphing utility to graph f and g in the same viewing window.

(b) Explain why g is nonnegative.

(c) Identify the points on the graph of g that correspond to the extrema of f.

(d) Does each of the zeros of f correspond to an extremum of g? Explain.

(e) Consider the function

$$h(t) = \int_{\pi/2}^t f(x)\, dx.$$

Use a graphing utility to graph h. What is the relationship between g and h? Verify your conjecture.

108. Finding a Limit Using a Definite Integral Find

$$\lim_{n\to\infty} \sum_{i=1}^{n} \frac{\sin(i\pi/n)}{n}$$

by evaluating an appropriate definite integral over the interval $[0, 1]$.

109. Rewriting Integrals

(a) Show that $\displaystyle\int_0^1 x^3(1 - x)^8\, dx = \int_0^1 x^8(1 - x)^3\, dx$.

(b) Show that $\displaystyle\int_0^1 x^a(1 - x)^b\, dx = \int_0^1 x^b(1 - x)^a\, dx$.

110. Rewriting Integrals

(a) Show that $\displaystyle\int_0^{\pi/2} \sin^2 x \, dx = \int_0^{\pi/2} \cos^2 x \, dx$.

(b) Show that

$$\int_0^{\pi/2} \sin^n x \, dx = \int_0^{\pi/2} \cos^n x \, dx$$

where n is a positive integer.

Probability In Exercises 111 and 112, the function

$$f(x) = kx^n(1 - x)^m, \quad 0 \le x \le 1$$

where $n > 0$, $m > 0$, and k is a constant, can be used to represent various probability distributions. If k is chosen such that

$$\int_0^1 f(x) \, dx = 1$$

then the probability that x will fall between a and b $(0 \le a \le b \le 1)$ is

$$P_{a,b} = \int_a^b f(x) \, dx.$$

111. The probability that a person will remember between $100a\%$ and $100b\%$ of material learned in an experiment is

$$P_{a,b} = \int_a^b \frac{15}{4} x \sqrt{1 - x} \, dx$$

where x represents the proportion remembered. (See figure.)

(a) For a randomly chosen individual, what is the probability that he or she will recall between 50% and 75% of the material?

(b) What is the median percent recall? That is, for what value of b is it true that the probability of recalling 0 to b is 0.5?

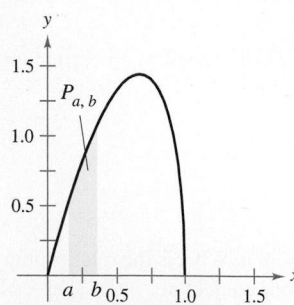

Figure for 111

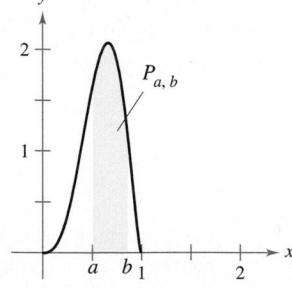

Figure for 112

112. The probability that ore samples taken from a region contain between $100a\%$ and $100b\%$ iron is

$$P_{a,b} = \int_a^b \frac{1155}{32} x^3 (1 - x)^{3/2} \, dx$$

where x represents the proportion of iron. (See figure.)

(a) What is the probability that a sample will contain between 0% and 25% iron?

(b) What is the probability that a sample will contain between 50% and 100% iron?

True or False? In Exercises 113–118, determine whether the statement is true or false. If it is false, explain why or give an example that shows it is false.

113. $\displaystyle\int 3x^2(x^3 + 5)^{-2} \, dx = -(x^3 + 5)^{-1} + C$

114. $\displaystyle\int x(x^2 + 1) \, dx = \frac{1}{2}x^2\left(\frac{1}{3}x^3 + x\right) + C$

115. $\displaystyle\int_{-10}^{10} (ax^3 + bx^2 + cx + d) \, dx = 2\int_0^{10} (bx^2 + d) \, dx$

116. $\displaystyle\int_a^b \sin x \, dx = \int_a^{b+2\pi} \sin x \, dx$

117. $\displaystyle 4\int \sin x \cos x \, dx = -\cos 2x + C$

118. $\displaystyle\int \sin^2 2x \cos 2x \, dx = \frac{1}{3} \sin^3 2x + C$

119. Rewriting Integrals Assume that f is continuous everywhere and that c is a constant. Show that

$$\int_{ca}^{cb} f(x) \, dx = c\int_a^b f(cx) \, dx.$$

120. Integration and Differentiation

(a) Verify that $\sin u - u \cos u + C = \displaystyle\int u \sin u \, du$.

(b) Use part (a) to show that $\displaystyle\int_0^{\pi^2} \sin\sqrt{x} \, dx = 2\pi$.

121. Proof Prove the second property of Theorem 5.16.

122. Rewriting Integrals Show that if f is continuous on the entire real number line, then

$$\int_a^b f(x + h) \, dx = \int_{a+h}^{b+h} f(x) \, dx.$$

PUTNAM EXAM CHALLENGE

123. If $a_0, a_1, \ldots, a_n$ are real numbers satisfying

$$\frac{a_0}{1} + \frac{a_1}{2} + \cdots + \frac{a_n}{n + 1} = 0,$$

show that the equation

$$a_0 + a_1 x + a_2 x^2 + \cdots + a_n x^n = 0$$

has at least one real root.

124. Find all the continuous positive functions $f(x)$, for $0 \le x \le 1$, such that

$$\int_0^1 f(x) \, dx = 1$$

$$\int_0^1 f(x)x \, dx = \alpha$$

$$\int_0^1 f(x)x^2 \, dx = \alpha^2$$

where α is a given real number.

These problems were composed by the Committee on the Putnam Prize Competition. © The Mathematical Association of America. All rights reserved.

5.6 Indeterminate Forms and L'Hôpital's Rule

■ Recognize limits that produce indeterminate forms.
■ Apply L'Hôpital's Rule to evaluate a limit.

Indeterminate Forms

Recall from Chapters 2 and 4 that the forms $0/0$ and ∞/∞ are called *indeterminate* because they do not guarantee that a limit exists, nor do they indicate what the limit is, if one does exist. When you encountered one of these indeterminate forms earlier in the text, you attempted to rewrite the expression by using various algebraic techniques.

Indeterminate Form	Limit	Algebraic Technique
$\dfrac{0}{0}$	$\displaystyle\lim_{x \to -1} \frac{2x^2 - 2}{x + 1} = \lim_{x \to -1} 2(x - 1)$ $= -4$	Divide numerator and denominator by $(x + 1)$.
$\dfrac{\infty}{\infty}$	$\displaystyle\lim_{x \to \infty} \frac{3x^2 - 1}{2x^2 + 1} = \lim_{x \to \infty} \frac{3 - (1/x^2)}{2 + (1/x^2)}$ $= \dfrac{3}{2}$	Divide numerator and denominator by x^2.

Occasionally, you can extend these algebraic techniques to find limits of transcendental functions. For instance, the limit

$$\lim_{x \to 0} \frac{e^{2x} - 1}{e^x - 1}$$

produces the indeterminate form $0/0$. Factoring and then dividing produces

$$\lim_{x \to 0} \frac{e^{2x} - 1}{e^x - 1} = \lim_{x \to 0} \frac{(e^x + 1)(e^x - 1)}{e^x - 1}$$
$$= \lim_{x \to 0} (e^x + 1)$$
$$= 2.$$

Not all indeterminate forms, however, can be evaluated by algebraic manipulation. This is often true when *both* algebraic and transcendental functions are involved. For instance, the limit

$$\lim_{x \to 0} \frac{e^{2x} - 1}{x}$$

produces the indeterminate form $0/0$. Rewriting the expression to obtain

$$\lim_{x \to 0} \left(\frac{e^{2x}}{x} - \frac{1}{x} \right)$$

merely produces another indeterminate form, $\infty - \infty$. Of course, you could use technology to estimate the limit, as shown in the table and in Figure 5.42. From the table and the graph, the limit appears to be 2. (This limit will be verified in Example 1.)

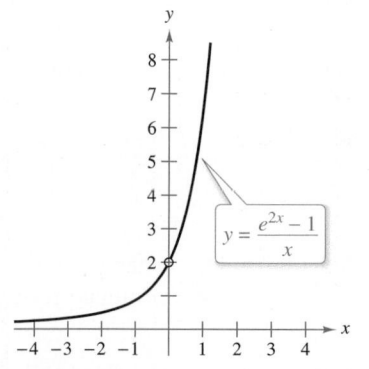

The limit as x approaches 0 appears to be 2.

Figure 5.42

x	-1	-0.1	-0.01	-0.001	0	0.001	0.01	0.1	1
$\dfrac{e^{2x} - 1}{x}$	0.865	1.813	1.980	1.998	?	2.002	2.020	2.214	6.389

L'Hôpital's Rule

To find the limit illustrated in Figure 5.42, you can use a theorem called **L'Hôpital's Rule.** This theorem states that under certain conditions, the limit of the quotient $f(x)/g(x)$ is determined by the limit of the quotient of the derivatives

$$\frac{f'(x)}{g'(x)}.$$

To prove this theorem, you can use a more general result called the **Extended Mean Value Theorem.**

GUILLAUME L'HÔPITAL
(1661–1704)

L'Hôpital's Rule is named after the French mathematician Guillaume François Antoine de L'Hôpital. L'Hôpital is credited with writing the first text on differential calculus (in 1696) in which the rule publicly appeared. It was recently discovered that the rule and its proof were written in a letter from John Bernoulli to L'Hôpital. "... I acknowledge that I owe very much to the bright minds of the Bernoulli brothers. ... I have made free use of their discoveries ...," said L'Hôpital.

See LarsonCalculus.com to read more of this biography.

THEOREM 5.17 The Extended Mean Value Theorem

If f and g are differentiable on an open interval (a, b) and continuous on $[a, b]$ such that $g'(x) \neq 0$ for any x in (a, b), then there exists a point c in (a, b) such that

$$\frac{f'(c)}{g'(c)} = \frac{f(b) - f(a)}{g(b) - g(a)}.$$

A proof of this theorem is given in Appendix A.

To see why Theorem 5.17 is called the Extended Mean Value Theorem, consider the special case in which $g(x) = x$. For this case, you obtain the "standard" Mean Value Theorem as presented in Section 4.2.

THEOREM 5.18 L'Hôpital's Rule

Let f and g be functions that are differentiable on an open interval (a, b) containing c, except possibly at c itself. Assume that $g'(x) \neq 0$ for all x in (a, b), except possibly at c itself. If the limit of $f(x)/g(x)$ as x approaches c produces the indeterminate form $0/0$, then

$$\lim_{x \to c} \frac{f(x)}{g(x)} = \lim_{x \to c} \frac{f'(x)}{g'(x)}$$

provided the limit on the right exists (or is infinite). This result also applies when the limit of $f(x)/g(x)$ as x approaches c produces any one of the indeterminate forms ∞/∞, $(-\infty)/\infty$, $\infty/(-\infty)$, or $(-\infty)/(-\infty)$.

A proof of this theorem is given in Appendix A.

■ **FOR FURTHER INFORMATION**
To enhance your understanding of the necessity of the restriction that $g'(x)$ be nonzero for all x in (a, b), except possibly at c, see the article "Counterexamples to L'Hôpital's Rule" by R. P. Boas in *The American Mathematical Monthly.* To view this article, go to *MathArticles.com.*

People occasionally use L'Hôpital's Rule incorrectly by applying the Quotient Rule to $f(x)/g(x)$. Be sure you see that the rule involves

$$\frac{f'(x)}{g'(x)}$$

not the derivative of $f(x)/g(x)$.

L'Hôpital's Rule can also be applied to one-sided limits. For instance, if the limit of $f(x)/g(x)$ as x approaches c *from the right* produces the indeterminate form $0/0$, then

$$\lim_{x \to c^+} \frac{f(x)}{g(x)} = \lim_{x \to c^+} \frac{f'(x)}{g'(x)}$$

provided the limit exists (or is infinite).

EXAMPLE 1 **Indeterminate Form 0/0**

Evaluate $\lim\limits_{x \to 0} \dfrac{e^{2x} - 1}{x}$.

Solution Because direct substitution results in the indeterminate form 0/0,

$$\lim_{x \to 0} \frac{e^{2x} - 1}{x} \nearrow \begin{array}{l} \lim\limits_{x \to 0} (e^{2x} - 1) = 0 \\ \\ \lim\limits_{x \to 0} x = 0 \end{array}$$

you can apply L'Hôpital's Rule, as shown below.

$$\lim_{x \to 0} \frac{e^{2x} - 1}{x} = \lim_{x \to 0} \frac{\dfrac{d}{dx}[e^{2x} - 1]}{\dfrac{d}{dx}[x]} \qquad \text{Apply L'Hôpital's Rule.}$$

$$= \lim_{x \to 0} \frac{2e^{2x}}{1} \qquad \text{Differentiate numerator and denominator.}$$

$$= 2 \qquad \text{Evaluate the limit.} \qquad \blacksquare$$

In the solution to Example 1, note that you actually do not know that the first limit is equal to the second limit until you have shown that the second limit exists. In other words, if the second limit had not existed, then it would not have been permissible to apply L'Hôpital's Rule.

Another form of L'Hôpital's Rule states that if the limit of $f(x)/g(x)$ as x approaches ∞ (or $-\infty$) produces the indeterminate form 0/0 or ∞/∞, then

$$\lim_{x \to \infty} \frac{f(x)}{g(x)} = \lim_{x \to \infty} \frac{f'(x)}{g'(x)}$$

provided the limit on the right exists.

EXAMPLE 2 **Indeterminate Form ∞/∞**

Evaluate $\lim\limits_{x \to \infty} \dfrac{\ln x}{x}$.

Solution Because direct substitution results in the indeterminate form ∞/∞, you can apply L'Hôpital's Rule to obtain

$$\lim_{x \to \infty} \frac{\ln x}{x} = \lim_{x \to \infty} \frac{\dfrac{d}{dx}[\ln x]}{\dfrac{d}{dx}[x]} \qquad \text{Apply L'Hôpital's Rule.}$$

$$= \lim_{x \to \infty} \frac{1}{x} \qquad \text{Differentiate numerator and denominator.}$$

$$= 0. \qquad \text{Evaluate the limit.} \qquad \blacksquare$$

▷ **TECHNOLOGY** Use a graphing utility to graph $y_1 = \ln x$ and $y_2 = x$ in the same viewing window. Which function grows faster as x approaches ∞? How is this observation related to Example 2?

Occasionally it is necessary to apply L'Hôpital's Rule more than once to remove an indeterminate form, as shown in Example 3.

■ **FOR FURTHER INFORMATION**
To read about the connection between Leonhard Euler and Guillaume L'Hôpital, see the article "When Euler Met l'Hôpital" by William Dunham in *Mathematics Magazine.* To view this article, go to *MathArticles.com.*

EXAMPLE 3   **Applying L'Hôpital's Rule More than Once**

Evaluate $\lim\limits_{x\to-\infty}\dfrac{x^2}{e^{-x}}$.

Solution   Because direct substitution results in the indeterminate form ∞/∞, you can apply L'Hôpital's Rule.

$$\lim_{x\to-\infty}\frac{x^2}{e^{-x}}=\lim_{x\to-\infty}\frac{\dfrac{d}{dx}[x^2]}{\dfrac{d}{dx}[e^{-x}]}=\lim_{x\to-\infty}\frac{2x}{-e^{-x}}$$

This limit yields the indeterminate form $(-\infty)/(-\infty)$, so you can apply L'Hôpital's Rule again to obtain

$$\lim_{x\to-\infty}\frac{2x}{-e^{-x}}=\lim_{x\to-\infty}\frac{\dfrac{d}{dx}[2x]}{\dfrac{d}{dx}[-e^{-x}]}=\lim_{x\to-\infty}\frac{2}{e^{-x}}=0.$$

In addition to the forms $0/0$ and ∞/∞, there are other indeterminate forms such as $0\cdot\infty$, 1^∞, ∞^0, 0^0, and $\infty-\infty$. For example, consider the following four limits that lead to the indeterminate form $0\cdot\infty$.

$$\underbrace{\lim_{x\to0}\left(\frac{1}{x}\right)(x)}_{\text{Limit is 1.}},\qquad\underbrace{\lim_{x\to0}\left(\frac{2}{x}\right)(x)}_{\text{Limit is 2.}},\qquad\underbrace{\lim_{x\to\infty}\left(\frac{1}{e^x}\right)(x)}_{\text{Limit is 0.}},\qquad\underbrace{\lim_{x\to\infty}\left(\frac{1}{x}\right)(e^x)}_{\text{Limit is }\infty.}$$

Because each limit is different, it is clear that the form $0\cdot\infty$ is indeterminate in the sense that it does not determine the value (or even the existence) of the limit. The remaining examples in this section show methods for evaluating these forms. Basically, you attempt to convert each of these forms to $0/0$ or ∞/∞ so that L'Hôpital's Rule can be applied.

EXAMPLE 4   **Indeterminate Form $0\cdot\infty$**

Evaluate $\lim\limits_{x\to\infty}e^{-x}\sqrt{x}$.

Solution   Because direct substitution produces the indeterminate form $0\cdot\infty$, you should try to rewrite the limit to fit the form $0/0$ or ∞/∞. In this case, you can rewrite the limit to fit the second form.

$$\lim_{x\to\infty}e^{-x}\sqrt{x}=\lim_{x\to\infty}\frac{\sqrt{x}}{e^x}$$

Now, by L'Hôpital's Rule, you have

$$\lim_{x\to\infty}\frac{\sqrt{x}}{e^x}=\lim_{x\to\infty}\frac{1/(2\sqrt{x})}{e^x}\qquad\text{Differentiate numerator and denominator.}$$

$$=\lim_{x\to\infty}\frac{1}{2\sqrt{x}\,e^x}\qquad\text{Simplify.}$$

$$=0.\qquad\text{Evaluate the limit.}$$

When rewriting a limit in one of the forms $0/0$ or ∞/∞ does not seem to work, try the other form. For instance, in Example 4, you can write the limit as

$$\lim_{x \to \infty} e^{-x}\sqrt{x} = \lim_{x \to \infty} \frac{e^{-x}}{x^{-1/2}}$$

which yields the indeterminate form $0/0$. As it happens, applying L'Hôpital's Rule to this limit produces

$$\lim_{x \to \infty} \frac{e^{-x}}{x^{-1/2}} = \lim_{x \to \infty} \frac{-e^{-x}}{-1/(2x^{3/2})}$$

which also yields the indeterminate form $0/0$.

The indeterminate forms 1^∞, ∞^0, and 0^0 arise from limits of functions that have variable bases and variable exponents. When you previously encountered this type of function, you used logarithmic differentiation to find the derivative. You can use a similar procedure when taking limits, as shown in the next example.

EXAMPLE 5 **Indeterminate Form 1^∞**

Evaluate $\displaystyle\lim_{x \to \infty} \left(1 + \frac{1}{x}\right)^x$.

Solution Because direct substitution yields the indeterminate form 1^∞, you can proceed as follows. To begin, assume that the limit exists and is equal to y.

$$y = \lim_{x \to \infty} \left(1 + \frac{1}{x}\right)^x$$

Taking the natural logarithm of each side produces

$$\ln y = \ln\left[\lim_{x \to \infty} \left(1 + \frac{1}{x}\right)^x\right].$$

Because the natural logarithmic function is continuous, you can write

$$\ln y = \lim_{x \to \infty} \left[x \ln\left(1 + \frac{1}{x}\right)\right] \qquad \text{Indeterminate form } \infty \cdot 0$$

$$= \lim_{x \to \infty} \left(\frac{\ln[1 + (1/x)]}{1/x}\right) \qquad \text{Indeterminate form } 0/0$$

$$= \lim_{x \to \infty} \left(\frac{(-1/x^2)\{1/[1 + (1/x)]\}}{-1/x^2}\right) \qquad \text{L'Hôpital's Rule}$$

$$= \lim_{x \to \infty} \frac{1}{1 + (1/x)}$$

$$= 1.$$

Now, because you have shown that

$$\ln y = 1$$

you can conclude that

$$y = e$$

and obtain

$$\lim_{x \to \infty} \left(1 + \frac{1}{x}\right)^x = e.$$

You can use a graphing utility to confirm this result, as shown in the figure at the right.

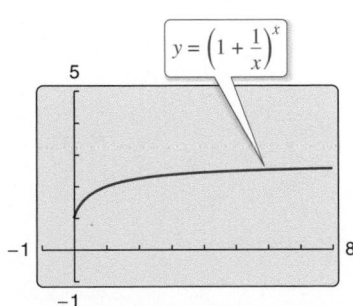

The limit of $[1 + (1/x)]^x$ as x approaches infinity is e.

L'Hôpital's Rule can also be applied to one-sided limits, as demonstrated in Examples 6 and 7.

EXAMPLE 6 **Indeterminate Form 0^0**

$\cdots \triangleright$ *See LarsonCalculus.com for an interactive version of this type of example.*

Evaluate $\lim\limits_{x \to 0^+} (\sin x)^x$.

Solution Because direct substitution produces the indeterminate form 0^0, you can proceed as shown below. To begin, assume that the limit exists and is equal to y.

$$y = \lim_{x \to 0^+} (\sin x)^x \qquad\qquad \text{Indeterminate form } 0^0$$

$$\ln y = \ln\left[\lim_{x \to 0^+} (\sin x)^x\right] \qquad \text{Take natural log of each side.}$$

$$= \lim_{x \to 0^+} \left[\ln(\sin x)^x\right] \qquad \text{Continuity}$$

$$= \lim_{x \to 0^+} \left[x \ln(\sin x)\right] \qquad \text{Indeterminate form } 0 \cdot (-\infty)$$

$$= \lim_{x \to 0^+} \frac{\ln(\sin x)}{1/x} \qquad\qquad \text{Indeterminate form } -\infty/\infty$$

$$= \lim_{x \to 0^+} \frac{\cot x}{-1/x^2} \qquad\qquad \text{L'Hôpital's Rule}$$

$$= \lim_{x \to 0^+} \frac{-x^2}{\tan x} \qquad\qquad \text{Indeterminate form } 0/0$$

$$= \lim_{x \to 0^+} \frac{-2x}{\sec^2 x} \qquad\qquad \text{L'Hôpital's Rule}$$

$$= 0$$

Now, because $\ln y = 0$, you can conclude that $y = e^0 = 1$, and it follows that

$$\lim_{x \to 0^+} (\sin x)^x = 1.$$

$\triangleright$ **TECHNOLOGY** When evaluating complicated limits such as the one in Example 6, it is helpful to check the reasonableness of the solution with a graphing utility. For instance, the calculations in the table and the graph in the figure (see below) are consistent with the conclusion that $(\sin x)^x$ approaches 1 as x approaches 0 from the right.

x	1	0.1	0.01	0.001	0.0001	0.00001
$(\sin x)^x$	0.8415	0.7942	0.9550	0.9931	0.9991	0.9999

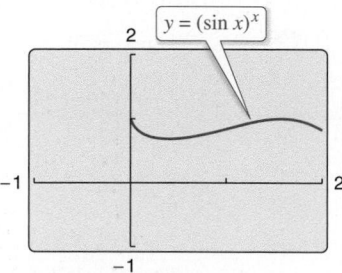

The limit of $(\sin x)^x$ is 1 as x approaches 0 from the right.

Use a graphing utility to estimate the limits $\lim\limits_{x \to 0} (1 - \cos x)^x$ and $\lim\limits_{x \to 0^+} (\tan x)^x$. Then try to verify your estimates analytically.

EXAMPLE 7 Indeterminate Form $\infty - \infty$

Evaluate $\displaystyle\lim_{x \to 1^+} \left(\frac{1}{\ln x} - \frac{1}{x-1} \right)$.

Solution Because direct substitution yields the indeterminate form $\infty - \infty$, you should try to rewrite the expression to produce a form to which you can apply L'Hôpital's Rule. In this case, you can combine the two fractions to obtain

$$\lim_{x \to 1^+} \left(\frac{1}{\ln x} - \frac{1}{x-1} \right) = \lim_{x \to 1^+} \frac{x - 1 - \ln x}{(x-1)\ln x}.$$

Now, because direct substitution produces the indeterminate form $0/0$, you can apply L'Hôpital's Rule to obtain

$$\lim_{x \to 1^+} \frac{x - 1 - \ln x}{(x-1)\ln x} = \lim_{x \to 1^+} \frac{\dfrac{d}{dx}[x - 1 - \ln x]}{\dfrac{d}{dx}[(x-1)\ln x]}$$

$$= \lim_{x \to 1^+} \frac{1 - (1/x)}{(x-1)(1/x) + \ln x}$$

$$= \lim_{x \to 1^+} \frac{x-1}{x - 1 + x \ln x}.$$

This limit also yields the indeterminate form $0/0$, so you can apply L'Hôpital's Rule again to obtain

$$\lim_{x \to 1^+} \frac{x-1}{x - 1 + x \ln x} = \lim_{x \to 1^+} \frac{1}{1 + x(1/x) + \ln x} = \frac{1}{2}.$$

You can check the reasonableness of this solution using a table, as shown at the left.

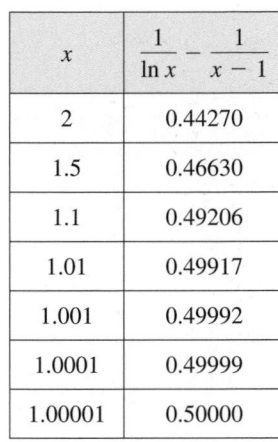

x	$\dfrac{1}{\ln x} - \dfrac{1}{x-1}$
2	0.44270
1.5	0.46630
1.1	0.49206
1.01	0.49917
1.001	0.49992
1.0001	0.49999
1.00001	0.50000

The forms $0/0$, ∞/∞, $\infty - \infty$, $0 \cdot \infty$, 0^0, 1^∞, and ∞^0 have been identified as *indeterminate*. There are similar forms that you should recognize as "determinate."

$\infty + \infty \to \infty$ Limit is positive infinity.

$-\infty - \infty \to -\infty$ Limit is negative infinity.

$0^\infty \to 0$ Limit is zero.

$0^{-\infty} \to \infty$ Limit is positive infinity.

As a final comment, remember that L'Hôpital's Rule can be applied only to quotients leading to the indeterminate forms $0/0$ and ∞/∞. For instance, the application of L'Hôpital's Rule shown below is *incorrect*.

$$\lim_{x \to 0} \frac{e^x}{x} \overset{?}{=} \lim_{x \to 0} \frac{e^x}{1} = 1 \qquad \text{Incorrect use of L'Hôpital's Rule}$$

REMARK You are asked to verify the last two forms in Exercises 110 and 111.

The reason this application is incorrect is that, even though the limit of the denominator is 0, the limit of the numerator is 1, which means that the hypotheses of L'Hôpital's Rule have not been satisfied.

Exploration

In each of the examples presented in this section, L'Hôpital's Rule is used to find a limit that exists. It can also be used to conclude that a limit is infinite. For instance, try using L'Hôpital's Rule to show that $\displaystyle\lim_{x \to \infty} e^x/x = \infty$.

5.6 Exercises

See **CalcChat.com** for tutorial help and worked-out solutions to odd-numbered exercises.

CONCEPT CHECK

1. **L'Hôpital's Rule** Explain the benefit of L'Hôpital's Rule.

2. **Indeterminate Forms** For each limit, use direct substitution. Then identify the form of the limit as either indeterminate or not.

(a) $\lim\limits_{x \to 0} \dfrac{x^2}{\sin 2x}$

(b) $\lim\limits_{x \to \infty} (e^x + x^2)$

(c) $\lim\limits_{x \to \infty} (\ln x - e^x)$

(d) $\lim\limits_{x \to 0^+} \left(\ln x^2 - \dfrac{1}{x}\right)$

Numerical and Graphical Analysis In Exercises 3–6, complete the table and use the result to estimate the limit. Use a graphing utility to graph the function to confirm your result.

3. $\lim\limits_{x \to 0} \dfrac{\sin 4x}{\sin 3x}$

x	-0.1	-0.01	-0.001	0	0.001	0.01	0.1
$f(x)$				?			

4. $\lim\limits_{x \to 0} \dfrac{1 - e^x}{x}$

x	-0.1	-0.01	-0.001	0	0.001	0.01	0.1
$f(x)$				?			

5. $\lim\limits_{x \to \infty} x^5 e^{-x/100}$

x	1	10	10^2	10^3	10^4	10^5
$f(x)$						

6. $\lim\limits_{x \to \infty} \dfrac{6x}{\sqrt{3x^2 - 2x}}$

x	1	10	10^2	10^3	10^4	10^5
$f(x)$						

Using Two Methods In Exercises 7–14, evaluate the limit (a) using techniques from Chapters 2 and 4 and (b) using L'Hôpital's Rule.

7. $\lim\limits_{x \to 4} \dfrac{3(x - 4)}{x^2 - 16}$

8. $\lim\limits_{x \to -4} \dfrac{2x^2 + 13x + 20}{x + 4}$

9. $\lim\limits_{x \to 6} \dfrac{\sqrt{x + 10} - 4}{x - 6}$

10. $\lim\limits_{x \to -1} \left(\dfrac{1 - \sqrt{x + 2}}{x + 1}\right)$

11. $\lim\limits_{x \to 0} \left(\dfrac{2 - 2\cos x}{6x}\right)$

12. $\lim\limits_{x \to 0} \dfrac{\sin 6x}{4x}$

13. $\lim\limits_{x \to \infty} \dfrac{5x^2 - 3x + 1}{3x^2 - 5}$

14. $\lim\limits_{x \to \infty} \dfrac{x^3 + 2x}{4 - x}$

Evaluating a Limit In Exercises 15–42, evaluate the limit, using L'Hôpital's Rule if necessary.

15. $\lim\limits_{x \to 3} \dfrac{x^2 - 2x - 3}{x - 3}$

16. $\lim\limits_{x \to -2} \dfrac{x^2 - 3x - 10}{x + 2}$

17. $\lim\limits_{x \to 0} \dfrac{\sqrt{25 - x^2} - 5}{x}$

18. $\lim\limits_{x \to 5^-} \dfrac{\sqrt{25 - x^2}}{x - 5}$

19. $\lim\limits_{x \to 0^+} \dfrac{e^x - (1 + x)}{x^3}$

20. $\lim\limits_{x \to 1} \dfrac{\ln x^3}{x^2 - 1}$

21. $\lim\limits_{x \to 1} \dfrac{x^{11} - 1}{x^4 - 1}$

22. $\lim\limits_{x \to 1} \dfrac{x^a - 1}{x^b - 1}$, where $a, b \neq 0$

23. $\lim\limits_{x \to 0} \dfrac{\sin 3x}{\sin 5x}$

24. $\lim\limits_{x \to 0} \dfrac{\sin ax}{\sin bx}$, where $a, b \neq 0$

25. $\lim\limits_{x \to \infty} \dfrac{7x^3 - 2x + 1}{6x^3 + 1}$

26. $\lim\limits_{x \to \infty} \dfrac{8 - x}{x^3}$

27. $\lim\limits_{x \to \infty} \dfrac{x^2 + 4x + 7}{x - 6}$

28. $\lim\limits_{x \to \infty} \dfrac{x^3}{x + 2}$

29. $\lim\limits_{x \to \infty} \dfrac{x^3}{e^{x/2}}$

30. $\lim\limits_{x \to \infty} \dfrac{e^{x^2}}{1 - x^3}$

31. $\lim\limits_{x \to \infty} \dfrac{x}{\sqrt{x^2 + 1}}$

32. $\lim\limits_{x \to \infty} \dfrac{x^2}{\sqrt{x^2 + 1}}$

33. $\lim\limits_{x \to \infty} \dfrac{\cos x}{x}$

34. $\lim\limits_{x \to \infty} \dfrac{\sin x}{x - \pi}$

35. $\lim\limits_{x \to \infty} \dfrac{\ln x}{x^2}$

36. $\lim\limits_{x \to \infty} \dfrac{\ln x^4}{x^3}$

37. $\lim\limits_{x \to \infty} \dfrac{e^x}{x^4}$

38. $\lim\limits_{x \to \infty} \dfrac{e^{2x - 9}}{3x}$

39. $\lim\limits_{x \to 0} \dfrac{\sin 5x}{\tan 9x}$

40. $\lim\limits_{x \to 1} \dfrac{\ln x}{\sin \pi x}$

41. $\lim\limits_{x \to -3^-} \dfrac{\int_x^{-3} \sin \theta \, d\theta}{x + 3}$

42. $\lim\limits_{x \to 1^+} \dfrac{\int_1^x \cos \theta \, d\theta}{x - 1}$

Evaluating a Limit In Exercises 43–62, (a) describe the type of indeterminate form (if any) that is obtained by direct substitution. (b) Evaluate the limit, using L'Hôpital's Rule if necessary. (c) Use a graphing utility to graph the function and verify the result in part (b).

43. $\lim\limits_{x \to \infty} x \ln x$

44. $\lim\limits_{x \to 0^+} x^3 \cot x$

45. $\lim\limits_{x \to \infty} x \sin \dfrac{1}{x}$

46. $\lim\limits_{x \to \infty} x \tan \dfrac{1}{x}$

47. $\lim\limits_{x \to 0^+} (e^x + x)^{2/x}$

48. $\lim\limits_{x \to 0^+} \left(1 + \dfrac{1}{x}\right)^x$

49. $\lim\limits_{x \to \infty} x^{1/x}$

50. $\lim\limits_{x \to 0^+} x^{1/x}$

51. $\lim\limits_{x \to 0^+} (1 + x)^{1/x}$

52. $\lim\limits_{x \to \infty} (1 + x)^{1/x}$

53. $\lim\limits_{x \to 0^+} 3x^{x/2}$

54. $\lim\limits_{x \to 4^+} [3(x - 4)]^{x - 4}$

55. $\lim\limits_{x\to 1^+} (\ln x)^{x-1}$

56. $\lim\limits_{x\to 0^+} \left[\cos\left(\dfrac{\pi}{2} - x\right)\right]^x$

57. $\lim\limits_{x\to 2^+} \left(\dfrac{8}{x^2 - 4} - \dfrac{x}{x - 2}\right)$

58. $\lim\limits_{x\to 2^+} \left(\dfrac{1}{x^2 - 4} - \dfrac{\sqrt{x - 1}}{x^2 - 4}\right)$

59. $\lim\limits_{x\to 1^+} \left(\dfrac{3}{\ln x} - \dfrac{2}{x - 1}\right)$

60. $\lim\limits_{x\to 0^+} \left(\dfrac{10}{x} - \dfrac{3}{x^2}\right)$

61. $\lim\limits_{x\to\infty} (e^x - x)$

62. $\lim\limits_{x\to\infty} \left(x - \sqrt{x^2 + 1}\right)$

EXPLORING CONCEPTS

63. Finding Functions Find differentiable functions f and g that satisfy the specified condition such that

$$\lim_{x\to 5} f(x) = 0 \quad \text{and} \quad \lim_{x\to 5} g(x) = 0.$$

Explain how you obtained your answers. (*Note:* There are many correct answers.)

(a) $\lim\limits_{x\to 5} \dfrac{f(x)}{g(x)} = 10$

(b) $\lim\limits_{x\to 5} \dfrac{f(x)}{g(x)} = 0$

(c) $\lim\limits_{x\to 5} \dfrac{f(x)}{g(x)} = \infty$

64. Finding Functions Find differentiable functions f and g such that

$$\lim_{x\to\infty} f(x) = \lim_{x\to\infty} g(x) = \infty \quad \text{and} \quad \lim_{x\to\infty} [f(x) - g(x)] = 25.$$

Explain how you obtained your answers. (*Note:* There are many correct answers.)

65. L'Hôpital's Rule Determine which of the following limits can be evaluated using L'Hôpital's Rule. Explain your reasoning. Do not evaluate the limit.

(a) $\lim\limits_{x\to 2} \dfrac{x - 2}{x^3 - x - 6}$

(b) $\lim\limits_{x\to 0} \dfrac{x^2 - 4x}{2x - 1}$

(c) $\lim\limits_{x\to\infty} \dfrac{x^3}{e^x}$

(d) $\lim\limits_{x\to 3} \dfrac{e^{x^2} - e^9}{x - 3}$

(e) $\lim\limits_{x\to 1} \dfrac{\cos \pi x}{\ln x}$

(f) $\lim\limits_{x\to 1} \dfrac{1 + x(\ln x - 1)}{(x - 1)\ln x}$

---

66. 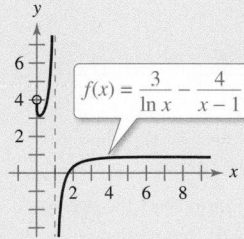 **HOW DO YOU SEE IT?** Use the graph of f to find each limit.

$f(x) = \dfrac{3}{\ln x} - \dfrac{4}{x - 1}$

(a) $\lim\limits_{x\to 1^-} f(x)$

(b) $\lim\limits_{x\to 1^+} f(x)$

(c) $\lim\limits_{x\to 1} f(x)$

---

67. Numerical Analysis Complete the table to show that x eventually "overpowers" $(\ln x)^4$.

x	10	10^2	10^4	10^6	10^8	10^{10}
$\dfrac{(\ln x)^4}{x}$						

68. Numerical Analysis Complete the table to show that e^x eventually "overpowers" x^5.

x	1	5	10	20	30	40	50	100
$\dfrac{e^x}{x^5}$								

Comparing Functions In Exercises 69–74, use L'Hôpital's Rule to determine the comparative rates of increase of the functions $f(x) = x^m$, $g(x) = e^{nx}$, and $h(x) = (\ln x)^n$, where $n > 0$, $m > 0$, and $x \to \infty$.

69. $\lim\limits_{x\to\infty} \dfrac{x^2}{e^{5x}}$

70. $\lim\limits_{x\to\infty} \dfrac{x^3}{e^{2x}}$

71. $\lim\limits_{x\to\infty} \dfrac{(\ln x)^3}{x}$

72. $\lim\limits_{x\to\infty} \dfrac{(\ln x)^2}{x^3}$

73. $\lim\limits_{x\to\infty} \dfrac{(\ln x)^n}{x^m}$

74. $\lim\limits_{x\to\infty} \dfrac{x^m}{e^{nx}}$

Asymptotes and Relative Extrema In Exercises 75–78, find any asymptotes and relative extrema that may exist and use a graphing utility to graph the function.

75. $y = x^{1/x}$, $x > 0$

76. $y = x^x$, $x > 0$

77. $y = 2xe^{-x}$

78. $y = \dfrac{\ln x}{x}$

Think About It In Exercises 79–82, L'Hôpital's Rule is used incorrectly. Describe the error.

79. $\lim\limits_{x\to 2} \dfrac{3x^2 + 4x + 1}{x^2 - x - 2} = \lim\limits_{x\to 2} \dfrac{6x + 4}{2x - 1} = \lim\limits_{x\to 2} \dfrac{6}{2} = 3$

80. $\lim\limits_{x\to 0} \dfrac{e^{2x} - 1}{e^x} = \lim\limits_{x\to 0} \dfrac{2e^{2x}}{e^x}$

$= \lim\limits_{x\to 0} 2e^x$

$= 2$

81. $\lim\limits_{x\to\infty} \dfrac{e^{-x}}{1 + e^{-x}} = \lim\limits_{x\to\infty} \dfrac{-e^{-x}}{-e^{-x}}$

$= \lim\limits_{x\to\infty} 1$

$= 1$

82. $\lim\limits_{x\to\infty} x \cos\dfrac{1}{x} = \lim\limits_{x\to\infty} \dfrac{\cos(1/x)}{1/x}$

$= \lim\limits_{x\to\infty} \dfrac{[-\sin(1/x)](1/x^2)}{-1/x^2}$

$= \lim\limits_{x\to\infty} \sin\dfrac{1}{x}$

$= 0$

Analytic and Graphical Analysis In Exercises 83 and 84, (a) explain why L'Hôpital's Rule cannot be used to find the limit, (b) find the limit analytically, and (c) use a graphing utility to graph the function and approximate the limit from the graph. Compare the result with that in part (b).

83. $\lim\limits_{x \to \infty} \dfrac{x}{\sqrt{x^2 + 1}}$

84. $\lim\limits_{x \to \pi/2^-} \dfrac{\tan x}{\sec x}$

Graphical Analysis In Exercises 85 and 86, graph $f(x)/g(x)$ and $f'(x)/g'(x)$ near $x = 0$. What do you notice about these ratios as $x \to 0$? How does this illustrate L'Hôpital's Rule?

85. $f(x) = \sin 3x$, $g(x) = \sin 4x$ **86.** $f(x) = e^{3x} - 1$, $g(x) = x$

87. Electric Circuit The diagram shows a simple electric circuit consisting of a power source, a resistor, and an inductor. If voltage V is first applied at time $t = 0$, then the current I flowing through the circuit at time t is given by

$$I = \frac{V}{R}(1 - e^{-Rt/L})$$

where L is the inductance and R is the resistance. Use L'Hôpital's Rule to find the formula for the current by fixing V and L and letting R approach 0 from the right.

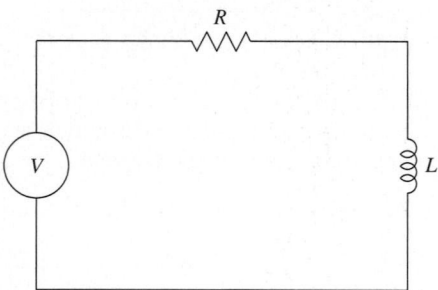

88. Velocity in a Resisting Medium The velocity v of an object falling through a resisting medium such as air or water is given by

$$v = \frac{32}{k}\left(1 - e^{-kt} + \frac{v_0 k e^{-kt}}{32}\right)$$

where v_0 is the initial velocity, t is the time in seconds, and k is the resistance constant of the medium. Use L'Hôpital's Rule to find the formula for the velocity of a falling body in a vacuum by fixing v_0 and t and letting k approach zero. (Assume that the downward direction is positive.)

89. The Gamma Function The Gamma Function $\Gamma(n)$ is defined in terms of the integral of the function given by $f(x) = x^{n-1}e^{-x}$, $n > 0$. Show that for any fixed value of n, the limit of $f(x)$ as x approaches infinity is zero.

90. Compound Interest The formula for the amount A in a savings account compounded n times per year for t years at an interest rate r and an initial deposit of P is given by

$$A = P\left(1 + \frac{r}{n}\right)^{nt}.$$

Use L'Hôpital's Rule to show that the limiting formula as the number of compoundings per year approaches infinity is given by $A = Pe^{rt}$.

Extended Mean Value Theorem In Exercises 91–94, verify that the Extended Mean Value Theorem can be applied to the functions f and g on the closed interval $[a, b]$. Then find all values c in the open interval (a, b) such that

$$\frac{f'(c)}{g'(c)} = \frac{f(b) - f(a)}{g(b) - g(a)}.$$

Functions	Interval
91. $f(x) = x^3$, $g(x) = x^2 + 1$	$[0, 1]$
92. $f(x) = \dfrac{1}{x}$, $g(x) = x^2 - 4$	$[1, 2]$
93. $f(x) = \sin x$, $g(x) = \cos x$	$\left[0, \dfrac{\pi}{2}\right]$
94. $f(x) = \ln x$, $g(x) = x^3$	$[1, 4]$

True or False? In Exercises 95–100, determine whether the statement is true or false. If it is false, explain why or give an example that shows it is false.

95. A limit of the form $\infty/0$ is indeterminate.

96. A limit of the form $\infty \cdot \infty$ is indeterminate.

97. An indeterminate form does not guarantee the existence of a limit.

98. $\lim\limits_{x \to 0} \dfrac{x^2 + x + 1}{x} = \lim\limits_{x \to 0} \dfrac{2x + 1}{1} = 1$

99. If $p(x)$ is a polynomial, then $\lim\limits_{x \to \infty} \dfrac{p(x)}{e^x} = 0$.

100. If $\lim\limits_{x \to \infty} \dfrac{f(x)}{g(x)} = 1$, then $\lim\limits_{x \to \infty} [f(x) - g(x)] = 0$.

101. Area Find the limit, as x approaches 0, of the ratio of the area of the triangle to the total shaded area in the figure.

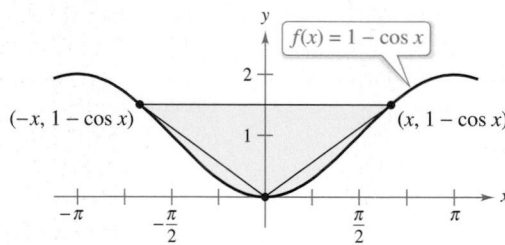

102. Finding a Limit In Section 2.3, a geometric argument (see figure) was used to prove that

$$\lim\limits_{\theta \to 0} \frac{\sin \theta}{\theta} = 1.$$

(a) Write the area of $\triangle ABD$ in terms of θ.

(b) Write the area of the shaded region in terms of θ.

(c) Write the ratio R of the area of $\triangle ABD$ to that of the shaded region.

(d) Find $\lim\limits_{\theta \to 0} R$.

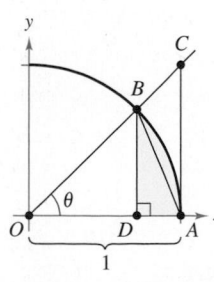

Continuous Function In Exercises 103 and 104, find the value of c that makes the function continuous at $x = 0$.

103. $f(x) = \begin{cases} \dfrac{4x - 2\sin 2x}{2x^3}, & x \neq 0 \\ c, & x = 0 \end{cases}$

104. $f(x) = \begin{cases} (e^x + x)^{1/x}, & x \neq 0 \\ c, & x = 0 \end{cases}$

105. Finding Values Find the values of a and b such that

$$\lim_{x \to 0} \frac{a - \cos bx}{x^2} = 2.$$

106. Evaluating a Limit Use a graphing utility to graph

$$f(x) = \frac{x^k - 1}{k}$$

for $k = 1, 0.1$, and 0.01. Then evaluate the limit

$$\lim_{k \to 0^+} \frac{x^k - 1}{k}.$$

107. Finding a Derivative

(a) Let $f'(x)$ be continuous. Show that

$$\lim_{h \to 0} \frac{f(x + h) - f(x - h)}{2h} = f'(x).$$

(b) Explain the result of part (a) graphically.

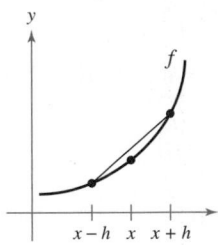

108. Finding a Second Derivative Let $f''(x)$ be continuous. Show that

$$\lim_{h \to 0} \frac{f(x + h) - 2f(x) + f(x - h)}{h^2} = f''(x).$$

109. Evaluating a Limit Consider the limit

$$\lim_{x \to 0^+} (-x \ln x).$$

(a) Describe the type of indeterminate form that is obtained by direct substitution.

(b) Evaluate the limit. Use a graphing utility to verify the result.

■ **FOR FURTHER INFORMATION** For a geometric approach to this exercise, see the article "A Geometric Proof of $\lim_{d \to 0^+} (-d \ln d) = 0$" by John H. Mathews in *The College Mathematics Journal*. To view this article, go to *MathArticles.com*.

110. Proof Prove that if $f(x) \geq 0$, $\lim_{x \to a} f(x) = 0$, and $\lim_{x \to a} g(x) = \infty$, then $\lim_{x \to a} f(x)^{g(x)} = 0$.

111. Proof Prove that if $f(x) \geq 0$, $\lim_{x \to a} f(x) = 0$, and $\lim_{x \to a} g(x) = -\infty$, then $\lim_{x \to a} f(x)^{g(x)} = \infty$.

112. Think About It Use two different methods to find the limit

$$\lim_{x \to \infty} \frac{\ln x^m}{\ln x^n}$$

where $m > 0$, $n > 0$, and $x > 0$.

113. Indeterminate Forms Show that the indeterminate forms 0^0, ∞^0, and 1^∞ do not always have a value of 1 by evaluating each limit.

(a) $\lim_{x \to 0^+} x^{(\ln 2)/(1 + \ln x)}$

(b) $\lim_{x \to \infty} x^{(\ln 2)/(1 + \ln x)}$

(c) $\lim_{x \to 0} (x + 1)^{(\ln 2)/x}$

114. Calculus History In L'Hôpital's 1696 calculus textbook, he illustrated his rule using the limit of the function

$$f(x) = \frac{\sqrt{2a^3 x - x^4} - a\sqrt[3]{a^2 x}}{a - \sqrt[4]{ax^3}}$$

as x approaches a, $a > 0$. Find this limit.

115. Finding a Limit Consider the function

$$h(x) = \frac{x + \sin x}{x}.$$

(a) Use a graphing utility to graph the function. Then use the *zoom* and *trace* features to investigate $\lim_{x \to \infty} h(x)$.

(b) Find $\lim_{x \to \infty} h(x)$ analytically by writing

$$h(x) = \frac{x}{x} + \frac{\sin x}{x}.$$

(c) Can you use L'Hôpital's Rule to find $\lim_{x \to \infty} h(x)$? Explain your reasoning.

116. Evaluating a Limit Let $f(x) = x + x \sin x$ and $g(x) = x^2 - 4$.

(a) Show that $\lim_{x \to \infty} \dfrac{f(x)}{g(x)} = 0$.

(b) Show that $\lim_{x \to \infty} f(x) = \infty$ and $\lim_{x \to \infty} g(x) = \infty$.

(c) Evaluate the limit

$$\lim_{x \to \infty} \frac{f'(x)}{g'(x)}.$$

What do you notice?

(d) Do your answers to parts (a) through (c) contradict L'Hôpital's Rule? Explain your reasoning.

PUTNAM EXAM CHALLENGE

117. Evaluate $\lim_{x \to \infty} \left[\dfrac{1}{x} \cdot \dfrac{a^x - 1}{a - 1} \right]^{1/x}$ where $a > 0$, $a \neq 1$.

5.7 The Natural Logarithmic Function: Integration

■ Use the Log Rule for Integration to integrate a rational function.
■ Integrate trigonometric functions.

Log Rule for Integration

In Chapter 3, you studied two differentiation rules for logarithms. The differentiation rule $d/dx[\ln x] = 1/x$ produces the Log Rule for Integration that you learned in Section 5.1. The differentiation rule $d/dx[\ln u] = u'/u$ produces the integration rule $\int 1/u = \ln|u| + C$. These rules are summarized below. (See Exercise 101.)

THEOREM 5.19 **Log Rule for Integration**

Let u be a differentiable function of x.

1. $\displaystyle\int \frac{1}{x}\, dx = \ln|x| + C$ **2.** $\displaystyle\int \frac{1}{u}\, du = \ln|u| + C$

Because $du = u'\, dx$, the second formula can also be written as

$$\int \frac{u'}{u}\, dx = \ln|u| + C.$$ Alternative form of Log Rule

EXAMPLE 1   Using the Log Rule for Integration

$$\int \frac{2}{x}\, dx = 2\int \frac{1}{x}\, dx$$ Constant Multiple Rule

$$= 2\ln|x| + C$$ Log Rule for Integration

$$= \ln x^2 + C$$ Property of logarithms

Because x^2 cannot be negative, the absolute value notation is unnecessary in the final form of the antiderivative.

EXAMPLE 2   Using the Log Rule with a Change of Variables

Find $\displaystyle\int \frac{1}{4x - 1}\, dx$.

Solution If you let $u = 4x - 1$, then $du = 4\, dx$.

$$\int \frac{1}{4x - 1}\, dx = \frac{1}{4}\int \left(\frac{1}{4x - 1}\right)4\, dx$$ Multiply and divide by 4.

$$= \frac{1}{4}\int \frac{1}{u}\, du$$ Substitute: $u = 4x - 1$.

$$= \frac{1}{4}\ln|u| + C$$ Apply Log Rule.

$$= \frac{1}{4}\ln|4x - 1| + C$$ Back-substitute.

Example 3 uses the alternative form of the Log Rule. To apply this rule, look for quotients in which the numerator is the derivative of the denominator.

EXAMPLE 3 **Finding Area with the Log Rule**

Find the area of the region bounded by the graph of

$$y = \frac{x}{x^2 + 1}$$

the x-axis, and the line $x = 3$.

Solution In Figure 5.43, you can see that the area of the region is given by the definite integral

$$\int_0^3 \frac{x}{x^2 + 1} \, dx.$$

If you let $u = x^2 + 1$, then $u' = 2x$. To apply the Log Rule, multiply and divide by 2 as shown.

$$\int_0^3 \frac{x}{x^2 + 1} \, dx = \frac{1}{2} \int_0^3 \frac{2x}{x^2 + 1} \, dx \qquad \text{Multiply and divide by 2.}$$

$$= \frac{1}{2} \left[\ln(x^2 + 1) \right]_0^3 \qquad \int \frac{u'}{u} \, dx = \ln|u| + C$$

$$= \frac{1}{2} (\ln 10 - \ln 1)$$

$$= \frac{1}{2} \ln 10 \qquad \ln 1 = 0$$

$$\approx 1.151$$

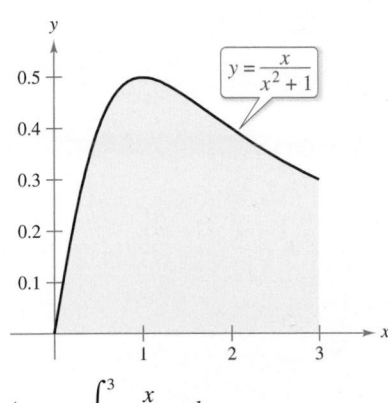

Area $= \displaystyle\int_0^3 \frac{x}{x^2 + 1} \, dx$

The area of the region bounded by the graph of y, the x-axis, and $x = 3$ is $\frac{1}{2} \ln 10$.

Figure 5.43

EXAMPLE 4 **Recognizing Quotient Forms of the Log Rule**

a. $\displaystyle\int \frac{3x^2 + 1}{x^3 + x} \, dx = \ln|x^3 + x| + C \qquad u = x^3 + x$

b. $\displaystyle\int \frac{\sec^2 x}{\tan x} \, dx = \ln|\tan x| + C \qquad u = \tan x$

c. $\displaystyle\int \frac{x + 1}{x^2 + 2x} \, dx = \frac{1}{2} \int \frac{2x + 2}{x^2 + 2x} \, dx \qquad u = x^2 + 2x$

$$= \frac{1}{2} \ln|x^2 + 2x| + C$$

d. $\displaystyle\int \frac{1}{3x + 2} \, dx = \frac{1}{3} \int \frac{3}{3x + 2} \, dx \qquad u = 3x + 2$

$$= \frac{1}{3} \ln|3x + 2| + C$$

With antiderivatives involving logarithms, it is easy to obtain forms that look quite different but are still equivalent. For instance, both

$$\ln|(3x + 2)^{1/3}| + C$$

and

$$\ln|3x + 2|^{1/3} + C$$

are equivalent to the antiderivative listed in Example 4(d).

Integrals to which the Log Rule can be applied often appear in disguised form. For instance, when a rational function has a *numerator of degree greater than or equal to that of the denominator*, division may reveal a form to which you can apply the Log Rule. This is shown in Example 5.

EXAMPLE 5 **Using Long Division Before Integrating**

$\cdots\cdots\triangleright$ *See LarsonCalculus.com for an interactive version of this type of example.*

Find the indefinite integral.

$$\int \frac{x^2 + x + 1}{x^2 + 1}\, dx$$

Solution Begin by using long division to rewrite the integrand.

$$\frac{x^2 + x + 1}{x^2 + 1} \quad\Longrightarrow\quad x^2 + 1\overline{)\,x^2 + x + 1\,} \atop \underline{x^2 \;+ 1} \atop x \quad\Longrightarrow\quad 1 + \frac{x}{x^2 + 1}$$

Now, you can integrate to obtain

$$\int \frac{x^2 + x + 1}{x^2 + 1}\, dx = \int \left(1 + \frac{x}{x^2 + 1}\right) dx \qquad \text{Rewrite using long division.}$$

$$= \int dx + \frac{1}{2}\int \frac{2x}{x^2 + 1}\, dx \qquad \text{Rewrite as two integrals.}$$

$$= x + \frac{1}{2}\ln(x^2 + 1) + C. \qquad \text{Integrate.}$$

Check this result by differentiating to obtain the original integrand. ∎

The next example presents another instance in which the use of the Log Rule is disguised. In this case, a change of variables helps you recognize the Log Rule.

EXAMPLE 6 **Change of Variables with the Log Rule**

Find the indefinite integral.

$$\int \frac{2x}{(x + 1)^2}\, dx$$

Solution If you let $u = x + 1$, then $du = dx$ and $x = u - 1$.

$$\int \frac{2x}{(x + 1)^2}\, dx = \int \frac{2(u - 1)}{u^2}\, du \qquad \text{Substitute.}$$

$$= 2\int \left(\frac{u}{u^2} - \frac{1}{u^2}\right) du \qquad \text{Rewrite as two fractions.}$$

$$= 2\int \frac{du}{u} - 2\int u^{-2}\, du \qquad \text{Rewrite as two integrals.}$$

$$= 2\ln|u| - 2\left(\frac{u^{-1}}{-1}\right) + C \qquad \text{Integrate.}$$

$$= 2\ln|u| + \frac{2}{u} + C \qquad \text{Simplify.}$$

$$= 2\ln|x + 1| + \frac{2}{x + 1} + C \qquad \text{Back-substitute.}$$

$\triangleright$ **TECHNOLOGY** If you have access to a computer algebra system, use it to find the indefinite integrals in Examples 5 and 6. How does the form of the antiderivative that it gives you compare with that given in Examples 5 and 6?

Check this result by differentiating to obtain the original integrand. ∎

As you study the methods shown in Examples 5 and 6, be aware that both methods involve rewriting a disguised integrand so that it fits one or more of the basic integration formulas. Throughout the remaining sections of Chapter 5 and in Chapter 8, much time will be devoted to integration techniques. To master these techniques, you must recognize the "form-fitting" nature of integration. In this sense, integration is not nearly as straightforward as differentiation. Differentiation takes the form

"Here is the question; what is the answer?"

Integration is more like

"Here is the answer; what is the question?"

Here are some guidelines you can use for integration.

GUIDELINES FOR INTEGRATION

1. Learn a basic list of integration formulas.

2. Find an integration formula that resembles all or part of the integrand and, by trial and error, find a choice of u that will make the integrand conform to the formula.

3. When you cannot find a u-substitution that works, try altering the integrand. You might try a trigonometric identity, multiplication and division by the same quantity, addition and subtraction of the same quantity, or long division. Be creative.

4. If you have access to computer software that will find antiderivatives symbolically, use it.

5. Check your result by differentiating to obtain the original integrand.

EXAMPLE 7 **u-Substitution and the Log Rule**

Solve the differential equation

$$\frac{dy}{dx} = \frac{1}{x \ln x}.$$

Solution The solution can be written as an indefinite integral.

$$y = \int \frac{1}{x \ln x}\, dx$$

Because the integrand is a quotient whose denominator is raised to the first power, you should try the Log Rule. There are three basic choices for u. The choices

$$u = x \quad \text{and} \quad u = x \ln x$$

fail to fit the u'/u form of the Log Rule. However, the third choice does fit. Letting $u = \ln x$ produces $u' = 1/x$, and you obtain the following.

$$\int \frac{1}{x \ln x}\, dx = \int \frac{1/x}{\ln x}\, dx \qquad \text{Divide numerator and denominator by } x.$$

$$= \int \frac{u'}{u}\, dx \qquad \text{Substitute: } u = \ln x.$$

$$= \ln|u| + C \qquad \text{Apply Log Rule.}$$

$$= \ln|\ln x| + C \qquad \text{Back-substitute.}$$

So, the solution is $y = \ln|\ln x| + C$.

· · REMARK Keep in mind that you can check your answer to an integration problem by differentiating the answer. For instance, in Example 7, the derivative of $y = \ln|\ln x| + C$ is $y' = 1/(x \ln x)$.

Integrals of Trigonometric Functions

In Section 5.1, you looked at six trigonometric integration rules—the six that correspond directly to differentiation rules. With the Log Rule, you can now complete the set of basic trigonometric integration formulas.

> **EXAMPLE 8**   **Using a Trigonometric Identity**

Find $\int \tan x \, dx$.

Solution   This integral does not seem to fit any formulas on our basic list. However, by using a trigonometric identity, you obtain

$$\int \tan x \, dx = \int \frac{\sin x}{\cos x} \, dx.$$

Knowing that $D_x[\cos x] = -\sin x$, you can let $u = \cos x$ and write

$$\int \tan x \, dx = -\int \frac{-\sin x}{\cos x} \, dx \qquad \text{Apply trigonometric identity and multiply and divide by } -1.$$

$$= -\int \frac{u'}{u} \, dx \qquad \text{Substitute: } u = \cos x.$$

$$= -\ln|u| + C \qquad \text{Apply Log Rule.}$$

$$= -\ln|\cos x| + C. \qquad \text{Back-substitute.} \qquad \blacksquare$$

Example 8 used a trigonometric identity to derive an integration rule for the tangent function. The next example takes a rather unusual step (multiplying and dividing by the same quantity) to derive an integration rule for the secant function.

> **EXAMPLE 9**   **Derivation of the Secant Formula**

Find $\int \sec x \, dx$.

Solution   Consider the following procedure.

$$\int \sec x \, dx = \int (\sec x)\left(\frac{\sec x + \tan x}{\sec x + \tan x}\right) dx \qquad \text{Multiply and divide by } \sec x + \tan x.$$

$$= \int \frac{\sec^2 x + \sec x \tan x}{\sec x + \tan x} \, dx$$

Letting u be the denominator of this quotient produces

$$u = \sec x + \tan x$$

and

$$u' = \sec x \tan x + \sec^2 x.$$

So, you can conclude that

$$\int \sec x \, dx = \int \frac{\sec^2 x + \sec x \tan x}{\sec x + \tan x} \, dx \qquad \text{Rewrite integrand.}$$

$$= \int \frac{u'}{u} \, dx \qquad \text{Substitute: } u = \sec x + \tan x.$$

$$= \ln|u| + C \qquad \text{Apply Log Rule.}$$

$$= \ln|\sec x + \tan x| + C. \qquad \text{Back-substitute.} \qquad \blacksquare$$

With the results of Examples 8 and 9, you now have integration formulas for $\sin x$, $\cos x$, $\tan x$, and $\sec x$. The integrals of the six basic trigonometric functions are summarized below. (For proofs of $\cot u$ and $\csc u$, see Exercises 85 and 86.)

> · · · · · · · · · · · · · · · · · ▷
>
> **·· REMARK** Using trigonometric identities and properties of logarithms, you could rewrite these six integration rules in other forms. For instance, you could write
>
> $$\int \csc u \, du$$
>
> $$= \ln|\csc u - \cot u| + C.$$
>
> (See Exercises 87–90.)

> ### INTEGRALS OF THE SIX BASIC TRIGONOMETRIC FUNCTIONS
>
> $$\int \sin u \, du = -\cos u + C \qquad \int \cos u \, du = \sin u + C$$
>
> $$\int \tan u \, du = -\ln|\cos u| + C \qquad \int \cot u \, du = \ln|\sin u| + C$$
>
> $$\int \sec u \, du = \ln|\sec u + \tan u| + C \qquad \int \csc u \, du = -\ln|\csc u + \cot u| + C$$

EXAMPLE 10 **Integrating Trigonometric Functions**

Evaluate $\displaystyle\int_0^{\pi/4} \sqrt{1 + \tan^2 x} \, dx$.

Solution Using $1 + \tan^2 x = \sec^2 x$, you can write

$$\int_0^{\pi/4} \sqrt{1 + \tan^2 x} \, dx = \int_0^{\pi/4} \sqrt{\sec^2 x} \, dx$$

$$= \int_0^{\pi/4} \sec x \, dx \qquad \sec x \geq 0 \text{ for } 0 \leq x \leq \frac{\pi}{4}.$$

$$= \ln|\sec x + \tan x| \Big]_0^{\pi/4}$$

$$= \ln(\sqrt{2} + 1) - \ln 1$$

$$\approx 0.881.$$

EXAMPLE 11 **Finding an Average Value**

Find the average value of

$$f(x) = \tan x$$

on the interval $[0, \pi/4]$.

Solution

$$\text{Average value} = \frac{1}{(\pi/4) - 0} \int_0^{\pi/4} \tan x \, dx \qquad \text{Average value} = \frac{1}{b - a}\int_a^b f(x)\, dx$$

$$= \frac{4}{\pi} \int_0^{\pi/4} \tan x \, dx \qquad \text{Simplify.}$$

$$= \frac{4}{\pi} \left[-\ln|\cos x| \right]_0^{\pi/4} \qquad \text{Integrate.}$$

$$= -\frac{4}{\pi} \left[\ln \frac{\sqrt{2}}{2} - \ln 1 \right]$$

$$= -\frac{4}{\pi} \ln \frac{\sqrt{2}}{2}$$

$$\approx 0.441$$

The average value is about 0.441, as shown in Figure 5.44.

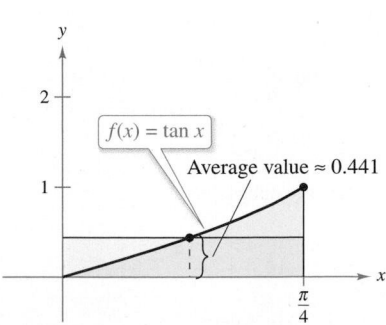

Figure 5.44

5.7 Exercises

CONCEPT CHECK

1. Log Rule for Integration   Can you use the Log Rule to find the integral below? Explain.

$$\int \frac{x}{(x^2 - 4)^3} dx$$

2. Long Division   Explain when to use long division before applying the Log Rule.

3. Guidelines for Integration   Describe two ways to alter an integrand so that it fits an integration formula.

4. Trigonometric Functions   Integrating which trigonometric function results in $\ln|\sin x| + C$?

Finding an Indefinite Integral   In Exercises 5–28, find the indefinite integral.

5. $\int \dfrac{5}{x} dx$

6. $\int \dfrac{1}{x - 5} dx$

7. $\int \dfrac{1}{2x + 5} dx$

8. $\int \dfrac{9}{5 - 4x} dx$

9. $\int \dfrac{x}{x^2 - 3} dx$

10. $\int \dfrac{x^2}{5 - x^3} dx$

11. $\int \dfrac{4x^3 + 3}{x^4 + 3x} dx$

12. $\int \dfrac{x^2 - 2x}{x^3 - 3x^2} dx$

13. $\int \dfrac{x^2 - 7}{7x} dx$

14. $\int \dfrac{x^3 - 8x}{x^2} dx$

15. $\int \dfrac{x^2 + 2x + 3}{x^3 + 3x^2 + 9x} dx$

16. $\int \dfrac{x^2 + 4x}{x^3 + 6x^2 + 5} dx$

17. $\int \dfrac{x^2 - 3x + 2}{x + 1} dx$

18. $\int \dfrac{2x^2 + 7x - 3}{x - 2} dx$

19. $\int \dfrac{x^3 - 3x^2 + 5}{x - 3} dx$

20. $\int \dfrac{x^3 - 6x - 20}{x + 5} dx$

21. $\int \dfrac{x^4 + x - 4}{x^2 + 2} dx$

22. $\int \dfrac{x^3 - 4x^2 - 4x + 20}{x^2 - 5} dx$

23. $\int \dfrac{(\ln x)^2}{x} dx$

24. $\int \dfrac{dx}{x(\ln x^2)^3}$

25. $\int \dfrac{1}{\sqrt{x}(1 - 3\sqrt{x})} dx$

26. $\int \dfrac{1}{x^{2/3}(1 + x^{1/3})} dx$

27. $\int \dfrac{6x}{(x - 5)^2} dx$

28. $\int \dfrac{x(x - 2)}{(x - 1)^3} dx$

Change of Variables   In Exercises 29–32, find the indefinite integral by making a change of variables (*Hint:* Let u be the denominator of the integrand.)

29. $\int \dfrac{1}{1 + \sqrt{2x}} dx$

30. $\int \dfrac{4}{1 + \sqrt{5x}} dx$

31. $\int \dfrac{\sqrt{x}}{\sqrt{x} - 3} dx$

32. $\int \dfrac{\sqrt[3]{x}}{\sqrt[3]{x} - 1} dx$

Finding an Indefinite Integral of a Trigonometric Function   In Exercises 33–42, find the indefinite integral.

33. $\int \cot \dfrac{\theta}{3} d\theta$

34. $\int \sec \dfrac{x}{2} dx$

35. $\int \csc 2x \, dx$

36. $\int \left(2 - \tan \dfrac{\theta}{4}\right) d\theta$

37. $\int \dfrac{\cos t}{1 + \sin t} dt$

38. $\int \dfrac{\csc^2 t}{\cot t} dt$

39. $\int \dfrac{\sec x \tan x}{\sec x - 1} dx$

40. $\int \dfrac{\sec(2 \ln t) + \tan(2 \ln t)}{t} dt$

41. $\int e^{-x} \tan(e^{-x}) dx$

42. $\int \sec t(\sec t + \tan t) dt$

Differential Equation   In Exercises 43–46, find the general solution of the differential equation. Use a graphing utility to graph three solutions, one of which passes through the given point.

43. $\dfrac{dy}{dx} = \dfrac{3}{2 - x}$, $(1, 0)$

44. $\dfrac{dy}{dx} = \dfrac{x - 2}{x}$, $(-1, 0)$

45. $\dfrac{dy}{dx} = \dfrac{2x}{x^2 - 9}$, $(0, 4)$

46. $\dfrac{dr}{dt} = \dfrac{\sec^2 t}{\tan t + 1}$, $(\pi, 4)$

Finding a Particular Solution   In Exercises 47 and 48, find the particular solution of the differential equation that satisfies the initial conditions.

47. $f''(x) = \dfrac{2}{x^2}$, $f'(1) = 1$, $f(1) = 1$, $x > 0$

48. $f''(x) = -\dfrac{4}{(x - 1)^2} - 2$, $f'(2) = 0$, $f(2) = 3$, $x > 1$

Slope Field   In Exercises 49 and 50, a differential equation, a point, and a slope field are given. (a) Sketch two approximate solutions of the differential equation on the slope field, one of which passes through the given point. (To print an enlarged copy of the graph, go to *MathGraphs.com*.) (b) Use integration and the given point to find the particular solution of the differential equation and use a graphing utility to graph the solution. Compare the result with the sketch in part (a) that passes through the given point.

49. $\dfrac{dy}{dx} = \dfrac{1}{x + 2}$, $(0, 1)$

50. $\dfrac{dy}{dx} = \dfrac{\ln x}{x}$, $(1, -2)$

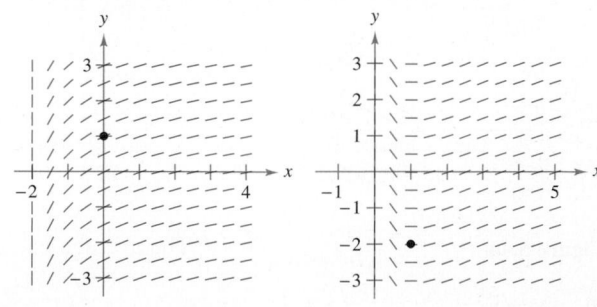

Evaluating a Definite Integral In Exercises 51–58, evaluate the definite integral. Use a graphing utility to verify your result.

51. $\int_0^4 \dfrac{5}{3x+1}\,dx$

52. $\int_{-1}^1 \dfrac{1}{2x+3}\,dx$

53. $\int_1^e \dfrac{(1+\ln x)^2}{x}\,dx$

54. $\int_e^{e^2} \dfrac{1}{x\ln x}\,dx$

55. $\int_0^2 \dfrac{x^2-2}{x+1}\,dx$

56. $\int_0^1 \dfrac{x-1}{x+1}\,dx$

57. $\int_1^2 \dfrac{1-\cos\theta}{\theta-\sin\theta}\,d\theta$

58. $\int_{\pi/8}^{\pi/4} (\csc 2\theta - \cot 2\theta)\,d\theta$

 **Finding an Integral Using Technology** In Exercises 59 and 60, use a computer algebra system to find or evaluate the integral.

59. $\displaystyle\int \dfrac{1-\sqrt{x}}{1+\sqrt{x}}\,dx$

60. $\displaystyle\int_{-\pi/4}^{\pi/4} \dfrac{\sin^2 x - \cos^2 x}{\cos x}\,dx$

Finding a Derivative In Exercises 61–64, find $F'(x)$.

61. $F(x) = \displaystyle\int_1^x \dfrac{1}{t}\,dt$

62. $F(x) = \displaystyle\int_0^x \tan t\,dt$

63. $F(x) = \displaystyle\int_1^{4x} \cot t\,dt$

64. $F(x) = \displaystyle\int_0^{x^2} \dfrac{3}{t+1}\,dt$

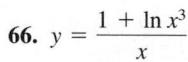 **Area** In Exercises 65–68, find the area of the given region. Use a graphing utility to verify your result.

65. $y = \dfrac{6}{x}$

66. $y = \dfrac{1+\ln x^3}{x}$

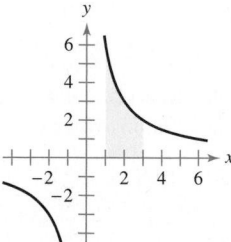

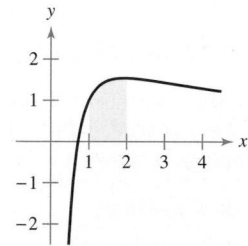

67. $y = \csc(x+1)$

68. $y = \dfrac{\sin x}{1+\cos x}$

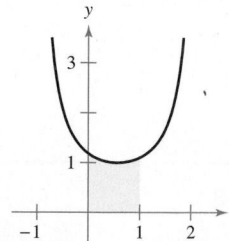

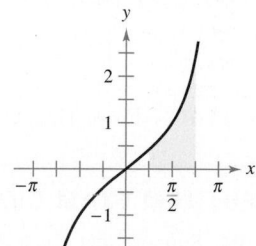

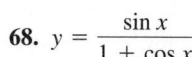 **Area** In Exercises 69–72, find the area of the region bounded by the graphs of the equations. Use a graphing utility to verify your result.

69. $y = \dfrac{x^2+4}{x}$, $\quad x = 1$, $\quad x = 4$, $\quad y = 0$

70. $y = \dfrac{5x}{x^2+2}$, $\quad x = 1$, $\quad x = 5$, $\quad y = 0$

71. $y = 2\sec\dfrac{\pi x}{6}$, $\quad x = 0$, $\quad x = 2$, $\quad y = 0$

72. $y = 2x - \tan 0.3x$, $\quad x = 1$, $\quad x = 4$, $\quad y = 0$

 **Finding the Average Value of a Function** In Exercises 73–76, find the average value of the function over the given interval.

73. $f(x) = \dfrac{8}{x^2}$, $\quad [2,4]$

74. $f(x) = \dfrac{4(x+1)}{x^2}$, $\quad [2,4]$

75. $f(x) = \dfrac{2\ln x}{x}$, $\quad [1,e]$

76. $f(x) = \sec\dfrac{\pi x}{6}$, $\quad [0,2]$

Midpoint Rule In Exercises 77 and 78, use the Midpoint Rule with $n = 4$ to approximate the value of the definite integral. Use a graphing utility to verify your result.

77. $\displaystyle\int_1^3 \dfrac{12}{x}\,dx$

78. $\displaystyle\int_0^{\pi/4} \sec x\,dx$

EXPLORING CONCEPTS

Approximation In Exercises 79 and 80, determine which value best approximates the area of the region between the x-axis and the graph of the function over the given interval. Make your selection on the basis of a sketch of the region, not by performing calculations.

79. $f(x) = \sec x$, $\quad [0,1]$

    (a) 6   (b) -6 (c) $\frac{1}{2}$ (d) 1.25   (e) 3

80. $f(x) = \dfrac{2x}{x^2+1}$, $\quad [0,4]$

    (a) 3   (b) 7   (c) -2 (d) 5   (e) 1

81. Napier's Inequality For $0 < x < y$, use the Mean Value Theorem to show that

$$\dfrac{1}{y} < \dfrac{\ln y - \ln x}{y - x} < \dfrac{1}{x}.$$

82. Think About It Is the function

$$F(x) = \int_x^{2x} \dfrac{1}{t}\,dt$$

constant, increasing, or decreasing on the interval $(0,\infty)$? Explain.

83. Finding a Value Find a value of x such that

$$\int_1^x \dfrac{3}{t}\,dt = \int_{1/4}^x \dfrac{1}{t}\,dt.$$

84. Finding a Value Find a value of x such that

$$\int_1^x \dfrac{1}{t}\,dt$$

is equal to (a) $\ln 5$ and (b) 1.

85. Proof Prove that

$$\int \cot u \, du = \ln|\sin u| + C.$$

86. Proof Prove that

$$\int \csc u \, du = -\ln|\csc u + \cot u| + C.$$

Using Properties of Logarithms and Trigonometric Identities In Exercises 87–90, show that the two formulas are equivalent.

87. $\int \tan x \, dx = -\ln|\cos x| + C$

$\int \tan x \, dx = \ln|\sec x| + C$

88. $\int \cot x \, dx = \ln|\sin x| + C$

$\int \cot x \, dx = -\ln|\csc x| + C$

89. $\int \sec x \, dx = \ln|\sec x + \tan x| + C$

$\int \sec x \, dx = -\ln|\sec x - \tan x| + C$

90. $\int \csc x \, dx = -\ln|\csc x + \cot x| + C$

$\int \csc x \, dx = \ln|\csc x - \cot x| + C$

91. Population Growth A population of bacteria P is changing at a rate of

$$\frac{dP}{dt} = \frac{3000}{1 + 0.25t}$$

where t is the time in days. The initial population (when $t = 0$) is 1000.

(a) Write an equation that gives the population at any time t.

(b) Find the population when $t = 3$ days.

92. Sales The rate of change in sales S is inversely proportional to time t $(t > 1)$, measured in weeks. Find S as a function of t when the sales after 2 and 4 weeks are 200 units and 300 units, respectively.

93. Heat Transfer

Find the time required for an object to cool from 300°F to 250°F by evaluating

$$t = \frac{10}{\ln 2} \int_{250}^{300} \frac{1}{T - 100} \, dT$$

where t is time in minutes.

94. Average Price The demand equation for a product is

$$p = \frac{90{,}000}{400 + 3x}$$

where p is the price (in dollars) and x is the number of units (in thousands). Find the average price p on the interval $40 \le x \le 50$.

95. Area and Slope Graph the function $f(x) = x/(1 + x^2)$ on the interval $[0, \infty)$.

(a) Find the area bounded by the graph of f and the line $y = \frac{1}{2}x$.

(b) Determine the values of the slope m such that the line $y = mx$ and the graph of f enclose a finite region.

(c) Calculate the area of this region as a function of m.

96. HOW DO YOU SEE IT? Use the graph of f' shown in the figure to answer the following.

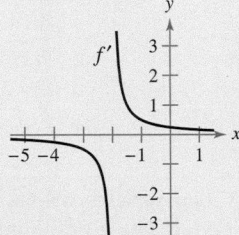

(a) Approximate the slope of f at $x = -1$. Explain.

(b) Approximate any open intervals on which the graph of f is increasing and any open intervals on which it is decreasing. Explain.

True or False? In Exercises 97–100, determine whether the statement is true or false. If it is false, explain why or give an example that shows it is false.

97. $\ln|x^4| = \ln x^4$

98. $\ln|\cos \theta^2| = \ln(\cos \theta^2)$

99. $\int \frac{1}{x} \, dx = \ln|cx|, \quad c \ne 0$

100. $\int_{-1}^{2} \frac{1}{x} \, dx = \left[\ln|x|\right]_{-1}^{2} = \ln 2 - \ln 1 = \ln 2$

101. Proof Prove Theorem 5.19.

PUTNAM EXAM CHALLENGE

102. Suppose that f is a function on the interval $[1, 3]$ such that $-1 \le f(x) \le 1$ for all x and $\int_{1}^{3} f(x) \, dx = 0$. How large can $\int_{1}^{3} \frac{f(x)}{x} \, dx$ be?

This problem was composed by the Committee on the Putnam Prize Competition.
© The Mathematical Association of America. All rights reserved.

5.8 Inverse Trigonometric Functions: Integration

- ■ Integrate functions whose antiderivatives involve inverse trigonometric functions.
- ■ Use the method of completing the square to integrate a function.
- ■ Review the basic integration rules involving elementary functions.

Integrals Involving Inverse Trigonometric Functions

The derivatives of the six inverse trigonometric functions fall into three pairs. In each pair, the derivative of one function is the negative of the other. For example,

$$\frac{d}{dx}[\arcsin x] = \frac{1}{\sqrt{1 - x^2}}$$

and

$$\frac{d}{dx}[\arccos x] = -\frac{1}{\sqrt{1 - x^2}}.$$

When listing the *antiderivative* that corresponds to each of the inverse trigonometric functions, you need to use only one member from each pair. It is conventional to use arcsin x as the antiderivative of $1/\sqrt{1 - x^2}$, rather than $-\arccos x$. The next theorem gives one antiderivative formula for each of the three pairs. The proofs of these integration rules are left to you (see Exercises 73–75).

■ **FOR FURTHER INFORMATION**
For a detailed proof of rule 2 of Theorem 5.20, see the article "A Direct Proof of the Integral Formula for Arctangent" by Arnold J. Insel in *The College Mathematics Journal*. To view this article, go to *MathArticles.com*.

> **THEOREM 5.20   Integrals Involving Inverse Trigonometric Functions**
>
> Let u be a differentiable function of x, and let $a > 0$.
>
> 1. $\displaystyle\int \frac{du}{\sqrt{a^2 - u^2}} = \arcsin \frac{u}{a} + C$
>
> 2. $\displaystyle\int \frac{du}{a^2 + u^2} = \frac{1}{a} \arctan \frac{u}{a} + C$
>
> 3. $\displaystyle\int \frac{du}{u\sqrt{u^2 - a^2}} = \frac{1}{a} \operatorname{arcsec} \frac{|u|}{a} + C$

EXAMPLE 1   Integration with Inverse Trigonometric Functions

a. $\displaystyle\int \frac{dx}{\sqrt{4 - x^2}} = \arcsin \frac{x}{2} + C$ $u = x, a = 2$

b. $\displaystyle\int \frac{dx}{2 + 9x^2} = \frac{1}{3}\int \frac{3\,dx}{(\sqrt{2})^2 + (3x)^2}$ $u = 3x, a = \sqrt{2}$

$\displaystyle\qquad\qquad = \frac{1}{3\sqrt{2}} \arctan \frac{3x}{\sqrt{2}} + C$

c. $\displaystyle\int \frac{dx}{x\sqrt{4x^2 - 9}} = \int \frac{2\,dx}{2x\sqrt{(2x)^2 - 3^2}}$ $u = 2x, a = 3$

$\displaystyle\qquad\qquad = \frac{1}{3} \operatorname{arcsec} \frac{|2x|}{3} + C$

The integrals in Example 1 are fairly straightforward applications of integration formulas. Unfortunately, this is not typical. The integration formulas for inverse trigonometric functions can be disguised in many ways.

EXAMPLE 2 **Integration by Substitution**

Find $\displaystyle\int \frac{dx}{\sqrt{e^{2x}-1}}$.

Solution As it stands, this integral does not fit any of the three inverse trigonometric formulas. Using the substitution $u = e^x$, however, produces

$$u = e^x \quad \Longrightarrow \quad du = e^x\, dx \quad \Longrightarrow \quad dx = \frac{du}{e^x} = \frac{du}{u}.$$

With this substitution, you can integrate as shown.

$$\int \frac{dx}{\sqrt{e^{2x}-1}} = \int \frac{dx}{\sqrt{(e^x)^2 - 1}} \qquad \text{Write } e^{2x} \text{ as } (e^x)^2.$$

$$= \int \frac{du/u}{\sqrt{u^2 - 1}} \qquad \text{Substitute.}$$

$$= \int \frac{du}{u\sqrt{u^2 - 1}} \qquad \text{Rewrite to fit Arcsecant Rule.}$$

$$= \text{arcsec}\,\frac{|u|}{1} + C \qquad \text{Apply Arcsecant Rule.}$$

$$= \text{arcsec}\, e^x + C \qquad \text{Back-substitute.} \qquad ■$$

▷ **TECHNOLOGY PITFALL** A symbolic integration utility can be useful for integrating functions such as the one in Example 2. In some cases, however, the utility may fail to find an antiderivative for two reasons. First, some elementary functions do not have antiderivatives that are elementary functions. Second, every utility has limitations—you might have entered a function that the utility was not programmed to handle. You should also remember that antiderivatives involving trigonometric functions or logarithmic functions can be written in many different forms. For instance, one utility found the integral in Example 2 to be

$$\int \frac{dx}{\sqrt{e^{2x}-1}} = \arctan \sqrt{e^{2x}-1} + C.$$

Try showing that this antiderivative is equivalent to the one found in Example 2.

EXAMPLE 3 **Rewriting as the Sum of Two Quotients**

Find $\displaystyle\int \frac{x+2}{\sqrt{4-x^2}}\, dx$.

Solution This integral does not appear to fit any of the basic integration formulas. By splitting the integrand into two parts, however, you can see that the first part can be found with the Power Rule and the second part yields an inverse sine function.

$$\int \frac{x+2}{\sqrt{4-x^2}}\, dx = \int \frac{x}{\sqrt{4-x^2}}\, dx + \int \frac{2}{\sqrt{4-x^2}}\, dx$$

$$= -\frac{1}{2}\int (4-x^2)^{-1/2}(-2x)\, dx + 2\int \frac{1}{\sqrt{4-x^2}}\, dx$$

$$= -\frac{1}{2}\left[\frac{(4-x^2)^{1/2}}{1/2}\right] + 2\,\text{arcsin}\,\frac{x}{2} + C$$

$$= -\sqrt{4-x^2} + 2\,\text{arcsin}\,\frac{x}{2} + C \qquad ■$$

Completing the Square

Completing the square helps when quadratic functions are involved in the integrand. For example, the quadratic $x^2 + bx + c$ can be written as the difference of two squares by adding and subtracting $(b/2)^2$.

$$x^2 + bx + c = x^2 + bx + \left(\frac{b}{2}\right)^2 - \left(\frac{b}{2}\right)^2 + c = \left(x + \frac{b}{2}\right)^2 - \left(\frac{b}{2}\right)^2 + c$$

EXAMPLE 4   Completing the Square

⋮ ⋯ ▷ *See LarsonCalculus.com for an interactive version of this type of example.*

Find $\displaystyle\int \frac{dx}{x^2 - 4x + 7}$.

Solution   You can write the denominator as the sum of two squares, as shown.

$$x^2 - 4x + 7 = (x^2 - 4x + 4) - 4 + 7 = (x - 2)^2 + 3 = u^2 + a^2$$

Now, in this completed square form, let $u = x - 2$ and $a = \sqrt{3}$.

$$\int \frac{dx}{x^2 - 4x + 7} = \int \frac{dx}{(x - 2)^2 + 3} = \frac{1}{\sqrt{3}} \arctan \frac{x - 2}{\sqrt{3}} + C$$

When the leading coefficient is not 1, it helps to factor before completing the square. For instance, you can complete the square of $2x^2 - 8x + 10$ by factoring first.

$$\begin{aligned}
2x^2 - 8x + 10 &= 2(x^2 - 4x + 5) \\
&= 2(x^2 - 4x + 4 - 4 + 5) \\
&= 2[(x - 2)^2 + 1]
\end{aligned}$$

To complete the square when the coefficient of x^2 is negative, use the same factoring process shown above. For instance, you can complete the square for $3x - x^2$ as shown.

$$3x - x^2 = -(x^2 - 3x) = -\left[x^2 - 3x + \left(\tfrac{3}{2}\right)^2 - \left(\tfrac{3}{2}\right)^2\right] = \left(\tfrac{3}{2}\right)^2 - \left(x - \tfrac{3}{2}\right)^2$$

EXAMPLE 5   Completing the Square

Find the area of the region bounded by the graph of

$$f(x) = \frac{1}{\sqrt{3x - x^2}}$$

the x-axis, and the lines $x = \frac{3}{2}$ and $x = \frac{9}{4}$.

Solution   In Figure 5.45, you can see that the area is

$$\begin{aligned}
\text{Area} &= \int_{3/2}^{9/4} \frac{1}{\sqrt{3x - x^2}}\, dx \\
&= \int_{3/2}^{9/4} \frac{dx}{\sqrt{(3/2)^2 - [x - (3/2)]^2}} \qquad \text{\footnotesize Use completed square form derived above.} \\
&= \arcsin \frac{x - (3/2)}{3/2} \Bigg]_{3/2}^{9/4} \\
&= \arcsin \frac{1}{2} - \arcsin 0 \\
&= \frac{\pi}{6} \\
&\approx 0.524.
\end{aligned}$$

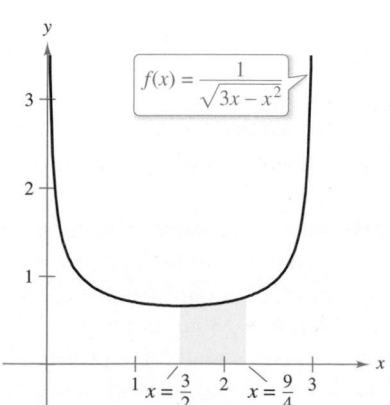

$$f(x) = \frac{1}{\sqrt{3x - x^2}}$$

The area of the region bounded by the graph of f, the x-axis, $x = \frac{3}{2}$, and $x = \frac{9}{4}$ is $\pi/6$.

Figure 5.45

Review of Basic Integration Rules

You have now completed the introduction of the **basic integration rules.** To be efficient at applying these rules, you should have practiced enough so that each rule is committed to memory.

BASIC INTEGRATION RULES ($a > 0$)

1. $\displaystyle \int k f(u)\, du = k \int f(u)\, du$

2. $\displaystyle \int [f(u) \pm g(u)]\, du = \int f(u)\, du \pm \int g(u)\, du$

3. $\displaystyle \int du = u + C$

4. $\displaystyle \int u^n\, du = \frac{u^{n+1}}{n+1} + C, \quad n \neq -1$

5. $\displaystyle \int \frac{du}{u} = \ln|u| + C$

6. $\displaystyle \int e^u\, du = e^u + C$

7. $\displaystyle \int a^u\, du = \left(\frac{1}{\ln a}\right) a^u + C$

8. $\displaystyle \int \sin u\, du = -\cos u + C$

9. $\displaystyle \int \cos u\, du = \sin u + C$

10. $\displaystyle \int \tan u\, du = -\ln|\cos u| + C$

11. $\displaystyle \int \cot u\, du = \ln|\sin u| + C$

12. $\displaystyle \int \sec u\, du = \ln|\sec u + \tan u| + C$

13. $\displaystyle \int \csc u\, du = -\ln|\csc u + \cot u| + C$

14. $\displaystyle \int \sec^2 u\, du = \tan u + C$

15. $\displaystyle \int \csc^2 u\, du = -\cot u + C$

16. $\displaystyle \int \sec u \tan u\, du = \sec u + C$

17. $\displaystyle \int \csc u \cot u\, du = -\csc u + C$

18. $\displaystyle \int \frac{du}{\sqrt{a^2 - u^2}} = \arcsin \frac{u}{a} + C$

19. $\displaystyle \int \frac{du}{a^2 + u^2} = \frac{1}{a} \arctan \frac{u}{a} + C$

20. $\displaystyle \int \frac{du}{u\sqrt{u^2 - a^2}} = \frac{1}{a} \operatorname{arcsec} \frac{|u|}{a} + C$

You can learn a lot about the nature of integration by comparing this list with the summary of differentiation rules given in Section 3.6. For differentiation, you now have rules that allow you to differentiate *any* elementary function. For integration, this is far from true.

The integration rules listed above are primarily those that were happened on during the development of differentiation rules. So far, you have not learned any rules or techniques for finding the antiderivative of a general product or quotient, the natural logarithmic function, or the inverse trigonometric functions. More important, you cannot apply any of the rules in this list unless you can create the proper *du* corresponding to the *u* in the formula. The point is that you need to work more on integration techniques, which you will do in Chapter 8. The next two examples should give you a better feeling for the integration problems that you *can* and *cannot* solve with the techniques and rules you now know.

EXAMPLE 6 **Comparing Integration Problems**

Find as many of the following integrals as you can using the formulas and techniques you have studied so far in the text.

a. $\displaystyle\int \frac{dx}{x\sqrt{x^2-1}}$

b. $\displaystyle\int \frac{x\,dx}{\sqrt{x^2-1}}$

c. $\displaystyle\int \frac{dx}{\sqrt{x^2-1}}$

Solution

a. You *can* find this integral (it fits the Arcsecant Rule).

$$\int \frac{dx}{x\sqrt{x^2-1}} = \text{arcsec}|x| + C$$

b. You *can* find this integral (it fits the Power Rule).

$$\int \frac{x\,dx}{\sqrt{x^2-1}} = \frac{1}{2}\int (x^2-1)^{-1/2}(2x)\,dx$$

$$= \frac{1}{2}\left[\frac{(x^2-1)^{1/2}}{1/2}\right] + C$$

$$= \sqrt{x^2-1} + C$$

c. You *cannot* find this integral using the techniques you have studied so far. (You should scan the list of basic integration rules to verify this conclusion.)

EXAMPLE 7 **Comparing Integration Problems**

Find as many of the following integrals as you can using the formulas and techniques you have studied so far in the text.

a. $\displaystyle\int \frac{dx}{x\ln x}$

b. $\displaystyle\int \frac{\ln x\,dx}{x}$

c. $\displaystyle\int \ln x\,dx$

Solution

a. You *can* find this integral (it fits the Log Rule).

$$\int \frac{dx}{x\ln x} = \int \frac{1/x}{\ln x}\,dx$$

$$= \ln|\ln x| + C$$

b. You *can* find this integral (it fits the Power Rule).

$$\int \frac{\ln x\,dx}{x} = \int \left(\frac{1}{x}\right)(\ln x)^1\,dx$$

$$= \frac{(\ln x)^2}{2} + C$$

• • REMARK Note in Examples 6 and 7 that the *simplest* functions are the ones that you cannot yet integrate.

c. You *cannot* find this integral using the techniques you have studied so far.

5.8 Exercises

See **CalcChat.com** for tutorial help and worked-out solutions to odd-numbered exercises.

CONCEPT CHECK

1. Integration Rules Decide whether you can find each integral using the formulas and techniques you have studied so far. Explain.

(a) $\int \dfrac{2\,dx}{\sqrt{x^2 + 4}}$

(b) $\int \dfrac{dx}{x\sqrt{x^2 - 9}}$

2. Completing the Square In your own words, describe the process of completing the square of a quadratic function. Explain when completing the square is useful for finding an integral.

Finding an Indefinite Integral In Exercises 3–22, find the indefinite integral.

3. $\int \dfrac{dx}{\sqrt{9 - x^2}}$

4. $\int \dfrac{dx}{\sqrt{1 - 4x^2}}$

5. $\int \dfrac{1}{x\sqrt{4x^2 - 1}}\,dx$

6. $\int \dfrac{12}{1 + 9x^2}\,dx$

7. $\int \dfrac{1}{\sqrt{1 - (x + 1)^2}}\,dx$

8. $\int \dfrac{7}{4 + (3 - x)^2}\,dx$

9. $\int \dfrac{t}{\sqrt{1 - t^4}}\,dt$

10. $\int \dfrac{1}{x\sqrt{x^4 - 4}}\,dx$

11. $\int \dfrac{t}{t^4 + 25}\,dt$

12. $\int \dfrac{1}{x\sqrt{1 - (\ln x)^2}}\,dx$

13. $\int \dfrac{e^{2x}}{4 + e^{4x}}\,dx$

14. $\int \dfrac{5}{x\sqrt{9x^2 - 11}}\,dx$

15. $\int \dfrac{-\csc x \cot x}{\sqrt{25 - \csc^2 x}}\,dx$

16. $\int \dfrac{\sin x}{7 + \cos^2 x}\,dx$

17. $\int \dfrac{1}{\sqrt{x}\sqrt{1 - x}}\,dx$

18. $\int \dfrac{3}{2\sqrt{x}(1 + x)}\,dx$

19. $\int \dfrac{x - 3}{x^2 + 1}\,dx$

20. $\int \dfrac{x^2 + 8}{x\sqrt{x^2 - 4}}\,dx$

21. $\int \dfrac{x + 5}{\sqrt{9 - (x - 3)^2}}\,dx$

22. $\int \dfrac{x - 2}{(x + 1)^2 + 4}\,dx$

Evaluating a Definite Integral In Exercises 23–34, evaluate the definite integral.

23. $\int_0^{1/6} \dfrac{3}{\sqrt{1 - 9x^2}}\,dx$

24. $\int_0^{\sqrt{2}} \dfrac{1}{\sqrt{4 - x^2}}\,dx$

25. $\int_0^{\sqrt{3}/2} \dfrac{1}{1 + 4x^2}\,dx$

26. $\int_{\sqrt{3}}^3 \dfrac{1}{x\sqrt{4x^2 - 9}}\,dx$

27. $\int_1^7 \dfrac{1}{9 + (x + 2)^2}\,dx$

28. $\int_1^4 \dfrac{1}{x\sqrt{16x^2 - 5}}\,dx$

29. $\int_0^{\ln 5} \dfrac{e^x}{1 + e^{2x}}\,dx$

30. $\int_{\ln 2}^{\ln 4} \dfrac{e^{-x}}{\sqrt{1 - e^{-2x}}}\,dx$

31. $\int_{\pi/2}^{\pi} \dfrac{\sin x}{1 + \cos^2 x}\,dx$

32. $\int_0^{\pi/2} \dfrac{\cos x}{1 + \sin^2 x}\,dx$

33. $\int_0^{1/\sqrt{2}} \dfrac{\arcsin x}{\sqrt{1 - x^2}}\,dx$

34. $\int_0^{1/\sqrt{2}} \dfrac{\arccos x}{\sqrt{1 - x^2}}\,dx$

Completing the Square In Exercises 35–42, find or evaluate the integral by completing the square.

35. $\int_0^2 \dfrac{dx}{x^2 - 2x + 2}$

36. $\int_{-2}^3 \dfrac{dx}{x^2 + 4x + 8}$

37. $\int \dfrac{dx}{\sqrt{-2x^2 + 8x + 4}}$

38. $\int \dfrac{dx}{3x^2 - 6x + 12}$

39. $\int \dfrac{1}{\sqrt{-x^2 - 4x}}\,dx$

40. $\int \dfrac{2}{\sqrt{-x^2 + 4x}}\,dx$

41. $\int_2^3 \dfrac{2x - 3}{\sqrt{4x - x^2}}\,dx$

42. $\int_3^4 \dfrac{1}{(x - 1)\sqrt{x^2 - 2x}}\,dx$

Integration by Substitution In Exercises 43–46, use the specified substitution to find or evaluate the integral.

43. $\int \sqrt{e^t - 3}\,dt$

$u = \sqrt{e^t - 3}$

44. $\int \dfrac{\sqrt{x - 2}}{x + 1}\,dx$

$u = \sqrt{x - 2}$

45. $\int_1^3 \dfrac{dx}{\sqrt{x}(1 + x)}$

$u = \sqrt{x}$

46. $\int_0^1 \dfrac{dx}{2\sqrt{3 - x}\sqrt{x + 1}}$

$u = \sqrt{x + 1}$

Comparing Integration Problems In Exercises 47–50, find the indefinite integrals, if possible, using the formulas and techniques you have studied so far in the text.

47. (a) $\int \dfrac{1}{\sqrt{1 - x^2}}\,dx$

(b) $\int \dfrac{x}{\sqrt{1 - x^2}}\,dx$

(c) $\int \dfrac{1}{x\sqrt{1 - x^2}}\,dx$

48. (a) $\int e^{x^2}\,dx$

(b) $\int xe^{x^2}\,dx$

(c) $\int \dfrac{1}{x^2} e^{1/x}\,dx$

49. (a) $\int \sqrt{x - 1}\,dx$

(b) $\int x\sqrt{x - 1}\,dx$

(c) $\int \dfrac{x}{\sqrt{x - 1}}\,dx$

50. (a) $\int \dfrac{1}{1 + x^4}\,dx$

(b) $\int \dfrac{x}{1 + x^4}\,dx$

(c) $\int \dfrac{x^3}{1 + x^4}\,dx$

EXPLORING CONCEPTS

Comparing Antiderivatives In Exercises 51 and 52, show that the antiderivatives are equivalent.

51. $\int \dfrac{3x^2}{\sqrt{1-x^6}}\,dx = \arcsin x^3 + C$ or $\arccos \sqrt{1-x^6} + C$

52. $\int \dfrac{6}{4+9x^2}\,dx = \arctan \dfrac{3x}{2} + C$ or $\operatorname{arccsc}\dfrac{\sqrt{4+9x^2}}{3x} + C$

53. Inverse Trigonometric Functions The antiderivative of

$$\int \frac{1}{\sqrt{1-x^2}}\,dx$$

can be either $\arcsin x + C$ or $-\arccos x + C$. Does this mean that $\arcsin x = -\arccos x$? Explain.

54. **HOW DO YOU SEE IT?** Using the graph, which value best approximates the area of the region between the x-axis and the function over the interval $\left[-\frac{1}{2}, \frac{1}{2}\right]$? Explain.

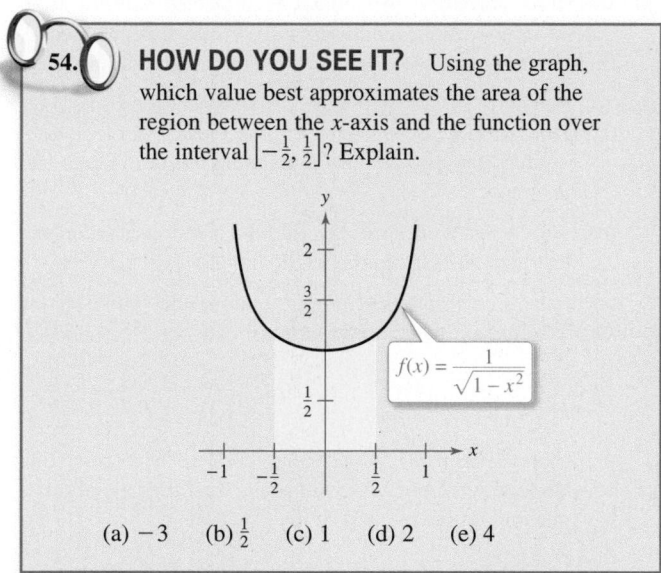

$f(x) = \dfrac{1}{\sqrt{1-x^2}}$

(a) -3 (b) $\frac{1}{2}$ (c) 1 (d) 2 (e) 4

Slope Field In Exercises 55 and 56, a differential equation, a point, and a slope field are given. (a) Sketch two approximate solutions of the differential equation on the slope field, one of which passes through the given point. (To print an enlarged copy of the graph, go to *MathGraphs.com*.) (b) Use integration and the given point to find the particular solution of the differential equation and use a graphing utility to graph the solution. Compare the result with the sketch in part (a) that passes through the given point.

55. $\dfrac{dy}{dx} = \dfrac{2}{9+x^2}$, $(0, 2)$ **56.** $\dfrac{dy}{dx} = \dfrac{2}{\sqrt{25-x^2}}$, $(5, \pi)$

Slope Field In Exercises 57–60, use a graphing utility to graph the slope field for the differential equation and graph the particular solution satisfying the specified initial condition.

57. $\dfrac{dy}{dx} = \dfrac{10}{x\sqrt{x^2-1}}$ **58.** $\dfrac{dy}{dx} = \dfrac{1}{12+x^2}$
$y(3) = 0$ $y(4) = 2$

59. $\dfrac{dy}{dx} = \dfrac{2y}{\sqrt{16-x^2}}$ **60.** $\dfrac{dy}{dx} = \dfrac{\sqrt{y}}{1+x^2}$
$y(0) = 2$ $y(0) = 4$

Differential Equation In Exercises 61 and 62, find the particular solution of the differential equation that satisfies the initial condition.

61. $\dfrac{dy}{dx} = \dfrac{1}{\sqrt{4-x^2}}$
$y(0) = \pi$

62. $\dfrac{dy}{dx} = \dfrac{1}{4+x^2}$
$y(2) = \pi$

Area In Exercises 63–66, find the area of the given region. Use a graphing utility to verify your result.

63. $y = \dfrac{2}{\sqrt{4-x^2}}$ **64.** $y = \dfrac{1}{x\sqrt{x^2-1}}$

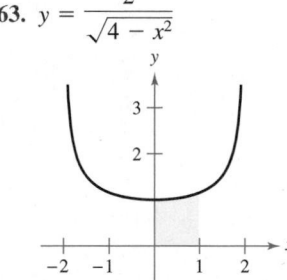 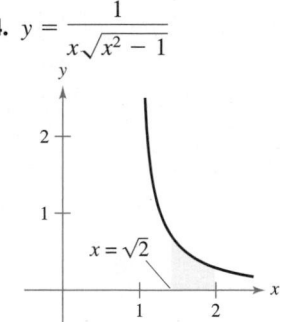

65. $y = \dfrac{3\cos x}{1+\sin^2 x}$ **66.** $y = \dfrac{4e^x}{1+e^{2x}}$

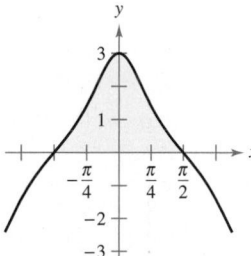 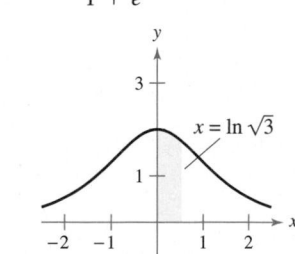

67. Area

(a) Sketch the region whose area is represented by

$$\int_0^1 \arcsin x\,dx.$$

(b) Use the integration capabilities of a graphing utility to approximate the area.

(c) Find the exact area analytically.

68. Approximating Pi

(a) Show that

$$\int_0^1 \frac{4}{1 + x^2}\, dx = \pi.$$

(b) Approximate the number π by using the integration capabilities of a graphing utility.

69. Investigation Consider the function

$$F(x) = \frac{1}{2}\int_x^{x+2} \frac{2}{t^2 + 1}\, dt.$$

(a) Write a short paragraph giving a geometric interpretation of the function $F(x)$ relative to the function

$$f(x) = \frac{2}{x^2 + 1}.$$

Use what you have written to guess the value of x that will make F maximum.

(b) Perform the specified integration to find an alternative form of $F(x)$. Use calculus to locate the value of x that will make F maximum and compare the result with your guess in part (a).

70. Comparing Integrals Consider the integral

$$\int \frac{1}{\sqrt{6x - x^2}}\, dx.$$

(a) Find the integral by completing the square of the radicand.

(b) Find the integral by making the substitution $u = \sqrt{x}$.

(c) The antiderivatives in parts (a) and (b) appear to be significantly different. Use a graphing utility to graph each antiderivative in the same viewing window and determine the relationship between them. Find the domain of each.

True or False? In Exercises 71 and 72, determine whether the statement is true or false. If it is false, explain why or give an example that shows it is false.

71. $\displaystyle\int \frac{dx}{3x\sqrt{9x^2 - 16}} = \frac{1}{4}\operatorname{arcsec}\frac{3x}{4} + C$

72. $\displaystyle\int \frac{dx}{25 + x^2} = \frac{1}{25}\arctan\frac{x}{25} + C$

Verifying an Integration Rule In Exercises 73–75, verify the rule by differentiating. Let $a > 0$.

73. $\displaystyle\int \frac{du}{\sqrt{a^2 - u^2}} = \arcsin\frac{u}{a} + C$

74. $\displaystyle\int \frac{du}{a^2 + u^2} = \frac{1}{a}\arctan\frac{u}{a} + C$

75. $\displaystyle\int \frac{du}{u\sqrt{u^2 - a^2}} = \frac{1}{a}\operatorname{arcsec}\frac{|u|}{a} + C$

76. Proof Graph $y_1 = \dfrac{x}{1 + x^2}$, $y_2 = \arctan x$, and $y_3 = x$ on $[0, 10]$. Prove that $\dfrac{x}{1 + x^2} < \arctan x < x$ for $x > 0$.

77. Numerical Integration

(a) Write an integral that represents the area of the region in the figure.

(b) Use the Midpoint Rule with $n = 8$ to estimate the area of the region.

(c) Explain how you can use the results of parts (a) and (b) to estimate π.

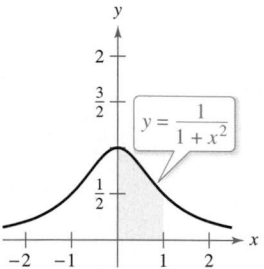

78. Vertical Motion An object is projected upward from ground level with an initial velocity of 500 feet per second. In this exercise, the goal is to analyze the motion of the object during its upward flight.

(a) If air resistance is neglected, find the velocity of the object as a function of time. Use a graphing utility to graph this function.

(b) Use the result of part (a) to find the position function and determine the maximum height attained by the object.

(c) If the air resistance is proportional to the square of the velocity, you obtain the equation

$$\frac{dv}{dt} = -(32 + kv^2)$$

where 32 feet per second per second is the acceleration due to gravity and k is a constant. Find the velocity as a function of time by solving the equation

$$\int \frac{dv}{32 + kv^2} = -\int dt.$$

(d) Use a graphing utility to graph the velocity function $v(t)$ in part (c) for $k = 0.001$. Use the graph to approximate the time t_0 at which the object reaches its maximum height.

(e) Use the integration capabilities of a graphing utility to approximate the integral

$$\int_0^{t_0} v(t)\, dt$$

where $v(t)$ and t_0 are those found in part (d). This is the approximation of the maximum height of the object.

(f) Explain the difference between the results in parts (b) and (e).

■ FOR FURTHER INFORMATION For more information on this topic, see the article "What Goes Up Must Come Down; Will Air Resistance Make It Return Sooner, or Later?" by John Lekner in *Mathematics Magazine*. To view this article, go to *MathArticles.com*.

5.9 Hyperbolic Functions

■ Develop properties of hyperbolic functions.
■ Differentiate and integrate hyperbolic functions.
■ Develop properties of inverse hyperbolic functions.
■ Differentiate and integrate functions involving inverse hyperbolic functions.

Hyperbolic Functions

In this section, you will look briefly at a special class of exponential functions called **hyperbolic functions.** The name *hyperbolic function* arose from comparison of the area of a semicircular region, as shown in Figure 5.46, with the area of a region under a hyperbola, as shown in Figure 5.47.

JOHANN HEINRICH LAMBERT (1728–1777)

The first person to publish a comprehensive study on hyperbolic functions was Johann Heinrich Lambert, a Swiss-German mathematician and colleague of Euler.
See LarsonCalculus.com to read more of this biography.

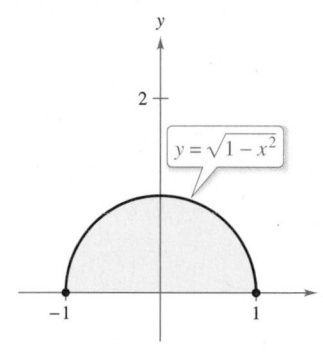

Circle: $x^2 + y^2 = 1$
Figure 5.46

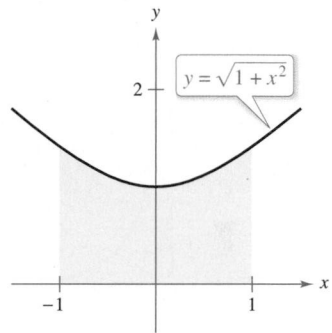

Hyperbola: $-x^2 + y^2 = 1$
Figure 5.47

The integral for the semicircular region involves an inverse trigonometric (circular) function:

$$\int_{-1}^{1} \sqrt{1 - x^2}\, dx = \frac{1}{2}\left[x\sqrt{1 - x^2} + \arcsin x \right]_{-1}^{1} = \frac{\pi}{2} \approx 1.571.$$

The integral for the hyperbolic region involves an inverse hyperbolic function:

$$\int_{-1}^{1} \sqrt{1 + x^2}\, dx = \frac{1}{2}\left[x\sqrt{1 + x^2} + \sinh^{-1} x \right]_{-1}^{1} \approx 2.296.$$

This is only one of many ways in which the hyperbolic functions are similar to the trigonometric functions.

REMARK The notation $\sinh x$ is read as "the hyperbolic sine of x," $\cosh x$ as "the hyperbolic cosine of x," and so on.

Definitions of the Hyperbolic Functions

$$\sinh x = \frac{e^x - e^{-x}}{2} \qquad\qquad \operatorname{csch} x = \frac{1}{\sinh x}, \quad x \neq 0$$

$$\cosh x = \frac{e^x + e^{-x}}{2} \qquad\qquad \operatorname{sech} x = \frac{1}{\cosh x}$$

$$\tanh x = \frac{\sinh x}{\cosh x} \qquad\qquad \coth x = \frac{1}{\tanh x}, \quad x \neq 0$$

■ **FOR FURTHER INFORMATION** For more information on the development of hyperbolic functions, see the article "An Introduction to Hyperbolic Functions in Elementary Calculus" by Jerome Rosenthal in *Mathematics Teacher*. To view this article, go to *MathArticles.com*.

The graphs of the six hyperbolic functions and their domains and ranges are shown in Figure 5.48. Note that the graph of sinh x can be obtained by adding the corresponding y-coordinates of the exponential functions $f(x) = \frac{1}{2}e^x$ and $g(x) = -\frac{1}{2}e^{-x}$. Likewise, the graph of cosh x can be obtained by adding the corresponding y-coordinates of the exponential functions $f(x) = \frac{1}{2}e^x$ and $h(x) = \frac{1}{2}e^{-x}$.

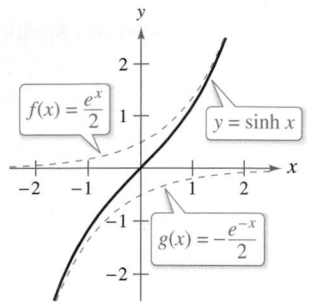

Domain: $(-\infty, \infty)$
Range: $(-\infty, \infty)$

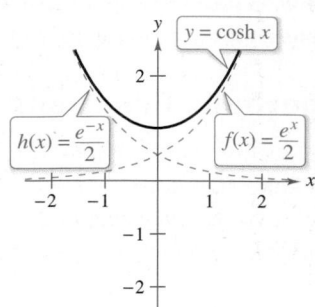

Domain: $(-\infty, \infty)$
Range: $[1, \infty)$

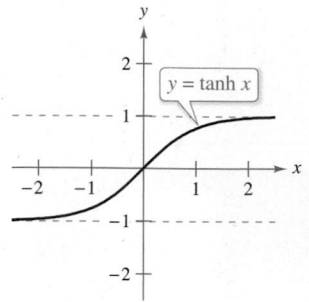

Domain: $(-\infty, \infty)$
Range: $(-1, 1)$

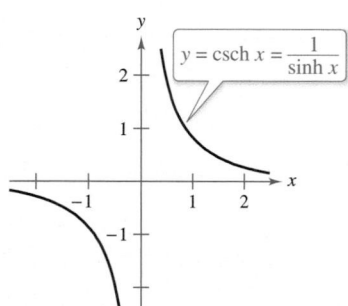

Domain: $(-\infty, 0) \cup (0, \infty)$
Range: $(-\infty, 0) \cup (0, \infty)$

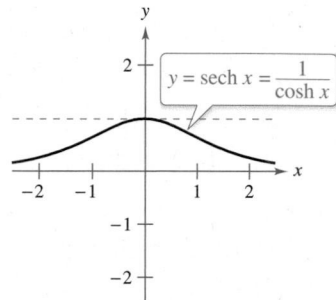

Domain: $(-\infty, \infty)$
Range: $(0, 1]$

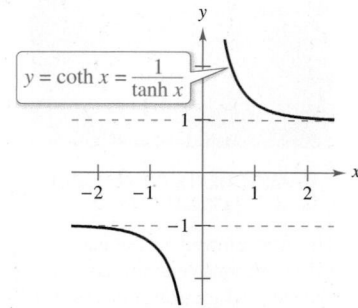

Domain: $(-\infty, 0) \cup (0, \infty)$
Range: $(-\infty, -1) \cup (1, \infty)$

Figure 5.48

Many of the trigonometric identities have corresponding *hyperbolic identities.* For instance,

$$\cosh^2 x - \sinh^2 x = \left(\frac{e^x + e^{-x}}{2}\right)^2 - \left(\frac{e^x - e^{-x}}{2}\right)^2$$

$$= \frac{e^{2x} + 2 + e^{-2x}}{4} - \frac{e^{2x} - 2 + e^{-2x}}{4}$$

$$= \frac{4}{4}$$

$$= 1.$$

■ **FOR FURTHER INFORMATION**
To understand geometrically the relationship between the hyperbolic and exponential functions, see the article "A Short Proof Linking the Hyperbolic and Exponential Functions" by Michael J. Seery in *The AMATYC Review.*

HYPERBOLIC IDENTITIES

$$\cosh^2 x - \sinh^2 x = 1 \qquad \sinh(x + y) = \sinh x \cosh y + \cosh x \sinh y$$

$$\tanh^2 x + \operatorname{sech}^2 x = 1 \qquad \sinh(x - y) = \sinh x \cosh y - \cosh x \sinh y$$

$$\coth^2 x - \operatorname{csch}^2 x = 1 \qquad \cosh(x + y) = \cosh x \cosh y + \sinh x \sinh y$$

$$\cosh(x - y) = \cosh x \cosh y - \sinh x \sinh y$$

$$\sinh^2 x = \frac{-1 + \cosh 2x}{2} \qquad \cosh^2 x = \frac{1 + \cosh 2x}{2}$$

$$\sinh 2x = 2 \sinh x \cosh x \qquad \cosh 2x = \cosh^2 x + \sinh^2 x$$

Differentiation and Integration of Hyperbolic Functions

Because the hyperbolic functions are written in terms of e^x and e^{-x}, you can easily derive rules for their derivatives. The next theorem lists these derivatives with the corresponding integration rules.

THEOREM 5.21   Derivatives and Integrals of Hyperbolic Functions

Let u be a differentiable function of x.

$$\frac{d}{dx}[\sinh u] = (\cosh u)u' \qquad \int \cosh u \, du = \sinh u + C$$

$$\frac{d}{dx}[\cosh u] = (\sinh u)u' \qquad \int \sinh u \, du = \cosh u + C$$

$$\frac{d}{dx}[\tanh u] = (\text{sech}^2 u)u' \qquad \int \text{sech}^2 u \, du = \tanh u + C$$

$$\frac{d}{dx}[\coth u] = -(\text{csch}^2 u)u' \qquad \int \text{csch}^2 u \, du = -\coth u + C$$

$$\frac{d}{dx}[\text{sech } u] = -(\text{sech } u \tanh u)u' \qquad \int \text{sech } u \tanh u \, du = -\text{sech } u + C$$

$$\frac{d}{dx}[\text{csch } u] = -(\text{csch } u \coth u)u' \qquad \int \text{csch } u \coth u \, du = -\text{csch } u + C$$

Proof   Here is a proof of two of the differentiation rules. (You are asked to prove some of the other differentiation rules in Exercises 99–101.)

$$\frac{d}{dx}[\sinh x] = \frac{d}{dx}\left[\frac{e^x - e^{-x}}{2}\right]$$

$$= \frac{e^x + e^{-x}}{2}$$

$$= \cosh x$$

$$\frac{d}{dx}[\tanh x] = \frac{d}{dx}\left[\frac{\sinh x}{\cosh x}\right]$$

$$= \frac{(\cosh x)(\cosh x) - (\sinh x)(\sinh x)}{\cosh^2 x}$$

$$= \frac{1}{\cosh^2 x}$$

$$= \text{sech}^2 x$$

EXAMPLE 1   **Differentiation of Hyperbolic Functions**

a. $\dfrac{d}{dx}[\sinh(x^2 - 3)] = 2x \cosh(x^2 - 3)$

b. $\dfrac{d}{dx}[\ln(\cosh x)] = \dfrac{\sinh x}{\cosh x} = \tanh x$

c. $\dfrac{d}{dx}[x \sinh x - \cosh x] = x \cosh x + \sinh x - \sinh x = x \cosh x$

d. $\dfrac{d}{dx}[(x - 1) \cosh x - \sinh x] = (x - 1) \sinh x + \cosh x - \cosh x = (x - 1) \sinh x$

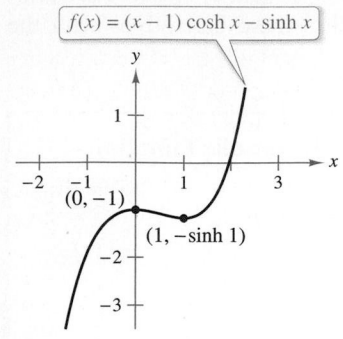

$$f(x) = (x-1)\cosh x - \sinh x$$

$f''(0) < 0$, so $(0, -1)$ is a relative maximum. $f''(1) > 0$, so $(1, -\sinh 1)$ is a relative minimum.
Figure 5.49

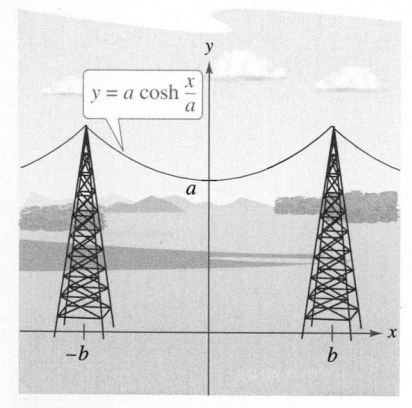

$$y = a\cosh\frac{x}{a}$$

Catenary
Figure 5.50

■ **FOR FURTHER INFORMATION**
In Example 3, the cable is a catenary between two supports at the same height. To learn about the shape of a cable hanging between supports of different heights, see the article "Reexamining the Catenary" by Paul Cella in *The College Mathematics Journal.* To view this article, go to *MathArticles.com.*

EXAMPLE 2 **Finding Relative Extrema**

Find the relative extrema of

$$f(x) = (x-1)\cosh x - \sinh x.$$

Solution Using the result of Example 1(d), set the first derivative of f equal to 0.

$$(x-1)\sinh x = 0$$

So, the critical numbers are $x = 1$ and $x = 0$. Using the Second Derivative Test, you can verify that the point $(0, -1)$ yields a relative maximum and the point $(1, -\sinh 1)$ yields a relative minimum, as shown in Figure 5.49. Try using a graphing utility to confirm this result. If your graphing utility does not have hyperbolic functions, you can use exponential functions, as shown.

$$f(x) = (x-1)\left(\frac{1}{2}\right)(e^x + e^{-x}) - \frac{1}{2}(e^x - e^{-x})$$

$$= \frac{1}{2}(xe^x + xe^{-x} - e^x - e^{-x} - e^x + e^{-x})$$

$$= \frac{1}{2}(xe^x + xe^{-x} - 2e^x)$$

When a uniform flexible cable, such as a telephone wire, is suspended from two points, it takes the shape of a *catenary*, as discussed in Example 3.

EXAMPLE 3 **Hanging Power Cables**

⋮
• • • ▷ *See LarsonCalculus.com for an interactive version of this type of example.*

Power cables are suspended between two towers, forming the catenary shown in Figure 5.50. The equation for this catenary is

$$y = a\cosh\frac{x}{a}.$$

The distance between the two towers is $2b$. Find the slope of the catenary at the point where the cable meets the right-hand tower.

Solution Differentiating produces

$$y' = a\left(\frac{1}{a}\right)\sinh\frac{x}{a} = \sinh\frac{x}{a}.$$

At the point $(b, a\cosh(b/a))$, the slope (from the left) is $m = \sinh\dfrac{b}{a}$.

EXAMPLE 4 **Integrating a Hyperbolic Function**

Find $\displaystyle\int \cosh 2x \sinh^2 2x\, dx.$

Solution

$$\int \cosh 2x \sinh^2 2x\, dx = \frac{1}{2}\int (\sinh 2x)^2(2\cosh 2x)\, dx \qquad u = \sinh 2x$$

$$= \frac{1}{2}\left[\frac{(\sinh 2x)^3}{3}\right] + C$$

$$= \frac{\sinh^3 2x}{6} + C$$

Inverse Hyperbolic Functions

Unlike trigonometric functions, hyperbolic functions are not periodic. In fact, by looking back at Figure 5.48, you can see that four of the six hyperbolic functions are actually one-to-one (the hyperbolic sine, tangent, cosecant, and cotangent). So, you can conclude that these four functions have inverse functions. The other two (the hyperbolic cosine and secant) are one-to-one when their domains are restricted to the positive real numbers, and for this restricted domain they also have inverse functions. Because the hyperbolic functions are defined in terms of exponential functions, it is not surprising to find that the inverse hyperbolic functions can be written in terms of logarithmic functions, as shown in the next theorem.

> **REMARK** Recall from Section 1.5 that a function has an inverse function if and only if it is one-to-one.

THEOREM 5.22 Inverse Hyperbolic Functions

Function	**Domain**
$\sinh^{-1} x = \ln\left(x + \sqrt{x^2 + 1}\right)$	$(-\infty, \infty)$
$\cosh^{-1} x = \ln\left(x + \sqrt{x^2 - 1}\right)$	$[1, \infty)$
$\tanh^{-1} x = \dfrac{1}{2} \ln \dfrac{1 + x}{1 - x}$	$(-1, 1)$
$\coth^{-1} x = \dfrac{1}{2} \ln \dfrac{x + 1}{x - 1}$	$(-\infty, -1) \cup (1, \infty)$
$\operatorname{sech}^{-1} x = \ln \dfrac{1 + \sqrt{1 - x^2}}{x}$	$(0, 1]$
$\operatorname{csch}^{-1} x = \ln\left(\dfrac{1}{x} + \dfrac{\sqrt{1 + x^2}}{|x|}\right)$	$(-\infty, 0) \cup (0, \infty)$

Proof The proof of this theorem is a straightforward application of the properties of the exponential and logarithmic functions. For example, for

$$f(x) = \sinh x = \frac{e^x - e^{-x}}{2}$$

and

$$g(x) = \ln\left(x + \sqrt{x^2 + 1}\right)$$

you can show that

$$f(g(x)) = x \quad \text{and} \quad g(f(x)) = x$$

which implies that g is the inverse function of f. ∎

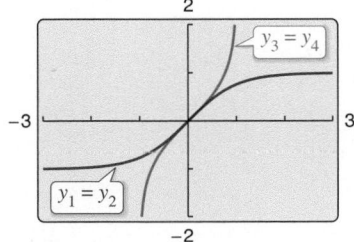

Graphs of the hyperbolic tangent function and the inverse hyperbolic tangent function

Figure 5.51

> **TECHNOLOGY** You can use a graphing utility to confirm graphically the results of Theorem 5.22. For instance, graph the following functions.

$y_1 = \tanh x$	Hyperbolic tangent
$y_2 = \dfrac{e^x - e^{-x}}{e^x + e^{-x}}$	Definition of hyperbolic tangent
$y_3 = \tanh^{-1} x$	Inverse hyperbolic tangent
$y_4 = \dfrac{1}{2} \ln \dfrac{1 + x}{1 - x}$	Definition of inverse hyperbolic tangent

The resulting display is shown in Figure 5.51. As you watch the graphs being traced out, notice that $y_1 = y_2$ and $y_3 = y_4$. Also notice that the graph of y_1 is the reflection of the graph of y_3 in the line $y = x$.

The graphs of the inverse hyperbolic functions are shown in Figure 5.52.

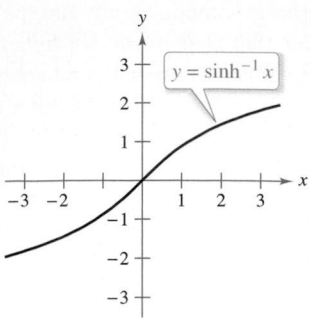

Domain: $(-\infty, \infty)$
Range: $(-\infty, \infty)$

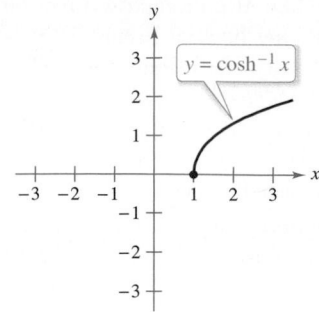

Domain: $[1, \infty)$
Range: $[0, \infty)$

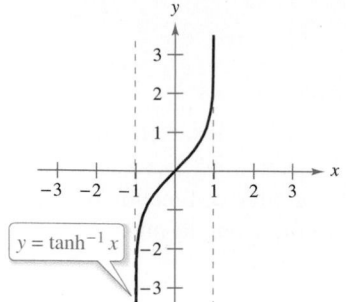

Domain: $(-1, 1)$
Range: $(-\infty, \infty)$

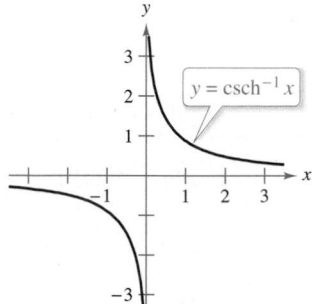

Domain: $(-\infty, 0) \cup (0, \infty)$
Range: $(-\infty, 0) \cup (0, \infty)$

Figure 5.52

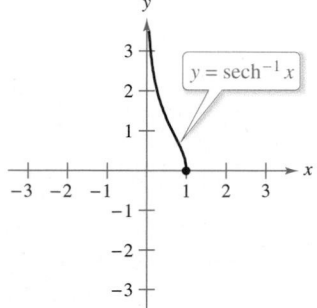

Domain: $(0, 1]$
Range: $[0, \infty)$

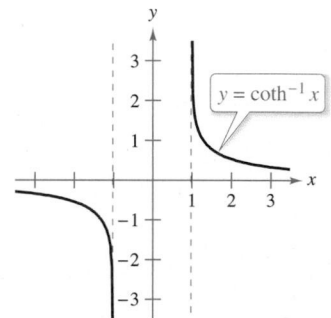

Domain: $(-\infty, -1) \cup (1, \infty)$
Range: $(-\infty, 0) \cup (0, \infty)$

The inverse hyperbolic secant can be used to define a curve called a *tractrix* or *pursuit curve*, as discussed in Example 5.

EXAMPLE 5 A Tractrix

A person is holding a rope that is tied to a boat, as shown in Figure 5.53. As the person walks along the dock, the boat travels along a **tractrix,** given by the equation

$$y = a \operatorname{sech}^{-1} \frac{x}{a} - \sqrt{a^2 - x^2}$$

where a is the length of the rope. For $a = 20$ feet, find the distance the person must walk to bring the boat to a position 5 feet from the dock.

Solution In Figure 5.53, notice that the distance the person has walked is

$$y_1 = y + \sqrt{20^2 - x^2}$$

$$= \left(20 \operatorname{sech}^{-1} \frac{x}{20} - \sqrt{20^2 - x^2}\right) + \sqrt{20^2 - x^2}$$

$$= 20 \operatorname{sech}^{-1} \frac{x}{20}.$$

When $x = 5$, this distance is

$$y_1 = 20 \operatorname{sech}^{-1} \frac{5}{20} = 20 \ln \frac{1 + \sqrt{1 - (1/4)^2}}{1/4} = 20 \ln\left(4 + \sqrt{15}\right) \approx 41.27 \text{ feet.}$$

So, the person must walk about 41.27 feet to bring the boat to a position 5 feet from the dock.

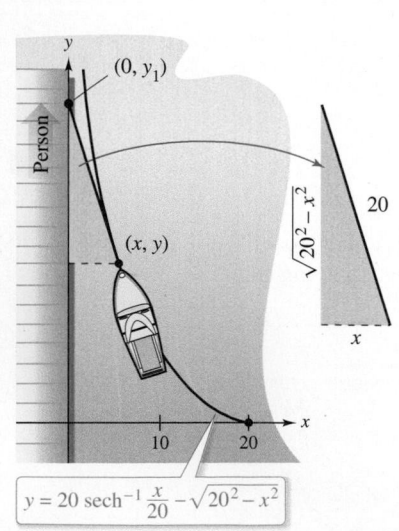

A person must walk about 41.27 feet to bring the boat to a position 5 feet from the dock.

Figure 5.53

Inverse Hyperbolic Functions: Differentiation and Integration

The derivatives of the inverse hyperbolic functions, which resemble the derivatives of the inverse trigonometric functions, are listed in Theorem 5.23 with the corresponding integration formulas (in logarithmic form). You can verify each of these formulas by applying the logarithmic definitions of the inverse hyperbolic functions. (See Exercises 102–104.)

THEOREM 5.23 **Differentiation and Integration Involving Inverse Hyperbolic Functions**

Let u be a differentiable function of x.

$$\frac{d}{dx}[\sinh^{-1} u] = \frac{u'}{\sqrt{u^2 + 1}} \qquad \frac{d}{dx}[\cosh^{-1} u] = \frac{u'}{\sqrt{u^2 - 1}}$$

$$\frac{d}{dx}[\tanh^{-1} u] = \frac{u'}{1 - u^2} \qquad \frac{d}{dx}[\coth^{-1} u] = \frac{u'}{1 - u^2}$$

$$\frac{d}{dx}[\operatorname{sech}^{-1} u] = \frac{-u'}{u\sqrt{1 - u^2}} \qquad \frac{d}{dx}[\operatorname{csch}^{-1} u] = \frac{-u'}{|u|\sqrt{1 + u^2}}$$

$$\int \frac{du}{\sqrt{u^2 \pm a^2}} = \ln\left(u + \sqrt{u^2 \pm a^2}\right) + C$$

$$\int \frac{du}{a^2 - u^2} = \frac{1}{2a} \ln\left|\frac{a + u}{a - u}\right| + C$$

$$\int \frac{du}{u\sqrt{a^2 \pm u^2}} = -\frac{1}{a} \ln \frac{a + \sqrt{a^2 \pm u^2}}{|u|} + C$$

EXAMPLE 6 **Differentiation of Inverse Hyperbolic Functions**

a. $\dfrac{d}{dx}\left[\sinh^{-1}(2x)\right] = \dfrac{2}{\sqrt{(2x)^2 + 1}}$

$$= \frac{2}{\sqrt{4x^2 + 1}}$$

b. $\dfrac{d}{dx}\left[\tanh^{-1}(x^3)\right] = \dfrac{3x^2}{1 - (x^3)^2}$

$$= \frac{3x^2}{1 - x^6}$$

EXAMPLE 7 **Integration Using Inverse Hyperbolic Functions**

a. $\displaystyle\int \frac{dx}{x\sqrt{4 - 9x^2}} = \int \frac{3\,dx}{(3x)\sqrt{2^2 - (3x)^2}}$ $\qquad \displaystyle\int \frac{du}{u\sqrt{a^2 - u^2}}$

$$= -\frac{1}{2} \ln \frac{2 + \sqrt{4 - 9x^2}}{|3x|} + C \qquad -\frac{1}{a} \ln \frac{a + \sqrt{a^2 - u^2}}{|u|} + C$$

• • • • • • • • • • • • • • • ▷
• • REMARK Let $a = 2$ and $u = 3x$.

b. $\displaystyle\int \frac{dx}{5 - 4x^2} = \frac{1}{2}\int \frac{2\,dx}{(\sqrt{5})^2 - (2x)^2}$ $\qquad \displaystyle\int \frac{du}{a^2 - u^2}$

$$= \frac{1}{2}\left(\frac{1}{2\sqrt{5}} \ln\left|\frac{\sqrt{5} + 2x}{\sqrt{5} - 2x}\right|\right) + C \qquad \frac{1}{2a} \ln\left|\frac{a + u}{a - u}\right| + C$$

$$= \frac{1}{4\sqrt{5}} \ln\left|\frac{\sqrt{5} + 2x}{\sqrt{5} - 2x}\right| + C$$

• • • • • • • • • • • • • • • ▷
• • REMARK Let $a = \sqrt{5}$ and $u = 2x$.

5.9 Exercises

See **CalcChat.com** for tutorial help and worked-out solutions to odd-numbered exercises.

CONCEPT CHECK

1. Hyperbolic Functions Describe how the name *hyperbolic function* arose.

2. Domains of Hyperbolic Functions Which hyperbolic functions have domains that are not all real numbers?

3. Hyperbolic Identities Which hyperbolic identity corresponds to the trigonometric identity

$$\sin^2 x = \frac{1 - \cos 2x}{2}?$$

4. Derivatives of Inverse Hyperbolic Functions What is the missing value?

$$\frac{d}{dx}[\text{sech}^{-1}(3x)] = \frac{}{3x\sqrt{1 - 9x^2}}$$

Evaluating a Function In Exercises 5–10, evaluate the function. If the value is not a rational number, round your answer to three decimal places.

5. (a) $\sinh 3$

   (b) $\tanh(-2)$

6. (a) $\cosh 0$

   (b) $\text{sech } 1$

7. (a) $\text{csch}(\ln 2)$

   (b) $\coth(\ln 5)$

8. (a) $\sinh^{-1} 0$

   (b) $\tanh^{-1} 0$

9. (a) $\cosh^{-1} 2$

   (b) $\text{sech}^{-1} \frac{2}{3}$

10. (a) $\text{csch}^{-1} 2$

   (b) $\coth^{-1} 3$

Verifying an Identity In Exercises 11–18, verify the identity.

11. $\sinh x + \cosh x = e^x$

12. $\cosh x - \sinh x = e^{-x}$

13. $\tanh^2 x + \text{sech}^2 x = 1$

14. $\coth^2 x - \text{csch}^2 x = 1$

15. $\cosh^2 x = \dfrac{1 + \cosh 2x}{2}$

16. $\sinh^2 x = \dfrac{-1 + \cosh 2x}{2}$

17. $\sinh 2x = 2 \sinh x \cosh x$

18. $\sinh(x + y) = \sinh x \cosh y + \cosh x \sinh y$

Finding Values of Hyperbolic Functions In Exercises 19 and 20, use the value of the given hyperbolic function to find the values of the other hyperbolic functions.

19. $\sinh x = \dfrac{3}{2}$

20. $\tanh x = \dfrac{1}{2}$

Finding a Limit In Exercises 21–24, find the limit.

21. $\lim\limits_{x \to \infty} \sinh x$

22. $\lim\limits_{x \to -\infty} \tanh x$

23. $\lim\limits_{x \to 0} \dfrac{\sinh x}{x}$

24. $\lim\limits_{x \to 0^-} \coth x$

 **Finding a Derivative** In Exercises 25–34, find the derivative of the function.

25. $f(x) = \sinh 9x$

26. $f(x) = \cosh(8x + 1)$

27. $y = \text{sech } 5x^2$

28. $f(x) = \tanh(4x^2 + 3x)$

29. $f(x) = \ln(\sinh x)$

30. $y = \ln\left(\tanh \dfrac{x}{2}\right)$

31. $h(t) = \dfrac{t}{6} \sinh(-3t)$

32. $y = (x^2 + 1) \coth \dfrac{x}{3}$

33. $f(t) = \arctan(\sinh t)$

34. $g(x) = \text{sech}^2 3x$

Finding an Equation of a Tangent Line In Exercises 35–38, find an equation of the tangent line to the graph of the function at the given point.

35. $y = \sinh(1 - x^2)$, $(1, 0)$

36. $y = x^{\cosh x}$, $(1, 1)$

37. $y = (\cosh x - \sinh x)^2$, $(0, 1)$

38. $y = e^{\sinh x}$, $(0, 1)$

 **Finding Relative Extrema** In Exercises 39–42, find the relative extrema of the function. Use a graphing utility to confirm your result.

39. $g(x) = x \text{ sech } x$

40. $h(x) = 2 \tanh x - x$

41. $f(x) = \sin x \sinh x - \cos x \cosh x$, $-4 \le x \le 4$

42. $f(x) = x \sinh(x - 1) - \cosh(x - 1)$

 **Catenary** In Exercises 43 and 44, a model for a power cable suspended between two towers is given. (a) Graph the model. (b) Find the heights of the cable at the towers and at the midpoint between the towers. (c) Find the slope of the cable at the point where the cable meets the right-hand tower.

43. $y = 10 + 15 \cosh \dfrac{x}{15}$, $-15 \le x \le 15$

44. $y = 18 + 25 \cosh \dfrac{x}{25}$, $-25 \le x \le 25$

 **Finding an Indefinite Integral** In Exercises 45–54, find the indefinite integral.

45. $\displaystyle \int \cosh 4x \, dx$

46. $\displaystyle \int \text{sech}^2 3x \, dx$

47. $\displaystyle \int \sinh(1 - 2x) \, dx$

48. $\displaystyle \int \dfrac{\cosh \sqrt{x}}{\sqrt{x}} \, dx$

49. $\displaystyle \int \cosh^2(x - 1) \sinh(x - 1) \, dx$

50. $\displaystyle \int \dfrac{\sinh x}{1 + \sinh^2 x} \, dx$

51. $\int \dfrac{\cosh x}{\sinh x}\, dx$

52. $\int \dfrac{\operatorname{csch}(1/x)\,\coth(1/x)}{x^2}\, dx$

53. $\int x\,\operatorname{csch}^2 \dfrac{x^2}{2}\, dx$

54. $\int \operatorname{sech}^3 x \tanh x\, dx$

Evaluating a Definite Integral In Exercises 55–60, evaluate the definite integral.

55. $\displaystyle\int_0^{\ln 2} \tanh x\, dx$

56. $\displaystyle\int_0^1 \cosh^2 x\, dx$

57. $\displaystyle\int_3^4 \operatorname{csch}^2 (x - 2)\, dx$

58. $\displaystyle\int_{1/2}^1 \operatorname{sech}^2 (2x - 1)\, dx$

59. $\displaystyle\int_{5/3}^2 \operatorname{csch}(3x - 4)\,\coth(3x - 4)\, dx$

60. $\displaystyle\int_0^{\ln 2} 2e^{-x} \cosh x\, dx$

EXPLORING CONCEPTS

61. Using a Graph Explain graphically why there is no solution to $\cosh x = \sinh x$.

62. Hyperbolic Functions Use the graphs on page 374 to determine whether each hyperbolic function is even, odd, or neither.

---

63. Think About It Verify the results of Exercise 62 algebraically.

64. **HOW DO YOU SEE IT?** Use the graphs of f and g shown in the figures to answer the following.

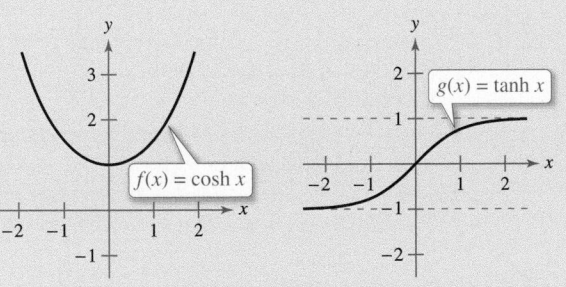

(a) Identify the open interval(s) on which the graphs of f and g are increasing or decreasing.

(b) Identify the open interval(s) on which the graphs of f and g are concave upward or concave downward.

Finding a Derivative In Exercises 65–74, find the derivative of the function.

65. $y = \cosh^{-1}(3x)$

66. $y = \operatorname{csch}^{-1}(1 - x)$

67. $y = \tanh^{-1}\sqrt{x}$

68. $f(x) = \coth^{-1}(x^2)$

69. $y = \sinh^{-1}(\tan x)$

70. $y = \tanh^{-1}(\sin 2x)$

71. $y = \operatorname{sech}^{-1}(\sin x),\ 0 < x < \pi/2$

72. $y = \coth^{-1}(e^{2x})$

73. $y = 2x\sinh^{-1}(2x) - \sqrt{1 + 4x^2}$

74. $y = x\tanh^{-1} x + \ln\sqrt{1 - x^2}$

Finding an Indefinite Integral In Exercises 75–82, find the indefinite integral using the formulas from Theorem 5.23.

75. $\int \dfrac{1}{3 - 9x^2}\, dx$

76. $\int \dfrac{1}{2x\sqrt{1 - 4x^2}}\, dx$

77. $\int \dfrac{1}{\sqrt{1 + e^{2x}}}\, dx$

78. $\int \dfrac{x}{9 - x^4}\, dx$

79. $\int \dfrac{1}{\sqrt{x}\sqrt{1 + x}}\, dx$

80. $\int \dfrac{\sqrt{x}}{\sqrt{1 + x^3}}\, dx$

81. $\int \dfrac{-1}{4x - x^2}\, dx$

82. $\int \dfrac{dx}{(x + 2)\sqrt{x^2 + 4x + 8}}$

Evaluating a Definite Integral In Exercises 83–86, evaluate the definite integral using the formulas from Theorem 5.23.

83. $\displaystyle\int_3^7 \dfrac{1}{\sqrt{x^2 - 4}}\, dx$

84. $\displaystyle\int_1^3 \dfrac{1}{x\sqrt{4 + x^2}}\, dx$

85. $\displaystyle\int_{-1}^1 \dfrac{1}{16 - 9x^2}\, dx$

86. $\displaystyle\int_0^1 \dfrac{1}{\sqrt{25x^2 + 1}}\, dx$

Differential Equation In Exercises 87 and 88, find the general solution of the differential equation.

87. $\dfrac{dy}{dx} = \dfrac{x^3 - 21x}{5 + 4x - x^2}$

88. $\dfrac{dy}{dx} = \dfrac{1 - 2x}{4x - x^2}$

Area In Exercises 89–92, find the area of the given region.

89. $y = \operatorname{sech}\dfrac{x}{2}$

90. $y = \tanh 2x$

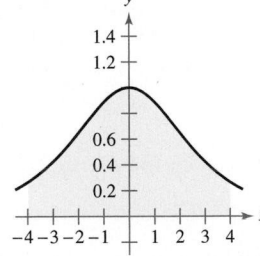

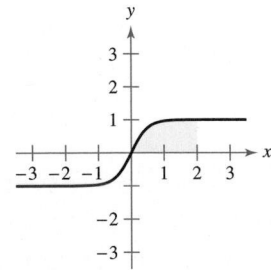
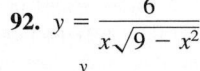

91. $y = \dfrac{5x}{\sqrt{x^4 + 1}}$

92. $y = \dfrac{6}{x\sqrt{9 - x^2}}$

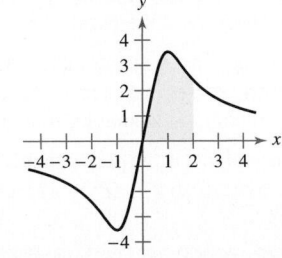

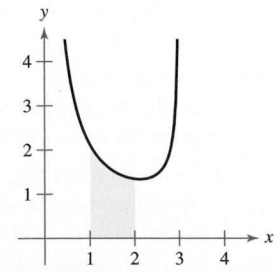

93. Tractrix Consider the equation of a tractrix

$$y = a \operatorname{sech}^{-1}\left(\frac{x}{a}\right) - \sqrt{a^2 - x^2}, \quad a > 0.$$

(a) Find dy/dx.

(b) Let L be the tangent line to the tractrix at the point P. When L intersects the y-axis at the point Q, show that the distance between P and Q is a.

94. Tractrix Show that the boat in Example 5 is always pointing toward the person.

95. Proof Prove that

$$\tanh^{-1} x = \frac{1}{2} \ln\left(\frac{1 + x}{1 - x}\right), \quad -1 < x < 1.$$

96. Proof Prove that

$$\sinh^{-1} t = \ln\left(t + \sqrt{t^2 + 1}\right).$$

97. Using a Right Triangle Show that

$$\arctan(\sinh x) = \arcsin(\tanh x).$$

98. Integration Let $x > 0$ and $b > 0$. Show that

$$\int_{-b}^{b} e^{xt}\, dt = \frac{2 \sinh bx}{x}.$$

Proof In Exercises 99–101, prove the differentiation formula.

99. $\dfrac{d}{dx}[\cosh x] = \sinh x$

100. $\dfrac{d}{dx}[\coth x] = -\operatorname{csch}^2 x$

101. $\dfrac{d}{dx}[\operatorname{sech} x] = -\operatorname{sech} x \tanh x$

Verifying a Differentiation Formula In Exercises 102–104, verify the differentiation formula.

102. $\dfrac{d}{dx}[\cosh^{-1} x] = \dfrac{1}{\sqrt{x^2 - 1}}$

103. $\dfrac{d}{dx}[\sinh^{-1} x] = \dfrac{1}{\sqrt{x^2 + 1}}$

104. $\dfrac{d}{dx}[\operatorname{sech}^{-1} x] = \dfrac{-1}{x\sqrt{1 - x^2}}$

PUTNAM EXAM CHALLENGE

105. From the vertex $(0, c)$ of the catenary $y = c \cosh(x/c)$ a line L is drawn perpendicular to the tangent to the catenary at point P. Prove that the length of L intercepted by the axes is equal to the ordinate y of the point P.

106. Prove or disprove: there is at least one straight line normal to the graph of $y = \cosh x$ at a point $(a, \cosh a)$ and also normal to the graph of $y = \sinh x$ at a point $(c, \sinh c)$.

[At a point on a graph, the normal line is the perpendicular to the tangent at that point. Also, $\cosh x = (e^x + e^{-x})/2$ and $\sinh x = (e^x - e^{-x})/2$.]

SECTION PROJECT

Mercator Map

When flying or sailing, pilots expect to be given a steady compass course to follow. On a standard flat map, this is difficult because a steady compass course results in a curved line, as shown below.

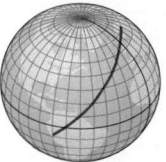
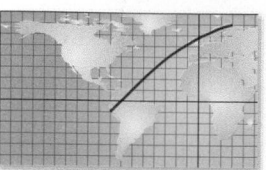

Globe: flight with constant 45° bearing

Standard flat map: flight with constant 45° bearing

For curved lines to appear as straight lines on a flat map, Flemish geographer Gerardus Mercator (1512-1594) realized that latitude lines must be stretched horizontally by a scaling factor of sec ϕ, where ϕ is the angle (in radians) of the latitude line. The Mercator map has latitude lines that are not equidistant, as shown at the right.

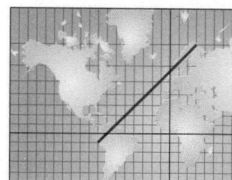

Mercator map: flight with constant 45° bearing

To calculate these vertical lengths, imagine a globe with radius R and latitude lines marked at angles of every $\Delta\phi$ radians, with $\Delta\phi = \phi_i - \phi_{i-1}$, as shown in the figure on the left below. The arc length of consecutive latitude lines is $R\Delta\phi$. On the corresponding Mercator map, the vertical distance between the ith and $(i-1)$st latitude lines is $R\Delta\phi \sec \phi_i$, and the total vertical distance from the equator to the nth latitude line is approximately $\sum_{i=1}^{n} R\Delta\phi \sec \phi_i$, as shown in the figure on the right below.

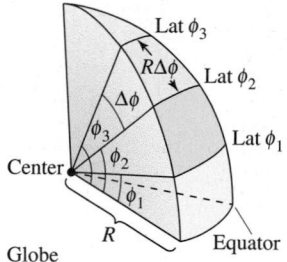
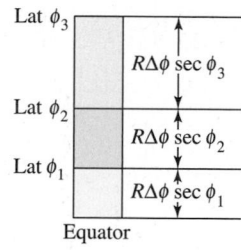

Globe

Mercator map

Mercator maps are still used by websites to display the world.

(a) Explain how to calculate the total vertical distance on a Mercator map from the equator to the nth latitude line using calculus.

(b) Using a globe radius of $R = 6$ inches, find the total vertical distances on a Mercator map from the equator to the latitude lines whose angles are 30°, 45°, and 60°.

(c) Explain what happens when you attempt to find the total vertical distance on a Mercator map from the equator to the North Pole.

(d) The *Gudermannian function* $\operatorname{gd}(y) = \displaystyle\int_0^y \frac{dt}{\cosh t}$ expresses the latitude $\phi(y) = \operatorname{gd}(y)$ in terms of the vertical position y on a Mercator map. Show that $\operatorname{gd}(y) = \arctan(\sinh y)$.

Finding an Indefinite Integral In Exercises 1–6, find the indefinite integral.

1. $\int (4x^2 + x + 3)\, dx$

2. $\int \dfrac{6}{\sqrt[3]{x}}\, dx$

3. $\int \dfrac{x^4 + 8}{x^3}\, dx$

4. $\int (5 \cos x - 2 \sec^2 x)\, dx$

5. $\int (5 - e^x)\, dx$

6. $\int \dfrac{10}{x}\, dx$

Finding a Particular Solution In Exercises 7–10, find the particular solution that satisfies the differential equation and the initial condition.

7. $f'(x) = -6x,\ f(1) = -2$ **8.** $f'(x) = 9x^2 + 1,\ f(0) = 7$

9. $f''(x) = 24x,\ f'(-1) = 7,\ f(1) = -4$

10. $f''(x) = 2 \cos x,\ f'(0) = 4,\ f(0) = -5$

11. Vertical Motion A ball is thrown vertically upward from ground level with an initial velocity of 96 feet per second. Assume the acceleration of the ball is $a(t) = -32$ feet per second per second. (Neglect air resistance.)

   (a) How long will it take the ball to rise to its maximum height? What is the maximum height?

   (b) After how many seconds is the velocity of the ball one-half the initial velocity?

   (c) What is the height of the ball when its velocity is one-half the initial velocity?

12. Vertical Motion With what initial velocity must an object be thrown upward (from a height of 3 meters) to reach a maximum height of 150 meters? Assume the acceleration of the object is $a(t) = -9.8$ meters per second per second. (Neglect air resistance.)

Finding a Sum In Exercises 13 and 14, find the sum. Use the summation capabilities of a graphing utility to verify your result.

13. $\displaystyle\sum_{i=1}^{5} (5i - 3)$

14. $\displaystyle\sum_{k=0}^{3} (k^2 + 1)$

Using Sigma Notation In Exercises 15 and 16, use sigma notation to write the sum.

15. $\dfrac{1}{5(3)} + \dfrac{2}{5(4)} + \dfrac{3}{5(5)} + \cdots + \dfrac{10}{5(12)}$

16. $\left(\dfrac{3}{n}\right)\left(\dfrac{1+1}{n}\right)^2 + \left(\dfrac{3}{n}\right)\left(\dfrac{2+1}{n}\right)^2 + \cdots + \left(\dfrac{3}{n}\right)\left(\dfrac{n+1}{n}\right)^2$

Evaluating a Sum In Exercises 17–20, use the properties of summation and Theorem 5.2 to evaluate the sum. Use the summation capabilities of a graphing utility to verify your result.

17. $\displaystyle\sum_{i=1}^{20} 2i$

18. $\displaystyle\sum_{i=1}^{30} (3i - 4)$

19. $\displaystyle\sum_{i=1}^{20} (i + 1)^2$

20. $\displaystyle\sum_{i=1}^{12} i(i^2 - 1)$

Finding Upper and Lower Sums for a Region In Exercises 21 and 22, find the upper and lower sums for the region bounded by the graph of the function and the x-axis on the given interval. Leave your answer in terms of n, the number of subintervals.

Function	Interval
21. $f(x) = 4x + 1$	$[2, 3]$
22. $f(x) = 7x^2$	$[0, 3]$

Finding Area by the Limit Definition In Exercises 23–26, use the limit process to find the area of the region bounded by the graph of the function and the x-axis over the given interval. Sketch the region.

23. $y = 8 - 2x,\ [0, 3]$ **24.** $y = x^2 + 3,\ [0, 2]$

25. $y = 5 - x^2,\ [-2, 1]$ **26.** $y = \frac{1}{4}x^3,\ [2, 4]$

Approximating Area with the Midpoint Rule In Exercises 27 and 28, use the Midpoint Rule with $n = 4$ to approximate the area of the region bounded by the graph of the function and the x-axis over the given interval.

27. $f(x) = 16 - x^2,\ [0, 4]$ **28.** $f(x) = \sin \pi x,\ [0, 1]$

Evaluating a Definite Integral as a Limit In Exercises 29 and 30, evaluate the definite integral by the limit definition.

29. $\displaystyle\int_{-3}^{5} 6x\, dx$

30. $\displaystyle\int_{0}^{3} (1 - 2x^2)\, dx$

Evaluating a Definite Integral Using a Geometric Formula In Exercises 31 and 32, sketch the region whose area is given by the definite integral. Then use a geometric formula to evaluate the integral.

31. $\displaystyle\int_{0}^{5} (5 - |x - 5|)\, dx$

32. $\displaystyle\int_{-6}^{6} \sqrt{36 - x^2}\, dx$

33. Using Properties of Definite Integrals Given

$$\int_{4}^{8} f(x)\, dx = 12 \quad \text{and} \quad \int_{4}^{8} g(x)\, dx = 5,\ \text{evaluate}$$

   (a) $\displaystyle\int_{4}^{8} [f(x) - g(x)]\, dx.$ (b) $\displaystyle\int_{4}^{8} [2f(x) - 3g(x)]\, dx.$

34. Using Properties of Definite Integrals Given

$$\int_{0}^{2} f(x)\, dx = 2 \quad \text{and} \quad \int_{2}^{5} f(x)\, dx = -5,\ \text{evaluate}$$

   (a) $\displaystyle\int_{0}^{5} f(x)\, dx.$ (b) $\displaystyle\int_{2}^{5} f(x)\, dx.$

Evaluating a Definite Integral In Exercises 35–40, use the Fundamental Theorem of Calculus to evaluate the definite integral.

35. $\displaystyle\int_{0}^{8} (3 + x)\, dx$

36. $\displaystyle\int_{2}^{3} (x^4 + 4x - 6)\, dx$

37. $\displaystyle\int_4^9 x\sqrt{x}\,dx$

38. $\displaystyle\int_{-\pi/4}^{\pi/4} \sec^2 t\,dt$

39. $\displaystyle\int_0^2 (x + e^x)\,dx$

40. $\displaystyle\int_1^6 \frac{3}{x}\,dx$

Finding the Area of a Region In Exercises 41–44, find the area of the region bounded by the graphs of the equations.

41. $y = 8 - x$, $\quad x = 0$, $\quad x = 6$, $\quad y = 0$

42. $y = \sqrt{x}(1 - x)$, $\quad y = 0$

43. $y = \dfrac{2}{x}$, $\quad y = 0$, $\quad x = 1$, $\quad x = 3$

44. $y = 1 + e^x$, $\quad y = 0$, $\quad x = 0$, $\quad x = 2$

Using the Mean Value Theorem for Integrals In Exercises 45 and 46, find the value(s) of c guaranteed by the Mean Value Theorem for Integrals for the function over the given interval.

45. $f(x) = 3x^2$, $\quad [1, 3]$

46. $f(x) = \sin x$, $\quad [0, \pi]$

Finding the Average Value of a Function In Exercises 47 and 48, find the average value of the function over the given interval and all values of x in the interval for which the function equals its average value.

47. $f(x) = \dfrac{1}{\sqrt{x}}$, $\quad [4, 9]$

48. $f(x) = x^3$, $\quad [0, 2]$

Using the Second Fundamental Theorem of Calculus In Exercises 49–52, use the Second Fundamental Theorem of Calculus to find $F'(x)$.

49. $F(x) = \displaystyle\int_0^x t^2\sqrt{1 + t^3}\,dt$

50. $F(x) = \displaystyle\int_1^x \frac{1}{t^2}\,dt$

51. $F(x) = \displaystyle\int_{-3}^x (t^2 + 3t + 2)\,dt$

52. $F(x) = \displaystyle\int_0^x \csc^2 t\,dt$

Finding an Indefinite Integral In Exercises 53–62, find the indefinite integral.

53. $\displaystyle\int \frac{x^2}{\sqrt{x^3 + 3}}\,dx$

54. $\displaystyle\int 6x^3\sqrt{3x^4 + 2}\,dx$

55. $\displaystyle\int x(1 - 3x^2)^4\,dx$

56. $\displaystyle\int \frac{x + 4}{(x^2 + 8x - 7)^2}\,dx$

57. $\displaystyle\int \frac{\cos\theta}{\sqrt{1 - \sin\theta}}\,d\theta$

58. $\displaystyle\int \sec 2x \tan 2x\,dx$

59. $\displaystyle\int xe^{-3x^2}\,dx$

60. $\displaystyle\int \frac{e^{1/x}}{x^2}\,dx$

61. $\displaystyle\int (x + 1)5^{(x+1)^2}\,dx$

62. $\displaystyle\int \frac{1}{t^2}(2^{-1/t})\,dt$

Evaluating a Definite Integral In Exercises 63–70, evaluate the definite integral. Use a graphing utility to verify your result.

63. $\displaystyle\int_0^1 (3x + 1)^5\,dx$

64. $\displaystyle\int_0^1 x^2(x^3 - 2)^3\,dx$

65. $\displaystyle\int_0^3 \frac{1}{\sqrt{1 + x}}\,dx$

66. $\displaystyle\int_3^6 \frac{x}{3\sqrt{x^2 - 8}}\,dx$

67. $2\pi\displaystyle\int_0^1 (y + 1)\sqrt{1 - y}\,dy$

68. $2\pi\displaystyle\int_{-1}^0 x^2\sqrt{x + 1}\,dx$

69. $\displaystyle\int_0^\pi \cos\frac{x}{2}\,dx$

70. $\displaystyle\int_{-\pi/4}^{\pi/4} \sin 2x\,dx$

Evaluating a Limit In Exercises 71–78, use L'Hôpital's Rule to evaluate the limit.

71. $\displaystyle\lim_{x\to 1} \frac{(\ln x)^2}{x - 1}$

72. $\displaystyle\lim_{x\to 0} \frac{\sin \pi x}{\sin 5\pi x}$

73. $\displaystyle\lim_{x\to\infty} \frac{e^{2x}}{x^2}$

74. $\displaystyle\lim_{x\to\infty} xe^{-x^2}$

75. $\displaystyle\lim_{x\to\infty} (\ln x)^{2/x}$

76. $\displaystyle\lim_{x\to 1^+} (x - 1)^{\ln x}$

77. $\displaystyle\lim_{n\to\infty} 1000\left(1 + \frac{0.09}{n}\right)^n$

78. $\displaystyle\lim_{x\to\infty} \left(1 + \frac{4}{x}\right)^x$

Finding an Indefinite Integral In Exercises 79–84, find the indefinite integral.

79. $\displaystyle\int \frac{1}{7x - 2}\,dx$

80. $\displaystyle\int \frac{x^2}{x^3 + 1}\,dx$

81. $\displaystyle\int \frac{\sin x}{1 + \cos x}\,dx$

82. $\displaystyle\int \frac{\ln\sqrt{x}}{x}\,dx$

83. $\displaystyle\int \frac{e^{2x} - e^{-2x}}{e^{2x} + e^{-2x}}\,dx$

84. $\displaystyle\int \frac{e^{2x}}{e^{2x} + 1}\,dx$

Evaluating a Definite Integral In Exercises 85–88, evaluate the definite integral.

85. $\displaystyle\int_1^4 \frac{2x + 1}{2x}\,dx$

86. $\displaystyle\int_1^e \frac{\ln x}{x}\,dx$

87. $\displaystyle\int_0^{\pi/3} \sec\theta\,d\theta$

88. $\displaystyle\int_0^\pi \tan\frac{\theta}{3}\,d\theta$

Finding an Indefinite Integral In Exercises 89–94, find the indefinite integral.

89. $\displaystyle\int \frac{1}{e^{2x} + e^{-2x}}\,dx$

90. $\displaystyle\int \frac{1}{3 + 25x^2}\,dx$

91. $\displaystyle\int \frac{x}{\sqrt{1 - x^4}}\,dx$

92. $\displaystyle\int \frac{1}{x\sqrt{9x^2 - 49}}\,dx$

93. $\displaystyle\int \frac{\arctan(x/2)}{4 + x^2}\,dx$

94. $\displaystyle\int \frac{\arcsin 2x}{\sqrt{1 - 4x^2}}x^2$

Finding a Derivative In Exercises 95–98, find the derivative of the function.

95. $y = \operatorname{sech}(4x - 1)$

96. $y = 2x - \cosh\sqrt{x}$

97. $y = \sinh^{-1}(4x)$

98. $y = x\tanh^{-1} 2x$

Finding an Indefinite Integral In Exercises 99–102, find the indefinite integral.

99. $\displaystyle\int x^2\operatorname{sech}^2 x^3\,dx$

100. $\displaystyle\int \sinh 6x\,dx$

101. $\displaystyle\int \frac{1}{9 - 4x^2}\,dx$

102. $\displaystyle\int \frac{x}{\sqrt{x^4 - 1}}\,dx$

P.S. Problem Solving

1. **Using a Function**   Let $L(x) = \int_1^x \frac{1}{t}\,dt$, $x > 0$.

   (a) Find $L(1)$.

   (b) Find $L'(x)$ and $L'(1)$.

   (c) Use a graphing utility to approximate the value of x (to three decimal places) for which $L(x) = 1$.

   (d) Prove that $L(x_1x_2) = L(x_1) + L(x_2)$ for all positive values of x_1 and x_2.

2. **Parabolic Arch**   Archimedes showed that the area of a parabolic arch is equal to $\frac{2}{3}$ the product of the base and the height (see figure).

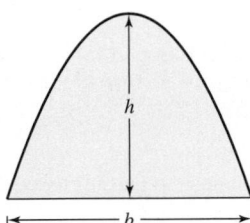

   (a) Graph the parabolic arch bounded by $y = 9 - x^2$ and the x-axis. Use an appropriate integral to find the area A.

   (b) Find the base and height of the arch and verify Archimedes' formula.

   (c) Prove Archimedes' formula for a general parabola.

3. **Using a Continuous Function**   Let f be continuous on the interval $[0, b]$, where $f(x) + f(b - x) \neq 0$ on $[0, b]$.

   (a) Show that $\int_0^b \frac{f(x)}{f(x) + f(b - x)}\,dx = \frac{b}{2}$.

   (b) Use the result in part (a) to evaluate

   $$\int_0^1 \frac{\sin x}{\sin(1 - x) + \sin x}\,dx.$$

   (c) Use the result in part (a) to evaluate

   $$\int_0^3 \frac{\sqrt{x}}{\sqrt{x} + \sqrt{3 - x}}\,dx.$$

4. **Fresnel Function**   The **Fresnel function** S is defined by the integral

   $$S(x) = \int_0^x \sin\!\left(\frac{\pi t^2}{2}\right)dt.$$

   (a) Graph the function $y = \sin\!\left(\frac{\pi x^2}{2}\right)$ on the interval $[0, 3]$.

   (b) Use the graph in part (a) to sketch the graph of S on the interval $[0, 3]$.

   (c) Locate all relative extrema of S on the interval $(0, 3)$.

   (d) Locate all points of inflection of S on the interval $(0, 3)$.

5. **Approximation**   The **Two-Point Gaussian Quadrature Approximation** for f is

   $$\int_{-1}^1 f(x)\,dx \approx f\!\left(-\frac{1}{\sqrt{3}}\right) + f\!\left(\frac{1}{\sqrt{3}}\right).$$

   (a) Use this formula to approximate

   $$\int_{-1}^1 \cos x\,dx.$$

   Find the error of the approximation.

   (b) Use this formula to approximate

   $$\int_{-1}^1 \frac{1}{1 + x^2}\,dx.$$

   (c) Prove that the Two-Point Gaussian Quadrature Approximation is exact for all polynomials of degree 3 or less.

6. **Extrema and Points of Inflection**   The graph of the function f consists of the three line segments joining the points $(0, 0)$, $(2, -2)$, $(6, 2)$, and $(8, 3)$. The function F is defined by the integral

   $$F(x) = \int_0^x f(t)\,dt.$$

   (a) Sketch the graph of f.

   (b) Complete the table.

x	0	1	2	3	4	5	6	7	8
$F(x)$									

   (c) Find the extrema of F on the interval $[0, 8]$.

   (d) Determine all points of inflection of F on the interval $(0, 8)$.

7. **Falling Objects**   Galileo Galilei (1564–1642) stated the following proposition concerning falling objects:

   The time in which any space is traversed by a uniformly accelerating body is equal to the time in which that same space would be traversed by the same body moving at a uniform speed whose value is the mean of the highest speed of the accelerating body and the speed just before acceleration began.

   Use the techniques of this chapter to verify this proposition.

8. **Proof**   Prove $\int_0^x f(t)(x - t)\,dt = \int_0^x \left(\int_0^t f(v)\,dv\right)dt$.

9. **Proof**   Prove $\int_a^b f(x)f'(x)\,dx = \frac{1}{2}([f(b)]^2 - [f(a)]^2)$.

10. **Riemann Sum**   Use an appropriate Riemann sum to evaluate the limit

    $$\lim_{n \to \infty} \frac{\sqrt{1} + \sqrt{2} + \sqrt{3} + \cdots + \sqrt{n}}{n^{3/2}}.$$

11. Riemann Sum Use an appropriate Riemann sum to evaluate the limit

$$\lim_{n \to \infty} \frac{1^5 + 2^5 + 3^5 + \cdots + n^5}{n^6}.$$

12. Proof Let f be integrable on $[a, b]$ and

$$0 < m \le f(x) \le M$$

for all x in the interval $[a, b]$. Prove that

$$m(a - b) \le \int_a^b f(x)\, dx \le M(b - a).$$

Use this result to estimate $\int_0^1 \sqrt{1 + x^4}\, dx$.

13. Velocity and Acceleration A car travels in a straight line for 1 hour. Its velocity v in miles per hour at six-minute intervals is shown in the table.

t (hours)	0	0.1	0.2	0.3	0.4	0.5
v (mi/h)	0	10	20	40	60	50

t (hours)	0.6	0.7	0.8	0.9	1.0
v (mi/h)	40	35	40	50	65

(a) Produce a reasonable graph of the velocity function v by graphing these points and connecting them with a smooth curve.

(b) Find the open intervals over which the acceleration a is positive.

(c) Find the average acceleration of the car (in miles per hour per hour) over the interval $[0, 0.4]$.

(d) What does the integral

$$\int_0^1 v(t)\, dt$$

signify? Approximate this integral using the Midpoint Rule with five subintervals.

(e) Approximate the acceleration at $t = 0.8$.

14. Proof Prove that if f is a continuous function on a closed interval $[a, b]$, then

$$\left| \int_a^b f(x)\, dx \right| \le \int_a^b |f(x)|\, dx.$$

15. Verifying a Sum Verify that

$$\sum_{i=1}^{n} i^2 = \frac{n(n + 1)(2n + 1)}{6}$$

by showing the following.

(a) $(1 + i)^3 - i^3 = 3i^2 + 3i + 1$

(b) $(n + 1)^3 = \sum_{i=1}^{n} (3i^2 + 3i + 1) + 1$

(c) $\sum_{i=1}^{n} i^2 = \frac{n(n + 1)(2n + 1)}{6}$

16. Area Consider the three regions A, B, and C determined by the graph of $f(x) = \arcsin x$, as shown in the figure.

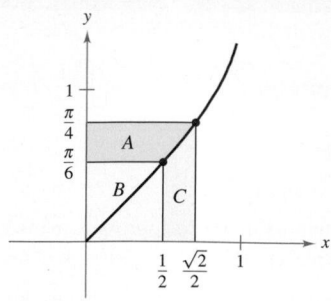

(a) Calculate the areas of regions A and B.

(b) Use your answers in part (a) to evaluate the integral

$$\int_{1/2}^{\sqrt{2}/2} \arcsin x\, dx.$$

(c) Use the methods in part (a) to evaluate the integral

$$\int_1^3 \ln x\, dx.$$

(d) Use the methods in part (a) to evaluate the integral

$$\int_1^{\sqrt{3}} \arctan x\, dx.$$

17. Area Use integration by substitution to find the area under the curve

$$y = \frac{1}{\sqrt{x} + x}$$

between $x = 1$ and $x = 4$.

18. Area Use integration by substitution to find the area under the curve

$$y = \frac{1}{\sin^2 x + 4 \cos^2 x}$$

between $x = 0$ and $x = \frac{\pi}{4}$.

19. Approximating a Function

(a) Use a graphing utility to compare the graph of the function $y = e^x$ with the graph of each given function.

(i) $y_1 = 1 + \frac{x}{1!}$

(ii) $y_2 = 1 + \frac{x}{1!} + \frac{x^2}{2!}$

(iii) $y_3 = 1 + \frac{x}{1!} + \frac{x^2}{2!} + \frac{x^3}{3!}$

(b) Identify the pattern of successive polynomials in part (a), extend the pattern one more term, and compare the graph of the resulting polynomial function with the graph of $y = e^x$.

(c) What do you think this pattern implies?

6 Differential Equations

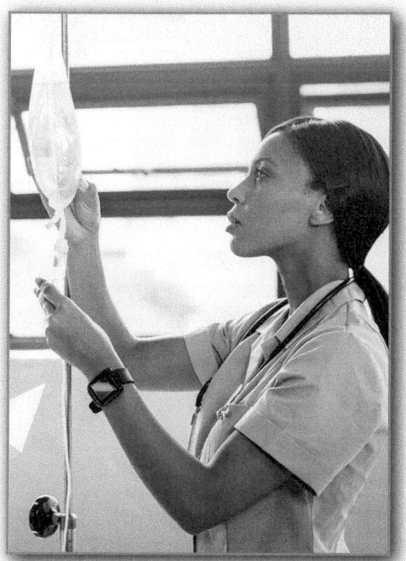

Intravenous Feeding
(Exercise 30, p. 429)

Sailing *(Exercise 85, p. 416)*

Wildlife Population *(Example 4, p. 407)*

Forestry
(Exercise 62, p. 404)

Radioactive Decay *(Example 3, p. 399)*

6.1 Slope Fields and Euler's Method

- Use initial conditions to find particular solutions of differential equations.
- Use slope fields to approximate solutions of differential equations.
- Use Euler's Method to approximate solutions of differential equations.

General and Particular Solutions

In this text, you will learn that physical phenomena can be described by differential equations. Recall that a **differential equation** in x and y is an equation that involves x, y, and derivatives of y. For example,

$$2xy' - 3y = 0 \qquad \text{Differential equation}$$

is a differential equation. In Section 6.2, you will see that problems involving radioactive decay, population growth, and Newton's Law of Cooling can be formulated in terms of differential equations.

A function $y = f(x)$ is called a **solution** of a differential equation if the equation is satisfied when y and its derivatives are replaced by $f(x)$ and its derivatives. For example, differentiation and substitution would show that $y = e^{-2x}$ is a solution of the differential equation $y' + 2y = 0$. It can be shown that every solution of this differential equation is of the form

$$y = Ce^{-2x} \qquad \text{General solution of } y' + 2y = 0$$

where C is any real number. This solution is called the **general solution.** Some differential equations have **singular solutions** that cannot be written as special cases of the general solution. Such solutions, however, are not considered in this text. The **order** of a differential equation is determined by the highest-order derivative in the equation. For instance, $y' = 4y$ is a first-order differential equation.

In Section 5.1, Example 9, you saw that the second-order differential equation $s''(t) = -32$ has the general solution

$$s(t) = -16t^2 + C_1 t + C_2 \qquad \text{General solution of } s''(t) = -32$$

which contains two arbitrary constants. It can be shown that a differential equation of order n has a general solution with n arbitrary constants.

EXAMPLE 1 **Determining Solutions**

Determine whether each function is a solution of the differential equation $y'' - y = 0$.

a. $y = \sin x$ **b.** $y = 4e^{-x}$ **c.** $y = Ce^x$

Solution

a. Because $y = \sin x$, $y' = \cos x$, and $y'' = -\sin x$, it follows that

$$y'' - y = -\sin x - \sin x = -2 \sin x \neq 0.$$

So, $y = \sin x$ is *not* a solution.

b. Because $y = 4e^{-x}$, $y' = -4e^{-x}$, and $y'' = 4e^{-x}$, it follows that

$$y'' - y = 4e^{-x} - 4e^{-x} = 0.$$

So, $y = 4e^{-x}$ is a solution.

c. Because $y = Ce^x$, $y' = Ce^x$, and $y'' = Ce^x$, it follows that

$$y'' - y = Ce^x - Ce^x = 0.$$

So, $y = Ce^x$ is a solution for any value of C.

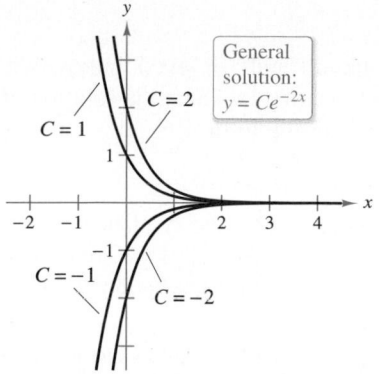

Several solution curves for $y' + 2y = 0$
Figure 6.1

Geometrically, the general solution of a first-order differential equation represents a family of curves known as **solution curves,** one for each value assigned to the arbitrary constant. For instance, you can verify that every function of the form

$$y = Ce^{-2x} \qquad \text{General solution of } y' + 2y = 0$$

is a solution of the differential equation

$$y' + 2y = 0.$$

Figure 6.1 shows four of the solution curves corresponding to different values of C.

As discussed in Section 5.1, **particular solutions** of a differential equation are obtained from **initial conditions** that give the values of the dependent variable or one of its derivatives for particular values of the independent variable. The term "initial condition" stems from the fact that, often in problems involving time, the value of the dependent variable or one of its derivatives is known at the *initial* time $t = 0$. For instance, the second-order differential equation

$$s''(t) = -32$$

having the general solution

$$s(t) = -16t^2 + C_1 t + C_2 \qquad \text{General solution of } s''(t) = -32$$

might have the following initial conditions.

$$s(0) = 80, \quad s'(0) = 64 \qquad \text{Initial conditions}$$

In this case, the initial conditions yield the particular solution

$$s(t) = -16t^2 + 64t + 80. \qquad \text{Particular solution}$$

EXAMPLE 2 Finding a Particular Solution

$\cdots\triangleright$ *See LarsonCalculus.com for an interactive version of this type of example.*

For the differential equation

$$xy' - 3y = 0$$

verify that $y = Cx^3$ is a solution. (Assume $x > 0$.) Then find the particular solution determined by the initial condition $y = 2$ when $x = 3$.

Solution You know that $y = Cx^3$ is a solution because $y' = 3Cx^2$ and

$$xy' - 3y = x(3Cx^2) - 3(Cx^3) = 0.$$

Furthermore, the initial condition $y = 2$ when $x = 3$ yields

$$
\begin{aligned}
y &= Cx^3 & \text{General solution}\\
2 &= C(3)^3 & \text{Substitute initial condition.}\\
\frac{2}{27} &= C & \text{Solve for } C.
\end{aligned}
$$

and you can conclude that the particular solution is

$$y = \frac{2x^3}{27}, \quad x > 0 \qquad \text{Particular solution}$$

as shown in Figure 6.2. Try checking this solution by substituting for y and y' in the original differential equation. ∎

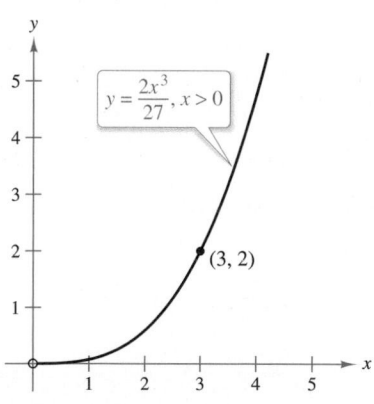

For the initial condition $y = 2$ when $x = 3$, the particular solution of the differential equation $xy' - 3y = 0$, $x > 0$, is $y = (2x^3)/27$.
Figure 6.2

Note that to determine a particular solution, the number of initial conditions must match the number of constants in the general solution.

Slope Fields

Solving a differential equation analytically can be difficult or even impossible. However, there is a graphical approach you can use to learn a lot about the solution of a differential equation. Consider a differential equation of the form

$$y' = F(x, y) \qquad \text{Differential equation}$$

where $F(x, y)$ is some expression in x and y. At each point (x, y) in the xy-plane where F is defined, the differential equation determines the slope $y' = F(x, y)$ of the solution at that point. If you draw short line segments with slope $F(x, y)$ at selected points (x, y) in the domain of F, then these line segments form a **slope field**, or a *direction field*, for the differential equation $y' = F(x, y)$. Each line segment has the same slope as the solution curve through that point. A slope field shows the general shape of all the solutions and can be helpful in getting a visual perspective of the directions of the solutions of a differential equation.

EXAMPLE 3   Sketching a Slope Field

Sketch a slope field for the differential equation $y' = x - y$ for the points $(-1, 1)$, $(0, 1)$, and $(1, 1)$.

Solution   The slope of the solution curve at any point (x, y) is

$$F(x, y) = x - y. \qquad \text{Slope at } (x, y)$$

So, the slope at each point can be found as shown.

Slope at $(-1, 1)$: $y' = -1 - 1 = -2$
Slope at $(0, 1)$: $y' = 0 - 1 = -1$
Slope at $(1, 1)$: $y' = 1 - 1 = 0$

Draw short line segments at the three points with their respective slopes, as shown in Figure 6.3.

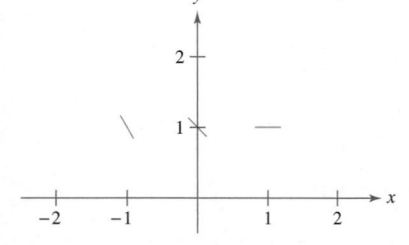

Figure 6.3

EXAMPLE 4   Identifying Slope Fields for Differential Equations

Match each slope field with its differential equation.

a. 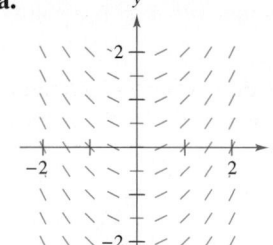   **b.** 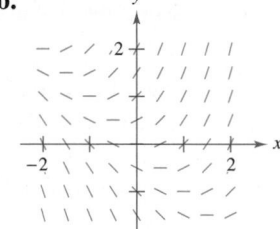   **c.**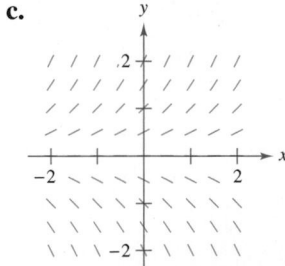

i. $y' = x + y$ **ii.** $y' = x$ **iii.** $y' = y$

Solution

a. You can see that the slope at any point along the y-axis is 0. The only equation that satisfies this condition is $y' = x$. So, the graph matches equation (ii).

b. You can see that the slope at the point $(1, -1)$ is 0. The only equation that satisfies this condition is $y' = x + y$. So, the graph matches equation (i).

c. You can see that the slope at any point along the x-axis is 0. The only equation that satisfies this condition is $y' = y$. So, the graph matches equation (iii).

A solution curve of a differential equation $y' = F(x, y)$ is simply a curve in the xy-plane whose tangent line at each point (x, y) has slope equal to $F(x, y)$. This is illustrated in Example 5.

EXAMPLE 5 Sketching a Solution Using a Slope Field

Sketch a slope field for the differential equation

$$y' = 2x + y.$$

Use the slope field to sketch the solution that passes through the point $(1, 1)$.

Solution Make a table showing the slopes at several points. The table shown is a small sample. The slopes at many other points should be calculated to get a representative slope field.

x	-2	-2	-1	-1	0	0	1	1	2	2
y	-1	1	-1	1	-1	1	-1	1	-1	1
$y' = 2x + y$	-5	-3	-3	-1	-1	1	1	3	3	5

Next, draw line segments at the points with their respective slopes, as shown in Figure 6.4.

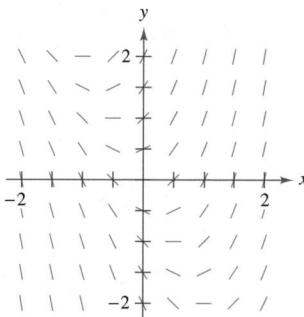

Slope field for $y' = 2x + y$
Figure 6.4

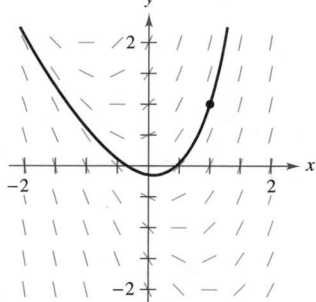

Particular solution for $y' = 2x + y$
passing through $(1, 1)$
Figure 6.5

After the slope field is drawn, start at the initial point $(1, 1)$ and move to the right in the direction of the line segment. Continue to draw the solution curve so that it moves parallel to the nearby line segments. Do the same to the left of $(1, 1)$. The resulting solution is shown in Figure 6.5.

In Example 5, note that the slope field shows that y' increases to infinity as x increases.

▷ **TECHNOLOGY** Drawing a slope field by hand is tedious. In practice, slope fields are usually drawn using a graphing utility. If you have access to a graphing utility that can graph slope fields, try graphing the slope field for the differential equation in Example 5. One example of a slope field drawn by a graphing utility is shown at the right.

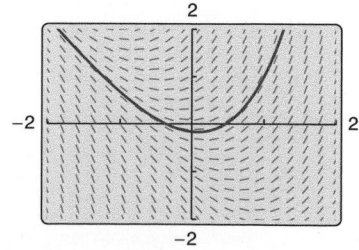

Euler's Method

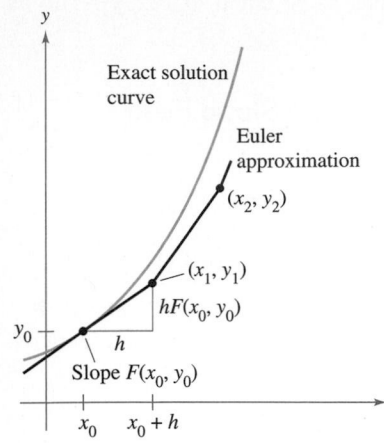

Figure 6.6

Euler's Method is a numerical approach to approximating the particular solution of the differential equation

$$y' = F(x, y)$$

that passes through the point (x_0, y_0). From the given information, you know that the graph of the solution passes through the point (x_0, y_0) and has a slope of $F(x_0, y_0)$ at this point. This gives you a "starting point" for approximating the solution.

From this starting point, you can proceed in the direction indicated by the slope. Using a small step h, move along the tangent line until you arrive at the point (x_1, y_1), where

$$x_1 = x_0 + h \quad \text{and} \quad y_1 = y_0 + hF(x_0, y_0)$$

as shown in Figure 6.6. Then, using (x_1, y_1) as a new starting point, you can repeat the process to obtain a second point (x_2, y_2). The values of x_i and y_i are shown below.

$$
\begin{aligned}
x_1 &= x_0 + h & y_1 &= y_0 + hF(x_0, y_0) \\
x_2 &= x_1 + h & y_2 &= y_1 + hF(x_1, y_1) \\
&\vdots & &\vdots \\
x_n &= x_{n-1} + h & y_n &= y_{n-1} + hF(x_{n-1}, y_{n-1})
\end{aligned}
$$

When using this method, note that you can obtain better approximations of the exact solution by choosing smaller and smaller step sizes.

EXAMPLE 6 **Approximating a Solution Using Euler's Method**

Use Euler's Method to approximate the particular solution of the differential equation

$$y' = x - y$$

passing through the point $(0, 1)$. Use a step of $h = 0.1$.

Solution Using $h = 0.1$, $x_0 = 0$, $y_0 = 1$, and $F(x, y) = x - y$, you have

$$x_0 = 0, \quad x_1 = 0.1, \quad x_2 = 0.2, \quad x_3 = 0.3$$

and the first three approximations are

$$y_1 = y_0 + hF(x_0, y_0) = 1 + (0.1)(0 - 1) = 0.9$$
$$y_2 = y_1 + hF(x_1, y_1) = 0.9 + (0.1)(0.1 - 0.9) = 0.82$$
$$y_3 = y_2 + hF(x_2, y_2) = 0.82 + (0.1)(0.2 - 0.82) = 0.758.$$

The first ten approximations are shown in the table. You can plot these values to see a graph of the approximate solution, as shown in Figure 6.7.

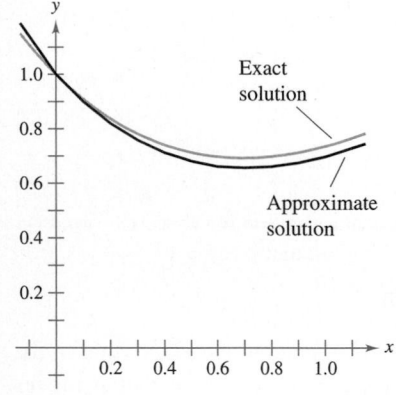

Figure 6.7

n	0	1	2	3	4	5	6	7	8	9	10
x_n	0	0.1	0.2	0.3	0.4	0.5	0.6	0.7	0.8	0.9	1
y_n	1	0.9	0.82	0.758	0.712	0.681	0.663	0.657	0.661	0.675	0.697

For the differential equation in Example 6, you can verify the exact solution to be the equation

$$y = x - 1 + 2e^{-x}.$$

Figure 6.7 compares this exact solution with the approximate solution obtained in Example 6.

6.1 Exercises

See **CalcChat.com** for tutorial help and worked-out solutions to odd-numbered exercises.

CONCEPT CHECK

1. Verifying a Solution Describe how to determine whether a function $y = f(x)$ is a solution of a differential equation.

2. General Solution What does the general solution of a first-order differential equation represent geometrically?

3. Slope Field What do the line segments on a slope field represent?

4. Euler's Method What does Euler's Method allow you to do?

 **Verifying a Solution** In Exercises 5–10, verify that the function is a solution of the differential equation.

Function	Differential Equation
5. $y = Ce^{5x}$	$y' = 5y$
6. $y = e^{-2x}$	$3y' + 5y = -e^{-2x}$
7. $y = C_1 \sin x - C_2 \cos x$	$y'' + y = 0$
8. $y = C_1 e^{-x} \cos x + C_2 e^{-x} \sin x$	$y'' + 2y' + 2y = 0$
9. $y = (-\cos x) \ln|\sec x + \tan x|$	$y'' + y = \tan x$
10. $y = \frac{2}{5}(e^{-4x} + e^x)$	$y'' + 4y' = 2e^x$

 **Verifying a Particular Solution** In Exercises 11–14, verify that the function is a particular solution of the differential equation.

Function	Differential Equation and Initial Condition
11. $y = \sin x \cos x - \cos^2 x$	$2y + y' = 2 \sin 2x - 1$
	$y\left(\dfrac{\pi}{4}\right) = 0$
12. $y = 6x - 4 \sin x + 1$	$y' = 6 - 4 \cos x$
	$y(0) = 1$
13. $y = 4e^{-6x^2}$	$y' = -12xy$
	$y(0) = 4$
14. $y = e^{-\cos x}$	$y' = y \sin x$
	$y\left(\dfrac{\pi}{2}\right) = 1$

 Determining a Solution In Exercises 15–22, determine whether the function is a solution of the differential equation $y^{(4)} - 16y = 0$.

15. $y = 3 \cos 2x$ **16.** $y = 3 \sin 2x$

17. $y = 3 \cos x$ **18.** $y = 2 \sin x$

19. $y = e^{-2x}$ **20.** $y = 5 \ln x$

21. $y = \ln x + e^{2x} + Cx^4$ **22.** $y = 3e^{2x} - 4 \sin 2x$

Determining a Solution In Exercises 23–30, determine whether the function is a solution of the differential equation $xy' - 2y = x^3 e^x$, $x > 0$.

23. $y = x^2 + e^x$

24. $y = x^3 - e^{-x}$

25. $y = x^2 e^x$

26. $y = x^2(2 + e^x)$

27. $y = e^x - \sin x$

28. $y = x^2 e^x + \sin x + \cos x$

29. $y = 2e^x \ln x$

30. $y = x^2 e^x - 5x^2$

 **Finding a Particular Solution** In Exercises 31–34, some of the curves corresponding to different values of C in the general solution of the differential equation are shown in the graph. Find the particular solution that passes through the point shown on the graph.

31. $y = Ce^{-x/2}$

$2y' + y = 0$

32. $y(x^2 + y) = C$

$2xy + (x^2 + 2y)y' = 0$

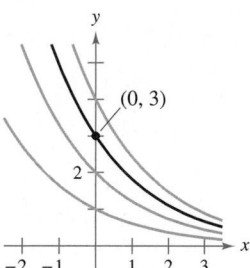

33. $y^2 = Cx^3$

$2xy' - 3y = 0$, $x > 0$

34. $2x^2 - y^2 = C$

$yy' - 2x = 0$

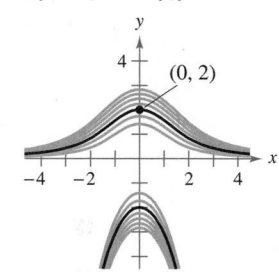

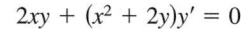 **Graphing Particular Solutions Using Technology** In Exercises 35 and 36, the general solution of the differential equation is given. Use a graphing utility to graph the particular solutions for the given values of C.

35. $4yy' - x = 0$

$4y^2 - x^2 = C$

$C = 0, C = \pm 1, C = \pm 4$

36. $yy' + x = 0$

$x^2 + y^2 = C$

$C = 0, C = 1, C = 4$

Finding a Particular Solution In Exercises 37–42, verify that the general solution satisfies the differential equation. Then find the particular solution that satisfies the initial condition(s).

37. $y = Ce^{-6x}$

$y' + 6y = 0$

$y = 3$ when $x = 0$

38. $3x^2 + 2y^2 = C$

$3x + 2yy' = 0$

$y = 3$ when $x = 1$

39. $y = C_1 \sin 3x + C_2 \cos 3x$

$y'' + 9y = 0$

$y = 2$ when $x = \dfrac{\pi}{6}$

$y' = 1$ when $x = \dfrac{\pi}{6}$

40. $y = C_1 + C_2 \ln x$

$xy'' + y' = 0, \; x > 0$

$y = 0$ when $x = 2$

$y' = \dfrac{1}{2}$ when $x = 2$

41. $y = C_1 x + C_2 x^3$

$x^2 y'' - 3xy' + 3y = 0, \; x > 0$

$y = 0$ when $x = 2$

$y' = 4$ when $x = 2$

42. $y = e^{2x/3}(C_1 + C_2 x)$

$9y'' - 12y' + 4y = 0$

$y = 4$ when $x = 0$

$y = 0$ when $x = 3$

Finding a General Solution In Exercises 43–52, use integration to find a general solution of the differential equation.

43. $\dfrac{dy}{dx} = 12x^2$

44. $\dfrac{dy}{dx} = 3x^8 - 2x$

45. $\dfrac{dy}{dx} = \dfrac{x}{1 + x^2}$

46. $\dfrac{dy}{dx} = \dfrac{e^x}{4 + e^x}$

47. $\dfrac{dy}{dx} = \sin 2x$

48. $\dfrac{dy}{dx} = \tan^2 x$

49. $\dfrac{dy}{dx} = x\sqrt{x - 6}$

50. $\dfrac{dy}{dx} = 2x\sqrt{4x^2 + 1}$

51. $\dfrac{dy}{dx} = xe^{x^2}$

52. $\dfrac{dy}{dx} = 5(\sin x)e^{\cos x}$

Slope Field In Exercises 53–56, a differential equation and its slope field are given. Complete the table by determining the slopes (if possible) in the slope field at the given points.

x	-4	-2	0	2	4	8
y	2	0	4	4	6	8
dy/dx						

53. $\dfrac{dy}{dx} = \dfrac{2x}{y}$

54. $\dfrac{dy}{dx} = y - x$

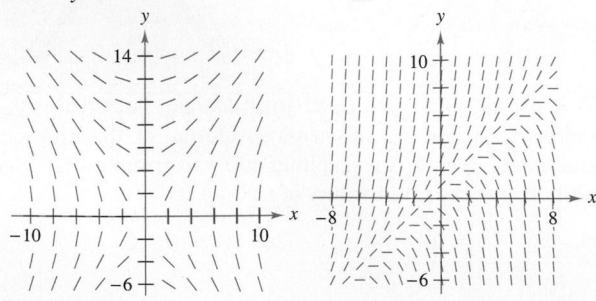

55. $\dfrac{dy}{dx} = x \cos \dfrac{\pi y}{8}$

56. $\dfrac{dy}{dx} = \tan \dfrac{\pi y}{6}$

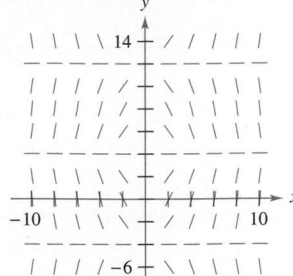

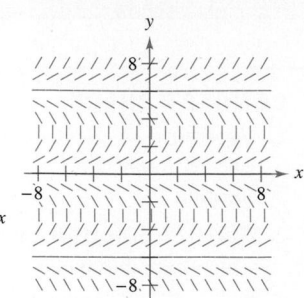

Matching In Exercises 57–60, match the differential equation with its slope field. [The slope fields are labeled (a), (b), (c), and (d).]

(a)

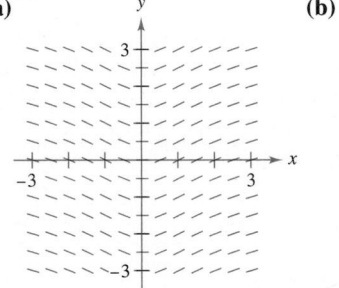

(b)

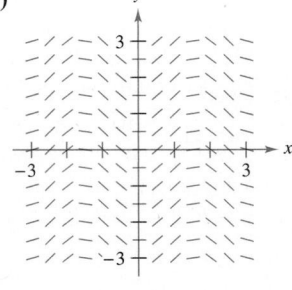

(c)

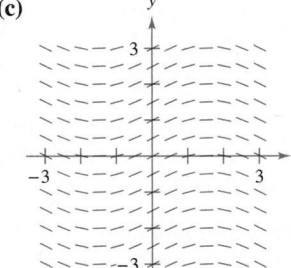

(d)

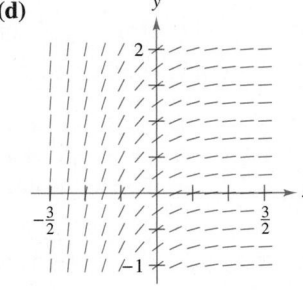

57. $\dfrac{dy}{dx} = \sin 2x$

58. $\dfrac{dy}{dx} = \dfrac{1}{2}\cos x$

59. $\dfrac{dy}{dx} = e^{-2x}$

60. $\dfrac{dy}{dx} = \dfrac{x}{x^2 + 1}$

Slope Field In Exercises 61–64, (a) sketch the slope field for the differential equation, (b) use the slope field to sketch the solution that passes through the given point, and (c) discuss the graph of the solution as $x \to \infty$ and $x \to -\infty$. Use a graphing utility to verify your results. To print a blank coordinate plane, go to *MathGraphs.com*.

61. $y' = 3 - x, \quad (4, 2)$

62. $y' = \frac{1}{3}x^2 - \frac{1}{2}x, \quad (1, 1)$

63. $y' = y - 4x, \quad (2, 2)$

64. $y' = y + xy, \quad (0, -4)$

65. Slope Field Use the slope field for the differential equation $y' = 1/x$, where $x > 0$, to sketch the graph of the solution that satisfies each given initial condition. Then make a conjecture about the behavior of a particular solution of $y' = 1/x$ as $x \to \infty$. To print an enlarged copy of the graph, go to *MathGraphs.com*.

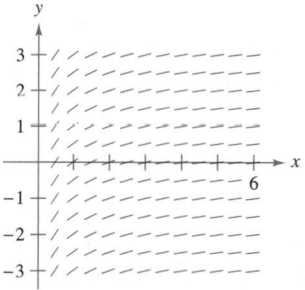

(a) $(1, 0)$

(b) $(2, -1)$

66. Slope Field Use the slope field for the differential equation $y' = 1/y$, where $y > 0$, to sketch the graph of the solution that satisfies each given initial condition. Then make a conjecture about the behavior of a particular solution of $y' = 1/y$ as $x \to \infty$. To print an enlarged copy of the graph, go to *MathGraphs.com*.

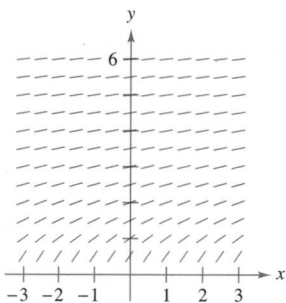

(a) $(0, 1)$

(b) $(1, 1)$

Slope Field In Exercises 67–72, use a computer algebra system to (a) graph the slope field for the differential equation and (b) graph the solution satisfying the specified initial condition.

67. $\dfrac{dy}{dx} = 0.25y, \quad y(0) = 4$

68. $\dfrac{dy}{dx} = 4 - y, \quad y(0) = 6$

69. $\dfrac{dy}{dx} = 0.02y(10 - y), \quad y(0) = 2$

70. $\dfrac{dy}{dx} = 0.2x(2 - y), \quad y(0) = 9$

71. $\dfrac{dy}{dx} = 0.4y(3 - x), \quad y(0) = 1$

72. $\dfrac{dy}{dx} = \dfrac{1}{2}e^{-x/8} \sin \dfrac{\pi y}{4}, \quad y(0) = 2$

Euler's Method In Exercises 73–78, use Euler's Method to make a table of values for the approximate solution of the differential equation with the specified initial value. Use n steps of size h.

73. $y' = x + y, \quad y(0) = 2, \quad n = 10, \quad h = 0.1$

74. $y' = x + y, \quad y(0) = 2, \quad n = 20, \quad h = 0.05$

75. $y' = 3x - 2y, \quad y(0) = 3, \quad n = 10, \quad h = 0.05$

76. $y' = 0.5x(3 - y), \quad y(0) = 1, \quad n = 5, \quad h = 0.4$

77. $y' = e^{xy}, \quad y(0) = 1, \quad n = 10, \quad h = 0.1$

78. $y' = \cos x + \sin y, \quad y(0) = 5, \quad n = 10, \quad h = 0.1$

Euler's Method In Exercises 79–81, complete the table using the exact solution of the differential equation and two approximations obtained using Euler's Method to approximate the particular solution of the differential equation. Use $h = 0.2$ and $h = 0.1$, and compute each approximation to four decimal places.

x	0	0.2	0.4	0.6	0.8	1
$y(x)$ (exact)						
$y(x)$ ($h = 0.2$)						
$y(x)$ ($h = 0.1$)						

	Differential Equation	Initial Condition	Exact Solution
79.	$\dfrac{dy}{dx} = y$	$(0, 3)$	$y = 3e^x$
80.	$\dfrac{dy}{dx} = \dfrac{2x}{y}$	$(0, 2)$	$y = \sqrt{2x^2 + 4}$
81.	$\dfrac{dy}{dx} = y + \cos x$	$(0, 0)$	$y = \dfrac{1}{2}(\sin x - \cos x + e^x)$

82. Euler's Method Compare the values of the approximations in Exercises 79–81 with the values given by the exact solution. How does the error change as h increases?

83. Temperature At time $t = 0$ minutes, the temperature of an object is 140°F. The temperature of the object is changing at the rate given by the differential equation

$$\frac{dy}{dt} = -\frac{1}{2}(y - 72).$$

(a) Use a graphing utility and Euler's Method to approximate the particular solutions of this differential equation at $t = 1, 2,$ and 3. Use a step size of $h = 0.1$. (A graphing utility program for Euler's Method is available at *LarsonCalculus.com*.)

(b) Compare your results with the exact solution

$$y = 72 + 68e^{-t/2}.$$

(c) Repeat parts (a) and (b) using a step size of $h = 0.05$. Compare the results.

84. **HOW DO YOU SEE IT?** The graph shows a solution of one of the following differential equations. Which differential equation was used? Explain your reasoning.

(a) $y' = xy$

(b) $y' = \dfrac{4x}{y}$

(c) $y' = -4xy$

(d) $y' = 4 - xy$

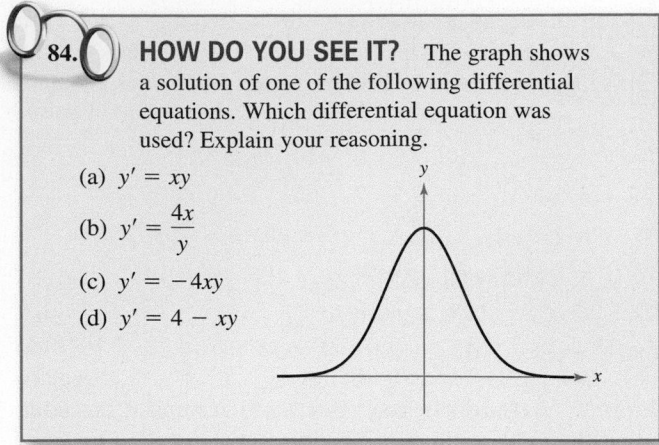

EXPLORING CONCEPTS

85. Euler's Method Explain when Euler's Method produces an exact particular solution of a differential equation.

86. Finding Values It is known that $y = Ce^{kx}$ is a solution of the differential equation $y' = 0.07y$. Is it possible to determine C or k from the information given? Explain.

True or False? In Exercises 87 and 88, determine whether the statement is true or false. If it is false, explain why or give an example that shows it is false.

87. If $y = f(x)$ is a solution of a first-order differential equation, then $y = f(x) + C$ is also a solution.

88. A slope field shows one particular solution of a differential equation.

89. Errors and Euler's Method The exact solution of the differential equation $y' = -2y$, where $y(0) = 4$, is $y = 4e^{-2x}$.

(a) Use a graphing utility to complete the table, where y is the exact value of the solution, y_1 is the approximate solution using Euler's Method with $h = 0.1$, y_2 is the approximate solution using Euler's Method with $h = 0.2$, e_1 is the absolute error $|y - y_1|$, e_2 is the absolute error $|y - y_2|$, and r is the ratio e_1/e_2.

x	0	0.2	0.4	0.6	0.8	1
y						
y_1						
y_2						
e_1						
e_2						
r						

(b) What can you conclude about the ratio r as h changes?

(c) Predict the absolute error when $h = 0.05$.

90. Errors and Euler's Method Repeat Exercise 89 for which the exact solution of the differential equation

$$\frac{dy}{dx} = x - y$$

where $y(0) = 1$, is $y = x - 1 + 2e^{-x}$.

91. Electric Circuit The diagram shows a simple electric circuit consisting of a power source, a resistor, and an inductor.

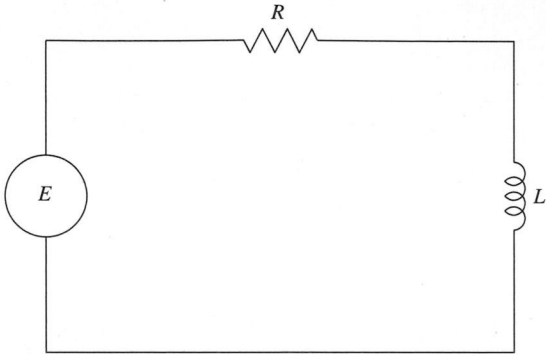

A model of the current I, in amperes (A), at time t is given by the first-order differential equation

$$L\frac{dI}{dt} + RI = E(t)$$

where $E(t)$ is the voltage (V) produced by the power source, R is the resistance, in ohms (Ω), and L is the inductance, in henrys (H). Suppose the electric circuit consists of a 24-V power source, a 12-Ω resistor, and a 4-H inductor.

(a) Sketch a slope field for the differential equation.

(b) What is the limiting value of the current? Explain.

92. Slope Field A slope field shows that the slope at the point $(1, 1)$ is 6. Does this slope field represent the family of solutions for the differential equation $y' = 4x + 2y$? Explain.

93. Think About It It is known that $y = A \sin \omega t$ is a solution of the differential equation $y'' + 16y = 0$. Find the value(s) of ω.

94. Think About It It is known that $y = e^{kt}$ is a solution of the differential equation $y'' - 16y = 0$. Find the value(s) of k.

PUTNAM EXAM CHALLENGE

95. Let f be a twice-differentiable real-valued function satisfying $f(x) + f''(x) = -xg(x)f'(x)$, where $g(x) \geq 0$ for all real x. Prove that $|f(x)|$ is bounded.

96. Prove that if the family of integral curves of the differential equation

$$\frac{dy}{dx} + p(x)y = q(x), \quad p(x) \cdot q(x) \neq 0$$

is cut by the line $x = k$, the tangents at the points of intersection are concurrent.

6.2   Growth and Decay

■ Use separation of variables to solve a simple differential equation.
■ Use exponential functions to model growth and decay in applied problems.

Differential Equations

In Section 6.1, you learned to analyze the solutions of differential equations visually using slope fields and to approximate solutions numerically using Euler's Method. Analytically, you have learned to solve only two types of differential equations—those of the forms $y' = f(x)$ and $y'' = f(x)$. In this section, you will learn how to solve a more general type of differential equation. The strategy is to rewrite the equation so that each variable occurs on only one side of the equation. This strategy is called *separation of variables*. (You will study this strategy in detail in Section 6.3.)

EXAMPLE 1   Solving a Differential Equation

$$y' = \frac{2x}{y} \qquad \text{Original equation}$$

$$yy' = 2x \qquad \text{Multiply each side by } y.$$

$$\int yy'\, dx = \int 2x\, dx \qquad \text{Integrate each side with respect to } x.$$

$$\int y\, dy = \int 2x\, dx \qquad dy = y'\, dx$$

$$\frac{1}{2}y^2 = x^2 + C_1 \qquad \text{Apply Power Rule.}$$

$$y^2 - 2x^2 = C \qquad \text{Rewrite, letting } C = 2C_1.$$

So, the general solution is $y^2 - 2x^2 = C$.

REMARK   You can use implicit differentiation to check the solution to Example 1.

When you integrate each side of the equation in Example 1, you do not need to add a constant of integration to each side. When you do, you still obtain the same result.

$$\int y\, dy = \int 2x\, dx$$

$$\frac{1}{2}y^2 + C_2 = x^2 + C_3$$

$$\frac{1}{2}y^2 = x^2 + (C_3 - C_2)$$

$$\frac{1}{2}y^2 = x^2 + C_1 \qquad \text{Rewrite, letting } C_1 = C_3 - C_2.$$

Some people prefer to use Leibniz notation and differentials when applying separation of variables. The solution to Example 1 is shown below using this notation.

$$\frac{dy}{dx} = \frac{2x}{y}$$

$$y\, dy = 2x\, dx$$

$$\int y\, dy = \int 2x\, dx$$

$$\frac{1}{2}y^2 = x^2 + C_1$$

$$y^2 - 2x^2 = C$$

Exploration

In Example 1, the general solution of the differential equation is

$$y^2 - 2x^2 = C.$$

Use a graphing utility to sketch the particular solutions for $C = \pm 2$, $C = \pm 1$, and $C = 0$. Describe the solutions graphically. Is the following statement true of each solution?

The slope of the graph at the point (x, y) is equal to twice the ratio of x and y.

Explain your reasoning. Are all curves for which this statement is true represented by the general solution?

Growth and Decay Models

In many applications, the rate of change of a variable y is proportional to the value of y. When y is a function of time t, the proportion can be written as shown.

Rate of change of y is   proportional to y.

$$\frac{dy}{dt} = ky$$

The general solution of this differential equation is given in the next theorem.

THEOREM 6.1 Exponential Growth and Decay

If y is a differentiable function of t such that $y > 0$ and $dy/dt = ky$ for some constant k, then

$$y = Ce^{kt}$$

where C is the **initial value** of y, and k is the **proportionality constant. Exponential growth** occurs when $k > 0$, and **exponential decay** occurs when $k < 0$.

Proof

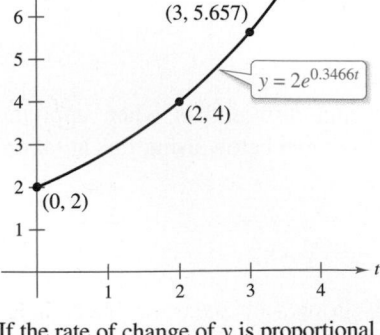

> • •**REMARK** Notice that you
> do not need to write $\ln|y|$
> because $y > 0$.
> • • • • • • • • • • • • • • • • ▷

$$\frac{dy}{dt} = ky \qquad\qquad \text{Write original equation.}$$

$$\frac{dy}{y} = k\, dt \qquad\qquad \text{Separate variables.}$$

$$\int \frac{dy}{y} = \int k\, dt \qquad\qquad \text{Integrate each side.}$$

$$\ln y = kt + C_1 \qquad\qquad \text{Find antiderivative of each side.}$$

$$y = e^{kt + C_1} \qquad\qquad \text{Exponentiate each side.}$$

$$y = e^{kt}e^{C_1} \qquad\qquad \text{Property of exponents}$$

$$y = Ce^{kt} \qquad\qquad \text{Let } C = e^{C_1}.$$

So, all solutions of $y' = ky$ are of the form $y = Ce^{kt}$. Remember that you can differentiate the function $y = Ce^{kt}$ with respect to t to verify that $y' = ky$. ∎

EXAMPLE 2   Using an Exponential Growth Model

The rate of change of y is proportional to y. When $t = 0$, $y = 2$, and when $t = 2$, $y = 4$. What is the value of y when $t = 3$?

Solution Because $y' = ky$, you know that y and t are related by the equation $y = Ce^{kt}$. You can find the values of the constants C and k by applying the initial conditions.

$$2 = Ce^0 \quad\Longrightarrow\quad C = 2 \qquad\qquad \text{When } t = 0, y = 2.$$

$$4 = 2e^{2k} \quad\Longrightarrow\quad k = \frac{1}{2}\ln 2 \approx 0.3466 \qquad \text{When } t = 2, y = 4.$$

So, the model is $y = 2e^{0.3466t}$. When $t = 3$, the value of y is $2e^{0.3466(3)} \approx 5.657$. See Figure 6.8. ∎

Using logarithmic properties, the value of k in Example 2 can also be written as $\ln\sqrt{2}$. So, the model becomes $y = 2e^{(\ln\sqrt{2})t}$, which can be rewritten as $y = 2\left(\sqrt{2}\right)^t$.

If the rate of change of y is proportional to y, then y follows an exponential model.

Figure 6.8

The graph shows points $(0, 2)$, $(2, 4)$, and $(3, 5.657)$ with the curve $y = 2e^{0.3466t}$.

▷ **TECHNOLOGY**   Most graphing utilities have curve-fitting capabilities that can be used to find models that represent data. Use the *exponential regression* feature of a graphing utility and the information in Example 2 to find a model for the data. How does your model compare with the given model?

Radioactive decay is measured in terms of *half-life*—the number of years required for half of the atoms in a sample of radioactive material to decay. The rate of decay is proportional to the amount present. The half-lives of some common radioactive isotopes are listed below.

Uranium (^{238}U)	4,470,000,000 years
Plutonium (^{239}Pu)	24,100 years
Carbon (^{14}C)	5715 years
Radium (^{226}Ra)	1599 years
Einsteinium (^{254}Es)	276 days
Radon (^{222}Rn)	3.82 days
Nobelium (^{257}No)	25 seconds

EXAMPLE 3   **Radioactive Decay**

Ten grams of the plutonium isotope ^{239}Pu were released in a nuclear accident. How long will it take for the 10 grams to decay to 1 gram?

Solution   Let y represent the mass (in grams) of the plutonium. Because the rate of decay is proportional to y, you know that

$$y = Ce^{kt}$$

where t is the time in years. To find the values of the constants C and k, apply the initial conditions. Using the fact that $y = 10$ when $t = 0$, you can write

$$10 = Ce^{k(0)} \implies 10 = Ce^0$$

which implies that $C = 10$. Next, using the fact that the half-life of ^{239}Pu is 24,100 years, you have $y = 10/2 = 5$ when $t = 24,100$. So, you can write

$$5 = 10e^{k(24,100)}$$

$$\frac{1}{2} = e^{24,100k}$$

$$\frac{1}{24,100} \ln \frac{1}{2} = k$$

$$-0.000028761 \approx k.$$

So, the model is

$$y = 10e^{-0.000028761t}. \qquad \text{Half-life model}$$

To find the time it would take for 10 grams to decay to 1 gram, you can solve for t in the equation

$$1 = 10e^{-0.000028761t}.$$

The solution is approximately 80,059 years.   ■

From Example 3, notice that in an exponential growth or decay problem, it is easy to solve for C when you are given the value of y at $t = 0$. The next example demonstrates a procedure for solving for C and k when you do not know the value of y at $t = 0$.

In a conventional nuclear reactor, 1 kilogram of ^{239}Pu can generate enough electricity to power about 900 homes for a year. *(Source: World Nuclear Association, U.S. Energy Information Administration)*

· · · · · · · · · · · · · · · ·▷

·· **REMARK**   The exponential decay model in Example 3 could also be written as $y = 10\left(\frac{1}{2}\right)^{t/24,100}$. This model is much easier to derive, but for some applications it is not as convenient to use.

| EXAMPLE 4 | **Population Growth** |

••••▷ *See LarsonCalculus.com for an interactive version of this type of example.*

An experimental population of fruit flies increases according to the law of exponential growth. There were 100 flies after the second day of the experiment and 300 flies after the fourth day. Approximately how many flies were in the original population?

Solution Let $y = Ce^{kt}$ be the number of flies at time t, where t is measured in days. Note that y is continuous, whereas the number of flies is discrete. Because $y = 100$ when $t = 2$ and $y = 300$ when $t = 4$, you can write

$$100 = Ce^{2k} \quad \text{and} \quad 300 = Ce^{4k}.$$

From the first equation, you know that

$$C = 100e^{-2k}.$$

Substituting this value into the second equation produces the following.

$$300 = 100e^{-2k}e^{4k}$$
$$300 = 100e^{2k}$$
$$3 = e^{2k}$$
$$\ln 3 = 2k$$
$$\frac{1}{2}\ln 3 = k$$
$$0.5493 \approx k$$

So, the exponential growth model is

$$y = Ce^{0.5493t}.$$

To solve for C, reapply the condition $y = 100$ when $t = 2$ and obtain

$$100 = Ce^{0.5493(2)}$$
$$C = 100e^{-1.0986}$$
$$C \approx 33.$$

So, the original population (when $t = 0$) consisted of approximately $y = C = 33$ flies, as shown in Figure 6.9.

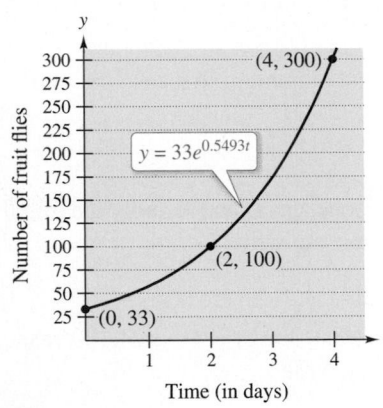

Figure 6.9

| EXAMPLE 5 | **Declining Sales** |

Four months after it stops advertising, a manufacturing company notices that its sales have dropped from 100,000 units per month to 80,000 units per month. The sales follow an exponential pattern of decline. What will the sales be after another 2 months?

Solution Use the exponential decay model $y = Ce^{kt}$, where t is measured in months. From the initial condition ($t = 0$), you know that $C = 100,000$. Moreover, because $y = 80,000$ when $t = 4$, you have

$$80,000 = 100,000e^{4k}$$
$$0.8 = e^{4k}$$
$$\ln(0.8) = 4k$$
$$-0.0558 \approx k.$$

So, after 2 more months ($t = 6$), you can expect the monthly sales to be

$$y = 100,000e^{-0.0558(6)}$$
$$\approx 71,500 \text{ units.}$$

See Figure 6.10.

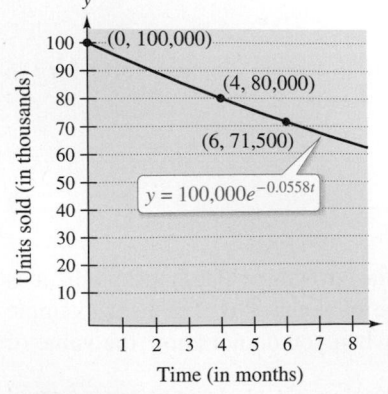

Figure 6.10

In Examples 2 through 5, you did not actually have to solve the differential equation $dy/dt = ky$. (This was done once in the proof of Theorem 6.1.) The next example demonstrates a problem whose solution involves the separation of variables technique. The example concerns **Newton's Law of Cooling,** which states that the rate of change of the temperature of an object is proportional to the difference between the object's temperature and the temperature of the surrounding medium.

EXAMPLE 6   Newton's Law of Cooling

Let y represent the temperature (in °F) of an object in a room whose temperature is kept at a constant 60°F. The object cools from 100°F to 90°F in 10 minutes. How much longer will it take for the temperature of the object to decrease to 80°F?

Solution   From Newton's Law of Cooling, you know that the rate of change of y is proportional to the difference between y and 60. This can be written as

$$\frac{dy}{dt} = k(y - 60), \quad 80 \le y \le 100.$$

To solve this differential equation, use separation of variables, as shown.

$$\frac{dy}{dt} = k(y - 60) \qquad \text{Differential equation}$$

$$\left(\frac{1}{y - 60}\right) dy = k \, dt \qquad \text{Separate variables.}$$

$$\int \frac{1}{y - 60} \, dy = \int k \, dt \qquad \text{Integrate each side.}$$

$$\ln|y - 60| = kt + C_1 \qquad \text{Find antiderivative of each side.}$$

Because $y > 60$, $|y - 60| = y - 60$, and you can omit the absolute value signs. Using exponential notation, you have

$$y - 60 = e^{kt + C_1}$$

$$y = 60 + Ce^{kt}. \qquad C = e^{C_1}$$

Using $y = 100$ when $t = 0$, you obtain

$$100 = 60 + Ce^{k(0)} = 60 + C$$

which implies that $C = 40$. Because $y = 90$ when $t = 10$,

$$90 = 60 + 40e^{k(10)}$$

$$30 = 40e^{10k}$$

$$k = \frac{1}{10} \ln \frac{3}{4}.$$

So, $k \approx -0.02877$ and the model is

$$y = 60 + 40e^{-0.02877t}. \qquad \text{Cooling model}$$

When $y = 80$, you obtain

$$80 = 60 + 40e^{-0.02877t}$$

$$20 = 40e^{-0.02877t}$$

$$\frac{1}{2} = e^{-0.02877t}$$

$$\ln \frac{1}{2} = -0.02877t$$

$$t \approx 24.09 \text{ minutes.}$$

So, it will require about 14.09 *more* minutes for the object to cool to a temperature of 80°F. See Figure 6.11.

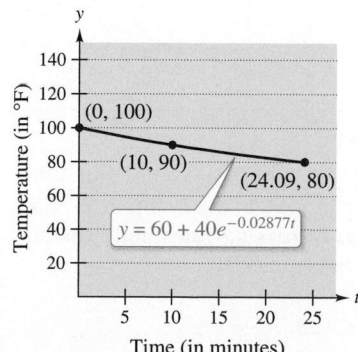

Figure 6.11

6.2 Exercises

See **CalcChat.com** for tutorial help and worked-out solutions to odd-numbered exercises.

CONCEPT CHECK

1. **Describing Values** Describe what the values of C and k represent in the exponential growth and decay model $y = Ce^{kt}$.

2. **Growth and Decay** For $y = Ce^{kt}$, explain why exponential growth occurs when $k > 0$ and exponential decay occurs when $k < 0$.

 **Solving a Differential Equation** In Exercises 3–12, find the general solution of the differential equation.

3. $\dfrac{dy}{dx} = x + 3$

4. $\dfrac{dy}{dx} = 5 - 8x$

5. $\dfrac{dy}{dx} = y + 3$

6. $\dfrac{dy}{dx} = 6 - y$

7. $y' = \dfrac{5x}{y}$

8. $y' = -\dfrac{\sqrt{x}}{4y}$

9. $y' = \sqrt{x}\, y$

10. $y' = x(1 + y)$

11. $(1 + x^2)y' - 2xy = 0$

12. $xy + y' = 100x$

Writing and Solving a Differential Equation In Exercises 13 and 14, write and find the general solution of the differential equation that models the verbal statement.

13. The rate of change of Q with respect to t is inversely proportional to the square of t.

14. The rate of change of P with respect to t is proportional to $25 - t$.

 **Slope Field** In Exercises 15 and 16, a differential equation, a point, and a slope field are given. (a) Sketch two approximate solutions of the differential equation on the slope field, one of which passes through the given point. (To print an enlarged copy of the graph, go to *MathGraphs.com.*) (b) Use integration and the given point to find the particular solution of the differential equation and use a graphing utility to graph the solution. Compare the result with the sketch in part (a) that passes through the given point.

15. $\dfrac{dy}{dx} = x(6 - y)$, $(0, 0)$

16. $\dfrac{dy}{dx} = xy$, $\left(0, \dfrac{1}{2}\right)$

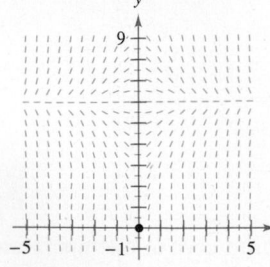

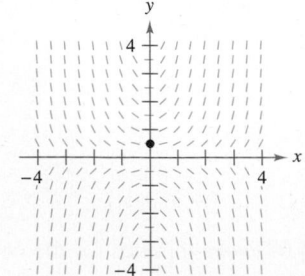

 **Finding a Particular Solution** In Exercises 17–20, find the function $y = f(t)$ passing through the point $(0, 10)$ with the given differential equation. Use a graphing utility to graph the solution.

17. $\dfrac{dy}{dt} = \dfrac{1}{2}t$

18. $\dfrac{dy}{dt} = -9\sqrt{t}$

19. $\dfrac{dy}{dt} = -\dfrac{1}{2}y$

20. $\dfrac{dy}{dt} = \dfrac{3}{4}y$

 **Writing and Solving a Differential Equation** In Exercises 21 and 22, write and find the general solution of the differential equation that models the verbal statement. Evaluate the solution at the specified value of the independent variable.

21. The rate of change of N is proportional to N. When $t = 0$, $N = 250$, and when $t = 1$, $N = 400$. What is the value of N when $t = 4$?

22. The rate of change of P is proportional to P. When $t = 0$, $P = 5000$, and when $t = 1$, $P = 4750$. What is the value of P when $t = 5$?

Finding an Exponential Function In Exercises 23–26, find the exponential function $y = Ce^{kt}$ that passes through the two given points.

23.

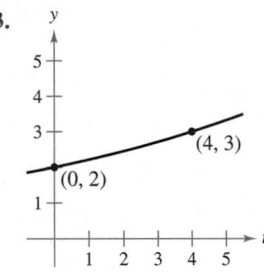

24.

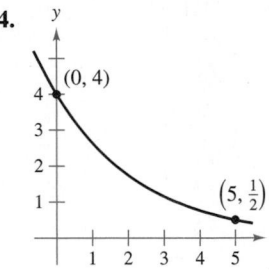

25.

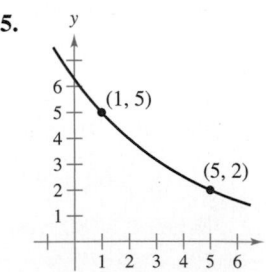

26.

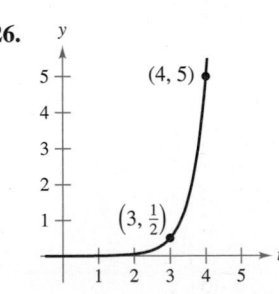

EXPLORING CONCEPTS

Increasing Function In Exercises 27 and 28, determine the quadrants in which the solution of the differential equation is an increasing function. Explain. (Do not solve the differential equation.)

27. $\dfrac{dy}{dx} = \dfrac{1}{2}xy$

28. $\dfrac{dy}{dx} = \dfrac{1}{2}x^2 y$

 **Radioactive Decay** In Exercises 29–36, complete the table for the radioactive isotope.

Isotope	Half-life (in years)	Initial Quantity	Amount After 1000 Years	Amount After 10,000 Years
29. ^{226}Ra	1599	20 g		
30. ^{226}Ra	1599		1.5 g	
31. ^{226}Ra	1599			0.1 g
32. ^{14}C	5715			3 g
33. ^{14}C	5715	5 g		
34. ^{14}C	5715		1.6 g	
35. ^{239}Pu	24,100		2.1 g	
36. ^{239}Pu	24,100			0.4 g

37. Radioactive Decay Radioactive radium has a half-life of approximately 1599 years. What percent of a given amount remains after 100 years?

38. Carbon Dating Carbon-14 dating assumes that the carbon dioxide on Earth today has the same radioactive content as it did centuries ago. If this is true, the amount of ^{14}C absorbed by a tree that grew several centuries ago should be the same as the amount of ^{14}C absorbed by a tree growing today. A piece of ancient charcoal contains only 15% as much of the radioactive carbon as a piece of modern charcoal. How long ago was the tree burned to make the ancient charcoal? (The half-life of ^{14}C is 5715 years.)

Compound Interest In Exercises 39–44, complete the table for a savings account in which interest is compounded continuously.

Initial Investment	Annual Rate	Time to Double	Amount After 10 Years
39. $1000	12%		
40. $28,000	2.5%		
41. $150		15 yr	
42. $31,000		8 yr	
43. $900			$1845.25
44. $6000			$6840

Compound Interest In Exercises 45–48, find the principal P that must be invested at rate r, compounded monthly, so that $1,000,000 will be available for retirement in t years.

45. $r = 7\frac{1}{2}\%$, $t = 20$ **46.** $r = 6\%$, $t = 40$

47. $r = 8\%$, $t = 35$ **48.** $r = 9\%$, $t = 25$

Compound Interest In Exercises 49 and 50, find the time necessary for $1000 to double when it is invested at rate r and compounded (a) annually, (b) monthly, (c) daily, and (d) continuously.

49. $r = 7\%$

50. $r = 5.5\%$

 **Population** In Exercises 51–54, the population (in millions) of a country in 2015 and the expected continuous annual rate of change k of the population are given. (*Source: U.S. Census Bureau, International Data Base*)

(a) Find the exponential growth model

$P = Ce^{kt}$

for the population by letting $t = 5$ correspond to 2015.

(b) Use the model to predict the population of the country in 2030.

(c) Discuss the relationship between the sign of k and the change in population for the country.

	Country	2015 Population	k
51.	Latvia	2.0	-0.011
52.	Canada	35.1	0.008
53.	Paraguay	6.8	0.012
54.	Ukraine	44.4	-0.006

55. Modeling Data One hundred bacteria are started in a culture and the number N of bacteria is counted each hour for 5 hours. The results are shown in the table, where t is the time in hours.

t	0	1	2	3	4	5
N	100	126	151	198	243	297

(a) Use the regression capabilities of a graphing utility to find an exponential model for the data.

(b) Use the model to estimate the time required for the population to quadruple in size.

56. Bacteria Growth The number of bacteria in a culture is increasing according to the law of exponential growth. There are 125 bacteria in the culture after 2 hours and 350 bacteria after 4 hours.

(a) Find the initial population.

(b) Write an exponential growth model for the bacteria population. Let t represent the time in hours.

(c) Use the model to determine the number of bacteria after 8 hours.

(d) After how many hours will the bacteria count be 25,000?

57. Learning Curve The management at a certain factory has found that a worker can produce at most 30 units in a day. The learning curve for the number of units N produced per day after a new employee has worked t days is

$N = 30(1 - e^{kt})$.

After 20 days on the job, a particular worker produces 19 units.

(a) Find the learning curve for this worker.

(b) How many days should pass before this worker is producing 25 units per day?

58. Learning Curve Suppose the management in Exercise 57 requires a new employee to produce at least 20 units per day after 30 days on the job.

(a) Find the learning curve that describes this minimum requirement.

(b) Find the number of days before a minimal achiever is producing 25 units per day.

59. Insect Population

(a) Suppose an insect population increases by a constant number each month. Explain why the number of insects can be represented by a linear function.

(b) Suppose an insect population increases by a constant percentage each month. Explain why the number of insects can be represented by an exponential function.

60. HOW DO YOU SEE IT? The functions f and g are both of the form $y = Ce^{kt}$.

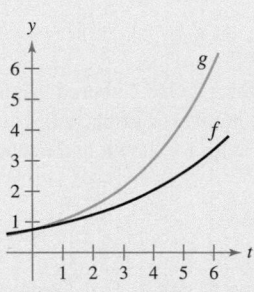

(a) Do the functions f and g represent exponential growth or exponential decay? Explain.

(b) Assume both functions have the same value of C. Which function has a greater value of k? Explain.

61. Modeling Data The table shows the cost of tuition and fees M (in dollars) at public four-year universities for selected years. (*Source: The College Board*)

Year	1980	1985	1990	1995
Cost, M	2320	2918	3492	4399

Year	2000	2005	2010	2015
Cost, M	4845	6708	8351	9410

(a) Use a graphing utility to find an exponential model M_1 for the data. Let $t = 0$ represent 1980.

(b) Use a graphing utility to find a linear model M_2 for the data. Let $t = 0$ represent 1980.

(c) Which model fits the data better? Explain.

(d) Use the exponential model to predict when the cost of tuition and fees will be $15,000. Does the result seem reasonable? Explain.

62. Forestry

The value of a tract of timber is

$$V(t) = 100,000e^{0.8\sqrt{t}}$$

where t is the time in years, with $t = 0$ corresponding to 2010. If money earns interest continuously at 10%, then the present value of the timber at any time t is

$$A(t) = V(t)e^{-0.10t}.$$

Find the year in which the timber should be harvested to maximize the present value function.

63. Sound Intensity The level of sound β (in decibels) with an intensity of I is $\beta(I) = 10 \log_{10}(I/I_0)$, where I_0 is an intensity of 10^{-16} watt per square centimeter, corresponding roughly to the faintest sound that can be heard. Determine $\beta(I)$ for the following.

(a) $I = 10^{-14}$ watt per square centimeter (whisper)

(b) $I = 10^{-9}$ watt per square centimeter (busy street corner)

(c) $I = 10^{-4}$ watt per square centimeter (threshold of pain)

64. Noise Level With the installation of noise suppression materials, the noise level in an auditorium was reduced from 93 to 80 decibels. Use the function in Exercise 63 to find the percent decrease in the intensity level of the noise as a result of the installation of these materials.

65. Newton's Law of Cooling When an object is removed from a furnace and placed in an environment with a constant temperature of 80°F, its core temperature is 1500°F. One hour after it is removed, the core temperature is 1120°F.

(a) Write an equation for the core temperature y of the object t hours after it is removed from the furnace.

(b) What is the core temperature of the object 6 hours after it is removed from the furnace?

66. Newton's Law of Cooling A container of hot liquid is placed in a freezer that is kept at a constant temperature of 20°F. The initial temperature of the liquid is 160°F. After 5 minutes, the liquid's temperature is 60°F.

(a) Write an equation for the temperature y of the liquid t minutes after it is placed in the freezer.

(b) How much longer will it take for the temperature of the liquid to decrease to 25°F?

True or False? In Exercises 67 and 68, determine whether the statement is true or false. If it is false, explain why or give an example that shows it is false.

67. Half of the atoms in a sample of radioactive radium decay in 799.5 years.

68. If prices are rising at a rate of 0.5% per month, then they are rising at a rate of 6% per year.

6.3 Separation of Variables

- Recognize and solve differential equations that can be solved by separation of variables.
- Use differential equations to model and solve applied problems.

Separation of Variables

Consider a differential equation that can be written in the form

$$M(x) + N(y)\frac{dy}{dx} = 0$$

where M is a continuous function of x alone and N is a continuous function of y alone. As you saw in Section 6.2, for this type of equation, all x-terms can be collected with dx and all y-terms with dy, and a solution can be obtained by integration. Such equations are said to be **separable,** and the solution procedure is called **separation of variables.** Below are some examples of differential equations that are separable.

Original Differential Equation	**Rewritten with Variables Separated**
$x^2 + 3y\dfrac{dy}{dx} = 0$	$3y\,dy = -x^2\,dx$
$(\sin x)y' = \cos x$	$dy = \cot x\,dx$
$\dfrac{xy'}{e^y + 1} = 2$	$\dfrac{1}{e^y + 1}\,dy = \dfrac{2}{x}\,dx$

EXAMPLE 1   Separation of Variables

· · · ·▷ *See LarsonCalculus.com for an interactive version of this type of example.*

Find the general solution of

$$(x^2 + 4)\frac{dy}{dx} = xy.$$

Solution   To begin, note that $y = 0$ is a solution. To find other solutions, assume that $y \neq 0$ and separate variables as shown.

$$(x^2 + 4)\,dy = xy\,dx \qquad \text{Differential form}$$

$$\frac{dy}{y} = \frac{x}{x^2 + 4}\,dx \qquad \text{Separate variables.}$$

Now, integrate to obtain

$$\int \frac{dy}{y} = \int \frac{x}{x^2 + 4}\,dx \qquad \text{Integrate.}$$

$$\ln|y| = \frac{1}{2}\ln(x^2 + 4) + C_1$$

$$\ln|y| = \ln\sqrt{x^2 + 4} + C_1$$

$$|y| = e^{\ln\sqrt{x^2+4}+C_1} \qquad \text{Exponentiate each side.}$$

$$|y| = e^{C_1}\sqrt{x^2 + 4} \qquad \text{Property of exponents}$$

$$y = \pm e^{C_1}\sqrt{x^2 + 4}.$$

Because $y = 0$ is also a solution, you can write the general solution as

$$y = C\sqrt{x^2 + 4}. \qquad \text{General solution}$$

··REMARK Be sure to check your solutions throughout this chapter. In Example 1, you can check the solution

$$y = C\sqrt{x^2 + 4}$$

by differentiating and substituting into the original equation.

$$(x^2 + 4)\frac{dy}{dx} = xy$$

$$(x^2 + 4)\frac{Cx}{\sqrt{x^2 + 4}} \overset{?}{=} x\left(C\sqrt{x^2 + 4}\right)$$

$$Cx\sqrt{x^2 + 4} = Cx\sqrt{x^2 + 4}$$

So, the solution checks.

In some cases, it is not feasible to write the general solution in the explicit form $y = f(x)$. The next example illustrates such a solution. Implicit differentiation can be used to verify this solution.

■ **FOR FURTHER INFORMATION**
For an example (from engineering) of a differential equation that is separable, see the article "Designing a Rose Cutter" by J. S. Hartzler in *The College Mathematics Journal*. To view this article, go to *MathArticles.com*.

EXAMPLE 2 Finding a Particular Solution

Given the initial condition $y(0) = 1$, find the particular solution of the equation

$$xy \, dx + e^{-x^2}(y^2 - 1) \, dy = 0.$$

Solution Note that $y = 0$ is a solution of the differential equation—but this solution does not satisfy the initial condition. So, you can assume that $y \neq 0$. To separate variables, you must rid the first term of y and the second term of e^{-x^2}. So, you should multiply by e^{x^2}/y and obtain the following.

$$xy \, dx + e^{-x^2}(y^2 - 1) \, dy = 0$$
$$e^{-x^2}(y^2 - 1) \, dy = -xy \, dx$$
$$\int \left(y - \frac{1}{y} \right) dy = \int -xe^{x^2} \, dx$$
$$\frac{y^2}{2} - \ln|y| = -\frac{1}{2}e^{x^2} + C$$

From the initial condition $y(0) = 1$, you have

$$\frac{1}{2} - 0 = -\frac{1}{2} + C$$

which implies that $C = 1$. So, the particular solution has the implicit form

$$\frac{y^2}{2} - \ln|y| = -\frac{1}{2}e^{x^2} + 1$$
$$y^2 - \ln y^2 + e^{x^2} = 2.$$

You can check this by differentiating and rewriting to get the original equation.

EXAMPLE 3 Finding a Particular Solution Curve

Find the equation of the curve that passes through the point $(1, 3)$ and has a slope of y/x^2 at any point (x, y).

Solution Because the slope of the curve is y/x^2, you have

$$\frac{dy}{dx} = \frac{y}{x^2}$$

with the initial condition $y(1) = 3$. Because the initial condition occurs in Quadrant I, assume $x > 0$. Then, separating variables and integrating produce

$$\int \frac{dy}{y} = \int \frac{dx}{x^2}, \quad y \neq 0, x > 0$$
$$\ln|y| = -\frac{1}{x} + C_1$$
$$y = e^{-(1/x)+C_1}$$
$$y = Ce^{-1/x}.$$

Because $y = 3$ when $x = 1$, it follows that $3 = Ce^{-1}$ and $C = 3e$. So, the equation of the specified curve is

$$y = (3e)e^{-1/x} \quad \implies \quad y = 3e^{(x-1)/x}, \quad x > 0.$$

See Figure 6.12.

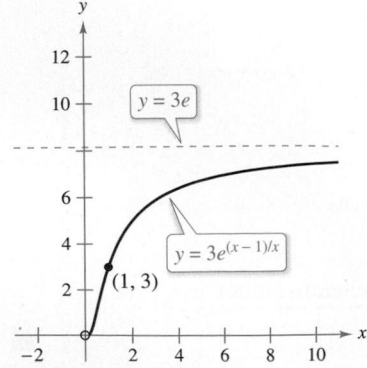

Figure 6.12

Applications

EXAMPLE 4   **Wildlife Population**

The rate of change of the number of coyotes $N(t)$ in a population is directly proportional to $650 - N(t)$, where t is the time in years. When $t = 0$, the population is 300, and when $t = 2$, the population has increased to 500. Find the population when $t = 3$.

Solution   Because the rate of change of the population is proportional to $650 - N(t)$, or $650 - N$, you can write the differential equation

$$\frac{dN}{dt} = k(650 - N).$$

You can solve this equation using separation of variables.

$$dN = k(650 - N)\, dt \qquad \text{Differential form}$$

$$\frac{dN}{650 - N} = k\, dt \qquad \text{Separate variables.}$$

$$-\ln|650 - N| = kt + C_1 \qquad \text{Integrate.}$$

$$\ln|650 - N| = -kt - C_1$$

$$650 - N = e^{-kt - C_1} \qquad \text{Exponentiate each side. (Assume } N < 650.)$$

$$650 - N = e^{-C_1}e^{-kt} \qquad \text{Property of exponents}$$

$$N = 650 - Ce^{-kt} \qquad \text{General solution}$$

Using $N = 300$ when $t = 0$, you can conclude that $C = 350$, which produces

$$N = 650 - 350e^{-kt}.$$

Then, using $N = 500$ when $t = 2$, it follows that

$$500 = 650 - 350e^{-2k} \quad \Longrightarrow \quad e^{-2k} = \frac{3}{7} \quad \Longrightarrow \quad k \approx 0.4236.$$

So, the model for the coyote population is

$$N = 650 - 350e^{-0.4236t}. \qquad \text{Model for population}$$

When $t = 3$, you can approximate the population to be

$$N = 650 - 350e^{-0.4236(3)}$$
$$\approx 552 \text{ coyotes.}$$

The model for the population is shown in Figure 6.13. Note that $N = 650$ is the horizontal asymptote of the graph and is the *carrying capacity* of the model. You will learn more about carrying capacity in the next section.

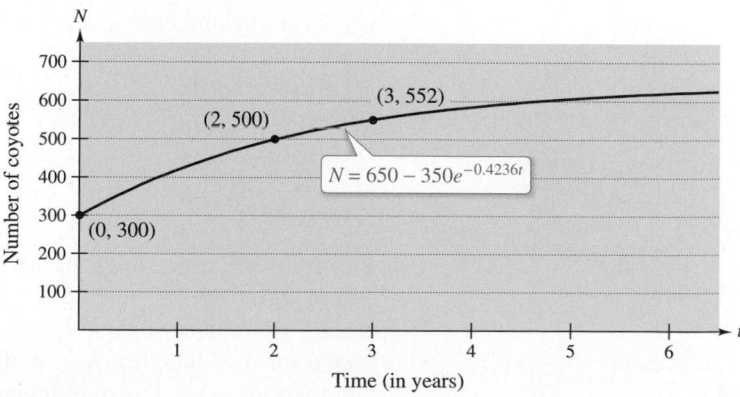

Figure 6.13

A common problem in electrostatics, thermodynamics, and hydrodynamics involves finding a family of curves, each of which is *orthogonal* to all members of a given family of curves. For example, the figure at the right shows a family of circles

$$x^2 + y^2 = C \qquad \text{Family of circles}$$

each of which intersects the lines in the family

$$y = Kx \qquad \text{Family of lines}$$

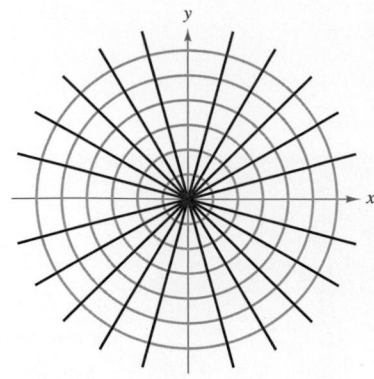

at right angles. Two such families of curves are said to be **mutually orthogonal,** and each curve in one of the families is called an **orthogonal trajectory** of the other family. In electrostatics, lines of force are orthogonal to the *equipotential curves.* In thermodynamics, the flow of heat across a plane surface is orthogonal to the *isothermal curves.* In hydrodynamics, the flow (stream) lines are orthogonal trajectories of the *velocity potential curves.*

Each line $y = Kx$ is an orthogonal trajectory of the family of circles $x^2 + y^2 = C$.

EXAMPLE 5   Finding Orthogonal Trajectories

Describe the orthogonal trajectories for the family of curves given by

$$y = \frac{C}{x}$$

for $C \neq 0$. Sketch several members of each family.

Solution   First, solve the given equation for C and write $xy = C$. Then, by differentiating implicitly with respect to x, you obtain the differential equation

$$x\frac{dy}{dx} + y = 0 \qquad \text{Differential equation}$$

$$x\frac{dy}{dx} = -y$$

$$\frac{dy}{dx} = -\frac{y}{x}. \qquad \text{Slope of given family}$$

Because dy/dx represents the slope of the given family of curves at (x, y), it follows that the orthogonal family has the negative reciprocal slope x/y. So,

$$\frac{dy}{dx} = \frac{x}{y}. \qquad \text{Slope of orthogonal family}$$

Now you can find the orthogonal family by separating variables and integrating.

$$\int y\, dy = \int x\, dx$$

$$\frac{y^2}{2} = \frac{x^2}{2} + C_1$$

$$y^2 - x^2 = K$$

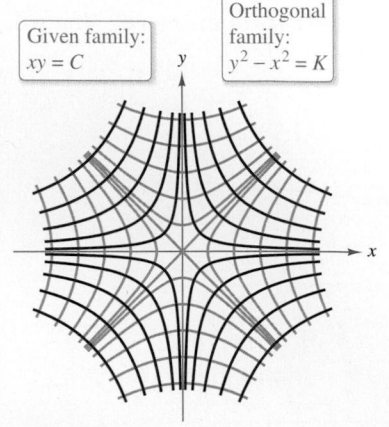

Given family: $xy = C$

Orthogonal family: $y^2 - x^2 = K$

Orthogonal trajectories
Figure 6.14

So, the orthogonal trajectories for the family of curves given by $y = C/x$ is the family of curves given by $y^2 - x^2 = K$. When $K \neq 0$, the orthogonal trajectories are hyperbolas with centers at the origin, and the transverse axes are vertical for $K > 0$ and horizontal for $K < 0$. When $K = 0$, the orthogonal trajectories are the lines $y = \pm x$. Several trajectories are shown in Figure 6.14.

EXAMPLE 6 **Modeling Advertising Awareness**

A new cereal product is introduced through an advertising campaign to a population of 1 million potential customers. The rate at which the population hears about the product is assumed to be proportional to the number of people who are not yet aware of the product. By the end of 1 year, half of the population has heard of the product. How many will have heard of it by the end of 2 years?

Solution Let y be the number of people (in millions) at time t who have heard of the product. This means that $(1 - y)$ is the number of people who have not heard of it, and dy/dt is the rate at which the population hears about the product. From the given assumption, you can write the differential equation as shown.

$$\frac{dy}{dt} = k(1 - y)$$

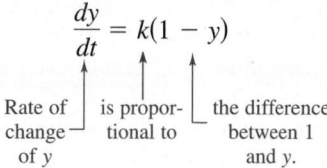

Rate of change of y | is proportional to | the difference between 1 and y.

You can solve this equation using separation of variables.

$dy = k(1 - y)\, dt$	Differential form
$\dfrac{dy}{1 - y} = k\, dt$	Separate variables.
$-\ln\lvert 1 - y \rvert = kt + C_1$	Integrate.
$\ln\lvert 1 - y \rvert = -kt - C_1$	Multiply each side by -1.
$1 - y = e^{-kt - C_1}$	Assume $y < 1$.
$y = 1 - Ce^{-kt}$	General solution

To solve for the constants C and k, use the initial conditions. That is, because $y = 0$ when $t = 0$, you can determine that $C = 1$. Similarly, because $y = 0.5$ when $t = 1$, it follows that $0.5 = 1 - e^{-k}$, which implies that

$$k = \ln 2 \approx 0.693.$$

So, the particular solution is

$$y = 1 - e^{-0.693t}. \qquad \text{Particular solution}$$

This model is shown in Figure 6.15. Using the model, you can determine that the number of people who have heard of the product after 2 years is

$$y = 1 - e^{-0.693(2)}$$

$$\approx 0.75 \text{ or } 750{,}000 \text{ people.}$$

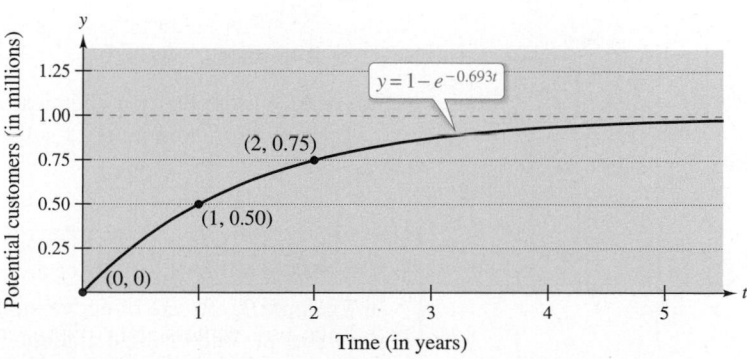

Figure 6.15

EXAMPLE 7   **Modeling a Chemical Reaction**

During a chemical reaction, substance A is converted into substance B at a rate that is proportional to the square of the amount of A. When $t = 0$, 60 grams of A is present, and after 1 hour $(t = 1)$, only 10 grams of A remains unconverted. How much of A is present after 2 hours?

Solution   Let Y be the amount of unconverted substance A at any time t. From the given assumption about the conversion rate, you can write the differential equation as shown.

$$\frac{dy}{dt} = ky^2$$

Rate of change of y is proportional to   the square of y.

You can solve this equation using separation of variables.

$$dy = ky^2\, dt \qquad \text{Differential form}$$

$$\frac{dy}{y^2} = k\, dt \qquad \text{Separate variables.}$$

$$-\frac{1}{y} = kt + C \qquad \text{Integrate.}$$

$$y = \frac{-1}{kt + C} \qquad \text{General solution}$$

To solve for the constants C and k, use the initial conditions. That is, because $y = 60$ when $t = 0$, you can determine that $C = -\frac{1}{60}$. Similarly, because $y = 10$ when $t = 1$, it follows that

$$10 = \frac{-1}{k - (1/60)}$$

which implies that $k = -\frac{1}{12}$. So, the particular solution is

$$y = \frac{-1}{(-1/12)t - (1/60)} \qquad \text{Substitute for } k \text{ and } C.$$

$$= \frac{60}{5t + 1}. \qquad \text{Particular solution}$$

Using the model, you can determine that the unconverted amount of substance A after 2 hours is

$$y = \frac{60}{5(2) + 1}$$

$$\approx 5.45 \text{ grams.}$$

In Figure 6.16, note that the chemical conversion is occurring rapidly during the first hour. Then, as more and more of substance A is converted, the conversion rate slows down. ∎

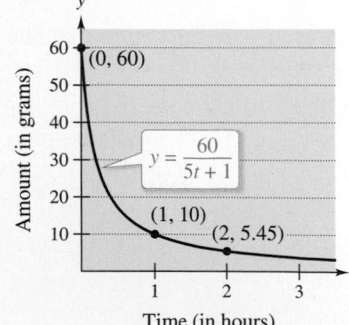

Figure 6.16

Exploration

In Example 7, the rate of conversion was assumed to be proportional to the *square* of the unconverted amount. How would the result change if the rate of conversion were assumed to be proportional to the unconverted amount?

The next example describes a growth model called a **Gompertz growth model.** This model assumes that the rate of change of y is proportional to the product of y and the natural log of L/y, where L is the population limit.

EXAMPLE 8 Modeling Population Growth

A population of 20 wolves has been introduced into a national park. The forest service estimates that the maximum population the park can sustain is 200 wolves. After 3 years, the population is estimated to be 40 wolves. According to a Gompertz growth model, how many wolves will there be 10 years after their introduction?

Solution Let y be the number of wolves at any time t. From the given assumption about the rate of growth of the population, you can write the differential equation as shown.

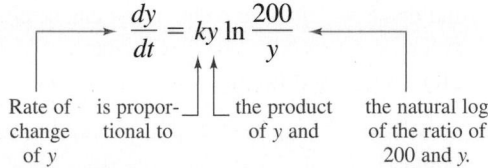

$$\frac{dy}{dt} = ky \ln \frac{200}{y}$$

Rate of change of y — is proportional to — the product of y and — the natural log of the ratio of 200 and y.

▷ **TECHNOLOGY** If you have access to a computer algebra system, try using it to find the general solution and the particular solution to Example 8.

Using separation of variables *or* a computer algebra system, you can find the general solution to be

$$y = 200e^{-Ce^{-kt}}. \qquad \text{General solution}$$

To solve for the constants C and k, use the initial conditions. That is, because $y = 20$ when $t = 0$, you can determine that

$$C = \ln 10$$
$$\approx 2.3026.$$

Similarly, because $y = 40$ when $t = 3$, it follows that

$$40 = 200e^{-2.3026e^{-3k}}$$

which implies that $k \approx 0.1194$. So, the particular solution is

$$y = 200e^{-2.3026e^{-0.1194t}}. \qquad \text{Particular solution}$$

Using the model, you can estimate the wolf population after 10 years to be

$$y = 200e^{-2.3026e^{-0.1194(10)}}$$
$$\approx 100 \text{ wolves.}$$

In Figure 6.17, note that after 10 years the population has reached about half of the estimated maximum population. Try checking the growth model to see that it yields $y = 20$ when $t = 0$ and $y = 40$ when $t = 3$.

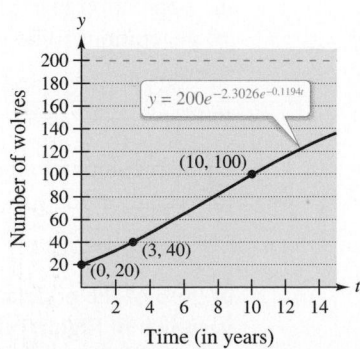

Figure 6.17

In genetics, a commonly used hybrid selection model is based on the differential equation

$$\frac{dy}{dt} = ky(1 - y)(a - by).$$

In this model, y represents the portion of the population that has a certain characteristic and t represents the time (measured in generations). The numbers a, b, and k are constants that depend on the genetic characteristic that is being studied.

EXAMPLE 9   Modeling Hybrid Selection

You are studying a population of beetles to determine how quickly characteristic D will pass from one generation to the next. At the beginning of your study ($t = 0$), you find that half the population has characteristic D. After four generations ($t = 4$), you find that 80% of the population has characteristic D. Use the hybrid selection model above with $a = 2$ and $b = 1$ to find the percent of the population that will have characteristic D after 10 generations.

Solution   Using $a = 2$ and $b = 1$, the differential equation for the hybrid selection model is

$$\frac{dy}{dt} = ky(1 - y)(2 - y).$$

Using separation of variables or a computer algebra system, you can find the general solution to be

$$\frac{y(2 - y)}{(1 - y)^2} = Ce^{2kt}. \qquad \text{General solution}$$

To solve for the constants C and k, use the initial conditions. That is, because $y = 0.5$ when $t = 0$, you can determine that $C = 3$. Similarly, because $y = 0.8$ when $t = 4$, it follows that

$$\frac{0.8(1.2)}{(0.2)^2} = 3e^{8k}$$

which implies that

$$k = \frac{1}{8} \ln 8 \approx 0.2599.$$

So, the particular solution is

$$\frac{y(2 - y)}{(1 - y)^2} = 3e^{0.5199t}. \qquad \text{Particular solution}$$

Using the model, you can estimate the percent of the population that will have characteristic D after 10 generations to be given by

$$\frac{y(2 - y)}{(1 - y)^2} = 3e^{0.5199(10)}.$$

Using a computer algebra system, you can solve this equation for y to obtain

$$y \approx 0.96$$

or 96% of the population. The graph of the model is shown in Figure 6.18.

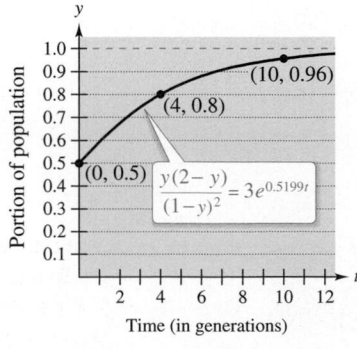

Figure 6.18

6.3 Exercises

See CalcChat.com for tutorial help and worked-out solutions to odd-numbered exercises.

CONCEPT CHECK

1. Separation of Variables Determine whether each differential equation is separable.

(a) $y = 2x^5 y' - y'$ (b) $\dfrac{y'}{x} = x^2 y + 1$

2. Mutually Orthogonal What does it mean for two families of curves to be mutually orthogonal?

Finding a General Solution Using Separation of Variables In Exercises 3–16, find the general solution of the differential equation.

3. $\dfrac{dy}{dx} = \dfrac{x}{y}$ **4.** $\dfrac{dy}{dx} = \dfrac{3x^2}{y^2}$

5. $\dfrac{dy}{dx} = \dfrac{x-1}{y^3}$ **6.** $\dfrac{dy}{dx} = \dfrac{6-x^2}{2y^3}$

7. $\dfrac{dr}{ds} = \dfrac{4}{9}r$ **8.** $\dfrac{dr}{ds} = \dfrac{9}{4}s$

9. $(2+x)y' = 3y$ **10.** $xy' = y$

11. $y^2 y' = \sin 9x$ **12.** $yy' = -8\cos(\pi x)$

13. $\sqrt{1-4x^2}\,y' = x$ **14.** $\sqrt{x^3-5}\,y' = x^2$

15. $y \ln x - xy' = 0$ **16.** $12yy' - 7e^x = 0$

Finding a Particular Solution Using Separation of Variables In Exercises 17–26, find the particular solution of the differential equation that satisfies the initial condition.

Differential Equation	Initial Condition
17. $yy' - 2e^x = 0$	$y(0) = 3$
18. $\sqrt{x} + \sqrt{y}\,y' = 0$	$y(1) = 9$
19. $y(x+1) + y' = 0$	$y(-2) = 1$
20. $2xy' - \ln x^2 = 0$	$y(1) = 2$
21. $y(1+x^2)y' - x(1+y^2) = 0$	$y(0) = \sqrt{3}$
22. $y\sqrt{1-x^2}\,y' - x\sqrt{1-y^2} = 0$	$y(\tfrac{1}{2}) = \tfrac{1}{2}$
23. $\dfrac{du}{dv} = uv \sin v^2$	$u(0) = e^2$
24. $\dfrac{dr}{ds} = e^{r-2s}$	$r(0) = 0$
25. $dP - kP\,dt = 0$	$P(0) = P_0$
26. $dT + k(T - 70)\,dt = 0$	$T(0) = 140$

Finding a Particular Solution Curve In Exercises 27–30, find an equation of the curve that passes through the point and has the given slope.

27. $(0, 2)$, $y' = \dfrac{x}{4y}$ **28.** $(1, 1)$, $y' = -\dfrac{9x}{16y}$

29. $(9, 1)$, $y' = \dfrac{y}{2x}$ **30.** $(8, 2)$, $y' = \dfrac{2y}{3x}$

Using Slope In Exercises 31 and 32, find all functions f having the indicated property.

31. The tangent to the graph of f at the point (x, y) intersects the x-axis at $(x + 2, 0)$.

32. All tangents to the graph of f pass through the origin.

Slope Field In Exercises 33–36, sketch a few solutions of the differential equation on the slope field and then find the general solution analytically. To print an enlarged copy of the graph, go to *MathGraphs.com*.

33. $\dfrac{dy}{dx} = x$ **34.** $\dfrac{dy}{dx} = -\dfrac{x}{y}$

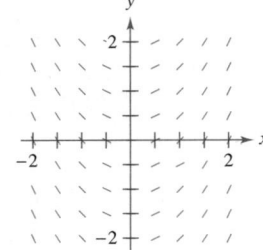 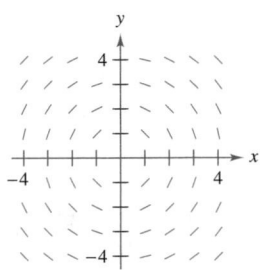

35. $\dfrac{dy}{dx} = 4 - y$ **36.** $\dfrac{dy}{dx} = 0.25x(4 - y)$

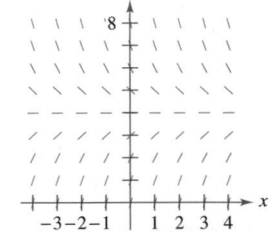 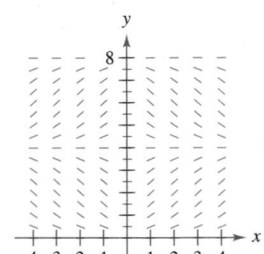

Euler's Method In Exercises 37 and 38, (a) use Euler's Method with a step size of $h = 0.1$ to approximate the particular solution of the initial value problem at the given x-value, (b) find the exact solution of the differential equation analytically, and (c) compare the solutions at the given x-value.

Differential Equation	Initial Condition	x-value
37. $\dfrac{dy}{dx} = -6xy$	$(0, 5)$	$x = 1$
38. $\dfrac{dy}{dx} = \dfrac{2x+12}{3y^2-4}$	$(1, 2)$	$x = 2$

39. Radioactive Decay The rate of decomposition of radioactive radium is proportional to the amount present at any time. The half-life of radioactive radium is 1599 years. What percent of a present amount will remain after 50 years?

40. Chemical Reaction In a chemical reaction, a certain compound changes into another compound at a rate proportional to the unchanged amount. There is 40 grams of the original compound initially and 35 grams after 1 hour. When will 75 percent of the compound be changed?

Slope Field **In Exercises 41–44, (a) write a differential equation for the statement, (b) match the differential equation with a possible slope field, and (c) verify your result by using a graphing utility to graph a slope field for the differential equation. [The slope fields are labeled (i), (ii), (iii), and (iv).]**

(i)

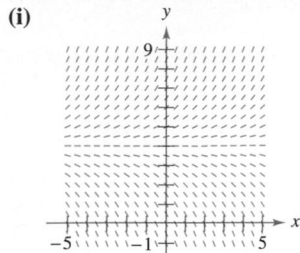

(ii)

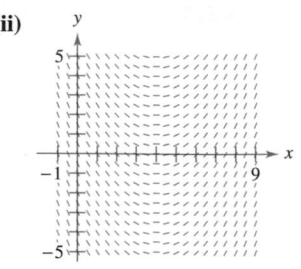

(iii)

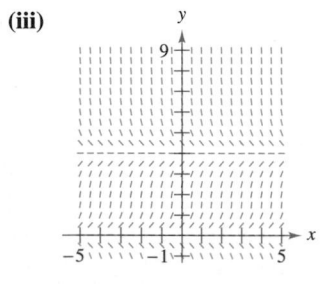

(iv)

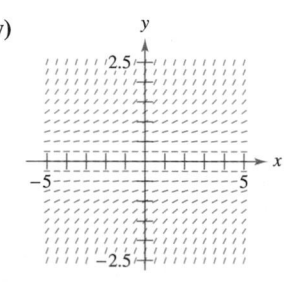

41. The rate of change of y with respect to x is proportional to the difference between y and 4.

42. The rate of change of y with respect to x is proportional to the difference between x and 4.

43. The rate of change of y with respect to x is proportional to the product of y and the difference between y and 4.

44. The rate of change of y with respect to x is proportional to y^2.

45. Weight Gain A calf that weighs 60 pounds at birth gains weight at the rate $dw/dt = k(1200 - w)$, where w is the weight in pounds and t is the time in years.

(a) Find the general solution of the differential equation.

(b) Use a graphing utility to graph the particular solutions for $k = 0.8, 0.9$, and 1.

(c) The animal is sold when its weight reaches 800 pounds. Find the time of sale for each of the models in part (b).

(d) What is the maximum weight of the animal for each of the models in part (b)?

46. Weight Gain A goat that weighs 7 pounds at birth gains weight at the rate $dw/dt = k(250 - w)$, where w is the weight in pounds and t is the time in years. Repeat Exercise 45 assuming that the goat is sold when its weight reaches 175 pounds.

 **Finding Orthogonal Trajectories In Exercises 47–52, find the orthogonal trajectories of the family. Use a graphing utility to graph several members of each family.**

47. $3x^2 - y^2 = C$ **48.** $x^2 - 2y^2 = C$

49. $x^2 = Cy$ **50.** $y^2 = 2Cx$

51. $y^2 = Cx^3$

52. $y = Ce^x$

53. Biology At any time t, the rate of growth of the population N of deer in a state park is proportional to the product of N and $L - N$, where $L = 500$ is the maximum number of deer the park can sustain. When $t = 0$, $N = 100$, and when $t = 4$, $N = 200$. Write N as a function of t.

54. Sales Growth The rate of change in sales S (in thousands of units) of a new product is proportional to the product of S and $L - S$, where L (in thousands of units) is the estimated maximum level of sales. When $t = 0$, $S = 10$. Write and solve the differential equation for this sales model.

Advertising Awareness **In Exercises 55 and 56, use the advertising awareness model described in Example 6 to find the number of people y (in millions) aware of the product as a function of time t (in years).**

55. $y = 0$ when $t = 0$; $y = 0.75$ when $t = 1$

56. $y = 0$ when $t = 0$; $y = 0.9$ when $t = 2$

Chemical Reaction **In Exercises 57 and 58, use the chemical reaction model given in Example 7 to find the amount y as a function of t, and use a graphing utility to graph the function.**

57. $y = 45$ grams when $t = 0$; $y = 4$ grams when $t = 2$

58. $y = 75$ grams when $t = 0$; $y = 12$ grams when $t = 1$

Using a Gompertz Growth Model **In Exercises 59 and 60, use the Gompertz growth model described on page 411 to find the growth function, and sketch its graph.**

59. $L = 500$; $y = 100$ when $t = 0$; $y = 150$ when $t = 2$

60. $L = 5000$; $y = 500$ when $t = 0$; $y = 625$ when $t = 1$

61. Biology A population of eight beavers has been introduced into a new wetlands area. Biologists estimate that the maximum population the wetlands can sustain is 60 beavers. After 3 years, the population is 15 beavers. According to a Gompertz growth model, how many beavers will be present in the wetlands after 10 years?

62. Biology A population of 30 rabbits has been introduced into a new region. It is estimated that the maximum population the region can sustain is 400 rabbits. After 1 year, the population is estimated to be 90 rabbits. According to a Gompertz growth model, how many rabbits will be present after 3 years?

Biology **In Exercises 63 and 64, use the hybrid selection model described on page 412 to find the percent of the population that has the given characteristic. (Assume $a = 2$ and $b = 1$.)**

63. You are studying a population of mayflies to determine how quickly characteristic A will pass from one generation to the next. At the start of the study, half the population has characteristic A. After four generations, 75% of the population has characteristic A. Find the percent of the population that will have characteristic A after 10 generations.

64. A research team is studying a population of snails to determine how quickly characteristic B will pass from one generation to the next. At the start of the study, 40% of the snails have characteristic B. After five generations, 80% of the population has characteristic B. Find the percent of the population that will have characteristic B after eight generations.

65. Chemical Mixture   A 100-gallon tank is full of a solution containing 25 pounds of a concentrate. Starting at time $t = 0$, distilled water is admitted to the tank at the rate of 5 gallons per minute, and the well-stirred solution is withdrawn at the same rate, as shown in the figure.

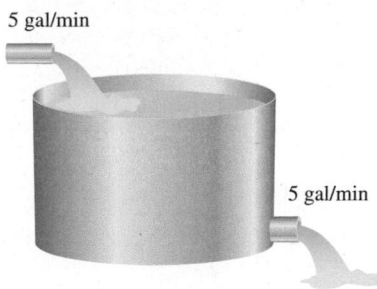

5 gal/min

5 gal/min

(a) Find the amount Q of the concentrate in the solution as a function of t. *(Hint: $Q' + Q/20 = 0$)*

(b) Find the time when the amount of concentrate in the tank reaches 15 pounds.

66. Chemical Mixture   A 200-gallon tank is half full of distilled water. At time $t = 0$, a solution containing 0.5 pound of concentrate per gallon enters the tank at the rate of 5 gallons per minute, and the well-stirred mixture is withdrawn at the same rate. Find the amount Q of concentrate in the tank after 30 minutes. *(Hint: $Q' + Q/20 = 5/2$)*

67. Chemical Reaction   In a chemical reaction, a compound changes into another compound at a rate proportional to the unchanged amount, according to the model

$$\frac{dy}{dt} = ky.$$

(a) Solve the differential equation.

(b) The initial amount of the original compound is 20 grams, and the amount remaining after 1 hour is 16 grams. When will 75% of the compound have been changed?

68. Snow Removal   The rate of change in the number of miles s of road cleared per hour by a snowplow is inversely proportional to the depth h of snow. That is,

$$\frac{ds}{dh} = \frac{k}{h}.$$

Find s as a function of h given that $s = 25$ miles when $h = 2$ inches and $s = 12$ miles when $h = 6$ inches ($2 \le h \le 15$).

69. Chemistry   A wet towel hung from a clothesline to dry loses moisture through evaporation at a rate proportional to its moisture content. After 1 hour, the towel has lost 40% of its original moisture content. After how long will it have lost 80%?

70. Biology   Let x and y be the sizes of two internal organs of a particular mammal at time t. Empirical data indicate that the relative growth rates of these two organs are equal, and can be modeled by

$$\frac{1}{x}\frac{dx}{dt} = \frac{1}{y}\frac{dy}{dt}.$$

Use this differential equation to write y as a function of x.

71. Population Growth   When predicting population growth, demographers must consider birth and death rates as well as the net change caused by the difference between the rates of immigration and emigration. Let P be the population at time t and let N be the net increase per unit time due to the difference between immigration and emigration. The rate of growth of the population is given by

$$\frac{dP}{dt} = kP + N, \quad N \text{ is constant.}$$

Solve this differential equation to find P as a function of time.

72. Meteorology   The barometric pressure y (in inches of mercury) at an altitude of x miles above sea level decreases at a rate proportional to the current pressure according to the model

$$\frac{dy}{dx} = -0.2y$$

where $y = 29.92$ inches when $x = 0$. Find the barometric pressure (a) at the top of Mount St. Helens (8364 feet) and (b) at the top of Mount Denali (20,310 feet).

73. Investment   A large corporation starts at time $t = 0$ to invest part of its receipts at a rate of P dollars per year in a fund for future corporate expansion. The fund earns r percent interest per year compounded continuously. The rate of growth of the amount A in the fund is given by

$$\frac{dA}{dt} = rA + P$$

where $A = 0$ when $t = 0$. Solve this differential equation for A as a function of t.

Investment   **In Exercises 74–76, use the result of Exercise 73.**

74. Find A for $P = \$275{,}000$, $r = 8\%$, and $t = 10$ years.

75. The corporation needs \$260,000,000 in 8 years and the fund earns $7\frac{1}{4}\%$ interest compounded continuously. Find P.

76. The corporation needs \$1,000,000 and it can invest \$125,000 per year in a fund earning 8% interest compounded continuously. Find t.

Using a Gompertz Growth Model   **In Exercises 77 and 78, use the Gompertz growth model described in Example 8.**

77. (a) Use a graphing utility to graph the slope field for the growth model when $k = 0.02$ and $L = 5000$.

(b) Describe the behavior of the graph as $t \to \infty$.

(c) Solve the growth model for $L = 5000$, $y_0 = 500$, and $k = 0.02$.

(d) Graph the equation you found in part (c). Determine the concavity of the graph.

78. (a) Use a graphing utility to graph the slope field for the growth model when $k = 0.05$ and $L = 1000$.

(b) Describe the behavior of the graph as $t \to \infty$.

(c) Solve the growth model for $L = 1000$, $y_0 = 100$, and $k = 0.05$.

(d) Graph the equation you found in part (c). Determine the concavity of the graph.

EXPLORING CONCEPTS

79. Separation of Variables Is an equation of the form

$$\frac{dy}{dx} = f(x)g(y) - f(x)h(y), \quad g(y) \neq h(y)$$

separable? Explain.

80. Finding a General Solution Find the general solution of $\frac{dy}{dx} = \frac{ay + b}{cy + d}$, where a, b, c, and d are nonzero constants.

Separation of Variables In Exercises 81–84, determine whether the differential equation is separable. If the equation is separable, rewrite it in the form $N(y)\,dy = M(x)\,dx$. (Do not solve the differential equation.)

81. $y(1 + x)\,dx + x\,dy = 0$ **82.** $y' = y^{1/2}$

83. $y' + xy = 5$ **84.** $y' = x - xy - y + 1$

85. Sailing

Ignoring resistance, a sailboat starting from rest accelerates (dv/dt) at a rate proportional to the difference between the velocities of the wind and the boat.

(a) The wind is blowing at 20 knots, and after 1 half-hour, the boat is moving at 10 knots. Write the velocity v as a function of time t.

(b) Use the result of part (a) to write the distance traveled by the boat as a function of time.

86. HOW DO YOU SEE IT? Recall from Example 1 that the general solution of

$$(x^2 + 4)\frac{dy}{dx} = xy$$

is $y = C\sqrt{x^2 + 4}$. The graphs below show the particular solutions for $C = 0.5$, 1, 2, and 3. Match the value of C with each graph. Explain your reasoning.

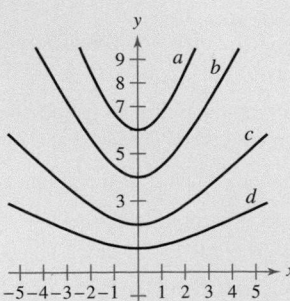

Determining If a Function Is Homogeneous In Exercises 87–94, determine whether the function is homogeneous, and if it is, determine its degree. A function $f(x, y)$ is *homogeneous of degree n* if $f(tx, ty) = t^n f(x, y)$.

87. $f(x, y) = x^3 - 4xy^2 + y^3$

88. $f(x, y) = x^4 + 2x^2y^2 + x + y$

89. $f(x, y) = e^{x/y}$

90. $f(x, y) = x^2 e^{y/x} + y^2$

91. $f(x, y) = 2 \ln xy$

92. $f(x, y) = \tan(x + y)$

93. $f(x, y) = 2 \ln \dfrac{x}{y}$

94. $f(x, y) = \tan \dfrac{y}{x}$

Solving a Homogeneous Differential Equation In Exercises 95–100, solve the homogeneous differential equation in terms of x and y. A *homogeneous differential equation* is an equation of the form

$$M(x, y)\,dx + N(x, y)\,dy = 0$$

where M and N are homogeneous functions of the same degree. To solve an equation of this form by the method of separation of variables, use the substitutions $y = vx$ and $dy = x\,dv + v\,dx$.

95. $(x + y)\,dx - 2x\,dy = 0$

96. $(x^3 + y^3)\,dx - xy^2\,dy = 0$

97. $(x - y)\,dx - (x + y)\,dy = 0$

98. $(x^2 + y^2)\,dx - 2xy\,dy = 0$

99. $xy\,dx + (y^2 - x^2)\,dy = 0$

100. $(2x + 3y)\,dx - x\,dy = 0$

True or False? In Exercises 101–103, determine whether the statement is true or false. If it is false, explain why or give an example that shows it is false.

101. The function $y = 0$ is always a solution of a differential equation that can be solved by separation of variables.

102. The differential equation $y' = xy - 2y + x - 2$ is separable.

103. The families $x^2 + y^2 = 2Cy$ and $x^2 + y^2 = 2Kx$ are mutually orthogonal.

PUTNAM EXAM CHALLENGE

104. A not uncommon calculus mistake is to believe that the product rule for derivatives says that $(fg)' = f'g'$. If

$$f(x) = e^{x^2}$$

determine, with proof, whether there exists an open interval (a, b) and a nonzero function g defined on (a, b) such that this wrong product rule is true for x in (a, b).

This problem was composed by the Committee on the Putnam Prize Competition.
© The Mathematical Association of America. All rights reserved.

6.4 The Logistic Equation

- Solve and analyze logistic differential equations.
- Use logistic differential equations to model and solve applied problems.

Logistic Differential Equation

In Section 6.2, the exponential growth model was derived from the fact that the rate of change of a variable y is proportional to the value of y. You observed that the differential equation $dy/dt = ky$ has the general solution $y = Ce^{kt}$. Exponential growth is unlimited, but when describing a population, there often exists some upper limit L past which growth cannot occur. This upper limit L is called the **carrying capacity,** which is the maximum population $y(t)$ that can be sustained or supported as time t increases. A model that is often used to describe this type of growth is the **logistic differential equation**

$$\frac{dy}{dt} = ky\left(1 - \frac{y}{L}\right) \qquad \text{Logistic differential equation}$$

where k and L are positive constants. A population that satisfies this equation does not grow without bound, but approaches the carrying capacity L as t increases.

From the equation, you can see that if y is between 0 and the carrying capacity L, then $dy/dt > 0$, and the population increases. If y is greater than L, then $dy/dt < 0$, and the population decreases. The general solution of the logistic differential equation is derived in the next example.

EXAMPLE 1 Deriving the General Solution

Solve the logistic differential equation $\dfrac{dy}{dt} = ky\left(1 - \dfrac{y}{L}\right)$.

Solution Begin by separating variables.

$$\frac{dy}{dt} = ky\left(1 - \frac{y}{L}\right) \qquad \text{Write differential equation.}$$

$$\frac{1}{y(1 - y/L)}\,dy = k\,dt \qquad \text{Separate variables.}$$

$$\int \frac{1}{y(1 - y/L)}\,dy = \int k\,dt \qquad \text{Integrate each side.}$$

$$\int \left(\frac{1}{y} + \frac{1}{L - y}\right) dy = \int k\,dt \qquad \text{Rewrite left side using partial fractions.}$$

$$\ln|y| - \ln|L - y| = kt + C \qquad \text{Find antiderivative of each side.}$$

$$\ln\left|\frac{L - y}{y}\right| = -kt - C \qquad \text{Multiply each side by } -1 \text{ and simplify.}$$

$$\left|\frac{L - y}{y}\right| = e^{-kt - C} \qquad \text{Exponentiate each side.}$$

$$\left|\frac{L - y}{y}\right| = e^{-C}e^{-kt} \qquad \text{Property of exponents}$$

$$\frac{L - y}{y} = be^{-kt} \qquad \text{Let } \pm e^{-C} = b.$$

> **REMARK** A review of the method of partial fractions is given in Section 8.5.

Solving this equation for y produces the general solution $y = \dfrac{L}{1 + be^{-kt}}$.

From Example 1, you can conclude that all solutions of the logistic differential equation are of the general form

$$y = \frac{L}{1 + be^{-kt}}.$$

The graph of the function y is called the *logistic curve*, as shown in Figure 6.19. In the next example, you will verify a particular solution of a logistic differential equation and find the initial condition.

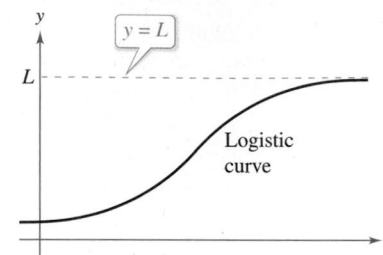

Note that as $t \to \infty$, $y \to L$.
Figure 6.19

EXAMPLE 2   Verifying a Particular Solution

Verify that the equation

$$y = \frac{4}{1 + 2e^{-3t}}$$

satisfies the logistic differential equation, and find the initial condition.

Solution   Comparing the given equation with the general form derived in Example 1, you know that $L = 4$, $b = 2$, and $k = 3$. You can verify that y satisfies the logistic differential equation as follows.

$$y = 4(1 + 2e^{-3t})^{-1} \qquad \text{Rewrite using negative exponent.}$$

$$y' = 4(-1)(1 + 2e^{-3t})^{-2}(-6e^{-3t}) \qquad \text{Apply Power Rule.}$$

$$= 3\left(\frac{4}{1 + 2e^{-3t}}\right)\left(\frac{2e^{-3t}}{1 + 2e^{-3t}}\right) \qquad \text{Rewrite.}$$

$$= 3y\left(\frac{2e^{-3t}}{1 + 2e^{-3t}}\right) \qquad \text{Rewrite using } y = \frac{4}{1 + 2e^{-3t}}.$$

$$= 3y\left(1 - \frac{1}{1 + 2e^{-3t}}\right) \qquad \text{Rewrite fraction using long division.}$$

$$= 3y\left(1 - \frac{4}{4(1 + 2e^{-3t})}\right) \qquad \text{Multiply fraction by } \frac{4}{4}.$$

$$= 3y\left(1 - \frac{y}{4}\right) \qquad \text{Rewrite using } y = \frac{4}{1 + 2e^{-3t}}.$$

So, y satisfies the logistic differential equation

$$y' = 3y\left(1 - \frac{y}{4}\right).$$

The initial condition can be found by letting $t = 0$ in the given equation.

$$y = \frac{4}{1 + 2e^{-3(0)}} = \frac{4}{3} \qquad \text{Let } t = 0 \text{ and simplify.}$$

So, the initial condition is $y(0) = \frac{4}{3}$.

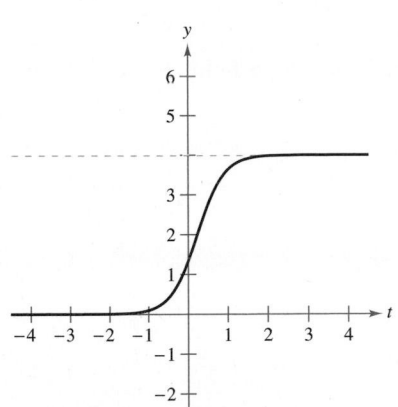

Figure 6.20

EXAMPLE 3 **Verifying the Upper Limit**

Verify that the upper limit of

$$y = \frac{4}{1 + 2e^{-3t}}$$

is 4.

Solution In Figure 6.20, you can see that the values of y appear to approach 4 as t increases without bound. You can come to this conclusion numerically, as shown in the table.

t	0	1	2	5	10	100
y	1.3333	3.6378	3.9803	4.0000	4.0000	4.0000

You can obtain the same results analytically, as follows.

$$\lim_{t \to \infty} y = \lim_{t \to \infty} \frac{4}{1 + 2e^{-3t}} = \frac{\lim_{t \to \infty} 4}{\lim_{t \to \infty} (1 + 2e^{-3t})} = \frac{4}{1 + 0} = 4$$

The upper limit of y is 4, which is also the carrying capacity $L = 4$.

EXAMPLE 4 **Determining the Point of Inflection**

Sketch a graph of

$$y = \frac{4}{1 + 2^{-3t}}.$$

Calculate y'' in terms of y and y'. Then determine the point of inflection.

Solution From Example 2, you know that

$$y' = 3y\left(1 - \frac{y}{4}\right).$$

Now calculate y'' in terms of y and y'.

$$y'' = 3y\left(-\frac{y'}{4}\right) + \left(1 - \frac{y}{4}\right)3y' \qquad \text{Differentiate using Product Rule.}$$

$$y'' = 3y'\left(1 - \frac{y}{2}\right) \qquad \text{Factor and simplify.}$$

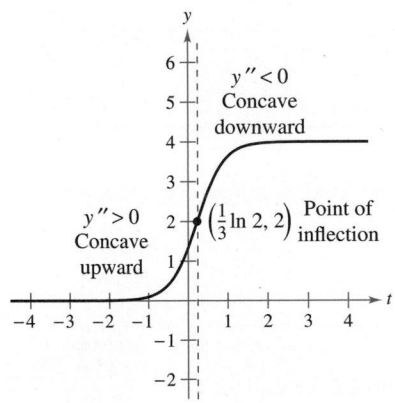

Figure 6.21

When $2 < y < 4$, $y'' < 0$ and the graph of y is concave downward. When $0 < y < 2$, $y'' > 0$ and the graph of y is concave upward. So, a point of inflection must occur at $y = 2$. The corresponding t-value is

$$2 = \frac{4}{1 + 2e^{-3t}} \implies 1 + 2e^{-3t} = 2 \implies e^{-3t} = \frac{1}{2} \implies t = \frac{1}{3}\ln 2.$$

The point of inflection is $\left(\frac{1}{3}\ln 2, 2\right)$, as shown in Figure 6.21. ∎

In Example 4, the point of inflection occurs at $y = L/2$. This is true for any logistic growth curve for which the solution starts below the carrying capacity L (see Exercise 31).

EXAMPLE 5   **Graphing a Slope Field and Solution Curves**

Graph a slope field for the logistic differential equation

$$y' = 0.05y\left(1 - \frac{y}{800}\right).$$

Then graph solution curves for the initial conditions $y(0) = 200$, $y(0) = 1200$, and $y(0) = 800$.

Solution   You can use a graphing utility to graph the slope field shown in Figure 6.22. The solution curves for the initial conditions

$$y(0) = 200, \quad y(0) = 1200, \quad \text{and} \quad y(0) = 800$$

are shown in Figures 6.23–6.25.

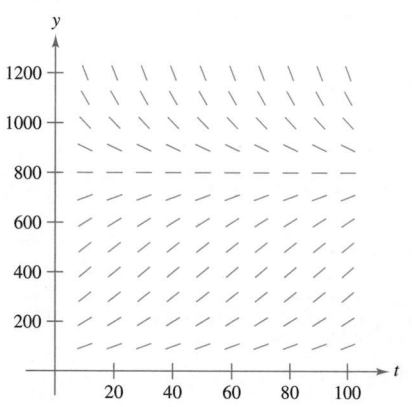

Slope field for

$$y' = 0.05y\left(1 - \frac{y}{800}\right)$$

Figure 6.22

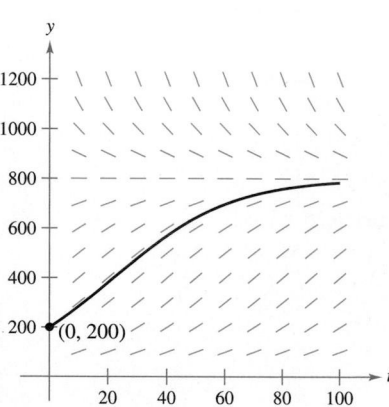

Particular solution for

$$y' = 0.05y\left(1 - \frac{y}{800}\right)$$

and initial condition $y(0) = 200$

Figure 6.23

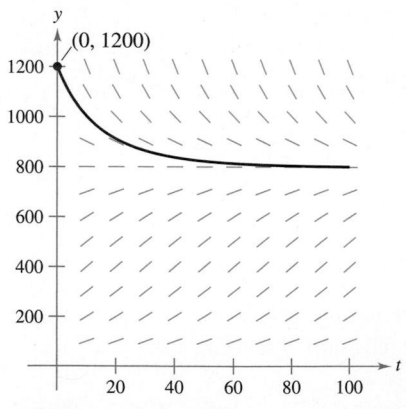

Particular solution for

$$y' = 0.05y\left(1 - \frac{y}{800}\right)$$

and initial condition $y(0) = 1200$

Figure 6.24

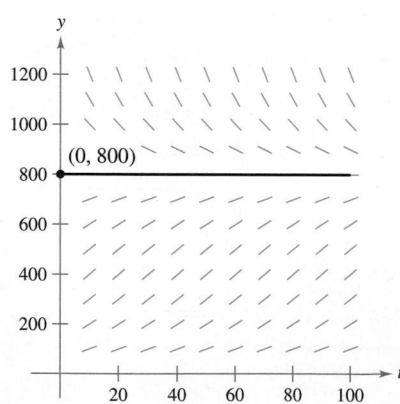

Particular solution for

$$y' = 0.05y\left(1 - \frac{y}{800}\right)$$

and initial condition $y(0) = 800$

Figure 6.25

Note that as t increases without bound, the solution curves in Figures 6.23–6.25 all tend to the same limit, which is the carrying capacity of 800.

Application

EXAMPLE 6   Solving a Logistic Differential Equation

A state game commission releases 40 elk into a game refuge. After 5 years, the elk population is 104. The commission believes that the environment can support no more than 4000 elk. The growth rate of the elk population p is

$$\frac{dp}{dt} = kp\left(1 - \frac{p}{4000}\right), \quad 40 \leq p \leq 4000$$

where t is the number of years.

a. Write a model for the elk population in terms of t.

b. Graph the slope field for the differential equation and the solution that passes through the point $(0, 40)$.

c. Use the model to estimate the elk population after 15 years.

d. Find the limit of the model as $t \to \infty$.

Solution

a. You know that $L = 4000$. So, the solution of the equation is of the form

$$p = \frac{4000}{1 + be^{-kt}}.$$

Because $p(0) = 40$, you can solve for b as follows.

$$40 = \frac{4000}{1 + be^{-k(0)}} \implies 40 = \frac{4000}{1 + b} \implies b = 99$$

Then, because $p = 104$ when $t = 5$, you can solve for k.

$$104 = \frac{4000}{1 + 99e^{-k(5)}} \implies k \approx 0.194$$

So, a model for the elk population is

$$p = \frac{4000}{1 + 99e^{-0.194t}}.$$

b. Using a graphing utility, you can graph the slope field of

$$\frac{dp}{dt} = 0.194p\left(1 - \frac{p}{4000}\right)$$

and the solution that passes through $(0, 40)$, as shown in Figure 6.26.

c. To estimate the elk population after 15 years, substitute 15 for t in the model.

$$p = \frac{4000}{1 + 99e^{-0.194(15)}} \qquad \text{Substitute 15 for } t.$$

$$= \frac{4000}{1 + 99e^{-2.91}} \qquad \text{Simplify.}$$

$$\approx 626 \qquad \text{Simplify.}$$

d. As t increases without bound, the denominator of

$$\frac{4000}{1 + 99e^{-0.194t}}$$

gets closer and closer to 1. So,

$$\lim_{t \to \infty} \frac{4000}{1 + 99e^{-0.194t}} = 4000.$$

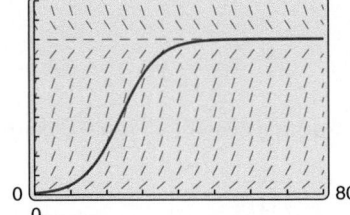

5000

0   0   80

Slope field for
$$\frac{dp}{dt} = 0.194p\left(1 - \frac{p}{4000}\right)$$
and the solution passing through $(0, 40)$
Figure 6.26

6.4 Exercises

See **CalcChat.com** for tutorial help and worked-out solutions to odd-numbered exercises.

CONCEPT CHECK

1. **Carrying Capacity** Describe carrying capacity in your own words.

2. **Logistic Curve** Write an equation that represents a logistic curve with a horizontal asymptote at $y = 3$. Explain.

Matching In Exercises 3–6, match the logistic equation with its graph. [The graphs are labeled (a), (b), (c), and (d).]

(a)

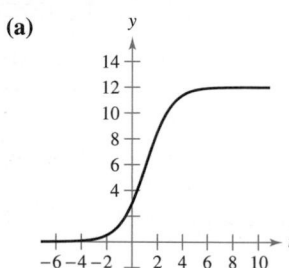

(b)

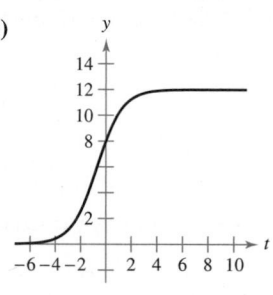

(c)

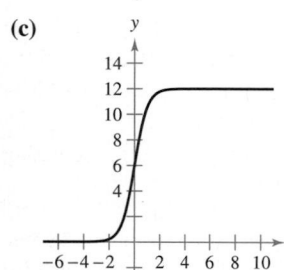

(d)

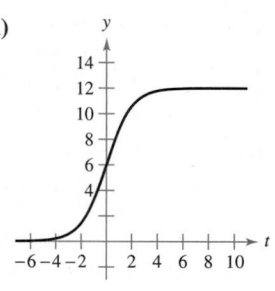

3. $y = \dfrac{12}{1 + e^{-t}}$

4. $y = \dfrac{12}{1 + 3e^{-t}}$

5. $y = \dfrac{12}{1 + \frac{1}{2}e^{-t}}$

6. $y = \dfrac{12}{1 + e^{-2t}}$

 **Verifying a Particular Solution** In Exercises 7–10, verify that the equation satisfies the logistic differential equation $\dfrac{dy}{dt} = ky\left(1 - \dfrac{y}{L}\right)$. Then find the initial condition.

7. $y = \dfrac{8}{1 + e^{-2t}}$

8. $y = \dfrac{12}{1 + 3e^{-t}}$

9. $y = \dfrac{12}{1 + 6e^{-t}}$

10. $y = \dfrac{12}{1 + e^{-2t}}$

 **Using a Logistic Equation** In Exercises 11–14, the logistic equation models the growth of a population. Use the equation to (a) find the value of k, (b) find the carrying capacity, (c) find the initial population, (d) determine when the population will reach 50% of its carrying capacity, and (e) write a logistic differential equation that has the solution $P(t)$.

11. $P(t) = \dfrac{2100}{1 + 29e^{-0.75t}}$

12. $P(t) = \dfrac{5000}{1 + 39e^{-0.2t}}$

13. $P(t) = \dfrac{6000}{1 + 4999e^{-0.8t}}$

14. $P(t) = \dfrac{1000}{1 + 8e^{-0.2t}}$

 **Using a Logistic Differential Equation** In Exercises 15–18, the logistic differential equation models the growth rate of a population. Use the equation to (a) find the value of k, (b) find the carrying capacity, (c) use a computer algebra system to graph a slope field, and (d) determine the value of P at which the population growth rate is the greatest.

15. $\dfrac{dP}{dt} = 3P\left(1 - \dfrac{P}{100}\right)$

16. $\dfrac{dP}{dt} = 0.5P\left(1 - \dfrac{P}{250}\right)$

17. $\dfrac{dP}{dt} = 0.1P - 0.0004P^2$

18. $\dfrac{dP}{dt} = 0.4P - 0.00025P^2$

Solving a Logistic Differential Equation In Exercises 19–22, find the logistic equation that satisfies the initial condition. Then use the logistic equation to find y when $t = 5$ and $t = 100$.

19. $\dfrac{dy}{dt} = y\left(1 - \dfrac{y}{36}\right)$, $(0, 4)$

20. $\dfrac{dy}{dt} = 2.8y\left(1 - \dfrac{y}{10}\right)$, $(0, 7)$

21. $\dfrac{dy}{dt} = \dfrac{4y}{5} - \dfrac{y^2}{150}$, $(0, 8)$

22. $\dfrac{dy}{dt} = \dfrac{3y}{20} - \dfrac{y^2}{1600}$, $(0, 15)$

Matching In Exercises 23–26, match the logistic differential equation and initial condition with the graph of its solution. [The graphs are labeled (a), (b), (c), and (d).]

(a)

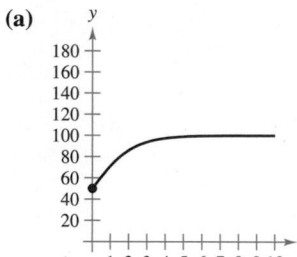

(b)

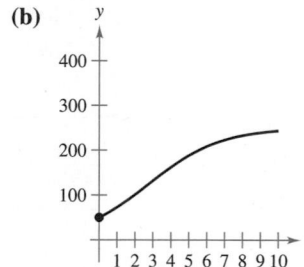

(c)

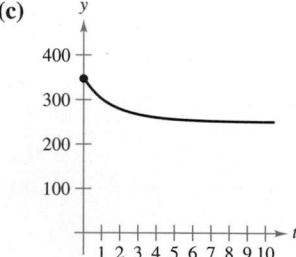

(d)

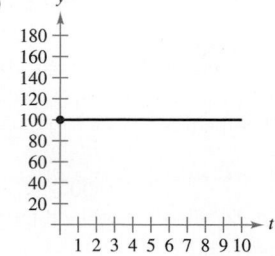

23. $\dfrac{dy}{dt} = 0.5y\left(1 - \dfrac{y}{250}\right)$, $(0, 350)$

24. $\dfrac{dy}{dt} = 0.9y\left(1 - \dfrac{y}{100}\right)$, $(0, 100)$

25. $\dfrac{dy}{dt} = 0.5y\left(1 - \dfrac{y}{250}\right)$, $(0, 50)$

26. $\dfrac{dy}{dt} = 0.9y\left(1 - \dfrac{y}{100}\right)$, $(0, 50)$

Slope Field In Exercises 27 and 28, a logistic differential equation, a point, and a slope field are given. (a) Sketch two approximate solutions of the differential equation on the slope field, one of which passes through the given point. (b) Find the particular solution of the differential equation and use a graphing utility to graph the solution. Compare the result with the sketch in part (a). To print an enlarged copy of the graph, go to *MathGraphs.com*.

27. $\dfrac{dy}{dt} = 0.2y\left(1 - \dfrac{y}{1000}\right)$, **28.** $\dfrac{dy}{dt} = 0.9y\left(1 - \dfrac{y}{200}\right)$,

(0, 105)   (0, 240)

EXPLORING CONCEPTS

29. Determining Values It is known that $y = \dfrac{L}{1 + be^{-kt}}$ is a solution of the logistic differential equation

$$\dfrac{dy}{dt} = 0.75y\left(1 - \dfrac{y}{2500}\right).$$

Is it possible to determine L, k, and b from the information given? If so, find their values. If not, which value(s) cannot be determined and what information do you need to determine the value(s)?

30. Concavity Describe the concavity of the logistic curve represented by $y = 4/(5 + 3e^{-6t})$. Explain your reasoning.

31. Point of Inflection For any logistic growth curve, show that the point of inflection occurs at $y = L/2$ when the solution starts below the carrying capacity L.

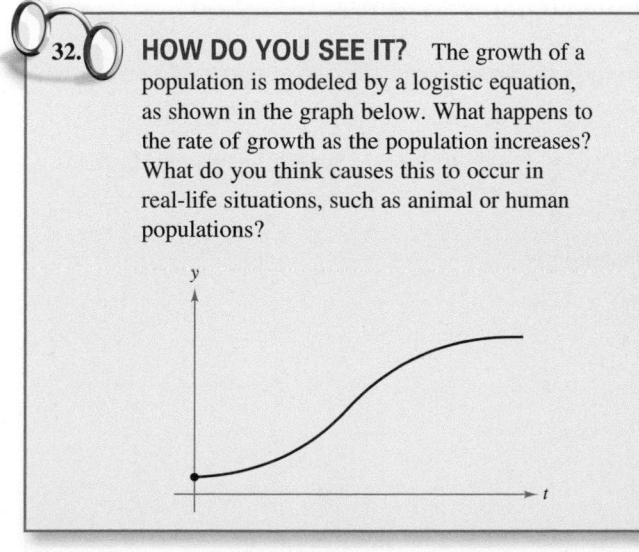

32. HOW DO YOU SEE IT? The growth of a population is modeled by a logistic equation, as shown in the graph below. What happens to the rate of growth as the population increases? What do you think causes this to occur in real-life situations, such as animal or human populations?

33. Endangered Species A conservation organization releases 25 Florida panthers into a game preserve. After 2 years, there are 39 panthers in the preserve. The Florida preserve has a carrying capacity of 200 panthers.

(a) Write a logistic equation that models the population of panthers in the preserve.

(b) Find the population after 5 years.

(c) When will the population reach 100?

(d) Write a logistic differential equation that models the growth rate of the panther population. Then repeat part (b) using Euler's Method with a step size of $h = 1$. Compare the approximation with the exact answer.

(e) After how many years is the panther population growing most rapidly? Explain.

34. Bacteria Growth At time $t = 0$, a bacterial culture weighs 1 gram. Two hours later, the culture weighs 4 grams. The maximum weight of the culture is 20 grams.

(a) Write a logistic equation that models the weight of the bacterial culture.

(b) Find the culture's weight after 5 hours.

(c) When will the culture's weight reach 18 grams?

(d) Write a logistic differential equation that models the growth rate of the culture's weight. Then repeat part (b) using Euler's Method with a step size of $h = 1$. Compare the approximation with the exact answer.

(e) After how many hours is the culture's weight increasing most rapidly? Explain.

True or False? In Exercises 35 and 36, determine whether the statement is true or false. If it is false, explain why or give an example that shows it is false.

35. For the logistic differential equation

$$\dfrac{dy}{dt} = ky\left(1 - \dfrac{y}{L}\right)$$

if $y > L$, then $dy/dt > 0$ and the population increases.

36. For the logistic differential equation

$$\dfrac{dy}{dt} = ky\left(1 - \dfrac{y}{L}\right)$$

if $0 < y < L$, then $dy/dt > 0$ and the population increases.

37. Logistic Differential Equation Identify two constant functions which are solutions to

$$\dfrac{dy}{dt} = ky\left(1 - \dfrac{y}{L}\right).$$

Assuming that the equation models a population, interpret your solutions.

38. Finding a Derivative Show that if $y = \dfrac{1}{1 + be^{-kt}}$, then

$$\dfrac{dy}{dt} = ky(1 - y).$$

6.5 First-Order Linear Differential Equations

■ Solve a first-order linear differential equation, and use linear differential equations to solve applied problems.

First-Order Linear Differential Equations

In this section, you will see how to solve a very important class of first-order differential equations—first-order linear differential equations.

Definition of First-Order Linear Differential Equation

A **first-order linear differential equation** is an equation of the form

$$\frac{dy}{dx} + P(x)y = Q(x)$$

where P and Q are continuous functions of x. This first-order linear differential equation is said to be in **standard form.**

ANNA JOHNSON PELL WHEELER
(1883–1966)

Anna Johnson Pell Wheeler was awarded a master's degree in 1904 from the University of Iowa for her thesis *The Extension of Galois Theory to Linear Differential Equations.* Influenced by David Hilbert, she worked on integral equations while studying infinite linear spaces.

To solve a linear differential equation, write it in standard form to identify the functions $P(x)$ and $Q(x)$. Then integrate $P(x)$ and form the expression

$$u(x) = e^{\int P(x)\,dx} \qquad \text{Integrating factor}$$

which is called an **integrating factor.** The general solution of the equation is

$$y = \frac{1}{u(x)} \int Q(x)u(x)\,dx. \qquad \text{General solution}$$

It is instructive to see why the integrating factor helps solve a linear differential equation of the form $y' + P(x)y = Q(x)$. When both sides of the equation are multiplied by the integrating factor $u(x) = e^{\int P(x)\,dx}$, the left side becomes the derivative of a product.

$$y'e^{\int P(x)\,dx} + P(x)ye^{\int P(x)\,dx} = Q(x)e^{\int P(x)\,dx}$$
$$\left[ye^{\int P(x)\,dx}\right]' = Q(x)e^{\int P(x)\,dx}$$

Integrating both sides of this second equation and dividing by $u(x)$ produce the general solution.

EXAMPLE 1   Solving a Linear Differential Equation

Find the general solution of

$$y' + y = e^x.$$

Solution   For this equation, $P(x) = 1$ and $Q(x) = e^x$. So, the integrating factor is

$$u(x) = e^{\int P(x)\,dx} = e^{\int dx} = e^x.$$

This implies that the general solution is

$$y = \frac{1}{u(x)} \int Q(x)u(x)\,dx$$
$$= \frac{1}{e^x} \int e^x(e^x)\,dx$$
$$= e^{-x}\left(\frac{1}{2}e^{2x} + C\right)$$
$$= \frac{1}{2}e^x + Ce^{-x}.$$

THEOREM 6.2 Solution of a First-Order Linear Differential Equation

An integrating factor for the first-order linear differential equation

$$y' + P(x)y = Q(x)$$

is $u(x) = e^{\int P(x)\,dx}$. The solution of the differential equation is

$$ye^{\int P(x)\,dx} = \int Q(x)e^{\int P(x)\,dx}\,dx + C.$$

EXAMPLE 2 **Solving a First-Order Linear Differential Equation**

⋯▷ *See LarsonCalculus.com for an interactive version of this type of example.*

Find the general solution of $xy' - 2y = x^2$, $x > 0$.

Solution The standard form of the equation is

$$y' + \left(-\frac{2}{x}\right)y = x, \quad x > 0. \qquad \text{Standard form}$$

So, $P(x) = -2/x$, and you have

$$\int P(x)\,dx = -\int \frac{2}{x}\,dx = -\ln x^2$$

which implies that the integrating factor is

$$e^{\int P(x)\,dx} = e^{-\ln x^2} = \frac{1}{e^{\ln x^2}} = \frac{1}{x^2}. \qquad \text{Integrating factor}$$

So, multiplying each side of the standard form by $1/x^2$ yields

$$\frac{y'}{x^2} - \frac{2y}{x^3} = \frac{1}{x}$$

$$\frac{d}{dx}\left[\frac{y}{x^2}\right] = \frac{1}{x}$$

$$\frac{y}{x^2} = \int \frac{1}{x}\,dx$$

$$\frac{y}{x^2} = \ln x + C$$

$$y = x^2(\ln x + C). \qquad \text{General solution}$$

Several solution curves (for $C = -2, -1, 0, 1, 2, 3,$ and 4) are shown in Figure 6.27. ∎

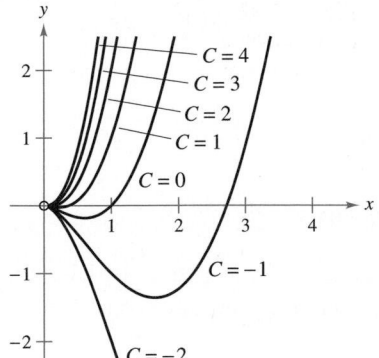

Figure 6.27

In most falling-body problems discussed so far in the text, air resistance has been neglected. The next example includes this factor. In the example, the air resistance on the falling object is assumed to be proportional to its velocity v. If g is the acceleration due to gravity, the downward force F on a falling object of mass m is given by $-mg - kv$. If a is the acceleration of the object, then by Newton's Second Law of Motion,

$$F = ma = m\frac{dv}{dt}$$

which yields the following differential equation.

$$m\frac{dv}{dt} = -mg - kv \quad \Longrightarrow \quad \frac{dv}{dt} + \frac{kv}{m} = -g$$

EXAMPLE 3 **A Falling Object with Air Resistance**

An object of mass m is dropped from a hovering helicopter. The air resistance is proportional to the velocity of the object. Find the velocity of the object as a function of time t.

Solution The velocity v satisfies the equation

$$\frac{dv}{dt} + \frac{kv}{m} = -g.$$

g = acceleration due to gravity,
k = constant of proportionality

Letting $b = k/m$, you can *separate variables* to obtain

$$dv = -(g + bv)\, dt$$

$$\int \frac{dv}{g + bv} = -\int dt$$

$$\frac{1}{b} \ln|g + bv| = -t + C_1$$

$$\ln|g + bv| = -bt + bC_1$$

$$g + bv = Ce^{-bt}. \qquad C = e^{bC_1}$$

> **••REMARK** Notice in Example 3 that the velocity approaches a limit of $-mg/k$ as a result of the air resistance. For falling-body problems in which air resistance is neglected, the velocity increases without bound.

Because the object was dropped, $v = 0$ when $t = 0$; so $g = C$, and it follows that

$$bv = -g + ge^{-bt} \implies v = \frac{-g(1 - e^{-bt})}{b} = -\frac{mg}{k}(1 - e^{-kt/m}).$$

A simple electric circuit consists of an electric current I (in amperes), a resistance R (in ohms), an inductance L (in henrys), and a constant electromotive force E (in volts), as shown in Figure 6.28. According to Kirchhoff's Second Law, if the switch S is closed when $t = 0$, then the applied electromotive force (voltage) is equal to the sum of the voltage drops in the rest of the circuit. This, in turn, means that the current I satisfies the differential equation

$$L\frac{dI}{dt} + RI = E.$$

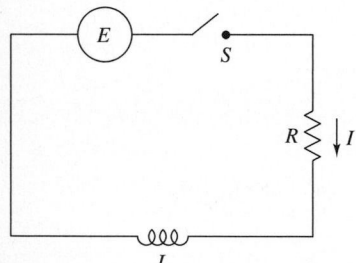

Figure 6.28

EXAMPLE 4 **An Electric Circuit Problem**

Find the current I as a function of time t (in seconds), given that I satisfies the differential equation $L(dI/dt) + RI = \sin 2t$, where R and L are nonzero constants.

Solution In standard form, the given linear equation is

$$\frac{dI}{dt} + \frac{R}{L}I = \frac{1}{L} \sin 2t.$$

Let $P(t) = R/L$, so that $e^{\int P(t)\, dt} = e^{(R/L)t}$, and by Theorem 6.2,

> **••REMARK** The integral in Example 4 was found using a computer algebra system. In Chapter 8, you will learn how to integrate functions of this type using integration by parts.

$$Ie^{(R/L)t} = \frac{1}{L} \int e^{(R/L)t} \sin 2t\, dt$$

$$= \frac{1}{4L^2 + R^2} e^{(R/L)t}(R \sin 2t - 2L \cos 2t) + C.$$

So, the general solution is

$$I = e^{-(R/L)t}\left[\frac{1}{4L^2 + R^2} e^{(R/L)t}(R \sin 2t - 2L \cos 2t) + C \right]$$

$$= \frac{1}{4L^2 + R^2}(R \sin 2t - 2L \cos 2t) + Ce^{-(R/L)t}.$$

One type of problem that can be described in terms of a differential equation involves chemical mixtures, as illustrated in the next example.

EXAMPLE 5   A Mixture Problem

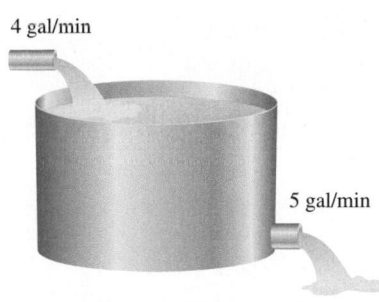

4 gal/min

5 gal/min

Figure 6.29

A tank contains 50 gallons of a solution composed of 90% water and 10% alcohol. A second solution containing 50% water and 50% alcohol is added to the tank at a rate of 4 gallons per minute. As the second solution is being added, the tank is being drained at a rate of 5 gallons per minute, as shown in Figure 6.29. The solution in the tank is stirred constantly. How much alcohol is in the tank after 10 minutes?

Solution   Let y be the number of gallons of alcohol in the tank at any time t. You know that $y = 5$ when $t = 0$. Because the number of gallons of solution in the tank at any time is $50 - t$, and the tank loses 5 gallons of solution per minute, it must lose

$$\left(\frac{5}{50 - t}\right)y$$

gallons of alcohol per minute. Furthermore, because the tank is gaining 2 gallons of alcohol per minute, the rate of change of alcohol in the tank is

$$\frac{dy}{dt} = 2 - \left(\frac{5}{50 - t}\right)y \quad \Longrightarrow \quad \frac{dy}{dt} + \left(\frac{5}{50 - t}\right)y = 2.$$

To solve this linear differential equation, let

$$P(t) = \frac{5}{50 - t}$$

and obtain

$$\int P(t)\, dt = \int \frac{5}{50 - t}\, dt = -5 \ln|50 - t|.$$

Because $t < 50$, you can drop the absolute value signs and conclude that

$$e^{\int P(t)\, dt} = e^{-5 \ln(50 - t)} = \frac{1}{(50 - t)^5}.$$

So, the general solution is

$$\frac{y}{(50 - t)^5} = \int \frac{2}{(50 - t)^5}\, dt$$

$$\frac{y}{(50 - t)^5} = \frac{1}{2(50 - t)^4} + C$$

$$y = \frac{50 - t}{2} + C(50 - t)^5.$$

Because $y = 5$ when $t = 0$, you have

$$5 = \frac{50}{2} + C(50)^5 \quad \Longrightarrow \quad -\frac{20}{50^5} = C$$

which means that the particular solution is

$$y = \frac{50 - t}{2} - 20\left(\frac{50 - t}{50}\right)^5.$$

Finally, when $t = 10$, the amount of alcohol in the tank is

$$y = \frac{50 - 10}{2} - 20\left(\frac{50 - 10}{50}\right)^5 \approx 13.45 \text{ gal}$$

which represents a solution containing 33.6% alcohol.

6.5 Exercises

See **CalcChat.com** for tutorial help and worked-out solutions to odd-numbered exercises.

CONCEPT CHECK

1. **First-Order** What does the term "first-order" refer to in a first-order linear differential equation?

2. **First-Order Linear Differential Equations** Describe how to solve a first-order linear differential equation.

Determining Whether a Differential Equation Is Linear In Exercises 3–6, determine whether the differential equation is linear. Explain your reasoning.

3. $x^3 y' + xy = e^x + 1$

4. $2xy - y' \ln x = y$

5. $y' - y \sin x = xy^2$

6. $\dfrac{2 - y'}{y} = 5x$

Solving a First-Order Linear Differential Equation In Exercises 7–14, find the general solution of the first-order linear differential equation for $x > 0$.

7. $\dfrac{dy}{dx} + \left(\dfrac{1}{x}\right)y = 6x + 2$

8. $\dfrac{dy}{dx} + \left(\dfrac{2}{x}\right)y = 3x - 5$

9. $y' + 2xy = 10x$

10. $y' + 3x^2 y = 6x^2$

11. $(y + 1)\cos x \, dx - dy = 0$

12. $(y - 1)\sin x \, dx - dy = 0$

13. $y' + 3y = e^{3x}$

14. $xy' + y = x^2 \ln x$

 **Slope Field** In Exercises 15 and 16, (a) sketch an approximate solution of the differential equation satisfying the given initial condition on the slope field, (b) find the particular solution that satisfies the given initial condition, and (c) use a graphing utility to graph the particular solution. Compare the graph with the sketch in part (a). To print an enlarged copy of the graph, go to *MathGraphs.com*.

15. $\dfrac{dy}{dx} = e^x - y,$

$(0, 1)$

16. $y' + \left(\dfrac{1}{x}\right)y = \sin x^2,$

$\left(\sqrt{\pi}, 0\right)$

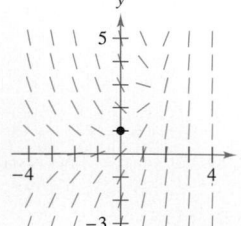
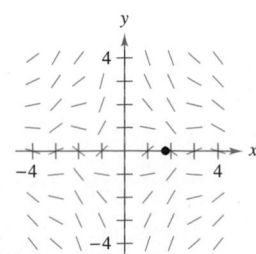

Finding a Particular Solution In Exercises 17–24, find the particular solution of the first-order linear differential equation for $x > 0$ that satisfies the initial condition.

Differential Equation	Initial Condition
17. $y' + y = 6e^x$	$y(0) = 3$
18. $x^3 y' + 2y = e^{1/x^2}$	$y(1) = e$
19. $y' + y \tan x = \sec x + \cos x$	$y(0) = 1$
20. $y' + y \sec x = \sec x$	$y(0) = 4$
21. $y' + \left(\dfrac{1}{x}\right)y = 0$	$y(2) = 2$
22. $y' + (2x - 1)y = 0$	$y(1) = 2$
23. $x \, dy = (x + y + 2) \, dx$	$y(1) = 10$
24. $2xy' - y = x^3 - x$	$y(4) = 2$

25. **Population Growth** When predicting population growth, demographers must consider birth and death rates as well as the net change caused by the difference between the rates of immigration and emigration. Let P be the population at time t and let N be the net increase per unit time resulting from the difference between immigration and emigration. So, the rate of growth of the population is given by

$$\frac{dP}{dt} = kP + N$$

where N is constant. Solve this differential equation to find P as a function of time, when at time $t = 0$ the size of the population is P_0.

26. **Investment Growth** A large corporation starts at time $t = 0$ to invest part of its receipts continuously at a rate of P dollars per year in a fund for future corporate expansion. Assume that the fund earns r percent interest per year compounded continuously. So, the rate of growth of the amount A in the fund is given by $dA/dt = rA + P$, where $A = 0$ when $t = 0$. Solve this differential equation for A as a function of t.

Investment Growth In Exercises 27 and 28, use the result of Exercise 26.

27. Find A for the following.

(a) $P = \$275{,}000$, $r = 8\%$, $t = 10$ years

(b) $P = \$550{,}000$, $r = 5.9\%$, $t = 25$ years

28. Find t if the corporation needs $\$1{,}000{,}000$ and it can invest $\$125{,}000$ per year in a fund earning 8% interest compounded continuously.

29. **Learning Curve** The management at a certain factory has found that the maximum number of units a worker can produce in a day is 75. The rate of increase in the number of units N produced with respect to time t in days by a new employee is proportional to $75 - N$.

(a) Determine the differential equation describing the rate of change of performance with respect to time.

(b) Solve the differential equation from part (a).

(c) Find the particular solution for a new employee who produced 20 units on the first day at the factory and 35 units on the twentieth day.

• **30. Intravenous Feeding** • • • • • • • • • • • • •

Glucose is added intravenously to the bloodstream at the rate of q units per minute, and the body removes glucose from the bloodstream at a rate proportional to the amount present. Assume that $Q(t)$ is the amount of glucose in the bloodstream at time t.

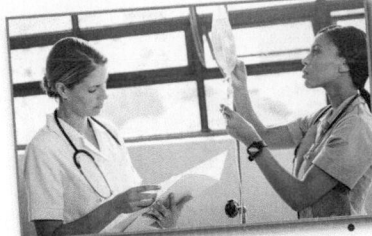

(a) Determine the differential equation describing the rate of change of glucose in the bloodstream with respect to time.

(b) Solve the differential equation from part (a), letting $Q = Q_0$ when $t = 0$.

(c) Find the limit of $Q(t)$ as $t \to \infty$.

Falling Object In Exercises 31 and 32, consider an object with a mass of 4 kilograms dropped from a height of 1500 meters, where the air resistance is proportional to the velocity.

31. Write the velocity of the object as a function of time t when the velocity after 5 seconds is approximately -31 meters per second. What is the limiting value of the velocity function?

32. Use the result of Exercise 31 to write the position of the object as a function of time t. Approximate the velocity of the object when it reaches ground level.

Electric Circuits In Exercises 33 and 34, use the differential equation for electric circuits given by

$$L\frac{dI}{dt} + RI + E.$$

In this equation, I is the current, R is the resistance, L is the inductance, and E is the electromotive force (voltage).

33. Solve the differential equation for the current given a constant voltage E_0.

34. Use the result of Exercise 33 to find the equation for the current when $I(0) = 0$, $E_0 = 120$ volts, $R = 600$ ohms, and $L = 4$ henrys. When does the current reach 90% of its limiting value?

Mixture In Exercises 35–38, consider a tank that at time $t = 0$ contains v_0 gallons of a solution of which, by weight, q_0 pounds is soluble concentrate. Another solution containing q_1 pounds of the concentrate per gallon is running into the tank at the rate of r_1 gallons per minute. The solution in the tank is kept well stirred and is withdrawn at the rate of r_2 gallons per minute.

35. Let Q be the amount of concentrate (in pounds) in the solution at any time t. Show that

$$\frac{dQ}{dt} + \frac{r_2 Q}{v_0 + (r_1 - r_2)t} = q_1 r_1.$$

36. Let Q be the amount of concentrate (in pounds) in the solution at any time t. Write the differential equation for the rate of change of Q with respect to t when $r_1 = r_2 = r$.

37. A 200-gallon tank is full of a solution containing 25 pounds of concentrate. Starting at time $t = 0$, distilled water is admitted to the tank at a rate of 10 gallons per minute, and the well-stirred solution is withdrawn at the same rate.

(a) Find the amount of concentrate Q (in pounds) in the solution as a function of t.

(b) Find the time at which the amount of concentrate in the tank reaches 15 pounds.

(c) Find the amount of concentrate (in pounds) in the solution as $t \to \infty$.

38. A 200-gallon tank is half full of distilled water. Starting at time $t = 0$, a solution containing 0.5 pound of concentrate per gallon is admitted to the tank at a rate of 5 gallons per minute, and the well-stirred mixture is withdrawn at a rate of 3 gallons per minute.

(a) At what time will the tank be full?

(b) At the time the tank is full, how many pounds of concentrate will it contain?

(c) Repeat parts (a) and (b), assuming that the solution entering the tank contains 1 pound of concentrate per gallon.

39. Using an Integrating Factor The expression $u(x)$ is an integrating factor for $y' + P(x)y = Q(x)$. Which of the following is equal to $u'(x)$? Verify your answer.

(a) $P(x)u(x)$ (b) $P'(x)u(x)$

(c) $Q(x)u(x)$ (d) $Q'(x)u(x)$

40. **HOW DO YOU SEE IT?** The graph shows the amount of concentrate Q (in pounds) in a solution in a tank at time t (in minutes) as a solution with concentrate enters the tank, is well stirred, and is withdrawn from the tank.

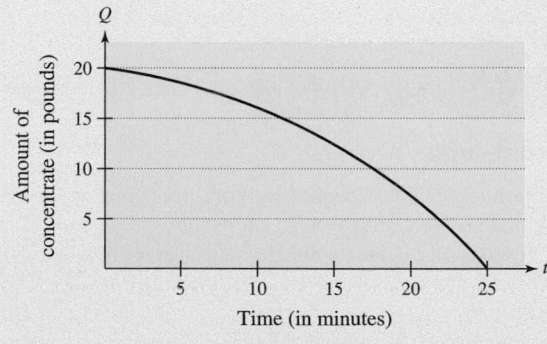

(a) How much concentrate is in the tank at time $t = 0$?

(b) Which is greater, the rate of solution into the tank or the rate of solution withdrawn from the tank? Explain.

(c) At what time is there no concentrate in the tank? What does this mean?

EXPLORING CONCEPTS

41. Using Different Methods   Describe two ways to find the general solution of

$$\frac{dy}{dx} + 3xy = x.$$

Verify that each method gives the same result.

42. Integrating Factor   Explain why you can omit the constant of integration when finding an integrating factor.

Matching   In Exercises 43–46, match the differential equation with its solution.

Differential Equation	Solution
43. $y' - 2x = 0$	(a) $y = Ce^{x^2}$
44. $y' - 2y = 0$	(b) $y = -\frac{1}{2} + Ce^{x^2}$
45. $y' - 2xy = 0$	(c) $y = x^2 + C$
46. $y' - 2xy = x$	(d) $y = Ce^{2x}$

Slope Field   In Exercises 47 and 48, (a) use a graphing utility to graph the slope field for the differential equation, (b) find the particular solutions of the differential equation passing through the given points, and (c) use a graphing utility to graph the particular solutions on the slope field in part (a).

Differential Equation	Points
47. $\dfrac{dy}{dx} - \dfrac{1}{x}y = x^2, \quad x > 0$	$(-2, 4), (2, 8)$
48. $\dfrac{dy}{dx} + 4x^3y = x^3$	$\left(0, \dfrac{7}{2}\right), \left(0, -\dfrac{1}{2}\right)$

Solving a First-Order Differential Equation   In Exercises 49–56, find the general solution of the first-order differential equation for $x > 0$ by any appropriate method.

49. $\dfrac{dy}{dx} = \dfrac{e^{2x+y}}{e^{x-y}}$

50. $y' \cos x^2 + \dfrac{y \cos x^2}{x} = \sec x^2$

51. $y \cos x - \cos x + \dfrac{dy}{dx} = 0$

52. $y' = 2x\sqrt{1 - y^2}$

53. $(2y - e^x)\, dx + x\, dy = 0$

54. $(x + y)\, dx - x\, dy = 0$

55. $3(y - 4x^2)\, dx + x\, dy = 0$

56. $x\, dx + (y + e^y)(x^2 + 1)\, dy = 0$

Solving a Bernoulli Differential Equation   In Exercises 57–64, solve the Bernoulli differential equation. The *Bernoulli equation* is a well-known nonlinear equation of the form

$$y' + P(x)y = Q(x)y^n$$

that can be reduced to a linear form by a substitution. The general solution of a Bernoulli equation is

$$y^{1-n}e^{\int (1-n)P(x)\, dx} = \int (1 - n)Q(x)e^{\int (1-n)P(x)\, dx}\, dx + C.$$

57. $y' + 3x^2y = x^2y^3$

58. $y' + xy = xy^{-1}$

59. $y' + \left(\dfrac{1}{x}\right)y = xy^2, \quad x > 0$

60. $y' + \left(\dfrac{1}{x}\right)y = x\sqrt{y}, \quad x > 0$

61. $xy' + y = xy^3, \quad x > 0$

62. $y' - y = y^3$

63. $y' - y = e^x\sqrt[3]{y}$

64. $yy' - 2y^2 = e^x$

True or False?   In Exercises 65 and 66, determine whether the statement is true or false. If it is false, explain why or give an example that shows it is false.

65. $y' + x\sqrt{y} = x^2$ is a first-order linear differential equation.

66. $y' + xy = e^xy$ is a first-order linear differential equation.

SECTION PROJECT

Weight Loss

A person's weight depends on both the number of calories consumed and the energy used. Moreover, the amount of energy used depends on a person's weight—the average amount of energy used by a person is 17.5 calories per pound per day. So, the more weight a person loses, the less energy a person uses (assuming that the person maintains a constant level of activity). An equation that can be used to model weight loss is

$$\frac{dw}{dt} = \frac{C}{3500} - \frac{17.5}{3500}w$$

where w is the person's weight (in pounds), t is the time in days, and C is the constant daily calorie consumption.

(a) Find the general solution of the differential equation.

(b) Consider a person who weighs 180 pounds and begins a diet of 2500 calories per day. How long will it take the person to lose 10 pounds? How long will it take the person to lose 35 pounds?

(c) Use a graphing utility to graph the particular solution from part (b). What is the "limiting" weight of the person?

(d) Repeat parts (b) and (c) for a person who weighs 200 pounds when the diet is started.

■ **FOR FURTHER INFORMATION**   For more information on modeling weight loss, see the article "A Linear Diet Model" by Arthur C. Segal in *The College Mathematics Journal.*

6.6 Predator-Prey Differential Equations

■ Analyze predator-prey differential equations.
■ Analyze competing-species differential equations.

Predator-Prey Differential Equations

ALFRED LOTKA (1880–1949)

VITO VOLTERRA (1860–1940)

Although Alfred Lotka and Vito Volterra both worked on other problems, they are most known for their work on predator-prey equations. Lotka was also a statistician, and Volterra did work in the development of integral equations and functional analysis.

In the 1920s, mathematicians Alfred Lotka (1880–1949) and Vito Volterra (1860–1940) independently developed mathematical models to represent many of the different ways in which two species can interact with each other. Two common ways in which species interact with each other are as predator and prey, and as competing species.

Consider a predator-prey relationship involving foxes (predators) and rabbits (prey). Assume that the rabbits are the primary food source for the foxes, the rabbits have an unlimited food supply, and there is no threat to the rabbits other than from the foxes. Let x represent the number of rabbits, let y represent the number of foxes, and let t represent time. When there are no foxes, the rabbit population grows according to the exponential growth model

$$\frac{dx}{dt} = ax, \quad a > 0.$$

When there are foxes but no rabbits, the foxes have no food and their population decays according to the exponential decay model

$$\frac{dy}{dt} = -my, \quad m > 0.$$

When both foxes and rabbits are present, there is an interaction rate of *decline* for the rabbit population given by $-bxy$, and an interaction rate of *increase* in the fox population given by nxy, where $b, n > 0$. So, the rates of change of each population can be modeled by the following predator-prey system of differential equations.

$$\frac{dx}{dt} = ax - bxy \qquad \text{Rate of change of prey}$$

$$\frac{dy}{dt} = -my + nxy \qquad \text{Rate of change of predators}$$

These equations are called **predator-prey equations** or **Lotka-Volterra equations.** The equations are **autonomous** because the rates of change do not depend explicitly on time t.

In general, it is not possible to solve the predator-prey equations explicitly for x and y. However, you can use techniques such as Euler's Method to approximate solutions. Also, you can discover properties of the solutions by analyzing the differential equations.

EXAMPLE 1 Analyzing Predator-Prey Equations

Write the predator-prey equations for $a = 0.04$, $b = 0.002$, $m = 0.08$, and $n = 0.0004$. Then find the values of x and y for which $dx/dt = dy/dt = 0$.

Solution For $a = 0.04$, $b = 0.002$, $m = 0.08$, and $n = 0.0004$, the predator-prey equations are shown below.

$$\frac{dx}{dt} = 0.04x - 0.002xy \qquad \text{Rate of change of prey}$$

$$\frac{dy}{dt} = -0.08y + 0.0004xy \qquad \text{Rate of change of predators}$$

Solving $dx/dt = x(0.04 - 0.002y) = 0$ and $dy/dt = y(-0.08 + 0.0004x) = 0$, you can see that $dx/dt = dy/dt = 0$ when $(x, y) = (0, 0)$ and when $(x, y) = (200, 20)$. ∎

There are two points of interest you should consider when analyzing predator-prey equations. Consider the predator-prey equations

$$\frac{dx}{dt} = ax - bxy \qquad \text{and} \qquad \frac{dy}{dt} = -my + nxy.$$

$\dfrac{dx}{dt} = 0$ when $x = 0$ or $y = \dfrac{a}{b}$, and $\dfrac{dy}{dt} = 0$ when $y = 0$ or $x = \dfrac{m}{n}$. So, at the points $(0, 0)$ and $\left(\dfrac{m}{n}, \dfrac{a}{b}\right)$, the prey and predator populations are constant. These points are called **critical points** or **equilibrium points** of the predator-prey equations.

EXAMPLE 2   **Analyzing Predator-Prey Equations Graphically**

Let the predator-prey equations from Example 1

$$\frac{dx}{dt} = 0.04x - 0.002xy \qquad \text{Rate of change of rabbit population}$$

and

$$\frac{dy}{dt} = -0.08y + 0.0004xy \qquad \text{Rate of change of fox population}$$

model a predator-prey relationship involving foxes and rabbits, where x is the number of rabbits and y is the number of foxes after t months. Use a graphing utility to graph the functions x and y when $0 \le t \le 240$ and the initial conditions are 200 rabbits and 10 foxes. What do you observe?

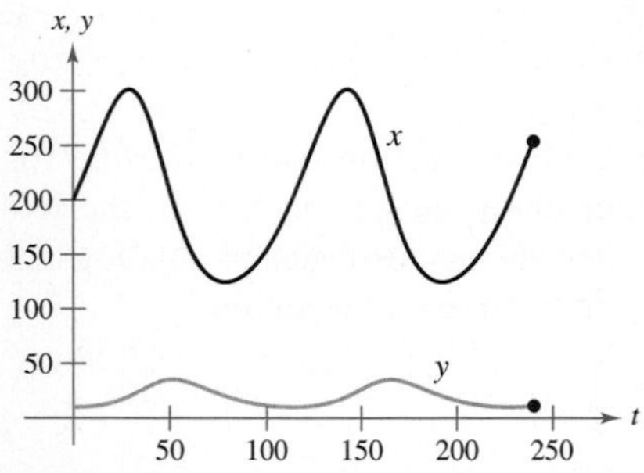

Solution   The graphs of x and y are shown in Figure 6.30. Here are some observations.

- The rabbit and fox populations oscillate periodically between their respective minimum and maximum values.
- The rabbit population oscillates from about 125 rabbits to about 300 rabbits.
- The fox population oscillates from about 10 foxes to about 35 foxes.
- About 20 months after the rabbit population peaks, the fox population peaks.
- The period of each population appears to be about 115 months.

Figure 6.30

In Example 2, the graph shows the curves plotted together with time t along the horizontal axis. You can also use the predator-prey equations dy/dt and dx/dt to graph a slope field. The slope field is graphed using the x-axis to represent the prey and the y-axis to represent the predators.

EXAMPLE 3   **Predator-Prey Equations and Slope Fields**

Use a graphing utility to graph the slope field of the predator-prey equations given in Example 2.

Solution   The slope field is shown in Figure 6.31. The x-axis represents the rabbit population, and the y-axis represents the fox population.

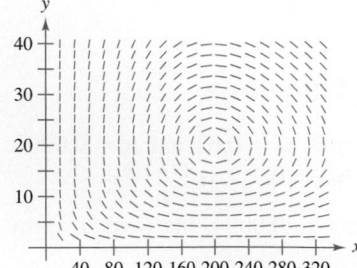

Figure 6.31

▷ **TECHNOLOGY**   If you are using a graphing utility, you may need to rewrite the equations as a function of x:

$$\frac{dy}{dx} = \frac{dy/dt}{dx/dt} = \frac{-my + nxy}{ax - bxy}.$$

EXAMPLE 4 **Graphing a Solution Curve**

Use the predator-prey equations

$$\frac{dx}{dt} = 0.04x - 0.002xy \quad \text{and} \quad \frac{dy}{dt} = -0.08y + 0.0004xy$$

and the slope field from Example 3 to graph the solution curve using the initial conditions of 200 rabbits and 10 foxes. Describe the changes in the populations as you trace the solution curve.

Solution The graph of the solution is a closed curve, as shown in Figures 6.32 and 6.33.

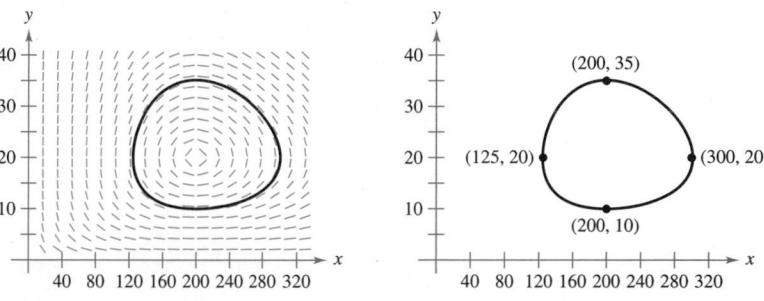

Figure 6.32   **Figure 6.33**

At $(200, 10)$, $dy/dt = 0$ and $dx/dt = 4$. So, the rabbit population is increasing at $(200, 10)$. This means that you should trace the curve counterclockwise as t increases. As you trace the curve, note the changes listed in Figure 6.34.

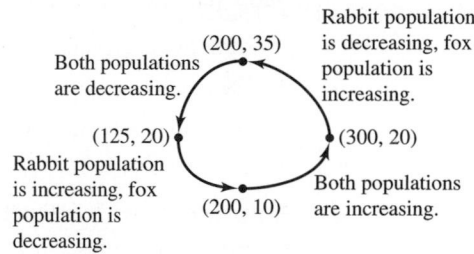

Figure 6.34

Although it is generally not possible to solve predator-prey equations explicitly for x and y, you can separate variables to derive an implicit solution. Begin by writing the equations dy/dt and dx/dt as a function of x.

$$\frac{dy}{dx} = \frac{y(-m + nx)}{x(a - by)} \qquad \text{Factor numerator and denominator.}$$

$$x(a - by)\, dy = y(-m + nx)\, dx \qquad \text{Differential form}$$

$$\frac{a - by}{y}\, dy = \frac{-m + nx}{x}\, dx \qquad \text{Separate variables.}$$

$$\int \frac{a - by}{y}\, dy = \int \frac{-m + nx}{x}\, dx \qquad \text{Integrate.}$$

$$a \ln y - by = -m \ln x + nx + C \qquad \text{Assume } x \text{ and } y \text{ are positive.}$$

$$a \ln y + m \ln x - by - nx = C \qquad \text{General solution}$$

The constant C is determined by the initial conditions.

REMARK The general solution

$$a \ln y + m \ln x - by - nx = C$$

can be rewritten as

$$\ln(y^a x^m) = C + by + nx$$

or as $y^a x^m = C_1 e^{by + nx}$.

Competing Species

Consider two species that compete with each other for the food available in their common environment. Assume that their populations are given by x and y at time t. When there is no interaction or competition between the species, the populations x and y each experience logistic growth. So, the populations of the first species x and the second species y can be modeled by the following differential equations.

$$\frac{dx}{dt} = ax - bx^2 \qquad \text{Rate of change of first species without interaction}$$

$$\frac{dy}{dt} = my - ny^2 \qquad \text{Rate of change of second species without interaction}$$

When the species interact, their competition for resources causes a rate of decline in each population proportional to the product xy. Using a negative interaction factor leads to the following **competing-species equations** (where a, b, c, m, n, and p are positive constants).

$$\frac{dx}{dt} = ax - bx^2 - cxy \qquad \text{Rate of change of first species with interaction}$$

$$\frac{dy}{dt} = my - ny^2 - pxy \qquad \text{Rate of change of second species with interaction}$$

In this text, it is assumed that competing-species equations have four critical points, as shown in Example 5.

EXAMPLE 5   Deriving the Critical Points

Show that the critical points of the competing-species equations

$$\frac{dx}{dt} = ax - bx^2 - cxy \qquad \text{and} \qquad \frac{dy}{dt} = my - ny^2 - pxy$$

are $(0, 0)$, $\left(0, \dfrac{m}{n}\right)$, $\left(\dfrac{a}{b}, 0\right)$, and $\left(\dfrac{an - mc}{bn - cp}, \dfrac{bm - ap}{bn - cp}\right)$.

Solution   Set dx/dt and dy/dt equal to 0 and then factor to obtain the following system of equations.

$$x(a - bx - cy) = 0 \qquad \text{Set } dx/dt \text{ equal to 0 and factor out } x.$$
$$y(m - ny - px) = 0 \qquad \text{Set } dy/dt \text{ equal to 0 and factor out } y.$$

If $x = 0$, then $y = 0$ or $y = m/n$. If $y = 0$, then $x = 0$ or $x = a/b$. So, three of the critical points are $(0, 0)$, $(0, m/n)$, and $(a/b, 0)$. At each of these critical points, one of the populations is 0. These points represent the possibility that both species cannot coexist.

The fourth critical point is obtained by solving the system

$$a - bx - cy = 0$$
$$m - ny - px = 0.$$

The solution of this system is

$$(x, y) = \left(\frac{an - mc}{bn - cp}, \frac{bm - ap}{bn - cp}\right).$$

Assuming this point exists and lies in Quadrant I of the xy-plane, the point represents the possibility that both species can coexist. ◼

EXAMPLE 6   Competing Species: One Species Survives

Consider the competing-species equations given by

$$\frac{dx}{dt} = 10x - x^2 - 2xy \qquad \text{and} \qquad \frac{dy}{dt} = 10y - y^2 - 2xy.$$

a. Find the critical points.

b. Use a graphing utility to graph the solution of the equations when $0 \le t \le 3$ and the initial conditions are $x(0) = 10$ and $y(0) = 15$. What do you observe?

Solution

a. Note that $a = 10$, $b = 1$, $c = 2$, $m = 10$, $n = 1$, and $p = 2$. So, the critical points are $(0, 0)$, $(0, 10)$, $(10, 0)$, and $\left(\dfrac{10 - 20}{1 - 4}, \dfrac{10 - 20}{1 - 4}\right) = \left(\dfrac{10}{3}, \dfrac{10}{3}\right)$.

b. The solution of the competing-species equations is shown in Figure 6.35. From the graph, it appears that one species survives. The population of the surviving species, represented by the graph of y, appears to remain constant at 10.

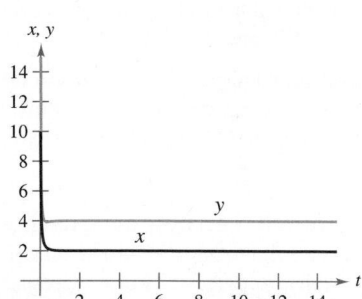

Figure 6.35

EXAMPLE 7   Competing Species: Both Species Survive

Consider the competing-species equations given by

$$\frac{dx}{dt} = 10x - 3x^2 - xy \qquad \text{and} \qquad \frac{dy}{dt} = 14y - 3y^2 - xy.$$

a. Find the critical points.

b. Use a graphing utility to graph the solution of the equations when $0 \le t \le 15$ and the initial conditions are $x(0) = 10$ and $y(0) = 15$. What do you observe?

Solution

a. Note that $a = 10$, $b = 3$, $c = 1$, $m = 14$, $n = 3$, and $p = 1$. So, the critical points are $(0, 0)$, $\left(0, \dfrac{14}{3}\right)$, $\left(\dfrac{10}{3}, 0\right)$, and $\left(\dfrac{30 - 14}{9 - 1}, \dfrac{42 - 10}{9 - 1}\right) = (2, 4)$.

b. The solution of the competing-species equations is shown in Figure 6.36. From the graph, it appears that both species survive. The population represented by y appears to remain constant at 4. The population represented by x appears to remain constant at 2.

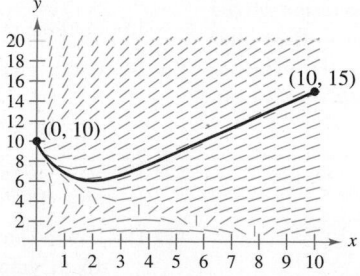 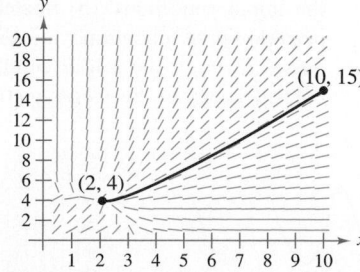

Figure 6.36

Examples 6 and 7 imply general conclusions about competing-species equations that have precisely four critical points. In general, it can be shown that when $bn > cp$, both species survive. When $bn < cp$, one species will survive and the other will not.

You can also use slope fields to analyze solutions of competing-species equations, as shown in Figures 6.37 (Example 6) and 6.38 (Example 7).

Figure 6.37   **Figure 6.38**

6.6 Exercises

See **CalcChat.com** for tutorial help and worked-out solutions to odd-numbered exercises.

CONCEPT CHECK

1. **Autonomous Differential Equation** What is an autonomous differential equation?

2. **Competing Species Equations** Which terms of the competing-species equations represent interaction between species? Explain.

Analyzing Predator-Prey Equations In Exercises 3 and 4, use the given values to write the predator-prey equations $dx/dt = ax - bxy$ and $dy/dt = -my + nxy$. Then find the values of x and y for which $dx/dt = dy/dt = 0$.

3. $a = 0.9$, $b = 0.05$, $m = 0.6$, $n = 0.008$

4. $a = 1.2$, $b = 0.04$, $m = 1.2$, $n = 0.02$

Predator-Prey Equations and Slope Fields In Exercises 5 and 6, predator-prey equations, a point, and a slope field are given. (a) Sketch a solution of the predator-prey equations on the slope field that passes through the given point. (b) Use a graphing utility to graph the solution. Compare the result with the sketch in part (a). To print an enlarged copy of the graph, go to *MathGraphs.com*.

5. $\dfrac{dx}{dt} = 0.04x - 0.002xy$

   $\dfrac{dy}{dt} = -0.08y + 0.0004xy$

6. $\dfrac{dx}{dt} = 0.03x - 0.006xy$

   $\dfrac{dy}{dt} = -0.04y + 0.004xy$

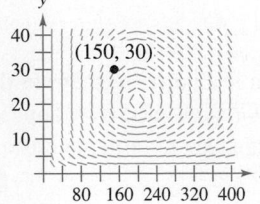

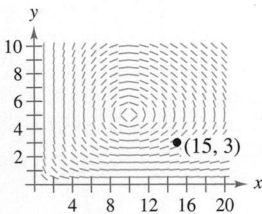

Predator-Prey Equations and Slope Fields In Exercises 7 and 8, two graphs are given. The first is a graph of the functions x and y of a set of predator-prey equations, where x is the number of prey and y is the number of predators at time t. The second graph is the corresponding slope field of the predator-prey equations. (a) Identify the initial conditions. (b) Sketch a solution of the predator-prey equations on the slope field that passes through the initial conditions. To print an enlarged copy of the graph, go to *MathGraphs.com*.

7.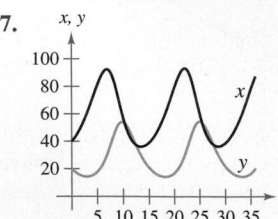

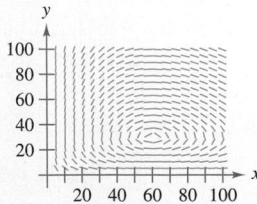

8.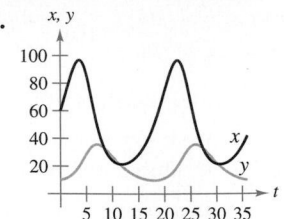

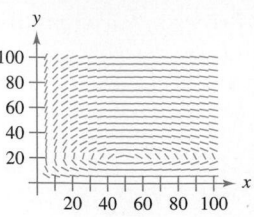

Rabbits and Foxes In Exercises 9–12, consider a predator-prey relationship involving foxes (predators) and rabbits (prey). Let x represent the number of rabbits, let y represent the number of foxes, and let t represent the time in months. Assume that the following predator-prey equations model the rates of change of each population.

$dx/dt = 0.8x - 0.04xy$ Rate of change of prey population

$dy/dt = -0.3y + 0.006xy$ Rate of change of predator population

When $t = 0$, $x = 55$ and $y = 10$.

9. Find the critical points of the predator-prey equations.

 10. Use a graphing utility to graph the functions x and y when $0 \le t \le 36$. Describe the behavior of each solution as t increases.

11. Use a graphing utility to graph a slope field of the predator-prey equations when $0 \le x \le 150$ and $0 \le y \le 50$.

12. Use the predator-prey equations and the slope field in Exercise 11 to graph the solution curve using the initial conditions. Describe the changes in the rabbit and fox populations as you trace the solution curve.

Prairie Dogs and Black-Footed Ferrets In Exercises 13–16, consider a predator-prey relationship involving black-footed ferrets (predators) and prairie dogs (prey). Let x represent the number of prairie dogs, let y represent the number of black-footed ferrets, and let t represent the time in months. Assume that the following predator-prey equations model the rates of change of each population.

$\dfrac{dx}{dt} = 0.1x - 0.00008xy$ Rate of change of prey population

$\dfrac{dy}{dt} = -0.4y + 0.00004xy$ Rate of change of predator population

When $t = 0$, $x = 4000$ and $y = 1000$.

13. Find the critical points of the predator-prey equations.

 14. Use a graphing utility to graph the functions x and y when $0 \le t \le 240$. Describe the behavior of each solution as t increases.

 15. Use a graphing utility to graph a slope field of the predator-prey equations when $0 \le x \le 25{,}000$ and $0 \le y \le 5000$.

16. Use the predator-prey equations and the slope field in Exercise 15 to graph the solution curve using the initial conditions. Describe the changes in the prairie dog and black-footed ferret populations as you trace the solution curve.

17. Critical Point as the Initial Condition In Exercise 9, you found the critical points of the predator-prey system. Assume that the critical point given by $(m/n, a/b)$ is the initial condition and repeat Exercises 10–12. Compare the results.

18. Critical Point as the Initial Condition In Exercise 13, you found the critical points of the predator-prey system. Assume that the critical point given by $(m/n, a/b)$ is the initial condition and repeat Exercises 14–16. Compare the results.

Analyzing Competing-Species Equations In Exercises 19–22, use the given values to write the competing-species equations $dx/dt = ax - bx^2 - cxy$ and $dy/dt = my - ny^2 - pxy$. Then find the values of x and y for which $dx/dt = dy/dt = 0$.

19. $a = 2, b = 3, c = 2, m = 2, n = 3, p = 2$

20. $a = 1, b = 0.5, c = 0.5, m = 2.5, n = 2, p = 0.5$

21. $a = 0.15, b = 0.6, c = 0.75, m = 0.15, n = 1.2, p = 0.45$

22. $a = 0.025, b = 0.1, c = 0.2, m = 0.3, n = 0.45, p = 0.1$

Bass and Trout In Exercises 23 and 24, consider a competing-species relationship involving bass and trout. Assume the bass and trout compete for the same resources. Let x represent the number of bass (in thousands), let y represent the number of trout (in thousands), and let t represent the time in months. Assume that the following competing-species equations model the rates of change of the two populations.

$$\frac{dx}{dt} = 0.8x - 0.4x^2 - 0.1xy \qquad \text{Rate of change of bass population}$$

$$\frac{dy}{dt} = 0.3y - 0.6y^2 - 0.1xy \qquad \text{Rate of change of trout population}$$

When $t = 0$, $x = 9$ and $y = 5$.

23. Find the critical points of the competing-species equations.

24. Use a graphing utility to graph the functions x and y when $0 \le t \le 36$. Describe the behavior of each solution as t increases.

Bass and Trout In Exercises 25 and 26, consider a competing-species relationship involving bass and trout. Assume the bass and trout compete for the same resources. Let x represent the number of bass (in thousands), let y represent the number of trout (in thousands), and let t represent the time in months. Assume that the following competing-species equations model the rates of change of the two populations.

$$\frac{dx}{dt} = 0.8x - 0.4x^2 - xy \qquad \text{Rate of change of bass population}$$

$$\frac{dy}{dt} = 0.3y - 0.6y^2 - xy \qquad \text{Rate of change of trout population}$$

When $t = 0$, $x = 7$ and $y = 6$.

25. Find the critical points of the competing-species equations.

26. Use a graphing utility to graph the functions x and y when $0 \le t \le 36$. Describe the behavior of each solution as t increases.

27. Critical Point as the Initial Condition In Exercise 23, you found the critical points of the competing-species system. Assume that the critical point given by $\left(\dfrac{an - mc}{bn - cp}, \dfrac{bm - ap}{bn - cp}\right)$ is the initial condition and repeat Exercise 24. Compare the results.

28. Critical Point as the Initial Condition In Exercise 25, you found the critical points of the competing-species system. Assume that the critical point given by $(0, m/n)$ is the initial condition and repeat Exercise 26. Compare the results.

EXPLORING CONCEPTS

29. Determining Initial Values Given a set of predator-prey equations, describe how to determine initial values so that both populations remain constant for all $t \ge 0$.

30. HOW DO YOU SEE IT? The populations of two species x and y are shown in the figure. Sketch the graph of the solution curve by hand for $0 \le t \le 20$.

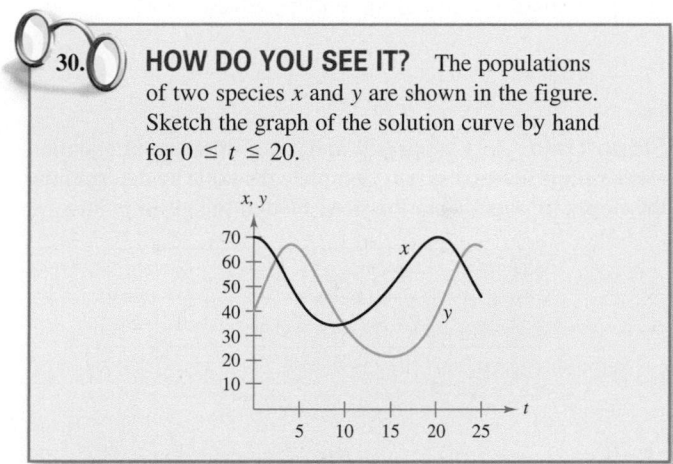

31. Revising the Predator-Prey Equations Consider a predator-prey relationship with x prey and y predators at time t. Assume both predator and prey are present. Then the rates of change of the two populations can be modeled by the following revised predator-prey system of differential equations.

$$\frac{dx}{dt} = ax\left(1 - \frac{x}{L}\right) - bxy \qquad \text{Rate of change of prey population}$$

$$\frac{dy}{dt} = -my + nxy \qquad \text{Rate of change of predator population}$$

(a) When there are no predators, the prey population will grow according to what model?

(b) Write the revised predator-prey equations for $a = 0.4$, $L = 100$, $b = 0.01$, $m = 0.3$, and $n = 0.005$. Find the critical numbers.

(c) Use a graphing utility to graph the functions x and y of the revised predator-prey equations when $0 \le t \le 72$ and the initial conditions are $x(0) = 40$ and $y(0) = 80$. Describe the behavior of each solution as t increases.

(d) Use a graphing utility to graph a slope field of the revised predator-prey equations when $0 \le x \le 100$ and $0 \le y \le 80$.

(e) Use the predator-prey equations and the slope field in part (d) to graph the solution curve using the initial conditions in part (c). Describe the changes in the prey and predator populations as you trace the solution curve.

Review Exercises

See **CalcChat.com** for tutorial help and worked-out solutions to odd-numbered exercises.

1. **Determining a Solution** Determine whether the function $y = x^3$ is a solution of the differential equation $2xy' + 4y = 10x^3$.

2. **Determining a Solution** Determine whether the function $y = 2 \sin 2x$ is a solution of the differential equation $y''' - 8y = 0$.

Finding a General Solution In Exercises 3–8, use integration to find a general solution of the differential equation.

3. $\dfrac{dy}{dx} = 4x^2 + 7$

4. $\dfrac{dy}{dx} = \dfrac{6 - x}{3x}, \quad x > 0$

5. $\dfrac{dy}{dx} = \cos 2x$

6. $\dfrac{dy}{dx} = 8 \csc x \cot x$

7. $\dfrac{dy}{dx} = e^{2-x}$

8. $\dfrac{dy}{dx} = 2e^{3x}$

Slope Field In Exercises 9 and 10, a differential equation and its slope field are given. Complete the table by determining the slopes (if possible) in the slope field at the given points.

x	-4	-2	0	2	4	8
y	2	0	4	4	6	8
dy/dx						

9. $\dfrac{dy}{dx} = 2x - y$

10. $\dfrac{dy}{dx} = x \sin \dfrac{\pi y}{4}$

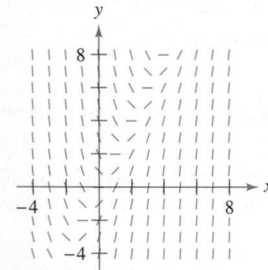 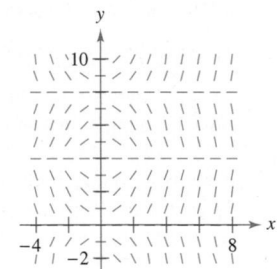

Slope Field In Exercises 11 and 12, (a) sketch the slope field for the differential equation, and (b) use the slope field to sketch the solution that passes through the given point. Use a graphing utility to verify your results. To print a blank coordinate plane, go to *MathGraphs.com*.

11. $y' = 2x^2 - x, \quad (0, 2)$

12. $y' = y + 4x, \quad (-1, 1)$

Euler's Method In Exercises 13 and 14, use Euler's Method to make a table of values for the approximate solution of the differential equation with the specified initial value. Use n steps of size h.

13. $y' = x - y, \quad y(0) = 4, \quad n = 10, \quad h = 0.05$

14. $y' = 5x - 2y, \quad y(0) = 2, \quad n = 10, \quad h = 0.1$

Solving a Differential Equation In Exercises 15–22, find the general solution of the differential equation.

15. $\dfrac{dy}{dx} = 6x - x^3$

16. $\dfrac{dy}{dx} = 3y + 5$

17. $\dfrac{dy}{dx} = (y - 1)^2$

18. $\dfrac{dy}{dx} = \dfrac{x}{x^2 + 2}$

19. $(2 + x)y' - xy = 0$

20. $xy' - (x + 1)y = 0$

21. $\sqrt{x + 1}\, y' - y = 0$

22. $y' + \sqrt{x}\, y = 9\sqrt{x}$

Writing and Solving a Differential Equation In Exercises 23 and 24, write and find the general solution of the differential equation that models the verbal statement.

23. The rate of change of y with respect to t is inversely proportional to the cube of t.

24. The rate of change of y with respect to t is proportional to $50 - t$.

Finding an Exponential Function In Exercises 25–28, find the exponential function $y = Ce^{kt}$ that passes through the two given points.

25.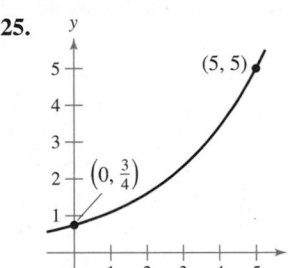

26.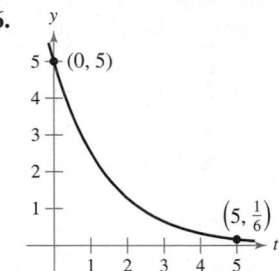

27.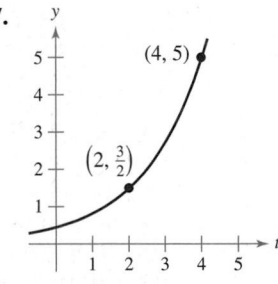

28.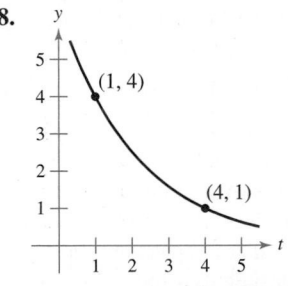

29. **Air Pressure** Under ideal conditions, air pressure decreases continuously with the height above sea level at a rate proportional to the pressure at that height. The barometer reads 30 inches at sea level and 15 inches at 18,000 feet. Find the barometric pressure at 35,000 feet.

30. **Radioactive Decay** Radioactive radium has a half-life of approximately 1599 years. The initial quantity is 15 grams. How much remains after 750 years?

31. **Population** A population grows exponentially at a rate of 1.85%. How long will it take the population to double?

32. **Compound Interest** Find the balance in an account when $400 is deposited for 11 years at an interest rate of 2% compounded continuously.

33. Sales The sales S (in thousands of units) of a new product after it has been on the market for t years is given by $S = Ce^{k/t}$. (a) Find S as a function of t when 5000 units have been sold after 1 year and the saturation point for the market is 30,000 units (that is, $\lim_{t \to \infty} S = 30$). (b) How many units will have been sold after 5 years?

34. Sales The sales S (in thousands of units) of a new product after it has been on the market for t years is given by $S = 25(1 - e^{kt})$. (a) Find S as a function of t when 4000 units have been sold after 1 year. (b) How many units will saturate this market?

Finding a General Solution Using Separation of Variables In Exercises 35–38, find the general solution of the differential equation.

35. $\dfrac{dy}{dx} = \dfrac{5x}{y}$

36. $\dfrac{dy}{dx} = \dfrac{x^3}{2y^2}$

37. $y'e^{y-3x} = e^{x+2y}$

38. $y' - e^y \sin x = 0$

Finding a Particular Solution Using Separation of Variables In Exercises 39–42, find the particular solution of the differential equation that satisfies the initial condition.

Differential Equation	Initial Condition
39. $y^3 y' - 3x = 0$	$y(2) = 2$
40. $yy' - 5e^{2x} = 0$	$y(0) = -3$
41. $y^3(x^4 + 1)y' - x^3(y^4 + 1) = 0$	$y(0) = 1$
42. $y' + \sin x \cos x = 0$	$y(\pi) = -2$

Slope Field In Exercises 43 and 44, sketch a few solutions of the differential equation on the slope field and then find the general solution analytically. To print an enlarged copy of the graph, go to *MathGraphs.com*.

43. $\dfrac{dy}{dx} = -\dfrac{4x}{y}$

44. $\dfrac{dy}{dx} = 3 - 2y$

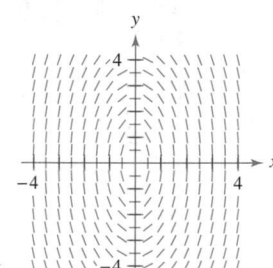

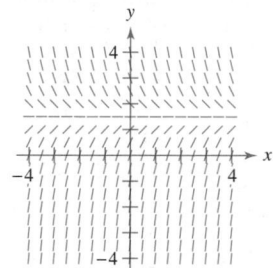

Finding a Particular Solution Curve In Exercises 45 and 46, find an equation of the curve that passes through the point and has the given slope.

45. $(1, 3)$, $y' = \dfrac{2x}{y}$

46. $(1, -2)$, $y' = \dfrac{y}{8x}$

Finding Orthogonal Trajectories In Exercises 47 and 48, find the orthogonal trajectories for the family of curves. Use a graphing utility to graph several members of each family.

47. $5x^2 - 4y^2 = C$

48. $x^3 = Cy$

Using a Logistic Equation In Exercises 49 and 50, the logistic equation models the growth of a population. Use the equation to (a) find the value of k, (b) find the carrying capacity, (c) find the initial population, (d) determine when the population will reach 50% of its carrying capacity, and (e) write a logistic differential equation that has the solution $P(t)$.

49. $P(t) = \dfrac{5250}{1 + 34e^{-0.55t}}$

50. $P(t) = \dfrac{4800}{1 + 14e^{-0.15t}}$

Solving a Logistic Differential Equation In Exercises 51 and 52, find the logistic equation that passes through the given point.

51. $\dfrac{dy}{dt} = y\left(1 - \dfrac{y}{80}\right)$, $(0, 8)$

52. $\dfrac{dy}{dt} = 1.76y\left(1 - \dfrac{y}{8}\right)$, $(0, 3)$

53. Wildlife Population The rate of change of the number of raccoons $N(t)$ in a population is directly proportional to $380 - N(t)$, where t is the time in years. When $t = 0$, the population is 110, and when $t = 4$, the population has increased to 150. Find the population when $t = 8$.

54. Environment A conservation department releases 1200 brook trout into a lake. It is estimated that the carrying capacity of the lake for the species is 20,400. After the first year, there are 2000 brook trout in the lake.

(a) Write a logistic equation that models the number of brook trout in the lake.

(b) Find the number of brook trout in the lake after 8 years.

(c) When will the number of brook trout reach 10,000?

(d) Write a logistic differential equation that models the growth rate of the brook trout population. Then repeat part (b) using Euler's method with a step size of $h = 1$. Compare the approximation with the exact answer.

(e) At what time is the brook trout population growing most rapidly? Explain.

55. Sales Growth The rate of change in sales S (in thousands of units) of a new product is proportional to $L - S$. L (in thousands of units) is the estimated maximum level of sales, and $S = 0$ when $t = 0$. Write and solve the differential equation for this sales model.

56. Sales Growth Use the result of Exercise 55 to write S as a function of t for (a) $L = 100$, $S = 25$ when $t = 2$, and (b) $L = 500$, $S = 50$ when $t = 1$.

Learning Theory In Exercises 57 and 58, assume that the rate of change in the proportion P of correct responses after n trials is proportional to the product of P and $L - P$, where L is the limiting proportion of correct responses.

57. Write and solve the differential equation for this learning theory model.

58. Use the solution of Exercise 57 to write P as a function of n, and then use a graphing utility to graph the solution.

(a) $L = 1.00$

(b) $L = 0.80$

$P = 0.50$ when $n = 0$ $P = 0.25$ when $n = 0$

$P = 0.85$ when $n = 4$ $P = 0.60$ when $n = 10$

Solving a First-Order Linear Differential Equation In Exercises 59–66, find the general solution of the first-order linear differential equation.

59. $y' - y = 10$

60. $e^x y' + 4e^x y = 1$

61. $4y' = e^{x/4} + y$

62. $\dfrac{dy}{dx} - \dfrac{5y}{x^2} = \dfrac{1}{x^2}$, $x > 0$

63. $(x - 2)y' + y = 1$, $x > 2$

64. $(x + 3)y' + 2y = 2(x + 3)^2$, $x > -3$

65. $y' + 5y = e^{5x}$

66. $xy' - ay = bx^4$

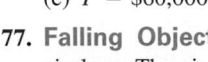 **Slope Field** In Exercises 67–70, (a) sketch an approximate solution of the differential equation satisfying the initial condition by hand on the slope field, (b) find the particular solution that satisfies the initial condition, and (c) use a graphing utility to graph the particular solution. Compare the graph with the hand-drawn graph in part (a). To print an enlarged copy of the graph, go to *MathGraphs.com*.

Differential Equation	Initial Condition
67. $\dfrac{dy}{dx} = e^{x/2} - y$	$(0, -1)$
68. $y' + 2y = \sin x$	$(0, 4)$
69. $y' = \csc x + y \cot x$	$(1, 1)$
70. $y' = \csc x - y \cot x$	$(1, 2)$

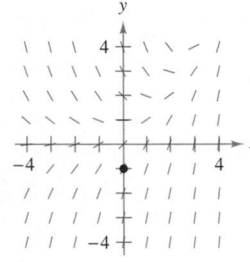

Figure for 67

Figure for 68

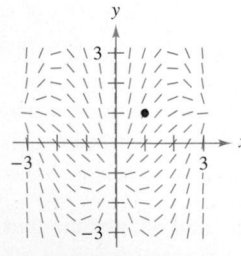

Figure for 69

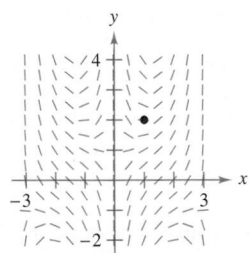

Figure for 70

Finding a Particular Solution In Exercises 71–74, find the particular solution of the first-order linear differential equation that satisfies the initial condition.

Differential Equation	Initial Condition
71. $y' + 5y = e^{5x}$	$y(0) = 3$
72. $y' - \left(\dfrac{3}{x}\right)y = 2x^3$	$y(1) = 1$
73. $(3y + 5)\cos x \, dx = dy$	$y(\pi) = 0$
74. $y' - 8x^3 y = e^{2x^4}$	$y(0) = 2$

75. Investment Let $A(t)$ be the amount in a fund earning interest at an annual rate r compounded continuously. When a continuous cash flow of P dollars per year is withdrawn from the fund, the rate of change of A is given by the differential equation

$$\frac{dA}{dt} = rA - P$$

where $A = A_0$ when $t = 0$. Solve this differential equation for A as a function of t.

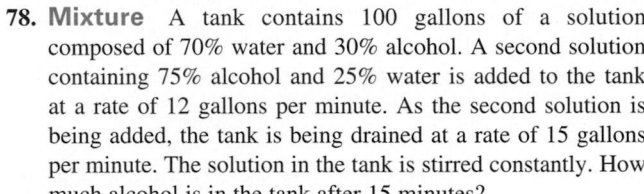 **76. Investment** A retired couple plans to withdraw P dollars per year from a retirement account of \$500,000 earning 10% interest compounded continuously. Use the result of Exercise 75 and a graphing utility to graph the function A for each of the following continuous annual cash flows. Use the graphs to describe what happens to the balance in the fund for each case.

(a) $P = \$40,000$

(b) $P = \$50,000$

(c) $P = \$60,000$

77. Falling Object A 12-pound object is dropped from an airplane. The air resistance is proportional to the velocity of the object. Write the velocity of the object as a function of time t when the velocity after 6 seconds is approximately 114 feet per second. What is the limiting value of the velocity function?

78. Mixture A tank contains 100 gallons of a solution composed of 70% water and 30% alcohol. A second solution containing 75% alcohol and 25% water is added to the tank at a rate of 12 gallons per minute. As the second solution is being added, the tank is being drained at a rate of 15 gallons per minute. The solution in the tank is stirred constantly. How much alcohol is in the tank after 15 minutes?

 Analyzing Predator-Prey Equations In Exercises 79 and 80, (a) use the given values to write a set of predator-prey equations, (b) find the values of x and y for which $x' = y' = 0$, and (c) use a graphing utility to graph the solutions x and y of the predator-prey equations for the given time frame. Describe the behavior of each solution as t increases.

79. Constants: $a = 0.3$, $b = 0.02$, $m = 0.4$, $n = 0.01$
Initial condition: $(20, 20)$
Time frame: $0 \le t \le 36$

80. Constants: $a = 0.4$, $b = 0.04$, $m = 0.6$, $n = 0.02$
Initial condition: $(30, 15)$
Time frame: $0 \le t \le 24$

Analyzing Competing-Species Equations In Exercises 81 and 82, (a) use the given values to write a set of competing-species equations, (b) find the values of x and y for which $x' = y' = 0$, and (c) use a graphing utility to graph the solutions x and y of the competing-species equations for the given time frame. Describe the behavior of each solution as t increases.

81. Constants: $a = 3$, $b = 1$, $c = 1$, $m = 2$, $n = 1$, $p = 0.5$
Initial condition: $(3, 2)$
Time frame: $0 \le t \le 6$

82. Constants: $a = 15$, $b = 2$, $c = 4$, $m = 17$, $n = 2$, $p = 4$
Initial condition: $(9, 10)$
Time frame: $0 \le t \le 4$

P.S. Problem Solving

See **CalcChat.com** for tutorial help and worked-out solutions to odd-numbered exercises.

1. Doomsday Equation The differential equation

$$\frac{dy}{dt} = ky^{1+\varepsilon}$$

where k and ε are positive constants, is called the **doomsday equation.**

(a) Solve the doomsday equation

$$\frac{dy}{dt} = y^{1.01}$$

given that $y(0) = 1$. Find the time T at which

$$\lim_{t \to T^-} y(t) = \infty.$$

(b) Solve the doomsday equation

$$\frac{dy}{dt} = ky^{1+\varepsilon}$$

given that $y(0) = y_0$. Explain why this equation is called the doomsday equation.

2. Sales Let S represent sales of a new product (in thousands of units), let L represent the maximum level of sales (in thousands of units), and let t represent time (in months). The rate of change of S with respect to t is proportional to the product of S and $L - S$.

(a) Write the differential equation for the sales model using these conditions.

When $t = 0$: $L = 100$, $S = 10$

When $t = 1$: $S = 20$

Verify that

$$S = \frac{L}{1 + Ce^{-kt}}.$$

(b) At what time is the growth in sales increasing most rapidly?

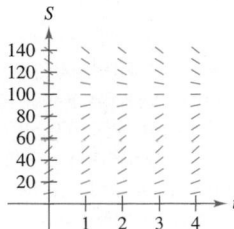 (c) Use a graphing utility to graph the sales function.

(d) Sketch the solution from part (a) on the slope field below. To print an enlarged copy of the graph, go to *MathGraphs.com*.

(e) Assume the estimated maximum level of sales is correct. Use the slope field to describe the shape of the solution curves for sales when, at some period of time, sales exceed L.

3. Modified Euler's Method Another numerical approach to approximating the particular solution of the differential equation $y' = F(x, y)$ is shown below.

$$x_n = x_{n-1} + h$$

$$y_n = y_{n-1} + hf\left(x_{n-1} + \frac{h}{2}, y_{n-1} + \frac{h}{2}f(x_{n-1}, y_{n-1})\right)$$

This approach is called **modified Euler's Method.**

(a) Use this method to approximate the solution of the differential equation $y' = x - y$ passing through the point $(0, 1)$. Use a step size of $h = 0.1$.

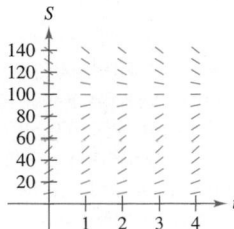 (b) Use a graphing utility to graph the exact solution and the approximations found using Euler's Method and modified Euler's Method (see Example 6, page 392). Compare the first 10 approximations found using modified Euler's Method to those found using Euler's Method and to the exact solution $y = x - 1 + 2e^{-x}$. Which approximation appears to be more accurate?

4. Error Using the Product Rule Although it is true for some functions f and g, a common mistake in calculus is to believe that the Product Rule for derivatives is $(fg)' = f'g'$.

(a) Given $g(x) = x$, find f such that $(fg)' = f'g'$.

(b) Given an arbitrary function g, find a function f such that $(fg)' = f'g'$.

(c) Describe what happens when $g(x) = e^x$.

5. Torricelli's Law **Torricelli's Law** states that water will flow from an opening at the bottom of a tank with the same speed that it would attain falling from the surface of the water to the opening. One of the forms of Torricelli's Law is

$$A(h)\frac{dh}{dt} = -k\sqrt{2gh}$$

where h is the height of the water in the tank, k is the area of the opening at the bottom of the tank, $A(h)$ is the horizontal cross-sectional area at height h, and g is the acceleration due to gravity ($g \approx 32$ feet per second per second). A hemispherical water tank has a radius of 6 feet. When the tank is full, a circular valve with a radius of 1 inch is opened at the bottom, as shown in the figure. How long will it take for the tank to drain completely?

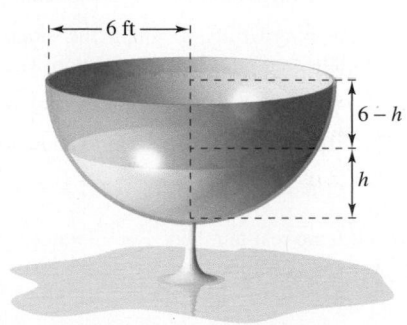

6. Torricelli's Law The cylindrical water tank shown in the figure has a height of 18 feet. When the tank is full, a circular valve is opened at the bottom of the tank. After 30 minutes, the depth of the water is 12 feet.

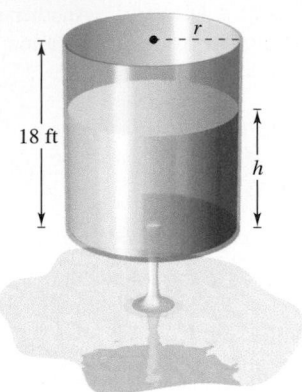

(a) Using Torricelli's Law, how long will it take for the tank to drain completely?

(b) What is the depth of the water in the tank after 1 hour?

7. Torricelli's Law A tank similar to the one in Exercise 6 has a height of 20 feet and a radius of 8 feet, and the valve is circular with a radius of 2 inches. The tank is full when the valve is opened. How long will it take for the tank to drain completely?

8. Rewriting the Logistic Equation Show that the logistic equation

$$y = \frac{L}{1 + be^{-kt}}$$

can be written as

$$y = \frac{1}{2}L\left[1 + \tanh\left(\frac{1}{2}k\left(t - \frac{\ln b}{k}\right)\right)\right].$$

What can you conclude about the graph of the logistic equation?

9. Biomass Biomass is a measure of the amount of living matter in an ecosystem. The biomass $s(t)$ in a given ecosystem increases at a rate of about 3.5 tons per year and decreases by about 1.9% per year. This situation can be modeled by the differential equation

$$\frac{ds}{dt} = 3.5 - 0.019s.$$

(a) Find the general solution of the differential equation.

(b) Use a graphing utility to graph the slope field for the differential equation. What do you notice?

(c) Explain what happens to the biomass as $t \to \infty$.

10. Finding a Function Consider a function f such that

$$f(0) = 1, \quad f'(0) = 1, \quad \text{and} \quad f(a + b) = f(a)f(b)$$

where a and b are real numbers. For all values of x, show that $f'(x) = f(x)$ and conclude that $f(x) = e^x$.

Medical Science In Exercises 11–13, a medical researcher wants to determine the concentration C (in moles per liter) of a tracer drug injected into a moving fluid. Solve this problem by considering a single-compartment dilution model (see figure). Assume that the fluid is continuously mixed and that the volume of the fluid in the compartment is constant.

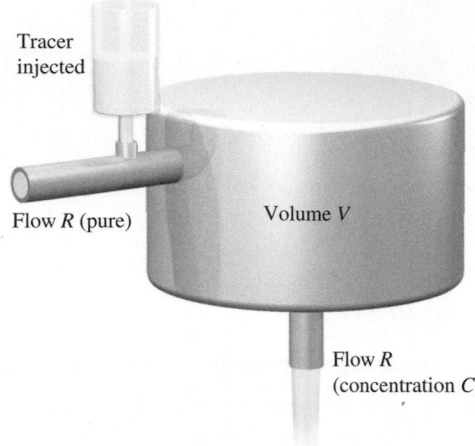

Tracer injected

Flow R (pure)

Volume V

Flow R (concentration C)

11. If the tracer is injected instantaneously at time $t = 0$, then the concentration of the fluid in the compartment begins diluting according to the differential equation

$$\frac{dC}{dt} = \left(-\frac{R}{V}\right)C$$

where $C = C_0$ when $t = 0$.

(a) Solve this differential equation to find the concentration C as a function of time t.

(b) Find the limit of C as $t \to \infty$.

12. Use the solution of the differential equation in Exercise 11 and the given values to find the concentration C as a function of time t, and use a graphing utility to graph the function.

(a) $V = 2$ liters

$R = 0.5$ liter per minute

$C_0 = 0.6$ mole per liter

(b) $V = 2$ liters

$R = 1.5$ liters per minute

$C_0 = 0.6$ mole per liter

13. In Exercises 11 and 12, it was assumed that there was a single initial injection of the tracer drug into the compartment. Now consider the case in which the tracer is continuously injected (beginning at $t = 0$) at the rate of Q moles per minute. Considering Q to be negligible compared with R, use the differential equation

$$\frac{dC}{dt} = \frac{Q}{V} - \left(\frac{R}{V}\right)C$$

where $C = 0$ when $t = 0$.

(a) Solve this differential equation to find the concentration C as a function of time t.

(b) Find the limit of C as $t \to \infty$.

7 Applications of Integration

Moving a Space Module into Orbit *(Example 3, p. 488)*

Pyramid of Khufu
(Section Project, p. 493)

Saturn *(Section Project, p. 473)*

Building Design *(Exercise 79, p. 453)*

3D Printing
(Exercise 68, p. 463)

443

7.1   Area of a Region Between Two Curves

■ Find the area of a region between two curves using integration.
■ Find the area of a region between intersecting curves using integration.
■ Describe integration as an accumulation process.

Area of a Region Between Two Curves

With a few modifications, you can extend the application of definite integrals from the area of a region *under* a curve to the area of a region *between* two curves. Consider two functions f and g that are continuous on the interval $[a, b]$. Also, the graphs of both f and g lie above the x-axis, and the graph of g lies below the graph of f, as shown in Figure 7.1. You can geometrically interpret the area of the region between the graphs as the area of the region under the graph of g subtracted from the area of the region under the graph of f, as shown in Figure 7.2.

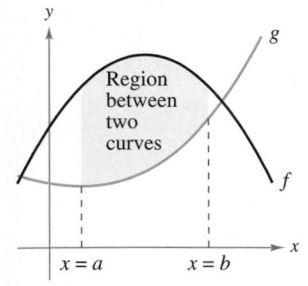

Figure 7.1

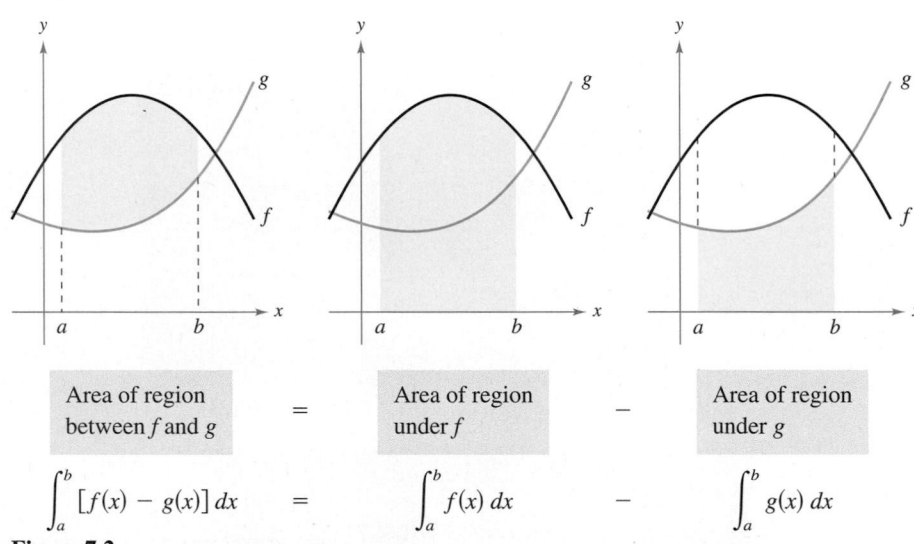

Area of region between f and g	=	Area of region under f	−	Area of region under g
$\displaystyle\int_a^b [f(x) - g(x)]\, dx$	=	$\displaystyle\int_a^b f(x)\, dx$	−	$\displaystyle\int_a^b g(x)\, dx$

Figure 7.2

To verify the reasonableness of the result shown in Figure 7.2, you can partition the interval $[a, b]$ into n subintervals, each of width Δx. Then, as shown in Figure 7.3, sketch a **representative rectangle** of width Δx and height $f(x_i) - g(x_i)$, where x_i is in the ith subinterval. The area of this representative rectangle is

$$\Delta A_i = (\text{height})(\text{width}) = [f(x_i) - g(x_i)]\Delta x.$$

By adding the areas of the n rectangles and taking the limit as $\|\Delta\| \to 0$ $(n \to \infty)$, you obtain

$$\lim_{n \to \infty} \sum_{i=1}^{n} [f(x_i) - g(x_i)]\Delta x.$$

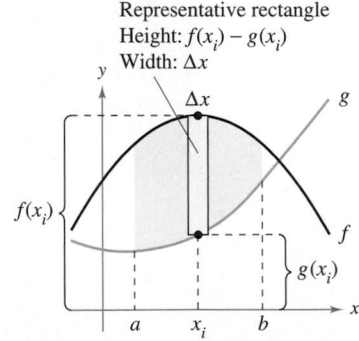

Representative rectangle
Height: $f(x_i) - g(x_i)$
Width: Δx

Figure 7.3

> **REMARK**   Recall from Section 5.3 that $\|\Delta\|$ is the norm of the partition. In a regular partition, the statements $\|\Delta\| \to 0$ and $n \to \infty$ are equivalent.

Because f and g are continuous on $[a, b]$, $f - g$ is also continuous on $[a, b]$ and the limit exists. So, the area of the region is

$$\text{Area} = \lim_{n \to \infty} \sum_{i=1}^{n} [f(x_i) - g(x_i)]\Delta x$$

$$= \int_a^b [f(x) - g(x)]\, dx.$$

Area of a Region Between Two Curves

If f and g are continuous on $[a, b]$ and $g(x) \le f(x)$ for all x in $[a, b]$, then the area of the region bounded by the graphs of f and g and the vertical lines $x = a$ and $x = b$ is

$$A = \int_a^b [f(x) - g(x)]\, dx.$$

In Figure 7.1, the graphs of f and g are shown above the x-axis. This, however, is not necessary. The same integrand $[f(x) - g(x)]$ can be used as long as f and g are continuous and $g(x) \le f(x)$ for all x in the interval $[a, b]$. This is summarized graphically in Figure 7.4. Notice in Figure 7.4 that the height of a representative rectangle is $f(x) - g(x)$ regardless of the relative position of the x-axis.

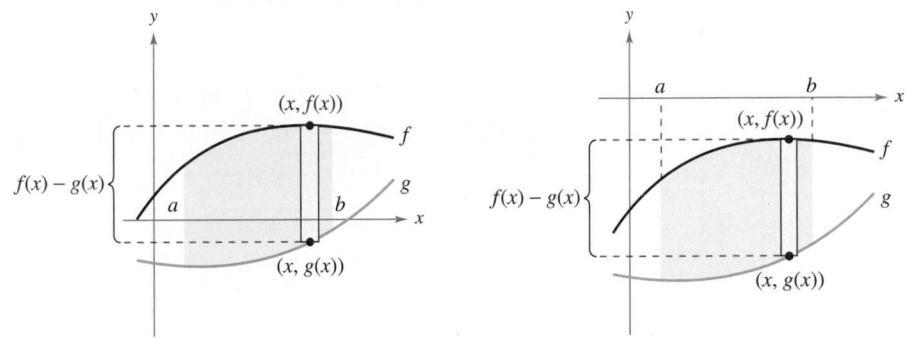

Figure 7.4

Representative rectangles are used throughout this chapter in various applications of integration. A vertical rectangle (of width Δx) implies integration with respect to x, whereas a horizontal rectangle (of width Δy) implies integration with respect to y.

EXAMPLE 1   Finding the Area of a Region Between Two Curves

Find the area of the region bounded by the graphs of $y = x^2 + 2$, $y = -x$, $x = 0$, and $x = 1$.

Solution   Let $g(x) = -x$ and $f(x) = x^2 + 2$. Then $g(x) \le f(x)$ for all x in $[0, 1]$, as shown in Figure 7.5. So, the area of the representative rectangle is

$$\Delta A = [f(x) - g(x)]\Delta x$$
$$= [(x^2 + 2) - (-x)]\Delta x$$

and the area of the region is

$$A = \int_a^b [f(x) - g(x)]\, dx$$
$$= \int_0^1 [(x^2 + 2) - (-x)]\, dx$$
$$= \left[\frac{x^3}{3} + \frac{x^2}{2} + 2x \right]_0^1$$
$$= \frac{1}{3} + \frac{1}{2} + 2$$
$$= \frac{17}{6}.$$

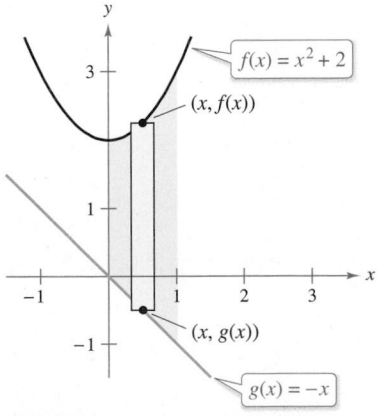

Region bounded by the graph of f, the graph of g, $x = 0$, and $x = 1$
Figure 7.5

Area of a Region Between Intersecting Curves

In Example 1, the graphs of $f(x) = x^2 + 2$ and $g(x) = -x$ do not intersect, and the values of a and b are given explicitly. A more common problem involves the area of a region bounded by two *intersecting* graphs, where the values of a and b must be calculated.

EXAMPLE 2 A Region Lying Between Two Intersecting Graphs

Find the area of the region bounded by the graphs of $f(x) = 2 - x^2$ and $g(x) = x$.

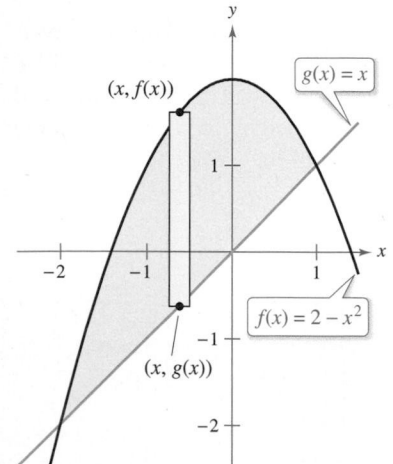

Region bounded by the graph of f and the graph of g
Figure 7.6

Solution In Figure 7.6, notice that the graphs of f and g have two points of intersection. To find the x-coordinates of these points, set $f(x)$ and $g(x)$ equal to each other and solve for x.

$$2 - x^2 = x \qquad \text{Set } f(x) \text{ equal to } g(x).$$
$$-x^2 - x + 2 = 0 \qquad \text{Write in general form.}$$
$$-(x + 2)(x - 1) = 0 \qquad \text{Factor.}$$
$$x = -2 \text{ or } 1 \qquad \text{Solve for } x.$$

So, $a = -2$ and $b = 1$. Because $g(x) \le f(x)$ for all x in the interval $[-2, 1]$, the representative rectangle has an area of

$$\Delta A = [f(x) - g(x)]\Delta x = [(2 - x^2) - x]\Delta x$$

and the area of the region is

$$A = \int_{-2}^{1} [(2 - x^2) - x]\, dx$$
$$= \left[-\frac{x^3}{3} - \frac{x^2}{2} + 2x \right]_{-2}^{1}$$
$$= \frac{9}{2}.$$

EXAMPLE 3 A Region Lying Between Two Intersecting Graphs

The sine and cosine curves intersect infinitely many times, bounding regions of equal areas, as shown in Figure 7.7. Find the area of one of these regions.

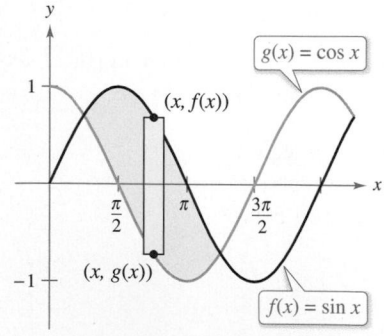

One of the regions bounded by the graphs of the sine and cosine functions
Figure 7.7

Solution Let $g(x) = \cos x$ and $f(x) = \sin x$. Then $g(x) \le f(x)$ for all x in the interval corresponding to the shaded region in Figure 7.7. To find the two points of intersection on this interval, set $f(x)$ and $g(x)$ equal to each other and solve for x.

$$\sin x = \cos x \qquad \text{Set } f(x) \text{ equal to } g(x).$$
$$\frac{\sin x}{\cos x} = 1 \qquad \text{Divide each side by } \cos x.$$
$$\tan x = 1 \qquad \text{Trigonometric identity}$$
$$x = \frac{\pi}{4} \text{ or } \frac{5\pi}{4}, \quad 0 \le x \le 2\pi \qquad \text{Solve for } x.$$

So, $a = \pi/4$ and $b = 5\pi/4$. Because $\sin x \ge \cos x$ for all x in the interval $[\pi/4, 5\pi/4]$, the area of the region is

$$A = \int_{\pi/4}^{5\pi/4} [\sin x - \cos x]\, dx$$
$$= \Big[-\cos x - \sin x \Big]_{\pi/4}^{5\pi/4}$$
$$= 2\sqrt{2}.$$

To find the area of the region between two curves that intersect at *more* than two points, first determine all points of intersection. Then check to see which curve is above the other in each interval determined by these points, as shown in Example 4.

EXAMPLE 4   Curves That Intersect at More than Two Points

•••▷ *See LarsonCalculus.com for an interactive version of this type of example.*

Find the area of the region between the graphs of

$$f(x) = 3x^3 - x^2 - 10x \quad \text{and} \quad g(x) = -x^2 + 2x.$$

Solution   Begin by setting $f(x)$ and $g(x)$ equal to each other and solving for x. This yields the x-values at all points of intersection of the two graphs.

$$3x^3 - x^2 - 10x = -x^2 + 2x \qquad \text{Set } f(x) \text{ equal to } g(x).$$
$$3x^3 - 12x = 0 \qquad \text{Write in general form.}$$
$$3x(x - 2)(x + 2) = 0 \qquad \text{Factor.}$$
$$x = -2, 0, 2 \qquad \text{Solve for } x.$$

So, the two graphs intersect when $x = -2$, 0, and 2. In Figure 7.8, notice that $g(x) \le f(x)$ on the interval $[-2, 0]$. The two graphs switch at the origin, however, and $f(x) \le g(x)$ on the interval $[0, 2]$. So, you need two integrals—one for the interval $[-2, 0]$ and one for the interval $[0, 2]$.

$$A = \int_{-2}^{0} [f(x) - g(x)]\,dx + \int_{0}^{2} [g(x) - f(x)]\,dx$$
$$= \int_{-2}^{0} (3x^3 - 12x)\,dx + \int_{0}^{2} (-3x^3 + 12x)\,dx$$
$$= \left[\frac{3x^4}{4} - 6x^2\right]_{-2}^{0} + \left[\frac{-3x^4}{4} + 6x^2\right]_{0}^{2}$$
$$= -(12 - 24) + (-12 + 24)$$
$$= 24$$

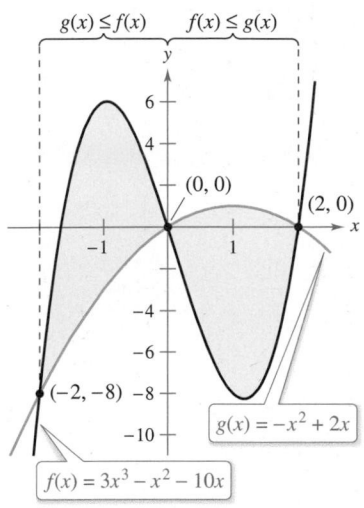

On $[-2, 0]$, $g(x) \le f(x)$, and on $[0, 2]$, $f(x) \le g(x)$.
Figure 7.8

REMARK   In Example 4, notice that you obtain an incorrect result when you integrate from -2 to 2. Such integration produces

$$\int_{-2}^{2} [f(x) - g(x)]\,dx = \int_{-2}^{2} (3x^3 - 12x)\,dx.$$
$$= 0.$$

When the graph of a function of y is a boundary of a region, it is often convenient to use representative rectangles that are *horizontal* and find the area by integrating with respect to y. In general, to determine the area between two curves, you can use

$$A = \int_{x_1}^{x_2} [(\text{top curve}) - (\text{bottom curve})]\,dx \qquad \text{Vertical rectangles}$$

$$\phantom{A = \int_{x_1}^{x_2}} \underbrace{\phantom{[(\text{top curve}) - (\text{bottom curve})]}}_{\text{in variable } x}$$

or

$$A = \int_{y_1}^{y_2} [(\text{right curve}) - (\text{left curve})]\,dy \qquad \text{Horizontal rectangles}$$

$$\phantom{A = \int_{y_1}^{y_2}} \underbrace{\phantom{[(\text{right curve}) - (\text{left curve})]}}_{\text{in variable } y}$$

where (x_1, y_1) and (x_2, y_2) are either adjacent points of intersection of the two curves involved or points on the specified boundary lines.

EXAMPLE 5   **Horizontal Representative Rectangles**

Find the area of the region bounded by the graphs of $x = 3 - y^2$ and $x = y + 1$.

Solution   Consider

$$g(y) = 3 - y^2 \quad \text{and} \quad f(y) = y + 1.$$

These two curves intersect when $y = -2$ and $y = 1$, as shown in Figure 7.9. Because $f(y) \le g(y)$ on this interval, you have

$$\Delta A = [g(y) - f(y)]\Delta y = [(3 - y^2) - (y + 1)]\Delta y.$$

So, the area is

$$
\begin{aligned}
A &= \int_{-2}^{1} [(3 - y^2) - (y + 1)]\, dy \\
&= \int_{-2}^{1} (-y^2 - y + 2)\, dy \\
&= \left[\frac{-y^3}{3} - \frac{y^2}{2} + 2y \right]_{-2}^{1} \\
&= \left(-\frac{1}{3} - \frac{1}{2} + 2 \right) - \left(\frac{8}{3} - 2 - 4 \right) \\
&= \frac{9}{2}.
\end{aligned}
$$

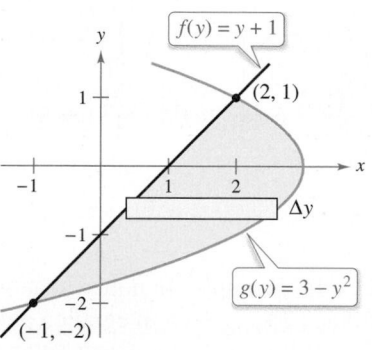

Horizontal rectangles (integration with respect to y)
Figure 7.9

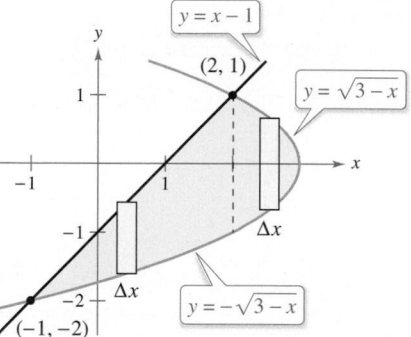

Vertical rectangles (integration with respect to x)
Figure 7.10

In Example 5, notice that by integrating with respect to y, you need only one integral. To integrate with respect to x, you would need two integrals because the upper boundary changes at $x = 2$, as shown in Figure 7.10.

$$
\begin{aligned}
A &= \int_{-1}^{2} \left[(x - 1) + \sqrt{3 - x} \right] dx + \int_{2}^{3} \left(\sqrt{3 - x} + \sqrt{3 - x} \right) dx \\
&= \int_{-1}^{2} \left[x - 1 + (3 - x)^{1/2} \right] dx + 2 \int_{2}^{3} (3 - x)^{1/2}\, dx \\
&= \left[\frac{x^2}{2} - x - \frac{(3 - x)^{3/2}}{3/2} \right]_{-1}^{2} - 2\left[\frac{(3 - x)^{3/2}}{3/2} \right]_{2}^{3} \\
&= \left(2 - 2 - \frac{2}{3} \right) - \left(\frac{1}{2} + 1 - \frac{16}{3} \right) - 2(0) + 2\left(\frac{2}{3} \right) \\
&= \frac{9}{2}
\end{aligned}
$$

Integration as an Accumulation Process

In this section, the integration formula for the area between two curves was developed by using a rectangle as the *representative element*. For each new application of integration in the remaining sections of this chapter, an appropriate representative element will be constructed using precalculus formulas you already know. Each integration formula will then be obtained by summing or accumulating these representative elements.

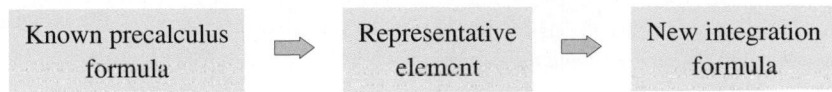

For example, the area formula in this section was developed as follows.

$$A = (\text{height})(\text{width}) \quad \Longrightarrow \quad \Delta A = [f(x) - g(x)]\Delta x \quad \Longrightarrow \quad A = \int_a^b [f(x) - g(x)]\, dx$$

EXAMPLE 6   **Integration as an Accumulation Process**

Find the area of the region bounded by the graph of $y = 4 - x^2$ and the x-axis. Describe the integration as an accumulation process.

Solution   The area of the region is

$$A = \int_{-2}^{2} (4 - x^2)\, dx.$$

You can think of the integration as an accumulation of the areas of the rectangles formed as the representative rectangle slides from $x = -2$ to $x = 2$, as shown in Figure 7.11.

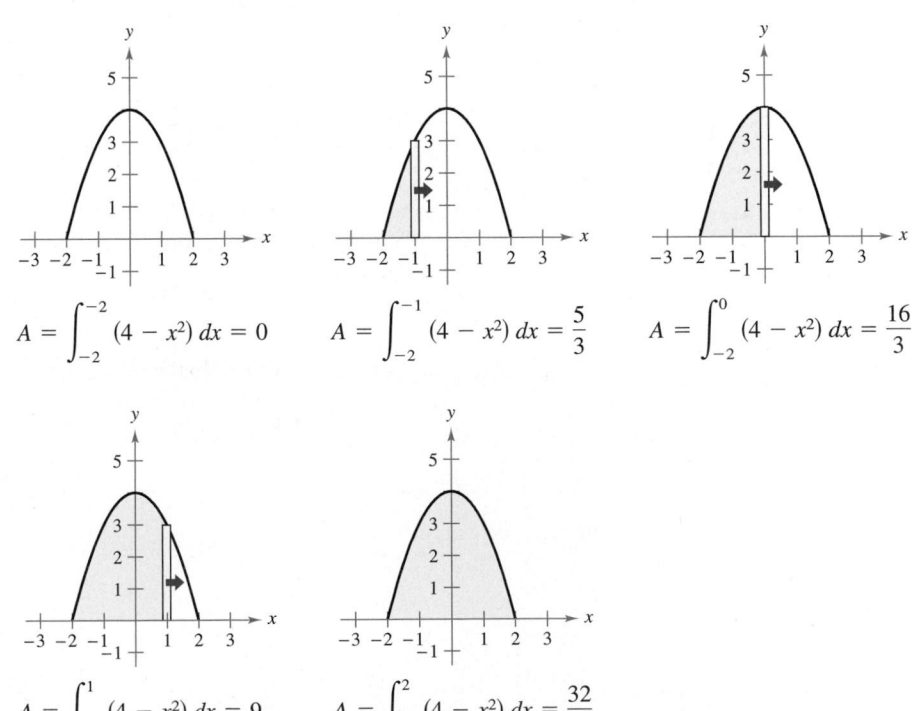

$$A = \int_{-2}^{-2} (4 - x^2)\, dx = 0 \qquad A = \int_{-2}^{-1} (4 - x^2)\, dx = \frac{5}{3} \qquad A = \int_{-2}^{0} (4 - x^2)\, dx = \frac{16}{3}$$

$$A = \int_{-2}^{1} (4 - x^2)\, dx = 9 \qquad A = \int_{-2}^{2} (4 - x^2)\, dx = \frac{32}{3}$$

Figure 7.11

7.1 Exercises

See **CalcChat.com** for tutorial help and worked-out solutions to odd-numbered exercises.

CONCEPT CHECK

1. Area What is the geometric interpretation of the area of the region between two curves?

2. Area Describe how to find the area of the region bounded by the graphs of $f(x)$ and $g(x)$ and the vertical lines $x = a$ and $x = b$, where f and g do not intersect on $[a, b]$.

3. Area Between Intersecting Curves Explain why it is important to determine all points of intersection of two curves when finding the area of the region between the curves.

4. Sketching a Region Sketch a region for which integration with respect to y is easier than integration with respect to x.

 **Writing a Definite Integral** In Exercises 5–10, write a definite integral that represents the area of the region. (Do not evaluate the integral.)

5. $y_1 = x^2 - 6x$
$y_2 = 0$

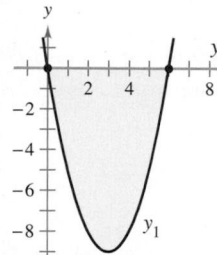

6. $y_1 = x^2 + 2x + 1$
$y_2 = 2x + 5$

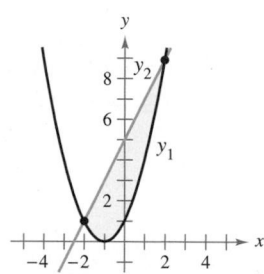

7. $y_1 = x^2 - 4x + 3$
$y_2 = -x^2 + 2x + 3$

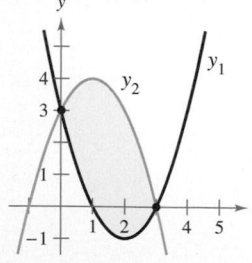

8. $y_1 = x^2$
$y_2 = x^3$

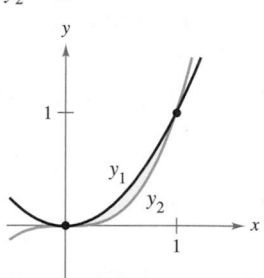

9. $y_1 = 3(x^3 - x)$
$y_2 = 0$

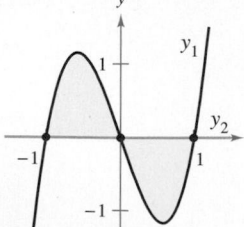

10. $y_1 = (x - 1)^3$
$y_2 = x - 1$

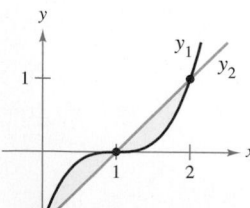

 **Finding a Region** In Exercises 11–14, the integrand of the definite integral is a difference of two functions. Sketch the graph of each function and shade the region whose area is represented by the integral.

11. $\int_0^4 \left[(x + 1) - \frac{x}{2} \right] dx$

12. $\int_2^3 \left[\left(\frac{x^3}{3} - x \right) - \frac{x}{3} \right] dx$

13. $\int_{-2}^1 \left[(2 - y) - y^2 \right] dy$

14. $\int_0^4 \left(2\sqrt{y} - y \right) dy$

 **Finding the Area of a Region** In Exercises 15–28, sketch the region bounded by the graphs of the equations and find the area of the region.

15. $y = x^2 - 1$, $y = -x + 2$, $x = 0$, $x = 1$

16. $y = -x^3 + 2$, $y = x - 3$, $x = -1$, $x = 1$

17. $f(x) = x^2 + 2x$, $g(x) = x + 2$

18. $y = -x^2 + 3x + 1$, $y = -x + 1$

19. $f(x) = \frac{1}{9x^2}$, $y = 1$, $x = 1$, $x = 2$

20. $f(x) = -\frac{4}{x^3}$, $y = 0$, $x = -3$, $x = -1$

21. $f(x) = x^5 + 2$, $g(x) = x + 2$

22. $f(x) = \sqrt[3]{x - 1}$, $g(x) = x - 1$

23. $f(y) = y^2$, $g(y) = y + 2$

24. $f(y) = y(2 - y)$, $g(y) = -y$

25. $f(y) = y^2 + 1$, $g(y) = 0$, $y = -1$, $y = 2$

26. $f(y) = \frac{y}{\sqrt{16 - y^2}}$, $g(y) = 0$, $y = 3$

27. $f(x) = \frac{10}{x}$, $x = 0$, $y = 2$, $y = 10$

28. $g(x) = \frac{4}{2 - x}$, $y = 4$, $x = 0$

 **Comparing Methods** In Exercises 29 and 30, find the area of the region by integrating (a) with respect to x and (b) with respect to y. (c) Compare your results. Which method is simpler? In general, will this method always be simpler than the other one? Why or why not?

29. $x = 4 - y^2$
$x = y - 2$

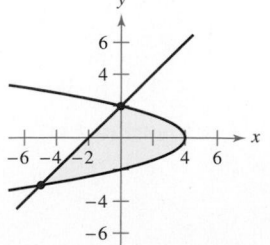

30. $y = x^2$
$y = 6 - x$

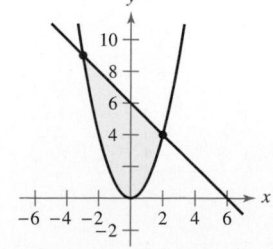

 **Finding the Area of a Region** In Exercises 31–36, (a) use a graphing utility to graph the region bounded by the graphs of the functions, (b) find the area of the region analytically, and (c) use the integration capabilities of the graphing utility to verify your results.

31. $f(x) = x(x^2 - 3x + 3)$, $g(x) = x^2$

32. $y = x^4 - 2x^2$, $y = 2x^2$

33. $f(x) = x^4 - 4x^2$, $g(x) = x^2 - 4$

34. $f(x) = x^4 - 9x^2$, $g(x) = x^3 - 9x$

35. $f(x) = \dfrac{1}{1 + x^2}$, $g(x) = \dfrac{1}{2}x^2$

36. $f(x) = \dfrac{6x}{x^2 + 1}$, $y = 0$, $0 \le x \le 3$

Finding the Area of a Region In Exercises 37–42, sketch the region bounded by the graphs of the equations and find the area of the region.

37. $f(x) = \cos x$, $g(x) = 2 - \cos x$, $0 \le x \le 2\pi$

38. $f(x) = \sin x$, $g(x) = \cos 2x$, $-\dfrac{\pi}{2} \le x \le \dfrac{\pi}{6}$

39. $f(x) = 2 \sin x$, $g(x) = \tan x$, $-\dfrac{\pi}{3} \le x \le \dfrac{\pi}{3}$

40. $f(x) = \sec \dfrac{\pi x}{4} \tan \dfrac{\pi x}{4}$, $g(x) = \left(\sqrt{2} - 4\right)x + 4$, $x = 0$

41. $f(x) = xe^{-x^2}$, $y = 0$, $0 \le x \le 1$

42. $f(x) = -2^x$, $g(x) = 1 - 3x$

 **Finding the Area of a Region** In Exercises 43–46, (a) use a graphing utility to graph the region bounded by the graphs of the equations, (b) find the area of the region analytically, and (c) use the integration capabilities of the graphing utility to verify your results.

43. $f(x) = 2 \sin x + \sin 2x$, $y = 0$, $0 \le x \le \pi$

44. $f(x) = 2 \sin x + \cos 2x$, $y = 0$, $0 < x \le \pi$

45. $f(x) = \dfrac{1}{x^2}e^{1/x}$, $y = 0$, $1 \le x \le 3$

46. $g(x) = \dfrac{4 \ln x}{x}$, $y = 0$, $x = 5$

 **Finding the Area of a Region** In Exercises 47–50, (a) use a graphing utility to graph the region bounded by the graphs of the equations, (b) explain why the area of the region is difficult to find analytically, and (c) use the integration capabilities of the graphing utility to approximate the area of the region to four decimal places.

47. $y = \sqrt{\dfrac{x^3}{4 - x}}$, $y = 0$, $x = 3$

48. $y = \sqrt{x}\, e^x$, $y = 0$, $x = 0$, $x = 1$

49. $y = x^2$, $y = 4 \cos x$

50. $y = x^2$, $y = \sqrt{3 + x}$

51. Finding the Area of a Region Find the area of the given region bounded by the graphs of y_1, y_2, and y_3, as shown in the figure.

$$y_1 = x^2 + 2, \quad y_2 = 4 - x^2, \quad y_3 = 2 - x$$

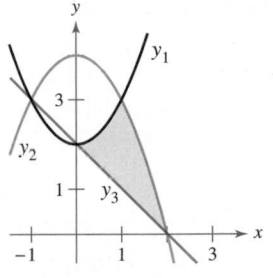 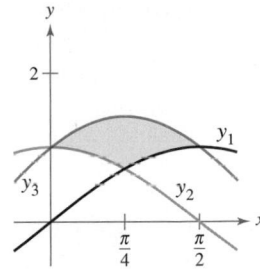

Figure for 51   Figure for 52

52. Finding the Area of a Region Find the area of the given region bounded by the graphs of y_1, y_2, and y_3, as shown in the figure.

$$y_1 = \sin x, \quad y_2 = \cos x, \quad y_3 = \sin x + \cos x$$

 **Integration as an Accumulation Process** In Exercises 53–56, find the accumulation function F. Then evaluate F at each value of the independent variable and graphically show the area given by each value of the independent variable.

53. $F(x) = \displaystyle\int_0^x \left(\dfrac{1}{2}t + 1\right) dt$ (a) $F(0)$ (b) $F(2)$ (c) $F(6)$

54. $F(x) = \displaystyle\int_0^x \left(\dfrac{1}{2}t^2 + 2\right) dt$ (a) $F(0)$ (b) $F(4)$ (c) $F(6)$

55. $F(\alpha) = \displaystyle\int_{-1}^{\alpha} \cos \dfrac{\pi\theta}{2}\, d\theta$ (a) $F(-1)$ (b) $F(0)$ (c) $F\left(\dfrac{1}{2}\right)$

56. $F(y) = \displaystyle\int_{-1}^{y} 4e^{x/2}\, dx$ (a) $F(-1)$ (b) $F(0)$ (c) $F(4)$

Finding the Area of a Figure In Exercises 57–60, use integration to find the area of the figure having the given vertices.

57. $(-1, -1), (1, 1), (2, -1)$

58. $(0, 0), (6, 0), (4, 3)$

59. $(0, 2), (4, 2), (0, -2), (-4, -2)$

60. $(0, 0), (1, 2), (3, -2), (1, -3)$

Using a Tangent Line In Exercises 61–64, write and evaluate the definite integral that represents the area of the region bounded by the graph of the function and the tangent line to the graph at the given point.

61. $f(x) = 2x^3 - 1$, $(1, 1)$

62. $f(x) = x - x^3$, $(-1, 0)$

63. $f(x) = \dfrac{1}{x^2 + 1}$, $\left(1, \dfrac{1}{2}\right)$

64. $y = \dfrac{2}{1 + 4x^2}$, $\left(\dfrac{1}{2}, 1\right)$

EXPLORING CONCEPTS

65. Area Between Curves The graphs of $y = 1 - x^2$ and $y = x^4 - 2x^2 + 1$ intersect at three points. However, the area between the curves *can* be found by a single integral. Explain why this is so, and write an integral that represents this area.

66. Using Symmetry The area of the region bounded by the graphs of $y = x^3$ and $y = x$ *cannot* be found by the single integral $\int_{-1}^{1} (x^3 - x)\, dx$. Explain why this is so. Use symmetry to write a single integral that does represent the area.

67. Interpreting Integrals Two cars with velocities $v_1(t)$ and $v_2(t)$ (in meters per second) are tested on a straight track. Consider the following integrals.

$$\int_0^5 [v_1(t) - v_2(t)]\, dt = 10 \quad \int_0^{10} [v_1(t) - v_2(t)]\, dt = 30$$

$$\int_{20}^{30} [v_1(t) - v_2(t)]\, dt = -5$$

(a) Write a verbal interpretation of each integral.

(b) Is it possible to determine the distance between the two cars when $t = 5$ seconds? Why or why not?

(c) Assume both cars start at the same time and place. Which car is ahead when $t = 10$ seconds? How far ahead is the car?

(d) Suppose Car 1 has velocity v_1 and is ahead of Car 2 by 13 meters when $t = 20$ seconds. How far ahead or behind is Car 1 when $t = 30$ seconds?

68. 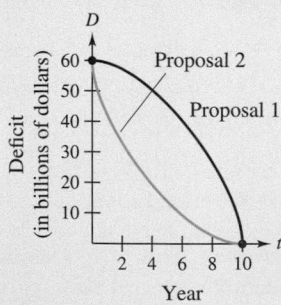 **HOW DO YOU SEE IT?** A state legislature is debating two proposals for eliminating the annual budget deficits after 10 years. The rate of decrease of the deficits for each proposal is shown in the figure.

(a) What does the area between the two curves represent?

(b) From the viewpoint of minimizing the cumulative state deficit, which is the better proposal? Explain.

Dividing a Region In Exercises 69 and 70, find b such that the line $y = b$ divides the region bounded by the graphs of the equations into two regions of equal area.

69. $y = 9 - x^2$, $\quad y = 0$ **70.** $y = 9 - |x|$, $\quad y = 0$

Dividing a Region In Exercises 71 and 72, find a such that the line $x = a$ divides the region bounded by the graphs of the equations into two regions of equal area.

71. $y = x$, $\quad y = 4$, $\quad x = 0$ **72.** $y^2 = 4 - x$, $\quad x = 0$

Limits and Integrals In Exercises 73 and 74, evaluate the limit and sketch the graph of the region whose area is represented by the limit.

73. $\displaystyle\lim_{\|\Delta\|\to 0} \sum_{i=1}^{n} (x_i - x_i^2)\Delta x$, where $x_i = \dfrac{i}{n}$ and $\Delta x = \dfrac{1}{n}$

74. $\displaystyle\lim_{\|\Delta\|\to 0} \sum_{i=1}^{n} (4 - x_i^2)\Delta x$, where $x_i = -2 + \dfrac{4i}{n}$ and $\Delta x = \dfrac{4}{n}$

Revenue In Exercises 75 and 76, two models R_1 and R_2 are given for revenue (in millions of dollars) for a corporation. Both models are estimates of revenues from 2020 through 2025, with $t = 0$ corresponding to 2020. Which model projects the greater revenue? How much more total revenue does that model project over the six-year period?

75. $R_1 = 7.21 + 0.58t$

$\quad\ \ R_2 = 7.21 + 0.45t$

76. $R_1 = 7.21 + 0.26t + 0.02t^2$

$\quad\ \ R_2 = 7.21 + 0.1t + 0.01t^2$

77. Lorenz Curve Economists use *Lorenz curves* to illustrate the distribution of income in a country. A Lorenz curve, $y = f(x)$, represents the actual income distribution in the country. In this model, x represents percents of families in the country from the poorest to the wealthiest and y represents percents of total income. The model $y = x$ represents a country in which each family has the same income. The area between these two models, where $0 \le x \le 100$, indicates a country's "income inequality." The table lists percents of income y for selected percents of families x in a country.

x	10	20	30	40	50
y	3.35	6.07	9.17	13.39	19.45

x	60	70	80	90
y	28.03	39.77	55.28	75.12

(a) Use a graphing utility to find a quadratic model for the Lorenz curve.

(b) Plot the data and graph the model.

(c) Graph the model $y = x$. How does this model compare with the model in part (a)?

(d) Use the integration capabilities of a graphing utility to approximate the "income inequality."

78. Profit The chief financial officer of a company reports that profits for the past fiscal year were $15.9 million. The officer predicts that profits for the next 5 years will grow at a continuous annual rate somewhere between $3\frac{1}{2}\%$ and 5%. Estimate the cumulative difference in total profit over the 5 years based on the predicted range of growth rates.

• • **79. Building Design** • • • • • • • • • • • • • • •

Concrete sections for a new building have the dimensions (in meters) and shape shown in the figure.

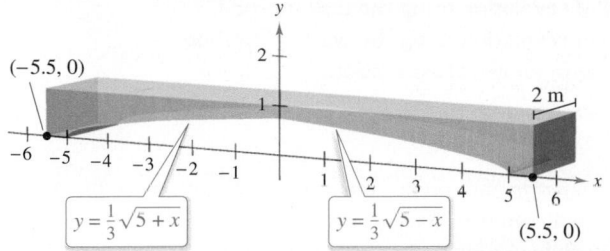

$(-5.5, 0)$

2 m

$y = \frac{1}{3}\sqrt{5 + x}$ $y = \frac{1}{3}\sqrt{5 - x}$ $(5.5, 0)$

(a) Find the area of the face of the section superimposed on the rectangular coordinate system.

(b) Find the volume of concrete in one of the sections by multiplying the area in part (a) by 2 meters.

(c) One cubic meter of concrete weighs 5000 pounds. Find the weight of the section.

80. Mechanical Design   The surface of a machine part is the region between the graphs of $y_1 = |x|$ and $y_2 = 0.08x^2 + k$ (see figure).

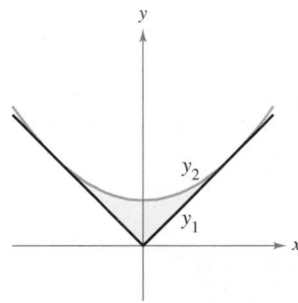

y_2

y_1

(a) Find k such that the parabola is tangent to the graph of y_1.

(b) Find the area of the surface of the machine part.

81. Area   Find the area between the graph of $y = \sin x$ and the line segment joining the points $(0, 0)$ and $\left(\frac{7\pi}{6}, -\frac{1}{2}\right)$, as shown in the figure.

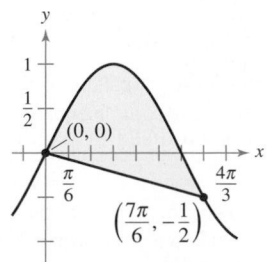

$(0, 0)$

$\frac{\pi}{6}$ $\frac{4\pi}{3}$

$\left(\frac{7\pi}{6}, -\frac{1}{2}\right)$

82. Area   Let $a > 0$ and $b > 0$. Show that the area of the ellipse $\frac{x^2}{a^2} + \frac{y^2}{b^2} = 1$ is πab (see figure).

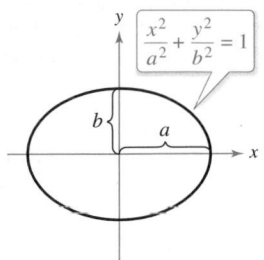

$\frac{x^2}{a^2} + \frac{y^2}{b^2} = 1$

b a

True or False?   **In Exercises 83–86, determine whether the statement is true or false. If it is false, explain why or give an example that shows it is false.**

83. If the area of the region bounded by the graphs of f and g is 1, then the area of the region bounded by the graphs of $h(x) = f(x) + C$ and $k(x) = g(x) + C$ is also 1.

84. If

$$\int_a^b [f(x) - g(x)]\, dx = A$$

then

$$\int_a^b [g(x) - f(x)]\, dx = -A.$$

85. If the graphs of f and g intersect midway between $x = a$ and $x = b$, then

$$\int_a^b [f(x) - g(x)]\, dx = 0.$$

86. The line

$$y = \left(1 - \sqrt[3]{0.5}\right)x$$

divides the region under the curve

$$f(x) = x(1 - x)$$

on $[0, 1]$ into two regions of equal area.

PUTNAM EXAM CHALLENGE

87. The horizontal line $y = c$ intersects the curve $y = 2x - 3x^3$ in the first quadrant as shown in the figure. Find c so that the areas of the two shaded regions are equal.

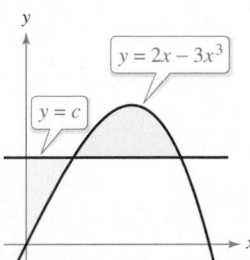

$y = 2x - 3x^3$

$y = c$

This problem was composed by the Committee on the Putnam Prize Competition.
© The Mathematical Association of America. All rights reserved.

7.2 Volume: The Disk Method

- Find the volume of a solid of revolution using the disk method.
- Find the volume of a solid of revolution using the washer method.
- Find the volume of a solid with known cross sections.

The Disk Method

You have already learned that area is only one of the *many* applications of the definite integral. Another important application is its use in finding the volume of a three-dimensional solid. In this section, you will study a particular type of three-dimensional solid—one whose cross sections are similar. Solids of revolution are used commonly in engineering and manufacturing. Some examples are axles, funnels, pills, bottles, and pistons, as shown in Figure 7.12.

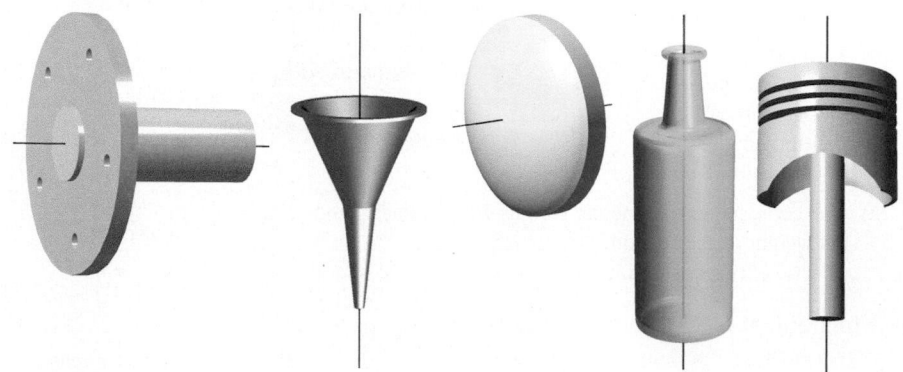

Solids of revolution
Figure 7.12

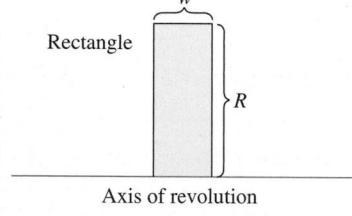

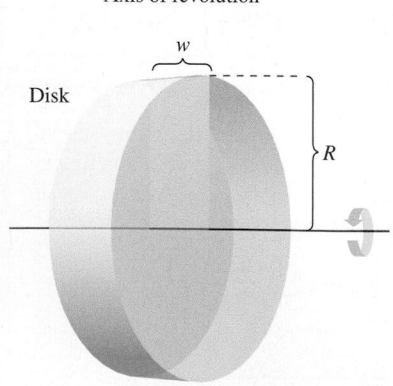

Volume of a disk: $\pi R^2 w$
Figure 7.13

When a region in the plane is revolved about a line, the resulting solid is a **solid of revolution,** and the line is called the **axis of revolution.** The simplest such solid is a right circular cylinder or **disk,** which is formed by revolving a rectangle about an axis adjacent to one side of the rectangle, as shown in Figure 7.13. The volume of such a disk is

Volume of disk = (area of disk)(width of disk)

$$= \pi R^2 w$$

where R is the radius of the disk and w is the width.

To see how to use the volume of a disk to find the volume of a general solid of revolution, consider a solid of revolution formed by revolving the plane region in Figure 7.14 (see next page) about the indicated axis. To determine the volume of this solid, consider a representative rectangle in the plane region. When this rectangle is revolved about the axis of revolution, it generates a representative disk whose volume is

$$\Delta V = \pi R^2 \, \Delta x.$$

Approximating the volume of the solid by n such disks of width Δx and radius $R(x_i)$ produces

$$\text{Volume of solid} \approx \sum_{i=1}^{n} \pi [R(x_i)]^2 \, \Delta x$$

$$= \pi \sum_{i=1}^{n} [R(x_i)]^2 \, \Delta x.$$

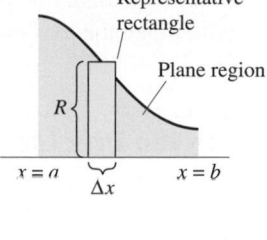

Disk method
Figure 7.14

This approximation appears to become better and better as $\|\Delta\| \to 0$ $(n \to \infty)$. So, you can define the volume of the solid as

$$\text{Volume of solid} = \lim_{\|\Delta\| \to 0} \pi \sum_{i=1}^{n} [R(x_i)]^2 \, \Delta x = \pi \int_a^b [R(x)]^2 \, dx.$$

Schematically, the disk method looks like this.

Known Precalculus Formula	**Representative Element**	**New Integration Formula**
Volume of disk $V = \pi R^2 w$	$\Delta V = \pi[R(x_i)]^2 \Delta x$	Solid of revolution $V = \pi \int_a^b [R(x)]^2 \, dx$

A similar formula can be derived when the axis of revolution is vertical.

THE DISK METHOD

To find the volume of a solid of revolution with the **disk method,** use one of the formulas below. (See Figure 7.15.)

Horizontal Axis of Revolution	**Vertical Axis of Revolution**
Volume $= V = \pi \int_a^b [R(x)]^2 \, dx$	Volume $= V = \pi \int_c^d [R(y)]^2 \, dy$

•• **REMARK** In Figure 7.15, note that you can determine the variable of integration by placing a representative rectangle in the plane region "perpendicular" to the axis of revolution. When the width of the rectangle is Δx, integrate with respect to x, and when the width of the rectangle is Δy, integrate with respect to y.

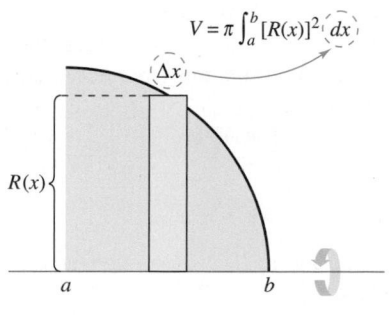

Horizontal axis of revolution
Figure 7.15

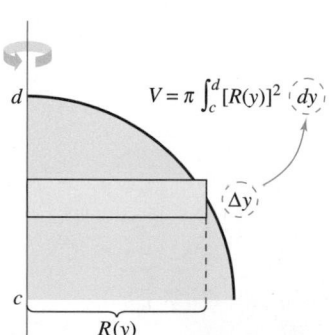

Vertical axis of revolution

The simplest application of the disk method involves a plane region bounded by the graph of f and the x-axis. When the axis of revolution is the x-axis, the radius $R(x)$ is simply $f(x)$.

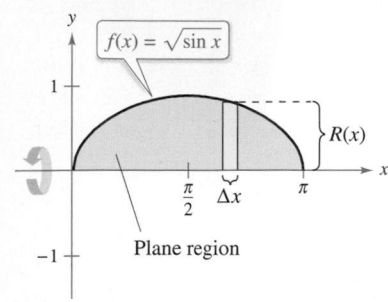

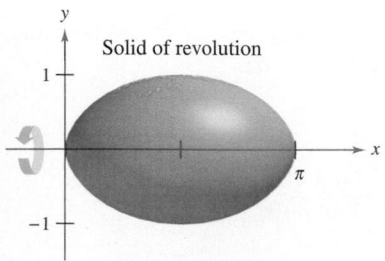

Figure 7.16

EXAMPLE 1 Using the Disk Method

Find the volume of the solid formed by revolving the region bounded by the graph of

$$f(x) = \sqrt{\sin x}$$

and the x-axis ($0 \le x \le \pi$) about the x-axis, as shown in Figure 7.16.

Solution From the representative rectangle in the upper graph in Figure 7.16, you can see that the radius of this solid is

$$R(x) = f(x)$$
$$= \sqrt{\sin x}.$$

So, the volume of the solid of revolution is

$$V = \pi \int_a^b [R(x)]^2 \, dx \qquad \text{Apply disk method.}$$

$$= \pi \int_0^\pi \left(\sqrt{\sin x} \right)^2 dx \qquad \text{Substitute } \sqrt{\sin x} \text{ for } R(x).$$

$$= \pi \int_0^\pi \sin x \, dx \qquad \text{Simplify.}$$

$$= \pi \left[-\cos x \right]_0^\pi \qquad \text{Integrate.}$$

$$= \pi(1 + 1)$$

$$= 2\pi.$$

EXAMPLE 2 Using a Line That Is Not a Coordinate Axis

Find the volume of the solid formed by revolving the region bounded by the graphs of

$$f(x) = 2 - x^2$$

and $g(x) = 1$ about the line $y = 1$, as shown in Figure 7.17.

Solution By equating $f(x)$ and $g(x)$, you can determine that the two graphs intersect when $x = \pm 1$. To find the radius, subtract $g(x)$ from $f(x)$.

$$R(x) = f(x) - g(x)$$
$$= (2 - x^2) - 1$$
$$= 1 - x^2$$

To find the volume, integrate between -1 and 1.

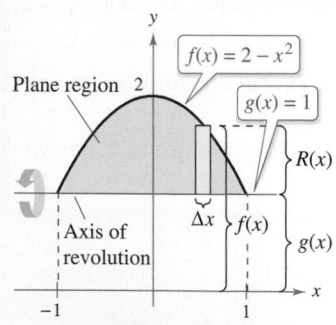

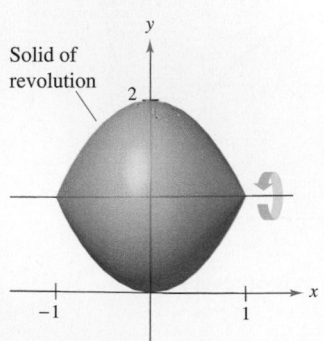

Figure 7.17

$$V = \pi \int_a^b [R(x)]^2 \, dx \qquad \text{Apply disk method.}$$

$$= \pi \int_{-1}^1 (1 - x^2)^2 \, dx \qquad \text{Substitute } 1 - x^2 \text{ for } R(x).$$

$$= \pi \int_{-1}^1 (1 - 2x^2 + x^4) \, dx \qquad \text{Simplify.}$$

$$= \pi \left[x - \frac{2x^3}{3} + \frac{x^5}{5} \right]_{-1}^1 \qquad \text{Integrate.}$$

$$= \frac{16\pi}{15}$$

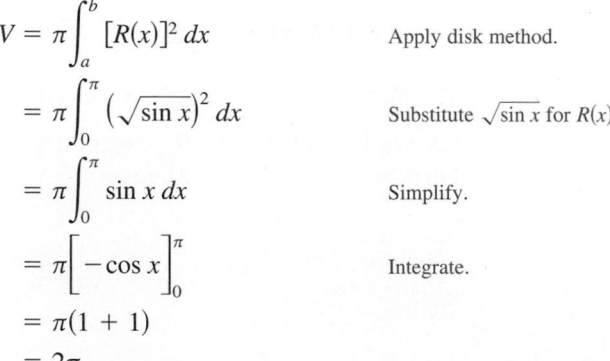

The Washer Method

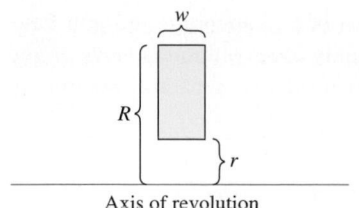

Axis of revolution

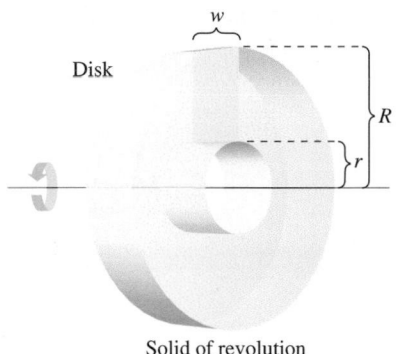

Disk

Solid of revolution

Figure 7.18

The disk method can be extended to cover solids of revolution with holes by replacing the representative disk with a representative **washer.** The washer is formed by revolving a rectangle about an axis, as shown in Figure 7.18. If r and R are the inner and outer radii of the washer, respectively, and w is the width of the washer, then the volume is

Volume of washer $= \pi(R^2 - r^2)w.$

To see how this concept can be used to find the volume of a solid of revolution, consider a region bounded by an **outer radius** $R(x)$ and an **inner radius** $r(x)$, as shown in Figure 7.19. If the region is revolved about its axis of revolution, then the volume of the resulting solid is

$$V = \pi \int_a^b \left([R(x)]^2 - [r(x)]^2\right) dx. \qquad \text{Washer method}$$

Note that the integral involving the inner radius represents the volume of the hole and is *subtracted* from the integral involving the outer radius.

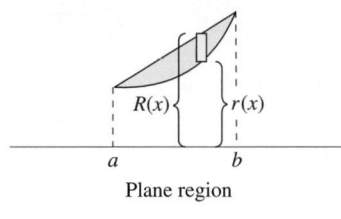

Plane region

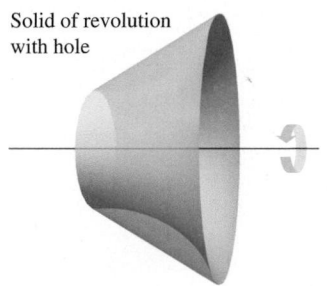

Solid of revolution with hole

Figure 7.19

EXAMPLE 3 Using the Washer Method

Find the volume of the solid formed by revolving the region bounded by the graphs of

$$y = \sqrt{x} \quad \text{and} \quad y = x^2$$

about the x-axis, as shown in Figure 7.20.

Solution In Figure 7.20, you can see that the outer and inner radii are as follows.

$R(x) = \sqrt{x}$ Outer radius

$r(x) = x^2$ Inner radius

Integrating between 0 and 1 produces

$$V = \pi \int_a^b \left([R(x)]^2 - [r(x)]^2\right) dx \qquad \text{Apply washer method.}$$

$$= \pi \int_0^1 \left[(\sqrt{x})^2 - (x^2)^2\right] dx \qquad \text{Substitute } \sqrt{x} \text{ for } R(x) \text{ and } x^2 \text{ for } r(x).$$

$$= \pi \int_0^1 (x - x^4) dx \qquad \text{Simplify.}$$

$$= \pi \left[\frac{x^2}{2} - \frac{x^5}{5}\right]_0^1 \qquad \text{Integrate.}$$

$$= \frac{3\pi}{10}.$$

Solid of revolution

Figure 7.20

In each example so far, the axis of revolution has been *horizontal* and you have integrated with respect to *x*. In the next example, the axis of revolution is *vertical* and you integrate with respect to *y*. In this example, you need two separate integrals to compute the volume.

EXAMPLE 4 Integrating with Respect to *y*: Two-Integral Case

Find the volume of the solid formed by revolving the region bounded by the graphs of

$$y = x^2 + 1, \quad y = 0, \quad x = 0, \quad \text{and} \quad x = 1$$

about the *y*-axis, as shown in Figure 7.21.

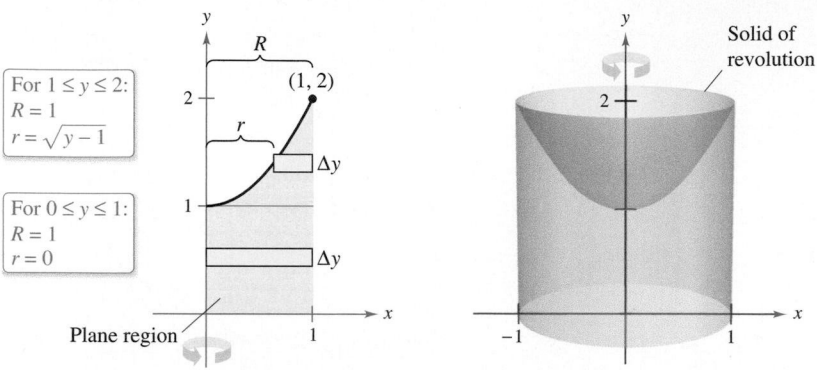

Figure 7.21

Solution For the region shown in Figure 7.21, the outer radius is simply $R = 1$. There is, however, no convenient formula that represents the inner radius. When $0 \le y \le 1$, $r = 0$, but when $1 \le y \le 2$, r is determined by the equation $y = x^2 + 1$, which implies that $r = \sqrt{y - 1}$.

$$r(y) = \begin{cases} 0, & 0 \le y \le 1 \\ \sqrt{y - 1}, & 1 \le y \le 2 \end{cases}$$

Using this definition of the inner radius, you can use two integrals to find the volume.

$$V = \pi \int_0^1 (1^2 - 0^2) \, dy + \pi \int_1^2 \left[1^2 - \left(\sqrt{y - 1} \right)^2 \right] dy \qquad \text{Apply washer method.}$$

$$= \pi \int_0^1 1 \, dy + \pi \int_1^2 (2 - y) \, dy \qquad \text{Simplify.}$$

$$= \pi \left[y \right]_0^1 + \pi \left[2y - \frac{y^2}{2} \right]_1^2 \qquad \text{Integrate.}$$

$$= \pi + \pi \left(4 - 2 - 2 + \frac{1}{2} \right)$$

$$= \frac{3\pi}{2}$$

Note that the first integral $\pi \int_0^1 1 \, dy$ represents the volume of a right circular cylinder of radius 1 and height 1. This portion of the volume could have been determined without using calculus. ∎

▷ **TECHNOLOGY** Some graphing utilities have the capability of generating (or have built-in software capable of generating) a solid of revolution. If you have access to such a utility, use it to graph some of the solids of revolution described in this section. For instance, the solid in Example 4 might appear like that shown in Figure 7.22.

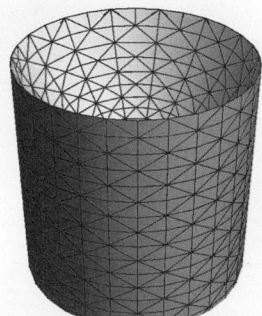

Generated by Mathematica

Figure 7.22

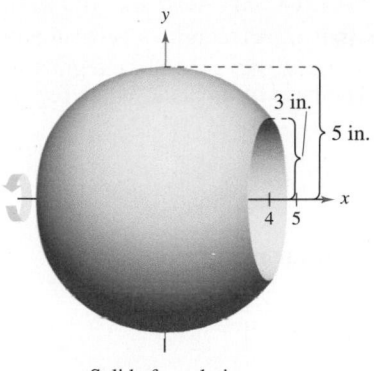

Solid of revolution

(a)

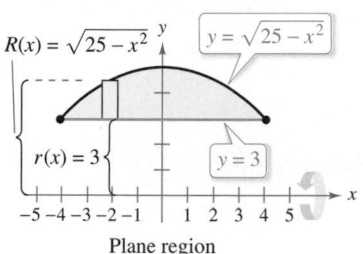

Plane region

(b)

Figure 7.23

Manufacturing

$\cdots \triangleright$ *See LarsonCalculus.com for an interactive version of this type of example.*

A manufacturer drills a hole through the center of a metal sphere of radius 5 inches, as shown in Figure 7.23(a). The hole has a radius of 3 inches. What is the volume of the resulting metal ring?

Solution You can imagine the ring to be generated by a segment of the circle whose equation is $x^2 + y^2 = 25$, as shown in Figure 7.23(b). Because the radius of the hole is 3 inches, you can let $y = 3$ and solve the equation $x^2 + y^2 - 25$ to determine that the limits of integration are $x = \pm 4$. So, the inner and outer radii are $r(x) = 3$ and $R(x) = \sqrt{25 - x^2}$, and the volume is

$$
\begin{aligned}
V &= \pi \int_a^b \left([R(x)]^2 - [r(x)]^2 \right) dx \\
&= \pi \int_{-4}^4 \left[\left(\sqrt{25 - x^2} \right)^2 - (3)^2 \right] dx \\
&= \pi \int_{-4}^4 (16 - x^2) \, dx \\
&= \pi \left[16x - \frac{x^3}{3} \right]_{-4}^4 \\
&= \frac{256\pi}{3} \text{ cubic inches.}
\end{aligned}
$$

Solids with Known Cross Sections

With the disk method, you can find the volume of a solid having a circular cross section whose area is $A = \pi R^2$. This method can be generalized to solids of any shape, as long as you know a formula for the area of an arbitrary cross section. Some common cross sections are squares, rectangles, triangles, semicircles, and trapezoids.

VOLUMES OF SOLIDS WITH KNOWN CROSS SECTIONS

1. For cross sections of area $A(x)$ taken perpendicular to the x-axis,

$$\text{Volume} = \int_a^b A(x) \, dx. \qquad \text{See Figure 7.24(a).}$$

2. For cross sections of area $A(y)$ taken perpendicular to the y-axis,

$$\text{Volume} = \int_c^d A(y) \, dy. \qquad \text{See Figure 7.24(b).}$$

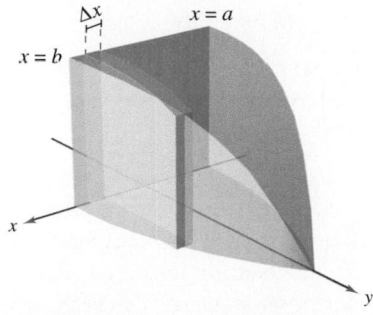

(a) Cross sections perpendicular to x-axis

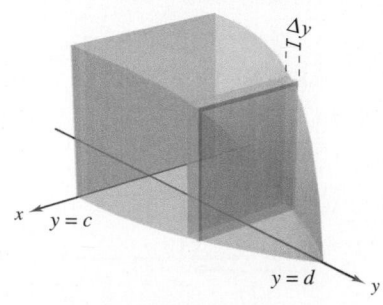

(b) Cross sections perpendicular to y-axis

Figure 7.24

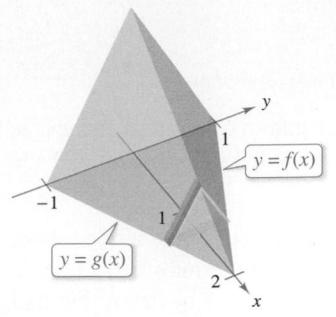

Cross sections are equilateral triangles.

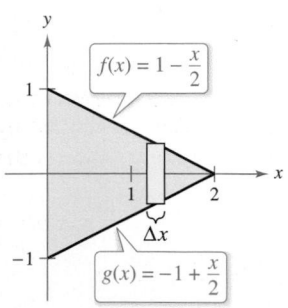

Triangular base in xy-plane
Figure 7.25

EXAMPLE 6 Triangular Cross Sections

Find the volume of the solid shown in Figure 7.25. The base of the solid is the region bounded by the lines

$$f(x) = 1 - \frac{x}{2}, \quad g(x) = -1 + \frac{x}{2}, \quad \text{and} \quad x = 0.$$

The cross sections perpendicular to the x-axis are equilateral triangles.

Solution The base and area of each triangular cross section are as follows.

$$\text{Base} = \left(1 - \frac{x}{2}\right) - \left(-1 + \frac{x}{2}\right) = 2 - x \qquad \text{Length of base}$$

$$\text{Area} = \frac{\sqrt{3}}{4}(\text{base})^2 \qquad \text{Area of equilateral triangle}$$

$$A(x) = \frac{\sqrt{3}}{4}(2 - x)^2 \qquad \text{Area of cross section}$$

Because x ranges from 0 to 2, the volume of the solid is

$$V = \int_a^b A(x)\, dx = \int_0^2 \frac{\sqrt{3}}{4}(2 - x)^2\, dx = -\frac{\sqrt{3}}{4}\left[\frac{(2 - x)^3}{3}\right]_0^2 = \frac{2\sqrt{3}}{3}.$$

EXAMPLE 7 An Application to Geometry

Prove that the volume of a pyramid with a square base is

$$V = \frac{1}{3}hB$$

where h is the height of the pyramid and B is the area of the base.

Solution As shown in Figure 7.26, you can intersect the pyramid with a plane parallel to the base at height y to form a square cross section whose sides are of length b'. Using similar triangles, you can show that

$$\frac{b'}{b} = \frac{h - y}{h} \quad \text{or} \quad b' = \frac{b}{h}(h - y)$$

where b is the length of the sides of the base of the pyramid. So,

$$A(y) = (b')^2 = \frac{b^2}{h^2}(h - y)^2.$$

Integrating between 0 and h produces

$$V = \int_0^h A(y)\, dy$$

$$= \int_0^h \frac{b^2}{h^2}(h - y)^2\, dy$$

$$= \frac{b^2}{h^2}\int_0^h (h - y)^2\, dy$$

$$= -\left(\frac{b^2}{h^2}\right)\left[\frac{(h - y)^3}{3}\right]_0^h$$

$$= \frac{b^2}{h^2}\left(\frac{h^3}{3}\right)$$

$$= \frac{1}{3}hB. \qquad \qquad B = b^2$$

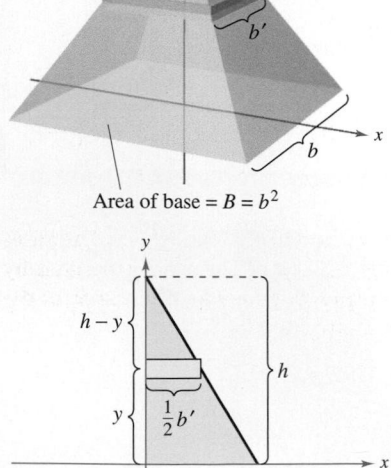

Figure 7.26

7.2 Exercises

See **CalcChat.com** for tutorial help and worked-out solutions to odd-numbered exercises.

CONCEPT CHECK

1. **Disk Method** Explain how to use the disk method to find the volume of a solid of revolution.

2. **Comparing Methods** What is the relationship between the disk method and the washer method?

3. **Finding the Volume of a Solid** In your own words, describe when it is necessary to use more than one integral to find the volume of a solid of revolution.

4. **Finding the Volume of a Solid** Explain how to find the volume of a solid with a known cross section.

 **Finding the Volume of a Solid** In Exercises 5–8, write and evaluate the definite integral that represents the volume of the solid formed by revolving the region about the **x-axis**.

5. $y = \sqrt{x}$

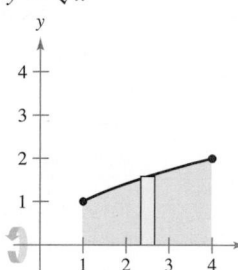

6. $y = -x + 1$

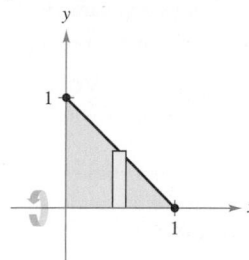

7. $y = x^2, \quad y = x^5$

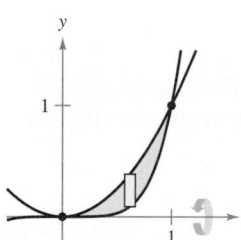

8. $y = 2, \quad y = 4 - \dfrac{x^2}{4}$

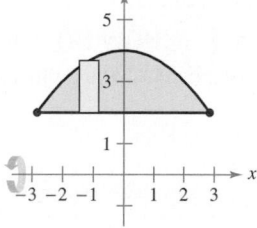

 **Finding the Volume of a Solid** In Exercises 9–12, write and evaluate the definite integral that represents the volume of the solid formed by revolving the region about the **y-axis**.

9. $y = x^2$

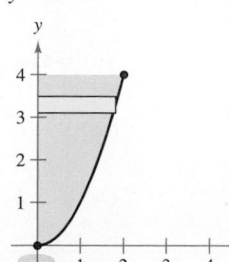

10. $y = \sqrt{16 - x^2}$

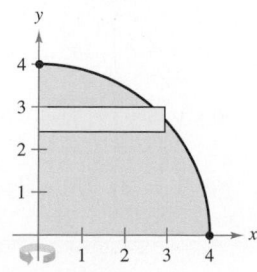

11. $y = x^{2/3}$

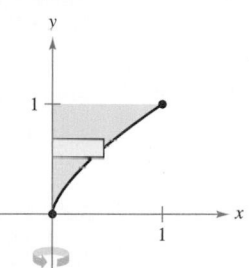

12. $x = -y^2 + 4y$

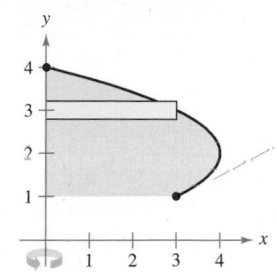

 Finding the Volume of a Solid In Exercises 13–16, find the volumes of the solids generated by revolving the region bounded by the graphs of the equations about the given lines.

13. $y = \sqrt{x}, \quad y = 0, \quad x = 3$

  (a) the x-axis   (b) the y-axis

  (c) the line $x = 3$ (d) the line $x = 6$

14. $y = 2x^2, \quad y = 0, \quad x = 2$

  (a) the y-axis   (b) the x-axis

  (c) the line $y = 8$ (d) the line $x = 2$

15. $y = x^2, \quad y = 4x - x^2$

  (a) the x-axis

  (b) the line $y = 6$

16. $y = 4 + 2x - x^2, \quad y = 4 - x$

  (a) the x-axis

  (b) the line $y = 1$

 **Finding the Volume of a Solid** In Exercises 17–20, find the volume of the solid generated by revolving the region bounded by the graphs of the equations about the line $y = 4$.

17. $y = x, \quad y = 3, \quad x = 0$

18. $y = \frac{1}{2}x^3, \quad y = 4, \quad x = 0$

19. $y = \dfrac{2}{1 + x}, \quad y = 0, \quad x = 0, \quad x = 4$

20. $y = \sqrt{1 - x}, \quad x = 0, \quad y = 0$

Finding the Volume of a Solid In Exercises 21–24, find the volume of the solid generated by revolving the region bounded by the graphs of the equations about the line $x = 5$.

21. $y = x, \quad y = 0, \quad y = 4, \quad x = 5$

22. $y = 2 - \dfrac{x}{2}, \quad y = 0, \quad y = 1, \quad x = 0$

23. $x = y^2, \quad x = 4$

24. $xy = 3, \quad y = 1, \quad y = 4, \quad x = 5$

Finding the Volume of a Solid In Exercises 25–32, find the volume of the solid generated by revolving the region bounded by the graphs of the equations about the *x*-axis.

25. $y = \dfrac{1}{\sqrt{3x+5}}, \quad y = 0, \quad x = 0, \quad x = 2$

26. $y = x\sqrt{4 - x^2}, \quad y = 0$

27. $y = \dfrac{6}{x}, \quad y = 0, \quad x = 1, \quad x = 3$

28. $y = \dfrac{2}{x+1}, \quad y = 0, \quad x = 0, \quad x = 6$

29. $y = e^{-3x}, \quad y = 0, \quad x = 0, \quad x = 2$

30. $y = e^{x/4}, \quad y = 0, \quad x = 0, \quad x = 6$

31. $y = x^2 + 1, \quad y = -x^2 + 2x + 5, \quad x = 0, \quad x = 3$

32. $y = \sqrt{x}, \quad y = -\frac{1}{2}x + 4, \quad x = 0, \quad x = 8$

Finding the Volume of a Solid In Exercises 33–36, find the volume of the solid generated by revolving the region bounded by the graphs of the equations about the *y*-axis.

33. $y = 3(2 - x), \quad y = 0, \quad x = 0$

34. $y = \sqrt{3x - 2}, \quad x = 0, \quad y = 0, \quad y = 1$

35. $y = 9 - x^2, \quad y = 0, \quad x = 2, \quad x = 3$

36. $y = \dfrac{x^3}{8}, \quad y = 0, \quad x = 4$

 **Finding the Volume of a Solid** In Exercises 37–40, find the volume of the solid generated by revolving the region bounded by the graphs of the equations about the *x*-axis. Verify your results using the integration capabilities of a graphing utility.

37. $y = \sin x, \quad y = 0, \quad x = 0, \quad x = \pi$

38. $y = \cos 2x, \quad y = 0, \quad x = 0, \quad x = \dfrac{\pi}{4}$

39. $y = e^{x-1}, \quad y = 0, \quad x = 1, \quad x = 2$

40. $y = e^{x/2} + e^{-x/2}, \quad y = 0, \quad x = -1, \quad x = 2$

Finding the Volume of a Solid In Exercises 41–48, find the volume of the solid generated by revolving the specified region about the given line.

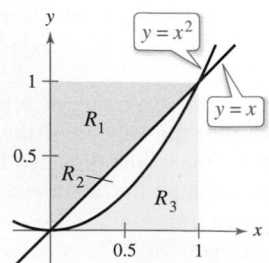

41. R_1 about $y = 0$ **42.** R_1 about $x = 1$

43. R_1 about $x = 0$ **44.** R_2 about $y = 1$

45. R_2 about $y = 0$ **46.** R_3 about $x = 1$

47. R_3 about $x = 0$ **48.** R_3 about $y = 1$

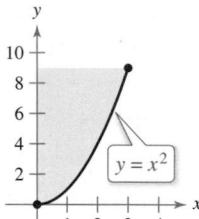 **Finding the Volume of a Solid Using Technology** In Exercises 49–52, use the integration capabilities of a graphing utility to approximate the volume of the solid generated by revolving the region bounded by the graphs of the equations about the *x*-axis.

49. $y = e^{-x^2}, \quad y = 0, \quad x = 0, \quad x = 2$

50. $y = \ln x, \quad y = 0, \quad x = 1, \quad x = 3$

51. $y = 2\arctan(0.2x), \quad y = 0, \quad x = 0, \quad x = 5$

52. $y = \sqrt{2x}, \quad y = x^2$

EXPLORING CONCEPTS

53. Describing a Solid Each integral represents the volume of a solid. Describe each solid.

(a) $\pi \displaystyle\int_0^{\pi/2} \sin^2 x \, dx$ (b) $\pi \displaystyle\int_2^4 y^4 \, dy$

54. Comparing Volumes A region bounded by the parabola $y = 4x - x^2$ and the *x*-axis is revolved about the *x*-axis. A second region bounded by the parabola $y = 4 - x^2$ and the *x*-axis is revolved about the *x*-axis. Without integrating, how do the volumes of the two solids compare? Explain.

55. Comparing Volumes The region in the figure is revolved about the indicated axes and line. Order the volumes of the resulting solids from least to greatest. Explain your reasoning.

(a) *x*-axis

(b) *y*-axis

(c) $x = 3$

56. 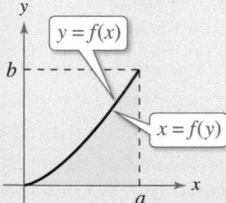 **HOW DO YOU SEE IT?** Use the graph to match the integral for the volume with the axis of rotation.

(a) $V = \pi \displaystyle\int_0^b \left(a^2 - [f(y)]^2\right) dy$ (i) *x*-axis

(b) $V = \pi \displaystyle\int_0^a \left(b^2 - [b - f(x)]^2\right) dx$ (ii) *y*-axis

(c) $V = \pi \displaystyle\int_0^a [f(x)]^2 \, dx$ (iii) $x = a$

(d) $V = \pi \displaystyle\int_0^b [a - f(y)]^2 \, dy$ (iv) $y = b$

Dividing a Solid In Exercises 57 and 58, consider the solid formed by revolving the region bounded by $y = \sqrt{x}$, $y = 0$, $x = 1$, and $x = 3$ about the x-axis.

57. Find the value of x in the interval $[1, 3]$ that divides the solid into two parts of equal volume.

58. Find the values of x in the interval $[1, 3]$ that divide the solid into three parts of equal volume.

59. Manufacturing A manufacturer drills a hole through the center of a metal sphere of radius R. The hole has a radius r. Find the volume of the resulting ring.

60. Manufacturing For the metal sphere in Exercise 59, let $R = 6$. What value of r will produce a ring whose volume is exactly half the volume of the sphere?

61. Volume of a Cone Use the disk method to verify that the volume of a right circular cone is $\frac{1}{3}\pi r^2 h$, where r is the radius of the base and h is the height.

62. Volume of a Sphere Use the disk method to verify that the volume of a sphere is $\frac{4}{3}\pi r^3$, where r is the radius.

63. Using a Cone A cone of height H with a base of radius r is cut by a plane parallel to and h units above the base, where $h < H$. Find the volume of the solid (frustum of a cone) below the plane.

64. Using a Sphere A sphere of radius r is cut by a plane h units above the equator, where $h < r$. Find the volume of the solid (spherical segment) above the plane.

65. Volume of a Fuel Tank A tank on the wing of a jet aircraft is formed by revolving the region bounded by the graph of $y = \frac{1}{8}x^2\sqrt{2 - x}$ and the x-axis ($0 \le x \le 2$) about the x-axis, where x and y are measured in meters. Use a graphing utility to graph the function. Find the volume of the tank analytically.

66. Volume of a Container A container can be modeled by revolving the graph of

$$y = \begin{cases} \sqrt{0.1x^3 - 2.2x^2 + 10.9x + 22.2}, & 0 \le x \le 11.5 \\ 2.95, & 11.5 < x \le 15 \end{cases}$$

about the x-axis, where x and y are measured in centimeters. Use a graphing utility to graph the function. Find the volume of the container analytically.

67. Finding Volumes of Solids Find the volumes of the solids (see figures) generated if the upper half of the ellipse $9x^2 + 25y^2 = 225$ is revolved about (a) the x-axis to form a prolate spheroid (shaped like a football) and (b) the y-axis to form an oblate spheroid (shaped like half of a candy).

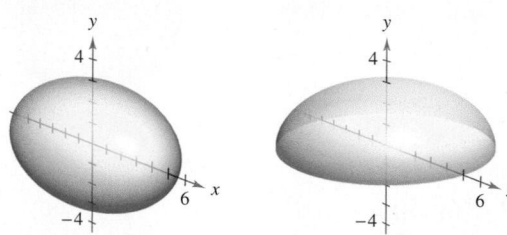

Figure for 67(a)   Figure for 67(b)

68. 3D Printing

A 3D printer is used to create a plastic drinking glass. The equations given to the printer for the inside of the glass are

$$x = \left(\frac{y}{4}\right)^{1/32} \quad \text{and} \quad y = 5$$

where x and y are measured in inches. What is the total volume that the drinking glass can hold when the region bounded by the graphs of the equations is revolved about the y-axis?

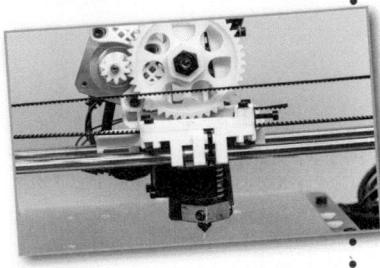

69. Minimum Volume The function $y = 4 - (x^2/4)$ on the interval $[0, 4]$ is revolved about the line $y = b$ (see figure).

(a) Find the volume of the resulting solid as a function of b.

(b) Use a graphing utility to graph the function in part (a), and use the graph to approximate the value of b that minimizes the volume of the solid.

(c) Use calculus to find the value of b that minimizes the volume of the solid, and compare the result with the answer to part (b).

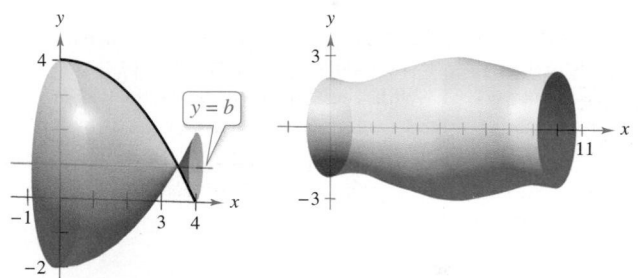

Figure for 69   Figure for 70

70. Modeling Data A draftsman is asked to determine the amount of material required to produce a machine part (see figure). The diameters d of the part at equally spaced points x are listed in the table. The measurements are listed in centimeters.

x	0	1	2	3	4	5
d	4.2	3.8	4.2	4.7	5.2	5.7

x	6	7	8	9	10
d	5.8	5.4	4.9	4.4	4.6

(a) Use the regression capabilities of a graphing utility to find a fourth-degree polynomial through the points representing the radius of the machine part. Plot the data and graph the model.

(b) Use the integration capabilities of a graphing utility to approximate the volume of the machine part.

71. Think About It Match each integral with the solid whose volume it represents, and give the dimensions of each solid.

(a) Right circular cylinder   (b) Ellipsoid

(c) Sphere   (d) Right circular cone   (e) Torus

(i) $\pi \int_0^h \left(\frac{rx}{h}\right)^2 dx$ (ii) $\pi \int_0^h r^2 \, dx$

(iii) $\pi \int_{-r}^r \left(\sqrt{r^2 - x^2}\right)^2 dx$

(iv) $\pi \int_{-b}^b \left(a\sqrt{1 - \frac{x^2}{b^2}}\right)^2 dx$

(v) $\pi \int_{-r}^r \left[(R + \sqrt{r^2 - x^2})^2 - (R - \sqrt{r^2 - x^2})^2\right] dx$

72. Cavalieri's Theorem Prove that if two solids have equal altitudes and all cross sections parallel to their bases and at equal distances from their bases have equal areas, then the solids have the same volume (see figure).

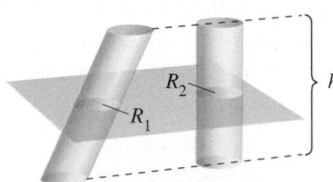

Area of R_1 = area of R_2

73. Using Cross Sections Find the volumes of the solids whose bases are bounded by the graphs of $y = x + 1$ and $y = x^2 - 1$, with the indicated cross sections taken perpendicular to the x-axis.

(a) Squares   (b) Rectangles of height 1

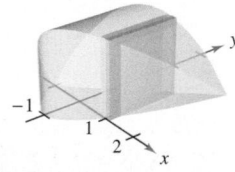   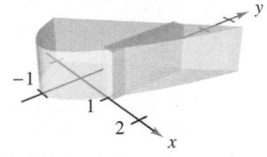

74. Using Cross Sections Find the volumes of the solids whose bases are bounded by the circle $x^2 + y^2 = 4$, with the indicated cross sections taken perpendicular to the x-axis.

(a) Squares   (b) Equilateral triangles

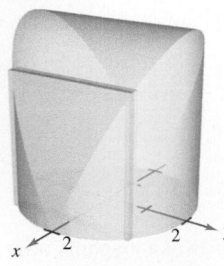   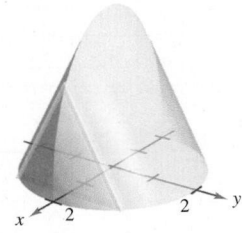

(c) Semicircles   (d) Isosceles right triangles

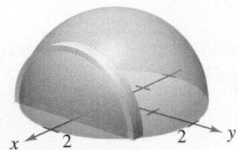   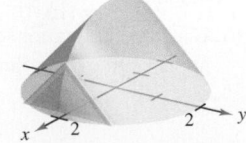

75. Using Cross Sections Find the volume of the solid of intersection (the solid common to both) of the two right circular cylinders of radius r whose axes meet at right angles (see figure).

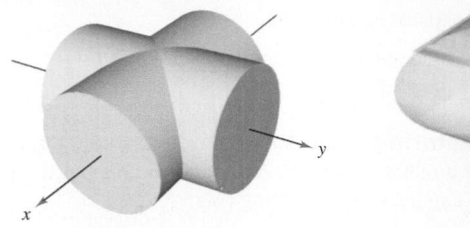

Two intersecting cylinders   Solid of intersection

■ **FOR FURTHER INFORMATION** For more information on this problem, see the article "Estimating the Volumes of Solid Figures with Curved Surfaces" by Donald Cohen in *Mathematics Teacher*. To view this article, go to *MathArticles.com*.

76. Using Cross Sections The solid shown in the figure has cross sections bounded by the graph of $|x|^a + |y|^a = 1$, where $1 \le a \le 2$.

(a) Describe the cross section when $a = 1$ and $a = 2$.

(b) Describe a procedure for approximating the volume of the solid.

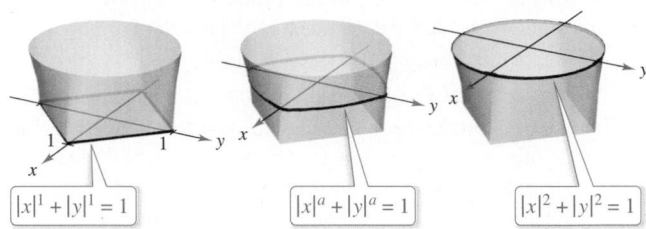

$|x|^1 + |y|^1 = 1$ $|x|^a + |y|^a = 1$ $|x|^2 + |y|^2 = 1$

77. Volume of a Wedge Two planes cut a right circular cylinder to form a wedge. One plane is perpendicular to the axis of the cylinder and the second makes an angle of θ degrees with the first (see figure).

(a) Find the volume of the wedge if $\theta = 45°$.

(b) Find the volume of the wedge for an arbitrary angle θ. Assuming that the cylinder has sufficient length, how does the volume of the wedge change as θ increases from $0°$ to $90°$?

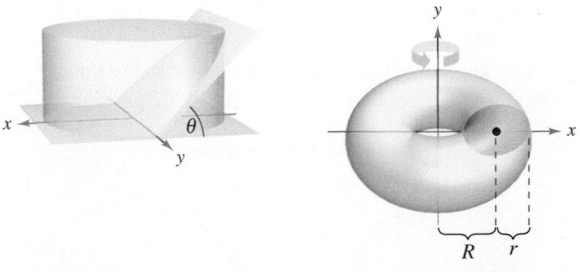

Figure for 77   Figure for 78

78. Volume of a Torus

(a) Show that the volume of the torus shown in the figure is given by the integral $8\pi R \int_0^r \sqrt{r^2 - y^2} \, dy$, where $R > r > 0$.

(b) Find the volume of the torus.

7.3 Volume: The Shell Method

■ Find the volume of a solid of revolution using the shell method.
■ Compare the uses of the disk method and the shell method.

The Shell Method

In this section, you will study an alternative method for finding the volume of a solid of revolution. This method is called the **shell method** because it uses cylindrical shells. A comparison of the advantages of the disk and shell methods is given later in this section.

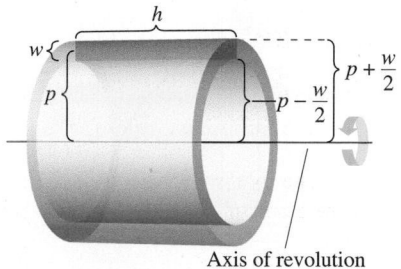

Figure 7.27

To begin, consider a representative rectangle as shown in Figure 7.27, where w is the width of the rectangle, h is the height of the rectangle, and p is the distance between the axis of revolution and the *center* of the rectangle. When this rectangle is revolved about its axis of revolution, it forms a cylindrical shell (or tube) of thickness w. To find the volume of this shell, consider two cylinders. The radius of the larger cylinder corresponds to the outer radius of the shell, and the radius of the smaller cylinder corresponds to the inner radius of the shell. Because p is the average radius of the shell, you know the outer radius is

$$p + \frac{w}{2} \qquad \text{Outer radius}$$

and the inner radius is

$$p - \frac{w}{2}. \qquad \text{Inner radius}$$

So, the volume of the shell is

$$\text{Volume of shell} = (\text{volume of cylinder}) - (\text{volume of hole})$$
$$= \pi\left(p + \frac{w}{2}\right)^2 h - \pi\left(p - \frac{w}{2}\right)^2 h$$
$$= 2\pi phw$$
$$= 2\pi(\text{average radius})(\text{height})(\text{thickness}).$$

You can use this formula to find the volume of a solid of revolution. For instance, the plane region in Figure 7.28 is revolved about a line to form the indicated solid. Consider a horizontal rectangle of width Δy. As the plane region is revolved about a line parallel to the x-axis, the rectangle generates a representative shell whose volume is

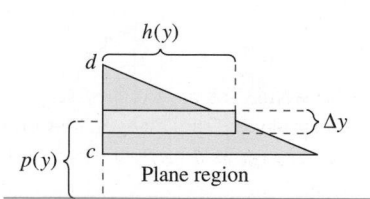

$$\Delta V = 2\pi[p(y)h(y)]\Delta y.$$

You can approximate the volume of the solid by n such shells of thickness Δy, height $h(y_i)$, and average radius $p(y_i)$.

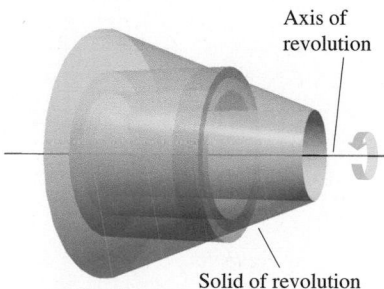

Figure 7.28

$$\text{Volume of solid} \approx \sum_{i=1}^{n} 2\pi[p(y_i)h(y_i)]\Delta y = 2\pi \sum_{i=1}^{n} [p(y_i)h(y_i)]\Delta y$$

This approximation appears to become better and better as $\|\Delta\| \to 0$ $(n \to \infty)$. So, the volume of the solid is

$$\text{Volume of solid} = \lim_{\|\Delta\| \to 0} 2\pi \sum_{i=1}^{n} [p(y_i)h(y_i)]\Delta y$$
$$= 2\pi \int_{c}^{d} [p(y)h(y)] \, dy.$$

THE SHELL METHOD

To find the volume of a solid of revolution with the **shell method,** use one of the formulas below. (See Figure 7.29.)

Horizontal Axis of Revolution   **Vertical Axis of Revolution**

$$\text{Volume} = V = 2\pi \int_c^d p(y)h(y)\, dy \qquad \text{Volume} = V = 2\pi \int_a^b p(x)h(x)\, dx$$

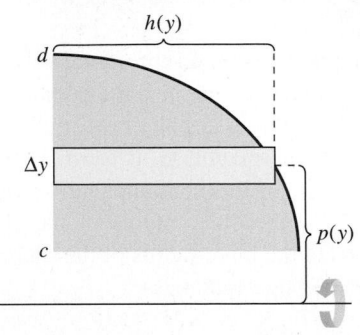

Horizontal axis of revolution
Figure 7.29   Vertical axis of revolution

EXAMPLE 1   **Using the Shell Method to Find Volume**

Find the volume of the solid formed by revolving the region bounded by

$$y = x - x^3$$

and the x-axis $(0 \le x \le 1)$ about the y-axis.

Solution   Because the axis of revolution is vertical, use a vertical representative rectangle, as shown in Figure 7.30. The width Δx indicates that x is the variable of integration. The distance from the center of the rectangle to the axis of revolution is $p(x) = x$, and the height of the rectangle is

$$h(x) = x - x^3.$$

Because x ranges from 0 to 1, apply the shell method to find the volume of the solid.

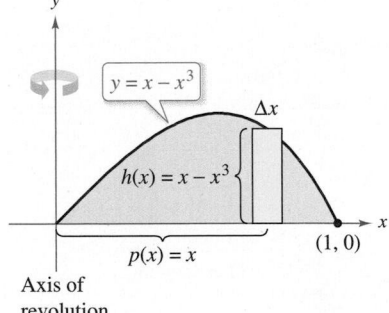

Axis of revolution

Figure 7.30

$$
\begin{aligned}
V &= 2\pi \int_a^b p(x)h(x)\, dx \\
&= 2\pi \int_0^1 x(x - x^3)\, dx \\
&= 2\pi \int_0^1 (-x^4 + x^2)\, dx \qquad \text{Simplify.} \\
&= 2\pi \left[-\frac{x^5}{5} + \frac{x^3}{3} \right]_0^1 \qquad \text{Integrate.} \\
&= 2\pi \left(-\frac{1}{5} + \frac{1}{3} \right) \\
&= \frac{4\pi}{15}
\end{aligned}
$$

EXAMPLE 2   **Using the Shell Method to Find Volume**

Find the volume of the solid formed by revolving the region bounded by the graph of

$$x = e^{-y^2}$$

and the y axis $(0 \le y \le 1)$ about the x-axis.

Solution   Because the axis of revolution is horizontal, use a horizontal representative rectangle, as shown in Figure 7.31. The width Δy indicates that y is the variable of integration. The distance from the center of the rectangle to the axis of revolution is $p(y) = y$, and the height of the rectangle is $h(y) = e^{-y^2}$. Because y ranges from 0 to 1, the volume of the solid is

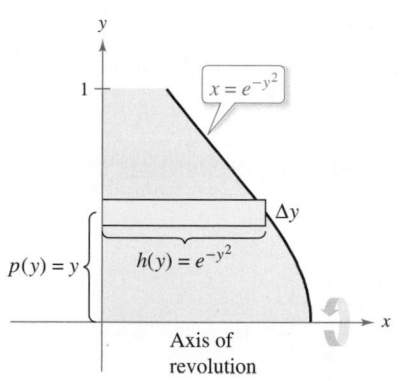

Figure 7.31

$$
\begin{aligned}
V &= 2\pi \int_c^d p(y)h(y)\, dy && \text{Apply shell method.}\\
&= 2\pi \int_0^1 y e^{-y^2}\, dy &&\\
&= -\pi \left[e^{-y^2} \right]_0^1 && \text{Integrate.}\\
&= \pi \left(1 - \frac{1}{e} \right)\\
&\approx 1.986.
\end{aligned}
$$

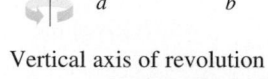

Exploration

To see the advantage of using the shell method in Example 2, solve the equation $x = e^{-y^2}$ for y.

$$
y = \begin{cases} 1, & 0 \le x \le 1/e \\ \sqrt{-\ln x}, & 1/e < x \le 1 \end{cases}
$$

Then use this equation to find the volume using the disk method.

Comparison of Disk and Shell Methods

The disk and shell methods can be distinguished as follows. For the disk method, the representative rectangle is always *perpendicular* to the axis of revolution, whereas for the shell method, the representative rectangle is always *parallel* to the axis of revolution, as shown in Figure 7.32.

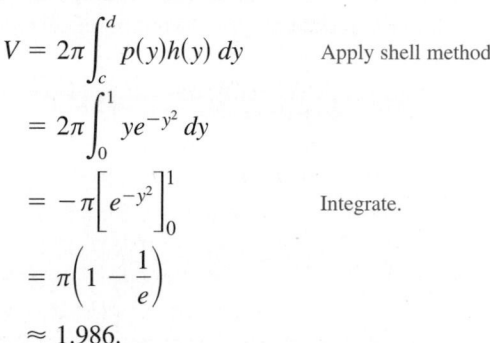

Vertical axis of revolution   Horizontal axis of revolution
        Disk method: Representative rectangle is
        perpendicular to the axis of revolution.

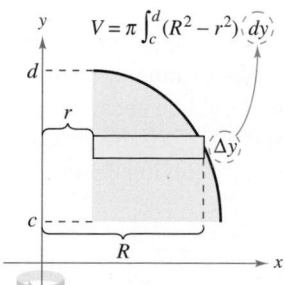

Vertical axis of revolution   Horizontal axis of revolution
        Shell method: Representative rectangle is
        parallel to the axis of revolution.

Figure 7.32

Often, one method is more convenient to use than the other. The next example illustrates a case in which the shell method is preferable.

EXAMPLE 3   Shell Method Preferable

⋮ ⋯▷ *See LarsonCalculus.com for an interactive version of this type of example.*

Find the volume of the solid formed by revolving the region bounded by the graphs of

$$y = x^2 + 1, \quad y = 0, \quad x = 0, \quad \text{and} \quad x = 1$$

about the *y*-axis.

Solution   In Example 4 in Section 7.2, you saw that the washer method requires two integrals to determine the volume of this solid. See Figure 7.33(a).

$$V = \pi \int_0^1 (1^2 - 0^2) \, dy + \pi \int_1^2 \left[1^2 - \left(\sqrt{y-1}\right)^2\right] dy \qquad \text{Apply washer method.}$$

$$= \pi \int_0^1 1 \, dy + \pi \int_1^2 (2 - y) \, dy \qquad \text{Simplify.}$$

$$= \pi \left[y\right]_0^1 + \pi \left[2y - \frac{y^2}{2}\right]_1^2 \qquad \text{Integrate.}$$

$$= \pi + \pi\left(4 - 2 - 2 + \frac{1}{2}\right)$$

$$= \frac{3\pi}{2}$$

In Figure 7.33(b), you can see that the shell method requires only one integral to find the volume.

$$V = 2\pi \int_a^b p(x)h(x) \, dx \qquad \text{Apply shell method.}$$

$$= 2\pi \int_0^1 x(x^2 + 1) \, dx$$

$$= 2\pi \left[\frac{x^4}{4} + \frac{x^2}{2}\right]_0^1 \qquad \text{Integrate.}$$

$$= 2\pi\left(\frac{3}{4}\right)$$

$$= \frac{3\pi}{2}$$

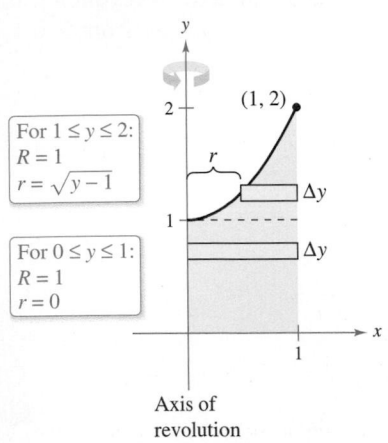

For $1 \le y \le 2$:
$R = 1$
$r = \sqrt{y-1}$

For $0 \le y \le 1$:
$R = 1$
$r = 0$

Axis of revolution

(a) Disk method

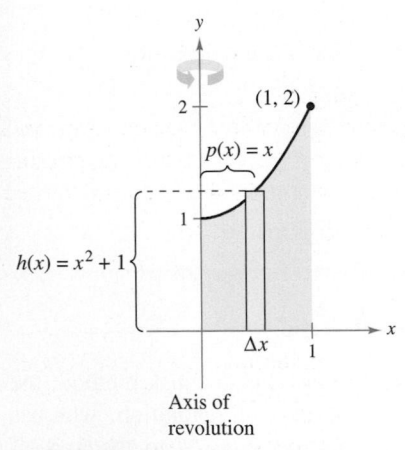

$p(x) = x$

$h(x) = x^2 + 1$

Axis of revolution

(b) Shell method

Figure 7.33

Consider the solid formed by revolving the region in Example 3 about the vertical line $x = 1$. Would the resulting solid of revolution have a greater volume or a smaller volume than the solid in Example 3? Without integrating, you should be able to reason that the resulting solid would have a smaller volume because "more" of the revolved region would be closer to the axis of revolution. To confirm this, try solving the integral

$$V = 2\pi \int_0^1 (1 - x)(x^2 + 1) \, dx \qquad p(x) = 1 - x$$

which gives the volume of the solid.

■ **FOR FURTHER INFORMATION**   To learn more about the disk and shell methods, see the article "The Disk and Shell Method" by Charles A. Cable in *The American Mathematical Monthly.* To view this article, go to *MathArticles.com.*

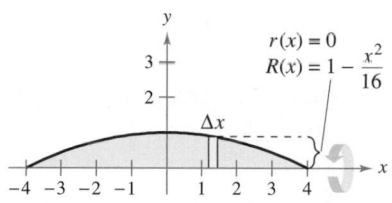

Figure 7.34

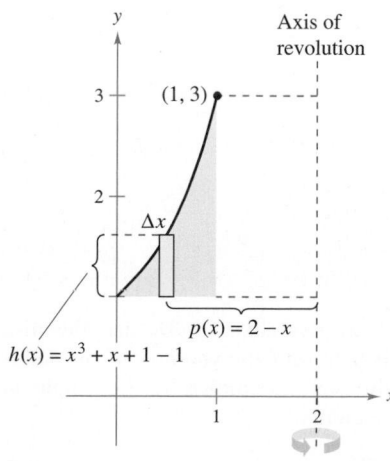

Disk method
Figure 7.35

EXAMPLE 4   **Volume of a Pontoon**

A pontoon is to be made in the shape shown in Figure 7.34. The pontoon is designed by rotating the graph of

$$y = 1 - \frac{x^2}{16}, \quad -4 \le x \le 4$$

about the x-axis, where x and y are measured in feet. Find the volume of the pontoon.

Solution   Refer to Figure 7.35 and use the disk method as shown.

$$V = \pi \int_{-4}^{4} \left(1 - \frac{x^2}{16}\right)^2 dx \qquad \text{Apply disk method.}$$

$$= \pi \int_{-4}^{4} \left(1 - \frac{x^2}{8} + \frac{x^4}{256}\right) dx \qquad \text{Simplify.}$$

$$= \pi \left[x - \frac{x^3}{24} + \frac{x^5}{1280}\right]_{-4}^{4} \qquad \text{Integrate.}$$

$$= \frac{64\pi}{15}$$

$$\approx 13.4 \text{ cubic feet}$$

To use the shell method in Example 4, you would have to solve for x in terms of y in the equation

$$y = 1 - \frac{x^2}{16}$$

and then evaluate an integral that requires a u-substitution.

Sometimes, solving for x is very difficult (or even impossible). In such cases, you must use a vertical rectangle (of width Δx), thus making x the variable of integration. The position (horizontal or vertical) of the axis of revolution then determines the method to be used. This is shown in Example 5.

EXAMPLE 5   **Shell Method Necessary**

Find the volume of the solid formed by revolving the region bounded by the graphs of $y = x^3 + x + 1$, $y = 1$, and $x = 1$ about the line $x = 2$, as shown in Figure 7.36.

Solution   In the equation $y = x^3 + x + 1$, you cannot easily solve for x in terms of y. (See the discussion at the end of Section 3.8.) Therefore, the variable of integration must be x, and you should choose a vertical representative rectangle. Because the rectangle is parallel to the axis of revolution, use the shell method.

$$V = 2\pi \int_{a}^{b} p(x)h(x) \, dx \qquad \text{Apply shell method.}$$

$$= 2\pi \int_{0}^{1} (2 - x)(x^3 + x + 1 - 1) \, dx$$

$$= 2\pi \int_{0}^{1} (-x^4 + 2x^3 - x^2 + 2x) \, dx \qquad \text{Simplify.}$$

$$= 2\pi \left[-\frac{x^5}{5} + \frac{x^4}{2} - \frac{x^3}{3} + x^2\right]_{0}^{1} \qquad \text{Integrate.}$$

$$= 2\pi \left(-\frac{1}{5} + \frac{1}{2} - \frac{1}{3} + 1\right)$$

$$= \frac{29\pi}{15}$$

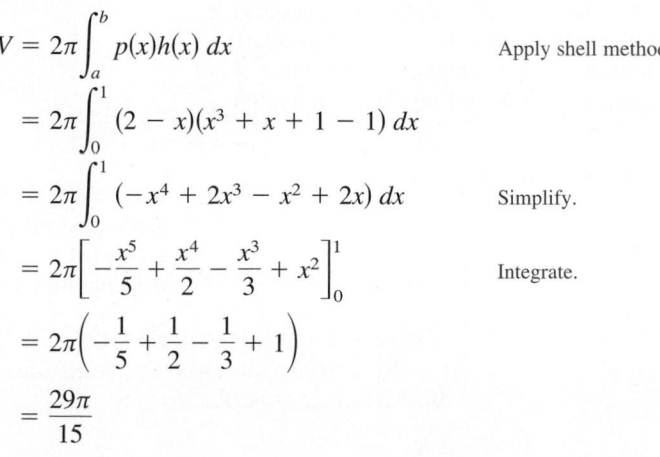

Figure 7.36

7.3 Exercises

See **CalcChat.com** for tutorial help and worked-out solutions to odd-numbered exercises.

CONCEPT CHECK

1. **Shell Method** Explain how to use the shell method to find the volume of a solid of revolution.

2. **Representative Rectangles** Compare the representative rectangles for the disk and shell methods.

Finding the Volume of a Solid In Exercises 3–12, use the shell method to write and evaluate the definite integral that represents the volume of the solid generated by revolving the plane region about the *y*-axis.

3. $y = x$

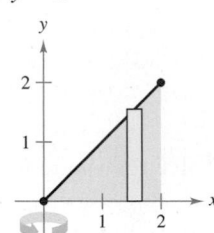

4. $y = 1 - x$

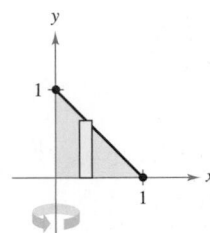

5. $y = \sqrt{x}$

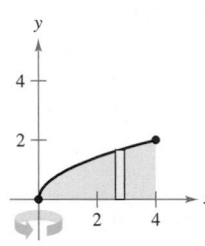

6. $y = \frac{1}{2}x^2 + 1$

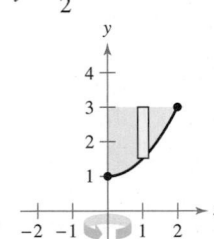

7. $y = \frac{1}{4}x^2$, $y = 0$, $x = 4$ **8.** $y = \frac{1}{2}x^3$, $y = 0$, $x = 3$

9. $y = x^2$, $y = 4x - x^2$ **10.** $y = x^3$, $y = 8$, $x = 0$

11. $y = \sqrt{2x - 5}$, $y = 0$, $x = 4$

12. $y = x^{3/2}$, $y = 8$, $x = 0$

Finding the Volume of a Solid In Exercises 13–22, use the shell method to write and evaluate the definite integral that represents the volume of the solid generated by revolving the plane region about the *x*-axis.

13. $y = x$

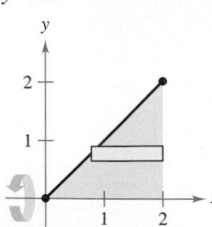

14. $y = 1 - x$

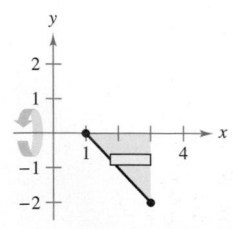

15. $y = \dfrac{1}{x}$

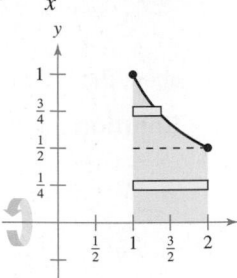

16. $x + y^2 = 4$

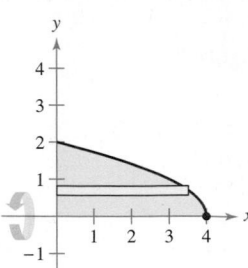

17. $y = x^3$, $x = 0$, $y = 8$ **18.** $y = 4x^2$, $x = 0$, $y = 4$

19. $x + y = 4$, $y = x$, $y = 0$ **20.** $y = 3 - x$, $y = 0$, $x = 6$

21. $y = 1 - \sqrt{x}$, $y = x + 1$, $y = 0$

22. $y = \sqrt{x + 2}$, $y = x$, $y = 0$

Finding the Volume of a Solid In Exercises 23–26, use the shell method to find the volume of the solid generated by revolving the region bounded by the graphs of the equations about the given line.

23. $y = 2x - x^2$, $y = 0$, about the line $x = 4$

24. $y = \sqrt{x}$, $y = 0$, $x = 4$, about the line $x = 6$

25. $y = 3x - x^2$, $y = x^2$, about the line $x = 2$

26. $y = \frac{1}{3}x^3$, $y = 6x - x^2$, about the line $x = 3$

Choosing a Method In Exercises 27 and 28, decide whether it is more convenient to use the disk method or the shell method to find the volume of the solid of revolution. Explain your reasoning. (Do not find the volume.)

27. $(y - 2)^2 = 4 - x$

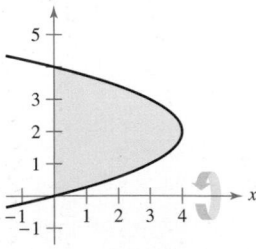

28. $y = 4 - e^x$

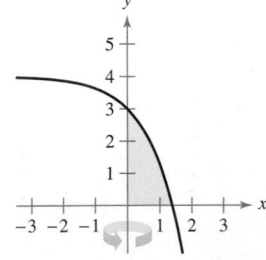

Choosing a Method In Exercises 29–32, use the disk method *or* the shell method to find the volumes of the solids generated by revolving the region bounded by the graphs of the equations about the given lines.

29. $y = x^3$, $y = 0$, $x = 2$

   (a) the *x*-axis   (b) the *y*-axis   (c) the line $x = 4$

30. $y = \dfrac{10}{x^2}$, $y = 0$, $x = 1$, $x = 5$

    (a) the x-axis   (b) the y-axis   (c) the line $y = 10$

31. $x^{1/2} + y^{1/2} = a^{1/2}$, $x = 0$, $y = 0$

    (a) the x-axis   (b) the y-axis   (c) the line $x = a$

32. $x^{2/3} + y^{2/3} = a^{2/3}$, $a > 0$ (hypocycloid)

    (a) the x-axis   (b) the y-axis

Finding the Volume of a Solid Using Technology In Exercises 33–36, (a) use a graphing utility to graph the region bounded by the graphs of the equations, and (b) use the integration capabilities of the graphing utility to approximate the volume of the solid generated by revolving the region about the y-axis.

33. $x^{4/3} + y^{4/3} = 1$, $x = 0$, $y = 0$, first quadrant

34. $y = \sqrt{1 - x^3}$, $y = 0$, $x = 0$

35. $y = \sqrt[3]{(x - 2)^2(x - 6)^2}$, $y = 0$, $x = 2$, $x = 6$

36. $y = \dfrac{2}{1 + e^{1/x}}$, $y = 0$, $x = 1$, $x = 3$

EXPLORING CONCEPTS

37. Describing Cylindrical Shells Consider the plane region bounded by the graphs of $y = k$, $y = 0$, $x = 0$, and $x = b$, where $k > 0$ and $b > 0$. What are the heights and radii of the cylinders generated when this region is revolved about (a) the x-axis and (b) the y-axis?

38. Think About It A solid is generated by revolving the region bounded by $y = 9 - x^2$ and $x = 0$ about the y-axis. Explain why you can use the shell method with limits of integration $x = 0$ and $x = 3$ to find the volume of the solid.

Comparing Integrals In Exercises 39 and 40, give a geometric argument that explains why the integrals have equal values.

39. $\pi \displaystyle\int_{1}^{5} (x - 1)\, dx = 2\pi \displaystyle\int_{0}^{2} y[5 - (y^2 + 1)]\, dy$

40. $\pi \displaystyle\int_{0}^{2} [16 - (2y)^2]\, dy = 2\pi \displaystyle\int_{0}^{4} x\left(\dfrac{x}{2}\right)\, dx$

---

41. Comparing Volumes The region in the figure is revolved about the indicated axes and line. Order the volumes of the resulting solids from least to greatest. Explain your reasoning.

    (a) x-axis   (b) y-axis   (c) $x = 4$

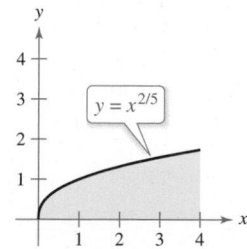

42. **HOW DO YOU SEE IT?** Use the graph to answer the following.

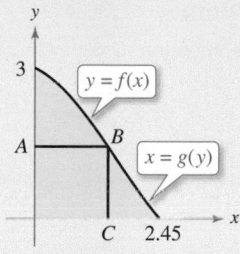

    (a) Describe the figure generated by revolving segment AB about the y-axis.

    (b) Describe the figure generated by revolving segment BC about the y-axis.

    (c) Assume the curve in the figure can be described as $y = f(x)$ or $x = g(y)$. A solid is generated by revolving the region bounded by the curve, $y = 0$, and $x = 0$ about the y-axis. Set up integrals to find the volume of this solid using the disk method and the shell method. (Do not integrate.)

Analyzing an Integral In Exercises 43–46, the integral represents the volume of a solid of revolution. Identify (a) the plane region that is revolved and (b) the axis of revolution.

43. $2\pi \displaystyle\int_{0}^{2} x^3\, dx$

44. $2\pi \displaystyle\int_{0}^{1} (y - y^{3/2})\, dy$

45. $2\pi \displaystyle\int_{0}^{6} (y + 2)\sqrt{6 - y}\, dy$

46. $2\pi \displaystyle\int_{0}^{1} (4 - x)e^x\, dx$

47. Machine Part A solid is generated by revolving the region bounded by $y = \frac{1}{2}x^2$ and $y = 2$ about the y-axis. A hole, centered along the axis of revolution, is drilled through this solid so that one-fourth of the volume is removed. Find the diameter of the hole.

48. Machine Part A solid is generated by revolving the region bounded by $y = \sqrt{9 - x^2}$ and $y = 0$ about the y-axis. A hole, centered along the axis of revolution, is drilled through this solid so that one-third of the volume is removed. Find the diameter of the hole.

49. Volume of a Torus A torus is formed by revolving the region bounded by the circle $x^2 + y^2 = 1$ about the line $x = 2$ (see figure). Find the volume of this "doughnut-shaped" solid. (*Hint:* The integral $\int_{-1}^{1} \sqrt{1 - x^2}\, dx$ represents the area of a semicircle.)

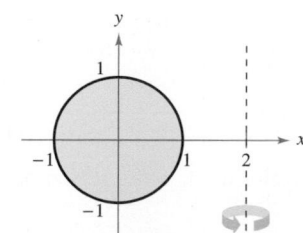

50. Volume of a Torus Repeat Exercise 49 for a torus formed by revolving the region bounded by the circle $x^2 + y^2 = r^2$ about the line $x = R$, where $r < R$.

51. Finding Volumes of Solids

(a) Use differentiation to verify that

$$\int x \sin x \, dx = \sin x - x \cos x + C.$$

(b) Use the result of part (a) to find the volume of the solid generated by revolving each plane region about the y-axis.

(i)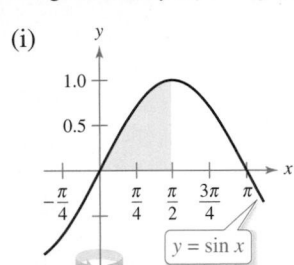

(ii)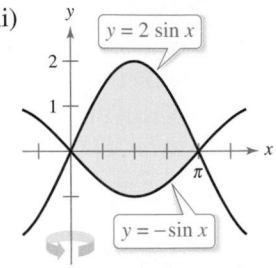

52. Finding Volumes of Solids

(a) Use differentiation to verify that

$$\int x \cos x \, dx = \cos x + x \sin x + C.$$

(b) Use the result of part (a) to find the volume of the solid generated by revolving each plane region about the y-axis. (*Hint:* Begin by approximating the points of intersection.)

(i)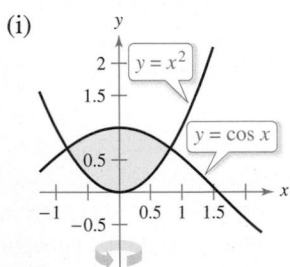

(ii)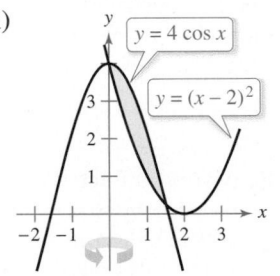

53. Volume of a Segment of a Sphere Let a sphere of radius r be cut by a plane, thereby forming a segment of height h. Show that the volume of this segment is

$$\frac{1}{3}\pi h^2 (3r - h).$$

54. Volume of an Ellipsoid Consider the plane region bounded by the graph of the ellipse

$$\left(\frac{x}{a}\right)^2 + \left(\frac{y}{b}\right)^2 = 1$$

where $a > 0$ and $b > 0$. Show that the volume of the ellipsoid formed when this region is revolved about the y-axis is

$$\frac{4}{3}\pi a^2 b.$$

What is the volume when the region is revolved about the x-axis?

55. Exploration Consider the region bounded by the graphs of $y = ax^n$, $y = ab^n$, and $x = 0$, as shown in the figure.

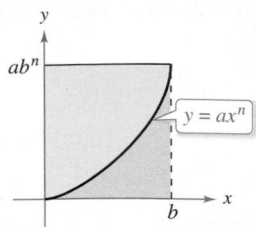

(a) Find the ratio $R_1(n)$ of the area of the region to the area of the circumscribed rectangle.

(b) Find $\lim\limits_{n \to \infty} R_1(n)$ and compare the result with the area of the circumscribed rectangle.

(c) Find the volume of the solid of revolution formed by revolving the region about the y-axis. Find the ratio $R_2(n)$ of this volume to the volume of the circumscribed right circular cylinder.

(d) Find $\lim\limits_{n \to \infty} R_2(n)$ and compare the result with the volume of the circumscribed cylinder.

(e) Use the results of parts (b) and (d) to make a conjecture about the shape of the graph of $y = ax^n$, $0 \le x \le b$, as $n \to \infty$.

56. Think About It Match each integral with the solid whose volume it represents and give the dimensions of each solid.

(a) Right circular cone   (b) Torus   (c) Sphere

(d) Right circular cylinder   (e) Ellipsoid

(i) $\displaystyle 2\pi \int_0^r hx \, dx$

(ii) $\displaystyle 2\pi \int_0^r hx\left(1 - \frac{x}{r}\right) dx$

(iii) $\displaystyle 2\pi \int_0^r 2x\sqrt{r^2 - x^2} \, dx$

(iv) $\displaystyle 2\pi \int_0^b 2ax\sqrt{1 - \frac{x^2}{b^2}} \, dx$

(v) $\displaystyle 2\pi \int_{-r}^r (R - x)\left(2\sqrt{r^2 - x^2}\right) dx$

57. Volume of a Storage Shed A storage shed has a circular base of diameter 80 feet. Starting at the center, the interior height is measured every 10 feet and recorded in the table (see figure). Find the volume of the shed.

x	Height
0	50
10	45
20	40
30	20
40	0

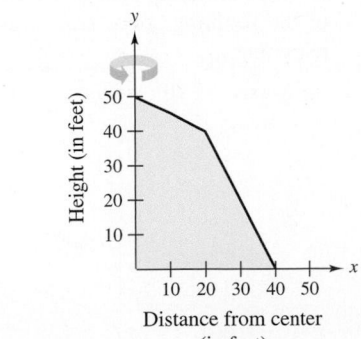

Distance from center
(in feet)

58. Modeling Data A pond is approximately circular, with a diameter of 400 feet. Starting at the center, the depth of the water is measured every 25 feet and recorded in the table (see figure).

x	0	25	50
Depth	20	19	19

x	75	100	125
Depth	17	15	14

x	150	175	200
Depth	10	6	0

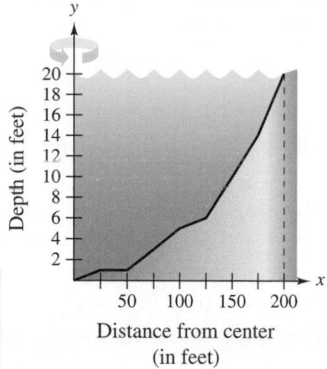

Distance from center (in feet)

(a) Use the regression capabilities of a graphing utility to find a quadratic model for the depths recorded in the table. Use the graphing utility to plot the depths and graph the model.

(b) Use the integration capabilities of a graphing utility and the model in part (a) to approximate the volume of water in the pond.

(c) Use the result of part (b) to approximate the number of gallons of water in the pond. (*Hint:* 1 cubic foot of water is approximately 7.48 gallons.)

59. Equal Volumes Let V_1 and V_2 be the volumes of the solids that result when the plane region bounded by

$$y = \frac{1}{x}, \quad y = 0, \quad x = \frac{1}{4}, \quad \text{and} \quad x = c, \quad c > \frac{1}{4}$$

is revolved about the x-axis and the y-axis, respectively. Find the value of c for which $V_1 = V_2$.

60. Volume of a Segment of a Paraboloid The region bounded by $y = r^2 - x^2$, $y = 0$, and $x = 0$ is revolved about the y-axis to form a paraboloid. A hole, centered along the axis of revolution, is drilled through this solid. The hole has a radius k, $0 < k < r$. Find the volume of the resulting ring (a) by integrating with respect to x and (b) by integrating with respect to y.

61. Finding Volumes of Solids Consider the graph of $y^2 = x(4 - x)^2$, as shown in the figure. Find the volumes of the solids that are generated when the loop of this graph is revolved about (a) the x-axis, (b) the y-axis, and (c) the line $x = 4$.

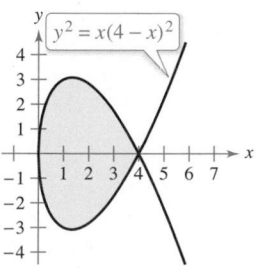

SECTION PROJECT

Saturn

The Oblateness of Saturn Saturn is the most oblate of the planets in our solar system. Its equatorial radius is 60,268 kilometers and its polar radius is 54,364 kilometers. The color-enhanced photograph of Saturn was taken by Voyager 1. In the photograph, the oblateness of Saturn is clearly visible.

(a) Find the ratio of the volumes of the sphere and the oblate ellipsoid shown below.

(b) If a planet were spherical and had the same volume as Saturn, what would its radius be?

Computer model of "spherical Saturn," whose equatorial radius is equal to its polar radius. The equation of the cross section passing through the pole is

$$x^2 + y^2 = 60,268^2.$$

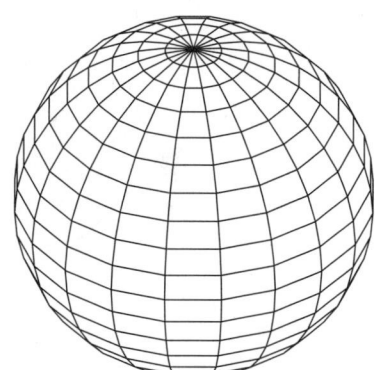

Computer model of "oblate Saturn," whose equatorial radius is greater than its polar radius. The equation of the cross section passing through the pole is

$$\frac{x^2}{60,268^2} + \frac{y^2}{54,364^2} = 1.$$

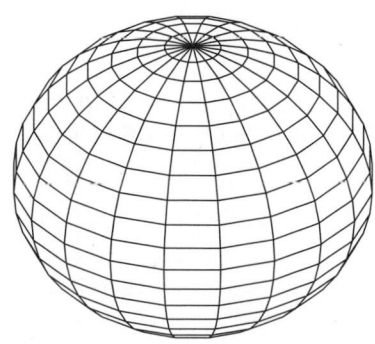

7.4 Arc Length and Surfaces of Revolution

■ Find the arc length of a smooth curve.
■ Find the area of a surface of revolution.

Arc Length

In this section, definite integrals are used to find the arc lengths of curves and the areas of surfaces of revolution. In either case, an arc (a segment of a curve) is approximated by straight line segments whose lengths are given by the familiar Distance Formula

$$d = \sqrt{(x_2 - x_1)^2 + (y_2 - y_1)^2}.$$

A **rectifiable** curve is one that has a finite arc length. You will see that a sufficient condition for the graph of a function f to be rectifiable between $(a, f(a))$ and $(b, f(b))$ is that f' be continuous on $[a, b]$. Such a function is **continuously differentiable** on $[a, b]$, and its graph on the interval $[a, b]$ is a **smooth curve.**

Consider a function $y = f(x)$ that is continuously differentiable on the interval $[a, b]$. You can approximate the graph of f by n line segments whose endpoints are determined by the partition

$$a = x_0 < x_1 < x_2 < \cdots < x_n = b$$

as shown in Figure 7.37. By letting $\Delta x_i = x_i - x_{i-1}$ and $\Delta y_i = y_i - y_{i-1}$, you can approximate the length of the graph by

$$s \approx \sum_{i=1}^{n} \sqrt{(x_i - x_{i-1})^2 + (y_i - y_{i-1})^2}$$

$$= \sum_{i=1}^{n} \sqrt{(\Delta x_i)^2 + (\Delta y_i)^2}$$

$$= \sum_{i=1}^{n} \sqrt{(\Delta x_i)^2 + \left(\frac{\Delta y_i}{\Delta x_i}\right)^2 (\Delta x_i)^2}$$

$$= \sum_{i=1}^{n} \sqrt{1 + \left(\frac{\Delta y_i}{\Delta x_i}\right)^2} (\Delta x_i).$$

This approximation appears to become better and better as $\|\Delta\| \to 0$ $(n \to \infty)$. So, the length of the graph is

$$s = \lim_{\|\Delta\| \to 0} \sum_{i=1}^{n} \sqrt{1 + \left(\frac{\Delta y_i}{\Delta x_i}\right)^2} (\Delta x_i).$$

Because $f'(x)$ exists for each x in (x_{i-1}, x_i), the Mean Value Theorem guarantees the existence of c_i in (x_{i-1}, x_i) such that

$$\frac{f(x_i) - f(x_{i-1})}{x_i - x_{i-1}} = f'(c_i)$$

$$\frac{\Delta y_i}{\Delta x_i} = f'(c_i).$$

Because f' is continuous on $[a, b]$, it follows that $\sqrt{1 + [f'(x)]^2}$ is also continuous (and therefore integrable) on $[a, b]$, which implies that

$$s = \lim_{\|\Delta\| \to 0} \sum_{i=1}^{n} \sqrt{1 + [f'(c_i)]^2} (\Delta x_i)$$

$$= \int_{a}^{b} \sqrt{1 + [f'(x)]^2} \, dx$$

where s is called the **arc length** of f between a and b.

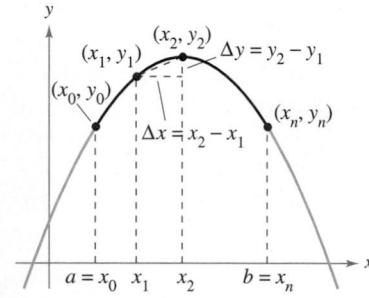

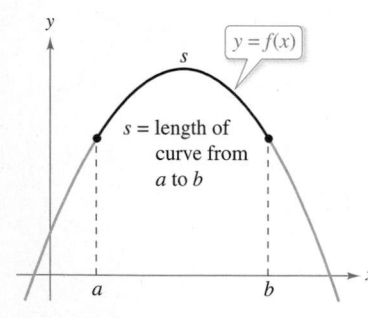

Figure 7.37

Definition of Arc Length

Let the function $y = f(x)$ represent a smooth curve on the interval $[a, b]$. The **arc length** of f between a and b is

$$s = \int_a^b \sqrt{1 + [f'(x)]^2}\, dx.$$

Similarly, for a smooth curve $x = g(y)$, the **arc length** of g between c and d is

$$s = \int_c^d \sqrt{1 + [g'(y)]^2}\, dy.$$

■ **FOR FURTHER INFORMATION** To see how arc length can be used to define trigonometric functions, see the article "Trigonometry Requires Calculus, Not Vice Versa" by Yves Nievergelt in *UMAP Modules*.

Because the definition of arc length can be applied to a linear function, you can check to see that this new definition agrees with the standard Distance Formula for the length of a line segment. This is shown in Example 1.

EXAMPLE 1 The Length of a Line Segment

Find the arc length from (x_1, y_1) to (x_2, y_2) on the graph of

$$f(x) = mx + b.$$

Solution Because

$$f'(x) = m = \frac{y_2 - y_1}{x_2 - x_1}$$

it follows that

$$
\begin{aligned}
s &= \int_{x_1}^{x_2} \sqrt{1 + [f'(x)]^2}\, dx && \text{Formula for arc length} \\
&= \int_{x_1}^{x_2} \sqrt{1 + \left(\frac{y_2 - y_1}{x_2 - x_1}\right)^2}\, dx \\
&= \sqrt{\frac{(x_2 - x_1)^2 + (y_2 - y_1)^2}{(x_2 - x_1)^2}}\,(x)\Bigg]_{x_1}^{x_2} && \text{Integrate and simplify.} \\
&= \sqrt{\frac{(x_2 - x_1)^2 + (y_2 - y_1)^2}{(x_2 - x_1)^2}}\,(x_2 - x_1) \\
&= \sqrt{(x_2 - x_1)^2 + (y_2 - y_1)^2}
\end{aligned}
$$

The formula for the arc length of the graph of f from (x_1, y_1) to (x_2, y_2) is the same as the standard Distance Formula.
Figure 7.38

which is the formula for the distance between two points in the plane, as shown in Figure 7.38. ■

▷ **TECHNOLOGY** Definite integrals representing arc length often are very difficult to evaluate. In this section, a few examples are presented. In the next chapter, with more advanced integration techniques, you will be able to tackle more difficult arc length problems. In the meantime, remember that you can always use a numerical integration program to approximate an arc length. For instance, use the *numerical integration* feature of a graphing utility to approximate the arc lengths in Examples 2 and 3.

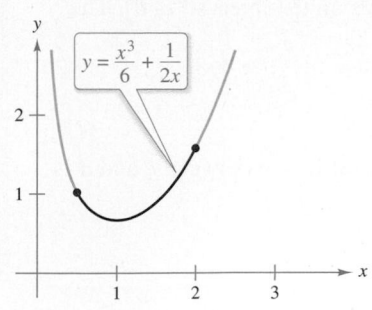

The arc length of the graph of y on $\left[\frac{1}{2}, 2\right]$

Figure 7.39

EXAMPLE 2 **Finding Arc Length**

Find the arc length of the graph of $y = \dfrac{x^3}{6} + \dfrac{1}{2x}$ on the interval $\left[\frac{1}{2}, 2\right]$, as shown in Figure 7.39.

Solution Using

$$\frac{dy}{dx} = \frac{3x^2}{6} - \frac{1}{2x^2} = \frac{1}{2}\left(x^2 - \frac{1}{x^2}\right)$$

yields an arc length of

$$
\begin{aligned}
s &= \int_a^b \sqrt{1 + \left(\frac{dy}{dx}\right)^2}\, dx && \text{Formula for arc length} \\
&= \int_{1/2}^2 \sqrt{1 + \left[\frac{1}{2}\left(x^2 - \frac{1}{x^2}\right)\right]^2}\, dx \\
&= \int_{1/2}^2 \sqrt{\frac{1}{4}\left(x^4 + 2 + \frac{1}{x^4}\right)}\, dx \\
&= \frac{1}{2}\int_{1/2}^2 \left(x^2 + \frac{1}{x^2}\right)\, dx && \text{Simplify.} \\
&= \frac{1}{2}\left[\frac{x^3}{3} - \frac{1}{x}\right]_{1/2}^2 && \text{Integrate.} \\
&= \frac{1}{2}\left(\frac{13}{6} + \frac{47}{24}\right) \\
&= \frac{33}{16}.
\end{aligned}
$$

EXAMPLE 3 **Finding Arc Length**

Find the arc length of the graph of $(y - 1)^3 = x^2$ on the interval $[0, 8]$, as shown in Figure 7.40.

Solution Solving for y yields $y = x^{2/3} + 1$ and $dy/dx = 2/(3x^{1/3})$. Because dy/dx is undefined when $x = 0$, the arc length formula with respect to x cannot be used. Solving for x in terms of y yields $x = \pm(y - 1)^{3/2}$. Choosing the positive value of x produces

$$\frac{dx}{dy} = \frac{3}{2}(y - 1)^{1/2}.$$

The x-interval $[0, 8]$ corresponds to the y-interval $[1, 5]$, and the arc length is

$$
\begin{aligned}
s &= \int_c^d \sqrt{1 + \left(\frac{dx}{dy}\right)^2}\, dy && \text{Formula for arc length} \\
&= \int_1^5 \sqrt{1 + \left[\frac{3}{2}(y - 1)^{1/2}\right]^2}\, dy \\
&= \int_1^5 \sqrt{\frac{9}{4}y - \frac{5}{4}}\, dy \\
&= \frac{1}{2}\int_1^5 \sqrt{9y - 5}\, dy && \text{Simplify.} \\
&= \frac{1}{18}\left[\frac{(9y - 5)^{3/2}}{3/2}\right]_1^5 && \text{Integrate.} \\
&= \frac{1}{27}(40^{3/2} - 4^{3/2}) \\
&\approx 9.073.
\end{aligned}
$$

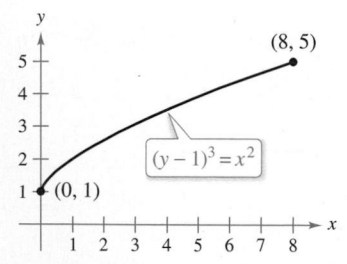

The arc length of the graph of y on $[0, 8]$

Figure 7.40

EXAMPLE 4 Finding Arc Length

····▷ *See LarsonCalculus.com for an interactive version of this type of example.*

Find the arc length of the graph of

$$y = \ln(\cos x)$$

from $x = 0$ to $x = \pi/4$, as shown in Figure 7.41.

Solution Using

$$\frac{dy}{dx} = -\frac{\sin x}{\cos x} = -\tan x$$

yields an arc length of

$$
\begin{aligned}
s &= \int_a^b \sqrt{1 + \left(\frac{dy}{dx}\right)^2}\, dx && \text{Formula for arc length}\\
&= \int_0^{\pi/4} \sqrt{1 + \tan^2 x}\, dx\\
&= \int_0^{\pi/4} \sqrt{\sec^2 x}\, dx && \text{Trigonometric identity}\\
&= \int_0^{\pi/4} \sec x\, dx && \text{Simplify.}\\
&= \Big[\ln|\sec x + \tan x|\Big]_0^{\pi/4} && \text{Integrate.}\\
&= \ln\left(\sqrt{2} + 1\right) - \ln 1\\
&\approx 0.881.
\end{aligned}
$$

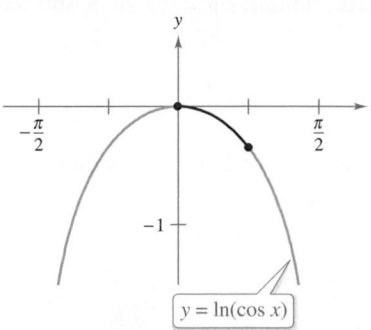

The arc length of the graph of y on $\left[0, \dfrac{\pi}{4}\right]$

Figure 7.41

EXAMPLE 5 Length of a Cable

An electric cable is hung between two towers that are 200 feet apart, as shown in Figure 7.42. The cable takes the shape of a catenary whose equation is

$$y = 75(e^{x/150} + e^{-x/150}) = 150 \cosh \frac{x}{150}.$$

Find the arc length of the cable between the two towers.

Solution Because $y' = \frac{1}{2}(e^{x/150} - e^{-x/150})$, you can write

$$(y')^2 = \frac{1}{4}(e^{x/75} - 2 + e^{-x/75})$$

and

$$1 + (y')^2 = \frac{1}{4}(e^{x/75} + 2 + e^{-x/75}) = \left[\frac{1}{2}(e^{x/150} + e^{-x/150})\right]^2.$$

Therefore, the arc length of the cable is

$$
\begin{aligned}
s &= \int_a^b \sqrt{1 + (y')^2}\, dx && \text{Formula for arc length}\\
&= \frac{1}{2}\int_{-100}^{100} (e^{x/150} + e^{-x/150})\, dx\\
&= 75\left[e^{x/150} - e^{-x/150}\right]_{-100}^{100} && \text{Integrate.}\\
&= 150(e^{2/3} - e^{-2/3})\\
&\approx 215 \text{ feet.}
\end{aligned}
$$

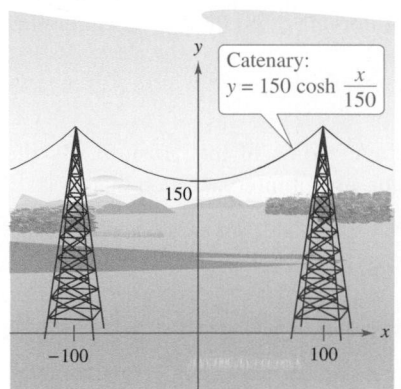

Catenary:
$y = 150 \cosh \dfrac{x}{150}$

Figure 7.42

Area of a Surface of Revolution

In Sections 7.2 and 7.3, integration was used to calculate the volume of a solid of revolution. You will now look at a procedure for finding the area of a surface of revolution.

> **Definition of Surface of Revolution**
>
> When the graph of a continuous function is revolved about a line, the resulting surface is a **surface of revolution.**

The area of a surface of revolution is derived from the formula for the lateral surface area of the frustum of a right circular cone. Consider the line segment in the figure at the right, where L is the length of the line segment, r_1 is the radius at the left end of the line segment, and r_2 is the radius at the right end of the line segment. When the line segment is revolved about its axis of revolution, it forms a frustum of a right circular cone, with

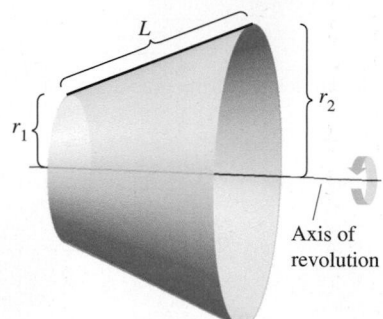

$$S = 2\pi r L \qquad \text{Lateral surface area of frustum}$$

where

$$r = \frac{1}{2}(r_1 + r_2). \qquad \text{Average radius of frustum}$$

(In Exercise 56, you are asked to verify the formula for S.)

Consider a function f that has a continuous derivative on the interval $[a, b]$. The graph of f is revolved about the x-axis to form a surface of revolution, as shown in Figure 7.43. Let Δ be a partition of $[a, b]$, with subintervals of width Δx_i. Then the line segment of length

$$\Delta L_i = \sqrt{(\Delta x_i)^2 + (\Delta y_i)^2}$$

generates a frustum of a cone. Let r_i be the average radius of this frustum. By the Intermediate Value Theorem, a point d_i exists (in the ith subinterval) such that

$$r_i = f(d_i).$$

The lateral surface area ΔS_i of the frustum is

$$\begin{aligned}
\Delta S_i &= 2\pi r_i \Delta L_i \\
&= 2\pi f(d_i)\sqrt{(\Delta x_i)^2 + (\Delta y_i)^2} \\
&= 2\pi f(d_i)\sqrt{1 + \left(\frac{\Delta y_i}{\Delta x_i}\right)^2}\, \Delta x_i.
\end{aligned}$$

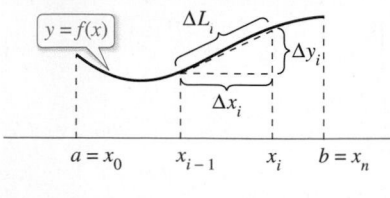

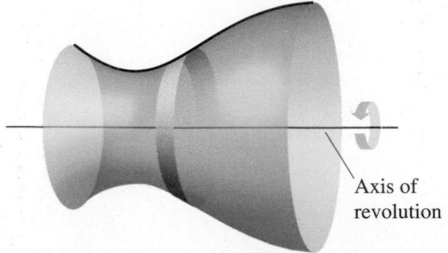

Figure 7.43

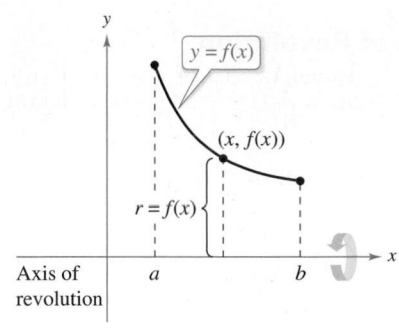

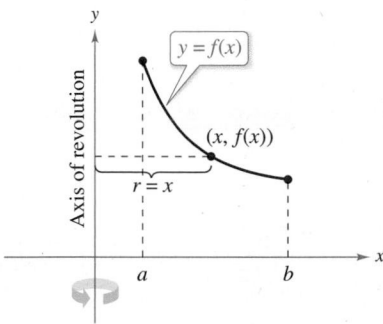

Figure 7.44

By the Mean Value Theorem, a number c_i exists in (x_{i-1}, x_i) such that

$$f'(c_i) = \frac{f(x_i) - f(x_{i-1})}{x_i - x_{i-1}}$$

$$= \frac{\Delta y_i}{\Delta x_i}.$$

So, $\Delta S_i = 2\pi f(d_i)\sqrt{1 + [f'(c_i)]^2}\, \Delta x_i$, and the total surface area can be approximated by

$$S \approx 2\pi \sum_{i=1}^{n} f(d_i)\sqrt{1 + [f'(c_i)]^2}\, \Delta x_i.$$

It can be shown that the limit of the right side as $\|\Delta\| \to 0$ $(n \to \infty)$ is

$$S = 2\pi \int_a^b f(x)\sqrt{1 + [f'(x)]^2}\, dx.$$

In a similar manner, if the graph of f is revolved about the y-axis, then S is

$$S = 2\pi \int_a^b x\sqrt{1 + [f'(x)]^2}\, dx.$$

In these two formulas for S, you can regard the products $2\pi f(x)$ and $2\pi x$ as the circumferences of the circles traced by a point (x, y) on the graph of f as it is revolved about the x-axis and the y-axis (Figure 7.44). In one case, the radius is $r = f(x)$, and in the other case, the radius is $r = x$. Moreover, by appropriately adjusting r, you can generalize the formula for surface area to cover *any* horizontal or vertical axis of revolution, as indicated in the next definition.

Definition of the Area of a Surface of Revolution

Let $y = f(x)$ have a continuous derivative on the interval $[a, b]$. The area S of the surface of revolution formed by revolving the graph of f about a horizontal or vertical axis is

$$S = 2\pi \int_a^b r(x)\sqrt{1 + [f'(x)]^2}\, dx \qquad \text{y is a function of x.}$$

where $r(x)$ is the distance between the graph of f and the axis of revolution. If $x = g(y)$ on the interval $[c, d]$, then the surface area is

$$S = 2\pi \int_c^d r(y)\sqrt{1 + [g'(y)]^2}\, dy \qquad \text{x is a function of y.}$$

where $r(y)$ is the distance between the graph of g and the axis of revolution.

The formulas in this definition are sometimes written as

$$S = 2\pi \int_a^b r(x)\, ds \qquad \text{y is a function of x.}$$

and

$$S = 2\pi \int_c^d r(y)\, ds \qquad \text{x is a function of y.}$$

where

$$ds = \sqrt{1 + [f'(x)]^2}\, dx \quad \text{and} \quad ds = \sqrt{1 + [g'(y)]^2}\, dy$$

respectively.

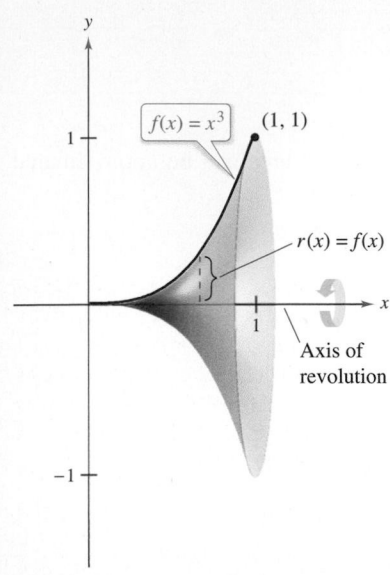

Figure 7.45

EXAMPLE 6 **The Area of a Surface of Revolution**

Find the area of the surface formed by revolving the graph of $f(x) = x^3$ on the interval $[0, 1]$ about the x-axis, as shown in Figure 7.45.

Solution The distance between the x-axis and the graph of f is $r(x) = f(x)$, and because $f'(x) = 3x^2$, the surface area is

$$
\begin{aligned}
S &= 2\pi \int_a^b r(x) \sqrt{1 + [f'(x)]^2}\, dx && \text{Formula for surface area} \\
&= 2\pi \int_0^1 x^3 \sqrt{1 + (3x^2)^2}\, dx \\
&= \frac{2\pi}{36} \int_0^1 (36x^3)(1 + 9x^4)^{1/2}\, dx && \text{Simplify.} \\
&= \frac{\pi}{18}\left[\frac{(1 + 9x^4)^{3/2}}{3/2}\right]_0^1 && \text{Integrate.} \\
&= \frac{\pi}{27}(10^{3/2} - 1) \\
&\approx 3.563.
\end{aligned}
$$

EXAMPLE 7 **The Area of a Surface of Revolution**

Find the area of the surface formed by revolving the graph of $f(x) = x^2$ on the interval $\left[0, \sqrt{2}\right]$ about the y-axis, as shown in the figure below.

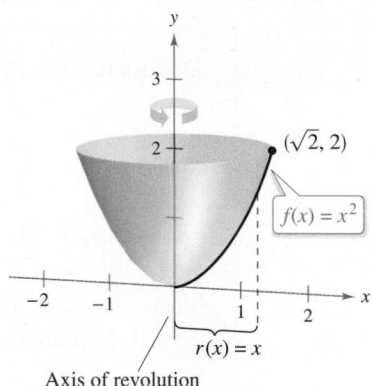

Solution In this case, the distance between the graph of f and the y-axis is $r(x) = x$. Using $f'(x) = 2x$ and the formula for surface area, you can determine that

$$
\begin{aligned}
S &= 2\pi \int_a^b r(x) \sqrt{1 + [f'(x)]^2}\, dx && \text{Formula for surface area} \\
&= 2\pi \int_0^{\sqrt{2}} x \sqrt{1 + (2x)^2}\, dx \\
&= \frac{2\pi}{8} \int_0^{\sqrt{2}} (1 + 4x^2)^{1/2}(8x)\, dx && \text{Simplify.} \\
&= \frac{\pi}{4}\left[\frac{(1 + 4x^2)^{3/2}}{3/2}\right]_0^{\sqrt{2}} && \text{Integrate.} \\
&= \frac{\pi}{6}[(1 + 8)^{3/2} - 1] \\
&= \frac{13\pi}{3} \\
&\approx 13.614.
\end{aligned}
$$

7.4 Exercises

See **CalcChat.com** for tutorial help and worked-out solutions to odd-numbered exercises.

CONCEPT CHECK

1. Rectifiable Curve Describe the condition for a curve to be rectifiable between two points.

2. Arc Length Explain how to find the arc length of a function that is a smooth curve on the interval $[a, b]$.

3. Arc Length Name a function for which the integral below represents the arc length of the function on the interval $[0, 2]$.

$$\int_0^2 \sqrt{1 + (4x)^2}\, dx$$

4. Surface of Revolution Describe a surface of revolution in your own words.

Finding Distance Using Two Methods In Exercises 5 and 6, find the distance between the points using (a) the Distance Formula and (b) integration.

5. $(2, 1), (5, 3)$ **6.** $(-2, 2), (4, -6)$

 **Finding Arc Length** In Exercises 7–20, find the arc length of the graph of the function over the indicated interval.

7. $y = \dfrac{2}{3}(x^2 + 1)^{3/2}$ **8.** $y = \dfrac{x^4}{8} + \dfrac{1}{4x^2}$, $[2, 3]$

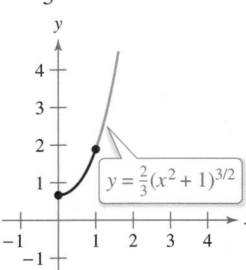   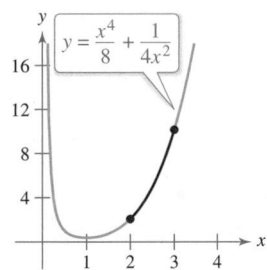

9. $y = \dfrac{2}{3}x^{3/2} + 1$ **10.** $y = 2x^{3/2} + 3$

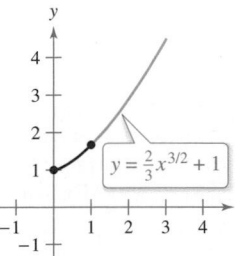   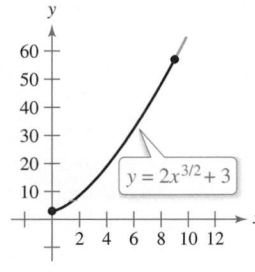

11. $y = \dfrac{3}{2}x^{2/3}$, $[1, 8]$ **12.** $y = \dfrac{3}{2}x^{2/3} + 4$, $[1, 27]$

13. $y = \dfrac{x^5}{10} + \dfrac{1}{6x^3}$, $[2, 5]$ **14.** $y = \dfrac{x^7}{14} + \dfrac{1}{10x^5}$, $[1, 2]$

15. $y = \ln(\sin x)$, $\left[\dfrac{\pi}{4}, \dfrac{3\pi}{4}\right]$ **16.** $y = \ln(\cos x)$, $\left[0, \dfrac{\pi}{3}\right]$

17. $y = \dfrac{1}{2}(e^x + e^{-x})$, $[0, 2]$

18. $y = \ln\left(\dfrac{e^x + 1}{e^x - 1}\right)$, $[\ln 6, \ln 8]$

19. $x = \dfrac{1}{3}(y^2 + 2)^{3/2}$, $0 \le y \le 4$

20. $x = \dfrac{1}{3}\sqrt{y}(y - 3)$, $1 \le y \le 4$

Finding Arc Length In Exercises 21–30, (a) sketch the graph of the function, highlighting the part indicated by the given interval, (b) write a definite integral that represents the arc length of the curve over the indicated interval and observe that the integral cannot be evaluated with the techniques studied so far, and (c) use the integration capabilities of a graphing utility to approximate the arc length.

21. $y = 4 - x^2$, $[0, 2]$ **22.** $y = x^2 + x - 2$, $[-2, 1]$

23. $y = \dfrac{1}{x}$, $[1, 3]$

24. $y = \dfrac{1}{x + 1}$, $[0, 1]$

25. $y = \sin x$, $[0, \pi]$

26. $y = \cos x$, $\left[-\dfrac{\pi}{2}, \dfrac{\pi}{2}\right]$

27. $y = 2 \arctan x$, $[0, 1]$

28. $y = \ln x$, $[1, 5]$

29. $x = e^{-y}$, $0 \le y \le 2$

30. $x = \sqrt{36 - y^2}$, $0 \le y \le 3$

Approximation In Exercises 31 and 32, approximate the arc length of the graph of the function over the interval $[0, 4]$ in three ways. (a) Use the Distance Formula to find the distance between the endpoints of the arc. (b) Use the Distance Formula to find the lengths of the four line segments connecting the points on the arc when $x = 0$, $x = 1$, $x = 2$, $x = 3$, and $x = 4$. Find the sum of the four lengths. (c) Use the integration capabilities of a graphing utility to approximate the integral yielding the indicated arc length.

31. $f(x) = x^3$ **32.** $f(x) = (x^2 - 4)^2$

33. Length of a Cable An electric cable is hung between two towers that are 40 meters apart (see figure). The cable takes the shape of a catenary whose equation is

$$y = 10(e^{x/20} + e^{-x/20}), \quad -20 \le x \le 20$$

where x and y are measured in meters. Find the arc length of the cable between the two towers.

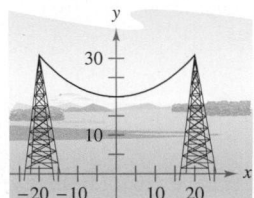

34. Roof Area A barn is 100 feet long and 40 feet wide (see figure). A cross section of the roof is the inverted catenary $y = 31 - 10(e^{x/20} + e^{-x/20})$. Find the number of square feet of roofing on the barn.

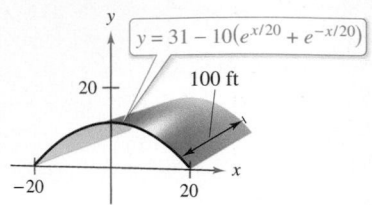

35. Length of Gateway Arch The Gateway Arch in St. Louis, Missouri, is closely approximated by the inverted catenary

$$y = 693.8597 - 68.7672 \cosh 0.0100333x,$$

$$-299.2239 \le x \le 299.2239.$$

Use the integration capabilities of a graphing utility to approximate the length of this curve (see figure).

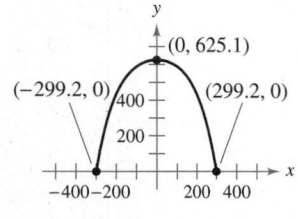   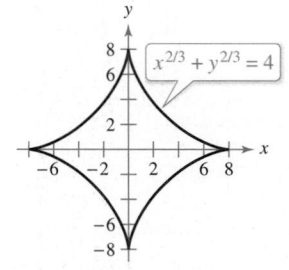

Figure for 35   Figure for 36

36. Astroid Find the total length of the graph of the astroid $x^{2/3} + y^{2/3} = 4$.

37. Arc Length of a Sector of a Circle Find the arc length from $(0, 3)$ clockwise to $(2, \sqrt{5})$ along the circle $x^2 + y^2 = 9$.

38. Arc Length of a Sector of a Circle Find the arc length from $(-3, 4)$ clockwise to $(4, 3)$ along the circle $x^2 + y^2 = 25$. Show that the result is one-fourth the circumference of the circle.

 **Finding the Area of a Surface of Revolution** In Exercises 39–44, write and evaluate the definite integral that represents the area of the surface generated by revolving the curve on the indicated interval about the x-axis.

39. $y = \frac{1}{3}x^3$

40. $y = 2\sqrt{x}$

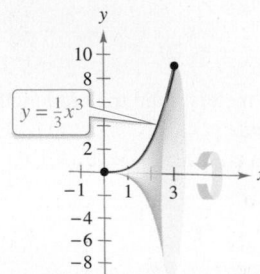

41. $y = \frac{x^3}{6} + \frac{1}{2x}$, $1 \le x \le 2$ **42.** $y = 3x$, $0 \le x \le 3$

43. $y = \sqrt{4 - x^2}$, $-1 \le x \le 1$

44. $y = \sqrt{9 - x^2}$, $-2 \le x \le 2$

 **Finding the Area of a Surface of Revolution** In Exercises 45–48, write and evaluate the definite integral that represents the area of the surface generated by revolving the curve on the indicated interval about the y-axis.

45. $y = \sqrt[3]{x} + 2$, $1 \le x \le 8$ **46.** $y = 9 - x^2$, $0 \le x \le 3$

47. $y = 1 - \frac{x^2}{4}$, $0 \le x \le 2$ **48.** $y = \frac{x}{2} + 3$, $1 \le x \le 5$

Finding the Area of a Surface of Revolution Using Technology In Exercises 49 and 50, use the integration capabilities of a graphing utility to approximate the area of the surface of revolution.

Function	Interval	Axis of Revolution
49. $y = \sin x$	$[0, \pi]$	x-axis
50. $y = \ln x$	$[1, e]$	y-axis

EXPLORING CONCEPTS

Approximation In Exercises 51 and 52, determine which value best approximates the length of the arc represented by the integral. Make your selection on the basis of a sketch of the arc, not by performing calculations.

51. $\displaystyle\int_0^2 \sqrt{1 + \left[\frac{d}{dx}\left(\frac{5}{x^2 + 1}\right)\right]^2}\, dx$

(a) 25 (b) 5 (c) 2 (d) −4 (e) 3

52. $\displaystyle\int_0^{\pi/4} \sqrt{1 + \left[\frac{d}{dx}(\tan x)\right]^2}\, dx$

(a) 3 (b) −2 (c) 4 (d) $\frac{4\pi}{3}$ (e) 1

53. Exploring Relationships Consider the function

$$f(x) = \frac{1}{4}e^x + e^{-x}.$$

Compare the definite integral of f on the interval $[a, b]$ with the arc length of f over the interval $[a, b]$.

54. **HOW DO YOU SEE IT?** The graphs of the functions f_1 and f_2 on the interval $[a, b]$ are shown in the figure. The graph of each function is revolved about the x-axis. Which surface of revolution has the greater surface area? Explain.

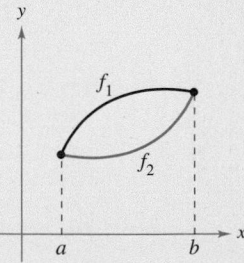

55. Think About It The figure shows the graphs of the functions $y_1 = x$, $y_2 = \frac{1}{2}x^{3/2}$, $y_3 = \frac{1}{4}x^2$, and $y_4 = \frac{1}{8}x^{5/2}$ on the interval $[0, 4]$. To print an enlarged copy of the graph, go to *MathGraphs.com*.

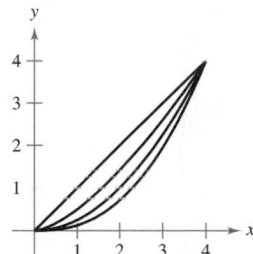

(a) Label the functions.

(b) Without calculating, list the functions in order of increasing arc length.

(c) Verify your answer in part (b) by using the integration capabilities of a graphing utility to approximate each arc length accurate to three decimal places.

56. Verifying a Formula

(a) Given a circular sector with radius L and central angle θ (see figure), show that the area of the sector is given by

$$S = \frac{1}{2}L^2\theta.$$

(b) By joining the straight-line edges of the sector in part (a), a right circular cone is formed (see figure) and the lateral surface area of the cone is the same as the area of the sector. Show that the area is $S = \pi rL$, where r is the radius of the base of the cone. (*Hint:* The arc length of the sector equals the circumference of the base of the cone.)

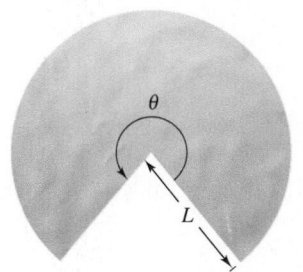

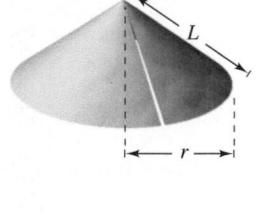

Figure for 56(a)   Figure for 56(b)

(c) Use the result of part (b) to verify that the formula for the lateral surface area of the frustum of a cone with slant height L and radii r_1 and r_2 (see figure) is $S = \pi(r_1 + r_2)L$. (*Note:* This formula was used to develop the integral for finding the surface area of a surface of revolution.)

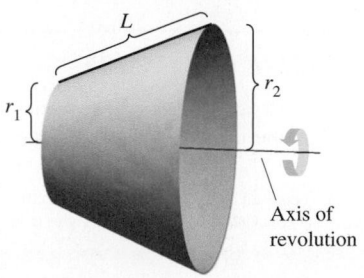

57. Lateral Surface Area of a Cone A right circular cone is generated by revolving the region bounded by $y = 3x/4$, $y = 3$, and $x = 0$ about the y-axis. Find the lateral surface area of the cone.

58. Lateral Surface Area of a Cone A right circular cone is generated by revolving the region bounded by $y = hx/r$, $y = h$, and $x - 0$ about the y-axis. Verify that the lateral surface area of the cone is $S = \pi r\sqrt{r^2 + h^2}$.

59. Using a Sphere Find the area of the segment of a sphere formed by revolving the graph of $y = \sqrt{9 - x^2}$, $0 \le x \le 2$, about the y-axis.

60. Using a Sphere Find the area of the segment of a sphere formed by revolving the graph of $y = \sqrt{r^2 - x^2}$, $0 \le x \le a$, about the y-axis. Assume that $a < r$.

61. Modeling Data The circumference C (in inches) of a vase is measured at three-inch intervals starting at its base. The measurements are shown in the table, where y is the vertical distance in inches from the base.

y	0	3	6	9	12	15	18
C	50	65.5	70	66	58	51	48

(a) Use the data to approximate the volume of the vase by summing the volumes of approximating disks.

(b) Use the data to approximate the outside surface area (excluding the base) of the vase by summing the outside surface areas of approximating frustums of right circular cones.

(c) Use the regression capabilities of a graphing utility to find a cubic model for the points (y, r), where $r = C/(2\pi)$. Use the graphing utility to plot the points and graph the model.

(d) Use the model in part (c) and the integration capabilities of a graphing utility to approximate the volume and outside surface area of the vase. Compare the results with your answers in parts (a) and (b).

62. Modeling Data Property bounded by two perpendicular roads and a stream is shown in the figure. All distances are measured in feet.

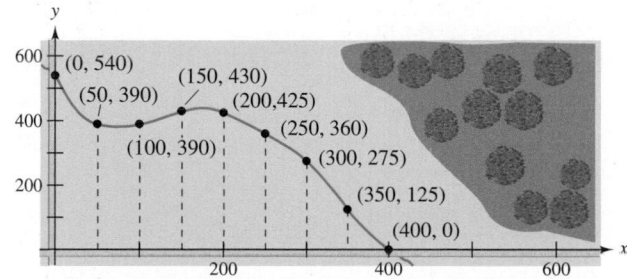

(a) Use the regression capabilities of a graphing utility to fit a fourth-degree polynomial to the path of the stream.

(b) Use the model in part (a) to approximate the area of the property in acres.

(c) Use the integration capabilities of a graphing utility to find the length of the stream that bounds the property.

63. Volume and Surface Area Let R be the region bounded by $y = 1/x$, the x-axis, $x = 1$, and $x = b$, where $b > 1$. Let D be the solid formed when R is revolved about the x-axis.

(a) Find the volume V of D.

(b) Write a definite integral that represents the surface area S of D.

(c) Show that V approaches a finite limit as $b \to \infty$.

(d) Show that $S \to \infty$ as $b \to \infty$.

64. Think About It Consider the equation

$$\frac{x^2}{9} + \frac{y^2}{4} = 1.$$

(a) Use a graphing utility to graph the equation.

(b) Write the definite integral for finding the first-quadrant arc length of the graph in part (a).

(c) Compare the interval of integration in part (b) and the domain of the integrand. Is it possible to evaluate the definite integral? Explain. (You will learn how to evaluate this type of integral in Section 8.8.)

Approximating Arc Length or Surface Area In Exercises 65–68, write the definite integral for finding the indicated arc length or surface area. Then use the integration capabilities of a graphing utility to approximate the arc length or surface area. (You will learn how to evaluate this type of integral in Section 8.8.)

65. Length of Pursuit A fleeing object leaves the origin and moves up the y-axis (see figure). At the same time, a pursuer leaves the point $(1, 0)$ and always moves toward the fleeing object. The pursuer's speed is twice that of the fleeing object. The equation of the path is modeled by

$$y = \frac{1}{3}(x^{3/2} - 3x^{1/2} + 2).$$

How far has the fleeing object traveled when it is caught? Show that the pursuer has traveled twice as far.

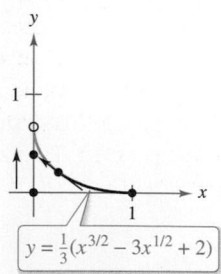

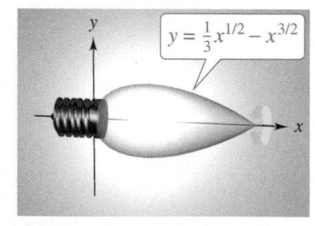

Figure for 65   Figure for 66

66. Bulb Design An ornamental light bulb is designed by revolving the graph of

$$y = \frac{1}{3}x^{1/2} - x^{3/2}, \quad 0 \le x \le \frac{1}{3}$$

about the x-axis, where x and y are measured in feet (see figure). Find the surface area of the bulb and use the result to approximate the amount of glass needed to make the bulb. Assume that the thickness of the glass is 0.015 inch.

67. Astroid Find the area of the surface formed by revolving the portion in the first quadrant of the graph of $x^{2/3} + y^{2/3} = 4$, $0 \le y \le 8$, about the y-axis.

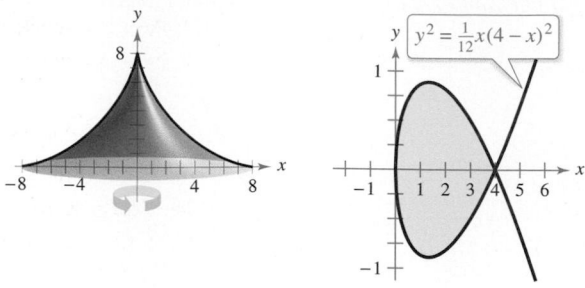

Figure for 67   Figure for 68

68. Using a Loop Consider the graph of

$$y^2 = \frac{1}{12}x(4 - x)^2$$

shown in the figure. Find the area of the surface formed when the loop of this graph is revolved about the x-axis.

69. Suspension Bridge A cable for a suspension bridge has the shape of a parabola with equation $y = kx^2$. Let h represent the height of the cable from its lowest point to its highest point and let $2w$ represent the total span of the bridge (see figure). Show that the length C of the cable is given by

$$C = 2\int_0^w \sqrt{1 + \left(\frac{4h^2}{w^2}\right)x^2}\, dx$$

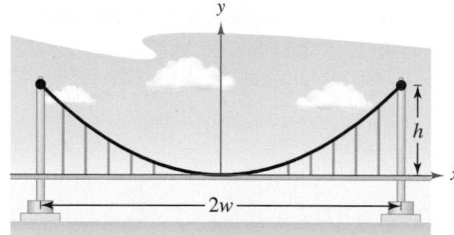

70. Suspension Bridge The Humber Bridge, located in the United Kingdom and opened in 1981, has a main span of about 1400 meters. Each of its towers has a height of about 155 meters. Use these dimensions, the integral in Exercise 69, and the integration capabilities of a graphing utility to approximate the length of a parabolic cable along the main span.

71. Arc Length and Area Let C be the curve given by $f(x) = \cosh x$ for $0 \le x \le t$, where $t > 0$. Show that the arc length of C is equal to the area bounded by C and the x-axis. Identify another curve on the interval $0 \le x \le t$ with this property.

PUTNAM EXAM CHALLENGE

72. Find the length of the curve $y^2 = x^3$ from the origin to the point where the tangent makes an angle of $45°$ with the x-axis.

7.5   Work

- ■ Find the work done by a constant force.
- ■ Find the work done by a variable force.

Work Done by a Constant Force

The concept of work is important to scientists and engineers for determining the energy needed to perform various jobs. For instance, it is useful to know the amount of work done when a crane lifts a steel girder, when a spring is compressed, when a rocket is propelled into the air, or when a truck pulls a load along a highway.

In general, **work** is done by a force when it moves an object. If the force applied to the object is *constant*, then the definition of work is as follows.

> ### Definition of Work Done by a Constant Force
>
> If an object is moved a distance D in the direction of an applied constant force F, then the **work** W done by the force is defined as $W = FD$.

There are four fundamental types of forces—gravitational, electromagnetic, strong nuclear, and weak nuclear. A **force** can be thought of as a *push* or a *pull*; a force changes the state of rest or state of motion of a body. For gravitational forces on Earth, it is common to use units of measure corresponding to the weight of an object.

EXAMPLE 1   Lifting an Object

Determine the work done in lifting a 50-pound object 4 feet.

Solution   The magnitude of the required force F is the weight of the object, as shown in Figure 7.46. So, the work done in lifting the object 4 feet is

$$W = FD = 50(4) = 200 \text{ foot-pounds.}$$ ■

In the U.S. measurement system, work is typically expressed in foot-pounds (ft-lb), inch-pounds, or foot-tons. In the International System of Units (SI), the basic unit of force is the **newton**—the force required to produce an acceleration of 1 meter per second per second on a mass of 1 kilogram. In this system, work is typically expressed in newton-meters, also called joules. In another system, the centimeter-gram-second (C-G-S) system, the basic unit of force is the **dyne**—the force required to produce an acceleration of 1 centimeter per second per second on a mass of 1 gram. In this system, work is typically expressed in dyne-centimeters, also called ergs, or in joules. The table below summarizes the units of measure that are commonly used to express the work done and lists several conversion factors.

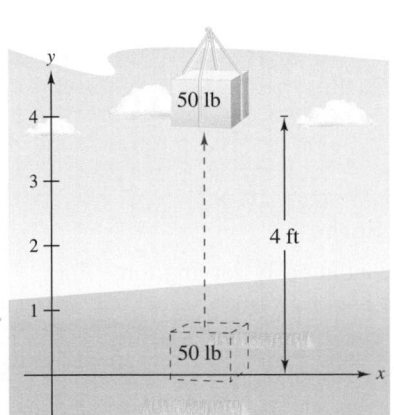

The work done in lifting a 50-pound object 4 feet is 200 foot-pounds.

Figure 7.46

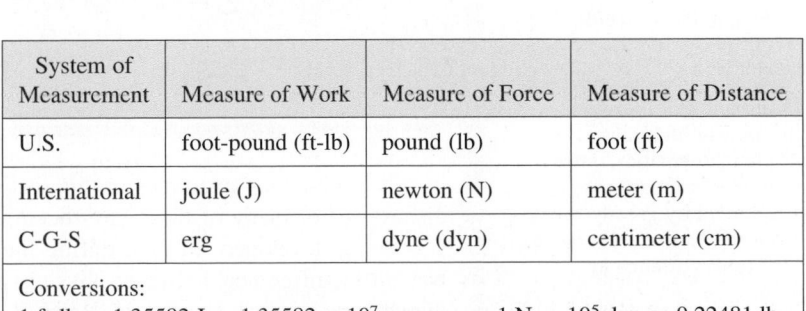

System of Measurement	Measure of Work	Measure of Force	Measure of Distance
U.S.	foot-pound (ft-lb)	pound (lb)	foot (ft)
International	joule (J)	newton (N)	meter (m)
C-G-S	erg	dyne (dyn)	centimeter (cm)
Conversions: 1 ft-lb $\approx$ 1.35582 J = 1.35582 × 10⁷ ergs 1 N = 10⁵ dyn $\approx$ 0.22481 lb 1 J = 10⁷ ergs $\approx$ 0.73756 ft-lb 1 lb $\approx$ 4.44822 N			

Work Done by a Variable Force

In Example 1, the force involved was *constant*. When a *variable* force is applied to an object, calculus is needed to determine the work done, because the amount of force changes as the object changes position. For instance, the force required to compress a spring increases as the spring is compressed.

Consider an object that is moved along a straight line from $x = a$ to $x = b$ by a continuously varying force $F(x)$. Let Δ be a partition that divides the interval $[a, b]$ into n subintervals determined by

$$a = x_0 < x_1 < x_2 < \cdots < x_n = b$$

and let $\Delta x_i = x_i - x_{i-1}$. For each i, choose c_i such that

$$x_{i-1} \le c_i \le x_i.$$

Then at c_i, the force is $F(c_i)$. Because F is continuous, you can approximate the work done in moving the object through the ith subinterval by the increment

$$\Delta W_i = F(c_i) \Delta x_i$$

as shown in Figure 7.47. So, the total work done as the object moves from $x = a$ to $x = b$ is approximated by

$$W \approx \sum_{i=1}^{n} \Delta W_i$$

$$= \sum_{i=1}^{n} F(c_i) \Delta x_i.$$

This approximation appears to become better and better as $\|\Delta\| \to 0$ $(n \to \infty)$. So, the work done is

$$W = \lim_{\|\Delta\| \to 0} \sum_{i=1}^{n} F(c_i) \Delta x_i$$

$$= \int_a^b F(x)\, dx.$$

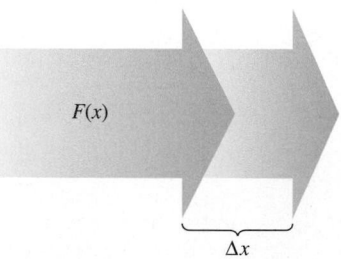

The amount of force changes as an object changes position (Δx).
Figure 7.47

Definition of Work Done by a Variable Force

If an object is moved along a straight line by a continuously varying force $F(x)$, then the **work** W done by the force as the object is moved from

$$x = a \quad \text{to} \quad x = b$$

is given by

$$W = \lim_{\|\Delta\| \to 0} \sum_{i=1}^{n} \Delta W_i$$

$$= \int_a^b F(x)\, dx.$$

**EMILIE DE BRETEUIL
(1706–1749)**

A major work by Breteuil was the translation of Newton's "Philosophiae Naturalis Principia Mathematica" into French. Her translation and commentary greatly contributed to the acceptance of Newtonian science in Europe.
See LarsonCalculus.com to read more of this biography.

The remaining examples in this section use some well-known physical laws. The discoveries of many of these laws occurred during the same period in which calculus was being developed. In fact, during the seventeenth and eighteenth centuries, there was little difference between physicists and mathematicians. One such physicist-mathematician was Emilie de Breteuil. Breteuil was instrumental in synthesizing the work of many other scientists, including Newton, Leibniz, Huygens, Kepler, and Descartes. Her physics text *Institutions* was widely used for many years.

The three laws of physics listed below were developed by Robert Hooke (1635–1703), Isaac Newton (1642–1727), and Charles Coulomb (1736–1806).

1. **Hooke's Law:** The force F required to compress or stretch a spring (within its elastic limits) is proportional to the distance d that the spring is compressed or stretched from its original length. That is,

$$F = kd$$

where the constant of proportionality k (the spring constant) depends on the specific nature of the spring.

2. **Newton's Law of Universal Gravitation:** The force F of attraction between two particles of masses m_1 and m_2 is proportional to the product of the masses and inversely proportional to the square of the distance d between the two particles. That is,

$$F = G\frac{m_1 m_2}{d^2}.$$

When m_1 and m_2 are in kilograms and d in meters, F will be in newtons for a value of $G = 6.67 \times 10^{-11}$ cubic meter per kilogram-second squared, where G is the **gravitational constant.**

3. **Coulomb's Law:** The force F between two charges q_1 and q_2 in a vacuum is proportional to the product of the charges and inversely proportional to the square of the distance d between the two charges. That is,

$$F = k\frac{q_1 q_2}{d^2}.$$

When q_1 and q_2 are given in electrostatic units and d in centimeters, F will be in dynes for a value of $k = 1$.

EXAMPLE 2   **Compressing a Spring**

$\vdots \cdots \triangleright$ *See LarsonCalculus.com for an interactive version of this type of example.*

A force of 30 newtons compresses a spring 0.3 meter from its natural length of 1.5 meters. Find the work done in compressing the spring an additional 0.3 meter.

Solution   By Hooke's Law, the force $F(x)$ required to compress the spring x units (from its natural length) is $F(x) = kx$. Because $F(0.3) = 30$, it follows that

$$F(0.3) = (k)(0.3) \quad\Longrightarrow\quad 30 = 0.3k \quad\Longrightarrow\quad 100 = k.$$

So, $F(x) = 100x$, as shown in Figure 7.48. To find the increment of work, assume that the force required to compress the spring over a small increment Δx is nearly constant. So, the increment of work is

$$\Delta W = (\text{force})(\text{distance increment}) = (100x)\Delta x.$$

Because the spring is compressed from $x = 0.3$ to $x = 0.6$ meter less than its natural length, the work required is

$$W = \int_a^b F(x)\,dx = \int_{0.3}^{0.6} 100x\,dx = 50x^2 \Big]_{0.3}^{0.6} = 18 - 4.5 = 13.5 \text{ joules.}$$

Note that you do *not* integrate from $x = 0$ to $x = 0.6$ because you were asked to determine the work done in compressing the spring an *additional* 0.3 meter (not including the first 0.3 meter). ■

Natural length: $F(0) = 0$

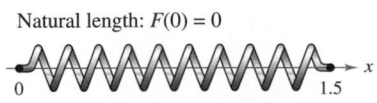

Compressed 0.3 meter: $F(0.3) = 30$

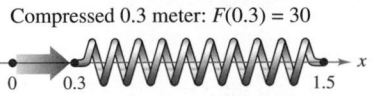

Compressed x meters: $F(x) = 100x$

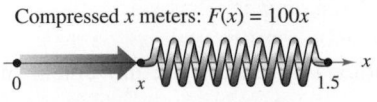

Figure 7.48

NASA's Space Launch System, or SLS, is a launch vehicle for exploration beyond Earth's orbit. NASA plans to use the SLS on missions to an asteroid and eventually to Mars. (*Source: NASA*)

EXAMPLE 3 **Moving a Space Module into Orbit**

A space module weighs 15 metric tons on the surface of Earth. How much work is done in propelling the module to a height of 800 miles above Earth, as shown in Figure 7.49? (Use 4000 miles as the radius of Earth. Do not consider the effect of air resistance or the weight of the propellant.)

Solution Because the weight of a body varies inversely as the square of its distance from the center of Earth, the force $F(x)$ exerted by gravity is

$$F(x) = \frac{C}{x^2}$$

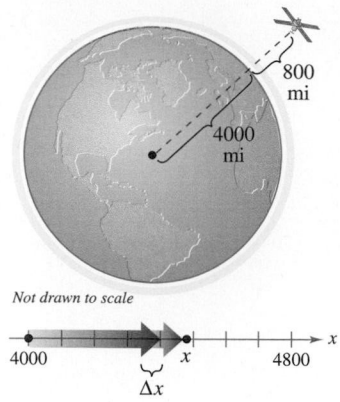

Not drawn to scale

Figure 7.49

where C is the constant of proportionality. Because the module weighs 15 metric tons on the surface of Earth and the radius of Earth is approximately 4000 miles, you have

$$15 = \frac{C}{(4000)^2} \quad \Longrightarrow \quad 240{,}000{,}000 = C.$$

So, the increment of work is

$$\Delta W = (\text{force})(\text{distance increment}) = \frac{240{,}000{,}000}{x^2} \Delta x.$$

Finally, because the module is propelled from $x = 4000$ to $x = 4800$ miles, the total work done is

$$
\begin{aligned}
W &= \int_a^b F(x)\, dx & \text{Formula for work} \\
&= \int_{4000}^{4800} \frac{240{,}000{,}000}{x^2}\, dx \\
&= \frac{-240{,}000{,}000}{x} \Bigg]_{4000}^{4800} & \text{Integrate.} \\
&= -50{,}000 + 60{,}000 \\
&= 10{,}000 \text{ mile-tons} \\
&\approx 1.164 \times 10^{11} \text{ foot-pounds.} & \text{1 mile = 5280 feet; 1 metric ton } \approx 2205 \text{ pounds}
\end{aligned}
$$

In SI units, using a conversion factor of 1 foot-pound ≈ 1.35582 joules, the work done is

$$W \approx 1.578 \times 10^{11} \text{ joules.} \qquad \blacksquare$$

The solutions to Examples 2 and 3 conform to our development of work as the summation of increments in the form

$$\Delta W = (\text{force})(\text{distance increment}) = (F)(\Delta x).$$

Another way to formulate the increment of work is

$$\Delta W = (\text{force increment})(\text{distance}) = (\Delta F)(x).$$

This second interpretation of ΔW is useful in problems involving the movement of nonrigid substances such as fluids and chains.

NASA

EXAMPLE 4 **Emptying a Tank of Oil**

A spherical tank of radius 8 feet is half full of oil that weighs 50 pounds per cubic foot. Find the work required to pump all of the oil out through a hole in the top of the tank.

Solution Consider the oil to be subdivided into disks of thickness Δy and radius x, as shown in Figure 7.50. Because the increment of force for each disk is given by its weight, you have

$$\Delta F = \text{weight}$$

$$= \left(\frac{50 \text{ pounds}}{\text{cubic root}}\right)(\text{volume})$$

$$= 50(\pi x^2 \Delta y) \text{ pounds.}$$

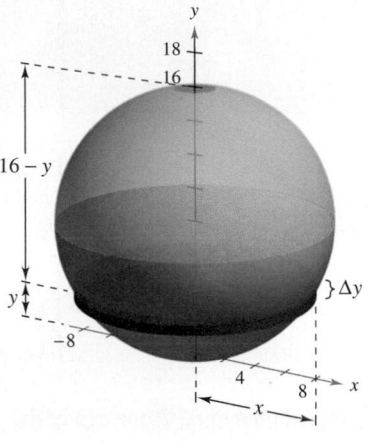

For a circle of radius 8 and center at $(0, 8)$, you have

Figure 7.50

$$x^2 + (y - 8)^2 = 8^2$$

$$x^2 = 16y - y^2$$

and you can write the force increment as

$$\Delta F = 50(\pi x^2 \Delta y)$$

$$= 50\pi(16y - y^2)\Delta y.$$

In Figure 7.50, note that a disk y feet from the bottom of the tank must be moved a distance of $(16 - y)$ feet. So, the increment of work is

$$\Delta W = (\Delta F)(16 - y)$$

$$= [50\pi(16y - y^2)\Delta y](16 - y)$$

$$= 50\pi(256y - 32y^2 + y^3)\Delta y.$$

Because the tank is half full, y ranges from 0 to 8 feet, and the work required to empty the tank is

$$W = \int_0^8 50\pi(256y - 32y^2 + y^3) \, dy$$

$$= 50\pi\left[128y^2 - \frac{32}{3}y^3 + \frac{y^4}{4}\right]_0^8$$

$$= 50\pi\left(\frac{11{,}264}{3}\right)$$

$$\approx 589{,}782 \text{ foot-pounds.}$$

To estimate the reasonableness of the result in Example 4, consider that the weight of the oil in the tank is

$$\left(\frac{1}{2}\right)(\text{volume})(\text{density}) = \frac{1}{2}\left(\frac{4}{3}\pi 8^3\right)(50) \approx 53{,}616.5 \text{ pounds.}$$

Lifting the entire half-tank of oil 8 feet would involve work of

$$W = FD \qquad \text{Formula for work done by a constant force}$$

$$\approx (53{,}616.5)(8)$$

$$= 428{,}932 \text{ foot-pounds.}$$

Because the oil is actually lifted between 8 and 16 feet, it seems reasonable that the work done is about 589,782 foot-pounds.

Work required to raise one end of the chain
Figure 7.51

| EXAMPLE 5 | **Lifting a Chain** |

A 20-foot chain weighing 5 pounds per foot is lying coiled on the ground. How much work is required to raise one end of the chain to a height of 20 feet so that it is fully extended, as shown in Figure 7.51?

Solution Imagine that the chain is divided into small sections, each of length Δy. Then the weight of each section is the increment of force

$$\Delta F = (\text{weight}) = \left(\frac{5 \text{ pounds}}{\text{foot}}\right)(\text{length}) = 5 \, \Delta y.$$

Because a typical section (initially on the ground) is raised to a height of y, the increment of work is

$$\Delta W = (\text{force increment})(\text{distance}) = (5 \, \Delta y)y = 5y \, \Delta y.$$

Because y ranges from 0 to 20 feet, the total work required to raise the chain is

$$W = \int_0^{20} 5y \, dy = \frac{5y^2}{2} \bigg]_0^{20} = \frac{5(400)}{2} = 1000 \text{ foot-pounds.} \quad ■$$

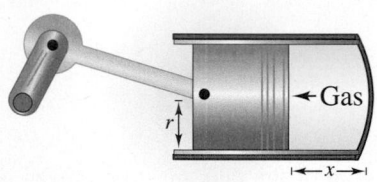

Work done by expanding gas
Figure 7.52

In the next example, you will consider a piston of radius r in a cylindrical casing, as shown in Figure 7.52. As the gas in the cylinder expands, the piston moves, and work is done. If p represents the pressure of the gas (in pounds per square foot) against the piston head and V represents the volume of the gas (in cubic feet), then the work increment involved in moving the piston Δx feet is

$$\Delta W = (\text{force})(\text{distance increment}) = F \, \Delta x = p(\pi r^2)\Delta x = p \, \Delta V.$$

So, as the volume of the gas expands from V_0 to V_1, the work done in moving the piston is

$$W = \int_{V_0}^{V_1} p \, dV.$$

Assuming the pressure of the gas to be inversely proportional to its volume, you have $p = k/V$ and the integral for work becomes

$$W = \int_{V_0}^{V_1} \frac{k}{V} \, dV.$$

| EXAMPLE 6 | **Work Done by an Expanding Gas** |

A quantity of gas with an initial volume of 1 cubic foot and a pressure of 500 pounds per square foot expands to a volume of 2 cubic feet. Find the work done by the gas. (Assume that the pressure is inversely proportional to the volume.)

Solution Because $p = k/V$ and $p = 500$ when $V = 1$, you have $k = 500$. So, the work done by the gas is

$$
\begin{aligned}
W &= \int_{V_0}^{V_1} \frac{k}{V} \, dV \\
&= \int_1^2 \frac{500}{V} \, dV \\
&= 500 \ln|V| \bigg]_1^2 \\
&\approx 346.6 \text{ foot-pounds.} \quad ■
\end{aligned}
$$

7.5 Exercises

See CalcChat.com for tutorial help and worked-out solutions to odd-numbered exercises.

CONCEPT CHECK

1. **Work** How do you know when work is done by a force?

2. **Comparing Methods** Describe the difference between finding the work done by a constant force and finding the work done by a variable force.

3. **Hooke's Law** Describe Hooke's Law in your own words.

4. **Work** What are two ways to write the increment of work?

 **Constant Force** In Exercises 5–8, determine the work done by the constant force.

5. A 1200-pound steel beam is lifted 40 feet.

6. An electric hoist lifts a 3000-pound car 6 feet.

7. A force of 112 newtons is required to slide a cement block 8 meters in a construction project.

8. The locomotive of a freight train pulls its cars with a constant force of 7 tons for a distance of one-quarter mile.

 **Hooke's Law** In Exercises 9–14, use Hooke's Law to determine the work done by the variable force in the spring problem.

9. A force of 5 pounds compresses a 15-inch spring a total of 3 inches. How much work is done in compressing the spring 7 inches?

10. A force of 250 newtons stretches a spring 30 centimeters. How much work is done in stretching the spring from 20 centimeters to 50 centimeters?

11. A force of 20 pounds stretches a spring 9 inches in an exercise machine. Find the work done in stretching the spring 1 foot from its natural position.

12. An overhead garage door has two springs, one on each side of the door. A force of 15 pounds is required to stretch each spring 1 foot. Because of the pulley system, the springs stretch only one-half the distance the door travels. The door moves a total of 8 feet, and the springs are at their natural length when the door is open. Find the work done by the pair of springs.

13. Eighteen foot-pounds of work is required to stretch a spring 4 inches from its natural length. Find the work required to stretch the spring an additional 3 inches.

14. Six joules of work is required to stretch a spring 0.5 meter from its natural length. Find the work required to stretch the spring an additional 0.25 meter.

15. **Propulsion** Neglecting air resistance and the weight of the propellant, determine the work done in propelling a five-metric-ton satellite to a height of (a) 100 miles above Earth and (b) 300 miles above Earth.

16. **Propulsion** Use the information in Exercise 15 to write the work W of the propulsion system as a function of the height h of the satellite above Earth. Find the limit (if it exists) of W as h approaches infinity.

17. **Propulsion** Neglecting air resistance and the weight of the propellant, determine the work done in propelling a 10-metric-ton satellite to a height of (a) 11,000 miles above Earth and (b) 22,000 miles above Earth.

18. **Propulsion** A lunar module weighs 12 metric tons on the surface of Earth. How much work is done in propelling the module from the surface of the moon to a height of 50 miles? Consider the radius of the moon to be 1100 miles and its force of gravity to be one-sixth that of Earth.

19. **Pumping Water** A rectangular tank with a base 4 feet by 5 feet and a height of 4 feet is full of water (see figure). The water weighs 62.4 pounds per cubic foot. How much work is done in pumping water out over the top edge in order to empty (a) half of the tank and (b) all of the tank?

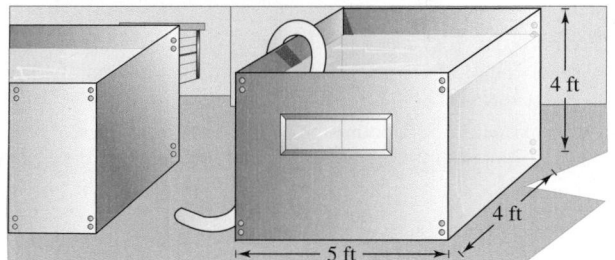

20. **Think About It** Explain why the answer in part (b) of Exercise 19 is not twice the answer in part (a).

21. **Pumping Water** A cylindrical water tank 4 meters high with a radius of 2 meters is buried so that the top of the tank is 1 meter below ground level (see figure). How much work is done in pumping a full tank of water up to ground level? The water weighs 9800 newtons per cubic meter.

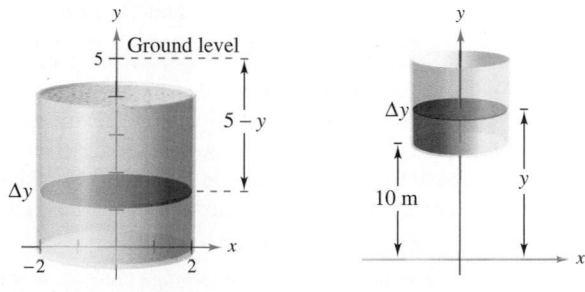

Figure for 21   Figure for 22

22. **Pumping Water** Suppose the tank in Exercise 21 is located on a tower so that the bottom of the tank is 10 meters above a stream (see figure). How much work is done in filling the tank half full of water through a hole in the bottom, using water from the stream?

23. Pumping Water An open tank has the shape of a right circular cone (see figure). The tank is 8 feet across the top and 6 feet high. How much work is done in emptying the tank by pumping the water over the top edge?

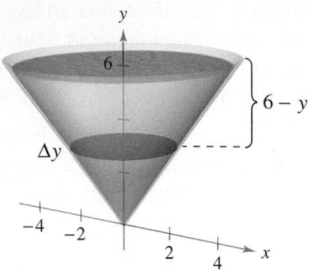

24. Pumping Water Water is pumped in through the bottom of the tank in Exercise 23. How much work is done to fill the tank

(a) to a depth of 2 feet?

(b) from a depth of 4 feet to a depth of 6 feet?

25. Pumping Water A hemispherical tank of radius 6 feet is positioned so that its base is circular. How much work is required to fill the tank with water through a hole in the base when the water source is at the base?

26. Pumping Diesel Fuel The fuel tank on a truck has trapezoidal cross sections with the dimensions (in feet) shown in the figure. Assume that the engine is approximately 3 feet above the top of the fuel tank and that diesel fuel weighs approximately 53.1 pounds per cubic foot. Find the work done by the fuel pump in raising a full tank of fuel to the level of the engine.

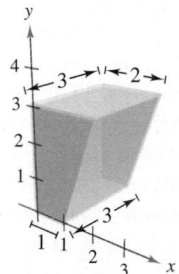

Pumping Gasoline In Exercises 27 and 28, find the work done in pumping gasoline that weighs 42 pounds per cubic foot.

27. A cylindrical gasoline tank 3 feet in diameter and 4 feet long is carried on the back of a truck and is used to fuel tractors. The axis of the tank is horizontal. The opening on the tractor tank is 5 feet above the top of the tank in the truck. Find the work done in pumping the entire contents of the fuel tank into the tractor.

28. The top of a cylindrical gasoline storage tank at a service station is 4 feet below ground level. The axis of the tank is horizontal and its diameter and length are 5 feet and 12 feet, respectively. Find the work done in pumping the entire contents of the full tank to a height of 3 feet above ground level.

Winding a Chain In Exercises 29–32, consider a 20-foot chain that weighs 3 pounds per foot hanging from a winch 20 feet above ground level. Find the work done by the winch in winding up the specified amount of chain.

29. Wind up the entire chain.

30. Wind up one-third of the chain.

31. Run the winch until the bottom of the chain is at the 10-foot level.

32. Wind up the entire chain with a 500-pound load attached to it.

Lifting a Chain In Exercises 33 and 34, consider a 15-foot hanging chain that weighs 3 pounds per foot. Find the work done in lifting the chain vertically to the indicated position.

33. Take the bottom of the chain and raise it to the 15-foot level, leaving the chain doubled and still hanging vertically (see figure).

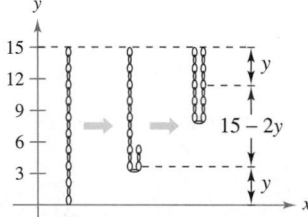

34. Repeat Exercise 33 raising the bottom of the chain to the 12-foot level.

EXPLORING CONCEPTS

35. Think About It Does it take any work to push an object that does not move? Explain.

36. Think About It In Example 1, 200 foot-pounds of work was needed to lift the 50-pound object 4 feet vertically off the ground. Does it take an additional 200 foot-pounds of work to lift the object another 4 feet vertically? Explain your reasoning.

37. Newton's Law of Universal Gravitation Consider two particles of masses m_1 and m_2. The position of the first particle is fixed, and the distance between the particles is a units. Using Newton's Law of Universal Gravitation, find the work needed to move the second particle so that the distance between the particles increases to b units.

38. Conjecture Use Newton's Law of Universal Gravitation to make a conjecture about what happens to the force of attraction between two particles when the distance between them is multiplied by a positive number n.

39. Electric Force Two electrons repel each other with a force that varies inversely as the square of the distance between them. One electron is fixed at the point $(2, 4)$. Find the work done in moving the second electron from $(-2, 4)$ to $(1, 4)$.

40. HOW DO YOU SEE IT? The graphs show the force F_i (in pounds) required to move an object 9 feet along the x-axis. Order the force functions from the one that yields the least work to the one that yields the most work without doing any calculations. Explain your reasoning.

(a)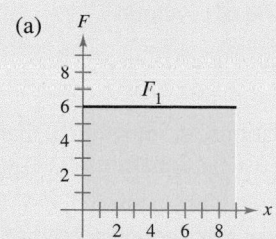

(b)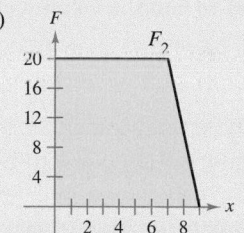

(c)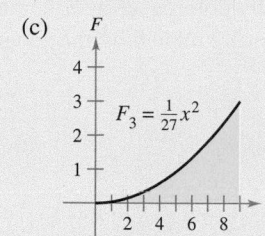

(d)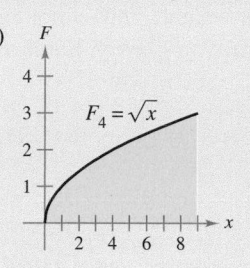

41. Ordering Forces Verify your answer to Exercise 40 by calculating the work for each force function.

42. Comparing Work Order the following from least to greatest in terms of total work done.

(a) A 60-pound box of books is lifted 3 feet.

(b) An 80-pound box of books is lifted 1 foot, and then a 40-pound box of books is lifted 1 foot.

(c) A 60-pound box is held 3 feet in the air for 3 minutes.

 **Boyle's Law** In Exercises 43 and 44, find the work done by the gas for the given volume and pressure. Assume that the pressure is inversely proportional to the volume. (See Example 6.)

43. A quantity of gas with an initial volume of 2 cubic feet and a pressure of 1000 pounds per square foot expands to a volume of 3 cubic feet.

44. A quantity of gas with an initial volume of 1 cubic foot and a pressure of 2500 pounds per square foot expands to a volume of 3 cubic feet.

Hydraulic Press In Exercises 45–48, use the integration capabilities of a graphing utility to approximate the work done by a press in a manufacturing process. A model for the variable force F (in pounds) and the distance x (in feet) the press moves is given.

Force	Interval
45. $F(x) = 1000[1.8 - \ln(x + 1)]$	$0 \le x \le 5$
46. $F(x) = \dfrac{e^{x^2} - 1}{100}$	$0 \le x \le 4$
47. $F(x) = 100x\sqrt{125 - x^3}$	$0 \le x \le 5$
48. $F(x) = 1000 \sinh x$	$0 \le x \le 2$

49. Modeling Data The hydraulic cylinder on a woodsplitter has a 4-inch bore (diameter) and a stroke of 2 feet. The hydraulic pump creates a maximum pressure of 2000 pounds per square inch. Therefore, the maximum force created by the cylinder is $2000(\pi \cdot 2^2) = 8000\pi$ pounds.

(a) Find the work done through one extension of the cylinder, given that the maximum force is required.

(b) The force exerted in splitting a piece of wood is variable. Measurements of the force obtained in splitting a piece of wood are shown in the table. The variable x measures the extension of the cylinder in feet, and F is the force in pounds. Use the regression capabilities of a graphing utility to find a fourth-degree polynomial model for the data. Plot the data and graph the model.

x	0	$\frac{1}{3}$	$\frac{2}{3}$	1	$\frac{4}{3}$	$\frac{5}{3}$	2
$F(x)$	0	20,000	22,000	15,000	10,000	5000	0

(c) Use the model in part (b) to approximate the extension of the cylinder when the force is maximum.

(d) Use the model in part (b) to approximate the work done in splitting the piece of wood.

SECTION PROJECT

Pyramid of Khufu

The Pyramid of Khufu (also known as the Great Pyramid of Giza) is the oldest of the Seven Wonders of the Ancient World. It is also the tallest of the three Giza pyramids in Egypt. The pyramid took 20 years to construct, ending around 2560 B.C. When it was built, it had a height of 481 feet and a square base with side lengths of 756 feet. Assume that the stone used to build it weighed 150 pounds per cubic foot.

(a) How much work was required to build the pyramid? Consider only vertical distance.

(b) Suppose that the pyramid builders worked 12 hours each day for 330 days a year for 20 years and that each worker did 200 foot-pounds of work per hour. Approximately how many workers were needed to build the pyramid?

7.6 Moments, Centers of Mass, and Centroids

- Understand the definition of mass.
- Find the center of mass in a one-dimensional system.
- Find the center of mass in a two-dimensional system.
- Find the center of mass of a planar lamina.
- Use the Theorem of Pappus to find the volume of a solid of revolution.

Mass

In this section, you will study several important applications of integration that are related to mass. Mass is a measure of a body's resistance to changes in motion, and is independent of the particular gravitational system in which the body is located. However, because so many applications involving mass occur on Earth's surface, an object's mass is sometimes equated with its weight. This is not technically correct. Weight is a type of force and as such is dependent on gravity. Force and mass are related by the equation

$$\text{Force} = (\text{mass})(\text{acceleration}).$$

The table below lists some commonly used measures of mass and force, together with their conversion factors.

System of Measurement	Measure of Mass	Measure of Force
U.S.	Slug	Pound = (slug)(ft/sec^2)
International	Kilogram	Newton = (kilogram)(m/sec^2)
C-G-S	Gram	Dyne = (gram)(cm/sec^2)

Conversions:

1 pound ≈ 4.44822 newtons   1 slug ≈ 14.59390 kilogram

1 newton ≈ 0.22481 pound   1 kilogram ≈ 0.0685218 slug

1 dyne ≈ 0.0000022481 pound   1 gram ≈ 0.0000685218 slug

1 dyne = 0.00001 newton   1 foot ≈ 0.30480 meter

EXAMPLE 1   Mass on the Surface of Earth

Find the mass (in slugs) of an object whose weight at sea level is 1 pound.

Solution   Use 32 feet per second per second as the acceleration due to gravity.

$$\text{Mass} = \frac{\text{force}}{\text{acceleration}}$$

$$= \frac{1 \text{ pound}}{32 \text{ feet per second per second}}$$

$$= 0.03125 \frac{\text{pound}}{\text{foot per second per second}}$$

$$= 0.03125 \text{ slug}$$

Force = (mass)(acceleration)

Because many applications involving mass occur on Earth's surface, this amount of mass is called a **pound mass.**

Center of Mass in a One-Dimensional System

You will now consider two types of moments of a mass—the **moment about a point** and the **moment about a line.** To define these two moments, consider an idealized situation in which a mass m is concentrated at a point. If x is the distance between this point mass and another point P, then the **moment of m about the point P is**

$$\text{Moment} = mx$$

and x is the **length of the moment arm.**

The concept of moment can be demonstrated simply by a seesaw, as shown in Figure 7.53. A child of mass 20 kilograms sits 2 meters to the left of fulcrum P, and an older child of mass 30 kilograms sits 2 meters to the right of P. From experience, you know that the seesaw will begin to rotate clockwise, moving the larger child down. This rotation occurs because the moment produced by the child on the left is less than the moment produced by the child on the right.

$$\text{Left moment} = (20)(2) = 40 \ \text{kilogram-meters}$$
$$\text{Right moment} = (30)(2) = 60 \ \text{kilogram-meters}$$

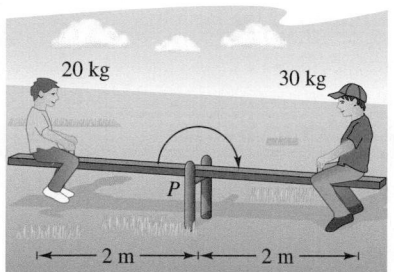

20 kg   30 kg

P

2 m   2 m

The seesaw will balance when the left and the right moments are equal.
Figure 7.53

To balance the seesaw, the two moments must be equal. For example, if the larger child moved to a position $\frac{4}{3}$ meters from the fulcrum, then the seesaw would balance, because each child would produce a moment of 40 kilogram-meters.

To generalize this, you can introduce a coordinate line on which the origin corresponds to the fulcrum, as shown in Figure 7.54. Several point masses are located on the x-axis. The measure of the tendency of this system to rotate about the origin is the **moment about the origin,** and it is defined as the sum of the n products $m_i x_i$. The moment about the origin is denoted by M_0 and can be written as

$$M_0 = m_1 x_1 + m_2 x_2 + \cdots + m_n x_n.$$

If M_0 is 0, then the system is said to be in **equilibrium.**

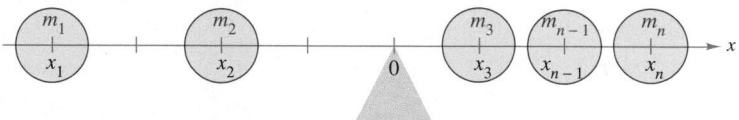

If $m_1 x_1 + m_2 x_2 + \cdots + m_n x_n = 0$, then the system is in equilibrium.
Figure 7.54

For a system that is not in equilibrium, the **center of mass** is defined as the point $\bar{x}$ at which the fulcrum could be relocated to attain equilibrium. If the system were translated $\bar{x}$ units, then each coordinate x_i would become

$$(x_i - \bar{x})$$

and because the moment of the translated system is 0, you have

$$\sum_{i=1}^{n} m_i(x_i - \bar{x}) = \sum_{i=1}^{n} m_i x_i - \sum_{i=1}^{n} m_i \bar{x} = 0.$$

Solving for $\bar{x}$ produces

$$\bar{x} = \frac{\displaystyle\sum_{i=1}^{n} m_i x_i}{\displaystyle\sum_{i=1}^{n} m_i} = \frac{\text{moment of system about origin}}{\text{total mass of system}}.$$

When $m_1 x_1 + m_2 x_2 + \cdots + m_n x_n = 0$, the system is in equilibrium.

Moment and Center of Mass: One-Dimensional System

Let the point masses $m_1, m_2, \ldots, m_n$ be located at $x_1, x_2, \ldots, x_n$.

1. The **moment about the origin** is

$$M_0 = m_1 x_1 + m_2 x_2 + \cdots + m_n x_n.$$

2. The **center of mass** is

$$\bar{x} = \frac{M_0}{m}$$

where $m = m_1 + m_2 + \cdots + m_n$ is the **total mass** of the system.

EXAMPLE 2 **The Center of Mass of a Linear System**

Find the center of mass of the linear system shown in Figure 7.55.

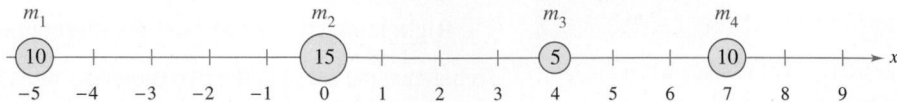

Figure 7.55

Solution The moment about the origin is

$$
\begin{aligned}
M_0 &= m_1 x_1 + m_2 x_2 + m_3 x_3 + m_4 x_4 \\
&= 10(-5) + 15(0) + 5(4) + 10(7) \\
&= -50 + 0 + 20 + 70 \\
&= 40. \qquad\qquad\qquad \text{Moment about origin}
\end{aligned}
$$

Because the total mass of the system is

$$m = 10 + 15 + 5 + 10 = 40 \qquad \text{Total mass}$$

the center of mass is

$$\bar{x} = \frac{M_0}{m} = \frac{40}{40} = 1. \qquad\qquad \text{Center of mass}$$

Note that the point masses will be in equilibrium when the fulcrum is located at $x = 1$.

Rather than defining the moment of a mass, you could define the moment of a *force.* In this context, the center of mass is called the **center of gravity.** Consider a system of point masses $m_1, m_2, \ldots, m_n$ that is located at $x_1, x_2, \ldots, x_n$. Then, because

$$\text{force} = (\text{mass})(\text{acceleration})$$

the total force of the system is

$$F = m_1 a + m_2 a + \cdots + m_n a = ma.$$

The **torque** (moment) about the origin is

$$T_0 = (m_1 a)x_1 + (m_2 a)x_2 + \cdots + (m_n a)x_n = M_0 a$$

and the **center of gravity** is

$$\frac{T_0}{F} = \frac{M_0 a}{ma} = \frac{M_0}{m} = \bar{x}.$$

So, the center of gravity and the center of mass have the same location.

Center of Mass in a Two-Dimensional System

You can extend the concept of moment to two dimensions by considering a system of masses located in the xy-plane at the points $(x_1, y_1), (x_2, y_2), \ldots, (x_n, y_n)$, as shown in Figure 7.56. Rather than defining a single moment (with respect to the origin), two moments are defined—one with respect to the x-axis and one with respect to the y-axis.

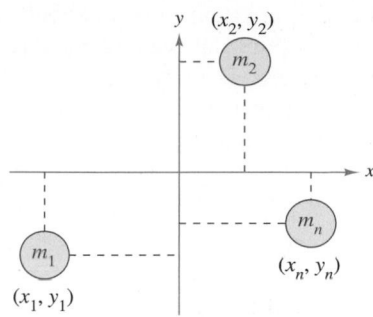

In a two-dimensional system, there is a moment about the y-axis M_y and a moment about the x-axis M_x.

Figure 7.56

Moments and Center of Mass: Two-Dimensional System

Let the point masses $m_1, m_2, \ldots, m_n$ be located at $(x_1, y_1), (x_2, y_2), \ldots, (x_n, y_n)$.

1. The **moment about the y-axis** is

$$M_y = m_1 x_1 + m_2 x_2 + \cdots + m_n x_n.$$

2. The **moment about the x-axis** is

$$M_x = m_1 y_1 + m_2 y_2 + \cdots + m_n y_n.$$

3. The **center of mass** $(\bar{x}, \bar{y})$ (or **center of gravity**) is

$$\bar{x} = \frac{M_y}{m} \quad \text{and} \quad \bar{y} = \frac{M_x}{m}$$

where

$$m = m_1 + m_2 + \cdots + m_n$$

is the **total mass** of the system.

The moment of a system of masses in the plane can be taken about any horizontal or vertical line. In general, the moment about a line is the sum of the product of the masses and the *directed distances* from the points to the line.

$$\text{Moment} = m_1(y_1 - b) + m_2(y_2 - b) + \cdots + m_n(y_n - b) \qquad \text{Horizontal line } y = b$$

$$\text{Moment} = m_1(x_1 - a) + m_2(x_2 - a) + \cdots + m_n(x_n - a) \qquad \text{Vertical line } x = a$$

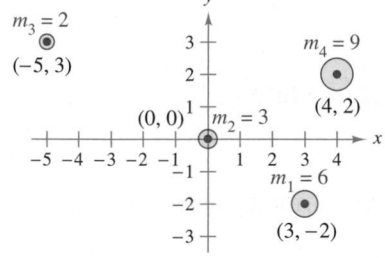

Figure 7.57

EXAMPLE 3 **The Center of Mass of a Two-Dimensional System**

Find the center of mass of a system of point masses $m_1 = 6$, $m_2 = 3$, $m_3 = 2$, and $m_4 = 9$, located at

$$(3, -2), \quad (0, 0), \quad (-5, 3), \quad \text{and} \quad (4, 2)$$

as shown in Figure 7.57.

Solution

$$
\begin{aligned}
M &= 6 \quad\;\; + 3 \;\;\; + 2 \;\;\;\;\; + 9 \;\;\;\; = 20 & \text{Mass} \\
M_y &= 6(3) \;\;\; + 3(0) + 2(-5) + 9(4) = 44 & \text{Moment about } y\text{-axis} \\
M_x &= 6(-2) + 3(0) + 2(3) \;\;\; + 9(2) = 12 & \text{Moment about } x\text{-axis}
\end{aligned}
$$

So,

$$\bar{x} = \frac{M_y}{m} = \frac{44}{20} = \frac{11}{5}$$

and

$$\bar{y} = \frac{M_x}{m} = \frac{12}{20} = \frac{3}{5}.$$

The center of mass is $\left(\frac{11}{5}, \frac{3}{5}\right)$.

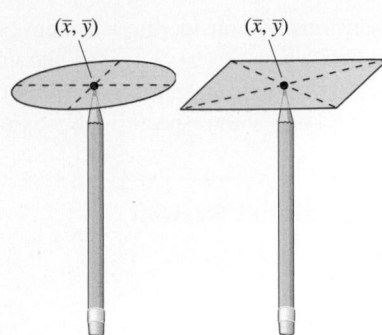

You can think of the center of mass $(\bar{x}, \bar{y})$ of a lamina as its balancing point. For a circular lamina, the center of mass is the center of the circle. For a rectangular lamina, the center of mass is the center of the rectangle.

Figure 7.58

Center of Mass of a Planar Lamina

So far in this section, you have assumed the total mass of a system to be distributed at discrete points in a plane or on a line. Now consider a thin, flat plate of material of constant density called a **planar lamina** (see Figure 7.58). **Density** is a measure of mass per unit of volume, such as grams per cubic centimeter. For planar laminas, however, density is considered to be a measure of mass per unit of area. Density is denoted by ρ, the lowercase Greek letter rho.

Consider an irregularly shaped planar lamina of uniform density ρ, bounded by the graphs of $y = f(x)$, $y = g(x)$, and $a \leq x \leq b$, as shown in Figure 7.59. The mass of this region is

$$
\begin{aligned}
m &= (\text{density})(\text{area}) \\
&= \rho \int_a^b [f(x) - g(x)]\, dx \\
&= \rho A
\end{aligned}
$$

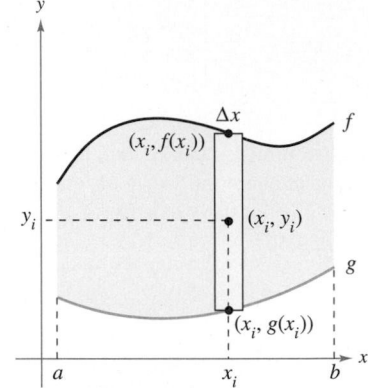

Planar lamina of uniform density ρ
Figure 7.59

where A is the area of the region. To find the center of mass of this lamina, partition the interval $[a, b]$ into n subintervals of equal width Δx. Let x_i be the center of the ith subinterval. You can approximate the portion of the lamina lying in the ith subinterval by a rectangle whose height is $h = f(x_i) - g(x_i)$. Because the density of the rectangle is ρ, its mass is

$$
m_i = (\text{density})(\text{area}) = \rho \underbrace{[f(x_i) - g(x_i)]}_{\text{Height}} \underbrace{\Delta x}_{\text{Width}}.
$$
$\underset{\text{Density}}{}$

Now, considering this mass to be located at the center (x_i, y_i) of the rectangle, the directed distance from the x-axis to (x_i, y_i) is $y_i = [f(x_i) + g(x_i)]/2$. So, the moment of m_i about the x-axis is

$$
\begin{aligned}
\text{Moment} &= (\text{mass})(\text{distance}) \\
&= m_i y_i \\
&= \rho [f(x_i) - g(x_i)] \Delta x \left[\frac{f(x_i) + g(x_i)}{2} \right].
\end{aligned}
$$

Summing the moments and taking the limit as $n \to \infty$ suggest the definitions below.

Moments and Center of Mass of a Planar Lamina

Let f and g be continuous functions such that $f(x) \geq g(x)$ on $[a, b]$, and consider the planar lamina of uniform density ρ bounded by the graphs of $y = f(x)$, $y = g(x)$, and $a \leq x \leq b$.

1. The **moments about the x- and y-axes** are

$$
M_x = \rho \int_a^b \left[\frac{f(x) + g(x)}{2} \right] [f(x) - g(x)]\, dx
$$

$$
M_y = \rho \int_a^b x[f(x) - g(x)]\, dx.
$$

2. The **center of mass** $(\bar{x}, \bar{y})$ is given by $\bar{x} = \dfrac{M_y}{m}$ and $\bar{y} = \dfrac{M_x}{m}$, where $m = \rho \int_a^b [f(x) - g(x)]\, dx$ is the mass of the lamina.

EXAMPLE 4 The Center of Mass of a Planar Lamina

•••▷ See LarsonCalculus.com for an interactive version of this type of example.

Find the center of mass of the lamina of uniform density ρ bounded by the graph of $f(x) = 4 - x^2$ and the x-axis.

Solution Because the center of mass lies on the axis of symmetry, you know that $\bar{x} = 0$. Moreover, the mass of the lamina is

$$m = \rho \int_{-2}^{2} (4 - x^2)\, dx$$

$$= \rho \left[4x - \frac{x^3}{3} \right]_{-2}^{2}$$

$$= \frac{32\rho}{3}.$$

To find the moment about the x-axis, place a representative rectangle in the region, as shown in the figure at the right. The distance from the x-axis to the center of this rectangle is

$$y_i = \frac{f(x)}{2} = \frac{4 - x^2}{2}.$$

Because the mass of the representative rectangle is

$$\rho f(x)\, \Delta x = \rho(4 - x^2)\, \Delta x$$

you have

$$M_x = \rho \int_{-2}^{2} \frac{4 - x^2}{2}(4 - x^2)\, dx$$

$$= \frac{\rho}{2} \int_{-2}^{2} (16 - 8x^2 + x^4)\, dx$$

$$= \frac{\rho}{2} \left[16x - \frac{8x^3}{3} + \frac{x^5}{5} \right]_{-2}^{2}$$

$$= \frac{256\rho}{15}$$

and $\bar{y}$ is

$$\bar{y} = \frac{M_x}{m} = \frac{256\rho/15}{32\rho/3} = \frac{8}{5}.$$

So, the center of mass (the balancing point) of the lamina is $\left(0, \frac{8}{5}\right)$, as shown in Figure 7.60.

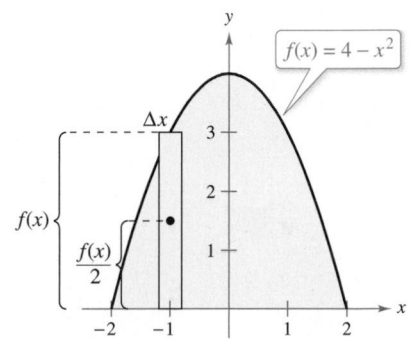

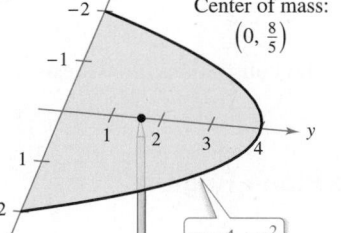

Center of mass: $\left(0, \frac{8}{5}\right)$

$y = 4 - x^2$

The center of mass is the balancing point.

Figure 7.60

The density ρ in Example 4 is a common factor of both the moments and the mass, and as such divides out of the quotients representing the coordinates of the center of mass. So, the center of mass of a lamina of *uniform* density depends only on the shape of the lamina and not on its density. For this reason, the point

$(\bar{x}, \bar{y})$ Center of mass or centroid

is sometimes called the center of mass of a *region* in the plane, or the **centroid** of the region. In other words, to find the centroid of a region in the plane, you can assume that the region has a constant density of $\rho = 1$ and therefore the mass of the region is equal to the area A, or $m = A$. Then you can calculate the corresponding center of mass, as shown in the next two examples.

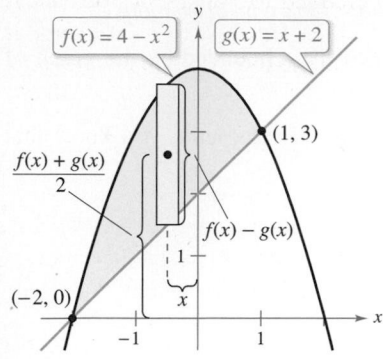

$f(x) = 4 - x^2$ $g(x) = x + 2$

$\dfrac{f(x) + g(x)}{2}$

$f(x) - g(x)$

$(1, 3)$

$(-2, 0)$

Figure 7.61

EXAMPLE 5 **The Centroid of a Plane Region**

Find the centroid of the region bounded by the graphs of $f(x) = 4 - x^2$ and $g(x) = x + 2$.

Solution The two graphs intersect at the points $(-2, 0)$ and $(1, 3)$, as shown in Figure 7.61. So, the area of the region is

$$A = \int_{-2}^{1} [f(x) - g(x)] \, dx = \int_{-2}^{1} (2 - x - x^2) \, dx = \frac{9}{2}.$$

The centroid $(\bar{x}, \bar{y})$ of the region has the following coordinates.

$$\bar{x} = \frac{1}{A} \int_{-2}^{1} x[(4 - x^2) - (x + 2)] \, dx$$

$$= \frac{2}{9} \int_{-2}^{1} (-x^3 - x^2 + 2x) \, dx$$

$$= \frac{2}{9} \left[-\frac{x^4}{4} - \frac{x^3}{3} + x^2 \right]_{-2}^{1}$$

$$= -\frac{1}{2}$$

$$\bar{y} = \frac{1}{A} \int_{-2}^{1} \left[\frac{(4 - x^2) + (x + 2)}{2} \right] [(4 - x^2) - (x + 2)] \, dx$$

$$= \frac{2}{9} \left(\frac{1}{2} \right) \int_{-2}^{1} (-x^2 + x + 6)(-x^2 - x + 2) \, dx$$

$$= \frac{1}{9} \int_{-2}^{1} (x^4 - 9x^2 - 4x + 12) \, dx$$

$$= \frac{1}{9} \left[\frac{x^5}{5} - 3x^3 - 2x^2 + 12x \right]_{-2}^{1}$$

$$= \frac{12}{5}$$

So, the centroid of the region is $(\bar{x}, \bar{y}) = \left(-\frac{1}{2}, \frac{12}{5} \right)$. ∎

For simple plane regions, you may be able to find the centroids without resorting to integration.

EXAMPLE 6 **The Centroid of a Simple Plane Region**

Find the centroid of the region shown in Figure 7.62(a).

Solution By superimposing a coordinate system on the region, as shown in Figure 7.62(b), you can locate the centroids of the three rectangles at

$$\left(\frac{1}{2}, \frac{3}{2} \right), \quad \left(\frac{5}{2}, \frac{1}{2} \right), \quad \text{and} \quad (5, 1).$$

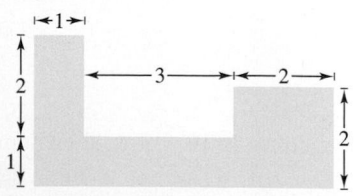

(a) Original region

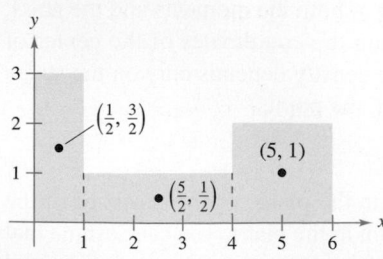

(b) The centroids of the three rectangles
Figure 7.62

Using these three points, you can find the centroid of the region.

$$A = \text{area of region} = 3 + 3 + 4 = 10$$

$$\bar{x} = \frac{(1/2)(3) + (5/2)(3) + (5)(4)}{10} = \frac{29}{10} = 2.9$$

$$\bar{y} = \frac{(3/2)(3) + (1/2)(3) + (1)(4)}{10} = \frac{10}{10} = 1$$

So, the centroid of the region is $(2.9, 1)$. Notice that $(2.9, 1)$ is not the "average" of $\left(\frac{1}{2}, \frac{3}{2} \right)$, $\left(\frac{5}{2}, \frac{1}{2} \right)$, and $(5, 1)$. ∎

Theorem of Pappus

The final topic in this section is a useful theorem credited to Pappus of Alexandria (ca. 300 A.D.), a Greek mathematician whose eight-volume *Mathematical Collection* is a record of much of classical Greek mathematics. You are asked to prove this theorem in Section 14.4.

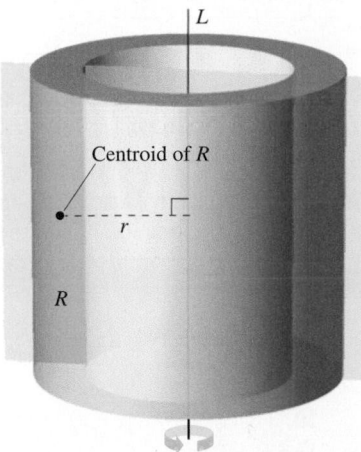

The volume V is $2\pi rA$, where A is the area of region R.

Figure 7.63

THEOREM 7.1 The Theorem of Pappus

Let R be a region in a plane and let L be a line in the same plane such that L does not intersect the interior of R, as shown in Figure 7.63. If r is the distance between the centroid of R and the line, then the volume V of the solid of revolution formed by revolving R about the line is

$$V = 2\pi rA$$

where A is the area of R. (Note that $2\pi r$ is the distance traveled by the centroid as the region is revolved about the line.)

The Theorem of Pappus can be used to find the volume of a torus, as shown in the next example. Recall that a torus is a doughnut-shaped solid formed by revolving a circular region about a line that lies in the same plane as the circle (but does not intersect the circle).

EXAMPLE 7   **Finding Volume by the Theorem of Pappus**

Find the volume of the torus shown in Figure 7.64(a), which was formed by revolving the circular region bounded by

$$(x - 2)^2 + y^2 = 1$$

about the y-axis, as shown in Figure 7.64(b).

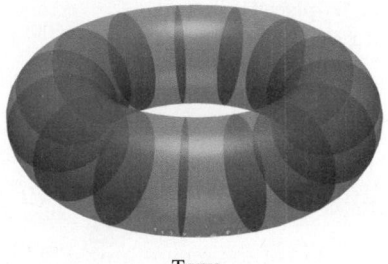

Torus

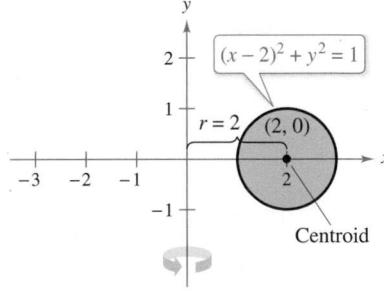

(a)   **(b)**

Figure 7.64

Solution In Figure 7.64(b), you can see that the centroid of the circular region is $(2, 0)$. So, the distance between the centroid and the axis of revolution is

$$r = 2.$$

Because the area of the circular region is $A = \pi$, the volume of the torus is

$$\begin{aligned} V &= 2\pi rA \\ &= 2\pi(2)(\pi) \\ &= 4\pi^2 \\ &\approx 39.5. \end{aligned}$$

Exploration

Use the shell method to show that the volume of the torus in Example 7 is

$$V = \int_1^3 4\pi x \sqrt{1 - (x - 2)^2}\, dx.$$

Evaluate this integral using a graphing utility. Does your answer agree with the one in Example 7?

7.6 Exercises

See **CalcChat.com** for tutorial help and worked-out solutions to odd-numbered exercises.

CONCEPT CHECK

1. Mass and Weight How are mass and weight related?

2. Moment The equation for the moment about the origin of a one-dimensional system is $M_0 = 5(-3) + 2(-1) + 1(1) + 5(2) + 1(6)$. Is the system in equilibrium? Explain.

3. Planar Lamina What is a planar lamina? Describe what the center of mass of a lamina represents.

4. Theorem of Pappus Explain why the Theorem of Pappus is useful.

 **Center of Mass of a Linear System** In Exercises 5–8, find the center of mass of the given system of point masses lying on the x-axis.

5. $m_1 = 7, m_2 = 3, m_3 = 5$
$x_1 = -5, x_2 = 0, x_3 = 3$

6. $m_1 = 0.1, m_2 = 0.2, m_3 = 0.2, m_4 = 0.5$
$x_1 = 1, x_2 = 2, x_3 = 3, x_4 = 4$

7. $m_1 = 1, m_2 = 3, m_3 = 2, m_4 = 9, m_5 = 5$
$x_1 = 6, x_2 = 10, x_3 = 3, x_4 = 2, x_5 = 4$

8. $m_1 = 8, m_2 = 5, m_3 = 5, m_4 = 12, m_5 = 2$
$x_1 = -2, x_2 = 6, x_3 = 0, x_4 = 3, x_5 = -5$

Equilibrium of a Linear System In Exercises 9 and 10, consider a beam of length L with a fulcrum x feet from one end (see figure). There are objects with weights W_1 and W_2 placed on opposite ends of the beam, where $W_1 < W_2$. Find x such that the system is in equilibrium.

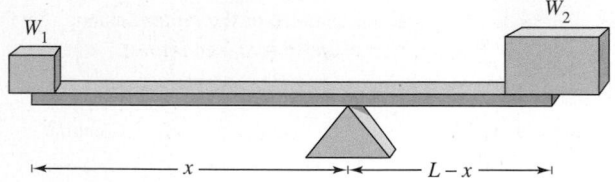

9. Two children weighing 48 pounds and 72 pounds are going to play on a seesaw that is 10 feet long.

10. In order to move a 600-pound rock, a person weighing 200 pounds wants to balance it on a beam that is 5 feet long.

 **Center of Mass of a Two-Dimensional System** In Exercises 11–14, find the center of mass of the given system of point masses.

11.

m_i	5	1	3
(x_i, y_i)	$(2, 2)$	$(-3, 1)$	$(1, -4)$

12.

m_i	8	1	4
(x_i, y_i)	$(-3, -1)$	$(0, 0)$	$(-1, 2)$

13.

m_i	12	6	4.5	15
(x_i, y_i)	$(2, 3)$	$(-1, 5)$	$(6, 8)$	$(2, -2)$

14.

m_i	3	4	2	1	6
(x_i, y_i)	$(-2, -3)$	$(5, 5)$	$(7, 1)$	$(0, 0)$	$(-3, 0)$

 **Center of Mass of a Planar Lamina** In Exercises 15–28, find M_x, M_y, and $(\bar{x}, \bar{y})$ for the lamina of uniform density ρ bounded by the graphs of the equations.

15. $y = \frac{1}{2}x, y = 0, x = 2$ **16.** $y = 6 - x, y = 0, x = 0$

17. $y = \sqrt{x}, y = 0, x = 4$ **18.** $y = \frac{1}{3}x^2, y = 0, x = 2$

19. $y = x^2, y = x^3$ **20.** $y = \sqrt{x}, y = \frac{1}{2}x$

21. $y = -x^2 + 4x + 2, y = x + 2$

22. $y = \sqrt{x} + 1, y = \frac{1}{3}x + 1$

23. $y = x^{2/3}, y = 0, x = 8$ **24.** $y = x^{2/3}, y = 4$

25. $x = 4 - y^2, x = 0$ **26.** $x = 3y - y^2, x = 0$

27. $x = -y, x = 2y - y^2$ **28.** $x = y + 2, x = y^2$

Approximating a Centroid Using Technology In Exercises 29 and 30, use a graphing utility to graph the region bounded by the graphs of the equations. Use the integration capabilities of the graphing utility to approximate the centroid of the region.

29. $y = 5\sqrt[3]{400 - x^2}, y = 0$

30. $y = \dfrac{8}{x^2 + 4}, y = 0, x = -2, x = 2$

 **Finding the Center of Mass** In Exercises 31–34, introduce an appropriate coordinate system and find the center of mass of the planar lamina. (The answer depends on the position of the coordinate system.)

31.

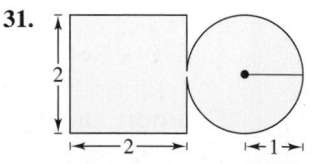

32.

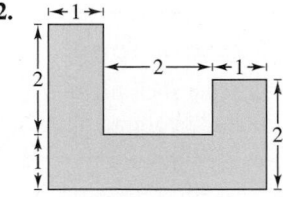

33.

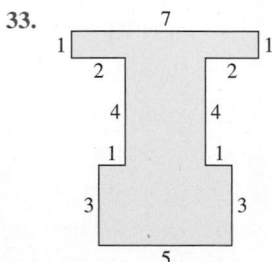

34.

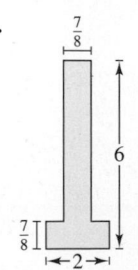

35. Finding the Center of Mass Find the center of mass of the lamina in Exercise 31 when the circular portion of the lamina has twice the density of the square portion of the lamina.

36. Finding the Center of Mass Find the center of mass of the lamina in Exercise 31 when the square portion of the lamina has twice the density of the circular portion of the lamina.

 **Finding Volume by the Theorem of Pappus** **In Exercises 37–40, use the Theorem of Pappus to find the volume of the solid of revolution.**

37. The torus formed by revolving the circular region bounded by

$$(x - 5)^2 + y^2 = 16$$

about the y-axis

38. The torus formed by revolving the circular region bounded by

$$x^2 + (y - 3)^2 = 4$$

about the x-axis

39. The solid formed by revolving the region bounded by the graphs of $y = x$, $y = 4$, and $x = 0$ about the x-axis

40. The solid formed by revolving the region bounded by the graphs of $y = 2\sqrt{x - 2}$, $y = 0$, and $x = 6$ about the y-axis

EXPLORING CONCEPTS

41. Center of Mass What happens to the center of mass of a linear system when each point mass is translated k units horizontally? Explain.

42. Centroid Explain why the centroid of a rectangle is the center of a rectangle.

43. Center of Mass Use rectangles to create a region such that the center of mass lies outside of the region. Verify algebraically that the center of mass lies outside of the region.

 **44.** **HOW DO YOU SEE IT?** The centroid of the plane region bounded by the graphs of $y = f(x)$, $y = 0$, $x = 0$, and $x = 3$ is $(1.2, 1.4)$. Without integrating, find the centroid of each of the regions bounded by the graphs of the following sets of equations. Explain your reasoning.

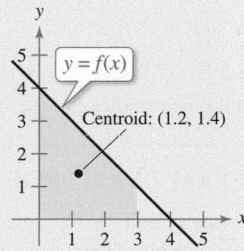

(a) $y = f(x) + 2$, $y = 2$, $x = 0$, and $x = 3$
(b) $y = f(x - 2)$, $y = 0$, $x = 2$, and $x = 5$
(c) $y = -f(x)$, $y = 0$, $x = 0$, and $x = 3$

Centroid of a Common Region **In Exercises 45–50, find and/or verify the centroid of the common region used in engineering.**

45. Triangle Show that the centroid of the triangle with vertices $(-a, 0)$, $(a, 0)$, and (b, c) is the point of intersection of the medians (see figure).

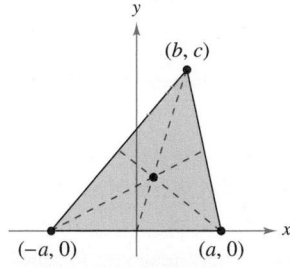

Figure for 45

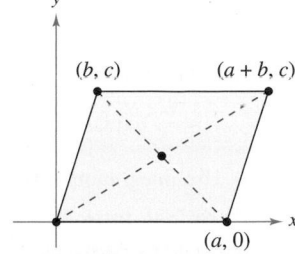

Figure for 46

46. Parallelogram Show that the centroid of the parallelogram with vertices $(0, 0)$, $(a, 0)$, (b, c), and $(a + b, c)$ is the point of intersection of the diagonals (see figure).

47. Trapezoid Find the centroid of the trapezoid with vertices $(0, 0)$, $(0, a)$, (c, b), and $(c, 0)$. Show that it is the intersection of the line connecting the midpoints of the parallel sides and the line connecting the extended parallel sides, as shown in the figure.

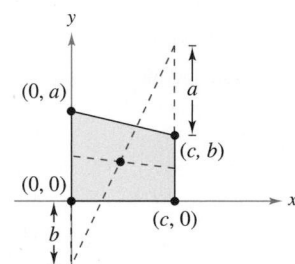

Figure for 47

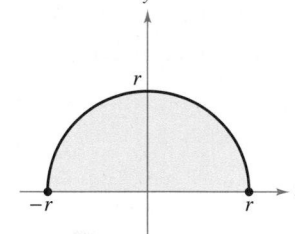

Figure for 48

48. Semicircle Find the centroid of the region bounded by the graphs of $y = \sqrt{r^2 - x^2}$ and $y = 0$ (see figure).

49. Semiellipse Find the centroid of the region bounded by the graphs of $y = \dfrac{b}{a}\sqrt{a^2 - x^2}$ and $y = 0$ (see figure).

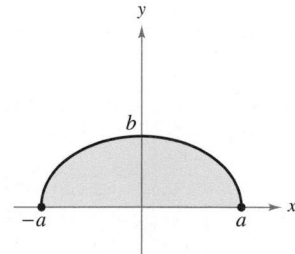

Figure for 49

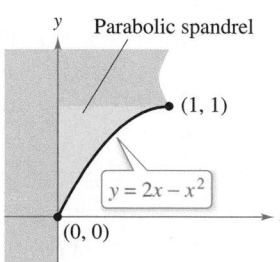

Figure for 50

50. Parabolic Spandrel Find the centroid of the **parabolic spandrel** shown in the figure.

51. Graphical Reasoning Consider the region bounded by the graphs of $y = x^2$ and $y = b$, where $b > 0$.

(a) Sketch a graph of the region.

(b) Set up the integral for finding M_y. Because of the form of the integrand, the value of the integral can be obtained without integrating. What is the form of the integrand? What is the value of the integral and what is the value of $\bar{x}$?

(c) Use the graph in part (a) to determine whether $\bar{y} > \dfrac{b}{2}$ or $\bar{y} < \dfrac{b}{2}$. Explain.

(d) Use integration to verify your answer in part (c).

52. Graphical and Numerical Reasoning Consider the region bounded by the graphs of $y = x^{2n}$ and $y = b$, where $b > 0$ and n is a positive integer.

(a) Sketch a graph of the region.

(b) Set up the integral for finding M_y. Because of the form of the integrand, the value of the integral can be obtained without integrating. What is the form of the integrand? What is the value of the integral and what is the value of $\bar{x}$?

(c) Use the graph in part (a) to determine whether $\bar{y} > \dfrac{b}{2}$ or $\bar{y} < \dfrac{b}{2}$. Explain.

(d) Use integration to find $\bar{y}$ as a function of n.

(e) Use the result of part (d) to complete the table.

n	1	2	3	4
$\bar{y}$				

(f) Find $\displaystyle\lim_{n\to\infty} \bar{y}$.

(g) Give a geometric explanation of the result in part (f).

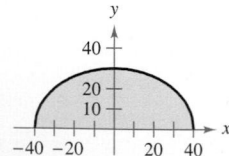 **53. Modeling Data** The manufacturer of glass for a window in a conversion van needs to approximate the center of mass of the glass. A coordinate system is superimposed on a prototype of the glass (see figure). The measurements (in centimeters) for the right half of the symmetric piece of glass are listed in the table.

x	0	10	20	30	40
y	30	29	26	20	0

(a) Use the regression capabilities of a graphing utility to find a fourth-degree polynomial model for the glass.

(b) Use the integration capabilities of a graphing utility and the model to approximate the center of mass of the glass.

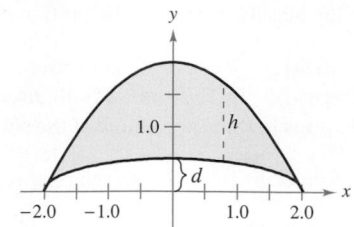 **54. Modeling Data** The manufacturer of a boat needs to approximate the center of mass of a section of the hull. A coordinate system is superimposed on a prototype (see figure). The measurements (in feet) for the right half of the symmetric prototype are listed in the table.

x	0	0.5	1.0	1.5	2
h	1.50	1.45	1.30	0.99	0
d	0.50	0.48	0.43	0.33	0

(a) Use the regression capabilities of a graphing utility to find fourth-degree polynomial models for both curves shown in the figure.

(b) Use the integration capabilities of a graphing utility and the models to approximate the center of mass of the hull section.

Second Theorem of Pappus In Exercises 55 and 56, use the *Second Theorem of Pappus*, which is stated as follows. If a segment of a plane curve C is revolved about an axis that does not intersect the curve (except possibly at its endpoints), then the area S of the resulting surface of revolution is equal to the product of the length of C times the distance d traveled by the centroid of C.

55. Find the area of the surface formed by revolving the graph of $y = 3 - x$, $0 \le x \le 3$, about the y-axis.

56. A torus is formed by revolving the graph of $(x - 1)^2 + y^2 = 1$ about the y-axis. Find the surface area of the torus.

57. Finding a Centroid Let $n \ge 1$ be constant, and consider the region bounded by $f(x) = x^n$, the x-axis, and $x = 1$. Find the centroid of this region. As $n \to \infty$, what does the region look like, and where is its centroid?

58. Finding a Centroid Consider the functions

$$f(x) = x^n \quad \text{and} \quad g(x) = x^m$$

on the interval $[0, 1]$, where m and n are positive integers and $n > m$. Find the centroid of the region bounded by f and g.

PUTNAM EXAM CHALLENGE

59. Let V be the region in the cartesian plane consisting of all points (x, y) satisfying the simultaneous conditions $|x| \le y \le |x| + 3$ and $y \le 4$. Find the centroid $(\bar{x}, \bar{y})$ of V.

7.7 Fluid Pressure and Fluid Force

■ Find fluid pressure and fluid force.

Fluid Pressure and Fluid Force

Swimmers know that the deeper an object is submerged in a fluid, the greater the pressure on the object. **Pressure** is defined as the force per unit of area over the surface of a body. For example, because a column of water that is 10 feet in height and 1 inch square weighs 4.3 pounds, the *fluid pressure* at a depth of 10 feet of water is 4.3 pounds per square inch.* At 20 feet, this would increase to 8.6 pounds per square inch, and in general the pressure is proportional to the depth of the object in the fluid.

BLAISE PASCAL (1623–1662)

Pascal is well known for his work in many areas of mathematics and physics, and also for his influence on Leibniz. Although much of Pascal's work in calculus was intuitive and lacked the rigor of modern mathematics, he nevertheless anticipated many important results.
See LarsonCalculus.com to read more of this biography.

Definition of Fluid Pressure

The **pressure** P on an object at depth h in a liquid is

$$P = wh$$

where w is the weight-density of the liquid per unit of volume.

Below are some common weight-densities of fluids in pounds per cubic foot.

Ethyl alcohol	49.4
Gasoline	41.0–43.0
Glycerin	78.6
Kerosene	51.2
Mercury	849.0
Seawater	64.0
Water	62.4

When calculating fluid pressure, you can use an important (and rather surprising) physical law called **Pascal's Principle,** named after the French mathematician Blaise Pascal. Pascal's Principle states that the pressure exerted by a fluid at a depth h is transmitted equally *in all directions*. For example, in Figure 7.65, the pressure at the indicated depth is the same for all three objects. Because fluid pressure is given in terms of force per unit area ($P = F/A$), the fluid force on a *submerged horizontal* surface of area A is

Fluid force $= F = PA =$ (pressure)(area).

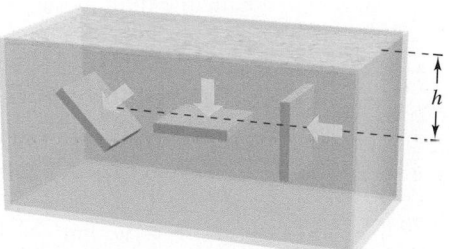

The pressure at h is the same for all three objects.
Figure 7.65

---

* The total pressure on an object in 10 feet of water would also include the pressure due to Earth's atmosphere. At sea level, atmospheric pressure is approximately 14.7 pounds per square inch.

The fluid force on a horizontal metal sheet is equal to the fluid pressure times the area.

Figure 7.66

EXAMPLE 1 **Fluid Force on a Submerged Sheet**

Find the fluid force on a rectangular metal sheet measuring 3 feet by 4 feet that is submerged in 6 feet of water, as shown in Figure 7.66.

Solution Because the weight-density of water is 62.4 pounds per cubic foot and the sheet is submerged in 6 feet of water, the fluid pressure is

$$P = (62.4)(6) \qquad\qquad P = wh$$
$$= 374.4 \text{ pounds per square foot.}$$

Because the total area of the sheet is $A = (3)(4) = 12$ square feet, the fluid force is

$$F = PA$$
$$= \left(374.4 \, \frac{\text{pounds}}{\text{square foot}}\right)(12 \text{ square feet})$$
$$= 4492.8 \text{ pounds.}$$

This result is independent of the size of the body of water. The fluid force would be the same in a swimming pool or lake. ◼

In Example 1, the fact that the sheet is rectangular and horizontal means that you do not need the methods of calculus to solve the problem. Consider a surface that is submerged vertically in a fluid. This problem is more difficult because the pressure is not constant over the surface.

Consider a vertical plate that is submerged in a fluid of weight-density w (per unit of volume), as shown in Figure 7.67. To determine the total force against *one side* of the region from depth c to depth d, you can subdivide the interval $[c, d]$ into n subintervals, each of width Δy. Next, consider the representative rectangle of width Δy and length $L(y_i)$, where y_i is in the ith subinterval. The force against this representative rectangle is

$$\Delta F_i = w(\text{depth})(\text{area})$$
$$= wh(y_i)L(y_i)\Delta y.$$

The force against n such rectangles is

$$\sum_{i=1}^{n} \Delta F_i = w \sum_{i=1}^{n} h(y_i)L(y_i)\Delta y.$$

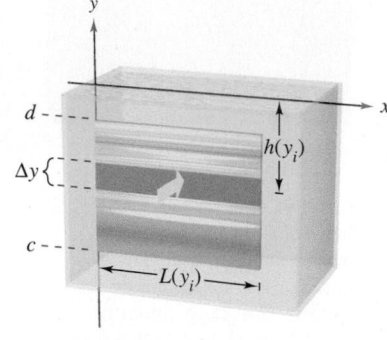

Calculus methods must be used to find the fluid force on a vertical metal plate.

Figure 7.67

Note that w is considered to be constant and is factored out of the summation. Therefore, taking the limit as $\|\Delta\| \to 0$ ($n \to \infty$) suggests the next definition.

Definition of Force Exerted by a Fluid

The **force F exerted by a fluid** of constant weight-density w (per unit of volume) against a submerged vertical plane region from $y = c$ to $y = d$ is

$$F = w \lim_{\|\Delta\|\to 0} \sum_{i=1}^{n} h(y_i)L(y_i)\Delta y$$
$$= w \int_{c}^{d} h(y)L(y) \, dy$$

where $h(y)$ is the depth of the fluid at y and $L(y)$ is the horizontal length of the region at y.

| EXAMPLE 2 | **Fluid Force on a Vertical Surface** |

⋮ ⊳ *See LarsonCalculus.com for an interactive version of this type of example.*

A vertical gate in a dam has the shape of an isosceles trapezoid 8 feet across the top and 6 feet across the bottom, with a height of 5 feet, as shown in Figure 7.68(a). What is the fluid force on the gate when the top of the gate is 4 feet below the surface of the water?

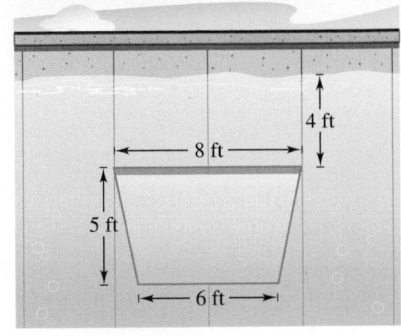

(a) Water gate in a dam

Solution In setting up a mathematical model for this problem, you are at liberty to locate the *x*- and *y*-axes in several different ways. A convenient approach is to let the *y*-axis bisect the gate and place the *x*-axis at the surface of the water, as shown in Figure 7.68(b). So, the depth of the water at *y* in feet is

$$\text{Depth} = h(y) = -y.$$

To find the length $L(y)$ of the region at *y*, find the equation of the line forming the right side of the gate. Because this line passes through the points $(3, -9)$ and $(4, -4)$, its equation is

$$y - (-9) = \frac{-4 - (-9)}{4 - 3}(x - 3)$$

$$y + 9 = 5(x - 3)$$

$$y = 5x - 24$$

$$x = \frac{y + 24}{5}.$$

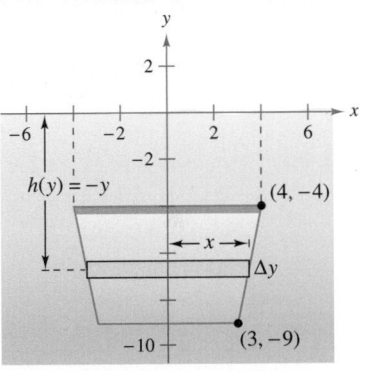

(b) The fluid force against the gate

Figure 7.68

In Figure 7.68(b), you can see that the length of the region at *y* is

$$\text{Length} = 2x = \frac{2}{5}(y + 24) = L(y).$$

Finally, by integrating from $y = -9$ to $y = -4$, you can calculate the fluid force to be

$$F = w \int_c^d h(y)L(y)\,dy$$

$$= 62.4 \int_{-9}^{-4} (-y)\left(\frac{2}{5}\right)(y + 24)\,dy$$

$$= -62.4\left(\frac{2}{5}\right)\int_{-9}^{-4} (y^2 + 24y)\,dy$$

$$= -62.4\left(\frac{2}{5}\right)\left[\frac{y^3}{3} + 12y^2\right]_{-9}^{-4}$$

$$= -62.4\left(\frac{2}{5}\right)\left(\frac{-1675}{3}\right)$$

$$= 13{,}936 \text{ pounds.}$$

In Example 2, the *x*-axis coincided with the surface of the water. This was convenient but arbitrary. In choosing a coordinate system to represent a physical situation, you should consider various possibilities. Often you can simplify the calculations in a problem by locating the coordinate system to take advantage of special characteristics of the problem, such as symmetry.

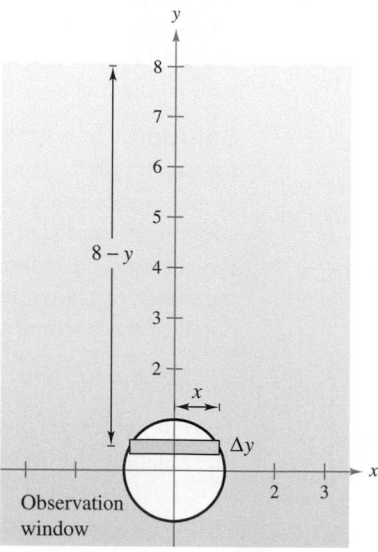

| EXAMPLE 3 | **Fluid Force on a Vertical Surface** |

A circular observation window at a seawater aquarium has a radius of 1 foot, and the center of the window is 8 feet below water level, as shown in Figure 7.69. What is the fluid force on the window?

The fluid force on the window
Figure 7.69

Solution To take advantage of symmetry, locate a coordinate system such that the origin coincides with the center of the window, as shown in Figure 7.69. The depth at y is then

$$\text{Depth} = h(y) = 8 - y.$$

The horizontal length of the window is $2x$, and you can use the equation for the circle, $x^2 + y^2 = 1$, to solve for x as shown.

$$\text{Length} = 2x = 2\sqrt{1 - y^2} = L(y)$$

Finally, because y ranges from -1 to 1, and using 64 pounds per cubic foot as the weight-density of seawater, you have

$$F = w \int_c^d h(y)L(y)\, dy = 64 \int_{-1}^{1} (8 - y)(2)\sqrt{1 - y^2}\, dy.$$

Initially it looks as though this integral would be difficult to solve. However, when you break the integral into two parts and apply symmetry, the solution is simpler.

$$F = 64(16) \int_{-1}^{1} \sqrt{1 - y^2}\, dy - 64(2) \int_{-1}^{1} y\sqrt{1 - y^2}\, dy$$

The second integral is 0 (because the integrand is odd and the limits of integration are symmetric with respect to the origin). Moreover, by recognizing that the first integral represents the area of a semicircle of radius 1, you obtain

$$F = 64(16)\left(\frac{\pi}{2}\right) - 64(2)(0)$$

$$= 512\pi$$

$$\approx 1608.5 \text{ pounds.}$$

So, the fluid force on the window is about 1608.5 pounds.

7.7 Exercises   See **CalcChat.com** for tutorial help and worked-out solutions to odd-numbered exercises.

CONCEPT CHECK

1. Fluid Pressure Describe fluid pressure in your own words.

2. Fluid Pressure Does fluid pressure change with depth?

Force on a Submerged Sheet **In Exercises 3–6, the area of the top side of a piece of sheet metal is given. The sheet metal is submerged horizontally in 8 feet of water. Find the fluid force on the top side.**

3. 3 square feet

4. 8 square feet

5. 10 square feet

6. 25 square feet

Force on a Submerged Sheet **In Exercises 7 and 8, the area of the top side of a piece of sheet metal is given. The sheet metal is submerged horizontally in 5 feet of ethyl alcohol. Find the fluid force on the top side.**

7. 9 square feet

8. 14 square feet

Fluid Force on a Tank Wall **In Exercises 9–14, find the fluid force on the vertical side of the tank, where the dimensions are given in feet. Assume that the tank is full of water.**

9. Rectangle

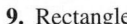
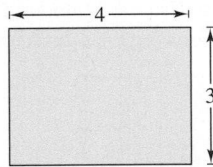

10. Triangle

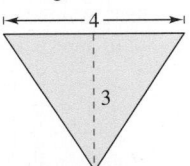

11. Trapezoid

12. Semicircle

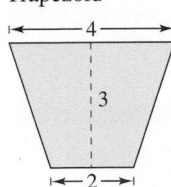
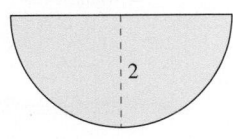

13. Parabola, $y = x^2$

14. Semiellipse,
$y = -\frac{1}{2}\sqrt{36 - 9x^2}$

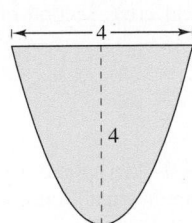
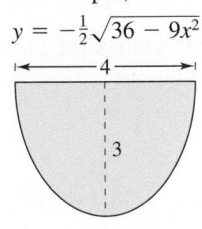

Fluid Force of Water **In Exercises 15–18, find the fluid force on the vertical plate submerged in water, where the dimensions are given in meters and the weight-density of water is 9800 newtons per cubic meter.**

15. Square

16. Rectangle

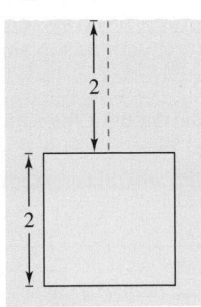
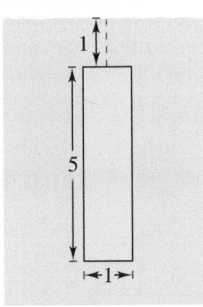

17. Triangle

18. Square

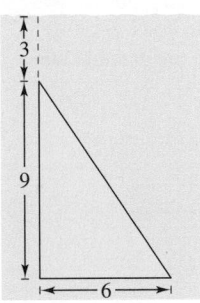
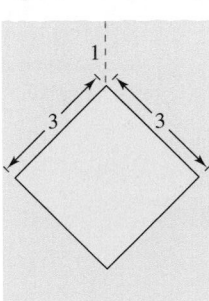

Force on a Concrete Form **In Exercises 19–22, the figure is the vertical side of a form for poured concrete that weighs 140.7 pounds per cubic foot. Determine the force on this part of the concrete form.**

19. Rectangle

20. Semiellipse,
$y = -\frac{3}{4}\sqrt{16 - x^2}$

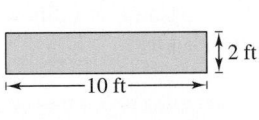
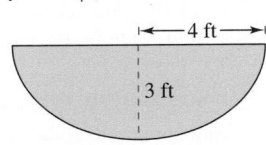

21. Rectangle

22. Triangle

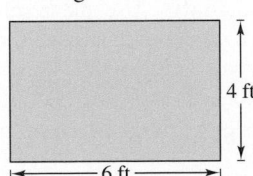
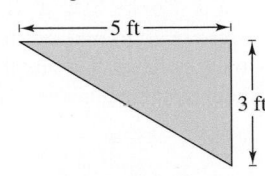

23. Fluid Force of Gasoline A cylindrical gasoline tank is placed so that the axis of the cylinder is horizontal. Find the fluid force on a circular end of the tank when the tank is half full, where the diameter is 3 feet and the gasoline weighs 42 pounds per cubic foot.

24. Fluid Force of Gasoline Repeat Exercise 23 for a tank that is full. (Evaluate one integral by a geometric formula and the other by observing that the integrand is an odd function.)

EXPLORING CONCEPTS

25. Fluid Pressure Explain why fluid pressure on a surface is calculated using horizontal representative rectangles instead of vertical representative rectangles.

26. Buoyant Force Buoyant force is the difference between the fluid forces on the top and bottom sides of a solid. Find an expression for the buoyant force of a rectangular solid submerged in a fluid with its top side parallel to the surface of the fluid.

27. Think About It Approximate the depth of the water in the tank in Exercise 9 if the fluid force is one-half as great as when the tank is full. Explain why the answer is not $\frac{3}{2}$.

---

28. HOW DO YOU SEE IT? Two identical semicircular windows are placed at the same depth in the vertical wall of an aquarium (see figure). Which is subjected to the greater fluid force? Explain.

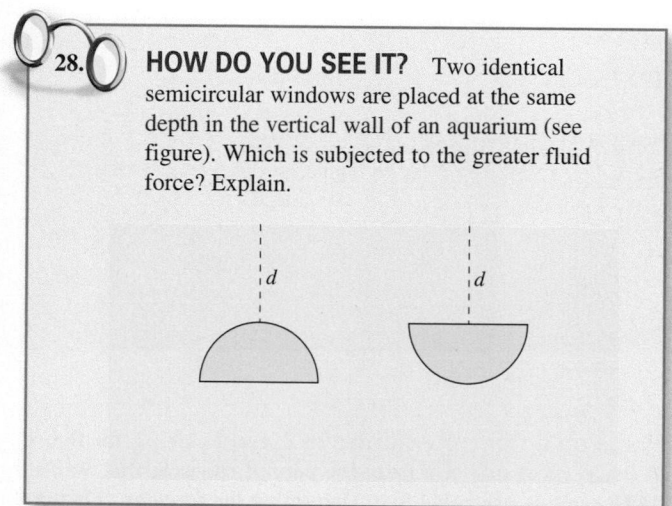

29. Fluid Force on a Circular Plate A circular plate of radius r feet is submerged vertically in a tank of fluid that weighs w pounds per cubic foot. The center of the circle is k feet below the surface of the fluid, where $k > r$. Show that the fluid force on the surface of the plate is

$$F = wk(\pi r^2).$$

(Evaluate one integral by a geometric formula and the other by observing that the integrand is an odd function.)

30. Fluid Force on a Circular Plate Use the result of Exercise 29 to find the fluid force on the circular plate shown in each figure. Assume that the tank is filled with water and the measurements are given in feet.

(a)   (b)

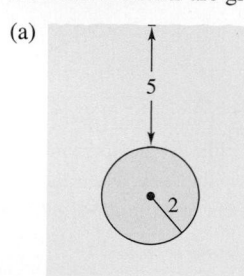   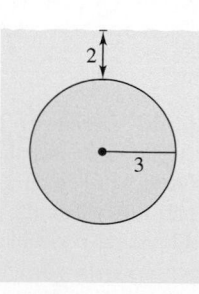

31. Fluid Force on a Rectangular Plate A rectangular plate of height h feet and base b feet is submerged vertically in a tank of fluid that weighs w pounds per cubic foot. The center of the rectangle is k feet below the surface of the fluid, where $k > h/2$. Show that the fluid force on the surface of the plate is

$$F = wkhb.$$

32. Fluid Force on a Rectangular Plate Use the result of Exercise 31 to find the fluid force on the rectangular plate shown in each figure. Assume that the tank is filled with water and the measurements are given in feet.

(a)   (b)

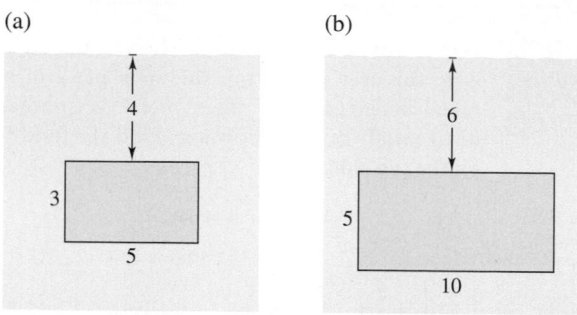

33. Submarine Porthole A square porthole on a vertical side of a submarine (submerged in seawater) has an area of 1 square foot. Find the fluid force on the porthole, assuming that the center of the square is 15 feet below the surface.

34. Submarine Porthole Repeat Exercise 33 for a circular porthole that has a diameter of 1 foot. The center of the circle is 15 feet below the surface.

35. Modeling Data The vertical stern of a boat partially submerged in seawater with a superimposed coordinate system is shown in the figure. The table shows the widths w of the stern (in feet) at indicated values of y. Use the Midpoint Rule with $n = 4$ to approximate the fluid force against the stern.

y	0	1	2	3	4
w	0	5	9	10.25	10.5

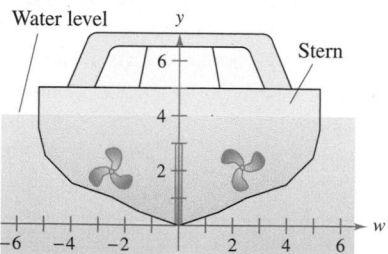

36. Irrigation Canal Gate The vertical cross section of an irrigation canal is modeled by

$$f(x) = \frac{5x^2}{x^2 + 4}$$

where x is measured in feet and $x = 0$ corresponds to the center of the canal. Use the integration capabilities of a graphing utility to approximate the fluid force against a vertical gate used to stop the flow of water when the water is 3 feet deep.

Review Exercises   See **CalcChat.com** for tutorial help and worked-out solutions to odd-numbered exercises.

Finding the Area of a Region   In Exercises 1–10, sketch the region bounded by the graphs of the equations and find the area of the region.

1. $y = 6 - \frac{1}{2}x^2$, $y = \frac{3}{4}x$, $x = -2$, $x = 2$

2. $y = \frac{1}{x^2}$, $y = 4$, $x = 5$

3. $y = \frac{1}{x^2 + 1}$, $y = 0$, $x = -1$, $x = 1$

4. $x = y^2 - 2y$, $x = -1$, $y = 0$

5. $y = x$, $y = x^3$

6. $x = y^2 + 1$, $x = y + 3$

7. $y = e^x$, $y = e^2$, $x = 0$

8. $y = \csc x$, $y = 2$, $\frac{\pi}{6} \le x \le \frac{5\pi}{6}$

9. $y = \sin x$, $y = \cos x$, $\frac{\pi}{4} \le x \le \frac{5\pi}{4}$

10. $x = \cos y$, $x = \frac{1}{2}$, $\frac{\pi}{3} \le y \le \frac{7\pi}{3}$

Finding the Area of a Region   In Exercises 11–14, (a) use a graphing utility to graph the region bounded by the graphs of the equations and (b) use the integration capabilities of the graphing utility to approximate the area of the region to four decimal places.

11. $y = x^2 - 8x + 3$, $y = 3 + 8x - x^2$

12. $y = x^2 - 4x + 3$, $y = x^3$, $x = 0$

13. $\sqrt{x} + \sqrt{y} = 1$, $y = 0$, $x = 0$

14. $y = x^4 - 2x^2$, $y = 2x^2$

Integration as an Accumulation Process   In Exercises 15 and 16, find the accumulation function F. Then evaluate F at each value of the independent variable and graphically show the area given by each value of the independent variable.

15. $F(x) = \int_0^x (3t + 1)\, dt$
   (a) $F(0)$ (b) $F(2)$ (c) $F(6)$

16. $F(x) = \int_{-\pi}^x (2 + \sin t)\, dt$
   (a) $F(-\pi)$ (b) $F(0)$ (c) $F(2\pi)$

Revenue   In Exercises 17 and 18, two models R_1 and R_2 are given for revenue (in millions of dollars) for a corporation. Both models are estimates of revenues from 2020 through 2025, with $t = 0$ corresponding to 2020. Which model projects the greater revenue? How much more total revenue does that model project over the six-year period?

17. $R_1 = 2.98 + 0.65t$
   $R_2 = 2.98 + 0.56t$

18. $R_1 = 4.87 + 0.55t + 0.01t^2$
   $R_2 = 4.87 + 0.61t + 0.07t^2$

Finding the Volume of a Solid   In Exercises 19 and 20, use the disk method to find the volume of the solid generated by revolving the region bounded by the graphs of the equations about the x-axis.

19. $y = \frac{1}{\sqrt{1 + x^2}}$, $y = 0$, $x = -1$, $x = 1$

20. $y = e^{-x}$, $y = 0$, $x = 0$, $x = 1$

Finding the Volume of a Solid   In Exercises 21 and 22, use the shell method to find the volume of the solid generated by revolving the region bounded by the graphs of the equations about the y-axis.

21. $y = \frac{1}{x^4 + 1}$, $y = 0$, $x = 0$, $x = 1$

22. $y = \frac{1}{x^2}$, $y = 0$, $x = 2$, $x = 5$

Finding the Volume of a Solid   In Exercises 23 and 24, use the disk method *or* the shell method to find the volumes of the solids generated by revolving the region bounded by the graphs of the equations about the given lines.

23. $y = x$, $y = 0$, $x = 3$
   (a) the x-axis
   (b) the y-axis
   (c) the line $x = 3$
   (d) the line $x = 6$

24. $y = \sqrt{x}$, $y = 2$, $x = 0$
   (a) the x-axis
   (b) the line $y = 2$
   (c) the y-axis
   (d) the line $x = -1$

25. Gasoline Tank   A gasoline tank is an oblate spheroid generated by revolving the region bounded by the graph of

$$\frac{x^2}{16} + \frac{y^2}{9} = 1$$

about the y-axis, where x and y are measured in feet. How much gasoline can the tank hold?

26. Using Cross Sections   Find the volume of the solid whose base is bounded by the circle $x^2 + y^2 = 9$ and whose cross sections perpendicular to the x-axis are equilateral triangles.

Finding Arc Length   In Exercises 27 and 28, find the arc length of the graph of the function over the indicated interval.

27. $f(x) = \frac{4}{5}x^{5/4}$, $[0, 4]$

28. $y = \frac{1}{3}x^{3/2} - 1$, $[2, 6]$

Finding the Area of a Surface of Revolution In Exercises 29 and 30, write and evaluate the definite integral that represents the area of the surface generated by revolving the curve on the indicated interval about the *x*-axis.

29. $y = \dfrac{x^3}{18}$, $3 \le x \le 6$

30. $y = \sqrt{25 - x^2}$, $-4 \le x \le 4$

Finding the Area of a Surface of Revolution In Exercises 31 and 32, write and evaluate the definite integral that represents the area of the surface generated by revolving the curve on the indicated interval about the *y*-axis.

31. $y = \dfrac{x^2}{2} + 4$, $0 \le x \le 2$

32. $y = \sqrt[3]{x}$, $1 \le x \le 2$

33. Hooke's Law A force of 5 pounds stretches a spring 1 inch from its natural position. Find the work done in stretching the spring from its natural length of 10 inches to a length of 15 inches.

34. Hooke's Law A force of 50 pounds stretches a spring 1 inch from its natural position. Find the work done in stretching the spring from its natural length of 10 inches to double that length.

35. Propulsion Neglecting air resistance and the weight of the propellant, determine the work done in propelling a five-metric-ton satellite to a height of 200 miles above Earth.

36. Pumping Water A water well has an 8-inch diameter and is 190 feet deep. The water is 25 feet from the top of the well. Determine the amount of work done in pumping the well dry.

37. Winding a Chain A chain 10 feet long weighs 4 pounds per foot and is hung from a platform 20 feet above the ground. How much work is required to raise the entire chain to the 20-foot level?

38. Winding a Cable A 200-foot cable weighing 5 pounds per foot is hanging from a winch 200 feet above ground level. Find the work done in winding up the cable when there is a 300-pound load attached to the end of the cable.

39. Boyle's Law A quantity of gas with an initial volume of 1 cubic foot and a pressure of 500 pounds per square foot expands to a volume of 4 cubic feet. Find the work done by the gas. Assume that the pressure is inversely proportional to the volume.

40. Boyle's Law A quantity of gas with an initial volume of 2 cubic feet and a pressure of 800 pounds per square foot expands to a volume of 3 cubic feet. Find the work done by the gas. Assume that the pressure is inversely proportional to the volume.

41. Center of Mass of a Linear System Find the center of mass of the given system of point masses lying on the *x*-axis.

$m_1 = 8$, $m_2 = 12$, $m_3 = 6$, $m_4 = 14$

$x_1 = -1$, $x_2 = 2$, $x_3 = 5$, $x_4 = 7$

42. Center of Mass of a Two-Dimensional System Find the center of mass of the given system of point masses.

m_i	3	2	6	9
(x_i, y_i)	$(2, 1)$	$(-3, 2)$	$(4, -1)$	$(6, 5)$

Center of Mass of a Planar Lamina In Exercises 43 and 44, find M_x, M_y, and $(\bar{x}, \bar{y})$ for the lamina of uniform density ρ bounded by the graphs of the equations.

43. $y = x^2$, $y = 2x + 3$

44. $y = x^{2/3}$, $y = \frac{1}{2}x$

45. Finding the Center of Mass Introduce an appropriate coordinate system and find the center of mass of the planar lamina. (The answer depends on the position of the coordinate system.)

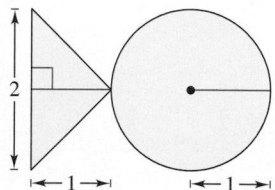

46. Finding Volume Use the Theorem of Pappus to find the volume of the torus formed by revolving the circular region bounded by $(x - 4)^2 + y^2 = 4$ about the *y*-axis.

Force on a Submerged Sheet In Exercises 47 and 48, the area of the top side of a piece of sheet metal is given. The sheet metal is submerged horizontally in 3 feet of water. Find the fluid force on the top side.

47. 2 square feet

48. 15 square feet

49. Fluid Force of Seawater Find the fluid force on the vertical plate submerged in seawater (see figure).

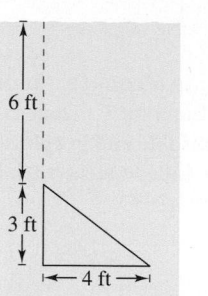   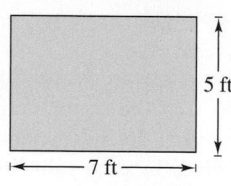

Figure for 49   Figure for 50

50. Force on a Concrete Form The vertical side of a form for poured concrete that weighs 140.7 pounds per cubic foot is shown in the figure. Determine the force on this part of the concrete form.

51. Submarine Porthole A circular porthole on a vertical side of a submarine (submerged in seawater) has a diameter of 3 feet. Find the fluid force on the porthole, assuming that the center of the circle is 1600 feet below the surface.

P.S. Problem Solving

See **CalcChat.com** for tutorial help and worked-out solutions to odd-numbered exercises.

1. Finding a Limit Let R be the area of the region in the first quadrant bounded by the parabola $y = x^2$ and the line $y = cx$, $c > 0$, as shown in the figure. Let T be the area of the triangle AOB. Calculate the limit

$$\lim_{c \to 0^+} \frac{T}{R}.$$

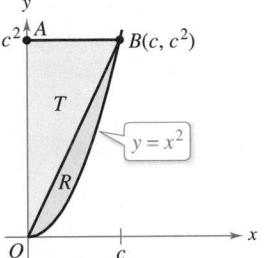

2. Center of Mass of a Lamina Let L be the lamina of uniform density $\rho = 1$ obtained by removing circle A of radius r from circle B of radius $2r$ (see figure).

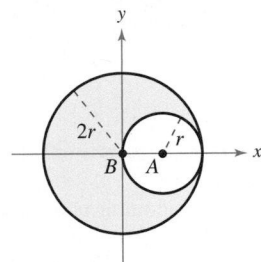

(a) Show that $M_x = 0$ for L.

(b) Show that M_y for L is equal to $(M_y$ for $B) - (M_y$ for $A)$.

(c) Find M_y for B and M_y for A. Then use part (b) to compute M_y for L.

(d) What is the center of mass of L?

3. Dividing a Region Let R be the region bounded by the parabola $y = x - x^2$ and the x-axis (see figure). Find the equation of the line $y = mx$ that divides this region into two regions of equal area.

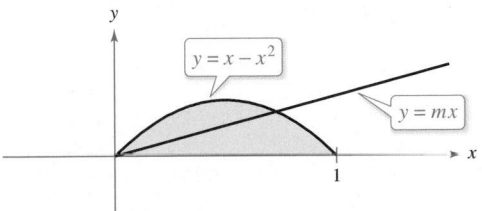

4. Surface Area Graph the curve

$$8y^2 = x^2(1 - x^2).$$

Use a computer algebra system to find the surface area of the solid of revolution obtained by revolving the curve about the y-axis.

5. Centroid A blade on an industrial fan has the configuration of a semicircle attached to a trapezoid (see figure). Find the centroid of the blade.

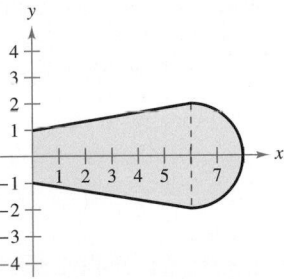

6. Volume A hole is cut through the center of a sphere of radius r (see figure). The height of the remaining spherical ring is h. Find the volume of the ring and show that it is independent of the radius of the sphere.

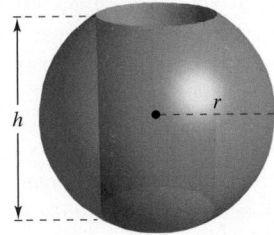

7. Volume A rectangle R of length ℓ and width w is revolved about the line L (see figure). Find the volume of the resulting solid of revolution.

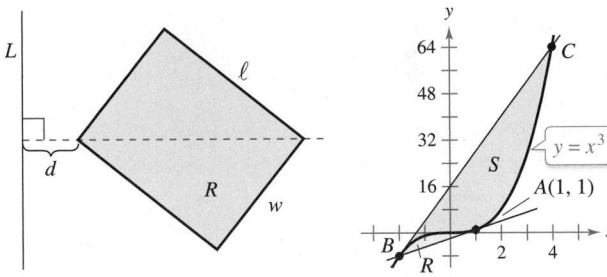

Figure for 7   Figure for 8

8. Comparing Areas of Regions

(a) The tangent line to the curve $y = x^3$ at the point $A(1, 1)$ intersects the curve at another point B. Let R be the area of the region bounded by the curve and the tangent line. The tangent line at B intersects the curve at another point C (see figure). Let S be the area of the region bounded by the curve and this second tangent line. How are the areas R and S related?

(b) Repeat the construction in part (a) by selecting an arbitrary point A on the curve $y = x^3$. Show that the two areas R and S are always related in the same way.

9. Using Arc Length The graph of $y = f(x)$ passes through the origin. The arc length of the curve from $(0, 0)$ to $(x, f(x))$ is given by

$$s(x) = \int_0^x \sqrt{1 + e^t}\, dt.$$

Identify the function f.

10. Using a Function Let f be rectifiable on the interval $[a, b]$, and let

$$s(x) = \int_a^x \sqrt{1 + [f'(t)]^2}\, dt.$$

(a) Find $\dfrac{ds}{dx}$.

(b) Find ds and $(ds)^2$.

(c) Find $s(x)$ on $[1, 3]$ when $f(t) = t^{3/2}$.

(d) Use the function and interval in part (c) to calculate $s(2)$ and describe what it signifies.

11. Archimedes' Principle **Archimedes' Principle** states that the upward or buoyant force on an object within a fluid is equal to the weight of the fluid that the object displaces. For a partially submerged object, you can obtain information about the relative densities of the floating object and the fluid by observing how much of the object is above and below the surface. You can also determine the size of a floating object if you know the amount that is above the surface and the relative densities. You can see the top of a floating iceberg (see figure). The density of ocean water is 1.03×10^3 kilograms per cubic meter, and that of ice is 0.92×10^3 kilograms per cubic meter. What percent of the total iceberg is below the surface?

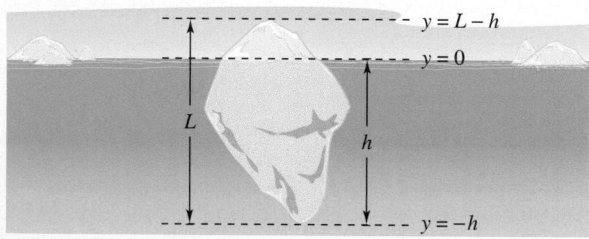

12. Finding a Centroid Sketch the region bounded on the left by $x = 1$, bounded above by $y = 1/x^3$, and bounded below by $y = -1/x^3$.

(a) Find the centroid of the region for $1 \le x \le 6$.

(b) Find the centroid of the region for $1 \le x \le b$.

(c) Where is the centroid as $b \to \infty$?

13. Finding a Centroid Sketch the region bounded on the left by $x = 1$, bounded above by $y = 1/x^4$, and bounded below by $y = -1/x^4$.

(a) Find the centroid of the region for $1 \le x \le 6$.

(b) Find the centroid of the region for $1 \le x \le b$.

(c) Where is the centroid as $b \to \infty$?

14. Work Find the work done by each force F.

(a)

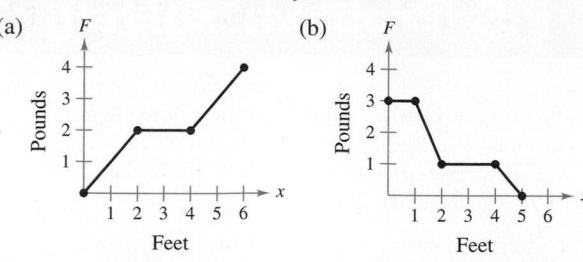

(b)

Consumer and Producer Surplus In Exercises 15 and 16, find the consumer surplus and producer surplus for the given demand $[p_1(x)]$ and supply $[p_2(x)]$ curves. The consumer surplus and producer surplus are represented by the areas shown in the figure.

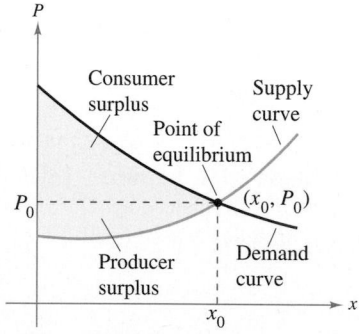

15. $p_1(x) = 50 - 0.5x$, $\quad p_2(x) = 0.125x$

16. $p_1(x) = 1000 - 0.4x^2$, $\quad p_2(x) = 42x$

17. Fluid Force A swimming pool is 20 feet wide, 40 feet long, 4 feet deep at one end, and 8 feet deep at the other end (see figures). The bottom is an inclined plane. Find the fluid force on each vertical wall when the pool is full of water.

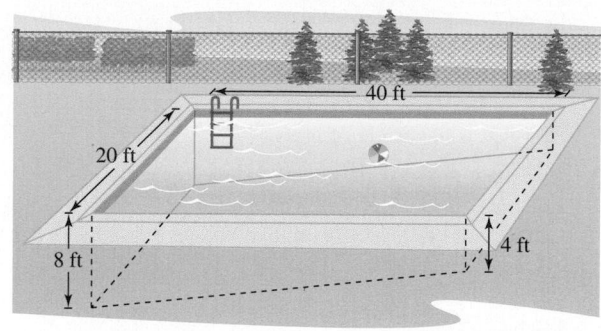

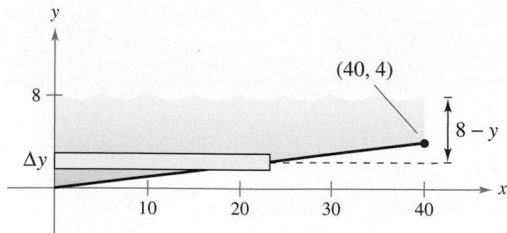

8 Integration Techniques and Improper Integrals

Sending a Space Module into Orbit
(Example 5, p. 575)

Chemical Reaction *(Exercise 50, p. 558)*

Fluid Force *(Exercise 63, p. 549)*

The Wallis Product
(Section Project, p. 540)

Memory Model *(Exercise 92, p. 531)*

Clockwise from top left, Dextroza/Shutterstock.com; Creations/Shutterstock.com;
agsandrew/Shutterstock.com; Juriah Mosin/Shutterstock.com; Andrea Izzotti/Shutterstock.com

8.1 Basic Integration Rules

■ Review procedures for fitting an integrand to one of the basic integration rules.

Fitting Integrands to Basic Integration Rules

In this chapter, you will study several integration techniques that greatly expand the set of integrals to which the basic integration rules can be applied. These rules are reviewed at the left. A major step in solving any integration problem is recognizing which basic integration rule to use.

REVIEW OF BASIC INTEGRATION RULES ($a > 0$)

1. $\int kf(u)\, du = k \int f(u)\, du$

2. $\int [f(u) \pm g(u)]\, du =$
   $\int f(u)\, du \pm \int g(u)\, du$

3. $\int du = u + C$

4. $\int u^n\, du = \dfrac{u^{n+1}}{n+1} + C$,
   $n \neq -1$

5. $\int \dfrac{du}{u} = \ln|u| + C$

6. $\int e^u\, du = e^u + C$

7. $\int a^u\, du = \left(\dfrac{1}{\ln a}\right) a^u + C$

8. $\int \sin u\, du = -\cos u + C$

9. $\int \cos u\, du = \sin u + C$

10. $\int \tan u\, du = -\ln|\cos u| + C$

11. $\int \cot u\, du = \ln|\sin u| + C$

12. $\int \sec u\, du =$
    $\ln|\sec u + \tan u| + C$

13. $\int \csc u\, du =$
    $-\ln|\csc u + \cot u| + C$

14. $\int \sec^2 u\, du = \tan u + C$

15. $\int \csc^2 u\, du = -\cot u + C$

16. $\int \sec u \tan u\, du = \sec u + C$

17. $\int \csc u \cot u\, du = -\csc u + C$

18. $\int \dfrac{du}{\sqrt{a^2 - u^2}} = \arcsin \dfrac{u}{a} + C$

19. $\int \dfrac{du}{a^2 + u^2} = \dfrac{1}{a} \arctan \dfrac{u}{a} + C$

20. $\int \dfrac{du}{u\sqrt{u^2 - a^2}} = \dfrac{1}{a}\text{arcsec}\dfrac{|u|}{a} + C$

EXAMPLE 1   A Comparison of Three Similar Integrals

⋯▷ *See LarsonCalculus.com for an interactive version of this type of example.*

Find each integral.

a. $\displaystyle\int \frac{4}{x^2 + 9}\, dx$ **b.** $\displaystyle\int \frac{4x}{x^2 + 9}\, dx$ **c.** $\displaystyle\int \frac{4x^2}{x^2 + 9}\, dx$

Solution

a. Use the Arctangent Rule and let $u = x$ and $a = 3$.

$$\int \frac{4}{x^2 + 9}\, dx = 4 \int \frac{1}{x^2 + 3^2}\, dx \qquad \text{Constant Multiple Rule}$$

$$= 4\left(\frac{1}{3} \arctan \frac{x}{3}\right) + C \qquad \text{Arctangent Rule}$$

$$= \frac{4}{3} \arctan \frac{x}{3} + C \qquad \text{Simplify.}$$

b. The Arctangent Rule does not apply because the numerator contains a factor of x. Consider the Log Rule and let $u = x^2 + 9$. Then $du = 2x\, dx$, and you have

$$\int \frac{4x}{x^2 + 9}\, dx = 2 \int \frac{2x\, dx}{x^2 + 9} \qquad \text{Constant Multiple Rule}$$

$$= 2 \int \frac{du}{u} \qquad \text{Substitute: } u = x^2 + 9.$$

$$= 2 \ln|u| + C \qquad \text{Log Rule}$$

$$= 2 \ln(x^2 + 9) + C. \qquad \text{Rewrite as a function of } x.$$

c. Because the degree of the numerator is equal to the degree of the denominator, you should first use division to rewrite the improper rational function as the sum of a polynomial and a proper rational function.

$$\int \frac{4x^2}{x^2 + 9}\, dx = \int \left(4 + \frac{-36}{x^2 + 9}\right) dx \qquad \text{Rewrite using long division.}$$

$$= \int 4\, dx - 36 \int \frac{1}{x^2 + 9}\, dx \qquad \text{Rewrite as two integrals.}$$

$$= 4x - 36\left(\frac{1}{3} \arctan \frac{x}{3}\right) + C \qquad \text{Integrate.}$$

$$= 4x - 12 \arctan \frac{x}{3} + C \qquad \text{Simplify.}$$

Note in Example 1(c) that some algebra is required before applying any integration rules, and more than one rule is needed to find the resulting integral.

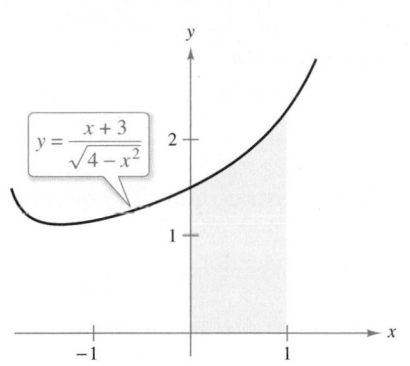

The area of the region is approximately 1.839.

Figure 8.1

EXAMPLE 2   **Using Two Basic Rules to Solve a Single Integral**

Evaluate $\displaystyle\int_0^1 \frac{x + 3}{\sqrt{4 - x^2}}\,dx$.

Solution   Begin by writing the integral as the sum of two integrals. Then apply the Power Rule and the Arcsine Rule.

$$\int_0^1 \frac{x + 3}{\sqrt{4 - x^2}}\,dx = \int_0^1 \frac{x}{\sqrt{4 - x^2}}\,dx + \int_0^1 \frac{3}{\sqrt{4 - x^2}}\,dx$$

$$= -\frac{1}{2}\int_0^1 (4 - x^2)^{-1/2}(-2x)\,dx + 3\int_0^1 \frac{1}{\sqrt{2^2 - x^2}}\,dx$$

$$= \left[-(4 - x^2)^{1/2} + 3\arcsin\frac{x}{2} \right]_0^1$$

$$= \left(-\sqrt{3} + \frac{\pi}{2} \right) - (-2 + 0)$$

$$\approx 1.839 \qquad\qquad \text{See Figure 8.1.}$$

▷ **TECHNOLOGY**   The Midpoint Rule can be used to give a good approximation of the value of the integral in Example 2 (for $n = 5$, the approximation is 1.837). When using numerical integration, however, you should be aware that the Midpoint Rule does not always give good approximations when one or both of the limits of integration are near a vertical asymptote. For instance, using the Fundamental Theorem of Calculus, you can obtain

$$\int_0^{1.99} \frac{x + 3}{\sqrt{4 - x^2}}\,dx \approx 6.213.$$

For $n = 5$, the Midpoint Rule gives an approximation of 5.667.

Rules 18, 19, and 20 of the basic integration rules on the preceding page all have expressions involving the sum or difference of two squares:

$$a^2 - u^2, \quad a^2 + u^2, \quad \text{and} \quad u^2 - a^2.$$

These expressions are often apparent after a u-substitution, as shown in Example 3.

EXAMPLE 3   **A Substitution Involving $a^2 - u^2$**

Find $\displaystyle\int \frac{x^2}{\sqrt{16 - x^6}}\,dx$.

Solution   Because the radical in the denominator can be written in the form

$$\sqrt{a^2 - u^2} = \sqrt{4^2 - (x^3)^2}$$

you can try the substitution $u = x^3$. Then $du = 3x^2\,dx$, and you have

$$\int \frac{x^2}{\sqrt{16 - x^6}}\,dx = \frac{1}{3}\int \frac{3x^2\,dx}{\sqrt{4^2 - (x^3)^2}} \qquad \text{Rewrite integral.}$$

$$= \frac{1}{3}\int \frac{du}{\sqrt{4^2 - u^2}} \qquad \text{Substitute: } u = x^3.$$

$$= \frac{1}{3}\arcsin\frac{u}{4} + C \qquad \text{Arcsine Rule}$$

$$= \frac{1}{3}\arcsin\frac{x^3}{4} + C. \qquad \text{Rewrite as a function of } x.$$

Two of the most commonly overlooked integration rules are the Log Rule and the Power Rule. Notice in the next two examples how these two integration rules can be disguised.

EXAMPLE 4 **A Disguised Form of the Log Rule**

Find $\displaystyle\int \frac{1}{1 + e^x}\, dx$.

Solution The integral does not appear to fit any of the basic rules. The quotient form, however, suggests the Log Rule. If you let $u = 1 + e^x$, then $du = e^x\, dx$. You can obtain the required du by adding and subtracting e^x in the numerator.

$$\int \frac{1}{1 + e^x}\, dx = \int \frac{1 + e^x - e^x}{1 + e^x}\, dx \qquad \text{Add and subtract } e^x \text{ in numerator.}$$

$$= \int \left(\frac{1 + e^x}{1 + e^x} - \frac{e^x}{1 + e^x}\right) dx \qquad \text{Rewrite as two fractions.}$$

$$= \int dx - \int \frac{e^x\, dx}{1 + e^x} \qquad \text{Rewrite as two integrals.}$$

$$= x - \ln(1 + e^x) + C \qquad \text{Integrate.} \quad\blacksquare$$

> **• • REMARK** Remember that you can separate numerators but not denominators. Watch out for this common error when fitting integrands to basic rules. For instance, you cannot separate denominators in Example 4.
>
> $$\frac{1}{1 + e^x} \neq \frac{1}{1} + \frac{1}{e^x}$$

There is usually more than one way to solve an integration problem. For instance, in Example 4, try integrating by multiplying the numerator and denominator by e^{-x} to obtain an integral of the form $-\int du/u$. See whether you can get the same answer by this procedure. (Be careful: the answer will appear in a different form.)

EXAMPLE 5 **A Disguised Form of the Power Rule**

Find $\displaystyle\int (\cot x)[\ln(\sin x)]\, dx$.

Solution Again, the integral does not appear to fit any of the basic rules. However, considering the two primary choices for u

$$u = \cot x \quad \text{or} \quad u = \ln(\sin x)$$

you can see that the second choice is the appropriate one because

$$u = \ln(\sin x) \quad \text{and} \quad du = \frac{\cos x}{\sin x}\, dx = \cot x\, dx.$$

So,

$$\int (\cot x)[\ln(\sin x)]\, dx = \int u\, du \qquad \text{Substitute: } u = \ln(\sin x).$$

$$= \frac{u^2}{2} + C \qquad \text{Integrate.}$$

$$= \frac{1}{2}[\ln(\sin x)]^2 + C. \qquad \text{Rewrite as a function of } x. \quad\blacksquare$$

In Example 5, try checking that the derivative of

$$\frac{1}{2}[\ln(\sin x)]^2 + C$$

is the integrand of the original integral.

Trigonometric identities can often be used to fit integrals to one of the basic integration rules.

EXAMPLE 6 Using Trigonometric Identities

Find $\displaystyle\int \tan^2 2x \, dx$.

▷ **TECHNOLOGY** If you have access to a computer algebra system, try using it to find the integrals in this section. Compare the *forms* of the antiderivatives given by the software with the forms obtained by hand. Sometimes the forms will be the same, but often they will differ. For instance, why is the antiderivative ln 2x + C equivalent to the antiderivative ln x + C?

Solution Note that $\tan^2 u$ is not in the list of basic integration rules. However, $\sec^2 u$ is in the list. This suggests the trigonometric identity $\tan^2 u = \sec^2 u - 1$. If you let $u = 2x$, then $du = 2 \, dx$ and

$$\int \tan^2 2x \, dx = \frac{1}{2} \int \tan^2 u \, du \qquad \text{Substitute: } u = 2x.$$

$$= \frac{1}{2} \int (\sec^2 u - 1) \, du \qquad \text{Trigonometric identity}$$

$$= \frac{1}{2} \int \sec^2 u \, du - \frac{1}{2} \int du \qquad \text{Rewrite as two integrals.}$$

$$= \frac{1}{2} \tan u - \frac{u}{2} + C \qquad \text{Integrate.}$$

$$= \frac{1}{2} \tan 2x - x + C. \qquad \text{Rewrite as a function of } x.$$ ■

This section concludes with a summary of the common procedures for fitting integrands to the basic integration rules.

PROCEDURES FOR FITTING INTEGRANDS TO BASIC INTEGRATION RULES

Technique	Example
Expand (numerator).	$(1 + e^x)^2 = 1 + 2e^x + e^{2x}$
Separate numerator.	$\dfrac{1 + x}{x^2 + 1} = \dfrac{1}{x^2 + 1} + \dfrac{x}{x^2 + 1}$
Complete the square.	$\dfrac{1}{\sqrt{2x - x^2}} = \dfrac{1}{\sqrt{1 - (x - 1)^2}}$
Divide improper rational function.	$\dfrac{x^2}{x^2 + 1} = 1 - \dfrac{1}{x^2 + 1}$
Add and subtract terms in numerator.	$\dfrac{2x}{x^2 + 2x + 1} = \dfrac{2x + 2 - 2}{x^2 + 2x + 1}$
	$= \dfrac{2x + 2}{x^2 + 2x + 1} - \dfrac{2}{(x + 1)^2}$
Use trigonometric identities.	$\cot^2 x = \csc^2 x - 1$
Multiply and divide by Pythagorean conjugate.	$\dfrac{1}{1 + \sin x} = \left(\dfrac{1}{1 + \sin x}\right)\left(\dfrac{1 - \sin x}{1 - \sin x}\right)$
	$= \dfrac{1 - \sin x}{1 - \sin^2 x}$
	$= \dfrac{1 - \sin x}{\cos^2 x}$
	$= \sec^2 x - \dfrac{\sin x}{\cos^2 x}$

8.1 Exercises

See **CalcChat.com** for tutorial help and worked-out solutions to odd-numbered exercises.

CONCEPT CHECK

1. Integration Technique Describe how to integrate a rational function with a numerator and denominator of the same degree.

2. Fitting Integrands to Basic Integration Rules What procedure should you use to fit each integrand to the basic integration rules? Do not integrate.

(a) $\int \dfrac{2 + x}{x^2 + 9}\, dx$ (b) $\int \cot^2 x\, dx$

Choosing an Antiderivative In Exercises 3 and 4, select the correct antiderivative.

3. $\int \dfrac{x}{\sqrt{x^2 + 1}}\, dx$

   (a) $2\sqrt{x^2 + 1} + C$ (b) $\sqrt{x^2 + 1} + C$

   (c) $\frac{1}{2}\sqrt{x^2 + 1} + C$ (d) $\ln(x^2 + 1) + C$

4. $\int \dfrac{1}{x^2 + 1}\, dx$

   (a) $\ln\sqrt{x^2 + 1} + C$ (b) $\dfrac{2x}{(x^2 + 1)^2} + C$

   (c) $\arctan x + C$ (d) $\ln(x^2 + 1) + C$

Choosing a Formula In Exercises 5–14, select the basic integration formula you can use to find the indefinite integral, and identify u and a when appropriate. Do not integrate.

5. $\int (5x - 3)^4\, dx$

6. $\int \dfrac{2t + 1}{t^2 + t - 4}\, dt$

7. $\int \dfrac{1}{\sqrt{x}\left(1 - 2\sqrt{x}\right)}\, dx$

8. $\int \dfrac{2}{(2t - 1)^2 + 4}\, dt$

9. $\int \dfrac{3}{\sqrt{1 - t^2}}\, dt$

10. $\int \dfrac{-2x}{\sqrt{x^2 - 4}}\, dx$

11. $\int t \sin t^2\, dt$

12. $\int \sec 5x \tan 5x\, dx$

13. $\int (\cos x)e^{\sin x}\, dx$

14. $\int \dfrac{1}{x\sqrt{x^2 - 4}}\, dx$

Finding an Indefinite Integral In Exercises 15–46, find the indefinite integral.

15. $\int 14(x - 5)^6\, dx$

16. $\int \dfrac{5}{(t + 6)^3}\, dt$

17. $\int \dfrac{7}{(z - 10)^7}\, dz$

18. $\int t^3\sqrt{t^4 + 1}\, dt$

19. $\int \left[z^2 + \dfrac{1}{(1 - z)^6}\right] dz$

20. $\int \left[4x - \dfrac{2}{(2x + 3)^2}\right] dx$

21. $\int \dfrac{t^2 - 3}{-t^3 + 9t + 1}\, dt$

22. $\int \dfrac{x + 1}{\sqrt{3x^2 + 6x}}\, dx$

23. $\int \dfrac{x^2}{x - 1}\, dx$

24. $\int \dfrac{3x}{x + 4}\, dx$

25. $\int \dfrac{x + 2}{x + 1}\, dx$

26. $\int \left(\dfrac{1}{9z - 5} - \dfrac{1}{9z + 5}\right) dz$

27. $\int (5 + 4x^2)^2\, dx$

28. $\int x\left(3 + \dfrac{2}{x}\right)^2 dx$

29. $\int x \cos 2\pi x^2\, dx$

30. $\int \csc \pi x \cot \pi x\, dx$

31. $\int \dfrac{\sin x}{\sqrt{\cos x}}\, dx$

32. $\int \dfrac{\csc^2 3t}{\cot 3t}\, dt$

33. $\int \dfrac{2}{e^{-x} + 1}\, dx$

34. $\int \dfrac{4}{3 - e^x}\, dx$

35. $\int \dfrac{\ln x^2}{x}\, dx$

36. $\int (\tan x)[\ln(\cos x)]\, dx$

37. $\int \dfrac{1 + \cos \alpha}{\sin \alpha}\, d\alpha$

38. $\int \dfrac{1}{\cos \theta - 1}\, d\theta$

39. $\int \dfrac{-1}{\sqrt{1 - (4t + 1)^2}}\, dt$

40. $\int \dfrac{1}{25 + 4x^2}\, dx$

41. $\int \dfrac{\tan(2/t)}{t^2}\, dt$

42. $\int \dfrac{e^{-1/t^3}}{t^4}\, dt$

43. $\int \dfrac{6}{z\sqrt{9z^2 - 25}}\, dz$

44. $\int \dfrac{1}{(x - 1)\sqrt{4x^2 - 8x + 3}}\, dx$

45. $\int \dfrac{4}{4x^2 + 4x + 65}\, dx$

46. $\int \dfrac{1}{x^2 - 4x + 9}\, dx$

Slope Field In Exercises 47 and 48, a differential equation, a point, and a slope field are given. (a) Sketch two approximate solutions of the differential equation on the slope field, one of which passes through the given point. (To print an enlarged copy of the graph, go to *MathGraphs.com*.) (b) Use integration and the given point to find the particular solution of the differential equation and use a graphing utility to graph the solution. Compare the result with the sketch in part (a) that passes through the given point.

47. $\dfrac{ds}{dt} = \dfrac{t}{\sqrt{1 - t^4}}$, $\left(0, -\dfrac{1}{2}\right)$

48. $\dfrac{dy}{dx} = \dfrac{1}{\sqrt{4x - x^2}}$, $\left(2, \dfrac{1}{2}\right)$

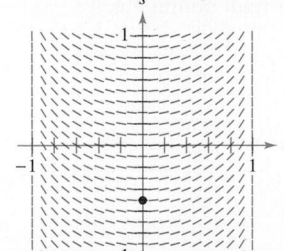

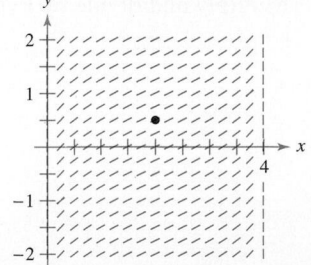

 **Slope Field** In Exercises 49 and 50, use a computer algebra system to graph the slope field for the differential equation and graph the solution satisfying the specified initial condition.

49. $\dfrac{dy}{dx} = 0.8y, \ y(0) = 4$

50. $\dfrac{dy}{dx} = 5 - y, \ y(0) = 1$

Differential Equation In Exercises 51–56, find the general solution of the differential equation.

51. $\dfrac{dy}{dx} = (e^x + 5)^2$

52. $\dfrac{dy}{dx} = (4 - e^{2x})^2$

53. $\dfrac{dr}{dt} = \dfrac{10e^t}{\sqrt{1 - e^{2t}}}$

54. $\dfrac{dr}{dt} = \dfrac{(1 + e^t)^2}{e^{3t}}$

55. $(4 + \tan^2 x)y' = \sec^2 x$

56. $y' = \dfrac{1}{x\sqrt{4x^2 - 9}}$

Evaluating a Definite Integral In Exercises 57–72, evaluate the definite integral. Use a graphing utility to verify your result.

57. $\displaystyle\int_{2/3}^{1} (2 - 3t)^4 \, dt$

58. $\displaystyle\int_{-1}^{0} \dfrac{5}{(t + 2)^{11}} \, dt$

59. $\displaystyle\int_{0}^{\pi/4} \cos 2x \, dx$

60. $\displaystyle\int_{0}^{\pi} \sin^2 t \cos t \, dt$

61. $\displaystyle\int_{0}^{1} xe^{-x^2} \, dx$

62. $\displaystyle\int_{1}^{e} \dfrac{1 - \ln x}{x} \, dx$

63. $\displaystyle\int_{2}^{3} \dfrac{\ln(x + 1)^3}{x + 1} \, dx$

64. $\displaystyle\int_{-3}^{1} \dfrac{e^x}{e^{2x} + 2e^x + 1} \, dx$

65. $\displaystyle\int_{0}^{8} \dfrac{2x}{\sqrt{x^2 + 36}} \, dx$

66. $\displaystyle\int_{1}^{3} \dfrac{2x^2 + 3x - 2}{x} \, dx$

67. $\displaystyle\int_{3}^{5} \dfrac{2t}{t^2 - 4t + 4} \, dt$

68. $\displaystyle\int_{2}^{4} \dfrac{4x^3}{x^4 - 6x^2 + 9} \, dx$

69. $\displaystyle\int_{0}^{2/\sqrt{3}} \dfrac{1}{4 + 9x^2} \, dx$

70. $\displaystyle\int_{0}^{7} \dfrac{1}{\sqrt{100 - x^2}} \, dx$

71. $\displaystyle\int_{-4}^{0} 3^{1-x} \, dx$

72. $\displaystyle\int_{0}^{1} 7^{x^2 + 2x}(x + 1) \, dx$

Area In Exercises 73–76, find the area of the given region.

73. $y = (-4x + 6)^{3/2}$

74. $y = \dfrac{3x + 2}{x^2 + 9}$

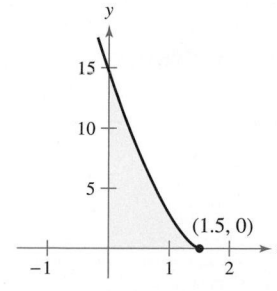

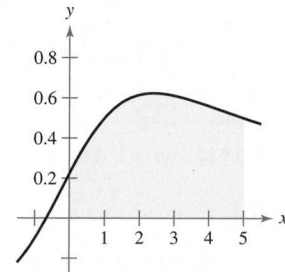

75. $y^2 = x^2(1 - x^2)$

76. $y = \sin 2x$

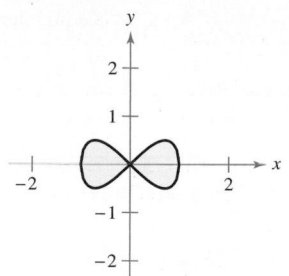

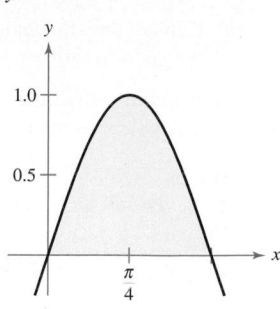

 **Finding an Integral Using Technology** In Exercises 77–80, use a computer algebra system to find the integral. Use the computer algebra system to graph two antiderivatives. Describe the relationship between the graphs of the two antiderivatives.

77. $\displaystyle\int \dfrac{1}{x^2 + 4x + 13} \, dx$

78. $\displaystyle\int \dfrac{x - 2}{x^2 + 4x + 13} \, dx$

79. $\displaystyle\int \dfrac{1}{1 + \sin \theta} \, d\theta$

80. $\displaystyle\int \left(\dfrac{e^x + e^{-x}}{2}\right)^3 \, dx$

EXPLORING CONCEPTS

81. Think About It When evaluating

$$\int_{-1}^{1} x^2 \, dx$$

is it appropriate to substitute

$$u = x^2, \quad x = \sqrt{u}, \quad \text{and} \quad dx = \dfrac{du}{2\sqrt{u}}$$

to obtain

$$\dfrac{1}{2}\int_{1}^{1} \sqrt{u} \, du = 0?$$

Explain.

82. Deriving a Rule Show that

$$\sec x = \dfrac{\sin x}{\cos x} + \dfrac{\cos x}{1 + \sin x}.$$

Then use this identity to derive the basic integration rule

$$\int \sec x \, dx = \ln|\sec x + \tan x| + C.$$

---

83. Finding Constants Determine the constants a and b such that

$$\sin x + \cos x = a \sin(x + b).$$

Use this result to integrate

$$\int \dfrac{dx}{\sin x + \cos x}.$$

84. Area The graphs of $f(x) = x$ and $g(x) = ax^2$ intersect at the points $(0, 0)$ and $(1/a, 1/a)$. Find $a \ (a > 0)$ such that the area of the region bounded by the graphs of these two functions is $\frac{2}{3}$.

85. Comparing Antiderivatives

(a) Explain why the antiderivative $y_1 = e^{x+C_1}$ is equivalent to the antiderivative $y_2 = Ce^x$.

(b) Explain why the antiderivative $y_1 = \sec^2 x + C_1$ is equivalent to the antiderivative $y_2 = \tan^2 x + C$.

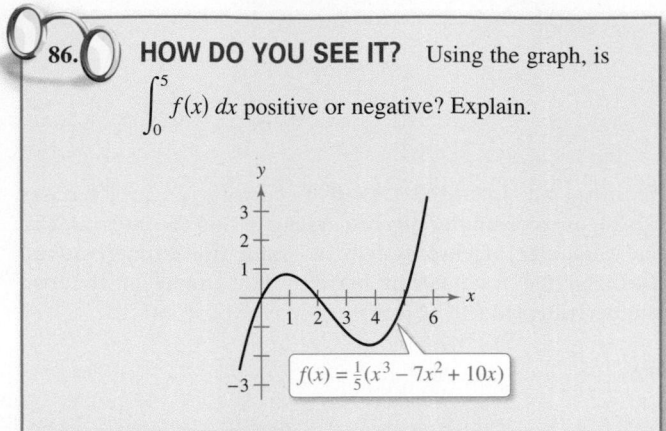

86. HOW DO YOU SEE IT? Using the graph, is $\int_0^5 f(x)\, dx$ positive or negative? Explain.

$$f(x) = \tfrac{1}{5}(x^3 - 7x^2 + 10x)$$

Approximation In Exercises 87 and 88, determine which value best approximates the area of the region between the x-axis and the graph of the function over the given interval. Make your selection on the basis of a sketch of the region, not by performing calculations.

87. $f(x) = \dfrac{4x}{x^2 + 1}$, $[0, 2]$

(a) 3 (b) 1 (c) -8 (d) 8 (e) 10

88. $f(x) = \dfrac{4}{x^2 + 1}$, $[0, 2]$

(a) 3 (b) 1 (c) -4 (d) 4 (e) 10

Interpreting Integrals In Exercises 89 and 90, (a) sketch the region whose area is given by the integral, (b) sketch the solid whose volume is given by the integral when the disk method is used, and (c) sketch the solid whose volume is given by the integral when the shell method is used. (There is more than one correct answer for each part.)

89. $\displaystyle\int_0^2 2\pi x^2\, dx$

90. $\displaystyle\int_0^4 \pi y\, dy$

91. Volume The region bounded by $y = e^{-x^2}$, $y = 0$, $x = 0$, and $x = b$ ($b > 0$) is revolved about the y-axis.

(a) Find the volume of the solid generated when $b = 1$.

(b) Find b such that the volume of the solid generated is $\tfrac{4}{3}$ cubic units.

92. Volume Consider the region bounded by the graphs of $x = 0$, $y = \cos x^2$, $y = \sin x^2$, and $x = \sqrt{x}/2$. Find the volume of the solid generated by revolving the region about the y-axis.

93. Arc Length Find the arc length of the graph of $y = \ln(\sin x)$ from $x = \pi/4$ to $x = \pi/2$.

94. Arc Length Find the arc length of the graph of $y = \ln(\cos x)$ from $x = 0$ to $x = \pi/3$.

95. Surface Area Find the area of the surface formed by revolving the graph of $y = 2\sqrt{x}$ on the interval $[0, 9]$ about the x-axis.

96. Centroid Find the centroid of the region bounded by the graphs of

$$y = \frac{1}{2x + 1}, \quad y = 0, \quad x = 0, \quad \text{and} \quad x = 2.$$

Average Value of a Function In Exercises 97 and 98, find the average value of the function over the given interval.

97. $f(x) = \dfrac{1}{1 + x^2}$, $-3 \le x \le 3$

98. $f(x) = \sin nx$, $0 \le x \le \pi/n$, n is a positive integer.

Arc Length In Exercises 99 and 100, use the integration capabilities of a graphing utility to approximate the arc length of the curve over the given interval.

99. $y = \tan \pi x$, $\left[0, \tfrac{1}{4}\right]$

100. $y = x^{2/3}$, $[1, 8]$

101. Finding a Pattern

(a) Find $\displaystyle\int \cos^3 x\, dx$.

(b) Find $\displaystyle\int \cos^5 x\, dx$.

(c) Find $\displaystyle\int \cos^7 x\, dx$.

(d) Explain how to find $\int \cos^{15} x\, dx$ without actually integrating.

102. Finding a Pattern

(a) Write $\int \tan^3 x\, dx$ in terms of $\int \tan x\, dx$. Then find $\int \tan^3 x\, dx$.

(b) Write $\int \tan^5 x\, dx$ in terms of $\int \tan^3 x\, dx$.

(c) Write $\int \tan^{2k+1} x\, dx$, where k is a positive integer, in terms of $\int \tan^{2k-1} x\, dx$.

(d) Explain how to find $\int \tan^{15} x\, dx$ without actually integrating.

103. Methods of Integration Show that the following results are equivalent. (You will learn about integration by tables in Section 8.7.)

Integration by tables:

$$\int \sqrt{x^2 + 1}\, dx = \frac{1}{2}\left(x\sqrt{x^2 + 1} + \ln\left|x + \sqrt{x^2 + 1}\right|\right) + C$$

Integration by computer algebra system:

$$\int \sqrt{x^2 + 1}\, dx = \frac{1}{2}\left[x\sqrt{x^2 + 1} + \operatorname{arcsinh}(x)\right] + C$$

PUTNAM EXAM CHALLENGE

104. Evaluate $\displaystyle\int_2^4 \frac{\sqrt{\ln(9 - x)}\, dx}{\sqrt{\ln(9 - x)} + \sqrt{\ln(x + 3)}}$.

8.2 Integration by Parts

■ Find an antiderivative using integration by parts.

Integration by Parts

In this section, you will study an important integration technique called **integration by parts.** This technique can be applied to a wide variety of functions and is particularly useful for integrands involving *products* of algebraic and transcendental functions. For instance, integration by parts works well with integrals such as

$$\int x \ln x \, dx, \quad \int x^2 e^x \, dx, \quad \text{and} \quad \int e^x \sin x \, dx.$$

Integration by parts is based on the formula for the derivative of a product

$$\frac{d}{dx}[uv] = u\frac{dv}{dx} + v\frac{du}{dx}$$

$$= uv' + vu'$$

where both u and v are differentiable functions of x. When u' and v' are continuous, you can integrate both sides of this equation to obtain

$$uv = \int uv' \, dx + \int vu' \, dx$$

$$= \int u \, dv + \int v \, du.$$

By rewriting this equation, you obtain the next theorem.

> **THEOREM 8.1 Integration by Parts**
>
> If u and v are functions of x and have continuous derivatives, then
>
> $$\int u \, dv = uv - \int v \, du.$$

This formula expresses the original integral in terms of another integral. Depending on the choices of u and dv, it may be easier to find the second integral than the original one. Because the choices of u and dv are critical in the integration by parts process, the guidelines below are provided.

> **GUIDELINES FOR INTEGRATION BY PARTS**
>
> 1. Try letting dv be the most complicated portion of the integrand that fits a basic integration rule. Then u will be the remaining factor(s) of the integrand.
> 2. Try letting u be the portion of the integrand whose derivative is a function simpler than u. Then dv will be the remaining factor(s) of the integrand.
>
> Note that dv always includes the dx of the original integrand.

When using integration by parts, note that you can first choose dv or first choose u. After you choose, however, the choice of the other factor is determined—it must be the remaining portion of the integrand. Also note that dv must contain the differential dx of the original integral.

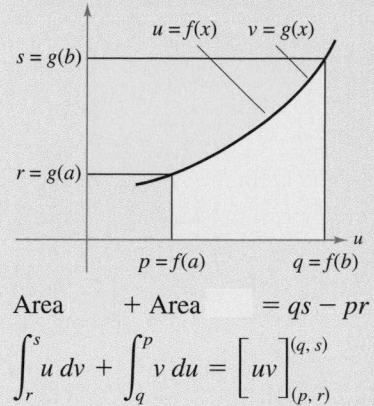

EXAMPLE 1 **Integration by Parts**

Find $\displaystyle\int xe^x\, dx$.

Solution To apply integration by parts, you need to write the integral in the form $\int u\, dv$. There are several ways to do this.

$$\int \underbrace{(x)}_{u}\underbrace{(e^x\, dx)}_{dv}, \quad \int \underbrace{(e^x)}_{u}\underbrace{(x\, dx)}_{dv}, \quad \int \underbrace{(1)}_{u}\underbrace{(xe^x\, dx)}_{dv}, \quad \int \underbrace{(xe^x)}_{u}\underbrace{(dx)}_{dv}$$

The guidelines on the preceding page suggest the first option because the derivative of $u = x$ is simpler than x, and $dv = e^x\, dx$ is the most complicated portion of the integrand that fits a basic integration formula.

$$dv = e^x\, dx \implies v = \int dv = \int e^x\, dx = e^x$$

$$u = x \implies du = dx$$

Now, integration by parts produces

$$\int u\, dv = uv - \int v\, du \qquad \text{Integration by parts formula}$$

$$\int xe^x\, dx = xe^x - \int e^x\, dx \qquad \text{Substitute.}$$

$$= xe^x - e^x + C. \qquad \text{Integrate.}$$

To check this, differentiate $xe^x - e^x + C$ to see that you obtain the original integrand.

> **REMARK** In Example 1, note that it is not necessary to include a constant of integration when solving
>
> $$v = \int e^x\, dx = e^x + C_1.$$
>
> To illustrate this, replace $v = e^x$ by $v = e^x + C_1$ and apply integration by parts to see that you obtain the same result.

EXAMPLE 2 **Integration by Parts**

Find $\displaystyle\int x^2 \ln x\, dx$.

Solution In this case, x^2 is more easily integrated than $\ln x$. Furthermore, the derivative of $\ln x$ is simpler than $\ln x$. So, you should let $dv = x^2\, dx$.

$$dv = x^2\, dx \implies v = \int x^2\, dx = \frac{x^3}{3}$$

$$u = \ln x \implies du = \frac{1}{x}\, dx$$

Integration by parts produces

$$\int u\, dv = uv - \int v\, du \qquad \text{Integration by parts formula}$$

$$\int x^2 \ln x\, dx = \frac{x^3}{3}\ln x - \int \left(\frac{x^3}{3}\right)\left(\frac{1}{x}\right) dx \qquad \text{Substitute.}$$

$$= \frac{x^3}{3}\ln x - \frac{1}{3}\int x^2\, dx \qquad \text{Simplify.}$$

$$= \frac{x^3}{3}\ln x - \frac{x^3}{9} + C. \qquad \text{Integrate.}$$

You can check this result by differentiating.

$$\frac{d}{dx}\left[\frac{x^3}{3}\ln x - \frac{x^3}{9} + C\right] = \frac{x^3}{3}\left(\frac{1}{x}\right) + (\ln x)(x^2) - \frac{x^2}{3} = x^2 \ln x$$

> ▷ **TECHNOLOGY**
> Try graphing
>
> $$f(x) = \int x^2 \ln x\, dx$$
>
> and
>
> $$g(x) = \frac{x^3}{3}\ln x - \frac{x^3}{9}$$
>
> on your graphing utility. Do you get the same graph?

One surprising application of integration by parts involves integrands consisting of single terms, such as

$$\int \ln x \, dx \quad \text{or} \quad \int \arcsin x \, dx.$$

In these cases, try letting $dv = dx$, as shown in the next example.

EXAMPLE 3 **An Integrand with a Single Term**

Evaluate $\displaystyle\int_0^1 \arcsin x \, dx$.

Solution Let $dv = dx$.

$$dv = dx \quad \Longrightarrow \quad v = \int dx = x$$

$$u = \arcsin x \quad \Longrightarrow \quad du = \frac{1}{\sqrt{1 - x^2}} \, dx$$

Integration by parts produces

$$\int u \, dv = uv - \int v \, du \qquad \text{Integration by parts formula}$$

$$\int \arcsin x \, dx = x \arcsin x - \int \frac{x}{\sqrt{1 - x^2}} \, dx \qquad \text{Substitute.}$$

$$= x \arcsin x + \frac{1}{2} \int (1 - x^2)^{-1/2}(-2x) \, dx \qquad \text{Rewrite.}$$

$$= x \arcsin x + \sqrt{1 - x^2} + C. \qquad \text{Integrate.}$$

Using this antiderivative, you can evaluate the definite integral as shown.

$$\int_0^1 \arcsin x \, dx = \left[x \arcsin x + \sqrt{1 - x^2} \right]_0^1$$

$$= \frac{\pi}{2} - 1$$

$$\approx 0.571$$

The area represented by this definite integral is shown in Figure 8.2.

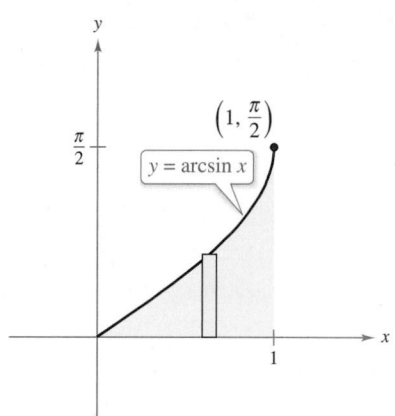

$\left(1, \dfrac{\pi}{2}\right)$

$y = \arcsin x$

The area of the region is approximately 0.571.

Figure 8.2

▷ **TECHNOLOGY** Remember that there are several ways to use technology to evaluate a definite integral: (1) use a numerical approximation such as the Midpoint Rule, or more advanced methods such as the Trapezoidal Rule and Simpson's Rule (see Section 8.6), (2) use a computer algebra system to find the antiderivative and then apply the Fundamental Theorem of Calculus, or (3) use the *numerical integration* feature of a graphing utility. However, these methods have shortcomings. For instance, to find the possible error when using Simpson's Rule, the integrand must have a continuous fourth derivative in the interval of integration (the integrand in Example 3 fails to meet this requirement). To apply the Fundamental Theorem of Calculus, the symbolic integration utility must be able to find the antiderivative. Often, for the *numerical integration* feature of a graphing utility, you are given no indication of the degree of accuracy of the approximation.

■ **FOR FURTHER INFORMATION** To see how integration by parts is used to prove Stirling's approximation $\ln(n!) = n \ln n - n$, see the article "The Validity of Stirling's Approximation: A Physical Chemistry Project" by A. S. Wallner and K. A. Brandt in *Journal of Chemical Education.*

Some integrals require repeated use of the integration by parts formula, as shown in the next example.

EXAMPLE 4   Repeated Use of Integration by Parts

Find $\int x^2 \sin x \, dx$.

Solution   The factors x^2 and $\sin x$ are equally easy to integrate. However, the derivative of x^2 becomes simpler, whereas the derivative of $\sin x$ does not. So, you should let $u = x^2$.

$$dv = \sin x \, dx \quad \Longrightarrow \quad v = \int \sin x \, dx = -\cos x$$

$$u = x^2 \qquad \Longrightarrow \quad du = 2x \, dx$$

Now, integration by parts produces

$$\int x^2 \sin x \, dx = -x^2 \cos x + \int 2x \cos x \, dx. \qquad \text{First use of integration by parts}$$

This first use of integration by parts has succeeded in simplifying the original integral, but the integral on the right still does not fit a basic integration rule. To find that integral, you can apply integration by parts again. This time, let $u = 2x$.

$$dv = \cos x \, dx \quad \Longrightarrow \quad v = \int \cos x \, dx = \sin x$$

$$u = 2x \qquad \Longrightarrow \quad du = 2 \, dx$$

Now, integration by parts produces

$$\int 2x \cos x \, dx = 2x \sin x - \int 2 \sin x \, dx \qquad \text{Second use of integration by parts}$$

$$= 2x \sin x + 2 \cos x + C.$$

Combining these two results, you can write

$$\int x^2 \sin x \, dx = -x^2 \cos x + 2x \sin x + 2 \cos x + C. \qquad \blacksquare$$

When making repeated applications of integration by parts, you need to be careful not to interchange the substitutions in successive applications. For instance, in Example 4, the first substitution was $u = x^2$ and $dv = \sin x \, dx$. If, in the second application, you had switched the substitution to $u = \cos x$ and $dv = 2x \, dx$, you would have obtained

$$\int x^2 \sin x \, dx = -x^2 \cos x + \int 2x \cos x \, dx$$

$$= -x^2 \cos x + x^2 \cos x + \int x^2 \sin x \, dx$$

$$= \int x^2 \sin x \, dx$$

thereby undoing the previous integration and returning to the *original* integral. When making repeated applications of integration by parts, you should also watch for the appearance of a *constant multiple* of the original integral. For instance, this occurs when you use integration by parts to find $\int e^x \cos 2x \, dx$, and it also occurs in Example 5 on the next page.

The integral in Example 5 is an important one. In Section 8.4 (Example 5), you will see that it is used to find the arc length of a parabolic segment.

EXAMPLE 5 Integration by Parts

Find $\displaystyle\int \sec^3 x \, dx$.

Solution The most complicated portion of the integrand that can be easily integrated is $\sec^2 x$, so you should let $dv = \sec^2 x \, dx$ and $u = \sec x$.

$$dv = \sec^2 x \, dx \implies v = \int \sec^2 x \, dx = \tan x$$

$$u = \sec x \implies du = \sec x \tan x \, dx$$

Integration by parts produces

$$\int u \, dv = uv - \int v \, du \qquad \text{Integration by parts formula}$$

$$\int \sec^3 x \, dx = \sec x \tan x - \int \sec x \tan^2 x \, dx \qquad \text{Substitute.}$$

$$\int \sec^3 x \, dx = \sec x \tan x - \int (\sec x)(\sec^2 x - 1) \, dx \qquad \text{Trigonometric identity}$$

$$\int \sec^3 x \, dx = \sec x \tan x - \int \sec^3 x \, dx + \int \sec x \, dx \qquad \text{Rewrite.}$$

$$2 \int \sec^3 x \, dx = \sec x \tan x + \int \sec x \, dx \qquad \text{Collect like integrals.}$$

$$2 \int \sec^3 x \, dx = \sec x \tan x + \ln|\sec x + \tan x| + C \qquad \text{Integrate.}$$

$$\int \sec^3 x \, dx = \frac{1}{2} \sec x \tan x + \frac{1}{2} \ln|\sec x + \tan x| + C. \qquad \text{Divide by 2.}$$

EXAMPLE 6 Finding a Centroid

A machine part is modeled by the region bounded by the graph of $y = \sin x$ and the x-axis, $0 \le x \le \pi/2$, as shown in Figure 8.3. Find the centroid of this region.

Solution Begin by finding the area of the region.

$$A = \int_0^{\pi/2} \sin x \, dx = \left[-\cos x \right]_0^{\pi/2} = 1$$

Now, you can find the coordinates of the centroid. To evaluate the integral for $\bar{y}$, first rewrite the integrand using the trigonometric identity $\sin^2 x = (1 - \cos 2x)/2$.

$$\bar{y} = \frac{1}{A} \int_0^{\pi/2} \frac{\sin x}{2}(\sin x) \, dx = \frac{1}{4} \int_0^{\pi/2} (1 - \cos 2x) \, dx = \frac{1}{4} \left[x - \frac{\sin 2x}{2} \right]_0^{\pi/2} = \frac{\pi}{8}$$

You can evaluate the integral for $\bar{x}$, $(1/A) \int_0^{\pi/2} x \sin x \, dx$, with integration by parts. To do this, let $dv = \sin x \, dx$ and $u = x$. This produces $v = -\cos x$ and $du = dx$, and you can write

$$\int x \sin x \, dx = -x \cos x + \int \cos x \, dx = -x \cos x + \sin x + C.$$

Finally, you can determine $\bar{x}$ to be

$$\bar{x} = \frac{1}{A} \int_0^{\pi/2} x \sin x \, dx = \left[-x \cos x + \sin x \right]_0^{\pi/2} = 1.$$

So, the centroid of the region is $(1, \pi/8)$.

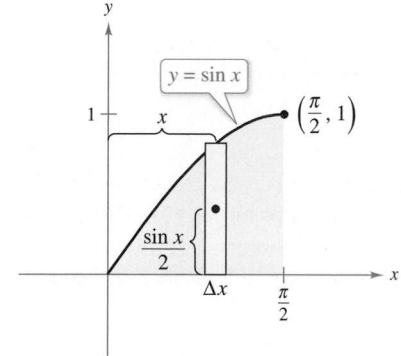

Figure 8.3

As you gain experience in using integration by parts, your skill in determining u and dv will increase. The next summary lists several common integrals with suggestions for the choices of u and dv.

REMARK You can use the acronym LIATE as a guideline for choosing u in integration by parts. In order, check the integrand for the following.

Is there a Logarithmic part?

Is there an Inverse trigonometric part?

Is there an Algebraic part?

Is there a Trigonometric part?

Is there an Exponential part?

SUMMARY: COMMON INTEGRALS USING INTEGRATION BY PARTS

1. For integrals of the form

$$\int x^n e^{ax}\, dx, \quad \int x^n \sin ax\, dx, \quad \text{or} \quad \int x^n \cos ax\, dx$$

let $u = x^n$ and let $dv = e^{ax}\, dx$, $\sin ax\, dx$, or $\cos ax\, dx$.

2. For integrals of the form

$$\int x^n \ln x\, dx, \quad \int x^n \arcsin ax\, dx, \quad \text{or} \quad \int x^n \arctan ax\, dx$$

let $u = \ln x$, $\arcsin ax$, or $\arctan ax$ and let $dv = x^n\, dx$.

3. For integrals of the form

$$\int e^{ax} \sin bx\, dx \quad \text{or} \quad \int e^{ax} \cos bx\, dx$$

let $u = \sin bx$ or $\cos bx$ and let $dv = e^{ax}\, dx$.

In problems involving repeated applications of integration by parts, a tabular method, illustrated in Example 7, can help to organize the work. This method works well for integrals of the form

$$\int x^n \sin ax\, dx, \quad \int x^n \cos ax\, dx, \quad \text{and} \quad \int x^n e^{ax}\, dx.$$

EXAMPLE 7 Using the Tabular Method

····▷ *See LarsonCalculus.com for an interactive version of this type of example.*

Find $\displaystyle\int x^2 \sin 4x\, dx$.

Solution Begin as usual by letting $u = x^2$ and $dv = v'\, dx = \sin 4x\, dx$. Next, create a table consisting of three columns, as shown.

Alternate Signs	u and Its Derivatives	v' and Its Antiderivatives
$+$	x^2	$\sin 4x$
$-$	$2x$	$-\frac{1}{4}\cos 4x$
$+$	2	$-\frac{1}{16}\sin 4x$
$-$	0	$\frac{1}{64}\cos 4x$

↑
Differentiate until you obtain
0 as a derivative.

The solution is obtained by adding the signed products of the diagonal entries.

$$\int x^2 \sin 4x\, dx = -\frac{1}{4}x^2 \cos 4x + \frac{1}{8}x \sin 4x + \frac{1}{32}\cos 4x + C$$

■ FOR FURTHER INFORMATION For more information on the tabular method, see the article "Tabular Integration by Parts" by David Horowitz in *The College Mathematics Journal*, and the article "More on Tabular Integration by Parts" by Leonard Gillman in *The College Mathematics Journal*. To view these articles, go to *MathArticles.com*.

8.2 Exercises

CONCEPT CHECK

1. **Integration by Parts** Integration by parts is based on what formula?

2. **Setting Up Integration by Parts** In your own words, describe how to choose u and dv when using integration by parts.

3. **Using Integration by Parts** How can you use integration by parts on an integrand with a single term that does not fit any of the basic integration rules?

4. **Using the Tabular Method** When is integrating using the tabular method useful?

 **Setting Up Integration by Parts** In Exercises 5–10, identify u and dv for finding the integral using integration by parts. Do not integrate.

5. $\int xe^{9x}\, dx$

6. $\int x^2 e^{2x}\, dx$

7. $\int (\ln x)^2\, dx$

8. $\int \ln 5x\, dx$

9. $\int x \sec^2 x\, dx$

10. $\int x^2 \cos x\, dx$

 Using Integration by Parts In Exercises 11–14, find the indefinite integral using integration by parts with the given choices of u and dv.

11. $\int x^3 \ln x\, dx;\ u = \ln x,\ dv = x^3\, dx$

12. $\int (7 - x)e^{x/2}\, dx;\ u = 7 - x,\ dv = e^{x/2}\, dx$

13. $\int (2x + 1)\sin 4x\, dx;\ u = 2x + 1,\ dv = \sin 4x\, dx$

14. $\int x \cos 4x\, dx;\ u = x,\ dv = \cos 4x\, dx$

 **Finding an Indefinite Integral** In Exercises 15–34, find the indefinite integral. (*Note:* Solve by the simplest method—not all require integration by parts.)

15. $\int xe^{4x}\, dx$

16. $\int \dfrac{5x}{e^{2x}}\, dx$

17. $\int x^3 e^x\, dx$

18. $\int \dfrac{e^{1/t}}{t^2}\, dt$

19. $\int t \ln(t + 1)\, dt$

20. $\int x^5 \ln 3x\, dx$

21. $\int \dfrac{(\ln x)^2}{x}\, dx$

22. $\int \dfrac{\ln x}{x^3}\, dx$

23. $\int \dfrac{xe^{2x}}{(2x + 1)^2}\, dx$

24. $\int \dfrac{x^3 e^{x^2}}{(x^2 + 1)^2}\, dx$

25. $\int x\sqrt{x - 5}\, dx$

26. $\int \dfrac{2x}{\sqrt{1 - 6x}}\, dx$

27. $\int x \csc^2 x\, dx$

28. $\int t \csc t \cot t\, dt$

29. $\int x^3 \sin x\, dx$

30. $\int x^2 \cos x\, dx$

31. $\int \arctan x\, dx$

32. $\int 4 \arccos x\, dx$

33. $\int e^{-3x} \sin 5x\, dx$

34. $\int e^{4x} \cos 2x\, dx$

Differential Equation In Exercises 35–38, find the general solution of the differential equation.

35. $y' = \ln x$

36. $y' = \arctan \dfrac{x}{2}$

37. $\dfrac{dy}{dt} = \dfrac{t^2}{\sqrt{3 + 5t}}$

38. $\dfrac{dy}{dx} = x^2 \sqrt{x - 3}$

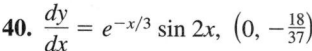 **Slope Field** In Exercises 39 and 40, a differential equation, a point, and a slope field are given. (a) Sketch two approximate solutions of the differential equation on the slope field, one of which passes through the given point. (To print an enlarged copy of the graph, go to *MathGraphs.com*.) (b) Use integration and the given point to find the particular solution of the differential equation and use a graphing utility to graph the solution. Compare the result with the sketch in part (a) that passes through the given point.

39. $\dfrac{dy}{dx} = x\sqrt{y}\cos x,\ (0, 4)$

40. $\dfrac{dy}{dx} = e^{-x/3}\sin 2x,\ \left(0, -\frac{18}{37}\right)$

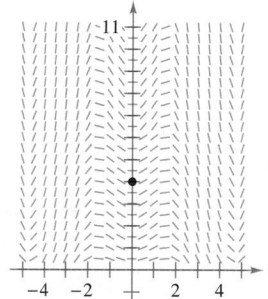

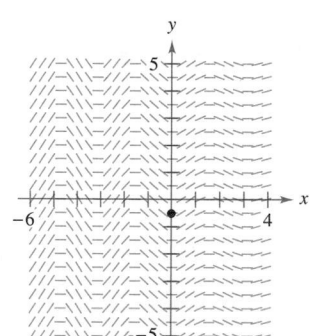

 **Slope Field** In Exercises 41 and 42, use a computer algebra system to graph the slope field for the differential equation and graph the solution satisfying the specified initial condition.

41. $\dfrac{dy}{dx} = \dfrac{x}{y}e^{x/8},\ y(0) = 2$

42. $\dfrac{dy}{dx} = \dfrac{x}{y}\sin x,\ y(0) = 4$

Evaluating a Definite Integral In Exercises 43–52, evaluate the definite integral. Use a graphing utility to verify your result.

43. $\displaystyle\int_0^3 xe^{x/2}\, dx$

44. $\displaystyle\int_0^2 x^2 e^{-2x}\, dx$

45. $\displaystyle\int_0^{\pi/4} x \cos 2x\, dx$

46. $\displaystyle\int_0^{\pi} x \sin 2x\, dx$

47. $\int_0^{1/2} \arccos x \, dx$

48. $\int_0^1 x \arcsin x^2 \, dx$

49. $\int_0^1 e^x \sin x \, dx$

50. $\int_0^1 \ln(4 + x^2) \, dx$

51. $\int_2^4 x \, \mathrm{arcsec}\, x \, dx$

52. $\int_0^{\pi/8} x \sec^2 2x \, dx$

 **Using the Tabular Method** In Exercises 53–58, use the tabular method to find the indefinite integral.

53. $\int x^2 e^{2x} \, dx$

54. $\int (1 - x)(e^{-x} + 1) \, dx$

55. $\int (x + 2)^2 \sin x \, dx$

56. $\int x^3 \cos 2x \, dx$

57. $\int (6 + x)\sqrt{4x + 9} \, dx$

58. $\int x^2(x - 2)^{3/2} \, dx$

EXPLORING CONCEPTS

59. Integration by Parts Write an integral that requires three applications of integration by parts. Explain why three applications are needed.

60. Integration by Parts When evaluating $\int x \sin x \, dx$, explain how letting $u = \sin x$ and $dv = x \, dx$ makes the solution more difficult to find.

61. Integration by Parts State whether you would use integration by parts to find each integral. If so, identify what you would use for u and dv. Explain your reasoning.

(a) $\int \dfrac{\ln x}{x} \, dx$ (b) $\int x \ln x \, dx$ (c) $\int x^2 e^{-3x} \, dx$

(d) $\int 2x e^{x^2} \, dx$ (e) $\int \dfrac{x}{\sqrt{x + 1}} \, dx$ (f) $\int \dfrac{x}{\sqrt{x^2 + 1}} \, dx$

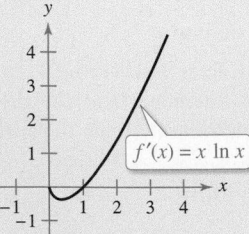

62. **HOW DO YOU SEE IT?** Use the graph of f' shown in the figure to answer the following.

$f'(x) = x \ln x$

(a) Approximate the slope of f at $x = 2$. Explain.

(b) Approximate any open intervals on which the graph of f is increasing and any open intervals on which it is decreasing. Explain.

Using Two Methods Together In Exercises 63–66, find the indefinite integral by using substitution followed by integration by parts.

63. $\int \sin \sqrt{x} \, dx$

64. $\int 2x^3 \cos x^2 \, dx$

65. $\int x^5 e^{x^2} \, dx$

66. $\int e^{\sqrt{2x}} \, dx$

67. Using Two Methods Integrate $\int \dfrac{x^3}{\sqrt{4 + x^2}} \, dx$

(a) by parts, letting $dv = \dfrac{x}{\sqrt{4 + x^2}} \, dx$.

(b) by substitution, letting $u = 4 + x^2$.

68. Using Two Methods Integrate $\int x\sqrt{4 - x} \, dx$

(a) by parts, letting $dv = \sqrt{4 - x} \, dx$.

(b) by substitution, letting $u = 4 - x$.

 **Finding a General Rule** In Exercises 69 and 70, use a computer algebra system to find the integrals for $n = 0, 1, 2,$ and 3. Use the result to obtain a general rule for the integrals for any positive integer n and test your results for $n = 4$.

69. $\int x^n \ln x \, dx$

70. $\int x^n e^x \, dx$

Proof In Exercises 71–76, use integration by parts to prove the formula. (For Exercises 71–74, assume that n is a positive integer.)

71. $\int x^n \sin x \, dx = -x^n \cos x + n \int x^{n-1} \cos x \, dx$

72. $\int x^n \cos x \, dx = x^n \sin x - n \int x^{n-1} \sin x \, dx$

73. $\int x^n \ln x \, dx = \dfrac{x^{n+1}}{(n + 1)^2}[-1 + (n + 1) \ln x] + C$

74. $\int x^n e^{ax} \, dx = \dfrac{x^n e^{ax}}{a} - \dfrac{n}{a} \int x^{n-1} e^{ax} \, dx$

75. $\int e^{ax} \sin bx \, dx = \dfrac{e^{ax}(a \sin bx - b \cos bx)}{a^2 + b^2} + C$

76. $\int e^{ax} \cos bx \, dx = \dfrac{e^{ax}(a \cos bx + b \sin bx)}{a^2 + b^2} + C$

Using Formulas In Exercises 77–82, find the indefinite integral by using the appropriate formula from Exercises 71–76.

77. $\int x^2 \sin x \, dx$

78. $\int x^2 \cos x \, dx$

79. $\int x^5 \ln x \, dx$

80. $\int x^3 e^{2x} \, dx$

81. $\int e^{-3x} \sin 4x \, dx$

82. $\int e^{2x} \cos 3x \, dx$

Area In Exercises 83–86, use a graphing utility to graph the region bounded by the graphs of the equations. Then find the area of the region analytically.

83. $y = 2xe^{-x}$, $y = 0$, $x = 3$

84. $y = \frac{1}{10}xe^{3x}$, $y = 0$, $x = 0$, $x = 2$

85. $y = e^{-x} \sin \pi x$, $y = 0$, $x = 1$

86. $y = x^3 \ln x$, $y = 0$, $x = 1$, $x = 3$

87. Area, Volume, and Centroid Given the region bounded by the graphs of $y = \ln x$, $y = 0$, and $x = e$, find

(a) the area of the region.

(b) the volume of the solid generated by revolving the region about the x-axis.

(c) the volume of the solid generated by revolving the region about the y-axis.

(d) the centroid of the region.

88. Area, Volume, and Centroid Given the region bounded by the graphs of $y = x \sin x$, $y = 0$, $x = 0$, and $x = \pi$, find

(a) the area of the region.

(b) the volume of the solid generated by revolving the region about the x-axis.

(c) the volume of the solid generated by revolving the region about the y-axis.

(d) the centroid of the region.

89. Centroid Find the centroid of the region bounded by the graphs of $y = \arcsin x$, $x = 0$, and $y = \pi/2$. How is this problem related to Example 6 in this section?

90. Centroid Find the centroid of the region bounded by the graphs of $f(x) = x^2$, $g(x) = 2^x$, $x = 2$, and $x = 4$.

91. Average Displacement A damping force affects the vibration of a spring so that the displacement of the spring is given by

$$y = e^{-4t}(\cos 2t + 5 \sin 2t).$$

Find the average value of y on the interval from $t = 0$ to $t = \pi$.

92. Memory Model

A model for the ability M of a child to memorize, measured on a scale from 0 to 10, is given by

$$M = 1 + 1.6t \ln t, \quad 0 < t \le 4$$

where t is the child's age in years. Find the average value of this model

(a) between the child's first and second birthdays.

(b) between the child's third and fourth birthdays.

Present Value In Exercises 93 and 94, find the present value P of a continuous income flow of $c(t)$ dollars per year using

$$P = \int_0^{t_1} c(t)e^{-rt}\, dt$$

where t_1 is the time in years and r is the annual interest rate compounded continuously.

93. $c(t) = 100,000 + 4000t$, $r = 5\%$, $t_1 = 10$

94. $c(t) = 1000 + 120t$, $r = 2\%$, $t_1 = 30$

Integrals Used to Find Fourier Coefficients In Exercises 95 and 96, verify the value of the definite integral, where n is a positive integer.

95. $\displaystyle\int_{-\pi}^{\pi} x \sin nx\, dx = \begin{cases} \dfrac{2\pi}{n}, & n \text{ is odd} \\[2mm] -\dfrac{2\pi}{n}, & n \text{ is even} \end{cases}$

96. $\displaystyle\int_{-\pi}^{\pi} x^2 \cos nx\, dx = \dfrac{(-1)^n\, 4\pi}{n^2}$

97. Vibrating String A string stretched between the two points $(0, 0)$ and $(2, 0)$ is plucked by displacing the string h units at its midpoint. The motion of the string is modeled by a **Fourier Sine Series** whose coefficients are given by

$$b_n = h \int_0^1 x \sin \frac{n\pi x}{2}\, dx + h \int_1^2 (-x + 2) \sin \frac{n\pi x}{2}\, dx.$$

Find b_n.

98. Finding a Pattern Find the area bounded by the graphs of $y = x \sin x$ and $y = 0$ over each interval.

(a) $[0, \pi]$ (b) $[\pi, 2\pi]$ (c) $[2\pi, 3\pi]$

Describe any patterns that you notice. What is the area between the graphs of $y = x \sin x$ and $y = 0$ over the interval $[n\pi, (n + 1)\pi]$, where n is any nonnegative integer? Explain.

99. Finding an Error Find the fallacy in the following argument that $0 = 1$.

$$dv = dx \implies v = \int dx = x$$

$$u = \frac{1}{x} \implies du = -\frac{1}{x^2}\, dx$$

$$0 + \int \frac{dx}{x} = \left(\frac{1}{x}\right)(x) - \int\left(-\frac{1}{x^2}\right)(x)\, dx$$

$$= 1 + \int \frac{dx}{x}$$

So, $0 = 1$.

PUTNAM EXAM CHALLENGE

100. Find a real number c and a positive number L for which

$$\lim_{r \to \infty} \frac{r^c \int_0^{\pi/2} x^r \sin x\, dx}{\int_0^{\pi/2} x^r \cos x\, dx} = L.$$

8.3 Trigonometric Integrals

■ Solve trigonometric integrals involving powers of sine and cosine.
■ Solve trigonometric integrals involving powers of secant and tangent.
■ Solve trigonometric integrals involving sine-cosine products.

Integrals Involving Powers of Sine and Cosine

SHEILA SCOTT MACINTYRE
(1910–1960)

Sheila Scott Macintyre published her first paper on the asymptotic periods of integral functions in 1935. She completed her doctorate work at Aberdeen University, where she taught. In 1958, she accepted a visiting research fellowship at the University of Cincinnati.

In this section, you will study techniques for evaluating integrals of the form

$$\int \sin^m x \cos^n x \, dx \quad \text{and} \quad \int \sec^m x \tan^n x \, dx$$

where either m or n is a positive integer. To find antiderivatives for these forms, try to break them into combinations of trigonometric integrals to which you can apply the Power Rule.

For instance, you can find

$$\int \sin^5 x \cos x \, dx$$

with the Power Rule by letting $u = \sin x$. Then, $du = \cos x \, dx$ and you have

$$\int \sin^5 x \cos x \, dx = \int u^5 \, du = \frac{u^6}{6} + C = \frac{\sin^6 x}{6} + C.$$

To break up $\int \sin^m x \cos^n x \, dx$ into forms to which you can apply the Power Rule, use these relationships.

$\sin^2 x + \cos^2 x = 1$	Pythagorean identity
$\sin^2 x = \dfrac{1 - \cos 2x}{2}$	Power-reducing formula for $\sin^2 x$
$\cos^2 x = \dfrac{1 + \cos 2x}{2}$	Power-reducing formula for $\cos^2 x$

GUIDELINES FOR EVALUATING INTEGRALS INVOLVING POWERS OF SINE AND COSINE

1. When the power of the sine is odd and positive, save one sine factor and convert the remaining factors to cosines. Then expand and integrate.

$$\int \overbrace{\sin^{2k+1}}^{\text{Odd}} x \cos^n x \, dx = \int \overbrace{(\sin^2 x)^k}^{\text{Convert to cosines}} \cos^n x \overbrace{\sin x \, dx}^{\text{Save for } du} = \int (1 - \cos^2 x)^k \cos^n x \sin x \, dx$$

2. When the power of the cosine is odd and positive, save one cosine factor and convert the remaining factors to sines. Then expand and integrate.

$$\int \sin^m x \overbrace{\cos^{2k+1}}^{\text{Odd}} x \, dx = \int (\sin^m x) \overbrace{(\cos^2 x)^k}^{\text{Convert to sines}} \overbrace{\cos x \, dx}^{\text{Save for } du} = \int (\sin^m x)(1 - \sin^2 x)^k \cos x \, dx$$

3. When the powers of both the sine and cosine are even and nonnegative, make repeated use of the formulas

$$\sin^2 x = \frac{1 - \cos 2x}{2} \quad \text{and} \quad \cos^2 x = \frac{1 + \cos 2x}{2}$$

to convert the integrand to odd powers of the cosine. Then proceed as in the second guideline.

EXAMPLE 1 **Power of Sine Is Odd and Positive**

Find $\displaystyle\int \sin^3 x \cos^4 x \, dx$.

Solution Because you expect to use the Power Rule with $u = \cos x$, *save one sine factor* to form du and convert the remaining sine factors to cosines.

$$\int \sin^3 x \cos^4 x \, dx = \int (\sin^2 x \cos^4 x)(\sin x) \, dx \qquad \text{Rewrite.}$$

$$= \int (1 - \cos^2 x) \cos^4 x \sin x \, dx \qquad \text{Trigonometric identity}$$

$$= \int (\cos^4 x - \cos^6 x) \sin x \, dx \qquad \text{Multiply.}$$

$$= \int \cos^4 x \sin x \, dx - \int \cos^6 x \sin x \, dx \qquad \text{Rewrite.}$$

$$= -\int (\cos^4 x)(-\sin x) \, dx + \int (\cos^6 x)(-\sin x) \, dx$$

$$= -\frac{\cos^5 x}{5} + \frac{\cos^7 x}{7} + C \qquad \text{Integrate.}$$

▷ **TECHNOLOGY** A computer algebra system used to find the integral in Example 1 yielded the following.

$$\int \sin^3 x \cos^4 x \, dx = (-\cos^5 x)\left(\frac{1}{7}\sin^2 x + \frac{2}{35}\right) + C$$

Is this equivalent to the result obtained in Example 1?

In Example 1, *both* of the powers m and n happened to be positive integers. This strategy will work as long as either m or n is odd and positive. For instance, in the next example, the power of the cosine is 3, but the power of the sine is $-\frac{1}{2}$.

EXAMPLE 2 **Power of Cosine Is Odd and Positive**

$\bullet \cdot \cdot \cdot \triangleright$ *See LarsonCalculus.com for an interactive version of this type of example.*

Evaluate $\displaystyle\int_{\pi/6}^{\pi/3} \frac{\cos^3 x}{\sqrt{\sin x}} \, dx$.

Solution Because you expect to use the Power Rule with $u = \sin x$, *save one cosine factor* to form du and convert the remaining cosine factors to sines.

$$\int_{\pi/6}^{\pi/3} \frac{\cos^3 x}{\sqrt{\sin x}} \, dx = \int_{\pi/6}^{\pi/3} \frac{\cos^2 x \cos x}{\sqrt{\sin x}} \, dx \qquad \text{Rewrite.}$$

$$= \int_{\pi/6}^{\pi/3} \frac{(1 - \sin^2 x)(\cos x)}{\sqrt{\sin x}} \, dx \qquad \text{Trigonometric identity}$$

$$= \int_{\pi/6}^{\pi/3} \left[(\sin x)^{-1/2} - (\sin x)^{3/2}\right] \cos x \, dx \qquad \text{Divide.}$$

$$= \left[\frac{(\sin x)^{1/2}}{1/2} - \frac{(\sin x)^{5/2}}{5/2}\right]_{\pi/6}^{\pi/3} \qquad \text{Integrate.}$$

$$= 2\left(\frac{\sqrt{3}}{2}\right)^{1/2} - \frac{2}{5}\left(\frac{\sqrt{3}}{2}\right)^{5/2} - \sqrt{2} + \frac{\sqrt{32}}{80}$$

$$\approx 0.239$$

Figure 8.4 shows the region whose area is represented by this integral.

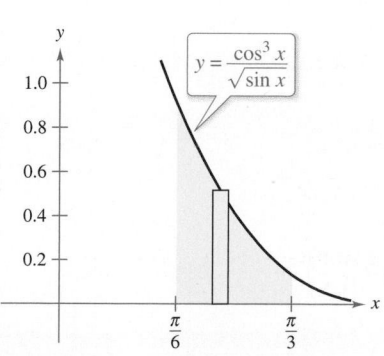

The area of the region is approximately 0.239.

Figure 8.4

EXAMPLE 3 **Power of Cosine Is Even and Nonnegative**

Find $\int \cos^4 x \, dx$.

Solution Because m and n are both even and nonnegative ($m = 0$), you can replace $\cos^4 x$ by

$$\left(\frac{1 + \cos 2x}{2}\right)^2.$$

So, you can integrate as shown.

$$\int \cos^4 x \, dx = \int \left(\frac{1 + \cos 2x}{2}\right)^2 dx \qquad \text{Power-reducing formula}$$

$$= \int \left(\frac{1}{4} + \frac{\cos 2x}{2} + \frac{\cos^2 2x}{4}\right) dx \qquad \text{Expand.}$$

$$= \int \left[\frac{1}{4} + \frac{\cos 2x}{2} + \frac{1}{4}\left(\frac{1 + \cos 4x}{2}\right)\right] dx \qquad \text{Power-reducing formula}$$

$$= \frac{3}{8} \int dx + \frac{1}{4} \int 2 \cos 2x \, dx + \frac{1}{32} \int 4 \cos 4x \, dx \qquad \text{Rewrite.}$$

$$= \frac{3x}{8} + \frac{\sin 2x}{4} + \frac{\sin 4x}{32} + C \qquad \text{Integrate.}$$

Use a symbolic differentiation utility to verify this. Can you simplify the derivative to obtain the original integrand? ■

In Example 3, when you evaluate the definite integral from 0 to $\pi/2$, you obtain

$$\int_0^{\pi/2} \cos^4 x \, dx = \left[\frac{3x}{8} + \frac{\sin 2x}{4} + \frac{\sin 4x}{32}\right]_0^{\pi/2}$$

$$= \left(\frac{3\pi}{16} + 0 + 0\right) - (0 + 0 + 0)$$

$$= \frac{3\pi}{16}.$$

Note that the only term that contributes to the solution is

$$\frac{3x}{8}.$$

This observation is generalized in the following formulas developed by John Wallis (1616–1703).

JOHN WALLIS (1616–1703)

Wallis did much of his work in calculus prior to Newton and Leibniz, and he influenced the thinking of both of these men. Wallis is also credited with introducing the present symbol (∞) for infinity.
See LarsonCalculus.com to read more of this biography.

Wallis's Formulas

1. If n is odd ($n \geq 3$), then

$$\int_0^{\pi/2} \cos^n x \, dx = \left(\frac{2}{3}\right)\left(\frac{4}{5}\right)\left(\frac{6}{7}\right) \cdots \left(\frac{n - 1}{n}\right).$$

2. If n is even ($n \geq 2$), then

$$\int_0^{\pi/2} \cos^n x \, dx = \left(\frac{1}{2}\right)\left(\frac{3}{4}\right)\left(\frac{5}{6}\right) \cdots \left(\frac{n - 1}{n}\right)\left(\frac{\pi}{2}\right).$$

These formulas are also valid when $\cos^n x$ is replaced by $\sin^n x$. (You are asked to prove both formulas in Exercise 87.)

Integrals Involving Powers of Secant and Tangent

The guidelines below can help you find integrals of the form

$$\int \sec^m x \tan^n x \, dx.$$

GUIDELINES FOR EVALUATING INTEGRALS INVOLVING POWERS OF SECANT AND TANGENT

1. When the power of the secant is even and positive, save a secant-squared factor and convert the remaining factors to tangents. Then expand and integrate.

$$\int \underset{\text{Even}}{\sec^{2k} x} \tan^n x \, dx = \int \overbrace{(\sec^2 x)^{k-1}}^{\text{Convert to tangents}} \tan^n x \overbrace{\sec^2 x \, dx}^{\text{Save for } du} = \int (1 + \tan^2 x)^{k-1} \tan^n x \sec^2 x \, dx$$

2. When the power of the tangent is odd and positive, save a secant-tangent factor and convert the remaining factors to secants. Then expand and integrate.

$$\int \sec^m x \overset{\text{Odd}}{\overbrace{\tan^{2k+1} x}} \, dx = \int \overbrace{(\sec^{m-1} x)(\tan^2 x)^k}^{\text{Convert to secants}} \overbrace{\sec x \tan x \, dx}^{\text{Save for } du} = \int (\sec^{m-1} x)(\sec^2 x - 1)^k \sec x \tan x \, dx$$

3. When there are no secant factors and the power of the tangent is even and positive, convert a tangent-squared factor to a secant-squared factor, then expand and repeat if necessary.

$$\int \tan^n x \, dx = \int (\tan^{n-2} x) \overbrace{(\tan^2 x)}^{\text{Convert to secants}} \, dx = \int (\tan^{n-2} x)(\sec^2 x - 1) \, dx$$

4. When the integral is of the form

$$\int \sec^m x \, dx$$

where m is odd and positive, use integration by parts, as illustrated in Example 5 in Section 8.2.

5. When the first four guidelines do not apply, try converting to sines and cosines.

EXAMPLE 4 **Power of Tangent Is Odd and Positive**

Find $\displaystyle\int \frac{\tan^3 x}{\sqrt{\sec x}} \, dx.$

Solution Because you expect to use the Power Rule with $u = \sec x$, *save a factor of (sec x tan x)* to form du and convert the remaining tangent factors to secants.

$$\int \frac{\tan^3 x}{\sqrt{\sec x}} \, dx = \int (\sec x)^{-1/2} \tan^3 x \, dx \qquad \text{Rewrite.}$$

$$= \int (\sec x)^{-3/2} (\tan^2 x)(\sec x \tan x) \, dx \qquad \text{Rewrite.}$$

$$= \int (\sec x)^{-3/2} (\sec^2 x - 1)(\sec x \tan x) \, dx \qquad \text{Trigonometric identity}$$

$$= \int [(\sec x)^{1/2} - (\sec x)^{-3/2}](\sec x \tan x) \, dx \qquad \text{Multiply.}$$

$$= \frac{2}{3}(\sec x)^{3/2} + 2(\sec x)^{-1/2} + C \qquad \text{Integrate.} \qquad \blacksquare$$

EXAMPLE 5 **Power of Secant Is Even and Positive**

Find $\int \sec^4 3x \tan^3 3x \, dx$.

Solution Let $u = \tan 3x$. Then $du = 3 \sec^2 3x \, dx$ and you can write

$$\int \sec^4 3x \tan^3 3x \, dx = \int (\sec^2 3x \tan^3 3x)(\sec^2 3x) \, dx \qquad \text{Rewrite.}$$

$$= \int (1 + \tan^2 3x)(\tan^3 3x)(\sec^2 3x) \, dx \qquad \text{Trigonometric identity}$$

$$= \frac{1}{3} \int (\tan^3 3x + \tan^5 3x)(3 \sec^2 3x) \, dx \qquad \text{Multiply.}$$

$$= \frac{1}{3} \left(\frac{\tan^4 3x}{4} + \frac{\tan^6 3x}{6} \right) + C \qquad \text{Integrate.}$$

$$= \frac{\tan^4 3x}{12} + \frac{\tan^6 3x}{18} + C.$$

In Example 5, the power of the tangent is odd and positive. So, you could also find the integral using the procedure described in the second guideline on the preceding page. In Exercises 67 and 68, you are asked to show that the results obtained by these two procedures differ only by a constant.

EXAMPLE 6 **Power of Tangent Is Even**

Evaluate $\displaystyle\int_0^{\pi/4} \tan^4 x \, dx$.

Solution Because there are no secant factors, you can begin by converting a tangent-squared factor to a secant-squared factor.

$$\int \tan^4 x \, dx = \int (\tan^2 x)(\tan^2 x) \, dx \qquad \text{Rewrite.}$$

$$= \int (\tan^2 x)(\sec^2 x - 1) \, dx \qquad \text{Trigonometric identity}$$

$$= \int \tan^2 x \sec^2 x \, dx - \int \tan^2 x \, dx \qquad \text{Rewrite.}$$

$$= \int \tan^2 x \sec^2 x \, dx - \int (\sec^2 x - 1) \, dx \qquad \text{Trigonometric identity}$$

$$= \frac{\tan^3 x}{3} - \tan x + x + C \qquad \text{Integrate.}$$

Next, evaluate the definite integral.

$$\int_0^{\pi/4} \tan^4 x \, dx = \left[\frac{\tan^3 x}{3} - \tan x + x \right]_0^{\pi/4}$$

$$= \frac{1}{3} - 1 + \frac{\pi}{4}$$

$$\approx 0.119$$

The area represented by the definite integral is shown in Figure 8.5. Try using the Midpoint Rule to approximate this integral. With $n = 15$, you should obtain an approximation that is within 0.001 of the actual value.

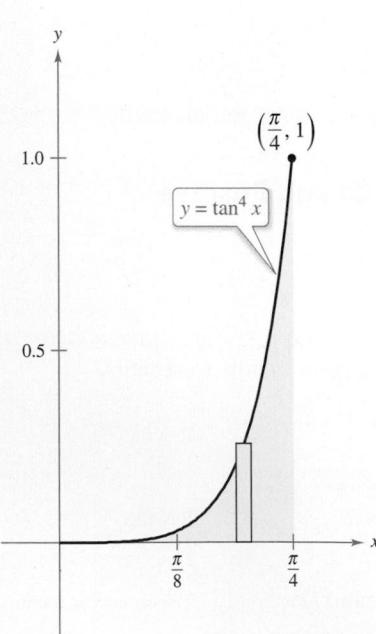

$\left(\frac{\pi}{4}, 1 \right)$

$y = \tan^4 x$

The area of the region is approximately 0.119.

Figure 8.5

For integrals involving powers of cotangents and cosecants, you can follow a strategy similar to that used for powers of tangents and secants. Also, when integrating trigonometric functions, remember that it sometimes helps to convert the entire integrand to powers of sines and cosines.

EXAMPLE 7   **Converting to Sines and Cosines**

Find $\displaystyle\int \frac{\sec x}{\tan^2 x}\, dx.$

Solution   Because the first four guidelines on page 535 do not apply, try converting the integrand to sines and cosines. In this case, you are able to integrate the resulting powers of sine and cosine as shown.

$$
\begin{aligned}
\int \frac{\sec x}{\tan^2 x}\, dx &= \int \left(\frac{1}{\cos x}\right)\left(\frac{\cos x}{\sin x}\right)^2 dx \\
&= \int (\sin x)^{-2}(\cos x)\, dx \\
&= -(\sin x)^{-1} + C \\
&= -\csc x + C
\end{aligned}
$$

Integrals Involving Sine-Cosine Products

Integrals involving the products of sines and cosines of two angles occur in many applications. You can evaluate these integrals using integration by parts. However, you may find it simpler to use the following product-to-sum formulas.

$$
\sin mx \sin nx = \frac{1}{2}(\cos[(m - n)x] - \cos[(m + n)x])
$$

$$
\sin mx \cos nx = \frac{1}{2}(\sin[(m - n)x] + \sin[(m + n)x])
$$

$$
\cos mx \cos nx = \frac{1}{2}(\cos[(m - n)x] + \cos[(m + n)x])
$$

EXAMPLE 8   **Using a Product-to-Sum Formula**

Find $\displaystyle\int \sin 5x \cos 4x\, dx.$

Solution   Considering the second product-to-sum formula above, you can write

$$
\begin{aligned}
\int \sin 5x \cos 4x\, dx &= \frac{1}{2}\int (\sin x + \sin 9x)\, dx \\
&= \frac{1}{2}\left(-\cos x - \frac{\cos 9x}{9}\right) + C \\
&= -\frac{\cos x}{2} - \frac{\cos 9x}{18} + C.
\end{aligned}
$$

■ **FOR FURTHER INFORMATION**   To learn more about integrals involving sine-cosine products with different angles, see the article "Integrals of Products of Sine and Cosine with Different Arguments" by Sherrie J. Nicol in *The College Mathematics Journal.* To view this article, go to *MathArticles.com.*

8.3 Exercises

See CalcChat.com for tutorial help and worked-out solutions to odd-numbered exercises.

CONCEPT CHECK

1. Analyzing Indefinite Integrals Which integral requires more steps to find? Explain. Do not integrate.

$$\int \sin^8 x \, dx \qquad \int \sin^8 x \cos x \, dx$$

2. Analyzing an Indefinite Integral Describe the technique for finding $\int \sec^5 x \tan^7 x \, dx$. Do not integrate.

Finding an Indefinite Integral Involving Sine and Cosine In Exercises 3–14, find the indefinite integral.

3. $\int \cos^5 x \sin x \, dx$

4. $\int \sin^7 2x \cos 2x \, dx$

5. $\int \cos^3 x \sin^4 x \, dx$

6. $\int \sin^3 3x \, dx$

7. $\int \sin^3 x \cos^2 x \, dx$

8. $\int \cos^3 \frac{x}{3} \, dx$

9. $\int \sin^3 2\theta \sqrt{\cos 2\theta} \, d\theta$

10. $\int \frac{\cos^5 t}{\sqrt{\sin t}} \, dt$

11. $\int \cos^2 3x \, dx$

12. $\int \sin^4 6\theta \, d\theta$

13. $\int 8x \cos^2 x \, dx$

14. $\int x^2 \sin^2 x \, dx$

Using Wallis's Formulas In Exercises 15–20, use Wallis's Formulas to evaluate the integral.

15. $\int_0^{\pi/2} \cos^3 x \, dx$

16. $\int_0^{\pi/2} \cos^6 x \, dx$

17. $\int_0^{\pi/2} \sin^2 x \, dx$

18. $\int_0^{\pi/2} \sin^9 x \, dx$

19. $\int_0^{\pi/2} \sin^{10} x \, dx$

20. $\int_0^{\pi/2} \cos^{11} x \, dx$

Finding an Indefinite Integral Involving Secant and Tangent In Exercises 21–34, find the indefinite integral.

21. $\int \sec 4x \, dx$

22. $\int \sec^4 x \, dx$

23. $\int \sec^3 \pi x \, dx$

24. $\int \tan^6 3x \, dx$

25. $\int \tan^5 \frac{x}{2} \, dx$

26. $\int \tan^3 \frac{\pi x}{2} \sec^2 \frac{\pi x}{2} \, dx$

27. $\int \tan^3 2t \sec^3 2t \, dt$

28. $\int \tan^5 x \sec^4 x \, dx$

29. $\int \sec^6 4x \tan 4x \, dx$

30. $\int \sec^2 \frac{x}{2} \tan \frac{x}{2} \, dx$

31. $\int \sec^5 x \tan^3 x \, dx$

32. $\int \tan^3 3x \, dx$

33. $\int \frac{\tan^2 x}{\sec x} \, dx$

34. $\int \frac{\tan^2 x}{\sec^5 x} \, dx$

Differential Equation In Exercises 35–38, find the general solution of the differential equation.

35. $\dfrac{dr}{d\theta} = \sin^4 \pi\theta$

36. $\dfrac{ds}{d\alpha} = \sin^2 \dfrac{\alpha}{2} \cos^2 \dfrac{\alpha}{2}$

37. $y' = \tan^3 3x \sec 3x$

38. $y' = \sqrt{\tan x} \sec^4 x$

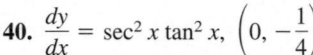 **Slope Field** In Exercises 39 and 40, a differential equation, a point, and a slope field are given. (a) Sketch two approximate solutions of the differential equation on the slope field, one of which passes through the given point. (To print an enlarged copy of the graph, go to *MathGraphs.com*.) (b) Use integration and the given point to find the particular solution of the differential equation and use a graphing utility to graph the solution. Compare the result with the sketch in part (a) that passes through the given point.

39. $\dfrac{dy}{dx} = \sin^2 x,\ (0, 0)$

40. $\dfrac{dy}{dx} = \sec^2 x \tan^2 x,\ \left(0, -\dfrac{1}{4}\right)$

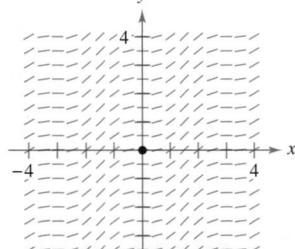

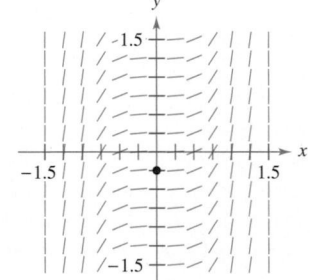

 **Slope Field** In Exercises 41 and 42, use a computer algebra system to graph the slope field for the differential equation and graph the solution satisfying the specified initial condition.

41. $\dfrac{dy}{dx} = \dfrac{3 \sin x}{y},\ y(0) = 2$

42. $\dfrac{dy}{dx} = 3\sqrt{y} \tan^2 x,\ y(0) = 3$

Using a Product-to-Sum Formula In Exercises 43–48, find the indefinite integral.

43. $\int \cos 2x \cos 6x \, dx$

44. $\int \cos 5\theta \cos 3\theta \, d\theta$

45. $\int \sin 2t \cos 9t \, dt$

46. $\int \sin 8x \cos 7x \, dx$

47. $\int \sin \theta \sin 3\theta \, d\theta$

48. $\int \sin 5x \sin 4x \, dx$

Finding an Indefinite Integral In Exercises 49–58, find the indefinite integral. Use a computer algebra system to confirm your result.

49. $\displaystyle\int \cot^3 2x \, dx$

50. $\displaystyle\int \tan^5 \frac{x}{4} \sec^4 \frac{x}{4} \, dx$

51. $\displaystyle\int \csc^4 3x \, dx$

52. $\displaystyle\int \cot^3 \frac{x}{2} \csc^4 \frac{x}{2} \, dx$

53. $\displaystyle\int \frac{\cot^2 t}{\csc t} \, dt$

54. $\displaystyle\int \frac{\cot^3 t}{\csc t} \, dt$

55. $\displaystyle\int \frac{1}{\sec x \tan x} \, dx$

56. $\displaystyle\int \frac{\sin^2 x - \cos^2 x}{\cos x} \, dx$

57. $\displaystyle\int (\tan^4 t - \sec^4 t) \, dt$

58. $\displaystyle\int \frac{1 - \sec t}{\cos t - 1} \, dt$

 **Evaluating a Definite Integral** In Exercises 59–66, evaluate the definite integral.

59. $\displaystyle\int_{-\pi}^{\pi} \sin^2 x \, dx$

60. $\displaystyle\int_{0}^{\pi/3} \tan^2 x \, dx$

61. $\displaystyle\int_{0}^{\pi/4} 6 \tan^3 x \, dx$

62. $\displaystyle\int_{0}^{\pi/3} \sec^{3/2} x \tan x \, dx$

63. $\displaystyle\int_{0}^{\pi/2} \frac{\cos(t)}{1 + \sin t} \, dt$

64. $\displaystyle\int_{\pi/6}^{\pi/3} \sin 6x \cos 4x \, dx$

65. $\displaystyle\int_{-\pi/2}^{\pi/2} 3 \cos^3 x \, dx$

66. $\displaystyle\int_{0}^{\pi} \sin^5 x \, dx$

EXPLORING CONCEPTS

Comparing Methods In Exercises 67 and 68, (a) find the indefinite integral in two different ways, (b) use a graphing utility to graph the antiderivative (without the constant of integration) obtained by each method to show that the results differ only by a constant, and (c) verify analytically that the results differ only by a constant.

67. $\displaystyle\int \sec^4 3x \tan^3 3x \, dx$

68. $\displaystyle\int \sec^2 x \tan x \, dx$

69. Comparing Methods Find the indefinite integral

$$\int \sin x \cos x \, dx$$

using the given method. Explain how your answers differ for each method.

(a) Substitution where $u = \sin x$

(b) Substitution where $u = \cos x$

(c) Integration by parts

(d) Using the identity $\sin 2x = 2 \sin x \cos x$

70. 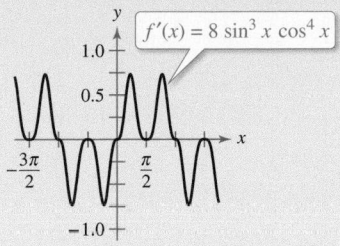 **HOW DO YOU SEE IT?** Use the graph of f' shown in the figure to answer the following.

(a) Using the interval shown in the graph, approximate the value(s) of x where f is maximum. Explain.

(b) Using the interval shown in the graph, approximate the value(s) of x where f is minimum. Explain.

Area In Exercises 71 and 72, find the area of the given region.

71. $y = \sin x, \ y = \sin^3 x$

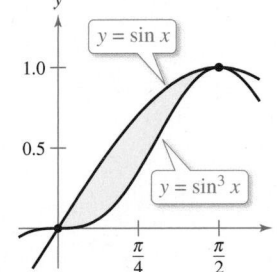

72. $y = \sin^2 \pi x$

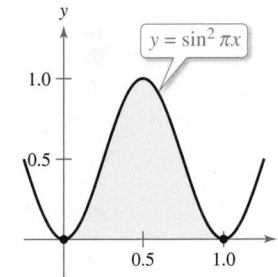

Area In Exercises 73 and 74, find the area of the region bounded by the graphs of the equations.

73. $y = \cos^2 x, \quad y = \sin^2 x, \quad x = -\dfrac{\pi}{4}, \quad x = \dfrac{\pi}{4}$

74. $y = \cos^2 x, \quad y = \sin x \cos x, \quad x = -\dfrac{\pi}{2}, \quad x = \dfrac{\pi}{4}$

Volume In Exercises 75 and 76, find the volume of the solid generated by revolving the region bounded by the graphs of the equations about the x-axis.

75. $y = \tan x, \quad y = 0, \quad x = -\dfrac{\pi}{4}, \quad x = \dfrac{\pi}{4}$

76. $y = \cos \dfrac{x}{2}, \quad y = \sin \dfrac{x}{2}, \quad x = 0, \quad x = \dfrac{\pi}{2}$

Volume and Centroid In Exercises 77 and 78, for the region bounded by the graphs of the equations, find (a) the volume of the solid generated by revolving the region about the x-axis and (b) the centroid of the region.

77. $y = \sin x, \quad y = 0, \quad x = 0, \quad x = \pi$

78. $y = \cos x, \quad y = 0, \quad x = 0, \quad x = \dfrac{\pi}{2}$

Verifying a Reduction Formula In Exercises 79–82, use integration by parts to verify the reduction formula. (A *reduction formula* reduces a given integral to the sum of a function and a simpler integral.)

79. $\displaystyle\int \sin^n x \, dx = -\frac{\sin^{n-1} x \cos x}{n} + \frac{n-1}{n} \int \sin^{n-2} x \, dx$

80. $\displaystyle\int \cos^n x \, dx = \frac{\cos^{n-1} x \sin x}{n} + \frac{n-1}{n} \int \cos^{n-2} x \, dx$

81. $\displaystyle\int \cos^m x \sin^n x \, dx$

$\qquad = -\frac{\cos^{m+1} x \sin^{n-1} x}{m+n} + \frac{n-1}{m+n} \int \cos^m x \sin^{n-2} x \, dx$

82. $\displaystyle\int \sec^n x \, dx = \frac{\sec^{n-2} x \tan x}{n-1} + \frac{n-2}{n-1} \int \sec^{n-2} x \, dx$

Using Formulas In Exercises 83–86, find the indefinite integral by using the appropriate formula from Exercises 79–82.

83. $\displaystyle\int \sin^5 x \, dx$

84. $\displaystyle\int \cos^4 x \, dx$

85. $\displaystyle\int \cos^2 x \sin^4 x \, dx$

86. $\displaystyle\int \sec^4 \frac{2\pi x}{5} \, dx$

87. Wallis's Formulas Use the result of Exercise 80 to prove the following versions of Wallis's Formulas.

(a) If n is odd ($n \geq 3$), then

$$\int_0^{\pi/2} \cos^n x \, dx = \left(\frac{2}{3}\right)\left(\frac{4}{5}\right)\left(\frac{6}{7}\right) \cdots \left(\frac{n-1}{n}\right).$$

(b) If n is even ($n \geq 2$), then

$$\int_0^{\pi/2} \cos^n x \, dx = \left(\frac{1}{2}\right)\left(\frac{3}{4}\right)\left(\frac{5}{6}\right) \cdots \left(\frac{n-1}{n}\right)\left(\frac{\pi}{2}\right).$$

88. Orthogonal Functions The **inner product** of two functions f and g on $[a, b]$ is given by

$$\langle f, g \rangle = \int_a^b f(x)g(x) \, dx.$$

Two distinct functions f and g are said to be **orthogonal** if $\langle f, g \rangle = 0$. Show that the following set of functions is orthogonal on $[-\pi, \pi]$.

$$\{\sin x, \sin 2x, \sin 3x, \ldots, \cos x, \cos 2x, \cos 3x, \ldots\}$$

89. Fourier Series The following sum is a *finite Fourier series*.

$$f(x) = \sum_{i=1}^{N} a_i \sin ix$$

$$= a_1 \sin x + a_2 \sin 2x + a_3 \sin 3x + \cdots + a_N \sin Nx$$

(a) Use Exercise 88 to show that the nth coefficient a_n is given by $a_n = \dfrac{1}{\pi} \displaystyle\int_{-\pi}^{\pi} f(x) \sin nx \, dx.$

(b) Let $f(x) = x$. Find a_1, a_2, and a_3.

SECTION PROJECT • • • • • • • • • • • •

The Wallis Product

The formula for π as an infinite product was derived by English mathematician John Wallis in 1655. This product, called the **Wallis Product,** appeared in his book *Arithmetica Infinitorum.*

$$\frac{\pi}{2} = \left(\frac{2 \cdot 2}{1 \cdot 3}\right)\left(\frac{4 \cdot 4}{3 \cdot 5}\right)\left(\frac{6 \cdot 6}{5 \cdot 7}\right) \cdots \left(\frac{(2n) \cdot (2n)}{(2n-1) \cdot (2n+1)}\right) \cdots$$

In 2015, physicists Carl Hagen and Tamar Friedmann (also a mathematician) stumbled upon a connection between quantum mechanics and the Wallis Product when they applied the variational principle to higher energy states of the hydrogen atom. This principle was previously used only on the ground energy state. The Wallis Product appeared naturally in the midst of their calculations involving gamma functions.

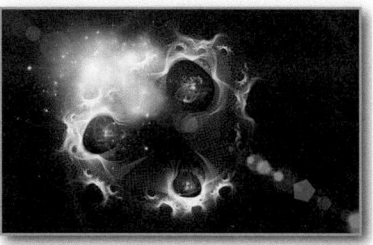

Quantum mechanics is the study of matter and light on the atomic and subatomic scale.

Consider Wallis's method of finding a formula for π. Let

$$I(n) = \int_0^{\pi/2} \sin^n x \, dx.$$

From Wallis's Formulas,

$$I(n) = \left(\frac{1}{2}\right)\left(\frac{3}{4}\right)\left(\frac{5}{6}\right) \cdots \left(\frac{n-1}{n}\right)\left(\frac{\pi}{2}\right), \quad n \text{ is even } (n \geq 2)$$

or

$$I(n) = \left(\frac{2}{3}\right)\left(\frac{4}{5}\right)\left(\frac{6}{7}\right) \cdots \left(\frac{n-1}{n}\right), \quad n \text{ is odd } (n \geq 3).$$

(a) Find $I(n)$ for $n = 2, 3, 4,$ and 5. What do you observe?

(b) Show that $I(n+1) \leq I(n)$ for $n \geq 2$.

(c) Show that

$$\lim_{n \to \infty} \frac{I(2n+1)}{I(2n)} = 1.$$

(Hint: Use the Squeeze Theorem.)

(d) Verify the Wallis Product using the limit in part (c).

■ **FOR FURTHER INFORMATION** For an alternative proof of the Wallis Product, see the article "An Elementary Proof of the Wallis Product Formula for pi" by Johan Wästlund in *The American Mathematical Monthly.* To view this article, go to *MathArticles.com.*

• •

8.4 Trigonometric Substitution

■ Use trigonometric substitution to find an integral.
■ Use integrals to model and solve real-life applications.

Trigonometric Substitution

Now that you can find integrals involving powers of trigonometric functions, you can use **trigonometric substitution** to find integrals involving the radicals

$$\sqrt{a^2 - u^2}, \quad \sqrt{a^2 + u^2}, \quad \text{and} \quad \sqrt{u^2 - a^2}.$$

The objective with trigonometric substitution is to eliminate the radical in the integrand. You do this by using the Pythagorean identities.

$$\cos^2 \theta = 1 - \sin^2 \theta$$
$$\sec^2 \theta = 1 + \tan^2 \theta$$
$$\tan^2 \theta = \sec^2 \theta - 1$$

For example, for $a > 0$, let $u = a \sin \theta$, where $-\pi/2 \le \theta \le \pi/2$. Then

$$
\begin{aligned}
\sqrt{a^2 - u^2} &= \sqrt{a^2 - a^2 \sin^2 \theta} \\
&= \sqrt{a^2(1 - \sin^2 \theta)} \\
&= \sqrt{a^2 \cos^2 \theta} \\
&= a \cos \theta.
\end{aligned}
$$

Note that $\cos \theta \ge 0$, because $-\pi/2 \le \theta \le \pi/2$.

Trigonometric Substitution ($a > 0$)

1. For integrals involving $\sqrt{a^2 - u^2}$, let

$$u = a \sin \theta.$$

Then $\sqrt{a^2 - u^2} = a \cos \theta$, where

$$-\pi/2 \le \theta \le \pi/2.$$

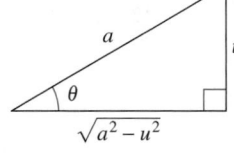

2. For integrals involving $\sqrt{a^2 + u^2}$, let

$$u = a \tan \theta.$$

Then $\sqrt{a^2 + u^2} = a \sec \theta$, where

$$-\pi/2 < \theta < \pi/2.$$

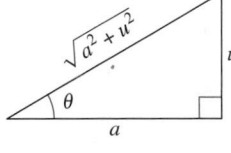

3. For integrals involving $\sqrt{u^2 - a^2}$, let

$$u = a \sec \theta.$$

Then

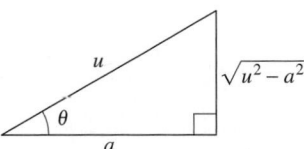

$$\sqrt{u^2 - a^2} = \begin{cases} a \tan \theta \text{ for } u > a, \text{ where } 0 \le \theta < \pi/2 \\ -a \tan \theta \text{ for } u < -a, \text{ where } \pi/2 < \theta \le \pi. \end{cases}$$

The restrictions on θ ensure that the function that defines the substitution is one-to-one. In fact, these are the same intervals over which the arcsine, arctangent, and arcsecant are defined.

EXAMPLE 1 **Trigonometric Substitution:** $u = a \sin \theta$

Find $\displaystyle\int \frac{dx}{x^2\sqrt{9 - x^2}}$.

Solution First, note that the basic integration rules do not apply. To use trigonometric substitution, you should observe that

$$\sqrt{9 - x^2}$$

is of the form $\sqrt{a^2 - u^2}$. So, you can use the substitution

$$x = a \sin \theta = 3 \sin \theta.$$

Using differentiation and the triangle shown in Figure 8.6, you obtain

$$dx = 3 \cos \theta \, d\theta, \quad \sqrt{9 - x^2} = 3 \cos \theta, \quad \text{and} \quad x^2 = 9 \sin^2 \theta.$$

So, trigonometric substitution yields

$$\int \frac{dx}{x^2\sqrt{9 - x^2}} = \int \frac{3 \cos \theta \, d\theta}{(9 \sin^2 \theta)(3 \cos \theta)} \qquad \text{Substitute.}$$

$$= \frac{1}{9}\int \frac{d\theta}{\sin^2 \theta} \qquad \text{Simplify.}$$

$$= \frac{1}{9}\int \csc^2 \theta \, d\theta \qquad \text{Trigonometric identity}$$

$$= -\frac{1}{9} \cot \theta + C \qquad \text{Apply Cosecant Rule.}$$

$$= -\frac{1}{9}\left(\frac{\sqrt{9 - x^2}}{x}\right) + C \qquad \text{Substitute for } \cot \theta.$$

$$= -\frac{\sqrt{9 - x^2}}{9x} + C.$$

Note that the triangle in Figure 8.6 can be used to convert the θ's back to x's, as shown.

$$\cot \theta = \frac{\text{adj.}}{\text{opp.}}$$

$$= \frac{\sqrt{9 - x^2}}{x}$$

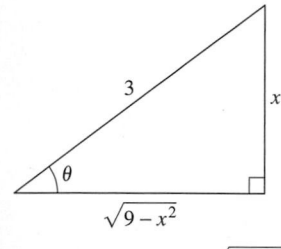

$$\sin \theta = \frac{x}{3}, \quad \cot \theta = \frac{\sqrt{9 - x^2}}{x}$$

Figure 8.6

▷ **TECHNOLOGY** Use a computer algebra system to find each indefinite integral.

$$\int \frac{dx}{\sqrt{9 - x^2}} \qquad \int \frac{dx}{x\sqrt{9 - x^2}}$$

$$\int \frac{dx}{x^2\sqrt{9 - x^2}} \qquad \int \frac{dx}{x^3\sqrt{9 - x^2}}$$

Then use trigonometric substitution to duplicate the results obtained with the computer algebra system.

In an earlier chapter, you saw how the inverse hyperbolic functions can be used to find the integrals

$$\int \frac{du}{\sqrt{u^2 \pm a^2}}, \quad \int \frac{du}{a^2 - u^2}, \quad \text{and} \quad \int \frac{du}{u\sqrt{a^2 \pm u^2}}.$$

You can also find these integrals using trigonometric substitution. This is shown in the next example.

EXAMPLE 2   **Trigonometric Substitution: $u = a \tan \theta$**

Find $\displaystyle\int \frac{dx}{\sqrt{4x^2 + 1}}$.

Solution   Let $u = 2x$, $a = 1$, and $2x = \tan\theta$, as shown in Figure 8.7. Then,

$$dx = \frac{1}{2}\sec^2\theta\, d\theta \quad \text{and} \quad \sqrt{4x^2 + 1} = \sec\theta.$$

Trigonometric substitution produces

$$\int \frac{dx}{\sqrt{4x^2 + 1}} = \frac{1}{2}\int \frac{\sec^2\theta\, d\theta}{\sec\theta} \qquad \text{Substitute.}$$

$$= \frac{1}{2}\int \sec\theta\, d\theta \qquad \text{Simplify.}$$

$$= \frac{1}{2}\ln|\sec\theta + \tan\theta| + C \qquad \text{Apply Secant Rule.}$$

$$= \frac{1}{2}\ln\left|\sqrt{4x^2 + 1} + 2x\right| + C. \qquad \text{Back-substitute.}$$

Try checking this result with a computer algebra system. Is the result given in this form or in the form of an inverse hyperbolic function?

You can extend the use of trigonometric substitution to cover integrals involving expressions such as $(a^2 - u^2)^{n/2}$ by writing the expression as

$$(a^2 - u^2)^{n/2} = \left(\sqrt{a^2 - u^2}\right)^n.$$

EXAMPLE 3   **Trigonometric Substitution: Rational Powers**

$\vdots \cdots \triangleright$ *See LarsonCalculus.com for an interactive version of this type of example.*

Find $\displaystyle\int \frac{dx}{(x^2 + 1)^{3/2}}$.

Solution   Begin by writing $(x^2 + 1)^{3/2}$ as

$$\left(\sqrt{x^2 + 1}\right)^3.$$

Then let $a = 1$ and $u = x = \tan\theta$, as shown in Figure 8.8. Using

$$dx = \sec^2\theta\, d\theta \quad \text{and} \quad \sqrt{x^2 + 1} = \sec\theta$$

you can apply trigonometric substitution, as shown.

$$\int \frac{dx}{(x^2 + 1)^{3/2}} = \int \frac{dx}{\left(\sqrt{x^2 + 1}\right)^3} \qquad \text{Rewrite denominator.}$$

$$= \int \frac{\sec^2\theta\, d\theta}{\sec^3\theta} \qquad \text{Substitute.}$$

$$= \int \frac{d\theta}{\sec\theta} \qquad \text{Simplify.}$$

$$= \int \cos\theta\, d\theta \qquad \text{Trigonometric identity}$$

$$= \sin\theta + C \qquad \text{Apply Cosine Rule.}$$

$$= \frac{x}{\sqrt{x^2 + 1}} + C \qquad \text{Back-substitute.}$$

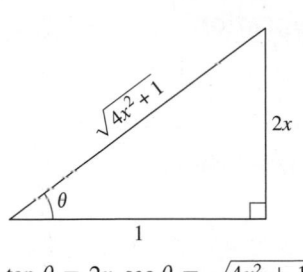

$\tan\theta = 2x$, $\sec\theta = \sqrt{4x^2 + 1}$
Figure 8.7

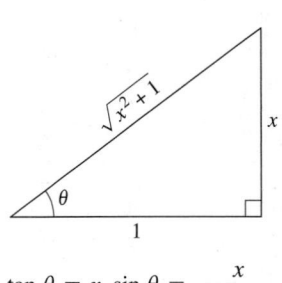

$\tan\theta = x$, $\sin\theta = \dfrac{x}{\sqrt{x^2 + 1}}$
Figure 8.8

For definite integrals, it is often convenient to determine integration limits for θ that avoid converting back to x. You might want to review this procedure in Section 5.5, Examples 8 and 9.

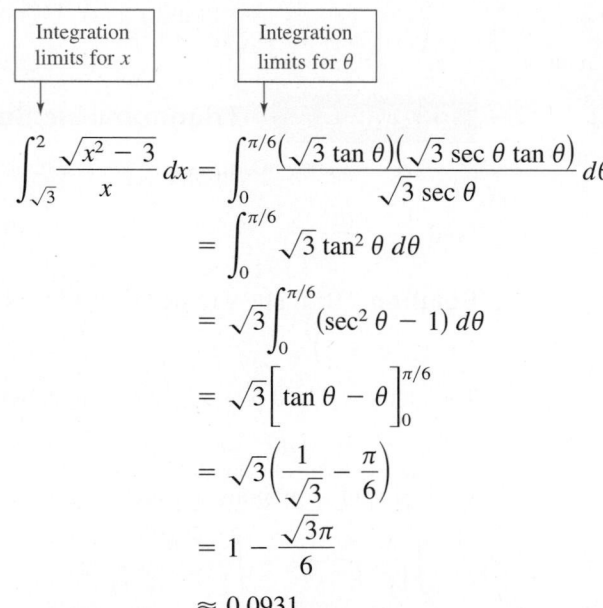

EXAMPLE 4 **Converting the Limits of Integration**

Evaluate $\displaystyle\int_{\sqrt{3}}^{2} \frac{\sqrt{x^2 - 3}}{x}\, dx.$

Solution Because $\sqrt{x^2 - 3}$ has the form $\sqrt{u^2 - a^2}$, you can consider

$$u = x, \quad a = \sqrt{3}, \quad \text{and} \quad x = \sqrt{3}\sec\theta$$

as shown in Figure 8.9. Then

$$dx = \sqrt{3}\sec\theta\tan\theta\, d\theta \quad \text{and} \quad \sqrt{x^2 - 3} = \sqrt{3}\tan\theta.$$

To determine the upper and lower limits of integration, use the substitution $x = \sqrt{3}\sec\theta$, as shown.

Lower Limit

When $x = \sqrt{3}$, $\sec\theta = 1$ and $\theta = 0$.

Upper Limit

When $x = 2$, $\sec\theta = \dfrac{2}{\sqrt{3}}$ and $\theta = \dfrac{\pi}{6}$.

So, you have

$$\int_{\sqrt{3}}^{2} \frac{\sqrt{x^2 - 3}}{x}\, dx = \int_{0}^{\pi/6} \frac{(\sqrt{3}\tan\theta)(\sqrt{3}\sec\theta\tan\theta)}{\sqrt{3}\sec\theta}\, d\theta$$

Integration limits for x · Integration limits for θ

$$= \int_{0}^{\pi/6} \sqrt{3}\tan^2\theta\, d\theta$$

$$= \sqrt{3}\int_{0}^{\pi/6} (\sec^2\theta - 1)\, d\theta$$

$$= \sqrt{3}\Big[\tan\theta - \theta\Big]_{0}^{\pi/6}$$

$$= \sqrt{3}\left(\frac{1}{\sqrt{3}} - \frac{\pi}{6}\right)$$

$$= 1 - \frac{\sqrt{3}\pi}{6}$$

$$\approx 0.0931.$$

In Example 4, try converting back to the variable x and evaluating the antiderivative at the original limits of integration. You should obtain

$$\int_{\sqrt{3}}^{2} \frac{\sqrt{x^2 - 3}}{x}\, dx = \sqrt{3}\left[\frac{\sqrt{x^2 - 3}}{\sqrt{3}} - \operatorname{arcsec}\frac{x}{\sqrt{3}}\right]_{\sqrt{3}}^{2}$$

$$= \sqrt{3}\left(\frac{1}{\sqrt{3}} - \frac{\pi}{6}\right)$$

$$\approx 0.0931.$$

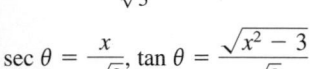

$\sec\theta = \dfrac{x}{\sqrt{3}}, \tan\theta = \dfrac{\sqrt{x^2 - 3}}{\sqrt{3}}$

Figure 8.9

When using trigonometric substitution to evaluate definite integrals, you must be careful to check that the values of θ lie in the intervals discussed at the beginning of this section. For instance, if in Example 4 you had been asked to evaluate the definite integral

$$\int_{-2}^{-\sqrt{3}} \frac{\sqrt{x^2 - 3}}{x} \, dx$$

then using $u = x$ and $a = \sqrt{3}$ in the interval $\left[-2, -\sqrt{3}\right]$ would imply that $u < -a$. So, when determining the upper and lower limits of integration, you would have to choose θ such that $\pi/2 < \theta \leq \pi$. In this case, the integral would be evaluated as shown.

$$\int_{-2}^{-\sqrt{3}} \frac{\sqrt{x^2 - 3}}{x} \, dx - \int_{5\pi/6}^{\pi} \frac{\left(-\sqrt{3} \tan \theta\right)\left(\sqrt{3} \sec \theta \tan \theta\right) d\theta}{\sqrt{3} \sec \theta}$$

$$= \int_{5\pi/6}^{\pi} -\sqrt{3} \tan^2 \theta \, d\theta$$

$$= -\sqrt{3} \int_{5\pi/6}^{\pi} (\sec^2 \theta - 1) \, d\theta$$

$$= -\sqrt{3} \left[\tan \theta - \theta \right]_{5\pi/6}^{\pi}$$

$$= -\sqrt{3} \left[(0 - \pi) - \left(-\frac{1}{\sqrt{3}} - \frac{5\pi}{6} \right) \right]$$

$$= -1 + \frac{\sqrt{3}\pi}{6}$$

$$\approx -0.0931$$

. . . . ▷

REMARK Recall from an earlier chapter that you used completing the square for integrands involving quadratic functions.

Trigonometric substitution can be used with completing the square. For instance, try finding the integral

$$\int \sqrt{x^2 - 2x} \, dx.$$

To begin, you could complete the square and write the integral as

$$\int \sqrt{(x - 1)^2 - 1^2} \, dx.$$

Because the integrand has the form

$$\sqrt{u^2 - a^2}$$

with $u = x - 1$ and $a = 1$, you can now use trigonometric substitution to find the integral.

Trigonometric substitution can be used to find the three integrals listed in the next theorem. These integrals will be encountered several times in the remainder of the text. When this happens, we will simply refer to this theorem. (In Exercise 65, you are asked to verify the formulas given in the theorem.)

THEOREM 8.2 Special Integration Formulas ($a > 0$)

1. $\displaystyle\int \sqrt{a^2 - u^2} \, du = \frac{1}{2}\left(u\sqrt{a^2 - u^2} + a^2 \arcsin \frac{u}{a} \right) + C$

2. $\displaystyle\int \sqrt{u^2 - a^2} \, du = \frac{1}{2}\left(u\sqrt{u^2 - a^2} - a^2 \ln\left| u + \sqrt{u^2 - a^2} \right| \right) + C, \quad u > a$

3. $\displaystyle\int \sqrt{u^2 + a^2} \, du = \frac{1}{2}\left(u\sqrt{u^2 + a^2} + a^2 \ln\left| u + \sqrt{u^2 + a^2} \right| \right) + C$

Applications

EXAMPLE 5 **Finding Arc Length**

Find the arc length of the graph of $f(x) = \frac{1}{2}x^2$ from $x = 0$ to $x = 1$ (see Figure 8.10).

Solution Refer to the arc length formula in Section 7.4.

$$
\begin{aligned}
s &= \int_0^1 \sqrt{1 + [f'(x)]^2}\, dx && \text{Formula for arc length} \\
&= \int_0^1 \sqrt{1 + x^2}\, dx && f'(x) = x \\
&= \int_0^{\pi/4} \sec^3 \theta\, d\theta && \text{Let } a = 1 \text{ and } x = \tan\theta. \\
&= \frac{1}{2}\Big[\sec\theta \tan\theta + \ln|\sec\theta + \tan\theta| \Big]_0^{\pi/4} && \text{Example 5, Section 8.2} \\
&= \frac{1}{2}\Big[\sqrt{2} + \ln\big(\sqrt{2} + 1\big) \Big] \\
&\approx 1.148
\end{aligned}
$$

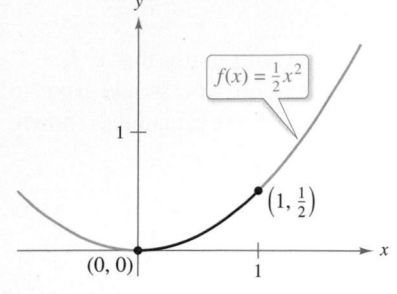

The arc length of the curve from $(0, 0)$ to $\left(1, \frac{1}{2}\right)$

Figure 8.10

EXAMPLE 6 **Comparing Two Fluid Forces**

A sealed barrel of oil (weighing 48 pounds per cubic foot) is floating in seawater (weighing 64 pounds per cubic foot), as shown in Figures 8.11 and 8.12. (The barrel is not completely full of oil. With the barrel lying on its side, the top 0.2 foot of the barrel is empty.) Compare the fluid forces against one end of the barrel from the inside and from the outside. (Assume the radius of the barrel is 1 foot and, with the barrel lying on its side, the top 0.6 foot of the barrel is above the water.)

Solution In Figure 8.12, locate the coordinate system with the origin at the center of the circle

$$x^2 + y^2 = 1.$$

To find the fluid force against an end of the barrel *from the inside*, integrate between -1 and 0.8 (using a weight of $w = 48$).

$$
\begin{aligned}
F &= w \int_c^d h(y)L(y)\, dy && \text{General equation (See Section 7.7.)} \\
F_{\text{inside}} &= 48 \int_{-1}^{0.8} (0.8 - y)(2)\sqrt{1 - y^2}\, dy \\
&= 76.8 \int_{-1}^{0.8} \sqrt{1 - y^2}\, dy - 96 \int_{-1}^{0.8} y\sqrt{1 - y^2}\, dy
\end{aligned}
$$

To find the fluid force *from the outside*, integrate between -1 and 0.4 (using a weight of $w = 64$).

$$
\begin{aligned}
F_{\text{outside}} &= 64 \int_{-1}^{0.4} (0.4 - y)(2)\sqrt{1 - y^2}\, dy \\
&= 51.2 \int_{-1}^{0.4} \sqrt{1 - y^2}\, dy - 128 \int_{-1}^{0.4} y\sqrt{1 - y^2}\, dy
\end{aligned}
$$

The details of integration are left for you to complete in Exercise 64. Intuitively, would you say that the force from the oil (the inside) or the force from the seawater (the outside) is greater? By evaluating these two integrals, you can determine that

$$F_{\text{inside}} \approx 121.3 \text{ pounds} \quad \text{and} \quad F_{\text{outside}} \approx 93.0 \text{ pounds.}$$

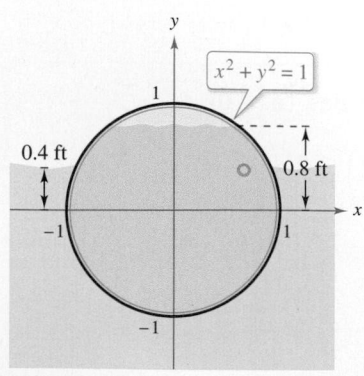

The barrel is not quite full of oil—the top 0.2 foot of the barrel is empty.

Figure 8.11

Figure 8.12

8.4 Exercises

See **CalcChat.com** for tutorial help and worked-out solutions to odd-numbered exercises.

CONCEPT CHECK

1. Trigonometric Substitution State the trigonometric substitution you would use to find the indefinite integral. Do not integrate.

(a) $\int (9 + x^2)^{-2}\, dx$

(b) $\int \sqrt{4 - x^2}\, dx$

(c) $\int \frac{x^2}{\sqrt{25 - x^2}}\, dx$

(d) $\int x^2(x^2 - 25)^{3/2}\, dx$

2. Trigonometric Substitution Why is completing the square useful when you are considering integration by trigonometric substitution?

Using Trigonometric Substitution In Exercises 3–6, find the indefinite integral using the substitution $x = 4\sin\theta$.

3. $\int \frac{1}{(16 - x^2)^{3/2}}\, dx$

4. $\int \frac{4}{x^2\sqrt{16 - x^2}}\, dx$

5. $\int \frac{\sqrt{16 - x^2}}{x}\, dx$

6. $\int \frac{x^3}{\sqrt{16 - x^2}}\, dx$

 **Using Trigonometric Substitution** In Exercises 7–10, find the indefinite integral using the substitution $x = 5\sec\theta$.

7. $\int \frac{1}{\sqrt{x^2 - 25}}\, dx$

8. $\int \frac{\sqrt{x^2 - 25}}{x}\, dx$

9. $\int x^3\sqrt{x^2 - 25}\, dx$

10. $\int \frac{x^3}{\sqrt{x^2 - 25}}\, dx$

Using Trigonometric Substitution In Exercises 11–14, find the indefinite integral using the substitution $x = 2\tan\theta$.

11. $\int \frac{x}{2}\sqrt{4 + x^2}\, dx$

12. $\int \frac{x^3}{4\sqrt{4 + x^2}}\, dx$

13. $\int \frac{4}{(4 + x^2)^2}\, dx$

14. $\int \frac{2x^2}{(4 + x^2)^2}\, dx$

Special Integration Formulas In Exercises 15–18, use the Special Integration Formulas (Theorem 8.2) to find the indefinite integral.

15. $\int \sqrt{49 - 16x^2}\, dx$

16. $\int \sqrt{5x^2 - 1}\, dx$

17. $\int \sqrt{36 - 5x^2}\, dx$

18. $\int \sqrt{9 + 4x^2}\, dx$

 **Finding an Indefinite Integral** In Exercises 19–32, find the indefinite integral.

19. $\int \sqrt{16 - 4x^2}\, dx$

20. $\int \frac{1}{\sqrt{x^2 - 4}}\, dx$

21. $\int \frac{\sqrt{1 - x^2}}{x^4}\, dx$

22. $\int \frac{\sqrt{25x^2 + 4}}{x^4}\, dx$

23. $\int \frac{1}{x\sqrt{4x^2 + 9}}\, dx$

24. $\int \frac{1}{x\sqrt{9x^2 + 1}}\, dx$

25. $\int \frac{-3}{(x^2 + 3)^{3/2}}\, dx$

26. $\int \frac{1}{(x^2 + 5)^{3/2}}\, dx$

27. $\int e^x\sqrt{1 - e^{2x}}\, dx$

28. $\int \frac{\sqrt{1 - x}}{\sqrt{x}}\, dx$

29. $\int \frac{1}{4 + 4x^2 + x^4}\, dx$

30. $\int \frac{x^3 + x + 1}{x^4 + 2x^2 + 1}\, dx$

31. $\int \operatorname{arcsec} 2x\, dx, \ x > \frac{1}{2}$

32. $\int x \arcsin x\, dx$

 **Completing the Square** In Exercises 33–36, complete the square and find the indefinite integral.

33. $\int \frac{x}{\sqrt{4x - x^2}}\, dx$

34. $\int \frac{x^2}{\sqrt{2x - x^2}}\, dx$

35. $\int \frac{x}{\sqrt{x^2 + 6x + 12}}\, dx$

36. $\int \frac{x}{\sqrt{x^2 - 6x + 5}}\, dx$

 **Converting the Limits of Integration** In Exercises 37–42, evaluate the definite integral using (a) the given integration limits and (b) the limits obtained by trigonometric substitution.

37. $\int_0^{\sqrt{3}/2} \frac{t^2}{(1 - t^2)^{3/2}}\, dt$

38. $\int_0^{\sqrt{3}/2} \frac{1}{(1 - t^2)^{5/2}}\, dt$

39. $\int_0^3 \frac{x^3}{\sqrt{x^2 + 9}}\, dx$

40. $\int_0^{3/5} \sqrt{9 - 25x^2}\, dx$

41. $\int_4^6 \frac{x^2}{\sqrt{x^2 - 9}}\, dx$

42. $\int_4^8 \frac{\sqrt{x^2 - 16}}{x^2}\, dx$

EXPLORING CONCEPTS

Choosing a Method In Exercises 43 and 44, state the method of integration you would use to find each integral. Explain why you chose that method. Do not integrate.

43. $\int x\sqrt{x^2 + 1}\, dx$

44. $\int x^2\sqrt{x^2 - 1}\, dx$

45. Comparing Methods

(a) Find the integral $\int \frac{x}{\sqrt{1 - x^2}}\, dx$ using u-substitution. Then find the integral using trigonometric substitution. Discuss the results.

(b) Find the integral $\int \frac{x^2}{x^2 + 9}\, dx$ algebraically using $x^2 = (x^2 + 9) - 9$. Then find the integral using trigonometric substitution. Discuss the results.

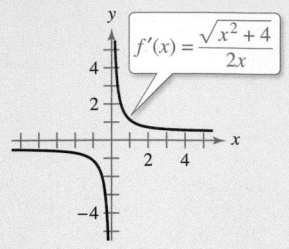

46. HOW DO YOU SEE IT? Use the graph of f' shown in the figure to answer the following.

$$f'(x) = \frac{\sqrt{x^2 + 4}}{2x}$$

(a) Identify the open interval(s) on which the graph of f is increasing or decreasing. Explain.

(b) Identify the open interval(s) on which the graph of f is concave upward or concave downward. Explain.

True or False? In Exercises 47–50, determine whether the statement is true or false. If it is false, explain why or give an example that shows it is false.

47. If $x = \sin \theta$, then

$$\int \frac{dx}{\sqrt{1 - x^2}} = \int d\theta.$$

48. If $x = \sec \theta$, then

$$\int \frac{\sqrt{x^2 - 1}}{x} \, dx = \int \sec \theta \tan \theta \, d\theta.$$

49. If $x = \tan \theta$, then

$$\int_0^{\sqrt{3}} \frac{dx}{(1 + x^2)^{3/2}} = \int_0^{4\pi/3} \cos \theta \, d\theta.$$

50. If $x = \sin \theta$, then

$$\int_{-1}^{1} x^2 \sqrt{1 - x^2} \, dx = 2 \int_0^{\pi/2} \sin^2 \theta \cos^2 \theta \, d\theta.$$

51. Area Find the area enclosed by the ellipse $\dfrac{x^2}{a^2} + \dfrac{y^2}{b^2} = 1$ shown in the figure.

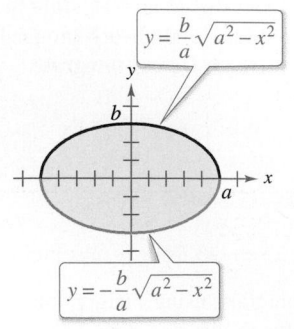

Figure for 51   **Figure for 52**

52. Area Find the area of the shaded region of the circle of radius a when the chord is h units $(0 < h < a)$ from the center of the circle (see figure).

Arc Length In Exercises 53 and 54, find the arc length of the graph of the function over the given interval.

53. $y = \ln x$, $[1, 5]$ **54.** $y = \dfrac{x^2}{4} - 2x$, $[4, 8]$

Volume of a Torus In Exercises 55 and 56, find the volume of the torus generated by revolving the region bounded by the graph of the circle about the y-axis.

55. $(x - 3)^2 + y^2 = 1$

56. $(x - h)^2 + y^2 = r^2$, $h > r$

Centroid In Exercises 57 and 58, find the centroid of the region bounded by the graphs of the inequalities.

57. $y \leq \dfrac{3}{\sqrt{x^2 + 9}}$, $y \geq 0$, $x \geq -4$, $x \leq 4$

58. $y \leq \frac{1}{4}x^2$, $(x - 4)^2 + y^2 \leq 16$, $y \geq 0$

59. Volume The axis of a storage tank in the form of a right circular cylinder is horizontal (see figure). The radius and length of the tank are 1 meter and 3 meters, respectively.

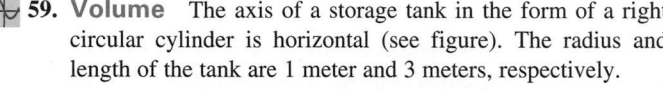
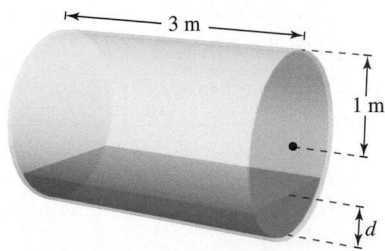

(a) Determine the volume of fluid in the tank as a function of its depth d.

(b) Use a graphing utility to graph the function in part (a).

(c) Design a dip stick for the tank with markings of $\frac{1}{4}$, $\frac{1}{2}$, and $\frac{3}{4}$.

(d) Fluid is entering the tank at a rate of $\frac{1}{4}$ cubic meter per minute. Determine the rate of change of the depth of the fluid as a function of its depth d.

(e) Use a graphing utility to graph the function in part (d). When will the rate of change of the depth be minimum? Does this agree with your intuition? Explain.

60. Field Strength The field strength H of a magnet of length $2L$ on a particle r units from the center of the magnet is

$$H = \frac{2mL}{(r^2 + L^2)^{3/2}}$$

where $\pm m$ are the poles of the magnet (see figure). Find the average field strength as the particle moves from 0 to R units from the center by evaluating the integral

$$\frac{1}{R} \int_0^R \frac{2mL}{(r^2 + L^2)^{3/2}} \, dr.$$

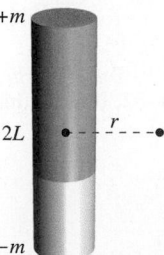

61. Tractrix A person moves from the origin along the positive y-axis pulling a weight at the end of a 12-meter rope (see figure). Initially, the weight is located at the point $(12, 0)$.

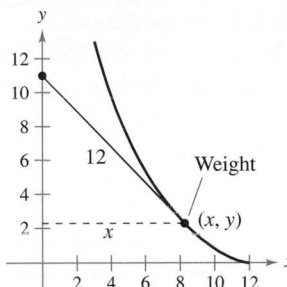

(a) Show that the slope of the tangent line of the path of the weight is

$$\frac{dy}{dx} = -\frac{\sqrt{144 - x^2}}{x}.$$

(b) Use the result of part (a) to find the equation of the path of the weight. Use a graphing utility to graph the path and compare it with the figure.

(c) Find any vertical asymptotes of the graphs in part (b).

(d) When the person has reached the point $(0, 12)$, how far has the weight moved?

62. Conjecture

(a) Find formulas for the distances between $(0, 0)$ and (a, a^2), $a > 0$, along the line between these points and along the parabola $y = x^2$.

(b) Use the formulas from part (a) to find the distances for $a = 1$, $a = 10$, and $a = 100$.

(c) Make a conjecture about the difference between the two distances as a increases.

• • **63. Fluid Force** • • • • • • • • • • • • • • • • • • •

Find the fluid force on a circular observation window of radius 1 foot in a vertical wall of a large water-filled tank at a fish hatchery when the center of the window is (a) 3 feet and (b) d feet ($d > 1$) below the water's surface (see figure). Use trigonometric substitution to evaluate the one integral. Water weighs 62.4 pounds per cubic foot. (Recall that in Section 7.7 in a similar problem, you evaluated one integral by a geometric formula and the other by observing that the integrand was odd.)

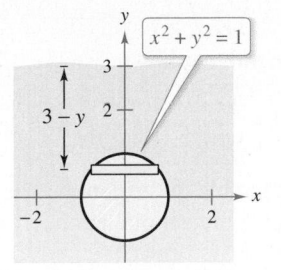

64. Fluid Force Evaluate the following two integrals, which yield the fluid forces given in Example 6.

(a) $F_{\text{inside}} = 48 \int_{-1}^{0.8} (0.8 - y)(2)\sqrt{1 - y^2}\, dy$

(b) $F_{\text{outside}} = 64 \int_{-1}^{0.4} (0.4 - y)(2)\sqrt{1 - y^2}\, dy$

65. Verifying Formulas Use trigonometric substitution to verify the integration formulas given in Theorem 8.2.

66. Arc Length Show that the arc length of the graph of $y = \sin x$ on the interval $[0, 2\pi]$ is equal to the circumference of the ellipse $x^2 + 2y^2 = 2$ (see figure).

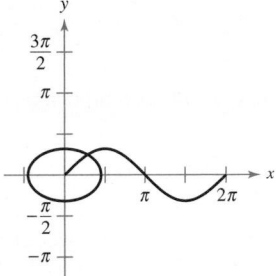

67. Area of a Lune The crescent-shaped region bounded by two circles forms a *lune* (see figure). Find the area of the lune given that the radius of the smaller circle is 3 and the radius of the larger circle is 5.

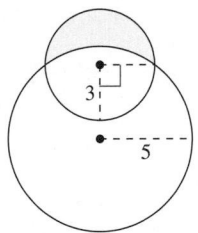

68. Area Two circles of radius 3, with centers at $(-2, 0)$ and $(2, 0)$, intersect as shown in the figure. Find the area of the shaded region.

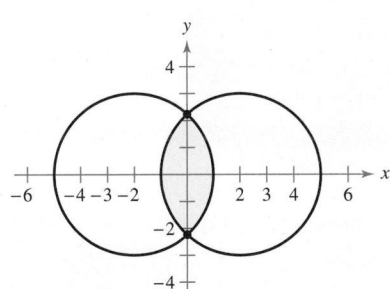

PUTNAM EXAM CHALLENGE

69. Evaluate

$$\int_0^1 \frac{\ln(x + 1)}{x^2 + 1}\, dx.$$

8.5 Partial Fractions

- Understand the concept of partial fraction decomposition.
- Use partial fraction decomposition with linear factors to integrate rational functions.
- Use partial fraction decomposition with quadratic factors to integrate rational functions.

Partial Fractions

This section examines a procedure for decomposing a rational function into simpler rational functions to which you can apply the basic integration formulas. This procedure is called the **method of partial fractions.** To see the benefit of the method of partial fractions, consider the integral

$$\int \frac{1}{x^2 - 5x + 6} \, dx.$$

To find this integral *without* partial fractions, you can complete the square and use trigonometric substitution (see Figure 8.13) to obtain

$$\int \frac{1}{x^2 - 5x + 6} \, dx = \int \frac{dx}{(x - 5/2)^2 - (1/2)^2} \qquad a = \tfrac{1}{2}, \; x - \tfrac{5}{2} = \tfrac{1}{2} \sec \theta$$

$$= \int \frac{(1/2) \sec \theta \tan \theta \, d\theta}{(1/4) \tan^2 \theta} \qquad dx = \tfrac{1}{2} \sec \theta \tan \theta \, d\theta$$

$$= 2 \int \csc \theta \, d\theta$$

$$= 2 \ln|\csc \theta - \cot \theta| + C$$

$$= 2 \ln\left| \frac{2x - 5}{2\sqrt{x^2 - 5x + 6}} - \frac{1}{2\sqrt{x^2 - 5x + 6}} \right| + C$$

$$= 2 \ln\left| \frac{x - 3}{\sqrt{x^2 - 5x + 6}} \right| + C$$

$$= \ln\left| \frac{(x - 3)^2}{x^2 - 5x + 6} \right| + C$$

$$= \ln\left| \frac{(x - 3)^2}{(x - 2)(x - 3)} \right| + C$$

$$= \ln\left| \frac{x - 3}{x - 2} \right| + C$$

$$= \ln|x - 3| - \ln|x - 2| + C.$$

$$\sec \theta = 2x - 5$$

Figure 8.13

Now, suppose you had observed that

$$\frac{1}{x^2 - 5x + 6} = \frac{1}{x - 3} - \frac{1}{x - 2}. \qquad \text{Partial fraction decomposition}$$

Then you could find the integral, as shown.

$$\int \frac{1}{x^2 - 5x + 6} \, dx = \int \left(\frac{1}{x - 3} - \frac{1}{x - 2} \right) dx = \ln|x - 3| - \ln|x - 2| + C$$

This method is clearly preferable to trigonometric substitution. Its use, however, depends on the ability to factor the denominator, $x^2 - 5x + 6$, and to find the **partial fractions**

$$\frac{1}{x - 3} \quad \text{and} \quad -\frac{1}{x - 2}.$$

In this section, you will study techniques for finding partial fraction decompositions.

JOHN BERNOULLI (1667–1748)

The method of partial fractions was introduced by John Bernoulli, a Swiss mathematician who was instrumental in the early development of calculus. John Bernoulli was a professor at the University of Basel and taught many outstanding students, the most famous of whom was Leonhard Euler.
See LarsonCalculus.com to read more of this biography.

Recall from algebra that every polynomial with real coefficients can be factored into linear and irreducible quadratic factors.* For instance, the polynomial

$$x^5 + x^4 - x - 1$$

can be written as

$$
\begin{aligned}
x^5 + x^4 - x - 1 &= x^4(x + 1) - (x + 1) \\
&= (x^4 - 1)(x + 1) \\
&= (x^2 + 1)(x^2 - 1)(x + 1) \\
&= (x^2 + 1)(x + 1)(x - 1)(x + 1) \\
&= (x - 1)(x + 1)^2(x^2 + 1)
\end{aligned}
$$

where $(x - 1)$ is a linear factor, $(x + 1)^2$ is a repeated linear factor, and $(x^2 + 1)$ is an irreducible quadratic factor. Using this factorization, you can write the partial fraction decomposition of the rational expression

$$\frac{N(x)}{x^5 + x^4 - x - 1}$$

where $N(x)$ is a polynomial of degree less than 5, as shown.

$$\frac{N(x)}{(x - 1)(x + 1)^2(x^2 + 1)} = \frac{A}{x - 1} + \frac{B}{x + 1} + \frac{C}{(x + 1)^2} + \frac{Dx + E}{x^2 + 1}$$

· · · · · · · · · · · · · · · ▷

· · REMARK In precalculus, you learned how to combine functions such as

$$\frac{1}{x - 2} + \frac{-1}{x + 3} = \frac{5}{(x - 2)(x + 3)}.$$

The method of partial fractions shows you how to reverse this process.

$$\frac{5}{(x - 2)(x + 3)} = \frac{?}{x - 2} + \frac{?}{x + 3}$$

Decomposition of $N(x)/D(x)$ into Partial Fractions

1. **Divide when improper:** When $N(x)/D(x)$ is an improper fraction (that is, when the degree of the numerator is greater than or equal to the degree of the denominator), divide the denominator into the numerator to obtain

$$\frac{N(x)}{D(x)} = (\text{a polynomial}) + \frac{N_1(x)}{D(x)}$$

where the degree of $N_1(x)$ is less than the degree of $D(x)$. Then apply Steps 2, 3, and 4 to the proper rational expression $N_1(x)/D(x)$.

2. **Factor denominator:** Completely factor the denominator into factors of the form

$$(px + q)^m \quad \text{and} \quad (ax^2 + bx + c)^n$$

where $ax^2 + bx + c$ is irreducible.

3. **Linear factors:** For each factor of the form $(px + q)^m$, the partial fraction decomposition must include the following sum of m fractions.

$$\frac{A_1}{(px + q)} + \frac{A_2}{(px + q)^2} + \cdots + \frac{A_m}{(px + q)^m}$$

4. **Quadratic factors:** For each factor of the form $(ax^2 + bx + c)^n$, the partial fraction decomposition must include the following sum of n fractions.

$$\frac{B_1x + C_1}{ax^2 + bx + c} + \frac{B_2x + C_2}{(ax^2 + bx + c)^2} + \cdots + \frac{B_nx + C_n}{(ax^2 + bx + c)^n}$$

---

* For a review of factorization techniques, see *Precalculus*, 10th edition, or *Precalculus: Real Mathematics, Real People*, 7th edition, both by Ron Larson (Boston, Massachusetts: Cengage Learning, 2018 and 2016, respectively).

Linear Factors

Algebraic techniques for determining the constants in the numerators of a partial fraction decomposition with linear or repeated linear factors are shown in Examples 1 and 2.

EXAMPLE 1 Distinct Linear Factors

Write the partial fraction decomposition for

$$\frac{1}{x^2 - 5x + 6}.$$

Solution Because $x^2 - 5x + 6 = (x - 3)(x - 2)$, you should include one partial fraction for each factor and write

$$\frac{1}{x^2 - 5x + 6} = \frac{A}{x - 3} + \frac{B}{x - 2}$$

where A and B are to be determined. Multiplying this equation by the least common denominator $(x - 3)(x - 2)$ yields the **basic equation**

$$1 = A(x - 2) + B(x - 3). \qquad \text{Basic equation}$$

Because this equation is to be true for all x, you can substitute any *convenient* values for x to obtain equations in A and B. The most convenient values are the ones that make particular factors equal to 0.

To solve for A, let $x = 3$.

$$1 = A(3 - 2) + B(3 - 3) \qquad \text{Let } x = 3 \text{ in basic equation.}$$
$$1 = A(1) + B(0)$$
$$1 = A$$

To solve for B, let $x = 2$.

$$1 = A(2 - 2) + B(2 - 3) \qquad \text{Let } x = 2 \text{ in basic equation.}$$
$$1 = A(0) + B(-1)$$
$$-1 = B$$

So, the decomposition is

$$\frac{1}{x^2 - 5x + 6} = \frac{1}{x - 3} - \frac{1}{x - 2}$$

as shown at the beginning of this section.

> **· · REMARK** Note that the substitutions for x in Example 1 are chosen for their convenience in determining values for A and B; $x = 3$ is chosen to eliminate the term $B(x - 3)$, and $x = 2$ is chosen to eliminate the term $A(x - 2)$. The goal is to make *convenient* substitutions whenever possible.

■ FOR FURTHER INFORMATION
To learn a different method for finding partial fraction decompositions, called the Heaviside Method, see the article "Calculus to Algebra Connections in Partial Fraction Decomposition" by Joseph Wiener and Will Watkins in *The AMATYC Review*.

Be sure you see that the method of partial fractions is practical only for integrals of rational functions whose denominators factor "nicely." For instance, when the denominator in Example 1 is changed to

$$x^2 - 5x + 5$$

its factorization as

$$x^2 - 5x + 5 = \left[x - \frac{5 + \sqrt{5}}{2} \right]\left[x - \frac{5 - \sqrt{5}}{2} \right]$$

would be too cumbersome to use with partial fractions. In such cases, you should use completing the square or a computer algebra system to perform the integration. When you do this, you should obtain

$$\int \frac{1}{x^2 - 5x + 5}\, dx = \frac{\sqrt{5}}{5} \ln\left|2x - \sqrt{5} - 5\right| - \frac{\sqrt{5}}{5} \ln\left|2x + \sqrt{5} - 5\right| + C.$$

EXAMPLE 2   Repeated Linear Factors

Find $\displaystyle\int \frac{5x^2 + 20x + 6}{x^3 + 2x^2 + x}\, dx$.

Solution   Because

$$x^3 + 2x^2 + x = x(x^2 + 2x + 1) = x(x + 1)^2$$

you should include one partial fraction for *each power* of x and $(x + 1)$ and write

$$\frac{5x^2 + 20x + 6}{x(x + 1)^2} = \frac{A}{x} + \frac{B}{x + 1} + \frac{C}{(x + 1)^2}.$$

Multiplying by the least common denominator $x(x + 1)^2$ yields the *basic equation*

$$5x^2 + 20x + 6 = A(x + 1)^2 + Bx(x + 1) + Cx. \qquad \text{Basic equation}$$

To solve for A, let $x = 0$. This eliminates the B and C terms and yields

$$6 = A(1) + 0 + 0$$
$$6 = A.$$

To solve for C, let $x = -1$. This eliminates the A and B terms and yields

$$5 - 20 + 6 = 0 + 0 - C$$
$$9 = C.$$

The most convenient choices for x have been used, so to find the value of B, you can use *any other value* of x along with the calculated values of A and C. Using $x = 1$, $A = 6$, and $C = 9$ produces

$$5 + 20 + 6 = A(4) + B(2) + C$$
$$31 = 6(4) + 2B + 9$$
$$-2 = 2B$$
$$-1 = B.$$

So, it follows that

$$\int \frac{5x^2 + 20x + 6}{x(x + 1)^2}\, dx = \int \left(\frac{6}{x} - \frac{1}{x + 1} + \frac{9}{(x + 1)^2} \right) dx$$
$$= 6 \ln|x| - \ln|x + 1| + 9\frac{(x + 1)^{-1}}{-1} + C$$
$$= \ln\left| \frac{x^6}{x + 1} \right| - \frac{9}{x + 1} + C.$$

Try checking this result by differentiating. Include algebra in your check, simplifying the derivative until you have obtained the original integrand. ∎

■ **FOR FURTHER INFORMATION**
For an alternative approach to using partial fractions, see the article "A Shortcut in Partial Fractions" by Xun-Cheng Huang in *The College Mathematics Journal*.

It is necessary to make as many substitutions for x as there are unknowns $(A, B, C, \ldots)$ to be determined. For instance, in Example 2, three substitutions $(x = 0, x = -1, \text{and } x = 1)$ were made to solve for A, B, and C.

▷ **TECHNOLOGY**   Most computer algebra systems, such as *Maple, Mathematica,* and the *TI-Nspire,* can be used to convert a rational function to its partial fraction decomposition. For instance, using *Mathematica,* you obtain the following.

Apart $[(5 * x \wedge 2 + 20 * x + 6)/ (x * (x + 1) \wedge 2), x]$

$$\frac{6}{x} + \frac{9}{(1 + x)^2} - \frac{1}{1 + x}$$

Quadratic Factors

When using the method of partial fractions with *linear* factors, a convenient choice of x immediately yields a value for one of the coefficients. With *quadratic* factors, a system of linear equations usually has to be solved, regardless of the choice of x.

EXAMPLE 3   **Distinct Linear and Quadratic Factors**

$\cdots\cdots\triangleright$ *See LarsonCalculus.com for an interactive version of this type of example.*

Find $\displaystyle\int \frac{2x^3 - 4x - 8}{(x^2 - x)(x^2 + 4)}\, dx.$

Solution   Because

$$(x^2 - x)(x^2 + 4) = x(x - 1)(x^2 + 4)$$

you should include one partial fraction for each factor and write

$$\frac{2x^3 - 4x - 8}{x(x - 1)(x^2 + 4)} = \frac{A}{x} + \frac{B}{x - 1} + \frac{Cx + D}{x^2 + 4}.$$

Multiplying by the least common denominator

$$x(x - 1)(x^2 + 4)$$

yields the *basic equation*

$$2x^3 - 4x - 8 = A(x - 1)(x^2 + 4) + Bx(x^2 + 4) + (Cx + D)(x)(x - 1).$$

To solve for A, let $x = 0$ and obtain

$$-8 = A(-1)(4) + 0 + 0$$
$$2 = A.$$

To solve for B, let $x = 1$ and obtain

$$-10 = 0 + B(5) + 0$$
$$-2 = B.$$

At this point, C and D are yet to be determined. You can find these remaining constants by choosing two other values for x and solving the resulting system of linear equations. Using $x = -1$, $A = 2$, and $B = -2$, you can write

$$-6 = (2)(-2)(5) + (-2)(-1)(5) + (-C + D)(-1)(-2)$$
$$2 = -C + D.$$

For $x = 2$, you have

$$0 = (2)(1)(8) + (-2)(2)(8) + (2C + D)(2)(1)$$
$$8 = 2C + D.$$

Solving the linear system by subtracting the first equation from the second

$$-C + D = 2$$
$$2C + D = 8$$

yields $C = 2$. Consequently, $D = 4$, and it follows that

$$\int \frac{2x^3 - 4x - 8}{x(x - 1)(x^2 + 4)}\, dx = \int \left(\frac{2}{x} - \frac{2}{x - 1} + \frac{2x}{x^2 + 4} + \frac{4}{x^2 + 4} \right) dx$$

$$= 2\ln|x| - 2\ln|x - 1| + \ln(x^2 + 4) + 2\arctan \frac{x}{2} + C.$$

In Examples 1, 2, and 3, the solution of the basic equation began with substituting values of x that made the linear factors equal to 0. This method works well when the partial fraction decomposition involves linear factors. When the decomposition involves only quadratic factors, however, an alternative procedure is often more convenient. For instance, try writing the right side of the basic equation in polynomial form and *equating the coefficients* of like terms. This method is shown in Example 4.

EXAMPLE 4   Repeated Quadratic Factors

Find $\displaystyle\int \frac{8x^3 + 13x}{(x^2 + 2)^2}\, dx.$

Solution   Include one partial fraction for each power of $(x^2 + 2)$ and write

$$\frac{8x^3 + 13x}{(x^2 + 2)^2} = \frac{Ax + B}{x^2 + 2} + \frac{Cx + D}{(x^2 + 2)^2}.$$

Multiplying by the least common denominator $(x^2 + 2)^2$ yields the *basic equation*

$$8x^3 + 13x = (Ax + B)(x^2 + 2) + Cx + D.$$

Expanding the basic equation and collecting like terms produce

$$8x^3 + 13x = Ax^3 + 2Ax + Bx^2 + 2B + Cx + D$$
$$8x^3 + 13x = Ax^3 + Bx^2 + (2A + C)x + (2B + D).$$

Now, you can equate the coefficients of like terms on opposite sides of the equation.

$$8 = A \qquad\qquad\qquad\qquad 0 = 2B + D$$

$$8x^3 + 0x^2 + 13x + 0 = Ax^3 + Bx^2 + (2A + C)x + (2B + D)$$

$$0 = B$$

$$13 = 2A + C$$

Using the known values $A = 8$ and $B = 0$, you can write

$$13 = 2A + C \quad\Longrightarrow\quad 13 = 2(8) + C \quad\Longrightarrow\quad -3 = C$$
$$0 = 2B + D \quad\Longrightarrow\quad 0 = 2(0) + D \quad\Longrightarrow\quad 0 = D.$$

Finally, you can conclude that

$$\int \frac{8x^3 + 13x}{(x^2 + 2)^2}\, dx = \int \left(\frac{8x}{x^2 + 2} + \frac{-3x}{(x^2 + 2)^2} \right) dx$$

$$= 4\ln(x^2 + 2) + \frac{3}{2(x^2 + 2)} + C.$$

▷ **TECHNOLOGY**   You can use a graphing utility to confirm the decomposition found in Example 4. To do this, graph

$$y_1 = \frac{8x^3 + 13x}{(x^2 + 2)^2}$$

and

$$y_2 = \frac{8x}{x^2 + 2} + \frac{-3x}{(x^2 + 2)^2}$$

in the same viewing window. The graphs should be identical, as shown at the right.

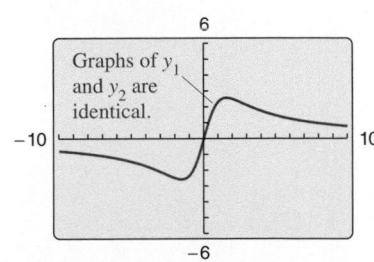

Graphs of y_1 and y_2 are identical.

When integrating rational expressions, keep in mind that for *improper* rational expressions such as

$$\frac{N(x)}{D(x)} = \frac{2x^3 + x^2 - 7x + 7}{x^2 + x - 2}$$

you must first divide to obtain

$$\frac{N(x)}{D(x)} = 2x - 1 + \frac{-2x + 5}{x^2 + x - 2}.$$

The proper rational expression is then decomposed into its partial fractions by the usual methods.

Here are some guidelines for solving the basic equation that is obtained in a partial fraction decomposition.

GUIDELINES FOR SOLVING THE BASIC EQUATION

Linear Factors

1. Substitute the roots of the distinct linear factors in the basic equation.

2. For repeated linear factors, use the coefficients determined in the first guideline to rewrite the basic equation. Then substitute other convenient values of x and solve for the remaining coefficients.

Quadratic Factors

1. Expand the basic equation.

2. Collect terms according to powers of x.

3. Equate the coefficients of like powers to obtain a system of linear equations involving A, B, C, and so on.

4. Solve the system of linear equations.

■ FOR FURTHER INFORMATION
To read about another method of evaluating integrals of rational functions, see the article "Alternate Approach to Partial Fractions to Evaluate Integrals of Rational Functions" by N. R. Nandakumar and Michael J. Bossé in *The Pi Mu Epsilon Journal*. To view this article, go to *MathArticles.com*.

Before concluding this section, here are a few things you should remember. First, it is not necessary to use the partial fractions technique on all rational functions. For instance, the following integral is found more easily by the Log Rule.

$$\int \frac{x^2 + 1}{x^3 + 3x - 4}\,dx = \frac{1}{3}\int \frac{3x^2 + 3}{x^3 + 3x - 4}\,dx$$

$$= \frac{1}{3}\ln|x^3 + 3x - 4| + C$$

Second, when the integrand is not in reduced form, reducing it may eliminate the need for partial fractions, as shown in the following integral.

$$\int \frac{x^2 - x - 2}{x^3 - 2x - 4}\,dx = \int \frac{(x + 1)(x - 2)}{(x - 2)(x^2 + 2x + 2)}\,dx$$

$$= \int \frac{x + 1}{x^2 + 2x + 2}\,dx$$

$$= \frac{1}{2}\ln|x^2 + 2x + 2| + C$$

Finally, partial fractions can be used with some quotients involving transcendental functions. For instance, the substitution $u = \sin x$ allows you to write

$$\int \frac{\cos x}{(\sin x)(\sin x - 1)}\,dx = \int \frac{du}{u(u - 1)}. \qquad u = \sin x,\ du = \cos x\,dx$$

8.5 Exercises

See **CalcChat.com** for tutorial help and worked-out solutions to odd-numbered exercises.

CONCEPT CHECK

1. Partial Fraction Decomposition Write the form of the partial fraction decomposition of each rational expression. Do not solve for the constants.

(a) $\dfrac{4}{x^2 - 8x}$

(h) $\dfrac{2x^2 + 1}{(x - 3)^3}$

(c) $\dfrac{2x - 3}{x^3 + 10x}$

(d) $\dfrac{2x - 1}{x(x^2 + 1)^2}$

2. Guidelines for Solving the Basic Equation In your own words, explain how to solve a basic equation obtained in a partial fraction decomposition that involves quadratic factors.

 **Using Partial Fractions** In Exercises 3–20, use partial fractions to find the indefinite integral.

3. $\displaystyle\int \frac{1}{x^2 - 9}\,dx$

4. $\displaystyle\int \frac{2}{9x^2 - 1}\,dx$

5. $\displaystyle\int \frac{5}{x^2 + 3x - 4}\,dx$

6. $\displaystyle\int \frac{3 - x}{3x^2 - 2x - 1}\,dx$

7. $\displaystyle\int \frac{x^2 + 12x + 12}{x^3 - 4x}\,dx$

8. $\displaystyle\int \frac{x^3 - x + 3}{x^2 + x - 2}\,dx$

9. $\displaystyle\int \frac{2x^3 - 4x^2 - 15x + 5}{x^2 - 2x - 8}\,dx$

10. $\displaystyle\int \frac{x + 2}{x^2 + 5x}\,dx$

11. $\displaystyle\int \frac{4x^2 + 2x - 1}{x^3 + x^2}\,dx$

12. $\displaystyle\int \frac{5x - 2}{(x - 2)^2}\,dx$

13. $\displaystyle\int \frac{x^2 - 6x + 2}{x^3 + 2x^2 + x}\,dx$

14. $\displaystyle\int \frac{8x}{x^3 + x^2 - x - 1}\,dx$

15. $\displaystyle\int \frac{9 - x^2}{7x^3 + x}\,dx$

16. $\displaystyle\int \frac{6x}{x^3 - 8}\,dx$

17. $\displaystyle\int \frac{x^2}{x^4 - 2x^2 - 8}\,dx$

18. $\displaystyle\int \frac{x}{16x^4 - 1}\,dx$

19. $\displaystyle\int \frac{x^2 + 5}{x^3 - x^2 + x + 3}\,dx$

20. $\displaystyle\int \frac{x^2 + 6x + 4}{x^4 + 8x^2 + 16}\,dx$

 **Evaluating a Definite Integral** In Exercises 21–24, use partial fractions to evaluate the definite integral. Use a graphing utility to verify your result.

21. $\displaystyle\int_0^2 \frac{3}{4x^2 + 5x + 1}\,dx$

22. $\displaystyle\int_1^5 \frac{x - 1}{x^2(x + 1)}\,dx$

23. $\displaystyle\int_1^2 \frac{x + 1}{x(x^2 + 1)}\,dx$

24. $\displaystyle\int_0^1 \frac{x^2 - x}{x^2 + x + 1}\,dx$

Finding an Indefinite Integral In Exercises 25–32, use substitution and partial fractions to find the indefinite integral.

25. $\displaystyle\int \frac{\sin x}{\cos x + \cos^2 x}\,dx$

26. $\displaystyle\int \frac{5\cos x}{\sin^2 x + 3\sin x - 4}\,dx$

27. $\displaystyle\int \frac{\sec^2 x}{\tan^2 x + 5\tan x + 6}\,dx$

28. $\displaystyle\int \frac{\sec^2 x}{(\tan x)(\tan x + 1)}\,dx$

29. $\displaystyle\int \frac{e^x}{(e^x - 1)(e^x + 4)}\,dx$

30. $\displaystyle\int \frac{e^x}{(e^{2x} + 1)(e^x - 1)}\,dx$

31. $\displaystyle\int \frac{\sqrt{x}}{x - 4}\,dx$

32. $\displaystyle\int \frac{1}{x(\sqrt{3} - \sqrt{x})}\,dx$

Verifying a Formula In Exercises 33–36, use the method of partial fractions to verify the integration formula.

33. $\displaystyle\int \frac{1}{x(a + bx)}\,dx = \frac{1}{a}\ln\left|\frac{x}{a + bx}\right| + C$

34. $\displaystyle\int \frac{1}{a^2 - x^2}\,dx = \frac{1}{2a}\ln\left|\frac{a + x}{a - x}\right| + C$

35. $\displaystyle\int \frac{x}{(a + bx)^2}\,dx = \frac{1}{b^2}\left(\frac{a}{a + bx} + \ln|a + bx|\right) + C$

36. $\displaystyle\int \frac{1}{x^2(a + bx)}\,dx = -\frac{1}{ax} - \frac{b}{a^2}\ln\left|\frac{x}{a + bx}\right| + C$

EXPLORING CONCEPTS

Choosing a Method In Exercises 37–39, state the method of integration you would use to find each integral. Explain why you chose that method. Do not integrate.

37. $\displaystyle\int \frac{x + 1}{x^2 + 2x - 8}\,dx$

38. $\displaystyle\int \frac{7x + 4}{x^2 + 2x - 8}\,dx$

39. $\displaystyle\int \frac{4}{x^2 + 2x + 5}\,dx$

 **40. HOW DO YOU SEE IT?** Use the graph of f' shown in the figure to answer the following.

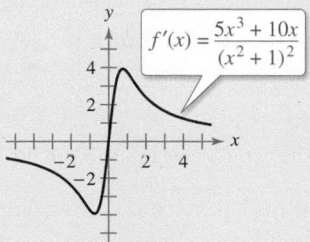

$f'(x) = \dfrac{5x^3 + 10x}{(x^2 + 1)^2}$

(a) Is $f(3) - f(2) > 0$? Explain.

(b) Which is greater, the area under the graph of f' from 1 to 2 or the area under the graph of f' from 3 to 4?

Area In Exercises 41–44, use partial fractions to find the area of the given region.

41. $y = \dfrac{12}{x^2 + 5x + 6}$

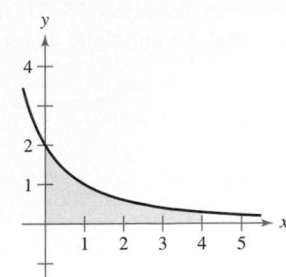

42. $y = \dfrac{15}{x^2 + 7x + 12}$

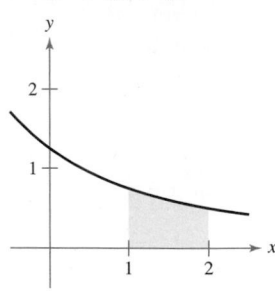
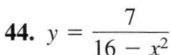

43. $y = \dfrac{15}{9 - x^2}$

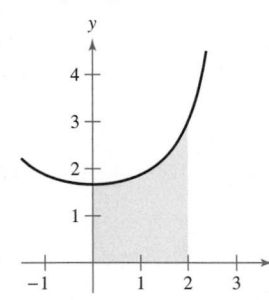

44. $y = \dfrac{7}{16 - x^2}$

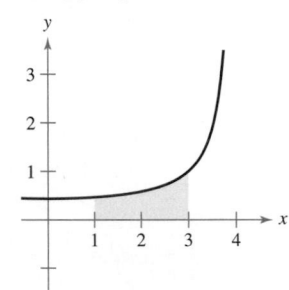

45. Modeling Data The predicted cost C (in hundreds of thousands of dollars) for a company to remove $p\%$ of a chemical from its waste water is shown in the table.

P	0	10	20	30	40
C	0	0.7	1.0	1.3	1.7

P	50	60	70	80	90
C	2.0	2.7	3.6	5.5	11.2

A model for the data is given by

$$C = \dfrac{124p}{(10 + p)(100 - p)}$$

for $0 \le p < 100$. Use the model to find the average cost of removing between 75% and 80% of the chemical.

46. Average Value of a Function Find the average value of

$$f(x) = \dfrac{1}{4x^2 - 1}$$

from $x = 1$ to $x = 4$.

47. Volume and Centroid Consider the region bounded by the graphs of

$$y = \dfrac{2x}{x^2 + 1}, \quad y = 0, \quad x = 0, \quad \text{and} \quad x = 3.$$

(a) Find the volume of the solid generated by revolving the region about the x-axis.

(b) Find the centroid of the region.

48. Volume Consider the region bounded by the graph of

$$y^2 = \dfrac{(2 - x)^2}{(1 + x)^2}$$

on the interval $[0, 1]$. Find the volume of the solid generated by revolving this region about the x-axis.

49. Epidemic Model A single infected individual enters a community of n susceptible individuals. Let x be the number of newly infected individuals at time t. The common epidemic model assumes that the disease spreads at a rate proportional to the product of the total number infected and the number not yet infected. So, $dx/dt = k(x + 1)(n - x)$ and you obtain

$$\int \dfrac{1}{(x + 1)(n - x)} \, dx = \int k \, dt.$$

Solve for x as a function of t.

• • 50. Chemical Reaction • • • • • • • • • • • • • •

In a chemical reaction, one unit of compound Y and one unit of compound Z are converted into a single unit of compound X. Let x be the amount of compound X formed. The rate of formation of X is proportional to the product of the amounts of unconverted compounds Y and Z. So, $dx/dt = k(y_0 - x)(z_0 - x)$, where y_0 and z_0 are the initial amounts of compounds Y and Z. From this equation, you obtain

$$\int \dfrac{1}{(y_0 - x)(z_0 - x)} \, dx = \int k \, dt.$$

(a) Solve for x as a function of t.

(b) Use the result of part (a) to find x as $t \to \infty$ for (1) $y_0 < z_0$, (2) $y_0 > z_0$, and (3) $y_0 = z_0$.

51. Using Two Methods Evaluate

$$\int_0^1 \dfrac{x}{1 + x^4} \, dx$$

in two different ways, one of which is partial fractions.

PUTNAM EXAM CHALLENGE

52. Prove $\dfrac{22}{7} - \pi = \displaystyle\int_0^1 \dfrac{x^4(1 - x)^4}{1 + x^2} \, dx$.

53. Let $p(x)$ be a nonzero polynomial of degree less than 1992 having no nonconstant factor in common with $x^3 - x$. Let

$$\dfrac{d^{1992}}{dx^{1992}} \left(\dfrac{p(x)}{x^3 - x} \right) = \dfrac{f(x)}{g(x)}$$

for polynomials $f(x)$ and $g(x)$. Find the smallest possible degree of $f(x)$.

These problems were composed by the Committee on the Putnam Prize Competition. © The Mathematical Association of America. All rights reserved.

8.6 Numerical Integration

■ Approximate a definite integral using the Trapezoidal Rule.
■ Approximate a definite integral using Simpson's Rule.
■ Analyze the approximate errors in the Trapezoidal Rule and Simpson's Rule.

The Trapezoidal Rule

Some elementary functions simply do not have antiderivatives that are elementary functions. For example, there is no elementary function that has any of the following functions as its derivative.

$$\sqrt[3]{x}\sqrt{1-x}, \qquad \sqrt{x}\cos x, \qquad \frac{\cos x}{x}, \qquad \sqrt{1-x^3}, \qquad \sin x^2$$

If you need to evaluate a definite integral involving a function whose antiderivative cannot be found, then while the Fundamental Theorem of Calculus is still true, it cannot be easily applied. In this case, it is easier to resort to an approximation technique. Two such techniques are described in this section.

One way to approximate a definite integral is to use n trapezoids, as shown in Figure 8.14. In the development of this method, assume that f is continuous and positive on the interval $[a, b]$. So, the definite integral

$$\int_a^b f(x)\, dx$$

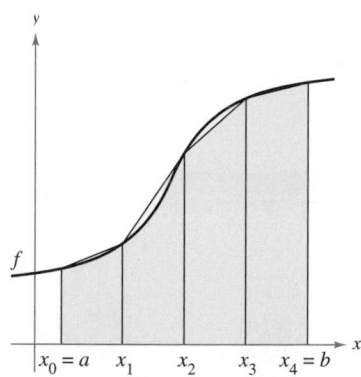

The area of the region can be approximated using four trapezoids.
Figure 8.14

represents the area of the region bounded by the graph of f and the x-axis, from $x = a$ to $x = b$. First, partition the interval $[a, b]$ into n subintervals, each of width $\Delta x = (b - a)/n$, such that

$$a = x_0 < x_1 < x_2 < \cdots < x_n = b.$$

Then form a trapezoid for each subinterval (see Figure 8.15). The area of the ith trapezoid is

$$\text{Area of } i\text{th trapezoid} = \left[\frac{f(x_{i-1}) + f(x_i)}{2}\right]\left(\frac{b-a}{n}\right).$$

This implies that the sum of the areas of the n trapezoids is

$$\text{Area} = \left(\frac{b-a}{n}\right)\left[\frac{f(x_0) + f(x_1)}{2} + \cdots + \frac{f(x_{n-1}) + f(x_n)}{2}\right]$$

$$= \left(\frac{b-a}{2n}\right)[f(x_0) + f(x_1) + f(x_1) + f(x_2) + \cdots + f(x_{n-1}) + f(x_n)]$$

$$= \left(\frac{b-a}{2n}\right)[f(x_0) + 2f(x_1) + 2f(x_2) + \cdots + 2f(x_{n-1}) + f(x_n)].$$

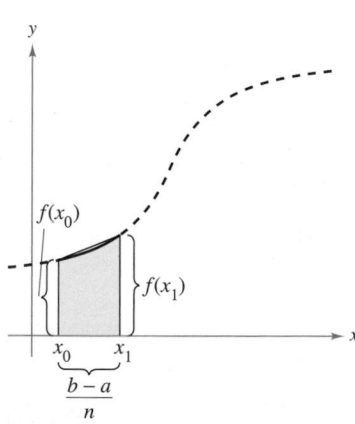

The area of the first trapezoid is
$$\left[\frac{f(x_0) + f(x_1)}{2}\right]\left(\frac{b-a}{n}\right).$$
Figure 8.15

Letting $\Delta x = (b - a)/n$, you can take the limit as $n \to \infty$ to obtain

$$\lim_{n\to\infty}\left(\frac{b-a}{2n}\right)[f(x_0) + 2f(x_1) + \cdots + 2f(x_{n-1}) + f(x_n)]$$

$$= \lim_{n\to\infty}\left[\frac{[f(a) - f(b)]\Delta x}{2} + \sum_{i=1}^{n} f(x_i)\, \Delta x\right]$$

$$= \lim_{n\to\infty}\frac{[f(a) - f(b)](b - a)}{2n} + \lim_{n\to\infty}\sum_{i=1}^{n} f(x_i)\, \Delta x$$

$$= 0 + \int_a^b f(x)\, dx.$$

The result is summarized in the next theorem.

> **THEOREM 8.3 The Trapezoidal Rule**
>
> Let f be continuous on $[a, b]$. The Trapezoidal Rule for approximating $\int_a^b f(x)\, dx$ is
>
> $$\int_a^b f(x)\, dx \approx \frac{b-a}{2n}[f(x_0) + 2f(x_1) + 2f(x_2) + \cdots + 2f(x_{n-1}) + f(x_n)].$$
>
> Moreover, as $n \to \infty$, the right-hand side approaches $\int_a^b f(x)\, dx$.

REMARK Observe that the coefficients in the Trapezoidal Rule have the following pattern.

$$1 \quad 2 \quad 2 \quad 2 \quad \cdots \quad 2 \quad 2 \quad 1$$

EXAMPLE 1 **Approximation with the Trapezoidal Rule**

Use the Trapezoidal Rule to approximate

$$\int_0^\pi \sin x\, dx.$$

Compare the results for $n = 4$ and $n = 8$, as shown in Figure 8.16.

Solution When $n = 4$, $\Delta x = \pi/4$, and you obtain

$$\int_0^\pi \sin x\, dx \approx \frac{\pi}{8}\left(\sin 0 + 2\sin\frac{\pi}{4} + 2\sin\frac{\pi}{2} + 2\sin\frac{3\pi}{4} + \sin\pi\right)$$

$$= \frac{\pi}{8}\left(0 + \sqrt{2} + 2 + \sqrt{2} + 0\right)$$

$$= \frac{\pi\left(1 + \sqrt{2}\right)}{4}$$

$$\approx 1.896.$$

When $n = 8$, $\Delta x = \pi/8$, and you obtain

$$\int_0^\pi \sin x\, dx \approx \frac{\pi}{16}\left(\sin 0 + 2\sin\frac{\pi}{8} + 2\sin\frac{\pi}{4} + 2\sin\frac{3\pi}{8} + 2\sin\frac{\pi}{2}\right.$$

$$\left. + 2\sin\frac{5\pi}{8} + 2\sin\frac{3\pi}{4} + 2\sin\frac{7\pi}{8} + \sin\pi\right)$$

$$= \frac{\pi}{16}\left(2 + 2\sqrt{2} + 4\sin\frac{\pi}{8} + 4\sin\frac{3\pi}{8}\right)$$

$$\approx 1.974.$$

For this particular integral, you could have found an antiderivative and determined that the exact area of the region is 2.

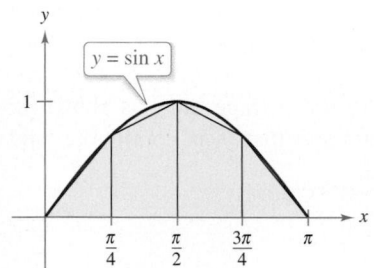

Four subintervals

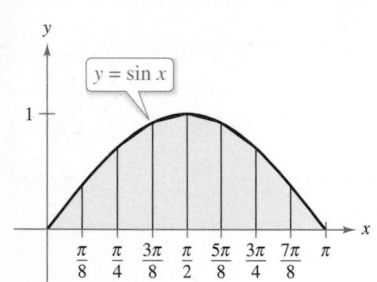

Eight subintervals

Trapezoidal approximations
Figure 8.16

▷ **TECHNOLOGY** Most graphing utilities have a *numerical integration* feature that can be used to approximate the value of a definite integral. Use this feature to approximate the integral in Example 1. How close is your approximation? When you use this feature, you need to be aware of its limitations. Often, you are given no indication of the degree of accuracy of the approximation. Other times, you may be given a result that is incorrect. For instance, use a graphing utility to evaluate

$$\int_{-1}^2 \frac{1}{x}\, dx.$$

Your graphing utility should give an error message. Does yours?

It is interesting to compare the Trapezoidal Rule with the Midpoint Rule given in Section 5.2. For the Trapezoidal Rule, you average the function values at the endpoints of the subintervals, but for the Midpoint Rule, you take the function values of the subinterval midpoints.

$$\int_a^b f(x)\,dx \approx \sum_{i=1}^{n} f\left(\frac{x_{i-1} + x_i}{2}\right) \Delta x \qquad \text{Midpoint Rule}$$

$$\int_a^b f(x)\,dx \approx \sum_{i=1}^{n} \left(\frac{f(x_{i-1}) + f(x_i)}{2}\right) \Delta x \qquad \text{Trapezoidal Rule}$$

There are two important points that should be made concerning the Trapezoidal Rule (or the Midpoint Rule). First, the approximation tends to become more accurate as n increases. For instance, in Example 1, when $n = 16$, the Trapezoidal Rule yields an approximation of 1.994. Second, although you could have used the Fundamental Theorem to evaluate the integral in Example 1, this theorem cannot be used to evaluate an integral as simple as $\int_0^{\pi} \sin x^2\,dx$ because $\sin x^2$ has no elementary antiderivative. Yet the Trapezoidal Rule can be applied to estimate this integral.

Simpson's Rule

One way to view the trapezoidal approximation of a definite integral is to say that on each subinterval, you approximate f by a *first*-degree polynomial. In Simpson's Rule, named after the English mathematician Thomas Simpson (1710–1761), you take this procedure one step further and approximate f by *second*-degree polynomials.

Before presenting Simpson's Rule, consider the next theorem for evaluating integrals of polynomials of degree 2 (or less).

THEOREM 8.4   Integral of $p(x) = Ax^2 + Bx + C$

If $p(x) = Ax^2 + Bx + C$, then

$$\int_a^b p(x)\,dx = \left(\frac{b - a}{6}\right)\left[p(a) + 4p\left(\frac{a + b}{2}\right) + p(b)\right].$$

Proof

$$\int_a^b p(x)\,dx = \int_a^b (Ax^2 + Bx + C)\,dx$$

$$= \left[\frac{Ax^3}{3} + \frac{Bx^2}{2} + Cx\right]_a^b$$

$$= \frac{A(b^3 - a^3)}{3} + \frac{B(b^2 - a^2)}{2} + C(b - a)$$

$$= \left(\frac{b - a}{6}\right)\left[2A(a^2 + ab + b^2) + 3B(b + a) + 6C\right]$$

By expansion and collection of terms, the expression inside the brackets becomes

$$\underbrace{(Aa^2 + Ba + C)}_{p(a)} + \underbrace{4\left[A\left(\frac{b + a}{2}\right)^2 + B\left(\frac{b + a}{2}\right) + C\right]}_{4p\left(\frac{a + b}{2}\right)} + \underbrace{(Ab^2 + Bb + C)}_{p(b)}$$

and you can write

$$\int_a^b p(x)\,dx = \left(\frac{b - a}{6}\right)\left[p(a) + 4p\left(\frac{a + b}{2}\right) + p(b)\right].$$

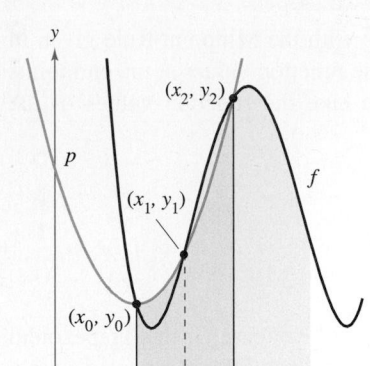

$$\int_{x_0}^{x_2} p(x)\, dx \approx \int_{x_0}^{x_2} f(x)\, dx$$

Figure 8.17

To develop Simpson's Rule for approximating a definite integral, you again partition the interval $[a, b]$ into n subintervals, each of width $\Delta x = (b - a)/n$. This time, however, n is required to be even, and the subintervals are grouped in pairs such that

$$a = x_0 < x_1 < \underbrace{x_2 < x_3 < x_4}_{} < \cdots < \underbrace{x_{n-2} < x_{n-1} < x_n}_{} = b.$$
$$\underbrace{\phantom{x_0 < x_1}}_{[x_0,\, x_2]} \quad \underbrace{\phantom{x_2 < x_4}}_{[x_2,\, x_4]} \qquad \underbrace{\phantom{x_{n-2} < x_n}}_{[x_{n-2},\, x_n]}$$

On each (double) subinterval $[x_{i-2}, x_i]$, you can approximate f by a polynomial p of degree less than or equal to 2. (See Exercise 47.) For example, on the subinterval $[x_0, x_2]$, choose the polynomial of least degree passing through the points (x_0, y_0), (x_1, y_1), and (x_2, y_2), as shown in Figure 8.17. Now, using p as an approximation of f on this subinterval, you have, by Theorem 8.4,

$$\int_{x_0}^{x_2} f(x)\, dx \approx \int_{x_0}^{x_2} p(x)\, dx$$

$$= \frac{x_2 - x_0}{6}\left[p(x_0) + 4p\left(\frac{x_0 + x_2}{2}\right) + p(x_2) \right]$$

$$= \frac{2[(b - a)/n]}{6}\left[p(x_0) + 4p(x_1) + p(x_2) \right]$$

$$= \frac{b - a}{3n}\left[f(x_0) + 4f(x_1) + f(x_2) \right].$$

Repeating this procedure on the entire interval $[a, b]$ produces the next theorem.

THEOREM 8.5 Simpson's Rule

Let f be continuous on $[a, b]$ and let n be an even integer. Simpson's Rule for approximating $\int_a^b f(x)\, dx$ is

$$\int_a^b f(x)\, dx \approx \frac{b - a}{3n}\left[f(x_0) + 4f(x_1) + 2f(x_2) + 4f(x_3) + \cdots \right.$$
$$\left. + 2f(x_{n-2}) + 4f(x_{n-1}) + f(x_n) \right].$$

Moreover, as $n \to \infty$, the right-hand side approaches $\int_a^b f(x)\, dx$.

In Example 1, the Trapezoidal Rule was used to estimate $\int_0^\pi \sin x\, dx$. In the next example, Simpson's Rule is applied to the same integral.

EXAMPLE 2   **Approximation with Simpson's Rule**

: • • • ▷ *See LarsonCalculus.com for an interactive version of this type of example.*

Use Simpson's Rule to approximate

$$\int_0^\pi \sin x\, dx.$$

Compare the results for $n = 4$ and $n = 8$.

Solution When $n = 4$, you have

$$\int_0^\pi \sin x\, dx \approx \frac{\pi}{12}\left(\sin 0 + 4\sin\frac{\pi}{4} + 2\sin\frac{\pi}{2} + 4\sin\frac{3\pi}{4} + \sin\pi \right) \approx 2.005.$$

When $n = 8$, you have $\displaystyle\int_0^\pi \sin x\, dx \approx 2.0003.$

■ **FOR FURTHER INFORMATION** For proofs of the formulas used for estimating the errors involved in the use of the Midpoint Rule and Simpson's Rule, see the article "Elementary Proofs of Error Estimates for the Midpoint and Simpson's Rules" by Edward C. Fazekas, Jr. and Peter R. Mercer in *Mathematics Magazine.* To view this article, go to *MathArticles.com.*

Error Analysis

When you use an approximation technique, it is important to know how accurate you can expect the approximation to be. The next theorem, which is listed without proof, gives the formulas for estimating the errors involved in the use of Simpson's Rule and the Trapezoidal Rule. In general, when using an approximation, you can think of the error E as the difference between $\int_a^b f(x)\,dx$ and the approximation.

THEOREM 8.6 **Errors in the Trapezoidal Rule and Simpson's Rule**

If f has a continuous second derivative on $[a, b]$, then the error E in approximating $\int_a^b f(x)\,dx$ by the Trapezoidal Rule is

$$|E| \le \frac{(b - a)^3}{12n^2}[\max |f''(x)|], \quad a \le x \le b. \qquad \text{Trapezoidal Rule}$$

Moreover, if f has a continuous fourth derivative on $[a, b]$, then the error E in approximating $\int_a^b f(x)\,dx$ by Simpson's Rule is

$$|E| \le \frac{(b - a)^5}{180n^4}[\max |f^{(4)}(x)|], \quad a \le x \le b. \qquad \text{Simpson's Rule}$$

Theorem 8.6 states that the errors generated by the Trapezoidal Rule and Simpson's Rule have upper bounds dependent on the extreme values of $f''(x)$ and $f^{(4)}(x)$ in the interval $[a, b]$. Furthermore, these errors can be made arbitrarily small by *increasing n*, provided that f'' and $f^{(4)}$ are continuous and therefore bounded in $[a, b]$.

▷ **TECHNOLOGY** If you have access to a computer algebra system, use it to evaluate the definite integral in Example 3. You should obtain a value of

$$\int_0^1 \sqrt{1 + x^2}\,dx$$

$$= \frac{1}{2}\left[\sqrt{2} + \ln\left(1 + \sqrt{2}\right)\right]$$

$$\approx 1.14779.$$

EXAMPLE 3 **The Approximate Error in the Trapezoidal Rule**

Determine a value of n such that the Trapezoidal Rule will approximate the value of

$$\int_0^1 \sqrt{1 + x^2}\,dx$$

with an error that is less than or equal to 0.01.

Solution Begin by letting $f(x) = \sqrt{1 + x^2}$ and finding the second derivative of f.

$$f'(x) = x(1 + x^2)^{-1/2} \quad \text{and} \quad f''(x) = (1 + x^2)^{-3/2}$$

The maximum value of $|f''(x)|$ on the interval $[0, 1]$ is $|f''(0)| = 1$. So, by Theorem 8.6, you can write

$$|E| \le \frac{(b - a)^3}{12n^2}|f''(0)| = \frac{1}{12n^2}(1) = \frac{1}{12n^2}.$$

To obtain an error E that is less than or equal to 0.01, you must choose n such that $1/(12n^2) \le 1/100$.

$$100 \le 12n^2 \quad \Longrightarrow \quad n \ge \sqrt{\tfrac{100}{12}} \approx 2.89$$

So, you can choose $n = 3$ (because n must be greater than or equal to 2.89) and apply the Trapezoidal Rule, as shown in Figure 8.18, to obtain

$$\int_0^1 \sqrt{1 + x^2}\,dx \approx \frac{1}{6}\left[\sqrt{1 + 0^2} + 2\sqrt{1 + \left(\tfrac{1}{3}\right)^2} + 2\sqrt{1 + \left(\tfrac{2}{3}\right)^2} + \sqrt{1 + 1^2}\right]$$

$$\approx 1.154.$$

So, by adding and subtracting the error from this estimate, you know that

$$1.144 \le \int_0^1 \sqrt{1 + x^2}\,dx \le 1.164.$$

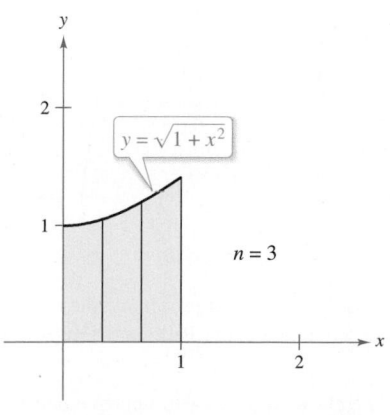

$$1.144 \le \int_0^1 \sqrt{1 + x^2}\,dx \le 1.164$$

Figure 8.18

8.6 Exercises

See **CalcChat.com** for tutorial help and worked-out solutions to odd-numbered exercises.

CONCEPT CHECK

1. Finding an Interval Would you use numerical integration to evaluate $\int_0^2 (e^x + 5x)\, dx$? Explain.

2. Errors in the Trapezoidal Rule and Simpson's Rule Describe how to decrease the error between an approximation and the exact value of an integral using the Trapezoidal Rule and Simpson's Rule.

Using the Trapezoidal Rule and Simpson's Rule In Exercises 3–14, use the Trapezoidal Rule and Simpson's Rule to approximate the value of the definite integral for the given value of n. Round your answer to four decimal places and compare the results with the exact value of the definite integral.

3. $\int_0^2 x^2\, dx, \quad n = 4$

4. $\int_1^2 \left(\frac{x^2}{4} + 1 \right) dx, \quad n = 4$

5. $\int_3^4 \frac{1}{x - 2}\, dx, \quad n = 4$

6. $\int_2^3 \frac{2}{x^2}\, dx, \quad n = 4$

7. $\int_1^3 x^3\, dx, \quad n = 6$

8. $\int_0^8 \sqrt[3]{x}\, dx, \quad n = 8$

9. $\int_4^9 \sqrt{x}\, dx, \quad n = 8$

10. $\int_1^4 (4 - x^2)\, dx, \quad n = 6$

11. $\int_0^1 \frac{2}{(x + 2)^2}\, dx, \quad n = 4$

12. $\int_0^2 x\sqrt{x^2 + 1}\, dx, \quad n = 4$

13. $\int_0^2 xe^{-x}\, dx, \quad n = 4$

14. $\int_0^2 x \ln(x + 1)\, dx, \quad n = 4$

Using the Trapezoidal Rule and Simpson's Rule In Exercises 15–24, approximate the definite integral using the Trapezoidal Rule and Simpson's Rule with $n = 4$. Compare these results with the approximation of the integral using a graphing utility.

15. $\int_0^2 \sqrt{1 + x^3}\, dx$

16. $\int_0^1 \sqrt{x}\,\sqrt{1 - x}\, dx$

17. $\int_0^1 \frac{1}{1 + x^2}\, dx$

18. $\int_0^2 \frac{1}{\sqrt{1 + x^3}}\, dx$

19. $\int_0^4 \sqrt{x}e^x\, dx$

20. $\int_1^3 \ln x\, dx$

21. $\int_0^{\sqrt{\pi/2}} \sin x^2\, dx$

22. $\int_{\pi/2}^{\pi} \sqrt{x} \sin x\, dx$

23. $\int_0^{\pi/4} x \tan x\, dx$

24. $\int_0^{\pi} f(x)\, dx, \quad f(x) = \begin{cases} \dfrac{\sin x}{x}, & x > 0 \\ 1, & x = 0 \end{cases}$

Estimating Errors In Exercises 25–28, use the error formulas in Theorem 8.6 to estimate the errors in approximating the integral, with $n = 4$, using (a) the Trapezoidal Rule and (b) Simpson's Rule.

25. $\int_0^2 (x^2 + 2x)\, dx$

26. $\int_1^3 2x^3\, dx$

27. $\int_2^4 \frac{1}{(x - 1)^2}\, dx$

28. $\int_0^1 e^{x^3}\, dx$

Estimating Errors In Exercises 29–32, use the error formulas in Theorem 8.6 to find n such that the error in the approximation of the definite integral is less than or equal to 0.00001 using (a) the Trapezoidal Rule and (b) Simpson's Rule.

29. $\int_1^3 \frac{1}{x}\, dx$

30. $\int_0^1 \frac{1}{1 + x}\, dx$

31. $\int_0^2 \sqrt{x + 2}\, dx$

32. $\int_1^3 e^{2x}\, dx$

Estimating Errors Using Technology In Exercises 33 and 34, use a computer algebra system and the error formulas to find n such that the error in the approximation of the definite integral is less than or equal to 0.00001 using (a) the Trapezoidal Rule and (b) Simpson's Rule.

33. $\int_0^1 \tan x^2\, dx$

34. $\int_0^2 (x + 1)^{2/3}\, dx$

35. Finding the Area of a Region Approximate the area of the shaded region using the Trapezoidal Rule and Simpson's Rule with $n = 4$.

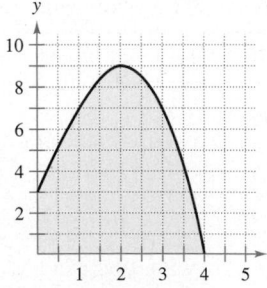

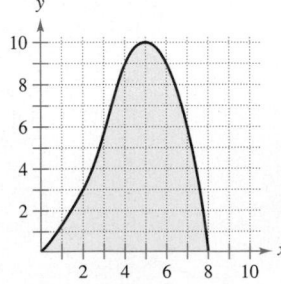

Figure for 35   Figure for 36

36. Finding the Area of a Region Approximate the area of the shaded region using the Trapezoidal Rule and Simpson's Rule with $n = 8$.

37. Area Use Simpson's Rule with $n = 14$ to approximate the area of the region bounded by the graphs of $y = \sqrt{x} \cos x$, $y = 0$, $x = 0$, and $x = \pi/2$.

38. **HOW DO YOU SEE IT?**   The function f is concave upward on the interval $[0, 2]$ and the function g is concave downward on the interval $[0, 2]$, as shown in the figure.

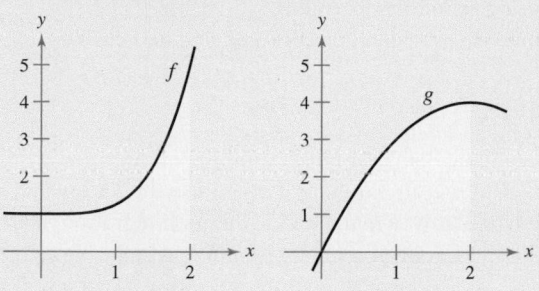

(a) Using the Trapezoidal Rule with $n = 4$, which integral would be overestimated, $\int_0^2 f(x)\, dx$ or $\int_0^2 g(x)\, dx$? Which integral would be underestimated? Explain your reasoning.

(b) Which rule would you use for more accurate approximations of $\int_0^2 f(x)\, dx$ and $\int_0^2 g(x)\, dx$, the Trapezoidal Rule or Simpson's Rule? Explain your reasoning.

EXPLORING CONCEPTS

39. Think About It   Explain how the Trapezoidal Rule is related to the approximations using left-hand and right-hand sums.

40. Describing an Error   Describe the size of the error when the Trapezoidal Rule is used to approximate $\int_a^b f(x)\, dx$ when $f(x)$ is a linear function. Use a graph to explain your answer.

41. Surveying   Use the Trapezoidal Rule to estimate the number of square meters of land, where x and y are measured in meters, as shown in the figure. The land is bounded by a stream and two straight roads that meet at right angles.

x	0	100	200	300	400	500
y	125	125	120	112	90	90

x	600	700	800	900	1000
y	95	88	75	35	0

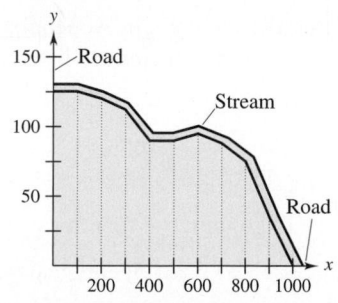

42. Circumference   The **elliptic integral**

$$8\sqrt{3} \int_0^{\pi/2} \sqrt{1 - \tfrac{2}{3} \sin^2 \theta}\; d\theta$$

gives the circumference of an ellipse. Use Simpson's Rule with $n = 8$ to approximate the circumference.

43. Work   To determine the size of the motor required to operate a press, a company must know the amount of work done when the press moves an object linearly 5 feet. The variable force to move the object is

$$F(x) = 100x\sqrt{125 - x^3}$$

where F is given in pounds and x gives the position of the unit in feet. Use Simpson's Rule with $n = 12$ to approximate the work W (in foot-pounds) done through one cycle when

$$W = \int_0^5 F(x)\, dx.$$

44. Approximating a Function   The table lists several measurements gathered in an experiment to approximate an unknown continuous function $y = f(x)$.

x	0.00	0.25	0.50	0.75	1.00
y	4.32	4.36	4.58	5.79	6.14

x	1.25	1.50	1.75	2.00
y	7.25	7.64	8.08	8.14

(a) Approximate the integral

$$\int_0^2 f(x)\, dx$$

using the Trapezoidal Rule and Simpson's Rule.

(b) Use a graphing utility to find a model of the form $y = ax^3 + bx^2 + cx + d$ for the data. Integrate the resulting polynomial over $[0, 2]$ and compare the result with the integral from part (a).

45. Using Simpson's Rule   Use Simpson's Rule with $n = 10$ and a computer algebra system to approximate t in the integral equation

$$\int_0^t \sin \sqrt{x}\, dx = 2.$$

46. Proof   Prove that Simpson's Rule is exact when approximating the integral of a cubic polynomial function, and demonstrate the result with $n = 4$ for

$$\int_0^1 x^3\, dx.$$

47. Proof   Prove that you can find a polynomial

$$p(x) = Ax^2 + Bx + C$$

that passes through any three points (x_1, y_1), (x_2, y_2), and (x_3, y_3), where the x_i's are distinct.

8.7   Integration by Tables and Other Integration Techniques

- Find an indefinite integral using a table of integrals.
- Find an indefinite integral using reduction formulas.
- Find an indefinite integral involving rational functions of sine and cosine.

Integration by Tables

So far in this chapter, you have studied several integration techniques that can be used with the basic integration rules. But merely knowing *how* to use the various techniques is not enough. You also need to know *when* to use them. Integration is first and foremost a problem of recognition. That is, you must recognize which rule or technique to apply to obtain an antiderivative. Frequently, a slight alteration of an integrand will require a different integration technique (or produce a function whose antiderivative is not an elementary function), as shown below.

$$\int x \ln x \, dx = \frac{x^2}{2} \ln x - \frac{x^2}{4} + C \qquad \text{Integration by parts}$$

$$\int \frac{\ln x}{x} \, dx = \frac{(\ln x)^2}{2} + C \qquad \text{Power Rule}$$

$$\int \frac{1}{x \ln x} \, dx = \ln|\ln x| + C \qquad \text{Log Rule}$$

$$\int \frac{x}{\ln x} \, dx = ? \qquad \text{Not an elementary function}$$

▷ **TECHNOLOGY** A computer algebra system consists, in part, of a database of integration formulas. The primary difference between using a computer algebra system and using tables of integrals is that with a computer algebra system, the computer searches through the database to find a fit. With integration tables, *you* must do the searching.

Many people find tables of integrals to be a valuable supplement to the integration techniques discussed in this chapter. Tables of common integrals can be found in Appendix B. **Integration by tables** is not a "cure-all" for all of the difficulties that can accompany integration—using tables of integrals requires considerable thought and insight and often involves substitution.

Each integration formula in Appendix B can be developed using one or more of the techniques in this chapter. You should try to verify several of the formulas. For instance, Formula 4

$$\int \frac{u}{(a + bu)^2} \, du = \frac{1}{b^2} \left(\frac{a}{a + bu} + \ln|a + bu| \right) + C \qquad \text{Formula 4}$$

can be verified using the method of partial fractions, Formula 19

$$\int \frac{\sqrt{a + bu}}{u} \, du = 2\sqrt{a + bu} + a \int \frac{du}{u\sqrt{a + bu}} \qquad \text{Formula 19}$$

can be verified using integration by parts, and Formula 84

$$\int \frac{1}{1 + e^u} \, du = u - \ln(1 + e^u) + C \qquad \text{Formula 84}$$

can be verified using substitution. Note that the integrals in Appendix B are classified according to the form of the integrand. Several of the forms are listed below.

u^n	$(a + bu)$
$(a + bu + cu^2)$	$\sqrt{a + bu}$
$(a^2 \pm u^2)$	$\sqrt{u^2 \pm a^2}$
$\sqrt{a^2 - u^2}$	Trigonometric functions
Inverse trigonometric functions	Exponential functions
Logarithmic functions	

EXAMPLE 1 Integration by Tables

Find $\displaystyle\int \frac{dx}{x\sqrt{x - 1}}$.

Solution Because the expression inside the radical is linear, you should consider forms involving $\sqrt{a + bu}$.

$$\int \frac{du}{u\sqrt{a + bu}} = \frac{2}{\sqrt{-a}}\arctan\sqrt{\frac{a + bu}{-a}} + C \qquad \text{Formula 17 } (a < 0)$$

Let $a = -1$, $b = 1$, and $u = x$. Then $du = dx$, and you can write

$$\int \frac{dx}{x\sqrt{x - 1}} = 2\arctan\sqrt{x - 1} + C.$$

EXAMPLE 2 Integration by Tables

⋯▷ *See LarsonCalculus.com for an interactive version of this type of example.*

Find $\displaystyle\int x\sqrt{x^4 - 9}\, dx$.

Solution Because the radical has the form $\sqrt{u^2 - a^2}$, you should consider Formula 26.

$$\int \sqrt{u^2 - a^2}\, du = \frac{1}{2}\left(u\sqrt{u^2 - a^2} - a^2\ln\left|u + \sqrt{u^2 - a^2}\right|\right) + C$$

Let $u = x^2$ and $a = 3$. Then $du = 2x\, dx$, and you have

$$\int x\sqrt{x^4 - 9}\, dx = \frac{1}{2}\int \sqrt{(x^2)^2 - 3^2}\,(2x)\, dx$$

$$= \frac{1}{4}\left(x^2\sqrt{x^4 - 9} - 9\ln\left|x^2 + \sqrt{x^4 - 9}\right|\right) + C.$$

EXAMPLE 3 Integration by Tables

Evaluate $\displaystyle\int_0^2 \frac{x}{1 + e^{-x^2}}\, dx$.

Solution Of the forms involving e^u, consider the formula

$$\int \frac{du}{1 + e^u} = u - \ln(1 + e^u) + C. \qquad \text{Formula 84}$$

Let $u = -x^2$. Then $du = -2x\, dx$, and you have

$$\int \frac{x}{1 + e^{-x^2}}\, dx = -\frac{1}{2}\int \frac{-2x\, dx}{1 + e^{-x^2}}$$

$$= -\frac{1}{2}\left[-x^2 - \ln\left(1 + e^{-x^2}\right)\right] + C$$

$$= \frac{1}{2}\left[x^2 + \ln\left(1 + e^{-x^2}\right)\right] + C.$$

So, the value of the definite integral is

$$\int_0^2 \frac{x}{1 + e^{-x^2}}\, dx = \frac{1}{2}\left[x^2 + \ln\left(1 + e^{-x^2}\right)\right]_0^2 = \frac{1}{2}\left[4 + \ln(1 + e^{-4}) - \ln 2\right] \approx 1.66.$$

Figure 8.19 shows the region whose area is represented by this integral. ◼

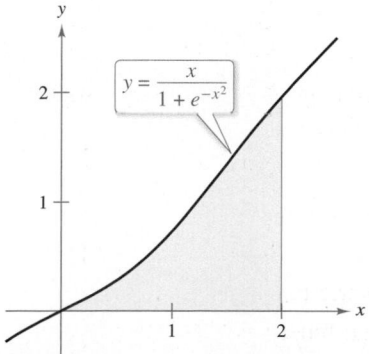

$$y = \frac{x}{1 + e^{-x^2}}$$

Figure 8.19

Reduction Formulas

Several of the integrals in the integration tables have the form

$$\int f(x)\,dx = g(x) + \int h(x)\,dx.$$

Such integration formulas are called **reduction formulas** because they reduce a given integral to the sum of a function and a simpler integral.

EXAMPLE 4 **Using a Reduction Formula**

Find $\displaystyle\int x^3 \sin x\,dx$.

Solution Consider the three formulas listed below.

$$\int u \sin u\,du = \sin u - u \cos u + C \qquad\qquad \text{Formula 52}$$

$$\int u^n \sin u\,du = -u^n \cos u + n\int u^{n-1} \cos u\,du \qquad \text{Formula 54}$$

$$\int u^n \cos u\,du = u^n \sin u - n\int u^{n-1} \sin u\,du \qquad \text{Formula 55}$$

Using Formula 54, Formula 55, and then Formula 52 produces

$$\int x^3 \sin x\,dx = -x^3 \cos x + 3\int x^2 \cos x\,dx$$

$$= -x^3 \cos x + 3\left(x^2 \sin x - 2\int x \sin x\,dx\right)$$

$$= -x^3 \cos x + 3x^2 \sin x + 6x \cos x - 6 \sin x + C.$$

EXAMPLE 5 **Using a Reduction Formula**

Find $\displaystyle\int \frac{\sqrt{3-5x}}{2x}\,dx$.

Solution Consider the two formulas listed below.

$$\int \frac{du}{u\sqrt{a+bu}} = \frac{1}{\sqrt{a}}\ln\left|\frac{\sqrt{a+bu}-\sqrt{a}}{\sqrt{a+bu}+\sqrt{a}}\right| + C \qquad \text{Formula 17 } (a > 0)$$

$$\int \frac{\sqrt{a+bu}}{u}\,du = 2\sqrt{a+bu} + a\int \frac{du}{u\sqrt{a+bu}} \qquad \text{Formula 19}$$

Using Formula 19, with $a = 3$, $b = -5$, and $u = x$, produces

$$\frac{1}{2}\int \frac{\sqrt{3-5x}}{x}\,dx = \frac{1}{2}\left(2\sqrt{3-5x} + 3\int \frac{dx}{x\sqrt{3-5x}}\right)$$

$$= \sqrt{3-5x} + \frac{3}{2}\int \frac{dx}{x\sqrt{3-5x}}.$$

Using Formula 17, with $a = 3$, $b = -5$, and $u = x$, produces

$$\int \frac{\sqrt{3-5x}}{2x}\,dx = \sqrt{3-5x} + \frac{3}{2}\left(\frac{1}{\sqrt{3}}\ln\left|\frac{\sqrt{3-5x}-\sqrt{3}}{\sqrt{3-5x}+\sqrt{3}}\right|\right) + C$$

$$= \sqrt{3-5x} + \frac{\sqrt{3}}{2}\ln\left|\frac{\sqrt{3-5x}-\sqrt{3}}{\sqrt{3-5x}+\sqrt{3}}\right| + C.$$

▷ **TECHNOLOGY** Sometimes when you use computer algebra systems, you obtain results that look very different, but are actually equivalent. Two different systems were used to find the integral in Example 5. The results are shown below.

Maple

$$\sqrt{3-5x} -$$
$$\sqrt{3}\,\mathrm{arctanh}\!\left(\tfrac{1}{3}\sqrt{3-5x}\sqrt{3}\right)$$

Mathematica

$$\sqrt{3-5x} -$$
$$\sqrt{3}\,\mathrm{ArcTanh}\left[\sqrt{1-\frac{5x}{3}}\right]$$

Notice that computer algebra systems do not include a constant of integration.

Rational Functions of Sine and Cosine

EXAMPLE 6   Integration by Tables

Find $\displaystyle\int \frac{\sin 2x}{2 + \cos x}\, dx.$

Solution   Substituting $2 \sin x \cos x$ for $\sin 2x$ produces

$$\int \frac{\sin 2x}{2 + \cos x}\, dx = 2 \int \frac{\sin x \cos x}{2 + \cos x}\, dx.$$

A check of the forms involving $\sin u$ or $\cos u$ in Appendix B shows that those listed do not apply. So, you can consider forms involving $a + bu$. For example,

$$\int \frac{u\, du}{a + bu} = \frac{1}{b^2}(bu - a \ln|a + bu|) + C. \qquad \text{Formula 3}$$

Let $a = 2$, $b = 1$, and $u = \cos x$. Then $du = -\sin x\, dx$, and you have

$$2 \int \frac{\sin x \cos x}{2 + \cos x}\, dx = -2 \int \frac{(\cos x)(-\sin x\, dx)}{2 + \cos x}$$
$$= -2(\cos x - 2 \ln|2 + \cos x|) + C$$
$$= -2 \cos x + 4 \ln|2 + \cos x| + C.$$

Example 6 involves a rational expression of $\sin x$ and $\cos x$. When you are unable to find an integral of this form in the integration tables, try using the following special substitution to convert the trigonometric expression to a standard rational expression.

Substitution for Rational Functions of Sine and Cosine

For integrals involving rational functions of sine and cosine, the substitution

$$u = \frac{\sin x}{1 + \cos x} = \tan \frac{x}{2}$$

yields

$$\cos x = \frac{1 - u^2}{1 + u^2}, \quad \sin x = \frac{2u}{1 + u^2}, \quad \text{and} \quad dx = \frac{2\, du}{1 + u^2}.$$

Proof   From the substitution for u, it follows that

$$u^2 = \frac{\sin^2 x}{(1 + \cos x)^2} = \frac{1 - \cos^2 x}{(1 + \cos x)^2} = \frac{1 - \cos x}{1 + \cos x}.$$

Solving for $\cos x$ produces

$$\cos x = \frac{1 - u^2}{1 + u^2}.$$

To find $\sin x$, write $u = (\sin x)/(1 + \cos x)$ as

$$\sin x = u(1 + \cos x) = u\left(1 + \frac{1 - u^2}{1 + u^2}\right) = \frac{2u}{1 + u^2}.$$

Finally, to find dx, consider $u = \tan(x/2)$. Then you have arctan $u = x/2$ and

$$dx = \frac{2\, du}{1 + u^2}.$$

8.7 Exercises

See **CalcChat.com** for tutorial help and worked-out solutions to odd-numbered exercises.

CONCEPT CHECK

1. Integration by Tables Which formula from the table of integrals would you use to find the integral below? Explain.

$$\int \frac{\sqrt{5 - 9x^2}}{x^2}\, dx$$

2. Reduction Formula Describe what is meant by a reduction formula. Give an example.

 **Integration by Tables** In Exercises 3 and 4, use a table of integrals with forms involving $a + bu$ to find the indefinite integral.

3. $\displaystyle\int \frac{x^2}{5 + x}\, dx$

4. $\displaystyle\int \frac{2}{x^2(4 + 3x)^2}\, dx$

Integration by Tables In Exercises 5 and 6, use a table of integrals with forms involving $\sqrt{a^2 - u^2}$ to find the indefinite integral.

5. $\displaystyle\int \frac{1}{x^2\sqrt{1 - x^2}}\, dx$

6. $\displaystyle\int \frac{\sqrt{64 - x^4}}{x}\, dx$

Integration by Tables In Exercises 7–10, use a table of integrals with forms involving the trigonometric functions to find the indefinite integral.

7. $\displaystyle\int \cos^4 3x\, dx$

8. $\displaystyle\int \frac{\sin^4 \sqrt{x}}{\sqrt{x}}\, dx$

9. $\displaystyle\int \frac{1}{\sqrt{x}\left(1 - \cos \sqrt{x}\right)}\, dx$

10. $\displaystyle\int \frac{1}{1 + \cot 4x}\, dx$

Integration by Tables In Exercises 11 and 12, use a table of integrals with forms involving e^u to find the indefinite integral.

11. $\displaystyle\int \frac{1}{1 + e^{2x}}\, dx$

12. $\displaystyle\int e^{-4x} \sin 3x\, dx$

Integration by Tables In Exercises 13 and 14, use a table of integrals with forms involving ln u to find the indefinite integral.

13. $\displaystyle\int x^6 \ln x\, dx$

14. $\displaystyle\int (\ln x)^3\, dx$

 **Using Two Methods** In Exercises 15–18, find the indefinite integral (a) using a table of integrals and (b) using the given method.

Integral	Method
15. $\displaystyle\int \ln \frac{x}{3}\, dx$	Integration by parts
16. $\displaystyle\int \sin^2 3x\, dx$	Power-reducing formula
17. $\displaystyle\int \frac{1}{x^2(x - 1)}\, dx$	Partial fractions
18. $\displaystyle\int \frac{dx}{(4 + x^2)^{3/2}}$	Trigonometric substitution

Finding an Indefinite Integral In Exercises 19–40, use a table of integrals to find the indefinite integral.

19. $\displaystyle\int x \, \text{arccsc}(x^2 + 1)\, dx$

20. $\displaystyle\int \text{arccot}(4x - 5)\, dx$

21. $\displaystyle\int \frac{2}{x^3\sqrt{x^4 - 1}}\, dx$

22. $\displaystyle\int \frac{1}{x^2 + 4x + 8}\, dx$

23. $\displaystyle\int \frac{x}{(7 - 6x)^2}\, dx$

24. $\displaystyle\int \frac{\theta^3}{1 + \sin \theta^4}\, d\theta$

25. $\displaystyle\int e^x \arccos e^x\, dx$

26. $\displaystyle\int \frac{e^x}{1 - \tan e^x}\, dx$

27. $\displaystyle\int \frac{x}{1 - \sec x^2}\, dx$

28. $\displaystyle\int \frac{1}{t[1 + (\ln t)^2]}\, dt$

29. $\displaystyle\int \frac{\cos \theta}{3 + 2 \sin \theta + \sin^2 \theta}\, d\theta$

30. $\displaystyle\int x^2\sqrt{3 + 25x^2}\, dx$

31. $\displaystyle\int \frac{1}{x^2\sqrt{2 + 9x^2}}\, dx$

32. $\displaystyle\int \sqrt{x} \arctan x^{3/2}\, dx$

33. $\displaystyle\int \frac{\ln x}{x(3 + 2 \ln x)}\, dx$

34. $\displaystyle\int \frac{e^x}{(1 - e^{2x})^{3/2}}\, dx$

35. $\displaystyle\int \frac{x}{(x^2 - 6x + 10)^2}\, dx$

36. $\displaystyle\int \sqrt{\frac{5 - x}{5 + x}}\, dx$

37. $\displaystyle\int \frac{x}{\sqrt{x^4 - 6x^2 + 5}}\, dx$

38. $\displaystyle\int \frac{\cos x}{\sqrt{\sin^2 x + 1}}\, dx$

39. $\displaystyle\int \frac{e^{3x}}{(1 + e^x)^3}\, dx$

40. $\displaystyle\int \cot^4 \theta\, d\theta$

 **Evaluating a Definite Integral** In Exercises 41–48, use a table of integrals to evaluate the definite integral.

41. $\displaystyle\int_0^1 \frac{x}{\sqrt{1 + x}}\, dx$

42. $\displaystyle\int_0^1 2x^3 e^{x^2}\, dx$

43. $\displaystyle\int_1^2 x^4 \ln x\, dx$

44. $\displaystyle\int_0^{\pi/2} x \sin 2x\, dx$

45. $\displaystyle\int_{-\pi/2}^{\pi/2} \frac{\cos x}{1 + \sin^2 x}\, dx$

46. $\displaystyle\int_0^5 \frac{x^2}{(5 + 2x)^2}\, dx$

47. $\displaystyle\int_0^{\pi/2} t^3 \cos t\, dt$

48. $\displaystyle\int_0^3 \sqrt{x^2 + 16}\, dx$

Verifying a Formula In Exercises 49–54, verify the integration formula.

49. $\displaystyle\int \frac{u^2}{(a + bu)^2}\, du = \frac{1}{b^3}\left(bu - \frac{a^2}{a + bu} - 2a \ln|a + bu|\right) + C$

50. $\displaystyle\int \frac{u^n}{\sqrt{a + bu}}\, du = \frac{2}{(2n + 1)b}\left(u^n\sqrt{a + bu} - na\int \frac{u^{n-1}}{\sqrt{a + bu}}\, du\right)$

51. $\displaystyle\int \frac{1}{(u^2 \pm a^2)^{3/2}} \, du = \frac{\pm u}{a^2 \sqrt{u^2 \pm a^2}} + C$

52. $\displaystyle\int u^n \cos u \, du = u^n \sin u - n \int u^{n-1} \sin u \, du$

53. $\displaystyle\int \arctan u \, du = u \arctan u - \ln \sqrt{1 + u^2} + C$

54. $\displaystyle\int (\ln u)^n \, du = u(\ln u)^n - n \int (\ln u)^{n-1} \, du$

Finding or Evaluating an Integral In Exercises 55–62, find or evaluate the integral.

55. $\displaystyle\int \frac{1}{2 - 3 \sin \theta} \, d\theta$

56. $\displaystyle\int \frac{\sin \theta}{1 + \cos^2 \theta} \, d\theta$

57. $\displaystyle\int_0^{\pi/2} \frac{1}{1 + \sin \theta + \cos \theta} \, d\theta$

58. $\displaystyle\int_0^{\pi/2} \frac{1}{3 - 2 \cos \theta} \, d\theta$

59. $\displaystyle\int \frac{\sin \theta}{3 - 2 \cos \theta} \, d\theta$

60. $\displaystyle\int \frac{\cos \theta}{1 + \cos \theta} \, d\theta$

61. $\displaystyle\int \frac{\sin \sqrt{\theta}}{\sqrt{\theta}} \, d\theta$

62. $\displaystyle\int \frac{4}{\csc \theta - \cot \theta} \, d\theta$

Area In Exercises 63 and 64, find the area of the region bounded by the graphs of the equations.

63. $y = \dfrac{x}{\sqrt{x + 3}}, \ y = 0, \ x = 6$

64. $y = \dfrac{x}{1 + e^{x^2}}, \ y = 0, \ x = 2$

EXPLORING CONCEPTS

65. Finding a Pattern

(a) Find $\int x^n \ln x \, dx$ for $n = 1$, 2, and 3. Describe any patterns you notice.

(b) Write a general rule for evaluating the integral in part (a) for an integer $n \geq 1$.

(c) Verify your rule from part (b) using integration by parts.

66. Choosing a Method State the method or integration formula you would use to find the antiderivative. Explain why you chose that method or formula. Do not integrate.

(a) $\displaystyle\int \frac{e^x}{e^{2x} + 1} \, dx$ (b) $\displaystyle\int \frac{e^x}{e^x + 1} \, dx$ (c) $\displaystyle\int x e^{x^2} \, dx$

(d) $\displaystyle\int x e^x \, dx$ (e) $\displaystyle\int e^{2x} \sqrt{e^{2x} + 1} \, dx$

---

67. Work A hydraulic cylinder on an industrial machine pushes a steel block a distance of x feet ($0 \leq x \leq 5$), where the variable force required is $F(x) = 2000xe^{-x}$ pounds. Find the work done in pushing the block the full 5 feet through the machine.

68. Work Repeat Exercise 67, using $F(x) = \dfrac{500x}{\sqrt{26 - x^2}}$ pounds.

69. Population A population is growing according to the logistic model

$$N = \frac{5000}{1 + e^{4.8 - 1.9t}}$$

where t is the time in days. Find the average population over the interval $[0, 2]$.

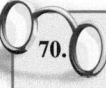

70. HOW DO YOU SEE IT? Use the graph of f' shown in the figure to answer the following.

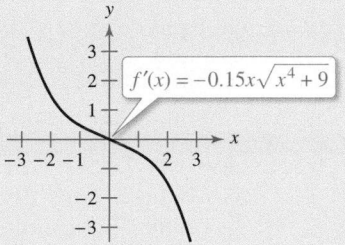

$f'(x) = -0.15x\sqrt{x^4 + 9}$

(a) Approximate the slope of f at $x = -1$. Explain.

(b) Approximate any open intervals on which the graph of f is increasing and any open intervals on which it is decreasing. Explain.

71. Volume Consider the region bounded by the graphs of

$$y = x\sqrt{16 - x^2}, \ y = 0, \ x = 0, \ \text{and} \ x = 4.$$

Find the volume of the solid generated by revolving the region about the y-axis.

72. Building Design The cross section of a precast concrete beam for a building is bounded by the graphs of the equations

$$x = \frac{2}{\sqrt{1 + y^2}}, \ x = \frac{-2}{\sqrt{1 + y^2}}, \ y = 0, \ \text{and} \ y = 3$$

where x and y are measured in feet. The length of the beam is 20 feet (see figure).

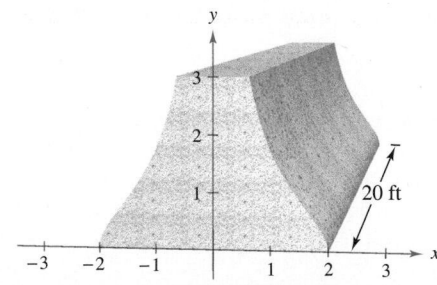

20 ft

(a) Find the volume V and the weight W of the beam. Assume the concrete weighs 148 pounds per cubic foot.

(b) Find the centroid of a cross section of the beam.

PUTNAM EXAM CHALLENGE

73. Evaluate $\displaystyle\int_0^{\pi/2} \frac{dx}{1 + (\tan x)^{\sqrt{2}}}$.

8.8 Improper Integrals

■ Evaluate an improper integral that has an infinite limit of integration.
■ Evaluate an improper integral that has an infinite discontinuity.

Improper Integrals with Infinite Limits of Integration

The definition of a definite integral

$$\int_a^b f(x)\, dx$$

requires that the interval $[a, b]$ be finite. Furthermore, the Fundamental Theorem of Calculus, by which you have been evaluating definite integrals, requires that f be continuous on $[a, b]$. In this section, you will study a procedure for evaluating integrals that do not satisfy these requirements—usually because either one or both of the limits of integration are infinite or because f has a finite number of infinite discontinuities in the interval $[a, b]$. Integrals that possess either property are **improper integrals.** Note that a function f is said to have an **infinite discontinuity** at c when, *from the right or left,*

$$\lim_{x \to c} f(x) = \infty \quad \text{or} \quad \lim_{x \to c} f(x) = -\infty.$$

To get an idea of how to evaluate an improper integral, consider the integral

$$\int_1^b \frac{dx}{x^2} = -\frac{1}{x} \Big]_1^b = -\frac{1}{b} + 1 = 1 - \frac{1}{b}$$

which can be interpreted as the area of the shaded region shown in Figure 8.20. Taking the limit as $b \to \infty$ produces

$$\int_1^\infty \frac{dx}{x^2} = \lim_{b \to \infty} \left(\int_1^b \frac{dx}{x^2} \right) = \lim_{b \to \infty} \left(1 - \frac{1}{b} \right) = 1.$$

This improper integral can be interpreted as the area of the *unbounded* region between the graph of $f(x) = 1/x^2$ and the x-axis (to the right of $x = 1$).

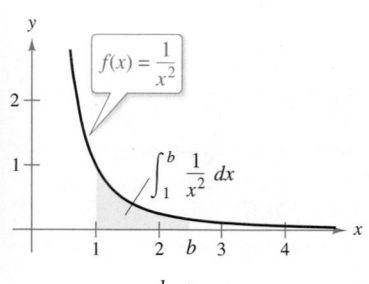

The unbounded region has an area of 1.
Figure 8.20

Definition of Improper Integrals with Infinite Integration Limits

1. If f is continuous on the interval $[a, \infty)$, then

$$\int_a^\infty f(x)\, dx = \lim_{b \to \infty} \int_a^b f(x)\, dx.$$

2. If f is continuous on the interval $(-\infty, b]$, then

$$\int_{-\infty}^b f(x)\, dx = \lim_{a \to -\infty} \int_a^b f(x)\, dx.$$

3. If f is continuous on the interval $(-\infty, \infty)$, then

$$\int_{-\infty}^\infty f(x)\, dx = \int_{-\infty}^c f(x)\, dx + \int_c^\infty f(x)\, dx$$

where c is any real number (see Exercise 107).

In the first two cases, the improper integral **converges** when the limit exists—otherwise, the improper integral **diverges.** In the third case, the improper integral on the left diverges when either of the improper integrals on the right diverges.

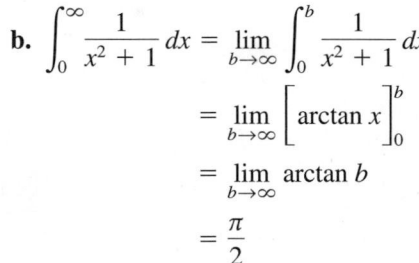

Diverges
(infinite area)

$y = \dfrac{1}{x}$

This unbounded region has an infinite area.

Figure 8.21

EXAMPLE 1 **An Improper Integral That Diverges**

Evaluate $\displaystyle\int_1^b \dfrac{dx}{x}$.

Solution

$$\int_1^\infty \frac{dx}{x} = \lim_{b \to \infty} \int_1^b \frac{dx}{x} \qquad \text{Take limit as } b \to \infty.$$

$$= \lim_{b \to \infty} \left[\ln x \right]_1^b \qquad \text{Apply Log Rule.}$$

$$= \lim_{b \to \infty} (\ln b - 0) \qquad \text{Apply Fundamental Theorem of Calculus.}$$

$$= \infty \qquad \text{Evaluate limit.}$$

The limit does not exist. So, you can conclude that the improper integral diverges. See Figure 8.21. ∎

Try comparing the regions shown in Figures 8.20 and 8.21. They look similar, yet the region in Figure 8.20 has a finite area of 1 and the region in Figure 8.21 has an infinite area.

EXAMPLE 2 **Improper Integrals That Converge**

Evaluate each improper integral.

a. $\displaystyle\int_0^\infty e^{-x}\, dx$

b. $\displaystyle\int_0^\infty \dfrac{1}{x^2 + 1}\, dx$

Solution

a. $\displaystyle\int_0^\infty e^{-x}\, dx = \lim_{b \to \infty} \int_0^b e^{-x}\, dx$

$$= \lim_{b \to \infty} \left[-e^{-x} \right]_0^b$$

$$= \lim_{b \to \infty} (-e^{-b} + 1)$$

$$= 1$$

See Figure 8.22.

b. $\displaystyle\int_0^\infty \dfrac{1}{x^2 + 1}\, dx = \lim_{b \to \infty} \int_0^b \dfrac{1}{x^2 + 1}\, dx$

$$= \lim_{b \to \infty} \left[\arctan x \right]_0^b$$

$$= \lim_{b \to \infty} \arctan b$$

$$= \frac{\pi}{2}$$

See Figure 8.23.

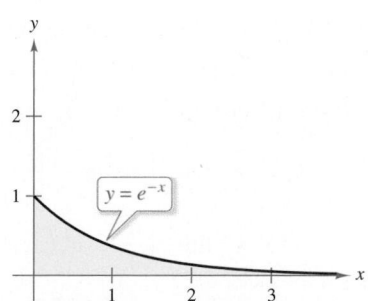

The area of the unbounded region is 1.

Figure 8.22

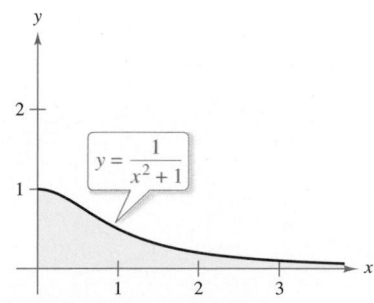

The area of the unbounded region is $\pi/2$.

Figure 8.23

In the next example, note how L'Hôpital's Rule can be used to evaluate an improper integral.

EXAMPLE 3 **Using L'Hôpital's Rule with an Improper Integral**

Evaluate $\displaystyle\int_1^\infty (1 - x)e^{-x}\, dx$.

Solution Use integration by parts, with $dv = e^{-x}\, dx$ and $u = (1 - x)$.

$$\int (1 - x)e^{-x}\, dx = -e^{-x}(1 - x) - \int e^{-x}\, dx$$
$$= -e^{-x} + xe^{-x} + e^{-x} + C$$
$$= xe^{-x} + C$$

Now, apply the definition of an improper integral.

$$\int_1^\infty (1 - x)e^{-x}\, dx = \lim_{b\to\infty} \Big[xe^{-x} \Big]_1^b$$
$$= \lim_{b\to\infty} \left(\frac{b}{e^b} - \frac{1}{e} \right)$$
$$= \lim_{b\to\infty} \frac{b}{e^b} - \lim_{b\to\infty} \frac{1}{e}$$

For the first limit, use L'Hôpital's Rule.

$$\lim_{b\to\infty} \frac{b}{e^b} = \lim_{b\to\infty} \frac{1}{e^b} = 0$$

So, you can conclude that

$$\int_1^\infty (1 - x)e^{-x}\, dx = \lim_{b\to\infty} \frac{b}{e^b} - \lim_{b\to\infty} \frac{1}{e}$$
$$= 0 - \frac{1}{e}$$
$$= -\frac{1}{e}. \qquad \text{See Figure 8.24.}$$

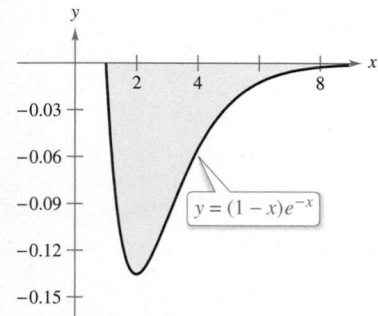

The area of the unbounded region is $|-1/e|$.
Figure 8.24

EXAMPLE 4 **Infinite Upper and Lower Limits of Integration**

Evaluate $\displaystyle\int_{-\infty}^\infty \frac{e^x}{1 + e^{2x}}\, dx$.

Solution Note that the integrand is continuous on $(-\infty, \infty)$. To evaluate the integral, you can break it into two parts, choosing $c = 0$ as a convenient value.

$$\int_{-\infty}^\infty \frac{e^x}{1 + e^{2x}}\, dx = \int_{-\infty}^0 \frac{e^x}{1 + e^{2x}}\, dx + \int_0^\infty \frac{e^x}{1 + e^{2x}}\, dx$$
$$= \lim_{a\to-\infty} \Big[\arctan e^x \Big]_a^0 + \lim_{b\to\infty} \Big[\arctan e^x \Big]_0^b$$
$$= \lim_{a\to-\infty} \left(\frac{\pi}{4} - \arctan e^a \right) + \lim_{b\to\infty} \left(\arctan e^b - \frac{\pi}{4} \right)$$
$$= \frac{\pi}{4} - 0 + \frac{\pi}{2} - \frac{\pi}{4}$$
$$= \frac{\pi}{2} \qquad \text{See Figure 8.25.}$$

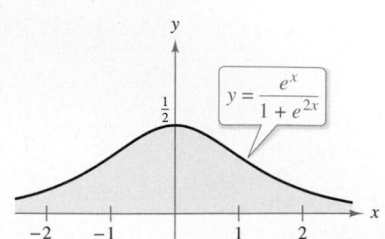

The area of the unbounded region is $\pi/2$.
Figure 8.25

The work required to move a 15-metric-ton space module an unlimited distance away from Earth is about 6.985×10^{11} foot-pounds.

EXAMPLE 5   **Sending a Space Module into Orbit**

In Example 3 in Section 7.5, you found that it would require 10,000 mile-tons of work to propel a 15-metric-ton space module to a height of 800 miles above Earth. How much work is required to propel the module an unlimited distance away from Earth's surface?

Solution   At first you might think that an infinite amount of work would be required. But if this were the case, it would be impossible to send rockets into outer space. Because this has been done, the work required must be finite. You can determine the work in the following manner. Using the integral in Example 3, Section 7.5, replace the upper bound of 4800 miles by ∞ and write

$$W = \int_{4000}^{\infty} \frac{240,000,000}{x^2}\, dx$$

$$= \lim_{b \to \infty} \left[-\frac{240,000,000}{x} \right]_{4000}^{b} \qquad \text{Integrate.}$$

$$= \lim_{b \to \infty} \left(-\frac{240,000,000}{b} + \frac{240,000,000}{4000} \right)$$

$$= 60,000 \text{ mile-tons}$$

$$= 6.985 \times 10^{11} \text{ foot-pounds.} \qquad \begin{array}{l} 1 \text{ mile} = 5280 \text{ feet;} \\ 1 \text{ metric ton} \approx 2205 \text{ pounds} \end{array}$$

In SI units, using a conversion factor of

$$1 \text{ foot-pound} \approx 1.35582 \text{ joules}$$

the work done is $W \approx 9.47 \times 10^{11}$ joules. ∎

Improper Integrals with Infinite Discontinuities

The second basic type of improper integral is one that has an infinite discontinuity *at or between* the limits of integration.

Definition of Improper Integrals with Infinite Discontinuities

1. If f is continuous on the interval $[a, b)$ and has an infinite discontinuity at b, then

$$\int_a^b f(x)\, dx = \lim_{c \to b^-} \int_a^c f(x)\, dx.$$

2. If f is continuous on the interval $(a, b]$ and has an infinite discontinuity at a, then

$$\int_a^b f(x)\, dx = \lim_{c \to a^+} \int_c^b f(x)\, dx.$$

3. If f is continuous on the interval $[a, b]$, except for some c in (a, b) at which f has an infinite discontinuity, then

$$\int_a^b f(x)\, dx = \int_a^c f(x)\, dx + \int_c^b f(x)\, dx.$$

In the first two cases, the improper integral **converges** when the limit exists— otherwise, the improper integral **diverges**. In the third case, the improper integral on the left diverges when either of the improper integrals on the right diverges.

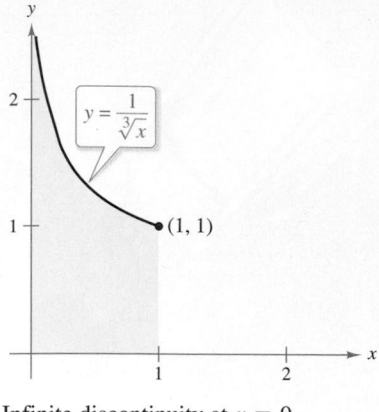

EXAMPLE 6 **An Improper Integral with an Infinite Discontinuity**

Evaluate $\int_0^1 \dfrac{dx}{\sqrt[3]{x}}$.

Solution The integrand has an infinite discontinuity at $x = 0$, as shown in the figure at the right. You can evaluate this integral as shown below.

$$\int_0^1 x^{-1/3}\, dx = \lim_{b \to 0^+} \left[\frac{x^{2/3}}{2/3} \right]_b^1$$

$$= \lim_{b \to 0^+} \frac{3}{2}(1 - b^{2/3})$$

$$= \frac{3}{2}$$

Infinite discontinuity at $x = 0$

EXAMPLE 7 **An Improper Integral That Diverges**

Evaluate $\int_0^2 \dfrac{dx}{x^3}$.

Solution Because the integrand has an infinite discontinuity at $x = 0$, you can write

$$\int_0^2 \frac{dx}{x^3} = \lim_{b \to 0^+} \left[-\frac{1}{2x^2} \right]_b^2$$

$$= \lim_{b \to 0^+} \left(-\frac{1}{8} + \frac{1}{2b^2} \right)$$

$$= \infty.$$

So, you can conclude that the improper integral diverges.

EXAMPLE 8 **An Improper Integral with an Interior Discontinuity**

Evaluate $\int_{-1}^2 \dfrac{dx}{x^3}$.

Solution This integral is improper because the integrand has an infinite discontinuity at the interior point $x = 0$, as shown in Figure 8.26. So, you can write

$$\int_{-1}^2 \frac{dx}{x^3} = \int_{-1}^0 \frac{dx}{x^3} + \int_0^2 \frac{dx}{x^3}.$$

From Example 7, you know that the second integral diverges. So, the original improper integral also diverges.

Remember to check for infinite discontinuities at interior points as well as at endpoints when determining whether an integral is improper. For instance, if you had not recognized that the integral in Example 8 was improper, you would have obtained the *incorrect* result

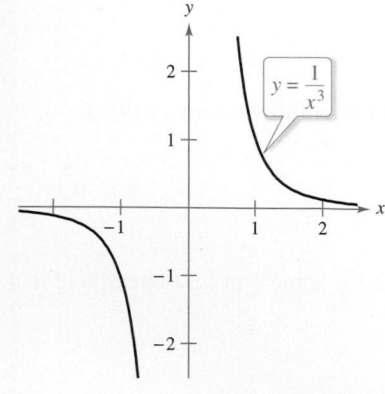

The improper integral $\int_{-1}^2 \dfrac{dx}{x^3}$ diverges.

Figure 8.26

$$\int_{-1}^2 \frac{dx}{x^3} \overset{?}{=} \left[\frac{-1}{2x^2} \right]_{-1}^2 = -\frac{1}{8} + \frac{1}{2} = \frac{3}{8}. \qquad \text{Incorrect evaluation}$$

The integral in the next example is improper for *two* reasons. One limit of integration is infinite, and the integrand has an infinite discontinuity at the other limit of integration.

EXAMPLE 9 **A Doubly Improper Integral**

:····▷ *See LarsonCalculus.com for an interactive version of this type of example.*

Evaluate $\displaystyle\int_0^\infty \frac{dx}{\sqrt{x}(x+1)}$.

Solution To evaluate this integral, split it at a convenient point (say, $x = 1$) and write

$$\int_0^\infty \frac{dx}{\sqrt{x}(x+1)} = \int_0^1 \frac{dx}{\sqrt{x}(x+1)} + \int_1^\infty \frac{dx}{\sqrt{x}(x+1)}$$

$$= \lim_{b\to0^+} \left[2\arctan\sqrt{x}\,\right]_b^1 + \lim_{c\to\infty} \left[2\arctan\sqrt{x}\,\right]_1^c$$

$$= \lim_{b\to0^+} \left(2\arctan 1 - 2\arctan\sqrt{b}\,\right) + \lim_{c\to\infty} \left(2\arctan\sqrt{c} - 2\arctan 1\right)$$

$$= 2\left(\frac{\pi}{4}\right) - 0 + 2\left(\frac{\pi}{2}\right) - 2\left(\frac{\pi}{4}\right)$$

$$= \pi.$$

See Figure 8.27.

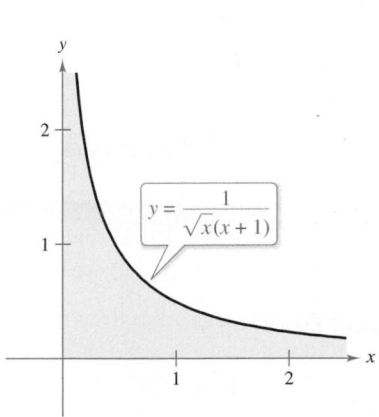

The area of the unbounded region is π.
Figure 8.27

EXAMPLE 10 **An Application Involving Arc Length**

Use the formula for arc length to show that the circumference of the circle $x^2 + y^2 = 1$ is 2π.

Solution To simplify the work, consider the quarter circle given by $y = \sqrt{1 - x^2}$, where $0 \le x \le 1$. The function y is differentiable for any x in this interval except $x = 1$. Therefore, the arc length of the quarter circle is given by the improper integral

$$s = \int_0^1 \sqrt{1 + (y')^2}\,dx$$

$$= \int_0^1 \sqrt{1 + \left(\frac{-x}{\sqrt{1-x^2}}\right)^2}\,dx$$

$$= \int_0^1 \frac{dx}{\sqrt{1-x^2}}.$$

This integral is improper because it has an infinite discontinuity at $x = 1$. So, you can write

$$s = \int_0^1 \frac{dx}{\sqrt{1-x^2}}$$

$$= \lim_{b\to1^-} \left[\arcsin x\right]_0^b$$

$$= \lim_{b\to1^-} (\arcsin b - \arcsin 0)$$

$$= \frac{\pi}{2} - 0$$

$$= \frac{\pi}{2}.$$

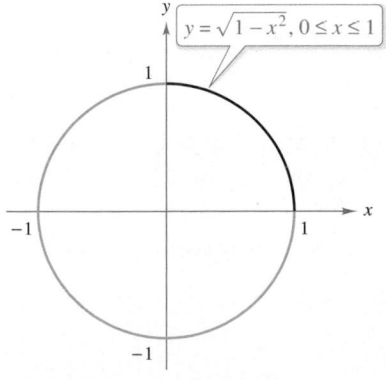

The circumference of the circle is 2π.
Figure 8.28

Finally, multiplying by 4, you can conclude that the circumference of the circle is $4s = 2\pi$, as shown in Figure 8.28.

This section concludes with a useful theorem describing the convergence or divergence of a common type of improper integral. The proof of this theorem is left as an exercise (see Exercise 49).

THEOREM 8.7   A Special Type of Improper Integral

$$\int_1^\infty \frac{dx}{x^p} = \begin{cases} \dfrac{1}{p-1}, & p > 1 \\ \text{diverges}, & p \le 1 \end{cases}$$

EXAMPLE 11   An Application Involving a Solid of Revolution

The solid formed by revolving (about the x-axis) the *unbounded* region lying between the graph of $f(x) = 1/x$ and the x-axis ($x \ge 1$) is called **Gabriel's Horn.** (See Figure 8.29.) Show that this solid has a finite volume and an infinite surface area.

Solution   Using the disk method and Theorem 8.7, you can determine the volume to be

$$V = \pi \int_1^\infty \left(\frac{1}{x}\right)^2 dx \qquad \text{Theorem 8.7, } p = 2 > 1$$

$$= \pi \left(\frac{1}{2-1}\right)$$

$$= \pi.$$

The surface area is given by

$$S = 2\pi \int_1^\infty f(x)\sqrt{1 + [f'(x)]^2}\, dx = 2\pi \int_1^\infty \frac{1}{x}\sqrt{1 + \frac{1}{x^4}}\, dx.$$

Because

$$\sqrt{1 + \frac{1}{x^4}} > 1$$

on the interval $[1, \infty)$, and the improper integral

$$\int_1^\infty \frac{1}{x}\, dx$$

diverges, you can conclude that the improper integral

$$\int_1^\infty \frac{1}{x}\sqrt{1 + \frac{1}{x^4}}\, dx$$

also diverges. (See Exercise 52.) So, the surface area is infinite.

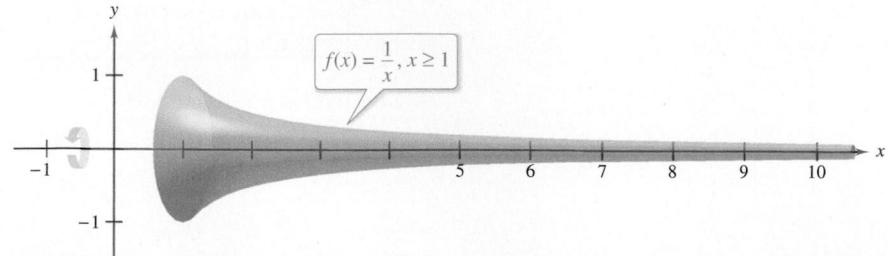

Gabriel's Horn has a finite volume and an infinite surface area.
Figure 8.29

■ **FOR FURTHER INFORMATION**
For further investigation of solids that have finite volumes and infinite surface areas, see the article "Supersolids: Solids Having Finite Volume and Infinite Surfaces" by William P. Love in *Mathematics Teacher*. To view this article, go to *MathArticles.com*.

■ **FOR FURTHER INFORMATION**
To learn about another function that has a finite volume and an infinite surface area, see the article "Gabriel's Wedding Cake" by Julian F. Fleron in *The College Mathematics Journal*. To view this article, go to *MathArticles.com*.

8.8 Exercises

See **CalcChat.com** for tutorial help and worked-out solutions to odd-numbered exercises.

CONCEPT CHECK

1. **Improper Integrals** Describe two ways for an integral to be improper.

2. **Improper Integrals** What does it mean for an improper integral to converge?

3. **Indefinite Integration Limits** Explain how to evaluate an improper integral that has an infinite limit of integration.

4. **Finding Values** For what values of a is each integral improper? Explain.

   (a) $\int_a^5 \frac{1}{x+2}\,dx$ (b) $\int_a^4 \frac{x}{3x-1}\,dx$

Determining Whether an Integral Is Improper In Exercises 5–12, decide whether the integral is improper. Explain your reasoning.

5. $\int_0^1 \frac{dx}{5x-3}$

6. $\int_1^2 \frac{dx}{x^3}$

7. $\int_0^1 \frac{2x-5}{x^2-5x+6}\,dx$

8. $\int_1^\infty \ln x^2\,dx$

9. $\int_0^2 e^{-x}\,dx$

10. $\int_0^\infty \cos x\,dx$

11. $\int_{-\infty}^\infty \frac{\sin x}{4+x^2}\,dx$

12. $\int_0^{\pi/4} \csc x\,dx$

Evaluating an Improper Integral In Exercises 13–16, explain why the integral is improper and determine whether it diverges or converges. Evaluate the integral if it converges.

13. $\int_0^4 \frac{1}{\sqrt{x}}\,dx$

14. $\int_3^4 \frac{1}{(x-3)^{3/2}}\,dx$

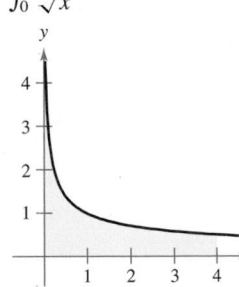

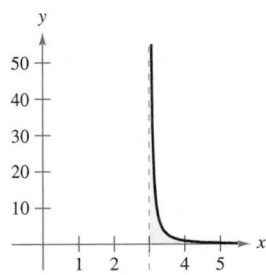

15. $\int_0^2 \frac{1}{(x-1)^2}\,dx$

16. $\int_{-\infty}^0 e^{3x}\,dx$

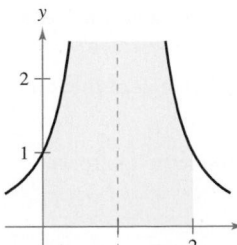

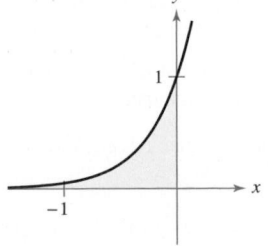

Evaluating an Improper Integral In Exercises 17–32, determine whether the improper integral diverges or converges. Evaluate the integral if it converges.

17. $\int_2^\infty \frac{1}{x^3}\,dx$

18. $\int_3^\infty \frac{1}{(x-1)^4}\,dx$

19. $\int_1^\infty \frac{3}{\sqrt[3]{x}}\,dx$

20. $\int_1^\infty \frac{4}{\sqrt[4]{x}}\,dx$

21. $\int_0^\infty e^{x/3}\,dx$

22. $\int_{-\infty}^0 xe^{-4x}\,dx$

23. $\int_0^\infty x^2 e^{-x}\,dx$

24. $\int_0^\infty e^{-x}\cos x\,dx$

25. $\int_4^\infty \frac{1}{x(\ln x)^3}\,dx$

26. $\int_1^\infty \frac{\ln x}{x}\,dx$

27. $\int_{-\infty}^\infty \frac{4}{16+x^2}\,dx$

28. $\int_0^\infty \frac{x^3}{(x^2+1)^2}\,dx$

29. $\int_0^\infty \frac{1}{e^x+e^{-x}}\,dx$

30. $\int_0^\infty \frac{e^x}{1+e^x}\,dx$

31. $\int_0^\infty \cos \pi x\,dx$

32. $\int_0^\infty \sin \frac{x}{2}\,dx$

Evaluating an Improper Integral In Exercises 33–48, determine whether the improper integral diverges or converges. Evaluate the integral if it converges, and check your results with the results obtained by using the integration capabilities of a graphing utility.

33. $\int_0^1 \frac{1}{x^2}\,dx$

34. $\int_0^5 \frac{10}{x}\,dx$

35. $\int_0^2 \frac{1}{\sqrt[3]{x-1}}\,dx$

36. $\int_0^8 \frac{3}{\sqrt{8-x}}\,dx$

37. $\int_0^e x \ln x\,dx$

38. $\int_0^e \ln x^2\,dx$

39. $\int_0^{\pi/2} \tan \theta\,d\theta$

40. $\int_0^{\pi/2} \sec \theta\,d\theta$

41. $\int_2^4 \frac{2}{x\sqrt{x^2-4}}\,dx$

42. $\int_3^6 \frac{1}{\sqrt{36-x^2}}\,dx$

43. $\int_3^5 \frac{1}{\sqrt{x^2-9}}\,dx$

44. $\int_0^5 \frac{1}{25-x^2}\,dx$

45. $\int_3^\infty \frac{1}{x\sqrt{x^2-9}}\,dx$

46. $\int_4^\infty \frac{\sqrt{x^2-16}}{x^2}\,dx$

47. $\int_0^\infty \frac{4}{\sqrt{x}(x+6)}\,dx$

48. $\int_1^\infty \frac{1}{x \ln x}\,dx$

Finding Values In Exercises 49 and 50, determine all values of p for which the improper integral converges.

49. $\int_1^\infty \frac{1}{x^p}\,dx$

50. $\int_0^1 \frac{1}{x^p}\,dx$

51. Mathematical Induction Use mathematical induction to verify that the following integral converges for any positive integer n.

$$\int_0^\infty x^n e^{-x}\,dx$$

52. Comparison Test for Improper Integrals In some cases, it is impossible to find the exact value of an improper integral, but it is important to determine whether the integral converges or diverges. Suppose the functions f and g are continuous and $0 \le g(x) \le f(x)$ on the interval $[a, \infty)$. It can be shown that if $\int_a^\infty f(x)\,dx$ converges, then $\int_a^\infty g(x)\,dx$ also converges, and if $\int_a^\infty g(x)\,dx$ diverges, then $\int_a^\infty f(x)\,dx$ also diverges. This is known as the Comparison Test for improper integrals.

(a) Use the Comparison Test to determine whether $\int_1^\infty e^{-x^2}\,dx$ converges or diverges. (*Hint:* Use the fact that $e^{-x^2} \le e^{-x}$ for $x \ge 1$.)

(b) Use the Comparison Test to determine whether $\int_1^\infty \frac{1}{x^5 + 1}\,dx$ converges or diverges. (*Hint:* Use the fact that $\frac{1}{x^5 + 1} \le \frac{1}{x^5}$ for $x \ge 1$.)

Convergence or Divergence In Exercises 53–60, use the results of Exercises 49–52 to determine whether the improper integral converges or diverges.

53. $\int_0^1 \frac{1}{\sqrt[6]{x}}\,dx$

54. $\int_0^1 \frac{1}{x^9}\,dx$

55. $\int_1^\infty \frac{1}{x^5}\,dx$

56. $\int_0^\infty x^4 e^{-x}\,dx$

57. $\int_1^\infty \frac{1}{x^2 + 5}\,dx$

58. $\int_2^\infty \frac{1}{\sqrt{x - 1}}\,dx$

59. $\int_1^\infty \frac{1 - \sin x}{x^2}\,dx$

60. $\int_0^\infty \frac{1}{e^x + x}\,dx$

EXPLORING CONCEPTS

61. Improper Integral Explain why $\int_{-1}^1 \frac{1}{x^3}\,dx \ne 0$.

62. Improper Integral Consider the integral

$$\int_0^3 \frac{10}{x^2 - 2x}\,dx.$$

To determine the convergence or divergence of the integral, how many improper integrals must be analyzed? What must be true of each of these integrals for the given integral to converge?

 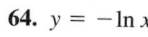 **Area** In Exercises 63–66, find the area of the unbounded shaded region.

63. $y = -\dfrac{7}{(x - 1)^3}$,

$-\infty < x \le -1$

64. $y = -\ln x$

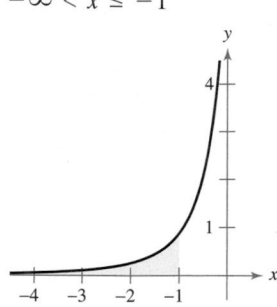
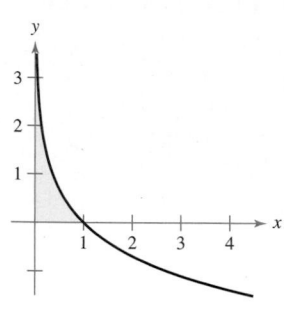

65. Witch of Agnesi:

$y = \dfrac{1}{x^2 + 1}$

66. Witch of Agnesi:

$y = \dfrac{8}{x^2 + 4}$

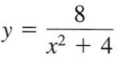
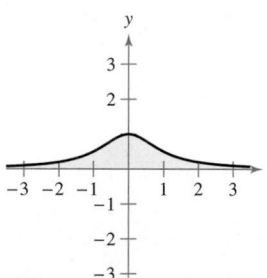
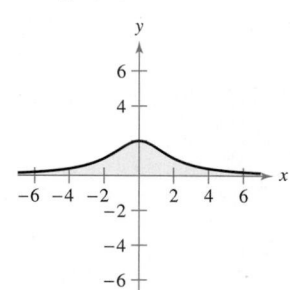

 **Area and Volume** In Exercises 67 and 68, consider the region satisfying the inequalities. (a) Find the area of the region. (b) Find the volume of the solid generated by revolving the region about the x-axis. (c) Find the volume of the solid generated by revolving the region about the y-axis.

67. $y \le e^{-x}$, $y \ge 0$, $x \ge 0$

68. $y \le \dfrac{1}{x^2}$, $y \ge 0$, $x \ge 1$

69. Arc Length Find the arc length of the graph of $y = \sqrt{16 - x^2}$ over the interval $[0, 4]$.

70. Surface Area Find the area of the surface formed by revolving the graph of $y = 2e^{-x}$ on the interval $[0, \infty)$ about the x-axis.

Propulsion In Exercises 71 and 72, use the weight of the rocket to answer each question. (Use 4000 miles as the radius of Earth and do not consider the effect of air resistance.)

(a) How much work is required to propel the rocket an unlimited distance away from Earth's surface?

(b) How far has the rocket traveled when half of the total work has occurred?

71. 5-metric-ton rocket

72. 10-metric-ton rocket

Probability A nonnegative function f is called a *probability density function* if

$$\int_{-\infty}^{\infty} f(t)\, dt = 1.$$

The probability that x lies between a and b is given by

$$P(a \le x \le b) = \int_{a}^{b} f(t)\, dt.$$

In Exercises 73 and 74, (a) show that the nonnegative function is a probability density function, and (b) find $P(0 \le x \le 6)$.

73. $f(t) = \begin{cases} \frac{1}{9}e^{-t/9}, & t \ge 0 \\ 0, & t < 0 \end{cases}$ 74. $f(t) = \begin{cases} \frac{5}{6}e^{-5t/6}, & t \ge 0 \\ 0, & t < 0 \end{cases}$

 75. **Normal Probability** The mean height of American men between 20 and 29 years old is 69 inches, and the standard deviation is 3 inches. A 20- to 29-year-old man is chosen at random from the population. The probability that he is 6 feet tall or taller is

$$P(72 \le x < \infty) = \int_{72}^{\infty} \frac{1}{3\sqrt{2\pi}} e^{-(x-69)^2/18}\, dx.$$

(*Source: National Center for Health Statistics*)

(a) Use a graphing utility to graph the integrand. Use the graphing utility to convince yourself that the area between the x-axis and the integrand is 1.

(b) Use a graphing utility to approximate $P(72 \le x < \infty)$.

(c) Approximate $0.5 - P(69 \le x \le 72)$ using a graphing utility. Use the graph in part (a) to explain why this result is the same as the answer in part (b).

76. **HOW DO YOU SEE IT?** The graph shows the probability density function for a car brand that has a mean fuel efficiency of 26 miles per gallon and a standard deviation of 2.4 miles per gallon.

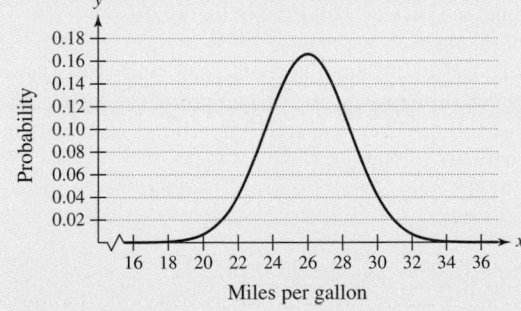

(a) Which is greater, the probability of choosing a car at random that gets between 26 and 28 miles per gallon or the probability of choosing a car at random that gets between 22 and 24 miles per gallon?

(b) Which is greater, the probability of choosing a car at random that gets between 20 and 22 miles per gallon or the probability of choosing a car at random that gets at least 30 miles per gallon?

Capitalized Cost In Exercises 77 and 78, find the capitalized cost C of an asset (a) for $n = 5$ years, (b) for $n = 10$ years, and (c) forever. The capitalized cost is given by

$$C = C_0 + \int_{0}^{n} c(t)e^{-rt}\, dt$$

where C_0 is the original investment, t is the time in years, r is the annual interest rate compounded continuously, and $c(t)$ is the annual cost of maintenance.

77. $C_0 = \$700{,}000$
    $c(t) = \$25{,}000$
    $r = 0.06$

78. $C_0 = \$700{,}000$
    $c(t) = \$25{,}000(1 + 0.08t)$
    $r = 0.06$

79. **Electromagnetic Theory** The magnetic potential P at a point on the axis of a circular coil is given by

$$P = \frac{2\pi N I r}{k} \int_{c}^{\infty} \frac{1}{(r^2 + x^2)^{3/2}}\, dx$$

where N, I, r, k, and c are constants. Find P.

80. **Gravitational Force** A "semi-infinite" uniform rod occupies the nonnegative x-axis. The rod has a linear density δ, which means that a segment of length dx has a mass of $\delta\, dx$. A particle of mass M is located at the point $(-a, 0)$. The gravitational force F that the rod exerts on the mass is given by

$$F = \int_{0}^{\infty} \frac{GM\delta}{(a + x)^2}\, dx$$

where G is the gravitational constant. Find F.

True or False? In Exercises 81–86, determine whether the statement is true or false. If it is false, explain why or give an example that shows it is false.

81. If f is continuous on $[0, \infty)$ and $\lim\limits_{x\to\infty} f(x) = 0$, then $\int_{0}^{\infty} f(x)\, dx$ converges.

82. If f is continuous on $[0, \infty)$ and $\int_{0}^{\infty} f(x)\, dx$ diverges, then $\lim\limits_{x\to\infty} f(x) \ne 0$.

83. If f' is continuous on $[0, \infty)$ and $\lim\limits_{x\to\infty} f(x) = 0$, then

$$\int_{0}^{\infty} f'(x)\, dx = -f(0).$$

84. If the graph of f is symmetric with respect to the origin or the y-axis, then $\int_{0}^{\infty} f(x)\, dx$ converges if and only if $\int_{-\infty}^{\infty} f(x)\, dx$ converges.

85. $\int_{0}^{\infty} e^{ax}\, dx$ converges for $a < 0$.

86. If $\lim\limits_{x\to\infty} f(x) = L$, then $\int_{0}^{\infty} f(x)\, dx$ converges.

87. **Comparing Integrals**

(a) Show that $\int_{-\infty}^{\infty} \sin x\, dx$ diverges.

(b) Show that $\lim\limits_{a\to\infty} \int_{-a}^{a} \sin x\, dx = 0$.

(c) What do parts (a) and (b) show about the definition of improper integrals?

88. Exploration Consider the integral

$$\int_0^{\pi/2} \frac{4}{1 + (\tan x)^n} \, dx$$

where n is a positive integer.

(a) Is the integral improper? Explain.

(b) Use a graphing utility to graph the integrand for $n = 2, 4, 8,$ and 12.

(c) Use the graphs to approximate the integral as $n \to \infty$.

(d) Use a computer algebra system to evaluate the integral for the values of n in part (b). Make a conjecture about the value of the integral for any positive integer n. Compare your results with your answer in part (c).

89. Comparing Integrals Let f be continuous on the interval $[a, \infty)$. Show that if the improper integral $\int_a^\infty |f(x)| \, dx$ converges, then the improper integral $\int_a^\infty f(x) \, dx$ also converges.

90. Writing

(a) The improper integrals

$$\int_1^\infty \frac{1}{x} \, dx \quad \text{and} \quad \int_1^\infty \frac{1}{x^2} \, dx$$

diverge and converge, respectively. Describe the essential difference between the integrands that cause one integral to converge and the other to diverge.

(b) Use a graphing utility to graph the function $y = (\sin x)/x$ over the interval $(1, \infty)$. Use your knowledge of the definite integral to make an inference as to whether the integral

$$\int_1^\infty \frac{\sin x}{x} \, dx$$

converges. Give reasons for your answer.

(c) Use one application of integration by parts and the result of Exercise 89 to determine the divergence or convergence of the integral in part (b).

Laplace Transforms Let $f(t)$ be a function defined for all positive values of t. **The Laplace Transform of $f(t)$ is defined by**

$$F(s) = \int_0^\infty e^{-st} f(t) \, dt$$

when the improper integral exists. Laplace Transforms are used to solve differential equations. In Exercises 91–98, find the Laplace Transform of the function.

91. $f(t) = 1$ **92.** $f(t) = t$

93. $f(t) = t^2$

94. $f(t) = e^{at}$

95. $f(t) = \cos at$

96. $f(t) = \sin at$

97. $f(t) = \cosh at$

98. $f(t) = \sinh at$

99. The Gamma Function The Gamma Function $\Gamma(n)$ is defined by

$$\Gamma(n) = \int_0^\infty x^{n-1} e^{-x} \, dx, \quad n > 0.$$

(a) Find $\Gamma(1)$, $\Gamma(2)$, and $\Gamma(3)$.

(b) Use integration by parts to show that $\Gamma(n + 1) = n\Gamma(n)$.

(c) Write $\Gamma(n)$ using factorial notation where n is a positive integer.

100. Proof Prove that $I_n = \left(\dfrac{n - 1}{n + 2} \right) I_{n-1}$, where

$$I_n = \int_0^\infty \frac{x^{2n-1}}{(x^2 + 1)^{n+3}} \, dx, \quad n \geq 1.$$

Then evaluate each integral.

(a) $\displaystyle\int_0^\infty \frac{x}{(x^2 + 1)^4} \, dx$ (b) $\displaystyle\int_0^\infty \frac{x^3}{(x^2 + 1)^5} \, dx$

(c) $\displaystyle\int_0^\infty \frac{x^5}{(x^2 + 1)^6} \, dx$

101. Finding a Value For what value of c does the integral

$$\int_0^\infty \left(\frac{1}{\sqrt{x^2 + 1}} - \frac{c}{x + 1} \right) dx$$

converge? Evaluate the integral for this value of c.

102. Finding a Value For what value of c does the integral

$$\int_1^\infty \left(\frac{cx}{x^2 + 2} - \frac{1}{3x} \right) dx$$

converge? Evaluate the integral for this value of c.

103. Volume Find the volume of the solid generated by revolving the region bounded by the graph of f about the x-axis.

$$f(x) = \begin{cases} x \ln x, & 0 < x \leq 2 \\ 0, & x = 0 \end{cases}$$

104. Volume Find the volume of the solid generated by revolving the unbounded region lying between $y = -\ln x$ and the y-axis $(y \geq 0)$ about the x-axis.

u-Substitution In Exercises 105 and 106, rewrite the improper integral as a proper integral using the given u-substitution. Then use the Trapezoidal Rule with $n = 5$ to approximate the integral.

105. $\displaystyle\int_0^1 \frac{\sin x}{\sqrt{x}} \, dx, \quad u = \sqrt{x}$

106. $\displaystyle\int_0^1 \frac{\cos x}{\sqrt{1 - x}} \, dx, \quad u = \sqrt{1 - x}$

107. Rewriting an Integral Let $\displaystyle\int_{-\infty}^\infty f(x) \, dx$ be convergent and let a and b be real numbers where $a \neq b$. Show that

$$\int_{-\infty}^a f(x) \, dx + \int_a^\infty f(x) \, dx = \int_{-\infty}^b f(x) \, dx + \int_b^\infty f(x) \, dx.$$

Review Exercises
See **CalcChat.com** for tutorial help and worked-out solutions to odd-numbered exercises.

Using Basic Integration Rules In Exercises 1–8, use the basic integration rules to find or evaluate the integral.

1. $\int x^2 \sqrt{x^3 - 27} \, dx$

2. $\int x e^{5-x^2} \, dx$

3. $\int \csc^2 \left(\frac{x+8}{4} \right) dx$

4. $\int \frac{x}{\sqrt[3]{4 - x^2}} \, dx$

5. $\int_1^e \frac{\ln 2x}{x} \, dx$

6. $\int_{3/2}^2 2x\sqrt{2x - 3} \, dx$

7. $\int \frac{100}{\sqrt{100 - x^2}} \, dx$

8. $\int \frac{2x}{x - 3} \, dx$

Using Integration by Parts In Exercises 9–16, use integration by parts to find the indefinite integral.

9. $\int x \, e^{1-x} \, dx$

10. $\int x^2 e^{x/2} \, dx$

11. $\int e^{2x} \sin 3x \, dx$

12. $\int x\sqrt{x - 1} \, dx$

13. $\int x \sec^2 x \, dx$

14. $\int \ln\sqrt{x^2 - 4} \, dx$

15. $\int x \arcsin 2x \, dx$

16. $\int \arctan 2x \, dx$

Finding a Trigonometric Integral In Exercises 17–26, find the trigonometric integral.

17. $\int \sin x \cos^4 x \, dx$

18. $\int \sin^2 x \cos^3 x \, dx$

19. $\int \cos^3(\pi x - 1) \, dx$

20. $\int \sin^2 \frac{\pi x}{2} \, dx$

21. $\int \sec^4 \frac{x}{2} \, dx$

22. $\int \tan \theta \sec^4 \theta \, d\theta$

23. $\int x \tan^4 x^2 \, dx$

24. $\int \frac{\tan^2 x}{\sec^3 x} \, dx$

25. $\int \frac{1}{1 - \sin \theta} \, d\theta$

26. $\int (\cos 2\theta)(\sin \theta + \cos \theta)^2 \, d\theta$

Area In Exercises 27 and 28, find the area of the given region.

27. $y = \sin^4 x$

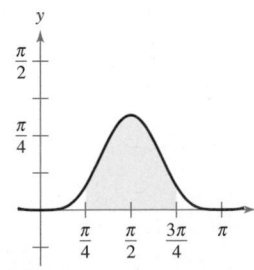

28. $y = \sin 3x \cos 2x$

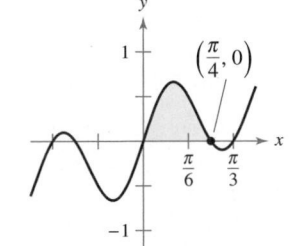

Using Trigonometric Substitution In Exercises 29–34, use trigonometric substitution to find or evaluate the integral.

29. $\int \frac{-12}{x^2\sqrt{4 - x^2}} \, dx$

30. $\int \frac{\sqrt{x^2 - 9}}{x} \, dx$

31. $\int \frac{x^3}{\sqrt{4 + x^2}} \, dx$

32. $\int \sqrt{25 - 9x^2} \, dx$

33. $\int_0^1 \frac{6x^3}{\sqrt{16 + x^2}} \, dx$

34. $\int_3^4 x^3\sqrt{x^2 - 9} \, dx$

Using Different Methods In Exercises 35 and 36, find the indefinite integral using each method.

35. $\int \frac{x^3}{\sqrt{4 + x^2}} \, dx$

  (a) Trigonometric substitution

  (b) Substitution: $u^2 = 4 + x^2$

  (c) Integration by parts: $dv = \frac{x}{\sqrt{4 + x^2}} \, dx$

36. $\int x\sqrt{4 + x} \, dx$

  (a) Trigonometric substitution

  (b) Substitution: $u^2 = 4 + x$

  (c) Substitution: $u = 4 + x$

  (d) Integration by parts: $dv = \sqrt{4 + x} \, dx$

Using Partial Fractions In Exercises 37–44, use partial fractions to find the indefinite integral.

37. $\int \frac{x - 8}{x^2 - x - 6} \, dx$

38. $\int \frac{5x - 2}{x^2 - x} \, dx$

39. $\int \frac{x^2 + 2x}{x^3 - x^2 + x - 1} \, dx$

40. $\int \frac{4x - 2}{3(x - 1)^2} \, dx$

41. $\int \frac{x^2}{x^2 - 2x + 1} \, dx$

42. $\int \frac{x^3 + 4}{x^2 - 4x} \, dx$

43. $\int \frac{4e^x}{(e^{2x} - 1)(e^x + 3)} \, dx$

44. $\int \frac{\sec^2 \theta}{(\tan \theta)(\tan \theta - 1)} \, d\theta$

Using the Trapezoidal Rule and Simpson's Rule In Exercises 45–48, approximate the definite integral using the Trapezoidal Rule and Simpson's Rule with $n = 4$. Compare these results with the approximation of the integral using a graphing utility.

45. $\int_2^3 \frac{2}{1 + x^2} \, dx$

46. $\int_0^1 \frac{x^{3/2}}{3 - x^2} \, dx$

47. $\int_0^{\pi/2} \sqrt{x} \cos x \, dx$

48. $\int_0^\pi \sqrt{1 + \sin^2 x} \, dx$

Integration by Tables   In Exercises 49–56, use integration tables to find or evaluate the integral.

49. $\displaystyle\int \frac{x}{(4+5x)^2}\, dx$

50. $\displaystyle\int \frac{x}{\sqrt{4+5x}}\, dx$

51. $\displaystyle\int_0^{\sqrt{\pi}/2} \frac{x}{1+\sin x^2}\, dx$

52. $\displaystyle\int_0^1 \frac{x}{1+e^{x^2}}\, dx$

53. $\displaystyle\int \frac{x}{x^2+4x+8}\, dx$

54. $\displaystyle\int \frac{3}{2x\sqrt{9x^2-1}}\, dx, \quad x > \frac{1}{3}$

55. $\displaystyle\int \frac{1}{\sin \pi x \cos \pi x}\, dx$

56. $\displaystyle\int \frac{1}{1+\tan \pi x}\, dx$

Finding an Indefinite Integral   In Exercises 57–64, find the indefinite integral using any method.

57. $\displaystyle\int \theta \sin \theta \cos \theta \, d\theta$

58. $\displaystyle\int \frac{\csc \sqrt{2x}}{\sqrt{x}}\, dx$

59. $\displaystyle\int \frac{x^{1/4}}{1+x^{1/2}}\, dx$

60. $\displaystyle\int \sqrt{1+\sqrt{x}}\, dx$

61. $\displaystyle\int \sqrt{1+\cos x}\, dx$

62. $\displaystyle\int \frac{3x^3+4x}{(x^2+1)^2}\, dx$

63. $\displaystyle\int \cos x \ln(\sin x)\, dx$

64. $\displaystyle\int (\sin \theta + \cos \theta)^2 \, d\theta$

Differential Equation   In Exercises 65–68, find the general solution of the differential equation using any method.

65. $\dfrac{dy}{dx} = \dfrac{25}{x^2-25}$

66. $\dfrac{dy}{dx} = \dfrac{\sqrt{4-x^2}}{2x}$

67. $y' = \ln(x^2+x)$

68. $y' = \sqrt{1-\cos \theta}$

Evaluating a Definite Integral   In Exercises 69–74, evaluate the definite integral using any method. Use a graphing utility to verify your result.

69. $\displaystyle\int_2^{\sqrt{5}} x(x^2-4)^{3/2}\, dx$

70. $\displaystyle\int_0^1 \frac{x}{(x-2)(x-4)}\, dx$

71. $\displaystyle\int_1^4 \frac{\ln x}{x}\, dx$

72. $\displaystyle\int_0^2 xe^{3x}\, dx$

73. $\displaystyle\int_2^{\pi} (x^2-4) \sin x \, dx$

74. $\displaystyle\int_0^5 \frac{x}{\sqrt{4+x}}\, dx$

Area   In Exercises 75 and 76, find the area of the given region using any method.

75. $y = x\sqrt{3-2x}$

76. $y = \dfrac{1}{25-x^2}$

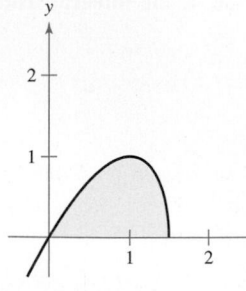

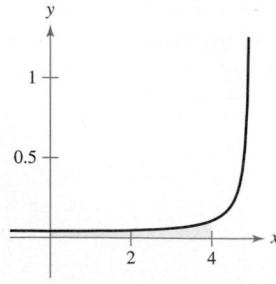

Centroid   In Exercises 77 and 78, find the centroid of the region bounded by the graphs of the equations using any method.

77. $y = \sqrt{1-x^2}, \quad y = 0$

78. $(x-1)^2 + y^2 = 1, \quad (x-4)^2 + y^2 = 4$

Evaluating an Improper Integral   In Exercises 79–86, determine whether the improper integral diverges or converges. Evaluate the integral if it converges.

79. $\displaystyle\int_0^{16} \frac{1}{\sqrt[4]{x}}\, dx$

80. $\displaystyle\int_0^2 \frac{7}{x-2}\, dx$

81. $\displaystyle\int_1^{\infty} x^2 \ln x \, dx$

82. $\displaystyle\int_0^{\infty} \frac{e^{-1/x}}{x^2}\, dx$

83. $\displaystyle\int_1^{\infty} \frac{\ln x}{x^2}\, dx$

84. $\displaystyle\int_1^{\infty} \frac{1}{\sqrt[4]{x}}\, dx$

85. $\displaystyle\int_2^{\infty} \frac{1}{x\sqrt{x^2-4}}\, dx$

86. $\displaystyle\int_0^{\infty} \frac{2}{\sqrt{x}(x+4)}\, dx$

87. Present Value   The board of directors of a corporation is calculating the price to pay for a business that is forecast to yield a continuous flow of profit of $500,000 per year. The money will earn a nominal rate of 5% per year compounded continuously. The present value of the business for t_0 years is

$$\text{Present value} = \int_0^{t_0} 500,000e^{-0.05t}\, dt.$$

(a) Find the present value of the business for 20 years.

(b) Find the present value of the business in perpetuity (forever).

88. Volume   Find the volume of the solid generated by revolving the region bounded by the graphs of $y \le xe^{-x}$, $y \ge 0$, and $x \ge 0$ about the x-axis.

89. Probability   The average lengths (from beak to tail) of different species of warblers in the eastern United States are approximately normally distributed with a mean of 12.9 centimeters and a standard deviation of 0.95 centimeter (see figure). The probability that a randomly selected warbler has a length between a and b centimeters is

$$P(a \le x \le b) = \frac{1}{0.95\sqrt{2\pi}} \int_a^b e^{-(x-12.9)^2/1.805}\, dx.$$

Use a graphing utility to approximate the probability that a randomly selected warbler has a length of (a) 13 centimeters or greater and (b) 15 centimeters or greater. (*Source: Peterson's Field Guide: Eastern Birds*)

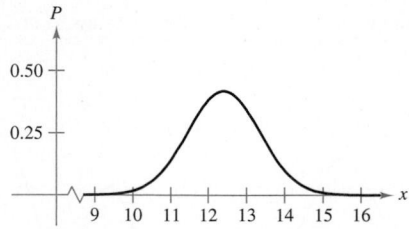

P.S. Problem Solving

1. Wallis's Formulas

(a) Evaluate the integrals

$$\int_{-1}^{1} (1 - x^2)\, dx \quad \text{and} \quad \int_{-1}^{1} (1 - x^2)^2\, dx.$$

(b) Use Wallis's Formulas to prove that

$$\int_{-1}^{1} (1 - x^2)^n\, dx = \frac{2^{2n+1}(n!)^2}{(2n + 1)!}$$

for all positive integers n.

2. Proof

(a) Evaluate the integrals

$$\int_{0}^{1} \ln x\, dx \quad \text{and} \quad \int_{0}^{1} (\ln x)^2\, dx.$$

(b) Prove that

$$\int_{0}^{1} (\ln x)^n\, dx = (-1)^n\, n!$$

for all positive integers n.

3. Comparing Methods Let $I = \int_{0}^{4} f(x)\, dx$, where f is shown in the figure. Let $L(n)$ and $R(n)$ represent the Riemann sums using the left-hand endpoints and right-hand endpoints of n subintervals of equal width. (Assume n is even.) Let $T(n)$ and $S(n)$ be the corresponding values of the Trapezoidal Rule and Simpson's Rule.

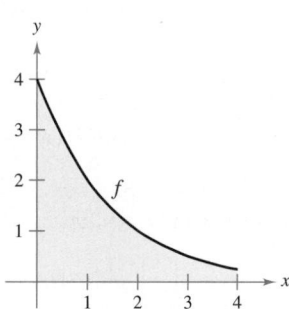

(a) For any n, list $L(n)$, $R(n)$, $T(n)$, and I in increasing order.

(b) Approximate $S(4)$.

4. Area Consider the problem of finding the area of the region bounded by the x-axis, the line $x = 4$, and the curve

$$y = \frac{x^2}{(x^2 + 9)^{3/2}}.$$

(a) Use a graphing utility to graph the region and approximate its area.

(b) Use an appropriate trigonometric substitution to find the exact area.

(c) Use the substitution $x = 3 \sinh u$ to find the exact area and verify that you obtain the same answer as in part (b).

5. Area Use the substitution

$$u = \tan \frac{x}{2}$$

to find the area of the shaded region under the graph of

$$y = \frac{1}{2 + \cos x}$$

for $0 \le x \le \pi/2$ (see figure).

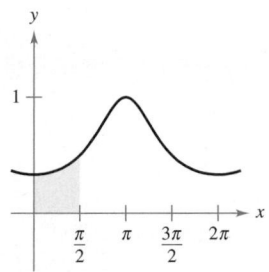

6. Arc Length Find the arc length of the graph of the function $y = \ln(1 - x^2)$ on the interval $0 \le x \le \frac{1}{2}$ (see figure).

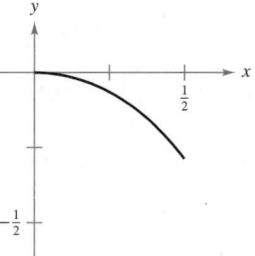

7. Centroid Find the centroid of the region bounded by the x-axis and the curve $y = e^{-c^2 x^2}$, where c is a positive constant (see figure).

$$\left(\text{Hint: Show that } \int_{0}^{\infty} e^{-c^2 x^2}\, dx = \frac{1}{c} \int_{0}^{\infty} e^{-x^2}\, dx.\right)$$

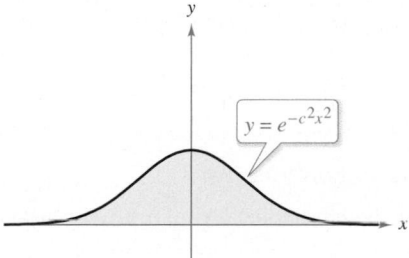

$$y = e^{-c^2 x^2}$$

8. Proof Prove the following generalization of the Mean Value Theorem. If f is twice differentiable on the closed interval $[a, b]$, then

$$f(b) - f(a) = f'(a)(b - a) - \int_{a}^{b} f''(t)(t - b)\, dt.$$

9. Inverse Function and Area

(a) Let $y = f^{-1}(x)$ be the inverse function of f. Use integration by parts to derive the formula

$$\int f^{-1}(x)\,dx = xf^{-1}(x) - \int f(y)\,dy.$$

(b) Use the formula in part (a) to find the integral

$$\int \arcsin x\,dx.$$

(c) Use the formula in part (a) to find the area under the graph of $y = \ln x$, $1 \le x \le e$ (see figure).

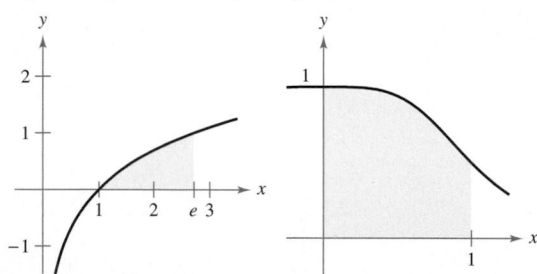

Figure for 9   Figure for 10

10. Area Factor the polynomial $p(x) = x^4 + 1$ and then find the area under the graph of

$$y = \frac{1}{x^4 + 1}, \quad 0 \le x \le 1 \text{ (see figure)}.$$

11. Partial Fraction Decomposition Suppose the denominator of a rational function can be factored into distinct linear factors

$$D(x) = (x - c_1)(x - c_2)\cdots(x - c_n)$$

for a positive integer n and distinct real numbers c_1, $c_2, \ldots, c_n$. If N is a polynomial of degree less than n, show that

$$\frac{N(x)}{D(x)} = \frac{P_1}{x - c_1} + \frac{P_2}{x - c_2} + \cdots + \frac{P_n}{x - c_n}$$

where $P_k = N(c_k)/D'(c_k)$ for $k = 1, 2, \ldots, n$. Note that this is the partial fraction decomposition of $N(x)/D(x)$.

12. Partial Fraction Decomposition Use the result of Exercise 11 to find the partial fraction decomposition of

$$\frac{x^3 - 3x^2 + 1}{x^4 - 13x^2 + 12x}.$$

13. Evaluating an Integral

(a) Use the substitution $u = \dfrac{\pi}{2} - x$ to evaluate the integral

$$\int_0^{\pi/2} \frac{\sin x}{\cos x + \sin x}\,dx.$$

(b) Let n be a positive integer. Evaluate the integral

$$\int_0^{\pi/2} \frac{\sin^n x}{\cos^n x + \sin^n x}\,dx.$$

14. Elementary Functions Some elementary functions, such as $f(x) = \sin(x^2)$, do not have antiderivatives that are elementary functions. Joseph Liouville proved that

$$\int \frac{e^x}{x}\,dx$$

does not have an elementary antiderivative. Use this fact to prove that

$$\int \frac{1}{\ln x}\,dx$$

does not have an elementary antiderivative.

15. Rocket The velocity v (in feet per second) of a rocket whose initial mass (including fuel) is m is given by

$$v = -gt + u \ln \frac{m}{m - rt}, \quad t < \frac{m}{r}$$

where u is the expulsion speed of the fuel, r is the rate at which the fuel is consumed, and $g = 32$ feet per second per second is the acceleration due to gravity. Find the position equation for a rocket for which $m = 50{,}000$ pounds, $u = 12{,}000$ feet per second, and $r = 400$ pounds per second. What is the height of the rocket when $t = 100$ seconds? (Assume that the rocket was fired from ground level and is moving straight upward.)

16. Proof Suppose that $f(a) = f(b) = g(a) = g(b) = 0$ and the second derivatives of f and g are continuous on the closed interval $[a, b]$. Prove that

$$\int_a^b f(x)g''(x)\,dx = \int_a^b f''(x)g(x)\,dx.$$

17. Proof Suppose that $f(a) = f(b) = 0$ and the second derivatives of f exist on the closed interval $[a, b]$. Prove that

$$\int_a^b (x - a)(x - b)f''(x)\,dx = 2\int_a^b f(x)\,dx.$$

18. Approximating an Integral Using the inequality

$$\frac{1}{x^5} + \frac{1}{x^{10}} + \frac{1}{x^{15}} < \frac{1}{x^5 - 1} < \frac{1}{x^5} + \frac{1}{x^{10}} + \frac{2}{x^{15}}$$

for $x \ge 2$, approximate $\displaystyle\int_2^\infty \frac{1}{x^5 - 1}\,dx.$

19. Volume Consider the shaded region between the graph of $y = \sin x$, where $0 \le x \le \pi$, and the line $y = c$, where $0 \le c \le 1$, as shown in the figure. A solid is formed by revolving the region about the line $y = c$.

(a) For what value of c does the solid have minimum volume?

(b) For what value of c does the solid have maximum volume?

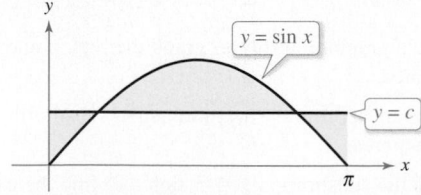

9 Infinite Series

Projectile Motion
(Exercise 80, p. 679)

Multiplier Effect *(Exercise 73, p. 606)*

Sphereflake *(Exercise 86, p. 607)*

Population
(Exercise 76, p. 598)

Government Expenditures *(Exercise 68, p. 597)*

9.1 Sequences

- Write the terms of a sequence.
- Determine whether a sequence converges or diverges.
- Write a formula for the *n*th term of a sequence.
- Use properties of monotonic sequences and bounded sequences.

Sequences

In mathematics, the word "sequence" is used in much the same way as it is in ordinary English. Saying that a collection of objects or events is *in sequence* usually means that the collection is ordered in such a way that it has an identified first member, second member, third member, and so on.

Mathematically, a **sequence** is defined as a function whose domain is the set of positive integers. Although a sequence is a function, it is common to represent sequences by subscript notation rather than by the standard function notation. For instance, in the sequence

$$1, \quad 2, \quad 3, \quad 4, \quad \ldots, \quad n, \quad \ldots$$

$$\downarrow \quad \downarrow \quad \downarrow \quad \downarrow \quad \quad \downarrow \quad \quad \downarrow \qquad \text{Sequence}$$

$$a_1, \quad a_2, \quad a_3, \quad a_4, \quad \ldots, \quad a_n, \quad \ldots$$

1 is mapped onto a_1, 2 is mapped onto a_2, and so on. The numbers

$$a_1, a_2, a_3, \ldots, a_n, \ldots$$

are the **terms** of the sequence. The number a_n is the **nth term** of the sequence, and the entire sequence is denoted by $\{a_n\}$. Occasionally, it is convenient to begin a sequence with a_0 so that the terms of the sequence become $a_0, a_1, a_2, a_3, \ldots, a_n, \ldots$ and the domain is the set of nonnegative integers.

EXAMPLE 1 Writing the Terms of a Sequence

a. The terms of the sequence $\{a_n\} = \{3 + (-1)^n\}$ are

$$3 + (-1)^1, \ 3 + (-1)^2, \ 3 + (-1)^3, \ 3 + (-1)^4, \ldots$$

$$2, \qquad\quad 4, \qquad\quad 2, \qquad\quad 4, \qquad \ldots.$$

b. The terms of the sequence $\{b_n\} = \left\{\dfrac{n}{1 - 2n}\right\}$ are

$$\frac{1}{1 - 2 \cdot 1}, \ \frac{2}{1 - 2 \cdot 2}, \ \frac{3}{1 - 2 \cdot 3}, \ \frac{4}{1 - 2 \cdot 4}, \ldots$$

$$-1, \qquad -\frac{2}{3}, \qquad -\frac{3}{5}, \qquad -\frac{4}{7}, \qquad \ldots.$$

c. The terms of the sequence $\{c_n\} = \left\{\dfrac{n^2}{2^n - 1}\right\}$ are

$$\frac{1^2}{2^1 - 1}, \ \frac{2^2}{2^2 - 1}, \ \frac{3^2}{2^3 - 1}, \ \frac{4^2}{2^4 - 1}, \ldots$$

$$\frac{1}{1}, \qquad \frac{4}{3}, \qquad \frac{9}{7}, \qquad \frac{16}{15}, \qquad \ldots.$$

• • **REMARK** Some sequences are defined recursively. To define a sequence recursively, you need to be given one or more of the first few terms. All other terms of the sequence are then defined using previous terms, as shown in Example 1(d).

▷ **d.** The terms of the **recursively defined** sequence $\{d_n\}$, where $d_1 = 25$ and $d_{n+1} = d_n - 5$, are

$$25, \quad 25 - 5 = 20, \quad 20 - 5 = 15, \quad 15 - 5 = 10, \ldots.$$

Limit of a Sequence

The primary focus of this chapter concerns sequences whose terms approach limiting values. Such sequences are said to **converge**. For instance, the sequence $\{1/2^n\}$

$$\frac{1}{2}, \frac{1}{4}, \frac{1}{8}, \frac{1}{16}, \frac{1}{32}, \cdots$$

converges to 0, as indicated in the next definition.

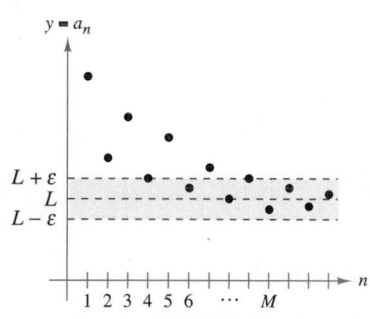

For $n > M$, the terms of the sequence all lie within ε units of L.

Figure 9.1

Definition of the Limit of a Sequence

Let L be a real number. The **limit** of a sequence $\{a_n\}$ is L, written as

$$\lim_{n \to \infty} a_n = L$$

if for each $\varepsilon > 0$, there exists $M > 0$ such that $|a_n - L| < \varepsilon$ whenever $n > M$. If the limit L of a sequence exists, then the sequence **converges** to L. If the limit of a sequence does not exist, then the sequence **diverges.**

Graphically, this definition says that eventually (for $n > M$ and $\varepsilon > 0$), the terms of a sequence that converges to L will lie within the band between the lines $y = L + \varepsilon$ and $y = L - \varepsilon$, as shown in Figure 9.1.

If a sequence $\{a_n\}$ agrees with a function f at every positive integer, and if $f(x)$ approaches a limit L as $x \to \infty$, then the sequence must converge to the same limit L.

REMARK The converse of Theorem 9.1 is not true (see Exercise 82).

THEOREM 9.1 Limit of a Sequence

Let L be a real number. Let f be a function of a real variable such that

$$\lim_{x \to \infty} f(x) = L.$$

If $\{a_n\}$ is a sequence such that $f(n) = a_n$ for every positive integer n, then

$$\lim_{n \to \infty} a_n = L.$$

EXAMPLE 2 Finding the Limit of a Sequence

Find the limit of the sequence whose nth term is $a_n = \left(1 + \dfrac{1}{n}\right)^n$.

Solution In Example 5, Section 5.6, you learned that

$$\lim_{x \to \infty} \left(1 + \frac{1}{x}\right)^x = e.$$

So, you can apply Theorem 9.1 to conclude that

$$\lim_{n \to \infty} a_n = \lim_{n \to \infty} \left(1 + \frac{1}{n}\right)^n = e. \qquad \blacksquare$$

There are different ways in which a sequence can fail to have a limit. One way is that the terms of the sequence increase without bound or decrease without bound. These cases are written symbolically, as shown below.

Terms increase without bound: $\displaystyle \lim_{n \to \infty} a_n = \infty$

Terms decrease without bound: $\displaystyle \lim_{n \to \infty} a_n = -\infty$

The properties of limits of sequences listed in the next theorem parallel those given for limits of functions of a real variable in Section 2.3.

THEOREM 9.2   Properties of Limits of Sequences

Let $\lim\limits_{n\to\infty} a_n = L$ and $\lim\limits_{n\to\infty} b_n = K$.

1. Scalar multiple: $\lim\limits_{n\to\infty} (ca_n) = cL$, c is any real number.

2. Sum or difference: $\lim\limits_{n\to\infty} (a_n \pm b_n) = L \pm K$

3. Product: $\lim\limits_{n\to\infty} (a_n b_n) = LK$

4. Quotient: $\lim\limits_{n\to\infty} \dfrac{a_n}{b_n} = \dfrac{L}{K}$, $b_n \neq 0$ and $K \neq 0$

EXAMPLE 3   Determining Convergence or Divergence

⋯▷ *See LarsonCalculus.com for an interactive version of this type of example.*

a. Because the sequence $\{a_n\} = \{3 + (-1)^n\}$ has terms

$$2, 4, 2, 4, \ldots \qquad \text{See Example 1(a).}$$

that alternate between 2 and 4, the limit

$$\lim\limits_{n\to\infty} a_n$$

does not exist. So, the sequence diverges.

b. For $\{b_n\} = \left\{\dfrac{n}{1 - 2n}\right\}$, divide the numerator and denominator by n to obtain

$$\lim\limits_{n\to\infty} \frac{n}{1 - 2n} = \lim\limits_{n\to\infty} \frac{1}{(1/n) - 2} = -\frac{1}{2} \qquad \text{See Example 1(b).}$$

which implies that the sequence converges to $-\frac{1}{2}$.

EXAMPLE 4   Using L'Hôpital's Rule to Determine Convergence

Show that the sequence whose nth term is $a_n = \dfrac{n^2}{2^n - 1}$ converges.

Solution   Consider the function of a real variable

$$f(x) = \frac{x^2}{2^x - 1}.$$

Applying L'Hôpital's Rule twice produces

$$\lim\limits_{x\to\infty} \frac{x^2}{2^x - 1} = \lim\limits_{x\to\infty} \frac{2x}{(\ln 2)2^x} = \lim\limits_{x\to\infty} \frac{2}{(\ln 2)^2 2^x} = 0.$$

Because $f(n) = a_n$ for every positive integer n, you can apply Theorem 9.1 to conclude that

$$\lim\limits_{n\to\infty} \frac{n^2}{2^n - 1} = 0. \qquad \text{See Example 1(c).}$$

So, the sequence converges to 0.

▷ **TECHNOLOGY** Use a graphing utility to graph the function in Example 4. Notice that as x approaches infinity, the value of the function gets closer and closer to 0. If you have access to a graphing utility that can generate terms of a sequence, try using it to calculate the first 20 terms of the sequence in Example 4. Then view the terms to observe numerically that the sequence converges to 0.

The symbol $n!$ (read "n factorial") is used to simplify some of the formulas developed in this chapter. Let n be a positive integer; then **n factorial** is defined as

$$n! = 1 \cdot 2 \cdot 3 \cdot 4 \cdots (n-1) \cdot n.$$

As a special case, **zero factorial** is defined as $0! = 1$. From this definition, you can see that $1! = 1$, $2! = 1 \cdot 2 = 2$, $3! = 1 \cdot 2 \cdot 3 = 6$, and so on. Factorials follow the same conventions for order of operations as exponents. That is, just as $2x^3$ and $(2x)^3$ imply different orders of operations, $2n!$ and $(2n)!$ imply the orders

$$2n! = 2(n!) = 2(1 \cdot 2 \cdot 3 \cdot 4 \cdots n)$$

and

$$(2n)! = 1 \cdot 2 \cdot 3 \cdot 4 \cdots n \cdot (n+1) \cdots 2n$$

respectively.

Another useful limit theorem that can be rewritten for sequences is the Squeeze Theorem from Section 2.3.

THEOREM 9.3 Squeeze Theorem for Sequences

If $\lim\limits_{n\to\infty} a_n = L = \lim\limits_{n\to\infty} b_n$ and there exists an integer N such that $a_n \le c_n \le b_n$ for all $n > N$, then $\lim\limits_{n\to\infty} c_n = L$.

EXAMPLE 5   Using the Squeeze Theorem

Show that the sequence $\{c_n\} = \left\{ (-1)^n \dfrac{1}{n!} \right\}$ converges, and find its limit.

Solution   To apply the Squeeze Theorem, you must find two convergent sequences that can be related to $\{c_n\}$. Two possibilities are $a_n = -1/2^n$ and $b_n = 1/2^n$, both of which converge to 0. By comparing the term $n!$ with 2^n, you can see that

$$n! = 1 \cdot 2 \cdot 3 \cdot 4 \cdot 5 \cdot 6 \cdots n = 24 \cdot \underbrace{5 \cdot 6 \cdots n}_{n-4 \text{ factors}} \qquad (n \ge 4)$$

and

$$2^n = 2 \cdot 2 \cdot 2 \cdot 2 \cdot 2 \cdot 2 \cdots 2 = 16 \cdot \underbrace{2 \cdot 2 \cdots 2}_{n-4 \text{ factors}}. \qquad (n \ge 4)$$

This implies that for $n \ge 4$, $2^n < n!$, and you have

$$\frac{-1}{2^n} \le (-1)^n \frac{1}{n!} \le \frac{1}{2^n}, \quad n \ge 4$$

as shown in Figure 9.2. So, by the Squeeze Theorem, it follows that

$$\lim_{n\to\infty} (-1)^n \frac{1}{n!} = 0.$$

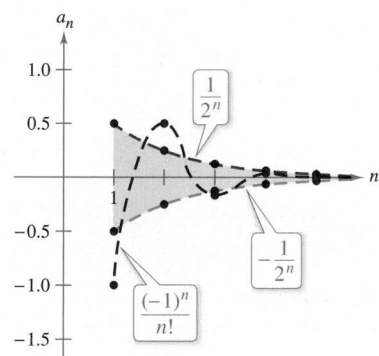

For $n \ge 4$, $(-1)^n/n!$ is squeezed between $-1/2^n$ and $1/2^n$.

Figure 9.2

Example 5 suggests something about the rate at which $n!$ increases as $n \to \infty$. As Figure 9.2 suggests, both $1/2^n$ and $1/n!$ approach 0 as $n \to \infty$. Yet $1/n!$ approaches 0 so much faster than $1/2^n$ does that

$$\lim_{n\to\infty} \frac{1/n!}{1/2^n} = \lim_{n\to\infty} \frac{2^n}{n!} = 0.$$

In fact, it can be shown that for any fixed number k, $\lim\limits_{n\to\infty} (k^n/n!) = 0$. This means that *the factorial function grows faster than any exponential function.*

In Example 5, the sequence $\{c_n\}$ has both positive and negative terms. For this sequence, it happens that the sequence of absolute values, $\{|c_n|\}$, also converges to 0. You can show this by the Squeeze Theorem using the inequality

$$0 \le \frac{1}{n!} \le \frac{1}{2^n}, \quad n \ge 4.$$

In such cases, it is often convenient to consider the sequence of absolute values and then apply Theorem 9.4, which states that if the absolute value sequence converges to 0, then the original signed sequence also converges to 0.

THEOREM 9.4   Absolute Value Theorem

For the sequence $\{a_n\}$, if

$$\lim_{n\to\infty} |a_n| = 0 \quad \text{then} \quad \lim_{n\to\infty} a_n = 0.$$

Proof   Consider the two sequences $\{|a_n|\}$ and $\{-|a_n|\}$. Because both of these sequences converge to 0 and

$$-|a_n| \le a_n \le |a_n|$$

you can use the Squeeze Theorem to conclude that $\{a_n\}$ converges to 0. ∎

Pattern Recognition for Sequences

Sometimes the terms of a sequence are generated by some rule that does not explicitly identify the nth term of the sequence. In such cases, you may be required to discover a *pattern* in the sequence and to describe the nth term. Once the nth term has been specified, you can investigate the convergence or divergence of the sequence.

EXAMPLE 6   **Finding the nth Term of a Sequence**

Find a sequence $\{a_n\}$ whose first five terms are

$$\frac{2}{1}, \frac{4}{3}, \frac{8}{5}, \frac{16}{7}, \frac{32}{9}, \dots$$

and then determine whether the sequence you have chosen converges or diverges.

Solution   First, note that the numerators are successive powers of 2, and the denominators form the sequence of positive odd integers. By comparing a_n with n, you have the following pattern.

$$\frac{2^1}{1}, \frac{2^2}{3}, \frac{2^3}{5}, \frac{2^4}{7}, \frac{2^5}{9}, \dots, \frac{2^n}{2n-1}, \dots$$

Consider the function of a real variable $f(x) = 2^x/(2x-1)$. Applying L'Hôpital's Rule produces

$$\lim_{x\to\infty} \frac{2^x}{2x-1} = \lim_{x\to\infty} \frac{2^x(\ln 2)}{2} = \infty.$$

Next, apply Theorem 9.1 to conclude that

$$\lim_{n\to\infty} \frac{2^n}{2n-1} = \infty.$$

So, the sequence diverges. ∎

Without a specific rule for generating the terms of a sequence or some knowledge of the context in which the terms of the sequence are obtained, it is not possible to determine the convergence or divergence of the sequence merely from its first several terms. For instance, although the first three terms of the following four sequences are identical, the first two sequences converge to 0, the third sequence converges to $\frac{1}{9}$, and the fourth sequence diverges.

$$\{a_n\}: \frac{1}{2}, \frac{1}{4}, \frac{1}{8}, \frac{1}{16}, \ldots, \frac{1}{2n}, \ldots$$

$$\{b_n\}: \frac{1}{2}, \frac{1}{4}, \frac{1}{8}, \frac{1}{15}, \ldots, \frac{6}{(n+1)(n^2-n+6)}, \ldots$$

$$\{c_n\}: \frac{1}{2}, \frac{1}{4}, \frac{1}{8}, \frac{7}{62}, \ldots, \frac{n^2-3n+3}{9n^2-25n+18}, \ldots$$

$$\{d_n\}: \frac{1}{2}, \frac{1}{4}, \frac{1}{8}, 0, \ldots, \frac{-n(n+1)(n-4)}{6(n^2+3n-2)}, \ldots$$

The process of determining an nth term from the pattern observed in the first several terms of a sequence is an example of *inductive reasoning*.

EXAMPLE 7   Finding the nth Term of a Sequence

Determine the nth term for a sequence whose first five terms are

$$-\frac{2}{1}, \frac{8}{2}, -\frac{26}{6}, \frac{80}{24}, -\frac{242}{120}, \ldots$$

and then decide whether the sequence converges or diverges.

Solution   Note that the numerators are 1 less than 3^n.

$$3^1 - 1 = 2 \qquad 3^2 - 1 = 8 \qquad 3^3 - 1 = 26 \qquad 3^4 - 1 = 80 \qquad 3^5 - 1 = 242$$

So, you can reason that the numerators are given by the rule

$$3^n - 1.$$

Factoring the denominators produces

$$1 = 1$$
$$2 = 1 \cdot 2$$
$$6 = 1 \cdot 2 \cdot 3$$
$$24 = 1 \cdot 2 \cdot 3 \cdot 4$$

and

$$120 = 1 \cdot 2 \cdot 3 \cdot 4 \cdot 5.$$

This suggests that the denominators are represented by $n!$. Finally, because the signs alternate, you can write the nth term as

$$a_n = (-1)^n \left(\frac{3^n - 1}{n!} \right).$$

From the discussion about the growth of $n!$ on the bottom of page 591, it follows that

$$\lim_{n \to \infty} |a_n| = \lim_{n \to \infty} \frac{3^n - 1}{n!} = 0.$$

Applying Theorem 9.4, you can conclude that

$$\lim_{n \to \infty} a_n = 0.$$

So, the sequence $\{a_n\}$ converges to 0.

Monotonic Sequences and Bounded Sequences

So far, you have determined the convergence of a sequence by finding its limit. Even when you cannot determine the limit of a particular sequence, it still may be useful to know whether the sequence converges. Theorem 9.5 (on the next page) provides a test for convergence of sequences without determining the limit. First, some preliminary definitions are given.

Definition of Monotonic Sequence

A sequence $\{a_n\}$ is **monotonic** when its terms are nondecreasing

$$a_1 \leq a_2 \leq a_3 \leq \cdots \leq a_n \leq \cdots$$

or when its terms are nonincreasing

$$a_1 \geq a_2 \geq a_3 \geq \cdots \geq a_n \geq \cdots .$$

EXAMPLE 8   **Determining Whether a Sequence Is Monotonic**

Determine whether each sequence having the given nth term is monotonic.

a. $a_n = 3 + (-1)^n$

b. $b_n = \dfrac{2n}{1 + n}$

c. $c_n = \dfrac{n^2}{2^n - 1}$

Solution

a. This sequence alternates between 2 and 4. So, it is not monotonic.

b. This sequence is monotonic because each successive term is greater than its predecessor. To see this, compare the terms b_n and b_{n+1}. [Note that, because n is positive, you can multiply each side of the inequality by $(1 + n)$ and $(2 + n)$ without reversing the inequality sign.]

$$b_n = \frac{2n}{1 + n} \overset{?}{<} \frac{2(n + 1)}{1 + (n + 1)} = b_{n+1}$$

$$2n(2 + n) \overset{?}{<} (1 + n)(2n + 2)$$

$$4n + 2n^2 \overset{?}{<} 2 + 4n + 2n^2$$

$$0 < 2$$

Starting with the final inequality, which is valid, you can reverse the steps to conclude that the original inequality is also valid.

c. This sequence is not monotonic because the second term is greater than both the first term and the third term. (Note that when you drop the first term, the remaining sequence $c_2, c_3, c_4, \ldots$ is monotonic.)

Figure 9.3 graphically illustrates these three sequences.   ∎

In Example 8(b), another way to see that the sequence is monotonic is to argue that the derivative of the corresponding differentiable function

$$f(x) = \frac{2x}{1 + x}$$

is positive for all x. This implies that f is increasing, which in turn implies that $\{b_n\}$ is increasing.

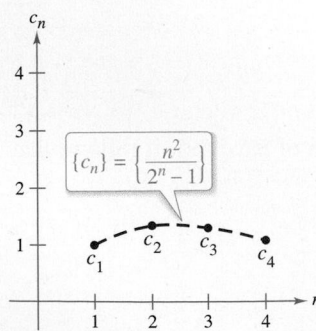

(a) Not monotonic

(b) Monotonic

(c) Not monotonic

Figure 9.3

Definition of Bounded Sequence

1. A sequence $\{a_n\}$ is **bounded above** when there is a real number M such that $a_n \leq M$ for all n. The number M is called an **upper bound** of the sequence.

2. A sequence $\{a_n\}$ is **bounded below** when there is a real number N such that $N \leq a_n$ for all n. The number N is called a **lower bound** of the sequence.

3. A sequence $\{a_n\}$ is **bounded** when it is bounded above and bounded below.

Note that all three sequences in Example 8 (and shown in Figure 9.3) are bounded. To see this, note that

$$2 \leq a_n \leq 4, \quad 1 \leq b_n \leq 2, \quad \text{and} \quad 0 \leq c_n \leq \frac{4}{3}.$$

One important property of the real numbers is that they are **complete**. Informally, this means that there are no holes or gaps on the real number line. (The set of rational numbers does not have the completeness property.) The completeness axiom for real numbers can be used to conclude that if a sequence has an upper bound, then it must have a **least upper bound** (an upper bound that is less than all other upper bounds for the sequence). For example, the least upper bound of the sequence $\{a_n\} = \{n/(n + 1)\}$,

$$\frac{1}{2}, \frac{2}{3}, \frac{3}{4}, \frac{4}{5}, \cdots, \frac{n}{n + 1}, \cdots$$

is 1. The completeness axiom is used in the proof of Theorem 9.5.

THEOREM 9.5 Bounded Monotonic Sequences

If a sequence $\{a_n\}$ is bounded and monotonic, then it converges.

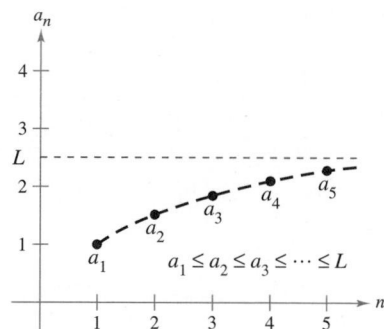

Every bounded, nondecreasing sequence converges.

Figure 9.4

Proof Assume that the sequence is nondecreasing, as shown in Figure 9.4. For the sake of simplicity, also assume that each term in the sequence is positive. Because the sequence is bounded, there must exist an upper bound M such that

$$a_1 \leq a_2 \leq a_3 \leq \cdots \leq a_n \leq \cdots \leq M.$$

From the completeness axiom, it follows that there is a least upper bound L such that

$$a_1 \leq a_2 \leq a_3 \leq \cdots \leq a_n \leq \cdots \leq L.$$

For $\varepsilon > 0$, it follows that $L - \varepsilon < L$, and therefore $L - \varepsilon$ cannot be an upper bound for the sequence. Consequently, at least one term of $\{a_n\}$ is greater than $L - \varepsilon$. That is, $L - \varepsilon < a_N$ for some positive integer N. Because the terms of $\{a_n\}$ are nondecreasing, it follows that $a_N \leq a_n$ for $n > N$. You now know that $L - \varepsilon < a_N \leq a_n \leq L < L + \varepsilon$, for every $n > N$. It follows that $|a_n - L| < \varepsilon$ for $n > N$, which by definition means that $\{a_n\}$ converges to L. The proof for a nonincreasing sequence is similar (see Exercise 89). ∎

EXAMPLE 9 Bounded and Monotonic Sequences

a. The sequence $\{a_n\} = \{1/n\}$ is both bounded and monotonic. So, by Theorem 9.5, it must converge.

b. The divergent sequence $\{b_n\} = \{n^2/(n + 1)\}$ is monotonic but not bounded. (It is bounded below.)

c. The divergent sequence $\{c_n\} = \{(-1)^n\}$ is bounded but not monotonic.

9.1 Exercises

See **CalcChat.com** for tutorial help and worked-out solutions to odd-numbered exercises.

CONCEPT CHECK

1. Recursively Defined Sequence What does it mean for a sequence to be defined recursively?

2. Properties of Limits of Sequences What is the value of L?

$\lim\limits_{n\to\infty} a_n = L$, $\lim\limits_{n\to\infty} b_n = 8$, and $\lim\limits_{n\to\infty} (a_n b_n) = 24$

3. Rate of Increase Which function grows faster as n approaches infinity? Explain.

$f(n) = 7^n$ $\qquad\qquad$ $g(n) = (n-1)!$

4. Bounded Monotonic Sequences A sequence $\{a_n\}$ is bounded below and nonincreasing. Does $\{a_n\}$ converge or diverge? Use a graph to support your conclusion.

 **Writing the Terms of a Sequence** In Exercises 5–10, write the first five terms of the sequence with the given nth term.

5. $a_n = 3^n$

6. $a_n = \left(-\dfrac{2}{5}\right)^n$

7. $a_n = \sin\dfrac{n\pi}{2}$

8. $a_n = \dfrac{3n}{n+4}$

9. $a_n = (-1)^{n+1}\left(\dfrac{2}{n}\right)$

10. $a_n = 2 + \dfrac{2}{n} - \dfrac{1}{n^2}$

Writing the Terms of a Sequence In Exercises 11 and 12, write the first five terms of the recursively defined sequence.

11. $a_1 = 3$, $a_{k+1} = 2(a_k - 1)$ $\qquad$ **12.** $a_1 = 6$, $a_{k+1} = \frac{1}{3}a_k^2$

Matching In Exercises 13–16, match the sequence with the given nth term with its graph. [The graphs are labeled (a), (b), (c), and (d).]

(a)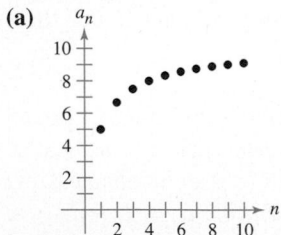

(b)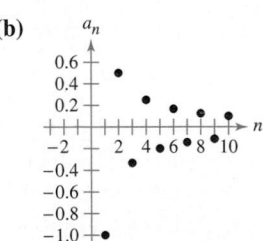

(c)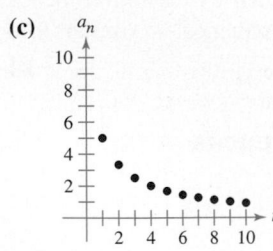

(d)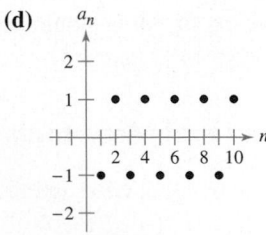

13. $a_n = \dfrac{10}{n+1}$ $\qquad\qquad$ **14.** $a_n = \dfrac{10n}{n+1}$

15. $a_n = (-1)^n$ $\qquad\qquad$ **16.** $a_n = \dfrac{(-1)^n}{n}$

Simplifying Factorials In Exercises 17–20, simplify the ratio of factorials.

17. $\dfrac{(n+1)!}{(n-1)!}$ $\qquad\qquad$ **18.** $\dfrac{(3n+1)!}{(3n)!}$

19. $\dfrac{n!}{(n-3)!}$ $\qquad\qquad$ **20.** $\dfrac{(4n+1)!}{(4n+3)!}$

 Finding the Limit of a Sequence In Exercises 21–24, find the limit of the sequence with the given nth term.

21. $a_n = \dfrac{n+1}{n}$ $\qquad\qquad$ **22.** $a_n = 6 + \dfrac{2}{n^2}$

23. $a_n = \dfrac{2n}{\sqrt{n^2+1}}$ $\qquad\qquad$ **24.** $a_n = \cos\dfrac{2}{n}$

Finding the Limit of a Sequence In Exercises 25–28, use a graphing utility to graph the first 10 terms of the sequence with the given nth term. Use the graph to make an inference about the convergence or divergence of the sequence. Verify your inference analytically and, if the sequence converges, find its limit.

25. $a_n = \dfrac{4n+1}{n}$ $\qquad\qquad$ **26.** $a_n = \dfrac{1}{n^{3/2}}$

27. $a_n = \sin\dfrac{n\pi}{2}$ $\qquad\qquad$ **28.** $a_n = 2 - \dfrac{1}{4^n}$

 **Determining Convergence or Divergence** In Exercises 29–44, determine the convergence or divergence of the sequence with the given nth term. If the sequence converges, find its limit.

29. $a_n = \dfrac{5}{n+2}$ $\qquad\qquad$ **30.** $a_n = n - \dfrac{1}{n!}$

31. $a_n = (-1)^n\left(\dfrac{n}{n+1}\right)$ $\qquad$ **32.** $a_n = \dfrac{1+(-1)^n}{n^2}$

33. $a_n = \dfrac{3n+\sqrt{n}}{4n}$ $\qquad\qquad$ **34.** $a_n = \dfrac{\sqrt[3]{n}}{\sqrt[3]{n}+1}$

35. $a_n = \dfrac{\ln(n^3)}{2n}$ $\qquad\qquad$ **36.** $a_n = \dfrac{e^n}{4^n}$

37. $a_n = \dfrac{(n+1)!}{n!}$ $\qquad\qquad$ **38.** $a_n = \dfrac{(n-2)!}{n!}$

39. $a_n = \dfrac{n^p}{e^n}$, $p > 0$ $\qquad$ **40.** $a_n = n \sin\dfrac{1}{n}$

41. $a_n = 2^{1/n}$ $\qquad\qquad$ **42.** $a_n = -3^{-n}$

43. $a_n = \dfrac{\sin n}{n}$ $\qquad\qquad$ **44.** $a_n = \dfrac{\cos 2n}{3^n}$

 **Finding the nth Term of a Sequence** In Exercises 45–52, write an expression for the nth term of the sequence and then determine whether the sequence you have chosen converges or diverges. (There is more than one correct answer.)

45. $2, 8, 14, 20, \ldots$ $\qquad\qquad$ **46.** $1, \frac{1}{2}, \frac{1}{6}, \frac{1}{24}, \frac{1}{120}, \ldots$

47. $-2, 1, 6, 13, 22, \ldots$ $\qquad\qquad$ **48.** $1, -\frac{1}{4}, \frac{1}{9}, -\frac{1}{16}, \ldots$

49. $\frac{2}{3}, \frac{3}{4}, \frac{4}{5}, \frac{5}{6}, \ldots$

50. 2, 24, 720, 40,320, 3,628,800, . . .

51. $2, 1 + \frac{1}{2}, 1 + \frac{1}{3}, 1 + \frac{1}{4}, 1 + \frac{1}{5}, \ldots$

52. $\frac{1}{2 \cdot 3}, \frac{2}{3 \cdot 4}, \frac{3}{4 \cdot 5}, \frac{4}{5 \cdot 6}, \ldots$

 **Monotonic and Bounded Sequences** In Exercises 53–60, determine whether the sequence with the given *n*th term is monotonic and whether it is bounded. Use a graphing utility to confirm your results.

53. $a_n = 4 - \dfrac{1}{n}$

54. $a_n = \dfrac{3n}{n + 2}$

55. $a_n = ne^{-n/2}$

56. $a_n = \left(-\dfrac{2}{3}\right)^n$

57. $a_n = \left(\dfrac{2}{3}\right)^n$

58. $a_n = \left(\dfrac{3}{2}\right)^n$

59. $a_n = \sin\dfrac{n\pi}{6}$

60. $a_n = \dfrac{\cos n}{n}$

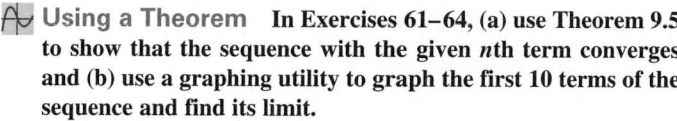 **Using a Theorem** In Exercises 61–64, (a) use Theorem 9.5 to show that the sequence with the given *n*th term converges and (b) use a graphing utility to graph the first 10 terms of the sequence and find its limit.

61. $a_n = 7 + \dfrac{1}{n}$

62. $a_n = 5 - \dfrac{2}{n}$

63. $a_n = \dfrac{1}{3}\left(1 - \dfrac{1}{3^n}\right)$

64. $a_n = 2 + \dfrac{1}{5^n}$

65. Compound Interest Consider the sequence $\{A_n\}$ whose *n*th term is given by

$$A_n = P\left(1 + \frac{r}{12}\right)^n$$

where P is the principal, A_n is the account balance after n months, and r is the interest rate compounded annually.

(a) Is $\{A_n\}$ a convergent sequence? Explain.

(b) Find the first 10 terms of the sequence when $P = \$10{,}000$ and $r = 0.055$.

66. Compound Interest A deposit of $100 is made in an account at the beginning of each month at an annual interest rate of 3% compounded monthly. The balance in the account after n months is $A_n = 100(401)[(1.0025)^n - 1]$.

(a) Compute the first six terms of the sequence $\{A_n\}$.

(b) Find the balance in the account after 5 years by computing the 60th term of the sequence.

(c) Find the balance in the account after 20 years by computing the 240th term of the sequence.

67. Inflation When the rate of inflation is $4\frac{1}{2}\%$ per year and the average price of a car is currently $25,000, the average price after n years is $P_n = \$25{,}000(1.045)^n$. Compute the average prices for the next 5 years.

· · 68. Government Expenditures · · · · · · · · ·

A government program that currently costs taxpayers $4.5 billion per year is cut back by 6% per year.

(a) Write an expression for the amount budgeted for this program after n years.

(b) Compute the budgets for the first 4 years.

(c) Determine the convergence or divergence of the sequence of reduced budgets. If the sequence converges, find its limit.

EXPLORING CONCEPTS

69. Writing a Sequence Give an example of a sequence satisfying the condition.

(a) A monotonically increasing sequence that converges to 10

(b) A sequence that converges to $\frac{3}{4}$

70. Writing a Sequence Give an example of a bounded sequence that has a limit and an example of a bounded sequence that does not have a limit.

71. Monotonic Sequence Let $\{a_n\}$ be a monotonic sequence such that $a_n \leq 1$. Discuss the convergence of $\{a_n\}$. When $\{a_n\}$ converges, what can you conclude about its limit?

72. HOW DO YOU SEE IT? The graphs of two sequences are shown in the figures. Which graph represents the sequence with alternating signs? Explain.

(a)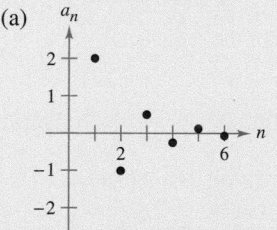

(b)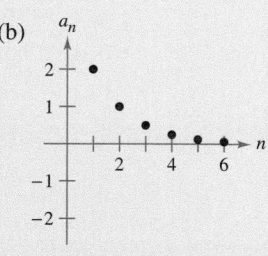

73. Using a Sequence Compute the first six terms of the sequence $\{a_n\} = \{\sqrt[n]{n}\}$. If the sequence converges, find its limit.

74. Using a Sequence Compute the first six terms of the sequence

$$\{a_n\} = \left\{\sqrt{n}\ln\left(1 + \frac{1}{n}\right)\right\}.$$

If the sequence converges, find its limit.

75. Proof Prove that if $\{s_n\}$ converges to L and $L > 0$, then there exists a number N such that $s_n > 0$ for $n > N$.

76. Population

The populations a_n (in millions) of Zimbabwe from 2000 through 2015 are given below as ordered pairs of the form (n, a_n), where n represents the year, with $n = 0$ corresponding to 2000. *(Source: U.S. Census Bureau)*

(0, 11.8), (1, 11.9),
(2, 11.9), (3, 11.8),
(4, 11.7), (5, 11.6),
(6, 11.5), (7, 11.4),
(8, 11.4), (9, 11.4),
(10, 11.7), (11, 12.1),
(12, 12.6), (13, 13.2),
(14, 13.8), (15, 14.2)

(a) Use the regression capabilities of a graphing utility to find a model of the form

$$a_n = bn^4 + cn^3 + dn^2 + en + f, \quad n = 0, 1, \ldots, 15$$

for the data. Use the graphing utility to plot the points and graph the model.

(b) Use the model to predict the population of Zimbabwe in 2020.

True or False? In Exercises 77–80, determine whether the statement is true or false. If it is false, explain why or give an example that shows it is false.

77. Sequences that diverge approach either ∞ or $-\infty$.

78. If $\{a_n\}$ converges, then $\lim\limits_{n \to \infty} (a_n - a_{n+1}) = 0$.

79. If $\{a_n\}$ converges, then $\{a_n/n\}$ converges to 0.

80. If $\{a_n\}$ diverges and $\{b_n\}$ diverges, then $\{a_n + b_n\}$ diverges.

81. Fibonacci Sequence In a study of the progeny of rabbits, Fibonacci (ca. 1170–ca. 1240) encountered the sequence now bearing his name. The sequence is defined recursively as $a_{n+2} = a_n + a_{n+1}$, where $a_1 = 1$ and $a_2 = 1$.

(a) Write the first 12 terms of the sequence.

(b) Write the first 10 terms of the sequence defined by
$$b_n = \frac{a_{n+1}}{a_n}, \quad n \geq 1.$$

(c) Using the definition in part (b), show that $b_n = 1 + \dfrac{1}{b_{n-1}}$.

(d) The **golden ratio** ρ can be defined by $\lim\limits_{n \to \infty} b_n = \rho$. Show that $\rho = 1 + (1/\rho)$ and solve this equation for ρ.

82. Using a Theorem Show that the converse of Theorem 9.1 is not true. [*Hint:* Find a function $f(x)$ such that $f(n) = a_n$ converges, but $\lim\limits_{x \to \infty} f(x)$ does not exist.]

83. Using a Sequence Consider the sequence
$$\sqrt{2}, \sqrt{2 + \sqrt{2}}, \sqrt{2 + \sqrt{2 + \sqrt{2}}}, \ldots.$$

(a) Compute the first five terms of this sequence.

(b) Write a recursion formula for a_n, for $n \geq 2$.

(c) Find $\lim\limits_{n \to \infty} a_n$.

84. Using a Sequence Consider the sequence $\{a_n\}$, where $a_1 = \sqrt{k}$, $a_{n+1} = \sqrt{k + a_n}$, and $k > 0$.

(a) Show that $\{a_n\}$ is increasing and bounded.

(b) Prove that $\lim\limits_{n \to \infty} a_n$ exists.

(c) Find $\lim\limits_{n \to \infty} a_n$.

85. Squeeze Theorem

(a) Show that $\int_1^n \ln x \, dx < \ln(n!)$ for $n \geq 2$.

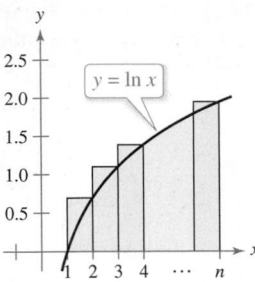

(b) Draw a graph similar to the one above that shows
$$\ln(n!) < \int_1^{n+1} \ln x \, dx.$$

(c) Use the results of parts (a) and (b) to show that
$$\frac{n^n}{e^{n-1}} < n! < \frac{(n+1)^{n+1}}{e^n}, \quad \text{for } n > 1.$$

(d) Use the Squeeze Theorem for Sequences and the result of part (c) to show that $\lim\limits_{n \to \infty} \left(\sqrt[n]{n!}/n\right) = 1/e$.

(e) Test the result of part (d) for $n = 20, 50,$ and 100.

86. Proof Prove, using the definition of the limit of a sequence, that
$$\lim\limits_{n \to \infty} \frac{1}{n^3} = 0.$$

87. Proof Prove, using the definition of the limit of a sequence, that $\lim\limits_{n \to \infty} r^n = 0$ for $-1 < r < 1$.

88. Using a Sequence Find a divergent sequence $\{a_n\}$ such that $\{a_{2n}\}$ converges.

89. Proof Prove Theorem 9.5 for a nonincreasing sequence.

PUTNAM EXAM CHALLENGE

90. Let $\{x_n\}$, $n \geq 0$, be a sequence of nonzero real numbers such that $x_n^2 - x_{n-1} x_{n+1} = 1$ for $n = 1, 2, 3, \ldots$. Prove there exists a real number a such that $x_{n+1} = ax_n - x_{n-1}$ for all $n \geq 1$.

91. Let $T_0 = 2$, $T_1 = 3$, $T_2 = 6$, and for $n \geq 3$,
$$T_n = (n + 4)T_{n-1} - 4nT_{n-2} + (4n - 8)T_{n-3}.$$

The first few terms are

2, 3, 6, 14, 40, 152, 784, 5168, 40576.

Find, with proof, a formula for T_n of the form $T_n = A_n + B_n$, where $\{A_n\}$ and $\{B_n\}$ are well-known sequences.

9.2 Series and Convergence

- Understand the definition of a convergent infinite series.
- Use properties of infinite geometric series.
- Use the *n*th-Term Test for Divergence of an infinite series.

Infinite Series

One important application of infinite sequences is in representing "infinite summations." Informally, if $\{a_n\}$ is an infinite sequence, then

$$\sum_{n=1}^{\infty} a_n = a_1 + a_2 + a_3 + \cdots + a_n + \cdots \qquad \text{Infinite series}$$

is an **infinite series** (or simply a **series**). The numbers a_1, a_2, a_3, and so on are the **terms** of the series. For some series, it is convenient to begin the index at $n = 0$ (or some other integer). As a typesetting convention, it is common to represent an infinite series as $\Sigma\, a_n$. In such cases, the starting value for the index must be taken from the context of the statement.

To find the sum of an infinite series, consider the **sequence of partial sums** listed below.

$$S_1 = a_1$$
$$S_2 = a_1 + a_2$$
$$S_3 = a_1 + a_2 + a_3$$
$$S_4 = a_1 + a_2 + a_3 + a_4$$
$$S_5 = a_1 + a_2 + a_3 + a_4 + a_5$$
$$\vdots$$
$$S_n = a_1 + a_2 + a_3 + a_4 + a_5 + \cdots + a_n$$

If this sequence of partial sums converges, then the series is said to converge and has the sum indicated in the next definition.

• • • • • • • • • • • • • • • • ▷
• • REMARK As you study this chapter, it is important to distinguish between an infinite series and a sequence. A sequence is an ordered collection of numbers

$$a_1, a_2, a_3, \ldots, a_n, \ldots$$

whereas a series is an infinite sum of terms from a sequence

$$a_1 + a_2 + a_3 + \cdots + a_n + \cdots.$$

INFINITE SERIES

The study of infinite series was considered a novelty in the fourteenth century. Logician Richard Suiseth, whose nickname was Calculator, solved this problem.

If throughout the first half of a given time interval a variation continues at a certain intensity, throughout the next quarter of the interval at double the intensity, throughout the following eighth at triple the intensity and so ad infinitum, then the average intensity for the whole interval will be the intensity of the variation during the second subinterval (or double the intensity).

This is the same as saying that the sum of the infinite series

$$\frac{1}{2} + \frac{2}{4} + \frac{3}{8} + \cdots + \frac{n}{2^n} + \cdots$$

is 2.

Definitions of Convergent and Divergent Series

For the infinite series $\displaystyle\sum_{n=1}^{\infty} a_n$, the **nth partial sum** is

$$S_n = a_1 + a_2 + \cdots + a_n.$$

If the sequence of partial sums $\{S_n\}$ converges to S, then the series $\displaystyle\sum_{n=1}^{\infty} a_n$ **converges.** The limit S is called the **sum of the series.**

$$S = a_1 + a_2 + \cdots + a_n + \cdots \qquad S = \sum_{n=1}^{\infty} a_n$$

If $\{S_n\}$ diverges, then the series **diverges.**

As you study this chapter, you will see that there are two basic questions involving infinite series.

- Does a series converge or does it diverge?
- When a series converges, what is its sum?

These questions are not always easy to answer, especially the second one.

EXAMPLE 1 **Convergent and Divergent Series**

a. The series

$$\sum_{n=1}^{\infty} \frac{1}{2^n} = \frac{1}{2} + \frac{1}{4} + \frac{1}{8} + \frac{1}{16} + \cdots$$

has the partial sums listed below.

$$S_1 = \frac{1}{2}$$

$$S_2 = \frac{1}{2} + \frac{1}{4} = \frac{3}{4}$$

$$S_3 = \frac{1}{2} + \frac{1}{4} + \frac{1}{8} = \frac{7}{8}$$

$$\vdots$$

$$S_n = \frac{1}{2} + \frac{1}{4} + \frac{1}{8} + \cdots + \frac{1}{2^n} = \frac{2^n - 1}{2^n}$$

Because

$$\lim_{n \to \infty} \frac{2^n - 1}{2^n} = 1$$

it follows that the series converges and its sum is 1. (You can also determine the partial sums of the series geometrically, as shown in Figure 9.5.)

b. The nth partial sum of the series

$$\sum_{n=1}^{\infty} \left(\frac{1}{n} - \frac{1}{n+1} \right) = \left(1 - \frac{1}{2} \right) + \left(\frac{1}{2} - \frac{1}{3} \right) + \left(\frac{1}{3} - \frac{1}{4} \right) + \cdots$$

is

$$S_n = 1 - \frac{1}{n+1}.$$

Because the limit of S_n is 1, the series converges and its sum is 1.

c. The series

$$\sum_{n=1}^{\infty} 1 = 1 + 1 + 1 + 1 + \cdots$$

diverges because $S_n = n$ and the sequence of partial sums diverges.

The series in Example 1(b) is a **telescoping series** of the form

$$(b_1 - b_2) + (b_2 - b_3) + (b_3 - b_4) + (b_4 - b_5) + \cdots. \qquad \text{Telescoping series}$$

Note that b_2 is canceled by the second term, b_3 is canceled by the third term, and so on. Because the nth partial sum of this series is

$$S_n = b_1 - b_{n+1}$$

it follows that a telescoping series will converge if and only if b_n approaches a finite number as $n \to \infty$. Moreover, if the series converges, then its sum is

$$S = b_1 - \lim_{n \to \infty} b_{n+1}.$$

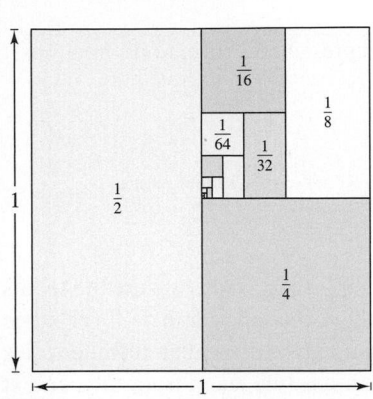

You can determine the partial sums of the series in Example 1(a) geometrically using this figure.
Figure 9.5

■ FOR FURTHER INFORMATION
To learn more about the partial sums of infinite series, see the article "Six Ways to Sum a Series" by Dan Kalman in *The College Mathematics Journal*. To view this article, go to *MathArticles.com.*

EXAMPLE 2 **Writing a Series in Telescoping Form**

Find the sum of the series $\displaystyle\sum_{n=1}^{\infty} \frac{2}{4n^2 - 1}$.

Solution

Using partial fractions, you can write

$$a_n = \frac{2}{4n^2 - 1} = \frac{2}{(2n - 1)(2n + 1)} = \frac{1}{2n - 1} - \frac{1}{2n + 1}.$$

From this telescoping form, you can see that the nth partial sum is

$$S_n = \left(\frac{1}{1} - \frac{1}{3}\right) + \left(\frac{1}{3} - \frac{1}{5}\right) + \cdots + \left(\frac{1}{2n - 1} - \frac{1}{2n + 1}\right) = 1 - \frac{1}{2n + 1}.$$

So, the series converges and its sum is 1. That is,

$$\sum_{n=1}^{\infty} \frac{2}{4n^2 - 1} = \lim_{n \to \infty} S_n = \lim_{n \to \infty} \left(1 - \frac{1}{2n + 1}\right) = 1.$$

Geometric Series

The series in Example 1(a) is a **geometric series.** In general, the series

$$\sum_{n=0}^{\infty} ar^n = a + ar + ar^2 + \cdots + ar^n + \cdots, \quad a \neq 0 \qquad \text{Geometric series}$$

is a **geometric series** with ratio r, $r \neq 0$.

THEOREM 9.6 Convergence of a Geometric Series

A geometric series with ratio r diverges when $|r| \geq 1$. If $|r| < 1$, then the series converges to the sum

$$\sum_{n=0}^{\infty} ar^n = \frac{a}{1 - r}, \quad |r| < 1.$$

Proof It is easy to see that the series diverges when $r = \pm 1$. If $r \neq \pm 1$, then

$$S_n = a + ar + ar^2 + \cdots + ar^{n-1}.$$

Multiplication by r yields

$$rS_n = ar + ar^2 + ar^3 + \cdots + ar^n.$$

Subtracting the second equation from the first produces $S_n - rS_n = a - ar^n$. Therefore, $S_n(1 - r) = a(1 - r^n)$, and the nth partial sum is

$$S_n = \frac{a}{1 - r}(1 - r^n).$$

When $|r| < 1$, it follows that $r^n \to 0$ as $n \to \infty$, and you obtain

$$\lim_{n \to \infty} S_n = \lim_{n \to \infty} \left[\frac{a}{1 - r}(1 - r^n)\right] = \frac{a}{1 - r}\left[\lim_{n \to \infty}(1 - r^n)\right] = \frac{a}{1 - r}$$

which means that the series *converges* and its sum is $a/(1 - r)$. It is left to you to show that the series diverges when $|r| > 1$.

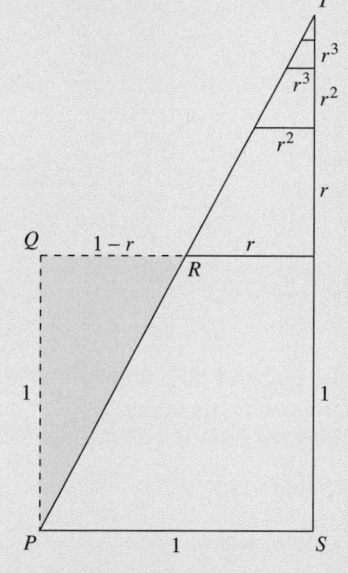

▷ TECHNOLOGY Figure 9.6 shows the first 20 partial sums of the infinite series in Example 3(a). Notice how the values appear to approach the line $y = 6$. Using a graphing utility to sum the first 20 terms, you should obtain a sum of about 5.999994.

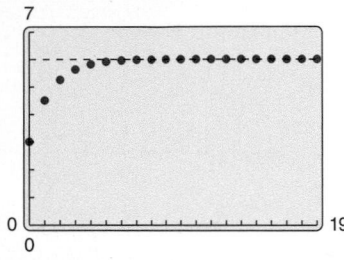

Figure 9.6

| EXAMPLE 3 | **Convergent and Divergent Geometric Series** |

a. The geometric series

$$\sum_{n=0}^{\infty} \frac{3}{2^n} = \sum_{n=0}^{\infty} 3\left(\frac{1}{2}\right)^n = 3(1) + 3\left(\frac{1}{2}\right) + 3\left(\frac{1}{2}\right)^2 + \cdots$$

has a ratio of $r = \frac{1}{2}$ with $a = 3$. Because $|r| < 1$, the series converges and its sum is

$$S = \frac{a}{1 - r} = \frac{3}{1 - (1/2)} = 6. \qquad \text{See Figure 9.6.}$$

b. The geometric series

$$\sum_{n=0}^{\infty} \left(\frac{3}{2}\right)^n = 1 + \frac{3}{2} + \frac{9}{4} + \frac{27}{8} + \cdots$$

has a ratio of $r = \frac{3}{2}$. Because $|r| \geq 1$, the series diverges. ∎

The formula for the sum of a geometric series can be used to write a repeating decimal as the ratio of two integers, as demonstrated in the next example.

| EXAMPLE 4 | **A Geometric Series for a Repeating Decimal** |

⋮ ⋯▷ *See LarsonCalculus.com for an interactive version of this type of example.*

Use a geometric series to write $0.\overline{08}$ as the ratio of two integers.

Solution For the repeating decimal $0.\overline{08}$, you can write

$$0.080808\ldots = \frac{8}{10^2} + \frac{8}{10^4} + \frac{8}{10^6} + \frac{8}{10^8} + \cdots$$

$$= \sum_{n=0}^{\infty} \left(\frac{8}{10^2}\right)\left(\frac{1}{10^2}\right)^n.$$

For this series, you have $a = 8/10^2$ and $r = 1/10^2$. So,

$$0.080808\ldots = \frac{a}{1 - r} = \frac{8/10^2}{1 - (1/10^2)} = \frac{8}{99}.$$

Try dividing 8 by 99 on a calculator to see that it produces $0.\overline{08}$. ∎

The convergence of a series is not affected by the removal of a finite number of terms from the beginning of the series. For instance, the geometric series

$$\sum_{n=4}^{\infty} \left(\frac{1}{2}\right)^n \quad \text{and} \quad \sum_{n=0}^{\infty} \left(\frac{1}{2}\right)^n$$

both converge. Furthermore, because the sum of the second series is

$$\frac{a}{1 - r} = \frac{1}{1 - (1/2)} = 2$$

you can conclude that the sum of the first series is

$$S = 2 - \left[\left(\frac{1}{2}\right)^0 + \left(\frac{1}{2}\right)^1 + \left(\frac{1}{2}\right)^2 + \left(\frac{1}{2}\right)^3\right]$$

$$= 2 - \frac{15}{8}$$

$$= \frac{1}{8}.$$

•• **REMARK** In words, Theorem 9.6 states that "the sum of a convergent geometric series is the first term of the series divided by the difference of 1 and the ratio r." For the series

$$\sum_{n=4}^{\infty} \left(\frac{1}{2}\right)^n$$

note that the first term is $(1/2)^4$ and $r = 1/2$. So, the sum is

$$S = \frac{(1/2)^4}{1 - (1/2)} = \frac{1}{8}.$$

The properties in the next theorem are direct consequences of the corresponding properties of limits of sequences.

THEOREM 9.7   Properties of Infinite Series

Let $\Sigma\, a_n$ and $\Sigma\, b_n$ be convergent series, and let A, B, and c be real numbers. If $\Sigma\, a_n = A$ and $\Sigma\, b_n = B$, then the following series converge to the indicated sums.

1. $\displaystyle\sum_{n=1}^{\infty} ca_n = cA$

2. $\displaystyle\sum_{n=1}^{\infty} (a_n + b_n) = A + B$

3. $\displaystyle\sum_{n=1}^{\infty} (a_n - b_n) = A - B$

---

nth-Term Test for Divergence

The next theorem states that when a series converges, the limit of its nth term must be 0.

THEOREM 9.8   Limit of the nth Term of a Convergent Series

If $\displaystyle\sum_{n=1}^{\infty} a_n$ converges, then $\displaystyle\lim_{n\to\infty} a_n = 0$.

Proof   Assume that

$$\sum_{n=1}^{\infty} a_n = \lim_{n\to\infty} S_n = L.$$

Then, because $S_n = S_{n-1} + a_n$ and

$$\lim_{n\to\infty} S_n = \lim_{n\to\infty} S_{n-1} = L$$

it follows that

$$
\begin{aligned}
L &= \lim_{n\to\infty} S_n \\
  &= \lim_{n\to\infty} (S_{n-1} + a_n) \\
  &= \lim_{n\to\infty} S_{n-1} + \lim_{n\to\infty} a_n \\
  &= L + \lim_{n\to\infty} a_n
\end{aligned}
$$

which implies that $\{a_n\}$ converges to 0. ∎

The contrapositive of Theorem 9.8 provides a useful test for *divergence*. This **nth-Term Test for Divergence** states that if the limit of the nth term of a series does *not* converge to 0, then the series must diverge.

THEOREM 9.9 nth-Term Test for Divergence

If $\displaystyle\lim_{n\to\infty} a_n \neq 0$ then $\displaystyle\sum_{n=1}^{\infty} a_n$ diverges.

EXAMPLE 5 **Using the *n*th-Term Test for Divergence**

a. For the series $\displaystyle\sum_{n=0}^{\infty} 2^n$, you have

$$\lim_{n\to\infty} 2^n = \infty.$$

So, the limit of the *n*th term is not 0, and the series diverges.

b. For the series $\displaystyle\sum_{n=1}^{\infty} \frac{n!}{2n! + 1}$, you have

$$\lim_{n\to\infty} \frac{n!}{2n! + 1} = \frac{1}{2}.$$

So, the limit of the *n*th term is not 0, and the series diverges.

c. For the series $\displaystyle\sum_{n=1}^{\infty} \frac{1}{n}$, you have

$$\lim_{n\to\infty} \frac{1}{n} = 0.$$

Because the limit of the *n*th term is 0, the *n*th-Term Test for Divergence does *not* apply and you can draw no conclusions about convergence or divergence. (In the next section, you will see that this particular series diverges.)

> **REMARK** The series in Example 5(c) will play an important role in this chapter.
>
> $$\sum_{n=1}^{\infty} \frac{1}{n} =$$
>
> $$1 + \frac{1}{2} + \frac{1}{3} + \frac{1}{4} + \cdots$$
>
> You will see that this series diverges even though the *n*th term approaches 0 as *n* approaches ∞.

EXAMPLE 6 **Bouncing Ball Problem**

A ball is dropped from a height of 6 feet and begins bouncing, as shown in Figure 9.7. The height of each bounce is three-fourths the height of the previous bounce. Find the total vertical distance traveled by the ball.

Solution When the ball hits the ground for the first time, it has traveled a distance of $D_1 = 6$ feet. For subsequent bounces, let D_i be the distance traveled up and down. For example, D_2 and D_3 are

$$D_2 = \underbrace{6\left(\frac{3}{4}\right)}_{\text{Up}} + \underbrace{6\left(\frac{3}{4}\right)}_{\text{Down}} = 12\left(\frac{3}{4}\right)$$

and

$$D_3 = \underbrace{6\left(\frac{3}{4}\right)\left(\frac{3}{4}\right)}_{\text{Up}} + \underbrace{6\left(\frac{3}{4}\right)\left(\frac{3}{4}\right)}_{\text{Down}} = 12\left(\frac{3}{4}\right)^2.$$

By continuing this process, it can be determined that the total vertical distance is

$$D = 6 + 12\left(\frac{3}{4}\right) + 12\left(\frac{3}{4}\right)^2 + 12\left(\frac{3}{4}\right)^3 + \cdots$$

$$= 6 + 12 \sum_{n=0}^{\infty} \left(\frac{3}{4}\right)^{n+1}$$

$$= 6 + 12\left(\frac{3}{4}\right) \sum_{n=0}^{\infty} \left(\frac{3}{4}\right)^n$$

$$= 6 + 9\left[\frac{1}{1 - (3/4)}\right]$$

$$= 6 + 9(4)$$

$$= 42 \text{ feet.}$$

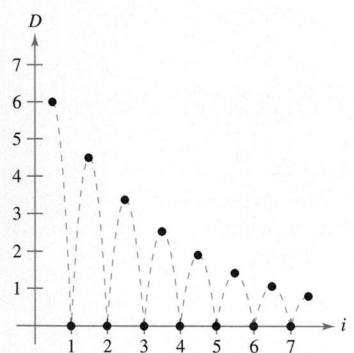

The height of each bounce is three-fourths the height of the preceding bounce.

Figure 9.7

9.2 Exercises

See **CalcChat.com** for tutorial help and worked-out solutions to odd-numbered exercises.

CONCEPT CHECK

1. **Sequence and Series** Describe the difference between $\lim\limits_{n\to\infty} a_n = 5$ and $\sum\limits_{n=1}^{\infty} a_n = 5$.

2. **Geometric Series** Define a geometric series, state when it converges, and give the formula for the sum of a convergent geometric series.

3. **Limit of the nth term of a Series** The limit of the nth term of a series converges to 0. What can you conclude about the convergence or divergence of the series?

4. **Limit of the nth Term of a Series** The limit of the nth term of a series does not converge to 0. What can you conclude about the convergence or divergence of the series?

Finding Partial Sums In Exercises 5–10, find the sequence of partial sums $S_1, S_2, S_3, S_4,$ and S_5.

5. $1 + \frac{1}{4} + \frac{1}{9} + \frac{1}{16} + \frac{1}{25} + \cdots$

6. $\frac{1}{2 \cdot 3} + \frac{2}{3 \cdot 4} + \frac{3}{4 \cdot 5} + \frac{4}{5 \cdot 6} + \frac{5}{6 \cdot 7} + \cdots$

7. $3 - \frac{9}{2} + \frac{27}{4} - \frac{81}{8} + \frac{243}{16} - \cdots$

8. $1 + \frac{1}{2} + \frac{1}{4} + \frac{1}{6} + \frac{1}{8} + \frac{1}{10} + \cdots$

9. $\sum\limits_{n=1}^{\infty} \frac{3}{2^{n-1}}$

10. $\sum\limits_{n=1}^{\infty} \frac{(-1)^{n+1}}{n!}$

Verifying Divergence In Exercises 11–18, verify that the infinite series diverges.

11. $\sum\limits_{n=0}^{\infty} 5\left(\frac{5}{2}\right)^n$

12. $\sum\limits_{n=0}^{\infty} 4(-1.05)^n$

13. $\sum\limits_{n=1}^{\infty} \frac{n}{n+1}$

14. $\sum\limits_{n=1}^{\infty} \frac{n}{2n+3}$

15. $\sum\limits_{n=1}^{\infty} \frac{n^3+1}{n^3+n^2}$

16. $\sum\limits_{n=1}^{\infty} \frac{2n}{\sqrt{n^2+1}}$

17. $\sum\limits_{n=1}^{\infty} \frac{4^n+3}{4^{n+1}}$

18. $\sum\limits_{n=1}^{\infty} \frac{(n+1)!}{5n!}$

Verifying Convergence In Exercises 19–24, verify that the infinite series converges.

19. $\sum\limits_{n=0}^{\infty} \left(\frac{5}{6}\right)^n$

20. $\sum\limits_{n=1}^{\infty} 2\left(-\frac{1}{2}\right)^n$

21. $\sum\limits_{n=0}^{\infty} (0.9)^n = 1 + 0.9 + 0.81 + 0.729 + \cdots$

22. $\sum\limits_{n=0}^{\infty} (-0.2)^n = 1 - 0.2 + 0.04 - 0.008 + \cdots$

23. $\sum\limits_{n=1}^{\infty} \frac{1}{n(n+1)}$ (*Hint:* Use partial fractions.)

24. $\sum\limits_{n=1}^{\infty} \frac{1}{n(n+2)}$ (*Hint:* Use partial fractions.)

Numerical, Graphical, and Analytic Analysis In Exercises 25–28, (a) find the sum of the series, (b) use a graphing utility to find the indicated partial sum S_n and complete the table, (c) use a graphing utility to graph the first 10 terms of the sequence of partial sums and a horizontal line representing the sum, and (d) explain the relationship between the magnitudes of the terms of the series and the rate at which the sequence of partial sums approaches the sum of the series.

n	5	10	20	50	100
S_n					

25. $\sum\limits_{n=1}^{\infty} \frac{6}{n(n+3)}$

26. $\sum\limits_{n=1}^{\infty} \frac{4}{n(n+4)}$

27. $\sum\limits_{n=1}^{\infty} 2(0.9)^{n-1}$

28. $\sum\limits_{n=1}^{\infty} 10\left(-\frac{1}{4}\right)^{n-1}$

Finding the Sum of a Convergent Series In Exercises 29–38, find the sum of the convergent series.

29. $\sum\limits_{n=0}^{\infty} 5\left(\frac{2}{3}\right)^n$

30. $\sum\limits_{n=0}^{\infty} \left(-\frac{1}{5}\right)^n$

31. $\sum\limits_{n=1}^{\infty} \frac{4}{n(n+2)}$

32. $\sum\limits_{n=1}^{\infty} \frac{1}{(2n+1)(2n+3)}$

33. $8 + 6 + \frac{9}{2} + \frac{27}{8} + \cdots$

34. $9 - 3 + 1 - \frac{1}{3} + \cdots$

35. $\sum\limits_{n=0}^{\infty} \left(\frac{1}{2^n} - \frac{1}{3^n}\right)$

36. $\sum\limits_{n=0}^{\infty} [(0.3)^n + (0.8)^n]$

37. $\sum\limits_{n=1}^{\infty} (\sin 1)^n$

38. $\sum\limits_{n=1}^{\infty} \frac{1}{9n^2 + 3n - 2}$

Using a Geometric Series In Exercises 39–44, (a) write the repeating decimal as a geometric series and (b) write the sum of the series as the ratio of two integers.

39. $0.\overline{4}$

40. $0.\overline{63}$

41. $0.\overline{12}$

42. $0.0\overline{1}$

43. $0.0\overline{75}$

44. $0.2\overline{15}$

Determining Convergence or Divergence In Exercises 45–58, determine the convergence or divergence of the series.

45. $\sum\limits_{n=0}^{\infty} (1.075)^n$

46. $\sum\limits_{n=0}^{\infty} \frac{6^n}{n+1}$

47. $\sum\limits_{n=1}^{\infty} \frac{n+1}{2n-1}$

48. $\sum\limits_{n=1}^{\infty} \frac{4n+1}{3n-1}$

49. $\sum\limits_{n=1}^{\infty} \left(\frac{1}{n} - \frac{1}{n+2}\right)$

50. $\sum\limits_{n=1}^{\infty} \left(\frac{1}{n+1} - \frac{1}{n+2}\right)$

51. $\sum\limits_{n=1}^{\infty} \frac{3^n}{n^3}$

52. $\sum\limits_{n=0}^{\infty} \frac{7}{5^n}$

53. $\displaystyle\sum_{n=2}^{\infty} \frac{n}{\ln n}$

54. $\displaystyle\sum_{n=1}^{\infty} \ln \frac{1}{n}$

55. $\displaystyle\sum_{n=1}^{\infty} \left(1 + \frac{k}{n}\right)^n$

56. $\displaystyle\sum_{n=1}^{\infty} e^{-n}$

57. $\displaystyle\sum_{n=1}^{\infty} \arctan n$

58. $\displaystyle\sum_{n=1}^{\infty} \ln\left(\frac{n+1}{n}\right)$

EXPLORING CONCEPTS

59. Using a Series You delete a finite number of terms from a divergent series. Will the new series still diverge? Explain your reasoning.

60. Using a Series You add a finite number of terms to a convergent series. Will the new series still converge? Explain your reasoning.

Making a Series Converge In Exercises 61–66, find all values of x for which the series converges. For these values of x, write the sum of the series as a function of x.

61. $\displaystyle\sum_{n=1}^{\infty} (3x)^n$

62. $\displaystyle\sum_{n=0}^{\infty} \left(\frac{2}{x}\right)^n$

63. $\displaystyle\sum_{n=1}^{\infty} (x-1)^n$

64. $\displaystyle\sum_{n=0}^{\infty} 5\left(\frac{x-2}{3}\right)^n$

65. $\displaystyle\sum_{n=0}^{\infty} (-1)^n x^n$

66. $\displaystyle\sum_{n=0}^{\infty} (-1)^n x^{2n}$

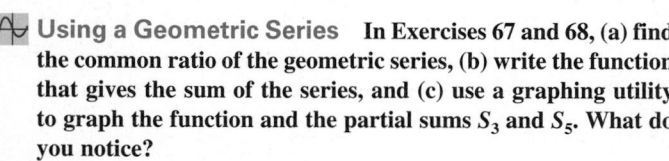 **Using a Geometric Series** In Exercises 67 and 68, (a) find the common ratio of the geometric series, (b) write the function that gives the sum of the series, and (c) use a graphing utility to graph the function and the partial sums S_3 and S_5. What do you notice?

67. $1 + x + x^2 + x^3 + \cdots$

68. $1 - \dfrac{x}{2} + \dfrac{x^2}{4} - \dfrac{x^3}{8} + \cdots$

Writing In Exercises 69 and 70, use a graphing utility to determine the first term that is less than 0.0001 in each of the convergent series. Note that the answers are very different. Explain how this will affect the rate at which the series converges.

69. $\displaystyle\sum_{n=1}^{\infty} \frac{1}{n(n+1)}, \quad \sum_{n=1}^{\infty} \left(\frac{1}{8}\right)^n$

70. $\displaystyle\sum_{n=1}^{\infty} \frac{1}{2^n}, \quad \sum_{n=1}^{\infty} (0.01)^n$

71. Marketing An electronic games manufacturer producing a new product estimates the annual sales to be 8000 units. Each year, 5% of the units that have been sold will become inoperative. So, 8000 units will be in use after 1 year, $[8000 + 0.95(8000)]$ units will be in use after 2 years, and so on. How many units will be in use after n years?

72. Depreciation A company buys a machine for $475,000 that depreciates at a rate of 30% per year. Find a formula for the value of the machine after n years. What is its value after 5 years?

---

73. Multiplier Effect

The total annual spending by tourists in a resort city is $200 million. Approximately 75% of that revenue is again spent in the resort city, and of that amount approximately 75% is again spent in the same city, and so on. Write the geometric series that gives the total amount of spending generated by the $200 million and find the sum of the series.

74. Multiplier Effect Repeat Exercise 73 when the percent of the revenue that is spent again in the city decreases to 60%.

75. Distance A ball is dropped from a height of 16 feet. Each time it drops h feet, it rebounds $0.81h$ feet. Find the total distance traveled by the ball.

76. Time The ball in Exercise 75 takes the following times for each fall.

$s_1 = -16t^2 + 16,$ $\qquad s_1 = 0$ when $t = 1$

$s_2 = -16t^2 + 16(0.81),$ $\qquad s_2 = 0$ when $t = 0.9$

$s_3 = -16t^2 + 16(0.81)^2,$ $\qquad s_3 = 0$ when $t = (0.9)^2$

$s_4 = -16t^2 + 16(0.81)^3,$ $\qquad s_4 = 0$ when $t = (0.9)^3$

$\qquad \vdots \qquad\qquad\qquad\qquad\qquad \vdots$

$s_n = -16t^2 + 16(0.81)^{n-1},$ $\quad s_n = 0$ when $t = (0.9)^{n-1}$

Beginning with s_2, the ball takes the same amount of time to bounce up as it does to fall, so the total time elapsed before it comes to rest is given by

$$t = 1 + 2\sum_{n=1}^{\infty} (0.9)^n.$$

Find this total time.

Probability In Exercises 77 and 78, the random variable n represents the number of units of a product sold per day in a store. The probability distribution of n is given by $P(n)$. Find the probability that two units are sold in a given day $[P(2)]$ and show that $P(0) + P(1) + P(2) + P(3) + \cdots = 1$.

77. $P(n) = \dfrac{1}{2}\left(\dfrac{1}{2}\right)^n$

78. $P(n) = \dfrac{1}{3}\left(\dfrac{2}{3}\right)^n$

79. Probability A fair coin is tossed repeatedly. The probability that the first head occurs on the nth toss is given by $P(n) = \left(\frac{1}{2}\right)^n$, where $n \geq 1$.

(a) Show that $\displaystyle\sum_{n=1}^{\infty} \left(\frac{1}{2}\right)^n = 1$.

(b) The expected number of tosses required until the first head occurs in the experiment is given by

$$\sum_{n=1}^{\infty} n\left(\frac{1}{2}\right)^n.$$

Is this series geometric?

(c) Use a computer algebra system to find the sum in part (b).

80. Probability In an experiment, three people toss a fair coin one at a time until one of them tosses a head. Determine, for each person, the probability that he or she tosses the first head. Verify that the sum of the three probabilities is 1.

81. Area The sides of a square are 16 inches in length. A new square is formed by connecting the midpoints of the sides of the original square, and two of the triangles outside the second square are shaded (see figure). Determine the area of the shaded regions (a) when this process is continued five more times and (b) when this pattern of shading is continued infinitely.

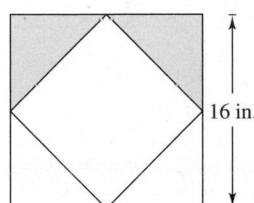

16 in.

Figure for 81   Figure for 82

82. Length A right triangle XYZ is shown above where $|XY| = z$ and $\angle X = \theta$. Line segments are continually drawn to be perpendicular to the triangle, as shown in the figure.

(a) Find the total length of the perpendicular line segments $|Yy_1| + |x_1y_1| + |x_1y_2| + \cdots$ in terms of z and θ.

(b) Find the total length of the perpendicular line segments when $z = 1$ and $\theta = \pi/6$.

Using a Geometric Series In Exercises 83–86, use the formula for the nth partial sum of a geometric series

$$\sum_{i=0}^{n-1} ar^i = \frac{a(1 - r^n)}{1 - r}.$$

83. Present Value The winner of a $2,000,000 sweepstakes will be paid $100,000 per year for 20 years. The money earns 6% interest per year. The present value of the winnings is

$$\sum_{n=1}^{20} 100,000 \left(\frac{1}{1.06} \right)^n.$$ Compute the present value and interpret its meaning.

84. Annuities When an employee receives a paycheck at the end of each month, P dollars is invested in a retirement account. These deposits are made each month for t years and the account earns interest at the annual percentage rate r. When the interest is compounded monthly, the amount A in the account at the end of t years is

$$A = P + P\left(1 + \frac{r}{12}\right) + \cdots + P\left(1 + \frac{r}{12}\right)^{12t - 1}$$

$$= P\left(\frac{12}{r}\right)\left[\left(1 + \frac{r}{12}\right)^{12t} - 1\right].$$

When the interest is compounded continuously, the amount A in the account after t years is

$$A = P + Pe^{r/12} + Pe^{2r/12} + \cdots + Pe^{(12t-1)r/12}$$

$$= \frac{P(e^{rt} - 1)}{e^{r/12} - 1}.$$

Verify the formulas for the sums given above.

85. Salary You go to work at a company that pays $0.01 for the first day, $0.02 for the second day, $0.04 for the third day, and so on. If the daily wage keeps doubling, what would your total income be for working (a) 29 days, (b) 30 days, and (c) 31 days?

• • 86. Sphereflake •

The sphereflake shown below is a computer-generated fractal that was created by Eric Haines. The radius of the large sphere is 1. To the large sphere, nine spheres of radius $\frac{1}{3}$ are attached. To each of these, nine spheres of radius $\frac{1}{9}$ are attached. This process is continued infinitely. Prove that the sphereflake has an infinite surface area.

Annuities In Exercises 87–90, consider making monthly deposits of P dollars in a savings account at an annual interest rate r. Use the results of Exercise 84 to find the balance A in the account after t years when the interest is compounded (a) monthly and (b) continuously.

87. $P = \$50$, $r = 2\%$, $t = 20$ years

88. $P = \$200$, $r = 5.5\%$, $t = 25$ years

89. $P = \$1050$, $r = 0.9\%$, $t = 35$ years

90. $P = \$175$, $r = 4\%$, $t = 50$ years

True or False? In Exercises 91–96, determine whether the statement is true or false. If it is false, explain why or give an example that shows it is false.

91. If $\lim_{n \to \infty} a_n = 0$, then $\sum_{n=1}^{\infty} a_n$ converges.

92. If $\sum_{n=1}^{\infty} a_n = L$, then $\sum_{n=0}^{\infty} a_n = L + a_0$.

93. If $|r| < 1$, then $\sum_{n=1}^{\infty} ar^n = \frac{a}{1 - r}$.

94. The series $\sum_{n=1}^{\infty} \frac{n}{1000(n + 1)}$ diverges.

95. $0.75 = 0.749999 \ldots$

96. Every decimal with a repeating pattern of digits is a rational number.

97. Using Divergent Series Find two divergent series $\Sigma\, a_n$ and $\Sigma\, b_n$ such that $\Sigma(a_n + b_n)$ converges.

98. Proof Given two infinite series $\Sigma\, a_n$ and $\Sigma\, b_n$ such that $\Sigma\, a_n$ converges and $\Sigma\, b_n$ diverges, prove that $\Sigma(a_n + b_n)$ diverges.

99. Fibonacci Sequence The Fibonacci sequence is defined recursively by $a_{n+2} = a_n + a_{n+1}$, where $a_1 = 1$ and $a_2 = 1$.

(a) Show that $\dfrac{1}{a_{n+1}a_{n+3}} = \dfrac{1}{a_{n+1}a_{n+2}} - \dfrac{1}{a_{n+2}a_{n+3}}$.

(b) Show that $\displaystyle\sum_{n=0}^{\infty} \dfrac{1}{a_{n+1}a_{n+3}} = 1$.

100. Remainder Let $\Sigma\, a_n$ be a convergent series, and let

$$R_N = a_{N+1} + a_{N+2} + \cdots$$

be the remainder of the series after the first N terms. Prove that $\displaystyle\lim_{N\to\infty} R_N = 0$.

101. Proof Prove that $\dfrac{1}{r} + \dfrac{1}{r^2} + \dfrac{1}{r^3} + \cdots = \dfrac{1}{r-1}$, for $|r| > 1$.

102. HOW DO YOU SEE IT? The figure below represents an informal way of showing that $\displaystyle\sum_{n=1}^{\infty} \dfrac{1}{n^2} < 2$. Explain how the figure implies this conclusion.

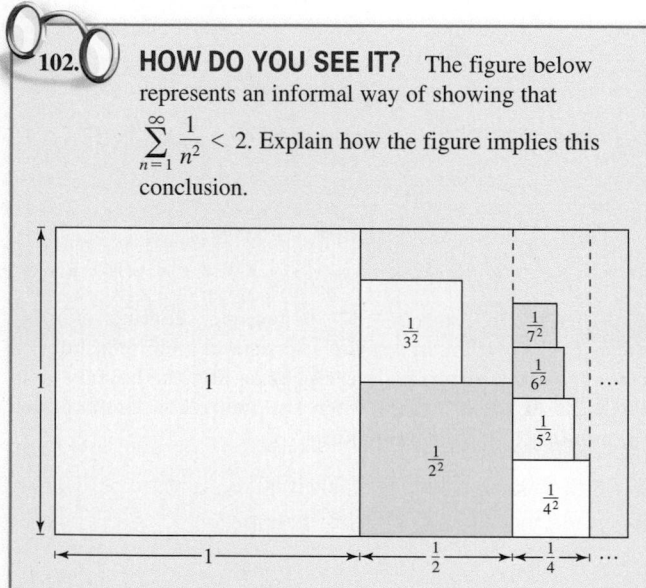

■ FOR FURTHER INFORMATION For more on this exercise, see the article "Convergence with Pictures" by P. J. Rippon in *American Mathematical Monthly*.

PUTNAM EXAM CHALLENGE

103. Express $\displaystyle\sum_{k=1}^{\infty} \dfrac{6^k}{(3^{k+1} - 2^{k+1})(3^k - 2^k)}$ as a rational number.

104. Let $f(n)$ be the sum of the first n terms of the sequence $0, 1, 1, 2, 2, 3, 3, 4, \ldots$, where the nth term is given by

$$a_n = \begin{cases} n/2, & \text{if } n \text{ is even} \\ (n-1)/2, & \text{if } n \text{ is odd} \end{cases}.$$

Show that if x and y are positive integers and $x > y$ then $xy = f(x + y) - f(x - y)$.

These problems were composed by the Committee on the Putnam Prize Competition. © The Mathematical Association of America. All rights reserved.

Cantor's Disappearing Table

The following procedure shows how to make a table disappear by removing only half of the table!

(a) Original table has a length of L.

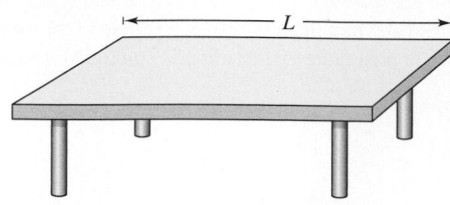

(b) Remove $\frac{1}{4}$ of the table centered at the midpoint. Each remaining piece has a length that is less than $\frac{1}{2}L$.

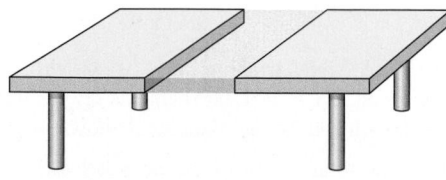

(c) Remove $\frac{1}{8}$ of the table by taking sections of length $\frac{1}{16}L$ from the centers of each of the two remaining pieces. Now you have removed $\frac{1}{4} + \frac{1}{8}$ of the table. Each remaining piece has a length that is less than $\frac{1}{4}L$.

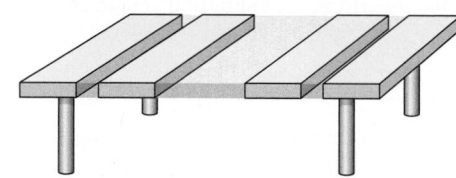

(d) Remove $\frac{1}{16}$ of the table by taking sections of length $\frac{1}{64}L$ from the centers of each of the four remaining pieces. Now you have removed $\frac{1}{4} + \frac{1}{8} + \frac{1}{16}$ of the table. Each remaining piece has a length that is less than $\frac{1}{8}L$.

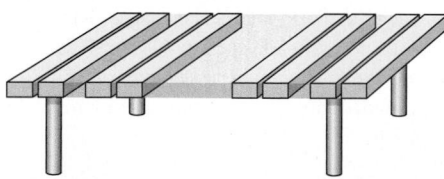

Will continuing this process cause the table to disappear, even though you have removed only half of the table? Why?

■ FOR FURTHER INFORMATION Read the article "Cantor's Disappearing Table" by Larry E. Knop in *The College Mathematics Journal*. To view this article, go to *MathArticles.com*.

9.3 The Integral Test and *p*-Series

■ Use the Integral Test to determine whether an infinite series converges or diverges.
■ Use properties of *p*-series and harmonic series.

The Integral Test

In this and the next section, you will study several convergence tests that apply to series with *positive* terms.

THEOREM 9.10   The Integral Test

If f is positive, continuous, and decreasing for $x \geq 1$ and $a_n = f(n)$, then

$$\sum_{n=1}^{\infty} a_n \quad \text{and} \quad \int_1^{\infty} f(x)\, dx$$

either both converge or both diverge.

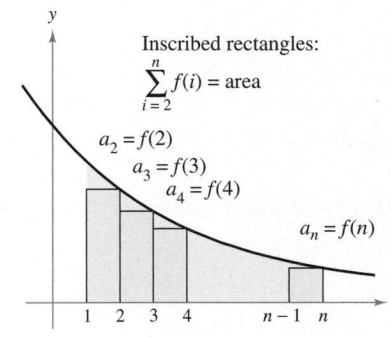

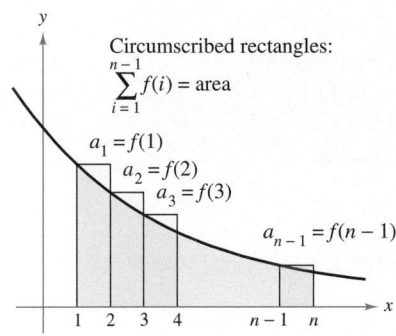

Figure 9.8

Proof   Begin by partitioning the interval $[1, n]$ into $(n - 1)$ unit intervals, as shown in Figure 9.8. The total areas of the inscribed rectangles and the circumscribed rectangles are

$$\sum_{i=2}^{n} f(i) = f(2) + f(3) + \cdots + f(n) \qquad \text{Inscribed area}$$

and

$$\sum_{i=1}^{n-1} f(i) = f(1) + f(2) + \cdots + f(n-1). \qquad \text{Circumscribed area}$$

The exact area under the graph of f from $x = 1$ to $x = n$ lies between the inscribed and circumscribed areas.

$$\sum_{i=2}^{n} f(i) \leq \int_1^n f(x)\, dx \leq \sum_{i=1}^{n-1} f(i)$$

Using the nth partial sum, $S_n = f(1) + f(2) + \cdots + f(n)$, you can write this inequality as

$$S_n - f(1) \leq \int_1^n f(x)\, dx \leq S_{n-1}.$$

Now, assuming that $\int_1^{\infty} f(x)\, dx$ converges to L, it follows that for $n \geq 1$,

$$S_n - f(1) \leq L \quad \Longrightarrow \quad S_n \leq L + f(1).$$

Consequently, $\{S_n\}$ is bounded and monotonic, and by Theorem 9.5 it converges. So, $\Sigma\, a_n$ converges. For the other direction of the proof, assume that the improper integral diverges. Then $\int_1^n f(x)\, dx$ approaches infinity as $n \to \infty$, and the inequality $S_{n-1} \geq \int_1^n f(x)\, dx$ implies that $\{S_n\}$ diverges. So, $\Sigma\, a_n$ diverges. ■

Remember that the convergence or divergence of $\Sigma\, a_n$ is not affected by deleting the first N terms. Similarly, when the conditions for the Integral Test are satisfied for all $x \geq N > 1$, you can simply use the integral $\int_N^{\infty} f(x)\, dx$ to test for convergence or divergence. (This is illustrated in Example 4.)

EXAMPLE 1 Using the Integral Test

Apply the Integral Test to the series $\displaystyle\sum_{n=1}^{\infty} \frac{n}{n^2 + 1}$.

Solution The function $f(x) = x/(x^2 + 1)$ is positive and continuous for $x \geq 1$. To determine whether f is decreasing, find the derivative.

$$f'(x) = \frac{(x^2 + 1)(1) - x(2x)}{(x^2 + 1)^2} = \frac{-x^2 + 1}{(x^2 + 1)^2}$$

So, $f'(x) < 0$ for $x > 1$ and it follows that f satisfies the conditions for the Integral Test. You can integrate to obtain

$$\int_1^{\infty} \frac{x}{x^2 + 1}\, dx = \frac{1}{2}\int_1^{\infty} \frac{2x}{x^2 + 1}\, dx$$

$$= \frac{1}{2}\lim_{b \to \infty} \int_1^b \frac{2x}{x^2 + 1}\, dx$$

$$= \frac{1}{2}\lim_{b \to \infty} \left[\ln(x^2 + 1) \right]_1^b$$

$$= \frac{1}{2}\lim_{b \to \infty} \left[\ln(b^2 + 1) - \ln 2 \right]$$

$$= \infty.$$

So, the series *diverges*.

> **••REMARK** Before applying the Integral Test, be sure to check that the function is positive, continuous, and decreasing for $x \geq 1$. When the function fails to satisfy one or more of these conditions, you cannot apply the Integral Test.

EXAMPLE 2 Using the Integral Test

•••▷ *See LarsonCalculus.com for an interactive version of this type of example.*

Apply the Integral Test to the series $\displaystyle\sum_{n=1}^{\infty} \frac{1}{n^2 + 1}$.

Solution Because $f(x) = 1/(x^2 + 1)$ satisfies the conditions for the Integral Test (check this), you can integrate to obtain

$$\int_1^{\infty} \frac{1}{x^2 + 1}\, dx = \lim_{b \to \infty} \int_1^b \frac{1}{x^2 + 1}\, dx$$

$$= \lim_{b \to \infty} \left[\arctan x \right]_1^b$$

$$= \lim_{b \to \infty} (\arctan b - \arctan 1)$$

$$= \frac{\pi}{2} - \frac{\pi}{4}$$

$$= \frac{\pi}{4}.$$

So, the series *converges* (see Figure 9.9).

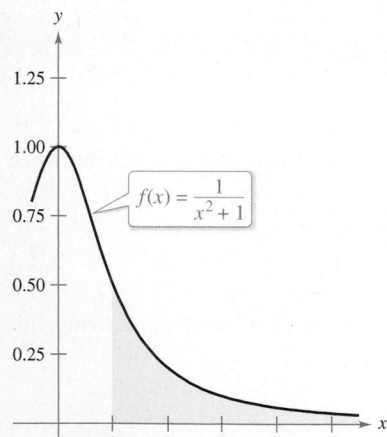

Because the improper integral converges, the infinite series also converges.
Figure 9.9

In Example 2, the fact that the improper integral converges to $\pi/4$ does not imply that the infinite series converges to $\pi/4$. To approximate the sum of the series, you can use the inequality

$$\sum_{n=1}^{N} \frac{1}{n^2 + 1} \leq \sum_{n=1}^{\infty} \frac{1}{n^2 + 1} \leq \sum_{n=1}^{N} \frac{1}{n^2 + 1} + \int_N^{\infty} \frac{1}{x^2 + 1}\, dx.$$

(See Exercise 52.) The larger the value of N, the better the approximation. For instance, using $N = 200$ produces $1.072 \leq \sum [1/(n^2 + 1)] \leq 1.077$.

p-Series and Harmonic Series

In the remainder of this section, you will investigate a second type of series that has a simple arithmetic test for convergence or divergence. A series of the form

$$\sum_{n=1}^{\infty} \frac{1}{n^p} = \frac{1}{1^p} + \frac{1}{2^p} + \frac{1}{3^p} + \cdots \qquad \text{\textit{p}-series}$$

is a **\textit{p}-series,** where p is a positive constant. For $p = 1$, the series

$$\sum_{n=1}^{\infty} \frac{1}{n} = 1 + \frac{1}{2} + \frac{1}{3} + \cdots \qquad \text{Harmonic series}$$

is the **harmonic series.** A **general harmonic series** is of the form $\sum [1/(an + b)]$. In music, strings of the same material, diameter, and tension, and whose lengths form a harmonic series, produce harmonic tones.

The Integral Test is convenient for establishing the convergence or divergence of *p*-series. This is shown in the proof of Theorem 9.11.

THEOREM 9.11   Convergence of *p*-Series

The *p*-series

$$\sum_{n=1}^{\infty} \frac{1}{n^p} = \frac{1}{1^p} + \frac{1}{2^p} + \frac{1}{3^p} + \frac{1}{4^p} + \cdots$$

converges for $p > 1$ and diverges for $0 < p \le 1$.

Proof   The proof follows from the Integral Test and from Theorem 8.7, which states that

$$\int_{1}^{\infty} \frac{1}{x^p}\, dx$$

converges for $p > 1$ and diverges for $0 < p \le 1$. ∎

EXAMPLE 3   **Convergent and Divergent *p*-Series**

Discuss the convergence or divergence of (a) the harmonic series and (b) the *p*-series with $p = 2$.

Solution

a. From Theorem 9.11, it follows that the harmonic series

$$\sum_{n=1}^{\infty} \frac{1}{n} = \frac{1}{1} + \frac{1}{2} + \frac{1}{3} + \cdots \qquad p = 1$$

diverges.

b. From Theorem 9.11, it follows that the *p*-series

$$\sum_{n=1}^{\infty} \frac{1}{n^2} = \frac{1}{1^2} + \frac{1}{2^2} + \frac{1}{3^2} + \cdots \qquad p = 2$$

converges. ∎

The sum of the series in Example 3(b) can be shown to be $\pi^2/6$. (This was proved by Leonhard Euler, but the proof is too difficult to present here.) Be sure you see that the Integral Test does not tell you that the sum of the series is equal to the value of the integral. For instance, the sum of the series in Example 3(b) is

$$\sum_{n=1}^{\infty} \frac{1}{n^2} = \frac{\pi^2}{6} \approx 1.645$$

whereas the value of the corresponding improper integral is

$$\int_1^{\infty} \frac{1}{x^2}\, dx = 1.$$

EXAMPLE 4 Testing a Series for Convergence

Determine whether the series

$$\sum_{n=2}^{\infty} \frac{1}{n \ln n}$$

converges or diverges.

Solution This series is similar to the divergent harmonic series. If its terms were greater than those of the harmonic series, you would expect it to diverge. However, because its terms are less than those of the harmonic series, you are not sure what to expect. The function

$$f(x) = \frac{1}{x \ln x}$$

is positive and continuous for $x \geq 2$. To determine whether f is decreasing, first rewrite f as

$$f(x) = (x \ln x)^{-1}$$

and then find its derivative.

$$f'(x) = (-1)(x \ln x)^{-2}(1 + \ln x) = -\frac{1 + \ln x}{x^2(\ln x)^2}$$

So, $f'(x) < 0$ for $x > 2$ and it follows that f satisfies the conditions for the Integral Test.

$$\int_2^{\infty} \frac{1}{x \ln x}\, dx = \int_2^{\infty} \frac{1/x}{\ln x}\, dx$$
$$= \lim_{b \to \infty} \left[\ln(\ln x) \right]_2^b$$
$$= \lim_{b \to \infty} \left[\ln(\ln b) - \ln(\ln 2) \right]$$
$$= \infty$$

The series diverges. ∎

Note that the infinite series in Example 4 diverges very slowly. For instance, as shown in the table, the sum of the first 10 terms is approximately 1.6878196, whereas the sum of the first 100 terms is just slightly greater: 2.3250871. In fact, the sum of the first 10,000 terms is approximately 3.0150217. You can see that although the infinite series "adds up to infinity," it does so very slowly.

n	11	101	1001	10,001	100,001
S_n	1.6878	2.3251	2.7275	3.0150	3.2382

9.3 Exercises

See **CalcChat.com** for tutorial help and worked-out solutions to odd-numbered exercises.

CONCEPT CHECK

1. Integral Test What conditions have to be satisfied to use the Integral Test?

2. *p*-Series Determine whether each series is a *p*-series.

(a) $\sum_{n=1}^{\infty} \frac{1}{n^{1.4}}$ (b) $\sum_{n=1}^{\infty} \frac{1}{n^{-2}}$ (c) $\sum_{n=1}^{\infty} \frac{1}{n^3}$

 **Using the Integral Test** In Exercises 3–22, confirm that the Integral Test can be applied to the series. Then use the Integral Test to determine the convergence or divergence of the series.

3. $\sum_{n=1}^{\infty} \frac{1}{n+3}$ **4.** $\sum_{n=1}^{\infty} \frac{2}{3n+5}$

5. $\sum_{n=1}^{\infty} \frac{1}{2^n}$ **6.** $\sum_{n=1}^{\infty} 3^{-n}$

7. $\sum_{n=1}^{\infty} e^{-n}$ **8.** $\sum_{n=1}^{\infty} ne^{-n/2}$

9. $\frac{\ln 2}{2} + \frac{\ln 3}{3} + \frac{\ln 4}{4} + \frac{\ln 5}{5} + \frac{\ln 6}{6} + \cdots$

10. $\frac{\ln 2}{\sqrt{2}} + \frac{\ln 3}{\sqrt{3}} + \frac{\ln 4}{\sqrt{4}} + \frac{\ln 5}{\sqrt{5}} + \frac{\ln 6}{\sqrt{6}} + \cdots$

11. $\frac{1}{3} + \frac{1}{5} + \frac{1}{7} + \frac{1}{9} + \frac{1}{11} + \cdots$

12. $\frac{1}{4} + \frac{2}{7} + \frac{3}{12} + \frac{4}{19} + \frac{5}{28} + \cdots$

13. $\sum_{n=1}^{\infty} \frac{\arctan n}{n^2 + 1}$ **14.** $\sum_{n=2}^{\infty} \frac{\ln n}{n^3}$

15. $\sum_{n=1}^{\infty} \frac{\ln n}{n^2}$ **16.** $\sum_{n=2}^{\infty} \frac{1}{n\sqrt{\ln n}}$

17. $\sum_{n=1}^{\infty} \frac{1}{(2n+3)^3}$ **18.** $\sum_{n=1}^{\infty} \frac{n+2}{n+1}$

19. $\sum_{n=1}^{\infty} \frac{4n}{2n^2 + 1}$ **20.** $\sum_{n=1}^{\infty} \frac{1}{\sqrt[3]{n+9}}$

21. $\sum_{n=1}^{\infty} \frac{n}{n^4 + 1}$ **22.** $\sum_{n=1}^{\infty} \frac{n}{n^4 + 2n^2 + 1}$

Using the Integral Test In Exercises 23 and 24, use the Integral Test to determine the convergence or divergence of the series, where *k* is a positive integer.

23. $\sum_{n=1}^{\infty} \frac{n^{k-1}}{n^k + c}$ **24.** $\sum_{n=1}^{\infty} n^k e^{-n}$

 **Conditions of the Integral Test** In Exercises 25–28, explain why the Integral Test does not apply to the series.

25. $\sum_{n=1}^{\infty} \frac{(-1)^n}{n}$ **26.** $\sum_{n=1}^{\infty} e^{-n} \cos n$

27. $\sum_{n=1}^{\infty} \frac{2 + \sin n}{n}$ **28.** $\sum_{n=1}^{\infty} \left(\frac{\sin n}{n}\right)^2$

Using the Integral Test In Exercises 29–32, use the Integral Test to determine the convergence or divergence of the *p*-series.

29. $\sum_{n=1}^{\infty} \frac{1}{n^7}$ **30.** $\sum_{n=1}^{\infty} \frac{1}{n^{1/2}}$

31. $\sum_{n=1}^{\infty} \frac{1}{n^{0.9}}$ **32.** $\sum_{n=1}^{\infty} \frac{1}{n^{1.001}}$

Using a *p*-Series In Exercises 33–38, use Theorem 9.11 to determine the convergence or divergence of the *p*-series.

33. $\sum_{n=1}^{\infty} \frac{1}{\sqrt[5]{n}}$

34. $\sum_{n=1}^{\infty} \frac{3}{n^{5/3}}$

35. $1 + \frac{1}{2\sqrt{2}} + \frac{1}{3\sqrt{3}} + \frac{1}{4\sqrt{4}} + \frac{1}{5\sqrt{5}} + \cdots$

36. $1 + \frac{1}{\sqrt[3]{4}} + \frac{1}{\sqrt[3]{9}} + \frac{1}{\sqrt[3]{16}} + \frac{1}{\sqrt[3]{25}} + \cdots$

37. $\sum_{n=1}^{\infty} \frac{1}{n^{1.03}}$

38. $\sum_{n=1}^{\infty} \frac{1}{n^{\pi}}$

39. Numerical and Graphical Analysis Use a graphing utility to find the indicated partial sum S_n and complete the table. Then use a graphing utility to graph the first 10 terms of the sequence of partial sums. For each series, compare the rate at which the sequence of partial sums approaches the sum of the series.

n	5	10	20	50	100
S_n					

(a) $\sum_{n=1}^{\infty} 3\left(\frac{1}{5}\right)^{n-1} = \frac{15}{4}$ (b) $\sum_{n=1}^{\infty} \frac{1}{n^2} = \frac{\pi^2}{6}$

40. Numerical Reasoning Because the harmonic series diverges, it follows that for any positive real number *M*, there exists a positive integer *N* such that the partial sum

$$\sum_{n=1}^{N} \frac{1}{n} > M.$$

(a) Use a graphing utility to complete the table.

M	2	4	6	8
N				

(b) As the real number *M* increases in equal increments, does the number *N* increase in equal increments? Explain.

EXPLORING CONCEPTS

41. Think About It Without performing any calculations, determine whether the following series converges. Explain.

$$\frac{1}{10{,}000} + \frac{1}{10{,}001} + \frac{1}{10{,}002} + \cdots$$

42. Using a Function Let f be a positive, continuous, and decreasing function for $x \geq 1$, such that $a_n = f(n)$. Use a graph to rank the following quantities in decreasing order. Explain your reasoning.

(a) $\displaystyle\sum_{n=2}^{7} a_n$ (b) $\displaystyle\int_{1}^{7} f(x)\, dx$ (c) $\displaystyle\sum_{n=1}^{6} a_n$

43. Using a Series Use a graph to show that the inequality is true. What can you conclude about the convergence or divergence of the series? Explain.

(a) $\displaystyle\sum_{n=1}^{\infty} \frac{1}{\sqrt{n}} > \int_{1}^{\infty} \frac{1}{\sqrt{x}}\, dx$

(b) $\displaystyle\sum_{n=2}^{\infty} \frac{1}{n^2} < \int_{1}^{\infty} \frac{1}{x^2}\, dx$

44. HOW DO YOU SEE IT? The graphs show the sequences of partial sums of the p-series

$$\sum_{n=1}^{\infty} \frac{1}{n^{0.4}} \quad \text{and} \quad \sum_{n=1}^{\infty} \frac{1}{n^{1.5}}.$$

Using Theorem 9.11, the first series diverges and the second series converges. Explain how the graphs show this.

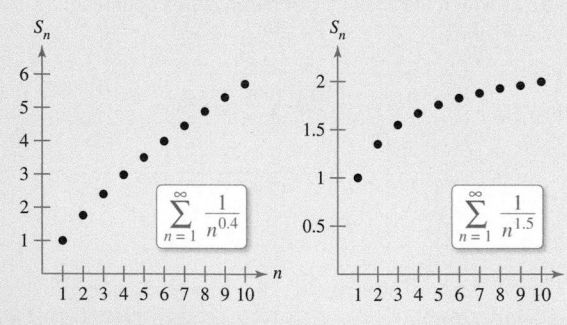

Finding Values In Exercises 45–50, find the positive values of p for which the series converges.

45. $\displaystyle\sum_{n=2}^{\infty} \frac{1}{n(\ln n)^p}$

46. $\displaystyle\sum_{n=2}^{\infty} \frac{\ln n}{n^p}$

47. $\displaystyle\sum_{n=1}^{\infty} \frac{n}{(1 + n^2)^p}$

48. $\displaystyle\sum_{n=1}^{\infty} n(1 + n^2)^p$

49. $\displaystyle\sum_{n=1}^{\infty} \left(\frac{3}{p}\right)^n$

50. $\displaystyle\sum_{n=3}^{\infty} \frac{1}{n(\ln n)[\ln(\ln n)]^p}$

51. Proof Let f be a positive, continuous, and decreasing function for $x \geq 1$, such that $a_n = f(n)$. Prove that if the series

$$\sum_{n=1}^{\infty} a_n$$

converges to S, then the remainder $R_N = S - S_N$ is bounded by

$$0 \leq R_N \leq \int_{N}^{\infty} f(x)\, dx.$$

52. Using a Remainder Show that the result of Exercise 51 can be written as

$$\sum_{n=1}^{N} a_n \leq \sum_{n=1}^{\infty} a_n \leq \sum_{n=1}^{N} a_n + \int_{N}^{\infty} f(x)\, dx.$$

Approximating a Sum In Exercises 53–58, use the result of Exercise 51 to approximate the sum of the convergent series using the indicated number of terms. Include an estimate of the maximum error for your approximation.

53. $\displaystyle\sum_{n=1}^{\infty} \frac{1}{n^4}$, three terms

54. $\displaystyle\sum_{n=1}^{\infty} \frac{1}{(n + 1)^3}$, six terms

55. $\displaystyle\sum_{n=1}^{\infty} \frac{1}{n^2 + 1}$, eight terms

56. $\displaystyle\sum_{n=1}^{\infty} \frac{1}{(n + 1)[\ln(n + 1)]^3}$, ten terms

57. $\displaystyle\sum_{n=1}^{\infty} n e^{-n^2}$, four terms

58. $\displaystyle\sum_{n=1}^{\infty} e^{-2n}$, five terms

Finding a Value In Exercises 59–62, use the result of Exercise 51 to find N such that $R_N \leq 0.001$ for the convergent series.

59. $\displaystyle\sum_{n=1}^{\infty} \frac{1}{n^4}$

60. $\displaystyle\sum_{n=1}^{\infty} \frac{1}{n^{3/2}}$

61. $\displaystyle\sum_{n=1}^{\infty} e^{-n/2}$

62. $\displaystyle\sum_{n=1}^{\infty} \frac{1}{n^2 + 1}$

63. Comparing Series

(a) Show that $\displaystyle\sum_{n=2}^{\infty} \frac{1}{n^{1.1}}$ converges and $\displaystyle\sum_{n=2}^{\infty} \frac{1}{n \ln n}$ diverges.

(b) Compare the first five terms of each series in part (a).

(c) Find $n > 3$ such that $\dfrac{1}{n^{1.1}} < \dfrac{1}{n \ln n}$.

64. Using a p-Series Ten terms are used to approximate a convergent p-series. Therefore, the remainder is a function of p and is

$$0 \leq R_{10}(p) \leq \int_{10}^{\infty} \frac{1}{x^p}\, dx, \quad p > 1.$$

(a) Perform the integration in the inequality.

(b) Use a graphing utility to represent the inequality graphically.

(c) Identify any asymptotes of the remainder function and interpret their meaning.

65. Euler's Constant   Let

$$S_n = \sum_{k=1}^{n} \frac{1}{k} = 1 + \frac{1}{2} + \cdots + \frac{1}{n}.$$

(a) Show that $\ln(n + 1) \le S_n \le 1 + \ln n$.

(b) Show that the sequence $\{a_n\} = \{S_n - \ln n\}$ is bounded.

(c) Show that the sequence $\{a_n\}$ is decreasing.

(d) Show that the sequence $\{a_n\}$ converges to a limit γ (called Euler's constant).

(e) Approximate γ using a_{100}.

66. Finding a Sum   Find the sum of the series

$$\sum_{n=2}^{\infty} \ln\left(1 - \frac{1}{n^2}\right).$$

67. Using a Series   Consider the series $\displaystyle\sum_{n=2}^{\infty} x^{\ln n}$.

(a) Determine the convergence or divergence of the series for $x = 1$.

(b) Determine the convergence or divergence of the series for $x = 1/e$.

(c) Find the positive values of x for which the series converges.

68. Riemann Zeta Function   The **Riemann zeta function** for real numbers is defined for all x for which the series

$$\zeta(x) = \sum_{n=1}^{\infty} n^{-x}$$

converges. Find the domain of the function.

Review   In Exercises 69–80, determine the convergence or divergence of the series.

69. $\displaystyle\sum_{n=1}^{\infty} \frac{1}{3n - 2}$

70. $\displaystyle\sum_{n=2}^{\infty} \frac{1}{n\sqrt{n^2 - 1}}$

71. $\displaystyle\sum_{n=1}^{\infty} \frac{1}{n\sqrt[4]{n}}$

72. $\displaystyle 3\sum_{n=1}^{\infty} \frac{1}{n^{0.95}}$

73. $\displaystyle\sum_{n=0}^{\infty} \left(\frac{2}{3}\right)^n$

74. $\displaystyle\sum_{n=0}^{\infty} \left(\frac{7}{5}\right)^n$

75. $\displaystyle\sum_{n=1}^{\infty} \frac{n}{\sqrt{3n^2 + 3}}$

76. $\displaystyle\sum_{n=1}^{\infty} \left(\frac{1}{n^2} - \frac{1}{n^3}\right)$

77. $\displaystyle\sum_{n=1}^{\infty} \left(1 + \frac{1}{n}\right)^n$

78. $\displaystyle\sum_{n=4}^{\infty} \ln\frac{n}{2}$

79. $\displaystyle\sum_{n=2}^{\infty} \frac{1}{n(\ln n)^3}$

80. $\displaystyle\sum_{n=3}^{\infty} \frac{1}{n(\ln n)[\ln(\ln n)]^4}$

SECTION PROJECT

The Harmonic Series

The harmonic series

$$\sum_{n=1}^{\infty} \frac{1}{n} = 1 + \frac{1}{2} + \frac{1}{3} + \frac{1}{4} + \cdots + \frac{1}{n} + \cdots$$

is one of the most important series in this chapter. Even though its terms tend to zero as n increases,

$$\lim_{n \to \infty} \frac{1}{n} = 0$$

the harmonic series diverges. In other words, even though the terms are getting smaller and smaller, the sum "adds up to infinity."

(a) One way to show that the harmonic series diverges is attributed to James Bernoulli. He grouped the terms of the harmonic series as follows:

$$1 + \frac{1}{2} + \underbrace{\frac{1}{3} + \frac{1}{4}}_{> \frac{1}{2}} + \underbrace{\frac{1}{5} + \cdots + \frac{1}{8}}_{> \frac{1}{2}} + \underbrace{\frac{1}{9} + \cdots + \frac{1}{16}}_{> \frac{1}{2}} +$$

$$\underbrace{\frac{1}{17} + \cdots + \frac{1}{32}}_{> \frac{1}{2}} + \cdots$$

Write a short paragraph explaining how you can use this grouping to show that the harmonic series diverges.

(b) Use the proof of the Integral Test, Theorem 9.10, to show that

$$\ln(n + 1) \le 1 + \frac{1}{2} + \frac{1}{3} + \frac{1}{4} + \cdots + \frac{1}{n} \le 1 + \ln n.$$

(c) Use part (b) to determine how many terms M you would need so that

$$\sum_{n=1}^{M} \frac{1}{n} > 50.$$

(d) Show that the sum of the first million terms of the harmonic series is less than 15.

(e) Show that the following inequalities are valid.

$$\ln\frac{21}{10} \le \frac{1}{10} + \frac{1}{11} + \cdots + \frac{1}{20} \le \ln\frac{20}{9}$$

$$\ln\frac{201}{100} \le \frac{1}{100} + \frac{1}{101} + \cdots + \frac{1}{200} \le \ln\frac{200}{99}$$

(f) Use the inequalities in part (e) to find the limit

$$\lim_{m \to \infty} \sum_{n=m}^{2m} \frac{1}{n}.$$

9.4 Comparisons of Series

■ Use the Direct Comparison Test to determine whether a series converges or diverges.
■ Use the Limit Comparison Test to determine whether a series converges or diverges.

Direct Comparison Test

For the convergence tests developed so far, the terms of the series have to be fairly simple and the series must have special characteristics in order for the convergence tests to be applied. A slight deviation from these special characteristics can make a test inapplicable. For example, in the pairs listed below, the second series cannot be tested by the same convergence test as the first series, even though it is similar to the first.

1. $\displaystyle\sum_{n=0}^{\infty} \frac{1}{2^n}$ is geometric, but $\displaystyle\sum_{n=0}^{\infty} \frac{n}{2^n}$ is not.

2. $\displaystyle\sum_{n=1}^{\infty} \frac{1}{n^3}$ is a p-series, but $\displaystyle\sum_{n=1}^{\infty} \frac{1}{n^3 + 1}$ is not.

3. $a_n = \dfrac{n}{(n^2 + 3)^2}$ is easily integrated, but $b_n = \dfrac{n^2}{(n^2 + 3)^2}$ is not.

In this section, you will study two additional tests for positive-term series. These two tests greatly expand the variety of series you are able to test for convergence or divergence. They allow you to *compare* a series having complicated terms with a simpler series whose convergence or divergence is known.

· · · · · · · · · · · · · · · ▷

· **REMARK** As stated, the Direct Comparison Test requires that $0 < a_n \le b_n$ for all n. Because the convergence of a series is not dependent on its first several terms, you could modify the test to require only that $0 < a_n \le b_n$ for all n greater than some integer N.

THEOREM 9.12 Direct Comparison Test

Let $0 < a_n \le b_n$ for all n.

1. If $\displaystyle\sum_{n=1}^{\infty} b_n$ converges, then $\displaystyle\sum_{n=1}^{\infty} a_n$ converges.

2. If $\displaystyle\sum_{n=1}^{\infty} a_n$ diverges, then $\displaystyle\sum_{n=1}^{\infty} b_n$ diverges.

Proof To prove the first property, let $L = \displaystyle\sum_{n=1}^{\infty} b_n$ and let

$$S_n = a_1 + a_2 + \cdots + a_n.$$

Because $0 < a_n \le b_n$, the sequence $S_1, S_2, S_3, \ldots$ is nondecreasing and bounded above by L. So, it must converge. Because

$$\lim_{n\to\infty} S_n = \sum_{n=1}^{\infty} a_n$$

it follows that $\displaystyle\sum_{n=1}^{\infty} a_n$ converges. The second property is logically equivalent to the first. ■

■ **FOR FURTHER INFORMATION** Is the Direct Comparison Test just for nonnegative series? To read about the generalization of this test to real series, see the article "The Comparison Test—Not Just for Nonnegative Series" by Michele Longo and Vincenzo Valori in *Mathematics Magazine*. To view this article, go to *MathArticles.com*.

Sequence of partial sums of $\sum\limits_{n=1}^{\infty} \dfrac{1}{3^n}$

Sequence of partial sums of $\sum\limits_{n=1}^{\infty} \dfrac{1}{2+3^n}$

For the given series in Example 1, the sequence of partial sums is less than the sequence of partial sums of the indicated convergent geometric series.
Figure 9.10

EXAMPLE 1 Using the Direct Comparison Test

Determine the convergence or divergence of

$$\sum_{n=1}^{\infty} \frac{1}{2+3^n}.$$

Solution This series resembles

$$\sum_{n=1}^{\infty} \frac{1}{3^n}. \qquad \text{Convergent geometric series}$$

Term-by-term comparison yields

$$a_n = \frac{1}{2+3^n} < \frac{1}{3^n} = b_n, \quad n \geq 1.$$

So, by the Direct Comparison Test, the given series converges. This conclusion is supported by Figure 9.10, which shows that the sequence of partial sums of $\Sigma\, a_n$ is less than the sequence of partial sums of the convergent geometric series $\Sigma\, b_n$.

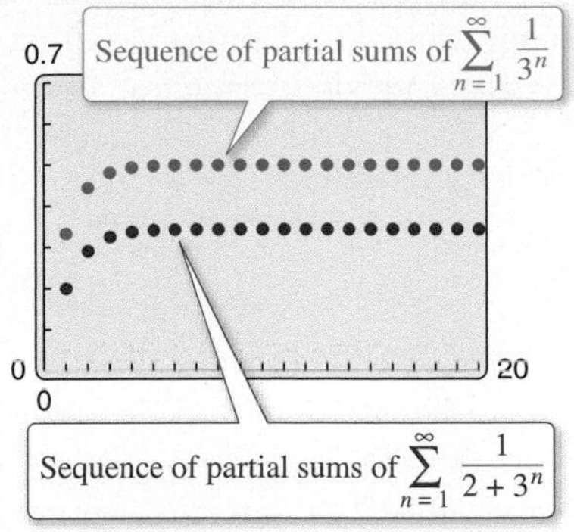

EXAMPLE 2 Using the Direct Comparison Test

•••▷ *See LarsonCalculus.com for an interactive version of this type of example.*

Determine the convergence or divergence of

$$\sum_{n=1}^{\infty} \frac{1}{2+\sqrt{n}}.$$

Solution This series resembles

$$\sum_{n=1}^{\infty} \frac{1}{n^{1/2}}. \qquad \text{Divergent } p\text{-series}$$

Term-by-term comparison yields

$$\frac{1}{2+\sqrt{n}} \leq \frac{1}{\sqrt{n}}, \quad n \geq 1$$

which *does not* meet the requirements for divergence. (Remember that when term-by-term comparison reveals a series that is *less* than a divergent series, the Direct Comparison Test tells you nothing.) Still expecting the series to diverge, you can compare the series with

$$\sum_{n=1}^{\infty} \frac{1}{n}. \qquad \text{Divergent harmonic series}$$

In this case, term-by-term comparison yields

$$a_n = \frac{1}{n} \leq \frac{1}{2+\sqrt{n}} = b_n, \quad n \geq 4$$

and, by the Direct Comparison Test, the given series diverges (see Figure 9.11). To verify the last inequality, try showing that

$$2 + \sqrt{n} \leq n$$

whenever $n \geq 4$.

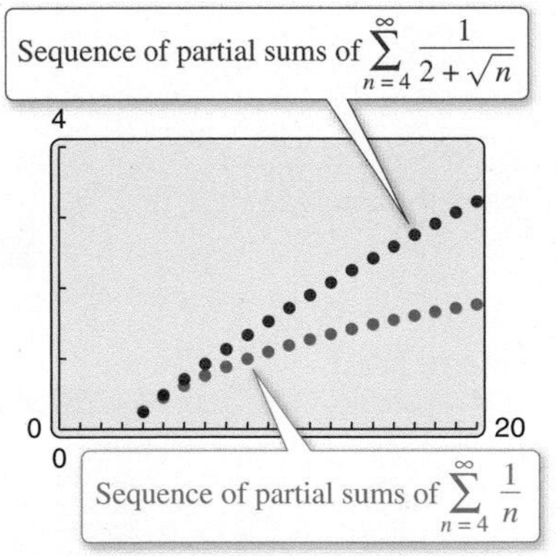

Sequence of partial sums of $\sum\limits_{n=4}^{\infty} \dfrac{1}{2+\sqrt{n}}$

Sequence of partial sums of $\sum\limits_{n=4}^{\infty} \dfrac{1}{n}$

For the given series in Example 2, the sequence of partial sums is greater than the sequence of partial sums of the divergent harmonic series.
Figure 9.11

Remember that both parts of the Direct Comparison Test require that $0 < a_n \leq b_n$. Informally, the test says the following about the two series with nonnegative terms.

1. If the "larger" series converges, then the "smaller" series must also converge.
2. If the "smaller" series diverges, then the "larger" series must also diverge.

Limit Comparison Test

Sometimes a series closely resembles a p-series or a geometric series, yet you cannot establish the term-by-term comparison necessary to apply the Direct Comparison Test. Under these circumstances, you may be able to apply a second comparison test, called the **Limit Comparison Test.**

· · · · · · · · · · · · · · · · · · ▷

· ·**REMARK** As with the Direct Comparison Test, the Limit Comparison Test could be modified to require only that a_n and b_n be positive for all n greater than some integer N.

THEOREM 9.13 Limit Comparison Test

If $a_n > 0$, $b_n > 0$, and

$$\lim_{n \to \infty} \frac{a_n}{b_n} = L$$

where L is *finite and positive*, then

$$\sum_{n=1}^{\infty} a_n \quad \text{and} \quad \sum_{n=1}^{\infty} b_n$$

either both converge or both diverge.

Proof Because $a_n > 0$, $b_n > 0$, and

$$\lim_{n \to \infty} \frac{a_n}{b_n} = L$$

there exists $N > 0$ such that

$$0 < \frac{a_n}{b_n} < L + 1, \quad \text{for } n \geq N.$$

This implies that

$$0 < a_n < (L + 1)b_n.$$

So, by the Direct Comparison Test, the convergence of $\Sigma\, b_n$ implies the convergence of $\Sigma\, a_n$. Similarly, the fact that

$$\lim_{n \to \infty} \frac{b_n}{a_n} = \frac{1}{L}$$

can be used to show that the convergence of $\Sigma\, a_n$ implies the convergence of $\Sigma\, b_n$. ∎

EXAMPLE 3 Using the Limit Comparison Test

Show that the general harmonic series below diverges.

$$\sum_{n=1}^{\infty} \frac{1}{an + b}, \quad a > 0, \quad b > 0$$

Solution By comparison with

$$\sum_{n=1}^{\infty} \frac{1}{n} \qquad \text{Divergent harmonic series}$$

you have

$$\lim_{n \to \infty} \frac{1/(an + b)}{1/n} = \lim_{n \to \infty} \frac{n}{an + b} = \frac{1}{a}.$$

Because this limit is greater than 0, you can conclude from the Limit Comparison Test that the series diverges. ∎

The Limit Comparison Test works well for comparing a "messy" algebraic series with a *p*-series. In choosing an appropriate *p*-series, you must choose one with an *n*th term of the same magnitude as the *n*th term of the given series.

Given Series	**Comparison Series**	**Conclusion**
$\displaystyle\sum_{n=1}^{\infty} \frac{1}{3n^2 - 4n + 5}$	$\displaystyle\sum_{n=1}^{\infty} \frac{1}{n^2}$	Both series converge.
$\displaystyle\sum_{n=1}^{\infty} \frac{1}{\sqrt{3n - 2}}$	$\displaystyle\sum_{n=1}^{\infty} \frac{1}{\sqrt{n}}$	Both series diverge.
$\displaystyle\sum_{n=1}^{\infty} \frac{n^2 - 10}{4n^5 + n^3}$	$\displaystyle\sum_{n=1}^{\infty} \frac{n^2}{n^5} = \sum_{n=1}^{\infty} \frac{1}{n^3}$	Both series converge.

In other words, when choosing a series for comparison, you can disregard all but the *highest powers of n* in both the numerator and the denominator.

EXAMPLE 4   Using the Limit Comparison Test

Determine the convergence or divergence of

$$\sum_{n=1}^{\infty} \frac{\sqrt{n}}{n^2 + 1}.$$

Solution Disregarding all but the highest powers of *n* in the numerator and the denominator, you can compare the series with

$$\sum_{n=1}^{\infty} \frac{\sqrt{n}}{n^2} = \sum_{n=1}^{\infty} \frac{1}{n^{3/2}}. \qquad \text{Convergent } p\text{-series}$$

Because

$$\lim_{n\to\infty} \frac{a_n}{b_n} = \lim_{n\to\infty} \left(\frac{\sqrt{n}}{n^2 + 1}\right)\left(\frac{n^{3/2}}{1}\right)$$

$$= \lim_{n\to\infty} \frac{n^2}{n^2 + 1}$$

$$= 1$$

you can conclude by the Limit Comparison Test that the series converges.

• • REMARK Recall when finding limits at $\pm\infty$ of a rational function that if the degree of the numerator is equal to the degree of the denominator, then the limit of the rational function is the ratio of the leading coefficients.

EXAMPLE 5   Using the Limit Comparison Test

Determine the convergence or divergence of

$$\sum_{n=1}^{\infty} \frac{n2^n}{4n^3 + 1}.$$

Solution A reasonable comparison would be with the series

$$\sum_{n=1}^{\infty} \frac{2^n}{n^2}. \qquad \text{Divergent series}$$

Note that this series diverges by the *n*th-Term Test. From the limit

$$\lim_{n\to\infty} \frac{a_n}{b_n} = \lim_{n\to\infty} \left(\frac{n2^n}{4n^3 + 1}\right)\left(\frac{n^2}{2^n}\right)$$

$$= \lim_{n\to\infty} \frac{n^3}{4n^3 + 1}$$

$$= \frac{1}{4}$$

you can conclude by the Limit Comparison Test that the series diverges.

9.4 Exercises

See **CalcChat.com** for tutorial help and worked-out solutions to odd-numbered exercises.

CONCEPT CHECK

1. **Direct Comparison Test** You want to compare the series $\Sigma\,a_n$ and $\Sigma\,b_n$, where $a_n > 0$, $b_n > 0$, and $\Sigma\,b_n$ converges. For $1 \le n \le 5$, $a_n > b_n$, and for $n \ge 6$, $a_n < b_n$. Explain whether the Direct Comparison Test can be used to compare the two series.

2. **Limit Comparison Test** When using the Limit Comparison Test, describe in your own words how to choose a series for comparison.

Graphical Analysis In Exercises 3 and 4, the figures show the graphs of the first 10 terms, and the graphs of the first 10 terms of the sequence of partial sums, of each series.

(a) Identify the series in each figure.

(b) Which series is a *p*-series? Does it converge or diverge?

(c) For the series that are not *p*-series, how do the magnitudes of the terms compare with the magnitudes of the terms of the *p*-series? What conclusion can you draw about the convergence or divergence of the series?

(d) Explain the relationship between the magnitudes of the terms of the series and the magnitudes of the terms of the partial sums.

3. $\displaystyle\sum_{n=1}^{\infty} \frac{6}{n^{3/2}}$, $\displaystyle\sum_{n=1}^{\infty} \frac{6}{n^{3/2} + 3}$, and $\displaystyle\sum_{n=1}^{\infty} \frac{6}{n\sqrt{n^2 + 0.5}}$

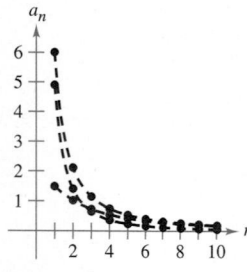
Graphs of terms

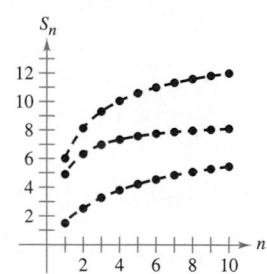
Graphs of partial sums

4. $\displaystyle\sum_{n=1}^{\infty} \frac{2}{\sqrt{n}}$, $\displaystyle\sum_{n=1}^{\infty} \frac{2}{\sqrt{n} - 0.5}$, and $\displaystyle\sum_{n=1}^{\infty} \frac{4}{\sqrt{n} + 0.5}$

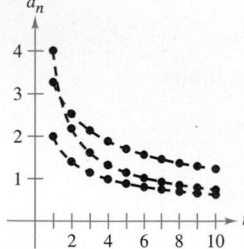
Graphs of terms

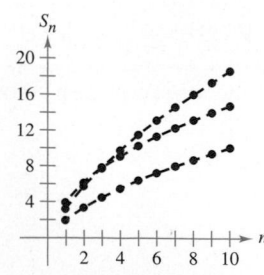
Graphs of partial sums

 Using the Direct Comparison Test In Exercises 5–16, use the Direct Comparison Test to determine the convergence or divergence of the series.

5. $\displaystyle\sum_{n=1}^{\infty} \frac{1}{2n - 1}$

6. $\displaystyle\sum_{n=1}^{\infty} \frac{1}{3n^2 + 2}$

7. $\displaystyle\sum_{n=2}^{\infty} \frac{1}{\sqrt{n} - 1}$

8. $\displaystyle\sum_{n=0}^{\infty} \frac{4^n}{5^n + 3}$

9. $\displaystyle\sum_{n=2}^{\infty} \frac{\ln n}{n + 1}$

10. $\displaystyle\sum_{n=1}^{\infty} \frac{1}{\sqrt{n^3 + 1}}$

11. $\displaystyle\sum_{n=0}^{\infty} \frac{1}{n!}$

12. $\displaystyle\sum_{n=1}^{\infty} \frac{1}{4\sqrt[3]{n} - 1}$

13. $\displaystyle\sum_{n=0}^{\infty} e^{-n^2}$

14. $\displaystyle\sum_{n=1}^{\infty} \frac{6^n + n}{5^n - 1}$

15. $\displaystyle\sum_{n=1}^{\infty} \frac{\sin^2 n}{n^3}$

16. $\displaystyle\sum_{n=1}^{\infty} \frac{\cos n + 2}{\sqrt{n}}$

 **Using the Limit Comparison Test** In Exercises 17–26, use the Limit Comparison Test to determine the convergence or divergence of the series.

17. $\displaystyle\sum_{n=1}^{\infty} \frac{n}{n^2 + 1}$

18. $\displaystyle\sum_{n=1}^{\infty} \frac{5}{4^n + 1}$

19. $\displaystyle\sum_{n=0}^{\infty} \frac{1}{\sqrt{n^2 + 1}}$

20. $\displaystyle\sum_{n=1}^{\infty} \frac{2^n + 1}{5^n + 1}$

21. $\displaystyle\sum_{n=1}^{\infty} \frac{2n^2 - 1}{3n^5 + 2n + 1}$

22. $\displaystyle\sum_{n=1}^{\infty} \frac{1}{n^2(n^2 + 4)}$

23. $\displaystyle\sum_{n=1}^{\infty} \frac{1}{n\sqrt{n^2 + 1}}$

24. $\displaystyle\sum_{n=1}^{\infty} \frac{n}{(n + 1)2^{n-1}}$

25. $\displaystyle\sum_{n=1}^{\infty} \frac{n^{k-1}}{n^k + 1}$, $k > 2$

26. $\displaystyle\sum_{n=1}^{\infty} \sin\frac{1}{n}$

Determining Convergence or Divergence In Exercises 27–34, test for convergence or divergence, using each test at least once. Identify which test was used.

(a) *n*th-Term Test
(b) Geometric Series Test
(c) *p*-Series Test
(d) Telescoping Series Test
(e) Integral Test
(f) Direct Comparison Test
(g) Limit Comparison Test

27. $\displaystyle\sum_{n=1}^{\infty} \frac{\sqrt[3]{n}}{n}$

28. $\displaystyle\sum_{n=0}^{\infty} 5\left(-\frac{4}{3}\right)^n$

29. $\displaystyle\sum_{n=1}^{\infty} \frac{1}{5^n + 1}$

30. $\displaystyle\sum_{n=3}^{\infty} \frac{1}{n^3 - 8}$

31. $\displaystyle\sum_{n=1}^{\infty} \frac{2n}{3n - 2}$

32. $\displaystyle\sum_{n=1}^{\infty} \left(\frac{1}{n + 1} - \frac{1}{n + 2}\right)$

33. $\displaystyle\sum_{n=1}^{\infty} \frac{n}{(n^2 + 1)^2}$

34. $\displaystyle\sum_{n=1}^{\infty} \frac{3}{n(n + 3)}$

35. Using the Limit Comparison Test Use the Limit Comparison Test with the harmonic series to show that the series $\Sigma\, a_n$ (where $0 < a_n < a_{n-1}$) diverges when $\lim\limits_{n\to\infty} na_n$ is finite and nonzero.

36. Proof Prove that, if $P(n)$ and $Q(n)$ are polynomials of degree j and k, respectively, then the series

$$\sum_{n=1}^{\infty} \frac{P(n)}{Q(n)}$$

converges if $j < k - 1$ and diverges if $j \geq k - 1$.

Determining Convergence or Divergence In Exercises 37–40, use the polynomial test given in Exercise 36 to determine whether the series converges or diverges.

37. $\frac{1}{2} + \frac{2}{5} + \frac{3}{10} + \frac{4}{17} + \frac{5}{26} + \cdots$

38. $\frac{1}{3} + \frac{1}{8} + \frac{1}{15} + \frac{1}{24} + \frac{1}{35} + \cdots$

39. $\sum_{n=1}^{\infty} \frac{1}{n^3 + 1}$

40. $\sum_{n=1}^{\infty} \frac{4n^5 + n^2 + 1}{n^4}$

Verifying Divergence In Exercises 41 and 42, use the divergence test given in Exercise 35 to show that the series diverges.

41. $\sum_{n=1}^{\infty} \frac{n^3}{5n^4 + 3}$ **42.** $\sum_{n=1}^{\infty} \frac{3n^2 + 1}{4n^3 + 2}$

Determining Convergence or Divergence In Exercises 43–46, determine the convergence or divergence of the series.

43. $\frac{1}{200} + \frac{1}{400} + \frac{1}{600} + \frac{1}{800} + \cdots$

44. $\frac{1}{200} + \frac{1}{208} + \frac{1}{216} + \frac{1}{224} + \cdots$

45. $\frac{1}{201} + \frac{1}{204} + \frac{1}{209} + \frac{1}{216} + \cdots$

46. $\frac{1}{201} + \frac{1}{208} + \frac{1}{227} + \frac{1}{264} + \cdots$

EXPLORING CONCEPTS

47. Using Series Review the results of Exercises 43–46. Explain why careful analysis is required to determine the convergence or divergence of a series and why considering only the magnitudes of the terms of a series could be misleading.

48. Comparing Series It appears that the terms of the series

$$\frac{1}{1000} + \frac{1}{1001} + \frac{1}{1002} + \frac{1}{1003} + \cdots$$

are less than the corresponding terms of the convergent series

$$1 + \frac{1}{4} + \frac{1}{9} + \frac{1}{16} + \cdots.$$

If the statement above is correct, then the first series converges. Is this correct? Why or why not? Make a statement about how the divergence or convergence of a series is affected by the inclusion or exclusion of the first finite number of terms.

---

49. Using a Series Consider the series $\sum_{n=1}^{\infty} \frac{1}{(2n-1)^2}$.

(a) Verify that the series converges.

(b) Use a graphing utility to complete the table.

n	5	10	20	50	100
S_n					

(c) The sum of the series is $\pi^2/8$. Find the sum of the series

$$\sum_{n=3}^{\infty} \frac{1}{(2n-1)^2}.$$

(d) Use a graphing utility to find the sum of the series

$$\sum_{n=10}^{\infty} \frac{1}{(2n-1)^2}.$$

50. Using a Series Consider the series $\sum_{n=1}^{\infty} \frac{1}{(n+2)^2}$.

(a) Verify that the series converges.

(b) Use a graphing utility to complete the table.

n	5	10	20	50	100
S_n					

(c) The sum of the series is $(\pi^2/6) - (5/4)$. Find the sum of the series

$$\sum_{n=6}^{\infty} \frac{1}{(n+2)^2}.$$

(d) Use a graphing utility to find the sum of the series

$$\sum_{n=15}^{\infty} \frac{1}{(n+2)^2}.$$

51. Decimal Representation of a Number Show that the series

$$\frac{x_1}{10} + \frac{x_2}{10^2} + \frac{x_3}{10^3} + \frac{x_4}{10^4} + \cdots$$

converges, where x_i is one of the numbers 0, 1, 2, . . ., 9.

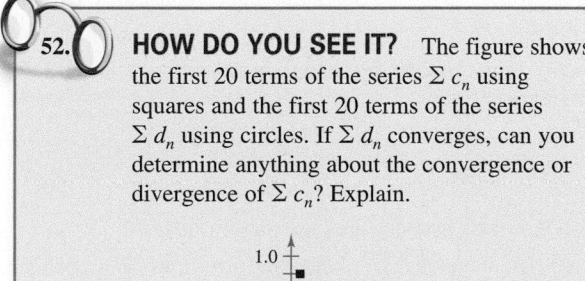

52. **HOW DO YOU SEE IT?** The figure shows the first 20 terms of the series $\Sigma\, c_n$ using squares and the first 20 terms of the series $\Sigma\, d_n$ using circles. If $\Sigma\, d_n$ converges, can you determine anything about the convergence or divergence of $\Sigma\, c_n$? Explain.

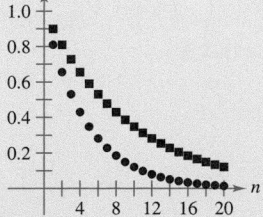

True or False? In Exercises 53–58, determine whether the statement is true or false. If it is false, explain why or give an example that shows it is false.

53. If $0 < a_n \le b_n$ and $\sum_{n=1}^{\infty} a_n$ converges, then $\sum_{n=1}^{\infty} b_n$ diverges.

54. If $0 < a_{n+10} \le b_n$ and $\sum_{n=1}^{\infty} b_n$ converges, then $\sum_{n=1}^{\infty} a_n$ converges.

55. If $a_n + b_n \le c_n$ and $\sum_{n=1}^{\infty} c_n$ converges, then the series $\sum_{n=1}^{\infty} a_n$ and $\sum_{n=1}^{\infty} b_n$ both converge. (Assume that the terms of all three series are positive.)

56. If $a_n \le b_n + c_n$ and $\sum_{n=1}^{\infty} a_n$ diverges, then the series $\sum_{n=1}^{\infty} b_n$ and $\sum_{n=1}^{\infty} c_n$ both diverge. (Assume that the terms of all three series are positive.)

57. If $0 < a_n \le b_n$ and $\sum_{n=1}^{\infty} a_n$ diverges, then $\sum_{n=1}^{\infty} b_n$ diverges.

58. If $0 < a_n \le b_n$ and $\sum_{n=1}^{\infty} b_n$ diverges, then $\sum_{n=1}^{\infty} a_n$ diverges.

59. Proof Prove that if the nonnegative series

$$\sum_{n=1}^{\infty} a_n \quad \text{and} \quad \sum_{n=1}^{\infty} b_n$$

converge, then so does the series $\sum_{n=1}^{\infty} a_n b_n$.

60. Proof Use the result of Exercise 59 to prove that if the nonnegative series

$$\sum_{n=1}^{\infty} a_n$$

converges, then so does the series

$$\sum_{n=1}^{\infty} a_n^2.$$

61. Finding Series Find two series that demonstrate the result of Exercise 59.

62. Finding Series Find two series that demonstrate the result of Exercise 60.

63. Proof Suppose that $\Sigma\, a_n$ and $\Sigma\, b_n$ are series with positive terms. Prove that if

$$\lim_{n \to \infty} \frac{a_n}{b_n} = 0$$

and $\Sigma\, b_n$ converges, then $\Sigma\, a_n$ also converges.

64. Proof Suppose that $\Sigma\, a_n$ and $\Sigma\, b_n$ are series with positive terms. Prove that if

$$\lim_{n \to \infty} \frac{a_n}{b_n} = \infty$$

and $\Sigma\, b_n$ diverges, then $\Sigma\, a_n$ also diverges.

65. Verifying Convergence Use the result of Exercise 63 to show that each series converges.

(a) $\sum_{n=1}^{\infty} \frac{1}{(n+1)^3}$

(b) $\sum_{n=1}^{\infty} \frac{1}{\sqrt{n}\pi^n}$

(c) $\sum_{n=2}^{\infty} \frac{\ln n}{n^3}$

(d) $\sum_{n=1}^{\infty} \frac{n^2+1}{e^n}$

66. Verifying Divergence Use the result of Exercise 64 to show that each series diverges.

(a) $\sum_{n=1}^{\infty} (n+2)^2$ (b) $\sum_{n=1}^{\infty} \frac{\ln n}{n}$

(c) $\sum_{n=2}^{\infty} \frac{1}{\ln n}$ (d) $\sum_{n=1}^{\infty} \frac{e^n}{\sqrt{n}}$

67. Proof Suppose that $\Sigma\, a_n$ is a series with positive terms. Prove that if $\Sigma\, a_n$ converges, then $\Sigma \sin a_n$ also converges.

68. Proof Prove that the series

$$\sum_{n=1}^{\infty} \frac{1}{1+2+3+\cdots+n}$$

converges.

69. Comparing Series Show that

$$\sum_{n=1}^{\infty} \frac{\ln n}{n\sqrt{n}}$$

converges by comparison with

$$\sum_{n=1}^{\infty} \frac{1}{n^{5/4}}.$$

70. Determining Convergence or Divergence Determine whether the every-other-term harmonic series

$$1 + \frac{1}{3} + \frac{1}{5} + \frac{1}{7} + \cdots$$

converges or diverges.

PUTNAM EXAM CHALLENGE

71. Is the infinite series

$$\sum_{n=1}^{\infty} \frac{1}{n^{(n+1)/n}}$$

convergent? Prove your statement.

72. Prove that if $\sum_{n=1}^{\infty} a_n$ is a convergent series of positive real numbers, then so is

$$\sum_{n=1}^{\infty} (a_n)^{n/(n+1)}.$$

These problems were composed by the Committee on the Putnam Prize Competition. © The Mathematical Association of America. All rights reserved.

9.5 Alternating Series

- Use the Alternating Series Test to determine whether an infinite series converges.
- Use the Alternating Series Remainder to approximate the sum of an alternating series.
- Classify a convergent series as absolutely or conditionally convergent.
- Rearrange an infinite series to obtain a different sum.

Alternating Series

So far, most series you have dealt with have had positive terms. In this section and the next section, you will study series that contain both positive and negative terms. The simplest such series is an **alternating series,** whose terms alternate in sign. For example, the geometric series

$$\sum_{n=0}^{\infty} \left(-\frac{1}{2} \right)^n = \sum_{n=0}^{\infty} (-1)^n \frac{1}{2^n}$$

$$= 1 - \frac{1}{2} + \frac{1}{4} - \frac{1}{8} + \frac{1}{16} - \cdots$$

is an *alternating geometric series* with $r = -\frac{1}{2}$. Alternating series occur in two ways: either the odd terms are negative or the even terms are negative.

THEOREM 9.14 Alternating Series Test

Let $a_n > 0$. The alternating series

$$\sum_{n=1}^{\infty} (-1)^n a_n \quad \text{and} \quad \sum_{n=1}^{\infty} (-1)^{n+1} a_n$$

converge when these two conditions are met.

1. $\lim\limits_{n \to \infty} a_n = 0$
2. $a_{n+1} \leq a_n$, for all n

> ••**REMARK** The second condition in the Alternating Series Test can be modified to require only that $0 < a_{n+1} \leq a_n$ for all n greater than some integer N.

Proof Consider the alternating series

$$\sum_{n=1}^{\infty} (-1)^{n+1} a_n.$$

For this series, the partial sum (where $2n$ is even)

$$S_{2n} = (a_1 - a_2) + (a_3 - a_4) + (a_5 - a_6) + \cdots + (a_{2n-1} - a_{2n})$$

has all nonnegative terms, and therefore $\{S_{2n}\}$ is a nondecreasing sequence. But you can also write

$$S_{2n} = a_1 - (a_2 - a_3) - (a_4 - a_5) - \cdots - (a_{2n-2} - a_{2n-1}) - a_{2n}$$

which implies that $S_{2n} \leq a_1$ for every integer n. So, $\{S_{2n}\}$ is a bounded, nondecreasing sequence that converges to some value L. Because $S_{2n-1} - a_{2n} = S_{2n}$ and $a_{2n} \to 0$, you have

$$\lim_{n \to \infty} S_{2n-1} = \lim_{n \to \infty} S_{2n} + \lim_{n \to \infty} a_{2n}$$

$$= L + \lim_{n \to \infty} a_{2n}$$

$$= L.$$

Because both S_{2n} and S_{2n-1} converge to the same limit L, it follows that $\{S_n\}$ also converges to L. Consequently, the given alternating series converges. ∎

REMARK The series in Example 1 is called the *alternating harmonic series.* More is said about this series in Example 8.

EXAMPLE 1   Using the Alternating Series Test

Determine the convergence or divergence of

$$\sum_{n=1}^{\infty} (-1)^{n+1} \frac{1}{n}.$$

Solution Note that $\lim\limits_{x \to \infty} a_n = \lim\limits_{n \to \infty} (1/n) = 0$. So, the first condition of Theorem 9.14 is satisfied. Also note that the second condition of Theorem 9.14 is satisfied because

$$a_{n+1} = \frac{1}{n+1} \leq \frac{1}{n} = a_n$$

for all n. So, applying the Alternating Series Test, you can conclude that the series converges.

EXAMPLE 2   Using the Alternating Series Test

Determine the convergence or divergence of

$$\sum_{n=1}^{\infty} \frac{n}{(-2)^{n-1}}.$$

Solution To apply the Alternating Series Test, note that, for $n \geq 1$,

$$\frac{1}{2} \leq \frac{n}{n+1}$$

$$\frac{2^{n-1}}{2^n} \leq \frac{n}{n+1}$$

$$(n+1)2^{n-1} \leq n2^n$$

$$\frac{n+1}{2^n} \leq \frac{n}{2^{n-1}}.$$

So, $a_{n+1} = (n+1)/2^n \leq n/2^{n-1} = a_n$ for all n. Furthermore, by L'Hôpital's Rule,

$$\lim_{x \to \infty} \frac{x}{2^{x-1}} = \lim_{x \to \infty} \frac{1}{2^{x-1}(\ln 2)} = 0 \implies \lim_{n \to \infty} \frac{n}{2^{n-1}} = 0.$$

Therefore, by the Alternating Series Test, the series converges.

EXAMPLE 3   When the Alternating Series Test Does Not Apply

REMARK In Example 3(a), remember that whenever a series does not pass the first condition of the Alternating Series Test, you can use the nth-Term Test for Divergence to conclude that the series diverges.

a. The alternating series

$$\sum_{n=1}^{\infty} \frac{(-1)^{n+1}(n+1)}{n} = \frac{2}{1} - \frac{3}{2} + \frac{4}{3} - \frac{5}{4} + \frac{6}{5} - \cdots$$

passes the second condition of the Alternating Series Test because $a_{n+1} \leq a_n$ for all n. You cannot apply the Alternating Series Test, however, because the series does not pass the first condition. In fact, the series diverges.

b. The alternating series

$$\frac{2}{1} - \frac{1}{1} + \frac{2}{2} - \frac{1}{2} + \frac{2}{3} - \frac{1}{3} + \frac{2}{4} - \frac{1}{4} + \cdots$$

passes the first condition because a_n approaches 0 as $n \to \infty$. You cannot apply the Alternating Series Test, however, because the series does not pass the second condition. To conclude that the series diverges, you can argue that S_{2N} equals the Nth partial sum of the divergent harmonic series. This implies that the sequence of partial sums diverges. So, the series diverges.

Alternating Series Remainder

For a convergent alternating series, the partial sum S_N can be a useful approximation for the sum S of the series. The error involved in using $S \approx S_N$ is the remainder $R_N = S - S_N$.

THEOREM 9.15   Alternating Series Remainder

If a convergent alternating series satisfies the condition $a_{n+1} \le a_n$, then the absolute value of the remainder R_N involved in approximating the sum S by S_N is less than (or equal to) the first neglected term. That is,

$$|S - S_N| = |R_N| \le a_{N+1}.$$

A proof of this theorem is given in Appendix A.

EXAMPLE 4   Approximating the Sum of an Alternating Series

∙∙∙∙▷ *See LarsonCalculus.com for an interactive version of this type of example.*

Approximate the sum of the series by its first six terms.

$$\sum_{n=1}^{\infty} (-1)^{n+1} \left(\frac{1}{n!} \right) = \frac{1}{1!} - \frac{1}{2!} + \frac{1}{3!} - \frac{1}{4!} + \frac{1}{5!} - \frac{1}{6!} + \cdots$$

Solution   The series converges by the Alternating Series Test because

$$\frac{1}{(n+1)!} \le \frac{1}{n!} \quad \text{and} \quad \lim_{n \to \infty} \frac{1}{n!} = 0.$$

The sum of the first six terms is

$$S_6 = 1 - \frac{1}{2} + \frac{1}{6} - \frac{1}{24} + \frac{1}{120} - \frac{1}{720} = \frac{91}{144} \approx 0.63194$$

and, by the Alternating Series Remainder, you have

$$|S - S_6| = |R_6| \le a_7 = \frac{1}{5040} \approx 0.0002.$$

So, the sum S lies between $0.63194 - 0.0002$ and $0.63194 + 0.0002$, and you have $0.63174 \le S \le 0.63214$.

▷ **TECHNOLOGY** Later, using the techniques in Section 9.10, you will be able to show that the series in Example 4 converges to

$$\frac{e-1}{e} \approx 0.63212.$$

(See Section 9.10, Exercise 58.) For now, try using a graphing utility to obtain an approximation of the sum of the series. How many terms do you need to obtain an approximation that is within 0.00001 of the actual sum?

EXAMPLE 5   Finding the Number of Terms

Determine the number of terms required to approximate the sum of the series with an error of less than 0.001.

$$\sum_{n=1}^{\infty} \frac{(-1)^{n+1}}{n^4}$$

Solution   By Theorem 9.15, you know that

$$|R_N| \le a_{N+1} = \frac{1}{(N+1)^4}.$$

For an error of less than 0.001, N must satisfy the inequality $1/(N+1)^4 < 0.001$.

$$\frac{1}{(N+1)^4} < 0.001 \quad \Longrightarrow \quad (N+1)^4 > 1000 \quad \Longrightarrow \quad N > \sqrt[4]{1000} - 1 \approx 4.6$$

So, you will need at least five terms. Using five terms, the sum is $S \approx S_5 \approx 0.94754$, which has an error of less than 0.001. ∎

Absolute and Conditional Convergence

Occasionally, a series may have both positive and negative terms and not be an alternating series. For instance, the series

$$\sum_{n=1}^{\infty} \frac{\sin n}{n^2} = \frac{\sin 1}{1} + \frac{\sin 2}{4} + \frac{\sin 3}{9} + \cdots$$

has both positive and negative terms, yet it is not an alternating series. One way to obtain some information about the convergence of this series is to investigate the convergence of the series

$$\sum_{n=1}^{\infty} \left| \frac{\sin n}{n^2} \right|.$$

By direct comparison, you have $|\sin n| \le 1$ for all n, so

$$\left| \frac{\sin n}{n^2} \right| \le \frac{1}{n^2}, \quad n \ge 1.$$

Therefore, by the Direct Comparison Test, the series $\sum |(\sin n)/n^2|$ converges. The next theorem tells you that the original series also converges.

THEOREM 9.16 Absolute Convergence

If the series $\sum |a_n|$ converges, then the series $\sum a_n$ also converges.

Proof Because $0 \le a_n + |a_n| \le 2|a_n|$ for all n, the series

$$\sum_{n=1}^{\infty} (a_n + |a_n|)$$

converges by comparison with the convergent series

$$\sum_{n=1}^{\infty} 2|a_n|.$$

Furthermore, because $a_n = (a_n + |a_n|) - |a_n|$, you can write

$$\sum_{n=1}^{\infty} a_n = \sum_{n=1}^{\infty} (a_n + |a_n|) - \sum_{n=1}^{\infty} |a_n|$$

where both series on the right converge. So, it follows that $\sum a_n$ converges.

The converse of Theorem 9.16 is not true. For instance, the **alternating harmonic series**

$$\sum_{n=1}^{\infty} \frac{(-1)^{n+1}}{n} = \frac{1}{1} - \frac{1}{2} + \frac{1}{3} - \frac{1}{4} + \cdots$$

converges by the Alternating Series Test. Yet the harmonic series diverges. This type of convergence is called **conditional.**

Definitions of Absolute and Conditional Convergence

1. The series $\sum a_n$ is **absolutely convergent** when $\sum |a_n|$ converges.
2. The series $\sum a_n$ is **conditionally convergent** when $\sum a_n$ converges but $\sum |a_n|$ diverges.

EXAMPLE 6   **Absolute and Conditional Convergence**

Determine whether each of the series is convergent or divergent. Classify any convergent series as absolutely or conditionally convergent.

a. $\displaystyle\sum_{n=0}^{\infty} \frac{(-1)^n n!}{2^n} = \frac{0!}{2^0} - \frac{1!}{2^1} + \frac{2!}{2^2} - \frac{3!}{2^3} + \cdots$

b. $\displaystyle\sum_{n=1}^{\infty} \frac{(-1)^n}{\sqrt{n}} = -\frac{1}{\sqrt{1}} + \frac{1}{\sqrt{2}} - \frac{1}{\sqrt{3}} + \frac{1}{\sqrt{4}} - \cdots$

Solution

a. This is an alternating series, but the Alternating Series Test does not apply because the limit of the nth term is not zero. By the nth-Term Test for Divergence, however, you can conclude that this series diverges.

b. This series can be shown to be convergent by the Alternating Series Test. Moreover, because the p-series

$$\sum_{n=1}^{\infty} \left| \frac{(-1)^n}{\sqrt{n}} \right| = \frac{1}{\sqrt{1}} + \frac{1}{\sqrt{2}} + \frac{1}{\sqrt{3}} + \frac{1}{\sqrt{4}} + \cdots$$

diverges, the given series is *conditionally* convergent.

EXAMPLE 7   **Absolute and Conditional Convergence**

Determine whether each of the series is convergent or divergent. Classify any convergent series as absolutely or conditionally convergent.

a. $\displaystyle\sum_{n=1}^{\infty} \frac{(-1)^{n(n+1)/2}}{3^n} = -\frac{1}{3} - \frac{1}{9} + \frac{1}{27} + \frac{1}{81} - \cdots$

b. $\displaystyle\sum_{n=1}^{\infty} \frac{(-1)^n}{\ln(n+1)} = -\frac{1}{\ln 2} + \frac{1}{\ln 3} - \frac{1}{\ln 4} + \frac{1}{\ln 5} - \cdots$

Solution

a. This is *not* an alternating series (the signs change in pairs). However, note that

$$\sum_{n=1}^{\infty} \left| \frac{(-1)^{n(n+1)/2}}{3^n} \right| = \sum_{n=1}^{\infty} \frac{1}{3^n}$$

is a convergent geometric series, with

$$r = \frac{1}{3}.$$

Consequently, by Theorem 9.16, you can conclude that the given series is *absolutely* convergent (and therefore convergent).

b. In this case, the Alternating Series Test indicates that the series converges. However, the series

$$\sum_{n=1}^{\infty} \left| \frac{(-1)^n}{\ln(n+1)} \right| = \frac{1}{\ln 2} + \frac{1}{\ln 3} + \frac{1}{\ln 4} + \cdots$$

diverges by direct comparison with the terms of the harmonic series. Therefore, the given series is *conditionally* convergent. ◼

◼ FOR FURTHER INFORMATION To read more about the convergence of alternating harmonic series, see the article "Almost Alternating Harmonic Series" by Curtis Feist and Ramin Naimi in *The College Mathematics Journal*. To view this article, go to *MathArticles.com*.

Rearrangement of Series

A finite sum such as

$$1 + 3 - 2 + 5 - 4$$

can be rearranged without changing the value of the sum. This is not necessarily true of an infinite series—it depends on whether the series is absolutely convergent or conditionally convergent.

1. If a series is *absolutely convergent*, then its terms can be rearranged in any order without changing the sum of the series.

2. If a series is *conditionally convergent*, then its terms can be rearranged to give a different sum.

The second case is illustrated in Example 8.

EXAMPLE 8 Rearrangement of a Series

■ **FOR FURTHER INFORMATION** Georg Friedrich Bernhard Riemann (1826–1866) proved that if $\Sigma\, a_n$ is conditionally convergent and S is any real number, then the terms of the series can be rearranged to converge to S. For more on this topic, see the article "Riemann's Rearrangement Theorem" by Stewart Galanor in *Mathematics Teacher*. To view this article, go to *MathArticles.com*.

The alternating harmonic series converges to ln 2. That is,

$$\sum_{n=1}^{\infty} (-1)^{n+1} \frac{1}{n} = \frac{1}{1} - \frac{1}{2} + \frac{1}{3} - \frac{1}{4} + \cdots = \ln 2. \qquad \text{See Exercise 55, Section 9.10.}$$

Rearrange the terms of the series to produce a different sum.

Solution Consider the rearrangement below.

$$1 - \frac{1}{2} - \frac{1}{4} + \frac{1}{3} - \frac{1}{6} - \frac{1}{8} + \frac{1}{5} - \frac{1}{10} - \frac{1}{12} + \frac{1}{7} - \frac{1}{14} - \cdots$$

$$= \left(1 - \frac{1}{2}\right) - \frac{1}{4} + \left(\frac{1}{3} - \frac{1}{6}\right) - \frac{1}{8} + \left(\frac{1}{5} - \frac{1}{10}\right) - \frac{1}{12} + \left(\frac{1}{7} - \frac{1}{14}\right) - \cdots$$

$$= \frac{1}{2} - \frac{1}{4} + \frac{1}{6} - \frac{1}{8} + \frac{1}{10} - \frac{1}{12} + \frac{1}{14} - \cdots$$

$$= \frac{1}{2}\left(1 - \frac{1}{2} + \frac{1}{3} - \frac{1}{4} + \frac{1}{5} - \frac{1}{6} + \frac{1}{7} - \cdots\right)$$

$$= \frac{1}{2}(\ln 2)$$

By rearranging the terms, you obtain a sum that is half the original sum. ■

Exploration

In Example 8, you learned that the alternating harmonic series

$$\sum_{n=1}^{\infty} (-1)^{n+1} \frac{1}{n} = 1 - \frac{1}{2} + \frac{1}{3} - \frac{1}{4} + \frac{1}{5} - \frac{1}{6} + \cdots$$

converges to ln 2 ≈ 0.693. Rearrangement of the terms of the series produces a different sum, $\frac{1}{2}$ ln 2 ≈ 0.347.

In this exploration, you will rearrange the terms of the alternating harmonic series in such a way that two positive terms follow each negative term. That is,

$$1 - \frac{1}{2} + \frac{1}{3} + \frac{1}{5} - \frac{1}{4} + \frac{1}{7} + \frac{1}{9} - \frac{1}{6} + \frac{1}{11} + \cdots.$$

Now calculate the partial sums S_4, S_7, S_{10}, S_{13}, S_{16}, and S_{19}. Then estimate the sum of this series to three decimal places.

9.5 Exercises

See **CalcChat.com** for tutorial help and worked-out solutions to odd-numbered exercises.

CONCEPT CHECK

1. Alternating Series An alternating series does not meet the first condition of the Alternating Series Test. What can you conclude about the convergence or divergence of the series? Explain.

2. Alternating Series Remainder What is the remainder of a convergent alternating series whose sum is approximated by the first N terms?

3. Absolute and Conditional Convergence In your own words, describe the difference between absolute and conditional convergence of an alternating series.

4. Rearrangement of Series Does rearranging the terms of a convergent series change the sum of the series? Explain.

 **Numerical and Graphical Analysis** In Exercises 5–8, explore the Alternating Series Remainder.

(a) Use a graphing utility to find the indicated partial sum S_n and complete the table.

n	1	2	3	4	5	6	7	8	9	10
S_n										

(b) Use a graphing utility to graph the first 10 terms of the sequence of partial sums and a horizontal line representing the sum.

(c) What pattern exists between the plot of the successive points in part (b) relative to the horizontal line representing the sum of the series? Do the distances between the successive points and the horizontal line increase or decrease?

(d) Discuss the relationship between the answers in part (c) and the Alternating Series Remainder as given in Theorem 9.15.

5. $\displaystyle\sum_{n=1}^{\infty} \frac{(-1)^{n-1}}{2n-1} = \frac{\pi}{4}$

6. $\displaystyle\sum_{n=1}^{\infty} \frac{(-1)^{n-1}}{(n-1)!} = \frac{1}{e}$

7. $\displaystyle\sum_{n=1}^{\infty} \frac{(-1)^{n-1}}{n^2} = \frac{\pi^2}{12}$

8. $\displaystyle\sum_{n=1}^{\infty} \frac{(-1)^{n-1}}{(2n-1)!} = \sin 1$

 **Determining Convergence or Divergence** In Exercises 9–30, determine the convergence or divergence of the series.

9. $\displaystyle\sum_{n=1}^{\infty} \frac{(-1)^{n+1}}{n+1}$

10. $\displaystyle\sum_{n=1}^{\infty} \frac{(-1)^{n+1} n}{3n+2}$

11. $\displaystyle\sum_{n=1}^{\infty} \frac{(-1)^n}{3^n}$

12. $\displaystyle\sum_{n=1}^{\infty} \frac{(-1)^n}{e^n}$

13. $\displaystyle\sum_{n=1}^{\infty} \frac{(-1)^n(5n-1)}{4n+1}$

14. $\displaystyle\sum_{n=1}^{\infty} \frac{(-1)^{n+1} n}{n^2+5}$

15. $\displaystyle\sum_{n=1}^{\infty} \frac{(-1)^n n}{\ln(n+1)}$

16. $\displaystyle\sum_{n=1}^{\infty} \frac{(-1)^n}{\ln(n+1)}$

17. $\displaystyle\sum_{n=1}^{\infty} \frac{(-1)^n}{\sqrt{n}}$

18. $\displaystyle\sum_{n=1}^{\infty} \frac{(-1)^{n+1} n^2}{n^2+4}$

19. $\displaystyle\sum_{n=1}^{\infty} \frac{(-1)^{n+1}(n+1)}{\ln(n+1)}$

20. $\displaystyle\sum_{n=1}^{\infty} \frac{(-1)^{n+1} \ln(n+1)}{n+1}$

21. $\displaystyle\sum_{n=1}^{\infty} \sin \frac{(2n-1)\pi}{2}$

22. $\displaystyle\sum_{n=1}^{\infty} \frac{1}{n} \cos n\pi$

23. $\displaystyle\sum_{n=0}^{\infty} \frac{(-1)^n}{n!}$

24. $\displaystyle\sum_{n=0}^{\infty} \frac{(-1)^n}{(2n+1)!}$

25. $\displaystyle\sum_{n=1}^{\infty} \frac{(-1)^{n+1} \sqrt{n}}{n+2}$

26. $\displaystyle\sum_{n=1}^{\infty} \frac{(-1)^{n+1} \sqrt{n}}{\sqrt[3]{n}}$

27. $\displaystyle\sum_{n=1}^{\infty} \frac{(-1)^{n+1} n!}{1 \cdot 3 \cdot 5 \cdots (2n-1)}$

28. $\displaystyle\sum_{n=1}^{\infty} (-1)^{n+1} \frac{1 \cdot 3 \cdot 5 \cdots (2n-1)}{1 \cdot 4 \cdot 7 \cdots (3n-2)}$

29. $\displaystyle\sum_{n=1}^{\infty} \frac{2(-1)^{n+1}}{e^n - e^{-n}} = \sum_{n=1}^{\infty} (-1)^{n+1} \operatorname{csch} n$

30. $\displaystyle\sum_{n=1}^{\infty} \frac{2(-1)^{n+1}}{e^n + e^{-n}} = \sum_{n=1}^{\infty} (-1)^{n+1} \operatorname{sech} n$

 **Approximating the Sum of an Alternating Series** In Exercises 31–34, approximate the sum of the series by using the first six terms. (See Example 4.)

31. $\displaystyle\sum_{n=0}^{\infty} \frac{(-1)^n 5}{n!}$

32. $\displaystyle\sum_{n=1}^{\infty} \frac{(-1)^{n+1} 4}{\ln(n+1)}$

33. $\displaystyle\sum_{n=1}^{\infty} \frac{(-1)^{n+1} 2}{n^3}$

34. $\displaystyle\sum_{n=1}^{\infty} \frac{(-1)^{n+1} n}{3^n}$

Finding the Number of Terms In Exercises 35–40, use Theorem 9.15 to determine the number of terms required to approximate the sum of the series with an error of less than 0.001.

35. $\displaystyle\sum_{n=1}^{\infty} \frac{(-1)^{n+1}}{n^3}$

36. $\displaystyle\sum_{n=1}^{\infty} \frac{(-1)^{n+1}}{n^2}$

37. $\displaystyle\sum_{n=1}^{\infty} \frac{(-1)^{n+1}}{2n^3 - 1}$

38. $\displaystyle\sum_{n=1}^{\infty} \frac{(-1)^{n+1}}{n^5}$

39. $\displaystyle\sum_{n=0}^{\infty} \frac{(-1)^n}{n!}$

40. $\displaystyle\sum_{n=0}^{\infty} \frac{(-1)^n}{(2n)!}$

 **Determining Absolute and Conditional Convergence** In Exercises 41–58, determine whether the series converges absolutely or conditionally, or diverges.

41. $\displaystyle\sum_{n=1}^{\infty} \frac{(-1)^n}{2^n}$

42. $\displaystyle\sum_{n=1}^{\infty} \frac{(-1)^{n+1}}{n^2}$

43. $\displaystyle\sum_{n=1}^{\infty} \frac{(-1)^n}{n!}$

44. $\displaystyle\sum_{n=1}^{\infty} \frac{(-1)^{n+1}}{n+3}$

45. $\sum_{n=1}^{\infty} \dfrac{(-1)^{n+1}}{\sqrt{n}}$

46. $\sum_{n=1}^{\infty} \dfrac{(-1)^{n+1}}{n\sqrt{n}}$

47. $\sum_{n=1}^{\infty} \dfrac{(-1)^{n+1}\,n^2}{(n+1)^2}$

48. $\sum_{n=1}^{\infty} \dfrac{(-1)^{n+1}n}{5n+1}$

49. $\sum_{n=2}^{\infty} \dfrac{(-1)^n}{n \ln n}$

50. $\sum_{n=0}^{\infty} (-1)^n\, e^{-n^2}$

51. $\sum_{n=2}^{\infty} \dfrac{(-1)^n\, n}{n^3 - 5}$

52. $\sum_{n=1}^{\infty} \dfrac{(-1)^{n+1}}{n^{4/3}}$

53. $\sum_{n=0}^{\infty} \dfrac{(-1)^n}{(2n+1)!}$

54. $\sum_{n=0}^{\infty} \dfrac{(-1)^n}{\sqrt{n+4}}$

55. $\sum_{n=0}^{\infty} \dfrac{\cos n\pi}{n+1}$

56. $\sum_{n=1}^{\infty} (-1)^{n+1} \arctan n$

57. $\sum_{n=1}^{\infty} \dfrac{\cos(n\pi/3)}{n^2}$

58. $\sum_{n=1}^{\infty} \dfrac{\sin[(2n-1)\pi/2]}{n}$

EXPLORING CONCEPTS

59. Alternating Series Determine whether S_{50} is an underestimate or an overestimate of the sum of the alternating series below. Explain.

$$\sum_{n=1}^{\infty} \dfrac{(-1)^n}{n}$$

60. Alternating Series Give an example of convergent alternating series $\Sigma\, a_n$ and $\Sigma\, b_n$ such that $\Sigma\, a_n b_n$ diverges.

61. Think About It Do you agree with the following statements? Why or why not?

(a) If both $\Sigma\, a_n$ and $\Sigma\,(-a_n)$ converge, then $\Sigma\, |a_n|$ converges.

(b) If $\Sigma\, a_n$ diverges, then $\Sigma\, |a_n|$ diverges.

---

62. **HOW DO YOU SEE IT?** The graphs of the sequences of partial sums of two series are shown in the figures. Which graph represents the partial sums of an alternating series? Explain.

(a)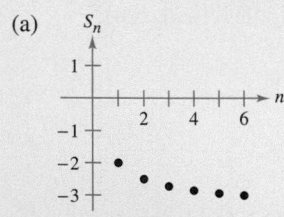

(b)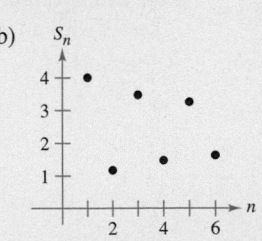

Finding Values In Exercises 63 and 64, find the values of p for which the series converges.

63. $\sum_{n=1}^{\infty} (-1)^n \left(\dfrac{1}{n^p} \right)$

64. $\sum_{n=1}^{\infty} (-1)^n \left(\dfrac{1}{n+p} \right)$

65. Proof Prove that if $\Sigma\, |a_n|$ converges, then $\Sigma\, a_n^2$ converges. Is the converse true? If not, give an example that shows it is false.

66. Finding a Series Use the result of Exercise 63 to give an example of an alternating p-series that converges but whose corresponding p-series diverges.

67. Finding a Series Give an example of a series that demonstrates the statement you proved in Exercise 65.

68. Finding Values Find all values of x for which the series $\Sigma\,(x^n/n)$ (a) converges absolutely and (b) converges conditionally.

Using a Series In Exercises 69 and 70, use the given series.

(a) Does the series meet the conditions of Theorem 9.14? Explain why or why not.

(b) Does the series converge? If so, what is the sum?

69. $\dfrac{1}{2} - \dfrac{1}{3} + \dfrac{1}{4} - \dfrac{1}{9} + \dfrac{1}{8} - \dfrac{1}{27} + \cdots + \dfrac{1}{2^n} - \dfrac{1}{3^n} + \cdots$

70. $\sum_{n=1}^{\infty} (-1)^{n+1} a_n,\ a_n = \begin{cases} \dfrac{1}{\sqrt{n}}, & \text{if } n \text{ is odd} \\ \dfrac{1}{n^3}, & \text{if } n \text{ is even} \end{cases}$

Review In Exercises 71–80, determine the convergence or divergence of the series and identify the test used.

71. $\sum_{n=1}^{\infty} \dfrac{8}{\sqrt[3]{n}}$

72. $\sum_{n=1}^{\infty} \dfrac{3n+5}{n^3 + 2n^2 + 4}$

73. $\sum_{n=1}^{\infty} \dfrac{3^n}{n^2}$

74. $\sum_{n=1}^{\infty} \dfrac{1}{6^n - 5}$

75. $\sum_{n=1}^{\infty} \left(\dfrac{9}{8} \right)^n$

76. $\sum_{n=1}^{\infty} \dfrac{2n^2}{(n+1)^2}$

77. $\sum_{n=1}^{\infty} 100 e^{-n/2}$

78. $\sum_{n=0}^{\infty} \dfrac{(-1)^n}{n+4}$

79. $\sum_{n=1}^{\infty} \dfrac{(-1)^{n+1}\,4}{3n^2 - 1}$

80. $\sum_{n=2}^{\infty} \dfrac{\ln n}{n}$

81. Describing an Error The following argument, that $0 = 1$, is *incorrect*. Describe the error.

$$0 = 0 + 0 + 0 + \cdots$$
$$= (1 - 1) + (1 - 1) + (1 - 1) + \cdots$$
$$= 1 + (-1 + 1) + (-1 + 1) + \cdots$$
$$= 1 + 0 + 0 + \cdots$$
$$= 1$$

PUTNAM EXAM CHALLENGE

82. Assume as known the (true) fact that the alternating harmonic series

(1) $1 - \dfrac{1}{2} + \dfrac{1}{3} - \dfrac{1}{4} + \dfrac{1}{5} - \dfrac{1}{6} + \dfrac{1}{7} - \dfrac{1}{8} + \cdots$

is convergent, and denote its sum by s. Rearrange the series (1) as follows:

(2) $1 + \dfrac{1}{3} - \dfrac{1}{2} + \dfrac{1}{5} + \dfrac{1}{7} - \dfrac{1}{4} + \dfrac{1}{9} + \dfrac{1}{11} - \dfrac{1}{6} + \cdots$

Assume as known the (true) fact that the series (2) is also convergent, and denote its sum by S. Denote by s_k, S_k the kth partial sum of the series (1) and (2), respectively. Prove the following statements.

(i) $S_{3n} = s_{4n} + \dfrac{1}{2} s_{2n}$,   (ii) $S \neq s$

9.6 The Ratio and Root Tests

- Use the Ratio Test to determine whether a series converges or diverges.
- Use the Root Test to determine whether a series converges or diverges.
- Review the tests for convergence and divergence of an infinite series.

The Ratio Test

This section begins with a test for absolute convergence—the **Ratio Test.**

THEOREM 9.17   Ratio Test

Let $\Sigma\, a_n$ be a series with nonzero terms.

1. The series $\Sigma\, a_n$ converges absolutely when $\lim\limits_{n\to\infty} \left| \dfrac{a_{n+1}}{a_n} \right| < 1.$

2. The series $\Sigma\, a_n$ diverges when $\lim\limits_{n\to\infty} \left| \dfrac{a_{n+1}}{a_n} \right| > 1$ or $\lim\limits_{n\to\infty} \left| \dfrac{a_{n+1}}{a_n} \right| = \infty.$

3. The Ratio Test is inconclusive when $\lim\limits_{n\to\infty} \left| \dfrac{a_{n+1}}{a_n} \right| = 1.$

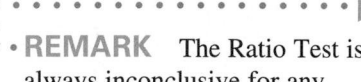

·· REMARK   The Ratio Test is always inconclusive for any p-series.

Proof   To prove Property 1, assume that

$$\lim_{n\to\infty} \left| \frac{a_{n+1}}{a_n} \right| = r < 1$$

and choose R such that $0 \le r < R < 1$. By the definition of the limit of a sequence, there exists some $N > 0$ such that $|a_{n+1}/a_n| < R$ for all $n > N$. Therefore, you can write the following inequalities.

$$|a_{N+1}| < |a_N|R$$
$$|a_{N+2}| < |a_{N+1}|R < |a_N|R^2$$
$$|a_{N+3}| < |a_{N+2}|R < |a_{N+1}|R^2 < |a_N|R^3$$
$$\vdots$$

The geometric series $\displaystyle\sum_{n=1}^{\infty} |a_N|R^n = |a_N|R + |a_N|R^2 + \cdots + |a_N|R^n + \cdots$ converges, and so, by the Direct Comparison Test, the series

$$\sum_{n=1}^{\infty} |a_{N+n}| = |a_{N+1}| + |a_{N+2}| + \cdots + |a_{N+n}| + \cdots$$

also converges. This in turn implies that the series $\Sigma\, |a_n|$ converges, because discarding a finite number of terms ($n = N - 1$) does not affect convergence. Consequently, by Theorem 9.16, the series $\Sigma\, a_n$ converges absolutely. The proof of Property 2 is similar and is left as an exercise (see Exercise 97). ∎

The fact that the Ratio Test is inconclusive when $|a_{n+1}/a_n| \to 1$ can be seen by comparing the two series $\Sigma\, (1/n)$ and $\Sigma\, (1/n^2)$. The first series diverges and the second one converges, but in both cases

$$\lim_{n\to\infty} \left| \frac{a_{n+1}}{a_n} \right| = 1.$$

Although the Ratio Test is not a cure for all ills related to testing for convergence, it is particularly useful for series that *converge rapidly*. Series involving factorials or exponentials are frequently of this type.

EXAMPLE 1 Using the Ratio Test

Determine the convergence or divergence of

$$\sum_{n=0}^{\infty} \frac{2^n}{n!}.$$

Solution Recall from Section 9.1 that the factorial function grows faster than any exponential function. So, you expect this series to converge. Because

$$a_n = \frac{2^n}{n!}$$

you can write the following.

$$\lim_{n\to\infty} \left| \frac{a_{n+1}}{a_n} \right| = \lim_{n\to\infty} \left[\frac{2^{n+1}}{(n+1)!} \div \frac{2^n}{n!} \right]$$

$$= \lim_{n\to\infty} \left[\frac{2^{n+1}}{(n+1)!} \cdot \frac{n!}{2^n} \right]$$

$$= \lim_{n\to\infty} \frac{2}{n+1}$$

$$= 0 < 1$$

This series converges because the limit of $|a_{n+1}/a_n|$ is less than 1.

> **REMARK** A step frequently used in applications of the Ratio Test involves simplifying quotients of factorials. In Example 1, for instance, notice that
>
> $$\frac{n!}{(n+1)!} = \frac{n!}{(n+1)n!} = \frac{1}{n+1}.$$

EXAMPLE 2 Using the Ratio Test

Determine whether each series converges or diverges.

a. $\displaystyle\sum_{n=0}^{\infty} \frac{n^2 2^{n+1}}{3^n}$ **b.** $\displaystyle\sum_{n=1}^{\infty} \frac{n^n}{n!}$

Solution

a. This series converges because the limit of $|a_{n+1}/a_n|$ is less than 1.

$$\lim_{n\to\infty} \left| \frac{a_{n+1}}{a_n} \right| = \lim_{n\to\infty} \left[(n+1)^2 \left(\frac{2^{n+2}}{3^{n+1}} \right) \left(\frac{3^n}{n^2 2^{n+1}} \right) \right]$$

$$= \lim_{n\to\infty} \frac{2(n+1)^2}{3n^2}$$

$$= \frac{2}{3} < 1$$

b. This series diverges because the limit of $|a_{n+1}/a_n|$ is greater than 1.

$$\lim_{n\to\infty} \left| \frac{a_{n+1}}{a_n} \right| = \lim_{n\to\infty} \left[\frac{(n+1)^{n+1}}{(n+1)!} \left(\frac{n!}{n^n} \right) \right]$$

$$= \lim_{n\to\infty} \left[\frac{(n+1)^{n+1}}{(n+1)} \left(\frac{1}{n^n} \right) \right]$$

$$= \lim_{n\to\infty} \frac{(n+1)^n}{n^n}$$

$$= \lim_{n\to\infty} \left(1 + \frac{1}{n} \right)^n$$

$$= e > 1$$

EXAMPLE 3 **A Failure of the Ratio Test**

••••▷ *See LarsonCalculus.com for an interactive version of this type of example.*

Determine the convergence or divergence of

$$\sum_{n=1}^{\infty} (-1)^n \frac{\sqrt{n}}{n+1}.$$

Solution The limit of $|a_{n+1}/a_n|$ is equal to 1.

$$\lim_{n\to\infty} \left| \frac{a_{n+1}}{a_n} \right| = \lim_{n\to\infty} \left[\left(\frac{\sqrt{n+1}}{n+2} \right) \left(\frac{n+1}{\sqrt{n}} \right) \right]$$

$$= \lim_{n\to\infty} \left[\sqrt{\frac{n+1}{n}} \left(\frac{n+1}{n+2} \right) \right]$$

$$= \sqrt{1}\,(1)$$

$$= 1$$

So, the Ratio Test is inconclusive. To determine whether the series converges, you need to try a different test. In this case, you can apply the Alternating Series Test. To show that $a_{n+1} \le a_n$, let

$$f(x) = \frac{\sqrt{x}}{x+1}.$$

Then the derivative is

$$f'(x) = \frac{-x+1}{2\sqrt{x}(x+1)^2}.$$

Because the derivative is negative for $x > 1$, you know that f is a decreasing function. Also, by L'Hôpital's Rule,

$$\lim_{x\to\infty} \frac{\sqrt{x}}{x+1} = \lim_{x\to\infty} \frac{1/(2\sqrt{x})}{1}$$

$$= \lim_{x\to\infty} \frac{1}{2\sqrt{x}}$$

$$= 0.$$

Therefore, by the Alternating Series Test, the series converges. ∎

The series in Example 3 is *conditionally convergent*. This follows from the fact that the series

$$\sum_{n=1}^{\infty} |a_n|$$

diverges $\left(\text{by the Limit Comparison Test with } \Sigma\, 1/\sqrt{n}\right)$, but the series

$$\sum_{n=1}^{\infty} a_n$$

converges.

▷ **TECHNOLOGY** A graphing utility can reinforce the conclusion that the series in Example 3 converges *conditionally*. By adding the first 100 terms of the series, you obtain a sum of about -0.2. (The sum of the first 100 terms of the series $\Sigma\, |a_n|$ is about 17.)

The Root Test

The next test for convergence or divergence of series works especially well for series involving nth powers. The proof of this theorem is similar to the proof given for the Ratio Test and is left as an exercise (see Exercise 98).

· · · · · · · · · · · · · · · ▷

· · REMARK The Root Test is always inconclusive for any p-series.

THEOREM 9.18 Root Test

1. The series $\Sigma \, a_n$ converges absolutely when $\lim\limits_{n\to\infty} \sqrt[n]{|a_n|} < 1$.

2. The series $\Sigma \, a_n$ diverges when $\lim\limits_{n\to\infty} \sqrt[n]{|a_n|} > 1$ or $\lim\limits_{n\to\infty} \sqrt[n]{|a_n|} = \infty$.

3. The Root Test is inconclusive when $\lim\limits_{n\to\infty} \sqrt[n]{|a_n|} = 1$.

EXAMPLE 4   **Using the Root Test**

Determine the convergence or divergence of

$$\sum_{n=1}^{\infty} \frac{e^{2n}}{n^n}.$$

Solution   You can apply the Root Test as follows.

$$\begin{aligned}
\lim_{n\to\infty} \sqrt[n]{|a_n|} &= \lim_{n\to\infty} \sqrt[n]{\frac{e^{2n}}{n^n}} \\
&= \lim_{n\to\infty} \frac{e^{2n/n}}{n^{n/n}} \\
&= \lim_{n\to\infty} \frac{e^2}{n} \\
&= 0 < 1
\end{aligned}$$

Because this limit is less than 1, you can conclude that the series converges absolutely (and therefore converges).

To see the usefulness of the Root Test for the series in Example 4, try applying the Ratio Test to that series. When you do this, you obtain the following.

$$\begin{aligned}
\lim_{n\to\infty} \left| \frac{a_{n+1}}{a_n} \right| &= \lim_{n\to\infty} \left[\frac{e^{2(n+1)}}{(n+1)^{n+1}} \div \frac{e^{2n}}{n^n} \right] \\
&= \lim_{n\to\infty} \left[\frac{e^{2(n+1)}}{(n+1)^{n+1}} \cdot \frac{n^n}{e^{2n}} \right] \\
&= \lim_{n\to\infty} e^2 \frac{n^n}{(n+1)^{n+1}} \\
&= \lim_{n\to\infty} e^2 \left(\frac{n}{n+1} \right)^n \left(\frac{1}{n+1} \right) \\
&= 0
\end{aligned}$$

Note that this limit is not as easily evaluated as the limit obtained by the Root Test in Example 4.

■ FOR FURTHER INFORMATION   For more information on the usefulness of the Root Test, see the article "*N*! and the Root Test" by Charles C. Mumma II in *The American Mathematical Monthly*. To view this article, go to *MathArticles.com*.

Strategies for Testing Series

You have now studied 10 tests for determining the convergence or divergence of an infinite series. (See the summary in the table on the next page.) Skill in choosing and applying the various tests will come only with practice. Below is a set of guidelines for choosing an appropriate test.

GUIDELINES FOR TESTING A SERIES FOR CONVERGENCE OR DIVERGENCE

1. Does the nth term approach 0? If not, the series diverges.
2. Is the series one of the special types—geometric, p-series, telescoping, or alternating?
3. Can the Integral Test, the Root Test, or the Ratio Test be applied?
4. Can the series be compared favorably to one of the special types?

In some instances, more than one test is applicable. However, your objective should be to learn to choose the most efficient test.

EXAMPLE 5 **Applying the Strategies for Testing Series**

Determine the convergence or divergence of each series.

a. $\displaystyle\sum_{n=1}^{\infty} \frac{n+1}{3n+1}$ **b.** $\displaystyle\sum_{n=1}^{\infty} \left(\frac{\pi}{6}\right)^n$ **c.** $\displaystyle\sum_{n=1}^{\infty} ne^{-n^2}$

d. $\displaystyle\sum_{n=1}^{\infty} \frac{1}{3n+1}$ **e.** $\displaystyle\sum_{n=1}^{\infty} (-1)^n \frac{3}{4n+1}$ **f.** $\displaystyle\sum_{n=1}^{\infty} \frac{n!}{10^n}$

g. $\displaystyle\sum_{n=1}^{\infty} \left(\frac{n+1}{2n+1}\right)^n$

Solution

a. For this series, the limit of the nth term is not 0 $\left(a_n \to \frac{1}{3} \text{ as } n \to \infty\right)$. So, by the nth-Term Test, the series diverges.

b. This series is geometric. Moreover, because the ratio of the terms

$$r = \frac{\pi}{6}$$

is less than 1 in absolute value, you can conclude that the series converges.

c. Because the function

$$f(x) = xe^{-x^2}$$

is easily integrated, you can use the Integral Test to conclude that the series converges.

d. The nth term of this series can be compared to the nth term of the harmonic series. After using the Limit Comparison Test, you can conclude that the series diverges.

e. This is an alternating series whose nth term approaches 0. Because $a_{n+1} \leq a_n$, you can use the Alternating Series Test to conclude that the series converges.

f. The nth term of this series involves a factorial, which indicates that the Ratio Test may work well. After applying the Ratio Test, you can conclude that the series diverges.

g. The nth term of this series involves a variable that is raised to the nth power, which indicates that the Root Test may work well. After applying the Root Test, you can conclude that the series converges.

SUMMARY OF TESTS FOR SERIES

Test	Series	Condition(s) of Convergence	Condition(s) of Divergence	Comment
nth-Term	$\displaystyle\sum_{n=1}^{\infty} a_n$		$\displaystyle\lim_{n\to\infty} a_n \neq 0$	This test cannot be used to show convergence.
Geometric Series $(r \neq 0)$	$\displaystyle\sum_{n=0}^{\infty} ar^n$	$\lvert r \rvert < 1$	$\lvert r \rvert \geq 1$	Sum: $S = \dfrac{a}{1-r}$
Telescoping Series	$\displaystyle\sum_{n=1}^{\infty} (b_n - b_{n+1})$	$\displaystyle\lim_{n\to\infty} b_n = L$		Sum: $S = b_1 - L$
p-Series	$\displaystyle\sum_{n=1}^{\infty} \dfrac{1}{n^p}$	$p > 1$	$0 < p \leq 1$	
Alternating Series $(a_n > 0)$	$\displaystyle\sum_{n=1}^{\infty} (-1)^{n-1} a_n$	$a_{n+1} \leq a_n$ and $\displaystyle\lim_{n\to\infty} a_n = 0$		Remainder: $\lvert R_N \rvert \leq a_{N+1}$
Integral (f is continuous, positive, and decreasing)	$\displaystyle\sum_{n=1}^{\infty} a_n,$ $a_n = f(n) \geq 0$	$\displaystyle\int_1^{\infty} f(x)\,dx$ converges	$\displaystyle\int_1^{\infty} f(x)\,dx$ diverges	Remainder: $0 < R_N < \displaystyle\int_N^{\infty} f(x)\,dx$
Root	$\displaystyle\sum_{n=1}^{\infty} a_n$	$\displaystyle\lim_{n\to\infty} \sqrt[n]{\lvert a_n \rvert} < 1$	$\displaystyle\lim_{n\to\infty} \sqrt[n]{\lvert a_n \rvert} > 1$ or $= \infty$	Test is inconclusive when $\displaystyle\lim_{n\to\infty} \sqrt[n]{\lvert a_n \rvert} = 1$.
Ratio	$\displaystyle\sum_{n=1}^{\infty} a_n$	$\displaystyle\lim_{n\to\infty} \left\lvert \dfrac{a_{n+1}}{a_n} \right\rvert < 1$	$\displaystyle\lim_{n\to\infty} \left\lvert \dfrac{a_{n+1}}{a_n} \right\rvert > 1$ or $= \infty$	Test is inconclusive when $\displaystyle\lim_{n\to\infty} \left\lvert \dfrac{a_{n+1}}{a_n} \right\rvert = 1$.
Direct Comparison $(a_n, b_n > 0)$	$\displaystyle\sum_{n=1}^{\infty} a_n$	$0 < a_n \leq b_n$ and $\displaystyle\sum_{n=1}^{\infty} b_n$ converges	$0 < b_n \leq a_n$ and $\displaystyle\sum_{n=1}^{\infty} b_n$ diverges	
Limit Comparison $(a_n, b_n > 0)$	$\displaystyle\sum_{n=1}^{\infty} a_n$	$\displaystyle\lim_{n\to\infty} \dfrac{a_n}{b_n} = L > 0$ and $\displaystyle\sum_{n=1}^{\infty} b_n$ converges	$\displaystyle\lim_{n\to\infty} \dfrac{a_n}{b_n} = L > 0$ and $\displaystyle\sum_{n=1}^{\infty} b_n$ diverges	

9.6 Exercises

See **CalcChat.com** for tutorial help and worked-out solutions to odd-numbered exercises.

CONCEPT CHECK

Ratio and Root Tests In Exercises 1–6, what can you conclude about the convergence or divergence of $\Sigma\, a_n$?

1. $\displaystyle\lim_{n\to\infty}\left|\frac{a_{n+1}}{a_n}\right| = 0$

2. $\displaystyle\lim_{n\to\infty}\left|\frac{a_{n+1}}{a_n}\right| = 1$

3. $\displaystyle\lim_{n\to\infty}\left|\frac{a_{n+1}}{a_n}\right| = \frac{3}{2}$

4. $\displaystyle\lim_{n\to\infty}\sqrt[n]{|a_n|} = 2$

5. $\displaystyle\lim_{n\to\infty}\sqrt[n]{|a_n|} = 1$

6. $\displaystyle\lim_{n\to\infty}\sqrt[n]{|a_n|} = e$

Verifying a Formula In Exercises 7 and 8, verify the formula.

7. $\dfrac{9^{n+1}(n-1)!}{9^n(n-2)!} = 9(n-1)$

8. $\dfrac{(2k-2)!}{(2k)!} = \dfrac{1}{(2k)(2k-1)}$

Matching In Exercises 9–14, match the series with the graph of its sequence of partial sums. [The graphs are labeled (a), (b), (c), (d), (e), and (f).]

(a)
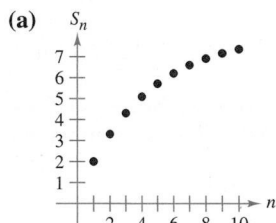

(b)
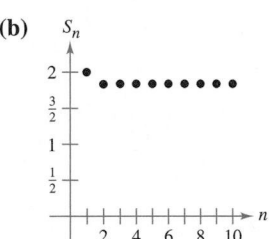

(c)
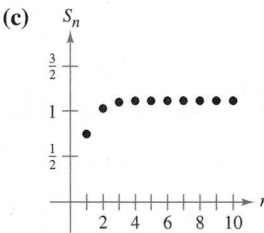

(d)
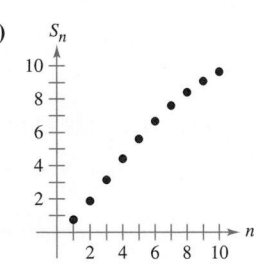

(e)
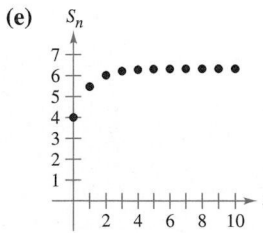

(f)
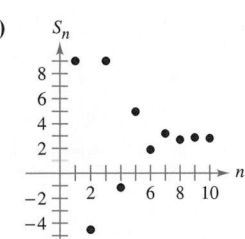

9. $\displaystyle\sum_{n=1}^{\infty} n\left(\frac{3}{4}\right)^n$

10. $\displaystyle\sum_{n=1}^{\infty} \left(\frac{3}{4}\right)^n\left(\frac{1}{n!}\right)$

11. $\displaystyle\sum_{n=1}^{\infty} \frac{(-3)^{n+1}}{n!}$

12. $\displaystyle\sum_{n=1}^{\infty} \frac{(-1)^{n-1}\,4}{(2n)!}$

13. $\displaystyle\sum_{n=1}^{\infty} \left(\frac{4n}{5n-3}\right)^n$

14. $\displaystyle\sum_{n=0}^{\infty} 4e^{-n}$

 **Numerical, Graphical, and Analytic Analysis** In Exercises 15 and 16, (a) use the Ratio Test to verify that the series converges, (b) use a graphing utility to find the indicated partial sum S_n and complete the table, (c) use a graphing utility to graph the first 10 terms of the sequence of partial sums, (d) use the table to estimate the sum of the series, and (e) explain the relationship between the magnitudes of the terms of the series and the rate at which the sequence of partial sums approaches the sum of the series.

n	5	10	15	20	25
S_n					

15. $\displaystyle\sum_{n=1}^{\infty} n^3\left(\frac{1}{2}\right)^n$

16. $\displaystyle\sum_{n=1}^{\infty} \frac{n^2+1}{n!}$

Using the Ratio Test In Exercises 17–38, use the Ratio Test to determine the convergence or divergence of the series. If the Ratio Test is inconclusive, determine the convergence or divergence of the series using other methods.

17. $\displaystyle\sum_{n=1}^{\infty} \frac{1}{8^n}$

18. $\displaystyle\sum_{n=1}^{\infty} \frac{5}{n!}$

19. $\displaystyle\sum_{n=1}^{\infty} \frac{(n-1)!}{4^n}$

20. $\displaystyle\sum_{n=0}^{\infty} \frac{2^n}{(n+2)!}$

21. $\displaystyle\sum_{n=0}^{\infty} (n+2)\left(\frac{9}{7}\right)^{n+1}$

22. $\displaystyle\sum_{n=1}^{\infty} n^2\left(\frac{5}{6}\right)^n$

23. $\displaystyle\sum_{n=1}^{\infty} \frac{9^n}{n^5}$

24. $\displaystyle\sum_{n=0}^{\infty} \frac{6^n}{(n+1)^3}$

25. $\displaystyle\sum_{n=1}^{\infty} \frac{n^3}{3^n}$

26. $\displaystyle\sum_{n=1}^{\infty} \frac{(-1)^{n+1}(n+2)}{n(n+1)}$

27. $\displaystyle\sum_{n=0}^{\infty} \frac{(-1)^n\,2^n}{n!}$

28. $\displaystyle\sum_{n=1}^{\infty} \frac{(-1)^{n-1}(3/2)^n}{n^2}$

29. $\displaystyle\sum_{n=1}^{\infty} \frac{n^2}{(n+1)(n^2+2)}$

30. $\displaystyle\sum_{n=1}^{\infty} \frac{(2n)!}{n^5}$

31. $\displaystyle\sum_{n=0}^{\infty} \frac{e^n}{n!}$

32. $\displaystyle\sum_{n=1}^{\infty} \frac{n!}{n^n}$

33. $\displaystyle\sum_{n=0}^{\infty} \frac{6^n}{(n+1)^n}$

34. $\displaystyle\sum_{n=0}^{\infty} \frac{(n!)^2}{(3n)!}$

35. $\displaystyle\sum_{n=0}^{\infty} \frac{5^n}{2^n+1}$

36. $\displaystyle\sum_{n=0}^{\infty} \frac{(-1)^n 2^{4n}}{(2n+1)!}$

37. $\displaystyle\sum_{n=0}^{\infty} \frac{(-1)^{n+1}n!}{1\cdot 3\cdot 5\cdots(2n+1)}$

38. $\displaystyle\sum_{n=1}^{\infty} \frac{(-1)^n[2\cdot 4\cdot 6\cdots(2n)]}{2\cdot 5\cdot 8\cdots(3n-1)}$

Using the Root Test In Exercises 39–52, use the Root Test to determine the convergence or divergence of the series.

39. $\displaystyle\sum_{n=1}^{\infty} \left(\frac{n}{2n+1}\right)^n$

40. $\displaystyle\sum_{n=1}^{\infty} \frac{1}{n^n}$

41. $\displaystyle\sum_{n=1}^{\infty} \left(\frac{3n+2}{n+3}\right)^n$

42. $\displaystyle\sum_{n=1}^{\infty} \left(\frac{n-2}{5n+1}\right)^n$

43. $\displaystyle\sum_{n=2}^{\infty} \frac{(-1)^n}{(\ln n)^n}$

44. $\displaystyle\sum_{n=1}^{\infty} \left(\frac{-3n}{2n+1}\right)^{3n}$

45. $\displaystyle\sum_{n=1}^{\infty} \left(2\sqrt[n]{n}+1\right)^n$

46. $\displaystyle\sum_{n=0}^{\infty} e^{-3n}$

47. $\displaystyle\sum_{n=1}^{\infty} \frac{n}{3^n}$

48. $\displaystyle\sum_{n=1}^{\infty} \left(\frac{n}{500}\right)^n$

49. $\displaystyle\sum_{n=1}^{\infty} \left(\frac{1}{n}-\frac{1}{n^2}\right)^n$

50. $\displaystyle\sum_{n=1}^{\infty} \left(\frac{\ln n}{n}\right)^n$

51. $\displaystyle\sum_{n=2}^{\infty} \frac{n}{(\ln n)^n}$

52. $\displaystyle\sum_{n=1}^{\infty} \frac{(n!)^n}{(n^n)^2}$

Review In Exercises 53–70, determine the convergence or divergence of the series using any appropriate test from this chapter. Identify the test used.

53. $\displaystyle\sum_{n=1}^{\infty} \frac{(-1)^{n+1} 5}{n}$

54. $\displaystyle\sum_{n=1}^{\infty} \frac{100}{n}$

55. $\displaystyle\sum_{n=1}^{\infty} \frac{3}{n\sqrt{n}}$

56. $\displaystyle\sum_{n=1}^{\infty} \left(\frac{2\pi}{3}\right)^n$

57. $\displaystyle\sum_{n=1}^{\infty} \frac{5n}{2n-1}$

58. $\displaystyle\sum_{n=1}^{\infty} \frac{n}{2n^2+1}$

59. $\displaystyle\sum_{n=1}^{\infty} \frac{(-1)^n 3^{n-2}}{2^n}$

60. $\displaystyle\sum_{n=1}^{\infty} \frac{10}{3\sqrt{n^3}}$

61. $\displaystyle\sum_{n=1}^{\infty} \frac{10n+3}{n2^n}$

62. $\displaystyle\sum_{n=1}^{\infty} \frac{2^n}{4n^2-1}$

63. $\displaystyle\sum_{n=1}^{\infty} \frac{\cos n}{3^n}$

64. $\displaystyle\sum_{n=2}^{\infty} \frac{(-1)^n}{n\ln n}$

65. $\displaystyle\sum_{n=1}^{\infty} \frac{n!}{n7^n}$

66. $\displaystyle\sum_{n=1}^{\infty} \frac{\ln n}{n^2}$

67. $\displaystyle\sum_{n=1}^{\infty} \frac{(-1)^n 3^{n-1}}{n!}$

68. $\displaystyle\sum_{n=1}^{\infty} \frac{(-1)^n 3^n}{n2^n}$

69. $\displaystyle\sum_{n=1}^{\infty} \frac{(-3)^n}{3\cdot5\cdot7\cdots(2n+1)}$

70. $\displaystyle\sum_{n=1}^{\infty} \frac{3\cdot5\cdot7\cdots(2n+1)}{18^n(2n-1)n!}$

Identifying Series In Exercises 71–74, identify the two series that are the same.

71. (a) $\displaystyle\sum_{n=1}^{\infty} \frac{n5^n}{n!}$

(b) $\displaystyle\sum_{n=0}^{\infty} \frac{n5^n}{(n+1)!}$

(c) $\displaystyle\sum_{n=0}^{\infty} \frac{(n+1)5^{n+1}}{(n+1)!}$

72. (a) $\displaystyle\sum_{n=4}^{\infty} n\left(\frac{3}{4}\right)^n$

(b) $\displaystyle\sum_{n=0}^{\infty} (n+1)\left(\frac{3}{4}\right)^n$

(c) $\displaystyle\sum_{n=1}^{\infty} n\left(\frac{3}{4}\right)^{n-1}$

73. (a) $\displaystyle\sum_{n=0}^{\infty} \frac{(-1)^n}{(2n+1)!}$

(b) $\displaystyle\sum_{n=1}^{\infty} \frac{(-1)^{n-1}}{(2n-1)!}$

(c) $\displaystyle\sum_{n=1}^{\infty} \frac{(-1)^{n-1}}{(2n+1)!}$

74. (a) $\displaystyle\sum_{n=2}^{\infty} \frac{(-1)^n}{(n-1)2^{n-1}}$

(b) $\displaystyle\sum_{n=1}^{\infty} \frac{(-1)^{n+1}}{n2^n}$

(c) $\displaystyle\sum_{n=0}^{\infty} \frac{(-1)^{n+1}}{(n+1)2^n}$

Writing an Equivalent Series In Exercises 75 and 76, write an equivalent series with the index of summation beginning at $n = 0$.

75. $\displaystyle\sum_{n=1}^{\infty} \frac{n}{7^n}$

76. $\displaystyle\sum_{n=2}^{\infty} \frac{4^{n+1}}{(n-2)!}$

Using a Recursively Defined Series In Exercises 77–82, the terms of a series $\displaystyle\sum_{n=1}^{\infty} a_n$ are defined recursively. Determine the convergence or divergence of the series. Explain your reasoning.

77. $a_1 = \dfrac{1}{2}, \ a_{n+1} = \dfrac{4n-1}{3n+2} a_n$

78. $a_1 = 2, \ a_{n+1} = \dfrac{2n+1}{5n-4} a_n$

79. $a_1 = 1, \ a_{n+1} = \dfrac{\sin n + 1}{\sqrt{n}} a_n$

80. $a_1 = \dfrac{1}{5}, \ a_{n+1} = \dfrac{\cos n + 1}{n} a_n$

81. $a_1 = \dfrac{1}{3}, \ a_{n+1} = \left(1 + \dfrac{1}{n}\right) a_n$

82. $a_1 = \dfrac{1}{4}, \ a_{n+1} = \sqrt[n]{a_n}$

Using the Ratio Test or Root Test In Exercises 83–86, use the Ratio Test or the Root Test to determine the convergence or divergence of the series.

83. $1 + \dfrac{1 \cdot 2}{1 \cdot 3} + \dfrac{1 \cdot 2 \cdot 3}{1 \cdot 3 \cdot 5} + \dfrac{1 \cdot 2 \cdot 3 \cdot 4}{1 \cdot 3 \cdot 5 \cdot 7} + \cdots$

84. $1 + \dfrac{2}{3} + \dfrac{3}{3^2} + \dfrac{4}{3^3} + \dfrac{5}{3^4} + \dfrac{6}{3^5} + \cdots$

85. $\dfrac{1}{(\ln 3)^3} + \dfrac{1}{(\ln 4)^4} + \dfrac{1}{(\ln 5)^5} + \dfrac{1}{(\ln 6)^6} + \cdots$

86. $1 + \dfrac{1 \cdot 3}{1 \cdot 2 \cdot 3} + \dfrac{1 \cdot 3 \cdot 5}{1 \cdot 2 \cdot 3 \cdot 4 \cdot 5} + \dfrac{1 \cdot 3 \cdot 5 \cdot 7}{1 \cdot 2 \cdot 3 \cdot 4 \cdot 5 \cdot 6 \cdot 7}$
$+ \cdots$

Finding Values In Exercises 87–92, find the values of x for which the series converges.

87. $\displaystyle\sum_{n=0}^{\infty} 2\left(\dfrac{x}{3}\right)^n$

88. $\displaystyle\sum_{n=0}^{\infty} \left(\dfrac{x-3}{5}\right)^n$

89. $\displaystyle\sum_{n=1}^{\infty} \dfrac{(-1)^n(x+1)^n}{n}$

90. $\displaystyle\sum_{n=0}^{\infty} 3(x-4)^n$

91. $\displaystyle\sum_{n=0}^{\infty} n!\left(\dfrac{x}{2}\right)^n$

92. $\displaystyle\sum_{n=0}^{\infty} \dfrac{(x+1)^n}{n!}$

EXPLORING CONCEPTS

93. Think About It What can you conclude about the convergence or divergence of $\Sigma\, a_n$ using the Ratio Test when a_n is a rational function of n? Explain.

94. Using Different Methods Describe two ways to show that the geometric series $\displaystyle\sum_{n=0}^{\infty} ar^n$, $r \neq 0$ converges when $|r| < 1$. Verify that both methods give the same result.

95. Think About It You are told that the terms of a positive series appear to approach zero rapidly as n approaches infinity. In fact, $a_7 \leq 0.0001$. Given no other information, does this imply that the series converges? Support your conclusion with examples.

---

96. HOW DO YOU SEE IT? The graphs show the sequences of partial sums of the series

$$\sum_{n=1}^{\infty} \dfrac{2^n}{n} \quad \text{and} \quad \sum_{n=1}^{\infty} \dfrac{n}{3^n}.$$

Using the Ratio Test, the first series diverges and the second series converges. Explain how the graphs show this.

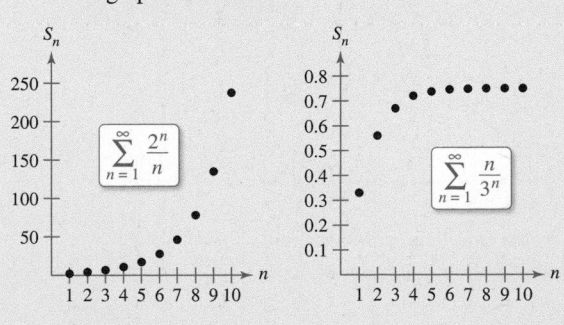

---

97. Proof Prove Property 2 of Theorem 9.17.

98. Proof Prove Theorem 9.18. (*Hint for Property 1:* If the limit equals $r < 1$, choose a real number R such that $r < R < 1$. By the definitions of the limit, there exists some $N > 0$ such that $\sqrt[n]{|a_n|} < R$ for $n > N$.)

Verifying an Inconclusive Test In Exercises 99–102, verify that the Ratio Test is inconclusive for the p-series.

99. $\displaystyle\sum_{n=1}^{\infty} \dfrac{1}{n^{3/2}}$

100. $\displaystyle\sum_{n=1}^{\infty} \dfrac{1}{n^{0.05}}$

101. $\displaystyle\sum_{n=1}^{\infty} \dfrac{1}{n^4}$

102. $\displaystyle\sum_{n=1}^{\infty} \dfrac{1}{n^p}$

103. Verifying an Inconclusive Test Show that the Root Test is inconclusive for the p-series

$$\sum_{n=1}^{\infty} \dfrac{1}{n^p}.$$

104. Verifying Inconclusive Tests Show that the Ratio Test and the Root Test are both inconclusive for the logarithmic p-series

$$\sum_{n=2}^{\infty} \dfrac{1}{n(\ln n)^p}.$$

105. Using Values Determine the convergence or divergence of the series

$$\sum_{n=1}^{\infty} \dfrac{(n!)^2}{(xn)!}$$

when (a) $x = 1$, (b) $x = 2$, (c) $x = 3$, and (d) x is a positive integer.

106. Using a Series Show that if

$$\sum_{n=1}^{\infty} a_n$$

is absolutely convergent, then

$$\left| \sum_{n=1}^{\infty} a_n \right| \leq \sum_{n=1}^{\infty} |a_n|.$$

PUTNAM EXAM CHALLENGE

107. Show that if the series

$$a_1 + a_2 + a_3 + \cdots + a_n + \cdots$$

converges, then the series

$$a_1 + \dfrac{a_2}{2} + \dfrac{a_3}{3} + \cdots + \dfrac{a_n}{n} + \cdots$$

converges also.

108. Is the following series convergent or divergent?

$$1 + \dfrac{1}{2} \cdot \dfrac{19}{7} + \dfrac{2!}{3^2}\left(\dfrac{19}{7}\right)^2 + \dfrac{3!}{4^3}\left(\dfrac{19}{7}\right)^3 + \dfrac{4!}{5^4}\left(\dfrac{19}{7}\right)^4 + \cdots$$

9.7 Taylor Polynomials and Approximations

■ Find polynomial approximations of elementary functions and compare them with the elementary functions.
■ Find Taylor and Maclaurin polynomial approximations of elementary functions.
■ Use the remainder of a Taylor polynomial.

Polynomial Approximations of Elementary Functions

The goal of this section is to show how polynomial functions can be used as approximations for other elementary functions. To find a polynomial function P that approximates another function f, begin by choosing a number c in the domain of f at which f and P have the same value. That is,

$$P(c) = f(c).$$ Graphs of f and P pass through $(c, f(c))$.

The approximating polynomial is said to be **expanded about** c or **centered at** c. Geometrically, the requirement that $P(c) = f(c)$ means that the graph of P passes through the point $(c, f(c))$. Of course, there are many polynomials whose graphs pass through the point $(c, f(c))$. Your task is to find a polynomial whose graph resembles the graph of f near this point. One way to do this is to impose the additional requirement that the slope of the polynomial function be the same as the slope of the graph of f at the point $(c, f(c))$.

$$P'(c) = f'(c)$$ Graphs of f and P have the same slope at $(c, f(c))$.

With these two requirements, you can obtain a simple linear approximation of f, as shown in Figure 9.12.

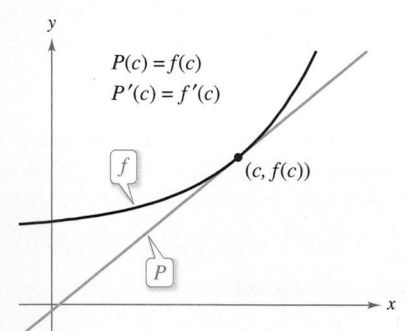

$P(c) = f(c)$
$P'(c) = f'(c)$

f

$(c, f(c))$

P

Near $(c, f(c))$, the graph of P can be used to approximate the graph of f.
Figure 9.12

▷ ·····•·•·•·✶·•·•·•·•·

• **REMARK** Example 1 is not the first time you have used a linear function to approximate another function. The same procedure was used as the basis for Newton's Method.

EXAMPLE 1   First-Degree Polynomial Approximation of $f(x) = e^x$

For the function $f(x) = e^x$, find a first-degree polynomial function $P_1(x) = a_0 + a_1x$ whose value and slope agree with the value and slope of f at $x = 0$.

Solution   Because $f(x) = e^x$ and $f'(x) = e^x$, the value and the slope of f at $x = 0$ are

$$f(0) = e^0 = 1$$ Value of f at $x = 0$

and

$$f'(0) = e^0 = 1.$$ Slope of f at $x = 0$

Because $P_1(x) = a_0 + a_1x$, you can use the condition that $P_1(0) = f(0)$ to conclude that $a_0 = 1$. Moreover, because $P_1'(x) = a_1$, you can use the condition that $P_1'(0) = f'(0)$ to conclude that $a_1 = 1$. Therefore, $P_1(x) = 1 + x$. The figure shows the graphs of $P_1(x) = 1 + x$ and $f(x) = e^x$.

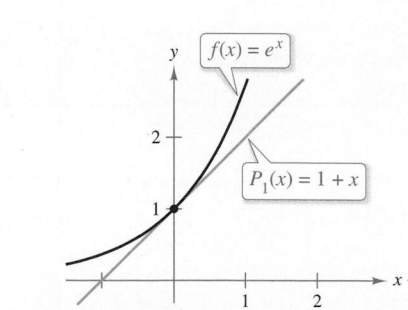

$f(x) = e^x$

$P_1(x) = 1 + x$

P_1 is the first-degree polynomial approximation of $f(x) = e^x$.

In Figure 9.13, you can see that, at points near $(0, 1)$, the graph of the first-degree polynomial function

$$P_1(x) = 1 + x \qquad \text{1st-degree approximation}$$

is reasonably close to the graph of $f(x) = e^x$. As you move away from $(0, 1)$, however, the graphs move farther and farther from each other and the accuracy of the approximation decreases. To improve the approximation, you can impose yet another requirement—that the values of the second derivatives of P and f agree when $x = 0$. The polynomial, P_2, of least degree that satisfies all three requirements $P_2(0) = f(0)$, $P_2'(0) = f'(0)$, and $P_2''(0) = f''(0)$ can be shown to be

$$P_2(x) = 1 + x + \frac{1}{2}x^2. \qquad \text{2nd-degree approximation}$$

Moreover, in Figure 9.13, you can see that P_2 is a better approximation of f than P_1. By requiring that the values of $P_n(x)$ and its first n derivatives match those of $f(x) = e^x$ at $x = 0$, you obtain the nth-degree approximation shown below.

$$P_n(x) = 1 + x + \frac{1}{2}x^2 + \frac{1}{3!}x^3 + \cdots + \frac{1}{n!}x^n \qquad \text{nth-degree approximation}$$

$$\approx e^x$$

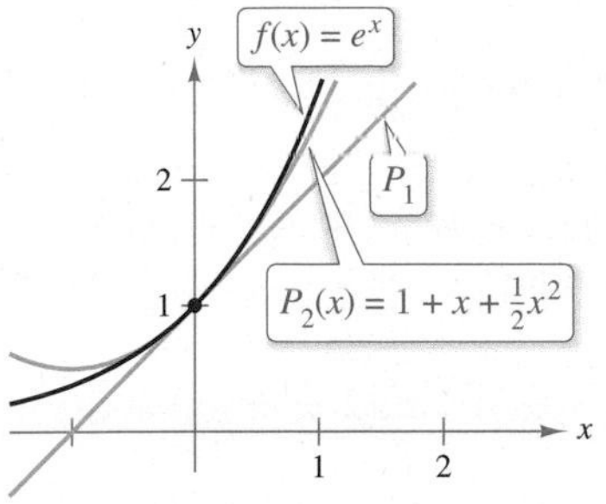

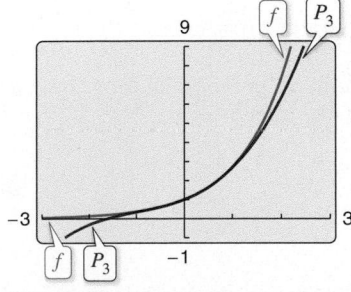

P_2 is the second-degree polynomial approximation of $f(x) = e^x$.
Figure 9.13

EXAMPLE 2   **Third-Degree Polynomial Approximation of $f(x) = e^x$**

Construct a table comparing the values of the polynomial

$$P_3(x) = 1 + x + \frac{1}{2}x^2 + \frac{1}{3!}x^3 \qquad \text{3rd-degree approximation}$$

with $f(x) = e^x$ for several values of x near 0.

Solution   Using a graphing utility, you can obtain the results shown in the table. Note that for $x = 0$, the two functions have the same value, but that as x moves farther away from 0, the accuracy of the approximating polynomial $P_3(x)$ decreases.

x	-1	-0.2	-0.1	0	0.1	0.2	1
e^x	0.3679	0.81873	0.904837	1	1.105171	1.22140	2.7183
$P_3(x)$	0.3333	0.81867	0.904833	1	1.105167	1.22133	2.6667

▷ **TECHNOLOGY**   A graphing utility can be used to compare the graph of the approximating polynomial with the graph of the function f. For instance, in Figure 9.14, the graph of

$$P_3(x) = 1 + x + \tfrac{1}{2}x^2 + \tfrac{1}{6}x^3 \qquad \text{3rd-degree approximation}$$

is compared with the graph of $f(x) = e^x$. Use a graphing utility to compare the graphs of

$$P_4(x) = 1 + x + \tfrac{1}{2}x^2 + \tfrac{1}{6}x^3 + \tfrac{1}{24}x^4 \qquad \text{4th-degree approximation}$$

$$P_5(x) = 1 + x + \tfrac{1}{2}x^2 + \tfrac{1}{6}x^3 + \tfrac{1}{24}x^4 + \tfrac{1}{120}x^5 \qquad \text{5th-degree approximation}$$

and

$$P_6(x) = 1 + x + \tfrac{1}{2}x^2 + \tfrac{1}{6}x^3 + \tfrac{1}{24}x^4 + \tfrac{1}{120}x^5 + \tfrac{1}{720}x^6 \qquad \text{6th-degree approximation}$$

with the graph of f. What do you notice?

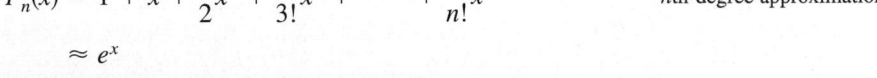

P_3 is the third-degree polynomial approximation of $f(x) = e^x$.
Figure 9.14

Taylor and Maclaurin Polynomials

The polynomial approximation of

$$f(x) = e^x$$

in Example 2 is expanded about $c = 0$. For expansions about an arbitrary value of c, it is convenient to write the polynomial in the form

$$P_n(x) = a_0 + a_1(x - c) + a_2(x - c)^2 + a_3(x - c)^3 + \cdots + a_n(x - c)^n.$$

In this form, repeated differentiation produces

$$P_n{'}(x) = a_1 + 2a_2(x - c) + 3a_3(x - c)^2 + \cdots + na_n(x - c)^{n-1}$$
$$P_n{''}(x) = 2a_2 + 2(3a_3)(x - c) + \cdots + n(n - 1)a_n(x - c)^{n-2}$$
$$P_n{'''}(x) = 2(3a_3) + \cdots + n(n - 1)(n - 2)a_n(x - c)^{n-3}$$
$$\vdots$$
$$P_n^{(n)}(x) = n(n - 1)(n - 2) \cdots (2)(1)a_n.$$

Letting $x = c$, you then obtain

$$P_n(c) = a_0, \quad P_n{'}(c) = a_1, \quad P_n{''}(c) = 2a_2, \ldots, \quad P_n^{(n)}(c) = n!a_n$$

and because the values of f and its first n derivatives must agree with the values of P_n and its first n derivatives at $x = c$, it follows that

$$f(c) = a_0, \quad f'(c) = a_1, \quad \frac{f''(c)}{2!} = a_2, \ldots, \quad \frac{f^{(n)}(c)}{n!} = a_n.$$

With these coefficients, you can obtain the following definition of **Taylor polynomials,** named after the English mathematician Brook Taylor, and **Maclaurin polynomials,** named after the Scottish mathematician Colin Maclaurin (1698–1746).

Definitions of nth Taylor Polynomial and nth Maclaurin Polynomial

If f has n derivatives at c, then the polynomial

$$P_n(x) = f(c) + f'(c)(x - c) + \frac{f''(c)}{2!}(x - c)^2 + \cdots + \frac{f^{(n)}(c)}{n!}(x - c)^n$$

is called the **nth Taylor polynomial for f at c.** If $c = 0$, then

$$P_n(x) = f(0) + f'(0)x + \frac{f''(0)}{2!}x^2 + \frac{f'''(0)}{3!}x^3 + \cdots + \frac{f^{(n)}(0)}{n!}x^n$$

is also called the **nth Maclaurin polynomial for f.**

· · REMARK Maclaurin polynomials are special types of Taylor polynomials for which $c = 0$.

EXAMPLE 3 **A Maclaurin Polynomial for $f(x) = e^x$**

From the discussion on the preceding page, the nth Maclaurin polynomial for $f(x) = e^x$ is given by

$$P_n(x) = 1 + x + \frac{1}{2!}x^2 + \frac{1}{3!}x^3 + \cdots + \frac{1}{n!}x^n.$$ ■

■ FOR FURTHER INFORMATION
To see how to use series to obtain other approximations to e, see the article "Novel Series-based Approximations to e" by John Knox and Harlan J. Brothers in *The College Mathematics Journal.* To view this article, go to *MathArticles.com.*

BROOK TAYLOR (1685–1731)

Although Taylor was not the first to seek polynomial approximations of transcendental functions, his account published in 1715 was one of the first comprehensive works on the subject.
See LarsonCalculus.com to read more of this biography.

EXAMPLE 4   **Finding Taylor Polynomials for ln x**

Find the Taylor polynomials P_0, P_1, P_2, P_3, and P_4 for

$$f(x) = \ln x$$

centered at $c = 1$.

Solution   Expanding about $c = 1$ yields the following.

$$
\begin{aligned}
f(x) &= \ln x & f(1) &= \ln 1 = 0 \\[4pt]
f'(x) &= \frac{1}{x} & f'(1) &= \frac{1}{1} = 1 \\[4pt]
f''(x) &= -\frac{1}{x^2} & f''(1) &= -\frac{1}{1^2} = -1 \\[4pt]
f'''(x) &= \frac{2!}{x^3} & f'''(1) &= \frac{2!}{1^3} = 2 \\[4pt]
f^{(4)}(x) &= -\frac{3!}{x^4} & f^{(4)}(1) &= -\frac{3!}{1^4} = -6
\end{aligned}
$$

Therefore, the Taylor polynomials are as follows.

$$P_0(x) = f(1) = 0$$
$$P_1(x) = f(1) + f'(1)(x - 1) = (x - 1)$$
$$P_2(x) = f(1) + f'(1)(x - 1) + \frac{f''(1)}{2!}(x - 1)^2$$

$$= (x - 1) - \frac{1}{2}(x - 1)^2$$

$$P_3(x) = f(1) + f'(1)(x - 1) + \frac{f''(1)}{2!}(x - 1)^2 + \frac{f'''(1)}{3!}(x - 1)^3$$

$$= (x - 1) - \frac{1}{2}(x - 1)^2 + \frac{1}{3}(x - 1)^3$$

$$P_4(x) = f(1) + f'(1)(x - 1) + \frac{f''(1)}{2!}(x - 1)^2 + \frac{f'''(1)}{3!}(x - 1)^3 + \frac{f^{(4)}(1)}{4!}(x - 1)^4$$

$$= (x - 1) - \frac{1}{2}(x - 1)^2 + \frac{1}{3}(x - 1)^3 - \frac{1}{4}(x - 1)^4$$

Figure 9.15 compares the graphs of P_1, P_2, P_3, and P_4 with the graph of $f(x) = \ln x$. Note that near $x = 1$, the graphs are nearly indistinguishable. For instance,

$$P_4(1.1) \approx 0.0953083$$

and

$$\ln(1.1) \approx 0.0953102.$$

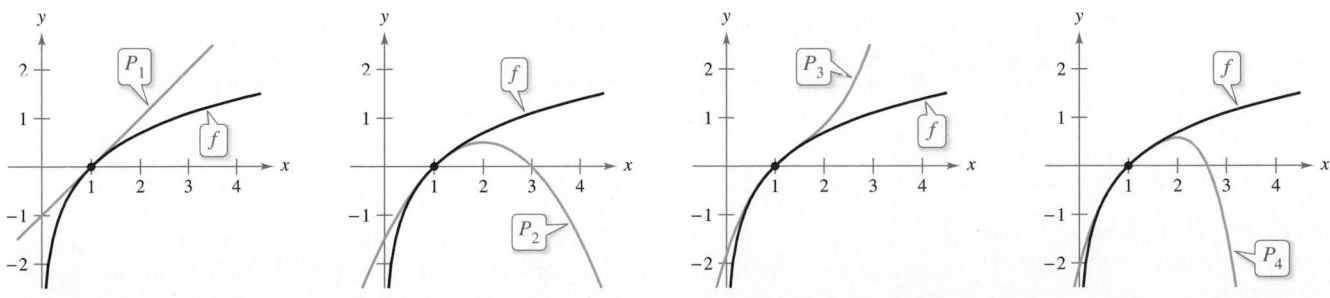

As n increases, the graph of P_n becomes a better and better approximation of the graph of $f(x) = \ln x$ near $x = 1$.
Figure 9.15

EXAMPLE 5 Finding Maclaurin Polynomials for cos x

Find the Maclaurin polynomials P_0, P_2, P_4, and P_6 for $f(x) = \cos x$. Use $P_6(x)$ to approximate the value of $\cos(0.1)$.

Solution Expanding about $c = 0$ yields the following.

$$\begin{aligned} f(x) &= \cos x & f(0) &= \cos 0 = 1 \\ f'(x) &= -\sin x & f'(0) &= -\sin 0 = 0 \\ f''(x) &= -\cos x & f''(0) &= -\cos 0 = -1 \\ f'''(x) &= \sin x & f'''(0) &= \sin 0 = 0 \end{aligned}$$

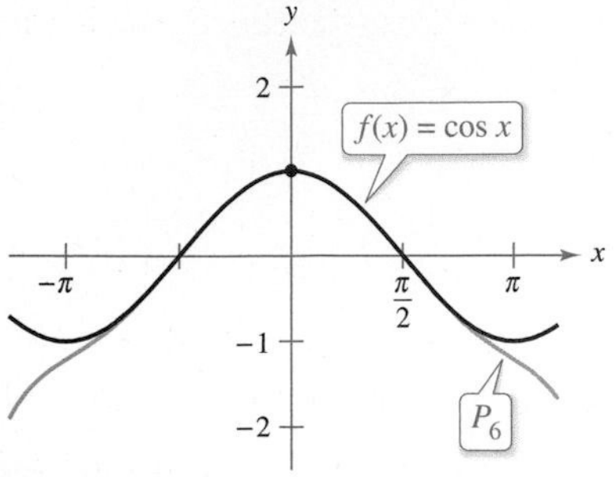

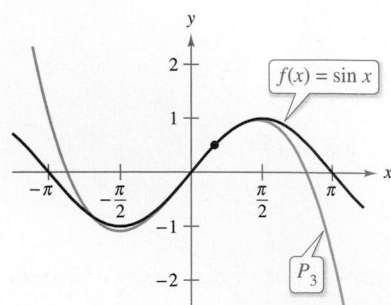

Near (0, 1), the graph of P_6 can be used to approximate the graph of $f(x) = \cos x$.
Figure 9.16

Through repeated differentiation, you can see that the pattern $1, 0, -1, 0$ continues, and you obtain the Maclaurin polynomials

$$P_0(x) = 1, \quad P_2(x) = 1 - \frac{1}{2!}x^2, \quad P_4(x) = 1 - \frac{1}{2!}x^2 + \frac{1}{4!}x^4,$$

and

$$P_6(x) = 1 - \frac{1}{2!}x^2 + \frac{1}{4!}x^4 - \frac{1}{6!}x^6.$$

To nine decimal places, the approximation

$$P_6(0.1) \approx 0.995004165$$

is the same as $\cos(0.1)$. Figure 9.16 compares the graphs of $f(x) = \cos x$ and P_6. ■

Note in Example 5 that the Maclaurin polynomials for $\cos x$ have only even powers of x. Similarly, the Maclaurin polynomials for $\sin x$ have only odd powers of x (see Exercise 19). This is not generally true of the Taylor polynomials for $\sin x$ and $\cos x$ expanded about $c \neq 0$, as shown in the next example.

EXAMPLE 6 Finding a Taylor Polynomial for sin x

See LarsonCalculus.com for an interactive version of this type of example.

Find the third Taylor polynomial for $f(x) = \sin x$, expanded about $c = \pi/6$.

Solution Expanding about $c = \pi/6$ yields the following.

$$\begin{aligned} f(x) &= \sin x & f\left(\frac{\pi}{6}\right) &= \sin\frac{\pi}{6} = \frac{1}{2} \\ f'(x) &= \cos x & f'\left(\frac{\pi}{6}\right) &= \cos\frac{\pi}{6} = \frac{\sqrt{3}}{2} \\ f''(x) &= -\sin x & f''\left(\frac{\pi}{6}\right) &= -\sin\frac{\pi}{6} = -\frac{1}{2} \\ f'''(x) &= -\cos x & f'''\left(\frac{\pi}{6}\right) &= -\cos\frac{\pi}{6} = -\frac{\sqrt{3}}{2} \end{aligned}$$

Near $(\pi/6, 1/2)$, the graph of P_3 can be used to approximate the graph of $f(x) = \sin x$.
Figure 9.17

So, the third Taylor polynomial for $f(x) = \sin x$, expanded about $c = \pi/6$, is

$$\begin{aligned} P_3(x) &= f\left(\frac{\pi}{6}\right) + f'\left(\frac{\pi}{6}\right)\left(x - \frac{\pi}{6}\right) + \frac{f''\left(\frac{\pi}{6}\right)}{2!}\left(x - \frac{\pi}{6}\right)^2 + \frac{f'''\left(\frac{\pi}{6}\right)}{3!}\left(x - \frac{\pi}{6}\right)^3 \\ &= \frac{1}{2} + \frac{\sqrt{3}}{2}\left(x - \frac{\pi}{6}\right) - \frac{1}{2(2!)}\left(x - \frac{\pi}{6}\right)^2 - \frac{\sqrt{3}}{2(3!)}\left(x - \frac{\pi}{6}\right)^3. \end{aligned}$$

Figure 9.17 compares the graphs of $f(x) = \sin x$ and P_3. ■

Taylor polynomials and Maclaurin polynomials can be used to approximate the value of a function at a specific point. For instance, to approximate the value of $\ln(1.1)$, you can use Taylor polynomials for $f(x) = \ln x$ expanded about $c = 1$, as shown in Example 4, or you can use Maclaurin polynomials, as shown in Example 7.

EXAMPLE 7   Approximation Using Maclaurin Polynomials

Use a fourth Maclaurin polynomial to approximate the value of $\ln(1.1)$.

Solution   Because 1.1 is closer to 1 than to 0, you should consider Maclaurin polynomials for the function $g(x) = \ln(1 + x)$.

$$g(x) = \ln(1 + x) \qquad\qquad g(0) = \ln(1 + 0) = 0$$
$$g'(x) = (1 + x)^{-1} \qquad\qquad g'(0) = (1 + 0)^{-1} = 1$$
$$g''(x) = -(1 + x)^{-2} \qquad\qquad g''(0) = -(1 + 0)^{-2} = -1$$
$$g'''(x) = 2(1 + x)^{-3} \qquad\qquad g'''(0) = 2(1 + 0)^{-3} = 2$$
$$g^{(4)}(x) = -6(1 + x)^{-4} \qquad\qquad g^{(4)}(0) = -6(1 + 0)^{-4} = -6$$

Note that you obtain the same coefficients as in Example 4. Therefore, the fourth Maclaurin polynomial for $g(x) = \ln(1 + x)$ is

$$P_4(x) = g(0) + g'(0)x + \frac{g''(0)}{2!}x^2 + \frac{g'''(0)}{3!}x^3 + \frac{g^{(4)}(0)}{4!}x^4$$

$$= x - \frac{1}{2}x^2 + \frac{1}{3}x^3 - \frac{1}{4}x^4.$$

Consequently,

$$\ln(1.1) = \ln(1 + 0.1) \approx P_4(0.1) \approx 0.0953083. \qquad\blacksquare$$

Exploration

Check to see that the fourth Taylor polynomial (from Example 4), evaluated at $x = 1.1$, yields the same result as the fourth Maclaurin polynomial in Example 7.

The table below illustrates the accuracy of the Maclaurin polynomial approximation of the calculator value of $\ln(1.1)$. You can see that as n increases, $P_n(0.1)$ approaches the value of $\ln(1.1) \approx 0.0953102$.

Maclaurin Polynomial Approximations of $\ln(1 + x)$ at $x = 0.1$

n	1	2	3	4
$P_n(0.1)$	0.1000000	0.0950000	0.0953333	0.0953083

On the other hand, the table below illustrates that as you move away from the expansion point $c = 0$, the accuracy of the approximation decreases.

Fourth Maclaurin Polynomial Approximation of $\ln(1 + x)$

x	0	0.1	0.5	0.75	1.0
$\ln(1 + x)$	0	0.0953102	0.4054651	0.5596158	0.6931472
$P_4(x)$	0	0.0953083	0.4010417	0.5302734	0.5833333

These two tables illustrate two very important points about the accuracy of Taylor (or Maclaurin) polynomials for use in approximations.

1. The approximation is usually better for higher-degree Taylor (or Maclaurin) polynomials than for those of lower degree.

2. The approximation is usually better at x-values close to c than at x-values far from c.

Remainder of a Taylor Polynomial

An approximation technique is of little value without some idea of its accuracy. To measure the accuracy of approximating a function value $f(x)$ by the Taylor polynomial $P_n(x)$, you can use the concept of a **remainder** $R_n(x)$, defined as follows.

$$f(x) = P_n(x) + R_n(x)$$

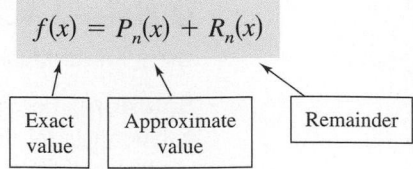

Exact value | Approximate value | Remainder

So, $R_n(x) = f(x) - P_n(x)$. The absolute value of $R_n(x)$ is called the **error** associated with the approximation. That is,

$$\text{Error} = |R_n(x)| = |f(x) - P_n(x)|.$$

The next theorem gives a general procedure for estimating the remainder associated with a Taylor polynomial. This important theorem is called **Taylor's Theorem,** and the remainder given in the theorem is called the **Lagrange form of the remainder.**

THEOREM 9.19 Taylor's Theorem

If a function f is differentiable through order $n + 1$ in an interval I containing c, then, for each x in I, there exists z between x and c such that

$$f(x) = f(c) + f'(c)(x - c) + \frac{f''(c)}{2!}(x - c)^2 + \cdots + \frac{f^{(n)}(c)}{n!}(x - c)^n + R_n(x)$$

where

$$R_n(x) = \frac{f^{(n+1)}(z)}{(n+1)!}(x - c)^{n+1}.$$

A proof of this theorem is given in Appendix A.

One useful consequence of Taylor's Theorem is that

$$|R_n(x)| \leq \frac{|x - c|^{n+1}}{(n+1)!} \max |f^{(n+1)}(z)|$$

where $\max |f^{(n+1)}(z)|$ is the maximum value of $f^{(n+1)}(z)$ between x and c.

For $n = 0$, Taylor's Theorem states that if f is differentiable in an interval I containing c, then, for each x in I, there exists z between x and c such that

$$f(x) = f(c) + f'(z)(x - c) \quad \text{or} \quad f'(z) = \frac{f(x) - f(c)}{x - c}.$$

Do you recognize this special case of Taylor's Theorem? (It is the Mean Value Theorem.)

When applying Taylor's Theorem, you should not expect to be able to find the exact value of z. (If you could do this, an approximation would not be necessary.) Rather, you are trying to find bounds for $f^{(n+1)}(z)$ from which you are able to tell how large the remainder $R_n(x)$ is.

EXAMPLE 8   Determining the Accuracy of an Approximation

The third Maclaurin polynomial for $\sin x$ is

$$P_3(x) = x - \frac{x^3}{3!}.$$

Use Taylor's Theorem to approximate $\sin(0.1)$ by $P_3(0.1)$ and determine the accuracy of the approximation.

Solution   Using Taylor's Theorem, you have

$$\sin x = x - \frac{x^3}{3!} + R_3(x) = x - \frac{x^3}{3!} + \frac{f^{(4)}(z)}{4!}x^4$$

where $0 < z < 0.1$. Therefore,

$$\sin(0.1) \approx 0.1 - \frac{(0.1)^3}{3!} \approx 0.1 - 0.000167 = 0.099833.$$

Because $f^{(4)}(z) = \sin z$, it follows that the error $|R_3(0.1)|$ can be bounded as follows.

$$0 < R_3(0.1) = \frac{\sin z}{4!}(0.1)^4 < \frac{0.0001}{4!} \approx 0.000004$$

This implies that

$$0.099833 < \sin(0.1) \approx 0.099833 + R_3(0.1) < 0.099833 + 0.000004$$

or

$$0.099833 < \sin(0.1) < 0.099837.$$

> **•• REMARK**   Note that when you use a calculator,
> $$\sin(0.1) \approx 0.0998334.$$

EXAMPLE 9   Approximating a Value to a Desired Accuracy

Determine the degree of the Taylor polynomial $P_n(x)$ expanded about $c = 1$ that should be used to approximate $\ln(1.2)$ so that the error is less than 0.001.

Solution   Following the pattern of Example 4, you can see that the $(n + 1)$st derivative of $f(x) = \ln x$ is

$$f^{(n+1)}(x) = (-1)^n \frac{n!}{x^{n+1}}.$$

Using Taylor's Theorem, you know that the error $|R_n(1.2)|$ is

$$|R_n(1.2)| = \left| \frac{f^{(n+1)}(z)}{(n + 1)!}(1.2 - 1)^{n+1} \right|$$

$$= \frac{n!}{z^{n+1}}\left[\frac{1}{(n + 1)!} \right](0.2)^{n+1}$$

$$= \frac{(0.2)^{n+1}}{z^{n+1}(n + 1)}$$

where $1 < z < 1.2$. In this interval, $(0.2)^{n+1}/[z^{n+1}(n + 1)]$ is less than $(0.2)^{n+1}/(n + 1)$. So, you are seeking a value of n such that

$$\frac{(0.2)^{n+1}}{(n + 1)} < 0.001 \quad \Longrightarrow \quad 1000 < (n + 1)5^{n+1}.$$

> **•• REMARK**   Note that when you use a calculator,
> $$P_3(1.2) \approx 0.1827$$
> and
> $$\ln(1.2) \approx 0.1823.$$

By trial and error, you can determine that the least value of n that satisfies this inequality is $n = 3$. So, you would need the third Taylor polynomial to achieve the desired accuracy in approximating $\ln(1.2)$.

9.7 Exercises

See **CalcChat.com** for tutorial help and worked-out solutions to odd-numbered exercises.

CONCEPT CHECK

1. **Polynomial Approximation** An elementary function is approximated by a polynomial. In your own words, describe what is meant by saying that the polynomial is *expanded about c* or *centered at c*.

2. **Taylor and Maclaurin Polynomials** How are Taylor polynomials and Maclaurin polynomials related?

3. **Accuracy of a Taylor Polynomial** Describe the accuracy of the nth-degree Taylor polynomial of f centered at c as the distance between c and x increases.

4. **Accuracy of a Taylor Polynomial** In general, how does the accuracy of a Taylor polynomial change as the degree of the polynomial increases? Explain your reasoning.

Matching In Exercises 5–8, match the Taylor polynomial approximation of the function $f(x) = e^{-x^2/2}$ with its graph. [The graphs are labeled (a), (b), (c), and (d).]

(a)
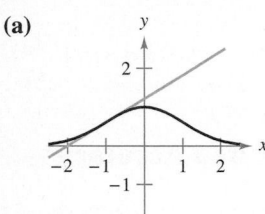

(b)
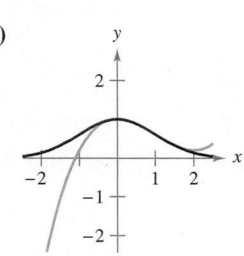

(c)
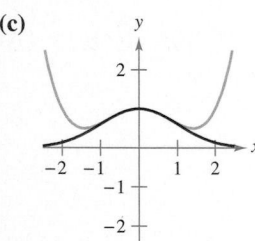

(d)
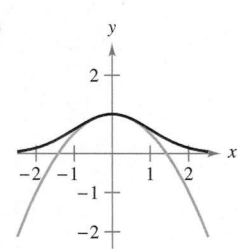

5. $g(x) = -\frac{1}{2}x^2 + 1$

6. $g(x) = \frac{1}{8}x^4 - \frac{1}{2}x^2 + 1$

7. $g(x) = e^{-1/2}[(x + 1) + 1]$

8. $g(x) = e^{-1/2}\left[\frac{1}{3}(x - 1)^3 - (x - 1) + 1\right]$

 **Finding a First-Degree Polynomial Approximation** In Exercises 9–12, find a first-degree polynomial function P_1 whose value and slope agree with the value and slope of f at $x = c$. Use a graphing utility to graph f and P_1.

9. $f(x) = \frac{\sqrt{x}}{4}, \quad c = 4$

10. $f(x) = \frac{6}{\sqrt[3]{x}}, \quad c = 8$

11. $f(x) = \sec x, \quad c = \frac{\pi}{6}$

12. $f(x) = \tan x, \quad c = \frac{\pi}{4}$

 **Graphical and Numerical Analysis** In Exercises 13 and 14, use a graphing utility to graph f and its second-degree polynomial approximation P_2 at $x = c$. Complete the table comparing the values of f and P_2.

13. $f(x) = \frac{4}{\sqrt{x}}, \quad c = 1$

$P_2(x) = 4 - 2(x - 1) + \frac{3}{2}(x - 1)^2$

x	0	0.8	0.9	1	1.1	1.2	2
$f(x)$							
$P_2(x)$							

14. $f(x) = \sec x, \quad c = \frac{\pi}{4}$

$P_2(x) = \sqrt{2} + \sqrt{2}\left(x - \frac{\pi}{4}\right) + \frac{3}{2}\sqrt{2}\left(x - \frac{\pi}{4}\right)^2$

x	-2.15	0.585	0.685	$\frac{\pi}{4}$	0.885	0.985	1.785
$f(x)$							
$P_2(x)$							

15. **Conjecture** Consider the function $f(x) = \cos x$ and its Maclaurin polynomials P_2, P_4, and P_6 (see Example 5).

 (a) Use a graphing utility to graph f and the indicated polynomial approximations.

(b) Evaluate and compare the values of $f^{(n)}(0)$ and $P_n^{(n)}(0)$ for $n = 2, 4$, and 6.

(c) Use the results in part (b) to make a conjecture about $f^{(n)}(0)$ and $P_n^{(n)}(0)$.

16. **Conjecture** Consider the function $f(x) = x^2 e^x$.

(a) Find the Maclaurin polynomials P_2, P_3, and P_4 for f.

(b) Use a graphing utility to graph f, P_2, P_3, and P_4.

(c) Evaluate and compare the values of $f^{(n)}(0)$ and $P_n^{(n)}(0)$ for $n = 2, 3$, and 4.

(d) Use the results in part (c) to make a conjecture about $f^{(n)}(0)$ and $P_n^{(n)}(0)$.

Finding a Maclaurin Polynomial In Exercises 17–26, find the nth Maclaurin polynomial for the function.

17. $f(x) = e^{4x}, \quad n = 4$

18. $f(x) = e^{-x}, \quad n = 5$

19. $f(x) = \sin x, \quad n = 5$

20. $f(x) = \cos \pi x, \quad n = 4$

21. $f(x) = xe^x, \quad n = 4$

22. $f(x) = x^2 e^{-x}, \quad n = 4$

23. $f(x) = \frac{1}{1 - x}, \quad n = 5$

24. $f(x) = \frac{x}{x + 1}, \quad n = 4$

25. $f(x) = \sec x, \quad n = 2$

26. $f(x) = \tan x, \quad n = 3$

 **Finding a Taylor Polynomial** In Exercises 27–32, find the nth Taylor polynomial for the function, centered at c.

27. $f(x) = \dfrac{2}{x}$, $n = 3$, $c = 1$

28. $f(x) = \dfrac{1}{x^2}$, $n = 4$, $c = -2$

29. $f(x) = \sqrt{x}$, $n = 2$, $c = 4$

30. $f(x) = \sqrt[3]{x}$, $n = 3$, $c = 8$

31. $f(x) = \ln x$, $n = 4$, $c = 2$

32. $f(x) = x^2 \cos x$, $n = 2$, $c = \pi$

Finding Taylor Polynomials Using Technology In Exercises 33 and 34, use a computer algebra system to find the indicated Taylor polynomials for the function f. Graph the function and the Taylor polynomials.

33. $f(x) = \tan \pi x$

    (a) $n = 3$, $c = 0$

    (b) $n = 3$, $c = 1/4$

34. $f(x) = \dfrac{1}{x^2 + 1}$

    (a) $n = 4$, $c = 0$

    (b) $n = 4$, $c = 1$

35. **Numerical and Graphical Approximations**

    (a) Use the Maclaurin polynomials $P_1(x)$, $P_3(x)$, and $P_5(x)$ for $f(x) = \sin x$ to complete the table.

x	0	0.25	0.50	0.75	1
$\sin x$	0	0.2474	0.4794	0.6816	0.8415
$P_1(x)$					
$P_3(x)$					
$P_5(x)$					

    (b) Use a graphing utility to graph $f(x) = \sin x$ and the Maclaurin polynomials in part (a).

    (c) Describe the change in accuracy of a polynomial approximation as the distance from the point where the polynomial is centered increases.

36. **Numerical and Graphical Approximations**

    (a) Use the Taylor polynomials $P_1(x)$, $P_2(x)$, and $P_4(x)$ for $f(x) = e^x$, centered at $c = 1$, to complete the table.

x	1	1.25	1.50	1.75	2
e^x	e	3.4903	4.4817	5.7546	7.3891
$P_1(x)$					
$P_2(x)$					
$P_4(x)$					

    (b) Use a graphing utility to graph $f(x) = e^x$ and the Taylor polynomials in part (a).

    (c) Describe the change in accuracy of polynomial approximations as the degree increases.

Numerical and Graphical Approximations In Exercises 37 and 38, (a) find the Maclaurin polynomial $P_3(x)$ for $f(x)$, (b) complete the table for $f(x)$ and $P_3(x)$, and (c) sketch the graphs of $f(x)$ and $P_3(x)$, on the same set of coordinate axes.

x	-0.75	-0.50	-0.25	0	0.25	0.50	0.75
$f(x)$							
$P_3(x)$							

37. $f(x) = \arcsin x$

38. $f(x) = \arctan x$

 **Approximating a Function Value** In Exercises 39–44, approximate the function at the given value of x, using the polynomial found in the indicated exercise.

39. $f(x) = e^{4x}$, $f\left(\frac{1}{4}\right)$, Exercise 17

40. $f(x) = x^2 e^{-x}$, $f\left(\frac{1}{5}\right)$, Exercise 22

41. $f(x) = \dfrac{1}{x^2}$, $f(-2.1)$, Exercise 28

42. $f(x) = \sqrt[3]{x}$, $f(8.05)$, Exercise 30

43. $f(x) = \ln x$, $f(2.1)$, Exercise 31

44. $f(x) = x^2 \cos x$, $f\left(\dfrac{7\pi}{8}\right)$, Exercise 32

 **Using Taylor's Theorem** In Exercises 45–50, use Taylor's Theorem to obtain an upper bound for the error of the approximation. Then calculate the exact value of the error.

45. $\cos(0.3) \approx 1 - \dfrac{(0.3)^2}{2!} + \dfrac{(0.3)^4}{4!}$

46. $\arccos(0.15) \approx \dfrac{\pi}{2} - 0.15$

47. $\sinh(0.2) \approx 0.2 + \dfrac{(0.2)^3}{3!} + \dfrac{(0.2)^5}{5!}$

48. $e \approx 1 + 1 + \dfrac{1^2}{2!} + \dfrac{1^3}{3!} + \dfrac{1^4}{4!} + \dfrac{1^5}{5!}$

49. $\arcsin(0.4) \approx 0.4 + \dfrac{(0.4)^3}{2 \cdot 3}$

50. $\arctan(0.4) \approx 0.4 - \dfrac{(0.4)^3}{3}$

Finding a Degree In Exercises 51–56, determine the degree of the Maclaurin polynomial required for the error in the approximation of the function at the indicated value of x to be less than 0.001.

51. $f(x) = \sin x$, approximate $f(0.3)$

52. $f(x) = \cos x$, approximate $f(0.4)$

53. $f(x) = e^x$, approximate $f(0.6)$

54. $f(x) = \ln(x + 1)$, approximate $f(1.25)$

55. $f(x) = \dfrac{1}{x - 2}$, approximate $f(0.15)$

56. $f(x) = \dfrac{1}{x + 1}$, approximate $f(0.2)$

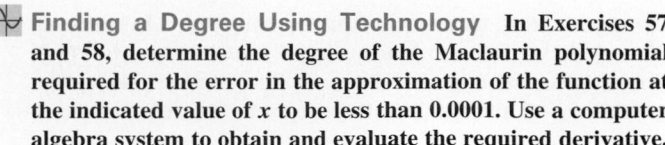

Finding a Degree Using Technology In Exercises 57 and 58, determine the degree of the Maclaurin polynomial required for the error in the approximation of the function at the indicated value of x to be less than 0.0001. Use a computer algebra system to obtain and evaluate the required derivative.

57. $f(x) = \ln(x + 1)$, approximate $f(0.5)$.

58. $f(x) = e^{-\pi x}$, approximate $f(1.3)$.

Finding Values In Exercises 59–62, determine the values of x for which the function can be replaced by the Taylor polynomial if the error cannot exceed 0.001.

59. $f(x) = e^x \approx 1 + x + \dfrac{x^2}{2!} + \dfrac{x^3}{3!}$, $x < 0$

60. $f(x) = \sin x \approx x - \dfrac{x^3}{3!}$

61. $f(x) = \cos x \approx 1 - \dfrac{x^2}{2!} + \dfrac{x^4}{4!}$

62. $f(x) = e^{-2x} \approx 1 - 2x + 2x^2 - \dfrac{4}{3}x^3$

EXPLORING CONCEPTS

63. Think About It What is the relationship between the equation of a tangent line to a differentiable function at a point and the first Taylor polynomial for that function centered at the point?

64. Maclaurin Polynomial Without performing any calculations, find the second Maclaurin polynomial for

$$f(x) = a + bx^2.$$

Explain your reasoning.

65. Maclaurin Polynomials Find the fourth Maclaurin polynomials for

$$f(x) = e^x \text{ and } g(x) = e^{2x}.$$

Explain how you can use the fourth Maclaurin polynomial for f to find the fourth Maclaurin polynomial for g.

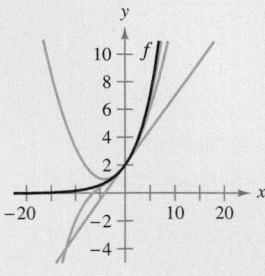

66. HOW DO YOU SEE IT? The figure shows the graphs of the first-, second-, and third-degree polynomial approximations P_1, P_2, and P_3 of a function f. Label the graphs of P_1, P_2, and P_3. To print an enlarged copy of the graph, go to *MathGraphs.com.*

67. Comparing Maclaurin Polynomials

(a) Compare the Maclaurin polynomials of degree 4 and degree 5, respectively, for the functions

$$f(x) = e^x \text{ and } g(x) = xe^x.$$

What is the relationship between them?

(b) Use the result in part (a) and the Maclaurin polynomial of degree 5 for $f(x) = \sin x$ to find a Maclaurin polynomial of degree 6 for the function $g(x) = x \sin x$.

(c) Use the result in part (a) and the Maclaurin polynomial of degree 5 for $f(x) = \sin x$ to find a Maclaurin polynomial of degree 4 for the function $g(x) = (\sin x)/x$.

68. Differentiating Maclaurin Polynomials

(a) Differentiate the Maclaurin polynomial of degree 5 for $f(x) = \sin x$ and compare the result with the Maclaurin polynomial of degree 4 for $g(x) = \cos x$.

(b) Differentiate the Maclaurin polynomial of degree 6 for $f(x) = \cos x$ and compare the result with the Maclaurin polynomial of degree 5 for $g(x) = \sin x$.

(c) Differentiate the Maclaurin polynomial of degree 4 for $f(x) = e^x$. Describe the relationship between the two series.

69. Graphical Reasoning The figure shows the graphs of the function $f(x) = \sin(\pi x/4)$ and the second-degree Taylor polynomial $P_2(x) = 1 - (\pi^2/32)(x - 2)^2$ centered at $x = 2$.

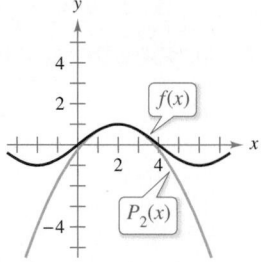

(a) Use the symmetry of the graph of f to write the second-degree Taylor polynomial $Q_2(x)$ for f centered at $x = -2$.

(b) Use a horizontal translation of the result in part (a) to find the second-degree Taylor polynomial $R_2(x)$ for f centered at $x = 6$.

(c) Is it possible to use a horizontal translation of the result in part (a) to write a second-degree Taylor polynomial for f centered at $x = 4$? Explain.

70. Proof Prove that if f is an odd function, then its nth Maclaurin polynomial contains only terms with odd powers of x.

71. Proof Prove that if f is an even function, then its nth Maclaurin polynomial contains only terms with even powers of x.

72. Proof Let $P_n(x)$ be the nth Taylor polynomial for f at c. Prove that $P_n(c) = f(c)$ and $P^{(k)}(c) = f^{(k)}(c)$ for $1 \le k \le n$.

73. Proof Consider a function f with continuous first and second derivatives at $x = c$. Prove that if f has a relative maximum at $x = c$, then the second Taylor polynomial centered at $x = c$ also has a relative maximum at $x = c$.

9.8 Power Series

■ Understand the definition of a power series.
■ Find the radius and interval of convergence of a power series.
■ Determine the endpoint convergence of a power series.
■ Differentiate and integrate a power series.

Power Series

In Section 9.7, you were introduced to the concept of approximating functions by Taylor polynomials. For instance, the function $f(x) = e^x$ can be *approximated* by its third-degree Maclaurin polynomial

$$e^x \approx 1 + x + \frac{x^2}{2!} + \frac{x^3}{3!}.$$

In that section, you saw that the higher the degree of the approximating polynomial, the better the approximation becomes.

In this and the next two sections, you will see that several important types of functions, including $f(x) = e^x$, can be represented *exactly* by an infinite series called a **power series.** For example, the power series representation for e^x is

$$e^x = 1 + x + \frac{x^2}{2!} + \frac{x^3}{3!} + \cdots + \frac{x^n}{n!} + \cdots.$$

For each real number x, it can be shown that the infinite series on the right converges to the number e^x. Before doing this, however, some preliminary results dealing with power series will be discussed—beginning with the next definition.

Exploration

Graphical Reasoning
Use a graphing utility to approximate the graph of each power series near $x = 0$. (Use the first several terms of each series.) Each series represents a well-known function. What is the function?

a. $\displaystyle\sum_{n=0}^{\infty} \frac{(-1)^n x^n}{n!}$

b. $\displaystyle\sum_{n=0}^{\infty} \frac{(-1)^n x^{2n}}{(2n)!}$

c. $\displaystyle\sum_{n=0}^{\infty} \frac{(-1)^n x^{2n+1}}{(2n+1)!}$

d. $\displaystyle\sum_{n=0}^{\infty} \frac{(-1)^n x^{2n+1}}{2n+1}$

e. $\displaystyle\sum_{n=0}^{\infty} \frac{2^n x^n}{n!}$

· ·REMARK To simplify the notation for power series, assume that $(x - c)^0 = 1$, even when $x = c$.

Definition of Power Series

If x is a variable, then an infinite series of the form

$$\sum_{n=0}^{\infty} a_n x^n = a_0 + a_1 x + a_2 x^2 + a_3 x^3 + \cdots + a_n x^n + \cdots$$

is called a **power series.** More generally, an infinite series of the form

$$\sum_{n=0}^{\infty} a_n (x - c)^n = a_0 + a_1(x - c) + a_2(x - c)^2 + \cdots + a_n(x - c)^n + \cdots$$

is called a **power series centered at c,** where c is a constant.

EXAMPLE 1　Power Series

a. The following power series is centered at 0.

$$\sum_{n=0}^{\infty} \frac{x^n}{n!} = 1 + x + \frac{x^2}{2} + \frac{x^3}{3!} + \cdots$$

b. The following power series is centered at -1.

$$\sum_{n=0}^{\infty} (-1)^n(x + 1)^n = 1 - (x + 1) + (x + 1)^2 - (x + 1)^3 + \cdots$$

c. The following power series is centered at 1.

$$\sum_{n=1}^{\infty} \frac{1}{n}(x - 1)^n = (x - 1) + \frac{1}{2}(x - 1)^2 + \frac{1}{3}(x - 1)^3 + \cdots$$

Radius and Interval of Convergence

A power series in x can be viewed as a function of x

$$f(x) = \sum_{n=0}^{\infty} a_n(x - c)^n$$

where the *domain of f* is the set of all x for which the power series converges. Determination of the domain of a power series is the primary concern in this section. Of course, every power series converges at its center c because

$$
\begin{aligned}
f(c) &= \sum_{n=0}^{\infty} a_n(c - c)^n \\
&= a_0(1) + 0 + 0 + \cdots + 0 + \cdots \\
&= a_0.
\end{aligned}
$$

So, c always lies in the domain of f. Theorem 9.20 (see below) states that the domain of a power series can take three basic forms: a single point, an interval centered at c, or the entire real number line, as shown in Figure 9.18.

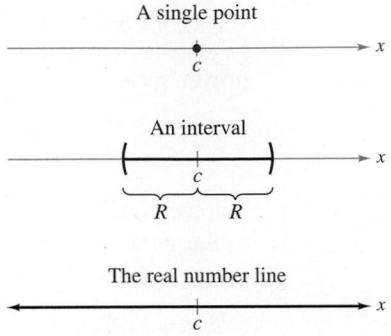

The domain of a power series has only three basic forms: a single point, an interval centered at c, or the entire real number line.
Figure 9.18

THEOREM 9.20 Convergence of a Power Series

For a power series centered at c, precisely one of the following is true.

1. The series converges only at c.

2. There exists a real number $R > 0$ such that the series converges absolutely for

$$|x - c| < R$$

and diverges for

$$|x - c| > R.$$

3. The series converges absolutely for all x.

The number R is the **radius of convergence** of the power series. If the series converges only at c, then the radius of convergence is $R = 0$. If the series converges for all x, then the radius of convergence is $R = \infty$. The set of all values of x for which the power series converges is the **interval of convergence** of the power series.

A proof of this theorem is given in Appendix A.

To determine the radius of convergence of a power series, use the Ratio Test, as demonstrated in Examples 2, 3, and 4.

EXAMPLE 2   Finding the Radius of Convergence

Find the radius of convergence of $\displaystyle\sum_{n=0}^{\infty} n!x^n$.

Solution For $x = 0$, you obtain

$$f(0) = \sum_{n=0}^{\infty} n!0^n = 1 + 0 + 0 + \ldots = 1.$$

For any fixed value of x such that $|x| > 0$, let $u_n = n!x^n$. Then

$$\lim_{n\to\infty} \left| \frac{u_{n+1}}{u_n} \right| = \lim_{n\to\infty} \left| \frac{(n+1)!x^{n+1}}{n!x^n} \right|$$

$$= |x| \lim_{n\to\infty} (n+1)$$

$$= \infty.$$

Therefore, by the Ratio Test, the series diverges for $|x| > 0$ and converges only at its center, 0. So, the radius of convergence is $R = 0$.

EXAMPLE 3   Finding the Radius of Convergence

Find the radius of convergence of

$$\sum_{n=0}^{\infty} 3(x-2)^n.$$

Solution For $x \neq 2$, let $u_n = 3(x-2)^n$. Then

$$\lim_{n\to\infty} \left| \frac{u_{n+1}}{u_n} \right| = \lim_{n\to\infty} \left| \frac{3(x-2)^{n+1}}{3(x-2)^n} \right|$$

$$= \lim_{n\to\infty} |x-2|$$

$$= |x-2|.$$

By the Ratio Test, the series converges for $|x-2| < 1$ and diverges for $|x-2| > 1$. Therefore, the radius of convergence of the series is $R = 1$.

EXAMPLE 4   Finding the Radius of Convergence

Find the radius of convergence of

$$\sum_{n=0}^{\infty} \frac{(-1)^n x^{2n+1}}{(2n+1)!}.$$

Solution Let $u_n = (-1)^n x^{2n+1}/(2n+1)!$. Then

$$\lim_{n\to\infty} \left| \frac{u_{n+1}}{u_n} \right| = \lim_{n\to\infty} \left| \frac{\dfrac{(-1)^{n+1} x^{2n+3}}{(2n+3)!}}{\dfrac{(-1)^n x^{2n+1}}{(2n+1)!}} \right|$$

$$= \lim_{n\to\infty} \frac{x^2}{(2n+3)(2n+2)}.$$

For any *fixed* value of x, this limit is 0. So, by the Ratio Test, the series converges for all x. Therefore, the radius of convergence is $R = \infty$.

Endpoint Convergence

Note that for a power series whose radius of convergence is a finite number R, Theorem 9.20 says nothing about the convergence at the *endpoints* of the interval of convergence. Each endpoint must be tested separately for convergence or divergence. As a result, the interval of convergence of a power series can take any one of the six forms shown in Figure 9.19.

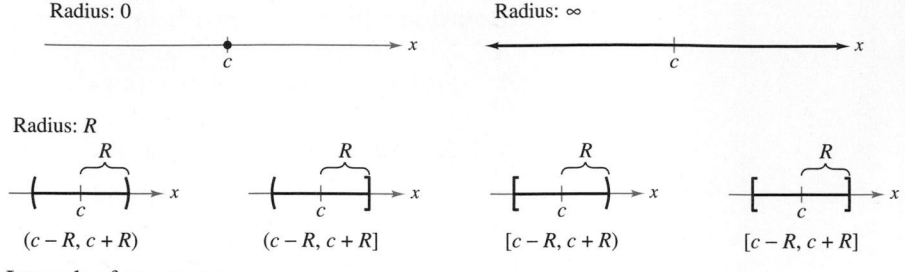

Intervals of convergence
Figure 9.19

EXAMPLE 5   Finding the Interval of Convergence

⋮ ⋯▷ *See LarsonCalculus.com for an interactive version of this type of example.*

Find the interval of convergence of

$$\sum_{n=1}^{\infty} \frac{x^n}{n}.$$

Solution   Letting $u_n = x^n/n$ produces

$$\lim_{n\to\infty} \left| \frac{u_{n+1}}{u_n} \right| = \lim_{n\to\infty} \left| \frac{\dfrac{x^{n+1}}{n+1}}{\dfrac{x^n}{n}} \right|$$

$$= \lim_{n\to\infty} \left| \frac{nx}{n+1} \right|$$

$$= |x|.$$

So, by the Ratio Test, the radius of convergence is $R = 1$. Moreover, because the series is centered at 0, it converges in the interval $(-1, 1)$. This interval, however, is not necessarily the *interval of convergence*. To determine this, you must test for convergence at each endpoint. When $x = 1$, you obtain the *divergent* harmonic series

$$\sum_{n=1}^{\infty} \frac{1}{n} = \frac{1}{1} + \frac{1}{2} + \frac{1}{3} + \cdots.$$ Diverges when $x = 1$.

When $x = -1$, you obtain the *convergent* alternating harmonic series

$$\sum_{n=1}^{\infty} \frac{(-1)^n}{n} = -1 + \frac{1}{2} - \frac{1}{3} + \frac{1}{4} - \cdots.$$ Converges when $x = -1$.

So, the interval of convergence for the series is $[-1, 1)$, as shown in Figure 9.20.

Interval: $[-1, 1)$
Radius: $R = 1$

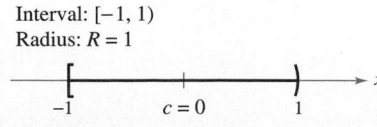

Figure 9.20

EXAMPLE 6 **Finding the Interval of Convergence**

Find the interval of convergence of $\displaystyle\sum_{n=0}^{\infty} \frac{(-1)^n(x+1)^n}{2^n}$.

Solution Letting $u_n = (-1)^n(x+1)^n/2^n$ produces

$$\lim_{n\to\infty}\left|\frac{u_{n+1}}{u_n}\right| = \lim_{n\to\infty}\left|\frac{\dfrac{(-1)^{n+1}(x+1)^{n+1}}{2^{n+1}}}{\dfrac{(-1)^n(x+1)^n}{2^n}}\right|$$

$$= \lim_{n\to\infty}\left|\frac{x+1}{2}\right|$$

$$= \left|\frac{x+1}{2}\right|.$$

By the Ratio Test, the series converges for

$$\left|\frac{x+1}{2}\right| < 1$$

or $|x+1| < 2$. So, the radius of convergence is $R = 2$. Because the series is centered at $x = -1$, it will converge in the interval $(-3, 1)$. Furthermore, at the endpoints, you have

$$\sum_{n=0}^{\infty}\frac{(-1)^n(-2)^n}{2^n} = \sum_{n=0}^{\infty}\frac{2^n}{2^n} = \sum_{n=0}^{\infty}1 \qquad \text{Diverges when } x = -3.$$

and

$$\sum_{n=0}^{\infty}\frac{(-1)^n(2)^n}{2^n} = \sum_{n=0}^{\infty}(-1)^n \qquad \text{Diverges when } x = 1.$$

both of which diverge. So, the interval of convergence is $(-3, 1)$, as shown in Figure 9.21.

Interval: $(-3, 1)$
Radius: $R = 2$

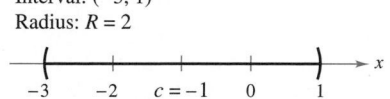

Figure 9.21

EXAMPLE 7 **Finding the Interval of Convergence**

Find the interval of convergence of

$$\sum_{n=1}^{\infty}\frac{x^n}{n^2}.$$

Solution Letting $u_n = x^n/n^2$ produces

$$\lim_{n\to\infty}\left|\frac{u_{n+1}}{u_n}\right| = \lim_{n\to\infty}\left|\frac{x^{n+1}/(n+1)^2}{x^n/n^2}\right|$$

$$= \lim_{n\to\infty}\left|\frac{n^2x}{(n+1)^2}\right|$$

$$= |x|.$$

So, the radius of convergence is $R = 1$. Because the series is centered at $x = 0$, it converges in the interval $(-1, 1)$. When $x = 1$, you obtain the convergent p-series

$$\sum_{n=1}^{\infty}\frac{1}{n^2} = \frac{1}{1^2} + \frac{1}{2^2} + \frac{1}{3^2} + \frac{1}{4^2} + \cdots . \qquad \text{Converges when } x = 1.$$

When $x = -1$, you obtain the convergent alternating series

$$\sum_{n=1}^{\infty}\frac{(-1)^n}{n^2} = -\frac{1}{1^2} + \frac{1}{2^2} - \frac{1}{3^2} + \frac{1}{4^2} - \cdots . \qquad \text{Converges when } x = -1.$$

Interval: $[-1, 1]$
Radius: $R = 1$

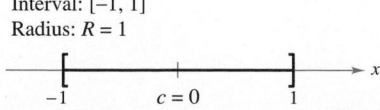

Figure 9.22

Therefore, the interval of convergence is $[-1, 1]$, as shown in Figure 9.22.

JAMES GREGORY (1638–1675)

One of the earliest mathematicians to work with power series was a Scotsman, James Gregory. He developed a power series method for interpolating table values—a method that was later used by Brook Taylor in the development of Taylor polynomials and Taylor series.

Differentiation and Integration of Power Series

Power series representation of functions has played an important role in the development of calculus. In fact, much of Newton's work with differentiation and integration was done in the context of power series—especially his work with complicated algebraic functions and transcendental functions. Euler, Lagrange, Leibniz, and the Bernoullis all used power series extensively in calculus.

Once you have defined a function with a power series, it is natural to wonder how you can determine the characteristics of the function. Is it continuous? Differentiable? Theorem 9.21, which is stated without proof, answers these questions.

THEOREM 9.21 **Properties of Functions Defined by Power Series**

If the function

$$f(x) = \sum_{n=0}^{\infty} a_n(x - c)^n$$
$$= a_0 + a_1(x - c) + a_2(x - c)^2 + a_3(x - c)^3 + \cdots$$

has a radius of convergence of $R > 0$, then, on the interval

$$(c - R, c + R)$$

f is differentiable (and therefore continuous). Moreover, the derivative and antiderivative of f are as follows.

1. $f'(x) = \sum_{n=1}^{\infty} na_n(x - c)^{n-1}$
$$= a_1 + 2a_2(x - c) + 3a_3(x - c)^2 + \cdots$$

2. $\int f(x)\, dx = C + \sum_{n=0}^{\infty} a_n \frac{(x - c)^{n+1}}{n + 1}$
$$= C + a_0(x - c) + a_1 \frac{(x - c)^2}{2} + a_2 \frac{(x - c)^3}{3} + \cdots$$

The *radius of convergence* of the series obtained by differentiating or integrating a power series is the same as that of the original power series. The *interval of convergence*, however, may differ as a result of the behavior at the endpoints.

Theorem 9.21 states that, in many ways, a function defined by a power series behaves like a polynomial. It is continuous in its interval of convergence, and both its derivative and its antiderivative can be determined by differentiating and integrating each term of the power series. For instance, the derivative of the power series

$$f(x) = \sum_{n=0}^{\infty} \frac{x^n}{n!}$$
$$= 1 + x + \frac{x^2}{2} + \frac{x^3}{3!} + \frac{x^4}{4!} + \cdots$$

is

$$f'(x) = 1 + (2)\frac{x}{2} + (3)\frac{x^2}{3!} + (4)\frac{x^3}{4!} + \cdots$$
$$= 1 + x + \frac{x^2}{2} + \frac{x^3}{3!} + \frac{x^4}{4!} + \cdots$$
$$= f(x).$$

Notice that $f'(x) = f(x)$. Do you recognize this function?

EXAMPLE 8 Intervals of Convergence for $f(x)$, $f'(x)$, and $\int f(x)\, dx$

Consider the function

$$f(x) = \sum_{n=1}^{\infty} \frac{x^n}{n} = x + \frac{x^2}{2} + \frac{x^3}{3} + \cdots.$$

Find the interval of convergence for each of the following.

a. $\int f(x)\, dx$ **b.** $f(x)$ **c.** $f'(x)$

Solution By Theorem 9.21, you have

$$f'(x) = \sum_{n=1}^{\infty} x^{n-1}$$
$$= 1 + x + x^2 + x^3 + \cdots$$

and

$$\int f(x)\, dx = C + \sum_{n=1}^{\infty} \frac{x^{n+1}}{n(n+1)}$$
$$= C + \frac{x^2}{1 \cdot 2} + \frac{x^3}{2 \cdot 3} + \frac{x^4}{3 \cdot 4} + \cdots.$$

• **REMARK** Notice in Example 8 that when differentiating the power series, differentiation is done on a *term-by-term* basis. Likewise, when integrating a series, integration is done on a *term-by-term* basis.

By the Ratio Test, you can show that each series has a radius of convergence of $R = 1$. Considering the interval $(-1, 1)$, you have the following.

a. For $\int f(x)\, dx$, the series

$$\sum_{n=1}^{\infty} \frac{x^{n+1}}{n(n+1)} \qquad \text{Interval of convergence: } [-1, 1]$$

converges for $x = \pm 1$, and its interval of convergence is $[-1, 1]$. See Figure 9.23(a).

b. For $f(x)$, the series

$$\sum_{n=1}^{\infty} \frac{x^n}{n} \qquad \text{Interval of convergence: } [-1, 1)$$

converges for $x = -1$ and diverges for $x = 1$. So, its interval of convergence is $[-1, 1)$. See Figure 9.23(b).

c. For $f'(x)$, the series

$$\sum_{n=1}^{\infty} x^{n-1} \qquad \text{Interval of convergence: } (-1, 1)$$

diverges for $x = \pm 1$, and its interval of convergence is $(-1, 1)$. See Figure 9.23(c).

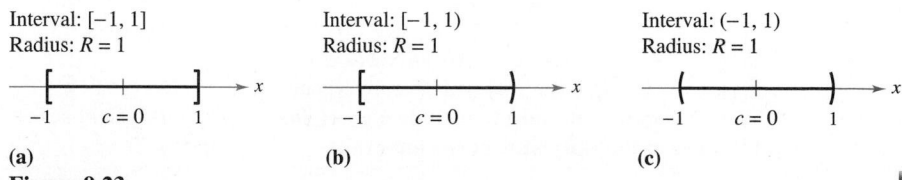

Interval: $[-1, 1]$
Radius: $R = 1$

Interval: $[-1, 1)$
Radius: $R = 1$

Interval: $(-1, 1)$
Radius: $R = 1$

(a)   **(b)**   **(c)**

Figure 9.23

From Example 8, it appears that of the three series, the one for the derivative, $f'(x)$, is the least likely to converge at the endpoints. In fact, it can be shown that if the series for $f'(x)$ converges at the endpoints

$$x = c \pm R$$

then the series for $f(x)$ will also converge there.

9.8 Exercises

See **CalcChat.com** for tutorial help and worked-out solutions to odd-numbered exercises.

CONCEPT CHECK

1. **Representing a Function** Explain how a Maclaurin polynomial and a power series centered at 0 for a function are different.

2. **Domain** What does the domain of
$$f(x) = \sum_{n=0}^{\infty} a_n(x - c)^n$$
represent?

3. **Radius of Convergence** Determine the radius of convergence for the power series $\sum_{n=0}^{\infty} a_n(x - 2)^n$ given the following result of the Ratio Test, where $u_n = a_n(x - 2)^n$.
$$\lim_{n \to \infty} \left| \frac{u_{n+1}}{u_n} \right| = \left| \frac{x - 2}{5} \right|$$

4. **Properties of Functions Defined by Power Series** In your own words, describe how a function defined by a power series behaves like a polynomial.

 **Finding the Center of a Power Series** In Exercises 5–8, state where the power series is centered.

5. $\sum_{n=0}^{\infty} nx^n$

6. $\sum_{n=1}^{\infty} \frac{(-1)^n(2n - 1)}{2^n n!} x^n$

7. $\sum_{n=1}^{\infty} \frac{(x - 2)^n}{n^3}$

8. $\sum_{n=0}^{\infty} \frac{(-1)^n(x - \pi)^{2n}}{(2n)!}$

 **Finding the Radius of Convergence** In Exercises 9–14, find the radius of convergence of the power series.

9. $\sum_{n=0}^{\infty} (-1)^n \frac{x^n}{n + 1}$

10. $\sum_{n=0}^{\infty} (3x)^n$

11. $\sum_{n=1}^{\infty} \frac{(4x)^n}{n^2}$

12. $\sum_{n=0}^{\infty} \frac{(-1)^n x^n}{5^n}$

13. $\sum_{n=0}^{\infty} \frac{x^{2n}}{(2n)!}$

14. $\sum_{n=0}^{\infty} \frac{(2n)! x^{3n}}{n!}$

 **Finding the Interval of Convergence** In Exercises 15–38, find the interval of convergence of the power series. (Be sure to include a check for convergence at the endpoints of the interval.)

15. $\sum_{n=0}^{\infty} \left(\frac{x}{4} \right)^n$

16. $\sum_{n=0}^{\infty} (2x)^n$

17. $\sum_{n=1}^{\infty} \frac{(-1)^n x^n}{n}$

18. $\sum_{n=0}^{\infty} (-1)^{n+1}(n + 1)x^n$

19. $\sum_{n=0}^{\infty} \frac{x^{5n}}{n!}$

20. $\sum_{n=0}^{\infty} \frac{(3x)^n}{(2n)!}$

21. $\sum_{n=0}^{\infty} (2n)! \left(\frac{x}{3} \right)^n$

22. $\sum_{n=0}^{\infty} \frac{(-1)^n x^n}{(n + 1)(n + 2)}$

23. $\sum_{n=1}^{\infty} \frac{(-1)^{n+1} x^n}{6^n}$

24. $\sum_{n=0}^{\infty} \frac{(-1)^n n!(x - 5)^n}{3^n}$

25. $\sum_{n=1}^{\infty} \frac{(-1)^{n+1}(x - 4)^n}{n9^n}$

26. $\sum_{n=0}^{\infty} \frac{(x - 3)^{n+1}}{(n + 1)4^{n+1}}$

27. $\sum_{n=0}^{\infty} \frac{(-1)^{n+1}(x - 1)^{n+1}}{n + 1}$

28. $\sum_{n=1}^{\infty} \frac{(-1)^{n+1}(x - 2)^n}{n2^n}$

29. $\sum_{n=1}^{\infty} \frac{(x - 3)^{n-1}}{3^{n-1}}$

30. $\sum_{n=0}^{\infty} \frac{(-1)^n x^{2n+1}}{2n + 1}$

31. $\sum_{n=1}^{\infty} \frac{n}{n + 1}(-2x)^{n-1}$

32. $\sum_{n=0}^{\infty} \frac{(-1)^n x^{2n}}{n!}$

33. $\sum_{n=0}^{\infty} \frac{x^{3n+1}}{(3n + 1)!}$

34. $\sum_{n=1}^{\infty} \frac{n! x^n}{(2n)!}$

35. $\sum_{n=1}^{\infty} \frac{2 \cdot 3 \cdot 4 \cdots (n + 1)x^n}{n!}$

36. $\sum_{n=1}^{\infty} \left[\frac{2 \cdot 4 \cdot 6 \cdots 2n}{3 \cdot 5 \cdot 7 \cdots (2n + 1)} \right] x^{2n+1}$

37. $\sum_{n=1}^{\infty} \frac{(-1)^{n+1} 3 \cdot 7 \cdot 11 \cdots (4n - 1)(x - 3)^n}{4^n}$

38. $\sum_{n=1}^{\infty} \frac{n!(x + 1)^n}{1 \cdot 3 \cdot 5 \cdots (2n - 1)}$

Finding the Radius of Convergence In Exercises 39 and 40, find the radius of convergence of the power series, where $c > 0$ and k is a positive integer.

39. $\sum_{n=1}^{\infty} \frac{(x - c)^{n-1}}{c^{n-1}}$

40. $\sum_{n=0}^{\infty} \frac{(n!)^k x^n}{(kn)!}$

Finding the Interval of Convergence In Exercises 41–44, find the interval of convergence of the power series, where $c > 0$ and k is a positive integer. (Be sure to include a check for convergence at the endpoints of the interval.)

41. $\sum_{n=0}^{\infty} \left(\frac{x}{k} \right)^n$

42. $\sum_{n=1}^{\infty} \frac{(-1)^{n+1}(x - c)^n}{nc^n}$

43. $\sum_{n=1}^{\infty} \frac{k(k + 1)(k + 2) \cdots (k + n - 1)x^n}{n!}$

44. $\sum_{n=1}^{\infty} \frac{n!(x - c)^n}{1 \cdot 3 \cdot 5 \cdots (2n - 1)}$

Writing an Equivalent Series In Exercises 45–48, write an equivalent series with the index of summation beginning at $n = 1$.

45. $\sum_{n=0}^{\infty} \frac{x^n}{n!}$

46. $\sum_{n=0}^{\infty} (-1)^{n+1}(n + 1)x^n$

47. $\sum_{n=2}^{\infty} \frac{x^{n-1}}{(7n - 1)!}$

48. $\sum_{n=2}^{\infty} \frac{x^{3n-1}}{(2n - 1)!}$

Finding Intervals of Convergence In Exercises 49–52, find the intervals of convergence of (a) $f(x)$, (b) $f'(x)$, (c) $f''(x)$, and (d) $\int f(x)\,dx$. (Be sure to include a check for convergence at the endpoints of the intervals.)

49. $f(x) = \sum_{n=0}^{\infty} \left(\frac{x}{3}\right)^n$

50. $f(x) = \sum_{n=1}^{\infty} \frac{(-1)^{n+1}(x-5)^n}{n5^n}$

51. $f(x) = \sum_{n=0}^{\infty} \frac{(-1)^{n+1}(x-1)^{n+1}}{n+1}$

52. $f(x) = \sum_{n=1}^{\infty} \frac{(-1)^{n+1}(x-2)^n}{n}$

EXPLORING CONCEPTS

Writing a Power Series In Exercises 53 and 54, write a power series that has the indicated interval of convergence. Explain your reasoning.

53. $(-3, 3)$ **54.** $[-3, 7]$

55. Conditional or Absolute Convergence Give examples that show that the convergence of a power series at an endpoint of its interval of convergence may be either conditional or absolute. Explain your reasoning.

56. HOW DO YOU SEE IT? Match the graph of the first 10 terms of the sequence of partial sums of the series

$$g(x) = \sum_{n=0}^{\infty} \left(\frac{x}{3}\right)^n$$

with the indicated value of the function. [The graphs are labeled (i), (ii), (iii), and (iv).] Explain how you made your choice.

(i)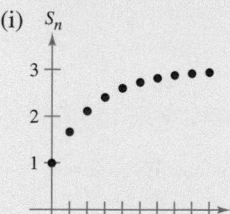

(ii)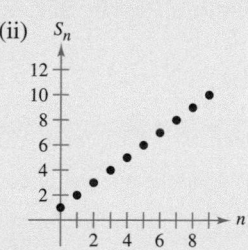

(iii)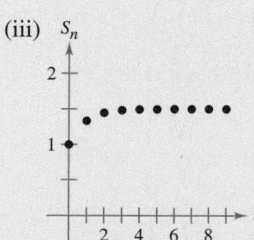

(iv)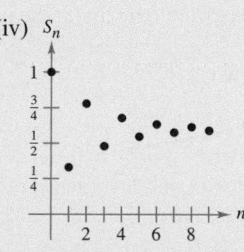

(a) $g(1)$ (b) $g(2)$
(c) $g(3)$ (d) $g(-2)$

---

57. Using Power Series Let $f(x) = \sum_{n=0}^{\infty} \frac{(-1)^n x^{2n+1}}{(2n+1)!}$ and $g(x) = \sum_{n=0}^{\infty} \frac{(-1)^n x^{2n}}{(2n)!}$.

(a) Find the intervals of convergence of f and g.

(b) Show that $f'(x) = g(x)$ and $g'(x) = -f(x)$.

(c) Identify the functions f and g.

58. Using a Power Series Let $f(x) = \sum_{n=0}^{\infty} \frac{x^{2n+1}}{(2n+1)!}$ and $g(x) = \sum_{n=0}^{\infty} \frac{x^{2n}}{(2n)!}$.

(a) Find the intervals of convergence of f and g.

(b) Show that $f'(x) = g(x)$ and $g'(x) = f(x)$.

(c) Identify the functions f and g.

Differential Equation In Exercises 59–64, show that the function represented by the power series is a solution of the differential equation.

59. $y = \sum_{n=0}^{\infty} \frac{(-1)^n + x^{2n+1}}{(2n+1)!}$, $y'' + y = 0$

60. $y = \sum_{n=0}^{\infty} \frac{(-1)^n x^{2n}}{(2n)!}$, $y'' + y = 0$

61. $y = \sum_{n=0}^{\infty} \frac{x^{2n+1}}{(2n+1)!}$, $y'' - y = 0$

62. $y = \sum_{n=0}^{\infty} \frac{x^{2n}}{(2n)!}$, $y'' - y = 0$

63. $y = \sum_{n=0}^{\infty} \frac{x^{2n}}{2^n n!}$, $y'' - xy' - y = 0$

64. $y = 1 + \sum_{n=1}^{\infty} \frac{(-1)^n x^{4n}}{2^{2n} n! \cdot 3 \cdot 7 \cdot 11 \cdots (4n-1)}$,
$y'' + x^2 y = 0$

65. Bessel Function The Bessel function of order 0 is

$$J_0(x) = \sum_{k=0}^{\infty} \frac{(-1)^k x^{2k}}{2^{2k}(k!)^2}.$$

(a) Show that the series converges for all x.

(b) Show that the series is a solution of the differential equation $x^2 J_0'' + x J_0' + x^2 J_0 = 0$.

(c) Use a graphing utility to graph the polynomial composed of the first four terms of J_0.

(d) Approximate $\int_0^1 J_0\,dx$ accurate to two decimal places.

66. Bessel Function The Bessel function of order 1 is

$$J_1(x) = x \sum_{k=0}^{\infty} \frac{(-1)^k x^{2k}}{2^{2k+1} k!(k+1)!}.$$

(a) Show that the series converges for all x.

(b) Show that the series is a solution of the differential equation $x^2 J_1'' + x J_1' + (x^2 - 1) J_1 = 0$.

(c) Use a graphing utility to graph the polynomial composed of the first four terms of J_1.

(d) Use J_0 from Exercise 65 to show that $J_0'(x) = -J_1(x)$.

67. Investigation The interval of convergence of the geometric series $\sum\limits_{n=0}^{\infty} \left(\dfrac{x}{4}\right)^n$ is $(-4, 4)$.

(a) Find the sum of the series when $x = \dfrac{5}{2}$. Use a graphing utility to graph the first six terms of the sequence of partial sums and the horizontal line representing the sum of the series.

(b) Repeat part (a) for $x = -\dfrac{5}{2}$.

(c) Write a short paragraph comparing the rates of convergence of the partial sums with the sums of the series in parts (a) and (b). How do the plots of the partial sums differ as they converge toward the sum of the series?

(d) Given any positive real number M, there exists a positive integer N such that the partial sum

$$\sum_{n=0}^{N} \left(\frac{5}{4}\right)^n > M.$$

Use a graphing utility to complete the table.

M	10	100	1000	10,000
N				

68. Investigation The interval of convergence of the series $\sum\limits_{n=0}^{\infty} (3x)^n$ is $\left(-\dfrac{1}{3}, \dfrac{1}{3}\right)$.

(a) Find the sum of the series when $x = \dfrac{1}{6}$. Use a graphing utility to graph the first six terms of the sequence of partial sums and the horizontal line representing the sum of the series.

(b) Repeat part (a) for $x = -\dfrac{1}{6}$.

(c) Write a short paragraph comparing the rates of convergence of the partial sums with the sums of the series in parts (a) and (b). How do the plots of the partial sums differ as they converge toward the sum of the series?

(d) Given any positive real number M, there exists a positive integer N such that the partial sum

$$\sum_{n=0}^{N} \left(3 \cdot \frac{2}{3}\right)^n > M.$$

Use a graphing utility to complete the table.

M	10	100	1000	10,000
N				

Identifying a Function In Exercises 69–72, the series represents a well-known function. Use a computer algebra system to graph the partial sum S_{10} and identify the function from the graph.

69. $f(x) = \sum\limits_{n=0}^{\infty} (-1)^n \dfrac{\pi^{2n} x^{2n}}{(2n)!}$

70. $f(x) = \sum\limits_{n=0}^{\infty} (-1)^{n+1} \dfrac{x^{2n+1}}{(2n+1)!}$

71. $f(x) = \sum\limits_{n=0}^{\infty} (-1)^n x^n, \quad -1 < x < 1$

72. $f(x) = \sum\limits_{n=0}^{\infty} (3x)^n$

True or False? In Exercises 73–76, determine whether the statement is true or false. If it is false, explain why or give an example that shows it is false.

73. If the power series $\sum\limits_{n=1}^{\infty} a_n x^n$ converges for $x = 2$, then it also converges for $x = -2$.

74. It is possible to find a power series whose interval of convergence is $[0, \infty)$.

75. If the interval of convergence for $\sum\limits_{n=0}^{\infty} a_n x^n$ is $(-1, 1)$, then the interval of convergence for $\sum\limits_{n=0}^{\infty} a_n (x - 1)^n$ is $(0, 2)$.

76. If $f(x) = \sum\limits_{n=0}^{\infty} a_n x^n$ converges for $|x| < 2$, then

$$\int_0^1 f(x) \, dx = \sum_{n=0}^{\infty} \frac{a_n}{n+1}.$$

77. Proof Prove that the power series

$$\sum_{n=0}^{\infty} \frac{(n+p)!}{n!(n+q)!} x^n$$

has a radius of convergence of $R = \infty$ when p and q are positive integers.

78. Using a Power Series Let

$$g(x) = 1 + 2x + x^2 + 2x^3 + x^4 + \cdots$$

where the coefficients are $c_{2n} = 1$ and $c_{2n+1} = 2$ for $n \geq 0$.

(a) Find the interval of convergence of the series.

(b) Find an explicit formula for $g(x)$.

79. Using a Power Series Let

$$f(x) = \sum_{n=0}^{\infty} c_n x^n$$

where $c_{n+3} = c_n$ for $n \geq 0$.

(a) Find the interval of convergence of the series.

(b) Find an explicit formula for $f(x)$.

80. Proof Prove that if the power series $\sum\limits_{n=0}^{\infty} c_n x^n$ has a radius of convergence of R, then $\sum\limits_{n=0}^{\infty} c_n x^{2n}$ has a radius of convergence of $\sqrt{R}$.

81. Proof For $n > 0$, let $R > 0$ and $c_n > 0$. Prove that if the interval of convergence of the series

$$\sum_{n=0}^{\infty} c_n (x - x_0)^n$$

is $[x_0 - R, x_0 + R]$, then the series converges conditionally at $x = x_0 - R$.

9.9 Representation of Functions by Power Series

■ Find a geometric power series that represents a function.
■ Construct a power series using series operations.

Geometric Power Series

In this section and the next, you will study several techniques for finding a power series that represents a function. Consider the function

$$f(x) = \frac{1}{1 - x}.$$

The form of f closely resembles the sum of a geometric series

$$\sum_{n=0}^{\infty} ar^n = \frac{a}{1 - r}, \quad |r| < 1.$$

In other words, when $a = 1$ and $r = x$, a power series representation for $1/(1 - x)$, centered at 0, is

$$\frac{1}{1 - x} = \sum_{n=0}^{\infty} ar^n$$

$$= \sum_{n=0}^{\infty} x^n$$

$$= 1 + x + x^2 + x^3 + \cdots, \quad |x| < 1.$$

Of course, this series represents $f(x) = 1/(1 - x)$ only on the interval $(-1, 1)$, whereas f is defined for all $x \neq 1$, as shown in Figure 9.24. To represent f in another interval, you must develop a different series. For instance, to obtain the power series centered at -1, you could write

$$\frac{1}{1 - x} = \frac{1}{2 - (x + 1)} = \frac{1/2}{1 - [(x + 1)/2]} = \frac{a}{1 - r}$$

which implies that $a = \frac{1}{2}$ and $r = (x + 1)/2$. So, for $|x + 1| < 2$, you have

$$\frac{1}{1 - x} = \sum_{n=0}^{\infty} \left(\frac{1}{2}\right)\left(\frac{x + 1}{2}\right)^n$$

$$= \frac{1}{2}\left[1 + \frac{(x + 1)}{2} + \frac{(x + 1)^2}{4} + \frac{(x + 1)^3}{8} + \cdots\right], \quad |x + 1| < 2$$

which converges on the interval $(-3, 1)$.

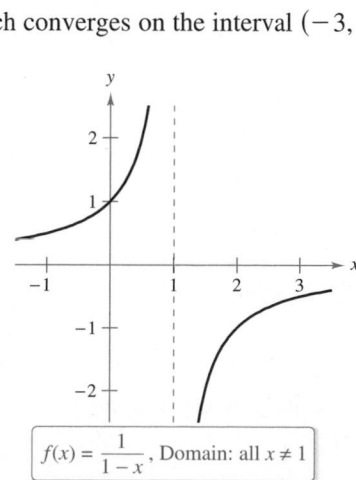
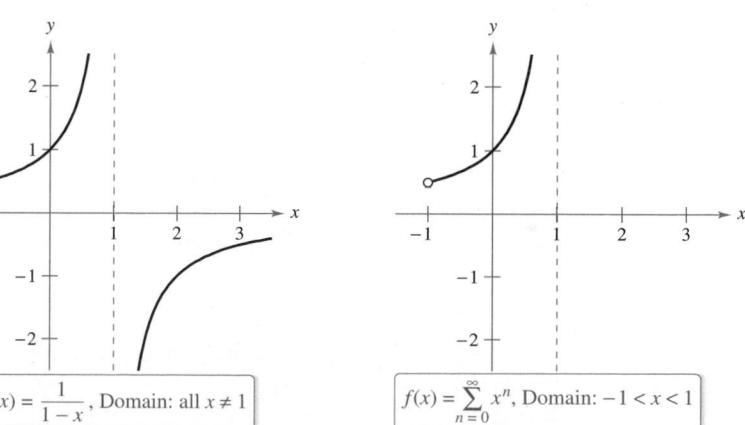

$f(x) = \dfrac{1}{1 - x}$, Domain: all $x \neq 1$

$f(x) = \displaystyle\sum_{n=0}^{\infty} x^n$, Domain: $-1 < x < 1$

Figure 9.24

EXAMPLE 1 **Finding a Geometric Power Series Centered at 0**

Find a power series for $f(x) = \dfrac{4}{x+2}$, centered at 0.

Solution Writing $f(x)$ in the form $a/(1-r)$ produces

$$\frac{4}{2+x} = \frac{2}{1-(-x/2)} = \frac{a}{1-r}$$

which implies that $a = 2$ and

$$r = -\frac{x}{2}.$$

So, the power series for $f(x)$ is

$$\frac{4}{x+2} = \sum_{n=0}^{\infty} ar^n$$
$$= \sum_{n=0}^{\infty} 2\left(-\frac{x}{2}\right)^n$$
$$= 2\left(1 - \frac{x}{2} + \frac{x^2}{4} - \frac{x^3}{8} + \cdots\right).$$

Long Division

$$
\begin{array}{r}
2 - x + \frac{1}{2}x^2 - \frac{1}{4}x^3 + \cdots \\
2+x\overline{)4} \\
\underline{4+2x} \\
-2x \\
\underline{-2x - x^2} \\
x^2 \\
\underline{x^2 + \frac{1}{2}x^3} \\
-\frac{1}{2}x^3 \\
\underline{-\frac{1}{2}x^3 - \frac{1}{4}x^4}
\end{array}
$$

This power series converges when

$$\left|-\frac{x}{2}\right| < 1$$

which implies that the interval of convergence is $(-2, 2)$. ∎

Another way to determine a power series for a rational function such as the one in Example 1 is to use long division. For instance, by dividing $2 + x$ into 4, you obtain the result shown at the left.

EXAMPLE 2 **Finding a Geometric Power Series Centered at 1**

Find a power series for $f(x) = \dfrac{1}{x}$, centered at 1.

Solution Writing $f(x)$ in the form $a/(1-r)$ produces

$$\frac{1}{x} = \frac{1}{1-(-x+1)} = \frac{a}{1-r}$$

which implies that $a = 1$ and $r = 1 - x = -(x-1)$. So, the power series for $f(x)$ is

$$\frac{1}{x} = \sum_{n=0}^{\infty} ar^n$$
$$= \sum_{n=0}^{\infty} [-(x-1)]^n$$
$$= \sum_{n=0}^{\infty} (-1)^n(x-1)^n$$
$$= 1 - (x-1) + (x-1)^2 - (x-1)^3 + \cdots.$$

This power series converges when

$$|x-1| < 1$$

which implies that the interval of convergence is $(0, 2)$. ∎

Operations with Power Series

The versatility of geometric power series will be shown later in this section, following a discussion of power series operations. These operations, used with differentiation and integration, provide a means of developing power series for a variety of elementary functions. (For simplicity, the operations are stated for a series centered at 0.)

Operations with Power Series

Let $f(x) = \displaystyle\sum_{n=0}^{\infty} a_n x^n$ and $g(x) = \displaystyle\sum_{n=0}^{\infty} b_n x^n$.

1. $f(kx) = \displaystyle\sum_{n=0}^{\infty} a_n k^n x^n$

2. $f(x^N) = \displaystyle\sum_{n=0}^{\infty} a_n x^{nN}$

3. $f(x) \pm g(x) = \displaystyle\sum_{n=0}^{\infty} (a_n \pm b_n) x^n$

The operations described above can change the interval of convergence for the resulting series. For example, in the addition shown below, the interval of convergence for the sum is the *intersection* of the intervals of convergence of the two original series.

$$\underbrace{\sum_{n=0}^{\infty} x^n}_{(-1,1)} + \underbrace{\sum_{n=0}^{\infty} \left(\frac{x}{2}\right)^n}_{(-2,2)} = \underbrace{\sum_{n=0}^{\infty} \left(1 + \frac{1}{2^n}\right) x^n}_{(-1,1)}$$

$$(-1,1) \ \cap \ (-2,2) \ = \ (-1,1)$$

EXAMPLE 3   Adding Two Power Series

Find a power series for

$$f(x) = \frac{3x - 1}{x^2 - 1}$$

centered at 0.

Solution   Using partial fractions, you can write $f(x)$ as

$$\frac{3x - 1}{x^2 - 1} = \frac{2}{x + 1} + \frac{1}{x - 1}.$$

By adding the two geometric power series

$$\frac{2}{x + 1} = \frac{2}{1 - (-x)} = \sum_{n=0}^{\infty} 2(-1)^n x^n, \quad |x| < 1$$

and

$$\frac{1}{x - 1} = \frac{-1}{1 - x} = -\sum_{n=0}^{\infty} x^n, \quad |x| < 1$$

you obtain the power series shown below.

$$\frac{3x - 1}{x^2 - 1} = \sum_{n=0}^{\infty} \left[2(-1)^n - 1 \right] x^n$$

$$= 1 - 3x + x^2 - 3x^3 + x^4 - \cdots$$

The interval of convergence for this power series is $(-1, 1)$.

EXAMPLE 4   Finding a Power Series by Integration

Find a power series for

$$f(x) = \ln x$$

centered at 1.

Solution   From Example 2, you know that

$$\frac{1}{x} = \sum_{n=0}^{\infty} (-1)^n (x-1)^n. \qquad \text{Interval of convergence: } (0, 2)$$

Integrating this series produces

$$\ln x = \int \frac{1}{x}\, dx + C$$

$$= C + \sum_{n=0}^{\infty} (-1)^n \frac{(x-1)^{n+1}}{n+1}.$$

By letting $x = 1$, you can conclude that $C = 0$. Therefore,

$$\ln x = \sum_{n=0}^{\infty} (-1)^n \frac{(x-1)^{n+1}}{n+1}$$

$$= \frac{(x-1)}{1} - \frac{(x-1)^2}{2} + \frac{(x-1)^3}{3} - \frac{(x-1)^4}{4} + \cdots. \qquad \begin{array}{l}\text{Interval of}\\\text{convergence: } (0, 2]\end{array}$$

Note that the series converges at $x = 2$. This is consistent with the observation in the preceding section that integration of a power series may alter the convergence at the endpoints of the interval of convergence.

■ **FOR FURTHER INFORMATION**   To read about finding a power series using integration by parts, see the article "Integration by Parts and Infinite Series" by Shelby J. Kilmer in *Mathematics Magazine*. To view this article, go to *MathArticles.com*.

In Section 9.7, Example 4, the fourth-degree Taylor polynomial (centered at $c = 1$) for the natural logarithmic function

$$\ln x \approx (x-1) - \frac{(x-1)^2}{2} + \frac{(x-1)^3}{3} - \frac{(x-1)^4}{4}$$

was used to approximate $\ln(1.1)$.

$$\ln(1.1) \approx (0.1) - \frac{1}{2}(0.1)^2 + \frac{1}{3}(0.1)^3 - \frac{1}{4}(0.1)^4$$

$$\approx 0.0953083$$

You now know from Example 4 in this section that this polynomial represents the first four terms of the power series for $\ln x$. Moreover, using the Alternating Series Remainder, you can determine that the error in this approximation is less than

$$|R_4| \le |a_5|$$

$$= \frac{1}{5}(0.1)^5$$

$$= 0.000002.$$

During the seventeenth and eighteenth centuries, mathematical tables for logarithms and values of other transcendental functions were computed in this manner. Such numerical techniques are far from outdated, because it is precisely by such means that many modern calculating devices are programmed to evaluate transcendental functions.

**SRINIVASA RAMANUJAN
(1887–1920)**

Series that can be used to approximate π have interested mathematicians for the past 300 years. An amazing series for approximating $1/\pi$ was discovered by the Indian mathematician Srinivasa Ramanujan in 1914 (see Exercise 57). Each successive term of Ramanujan's series adds roughly eight more correct digits to the value of $1/\pi$. For more information about Ramanujan's work, see the article "Ramanujan and Pi" by Jonathan M. Borwein and Peter B. Borwein in *Scientific American*.
See LarsonCalculus.com to read more of this biography.

EXAMPLE 5 **Finding a Power Series by Integration**

⋮⋯▷ *See LarsonCalculus.com for an interactive version of this type of example.*

Find a power series for

$$g(x) = \arctan x$$

centered at 0.

Solution Because $D_x[\arctan x] = 1/(1 + x^2)$, you can use the series

$$f(x) = \frac{1}{1+x} = \sum_{n=0}^{\infty} (-1)^n x^n. \qquad \text{Interval of convergence: } (-1, 1)$$

Substituting x^2 for x produces

$$f(x^2) = \frac{1}{1+x^2} = \sum_{n=0}^{\infty} (-1)^n x^{2n}.$$

Finally, by integrating, you obtain

$$
\begin{aligned}
\arctan x &= \int \frac{1}{1+x^2}\, dx + C \\
&= C + \sum_{n=0}^{\infty} (-1)^n \frac{x^{2n+1}}{2n+1} \\
&= \sum_{n=0}^{\infty} (-1)^n \frac{x^{2n+1}}{2n+1} \qquad \text{Let } x = 0, \text{ then } C = 0. \\
&= x - \frac{x^3}{3} + \frac{x^5}{5} - \frac{x^7}{7} + \cdots. \qquad \text{Interval of convergence: } (-1, 1)
\end{aligned}
$$

It can be shown that the power series developed for $\arctan x$ in Example 5 also converges (to $\arctan x$) for $x = \pm 1$. For instance, when $x = 1$, you can write

$$
\begin{aligned}
\arctan 1 &= 1 - \frac{1}{3} + \frac{1}{5} - \frac{1}{7} + \cdots \\
&= \frac{\pi}{4}.
\end{aligned}
$$

However, this series (developed by James Gregory in 1671) is not a practical way of approximating π because it converges so slowly that hundreds of terms would have to be used to obtain reasonable accuracy. Example 6 shows how to use *two* different arctangent series to obtain a very good approximation of π using only a few terms. This approximation was developed by John Machin in 1706.

FOR FURTHER INFORMATION
To read about other methods for approximating π, see the article "Two Methods for Approximating π" by Chien-Lih Hwang in *Mathematics Magazine*. To view this article, go to *MathArticles.com*.

EXAMPLE 6 **Approximating π with a Series**

Use the trigonometric identity

$$4 \arctan \frac{1}{5} - \arctan \frac{1}{239} = \frac{\pi}{4}$$

to approximate the number π. [See Exercise 44(b).]

Solution By using only five terms from each of the series for $\arctan(1/5)$ and $\arctan(1/239)$, you obtain

$$4\left(4 \arctan \frac{1}{5} - \arctan \frac{1}{239}\right) \approx 3.1415927$$

which agrees with the exact value of π with an error of less than 0.0000001.

9.9 Exercises

See **CalcChat.com** for tutorial help and worked-out solutions to odd-numbered exercises.

CONCEPT CHECK

1. Using Power Series Explain how to use a geometric power series to represent a function of the form $f(x) = \dfrac{b}{c-x}$.

2. Power Series Operations Consider $f(x) = \displaystyle\sum_{n=0}^{\infty} 5x^{2n}$.

What are the values of a and b in terms of n?

$$f\left(\frac{x^3}{5}\right) = \sum_{n=0}^{\infty} \frac{x^a}{5^b}$$

 **Finding a Geometric Power Series** In Exercises 3–6, find a geometric power series for the function, centered at 0, (a) by the technique shown in Examples 1 and 2 and (b) by long division.

3. $f(x) = \dfrac{1}{4-x}$

4. $f(x) = \dfrac{1}{2+x}$

5. $f(x) = \dfrac{4}{3+x}$

6. $f(x) = \dfrac{2}{5-x}$

 **Finding a Power Series** In Exercises 7–18, find a power series for the function, centered at c, and determine the interval of convergence.

7. $f(x) = \dfrac{1}{6-x}$, $\quad c = 1$

8. $f(x) = \dfrac{2}{6-x}$, $\quad c = -2$

9. $f(x) = \dfrac{1}{1-3x}$, $\quad c = 0$

10. $h(x) = \dfrac{1}{1-4x}$, $\quad c = 0$

11. $g(x) = \dfrac{5}{2x-3}$, $\quad c = -3$

12. $f(x) = \dfrac{3}{2x-1}$, $\quad c = 2$

13. $f(x) = \dfrac{2}{5x+4}$, $\quad c = -1$

14. $f(x) = \dfrac{4}{3x+2}$, $\quad c = 3$

15. $g(x) = \dfrac{4x}{x^2+2x-3}$, $\quad c = 0$

16. $g(x) = \dfrac{3x-8}{3x^2+5x-2}$, $\quad c = 0$

17. $f(x) = \dfrac{2}{1-x^2}$, $\quad c = 0$

18. $f(x) = \dfrac{5}{4-x^2}$, $\quad c = 0$

 **Using a Power Series** In Exercises 19–28, use the power series

$$\frac{1}{1+x} = \sum_{n=0}^{\infty} (-1)^n x^n, \quad |x| < 1$$

to find a power series for the function, centered at 0, and determine the interval of convergence.

19. $h(x) = \dfrac{-2}{x^2-1} = \dfrac{1}{1+x} + \dfrac{1}{1-x}$

20. $h(x) = \dfrac{x}{x^2-1} = \dfrac{1}{2(1+x)} - \dfrac{1}{2(1-x)}$

21. $f(x) = -\dfrac{1}{(x+1)^2} = \dfrac{d}{dx}\left[\dfrac{1}{x+1}\right]$

22. $f(x) = \dfrac{2}{(x+1)^3} = \dfrac{d^2}{dx^2}\left[\dfrac{1}{x+1}\right]$

23. $f(x) = \ln(x+1) = \displaystyle\int \dfrac{1}{x+1}\,dx$

24. $f(x) = \ln(1-x^2) = \displaystyle\int \dfrac{1}{1+x}\,dx - \int \dfrac{1}{1-x}\,dx$

25. $g(x) = \dfrac{1}{x^2+1}$

26. $f(x) = \ln(x^2+1)$

27. $h(x) = \dfrac{1}{4x^2+1}$ **28.** $f(x) = \arctan 2x$

Graphical and Numerical Analysis In Exercises 29 and 30, let

$$S_n = x - \frac{x^2}{2} + \frac{x^3}{3} - \frac{x^4}{4} + \cdots \pm \frac{x^n}{n}.$$

Use a graphing utility to confirm the inequality graphically. Then complete the table to confirm the inequality numerically.

x	0.0	0.2	0.4	0.6	0.8	1.0
S_n						
$\ln(x+1)$						
S_{n+1}						

29. $S_2 \le \ln(x+1) \le S_3$

30. $S_4 \le \ln(x+1) \le S_5$

Approximating a Sum In Exercises 31 and 32, (a) use a graphing utility to graph several partial sums of the series, (b) find the sum of the series and its radius of convergence, (c) use a graphing utility and 50 terms of the series to approximate the sum when $x = 0.5$, and (d) determine what the approximation represents and how good the approximation is.

31. $\displaystyle\sum_{n=1}^{\infty} \dfrac{(-1)^{n+1}(x-1)^n}{n}$

32. $\displaystyle\sum_{n=0}^{\infty} \dfrac{(-1)^n x^{2n+1}}{(2n+1)!}$

 **Approximating a Value** In Exercises 33–36, use the power series for $f(x) = \arctan x$ to approximate the value, using $R_N \le 0.001$.

33. $\arctan \dfrac{1}{4}$

34. $\displaystyle\int_{0}^{3/4} \arctan x^2\,dx$

35. $\displaystyle\int_{0}^{1/2} \dfrac{\arctan x^2}{x}\,dx$

36. $\displaystyle\int_{0}^{1/2} x^2 \arctan x\,dx$

Using a Power Series In Exercises 37–40, use the power series

$$\frac{1}{1-x} = \sum_{n=0}^{\infty} x^n, \quad |x| < 1$$

to find a power series for the function, centered at 0, and determine the interval of convergence.

37. $f(x) = \dfrac{1}{(1-x)^2}$ **38.** $f(x) = \dfrac{x}{(1-x)^2}$

39. $f(x) = \dfrac{1+x}{(1-x)^2}$ **40.** $f(x) = \dfrac{x(1+x)}{(1-x)^2}$

41. Probability A fair coin is tossed repeatedly. The probability that the first head occurs on the nth toss is $P(n) = \left(\frac{1}{2}\right)^n$. When this game is repeated many times, the average number of tosses required until the first head occurs is

$$E(n) = \sum_{n=1}^{\infty} nP(n).$$

(This value is called the *expected value of n.*) Use the results of Exercises 37–40 to find $E(n)$. Is the answer what you expected? Why or why not?

42. Finding the Sum of a Series Use the results of Exercises 37–40 to find the sum of each series.

(a) $\dfrac{1}{3} \displaystyle\sum_{n=1}^{\infty} n\left(\dfrac{2}{3}\right)^n$ (b) $\dfrac{1}{10} \displaystyle\sum_{n=1}^{\infty} n\left(\dfrac{9}{10}\right)^n$

43. Proof Prove that

$$\arctan x + \arctan y = \arctan \frac{x+y}{1-xy}$$

for $xy \neq 1$ provided the value of the left side of the equation is between $-\pi/2$ and $\pi/2$.

44. Verifying an Identity Use the result of Exercise 43 to verify each identity.

(a) $\arctan \dfrac{120}{119} - \arctan \dfrac{1}{239} = \dfrac{\pi}{4}$

(b) $4 \arctan \dfrac{1}{5} - \arctan \dfrac{1}{239} = \dfrac{\pi}{4}$

[*Hint:* Use Exercise 43 twice to find $4 \arctan \frac{1}{5}$. Then use part (a).]

Approximating Pi In Exercises 45 and 46, (a) use the result of Exercise 43 to verify the given identity and (b) use the identity and the series for the arctangent to approximate π by using four terms of each series.

45. $2 \arctan \dfrac{1}{2} - \arctan \dfrac{1}{7} = \dfrac{\pi}{4}$

46. $\arctan \dfrac{1}{2} + \arctan \dfrac{1}{3} = \dfrac{\pi}{4}$

Finding the Sum of a Series In Exercises 47–52, find the sum of the convergent series by using a well-known function. Identify the function and explain how you obtained the sum.

47. $\displaystyle\sum_{n=1}^{\infty} (-1)^{n+1} \dfrac{1}{2^n n}$ **48.** $\displaystyle\sum_{n=1}^{\infty} (-1)^{n+1} \dfrac{1}{3^n n}$

49. $\displaystyle\sum_{n=1}^{\infty} (-1)^{n+1} \dfrac{2^n}{5^n n}$

50. $\displaystyle\sum_{n=0}^{\infty} (-1)^n \dfrac{1}{2n+1}$

51. $\displaystyle\sum_{n=0}^{\infty} (-1)^n \dfrac{1}{2^{2n+1}(2n+1)}$

52. $\displaystyle\sum_{n=1}^{\infty} (-1)^{n+1} \dfrac{1}{3^{2n-1}(2n-1)}$

EXPLORING CONCEPTS

53. Using Series One of the series in Exercises 47–52 converges to its sum at a much lower rate than the other five series. Which is it? Explain why this series converges so slowly. Use a graphing utility to illustrate the rate of convergence.

54. Radius of Convergence The radius of convergence of the power series $\displaystyle\sum_{n=0}^{\infty} a_n x^n$ is 3. What is the radius of convergence of the series

$$\sum_{n=1}^{\infty} na_n x^{n-1}?$$

Explain.

55. Convergence of a Power Series The power series $\displaystyle\sum_{n=0}^{\infty} a_n x^n$ converges for $|x+1| < 4$. What can you conclude about the convergence of the series

$$\sum_{n=0}^{\infty} a_n \frac{x^{n+1}}{n+1}?$$

Explain.

 **56.** **HOW DO YOU SEE IT?** The figure on the left shows the graph of a function. The figure on the right shows the graph of a power series representation of the function.

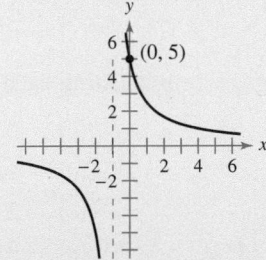 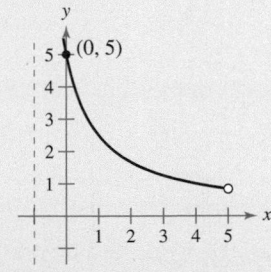

(a) Identify the function.

(b) What are the center and interval of convergence of the power series?

57. Ramanujan and Pi Use a graphing utility to show that

$$\frac{\sqrt{8}}{9801} \sum_{n=0}^{\infty} \frac{(4n)!(1103 + 26{,}390n)}{(n!)396^{4n}} = \frac{1}{\pi}.$$

9.10 Taylor and Maclaurin Series

■ Find a Taylor or Maclaurin series for a function.
■ Find a binomial series.
■ Use a basic list of Taylor series to find other Taylor series.

Taylor Series and Maclaurin Series

In Section 9.9, you derived power series for several functions using geometric series with term-by-term differentiation or integration. In this section, you will study a *general* procedure for deriving the power series for a function that has derivatives of all orders. The next theorem gives the form that *every* convergent power series must take.

> **REMARK** Be sure you understand Theorem 9.22. The theorem says that *if a power series converges to $f(x)$, then the series must be a Taylor series.* The theorem does *not* say that every series formed with the Taylor coefficients $a_n = f^{(n)}(c)/n!$ will converge to $f(x)$.

THEOREM 9.22 The Form of a Convergent Power Series

If f is represented by a power series $f(x) = \sum a_n(x - c)^n$ for all x in an open interval I containing c, then

$$a_n = \frac{f^{(n)}(c)}{n!}$$

and

$$f(x) = f(c) + f'(c)(x - c) + \frac{f''(c)}{2!}(x - c)^2 + \cdots$$
$$+ \frac{f^{(n)}(c)}{n!}(x - c)^n + \cdots.$$

Proof Consider a power series $\sum a_n(x - c)^n$ that has a radius of convergence R. Then, by Theorem 9.21, you know that the nth derivative of f exists for $|x - c| < R$, and by successive differentiation you obtain the following.

$$f^{(0)}(x) = a_0 + a_1(x - c) + a_2(x - c)^2 + a_3(x - c)^3 + a_4(x - c)^4 + \cdots$$
$$f^{(1)}(x) = a_1 + 2a_2(x - c) + 3a_3(x - c)^2 + 4a_4(x - c)^3 + \cdots$$
$$f^{(2)}(x) = 2a_2 + 3!a_3(x - c) + 4 \cdot 3a_4(x - c)^2 + \cdots$$
$$f^{(3)}(x) = 3!a_3 + 4!a_4(x - c) + \cdots$$
$$\vdots$$
$$f^{(n)}(x) = n!a_n + (n + 1)!a_{n+1}(x - c) + \cdots$$

Evaluating each of these derivatives at $x = c$ yields

$$f^{(0)}(c) = 0!a_0$$
$$f^{(1)}(c) = 1!a_1$$
$$f^{(2)}(c) = 2!a_2$$
$$f^{(3)}(c) = 3!a_3$$

and, in general, $f^{(n)}(c) = n!a_n$. By solving for a_n, you find that the coefficients of the power series representation of $f(x)$ are

$$a_n = \frac{f^{(n)}(c)}{n!}.$$ ∎

Notice that the coefficients of the power series in Theorem 9.22 are precisely the coefficients of the Taylor polynomials for f at c as defined in Section 9.7. For this reason, the series is called the **Taylor series** for f at c.

COLIN MACLAURIN (1698–1746)

The development of power series to represent functions is credited to the combined work of many seventeenth- and eighteenth-century mathematicians. Gregory, Newton, John and James Bernoulli, Leibniz, Euler, Lagrange, Wallis, and Fourier all contributed to this work. However, the two names that are most commonly associated with power series are Brook Taylor and Colin Maclaurin.
See LarsonCalculus.com to read more of this biography.

Definition of Taylor and Maclaurin Series

If a function f has derivatives of all orders at $x = c$, then the series

$$\sum_{n=0}^{\infty} \frac{f^{(n)}(c)}{n!}(x - c)^n = f(c) + f'(c)(x - c) + \cdots + \frac{f^{(n)}(c)}{n!}(x - c)^n + \cdots$$

is called the **Taylor series for f at c.** Moreover, if $c = 0$, then the series is the **Maclaurin series for f.**

When you know the pattern for the coefficients of the Taylor polynomials for a function, you can extend the pattern to form the corresponding Taylor series. For instance, in Example 4 in Section 9.7, you found the fourth Taylor polynomial for $\ln x$, centered at 1, to be

$$P_4(x) = (x - 1) - \frac{1}{2}(x - 1)^2 + \frac{1}{3}(x - 1)^3 - \frac{1}{4}(x - 1)^4.$$

From this pattern, you can obtain the Taylor series for $\ln x$ centered at $c = 1$,

$$(x - 1) - \frac{1}{2}(x - 1)^2 + \cdots + \frac{(-1)^{n+1}}{n}(x - 1)^n + \cdots.$$

EXAMPLE 1   **Forming a Power Series**

Use the function

$$f(x) = \sin x$$

to form the Maclaurin series

$$\sum_{n=0}^{\infty} \frac{f^{(n)}(0)}{n!}x^n = f(0) + f'(0)x + \frac{f''(0)}{2!}x^2 + \frac{f'''(0)}{3!}x^3 + \frac{f^{(4)}(0)}{4!}x^4 + \cdots$$

and determine the interval of convergence.

Solution   Taking successive derivatives of f yields

$$
\begin{aligned}
f(x) &= \sin x & f(0) &= \sin 0 = 0 \\
f'(x) &= \cos x & f'(0) &= \cos 0 = 1 \\
f''(x) &= -\sin x & f''(0) &= -\sin 0 = 0 \\
f'''(x) &= -\cos x & f'''(0) &= -\cos 0 = -1 \\
f^{(4)}(x) &= \sin x & f^{(4)}(0) &= \sin 0 = 0 \\
f^{(5)}(x) &= \cos x & f^{(5)}(0) &= \cos 0 = 1
\end{aligned}
$$

and so on. The pattern repeats after the third derivative. So, the power series is as follows.

$$
\begin{aligned}
\sum_{n=0}^{\infty} \frac{f^{(n)}(0)}{n!}x^n &= f(0) + f'(0)x + \frac{f''(0)}{2!}x^2 + \frac{f'''(0)}{3!}x^3 + \frac{f^{(4)}(0)}{4!}x^4 + \cdots \\
&= 0 + (1)x + \frac{0}{2!}x^2 + \frac{(-1)}{3!}x^3 + \frac{0}{4!}x^4 + \frac{1}{5!}x^5 + \frac{0}{6!}x^6 + \frac{(-1)}{7!}x^7 + \cdots \\
&= x - \frac{x^3}{3!} + \frac{x^5}{5!} - \frac{x^7}{7!} + \cdots \\
&= \sum_{n=0}^{\infty} \frac{(-1)^n x^{2n+1}}{(2n+1)!}
\end{aligned}
$$

By the Ratio Test, you can conclude that this series converges for all x.

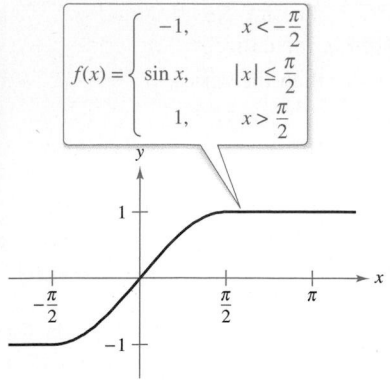

$$f(x) = \begin{cases} -1, & x < -\dfrac{\pi}{2} \\ \sin x, & |x| \le \dfrac{\pi}{2} \\ 1, & x > \dfrac{\pi}{2} \end{cases}$$

Figure 9.25

Notice that in Example 1, you cannot conclude that the power series converges to $\sin x$ for all x. You can simply conclude that the power series converges to some function, but you are not sure what function it is. This is a subtle, but important, point in dealing with Taylor or Maclaurin series. To persuade yourself that the series

$$f(c) + f'(c)(x - c) + \frac{f''(c)}{2!}(x - c)^2 + \cdots + \frac{f^{(n)}(c)}{n!}(x - c)^n + \cdots$$

might converge to a function other than f, remember that the derivatives are being evaluated at a single point. It can easily happen that another function will agree with the values of $f^{(n)}(x)$ when $x = c$ and disagree at other x-values. For instance, the power series (centered at 0) for the function f shown in Figure 9.25 is the same series as in Example 1. You know that the series converges for all x, and yet it obviously cannot converge to both $f(x)$ and $\sin x$ for all x.

Let f have derivatives of all orders in an open interval I centered at c. The Taylor series for f may fail to converge for some x in I. Even when it is convergent, it may fail to have $f(x)$ as its sum. Nevertheless, Theorem 9.19 tells us that for each n,

$$f(x) = f(c) + f'(c)(x - c) + \frac{f''(c)}{2!}(x - c)^2 + \cdots + \frac{f^{(n)}(c)}{n!}(x - c)^n + R_n(x)$$

where

$$R_n(x) = \frac{f^{(n+1)}(z)}{(n + 1)!}(x - c)^{n+1}.$$

Note that in this remainder formula, the particular value of z that makes the remainder formula true depends on the values of x and n. If $R_n \to 0$, then the next theorem tells us that the Taylor series for f actually converges to $f(x)$ for all x in I.

THEOREM 9.23   Convergence of Taylor Series

If $\lim\limits_{n \to \infty} R_n = 0$ for all x in the interval I, then the Taylor series for f converges and equals $f(x)$,

$$f(x) = \sum_{n=0}^{\infty} \frac{f^{(n)}(c)}{n!}(x - c)^n.$$

Proof For a Taylor series, the nth partial sum coincides with the nth Taylor polynomial. That is, $S_n(x) = P_n(x)$. Moreover, because

$$P_n(x) = f(x) - R_n(x)$$

it follows that

$$\begin{aligned} \lim_{n \to \infty} S_n(x) &= \lim_{n \to \infty} P_n(x) \\ &= \lim_{n \to \infty} [f(x) - R_n(x)] \\ &= f(x) - \lim_{n \to \infty} R_n(x). \end{aligned}$$

So, for a given x, the Taylor series (the sequence of partial sums) converges to $f(x)$ if and only if $R_n(x) \to 0$ as $n \to \infty$. ■

Stated another way, Theorem 9.23 says that a power series formed with Taylor coefficients

$$a_n = \frac{f^{(n)}(c)}{n!}$$

converges to the function from which it was derived at precisely those values for which the remainder approaches 0 as $n \to \infty$.

In Example 1, you derived the power series from the sine function and you also concluded that the series converges to some function on the entire real number line. In Example 2, you will see that the series actually converges to $\sin x$. The key observation is that although the value of z is not known, it is possible to obtain an upper bound for

$$\left| f^{(n+1)}(z) \right|.$$

EXAMPLE 2   A Convergent Maclaurin Series

Show that the Maclaurin series for

$$f(x) = \sin x$$

converges to $\sin x$ for all x.

Solution   Using the result in Example 1, you need to show that

$$\sin x = x - \frac{x^3}{3!} + \frac{x^5}{5!} - \frac{x^7}{7!} + \cdots + \frac{(-1)^n x^{2n+1}}{(2n+1)!} + \cdots$$

is true for all x. Because

$$f^{(n+1)}(x) = \pm \sin x$$

or

$$f^{(n+1)}(x) = \pm \cos x$$

you know that

$$\left| f^{(n+1)}(z) \right| \le 1$$

for every real number z. Therefore, for any fixed x, you can apply Taylor's Theorem (Theorem 9.19) to conclude that

$$0 \le \left| R_n(x) \right| = \left| \frac{f^{(n+1)}(z)}{(n+1)!} x^{n+1} \right| \le \frac{|x|^{n+1}}{(n+1)!}.$$

From the discussion in Section 9.1 regarding the relative rates of convergence of exponential and factorial sequences, it follows that for a fixed x,

$$\lim_{n \to \infty} \frac{|x|^{n+1}}{(n+1)!} = 0.$$

Finally, by the Squeeze Theorem, it follows that for all x, $R_n(x) \to 0$ as $n \to \infty$. So, by Theorem 9.23, the Maclaurin series for $\sin x$ converges to $\sin x$ for all x.

Figure 9.26 visually illustrates the convergence of the Maclaurin series for $\sin x$ by comparing the graphs of the Maclaurin polynomials P_1, P_3, P_5, and P_7 with the graph of the sine function. Notice that as the degree of the polynomial increases, its graph more closely resembles that of the sine function.

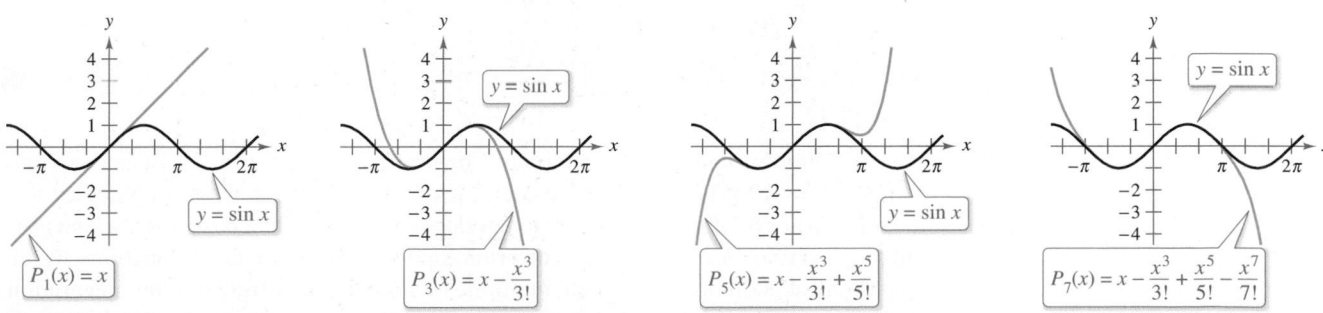

As n increases, the graph of P_n more closely resembles the sine function.
Figure 9.26

The guidelines for finding a Taylor series for f at c are summarized below.

GUIDELINES FOR FINDING A TAYLOR SERIES

1. Differentiate f with respect to x several times and evaluate each derivative at c.

$$f(c), f'(c), f''(c), f'''(c), \cdots, f^{(n)}(c), \cdots$$

Try to recognize a pattern in these numbers.

2. Use the sequence developed in the first step to form the Taylor coefficients $a_n = f^{(n)}(c)/n!$ and determine the interval of convergence for the resulting power series

$$f(c) + f'(c)(x - c) + \frac{f''(c)}{2!}(x - c)^2 + \cdots + \frac{f^{(n)}(c)}{n!}(x - c)^n + \cdots.$$

3. Within this interval of convergence, determine whether the series converges to $f(x)$.

▷

· · REMARK When you have difficulty recognizing a pattern, remember that you can use Theorem 9.22 to find the Taylor series. Also, you can try using the coefficients of a known Taylor or Maclaurin series, as shown in Example 3.

The direct determination of Taylor or Maclaurin coefficients using successive differentiation can be difficult, and the next example illustrates a shortcut for finding the coefficients indirectly—using the coefficients of a known Taylor or Maclaurin series.

EXAMPLE 3 **Maclaurin Series for a Composite Function**

Find the Maclaurin series for

$$f(x) = \sin x^2.$$

Solution To find the coefficients for this Maclaurin series directly, you must calculate successive derivatives of $f(x) = \sin x^2$. By calculating just the first two,

$$f'(x) = 2x \cos x^2$$

and

$$f''(x) = -4x^2 \sin x^2 + 2 \cos x^2$$

you can see that this task would be quite cumbersome. Fortunately, there is an alternative. First, consider the Maclaurin series for $\sin x$ found in Example 1.

$$g(x) = \sin x$$

$$= x - \frac{x^3}{3!} + \frac{x^5}{5!} - \frac{x^7}{7!} + \cdots$$

Now, because $\sin x^2 = g(x^2)$, you can substitute x^2 for x in the series for $\sin x$ to obtain

$$\sin x^2 = g(x^2)$$

$$= x^2 - \frac{x^6}{3!} + \frac{x^{10}}{5!} - \frac{x^{14}}{7!} + \cdots.$$ ■

Be sure to understand the point illustrated in Example 3. Because direct computation of Taylor or Maclaurin coefficients can be tedious, the most practical way to find a Taylor or Maclaurin series is to develop power series for a *basic list* of elementary functions. From this list, you can determine power series for other functions by the operations of addition, subtraction, multiplication, division, differentiation, integration, and composition with known power series.

Binomial Series

Before presenting the basic list for elementary functions, you will develop one more series—for a function of the form $f(x) = (1 + x)^k$. This produces the **binomial series.**

EXAMPLE 4 Binomial Series

Find the Maclaurin series for $f(x) = (1 + x)^k$ and determine its radius of convergence. Assume that k is not a positive integer and $k \neq 0$.

Solution By successive differentiation, you have

$$f(x) = (1 + x)^k \qquad\qquad f(0) = 1$$
$$f'(x) = k(1 + x)^{k-1} \qquad\qquad f'(0) = k$$
$$f''(x) = k(k - 1)(1 + x)^{k-2} \qquad\qquad f''(0) = k(k - 1)$$
$$f'''(x) = k(k - 1)(k - 2)(1 + x)^{k-3} \qquad f'''(0) = k(k - 1)(k - 2)$$
$$\vdots \qquad\qquad\qquad\qquad \vdots$$
$$f^{(n)}(x) = k \cdots (k - n + 1)(1 + x)^{k-n} \qquad f^{(n)}(0) = k(k - 1) \cdots (k - n + 1)$$

which produces the series

$$1 + kx + \frac{k(k - 1)x^2}{2} + \cdots + \frac{k(k - 1) \cdots (k - n + 1)x^n}{n!} + \cdots.$$

By the Ratio Test, you can conclude that the radius of convergence is $R = 1$. So, the series converges to some function in the interval $(-1, 1)$. ■

Note that Example 4 shows that the Taylor series for $(1 + x)^k$ converges to some function in the interval $(-1, 1)$. However, the example does not show that the series actually converges to $(1 + x)^k$. To do this, you could show that the remainder $R_n(x)$ converges to 0, as illustrated in Example 2. You now have enough information to find a binomial series for a function, as shown in the next example.

EXAMPLE 5 Finding a Binomial Series

Find the power series for

$$f(x) = \sqrt[3]{1 + x}.$$

Solution Using the binomial series

$$(1 + x)^k = 1 + kx + \frac{k(k - 1)x^2}{2!} + \frac{k(k - 1)(k - 2)x^3}{3!} + \cdots$$

let $k = \frac{1}{3}$ and write

$$(1 + x)^{1/3} = 1 + \frac{x}{3} - \frac{2x^2}{3^2 2!} + \frac{2 \cdot 5 x^3}{3^3 3!} - \frac{2 \cdot 5 \cdot 8 x^4}{3^4 4!} + \cdots. \qquad ■$$

▷ **TECHNOLOGY** Use a graphing utility to confirm the result in Example 5. When you graph the functions

$$f(x) = (1 + x)^{1/3}$$

and

$$P_4(x) = 1 + \frac{x}{3} - \frac{x^2}{9} + \frac{5x^3}{81} - \frac{10x^4}{243}$$

in the same viewing window, you should obtain the result shown in Figure 9.27.

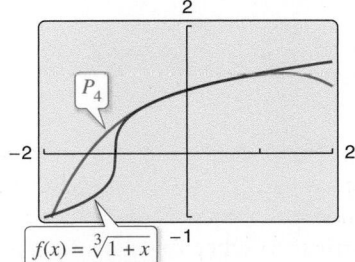

$f(x) = \sqrt[3]{1 + x}$

Figure 9.27

Deriving Taylor Series from a Basic List

The list below provides the power series for several elementary functions with the corresponding intervals of convergence.

POWER SERIES FOR ELEMENTARY FUNCTIONS

Function	Interval of Convergence

$$\frac{1}{x} = 1 - (x - 1) + (x - 1)^2 - (x - 1)^3 + (x - 1)^4 - \cdots + (-1)^n(x - 1)^n + \cdots \qquad 0 < x < 2$$

$$\frac{1}{1 + x} = 1 - x + x^2 - x^3 + x^4 - x^5 + \cdots + (-1)^n x^n + \cdots \qquad -1 < x < 1$$

$$\ln x = (x - 1) - \frac{(x - 1)^2}{2} + \frac{(x - 1)^3}{3} - \frac{(x - 1)^4}{4} + \cdots + \frac{(-1)^{n-1}(x - 1)^n}{n} + \cdots \qquad 0 < x \le 2$$

$$e^x = 1 + x + \frac{x^2}{2!} + \frac{x^3}{3!} + \frac{x^4}{4!} + \frac{x^5}{5!} + \cdots + \frac{x^n}{n!} + \cdots \qquad -\infty < x < \infty$$

$$\sin x = x - \frac{x^3}{3!} + \frac{x^5}{5!} - \frac{x^7}{7!} + \frac{x^9}{9!} - \cdots + \frac{(-1)^n x^{2n+1}}{(2n + 1)!} + \cdots \qquad -\infty < x < \infty$$

$$\cos x = 1 - \frac{x^2}{2!} + \frac{x^4}{4!} - \frac{x^6}{6!} + \frac{x^8}{8!} - \cdots + \frac{(-1)^n x^{2n}}{(2n)!} + \cdots \qquad -\infty < x < \infty$$

$$\arctan x = x - \frac{x^3}{3} + \frac{x^5}{5} - \frac{x^7}{7} + \frac{x^9}{9} - \cdots + \frac{(-1)^n x^{2n+1}}{2n + 1} + \cdots \qquad -1 \le x \le 1$$

$$\arcsin x = x + \frac{x^3}{2 \cdot 3} + \frac{1 \cdot 3x^5}{2 \cdot 4 \cdot 5} + \frac{1 \cdot 3 \cdot 5x^7}{2 \cdot 4 \cdot 6 \cdot 7} + \cdots + \frac{(2n)! x^{2n+1}}{(2^n n!)^2 (2n + 1)} + \cdots \qquad -1 \le x \le 1$$

$$(1 + x)^k = 1 + kx + \frac{k(k - 1)x^2}{2!} + \frac{k(k - 1)(k - 2)x^3}{3!} + \cdots + \frac{k(k - 1) \cdots (k - n + 1)x^n}{n!} + \cdots \qquad -1 < x < 1^*$$

---

* The convergence at $x = \pm 1$ depends on the value of k.

Note that the binomial series is valid for noninteger values of k. Also, when k is a positive integer, the binomial series reduces to a simple binomial expansion.

EXAMPLE 6   **Deriving a Power Series from a Basic List**

Find the power series for

$$f(x) = \cos \sqrt{x}.$$

Solution   Using the power series

$$\cos x = 1 - \frac{x^2}{2!} + \frac{x^4}{4!} - \frac{x^6}{6!} + \frac{x^8}{8!} - \cdots$$

you can replace x by

$$\sqrt{x}$$

to obtain the series

$$\cos \sqrt{x} = 1 - \frac{x}{2!} + \frac{x^2}{4!} - \frac{x^3}{6!} + \frac{x^4}{8!} - \cdots.$$

This series converges for all x in the domain of $\cos \sqrt{x}$—that is, for $x \ge 0$.

Power series can be multiplied and divided like polynomials. After finding the first few terms of the product (or quotient), you may be able to recognize a pattern.

EXAMPLE 7 **Multiplication of Power Series**

Find the first three nonzero terms in the Maclaurin series $e^x \arctan x$.

Solution Using the Maclaurin series for e^x and $\arctan x$ in the table, you have

$$e^x \arctan x = \left(1 + x + \frac{x^2}{2!} + \frac{x^3}{3!} + \frac{x^4}{4!} + \cdots\right)\left(x - \frac{x^3}{3} + \frac{x^5}{5} - \cdots\right).$$

Multiply these expressions and collect like terms as you would in multiplying polynomials.

$$1 + x + \frac{1}{2}x^2 + \frac{1}{6}x^3 + \frac{1}{24}x^4 + \cdots$$

$$x \qquad\qquad - \frac{1}{3}x^3 \qquad\qquad + \frac{1}{5}x^5 - \cdots$$

$$\overline{}$$

$$x + \quad x^2 + \frac{1}{2}x^3 + \frac{1}{6}x^4 + \frac{1}{24}x^5 + \cdots$$

$$- \frac{1}{3}x^3 - \frac{1}{3}x^4 - \frac{1}{6}x^5 - \cdots$$

$$+ \frac{1}{5}x^5 + \cdots$$

$$\overline{}$$

$$x + \quad x^2 + \frac{1}{6}x^3 - \frac{1}{6}x^4 + \frac{3}{40}x^5 + \cdots$$

So, $e^x \arctan x = x + x^2 + \frac{1}{6}x^3 + \cdots$.

EXAMPLE 8 **Division of Power Series**

Find the first three nonzero terms in the Maclaurin series $\tan x$.

Solution Using the Maclaurin series for $\sin x$ and $\cos x$ in the table, you have

$$\tan x = \frac{\sin x}{\cos x} = \frac{x - \dfrac{x^3}{3!} + \dfrac{x^5}{5!} - \cdots}{1 - \dfrac{x^2}{2!} + \dfrac{x^4}{4!} - \cdots}.$$

Divide using long division.

$$x + \frac{1}{3}x^3 + \frac{2}{15}x^5 + \cdots$$

$$1 - \frac{1}{2}x^2 + \frac{1}{24}x^4 - \cdots \overline{)\, x - \frac{1}{6}x^3 + \frac{1}{120}x^5 - \cdots}$$

$$x - \frac{1}{2}x^3 + \frac{1}{24}x^5 - \cdots$$

$$\overline{}$$

$$\frac{1}{3}x^3 - \frac{1}{30}x^5 + \cdots$$

$$\frac{1}{3}x^3 - \frac{1}{6}x^5 + \cdots$$

$$\overline{}$$

$$\frac{2}{15}x^5 + \cdots$$

So, $\tan x = x + \frac{1}{3}x^3 + \frac{2}{15}x^5 + \cdots$.

EXAMPLE 9 **A Power Series for sin² x**

Find the power series for

$$f(x) = \sin^2 x.$$

Solution Consider rewriting $\sin^2 x$ as

$$\sin^2 x = \frac{1 - \cos 2x}{2} = \frac{1}{2} - \frac{1}{2}\cos 2x.$$

Now, use the series for $\cos x$.

$$\cos x = 1 - \frac{x^2}{2!} + \frac{x^4}{4!} - \frac{x^6}{6!} + \frac{x^8}{8!} - \cdots$$

$$\cos 2x = 1 - \frac{2^2}{2!}x^2 + \frac{2^4}{4!}x^4 - \frac{2^6}{6!}x^6 + \frac{2^8}{8!}x^8 - \cdots$$

$$-\frac{1}{2}\cos 2x = -\frac{1}{2} + \frac{2}{2!}x^2 - \frac{2^3}{4!}x^4 + \frac{2^5}{6!}x^6 - \frac{2^7}{8!}x^8 + \cdots$$

$$\frac{1}{2} - \frac{1}{2}\cos 2x = \frac{1}{2} - \frac{1}{2} + \frac{2}{2!}x^2 - \frac{2^3}{4!}x^4 + \frac{2^5}{6!}x^6 - \frac{2^7}{8!}x^8 + \cdots$$

So, the series for $f(x) = \sin^2 x$ is

$$\sin^2 x = \frac{2}{2!}x^2 - \frac{2^3}{4!}x^4 + \frac{2^5}{6!}x^6 - \frac{2^7}{8!}x^8 + \cdots.$$

This series converges for $-\infty < x < \infty$. ∎

As mentioned in the preceding section, power series can be used to obtain tables of values of transcendental functions. They are also useful for estimating the values of definite integrals for which antiderivatives cannot be found. The next example demonstrates this use.

EXAMPLE 10 **Power Series Approximation of a Definite Integral**

⋅⋅⋅▷ *See LarsonCalculus.com for an interactive version of this type of example.*

Use a power series to approximate

$$\int_0^1 e^{-x^2}\, dx$$

with an error of less than 0.01.

Solution Replacing x with $-x^2$ in the series for e^x produces the following.

$$e^{-x^2} = 1 - x^2 + \frac{x^4}{2!} - \frac{x^6}{3!} + \frac{x^8}{4!} - \cdots$$

$$\int_0^1 e^{-x^2}\, dx = \left[x - \frac{x^3}{3} + \frac{x^5}{5 \cdot 2!} - \frac{x^7}{7 \cdot 3!} + \frac{x^9}{9 \cdot 4!} - \cdots \right]_0^1$$

$$= 1 - \frac{1}{3} + \frac{1}{10} - \frac{1}{42} + \frac{1}{216} - \cdots$$

Summing the first four terms, you have

$$\int_0^1 e^{-x^2}\, dx \approx 0.74$$

which, by the Alternating Series Test, has an error of less than $\frac{1}{216} \approx 0.005$. ∎

9.10 Exercises

See CalcChat.com for tutorial help and worked-out solutions to odd-numbered exercises.

CONCEPT CHECK

1. Convergence of a Taylor Series Explain how to determine whether a Taylor series for a function f converges to f.

2. Binomial Series The binomial series is used to represent a function of what form? What is the radius of convergence for the binomial series?

3. Power Series How can you multiply and divide power series?

4. Finding a Taylor Series Explain how to use the series $g(x) = e^x = \sum_{n=0}^{\infty} \frac{x^n}{n!}$ to find the series for $f(x) = x^2 e^{-3x}$. Do not find the series.

 **Finding a Taylor Series** In Exercises 5–16, use the definition of Taylor series to find the Taylor series, centered at c, for the function.

5. $f(x) = e^{2x}, \quad c = 0$

6. $f(x) = e^{-4x}, \quad c = 0$

7. $f(x) = \cos x, \quad c = \dfrac{\pi}{4}$

8. $f(x) = \sin x, \quad c = \dfrac{\pi}{4}$

9. $f(x) = \dfrac{1}{x}, \quad c = 1$

10. $f(x) = \dfrac{1}{1-x}, \quad c = 2$

11. $f(x) = \ln x, \quad c = 1$

12. $f(x) = e^x, \quad c = 1$

13. $f(x) = \sin 3x, \quad c = 0$

14. $f(x) = \ln(x^2 + 1), \quad c = 0$

15. $f(x) = \sec x, \quad c = 0$ (first three nonzero terms)

16. $f(x) = \tan x, \quad c = 0$ (first three nonzero terms)

 **Proof** In Exercises 17–20, prove that the Maclaurin series for the function converges to the function for all x.

17. $f(x) = \cos x$

18. $f(x) = e^{-2x}$

19. $f(x) = \sinh x$

20. $f(x) = \cosh x$

 **Using a Binomial Series** In Exercises 21–26, use the binomial series to find the Maclaurin series for the function.

21. $f(x) = \dfrac{1}{\sqrt{1-x}}$

22. $f(x) = \dfrac{1}{(1+x)^4}$

23. $f(x) = \dfrac{1}{\sqrt{1-x^2}}$

24. $f(x) = \dfrac{1}{(2+x)^3}$

25. $f(x) = \sqrt[4]{1+x}$

26. $f(x) = \sqrt{1+x^3}$

 **Finding a Maclaurin Series** In Exercises 27–40, find the Maclaurin series for the function. Use the table of power series for elementary functions on page 674.

27. $f(x) = e^{x^2/2}$

28. $g(x) = e^{-x/3}$

29. $f(x) = \ln(1+x)$

30. $f(x) = \ln(1+x^3)$

31. $f(x) = \cos 4x$

32. $f(x) = \sin \pi x$

33. $g(x) = \arctan 5x$

34. $f(x) = \arcsin \pi x$

35. $f(x) = \cos x^{3/2}$

36. $g(x) = 2 \sin x^3$

37. $f(x) = \frac{1}{2}(e^x - e^{-x}) = \sinh x$

38. $f(x) = e^x + e^{-x} - 2 \cosh x$

39. $f(x) = \cos^2 x$

40. $f(x) = \sinh^{-1} x = \ln\left(x + \sqrt{x^2 + 1}\right)$

$\left(\text{\textit{Hint:} Integrate the series for } \dfrac{1}{\sqrt{x^2 + 1}}.\right)$

Verifying a Formula In Exercises 41 and 42, use a power series and the fact that $i^2 = -1$ to verify the formula.

41. $g(x) = \dfrac{1}{2i}(e^{ix} - e^{-ix}) = \sin x$

42. $g(x) = \dfrac{1}{2}(e^{ix} + e^{-ix}) = \cos x$

 **Finding a Maclaurin Series** In Exercises 43–46, find the Maclaurin series for the function.

43. $f(x) = x \sin x$

44. $h(x) = x \cos x$

45. $g(x) = \begin{cases} \dfrac{\sin x}{x}, & x \neq 0 \\ 1, & x = 0 \end{cases}$

46. $f(x) = \begin{cases} \dfrac{\arcsin x}{x}, & x \neq 0 \\ 1, & x = 0 \end{cases}$

 **Finding Terms of a Maclaurin Series** In Exercises 47–52, find the first four nonzero terms of the Maclaurin series for the function by multiplying or dividing the appropriate power series. Use the table of power series for elementary functions on page 674. Use a graphing utility to graph the function and its corresponding polynomial approximation.

47. $f(x) = e^x \sin x$

48. $g(x) = e^x \cos x$

49. $h(x) = (\cos x) \ln(1 + x)$

50. $f(x) = e^x \ln(1 + x)$

51. $g(x) = \dfrac{\sin x}{1 + x}$

52. $f(x) = \dfrac{e^x}{1 + x}$

Finding a Maclaurin Series In Exercises 53 and 54, find a Maclaurin series for $f(x)$.

53. $f(x) = \displaystyle\int_0^x (e^{-t^2} - 1) \, dt$

54. $f(x) = \displaystyle\int_0^x \sqrt{1 + t^3} \, dt$

 **Verifying a Sum** In Exercises 55–58, verify the sum. Then use a graphing utility to approximate the sum with an error of less than 0.0001.

55. $\displaystyle\sum_{n=1}^{\infty} (-1)^{n+1} \frac{1}{n} = \ln 2$

56. $\displaystyle\sum_{n=0}^{\infty} (-1)^n \left[\frac{1}{(2n+1)!} \right] = \sin 1$

57. $\displaystyle\sum_{n=0}^{\infty} \frac{2^n}{n!} = e^2$

58. $\displaystyle\sum_{n=1}^{\infty} (-1)^{n-1} \left(\frac{1}{n!} \right) = \frac{e-1}{e}$

Finding a Limit In Exercises 59–62, use the series representation of the function f to find $\lim_{x \to 0} f(x)$, if it exists.

59. $f(x) = \dfrac{1 - \cos x}{x}$

60. $f(x) = \dfrac{\sin x}{x}$

61. $f(x) = \dfrac{e^x - 1}{x}$

62. $f(x) = \dfrac{\ln(x + 1)}{x}$

Approximating an Integral In Exercises 63–70, use a power series to approximate the value of the definite integral with an error of less than 0.0001. (In Exercises 65 and 67, assume that the integrand is defined as 1 when $x = 0$.)

63. $\displaystyle\int_0^1 e^{-x^3} \, dx$

64. $\displaystyle\int_0^{1/4} x \ln(x + 1) \, dx$

65. $\displaystyle\int_0^1 \frac{\sin x}{x} \, dx$

66. $\displaystyle\int_0^1 \cos x^2 \, dx$

67. $\displaystyle\int_0^{1/2} \frac{\arctan x}{x} \, dx$

68. $\displaystyle\int_0^{1/2} \arctan x^2 \, dx$

69. $\displaystyle\int_{0.1}^{0.3} \sqrt{1 + x^3} \, dx$

70. $\displaystyle\int_0^{0.2} \sqrt{1 + x^2} \, dx$

Area In Exercises 71 and 72, use a power series to approximate the area of the region with an error of less than 0.0001. Use a graphing utility to verify the result.

71. $\displaystyle\int_0^{\pi/2} \sqrt{x} \cos x \, dx$

72. $\displaystyle\int_{0.5}^1 \cos \sqrt{x} \, dx$

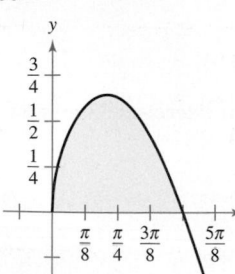

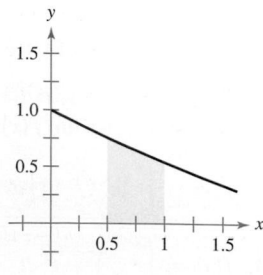

Probability In Exercises 73 and 74, approximate the probability with an error of less than 0.0001, where the probability is given by

$$P(a < x < b) = \frac{1}{\sqrt{2\pi}} \int_a^b e^{-x^2/2} \, dx.$$

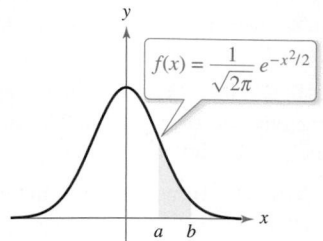

73. $P(0 < x < 1)$

74. $P(1 < x < 2)$

EXPLORING CONCEPTS

75. Comparing Methods Describe three ways to find the Maclaurin series for $\cos^2 x$. Show that each method produces the same first three terms.

76. Maclaurin Series Explain how to use the power series for $f(x) = \arctan x$ to find the Maclaurin series for

$$g(x) = \frac{1}{1 + x^2}.$$

What is another way to find the Maclaurin series for g using a power series for an elementary function?

77. Finding a Function Which function has the Maclaurin series

$$\sum_{n=0}^{\infty} \frac{(-1)^n (x + 3)^{2n+1}}{2^2 (2n + 1)!}?$$

Explain your reasoning.

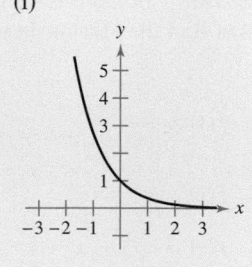 **78.** **HOW DO YOU SEE IT?** Identify the function represented by each power series and match the function with its graph. [The graphs are labeled (i) and (ii).]

(i)

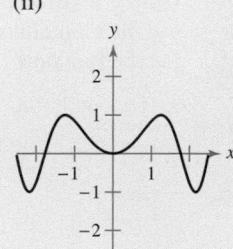

(ii)

(a) $\displaystyle\sum_{n=0}^{\infty} \frac{(-1)^n x^{4n+2}}{(2n+1)!}$

(b) $\displaystyle\sum_{n=0}^{\infty} \frac{(-1)^n x^n}{n!}$

79. Projectile Motion A projectile fired from the ground follows the trajectory given by

$$y = \left(\tan \theta + \frac{g}{kv_0 \cos \theta}\right) x + \frac{g}{k^2} \ln\left(1 - \frac{kx}{v_0 \cos \theta}\right)$$

where v_0 is the initial speed, θ is the angle of projection, g is the acceleration due to gravity, and k is the drag factor caused by air resistance. Using the power series representation

$$\ln(1 + x) = x - \frac{x^2}{2} + \frac{x^3}{3} - \frac{x^4}{4} + \cdots, \quad -1 < x < 1$$

verify that the trajectory can be rewritten as

$$y = (\tan \theta)x - \frac{gx^2}{2v_0^2 \cos^2 \theta} - \frac{kgx^3}{3v_0^3 \cos^3 \theta} - \frac{k^2 gx^4}{4v_0^4 \cos^4 \theta} - \cdots.$$

80. Projectile Motion

Use the result of Exercise 79 to determine the series for the path of a projectile launched from ground level at an angle of $\theta = 60°$, with an initial speed of $v_0 = 64$ feet per second and a drag factor of $k = \frac{1}{16}$.

81. Investigation Consider the function f defined by

$$f(x) = \begin{cases} e^{-1/x^2}, & x \neq 0 \\ 0, & x = 0. \end{cases}$$

(a) Sketch a graph of the function.

(b) Use the alternative form of the definition of the derivative (Section 3.1) and L'Hôpital's Rule to show that $f'(0) = 0$. [By continuing this process, it can be shown that $f^{(n)}(0) = 0$ for $n > 1$.]

(c) Using the result in part (b), find the Maclaurin series for f. Does the series converge to f?

82. Investigation

(a) Find the power series centered at 0 for the function

$$f(x) = \frac{\ln(x^2 + 1)}{x^2}.$$

(b) Use a graphing utility to graph f and the eighth-degree Taylor polynomial $P_8(x)$ for f.

(c) Use a graphing utility to complete the table, where

$$F(x) = \int_0^x \frac{\ln(t^2 + 1)}{t^2}\, dt \quad \text{and} \quad G(x) = \int_0^x P_8(t)\, dt.$$

x	0.25	0.50	0.75	1.00	1.50	2.00
$F(x)$						
$G(x)$						

(d) Describe the relationship between the graphs of f and P_8 and the results given in the table in part (c).

83. Proof Prove that $\lim_{n \to \infty} \dfrac{x^n}{n!} = 0$ for any real x.

84. Finding a Maclaurin Series Find the Maclaurin series for

$$f(x) = \ln \frac{1 + x}{1 - x}$$

and determine its radius of convergence. Use the first four terms of the series to approximate $\ln 3$.

Evaluating a Binomial Coefficient In Exercises 85–88, evaluate the binomial coefficient using the formula

$$\binom{k}{n} = \frac{k(k - 1)(k - 2)(k - 3) \cdots (k - n + 1)}{n!}$$

where k is a real number, n is a positive integer, and $\binom{k}{0} = 1$.

85. $\binom{6}{3}$ **86.** $\binom{0.25}{2}$

87. $\binom{-0.8}{5}$ **88.** $\binom{-5}{6}$

89. Writing a Power Series Write the power series for $(1 + x)^k$ in terms of binomial coefficients.

90. Proof Prove that the Taylor series for e^x, centered at $x = a$, is given by

$$e^a\left[1 + (x - a) + \frac{(x - a)^2}{2!} + \cdots\right].$$

91. Proof Prove that e is irrational. [*Hint:* Assume that $e = p/q$ is rational (p and q are integers) and consider

$$e = 1 + 1 + \frac{1}{2!} + \cdots + \frac{1}{n!} + \cdots.\Bigg]$$

92. Using Fibonacci Numbers Show that the Maclaurin series for the function

$$g(x) = \frac{x}{1 - x - x^2}$$

is

$$\sum_{n=1}^{\infty} F_n x^n$$

where F_n is the nth Fibonacci number with $F_1 = F_2 = 1$ and $F_n = F_{n-2} + F_{n-1}$, for $n \geq 3$. (*Hint:* Write

$$\frac{x}{1 - x - x^2} = a_0 + a_1 x + a_2 x^2 + \cdots$$

and multiply each side of this equation by $1 - x - x^2$.)

PUTNAM EXAM CHALLENGE

93. Assume that $|f(x)| \leq 1$ and $|f''(x)| \leq 1$ for all x on an interval of length at least 2. Show that $|f'(x)| \leq 2$ on the interval.

Review Exercises

See **CalcChat.com** for tutorial help and worked-out solutions to odd-numbered exercises.

Writing the Terms of a Sequence In Exercises 1–4, write the first five terms of the sequence with the given nth term.

1. $a_n = 6^n - 2$

2. $a_n = \dfrac{5^{n-1}}{n!}$

3. $a_n = \left(-\dfrac{1}{4}\right)^n$

4. $a_n = \dfrac{2n}{n+5}$

Matching In Exercises 5–8, match the sequence with the given nth term with its graph. [The graphs are labeled (a), (b), (c), and (d).]

(a)

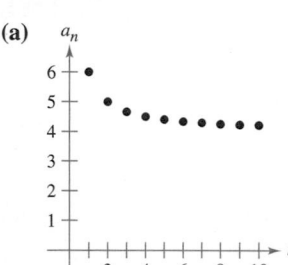

(b)

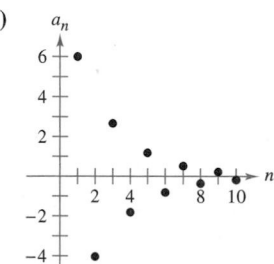

(c)

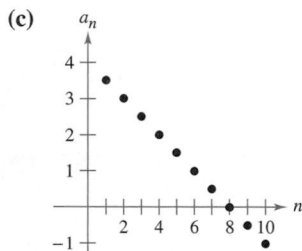

(d)

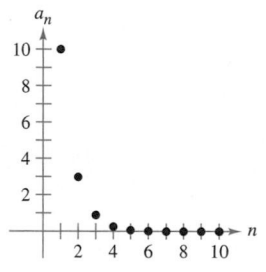

5. $a_n = 4 + \dfrac{2}{n}$

6. $a_n = 4 - \dfrac{n}{2}$

7. $a_n = 10(0.3)^{n-1}$

8. $a_n = 6\left(-\dfrac{2}{3}\right)^{n-1}$

Finding the Limit of a Sequence In Exercises 9 and 10, use a graphing utility to graph the first 10 terms of the sequence with the given nth term. Use the graph to make an inference about the convergence or divergence of the sequence. Verify your inference analytically and, if the sequence converges, find its limit.

9. $a_n = \dfrac{5n + 2}{n}$

10. $a_n = \cos \dfrac{n\pi}{3}$

Determining Convergence or Divergence In Exercises 11–18, determine the convergence or divergence of the sequence with the given nth term. If the sequence converges, find its limit.

11. $a_n = \dfrac{1}{\sqrt{n}}$

12. $a_n = \dfrac{n}{n^2 + 1}$

13. $a_n = \left(\dfrac{2}{5}\right)^n + 5$

14. $a_n = \dfrac{2n^3 - 1}{3n + 4}$

15. $a_n = \dfrac{(4n)!}{(4n - 1)!}$

16. $a_n = \dfrac{n}{\ln n}$

17. $a_n = \dfrac{e^{2n}}{\ln n}$

18. $a_n = \dfrac{\sin \sqrt{n}}{\sqrt{n}}$

Finding the nth Term of a Sequence In Exercises 19–22, write an expression for the nth term of the sequence and then determine whether the sequence you have chosen converges or diverges. (There is more than one correct answer.)

19. $3, 8, 13, 18, 23, \ldots$

20. $-5, -2, 3, 10, 19, \ldots$

21. $\dfrac{1}{2}, \dfrac{1}{3}, \dfrac{1}{7}, \dfrac{1}{25}, \dfrac{1}{121}, \ldots$

22. $\dfrac{1}{2}, \dfrac{2}{5}, \dfrac{3}{10}, \dfrac{4}{17}, \ldots$

Monotonic and Bounded Sequences In Exercises 23 and 24, determine whether the sequence with the given nth term is monotonic and whether it is bounded. Use a graphing utility to confirm your results.

23. $a_n = 3 - \dfrac{1}{2n}$

24. $a_n = \left(\dfrac{4}{3}\right)^n$

25. Compound Interest A deposit of $8000 is made in an account that earns 5% interest compounded quarterly. The balance in the account after n quarters is

$$A_n = 8000(1.0125)^n, \quad n = 1, 2, 3, \ldots .$$

(a) Compute the first eight terms of the sequence $\{A_n\}$.

(b) Find the balance in the account after 10 years by computing the 40th term of the sequence.

26. Depreciation A company buys a machine for $175,000. During the next 5 years, the machine will depreciate at a rate of 30% per year. (That is, at the end of each year, the depreciated value will be 70% of what it was at the beginning of the year.)

(a) Write an expression for the value of the machine after n years.

(b) Compute the depreciated values of the machine for the first 5 years.

Finding Partial Sums In Exercises 27 and 28, find the sequence of partial sums $S_1, S_2, S_3, S_4,$ and S_5.

27. $3 + \dfrac{3}{2} + 1 + \dfrac{3}{4} + \dfrac{3}{5} + \cdots$

28. $-7 + 1 - \dfrac{1}{7} + \dfrac{1}{49} - \dfrac{1}{343} + \cdots$

Numerical, Graphical, and Analytic Analysis In Exercises 29 and 30, (a) use a graphing utility to find the indicated partial sum S_n and complete the table, and (b) use a graphing utility to graph the first 10 terms of the sequence of partial sums.

n	5	10	15	20	25
S_n					

29. $\displaystyle\sum_{n=1}^{\infty} \left(\dfrac{3}{2}\right)^{n-1}$

30. $\displaystyle\sum_{n=1}^{\infty} \dfrac{1}{n(n+1)}$

Finding the Sum of a Convergent Series   In Exercises 31–34, find the sum of the convergent series.

31. $\displaystyle\sum_{n=0}^{\infty} \left(\frac{2}{5}\right)^n$

32. $\displaystyle\sum_{n=0}^{\infty} \frac{3^{n+2}}{7^n}$

33. $\displaystyle\sum_{n=0}^{\infty} [(0.4)^n + (0.9)^n]$

34. $\displaystyle\sum_{n=0}^{\infty} \left[\left(\frac{3}{4}\right)^n - \frac{1}{(n+1)(n+2)}\right]$

Using a Geometric Series   In Exercises 35 and 36, (a) write the repeating decimal as a geometric series and (b) write the sum of the series as the ratio of two integers.

35. $0.\overline{09}$

36. $0.\overline{64}$

Using a Geometric Series or the nth-Term Test   In Exercises 37–40, use a geometric series or the nth-Term Test to determine the convergence or divergence of the series.

37. $\displaystyle\sum_{n=0}^{\infty} (1.67)^n$

38. $\displaystyle\sum_{n=0}^{\infty} 9^{-n}$

39. $\displaystyle\sum_{n=0}^{\infty} \frac{2n+1}{3n+2}$

40. $\displaystyle\sum_{n=1}^{\infty} \frac{5n! + 6}{n! + 1}$

41. Marketing   A manufacturer producing a new product estimates the annual sales to be 9600 units. Each year, 8% of the units that have been sold will become inoperative. So, 9600 units will be in use after 1 year, $[9600 + 0.92(9600)]$ units will be in use after 2 years, and so on. How many units will be in use after n years?

42. Distance   A ball is dropped from a height of 8 meters. Each time it drops h meters, it rebounds $0.7h$ meters. Find the total distance traveled by the ball.

Using the Integral Test or a p-Series   In Exercises 43–48, use the Integral Test or a p-series to determine the convergence or divergence of the series.

43. $\displaystyle\sum_{n=1}^{\infty} \frac{2}{6n+1}$

44. $\displaystyle\sum_{n=1}^{\infty} \frac{1}{\sqrt[4]{n^3}}$

45. $\displaystyle\sum_{n=1}^{\infty} \frac{1}{n^{5/2}}$

46. $\displaystyle\sum_{n=1}^{\infty} \frac{1}{5^n}$

47. $\displaystyle\sum_{n=1}^{\infty} \left(\frac{1}{n^2} - \frac{1}{n}\right)$

48. $\displaystyle\sum_{n=1}^{\infty} \frac{\ln n}{n^4}$

Using the Direct Comparison Test or the Limit Comparison Test   In Exercises 49–54, use the Direct Comparison Test or the Limit Comparison Test to determine the convergence or divergence of the series.

49. $\displaystyle\sum_{n=2}^{\infty} \frac{1}{\sqrt[3]{n}-1}$

50. $\displaystyle\sum_{n=0}^{\infty} \frac{7^n}{8^n+5}$

51. $\displaystyle\sum_{n=1}^{\infty} \frac{1}{\sqrt{n^3+2n}}$

52. $\displaystyle\sum_{n=1}^{\infty} \frac{n+1}{n(n+2)}$

53. $\displaystyle\sum_{n=1}^{\infty} \frac{1\cdot 3\cdot 5\cdots(2n-1)}{2\cdot 4\cdot 6\cdots(2n)}$

54. $\displaystyle\sum_{n=1}^{\infty} \frac{1}{3^n-5}$

Using the Alternating Series Test   In Exercises 55–60, use the Alternating Series Test, if applicable, to determine the convergence or divergence of the series.

55. $\displaystyle\sum_{n=1}^{\infty} \frac{(-1)^n}{n^5}$

56. $\displaystyle\sum_{n=1}^{\infty} \frac{(-1)^n(n+1)}{n^2+1}$

57. $\displaystyle\sum_{n=2}^{\infty} \frac{(-1)^n n}{n^2-3}$

58. $\displaystyle\sum_{n=4}^{\infty} \frac{(-1)^n n}{n-3}$

59. $\displaystyle\sum_{n=1}^{\infty} \frac{(-1)^{n+1}\sqrt{n}}{\sqrt[4]{n}+2}$

60. $\displaystyle\sum_{n=2}^{\infty} \frac{(-1)^n \ln n^3}{n}$

Finding the Number of Terms   In Exercises 61 and 62, use Theorem 9.15 to determine the number of terms required to approximate the sum of the series with an error of less than 0.0001.

61. $\displaystyle\sum_{n=1}^{\infty} \frac{(-1)^n}{n^4}$

62. $\displaystyle\sum_{n=1}^{\infty} \frac{(-1)^{n+1}}{3n^3-2}$

Using the Ratio Test or the Root Test   In Exercises 63–68, use the Ratio Test or the Root Test to determine the convergence or divergence of the series.

63. $\displaystyle\sum_{n=1}^{\infty} \left(\frac{3n-1}{2n+5}\right)^n$

64. $\displaystyle\sum_{n=1}^{\infty} \left(\frac{4n}{7n-1}\right)^n$

65. $\displaystyle\sum_{n=1}^{\infty} \frac{2^n}{n^3}$

66. $\displaystyle\sum_{n=0}^{\infty} \frac{7^n}{(2n+3)^n}$

67. $\displaystyle\sum_{n=1}^{\infty} \frac{n}{e^{n^2}}$

68. $\displaystyle\sum_{n=1}^{\infty} \frac{n!}{e^{2n}}$

Numerical, Graphical, and Analytic Analysis   In Exercises 69 and 70, (a) verify that the series converges, (b) use a graphing utility to find the indicated partial sum S_n and complete the table, (c) use a graphing utility to graph the first 10 terms of the sequence of partial sums, and (d) use the table to estimate the sum of the series.

n	5	10	15	20	25
S_n					

69. $\displaystyle\sum_{n=1}^{\infty} n\left(\frac{3}{5}\right)^n$

70. $\displaystyle\sum_{n=1}^{\infty} \frac{(-1)^{n-1}n}{n^3+5}$

Review   In Exercises 71–76, determine the convergence or divergence of the series using any appropriate test from this chapter. Identify the test used.

71. $\displaystyle\sum_{n=1}^{\infty} \frac{4}{n^{2\pi}}$

72. $\displaystyle\sum_{n=0}^{\infty} \frac{7^{n+1}}{8^n}$

73. $\displaystyle\sum_{n=1}^{\infty} \frac{5n^3+6}{7n^3+2n}$

74. $\displaystyle\sum_{n=1}^{\infty} e^{-n/3}$

75. $\displaystyle\sum_{n=1}^{\infty} \frac{10^n}{4+9^n}$

76. $\displaystyle\sum_{n=1}^{\infty} \frac{(n!)^n}{3^n}$

Finding a Maclaurin Polynomial   In Exercises 77 and 78, find the nth Maclaurin polynomial for the function.

77. $f(x) = e^{-2x}$, $n = 3$

78. $f(x) = \cos 3x$, $n = 4$

Finding a Taylor Polynomial In Exercises 79 and 80, find the third Taylor polynomial for the function, centered at c.

79. $f(x) = \dfrac{1}{x^3}, \quad c = 1$

80. $f(x) = \tan x, \quad c = -\dfrac{\pi}{4}$

Finding a Degree In Exercises 81 and 82, determine the degree of the Maclaurin polynomial required for the error in the approximation of the function at the indicated value of x to be less than 0.001.

81. $f(x) = \cos x, \quad$ approximate $f(0.75)$

82. $f(x) = e^x, \quad$ approximate $f(-0.25)$

Finding the Interval of Convergence In Exercises 83–88, find the interval of convergence of the power series. (Be sure to include a check for convergence at the endpoints of the interval.)

83. $\displaystyle\sum_{n=0}^{\infty} \left(\dfrac{x}{10}\right)^n$

84. $\displaystyle\sum_{n=0}^{\infty} (5x)^n$

85. $\displaystyle\sum_{n=0}^{\infty} \dfrac{(-1)^n (x-2)^n}{(n+1)^2}$

86. $\displaystyle\sum_{n=1}^{\infty} \dfrac{4^n (x-1)^n}{n}$

87. $\displaystyle\sum_{n=0}^{\infty} n!(x-2)^n$

88. $\displaystyle\sum_{n=0}^{\infty} \dfrac{(x-3)^n}{3^n}$

Finding Intervals of Convergence In Exercises 89 and 90, find the intervals of convergence of (a) $f(x)$, (b) $f'(x)$, (c) $f''(x)$, and (d) $\int f(x)\,dx$. (Be sure to include a check for convergence at the endpoints of the intervals.)

89. $f(x) = \displaystyle\sum_{n=0}^{\infty} \left(\dfrac{x}{5}\right)^n$

90. $f(x) = \displaystyle\sum_{n=1}^{\infty} \dfrac{(-1)^{n+1}(x-4)^n}{n}$

Differential Equation In Exercises 91 and 92, show that the function represented by the power series is a solution of the differential equation.

91. $y = \displaystyle\sum_{n=0}^{\infty} (-1)^n \dfrac{x^{2n}}{4^n (n!)^2}, \quad x^2 y'' + xy' + x^2 y = 0$

92. $y = \displaystyle\sum_{n=0}^{\infty} \dfrac{(-3)^n x^{2n}}{2^n n!}, \quad y'' + 3xy' + 3y = 0$

Finding a Geometric Power Series In Exercises 93 and 94, find a geometric power series for the function, centered at 0.

93. $g(x) = \dfrac{2}{3-x}$

94. $h(x) = \dfrac{3}{2+x}$

Finding a Power Series In Exercises 95 and 96, find a power series for the function, centered at c, and determine the interval of convergence.

95. $f(x) = \dfrac{6}{4-x}, \quad c = 1$

96. $f(x) = \dfrac{6x}{x^2 + 4x - 5}, \quad c = 0$

Finding the Sum of a Series In Exercises 97–102, find the sum of the convergent series by using a well-known function. Identify the function and explain how you obtained the sum.

97. $\displaystyle\sum_{n=1}^{\infty} (-1)^{n+1} \dfrac{1}{4^n n}$

98. $\displaystyle\sum_{n=1}^{\infty} (-1)^{n+1} \dfrac{1}{5^n n}$

99. $\displaystyle\sum_{n=0}^{\infty} \dfrac{1}{2^n n!}$

100. $\displaystyle\sum_{n=0}^{\infty} \dfrac{2^n}{3^n n!}$

101. $\displaystyle\sum_{n=0}^{\infty} (-1)^n \dfrac{2^{2n}}{3^{2n} (2n)!}$

102. $\displaystyle\sum_{n=0}^{\infty} (-1)^n \dfrac{1}{3^{2n+1}(2n+1)!}$

Finding a Taylor Series In Exercises 103–110, use the definition of Taylor series to find the Taylor series, centered at c, for the function.

103. $f(x) = \sin x, \quad c = \dfrac{3\pi}{4}$

104. $f(x) = \cos x, \quad c = -\dfrac{\pi}{4}$

105. $f(x) = 3^x, \quad c = 0$

106. $f(x) = \csc x, \quad c = \dfrac{\pi}{2}$ (first three nonzero terms)

107. $f(x) = \dfrac{1}{x}, \quad c = -1$

108. $f(x) = \sqrt{x}, \quad c = 4$

109. $g(x) = \sqrt[5]{1+x}, \quad c = 0$

110. $h(x) = \dfrac{1}{(1+x)^3}, \quad c = 0$

111. Forming Maclaurin Series Determine the first four terms of the Maclaurin series for e^{2x}

(a) by using the definition of Maclaurin series.

(b) by replacing x by $2x$ in the series for e^x.

(c) by multiplying the series for e^x by itself, because $e^{2x} = e^x \cdot e^x$.

112. Forming Maclaurin Series Determine the first four terms of the Maclaurin series for $\sin 2x$

(a) by using the definition of Maclaurin series.

(b) by replacing x by $2x$ in the series for $\sin 2x$.

(c) by multiplying 2 by the series for $\sin x$ by the series for $\cos x$, because $\sin 2x = 2 \sin x \cos x$.

Finding a Maclaurin Series In Exercises 113–116, find the Maclaurin series for the function. Use the table of power series for elementary functions on page 674.

113. $f(x) = e^{6x}$

114. $f(x) = \ln(x-1)$

115. $f(x) = \sin 5x$

116. $f(x) = \cos 3x$

Approximating an Integral In Exercises 117 and 118, use a power series to approximate the value of the definite integral with an error of less than 0.01.

117. $\displaystyle\int_0^{0.5} \cos x^3 \, dx$

118. $\displaystyle\int_0^1 e^{-x^4} \, dx$

P.S. Problem Solving

See **CalcChat.com** for tutorial help and worked-out solutions to odd-numbered exercises.

1. Cantor Set The **Cantor set** (Georg Cantor, 1845–1918) is a subset of the unit interval $[0, 1]$. To construct the Cantor set, first remove the middle third $\left(\frac{1}{3}, \frac{2}{3}\right)$ of the interval, leaving two line segments. For the second step, remove the middle third of each of the two remaining segments, leaving four line segments. Continue this procedure indefinitely, as shown in the figure. The Cantor set consists of all numbers in the unit interval $[0, 1]$ that still remain.

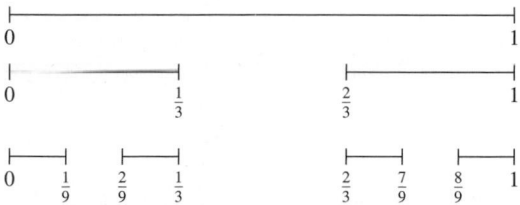

(a) Find the total length of all the line segments that are removed.

(b) Write down three numbers that are in the Cantor set.

(c) Let C_n denote the total length of the remaining line segments after n steps. Find $\lim\limits_{n \to \infty} C_n$.

2. Using Sequences

(a) Given that $\lim\limits_{n \to \infty} a_{2n} = L$ and $\lim\limits_{n \to \infty} a_{2n+1} = L$, show that $\{a_n\}$ is convergent and $\lim\limits_{n \to \infty} a_n = L$.

(b) Let $a_1 = 1$ and $a_{n+1} = 1 + \dfrac{1}{1 + a_n}$. Write out the first eight terms of $\{a_n\}$. Use part (a) to show that $\lim\limits_{n \to \infty} a_n = \sqrt{2}$.

This gives the **continued fraction expansion**

$$\sqrt{2} = 1 + \cfrac{1}{2 + \cfrac{1}{2 + \cdots}}.$$

3. Using a Series It can be shown that

$$\sum_{n=1}^{\infty} \frac{1}{n^2} = \frac{\pi^2}{6} \text{ [see Section 9.3, page 612]}.$$

Use this fact to show that $\displaystyle\sum_{n=1}^{\infty} \frac{1}{(2n-1)^2} = \frac{\pi^2}{8}$.

4. Finding a Limit Let T be an equilateral triangle with sides of length 1. Let a_n be the number of circles that can be packed tightly in n rows inside the triangle. For example, $a_1 = 1$, $a_2 = 3$, and $a_3 = 6$, as shown in the figure. Let A_n be the combined area of the a_n circles. Find $\lim\limits_{n \to \infty} A_n$.

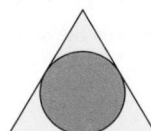 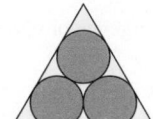 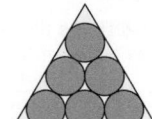

5. Using Center of Gravity Identical blocks of unit length are stacked on top of each other at the edge of a table. The center of gravity of the top block must lie over the block below it, the center of gravity of the top two blocks must lie over the block below them, and so on (see figure).

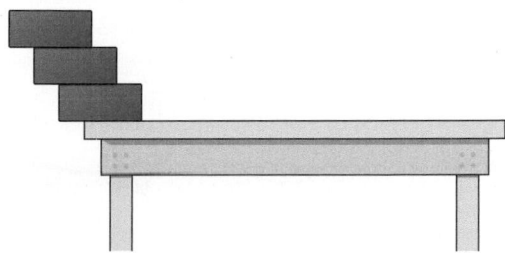

(a) When there are three blocks, show that it is possible to stack them so that the left edge of the top block extends $\frac{11}{12}$ unit beyond the edge of the table.

(b) Is it possible to stack the blocks so that the right edge of the top block extends beyond the edge of the table?

(c) How far beyond the table can the blocks be stacked?

6. Using Power Series

(a) Consider the power series

$$\sum_{n=0}^{\infty} a_n x^n = 1 + 2x + 3x^2 + x^3 + 2x^4 + 3x^5 + x^6 + \cdots$$

in which the coefficients $a_n = 1, 2, 3, 1, 2, 3, 1, \ldots$ are periodic of period $p = 3$. Find the radius of convergence and the sum of this power series.

(b) Consider a power series

$$\sum_{n=0}^{\infty} a_n x^n$$

in which the coefficients are periodic, $(a_{n+p} = a_p)$, and $a_n > 0$. Find the radius of convergence and the sum of this power series.

7. Finding Sums of Series

(a) Find a power series for the function

$$f(x) = xe^x$$

centered at 0. Use this representation to find the sum of the infinite series

$$\sum_{n=1}^{\infty} \frac{1}{n!(n+2)}.$$

(b) Differentiate the power series for $f(x) = xe^x$. Use the result to find the sum of the infinite series

$$\sum_{n=0}^{\infty} \frac{n+1}{n!}.$$

8. **Using the Alternating Series Test** The graph of the function

$$f(x) = \begin{cases} 1, & x = 0 \\ \dfrac{\sin x}{x}, & x > 0 \end{cases}$$

is shown below. Use the Alternating Series Test to show that the improper integral $\displaystyle\int_1^\infty f(x)\,dx$ converges.

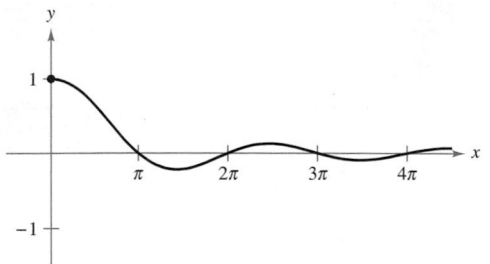

9. **Conditional and Absolute Convergence** For what values of the positive constants a and b does the following series converge absolutely? For what values does it converge conditionally?

$$a - \frac{b}{2} + \frac{a}{3} - \frac{b}{4} + \frac{a}{5} - \frac{b}{6} + \frac{a}{7} - \frac{b}{8} + \cdots$$

10. **Proof**

   (a) Consider the following sequence of numbers defined recursively.

   $$a_1 = 3$$
   $$a_2 = \sqrt{3}$$
   $$a_3 = \sqrt{3 + \sqrt{3}}$$
   $$\vdots$$
   $$a_{n+1} = \sqrt{3 + a_n}$$

   Write the decimal approximations for the first six terms of this sequence. Prove that the sequence converges and find its limit.

   (b) Consider the following sequence defined recursively by $a_1 = \sqrt{a}$ and $a_{n+1} = \sqrt{a + a_n}$, where $a > 2$.

   $$\sqrt{a},\quad \sqrt{a + \sqrt{a}},\quad \sqrt{a + \sqrt{a + \sqrt{a}}},\ldots$$

   Prove that this sequence converges and find its limit.

11. **Proof** Let $\{a_n\}$ be a sequence of positive numbers satisfying $\displaystyle\lim_{n\to\infty}(a_n)^{1/n} = L < \frac{1}{r}$, $r > 0$. Prove that the series $\displaystyle\sum_{n=1}^\infty a_n r^n$ converges.

12. **Using a Series** Consider the infinite series $\displaystyle\sum_{n=1}^\infty \frac{1}{2^{n+(-1)^n}}$.

   (a) Find the first five terms of the sequence of partial sums.

   (b) Show that the Ratio Test is inconclusive for this series.

   (c) Use the Root Test to determine the convergence or divergence of this series.

13. **Deriving Identities** Derive each identity using the appropriate geometric series.

   (a) $\dfrac{1}{0.99} = 1.01010101\ldots$

   (b) $\dfrac{1}{0.98} = 1.0204081632\ldots$

14. **Population** Consider an idealized population with the characteristic that each member of the population produces one offspring at the end of every time period. Each member has a life span of three time periods and the population begins with 10 newborn members. The following table shows the population during the first five time periods.

	Time Period				
Age Bracket	1	2	3	4	5
0–1	10	10	20	40	70
1–2		10	10	20	40
2–3			10	10	20
Total	10	20	40	70	130

The sequence for the total population has the property that

$$S_n = S_{n-1} + S_{n-2} + S_{n-3}, \quad n > 3.$$

Find the total population during each of the next five time periods.

15. **Spheres** Imagine you are stacking an infinite number of spheres of decreasing radii on top of each other, as shown in the figure. The radii of the spheres are 1 meter, $1/\sqrt{2}$ meter, $1/\sqrt{3}$ meter, and so on. The spheres are made of a material that weighs 1 newton per cubic meter.

   (a) How high is this infinite stack of spheres?

   (b) What is the total surface area of all the spheres in the stack?

   (c) Show that the weight of the stack is finite.

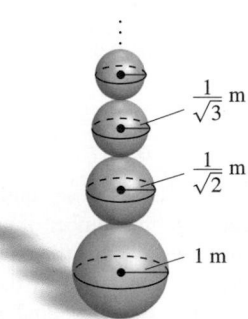

16. **Determining Convergence or Divergence**

   Determine the convergence or divergence of the series

   $$\sum_{n=1}^\infty \left(\sin \frac{1}{2n} - \sin \frac{1}{2n+1} \right).$$

10 Conics, Parametric Equations, and Polar Coordinates

Antenna Radiation *(Exercise 49, p. 736)*

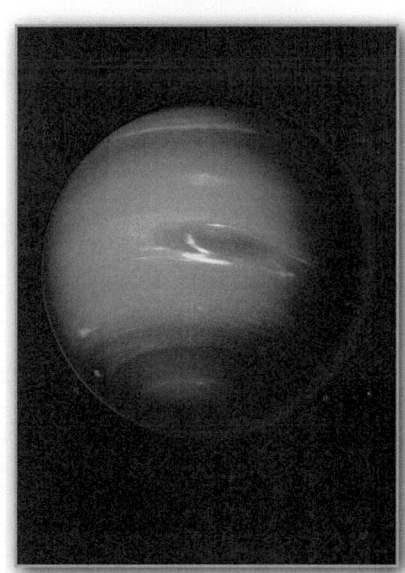

Planetary Motion
(Exercise 65, p. 745)

Baseball *(Exercise 81, p. 709)*

Halley's Comet
(Exercise 73, p. 698)

Architecture *(Exercise 67, p. 698)*

10.1 Conics and Calculus

- Understand the definition of a conic section.
- Analyze and write equations of parabolas using properties of parabolas.
- Analyze and write equations of ellipses using properties of ellipses.
- Analyze and write equations of hyperbolas using properties of hyperbolas.

Conic Sections

Each **conic section** (or simply **conic**) can be described as the intersection of a plane and a double-napped cone. Notice in Figure 10.1 that for the four basic conics, the intersecting plane does not pass through the vertex of the cone. When the plane passes through the vertex, the resulting figure is a **degenerate conic,** as shown in Figure 10.2.

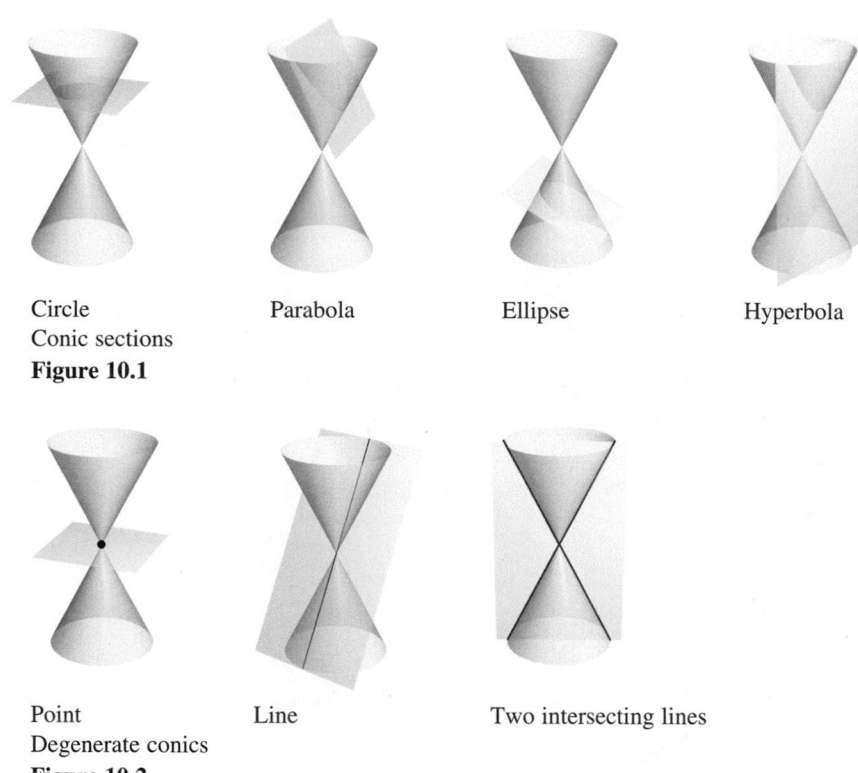

Circle
Conic sections
Figure 10.1

Parabola

Ellipse

Hyperbola

Point
Degenerate conics
Figure 10.2

Line

Two intersecting lines

HYPATIA (370–415 A.D.)

The Greeks discovered conic sections sometime between 600 and 300 B.C. By the beginning of the Alexandrian period, enough was known about conics for Apollonius (262–190 B.C.) to produce an eight-volume work on the subject. Later, toward the end of the Alexandrian period, Hypatia wrote a textbook entitled *On the Conics of Apollonius*. Her death marked the end of major mathematical discoveries in Europe for several hundred years.

The early Greeks were largely concerned with the geometric properties of conics. It was not until 1900 years later, in the early seventeenth century, that the broader applicability of conics became apparent. Conics then played a prominent role in the development of calculus.

See LarsonCalculus.com to read more of this biography.

There are several ways to study conics. You could begin as the Greeks did, by defining the conics in terms of the intersections of planes and cones, or you could define them algebraically in terms of the general second-degree equation

$$Ax^2 + Bxy + Cy^2 + Dx + Ey + F = 0. \qquad \text{General second-degree equation}$$

However, a third approach, in which each of the conics is defined as a **locus** (collection) of points satisfying a certain geometric property, works best. For example, a circle can be defined as the collection of all points (x, y) that are equidistant from a fixed point (h, k). This locus definition easily produces the standard equation of a circle

$$(x - h)^2 + (y - k)^2 = r^2. \qquad \text{Standard equation of a circle}$$

For information about rotating second-degree equations in two variables, see Appendix D.

■ **FOR FURTHER INFORMATION**
To learn more about the mathematical activities of Hypatia, see the article "Hypatia and Her Mathematics" by Michael A. B. Deakin in *The American Mathematical Monthly*. To view this article, go to *MathArticles.com*.

Parabolas

A **parabola** is the set of all points (x, y) that are equidistant from a fixed line, the **directrix,** and a fixed point, the **focus,** not on the line. The midpoint between the focus and the directrix is the **vertex,** and the line passing through the focus and the vertex is the **axis** of the parabola. Note in Figure 10.3 that a parabola is symmetric with respect to its axis.

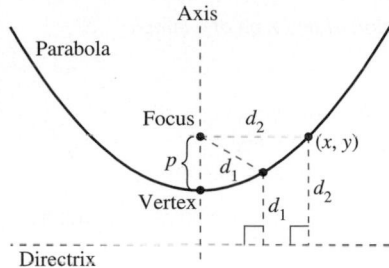

Figure 10.3

THEOREM 10.1   Standard Equation of a Parabola

The **standard form** of the equation of a parabola with vertex (h, k) and directrix $y = k - p$ is

$$(x - h)^2 = 4p(y - k). \qquad \text{Vertical axis}$$

For directrix $x = h - p$, the equation is

$$(y - k)^2 = 4p(x - h). \qquad \text{Horizontal axis}$$

The focus lies on the axis p units (*directed distance*) from the vertex. The coordinates of the focus are as follows.

$$(h, k + p) \qquad \text{Vertical axis}$$
$$(h + p, k) \qquad \text{Horizontal axis}$$

EXAMPLE 1   **Finding the Focus of a Parabola**

Find the focus of the parabola

$$y = \frac{1}{2} - x - \frac{1}{2}x^2.$$

Solution   To find the focus, convert to standard form by completing the square.

$$
\begin{aligned}
y &= \frac{1}{2} - x - \frac{1}{2}x^2 & \text{Write original equation.}\\
2y &= 1 - 2x - x^2 & \text{Multiply each side by 2.}\\
2y &= 1 - (x^2 + 2x) & \text{Group terms.}\\
2y &= 2 - (x^2 + 2x + 1) & \text{Add and subtract 1 on right side.}\\
x^2 + 2x + 1 &= -2y + 2 \\
(x + 1)^2 &= -2(y - 1) & \text{Write in standard form.}
\end{aligned}
$$

Comparing this equation with

$$(x - h)^2 = 4p(y - k)$$

you can conclude that

$$h = -1, \quad k = 1, \quad \text{and} \quad p = -\frac{1}{2}.$$

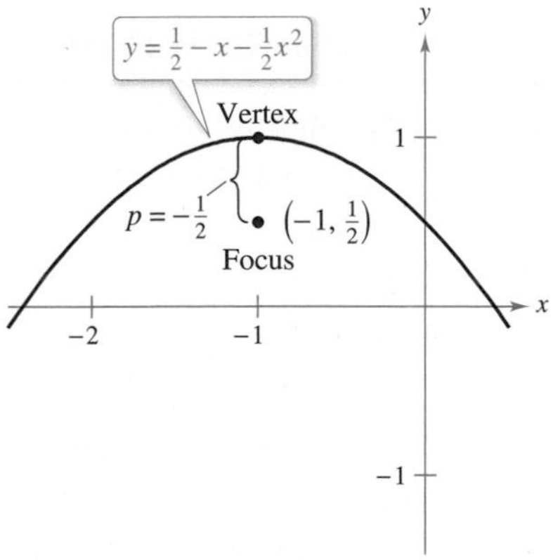

Because p is negative, the parabola opens downward, as shown in Figure 10.4. So, the focus of the parabola is p units from the vertex, or

$$(h, k + p) = \left(-1, \frac{1}{2}\right). \qquad \text{Focus}$$

Parabola with a vertical axis, $p < 0$
Figure 10.4

A line segment that passes through the focus of a parabola and has endpoints on the parabola is called a **focal chord.** The specific focal chord perpendicular to the axis of the parabola is the **latus rectum.** The next example shows how to determine the length of the latus rectum and the length of the corresponding intercepted arc.

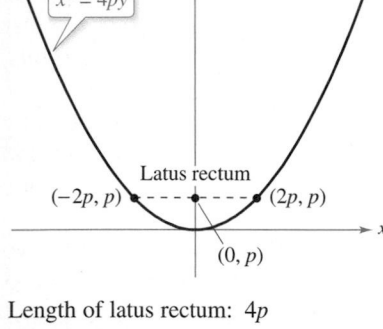

Length of latus rectum: $4p$
Figure 10.5

EXAMPLE 2 **Focal Chord Length and Arc Length**

⋯▷ *See LarsonCalculus.com for an interactive version of this type of example.*

Find the length of the latus rectum of the parabola

$$x^2 = 4py.$$

Then find the length of the parabolic arc intercepted by the latus rectum.

Solution Because the latus rectum passes through the focus $(0, p)$ and is perpendicular to the y-axis, the coordinates of its endpoints are

$$(-x, p) \quad \text{and} \quad (x, p).$$

Substituting p for y in the equation of the parabola produces

$$x^2 = 4p(p) \quad \Longrightarrow \quad x = \pm 2p.$$

So, the endpoints of the latus rectum are $(-2p, p)$ and $(2p, p)$, and you can conclude that its length is $4p$, as shown in Figure 10.5. In contrast, the length of the intercepted arc is

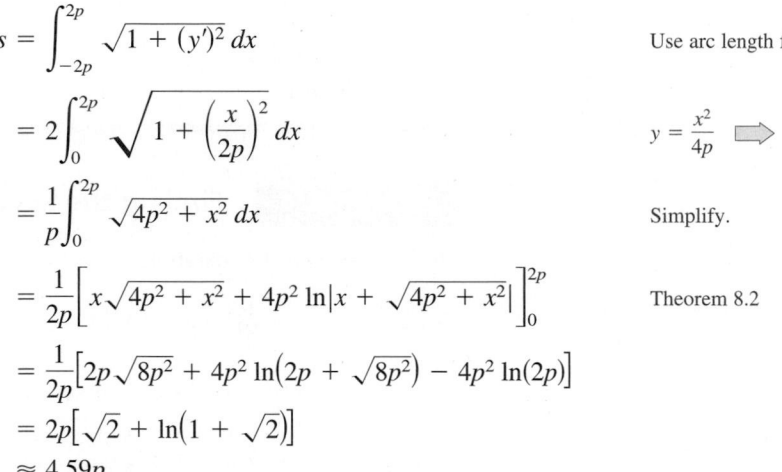

$$s = \int_{-2p}^{2p} \sqrt{1 + (y')^2} \, dx \qquad \text{Use arc length formula.}$$

$$= 2 \int_{0}^{2p} \sqrt{1 + \left(\frac{x}{2p}\right)^2} \, dx \qquad y = \frac{x^2}{4p} \ \Longrightarrow \ y' = \frac{x}{2p}$$

$$= \frac{1}{p} \int_{0}^{2p} \sqrt{4p^2 + x^2} \, dx \qquad \text{Simplify.}$$

$$= \frac{1}{2p} \left[x\sqrt{4p^2 + x^2} + 4p^2 \ln\left|x + \sqrt{4p^2 + x^2}\right| \right]_{0}^{2p} \qquad \text{Theorem 8.2}$$

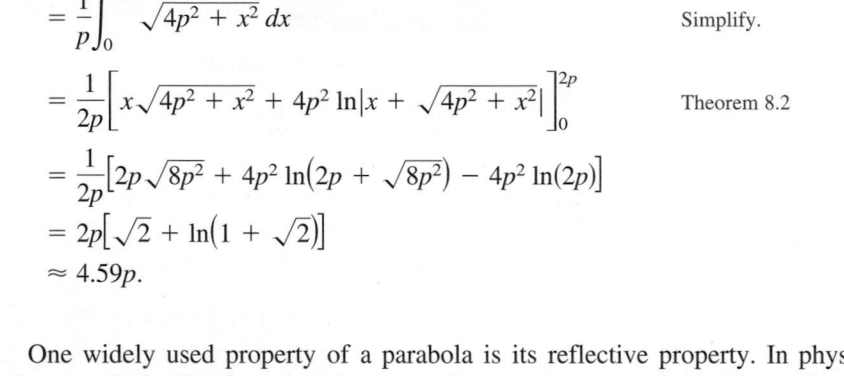

$$= \frac{1}{2p}\left[2p\sqrt{8p^2} + 4p^2 \ln\left(2p + \sqrt{8p^2}\right) - 4p^2 \ln(2p)\right]$$

$$= 2p\left[\sqrt{2} + \ln\left(1 + \sqrt{2}\right)\right]$$

$$\approx 4.59p. \qquad \blacksquare$$

One widely used property of a parabola is its reflective property. In physics, a surface is called **reflective** when the tangent line at any point on the surface makes equal angles with an incoming ray and the resulting outgoing ray. The angle corresponding to the incoming ray is the **angle of incidence,** and the angle corresponding to the outgoing ray is the **angle of reflection.** One example of a reflective surface is a flat mirror.

Another type of reflective surface is that formed by revolving a parabola about its axis. The resulting surface has the property that all incoming rays parallel to the axis are directed through the focus of the parabola. This is the principle behind the design of the parabolic mirrors used in reflecting telescopes. Conversely, all light rays emanating from the focus of a parabolic reflector used in a flashlight are parallel, as shown in Figure 10.6.

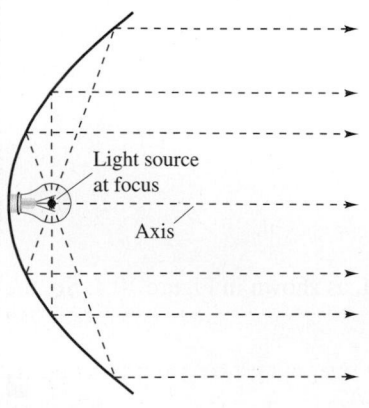

Parabolic reflector: light is reflected in parallel rays.
Figure 10.6

THEOREM 10.2 **Reflective Property of a Parabola**

Let P be a point on a parabola. The tangent line to the parabola at point P makes equal angles with the following two lines.

1. The line passing through P and the focus

2. The line passing through P parallel to the axis of the parabola

Ellipses

More than a thousand years after the close of the Alexandrian period of Greek mathematics, Western civilization finally began a Renaissance of mathematical and scientific discovery. One of the principal figures in this rebirth was the Polish astronomer Nicolaus Copernicus (1473–1543). In his work *On the Revolutions of the Heavenly Spheres,* Copernicus claimed that all of the planets, including Earth, revolved about the sun in circular orbits. Although some of Copernicus's claims were invalid, the controversy set off by his heliocentric theory motivated astronomers to search for a mathematical model to explain the observed movements of the sun and planets. The first to find an accurate model was the German astronomer Johannes Kepler (1571–1630). Kepler discovered that the planets move about the sun in elliptical orbits, with the sun not as the center but as a focal point of the orbit.

The use of ellipses to explain the movements of the planets is only one of many practical and aesthetic uses. As with parabolas, you will begin your study of this second type of conic by defining it as a locus of points. Now, however, *two* focal points are used rather than one.

An **ellipse** is the set of all points (x, y) the sum of whose distances from two distinct fixed points called **foci** is constant. (See Figure 10.7.) The line through the foci intersects the ellipse at two points, called the **vertices.** The chord joining the vertices is the **major axis,** and its midpoint is the **center** of the ellipse. The chord perpendicular to the major axis at the center is the **minor axis** of the ellipse. (See Figure 10.8.)

**NICOLAUS COPERNICUS
(1473–1543)**

Copernicus began to study planetary motion when he was asked to revise the calendar. At that time, the exact length of the year could not be accurately predicted using the theory that Earth was the center of the universe.
See LarsonCalculus.com to read more of this biography.

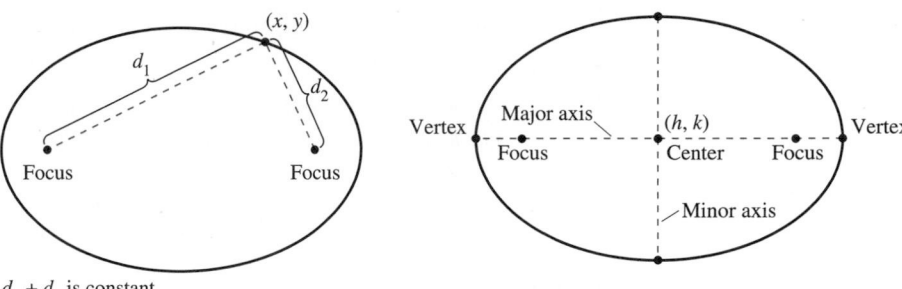

$d_1 + d_2$ is constant.

Figure 10.7

Figure 10.8

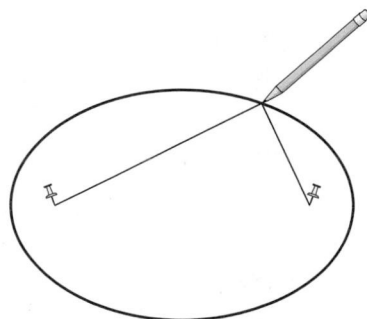

If the ends of a fixed length of string are fastened to the thumbtacks and the string is drawn taut with a pencil, then the path traced by the pencil will be an ellipse.

Figure 10.9

THEOREM 10.3   Standard Equation of an Ellipse

The **standard form** of the equation of an ellipse with center (h, k) and major and minor axes of lengths $2a$ and $2b$, respectively, where $a > b$, is

$$\frac{(x - h)^2}{a^2} + \frac{(y - k)^2}{b^2} = 1 \qquad \text{Major axis is horizontal.}$$

or

$$\frac{(x - h)^2}{b^2} + \frac{(y - k)^2}{a^2} = 1. \qquad \text{Major axis is vertical.}$$

The foci lie on the major axis, c units from the center, with

$$c^2 = a^2 - b^2.$$

You can visualize the definition of an ellipse by imagining two thumbtacks placed at the foci, as shown in Figure 10.9.

■ **FOR FURTHER INFORMATION** To learn about how an ellipse may be "exploded" into a parabola, see the article "Exploding the Ellipse" by Arnold Good in *Mathematics Teacher.* To view this article, go to *MathArticles.com.*

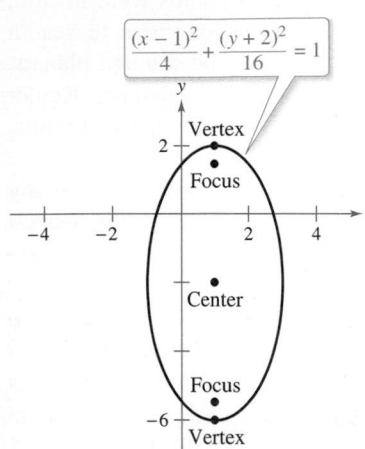

$$\frac{(x-1)^2}{4} + \frac{(y+2)^2}{16} = 1$$

Ellipse with a vertical major axis.
Figure 10.10

EXAMPLE 3 Analyzing an Ellipse

Find the center, vertices, and foci of the ellipse

$$4x^2 + y^2 - 8x + 4y - 8 = 0.$$ General second-degree equation

Solution Complete the square to write the original equation in standard form.

$$4x^2 + y^2 - 8x + 4y - 8 = 0$$ Write original equation.

$$4x^2 - 8x + y^2 + 4y = 8$$

$$4(x^2 - 2x + 1) + (y^2 + 4y + 4) = 8 + 4 + 4$$

$$4(x-1)^2 + (y+2)^2 = 16$$

$$\frac{(x-1)^2}{4} + \frac{(y+2)^2}{16} = 1$$ Write in standard form.

So, the major axis is vertical, where $h = 1$, $k = -2$, $a = 4$, $b = 2$, and

$$c = \sqrt{16 - 4} = 2\sqrt{3}.$$

So, you obtain the following.

Center: $(1, -2)$	(h, k)
Vertices: $(1, -6)$ and $(1, 2)$	$(h, k \pm a)$
Foci: $\left(1, -2 - 2\sqrt{3}\right)$ and $\left(1, -2 + 2\sqrt{3}\right)$	$(h, k \pm c)$

The graph of the ellipse is shown in Figure 10.10.

In Example 3, the constant term in the general second-degree equation is $F = -8$. For a constant term greater than or equal to 8, you would obtain one of these degenerate cases.

1. $F = 8$, single point, $(1, -2)$: $\dfrac{(x-1)^2}{4} + \dfrac{(y+2)^2}{16} = 0$

2. $F > 8$, no solution points: $\dfrac{(x-1)^2}{4} + \dfrac{(y+2)^2}{16} < 0$

EXAMPLE 4 The Orbit of the Moon

The moon orbits Earth in an elliptical path with the center of Earth at one focus, as shown in Figure 10.11. The major and minor axes of the orbit have lengths of 768,800 kilometers and 767,641 kilometers, respectively. Find the greatest and least distances (the apogee and perigee) from Earth's center to the moon's center.

Solution Begin by solving for a and b.

$2a = 768,800$	Length of major axis
$a = 384,400$	Solve for a.
$2b = 767,641$	Length of minor axis
$b = 383,820.5$	Solve for b.

Now, using these values, you can solve for c as follows.

$$c = \sqrt{a^2 - b^2} \approx 21,099$$

The greatest distance between the center of Earth and the center of the moon is

$$a + c \approx 405,499 \text{ kilometers}$$

and the least distance is

$$a - c \approx 363,301 \text{ kilometers}.$$

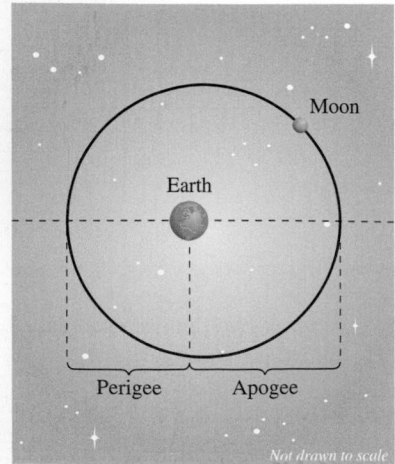

Figure 10.11

■ **FOR FURTHER INFORMATION**
For more information on some uses of the reflective properties of conics, see the article "Parabolic Mirrors, Elliptic and Hyperbolic Lenses" by Mohsen Maesumi in *The American Mathematical Monthly*. Also see the article "The Geometry of Microwave Antennas" by William R. Parzynski in *Mathematics Teacher*.

Theorem 10.2 presented a reflective property of parabolas. Ellipses have a similar reflective property. You are asked to prove the next theorem in Exercise 84.

THEOREM 10.4   Reflective Property of an Ellipse

Let P be a point on an ellipse. The tangent line to the ellipse at point P makes equal angles with the lines through P and the foci.

One of the reasons that astronomers had difficulty detecting that the orbits of the planets are ellipses is that the foci of the planetary orbits are relatively close to the center of the sun, making the orbits nearly circular. To measure the ovalness of an ellipse, you can use the concept of **eccentricity.**

Definition of Eccentricity of an Ellipse

The **eccentricity** e of an ellipse is given by the ratio

$$e = \frac{c}{a}.$$

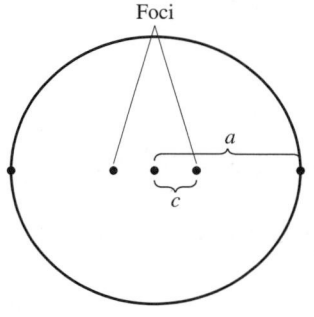

(a) $\dfrac{c}{a}$ is small.

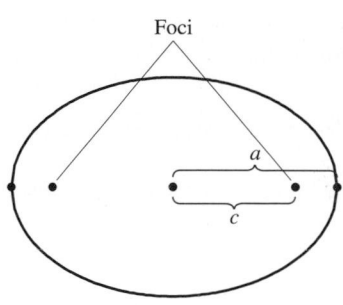

(b) $\dfrac{c}{a}$ is close to 1.

Eccentricity is the ratio $\dfrac{c}{a}$.

Figure 10.12

To see how this ratio is used to describe the shape of an ellipse, note that because the foci of an ellipse are located along the major axis between the vertices and the center, it follows that

$$0 < c < a.$$

For an ellipse that is nearly circular, the foci are close to the center and the ratio c/a is close to 0, and for an elongated ellipse, the foci are close to the vertices and the ratio c/a is close to 1, as shown in Figure 10.12. Note that

$$0 < e < 1$$

for every ellipse.

The orbit of the moon has an eccentricity of $e \approx 0.0549$, and the eccentricities of the eight planetary orbits are listed below.

Mercury:	$e \approx 0.2056$	Jupiter:	$e \approx 0.0489$
Venus:	$e \approx 0.0067$	Saturn:	$e \approx 0.0565$
Earth:	$e \approx 0.0167$	Uranus:	$e \approx 0.0457$
Mars:	$e \approx 0.0935$	Neptune:	$e \approx 0.0113$

You can use integration to show that the area of an ellipse is $A = \pi ab$. For instance, the area of the ellipse

$$\frac{x^2}{a^2} + \frac{y^2}{b^2} = 1$$

is

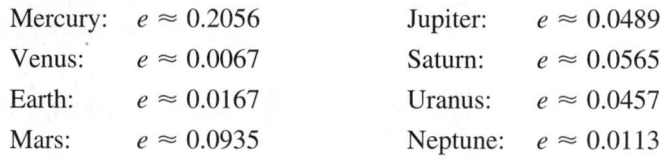

$$A = 4 \int_0^a \frac{b}{a} \sqrt{a^2 - x^2}\, dx$$

$$= \frac{4b}{a} \int_0^{\pi/2} a^2 \cos^2 \theta\, d\theta. \qquad \text{Trigonometric substitution } x = a \sin \theta$$

However, it is not so simple to find the *circumference* of an ellipse. The next example shows how to use eccentricity to set up an "elliptic integral" for the circumference of an ellipse.

EXAMPLE 5 Finding the Circumference of an Ellipse

$\cdots \triangleright$ *See LarsonCalculus.com for an interactive version of this type of example.*

Show that the circumference of the ellipse $(x^2/a^2) + (y^2/b^2) = 1$ is

$$4a \int_0^{\pi/2} \sqrt{1 - e^2 \sin^2 \theta} \, d\theta. \qquad e = \frac{c}{a}$$

Solution Because the ellipse is symmetric with respect to both the x-axis and the y-axis, you know that its circumference C is four times the arc length of

$$y = \frac{b}{a}\sqrt{a^2 - x^2}$$

in the first quadrant. The function y is differentiable for all x in the interval $[0, a]$ except at $x = a$. So, the circumference is given by the improper integral

$$C = \lim_{d \to a^-} 4 \int_0^d \sqrt{1 + (y')^2} \, dx = 4 \int_0^a \sqrt{1 + (y')^2} \, dx = 4 \int_0^a \sqrt{1 + \frac{b^2 x^2}{a^2(a^2 - x^2)}} \, dx.$$

Using the trigonometric substitution $x = a \sin \theta$, you obtain

$$C = 4 \int_0^{\pi/2} \sqrt{1 + \frac{b^2 \sin^2 \theta}{a^2 \cos^2 \theta}} \, (a \cos \theta) \, d\theta$$

$$= 4 \int_0^{\pi/2} \sqrt{a^2 \cos^2 \theta + b^2 \sin^2 \theta} \, d\theta$$

$$= 4 \int_0^{\pi/2} \sqrt{a^2(1 - \sin^2 \theta) + b^2 \sin^2 \theta} \, d\theta$$

$$= 4 \int_0^{\pi/2} \sqrt{a^2 - (a^2 - b^2) \sin^2 \theta} \, d\theta.$$

Because $e^2 = c^2/a^2 = (a^2 - b^2)/a^2$, you can rewrite this integral as

$$C = 4a \int_0^{\pi/2} \sqrt{1 - e^2 \sin^2 \theta} \, d\theta.$$

A great deal of time has been devoted to the study of elliptic integrals. Such integrals generally do not have elementary antiderivatives. To find the circumference of an ellipse, you must usually resort to an approximation technique.

EXAMPLE 6 Approximating the Value of an Elliptic Integral

Use the elliptic integral in Example 5 to approximate the circumference of the ellipse

$$\frac{x^2}{25} + \frac{y^2}{16} = 1.$$

Solution Because $e^2 = c^2/a^2 = (a^2 - b^2)/a^2 = 9/25$, you have

$$C = (4)(5) \int_0^{\pi/2} \sqrt{1 - \frac{9 \sin^2 \theta}{25}} \, d\theta.$$

Applying Simpson's Rule with $n = 4$ produces

$$C \approx 20 \left[\frac{\pi/2}{3(4)} \right] [1 + 4(0.9733) + 2(0.9055) + 4(0.8323) + 0.8]$$

$$\approx 28.36.$$

So, the ellipse has a circumference of about 28.36 units, as shown in Figure 10.13.

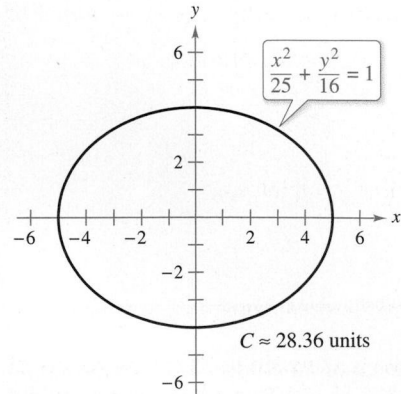

$C \approx 28.36$ units

Figure 10.13

Hyperbolas

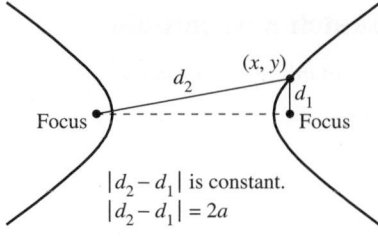

$|d_2 - d_1|$ is constant.

$|d_2 - d_1| = 2a$

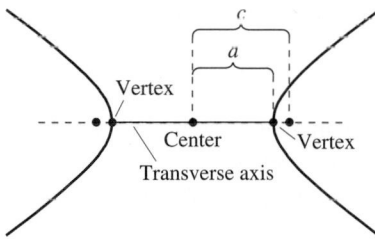

Figure 10.14

The definition of a hyperbola is similar to that of an ellipse. For an ellipse, the *sum* of the distances between the foci and a point on the ellipse is fixed, whereas for a hyperbola, the absolute value of the *difference* between these distances is fixed.

A **hyperbola** is the set of all points (x, y) for which the absolute value of the difference between the distances from two distinct fixed points called **foci** is constant. (See Figure 10.14.) The line through the two foci intersects a hyperbola at two points called the **vertices**. The line segment connecting the vertices is the **transverse axis,** and the midpoint of the transverse axis is the **center** of the hyperbola. One distinguishing feature of a hyperbola is that its graph has two separate *branches*.

THEOREM 10.5 Standard Equation of a Hyperbola

The **standard form** of the equation of a hyperbola with center at (h, k) is

$$\frac{(x - h)^2}{a^2} - \frac{(y - k)^2}{b^2} = 1 \qquad \text{Transverse axis is horizontal.}$$

or

$$\frac{(y - k)^2}{a^2} - \frac{(x - h)^2}{b^2} = 1. \qquad \text{Transverse axis is vertical.}$$

The vertices are a units from the center, and the foci are c units from the center, where $c^2 = a^2 + b^2$.

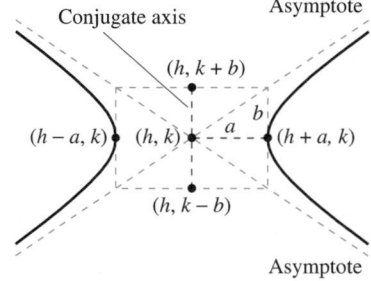

Figure 10.15

Note that the constants a, b, and c do not have the same relationship for hyperbolas as they do for ellipses. For hyperbolas, $c^2 = a^2 + b^2$, but for ellipses, $c^2 = a^2 - b^2$.

An important aid in sketching the graph of a hyperbola is the determination of its **asymptotes,** as shown in Figure 10.15. Each hyperbola has two asymptotes that intersect at the center of the hyperbola. The asymptotes pass through the vertices of a rectangle of dimensions $2a$ by $2b$, with its center at (h, k). The line segment of length $2b$ joining

$$(h, k + b)$$

and

$$(h, k - b)$$

is referred to as the **conjugate axis** of the hyperbola.

THEOREM 10.6 Asymptotes of a Hyperbola

For a *horizontal* transverse axis, the equations of the asymptotes are

$$y = k + \frac{b}{a}(x - h) \quad \text{and} \quad y = k - \frac{b}{a}(x - h).$$

For a *vertical* transverse axis, the equations of the asymptotes are

$$y = k + \frac{a}{b}(x - h) \quad \text{and} \quad y = k - \frac{a}{b}(x - h).$$

In Figure 10.15, you can see that the asymptotes coincide with the diagonals of the rectangle with dimensions $2a$ and $2b$, centered at (h, k). This provides you with a quick means of sketching the asymptotes, which in turn aids in sketching the hyperbola.

> **EXAMPLE 7**   **Using Asymptotes to Sketch a Hyperbola**

:····▷ *See LarsonCalculus.com for an interactive version of this type of example.*

Sketch the graph of the hyperbola

$$4x^2 - y^2 = 16.$$

Solution   Begin by rewriting the equation in standard form.

$$\frac{x^2}{4} - \frac{y^2}{16} = 1$$

The transverse axis is horizontal and the vertices occur at $(-2, 0)$ and $(2, 0)$. The ends of the conjugate axis occur at $(0, -4)$ and $(0, 4)$. Using these four points, you can sketch the rectangle shown in Figure 10.16(a). By drawing the asymptotes through the corners of this rectangle, you can complete the sketch as shown in Figure 10.16(b).

▷ **TECHNOLOGY** You can use a graphing utility to verify the graph obtained in Example 7 by solving the original equation for y and graphing the following equations.

$$y_1 = \sqrt{4x^2 - 16}$$
$$y_2 = -\sqrt{4x^2 - 16}$$

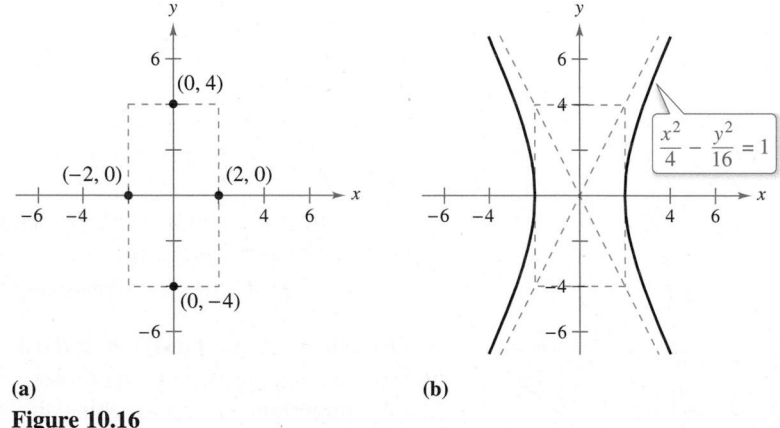

(a)   (b)

Figure 10.16

Definition of Eccentricity of a Hyperbola

The **eccentricity** e of a hyperbola is given by the ratio

$$e = \frac{c}{a}.$$

■ **FOR FURTHER INFORMATION** To read about using a string that traces both elliptic and hyperbolic arcs having the same foci, see the article "Ellipse to Hyperbola: 'With This String I Thee Wed'" by Tom M. Apostol and Mamikon A. Mnatsakanian in *Mathematics Magazine.* To view this article, go to *MathArticles.com.*

As with an ellipse, the **eccentricity** of a hyperbola is $e = c/a$. Because $c > a$ for hyperbolas, it follows that $e > 1$ for hyperbolas. If the eccentricity is large, then the branches of the hyperbola are nearly flat. If the eccentricity is close to 1, then the branches of the hyperbola are more pointed, as shown in Figure 10.17.

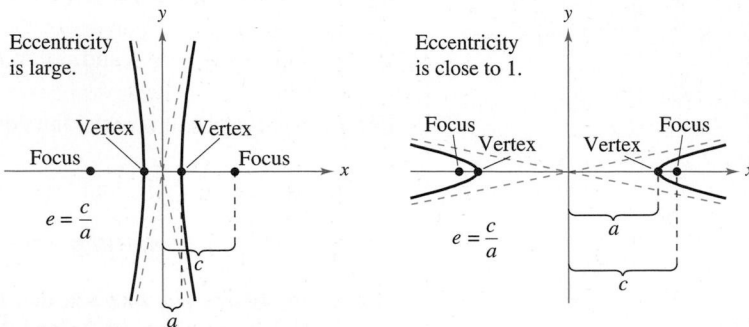

Figure 10.17

The application in Example 8 was developed during World War II. It shows how the properties of hyperbolas can be used in radar and other detection systems.

EXAMPLE 8 **A Hyperbolic Detection System**

Two microphones, 1 mile apart, record an explosion. Microphone A receives the sound 2 seconds before microphone B. Where was the explosion?

Solution   Assuming that sound travels at 1100 feet per second, you know that the explosion took place 2200 feet farther from B than from A, as shown in Figure 10.18. The locus of all points that are 2200 feet closer to A than to B is one branch of the hyperbola

$$\frac{x^2}{a^2} - \frac{y^2}{b^2} = 1$$

where

$$c = \frac{1 \text{ mile}}{2} = \frac{5280 \text{ ft}}{2} = 2640 \text{ feet}$$

and

$$a = \frac{2200 \text{ ft}}{2} = 1100 \text{ feet.}$$

Because $c^2 = a^2 + b^2$, it follows that

$$
\begin{aligned}
b^2 &= c^2 - a^2 \\
&= (2640)^2 - (1100)^2 \\
&= 5,759,600
\end{aligned}
$$

and you can conclude that the explosion occurred somewhere on the right branch of the hyperbola

$$\frac{x^2}{1,210,000} - \frac{y^2}{5,759,600} = 1.$$

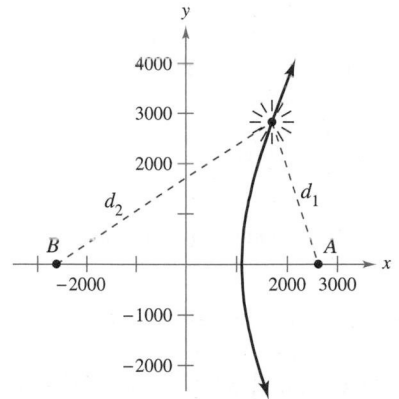

$2c = 5280$
$d_2 - d_1 = 2a = 2200$
Figure 10.18

In Example 8, you were able to determine only the hyperbola on which the explosion occurred, but not the exact location of the explosion. If, however, you had received the sound at a third position C, then two other hyperbolas would be determined. The exact location of the explosion would be the point at which these three hyperbolas intersect.

Another interesting application of conics involves the orbits of comets in our solar system. Of the 610 comets identified prior to 1970, 245 have elliptical orbits, 295 have parabolic orbits, and 70 have hyperbolic orbits. The center of the sun is a focus of each orbit, and each orbit has a vertex at the point at which the comet is closest to the sun. Undoubtedly, many comets with parabolic or hyperbolic orbits have not been identified—such comets pass through our solar system only once. Only comets with elliptical orbits, such as Halley's comet, remain in our solar system.

The type of orbit for a comet can be determined as follows.

1. Ellipse: $v < \sqrt{2GM/p}$
2. Parabola: $v = \sqrt{2GM/p}$
3. Hyperbola: $v > \sqrt{2GM/p}$

In each of the above, p is the distance between one vertex and one focus of the comet's orbit (in meters), v is the velocity of the comet at the vertex (in meters per second), $M \approx 1.989 \times 10^{30}$ kilograms is the mass of the sun, and $G \approx 6.67 \times 10^{-11}$ cubic meter per kilogram-second squared is the gravitational constant.

CAROLINE HERSCHEL
(1750–1848)

The first woman to be credited with detecting a new comet was the English astronomer Caroline Herschel. During her life, Caroline Herschel discovered a total of eight new comets.
See LarsonCalculus.com to read more of this biography.

10.1 Exercises

See **CalcChat.com** for tutorial help and worked-out solutions to odd-numbered exercises.

CONCEPT CHECK

1. Conic Sections State the definitions of parabola, ellipse, and hyperbola in your own words.

2. Reflective Property Use a sketch to illustrate the reflective property of an ellipse.

3. Eccentricity Consider an ellipse with eccentricity e.

(a) What are the possible values of e?

(b) What happens to the graph of the ellipse as e increases?

4. Hyperbola Explain how to sketch a hyperbola with a vertical transverse axis.

Matching In Exercises 5–10, match the equation with its graph. [The graphs are labeled (a), (b), (c), (d), (e), and (f).]

(a)
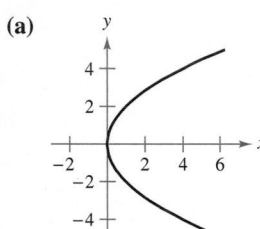

(b)
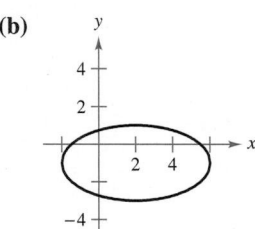

(c)
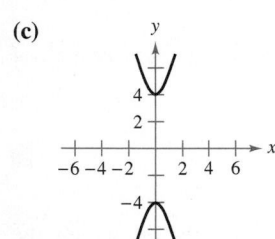

(d)
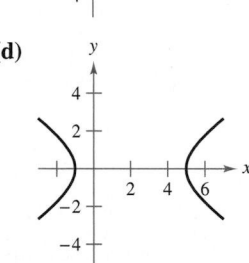

(e)
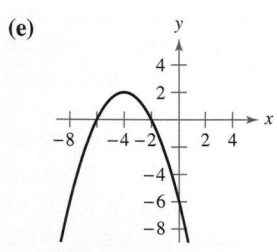

(f)
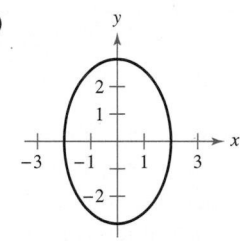

5. $y^2 = 4x$

6. $(x + 4)^2 = -2(y - 2)$

7. $\dfrac{y^2}{16} - \dfrac{x^2}{1} = 1$

8. $\dfrac{(x - 2)^2}{16} + \dfrac{(y + 1)^2}{4} = 1$

9. $\dfrac{x^2}{4} + \dfrac{y^2}{9} = 1$

10. $\dfrac{(x - 2)^2}{9} - \dfrac{y^2}{4} = 1$

 **Sketching a Parabola** In Exercises 11–16, find the vertex, focus, and directrix of the parabola, and sketch its graph.

11. $(x + 5) + (y - 3)^2 = 0$

12. $(x - 6)^2 - 2(y + 7) = 0$

13. $y^2 - 4y - 4x = 0$

14. $y^2 + 6y + 8x + 25 = 0$

15. $x^2 + 4x + 4y - 4 = 0$

16. $x^2 - 2x - 4y - 7 = 0$

 **Finding the Standard Equation of a Parabola** In Exercises 17–24, find the standard form of the equation of the parabola with the given characteristics.

17. Vertex: $(5, 4)$
Focus: $(3, 4)$

18. Vertex: $(-3, -1)$
Focus: $(-3, 1)$

19. Vertex: $(0, 5)$
Directrix: $y = -3$

20. Focus: $(2, 2)$
Directrix: $x = -2$

21. Vertex: $(1, -1)$
Points on the parabola:
$(-1, -4), (3, -4)$

22. Vertex: $(2, 4)$
Points on the parabola:
$(0, 0), (4, 0)$

23. Axis is parallel to y-axis; graph passes through $(0, 3)$, $(3, 4)$, and $(4, 11)$.

24. Directrix: $y = -2$; endpoints of latus rectum are $(0, 2)$ and $(8, 2)$.

 Sketching an Ellipse In Exercises 25–30, find the center, foci, vertices, and eccentricity of the ellipse, and sketch its graph.

25. $16x^2 + y^2 = 16$

26. $3x^2 + 7y^2 = 63$

27. $\dfrac{(x - 3)^2}{16} + \dfrac{(y - 1)^2}{25} = 1$

28. $(x + 4)^2 + \dfrac{(y + 6)^2}{1/4} = 1$

29. $9x^2 + 4y^2 + 36x - 24y - 36 = 0$

30. $x^2 + 10y^2 - 6x + 20y + 18 = 0$

 **Finding the Standard Equation of an Ellipse** In Exercises 31–36, find the standard form of the equation of the ellipse with the given characteristics.

31. Center: $(0, 0)$
Focus: $(5, 0)$
Vertex: $(6, 0)$

32. Vertices: $(0, 3), (8, 3)$
Eccentricity: $\frac{3}{4}$

33. Vertices: $(3, 1), (3, 9)$
Minor axis length: 6

34. Foci: $(0, \pm 9)$
Major axis length: 22

35. Center: $(0, 0)$
Major axis: horizontal
Points on the ellipse:
$(3, 1), (4, 0)$

36. Center: $(1, 2)$
Major axis: vertical
Points on the ellipse:
$(1, 6), (3, 2)$

 **Sketching a Hyperbola** In Exercises 37–40, find the center, foci, vertices, and eccentricity of the hyperbola, and sketch its graph using asymptotes as an aid.

37. $\dfrac{x^2}{25} - \dfrac{y^2}{16} = 1$

38. $\dfrac{(y + 3)^2}{225} - \dfrac{(x - 5)^2}{64} = 1$

39. $9x^2 - y^2 - 36x - 6y + 18 = 0$

40. $y^2 - 16x^2 + 64x - 208 = 0$

Finding the Standard Equation of a Hyperbola In Exercises 41–48, find the standard form of the equation of the hyperbola with the given characteristics.

41. Vertices: $(\pm 1, 0)$
Asymptotes: $y = \pm 5x$

42. Vertices: $(0, \pm 4)$
Asymptotes: $y = \pm 2x$

43. Vertices: $(2, \pm 3)$
Point on graph: $(0, 5)$

44. Vertices: $(2, \pm 3)$
Foci: $(2, \pm 5)$

45. Center: $(0, 0)$
Vertex: $(0, 2)$
Focus: $(0, 4)$

46. Center: $(0, 0)$
Vertex: $(6, 0)$
Focus: $(10, 0)$

47. Vertices: $(0, 2)$, $(6, 2)$
Asymptotes: $y = \frac{2}{3}x$
$y = 4 - \frac{2}{3}x$

48. Focus: $(20, 0)$
Asymptotes: $y = \pm\frac{3}{4}x$

Finding Equations of Tangent Lines and Normal Lines In Exercises 49 and 50, find equations for (a) the tangent lines and (b) the normal lines to the hyperbola for the given value of x. (The *normal line* at a point is perpendicular to the tangent line at the point.)

49. $\dfrac{x^2}{9} - y^2 = 1$, $x = 6$

50. $\dfrac{y^2}{4} - \dfrac{x^2}{2} = 1$, $x = 4$

Classifying the Graph of an Equation In Exercises 51–56, classify the graph of the equation as a circle, a parabola, an ellipse, or a hyperbola.

51. $25x^2 - 10x - 200y - 119 = 0$

52. $4x^2 - y^2 - 4x - 3 = 0$

53. $3(x - 1)^2 = 6 + 2(y + 1)^2$ **54.** $9(x + 3)^2 = 36 - 4(y - 2)^2$

55. $9x^2 + 9y^2 - 36x + 6y + 34 = 0$

56. $y^2 - 4y = x + 5$

EXPLORING CONCEPTS

57. Using an Equation Consider the equation $9x^2 + 4y^2 - 36x - 24y - 36 = 0$.

(a) Classify the graph of the equation as a circle, a parabola, an ellipse, or a hyperbola.

(b) Change the $4y^2$-term in the equation to $-4y^2$. Classify the graph of the new equation.

(c) Change the $9x^2$-term in the original equation to $4x^2$. Classify the graph of the new equation.

(d) Describe one way you could change the original equation so that its graph is a parabola.

58. Investigation Sketch the graphs of $x^2 = 4py$ for $p = \frac{1}{4}, \frac{1}{2}, 1, \frac{3}{2}$, and 2 on the same coordinate axes. Discuss the change in the graphs as p increases.

59. Ellipse Let C be the circumference of the ellipse $(x^2/a^2) + (y^2/b^2) = 1$, $b < a$. Explain why $2\pi b < C < 2\pi a$. Use a graph to support your explanation.

---

60. **HOW DO YOU SEE IT?** Describe in words how a plane could intersect with the double-napped cone to form each conic section (see figure).

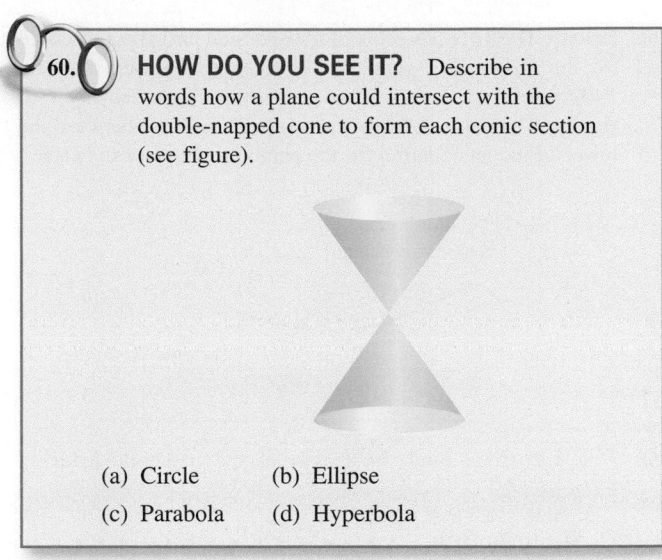

(a) Circle (b) Ellipse

(c) Parabola (d) Hyperbola

61. Solar Collector A solar collector for heating water is constructed with a sheet of stainless steel that is formed into the shape of a parabola (see figure). The water will flow through a pipe that is located at the focus of the parabola. At what distance from the vertex is the pipe?

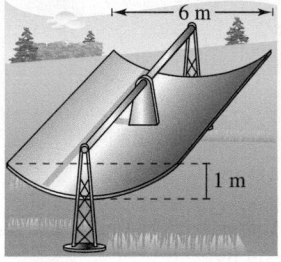

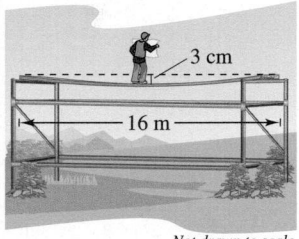

Not drawn to scale

Figure for 61　　　　Figure for 62

62. Beam Deflection A simply supported beam that is 16 meters long has a load concentrated at the center (see figure). The deflection of the beam at its center is 3 centimeters. Assume that the shape of the deflected beam is parabolic.

(a) Find an equation of the parabola. (Assume that the origin is at the center of the beam.)

(b) How far from the center of the beam is the deflection 1 centimeter?

63. Proof

(a) Prove that any two distinct tangent lines to a parabola intersect.

(b) Demonstrate the result of part (a) by finding the point of intersection of the tangent lines to the parabola $x^2 - 4x - 4y = 0$ at the points $(0, 0)$ and $(6, 3)$.

64. Proof

(a) Prove that if any two tangent lines to a parabola intersect at right angles, then their point of intersection must lie on the directrix.

(b) Demonstrate the result of part (a) by showing that the tangent lines to the parabola $x^2 - 4x - 4y + 8 = 0$ at the points $(-2, 5)$ and $\left(3, \frac{5}{4}\right)$ intersect at right angles and that their point of intersection lies on the directrix.

65. Bridge Design A cable of a suspension bridge is suspended (in the shape of a parabola) between two towers that are 120 meters apart and 20 meters above the roadway (see figure). The cable touches the roadway midway between the towers. Find an equation for the parabolic shape of the cable.

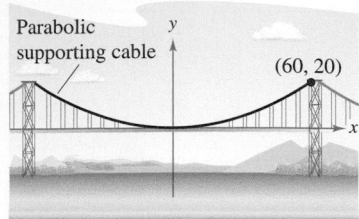

Parabolic supporting cable

(60, 20)

66. Arc Length Find the length of the parabolic cable in Exercise 65.

67. Architecture

A church window is bounded above by a parabola and below by the arc of a circle (see figure). Find the area of the window.

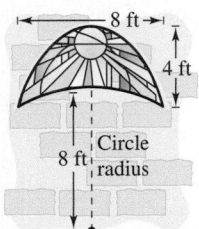

8 ft

4 ft

8 ft Circle radius

68. Surface Area A satellite signal receiving dish is formed by revolving the parabola given by

$$x^2 = 20y$$

about the y-axis. The radius of the dish is r feet. Verify that the surface area of the dish is given by

$$2\pi \int_0^r x \sqrt{1 + \left(\frac{x}{10}\right)^2}\, dx = \frac{\pi}{15}[(100 + r^2)^{3/2} - 1000].$$

69. Orbit of Earth Earth moves in an elliptical orbit with the sun at one of the foci. The length of half of the major axis is 149,598,000 kilometers, and the eccentricity is 0.0167. Find the minimum distance (*perihelion*) and the maximum distance (*aphelion*) of Earth from the sun.

70. Satellite Orbit The *apogee* (the point in orbit farthest from Earth) and the *perigee* (the point in orbit closest to Earth) of an elliptical orbit of an Earth satellite are given by A and P, respectively. Show that the eccentricity of the orbit is

$$e = \frac{A - P}{A + P}.$$

71. Explorer 1 On January 31, 1958, the United States launched the research satellite Explorer 1. Its low and high points above the surface of Earth were 220 miles and 1563 miles. Find the eccentricity of its elliptical orbit. (Use 4000 miles as the radius of Earth.)

72. Explorer 55 On November 20, 1975, the United States launched the research satellite Explorer 55. Its low and high points above the surface of Earth were 96 miles and 1865 miles. Find the eccentricity of its elliptical orbit. (Use 4000 miles as the radius of Earth.)

73. Halley's Comet

Probably the most famous of all comets, Halley's comet, has an elliptical orbit with the sun at one focus. Its maximum distance from the sun is approximately 35.29 AU (1 astronomical unit is approximately 92.956×10^6 miles), and its minimum distance is approximately 0.59 AU. Find the eccentricity of the orbit.

74. Particle Motion Consider a particle traveling clockwise on the elliptical path

$$\frac{x^2}{100} + \frac{y^2}{25} = 1.$$

The particle leaves the orbit at the point $(-8, 3)$ and travels in a straight line tangent to the ellipse. At what point will the particle cross the y-axis?

Area, Volume, and Surface Area In Exercises 75 and 76, find (a) the area of the region bounded by the ellipse, (b) the volume and surface area of the solid generated by revolving the region about its major axis (prolate spheroid), and (c) the volume and surface area of the solid generated by revolving the region about its minor axis (oblate spheroid).

75. $\dfrac{x^2}{4} + \dfrac{y^2}{1} = 1$

76. $\dfrac{x^2}{16} + \dfrac{y^2}{9} = 1$

77. Arc Length Use the integration capabilities of a graphing utility to approximate to two-decimal-place accuracy the elliptical integral representing the circumference of the ellipse

$$\frac{x^2}{25} + \frac{y^2}{49} = 1.$$

78. Conjecture

(a) Show that the equation of an ellipse can be written as

$$\frac{(x - h)^2}{a^2} + \frac{(y - k)^2}{a^2(1 - e^2)} = 1.$$

(b) Use a graphing utility to graph the ellipse

$$\frac{(x - 2)^2}{4} + \frac{(y - 3)^2}{4(1 - e^2)} = 1$$

for $e = 0.95$, $e = 0.75$, $e = 0.5$, $e = 0.25$, and $e = 0$.

(c) Use the results of part (b) to make a conjecture about the change in the shape of the ellipse as e approaches 0.

79. Geometry The area of the ellipse in the figure is twice the area of the circle. What is the length of the major axis?

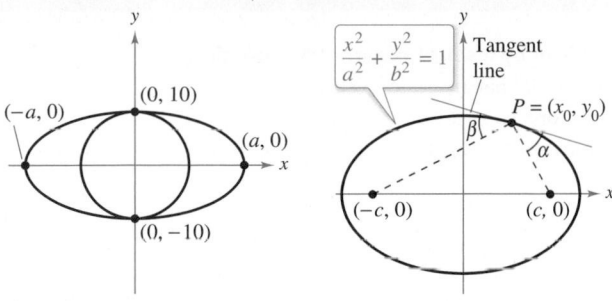

Figure for 79   Figure for 80

80. Proof Prove Theorem 10.4 by showing that the tangent line to an ellipse at a point P makes equal angles with lines through P and the foci (see figure). [*Hint:* (1) Find the slope of the tangent line at P, (2) find the slopes of the lines through P and each focus, and (3) use the formula for the tangent of the angle θ between two lines with slopes m_1 and m_2,

$$\tan \theta = \left| \frac{m_1 - m_2}{1 + m_1 m_2} \right|.]$$

81. Finding an Equation of a Hyperbola Find an equation of the hyperbola such that for any point on the hyperbola, the difference between its distances from the points $(2, 2)$ and $(10, 2)$ is 6.

82. Hyperbola Consider a hyperbola centered at the origin with a horizontal transverse axis. Use the definition of a hyperbola to derive its standard form

$$\frac{x^2}{a^2} - \frac{y^2}{b^2} = 1.$$

83. Navigation LORAN (long distance radio navigation) for aircraft and ships uses synchronized pulses transmitted by widely separated transmitting stations. These pulses travel at the speed of light (186,000 miles per second). The difference in the times of arrival of these pulses at an aircraft or ship is constant on a hyperbola having the transmitting stations as foci. Assume that two stations, 300 miles apart, are positioned on a rectangular coordinate system at $(-150, 0)$ and $(150, 0)$ and that a ship is traveling on a path with coordinates $(x, 75)$ (see figure). Find the x-coordinate of the position of the ship when the time difference between the pulses from the transmitting stations is 1000 microseconds (0.001 second).

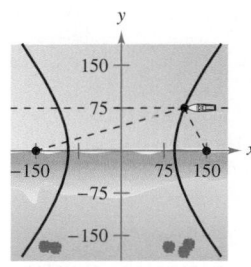

Figure for 83

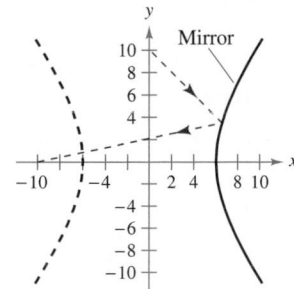

Figure for 84

84. Hyperbolic Mirror A hyperbolic mirror (used in some telescopes) has the property that a light ray directed at the focus will be reflected to the other focus. The mirror in the figure has the equation

$$\frac{x^2}{36} - \frac{y^2}{64} = 1.$$

At which point on the mirror will light from the point $(0, 10)$ be reflected to the other focus?

85. Tangent Line Show that the equation of the tangent line to $\dfrac{x^2}{a^2} - \dfrac{y^2}{b^2} = 1$ at the point (x_0, y_0) is $\left(\dfrac{x_0}{a^2}\right)x - \left(\dfrac{y_0}{b^2}\right)y = 1.$

86. Proof Prove that the graph of the equation

$$Ax^2 + Cy^2 + Dx + Ey + F = 0$$

is one of the following (except in degenerate cases).

Conic	Condition
(a) Circle	$A = C$
(b) Parabola	$A = 0$ or $C = 0$ (but not both)
(c) Ellipse	$AC > 0$
(d) Hyperbola	$AC < 0$

True or False? **In Exercises 87–92, determine whether the statement is true or false. If it is false, explain why or give an example that shows it is false.**

87. It is possible for a parabola to intersect its directrix.

88. The point on a parabola closest to its focus is its vertex.

89. The eccentricity of a hyperbola with a horizontal transverse axis is $e = \sqrt{1 + m^2}$, where m and $-m$ the slopes of the asymptotes.

90. If $D \neq 0$ or $E \neq 0$, then the graph of

$$y^2 - x^2 + Dx + Ey = 0$$

is a hyperbola.

91. If the asymptotes of the hyperbola $(x^2/a^2) - (y^2/b^2) = 1$ intersect at right angles, then $a = b$.

92. Every tangent line to a hyperbola intersects the hyperbola only at the point of tangency.

PUTNAM EXAM CHALLENGE

93. For a point P on an ellipse, let d be the distance from the center of the ellipse to the line tangent to the ellipse at P. Prove that $(PF_1)(PF_2)d^2$ is constant as P varies on the ellipse, where PF_1 and PF_2 are the distances from P to the foci F_1 and F_2 of the ellipse.

94. Find the minimum value of

$$(u - v)^2 + \left(\sqrt{2 - u^2} - \frac{9}{v} \right)^2$$

for $0 < u < \sqrt{2}$ and $v > 0$.

10.2 Plane Curves and Parametric Equations

■ Sketch the graph of a curve given by a set of parametric equations.
■ Eliminate the parameter in a set of parametric equations.
■ Find a set of parametric equations to represent a curve.
■ Understand two classic calculus problems, the tautochrone and brachistochrone problems.

Plane Curves and Parametric Equations

Until now, you have been representing a graph by a single equation involving *two* variables. In this section, you will study situations in which *three* variables are used to represent a curve in the plane.

Consider the path followed by an object that is propelled into the air at an angle of 45°. For an initial velocity of 48 feet per second, the object travels the parabolic path given by the rectangular equation

$$y = -\frac{x^2}{72} + x$$

as shown in Figure 10.19. This equation, however, does not tell the whole story. Although it does tell you *where* the object has been, it does not tell you *when* the object was at a given point (x, y). To determine this time, you can introduce a third variable t, called a **parameter.** By writing both x and y as functions of t, you obtain the **parametric equations**

$$x = 24\sqrt{2}t \qquad \text{Parametric equation for } x$$

and

$$y = -16t^2 + 24\sqrt{2}t. \qquad \text{Parametric equation for } y$$

From this set of equations, you can determine that at time $t = 0$, the object is at the point $(0, 0)$. Similarly, at time $t = 1$, the object is at the point

$$\left(24\sqrt{2}, 24\sqrt{2} - 16\right)$$

and so on. (You will learn a method for determining this particular set of parametric equations—the equations of motion—later, in Section 12.3.)

For this particular motion problem, x and y are continuous functions of t, and the resulting path is called a **plane curve.**

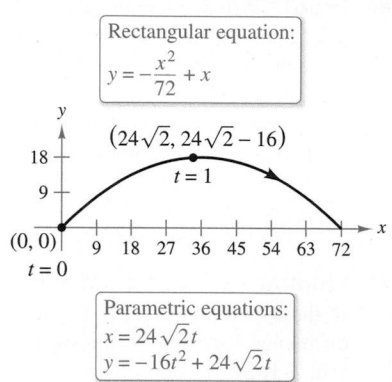

Rectangular equation:
$$y = -\frac{x^2}{72} + x$$

Parametric equations:
$$x = 24\sqrt{2}t$$
$$y = -16t^2 + 24\sqrt{2}t$$

Curvilinear motion: two variables for position, one variable for time
Figure 10.19

•• **REMARK** At times, it is important to distinguish between a graph (the set of points) and a curve (the points together with their defining parametric equations). When it is important, the distinction will be explicit. When it is not important, C will be used to represent either the graph or the curve.

Definition of a Plane Curve

If f and g are continuous functions of t on an interval I, then the equations

$$x = f(t) \quad \text{and} \quad y = g(t)$$

are **parametric equations** and t is the **parameter.** The set of points (x, y) obtained as t varies over the interval I is the **graph** of the parametric equations. Taken together, the parametric equations and the graph are a **plane curve,** denoted by C.

When sketching a curve represented by a set of parametric equations, you can plot points in the xy-plane. Each set of coordinates (x, y) is determined from a value chosen for the parameter t. By plotting the resulting points in order of increasing values of t, the curve is traced out in a specific direction. This is called the **orientation** of the curve.

EXAMPLE 1 Sketching a Curve

Sketch the curve described by the parametric equations

$$x = f(t) = t^2 - 4$$

and

$$y = g(t) = \frac{t}{2}$$

where $-2 \leq t \leq 3$.

Solution For values of t on the given interval, the parametric equations yield the points (x, y) shown in the table.

t	-2	-1	0	1	2	3
x	0	-3	-4	-3	0	5
y	-1	$-\frac{1}{2}$	0	$\frac{1}{2}$	1	$\frac{3}{2}$

By plotting these points in order of increasing values of t and using the continuity of f and g, you obtain the curve C shown in Figure 10.20. Note that the arrows on the curve indicate its orientation as t increases from -2 to 3.

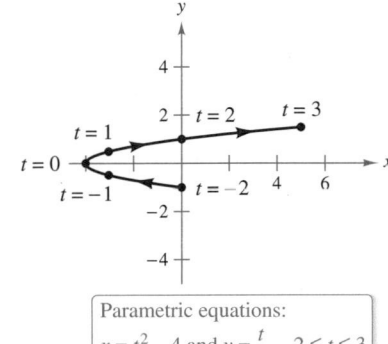

Parametric equations:
$x = t^2 - 4$ and $y = \dfrac{t}{2}, -2 \leq t \leq 3$

Figure 10.20

According to the Vertical Line Test, the graph shown in Figure 10.20 does not define y as a function of x. This points out one benefit of parametric equations—they can be used to represent graphs that are more general than graphs of functions.

It often happens that two different sets of parametric equations have the same graph. For instance, the set of parametric equations

$$x = 4t^2 - 4 \quad \text{and} \quad y = t, \quad -1 \leq t \leq \frac{3}{2}$$

has the same graph as the set given in Example 1. (See Figure 10.21.) However, comparing the values of t in Figures 10.20 and 10.21, you can see that the second graph is traced out more *rapidly* (considering t as time) than the first graph. So, in applications, different parametric representations can be used to represent various *speeds* at which objects travel along a given path.

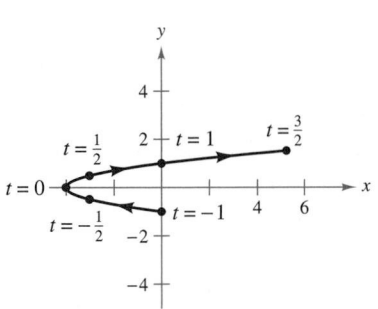

Parametric equations:
$x = 4t^2 - 4$ and $y = t, -1 \leq t \leq \frac{3}{2}$

Figure 10.21

▷ **TECHNOLOGY** Most graphing utilities have a *parametric* graphing mode. If you have access to such a utility, use it to confirm the graphs shown in Figures 10.20 and 10.21. Does the curve given by the parametric equations

$$x = 4t^2 - 8t \quad \text{and} \quad y = 1 - t, \quad -\frac{1}{2} \leq t \leq 2$$

represent the same graph as that shown in Figures 10.20 and 10.21? What do you notice about the *orientation* of this curve?

Eliminating the Parameter

Finding a rectangular equation that represents the graph of a set of parametric equations is called **eliminating the parameter.** For instance, you can eliminate the parameter from the set of parametric equations in Example 1 as follows.

Parametric equations	⇨	Solve for t in one equation.	⇨	Substitute into second equation.	⇨	Rectangular equation
$x = t^2 - 4$ $y = t/2$		$t = 2y$		$x = (2y)^2 - 4$		$x = 4y^2 - 4$

Once you have eliminated the parameter, you can recognize that the equation $x = 4y^2 - 4$ represents a parabola with a horizontal axis and vertex at $(-4, 0)$, as shown in Figure 10.20.

The range of x and y implied by the parametric equations may be altered by the change to rectangular form. In such instances, the domain of the rectangular equation must be adjusted so that its graph matches the graph of the parametric equations. Such a situation is demonstrated in the next example.

EXAMPLE 2 Adjusting the Domain

Sketch the curve represented by the equations

$$x = \frac{1}{\sqrt{t + 1}} \quad \text{and} \quad y = \frac{t}{t + 1}, \quad t > -1$$

by eliminating the parameter and adjusting the domain of the resulting rectangular equation.

Solution Begin by solving one of the parametric equations for t. For instance, you can solve the first equation for t as follows.

$$x = \frac{1}{\sqrt{t + 1}} \qquad \text{Parametric equation for } x$$

$$x^2 = \frac{1}{t + 1} \qquad \text{Square each side.}$$

$$t + 1 = \frac{1}{x^2}$$

$$t = \frac{1}{x^2} - 1$$

$$t = \frac{1 - x^2}{x^2} \qquad \text{Solve for } t.$$

Now, substituting into the parametric equation for y produces

$$y = \frac{t}{t + 1} \qquad \text{Parametric equation for } y$$

$$y = \frac{(1 - x^2)/x^2}{[(1 - x^2)/x^2] + 1} \qquad \text{Substitute } (1 - x^2)/x^2 \text{ for } t.$$

$$y = 1 - x^2. \qquad \text{Simplify.}$$

The rectangular equation, $y = 1 - x^2$, is defined for all values of x, but from the parametric equation for x, you can see that the curve is defined only when $t > -1$. This implies that you should restrict the domain of x to positive values, as shown in Figure 10.22.

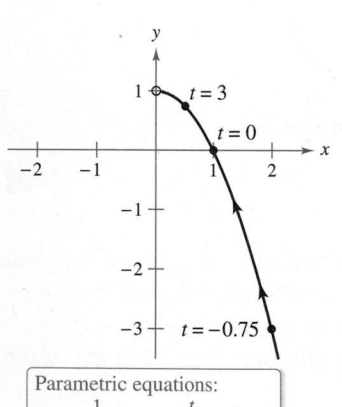

Parametric equations:
$x = \dfrac{1}{\sqrt{t + 1}}, \; y = \dfrac{t}{t + 1}, \; t > -1$

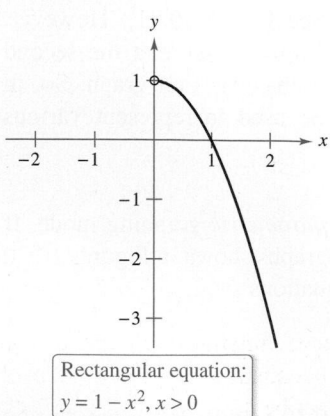

Rectangular equation:
$y = 1 - x^2, \; x > 0$

Figure 10.22

It is not necessary for the parameter in a set of parametric equations to represent time. The next example uses an *angle* as the parameter.

EXAMPLE 3 **Using Trigonometry to Eliminate a Parameter**

⋯▷ *See LarsonCalculus.com for an interactive version of this type of example.*

Sketch the curve represented by

$$x = 3 \cos \theta \quad \text{and} \quad y = 4 \sin \theta, \quad 0 \le \theta \le 2\pi$$

by eliminating the parameter and finding the corresponding rectangular equation.

Solution Begin by solving for $\cos \theta$ and $\sin \theta$ in the given equations.

$$\cos \theta = \frac{x}{3} \qquad \text{Solve for } \cos \theta.$$

and

$$\sin \theta = \frac{y}{4} \qquad \text{Solve for } \sin \theta.$$

Next, make use of the identity

$$\sin^2 \theta + \cos^2 \theta = 1$$

to form an equation involving only x and y.

$$\cos^2 \theta + \sin^2 \theta = 1 \qquad \text{Trigonometric identity}$$

$$\left(\frac{x}{3}\right)^2 + \left(\frac{y}{4}\right)^2 = 1 \qquad \text{Substitute.}$$

$$\frac{x^2}{9} + \frac{y^2}{16} = 1 \qquad \text{Rectangular equation}$$

From this rectangular equation, you can see that the graph is an ellipse centered at $(0, 0)$, with vertices at $(0, 4)$ and $(0, -4)$ and minor axis of length $2b = 6$, as shown in Figure 10.23. Note that the ellipse is traced out *counterclockwise* as θ varies from 0 to 2π.

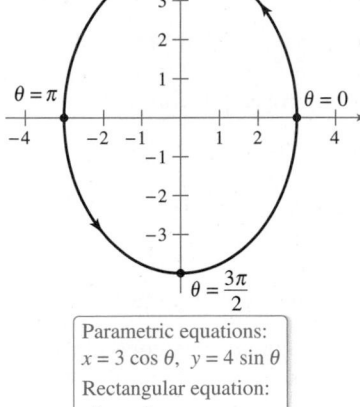

Parametric equations:
$x = 3 \cos \theta, \ y = 4 \sin \theta$

Rectangular equation:
$\dfrac{x^2}{9} + \dfrac{y^2}{16} = 1$

Figure 10.23

Using the technique shown in Example 3, you can conclude that the graph of the parametric equations

$$x = h + a \cos \theta \quad \text{and} \quad y = k + b \sin \theta, \quad 0 \le \theta \le 2\pi$$

is the ellipse (traced counterclockwise) given by

$$\frac{(x - h)^2}{a^2} + \frac{(y - k)^2}{b^2} = 1.$$

The graph of the parametric equations

$$x = h + a \sin \theta \quad \text{and} \quad y = k + b \cos \theta, \quad 0 \le \theta \le 2\pi$$

is also the ellipse (traced clockwise) given by

$$\frac{(x - h)^2}{a^2} + \frac{(y - k)^2}{b^2} = 1.$$

In Examples 2 and 3, it is important to realize that eliminating the parameter is primarily an *aid to curve sketching*. When the parametric equations represent the path of a moving object, the graph alone is not sufficient to describe the motion of the object. You still need the parametric equations to tell you the *position, direction,* and *speed* at a given time.

▷ **TECHNOLOGY** Use a graphing utility in *parametric* mode to graph several ellipses.

Finding Parametric Equations

The first three examples in this section illustrate techniques for sketching the graph represented by a set of parametric equations. You will now investigate the reverse problem. How can you determine a set of parametric equations for a given graph or a given physical description? From the discussion following Example 1, you know that such a representation is not unique. This is demonstrated further in the next example, which finds two different parametric representations for a given graph.

EXAMPLE 4   Finding Parametric Equations for a Given Graph

Find a set of parametric equations that represents the graph of $y = 1 - x^2$, using each of the following parameters.

a. $t = x$ **b.** The slope $m = \dfrac{dy}{dx}$ at the point (x, y)

Solution

a. Letting $x = t$ produces the parametric equations

$$x = t \quad \text{and} \quad y = 1 - x^2 = 1 - t^2.$$

b. To write x and y in terms of the parameter m, you can proceed as follows.

$$m = \frac{dy}{dx}$$

$$m = -2x \qquad \text{Differentiate } y = 1 - x^2.$$

$$x = -\frac{m}{2} \qquad \text{Solve for } x.$$

This produces a parametric equation for x. To obtain a parametric equation for y, substitute $-m/2$ for x in the original equation.

$$y = 1 - x^2 \qquad \text{Write original rectangular equation.}$$

$$y = 1 - \left(-\frac{m}{2}\right)^2 \qquad \text{Substitute } -m/2 \text{ for } x.$$

$$y = 1 - \frac{m^2}{4} \qquad \text{Simplify.}$$

So, the parametric equations are

$$x = -\frac{m}{2} \quad \text{and} \quad y = 1 - \frac{m^2}{4}.$$

In Figure 10.24, note that the resulting curve has a right-to-left orientation as determined by the increasing values of slope m. For part (a), the curve would have the opposite orientation. ∎

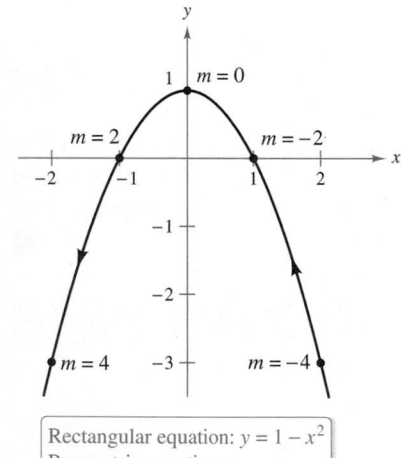

Rectangular equation: $y = 1 - x^2$
Parametric equations:

$$x = -\frac{m}{2}, \ y = 1 - \frac{m^2}{4}$$

Figure 10.24

▷ **TECHNOLOGY** To be efficient at using a graphing utility, it is important that you develop skill in representing a graph by a set of parametric equations. The reason for this is that many graphing utilities have only three graphing modes—(1) functions, (2) parametric equations, and (3) polar equations. Most graphing utilities are not programmed to graph a general equation. For instance, suppose you want to graph the hyperbola $x^2 - y^2 = 1$. To graph the hyperbola in *function* mode, you need two equations

$$y = \sqrt{x^2 - 1} \quad \text{and} \quad y = -\sqrt{x^2 - 1}.$$

In *parametric* mode, you can represent the graph by $x = \sec t$ and $y = \tan t$.

■ **FOR FURTHER INFORMATION**
To read about other methods for finding parametric equations, see the article "Finding Rational Parametric Curves of Relative Degree One or Two" by Dave Boyles in *The College Mathematics Journal*. To view this article, go to *MathArticles.com*.

■ **FOR FURTHER INFORMATION**
For more information on cycloids, see the article "The Geometry of Rolling Curves" by John Bloom and Lee Whitt in *The American Mathematical Monthly.* To view this article, go to *MathArticles.com.*

EXAMPLE 5 **Parametric Equations for a Cycloid**

Determine the curve traced by a point P on the circumference of a circle of radius a rolling along a straight line in a plane. Such a curve is called a **cycloid.**

Solution Let the parameter θ be the measure of the circle's rotation, and let the point $P = (x, y)$ begin at the origin. When $\theta = 0$, P is at the origin. When $\theta = \pi$, P is at a maximum point $(\pi a, 2a)$. When $\theta = 2\pi$, P is back on the x-axis at $(2\pi a, 0)$. From Figure 10.25, you can see that $\angle APC = 180° - \theta$. So,

$$\sin \theta = \sin(180° - \theta) = \sin(\angle APC) - \frac{AC}{a} = \frac{BD}{a}$$

$$\cos \theta = -\cos(180° - \theta) = -\cos(\angle APC) = \frac{AP}{-a}$$

which implies that $AP = -a \cos \theta$ and $BD = a \sin \theta$.

Because the circle rolls along the x-axis, you know that $OD = \overset{\frown}{PD} = a\theta$. Furthermore, because $BA = DC = a$, you have

$$x = OD - BD = a\theta - a \sin \theta$$
$$y = BA + AP = a - a \cos \theta.$$

So, the parametric equations are

$$x = a(\theta - \sin \theta) \quad \text{and} \quad y = a(1 - \cos \theta).$$

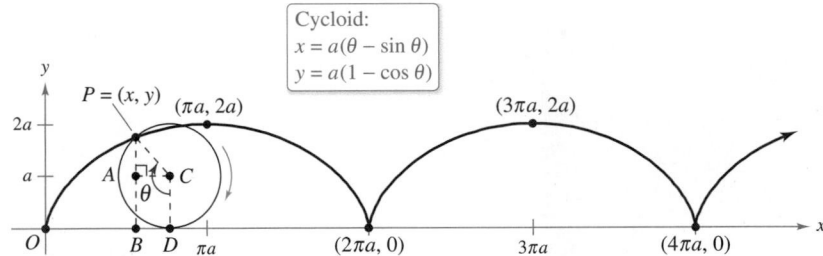

Cycloid:
$x = a(\theta - \sin \theta)$
$y = a(1 - \cos \theta)$

Figure 10.25

▷ **TECHNOLOGY** Some graphing utilities allow you to simulate the motion of an object that is moving in the plane or in space. If you have access to such a utility, use it to trace out the path of the cycloid shown in Figure 10.25.

The cycloid in Figure 10.25 has sharp corners called **cusps** at the values $x = 2n\pi a$. Notice that the derivatives $x'(\theta)$ and $y'(\theta)$ are both zero at the points for which $\theta = 2n\pi$.

$$x(\theta) = a(\theta - \sin \theta) \qquad y(\theta) = a(1 - \cos \theta)$$
$$x'(\theta) = a - a \cos \theta \qquad y'(\theta) = a \sin \theta$$
$$x'(2n\pi) = 0 \qquad y'(2n\pi) = 0$$

Between these points, the cycloid is called **smooth.**

Definition of a Smooth Curve

A curve C represented by $x = f(t)$ and $y = g(t)$ on an interval I is called **smooth** when f' and g' are continuous on I and not simultaneously 0, except possibly at the endpoints of I. The curve C is called **piecewise smooth** when it is smooth on each subinterval of some partition of I.

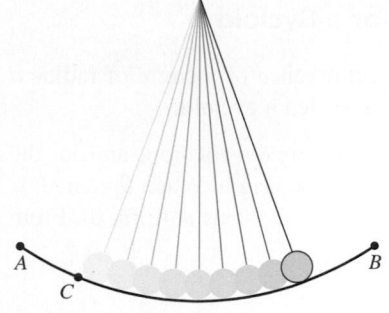

The time required to complete a full swing of the pendulum when starting from point *C* is only approximately the same as the time required when starting from point *A*.

Figure 10.26

JAMES BERNOULLI (1654–1705)

James Bernoulli, also called Jacques, was the older brother of John. He was one of several accomplished mathematicians of the Swiss Bernoulli family. James's mathematical accomplishments have given him a prominent place in the early development of calculus. *See LarsonCalculus.com to read more of this biography.*

The Tautochrone and Brachistochrone Problems

The curve described in Example 5 is related to one of the most famous pairs of problems in the history of calculus. The first problem (called the **tautochrone problem**) began with Galileo's discovery that the time required to complete a full swing of a pendulum is *approximately* the same whether it makes a large movement at high speed or a small movement at lower speed (see Figure 10.26). Late in his life, Galileo realized that he could use this principle to construct a clock. However, he was not able to conquer the mechanics of actual construction. Christian Huygens (1629–1695) was the first to design and construct a working model. In his work with pendulums, Huygens realized that a pendulum does not take exactly the same time to complete swings of varying lengths. (This does not affect a pendulum clock, because the length of the circular arc is kept constant by giving the pendulum a slight boost each time it passes its lowest point.) But, in studying the problem, Huygens discovered that a ball rolling back and forth on an inverted cycloid does complete each cycle in exactly the same time.

The second problem, which was posed by John Bernoulli in 1696, is called the **brachistochrone problem**—in Greek, *brachys* means short and *chronos* means time. The problem was to determine the path down which a particle (such as a ball) will slide from point *A* to point *B* in the *shortest time*. Several mathematicians took up the challenge, and the following year the problem was solved by Newton, Leibniz, L'Hôpital, John Bernoulli, and James Bernoulli. As it turns out, the solution is not a straight line from *A* to *B*, but an inverted cycloid passing through the points *A* and *B*, as shown in Figure 10.27.

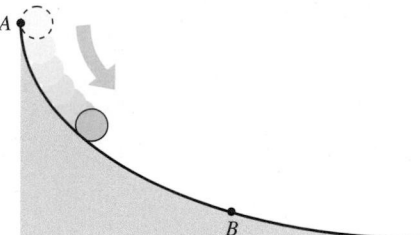

An inverted cycloid is the path down which a ball will roll in the shortest time.
Figure 10.27

The amazing part of the solution to the brachistochrone problem is that a particle starting at rest at *any* point *C* of the cycloid between *A* and *B* will take exactly the same time to reach *B*, as shown in Figure 10.28.

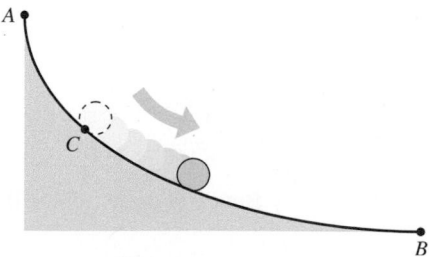

A ball starting at point *C* takes the same time to reach point *B* as one that starts at point *A*.
Figure 10.28

■ **FOR FURTHER INFORMATION** To see a proof of the famous brachistochrone problem, see the article "A New Minimization Proof for the Brachistochrone" by Gary Lawlor in *The American Mathematical Monthly*. To view this article, go to *MathArticles.com*.

10.2 Exercises

See **CalcChat.com** for tutorial help and worked-out solutions to odd-numbered exercises.

CONCEPT CHECK

1. **Parametric Equations** What information does a set of parametric equations provide that is lacking in a rectangular equation for describing the motion of an object?

2. **Plane Curve** Explain the process of sketching a plane curve given by parametric equations. What is meant by the orientation of the curve?

3. **Think About It** How can two sets of parametric equations represent the same graph but different curves?

4. **Adjusting a Domain** Consider the parametric equations

$$x = \sqrt{t - 2} \quad \text{and} \quad y = \frac{1}{2}t + 1, \quad t \geq 2.$$

What is implied about the domain of the resulting rectangular equation?

 **Using Parametric Equations** In Exercises 5–22, sketch the curve represented by the parametric equations (indicate the orientation of the curve), and write the corresponding rectangular equation by eliminating the parameter.

5. $x = 2t - 3, \quad y = 3t + 1$

6. $x = 5 - 4t, \quad y = 2 + 5t$

7. $x = t + 1, \quad y = t^2$

8. $x = 2t^2, \quad y = t^4 + 1$

9. $x = t^3, \quad y = \dfrac{t^2}{2}$

10. $x = t^2 + t, \quad y = t^2 - t$

11. $x = \sqrt{t}, \quad y = t - 5$

12. $x = \sqrt[4]{t}, \quad y = 8 - t$

13. $x = t - 3, \quad y = \dfrac{t}{t - 3}$

14. $x = 1 + \dfrac{1}{t}, \quad y = t - 1$

15. $x = 2t, \quad y = |t - 2|$

16. $x = |t - 1|, \quad y = t + 2$

17. $x = e^t, \quad y = e^{3t} + 1$

18. $x = e^{-t}, \quad y = e^{2t} - 1$

19. $x = 8 \cos \theta, \quad y = 8 \sin \theta$

20. $x = 3 \cos \theta, \quad y = 7 \sin \theta$

21. $x = \sec \theta, \quad y = \cos \theta, \quad 0 \leq \theta < \pi/2, \quad \pi/2 < \theta \leq \pi$

22. $x = \tan^2 \theta, \quad y = \sec^2 \theta$

 **Using Parametric Equations** In Exercises 23–34, use a graphing utility to graph the curve represented by the parametric equations (indicate the orientation of the curve). Eliminate the parameter and write the corresponding rectangular equation.

23. $x = 6 \sin 2\theta$
    $y = 4 \cos 2\theta$

24. $x = \cos \theta$
    $y = 2 \sin 2\theta$

25. $x = 4 + 2 \cos \theta$
    $y = -1 + \sin \theta$

26. $x = -2 + 3 \cos \theta$
    $y = -5 + 3 \sin \theta$

27. $x = -3 + 4 \cos \theta$
    $y = 2 + 5 \sin \theta$

28. $x = \sec \theta$
    $y = \tan \theta$

29. $x = 4 \sec \theta$
    $y = 3 \tan \theta$

30. $x = \cos^3 \theta$
    $y = \sin^3 \theta$

31. $x = t^3, \quad y = 3 \ln t$

32. $x = \ln 2t, \quad y = t^2$

33. $x = e^{-t}, \quad y = e^{3t}$

34. $x = e^{2t}, \quad y = e^t$

Comparing Plane Curves In Exercises 35–38, determine any differences between the curves of the parametric equations. Are the graphs the same? Are the orientations the same? Are the curves smooth? Explain.

35. (a) $x = t, y = t^2$ (b) $x = -t, y = t^2$

36. (a) $x = t + 1, y = t^3$ (b) $x = -t + 1, y = (-t)^3$

37. (a) $x = t$ (b) $x = \cos \theta$
    $\quad y = 2t + 1$ $y = 2 \cos \theta + 1$
    (c) $x = e^{-t}$ (d) $x = e^t$
    $\quad y = 2e^{-t} + 1$ $y = 2e^t + 1$

38. (a) $x = 2 \cos \theta$ (b) $x = \sqrt{4t^2 - 1}/|t|$
    $\quad y = 2 \sin \theta$ $y = 1/t$
    (c) $x = \sqrt{t}$ (d) $x = -\sqrt{4 - e^{2t}}$
    $\quad y = \sqrt{4 - t}$ $y = e^t$

Eliminating a Parameter In Exercises 39–42, eliminate the parameter and obtain the standard form of the rectangular equation.

39. Line through (x_1, y_1) and (x_2, y_2):
    $x = x_1 + t(x_2 - x_1), \quad y = y_1 + t(y_2 - y_1)$

40. Circle: $x = h + r \cos \theta, \quad y = k + r \sin \theta$

41. Ellipse: $x = h + a \cos \theta, \quad y = k + b \sin \theta$

42. Hyperbola: $x = h + a \sec \theta, \quad y = k + b \tan \theta$

Writing a Set of Parametric Equations In Exercises 43–50, use the results of Exercises 39–42 to find a set of parametric equations for the line or conic.

43. Line: passes through $(0, 0)$ and $(4, -7)$

44. Line: passes through $(-3, 1)$ and $(1, 9)$

45. Circle: center: $(1, 1)$; radius: 2

46. Circle: center: $\left(-\frac{1}{2}, -4\right)$; radius: $\frac{1}{2}$

47. Ellipse: vertices: $(-3, 0), (7, 0)$; foci: $(-1, 0), (5, 0)$

48. Ellipse: vertices: $(-1, 8), (-1, -12)$; foci: $(-1, 4), (-1, -8)$

49. Hyperbola: vertices: $(0, \pm 1)$; foci: $\left(0, \pm \sqrt{5}\right)$

50. Hyperbola: vertices: $(-2, 1), (0, 1)$; foci: $(-3, 1), (1, 1)$

 **Finding Parametric Equations** In Exercises 51–54, find two different sets of parametric equations for the rectangular equation.

51. $y = 6x - 5$

52. $y = 4/(x - 1)$

53. $y = x^3$

54. $y = x^2$

Finding Parametric Equations In Exercises 55–58, find a set of parametric equations for the rectangular equation that satisfies the given condition.

55. $y = 2x - 5$, $t = 0$ at the point $(3, 1)$

56. $y = 4x + 1$, $t = -1$ at the point $(-2, -7)$

57. $y = x^2$, $t = 4$ at the point $(4, 16)$

58. $y = 4 - x^2$, $t = 1$ at the point $(1, 3)$

Graphing a Plane Curve In Exercises 59–66, use a graphing utility to graph the curve represented by the parametric equations. Indicate the orientation of the curve. Identify any points at which the curve is not smooth.

59. Cycloid: $x = 2(\theta - \sin \theta)$, $y = 2(1 - \cos \theta)$

60. Cycloid: $x = \theta + \sin \theta$, $y = 1 - \cos \theta$

61. Prolate cycloid: $x = \theta - \frac{3}{2}\sin \theta$, $y = 1 - \frac{3}{2}\cos \theta$

62. Prolate cycloid: $x = 2\theta - 4\sin \theta$, $y = 2 - 4\cos \theta$

63. Hypocycloid: $x = 3\cos^3 \theta$, $y = 3\sin^3 \theta$

64. Curtate cycloid: $x = 2\theta - \sin \theta$, $y = 2 - \cos \theta$

65. Witch of Agnesi: $x = 2\cot \theta$, $y = 2\sin^2 \theta$

66. Folium of Descartes: $x = 3t/(1 + t^3)$, $y = 3t^2/(1 + t^3)$

EXPLORING CONCEPTS

67. Orientation Describe the orientation of the parametric equations $x = t^2$ and $y = t^4$ for $-1 \le t \le 1$.

68. Conjecture Make a conjecture about the change in the graph of parametric equations when the sign of the parameter is changed. Explain your reasoning using examples to support your conjecture.

69. Think About It The following sets of parametric equations have the same graph. Does this contradict your conjecture from Exercise 68? Explain.

$x = \cos \theta$, $y = \sin^2 \theta$, $0 < \theta < \pi$

$x = \cos(-\theta)$, $y = \sin^2(-\theta)$, $0 < \theta < \pi$

70. **HOW DO YOU SEE IT?** Which set of parametric equations is shown in the graph below? Explain your reasoning.

(a) $x = t$ (b) $x = t^2$

    $y = t^2$ $y = t$

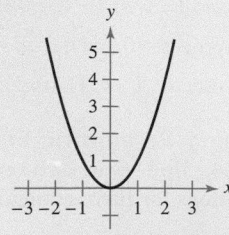

Matching In Exercises 71–74, match the set of parametric equations with its graph. [The graphs are labeled (a), (b), (c), and (d).] Explain your reasoning.

(a)
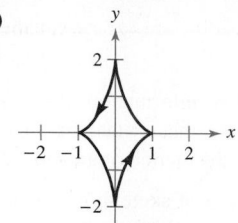

(b)
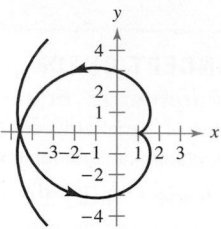

(c)
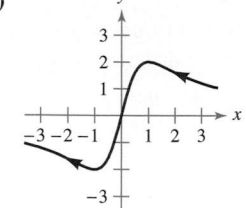

(d)
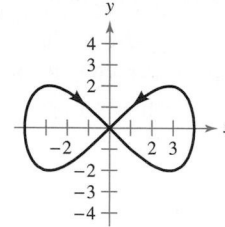

71. Lissajous curve: $x = 4\cos \theta$, $y = 2\sin 2\theta$

72. Evolute of ellipse: $x = \cos^3 \theta$, $y = 2\sin^3 \theta$

73. Involute of circle: $x = \cos \theta + \theta \sin \theta$, $y = \sin \theta - \theta \cos \theta$

74. Serpentine curve: $x = \cot \theta$, $y = 4\sin \theta \cos \theta$

75. Curtate Cycloid A wheel of radius a rolls along a line without slipping. The curve traced by a point P that is b units from the center ($b < a$) is called a **curtate cycloid** (see figure). Use the angle θ to find a set of parametric equations for this curve.

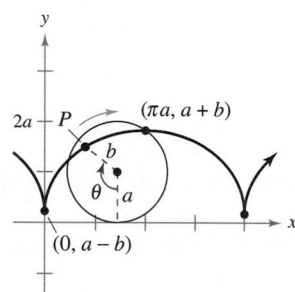

Figure for 75   Figure for 76

76. Epicycloid A circle of radius 1 rolls around the outside of a circle of radius 2 without slipping. The curve traced by a point on the circumference of the smaller circle is called an **epicycloid** (see figure). Use the angle θ to find a set of parametric equations for this curve.

True or False? In Exercises 77–79, determine whether the statement is true or false. If it is false, explain why or give an example that shows it is false.

77. The graph of the parametric equations $x = t^2$ and $y = t^2$ is the line $y = x$.

78. If y is a function of t and x is a function of t, then y is a function of x.

79. The curve represented by the parametric equations $x = t$ and $y = \cos t$ can be written as an equation of the form $y = f(x)$.

80. Translation of a Plane Curve   Consider the parametric equations $x = 8 \cos t$ and $y = 8 \sin t$.

(a) Describe the curve represented by the parametric equations.

(b) How does the curve represented by the parametric equations

$$x = 8 \cos t + 3 \text{ and } y = 8 \sin t + 6$$

compare to the curve described in part (a)?

(c) How does the original curve change when cosine and sine are interchanged?

Projectile Motion   **In Exercises 81 and 82, consider a projectile launched at a height h feet above the ground and at an angle θ with the horizontal. When the initial velocity is v_0 feet per second, the path of the projectile is modeled by the parametric equations**

$$x = (v_0 \cos \theta)t$$

and

$$y = h + (v_0 \sin \theta)t - 16t^2.$$

• **81. Baseball** • • • • • • • • • • • • • • • • • • •

The center field fence in a ballpark is 10 feet high and 400 feet from home plate. The ball is hit 3 feet above the ground. It leaves the bat at an angle of θ degrees with the horizontal at a speed of 100 miles per hour.

(a) Write a set of parametric equations for the path of the ball.

(b) Use a graphing utility to graph the path of the ball when $\theta = 15°$. Is the hit a home run?

(c) Use a graphing utility to graph the path of the ball when $\theta = 23°$. Is the hit a home run?

(d) Find the minimum angle at which the ball must leave the bat in order for the hit to be a home run.

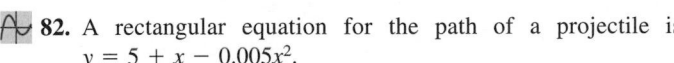

82. A rectangular equation for the path of a projectile is $y = 5 + x - 0.005x^2$.

(a) Eliminate the parameter t from the position function for the motion of a projectile to show that the rectangular equation is

$$y = -\frac{16 \sec^2 \theta}{v_0^2}x^2 + (\tan \theta)x + h.$$

(b) Use the result of part (a) to find h, v_0, and θ. Find the parametric equations of the path.

(c) Use a graphing utility to graph the rectangular equation for the path of the projectile. Confirm your answer in part (b) by sketching the curve represented by the parametric equations.

(d) Use a graphing utility to approximate the maximum height of the projectile and its range.

SECTION PROJECT

Cycloids

In Greek, the word *cycloid* means *wheel*, the word *hypocycloid* means *under the wheel*, and the word *epicycloid* means *upon the wheel*. Match the hypocycloid or epicycloid with its graph. [The graphs are labeled (a), (b), (c), (d), (e), and (f).]

Hypocycloid, H(A, B)

The path traced by a fixed point on a circle of radius B as it rolls around the *inside* of a circle of radius A

$$x = (A - B) \cos t + B \cos\left(\frac{A - B}{B}\right)t$$

$$y = (A - B) \sin t - B \sin\left(\frac{A - B}{B}\right)t$$

Epicycloid, E(A, B)

The path traced by a fixed point on a circle of radius B as it rolls around the *outside* of a circle of radius A

$$x = (A + B) \cos t - B \cos\left(\frac{A + B}{B}\right)t$$

$$y = (A + B) \sin t - B \sin\left(\frac{A + B}{B}\right)t$$

I. $H(8, 3)$	II. $E(8, 3)$	III. $H(8, 7)$
IV. $E(24, 3)$	V. $H(24, 7)$	VI. $E(24, 7)$

(a)
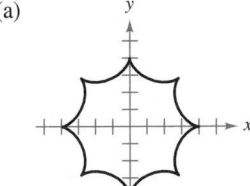

(b)
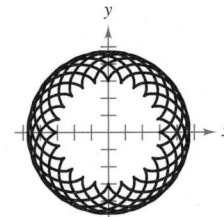

(c)
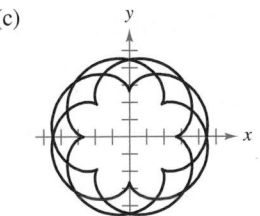

(d)
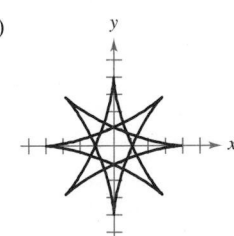

(e)
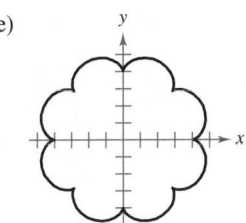

(f)
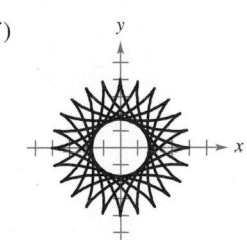

Exercises based on "Mathematical Discovery via Computer Graphics: Hypocycloids and Epicycloids" by Florence S. Gordon and Sheldon P. Gordon, *College Mathematics Journal*, November 1984, p. 441. Used by permission of the authors.

10.3 Parametric Equations and Calculus

- Find the slope of a tangent line to a curve given by a set of parametric equations.
- Find the arc length of a curve given by a set of parametric equations.
- Find the area of a surface of revolution (parametric form).

Slope and Tangent Lines

Now that you can represent a graph in the plane by a set of parametric equations, it is natural to ask how to use calculus to study plane curves. Consider the projectile represented by the parametric equations

$$x = 24\sqrt{2}t \quad \text{and} \quad y = -16t^2 + 24\sqrt{2}t$$

as shown in Figure 10.29. From the discussion at the beginning of Section 10.2, you know that these equations enable you to locate the position of the projectile at a given time. You also know that the object is initially projected at an angle of 45°, or a slope of $m = \tan 45° = 1$. But how can you find the slope at some other time t? The next theorem answers this question by giving a formula for the slope of the tangent line as a function of t.

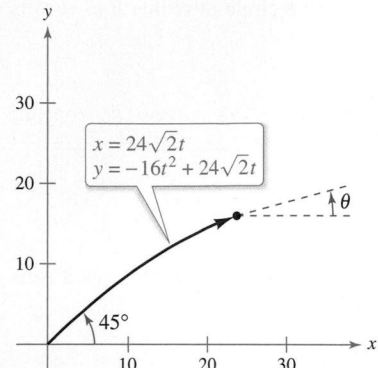

At time t, the angle of elevation of the projectile is θ.
Figure 10.29

THEOREM 10.7 Parametric Form of the Derivative

If a smooth curve C is given by the equations

$$x = f(t) \quad \text{and} \quad y = g(t)$$

then the slope of C at (x, y) is

$$\frac{dy}{dx} = \frac{dy/dt}{dx/dt}, \quad \frac{dx}{dt} \neq 0.$$

Proof In Figure 10.30, consider $\Delta t > 0$ and let

$$\Delta y = g(t + \Delta t) - g(t) \quad \text{and} \quad \Delta x = f(t + \Delta t) - f(t).$$

Because $\Delta x \to 0$ as $\Delta t \to 0$, you can write

$$\frac{dy}{dx} = \lim_{\Delta x \to 0} \frac{\Delta y}{\Delta x}$$

$$= \lim_{\Delta t \to 0} \frac{g(t + \Delta t) - g(t)}{f(t + \Delta t) - f(t)}.$$

Dividing both the numerator and denominator by Δt, you can use the differentiability of f and g to conclude that

$$\frac{dy}{dx} = \lim_{\Delta t \to 0} \frac{[g(t + \Delta t) - g(t)]/\Delta t}{[f(t + \Delta t) - f(t)]/\Delta t}$$

$$= \frac{\displaystyle\lim_{\Delta t \to 0} \frac{g(t + \Delta t) - g(t)}{\Delta t}}{\displaystyle\lim_{\Delta t \to 0} \frac{f(t + \Delta t) - f(t)}{\Delta t}}$$

$$= \frac{g'(t)}{f'(t)}$$

$$= \frac{dy/dt}{dx/dt}.$$

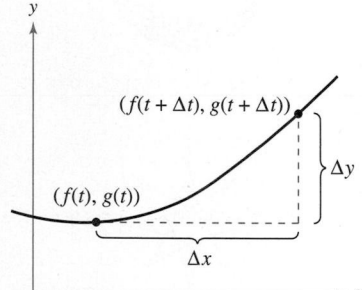

The slope of the secant line through the points $(f(t), g(t))$ and $(f(t + \Delta t), g(t + \Delta t))$ is $\Delta y/\Delta x$.
Figure 10.30

EXAMPLE 1   Differentiation and Parametric Form

Find dy/dx for the curve given by $x = \sin t$ and $y = \cos t$.

Solution

$$\frac{dy}{dx} = \frac{dy/dt}{dx/dt}$$

$$= \frac{-\sin t}{\cos t}$$

$$= -\tan t$$

Because dy/dx is a function of t, you can use Theorem 10.7 repeatedly to find *higher-order* derivatives. For instance,

$$\frac{d^2y}{dx^2} = \frac{d}{dx}\left[\frac{dy}{dx}\right] = \frac{\frac{d}{dt}\left[\frac{dy}{dx}\right]}{dx/dt} \qquad \text{Second derivative}$$

$$\frac{d^3y}{dx^3} = \frac{d}{dx}\left[\frac{d^2y}{dx^2}\right] = \frac{\frac{d}{dt}\left[\frac{d^2y}{dx^2}\right]}{dx/dt}. \qquad \text{Third derivative}$$

EXAMPLE 2   Finding Slope and Concavity

For the curve given by

$$x = \sqrt{t} \quad \text{and} \quad y = \frac{1}{4}(t^2 - 4), \quad t \ge 0$$

find the slope and concavity at the point $(2, 3)$.

Solution   Because

$$\frac{dy}{dx} = \frac{dy/dt}{dx/dt} = \frac{(1/2)t}{(1/2)t^{-1/2}} = t^{3/2} \qquad \text{Parametric form of first derivative}$$

you can find the second derivative to be

$$\frac{d^2y}{dx^2} = \frac{\frac{d}{dt}\left[\frac{dy}{dx}\right]}{dx/dt} = \frac{\frac{d}{dt}[t^{3/2}]}{dx/dt} = \frac{(3/2)t^{1/2}}{(1/2)t^{-1/2}} = 3t. \qquad \text{Parametric form of second derivative}$$

At $(x, y) = (2, 3)$, it follows that $t = 4$, and the slope is

$$\frac{dy}{dx} = (4)^{3/2} = 8.$$

Moreover, when $t = 4$, the second derivative is

$$\frac{d^2y}{dx^2} = 3(4) = 12 > 0$$

and you can conclude that the graph is concave upward at $(2, 3)$, as shown in Figure 10.31.

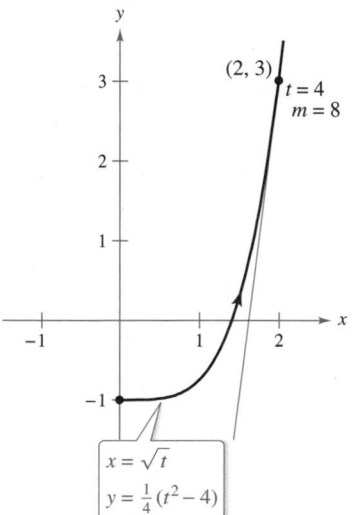

The graph is concave upward at $(2, 3)$ when $t = 4$.
Figure 10.31

Because the parametric equations $x = f(t)$ and $y = g(t)$ need not define y as a function of x, it is possible for a plane curve to loop around and cross itself. At such points, the curve may have more than one tangent line, as shown in the next example.

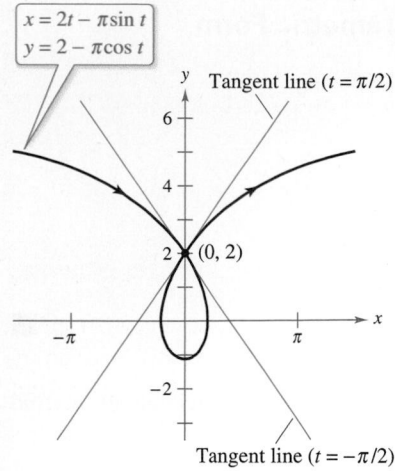

$x = 2t - \pi \sin t$
$y = 2 - \pi \cos t$

Tangent line $(t = \pi/2)$

Tangent line $(t = -\pi/2)$

This prolate cycloid has two tangent lines at the point $(0, 2)$.
Figure 10.32

EXAMPLE 3 **A Curve with Two Tangent Lines at a Point**

· · · · ▷ *See LarsonCalculus.com for an interactive version of this type of example.*

The **prolate cycloid** given by

$$x = 2t - \pi \sin t \quad \text{and} \quad y = 2 - \pi \cos t$$

crosses itself at the point $(0, 2)$, as shown in Figure 10.32. Find the equations of both tangent lines at this point.

Solution Because $x = 0$ and $y = 2$ when $t = \pm\pi/2$, and

$$\frac{dy}{dx} = \frac{dy/dt}{dx/dt} = \frac{\pi \sin t}{2 - \pi \cos t}$$

you have $dy/dx = -\pi/2$ when $t = -\pi/2$ and $dy/dx = \pi/2$ when $t = \pi/2$. So, the two tangent lines at $(0, 2)$ are

$$y - 2 = -\frac{\pi}{2}x \qquad \text{Tangent line when } t = -\frac{\pi}{2}$$

and

$$y - 2 = \frac{\pi}{2}x. \qquad \text{Tangent line when } t = \frac{\pi}{2}$$

If $dy/dt = 0$ and $dx/dt \neq 0$ when $t = t_0$, then the curve represented by $x = f(t)$ and $y = g(t)$ has a horizontal tangent at $(f(t_0), g(t_0))$. For instance, in Example 3, the given curve has a horizontal tangent at the point $(0, 2 - \pi)$ (when $t = 0$). Similarly, if $dx/dt = 0$ and $dy/dt \neq 0$ when $t = t_0$, then the curve represented by $x = f(t)$ and $y = g(t)$ has a vertical tangent at $(f(t_0), g(t_0))$. If dy/dt and dx/dt are *simultaneously* 0, then no conclusion can be drawn about tangent lines.

Arc Length

You have seen how parametric equations can be used to describe the path of a particle moving in the plane. You will now develop a formula for determining the *distance* traveled by the particle along its path.

Recall from Section 7.4 that the formula for the arc length of a curve C given by $y = h(x)$ over the interval $[x_0, x_1]$ is

$$s = \int_{x_0}^{x_1} \sqrt{1 + [h'(x)]^2} \, dx$$

$$= \int_{x_0}^{x_1} \sqrt{1 + \left(\frac{dy}{dx}\right)^2} \, dx.$$

If C is represented by the parametric equations $x = f(t)$ and $y = g(t)$, $a \leq t \leq b$, and if $dx/dt = f'(t) > 0$, then

$$s = \int_{x_0}^{x_1} \sqrt{1 + \left(\frac{dy}{dx}\right)^2} \, dx$$

$$= \int_{x_0}^{x_1} \sqrt{1 + \left(\frac{dy/dt}{dx/dt}\right)^2} \, dx$$

$$= \int_{a}^{b} \sqrt{\frac{(dx/dt)^2 + (dy/dt)^2}{(dx/dt)^2}} \frac{dx}{dt} \, dt$$

$$= \int_{a}^{b} \sqrt{\left(\frac{dx}{dt}\right)^2 + \left(\frac{dy}{dt}\right)^2} \, dt$$

$$= \int_{a}^{b} \sqrt{[f'(t)]^2 + [g'(t)]^2} \, dt.$$

> **THEOREM 10.8   Arc Length in Parametric Form**
>
> If a smooth curve C is given by $x = f(t)$ and $y = g(t)$ such that C does not intersect itself on the interval $a \le t \le b$ (except possibly at the endpoints), then the arc length of C over the interval is given by
>
> $$s = \int_a^b \sqrt{\left(\frac{dx}{dt}\right)^2 + \left(\frac{dy}{dt}\right)^2} \, dt = \int_a^b \sqrt{[f'(t)]^2 + [g'(t)]^2} \, dt.$$

REMARK   When applying the arc length formula to a curve, be sure that the curve is traced out only once on the interval of integration. For instance, the circle given by $x = \cos t$ and $y = \sin t$ is traced out once on the interval $0 \le t \le 2\pi$ but is traced out twice on the interval $0 \le t \le 4\pi$.

In the preceding section, you saw that if a circle rolls along a line, then a point on its circumference will trace a path called a cycloid. If the circle rolls around the circumference of another circle, then the path of the point is an **epicycloid**. The next example shows how to find the arc length of an epicycloid.

ARCH OF A CYCLOID

The arc length of an arch of a cycloid was first calculated in 1658 by British architect and mathematician Christopher Wren, famous for rebuilding many buildings and churches in London, including St. Paul's Cathedral.

EXAMPLE 4   Finding Arc Length

A circle of radius 1 rolls around the circumference of a larger circle of radius 4, as shown in Figure 10.33. The epicycloid traced by a point on the circumference of the smaller circle is given by

$$x = 5 \cos t - \cos 5t \quad \text{and} \quad y = 5 \sin t - \sin 5t.$$

Find the distance traveled by the point in one complete trip about the larger circle.

Solution   Before applying Theorem 10.8, note in Figure 10.33 that the curve has sharp points when $t = 0$ and $t = \pi/2$. Between these two points, dx/dt and dy/dt are not simultaneously 0. So, the portion of the curve generated from $t = 0$ to $t = \pi/2$ is smooth. To find the total distance traveled by the point, you can find the arc length of that portion lying in the first quadrant and multiply by 4.

$$s = 4\int_0^{\pi/2} \sqrt{\left(\frac{dx}{dt}\right)^2 + \left(\frac{dy}{dt}\right)^2} \, dt \qquad \text{Parametric form for arc length}$$

$$= 4\int_0^{\pi/2} \sqrt{(-5\sin t + 5\sin 5t)^2 + (5\cos t - 5\cos 5t)^2} \, dt$$

$$= 20\int_0^{\pi/2} \sqrt{2 - 2\sin t \sin 5t - 2\cos t \cos 5t} \, dt$$

$$= 20\int_0^{\pi/2} \sqrt{2 - 2\cos 4t} \, dt \qquad \text{Difference formula for cosine}$$

$$= 20\int_0^{\pi/2} \sqrt{4\sin^2 2t} \, dt \qquad \text{Double-angle formula}$$

$$= 40\int_0^{\pi/2} \sin 2t \, dt$$

$$= -20\left[\cos 2t\right]_0^{\pi/2}$$

$$= 40$$

For the epicycloid shown in Figure 10.33, an arc length of 40 seems about right because the circumference of a circle of radius 6 is

$$2\pi r = 12\pi \approx 37.7.$$

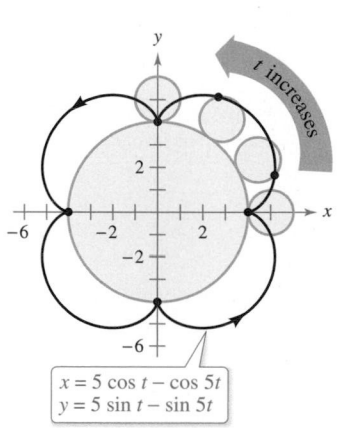

$x = 5 \cos t - \cos 5t$
$y = 5 \sin t - \sin 5t$

An epicycloid is traced by a point on the smaller circle as it rolls around the larger circle.
Figure 10.33

Area of a Surface of Revolution

You can use the formula for the area of a surface of revolution in rectangular form to develop a formula for surface area in parametric form.

THEOREM 10.9 Area of a Surface of Revolution

If a smooth curve C given by $x = f(t)$ and $y = g(t)$ does not cross itself on an interval $a \leq t \leq b$, then the area S of the surface of revolution formed by revolving C about the coordinate axes is given by the following.

1. $S = 2\pi \displaystyle\int_a^b g(t) \sqrt{\left(\dfrac{dx}{dt}\right)^2 + \left(\dfrac{dy}{dt}\right)^2}\, dt$ Revolution about the x-axis: $g(t) \geq 0$

2. $S = 2\pi \displaystyle\int_a^b f(t) \sqrt{\left(\dfrac{dx}{dt}\right)^2 + \left(\dfrac{dy}{dt}\right)^2}\, dt$ Revolution about the y-axis: $f(t) \geq 0$

These formulas may be easier to remember if you think of the differential of arc length as

$$ds = \sqrt{\left(\frac{dx}{dt}\right)^2 + \left(\frac{dy}{dt}\right)^2}\, dt.$$ Differential of arc length

Then the formulas in Theorem 10.9 can be written as follows.

1. $S = 2\pi \displaystyle\int_a^b g(t)\, ds$ 2. $S = 2\pi \displaystyle\int_a^b f(t)\, ds$

EXAMPLE 5 **Finding the Area of a Surface of Revolution**

Let C be the arc of the circle $x^2 + y^2 = 9$ from $(3, 0)$ to

$$\left(\frac{3}{2}, \frac{3\sqrt{3}}{2}\right)$$

as shown in Figure 10.34. Find the area of the surface formed by revolving C about the x-axis.

Solution You can represent C parametrically by the equations

$$x = 3\cos t \quad \text{and} \quad y = 3\sin t, \quad 0 \leq t \leq \pi/3.$$

(Note that you can determine the interval for t by observing that $t = 0$ when $x = 3$ and $t = \pi/3$ when $x = 3/2$.) On this interval, C is smooth and y is nonnegative, and you can apply Theorem 10.9 to obtain a surface area of

$$S = 2\pi \int_0^{\pi/3} (3\sin t)\sqrt{(-3\sin t)^2 + (3\cos t)^2}\, dt$$ Apply formula for area of a surface of revolution.

$$= 6\pi \int_0^{\pi/3} \sin t \sqrt{9(\sin^2 t + \cos^2 t)}\, dt$$

$$= 6\pi \int_0^{\pi/3} 3\sin t\, dt$$ Trigonometric identity

$$= -18\pi \Big[\cos t\Big]_0^{\pi/3}$$

$$= -18\pi \left(\frac{1}{2} - 1\right)$$

$$= 9\pi.$$

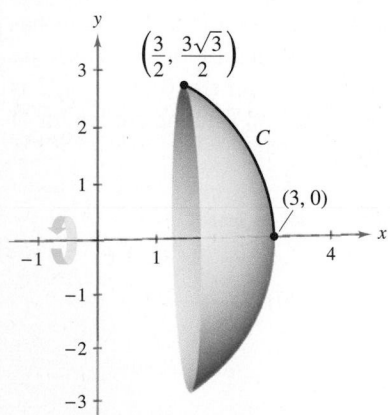

The surface of revolution has a surface area of 9π.

Figure 10.34

10.3   Exercises

CONCEPT CHECK

1. **Parametric Form of the Derivative**   What does the parametric form of the derivative represent?

2. **Tangent Lines**   Under what circumstances can a graph that represents a set of parametric equations have more than one tangent line at a given point?

3. **Tangent Lines**   Consider a curve represented by the parametric equations $x = f(t)$ and $y = g(t)$. When does the graph have horizontal tangent lines? Vertical tangent lines?

4. **Arc Length**   Why does the arc length formula require that the curve not intersect itself on an interval, except possibly at the endpoints?

 **Finding a Derivative**   In Exercises 5–8, find dy/dx.

5. $x = t^2$, $y = 7 - 6t$ 6. $x = \sqrt[3]{t}$, $y = 4 - t$

7. $x = \sin^2 \theta$, $y = \cos^2 \theta$ 8. $x = 2e^\theta$, $y = e^{-\theta/2}$

 **Finding Slope and Concavity**   In Exercises 9–18, find dy/dx and d^2y/dx^2, and find the slope and concavity (if possible) at the given value of the parameter.

Parametric Equations	Parameter
9. $x = 4t$, $y = 3t - 2$	$t = 3$
10. $x = \sqrt{t}$, $y = 3t - 1$	$t = 1$
11. $x = t + 1$, $y = t^2 + 3t$	$t = -2$
12. $x = t^2 + 5t + 4$, $y = 4t$	$t = 0$
13. $x = 4\cos\theta$, $y = 4\sin\theta$	$\theta = \dfrac{\pi}{4}$
14. $x = \cos\theta$, $y = 3\sin\theta$	$\theta = 0$
15. $x = 2 + \sec\theta$, $y = 1 + 2\tan\theta$	$\theta = -\dfrac{\pi}{3}$
16. $x = \sqrt{t}$, $y = \sqrt{t-1}$	$t = 5$
17. $x = \cos^3\theta$, $y = \sin^3\theta$	$\theta = \dfrac{\pi}{4}$
18. $x = \theta - \sin\theta$, $y = 1 - \cos\theta$	$\theta = \pi$

 **Finding Equations of Tangent Lines**   In Exercises 19–22, find an equation of the tangent line to the curve at each given point.

19. $x = 2\cot\theta$, $y = 2\sin^2\theta$, $\left(-\dfrac{2}{\sqrt{3}}, \dfrac{3}{2}\right)$, $(0, 2)$, $\left(2\sqrt{3}, \dfrac{1}{2}\right)$

20. $x = 2 - 3\cos\theta$, $y = 3 + 2\sin\theta$,

   $(-1, 3)$, $(2, 5)$, $\left(\dfrac{4 + 3\sqrt{3}}{2}, 2\right)$

21. $x = t^2 - 4$, $y = t^2 - 2t$, $(0, 0)$, $(-3, -1)$, $(-3, 3)$

22. $x = t^4 + 2$, $y = t^3 + t$, $(2, 0)$, $(3, -2)$, $(18, 10)$

 **Finding an Equation of a Tangent Line**   In Exercises 23–26, (a) use a graphing utility to graph the curve represented by the parametric equations, (b) use a graphing utility to find dx/dt, dy/dt, and dy/dx at the given value of the parameter, (c) find an equation of the tangent line to the curve at the given value of the parameter, and (d) use a graphing utility to graph the curve and the tangent line from part (c).

Parametric Equations	Parameter
23. $x = 6t$, $y = 1 - 4t^2$	$t = \dfrac{1}{2}$
24. $x = t - 2$, $y = \dfrac{1}{t} + 3$	$t = 1$
25. $x = t^2 - t + 2$, $y = t^3 - 3t$	$t = -1$
26. $x = 3t - t^2$, $y = 2t^{3/2}$	$t = \dfrac{1}{4}$

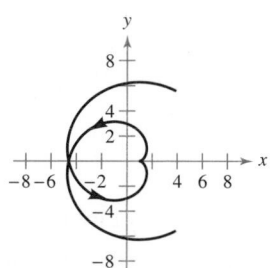 **Finding Equations of Tangent Lines**   In Exercises 27–30, find the equations of the tangent lines at the point where the curve crosses itself.

27. $x = 2\sin 2t$, $y = 3\sin t$

28. $x = 2 - \pi\cos t$, $y = 2t - \pi\sin t$

29. $x = t^2 - t$, $y = t^3 - 3t - 1$

30. $x = t^3 - 6t$, $y = t^2$

Horizontal and Vertical Tangency   In Exercises 31 and 32, find all points (if any) of horizontal and vertical tangency to the curve on the given interval.

31. $x = \cos\theta + \theta\sin\theta$
    $y = \sin\theta - \theta\cos\theta$
    $-2\pi \le \theta \le 2\pi$

32. $x = 2\theta$
    $y = 2(1 - \cos\theta)$
    $0 \le \theta \le 2\pi$

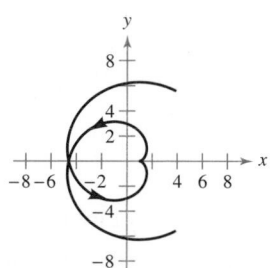
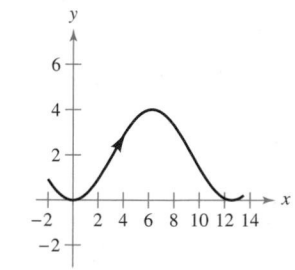

 **Horizontal and Vertical Tangency**   In Exercises 33–42, find all points (if any) of horizontal and vertical tangency to the curve. Use a graphing utility to confirm your results.

33. $x = 9 - t$, $y = -t^2$ 34. $x = t + 1$, $y = t^2 + 3t$

35. $x = t + 4$, $y = t^3 - 12t + 6$

36. $x = t^2 - t + 2$, $y = t^3 - 3t$

37. $x = 7\cos\theta$, $y = 7\sin\theta$ 38. $x = \cos\theta$, $y = 2\sin 2\theta$

39. $x = 5 + 3\cos\theta$, $y = -2 + \sin\theta$

40. $x = \sec\theta$, $y = \tan\theta$

41. $x = 4\cos^2\theta$, $y = 2\sin\theta$ **42.** $x = \cos^2\theta$, $y = \cos\theta$

 **Determining Concavity** In Exercises 43–48, determine the open t-intervals on which the curve is concave downward or concave upward.

43. $x = 3t^2$, $y = t^3 - t$ **44.** $x = 2 + t^2$, $y = t^2 + t^3$

45. $x = 2t + \ln t$, $y = 2t - \ln t$

46. $x = t^2$, $y = \ln t$

47. $x = \sin t$, $y = \cos t$, $0 < t < \pi$

48. $x = 4\cos t$, $y = 2\sin t$, $0 < t < 2\pi$

 **Arc Length** In Exercises 49–54, find the arc length of the curve on the given interval.

Parametric Equations	Interval
49. $x = 3t + 5$, $y = 7 - 2t$	$-1 \le t \le 3$
50. $x = 6t^2$, $y = 2t^3$	$1 \le t \le 4$
51. $x = e^{-t}\cos t$, $y = e^{-t}\sin t$	$0 \le t \le \dfrac{\pi}{2}$
52. $x = \arcsin t$, $y = \ln\sqrt{1 - t^2}$	$0 \le t \le \frac{1}{2}$
53. $x = \sqrt{t}$, $y = 3t - 1$	$0 \le t \le 1$
54. $x = t$, $y = \dfrac{t^5}{10} + \dfrac{1}{6t^3}$	$1 \le t \le 2$

Arc Length In Exercises 55–58, find the arc length of the curve on the interval $[0, 2\pi]$.

55. Hypocycloid perimeter: $x = a\cos^3\theta$, $y = a\sin^3\theta$

56. Involute of a circle: $x = \cos\theta + \theta\sin\theta$
$$y = \sin\theta - \theta\cos\theta$$

57. Cycloid arch: $x = a(\theta - \sin\theta)$, $y = a(1 - \cos\theta)$

58. Nephroid perimeter: $x = a(3\cos t - \cos 3t)$
$$y = a(3\sin t - \sin 3t)$$

 **59. Path of a Projectile** The path of a projectile is modeled by the parametric equations
$$x = (90\cos 30°)t \quad \text{and} \quad y = (90\sin 30°)t - 16t^2$$
where x and y are measured in feet.

(a) Use a graphing utility to graph the path of the projectile.

(b) Use a graphing utility to approximate the range of the projectile.

(c) Use the integration capabilities of a graphing utility to approximate the arc length of the path. Compare this result with the range of the projectile.

60. Path of a Projectile When the projectile in Exercise 59 is launched at an angle θ with the horizontal, its parametric equations are $x = (90\cos\theta)t$ and $y = (90\sin\theta)t - 16t^2$. Find the angle that maximizes the range of the projectile. Use a graphing utility to find the angle that maximizes the arc length of the trajectory.

61. Folium of Descartes Consider the parametric equations
$$x = \frac{4t}{1 + t^3} \quad \text{and} \quad y = \frac{4t^2}{1 + t^3}.$$

(a) Use a graphing utility to graph the curve represented by the parametric equations.

(b) Use a graphing utility to find the points of horizontal tangency to the curve.

(c) Use the integration capabilities of a graphing utility to approximate the arc length of the closed loop. (*Hint:* Use symmetry and integrate over the interval $0 \le t \le 1$.)

62. Witch of Agnesi Consider the parametric equations
$$x = 4\cot\theta \quad \text{and} \quad y = 4\sin^2\theta, \quad -\frac{\pi}{2} \le \theta \le \frac{\pi}{2}.$$

(a) Use a graphing utility to graph the curve represented by the parametric equations.

(b) Use a graphing utility to find the points of horizontal tangency to the curve.

(c) Use the integration capabilities of a graphing utility to approximate the arc length over the interval $\pi/4 \le \theta \le \pi/2$.

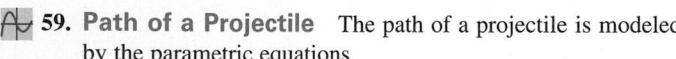 **Surface Area** In Exercises 63–68, find the area of the surface generated by revolving the curve about each given axis.

63. $x = 2t$, $y = 3t$, $0 \le t \le 3$
(a) x-axis
(b) y-axis

64. $x = t$, $y = 4 - 2t$, $0 \le t \le 2$
(a) x-axis
(b) y-axis

65. $x = 5\cos\theta$, $y = 5\sin\theta$, $0 \le \theta \le \dfrac{\pi}{2}$, y-axis

66. $x = \frac{1}{3}t^3$, $y = t + 1$, $1 \le t \le 2$, y-axis

67. $x = a\cos^3\theta$, $y = a\sin^3\theta$, $0 \le \theta \le \pi$, x-axis

68. $x = a\cos\theta$, $y = b\sin\theta$, $0 \le \theta \le 2\pi$
(a) x-axis
(b) y-axis

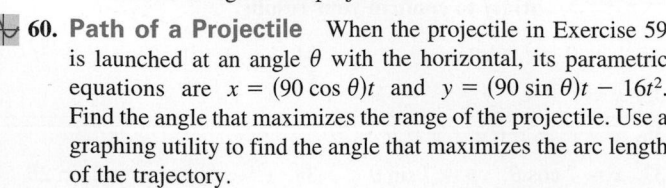 **Surface Area** In Exercises 69–72, write an integral that represents the area of the surface generated by revolving the curve about the x-axis. Use a graphing utility to approximate the integral.

Parametric Equations	Interval
69. $x = t^3$, $y = t + 2$	$0 \le t \le 2$
70. $x = t^2$, $y = \sqrt{t}$	$1 \le t \le 3$
71. $x = \cos^2\theta$, $y = \cos\theta$	$0 \le \theta \le \dfrac{\pi}{2}$
72. $x = \theta + \sin\theta$, $y = \theta + \cos\theta$	$0 \le \theta \le \dfrac{\pi}{2}$

EXPLORING CONCEPTS

73. Writing

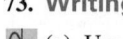 (a) Use a graphing utility to graph each set of parametric equations.

$$x = t - \sin t, \quad y = 1 - \cos t, \quad 0 \le t \le 2\pi$$

$$x = 2t - \sin(2t), \quad y = 1 - \cos(2t), \quad 0 \le t \le \pi$$

(b) Compare the graphs of the two sets of parametric equations in part (a). When the curve represents the motion of a particle and t is time, what can you infer about the average speeds of the particle on the paths represented by the two sets of parametric equations?

(c) Without graphing the curve, determine the time required for a particle to traverse the same path as in parts (a) and (b) when the path is modeled by

$$x = \tfrac{1}{2}t - \sin\left(\tfrac{1}{2}t\right) \quad \text{and} \quad y = 1 - \cos\left(\tfrac{1}{2}t\right).$$

74. Writing

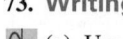 (a) Each set of parametric equations represents the motion of a particle. Use a graphing utility to graph each set.

First Particle: $x = 3 \cos t, \quad y = 4 \sin t, \quad 0 \le t \le 2\pi$

Second Particle: $x = 4 \sin t, \quad y = 3 \cos t,$
$$0 \le t \le 2\pi$$

(b) Determine the number of points of intersection.

(c) Will the particles ever be at the same place at the same time? If so, identify the point(s).

75. Sketching a Graph Find a set of parametric equations $x = f(t)$ and $y = g(t)$ such that $dx/dt < 0$ and $dy/dt < 0$ for all real numbers t. Then sketch a graph of the curve.

 **76. HOW DO YOU SEE IT?** Using the graph of f, (a) determine whether dy/dt is positive or negative given that dx/dt is negative and (b) determine whether dx/dt is positive or negative given that dy/dt is positive. Explain your reasoning.

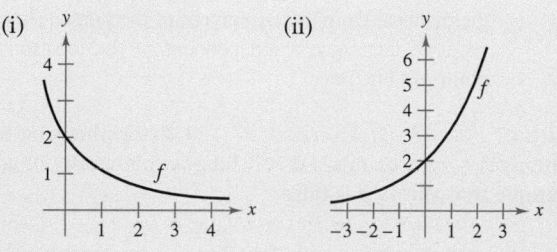

77. Integration by Substitution Use integration by substitution to show that if y is a continuous function of x on the interval $a \le x \le b$, where $x = f(t)$ and $y = g(t)$, then

$$\int_a^b y \, dx = \int_{t_1}^{t_2} g(t) f'(t) \, dt$$

where $f(t_1) = a$, $f(t_2) = b$, and both g and f' are continuous on $[t_1, t_2]$.

78. Surface Area A portion of a sphere of radius r is removed by cutting out a circular cone with its vertex at the center of the sphere. The vertex of the cone forms an angle of 2θ. Find the surface area removed from the sphere.

Area In Exercises 79 and 80, find the area of the region. (Use the result of Exercise 77.)

79. $x = 2 \sin^2 \theta$

$y = 2 \sin^2 \theta \tan \theta$

$0 \le \theta < \dfrac{\pi}{2}$

80. $x = 2 \cot \theta$

$y = 2 \sin^2 \theta$

$0 < \theta < \pi$

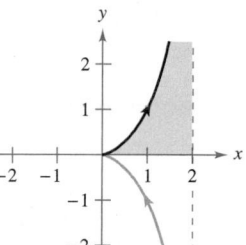

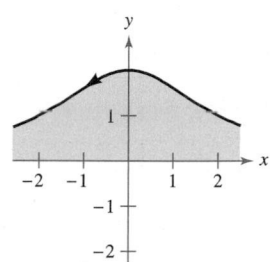

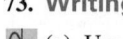 **Areas of Simple Closed Curves** In Exercises 81–86, use a computer algebra system and the result of Exercise 77 to match the closed curve with its area. (These exercises were based on "The Surveyor's Area Formula" by Bart Braden, *College Mathematics Journal*, September 1986, pp. 335–337, by permission of the author.)

(a) $\tfrac{8}{3}ab$ (b) $\tfrac{3}{8}\pi a^2$

(c) $2\pi a^2$ (d) πab

(e) $2\pi ab$ (f) $6\pi a^2$

81. Ellipse: $(0 \le t \le 2\pi)$

$x = b \cos t$

$y = a \sin t$

82. Astroid: $(0 \le t \le 2\pi)$

$x = a \cos^3 t$

$y = a \sin^3 t$

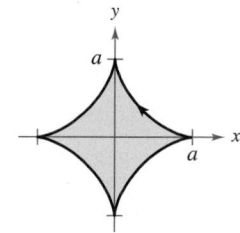

83. Cardioid: $(0 \le t \le 2\pi)$

$x = 2a \cos t - a \cos 2t$

$y = 2a \sin t - a \sin 2t$

84. Deltoid: $(0 \le t \le 2\pi)$

$x = 2a \cos t + a \cos 2t$

$y = 2a \sin t - a \sin 2t$

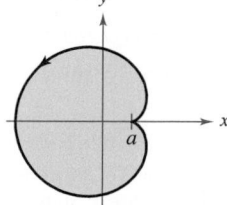

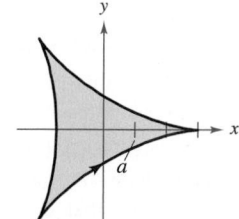

85. Hourglass: $(0 \le t \le 2\pi)$ **86. Teardrop:** $(0 \le t \le 2\pi)$

$x = a \sin 2t$ $x = 2a \cos t - a \sin 2t$

$y = b \sin t$ $y = b \sin t$

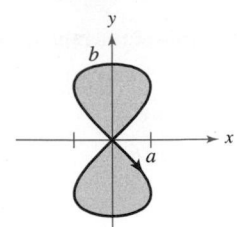   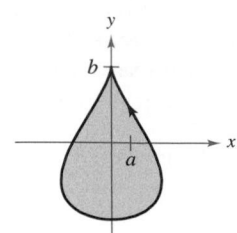

Centroid **In Exercises 87 and 88, find the centroid of the region bounded by the graph of the parametric equations and the coordinate axes. (Use the result of Exercise 77.)**

87. $x = \sqrt{t}, \quad y = 4 - t$

88. $x = \sqrt{4 - t}, \quad y = \sqrt{t}$

Volume **In Exercises 89 and 90, find the volume of the solid formed by revolving the region bounded by the graph of the parametric equations about the x-axis. (Use the result of Exercise 77.)**

89. $x = 6 \cos \theta, \quad y = 6 \sin \theta$

90. $x = \cos \theta, \quad y = 3 \sin \theta, \quad a > 0$

91. Cycloid Use the parametric equations

$$x = a(\theta - \sin \theta) \quad \text{and} \quad y = a(1 - \cos \theta), \, a > 0$$

to answer the following.

(a) Find dy/dx and d^2y/dx^2.

(b) Find the equation of the tangent line at the point where $\theta = \pi/6$.

(c) Find all points of horizontal tangency.

(d) Determine where the curve is concave upward or concave downward.

(e) Find the length of one arc of the curve.

92. Using Parametric Equations Use the parametric equations

$$x = t^2\sqrt{3} \quad \text{and} \quad y = 3t - \frac{1}{3}t^3$$

to answer the following.

(a) Use a graphing utility to graph the curve on the interval $-3 \le t \le 3$.

(b) Find dy/dx and d^2y/dx^2.

(c) Find the equation of the tangent line at the point $\left(\sqrt{3}, \frac{8}{3}\right)$.

(d) Find the length of the curve on the interval $-3 \le t \le 3$.

(e) Find the area of the surface generated by revolving the curve about the x-axis.

93. Involute of a Circle The involute of a circle is described by the endpoint P of a string that is held taut as it is unwound from a spool that does not turn (see figure). Show that a parametric representation of the involute is

$$x = r(\cos \theta + \theta \sin \theta) \quad \text{and} \quad y = r(\sin \theta - \theta \cos \theta).$$

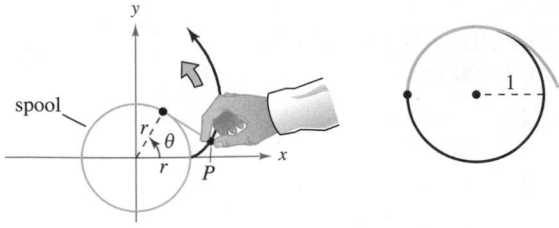

Figure for 93   Figure for 94

94. Involute of a Circle The figure shows a piece of string tied to a circle with a radius of one unit. The string is just long enough to reach the opposite side of the circle. Find the area that is covered when the string is unwound counterclockwise.

95. Using Parametric Equations

(a) Use a graphing utility to graph the curve given by

$$x = \frac{1 - t^2}{1 + t^2} \quad \text{and} \quad y = \frac{2t}{1 + t^2}, \quad -20 \le t \le 20.$$

(b) Describe the graph and confirm your result analytically.

(c) Discuss the speed at which the curve is traced as t increases from -20 to 20.

96. Tractrix A person moves from the origin along the positive y-axis pulling a weight at the end of a 12-meter rope. Initially, the weight is located at the point $(12, 0)$.

(a) In Exercise 61 of Section 8.4, it was shown that the path of the weight is modeled by the rectangular equation

$$y = -12 \ln \frac{12 - \sqrt{144 - x^2}}{x} - \sqrt{144 - x^2}$$

where $0 < x \le 12$. Use a graphing utility to graph the rectangular equation.

(b) Use a graphing utility to graph the parametric equations

$$y = 12 \operatorname{sech} \frac{t}{12} \quad \text{and} \quad y = t - 12 \tanh \frac{t}{12}$$

where $t \ge 0$. How does this graph compare with the graph in part (a)? Which graph (if either) do you think is a better representation of the path?

(c) Use the parametric equations for the tractrix to verify that the distance from the y-intercept of the tangent line to the point of tangency is independent of the location of the point of tangency.

True or False? **In Exercises 97–100, determine whether the statement is true or false. If it is false, explain why or give an example that shows it is false.**

97. If $x = f(t)$ and $y = g(t)$, then $\dfrac{d^2y}{dx^2} = \dfrac{g''(t)}{f''(t)}$.

98. The curve given by $x = t^3$ and $y = t^2$ has a horizontal tangent at the origin because $dy/dt = 0$ when $t = 0$.

99. The curve given by $x = x_1 + t(x_2 - x_1)$ and $y = y_1 + t(y_2 - y_1)$, $y_1 \ne y_2$, has at least one horizontal asymptote.

100. The curve given by $x = h + a \cos \theta$ and $y = k + b \sin \theta$ has two horizontal asymptotes and two vertical asymptotes.

10.4 Polar Coordinates and Polar Graphs

- Understand the polar coordinate system.
- Rewrite rectangular coordinates and equations in polar form and vice versa.
- Sketch the graph of an equation given in polar form.
- Find the slope of a tangent line to a polar graph.
- Identify several types of special polar graphs.

Polar Coordinates

So far, you have been representing graphs as collections of points (x, y) on the rectangular coordinate system. The corresponding equations for these graphs have been in either rectangular or parametric form. In this section, you will study a coordinate system called the **polar coordinate system.**

To form the polar coordinate system in the plane, fix a point O, called the **pole** (or **origin**), and construct from O an initial ray called the **polar axis,** as shown in Figure 10.35. Then each point P in the plane can be assigned **polar coordinates** (r, θ), as follows.

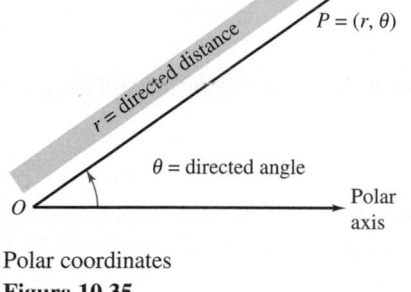

Polar coordinates
Figure 10.35

$r = $ *directed distance* from O to P

$\theta = $ *directed angle*, counterclockwise from polar axis to segment $\overline{OP}$

Figure 10.36 shows three points on the polar coordinate system. Notice that in this system, it is convenient to locate points with respect to a grid of concentric circles intersected by **radial lines** through the pole.

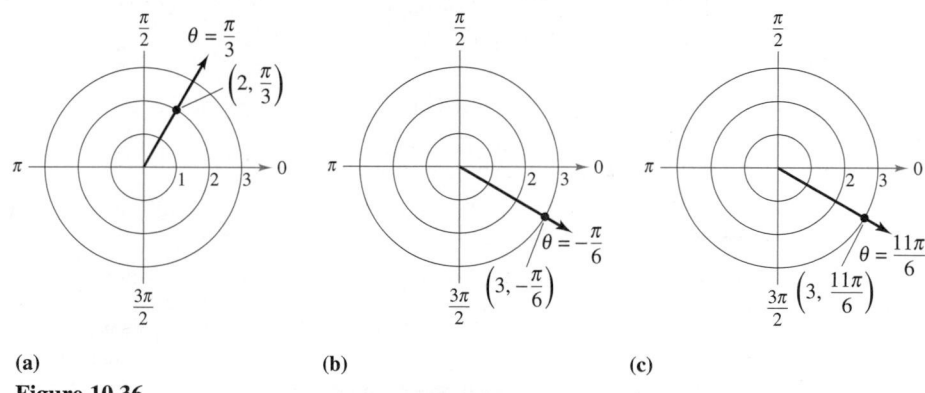

(a)   (b)   (c)

Figure 10.36

With rectangular coordinates, each point (x, y) has a unique representation. This is not true with polar coordinates. For instance, the coordinates

$$(r, \theta) \quad \text{and} \quad (r, 2\pi + \theta)$$

represent the same point [see parts (b) and (c) in Figure 10.36]. Also, because r is a *directed distance,* the coordinates

$$(r, \theta) \quad \text{and} \quad (-r, \theta + \pi)$$

represent the same point. In general, the point (r, θ) can be written as

$$(r, \theta) = (r, \theta + 2n\pi)$$

or

$$(r, \theta) = (-r, \theta + (2n + 1)\pi)$$

where n is any integer. Moreover, the pole is represented by $(0, \theta)$, where θ is any angle.

POLAR COORDINATES

The mathematician credited with first using polar coordinates was James Bernoulli, who introduced them in 1691. However, there is some evidence that it may have been Isaac Newton who first used them.

Coordinate Conversion

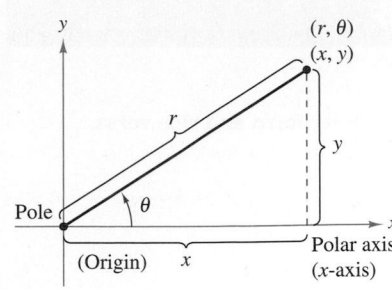

Relating polar and rectangular coordinates
Figure 10.37

To establish the relationship between polar and rectangular coordinates, let the polar axis coincide with the positive x-axis and the pole with the origin, as shown in Figure 10.37. Because (x, y) lies on a circle of radius r, it follows that

$$r^2 = x^2 + y^2.$$

Moreover, for $r > 0$, the definitions of the trigonometric functions imply that

$$\tan \theta = \frac{y}{x}, \quad \cos \theta = \frac{x}{r}, \quad \text{and} \quad \sin \theta = \frac{y}{r}.$$

You can show that the same relationships hold for $r < 0$.

THEOREM 10.10 Coordinate Conversion

The polar coordinates (r, θ) of a point are related to the rectangular coordinates (x, y) of the point as follows.

Polar-to-Rectangular	**Rectangular-to-Polar**
$x = r \cos \theta$	$\tan \theta = \dfrac{y}{x}$
$y = r \sin \theta$	$r^2 = x^2 + y^2$

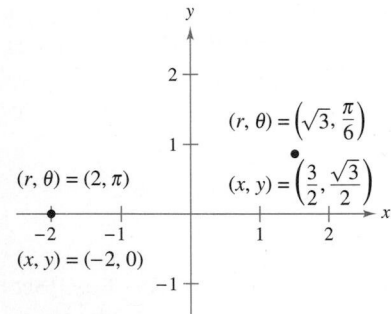

To convert from polar to rectangular coordinates, let $x = r \cos \theta$ and $y = r \sin \theta$.
Figure 10.38

EXAMPLE 1   **Polar-to-Rectangular Conversion**

a. For the point $(r, \theta) = (2, \pi)$,

$$x = r \cos \theta = 2 \cos \pi = -2 \quad \text{and} \quad y = r \sin \theta = 2 \sin \pi = 0.$$

So, the rectangular coordinates are $(x, y) = (-2, 0)$.

b. For the point $(r, \theta) = (\sqrt{3}, \pi/6)$,

$$x = \sqrt{3} \cos \frac{\pi}{6} = \frac{3}{2} \quad \text{and} \quad y = \sqrt{3} \sin \frac{\pi}{6} = \frac{\sqrt{3}}{2}.$$

So, the rectangular coordinates are $(x, y) = (3/2, \sqrt{3}/2)$.

See Figure 10.38.

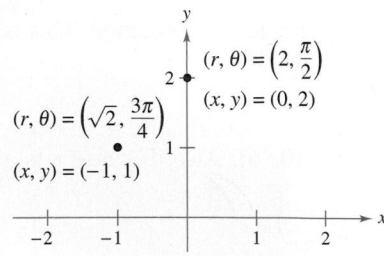

To convert from rectangular to polar coordinates, let $\tan \theta = y/x$ and $r = \sqrt{x^2 + y^2}$.
Figure 10.39

EXAMPLE 2   **Rectangular-to-Polar Conversion**

a. For the second-quadrant point $(x, y) = (-1, 1)$,

$$\tan \theta = \frac{y}{x} = -1 \quad \Longrightarrow \quad \theta = \frac{3\pi}{4}.$$

Because θ was chosen to be in the same quadrant as (x, y), use a positive value of r.

$$r = \sqrt{x^2 + y^2} = \sqrt{(-1)^2 + (1)^2} = \sqrt{2}$$

This implies that *one* set of polar coordinates is $(r, \theta) = (\sqrt{2}, 3\pi/4)$.

b. Because the point $(x, y) = (0, 2)$ lies on the positive y-axis, choose $\theta = \pi/2$ and $r = 2$, and one set of polar coordinates is $(r, \theta) = (2, \pi/2)$.

See Figure 10.39.

Note that you can also use Theorem 10.10 to convert a polar equation to a rectangular equation (and vice versa), as shown in Example 3.

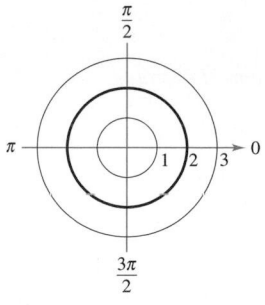

(a) Circle: $r = 2$

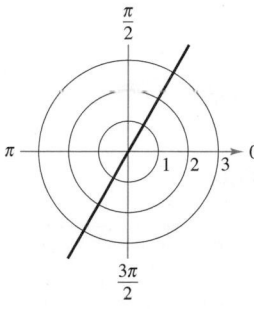

(b) Radial line: $\theta = \dfrac{\pi}{3}$

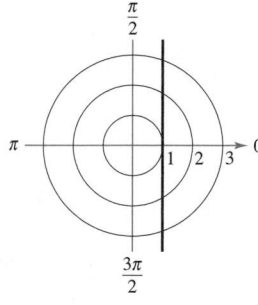

(c) Vertical line: $r = \sec \theta$
Figure 10.40

Polar Graphs

One way to sketch the graph of a polar equation is to convert to rectangular coordinates and then sketch the graph of the rectangular equation.

EXAMPLE 3   Graphing Polar Equations

Describe the graph of each polar equation. Confirm each description by converting to a rectangular equation.

a. $r = 2$ **b.** $\theta = \dfrac{\pi}{3}$ **c.** $r = \sec \theta$

Solution

a. The graph of the polar equation $r = 2$ consists of all points that are two units from the pole. So, this graph is a circle centered at the origin with a radius of 2. [See Figure 10.40(a).] You can confirm this by using the relationship $r^2 = x^2 + y^2$ to obtain the rectangular equation

$$x^2 + y^2 = 2^2. \qquad \text{Rectangular equation}$$

b. The graph of the polar equation $\theta = \pi/3$ consists of all points on the line that makes an angle of $\pi/3$ with the positive x-axis. [See Figure 10.40(b).] You can confirm this by using the relationship $\tan \theta = y/x$ to obtain the rectangular equation

$$y = \sqrt{3}x. \qquad \text{Rectangular equation}$$

c. The graph of the polar equation $r = \sec \theta$ is not evident by simple inspection, so you can begin by converting to rectangular form using the relationship $r \cos \theta = x$.

$$r = \sec \theta \qquad \text{Polar equation}$$
$$r \cos \theta = 1$$
$$x = 1 \qquad \text{Rectangular equation}$$

From the rectangular equation, you can see that the graph is a vertical line. [See Figure 10.40(c).]

▷ **TECHNOLOGY**   Sketching the graphs of complicated polar equations *by hand* can be tedious. With technology, however, the task is not difficult. Use a graphing utility in *polar* mode to graph the equations in the exercise set. If your graphing utility does not have a *polar* mode but does have a *parametric* mode, you can graph $r = f(\theta)$ by writing the equation as

$$x = f(\theta) \cos \theta$$
$$y = f(\theta) \sin \theta.$$

For instance, the graph of $r = \frac{1}{2}\theta$ shown in Figure 10.41 was produced with a graphing utility in parametric mode. This equation was graphed using the parametric equations

$$x = \frac{1}{2}\theta \cos \theta$$

$$y = \frac{1}{2}\theta \sin \theta$$

with the values of θ varying from -4π to 4π. This curve is of the form $r = a\theta$ and is called a **spiral of Archimedes.**

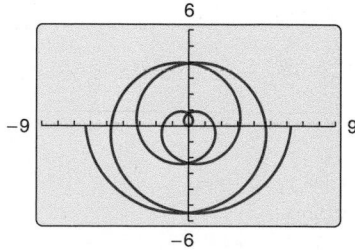

Spiral of Archimedes
Figure 10.41

EXAMPLE 4 **Sketching a Polar Graph**

∙∙∙∙▷ *See LarsonCalculus.com for an interactive version of this type of example.*

Sketch the graph of $r = 2 \cos 3\theta$.

Solution Begin by writing the polar equation in parametric form.

$$x = 2 \cos 3\theta \cos \theta \quad \text{and} \quad y = 2 \cos 3\theta \sin \theta$$

After some experimentation, you will find that the entire curve, which is called a **rose curve**, can be sketched by letting θ vary from 0 to π, as shown in Figure 10.42. If you try duplicating this graph with a graphing utility, you will find that by letting θ vary from 0 to 2π, you will actually trace the entire curve *twice*.

REMARK One way to sketch the graph of $r = 2 \cos 3\theta$ by hand is to make a table of values.

θ	0	$\dfrac{\pi}{6}$	$\dfrac{\pi}{3}$	$\dfrac{\pi}{2}$	$\dfrac{2\pi}{3}$
r	2	0	-2	0	2

By extending the table and plotting the points, you will obtain the curve shown in Example 4.

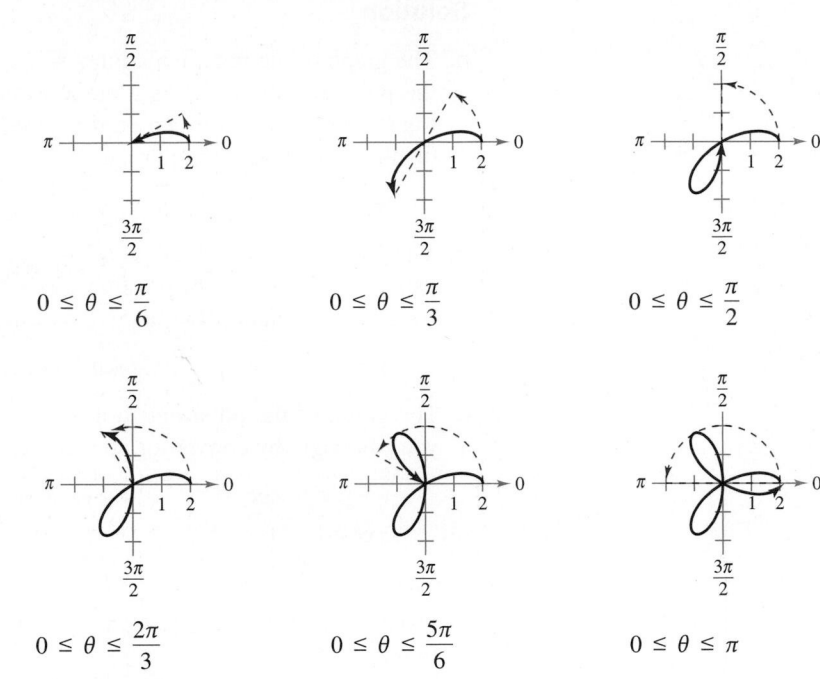

$$0 \le \theta \le \frac{\pi}{6} \qquad 0 \le \theta \le \frac{\pi}{3} \qquad 0 \le \theta \le \frac{\pi}{2}$$

$$0 \le \theta \le \frac{2\pi}{3} \qquad 0 \le \theta \le \frac{5\pi}{6} \qquad 0 \le \theta \le \pi$$

Figure 10.42

Use a graphing utility to experiment with other rose curves. Note that rose curves are of the form

$$r = a \cos n\theta \quad \text{or} \quad r = a \sin n\theta.$$

For instance, Figure 10.43 shows the graphs of two other rose curves.

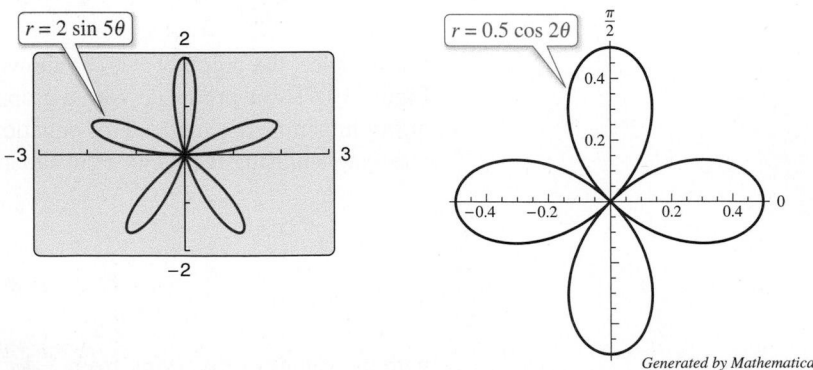

$r = 2 \sin 5\theta$

$r = 0.5 \cos 2\theta$

Generated by Mathematica

Rose curves
Figure 10.43

Slope and Tangent Lines

To find the slope of a tangent line to a polar graph, consider a differentiable function given by $r = f(\theta)$. To find the slope in polar form, use the parametric equations

$$x = r \cos \theta = f(\theta) \cos \theta \quad \text{and} \quad y = r \sin \theta = f(\theta) \sin \theta.$$

Using the parametric form of dy/dx given in Theorem 10.7, you have

$$\frac{dy}{dx} = \frac{dy/d\theta}{dx/d\theta} = \frac{f(\theta) \cos \theta + f'(\theta) \sin \theta}{-f(\theta) \sin \theta + f'(\theta) \cos \theta}$$

which establishes the next theorem.

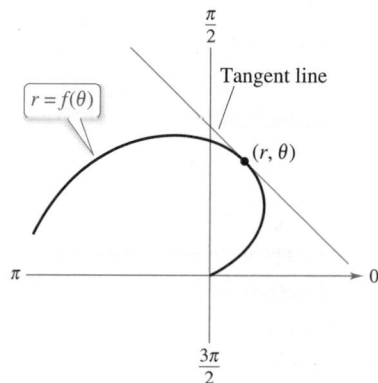

Tangent line to polar curve
Figure 10.44

THEOREM 10.11 Slope in Polar Form

If f is a differentiable function of θ, then the *slope* of the tangent line to the graph of $r = f(\theta)$ at the point (r, θ) is

$$\frac{dy}{dx} = \frac{dy/d\theta}{dx/d\theta} = \frac{f(\theta) \cos \theta + f'(\theta) \sin \theta}{-f(\theta) \sin \theta + f'(\theta) \cos \theta}$$

provided that $dx/d\theta \neq 0$ at (r, θ). (See Figure 10.44.)

From Theorem 10.11, you can make the following observations.

1. Solutions of $\dfrac{dy}{d\theta} = 0$ yield horizontal tangents, provided that $\dfrac{dx}{d\theta} \neq 0$.

2. Solutions of $\dfrac{dx}{d\theta} = 0$ yield vertical tangents, provided that $\dfrac{dy}{d\theta} \neq 0$.

If $dy/d\theta$ and $dx/d\theta$ are *simultaneously* 0, then no conclusion can be drawn about tangent lines.

EXAMPLE 5   **Finding Horizontal and Vertical Tangent Lines**

Find the horizontal and vertical tangent lines of $r = \sin \theta$, where $0 \leq \theta < \pi$.

Solution   Begin by writing the equation in parametric form.

$$x = r \cos \theta = \sin \theta \cos \theta$$

and

$$y = r \sin \theta = \sin \theta \sin \theta = \sin^2 \theta$$

Next, differentiate x and y with respect to θ and set each derivative equal to 0.

$$\frac{dx}{d\theta} = \cos^2 \theta - \sin^2 \theta = \cos 2\theta = 0 \quad \Longrightarrow \quad \theta = \frac{\pi}{4}, \frac{3\pi}{4}$$

$$\frac{dy}{d\theta} = 2 \sin \theta \cos \theta = \sin 2\theta = 0 \quad \Longrightarrow \quad \theta = 0, \frac{\pi}{2}$$

So, the graph has vertical tangent lines at

$$\left(\frac{\sqrt{2}}{2}, \frac{\pi}{4}\right) \quad \text{and} \quad \left(\frac{\sqrt{2}}{2}, \frac{3\pi}{4}\right)$$

and it has horizontal tangent lines at

$$(0, 0) \quad \text{and} \quad \left(1, \frac{\pi}{2}\right)$$

as shown in Figure 10.45.

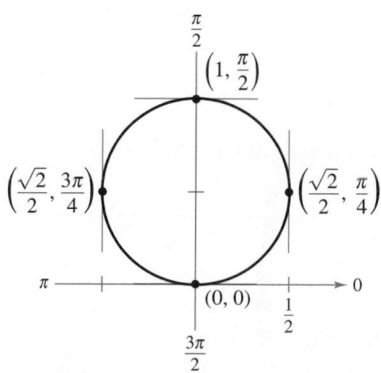

Horizontal and vertical tangent lines of $r = \sin \theta$
Figure 10.45

EXAMPLE 6 Finding Horizontal and Vertical Tangent Lines

Find the horizontal and vertical tangent lines to the graph of $r = 2(1 - \cos \theta)$, where $0 \le \theta < 2\pi$.

Solution Let $y = r \sin \theta$ and then differentiate with respect to θ.

$$y = r \sin \theta \qquad \text{Parametric equation for } y$$

$$= 2(1 - \cos \theta) \sin \theta \qquad \text{Substitute for } r.$$

$$\frac{dy}{d\theta} = 2[(1 - \cos \theta)(\cos \theta) + (\sin \theta)(\sin \theta)] \qquad \text{Derivative of } y \text{ with respect to } \theta$$

$$= 2(\cos \theta - \cos^2 \theta + \sin^2 \theta) \qquad \text{Multiply.}$$

$$= 2(\cos \theta - \cos^2 \theta + 1 - \cos^2 \theta) \qquad \text{Pythagorean identity}$$

$$= -2(2 \cos^2 \theta - \cos \theta - 1) \qquad \text{Combine like terms; factor out } -1$$

$$= -2(2 \cos \theta + 1)(\cos \theta - 1) \qquad \text{Factor.}$$

Setting $dy/d\theta$ equal to 0, you can see that $\cos \theta = -\frac{1}{2}$ and $\cos \theta = 1$. So, $dy/d\theta = 0$ when $\theta = 2\pi/3, 4\pi/3$, and 0. Similarly, using $x = r \cos \theta$, you have

$$x = r \cos \theta \qquad \text{Parametric equation for } x$$

$$= 2(1 - \cos \theta) \cos \theta \qquad \text{Substitute for } r.$$

$$= 2 \cos \theta - 2 \cos^2 \theta \qquad \text{Multiply.}$$

$$\frac{dx}{d\theta} = -2 \sin \theta + 4 \cos \theta \sin \theta \qquad \text{Derivative of } x \text{ with respect to } \theta$$

$$= (2 \sin \theta)(2 \cos \theta - 1). \qquad \text{Factor.}$$

Setting $dx/d\theta$ equal to 0, you can see that $\sin \theta = 0$ and $\cos \theta = \frac{1}{2}$. So, you can conclude that $dx/d\theta = 0$ when $\theta = 0, \pi, \pi/3$, and $5\pi/3$. From these results and from the graph shown in Figure 10.46, you can conclude that the graph has horizontal tangents at $(3, 2\pi/3)$ and $(3, 4\pi/3)$ and has vertical tangents at $(1, \pi/3)$, $(1, 5\pi/3)$, and $(4, \pi)$. This graph is called a **cardioid.** Note that both derivatives $(dy/d\theta$ and $dx/d\theta)$ are 0 when $\theta = 0$. Using this information alone, you do not know whether the graph has a horizontal or vertical tangent line at the pole. From Figure 10.46, however, you can see that the graph has a cusp at the pole. ∎

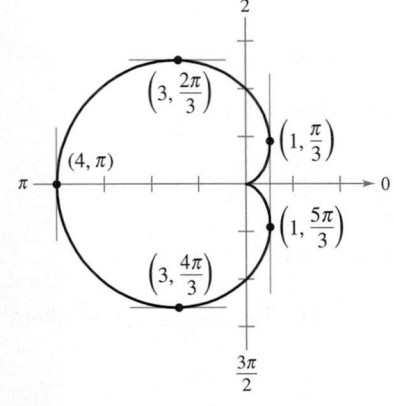

Horizontal and vertical tangent lines of $r = 2(1 - \cos \theta)$

Figure 10.46

Theorem 10.11 has an important consequence. If the graph of $r = f(\theta)$ passes through the pole when $\theta = \alpha$ and $f'(\alpha) \neq 0$, then the formula for dy/dx simplifies as follows.

$$\frac{dy}{dx} = \frac{f'(\alpha) \sin \alpha + f(\alpha) \cos \alpha}{f'(\alpha) \cos \alpha - f(\alpha) \sin \alpha} = \frac{f'(\alpha) \sin \alpha + 0}{f'(\alpha) \cos \alpha - 0} = \frac{\sin \alpha}{\cos \alpha} = \tan \alpha$$

So, the line $\theta = \alpha$ is tangent to the graph at the pole, $(0, \alpha)$.

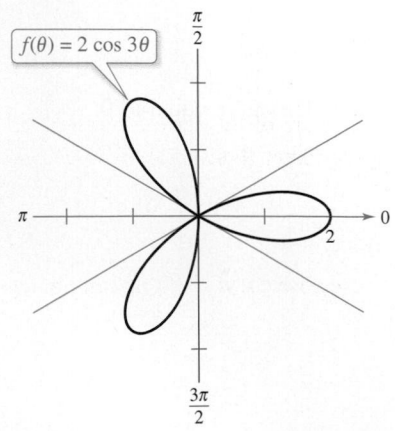

$f(\theta) = 2 \cos 3\theta$

This rose curve has three tangent lines ($\theta = \pi/6$, $\theta = \pi/2$, and $\theta = 5\pi/6$) at the pole.

Figure 10.47

> **THEOREM 10.12 Tangent Lines at the Pole**
>
> If $f(\alpha) = 0$ and $f'(\alpha) \neq 0$, then the line $\theta = \alpha$ is tangent at the pole to the graph of $r = f(\theta)$.

Theorem 10.12 is useful because it states that the zeros of $r = f(\theta)$ can be used to find the tangent lines at the pole. Note that because a polar curve can cross the pole more than once, it can have more than one tangent line at the pole. For example, the rose curve $f(\theta) = 2 \cos 3\theta$ has three tangent lines at the pole, as shown in Figure 10.47. For this curve, $f(\theta) = 2 \cos 3\theta$ is 0 when θ is $\pi/6$, $\pi/2$, and $5\pi/6$. Moreover, the derivative $f'(\theta) = -6 \sin 3\theta$ is not 0 for these values of θ.

Special Polar Graphs

Several important types of graphs have equations that are simpler in polar form than in rectangular form. For example, the polar equation of a circle having a radius of a and centered at the origin is simply $r = a$. Later in the text, you will come to appreciate this benefit. For now, several other types of graphs that have simpler equations in polar form are shown below. (Conics are considered in Section 10.6.)

Limaçons

$r = a \pm b \cos \theta$
$r = a \pm b \sin \theta$
$(a > 0, b > 0)$

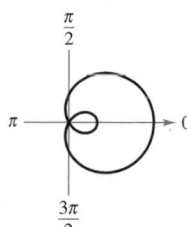

$\dfrac{a}{b} < 1$

Limaçon with inner loop

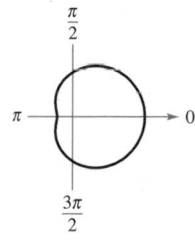

$\dfrac{a}{b} = 1$

Cardioid (heart-shaped)

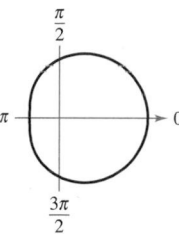

$1 < \dfrac{a}{b} < 2$

Dimpled limaçon

$\dfrac{a}{b} \geq 2$

Convex limaçon

Rose Curves

n petals when n is odd
$2n$ petals when n is even
$(n \geq 2)$

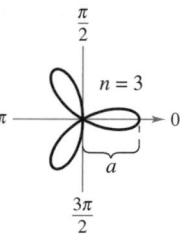

$r = a \cos n\theta$
Rose curve

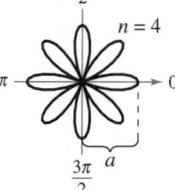

$r = a \cos n\theta$
Rose curve

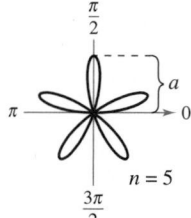

$r = a \sin n\theta$
Rose curve

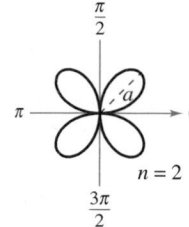

$r = a \sin n\theta$
Rose curve

Circles and Lemniscates

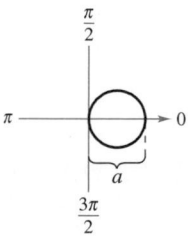
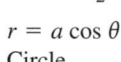

$r = a \cos \theta$
Circle

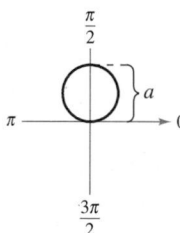

$r = a \sin \theta$
Circle

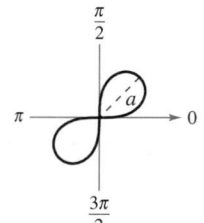

$r^2 = a^2 \sin 2\theta$
Lemniscate

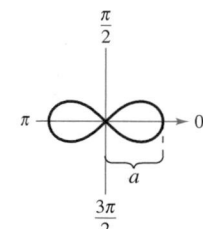

$r^2 = a^2 \cos 2\theta$
Lemniscate

▷ **TECHNOLOGY**   The rose curves described above are of the form $r = a \cos n\theta$ or $r = a \sin n\theta$, where n is a positive integer that is greater than or equal to 2. Use a graphing utility to graph

$$r = a \cos n\theta \quad \text{or} \quad r = a \sin n\theta$$

for some noninteger values of n. Are these graphs also rose curves? For example, try sketching the graph of

$$r = \cos \frac{2}{3}\theta, \quad 0 \leq \theta \leq 6\pi.$$

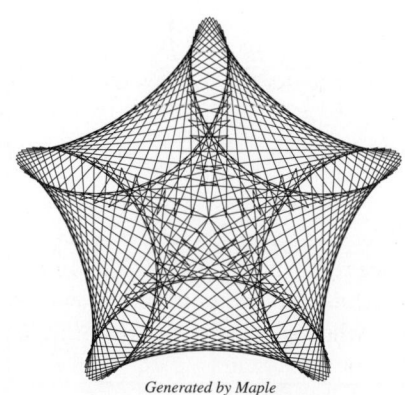

Generated by Maple

■ **FOR FURTHER INFORMATION**   For more information on rose curves and related curves, see the article "A Rose is a Rose . . ." by Peter M. Maurer in *The American Mathematical Monthly.* The computer-generated graph at the left is the result of an algorithm that Maurer calls "The Rose." To view this article, go to *MathArticles.com.*

10.4 Exercises

See **CalcChat.com** for tutorial help and worked-out solutions to odd-numbered exercises.

CONCEPT CHECK

1. Polar Coordinates Consider the polar coordinates (r, θ). What does r represent? What does θ represent?

2. Plotting Points Plot the points below on the same set of coordinate axes.

$$(r, \theta) = \left(2, \frac{\pi}{2}\right) \quad \text{and} \quad (x, y) = \left(2, \frac{\pi}{2}\right)$$

3. Comparing Coordinate Systems Describe the differences between the rectangular coordinate system and the polar coordinate system.

4. Parametric Form of a Polar Equation Explain how to write a polar equation in parametric form.

 **Polar-to-Rectangular Conversion In Exercises 5–14, the polar coordinates of a point are given. Plot the point and find the corresponding rectangular coordinates for the point.**

5. $\left(8, \frac{\pi}{2}\right)$ **6.** $\left(-2, \frac{5\pi}{3}\right)$

7. $\left(-4, -\frac{3\pi}{4}\right)$ **8.** $\left(0, -\frac{7\pi}{6}\right)$

9. $\left(7, \frac{5\pi}{4}\right)$ **10.** $\left(-2, \frac{11\pi}{6}\right)$

11. $\left(\sqrt{2}, 2.36\right)$ **12.** $(-3, -1.57)$

13. $(-8, 0.75)$ **14.** $(1.25, -5)$

 **Rectangular-to-Polar Conversion In Exercises 15–24, the rectangular coordinates of a point are given. Plot the point and find *two* sets of polar coordinates for the point for $0 \leq \theta < 2\pi$.**

15. $(1, 0)$ **16.** $(0, -9)$

17. $(-3, 4)$ **18.** $(6, -2)$

19. $\left(-5, -5\sqrt{3}\right)$ **20.** $\left(3, -\sqrt{3}\right)$

21. $\left(\sqrt{7}, -\sqrt{7}\right)$ **22.** $\left(-2\sqrt{2}, -2\sqrt{2}\right)$

23. $(4, 5)$ **24.** $(1, 8)$

 **Rectangular-to-Polar Conversion In Exercises 25–34, convert the rectangular equation to polar form and sketch its graph.**

25. $x^2 + y^2 = 9$ **26.** $x^2 - y^2 = 9$

27. $x^2 + y^2 = a^2$ **28.** $x^2 + y^2 - 2ax = 0$

29. $y = 8$ **30.** $x = 12$

31. $3x - y + 2 = 0$ **32.** $xy = 4$

33. $y^2 = 9x$

34. $(x^2 + y^2)^2 - 9(x^2 - y^2) = 0$

 **Polar-to-Rectangular Conversion In Exercises 35–44, convert the polar equation to rectangular form and sketch its graph.**

35. $r = 4$ **36.** $r = -1$

37. $r = 3 \sin \theta$ **38.** $r = 5 \cos \theta$

39. $r = \theta$ **40.** $\theta = \frac{5\pi}{6}$

41. $r = 3 \sec \theta$ **42.** $r = -6 \csc \theta$

43. $r = \sec \theta \tan \theta$ **44.** $r = \cot \theta \csc \theta$

Graphing a Polar Equation In Exercises 45–54, use a graphing utility to graph the polar equation. Find an interval for θ over which the graph is traced *only once*.

45. $r = 2 - 5 \cos \theta$ **46.** $r = 3(1 - 4 \cos \theta)$

47. $r = -1 + \sin \theta$ **48.** $r = 4 + 3 \cos \theta$

49. $r = \dfrac{2}{1 + \cos \theta}$ **50.** $r = \dfrac{1}{4 - 3 \sin \theta}$

51. $r = 5 \cos \dfrac{3\theta}{2}$ **52.** $r = 3 \sin \dfrac{5\theta}{2}$

53. $r^2 = 4 \sin 2\theta$ **54.** $r^2 = \dfrac{1}{\theta}$

55. Verifying a Polar Equation Convert the equation

$$r = 2(h \cos \theta + k \sin \theta)$$

to rectangular form and verify that it is the equation of a circle. Find the radius and the rectangular coordinates of the center of the circle.

56. HOW DO YOU SEE IT? Identify each special polar graph and write its equation.

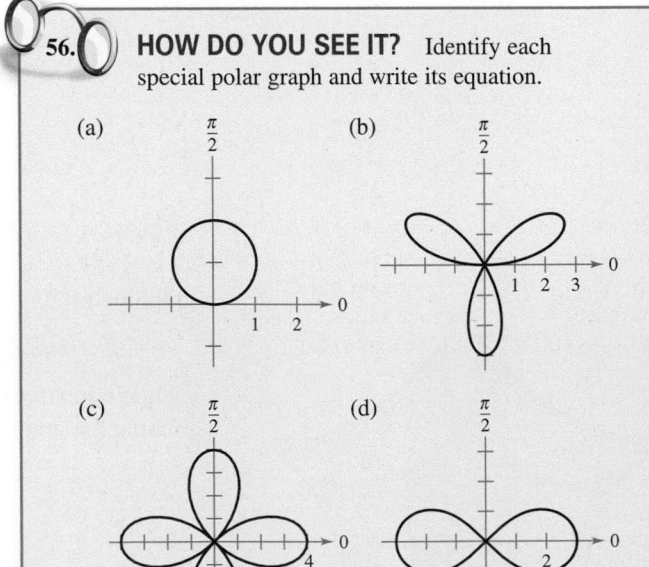

57. Sketching a Graph   Sketch the graph of $r = 4 \sin \theta$ over each interval.

(a) $0 \leq \theta \leq \dfrac{\pi}{2}$ (b) $\dfrac{\pi}{2} \leq \theta \leq \pi$ (c) $-\dfrac{\pi}{2} \leq \theta \leq \dfrac{\pi}{2}$

58. Distance Formula

(a) Verify that the Distance Formula for the distance between the two points (r_1, θ_1) and (r_2, θ_2) in polar coordinates is

$$d = \sqrt{r_1^2 + r_2^2 - 2r_1 r_2 \cos(\theta_1 - \theta_2)}.$$

(b) Describe the positions of the points relative to each other for $\theta_1 = \theta_2$. Simplify the Distance Formula for this case. Is the simplification what you expected? Explain.

(c) Simplify the Distance Formula for $\theta_1 - \theta_2 = 90°$. Is the simplification what you expected? Explain.

(d) Choose two points on the polar coordinate system and find the distance between them. Then choose different polar representations of the same two points and apply the Distance Formula again. Discuss the result.

Distance Formula   In Exercises 59–62, use the result of Exercise 58 to find the distance between the two points in polar coordinates.

59. $\left(1, \dfrac{5\pi}{6}\right)$, $\left(4, \dfrac{\pi}{3}\right)$ **60.** $\left(8, \dfrac{7\pi}{4}\right)$, $(5, \pi)$

61. $(2, 0.5)$, $(7, 1.2)$ **62.** $(4, 2.5)$, $(12, 1)$

 **Finding Slopes of Tangent Lines**   In Exercises 63 and 64, find dy/dx and the slopes of the tangent lines shown on the graph of the polar equation.

63. $r = 2(1 - \sin \theta)$ **64.** $r = 2 + 3 \sin \theta$

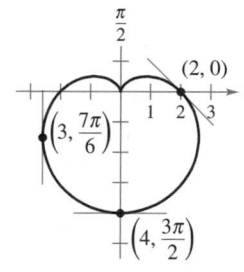

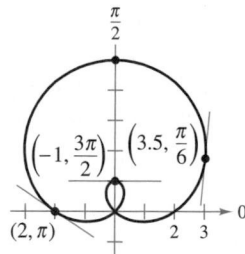

 **Finding Slopes of Tangent Lines Using Technology**   In Exercises 65–68, use a graphing utility to (a) graph the polar equation, (b) draw the tangent line at the given value of θ, and (c) find dy/dx at the given value of θ. (*Hint:* Let the increment between the values of θ equal $\pi/24$.)

65. $r = 3(1 - \cos \theta)$, $\theta = \dfrac{\pi}{2}$ **66.** $r = 3 - 2 \cos \theta$, $\theta = 0$

67. $r = 3 \sin \theta$, $\theta = \dfrac{\pi}{3}$ **68.** $r = 4$, $\theta = \dfrac{\pi}{4}$

 **Horizontal and Vertical Tangency**   In Exercises 69 and 70, find the points of horizontal and vertical tangency to the polar curve.

69. $r = 1 - \sin \theta$ **70.** $r = a \sin \theta$

Horizontal Tangency   In Exercises 71 and 72, find the points of horizontal tangency to the polar curve.

71. $r = 2 \csc \theta + 3$ **72.** $r = a \sin \theta \cos^2 \theta$

 **Tangent Lines at the Pole**   In Exercises 73–80, sketch a graph of the polar equation and find the tangent line(s) at the pole (if any).

73. $r = 5 \sin \theta$ **74.** $r = 5 \cos \theta$

75. $r = 4(1 - \sin \theta)$ **76.** $r = 2(1 - \cos \theta)$

77. $r = 4 \cos 3\theta$ **78.** $r = -\sin 5\theta$

79. $r = 3 \sin 2\theta$ **80.** $r = 3 \cos 2\theta$

Sketching a Polar Graph   In Exercises 81–92, sketch a graph of the polar equation.

81. $r = 8$ **82.** $r = 1$

83. $r = 4(1 + \cos \theta)$ **84.** $r = 1 + \sin \theta$

85. $r = 3 - 2 \cos \theta$ **86.** $r = 5 - 4 \sin \theta$

87. $r = -7 \csc \theta$ **88.** $r = \dfrac{6}{2 \sin \theta - 3 \cos \theta}$

89. $r = 3\theta$ **90.** $r = \dfrac{1}{\theta}$

91. $r^2 = 4 \cos 2\theta$ **92.** $r^2 = 4 \sin \theta$

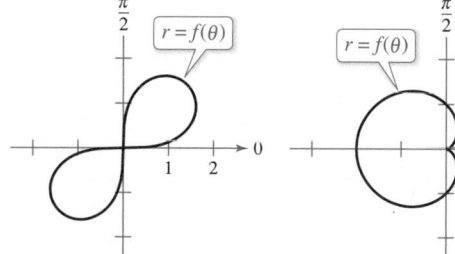 **Asymptote**   In Exercises 93–96, use a graphing utility to graph the equation and show that the given line is an asymptote of the graph.

	Name of Graph	Polar Equation	Asymptote
93.	Conchoid	$r = 2 - \sec \theta$	$x = -1$
94.	Conchoid	$r = 2 + \csc \theta$	$y = 1$
95.	Hyperbolic spiral	$r = 2/\theta$	$y = 2$
96.	Strophoid	$r = 2 \cos 2\theta \sec \theta$	$x = -2$

EXPLORING CONCEPTS

Transformations of Polar Graphs   In Exercises 97 and 98, use the graph of $r = f(\theta)$ to sketch a graph of the transformation.

97. $r = f(-\theta)$ **98.** $r = -f(\theta)$

99. Symmetry of Polar Graphs   Describe how to test whether a polar graph is symmetric about (a) the x-axis and (b) the y-axis.

100. Think About It Use a graphing utility to graph the polar equation $r = 6[1 + \cos(\theta - \phi)]$ for (a) $\phi = 0$, (b) $\phi = \pi/4$, and (c) $\phi = \pi/2$. Use the graphs to describe the effect of the angle ϕ. Write the equation as a function of $\sin \theta$ for part (c).

101. Rotated Curve Verify that if the curve whose polar equation is $r = f(\theta)$ is rotated about the pole through an angle ϕ, then an equation for the rotated curve is $r = f(\theta - \phi)$.

102. Rotated Curve The polar form of an equation of a curve is $r = f(\sin \theta)$. Show that the form becomes

(a) $r = f(-\cos \theta)$ if the curve is rotated counterclockwise $\pi/2$ radians about the pole.

(b) $r = f(-\sin \theta)$ if the curve is rotated counterclockwise π radians about the pole.

(c) $r = f(\cos \theta)$ if the curve is rotated counterclockwise $3\pi/2$ radians about the pole.

Rotated Curve **In Exercises 103–105, use the results of Exercises 101 and 102.**

 103. Write an equation for the limaçon $r = 2 - \sin \theta$ after it has been rotated counterclockwise by an angle of (a) $\theta = \pi/4$, (b) $\theta = \pi/2$, (c) $\theta = \pi$, and (d) $\theta = 3\pi/2$. Use a graphing utility to graph each rotated limaçon.

 104. Write an equation for the rose curve $r = 2 \sin 2\theta$ after it has been rotated counterclockwise by an angle of (a) $\theta = \pi/6$, (b) $\theta = \pi/2$, (c) $\theta = 2\pi/3$, and (d) $\theta = \pi$. Use a graphing utility to graph each rotated rose curve.

105. Sketch the graph of each equation.

(a) $r = 1 - \sin \theta$ (b) $r = 1 - \sin\left(\theta - \dfrac{\pi}{4}\right)$

106. Proof Prove that the tangent of the angle ψ ($0 \le \psi \le \pi/2$) between the radial line and the tangent line at the point (r, θ) on the graph of $r = f(\theta)$ (see figure) is given by

$$\tan \psi = \left| \frac{r}{dr/d\theta} \right|.$$

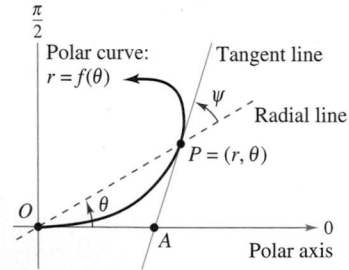

 Finding an Angle **In Exercises 107–112, use the result of Exercise 106 to find the angle ψ between the radial and tangent lines to the graph for the indicated value of θ. Use a graphing utility to graph the polar equation, the radial line, and the tangent line for the indicated value of θ. Identify the angle ψ.**

Polar Equation	Value of θ
107. $r = 2(1 - \cos \theta)$	$\theta = \pi$
108. $r = 3(1 - \cos \theta)$	$\theta = \dfrac{3\pi}{4}$
109. $r = 2 \cos 3\theta$	$\theta = \dfrac{\pi}{4}$
110. $r = 4 \sin 2\theta$	$\theta = \dfrac{\pi}{6}$
111. $r = \dfrac{6}{1 - \cos \theta}$	$\theta = \dfrac{2\pi}{3}$
112. $r = 5$	$\theta = \dfrac{\pi}{6}$

True or False? **In Exercises 113–116, determine whether the statement is true or false. If it is false, explain why or give an example that shows it is false.**

113. If (r_1, θ_1) and (r_2, θ_2) represent the same point on the polar coordinate system, then $|r_1| = |r_2|$.

114. If (r, θ_1) and (r, θ_2) represent the same point on the polar coordinate system, then $\theta_1 = \theta_2 + 2n\pi$ for some integer n.

115. If $x > 0$, then the point (x, y) on the rectangular coordinate system can be represented by (r, θ) on the polar coordinate system, where $r = \sqrt{x^2 + y^2}$ and $\theta = \arctan(y/x)$.

116. The polar equations $r = \sin 2\theta$, $r = -\sin 2\theta$, and $r = \sin(-2\theta)$ all have the same graph.

SECTION PROJECT

Cassini Oval

A Cassini oval is defined as the set of all points the product of whose distances from two fixed points is constant. These curves are named after the astronomer Giovanni Domenico Cassini (1625–1712). He suspected that these curves could model planetary motion. However, as you saw in Section 10.1, Kepler used ellipses to describe planetary motion. You will learn more about Kepler's Laws of planetary motion in Section 10.6.

Let $(-c, 0)$ and $(c, 0)$ be two fixed points in the plane. A point (x, y) lies on a Cassini oval when the distance between (x, y) and $(-c, 0)$ times the distance between (x, y) and $(c, 0)$ is b^2, where b is a constant.

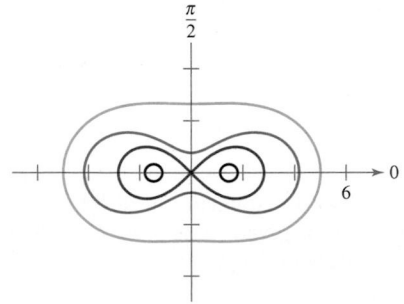

Four different types of Cassini ovals

(a) Show that $(x^2 + y^2)^2 - 2c^2(x^2 - y^2) + c^4 = b^4$.

(b) Convert the equation in part (a) to polar coordinates.

(c) Show that if $b = c$, then the Cassini oval is a lemniscate.

 (d) Use a graphing utility to graph the Cassini oval for $c = 1$ and $b = 2$.

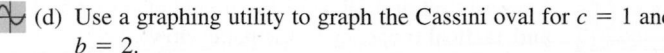

10.5   Area and Arc Length in Polar Coordinates

■ Find the area of a region bounded by a polar graph.
■ Find the points of intersection of two polar graphs.
■ Find the arc length of a polar graph.
■ Find the area of a surface of revolution (polar form).

Area of a Polar Region

The development of a formula for the area of a polar region parallels that for the area of a region on the rectangular coordinate system but uses sectors of a circle instead of rectangles as the basic elements of area. In Figure 10.48, note that the area of a circular sector of radius r is $\frac{1}{2}\theta r^2$, provided θ is measured in radians.

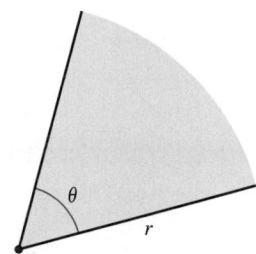

The area of a sector of a circle is $A = \frac{1}{2}\theta r^2$.
Figure 10.48

Consider the function $r = f(\theta)$, where f is continuous and nonnegative on the interval $\alpha \le \theta \le \beta$. The region bounded by the graph of f and the radial lines $\theta = \alpha$ and $\theta = \beta$ is shown in Figure 10.49(a). To find the area of this region, partition the interval $[\alpha, \beta]$ into n equal subintervals

$$\alpha = \theta_0 < \theta_1 < \theta_2 < \cdots < \theta_{n-1} < \theta_n = \beta.$$

Then approximate the area of the region by the sum of the areas of the n sectors, as shown in Figure 10.49(b).

$$\text{Radius of } i\text{th sector} = f(\theta_i)$$

$$\text{Central angle of } i\text{th sector} = \frac{\beta - \alpha}{n} = \Delta\theta$$

$$A \approx \sum_{i=1}^{n} \left(\frac{1}{2}\right)\Delta\theta[f(\theta_i)]^2$$

Taking the limit as $n \to \infty$ produces

$$A = \lim_{n \to \infty} \frac{1}{2} \sum_{i=1}^{n} [f(\theta_i)]^2 \Delta\theta$$

$$= \frac{1}{2}\int_{\alpha}^{\beta} [f(\theta)]^2 \, d\theta$$

which leads to the next theorem.

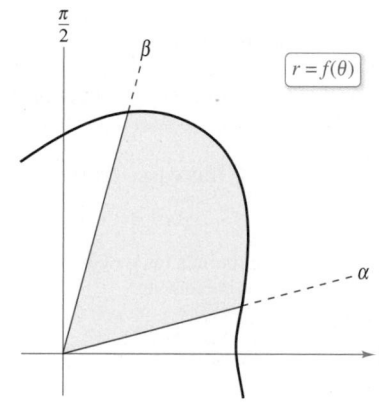

(a)

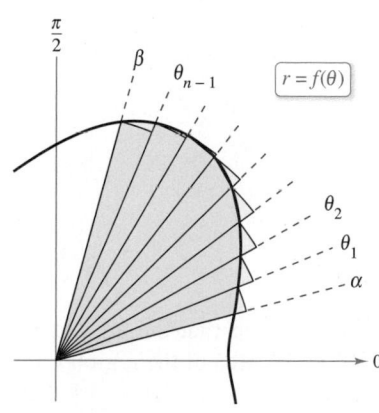

(b)
Figure 10.49

THEOREM 10.13   Area in Polar Coordinates

If f is continuous and nonnegative on the interval $[\alpha, \beta]$, $0 < \beta - \alpha \le 2\pi$, then the area of the region bounded by the graph of $r = f(\theta)$ between the radial lines $\theta = \alpha$ and $\theta = \beta$ is

$$A = \frac{1}{2}\int_{\alpha}^{\beta} [f(\theta)]^2 \, d\theta$$

$$= \frac{1}{2}\int_{\alpha}^{\beta} r^2 \, d\theta. \qquad 0 < \beta - \alpha \le 2\pi$$

You can use the formula in Theorem 10.13 to find the area of a region bounded by the graph of a continuous *nonpositive* function. The formula is not necessarily valid, however, when f takes on both positive *and* negative values in the interval $[\alpha, \beta]$.

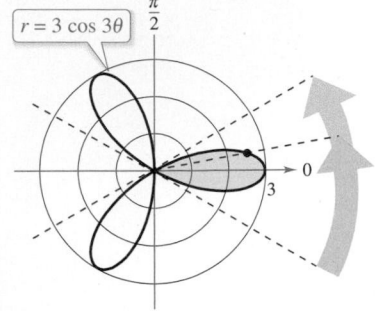

The area of one petal of the rose curve that lies between the radial lines $\theta = -\pi/6$ and $\theta = \pi/6$ is $3\pi/4$.

Figure 10.50

EXAMPLE 1 **Finding the Area of a Polar Region**

$\vdots \cdots \triangleright$ *See LarsonCalculus.com for an interactive version of this type of example.*

Find the area of one petal of the rose curve $r = 3 \cos 3\theta$.

Solution In Figure 10.50, you can see that the petal on the right is traced as θ increases from $-\pi/6$ to $\pi/6$. So, the area is

$$A = \frac{1}{2}\int_\alpha^\beta r^2 \, d\theta = \frac{1}{2}\int_{-\pi/6}^{\pi/6} (3 \cos 3\theta)^2 \, d\theta \qquad \text{Use formula for area in polar coordinates.}$$

$$= \frac{9}{2}\int_{-\pi/6}^{\pi/6} \frac{1 + \cos 6\theta}{2} \, d\theta \qquad \text{Power-reducing formula}$$

$$= \frac{9}{4}\left[\theta + \frac{\sin 6\theta}{6}\right]_{-\pi/6}^{\pi/6}$$

$$= \frac{9}{4}\left(\frac{\pi}{6} + \frac{\pi}{6}\right)$$

$$= \frac{3\pi}{4}.$$

To find the area of the region lying inside all three petals of the rose curve in Example 1, you could *not* simply integrate between 0 and 2π. By doing this, you would obtain $9\pi/2$, which is twice the area of the three petals. The duplication occurs because the rose curve is traced twice as θ increases from 0 to 2π.

EXAMPLE 2 **Finding the Area Bounded by a Single Curve**

Find the area of the region lying between the inner and outer loops of the limaçon $r = 1 - 2 \sin \theta$.

Solution In Figure 10.51, note that the inner loop is traced as θ increases from $\pi/6$ to $5\pi/6$. So, the area inside the *inner loop* is

$$A_1 = \frac{1}{2}\int_{\pi/6}^{5\pi/6} (1 - 2 \sin \theta)^2 \, d\theta \qquad \text{Use formula for area in polar coordinates.}$$

$$= \frac{1}{2}\int_{\pi/6}^{5\pi/6} (1 - 4 \sin \theta + 4 \sin^2 \theta) \, d\theta$$

$$= \frac{1}{2}\int_{\pi/6}^{5\pi/6} \left[1 - 4 \sin \theta + 4\left(\frac{1 - \cos 2\theta}{2}\right)\right] d\theta \qquad \text{Power-reducing formula}$$

$$= \frac{1}{2}\int_{\pi/6}^{5\pi/6} (3 - 4 \sin \theta - 2 \cos 2\theta) \, d\theta \qquad \text{Simplify.}$$

$$= \frac{1}{2}\left[3\theta + 4 \cos \theta - \sin 2\theta\right]_{\pi/6}^{5\pi/6}$$

$$= \frac{1}{2}\left(2\pi - 3\sqrt{3}\right)$$

$$= \pi - \frac{3\sqrt{3}}{2}.$$

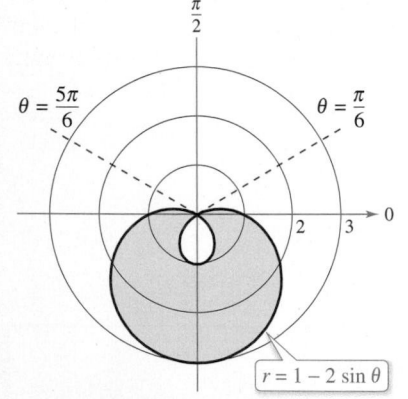

The area between the inner and outer loops is approximately 8.34.

Figure 10.51

In a similar way, you can integrate from $5\pi/6$ to $13\pi/6$ to find that the area of the region lying inside the *outer loop* is $A_2 = 2\pi + \left(3\sqrt{3}/2\right)$. The area of the region lying between the two loops is the difference of A_2 and A_1.

$$A = A_2 - A_1 = \left(2\pi + \frac{3\sqrt{3}}{2}\right) - \left(\pi - \frac{3\sqrt{3}}{2}\right) = \pi + 3\sqrt{3} \approx 8.34$$

Points of Intersection of Polar Graphs

Because a point may be represented in different ways in polar coordinates, care must be taken in determining the points of intersection of two polar graphs. For example, consider the points of intersection of the graphs of

$$r = 1 - 2 \cos \theta \quad \text{and} \quad r = 1$$

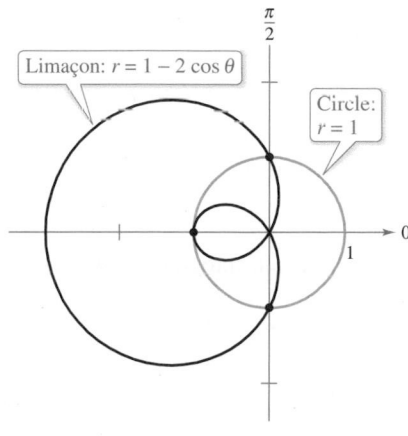

Three points of intersection: $(1, \pi/2)$, $(-1, 0)$, and $(1, 3\pi/2)$
Figure 10.52

as shown in Figure 10.52. As with rectangular equations, you can attempt to find the points of intersection by solving the two equations simultaneously, as shown.

$$r = 1 - 2 \cos \theta \qquad \text{First equation}$$
$$1 = 1 - 2 \cos \theta \qquad \text{Substitute } r = 1 \text{ from second equation into first equation.}$$
$$\cos \theta = 0 \qquad \text{Simplify.}$$
$$\theta = \frac{\pi}{2}, \ \frac{3\pi}{2} \qquad \text{Solve for } \theta.$$

The corresponding points of intersection are $(1, \pi/2)$ and $(1, 3\pi/2)$. From Figure 10.52, however, you can see that there is a *third* point of intersection that did not show up when the two polar equations were solved simultaneously. (This is one reason why you should sketch a graph when finding the area of a polar region.) The reason the third point was not found is that it does not occur with the same coordinates in the two graphs. On the graph of $r = 1$, the point occurs with coordinates $(1, \pi)$, but on the graph of

$$r = 1 - 2 \cos \theta$$

the point occurs with coordinates $(-1, 0)$.

In addition to solving equations simultaneously and sketching a graph, note that because the pole can be represented by $(0, \theta)$, where θ is *any* angle, you should check separately for the pole when finding points of intersection.

You can compare the problem of finding points of intersection of two polar graphs with that of finding collision points of two satellites in intersecting orbits about Earth, as shown in Figure 10.53. The satellites will not collide as long as they reach the points of intersection at different times (θ-values). Collisions will occur only at the points of intersection that are "simultaneous points"—those that are reached at the same time (θ-value).

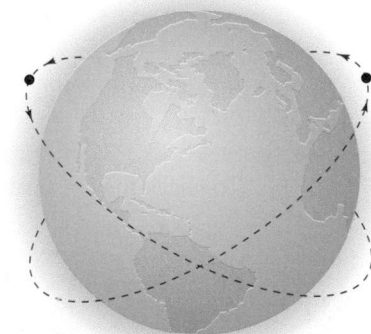

The paths of satellites can cross without causing a collision.
Figure 10.53

■ **FOR FURTHER INFORMATION** For more information on using technology to find points of intersection, see the article "Finding Points of Intersection of Polar-Coordinate Graphs" by Warren W. Esty in *Mathematics Teacher*. To view this article, go to *MathArticles.com*.

| EXAMPLE 3 | **Finding the Area of a Region Between Two Curves** |

Find the area of the region common to the two regions bounded by the curves

$$r = -6 \cos \theta \qquad \text{Circle}$$

and

$$r = 2 - 2 \cos \theta. \qquad \text{Cardioid}$$

Solution Because both curves are symmetric with respect to the x-axis, you can work with the upper half-plane, as shown in Figure 10.54. The blue shaded region lies between the circle and the radial line

$$\theta = \frac{2\pi}{3}.$$

Because the circle has coordinates $(0, \pi/2)$ at the pole, you can integrate between $\pi/2$ and $2\pi/3$ to obtain the area of this region. The region that is shaded red is bounded by the cardioid and the radial lines $\theta = 2\pi/3$ and $\theta = \pi$. So, you can find the area of this second region by integrating between $2\pi/3$ and π. The sum of these two integrals gives the area of the common region lying *above* the radial line $\theta = \pi$.

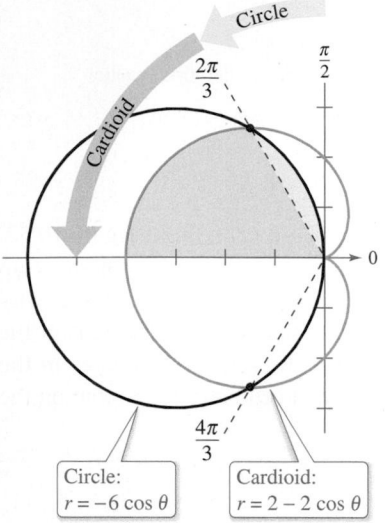

Circle: $r = -6 \cos \theta$ Cardioid: $r = 2 - 2 \cos \theta$

Figure 10.54

$$\overbrace{\text{Region between circle and radial line } \theta = 2\pi/3}^{} \qquad \overbrace{\text{Region between cardioid and radial lines } \theta = 2\pi/3 \text{ and } \theta = \pi}^{}$$

$$\frac{A}{2} = \frac{1}{2} \int_{\pi/2}^{2\pi/3} (-6 \cos \theta)^2 \, d\theta + \frac{1}{2} \int_{2\pi/3}^{\pi} (2 - 2 \cos \theta)^2 \, d\theta$$

$$= 18 \int_{\pi/2}^{2\pi/3} \cos^2 \theta \, d\theta + \frac{1}{2} \int_{2\pi/3}^{\pi} (4 - 8 \cos \theta + 4 \cos^2 \theta) \, d\theta$$

$$= 9 \int_{\pi/2}^{2\pi/3} (1 + \cos 2\theta) \, d\theta + \int_{2\pi/3}^{\pi} (3 - 4 \cos \theta + \cos 2\theta) \, d\theta$$

$$= 9 \left[\theta + \frac{\sin 2\theta}{2} \right]_{\pi/2}^{2\pi/3} + \left[3\theta - 4 \sin \theta + \frac{\sin 2\theta}{2} \right]_{2\pi/3}^{\pi}$$

$$= 9 \left(\frac{2\pi}{3} - \frac{\sqrt{3}}{4} - \frac{\pi}{2} \right) + \left(3\pi - 2\pi + 2\sqrt{3} + \frac{\sqrt{3}}{4} \right)$$

$$= \frac{5\pi}{2}$$

Finally, multiplying by 2, you can conclude that the total area is

$$5\pi \approx 15.7. \qquad \text{Area of region inside circle and cardioid}$$

To check the reasonableness of this result, note that the area of the circular region is

$$\pi r^2 = 9\pi. \qquad \text{Area of circle}$$

So, it seems reasonable that the area of the region lying inside the circle and the cardioid is 5π.   ■

To see the benefit of polar coordinates for finding the area in Example 3, consider the integral below, which gives the comparable area in rectangular coordinates.

$$\frac{A}{2} = \int_{-4}^{-3/2} \sqrt{2\sqrt{1 - 2x} - x^2 - 2x + 2} \, dx + \int_{-3/2}^{0} \sqrt{-x^2 - 6x} \, dx$$

Use the integration capabilities of a graphing utility to show that you obtain the same area as that found in Example 3.

Arc Length in Polar Form

The formula for the length of a polar arc can be obtained from the arc length formula for a curve described by parametric equations. (See Exercise 84.)

• • • • • • • • • • • • • • ▷
•• **REMARK** When applying the arc length formula to a polar curve, be sure that the curve is traced out only once on the interval of integration. For instance, the rose curve $r = \cos 3\theta$ is traced out once on the interval $0 \le \theta \le \pi$ but is traced out twice on the interval $0 \le \theta \le 2\pi$.

> **THEOREM 10.14   Arc Length of a Polar Curve**
>
> Let f be a function whose derivative is continuous on an interval $\alpha \le \theta \le \beta$. The length of the graph of $r = f(\theta)$ from $\theta = \alpha$ to $\theta = \beta$ is
>
> $$s = \int_\alpha^\beta \sqrt{[f(\theta)]^2 + [f'(\theta)]^2}\, d\theta = \int_\alpha^\beta \sqrt{r^2 + \left(\frac{dr}{d\theta}\right)^2}\, d\theta.$$

EXAMPLE 4   **Finding the Length of a Polar Curve**

Find the length of the arc from $\theta = 0$ to $\theta = 2\pi$ for the cardioid

$$r = f(\theta) = 2 - 2\cos\theta$$

as shown in Figure 10.55.

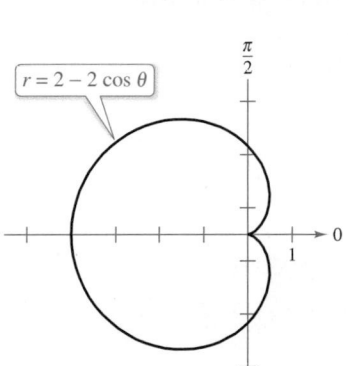

$r = 2 - 2\cos\theta$

Figure 10.55

Solution   Because $f'(\theta) = 2\sin\theta$, you can find the arc length as follows.

$$s = \int_\alpha^\beta \sqrt{[f(\theta)]^2 + [f'(\theta)]^2}\, d\theta \qquad \text{Formula for arc length of a polar curve}$$

$$= \int_0^{2\pi} \sqrt{(2 - 2\cos\theta)^2 + (2\sin\theta)^2}\, d\theta$$

$$= 2\sqrt{2} \int_0^{2\pi} \sqrt{1 - \cos\theta}\, d\theta \qquad \text{Simplify.}$$

$$= 2\sqrt{2} \int_0^{2\pi} \sqrt{2\sin^2\frac{\theta}{2}}\, d\theta \qquad \text{Trigonometric identity}$$

$$= 4 \int_0^{2\pi} \sin\frac{\theta}{2}\, d\theta \qquad \sin\frac{\theta}{2} \ge 0 \text{ for } 0 \le \theta \le 2\pi$$

$$= 8 \left[-\cos\frac{\theta}{2} \right]_0^{2\pi}$$

$$= 8(1 + 1)$$

$$= 16$$

Using Figure 10.55, you can determine the reasonableness of this answer by comparing it with the circumference of a circle. For example, a circle of radius $\frac{5}{2}$ has a circumference of

$$5\pi \approx 15.7.$$

Note that in the fifth step of the solution, it is legitimate to write

$$\sqrt{2\sin^2\frac{\theta}{2}} = \sqrt{2}\sin\frac{\theta}{2}$$

rather than

$$\sqrt{2\sin^2\frac{\theta}{2}} = \sqrt{2}\left|\sin\frac{\theta}{2}\right|$$

because $\sin(\theta/2) \ge 0$ for $0 \le \theta \le 2\pi$. ∎

Area of a Surface of Revolution

The polar coordinate versions of the formulas for the area of a surface of revolution can be obtained from the parametric versions given in Theorem 10.9, using the equations $x = r \cos \theta$ and $y = r \sin \theta$.

················▷

··REMARK When using Theorem 10.15, check to see that the graph of $r = f(\theta)$ is traced only once on the interval $\alpha \le \theta \le \beta$. For example, the circle $r = \cos \theta$ is traced only once on the interval $0 \le \theta \le \pi$.

> **THEOREM 10.15   Area of a Surface of Revolution**
>
> Let f be a function whose derivative is continuous on an interval $\alpha \le \theta \le \beta$. The area of the surface formed by revolving the graph of $r = f(\theta)$ from $\theta = \alpha$ to $\theta = \beta$ about the indicated line is as follows.
>
> **1.** $S = 2\pi \displaystyle\int_{\alpha}^{\beta} f(\theta) \sin \theta \sqrt{[f(\theta)]^2 + [f'(\theta)]^2}\, d\theta$ About the polar axis
>
> **2.** $S = 2\pi \displaystyle\int_{\alpha}^{\beta} f(\theta) \cos \theta \sqrt{[f(\theta)]^2 + [f'(\theta)]^2}\, d\theta$ About the line $\theta = \dfrac{\pi}{2}$

EXAMPLE 5   Finding the Area of a Surface of Revolution

Find the area of the surface formed by revolving the circle $r = f(\theta) = \cos \theta$ about the line $\theta = \pi/2$, as shown in Figure 10.56.

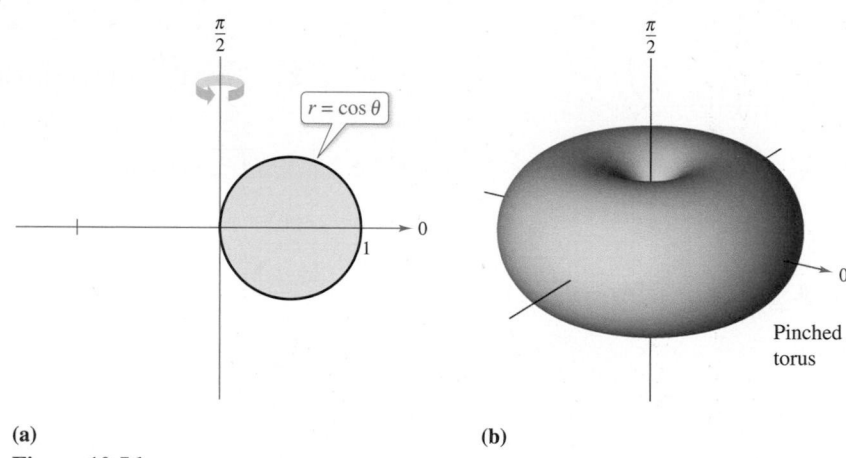

(a)   (b)

Figure 10.56

Solution   Use the second formula in Theorem 10.15 with $f'(\theta) = -\sin \theta$. Because the circle is traced once as θ increases from 0 to π, you have

$$S = 2\pi \int_{\alpha}^{\beta} f(\theta) \cos \theta \sqrt{[f(\theta)]^2 + [f'(\theta)]^2}\, d\theta \qquad \text{Formula for area of a surface of revolution}$$

$$= 2\pi \int_{0}^{\pi} (\cos \theta)(\cos \theta) \sqrt{\cos^2 \theta + \sin^2 \theta}\, d\theta$$

$$= 2\pi \int_{0}^{\pi} \cos^2 \theta\, d\theta \qquad \text{Trigonometric identity}$$

$$= \pi \int_{0}^{\pi} (1 + \cos 2\theta)\, d\theta \qquad \text{Trigonometric identity}$$

$$= \pi \left[\theta + \frac{\sin 2\theta}{2} \right]_{0}^{\pi}$$

$$= \pi^2.$$

10.5 Exercises

See **CalcChat.com** for tutorial help and worked-out solutions to odd-numbered exercises.

CONCEPT CHECK

1. **Area of a Polar Region** What should you check before applying Theorem 10.13 to find the area of the region bounded by the graph of $r = f(\theta)$?

2. **Points of Intersection** Explain why finding points of intersection of polar graphs may require further analysis beyond solving two equations simultaneously.

 **Area of a Polar Region** In Exercises 3–6, write an integral that represents the area of the shaded region of the figure. Do not evaluate the integral.

3. $r = 4 \sin \theta$

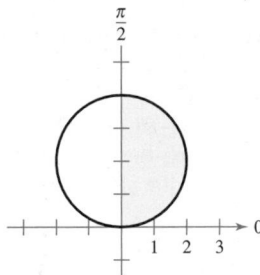

4. $r = \cos 2\theta$

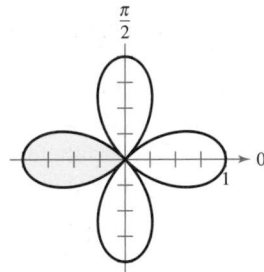

5. $r = 3 - 2 \sin \theta$

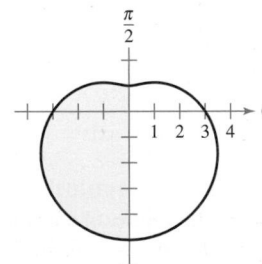

6. $r = 1 - \cos 2\theta$

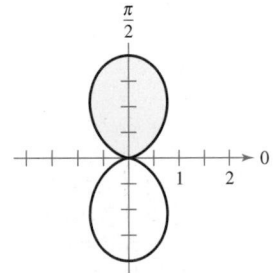

 **Finding the Area of a Polar Region** In Exercises 7–18, find the area of the region.

7. Interior of $r = 6 \sin \theta$

8. Interior of $r = 3 \cos \theta$

9. One petal of $r = 2 \cos 3\theta$

10. Two petals of $r = 4 \sin 3\theta$

11. Two petals of $r = \sin 8\theta$

12. Three petals of $r = \cos 5\theta$

13. Interior of $r = 6 + 5 \sin \theta$ (below the polar axis)

14. Interior of $r = 9 - \sin \theta$ (above the polar axis)

15. Interior of $r = 4 + \sin \theta$

16. Interior of $r = 1 - \cos \theta$

17. Interior of $r^2 = 4 \cos 2\theta$

18. Interior of $r^2 = 6 \sin 2\theta$

 **Finding the Area of a Polar Region** In Exercises 19–26, use a graphing utility to graph the polar equation. Find the area of the given region analytically.

19. Inner loop of $r = 1 + 2 \cos \theta$

20. Inner loop of $r = 2 - 4 \cos \theta$

21. Inner loop of $r = 1 + 2 \sin \theta$

22. Inner loop of $r = 4 - 6 \sin \theta$

23. Between the loops of $r = 1 + 2 \cos \theta$

24. Between the loops of $r = 2(1 + 2 \sin \theta)$

25. Between the loops of $r = 3 - 6 \sin \theta$

26. Between the loops of $r = \frac{1}{2} + \cos \theta$

 **Finding Points of Intersection** In Exercises 27–34, find the points of intersection of the graphs of the equations.

27. $r = 1 + \cos \theta$
    $r = 1 - \cos \theta$

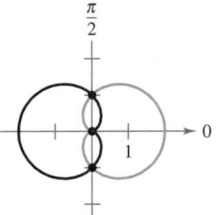

28. $r = 3(1 + \sin \theta)$
    $r = 3(1 - \sin \theta)$

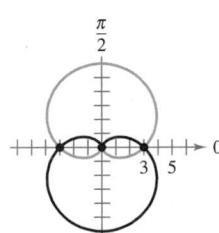

29. $r = 1 + \cos \theta$
    $r = 1 - \sin \theta$

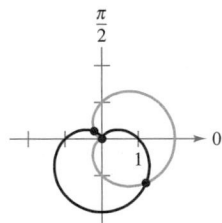

30. $r = 2 - 3 \cos \theta$
    $r = \cos \theta$

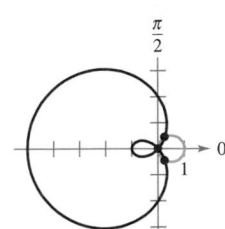

31. $r = 4 - 5 \sin \theta$
    $r = 3 \sin \theta$

32. $r = 3 + \sin \theta$
    $r = 2 \csc \theta$

33. $r = \dfrac{\theta}{2}$
    $r = 2$

34. $\theta = \dfrac{\pi}{4}$
    $r = 2$

Writing In Exercises 35 and 36, use a graphing utility to graph the polar equations and approximate the points of intersection of the graphs. Watch the graphs as they are traced in the viewing window. Explain why the pole is not a point of intersection obtained by solving the equations simultaneously.

35. $r = \cos \theta$
    $r = 2 - 3 \sin \theta$

36. $r = 4 \sin \theta$
    $r = 2(1 + \sin \theta)$

Finding the Area of a Polar Region Between Two Curves In Exercises 37–44, use a graphing utility to graph the polar equations. Find the area of the given region analytically.

37. Common interior of $r = 4 \sin 2\theta$ and $r = 2$

38. Common interior of $r = 2(1 + \cos \theta)$ and $r = 2(1 - \cos \theta)$

39. Common interior of $r = 3 - 2 \sin \theta$ and $r = -3 + 2 \sin \theta$

40. Common interior of $r = 5 - 3 \sin \theta$ and $r = 5 - 3 \cos \theta$

41. Common interior of $r = 4 \sin \theta$ and $r = 2$

42. Common interior of $r = 2 \cos \theta$ and $r = 2 \sin \theta$

43. Inside $r = 2 \cos \theta$ and outside $r = 1$

44. Inside $r = 3 \sin \theta$ and outside $r = 1 + \sin \theta$

Finding the Area of a Polar Region Between Two Curves In Exercises 45–48, find the area of the region.

45. Inside $r = a(1 + \cos \theta)$ and outside $r = a \cos \theta$

46. Inside $r = 2a \cos \theta$ and outside $r = a$

47. Common interior of $r = a(1 + \cos \theta)$ and $r = a \sin \theta$

48. Common interior of $r = a \cos \theta$ and $r = a \sin \theta$, where $a > 0$

• • **49. Antenna Radiation** • • • • • • • • • • • • • •

The radiation from a transmitting antenna is not uniform in all directions. The intensity from a particular antenna is modeled by $r = a \cos^2 \theta$.

(a) Convert the polar equation to rectangular form.

(b) Use a graphing utility to graph the model for $a = 4$ and $a = 6$.

(c) Find the area of the geographical region between the two curves in part (b).

50. **Area** The area inside one or more of the three interlocking circles $r = 2a \cos \theta$, $r = 2a \sin \theta$, and $r = a$ is divided into seven regions. Find the area of each region.

51. **Conjecture** Find the area of the region enclosed by

$$r = a \cos(n\theta)$$

for $n = 1, 2, 3, \ldots$. Use the results to make a conjecture about the area enclosed by the function when n is even and when n is odd.

52. **Area** Sketch the strophoid

$$r = \sec \theta - 2 \cos \theta, \quad -\frac{\pi}{2} < \theta < \frac{\pi}{2}.$$

Convert this equation to rectangular coordinates. Find the area enclosed by the loop.

Finding the Arc Length of a Polar Curve In Exercises 53–58, find the length of the curve over the given interval.

53. $r = 8$, $\left[0, \dfrac{\pi}{6}\right]$ 54. $r = a$, $[0, 2\pi]$

55. $r = 4 \sin \theta$, $[0, \pi]$ 56. $r = 2a \cos \theta$, $\left[-\dfrac{\pi}{4}, \dfrac{\pi}{4}\right]$

57. $r = 1 + \sin \theta$, $[0, 2\pi]$ 58. $r = 8(1 + \cos \theta)$, $\left[0, \dfrac{\pi}{3}\right]$

 **Finding the Arc Length of a Polar Curve** In Exercises 59–64, use a graphing utility to graph the polar equation over the given interval. Use the integration capabilities of the graphing utility to approximate the length of the curve.

59. $r = 2\theta$, $\left[0, \dfrac{\pi}{2}\right]$ 60. $r = \sec \theta$, $\left[0, \dfrac{\pi}{3}\right]$

61. $r = \dfrac{1}{\theta}$, $[\pi, 2\pi]$ 62. $r = e^{\theta}$, $[0, \pi]$

63. $r = \sin(3 \cos \theta)$, $[0, \pi]$ 64. $r = 2 \sin(2 \cos \theta)$, $[0, \pi]$

Finding the Area of a Surface of Revolution In Exercises 65–68, find the area of the surface formed by revolving the polar equation over the given interval about the given line.

Polar Equation	Interval	Axis of Revolution
65. $r = 6 \cos \theta$	$0 \le \theta \le \dfrac{\pi}{2}$	Polar axis
66. $r = a \cos \theta$	$0 \le \theta \le \dfrac{\pi}{2}$	$\theta = \dfrac{\pi}{2}$
67. $r = e^{a\theta}$	$0 \le \theta \le \dfrac{\pi}{2}$	$\theta = \dfrac{\pi}{2}$
68. $r = a(1 + \cos \theta)$	$0 \le \theta \le \pi$	Polar axis

Finding the Area of a Surface of Revolution In Exercises 69 and 70, use the integration capabilities of a graphing utility to approximate the area of the surface formed by revolving the polar equation over the given interval about the polar axis.

69. $r = 4 \cos 2\theta$, $\left[0, \dfrac{\pi}{4}\right]$ 70. $r = \theta$, $[0, \pi]$

EXPLORING CONCEPTS

Using Different Methods In Exercises 71 and 72, (a) sketch the graph of the polar equation, (b) determine the interval that traces the graph only once, (c) find the area of the region bounded by the graph using a geometric formula, and (d) find the area of the region bounded by the graph using integration.

71. $r = 10 \cos \theta$ 72. $r = 5 \sin \theta$

73. **Think About It** Let $f(\theta) > 0$ for all θ and let $g(\theta) < 0$ for all θ. Find polar equations $r = f(\theta)$ and $r = g(\theta)$ such that their graphs intersect.

74. HOW DO YOU SEE IT? Which graph, traced out only once, has a larger arc length? Explain your reasoning.

(a)

$\frac{\pi}{2}$

(b)

$\frac{\pi}{2}$

75. Surface Area of a Torus Find the surface area of the torus generated by revolving the circle given by $r = 2$ about the line $r = 5 \sec \theta$.

76. Surface Area of a Torus Find the surface area of the torus generated by revolving the circle given by $r = a$ about the line $r = b \sec \theta$, where $0 < a < b$.

77. Approximating Area Consider the circle

$r = 8 \cos \theta.$

(a) Find the area of the circle.

(b) Complete the table for the areas A of the sectors of the circle between $\theta = 0$ and the values of θ in the table.

θ	0.2	0.4	0.6	0.8	1.0	1.2	1.4
A							

(c) Use the table in part (b) to approximate the values of θ for which the sector of the circle composes $\frac{1}{4}$, $\frac{1}{2}$, and $\frac{3}{4}$ of the total area of the circle.

(d) Use a graphing utility to approximate, to two decimal places, the angles θ for which the sector of the circle composes $\frac{1}{4}$, $\frac{1}{2}$, and $\frac{3}{4}$ of the total area of the circle.

(e) Do the results of part (d) depend on the radius of the circle? Explain.

78. Approximating Area Consider the circle

$r = 3 \sin \theta.$

(a) Find the area of the circle.

(b) Complete the table for the areas A of the sectors of the circle between $\theta = 0$ and the values of θ in the table.

θ	0.2	0.4	0.6	0.8	1.0	1.2	1.4
A							

(c) Use the table in part (b) to approximate the values of θ for which the sector of the circle composes $\frac{1}{8}$, $\frac{1}{4}$, and $\frac{1}{2}$ of the total area of the circle.

(d) Use a graphing utility to approximate, to two decimal places, the angles θ for which the sector of the circle composes $\frac{1}{8}$, $\frac{1}{4}$, and $\frac{1}{2}$ of the total area of the circle.

79. Spiral of Archimedes The curve represented by the equation $r = a\theta$, where a is a constant, is called the **spiral of Archimedes**.

(a) Use a graphing utility to graph $r = \theta$, where $\theta \geq 0$. What happens to the graph of $r = a\theta$ as a increases? What happens if $\theta \leq 0$?

(b) Determine the points on the spiral $r = a\theta$ ($a > 0, \theta \geq 0$), where the curve crosses the polar axis.

(c) Find the length of $r = \theta$ over the interval $0 \leq \theta \leq 2\pi$.

(d) Find the area under the curve $r = \theta$ for $0 \leq \theta \leq 2\pi$.

80. Logarithmic Spiral The curve represented by the equation $r = ae^{b\theta}$, where a and b are constants, is called a **logarithmic spiral**. The figure shows the graph of $r = e^{\theta/6}$, $-2\pi \leq \theta \leq 2\pi$. Find the area of the shaded region.

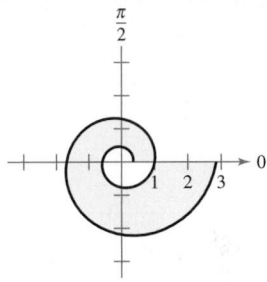

81. Area The larger circle in the figure is the graph of $r = 1$. Find the polar equation of the smaller circle such that the shaded regions are equal.

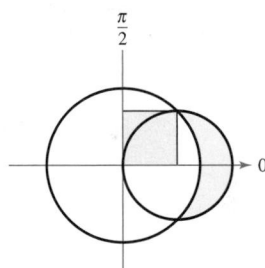

82. Area Find the area of the circle given by

$r = \sin \theta + \cos \theta.$

Check your result by converting the polar equation to rectangular form, then using the formula for the area of a circle.

83. Folium of Descartes A curve called the **folium of Descartes** can be represented by the parametric equations

$$x = \frac{3t}{1 + t^3} \quad \text{and} \quad y = \frac{3t^2}{1 + t^3}.$$

(a) Convert the parametric equations to polar form.

(b) Sketch the graph of the polar equation from part (a).

(c) Use a graphing utility to approximate the area enclosed by the loop of the curve.

84. Arc Length in Polar Form Use the formula for the arc length of a curve in parametric form to derive the formula for the arc length of a polar curve.

10.6 Polar Equations of Conics and Kepler's Laws

■ Analyze and write polar equations of conics.
■ Understand and use Kepler's Laws of planetary motion.

Polar Equations of Conics

In this chapter, you have seen that the rectangular equations of ellipses and hyperbolas take simple forms when the origin lies at their *centers*. As it happens, there are many important applications of conics in which it is more convenient to use one of the foci as the reference point (the origin) for the coordinate system. Here are two examples.

1. The sun lies at a focus of Earth's orbit.

2. The light source of a parabolic reflector lies at its focus.

In this section, you will see that the polar equations of conics take simpler forms when one of the foci lies at the pole.

The next theorem uses the concept of *eccentricity*, as defined in Section 10.1, to classify the three basic types of conics.

> **THEOREM 10.16 Classification of Conics by Eccentricity**
>
> Let F be a fixed point (*focus*) and let D be a fixed line (*directrix*) in the plane. Let P be another point in the plane and let e (*eccentricity*) be the ratio of the distance between P and F to the distance between P and D. The collection of all points P with a given eccentricity is a conic.
>
> 1. The conic is an ellipse for $0 < e < 1$.
>
> 2. The conic is a parabola for $e = 1$.
>
> 3. The conic is a hyperbola for $e > 1$.
>
> A proof of this theorem is given in Appendix A.

In Figure 10.57, note that for each type of conic, the pole corresponds to the fixed point (focus) given in the definition.

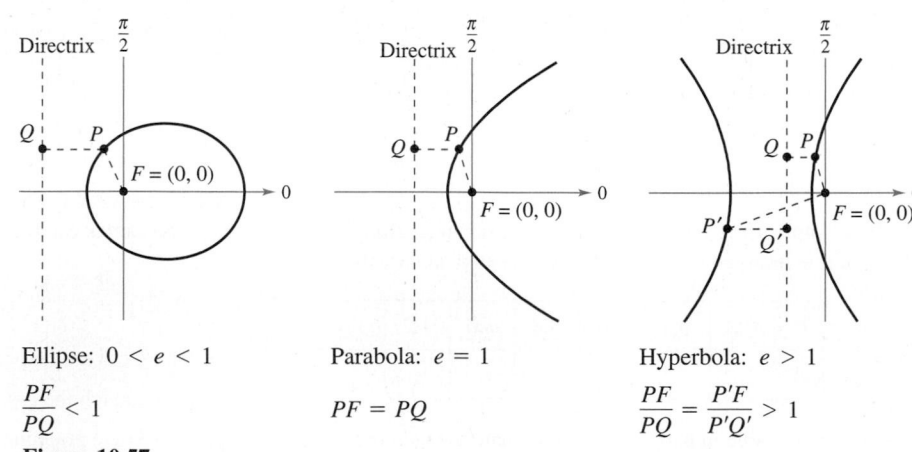

Ellipse: $0 < e < 1$
$$\frac{PF}{PQ} < 1$$

Parabola: $e = 1$
$$PF = PQ$$

Hyperbola: $e > 1$
$$\frac{PF}{PQ} = \frac{P'F}{P'Q'} > 1$$

Figure 10.57

The benefit of locating a focus of a conic at the pole is that the equation of the conic becomes simpler, as seen in the proof of the next theorem.

THEOREM 10.17   Polar Equations of Conics

The graph of a polar equation of the form

$$r = \frac{ed}{1 \pm e \cos \theta} \quad \text{or} \quad r = \frac{ed}{1 \pm e \sin \theta}$$

is a conic, where $e > 0$ is the eccentricity and $|d|$ is the distance between the focus at the pole and its corresponding directrix.

Proof   This is a proof for $r = ed/(1 + e \cos \theta)$ with $d > 0$. In Figure 10.58, consider a vertical directrix d units to the right of the focus $F = (0, 0)$. If $P = (r, \theta)$ is a point on the graph of $r = ed/(1 + e \cos \theta)$, then the distance between P and the directrix can be shown to be

$$PQ = |d - x| = |d - r \cos \theta| = \left| \frac{r(1 + e \cos \theta)}{e} - r \cos \theta \right| = \left| \frac{r}{e} \right|.$$

Because the distance between P and the pole is simply $PF = |r|$, the ratio of PF to PQ is

$$\frac{PF}{PQ} = \frac{|r|}{|r/e|} = |e| = e$$

and, by Theorem 10.16, the graph of the equation must be a conic. The proofs of the other cases are similar. ∎

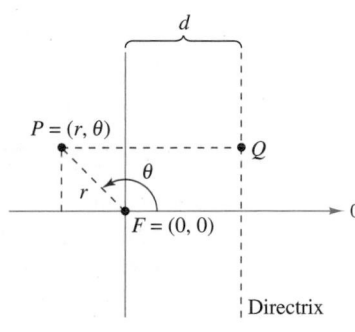

Figure 10.58

The four types of equations indicated in Theorem 10.17 can be classified as follows, where $d > 0$.

a. Horizontal directrix above the pole: $r = \dfrac{ed}{1 + e \sin \theta}$

b. Horizontal directrix below the pole: $r = \dfrac{ed}{1 - e \sin \theta}$

c. Vertical directrix to the right of the pole: $r = \dfrac{ed}{1 + e \cos \theta}$

d. Vertical directrix to the left of the pole: $r = \dfrac{ed}{1 - e \cos \theta}$

Figure 10.59 illustrates these four possibilities for a parabola. Note that for convenience, the equation for the directrix is shown in rectangular form.

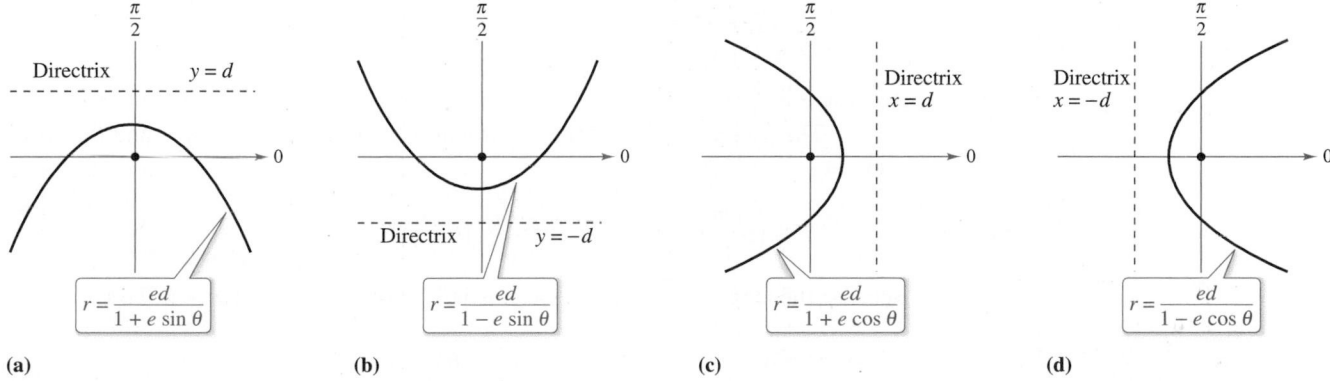

(a)   (b)   (c)   (d)

The four types of polar equations for a parabola

Figure 10.59

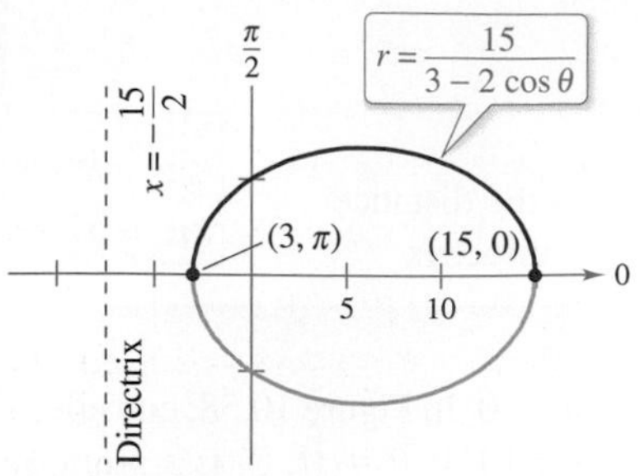

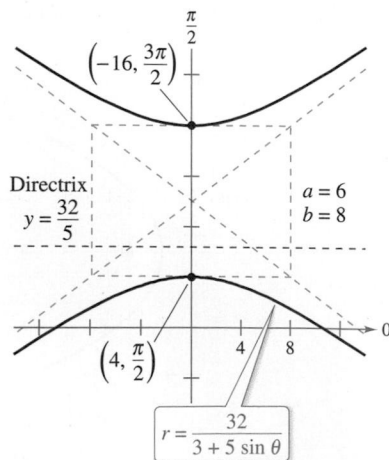

The graph of the conic is an ellipse with $e = \frac{2}{3}$.

Figure 10.60

EXAMPLE 1 **Determining a Conic from Its Equation**

Sketch the graph of the conic $r = \dfrac{15}{3 - 2 \cos \theta}$.

Solution To determine the type of conic, rewrite the equation as

$$r = \frac{15}{3 - 2 \cos \theta} \qquad \text{Write original equation.}$$

$$= \frac{5}{1 - (2/3) \cos \theta}. \qquad \text{Divide numerator and denominator by 3.}$$

So, the graph is an ellipse with $e = \frac{2}{3}$. You can sketch the upper half of the ellipse by plotting points from $\theta = 0$ to $\theta = \pi$, as shown in Figure 10.60. Then, using symmetry with respect to the polar axis, you can sketch the lower half.

For the ellipse in Figure 10.60, the major axis is horizontal and the vertices lie at $(15, 0)$ and $(3, \pi)$. So, the length of the *major* axis is $2a = 18$. To find the length of the minor axis, you can use the equations $e = c/a$ and $b^2 = a^2 - c^2$ to conclude that

$$b^2 = a^2 - c^2 = a^2 - (ea)^2 = a^2(1 - e^2). \qquad \text{Ellipse}$$

Because $e = \frac{2}{3}$, you have

$$b^2 = 9^2\left[1 - \left(\tfrac{2}{3}\right)^2\right] = 45$$

which implies that $b = \sqrt{45} = 3\sqrt{5}$. So, the length of the minor axis is $2b = 6\sqrt{5}$. A similar analysis for hyperbolas yields

$$b^2 = c^2 - a^2 = (ea)^2 - a^2 = a^2(e^2 - 1). \qquad \text{Hyperbola}$$

EXAMPLE 2 **Sketching a Conic from Its Polar Equation**

∙∙∙∙▷ *See LarsonCalculus.com for an interactive version of this type of example.*

Sketch the graph of the polar equation $r = \dfrac{32}{3 + 5 \sin \theta}$.

Solution Dividing the numerator and denominator by 3 produces

$$r = \frac{32/3}{1 + (5/3) \sin \theta}.$$

Because $e = \frac{5}{3} > 1$, the graph is a hyperbola. Because $d = \frac{32}{5}$, the directrix is the line $y = \frac{32}{5}$. The transverse axis of the hyperbola lies on the line $\theta = \pi/2$, and the vertices occur at

$$(r, \theta) = \left(4, \frac{\pi}{2}\right) \quad \text{and} \quad (r, \theta) = \left(-16, \frac{3\pi}{2}\right).$$

Because the length of the transverse axis is 12, you can see that $a = 6$. To find b, write

$$b^2 = a^2(e^2 - 1) = 6^2\left[\left(\frac{5}{3}\right)^2 - 1\right] = 64.$$

Therefore, $b = 8$. Finally, you can use a and b to determine the asymptotes of the hyperbola and obtain the sketch shown in Figure 10.61.

The graph of the conic is a hyperbola with $e = \frac{5}{3}$.

Figure 10.61

JOHANNES KEPLER (1571–1630)

Kepler formulated his three
laws from the extensive
data recorded by Danish
astronomer Tycho Brahe and
from direct observation of the
orbit of Mars.
*See LarsonCalculus.com to read
more of this biography.*

Kepler's Laws

Kepler's Laws, named after the German astronomer Johannes Kepler, can be used to describe the orbits of the planets about the sun.

1. Each planet moves in an elliptical orbit with the sun as a focus.
2. A ray from the sun to the planet sweeps out equal areas of the ellipse in equal times.
3. The square of the period is proportional to the cube of the mean distance between the planet and the sun.*

Although Kepler derived these laws empirically, they were later validated by Newton. In fact, Newton was able to show that each law can be deduced from a set of universal laws of motion and gravitation that govern the movement of all heavenly bodies, including comets and satellites. This is shown in the next example, involving the comet named after the English mathematician and physicist Edmund Halley (1656–1742).

EXAMPLE 3 Halley's Comet

Halley's comet has an elliptical orbit with the sun at one focus and has an eccentricity of $e \approx 0.967$. The length of the major axis of the orbit is approximately 35.88 astronomical units (AU). (An astronomical unit is defined as the mean distance between Earth and the sun, which is 93 million miles.) Find a polar equation for the orbit. How close does Halley's comet come to the sun?

Solution Using a vertical axis, you can choose an equation of the form

$$r = \frac{ed}{1 + e \sin \theta}.$$

Because the vertices of the ellipse occur when $\theta = \pi/2$ and $\theta = 3\pi/2$, you can determine the length of the major axis to be the sum of the r-values of the vertices, as shown in Figure 10.62. That is,

$$2a = \frac{0.967d}{1 + 0.967} + \frac{0.967d}{1 - 0.967}$$

$$35.88 \approx 29.79d. \qquad\qquad 2a \approx 35.88$$

So, $d \approx 1.204$ and

$$ed \approx (0.967)(1.204) \approx 1.164.$$

Using this value in the equation produces

$$r = \frac{1.164}{1 + 0.967 \sin \theta}$$

where r is measured in astronomical units. To find the closest point to the sun (the focus), you can write

$$c = ea \approx (0.967)(17.94) \approx 17.35.$$

Because c is the distance between the focus and the center, the closest point is

$$a - c \approx 17.94 - 17.35$$
$$= 0.59 \text{ AU}$$
$$\approx 55,000,000 \text{ miles.}$$

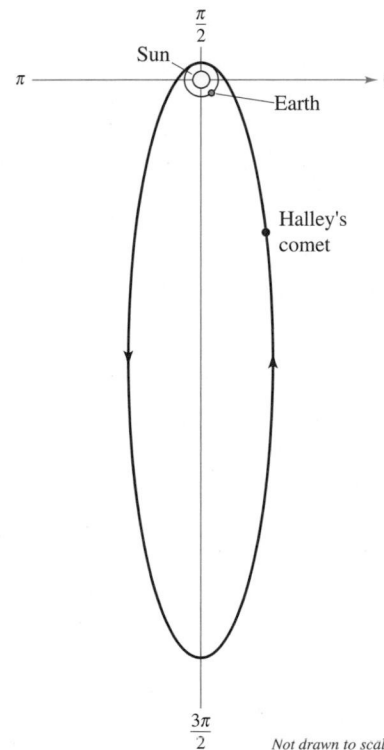

Figure 10.62

Not drawn to scale

---

* If Earth is used as a reference with a period of 1 year and a distance of 1 astronomical unit, then the proportionality constant is 1. For example, because Mars has a mean distance to the sun of $D \approx 1.524$ AU, its period P is $D^3 = P^2$. So, the period for Mars is $P \approx 1.88$ years.

Kepler's Second Law states that as a planet moves about the sun, a ray from the sun to the planet sweeps out equal areas in equal times. This law can also be applied to comets or asteroids with elliptical orbits. For example, Figure 10.63 shows the orbit of the asteroid Apollo about the sun. Applying Kepler's Second Law to this asteroid, you know that the closer it is to the sun, the greater its velocity, because a short ray must be moving quickly to sweep out as much area as a long ray.

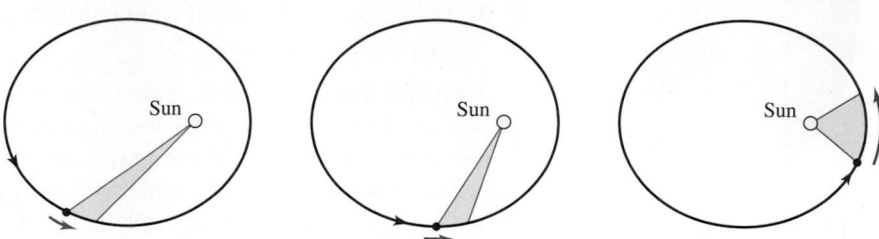

A ray from the sun to the asteroid Apollo sweeps out equal areas in equal times.
Figure 10.63

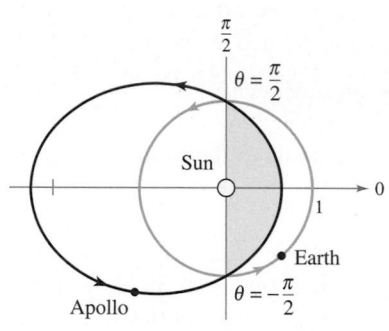

Figure 10.64

EXAMPLE 4   The Asteroid Apollo

The asteroid Apollo has a period of about 661 Earth days, and its orbit is approximated by the ellipse

$$r = \frac{1}{1 + (5/9)\cos\theta} = \frac{9}{9 + 5\cos\theta}$$

where r is measured in astronomical units. How long does it take Apollo to move from the position $\theta = -\pi/2$ to $\theta = \pi/2$, as shown in Figure 10.64?

Solution   Begin by finding the area swept out as θ increases from $-\pi/2$ to $\pi/2$.

$$A = \frac{1}{2}\int_{\alpha}^{\beta} r^2 \, d\theta \qquad \text{Formula for area of a polar graph}$$

$$= \frac{1}{2}\int_{-\pi/2}^{\pi/2} \left(\frac{9}{9 + 5\cos\theta}\right)^2 d\theta$$

Using the substitution $u = \tan(\theta/2)$, as discussed in Section 8.7, you obtain

$$A = \frac{81}{112}\left[\frac{-5\sin\theta}{9 + 5\cos\theta} + \frac{18}{\sqrt{56}}\arctan\frac{\sqrt{56}\tan(\theta/2)}{14}\right]_{-\pi/2}^{\pi/2} \approx 0.90429.$$

Because the major axis of the ellipse has length $2a = 81/28$ and the eccentricity is $e = 5/9$, you can determine that

$$b = a\sqrt{1 - e^2} = \frac{9}{\sqrt{56}}.$$

So, the area of the ellipse is

$$\text{Area of ellipse} = \pi ab = \pi\left(\frac{81}{56}\right)\left(\frac{9}{\sqrt{56}}\right) \approx 5.46507.$$

Because the time required to complete the orbit is 661 days, you can apply Kepler's Second Law to conclude that the time t required to move from the position $\theta = -\pi/2$ to $\theta = \pi/2$ is

$$\frac{t}{661} = \frac{\text{area of elliptical segment}}{\text{area of ellipse}} \approx \frac{0.90429}{5.46507}$$

which implies that $t \approx 109$ days.

10.6 Exercises

See **CalcChat.com** for tutorial help and worked-out solutions to odd-numbered exercises.

CONCEPT CHECK

1. Classification of Conics Identify each conic using eccentricity.

(a) $r = \dfrac{4}{1 + 3 \sin \theta}$

(b) $r = \dfrac{7}{1 - \cos \theta}$

(c) $r = \dfrac{8}{6 + 5 \cos \theta}$

(d) $r = \dfrac{3}{2 - 3 \sin \theta}$

2. Comparing Conics Without graphing, how are the graphs of the following conics different? Explain.

$$r = \frac{1}{1 + \sin \theta} \quad \text{and} \quad r = \frac{1}{1 - \sin \theta}$$

 Graphing a Conic In Exercises 3 and 4, use a graphing utility to graph the polar equation when (a) $e = 1$, (b) $e = 0.5$, and (c) $e = 1.5$. Identify the conic.

3. $r = \dfrac{2e}{1 + e \cos \theta}$

4. $r = \dfrac{2e}{1 - e \sin \theta}$

Writing In Exercises 5 and 6, consider the polar equation

$$r = \frac{4}{1 + e \sin \theta}.$$

5. Use a graphing utility to graph the equation for $e = 0.1$, $e = 0.25$, $e = 0.5$, $e = 0.75$, and $e = 0.9$. Identify the conic and discuss the change in its shape as $e \to 1^-$ and $e \to 0^+$.

6. Use a graphing utility to graph the equation for $e = 1.1$, $e = 1.5$, and $e = 2$. Identify the conic and discuss the change in its shape as $e \to 1^+$ and $e \to \infty$.

Matching In Exercises 7–12, match the polar equation with its correct graph. [The graphs are labeled (a), (b), (c), (d), (e), and (f).]

(a)
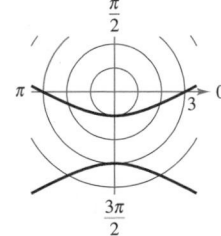

(b)
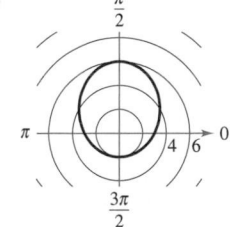

(c)
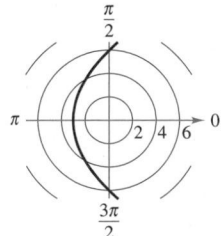

(d)
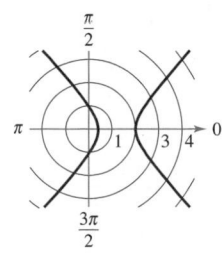

(e)
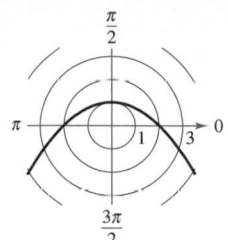

(f)
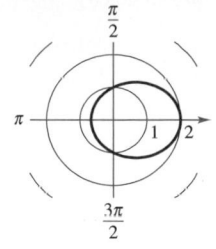

7. $r = \dfrac{6}{1 - \cos \theta}$

8. $r = \dfrac{2}{2 - \cos \theta}$

9. $r = \dfrac{3}{1 - 2 \sin \theta}$

10. $r = \dfrac{2}{1 + \sin \theta}$

11. $r = \dfrac{6}{2 - \sin \theta}$

12. $r = \dfrac{2}{2 + 3 \cos \theta}$

 Identifying and Sketching a Conic In Exercises 13–22, find the eccentricity and the distance from the pole to the directrix of the conic. Then identify the conic and sketch its graph. Use a graphing utility to confirm your results.

13. $r = \dfrac{1}{1 - \cos \theta}$

14. $r = \dfrac{5}{5 - 3 \cos \theta}$

15. $r = \dfrac{7}{4 + 8 \sin \theta}$

16. $r = \dfrac{4}{1 + \cos \theta}$

17. $r = \dfrac{6}{-2 + 3 \cos \theta}$

18. $r = \dfrac{10}{5 + 4 \sin \theta}$

19. $r = \dfrac{6}{2 + \cos \theta}$

20. $r = \dfrac{-6}{3 + 7 \sin \theta}$

21. $r = \dfrac{300}{-12 + 6 \sin \theta}$

22. $r = \dfrac{24}{25 + 25 \cos \theta}$

Identifying a Conic In Exercises 23–26, use a graphing utility to graph the polar equation. Identify the graph and find its eccentricity.

23. $r = \dfrac{3}{-4 + 2 \sin \theta}$

24. $r = \dfrac{-15}{2 + 8 \sin \theta}$

25. $r = \dfrac{-10}{1 - \cos \theta}$

26. $r = \dfrac{6}{6 + 7 \cos \theta}$

Comparing Graphs In Exercises 27–30, use a graphing utility to graph the conic. Describe how the graph differs from the graph in the indicated exercise.

27. $r = \dfrac{4}{1 + \cos(\theta - \pi/3)}$ (See Exercise 16.)

28. $r = \dfrac{10}{5 + 4 \sin(\theta - \pi/4)}$ (See Exercise 18.)

29. $r = \dfrac{6}{2 + \cos(\theta + \pi/6)}$ (See Exercise 19.)

30. $r = \dfrac{-6}{3 + 7 \sin(\theta + 2\pi/3)}$ (See Exercise 20.)

31. Rotated Ellipse Write the equation for the ellipse rotated $\pi/6$ radian clockwise from the ellipse

$$r = \frac{8}{8 + 5 \cos \theta}.$$

32. Rotated Parabola Write the equation for the parabola rotated $\pi/4$ radian counterclockwise from the parabola

$$r = \frac{9}{1 + \sin \theta}.$$

Finding a Polar Equation In Exercises 33–38, find a polar equation for the conic with its focus at the pole and the given eccentricity and directrix. (For convenience, the equation for the directrix is given in rectangular form.)

	Conic	Eccentricity	Directrix
33.	Parabola	$e = 1$	$x = -3$
34.	Parabola	$e = 1$	$y = 4$
35.	Ellipse	$e = \frac{1}{4}$	$y = 1$
36.	Ellipse	$e = \frac{5}{6}$	$y = -2$
37.	Hyperbola	$e = \frac{4}{3}$	$x = 2$
38.	Hyperbola	$e = \frac{3}{2}$	$x = -1$

Finding a Polar Equation In Exercises 39–44, find a polar equation for the conic with its focus at the pole and the given vertex or vertices.

	Conic	Vertex or Vertices
39.	Parabola	$\left(1, -\frac{\pi}{2}\right)$
40.	Parabola	$(5, \pi)$
41.	Ellipse	$(2, 0), \quad (8, \pi)$
42.	Ellipse	$\left(2, \frac{\pi}{2}\right), \quad \left(4, \frac{3\pi}{2}\right)$
43.	Hyperbola	$\left(1, \frac{3\pi}{2}\right), \quad \left(9, \frac{3\pi}{2}\right)$
44.	Hyperbola	$(2, 0), \quad (10, 0)$

EXPLORING CONCEPTS

45. Eccentricity Consider two ellipses, where the foci of the first ellipse are farther apart than the foci of the second ellipse. Is the eccentricity of the first ellipse always greater than the eccentricity of the second ellipse? Explain.

46. Distance Describe what happens to the distance between the directrix and the center of an ellipse when the foci remain fixed and e approaches 0.

47. Finding a Polar Equation Find a polar equation for the ellipse with the following characteristics.

Focus: $(0, 0)$

Eccentricity: $e = \frac{1}{2}$

Directrix: $r = 4 \sec \theta$

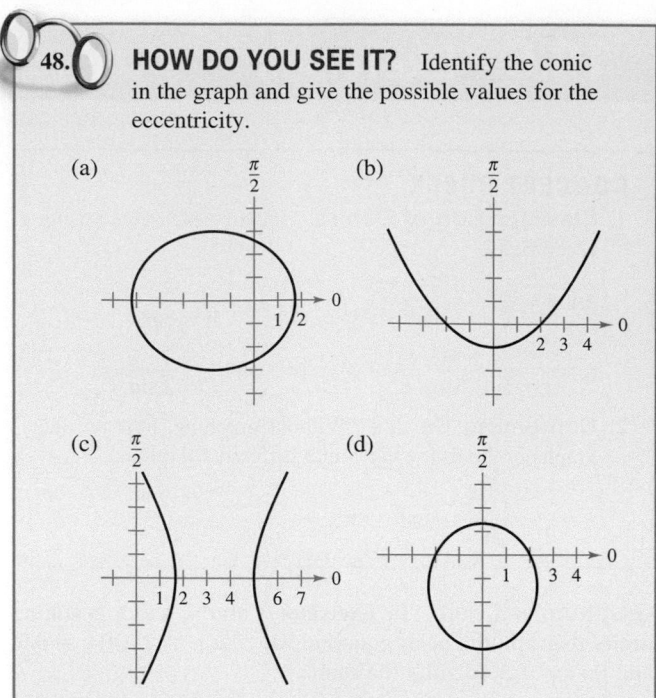

48. HOW DO YOU SEE IT? Identify the conic in the graph and give the possible values for the eccentricity.

(a)   (b)   (c)   (d)

49. Ellipse Show that the polar equation for $\dfrac{x^2}{a^2} + \dfrac{y^2}{b^2} = 1$ is

$$r^2 = \frac{b^2}{1 - e^2 \cos^2 \theta}. \qquad \text{Ellipse}$$

50. Hyperbola Show that the polar equation for $\dfrac{x^2}{a^2} - \dfrac{y^2}{b^2} = 1$ is

$$r^2 = \frac{-b^2}{1 - e^2 \cos^2 \theta}. \qquad \text{Hyperbola}$$

Finding a Polar Equation In Exercises 51–54, use the results of Exercises 49 and 50 to write the polar form of the equation of the conic.

51. Ellipse: focus at $(4, 0)$; vertices at $(5, 0), (5, \pi)$

52. Hyperbola: focus at $(5, 0)$; vertices at $(4, 0), (4, \pi)$

53. $\dfrac{x^2}{9} - \dfrac{y^2}{16} = 1$

54. $\dfrac{x^2}{4} + y^2 = 1$

Area of a Region In Exercises 55–58, use the integration capabilities of a graphing utility to approximate the area of the region bounded by the graph of the polar equation.

55. $r = \dfrac{3}{2 - \cos \theta}$

56. $r = \dfrac{9}{4 + \cos \theta}$

57. $r = \dfrac{2}{7 - 6 \sin \theta}$

58. $r = \dfrac{3}{6 + 5 \sin \theta}$

59. Explorer 18 On November 27, 1963, the United States launched Explorer 18. Its low and high points above the surface of Earth were approximately 119 miles and 123,000 miles (see figure). The center of Earth is a focus of the orbit. Find the polar equation for the orbit and find the distance between the surface of Earth and the satellite when $\theta = 60°$. (Assume that the radius of Earth is 4000 miles.)

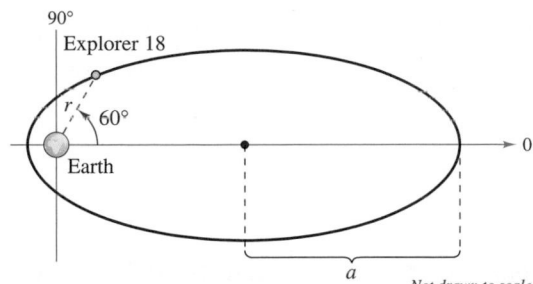

60. Planetary Motion The planets travel in elliptical orbits with the sun as a focus, as shown in the figure.

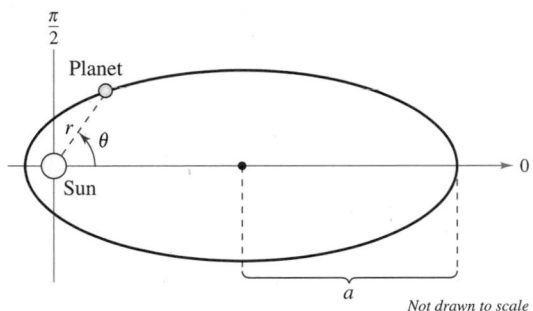

(a) Show that the polar equation of the orbit is given by

$$r = \frac{(1 - e^2)a}{1 - e\cos\theta}$$

where e is the eccentricity.

(b) Show that the minimum distance (*perihelion*) from the sun to the planet is $r = a(1 - e)$ and the maximum distance (*aphelion*) is $r = a(1 + e)$.

Planetary Motion In Exercises 61–64, use Exercise 60 to find the polar equation of the elliptical orbit of the planet and the perihelion and aphelion distances.

61. Earth

$a = 1.496 \times 10^8$ kilometers

$e = 0.0167$

62. Saturn

$a = 1.434 \times 10^9$ kilometers

$e = 0.0565$

63. Neptune

$a = 4.495 \times 10^9$ kilometers

$e = 0.0113$

64. Mercury

$a = 5.791 \times 10^7$ kilometers

$e = 0.2056$

NASA

65. Planetary Motion

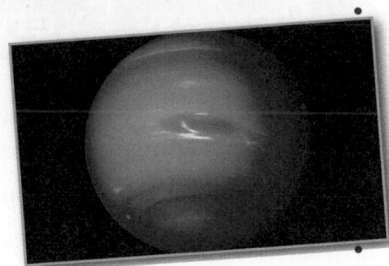

In Exercise 63, the polar equation for the elliptical orbit of Neptune was found. Use the equation and a computer algebra system to perform each of the following.

(a) Approximate the area swept out by a ray from the sun to the planet as θ increases from 0 to $\pi/9$. Use this result to determine the number of years required for the planet to move through this arc when the period of one revolution around the sun is 165 years.

(b) By trial and error, approximate the angle α such that the area swept out by a ray from the sun to the planet as θ increases from π to α equals the area found in part (a) (see figure). Does the ray sweep through a larger or smaller angle than in part (a) to generate the same area? Why is this the case?

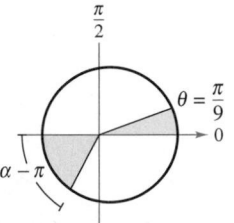

(c) Approximate the distances the planet traveled in parts (a) and (b). Use these distances to approximate the average number of kilometers per year the planet traveled in the two cases.

66. Comet Hale-Bopp The comet Hale-Bopp has an elliptical orbit with the sun at one focus and has an eccentricity of $e \approx 0.995$. The length of the major axis of the orbit is approximately 500 astronomical units. (a) Find the length of its minor axis. (b) Find a polar equation for the orbit. (c) Find the perihelion and aphelion distances.

Eccentricity In Exercises 67 and 68, let r_0 represent the distance from a focus to the nearest vertex, and let r_1 represent the distance from the focus to the farthest vertex.

67. Show that the eccentricity of an ellipse can be written as

$$e = \frac{r_1 - r_0}{r_1 + r_0}.$$

Then show that $\frac{r_1}{r_0} = \frac{1 + e}{1 - e}.$

68. Show that the eccentricity of a hyperbola can be written as

$$e = \frac{r_1 + r_0}{r_1 - r_0}.$$

Then show that $\frac{r_1}{r_0} = \frac{e + 1}{e - 1}.$

Review Exercises

See CalcChat.com for tutorial help and worked-out solutions to odd-numbered exercises.

Matching In Exercises 1–6, match the equation with its graph. [The graphs are labeled (a), (b), (c), (d), (e), and (f).]

(a)

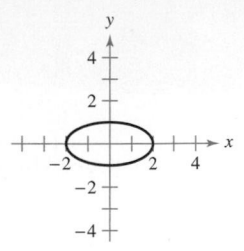

(b)

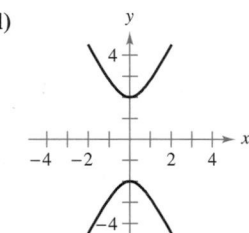

(c)

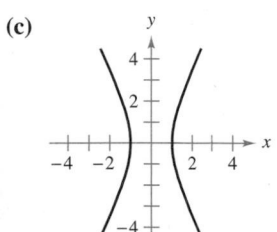

(d)

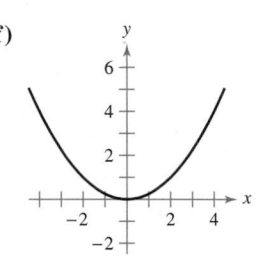

(e)

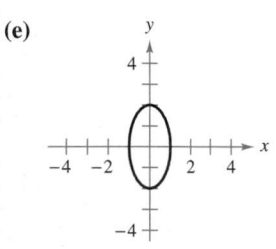

(f)

1. $4x^2 + y^2 = 4$

2. $4x^2 - y^2 = 4$

3. $y^2 = -4x$

4. $y^2 - 4x^2 = 4$

5. $x^2 + 4y^2 = 4$

6. $x^2 = 4y$

Identifying a Conic In Exercises 7–14, identify the conic, analyze the equation (center, radius, vertices, foci, eccentricity, directrix, and asymptotes, if possible), and sketch its graph. Use a graphing utility to confirm your results.

7. $x^2 + y^2 - 2x - 8y - 8 = 0$

8. $y^2 - 12y - 8x + 20 = 0$

9. $3x^2 - 2y^2 + 24x + 12y + 24 = 0$

10. $5x^2 + y^2 - 20x + 19 = 0$

11. $16x^2 + 16y^2 - 16x + 24y - 3 = 0$

12. $-4x^2 + 3y^2 - 16x - 18y + 10 = 0$

13. $x^2 + 10x - 12y + 13 = 0$

14. $9x^2 + 25y^2 + 18x - 100y - 116 = 0$

Finding the Standard Equation of a Parabola In Exercises 15 and 16, find the standard form of the equation of the parabola with the given characteristics.

15. Vertex: $(7, 0)$

 Directrix: $x = 5$

16. Vertex: $(2, 6)$

 Focus: $(2, 4)$

Finding the Standard Equation of an Ellipse In Exercises 17–20, find the standard form of the equation of the ellipse with the given characteristics.

17. Center: $(0, 1)$

 Focus: $(4, 1)$

 Vertex: $(6, 1)$

18. Center: $(0, 0)$

 Major axis: vertical

 Points on the ellipse: $(1, 2), (2, 0)$

19. Vertices: $(3, 1), (3, 7)$

 Eccentricity: $\frac{2}{3}$

20. Foci: $(0, \pm 7)$

 Major axis length: 20

Finding the Standard Equation of a Hyperbola In Exercises 21–24, find the standard form of the equation of the hyperbola with the given characteristics.

21. Vertices: $(0, \pm 8)$

 Asymptotes: $y = \pm 2x$

22. Vertices: $(\pm 2, 0)$

 Asymptotes: $y = \pm 32x$

23. Vertices: $(\pm 7, -1)$

 Foci: $(\pm 9, -1)$

24. Center: $(0, 0)$

 Vertex: $(0, 3)$

 Focus: $(0, 6)$

25. Satellite Antenna A cross section of a large parabolic antenna is modeled by the graph of $y = x^2/200$, $-100 \le x \le 100$. The receiving and transmitting equipment is positioned at the focus.

 (a) Find the coordinates of the focus.

 (b) Find the surface area of the antenna.

26. Using an Ellipse Consider the ellipse $\dfrac{x^2}{25} + \dfrac{y^2}{9} = 1$.

 (a) Find the area of the region bounded by the ellipse.

 (b) Find the volume of the solid generated by revolving the region about its major axis.

Using Parametric Equations In Exercises 27–34, sketch the curve represented by the parametric equations (indicate the orientation of the curve), and write the corresponding rectangular equation by eliminating the parameter.

27. $x = 1 + 8t$, $y = 3 - 4t$

28. $x = t - 2$, $y = t^2 - 1$

29. $x = \sqrt{t} + 1$, $y = t - 3$

30. $x = e^t - 1$, $y = e^{3t}$

31. $x = 6 \cos \theta$, $y = 6 \sin \theta$

32. $x = 2 + 5 \cos t$, $y = 3 + 2 \sin t$

33. $x = 2 + \sec \theta$, $y = 3 + \tan \theta$

34. $x = 5 \sin^3 \theta$, $y = 5 \cos^3 \theta$

Finding Parametric Equations In Exercises 35 and 36, find two different sets of parametric equations for the rectangular equation.

35. $y = 4x + 3$

36. $y = x^2 - 2$

37. Rotary Engine The rotary engine was developed by Felix Wankel in the 1950s. It features a rotor that is a modified equilateral triangle. The rotor moves in a chamber that, in two dimensions, is an epitrochoid. Use a graphing utility to graph the chamber modeled by the parametric equations

$$x = \cos 3\theta + 5\cos\theta$$

and

$$y = \sin 3\theta + 5\sin\theta.$$

38. Serpentine Curve Consider the parametric equations $x = 2\cot\theta$ and $y = 4\sin\theta\cos\theta$, $0 < \theta < \pi$.

(a) Use a graphing utility to graph the curve.

(b) Eliminate the parameter to show that the rectangular equation of the serpentine curve is $(4 + x^2)y = 8x$.

Finding Slope and Concavity In Exercises 39–46, find dy/dx and d^2y/dx^2, and find the slope and concavity (if possible) at the given value of the parameter.

Parametric Equations	Parameter
39. $x = 1 + 6t$, $y = 4 - 5t$	$t = 3$
40. $x = t - 6$, $y = t^2$	$t = 5$
41. $x = \dfrac{1}{t}$, $y = t^2$	$t = -2$
42. $x = \dfrac{1}{\sqrt{t}} + 1$, $y = 3 - 2t$	$t = 4$
43. $x = e^t$, $y = e^{-t}$	$t = 1$
44. $x = 5 + \cos\theta$, $y = 3 + 4\sin\theta$	$\theta = \dfrac{\pi}{6}$
45. $x = 10\cos\theta$, $y = 10\sin\theta$	$\theta = \dfrac{\pi}{4}$
46. $x = \cos^4\theta$, $y = \sin^4\theta$	$\theta = -\dfrac{\pi}{3}$

Finding an Equation of a Tangent Line In Exercises 47 and 48, (a) use a graphing utility to graph the curve represented by the parametric equations, (b) use a graphing utility to find $dx/d\theta$, $dy/d\theta$, and dy/dx at the given value of the parameter, (c) find an equation of the tangent line to the curve at the given value of the parameter, and (d) use a graphing utility to graph the curve and the tangent line from part (c).

Parametric Equations	Parameter
47. $x = \cot\theta$, $y = \sin 2\theta$	$\theta = \dfrac{\pi}{6}$
48. $x = \dfrac{1}{4}\tan\theta$, $y = 6\sin\theta$	$\theta = \dfrac{\pi}{3}$

Horizontal and Vertical Tangency In Exercises 49–52, find all points (if any) of horizontal and vertical tangency to the curve. Use a graphing utility to confirm your results.

49. $x = 5 - t$, $y = 2t^2$

50. $x = t + 2$, $y = t^3 - 2t$

51. $x = 2 + 2\sin\theta$, $y = 1 + \cos\theta$

52. $x = 2 - 2\cos\theta$, $y = 2\sin 2\theta$

Arc Length In Exercises 53 and 54, find the arc length of the curve on the given interval.

Parametric Equations	Interval
53. $x = t^2 + 1$, $y = 4t^3 + 3$	$0 \le t \le 2$
54. $x = 7\cos\theta$, $y = 7\sin\theta$	$0 \le \theta \le \pi$

Surface Area In Exercises 55 and 56, find the area of the surface generated by revolving the curve about (a) the x-axis and (b) the y-axis.

55. $x = 4t$, $y = 3t + 1$, $0 \le t \le 1$

56. $x = 2\cos\theta$, $y = 2\sin\theta$, $0 \le \theta \le \dfrac{\pi}{2}$

Area In Exercises 57 and 58, find the area of the region.

57. $x = 3\sin\theta$
$y = 2\cos\theta$
$-\dfrac{\pi}{2} \le \theta \le \dfrac{\pi}{2}$

58. $x = 2\cos\theta$
$y = \sin\theta$
$0 \le \theta \le \pi$

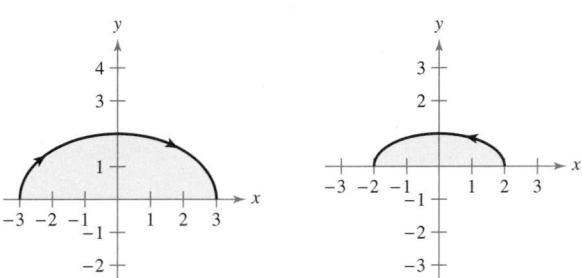

Polar-to-Rectangular Conversion In Exercises 59–62, the polar coordinates of a point are given. Plot the point and find the corresponding rectangular coordinates for the point.

59. $\left(5, \dfrac{3\pi}{2}\right)$

60. $\left(-6, \dfrac{5\pi}{6}\right)$

61. $\left(\sqrt{7}, 3.25\right)$

62. $(-2, -2.45)$

Rectangular-to-Polar Conversion In Exercises 63–66, the rectangular coordinates of a point are given. Plot the point and find *two* sets of polar coordinates for the point for $0 \le \theta < 2\pi$.

63. $(4, -4)$

64. $(0, -7)$

65. $(-1, 3)$

66. $\left(-\sqrt{3}, -\sqrt{3}\right)$

Rectangular-to-Polar Conversion In Exercises 67–72, convert the rectangular equation to polar form and sketch its graph.

67. $x^2 + y^2 = 25$

68. $x^2 - y^2 = 4$

69. $y = 9$

70. $x = 6$

71. $-x + 4y - 3 = 0$

72. $x^2 = 4y$

Polar-to-Rectangular Conversion In Exercises 73–78, convert the polar equation to rectangular form and sketch its graph.

73. $r = 6 \cos \theta$

74. $r = 10$

75. $r = -4 \sec \theta$

76. $r = 3 \csc \theta$

77. $\theta = \dfrac{3\pi}{4}$

78. $r = -2 \sec \theta \tan \theta$

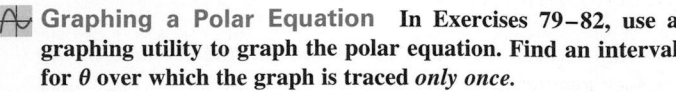 **Graphing a Polar Equation** In Exercises 79–82, use a graphing utility to graph the polar equation. Find an interval for θ over which the graph is traced *only once*.

79. $r = \dfrac{3\pi}{2} \sin 3\theta$

80. $r = 2 \sin \theta \cos^2 \theta$

81. $r = 4 \cos 2\theta \sec \theta$

82. $r = 4(\sec \theta - \cos \theta)$

Horizontal and Vertical Tangency In Exercises 83 and 84, find the points of horizontal and vertical tangency (if any) to the polar curve.

83. $r = 1 - \cos \theta$

84. $r = 3 \tan \theta$

Tangent Lines at the Pole In Exercises 85 and 86, sketch a graph of the polar equation and find the tangent lines at the pole.

85. $r = 4 \sin 3\theta$

86. $r = 3 \cos 4\theta$

Sketching a Polar Graph In Exercises 87–96, sketch a graph of the polar equation.

87. $r = 6$

88. $\theta = \dfrac{\pi}{10}$

89. $r = -\sec \theta$

90. $r = 5 \csc \theta$

91. $r = 4 - 3 \cos \theta$

92. $r = 3 + 2 \sin \theta$

93. $r = 4\theta$

94. $r = -3 \cos 2\theta$

95. $r^2 = 4 \sin 2\theta$

96. $r^2 = 9 \cos 2\theta$

Finding the Area of a Polar Region In Exercises 97–100, find the area of the region.

97. One petal of $r = 3 \cos 5\theta$

98. One petal of $r = 2 \sin 6\theta$

99. Interior of $r = 2 + \cos \theta$

100. Interior of $r = 5(1 - \sin \theta)$

Finding Points of Intersection In Exercises 101 and 102, find the points of intersection of the graphs of the equations.

101. $r = 1 - \cos \theta$
$r = 1 + \sin \theta$

102. $r = 1 + \sin \theta$
$r = 3 \sin \theta$

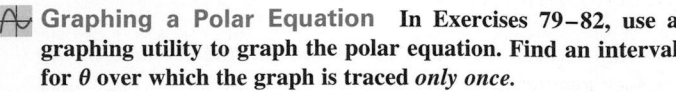 **Finding the Area of a Polar Region** In Exercises 103–108, use a graphing utility to graph the polar equation. Find the area of the given region analytically.

103. Inner loop of $r = 3 - 6 \cos \theta$

104. Inner loop of $r = 4 + 8 \sin \theta$

105. Between the loops of $r = 3 - 6 \cos \theta$

106. Between the loops of $r = 4 + 8 \sin \theta$

107. Common interior of $r = 5 - 2 \sin \theta$ and $r = -5 + 2 \sin \theta$

108. Common interior of $r = 4 \cos \theta$ and $r = 2$

Finding the Arc Length of a Polar Curve In Exercises 109 and 110, find the length of the curve over the given interval.

Polar Equation	Interval
109. $r = 5 \cos \theta$	$\left[\dfrac{\pi}{2}, \pi\right]$
110. $r = 3(1 - \cos \theta)$	$[0, \pi]$

Finding the Area of a Surface of Revolution In Exercises 111 and 112, find the area of the surface formed by revolving the polar equation over the given interval about the given line.

Polar Equation	Interval	Axis of Revolution
111. $r = 2 \sin \theta$	$0 \le \theta \le \pi$	Polar axis
112. $r = 2 \sin \theta$	$0 \le \theta \le \dfrac{\pi}{2}$	$\theta = \dfrac{\pi}{2}$

Identifying and Sketching a Conic In Exercises 113–118, find the eccentricity and the distance from the pole to the directrix of the conic. Then identify the conic and sketch its graph. Use a graphing utility to confirm your results.

113. $r = \dfrac{6}{1 - \sin \theta}$

114. $r = \dfrac{2}{1 + \cos \theta}$

115. $r = \dfrac{6}{3 + 2 \cos \theta}$

116. $r = \dfrac{4}{5 - 3 \sin \theta}$

117. $r = \dfrac{4}{2 - 3 \sin \theta}$

118. $r = \dfrac{8}{2 - 5 \cos \theta}$

Finding a Polar Equation In Exercises 119–122, find a polar equation for the conic with its focus at the pole and the given eccentricity and directrix. (For convenience, the equation for the directrix is given in rectangular form.)

	Conic	Eccentricity	Directrix
119.	Parabola	$e = 1$	$x = 5$
120.	Ellipse	$e = \dfrac{3}{4}$	$y = -2$
121.	Hyperbola	$e = 3$	$y = 3$
122.	Hyperbola	$e = \dfrac{5}{2}$	$x = -1$

Finding a Polar Equation In Exercises 123–126, find a polar equation for the conic with its focus at the pole and the given vertex or vertices.

	Conic	Vertex or Vertices
123.	Parabola	$\left(2, \dfrac{\pi}{2}\right)$
124.	Parabola	$(3, \pi)$
125.	Ellipse	$(5, 0), (1, \pi)$
126.	Hyperbola	$(1, 0), (7, 0)$

P.S. Problem Solving

See **CalcChat.com** for tutorial help and worked-out solutions to odd-numbered exercises.

1. **Using a Parabola** Consider the parabola

   $$x^2 = 4y$$

   and the focal chord

   $$y = \tfrac{3}{4}x + 1.$$

   (a) Sketch the graph of the parabola and the focal chord.

   (b) Show that the tangent lines to the parabola at the endpoints of the focal chord intersect at right angles.

   (c) Show that the tangent lines to the parabola at the endpoints of the focal chord intersect on the directrix of the parabola.

2. **Using a Parabola** Consider the parabola $x^2 = 4py$ and one of its focal chords.

   (a) Show that the tangent lines to the parabola at the endpoints of the focal chord intersect at right angles.

   (b) Show that the tangent lines to the parabola at the endpoints of the focal chord intersect on the directrix of the parabola.

3. **Proof** Prove Theorem 10.2, Reflective Property of a Parabola, as shown in the figure.

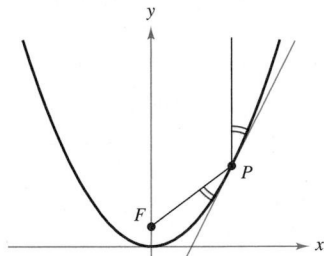

4. **Flight Paths** An air traffic controller spots two planes at the same altitude flying toward each other (see figure). Their flight paths are 20° and 315°. One plane is 150 miles from point P with a speed of 375 miles per hour. The other is 190 miles from point P with a speed of 450 miles per hour.

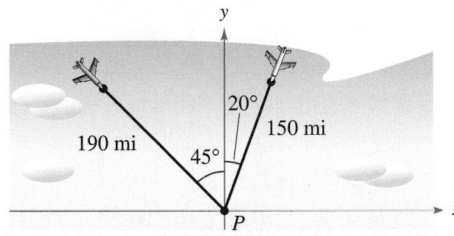

   (a) Find parametric equations for the path of each plane where t is the time in hours, with $t = 0$ corresponding to the time at which the air traffic controller spots the planes.

   (b) Use the result of part (a) to write the distance between the planes as a function of t.

   (c) Use a graphing utility to graph the function in part (b). When will the distance between the planes be minimum? If the planes must keep a separation of at least 3 miles, is the requirement met?

5. **Strophoid** The curve given by the parametric equations

   $$x(t) = \frac{1 - t^2}{1 + t^2} \quad \text{and} \quad y(t) = \frac{t(1 - t^2)}{1 + t^2}$$

   is called a **strophoid.**

   (a) Find a rectangular equation of the strophoid.

   (b) Find a polar equation of the strophoid.

   (c) Sketch a graph of the strophoid.

   (d) Find the equations of the two tangent lines at the origin.

   (e) Find the points on the graph at which the tangent lines are horizontal.

6. **Finding a Rectangular Equation** Find a rectangular equation of the portion of the cycloid given by the parametric equations $x = a(\theta - \sin \theta)$ and $y = a(1 - \cos \theta)$, $0 \le \theta \le \pi$, as shown in the figure.

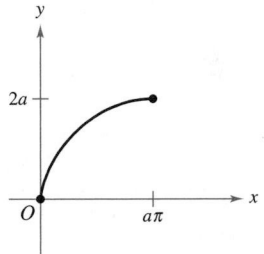

7. **Cornu Spiral** Consider the **cornu spiral** given by

   $$x(t) = \int_0^t \cos \frac{\pi u^2}{2} \, du \quad \text{and} \quad y(t) = \int_0^t \sin \frac{\pi u^2}{2} \, du.$$

   (a) Use a graphing utility to graph the spiral over the interval $-\pi \le t \le \pi$.

   (b) Show that the cornu spiral is symmetric with respect to the origin.

   (c) Find the length of the cornu spiral from $t = 0$ to $t = a$. What is the length of the spiral from $t = -\pi$ to $t = \pi$?

8. **Using an Ellipse** Consider the region bounded by the ellipse

   $$\frac{x^2}{a^2} + \frac{y^2}{b^2} = 1$$

   with eccentricity $e = c/a$.

   (a) Show that the solid (oblate spheroid) generated by revolving the region about the minor axis of the ellipse has a volume of $V = 4\pi^2 b/3$ and a surface area of

   $$S = 2\pi a^2 + \pi \left(\frac{b^2}{e}\right) \ln\left(\frac{1 + e}{1 - e}\right).$$

   (b) Show that the solid (prolate spheroid) generated by revolving the region about the major axis of the ellipse has a volume of $V = 4\pi ab^2/3$ and a surface area of

   $$S = 2\pi b^2 + 2\pi \left(\frac{ab}{e}\right) \arcsin e.$$

9. Area Let a and b be positive constants. Find the area of the region in the first quadrant bounded by the graph of the polar equation

$$r = \frac{ab}{(a \sin \theta + b \cos \theta)}, \quad 0 \le \theta \le \frac{\pi}{2}.$$

10. Arc Length Consider the **logarithmic spiral**

$$r = e^{a\theta}$$

where a is a constant greater than 0 (see figure). Find the arc length from the point $(1, 0)$ to the pole. Notice that the graph of the curve makes infinitely many rotations to reach the pole.

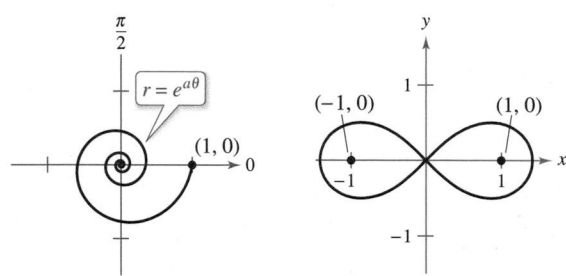

Figure for 10   Figure for 11

11. Finding a Polar Equation Determine the polar equation of the set of all points (r, θ), the product of whose distances from the points $(1, 0)$ and $(-1, 0)$ is equal to 1, as shown in the figure.

12. Arc Length A particle is moving along the path described by the parametric equations

$$x = \frac{1}{t} \quad \text{and} \quad y = \frac{\sin t}{t}$$

for $1 \le t < \infty$, as shown in the figure. Find the length of this path.

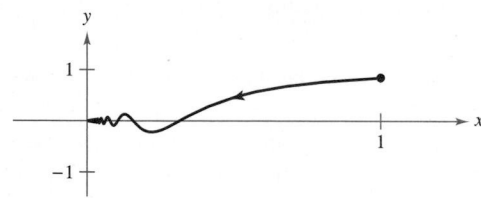

13. Finding a Polar Equation Four dogs are located at the corners of a square with sides of length d. The dogs all move counterclockwise at the same speed directly toward the next dog, as shown in the figure. Find the polar equation of a dog's path as it spirals toward the center of the square.

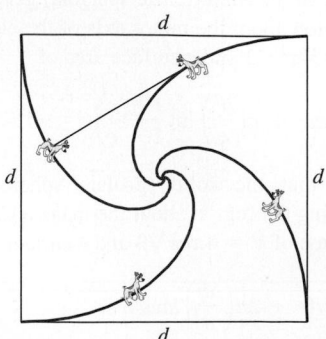

14. Using a Hyperbola Consider the hyperbola

$$\frac{x^2}{a^2} - \frac{y^2}{b^2} = 1$$

with foci F_1 and F_2, as shown in the figure. Let T be the tangent line at a point M on the hyperbola. Show that incoming rays of light aimed at one focus are reflected by a hyperbolic mirror toward the other focus.

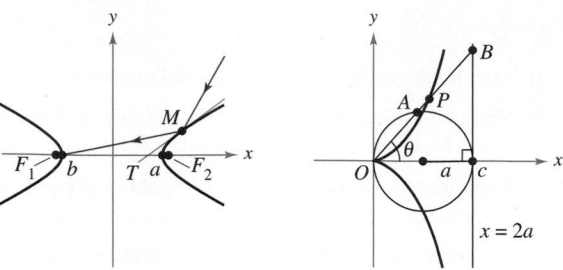

Figure for 14   Figure for 15

15. Cissoid of Diocles Consider a circle of radius a tangent to the y-axis and the line $x = 2a$, as shown in the figure. Let A be the point where the segment OB intersects the circle, where point B lies on the line $x = 2a$. The **cissoid of Diocles** consists of all points P such that $OP = AB$.

(a) Find a polar equation of the cissoid.

(b) Find a set of parametric equations for the cissoid that does not contain trigonometric functions.

(c) Find a rectangular equation of the cissoid.

16. Butterfly Curve Use a graphing utility to graph the curve shown in the figure below. The curve is given by

$$r = e^{\cos \theta} - 2 \cos 4\theta + \sin^5 \frac{\theta}{12}.$$

Over what interval must θ vary to produce the curve?

FOR FURTHER INFORMATION For more information on this curve, see the article "A Study in Step Size" by Temple H. Fay in *Mathematics Magazine*. To view this article, go to *MathArticles.com*.

17. Graphing Polar Equations Use a graphing utility to graph the polar equation $r = \cos 5\theta + n \cos \theta$ for $0 \le \theta < \pi$ and for the integers $n = -5$ to $n = 5$. What values of n produce the "heart" portion of the curve? What values of n produce the "bell" portion? (This curve, created by Michael W. Chamberlin, appeared in *The College Mathematics Journal*.)

11 Vectors and the Geometry of Space

Geography *(Exercise 47, p. 807)*

Modeling Data
(Exercise 105, p. 796)

Work *(Exercise 62, p. 778)*

Auditorium Lights
(Exercise 99, p. 769)

Navigation *(Exercise 84, p. 761)*

11.1   Vectors in the Plane

- Write the component form of a vector.
- Perform vector operations and interpret the results geometrically.
- Write a vector as a linear combination of standard unit vectors.

Component Form of a Vector

Many quantities in geometry and physics, such as area, volume, temperature, mass, and time, can be characterized by a single real number that is scaled to appropriate units of measure. These are called **scalar quantities,** and the real number associated with each is called a **scalar.**

Other quantities, such as force, velocity, and acceleration, involve both magnitude and direction and cannot be characterized completely by a single real number. A **directed line segment** is used to represent such a quantity, as shown in Figure 11.1. The directed line segment $\overrightarrow{PQ}$ has **initial point** P and **terminal point** Q, and its **length** (or **magnitude**) is denoted by $\|\overrightarrow{PQ}\|$. Directed line segments that have the same length and direction are **equivalent,** as shown in Figure 11.2. The set of all directed line segments that are equivalent to a given directed line segment $\overrightarrow{PQ}$ is a **vector in the plane** and is denoted by

$$\mathbf{v} = \overrightarrow{PQ}.$$

In typeset material, vectors are usually denoted by lowercase, boldface letters such as **u**, **v**, and **w**. When written by hand, however, vectors are often denoted by letters with arrows above them, such as $\vec{u}$, $\vec{v}$, and $\vec{w}$.

Be sure you understand that a vector represents a *set* of directed line segments (each having the same length and direction). In practice, however, it is common not to distinguish between a vector and one of its representatives.

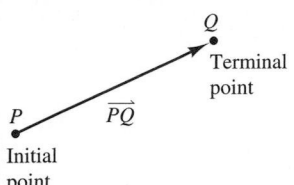

A directed line segment

Figure 11.1

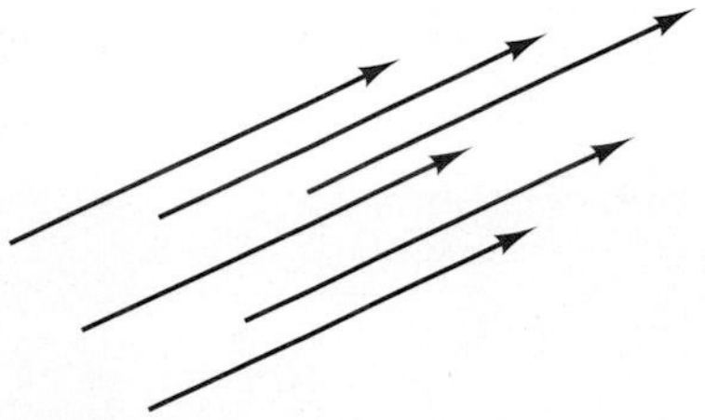

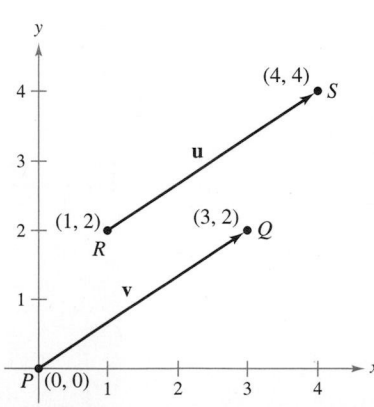

Equivalent directed line segments

Figure 11.2

EXAMPLE 1   **Vector Representation: Directed Line Segments**

Let **v** be represented by the directed line segment from $(0, 0)$ to $(3, 2)$, and let **u** be represented by the directed line segment from $(1, 2)$ to $(4, 4)$. Show that **v** and **u** are equivalent.

Solution   Let $P(0, 0)$ and $Q(3, 2)$ be the initial and terminal points of **v**, and let $R(1, 2)$ and $S(4, 4)$ be the initial and terminal points of **u**, as shown in Figure 11.3. You can use the Distance Formula to show that $\overrightarrow{PQ}$ and $\overrightarrow{RS}$ have the *same length*.

$$\|\overrightarrow{PQ}\| = \sqrt{(3 - 0)^2 + (2 - 0)^2} = \sqrt{13}$$
$$\|\overrightarrow{RS}\| = \sqrt{(4 - 1)^2 + (4 - 2)^2} = \sqrt{13}$$

Both line segments have the *same direction,* because they both are directed toward the upper right on lines having the same slope.

$$\text{Slope of } \overrightarrow{PQ} = \frac{2 - 0}{3 - 0} = \frac{2}{3}$$

and

$$\text{Slope of } \overrightarrow{RS} = \frac{4 - 2}{4 - 1} = \frac{2}{3}$$

Because $\overrightarrow{PQ}$ and $\overrightarrow{RS}$ have the same length and direction, you can conclude that the two vectors are equivalent. That is, **v** and **u** are equivalent.

The vectors **u** and **v** are equivalent.

Figure 11.3

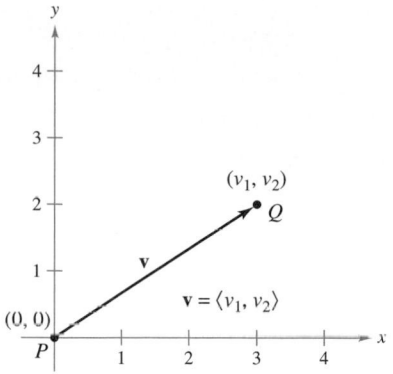

A vector in standard position

Figure 11.4

The directed line segment whose initial point is the origin is often the most convenient representative of a set of equivalent directed line segments such as those shown in Figure 11.3. This representation of **v** is said to be in **standard position.** A directed line segment whose initial point is the origin can be uniquely represented by the coordinates of its terminal point $Q(v_1, v_2)$, as shown in Figure 11.4. In the next definition, note the difference in the notation between the *component form* of a vector $\mathbf{v} = \langle v_1, v_2 \rangle$ and the point (v_1, v_2).

Definition of Component Form of a Vector in the Plane

If **v** is a vector in the plane whose initial point is the origin and whose terminal point is (v_1, v_2), then the **component form of v** is $\mathbf{v} = \langle v_1, v_2 \rangle$. The coordinates v_1 and v_2 are called the **components of v.** If both the initial point and the terminal point lie at the origin, then **v** is called the **zero vector** and is denoted by $\mathbf{0} = \langle 0, 0 \rangle$.

This definition implies that two vectors $\mathbf{u} = \langle u_1, u_2 \rangle$ and $\mathbf{v} = \langle v_1, v_2 \rangle$ are **equal** if and only if $u_1 = v_1$ and $u_2 = v_2$.

The procedures listed below can be used to convert directed line segments to component form or vice versa.

1. If $P(p_1, p_2)$ and $Q(q_1, q_2)$ are the initial and terminal points of a directed line segment, then the component form of the vector **v** represented by $\overrightarrow{PQ}$ is

$$\langle v_1, v_2 \rangle = \langle q_1 - p_1, q_2 - p_2 \rangle.$$

Moreover, from the Distance Formula, you can see that the **length** (or **magnitude**) **of v** is

$$\|\mathbf{v}\| = \sqrt{(q_1 - p_1)^2 + (q_2 - p_2)^2} \qquad \text{Length of a vector}$$
$$= \sqrt{v_1^2 + v_2^2}.$$

2. If $\mathbf{v} = \langle v_1, v_2 \rangle$, then **v** can be represented by the directed line segment, in standard position, from $P(0, 0)$ to $Q(v_1, v_2)$.

The length of **v** is also called the **norm of v.** If $\|\mathbf{v}\| = 1$, then **v** is a **unit vector.** Moreover, $\|\mathbf{v}\| = 0$ if and only if **v** is the zero vector **0**.

EXAMPLE 2   **Component Form and Length of a Vector**

Find the component form and length of the vector **v** that has initial point $(3, -7)$ and terminal point $(-2, 5)$.

Solution   Let $P(3, -7) = (p_1, p_2)$ and $Q(-2, 5) = (q_1, q_2)$. Then the components of $\mathbf{v} = \langle v_1, v_2 \rangle$ are

$$v_1 = q_1 - p_1 = -2 - 3 = -5$$

and

$$v_2 = q_2 - p_2 = 5 - (-7) = 12.$$

So, as shown in Figure 11.5, $\mathbf{v} = \langle -5, 12 \rangle$, and the length of **v** is

$$\|\mathbf{v}\| = \sqrt{(-5)^2 + 12^2}$$
$$= \sqrt{169}$$
$$= 13.$$

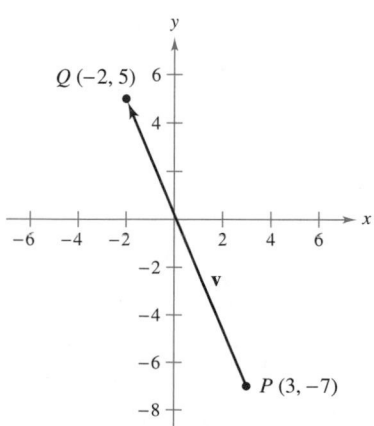

Component form of **v**: $\mathbf{v} = \langle -5, 12 \rangle$

Figure 11.5

Vector Operations

> **Definitions of Vector Addition and Scalar Multiplication**
>
> Let $\mathbf{u} = \langle u_1, u_2 \rangle$ and $\mathbf{v} = \langle v_1, v_2 \rangle$ be vectors and let c be a scalar.
>
> 1. The **vector sum** of $\mathbf{u}$ and $\mathbf{v}$ is the vector $\mathbf{u} + \mathbf{v} = \langle u_1 + v_1, u_2 + v_2 \rangle$.
>
> 2. The **scalar multiple** of c and $\mathbf{u}$ is the vector
> $$c\mathbf{u} = \langle cu_1, cu_2 \rangle.$$
>
> 3. The **negative** of $\mathbf{v}$ is the vector
> $$-\mathbf{v} = (-1)\mathbf{v} = \langle -v_1, -v_2 \rangle.$$
>
> 4. The **difference** of $\mathbf{u}$ and $\mathbf{v}$ is
> $$\mathbf{u} - \mathbf{v} = \mathbf{u} + (-\mathbf{v}) = \langle u_1 - v_1, u_2 - v_2 \rangle.$$

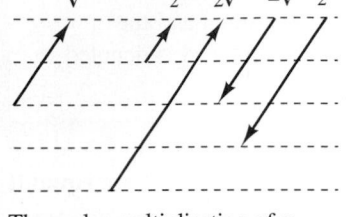

The scalar multiplication of $\mathbf{v}$

Figure 11.6

Geometrically, the scalar multiple of a vector $\mathbf{v}$ and a scalar c is the vector that is $|c|$ times as long as $\mathbf{v}$, as shown in Figure 11.6. If c is positive, then $c\mathbf{v}$ has the same direction as $\mathbf{v}$. If c is negative, then $c\mathbf{v}$ has the opposite direction.

The sum of two vectors can be represented geometrically by positioning the vectors (without changing their magnitudes or directions) so that the initial point of one coincides with the terminal point of the other, as shown in Figure 11.7. The vector $\mathbf{u} + \mathbf{v}$, called the **resultant vector,** is the diagonal of a parallelogram having $\mathbf{u}$ and $\mathbf{v}$ as its adjacent sides.

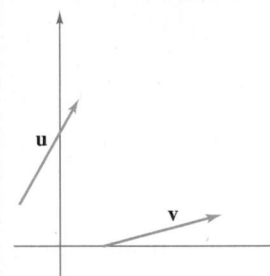 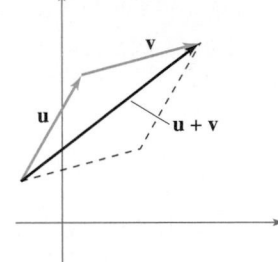 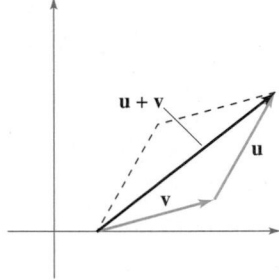

| To find $\mathbf{u} + \mathbf{v}$, | (1) move the initial point of $\mathbf{v}$ to the terminal point of $\mathbf{u}$, or | (2) move the initial point of $\mathbf{u}$ to the terminal point of $\mathbf{v}$. |

Figure 11.7

Figure 11.8 shows the equivalence of the geometric and algebraic definitions of vector addition and scalar multiplication and presents (at far right) a geometric interpretation of $\mathbf{u} - \mathbf{v}$.

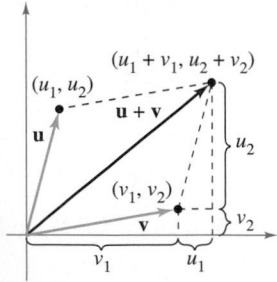 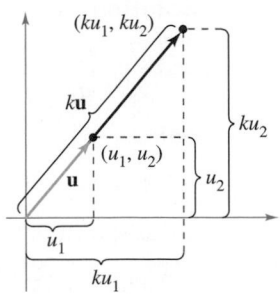 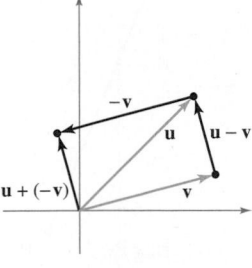

Vector addition   Scalar multiplication   Vector subtraction

Figure 11.8

EXAMPLE 3 **Vector Operations**

For $\mathbf{v} = \langle -2, 5 \rangle$ and $\mathbf{w} = \langle 3, 4 \rangle$, find each of the vectors.

a. $\frac{1}{2}\mathbf{v}$ **b.** $\mathbf{w} - \mathbf{v}$ **c.** $\mathbf{v} + 2\mathbf{w}$

Solution

a. $\frac{1}{2}\mathbf{v} = \langle \frac{1}{2}(-2), \frac{1}{2}(5) \rangle = \langle -1, \frac{5}{2} \rangle$

b. $\mathbf{w} - \mathbf{v} = \langle w_1 - v_1, w_2 - v_2 \rangle$

$\qquad = \langle 3 - (-2), 4 - 5 \rangle$

$\qquad = \langle 5, -1 \rangle$

c. Using $2\mathbf{w} = \langle 6, 8 \rangle$, you have

$$\mathbf{v} + 2\mathbf{w} = \langle -2, 5 \rangle + \langle 6, 8 \rangle$$
$$= \langle -2 + 6, 5 + 8 \rangle$$
$$= \langle 4, 13 \rangle.$$

Vector addition and scalar multiplication share many properties of ordinary arithmetic, as shown in the next theorem.

EMMY NOETHER (1882–1935)

One person who contributed to our knowledge of axiomatic systems was the German mathematician Emmy Noether. Noether is generally recognized as the leading woman mathematician in recent history.

THEOREM 11.1 Properties of Vector Operations

Let $\mathbf{u}$, $\mathbf{v}$, and $\mathbf{w}$ be vectors in the plane, and let c and d be scalars.

1. $\mathbf{u} + \mathbf{v} = \mathbf{v} + \mathbf{u}$		Commutative Property
2. $(\mathbf{u} + \mathbf{v}) + \mathbf{w} = \mathbf{u} + (\mathbf{v} + \mathbf{w})$		Associative Property
3. $\mathbf{u} + \mathbf{0} = \mathbf{u}$		Additive Identity Property
4. $\mathbf{u} + (-\mathbf{u}) = \mathbf{0}$		Additive Inverse Property
5. $c(d\mathbf{u}) = (cd)\mathbf{u}$		
6. $(c + d)\mathbf{u} = c\mathbf{u} + d\mathbf{u}$		Distributive Property
7. $c(\mathbf{u} + \mathbf{v}) = c\mathbf{u} + c\mathbf{v}$		Distributive Property
8. $1(\mathbf{u}) = \mathbf{u},\ 0(\mathbf{u}) = \mathbf{0}$		

Proof The proof of the *Associative Property* of vector addition uses the Associative Property of addition of real numbers.

$$(\mathbf{u} + \mathbf{v}) + \mathbf{w} = [\langle u_1, u_2 \rangle + \langle v_1, v_2 \rangle] + \langle w_1, w_2 \rangle$$
$$= \langle u_1 + v_1, u_2 + v_2 \rangle + \langle w_1, w_2 \rangle$$
$$= \langle (u_1 + v_1) + w_1, (u_2 + v_2) + w_2 \rangle$$
$$= \langle u_1 + (v_1 + w_1), u_2 + (v_2 + w_2) \rangle$$
$$= \langle u_1, u_2 \rangle + \langle v_1 + w_1, v_2 + w_2 \rangle$$
$$= \mathbf{u} + (\mathbf{v} + \mathbf{w})$$

The other properties can be proved in a similar manner.

Any set of vectors (with an accompanying set of scalars) that satisfies the eight properties listed in Theorem 11.1 is a **vector space.*** The eight properties are the *vector space axioms*. So, this theorem states that the set of vectors in the plane (with the set of real numbers) forms a vector space.

■ **FOR FURTHER INFORMATION**
For more information on Emmy Noether, see the article "Emmy Noether, Greatest Woman Mathematician" by Clark Kimberling in *Mathematics Teacher*. To view this article, go to *MathArticles.com*.

---

* For more information about vector spaces, see *Elementary Linear Algebra*, Eight Edition, by Ron Larson (Boston, Massachusetts: Cengage Learning, 2017).

> **THEOREM 11.2 Length of a Scalar Multiple**
>
> Let $\mathbf{v}$ be a vector and let c be a scalar. Then
>
> $$\|c\mathbf{v}\| = |c|\,\|\mathbf{v}\|. \qquad \text{$|c|$ is the absolute value of c.}$$

Proof Because $c\mathbf{v} = \langle cv_1, cv_2 \rangle$, it follows that

$$\begin{aligned}
\|c\mathbf{v}\| &= \|\langle cv_1, cv_2 \rangle\| \\
&= \sqrt{(cv_1)^2 + (cv_2)^2} \\
&= \sqrt{c^2 v_1^2 + c^2 v_2^2} \\
&= \sqrt{c^2(v_1^2 + v_2^2)} \\
&= |c|\sqrt{v_1^2 + v_2^2} \\
&= |c|\,\|\mathbf{v}\|.
\end{aligned}$$

In many applications of vectors, it is useful to find a unit vector that has the same direction as a given vector. The next theorem gives a procedure for doing this.

> **THEOREM 11.3 Unit Vector in the Direction of v**
>
> If $\mathbf{v}$ is a nonzero vector in the plane, then the vector
>
> $$\mathbf{u} = \frac{\mathbf{v}}{\|\mathbf{v}\|} = \frac{1}{\|\mathbf{v}\|}\mathbf{v}$$
>
> has length 1 and the same direction as $\mathbf{v}$.

Proof Because $1/\|\mathbf{v}\|$ is positive and $\mathbf{u} = (1/\|\mathbf{v}\|)\mathbf{v}$, you can conclude that $\mathbf{u}$ has the same direction as $\mathbf{v}$. To see that $\|\mathbf{u}\| = 1$, note that

$$\|\mathbf{u}\| = \left\| \left(\frac{1}{\|\mathbf{v}\|} \right) \mathbf{v} \right\| = \left| \frac{1}{\|\mathbf{v}\|} \right| \|\mathbf{v}\| = \frac{1}{\|\mathbf{v}\|} \|\mathbf{v}\| = 1.$$

So, $\mathbf{u}$ has length 1 and the same direction as $\mathbf{v}$.

In Theorem 11.3, $\mathbf{u}$ is called a **unit vector in the direction of v.** The process of multiplying $\mathbf{v}$ by $1/\|\mathbf{v}\|$ to get a unit vector is called **normalization of v.**

EXAMPLE 4 **Finding a Unit Vector**

Find a unit vector in the direction of $\mathbf{v} = \langle -2, 5 \rangle$ and verify that it has length 1.

Solution From Theorem 11.3, the unit vector in the direction of $\mathbf{v}$ is

$$\begin{aligned}
\frac{\mathbf{v}}{\|\mathbf{v}\|} &= \frac{\langle -2, 5 \rangle}{\sqrt{(-2)^2 + (5)^2}} \\
&= \frac{1}{\sqrt{29}} \langle -2, 5 \rangle \\
&= \left\langle \frac{-2}{\sqrt{29}}, \frac{5}{\sqrt{29}} \right\rangle.
\end{aligned}$$

This vector has length 1, because

$$\sqrt{\left(\frac{-2}{\sqrt{29}} \right)^2 + \left(\frac{5}{\sqrt{29}} \right)^2} = \sqrt{\frac{4}{29} + \frac{25}{29}} = \sqrt{\frac{29}{29}} = 1.$$

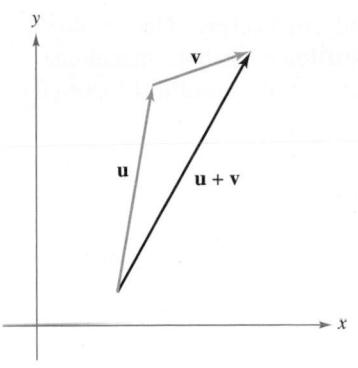

Triangle inequality
Figure 11.9

Generally, the length of the sum of two vectors is not equal to the sum of their lengths. To see this, consider the vectors **u** and **v** as shown in Figure 11.9. With **u** and **v** as two sides of a triangle, the length of the third side is $\|\mathbf{u} + \mathbf{v}\|$, and

$$\|\mathbf{u} + \mathbf{v}\| \le \|\mathbf{u}\| + \|\mathbf{v}\|.$$

Equality occurs only when the vectors **u** and **v** have the *same direction*. This result is called the **triangle inequality** for vectors. (You are asked to prove this in Exercise 73, Section 11.3.)

Standard Unit Vectors

The unit vectors $\langle 1, 0 \rangle$ and $\langle 0, 1 \rangle$ are called the **standard unit vectors** in the plane and are denoted by

$$\mathbf{i} = \langle 1, 0 \rangle \quad \text{and} \quad \mathbf{j} = \langle 0, 1 \rangle \qquad \text{Standard unit vectors}$$

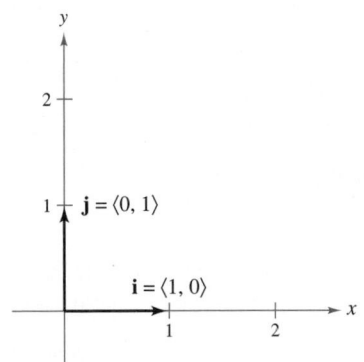

Standard unit vectors **i** and **j**
Figure 11.10

as shown in Figure 11.10. These vectors can be used to represent any vector uniquely, as follows.

$$\mathbf{v} = \langle v_1, v_2 \rangle = \langle v_1, 0 \rangle + \langle 0, v_2 \rangle = v_1 \langle 1, 0 \rangle + v_2 \langle 0, 1 \rangle = v_1 \mathbf{i} + v_2 \mathbf{j}$$

The vector $\mathbf{v} = v_1 \mathbf{i} + v_2 \mathbf{j}$ is called a **linear combination** of **i** and **j**. The scalars v_1 and v_2 are called the **horizontal** and **vertical components of v.**

EXAMPLE 5   **Writing a Linear Combination of Unit Vectors**

Let **u** be the vector with initial point $(2, -5)$ and terminal point $(-1, 3)$, and let $\mathbf{v} = 2\mathbf{i} - \mathbf{j}$. Write each vector as a linear combination of **i** and **j**.

a. u

b. w $= 2\mathbf{u} - 3\mathbf{v}$

Solution

a. $\mathbf{u} = \langle q_1 - p_1, q_2 - p_2 \rangle = \langle -1 - 2, 3 - (-5) \rangle = \langle -3, 8 \rangle = -3\mathbf{i} + 8\mathbf{j}$

b. $\mathbf{w} = 2\mathbf{u} - 3\mathbf{v} = 2(-3\mathbf{i} + 8\mathbf{j}) - 3(2\mathbf{i} - \mathbf{j}) = -6\mathbf{i} + 16\mathbf{j} - 6\mathbf{i} + 3\mathbf{j} = -12\mathbf{i} + 19\mathbf{j}$

If **u** is a unit vector and θ is the angle (measured counterclockwise) from the positive *x*-axis to **u**, then the terminal point of **u** lies on the unit circle, and you have

$$\mathbf{u} = \langle \cos \theta, \sin \theta \rangle = \cos \theta \mathbf{i} + \sin \theta \mathbf{j} \qquad \text{Unit vector}$$

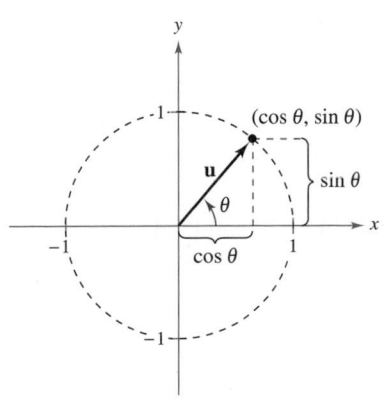

The angle θ from the positive *x*-axis to the vector **u**
Figure 11.11

as shown in Figure 11.11. Moreover, it follows that any other nonzero vector **v** making an angle θ with the positive *x*-axis has the same direction as **u**, and you can write

$$\mathbf{v} = \|\mathbf{v}\| \langle \cos \theta, \sin \theta \rangle = \|\mathbf{v}\| \cos \theta \mathbf{i} + \|\mathbf{v}\| \sin \theta \mathbf{j}.$$

EXAMPLE 6   **Writing a Vector of Given Magnitude and Direction**

The vector **v** has a magnitude of 3 and makes an angle of $30° = \pi/6$ with the positive *x*-axis. Write **v** as a linear combination of the unit vectors **i** and **j**.

Solution   Because the angle between **v** and the positive *x*-axis is $\theta = \pi/6$, you can write

$$\mathbf{v} = \|\mathbf{v}\| \cos \theta \mathbf{i} + \|\mathbf{v}\| \sin \theta \mathbf{j} = 3 \cos \frac{\pi}{6} \mathbf{i} + 3 \sin \frac{\pi}{6} \mathbf{j} = \frac{3\sqrt{3}}{2} \mathbf{i} + \frac{3}{2} \mathbf{j}.$$

Vectors have many applications in physics and engineering. One example is force. A vector can be used to represent force, because force has both magnitude and direction. If two or more forces are acting on an object, then the **resultant force** on the object is the vector sum of the vector forces.

EXAMPLE 7 Finding the Resultant Force

Two tugboats are pushing an ocean liner, as shown in Figure 11.12. Each boat is exerting a force of 400 pounds. What is the resultant force on the ocean liner?

Solution Using Figure 11.12, you can represent the forces exerted by the first and second tugboats as

$$\mathbf{F}_1 = 400\langle\cos 20°, \sin 20°\rangle = 400 \cos(20°)\mathbf{i} + 400 \sin(20°)\mathbf{j}$$
$$\mathbf{F}_2 = 400\langle\cos(-20°), \sin(-20°)\rangle = 400 \cos(20°)\mathbf{i} - 400 \sin(20°)\mathbf{j}.$$

The resultant force on the ocean liner is

$$\mathbf{F} = \mathbf{F}_1 + \mathbf{F}_2$$
$$= [400 \cos(20°)\mathbf{i} + 400 \sin(20°)\mathbf{j}] + [400 \cos(20°)\mathbf{i} - 400 \sin(20°)\mathbf{j}]$$
$$= 800 \cos(20°)\mathbf{i}$$
$$\approx 752\mathbf{i}.$$

So, the resultant force on the ocean liner is approximately 752 pounds in the direction of the positive x-axis.

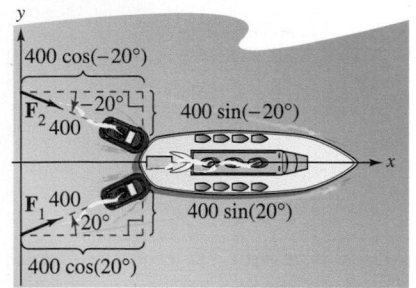

The resultant force on the ocean liner that is exerted by the two tugboats
Figure 11.12

In surveying and navigation, a **bearing** is a direction that measures the acute angle that a path or line of sight makes with a fixed north-south line. In air navigation, bearings are measured in degrees clockwise from north.

EXAMPLE 8 Finding a Velocity

⋮····▷ *See LarsonCalculus.com for an interactive version of this type of example.*

An airplane is traveling at a fixed altitude with a negligible wind factor. The airplane is traveling at a speed of 500 miles per hour with a bearing of 330°, as shown in Figure 11.13(a). As the airplane reaches a certain point, it encounters wind with a velocity of 70 miles per hour in the direction N 45° E (45° east of north), as shown in Figure 11.13(b). What are the resultant speed and direction of the airplane?

Solution Using Figure 11.13(a), represent the velocity of the airplane (alone) as

$$\mathbf{v}_1 = 500 \cos(120°)\mathbf{i} + 500 \sin(120°)\mathbf{j}.$$

The velocity of the wind is represented by the vector

$$\mathbf{v}_2 = 70 \cos(45°)\mathbf{i} + 70 \sin(45°)\mathbf{j}.$$

The resultant velocity of the airplane (in the wind) is

$$\mathbf{v} = \mathbf{v}_1 + \mathbf{v}_2$$
$$= 500 \cos(120°)\mathbf{i} + 500 \sin(120°)\mathbf{j} + 70 \cos(45°)\mathbf{i} + 70 \sin(45°)\mathbf{j}$$
$$\approx -200.5\mathbf{i} + 482.5\mathbf{j}.$$

To find the resultant speed and direction, write $\mathbf{v} = \|\mathbf{v}\|(\cos\theta\mathbf{i} + \sin\theta\mathbf{j})$. Because $\|\mathbf{v}\| \approx \sqrt{(-200.5)^2 + (482.5)^2} \approx 522.5$, you can write

$$\mathbf{v} \approx 522.5\left(\frac{-200.5}{522.5}\mathbf{i} + \frac{482.5}{522.5}\mathbf{j}\right) \approx 522.5[\cos(112.6°)\mathbf{i} + \sin(112.6°)\mathbf{j}].$$

The new speed of the airplane, as altered by the wind, is approximately 522.5 miles per hour in a path that makes an angle of 112.6° with the positive x-axis.

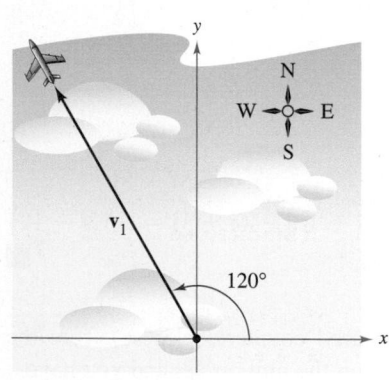

(a) Direction without wind

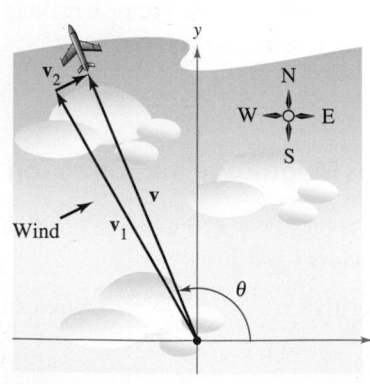

(b) Direction with wind
Figure 11.13

11.1 Exercises

See **CalcChat.com** for tutorial help and worked-out solutions to odd-numbered exercises.

CONCEPT CHECK

1. **Scalar and Vector** Describe the difference between a scalar and a vector. Give examples of each.

2. **Vector** Two points and a vector are given. Determine which point is the initial point and which point is the terminal point. Explain.

$$P(2, -1), \quad Q(-4, 6), \quad \text{and} \quad \mathbf{v} = \langle 6, -7 \rangle$$

Sketching a Vector In Exercises 3 and 4, (a) find the component form of the vector v and (b) sketch the vector with its initial point at the origin.

3.

4.

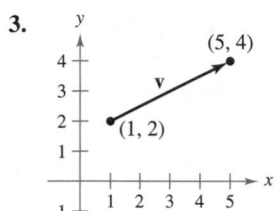 **Equivalent Vectors** In Exercises 5–8, find the vectors u and v whose initial and terminal points are given. Show that u and v are equivalent.

	Terminal		Terminal
Initial Point	Point	Initial Point	Point
5. u: $(3, 2)$	$(5, 6)$	**6. u:** $(-4, 0)$	$(1, 8)$
v: $(1, 4)$	$(3, 8)$	**v:** $(2, -1)$	$(7, 7)$
7. u: $(0, 3)$	$(6, -2)$	**8. u:** $(-4, -1)$	$(11, -4)$
v: $(3, 10)$	$(9, 5)$	**v:** $(10, 13)$	$(25, 10)$

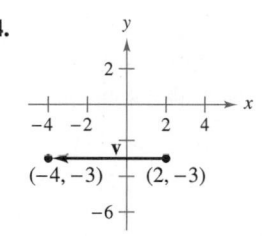 **Writing a Vector in Different Forms** In Exercises 9–16, the initial and terminal points of a vector v are given. (a) Sketch the given directed line segment. (b) Write the vector in component form. (c) Write the vector as the linear combination of the standard unit vectors i and j. (d) Sketch the vector with its initial point at the origin.

	Terminal		Terminal
Initial Point	Point	Initial Point	Point
9. $(2, 0)$	$(5, 5)$	**10.** $(4, -6)$	$(3, 6)$
11. $(8, 3)$	$(6, -1)$	**12.** $(0, -4)$	$(-5, -1)$
13. $(6, 2)$	$(6, 6)$	**14.** $(7, -1)$	$(-3, -1)$
15. $\left(\frac{3}{2}, \frac{4}{3}\right)$	$\left(\frac{1}{2}, 3\right)$	**16.** $(0.12, 0.60)$	$(0.84, 1.25)$

Finding a Terminal Point In Exercises 17 and 18, the vector v and its initial point are given. Find the terminal point.

17. $\mathbf{v} = \langle -1, 3 \rangle$; Initial point: $(4, 2)$

18. $\mathbf{v} = \langle 4, -9 \rangle$; Initial point: $(5, 3)$

 Finding a Magnitude of a Vector In Exercises 19–24, find the magnitude of v.

19. $\mathbf{v} = 4\mathbf{i}$

20. $\mathbf{v} = -9\mathbf{j}$

21. $\mathbf{v} = \langle 8, 15 \rangle$

22. $\mathbf{v} = \langle -24, 7 \rangle$

23. $\mathbf{v} = -\mathbf{i} - 5\mathbf{j}$

24. $\mathbf{v} = 3\mathbf{i} + 3\mathbf{j}$

Sketching Scalar Multiples In Exercises 25 and 26, sketch each scalar multiple of v.

25. $\mathbf{v} = \langle 3, 5 \rangle$ (a) $2\mathbf{v}$ (b) $-3\mathbf{v}$ (c) $\frac{7}{2}\mathbf{v}$ (d) $\frac{2}{3}\mathbf{v}$

26. $\mathbf{v} = \langle -2, 3 \rangle$ (a) $4\mathbf{v}$ (b) $-\frac{1}{2}\mathbf{v}$ (c) $0\mathbf{v}$ (d) $-6\mathbf{v}$

 **Using Vector Operations** In Exercises 27 and 28, find (a) $\frac{2}{3}\mathbf{u}$, (b) $3\mathbf{v}$, (c) $\mathbf{v} - \mathbf{u}$, and (d) $2\mathbf{u} + 5\mathbf{v}$.

27. $\mathbf{u} = \langle 4, 9 \rangle$, $\mathbf{v} = \langle 2, -5 \rangle$ 28. $\mathbf{u} = \langle -3, -8 \rangle$, $\mathbf{v} = \langle 8, 7 \rangle$

Sketching a Vector In Exercises 29–34, use the figure to sketch a graph of the vector. To print an enlarged copy of the graph, go to *MathGraphs.com*.

29. $-\mathbf{u}$

30. $2\mathbf{u}$

31. $-\mathbf{v}$

32. $\frac{1}{2}\mathbf{v}$

33. $\mathbf{u} - \mathbf{v}$

34. $\mathbf{u} + 2\mathbf{v}$

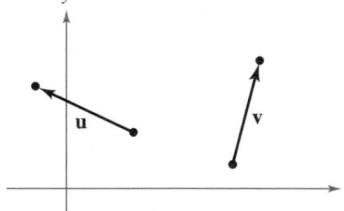

 **Finding a Unit Vector** In Exercises 35–38, find the unit vector in the direction of v and verify that it has length 1.

35. $\mathbf{v} = \langle 3, 12 \rangle$

36. $\mathbf{v} = \langle -5, 15 \rangle$

37. $\mathbf{v} = \left\langle \frac{3}{2}, \frac{5}{2} \right\rangle$

38. $\mathbf{v} = \langle -6.2, 3.4 \rangle$

Finding Magnitudes In Exercises 39–42, find the following.

(a) $\|\mathbf{u}\|$ (b) $\|\mathbf{v}\|$ (c) $\|\mathbf{u} + \mathbf{v}\|$

(d) $\left\| \dfrac{\mathbf{u}}{\|\mathbf{u}\|} \right\|$ (e) $\left\| \dfrac{\mathbf{v}}{\|\mathbf{v}\|} \right\|$ (f) $\left\| \dfrac{\mathbf{u} + \mathbf{v}}{\|\mathbf{u} + \mathbf{v}\|} \right\|$

39. $\mathbf{u} = \langle 1, -1 \rangle$, $\mathbf{v} = \langle -1, 2 \rangle$

40. $\mathbf{u} = \langle 0, 1 \rangle$, $\mathbf{v} = \langle 3, -3 \rangle$

41. $\mathbf{u} = \left\langle 1, \frac{1}{2} \right\rangle$, $\mathbf{v} = \langle 2, 3 \rangle$

42. $\mathbf{u} = \langle 2, -4 \rangle$, $\mathbf{v} = \langle 5, 5 \rangle$

Using the Triangle Inequality In Exercises 43 and 44, sketch a graph of u, v, and u + v. Then demonstrate the triangle inequality using the vectors u and v.

43. $\mathbf{u} = \langle 2, 1 \rangle$, $\mathbf{v} = \langle 5, 4 \rangle$

44. $\mathbf{u} = \langle -3, 2 \rangle$, $\mathbf{v} = \langle 1, -2 \rangle$

Finding a Vector In Exercises 45–48, find the vector **v** with the given magnitude and the same direction as **u**.

Magnitude	Direction
45. $\|\mathbf{v}\| = 6$	$\mathbf{u} = \langle 0, 3 \rangle$
46. $\|\mathbf{v}\| = 4$	$\mathbf{u} = \langle 1, 1 \rangle$
47. $\|\mathbf{v}\| = 5$	$\mathbf{u} = \langle -1, 2 \rangle$
48. $\|\mathbf{v}\| = 2$	$\mathbf{u} = \langle \sqrt{3}, 3 \rangle$

Finding a Vector In Exercises 49–52, find the component form of **v** given its magnitude and the angle it makes with the positive x-axis.

49. $\|\mathbf{v}\| = 3$, $\theta = 0°$ **50.** $\|\mathbf{v}\| = 5$, $\theta = 120°$

51. $\|\mathbf{v}\| = 2$, $\theta = 150°$ **52.** $\|\mathbf{v}\| = 4$, $\theta = 3.5°$

Finding a Vector In Exercises 53–56, find the component form of **u** + **v** given the lengths of **u** and **v** and the angles that **u** and **v** make with the positive x-axis.

53. $\|\mathbf{u}\| = 1$, $\theta_{\mathbf{u}} = 0°$ **54.** $\|\mathbf{u}\| = 4$, $\theta_{\mathbf{u}} = 0°$

     $\|\mathbf{v}\| = 3$, $\theta_{\mathbf{v}} = 45°$ $\|\mathbf{v}\| = 2$, $\theta_{\mathbf{v}} = 60°$

55. $\|\mathbf{u}\| = 2$, $\theta_{\mathbf{u}} = 4$ **56.** $\|\mathbf{u}\| = 5$, $\theta_{\mathbf{u}} = -0.5$

     $\|\mathbf{v}\| = 1$, $\theta_{\mathbf{v}} = 2$ $\|\mathbf{v}\| = 5$, $\theta_{\mathbf{v}} = 0.5$

EXPLORING CONCEPTS

Think About It In Exercises 57 and 58, consider two forces of equal magnitude acting on a point.

57. When the magnitude of the resultant is the sum of the magnitudes of the two forces, make a conjecture about the angle between the forces.

58. When the resultant of the forces is 0, make a conjecture about the angle between the forces.

59. Triangle Consider a triangle with vertices X, Y, and Z. What is $\overrightarrow{XY} + \overrightarrow{YZ} + \overrightarrow{ZX}$? Explain.

---

60. **HOW DO YOU SEE IT?** Use the figure to determine whether each statement is true or false. Justify your answer.

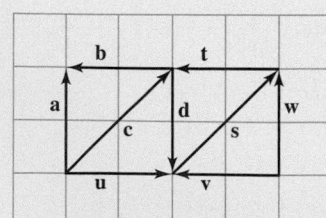

  (a) $\mathbf{a} = -\mathbf{d}$ (b) $\mathbf{c} = \mathbf{s}$

  (c) $\mathbf{a} + \mathbf{u} = \mathbf{c}$ (d) $\mathbf{v} + \mathbf{w} = -\mathbf{s}$

  (e) $\mathbf{a} + \mathbf{d} = \mathbf{0}$ (f) $\mathbf{u} - \mathbf{v} = -2(\mathbf{b} + \mathbf{t})$

Finding Values In Exercises 61–66, find a and b such that $\mathbf{v} = a\mathbf{u} + b\mathbf{w}$, where $\mathbf{u} = \langle 1, 2 \rangle$ and $\mathbf{w} = \langle 1, -1 \rangle$.

61. $\mathbf{v} = \langle 4, 5 \rangle$ **62.** $\mathbf{v} = \langle -7, -2 \rangle$

63. $\mathbf{v} = \langle -6, 0 \rangle$ **64.** $\mathbf{v} = \langle 0, 6 \rangle$

65. $\mathbf{v} = \langle 1, -3 \rangle$ **66.** $\mathbf{v} = \langle -1, 8 \rangle$

Finding Unit Vectors In Exercises 67–72, find a unit vector (a) parallel to and (b) perpendicular to the graph of f at the given point. Then sketch the graph of f and sketch the vectors at the given point.

67. $f(x) = x^2$, $(3, 9)$ **68.** $f(x) = -x^2 + 5$, $(1, 4)$

69. $f(x) = x^3$, $(1, 1)$ **70.** $f(x) = x^3$, $(-2, -8)$

71. $f(x) = \sqrt{25 - x^2}$, $(3, 4)$

72. $f(x) = \tan x$, $\left(\dfrac{\pi}{4}, 1\right)$

Finding a Vector In Exercises 73 and 74, find the component form of **v** given the magnitudes of **u** and **u** + **v** and the angles that **u** and **u** + **v** make with the positive x-axis.

73. $\|\mathbf{u}\| = 1$, $\theta = 45°$ **74.** $\|\mathbf{u}\| = 4$, $\theta = 30°$

     $\|\mathbf{u} + \mathbf{v}\| = \sqrt{2}$, $\theta = 90°$ $\|\mathbf{u} + \mathbf{v}\| = 6$, $\theta = 120°$

75. Resultant Force Forces with magnitudes of 500 pounds and 200 pounds act on a machine part at angles of 30° and $-45°$, respectively, with the x-axis (see figure). Find the direction and magnitude of the resultant force.

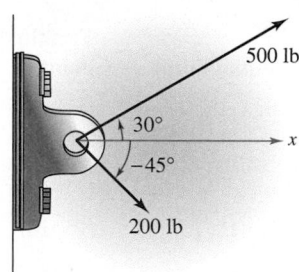

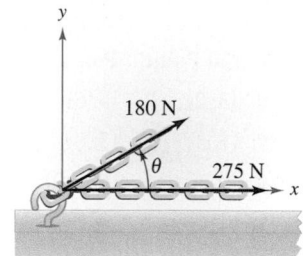

Figure for 75   Figure for 76

76. Numerical and Graphical Analysis Forces with magnitudes of 180 newtons and 275 newtons act on a hook (see figure). The angle between the two forces is θ degrees.

(a) When $\theta = 30°$, find the direction and magnitude of the resultant force.

(b) Write the magnitude M and direction α of the resultant force as functions of θ, where $0° \le \theta \le 180°$.

(c) Use a graphing utility to complete the table.

θ	0°	30°	60°	90°	120°	150°	180°
M							
α							

(d) Use a graphing utility to graph the two functions M and α.

(e) Explain why one of the functions decreases for increasing values of θ, whereas the other does not.

77. Resultant Force Three forces with magnitudes of 75 pounds, 100 pounds, and 125 pounds act on an object at angles of 30°, 45°, and 120°, respectively, with the positive *x*-axis. Find the direction and magnitude of the resultant force.

78. Resultant Force Three forces with magnitudes of 400 newtons, 280 newtons, and 350 newtons act on an object at angles of −30°, 45°, and 135°, respectively, with the positive *x*-axis. Find the direction and magnitude of the resultant force.

Cable Tension **In Exercises 79 and 80, determine the tension in the cable supporting the given load.**

79.

80.

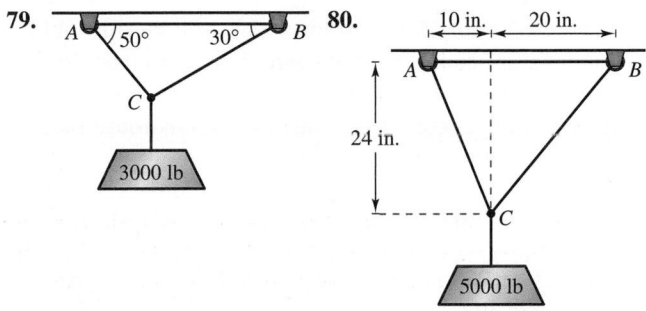

81. Projectile Motion A gun with a muzzle velocity of 1200 feet per second is fired at an angle of 6° above the horizontal. Find the vertical and horizontal components of the velocity.

82. Shared Load To carry a 100-pound cylindrical weight, two workers lift on the ends of short ropes tied to an eyelet on the top center of the cylinder. One rope makes a 20° angle away from the vertical and the other makes a 30° angle (see figure).

(a) Find each rope's tension when the resultant force is vertical.

(b) Find the vertical component of each worker's force.

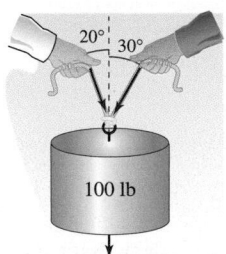

83. Navigation A plane is flying with a bearing of 302°. Its speed with respect to the air is 900 kilometers per hour. The wind at the plane's altitude is from the southwest at 100 kilometers per hour (see figure). What is the true direction of the plane, and what is its speed with respect to the ground?

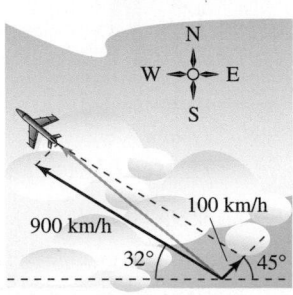

• • 84. Navigation • • • • • • • • • • • • •

A plane flies at a constant groundspeed of 400 miles per hour due east and encounters a 50-mile-per-hour wind from the northwest. Find the airspeed and compass direction that will allow the plane to maintain its groundspeed and eastward direction.

True or False? **In Exercises 85–94, determine whether the statement is true or false. If it is false, explain why or give an example that shows it is false.**

85. The weight of a car is a scalar.

86. The mass of a book is a scalar.

87. The temperature of your blood is a scalar.

88. The velocity of a bicycle is a vector.

89. If **u** and **v** have the same magnitude and direction, then **u** and **v** are equivalent.

90. If **u** is a unit vector in the direction of **v**, then $\mathbf{v} = \|\mathbf{v}\|\mathbf{u}$.

91. If $\mathbf{u} = a\mathbf{i} + b\mathbf{j}$ is a unit vector, then $a^2 + b^2 = 1$.

92. If $\mathbf{v} = a\mathbf{i} + b\mathbf{j} = \mathbf{0}$, then $a = -b$.

93. If $a = b$, then $\|a\mathbf{i} + b\mathbf{j}\| = \sqrt{2}\,a$.

94. If **u** and **v** have the same magnitude but opposite directions, then $\mathbf{u} + \mathbf{v} = \mathbf{0}$.

95. Proof Prove that

$$\mathbf{u} = (\cos\theta)\mathbf{i} - (\sin\theta)\mathbf{j} \quad \text{and} \quad \mathbf{v} = (\sin\theta)\mathbf{i} + (\cos\theta)\mathbf{j}$$

are unit vectors for any angle θ.

96. Geometry Using vectors, prove that the line segment joining the midpoints of two sides of a triangle is parallel to, and one-half the length of, the third side.

97. Geometry Using vectors, prove that the diagonals of a parallelogram bisect each other.

98. Proof Prove that the vector $\mathbf{w} = \|\mathbf{u}\|\mathbf{v} + \|\mathbf{v}\|\mathbf{u}$ bisects the angle between **u** and **v**.

99. Using a Vector Consider the vector $\mathbf{u} = \langle x, y \rangle$. Describe the set of all points (x, y) such that $\|\mathbf{u}\| = 5$.

PUTNAM EXAM CHALLENGE

100. A coast artillery gun can fire at any angle of elevation between 0° and 90° in a fixed vertical plane. If air resistance is neglected and the muzzle velocity is constant $(= v_0)$, determine the set H of points in the plane and above the horizontal which can be hit.

11.2 Space Coordinates and Vectors in Space

■ Understand the three-dimensional rectangular coordinate system.
■ Analyze vectors in space.

Coordinates in Space

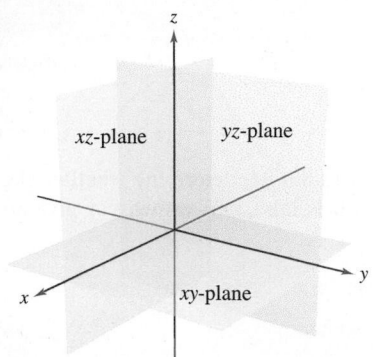

The three-dimensional coordinate system
Figure 11.14

Up to this point in the text, you have been primarily concerned with the two-dimensional coordinate system. Much of the remaining part of your study of calculus will involve the three-dimensional coordinate system.

Before extending the concept of a vector to three dimensions, you must be able to identify points in the **three-dimensional coordinate system.** You can construct this system by passing a z-axis perpendicular to both the x- and y-axes at the origin, as shown in Figure 11.14. Taken as pairs, the axes determine three **coordinate planes:** the **xy-plane,** the **xz-plane,** and the **yz-plane.** These three coordinate planes separate three-space into eight **octants.** The first octant is the one for which all three coordinates are positive. In this three-dimensional system, a point P in space is determined by an ordered triple (x, y, z), where x, y, and z are as follows.

x = directed distance from yz-plane to P

y = directed distance from xz-plane to P

z = directed distance from xy-plane to P

Several points are shown in Figure 11.15.

• **REMARK** The
three-dimensional rotatable
graphs that are available at
LarsonCalculus.com can help
you visualize points or objects
in a three-dimensional
coordinate system.

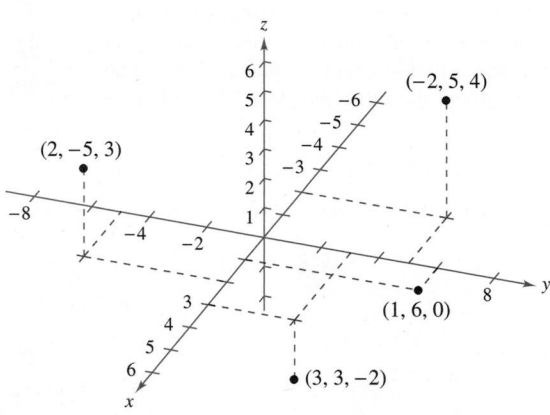

Points in the three-dimensional coordinate system are represented by ordered triples.
Figure 11.15

A three-dimensional coordinate system can have either a **right-handed** or a **left-handed** orientation. To determine the orientation of a system, imagine that you are standing at the origin, with your arms pointing in the direction of the positive x- and y-axes and with the positive z-axis pointing up, as shown in Figure 11.16. The system is right-handed or left-handed depending on which hand points along the x-axis. In this text, you will work exclusively with the right-handed system.

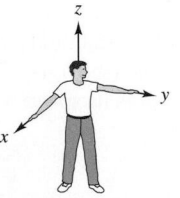

Right-handed
system

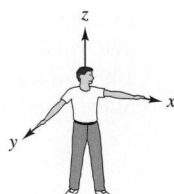

Left-handed
system
Figure 11.16

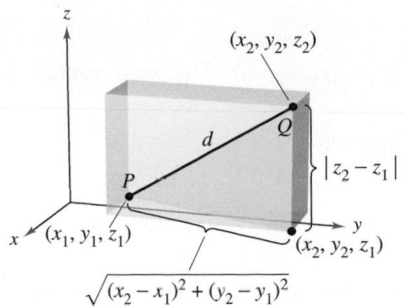

The distance between two points in space

Figure 11.17

Many of the formulas established for the two-dimensional coordinate system can be extended to three dimensions. For example, to find the distance between two points in space, you can use the Pythagorean Theorem twice, as shown in Figure 11.17. By doing this, you will obtain the formula for the distance between the points (x_1, y_1, z_1) and (x_2, y_2, z_2).

$$d = \sqrt{(x_2 - x_1)^2 + (y_2 - y_1)^2 + (z_2 - z_1)^2} \qquad \text{Distance Formula}$$

EXAMPLE 1   **Finding the Distance Between Two Points in Space**

Find the distance between the points $(2, -1, 3)$ and $(1, 0, -2)$.

Solution

$$\begin{aligned}
d &= \sqrt{(1 - 2)^2 + (0 + 1)^2 + (-2 - 3)^2} \qquad \text{Distance Formula} \\
&= \sqrt{1 + 1 + 25} \\
&= \sqrt{27} \\
&= 3\sqrt{3}
\end{aligned}$$

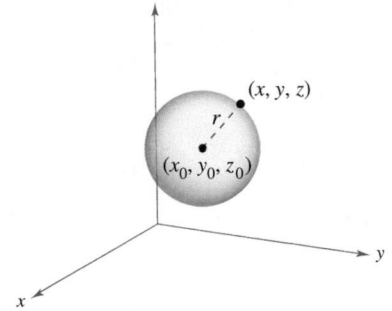

Figure 11.18

A **sphere** with center at (x_0, y_0, z_0) and radius r is defined to be the set of all points (x, y, z) such that the distance between (x, y, z) and (x_0, y_0, z_0) is r. You can use the Distance Formula to find the **standard equation of a sphere** of radius r, centered at (x_0, y_0, z_0). If (x, y, z) is an arbitrary point on the sphere, then the equation of the sphere is

$$(x - x_0)^2 + (y - y_0)^2 + (z - z_0)^2 = r^2 \qquad \text{Equation of sphere}$$

as shown in Figure 11.18. Moreover, the midpoint of the line segment joining the points (x_1, y_1, z_1) and (x_2, y_2, z_2) has coordinates

$$\left(\frac{x_1 + x_2}{2}, \frac{y_1 + y_2}{2}, \frac{z_1 + z_2}{2} \right). \qquad \text{Midpoint Formula}$$

EXAMPLE 2   **Finding the Equation of a Sphere**

Find the standard equation of the sphere that has the points

$$(5, -2, 3) \quad \text{and} \quad (0, 4, -3)$$

as endpoints of a diameter.

Solution   Using the Midpoint Formula, the center of the sphere is

$$\left(\frac{5 + 0}{2}, \frac{-2 + 4}{2}, \frac{3 - 3}{2} \right) = \left(\frac{5}{2}, 1, 0 \right). \qquad \text{Midpoint Formula}$$

By the Distance Formula, the radius is

$$r = \sqrt{ \left(0 - \frac{5}{2} \right)^2 + (4 - 1)^2 + (-3 - 0)^2 } = \sqrt{ \frac{97}{4} } = \frac{\sqrt{97}}{2}.$$

Therefore, the standard equation of the sphere is

$$\left(x - \frac{5}{2} \right)^2 + (y - 1)^2 + z^2 = \frac{97}{4}. \qquad \text{Equation of sphere}$$

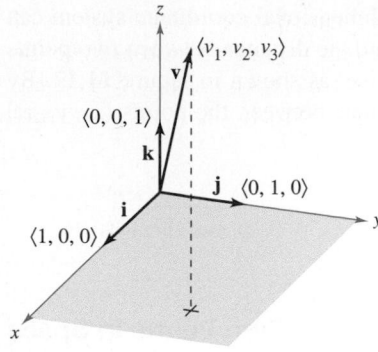

The standard unit vectors in space
Figure 11.19

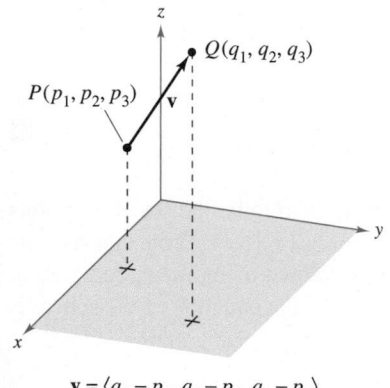

$$v = \langle q_1 - p_1, q_2 - p_2, q_3 - p_3 \rangle$$

Figure 11.20

Vectors in Space

In space, vectors are denoted by ordered triples $\mathbf{v} = \langle v_1, v_2, v_3 \rangle$. The **zero vector** is denoted by $\mathbf{0} = \langle 0, 0, 0 \rangle$. Using the unit vectors

$$\mathbf{i} = \langle 1, 0, 0 \rangle, \quad \mathbf{j} = \langle 0, 1, 0 \rangle, \quad \text{and} \quad \mathbf{k} = \langle 0, 0, 1 \rangle$$

the **standard unit vector notation** for $\mathbf{v}$ is

$$\mathbf{v} = v_1 \mathbf{i} + v_2 \mathbf{j} + v_3 \mathbf{k}$$

as shown in Figure 11.19. If $\mathbf{v}$ is represented by the directed line segment from $P(p_1, p_2, p_3)$ to $Q(q_1, q_2, q_3)$, as shown in Figure 11.20, then the component form of $\mathbf{v}$ is written by subtracting the coordinates of the initial point from the coordinates of the terminal point, as follows.

$$\mathbf{v} = \langle v_1, v_2, v_3 \rangle = \langle q_1 - p_1, q_2 - p_2, q_3 - p_3 \rangle$$

Vectors in Space

Let $\mathbf{u} = \langle u_1, u_2, u_3 \rangle$ and $\mathbf{v} = \langle v_1, v_2, v_3 \rangle$ be vectors in space and let c be a scalar.

1. *Equality of Vectors:* $\mathbf{u} = \mathbf{v}$ if and only if $u_1 = v_1$, $u_2 = v_2$, and $u_3 = v_3$.
2. *Component Form:* If $\mathbf{v}$ is represented by the directed line segment from $P(p_1, p_2, p_3)$ to $Q(q_1, q_2, q_3)$, then

$$\mathbf{v} = \langle v_1, v_2, v_3 \rangle = \langle q_1 - p_1, q_2 - p_2, q_3 - p_3 \rangle.$$

3. *Length:* $\|\mathbf{v}\| = \sqrt{v_1^2 + v_2^2 + v_3^2}$
4. *Unit Vector in the Direction of* $\mathbf{v}$: $\dfrac{\mathbf{v}}{\|\mathbf{v}\|} = \left(\dfrac{1}{\|\mathbf{v}\|}\right)\langle v_1, v_2, v_3 \rangle, \quad \mathbf{v} \neq \mathbf{0}$
5. *Vector Addition:* $\mathbf{v} + \mathbf{u} = \langle v_1 + u_1, v_2 + u_2, v_3 + u_3 \rangle$
6. *Scalar Multiplication:* $c\mathbf{v} = \langle cv_1, cv_2, cv_3 \rangle$

Note that the properties of vector operations listed in Theorem 11.1 (see Section 11.1) are also valid for vectors in space.

EXAMPLE 3   Finding the Component Form of a Vector in Space

⋮ ⋯▷ *See LarsonCalculus.com for an interactive version of this type of example.*

Find the component form and magnitude of the vector $\mathbf{v}$ having initial point $(-2, 3, 1)$ and terminal point $(0, -4, 4)$. Then find a unit vector in the direction of $\mathbf{v}$.

Solution   The component form of $\mathbf{v}$ is

$$\mathbf{v} = \langle q_1 - p_1, q_2 - p_2, q_3 - p_3 \rangle = \langle 0 - (-2), -4 - 3, 4 - 1 \rangle = \langle 2, -7, 3 \rangle$$

which implies that its magnitude is

$$\|\mathbf{v}\| = \sqrt{2^2 + (-7)^2 + 3^2} = \sqrt{62}.$$

The unit vector in the direction of $\mathbf{v}$ is

$$\begin{aligned}
\mathbf{u} &= \frac{\mathbf{v}}{\|\mathbf{v}\|} \\
&= \frac{1}{\sqrt{62}}\langle 2, -7, 3 \rangle \\
&= \left\langle \frac{2}{\sqrt{62}}, \frac{-7}{\sqrt{62}}, \frac{3}{\sqrt{62}} \right\rangle.
\end{aligned}$$

∎

Recall from the definition of scalar multiplication that positive scalar multiples of a nonzero vector **v** have the same direction as **v**, whereas negative multiples have the direction opposite of **v**. In general, two nonzero vectors **u** and **v** are **parallel** when there is some scalar c such that $\mathbf{u} = c\mathbf{v}$. For example, in Figure 11.21, the vectors **u**, **v**, and **w** are parallel because

$$\mathbf{u} = 2\mathbf{v} \quad \text{and} \quad \mathbf{w} = -\mathbf{v}.$$

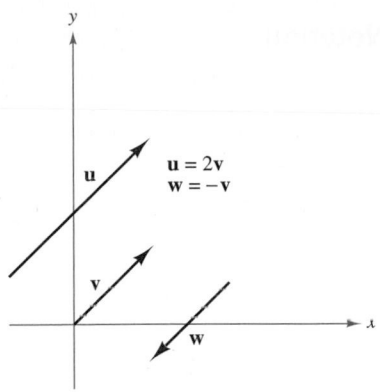

$\mathbf{u} = 2\mathbf{v}$
$\mathbf{w} = -\mathbf{v}$

Parallel vectors
Figure 11.21

> **Definition of Parallel Vectors**
>
> Two nonzero vectors **u** and **v** are **parallel** when there is some scalar c such that $\mathbf{u} = c\mathbf{v}$.

EXAMPLE 4 Parallel Vectors

Vector **w** has initial point $(2, -1, 3)$ and terminal point $(-4, 7, 5)$. Which of the following vectors is parallel to **w**?

a. $\mathbf{u} = \langle 3, -4, -1 \rangle$

b. $\mathbf{v} = \langle 12, -16, 4 \rangle$

Solution Begin by writing **w** in component form.

$$\mathbf{w} = \langle -4 - 2, 7 - (-1), 5 - 3 \rangle = \langle -6, 8, 2 \rangle$$

a. Because $\mathbf{u} = \langle 3, -4, -1 \rangle = -\frac{1}{2}\langle -6, 8, 2 \rangle = -\frac{1}{2}\mathbf{w}$, you can conclude that **u** is parallel to **w**.

b. In this case, you want to find a scalar c such that

$$\langle 12, -16, 4 \rangle = c\langle -6, 8, 2 \rangle.$$

To find c, equate the corresponding components and solve as shown.

$$12 = -6c \implies c = -2$$
$$-16 = 8c \implies c = -2$$
$$4 = 2c \implies c = 2$$

Note that $c = -2$ for the first two components and $c = 2$ for the third component. This means that the equation $\langle 12, -16, 4 \rangle = c\langle -6, 8, 2 \rangle$ has no solution, and the vectors are not parallel.

EXAMPLE 5 Using Vectors to Determine Collinear Points

Determine whether the points

$$P(1, -2, 3), \quad Q(2, 1, 0), \quad \text{and} \quad R(4, 7, -6)$$

are collinear.

Solution The component forms of $\overrightarrow{PQ}$ and $\overrightarrow{PR}$ are

$$\overrightarrow{PQ} = \langle 2 - 1, 1 - (-2), 0 - 3 \rangle = \langle 1, 3, -3 \rangle$$

and

$$\overrightarrow{PR} = \langle 4 - 1, 7 - (-2), -6 - 3 \rangle = \langle 3, 9, -9 \rangle.$$

These two vectors have a common initial point. So, P, Q, and R lie on the same line if and only if $\overrightarrow{PQ}$ and $\overrightarrow{PR}$ are parallel—which they are because $\overrightarrow{PR} = 3\,\overrightarrow{PQ}$, as shown in Figure 11.22.

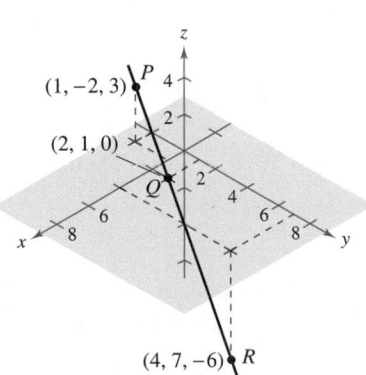

The points P, Q, and R lie on the same line.
Figure 11.22

EXAMPLE 6 **Standard Unit Vector Notation**

a. Write the vector $\mathbf{v} = 4\mathbf{i} - 5\mathbf{k}$ in component form.

b. Find the terminal point of the vector $\mathbf{v} = 7\mathbf{i} - \mathbf{j} + 3\mathbf{k}$, given that the initial point is $P(-2, 3, 5)$.

c. Find the magnitude of the vector $\mathbf{v} = -6\mathbf{i} + 2\mathbf{j} - 3\mathbf{k}$. Then find a unit vector in the direction of $\mathbf{v}$.

Solution

a. Because $\mathbf{j}$ is missing, its component is 0 and

$$\mathbf{v} = 4\mathbf{i} - 5\mathbf{k} = \langle 4, 0, -5 \rangle.$$

b. You need to find $Q(q_1, q_2, q_3)$ such that

$$\mathbf{v} = \overrightarrow{PQ} = 7\mathbf{i} - \mathbf{j} + 3\mathbf{k}.$$

This implies that $q_1 - (-2) = 7$, $q_2 - 3 = -1$, and $q_3 - 5 = 3$. The solution of these three equations is $q_1 = 5$, $q_2 = 2$, and $q_3 = 8$. Therefore, Q is $(5, 2, 8)$.

c. Note that $v_1 = -6$, $v_2 = 2$, and $v_3 = -3$. So, the magnitude of $\mathbf{v}$ is

$$\|\mathbf{v}\| = \sqrt{(-6)^2 + 2^2 + (-3)^2} = \sqrt{49} = 7.$$

The unit vector in the direction of $\mathbf{v}$ is

$$\tfrac{1}{7}(-6\mathbf{i} + 2\mathbf{j} - 3\mathbf{k}) = -\tfrac{6}{7}\mathbf{i} + \tfrac{2}{7}\mathbf{j} - \tfrac{3}{7}\mathbf{k}.$$

EXAMPLE 7 **Measuring Force**

A television camera weighing 120 pounds is supported by a tripod, as shown in Figure 11.23. Represent the force exerted on each leg of the tripod as a vector.

Solution Let the vectors $\mathbf{F}_1$, $\mathbf{F}_2$, and $\mathbf{F}_3$ represent the forces exerted on the three legs. From Figure 11.23, you can determine the directions of $\mathbf{F}_1$, $\mathbf{F}_2$, and $\mathbf{F}_3$ to be as follows.

$$\mathbf{F}_1 = \overrightarrow{PQ_1} = \langle 0 - 0, -1 - 0, 0 - 4 \rangle = \langle 0, -1, -4 \rangle$$

$$\mathbf{F}_2 = \overrightarrow{PQ_2} = \left\langle \frac{\sqrt{3}}{2} - 0, \frac{1}{2} - 0, 0 - 4 \right\rangle = \left\langle \frac{\sqrt{3}}{2}, \frac{1}{2}, -4 \right\rangle$$

$$\mathbf{F}_3 = \overrightarrow{PQ_3} = \left\langle -\frac{\sqrt{3}}{2} - 0, \frac{1}{2} - 0, 0 - 4 \right\rangle = \left\langle -\frac{\sqrt{3}}{2}, \frac{1}{2}, -4 \right\rangle.$$

Because all three legs have the same length and the total force is distributed equally among the three legs, you know that $\|\mathbf{F}_1\| = \|\mathbf{F}_2\| = \|\mathbf{F}_3\|$. So, there exists a constant c such that

$$\mathbf{F}_1 = c\langle 0, -1, -4 \rangle, \quad \mathbf{F}_2 = c\left\langle \frac{\sqrt{3}}{2}, \frac{1}{2}, -4 \right\rangle, \quad \text{and} \quad \mathbf{F}_3 = c\left\langle -\frac{\sqrt{3}}{2}, \frac{1}{2}, -4 \right\rangle.$$

Let the total force exerted by the object be given by $\mathbf{F} = \langle 0, 0, -120 \rangle$. Then, using the fact that

$$\mathbf{F} = \mathbf{F}_1 + \mathbf{F}_2 + \mathbf{F}_3$$

you can conclude that $\mathbf{F}_1$, $\mathbf{F}_2$, and $\mathbf{F}_3$ all have a vertical component of -40. This implies that $c(-4) = -40$ and $c = 10$. Therefore, the forces exerted on the legs can be represented by

$$\mathbf{F}_1 = \langle 0, -10, -40 \rangle,$$
$$\mathbf{F}_2 = \langle 5\sqrt{3}, 5, -40 \rangle,$$

and

$$\mathbf{F}_3 = \langle -5\sqrt{3}, 5, -40 \rangle.$$

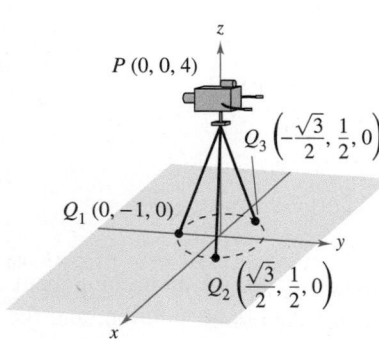

Figure 11.23

11.2 Exercises

See **CalcChat.com** for tutorial help and worked-out solutions to odd-numbered exercises.

CONCEPT CHECK

1. **Describing Coordinates** A point in the three-dimensional coordinate system has coordinates (x_0, y_0, z_0). Describe what each coordinate measures.

2. **Coordinates in Space** What is the y coordinate of any point in the xz-plane?

3. **Comparing Graphs** Describe the graph of $x = 4$ on (a) the number line, (b) the two-dimensional coordinate system, and (c) the three-dimensional coordinate system.

4. **Parallel Vectors** Explain how to determine whether two nonzero vectors **u** and **v** are parallel.

Plotting Points In Exercises 5–8, plot the points in the same three-dimensional coordinate system.

5. (a) $(2, 1, 3)$ (b) $(-1, 2, 1)$

6. (a) $(3, -2, 5)$ (b) $\left(\frac{3}{2}, 4, -2\right)$

7. (a) $(5, -2, 2)$ (b) $(5, -2, -2)$

8. (a) $(0, 4, -5)$ (b) $(4, 0, 5)$

 **Finding Coordinates of a Point** In Exercises 9–12, find the coordinates of the point.

9. The point is located three units behind the yz-plane, four units to the right of the xz-plane, and five units above the xy-plane.

10. The point is located seven units in front of the yz-plane, two units to the left of the xz-plane, and one unit below the xy-plane.

11. The point is located on the x-axis, 12 units in front of the yz-plane.

12. The point is located in the yz-plane, three units to the right of the xz-plane, and two units above the xy-plane.

Using the Three-Dimensional Coordinate System In Exercises 13–24, determine the location of a point (x, y, z) that satisfies the condition(s).

13. $z = 1$ 14. $y = 6$

15. $x = -3$ 16. $z = -5$

17. $y < 0$ 18. $x > 0$

19. $|y| \le 3$ 20. $|x| > 4$

21. $xy > 0, \quad z = -3$ 22. $xy < 0, \quad z = 4$

23. $xyz < 0$ 24. $xyz > 0$

 **Finding the Distance Between Two Points in Space** In Exercises 25–28, find the distance between the points.

25. $(4, 1, 5), (8, 2, 6)$ 26. $(-1, 1, 1), (-3, 5, -3)$

27. $(0, 2, 4), (3, 2, 8)$ 28. $(-3, 7, 1), (-5, 8, -4)$

Classifying a Triangle In Exercises 29–32, find the lengths of the sides of the triangle with the indicated vertices, and determine whether the triangle is a right triangle, an isosceles triangle, or neither.

29. $(0, 0, 4), (2, 6, 7), (6, 4, -8)$

30. $(3, 4, 1), (0, 6, 2), (3, 5, 6)$

31. $(-1, 0, -2), (-1, 5, 2), (-3, -1, 1)$

32. $(4, -1, -1), (2, 0, -4), (3, 5, -1)$

 **Finding the Midpoint** In Exercises 33–36, find the coordinates of the midpoint of the line segment joining the points.

33. $(4, 0, -6), (8, 8, 20)$

34. $(7, 2, 2), (-5, -2, -3)$

35. $(3, 4, 6), (1, 8, 0)$

36. $(5, -9, 7), (-2, 3, 3)$

 Finding the Equation of a Sphere In Exercises 37–42, find the standard equation of the sphere with the given characteristics.

37. Center: $(7, 1, -2)$; Radius: 1

38. Center: $(-1, -5, 8)$; Radius: 5

39. Endpoints of a diameter: $(2, 1, 3), (1, 3, -1)$

40. Endpoints of a diameter: $(-2, 4, -5), (-4, 0, 3)$

41. Center: $(-7, 7, 6)$, tangent to the xy-plane

42. Center: $(-4, 0, 0)$, tangent to the yz-plane

Finding the Equation of a Sphere In Exercises 43–46, complete the square to write the equation of the sphere in standard form. Find the center and radius.

43. $x^2 + y^2 + z^2 - 2x + 6y + 8z + 1 = 0$

44. $x^2 + y^2 + z^2 + 9x - 2y + 10z + 19 = 0$

45. $9x^2 + 9y^2 + 9z^2 - 6x + 18y + 1 = 0$

46. $4x^2 + 4y^2 + 4z^2 - 24x - 4y + 8z - 23 = 0$

Finding the Component Form of a Vector in Space In Exercises 47 and 48, (a) find the component form of the vector **v**, (b) write the vector using standard unit vector notation, and (c) sketch the vector with its initial point at the origin.

47.   48.

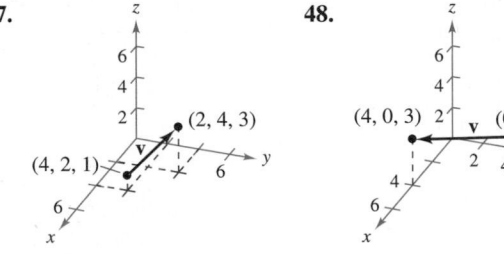

Writing a Vector in Different Forms In Exercises 49 and 50, the initial and terminal points of a vector v are given. (a) Sketch the directed line segment. (b) Find the component form of the vector. (c) Write the vector using standard unit vector notation. (d) Sketch the vector with its initial point at the origin.

49. Initial point: $(-1, 2, 3)$

Terminal point: $(3, 3, 4)$

50. Initial point: $(2, -1, -2)$

Terminal point: $(-4, 3, 7)$

 **Finding the Component Form of a Vector in Space** In Exercises 51–54, find the component form and magnitude of the vector v with the given initial and terminal points. Then find a unit vector in the direction of v.

51. Initial point: $(3, 2, 0)$

Terminal point: $(4, 1, 6)$

52. Initial point: $(1, -2, 4)$

Terminal point: $(2, 4, -2)$

53. Initial point: $(4, 2, 0)$

Terminal point: $(0, 5, 2)$

54. Initial point: $(1, -2, 0)$

Terminal point: $(1, -2, -3)$

Finding a Terminal Point In Exercises 55 and 56, the vector v and its initial point are given. Find the terminal point.

55. $\mathbf{v} = \langle 3, -5, 6 \rangle$

Initial point: $(0, 6, 2)$

56. $\mathbf{v} = \left\langle 1, -\frac{2}{3}, \frac{1}{2} \right\rangle$

Initial point: $\left(0, 2, \frac{5}{2}\right)$

Finding Scalar Multiples In Exercises 57 and 58, find each scalar multiple of v and sketch its graph.

57. $\mathbf{v} = \langle 1, 2, 2 \rangle$

(a) $2\mathbf{v}$ (b) $-\mathbf{v}$

(c) $\frac{3}{2}\mathbf{v}$ (d) $0\mathbf{v}$

58. $\mathbf{v} = \langle 2, -2, 1 \rangle$

(a) $-\mathbf{v}$ (b) $2\mathbf{v}$

(c) $\frac{1}{2}\mathbf{v}$ (d) $\frac{5}{2}\mathbf{v}$

Finding a Vector In Exercises 59–62, find the vector z, given that $\mathbf{u} = \langle 1, 2, 3 \rangle$, $\mathbf{v} = \langle 2, 2, -1 \rangle$, and $\mathbf{w} = \langle 4, 0, -4 \rangle$.

59. $\mathbf{z} = \mathbf{u} - \mathbf{v} + \mathbf{w}$

60. $\mathbf{z} = 5\mathbf{u} - 3\mathbf{v} - \frac{1}{2}\mathbf{w}$

61. $\frac{1}{3}\mathbf{z} - 3\mathbf{u} = \mathbf{w}$

62. $2\mathbf{u} + \mathbf{v} - \mathbf{w} + 3\mathbf{z} = 0$

 Parallel Vectors In Exercises 63–66, determine which of the vectors is/are parallel to z. Use a graphing utility to confirm your results.

63. $\mathbf{z} = \langle 3, 2, -5 \rangle$

(a) $\langle -6, -4, 10 \rangle$

(b) $\left\langle 2, \frac{4}{3}, -\frac{10}{3} \right\rangle$

(c) $\langle 6, 4, 10 \rangle$

(d) $\langle 1, -4, 2 \rangle$

64. $\mathbf{z} = \frac{1}{2}\mathbf{i} - \frac{2}{3}\mathbf{j} + \frac{3}{4}\mathbf{k}$

(a) $6\mathbf{i} - 4\mathbf{j} + 9\mathbf{k}$

(b) $-\mathbf{i} + \frac{4}{3}\mathbf{j} - \frac{3}{2}\mathbf{k}$

(c) $12\mathbf{i} + 9\mathbf{k}$

(d) $\frac{3}{4}\mathbf{i} - \mathbf{j} + \frac{9}{8}\mathbf{k}$

65. z has initial point $(1, -1, 3)$ and terminal point $(-2, 3, 5)$.

(a) $-6\mathbf{i} + 8\mathbf{j} + 4\mathbf{k}$ (b) $4\mathbf{j} + 2\mathbf{k}$

66. z has initial point $(5, 4, 1)$ and terminal point $(-2, -4, 4)$.

(a) $\langle 7, 6, 2 \rangle$ (b) $\langle 14, 16, -6 \rangle$

 **Using Vectors to Determine Collinear Points** In Exercises 67–70, use vectors to determine whether the points are collinear.

67. $(0, -2, -5), (3, 4, 4), (2, 2, 1)$

68. $(4, -2, 7), (-2, 0, 3), (7, -3, 9)$

69. $(1, 2, 4), (2, 5, 0), (0, 1, 5)$

70. $(0, 0, 0), (1, 3, -2), (2, -6, 4)$

Verifying a Parallelogram In Exercises 71 and 72, use vectors to show that the points form the vertices of a parallelogram.

71. $(2, 9, 1), (3, 11, 4), (0, 10, 2), (1, 12, 5)$

72. $(1, 1, 3), (9, -1, -2), (11, 2, -9), (3, 4, -4)$

Finding the Magnitude In Exercises 73–78, find the magnitude of v.

73. $\mathbf{v} = \langle -1, 0, 1 \rangle$

74. $\mathbf{v} = \langle -5, -3, -4 \rangle$

75. $\mathbf{v} = 3\mathbf{j} - 5\mathbf{k}$

76. $\mathbf{v} = 2\mathbf{i} + 5\mathbf{j} - \mathbf{k}$

77. $\mathbf{v} = \mathbf{i} - 2\mathbf{j} - 3\mathbf{k}$

78. $\mathbf{v} = -4\mathbf{i} + 3\mathbf{j} + 7\mathbf{k}$

 **Finding Unit Vectors** In Exercises 79–82, find a unit vector (a) in the direction of v and (b) in the direction opposite of v.

79. $\mathbf{v} = \langle 2, -1, 2 \rangle$

80. $\mathbf{v} = \langle 6, 0, 8 \rangle$

81. $\mathbf{v} = 4\mathbf{i} - 5\mathbf{j} + 3\mathbf{k}$

82. $\mathbf{v} = 5\mathbf{i} + 3\mathbf{j} - \mathbf{k}$

Finding a Vector In Exercises 83–86, find the vector v with the given magnitude and the same direction as u.

Magnitude	Direction
83. $\|\mathbf{v}\| = 10$	$\mathbf{u} = \langle 0, 3, 3 \rangle$
84. $\|\mathbf{v}\| = 3$	$\mathbf{u} = \langle 1, 1, 1 \rangle$
85. $\|\mathbf{v}\| = \frac{3}{2}$	$\mathbf{u} = \langle 2, -2, 1 \rangle$
86. $\|\mathbf{v}\| = 7$	$\mathbf{u} = \langle -4, 6, 2 \rangle$

Sketching a Vector In Exercises 87 and 88, sketch the vector v and write its component form.

87. v lies in the yz-plane, has magnitude 2, and makes an angle of $30°$ with the positive y-axis.

88. v lies in the xz-plane, has magnitude 5, and makes an angle of $45°$ with the positive z-axis.

Finding a Point Using Vectors In Exercises 89 and 90, use vectors to find the point that lies two-thirds of the way from P to Q.

89. $P(4, 3, 0), Q(1, -3, 3)$

90. $P(1, 2, 5), Q(6, 8, 2)$

EXPLORING CONCEPTS

91. Writing The initial and terminal points of the vector **v** are (x_1, y_1, z_1) and (x, y, z). Describe the set of all points (x, y, z) such that $\|\mathbf{v}\| = 4$.

92. Writing Let $\mathbf{r} = \langle x, y, z \rangle$ and $\mathbf{r}_0 = \langle 1, 1, 1 \rangle$. Describe the set of all points (x, y, z) such that $\|\mathbf{r} - \mathbf{r}_0\| = 2$.

93. Writing Let $\mathbf{r} = \langle x, y, z \rangle$. Describe the set of all points (x, y, z) such that $\|\mathbf{r}\| > 1$.

94. HOW DO YOU SEE IT? Determine (x, y, z) for each figure. Then find the component form of the vector from the point on the x-axis to the point (x, y, z).

(a)   (b)

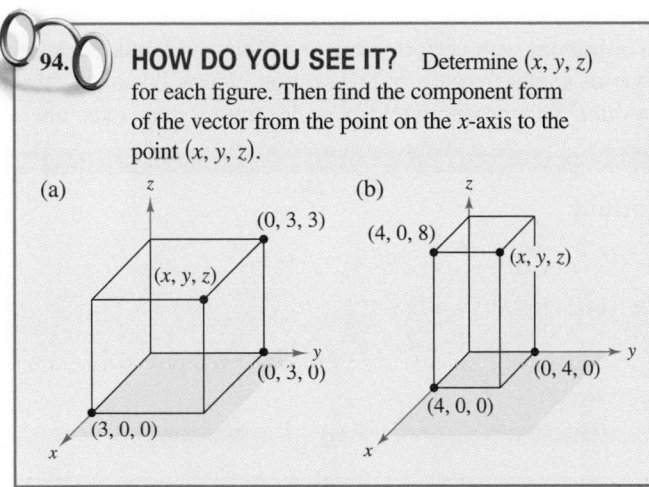

95. Using Vectors Consider two nonzero vectors **u** and **v**, and let s and t be real numbers. Describe the geometric figure generated by connecting the terminal points of the three vectors $t\mathbf{v}$, $\mathbf{u} + t\mathbf{v}$, and $s\mathbf{u} + t\mathbf{v}$.

96. Using Vectors Let $\mathbf{u} = \mathbf{i} + \mathbf{j}$, $\mathbf{v} = \mathbf{j} + \mathbf{k}$, and $\mathbf{w} = a\mathbf{u} + b\mathbf{v}$.

(a) Sketch **u** and **v**.

(b) If $\mathbf{w} = \mathbf{0}$, show that a and b must both be zero.

(c) Find a and b such that $\mathbf{w} = \mathbf{i} + 2\mathbf{j} + \mathbf{k}$.

(d) Show that no choice of a and b yields $\mathbf{w} = \mathbf{i} + 2\mathbf{j} + 3\mathbf{k}$.

97. Diagonal of a Cube Find the component form of the unit vector **v** in the direction of the diagonal of the cube shown in the figure.

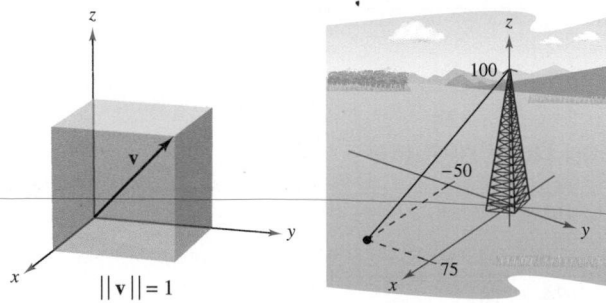

Figure for 97   Figure for 98

98. Tower Guy Wire The guy wire supporting a 100-foot tower has a tension of 550 pounds. Using the distance shown in the figure, write the component form of the vector **F** representing the tension in the wire.

• 99. Auditorium Lights •

The lights in an auditorium are 24-pound discs of radius 18 inches. Each disc is supported by three equally spaced cables that are L inches long (see figure).

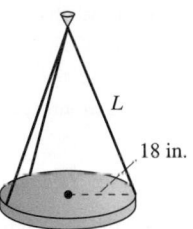 

(a) Write the tension T in each cable as a function of L. Determine the domain of the function.

(b) Use a graphing utility and the function in part (a) to complete the table.

L	20	25	30	35	40	45	50
T							

(c) Use a graphing utility to graph the function in part (a). Determine the asymptotes of the graph.

(d) Confirm the asymptotes of the graph in part (c) analytically.

(e) Determine the minimum length of each cable when a cable is designed to carry a maximum load of 10 pounds.

100. Think About It Suppose the length of each cable in Exercise 99 has a fixed length $L = a$ and the radius of each disc is r_0 inches. Make a conjecture about the limit $\lim\limits_{r_0 \to a^-} T$ and give a reason for your answer.

101. Load Supports Find the tension in each of the supporting cables in the figure when the weight of the crate is 500 newtons.

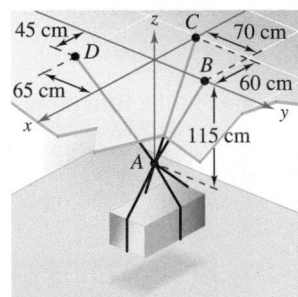 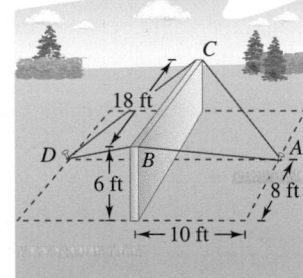

Figure for 101   Figure for 102

102. Construction A precast concrete wall is temporarily kept in its vertical position by ropes (see figure). Find the total force exerted on the pin at position A. The tensions in AB and AC are 420 pounds and 650 pounds, respectively.

103. Geometry Write an equation whose graph consists of the set of points $P(x, y, z)$ that are twice as far from $A(0, -1, 1)$ as from $B(1, 2, 0)$. Describe the geometric figure represented by the equation.

11.3 The Dot Product of Two Vectors

- Use properties of the dot product of two vectors.
- Find the angle between two vectors using the dot product.
- Find the direction cosines of a vector in space.
- Find the projection of a vector onto another vector.
- Use vectors to find the work done by a constant force.

The Dot Product

So far, you have studied two operations with vectors—vector addition and multiplication by a scalar—each of which yields another vector. In this section, you will study a third vector operation, the **dot product.** This product yields a scalar, rather than a vector.

> **REMARK** Because the dot product of two vectors yields a scalar, it is also called the *scalar product* (or *inner product*) of the two vectors.

Definition of Dot Product

The **dot product** of $\mathbf{u} = \langle u_1, u_2 \rangle$ and $\mathbf{v} = \langle v_1, v_2 \rangle$ is

$$\mathbf{u} \cdot \mathbf{v} = u_1v_1 + u_2v_2.$$

The **dot product** of $\mathbf{u} = \langle u_1, u_2, u_3 \rangle$ and $\mathbf{v} = \langle v_1, v_2, v_3 \rangle$ is

$$\mathbf{u} \cdot \mathbf{v} = u_1v_1 + u_2v_2 + u_3v_3.$$

Exploration

Interpreting a Dot Product
Several vectors are shown below on the unit circle. Find the dot products of several pairs of vectors. Then find the angle between each pair that you used. Make a conjecture about the relationship between the dot product of two vectors and the angle between the vectors.

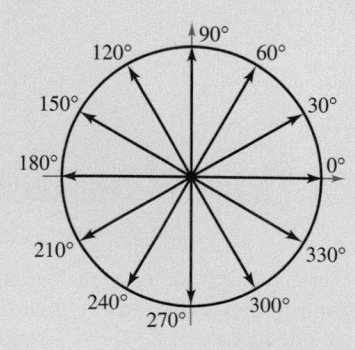

THEOREM 11.4 Properties of the Dot Product

Let $\mathbf{u}$, $\mathbf{v}$, and $\mathbf{w}$ be vectors in the plane or in space and let c be a scalar.

1. $\mathbf{u} \cdot \mathbf{v} = \mathbf{v} \cdot \mathbf{u}$ Commutative Property
2. $\mathbf{u} \cdot (\mathbf{v} + \mathbf{w}) = \mathbf{u} \cdot \mathbf{v} + \mathbf{u} \cdot \mathbf{w}$ Distributive Property
3. $c(\mathbf{u} \cdot \mathbf{v}) = c\mathbf{u} \cdot \mathbf{v} = \mathbf{u} \cdot c\mathbf{v}$ Associative Property
4. $\mathbf{0} \cdot \mathbf{v} = 0$
5. $\mathbf{v} \cdot \mathbf{v} = \|\mathbf{v}\|^2$

Proof To prove the first property, let $\mathbf{u} = \langle u_1, u_2, u_3 \rangle$ and $\mathbf{v} = \langle v_1, v_2, v_3 \rangle$. Then

$$\mathbf{u} \cdot \mathbf{v} = u_1v_1 + u_2v_2 + u_3v_3 = v_1u_1 + v_2u_2 + v_3u_3 = \mathbf{v} \cdot \mathbf{u}.$$

For the fifth property, let $\mathbf{v} = \langle v_1, v_2, v_3 \rangle$. Then

$$\mathbf{v} \cdot \mathbf{v} = v_1^2 + v_2^2 + v_3^2 = \left(\sqrt{v_1^2 + v_2^2 + v_3^2}\right)^2 = \|\mathbf{v}\|^2.$$

Proofs of the other properties are left to you. ∎

EXAMPLE 1   Finding Dot Products

Let $\mathbf{u} = \langle 2, -2 \rangle$, $\mathbf{v} = \langle 5, 8 \rangle$, and $\mathbf{w} = \langle -4, 3 \rangle$.

a. $\mathbf{u} \cdot \mathbf{v} = \langle 2, -2 \rangle \cdot \langle 5, 8 \rangle = 2(5) + (-2)(8) = -6$

b. $(\mathbf{u} \cdot \mathbf{v})\mathbf{w} = -6\langle -4, 3 \rangle = \langle 24, -18 \rangle$

c. $\mathbf{u} \cdot (2\mathbf{v}) = 2(\mathbf{u} \cdot \mathbf{v}) = 2(-6) = -12$

d. $\|\mathbf{w}\|^2 = \mathbf{w} \cdot \mathbf{w} = \langle -4, 3 \rangle \cdot \langle -4, 3 \rangle = (-4)(-4) + (3)(3) = 25$

Notice that the result of part (b) is a *vector* quantity, whereas the results of the other three parts are *scalar* quantities.

Angle Between Two Vectors

The **angle between two nonzero vectors** is the angle θ, $0 \le \theta \le \pi$, between their respective standard position vectors, as shown in Figure 11.24. The next theorem shows how to find this angle using the dot product. (Note that the angle between the zero vector and another vector is not defined here.)

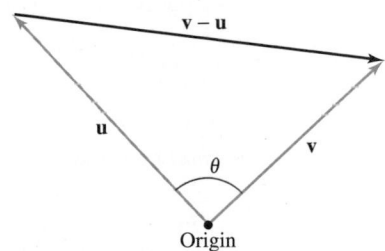

The angle between two vectors
Figure 11.24

THEOREM 11.5   Angle Between Two Vectors

If θ is the angle between two nonzero vectors $\mathbf{u}$ and $\mathbf{v}$, where $0 \le \theta \le \pi$, then

$$\cos \theta = \frac{\mathbf{u} \cdot \mathbf{v}}{\|\mathbf{u}\|\,\|\mathbf{v}\|}.$$

Proof   Consider the triangle determined by vectors $\mathbf{u}$, $\mathbf{v}$, and $\mathbf{v} - \mathbf{u}$, as shown in Figure 11.24. By the Law of Cosines, you can write

$$\|\mathbf{v} - \mathbf{u}\|^2 = \|\mathbf{u}\|^2 + \|\mathbf{v}\|^2 - 2\|\mathbf{u}\|\,\|\mathbf{v}\| \cos \theta.$$

Using the properties of the dot product, the left side can be rewritten as

$$
\begin{aligned}
\|\mathbf{v} - \mathbf{u}\|^2 &= (\mathbf{v} - \mathbf{u}) \cdot (\mathbf{v} - \mathbf{u}) \\
&= (\mathbf{v} - \mathbf{u}) \cdot \mathbf{v} - (\mathbf{v} - \mathbf{u}) \cdot \mathbf{u} \\
&= \mathbf{v} \cdot \mathbf{v} - \mathbf{u} \cdot \mathbf{v} - \mathbf{v} \cdot \mathbf{u} + \mathbf{u} \cdot \mathbf{u} \\
&= \|\mathbf{v}\|^2 - 2\mathbf{u} \cdot \mathbf{v} + \|\mathbf{u}\|^2
\end{aligned}
$$

and substitution back into the Law of Cosines yields

$$
\begin{aligned}
\|\mathbf{v}\|^2 - 2\mathbf{u} \cdot \mathbf{v} + \|\mathbf{u}\|^2 &= \|\mathbf{u}\|^2 + \|\mathbf{v}\|^2 - 2\|\mathbf{u}\|\,\|\mathbf{v}\| \cos \theta \\
-2\mathbf{u} \cdot \mathbf{v} &= -2\|\mathbf{u}\|\,\|\mathbf{v}\| \cos \theta \\
\cos \theta &= \frac{\mathbf{u} \cdot \mathbf{v}}{\|\mathbf{u}\|\,\|\mathbf{v}\|}.
\end{aligned}
$$

Note in Theorem 11.5 that because $\|\mathbf{u}\|$ and $\|\mathbf{v}\|$ are always positive, $\mathbf{u} \cdot \mathbf{v}$ and $\cos \theta$ will always have the same sign. Figure 11.25 shows the possible orientations of two vectors.

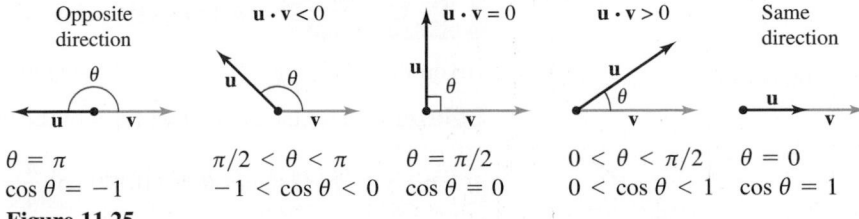

Opposite direction	$\mathbf{u} \cdot \mathbf{v} < 0$	$\mathbf{u} \cdot \mathbf{v} = 0$	$\mathbf{u} \cdot \mathbf{v} > 0$	Same direction
$\theta = \pi$	$\pi/2 < \theta < \pi$	$\theta = \pi/2$	$0 < \theta < \pi/2$	$\theta = 0$
$\cos \theta = -1$	$-1 < \cos \theta < 0$	$\cos \theta = 0$	$0 < \cos \theta < 1$	$\cos \theta = 1$

Figure 11.25

From Theorem 11.5, you can see that two nonzero vectors meet at a right angle if and only if their dot product is zero. Two such vectors are said to be **orthogonal.**

> **Definition of Orthogonal Vectors**
>
> The vectors **u** and **v** are orthogonal when $\mathbf{u} \cdot \mathbf{v} = 0$.

REMARK   The terms "perpendicular," "orthogonal," and "normal" all mean essentially the same thing—meeting at right angles. It is common, however, to say that two vectors are *orthogonal*, two lines or planes are *perpendicular*, and a vector is *normal* to a line or plane.

From this definition, it follows that the zero vector is orthogonal to every vector **u**, because $\mathbf{0} \cdot \mathbf{u} = 0$. Moreover, for $0 \leq \theta \leq \pi$, you know that $\cos \theta = 0$ if and only if $\theta = \pi/2$. So, you can use Theorem 11.5 to conclude that two *nonzero* vectors are orthogonal if and only if the angle between them is $\pi/2$.

EXAMPLE 2   Finding the Angle Between Two Vectors

▷ *See LarsonCalculus.com for an interactive version of this type of example.*

For $\mathbf{u} = \langle 3, -1, 2 \rangle$, $\mathbf{v} = \langle -4, 0, 2 \rangle$, $\mathbf{w} = \langle 1, -1, -2 \rangle$, and $\mathbf{z} = \langle 2, 0, -1 \rangle$, find the angle between each pair of vectors.

a. **u** and **v**   **b.** **u** and **w**   **c.** **v** and **z**

Solution

a. $\cos \theta = \dfrac{\mathbf{u} \cdot \mathbf{v}}{\|\mathbf{u}\| \|\mathbf{v}\|} = \dfrac{-12 + 0 + 4}{\sqrt{14}\sqrt{20}} = \dfrac{-8}{2\sqrt{14}\sqrt{5}} = \dfrac{-4}{\sqrt{70}}$

Because $\mathbf{u} \cdot \mathbf{v} < 0$, $\theta = \arccos \dfrac{-4}{\sqrt{70}} \approx 2.069$ radians.

b. $\cos \theta = \dfrac{\mathbf{u} \cdot \mathbf{w}}{\|\mathbf{u}\| \|\mathbf{w}\|} = \dfrac{3 + 1 - 4}{\sqrt{14}\sqrt{6}} = \dfrac{0}{\sqrt{84}} = 0$

Because $\mathbf{u} \cdot \mathbf{w} = 0$, **u** and **w** are *orthogonal*. So, $\theta = \pi/2$.

c. $\cos \theta = \dfrac{\mathbf{v} \cdot \mathbf{z}}{\|\mathbf{v}\| \|\mathbf{z}\|} = \dfrac{-8 + 0 - 2}{\sqrt{20}\sqrt{5}} = \dfrac{-10}{\sqrt{100}} = -1$

Consequently, $\theta = \pi$. Note that **v** and **z** are parallel, with $\mathbf{v} = -2\mathbf{z}$. ∎

REMARK   The angle between **u** and **v** in Example 3(a) can also be written as approximately 118.561°.

When the angle between two vectors is known, rewriting Theorem 11.5 in the form

$$\mathbf{u} \cdot \mathbf{v} = \|\mathbf{u}\| \|\mathbf{v}\| \cos \theta \qquad \text{Alternative form of dot product}$$

produces an alternative way to calculate the dot product.

EXAMPLE 3   Alternative Form of the Dot Product

Given that $\|\mathbf{u}\| = 10$, $\|\mathbf{v}\| = 7$, and the angle between **u** and **v** is $\pi/4$, find $\mathbf{u} \cdot \mathbf{v}$.

Solution   Use the alternative form of the dot product as shown.

$$\mathbf{u} \cdot \mathbf{v} = \|\mathbf{u}\| \|\mathbf{v}\| \cos \theta = (10)(7) \cos \frac{\pi}{4} = 35\sqrt{2}$$ ∎

Direction Cosines

For a vector in the plane, you have seen that it is convenient to measure direction in terms of the angle, measured counterclockwise, *from* the positive x-axis *to* the vector. In space, it is more convenient to measure direction in terms of the angles *between* the nonzero vector $\mathbf{v}$ and the three unit vectors $\mathbf{i}$, $\mathbf{j}$, and $\mathbf{k}$, as shown in Figure 11.26. The angles α, β, and γ are the **direction angles of v**, and $\cos \alpha$, $\cos \beta$, and $\cos \gamma$ are the **direction cosines of v.** Because

$$\mathbf{v} \cdot \mathbf{i} = \|\mathbf{v}\| \|\mathbf{i}\| \cos \alpha = \|\mathbf{v}\| \cos \alpha$$

and

$$\mathbf{v} \cdot \mathbf{i} = \langle v_1, v_2, v_3 \rangle \cdot \langle 1, 0, 0 \rangle = v_1$$

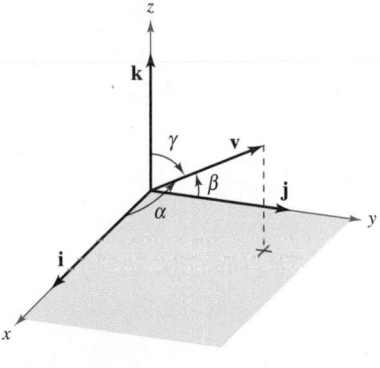

Direction angles
Figure 11.26

it follows that $\cos \alpha = v_1 / \|\mathbf{v}\|$. By similar reasoning with the unit vectors $\mathbf{j}$ and $\mathbf{k}$, you have

$$\cos \alpha = \frac{v_1}{\|\mathbf{v}\|} \qquad \text{α is the angle between $\mathbf{v}$ and $\mathbf{i}$.}$$

$$\cos \beta = \frac{v_2}{\|\mathbf{v}\|} \qquad \text{β is the angle between $\mathbf{v}$ and $\mathbf{j}$.}$$

$$\cos \gamma = \frac{v_3}{\|\mathbf{v}\|}. \qquad \text{γ is the angle between $\mathbf{v}$ and $\mathbf{k}$.}$$

Consequently, any nonzero vector $\mathbf{v}$ in space has the normalized form

$$\frac{\mathbf{v}}{\|\mathbf{v}\|} = \frac{v_1}{\|\mathbf{v}\|}\mathbf{i} + \frac{v_2}{\|\mathbf{v}\|}\mathbf{j} + \frac{v_3}{\|\mathbf{v}\|}\mathbf{k} = \cos \alpha \mathbf{i} + \cos \beta \mathbf{j} + \cos \gamma \mathbf{k}$$

and because $\mathbf{v}/\|\mathbf{v}\|$ is a unit vector, it follows that

$$\cos^2 \alpha + \cos^2 \beta + \cos^2 \gamma = 1.$$

REMARK Recall that α, β, and γ are the Greek letters alpha, beta, and gamma, respectively.

EXAMPLE 4   Finding Direction Angles

Find the direction cosines and angles for the vector $\mathbf{v} = 2\mathbf{i} + 3\mathbf{j} + 4\mathbf{k}$, and show that $\cos^2 \alpha + \cos^2 \beta + \cos^2 \gamma = 1$.

Solution   Because $\|\mathbf{v}\| = \sqrt{2^2 + 3^2 + 4^2} = \sqrt{29}$, you can write the following.

$$\cos \alpha = \frac{v_1}{\|\mathbf{v}\|} = \frac{2}{\sqrt{29}} \implies \alpha \approx 68.2° \qquad \text{Angle between $\mathbf{v}$ and $\mathbf{i}$}$$

$$\cos \beta = \frac{v_2}{\|\mathbf{v}\|} = \frac{3}{\sqrt{29}} \implies \beta \approx 56.1° \qquad \text{Angle between $\mathbf{v}$ and $\mathbf{j}$}$$

$$\cos \gamma = \frac{v_3}{\|\mathbf{v}\|} = \frac{4}{\sqrt{29}} \implies \gamma \approx 42.0° \qquad \text{Angle between $\mathbf{v}$ and $\mathbf{k}$}$$

Furthermore, the sum of the squares of the direction cosines is

$$\cos^2 \alpha + \cos^2 \beta + \cos^2 \gamma = \frac{4}{29} + \frac{9}{29} + \frac{16}{29}$$

$$= \frac{29}{29}$$

$$= 1.$$

See Figure 11.27.

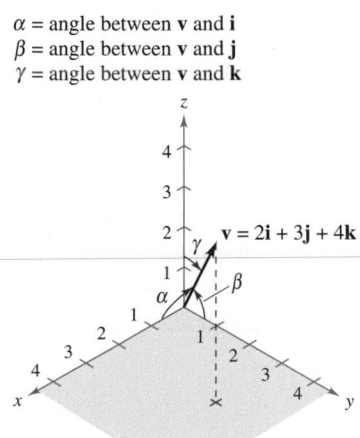

α = angle between $\mathbf{v}$ and $\mathbf{i}$
β = angle between $\mathbf{v}$ and $\mathbf{j}$
γ = angle between $\mathbf{v}$ and $\mathbf{k}$

The direction angles of $\mathbf{v}$
Figure 11.27

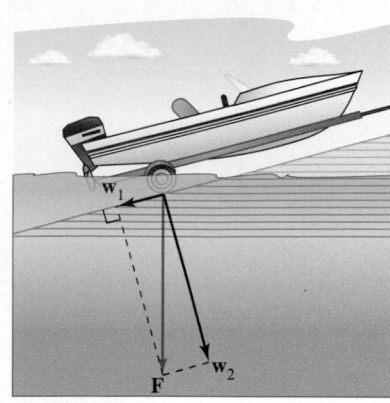

The force due to gravity pulls the boat against the ramp and down the ramp.
Figure 11.28

Projections and Vector Components

You have already seen applications in which two vectors are added to produce a resultant vector. Many applications in physics and engineering pose the reverse problem—decomposing a vector into the sum of two **vector components.** The following physical example enables you to see the usefulness of this procedure.

Consider a boat on an inclined ramp, as shown in Figure 11.28. The force $\mathbf{F}$ due to gravity pulls the boat *down* the ramp and *against* the ramp. These two forces, $\mathbf{w}_1$ and $\mathbf{w}_2$, are orthogonal—they are called the vector components of $\mathbf{F}$.

$$\mathbf{F} = \mathbf{w}_1 + \mathbf{w}_2 \qquad \text{Vector components of } \mathbf{F}$$

The forces $\mathbf{w}_1$ and $\mathbf{w}_2$ help you analyze the effect of gravity on the boat. For example, $\mathbf{w}_1$ indicates the force necessary to keep the boat from rolling down the ramp, whereas $\mathbf{w}_2$ indicates the force that the tires must withstand.

Definitions of Projection and Vector Components

Let $\mathbf{u}$ and $\mathbf{v}$ be nonzero vectors. Moreover, let

$$\mathbf{u} = \mathbf{w}_1 + \mathbf{w}_2$$

where $\mathbf{w}_1$ is parallel to $\mathbf{v}$ and $\mathbf{w}_2$ is orthogonal to $\mathbf{v}$, as shown in Figure 11.29.

1. $\mathbf{w}_1$ is called the **projection of u onto v** or the **vector component of u along v,** and is denoted by $\mathbf{w}_1 = \text{proj}_{\mathbf{v}}\mathbf{u}$.

2. $\mathbf{w}_2 = \mathbf{u} - \mathbf{w}_1$ is called the **vector component of u orthogonal to v.**

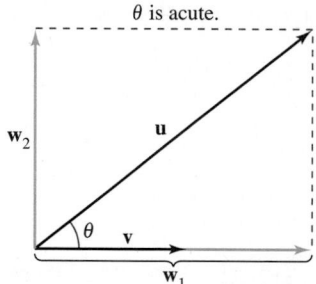
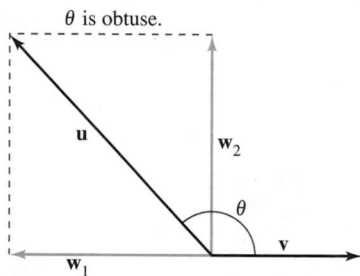

$\mathbf{w}_1 = \text{proj}_{\mathbf{v}}\mathbf{u} = $ projection of $\mathbf{u}$ onto $\mathbf{v} = $ vector component of $\mathbf{u}$ along $\mathbf{v}$
$\mathbf{w}_2 = $ vector component of $\mathbf{u}$ orthogonal to $\mathbf{v}$
Figure 11.29

EXAMPLE 5   **Finding a Vector Component of u Orthogonal to v**

Find the vector component of $\mathbf{u} = \langle 5, 10 \rangle$ that is orthogonal to $\mathbf{v} = \langle 4, 3 \rangle$, given that

$$\mathbf{w}_1 = \text{proj}_{\mathbf{v}}\mathbf{u} = \langle 8, 6 \rangle$$

and

$$\mathbf{u} = \langle 5, 10 \rangle = \mathbf{w}_1 + \mathbf{w}_2.$$

Solution   Because $\mathbf{u} = \mathbf{w}_1 + \mathbf{w}_2$, where $\mathbf{w}_1$ is parallel to $\mathbf{v}$, it follows that $\mathbf{w}_2$ is the vector component of $\mathbf{u}$ orthogonal to $\mathbf{v}$. So, you have

$$\begin{aligned}\mathbf{w}_2 &= \mathbf{u} - \mathbf{w}_1 \\ &= \langle 5, 10 \rangle - \langle 8, 6 \rangle \\ &= \langle -3, 4 \rangle.\end{aligned}$$

Check to see that $\mathbf{w}_2$ is orthogonal to $\mathbf{v}$, as shown in Figure 11.30.

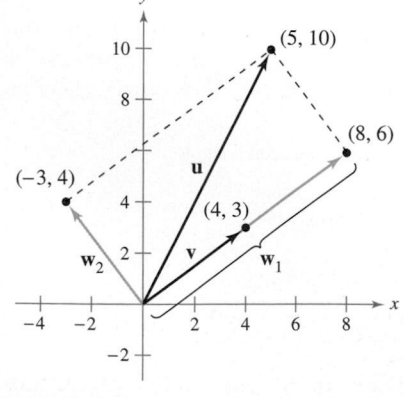

$\mathbf{u} = \mathbf{w}_1 + \mathbf{w}_2$
Figure 11.30

From Example 5, you can see that it is easy to find the vector component $\mathbf{w}_2$ once you have found the projection, $\mathbf{w}_1$, of $\mathbf{u}$ onto $\mathbf{v}$. To find this projection, use the dot product in the next theorem, which you will prove in Exercise 74.

THEOREM 11.6   Projection Using the Dot Product

If $\mathbf{u}$ and $\mathbf{v}$ are nonzero vectors, then the projection of $\mathbf{u}$ onto $\mathbf{v}$ is

$$\text{proj}_{\mathbf{v}}\mathbf{u} = \left(\frac{\mathbf{u} \cdot \mathbf{v}}{\|\mathbf{v}\|^2}\right)\mathbf{v}.$$

• • **REMARK** Note the distinction between the terms "component" and "vector component." For example, using the standard unit vectors with $\mathbf{u} = u_1\mathbf{i} + u_2\mathbf{j}$, u_1 is the *component* of $\mathbf{u}$ in the direction of $\mathbf{i}$, and $u_1\mathbf{i}$ is the *vector component* in the direction of $\mathbf{i}$.

The projection of $\mathbf{u}$ onto $\mathbf{v}$ can be written as a scalar multiple of a unit vector in the direction of $\mathbf{v}$. That is,

$$\left(\frac{\mathbf{u} \cdot \mathbf{v}}{\|\mathbf{v}\|^2}\right)\mathbf{v} = \left(\frac{\mathbf{u} \cdot \mathbf{v}}{\|\mathbf{v}\|}\right)\frac{\mathbf{v}}{\|\mathbf{v}\|} = (k)\frac{\mathbf{v}}{\|\mathbf{v}\|}.$$

The scalar k is called the **component of u in the direction of v.** So,

$$k = \frac{\mathbf{u} \cdot \mathbf{v}}{\|\mathbf{v}\|} = \|\mathbf{u}\| \cos \theta.$$

EXAMPLE 6   **Decomposing a Vector into Vector Components**

Find the projection of $\mathbf{u}$ onto $\mathbf{v}$ and the vector component of $\mathbf{u}$ orthogonal to $\mathbf{v}$ for

$$\mathbf{u} = 3\mathbf{i} - 5\mathbf{j} + 2\mathbf{k} \quad \text{and} \quad \mathbf{v} = 7\mathbf{i} + \mathbf{j} - 2\mathbf{k}.$$

Solution   The projection of $\mathbf{u}$ onto $\mathbf{v}$ is

$$\mathbf{w}_1 = \text{proj}_{\mathbf{v}}\mathbf{u} = \left(\frac{\mathbf{u} \cdot \mathbf{v}}{\|\mathbf{v}\|^2}\right)\mathbf{v} = \left(\frac{12}{54}\right)(7\mathbf{i} + \mathbf{j} - 2\mathbf{k}) = \frac{14}{9}\mathbf{i} + \frac{2}{9}\mathbf{j} - \frac{4}{9}\mathbf{k}.$$

The vector component of $\mathbf{u}$ orthogonal to $\mathbf{v}$ is the vector

$$\mathbf{w}_2 = \mathbf{u} - \mathbf{w}_1 = (3\mathbf{i} - 5\mathbf{j} + 2\mathbf{k}) - \left(\frac{14}{9}\mathbf{i} + \frac{2}{9}\mathbf{j} - \frac{4}{9}\mathbf{k}\right) = \frac{13}{9}\mathbf{i} - \frac{47}{9}\mathbf{j} + \frac{22}{9}\mathbf{k}.$$

See Figure 11.31.

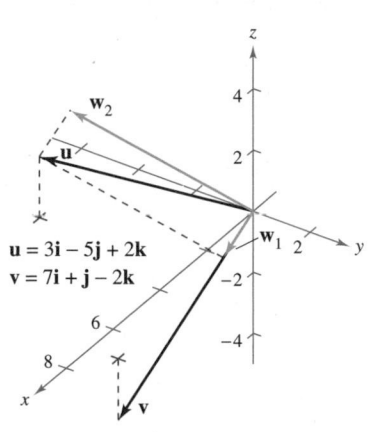

$\mathbf{u} = 3\mathbf{i} - 5\mathbf{j} + 2\mathbf{k}$
$\mathbf{v} = 7\mathbf{i} + \mathbf{j} - 2\mathbf{k}$

$\mathbf{u} = \mathbf{w}_1 + \mathbf{w}_2$
Figure 11.31

EXAMPLE 7   **Finding a Force**

A 600-pound boat sits on a ramp inclined at 30°, as shown in Figure 11.32. What force is required to keep the boat from rolling down the ramp?

Solution   Because the force due to gravity is vertical and downward, you can represent the gravitational force by the vector $\mathbf{F} = -600\mathbf{j}$. To find the force required to keep the boat from rolling down the ramp, project $\mathbf{F}$ onto a unit vector $\mathbf{v}$ in the direction of the ramp, as follows.

$$\mathbf{v} = \cos 30°\mathbf{i} + \sin 30°\mathbf{j} = \frac{\sqrt{3}}{2}\mathbf{i} + \frac{1}{2}\mathbf{j} \qquad \text{Unit vector along ramp}$$

Therefore, the projection of $\mathbf{F}$ onto $\mathbf{v}$ is

$$\mathbf{w}_1 = \text{proj}_{\mathbf{v}}\mathbf{F} = \left(\frac{\mathbf{F} \cdot \mathbf{v}}{\|\mathbf{v}\|^2}\right)\mathbf{v} = (\mathbf{F} \cdot \mathbf{v})\mathbf{v} = (-600)\left(\frac{1}{2}\right)\mathbf{v} = -300\left(\frac{\sqrt{3}}{2}\mathbf{i} + \frac{1}{2}\mathbf{j}\right).$$

The magnitude of this force is 300, so a force of 300 pounds is required to keep the boat from rolling down the ramp.

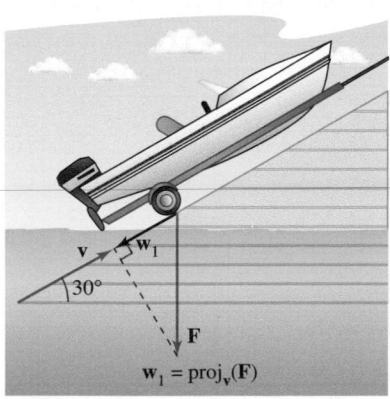

Figure 11.32

Work

The work W done by the constant force $\mathbf{F}$ acting along the line of motion of an object is given by

$$W = (\text{magnitude of force})(\text{distance}) = \|\mathbf{F}\|\|\overrightarrow{PQ}\|$$

as shown in Figure 11.33(a). When the constant force $\mathbf{F}$ is not directed along the line of motion, you can see from Figure 11.33(b) that the work W done by the force is

$$W = \|\text{proj}_{\overrightarrow{PQ}}\mathbf{F}\|\|\overrightarrow{PQ}\| = (\cos\theta)\|\mathbf{F}\|\|\overrightarrow{PQ}\| = \mathbf{F}\cdot\overrightarrow{PQ}.$$

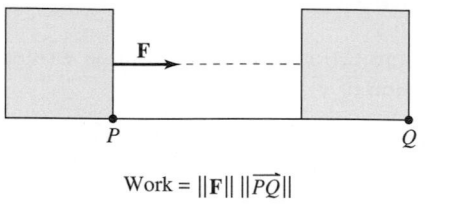

Work $= \|\mathbf{F}\|\|\overrightarrow{PQ}\|$

(a) Force acts along the line of motion.

Figure 11.33

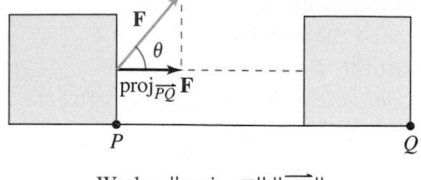

Work $= \|\text{proj}_{\overrightarrow{PQ}}\mathbf{F}\|\|\overrightarrow{PQ}\|$

(b) Force acts at angle θ with the line of motion.

This notion of work is summarized in the next definition.

Definition of Work

The work W done by a constant force $\mathbf{F}$ as its point of application moves along the vector $\overrightarrow{PQ}$ is one of the following.

1. $W = \|\text{proj}_{\overrightarrow{PQ}}\mathbf{F}\|\|\overrightarrow{PQ}\|$ Projection form

2. $W = \mathbf{F}\cdot\overrightarrow{PQ}$ Dot product form

EXAMPLE 8   **Finding Work**

To close a sliding door, a person pulls on a rope with a constant force of 50 pounds at a constant angle of 60°, as shown in Figure 11.34. Find the work done in moving the door 12 feet to its closed position.

Figure 11.34

Solution   Using a projection, you can calculate the work as follows.

$$W = \|\text{proj}_{\overrightarrow{PQ}}\mathbf{F}\|\|\overrightarrow{PQ}\| = \cos(60°)\|\mathbf{F}\|\|\overrightarrow{PQ}\| = \frac{1}{2}(50)(12) = 300 \text{ foot-pounds} \quad \blacksquare$$

11.3 Exercises

See **CalcChat.com** for tutorial help and worked-out solutions to odd-numbered exercises.

CONCEPT CHECK

1. Dot Product What can you say about the relative position of two nonzero vectors if their dot product is zero?

2. Direction Cosines Consider the vector

$$\mathbf{v} = \langle v_1, v_2, v_3 \rangle.$$

What is the meaning of $\arccos \dfrac{v_2}{\|\mathbf{v}\|} = 30°$?

 **Finding Dot Products** In Exercises 3–10, find (a) $\mathbf{u} \cdot \mathbf{v}$, (b) $\mathbf{u} \cdot \mathbf{u}$, (c) $\|\mathbf{v}\|^2$, (d) $(\mathbf{u} \cdot \mathbf{v})\mathbf{v}$, and (e) $\mathbf{u} \cdot (3\mathbf{v})$.

3. $\mathbf{u} = \langle 3, 4 \rangle$, $\mathbf{v} = \langle -1, 5 \rangle$ **4.** $\mathbf{u} = \langle 4, 10 \rangle$, $\mathbf{v} = \langle -2, 3 \rangle$

5. $\mathbf{u} = \langle 6, -4 \rangle$, $\mathbf{v} = \langle -3, 2 \rangle$ **6.** $\mathbf{u} = \langle -7, -1 \rangle$, $\mathbf{v} = \langle -4, -1 \rangle$

7. $\mathbf{u} = \langle 2, -3, 4 \rangle$, $\mathbf{v} = \langle 0, 6, 5 \rangle$

8. $\mathbf{u} = \langle -5, 0, 5 \rangle$, $\mathbf{v} = \langle -1, 2, 1 \rangle$

9. $\mathbf{u} = 2\mathbf{i} - \mathbf{j} + \mathbf{k}$ **10.** $\mathbf{u} = 2\mathbf{i} + \mathbf{j} - 2\mathbf{k}$

$\quad\ \mathbf{v} = \mathbf{i} - \mathbf{k}$ $\quad\ \ \mathbf{v} = \mathbf{i} - 3\mathbf{j} + 2\mathbf{k}$

 **Finding the Angle Between Two Vectors** In Exercises 11–18, find the angle θ between the vectors (a) in radians and (b) in degrees.

11. $\mathbf{u} = \langle 1, 1 \rangle$, $\mathbf{v} = \langle 2, -2 \rangle$

12. $\mathbf{u} = \langle 3, 1 \rangle$, $\mathbf{v} = \langle 2, -1 \rangle$

13. $\mathbf{u} = 3\mathbf{i} + \mathbf{j}$, $\mathbf{v} = -2\mathbf{i} + 4\mathbf{j}$

14. $\mathbf{u} = \cos\left(\dfrac{\pi}{6}\right)\mathbf{i} + \sin\left(\dfrac{\pi}{6}\right)\mathbf{j}$, $\mathbf{v} = \cos\left(\dfrac{3\pi}{4}\right)\mathbf{i} + \sin\left(\dfrac{3\pi}{4}\right)\mathbf{j}$

15. $\mathbf{u} = \langle 1, 1, 1 \rangle$, $\mathbf{v} = \langle 2, 1, -1 \rangle$

16. $\mathbf{u} = 3\mathbf{i} + 2\mathbf{j} + \mathbf{k}$, $\mathbf{v} = 2\mathbf{i} - 3\mathbf{j}$

17. $\mathbf{u} = 3\mathbf{i} + 4\mathbf{j}$, $\mathbf{v} = -2\mathbf{j} + 3\mathbf{k}$

18. $\mathbf{u} = 2\mathbf{i} - 3\mathbf{j} + \mathbf{k}$, $\mathbf{v} = \mathbf{i} - 2\mathbf{j} + \mathbf{k}$

 **Alternative Form of Dot Product** In Exercises 19 and 20, use the alternative form of the dot product to find $\mathbf{u} \cdot \mathbf{v}$.

19. $\|\mathbf{u}\| = 8$, $\|\mathbf{v}\| = 5$, and the angle between $\mathbf{u}$ and $\mathbf{v}$ is $\pi/3$.

20. $\|\mathbf{u}\| = 40$, $\|\mathbf{v}\| = 25$, and the angle between $\mathbf{u}$ and $\mathbf{v}$ is $5\pi/6$.

Comparing Vectors In Exercises 21–26, determine whether $\mathbf{u}$ and $\mathbf{v}$ are orthogonal, parallel, or neither.

21. $\mathbf{u} = \langle 4, 3 \rangle$ **22.** $\mathbf{u} = -\frac{1}{3}(\mathbf{i} - 2\mathbf{j})$

$\quad\ \mathbf{v} = \langle \frac{1}{2}, -\frac{2}{3} \rangle$ $\quad\ \ \mathbf{v} = 2\mathbf{i} - 4\mathbf{j}$

23. $\mathbf{u} = \mathbf{j} + 6\mathbf{k}$ **24.** $\mathbf{u} = -2\mathbf{i} + 3\mathbf{j} - \mathbf{k}$

$\quad\ \mathbf{v} = \mathbf{i} - 2\mathbf{j} - \mathbf{k}$ $\quad\ \ \mathbf{v} = 2\mathbf{i} + \mathbf{j} - \mathbf{k}$

25. $\mathbf{u} = \langle 2, -3, 1 \rangle$ **26.** $\mathbf{u} = \langle \cos\theta, \sin\theta, -1 \rangle$

$\quad\ \mathbf{v} = \langle -1, -1, -1 \rangle$ $\quad\ \ \mathbf{v} = \langle \sin\theta, -\cos\theta, 0 \rangle$

Classifying a Triangle In Exercises 27–30, the vertices of a triangle are given. Determine whether the triangle is an acute triangle, an obtuse triangle, or a right triangle. Explain your reasoning.

27. $(1, 2, 0), (0, 0, 0), (-2, 1, 0)$

28. $(-3, 0, 0), (0, 0, 0), (1, 2, 3)$

29. $(2, 0, 1), (0, 1, 2), (-0.5, 1.5, 0)$

30. $(2, -7, 3), (-1, 5, 8), (4, 6, -1)$

 **Finding Direction Angles** In Exercises 31–36, find the direction cosines and angles of $\mathbf{u}$ and show that $\cos^2\alpha + \cos^2\beta + \cos^2\gamma = 1$.

31. $\mathbf{u} = \mathbf{i} + 2\mathbf{j} + 2\mathbf{k}$ **32.** $\mathbf{u} = 5\mathbf{i} + 3\mathbf{j} - \mathbf{k}$

33. $\mathbf{u} = 7\mathbf{i} + \mathbf{j} - \mathbf{k}$ **34.** $\mathbf{u} = -4\mathbf{i} + 3\mathbf{j} + 5\mathbf{k}$

35. $\mathbf{u} = \langle 0, 6, -4 \rangle$ **36.** $\mathbf{u} = \langle -1, 5, 2 \rangle$

 Finding the Projection of u onto v In Exercises 37–44, (a) find the projection of $\mathbf{u}$ onto $\mathbf{v}$ and (b) find the vector component of $\mathbf{u}$ orthogonal to $\mathbf{v}$.

37. $\mathbf{u} = \langle 6, 7 \rangle$, $\mathbf{v} = \langle 1, 4 \rangle$ **38.** $\mathbf{u} = \langle 9, 7 \rangle$, $\mathbf{v} = \langle 1, 3 \rangle$

39. $\mathbf{u} = 2\mathbf{i} + 3\mathbf{j}$, $\mathbf{v} = 5\mathbf{i} + \mathbf{j}$

40. $\mathbf{u} = 2\mathbf{i} - 3\mathbf{j}$, $\mathbf{v} = 3\mathbf{i} + 2\mathbf{j}$

41. $\mathbf{u} = \langle 0, 3, 3 \rangle$, $\mathbf{v} = \langle -1, 1, 1 \rangle$

42. $\mathbf{u} = \langle 8, 2, 0 \rangle$, $\mathbf{v} = \langle 2, 1, -1 \rangle$

43. $\mathbf{u} = -9\mathbf{i} - 2\mathbf{j} - 4\mathbf{k}$, $\mathbf{v} = 4\mathbf{j} + 4\mathbf{k}$

44. $\mathbf{u} = 5\mathbf{i} - \mathbf{j} - \mathbf{k}$, $\mathbf{v} = -\mathbf{i} + 5\mathbf{j} + 8\mathbf{k}$

EXPLORING CONCEPTS

45. Using Vectors Explain why $\mathbf{u} + \mathbf{v} \cdot \mathbf{w}$ is not defined, where $\mathbf{u}$, $\mathbf{v}$, and $\mathbf{w}$ are nonzero vectors.

46. Projection What can be said about the vectors $\mathbf{u}$ and $\mathbf{v}$ when the projection of $\mathbf{u}$ onto $\mathbf{v}$ equals $\mathbf{u}$?

47. Projection When the projection of $\mathbf{u}$ onto $\mathbf{v}$ has the same magnitude as the projection of $\mathbf{v}$ onto $\mathbf{u}$, can you conclude that $\|\mathbf{u}\| = \|\mathbf{v}\|$? Explain.

 48. **HOW DO YOU SEE IT?** What is known about θ, the angle between two nonzero vectors $\mathbf{u}$ and $\mathbf{v}$, when

(a) $\mathbf{u} \cdot \mathbf{v} = 0$?   (b) $\mathbf{u} \cdot \mathbf{v} > 0$?   (c) $\mathbf{u} \cdot \mathbf{v} < 0$?

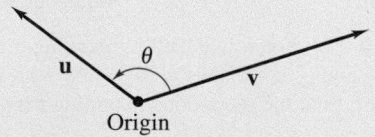

49. Revenue The vector $\mathbf{u} = \langle 3240, 1450, 2235 \rangle$ gives the numbers of hamburgers, chicken sandwiches, and cheeseburgers, respectively, sold at a fast-food restaurant in one week. The vector $\mathbf{v} = \langle 2.25, 2.95, 2.65 \rangle$ gives the prices (in dollars) per unit for the three food items. Find the dot product $\mathbf{u} \cdot \mathbf{v}$ and explain what information it gives.

50. Revenue Repeat Exercise 49 after decreasing the prices by 2%. Identify the vector operation used to decrease the prices by 2%.

Orthogonal Vectors In Exercises 51–54, find two vectors in opposite directions that are orthogonal to the vector u. (The answers are not unique.)

51. $\mathbf{u} = -\frac{1}{4}\mathbf{i} + \frac{3}{2}\mathbf{j}$

52. $\mathbf{u} = 9\mathbf{i} - 4\mathbf{j}$

53. $\mathbf{u} = \langle 3, 1, -2 \rangle$

54. $\mathbf{u} = \langle 4, -3, 6 \rangle$

55. Finding an Angle Find the angle between a cube's diagonal and one of its edges.

56. Finding an Angle Find the angle between the diagonal of a cube and the diagonal of one of its sides.

57. Braking Load A 48,000-pound truck is parked on a 10° slope (see figure). Assume the only force to overcome is that due to gravity. Find (a) the force required to keep the truck from rolling down the hill and (b) the force perpendicular to the hill.

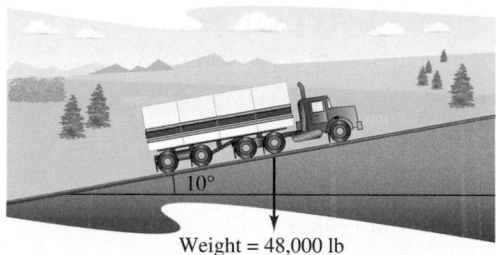

Weight = 48,000 lb

58. Braking Load A 5400-pound sport utility vehicle is parked on an 18° slope. Assume the only force to overcome is that due to gravity. Find (a) the force required to keep the vehicle from rolling down the hill and (b) the force perpendicular to the hill.

59. Work An object is pulled 10 feet across a floor using a force of 85 pounds. The direction of the force is 60° above the horizontal (see figure). Find the work done.

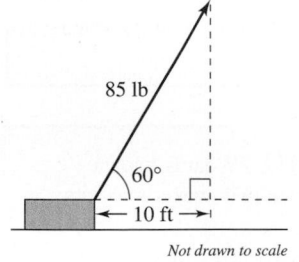

85 lb

60°

← 10 ft →

Not drawn to scale

Figure for 59

Figure for 60

60. Work A wagon is pulled by exerting a force of 65 pounds on a handle that makes a 20° angle with the horizontal (see figure). Find the work done in pulling the wagon 50 feet.

61. Work A car is towed using a force of 1600 newtons. The chain used to pull the car makes a 25° angle with the horizontal. Find the work done in towing the car 2 kilometers.

62. Work

A pallet truck is pulled by exerting a force of 400 newtons on a handle that makes a 60° angle with the horizontal. Find the work done in pulling the truck 40 meters.

True or False? In Exercises 63 and 64, determine whether the statement is true or false. If it is false, explain why or give an example that shows it is false.

63. If $\mathbf{u} \cdot \mathbf{v} = \mathbf{u} \cdot \mathbf{w}$ and $\mathbf{u} \neq \mathbf{0}$, then $\mathbf{v} = \mathbf{w}$.

64. If $\mathbf{u}$ and $\mathbf{v}$ are orthogonal to $\mathbf{w}$, then $\mathbf{u} + \mathbf{v}$ is orthogonal to $\mathbf{w}$.

Using Points of Intersection In Exercises 65–68, (a) find all points of intersection of the graphs of the two equations, (b) find the unit tangent vectors to each curve at their points of intersection, and (c) find the angles ($0° \leq \theta \leq 90°$) between the curves at their points of intersection.

65. $y = x^2$, $y = x^{1/3}$

66. $y = x^3$, $y = x^{1/3}$

67. $y = 1 - x^2$, $y = x^2 - 1$

68. $(y + 1)^2 = x$, $y = x^3 - 1$

69. Proof Use vectors to prove that the diagonals of a rhombus are perpendicular.

70. Proof Use vectors to prove that a parallelogram is a rectangle if and only if its diagonals are equal in length.

71. Bond Angle Consider a regular tetrahedron with vertices $(0, 0, 0)$, $(k, k, 0)$, $(k, 0, k)$, and $(0, k, k)$, where k is a positive real number.

(a) Sketch the graph of the tetrahedron.

(b) Find the length of each edge.

(c) Find the angle between any two edges.

(d) Find the angle between the line segments from the centroid $(k/2, k/2, k/2)$ to two vertices. This is the bond angle for a molecule, such as CH_4 (methane) or $PbCl_4$ (lead tetrachloride), where the structure of the molecule is a tetrahedron.

72. Proof Consider the vectors $\mathbf{u} = \langle \cos \alpha, \sin \alpha, 0 \rangle$ and $\mathbf{v} = \langle \cos \beta, \sin \beta, 0 \rangle$, where $\alpha > \beta$. Find the dot product of the vectors and use the result to prove the identity

$$\cos(\alpha - \beta) = \cos \alpha \cos \beta + \sin \alpha \sin \beta.$$

73. Proof Prove the triangle inequality $\|\mathbf{u} + \mathbf{v}\| \leq \|\mathbf{u}\| + \|\mathbf{v}\|$.

74. Proof Prove Theorem 11.6.

75. Proof Prove the **Cauchy-Schwarz Inequality**,

$$|\mathbf{u} \cdot \mathbf{v}| \leq \|\mathbf{u}\| \|\mathbf{v}\|.$$

11.4 The Cross Product of Two Vectors in Space

■ Find the cross product of two vectors in space.
■ Use the triple scalar product of three vectors in space.

The Cross Product

Many applications in physics, engineering, and geometry involve finding a vector in space that is orthogonal to two given vectors. In this section, you will study a product that will yield such a vector. It is called the **cross product,** and it is most conveniently defined and calculated using the standard unit vector form. Because the cross product yields a vector, it is also called the **vector product.**

Definition of Cross Product of Two Vectors in Space

Let

$$\mathbf{u} = u_1\mathbf{i} + u_2\mathbf{j} + u_3\mathbf{k} \quad \text{and} \quad \mathbf{v} = v_1\mathbf{i} + v_2\mathbf{j} + v_3\mathbf{k}$$

be vectors in space. The **cross product** of $\mathbf{u}$ and $\mathbf{v}$ is the vector

$$\mathbf{u} \times \mathbf{v} = (u_2v_3 - u_3v_2)\mathbf{i} - (u_1v_3 - u_3v_1)\mathbf{j} + (u_1v_2 - u_2v_1)\mathbf{k}.$$

It is important to note that this definition applies only to three-dimensional vectors. The cross product is not defined for two-dimensional vectors.

A convenient way to calculate $\mathbf{u} \times \mathbf{v}$ is to use the *determinant form* with cofactor expansion shown below. (This 3×3 determinant form is used simply to help remember the formula for the cross product. The corresponding array is technically not a matrix because its entries are not all numbers.)

$$\mathbf{u} \times \mathbf{v} = \begin{vmatrix} \mathbf{i} & \mathbf{j} & \mathbf{k} \\ u_1 & u_2 & u_3 \\ v_1 & v_2 & v_3 \end{vmatrix} \quad \begin{matrix} \longleftarrow \text{Put "}\mathbf{u}\text{" in Row 2.} \\ \longleftarrow \text{Put "}\mathbf{v}\text{" in Row 3.} \end{matrix}$$

$$= \begin{vmatrix} \mathbf{i} & \mathbf{j} & \mathbf{k} \\ u_1 & u_2 & u_3 \\ v_1 & v_2 & v_3 \end{vmatrix}\mathbf{i} - \begin{vmatrix} \mathbf{i} & \mathbf{j} & \mathbf{k} \\ u_1 & u_2 & u_3 \\ v_1 & v_2 & v_3 \end{vmatrix}\mathbf{j} + \begin{vmatrix} \mathbf{i} & \mathbf{j} & \mathbf{k} \\ u_1 & u_2 & u_3 \\ v_1 & v_2 & v_3 \end{vmatrix}\mathbf{k}$$

$$= \begin{vmatrix} u_2 & u_3 \\ v_2 & v_3 \end{vmatrix}\mathbf{i} - \begin{vmatrix} u_1 & u_3 \\ v_1 & v_3 \end{vmatrix}\mathbf{j} + \begin{vmatrix} u_1 & u_2 \\ v_1 & v_2 \end{vmatrix}\mathbf{k}$$

$$= (u_2v_3 - u_3v_2)\mathbf{i} - (u_1v_3 - u_3v_1)\mathbf{j} + (u_1v_2 - u_2v_1)\mathbf{k}$$

Note the minus sign in front of the $\mathbf{j}$-component. Each of the three 2×2 determinants can be evaluated by using the diagonal pattern

$$\begin{vmatrix} a & b \\ c & d \end{vmatrix} = ad - bc.$$

Here are a couple of examples.

$$\begin{vmatrix} 2 & 4 \\ 3 & -1 \end{vmatrix} = (2)(-1) - (4)(3) = -2 - 12 = -14$$

and

$$\begin{vmatrix} 4 & 0 \\ -6 & 3 \end{vmatrix} = (4)(3) - (0)(-6) = 12$$

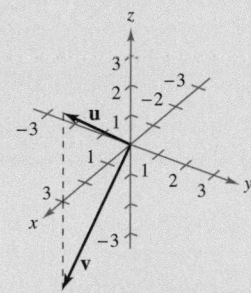
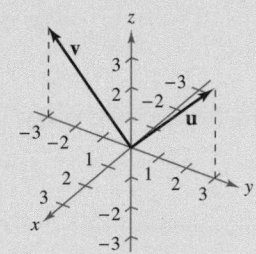
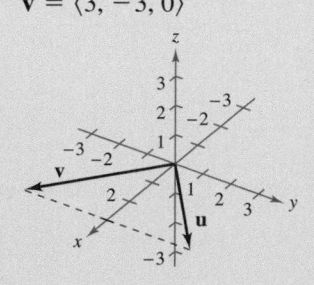

EXAMPLE 1 **Finding the Cross Product**

For $\mathbf{u} = \mathbf{i} - 2\mathbf{j} + \mathbf{k}$ and $\mathbf{v} = 3\mathbf{i} + \mathbf{j} - 2\mathbf{k}$, find each of the following.

a. $\mathbf{u} \times \mathbf{v}$ **b.** $\mathbf{v} \times \mathbf{u}$ **c.** $\mathbf{v} \times \mathbf{v}$

Solution

a. $\mathbf{u} \times \mathbf{v} = \begin{vmatrix} \mathbf{i} & \mathbf{j} & \mathbf{k} \\ 1 & -2 & 1 \\ 3 & 1 & -2 \end{vmatrix}$

$= \begin{vmatrix} -2 & 1 \\ 1 & -2 \end{vmatrix}\mathbf{i} - \begin{vmatrix} 1 & 1 \\ 3 & -2 \end{vmatrix}\mathbf{j} + \begin{vmatrix} 1 & -2 \\ 3 & 1 \end{vmatrix}\mathbf{k}$

$= (4 - 1)\mathbf{i} - (-2 - 3)\mathbf{j} + (1 + 6)\mathbf{k}$

$= 3\mathbf{i} + 5\mathbf{j} + 7\mathbf{k}$

b. $\mathbf{v} \times \mathbf{u} = \begin{vmatrix} \mathbf{i} & \mathbf{j} & \mathbf{k} \\ 3 & 1 & -2 \\ 1 & -2 & 1 \end{vmatrix}$

$= \begin{vmatrix} 1 & -2 \\ -2 & 1 \end{vmatrix}\mathbf{i} - \begin{vmatrix} 3 & -2 \\ 1 & 1 \end{vmatrix}\mathbf{j} + \begin{vmatrix} 3 & 1 \\ 1 & -2 \end{vmatrix}\mathbf{k}$

$= (1 - 4)\mathbf{i} - (3 + 2)\mathbf{j} + (-6 - 1)\mathbf{k}$

$= -3\mathbf{i} - 5\mathbf{j} - 7\mathbf{k}$

c. $\mathbf{v} \times \mathbf{v} = \begin{vmatrix} \mathbf{i} & \mathbf{j} & \mathbf{k} \\ 3 & 1 & -2 \\ 3 & 1 & -2 \end{vmatrix} = \mathbf{0}$

> **•• REMARK** Note that this result is the negative of that in part (a).

The results obtained in Example 1 suggest some interesting *algebraic* properties of the cross product. For instance, $\mathbf{u} \times \mathbf{v} = -(\mathbf{v} \times \mathbf{u})$, and $\mathbf{v} \times \mathbf{v} = \mathbf{0}$. These properties, and several others, are summarized in the next theorem.

THEOREM 11.7 **Algebraic Properties of the Cross Product**

Let $\mathbf{u}$, $\mathbf{v}$, and $\mathbf{w}$ be vectors in space, and let c be a scalar.

1. $\mathbf{u} \times \mathbf{v} = -(\mathbf{v} \times \mathbf{u})$
2. $\mathbf{u} \times (\mathbf{v} + \mathbf{w}) = (\mathbf{u} \times \mathbf{v}) + (\mathbf{u} \times \mathbf{w})$
3. $c(\mathbf{u} \times \mathbf{v}) = (c\mathbf{u}) \times \mathbf{v} = \mathbf{u} \times (c\mathbf{v})$
4. $\mathbf{u} \times \mathbf{0} = \mathbf{0} \times \mathbf{u} = \mathbf{0}$
5. $\mathbf{u} \times \mathbf{u} = \mathbf{0}$
6. $\mathbf{u} \cdot (\mathbf{v} \times \mathbf{w}) = (\mathbf{u} \times \mathbf{v}) \cdot \mathbf{w}$

Proof To prove Property 1, let $\mathbf{u} = u_1\mathbf{i} + u_2\mathbf{j} + u_3\mathbf{k}$ and $\mathbf{v} = v_1\mathbf{i} + v_2\mathbf{j} + v_3\mathbf{k}$. Then

$$\mathbf{u} \times \mathbf{v} = (u_2v_3 - u_3v_2)\mathbf{i} - (u_1v_3 - u_3v_1)\mathbf{j} + (u_1v_2 - u_2v_1)\mathbf{k}$$

and

$$\mathbf{v} \times \mathbf{u} = (v_2u_3 - v_3u_2)\mathbf{i} - (v_1u_3 - v_3u_1)\mathbf{j} + (v_1u_2 - v_2u_1)\mathbf{k}$$

which implies that $\mathbf{u} \times \mathbf{v} = -(\mathbf{v} \times \mathbf{u})$. Proofs of Properties 2, 3, 5, and 6 are left as exercises (see Exercises 47–50).

Note that Property 1 of Theorem 11.7 indicates that the cross product is *not commutative*. In particular, this property indicates that the vectors $\mathbf{u} \times \mathbf{v}$ and $\mathbf{v} \times \mathbf{u}$ have equal lengths but opposite directions. The next theorem lists some other *geometric* properties of the cross product of two vectors.

> **THEOREM 11.8 Geometric Properties of the Cross Product**
>
> Let $\mathbf{u}$ and $\mathbf{v}$ be nonzero vectors in space, and let θ be the angle between $\mathbf{u}$ and $\mathbf{v}$.
>
> 1. $\mathbf{u} \times \mathbf{v}$ is orthogonal to both $\mathbf{u}$ and $\mathbf{v}$.
> 2. $\|\mathbf{u} \times \mathbf{v}\| - \|\mathbf{u}\|\|\mathbf{v}\| \sin \theta$
> 3. $\mathbf{u} \times \mathbf{v} = \mathbf{0}$ if and only if $\mathbf{u}$ and $\mathbf{v}$ are scalar multiples of each other.
> 4. $\|\mathbf{u} \times \mathbf{v}\| =$ area of parallelogram having $\mathbf{u}$ and $\mathbf{v}$ as adjacent sides.

• • • • • • • • • • • • • • • • • • ▷
• **REMARK** It follows from Properties 1 and 2 in Theorem 11.8 that if $\mathbf{n}$ is a unit vector orthogonal to both $\mathbf{u}$ and $\mathbf{v}$, then

$$\mathbf{u} \times \mathbf{v} = \pm(\|\mathbf{u}\|\|\mathbf{v}\| \sin \theta)\mathbf{n}.$$

Proof To prove Property 2, note because $\cos \theta = (\mathbf{u} \cdot \mathbf{v})/(\|\mathbf{u}\|\|\mathbf{v}\|)$, it follows that

$$\|\mathbf{u}\|\|\mathbf{v}\| \sin \theta = \|\mathbf{u}\|\|\mathbf{v}\|\sqrt{1 - \cos^2 \theta}$$

$$= \|\mathbf{u}\|\|\mathbf{v}\|\sqrt{1 - \frac{(\mathbf{u} \cdot \mathbf{v})^2}{\|\mathbf{u}\|^2\|\mathbf{v}\|^2}}$$

$$= \sqrt{\|\mathbf{u}\|^2\|\mathbf{v}\|^2 - (\mathbf{u} \cdot \mathbf{v})^2}$$

$$= \sqrt{(u_1^2 + u_2^2 + u_3^2)(v_1^2 + v_2^2 + v_3^2) - (u_1v_1 + u_2v_2 + u_3v_3)^2}$$

$$= \sqrt{(u_2v_3 - u_3v_2)^2 + (u_1v_3 - u_3v_1)^2 + (u_1v_2 - u_2v_1)^2}$$

$$= \|\mathbf{u} \times \mathbf{v}\|.$$

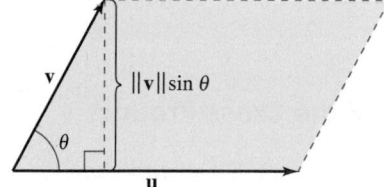

The vectors $\mathbf{u}$ and $\mathbf{v}$ form adjacent sides of a parallelogram.
Figure 11.35

To prove Property 4, refer to Figure 11.35, which is a parallelogram having $\mathbf{v}$ and $\mathbf{u}$ as adjacent sides. Because the height of the parallelogram is $\|\mathbf{v}\| \sin \theta$, the area is

$$\text{Area} = (\text{base})(\text{height})$$

$$= \|\mathbf{u}\|\|\mathbf{v}\| \sin \theta$$

$$= \|\mathbf{u} \times \mathbf{v}\|.$$

Proofs of Properties 1 and 3 are left as exercises (see Exercises 51 and 52). ∎

Both $\mathbf{u} \times \mathbf{v}$ and $\mathbf{v} \times \mathbf{u}$ are perpendicular to the plane determined by $\mathbf{u}$ and $\mathbf{v}$. One way to remember the orientations of the vectors $\mathbf{u}$, $\mathbf{v}$, and $\mathbf{u} \times \mathbf{v}$ is to compare them with the unit vectors $\mathbf{i}$, $\mathbf{j}$, and $\mathbf{k} = \mathbf{i} \times \mathbf{j}$, as shown in Figure 11.36. The three vectors $\mathbf{u}$, $\mathbf{v}$, and $\mathbf{u} \times \mathbf{v}$ form a *right-handed system*, whereas the three vectors $\mathbf{u}$, $\mathbf{v}$, and $\mathbf{v} \times \mathbf{u}$ form a *left-handed system*.

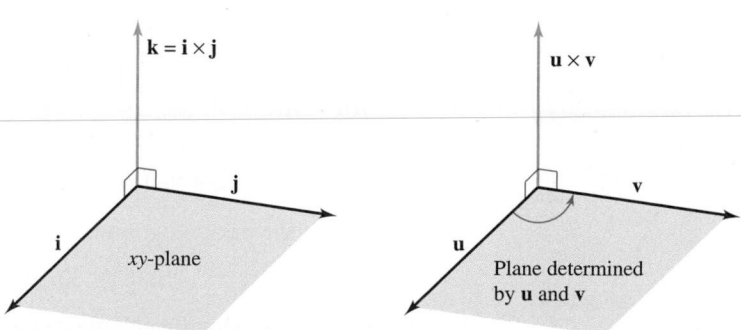

Right-handed systems
Figure 11.36

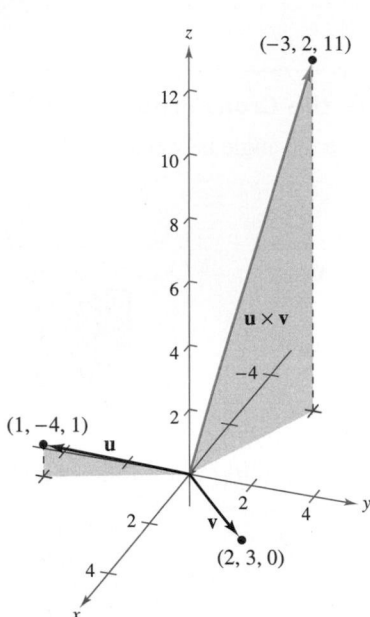

The vector $\mathbf{u} \times \mathbf{v}$ is orthogonal to both $\mathbf{u}$ and $\mathbf{v}$.
Figure 11.37

EXAMPLE 2 **Using the Cross Product**

⋮ ⋯▷ *See LarsonCalculus.com for an interactive version of this type of example.*

Find a unit vector that is orthogonal to both

$$\mathbf{u} = \mathbf{i} - 4\mathbf{j} + \mathbf{k}$$

and

$$\mathbf{v} = 2\mathbf{i} + 3\mathbf{j}.$$

Solution The cross product $\mathbf{u} \times \mathbf{v}$, as shown in Figure 11.37, is orthogonal to both $\mathbf{u}$ and $\mathbf{v}$.

$$\mathbf{u} \times \mathbf{v} = \begin{vmatrix} \mathbf{i} & \mathbf{j} & \mathbf{k} \\ 1 & -4 & 1 \\ 2 & 3 & 0 \end{vmatrix} \qquad \text{Cross product}$$

$$= -3\mathbf{i} + 2\mathbf{j} + 11\mathbf{k}$$

Because

$$\|\mathbf{u} \times \mathbf{v}\| = \sqrt{(-3)^2 + 2^2 + 11^2} = \sqrt{134}$$

a unit vector orthogonal to both $\mathbf{u}$ and $\mathbf{v}$ is

$$\frac{\mathbf{u} \times \mathbf{v}}{\|\mathbf{u} \times \mathbf{v}\|} = -\frac{3}{\sqrt{134}}\mathbf{i} + \frac{2}{\sqrt{134}}\mathbf{j} + \frac{11}{\sqrt{134}}\mathbf{k}. \qquad ■$$

In Example 2, note that you could have used the cross product $\mathbf{v} \times \mathbf{u}$ to form a unit vector that is orthogonal to both $\mathbf{u}$ and $\mathbf{v}$. With that choice, you would have obtained the negative of the unit vector found in the example.

EXAMPLE 3 **Geometric Application of the Cross Product**

The vertices of a quadrilateral are listed below. Show that the quadrilateral is a parallelogram and find its area.

$$A = (5, 2, 0) \qquad B = (2, 6, 1)$$
$$C = (2, 4, 7) \qquad D = (5, 0, 6)$$

Solution From Figure 11.38, you can see that the sides of the quadrilateral correspond to the following four vectors.

$$\overrightarrow{AB} = -3\mathbf{i} + 4\mathbf{j} + \mathbf{k} \qquad \overrightarrow{CD} = 3\mathbf{i} - 4\mathbf{j} - \mathbf{k} = -\overrightarrow{AB}$$
$$\overrightarrow{AD} = 0\mathbf{i} - 2\mathbf{j} + 6\mathbf{k} \qquad \overrightarrow{CB} = 0\mathbf{i} + 2\mathbf{j} - 6\mathbf{k} = -\overrightarrow{AD}$$

So, $\overrightarrow{AB}$ is parallel to $\overrightarrow{CD}$ and $\overrightarrow{AD}$ is parallel to $\overrightarrow{CB}$, and you can conclude that the quadrilateral is a parallelogram with $\overrightarrow{AB}$ and $\overrightarrow{AD}$ as adjacent sides. Moreover, because

$$\overrightarrow{AB} \times \overrightarrow{AD} = \begin{vmatrix} \mathbf{i} & \mathbf{j} & \mathbf{k} \\ -3 & 4 & 1 \\ 0 & -2 & 6 \end{vmatrix} \qquad \text{Cross product}$$

$$= 26\mathbf{i} + 18\mathbf{j} + 6\mathbf{k}$$

the area of the parallelogram is

$$\|\overrightarrow{AB} \times \overrightarrow{AD}\| = \sqrt{1036} \approx 32.19.$$

Is the parallelogram a rectangle? You can determine whether it is by finding the angle between the vectors $\overrightarrow{AB}$ and $\overrightarrow{AD}$. ■

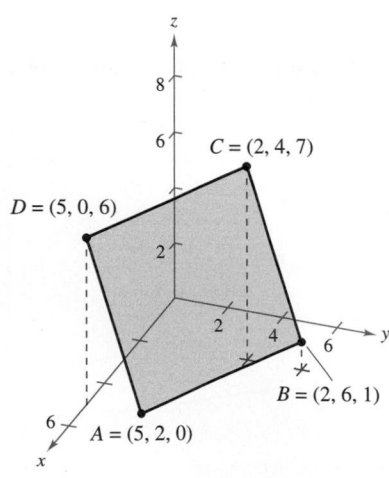

The area of the parallelogram is approximately 32.19.
Figure 11.38

In physics, the cross product can be used to measure **torque**—the **moment M of a force F about a point P,** as shown in Figure 11.39. If the point of application of the force is Q, then the moment of **F** about P is

$$\mathbf{M} = \overrightarrow{PQ} \times \mathbf{F}. \qquad \text{Moment of } \mathbf{F} \text{ about } P$$

The magnitude of the moment **M** measures the tendency of the vector $\overrightarrow{PQ}$ to rotate counterclockwise (using the right-hand rule) about an axis directed along the vector **M**.

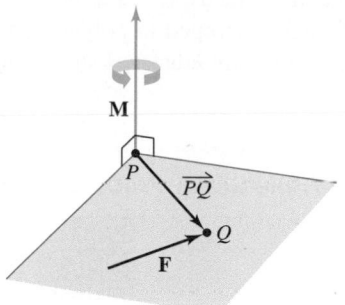

The moment of **F** about P

Figure 11.39

EXAMPLE 4 An Application of the Cross Product

A vertical force of 50 pounds is applied to the end of a one-foot lever that is attached to an axle at point P, as shown in Figure 11.40. Find the moment of this force about the point P when $\theta = 60°$.

Solution Represent the 50-pound force as

$$\mathbf{F} = -50\mathbf{k}$$

and the lever as

$$\overrightarrow{PQ} = \cos(60°)\mathbf{j} + \sin(60°)\mathbf{k} = \frac{1}{2}\mathbf{j} + \frac{\sqrt{3}}{2}\mathbf{k}.$$

The moment of **F** about P is

$$\mathbf{M} = \overrightarrow{PQ} \times \mathbf{F} = \begin{vmatrix} \mathbf{i} & \mathbf{j} & \mathbf{k} \\ 0 & \dfrac{1}{2} & \dfrac{\sqrt{3}}{2} \\ 0 & 0 & -50 \end{vmatrix} = -25\mathbf{i}. \qquad \text{Moment of } \mathbf{F} \text{ about } P$$

The magnitude of this moment is 25 foot-pounds. ■

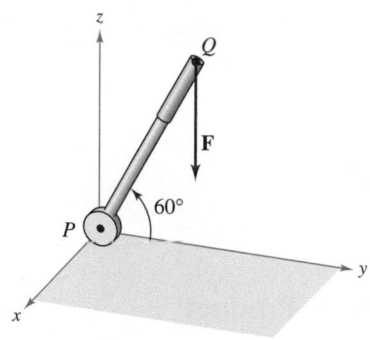

A vertical force of 50 pounds is applied at point Q.

Figure 11.40

In Example 4, note that the moment (the tendency of the lever to rotate about its axle) is dependent on the angle θ. When $\theta = \pi/2$, the moment is 0. The moment is greatest when $\theta = 0$.

The Triple Scalar Product

For vectors **u**, **v**, and **w** in space, the dot product of **u** and $\mathbf{v} \times \mathbf{w}$

$$\mathbf{u} \cdot (\mathbf{v} \times \mathbf{w})$$

is called the **triple scalar product,** as defined in Theorem 11.9. The proof of this theorem is left as an exercise (see Exercise 55).

THEOREM 11.9 The Triple Scalar Product

For $\mathbf{u} = u_1\mathbf{i} + u_2\mathbf{j} + u_3\mathbf{k}$, $\mathbf{v} = v_1\mathbf{i} + v_2\mathbf{j} + v_3\mathbf{k}$, and $\mathbf{w} = w_1\mathbf{i} + w_2\mathbf{j} + w_3\mathbf{k}$, the triple scalar product is

$$\mathbf{u} \cdot (\mathbf{v} \times \mathbf{w}) = \begin{vmatrix} u_1 & u_2 & u_3 \\ v_1 & v_2 & v_3 \\ w_1 & w_2 & w_3 \end{vmatrix}.$$

Note that the value of a determinant is multiplied by -1 when two rows are interchanged. After two such interchanges, the value of the determinant will be unchanged. So, the following triple scalar products are equivalent.

$$\mathbf{u} \cdot (\mathbf{v} \times \mathbf{w}) = \mathbf{v} \cdot (\mathbf{w} \times \mathbf{u}) = \mathbf{w} \cdot (\mathbf{u} \times \mathbf{v})$$

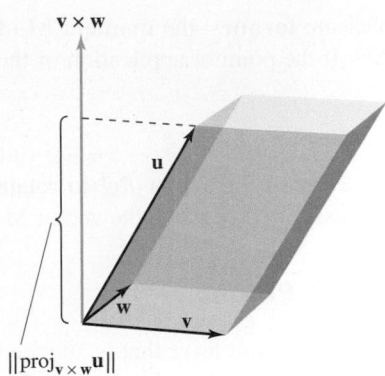

$\|proj_{v \times w}u\|$

Area of base = $\|v \times w\|$
Volume of parallelepiped = $|u \cdot (v \times w)|$
Figure 11.41

If the vectors **u**, **v**, and **w** do not lie in the same plane, then the triple scalar product **u** · (**v** × **w**) can be used to determine the volume of the parallelepiped (a polyhedron, all of whose faces are parallelograms) with **u**, **v**, and **w** as adjacent edges, as shown in Figure 11.41. This is established in the next theorem.

THEOREM 11.10 Geometric Property of the Triple Scalar Product

The volume V of a parallelepiped with vectors **u**, **v**, and **w** as adjacent edges is

$$V = |\mathbf{u} \cdot (\mathbf{v} \times \mathbf{w})|.$$

Proof In Figure 11.41, note that the area of the base is $\|\mathbf{v} \times \mathbf{w}\|$ and the height of the parallelepiped is $\|proj_{\mathbf{v} \times \mathbf{w}}\mathbf{u}\|$. Therefore, the volume is

$$V = (\text{height})(\text{area of base})$$
$$= \|proj_{\mathbf{v} \times \mathbf{w}}\mathbf{u}\| \, \|\mathbf{v} \times \mathbf{w}\|$$
$$= \left| \frac{\mathbf{u} \cdot (\mathbf{v} \times \mathbf{w})}{\|\mathbf{v} \times \mathbf{w}\|} \right| \|\mathbf{v} \times \mathbf{w}\|$$
$$= |\mathbf{u} \cdot (\mathbf{v} \times \mathbf{w})|. \qquad ■$$

EXAMPLE 5 **Volume by the Triple Scalar Product**

Find the volume of the parallelepiped shown in Figure 11.42 having

$$\mathbf{u} = 3\mathbf{i} - 5\mathbf{j} + \mathbf{k}$$
$$\mathbf{v} = 2\mathbf{j} - 2\mathbf{k}$$

and

$$\mathbf{w} = 3\mathbf{i} + \mathbf{j} + \mathbf{k}$$

as adjacent edges.

The parallelepiped has a volume of 36.
Figure 11.42

Solution By Theorem 11.10, you have

$$V = |\mathbf{u} \cdot (\mathbf{v} \times \mathbf{w})| \qquad \text{Triple scalar product}$$
$$= \begin{vmatrix} 3 & -5 & 1 \\ 0 & 2 & -2 \\ 3 & 1 & 1 \end{vmatrix}$$
$$= 3 \begin{vmatrix} 2 & -2 \\ 1 & 1 \end{vmatrix} - (-5) \begin{vmatrix} 0 & -2 \\ 3 & 1 \end{vmatrix} + (1) \begin{vmatrix} 0 & 2 \\ 3 & 1 \end{vmatrix}$$
$$= 3(4) + 5(6) + 1(-6)$$
$$= 36. \qquad ■$$

A natural consequence of Theorem 11.10 is that the volume of the parallelepiped is 0 if and only if the three vectors are coplanar. That is, when the vectors $\mathbf{u} = \langle u_1, u_2, u_3 \rangle$, $\mathbf{v} = \langle v_1, v_2, v_3 \rangle$, and $\mathbf{w} = \langle w_1, w_2, w_3 \rangle$ have the same initial point, they lie in the same plane if and only if

$$\mathbf{u} \cdot (\mathbf{v} \times \mathbf{w}) = \begin{vmatrix} u_1 & u_2 & u_3 \\ v_1 & v_2 & v_3 \\ w_1 & w_2 & w_3 \end{vmatrix} = 0.$$

11.4 Exercises

See **CalcChat.com** for tutorial help and worked-out solutions to odd-numbered exercises.

CONCEPT CHECK

1. Vectors Explain what $\mathbf{u} \times \mathbf{v}$ represents geometrically.

2. Area Explain how to find the area of a parallelogram using vectors.

Cross Product of Unit Vectors In Exercises 3–6, find the cross product of the unit vectors and sketch your result.

3. $\mathbf{j} \times \mathbf{i}$

4. $\mathbf{j} \times \mathbf{k}$

5. $\mathbf{i} \times \mathbf{k}$

6. $\mathbf{k} \times \mathbf{i}$

 **Finding Cross Products** In Exercises 7–10, find (a) $\mathbf{u} \times \mathbf{v}$, (b) $\mathbf{v} \times \mathbf{u}$, and (c) $\mathbf{v} \times \mathbf{v}$.

7. $\mathbf{u} = -2\mathbf{i} + 4\mathbf{j}$
   $\mathbf{v} = 3\mathbf{i} + 2\mathbf{j} + 5\mathbf{k}$

8. $\mathbf{u} = 3\mathbf{i} + 5\mathbf{k}$
   $\mathbf{v} = 2\mathbf{i} + 3\mathbf{j} - 2\mathbf{k}$

9. $\mathbf{u} = \langle 7, 3, 2 \rangle$
   $\mathbf{v} = \langle 1, -1, 5 \rangle$

10. $\mathbf{u} = \langle 2, 1, -9 \rangle$
   $\mathbf{v} = \langle -6, -2, -1 \rangle$

Finding a Cross Product In Exercises 11–14, find $\mathbf{u} \times \mathbf{v}$ and show that it is orthogonal to both $\mathbf{u}$ and $\mathbf{v}$.

11. $\mathbf{u} = \langle 4, -1, 0 \rangle$
   $\mathbf{v} = \langle -6, 3, 0 \rangle$

12. $\mathbf{u} = \langle -5, 2, 2 \rangle$
   $\mathbf{v} = \langle 0, 1, 8 \rangle$

13. $\mathbf{u} = \mathbf{i} + \mathbf{j} + \mathbf{k}$
   $\mathbf{v} = 2\mathbf{i} + \mathbf{j} - \mathbf{k}$

14. $\mathbf{u} = \mathbf{i} + 6\mathbf{j}$
   $\mathbf{v} = -2\mathbf{i} + \mathbf{j} + \mathbf{k}$

 Finding a Unit Vector In Exercises 15–18, find a unit vector that is orthogonal to both $\mathbf{u}$ and $\mathbf{v}$.

15. $\mathbf{u} = \langle 4, -3, 1 \rangle$
   $\mathbf{v} = \langle 2, 5, 3 \rangle$

16. $\mathbf{u} = \langle -8, -6, 4 \rangle$
   $\mathbf{v} = \langle 10, -12, -2 \rangle$

17. $\mathbf{u} = -3\mathbf{i} + 2\mathbf{j} - 5\mathbf{k}$
   $\mathbf{v} = \mathbf{i} - \mathbf{j} + 4\mathbf{k}$

18. $\mathbf{u} = 2\mathbf{k}$
   $\mathbf{v} = 4\mathbf{i} + 6\mathbf{k}$

Area In Exercises 19–22, find the area of the parallelogram that has the given vectors as adjacent sides. Use a computer algebra system or a graphing utility to verify your result.

19. $\mathbf{u} = \mathbf{j}$
   $\mathbf{v} = \mathbf{j} + \mathbf{k}$

20. $\mathbf{u} = \mathbf{i} + \mathbf{j} + \mathbf{k}$
   $\mathbf{v} = \mathbf{j} + \mathbf{k}$

21. $\mathbf{u} = \langle 3, 2, -1 \rangle$
   $\mathbf{v} = \langle 1, 2, 3 \rangle$

22. $\mathbf{u} = \langle 2, -1, 0 \rangle$
   $\mathbf{v} = \langle -1, 2, 0 \rangle$

 **Area** In Exercises 23 and 24, verify that the points are the vertices of a parallelogram, and find its area.

23. $A(0, 3, 2)$, $B(1, 5, 5)$, $C(6, 9, 5)$, $D(5, 7, 2)$

24. $A(2, -3, 1)$, $B(6, 5, -1)$, $C(7, 2, 2)$, $D(3, -6, 4)$

Area In Exercises 25 and 26, find the area of the triangle with the given vertices. $\left(\textit{Hint: } \frac{1}{2}\|\mathbf{u} \times \mathbf{v}\| \text{ is the area of the triangle having u and v as adjacent sides.}\right)$

25. $A(0, 0, 0)$, $B(1, 0, 3)$, $C(-3, 2, 0)$

26. $A(2, -3, 4)$, $B(0, 1, 2)$, $C(-1, 2, 0)$

27. Torque The brakes on a bicycle are applied using a downward force of 20 pounds on the pedal when the crank makes a 40° angle with the horizontal (see figure). The crank is 6 inches in length. Find the torque at P.

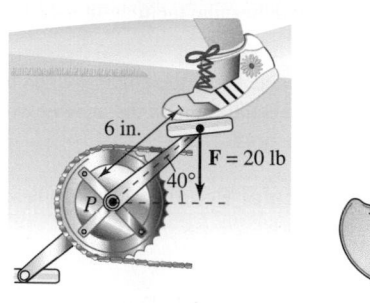

Figure for 27   Figure for 28

28. Torque Both the magnitude and the direction of the force on a crankshaft change as the crankshaft rotates. Find the torque on the crankshaft using the position and data shown in the figure.

29. Optimization A force of 180 pounds acts on the bracket shown in the figure.

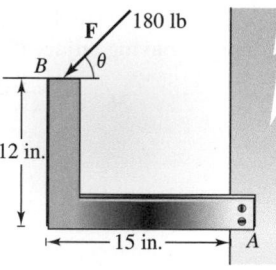

(a) Determine the vector $\overrightarrow{AB}$ and the vector $\mathbf{F}$ representing the force. ($\mathbf{F}$ will be in terms of θ.)

(b) Find the magnitude of the moment about A by evaluating $\|\overrightarrow{AB} \times \mathbf{F}\|$.

(c) Use the result of part (b) to determine the magnitude of the moment when $\theta = 30°$.

(d) Use the result of part (b) to determine the angle θ when the magnitude of the moment is maximum. At that angle, what is the relationship between the vectors $\mathbf{F}$ and $\overrightarrow{AB}$? Is it what you expected? Why or why not?

(e) Use a graphing utility to graph the function for the magnitude of the moment about A for $0° \leq \theta \leq 180°$. Find the zero of the function in the given domain. Interpret the meaning of the zero in the context of the problem.

30. Optimization A force of 56 pounds acts on the pipe wrench shown in the figure.

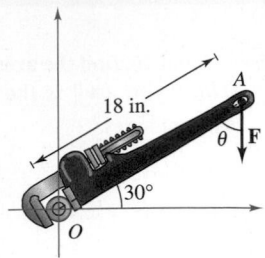

 (a) Find the magnitude of the moment about O by evaluating $\|\overrightarrow{OA} \times \mathbf{F}\|$. Use a graphing utility to graph the resulting function of θ.

(b) Use the result of part (a) to determine the magnitude of the moment when $\theta = 45°$.

(c) Use the result of part (a) to determine the angle θ when the magnitude of the moment is maximum. Is the answer what you expected? Why or why not?

Finding a Triple Scalar Product In Exercises 31–34, find $\mathbf{u} \cdot (\mathbf{v} \times \mathbf{w})$.

31. $\mathbf{u} = \mathbf{i}$
$\mathbf{v} = \mathbf{j}$
$\mathbf{w} = \mathbf{k}$

32. $\mathbf{u} = \langle 1, 1, 1 \rangle$
$\mathbf{v} = \langle 2, 1, 0 \rangle$
$\mathbf{w} = \langle 0, 0, 1 \rangle$

33. $\mathbf{u} = \langle 2, 0, 1 \rangle$
$\mathbf{v} = \langle 0, 3, 0 \rangle$
$\mathbf{w} = \langle 0, 0, 1 \rangle$

34. $\mathbf{u} = \langle 2, 0, 0 \rangle$
$\mathbf{v} = \langle 1, 1, 1 \rangle$
$\mathbf{w} = \langle 0, 2, 2 \rangle$

Volume In Exercises 35 and 36, use the triple scalar product to find the volume of the parallelepiped having adjacent edges u, v, and w.

35. $\mathbf{u} = \mathbf{i} + \mathbf{j}$
$\mathbf{v} = \mathbf{j} + \mathbf{k}$
$\mathbf{w} = \mathbf{i} + \mathbf{k}$

36. $\mathbf{u} = \langle 1, 3, 1 \rangle$
$\mathbf{v} = \langle 0, 6, 6 \rangle$
$\mathbf{w} = \langle -4, 0, -4 \rangle$

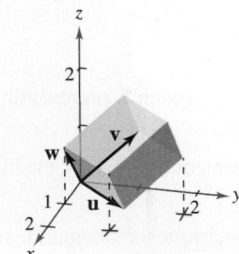 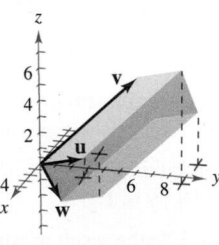

Volume In Exercises 37 and 38, find the volume of the parallelepiped with the given vertices.

37. $(0, 0, 0), (3, 0, 0), (0, 5, 1), (2, 0, 5),$
$(3, 5, 1), (5, 0, 5), (2, 5, 6), (5, 5, 6)$

38. $(0, 0, 0), (0, 4, 0), (-3, 0, 0), (-1, 1, 5),$
$(-3, 4, 0), (-1, 5, 5), (-4, 1, 5), (-4, 5, 5)$

EXPLORING CONCEPTS

39. Comparing Dot Products Identify the dot products that are equal. Explain your reasoning. (Assume **u**, **v**, and **w** are nonzero vectors.)

(a) $\mathbf{u} \cdot (\mathbf{v} \times \mathbf{w})$ (b) $(\mathbf{v} \times \mathbf{w}) \cdot \mathbf{u}$

(c) $(\mathbf{u} \times \mathbf{v}) \cdot \mathbf{w}$ (d) $(\mathbf{u} \times -\mathbf{w}) \cdot \mathbf{v}$

(e) $\mathbf{u} \cdot (\mathbf{w} \times \mathbf{v})$ (f) $\mathbf{w} \cdot (\mathbf{v} \times \mathbf{u})$

(g) $(-\mathbf{u} \times \mathbf{v}) \cdot \mathbf{w}$ (h) $(\mathbf{w} \times \mathbf{u}) \cdot \mathbf{v}$

40. Using Dot and Cross Products When $\mathbf{u} \times \mathbf{v} = \mathbf{0}$ and $\mathbf{u} \cdot \mathbf{v} = 0$, what can you conclude about **u** and **v**?

41. Cross Product Two nonzero vectors lie in the yz-plane. Where does the cross product of the vectors lie? Explain.

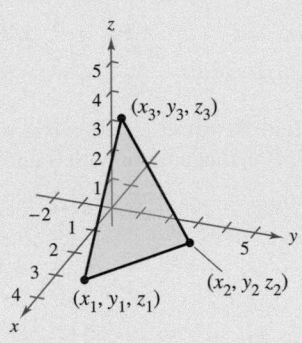

42. HOW DO YOU SEE IT? The vertices of a triangle in space are $(x_1, y_1, z_1), (x_2, y_2, z_2),$ and (x_3, y_3, z_3). Explain how to find a vector perpendicular to the triangle.

True or False? In Exercises 43–46, determine whether the statement is true or false. If it is false, explain why or give an example that shows it is false.

43. It is possible to find the cross product of two vectors in a two-dimensional coordinate system.

44. The cross product of two nonzero vectors is a nonzero vector.

45. If $\mathbf{u} \neq \mathbf{0}$ and $\mathbf{u} \times \mathbf{v} = \mathbf{u} \times \mathbf{w}$, then $\mathbf{v} = \mathbf{w}$.

46. If $\mathbf{u} \neq \mathbf{0}, \mathbf{u} \cdot \mathbf{v} = \mathbf{u} \cdot \mathbf{w}$, and $\mathbf{u} \times \mathbf{v} = \mathbf{u} \times \mathbf{w}$, then $\mathbf{v} = \mathbf{w}$.

Proof In Exercises 47–52, prove the property of the cross product.

47. $\mathbf{u} \times (\mathbf{v} + \mathbf{w}) = (\mathbf{u} \times \mathbf{v}) + (\mathbf{u} \times \mathbf{w})$

48. $c(\mathbf{u} \times \mathbf{v}) = (c\mathbf{u}) \times \mathbf{v} = \mathbf{u} \times (c\mathbf{v})$

49. $\mathbf{u} \times \mathbf{u} = \mathbf{0}$ **50.** $\mathbf{u} \cdot (\mathbf{v} \times \mathbf{w}) = (\mathbf{u} \times \mathbf{v}) \cdot \mathbf{w}$

51. $\mathbf{u} \times \mathbf{v}$ is orthogonal to both **u** and **v**.

52. $\mathbf{u} \times \mathbf{v} = \mathbf{0}$ if and only if **u** and **v** are scalar multiples of each other.

53. Proof Prove that $\|\mathbf{u} \times \mathbf{v}\| = \|\mathbf{u}\| \|\mathbf{v}\|$ if **u** and **v** are orthogonal.

54. Proof Prove that $\mathbf{u} \times (\mathbf{v} \times \mathbf{w}) = (\mathbf{u} \cdot \mathbf{w})\mathbf{v} - (\mathbf{u} \cdot \mathbf{v})\mathbf{w}$.

55. Proof Prove Theorem 11.9.

11.5 Lines and Planes in Space

■ Write a set of parametric equations for a line in space.
■ Write a linear equation to represent a plane in space.
■ Sketch the plane given by a linear equation.
■ Find the distances between points, planes, and lines in space.

Lines in Space

In the plane, *slope* is used to determine the equation of a line. In space, it is more convenient to use *vectors* to determine the equation of a line.

In Figure 11.43, consider the line L through the point $P(x_1, y_1, z_1)$ and parallel to the vector $\mathbf{v} = \langle a, b, c \rangle$. The vector $\mathbf{v}$ is a **direction vector** for the line L, and a, b, and c are **direction numbers.** One way of describing the line L is to say that it consists of all points $Q(x, y, z)$ for which the vector $\overrightarrow{PQ}$ is parallel to $\mathbf{v}$. This means that $\overrightarrow{PQ}$ is a scalar multiple of $\mathbf{v}$ and you can write $\overrightarrow{PQ} = t\mathbf{v}$, where t is a scalar (a real number).

$$\overrightarrow{PQ} = \langle x - x_1, y - y_1, z - z_1 \rangle = \langle at, bt, ct \rangle = t\mathbf{v}$$

By equating corresponding components, you can obtain **parametric equations** of a line in space.

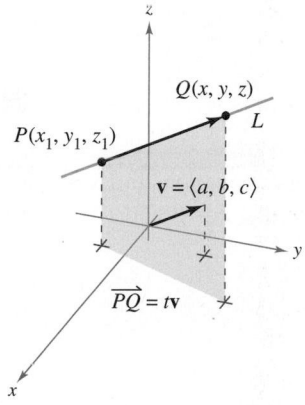

Line L and its direction vector $\mathbf{v}$
Figure 11.43

THEOREM 11.11 Parametric Equations of a Line in Space

A line L parallel to the vector $\mathbf{v} = \langle a, b, c \rangle$ and passing through the point $P(x_1, y_1, z_1)$ is represented by the **parametric equations**

$$x = x_1 + at, \quad y = y_1 + bt, \quad \text{and} \quad z = z_1 + ct.$$

If the direction numbers a, b, and c are all nonzero, then you can eliminate the parameter t in the parametric equations to obtain **symmetric equations** of the line.

$$\frac{x - x_1}{a} = \frac{y - y_1}{b} = \frac{z - z_1}{c} \qquad \text{Symmetric equations}$$

EXAMPLE 1 Finding Parametric and Symmetric Equations

Find parametric and symmetric equations of the line L that passes through the point $(1, -2, 4)$ and is parallel to $\mathbf{v} = \langle 2, 4, -4 \rangle$, as shown in Figure 11.44.

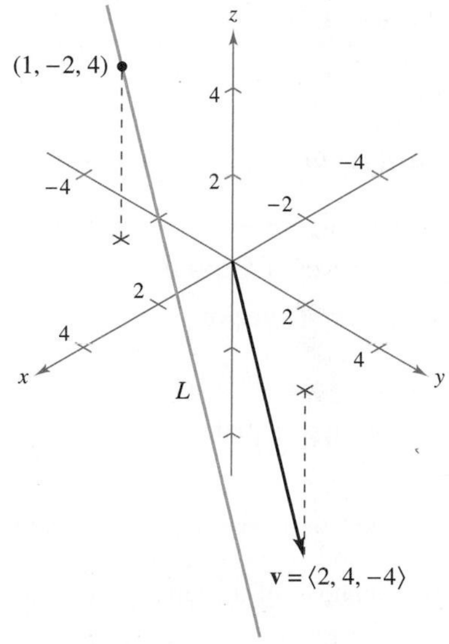

Solution To find a set of parametric equations of the line, use the coordinates $x_1 = 1$, $y_1 = -2$, and $z_1 = 4$ and direction numbers $a = 2$, $b = 4$, and $c = -4$.

$$x = 1 + 2t, \quad y = -2 + 4t, \quad z = 4 - 4t \qquad \text{Parametric equations}$$

Because a, b, and c are all nonzero, a set of symmetric equations is

$$\frac{x - 1}{2} = \frac{y + 2}{4} = \frac{z - 4}{-4}. \qquad \text{Symmetric equations}$$

Neither parametric equations nor symmetric equations of a given line are unique. For instance, in Example 1, by letting $t = 1$ in the parametric equations, you would obtain the point $(3, 2, 0)$. Using this point with the direction numbers $a = 2$, $b = 4$, and $c = -4$ would produce a different set of parametric equations

$$x = 3 + 2t, \quad y = 2 + 4t, \quad \text{and} \quad z = -4t.$$

The vector $\mathbf{v}$ is parallel to the line L.
Figure 11.44

> **EXAMPLE 2** **Parametric Equations of a Line Through Two Points**
>
> ⋅ ⋅ ⋅ ▷ *See LarsonCalculus.com for an interactive version of this type of example.*

Find a set of parametric equations of the line that passes through the points

$$(-2, 1, 0) \quad \text{and} \quad (1, 3, 5).$$

Solution Begin by using the points $P(-2, 1, 0)$ and $Q(1, 3, 5)$ to find a direction vector for the line passing through P and Q.

$$\mathbf{v} = \overrightarrow{PQ} = \langle 1 - (-2), 3 - 1, 5 - 0 \rangle = \langle 3, 2, 5 \rangle = \langle a, b, c \rangle$$

Using the direction numbers $a = 3$, $b = 2$, and $c = 5$ with the point $P(-2, 1, 0)$, you obtain the parametric equations

$$x = -2 + 3t, \quad y = 1 + 2t, \quad \text{and} \quad z = 5t. \qquad \blacksquare$$

⋅ ⋅ ⋅ ⋅ ⋅ ⋅ **REMARK** As t varies over all real numbers, the parametric equations in Example 2 determine the points (x, y, z) on the line. In particular, note that $t = 0$ and $t = 1$ give the original points $(-2, 1, 0)$ and $(1, 3, 5)$.

Planes in Space

You have seen how an equation of a line in space can be obtained from a point on the line and a vector *parallel* to it. You will now see that an equation of a plane in space can be obtained from a point in the plane and a vector *normal* (perpendicular) to the plane.

Consider the plane containing the point $P(x_1, y_1, z_1)$ having a nonzero normal vector

$$\mathbf{n} = \langle a, b, c \rangle$$

as shown in Figure 11.45. This plane consists of all points $Q(x, y, z)$ for which vector $\overrightarrow{PQ}$ is orthogonal to $\mathbf{n}$. Using the dot product, you can write the following.

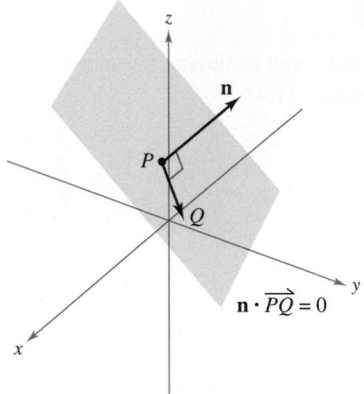

The normal vector $\mathbf{n}$ is orthogonal to each vector $\overrightarrow{PQ}$ in the plane.
Figure 11.45

$$\mathbf{n} \cdot \overrightarrow{PQ} = 0$$
$$\langle a, b, c \rangle \cdot \langle x - x_1, y - y_1, z - z_1 \rangle = 0$$
$$a(x - x_1) + b(y - y_1) + c(z - z_1) = 0$$

The third equation of the plane is said to be in **standard form.**

THEOREM 11.12 **Standard Equation of a Plane in Space**

The plane containing the point (x_1, y_1, z_1) and having normal vector

$$\mathbf{n} = \langle a, b, c \rangle$$

can be represented by the **standard form** of the equation of a plane

$$a(x - x_1) + b(y - y_1) + c(z - z_1) = 0.$$

By regrouping terms in the standard form of the equation of a plane, you obtain the **general form.**

$$ax + by + cz + d = 0 \qquad \text{General form of equation of plane}$$

Given the general form of the equation of a plane, it is easy to find a normal vector to the plane. Simply use the coefficients of x, y, and z and write $\mathbf{n} = \langle a, b, c \rangle$.

EXAMPLE 3   Finding an Equation of a Plane in Three-Space

Find an equation (in standard form and in general form) of the plane containing the points

$$(2, 1, 1), \quad (1, 4, 1), \quad \text{and} \quad (-2, 0, 4).$$

Solution   To apply Theorem 11.12, you need a point in the plane and a vector that is normal to the plane. There are three choices for the point, but no normal vector is given. To obtain a normal vector, use the cross product of vectors $\mathbf{u}$ and $\mathbf{v}$ extending from the point $(2, 1, 1)$ to the points $(1, 4, 1)$ and $(-2, 0, 4)$, as shown in Figure 11.46. The component forms of $\mathbf{u}$ and $\mathbf{v}$ are

$$\mathbf{u} = \langle 1 - 2, 4 - 1, 1 - 1 \rangle = \langle -1, 3, 0 \rangle$$

and

$$\mathbf{v} = \langle -2 - 2, 0 - 1, 4 - 1 \rangle = \langle -4, -1, 3 \rangle.$$

So, it follows that a vector normal to the given plane is

$$\mathbf{n} = \mathbf{u} \times \mathbf{v}$$

$$= \begin{vmatrix} \mathbf{i} & \mathbf{j} & \mathbf{k} \\ -1 & 3 & 0 \\ -4 & -1 & 3 \end{vmatrix}$$

$$= 9\mathbf{i} + 3\mathbf{j} + 13\mathbf{k}$$

$$= \langle a, b, c \rangle.$$

Using the direction numbers for $\mathbf{n}$ and the point $(x_1, y_1, z_1) = (2, 1, 1)$, you can determine an equation of the plane in standard form to be

$$a(x - x_1) + b(y - y_1) + c(z - z_1) = 0$$

$$9(x - 2) + 3(y - 1) + 13(z - 1) = 0. \qquad \text{Standard form}$$

By regrouping terms, the general form is

$$9x - 18 + 3y - 3 + 13z - 13 = 0$$

$$9x + 3y + 13z - 34 = 0. \qquad \text{General form}$$

REMARK   In Example 3, check to see that each of the three original points satisfies the equation $9x + 3y + 13z - 34 = 0$.

Two distinct planes in three-space either are parallel or intersect in a line. For two planes that intersect, you can determine the angle $(0 \le \theta \le \pi/2)$ between them from the angle between their normal vectors, as shown in Figure 11.47. Specifically, if vectors $\mathbf{n}_1$ and $\mathbf{n}_2$ are normal to two intersecting planes, then the angle θ between the normal vectors is equal to the angle between the two planes and is

$$\cos \theta = \frac{|\mathbf{n}_1 \cdot \mathbf{n}_2|}{\|\mathbf{n}_1\| \|\mathbf{n}_2\|}. \qquad \text{Angle between two planes}$$

Consequently, two planes with normal vectors $\mathbf{n}_1$ and $\mathbf{n}_2$ are

1. *perpendicular* when $\mathbf{n}_1 \cdot \mathbf{n}_2 = 0$.
2. *parallel* when $\mathbf{n}_1$ is a scalar multiple of $\mathbf{n}_2$.

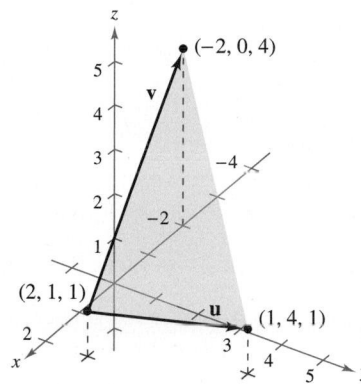

A plane determined by $\mathbf{u}$ and $\mathbf{v}$
Figure 11.46

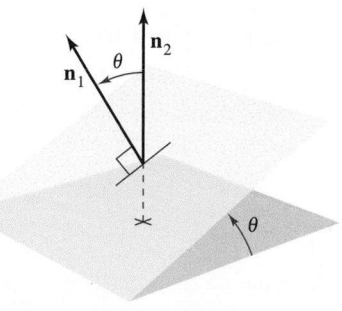

The angle θ between two planes
Figure 11.47

EXAMPLE 4   Finding the Line of Intersection of Two Planes

Find the angle between the two planes $x - 2y + z = 0$ and $2x + 3y - 2z = 0$. Then find parametric equations of their line of intersection (see Figure 11.48).

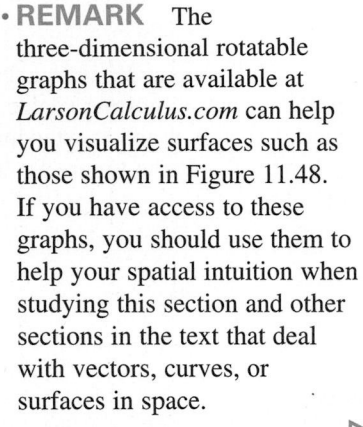

• • REMARK   The three-dimensional rotatable graphs that are available at *LarsonCalculus.com* can help you visualize surfaces such as those shown in Figure 11.48. If you have access to these graphs, you should use them to help your spatial intuition when studying this section and other sections in the text that deal with vectors, curves, or surfaces in space.

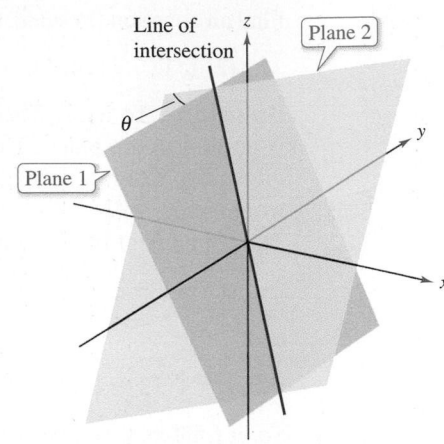

Figure 11.48

Solution   Normal vectors for the planes are $\mathbf{n}_1 = \langle 1, -2, 1 \rangle$ and $\mathbf{n}_2 = \langle 2, 3, -2 \rangle$. Consequently, the angle between the two planes is determined as follows.

$$\cos \theta = \frac{|\mathbf{n}_1 \cdot \mathbf{n}_2|}{\|\mathbf{n}_1\| \|\mathbf{n}_2\|} = \frac{|-6|}{\sqrt{6}\sqrt{17}} = \frac{6}{\sqrt{102}} \approx 0.59409$$

This implies that the angle between the two planes is $\theta \approx 53.55°$. You can find the line of intersection of the two planes by simultaneously solving the two linear equations representing the planes. One way to do this is to multiply the first equation by -2 and add the result to the second equation.

$$
\begin{array}{ll}
x - 2y + z = 0 \implies & -2x + 4y - 2z = 0 \qquad \text{Multiply Equation 1 by } -2. \\
2x + 3y - 2z = 0 \implies & \underline{2x + 3y - 2z = 0} \qquad \text{Write Equation 2.} \\
& 7y - 4z = 0 \qquad \text{Add equations.} \\
& y = \dfrac{4z}{7} \qquad \text{Solve for } y.
\end{array}
$$

Substituting $y = 4z/7$ back into one of the original equations, you can determine that $x = z/7$. Finally, by letting $t = z/7$, you obtain the parametric equations

$$x = t, \quad y = 4t, \quad \text{and} \quad z = 7t \qquad \text{Line of intersection}$$

which indicate that 1, 4, and 7 are direction numbers for the line of intersection. ∎

Note that the direction numbers in Example 4 can be obtained from the cross product of the two normal vectors as follows.

$$
\mathbf{n}_1 \times \mathbf{n}_2 = \begin{vmatrix} \mathbf{i} & \mathbf{j} & \mathbf{k} \\ 1 & -2 & 1 \\ 2 & 3 & -2 \end{vmatrix}
$$

$$
= \begin{vmatrix} -2 & 1 \\ 3 & -2 \end{vmatrix} \mathbf{i} - \begin{vmatrix} 1 & 1 \\ 2 & -2 \end{vmatrix} \mathbf{j} + \begin{vmatrix} 1 & -2 \\ 2 & 3 \end{vmatrix} \mathbf{k}
$$

$$
= \mathbf{i} + 4\mathbf{j} + 7\mathbf{k}
$$

This means that the line of intersection of the two planes is parallel to the cross product of their normal vectors.

Sketching Planes in Space

If a plane in space intersects one of the coordinate planes, then the line of intersection is called the **trace** of the given plane in the coordinate plane. To sketch a plane in space, it is helpful to find its points of intersection with the coordinate axes and its traces in the coordinate planes. For example, consider the plane

$$3x + 2y + 4z = 12. \qquad \text{Equation of plane}$$

You can find the xy-trace by letting $z = 0$ and sketching the line

$$3x + 2y = 12 \qquad \text{xy-trace}$$

in the xy-plane. This line intersects the x-axis at $(4, 0, 0)$ and the y-axis at $(0, 6, 0)$. In Figure 11.49, this process is continued by finding the yz-trace and the xz-trace and then shading the triangular region lying in the first octant.

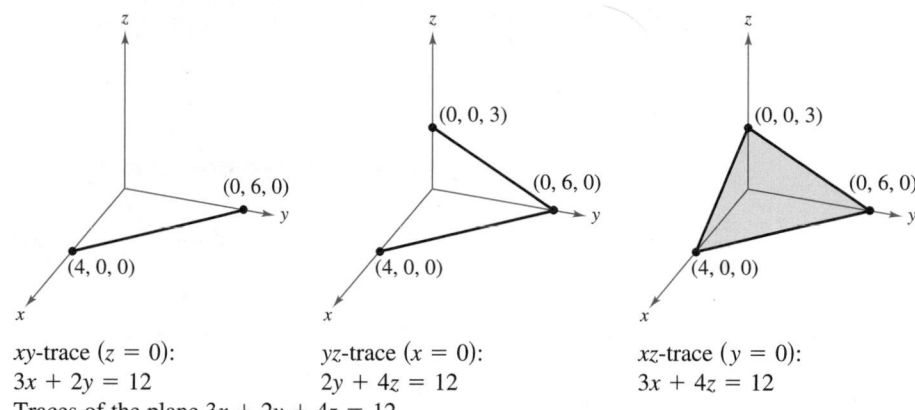

xy-trace $(z = 0)$:
$3x + 2y = 12$
Traces of the plane $3x + 2y + 4z = 12$

yz-trace $(x = 0)$:
$2y + 4z = 12$

xz-trace $(y = 0)$:
$3x + 4z = 12$

Figure 11.49

If an equation of a plane has a missing variable, such as

$$2x + z = 1$$

then the plane must be *parallel to the axis* represented by the missing variable, as shown in Figure 11.50. If two variables are missing from an equation of a plane, such as

$$ax + d = 0$$

then it is *parallel to the coordinate plane* represented by the missing variables, as shown in Figure 11.51.

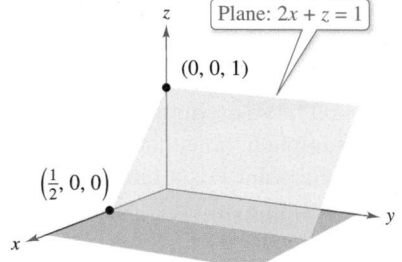

Plane $2x + z = 1$ is parallel to the y-axis.

Figure 11.50

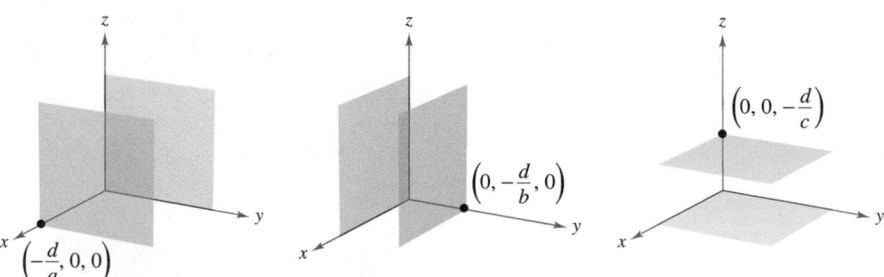

Plane $ax + d = 0$ is parallel to the yz-plane.

Plane $by + d = 0$ is parallel to the xz-plane.

Plane $cz + d = 0$ is parallel to the xy-plane.

Figure 11.51

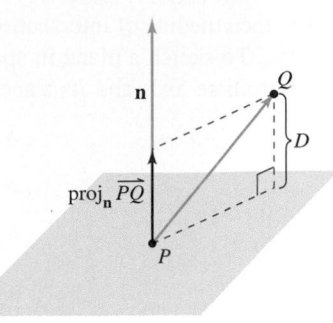

$$D = \|\text{proj}_\mathbf{n} \overrightarrow{PQ}\|$$

The distance between a point and a plane

Figure 11.52

Distances Between Points, Planes, and Lines

Consider two types of problems involving distance in space: (1) finding the distance between a point and a plane and (2) finding the distance between a point and a line. The solutions of these problems illustrate the versatility and usefulness of vectors in coordinate geometry: the first problem uses the *dot product* of two vectors, and the second problem uses the *cross product*.

The distance D between a point Q and a plane is the length of the shortest line segment connecting Q to the plane, as shown in Figure 11.52. For *any* point P in the plane, you can find this distance by projecting the vector $\overrightarrow{PQ}$ onto the normal vector $\mathbf{n}$. The length of this projection is the desired distance.

THEOREM 11.13   Distance Between a Point and a Plane

The distance between a plane and a point Q (not in the plane) is

$$D = \|\text{proj}_\mathbf{n}\overrightarrow{PQ}\| = \frac{|\overrightarrow{PQ} \cdot \mathbf{n}|}{\|\mathbf{n}\|}$$

where P is a point in the plane and $\mathbf{n}$ is normal to the plane.

To find a point in the plane $ax + by + cz + d = 0$, where $a \neq 0$, let $y = 0$ and $z = 0$. Then, from the equation $ax + d = 0$, you can conclude that the point

$$\left(-\frac{d}{a}, 0, 0\right)$$

lies in the plane.

EXAMPLE 5   **Finding the Distance Between a Point and a Plane**

Find the distance between the point $Q(1, 5, -4)$ and the plane $3x - y + 2z = 6$.

Solution   You know that $\mathbf{n} = \langle 3, -1, 2 \rangle$ is normal to the plane. To find a point in the plane, let $y = 0$ and $z = 0$, and obtain the point $P(2, 0, 0)$. The vector from P to Q is

$$\overrightarrow{PQ} = \langle 1 - 2, 5 - 0, -4 - 0 \rangle$$
$$= \langle -1, 5, -4 \rangle.$$

Using the Distance Formula given in Theorem 11.13 produces

$$D = \frac{|\overrightarrow{PQ} \cdot \mathbf{n}|}{\|\mathbf{n}\|} = \frac{|\langle -1, 5, -4 \rangle \cdot \langle 3, -1, 2 \rangle|}{\sqrt{9 + 1 + 4}} = \frac{|-3 - 5 - 8|}{\sqrt{14}} = \frac{16}{\sqrt{14}} \approx 4.28.$$

> **REMARK**   In the solution to Example 5, note that the choice of the point P is arbitrary. Try choosing a different point in the plane to verify that you obtain the same distance.

From Theorem 11.13, you can determine that the distance between the point $Q(x_0, y_0, z_0)$ and the plane $ax + by + cz + d = 0$ is

$$D = \frac{|a(x_0 - x_1) + b(y_0 - y_1) + c(z_0 - z_1)|}{\sqrt{a^2 + b^2 + c^2}}$$

or

$$D = \frac{|ax_0 + by_0 + cz_0 + d|}{\sqrt{a^2 + b^2 + c^2}} \qquad \text{Distance between a point and a plane}$$

where $P(x_1, y_1, z_1)$ is a point in the plane and $d = -(ax_1 + by_1 + cz_1)$.

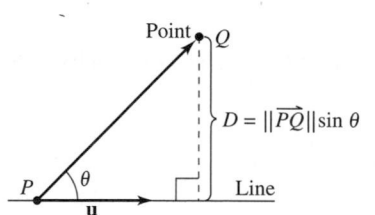

The distance between the parallel
planes is approximately 2.14.
Figure 11.53

EXAMPLE 6 **Finding the Distance Between Two Parallel Planes**

Two parallel planes, $3x - y + 2z - 6 = 0$ and $6x - 2y + 4z + 4 = 0$, are shown in
Figure 11.53. To find the distance between the planes, choose a point in the first plane,
such as $(x_0, y_0, z_0) = (2, 0, 0)$. Then, from the second plane, you can determine that
$a = 6$, $b = -2$, $c = 4$, and $d = 4$ and conclude that the distance is

$$
\begin{aligned}
D &= \frac{|ax_0 + by_0 + cz_0 + d|}{\sqrt{a^2 + b^2 + c^2}} \\
&= \frac{|6(2) + (-2)(0) + (4)(0) + 4|}{\sqrt{6^2 + (-2)^2 + 4^2}} \\
&= \frac{16}{\sqrt{56}} = \frac{8}{\sqrt{14}} \approx 2.14.
\end{aligned}
$$

The formula for the distance between a point and a line in space resembles that
for the distance between a point and a plane—except that you replace the dot product
with the length of the cross product and the normal vector **n** with a direction vector for
the line.

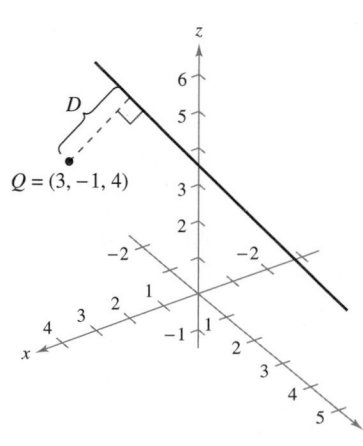

The distance between a point and a line
Figure 11.54

THEOREM 11.14 **Distance Between a Point and a Line in Space**

The distance between a point Q and a line in space is

$$ D = \frac{\|\overrightarrow{PQ} \times \mathbf{u}\|}{\|\mathbf{u}\|} $$

where **u** is a direction vector for the line and P is a point
on the line.

Proof In Figure 11.54, let D be the distance between the point Q and the line. Then
$D = \|\overrightarrow{PQ}\| \sin \theta$, where θ is the angle between **u** and $\overrightarrow{PQ}$. By Property 2 of Theorem 11.8,
you have $\|\mathbf{u}\| \|\overrightarrow{PQ}\| \sin \theta = \|\mathbf{u} \times \overrightarrow{PQ}\| = \|\overrightarrow{PQ} \times \mathbf{u}\|$. Consequently,

$$ D = \|\overrightarrow{PQ}\| \sin \theta = \frac{\|\overrightarrow{PQ} \times \mathbf{u}\|}{\|\mathbf{u}\|}. $$

EXAMPLE 7 **Finding the Distance Between a Point and a Line**

Find the distance between the point $Q(3, -1, 4)$ and the line

$$ x = -2 + 3t, \quad y = -2t, \quad \text{and} \quad z = 1 + 4t. $$

Solution Using the direction numbers 3, -2, and 4, a direction vector for the line
is $\mathbf{u} = \langle 3, -2, 4 \rangle$. To find a point on the line, let $t = 0$ and obtain $P = (-2, 0, 1)$. So,

$$ \overrightarrow{PQ} = \langle 3 - (-2), -1 - 0, 4 - 1 \rangle = \langle 5, -1, 3 \rangle $$

and you can form the cross product

$$ \overrightarrow{PQ} \times \mathbf{u} = \begin{vmatrix} \mathbf{i} & \mathbf{j} & \mathbf{k} \\ 5 & -1 & 3 \\ 3 & -2 & 4 \end{vmatrix} = 2\mathbf{i} - 11\mathbf{j} - 7\mathbf{k} = \langle 2, -11, -7 \rangle. $$

The distance between the point Q and
the line is $\sqrt{6} \approx 2.45$.
Figure 11.55

Finally, using Theorem 11.14, you can find the distance to be

$$ D = \frac{\|\overrightarrow{PQ} \times \mathbf{u}\|}{\|\mathbf{u}\|} = \frac{\sqrt{174}}{\sqrt{29}} = \sqrt{6} \approx 2.45. \qquad \text{See Figure 11.55.} $$

11.5 Exercises

See **CalcChat.com** for tutorial help and worked-out solutions to odd-numbered exercises.

CONCEPT CHECK

1. **Parametric and Symmetric Equations** Give the parametric equations and the symmetric equations of a line in space. Describe what is required to find these equations.

2. **Normal Vector** The equation of a plane in space is $2(x - 1) + 4(y - 3) - (z + 5) = 0$. What is the normal vector to this plane?

3. **Plane** Write an equation of a plane in space that is parallel to the x-axis.

4. **Parallel Planes** Explain how to find the distance between two parallel planes.

Checking Points on a Line In Exercises 5 and 6, determine whether each point lies on the line.

5. $x = -2 + t, \ y = 3t, \ z = 4 + t$
   (a) $(0, 6, 6)$ (b) $(2, 3, 5)$ (c) $(-4, -6, 2)$

6. $\dfrac{x - 3}{2} = \dfrac{y - 7}{8} = z + 2$
   (a) $(7, 23, 0)$ (b) $(1, -1, -3)$ (c) $(-7, 47, -7)$

Finding Parametric and Symmetric Equations In Exercises 7–12, find sets of (a) parametric equations and (b) symmetric equations of the line that passes through the given point and is parallel to the given vector or line. (For each line, write the direction numbers as integers.)

Point	Parallel to
7. $(0, 0, 0)$	$\mathbf{v} = \langle 3, 1, 5 \rangle$
8. $(0, 0, 0)$	$\mathbf{v} = \langle -2, \frac{5}{2}, 1 \rangle$
9. $(-2, 0, 3)$	$\mathbf{v} = 2\mathbf{i} + 4\mathbf{j} - 2\mathbf{k}$
10. $(-3, 0, 2)$	$\mathbf{v} = 6\mathbf{j} + 3\mathbf{k}$
11. $(1, 0, 1)$	$x = 3 + 3t, y = 5 - 2t, z = -7 + t$
12. $(-3, 5, 4)$	$\dfrac{x - 1}{3} = \dfrac{y + 1}{-2} = z - 3$

Finding Parametric and Symmetric Equations In Exercises 13–16, find sets of (a) parametric equations and (b) symmetric equations of the line that passes through the two points (if possible). (For each line, write the direction numbers as integers.)

13. $(5, -3, -2), \left(-\frac{2}{3}, \frac{2}{3}, 1\right)$ 14. $(0, 4, 3), (-1, 2, 5)$

15. $(7, -2, 6), (-3, 0, 6)$ 16. $(0, 0, 25), (10, 10, 0)$

Finding Parametric Equations In Exercises 17–24, find a set of parametric equations of the line with the given characteristics.

17. The line passes through the point $(2, 3, 4)$ and is parallel to the xz-plane and the yz-plane.

18. The line passes through the point $(-4, 5, 2)$ and is parallel to the xy-plane and the yz-plane.

19. The line passes through the point $(2, 3, 4)$ and is perpendicular to the plane given by $3x + 2y - z = 6$.

20. The line passes through the point $(-4, 5, 2)$ and is perpendicular to the plane given by $-x + 2y + z = 5$.

21. The line passes through the point $(5, -3, -4)$ and is parallel to $\mathbf{v} = \langle 2, -1, 3 \rangle$.

22. The line passes through the point $(-1, 4, -3)$ and is parallel to $\mathbf{v} = 5\mathbf{i} - \mathbf{j}$.

23. The line passes through the point $(2, 1, 2)$ and is parallel to the line $x = -t, y = 1 + t, z = -2 + t$.

24. The line passes through the point $(-6, 0, 8)$ and is parallel to the line $x = 5 - 2t, y = -4 + 2t, z = 0$.

Using Parametric and Symmetric Equations In Exercises 25–28, find the coordinates of a point P on the line and a vector $\mathbf{v}$ parallel to the line.

25. $x = 3 - t, \quad y = -1 + 2t, \quad z = -2$

26. $x = 4t, \quad y = 5 - t, \quad z = 4 + 3t$

27. $\dfrac{x - 7}{4} = \dfrac{y + 6}{2} = z + 2$ 28. $\dfrac{x + 3}{5} = \dfrac{y}{8} = \dfrac{z - 3}{6}$

Determining Parallel Lines In Exercises 29–32, determine whether the lines are parallel or identical.

29. $x = 6 - 3t, \quad y = -2 + 2t, \quad z = 5 + 4t$
   $x = 6t, \quad y = 2 - 4t, \quad z = 13 - 8t$

30. $x = 1 + 2t, \quad y = -1 - t, \quad z = 3t$
   $x = 5 + 2t, \quad y = 1 - t, \quad z = 8 + 3t$

31. $\dfrac{x - 8}{4} = \dfrac{y + 5}{-2} = \dfrac{z + 9}{3}$
   $\dfrac{x + 4}{-8} = \dfrac{y - 1}{4} = \dfrac{z + 18}{-6}$

32. $\dfrac{x - 1}{4} = \dfrac{y - 1}{2} = \dfrac{z + 3}{4}$
   $\dfrac{x + 2}{1} = \dfrac{y - 1}{0.5} = \dfrac{z - 3}{1}$

Finding a Point of Intersection In Exercises 33–36, determine whether the lines intersect, and if so, find the point of intersection and the angle between the lines.

33. $x = 4t + 2, \quad y = 3, \quad z = -t + 1$
   $x = 2s + 2, \quad y = 2s + 3, \quad z = s + 1$

34. $x = -3t + 1, \quad y = 4t + 1, \quad z = 2t + 4$
   $x = 3s + 1, \quad y = 2s + 4, \quad z = -s + 1$

35. $\dfrac{x}{3} = \dfrac{y - 2}{-1} = z + 1, \quad \dfrac{x - 1}{4} = y + 2 = \dfrac{z + 3}{-3}$

36. $\dfrac{x - 2}{-3} = \dfrac{y - 2}{6} = z - 3, \quad \dfrac{x - 3}{2} = y + 5 = \dfrac{z + 2}{4}$

Checking Points in a Plane In Exercises 37 and 38, determine whether each point lies in the plane.

37. $x + 2y - 4z - 1 = 0$

   (a) $(-7, 2, -1)$ (b) $(5, 2, 2)$ (c) $(-6, 1, -1)$

38. $2x + y + 3z - 6 = 0$

   (a) $(3, 6, -2)$ (b) $(-1, 5, -1)$ (c) $(2, 1, 0)$

 **Finding an Equation of a Plane** In Exercises 39–44, find an equation of the plane that passes through the given point and is perpendicular to the given vector or line.

Point	Perpendicular to
39. $(1, 3, -7)$	$\mathbf{n} = \mathbf{j}$
40. $(0, -1, 4)$	$\mathbf{n} = \mathbf{k}$
41. $(3, 2, 2)$	$\mathbf{n} = 2\mathbf{i} + 3\mathbf{j} - \mathbf{k}$
42. $(0, 0, 0)$	$\mathbf{n} = -3\mathbf{i} + 2\mathbf{k}$
43. $(-1, 4, 0)$	$x = -1 + 2t, y = 5 - t, z = 3 - 2t$
44. $(3, 2, 2)$	$\dfrac{x - 1}{4} = y + 2 = \dfrac{z + 3}{-3}$

 **Finding an Equation of a Plane** In Exercises 45–56, find an equation of the plane with the given characteristics.

45. The plane passes through $(0, 0, 0)$, $(2, 0, 3)$, and $(-3, -1, 5)$.

46. The plane passes through $(3, -1, 2)$, $(2, 1, 5)$, and $(1, -2, -2)$.

47. The plane passes through $(1, 2, 3)$, $(3, 2, 1)$, and $(-1, -2, 2)$.

48. The plane passes through the point $(1, 2, 3)$ and is parallel to the yz-plane.

49. The plane passes through the point $(1, 2, 3)$ and is parallel to the xy-plane.

50. The plane contains the y-axis and makes an angle of $\pi/6$ with the positive x-axis.

51. The plane contains the lines given by

$$\frac{x - 1}{-2} = y - 4 = z \quad \text{and} \quad \frac{x - 2}{-3} = \frac{y - 1}{4} = \frac{z - 2}{-1}.$$

52. The plane passes through the point $(2, 2, 1)$ and contains the line given by

$$\frac{x}{2} = \frac{y - 4}{-1} = z.$$

53. The plane passes through the points $(2, 2, 1)$ and $(-1, 1, -1)$ and is perpendicular to the plane

$$2x - 3y + z = 3.$$

54. The plane passes through the points $(3, 2, 1)$ and $(3, 1, -5)$ and is perpendicular to the plane

$$6x + 7y + 2z = 10.$$

55. The plane passes through the points $(1, -2, -1)$ and $(2, 5, 6)$ and is parallel to the x-axis.

56. The plane passes through the points $(4, 2, 1)$ and $(-3, 5, 7)$ and is parallel to the z-axis.

Finding an Equation of a Plane In Exercises 57–60, find an equation of the plane that contains all the points that are equidistant from the given points.

57. $(2, 2, 0)$, $(0, 2, 2)$

58. $(1, 0, 2)$, $(2, 0, 1)$

59. $(-3, 1, 2)$, $(6, -2, 4)$

60. $(-5, 1, -3)$, $(2, -1, 6)$

Parallel Planes In Exercises 61–64, determine whether the planes are parallel or identical.

61. $-5x + 2y - 8z = 6$ **62.** $2x - y + 3z = 8$

    $15x - 6y + 24z = 17$ $8x - 4y + 12z = 5$

63. $3x - 2y + 5z = 10$

    $75x - 50y + 125z = 250$

64. $-x + 4y - z = 6$

    $-\frac{5}{2}x + 10y - \frac{5}{2}z = 15$

 **Intersection of Planes** In Exercises 65–68, (a) find the angle between the two planes and (b) find a set of parametric equations for the line of intersection of the planes.

65. $3x + 2y - z = 7$ **66.** $-2x + y + z = 2$

    $x - 4y + 2z = 0$ $6x - 3y + 2z = 4$

67. $3x - y + z = 7$ **68.** $6x - 3y + z = 5$

    $4x + 6y + 3z = 2$ $-x + y + 5z = 5$

Comparing Planes In Exercises 69–74, determine whether the planes are parallel, orthogonal, or neither. If they are neither parallel nor orthogonal, find the angle between the planes.

69. $5x - 3y + z = 4$ **70.** $3x + y - 4z = 3$

    $x + 4y + 7z = 1$ $-9x - 3y + 12z = 4$

71. $x - 3y + 6z = 4$ **72.** $3x + 2y - z = 7$

    $5x + y - z = 4$ $x - 4y + 2z = 0$

73. $x - 5y - z = 1$ **74.** $2x - z = 1$

    $5x - 25y - 5z = -3$ $4x + y + 8z = 10$

 **Sketching a Graph of a Plane** In Exercises 75–82, sketch a graph of the plane and label any intercepts.

75. $y = -2$ **76.** $z = 1$

77. $x + z = 6$ **78.** $2x + y = 8$

79. $4x + 2y + 6z = 12$ **80.** $3x + 6y + 2z = 6$

81. $2x - y + 3z = 4$ **82.** $2x - y + z = 4$

Intersection of a Plane and a Line In Exercises 83–86, find the point(s) of intersection (if any) of the plane and the line. Also, determine whether the line lies in the plane.

83. $x + 3y - z = 6$, $\dfrac{x + 7}{2} = y - 4 = \dfrac{z + 1}{5}$

84. $2x + 3y = -5$, $\dfrac{x - 1}{4} = \dfrac{y}{2} = \dfrac{z - 3}{6}$

85. $2x + 3y = 10$, $\dfrac{x-1}{3} = \dfrac{y+1}{-2} = z - 3$

86. $5x + 3y = 17$, $\dfrac{x-4}{2} = \dfrac{y+1}{-3} = \dfrac{z+2}{5}$

Finding the Distance Between a Point and a Plane In Exercises 87–90, find the distance between the point and the plane.

87. $(0, 0, 0)$
$2x + 3y + z = 12$

88. $(0, 0, 0)$
$5x + y - z = 9$

89. $(2, 8, 4)$
$2x + y + z = 5$

90. $(1, 3, -1)$
$3x - 4y + 5z = 6$

Finding the Distance Between Two Parallel Planes In Exercises 91–94, verify that the two planes are parallel and find the distance between the planes.

91. $x - 3y + 4z = 10$
$x - 3y + 4z = 6$

92. $2x + 7y + z = 13$
$2x + 7y + z = 9$

93. $-3x + 6y + 7z = 1$
$6x - 12y - 14z = 25$

94. $-x + 6y + 2z = 3$
$-\frac{1}{2}x + 3y + z = 4$

Finding the Distance Between a Point and a Line In Exercises 95–98, find the distance between the point and the line given by the set of parametric equations.

95. $(1, 5, -2)$; $x = 4t - 2$, $y = 3$, $z = -t + 1$

96. $(1, -2, 4)$; $x = 2t$, $y = t - 3$, $z = 2t + 2$

97. $(-2, 1, 3)$; $x = 1 - t$, $y = 2 + t$, $z = -2t$

98. $(4, -1, 5)$; $x = 3$, $y = 1 + 3t$, $z = 1 + t$

Finding the Distance Between Two Parallel Lines In Exercises 99 and 100, verify that the two lines are parallel and find the distance between the lines.

99. L_1: $x = 2 - t$, $y = 3 + 2t$, $z = 4 + t$
   L_2: $x = 3t$, $y = 1 - 6t$, $z = 4 - 3t$

100. L_1: $x = 3 + 6t$, $y = -2 + 9t$, $z = 1 - 12t$
   L_2: $x = -1 + 4t$, $y = 3 + 6t$, $z = -8t$

EXPLORING CONCEPTS

101. Planes Consider a line and a point not on the line. How many planes contain the line and the point? Explain.

102. Planes How many planes are orthogonal to a given plane in space? Explain.

103. Think About It Do two distinct lines in space determine a unique plane? Explain.

104. **HOW DO YOU SEE IT?** Match the general equation with its graph. Then state what axis or plane the equation is parallel to.

(a) $ax + by + d = 0$

(b) $ax + d = 0$

(c) $cz + d = 0$

(d) $ax + cz + d = 0$

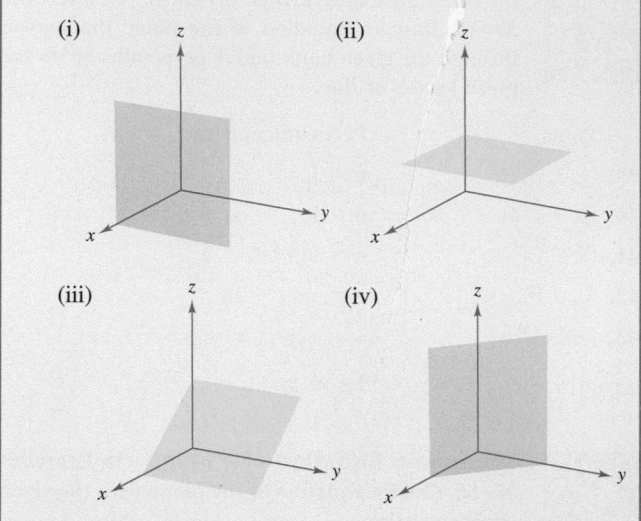

105. Modeling Data

Personal consumption expenditures (in billions of dollars) for several types of recreation from 2009 through 2014 are shown in the table, where x is the expenditures on amusement parks and campgrounds, y is the expenditures on live entertainment (excluding sports), and z is the expenditures on spectator sports. *(Source: U.S. Bureau of Economic Analysis)*

Year	2009	2010	2011	2012	2013	2014
x	37.2	38.8	41.3	44.6	47.0	50.3
y	25.2	26.3	28.3	28.5	28.0	30.0
z	18.8	19.2	20.4	20.6	21.6	22.4

A model for the data is given by

$0.23x + 0.14y - z = -6.85$.

(a) Complete a fourth row in the table using the model to approximate z for the given values of x and y. Compare the approximations with the actual values of z.

(b) According to this model, increases in expenditures on recreation types x and y would correspond to what kind of change in expenditures on recreation type z?

106. Mechanical Design The figure shows a chute at the top of a grain elevator of a combine that funnels the grain into a bin. Find the angle between two adjacent sides.

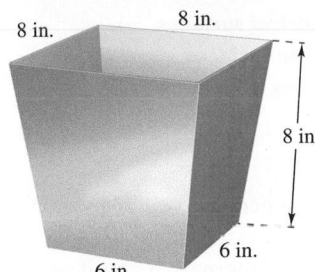

8 in. 8 in.

8 in.

6 in.

6 in.

107. Distance Two insects are crawling along different lines in three-space. At time t (in minutes), the first insect is at the point (x, y, z) on the line $x = 6 + t$, $y = 8 - t$, $z = 3 + t$. Also, at time t, the second insect is at the point (x, y, z) on the line $x = 1 + t$, $y = 2 + t$, $z = 2t$. Assume that distances are given in inches.

(a) Find the distance between the two insects at time $t = 0$.

(b) Use a graphing utility to graph the distance between the insects from $t = 0$ to $t = 10$.

(c) Using the graph from part (b), what can you conclude about the distance between the insects?

(d) How close to each other do the insects get?

108. Finding an Equation of a Sphere Find the standard equation of the sphere with center $(-3, 2, 4)$ that is tangent to the plane given by $2x + 4y - 3z = 8$.

109. Finding a Point of Intersection Find the point of intersection of the plane $3x - y + 4z = 7$ and the line through $(5, 4, -3)$ that is perpendicular to this plane.

110. Finding the Distance Between a Plane and a Line Show that the plane $2x - y - 3z = 4$ is parallel to the line $x = -2 + 2t$, $y = -1 + 4t$, $z = 4$, and find the distance between them.

111. Finding Parametric Equations Find a set of parametric equations for the line passing through the point $(0, 1, 4)$ that is perpendicular to $\mathbf{u} = \langle 2, -5, 1 \rangle$ and $\mathbf{v} = \langle -3, 1, 4 \rangle$.

112. Finding Parametric Equations Find a set of parametric equations for the line passing through the point $(1, 0, 2)$ that is parallel to the plane given by $x + y + z = 5$ and perpendicular to the line $x = t$, $y = 1 + t$, $z = 1 + t$.

True or False? In Exercises 113–118, determine whether the statement is true or false. If it is false, explain why or give an example that shows it is false.

113. If $\mathbf{v} = a_1\mathbf{i} + b_1\mathbf{j} + c_1\mathbf{k}$ is any vector in the plane given by $a_2 x + b_2 y + c_2 z + d_2 = 0$, then $a_1 a_2 + b_1 b_2 + c_1 c_2 = 0$.

114. Two lines in space are either intersecting or parallel.

115. Two planes in space are either intersecting or parallel.

116. If two lines L_1 and L_2 are each parallel to a plane, then L_1 and L_2 are parallel.

117. If two planes P_1 and P_2 are each perpendicular to a third plane in space, then P_1 and P_2 are parallel.

118. A plane and a line in space are either intersecting or parallel.

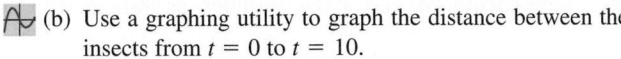

SECTION PROJECT

Distances in Space

You have learned two distance formulas in this section—one for the distance between a point and a plane, and one for the distance between a point and a line. In this project, you will study a third distance problem—the distance between two skew lines. Two lines in space are *skew* if they are neither parallel nor intersecting (see figure).

(a) Consider the following two lines in space.

L_1: $x = 4 + 5t$, $y = 5 + 5t$, $z = 1 - 4t$
L_2: $x = 4 + s$, $y = -6 + 8s$, $z = 7 - 3s$

(i) Show that these lines are not parallel.

(ii) Show that these lines do not intersect and therefore are skew lines.

(iii) Show that the two lines lie in parallel planes.

(iv) Find the distance between the parallel planes from part (iii). This is the distance between the original skew lines.

(b) Use the procedure in part (a) to find the distance between the lines.

L_1: $x = 2t$, $y = 4t$, $z = 6t$
L_2: $x = 1 - s$, $y = 4 + s$, $z = -1 + s$

(c) Use the procedure in part (a) to find the distance between the lines.

L_1: $x = 3t$, $y = 2 - t$, $z = -1 + t$
L_2: $x = 1 + 4s$, $y = -2 + s$, $z = -3 - 3s$

(d) Develop a formula for finding the distance between the skew lines.

L_1: $x = x_1 + a_1 t$, $y = y_1 + b_1 t$, $z = z_1 + c_1 t$
L_2: $x = x_2 + a_2 s$, $y = y_2 + b_2 s$, $z = z_2 + c_2 s$

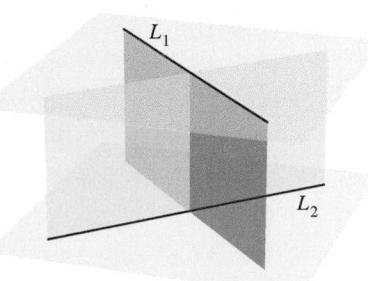

11.6 Surfaces in Space

- Recognize and write equations of cylindrical surfaces.
- Recognize and write equations of quadric surfaces.
- Recognize and write equations of surfaces of revolution.

Cylindrical Surfaces

The first five sections of this chapter contained the vector portion of the preliminary work necessary to study vector calculus and the calculus of space. In this and the next section, you will study surfaces in space and alternative coordinate systems for space. You have already studied two special types of surfaces.

1. Spheres: $(x - x_0)^2 + (y - y_0)^2 + (z - z_0)^2 = r^2$ Section 11.2

2. Planes: $ax + by + cz + d = 0$ Section 11.5

A third type of surface in space is a **cylindrical surface,** or simply a **cylinder.** To define a cylinder, consider the familiar right circular cylinder shown in Figure 11.56. The cylinder was generated by a vertical line moving around the circle $x^2 + y^2 = a^2$ in the xy-plane. This circle is a **generating curve** for the cylinder, as indicated in the next definition.

Definition of a Cylinder

Let C be a curve in a plane and let L be a line not in a parallel plane. The set of all lines parallel to L and intersecting C is a **cylinder.** The curve C is the **generating curve** (or **directrix**) of the cylinder, and the parallel lines are **rulings.**

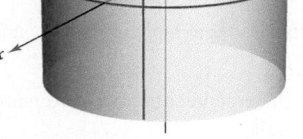

Right circular cylinder:
$x^2 + y^2 = a^2$

Rulings are parallel to z-axis
Figure 11.56

Without loss of generality, you can assume that C lies in one of the three coordinate planes. Moreover, this text restricts the discussion to *right* cylinders—cylinders whose rulings are perpendicular to the coordinate plane containing C, as shown in Figure 11.57. Note that the rulings intersect C and are parallel to the line L.

For the right circular cylinder shown in Figure 11.56, the equation of the generating curve in the xy-plane is

$$x^2 + y^2 = a^2.$$

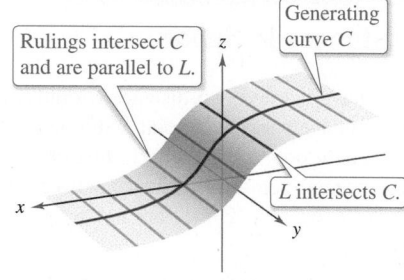

Right cylinder: A cylinder whose rulings are perpendicular to the coordinate plane containing C
Figure 11.57

To find an equation of the cylinder, note that you can generate any one of the rulings by fixing the values of x and y and then allowing z to take on all real values. In this sense, the value of z is arbitrary and is, therefore, not included in the equation. In other words, the equation of this cylinder is simply the equation of its generating curve.

$$x^2 + y^2 = a^2$$ Equation of cylinder in space

Equations of Cylinders

The equation of a cylinder whose rulings are parallel to one of the coordinate axes contains only the variables corresponding to the other two axes.

EXAMPLE 1   **Sketching a Cylinder**

Sketch the surface represented by each equation.

a. $z = y^2$ **b.** $z = \sin x, \quad 0 \le x \le 2\pi$

Solution

a. The graph is a cylinder whose generating curve, $z = y^2$, is a parabola in the yz-plane. The rulings of the cylinder are parallel to the x-axis, as shown in Figure 11.58(a).

b. The graph is a cylinder generated by the sine curve in the xz-plane. The rulings are parallel to the y axis, as shown in Figure 11.58(b).

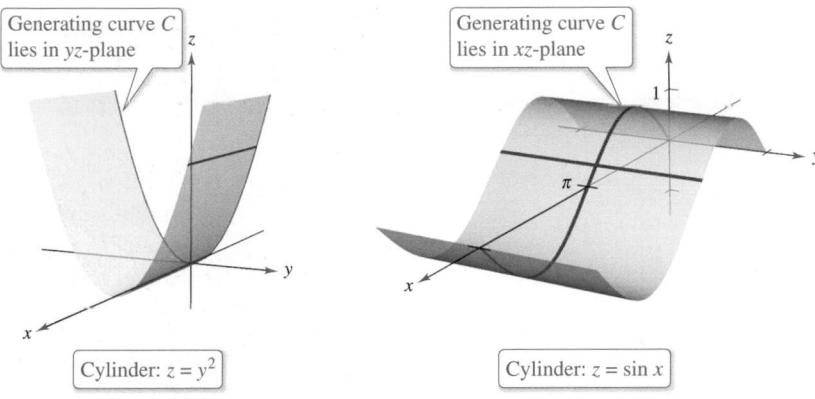

(a) Rulings are parallel to x-axis.	**(b)** Rulings are parallel to y-axis.

Figure 11.58

Quadric Surfaces

The fourth basic type of surface in space is a **quadric surface.** Quadric surfaces are the three-dimensional analogs of conic sections.

Quadric Surface

The equation of a **quadric surface** in space is a second-degree equation in three variables. The **general form** of the equation is

$$Ax^2 + By^2 + Cz^2 + Dxy + Exz + Fyz + Gx + Hy + Iz + J = 0.$$

There are six basic types of quadric surfaces: **ellipsoid, hyperboloid of one sheet, hyperboloid of two sheets, elliptic cone, elliptic paraboloid,** and **hyperbolic paraboloid.**

The intersection of a surface with a plane is called the **trace of the surface** in the plane. To visualize a surface in space, it is helpful to determine its traces in some well-chosen planes. The traces of quadric surfaces are conics. These traces, together with the **standard form** of the equation of each quadric surface, are shown in the table on the next two pages.

In the table on the next two pages, only one of several orientations of each quadric surface is shown. When the surface is oriented along a different axis, its standard equation will change accordingly, as illustrated in Examples 2 and 3. The fact that the two types of paraboloids have one variable raised to the first power can be helpful in classifying quadric surfaces. The other four types of basic quadric surfaces have equations that are of *second degree* in all three variables.

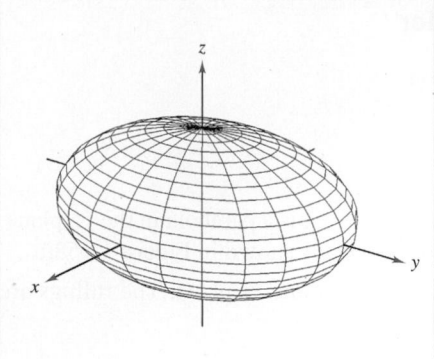

Ellipsoid

$$\frac{x^2}{a^2} + \frac{y^2}{b^2} + \frac{z^2}{c^2} = 1$$

Trace	Plane
Ellipse	Parallel to xy-plane
Ellipse	Parallel to xz-plane
Ellipse	Parallel to yz-plane

The surface is a sphere when $a = b = c \neq 0$.

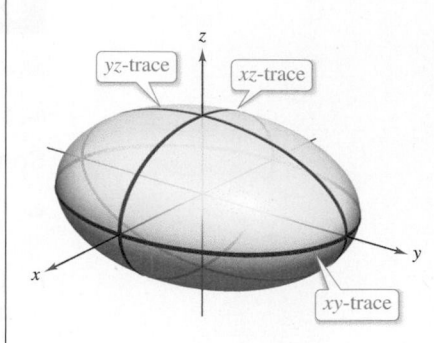

Hyperboloid of One Sheet

$$\frac{x^2}{a^2} + \frac{y^2}{b^2} - \frac{z^2}{c^2} = 1$$

Trace	Plane
Ellipse	Parallel to xy-plane
Hyperbola	Parallel to xz-plane
Hyperbola	Parallel to yz-plane

The axis of the hyperboloid corresponds to the variable whose coefficient is negative.

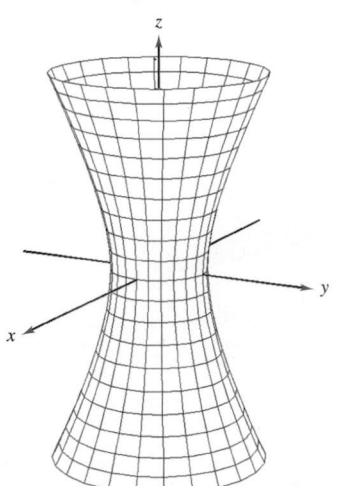

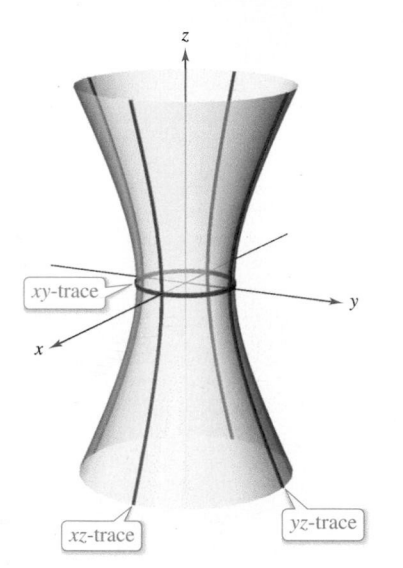

Hyperboloid of Two Sheets

$$\frac{z^2}{c^2} - \frac{x^2}{a^2} - \frac{y^2}{b^2} = 1$$

Trace	Plane
Ellipse	Parallel to xy-plane
Hyperbola	Parallel to xz-plane
Hyperbola	Parallel to yz-plane

The axis of the hyperboloid corresponds to the variable whose coefficient is positive. There is no trace in the coordinate plane perpendicular to this axis.

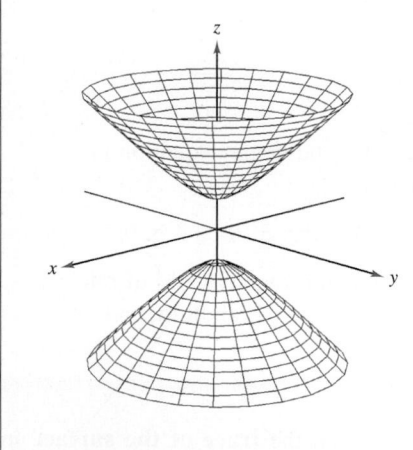

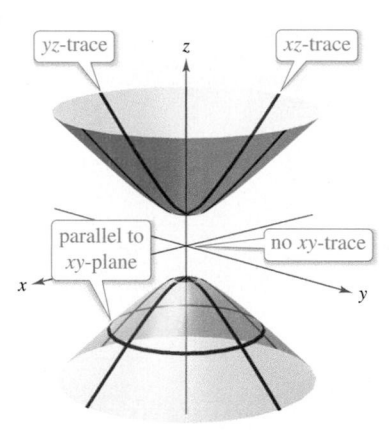

Elliptic Cone

$$\frac{x^2}{a^2} + \frac{y^2}{b^2} - \frac{z^2}{c^2} = 0$$

Trace	Plane
Ellipse	Parallel to xy-plane
Hyperbola	Parallel to xz-plane
Hyperbola	Parallel to yz-plane

The axis of the cone corresponds to the variable whose coefficient is negative. The traces in the coordinate planes parallel to this axis are intersecting lines.

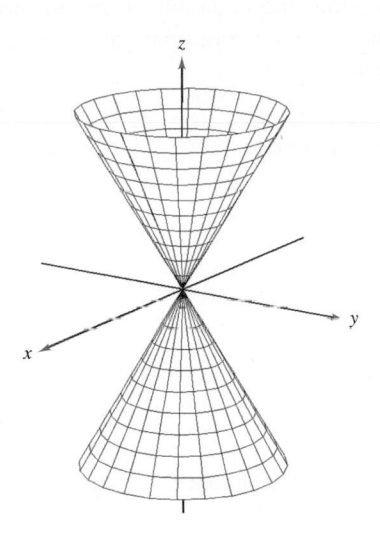

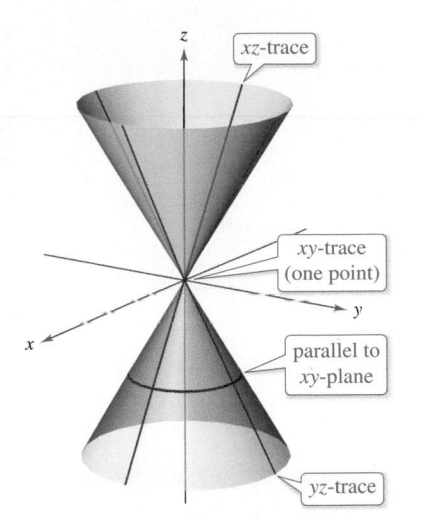

Elliptic Paraboloid

$$z = \frac{x^2}{a^2} + \frac{y^2}{b^2}$$

Trace	Plane
Ellipse	Parallel to xy-plane
Parabola	Parallel to xz-plane
Parabola	Parallel to yz-plane

The axis of the paraboloid corresponds to the variable raised to the first power.

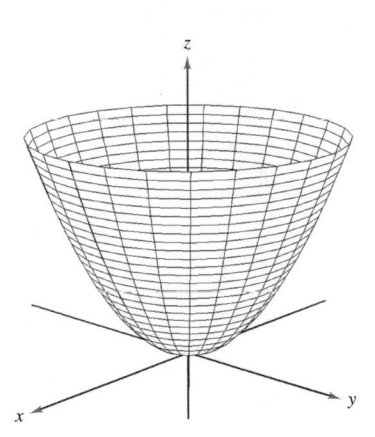

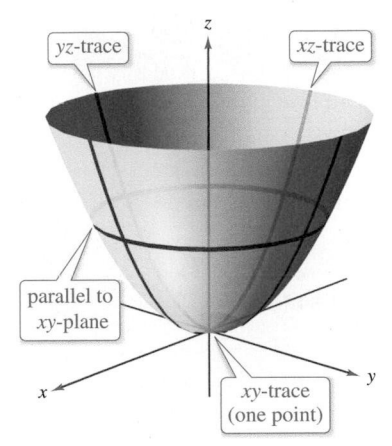

Hyperbolic Paraboloid

$$z = \frac{y^2}{b^2} - \frac{x^2}{a^2}$$

Trace	Plane
Hyperbola	Parallel to xy-plane
Parabola	Parallel to xz-plane
Parabola	Parallel to yz-plane

The axis of the paraboloid corresponds to the variable raised to the first power.

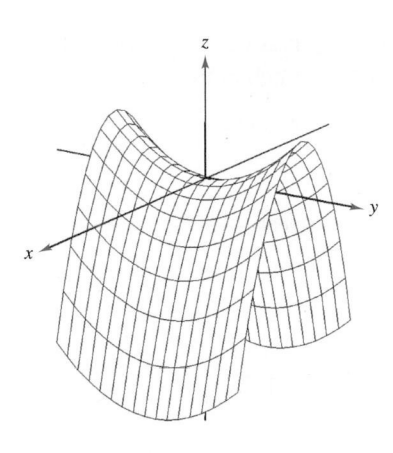

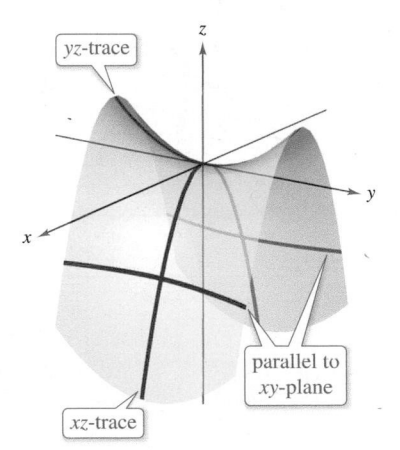

To classify a quadric surface, begin by writing the equation of the surface in standard form. Then, determine several traces taken in the coordinate planes *or* taken in planes that are parallel to the coordinate planes.

EXAMPLE 2   **Sketching a Quadric Surface**

Classify and sketch the surface

$$4x^2 - 3y^2 + 12z^2 + 12 = 0.$$

Solution   Begin by writing the equation in standard form.

$$4x^2 - 3y^2 + 12z^2 + 12 = 0 \qquad \text{Write original equation.}$$

$$\frac{x^2}{-3} + \frac{y^2}{4} - z^2 - 1 = 0 \qquad \text{Divide by } -12.$$

$$\frac{y^2}{4} - \frac{x^2}{3} - z^2 = 1 \qquad \text{Standard form}$$

From the table on pages 800 and 801, you can conclude that the surface is a hyperboloid of two sheets with the *y*-axis as its axis. To sketch the graph of this surface, it helps to find the traces in the coordinate planes.

$$xy\text{-trace } (z = 0): \quad \frac{y^2}{4} - \frac{x^2}{3} = 1 \qquad \text{Hyperbola}$$

$$xz\text{-trace } (y = 0): \quad \frac{x^2}{3} + \frac{z^2}{1} = -1 \qquad \text{No trace}$$

$$yz\text{-trace } (x = 0): \quad \frac{y^2}{4} - \frac{z^2}{1} = 1 \qquad \text{Hyperbola}$$

The graph is shown in Figure 11.59.

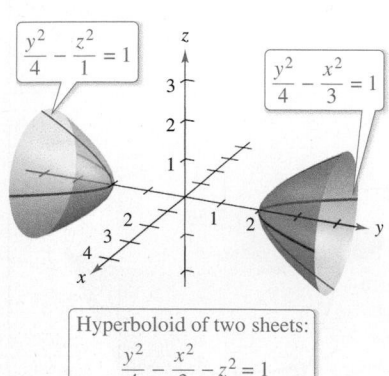

$$\frac{y^2}{4} - \frac{z^2}{1} = 1$$

$$\frac{y^2}{4} - \frac{x^2}{3} = 1$$

Hyperboloid of two sheets:
$$\frac{y^2}{4} - \frac{x^2}{3} - z^2 = 1$$

Figure 11.59

EXAMPLE 3   **Sketching a Quadric Surface**

Classify and sketch the surface

$$x - y^2 - 4z^2 = 0.$$

Solution   Because *x* is raised only to the first power, the surface is a paraboloid. The axis of the paraboloid is the *x*-axis. In standard form, the equation is

$$x = y^2 + 4z^2. \qquad \text{Standard form}$$

Some convenient traces are listed below.

xy-trace $(z = 0)$:	$x = y^2$	Parabola
xz-trace $(y = 0)$:	$x = 4z^2$	Parabola
parallel to yz-plane $(x = 4)$:	$\dfrac{y^2}{4} + \dfrac{z^2}{1} = 1$	Ellipse

The surface is an *elliptic* paraboloid, as shown in Figure 11.60.

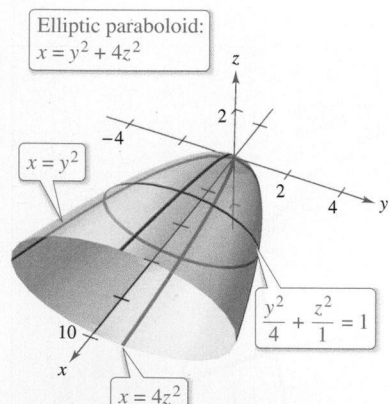

Elliptic paraboloid:
$$x = y^2 + 4z^2$$

$$x = y^2$$

$$\frac{y^2}{4} + \frac{z^2}{1} = 1$$

$$x = 4z^2$$

Figure 11.60

Some second-degree equations in *x*, *y*, and *z* do not represent any of the basic types of quadric surfaces. For example, the graph of

$$x^2 + y^2 + z^2 = 0 \qquad \text{Single point}$$

is a single point, and the graph of

$$x^2 + y^2 = 1 \qquad \text{Right circular cylinder}$$

is a right circular cylinder.

For a quadric surface not centered at the origin, you can form the standard equation by completing the square, as demonstrated in Example 4.

EXAMPLE 4 **A Quadric Surface Not Centered at the Origin**

···▷ *See LarsonCalculus.com for an interactive version of this type of example.*

Classify and sketch the surface

$$x^2 + 2y^2 + z^2 - 4x + 4y - 2z + 3 = 0.$$

Solution Begin by grouping terms and factoring where possible.

$$x^2 - 4x + 2(y^2 + 2y) + z^2 - 2z = -3$$

Next, complete the square for each variable and write the equation in standard form.

$$(x^2 - 4x + \quad) + 2(y^2 + 2y + \quad) + (z^2 - 2z + \quad) = -3$$
$$(x^2 - 4x + 4) + 2(y^2 + 2y + 1) + (z^2 - 2z + 1) = -3 + 4 + 2 + 1$$
$$(x - 2)^2 + 2(y + 1)^2 + (z - 1)^2 = 4$$
$$\frac{(x - 2)^2}{4} + \frac{(y + 1)^2}{2} + \frac{(z - 1)^2}{4} = 1$$

From this equation, you can see that the quadric surface is an ellipsoid that is centered at $(2, -1, 1)$. Its graph is shown in Figure 11.61. ∎

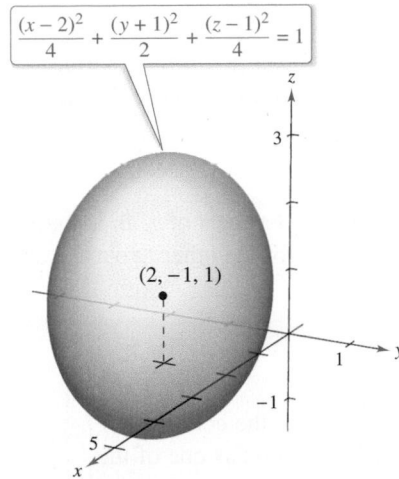

$$\frac{(x - 2)^2}{4} + \frac{(y + 1)^2}{2} + \frac{(z - 1)^2}{4} = 1$$

$(2, -1, 1)$

An ellipsoid centered at $(2, -1, 1)$
Figure 11.61

▷ **TECHNOLOGY** A 3-D graphing utility can help you visualize a surface in space.* Such a graphing utility may create a three-dimensional graph by sketching several traces of the surface and then applying a "hidden-line" routine that blocks out portions of the surface that lie behind other portions of the surface. Two examples of figures that were generated by *Mathematica* are shown below.

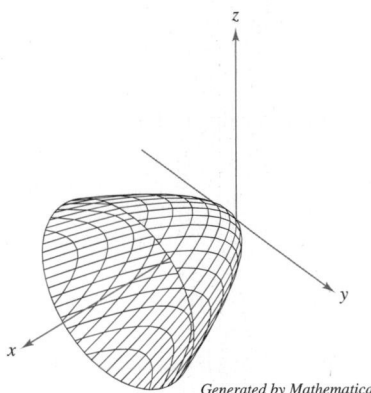

Generated by Mathematica

Elliptic paraboloid
$$x = \frac{y^2}{2} + \frac{z^2}{2}$$

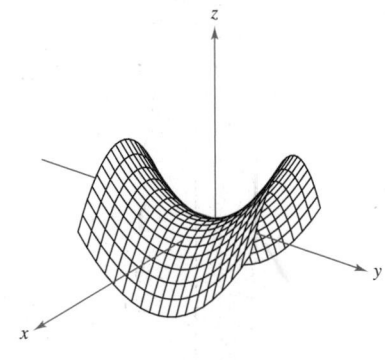

Generated by Mathematica

Hyperbolic paraboloid
$$z = \frac{y^2}{16} - \frac{x^2}{16}$$

Using a graphing utility to graph a surface in space requires practice. For one thing, you must know enough about the surface to be able to specify a *viewing window* that gives a representative view of the surface. Also, you can often improve the view of a surface by rotating the axes. For instance, note that the elliptic paraboloid in the figure is seen from a line of sight that is "higher" than the line of sight used to view the hyperbolic paraboloid.

* Some 3-D graphing utilities require surfaces to be entered with parametric equations. For a discussion of this technique, see Section 15.5.

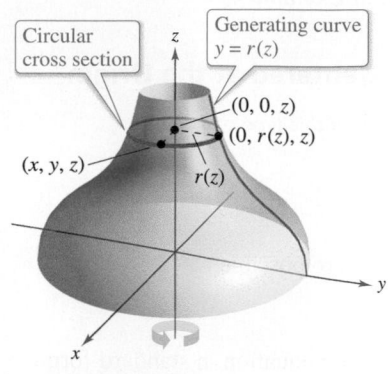

Circular cross section

Generating curve $y = r(z)$

$(0, 0, z)$

$(0, r(z), z)$

(x, y, z)

$r(z)$

Figure 11.62

Surfaces of Revolution

The fifth special type of surface you will study is a **surface of revolution.** In Section 7.4, you studied a method for finding the *area* of such a surface. You will now look at a procedure for finding its *equation.* Consider the graph of the **radius function**

$$y = r(z) \qquad \text{Generating curve}$$

in the yz-plane. When this graph is revolved about the z-axis, it forms a surface of revolution, as shown in Figure 11.62. The trace of the surface in the plane $z = z_0$ is a circle whose radius is $r(z_0)$ and whose equation is

$$x^2 + y^2 = [r(z_0)]^2. \qquad \text{Circular trace in plane: } z = z_0$$

Replacing z_0 with z produces an equation that is valid for all values of z. In a similar manner, you can obtain equations for surfaces of revolution for the other two axes, and the results are summarized as follows.

Surface of Revolution

If the graph of a radius function r is revolved about one of the coordinate axes, then the equation of the resulting surface of revolution has one of the forms listed below.

1. Revolved about the x-axis: $y^2 + z^2 = [r(x)]^2$
2. Revolved about the y-axis: $x^2 + z^2 = [r(y)]^2$
3. Revolved about the z-axis: $x^2 + y^2 = [r(z)]^2$

EXAMPLE 5 Finding an Equation for a Surface of Revolution

Find an equation for the surface of revolution formed by revolving (a) the graph of $y = 1/z$ about the z-axis and (b) the graph of $9x^2 = y^3$ about the y-axis.

Solution

a. An equation for the surface of revolution formed by revolving the graph of

$$y = \frac{1}{z} \qquad \text{Radius function}$$

about the z-axis is

$$x^2 + y^2 = [r(z)]^2 \qquad \text{Revolved about the } z\text{-axis}$$

$$x^2 + y^2 = \left(\frac{1}{z}\right)^2. \qquad \text{Substitute } 1/z \text{ for } r(z).$$

b. To find an equation for the surface formed by revolving the graph of $9x^2 = y^3$ about the y-axis, solve for x in terms of y to obtain

$$x = \frac{1}{3}y^{3/2} = r(y). \qquad \text{Radius function}$$

So, the equation for this surface is

$$x^2 + z^2 = [r(y)]^2 \qquad \text{Revolved about the } y\text{-axis}$$

$$x^2 + z^2 = \left(\frac{1}{3}y^{3/2}\right)^2 \qquad \text{Substitute } \tfrac{1}{3}y^{3/2} \text{ for } r(y).$$

$$x^2 + z^2 = \frac{1}{9}y^3. \qquad \text{Equation of surface}$$

The graph is shown in Figure 11.63.

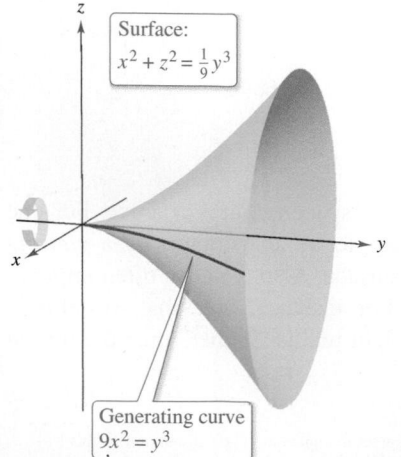

Surface: $x^2 + z^2 = \frac{1}{9}y^3$

Generating curve $9x^2 = y^3$

Figure 11.63

The generating curve for a surface of revolution is not unique. For instance, the surface

$$x^2 + z^2 = e^{-2y}$$

can be formed by revolving either the graph of

$$x = e^{-y}$$

about the y-axis or the graph of

$$z = e^{-y}$$

about the y-axis, as shown in Figure 11.64.

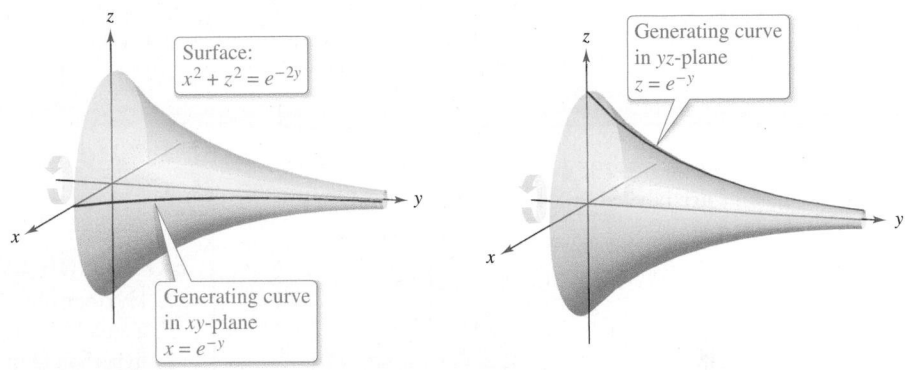

Surface:
$x^2 + z^2 = e^{-2y}$

Generating curve
in xy-plane
$x = e^{-y}$

Generating curve
in yz-plane
$z = e^{-y}$

Figure 11.64

EXAMPLE 6 **Finding a Generating Curve**

Find a generating curve and the axis of revolution for the surface

$$x^2 + 3y^2 + z^2 = 9.$$

Solution The equation has one of the forms listed below.

$$x^2 + y^2 = [r(z)]^2 \qquad \text{Revolved about } z\text{-axis}$$
$$y^2 + z^2 = [r(x)]^2 \qquad \text{Revolved about } x\text{-axis}$$
$$x^2 + z^2 = [r(y)]^2 \qquad \text{Revolved about } y\text{-axis}$$

Because the coefficients of x^2 and z^2 are equal, you should choose the third form and write

$$x^2 + z^2 = 9 - 3y^2.$$

The y-axis is the axis of revolution. You can choose a generating curve from either of the traces

$$x^2 = 9 - 3y^2 \qquad \text{Trace in } xy\text{-plane}$$

or

$$z^2 = 9 - 3y^2. \qquad \text{Trace in } yz\text{-plane}$$

For instance, using the first trace, the generating curve is the semiellipse

$$x = \sqrt{9 - 3y^2}. \qquad \text{Generating curve}$$

The graph of this surface is shown in Figure 11.65.

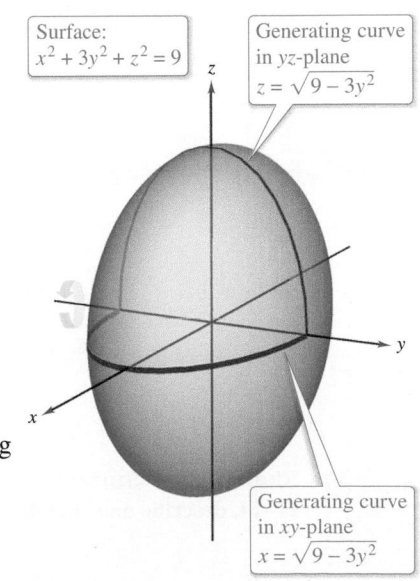

Surface:
$x^2 + 3y^2 + z^2 = 9$

Generating curve
in yz-plane
$z = \sqrt{9 - 3y^2}$

Generating curve
in xy-plane
$x = \sqrt{9 - 3y^2}$

Figure 11.65

11.6 Exercises

See **CalcChat.com** for tutorial help and worked-out solutions to odd-numbered exercises.

CONCEPT CHECK

1. **Quadric Surfaces** How are quadric surfaces and conic sections related?

2. **Classifying an Equation** What does the equation $z = x^2$ represent in the xz-plane? What does it represent in three-space?

3. **Trace of a Surface** What is meant by the trace of a surface? How do you find a trace?

4. **Think About It** Does every second-degree equation in x, y, and z represent a quadric surface? Explain.

Matching In Exercises 5–10, match the equation with its graph. [The graphs are labeled (a), (b), (c), (d), (e), and (f).]

(a)

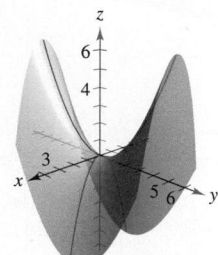

(b)

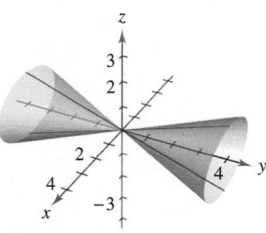

(c)

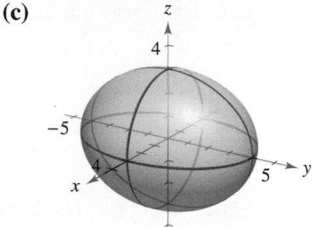

(d)

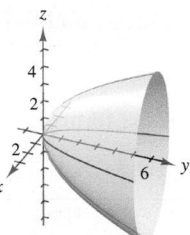

(e)

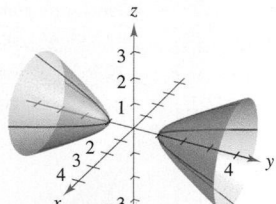

(f)

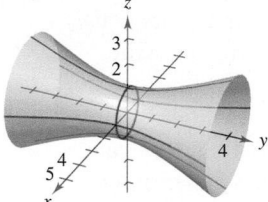

5. $\dfrac{x^2}{9} + \dfrac{y^2}{16} + \dfrac{z^2}{9} = 1$

6. $15x^2 - 4y^2 + 15z^2 = -4$

7. $4x^2 - y^2 + 4z^2 = 4$

8. $y^2 = 4x^2 + 9z^2$

9. $4x^2 - 4y + z^2 = 0$

10. $4x^2 - y^2 + 4z = 0$

Sketching a Surface in Space In Exercises 11–14, describe and sketch the surface.

11. $y^2 + z^2 = 9$

12. $y^2 + z = 6$

13. $4x^2 + y^2 = 4$

14. $y^2 - z^2 = 25$

Sketching a Quadric Surface In Exercises 15–26, classify and sketch the quadric surface. Use a computer algebra system or a graphing utility to confirm your sketch.

15. $4x^2 - y^2 - z^2 = 1$

16. $\dfrac{x^2}{16} + \dfrac{y^2}{25} + \dfrac{z^2}{25} = 1$

17. $16x^2 - y^2 + 16z^2 = 4$

18. $z = x^2 + 4y^2$

19. $x^2 + \dfrac{y^2}{4} + z^2 = 1$

20. $z^2 - x^2 - \dfrac{y^2}{4} = 1$

21. $z^2 = x^2 + \dfrac{y^2}{9}$

22. $3z = -y^2 + x^2$

23. $x^2 - y^2 + z = 0$

24. $x^2 = 2y^2 + 2z^2$

25. $x^2 - y + z^2 = 0$

26. $-8x^2 + 18y^2 + 18z^2 = 2$

EXPLORING CONCEPTS

27. **Hyperboloid** Explain how to determine whether a quadric surface is a hyperboloid of one sheet or a hyperboloid of two sheets.

28. **Ellipsoid** Is every trace of an ellipsoid an ellipse? Explain.

29. **Quadric Surface** Is there a quadric surface whose traces are all parabolas? Explain.

30. **HOW DO YOU SEE IT?** The four figures below are graphs of the quadric surface $z = x^2 + y^2$. Match each of the four graphs with the point in space from which the paraboloid is viewed.

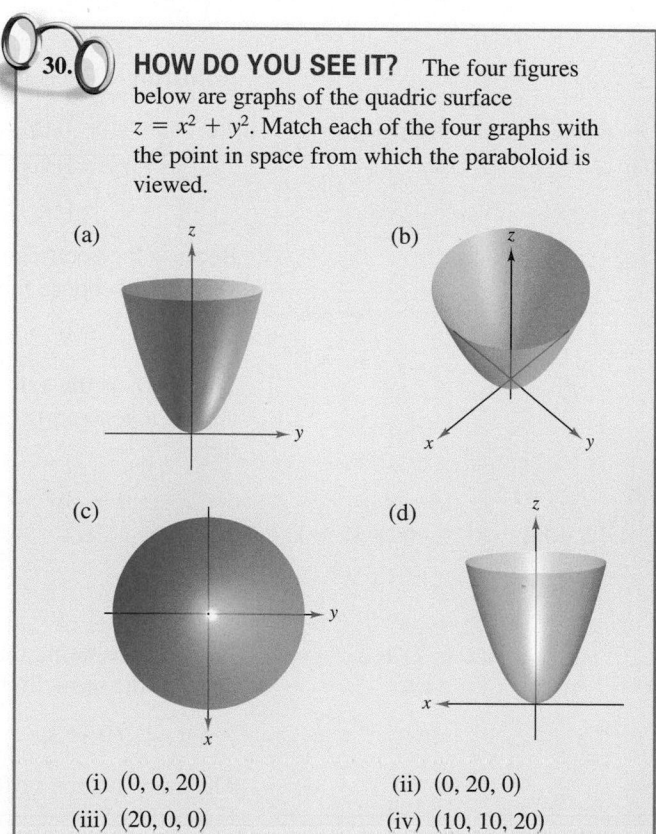

(i) $(0, 0, 20)$

(ii) $(0, 20, 0)$

(iii) $(20, 0, 0)$

(iv) $(10, 10, 20)$

 **Finding an Equation for a Surface of Revolution** In Exercises 31–36, find an equation for the surface of revolution formed by revolving the curve in the indicated coordinate plane about the given axis.

Equation of Curve	Coordinate Plane	Axis of Revolution
31. $z = 5y$	yz-plane	y-axis
32. $z^2 = 9y$	yz-plane	y-axis
33. $y^3 = 8z$	yz plane	z-axis
34. $z = \ln x$	xz-plane	z-axis
35. $xy = 2$	xy-plane	x-axis
36. $2z = \sqrt{4 - x^2}$	xz-plane	x-axis

 **Finding a Generating Curve** In Exercises 37–40, find an equation of a generating curve given the equation of its surface of revolution.

37. $x^2 + y^2 - 2z = 0$

38. $x^2 + z^2 = \cos^2 y$

39. $8x^2 + y^2 + z^2 = 5$

40. $6x^2 + 2y^2 + 2z^2 = 1$

Finding the Volume of a Solid In Exercises 41 and 42, use the shell method to find the volume of the solid below the surface of revolution and above the xy-plane.

41. The curve $z = 4x - x^2$ in the xz-plane is revolved about the z-axis.

42. The curve

$$z = \sin y, \quad 0 \le y \le \pi$$

in the yz-plane is revolved about the z-axis.

Analyzing a Trace In Exercises 43 and 44, analyze the trace when the surface

$$z = \tfrac{1}{2}x^2 + \tfrac{1}{4}y^2$$

is intersected by the indicated planes.

43. Find the lengths of the major and minor axes and the coordinates of the foci of the ellipse generated when the surface is intersected by the planes given by

(a) $z = 2$ and   (b) $z = 8$.

44. Find the coordinates of the focus of the parabola formed when the surface is intersected by the planes given by

(a) $y = 4$ and   (b) $x = 2$.

Finding an Equation of a Surface In Exercises 45 and 46, find an equation of the surface satisfying the conditions, and identify the surface.

45. The set of all points equidistant from the point $(0, 2, 0)$ and the plane $y = -2$

46. The set of all points equidistant from the point $(0, 0, 4)$ and the xy-plane

47. Geography

Because of the forces caused by its rotation, Earth is an oblate ellipsoid rather than a sphere. The equatorial radius is 3963 miles and the polar radius is 3950 miles. Find an equation of the ellipsoid. (Assume that the center of Earth is at the origin and that the trace formed by the plane $z = 0$ corresponds to the equator.)

48. Machine Design The top of a rubber bushing designed to absorb vibrations in an automobile is the surface of revolution generated by revolving the curve

$$z = \frac{1}{2}y^2 + 1$$

for $0 \le y \le 2$ in the yz-plane about the z-axis.

(a) Find an equation for the surface of revolution.

(b) All measurements are in centimeters and the bushing is set on the xy-plane. Use the shell method to find its volume.

(c) The bushing has a hole of diameter 1 centimeter through its center and parallel to the axis of revolution. Find the volume of the rubber bushing.

49. Using a Hyperbolic Paraboloid Determine the intersection of the hyperbolic paraboloid

$$z = \frac{y^2}{b^2} - \frac{x^2}{a^2}$$

with the plane $bx + ay - z = 0$. (Assume $a, b > 0$.)

50. Intersection of Surfaces Explain why the curve of intersection of the surfaces $x^2 + 3y^2 - 2z^2 + 2y = 4$ and $2x^2 + 6y^2 - 4z^2 - 3x = 2$ lies in a plane.

51. Think About It Three types of classic *topological surfaces* are shown below. The sphere and torus have both an "inside" and an "outside." Does the Klein bottle have both an "inside" and an "outside?" Explain.

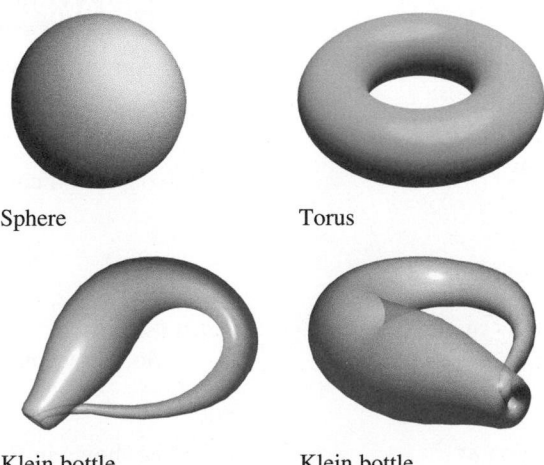

Sphere   Torus

Klein bottle   Klein bottle

11.7 Cylindrical and Spherical Coordinates

■ Use cylindrical coordinates to represent surfaces in space.
■ Use spherical coordinates to represent surfaces in space.

Cylindrical Coordinates

You have already seen that some two-dimensional graphs are easier to represent in polar coordinates than in rectangular coordinates. A similar situation exists for surfaces in space. In this section, you will study two alternative space-coordinate systems. The first, the **cylindrical coordinate system,** is an extension of polar coordinates in the plane to three-dimensional space.

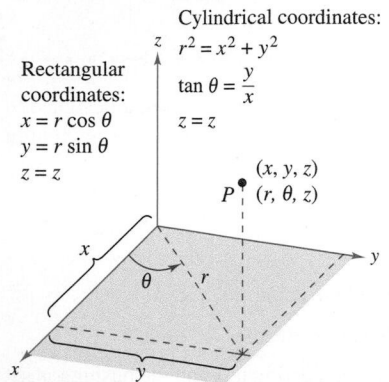

Rectangular coordinates:
$x = r \cos \theta$
$y = r \sin \theta$
$z = z$

Cylindrical coordinates:
$r^2 = x^2 + y^2$
$\tan \theta = \dfrac{y}{x}$
$z = z$

Figure 11.66

The Cylindrical Coordinate System

In a **cylindrical coordinate system,** a point P in space is represented by an ordered triple (r, θ, z).

1. (r, θ) is a polar representation of the projection of P in the xy-plane.
2. z is the directed distance from (r, θ) to P.

To convert from rectangular to cylindrical coordinates (or vice versa), use the conversion guidelines for polar coordinates listed below and illustrated in Figure 11.66.

Cylindrical to rectangular:

$$x = r \cos \theta, \quad y = r \sin \theta, \quad z = z$$

Rectangular to cylindrical:

$$r^2 = x^2 + y^2, \quad \tan \theta = \frac{y}{x}, \quad z = z$$

The point $(0, 0, 0)$ is called the **pole.** Moreover, because the representation of a point in the polar coordinate system is not unique, it follows that the representation in the cylindrical coordinate system is also not unique.

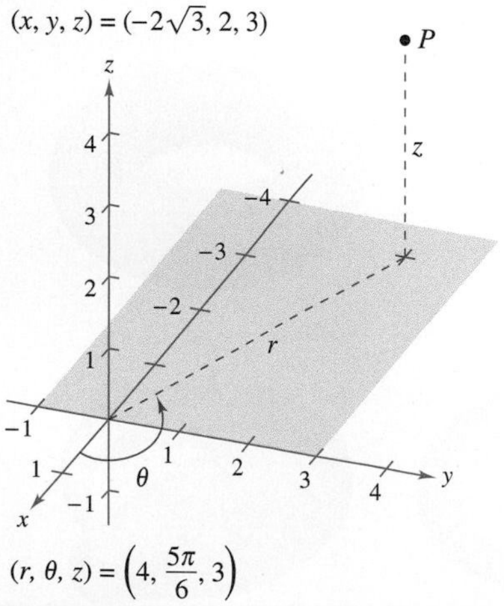

$(x, y, z) = (-2\sqrt{3}, 2, 3)$

$(r, \theta, z) = \left(4, \dfrac{5\pi}{6}, 3\right)$

Figure 11.67

EXAMPLE 1   Cylindrical-to-Rectangular Conversion

Convert the point $(r, \theta, z) = (4, 5\pi/6, 3)$ to rectangular coordinates.

Solution   Using the cylindrical-to-rectangular conversion equations produces

$$x = 4 \cos \frac{5\pi}{6} = 4\left(-\frac{\sqrt{3}}{2}\right) = -2\sqrt{3}$$

$$y = 4 \sin \frac{5\pi}{6} = 4\left(\frac{1}{2}\right) = 2$$

$$z = 3.$$

So, in rectangular coordinates, the point is

$$(x, y, z) = \left(-2\sqrt{3}, 2, 3\right)$$

as shown in Figure 11.67.

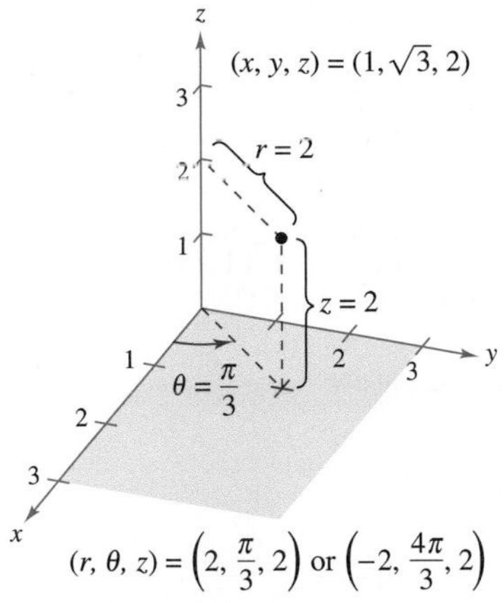

EXAMPLE 2   **Rectangular-to-Cylindrical Conversion**

Convert the point

$$(x, y, z) = \left(1, \sqrt{3}, 2\right)$$

to cylindrical coordinates.

Solution   Use the rectangular-to-cylindrical conversion equations.

$$r = \pm\sqrt{1 + 3} = \pm 2$$

$$\tan \theta = \sqrt{3} \implies \theta = \arctan\sqrt{3} + n\pi = \frac{\pi}{3} + n\pi$$

$$z = 2$$

You have two choices for r and infinitely many choices for θ. As shown in Figure 11.68, two convenient representations of the point are

$$\left(2, \frac{\pi}{3}, 2\right) \qquad\qquad r > 0 \text{ and } \theta \text{ in Quadrant I}$$

and

$$\left(-2, \frac{4\pi}{3}, 2\right). \qquad r < 0 \text{ and } \theta \text{ in Quadrant III}$$

(Figure 11.68 label:) $(r, \theta, z) = \left(2, \frac{\pi}{3}, 2\right)$ or $\left(-2, \frac{4\pi}{3}, 2\right)$

Figure 11.68

Cylindrical coordinates are especially convenient for representing cylindrical surfaces and surfaces of revolution with the z-axis as the axis of symmetry, as shown in Figure 11.69.

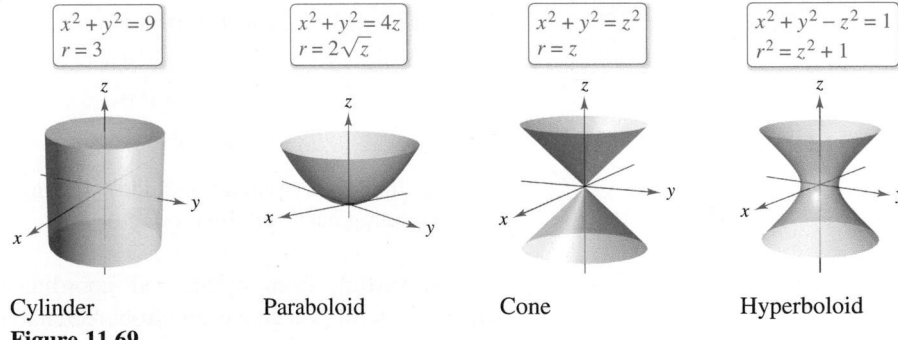

| $x^2 + y^2 = 9$
 $r = 3$ | $x^2 + y^2 = 4z$
 $r = 2\sqrt{z}$ | $x^2 + y^2 = z^2$
 $r = z$ | $x^2 + y^2 - z^2 = 1$
 $r^2 = z^2 + 1$ |

Cylinder   Paraboloid   Cone   Hyperboloid

Figure 11.69

Vertical planes containing the z-axis and horizontal planes also have simple cylindrical coordinate equations, as shown in Figure 11.70.

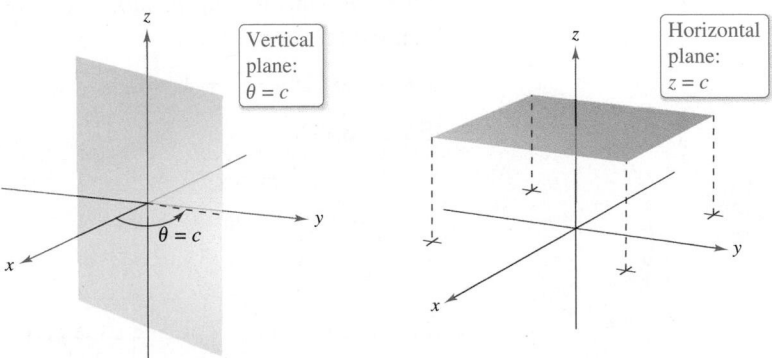

Vertical plane: $\theta = c$

Horizontal plane: $z = c$

Figure 11.70

EXAMPLE 3 **EXAMPLE 3** **Rectangular-to-Cylindrical Conversion**

Find an equation in cylindrical coordinates for the surface represented by each rectangular equation.

a. $x^2 + y^2 = 4z^2$

b. $y^2 = x$

Solution

a. From Section 11.6, you know that the graph of

$$x^2 + y^2 = 4z^2$$

is an elliptic cone with its axis along the z-axis, as shown in Figure 11.71. When you replace $x^2 + y^2$ with r^2, the equation in cylindrical coordinates is

| $x^2 + y^2 = 4z^2$ | Rectangular equation |
| $r^2 = 4z^2$. | Cylindrical equation |

b. The graph of the surface

$$y^2 = x$$

is a parabolic cylinder with rulings parallel to the z-axis, as shown in Figure 11.72. To obtain the equation in cylindrical coordinates, replace y^2 with $r^2 \sin^2 \theta$ and x with $r \cos \theta$, as shown.

$y^2 = x$	Rectangular equation
$r^2 \sin^2 \theta = r \cos \theta$	Substitute $r \sin \theta$ for y and $r \cos \theta$ for x.
$r(r \sin^2 \theta - \cos \theta) = 0$	Collect terms and factor.
$r \sin^2 \theta - \cos \theta = 0$	Divide each side by r.
$r = \dfrac{\cos \theta}{\sin^2 \theta}$	Solve for r.
$r = \csc \theta \cot \theta$	Cylindrical equation

Note that this equation includes a point for which $r = 0$, so nothing was lost by dividing each side by the factor r. ■

Converting from cylindrical coordinates to rectangular coordinates is less straightforward than converting from rectangular coordinates to cylindrical coordinates, as demonstrated in Example 4.

EXAMPLE 4 **EXAMPLE 4** **Cylindrical-to-Rectangular Conversion**

Find an equation in rectangular coordinates for the surface represented by the cylindrical equation

$$r^2 \cos 2\theta + z^2 + 1 = 0.$$

Solution

$r^2 \cos 2\theta + z^2 + 1 = 0$	Cylindrical equation
$r^2(\cos^2 \theta - \sin^2 \theta) + z^2 + 1 = 0$	Trigonometric identity
$r^2 \cos^2 \theta - r^2 \sin^2 \theta + z^2 = -1$	
$x^2 - y^2 + z^2 = -1$	Replace $r \cos \theta$ with x and $r \sin \theta$ with y.
$y^2 - x^2 - z^2 = 1$	Rectangular equation

This is a hyperboloid of two sheets whose axis lies along the y-axis, as shown in Figure 11.73. ■

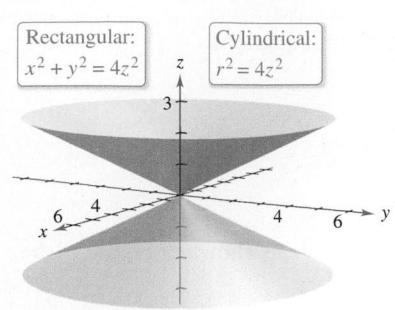

Rectangular: $x^2 + y^2 = 4z^2$ Cylindrical: $r^2 = 4z^2$

Figure 11.71

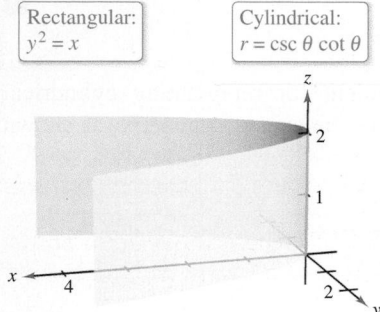

Rectangular: $y^2 = x$ Cylindrical: $r = \csc \theta \cot \theta$

Figure 11.72

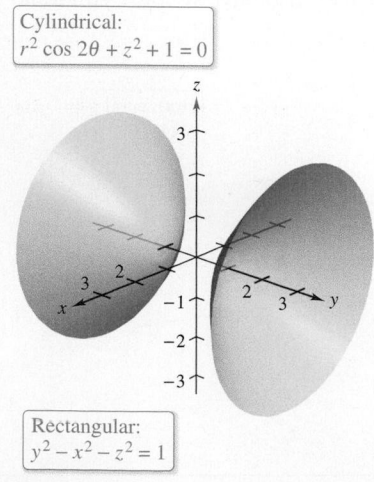

Cylindrical: $r^2 \cos 2\theta + z^2 + 1 = 0$

Rectangular: $y^2 - x^2 - z^2 = 1$

Figure 11.73

Spherical Coordinates

In the **spherical coordinate system,** each point is represented by an ordered triple: the first coordinate is a distance, and the second and third coordinates are angles. This system is similar to the latitude-longitude system used to identify points on the surface of Earth. For example, the point on the surface of Earth whose latitude is 40° North (of the equator) and whose longitude is 80° West (of the prime meridian) is shown in Figure 11.74. Assuming that Earth is spherical and has a radius of 4000 miles, you would label this point as

$$(4000, -80°, 50°).$$

Radius   80° clockwise from   50° down from
prime meridian   North Pole

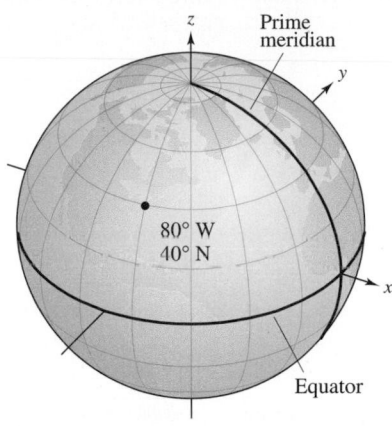

Figure 11.74

The Spherical Coordinate System

In a **spherical coordinate system,** a point P in space is represented by an ordered triple (ρ, θ, ϕ), where ρ is the lowercase Greek letter rho and ϕ is the lowercase Greek letter phi.

1. ρ is the distance between P and the origin, $\rho \geq 0$.
2. θ is the same angle used in cylindrical coordinates for $r \geq 0$.
3. ϕ is the angle *between* the positive z-axis and the line segment $\overrightarrow{OP}$, $0 \leq \phi \leq \pi$.

Note that the first and third coordinates, ρ and ϕ, are nonnegative.

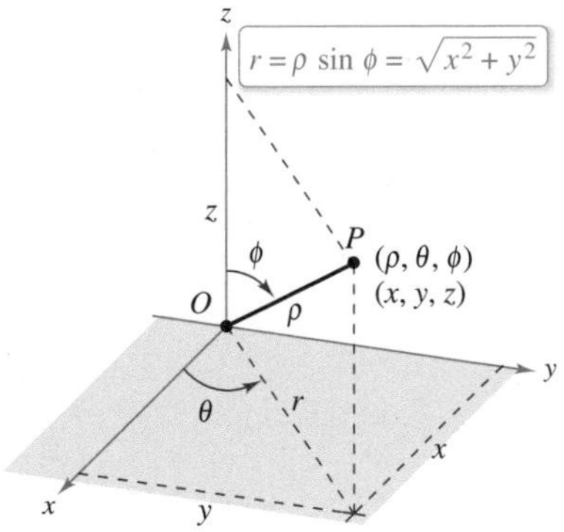

$$r = \rho \sin \phi = \sqrt{x^2 + y^2}$$

P (ρ, θ, ϕ)
(x, y, z)

Spherical coordinates
Figure 11.75

The relationship between rectangular and spherical coordinates is illustrated in Figure 11.75. To convert from one system to the other, use the conversion guidelines listed below.

Spherical to rectangular:

$$x = \rho \sin \phi \cos \theta, \quad y = \rho \sin \phi \sin \theta, \quad z = \rho \cos \phi$$

Rectangular to spherical:

$$\rho^2 = x^2 + y^2 + z^2, \quad \tan \theta = \frac{y}{x}, \quad \phi = \arccos \frac{z}{\sqrt{x^2 + y^2 + z^2}}$$

To change coordinates between the cylindrical and spherical systems, use the conversion guidelines listed below.

Spherical to cylindrical ($r \geq 0$):

$$r^2 = \rho^2 \sin^2 \phi, \quad \theta = \theta, \quad z = \rho \cos \phi$$

Cylindrical to spherical ($r \geq 0$):

$$\rho = \sqrt{r^2 + z^2}, \quad \theta = \theta, \quad \phi = \arccos \frac{z}{\sqrt{r^2 + z^2}}$$

The spherical coordinate system is useful primarily for surfaces in space that have a *point* or *center* of symmetry. For example, Figure 11.76 shows three surfaces with simple spherical equations.

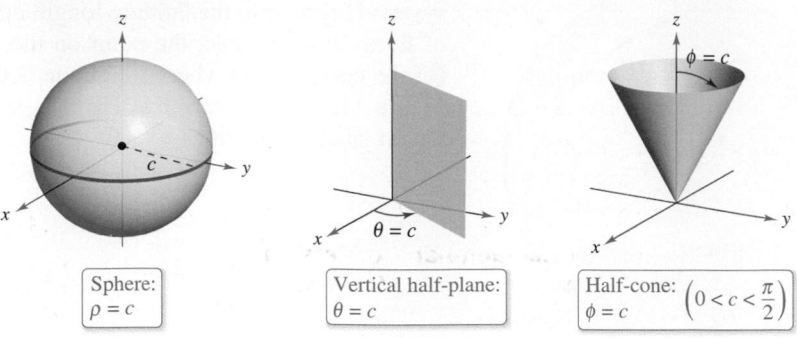

Sphere:
$\rho = c$

Vertical half-plane:
$\theta = c$

Half-cone: $\left(0 < c < \dfrac{\pi}{2}\right)$
$\phi = c$

Figure 11.76

EXAMPLE 5   **Rectangular-to-Spherical Conversion**

⋮ ⋯▷ *See LarsonCalculus.com for an interactive version of this type of example.*

Find an equation in spherical coordinates for the surface represented by each rectangular equation.

a. Cone: $x^2 + y^2 = z^2$ **b.** Sphere: $x^2 + y^2 + z^2 - 4z = 0$

Solution

a. Use the spherical-to-rectangular equations

$$x = \rho \sin \phi \cos \theta, \quad y = \rho \sin \phi \sin \theta, \quad \text{and} \quad z = \rho \cos \phi$$

and substitute in the rectangular equation as shown.

$$x^2 + y^2 = z^2$$
$$\rho^2 \sin^2 \phi \cos^2 \theta + \rho^2 \sin^2 \phi \sin^2 \theta = \rho^2 \cos^2 \phi$$
$$\rho^2 \sin^2 \phi(\cos^2 \theta + \sin^2 \theta) = \rho^2 \cos^2 \phi$$
$$\rho^2 \sin^2 \phi = \rho^2 \cos^2 \phi$$
$$\frac{\sin^2 \phi}{\cos^2 \phi} = 1 \qquad\qquad \rho \geq 0$$
$$\tan^2 \phi = 1$$
$$\tan \phi = \pm 1$$

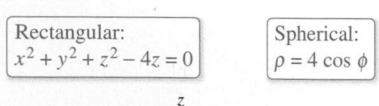

Rectangular:
$x^2 + y^2 + z^2 - 4z = 0$

Spherical:
$\rho = 4 \cos \phi$

So, you can conclude that

$$\phi = \frac{\pi}{4} \quad \text{or} \quad \phi = \frac{3\pi}{4}.$$

The equation $\phi = \pi/4$ represents the *upper* half-cone, and the equation $\phi = 3\pi/4$ represents the *lower* half-cone.

b. Because $\rho^2 = x^2 + y^2 + z^2$ and $z = \rho \cos \phi$, the rectangular equation has the following spherical form.

$$\rho^2 - 4\rho \cos \phi = 0 \quad \Longrightarrow \quad \rho(\rho - 4 \cos \phi) = 0$$

Temporarily discarding the possibility that $\rho = 0$, you have the spherical equation

$$\rho - 4 \cos \phi = 0 \quad \text{or} \quad \rho = 4 \cos \phi.$$

Note that the solution set for this equation includes a point for which $\rho = 0$, so nothing is lost by discarding the factor ρ. The sphere represented by the equation $\rho = 4 \cos \phi$ is shown in Figure 11.77. ◼

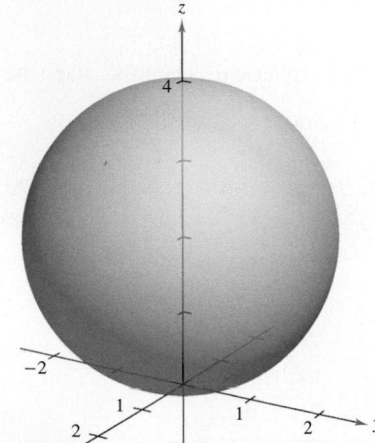

Figure 11.77

11.7 Exercises

See CalcChat.com for tutorial help and worked-out solutions to odd-numbered exercises.

CONCEPT CHECK

1. **Cylindrical Coordinates** Describe the cylindrical coordinate system in your own words.

2. **Spherical Coordinates** Describe the position of the point $(2, 0°, 30°)$ given in spherical coordinates.

 **Cylindrical-to-Rectangular Conversion** In Exercises 3–8, convert the point from cylindrical coordinates to rectangular coordinates.

3. $(-7, 0, 5)$

4. $(2, -\pi, -4)$

5. $\left(3, \dfrac{\pi}{4}, 1\right)$

6. $\left(6, -\dfrac{3\pi}{2}, 2\right)$

7. $\left(4, \dfrac{7\pi}{6}, -3\right)$

8. $\left(-\dfrac{2}{3}, \dfrac{4\pi}{3}, 8\right)$

 **Rectangular-to-Cylindrical Conversion** In Exercises 9–14, convert the point from rectangular coordinates to cylindrical coordinates.

9. $(0, 5, 1)$

10. $\left(6, 2\sqrt{3}, -1\right)$

11. $(2, -2, -4)$

12. $(3, -3, 7)$

13. $\left(1, \sqrt{3}, 4\right)$

14. $\left(2\sqrt{3}, -2, 6\right)$

 **Rectangular-to-Cylindrical Conversion** In Exercises 15–22, find an equation in cylindrical coordinates for the surface represented by the rectangular equation.

15. $z = 4$

16. $x = 9$

17. $x^2 + y^2 - 2z^2 = 5$

18. $z = x^2 + y^2 - 11$

19. $y = x^2$

20. $x^2 + y^2 = 8x$

21. $y^2 = 10 - z^2$

22. $x^2 + y^2 + z^2 - 3z = 0$

 **Cylindrical-to-Rectangular Conversion** In Exercises 23–30, find an equation in rectangular coordinates for the surface represented by the cylindrical equation, and sketch its graph.

23. $r = 3$

24. $z = -2$

25. $\theta = \dfrac{\pi}{6}$

26. $r = \dfrac{1}{2}z$

27. $r^2 + z^2 = 5$

28. $z = r^2 \cos^2 \theta$

29. $r = 4 \sin \theta$

30. $r = 2 \cos \theta$

Rectangular-to-Spherical Conversion In Exercises 31–36, convert the point from rectangular coordinates to spherical coordinates.

31. $(4, 0, 0)$

32. $(-4, 0, 0)$

33. $\left(-2, 2\sqrt{3}, 4\right)$

34. $\left(-5, -5, \sqrt{2}\right)$

35. $\left(\sqrt{3}, 1, 2\sqrt{3}\right)$

36. $(-1, 2, 1)$

 Spherical-to-Rectangular Conversion In Exercises 37–42, convert the point from spherical coordinates to rectangular coordinates.

37. $\left(4, \dfrac{\pi}{6}, \dfrac{\pi}{4}\right)$

38. $\left(6, \pi, \dfrac{\pi}{2}\right)$

39. $\left(12, -\dfrac{\pi}{4}, 0\right)$

40. $\left(9, \dfrac{\pi}{4}, \pi\right)$

41. $\left(5, \dfrac{\pi}{4}, \dfrac{\pi}{12}\right)$

42. $\left(7, \dfrac{3\pi}{4}, \dfrac{\pi}{9}\right)$

 **Rectangular-to-Spherical Conversion** In Exercises 43–50, find an equation in spherical coordinates for the surface represented by the rectangular equation.

43. $y = 2$

44. $z = 6$

45. $x^2 + y^2 + z^2 = 49$

46. $x^2 + y^2 - 3z^2 = 0$

47. $x^2 + y^2 = 16$

48. $x = 13$

49. $x^2 + y^2 = 2z^2$

50. $x^2 + y^2 + z^2 - 9z = 0$

Spherical-to-Rectangular Conversion In Exercises 51–58, find an equation in rectangular coordinates for the surface represented by the spherical equation, and sketch its graph.

51. $\rho = 1$

52. $\theta = \dfrac{3\pi}{4}$

53. $\phi = \dfrac{\pi}{6}$

54. $\phi = \dfrac{\pi}{2}$

55. $\rho = 4 \cos \phi$

56. $\rho = 2 \sec \phi$

57. $\rho = \csc \phi$

58. $\rho = 4 \csc \phi \sec \theta$

Cylindrical-to-Spherical Conversion In Exercises 59–64, convert the point from cylindrical coordinates to spherical coordinates.

59. $\left(4, \dfrac{\pi}{4}, 0\right)$

60. $\left(3, -\dfrac{\pi}{4}, 0\right)$

61. $\left(6, \dfrac{\pi}{2}, -6\right)$

62. $\left(-4, \dfrac{\pi}{3}, 4\right)$

63. $(12, \pi, 5)$

64. $\left(4, \dfrac{\pi}{2}, 3\right)$

Spherical-to-Cylindrical Conversion In Exercises 65–70, convert the point from spherical coordinates to cylindrical coordinates.

65. $\left(10, \dfrac{\pi}{6}, \dfrac{\pi}{2}\right)$

66. $\left(4, \dfrac{\pi}{18}, \dfrac{\pi}{2}\right)$

67. $\left(6, -\dfrac{\pi}{6}, \dfrac{\pi}{3}\right)$

68. $\left(5, -\dfrac{5\pi}{6}, \pi\right)$

69. $\left(8, \dfrac{7\pi}{6}, \dfrac{\pi}{6}\right)$

70. $\left(7, \dfrac{\pi}{4}, \dfrac{3\pi}{4}\right)$

Matching In Exercises 71–76, match the equation (written in terms of cylindrical or spherical coordinates) with its graph. [The graphs are labeled (a), (b), (c), (d), (e), and (f).]

(a)

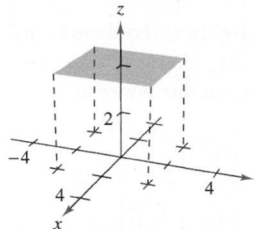

(b)

(c)

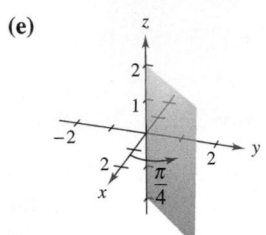

(d)

(e)

(f)

71. $r = 5$

72. $\theta = \dfrac{\pi}{4}$

73. $\rho = 5$

74. $\phi = \dfrac{\pi}{4}$

75. $r^2 = z$

76. $\rho = 4 \sec \phi$

77. Spherical Coordinates Explain why in spherical coordinates the graph of $\theta = c$ is a half-plane and not an entire plane.

78. HOW DO YOU SEE IT? Identify the surface graphed and match the graph with its rectangular equation. Then find an equation in cylindrical coordinates for the equation given in rectangular coordinates.

(a)

(b)

(i) $x^2 + y^2 = \dfrac{4}{9}z^2$

(ii) $x^2 + y^2 - z^2 = 2$

Converting a Rectangular Equation In Exercises 79–86, convert the rectangular equation to an equation in (a) cylindrical coordinates and (b) spherical coordinates.

79. $x^2 + y^2 + z^2 = 27$

80. $4(x^2 + y^2) = z^2$

81. $x^2 + y^2 + z^2 - 2z = 0$

82. $x^2 + y^2 = z$

83. $x^2 + y^2 = 4y$

84. $x^2 + y^2 = 45$

85. $x^2 - y^2 = 9$

86. $y = 4$

Sketching a Solid In Exercises 87–90, sketch the solid that has the given description in cylindrical coordinates.

87. $0 \le \theta \le \pi/2, 0 \le r \le 2, 0 \le z \le 4$

88. $-\pi/2 \le \theta \le \pi/2, 0 \le r \le 3, 0 \le z \le r \cos \theta$

89. $0 \le \theta \le 2\pi, 0 \le r \le a, r \le z \le a$

90. $0 \le \theta \le 2\pi, 2 \le r \le 4, z^2 \le -r^2 + 6r - 8$

Sketching a Solid In Exercises 91–94, sketch the solid that has the given description in spherical coordinates.

91. $0 \le \theta \le 2\pi, 0 \le \phi \le \pi/6, 0 \le \rho \le a \sec \phi$

92. $0 \le \theta \le 2\pi, \pi/4 \le \phi \le \pi/2, 0 \le \rho \le 1$

93. $0 \le \theta \le \pi/2, 0 \le \phi \le \pi/2, 0 \le \rho \le 2$

94. $0 \le \theta \le \pi, 0 \le \phi \le \pi/2, 1 \le \rho \le 3$

EXPLORING CONCEPTS

Think About It In Exercises 95–100, find inequalities that describe the solid and state the coordinate system used. Position the solid on the coordinate system such that the inequalities are as simple as possible.

95. A cube with each edge 10 centimeters long

96. A cylindrical shell 8 meters long with an inside diameter of 0.75 meter and an outside diameter of 1.25 meters

97. A spherical shell with inside and outside radii of 4 inches and 6 inches, respectively

98. The solid that remains after a hole 1 inch in diameter is drilled through the center of a sphere 6 inches in diameter

99. The solid inside both $x^2 + y^2 + z^2 = 9$ and $\left(x - \frac{3}{2}\right)^2 + y^2 = \frac{9}{4}$

100. The solid between the spheres $x^2 + y^2 + z^2 = 4$ and $x^2 + y^2 + z^2 = 9$, and inside the cone $z^2 = x^2 + y^2$

True or False? In Exercises 101 and 102, determine whether the statement is true or false. If it is false, explain why or give an example that shows it is false.

101. The cylindrical coordinates of a point (x, y, z) are unique.

102. The spherical coordinates of a point (x, y, z) are unique.

103. Intersection of Surfaces Identify the curve of intersection of the surfaces (in cylindrical coordinates) $z = \sin \theta$ and $r = 1$.

104. Intersection of Surfaces Identify the curve of intersection of the surfaces (in spherical coordinates) $\rho = 2 \sec \phi$ and $\rho = 4$.

Review Exercises   See CalcChat.com for tutorial help and worked-out solutions to odd-numbered exercises.

Writing Vectors in Different Forms In Exercises 1 and 2, let $u = \overrightarrow{PQ}$ and $v = \overrightarrow{PR}$ and (a) write u and v in component form, (b) write u and v as the linear combination of the standard unit vectors i and j, (c) find the magnitudes of u and v, and (d) find $-3u + v$.

1. $P - (1, 2)$, $Q = (4, 1)$, $R - (5, 4)$

2. $P = (-2, -1)$, $Q = (5, -1)$, $R = (2, 4)$

Finding a Vector In Exercises 3 and 4, find the component form of v given its magnitude and the angle it makes with the positive x-axis.

3. $\|v\| = 8$, $\theta = 60°$ 4. $\|v\| = \frac{1}{2}$, $\theta = 225°$

5. **Finding Coordinates of a Point** Find the coordinates of the point located in the xy-plane, four units to the right of the xz-plane, and five units behind the yz-plane.

6. **Using the Three-Dimensional Coordinate System** Determine the location of a point (x, y, z) that satisfies the condition $y = 3$.

Finding the Distance Between Two Points in Space In Exercises 7 and 8, find the distance between the points.

7. $(1, 6, 3)$, $(-2, 3, 5)$

8. $(-2, 1, -5)$, $(4, -1, -1)$

Finding the Equation of a Sphere In Exercises 9 and 10, find the standard equation of the sphere with the given characteristics.

9. Center: $(3, -2, 6)$; Radius: 4

10. Endpoints of a diameter: $(0, 0, 4)$, $(4, 6, 0)$

Finding the Equation of a Sphere In Exercises 11 and 12, complete the square to write the equation of the sphere in standard form. Find the center and radius.

11. $x^2 + y^2 + z^2 - 4x - 6y + 4 = 0$

12. $x^2 + y^2 + z^2 - 10x + 6y - 4z + 34 = 0$

Writing a Vector in Different Forms In Exercises 13 and 14, the initial and terminal points of a vector are given. (a) Sketch the directed line segment. (b) Find the component form of the vector. (c) Write the vector using standard unit vector notation. (d) Sketch the vector with its initial point at the origin.

13. Initial point: $(2, -1, 3)$ 14. Initial point: $(6, 2, 0)$
    Terminal point: $(4, 4, -7)$ Terminal point: $(3, -3, 8)$

Finding a Vector In Exercises 15 and 16, find the vector z, given that $u = \langle 5, -2, 3 \rangle$, $v = \langle 0, 2, 1 \rangle$, and $w = \langle -6, -6, 2 \rangle$.

15. $z = -u + 3v + \frac{1}{2}w$

16. $u - v + w - 2z = 0$

Using Vectors to Determine Collinear Points In Exercises 17 and 18, use vectors to determine whether the points are collinear.

17. $(3, 4, -1)$, $(-1, 6, 9)$, $(5, 3, -6)$

18. $(5, -4, 7)$, $(8, -5, 5)$, $(11, 6, 3)$

19. **Finding a Unit Vector** Find a unit vector in the direction of $u = \langle 2, 3, 5 \rangle$.

20. **Finding a Vector** Find the vector v of magnitude 8 in the direction $\langle 6, -3, 2 \rangle$.

Finding Dot Products In Exercises 21 and 22, let $u = \overrightarrow{PQ}$ and $v = \overrightarrow{PR}$, and find (a) the component forms of u and v, (b) $u \cdot v$, and (c) $v \cdot v$.

21. $P = (5, 0, 0)$, $Q = (4, 4, 0)$, $R = (2, 0, 6)$

22. $P = (2, -1, 3)$, $Q = (0, 5, 1)$, $R = (5, 5, 0)$

Finding the Angle Between Two Vectors In Exercises 23 and 24, find the angle θ between the vectors (a) in radians and (b) in degrees.

23. $u = 5[\cos(3\pi/4)i + \sin(3\pi/4)j]$
    $v = 2[\cos(2\pi/3)i + \sin(2\pi/3)j]$

24. $u = \langle 1, 0, -3 \rangle$, $v = \langle 2, -2, 1 \rangle$

Comparing Vectors In Exercises 25 and 26, determine whether u and v are orthogonal, parallel, or neither.

25. $u = \langle 7, -2, 3 \rangle$ 26. $u = \langle -3, 0, 9 \rangle$
    $v = \langle -1, 4, 5 \rangle$ $v = \langle 1, 0, -3 \rangle$

Finding the Projection of u onto v In Exercises 27 and 28, (a) find the projection of u onto v, and (b) find the vector component of u orthogonal to v.

27. $u = 4i + 2j$, $v = 3i + 4j$

28. $u = \langle 1, -1, 1 \rangle$, $v = \langle 2, 0, 2 \rangle$

29. **Orthogonal Vectors** Find two vectors in opposite directions that are orthogonal to the vector $u = \langle 5, 6, -3 \rangle$.

30. **Work** An object is pulled 8 feet across a floor using a force of 75 pounds. The direction of the force is 30° above the horizontal. Find the work done.

Finding Cross Products In Exercises 31 and 32, find (a) $u \times v$, (b) $v \times u$, and (c) $v \times v$.

31. $u = 4i + 3j + 6k$ 32. $u = \langle 0, 2, 1 \rangle$
    $v = 5i + 2j + k$ $v = \langle 1, -3, 4 \rangle$

33. **Finding a Unit Vector** Find a unit vector that is orthogonal to both $u = \langle 2, -10, 8 \rangle$ and $v = \langle 4, 6, -8 \rangle$.

34. **Area** Find the area of the parallelogram that has the vectors $u = \langle 3, -1, 5 \rangle$ and $v = \langle 2, -4, 1 \rangle$ as adjacent sides.

35. Torque A vertical force of 40 pounds acts on a wrench, as shown in the figure. Find the torque at P.

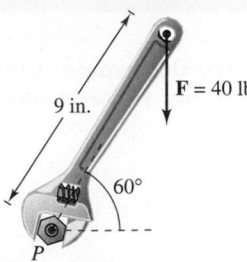

9 in.

$\mathbf{F} = 40$ lb

$60°$

P

36. Volume Use the triple scalar product to find the volume of the parallelepiped having adjacent edges $\mathbf{u} = 2\mathbf{i} + \mathbf{j}$, $\mathbf{v} = 2\mathbf{j} + \mathbf{k}$, and $\mathbf{w} = -\mathbf{j} + 2\mathbf{k}$.

Finding Parametric and Symmetric Equations In Exercises 37 and 38, find sets of (a) parametric equations and (b) symmetric equations of the line that passes through the two points. (For each line, write the direction numbers as integers.)

37. $(3, 0, 2)$, $(9, 11, 6)$ **38.** $(-1, 4, 3)$, $(8, 10, 5)$

Finding Parametric Equations In Exercises 39 and 40, find a set of parametric equations of the line with the given characteristics.

39. The line passes through the point $(-6, -8, 2)$ and is perpendicular to the xz-plane.

40. The line passes through the point $(1, 2, 3)$ and is parallel to the line given by $x = y = z$.

Finding an Equation of a Plane In Exercises 41–44, find an equation of the plane with the given characteristics.

41. The plane passes through $(-3, -4, 2)$, $(-3, 4, 1)$, and $(1, 1, -2)$.

42. The plane passes through the point $(-2, 3, 1)$ and is perpendicular to $\mathbf{n} = 3\mathbf{i} - \mathbf{j} + \mathbf{k}$.

43. The plane contains the lines given by

$$\frac{x - 1}{-2} = y = z + 1$$

and

$$\frac{x + 1}{-2} = y - 1 = z - 2.$$

44. The plane passes through the points $(5, 1, 3)$ and $(2, -2, 1)$ and is perpendicular to the plane $2x + y - z = 4$.

45. Distance Find the distance between the point $(1, 0, 2)$ and the plane $2x - 3y + 6z = 6$.

46. Distance Find the distance between the point $(3, -2, 4)$ and the plane $2x - 5y + z = 10$.

47. Distance Find the distance between the planes $5x - 3y + z = 2$ and $5x - 3y + z = -3$.

48. Distance Find the distance between the point $(-5, 1, 3)$ and the line given by $x = 1 + t$, $y = 3 - 2t$, and $z = 5 - t$.

Sketching a Surface in Space In Exercises 49–58, describe and sketch the surface.

49. $x + 2y + 3z = 6$ **50.** $y = z^2$

51. $y = \frac{1}{2}z$ **52.** $y = \cos z$

53. $\dfrac{x^2}{16} + \dfrac{y^2}{9} + z^2 = 1$ **54.** $16x^2 + 16y^2 - 9z^2 = 0$

55. $\dfrac{x^2}{16} - \dfrac{y^2}{9} + z^2 = -1$ **56.** $\dfrac{x^2}{25} + \dfrac{y^2}{4} - \dfrac{z^2}{100} = 1$

57. $x^2 + z^2 = 4$ **58.** $y^2 + z^2 = 16$

59. Surface of Revolution Find an equation for the surface of revolution formed by revolving the curve $z^2 = 2y$ in the yz-plane about the y-axis.

60. Surface of Revolution Find an equation for the surface of revolution formed by revolving the curve $2x + 3z = 1$ in the xz-plane about the x-axis.

Converting Rectangular Coordinates In Exercises 61 and 62, convert the point from rectangular coordinates to (a) cylindrical coordinates and (b) spherical coordinates.

61. $\left(-\sqrt{3}, 3, -5\right)$ **62.** $(8, 8, 1)$

Cylindrical-to-Rectangular Conversion In Exercises 63 and 64, convert the point from cylindrical coordinates to rectangular coordinates.

63. $(5, \pi, 1)$ **64.** $\left(-2, \dfrac{\pi}{3}, 3\right)$

Spherical-to-Rectangular Conversion In Exercises 65 and 66, convert the point from spherical coordinates to rectangular coordinates.

65. $\left(4, \pi, \dfrac{\pi}{4}\right)$ **66.** $\left(8, -\dfrac{\pi}{6}, \dfrac{\pi}{3}\right)$

Converting a Rectangular Equation In Exercises 67 and 68, convert the rectangular equation to an equation in (a) cylindrical coordinates and (b) spherical coordinates.

67. $x^2 - y^2 = 2z$ **68.** $x^2 + y^2 + z^2 = 16$

Cylindrical-to-Rectangular Conversion In Exercises 69 and 70, find an equation in rectangular coordinates for the surface represented by the cylindrical equation, and sketch its graph.

69. $z = r^2 \sin^2 \theta + 3r \cos \theta$

70. $r = -5z$

Spherical-to-Rectangular Conversion In Exercises 71 and 72, find an equation in rectangular coordinates for the surface represented by the spherical equation, and sketch its graph.

71. $\phi = \dfrac{\pi}{4}$

72. $\rho = 9 \sec \theta$

P.S. Problem Solving

See **CalcChat.com** for tutorial help and worked-out solutions to odd-numbered exercises.

1. Proof Using vectors, prove the Law of Sines: If **a**, **b**, and **c** are the three sides of the triangle shown in the figure, then

$$\frac{\sin A}{\|\mathbf{a}\|} = \frac{\sin B}{\|\mathbf{b}\|} = \frac{\sin C}{\|\mathbf{c}\|}.$$

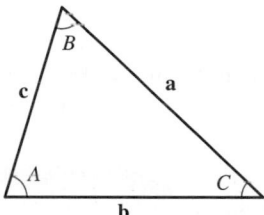

2. Using an Equation Consider the function

$$f(x) = \int_0^x \sqrt{t^4 + 1}\, dt.$$

(a) Use a graphing utility to graph the function on the interval $-2 \le x \le 2$.

(b) Find a unit vector parallel to the graph of f at the point $(0, 0)$.

(c) Find a unit vector perpendicular to the graph of f at the point $(0, 0)$.

(d) Find the parametric equations of the tangent line to the graph of f at the point $(0, 0)$.

3. Proof Using vectors, prove that the line segments joining the midpoints of the sides of a parallelogram form a parallelogram (see figure).

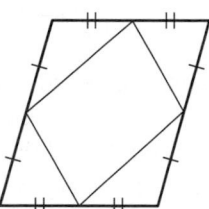

4. Proof Using vectors, prove that the diagonals of a rhombus are perpendicular (see figure).

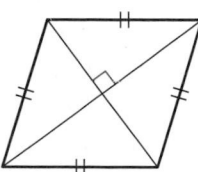

5. Distance

(a) Find the shortest distance between the point $Q(2, 0, 0)$ and the line determined by the points $P_1(0, 0, 1)$ and $P_2(0, 1, 2)$.

(b) Find the shortest distance between the point $Q(2, 0, 0)$ and the line segment joining the points $P_1(0, 0, 1)$ and $P_2(0, 1, 2)$.

6. Orthogonal Vectors Let P_0 be a point in the plane with normal vector **n**. Describe the set of points P in the plane for which $\left(\mathbf{n} + \overrightarrow{PP_0}\right)$ is orthogonal to $\left(\mathbf{n} - \overrightarrow{PP_0}\right)$.

7. Volume

(a) Find the volume of the solid bounded below by the paraboloid

$$z = x^2 + y^2$$

and above by the plane $z = 1$.

(b) Find the volume of the solid bounded below by the elliptic paraboloid

$$z = \frac{x^2}{a^2} + \frac{y^2}{b^2}$$

and above by the plane $z = k$, where $k > 0$.

(c) Show that the volume of the solid in part (b) is equal to one-half the product of the area of the base times the altitude, as shown in the figure.

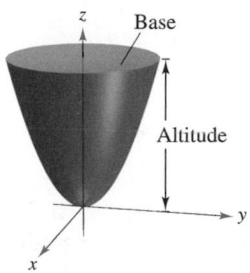

8. Volume

(a) Use the disk method to find the volume of the sphere $x^2 + y^2 + z^2 = r^2$.

(b) Find the volume of the ellipsoid $\dfrac{x^2}{a^2} + \dfrac{y^2}{b^2} + \dfrac{z^2}{c^2} = 1$.

9. Proof Prove the following property of the cross product.

$$(\mathbf{u} \times \mathbf{v}) \times (\mathbf{w} \times \mathbf{z}) = [(\mathbf{u} \times \mathbf{v}) \cdot \mathbf{z}]\mathbf{w} - [(\mathbf{u} \times \mathbf{v}) \cdot \mathbf{w}]\mathbf{z}$$

10. Using Parametric Equations Consider the line given by the parametric equations

$$x = -t + 3, \quad y = \tfrac{1}{2}t + 1, \quad z = 2t - 1$$

and the point $(4, 3, s)$ for any real number s.

(a) Write the distance between the point and the line as a function of s.

(b) Use a graphing utility to graph the function in part (a). Use the graph to find the value of s such that the distance between the point and the line is minimum.

(c) Use the *zoom* feature of a graphing utility to zoom out several times on the graph in part (b). Does it appear that the graph has slant asymptotes? Explain. If it appears to have slant asymptotes, find them.

11. Sketching Graphs Sketch the graph of each equation given in spherical coordinates.

(a) $\rho = 2 \sin \phi$ (b) $\rho = 2 \cos \phi$

12. Sketching Graphs Sketch the graph of each equation given in cylindrical coordinates.

(a) $r = 2 \cos \theta$ (b) $z = r^2 \cos 2\theta$

13. Tetherball A tetherball weighing 1 pound is pulled outward from the pole by a horizontal force $\mathbf{u}$ until the rope makes an angle of θ degrees with the pole (see figure).

(a) Determine the resulting tension in the rope and the magnitude of $\mathbf{u}$ when $\theta = 30°$.

(b) Write the tension T in the rope and the magnitude of $\mathbf{u}$ as functions of θ. Determine the domains of the functions.

(c) Use a graphing utility to complete the table.

θ	0°	10°	20°	30°	40°	50°	60°
T							
$\|\mathbf{u}\|$							

(d) Use a graphing utility to graph the two functions for $0° \leq \theta \leq 60°$.

(e) Compare T and $\|\mathbf{u}\|$ as θ increases.

(f) Find (if possible)

$$\lim_{\theta \to \pi/2^-} T \text{ and } \lim_{\theta \to \pi/2^-} \|\mathbf{u}\|.$$

Are the results what you expected? Explain.

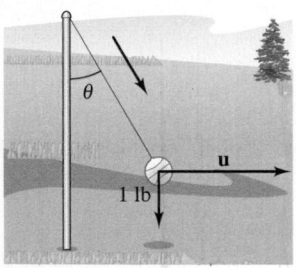 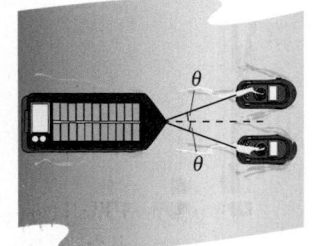

Figure for 13 Figure for 14

14. Towing A loaded barge is being towed by two tugboats, and the magnitude of the resultant is 6000 pounds directed along the axis of the barge (see figure). Each towline makes an angle of θ degrees with the axis of the barge.

(a) Find the tension in the towlines when $\theta = 20°$.

(b) Write the tension T of each line as a function of θ. Determine the domain of the function.

(c) Use a graphing utility to complete the table.

θ	10°	20°	30°	40°	50°	60°
T						

(d) Use a graphing utility to graph the tension function.

(e) Explain why the tension increases as θ increases.

15. Proof Consider the vectors

$$\mathbf{u} = \langle \cos \alpha, \sin \alpha, 0 \rangle \quad \text{and} \quad \mathbf{v} = \langle \cos \beta, \sin \beta, 0 \rangle$$

where $\alpha > \beta$. Find the cross product of the vectors and use the result to prove the identity

$$\sin(\alpha - \beta) = \sin \alpha \cos \beta - \cos \alpha \sin \beta.$$

16. Latitude-Longitude System Los Angeles is located at 34.05° North latitude and 118.24° West longitude, and Rio de Janeiro, Brazil, is located at 22.90° South latitude and 43.23° West longitude (see figure). Assume that Earth is spherical and has a radius of 4000 miles.

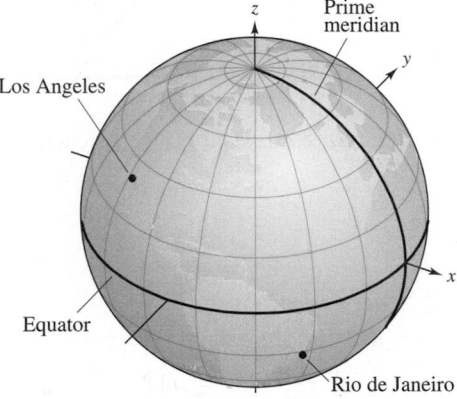

(a) Find the spherical coordinates for the location of each city.

(b) Find the rectangular coordinates for the location of each city.

(c) Find the angle (in radians) between the vectors from the center of Earth to the two cities.

(d) Find the great-circle distance s between the cities. (*Hint:* $s = r\theta$)

(e) Repeat parts (a)–(d) for the cities of Boston, located at 42.36° North latitude and 71.06° West longitude, and Honolulu, located at 21.31° North latitude and 157.86° West longitude.

17. Distance Between a Point and a Plane Consider the plane that passes through the points P, R, and S. Show that the distance from a point Q to this plane is

$$\text{Distance} = \frac{|\mathbf{u} \cdot (\mathbf{v} \times \mathbf{w})|}{\|\mathbf{u} \times \mathbf{v}\|}$$

where $\mathbf{u} = \overrightarrow{PR}$, $\mathbf{v} = \overrightarrow{PS}$, and $\mathbf{w} = \overrightarrow{PQ}$.

18. Distance Between Parallel Planes Show that the distance between the parallel planes

$$ax + by + cz + d_1 = 0 \quad \text{and} \quad ax + by + cz + d_2 = 0$$

is

$$\text{Distance} = \frac{|d_1 - d_2|}{\sqrt{a^2 + b^2 + c^2}}.$$

19. Intersection of Planes Show that the curve of intersection of the plane $z = 2y$ and the cylinder $x^2 + y^2 = 1$ is an ellipse.

12 Vector-Valued Functions

Speed *(Exercise 66, p. 865)*

Air Traffic Control
(Exercise 61, p. 854)

Football *(Exercise 34, p. 843)*

Shot-Put Throw
(Exercise 44, p. 843)

Staircase *(Exercise 81, p. 827)*

12.1 Vector-Valued Functions

■ Analyze and sketch a space curve given by a vector-valued function.
■ Extend the concepts of limits and continuity to vector-valued functions.

Space Curves and Vector-Valued Functions

In Section 10.2, a *plane curve* was defined as the set of ordered pairs $(f(t), g(t))$ together with their defining parametric equations $x = f(t)$ and $y = g(t)$, where f and g are continuous functions of t on an interval I. This definition can be extended naturally to three-dimensional space. A **space curve** C is the set of all ordered triples $(f(t), g(t), h(t))$ together with their defining parametric equations

$$x = f(t), \quad y = g(t), \quad \text{and} \quad z = h(t)$$

where f, g, and h are continuous functions of t on an interval I.

Before looking at examples of space curves, a new type of function, called a **vector-valued function,** is introduced. This type of function maps real numbers to vectors.

Definition of Vector-Valued Function

A function of the form

$$\mathbf{r}(t) = f(t)\mathbf{i} + g(t)\mathbf{j} \qquad \text{Plane}$$

or

$$\mathbf{r}(t) = f(t)\mathbf{i} + g(t)\mathbf{j} + h(t)\mathbf{k} \qquad \text{Space}$$

is a **vector-valued function,** where the **component functions** f, g, and h are real-valued functions of the parameter t. Vector-valued functions are sometimes denoted as

$$\mathbf{r}(t) = \langle f(t), g(t) \rangle \qquad \text{Plane}$$

or

$$\mathbf{r}(t) = \langle f(t), g(t), h(t) \rangle. \qquad \text{Space}$$

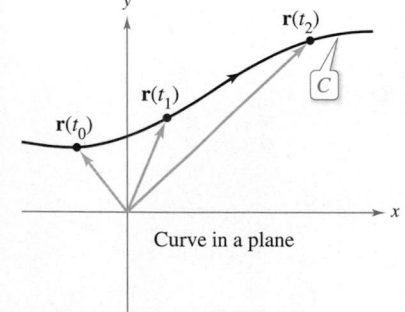

Curve in a plane

Technically, a curve in a plane or in space consists of a collection of points and the defining parametric equations. Two different curves can have the same graph. For instance, each of the curves

$$\mathbf{r}(t) = \sin t\,\mathbf{i} + \cos t\,\mathbf{j} \quad \text{and} \quad \mathbf{r}(t) = \sin t^2\mathbf{i} + \cos t^2\mathbf{j}$$

has the unit circle as its graph, but these equations do not represent the same curve—because the circle is traced out in different ways on the graphs.

Be sure you see the distinction between the vector-valued function $\mathbf{r}$ and the real-valued functions f, g, and h. All are functions of the real variable t, but $\mathbf{r}(t)$ is a vector, whereas $f(t)$, $g(t)$, and $h(t)$ are real numbers (for each specific value of t). Real-valued functions are sometimes called **scalar functions** to distinguish them from vector-valued functions.

Vector-valued functions serve dual roles in the representation of curves. By letting the parameter t represent time, you can use a vector-valued function to represent *motion* along a curve. Or, in the more general case, you can use a vector-valued function to *trace the graph* of a curve. In either case, the terminal point of the position vector $\mathbf{r}(t)$ coincides with the point (x, y) or (x, y, z) on the curve given by the parametric equations, as shown in Figure 12.1. The arrowhead on the curve indicates the curve's *orientation* by pointing in the direction of increasing values of t.

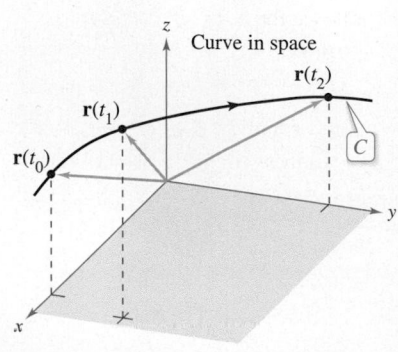

Curve in space

Curve C is traced out by the terminal point of position vector $\mathbf{r}(t)$.
Figure 12.1

Unless stated otherwise, the **domain** of a vector-valued function **r** is considered to be the intersection of the domains of the component functions f, g, and h. For instance, the domain of $\mathbf{r}(t) = \ln t \mathbf{i} + \sqrt{1 - t}\mathbf{j} + t\mathbf{k}$ is the interval $(0, 1]$.

EXAMPLE 1   Sketching a Plane Curve

Sketch the plane curve represented by the vector-valued function

$$\mathbf{r}(t) = 2 \cos t \mathbf{i} - 3 \sin t \mathbf{j}, \quad 0 \le t \le 2\pi. \qquad \text{Vector-valued function}$$

Solution   From the position vector $\mathbf{r}(t)$, you can write the parametric equations

$$x = 2 \cos t \quad \text{and} \quad y = -3 \sin t.$$

Solving for $\cos t$ and $\sin t$ and using the identity $\cos^2 t + \sin^2 t = 1$, you get the rectangular equation

$$\frac{x^2}{2^2} + \frac{y^2}{3^2} = 1. \qquad \text{Rectangular equation}$$

The graph of this rectangular equation is the ellipse shown in Figure 12.2. The curve has a *clockwise* orientation. That is, as t increases from 0 to 2π, the position vector $\mathbf{r}(t)$ moves clockwise, and its terminal point traces the ellipse.

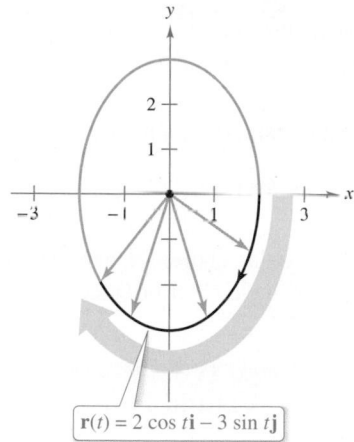

$\mathbf{r}(t) = 2 \cos t \mathbf{i} - 3 \sin t \mathbf{j}$

The ellipse is traced clockwise as t increases from 0 to 2π.
Figure 12.2

EXAMPLE 2   Sketching a Space Curve

⋅⋅⋅▷ *See LarsonCalculus.com for an interactive version of this type of example.*

Sketch the space curve represented by the vector-valued function

$$\mathbf{r}(t) = 4 \cos t \mathbf{i} + 4 \sin t \mathbf{j} + t\mathbf{k}, \quad 0 \le t \le 4\pi. \qquad \text{Vector-valued function}$$

Solution   From the first two parametric equations

$$x = 4 \cos t \quad \text{and} \quad y = 4 \sin t$$

you can obtain

$$x^2 + y^2 = 16. \qquad \text{Rectangular equation}$$

This means that the curve lies on a right circular cylinder of radius 4, centered about the z-axis. To locate the curve on this cylinder, you can use the third parametric equation

$$z = t.$$

In Figure 12.3, note that as t increases from 0 to 4π, the point (x, y, z) spirals up the cylinder to produce a **helix**. A real-life example of a helix is shown in the drawing at the left.

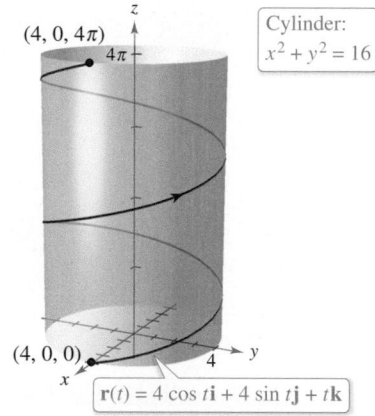

Cylinder: $x^2 + y^2 = 16$

$\mathbf{r}(t) = 4 \cos t \mathbf{i} + 4 \sin t \mathbf{j} + t\mathbf{k}$

As t increases from 0 to 4π, two spirals on the helix are traced out.
Figure 12.3 ▪

In 1953, Francis Crick and James D. Watson discovered the double helix structure of DNA.

In Examples 1 and 2, you were given a vector-valued function and were asked to sketch the corresponding curve. The next two examples address the reverse problem—finding a vector-valued function to represent a given graph. Of course, when the graph is described parametrically, representation by a vector-valued function is straightforward. For instance, to represent the line in space given by $x = 2 + t$, $y = 3t$, and $z = 4 - t$, you can simply use the vector-valued function

$$\mathbf{r}(t) = (2 + t)\mathbf{i} + 3t\mathbf{j} + (4 - t)\mathbf{k}.$$

When a set of parametric equations for the graph is not given, the problem of representing the graph by a vector-valued function boils down to finding a set of parametric equations.

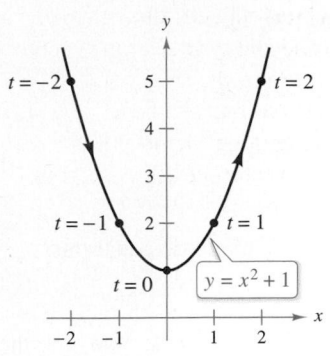

There are many ways to parametrize this graph. One way is to let $x = t$.

Figure 12.4

EXAMPLE 3 **Representing a Graph: Vector-Valued Function**

Represent the parabola

$$y = x^2 + 1$$

by a vector-valued function.

Solution Although there are many ways to choose the parameter t, a natural choice is to let $x = t$. Then $y = t^2 + 1$ and you have

$$\mathbf{r}(t) = t\mathbf{i} + (t^2 + 1)\mathbf{j}. \qquad \text{Vector-valued function}$$

Note in Figure 12.4 the orientation produced by this particular choice of parameter. Had you chosen $x = -t$ as the parameter, the curve would have been oriented in the opposite direction.

EXAMPLE 4 **Representing a Graph: Vector-Valued Function**

Sketch the space curve C represented by the intersection of the semiellipsoid

$$\frac{x^2}{12} + \frac{y^2}{24} + \frac{z^2}{4} = 1, \quad z \geq 0$$

and the parabolic cylinder $y = x^2$. Then find a vector-valued function to represent the graph.

Solution The intersection of the two surfaces is shown in Figure 12.5. As in Example 3, a natural choice of parameter is $x = t$. For this choice, you can use the given equation $y = x^2$ to obtain $y = t^2$. Then it follows that

$$\frac{z^2}{4} = 1 - \frac{x^2}{12} - \frac{y^2}{24} = 1 - \frac{t^2}{12} - \frac{t^4}{24} = \frac{24 - 2t^2 - t^4}{24} = \frac{(6 + t^2)(4 - t^2)}{24}.$$

Because the curve lies above the xy-plane, you should choose the positive square root for z and obtain the parametric equations

$$x = t, \quad y = t^2, \quad \text{and} \quad z = \sqrt{\frac{(6 + t^2)(4 - t^2)}{6}}.$$

The resulting vector-valued function is

$$\mathbf{r}(t) = t\mathbf{i} + t^2\mathbf{j} + \sqrt{\frac{(6 + t^2)(4 - t^2)}{6}}\mathbf{k}, \quad -2 \leq t \leq 2. \qquad \text{Vector-valued function}$$

(Note that the $\mathbf{k}$-component of $\mathbf{r}(t)$ implies $-2 \leq t \leq 2$.) From the points $(-2, 4, 0)$ and $(2, 4, 0)$ shown in Figure 12.5, you can see that the curve is traced as t increases from -2 to 2.

• • REMARK Curves in space can be specified in various ways. For instance, the curve in Example 4 is described as the intersection of two surfaces in space.

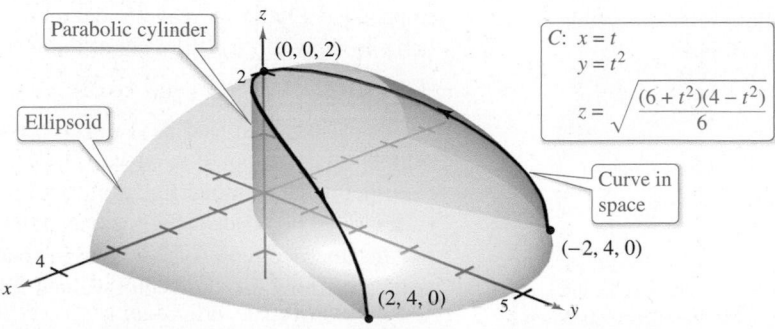

The curve C is the intersection of the semiellipsoid and the parabolic cylinder.

Figure 12.5

Limits and Continuity

Many techniques and definitions used in the calculus of real-valued functions can be applied to vector-valued functions. For instance, you can add and subtract vector-valued functions, multiply a vector-valued function by a scalar, take the limit of a vector-valued function, differentiate a vector-valued function, and so on. The basic approach is to capitalize on the linearity of vector operations by extending the definitions on a component-by-component basis. For example, to add two vector-valued functions (in the plane), you can write

$$\mathbf{r}_1(t) + \mathbf{r}_2(t) = [f_1(t)\mathbf{i} + g_1(t)\mathbf{j}] + [f_2(t)\mathbf{i} + g_2(t)\mathbf{j}] \qquad \text{Sum}$$
$$= [f_1(t) + f_2(t)]\mathbf{i} + [g_1(t) + g_2(t)]\mathbf{j}.$$

To subtract two vector-valued functions, you can write

$$\mathbf{r}_1(t) - \mathbf{r}_2(t) = [f_1(t)\mathbf{i} + g_1(t)\mathbf{j}] - [f_2(t)\mathbf{i} + g_2(t)\mathbf{j}] \qquad \text{Difference}$$
$$= [f_1(t) - f_2(t)]\mathbf{i} + [g_1(t) - g_2(t)]\mathbf{j}.$$

Similarly, to multiply a vector-valued function by a scalar, you can write

$$c\mathbf{r}(t) = c[f_1(t)\mathbf{i} + g_1(t)\mathbf{j}] \qquad \text{Scalar multiplication}$$
$$= cf_1(t)\mathbf{i} + cg_1(t)\mathbf{j}.$$

To divide a vector-valued function by a scalar, you can write

$$\frac{\mathbf{r}(t)}{c} = \frac{[f_1(t)\mathbf{i} + g_1(t)\mathbf{j}]}{c}, \quad c \neq 0 \qquad \text{Scalar division}$$
$$= \frac{f_1(t)}{c}\mathbf{i} + \frac{g_1(t)}{c}\mathbf{j}.$$

This component-by-component extension of operations with real-valued functions to vector-valued functions is further illustrated in the definition of the limit of a vector-valued function.

Definition of the Limit of a Vector-Valued Function

1. If $\mathbf{r}$ is a vector-valued function such that $\mathbf{r}(t) = f(t)\mathbf{i} + g(t)\mathbf{j}$, then

$$\lim_{t \to a} \mathbf{r}(t) = \left[\lim_{t \to a} f(t)\right]\mathbf{i} + \left[\lim_{t \to a} g(t)\right]\mathbf{j} \qquad \text{Plane}$$

provided f and g have limits as $t \to a$.

2. If $\mathbf{r}$ is a vector-valued function such that $\mathbf{r}(t) = f(t)\mathbf{i} + g(t)\mathbf{j} + h(t)\mathbf{k}$, then

$$\lim_{t \to a} \mathbf{r}(t) = \left[\lim_{t \to a} f(t)\right]\mathbf{i} + \left[\lim_{t \to a} g(t)\right]\mathbf{j} + \left[\lim_{t \to a} h(t)\right]\mathbf{k} \qquad \text{Space}$$

provided f, g, and h have limits as $t \to a$.

If $\mathbf{r}(t)$ approaches the vector $\mathbf{L}$ as $t \to a$, then the length of the vector $\mathbf{r}(t) - \mathbf{L}$ approaches 0. That is,

$$\|\mathbf{r}(t) - \mathbf{L}\| \to 0 \quad \text{as} \quad t \to a.$$

This is illustrated graphically in Figure 12.6. With this definition of the limit of a vector-valued function, you can develop vector versions of most of the limit theorems given in Chapter 2. For example, the limit of the sum of two vector-valued functions is the sum of their individual limits. Also, you can use the orientation of the curve $\mathbf{r}(t)$ to define one-sided limits of vector-valued functions. The next definition extends the notion of continuity to vector-valued functions.

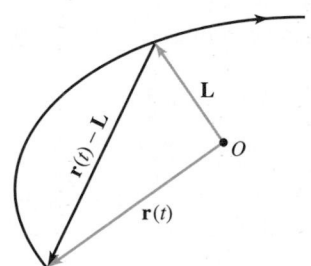

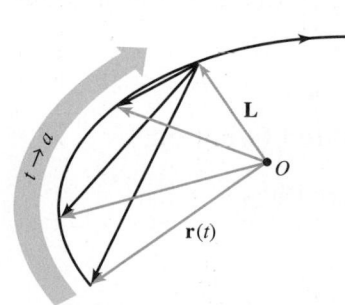

As t approaches a, $\mathbf{r}(t)$ approaches the limit $\mathbf{L}$. For the limit $\mathbf{L}$ to exist, it is not necessary that $\mathbf{r}(a)$ be defined or that $\mathbf{r}(a)$ be equal to $\mathbf{L}$.

Figure 12.6

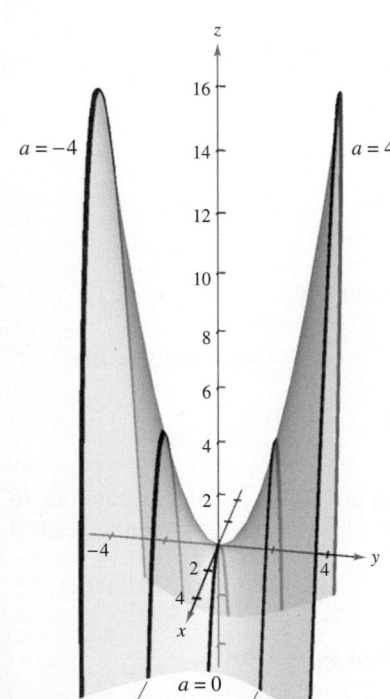

For each value of a, the curve represented by the vector-valued function $\mathbf{r}(t) = t\mathbf{i} + a\mathbf{j} + (a^2 - t^2)\mathbf{k}$ is a parabola.
Figure 12.7

> **Definition of Continuity of a Vector-Valued Function**
>
> A vector-valued function $\mathbf{r}$ is **continuous at the point** given by $t = a$ when the limit of $\mathbf{r}(t)$ exists as $t \to a$ and
>
> $$\lim_{t \to a} \mathbf{r}(t) = \mathbf{r}(a).$$
>
> A vector-valued function $\mathbf{r}$ is **continuous on an interval** I when it is continuous at every point in the interval.

From this definition, it follows that a vector-valued function is continuous at $t = a$ if and only if each of its component functions is continuous at $t = a$.

EXAMPLE 5 **Continuity of a Vector-Valued Function**

Discuss the continuity of the vector-valued function

$$\mathbf{r}(t) = t\mathbf{i} + a\mathbf{j} + (a^2 - t^2)\mathbf{k} \qquad a \text{ is a constant.}$$

at $t = 0$.

Solution As t approaches 0, the limit is

$$\lim_{t \to 0} \mathbf{r}(t) = \left[\lim_{t \to 0} t\right]\mathbf{i} + \left[\lim_{t \to 0} a\right]\mathbf{j} + \left[\lim_{t \to 0} (a^2 - t^2)\right]\mathbf{k}$$
$$= 0\mathbf{i} + a\mathbf{j} + a^2\mathbf{k}$$
$$= a\mathbf{j} + a^2\mathbf{k}.$$

Because

$$\mathbf{r}(0) = (0)\mathbf{i} + (a)\mathbf{j} + (a^2)\mathbf{k}$$
$$= a\mathbf{j} + a^2\mathbf{k}$$

you can conclude that $\mathbf{r}$ is continuous at $t = 0$. By similar reasoning, you can conclude that the vector-valued function $\mathbf{r}$ is continuous at all real-number values of t.

For each value of a, the curve represented by the vector-valued function in Example 5

$$\mathbf{r}(t) = t\mathbf{i} + a\mathbf{j} + (a^2 - t^2)\mathbf{k} \qquad a \text{ is a constant.}$$

is a parabola. You can think of each parabola as the intersection of the vertical plane $y = a$ and the hyperbolic paraboloid

$$y^2 - x^2 = z$$

as shown in Figure 12.7.

▷ **TECHNOLOGY** Almost any type of three-dimensional sketch is difficult to do by hand, but sketching curves in space is especially difficult. The problem is trying to create the illusion of three dimensions. Graphing utilities use a variety of techniques to add "three-dimensionality" to graphs of space curves. One way is to show the curve on a surface, as in Figure 12.7.

EXAMPLE 6 **Continuity of a Vector-Valued Function**

Determine the interval(s) on which the vector-valued function

$$\mathbf{r}(t) = t\mathbf{i} + \sqrt{t + 1}\mathbf{j} + (t^2 + 1)\mathbf{k}$$

is continuous.

Solution The component functions are

$$f(t) = t, \quad g(t) = \sqrt{t + 1}, \quad \text{and} \quad h(t) = (t^2 + 1).$$

Both f and h are continuous for all real-number values of t. The function g, however, is continuous only for $t \geq -1$. So, $\mathbf{r}$ is continuous on the interval $[-1, \infty)$.

12.1   Exercises

See **CalcChat.com** for tutorial help and worked-out solutions to odd-numbered exercises.

CONCEPT CHECK

1. **Vector-Valued Function**   Describe how you can use a vector-valued function to represent a curve.

2. **Continuity of a Vector-Valued Function**   Describe what it means for a vector-valued function $\mathbf{r}(t)$ to be continuous at a point.

Finding the Domain   In Exercises 3–10, find the domain of the vector-valued function.

3. $\mathbf{r}(t) = \dfrac{1}{t+1}\mathbf{i} + \dfrac{t}{2}\mathbf{j} - 3t\mathbf{k}$

4. $\mathbf{r}(t) = \sqrt{4 - t^2}\,\mathbf{i} + t^2\mathbf{j} - 6t\mathbf{k}$

5. $\mathbf{r}(t) = \ln t\,\mathbf{i} - e^t\mathbf{j} - t\mathbf{k}$

6. $\mathbf{r}(t) = \sin t\,\mathbf{i} + 4\cos t\,\mathbf{j} + t\mathbf{k}$

7. $\mathbf{r}(t) = \mathbf{F}(t) + \mathbf{G}(t)$, where
   $\mathbf{F}(t) = \cos t\,\mathbf{i} - \sin t\,\mathbf{j} + \sqrt{t}\,\mathbf{k}$, $\mathbf{G}(t) = \cos t\,\mathbf{i} + \sin t\,\mathbf{j}$

8. $\mathbf{r}(t) = \mathbf{F}(t) - \mathbf{G}(t)$, where
   $\mathbf{F}(t) = \ln t\,\mathbf{i} + 5t\mathbf{j} - 3t^2\mathbf{k}$, $\mathbf{G}(t) = \mathbf{i} + 4t\mathbf{j} - 3t^2\mathbf{k}$

9. $\mathbf{r}(t) = \mathbf{F}(t) \times \mathbf{G}(t)$, where
   $\mathbf{F}(t) = \sin t\,\mathbf{i} + \cos t\,\mathbf{j}$, $\mathbf{G}(t) = \sin t\,\mathbf{j} + \cos t\,\mathbf{k}$

10. $\mathbf{r}(t) = \mathbf{F}(t) \times \mathbf{G}(t)$, where
    $\mathbf{F}(t) = t^3\mathbf{i} - t\mathbf{j} + t\mathbf{k}$, $\mathbf{G}(t) = \sqrt[3]{t}\,\mathbf{i} + \dfrac{1}{t+1}\mathbf{j} + (t+2)\mathbf{k}$

Evaluating a Function   In Exercises 11 and 12, evaluate the vector-valued function at each given value of t.

11. $\mathbf{r}(t) = \frac{1}{2}t^2\mathbf{i} - (t-1)\mathbf{j}$

   (a) $\mathbf{r}(1)$ (b) $\mathbf{r}(0)$ (c) $\mathbf{r}(s+1)$

   (d) $\mathbf{r}(2 + \Delta t) - \mathbf{r}(2)$

12. $\mathbf{r}(t) = \cos t\,\mathbf{i} + 2\sin t\,\mathbf{j}$

   (a) $\mathbf{r}(0)$ (b) $\mathbf{r}(\pi/4)$ (c) $\mathbf{r}(\theta - \pi)$

   (d) $\mathbf{r}(\pi/6 + \Delta t) - \mathbf{r}(\pi/6)$

Writing a Vector-Valued Function   In Exercises 13–16, represent the line segment from P to Q by a vector-valued function and by a set of parametric equations.

13. $P(0, 0, 0)$, $Q(5, 2, 2)$ 14. $P(0, 2, -1)$, $Q(4, 7, 2)$

15. $P(-3, -6, -1)$, $Q(-1, -9, -8)$

16. $P(1, -6, 8)$, $Q(-3, -2, 5)$

Think About It   In Exercises 17 and 18, find $\mathbf{r}(t) \cdot \mathbf{u}(t)$. Is the result a vector-valued function? Explain.

17. $\mathbf{r}(t) = (3t - 1)\mathbf{i} + \frac{1}{4}t^3\mathbf{j} + 4\mathbf{k}$, $\mathbf{u}(t) = t^2\mathbf{i} - 8\mathbf{j} + t^3\mathbf{k}$

18. $\mathbf{r}(t) = \langle 3\cos t, 2\sin t, t - 2 \rangle$, $\mathbf{u}(t) = \langle 4\sin t, -6\cos t, t^2 \rangle$

Matching   In Exercises 19–22, match the equation with its graph. [The graphs are labeled (a), (b), (c), and (d).]

(a)
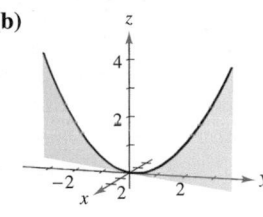

(b)
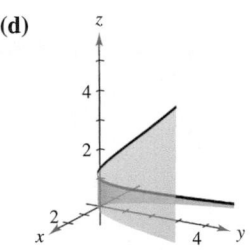

(c)
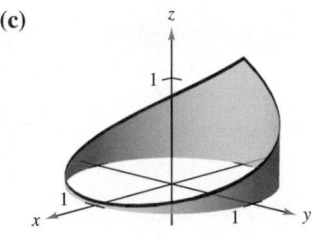

(d)

19. $\mathbf{r}(t) = t\mathbf{i} + 2t\mathbf{j} + t^2\mathbf{k}$, $-2 \le t \le 2$

20. $\mathbf{r}(t) = \cos(\pi t)\mathbf{i} + \sin(\pi t)\mathbf{j} + t^2\mathbf{k}$, $-1 \le t \le 1$

21. $\mathbf{r}(t) = t\mathbf{i} + t^2\mathbf{j} + e^{0.75t}\mathbf{k}$, $-2 \le t \le 2$

22. $\mathbf{r}(t) = t\mathbf{i} + \ln t\,\mathbf{j} + \dfrac{2t}{3}\mathbf{k}$, $0.1 < t < 5$

Sketching a Plane Curve   In Exercises 23–30, sketch the plane curve represented by the vector-valued function and give the orientation of the curve.

23. $\mathbf{r}(t) = \dfrac{t}{4}\mathbf{i} + (t-1)\mathbf{j}$ 24. $\mathbf{r}(t) = (5 - t)\mathbf{i} + \sqrt{t}\,\mathbf{j}$

25. $\mathbf{r}(t) = t^3\mathbf{i} + t^2\mathbf{j}$

26. $\mathbf{r}(t) = (t^2 + t)\mathbf{i} + (t^2 - t)\mathbf{j}$

27. $\mathbf{r}(\theta) = \cos\theta\,\mathbf{i} + 3\sin\theta\,\mathbf{j}$

28. $\mathbf{r}(t) = 2\cos t\,\mathbf{i} + 2\sin t\,\mathbf{j}$

29. $\mathbf{r}(\theta) = 3\sec\theta\,\mathbf{i} + 2\tan\theta\,\mathbf{j}$

30. $\mathbf{r}(t) = 2\cos^3 t\,\mathbf{i} + 2\sin^3 t\,\mathbf{j}$

Sketching a Space Curve   In Exercises 31–38, sketch the space curve represented by the vector-valued function and give the orientation of the curve.

31. $\mathbf{r}(t) = (-t + 1)\mathbf{i} + (4t + 2)\mathbf{j} + (2t + 3)\mathbf{k}$

32. $\mathbf{r}(t) = t\mathbf{i} + (2t - 5)\mathbf{j} + 3t\mathbf{k}$

33. $\mathbf{r}(t) = 2\cos t\,\mathbf{i} + 2\sin t\,\mathbf{j} + t\mathbf{k}$

34. $\mathbf{r}(t) = t\mathbf{i} + 3\cos t\,\mathbf{j} + 3\sin t\,\mathbf{k}$

35. $\mathbf{r}(t) = 2\sin t\,\mathbf{i} + 2\cos t\,\mathbf{j} + e^{-t}\mathbf{k}$

36. $\mathbf{r}(t) = t^2\mathbf{i} + 2t\mathbf{j} + \frac{3}{2}t\mathbf{k}$

37. $\mathbf{r}(t) = \left\langle t, t^2, \frac{2}{3}t^3 \right\rangle$

38. $\mathbf{r}(t) = \langle \cos t + t\sin t, \sin t - t\cos t, t \rangle$

 **Identifying a Common Curve** In Exercises 39 and 40, use a computer algebra system to graph the vector-valued function and identify the common curve.

39. $\mathbf{r}(t) = -\frac{1}{2}t^2\mathbf{i} + t\mathbf{j} - \frac{\sqrt{3}}{2}t^2\mathbf{k}$

40. $\mathbf{r}(t) = -\sqrt{2}\sin t\mathbf{i} + 2\cos t\mathbf{j} + \sqrt{2}\sin t\mathbf{k}$

 **Transformations of Vector-Valued Functions** In Exercises 41 and 42, use a computer algebra system to graph the vector-valued function $\mathbf{r}(t)$. For each $\mathbf{u}(t)$, make a conjecture about the transformation (if any) of the graph of $\mathbf{r}(t)$. Use a computer algebra system to verify your conjecture.

41. $\mathbf{r}(t) = 2\cos t\mathbf{i} + 2\sin t\mathbf{j} + \frac{1}{2}t\mathbf{k}$

(a) $\mathbf{u}(t) = 2(\cos t - 1)\mathbf{i} + 2\sin t\mathbf{j} + \frac{1}{2}t\mathbf{k}$

(b) $\mathbf{u}(t) = 2\cos t\mathbf{i} + 2\sin t\mathbf{j} + 2t\mathbf{k}$

(c) $\mathbf{u}(t) = 2\cos(-t)\mathbf{i} + 2\sin(-t)\mathbf{j} + \frac{1}{2}(-t)\mathbf{k}$

(d) $\mathbf{u}(t) = 6\cos t\mathbf{i} + 6\sin t\mathbf{j} + \frac{1}{2}t\mathbf{k}$

42. $\mathbf{r}(t) = t\mathbf{i} + t^2\mathbf{j} + \frac{1}{2}t^3\mathbf{k}$

(a) $\mathbf{u}(t) = (-t)\mathbf{i} + (-t)^2\mathbf{j} + \frac{1}{2}(-t)^3\mathbf{k}$

(b) $\mathbf{u}(t) = t^2\mathbf{i} + t\mathbf{j} + \frac{1}{2}t^3\mathbf{k}$

(c) $\mathbf{u}(t) = t\mathbf{i} + t^2\mathbf{j} + \left(\frac{1}{2}t^3 + 4\right)\mathbf{k}$

(d) $\mathbf{u}(t) = t\mathbf{i} + t^2\mathbf{j} + \frac{1}{8}t^3\mathbf{k}$

Writing a Transformation In Exercises 43–46, consider the vector-valued function $\mathbf{r}(t) = 3t^2\mathbf{i} + (t-1)\mathbf{j} + t\mathbf{k}$. Write a vector-valued function $\mathbf{u}(t)$ that is the specified transformation of $\mathbf{r}$.

43. A vertical translation two units upward

44. A horizontal translation one unit in the direction of the positive x-axis

45. The y-value increases by a factor of two

46. The z-value increases by a factor of three

 **Representing a Graph by a Vector-Valued Function** In Exercises 47–54, represent the plane curve by a vector-valued function. (There are many correct answers.)

47. $y = x + 5$

48. $2x - 3y + 5 = 0$

49. $y = (x - 2)^2$

50. $y = 4 - x^2$

51. $x^2 + y^2 = 25$

52. $(x - 2)^2 + y^2 = 4$

53. $\frac{x^2}{16} - \frac{y^2}{4} = 1$

54. $\frac{x^2}{9} + \frac{y^2}{16} = 1$

 **Representing a Graph by a Vector-Valued Function** In Exercises 55–62, sketch the space curve represented by the intersection of the surfaces. Then represent the curve by a vector-valued function using the given parameter.

Surfaces	Parameter
55. $z = x^2 + y^2$, $x + y = 0$	$x = t$
56. $z = x^2 + y^2$, $z = 4$	$x = 2\cos t$

Surfaces	Parameter
57. $x^2 + y^2 = 4$, $z = x^2$	$x = 2\sin t$
58. $4x^2 + 4y^2 + z^2 = 16$, $x = z^2$	$z = t$
59. $x^2 + y^2 + z^2 = 4$, $x + z = 2$	$x = 1 + \sin t$
60. $x^2 + y^2 + z^2 = 10$, $x + y = 4$	$x = 2 + \sin t$
61. $x^2 + z^2 = 4$, $y^2 + z^2 = 4$	$x = t$ (first octant)
62. $x^2 + y^2 + z^2 = 16$, $xy = 4$	$x = t$ (first octant)

63. Sketching a Curve Show that the vector-valued function $\mathbf{r}(t) = t\mathbf{i} + 2t\cos t\mathbf{j} + 2t\sin t\mathbf{k}$ lies on the cone $4x^2 = y^2 + z^2$. Sketch the curve.

64. Sketching a Curve Show that the vector-valued function $\mathbf{r}(t) = e^{-t}\cos t\mathbf{i} + e^{-t}\sin t\mathbf{j} + e^{-t}\mathbf{k}$ lies on the cone $z^2 = x^2 + y^2$. Sketch the curve.

 **Finding a Limit** In Exercises 65–70, find the limit (if it exists).

65. $\lim\limits_{t \to \pi} (t\mathbf{i} + \cos t\mathbf{j} + \sin t\mathbf{k})$

66. $\lim\limits_{t \to 2} \left(3t\mathbf{i} + \frac{2}{t^2 - 1}\mathbf{j} + \frac{1}{t}\mathbf{k}\right)$

67. $\lim\limits_{t \to 0} \left(t^2\mathbf{i} + 3t\mathbf{j} + \frac{1 - \cos t}{t}\mathbf{k}\right)$

68. $\lim\limits_{t \to 1} \left(\sqrt{t}\mathbf{i} + \frac{\ln t}{t^2 - 1}\mathbf{j} + \frac{1}{t - 1}\mathbf{k}\right)$

69. $\lim\limits_{t \to 0} \left(e^t\mathbf{i} + \frac{\sin t}{t}\mathbf{j} + e^{-t}\mathbf{k}\right)$

70. $\lim\limits_{t \to \infty} \left(e^{-t}\mathbf{i} + \frac{1}{t}\mathbf{j} + t^{1/t}\mathbf{k}\right)$

 **Continuity of a Vector-Valued Function** In Exercises 71–76, determine the interval(s) on which the vector-valued function is continuous.

71. $\mathbf{r}(t) = \frac{1}{2t + 1}\mathbf{i} + \frac{1}{t}\mathbf{j}$

72. $\mathbf{r}(t) = \sqrt{t}\mathbf{i} + \sqrt{t - 1}\mathbf{j}$

73. $\mathbf{r}(t) = t\mathbf{i} + \arcsin t\mathbf{j} + (t - 1)\mathbf{k}$

74. $\mathbf{r}(t) = 2e^{-t}\mathbf{i} + e^{-t}\mathbf{j} + \ln(t - 1)\mathbf{k}$

75. $\mathbf{r}(t) = \langle e^{-t}, t^2, \tan t \rangle$

76. $\mathbf{r}(t) = \langle 8, \sqrt{t}, \sqrt[3]{t} \rangle$

EXPLORING CONCEPTS

77. Think About It Consider first-degree polynomial functions $f(t)$, $g(t)$, and $h(t)$. Determine whether the curve represented by $\mathbf{r}(t) = f(t)\mathbf{i} + g(t)\mathbf{j} + h(t)\mathbf{k}$ is a line. Explain.

78. Think About It The curve represented by $\mathbf{r}(t) = f(t)\mathbf{i} + g(t)\mathbf{j} + h(t)\mathbf{k}$ is a line. Are f, g, and h first-degree polynomial functions of t? Explain.

79. Continuity of a Vector-Valued Function Give an example of a vector-valued function that is defined but not continuous at $t = 3$.

80. Comparing Functions Which of the following vector-valued functions represent the same graph?

(a) $\mathbf{r}(t) = (-3 \cos t + 1)\mathbf{i} + (5 \sin t + 2)\mathbf{j} + 4\mathbf{k}$

(b) $\mathbf{r}(t) = 4\mathbf{i} + (-3 \cos t + 1)\mathbf{j} + (5 \sin t + 2)\mathbf{k}$

(c) $\mathbf{r}(t) = (3 \cos t - 1)\mathbf{i} + (-5 \sin t - 2)\mathbf{j} + 4\mathbf{k}$

(d) $\mathbf{r}(t) = (-3 \cos 2t + 1)\mathbf{i} + (5 \sin 2t + 2)\mathbf{j} + 4\mathbf{k}$

81. Staircase

The outer bottom edge of a staircase is in the shape of a helix of radius 1 meter. The staircase has a height of 4 meters and makes two complete revolutions from top to bottom. Find a vector-valued function for the staircase. Use a computer algebra system to graph your function. (There are many correct answers.)

82. **HOW DO YOU SEE IT?** The four figures below are graphs of the vector-valued function

$$\mathbf{r}(t) = 4 \cos t\,\mathbf{i} + 4 \sin t\,\mathbf{j} + \frac{t}{4}\mathbf{k}.$$

Match each of the four graphs with the point in space from which the helix is viewed.

(a)

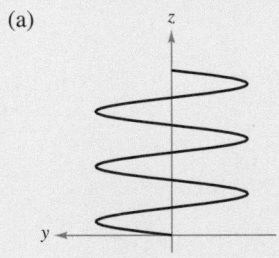

Generated by Mathematica

(b)

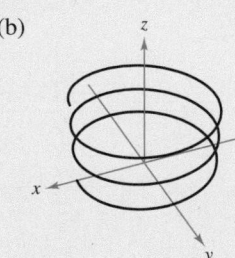

Generated by Mathematica

(c)

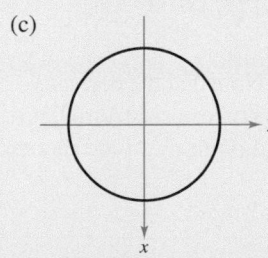

Generated by Mathematica

(d)

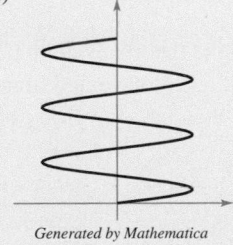

Generated by Mathematica

(i) $(0, 0, 20)$

(ii) $(20, 0, 0)$

(iii) $(-20, 0, 0)$

(iv) $(10, 20, 10)$

83. Proof Let $\mathbf{r}(t)$ and $\mathbf{u}(t)$ be vector-valued functions whose limits exist as $t \to c$. Prove that

$$\lim_{t \to c} [\mathbf{r}(t) \times \mathbf{u}(t)] = \lim_{t \to c} \mathbf{r}(t) \times \lim_{t \to c} \mathbf{u}(t).$$

84. Proof Let $\mathbf{r}(t)$ and $\mathbf{u}(t)$ be vector-valued functions whose limits exist as $t \to c$. Prove that

$$\lim_{t \to c} [\mathbf{r}(t) \cdot \mathbf{u}(t)] = \lim_{t \to c} \mathbf{r}(t) \cdot \lim_{t \to c} \mathbf{u}(t).$$

85. Proof Prove that if $\mathbf{r}$ is a vector-valued function that is continuous at c, then $\|\mathbf{r}\|$ is continuous at c.

86. Verifying a Converse Verify that the converse of Exercise 85 is not true by finding a vector-valued function $\mathbf{r}$ such that $\|\mathbf{r}\|$ is continuous at c but $\mathbf{r}$ is not continuous at c.

Think About It **In Exercises 87 and 88, two particles travel along the space curves $\mathbf{r}(t)$ and $\mathbf{u}(t)$.**

87. If $\mathbf{r}(t)$ and $\mathbf{u}(t)$ intersect, will the particles collide?

88. If the particles collide, do their paths $\mathbf{r}(t)$ and $\mathbf{u}(t)$ intersect?

Particle Motion **In Exercises 89 and 90, two particles travel along the space curves $\mathbf{r}(t)$ and $\mathbf{u}(t)$. Do the particles collide? Do their paths intersect?**

89. $\mathbf{r}(t) = t^2\mathbf{i} + (9t - 20)\mathbf{j} + t^2\mathbf{k}$

$\mathbf{u}(t) = (3t + 4)\mathbf{i} + t^2\mathbf{j} + (5t - 4)\mathbf{k}$

90. $\mathbf{r}(t) = t\mathbf{i} + t^2\mathbf{j} + t^3\mathbf{k}$

$\mathbf{u}(t) = (-2t + 3)\mathbf{i} + 8t\mathbf{j} + (12t + 2)\mathbf{k}$

SECTION PROJECT

Witch of Agnesi

In Section 4.5, you studied a famous curve called the **Witch of Agnesi**. In this project, you will take a closer look at this function.

Consider a circle of radius a centered on the y-axis at $(0, a)$. Let A be a point on the horizontal line $y = 2a$, let O be the origin, and let B be the point where the segment OA intersects the circle. A point P is on the Witch of Agnesi when P lies on the horizontal line through B and on the vertical line through A.

(a) Show that the point A is traced out by the vector-valued function

$$\mathbf{r}_A(\theta) = 2a \cot \theta\,\mathbf{i} + 2a\mathbf{j}, \quad 0 < \theta < \pi$$

where θ is the angle that OA makes with the positive x-axis.

(b) Show that the point B is traced out by the vector-valued function $\mathbf{r}_B(\theta) = a \sin 2\theta\,\mathbf{i} + a(1 - \cos 2\theta)\mathbf{j}, 0 < \theta < \pi$.

(c) Combine the results of parts (a) and (b) to find the vector-valued function $\mathbf{r}(\theta)$ for the Witch of Agnesi. Use a graphing utility to graph this curve for $a = 1$.

(d) Describe the limits $\lim_{\theta \to 0^+} \mathbf{r}(\theta)$ and $\lim_{\theta \to \pi^-} \mathbf{r}(\theta)$.

(e) Eliminate the parameter θ and determine the rectangular equation of the Witch of Agnesi. Use a graphing utility to graph this function for $a = 1$ and compare your graph with that obtained in part (c).

12.2 Differentiation and Integration of Vector-Valued Functions

■ Differentiate a vector-valued function.
■ Integrate a vector-valued function.

Differentiation of Vector-Valued Functions

In Sections 12.3–12.5, you will study several important applications involving the calculus of vector-valued functions. In preparation for that study, this section is devoted to the mechanics of differentiation and integration of vector-valued functions.

The definition of the derivative of a vector-valued function parallels the definition for real-valued functions.

Definition of the Derivative of a Vector-Valued Function

The **derivative of a vector-valued function r** is

$$\mathbf{r}'(t) = \lim_{\Delta t \to 0} \frac{\mathbf{r}(t + \Delta t) - \mathbf{r}(t)}{\Delta t}$$

for all t for which the limit exists. If $\mathbf{r}'(t)$ exists, then **r** is **differentiable at t.** If $\mathbf{r}'(t)$ exists for all t in an open interval I, then **r** is **differentiable on the interval I.** Differentiability of vector-valued functions can be extended to closed intervals by considering one-sided limits.

Differentiation of vector-valued functions can be done on a *component-by-component basis.* To see why this is true, consider the function $\mathbf{r}(t) = f(t)\mathbf{i} + g(t)\mathbf{j}$. Applying the definition of the derivative produces the following.

$$\mathbf{r}'(t) = \lim_{\Delta t \to 0} \frac{\mathbf{r}(t + \Delta t) - \mathbf{r}(t)}{\Delta t}$$

$$= \lim_{\Delta t \to 0} \frac{f(t + \Delta t)\mathbf{i} + g(t + \Delta t)\mathbf{j} - f(t)\mathbf{i} - g(t)\mathbf{j}}{\Delta t}$$

$$= \lim_{\Delta t \to 0} \left\{ \left[\frac{f(t + \Delta t) - f(t)}{\Delta t} \right]\mathbf{i} + \left[\frac{g(t + \Delta t) - g(t)}{\Delta t} \right]\mathbf{j} \right\}$$

$$= \left\{ \lim_{\Delta t \to 0} \left[\frac{f(t + \Delta t) - f(t)}{\Delta t} \right] \right\}\mathbf{i} + \left\{ \lim_{\Delta t \to 0} \left[\frac{g(t + \Delta t) - g(t)}{\Delta t} \right] \right\}\mathbf{j}$$

$$= f'(t)\mathbf{i} + g'(t)\mathbf{j}$$

This important result is listed in the theorem shown below. Note that the derivative of the vector-valued function **r** is itself a vector-valued function. You can see from Figure 12.8 that $\mathbf{r}'(t)$ is a vector tangent to the curve given by $\mathbf{r}(t)$ and pointing in the direction of increasing t-values.

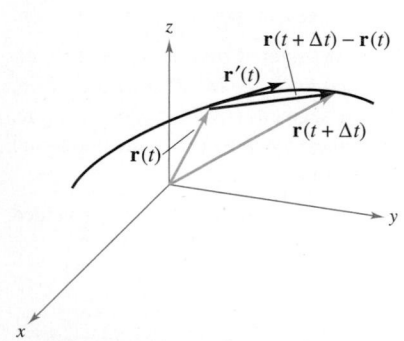

Figure 12.8

THEOREM 12.1 Differentiation of Vector-Valued Functions

1. If $\mathbf{r}(t) = f(t)\mathbf{i} + g(t)\mathbf{j}$, where f and g are differentiable functions of t, then

$$\mathbf{r}'(t) = f'(t)\mathbf{i} + g'(t)\mathbf{j}. \qquad \text{Plane}$$

2. If $\mathbf{r}(t) = f(t)\mathbf{i} + g(t)\mathbf{j} + h(t)\mathbf{k}$, where f, g, and h are differentiable functions of t, then

$$\mathbf{r}'(t) = f'(t)\mathbf{i} + g'(t)\mathbf{j} + h'(t)\mathbf{k}. \qquad \text{Space}$$

EXAMPLE 1 **Differentiation of a Vector-Valued Function**

∵ ▷ *See LarsonCalculus.com for an interactive version of this type of example.*

For the vector-valued function

$$\mathbf{r}(t) = t\mathbf{i} + (t^2 + 2)\mathbf{j}$$

find $\mathbf{r}'(t)$. Then sketch the plane curve represented by $\mathbf{r}(t)$ and the graphs of $\mathbf{r}(1)$ and $\mathbf{r}'(1)$.

Solution Differentiate on a component-by-component basis to obtain

$$\mathbf{r}'(t) = \mathbf{i} + 2t\mathbf{j}. \qquad \text{Derivative}$$

From the position vector $\mathbf{r}(t)$, you can write the parametric equations $x = t$ and $y = t^2 + 2$. The corresponding rectangular equation is $y = x^2 + 2$. When $t = 1$,

$$\mathbf{r}(1) = \mathbf{i} + 3\mathbf{j}$$

and

$$\mathbf{r}'(1) = \mathbf{i} + 2\mathbf{j}.$$

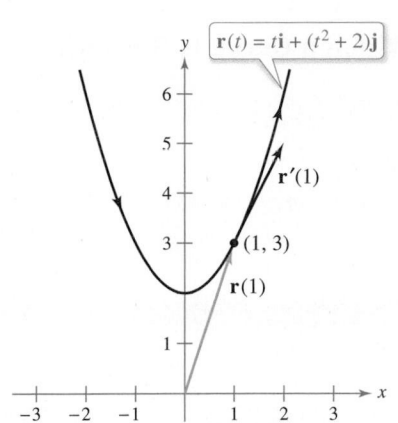

Figure 12.9

In Figure 12.9, $\mathbf{r}(1)$ is drawn starting at the origin, and $\mathbf{r}'(1)$ is drawn starting at the terminal point of $\mathbf{r}(1)$. Note that at $(1, 3)$, the vector $\mathbf{r}'(1)$ is tangent to the curve given by $\mathbf{r}(t)$ and is pointing in the direction of increasing t-values. ■

Higher-order derivatives of vector-valued functions are obtained by successive differentiation of each component function.

EXAMPLE 2 **Higher-Order Differentiation**

For the vector-valued function

$$\mathbf{r}(t) = \cos t\mathbf{i} + \sin t\mathbf{j} + 2t\mathbf{k}$$

find each of the following.

a. $\mathbf{r}'(t)$

b. $\mathbf{r}''(t)$

c. $\mathbf{r}'(t) \cdot \mathbf{r}''(t)$

d. $\mathbf{r}'(t) \times \mathbf{r}''(t)$

Solution

a. $\mathbf{r}'(t) = -\sin t\mathbf{i} + \cos t\mathbf{j} + 2\mathbf{k}$ First derivative

b. $\mathbf{r}''(t) = -\cos t\mathbf{i} - \sin t\mathbf{j} + 0\mathbf{k}$

       $= -\cos t\mathbf{i} - \sin t\mathbf{j}$ Second derivative

c. $\mathbf{r}'(t) \cdot \mathbf{r}''(t) = \sin t \cos t - \sin t \cos t = 0$ Dot product

d. $\mathbf{r}'(t) \times \mathbf{r}''(t) = \begin{vmatrix} \mathbf{i} & \mathbf{j} & \mathbf{k} \\ -\sin t & \cos t & 2 \\ -\cos t & -\sin t & 0 \end{vmatrix}$ Cross product

$$= \begin{vmatrix} \cos t & 2 \\ -\sin t & 0 \end{vmatrix}\mathbf{i} - \begin{vmatrix} -\sin t & 2 \\ -\cos t & 0 \end{vmatrix}\mathbf{j} + \begin{vmatrix} -\sin t & \cos t \\ -\cos t & -\sin t \end{vmatrix}\mathbf{k}$$

$$= 2\sin t\mathbf{i} - 2\cos t\mathbf{j} + \mathbf{k}$$ ■

In Example 2(c), note that the dot product is a real-valued function, not a vector-valued function.

The parametrization of the curve represented by the vector-valued function

$$\mathbf{r}(t) = f(t)\mathbf{i} + g(t)\mathbf{j} + h(t)\mathbf{k}$$

is **smooth on an open interval** I when f', g', and h' are continuous on I and $\mathbf{r}'(t) \neq \mathbf{0}$ for any value of t in the interval I.

EXAMPLE 3 Finding Intervals on Which a Curve Is Smooth

Find the intervals on which the epicycloid C given by

$$\mathbf{r}(t) = (5 \cos t - \cos 5t)\mathbf{i} + (5 \sin t - \sin 5t)\mathbf{j}, \quad 0 \leq t \leq 2\pi$$

is smooth.

Solution The derivative of $\mathbf{r}$ is

$$\mathbf{r}'(t) = (-5 \sin t + 5 \sin 5t)\mathbf{i} + (5 \cos t - 5 \cos 5t)\mathbf{j}.$$

In the interval $[0, 2\pi]$, the only values of t for which

$$\mathbf{r}'(t) = 0\mathbf{i} + 0\mathbf{j}$$

are $t = 0, \pi/2, \pi, 3\pi/2$, and 2π. Therefore, you can conclude that C is smooth on the intervals

$$\left(0, \frac{\pi}{2}\right), \quad \left(\frac{\pi}{2}, \pi\right), \quad \left(\pi, \frac{3\pi}{2}\right), \quad \text{and} \quad \left(\frac{3\pi}{2}, 2\pi\right)$$

as shown in Figure 12.10. ■

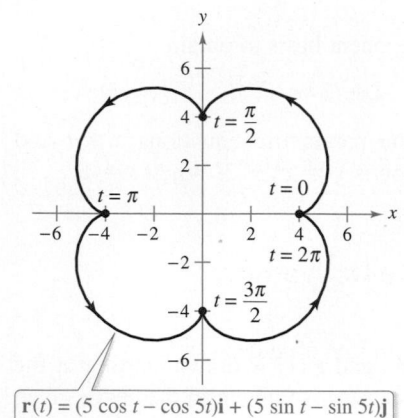

$\mathbf{r}(t) = (5 \cos t - \cos 5t)\mathbf{i} + (5 \sin t - \sin 5t)\mathbf{j}$

The epicycloid is not smooth at the points where it intersects the axes.
Figure 12.10

In Figure 12.10, note that the curve is not smooth at points at which the curve makes abrupt changes in direction. Such points are called *cusps* or **nodes.**

Most of the differentiation rules in Chapter 3 have counterparts for vector-valued functions, and several of these are listed in the next theorem. Note that the theorem contains three versions of "product rules." Property 3 gives the derivative of the product of a real-valued function w and a vector-valued function $\mathbf{r}$, Property 4 gives the derivative of the dot product of two vector-valued functions, and Property 5 gives the derivative of the cross product of two vector-valued functions (in space).

THEOREM 12.2 Properties of the Derivative

Let $\mathbf{r}$ and $\mathbf{u}$ be differentiable vector-valued functions of t, let w be a differentiable real-valued function of t, and let c be a scalar.

1. $\dfrac{d}{dt}[c\mathbf{r}(t)] = c\mathbf{r}'(t)$

2. $\dfrac{d}{dt}[\mathbf{r}(t) \pm \mathbf{u}(t)] = \mathbf{r}'(t) \pm \mathbf{u}'(t)$

3. $\dfrac{d}{dt}[w(t)\mathbf{r}(t)] = w(t)\mathbf{r}'(t) + w'(t)\mathbf{r}(t)$

4. $\dfrac{d}{dt}[\mathbf{r}(t) \cdot \mathbf{u}(t)] = \mathbf{r}(t) \cdot \mathbf{u}'(t) + \mathbf{r}'(t) \cdot \mathbf{u}(t)$

5. $\dfrac{d}{dt}[\mathbf{r}(t) \times \mathbf{u}(t)] = \mathbf{r}(t) \times \mathbf{u}'(t) + \mathbf{r}'(t) \times \mathbf{u}(t)$

6. $\dfrac{d}{dt}[\mathbf{r}(w(t))] = \mathbf{r}'(w(t))w'(t)$

7. If $\mathbf{r}(t) \cdot \mathbf{r}(t) = c$, then $\mathbf{r}(t) \cdot \mathbf{r}'(t) = 0$.

• • REMARK Note that Property 5 applies only to three-dimensional vector-valued functions because the cross product is not defined for two-dimensional vectors.

Proof To prove Property 4, let

$$\mathbf{r}(t) = f_1(t)\mathbf{i} + g_1(t)\mathbf{j} \quad \text{and} \quad \mathbf{u}(t) = f_2(t)\mathbf{i} + g_2(t)\mathbf{j}$$

where f_1, f_2, g_1, and g_2 are differentiable functions of t. Then

$$\mathbf{r}(t) \cdot \mathbf{u}(t) = f_1(t)f_2(t) + g_1(t)g_2(t)$$

and it follows that

$$\frac{d}{dt}[\mathbf{r}(t) \cdot \mathbf{u}(t)] = f_1(t)f_2'(t) + f_1'(t)f_2(t) + g_1(t)g_2'(t) + g_1'(t)g_2(t)$$

$$= [f_1(t)f_2'(t) + g_1(t)g_2'(t)] + [f_1'(t)f_2(t) + g_1'(t)g_2(t)]$$

$$= \mathbf{r}(t) \cdot \mathbf{u}'(t) + \mathbf{r}'(t) \cdot \mathbf{u}(t).$$

Proofs of the other properties are left as exercises (see Exercises 61–65 and Exercise 68). ∎

Exploration

Let $\mathbf{r}(t) = \cos t\mathbf{i} + \sin t\mathbf{j}$. Sketch the graph of $\mathbf{r}(t)$. Explain why the graph is a circle of radius 1 centered at the origin. Calculate $\mathbf{r}(\pi/4)$ and $\mathbf{r}'(\pi/4)$. Position the vector $\mathbf{r}'(\pi/4)$ so that its initial point is at the terminal point of $\mathbf{r}(\pi/4)$. What do you observe? Show that $\mathbf{r}(t) \cdot \mathbf{r}(t)$ is constant and that $\mathbf{r}(t) \cdot \mathbf{r}'(t) = 0$ for all t. How does this example relate to Property 7 of Theorem 12.2?

EXAMPLE 4   Using Properties of the Derivative

For $\mathbf{r}(t) = \dfrac{1}{t}\mathbf{i} - \mathbf{j} + \ln t\mathbf{k}$ and $\mathbf{u}(t) = t^2\mathbf{i} - 2t\mathbf{j} + \mathbf{k}$, find each derivative.

a. $\dfrac{d}{dt}[\mathbf{r}(t) \cdot \mathbf{u}(t)]$

b. $\dfrac{d}{dt}[\mathbf{u}(t) \times \mathbf{u}'(t)]$

Solution

a. Because $\mathbf{r}'(t) = -\dfrac{1}{t^2}\mathbf{i} + \dfrac{1}{t}\mathbf{k}$ and $\mathbf{u}'(t) = 2t\mathbf{i} - 2\mathbf{j}$, you have

$$\frac{d}{dt}[\mathbf{r}(t) \cdot \mathbf{u}(t)]$$

$$= \mathbf{r}(t) \cdot \mathbf{u}'(t) + \mathbf{r}'(t) \cdot \mathbf{u}(t)$$

$$= \left(\frac{1}{t}\mathbf{i} - \mathbf{j} + \ln t\mathbf{k}\right) \cdot (2t\mathbf{i} - 2\mathbf{j}) + \left(-\frac{1}{t^2}\mathbf{i} + \frac{1}{t}\mathbf{k}\right) \cdot (t^2\mathbf{i} - 2t\mathbf{j} + \mathbf{k})$$

$$= 2 + 2 + (-1) + \frac{1}{t}$$

$$= 3 + \frac{1}{t}.$$

b. Because $\mathbf{u}'(t) = 2t\mathbf{i} - 2\mathbf{j}$ and $\mathbf{u}''(t) = 2\mathbf{i}$, you have

$$\frac{d}{dt}[\mathbf{u}(t) \times \mathbf{u}'(t)] = [\mathbf{u}(t) \times \mathbf{u}''(t)] + [\mathbf{u}'(t) \times \mathbf{u}'(t)]$$

$$= \begin{vmatrix} \mathbf{i} & \mathbf{j} & \mathbf{k} \\ t^2 & -2t & 1 \\ 2 & 0 & 0 \end{vmatrix} + \mathbf{0}$$

$$= \begin{vmatrix} -2t & 1 \\ 0 & 0 \end{vmatrix}\mathbf{i} - \begin{vmatrix} t^2 & 1 \\ 2 & 0 \end{vmatrix}\mathbf{j} + \begin{vmatrix} t^2 & -2t \\ 2 & 0 \end{vmatrix}\mathbf{k}$$

$$= 0\mathbf{i} - (-2)\mathbf{j} + 4t\mathbf{k}$$

$$= 2\mathbf{j} + 4t\mathbf{k}.$$ ∎

Try reworking parts (a) and (b) in Example 4 by first forming the dot and cross products and then differentiating to see that you obtain the same results.

Integration of Vector-Valued Functions

The next definition is a consequence of the definition of the derivative of a vector-valued function.

Definition of Integration of Vector-Valued Functions

1. If $\mathbf{r}(t) = f(t)\mathbf{i} + g(t)\mathbf{j}$, where f and g are continuous on $[a, b]$, then the **indefinite integral (antiderivative)** of $\mathbf{r}$ is

$$\int \mathbf{r}(t)\, dt = \left[\int f(t)\, dt \right]\mathbf{i} + \left[\int g(t)\, dt \right]\mathbf{j} \qquad \text{Plane}$$

and its **definite integral** over the interval $a \leq t \leq b$ is

$$\int_a^b \mathbf{r}(t)\, dt = \left[\int_a^b f(t)\, dt \right]\mathbf{i} + \left[\int_a^b g(t)\, dt \right]\mathbf{j}.$$

2. If $\mathbf{r}(t) = f(t)\mathbf{i} + g(t)\mathbf{j} + h(t)\mathbf{k}$, where f, g, and h are continuous on $[a, b]$, then the **indefinite integral (antiderivative)** of $\mathbf{r}$ is

$$\int \mathbf{r}(t)\, dt = \left[\int f(t)\, dt \right]\mathbf{i} + \left[\int g(t)\, dt \right]\mathbf{j} + \left[\int h(t)\, dt \right]\mathbf{k} \qquad \text{Space}$$

and its **definite integral** over the interval $a \leq t \leq b$ is

$$\int_a^b \mathbf{r}(t)\, dt = \left[\int_a^b f(t)\, dt \right]\mathbf{i} + \left[\int_a^b g(t)\, dt \right]\mathbf{j} + \left[\int_a^b h(t)\, dt \right]\mathbf{k}.$$

The antiderivative of a vector-valued function is a family of vector-valued functions all differing by a constant vector $\mathbf{C}$. For instance, if $\mathbf{r}(t)$ is a three-dimensional vector-valued function, then for the indefinite integral $\int \mathbf{r}(t)\, dt$, you obtain three constants of integration

$$\int f(t)\, dt = F(t) + C_1, \quad \int g(t)\, dt = G(t) + C_2, \quad \int h(t)\, dt = H(t) + C_3$$

where $F'(t) = f(t)$, $G'(t) = g(t)$, and $H'(t) = h(t)$. These three *scalar* constants produce one *vector* constant of integration

$$\int \mathbf{r}(t)\, dt = [F(t) + C_1]\mathbf{i} + [G(t) + C_2]\mathbf{j} + [H(t) + C_3]\mathbf{k}$$

$$= [F(t)\mathbf{i} + G(t)\mathbf{j} + H(t)\mathbf{k}] + [C_1\mathbf{i} + C_2\mathbf{j} + C_3\mathbf{k}]$$

$$= \mathbf{R}(t) + \mathbf{C}$$

where $\mathbf{R}'(t) = \mathbf{r}(t)$.

EXAMPLE 5   **Integrating a Vector-Valued Function**

Find the indefinite integral

$$\int (t\mathbf{i} + 3\mathbf{j})\, dt.$$

Solution   Integrating on a component-by-component basis produces

$$\int (t\mathbf{i} + 3\mathbf{j})\, dt = \frac{t^2}{2}\mathbf{i} + 3t\mathbf{j} + \mathbf{C}.$$

Example 6 shows how to evaluate the definite integral of a vector-valued function.

EXAMPLE 6 **Definite Integral of a Vector-Valued Function**

Evaluate the integral

$$\int_0^1 \mathbf{r}(t)\, dt = \int_0^1 \left(\sqrt[3]{t}\,\mathbf{i} + \frac{1}{t+1}\mathbf{j} + e^{-t}\mathbf{k} \right) dt.$$

Solution

$$\int_0^1 \mathbf{r}(t)\, dt = \left(\int_0^1 t^{1/3}\, dt \right)\mathbf{i} + \left(\int_0^1 \frac{1}{t+1}\, dt \right)\mathbf{j} + \left(\int_0^1 e^{-t}\, dt \right)\mathbf{k}$$

$$= \left[\left(\frac{3}{4} \right)t^{4/3} \right]_0^1 \mathbf{i} + \left[\ln|t+1| \right]_0^1 \mathbf{j} + \left[-e^{-t} \right]_0^1 \mathbf{k}$$

$$= \frac{3}{4}\mathbf{i} + (\ln 2)\mathbf{j} + \left(1 - \frac{1}{e} \right)\mathbf{k}$$ ∎

As with real-valued functions, you can narrow the family of antiderivatives of a vector-valued function $\mathbf{r}'$ down to a single antiderivative by imposing an initial condition on the vector-valued function $\mathbf{r}$. This is demonstrated in the next example.

EXAMPLE 7 **The Antiderivative of a Vector-Valued Function**

Find the antiderivative of

$$\mathbf{r}'(t) = \cos 2t\mathbf{i} - 2\sin t\mathbf{j} + \frac{1}{1+t^2}\mathbf{k}$$

that satisfies the initial condition

$$\mathbf{r}(0) = 3\mathbf{i} - 2\mathbf{j} + \mathbf{k}.$$

Solution

$$\mathbf{r}(t) = \int \mathbf{r}'(t)\, dt$$

$$= \left(\int \cos 2t\, dt \right)\mathbf{i} + \left(\int -2\sin t\, dt \right)\mathbf{j} + \left(\int \frac{1}{1+t^2}\, dt \right)\mathbf{k}$$

$$= \left(\frac{1}{2}\sin 2t + C_1 \right)\mathbf{i} + (2\cos t + C_2)\mathbf{j} + (\arctan t + C_3)\mathbf{k}$$

Letting $t = 0$, you can write

$$\mathbf{r}(0) = (0 + C_1)\mathbf{i} + (2 + C_2)\mathbf{j} + (0 + C_3)\mathbf{k}.$$

Using the fact that $\mathbf{r}(0) = 3\mathbf{i} - 2\mathbf{j} + \mathbf{k}$, you have

$$(0 + C_1)\mathbf{i} + (2 + C_2)\mathbf{j} + (0 + C_3)\mathbf{k} = 3\mathbf{i} - 2\mathbf{j} + \mathbf{k}.$$

Equating corresponding components produces

$$C_1 = 3, \quad 2 + C_2 = -2, \quad \text{and} \quad C_3 = 1.$$

So, the antiderivative that satisfies the initial condition is

$$\mathbf{r}(t) = \left(\frac{1}{2}\sin 2t + 3 \right)\mathbf{i} + (2\cos t - 4)\mathbf{j} + (\arctan t + 1)\mathbf{k}.$$ ∎

12.2 Exercises

See CalcChat.com for tutorial help and worked-out solutions to odd-numbered exercises.

CONCEPT CHECK

1. Derivative Describe the relationship between the graph of $\mathbf{r}'(t_0)$ and the curve represented by $\mathbf{r}(t)$.

2. Integration Explain why the family of vector-valued functions that are the antiderivatives of a vector-valued function differ by a constant vector.

Differentiation of Vector-Valued Functions In Exercises 3–10, find $\mathbf{r}'(t)$, $\mathbf{r}(t_0)$, and $\mathbf{r}'(t_0)$ for the given value of t_0. Then sketch the curve represented by the vector-valued function and sketch the vectors $\mathbf{r}(t_0)$ and $\mathbf{r}'(t_0)$.

3. $\mathbf{r}(t) = (1 - t^2)\mathbf{i} + t\mathbf{j}$, $t_0 = 3$

4. $\mathbf{r}(t) = (1 + t)\mathbf{i} + t^3\mathbf{j}$, $t_0 = 1$

5. $\mathbf{r}(t) = \cos t\mathbf{i} + \sin t\mathbf{j}$, $t_0 = \dfrac{\pi}{2}$

6. $\mathbf{r}(t) = 3\sin t\mathbf{i} + 4\cos t\mathbf{j}$, $t_0 = \dfrac{\pi}{2}$

7. $\mathbf{r}(t) = \langle e^t, e^{2t} \rangle$, $t_0 = 0$

8. $\mathbf{r}(t) = \langle e^{-t}, e^t \rangle$, $t_0 = 0$

9. $\mathbf{r}(t) = 2\cos t\mathbf{i} + 2\sin t\mathbf{j} + t\mathbf{k}$, $t_0 = \dfrac{3\pi}{2}$

10. $\mathbf{r}(t) = t\mathbf{i} + t^2\mathbf{j} + \dfrac{3}{2}\mathbf{k}$, $t_0 = 2$

Finding a Derivative In Exercises 11–18, find $\mathbf{r}'(t)$.

11. $\mathbf{r}(t) = t^4\mathbf{i} - 5t\mathbf{j}$

12. $\mathbf{r}(t) = \sqrt{t}\mathbf{i} + (1 - t^3)\mathbf{j}$

13. $\mathbf{r}(t) = 3\cos^3 t\mathbf{i} + 2\sin^3 t\mathbf{j} + \mathbf{k}$

14. $\mathbf{r}(t) = 4\sqrt{t}\mathbf{i} + t^2\sqrt{t}\mathbf{j} + \ln t^2\mathbf{k}$

15. $\mathbf{r}(t) = e^{-t}\mathbf{i} + 4\mathbf{j} + 5te^t\mathbf{k}$

16. $\mathbf{r}(t) = \langle t^3, \cos 3t, \sin 3t \rangle$

17. $\mathbf{r}(t) = \langle t\sin t, t\cos t, t \rangle$

18. $\mathbf{r}(t) = \langle \arcsin t, \arccos t, 0 \rangle$

Higher-Order Differentiation In Exercises 19–22, find (a) $\mathbf{r}'(t)$, (b) $\mathbf{r}''(t)$, and (c) $\mathbf{r}'(t) \cdot \mathbf{r}''(t)$.

19. $\mathbf{r}(t) = t^3\mathbf{i} + \frac{1}{2}t^2\mathbf{j}$ **20.** $\mathbf{r}(t) = (t^2 + t)\mathbf{i} + (t^2 - t)\mathbf{j}$

21. $\mathbf{r}(t) = 4\cos t\mathbf{i} + 4\sin t\mathbf{j}$ **22.** $\mathbf{r}(t) = 8\cos t\mathbf{i} + 3\sin t\mathbf{j}$

Higher-Order Differentiation In Exercises 23–26, find (a) $\mathbf{r}'(t)$, (b) $\mathbf{r}''(t)$, (c) $\mathbf{r}'(t) \cdot \mathbf{r}''(t)$, and (d) $\mathbf{r}'(t) \times \mathbf{r}''(t)$.

23. $\mathbf{r}(t) = \frac{1}{2}t^2\mathbf{i} - t\mathbf{j} + \frac{1}{6}t^3\mathbf{k}$

24. $\mathbf{r}(t) = t^3\mathbf{i} + (2t^2 + 3)\mathbf{j} + (3t - 5)\mathbf{k}$

25. $\mathbf{r}(t) = \langle \cos t + t\sin t, \sin t - t\cos t, t \rangle$

26. $\mathbf{r}(t) = \langle e^{-t}, t^2, \tan t \rangle$

Finding Intervals on Which a Curve Is Smooth In Exercises 27–34, find the open interval(s) on which the curve given by the vector-valued function is smooth.

27. $\mathbf{r}(t) = t^2\mathbf{i} + t^3\mathbf{j}$ **28.** $\mathbf{r}(t) = 5t^5\mathbf{i} - t^4\mathbf{j}$

29. $\mathbf{r}(\theta) = 2\cos^3\theta\mathbf{i} + 3\sin^3\theta\mathbf{j}$, $0 \le \theta \le 2\pi$

30. $\mathbf{r}(\theta) = (\theta + \sin\theta)\mathbf{i} + (1 - \cos\theta)\mathbf{j}$, $0 \le \theta \le 2\pi$

31. $\mathbf{r}(t) = \dfrac{2t}{8 + t^3}\mathbf{i} + \dfrac{2t^2}{8 + t^3}\mathbf{j}$

32. $\mathbf{r}(t) = e^t\mathbf{i} - e^{-t}\mathbf{j} + 3t\mathbf{k}$

33. $\mathbf{r}(t) = t\mathbf{i} - 3t\mathbf{j} + \tan t\mathbf{k}$

34. $\mathbf{r}(t) = \sqrt{t}\mathbf{i} + (t^2 - 1)\mathbf{j} + \frac{1}{4}t\mathbf{k}$

Using Properties of the Derivative In Exercises 35 and 36, use the properties of the derivative to find the following.

(a) $\mathbf{r}'(t)$ (b) $\dfrac{d}{dt}[3\mathbf{r}(t) - \mathbf{u}(t)]$ (c) $\dfrac{d}{dt}[(5t)\mathbf{u}(t)]$

(d) $\dfrac{d}{dt}[\mathbf{r}(t) \cdot \mathbf{u}(t)]$ (e) $\dfrac{d}{dt}[\mathbf{r}(t) \times \mathbf{u}(t)]$ (f) $\dfrac{d}{dt}[\mathbf{r}(2t)]$

35. $\mathbf{r}(t) = t\mathbf{i} + 3t\mathbf{j} + t^2\mathbf{k}$, $\mathbf{u}(t) = 4t\mathbf{i} + t^2\mathbf{j} + t^3\mathbf{k}$

36. $\mathbf{r}(t) = \langle t, 2\sin t, 2\cos t \rangle$, $\mathbf{u}(t) = \left\langle \dfrac{1}{t}, 2\sin t, 2\cos t \right\rangle$

Using Two Methods In Exercises 37 and 38, find (a) $\dfrac{d}{dt}[\mathbf{r}(t) \cdot \mathbf{u}(t)]$ and (b) $\dfrac{d}{dt}[\mathbf{r}(t) \times \mathbf{u}(t)]$ in two different ways.

 (i) **Find the product first, then differentiate.**

(ii) **Apply the properties of Theorem 12.2.**

37. $\mathbf{r}(t) = t\mathbf{i} + 2t^2\mathbf{j} + t^3\mathbf{k}$, $\mathbf{u}(t) = t^4\mathbf{k}$

38. $\mathbf{r}(t) = \cos t\mathbf{i} + \sin t\mathbf{j} + t\mathbf{k}$, $\mathbf{u}(t) = \mathbf{j} + t\mathbf{k}$

Finding an Indefinite Integral In Exercises 39–46, find the indefinite integral.

39. $\displaystyle\int (2t\mathbf{i} + \mathbf{j} + 9\mathbf{k})\,dt$ **40.** $\displaystyle\int \left(4t^3\mathbf{i} + 6t\mathbf{j} - 4\sqrt{t}\mathbf{k}\right) dt$

41. $\displaystyle\int \left(\frac{1}{t}\mathbf{i} + \mathbf{j} - t^{3/2}\mathbf{k}\right) dt$ **42.** $\displaystyle\int \left(\ln t\mathbf{i} + \frac{1}{t}\mathbf{j} + \mathbf{k}\right) dt$

43. $\displaystyle\int (\mathbf{i} + 4t^3\mathbf{j} + 5^t\mathbf{k})\,dt$

44. $\displaystyle\int \left(\sec^2 t\mathbf{i} + \frac{1}{1 + t^2}\mathbf{j}\right) dt$

45. $\displaystyle\int (e^t\mathbf{i} + \mathbf{j} + t\cos t\mathbf{k})\,dt$

46. $\displaystyle\int (e^{-t}\sin t\mathbf{i} + \cot t\mathbf{j})\,dt$

Evaluating a Definite Integral In Exercises 47–52, evaluate the definite integral.

47. $\int_0^1 (8t\mathbf{i} + t\mathbf{j} - \mathbf{k}) \, dt$

48. $\int_{-1}^1 \left(t\mathbf{i} + t^3\mathbf{j} + \sqrt[3]{t}\,\mathbf{k}\right) dt$

49. $\int_0^{\pi/2} [(5\cos t)\mathbf{i} + (6\sin t)\mathbf{j} + \mathbf{k}] \, dt$

50. $\int_0^{\pi/4} [(\sec t \tan t)\mathbf{i} + (\tan t)\mathbf{j} + (2\sin t \cos t)\mathbf{k}] \, dt$

51. $\int_0^2 (t\mathbf{i} + e^t\mathbf{j} - te^t\mathbf{k}) \, dt$ 52. $\int_0^3 \|t\mathbf{i} + t^2\mathbf{j}\| \, dt$

Finding an Antiderivative In Exercises 53–58, find $\mathbf{r}(t)$ that satisfies the initial condition(s).

53. $\mathbf{r}'(t) = 4e^{2t}\mathbf{i} + 3e^t\mathbf{j}, \quad \mathbf{r}(0) = 2\mathbf{i}$

54. $\mathbf{r}'(t) = 3t^2\mathbf{j} + 6\sqrt{t}\,\mathbf{k}, \quad \mathbf{r}(0) = \mathbf{i} + 2\mathbf{j}$

55. $\mathbf{r}''(t) = -32\mathbf{j}, \quad \mathbf{r}'(0) = 600\sqrt{3}\,\mathbf{i} + 600\mathbf{j}, \quad \mathbf{r}(0) = \mathbf{0}$

56. $\mathbf{r}''(t) = -4\cos t\mathbf{j} - 3\sin t\mathbf{k}, \quad \mathbf{r}'(0) = 3\mathbf{k}, \quad \mathbf{r}(0) = 4\mathbf{j}$

57. $\mathbf{r}'(t) = te^{-t^2}\mathbf{i} - e^{-t}\mathbf{j} + \mathbf{k}, \quad \mathbf{r}(0) = \tfrac{1}{2}\mathbf{i} - \mathbf{j} + \mathbf{k}$

58. $\mathbf{r}'(t) = \dfrac{1}{1 + t^2}\mathbf{i} + \dfrac{1}{t^2}\mathbf{j} + \dfrac{1}{t}\mathbf{k}, \quad \mathbf{r}(1) = 2\mathbf{i}$

EXPLORING CONCEPTS

59. **Using a Derivative** The three components of the derivative of the vector-valued function $\mathbf{u}$ are positive at $t = t_0$. Describe the behavior of $\mathbf{u}$ at $t = t_0$.

60. **Think About It** Find two vector-valued functions $\mathbf{f}(t)$ and $\mathbf{g}(t)$ such that

$$\int_a^b [\mathbf{f}(t) \cdot \mathbf{g}(t)] \, dt \neq \left[\int_a^b \mathbf{f}(t) \, dt \right] \cdot \left[\int_a^b \mathbf{g}(t) \, dt \right].$$

Proof In Exercises 61–68, prove the property. In each case, assume r, u, and v are differentiable vector-valued functions of t in space, w is a differentiable real-valued function of t, and c is a scalar.

61. $\dfrac{d}{dt}[c\mathbf{r}(t)] = c\mathbf{r}'(t)$

62. $\dfrac{d}{dt}[\mathbf{r}(t) \pm \mathbf{u}(t)] = \mathbf{r}'(t) \pm \mathbf{u}'(t)$

63. $\dfrac{d}{dt}[w(t)\mathbf{r}(t)] = w(t)\mathbf{r}'(t) + w'(t)\mathbf{r}(t)$

64. $\dfrac{d}{dt}[\mathbf{r}(t) \times \mathbf{u}(t)] = \mathbf{r}(t) \times \mathbf{u}'(t) + \mathbf{r}'(t) \times \mathbf{u}(t)$

65. $\dfrac{d}{dt}[\mathbf{r}(w(t))] = \mathbf{r}'(w(t))w'(t)$

66. $\dfrac{d}{dt}[\mathbf{r}(t) \times \mathbf{r}'(t)] = \mathbf{r}(t) \times \mathbf{r}''(t)$

67. $\dfrac{d}{dt}\{\mathbf{r}(t) \cdot [\mathbf{u}(t) \times \mathbf{v}(t)]\} = \mathbf{r}'(t) \cdot [\mathbf{u}(t) \times \mathbf{v}(t)] +$
$\mathbf{r}(t) \cdot [\mathbf{u}'(t) \times \mathbf{v}(t)] +$
$\mathbf{r}(t) \cdot [\mathbf{u}(t) \times \mathbf{v}'(t)]$

68. If $\mathbf{r}(t) \cdot \mathbf{r}(t)$ is a constant, then $\mathbf{r}(t) \cdot \mathbf{r}'(t) = 0$.

69. **Particle Motion** A particle moves in the xy-plane along the curve represented by the vector-valued function $\mathbf{r}(t) = (t - \sin t)\mathbf{i} + (1 - \cos t)\mathbf{j}$.

(a) Use a graphing utility to graph $\mathbf{r}$. Describe the curve.

(b) Find the minimum and maximum values of $\|\mathbf{r}'\|$ and $\|\mathbf{r}''\|$.

70. **Particle Motion** A particle moves in the yz-plane along the curve represented by the vector-valued function $\mathbf{r}(t) = (2\cos t)\mathbf{j} + (3\sin t)\mathbf{k}$.

(a) Describe the curve.

(b) Find the minimum and maximum values of $\|\mathbf{r}'\|$ and $\|\mathbf{r}''\|$.

71. **Perpendicular Vectors** Consider the vector-valued function $\mathbf{r}(t) = (e^t \sin t)\mathbf{i} + (e^t \cos t)\mathbf{j}$. Show that $\mathbf{r}(t)$ and $\mathbf{r}''(t)$ are always perpendicular to each other.

72. **HOW DO YOU SEE IT?** The graph shows a vector-valued function $\mathbf{r}(t)$ for $0 \le t \le 2\pi$ and its derivative $\mathbf{r}'(t)$ for several values of t.

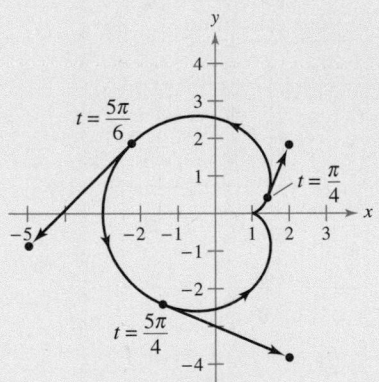

(a) For each derivative shown in the graph, determine whether each component is positive or negative.

(b) Is the curve smooth on the interval $[0, 2\pi]$? Explain.

True or False? In Exercises 73–76, determine whether the statement is true or false. If it is false, explain why or give an example that shows it is false.

73. If a particle moves along a sphere centered at the origin, then its derivative vector is always tangent to the sphere.

74. The definite integral of a vector-valued function is a real number.

75. $\dfrac{d}{dt}[\|\mathbf{r}(t)\|] = \|\mathbf{r}'(t)\|$

76. If $\mathbf{r}$ and $\mathbf{u}$ are differentiable vector-valued functions of t, then

$$\dfrac{d}{dt}[\mathbf{r}(t) \cdot \mathbf{u}(t)] = \mathbf{r}'(t) \cdot \mathbf{u}'(t).$$

12.3 Velocity and Acceleration

■ Describe the velocity and acceleration associated with a vector-valued function.
■ Use a vector-valued function to analyze projectile motion.

Velocity and Acceleration

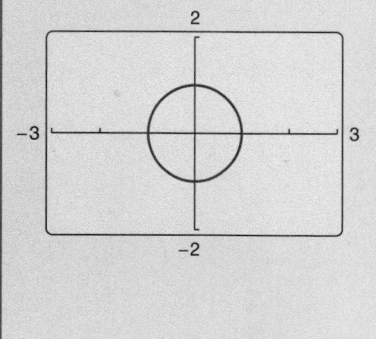
You are now ready to combine your study of parametric equations, curves, vectors, and vector-valued functions to form a model for motion along a curve. You will begin by looking at the motion of an object in the plane. (The motion of an object in space can be developed similarly.)

As an object moves along a curve in the plane, the coordinates x and y of its center of mass are each functions of time t. Rather than using the letters f and g to represent these two functions, it is convenient to write $x = x(t)$ and $y = y(t)$. So, the position vector $\mathbf{r}(t)$ takes the form

$$\mathbf{r}(t) = x(t)\mathbf{i} + y(t)\mathbf{j}. \qquad \text{Position vector}$$

The beauty of this vector model for representing motion is that you can use the first and second derivatives of the vector-valued function $\mathbf{r}$ to find the object's velocity and acceleration. (Recall from the preceding chapter that velocity and acceleration are both vector quantities having magnitude and direction.) To find the velocity and acceleration vectors at a given time t, consider a point $Q(x(t + \Delta t), y(t + \Delta t))$ that is approaching the point $P(x(t), y(t))$ along the curve C given by $\mathbf{r}(t) = x(t)\mathbf{i} + y(t)\mathbf{j}$, as shown in Figure 12.11. As $\Delta t \to 0$, the direction of the vector $\overrightarrow{PQ}$ (denoted by $\Delta \mathbf{r}$) approaches the *direction of motion* at time t.

$$\Delta \mathbf{r} = \mathbf{r}(t + \Delta t) - \mathbf{r}(t)$$

$$\frac{\Delta \mathbf{r}}{\Delta t} = \frac{\mathbf{r}(t + \Delta t) - \mathbf{r}(t)}{\Delta t}$$

$$\lim_{\Delta t \to 0} \frac{\Delta \mathbf{r}}{\Delta t} = \lim_{\Delta t \to 0} \frac{\mathbf{r}(t + \Delta t) - \mathbf{r}(t)}{\Delta t}$$

When this limit exists, it is defined as the **velocity vector** or **tangent vector** to the curve at point P. Note that this is the same limit used to define $\mathbf{r}'(t)$. So, the direction of $\mathbf{r}'(t)$ gives the direction of motion at time t. Moreover, the magnitude of the vector $\mathbf{r}'(t)$

$$\|\mathbf{r}'(t)\| = \|x'(t)\mathbf{i} + y'(t)\mathbf{j}\| = \sqrt{[x'(t)]^2 + [y'(t)]^2}$$

gives the **speed** of the object at time t.

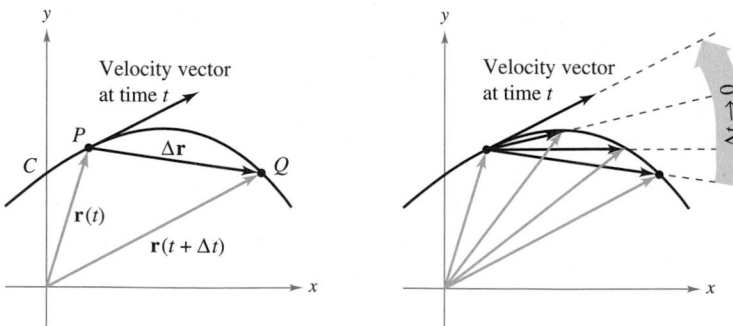

As $\Delta t \to 0$, $\dfrac{\Delta \mathbf{r}}{\Delta t}$ approaches the velocity vector.

Figure 12.11

Similar to how $\mathbf{r}'(t)$ is used to find velocity, you can use $\mathbf{r}''(t)$ to find acceleration, as indicated in the definitions at the top of the next page.

Definitions of Velocity and Acceleration

If x and y are twice-differentiable functions of t, and $\mathbf{r}$ is a vector-valued function given by $\mathbf{r}(t) = x(t)\mathbf{i} + y(t)\mathbf{j}$, then the velocity vector, acceleration vector, and speed at time t are as follows.

$$\textbf{Velocity} = \mathbf{v}(t) = \mathbf{r}'(t) = x'(t)\mathbf{i} + y'(t)\mathbf{j}$$

$$\textbf{Acceleration} = \mathbf{a}(t) = \mathbf{r}''(t) = x''(t)\mathbf{i} + y''(t)\mathbf{j}$$

$$\textbf{Speed} = \|\mathbf{v}(t)\| = \|\mathbf{r}'(t)\| = \sqrt{[x'(t)]^2 + [y'(t)]^2}$$

For motion along a space curve, the definitions are similar. That is, for

$$\mathbf{r}(t) = x(t)\mathbf{i} + y(t)\mathbf{j} + z(t)\mathbf{k}$$

you have the following.

$$\textbf{Velocity} = \mathbf{v}(t) = \mathbf{r}'(t) = x'(t)\mathbf{i} + y'(t)\mathbf{j} + z'(t)\mathbf{k}$$

$$\textbf{Acceleration} = \mathbf{a}(t) = \mathbf{r}''(t) = x''(t)\mathbf{i} + y''(t)\mathbf{j} + z''(t)\mathbf{k}$$

$$\textbf{Speed} = \|\mathbf{v}(t)\| = \|\mathbf{r}'(t)\| = \sqrt{[x'(t)]^2 + [y'(t)]^2 + [z'(t)]^2}$$

EXAMPLE 1   **Velocity and Acceleration Along a Plane Curve**

Find the (a) velocity vector, (b) speed, and (c) acceleration vector for the particle that moves along the plane curve C described by

$$\mathbf{r}(t) = 2\sin\frac{t}{2}\mathbf{i} + 2\cos\frac{t}{2}\mathbf{j}. \qquad \text{Position vector}$$

Solution

a. $\mathbf{v}(t) = \mathbf{r}'(t) = \cos\dfrac{t}{2}\mathbf{i} - \sin\dfrac{t}{2}\mathbf{j}$ Velocity vector

b. $\|\mathbf{r}'(t)\| = \sqrt{\cos^2\dfrac{t}{2} + \sin^2\dfrac{t}{2}} = 1$ Speed (at any time)

c. $\mathbf{a}(t) = \mathbf{r}''(t) = -\dfrac{1}{2}\sin\dfrac{t}{2}\mathbf{i} - \dfrac{1}{2}\cos\dfrac{t}{2}\mathbf{j}$ Acceleration vector

• • • • • • • • • • • • • • • • ▷

•• **REMARK** In Example 1, note that the velocity and acceleration vectors are orthogonal at any point in time (see Figure 12.12). This is characteristic of motion at a constant speed. (See Exercise 59.)

The parametric equations for the curve in Example 1 are

$$x = 2\sin\frac{t}{2} \quad \text{and} \quad y = 2\cos\frac{t}{2}.$$

By eliminating the parameter t, you obtain the rectangular equation

$$x^2 + y^2 = 4. \qquad \text{Rectangular equation}$$

So, the curve is a circle of radius 2 centered at the origin, as shown in Figure 12.12. Because the velocity vector

$$\mathbf{v}(t) = \cos\frac{t}{2}\mathbf{i} - \sin\frac{t}{2}\mathbf{j}$$

has a constant magnitude but a changing direction as t increases, the particle moves around the circle at a constant speed.

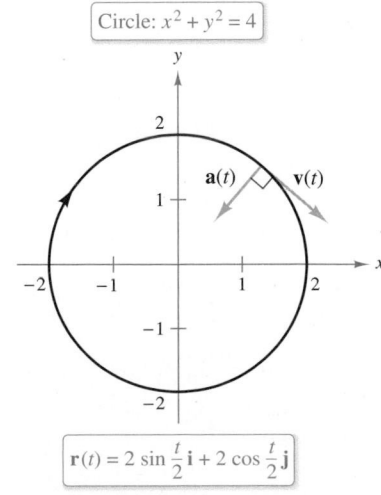

Circle: $x^2 + y^2 = 4$

$\mathbf{r}(t) = 2\sin\dfrac{t}{2}\mathbf{i} + 2\cos\dfrac{t}{2}\mathbf{j}$

The particle moves around the circle at a constant speed.
Figure 12.12

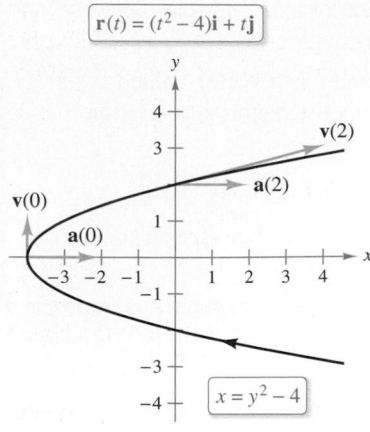

At each point on the curve, the acceleration vector points to the right.
Figure 12.13

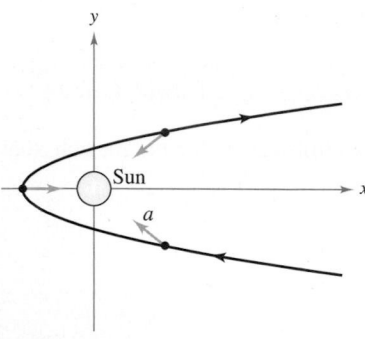

At each point in the comet's orbit, the acceleration vector points toward the sun.
Figure 12.14

EXAMPLE 2 **Velocity and Acceleration Vectors in the Plane**

Sketch the path of an object moving along the plane curve given by

$$\mathbf{r}(t) = (t^2 - 4)\mathbf{i} + t\mathbf{j} \qquad \text{Position vector}$$

and find the velocity and acceleration vectors when $t = 0$ and $t = 2$.

Solution Using the parametric equations $x = t^2 - 4$ and $y = t$, you can determine that the curve is a parabola given by

$$x = y^2 - 4 \qquad \text{Rectangular equation}$$

as shown in Figure 12.13. The velocity vector (at any time) is

$$\mathbf{v}(t) = \mathbf{r}'(t) = 2t\mathbf{i} + \mathbf{j} \qquad \text{Velocity vector}$$

and the acceleration vector (at any time) is

$$\mathbf{a}(t) = \mathbf{r}''(t) = 2\mathbf{i}. \qquad \text{Acceleration vector}$$

When $t = 0$, the velocity and acceleration vectors are

$$\mathbf{v}(0) = 2(0)\mathbf{i} + \mathbf{j} = \mathbf{j} \quad \text{and} \quad \mathbf{a}(0) = 2\mathbf{i}.$$

When $t = 2$, the velocity and acceleration vectors are

$$\mathbf{v}(2) = 2(2)\mathbf{i} + \mathbf{j} = 4\mathbf{i} + \mathbf{j} \quad \text{and} \quad \mathbf{a}(2) = 2\mathbf{i}. \qquad \blacksquare$$

For the object moving along the path shown in Figure 12.13, note that the acceleration vector is constant (it has a magnitude of 2 and points to the right). This implies that the speed of the object is decreasing as the object moves toward the vertex of the parabola, and the speed is increasing as the object moves away from the vertex of the parabola.

This type of motion is *not* characteristic of comets that travel on parabolic paths through our solar system. For such comets, the acceleration vector always points to the origin (the sun), which implies that the comet's speed increases as it approaches the vertex of the path and decreases as it moves away from the vertex. (See Figure 12.14.)

EXAMPLE 3 **Velocity and Acceleration Vectors in Space**

$\vdots \cdots \triangleright$ *See LarsonCalculus.com for an interactive version of this type of example.*

Sketch the path of an object moving along the space curve C given by

$$\mathbf{r}(t) = t\mathbf{i} + t^3\mathbf{j} + 3t\mathbf{k}, \quad t \geq 0 \qquad \text{Position vector}$$

and find the velocity and acceleration vectors when $t = 1$.

Solution Using the parametric equations $x = t$ and $y = t^3$, you can determine that the path of the object lies on the cubic cylinder given by

$$y = x^3. \qquad \text{Rectangular equation}$$

Moreover, because $z = 3t$, the object starts at $(0, 0, 0)$ and moves upward as t increases, as shown in Figure 12.15. Because $\mathbf{r}(t) = t\mathbf{i} + t^3\mathbf{j} + 3t\mathbf{k}$, you have

$$\mathbf{v}(t) = \mathbf{r}'(t) = \mathbf{i} + 3t^2\mathbf{j} + 3\mathbf{k} \qquad \text{Velocity vector}$$

and

$$\mathbf{a}(t) = \mathbf{r}''(t) = 6t\mathbf{j}. \qquad \text{Acceleration vector}$$

When $t = 1$, the velocity and acceleration vectors are

$$\mathbf{v}(1) = \mathbf{r}'(1) = \mathbf{i} + 3\mathbf{j} + 3\mathbf{k} \quad \text{and} \quad \mathbf{a}(1) = \mathbf{r}''(1) = 6\mathbf{j}. \qquad \blacksquare$$

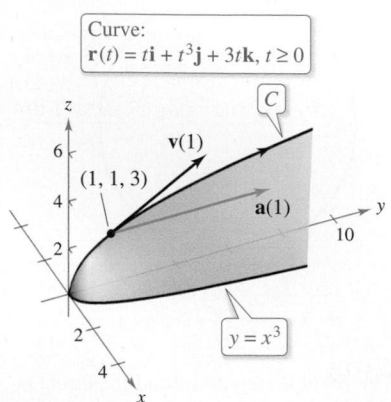

Figure 12.15

So far in this section, you have concentrated on finding the velocity and acceleration by differentiating the position vector. Many practical applications involve the reverse problem—finding the position vector for a given velocity or acceleration. This is demonstrated in the next example.

EXAMPLE 4   **Finding a Position Vector by Integration**

An object starts from rest at the point $(1, 2, 0)$ and moves with an acceleration of

$$\mathbf{a}(t) = \mathbf{j} + 2\mathbf{k} \qquad\qquad \text{Acceleration vector}$$

where $\|\mathbf{a}(t)\|$ is measured in feet per second per second. Find the location of the object after $t = 2$ seconds.

Solution   From the description of the object's motion, you can deduce the following *initial conditions*. Because the object starts from rest, you have

$$\mathbf{v}(0) = \mathbf{0}.$$

Moreover, because the object starts at the point $(x, y, z) = (1, 2, 0)$, you have

$$\mathbf{r}(0) = x(0)\mathbf{i} + y(0)\mathbf{j} + z(0)\mathbf{k} = 1\mathbf{i} + 2\mathbf{j} + 0\mathbf{k} = \mathbf{i} + 2\mathbf{j}.$$

To find the position vector, you should integrate twice, each time using one of the initial conditions to solve for the constant of integration. The velocity vector is

$$\mathbf{v}(t) = \int \mathbf{a}(t)\,dt$$

$$= \int (\mathbf{j} + 2\mathbf{k})\,dt$$

$$= t\mathbf{j} + 2t\mathbf{k} + \mathbf{C}$$

where $\mathbf{C} = C_1\mathbf{i} + C_2\mathbf{j} + C_3\mathbf{k}$. Letting $t = 0$ and applying the initial condition $\mathbf{v}(0) = \mathbf{0}$, you obtain

$$\mathbf{v}(0) = C_1\mathbf{i} + C_2\mathbf{j} + C_3\mathbf{k} = \mathbf{0} \implies C_1 = C_2 = C_3 = 0.$$

So, the *velocity* at any time t is

$$\mathbf{v}(t) = t\mathbf{j} + 2t\mathbf{k}. \qquad\qquad \text{Velocity vector}$$

Integrating once more produces

$$\mathbf{r}(t) = \int \mathbf{v}(t)\,dt$$

$$= \int (t\mathbf{j} + 2t\mathbf{k})\,dt$$

$$= \frac{t^2}{2}\mathbf{j} + t^2\mathbf{k} + \mathbf{C}$$

where $\mathbf{C} = C_4\mathbf{i} + C_5\mathbf{j} + C_6\mathbf{k}$. Letting $t = 0$ and applying the initial condition $\mathbf{r}(0) = \mathbf{i} + 2\mathbf{j}$, you have

$$\mathbf{r}(0) = C_4\mathbf{i} + C_5\mathbf{j} + C_6\mathbf{k} = \mathbf{i} + 2\mathbf{j} \implies C_4 = 1, C_5 = 2, C_6 = 0.$$

So, the *position* vector is

$$\mathbf{r}(t) = \mathbf{i} + \left(\frac{t^2}{2} + 2\right)\mathbf{j} + t^2\mathbf{k}. \qquad\qquad \text{Position vector}$$

The location of the object after $t = 2$ seconds is given by

$$\mathbf{r}(2) = \mathbf{i} + 4\mathbf{j} + 4\mathbf{k}$$

as shown in Figure 12.16.

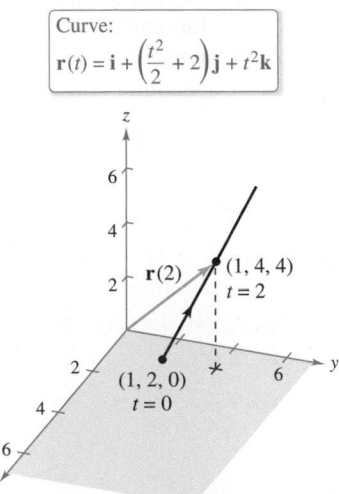

Curve:
$$\mathbf{r}(t) = \mathbf{i} + \left(\frac{t^2}{2} + 2\right)\mathbf{j} + t^2\mathbf{k}$$

The object takes 2 seconds to move from point $(1, 2, 0)$ to point $(1, 4, 4)$ along the curve.

Figure 12.16

Projectile Motion

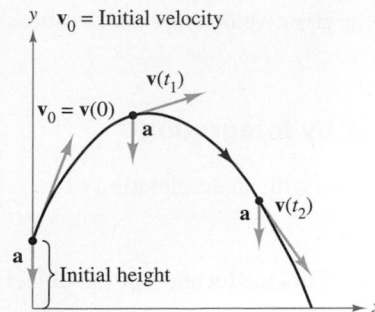

y $\mathbf{v}_0$ = Initial velocity

$\mathbf{v}(t_1)$

$\mathbf{v}_0 = \mathbf{v}(0)$

$\mathbf{a}$

$\mathbf{a}$ $\mathbf{v}(t_2)$

$\mathbf{a}$ Initial height

Figure 12.17

To derive the parametric equations for the path of a projectile, assume that gravity is the only force acting on the projectile after it is launched. So, the motion occurs in a vertical plane, which can be represented by the xy-coordinate system with the origin as a point on Earth's surface, as shown in Figure 12.17. For a projectile of mass m, the force due to gravity is

$$\mathbf{F} = -mg\mathbf{j} \qquad \text{Force due to gravity}$$

where the acceleration due to gravity is $g = 32$ feet per second per second, or 9.8 meters per second per second. By **Newton's Second Law of Motion,** this same force produces an acceleration $\mathbf{a} = \mathbf{a}(t)$ and satisfies the equation $\mathbf{F} = m\mathbf{a}$. Consequently, the acceleration of the projectile is given by $m\mathbf{a} = -mg\mathbf{j}$, which implies that

$$\mathbf{a} = -g\mathbf{j}. \qquad \text{Acceleration of projectile}$$

EXAMPLE 5   **Derivation of the Position Vector for a Projectile**

A projectile of mass m is launched from an initial position $\mathbf{r}_0$ with an initial velocity $\mathbf{v}_0$. Find its position vector as a function of time.

Solution   Begin with the acceleration $\mathbf{a}(t) = -g\mathbf{j}$ and integrate twice.

$$\mathbf{v}(t) = \int \mathbf{a}(t)\, dt = \int -g\mathbf{j}\, dt = -gt\mathbf{j} + \mathbf{C}_1$$

$$\mathbf{r}(t) = \int \mathbf{v}(t)\, dt = \int (-gt\mathbf{j} + \mathbf{C}_1)\, dt = -\frac{1}{2}gt^2\mathbf{j} + \mathbf{C}_1 t + \mathbf{C}_2$$

You can use the initial conditions $\mathbf{v}(0) = \mathbf{v}_0$ and $\mathbf{r}(0) = \mathbf{r}_0$ to solve for the constant vectors $\mathbf{C}_1$ and $\mathbf{C}_2$. Doing this produces

$$\mathbf{C}_1 = \mathbf{v}_0 \quad \text{and} \quad \mathbf{C}_2 = \mathbf{r}_0.$$

Therefore, the position vector is

$$\mathbf{r}(t) = -\frac{1}{2}gt^2\mathbf{j} + t\mathbf{v}_0 + \mathbf{r}_0. \qquad \text{Position vector}$$

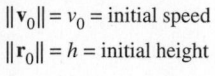

$\|\mathbf{v}_0\| = v_0$ = initial speed

$\|\mathbf{r}_0\| = h$ = initial height

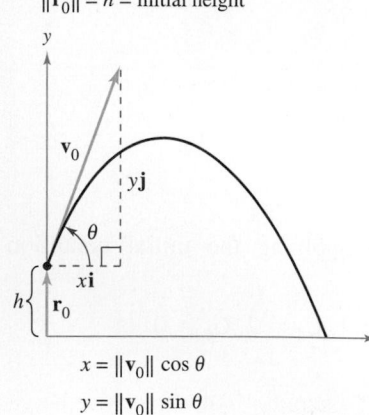

y

$\mathbf{v}_0$

$y\mathbf{j}$

θ

$x\mathbf{i}$

h $\mathbf{r}_0$

$x = \|\mathbf{v}_0\| \cos\theta$

$y = \|\mathbf{v}_0\| \sin\theta$

Figure 12.18

In many projectile problems, the constant vectors $\mathbf{r}_0$ and $\mathbf{v}_0$ are not given explicitly. Often you are given the initial height h, the initial speed v_0, and the angle θ at which the projectile is launched, as shown in Figure 12.18. From the given height, you can deduce that $\mathbf{r}_0 = h\mathbf{j}$. Because the speed gives the magnitude of the initial velocity, it follows that $v_0 = \|\mathbf{v}_0\|$ and you can write

$$\mathbf{v}_0 = x\mathbf{i} + y\mathbf{j}$$
$$= \left(\|\mathbf{v}_0\| \cos\theta\right)\mathbf{i} + \left(\|\mathbf{v}_0\| \sin\theta\right)\mathbf{j}$$
$$= v_0 \cos\theta\mathbf{i} + v_0 \sin\theta\mathbf{j}.$$

So, the position vector can be written in the form

$$\mathbf{r}(t) = -\frac{1}{2}gt^2\mathbf{j} + t\mathbf{v}_0 + \mathbf{r}_0 \qquad \text{Position vector}$$

$$= -\frac{1}{2}gt^2\mathbf{j} + tv_0 \cos\theta\mathbf{i} + tv_0 \sin\theta\mathbf{j} + h\mathbf{j}$$

$$= (v_0 \cos\theta)t\mathbf{i} + \left[h + (v_0 \sin\theta)t - \frac{1}{2}gt^2\right]\mathbf{j}.$$

THEOREM 12.3 Position Vector for a Projectile

Neglecting air resistance, the path of a projectile launched from an initial height h with initial speed v_0 and angle of elevation θ is described by the vector function

$$\mathbf{r}(t) = (v_0 \cos \theta)t\mathbf{i} + \left[h + (v_0 \sin \theta)t - \tfrac{1}{2}gt^2\right]\mathbf{j}$$

where g is the acceleration due to gravity.

EXAMPLE 6 **Describing the Path of a Baseball**

A baseball is hit 3 feet above ground level at 100 feet per second and at an angle of 45° with respect to the ground, as shown in Figure 12.19. Find the maximum height reached by the baseball. Will it clear a 10-foot-high fence located 300 feet from home plate?

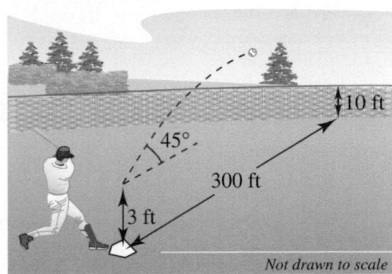

Figure 12.19

Solution You are given

$$h = 3, \quad v_0 = 100, \quad \text{and} \quad \theta = 45°.$$

So, using Theorem 12.3 with $g = 32$ feet per second per second produces

$$\mathbf{r}(t) = \left(100 \cos \frac{\pi}{4}\right)t\mathbf{i} + \left[3 + \left(100 \sin \frac{\pi}{4}\right)t - 16t^2\right]\mathbf{j}$$
$$= \left(50\sqrt{2}\,t\right)\mathbf{i} + \left(3 + 50\sqrt{2}\,t - 16t^2\right)\mathbf{j}.$$

The velocity vector is

$$\mathbf{v}(t) = \mathbf{r}'(t) = 50\sqrt{2}\,\mathbf{i} + \left(50\sqrt{2} - 32t\right)\mathbf{j}.$$

The maximum height occurs when

$$y'(t) = 50\sqrt{2} - 32t$$

is equal to 0, which implies that

$$t = \frac{25\sqrt{2}}{16} \approx 2.21 \text{ seconds.}$$

So, the maximum height reached by the ball is

$$y = 3 + 50\sqrt{2}\left(\frac{25\sqrt{2}}{16}\right) - 16\left(\frac{25\sqrt{2}}{16}\right)^2$$

$$= \frac{649}{8}$$

$$\approx 81 \text{ feet.} \qquad \text{Maximum height when } t \approx 2.21 \text{ seconds}$$

The ball is 300 feet from where it was hit when

$$x(t) = 300 \quad \Longrightarrow \quad 50\sqrt{2}\,t = 300.$$

Solving this equation for t produces $t = 3\sqrt{2} \approx 4.24$ seconds. At this time, the height of the ball is

$$y = 3 + 50\sqrt{2}\left(3\sqrt{2}\right) - 16\left(3\sqrt{2}\right)^2$$
$$= 303 - 288$$
$$= 15 \text{ feet.} \qquad \text{Height when } t \approx 4.24 \text{ seconds}$$

Therefore, the ball clears the 10-foot fence for a home run.

12.3 Exercises

See **CalcChat.com** for tutorial help and worked-out solutions to odd-numbered exercises.

CONCEPT CHECK

1. **Velocity Vector** An object moves along a curve in the plane. What information do you gain about the motion of the object from the velocity vector to the curve at time t?

2. **Acceleration Vectors** For each scenario, describe the direction of the acceleration vectors. Explain your reasoning.

   (a) A comet traveling through our solar system in a parabolic path

   (b) An object thrown on Earth's surface

 **Finding Velocity and Acceleration Along a Plane Curve** In Exercises 3–10, the position vector r describes the path of an object moving in the xy-plane.

(a) Find the velocity vector, speed, and acceleration vector of the object.

(b) Evaluate the velocity vector and acceleration vector of the object at the given point.

(c) Sketch a graph of the path and sketch the velocity and acceleration vectors at the given point.

Position Vector	Point
3. $\mathbf{r}(t) = 3t\mathbf{i} + (t-1)\mathbf{j}$	$(3, 0)$
4. $\mathbf{r}(t) = t\mathbf{i} + (-t^2 + 4)\mathbf{j}$	$(1, 3)$
5. $\mathbf{r}(t) = t^2\mathbf{i} + t\mathbf{j}$	$(4, 2)$
6. $\mathbf{r}(t) = \left(\frac{1}{4}t^3 + 1\right)\mathbf{i} + t\mathbf{j}$	$(3, 2)$
7. $\mathbf{r}(t) = 2\cos t\mathbf{i} + 2\sin t\mathbf{j}$	$(\sqrt{2}, \sqrt{2})$
8. $\mathbf{r}(t) = 3\cos t\mathbf{i} + 2\sin t\mathbf{j}$	$(3, 0)$
9. $\mathbf{r}(t) = \langle t - \sin t, 1 - \cos t \rangle$	$(\pi, 2)$
10. $\mathbf{r}(t) = \langle e^{-t}, e^t \rangle$	$(1, 1)$

 **Finding Velocity and Acceleration Vectors in Space** In Exercises 11–20, the position vector r describes the path of an object moving in space.

(a) Find the velocity vector, speed, and acceleration vector of the object.

(b) Evaluate the velocity vector and acceleration vector of the object at the given value of t.

Position Vector	Time
11. $\mathbf{r}(t) = t\mathbf{i} + 5t\mathbf{j} + 3t\mathbf{k}$	$t = 1$
12. $\mathbf{r}(t) = 4t\mathbf{i} + 4t\mathbf{j} - 2t\mathbf{k}$	$t = 3$
13. $\mathbf{r}(t) = t\mathbf{i} + t^2\mathbf{j} + \frac{1}{2}t^2\mathbf{k}$	$t = 4$
14. $\mathbf{r}(t) = 3t\mathbf{i} + t\mathbf{j} + \frac{1}{4}t^2\mathbf{k}$	$t = 2$
15. $\mathbf{r}(t) = t\mathbf{i} - t\mathbf{j} + \sqrt{9 - t^2}\,\mathbf{k}$	$t = 0$
16. $\mathbf{r}(t) = t^2\mathbf{i} + t\mathbf{j} + 2t^{3/2}\mathbf{k}$	$t = 4$

Position Vector	Time
17. $\mathbf{r}(t) = \langle 4t, 3\cos t, 3\sin t \rangle$	$t = \pi$
18. $\mathbf{r}(t) = \langle 2\cos t, \sin 3t, t^2 \rangle$	$t = \dfrac{\pi}{4}$
19. $\mathbf{r}(t) = \langle e^t \cos t, e^t \sin t, e^t \rangle$	$t = 0$
20. $\mathbf{r}(t) = \left\langle \ln t, \dfrac{1}{t^2}, t^4 \right\rangle$	$t = \sqrt{3}$

 **Finding a Position Vector by Integration** In Exercises 21–26, use the given acceleration vector and initial conditions to find the velocity and position vectors. Then find the position at time $t = 2$.

21. $\mathbf{a}(t) = \mathbf{i} + \mathbf{j} + \mathbf{k}$, $\mathbf{v}(0) = \mathbf{0}$, $\mathbf{r}(0) = \mathbf{0}$

22. $\mathbf{a}(t) = 2\mathbf{i} + 3\mathbf{k}$, $\mathbf{v}(0) = 4\mathbf{j}$, $\mathbf{r}(0) = \mathbf{0}$

23. $\mathbf{a}(t) = t\mathbf{j} + t\mathbf{k}$, $\mathbf{v}(1) = 5\mathbf{j}$, $\mathbf{r}(1) = \mathbf{0}$

24. $\mathbf{a}(t) = -32\mathbf{k}$, $\mathbf{v}(0) = 3\mathbf{i} - 2\mathbf{j} + \mathbf{k}$, $\mathbf{r}(0) = 5\mathbf{j} + 2\mathbf{k}$

25. $\mathbf{a}(t) = -\cos t\mathbf{i} - \sin t\mathbf{j}$, $\mathbf{v}(0) = \mathbf{j} + \mathbf{k}$, $\mathbf{r}(0) = \mathbf{i}$

26. $\mathbf{a}(t) = e^t\mathbf{i} - 8\mathbf{k}$, $\mathbf{v}(0) = 2\mathbf{i} + 3\mathbf{j} + \mathbf{k}$, $\mathbf{r}(0) = \mathbf{0}$

Projectile Motion In Exercises 27–40, use the model for projectile motion, assuming there is no air resistance and $g = 32$ feet per second per second.

27. A baseball is hit from a height of 2.5 feet above the ground with an initial speed of 140 feet per second and at an angle of 22° above the horizontal. Find the maximum height reached by the baseball. Determine whether it will clear a 10-foot-high fence located 375 feet from home plate.

28. Determine the maximum height and range of a projectile fired at a height of 3 feet above the ground with an initial speed of 900 feet per second and at an angle of 45° above the horizontal.

29. A baseball, hit 3 feet above the ground, leaves the bat at an angle of 45° and is caught by an outfielder 3 feet above the ground and 300 feet from home plate. What is the initial speed of the ball, and how high does it rise?

30. A baseball player at second base throws a ball 90 feet to the player at first base. The ball is released at a point 5 feet above the ground with an initial speed of 50 miles per hour and at an angle of 15° above the horizontal. At what height does the player at first base catch the ball?

31. Eliminate the parameter t from the position vector for the motion of a projectile to show that the rectangular equation is

$$y = -\frac{g\sec^2\theta}{2v_0^2}x^2 + (\tan\theta)x + h.$$

32. The path of a ball is given by the rectangular equation $y = x - 0.005x^2$. Use the result of Exercise 31 to find the position vector. Then find the speed and direction of the ball at the point at which it has traveled 60 feet horizontally.

33. The Rogers Centre in Toronto, Ontario, has a center field fence that is 10 feet high and 400 feet from home plate. A ball is hit 3 feet above the ground and leaves the bat at a speed of 100 miles per hour.

(a) The ball leaves the bat at an angle of $\theta = \theta_0$ with the horizontal. Write the vector-valued function for the path of the ball.

(b) Use a graphing utility to graph the vector-valued function for $\theta_0 = 10°$, $\theta_0 = 15°$, $\theta_0 = 20°$, and $\theta_0 = 25°$. Use the graphs to approximate the minimum angle required for the hit to be a home run.

(c) Determine analytically the minimum angle required for the hit to be a home run.

•• **34. Football** ••••••••••••••••••

The quarterback of a football team releases a pass at a height of 7 feet above the playing field, and the football is caught by a receiver 30 yards directly downfield at a height of 4 feet. The pass is released at an angle of 35° with the horizontal.

(a) Find the speed of the football when it is released.

(b) Find the maximum height of the football.

(c) Find the time the receiver has to reach the proper position after the quarterback releases the football.

••••••••••••••••••••••••••••

35. A bale ejector consists of two variable-speed belts at the end of a baler. Its purpose is to toss bales into a trailing wagon. In loading the back of a wagon, a bale must be thrown to a position 8 feet above and 16 feet behind the ejector.

(a) Find the minimum initial speed of the bale and the corresponding angle at which it must be ejected from the baler.

(b) The ejector has a fixed angle of 45°. Find the initial speed required.

36. A bomber is flying horizontally at an altitude of 30,000 feet with a speed of 540 miles per hour (see figure). When should the bomb be released for it to hit the target? (Give your answer in terms of the angle of depression from the plane to the target.) What is the speed of the bomb at the time of impact?

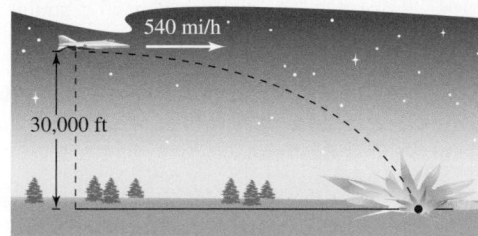

540 mi/h

30,000 ft

37. A shot fired from a gun with a muzzle speed of 1200 feet per second is to hit a target 3000 feet away. Determine the minimum angle of elevation of the gun.

38. A projectile is fired from ground level at an angle of 12° with the horizontal. The projectile is to have a range of 200 feet. Find the minimum initial speed necessary.

39. Use a graphing utility to graph the paths of a projectile for the given values of θ and v_0. For each case, use the graph to approximate the maximum height and range of the projectile. (Assume that the projectile is launched from ground level.)

(a) $\theta = 10°$, $v_0 = 66$ ft/sec

(b) $\theta = 10°$, $v_0 = 146$ ft/sec

(c) $\theta = 45°$, $v_0 = 66$ ft/sec

(d) $\theta = 45°$, $v_0 = 146$ ft/sec

(e) $\theta = 60°$, $v_0 = 66$ ft/sec

(f) $\theta = 60°$, $v_0 = 146$ ft/sec

40. Find the angles at which an object must be thrown to obtain (a) the maximum range and (b) the maximum height.

Projectile Motion **In Exercises 41 and 42, use the model for projectile motion, assuming there is no air resistance and $g = 9.8$ meters per second per second.**

41. Determine the maximum height and range of a projectile fired at a height of 1.5 meters above the ground with an initial speed of 100 meters per second and at an angle of 30° above the horizontal.

42. A projectile is fired from ground level at an angle of 8° with the horizontal. The projectile is to have a range of 50 meters. Find the minimum initial speed necessary.

43. Shot-Put Throw The path of a shot thrown at an angle θ is

$$\mathbf{r}(t) = (v_0 \cos \theta)t\,\mathbf{i} + \left[h + (v_0 \sin \theta)t - \frac{1}{2}gt^2\right]\mathbf{j}$$

where v_0 is the initial speed, h is the initial height, t is the time in seconds, and g is the acceleration due to gravity. Verify that the shot will remain in the air for a total of

$$t = \frac{v_0 \sin \theta + \sqrt{v_0^2 \sin^2 \theta + 2gh}}{g} \text{ seconds}$$

and will travel a horizontal distance of

$$\frac{v_0^2 \cos \theta}{g}\left(\sin \theta + \sqrt{\sin^2 \theta + \frac{2gh}{v_0^2}}\right) \text{ feet.}$$

•• **44. Shot-Put Throw** •••••••••••••••••

A shot is thrown from a height of $h = 5.75$ feet with an initial speed of $v_0 = 41$ feet per second and at an angle of $\theta = 42.5°$ with the horizontal. Use the result of Exercise 43 to find the total time of travel and the total horizontal distance traveled.

••••••••••••••••••••••••••••

Cycloidal Motion In Exercises 45 and 46, consider the motion of a point (or particle) on the circumference of a rolling circle. As the circle rolls, it generates the cycloid

$$\mathbf{r}(t) = b(\omega t - \sin \omega t)\mathbf{i} + b(1 - \cos \omega t)\mathbf{j}$$

where ω is the constant angular speed of the circle and b is the radius of the circle.

45. Find the velocity and acceleration vectors of the particle. Use the results to determine the times at which the speed of the particle will be (a) zero and (b) maximized.

46. Find the maximum speed of a point on the circumference of an automobile tire of radius 1 foot when the automobile is traveling at 60 miles per hour. Compare this speed with the speed of the automobile.

Circular Motion In Exercises 47–50, consider a particle moving on a circular path of radius b described by

$$\mathbf{r}(t) = b \cos \omega t\mathbf{i} + b \sin \omega t\mathbf{j}$$

where $\omega = du/dt$ is the constant angular speed.

47. Find the velocity vector and show that it is orthogonal to $\mathbf{r}(t)$.

48. (a) Show that the speed of the particle is $b\omega$.

    (b) Use a graphing utility in *parametric* mode to graph the circle for $b = 6$. Try different values of ω. Does the graphing utility draw the circle faster for greater values of ω?

49. Find the acceleration vector and show that its direction is always toward the center of the circle.

50. Show that the magnitude of the acceleration vector is $b\omega^2$.

Circular Motion In Exercises 51 and 52, use the results of Exercises 47–50.

51. A psychrometer (an instrument used to measure humidity) weighing 4 ounces is whirled horizontally using a 6-inch string (see figure). The string will break under a force of 2 pounds. Find the maximum speed the instrument can attain without breaking the string. (Use $\mathbf{F} = m\mathbf{a}$, where $m = 1/128$.)

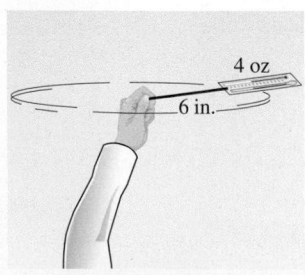

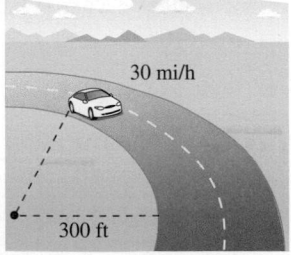

Figure for 51   Figure for 52

52. A 3400-pound automobile is negotiating a circular interchange of radius 300 feet at 30 miles per hour (see figure). Assuming the roadway is level, find the force between the tires and the road such that the car stays on the circular path and does not skid. (Use $\mathbf{F} = m\mathbf{a}$, where $m = 3400/32$.) Find the angle at which the roadway should be banked so that no lateral frictional force is exerted on the tires of the automobile.

EXPLORING CONCEPTS

53. Constant Speed Explain how a particle can be accelerating even though its speed is constant.

54. Think About It Consider a particle that is moving along the space curve given by $\mathbf{r}_1(t) = t^3\mathbf{i} + (3 - t)\mathbf{j} + 2t^2\mathbf{k}$. Write a vector-valued function $\mathbf{r}_2$ for a particle that moves four times as fast as the particle represented by $\mathbf{r}_1$. Explain how you found the function.

55. Circular Motion Consider a particle that moves around a circle. Is the velocity vector of the particle always orthogonal to the acceleration vector of the particle? Explain.

56. Particle Motion Consider a particle moving on an elliptical path described by $\mathbf{r}(t) = a \cos \omega t\mathbf{i} + b \sin \omega t\mathbf{j}$, where $\omega = d\theta/dt$ is the constant angular speed.

    (a) Find the velocity vector. What is the speed of the particle?

    (b) Find the acceleration vector and show that its direction is always toward the center of the ellipse.

57. Path of an Object When $t = 0$, an object is at the point $(0, 1)$ and has a velocity vector $\mathbf{v}(0) = -\mathbf{i}$. It moves with an acceleration of $\mathbf{a}(t) = \sin t\mathbf{i} - \cos t\mathbf{j}$. Show that the path of the object is a circle.

58. HOW DO YOU SEE IT? The graph shows the path of a projectile and the velocity and acceleration vectors at times t_1 and t_2. Classify the angle between the velocity vector and the acceleration vector at times t_1 and t_2. Using the vectors, is the speed increasing or decreasing at times t_1 and t_2? Explain your reasoning.

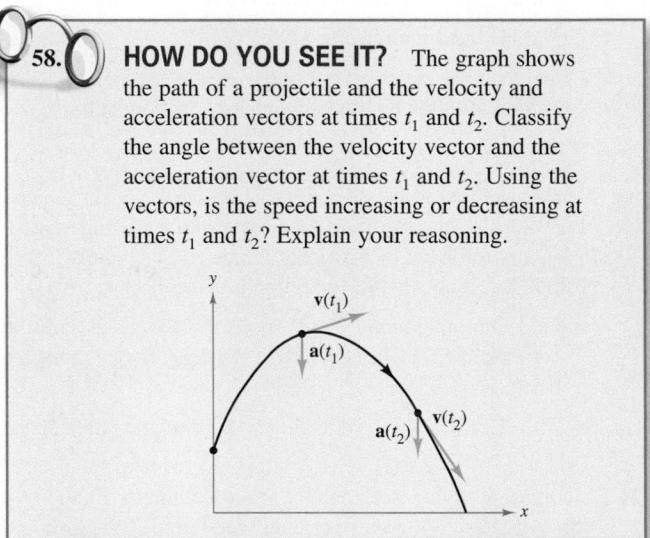

59. Proof Prove that when an object is traveling at a constant speed, its velocity and acceleration vectors are orthogonal.

60. Proof Prove that an object moving in a straight line at a constant speed has an acceleration of 0.

True or False? In Exercises 61–63, determine whether the statement is true or false. If it is false, explain why or give an example that shows it is false.

61. The velocity vector points in the direction of motion.

62. If a particle moves along a straight line, then the velocity and acceleration vectors are orthogonal.

63. A velocity vector of variable magnitude cannot have a constant direction.

12.4   Tangent Vectors and Normal Vectors

■ Find a unit tangent vector and a principal unit normal vector at a point on a space curve.
■ Find the tangential and normal components of acceleration.

Tangent Vectors and Normal Vectors

In the preceding section, you learned that the velocity vector points in the direction of motion. This observation leads to the next definition, which applies to any smooth curve—not just to those for which the parameter represents time.

> ### Definition of Unit Tangent Vector
>
> Let C be a smooth curve represented by **r** on an open interval I. The **unit tangent vector** $\mathbf{T}(t)$ at t is defined as
>
> $$\mathbf{T}(t) = \frac{\mathbf{r}'(t)}{\|\mathbf{r}'(t)\|}, \quad \mathbf{r}'(t) \neq \mathbf{0}.$$

Recall that a curve is *smooth* on an interval when **r**′ is continuous and nonzero on the interval. So, "smoothness" is sufficient to guarantee that a curve has a unit tangent vector.

EXAMPLE 1   Finding the Unit Tangent Vector

Find the unit tangent vector to the curve given by

$$\mathbf{r}(t) = t\mathbf{i} + t^2\mathbf{j}$$

when $t = 1$.

Solution   The derivative of $\mathbf{r}(t)$ is

$$\mathbf{r}'(t) = \mathbf{i} + 2t\mathbf{j}. \qquad \text{Derivative of } \mathbf{r}(t)$$

So, the unit tangent vector is

$$\mathbf{T}(t) = \frac{\mathbf{r}'(t)}{\|\mathbf{r}'(t)\|} \qquad \text{Definition of } \mathbf{T}(t)$$

$$= \frac{1}{\sqrt{1 + 4t^2}}(\mathbf{i} + 2t\mathbf{j}). \qquad \text{Substitute for } \mathbf{r}'(t).$$

When $t = 1$, the unit tangent vector is

$$\mathbf{T}(1) = \frac{1}{\sqrt{5}}(\mathbf{i} + 2\mathbf{j})$$

as shown in Figure 12.20.

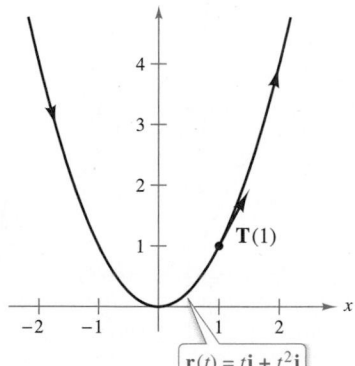

The direction of the unit tangent vector depends on the orientation of the curve.
Figure 12.20

In Example 1, note that the direction of the unit tangent vector depends on the orientation of the curve. For the parabola described by

$$\mathbf{r}(t) = -(t - 2)\mathbf{i} + (t - 2)^2\mathbf{j}$$

$\mathbf{T}(1)$ would still represent the unit tangent vector at the point $(1, 1)$, but it would point in the opposite direction. Try verifying this.

The **tangent line to a curve** at a point is the line that passes through the point and is parallel to the unit tangent vector. In Example 2, the unit tangent vector is used to find the tangent line at a point on a helix.

EXAMPLE 2   Finding the Tangent Line at a Point on a Curve

Find $\mathbf{T}(t)$ and then find a set of parametric equations for the tangent line to the helix given by

$$\mathbf{r}(t) = 2 \cos t\mathbf{i} + 2 \sin t\mathbf{j} + t\mathbf{k}$$

at the point $\left(\sqrt{2}, \sqrt{2}, \dfrac{\pi}{4} \right)$.

Solution   The derivative of $\mathbf{r}(t)$ is

$$\mathbf{r}'(t) = -2 \sin t\mathbf{i} + 2 \cos t\mathbf{j} + \mathbf{k}$$

which implies that $\|\mathbf{r}'(t)\| = \sqrt{4 \sin^2 t + 4 \cos^2 t + 1} = \sqrt{5}$. Therefore, the unit tangent vector is

$$\mathbf{T}(t) = \frac{\mathbf{r}'(t)}{\|\mathbf{r}'(t)\|}$$

$$= \frac{1}{\sqrt{5}}(-2 \sin t\mathbf{i} + 2 \cos t\mathbf{j} + \mathbf{k}). \qquad \text{Unit tangent vector}$$

At the point $\left(\sqrt{2}, \sqrt{2}, \pi/4 \right)$, $t = \pi/4$ and the unit tangent vector is

$$\mathbf{T}\!\left(\frac{\pi}{4}\right) = \frac{1}{\sqrt{5}}\!\left(-2\frac{\sqrt{2}}{2}\mathbf{i} + 2\frac{\sqrt{2}}{2}\mathbf{j} + \mathbf{k}\right)$$

$$= \frac{1}{\sqrt{5}}(-\sqrt{2}\mathbf{i} + \sqrt{2}\mathbf{j} + \mathbf{k}).$$

Using the direction numbers $a = -\sqrt{2}$, $b = \sqrt{2}$, and $c = 1$, and the point $(x_1, y_1, z_1) = \left(\sqrt{2}, \sqrt{2}, \pi/4 \right)$, you can obtain the parametric equations (given with parameter s) listed below.

$$x = x_1 + as = \sqrt{2} - \sqrt{2}s$$
$$y = y_1 + bs = \sqrt{2} + \sqrt{2}s$$
$$z = z_1 + cs = \frac{\pi}{4} + s$$

This tangent line is shown in Figure 12.21.

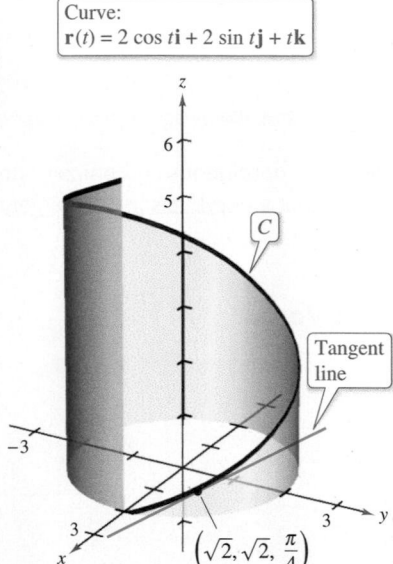

Curve:
$\mathbf{r}(t) = 2 \cos t\mathbf{i} + 2 \sin t\mathbf{j} + t\mathbf{k}$

The tangent line to a curve at a point is determined by the unit tangent vector at the point.

Figure 12.21

In Example 2, there are infinitely many vectors that are orthogonal to the tangent vector $\mathbf{T}(t)$. One of these is the vector $\mathbf{T}'(t)$. This follows from Property 7 of Theorem 12.2. That is,

$$\mathbf{T}(t) \cdot \mathbf{T}(t) = \|\mathbf{T}(t)\|^2 = 1 \quad \Longrightarrow \quad \mathbf{T}(t) \cdot \mathbf{T}'(t) = 0.$$

By normalizing the vector $\mathbf{T}'(t)$, you obtain a special vector called the **principal unit normal vector,** as indicated in the next definition.

> ### Definition of Principal Unit Normal Vector
>
> Let C be a smooth curve represented by $\mathbf{r}$ on an open interval I. If $\mathbf{T}'(t) \neq \mathbf{0}$, then the **principal unit normal vector** at t is defined as
>
> $$\mathbf{N}(t) = \frac{\mathbf{T}'(t)}{\|\mathbf{T}'(t)\|}.$$

EXAMPLE 3 **Finding the Principal Unit Normal Vector**

Find $\mathbf{N}(t)$ and $\mathbf{N}(1)$ for the curve represented by $\mathbf{r}(t) = 3t\mathbf{i} + 2t^2\mathbf{j}$.

Solution By differentiating, you obtain

$$\mathbf{r}'(t) = 3\mathbf{i} + 4t\mathbf{j}$$

which implies that

$$\|\mathbf{r}'(t)\| = \sqrt{9 + 16t^2}.$$

So, the unit tangent vector is

$$\mathbf{T}(t) = \frac{\mathbf{r}'(t)}{\|\mathbf{r}'(t)\|}$$

$$= \frac{1}{\sqrt{9 + 16t^2}}(3\mathbf{i} + 4t\mathbf{j}). \qquad \text{Unit tangent vector}$$

Using Theorem 12.2, differentiate $\mathbf{T}(t)$ with respect to t to obtain

$$\mathbf{T}'(t) = \frac{1}{\sqrt{9 + 16t^2}}(4\mathbf{j}) - \frac{16t}{(9 + 16t^2)^{3/2}}(3\mathbf{i} + 4t\mathbf{j})$$

$$= \frac{12}{(9 + 16t^2)^{3/2}}(-4t\mathbf{i} + 3\mathbf{j})$$

which implies that

$$\|\mathbf{T}'(t)\| = 12\sqrt{\frac{9 + 16t^2}{(9 + 16t^2)^3}} = \frac{12}{9 + 16t^2}.$$

Therefore, the principal unit normal vector is

$$\mathbf{N}(t) = \frac{\mathbf{T}'(t)}{\|\mathbf{T}'(t)\|}$$

$$= \frac{1}{\sqrt{9 + 16t^2}}(-4t\mathbf{i} + 3\mathbf{j}). \qquad \text{Principal unit normal vector}$$

When $t = 1$, the principal unit normal vector is

$$\mathbf{N}(1) = \frac{1}{5}(-4\mathbf{i} + 3\mathbf{j})$$

as shown in Figure 12.22.

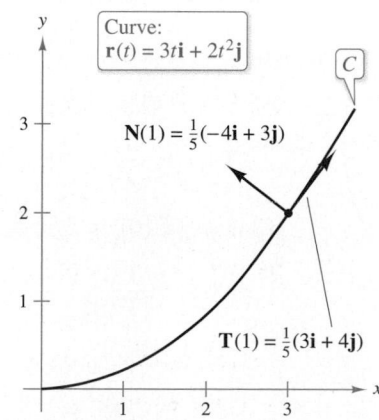

The principal unit normal vector points toward the concave side of the curve.

Figure 12.22

The principal unit normal vector can be difficult to evaluate algebraically. For plane curves, you can simplify the algebra by finding

$$\mathbf{T}(t) = x(t)\mathbf{i} + y(t)\mathbf{j}$$ Unit tangent vector

and observing that $\mathbf{N}(t)$ must be either

$$\mathbf{N}_1(t) = y(t)\mathbf{i} - x(t)\mathbf{j} \quad \text{or} \quad \mathbf{N}_2(t) = -y(t)\mathbf{i} + x(t)\mathbf{j}.$$

Because $\sqrt{[x(t)]^2 + [y(t)]^2} = 1$, it follows that both $\mathbf{N}_1(t)$ and $\mathbf{N}_2(t)$ are unit normal vectors. The *principal* unit normal vector $\mathbf{N}$ is the one that points toward the concave side of the curve, as shown in Figure 12.22 (see Exercise 72). This also holds for curves in space. That is, for an object moving along a curve C in space, the vector $\mathbf{T}(t)$ points in the direction the object is moving, whereas the vector $\mathbf{N}(t)$ is orthogonal to $\mathbf{T}(t)$ and points in the direction in which the object is turning, as shown in Figure 12.23.

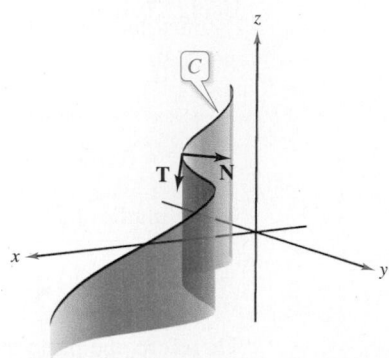

At any point on a curve, a unit normal vector is orthogonal to the unit tangent vector. The *principal* unit normal vector points in the direction in which the curve is turning.
Figure 12.23

EXAMPLE 4   **Finding the Principal Unit Normal Vector**

Find the principal unit normal vector for the helix $\mathbf{r}(t) = 2 \cos t\mathbf{i} + 2 \sin t\mathbf{j} + t\mathbf{k}$.

Solution   From Example 2, you know that the unit tangent vector is

$$\mathbf{T}(t) = \frac{1}{\sqrt{5}}(-2 \sin t\mathbf{i} + 2 \cos t\mathbf{j} + \mathbf{k}).$$ Unit tangent vector

So, $\mathbf{T}'(t)$ is given by

$$\mathbf{T}'(t) = \frac{1}{\sqrt{5}}(-2 \cos t\mathbf{i} - 2 \sin t\mathbf{j}).$$

Because $\|\mathbf{T}'(t)\| = 2/\sqrt{5}$, it follows that the principal unit normal vector is

$$\mathbf{N}(t) = \frac{\mathbf{T}'(t)}{\|\mathbf{T}'(t)\|}$$

$$= \frac{1}{2}(-2 \cos t\mathbf{i} - 2 \sin t\mathbf{j})$$

$$= -\cos t\mathbf{i} - \sin t\mathbf{j}.$$ Principal unit normal vector

Note that this vector is horizontal and points toward the z-axis, as shown in Figure 12.24.

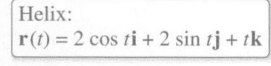

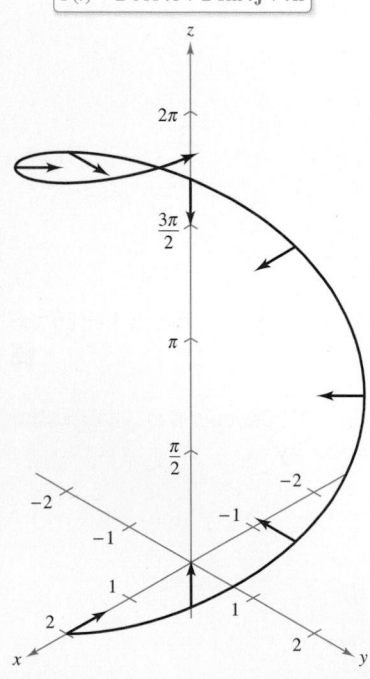

$\mathbf{N}(t)$ is horizontal and points toward the z-axis.
Figure 12.24

Tangential and Normal Components of Acceleration

In the preceding section, you considered the problem of describing the motion of an object along a curve. You saw that for an object traveling at a *constant speed,* the velocity and acceleration vectors are perpendicular. This seems reasonable, because the speed would not be constant if any acceleration were acting in the direction of motion. You can verify this observation by noting that

$$\mathbf{r}''(t) \cdot \mathbf{r}'(t) = 0$$

when $\|\mathbf{r}'(t)\|$ is a constant. (See Property 7 of Theorem 12.2.)

For an object traveling at a *variable speed,* however, the velocity and acceleration vectors are not necessarily perpendicular. For instance, you saw that the acceleration vector for a projectile always points down, regardless of the direction of motion.

In general, part of the acceleration (the tangential component) acts in the line of motion, and part of it (the normal component) acts perpendicular to the line of motion. In order to determine these two components, you can use the unit vectors $\mathbf{T}(t)$ and $\mathbf{N}(t)$, which serve in much the same way as do $\mathbf{i}$ and $\mathbf{j}$ in representing vectors in the plane. The next theorem states that the acceleration vector lies in the plane determined by $\mathbf{T}(t)$ and $\mathbf{N}(t)$.

THEOREM 12.4 Acceleration Vector

If $\mathbf{r}(t)$ is the position vector for a smooth curve C and $\mathbf{N}(t)$ exists, then the acceleration vector $\mathbf{a}(t)$ lies in the plane determined by $\mathbf{T}(t)$ and $\mathbf{N}(t)$.

Proof To simplify the notation, write $\mathbf{T}$ for $\mathbf{T}(t)$, $\mathbf{T}'$ for $\mathbf{T}'(t)$, and so on. Because $\mathbf{T} = \mathbf{r}'/\|\mathbf{r}'\| = \mathbf{v}/\|\mathbf{v}\|$, it follows that

$$\mathbf{v} = \|\mathbf{v}\|\mathbf{T}.$$

By differentiating, you obtain

$$\begin{aligned}
\mathbf{a} &= \mathbf{v}' \\
&= \frac{d}{dt}\big[\|\mathbf{v}\|\big]\mathbf{T} + \|\mathbf{v}\|\mathbf{T}' && \text{Product Rule} \\
&= \frac{d}{dt}\big[\|\mathbf{v}\|\big]\mathbf{T} + \|\mathbf{v}\|\mathbf{T}'\left(\frac{\|\mathbf{T}'\|}{\|\mathbf{T}'\|}\right) \\
&= \frac{d}{dt}\big[\|\mathbf{v}\|\big]\mathbf{T} + \|\mathbf{v}\|\,\|\mathbf{T}'\|\mathbf{N}. && \mathbf{N} = \mathbf{T}'/\|\mathbf{T}'\|
\end{aligned}$$

Because $\mathbf{a}$ is written as a linear combination of $\mathbf{T}$ and $\mathbf{N}$, it follows that $\mathbf{a}$ lies in the plane determined by $\mathbf{T}$ and $\mathbf{N}$. ■

The coefficients of $\mathbf{T}$ and $\mathbf{N}$ in the proof of Theorem 12.4 are called the **tangential and normal components of acceleration** and are denoted by

$$a_{\mathbf{T}} = \frac{d}{dt}\big[\|\mathbf{v}\|\big] \quad \text{and} \quad a_{\mathbf{N}} = \|\mathbf{v}\|\,\|\mathbf{T}'\|.$$

So, you can write

$$\mathbf{a}(t) = a_{\mathbf{T}}\mathbf{T}(t) + a_{\mathbf{N}}\mathbf{N}(t).$$

The next theorem lists some convenient formulas for $a_{\mathbf{T}}$ and $a_{\mathbf{N}}$.

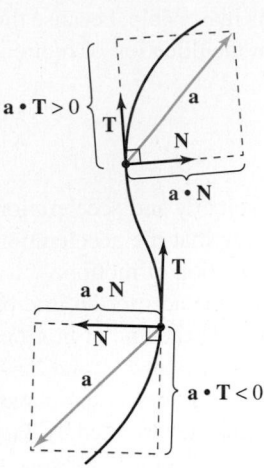

The tangential and normal components of acceleration are obtained by projecting **a** onto **T** and **N**.

Figure 12.25

THEOREM 12.5 Tangential and Normal Components of Acceleration

If $\mathbf{r}(t)$ is the position vector for a smooth curve C [for which $\mathbf{N}(t)$ exists], then the tangential and normal components of acceleration are as follows.

$$a_{\mathbf{T}} = \frac{d}{dt}\big[\|\mathbf{v}\|\big] = \mathbf{a} \cdot \mathbf{T} = \frac{\mathbf{v} \cdot \mathbf{a}}{\|\mathbf{v}\|}$$

$$a_{\mathbf{N}} = \|\mathbf{v}\|\,\|\mathbf{T}'\| = \mathbf{a} \cdot \mathbf{N} = \frac{\|\mathbf{v} \times \mathbf{a}\|}{\|\mathbf{v}\|} = \sqrt{\|\mathbf{a}\|^2 - a_{\mathbf{T}}^2}$$

Note that $a_{\mathbf{N}} \geq 0$. The normal component of acceleration is also called the **centripetal component of acceleration.**

Proof Note that **a** lies in the plane of **T** and **N**. So, you can use Figure 12.25 to conclude that, for any time t, the components of the projection of the acceleration vector onto **T** and onto **N** are given by $a_{\mathbf{T}} = \mathbf{a} \cdot \mathbf{T}$ and $a_{\mathbf{N}} = \mathbf{a} \cdot \mathbf{N}$, respectively. Moreover, because $\mathbf{a} = \mathbf{v}'$ and $\mathbf{T} = \mathbf{v}/\|\mathbf{v}\|$, you have

$$a_{\mathbf{T}} = \mathbf{a} \cdot \mathbf{T} = \mathbf{T} \cdot \mathbf{a} = \frac{\mathbf{v}}{\|\mathbf{v}\|} \cdot \mathbf{a} = \frac{\mathbf{v} \cdot \mathbf{a}}{\|\mathbf{v}\|}.$$

In Exercises 74 and 75, you are asked to prove the other parts of the theorem. ■

EXAMPLE 5 **Tangential and Normal Components of Acceleration**

$\vdots \cdots \triangleright$ *See LarsonCalculus.com for an interactive version of this type of example.*

Find the tangential and normal components of acceleration for the position vector given by $\mathbf{r}(t) = 3t\mathbf{i} - t\mathbf{j} + t^2\mathbf{k}$.

Solution Begin by finding the velocity, speed, and acceleration.

$$\mathbf{v}(t) = \mathbf{r}'(t) = 3\mathbf{i} - \mathbf{j} + 2t\mathbf{k} \qquad \text{Velocity vector}$$

$$\|\mathbf{v}(t)\| = \sqrt{9 + 1 + 4t^2} = \sqrt{10 + 4t^2} \qquad \text{Speed}$$

$$\mathbf{a}(t) = \mathbf{r}''(t) = 2\mathbf{k} \qquad \text{Acceleration vector}$$

By Theorem 12.5, the tangential component of acceleration is

$$a_{\mathbf{T}} = \frac{\mathbf{v} \cdot \mathbf{a}}{\|\mathbf{v}\|} = \frac{4t}{\sqrt{10 + 4t^2}} \qquad \text{Tangential component of acceleration}$$

and because

$$\mathbf{v} \times \mathbf{a} = \begin{vmatrix} \mathbf{i} & \mathbf{j} & \mathbf{k} \\ 3 & -1 & 2t \\ 0 & 0 & 2 \end{vmatrix} = -2\mathbf{i} - 6\mathbf{j}$$

the normal component of acceleration is

$$a_{\mathbf{N}} = \frac{\|\mathbf{v} \times \mathbf{a}\|}{\|\mathbf{v}\|} = \frac{\sqrt{4 + 36}}{\sqrt{10 + 4t^2}} = \frac{2\sqrt{10}}{\sqrt{10 + 4t^2}}. \qquad \text{Normal component of acceleration}$$

In Example 5, you could have used the alternative formula for $a_{\mathbf{N}}$ as follows.

$$a_{\mathbf{N}} = \sqrt{\|\mathbf{a}\|^2 - a_{\mathbf{T}}^2} = \sqrt{(2)^2 - \frac{16t^2}{10 + 4t^2}} = \frac{2\sqrt{10}}{\sqrt{10 + 4t^2}}$$

EXAMPLE 6 Finding a_T and a_N for a Circular Helix

Find the tangential and normal components of acceleration for the helix given by

$$\mathbf{r}(t) = b\cos t\mathbf{i} + b\sin t\mathbf{j} + ct\mathbf{k}, \quad b > 0.$$

Solution

$$\mathbf{v}(t) = \mathbf{r}'(t) = -b\sin t\mathbf{i} + b\cos t\mathbf{j} + c\mathbf{k} \qquad \text{Velocity vector}$$

$$\|\mathbf{v}(t)\| = \sqrt{b^2\sin^2 t + b^2\cos^2 t + c^2} = \sqrt{b^2 + c^2} \qquad \text{Speed}$$

$$\mathbf{a}(t) = \mathbf{r}''(t) = -b\cos t\mathbf{i} - b\sin t\mathbf{j} \qquad \text{Acceleration vector}$$

By Theorem 12.5, the tangential component of acceleration is

$$a_T = \frac{\mathbf{v}\cdot\mathbf{a}}{\|\mathbf{v}\|} = \frac{b^2\sin t\cos t - b^2\sin t\cos t + 0}{\sqrt{b^2+c^2}} = 0. \qquad \begin{array}{l}\text{Tangential component}\\\text{of acceleration}\end{array}$$

Moreover, because

$$\|\mathbf{a}\| = \sqrt{b^2\cos^2 t + b^2\sin^2 t} = b$$

you can use the alternative formula for the normal component of acceleration to obtain

$$a_N = \sqrt{\|\mathbf{a}\|^2 - a_T^2} = \sqrt{b^2 - 0^2} = b. \qquad \begin{array}{l}\text{Normal component}\\\text{of acceleration}\end{array}$$

Note that the normal component of acceleration is equal to the magnitude of the acceleration. In other words, because the speed is constant, the acceleration is perpendicular to the velocity. See Figure 12.26.

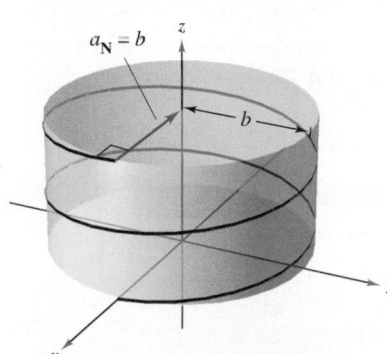

$a_N = b$

The normal component of acceleration is equal to the radius of the cylinder around which the helix is spiraling.
Figure 12.26

EXAMPLE 7 Projectile Motion

The position vector for the projectile shown in Figure 12.27 is

$$\mathbf{r}(t) = (50\sqrt{2}t)\mathbf{i} + (50\sqrt{2}t - 16t^2)\mathbf{j}. \qquad \text{Position vector}$$

Find the tangential components of acceleration when $t = 0$, 1, and $25\sqrt{2}/16$.

Solution

$$\mathbf{v}(t) = 50\sqrt{2}\mathbf{i} + (50\sqrt{2} - 32t)\mathbf{j} \qquad \text{Velocity vector}$$

$$\|\mathbf{v}(t)\| = 2\sqrt{50^2 - 16(50)\sqrt{2}t + 16^2t^2} \qquad \text{Speed}$$

$$\mathbf{a}(t) = -32\mathbf{j} \qquad \text{Acceleration vector}$$

The tangential component of acceleration is

$$a_T(t) = \frac{\mathbf{v}(t)\cdot\mathbf{a}(t)}{\|\mathbf{v}(t)\|} = \frac{-32(50\sqrt{2} - 32t)}{2\sqrt{50^2 - 16(50)\sqrt{2}t + 16^2t^2}}. \qquad \begin{array}{l}\text{Tangential component}\\\text{of acceleration}\end{array}$$

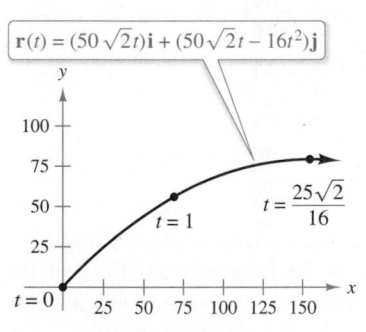

$\mathbf{r}(t) = (50\sqrt{2}t)\mathbf{i} + (50\sqrt{2}t - 16t^2)\mathbf{j}$

The path of a projectile
Figure 12.27

At the specified times, you have

$$a_T(0) = \frac{-32(50\sqrt{2})}{100} = -16\sqrt{2} \approx -22.6$$

$$a_T(1) = \frac{-32(50\sqrt{2} - 32)}{2\sqrt{50^2 - 16(50)\sqrt{2} + 16^2}} \approx -15.4$$

$$a_T\left(\frac{25\sqrt{2}}{16}\right) = \frac{-32(50\sqrt{2} - 50\sqrt{2})}{50\sqrt{2}} = 0.$$

You can see from Figure 12.27 that at the maximum height, when $t = 25\sqrt{2}/16$, the tangential component is 0. This is reasonable because the direction of motion is horizontal at the point and the tangential component of the acceleration is equal to the horizontal component of the acceleration.

12.4 Exercises

See **CalcChat.com** for tutorial help and worked-out solutions to odd-numbered exercises.

CONCEPT CHECK

1. Unit Tangent Vector How is the unit tangent vector related to the orientation of a curve? Explain.

2. Principal Unit Normal Vector In what direction does the principal unit normal vector point?

 **Finding the Unit Tangent Vector** In Exercises 3–8, find the unit tangent vector to the curve at the specified value of the parameter.

3. $\mathbf{r}(t) = t^2\mathbf{i} + 2t\mathbf{j}, \quad t = 1$

4. $\mathbf{r}(t) = t^3\mathbf{i} + 2t^2\mathbf{j}, \quad t = 1$

5. $\mathbf{r}(t) = 5\cos t\mathbf{i} + 5\sin t\mathbf{j}, \quad t = \dfrac{\pi}{3}$

6. $\mathbf{r}(t) = 6\sin t\mathbf{i} - 2\cos t\mathbf{j}, \quad t = \dfrac{\pi}{6}$

7. $\mathbf{r}(t) = 3t\mathbf{i} - \ln t\mathbf{j}, \quad t = e$

8. $\mathbf{r}(t) = e^t\cos t\mathbf{i} + e^t\mathbf{j}, \quad t = 0$

 **Finding a Tangent Line** In Exercises 9–14, find the unit tangent vector $\mathbf{T}(t)$ and a set of parametric equations for the tangent line to the space curve at point P.

9. $\mathbf{r}(t) = t\mathbf{i} + t^2\mathbf{j} + t\mathbf{k}, \quad P(0, 0, 0)$

10. $\mathbf{r}(t) = t^2\mathbf{i} + t\mathbf{j} + \frac{4}{3}\mathbf{k}, \quad P\left(1, 1, \frac{4}{3}\right)$

11. $\mathbf{r}(t) = \cos t\mathbf{i} + 3\sin t\mathbf{j} + (3t - 4)\mathbf{k}, \quad P(1, 0, -4)$

12. $\mathbf{r}(t) = \left\langle t, t, \sqrt{4 - t^2} \right\rangle, \quad P(1, 1, \sqrt{3})$

13. $\mathbf{r}(t) = \langle 2\cos t, 2\sin t, 4 \rangle, \quad P(\sqrt{2}, \sqrt{2}, 4)$

14. $\mathbf{r}(t) = \langle 2\sin t, 2\cos t, 4\sin^2 t \rangle, \quad P(1, \sqrt{3}, 1)$

 **Finding the Principal Unit Normal Vector** In Exercises 15–20, find the principal unit normal vector to the curve at the specified value of the parameter.

15. $\mathbf{r}(t) = t\mathbf{i} + \frac{1}{2}t^2\mathbf{j}, \quad t = 2$ 16. $\mathbf{r}(t) = t\mathbf{i} + \dfrac{6}{t}\mathbf{j}, \quad t = 3$

17. $\mathbf{r}(t) = t\mathbf{i} + t^2\mathbf{j} + \ln t\mathbf{k}, \quad t = 1$

18. $\mathbf{r}(t) = \sqrt{2}t\mathbf{i} + e^t\mathbf{j} + e^{-t}\mathbf{k}, \quad t = 0$

19. $\mathbf{r}(t) = 6\cos t\mathbf{i} + 6\sin t\mathbf{j} + \mathbf{k}, \quad t = \dfrac{3\pi}{4}$

20. $\mathbf{r}(t) = \cos 3t\mathbf{i} + 2\sin 3t\mathbf{j} + \mathbf{k}, \quad t = \pi$

Sketching a Graph and Vectors In Exercises 21–24, sketch the graph of the plane curve $\mathbf{r}(t)$ and sketch the vectors $\mathbf{T}(t)$ and $\mathbf{N}(t)$ at the given value of t.

21. $\mathbf{r}(t) = t\mathbf{i} + \dfrac{1}{t}\mathbf{j}, \quad t = 2$

22. $\mathbf{r}(t) = t\mathbf{i} - t^3\mathbf{j}, \quad t = 1$

23. $\mathbf{r}(t) = (2t + 1)\mathbf{i} - t^2\mathbf{j}, \quad t = 2$

24. $\mathbf{r}(t) = 2\cos t\mathbf{i} + 2\sin t\mathbf{j}, \quad t = \dfrac{7\pi}{6}$

 **Finding Tangential and Normal Components of Acceleration** In Exercises 25–30, find the tangential and normal components of acceleration at the given time t for the plane curve $\mathbf{r}(t)$.

25. $\mathbf{r}(t) = t\mathbf{i} + \dfrac{1}{t}\mathbf{j}, \quad t = 1$

26. $\mathbf{r}(t) = t^2\mathbf{i} + 2t\mathbf{j}, \quad t = 1$

27. $\mathbf{r}(t) = e^t\mathbf{i} + e^{-2t}\mathbf{j}, \quad t = 0$

28. $\mathbf{r}(t) = e^t\mathbf{i} + e^{-t}\mathbf{j}, \quad t = 0$

29. $\mathbf{r}(t) = e^t\cos t\mathbf{i} + e^t\sin t\mathbf{j}, \quad t = \dfrac{\pi}{2}$

30. $\mathbf{r}(t) = 4\cos 3t\mathbf{i} + 4\sin 3t\mathbf{j}, \quad t = \pi$

Circular Motion In Exercises 31–34, consider an object moving according to the position vector

$$\mathbf{r}(t) = a\cos \omega t\mathbf{i} + a\sin \omega t\mathbf{j}.$$

31. Find $\mathbf{T}(t)$, $\mathbf{N}(t)$, $a_\mathbf{T}$, and $a_\mathbf{N}$.

32. Determine the directions of $\mathbf{T}$ and $\mathbf{N}$ relative to the position vector $\mathbf{r}$.

33. Determine the speed of the object at any time t and explain its value relative to the value of $a_\mathbf{T}$.

34. When the angular speed ω is halved, by what factor is $a_\mathbf{N}$ changed?

 **Finding Tangential and Normal Components of Acceleration** In Exercises 35–40, find the tangential and normal components of acceleration at the given time t for the space curve $\mathbf{r}(t)$.

35. $\mathbf{r}(t) = t\mathbf{i} + 2t\mathbf{j} - 3t\mathbf{k}, \quad t = 1$

36. $\mathbf{r}(t) = \cos t\mathbf{i} + \sin t\mathbf{j} + 2t\mathbf{k}, \quad t = \dfrac{\pi}{3}$

37. $\mathbf{r}(t) = t\mathbf{i} + t^2\mathbf{j} + \dfrac{t^2}{2}\mathbf{k}, \quad t = 1$

38. $\mathbf{r}(t) = (2t - 1)\mathbf{i} + t^2\mathbf{j} - 4t\mathbf{k}, \quad t = 2$

39. $\mathbf{r}(t) = e^t\sin t\mathbf{i} + e^t\cos t\mathbf{j} + e^t\mathbf{k}, \quad t = 0$

40. $\mathbf{r}(t) = e^t\mathbf{i} + 2t\mathbf{j} + e^{-t}\mathbf{k}, \quad t = 0$

EXPLORING CONCEPTS

41. Acceleration Describe the motion of a particle when the normal component of acceleration is 0.

42. Acceleration Describe the motion of a particle when the tangential component of acceleration is 0.

43. Finding Vectors An object moves along the path given by

$$\mathbf{r}(t) = 3t\mathbf{i} + 4t\mathbf{j}.$$

Find $\mathbf{v}(t)$, $\mathbf{a}(t)$, $\mathbf{T}(t)$, and $\mathbf{N}(t)$ (if it exists). What is the form of the path? Is the speed of the object constant or changing?

44. **HOW DO YOU SEE IT?** The figures show the paths of two particles.

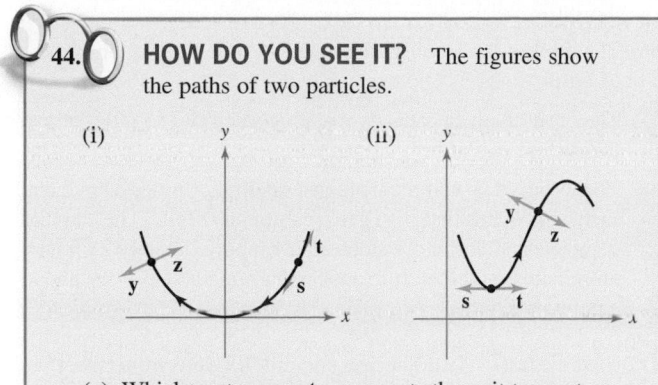

(i) (ii)

(a) Which vector, **s** or **t**, represents the unit tangent vector? Explain.

(b) Which vector, **y** or **z**, represents the principal unit normal vector? Explain.

45. Cycloidal Motion The figure shows the path of a particle modeled by the vector-valued function

$$\mathbf{r}(t) = \langle \pi t - \sin \pi t, 1 - \cos \pi t \rangle.$$

The figure also shows the vectors $\mathbf{v}(t)/\|\mathbf{v}(t)\|$ and $\mathbf{a}(t)/\|\mathbf{a}(t)\|$ at the indicated values of t.

(a) Find $a_{\mathbf{T}}$ and $a_{\mathbf{N}}$ at $t = \frac{1}{2}$, $t = 1$, and $t = \frac{3}{2}$.

(b) Determine whether the speed of the particle is increasing or decreasing at each of the indicated values of t. Give reasons for your answers.

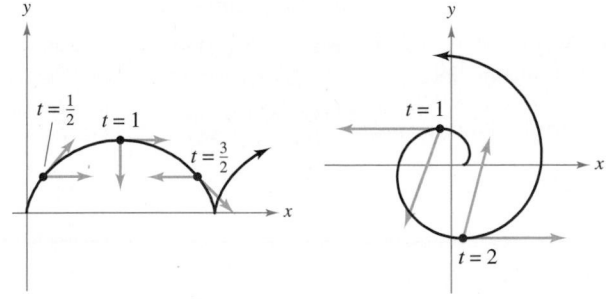

Figure for 45   Figure for 46

46. Motion Along an Involute of a Circle The figure shows a particle moving along a path modeled by

$$\mathbf{r}(t) = \langle \cos \pi t + \pi t \sin \pi t, \sin \pi t - \pi t \cos \pi t \rangle.$$

The figure also shows the vectors $\mathbf{v}(t)$ and $\mathbf{a}(t)$ for $t = 1$ and $t = 2$.

(a) Find $a_{\mathbf{T}}$ and $a_{\mathbf{N}}$ at $t = 1$ and $t = 2$.

(b) Determine whether the speed of the particle is increasing or decreasing at each of the indicated values of t. Give reasons for your answers.

Finding a Binormal Vector In Exercises 47–52, find the vectors **T** and **N** and the binormal vector $\mathbf{B} = \mathbf{T} \times \mathbf{N}$ for the vector-valued function $\mathbf{r}(t)$ at the given value of t.

47. $\mathbf{r}(t) = 2 \cos t\mathbf{i} + 2 \sin t\mathbf{j} + \dfrac{t}{2}\mathbf{k}$, $\quad t = \dfrac{\pi}{2}$

48. $\mathbf{r}(t) = t\mathbf{i} + t^2\mathbf{j} + \dfrac{t^3}{3}\mathbf{k}$, $\quad t = 1$

49. $\mathbf{r}(t) = \mathbf{i} + \sin t\mathbf{j} + \cos t\mathbf{k}$, $\quad t = \dfrac{\pi}{4}$

50. $\mathbf{r}(t) = 2e^t\mathbf{i} + e^t \cos t\mathbf{j} + e^t \sin t\mathbf{k}$, $\quad t = 0$

51. $\mathbf{r}(t) = 4 \sin t\mathbf{i} + 4 \cos t\mathbf{j} + 2t\mathbf{k}$, $\quad t = \dfrac{\pi}{3}$

52. $\mathbf{r}(t) = 3 \cos 2t\mathbf{i} + 3 \sin 2t\mathbf{j} + t\mathbf{k}$, $\quad t = \dfrac{\pi}{4}$

Alternative Formula for the Principal Unit Normal Vector In Exercises 53–56, use the vector-valued function $\mathbf{r}(t)$ to find the principal unit normal vector $\mathbf{N}(t)$ using the alternative formula

$$\mathbf{N} = \frac{(\mathbf{v} \cdot \mathbf{v})\mathbf{a} - (\mathbf{v} \cdot \mathbf{a})\mathbf{v}}{\|(\mathbf{v} \cdot \mathbf{v})\mathbf{a} - (\mathbf{v} \cdot \mathbf{a})\mathbf{v}\|}.$$

53. $\mathbf{r}(t) = 3t\mathbf{i} + 2t^2\mathbf{j}$

54. $\mathbf{r}(t) = 3 \cos 2t\mathbf{i} + 3 \sin 2t\mathbf{j}$

55. $\mathbf{r}(t) = 2t\mathbf{i} + 4t\mathbf{j} + t^2\mathbf{k}$

56. $\mathbf{r}(t) = 5 \cos t\mathbf{i} + 5 \sin t\mathbf{j} + 3t\mathbf{k}$

57. Projectile Motion Find the tangential and normal components of acceleration for a projectile fired at an angle θ with the horizontal at an initial speed of v_0. What are the components when the projectile is at its maximum height?

58. Projectile Motion Use your results from Exercise 57 to find the tangential and normal components of acceleration for a projectile fired at an angle of $45°$ with the horizontal at an initial speed of 150 feet per second. What are the components when the projectile is at its maximum height?

59. Projectile Motion A projectile is launched with an initial speed of 120 feet per second at a height of 5 feet and at an angle of $30°$ with the horizontal.

(a) Determine the vector-valued function for the path of the projectile.

(b) Use a graphing utility to graph the path and approximate the maximum height and range of the projectile.

(c) Find $\mathbf{v}(t)$, $\|\mathbf{v}(t)\|$, and $\mathbf{a}(t)$.

(d) Use a graphing utility to complete the table.

t	0.5	1.0	1.5	2.0	2.5	3.0
Speed						

(e) Use a graphing utility to graph the scalar functions $a_{\mathbf{T}}$ and $a_{\mathbf{N}}$. How is the speed of the projectile changing when $a_{\mathbf{T}}$ and $a_{\mathbf{N}}$ have opposite signs?

60. Projectile Motion A projectile is launched with an initial speed of 220 feet per second at a height of 4 feet and at an angle of 45° with the horizontal.

(a) Determine the vector-valued function for the path of the projectile.

(b) Use a graphing utility to graph the path and approximate the maximum height and range of the projectile.

(c) Find $\mathbf{v}(t)$, $\|\mathbf{v}(t)\|$, and $\mathbf{a}(t)$.

(d) Use a graphing utility to complete the table.

t	0.5	1.0	1.5	2.0	2.5	3.0
Speed						

61. Air Traffic Control

Because of a storm, ground controllers instruct the pilot of a plane flying at an altitude of 4 miles to make a 90° turn and climb to an altitude of 4.2 miles. The model for the path of the plane during this maneuver is

$$\mathbf{r}(t) = \langle 10 \cos 10\pi t, 10 \sin 10\pi t, 4 + 4t \rangle, \quad 0 \le t \le \tfrac{1}{20}$$

where t is the time in hours and $\mathbf{r}$ is the distance in miles.

(a) Determine the speed of the plane.

(b) Calculate $a_\mathbf{T}$ and $a_\mathbf{N}$. Why is one of these equal to 0?

62. Projectile Motion A plane flying at an altitude of 36,000 feet at a speed of 600 miles per hour releases a bomb. Find the tangential and normal components of acceleration acting on the bomb.

63. Centripetal Acceleration An object is spinning at a constant speed on the end of a string, according to the position vector $\mathbf{r}(t) = a \cos \omega t \mathbf{i} + a \sin \omega t \mathbf{j}$.

(a) When the angular speed ω is doubled, how is the centripetal component of acceleration changed?

(b) When the angular speed is unchanged but the length of the string is halved, how is the centripetal component of acceleration changed?

64. Centripetal Force An object of mass m moves at a constant speed v in a circular path of radius r, according to the position vector $\mathbf{r}(t) = r \cos \omega t \mathbf{i} + r \sin \omega t \mathbf{j}$.

(a) The force required to produce the centripetal component of acceleration is called the *centripetal force* and is given by $F = mv^2/r$. Use $\mathbf{F} = m\mathbf{a}$ to verify the centripetal force.

(b) Newton's Law of Universal Gravitation is given by $F = GMm/d^2$, where d is the distance between the centers of the two bodies of masses M and m, and G is the gravitational constant. Use this law to show that the speed required for circular motion is $v = \sqrt{GM/r}$.

Orbital Speed In Exercises 65–68, use the result of Exercise 64 to find the speed necessary for the given circular orbit around Earth. Let $GM = 9.56 \times 10^4$ cubic miles per second per second, and assume the radius of Earth is 4000 miles.

65. The orbit of the International Space Station 255 miles above the surface of Earth

66. The orbit of the Hubble telescope 340 miles above the surface of Earth

67. The orbit of a heat capacity mapping satellite 385 miles above the surface of Earth

68. The orbit of a communications satellite r miles above the surface of Earth that is in geosynchronous orbit. [The satellite completes one orbit per sidereal day (approximately 23 hours, 56 minutes) and therefore appears to remain stationary above a point on Earth.]

True or False? In Exercises 69 and 70, determine whether the statement is true or false. If it is false, explain why or give an example that shows it is false.

69. The velocity and acceleration vectors of a moving object are always perpendicular.

70. If $a_\mathbf{N} = 0$ for a moving object, then the object is moving in a straight line.

71. Motion of a Particle A particle moves along a path modeled by

$$\mathbf{r}(t) = \cosh(bt)\mathbf{i} + \sinh(bt)\mathbf{j}$$

where b is a positive constant.

(a) Show that the path of the particle is a hyperbola.

(b) Show that $\mathbf{a}(t) = b^2\mathbf{r}(t)$.

72. Proof Prove that the principal unit normal vector $\mathbf{N}$ points toward the concave side of a plane curve.

73. Proof Prove that the vector $\mathbf{T}'(t)$ is 0 for an object moving in a straight line.

74. Proof Prove that $a_\mathbf{N} = \dfrac{\|\mathbf{v} \times \mathbf{a}\|}{\|\mathbf{v}\|}$.

75. Proof Prove that $a_\mathbf{N} = \sqrt{\|\mathbf{a}\|^2 - a_\mathbf{T}^2}$.

PUTNAM EXAM CHALLENGE

76. A particle of unit mass moves on a straight line under the action of a force which is a function $f(v)$ of the velocity v of the particle, but the form of this function is not known. A motion is observed, and the distance x covered in time t is found to be connected with t by the formula

$$x = at + bt^2 + ct^3$$

where a, b, and c have numerical values determined by observation of the motion. Find the function $f(v)$ for the range of v covered by the experiment.

This problem was composed by the Committee on the Putnam Prize Competition.
© The Mathematical Association of America. All rights reserved.

12.5 Arc Length and Curvature

■ Find the arc length of a space curve.
■ Use the arc length parameter to describe a plane curve or space curve.
■ Find the curvature of a curve at a point on the curve.
■ Use a vector-valued function to find frictional force.

Arc Length

In Section 10.3, you saw that the arc length of a smooth *plane* curve C given by the parametric equations $x = x(t)$ and $y = y(t)$, $a \le t \le b$, is

$$s = \int_a^b \sqrt{[x'(t)]^2 + [y'(t)]^2}\, dt.$$

In vector form, where C is given by $\mathbf{r}(t) = x(t)\mathbf{i} + y(t)\mathbf{j}$, you can rewrite this equation for arc length as

$$s = \int_a^b \|\mathbf{r}'(t)\|\, dt.$$

The formula for the arc length of a plane curve has a natural extension to a smooth curve in *space,* as stated in the next theorem.

> **THEOREM 12.6 Arc Length of a Space Curve**
>
> If C is a smooth curve given by $\mathbf{r}(t) = x(t)\mathbf{i} + y(t)\mathbf{j} + z(t)\mathbf{k}$ on an interval $[a, b]$, then the arc length of C on the interval is
>
> $$s = \int_a^b \sqrt{[x'(t)]^2 + [y'(t)]^2 + [z'(t)]^2}\, dt = \int_a^b \|\mathbf{r}'(t)\|\, dt.$$

EXAMPLE 1 **Finding the Arc Length of a Curve in Space**

• • • ▷ *See LarsonCalculus.com for an interactive version of this type of example.*

Find the arc length of the curve given by

$$\mathbf{r}(t) = t\mathbf{i} + \frac{4}{3}t^{3/2}\mathbf{j} + \frac{1}{2}t^2\mathbf{k}$$

from $t = 0$ to $t = 2$, as shown in Figure 12.28.

Solution Using $x(t) = t$, $y(t) = \frac{4}{3}t^{3/2}$, and $z(t) = \frac{1}{2}t^2$, you obtain $x'(t) = 1$, $y'(t) = 2t^{1/2}$, and $z'(t) = t$. So, the arc length from $t = 0$ to $t = 2$ is given by

$$s = \int_0^2 \sqrt{[x'(t)]^2 + [y'(t)]^2 + [z'(t)]^2}\, dt \qquad \text{Formula for arc length}$$

$$= \int_0^2 \sqrt{1 + 4t + t^2}\, dt$$

$$= \int_0^2 \sqrt{(t+2)^2 - 3}\, dt \qquad \begin{matrix}\text{Integration tables}\\\text{(Appendix B), Formula 26}\end{matrix}$$

$$= \left[\frac{t+2}{2}\sqrt{(t+2)^2 - 3} - \frac{3}{2}\ln|(t+2) + \sqrt{(t+2)^2 - 3}| \right]_0^2$$

$$= 2\sqrt{13} - \frac{3}{2}\ln(4 + \sqrt{13}) - 1 + \frac{3}{2}\ln 3$$

$$\approx 4.816.$$

Exploration

Arc Length Formula The formula for the arc length of a space curve is given in terms of the parametric equations used to represent the curve. Does this mean that the arc length of the curve depends on the parameter being used? Would you want this to be true? Explain your reasoning.

Here is a different parametric representation of the curve in Example 1.

$$\mathbf{r}(t) = t^2\mathbf{i} + \frac{4}{3}t^3\mathbf{j} + \frac{1}{2}t^4\mathbf{k}$$

Find the arc length from $t = 0$ to $t = \sqrt{2}$ and compare the result with that found in Example 1.

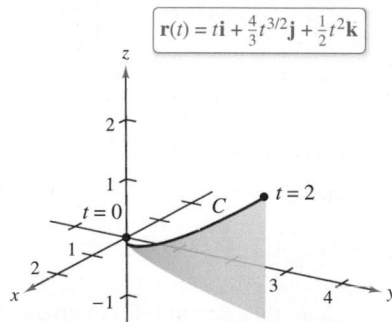

$$\mathbf{r}(t) = t\mathbf{i} + \frac{4}{3}t^{3/2}\mathbf{j} + \frac{1}{2}t^2\mathbf{k}$$

As t increases from 0 to 2, the vector $\mathbf{r}(t)$ traces out a curve.
Figure 12.28

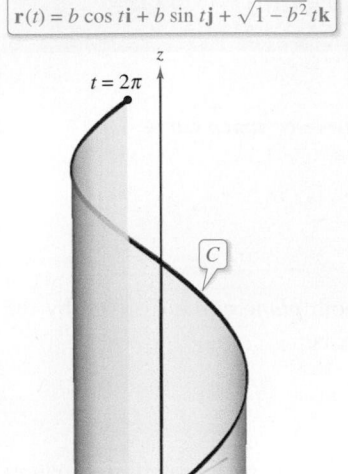

Curve:
$\mathbf{r}(t) = b \cos t\mathbf{i} + b \sin t\mathbf{j} + \sqrt{1 - b^2}\,t\mathbf{k}$

One turn of a helix
Figure 12.29

EXAMPLE 2   **Finding the Arc Length of a Helix**

Find the length of one turn of the helix given by

$$\mathbf{r}(t) = b \cos t\mathbf{i} + b \sin t\mathbf{j} + \sqrt{1 - b^2}\,t\mathbf{k}$$

as shown in Figure 12.29.

Solution   Begin by finding the derivative.

$$\mathbf{r}'(t) = -b \sin t\mathbf{i} + b \cos t\mathbf{j} + \sqrt{1 - b^2}\,\mathbf{k} \qquad \text{Derivative}$$

Now, using the formula for arc length, you can find the length of one turn of the helix by integrating $\|\mathbf{r}'(t)\|$ from 0 to 2π.

$$\begin{aligned}
s &= \int_0^{2\pi} \|\mathbf{r}'(t)\|\, dt && \text{Formula for arc length} \\
&= \int_0^{2\pi} \sqrt{b^2(\sin^2 t + \cos^2 t) + (1 - b^2)}\, dt \\
&= \int_0^{2\pi} dt \\
&= t\, \Big]_0^{2\pi} \\
&= 2\pi
\end{aligned}$$

So, the length is 2π units.   ∎

Arc Length Parameter

You have seen that curves can be represented by vector-valued functions in different ways, depending on the choice of parameter. For *motion* along a curve, the convenient parameter is time t. For studying the *geometric properties* of a curve, however, the convenient parameter is often arc length s.

$s(t) = \int_a^t \sqrt{[x'(u)]^2 + [y'(u)]^2 + [z'(u)]^2}\, du$

Figure 12.30

> **Definition of Arc Length Function**
>
> Let C be a smooth curve given by $\mathbf{r}(t)$ defined on the closed interval $[a, b]$. For $a \le t \le b$, the **arc length function** is
>
> $$s(t) = \int_a^t \|\mathbf{r}'(u)\|\, du = \int_a^t \sqrt{[x'(u)]^2 + [y'(u)]^2 + [z'(u)]^2}\, du.$$
>
> The arc length s is called the **arc length parameter.** (See Figure 12.30.)

Note that the arc length function s is *nonnegative*. It measures the distance along C from the initial point $(x(a), y(a), z(a))$ to the point $(x(t), y(t), z(t))$.

Using the definition of the arc length function and the Second Fundamental Theorem of Calculus, you can conclude that

$$\frac{ds}{dt} = \|\mathbf{r}'(t)\|. \qquad \text{Derivative of arc length function}$$

In differential form, you can write

$$ds = \|\mathbf{r}'(t)\|\, dt.$$

$\mathbf{r}(t) = (3 - 3t)\mathbf{i} + 4t\mathbf{j}$
$0 \le t \le 1$

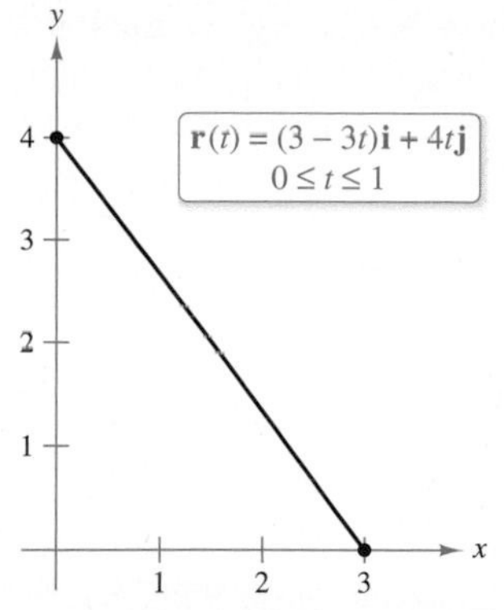

The line segment from $(3, 0)$ to $(0, 4)$ can be parametrized using the arc length parameter s.

Figure 12.31

EXAMPLE 3 **Finding the Arc Length Function for a Line**

Find the arc length function $s(t)$ for the line segment given by

$$\mathbf{r}(t) = (3 - 3t)\mathbf{i} + 4t\mathbf{j}, \quad 0 \le t \le 1$$

and write $\mathbf{r}$ as a function of the parameter s. (See Figure 12.31.)

Solution Because $\mathbf{r}'(t) = -3\mathbf{i} + 4\mathbf{j}$ and

$$\|\mathbf{r}'(t)\| = \sqrt{(-3)^2 + 4^2} = 5$$

you have

$$
\begin{aligned}
s(t) &= \int_0^t \|\mathbf{r}'(u)\| \, du \\
&= \int_0^t 5 \, du \\
&= 5t.
\end{aligned}
$$

Using $s = 5t$ (or $t = s/5$), you can rewrite $\mathbf{r}$ using the arc length parameter as follows.

$$\mathbf{r}(s) = \left(3 - \frac{3}{5}s\right)\mathbf{i} + \frac{4}{5}s\mathbf{j}, \quad 0 \le s \le 5$$

One of the advantages of writing a vector-valued function in terms of the arc length parameter is that $\|\mathbf{r}'(s)\| = 1$. For instance, in Example 3, you have

$$\|\mathbf{r}'(s)\| = \sqrt{\left(-\frac{3}{5}\right)^2 + \left(\frac{4}{5}\right)^2} = 1.$$

So, for a smooth curve C represented by $\mathbf{r}(s)$, where s is the arc length parameter, the arc length between a and b is

$$
\begin{aligned}
\text{Length of arc} &= \int_a^b \|\mathbf{r}'(s)\| \, ds \\
&= \int_a^b ds \\
&= b - a \\
&= \text{length of interval.}
\end{aligned}
$$

Furthermore, if t is *any* parameter such that $\|\mathbf{r}'(t)\| = 1$, then t must be the arc length parameter. These results are summarized in the next theorem, which is stated without proof.

THEOREM 12.7 **Arc Length Parameter**

If C is a smooth curve given by

$$\mathbf{r}(s) = x(s)\mathbf{i} + y(s)\mathbf{j} \qquad \text{Plane curve}$$

or

$$\mathbf{r}(s) = x(s)\mathbf{i} + y(s)\mathbf{j} + z(s)\mathbf{k} \qquad \text{Space curve}$$

where s is the arc length parameter, then

$$\|\mathbf{r}'(s)\| = 1.$$

Moreover, if t is *any* parameter for the vector-valued function $\mathbf{r}$ such that $\|\mathbf{r}'(t)\| = 1$, then t must be the arc length parameter.

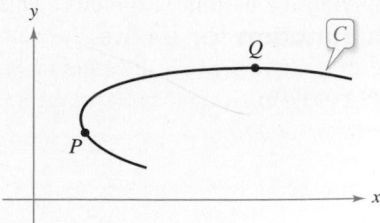

Curvature at P is greater than at Q.
Figure 12.32

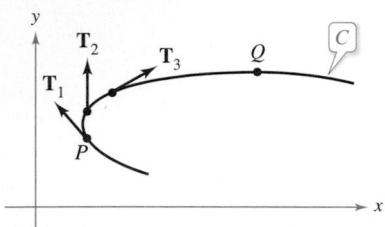

The magnitude of the rate of change of **T** with respect to the arc length is the curvature of a curve.
Figure 12.33

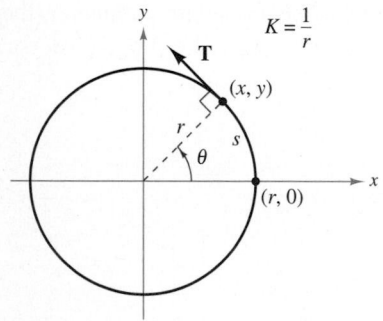

The curvature of a circle is constant.
Figure 12.34

Curvature

An important use of the arc length parameter is to find **curvature**—the measure of how sharply a curve bends. For instance, in Figure 12.32, the curve bends more sharply at P than at Q, and you can say that the curvature is greater at P than at Q. You can calculate curvature by calculating the magnitude of the rate of change of the unit tangent vector **T** with respect to the arc length s, as shown in Figure 12.33.

Definition of Curvature

Let C be a smooth curve (in the plane *or* in space) given by $\mathbf{r}(s)$, where s is the arc length parameter. The **curvature** K at s is

$$K = \left\| \frac{d\mathbf{T}}{ds} \right\| = \| \mathbf{T}'(s) \|.$$

A circle has the same curvature at any point. Moreover, the curvature and the radius of the circle are inversely related. That is, a circle with a large radius has a small curvature, and a circle with a small radius has a large curvature. This inverse relationship is made explicit in the next example.

EXAMPLE 4 **Finding the Curvature of a Circle**

Show that the curvature of a circle of radius r is

$$K = \frac{1}{r}.$$

Solution Without loss of generality, you can consider the circle to be centered at the origin. Let (x, y) be any point on the circle and let s be the length of the arc from $(r, 0)$ to (x, y), as shown in Figure 12.34. By letting θ be the central angle of the circle, you can represent the circle by

$$\mathbf{r}(\theta) = r \cos \theta \mathbf{i} + r \sin \theta \mathbf{j}. \qquad \theta \text{ is the parameter.}$$

Using the formula for the length of a circular arc $s = r\theta$, you can rewrite $\mathbf{r}(\theta)$ in terms of the arc length parameter as follows.

$$\mathbf{r}(s) = r \cos \frac{s}{r} \mathbf{i} + r \sin \frac{s}{r} \mathbf{j} \qquad \text{Arc length } s \text{ is the parameter.}$$

So, $\mathbf{r}'(s) = -\sin \frac{s}{r} \mathbf{i} + \cos \frac{s}{r} \mathbf{j}$, and it follows that $\| \mathbf{r}'(s) \| = 1$, which implies that the unit tangent vector is

$$\mathbf{T}(s) = \frac{\mathbf{r}'(s)}{\| \mathbf{r}'(s) \|} = -\sin \frac{s}{r} \mathbf{i} + \cos \frac{s}{r} \mathbf{j}$$

and the curvature is

$$K = \| \mathbf{T}'(s) \| = \left\| -\frac{1}{r} \cos \frac{s}{r} \mathbf{i} - \frac{1}{r} \sin \frac{s}{r} \mathbf{j} \right\| = \frac{1}{r}$$

at every point on the circle. ∎

Because a straight line does not curve, you would expect its curvature to be 0. Try checking this by finding the curvature of the line given by

$$\mathbf{r}(s) = \left(3 - \frac{3}{5} s \right) \mathbf{i} + \frac{4}{5} s \mathbf{j}.$$

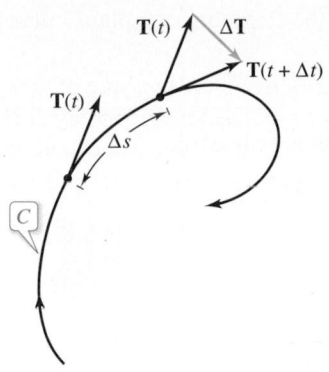

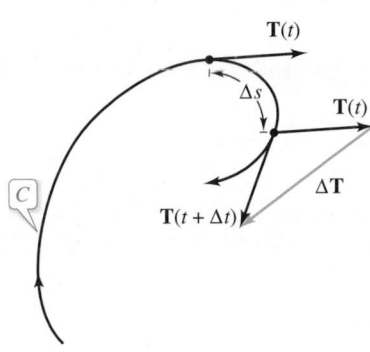

Figure 12.35

In Example 4, the curvature was found by applying the definition directly. This requires that the curve be written in terms of the arc length parameter s. The next theorem gives two other formulas for finding the curvature of a curve written in terms of an arbitrary parameter t. The proof of this theorem is left as an exercise [see Exercise 82, parts (a) and (b)].

THEOREM 12.8 Formulas for Curvature

If C is a smooth curve given by $\mathbf{r}(t)$, then the curvature K of C at t is

$$K = \frac{\|\mathbf{T}'(t)\|}{\|\mathbf{r}'(t)\|} = \frac{\|\mathbf{r}'(t) \times \mathbf{r}''(t)\|}{\|\mathbf{r}'(t)\|^3}.$$

Because $\|\mathbf{r}'(t)\| = ds/dt$, the first formula implies that curvature is the ratio of the rate of change of the unit tangent vector $\mathbf{T}$ to the rate of change of the arc length. To see that this is reasonable, let Δt be a "small number." Then,

$$\frac{\mathbf{T}'(t)}{ds/dt} \approx \frac{[\mathbf{T}(t + \Delta t) - \mathbf{T}(t)]/\Delta t}{[s(t + \Delta t) - s(t)]/\Delta t} = \frac{\mathbf{T}(t + \Delta t) - \mathbf{T}(t)}{s(t + \Delta t) - s(t)} = \frac{\Delta \mathbf{T}}{\Delta s}.$$

In other words, for a given Δs, the greater the length of $\Delta \mathbf{T}$, the more the curve bends at t, as shown in Figure 12.35.

EXAMPLE 5 Finding the Curvature of a Space Curve

Find the curvature of the curve given by

$$\mathbf{r}(t) = 2t\mathbf{i} + t^2\mathbf{j} - \frac{1}{3}t^3\mathbf{k}.$$

Solution It is not apparent whether this parameter represents arc length, so you should use the formula $K = \|\mathbf{T}'(t)\|/\|\mathbf{r}'(t)\|$.

$$\mathbf{r}'(t) = 2\mathbf{i} + 2t\mathbf{j} - t^2\mathbf{k}$$

$$\|\mathbf{r}'(t)\| = \sqrt{4 + 4t^2 + t^4} \qquad\qquad \text{Length of } \mathbf{r}'(t)$$

$$= t^2 + 2$$

$$\mathbf{T}(t) = \frac{\mathbf{r}'(t)}{\|\mathbf{r}'(t)\|}$$

$$= \frac{2\mathbf{i} + 2t\mathbf{j} - t^2\mathbf{k}}{t^2 + 2}$$

$$\mathbf{T}'(t) = \frac{(t^2 + 2)(2\mathbf{j} - 2t\mathbf{k}) - (2t)(2\mathbf{i} + 2t\mathbf{j} - t^2\mathbf{k})}{(t^2 + 2)^2}$$

$$= \frac{-4t\mathbf{i} + (4 - 2t^2)\mathbf{j} - 4t\mathbf{k}}{(t^2 + 2)^2}$$

$$\|\mathbf{T}'(t)\| = \frac{\sqrt{16t^2 + 16 - 16t^2 + 4t^4 + 16t^2}}{(t^2 + 2)^2}$$

$$= \frac{2(t^2 + 2)}{(t^2 + 2)^2}$$

$$= \frac{2}{t^2 + 2} \qquad\qquad \text{Length of } \mathbf{T}'(t)$$

Therefore,

$$K = \frac{\|\mathbf{T}'(t)\|}{\|\mathbf{r}'(t)\|} = \frac{2}{(t^2 + 2)^2}. \qquad\qquad \text{Curvature}$$

The next theorem presents a formula for calculating the curvature of a plane curve given by $y = f(x)$.

> **THEOREM 12.9 Curvature in Rectangular Coordinates**
>
> If C is the graph of a twice-differentiable function given by $y = f(x)$, then the curvature K at the point (x, y) is
>
> $$K = \frac{|y''|}{[1 + (y')^2]^{3/2}}.$$

Proof By representing the curve C by $\mathbf{r}(x) = x\mathbf{i} + f(x)\mathbf{j} + 0\mathbf{k}$, where x is the parameter, you obtain $\mathbf{r}'(x) = \mathbf{i} + f'(x)\mathbf{j}$,

$$\|\mathbf{r}'(x)\| = \sqrt{1 + [f'(x)]^2}$$

and $\mathbf{r}''(x) = f''(x)\mathbf{j}$. Because $\mathbf{r}'(x) \times \mathbf{r}''(x) = f''(x)\mathbf{k}$, it follows that the curvature is

$$K = \frac{\|\mathbf{r}'(x) \times \mathbf{r}''(x)\|}{\|\mathbf{r}'(x)\|^3}$$

$$= \frac{|f''(x)|}{\{1 + [f'(x)]^2\}^{3/2}}$$

$$= \frac{|y''|}{[1 + (y')^2]^{3/2}}.$$

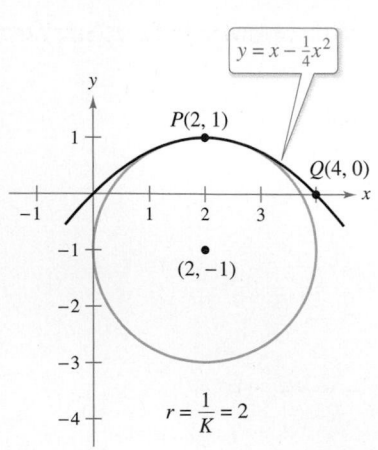

r = radius of curvature

$K = \dfrac{1}{r}$

P

Center of curvature

C

The circle of curvature

Figure 12.36

Let C be a curve with curvature K at point P. The circle passing through point P with radius $r = 1/K$ is called the **circle of curvature** when the circle lies on the concave side of the curve and shares a common tangent line with the curve at point P. The radius is called the **radius of curvature** at P, and the center of the circle is called the **center of curvature.**

The circle of curvature gives you a nice way to estimate the curvature K at a point P on a curve graphically. Using a compass, you can sketch a circle that lies against the concave side of the curve at point P, as shown in Figure 12.36. If the circle has a radius of r, then you can estimate the curvature to be $K = 1/r$.

EXAMPLE 6 **Finding Curvature in Rectangular Coordinates**

Find the curvature of the parabola given by

$$y = x - \frac{1}{4}x^2$$

at $x = 2$. Sketch the circle of curvature at $(2, 1)$.

Solution The curvature at $x = 2$ is as follows.

$$y' = 1 - \frac{x}{2} \qquad\qquad y' = 0$$

$$y'' = -\frac{1}{2} \qquad\qquad y'' = -\frac{1}{2}$$

$$K = \frac{|y''|}{[1 + (y')^2]^{3/2}} \qquad K = \frac{1}{2}$$

The circle of curvature

Figure 12.37

Because the curvature at $P(2, 1)$ is $\frac{1}{2}$, it follows that the radius of the circle of curvature at that point is 2. So, the center of curvature is $(2, -1)$, as shown in Figure 12.37. [In the figure, note that the curve has the greatest curvature at P. Try showing that the curvature at $Q(4, 0)$ is $1/2^{5/2} \approx 0.177$.]

The amount of thrust felt by passengers in a car that is turning depends on two things—the speed of the car and the sharpness of the turn.
Figure 12.38

Arc length and curvature are closely related to the tangential and normal components of acceleration. The tangential component of acceleration is the rate of change of the speed, which in turn is the rate of change of the arc length. This component is negative as a moving object slows down and positive as it speeds up—regardless of whether the object is turning or traveling in a straight line. So, the tangential component is solely a function of the arc length and is independent of the curvature.

On the other hand, the normal component of acceleration is a function of *both* speed and curvature. This component measures the acceleration acting perpendicular to the direction of motion. To see why the normal component is affected by both speed and curvature, imagine that you are driving a car around a turn, as shown in Figure 12.38. When your speed is high and the turn is sharp, you feel yourself thrown against the car door. By lowering your speed *or* taking a more gentle turn, you are able to lessen this sideways thrust.

The next theorem explicitly states the relationships among speed, curvature, and the components of acceleration.

· · · · · · · · · · · · · ▷

· ·REMARK Note that Theorem 12.10 gives additional formulas for $a_\mathbf{T}$ and $a_\mathbf{N}$.

> **THEOREM 12.10 Acceleration, Speed, and Curvature**
>
> If $\mathbf{r}(t)$ is the position vector for a smooth curve C, then the acceleration vector is given by
>
> $$\mathbf{a}(t) = \frac{d^2 s}{dt^2}\mathbf{T} + K\left(\frac{ds}{dt}\right)^2 \mathbf{N}$$
>
> where K is the curvature of C and ds/dt is the speed.

Proof For the position vector $\mathbf{r}(t)$, you have

$$\mathbf{a}(t) = a_\mathbf{T}\mathbf{T} + a_\mathbf{N}\mathbf{N}$$

$$= \frac{d}{dt}\big[\|\mathbf{v}\|\big]\mathbf{T} + \|\mathbf{v}\|\,\|\mathbf{T}'\|\mathbf{N}$$

$$= \frac{d^2 s}{dt^2}\mathbf{T} + \frac{ds}{dt}(\|\mathbf{v}\|K)\mathbf{N}$$

$$= \frac{d^2 s}{dt^2}\mathbf{T} + K\left(\frac{ds}{dt}\right)^2 \mathbf{N}.$$

EXAMPLE 7   **Tangential and Normal Components of Acceleration**

Find $a_\mathbf{T}$ and $a_\mathbf{N}$ for the curve given by

$$\mathbf{r}(t) = 2t\mathbf{i} + t^2\mathbf{j} - \tfrac{1}{3}t^3\mathbf{k}.$$

Solution From Example 5, you know that

$$\frac{ds}{dt} = \|\mathbf{r}'(t)\| = t^2 + 2 \quad \text{and} \quad K = \frac{2}{(t^2 + 2)^2}.$$

Therefore,

$$a_\mathbf{T} = \frac{d^2 s}{dt^2} = 2t \qquad\qquad \text{Tangential component}$$

and

$$a_\mathbf{N} = K\left(\frac{ds}{dt}\right)^2 = \frac{2}{(t^2 + 2)^2}(t^2 + 2)^2 = 2. \qquad \text{Normal component}$$

Application

There are many applications in physics and engineering dynamics that involve the relationships among speed, arc length, curvature, and acceleration. One such application concerns frictional force.

A moving object with mass m is in contact with a stationary object. The total force required to produce an acceleration $\mathbf{a}$ along a given path is

$$\mathbf{F} = m\mathbf{a}$$

$$= m\left(\frac{d^2s}{dt^2}\right)\mathbf{T} + mK\left(\frac{ds}{dt}\right)^2\mathbf{N}$$

$$= ma_{\mathbf{T}}\mathbf{T} + ma_{\mathbf{N}}\mathbf{N}.$$

The portion of this total force that is supplied by the stationary object is called the **force of friction.** For example, when a car moving with constant speed is rounding a turn, the roadway exerts a frictional force that keeps the car from sliding off the road. If the car is not sliding, the frictional force is perpendicular to the direction of motion and has magnitude equal to the normal component of acceleration, as shown in Figure 12.39. The potential frictional force of a road around a turn can be increased by banking the roadway.

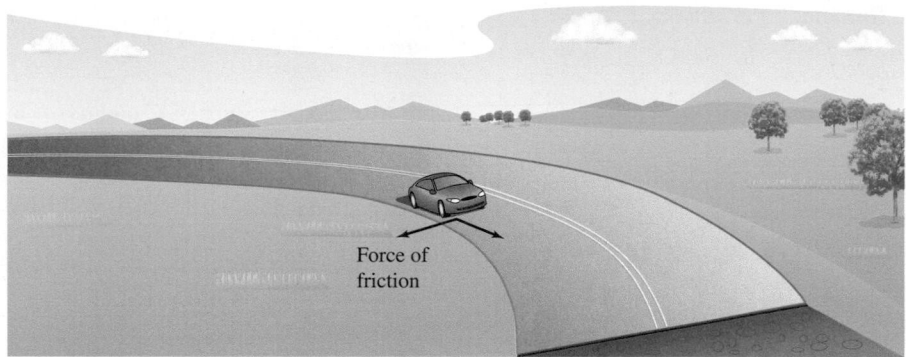

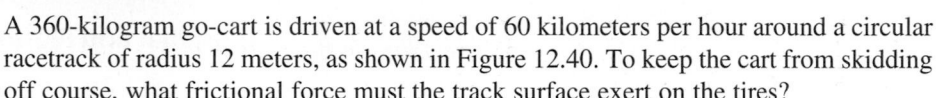
Force of friction

The force of friction is perpendicular to the direction of motion.
Figure 12.39

EXAMPLE 8 **Frictional Force**

A 360-kilogram go-cart is driven at a speed of 60 kilometers per hour around a circular racetrack of radius 12 meters, as shown in Figure 12.40. To keep the cart from skidding off course, what frictional force must the track surface exert on the tires?

Solution The frictional force must equal the mass times the normal component of acceleration. For this circular path, you know that the curvature is

$$K = \frac{1}{12}. \qquad \text{Curvature of circular racetrack}$$

Therefore, the frictional force is

$$ma_{\mathbf{N}} = mK\left(\frac{ds}{dt}\right)^2$$

$$= (360 \text{ kg})\left(\frac{1}{12 \text{ m}}\right)\left(\frac{60,000 \text{ m}}{3600 \text{ sec}}\right)^2$$

$$\approx 8333 \text{ (kg)(m)/sec}^2.$$

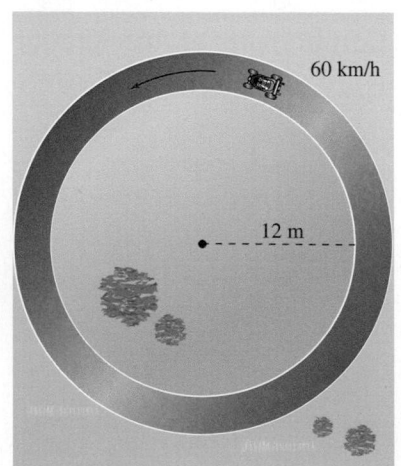

60 km/h

12 m

Figure 12.40

SUMMARY OF VELOCITY, ACCELERATION, AND CURVATURE

Unless noted otherwise, let C be a curve (in the plane or in space) given by the position vector

$$\mathbf{r}(t) = x(t)\mathbf{i} + y(t)\mathbf{j} \qquad \text{Curve in the plane}$$

or

$$\mathbf{r}(t) = x(t)\mathbf{i} + y(t)\mathbf{j} + z(t)\mathbf{k} \qquad \text{Curve in space}$$

where x, y, and z are twice-differentiable functions of t.

Velocity vector, speed, and acceleration vector

$$\mathbf{v}(t) = \mathbf{r}'(t) \qquad \text{Velocity vector}$$

$$\|\mathbf{v}(t)\| = \|\mathbf{r}'(t)\| = \frac{ds}{dt} \qquad \text{Speed}$$

$$\mathbf{a}(t) = \mathbf{r}''(t) \qquad \text{Acceleration vector}$$

$$= a_{\mathbf{T}}\mathbf{T}(t) + a_{\mathbf{N}}\mathbf{N}(t)$$

$$= \frac{d^2s}{dt^2}\mathbf{T}(t) + K\left(\frac{ds}{dt}\right)^2\mathbf{N}(t) \qquad \text{K is curvature and $\frac{ds}{dt}$ is speed.}$$

Unit tangent vector and principal unit normal vector

$$\mathbf{T}(t) = \frac{\mathbf{r}'(t)}{\|\mathbf{r}'(t)\|} \qquad \text{Unit tangent vector}$$

$$\mathbf{N}(t) = \frac{\mathbf{T}'(t)}{\|\mathbf{T}'(t)\|} \qquad \text{Principal unit normal vector}$$

Components of acceleration

$$a_{\mathbf{T}} = \mathbf{a} \cdot \mathbf{T} = \frac{\mathbf{v} \cdot \mathbf{a}}{\|\mathbf{v}\|} = \frac{d^2s}{dt^2} \qquad \text{Tangential component of acceleration}$$

$$a_{\mathbf{N}} = \mathbf{a} \cdot \mathbf{N} \qquad \text{Normal component of acceleration}$$

$$= \frac{\|\mathbf{v} \times \mathbf{a}\|}{\|\mathbf{v}\|}$$

$$= \sqrt{\|\mathbf{a}\|^2 - a_{\mathbf{T}}^2}$$

$$= K\left(\frac{ds}{dt}\right)^2 \qquad \text{K is curvature and $\frac{ds}{dt}$ is speed.}$$

Formulas for curvature in the plane

$$K = \frac{|y''|}{[1 + (y')^2]^{3/2}} \qquad \text{C given by $y = f(x)$}$$

$$K = \frac{|x'y'' - y'x''|}{[(x')^2 + (y')^2]^{3/2}} \qquad \text{C given by $x = x(t)$, $y = y(t)$}$$

Formulas for curvature in the plane or in space

$$K = \|\mathbf{T}'(s)\| = \|\mathbf{r}''(s)\| \qquad \text{s is arc length parameter.}$$

$$K = \frac{\|\mathbf{T}'(t)\|}{\|\mathbf{r}'(t)\|} = \frac{\|\mathbf{r}'(t) \times \mathbf{r}''(t)\|}{\|\mathbf{r}'(t)\|^3} \qquad \text{t is general parameter.}$$

$$K = \frac{\mathbf{a}(t) \cdot \mathbf{N}(t)}{\|\mathbf{v}(t)\|^2}$$

Cross product formulas apply only to curves in space.

12.5 Exercises

See **CalcChat.com** for tutorial help and worked-out solutions to odd-numbered exercises.

CONCEPT CHECK

1. Curvature Consider points P and Q on a curve. What does it mean for the curvature at P to be less than the curvature at Q?

2. Arc Length Parameter Let $\mathbf{r}(t)$ be a space curve. How can you determine whether t is the arc length parameter?

Finding the Arc Length of a Plane Curve In Exercises 3–8, sketch the plane curve and find its length over the given interval.

3. $\mathbf{r}(t) = 3t\mathbf{i} - t\mathbf{j}$, $[0, 3]$ **4.** $\mathbf{r}(t) = t\mathbf{i} + t^2\mathbf{j}$, $[0, 4]$

5. $\mathbf{r}(t) = t^3\mathbf{i} + t^2\mathbf{j}$, $[0, 1]$

6. $\mathbf{r}(t) = t^2\mathbf{i} - 4t\mathbf{j}$, $[0, 5]$

7. $\mathbf{r}(t) = a\cos^3 t\mathbf{i} + a\sin^3 t\mathbf{j}$, $[0, 2\pi]$

8. $\mathbf{r}(t) = a\cos t\mathbf{i} + a\sin t\mathbf{j}$, $[0, 2\pi]$

9. Projectile Motion The position of a baseball is represented by $\mathbf{r}(t) = 50\sqrt{2}\,t\mathbf{i} + (3 + 50\sqrt{2}\,t - 16t^2)\mathbf{j}$. Find the arc length of the trajectory of the baseball.

10. Projectile Motion The position of a baseball is represented by $\mathbf{r}(t) = 40\sqrt{3}\,t\mathbf{i} + (4 + 40t - 16t^2)\mathbf{j}$. Find the arc length of the trajectory of the baseball.

 **Finding the Arc Length of a Curve in Space** In Exercises 11–16, sketch the space curve and find its length over the given interval.

11. $\mathbf{r}(t) = -t\mathbf{i} + 4t\mathbf{j} + 3t\mathbf{k}$, $[0, 1]$

12. $\mathbf{r}(t) = \mathbf{i} + t^2\mathbf{j} + t^3\mathbf{k}$, $[0, 2]$

13. $\mathbf{r}(t) = \langle 4t, -\cos t, \sin t \rangle$, $\left[0, \dfrac{3\pi}{2}\right]$

14. $\mathbf{r}(t) = \langle 2\sin t, 5t, 2\cos t \rangle$, $[0, \pi]$

15. $\mathbf{r}(t) = a\cos t\mathbf{i} + a\sin t\mathbf{j} + bt\mathbf{k}$, $[0, 2\pi]$

16. $\mathbf{r}(t) = \langle \cos t + t\sin t, \sin t - t\cos t, t^2 \rangle$, $\left[0, \dfrac{\pi}{2}\right]$

17. Investigation Consider the graph of the vector-valued function $\mathbf{r}(t) = t\mathbf{i} + (4 - t^2)\mathbf{j} + t^3\mathbf{k}$ on the interval $[0, 2]$.

(a) Approximate the length of the curve by finding the length of the line segment connecting its endpoints.

(b) Approximate the length of the curve by summing the lengths of the line segments connecting the terminal points of the vectors $\mathbf{r}(0)$, $\mathbf{r}(0.5)$, $\mathbf{r}(1)$, $\mathbf{r}(1.5)$, and $\mathbf{r}(2)$.

(c) Describe how you could obtain a more accurate approximation by continuing the processes in parts (a) and (b).

(d) Use the integration capabilities of a graphing utility to approximate the length of the curve. Compare this result with the answers in parts (a) and (b).

18. Investigation Consider the helix represented by the vector-valued function $\mathbf{r}(t) = \langle 2\cos t, 2\sin t, t \rangle$.

(a) Write the length of the arc s on the helix as a function of t by evaluating the integral

$$s = \int_0^t \sqrt{[x'(u)]^2 + [y'(u)]^2 + [z'(u)]^2}\; du.$$

(b) Solve for t in the relationship derived in part (a), and substitute the result into the original vector-valued function. This yields a parametrization of the curve in terms of the arc length parameter s.

(c) Find the coordinates of the point on the helix for arc lengths $s = \sqrt{5}$ and $s = 4$.

(d) Verify that $\|\mathbf{r}'(s)\| = 1$.

 **Finding Curvature** In Exercises 19–22, find the curvature of the curve, where s is the arc length parameter.

19. $\mathbf{r}(s) = \left(1 + \dfrac{\sqrt{2}}{2}s\right)\mathbf{i} + \left(1 - \dfrac{\sqrt{2}}{2}s\right)\mathbf{j}$

20. $\mathbf{r}(s) = (3 + s)\mathbf{i} + \mathbf{j}$

21. $\mathbf{r}(s) = \cos\dfrac{1}{2}s\mathbf{i} + \dfrac{\sqrt{3}}{2}s\mathbf{j} + \sin\dfrac{1}{2}s\mathbf{k}$

22. $\mathbf{r}(s) = \cos s\mathbf{i} + \sin s\mathbf{j} + 5\mathbf{k}$

Finding Curvature In Exercises 23–28, find the curvature of the plane curve at the given value of the parameter.

23. $\mathbf{r}(t) = 4t\mathbf{i} - 2t\mathbf{j}$, $t = 1$ **24.** $\mathbf{r}(t) = t^2\mathbf{i} + \mathbf{j}$, $t = 2$

25. $\mathbf{r}(t) = t\mathbf{i} + \dfrac{1}{t}\mathbf{j}$, $t = 1$ **26.** $\mathbf{r}(t) = t\mathbf{i} + \dfrac{1}{9}t^3\mathbf{j}$, $t = 2$

27. $\mathbf{r}(t) = \langle t, \sin t \rangle$, $t = \dfrac{\pi}{2}$

28. $\mathbf{r}(t) = \langle 5\cos t, 4\sin t \rangle$, $t = \dfrac{\pi}{3}$

 **Finding Curvature** In Exercises 29–36, find the curvature of the curve.

29. $\mathbf{r}(t) = 4\cos 2\pi t\mathbf{i} + 4\sin 2\pi t\mathbf{j}$

30. $\mathbf{r}(t) = 2\cos \pi t\mathbf{i} + \sin \pi t\mathbf{j}$

31. $\mathbf{r}(t) = a\cos \omega t\mathbf{i} + a\sin \omega t\mathbf{j}$

32. $\mathbf{r}(t) = a\cos \omega t\mathbf{i} + b\sin \omega t\mathbf{j}$

33. $\mathbf{r}(t) = t\mathbf{i} + t^2\mathbf{j} + \dfrac{t^2}{2}\mathbf{k}$

34. $\mathbf{r}(t) = 2t^2\mathbf{i} + t\mathbf{j} + \dfrac{1}{2}t^2\mathbf{k}$

35. $\mathbf{r}(t) = 4t\mathbf{i} + 3\cos t\mathbf{j} + 3\sin t\mathbf{k}$

36. $\mathbf{r}(t) = e^{2t}\mathbf{i} + e^{2t}\cos t\mathbf{j} + e^{2t}\sin t\mathbf{k}$

Finding Curvature In Exercises 37–40, find the curvature of the curve at the point P.

37. $\mathbf{r}(t) = 3t\mathbf{i} + 2t^2\mathbf{j}, \quad P(-3, 2)$

38. $\mathbf{r}(t) = e^t\mathbf{i} + 4t\mathbf{j}, \quad P(1, 0)$

39. $\mathbf{r}(t) = t\mathbf{i} + t^2\mathbf{j} + \dfrac{t^3}{4}\mathbf{k}, \quad P(2, 4, 2)$

40. $\mathbf{r}(t) = e^t\cos t\mathbf{i} + e^t\sin t\mathbf{j} + e^t\mathbf{k}, \quad P(1, 0, 1)$

 **Finding Curvature in Rectangular Coordinates** In Exercises 41–48, find the curvature and radius of curvature of the plane curve at the given value of x.

41. $y = 6x, \quad x = 3$

42. $y = x - \dfrac{4}{x}, \quad x = 2$

43. $y = 5x^2 + 7, \quad x = -1$

44. $y = 2\sqrt{9 - x^2}, \quad x = 0$

45. $y = \sin 2x, \quad x = \dfrac{\pi}{4}$

46. $y = e^{-x/4}, \quad x = 8$

47. $y = x^3, \quad x = 2$

48. $y = x^n, \quad x = 1, \quad n \geq 2$

Maximum Curvature In Exercises 49–54, (a) find the point on the curve at which the curvature is a maximum and (b) find the limit of the curvature as $x \to \infty$.

49. $y = (x - 1)^2 + 3$

50. $y = x^3$

51. $y = x^{2/3}$

52. $y = \dfrac{1}{x}$

53. $y = \ln x$

54. $y = e^x$

Curvature In Exercises 55–58, find all points on the graph of the function such that the curvature is zero.

55. $y = 1 - x^4$

56. $y = (x - 2)^6 + 3x$

57. $y = \cos\dfrac{x}{2}$

58. $y = \sin x$

EXPLORING CONCEPTS

59. Curvature Consider the function $f(x) = e^{cx}$. What value(s) of c produce a maximum curvature at $x = 0$?

60. Curvature Given a twice-differentiable function $y = f(x)$, determine its curvature at a relative extremum. Can the curvature ever be greater than it is at a relative extremum? Why or why not?

---

61. Investigation Consider the function $f(x) = x^4 - x^2$.

(a) Use a computer algebra system to find the curvature K of the curve as a function of x.

(b) Use the result of part (a) to find the circles of curvature to the graph of f when $x = 0$ and $x = 1$. Use a computer algebra system to graph the function and the two circles of curvature.

(c) Graph the function $K(x)$ and compare it with the graph of $f(x)$. For example, do the extrema of f and K occur at the same critical numbers? Explain your reasoning.

62. Motion of a Particle A particle moves along the plane curve C described by $\mathbf{r}(t) = t\mathbf{i} + t^2\mathbf{j}$.

(a) Find the length of C on the interval $0 \leq t \leq 2$.

(b) Find the curvature of C at $t = 0$, $t = 1$, and $t = 2$.

(c) Describe the curvature of C as t changes from $t = 0$ to $t = 2$.

63. Investigation Find all a and b such that the two curves given by

$$y_1 = ax(b - x) \quad \text{and} \quad y_2 = \frac{x}{x + 2}$$

intersect at only one point and have a common tangent line and equal curvature at that point. Sketch a graph for each set of values of a and b.

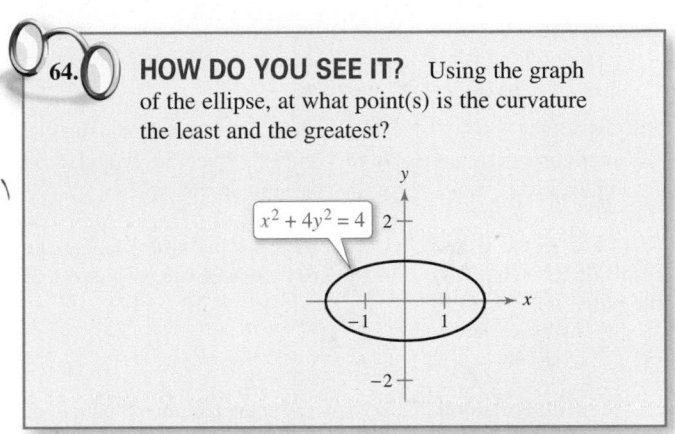

64. HOW DO YOU SEE IT? Using the graph of the ellipse, at what point(s) is the curvature the least and the greatest?

$x^2 + 4y^2 = 4$

65. Sphere and Paraboloid A sphere of radius 4 is dropped into the paraboloid given by $z = x^2 + y^2$.

(a) How close will the sphere come to the vertex of the paraboloid?

(b) What is the radius of the largest sphere that will touch the vertex?

66. Speed

The smaller the curvature of a bend in a road, the faster a car can travel. Assume that the maximum speed around a turn is inversely proportional to the square root of the curvature. A car moving on the path $y = \frac{1}{3}x^3$, where x and y are measured in miles, can safely go 30 miles per hour at $\left(1, \frac{1}{3}\right)$. How fast can it go at $\left(\frac{3}{2}, \frac{9}{8}\right)$?

67. Center of Curvature Let C be a curve given by $y = f(x)$. Let K be the curvature ($K \neq 0$) at the point $P(x_0, y_0)$ and let

$$z = \frac{1 + f'(x_0)^2}{f''(x_0)}.$$

Show that the coordinates (α, β) of the center of curvature at P are $(\alpha, \beta) = (x_0 - f'(x_0)z, y_0 + z)$.

68. Center of Curvature Use the result of Exercise 67 to find the center of curvature for the curve at the given point.

(a) $y = e^x$, $(0, 1)$ (b) $y = \dfrac{x^2}{2}$, $\left(1, \dfrac{1}{2}\right)$ (c) $y = x^2$, $(0, 0)$

69. Curvature A curve C is given by the polar equation $r = f(\theta)$. Show that the curvature K at the point (r, θ) is

$$K = \dfrac{|2(r')^2 - rr'' + r^2|}{[(r')^2 + r^2]^{3/2}}.$$

[*Hint:* Represent the curve by $\mathbf{r}(\theta) = r\cos\theta\mathbf{i} + r\sin\theta\mathbf{j}$.]

70. Curvature Use the result of Exercise 69 to find the curvature of each polar curve.

(a) $r = 1 + \sin\theta$ (b) $r = \theta$

(c) $r = a\sin\theta$ (d) $r = e^\theta$

71. Curvature Given the polar curve $r = e^{a\theta}$, $a > 0$, use the result of Exercise 69 to find the curvature K and determine the limit of K as (a) $\theta \to \infty$ and (b) $a \to \infty$.

72. Curvature at the Pole Show that the formula for the curvature of a polar curve $r = f(\theta)$ given in Exercise 69 reduces to $K = 2/|r'|$ for the curvature at the pole.

Curvature at the Pole In Exercises 73 and 74, use the result of Exercise 72 to find the curvature of the rose curve at the pole.

73. $r = 4\sin 2\theta$ **74.** $r = \cos 3\theta$

75. Proof For a smooth curve given by the parametric equations $x = f(t)$ and $y = g(t)$, prove that the curvature is given by

$$K = \dfrac{|f'(t)g''(t) - g'(t)f''(t)|}{\{[f'(t)]^2 + [g'(t)]^2\}^{3/2}}.$$

 76. Horizontal Asymptotes Use the result of Exercise 75 to find the curvature K of the curve represented by the parametric equations $x(t) = t^3$ and $y(t) = \frac{1}{2}t^2$. Use a graphing utility to graph K and determine any horizontal asymptotes. Interpret the asymptotes in the context of the problem.

77. Curvature of a Cycloid Use the result of Exercise 75 to find the curvature K of the cycloid represented by the parametric equations

$$x(\theta) = a(\theta - \sin\theta) \quad \text{and} \quad y(\theta) = a(1 - \cos\theta).$$

What are the minimum and maximum values of K?

78. Tangential and Normal Components of Acceleration Use Theorem 12.10 to find $a_{\mathbf{T}}$ and $a_{\mathbf{N}}$ for each curve given by the vector-valued function.

(a) $\mathbf{r}(t) = 3t^2\mathbf{i} + (3t - t^3)\mathbf{j}$

(b) $\mathbf{r}(t) = t\mathbf{i} + t^2\mathbf{j} + \frac{1}{2}t^2\mathbf{k}$

79. Frictional Force A 5500-pound vehicle is driven at a speed of 30 miles per hour on a circular interchange of radius 100 feet. To keep the vehicle from skidding off course, what frictional force must the road surface exert on the tires?

80. Frictional Force A 6400-pound vehicle is driven at a speed of 35 miles per hour on a circular interchange of radius 250 feet. To keep the vehicle from skidding off course, what frictional force must the road surface exert on the tires?

81. Curvature Verify that the curvature at any point (x, y) on the graph of $y = \cosh x$ is $1/y^2$.

82. Formulas for Curvature Use the definition of curvature in space, $K = \|\mathbf{T}'(s)\| = \|\mathbf{r}''(s)\|$, to verify each formula.

(a) $K = \dfrac{\|\mathbf{T}'(t)\|}{\|\mathbf{r}'(t)\|}$

(b) $K = \dfrac{\|\mathbf{r}'(t) \times \mathbf{r}''(t)\|}{\|\mathbf{r}'(t)\|^3}$

(c) $K = \dfrac{\mathbf{a}(t) \cdot \mathbf{N}(t)}{\|\mathbf{v}(t)\|^2}$

True or False? In Exercises 83–86, determine whether the statement is true or false. If it is false, explain why or give an example that shows it is false.

83. The arc length of a space curve depends on the parametrization.

84. The curvature of a plane curve at an inflection point is zero.

85. The curvature of a parabola is a maximum at its vertex.

86. The normal component of acceleration is a function of both speed and curvature.

Kepler's Laws In Exercises 87–94, you are asked to verify Kepler's Laws of Planetary Motion. For these exercises, assume that each planet moves in an orbit given by the vector-valued function **r**. Let $r = \|\mathbf{r}\|$, let G represent the universal gravitational constant, let M represent the mass of the sun, and let m represent the mass of the planet.

87. Prove that $\mathbf{r} \cdot \mathbf{r}' = r\dfrac{dr}{dt}$.

88. Using Newton's Second Law of Motion, $\mathbf{F} = m\mathbf{a}$, and Newton's Second Law of Gravitation

$$\mathbf{F} = -\dfrac{GmM}{r^3}\mathbf{r}$$

show that **a** and **r** are parallel, and that $\mathbf{r}(t) \times \mathbf{r}'(t) = \mathbf{L}$ is a constant vector. So, $\mathbf{r}(t)$ moves *in a fixed plane*, orthogonal to **L**.

89. Prove that $\dfrac{d}{dt}\left[\dfrac{\mathbf{r}}{r}\right] = \dfrac{1}{r^3}[(\mathbf{r} \times \mathbf{r}') \times \mathbf{r}]$.

90. Show that $\dfrac{\mathbf{r}'}{GM} \times \mathbf{L} - \dfrac{\mathbf{r}}{r} = \mathbf{e}$ is a constant vector.

91. Prove Kepler's First Law: Each planet moves in an elliptical orbit with the sun as a focus.

92. Assume that the elliptical orbit

$$r = \dfrac{ed}{1 + e\cos\theta}$$

is in the *xy*-plane, with **L** along the *z*-axis. Prove that

$$\|\mathbf{L}\| = r^2\dfrac{d\theta}{dt}.$$

93. Prove Kepler's Second Law: Each ray from the sun to a planet sweeps out equal areas of the ellipse in equal times.

94. Prove Kepler's Third Law: The square of the period of a planet's orbit is proportional to the cube of the mean distance between the planet and the sun.

Domain and Continuity In Exercises 1–4, (a) find the domain of **r**, and (b) determine the interval(s) on which the function is continuous.

1. $\mathbf{r}(t) = \tan t\mathbf{i} + \mathbf{j} + t\mathbf{k}$

2. $\mathbf{r}(t) = \sqrt{t}\mathbf{i} + \dfrac{1}{t-4}\mathbf{j} + \mathbf{k}$

3. $\mathbf{r}(t) = \sqrt{t^2 - 9}\mathbf{i} - \mathbf{j} + \ln(t-1)\mathbf{k}$

4. $\mathbf{r}(t) = (2t+1)\mathbf{i} + t^2\mathbf{j} + t\mathbf{k}$

Evaluating a Function In Exercises 5 and 6, evaluate the vector-valued function at each given value of t.

5. $\mathbf{r}(t) = (2t+1)\mathbf{i} + t^2\mathbf{j} - \sqrt{t+2}\mathbf{k}$

   (a) $\mathbf{r}(0)$ (b) $\mathbf{r}(-2)$ (c) $\mathbf{r}(c-1)$

   (d) $\mathbf{r}(1 + \Delta t) - \mathbf{r}(1)$

6. $\mathbf{r}(t) = 3\cos t\mathbf{i} + (1 - \sin t)\mathbf{j} - t\mathbf{k}$

   (a) $\mathbf{r}(0)$ (b) $\mathbf{r}\left(\dfrac{\pi}{2}\right)$ (c) $\mathbf{r}(s - \pi)$

   (d) $\mathbf{r}(\pi + \Delta t) - \mathbf{r}(\pi)$

Writing a Vector-Valued Function In Exercises 7 and 8, represent the line segment from P to Q by a vector-valued function and by a set of parametric equations.

7. $P(3, 0, 5), \quad Q(2, -2, 3)$

8. $P(-2, -3, 8), \quad Q(5, 1, -2)$

Sketching a Curve In Exercises 9–12, sketch the curve represented by the vector-valued function and give the orientation of the curve.

9. $\mathbf{r}(t) = \langle \pi \cos t, \pi \sin t \rangle$

10. $\mathbf{r}(t) = \langle t + 2, t^2 - 1 \rangle$

11. $\mathbf{r}(t) = (t + 1)\mathbf{i} + (3t - 1)\mathbf{j} + 2t\mathbf{k}$

12. $\mathbf{r}(t) = 2\cos t\mathbf{i} + t\mathbf{j} + 2\sin t\mathbf{k}$

Representing a Graph by a Vector-Valued Function In Exercises 13 and 14, represent the plane curve by a vector-valued function. (There are many correct answers.)

13. $3x + 4y - 12 = 0$ 14. $y = 9 - x^2$

Representing a Graph by a Vector-Valued Function In Exercises 15 and 16, sketch the space curve represented by the intersection of the surfaces. Then use the parameter $x = t$ to find a vector-valued function for the space curve.

15. $z = x^2 + y^2, \quad y = 2$

16. $x^2 + z^2 = 4, \quad x - y = 0$

Finding a Limit In Exercises 17 and 18, find the limit.

17. $\lim\limits_{t \to 3} \left(\sqrt{3 - t}\mathbf{i} + \ln t\mathbf{j} - \dfrac{1}{t}\mathbf{k} \right)$

18. $\lim\limits_{t \to 0} \left(\dfrac{\sin 2t}{t}\mathbf{i} + e^{-t}\mathbf{j} + 4\mathbf{k} \right)$

Higher-Order Differentiation In Exercises 19 and 20, find (a) $\mathbf{r}'(t)$, (b) $\mathbf{r}''(t)$, and (c) $\mathbf{r}'(t) \cdot \mathbf{r}''(t)$.

19. $\mathbf{r}(t) = (t^2 + 4t)\mathbf{i} - 3t^2\mathbf{j}$

20. $\mathbf{r}(t) = 5\cos t\mathbf{i} + 2\sin t\mathbf{j}$

Higher-Order Differentiation In Exercises 21 and 22, find (a) $\mathbf{r}'(t)$, (b) $\mathbf{r}''(t)$, (c) $\mathbf{r}'(t) \cdot \mathbf{r}''(t)$, and (d) $\mathbf{r}'(t) \times \mathbf{r}''(t)$.

21. $\mathbf{r}(t) = 2t^3\mathbf{i} + 4t\mathbf{j} - t^2\mathbf{k}$

22. $\mathbf{r}(t) = (4t + 3)\mathbf{i} + t^2\mathbf{j} + (2t^2 + 4)\mathbf{k}$

Finding Intervals on Which a Curve is Smooth In Exercises 23 and 24, find the open interval(s) on which the curve given by the vector-valued function is smooth.

23. $\mathbf{r}(t) = (t - 1)^3\mathbf{i} + (t - 1)^4\mathbf{j}$

24. $\mathbf{r}(t) = \dfrac{t}{t - 2}\mathbf{i} + t\mathbf{j} + \sqrt{1 + t}\mathbf{k}$

Using Properties of the Derivative In Exercises 25 and 26, use the properties of the derivative to find the following.

  (a) $\mathbf{r}'(t)$ (b) $\dfrac{d}{dt}[\mathbf{u}(t) - 2\mathbf{r}(t)]$ (c) $\dfrac{d}{dt}[(3t)\mathbf{r}(t)]$

  (d) $\dfrac{d}{dt}[\mathbf{r}(t) \cdot \mathbf{u}(t)]$ (e) $\dfrac{d}{dt}[\mathbf{r}(t) \times \mathbf{u}(t)]$ (f) $\dfrac{d}{dt}[\mathbf{u}(2t)]$

25. $\mathbf{r}(t) = 3t\mathbf{i} + (t - 1)\mathbf{j}, \quad \mathbf{u}(t) = t\mathbf{i} + t^2\mathbf{j} + \frac{2}{3}t^3\mathbf{k}$

26. $\mathbf{r}(t) = \sin t\mathbf{i} + \cos t\mathbf{j} + t\mathbf{k}, \quad \mathbf{u}(t) = \sin t\mathbf{i} + \cos t\mathbf{j} + \dfrac{1}{t}\mathbf{k}$

Finding an Indefinite Integral In Exercises 27–30, find the indefinite integral.

27. $\displaystyle\int (t^2\mathbf{i} + 5t\mathbf{j} + 8t^3\mathbf{k})\, dt$ 28. $\displaystyle\int (6\mathbf{i} - 2t\mathbf{j} + \ln t\mathbf{k})\, dt$

29. $\displaystyle\int \left(3\sqrt{t}\mathbf{i} + \dfrac{2}{t}\mathbf{j} + \mathbf{k} \right) dt$ 30. $\displaystyle\int (\sin t\mathbf{i} + \mathbf{j} + e^{2t}\mathbf{k})\, dt$

Evaluating a Definite Integral In Exercises 31–34, evaluate the definite integral.

31. $\displaystyle\int_{-2}^{2} (3t\mathbf{i} + 2t^2\mathbf{j} - t^3\mathbf{k})\, dt$

32. $\displaystyle\int_{0}^{3} (t\mathbf{i} + \sqrt{t}\mathbf{j} + 4t\mathbf{k})\, dt$

33. $\displaystyle\int_{0}^{2} (e^{t/2}\mathbf{i} - 3t^2\mathbf{j} - \mathbf{k})\, dt$

34. $\displaystyle\int_{0}^{\pi/3} (2\cos t\mathbf{i} + \sin t\mathbf{j} + 3\mathbf{k})\, dt$

Finding an Antiderivative In Exercises 35 and 36, find $\mathbf{r}(t)$ that satisfies the initial condition(s).

35. $\mathbf{r}'(t) = 2t\mathbf{i} + e^t\mathbf{j} + e^{-t}\mathbf{k}, \quad \mathbf{r}(0) = \mathbf{i} + 3\mathbf{j} - 5\mathbf{k}$

36. $\mathbf{r}'(t) = \sec t\mathbf{i} + \tan t\mathbf{j} + t^2\mathbf{k}, \quad r(0) = 3\mathbf{k}$

Finding Velocity and Acceleration Vectors in Space In Exercises 37–40, the position vector r describes the path of an object moving in space. (a) Find the velocity vector, speed, and acceleration vector of the object. (b) Evaluate the velocity vector and acceleration vector of the object at the given value of t.

Position Vector	Time
37. $\mathbf{r}(t) = 4t\mathbf{i} + t^3\mathbf{j} - t\mathbf{k}$	$t = 1$
38. $\mathbf{r}(t) = \sqrt{t}\mathbf{i} + 5t\mathbf{j} + 2t^2\mathbf{k}$	$t = 4$
39. $\mathbf{r}(t) = \langle \cos^3 t, \sin^3 t, 3t \rangle$	$t = \pi$
40. $\mathbf{r}(t) = \langle t, -\tan t, e^t \rangle$	$t = 0$

Projectile Motion In Exercises 41 and 42, use the model for projectile motion, assuming there is no air resistance and $g = 32$ feet per second per second.

41. A baseball is hit from a height of 3.5 feet above the ground with an initial speed of 120 feet per second and at an angle of 30° above the horizontal. Find the maximum height reached by the baseball. Determine whether it will clear an 8-foot-high fence located 375 feet from home plate.

42. Determine the maximum height and range of a projectile fired at a height of 6 feet above the ground with an initial speed of 400 feet per second and an angle of 60° above the horizontal.

Finding the Unit Tangent Vector In Exercises 43 and 44, find the unit tangent vector to the curve at the specified value of the parameter.

43. $\mathbf{r}(t) = 6t\mathbf{i} - t^2\mathbf{j}, \quad t = 2$

44. $\mathbf{r}(t) = 2 \sin t\mathbf{i} + 4 \cos t\mathbf{j}, \quad t = \dfrac{\pi}{6}$

Finding a Tangent Line In Exercises 45 and 46, find the unit tangent vector $\mathbf{T}(t)$ and a set of parametric equations for the tangent line to the space curve at point P.

45. $\mathbf{r}(t) = e^{2t}\mathbf{i} + \cos t\mathbf{j} - \sin 3t\mathbf{k}, \quad P(1, 1, 0)$

46. $\mathbf{r}(t) = t\mathbf{i} + t^2\mathbf{j} + \frac{2}{3}t^3\mathbf{k}, \quad P\left(2, 4, \frac{16}{3}\right)$

Finding the Principal Unit Normal Vector In Exercises 47–50, find the principal unit normal vector to the curve at the specified value of the parameter.

47. $\mathbf{r}(t) = 2t\mathbf{i} + 3t^2\mathbf{j}, \quad t = 1$ **48.** $\mathbf{r}(t) = t\mathbf{i} + \ln t\mathbf{j}, \quad t = 2$

49. $\mathbf{r}(t) = 3 \cos 2t\mathbf{i} + 3 \sin 2t\mathbf{j} + 3\mathbf{k}, \quad t = \dfrac{\pi}{4}$

50. $\mathbf{r}(t) = 4 \cos t\mathbf{i} + 4 \sin t\mathbf{j} + \mathbf{k}, \quad t = \dfrac{2\pi}{3}$

Finding Tangential and Normal Components of Acceleration In Exercises 51 and 52, find the tangential and normal components of acceleration at the given time t for the plane curve $\mathbf{r}(t)$.

51. $\mathbf{r}(t) = \dfrac{3}{t}\mathbf{i} - 6t\mathbf{j}, \quad t = 3$

52. $\mathbf{r}(t) = 3 \cos 2t\mathbf{i} + 3 \sin 2t\mathbf{j}, \quad t = \dfrac{\pi}{6}$

Finding Tangential and Normal Components of Acceleration In Exercises 53 and 54, find the tangential and normal components of acceleration at the given time t for the space curve $\mathbf{r}(t)$.

53. $\mathbf{r}(t) = \sin t\mathbf{i} - 3t\mathbf{j} + \cos t\mathbf{k}, \quad t = \dfrac{\pi}{6}$

54. $\mathbf{r}(t) = \dfrac{t^3}{3}\mathbf{i} - 6t\mathbf{j} + t^2\mathbf{k}, \quad t = 2$

Finding the Arc Length of a Plane Curve In Exercises 55–58, sketch the plane curve and find its length over the given interval.

55. $\mathbf{r}(t) = 2t\mathbf{i} - 3t\mathbf{j}, \quad [0, 5]$

56. $\mathbf{r}(t) = t^2\mathbf{i} + 2t\mathbf{k}, \quad [0, 3]$

57. $\mathbf{r}(t) = 2 \sin t\mathbf{i} + \mathbf{j}, \quad \left[\dfrac{\pi}{2}, \pi\right]$

58. $\mathbf{r}(t) = 10 \cos t\mathbf{i} + 10 \sin t\mathbf{j}, \quad [0, 2\pi]$

Finding the Arc Length of a Curve in Space In Exercises 59–62, sketch the space curve and find its length over the given interval.

59. $\mathbf{r}(t) = -3t\mathbf{i} + 2t\mathbf{j} + 4t\mathbf{k}, \quad [0, 3]$

60. $\mathbf{r}(t) = t\mathbf{i} + t^2\mathbf{j} + 2t\mathbf{k}, \quad [0, 2]$

61. $\mathbf{r}(t) = \langle 8 \cos t, 8 \sin t, t \rangle, \quad \left[0, \dfrac{\pi}{2}\right]$

62. $\mathbf{r}(t) = \langle 2(\sin t - t \cos t), 2(\cos t + t \sin t), t \rangle, \quad \left[0, \dfrac{\pi}{2}\right]$

Finding Curvature In Exercises 63–66, find the curvature of the curve.

63. $\mathbf{r}(t) = 3t\mathbf{i} + 2t\mathbf{j}$

64. $\mathbf{r}(t) = 2\sqrt{t}\mathbf{i} + 3t\mathbf{j}$

65. $\mathbf{r}(t) = 2t\mathbf{i} + \frac{1}{2}t^2\mathbf{j} + t^2\mathbf{k}$

66. $\mathbf{r}(t) = 2t\mathbf{i} + 5 \cos t\mathbf{j} + 5 \sin t\mathbf{k}$

Finding Curvature In Exercises 67 and 68, find the curvature of the curve at the point P.

67. $\mathbf{r}(t) = \frac{1}{2}t^2\mathbf{i} + t\mathbf{j} + \frac{1}{3}t^3\mathbf{k}, \quad P\left(\frac{1}{2}, 1, \frac{1}{3}\right)$

68. $\mathbf{r}(t) = 4 \cos t\mathbf{i} + 3 \sin t\mathbf{j} + t\mathbf{k}, \quad P(-4, 0, \pi)$

Finding Curvature in Rectangular Coordinates In Exercises 69–72, find the curvature and radius of curvature of the plane curve at the given value of x.

69. $y = \frac{1}{2}x^2 + x, \quad x = 4$ **70.** $y = e^{-x/2}, \quad x = 0$

71. $y = \ln x, \quad x = 1$

72. $y = \tan x, \quad x = \dfrac{\pi}{4}$

73. Frictional Force A 7200-pound vehicle is driven at a speed of 25 miles per hour on a circular interchange of radius 150 feet. To keep the vehicle from skidding off course, what frictional force must the road surface exert on the tires?

P.S. Problem Solving

See **CalcChat.com** for tutorial help and worked-out solutions to odd-numbered exercises.

1. **Cornu Spiral** The **cornu spiral** is given by

$$x(t) = \int_0^t \cos\left(\frac{\pi u^2}{2}\right) du \quad \text{and} \quad y(t) = \int_0^t \sin\left(\frac{\pi u^2}{2}\right) du.$$

The spiral shown in the figure was plotted over the interval $-\pi \le t \le \pi$.

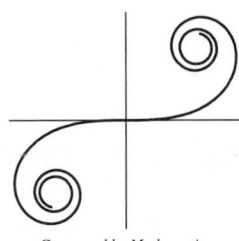

Generated by Mathematica

(a) Find the arc length of this curve from $t = 0$ to $t = a$.

(b) Find the curvature of the graph when $t = a$.

(c) The cornu spiral was discovered by James Bernoulli. He found that the spiral has an amazing relationship between curvature and arc length. What is this relationship?

2. **Radius of Curvature** Let T be the tangent line at the point $P(x, y)$ on the graph of the curve $x^{2/3} + y^{2/3} = a^{2/3}$, $a > 0$, as shown in the figure. Show that the radius of curvature at P is three times the distance from the origin to the tangent line T.

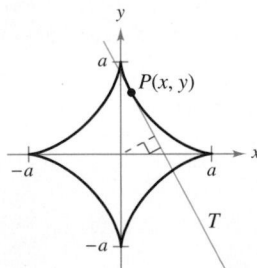

3. **Projectile Motion** A bomber is flying horizontally at an altitude of 3200 feet with a speed of 400 feet per second when it releases a bomb. A projectile is launched 5 seconds later from a cannon at a site facing the bomber and 5000 feet from the point that was directly beneath the bomber when the bomb was released, as shown in the figure. The projectile is to intercept the bomb at an altitude of 1600 feet. Determine the required initial speed and angle of inclination of the projectile. (Ignore air resistance.)

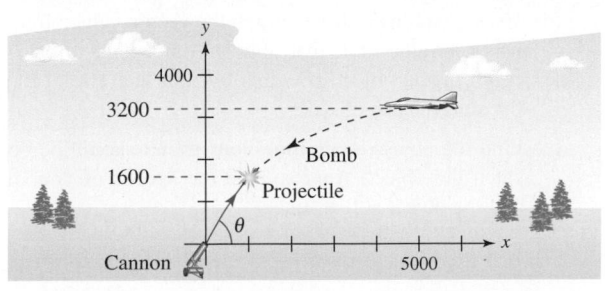

4. **Projectile Motion** Repeat Exercise 3 for the case in which the bomber is facing *away* from the launch site, as shown in the figure.

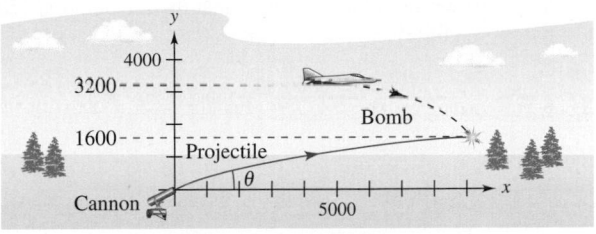

5. **Cycloid** Consider one arch of the cycloid

$$\mathbf{r}(\theta) = (\theta - \sin\theta)\mathbf{i} + (1 - \cos\theta)\mathbf{j}, \quad 0 \le \theta \le 2\pi$$

as shown in the figure. Let $s(\theta)$ be the arc length from the highest point on the arch to the point $(x(\theta), y(\theta))$, and let $\rho(\theta) = 1/K$ be the radius of curvature at the point $(x(\theta), y(\theta))$. Show that s and ρ are related by the equation $s^2 + \rho^2 = 16$. (This equation is called a *natural equation* for the curve.)

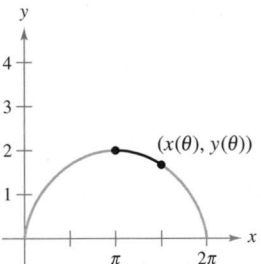

6. **Cardioid** Consider the cardioid

$$r = 1 - \cos\theta, \quad 0 \le \theta \le 2\pi$$

as shown in the figure. Let $s(\theta)$ be the arc length from the point $(2, \pi)$ on the cardioid to the point (r, θ), and let $\rho(\theta) = 1/K$ be the radius of curvature at the point (r, θ). Show that s and ρ are related by the equation $s^2 + 9\rho^2 = 16$. (This equation is called a *natural equation* for the curve.)

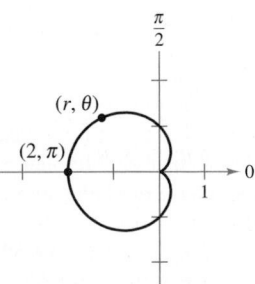

7. **Proof** If $\mathbf{r}(t)$ is a nonzero differentiable function of t, prove that

$$\frac{d}{dt}\big[\|\mathbf{r}(t)\|\big] = \frac{\mathbf{r}(t) \cdot \mathbf{r}'(t)}{\|\mathbf{r}(t)\|}.$$

8. Satellite A communications satellite moves in a circular orbit around Earth at a distance of 42,000 kilometers from the center of Earth. The angular speed

$$\frac{d\theta}{dt} = \omega = \frac{\pi}{12} \text{ radian per hour}$$

is constant.

(a) Use polar coordinates to show that the acceleration vector is given by

$$\mathbf{a} = \frac{d^2\mathbf{r}}{dt^2} = \left[\frac{d^2r}{dt^2} - r\left(\frac{d\theta}{dt}\right)^2\right]\mathbf{u_r} + \left[r\frac{d^2\theta}{dt^2} + 2\frac{dr}{dt}\frac{d\theta}{dt}\right]\mathbf{u_\theta}$$

where $\mathbf{u_r} = \cos\theta\mathbf{i} + \sin\theta\mathbf{j}$ is the unit vector in the radial direction and $\mathbf{u_\theta} = -\sin\theta\mathbf{i} + \cos\theta\mathbf{j}$.

(b) Find the radial and angular components of acceleration for the satellite.

Binormal Vector **In Exercises 9–11, use the binormal vector defined by the equation $\mathbf{B} = \mathbf{T} \times \mathbf{N}$.**

9. Find the unit tangent, principal unit normal, and binormal vectors for the helix

$$\mathbf{r}(t) = 4\cos t\mathbf{i} + 4\sin t\mathbf{j} + 3t\mathbf{k}$$

at $t = \pi/2$. Sketch the helix together with these three mutually orthogonal unit vectors.

10. Find the unit tangent, principal unit normal, and binormal vectors for the curve

$$\mathbf{r}(t) = \cos t\mathbf{i} + \sin t\mathbf{j} - \mathbf{k}$$

at $t = \pi/4$. Sketch the curve together with these three mutually orthogonal unit vectors.

11. (a) Prove that there exists a scalar τ, called the **torsion,** such that $d\mathbf{B}/ds = -\tau\mathbf{N}$.

(b) Prove that $\dfrac{d\mathbf{N}}{ds} = -K\mathbf{T} + \tau\mathbf{B}$.

(The three equations $d\mathbf{T}/ds = K\mathbf{N}$, $d\mathbf{N}/ds = -K\mathbf{T} + \tau\mathbf{B}$, and $d\mathbf{B}/ds = -\tau\mathbf{N}$ are called the *Frenet-Serret formulas.*)

12. Exit Ramp A highway has an exit ramp that begins at the origin of a coordinate system and follows the curve

$$y = \frac{1}{32}x^{5/2}$$

to the point $(4, 1)$ (see figure). Then it follows a circular path whose curvature is that given by the curve at $(4, 1)$. What is the radius of the circular arc? Explain why the curve and the circular arc should have the same curvature at $(4, 1)$.

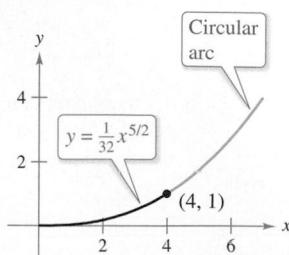

13. Arc Length and Curvature Consider the vector-valued function

$$\mathbf{r}(t) = \langle t\cos\pi t, t\sin\pi t\rangle, \quad 0 \le t \le 2.$$

(a) Use a graphing utility to graph the function.

(b) Find the length of the arc in part (a).

(c) Find the curvature K as a function of t. Find the curvature at $t = 0$, $t = 1$, and $t = 2$.

(d) Use a graphing utility to graph the function K.

(e) Find (if possible) $\lim\limits_{t\to\infty} K$.

(f) Using the result of part (e), make a conjecture about the graph of $\mathbf{r}$ as $t \to \infty$.

14. Ferris Wheel You want to toss an object to a friend who is riding a Ferris wheel (see figure). The following parametric equations give the path of the friend $\mathbf{r_1}(t)$ and the path of the object $\mathbf{r_2}(t)$. Distance is measured in meters, and time is measured in seconds.

$$\mathbf{r_1}(t) = 15\left(\sin\frac{\pi t}{10}\right)\mathbf{i} + \left(16 - 15\cos\frac{\pi t}{10}\right)\mathbf{j}$$

$$\mathbf{r_2}(t) = [22 - 8.03(t - t_0)]\mathbf{i} + [1 + 11.47(t - t_0) - 4.9(t - t_0)^2]\mathbf{j}$$

(a) Locate your friend's position on the Ferris wheel at time $t = 0$.

(b) Determine the number of revolutions per minute of the Ferris wheel.

(c) What are the speed and angle of inclination (in degrees) at which the object is thrown at time $t = t_0$?

(d) Use a graphing utility to graph the vector-valued functions using a value of t_0 that allows your friend to be within reach of the object. (Do this by trial and error.) Explain the significance of t_0.

(e) Find the approximate time your friend should be able to catch the object. Approximate the speeds of your friend and the object at that time.

13 Functions of Several Variables

Ocean Floor *(Exercise 66, p. 930)*

Manufacturing
(Example 2, p. 959)

Wind Chill *(Exercise 27, p. 910)*

Marginal Costs
(Exercise 118, p. 902)

Forestry *(Exercise 77, p. 882)*

13.1 Introduction to Functions of Several Variables

- Understand the notation for a function of several variables.
- Sketch the graph of a function of two variables.
- Sketch level curves for a function of two variables.
- Sketch level surfaces for a function of three variables.
- Use computer graphics to graph a function of two variables.

Functions of Several Variables

So far in this text, you have dealt only with functions of a single (independent) variable. Many familiar quantities, however, are functions of two or more variables. Here are three examples.

1. The work done by a force, $W = FD$, is a function of two variables.
2. The volume of a right circular cylinder, $V = \pi r^2 h$, is a function of two variables.
3. The volume of a rectangular solid, $V = lwh$, is a function of three variables.

The notation for a function of two or more variables is similar to that for a function of a single variable. Here are two examples.

$$z = f(\underbrace{x, y}_{\text{2 variables}}) = x^2 + xy \qquad \text{Function of two variables}$$

and

$$w = f(\underbrace{x, y, z}_{\text{3 variables}}) = x + 2y - 3z \qquad \text{Function of three variables}$$

Exploration

Without using a graphing utility, describe the graph of each function of two variables.

a. $z = x^2 + y^2$
b. $z = x + y$
c. $z = x^2 + y$
d. $z = \sqrt{x^2 + y^2}$
e. $z = \sqrt{1 - x^2 + y^2}$

MARY FAIRFAX SOMERVILLE (1780–1872)

Somerville was interested in the problem of creating geometric models for functions of several variables. Her most well-known book, *The Mechanics of the Heavens*, was published in 1831.
See LarsonCalculus.com to read more of this biography.

Definition of a Function of Two Variables

Let D be a set of ordered pairs of real numbers. If to each ordered pair (x, y) in D there corresponds a unique real number $f(x, y)$, then f is a **function of x and y**. The set D is the **domain** of f, and the corresponding set of values for $f(x, y)$ is the **range** of f. For the function

$$z = f(x, y)$$

x and y are called the **independent variables** and z is called the **dependent variable**.

Similar definitions can be given for functions of three, four, or n variables, where the domains consist of ordered triples (x_1, x_2, x_3), quadruples (x_1, x_2, x_3, x_4), and n-tuples $(x_1, x_2, \ldots, x_n)$. In all cases, the range is a set of real numbers. In this chapter, you will study only functions of two or three variables.

As with functions of one variable, the most common way to describe a function of several variables is with an *equation*, and unless it is otherwise restricted, you can assume that the domain is the set of all points for which the equation is defined. For instance, the domain of the function

$$f(x, y) = x^2 + y^2$$

is the entire xy-plane. Similarly, the domain of

$$f(x, y) = \ln xy$$

is the set of all points (x, y) in the plane for which $xy > 0$. This consists of all points in the first and third quadrants.

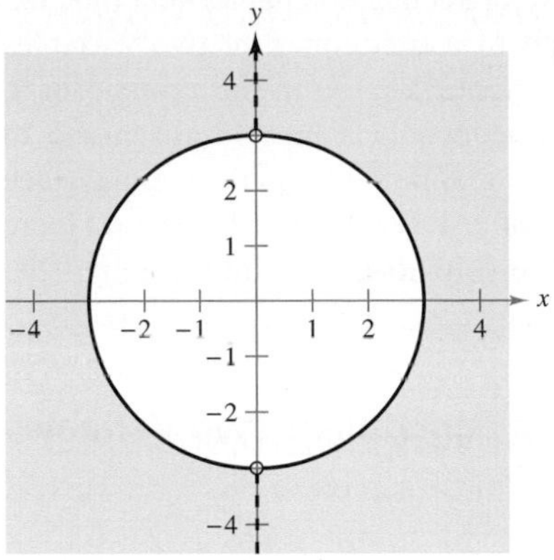

Domain of
$$f(x, y) = \frac{\sqrt{x^2 + y^2 - 9}}{x}$$

Figure 13.1

EXAMPLE 1 Domains of Functions of Several Variables

Find the domain of each function.

a. $f(x, y) = \dfrac{\sqrt{x^2 + y^2 - 9}}{x}$ **b.** $g(x, y, z) = \dfrac{x}{\sqrt{9 - x^2 - y^2 - z^2}}$

Solution

a. The function f is defined for all points (x, y) such that $x \neq 0$ and
$$x^2 + y^2 \geq 9.$$
So, the domain is the set of all points lying on or outside the circle $x^2 + y^2 = 9$ *except* those points on the y-axis, as shown in Figure 13.1.

b. The function g is defined for all points (x, y, z) such that
$$x^2 + y^2 + z^2 < 9.$$
Consequently, the domain is the set of all points (x, y, z) lying inside a sphere of radius 3 that is centered at the origin. ■

Functions of several variables can be combined in the same ways as functions of single variables. For instance, you can form the sum, difference, product, and quotient of two functions of two variables as follows.

$$(f \pm g)(x, y) = f(x, y) \pm g(x, y) \qquad \text{Sum or difference}$$
$$(fg)(x, y) = f(x, y)g(x, y) \qquad \text{Product}$$
$$\frac{f}{g}(x, y) = \frac{f(x, y)}{g(x, y)}, \quad g(x, y) \neq 0 \qquad \text{Quotient}$$

You cannot form the composite of two functions of several variables. You can, however, form the **composite** function $(g \circ h)(x, y)$, where g is a function of a single variable and h is a function of two variables.

$$(g \circ h)(x, y) = g(h(x, y)) \qquad \text{Composition}$$

The domain of this composite function consists of all (x, y) in the domain of h such that $h(x, y)$ is in the domain of g. For example, the function
$$f(x, y) = \sqrt{16 - 4x^2 - y^2}$$
can be viewed as the composite of the function of two variables given by
$$h(x, y) = 16 - 4x^2 - y^2$$
and the function of a single variable given by
$$g(u) = \sqrt{u}.$$
The domain of this function is the set of all points lying on or inside the ellipse $4x^2 + y^2 = 16$.

A function that can be written as a sum of functions of the form $cx^m y^n$ (where c is a real number and m and n are nonnegative integers) is called a **polynomial function** of two variables. For instance, the functions
$$f(x, y) = x^2 + y^2 - 2xy + x + 2 \quad \text{and} \quad g(x, y) = 3xy^2 + x - 2$$
are polynomial functions of two variables. A **rational function** is the quotient of two polynomial functions. Similar terminology is used for functions of more than two variables.

The Graph of a Function of Two Variables

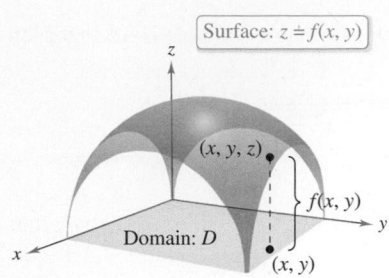

Figure 13.2

As with functions of a single variable, you can learn a lot about the behavior of a function of two variables by sketching its graph. The **graph** of a function f of two variables is the set of all points (x, y, z) for which $z = f(x, y)$ and (x, y) is in the domain of f. This graph can be interpreted geometrically as a *surface in space,* as discussed in Sections 11.5 and 11.6. In Figure 13.2, note that the graph of $z = f(x, y)$ is a surface whose projection onto the xy-plane is D, the domain of f. To each point (x, y) in D there corresponds a point (x, y, z) on the surface, and, conversely, to each point (x, y, z) on the surface there corresponds a point (x, y) in D.

EXAMPLE 2   Describing the Graph of a Function of Two Variables

Consider the function given by

$$f(x, y) = \sqrt{16 - 4x^2 - y^2}.$$

a. Find the domain and range of the function.

b. Describe the graph of f.

Solution

a. The domain D implied by the equation of f is the set of all points (x, y) such that

$$16 - 4x^2 - y^2 \geq 0.$$

So, D is the set of all points lying on or inside the ellipse

$$\frac{x^2}{4} + \frac{y^2}{16} = 1. \qquad \text{Ellipse in the } xy\text{-plane}$$

The range of f is all values $z = f(x, y)$ such that $0 \leq z \leq \sqrt{16}$, or

$$0 \leq z \leq 4. \qquad \text{Range of } f$$

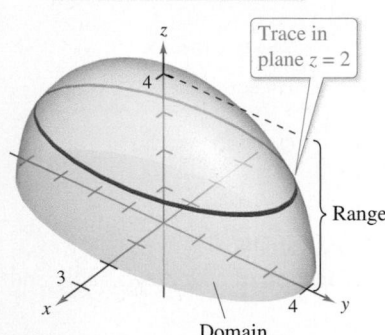

The graph of
$f(x, y) = \sqrt{16 - 4x^2 - y^2}$ is the
upper half of an ellipsoid.
Figure 13.3

b. A point (x, y, z) is on the graph of f if and only if

$$z = \sqrt{16 - 4x^2 - y^2}$$
$$z^2 = 16 - 4x^2 - y^2$$
$$4x^2 + y^2 + z^2 = 16$$
$$\frac{x^2}{4} + \frac{y^2}{16} + \frac{z^2}{16} = 1, \quad 0 \leq z \leq 4.$$

From Section 11.6, you know that the graph of f is the upper half of an ellipsoid, as shown in Figure 13.3.

To sketch a surface in space *by hand,* it helps to use traces in planes parallel to the coordinate planes, as shown in Figure 13.3. For example, to find the trace of the surface in the plane $z = 2$, substitute $z = 2$ in the equation $z = \sqrt{16 - 4x^2 - y^2}$ and obtain

$$2 = \sqrt{16 - 4x^2 - y^2} \implies \frac{x^2}{3} + \frac{y^2}{12} = 1.$$

So, the trace is an ellipse centered at the point $(0, 0, 2)$ with major and minor axes of lengths $4\sqrt{3}$ and $2\sqrt{3}$.

Traces are also used with most three-dimensional graphing utilities. For instance, Figure 13.4 shows a computer-generated version of the surface given in Example 2. For this graph, the computer took 25 traces parallel to the xy-plane and 12 traces in vertical planes.

If you have access to a three-dimensional graphing utility, use it to graph several surfaces.

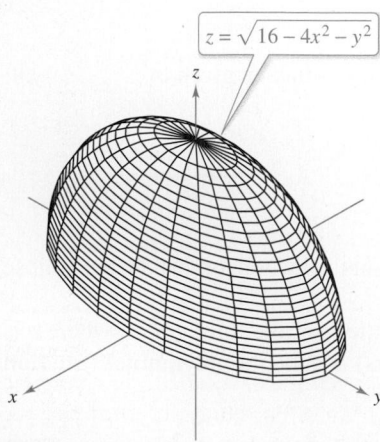

Figure 13.4

Level Curves

A second way to visualize a function of two variables is to use a **scalar field** in which the scalar

$$z = f(x, y)$$

is assigned to the point (x, y). A scalar field can be characterized by **level curves** (or **contour lines**) along which the value of $f(x, y)$ is constant. For instance, the weather map in Figure 13.5 shows level curves of equal pressure called **isobars.** In weather maps for which the level curves represent points of equal temperature, the level curves are called **isotherms,** as shown in Figure 13.6. Another common use of level curves is in representing electric potential fields. In this type of map, the level curves are called **equipotential lines.**

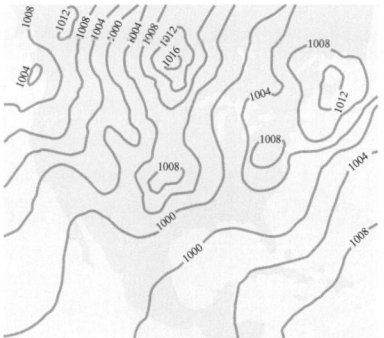

Level curves show the lines of equal pressure (isobars), measured in millibars.

Figure 13.5

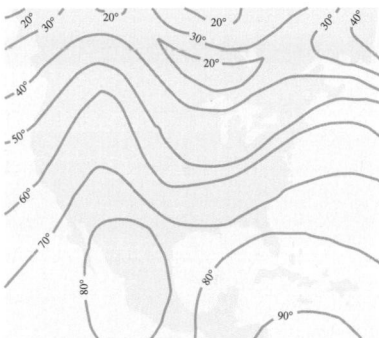

Level curves show the lines of equal temperature (isotherms), measured in degrees Fahrenheit.

Figure 13.6

Contour maps are commonly used to show regions on Earth's surface, with the level curves representing the height above sea level. This type of map is called a **topographic map.** For example, the mountain shown in Figure 13.7 is represented by the topographic map in Figure 13.8.

Figure 13.7

Figure 13.8

A contour map depicts the variation of z with respect to x and y by the spacing between level curves. Much space between level curves indicates that z is changing slowly, whereas little space indicates a rapid change in z. Furthermore, to produce a good three-dimensional illusion in a contour map, it is important to choose c-values that are *evenly spaced.*

EXAMPLE 3 **Sketching a Contour Map**

The hemisphere

$$f(x, y) = \sqrt{64 - x^2 - y^2}$$

is shown in Figure 13.9. Sketch a contour map of this surface using level curves corresponding to $c = 0, 1, 2, \ldots, 8$.

Solution For each value of c, the equation $f(x, y) = c$ is a circle (or point) in the xy-plane. For example, when $c_1 = 0$, the level curve is

$$x^2 + y^2 = 64 \qquad \text{Circle of radius 8}$$

which is a circle of radius 8. Figure 13.10 shows the nine level curves for the hemisphere.

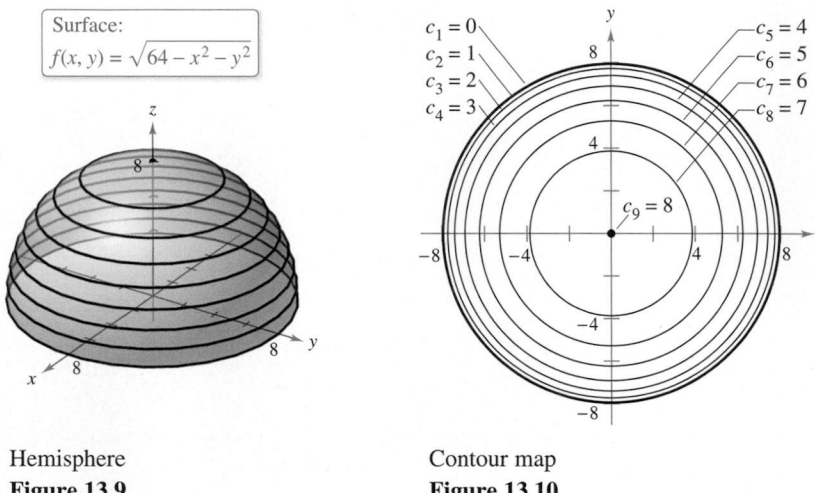

Hemisphere
Figure 13.9

Contour map
Figure 13.10

EXAMPLE 4 **Sketching a Contour Map**

$\vdots \cdots \triangleright$ *See LarsonCalculus.com for an interactive version of this type of example.*

The hyperbolic paraboloid

$$z = y^2 - x^2$$

is shown in Figure 13.11. Sketch a contour map of this surface.

Solution For each value of c, let $f(x, y) = c$ and sketch the resulting level curve in the xy-plane. For this function, each of the level curves ($c \neq 0$) is a hyperbola whose asymptotes are the lines $y = \pm x$. When $c < 0$, the transverse axis is horizontal. For instance, the level curve for $c = -4$ is

$$\frac{x^2}{2^2} - \frac{y^2}{2^2} = 1.$$

When $c > 0$, the transverse axis is vertical. For instance, the level curve for $c = 4$ is

$$\frac{y^2}{2^2} - \frac{x^2}{2^2} = 1.$$

When $c = 0$, the level curve is the degenerate conic representing the intersecting asymptotes, as shown in Figure 13.12.

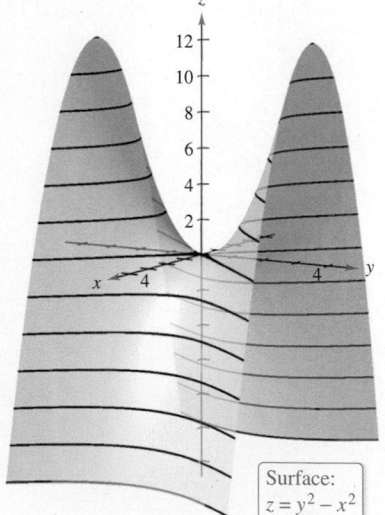

Hyperbolic paraboloid
Figure 13.11

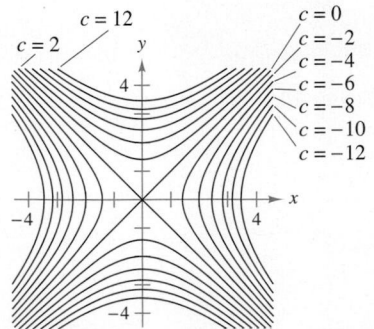

Hyperbolic level curves (at increments of 2)
Figure 13.12

One example of a function of two variables used in economics is the **Cobb-Douglas production function.** This function is used as a model to represent the numbers of units produced by varying amounts of labor and capital. If x measures the units of labor and y measures the units of capital, then the number of units produced is

$$f(x, y) = Cx^a y^{1-a}$$

where C and a are constants with $0 < a < 1$.

EXAMPLE 5   **The Cobb-Douglas Production Function**

A manufacturer estimates a production function to be

$$f(x, y) = 100x^{0.6}y^{0.4}$$

where x is the number of units of labor and y is the number of units of capital. Compare the production level when $x = 1000$ and $y = 500$ with the production level when $x = 2000$ and $y = 1000$.

Solution   When $x = 1000$ and $y = 500$, the production level is

$$f(1000, 500) = 100(1000^{0.6})(500^{0.4})$$
$$\approx 75{,}786.$$

When $x = 2000$ and $y = 1000$, the production level is

$$f(2000, 1000) = 100(2000^{0.6})(1000^{0.4})$$
$$\approx 151{,}572.$$

The level curves of $z = f(x, y)$ are shown in Figure 13.13. Note that by doubling both x and y, you double the production level (see Exercise 83). ■

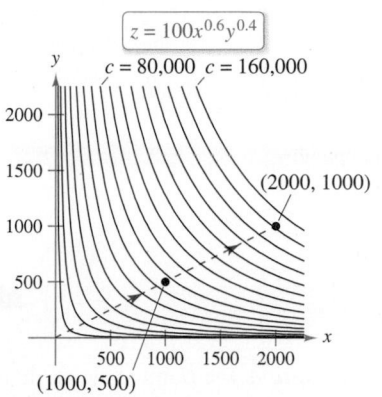

$z = 100x^{0.6}y^{0.4}$

$c = 80{,}000 \quad c = 160{,}000$

$(2000, 1000)$

$(1000, 500)$

Level curves (at increments of 10,000)
Figure 13.13

Level Surfaces

The concept of a level curve can be extended by one dimension to define a **level surface.** If f is a function of three variables and c is a constant, then the graph of the equation

$$f(x, y, z) = c$$

is a **level surface** of f, as shown in Figure 13.14.

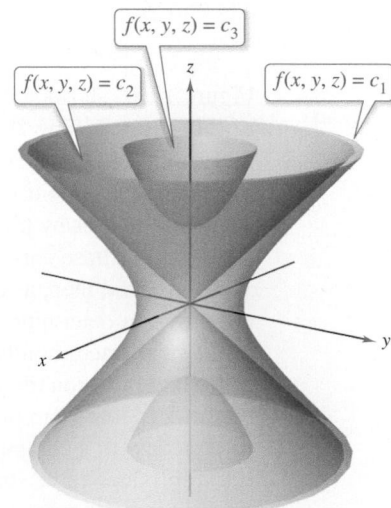

$f(x, y, z) = c_3$

$f(x, y, z) = c_2$

$f(x, y, z) = c_1$

Level surfaces of f
Figure 13.14

EXAMPLE 6 Level Surfaces

Describe the level surfaces of

$$f(x, y, z) = 4x^2 + y^2 + z^2.$$

Solution Each level surface has an equation of the form

$$4x^2 + y^2 + z^2 = c. \qquad \text{Equation of level surface}$$

So, the level surfaces are ellipsoids (whose cross sections parallel to the yz-plane are circles). As c increases, the radii of the circular cross sections increase according to the square root of c. For example, the level surfaces corresponding to the values $c = 0$, $c = 4$, and $c = 16$ are as follows.

$$4x^2 + y^2 + z^2 = 0 \qquad \text{Level surface for } c = 0 \text{ (single point)}$$

$$\frac{x^2}{1} + \frac{y^2}{4} + \frac{z^2}{4} = 1 \qquad \text{Level surface for } c = 4 \text{ (ellipsoid)}$$

$$\frac{x^2}{4} + \frac{y^2}{16} + \frac{z^2}{16} = 1 \qquad \text{Level surface for } c = 16 \text{ (ellipsoid)}$$

These level surfaces are shown in Figure 13.15. ■

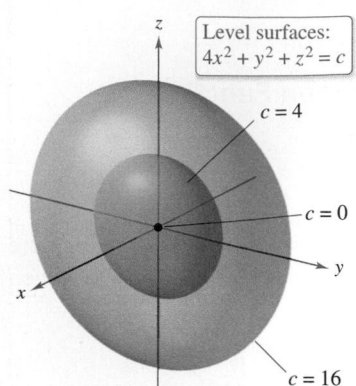

Level surfaces:
$4x^2 + y^2 + z^2 = c$

$c = 4$

$c = 0$

$c = 16$

Figure 13.15

If the function in Example 6 represented the *temperature* at the point (x, y, z), then the level surfaces shown in Figure 13.15 would be called **isothermal surfaces.**

Computer Graphics

The problem of sketching the graph of a surface in space can be simplified by using a computer. Although there are several types of three-dimensional graphing utilities, most use some form of trace analysis to give the illusion of three dimensions. To use such a graphing utility, you usually need to enter the equation of the surface and the region in the xy-plane over which the surface is to be plotted. (You might also need to enter the number of traces to be taken.) For instance, to graph the surface

$$f(x, y) = (x^2 + y^2)e^{1 - x^2 - y^2}$$

you might choose the following bounds for x, y, and z.

$$-3 \le x \le 3 \qquad \text{Bounds for } x$$

$$-3 \le y \le 3 \qquad \text{Bounds for } y$$

$$0 \le z \le 3 \qquad \text{Bounds for } z$$

Figure 13.16 shows a computer-generated graph of this surface using 26 traces taken parallel to the yz-plane. To heighten the three-dimensional effect, the program uses a "hidden line" routine. That is, it begins by plotting the traces in the foreground (those corresponding to the largest x-values), and then, as each new trace is plotted, the program determines whether all or only part of the next trace should be shown.

The graphs on the next page show a variety of surfaces that were plotted by computer. If you have access to a computer drawing program, use it to reproduce these surfaces. Remember also that the three-dimensional graphics in this text can be viewed and rotated. These rotatable graphs are available at *LarsonCalculus.com*.

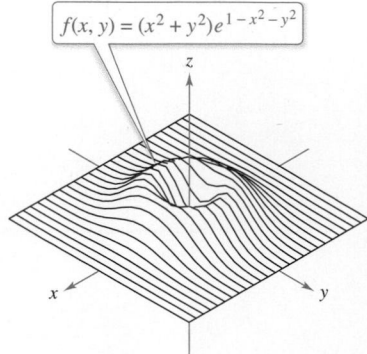

$f(x, y) = (x^2 + y^2)e^{1 - x^2 - y^2}$

Figure 13.16

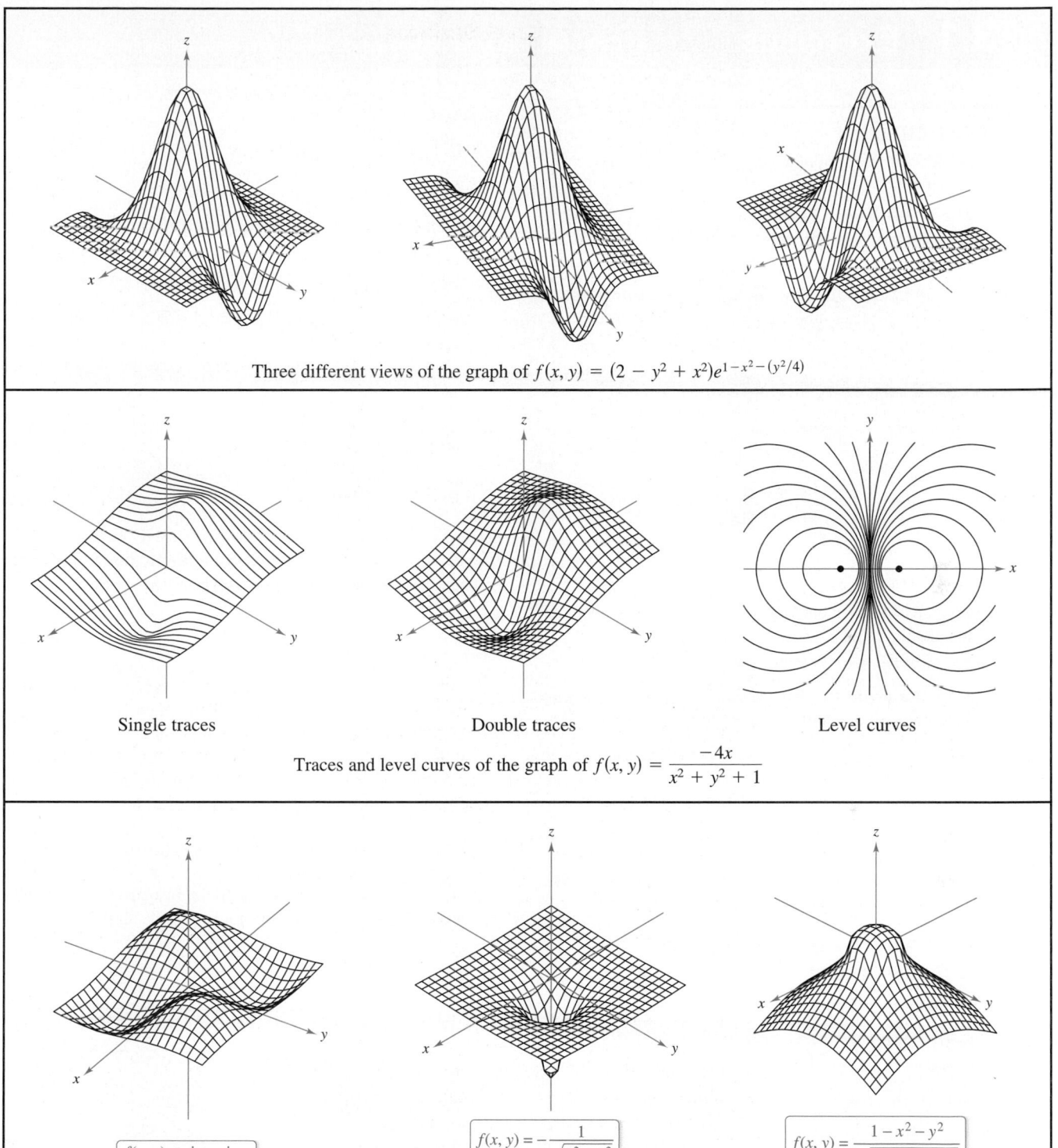

Three different views of the graph of $f(x, y) = (2 - y^2 + x^2)e^{1-x^2-(y^2/4)}$

Single traces   Double traces   Level curves

Traces and level curves of the graph of $f(x, y) = \dfrac{-4x}{x^2 + y^2 + 1}$

$f(x, y) = \sin x \sin y$

$f(x, y) = -\dfrac{1}{\sqrt{x^2 + y^2}}$

$f(x, y) = \dfrac{1 - x^2 - y^2}{\sqrt{|1 - x^2 - y^2|}}$

13.1 Exercises

See **CalcChat.com** for tutorial help and worked-out solutions to odd-numbered exercises.

CONCEPT CHECK

1. Think About It Explain why $z^2 = x + 3y$ is not a function of x and y.

2. Function of Two Variables What is a graph of a function of two variables? How is it interpreted geometrically?

3. Determining Whether a Graph Is a Function Use the graph to determine whether z is a function of x and y. Explain.

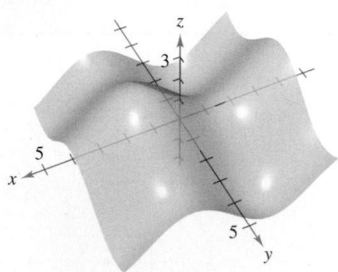

4. Contour Map Explain how to sketch a contour map of a function of x and y.

Determining Whether an Equation Is a Function In Exercises 5–8, determine whether z is a function of x and y.

5. $x^2z + 3y^2 - xy = 10$ **6.** $xz^2 + 2xy - y^2 = 4$

7. $\dfrac{x^2}{4} + \dfrac{y^2}{9} + z^2 = 1$ **8.** $z + x \ln y - 8yz = 0$

 Evaluating a Function In Exercises 9–20, evaluate the function at the given values of the independent variables. Simplify the results.

9. $f(x, y) = 2x - y + 3$ **10.** $f(x, y) = 4 - x^2 - 4y^2$

(a) $f(0, 2)$ (b) $f(-1, 0)$ (a) $f(0, 0)$ (b) $f(0, 1)$

(c) $f(5, 30)$ (d) $f(3, y)$ (c) $f(2, 3)$ (d) $f(1, y)$

(e) $f(x, 4)$ (f) $f(5, t)$ (e) $f(x, 0)$ (f) $f(t, 1)$

11. $f(x, y) = xe^y$ **12.** $g(x, y) = \ln|x + y|$

(a) $f(-1, 0)$ (b) $f(0, 2)$ (a) $g(1, 0)$ (b) $g(0, -t^2)$

(c) $f(x, 3)$ (d) $f(t, -y)$ (c) $g(e, 0)$ (d) $g(e, e)$

13. $h(x, y, z) = \dfrac{xy}{z}$ **14.** $f(x, y, z) = \sqrt{x + y + z}$

(a) $h(-1, 3, -1)$ (a) $f(2, 2, 5)$

(b) $h(2, 2, 2)$ (b) $f(0, 6, -2)$

(c) $h(4, 4t, t^2)$ (c) $f(8, -7, 2)$

(d) $h(-3, 2, 5)$ (d) $f(0, 1, -1)$

15. $f(x, y) - x \sin y$

(a) $f(2, \pi/4)$ (b) $f(3, 1)$ (c) $f(-3, 0)$ (d) $f(4, \pi/2)$

16. $V(r, h) = \pi r^2 h$

(a) $V(3, 10)$ (b) $V(5, 2)$ (c) $V(4, 8)$ (d) $V(6, \pi)$

17. $g(x, y) = \displaystyle\int_x^y (2t - 3)\, dt$

(a) $g(4, 0)$ (b) $g(4, 1)$ (c) $g\left(4, \frac{3}{2}\right)$ (d) $g\left(\frac{3}{2}, 0\right)$

18. $g(x, y) = \displaystyle\int_x^y \dfrac{1}{t}\, dt$

(a) $g(4, 1)$ (b) $g(6, 3)$ (c) $g(2, 5)$ (d) $g\left(\frac{1}{2}, 7\right)$

19. $f(x, y) = 2x + y^2$

(a) $\dfrac{f(x + \Delta x, y) - f(x, y)}{\Delta x}$ (b) $\dfrac{f(x, y + \Delta y) - f(x, y)}{\Delta y}$

20. $f(x, y) = 3x^2 - 2y$

(a) $\dfrac{f(x + \Delta x, y) - f(x, y)}{\Delta x}$ (b) $\dfrac{f(x, y + \Delta y) - f(x, y)}{\Delta y}$

Finding the Domain and Range of a Function In Exercises 21–32, find the domain and range of the function.

21. $f(x, y) = 3x^2 - y$ **22.** $f(x, y) = e^{xy}$

23. $g(x, y) = x\sqrt{y}$ **24.** $g(x, y) = \dfrac{y}{\sqrt{x}}$

25. $z = \dfrac{x + y}{xy}$ **26.** $z = \dfrac{xy}{x + y}$

27. $f(x, y) = \sqrt{4 - x^2 - y^2}$ **28.** $f(x, y) = \sqrt{9 - 6x^2 + y^2}$

29. $f(x, y) = \arccos(x + y)$ **30.** $f(x, y) = \arcsin(y/x)$

31. $f(x, y) = \ln(5 - x - y)$ **32.** $f(x, y) = \ln(xy - 6)$

33. Think About It The graphs labeled (a), (b), (c), and (d) are graphs of the function $f(x, y) = -4x/(x^2 + y^2 + 1)$. Match each of the four graphs with the point in space from which the surface is viewed. The four points are $(20, 15, 25)$, $(-15, 10, 20)$, $(20, 20, 0)$, and $(20, 0, 0)$.

(a)
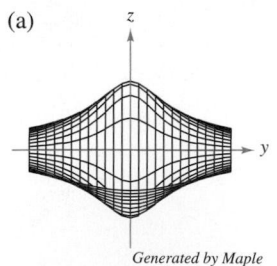
Generated by Maple

(b)
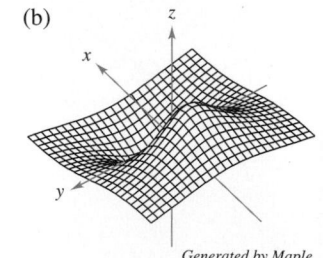
Generated by Maple

(c)
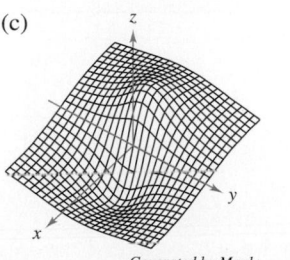
Generated by Maple

(d)
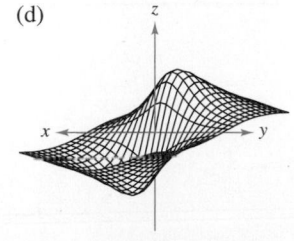
Generated by Maple

34. Think About It Use the function given in Exercise 33.

(a) Find the domain and range of the function.

(b) Identify the points in the xy-plane at which the function value is 0.

(c) Does the surface pass through all the octants of the rectangular coordinate system? Give reasons for your answer.

 **Sketching a Surface** In Exercises 35–42, describe and sketch the surface given by the function.

35. $f(x, y) = 4$

36. $f(x, y) = 6 - 2x - 3y$

37. $f(x, y) = y^2$

38. $g(x, y) = \frac{1}{2}y$

39. $z = -x^2 - y^2$

40. $z = \frac{1}{2}\sqrt{x^2 + y^2}$

41. $f(x, y) = e^{-x}$

42. $f(x, y) = \begin{cases} xy, & x \geq 0, y \geq 0 \\ 0, & x < 0 \text{ or } y < 0 \end{cases}$

Graphing a Function Using Technology In Exercises 43–46, use a computer algebra system to graph the function.

43. $z = y^2 - x^2 + 1$

44. $z = \frac{1}{12}\sqrt{144 - 16x^2 - 9y^2}$

45. $f(x, y) = x^2 e^{(-xy/2)}$

46. $f(x, y) = x \sin y$

Matching In Exercises 47–50, match the graph of the surface with one of the contour maps. [The contour maps are labeled (a), (b), (c), and (d).]

(a)
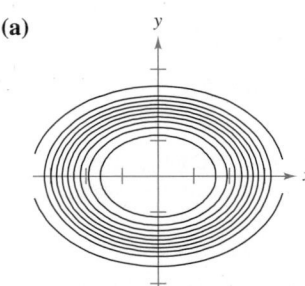

(b)
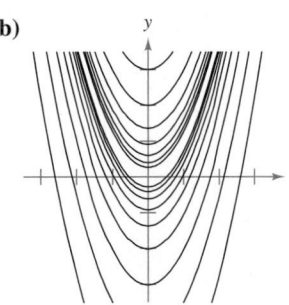

(c)
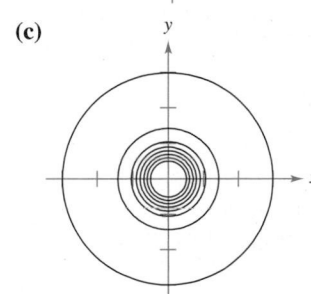

(d)
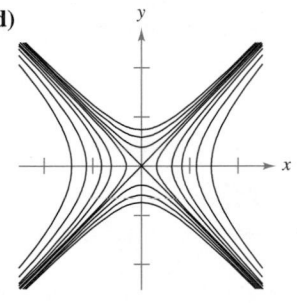

47. $f(x, y) = e^{1-x^2-y^2}$

48. $f(x, y) = e^{1-x^2+y^2}$

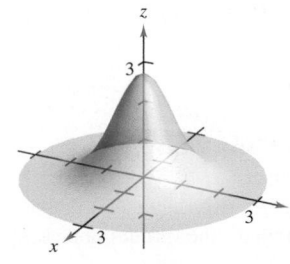

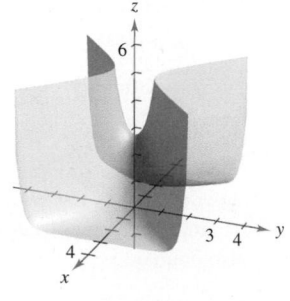

49. $f(x, y) = \ln|y - x^2|$

50. $f(x, y) = \cos\left(\dfrac{x^2 + 2y^2}{4}\right)$

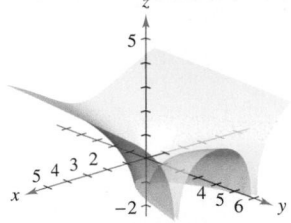

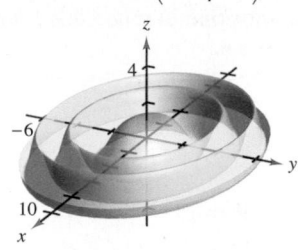

 **Sketching a Contour Map** In Exercises 51–58, describe the level curves of the function. Sketch a contour map of the surface using level curves for the given c-values.

51. $z = x + y$, $c = -1, 0, 2, 4$

52. $z = 6 - 2x - 3y$, $c = 0, 2, 4, 6, 8, 10$

53. $z = x^2 + 4y^2$, $c = 0, 1, 2, 3, 4$

54. $f(x, y) = \sqrt{9 - x^2 - y^2}$, $c = 0, 1, 2, 3$

55. $f(x, y) = xy$, $c = \pm 1, \pm 2, \ldots, \pm 6$

56. $f(x, y) = e^{xy/2}$, $c = 2, 3, 4, \frac{1}{2}, \frac{1}{3}, \frac{1}{4}$

57. $f(x, y) = x/(x^2 + y^2)$, $c = \pm\frac{1}{2}, \pm 1, \pm\frac{3}{2}, \pm 2$

58. $f(x, y) = \ln(x - y)$, $c = 0, \pm\frac{1}{2}, \pm 1, \pm\frac{3}{2}, \pm 2$

Graphing Level Curves Using Technology In Exercises 59–62, use a graphing utility to graph six level curves of the function.

59. $f(x, y) = x^2 - y^2 + 2$

60. $f(x, y) = |xy|$

61. $g(x, y) = \dfrac{8}{1 + x^2 + y^2}$

62. $h(x, y) = 3\sin(|x| + |y|)$

EXPLORING CONCEPTS

63. Vertical Line Test Does the Vertical Line Test apply to functions of two variables? Explain your reasoning.

64. Using Level Curves All of the level curves of the surface given by $z = f(x, y)$ are concentric circles. Does this imply that the graph of f is a hemisphere? Illustrate your answer with an example.

65. Creating a Function Construct a function whose level curves are lines passing through the origin.

66. Conjecture Consider the function $f(x, y) = xy$, for $x \geq 0$ and $y \geq 0$.

(a) Sketch the graph of the surface given by f.

(b) Make a conjecture about the relationship between the graphs of f and $g(x, y) = f(x, y) - 3$. Explain your reasoning.

(c) Repeat part (b) for $g(x, y) = -f(x, y)$.

(d) Repeat part (b) for $g(x, y) = \frac{1}{2}f(x, y)$.

(e) On the surface in part (a), sketch the graph of $z = f(x, x)$.

Writing In Exercises 67 and 68, use the graphs of the level curves (*c*-values evenly spaced) of the function f to write a description of a possible graph of f. Is the graph of f unique? Explain.

67.

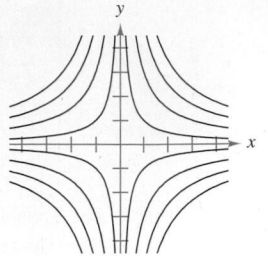

68.

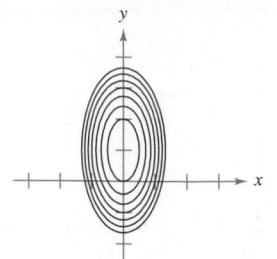

69. Investment In 2016, an investment of $1000 was made in a bond earning 6% compounded annually. Assume that the buyer pays tax at rate R and the annual rate of inflation is I. In the year 2026, the value V of the investment in constant 2016 dollars is

$$V(I, R) = 1000\left[\frac{1 + 0.06(1 - R)}{1 + I}\right]^{10}.$$

Use this function of two variables to complete the table.

	Inflation Rate		
Tax Rate	0	0.03	0.05
0			
0.28			
0.35			

70. Investment A principal of $5000 is deposited in a savings account that earns interest at a rate of r (written as a decimal), compounded continuously. The amount $A(r, t)$ after t years is

$$A(r, t) = 5000e^{rt}.$$

Use this function of two variables to complete the table.

	Number of Years			
Rate	5	10	15	20
0.02				
0.03				
0.04				
0.05				

 Sketching a Level Surface In Exercises 71–76, describe and sketch the graph of the level surface $f(x, y, z) = c$ at the given value of c.

71. $f(x, y, z) = x - y + z, \quad c = 1$

72. $f(x, y, z) = 4x + y + 2z, \quad c = 4$

73. $f(x, y, z) = x^2 + y^2 + z^2, \quad c = 9$

74. $f(x, y, z) = x^2 + \frac{1}{4}y^2 - z, \quad c = 1$

75. $f(x, y, z) = 4x^2 + 4y^2 - z^2, \quad c = 0$

76. $f(x, y, z) = \sin x - z, \quad c = 0$

77. Forestry

The *Doyle Log Rule* is one of several methods used to determine the lumber yield of a log (in board-feet) in terms of its diameter d (in inches) and its length L (in feet). The number of board-feet is

$$N(d, L) = \left(\frac{d - 4}{4}\right)^2 L.$$

(a) Find the number of board-feet of lumber in a log 22 inches in diameter and 12 feet in length.

(b) Find $N(30, 12)$.

78. Queuing Model The average length of time that a customer waits in line for service is

$$W(x, y) = \frac{1}{x - y}, \quad x > y$$

where y is the average arrival rate, written as the number of customers per unit of time, and x is the average service rate, written in the same units. Evaluate each of the following.

(a) $W(15, 9)$ (b) $W(15, 13)$

(c) $W(12, 7)$ (d) $W(5, 2)$

79. Temperature Distribution The temperature T (in degrees Celsius) at any point (x, y) on a circular steel plate of radius 10 meters is

$$T = 600 - 0.75x^2 - 0.75y^2$$

where x and y are measured in meters. Sketch the isothermal curves for $T = 0, 100, 200, \ldots, 600$.

80. Electric Potential The electric potential V at any point (x, y) is

$$V(x, y) = \frac{5}{\sqrt{25 + x^2 + y^2}}.$$

Sketch the equipotential curves for $V = \frac{1}{2}$, $V = \frac{1}{3}$, and $V = \frac{1}{4}$.

 **Cobb-Douglas Production Function** In Exercises 81 and 82, use the Cobb-Douglas production function to find the production level when $x = 600$ units of labor and $y = 350$ units of capital.

81. $f(x, y) = 80x^{0.5}y^{0.5}$ **82.** $f(x, y) = 100x^{0.65}y^{0.35}$

83. Cobb-Douglas Production Function Use the Cobb-Douglas production function, $f(x, y) = Cx^a y^{1-a}$, to show that when the number of units of labor and the number of units of capital are doubled, the production level is also doubled.

84. Cobb-Douglas Production Function Show that the Cobb-Douglas production function $z = Cx^a y^{1-a}$ can be rewritten as

$$\ln \frac{z}{y} = \ln C + a \ln \frac{x}{y}.$$

85. Ideal Gas Law According to the Ideal Gas Law, $PV = kT$, where P is pressure, V is volume, T is temperature (in kelvins), and k is a constant of proportionality. A tank contains 2000 cubic inches of nitrogen at a pressure of 26 pounds per square inch and a temperature of 300 K.

(a) Determine k.

(b) Write P as a function of V and T and describe the level curves.

86. Modeling Data The table shows the net sales x (in billions of dollars), the total assets y (in billions of dollars), and the shareholder's equity z (in billions of dollars) for Walmart for the years 2010 through 2015. *(Source: Wal-Mart Stores, Inc.)*

Year	2010	2011	2012	2013	2014	2015
x	405.0	418.5	443.4	465.6	473.1	482.2
y	170.7	180.8	193.4	203.1	204.8	203.7
z	70.7	68.5	71.3	76.3	76.3	81.4

A model for the data is $z = f(x, y) = 0.428x - 0.653y + 8.172$.

(a) Complete a fourth row in the table using the model to approximate z for the given values of x and y. Compare the approximations with the actual values of z.

(b) Which of the two variables in this model has more influence on shareholder's equity? Explain.

(c) Simplify the expression for $f(x, 150)$ and interpret its meaning in the context of the problem.

87. Meteorology Meteorologists measure the atmospheric pressure in millibars. From these observations, they create weather maps on which the curves of equal atmospheric pressure (isobars) are drawn (see figure). On the map, the closer the isobars, the higher the wind speed. Match points A, B, and C with (a) highest pressure, (b) lowest pressure, and (c) highest wind velocity.

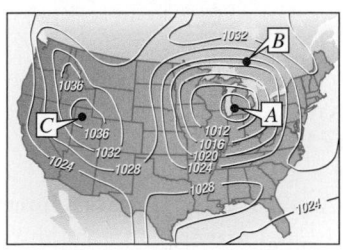

Figure for 87

Figure for 88

88. Acid Rain The acidity of rainwater is measured in units called pH. A pH of 7 is neutral, smaller values are increasingly acidic, and larger values are increasingly alkaline. The map shows curves of equal pH and gives evidence that downwind of heavily industrialized areas, the acidity has been increasing. Using the level curves on the map, determine the direction of the prevailing winds in the northeastern United States.

89. Construction Cost A rectangular storage box with an open top has a length of x feet, a width of y feet, and a height of z feet. It costs \$4.50 per square foot to build the base and \$2.50 per square foot to build the sides. Write the cost C of constructing the box as a function of x, y, and z.

90. HOW DO YOU SEE IT? The contour map of the Southern Hemisphere shown in the figure was computer generated using data collected by satellite instrumentation. Color is used to show the "ozone hole" in Earth's atmosphere. The purple and blue areas represent the lowest levels of ozone, and the green areas represent the highest levels. *(Source: NASA)*

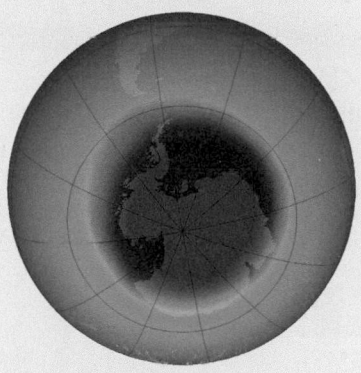

(a) Do the level curves correspond to equally spaced ozone levels? Explain.

(b) Describe how to obtain a more detailed contour map.

True or False? In Exercises 91–94, determine whether the statement is true or false. If it is false, explain why or give an example that shows it is false.

91. If $f(x_0, y_0) = f(x_1, y_1)$, then $x_0 = x_1$ and $y_0 = y_1$.

92. If f is a function, then $f(ax, ay) = a^2 f(x, y)$.

93. The equation for a sphere is a function of three variables.

94. Two different level curves of the graph of $z = f(x, y)$ can intersect.

PUTNAM EXAM CHALLENGE

95. Let $f \colon \mathbb{R}^2 \to \mathbb{R}$ be a function such that

$$f(x, y) + f(y, z) + f(z, x) = 0$$

for all real numbers x, y, and z. Prove that there exists a function $g \colon \mathbb{R} \to \mathbb{R}$ such that

$$f(x, y) = g(x) - g(y)$$

for all real numbers x and y.

This problem was composed by the Committee on the Putnam Prize Competition. © The Mathematical Association of America. All rights reserved.

13.2 Limits and Continuity

- Understand the definition of a neighborhood in the plane.
- Understand and use the definition of the limit of a function of two variables.
- Extend the concept of continuity to a function of two variables.
- Extend the concept of continuity to a function of three variables.

Neighborhoods in the Plane

In this section, you will study limits and continuity involving functions of two or three variables. The section begins with functions of two variables. At the end of the section, the concepts are extended to functions of three variables.

Your study of the limit of a function of two variables begins by defining a two-dimensional analog to an interval on the real number line. Using the formula for the distance between two points

$$(x, y) \quad \text{and} \quad (x_0, y_0)$$

in the plane, you can define the **δ-neighborhood** about (x_0, y_0) to be the **disk** centered at (x_0, y_0) with radius $\delta > 0$

$$\{(x, y): \sqrt{(x - x_0)^2 + (y - y_0)^2} < \delta\} \qquad \text{Open disk}$$

as shown in Figure 13.17. When this formula contains the *less than* inequality sign, $<$, the disk is called **open,** and when it contains the *less than or equal to* inequality sign, $\leq$, the disk is called **closed.** This corresponds to the use of $<$ and $\leq$ to define open and closed intervals.

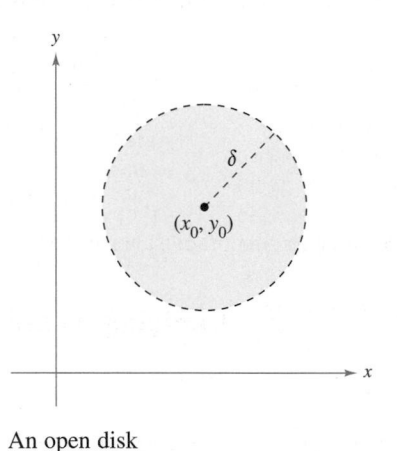

An open disk

Figure 13.17

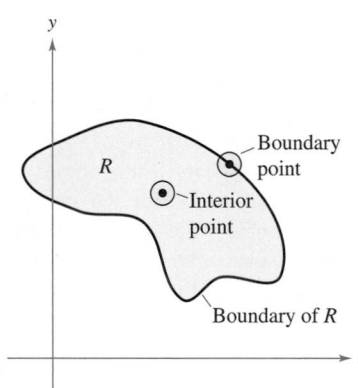

The boundary and interior points of a region R

Figure 13.18

Let the region R be a set of points in the plane. A point (x_0, y_0) in R is an **interior point** of R if there exists a δ-neighborhood about (x_0, y_0) that lies entirely in R, as shown in Figure 13.18. If every point in R is an interior point, then R is an **open region.** A point (x_0, y_0) is a **boundary point** of R if every open disk centered at (x_0, y_0) contains points inside R *and* points outside R. If R contains all its boundary points, then R is a **closed region.**

■ **FOR FURTHER INFORMATION** For more information on Sonya Kovalevsky, see the article "S. Kovalevsky: A Mathematical Lesson" by Karen D. Rappaport in *The American Mathematical Monthly.* To view this article, go to *MathArticles.com.*

SONYA KOVALEVSKY (1850–1891)

Much of the terminology used to define limits and continuity of a function of two or three variables was introduced by the German mathematician Karl Weierstrass (1815–1897). Weierstrass's rigorous approach to limits and other topics in calculus gained him the reputation as the "father of modern analysis." Weierstrass was a gifted teacher. One of his best-known students was the Russian mathematician Sonya Kovalevsky, who applied many of Weierstrass's techniques to problems in mathematical physics and became one of the first women to gain acceptance as a research mathematician.

Limit of a Function of Two Variables

> **Definition of the Limit of a Function of Two Variables**
>
> Let f be a function of two variables defined, except possibly at (x_0, y_0), on an open disk centered at (x_0, y_0), and let L be a real number. Then
>
> $$\lim_{(x, y) \to (x_0, y_0)} f(x, y) = L$$
>
> if for each $\varepsilon > 0$ there corresponds a $\delta > 0$ such that
>
> $$|f(x, y) - L| < \varepsilon \quad \text{whenever} \quad 0 < \sqrt{(x - x_0)^2 + (y - y_0)^2} < \delta.$$

Graphically, the definition of the limit of a function of two variables implies that for any point $(x, y) \neq (x_0, y_0)$ in the disk of radius δ, the value $f(x, y)$ lies between $L + \varepsilon$ and $L - \varepsilon$, as shown in Figure 13.19.

The definition of the limit of a function of two variables is similar to the definition of the limit of a function of a single variable, yet there is a critical difference. To determine whether a function of a single variable has a limit, you need only test the approach from two directions—from the right and from the left. When the function approaches the same limit from the right and from the left, you can conclude that the limit exists. For a function of two variables, however, the statement

$$(x, y) \to (x_0, y_0)$$

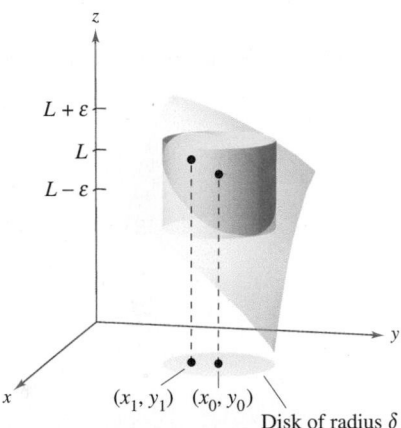

For any (x, y) in the disk of radius δ, the value $f(x, y)$ lies between $L + \varepsilon$ and $L - \varepsilon$.

Figure 13.19

means that the point (x, y) is allowed to approach (x_0, y_0) from any direction. If the value of

$$\lim_{(x, y) \to (x_0, y_0)} f(x, y)$$

is not the same for all possible approaches, or **paths,** to (x_0, y_0), then the limit does not exist.

EXAMPLE 1 Verifying a Limit by the Definition

Show that $\displaystyle\lim_{(x, y) \to (a, b)} x = a.$

Solution Let $f(x, y) = x$ and $L = a$. You need to show that for each $\varepsilon > 0$, there exists a δ-neighborhood about (a, b) such that

$$|f(x, y) - L| = |x - a| < \varepsilon$$

whenever $(x, y) \neq (a, b)$ lies in the neighborhood. You can first observe that from

$$0 < \sqrt{(x - a)^2 + (y - b)^2} < \delta$$

it follows that

$$\begin{aligned} |f(x, y) - L| &= |x - a| \\ &= \sqrt{(x - a)^2} \\ &\leq \sqrt{(x - a)^2 + (y - b)^2} \\ &< \delta. \end{aligned}$$

So, you can choose $\delta = \varepsilon$, and the limit is verified.

Limits of functions of several variables have the same properties regarding sums, differences, products, and quotients as do limits of functions of single variables. (See Theorem 2.2 in Section 2.3.) Some of these properties are used in the next example.

EXAMPLE 2 **Finding a Limit**

Find the limit.

$$\lim_{(x,\,y)\to(1,\,2)} \frac{5x^2y}{x^2 + y^2}$$

Solution By using the properties of limits of products and sums, you obtain

$$\lim_{(x,\,y)\to(1,\,2)} 5x^2y = 5(1^2)(2) = 10$$

and

$$\lim_{(x,\,y)\to(1,\,2)} (x^2 + y^2) = (1^2 + 2^2) = 5.$$

Because the limit of a quotient is equal to the quotient of the limits (and the denominator is not 0), you have

$$\lim_{(x,\,y)\to(1,\,2)} \frac{5x^2y}{x^2 + y^2} = \frac{10}{5} = 2.$$

EXAMPLE 3 **Finding a Limit**

Find the limit: $\displaystyle \lim_{(x,\,y)\to(0,\,0)} \frac{5x^2y}{x^2 + y^2}$.

Solution In this case, the limits of the numerator and of the denominator are both 0, so you cannot determine the existence (or nonexistence) of a limit by taking the limits of the numerator and denominator separately and then dividing. From the graph of f in Figure 13.20, however, it seems reasonable that the limit might be 0. So, you can try applying the definition to $L = 0$. First, note that

$$|y| \le \sqrt{x^2 + y^2}$$

and

$$\frac{x^2}{x^2 + y^2} \le 1.$$

Then, in a δ-neighborhood about $(0, 0)$, you have

$$0 < \sqrt{x^2 + y^2} < \delta$$

and it follows that, for $(x, y) \ne (0, 0)$,

$$\begin{aligned}
|f(x, y) - 0| &= \left| \frac{5x^2y}{x^2 + y^2} \right| \\
&= 5|y|\left(\frac{x^2}{x^2 + y^2} \right) \\
&\le 5|y| \\
&\le 5\sqrt{x^2 + y^2} \\
&< 5\delta.
\end{aligned}$$

So, you can choose $\delta = \varepsilon/5$ and conclude that

$$\lim_{(x,\,y)\to(0,\,0)} \frac{5x^2y}{x^2 + y^2} = 0.$$

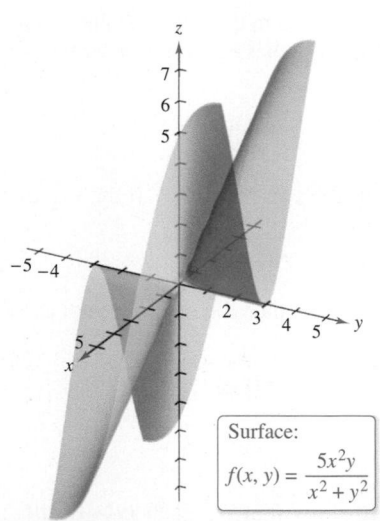

Surface:

$$f(x, y) = \frac{5x^2y}{x^2 + y^2}$$

Figure 13.20

For some functions, it is easy to recognize that a limit does not exist. For instance, it is clear that the limit

$$\lim_{(x, y) \to (0, 0)} \frac{1}{x^2 + y^2}$$

does not exist because the values of $f(x, y)$ increase without bound as (x, y) approaches $(0, 0)$ along *any* path (see Figure 13.21).

For other functions, it is not so easy to recognize that a limit does not exist. For instance, the next example describes a limit that does not exist because the function approaches different values along different paths.

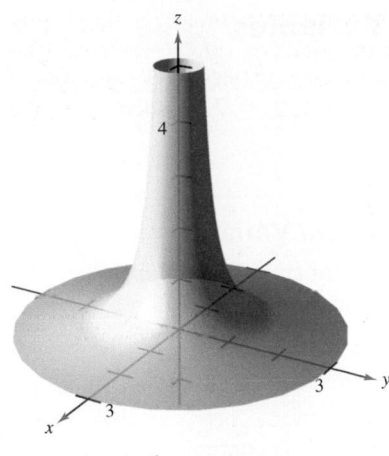

$$\lim_{(x, y) \to (0, 0)} \frac{1}{x^2 + y^2}$$ does not exist.

Figure 13.21

EXAMPLE 4 A Limit That Does Not Exist

⋮ ⋯ ▷ *See LarsonCalculus.com for an interactive version of this type of example.*

Show that the limit does not exist.

$$\lim_{(x, y) \to (0, 0)} \left(\frac{x^2 - y^2}{x^2 + y^2} \right)^2$$

Solution The domain of the function

$$f(x, y) = \left(\frac{x^2 - y^2}{x^2 + y^2} \right)^2$$

consists of all points in the *xy*-plane except for the point $(0, 0)$. To show that the limit as (x, y) approaches $(0, 0)$ does not exist, consider approaching $(0, 0)$ along two different "paths," as shown in Figure 13.22. Along the *x*-axis, every point is of the form

$$(x, 0)$$

and the limit along this approach is

$$\lim_{(x, 0) \to (0, 0)} \left(\frac{x^2 - 0^2}{x^2 + 0^2} \right)^2 = \lim_{(x, 0) \to (0, 0)} 1^2 = 1. \qquad \text{Limit along } x\text{-axis}$$

However, when (x, y) approaches $(0, 0)$ along the line $y = x$, you obtain

$$\lim_{(x, x) \to (0, 0)} \left(\frac{x^2 - x^2}{x^2 + x^2} \right)^2 = \lim_{(x, x) \to (0, 0)} \left(\frac{0}{2x^2} \right)^2 = 0. \qquad \text{Limit along line } y = x$$

This means that in any open disk centered at $(0, 0)$, there are points (x, y) at which f takes on the value 1 and other points at which f takes on the value 0. For instance,

$$f(x, y) = 1$$

at $(1, 0)$, $(0.1, 0)$, $(0.01, 0)$, and $(0.001, 0)$, and

$$f(x, y) = 0$$

at $(1, 1)$, $(0.1, 0.1)$, $(0.01, 0.01)$, and $(0.001, 0.001)$. So, f does not have a limit as (x, y) approaches $(0, 0)$.

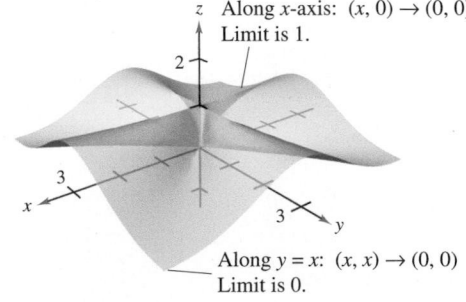

Along *x*-axis: $(x, 0) \to (0, 0)$
Limit is 1.

Along $y = x$: $(x, x) \to (0, 0)$
Limit is 0.

$$\lim_{(x, y) \to (0, 0)} \left(\frac{x^2 - y^2}{x^2 + y^2} \right)^2$$ does not exist.

Figure 13.22

In Example 4, you could conclude that the limit does not exist because you found two approaches that produced different limits. Be sure you understand that when two approaches produce the same limit, you *cannot* conclude that the limit exists. To form such a conclusion, you must show that the limit is the same along *all* possible approaches.

Continuity of a Function of Two Variables

Notice in Example 2 that the limit of $f(x, y) = 5x^2y/(x^2 + y^2)$ as $(x, y) \to (1, 2)$ can be evaluated by direct substitution. That is, the limit is $f(1, 2) = 2$. In such cases, the function f is said to be **continuous** at the point $(1, 2)$.

· · REMARK This definition of continuity can be extended to *boundary points* of the open region R by considering a special type of limit in which (x, y) is allowed to approach (x_0, y_0) along paths lying in the region R. This notion is similar to that of one-sided limits, as discussed in Chapter 1.

Definition of Continuity of a Function of Two Variables

A function f of two variables is **continuous at a point** (x_0, y_0) in an open region R if $f(x_0, y_0)$ is defined and is equal to the limit of $f(x, y)$ as (x, y) approaches (x_0, y_0). That is,

$$\lim_{(x, y) \to (x_0, y_0)} f(x, y) = f(x_0, y_0).$$

The function f is **continuous in the open region R** if it is continuous at every point in R.

In Example 3, it was shown that the function

$$f(x, y) = \frac{5x^2y}{x^2 + y^2}$$

is not continuous at $(0, 0)$. Because the limit at this point exists, however, you can remove the discontinuity by defining f at $(0, 0)$ as being equal to its limit there. Such a discontinuity is called **removable**. In Example 4, the function

$$f(x, y) = \left(\frac{x^2 - y^2}{x^2 + y^2}\right)^2$$

was also shown not to be continuous at $(0, 0)$, but this discontinuity is **nonremovable.**

THEOREM 13.1 Continuous Functions of Two Variables

If k is a real number and $f(x, y)$ and $g(x, y)$ are continuous at (x_0, y_0), then the following functions are also continuous at (x_0, y_0).

1. Scalar multiple: kf
2. Sum or difference: $f \pm g$
3. Product: fg
4. Quotient: f/g, $g(x_0, y_0) \neq 0$

Theorem 13.1 establishes the continuity of *polynomial* and *rational* functions at every point in their domains. Furthermore, the continuity of other types of functions can be extended naturally from one to two variables. For instance, the functions whose graphs are shown in Figures 13.23 and 13.24 are continuous at every point in the plane.

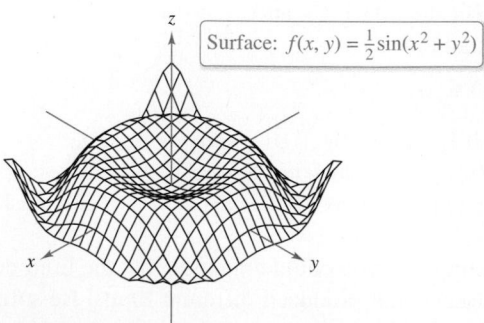

The function f is continuous at every point in the plane.
Figure 13.23

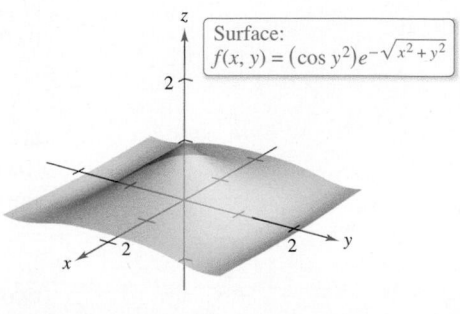

The function f is continuous at every point in the plane.
Figure 13.24

The next theorem states conditions under which a composite function is continuous.

THEOREM 13.2   Continuity of a Composite Function

If h is continuous at (x_0, y_0) and g is continuous at $h(x_0, y_0)$, then the composite function given by $(g \circ h)(x, y) = g(h(x, y))$ is continuous at (x_0, y_0). That is,

$$\lim_{(x, y) \to (x_0, y_0)} g(h(x, y)) = g(h(x_0, y_0)).$$

Note in Theorem 13.2 that h is a function of two variables and g is a function of one variable.

EXAMPLE 5   **Testing for Continuity**

Discuss the continuity of each function.

a. $f(x, y) = \dfrac{x - 2y}{x^2 + y^2}$ **b.** $g(x, y) = \dfrac{2}{y - x^2}$

Solution

a. Because a rational function is continuous at every point in its domain, you can conclude that f is continuous at each point in the xy-plane except at $(0, 0)$, as shown in Figure 13.25.

b. The function

$$g(x, y) = \frac{2}{y - x^2}$$

is continuous except at the points at which the denominator is 0. These points are given by the equation

$$y - x^2 = 0.$$

So, you can conclude that the function is continuous at all points except those lying on the parabola $y = x^2$. Inside this parabola, you have $y > x^2$, and the surface represented by the function lies above the xy-plane, as shown in Figure 13.26. Outside the parabola, $y < x^2$, and the surface lies below the xy-plane.

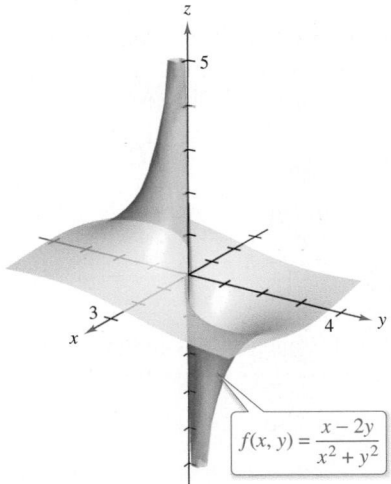

The function f is not continuous at $(0, 0)$.

Figure 13.25

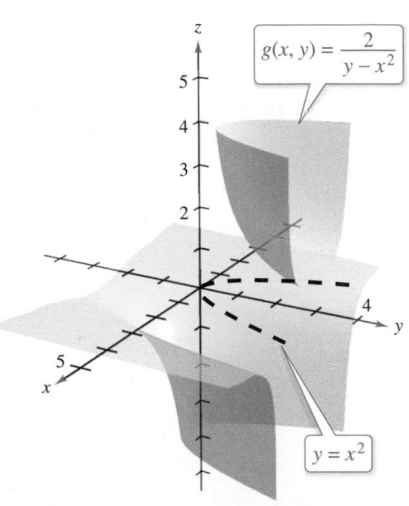

The function g is not continuous on the parabola $y = x^2$.

Figure 13.26

Continuity of a Function of Three Variables

The preceding definitions of limits and continuity can be extended to functions of three variables by considering points (x, y, z) within the *open sphere*

$$(x - x_0)^2 + (y - y_0)^2 + (z - z_0)^2 < \delta^2. \qquad \text{Open sphere}$$

The radius of this sphere is δ, and the sphere is centered at (x_0, y_0, z_0), as shown in Figure 13.27.

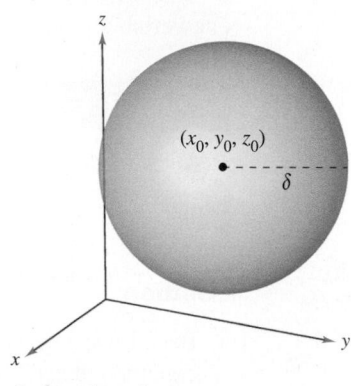

Open sphere in space

Figure 13.27

A point (x_0, y_0, z_0) in a region R in space is an **interior point** of R if there exists a δ-sphere about (x_0, y_0, z_0) that lies entirely in R. If every point in R is an interior point, then R is called **open.**

Definition of Continuity of a Function of Three Variables

A function f of three variables is **continuous at a point (x_0, y_0, z_0)** in an open region R if $f(x_0, y_0, z_0)$ is defined and is equal to the limit of $f(x, y, z)$ as (x, y, z) approaches (x_0, y_0, z_0). That is,

$$\lim_{(x, y, z) \to (x_0, y_0, z_0)} f(x, y, z) = f(x_0, y_0, z_0).$$

The function f is **continuous in the open region R** if it is continuous at every point in R.

EXAMPLE 6 **Testing Continuity of a Function of Three Variables**

Discuss the continuity of

$$f(x, y, z) = \frac{1}{x^2 + y^2 - z}.$$

Solution The function f is continuous except at the points at which the denominator is 0, which are given by the equation

$$x^2 + y^2 - z = 0.$$

So, f is continuous at each point in space except at the points on the paraboloid

$$z = x^2 + y^2.$$

13.2 Exercises

See **CalcChat.com** for tutorial help and worked-out solutions to odd-numbered exercises.

CONCEPT CHECK

1. Describing Notation Write a brief description of the meaning of the notation $\lim\limits_{(x, y) \to (-1, 3)} f(x, y) = 1$.

2. Limits Explain how examining limits along different paths might show that a limit does not exist. Does this type of examination show that a limit does exist? Explain.

 **Verifying a Limit by the Definition** In Exercises 3–6, use the definition of the limit of a function of two variables to verify the limit.

3. $\lim\limits_{(x, y) \to (1, 0)} x = 1$

4. $\lim\limits_{(x, y) \to (4, -1)} x = 4$

5. $\lim\limits_{(x, y) \to (1, -3)} y = -3$

6. $\lim\limits_{(x, y) \to (a, b)} y = b$

Using Properties of Limits In Exercises 7–10, find the indicated limit by using the limits

$$\lim\limits_{(x, y) \to (a, b)} f(x, y) = 4 \quad \text{and} \quad \lim\limits_{(x, y) \to (a, b)} g(x, y) = -5.$$

7. $\lim\limits_{(x, y) \to (a, b)} [f(x, y) - g(x, y)]$

8. $\lim\limits_{(x, y) \to (a, b)} \left[\dfrac{3f(x, y)}{g(x, y)} \right]$

9. $\lim\limits_{(x, y) \to (a, b)} [f(x, y)g(x, y)]$

10. $\lim\limits_{(x, y) \to (a, b)} \left[\dfrac{f(x, y) + g(x, y)}{f(x, y)} \right]$

 **Limit and Continuity** In Exercises 11–24, find the limit and discuss the continuity of the function.

11. $\lim\limits_{(x, y) \to (3, 1)} (x^2 - 2y)$

12. $\lim\limits_{(x, y) \to (-1, 1)} (x + 4y^2 + 5)$

13. $\lim\limits_{(x, y) \to (1, 2)} e^{xy}$

14. $\lim\limits_{(x, y) \to (2, 4)} \dfrac{x + y}{x^2 + 1}$

15. $\lim\limits_{(x, y) \to (0, 2)} \dfrac{x}{y}$

16. $\lim\limits_{(x, y) \to (-1, 2)} \dfrac{x + y}{x - y}$

17. $\lim\limits_{(x, y) \to (1, 1)} \dfrac{xy}{x^2 + y^2}$

18. $\lim\limits_{(x, y) \to (1, 1)} \dfrac{x}{\sqrt{x + y}}$

19. $\lim\limits_{(x, y) \to (\pi/3, 2)} y \cos xy$

20. $\lim\limits_{(x, y) \to (\pi, -4)} \sin\dfrac{x}{y}$

21. $\lim\limits_{(x, y) \to (0, 1)} \dfrac{\arcsin xy}{1 - xy}$

22. $\lim\limits_{(x, y) \to (0, 1)} \dfrac{\arccos(x/y)}{1 + xy}$

23. $\lim\limits_{(x, y, z) \to (1, 3, 4)} \sqrt{x + y + z}$

24. $\lim\limits_{(x, y, z) \to (-2, 1, 0)} xe^{yz}$

 **Finding a Limit** In Exercises 25–36, find the limit (if it exists). If the limit does not exist, explain why.

25. $\lim\limits_{(x, y) \to (1, 1)} \dfrac{xy - 1}{1 + xy}$

26. $\lim\limits_{(x, y) \to (1, -1)} \dfrac{x^2 y}{1 + xy^2}$

27. $\lim\limits_{(x, y) \to (0, 0)} \dfrac{1}{x + y}$

28. $\lim\limits_{(x, y) \to (0, 0)} \dfrac{1}{x^2 y^2}$

29. $\lim\limits_{(x, y) \to (0, 0)} \dfrac{x - y}{\sqrt{x} - \sqrt{y}}$

30. $\lim\limits_{(x, y) \to (2, 1)} \dfrac{x - y - 1}{\sqrt{x - y} - 1}$

31. $\lim\limits_{(x, y) \to (0, 0)} \dfrac{x + y}{x^2 + y}$

32. $\lim\limits_{(x, y) \to (0, 0)} \dfrac{x}{x^2 - y^2}$

33. $\lim\limits_{(x, y) \to (0, 0)} \dfrac{x^2}{(x^2 + 1)(y^2 + 1)}$

34. $\lim\limits_{(x, y) \to (0, 0)} \ln(x^2 + y^2)$

35. $\lim\limits_{(x, y, z) \to (0, 0, 0)} \dfrac{xy + yz + xz}{x^2 + y^2 + z^2}$

36. $\lim\limits_{(x, y, z) \to (0, 0, 0)} \dfrac{xy + yz^2 + xz^2}{x^2 + y^2 + z^2}$

EXPLORING CONCEPTS

37. Limits If $f(2, 3) = 4$, can you conclude anything about $\lim\limits_{(x, y) \to (2, 3)} f(x, y)$? Explain.

38. Limits If $\lim\limits_{(x, y) \to (2, 3)} f(x, y) = 4$, can you conclude anything about $f(2, 3)$? Explain.

39. Think About It Given that $\lim\limits_{(x, y) \to (0, 0)} f(x, y) = 0$, does $\lim\limits_{(x, 0) \to (0, 0)} f(x, 0) = 0$? Explain.

40. **HOW DO YOU SEE IT?** The figure shows the graph of $f(x, y) = \ln(x^2 + y^2)$. From the graph, does it appear that the limit at each point exists?

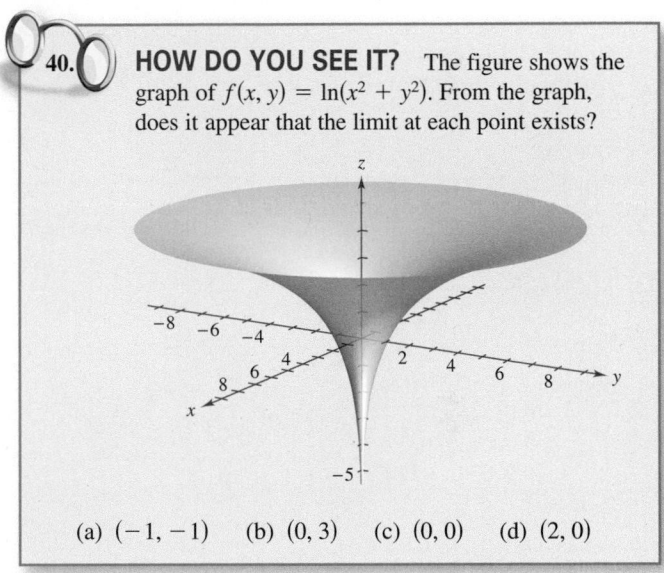

    (a) $(-1, -1)$ (b) $(0, 3)$ (c) $(0, 0)$ (d) $(2, 0)$

Continuity In Exercises 41 and 42, discuss the continuity of the function and evaluate the limit of $f(x, y)$ (if it exists) as $(x, y) \to (0, 0)$.

41. $f(x, y) = e^{xy}$

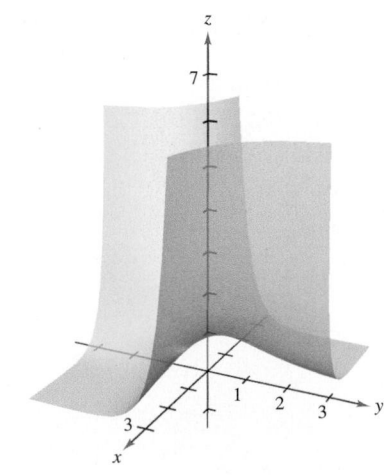

42. $f(x, y) = 1 - \dfrac{\cos(x^2 + y^2)}{x^2 + y^2}$

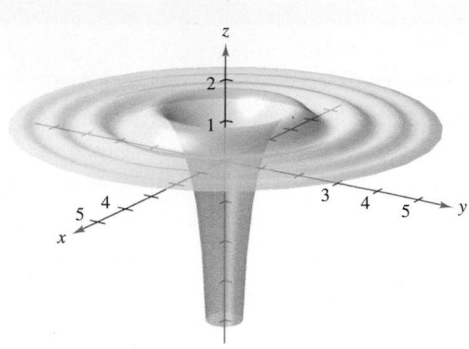

![Limit and Continuity icon] **Limit and Continuity** In Exercises 43–46, use a graphing utility to make a table showing the values of $f(x, y)$ at the given points for each path. Use the result to make a conjecture about the limit of $f(x, y)$ as $(x, y) \to (0, 0)$. Determine analytically whether the limit exists and discuss the continuity of the function.

43. $f(x, y) = \dfrac{xy}{x^2 + y^2}$

Path: $y = 0$

Points: $(1, 0)$, $(0.5, 0)$,
$(0.1, 0)$, $(0.01, 0)$,
$(0.001, 0)$

Path: $y = x$

Points: $(1, 1)$, $(0.5, 0.5)$,
$(0.1, 0.1)$, $(0.01, 0.01)$,
$(0.001, 0.001)$

44. $f(x, y) = -\dfrac{xy^2}{x^2 + y^4}$

Path: $x = y^2$

Points: $(1, 1)$, $(0.25, 0.5)$,
$(0.01, 0.1)$,
$(0.0001, 0.01)$,
$(0.000001, 0.001)$

Path: $x = -y^2$

Points: $(-1, 1)$, $(-0.25, 0.5)$,
$(-0.01, 0.1)$,
$(-0.0001, 0.01)$,
$(-0.000001, 0.001)$

45. $f(x, y) = \dfrac{y}{x^2 + y^2}$

Path: $y = 0$

Points: $(1, 0)$, $(0.5, 0)$,
$(0.1, 0)$, $(0.01, 0)$,
$(0.001, 0)$

Path: $y = x$

Points: $(1, 1)$, $(0.5, 0.5)$,
$(0.1, 0.1)$,
$(0.01, 0.01)$,
$(0.001, 0.001)$

46. $f(x, y) = \dfrac{2x - y^2}{2x^2 + y}$

Path: $y = 0$

Points: $(1, 0)$, $(0.25, 0)$,
$(0.01, 0)$,
$(0.001, 0)$,
$(0.000001, 0)$

Path: $y = x$

Points: $(1, 1)$, $(0.25, 0.25)$,
$(0.01, 0.01)$,
$(0.001, 0.001)$,
$(0.0001, 0.0001)$

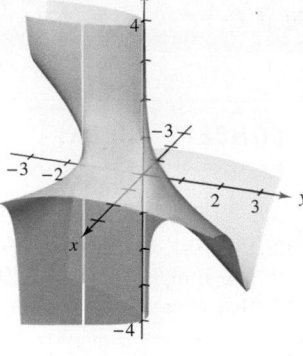

47. Limit Consider $\displaystyle\lim_{(x, y) \to (0, 0)} \dfrac{x^2 + y^2}{xy}$ (see figure).

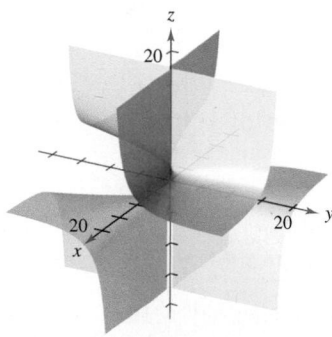

(a) Determine (if possible) the limit along any line of the form $y = ax$.

(b) Determine (if possible) the limit along the parabola $y = x^2$.

(c) Does the limit exist? Explain.

48. Limit Consider $\displaystyle\lim_{(x, y) \to (0, 0)} \dfrac{x^2 y}{x^4 + y^2}$ (see figure).

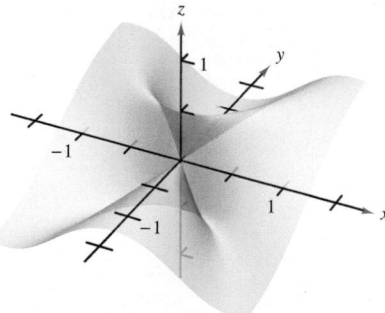

Comparing Continuity In Exercises 49 and 50, discuss the continuity of the functions f and g. Explain any differences.

49. $f(x, y) = \begin{cases} \dfrac{x^4 - y^4}{x^2 + y^2}, & (x, y) \neq (0, 0) \\ 0, & (x, y) = (0, 0) \end{cases}$

$g(x, y) = \begin{cases} \dfrac{x^4 - y^4}{x^2 + y^2}, & (x, y) \neq (0, 0) \\ 1, & (x, y) = (0, 0) \end{cases}$

50. $f(x, y) = \begin{cases} \dfrac{x^2 + 2xy^2 + y^2}{x^2 + y^2}, & (x, y) \ne (0, 0) \\ 0, & (x, y) = (0, 0) \end{cases}$

$g(x, y) = \begin{cases} \dfrac{x^2 + 2xy^2 + y^2}{x^2 + y^2}, & (x, y) \ne (0, 0) \\ 1, & (x, y) = (0, 0) \end{cases}$

Finding a Limit Using Polar Coordinates   In Exercises 51–56, use polar coordinates to find the limit. [*Hint:* Let $x = r \cos \theta$ and $y = r \sin \theta$, and note that $(x, y) \to (0, 0)$ implies $r \to 0$.]

51. $\lim\limits_{(x, y) \to (0, 0)} \dfrac{xy^2}{x^2 + y^2}$

52. $\lim\limits_{(x, y) \to (0, 0)} \dfrac{x^3 + y^3}{x^2 + y^2}$

53. $\lim\limits_{(x, y) \to (0, 0)} \dfrac{x^2 y^2}{x^2 + y^2}$

54. $\lim\limits_{(x, y) \to (0, 0)} \dfrac{x^2 - y^2}{\sqrt{x^2 + y^2}}$

55. $\lim\limits_{(x, y) \to (0, 0)} \cos(x^2 + y^2)$

56. $\lim\limits_{(x, y) \to (0, 0)} \sin\sqrt{x^2 + y^2}$

Finding a Limit Using Polar Coordinates   In Exercises 57–60, use polar coordinates and L'Hôpital's Rule to find the limit.

57. $\lim\limits_{(x, y) \to (0, 0)} \dfrac{\sin\sqrt{x^2 + y^2}}{\sqrt{x^2 + y^2}}$

58. $\lim\limits_{(x, y) \to (0, 0)} \dfrac{\sin(x^2 + y^2)}{x^2 + y^2}$

59. $\lim\limits_{(x, y) \to (0, 0)} \dfrac{1 - \cos(x^2 + y^2)}{x^2 + y^2}$

60. $\lim\limits_{(x, y) \to (0, 0)} (x^2 + y^2) \ln(x^2 + y^2)$

Continuity   In Exercises 61–66, discuss the continuity of the function.

61. $f(x, y, z) = \dfrac{1}{\sqrt{x^2 + y^2 + z^2}}$

62. $f(x, y, z) = \dfrac{z}{x^2 + y^2 - 4}$

63. $f(x, y, z) = \dfrac{\sin z}{e^x + e^y}$

64. $f(x, y, z) = xy \sin z$

65. $f(x, y) = \begin{cases} \dfrac{\sin xy}{xy}, & xy \ne 0 \\ 1, & xy = 0 \end{cases}$

66. $f(x, y) = \begin{cases} \dfrac{\sin(x^2 - y^2)}{x^2 - y^2}, & x^2 \ne y^2 \\ 1, & x^2 = y^2 \end{cases}$

Continuity of a Composite Function   In Exercises 67–70, discuss the continuity of the composite function $f \circ g$.

67. $f(t) = t^2$

$g(x, y) = 2x - 3y$

68. $f(t) = \dfrac{1}{t}$

$g(x, y) = x^2 + y^2$

69. $f(t) = \dfrac{1}{t}$

$g(x, y) = 2x - 3y$

70. $f(t) = \dfrac{1}{1 - t}$

$g(x, y) = x^2 + y^2$

Finding a Limit   In Exercises 71–76, find each limit.

(a) $\lim\limits_{\Delta x \to 0} \dfrac{f(x + \Delta x, y) - f(x, y)}{\Delta x}$

(b) $\lim\limits_{\Delta y \to 0} \dfrac{f(x, y + \Delta y) - f(x, y)}{\Delta y}$

71. $f(x, y) = x^2 - 4y$

72. $f(x, y) = 3x^2 + y^2$

73. $f(x, y) = \dfrac{x}{y}$

74. $f(x, y) = \dfrac{1}{x + y}$

75. $f(x, y) = 3x + xy - 2y$

76. $f(x, y) = \sqrt{y}(y + 1)$

Finding a Limit Using Spherical Coordinates   In Exercises 77 and 78, use spherical coordinates to find the limit. [*Hint:* Let $x = \rho \sin \phi \cos \theta$, $y = \rho \sin \phi \sin \theta$, and $z = \rho \cos \phi$, and note that $(x, y, z) \to (0, 0, 0)$ implies $\rho \to 0^+$.]

77. $\lim\limits_{(x, y) \to (0, 0, 0)} \dfrac{xyz}{x^2 + y^2 + z^2}$

78. $\lim\limits_{(x, y) \to (0, 0, 0)} \tan^{-1}\left(\dfrac{1}{x^2 + y^2 + z^2}\right)$

True or False?   In Exercises 79–82, determine whether the statement is true or false. If it is false, explain why or give an example that shows it is false.

79. A closed region contains all of its boundary points.

80. Every point in an open region is an interior point.

81. If f is continuous for all nonzero x and y, and $f(0, 0) = 0$, then $\lim\limits_{(x, y) \to (0, 0)} f(x, y) = 0$.

82. If g is a continuous function of x, h is a continuous function of y, and $f(x, y) = g(x) + h(y)$, then f is continuous.

83. Finding a Limit   Find the following limit.

$$\lim\limits_{(x, y) \to (0, 1)} \tan^{-1}\left[\dfrac{x^2 + 1}{x^2 + (y - 1)^2}\right]$$

84. Continuity   For the function

$$f(x, y) = xy\left(\dfrac{x^2 - y^2}{x^2 + y^2}\right)$$

define $f(0, 0)$ such that f is continuous at the origin.

85. Proof   Prove that

$$\lim\limits_{(x, y) \to (a, b)} [f(x, y) + g(x, y)] = L_1 + L_2$$

where $f(x, y)$ approaches L_1 and $g(x, y)$ approaches L_2 as $(x, y) \to (a, b)$.

86. Proof   Prove that if f is continuous and $f(a, b) < 0$, then there exists a δ-neighborhood about (a, b) such that $f(x, y) < 0$ for every point (x, y) in the neighborhood.

13.3 Partial Derivatives

- Find and use partial derivatives of a function of two variables.
- Find and use partial derivatives of a function of three or more variables.
- Find higher-order partial derivatives of a function of two or three variables.

Partial Derivatives of a Function of Two Variables

In applications of functions of several variables, the question often arises, "How will the value of a function be affected by a change in one of its independent variables?" You can answer this by considering the independent variables one at a time. For example, to determine the effect of a catalyst in an experiment, a chemist could conduct the experiment several times using varying amounts of the catalyst while keeping constant other variables such as temperature and pressure. You can use a similar procedure to determine the rate of change of a function f with respect to one of its several independent variables. This process is called **partial differentiation,** and the result is referred to as the **partial derivative** of f with respect to the chosen independent variable.

JEAN LE ROND D'ALEMBERT (1717–1783)

The introduction of partial derivatives followed Newton's and Leibniz's work in calculus by several years. Between 1730 and 1760, Leonhard Euler and Jean Le Rond d'Alembert separately published several papers on dynamics, in which they established much of the theory of partial derivatives. These papers used functions of two or more variables to study problems involving equilibrium, fluid motion, and vibrating strings.
See LarsonCalculus.com to read more of this biography.

Definition of Partial Derivatives of a Function of Two Variables

If $z = f(x, y)$, then the **first partial derivatives** of f with respect to x and y are the functions f_x and f_y defined by

$$f_x(x, y) = \lim_{\Delta x \to 0} \frac{f(x + \Delta x, y) - f(x, y)}{\Delta x}$$ Partial derivative with respect to x

and

$$f_y(x, y) = \lim_{\Delta y \to 0} \frac{f(x, y + \Delta y) - f(x, y)}{\Delta y}$$ Partial derivative with respect to y

provided the limits exist.

This definition indicates that if $z = f(x, y)$, then to find f_x, you *consider y constant* and differentiate with respect to x. Similarly, to find f_y, you *consider x constant* and differentiate with respect to y.

EXAMPLE 1   **Finding Partial Derivatives**

a. To find f_x for $f(x, y) = 3x - x^2y^2 + 2x^3y$, consider y to be constant and differentiate with respect to x.

$$f_x(x, y) = 3 - 2xy^2 + 6x^2y$$ Partial derivative with respect to x

To find f_y, consider x to be constant and differentiate with respect to y.

$$f_y(x, y) = -2x^2y + 2x^3$$ Partial derivative with respect to y

b. To find f_x for $f(x, y) = (\ln x)(\sin x^2y)$, consider y to be constant and differentiate with respect to x.

$$f_x(x, y) = (\ln x)(\cos x^2y)(2xy) + \frac{\sin x^2y}{x}$$ Partial derivative with respect to x

To find f_y, consider x to be constant and differentiate with respect to y.

$$f_y(x, y) = (\ln x)(\cos x^2y)(x^2)$$ Partial derivative with respect to y

· · · · · · · · · · · · · · · · · · ▷

· · REMARK   The notation $\partial z / \partial x$ is read as "the partial derivative of z with respect to x," and $\partial z / \partial y$ is read as "the partial derivative of z with respect to y."

Notation for First Partial Derivatives

For $z = f(x, y)$, the partial derivatives f_x and f_y are denoted by

$$\frac{\partial}{\partial x} f(x, y) = f_x(x, y) = z_x = \frac{\partial z}{\partial x} \qquad \text{Partial derivative with respect to } x$$

and

$$\frac{\partial}{\partial y} f(x, y) = f_y(x, y) = z_y = \frac{\partial z}{\partial y}. \qquad \text{Partial derivative with respect to } y$$

The first partials evaluated at the point (a, b) are denoted by

$$\left. \frac{\partial z}{\partial x} \right|_{(a, b)} = f_x(a, b)$$

and

$$\left. \frac{\partial z}{\partial y} \right|_{(a, b)} = f_y(a, b).$$

EXAMPLE 2   Finding and Evaluating Partial Derivatives

For $f(x, y) = xe^{x^2 y}$, find f_x and f_y, and evaluate each at the point $(1, \ln 2)$.

Solution   Because

$$f_x(x, y) = xe^{x^2 y}(2xy) + e^{x^2 y} \qquad \text{Partial derivative with respect to } x$$

the partial derivative of f with respect to x at $(1, \ln 2)$ is

$$f_x(1, \ln 2) = e^{\ln 2}(2 \ln 2) + e^{\ln 2}$$
$$= 4 \ln 2 + 2.$$

Because

$$f_y(x, y) = xe^{x^2 y}(x^2)$$
$$= x^3 e^{x^2 y} \qquad \text{Partial derivative with respect to } y$$

the partial derivative of f with respect to y at $(1, \ln 2)$ is

$$f_y(1, \ln 2) = e^{\ln 2}$$
$$= 2. \qquad ■$$

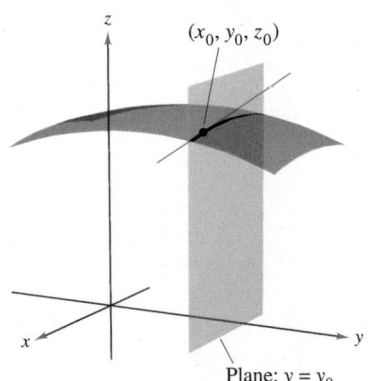

$\frac{\partial f}{\partial x} = $ slope in x-direction

Figure 13.28

The partial derivatives of a function of two variables, $z = f(x, y)$, have a useful geometric interpretation. If $y = y_0$, then $z = f(x, y_0)$ represents the curve formed by intersecting the surface $z = f(x, y)$ with the plane $y = y_0$, as shown in Figure 13.28. Therefore,

$$f_x(x_0, y_0) = \lim_{\Delta x \to 0} \frac{f(x_0 + \Delta x, y_0) - f(x_0, y_0)}{\Delta x}$$

represents the slope of this curve at the point $(x_0, y_0, f(x_0, y_0))$. Note that both the curve and the tangent line lie in the plane $y = y_0$. Similarly,

$$f_y(x_0, y_0) = \lim_{\Delta y \to 0} \frac{f(x_0, y_0 + \Delta y) - f(x_0, y_0)}{\Delta y}$$

represents the slope of the curve given by the intersection of $z = f(x, y)$ and the plane $x = x_0$ at $(x_0, y_0, f(x_0, y_0))$, as shown in Figure 13.29.

Informally, the values of $\partial f / \partial x$ and $\partial f / \partial y$ at the point (x_0, y_0, z_0) denote the **slopes of the surface in the x- and y-directions,** respectively.

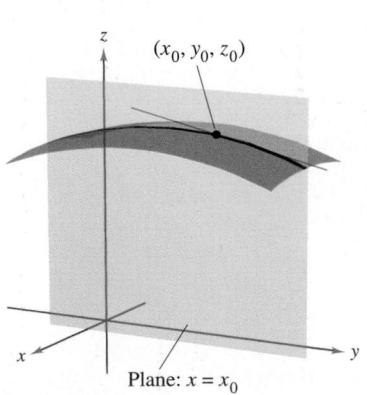

$\frac{\partial f}{\partial y} = $ slope in y-direction

Figure 13.29

EXAMPLE 3 Finding the Slopes of a Surface

: · · · ▷ *See LarsonCalculus.com for an interactive version of this type of example.*

Find the slopes in the x-direction and in the y-direction of the surface

$$f(x, y) = -\frac{x^2}{2} - y^2 + \frac{25}{8}$$

at the point $\left(\frac{1}{2}, 1, 2\right)$.

Solution The partial derivatives of f with respect to x and y are

$$f_x(x, y) = -x \quad \text{and} \quad f_y(x, y) = -2y. \qquad \text{Partial derivatives}$$

So, in the x-direction, the slope is

$$f_x\left(\frac{1}{2}, 1\right) = -\frac{1}{2} \qquad \qquad \text{Figure 13.30}$$

and in the y-direction, the slope is

$$f_y\left(\frac{1}{2}, 1\right) = -2. \qquad \qquad \text{Figure 13.31}$$

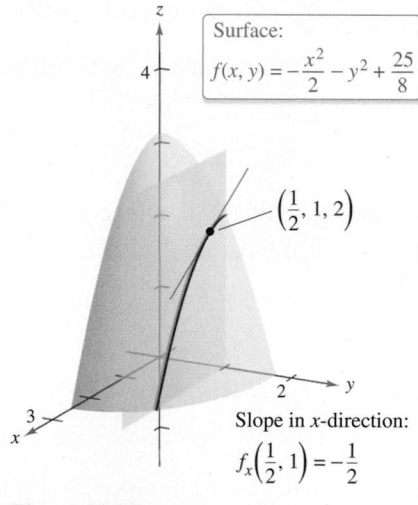

Figure 13.30

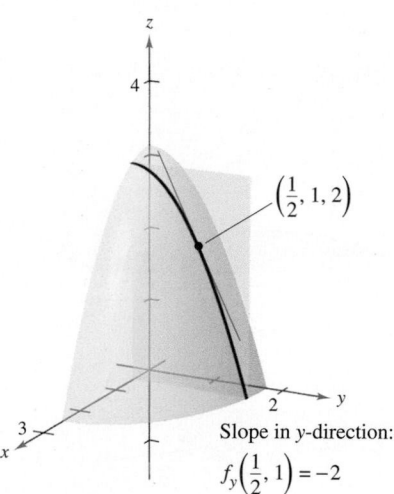

Figure 13.31

EXAMPLE 4 Finding the Slopes of a Surface

Find the slopes of the surface

$$f(x, y) = 1 - (x - 1)^2 - (y - 2)^2$$

at the point $(1, 2, 1)$ in the x-direction and in the y-direction.

Solution The partial derivatives of f with respect to x and y are

$$f_x(x, y) = -2(x - 1) \quad \text{and} \quad f_y(x, y) = -2(y - 2). \qquad \text{Partial derivatives}$$

So, at the point $(1, 2, 1)$, the slope in the x-direction is

$$f_x(1, 2) = -2(1 - 1) = 0$$

and the slope in the y-direction is

$$f_y(1, 2) = -2(2 - 2) = 0$$

as shown in Figure 13.32.

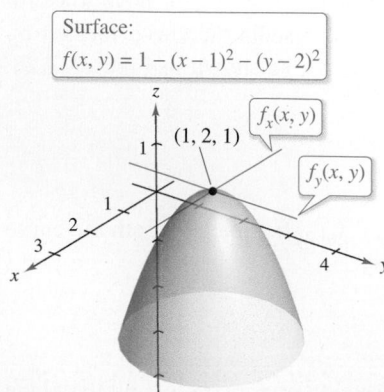

Figure 13.32

No matter how many variables are involved, partial derivatives can be interpreted as *rates of change.*

EXAMPLE 5 **Using Partial Derivatives to Find Rates of Change**

The area of a parallelogram with adjacent sides a and b and included angle θ is given by $A = ab \sin \theta$, as shown in Figure 13.33.

a. Find the rate of change of A with respect to a for $a = 10$, $b = 20$, and $\theta = \pi/6$.

b. Find the rate of change of A with respect to θ for $a = 10$, $b = 20$, and $\theta = \pi/6$.

Solution

a. To find the rate of change of the area with respect to a, hold b and θ constant and differentiate with respect to a to obtain

$$\frac{\partial A}{\partial a} = b \sin \theta. \qquad \text{Find partial derivative with respect to } a.$$

For $a = 10$, $b = 20$, and $\theta = \pi/6$, the rate of change of the area with respect to a is

$$\frac{\partial A}{\partial a} = 20 \sin \frac{\pi}{6} = 10. \qquad \text{Substitute for } b \text{ and } \theta.$$

b. To find the rate of change of the area with respect to θ, hold a and b constant and differentiate with respect to θ to obtain

$$\frac{\partial A}{\partial \theta} = ab \cos \theta. \qquad \text{Find partial derivative with respect to } \theta.$$

For $a = 10$, $b = 20$, and $\theta = \pi/6$, the rate of change of the area with respect to θ is

$$\frac{\partial A}{\partial \theta} = 200 \cos \frac{\pi}{6} = 100\sqrt{3}. \qquad \text{Substitute for } a, b, \text{ and } \theta. \qquad \blacksquare$$

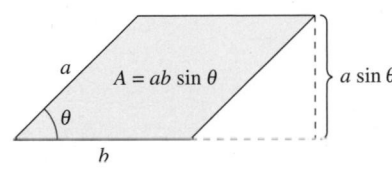

The area of the parallelogram is $ab \sin \theta$.

Figure 13.33

Partial Derivatives of a Function of Three or More Variables

The concept of a partial derivative can be extended naturally to functions of three or more variables. For instance, if $w = f(x, y, z)$, then there are three partial derivatives, each of which is formed by holding two of the variables constant. That is, to define the partial derivative of w with respect to x, consider y and z to be constant and differentiate with respect to x. A similar process is used to find the derivatives of w with respect to y and with respect to z.

$$\frac{\partial w}{\partial x} = f_x(x, y, z) = \lim_{\Delta x \to 0} \frac{f(x + \Delta x, y, z) - f(x, y, z)}{\Delta x}$$

$$\frac{\partial w}{\partial y} = f_y(x, y, z) = \lim_{\Delta y \to 0} \frac{f(x, y + \Delta y, z) - f(x, y, z)}{\Delta y}$$

$$\frac{\partial w}{\partial z} = f_z(x, y, z) = \lim_{\Delta z \to 0} \frac{f(x, y, z + \Delta z) - f(x, y, z)}{\Delta z}$$

In general, if $w = f(x_1, x_2, \ldots, x_n)$, then there are n partial derivatives denoted by

$$\frac{\partial w}{\partial x_k} = f_{x_k}(x_1, x_2, \ldots, x_n), \quad k = 1, 2, \ldots, n.$$

To find the partial derivative with respect to one of the variables, hold the other variables constant and differentiate with respect to the given variable.

EXAMPLE 6   **Finding Partial Derivatives**

a. To find the partial derivative of $f(x, y, z) = xy + yz^2 + xz$ with respect to z, consider x and y to be constant and obtain

$$\frac{\partial}{\partial z}[xy + yz^2 + xz] = 2yz + x.$$

b. To find the partial derivative of $f(x, y, z) = z \sin(xy^2 + 2z)$ with respect to z, consider x and y to be constant. Then, using the Product Rule, you obtain

$$\frac{\partial}{\partial z}[z \sin(xy^2 + 2z)] = (z)\frac{\partial}{\partial z}[\sin(xy^2 + 2z)] + \sin(xy^2 + 2z)\frac{\partial}{\partial z}[z]$$

$$= (z)[\cos(xy^2 + 2z)](2) + \sin(xy^2 + 2z)$$

$$= 2z \cos(xy^2 + 2z) + \sin(xy^2 + 2z).$$

c. To find the partial derivative of

$$f(x, y, z, w) = \frac{x + y + z}{w}$$

with respect to w, consider x, y, and z to be constant and obtain

$$\frac{\partial}{\partial w}\left[\frac{x + y + z}{w}\right] = -\frac{x + y + z}{w^2}.$$

Higher-Order Partial Derivatives

As is true for ordinary derivatives, it is possible to take second, third, and higher-order partial derivatives of a function of several variables, provided such derivatives exist. Higher-order derivatives are denoted by the order in which the differentiation occurs. For instance, the function $z = f(x, y)$ has the following second partial derivatives.

1. Differentiate twice with respect to x:

$$\frac{\partial}{\partial x}\left(\frac{\partial f}{\partial x}\right) = \frac{\partial^2 f}{\partial x^2} = f_{xx}.$$

2. Differentiate twice with respect to y:

$$\frac{\partial}{\partial y}\left(\frac{\partial f}{\partial y}\right) = \frac{\partial^2 f}{\partial y^2} = f_{yy}.$$

3. Differentiate first with respect to x and then with respect to y:

$$\frac{\partial}{\partial y}\left(\frac{\partial f}{\partial x}\right) = \frac{\partial^2 f}{\partial y \partial x} = f_{xy}.$$

4. Differentiate first with respect to y and then with respect to x:

$$\frac{\partial}{\partial x}\left(\frac{\partial f}{\partial y}\right) = \frac{\partial^2 f}{\partial x \partial y} = f_{yx}.$$

The third and fourth cases are called **mixed partial derivatives.**

• • REMARK Note that the two types of notation for mixed partials have different conventions for indicating the order of differentiation.

$$\frac{\partial}{\partial y}\left(\frac{\partial f}{\partial x}\right) = \frac{\partial^2 f}{\partial y \partial x} \quad \text{Right-to-left order}$$

$$(f_x)_y = f_{xy} \quad \text{Left-to-right order}$$

You can remember the order by observing that in both notations you differentiate first with respect to the variable "nearest" f.

EXAMPLE 7 **Finding Second Partial Derivatives**

Find the second partial derivatives of

$$f(x, y) = 3xy^2 - 2y + 5x^2y^2$$

and determine the value of $f_{xy}(-1, 2)$.

Solution Begin by finding the first partial derivatives with respect to x and y.

$$f_x(x, y) = 3y^2 + 10xy^2 \quad \text{and} \quad f_y(x, y) = 6xy - 2 + 10x^2y$$

Then, differentiate each of these with respect to x and y.

$$f_{xx}(x, y) = 10y^2 \qquad \text{and} \quad f_{yy}(x, y) = 6x + 10x^2$$
$$f_{xy}(x, y) = 6y + 20xy \quad \text{and} \quad f_{yx}(x, y) = 6y + 20xy$$

At $(-1, 2)$, the value of f_{xy} is

$$f_{xy}(-1, 2) = 12 - 40 = -28.$$

Notice in Example 7 that the two mixed partials are equal. Sufficient conditions for this occurrence are given in Theorem 13.3.

THEOREM 13.3 **Equality of Mixed Partial Derivatives**

If f is a function of x and y such that f_{xy} and f_{yx} are continuous on an open disk R, then, for every (x, y) in R,

$$f_{xy}(x, y) = f_{yx}(x, y).$$

Theorem 13.3 also applies to a function f of *three or more variables* as long as all second partial derivatives are continuous. For example, if

$$w = f(x, y, z) \qquad \text{Function of three variables}$$

and all the second partial derivatives are continuous in an open region R, then at each point in R, the order of differentiation in the mixed second partial derivatives is irrelevant. If the third partial derivatives of f are also continuous, then the order of differentiation of the mixed third partial derivatives is irrelevant.

EXAMPLE 8 **Finding Higher-Order Partial Derivatives**

Show that $f_{xz} = f_{zx}$ and $f_{xzz} = f_{zxz} = f_{zzx}$ for the function

$$f(x, y, z) = ye^x + x \ln z.$$

Solution

First partials:

$$f_x(x, y, z) = ye^x + \ln z, \quad f_z(x, y, z) = \frac{x}{z}$$

Second partials (note that the first two are equal):

$$f_{xz}(x, y, z) = \frac{1}{z}, \quad f_{zx}(x, y, z) = \frac{1}{z}, \quad f_{zz}(x, y, z) = -\frac{x}{z^2}$$

Third partials (note that all three are equal):

$$f_{xzz}(x, y, z) = -\frac{1}{z^2}, \quad f_{zxz}(x, y, z) = -\frac{1}{z^2}, \quad f_{zzx}(x, y, z) = -\frac{1}{z^2}$$

13.3 Exercises

See CalcChat.com for tutorial help and worked-out solutions to odd-numbered exercises.

CONCEPT CHECK

1. **First Partial Derivatives** List three ways of writing the first partial derivative with respect to x of $z = f(x, y)$.

2. **First Partial Derivatives** Sketch a surface representing a function f of two variables x and y. Use the sketch to give geometric interpretations of $\partial f/\partial x$ and $\partial f/\partial y$.

3. **Higher-Order Partial Derivatives** Describe the order in which the differentiation of $f(x, y, z)$ occurs for (a) f_{yxz} and (b) $\partial^2 f/\partial x \partial z$.

4. **Mixed Partial Derivatives** If f is a function of x and y such that f_{xy} and f_{yx} are continuous, what is the relationship between the mixed partial derivatives?

Examining a Partial Derivative In Exercises 5–10, explain whether the Quotient Rule should be used to find the partial derivative. Do not differentiate.

5. $\dfrac{\partial}{\partial x}\left(\dfrac{x^2 y}{y^2 - 3}\right)$

6. $\dfrac{\partial}{\partial y}\left(\dfrac{x^2 y}{y^2 - 3}\right)$

7. $\dfrac{\partial}{\partial y}\left(\dfrac{x - y}{x^2 + 1}\right)$

8. $\dfrac{\partial}{\partial x}\left(\dfrac{x - y}{x^2 + 1}\right)$

9. $\dfrac{\partial}{\partial x}\left(\dfrac{xy}{x^2 + 1}\right)$

10. $\dfrac{\partial}{\partial y}\left(\dfrac{xy}{x^2 + 1}\right)$

 **Finding Partial Derivatives** In Exercises 11–40, find both first partial derivatives.

11. $f(x, y) = 2x - 5y + 3$

12. $f(x, y) = x^2 - 2y^2 + 4$

13. $z = 6x - x^2 y + 8y^2$

14. $f(x, y) = 4x^3 y^{-2}$

15. $z = x\sqrt{y}$

16. $z = 2y^2\sqrt{x}$

17. $z = e^{xy}$

18. $z = e^{x/y}$

19. $z = x^2 e^{2y}$

20. $z = 7y e^{y/x}$

21. $z = \ln \dfrac{x}{y}$

22. $z = \ln\sqrt{xy}$

23. $z = \ln(x^2 + y^2)$

24. $z = \ln\dfrac{x + y}{x - y}$

25. $z = \dfrac{x^2}{2y} + \dfrac{3y^2}{x}$

26. $z = \dfrac{xy}{x^2 + y^2}$

27. $h(x, y) = e^{-(x^2 + y^2)}$

28. $g(x, y) = \ln\sqrt{x^2 + y^2}$

29. $f(x, y) = \sqrt{x^2 + y^2}$

30. $f(x, y) = \sqrt{2x + y^3}$

31. $z = \cos xy$

32. $z = \sin(x + 2y)$

33. $z = \tan(2x - y)$

34. $z = \sin 5x \cos 5y$

35. $z = e^y \sin 8xy$

36. $z = \cos(x^2 + y^2)$

37. $z = \sinh(2x + 3y)$

38. $z = \cosh xy^2$

39. $f(x, y) = \displaystyle\int_x^y (t^2 - 1)\, dt$

40. $f(x, y) = \displaystyle\int_x^y (2t + 1)\, dt + \int_y^x (2t - 1)\, dt$

 **Finding Partial Derivatives** In Exercises 41–44, use the limit definition of partial derivatives to find $f_x(x, y)$ and $f_y(x, y)$.

41. $f(x, y) = 3x + 2y$

42. $f(x, y) = x^2 - 2xy + y^2$

43. $f(x, y) = \sqrt{x + y}$

44. $f(x, y) = \dfrac{1}{x + y}$

 **Finding and Evaluating Partial Derivatives** In Exercises 45–52, find f_x and f_y, and evaluate each at the given point.

45. $f(x, y) = e^x y^2$, $(\ln 3, 2)$

46. $f(x, y) = x^3 \ln 5y$, $(1, 1)$

47. $f(x, y) = \cos(2x - y)$, $\left(\dfrac{\pi}{4}, \dfrac{\pi}{3}\right)$

48. $f(x, y) = \sin xy$, $\left(2, \dfrac{\pi}{4}\right)$

49. $f(x, y) = \arctan\dfrac{y}{x}$, $(2, -2)$

50. $f(x, y) = \arccos xy$, $(1, 1)$

51. $f(x, y) = \dfrac{xy}{x - y}$, $(2, -2)$

52. $f(x, y) = \dfrac{2xy}{\sqrt{4x^2 + 5y^2}}$, $(1, 1)$

Finding the Slopes of a Surface In Exercises 53–56, find the slopes of the surface in the x- and y-directions at the given point.

53. $z = xy$

$(1, 2, 2)$

54. $z = \sqrt{25 - x^2 - y^2}$

$(3, 0, 4)$

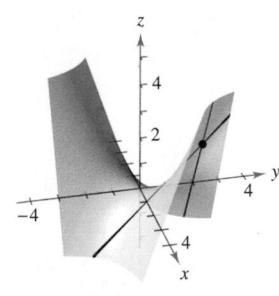

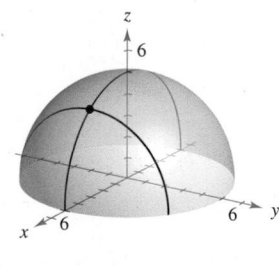

55. $g(x, y) = 4 - x^2 - y^2$

$(1, 1, 2)$

56. $h(x, y) = x^2 - y^2$

$(-2, 1, 3)$

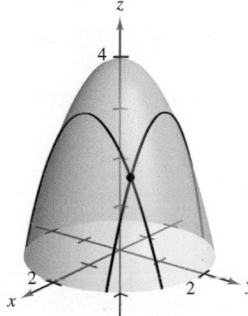

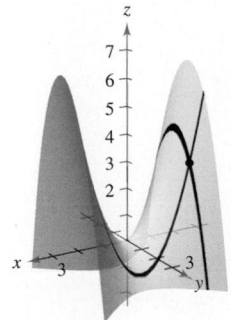

 **Finding Partial Derivatives** In Exercises 57–62, find the first partial derivatives with respect to x, y, and z.

57. $H(x, y, z) = \sin(x + 2y + 3z)$

58. $f(x, y, z) = 3x^2y - 5xyz + 10yz^2$

59. $w = \sqrt{x^2 + y^2 + z^2}$

60. $w = \dfrac{7xz}{x + y}$

61. $F(x, y, z) = \ln\sqrt{x^2 + y^2 + z^2}$

62. $G(x, y, z) = \dfrac{1}{\sqrt{1 - x^2 - y^2 - z^2}}$

Finding and Evaluating Partial Derivatives In Exercises 63–68, find f_x, f_y, and f_z, and evaluate each at the given point.

63. $f(x, y, z) = x^3yz^2$, $(1, 1, 1)$

64. $f(x, y, z) = x^2y^3 + 2xyz - 3yz$, $(-2, 1, 2)$

65. $f(x, y, z) = \dfrac{\ln x}{yz}$, $(1, -1, -1)$

66. $f(x, y, z) = \dfrac{xy}{x + y + z}$, $(3, 1, -1)$

67. $f(x, y, z) = z \sin(x + 6y)$, $\left(0, \dfrac{\pi}{2}, -4\right)$

68. $f(x, y, z) = \sqrt{3x^2 + y^2 - 2z^2}$, $(1, -2, 1)$

Using First Partial Derivatives In Exercises 69–76, find all values of x and y such that $f_x(x, y) = 0$ and $f_y(x, y) = 0$ simultaneously.

69. $f(x, y) = x^2 + xy + y^2 - 2x + 2y$

70. $f(x, y) = x^2 - xy + y^2 - 5x + y$

71. $f(x, y) = x^2 + 4xy + y^2 - 4x + 16y + 3$

72. $f(x, y) = x^2 - xy + y^2$

73. $f(x, y) = \dfrac{1}{x} + \dfrac{1}{y} + xy$

74. $f(x, y) = 3x^3 - 12xy + y^3$

75. $f(x, y) = e^{x^2 + xy + y^2}$

76. $f(x, y) = \ln(x^2 + y^2 + 1)$

 **Finding Second Partial Derivatives** In Exercises 77–86, find the four second partial derivatives. Observe that the second mixed partials are equal.

77. $z = 3xy^2$ **78.** $z = x^2 + 3y^2$

79. $z = x^4 - 2xy + 3y^3$ **80.** $z = x^4 - 3x^2y^2 + y^4$

81. $z = \sqrt{x^2 + y^2}$

82. $z = \ln(x - y)$

83. $z = e^x \tan y$

84. $z = 2xe^y - 3ye^{-x}$

85. $z = \cos xy$

86. $z = \arctan \dfrac{y}{x}$

 **Finding Partial Derivatives Using Technology** In Exercises 87–90, use a computer algebra system to find the first and second partial derivatives of the function. Determine whether there exist values of x and y such that $f_x(x, y) = 0$ and $f_y(x, y) = 0$ simultaneously.

87. $f(x, y) = x \sec y$ **88.** $f(x, y) = \sqrt{25 - x^2 - y^2}$

89. $f(x, y) = \ln \dfrac{x}{x^2 + y^2}$ **90.** $f(x, y) = \dfrac{xy}{x - y}$

Finding Higher-Order Partial Derivatives In Exercises 91–94, show that the mixed partial derivatives f_{xyy}, f_{yxy}, and f_{yyx} are equal.

91. $f(x, y, z) = xyz$

92. $f(x, y, z) = x^2 - 3xy + 4yz + z^3$

93. $f(x, y, z) = e^{-x} \sin yz$

94. $f(x, y, z) = \dfrac{2z}{x + y}$

Laplace's Equation In Exercises 95–98, show that the function satisfies Laplace's equation $\partial^2 z / \partial x^2 + \partial^2 z / \partial y^2 = 0$.

95. $z = 5xy$ **96.** $z = \frac{1}{2}(e^y - e^{-y}) \sin x$

97. $z = e^x \sin y$ **98.** $z = \arctan \dfrac{y}{x}$

Wave Equation In Exercises 99–102, show that the function satisfies the wave equation $\partial^2 z / \partial t^2 = c^2(\partial^2 z / \partial x^2)$.

99. $z = \sin(x - ct)$

100. $z = \cos(4x + 4ct)$

101. $z = \ln(x + ct)$

102. $z = \sin \omega ct \sin \omega x$

Heat Equation In Exercises 103 and 104, show that the function satisfies the heat equation $\partial z / \partial t = c^2(\partial^2 z / \partial x^2)$.

103. $z = e^{-t} \cos \dfrac{x}{c}$

104. $z = e^{-t} \sin \dfrac{x}{c}$

Cauchy-Riemann Equations In Exercises 105 and 106, show that the functions u and v satisfy the Cauchy-Riemann equations

$$\dfrac{\partial u}{\partial x} = \dfrac{\partial v}{\partial y} \quad \text{and} \quad \dfrac{\partial u}{\partial y} = -\dfrac{\partial v}{\partial x}.$$

105. $u = x^2 - y^2$, $v = 2xy$

106. $u = e^x \cos y$, $v = e^x \sin y$

Using First Partial Derivatives In Exercises 107 and 108, determine whether there exists a function $f(x, y)$ with the given partial derivatives. Explain your reasoning. If such a function exists, give an example.

107. $f_x(x, y) = -3 \sin(3x - 2y)$, $f_y(x, y) = 2 \sin(3x - 2y)$

108. $f_x(x, y) = 2x + y$, $f_y(x, y) = x - 4y$

EXPLORING CONCEPTS

109. Think About It Consider $z = f(x, y)$ such that $z_x = z_y$. Does $z = c(x + y)$? Explain.

110. First Partial Derivatives Given $z = f(x)g(y)$, find $z_x + z_y$.

111. Sketching a Graph Sketch the graph of a function $z = f(x, y)$ whose derivative f_x is always negative and whose derivative f_y is always positive.

112. Sketching a Graph Sketch the graph of a function $z = f(x, y)$ whose derivatives f_x and f_y are always positive.

113. Think About It The price P (in dollars) of a used car is a function of its initial cost C (in dollars) and its age A (in years). What are the units of $\partial P/\partial A$? Is $\partial P/\partial A$ positive or negative? Explain.

---

114. **HOW DO YOU SEE IT?** Use the graph of the surface to determine the sign of each partial derivative. Explain your reasoning.

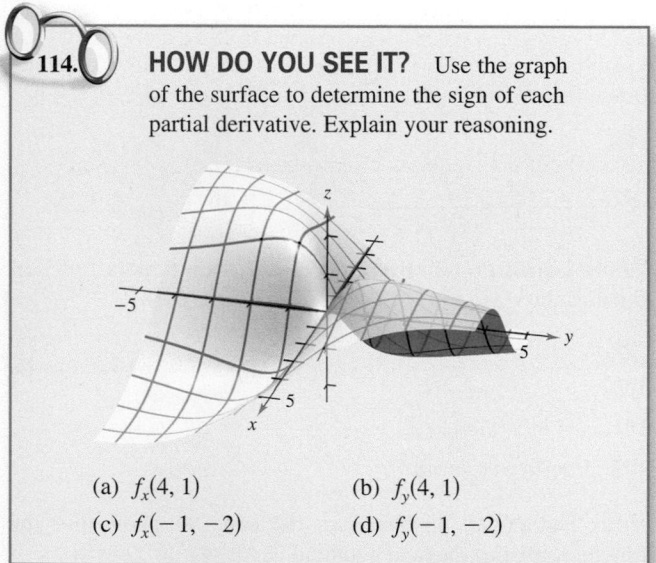

(a) $f_x(4, 1)$ (b) $f_y(4, 1)$

(c) $f_x(-1, -2)$ (d) $f_y(-1, -2)$

---

115. Area The area of a triangle is represented by $A = \frac{1}{2}ab \sin \theta$, where a and b are two of the side lengths and θ is the angle between a and b.

(a) Find the rate of change of A with respect to b for $a = 4$, $b = 1$, and $\theta = \pi/4$.

(b) Find the rate of change of A with respect to θ for $a = 2$, $b = 5$, and $\theta = \pi/3$.

116. Volume The volume of a right-circular cone of radius r and height h is represented by $V = \frac{1}{3}\pi r^2 h$.

(a) Find the rate of change of V with respect to r for $r = 2$ and $h = 2$.

(b) Find the rate of change of V with respect to h for $r = 2$ and $h = 2$.

117. Marginal Revenue A pharmaceutical corporation has two plants that produce the same over-the-counter medicine. If x_1 and x_2 are the numbers of units produced at plant 1 and plant 2, respectively, then the total revenue for the product is given by $R = 200x_1 + 200x_2 - 4x_1^2 - 8x_1x_2 - 4x_2^2$. When $x_1 = 4$ and $x_2 = 12$, find (a) the marginal revenue for plant 1, $\partial R/\partial x_1$, and (b) the marginal revenue for plant 2, $\partial R/\partial x_2$.

118. Marginal Costs

A company manufactures two types of wood-burning stoves: a freestanding model and a fireplace-insert model. The cost function for producing x freestanding and y fireplace-insert stoves is

$$C = 32\sqrt{xy} + 175x + 205y + 1050.$$

(a) Find the marginal costs ($\partial C/\partial x$ and $\partial C/\partial y$) when $x = 80$ and $y = 20$.

(b) When additional production is required, which model of stove results in the cost increasing at a higher rate? How can this be determined from the cost model?

119. Psychology Early in the twentieth century, an intelligence test called the *Stanford-Binet Test* (more commonly known as the IQ test) was developed. In this test, an individual's mental age M is divided by the individual's chronological age C and then the quotient is multiplied by 100. The result is the individual's *IQ*.

$$IQ(M, C) = \frac{M}{C} \times 100$$

Find the partial derivatives of *IQ* with respect to M and with respect to C. Evaluate the partial derivatives at the point $(12, 10)$ and interpret the result. *(Source: Adapted from Bernstein/Clark-Stewart/Roy/Wickens, Psychology, Fourth Edition)*

120. Marginal Productivity Consider the Cobb-Douglas production function $f(x, y) = 200x^{0.7}y^{0.3}$. When $x = 1000$ and $y = 500$, find (a) the marginal productivity of labor, $\partial f/\partial x$, and (b) the marginal productivity of capital, $\partial f/\partial y$.

121. Think About It Let N be the number of applicants to a university, p the charge for food and housing at the university, and t the tuition. Suppose that N is a function of p and t such that $\partial N/\partial p < 0$ and $\partial N/\partial t < 0$. What information is gained by noticing that both partials are negative?

122. Investment The value of an investment of \$1000 earning 6% compounded annually is

$$V(I, R) = 1000\left[\frac{1 + 0.06(1 - R)}{1 + I}\right]^{10}$$

where I is the annual rate of inflation and R is the tax rate for the person making the investment. Calculate $V_I(0.03, 0.28)$ and $V_R(0.03, 0.28)$. Determine whether the tax rate or the rate of inflation is the greater "negative" factor in the growth of the investment.

123. Temperature Distribution The temperature at any point (x, y) on a steel plate is $T = 500 - 0.6x^2 - 1.5y^2$, where x and y are measured in meters. At the point $(2, 3)$, find the rates of change of the temperature with respect to the distances moved along the plate in the directions of the x- and y-axes.

124. Apparent Temperature A measure of how hot weather feels to an average person is the Apparent Temperature Index. A model for this index is

$$A = 0.885t - 22.4h + 1.20th - 0.544$$

where A is the apparent temperature in degrees Celsius, t is the air temperature, and h is the relative humidity in decimal form. *(Source: The UMAP Journal)*

(a) Find $\dfrac{\partial A}{\partial t}$ and $\dfrac{\partial A}{\partial h}$ when $t = 30°$ and $h = 0.80$.

(b) Which has a greater effect on A, air temperature or humidity? Explain.

125. Ideal Gas Law The Ideal Gas Law states that

$$PV = nRT$$

where P is pressure, V is volume, n is the number of moles of gas, R is a fixed constant (the gas constant), and T is absolute temperature. Show that

$$\frac{\partial T}{\partial P} \cdot \frac{\partial P}{\partial V} \cdot \frac{\partial V}{\partial T} = -1.$$

126. Marginal Utility The utility function $U = f(x, y)$ is a measure of the utility (or satisfaction) derived by a person from the consumption of two products x and y. The utility function for two products is

$$U = -5x^2 + xy - 3y^2.$$

(a) Determine the marginal utility of product x.

(b) Determine the marginal utility of product y.

(c) When $x = 2$ and $y = 3$, should a person consume one more unit of product x or one more unit of product y? Explain your reasoning.

(d) Use a computer algebra system to graph the function. Interpret the marginal utilities of products x and y graphically.

127. Modeling Data Personal consumption expenditures (in billions of dollars) for several types of recreation from 2009 through 2014 are shown in the table, where x is the expenditures on amusement parks and campgrounds, y is the expenditures on live entertainment (excluding sports), and z is the expenditures on spectator sports. *(Source: U.S. Bureau of Economic Analysis)*

Year	2009	2010	2011	2012	2013	2014
x	37.2	38.8	41.3	44.6	47.0	50.3
y	25.2	26.3	28.3	28.5	28.0	30.0
z	18.8	19.2	20.4	20.6	21.6	22.4

A model for the data is given by

$$z = 0.23x + 0.14y + 6.85.$$

(a) Find $\dfrac{\partial z}{\partial x}$ and $\dfrac{\partial z}{\partial y}$.

(b) Interpret the partial derivatives in the context of the problem.

128. Modeling Data The table shows the national health expenditures (in billions of dollars) for the Department of Veterans Affairs x, workers' compensation y, and Medicaid z from 2009 through 2014. *(Source: Centers for Medicare and Medicaid Services)*

Year	2009	2010	2011	2012	2013	2014
x	42.5	45.7	48.2	49.8	52.8	57.2
y	36.0	36.1	39.1	41.7	44.1	47.3
z	374.5	397.2	406.4	422.0	446.7	495.8

A model for the data is given by

$$z = -0.120x^2 + 0.657y^2 + 17.70x - 51.53y + 842.5.$$

(a) Find $\dfrac{\partial^2 z}{\partial x^2}$ and $\dfrac{\partial^2 z}{\partial y^2}$.

(b) Determine the concavity of traces parallel to the xz-plane. Interpret the result in the context of the problem.

(c) Determine the concavity of traces parallel to the yz-plane. Interpret the result in the context of the problem.

129. Using a Function Consider the function defined by

$$f(x, y) = \begin{cases} \dfrac{xy(x^2 - y^2)}{x^2 + y^2}, & (x, y) \neq (0, 0) \\ 0, & (x, y) = (0, 0) \end{cases}.$$

(a) Find $f_x(x, y)$ and $f_y(x, y)$ for $(x, y) \neq (0, 0)$.

(b) Use the definition of partial derivatives to find $f_x(0, 0)$ and $f_y(0, 0)$.

$$\left[\textit{Hint: } f_x(0, 0) = \lim_{\Delta x \to 0} \frac{f(\Delta x, 0) - f(0, 0)}{\Delta x} \right]$$

(c) Use the definition of partial derivatives to find $f_{xy}(0, 0)$ and $f_{yx}(0, 0)$.

(d) Using Theorem 13.3 and the result of part (c), what can be said about f_{xy} or f_{yx}?

130. Using a Function Consider the function

$$f(x, y) = (x^3 + y^3)^{1/3}.$$

(a) Find $f_x(0, 0)$ and $f_y(0, 0)$.

(b) Determine the points (if any) at which $f_x(x, y)$ or $f_y(x, y)$ fails to exist.

131. Using a Function Consider the function

$$f(x, y) = (x^2 + y^2)^{2/3}.$$

Show that

$$f_x(x, y) = \begin{cases} \dfrac{4x}{3(x^2 + y^2)^{1/3}}, & (x, y) \neq (0, 0) \\ 0, & (x, y) = (0, 0) \end{cases}.$$

■ **FOR FURTHER INFORMATION** For more information about this problem, see the article "A Classroom Note on a Naturally Occurring Piecewise Defined Function" by Don Cohen in *Mathematics and Computer Education.*

13.4 Differentials

■ Understand the concepts of increments and differentials.
■ Extend the concept of differentiability to a function of two variables.
■ Use a differential as an approximation.

Increments and Differentials

In this section, the concepts of increments and differentials are generalized to functions of two or more variables. Recall from Section 4.8 that for $y = f(x)$, the differential of y was defined as

$$dy = f'(x)\, dx.$$

Similar terminology is used for a function of two variables, $z = f(x, y)$. That is, Δx and Δy are the **increments of x and y,** and the **increment of z** is

$$\Delta z = f(x + \Delta x, y + \Delta y) - f(x, y). \qquad \text{Increment of } z$$

Definition of Total Differential

If $z = f(x, y)$ and Δx and Δy are increments of x and y, then the **differentials** of the independent variables x and y are

$$dx = \Delta x \quad \text{and} \quad dy = \Delta y$$

and the **total differential** of the dependent variable z is

$$dz = \frac{\partial z}{\partial x}\, dx + \frac{\partial z}{\partial y}\, dy = f_x(x, y)\, dx + f_y(x, y)\, dy.$$

This definition can be extended to a function of three or more variables. For instance, if $w = f(x, y, z, u)$, then $dx = \Delta x$, $dy = \Delta y$, $dz = \Delta z$, $du = \Delta u$, and the total differential of w is

$$dw = \frac{\partial w}{\partial x}\, dx + \frac{\partial w}{\partial y}\, dy + \frac{\partial w}{\partial z}\, dz + \frac{\partial w}{\partial u}\, du.$$

EXAMPLE 1 Finding the Total Differential

Find the total differential for each function.

a. $z = 2x \sin y - 3x^2 y^2$ **b.** $w = x^2 + y^2 + z^2$

Solution

a. The total differential dz for $z = 2x \sin y - 3x^2 y^2$ is

$$dz = \frac{\partial z}{\partial x}\, dx + \frac{\partial z}{\partial y}\, dy \qquad \text{Total differential } dz$$

$$= (2 \sin y - 6xy^2)\, dx + (2x \cos y - 6x^2 y)\, dy.$$

b. The total differential dw for $w = x^2 + y^2 + z^2$ is

$$dw = \frac{\partial w}{\partial x}\, dx + \frac{\partial w}{\partial y}\, dy + \frac{\partial w}{\partial z}\, dz \qquad \text{Total differential } dw$$

$$= 2x\, dx + 2y\, dy + 2z\, dz.$$

Differentiability

In Section 4.8, you learned that for a *differentiable* function given by $y = f(x)$, you can use the differential $dy = f'(x)\,dx$ as an approximation (for small Δx) of the value $\Delta y = f(x + \Delta x) - f(x)$. When a similar approximation is possible for a function of two variables, the function is said to be **differentiable.** This is stated explicitly in the next definition.

Definition of Differentiability

A function f given by $z = f(x, y)$ is **differentiable** at (x_0, y_0) if Δz can be written in the form

$$\Delta z = f_x(x_0, y_0)\,\Delta x + f_y(x_0, y_0)\,\Delta y + \varepsilon_1 \Delta x + \varepsilon_2 \Delta y$$

where both ε_1 and $\varepsilon_2 \to 0$ as

$$(\Delta x, \Delta y) \to (0, 0).$$

The function f is **differentiable in a region R** if it is differentiable at each point in R.

EXAMPLE 2   **Showing that a Function Is Differentiable**

Show that the function

$$f(x, y) = x^2 + 3y$$

is differentiable at every point in the plane.

Solution   Letting $z = f(x, y)$, the increment of z at an arbitrary point (x, y) in the plane is

$$\begin{aligned}
\Delta z &= f(x + \Delta x, y + \Delta y) - f(x, y) &&\text{Increment of } z \\
&= (x + \Delta x)^2 + 3(y + \Delta y) - (x^2 + 3y) \\
&= x^2 + 2x\Delta x + (\Delta x)^2 + 3y + 3\Delta y - x^2 - 3y \\
&= 2x\Delta x + (\Delta x)^2 + 3\Delta y \\
&= 2x(\Delta x) + 3(\Delta y) + \Delta x(\Delta x) + 0(\Delta y) \\
&= f_x(x, y)\Delta x + f_y(x, y)\Delta y + \varepsilon_1 \Delta x + \varepsilon_2 \Delta y
\end{aligned}$$

where $\varepsilon_1 = \Delta x$ and $\varepsilon_2 = 0$. Because $\varepsilon_1 \to 0$ and $\varepsilon_2 \to 0$ as $(\Delta x, \Delta y) \to (0, 0)$, it follows that f is differentiable at every point in the plane. The graph of f is shown in Figure 13.34.

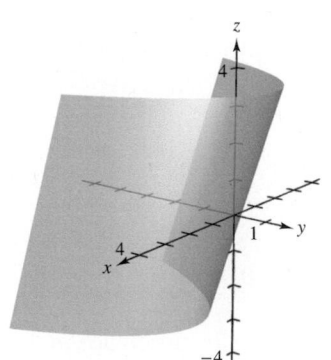

Figure 13.34

Be sure you see that the term "differentiable" is used differently for functions of two variables than for functions of one variable. A function of one variable is differentiable at a point when its derivative exists at the point. For a function of two variables, however, the existence of the partial derivatives f_x and f_y does not guarantee that the function is differentiable (see Example 5). The next theorem gives a *sufficient* condition for differentiability of a function of two variables.

THEOREM 13.4   Sufficient Condition for Differentiability

If f is a function of x and y, where f_x and f_y are continuous in an open region R, then f is differentiable on R.
A proof of this theorem is given in Appendix A.

Approximation by Differentials

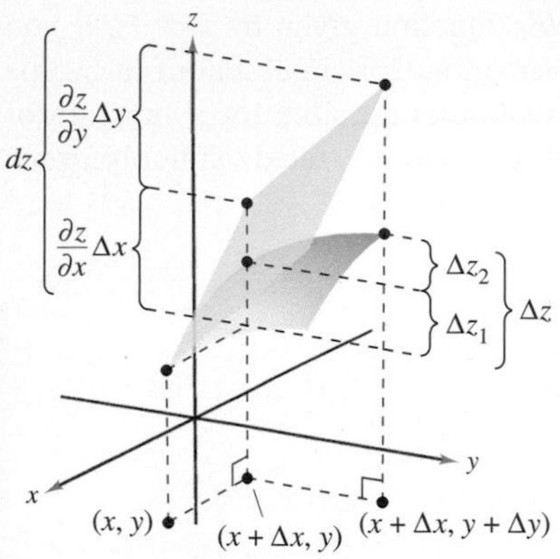

Theorem 13.4 tells you that you can choose $(x + \Delta x, y + \Delta y)$ close enough to (x, y) to make $\varepsilon_1 \Delta x$ and $\varepsilon_2 \Delta y$ insignificant. In other words, for small Δx and Δy, you can use the approximation

$$\Delta z \approx dz. \qquad \text{Approximate change in } z$$

This approximation is illustrated graphically in Figure 13.35. Recall that the partial derivatives $\partial z / \partial x$ and $\partial z / \partial y$ can be interpreted as the slopes of the surface in the x- and y-directions. This means that

$$dz = \frac{\partial z}{\partial x} \Delta x + \frac{\partial z}{\partial y} \Delta y$$

represents the change in height of a plane that is tangent to the surface at the point $(x, y, f(x, y))$. Because a plane in space is represented by a linear equation in the variables x, y, and z, the approximation of Δz by dz is called a **linear approximation.** You will learn more about this geometric interpretation in Section 13.7.

The exact change in z is Δz. This change can be approximated by the differential dz.

Figure 13.35

EXAMPLE 3   **Using a Differential as an Approximation**

 · · · ▷ *See LarsonCalculus.com for an interactive version of this type of example.*

Use the differential dz to approximate the change in

$$z = \sqrt{4 - x^2 - y^2}$$

as (x, y) moves from the point $(1, 1)$ to the point $(1.01, 0.97)$. Compare this approximation with the exact change in z.

Solution   Letting $(x, y) = (1, 1)$ and $(x + \Delta x, y + \Delta y) = (1.01, 0.97)$ produces

$$dx = \Delta x = 0.01 \quad \text{and} \quad dy = \Delta y = -0.03.$$

So, the change in z can be approximated by

$$\Delta z \approx dz = \frac{\partial z}{\partial x} dx + \frac{\partial z}{\partial y} dy = \frac{-x}{\sqrt{4 - x^2 - y^2}} \Delta x + \frac{-y}{\sqrt{4 - x^2 - y^2}} \Delta y.$$

When $x = 1$ and $y = 1$, you have

$$\Delta z \approx -\frac{1}{\sqrt{2}}(0.01) - \frac{1}{\sqrt{2}}(-0.03) = \frac{0.02}{\sqrt{2}} = \sqrt{2}(0.01) \approx 0.0141.$$

In Figure 13.36, you can see that the exact change corresponds to the difference in the heights of two points on the surface of a hemisphere. This difference is given by

$$\begin{aligned} \Delta z &= f(1.01, 0.97) - f(1, 1) \\ &= \sqrt{4 - (1.01)^2 - (0.97)^2} - \sqrt{4 - 1^2 - 1^2} \\ &\approx 0.0137. \end{aligned}$$

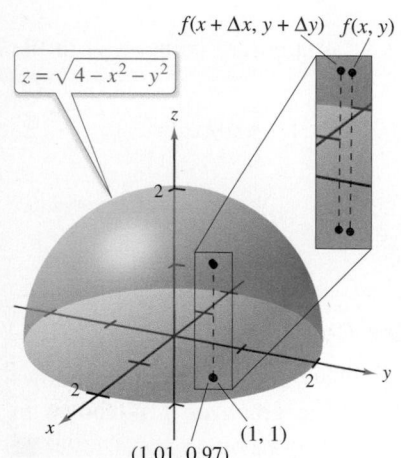

As (x, y) moves from the point $(1, 1)$ to the point $(1.01, 0.97)$, the value of $f(x, y)$ changes by about 0.0137.

Figure 13.36

A function of three variables $w = f(x, y, z)$ is **differentiable** at (x, y, z) provided that $\Delta w = f(x + \Delta x, y + \Delta y, z + \Delta z) - f(x, y, z)$ can be written in the form

$$\Delta w = f_x \Delta x + f_y \Delta y + f_z \Delta z + \varepsilon_1 \Delta x + \varepsilon_2 \Delta y + \varepsilon_3 \Delta z$$

where ε_1, ε_2, and $\varepsilon_3 \to 0$ as $(\Delta x, \Delta y, \Delta z) \to (0, 0, 0)$. With this definition of differentiability, Theorem 13.4 has the following extension for functions of three variables: If f is a function of x, y, and z, where f_x, f_y, and f_z are continuous in an open region R, then f is differentiable on R.

In Section 4.8, you used differentials to approximate the propagated error introduced by an error in measurement. This application of differentials is further illustrated in Example 4.

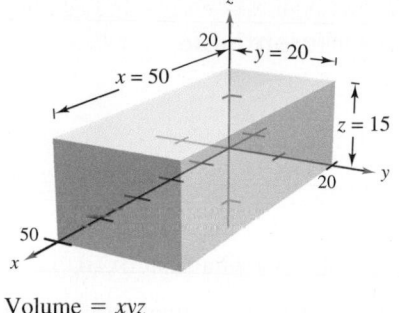

Volume = xyz
Figure 13.37

EXAMPLE 4 **Error Analysis**

The possible error involved in measuring each dimension of a rectangular box is ± 0.1 millimeter. The dimensions of the box are $x = 50$ centimeters, $y = 20$ centimeters, and $z = 15$ centimeters, as shown in Figure 13.37. Use dV to estimate the propagated error and the relative error in the calculated volume of the box.

Solution   The volume of the box is $V = xyz$, so

$$dV = \frac{\partial V}{\partial x}\,dx + \frac{\partial V}{\partial y}\,dy + \frac{\partial V}{\partial z}\,dz$$
$$= yz\,dx + xz\,dy + xy\,dz.$$

Using 0.1 millimeter $= 0.01$ centimeter, you have

$$dx = dy = dz = \pm 0.01$$

and the propagated error is approximately

$$dV = (20)(15)(\pm 0.01) + (50)(15)(\pm 0.01) + (50)(20)(\pm 0.01)$$
$$= 300(\pm 0.01) + 750(\pm 0.01) + 1000(\pm 0.01)$$
$$= 2050(\pm 0.01)$$
$$= \pm 20.5 \text{ cubic centimeters.}$$

Because the measured volume is

$$V = (50)(20)(15) = 15{,}000 \text{ cubic centimeters}$$

the relative error, $\Delta V/V$, is approximately

$$\frac{\Delta V}{V} \approx \frac{dV}{V} = \frac{\pm 20.5}{15{,}000} \approx \pm 0.0014$$

which is a percent error of about 0.14%.

As is true for a function of a single variable, when a function in two or more variables is differentiable at a point, it is also continuous there.

THEOREM 13.5   **Differentiability Implies Continuity**

If a function of x and y is differentiable at (x_0, y_0), then it is continuous at (x_0, y_0).

Proof   Let f be differentiable at (x_0, y_0), where $z = f(x, y)$. Then

$$\Delta z = [f_x(x_0, y_0) + \varepsilon_1]\,\Delta x + [f_y(x_0, y_0) + \varepsilon_2]\,\Delta y$$

where both ε_1 and $\varepsilon_2 \to 0$ as $(\Delta x, \Delta y) \to (0, 0)$. However, by definition, you know that Δz is

$$\Delta z = f(x_0 + \Delta x, y_0 + \Delta y) - f(x_0, y_0).$$

Letting $x = x_0 + \Delta x$ and $y = y_0 + \Delta y$ produces

$$f(x, y) - f(x_0, y_0) = [f_x(x_0, y_0) + \varepsilon_1]\,\Delta x + [f_y(x_0, y_0) + \varepsilon_2]\,\Delta y$$
$$= [f_x(x_0, y_0) + \varepsilon_1](x - x_0) + [f_y(x_0, y_0) + \varepsilon_2](y - y_0).$$

Taking the limit as $(x, y) \to (x_0, y_0)$, you have

$$\lim_{(x,\, y) \to (x_0,\, y_0)} f(x, y) = f(x_0, y_0)$$

which means that f is continuous at (x_0, y_0).

Remember that the existence of f_x and f_y is not sufficient to guarantee differentiability, as illustrated in the next example.

EXAMPLE 5 **A Function That Is Not Differentiable**

For the function

$$f(x, y) = \begin{cases} \dfrac{-3xy}{x^2 + y^2}, & (x, y) \neq (0, 0) \\ 0, & (x, y) = (0, 0) \end{cases}$$

show that $f_x(0, 0)$ and $f_y(0, 0)$ both exist but that f is not differentiable at $(0, 0)$.

Solution You can show that f is not differentiable at $(0, 0)$ by showing that it is not continuous at this point. To see that f is not continuous at $(0, 0)$, look at the values of $f(x, y)$ along two different approaches to $(0, 0)$, as shown in Figure 13.38. Along the line $y = x$, the limit is

$$\lim_{(x, x) \to (0, 0)} f(x, y) = \lim_{(x, x) \to (0, 0)} \frac{-3x^2}{2x^2} = -\frac{3}{2}$$

whereas along $y = -x$, you have

$$\lim_{(x, -x) \to (0, 0)} f(x, y) = \lim_{(x, -x) \to (0, 0)} \frac{3x^2}{2x^2} = \frac{3}{2}.$$

So, the limit of $f(x, y)$ as $(x, y) \to (0, 0)$ does not exist, and you can conclude that f is not continuous at $(0, 0)$. Therefore, by Theorem 13.5, you know that f is not differentiable at $(0, 0)$. On the other hand, by the definition of the partial derivatives f_x and f_y, you have

$$f_x(0, 0) = \lim_{\Delta x \to 0} \frac{f(\Delta x, 0) - f(0, 0)}{\Delta x} = \lim_{\Delta x \to 0} \frac{0 - 0}{\Delta x} = 0$$

and

$$f_y(0, 0) = \lim_{\Delta y \to 0} \frac{f(0, \Delta y) - f(0, 0)}{\Delta y} = \lim_{\Delta y \to 0} \frac{0 - 0}{\Delta y} = 0.$$

So, the partial derivatives at $(0, 0)$ exist.

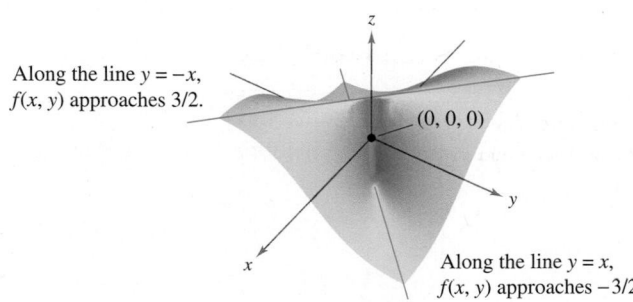

$$f(x, y) = \begin{cases} \dfrac{-3xy}{x^2 + y^2}, & (x, y) \neq (0, 0) \\ 0, & (x, y) = (0, 0) \end{cases}$$

Along the line $y = -x$,
$f(x, y)$ approaches 3/2.

$(0, 0, 0)$

Along the line $y = x$,
$f(x, y)$ approaches $-3/2$.

Figure 13.38

Generated by Mathematica

▷ **TECHNOLOGY** A graphing utility can be used to graph piecewise-defined functions like the one given in Example 5. For instance, the graph shown at the left was generated by *Mathematica*.

13.4 Exercises

See **CalcChat.com** for tutorial help and worked-out solutions to odd-numbered exercises.

CONCEPT CHECK

1. **Approximation** Describe the change in accuracy of dz as an approximation of Δz as Δx and Δy increase.

2. **Linear Approximation** What is meant by a linear approximation of $z = f(x, y)$ at the point $P(x_0, y_0)$?

 **Finding a Total Differential** In Exercises 3–8, find the total differential.

3. $z = 5x^3y^2$

4. $z = 2x^3y - 8xy^4$

5. $z = \frac{1}{2}\left(e^{x^2+y^2} - e^{-x^2-y^2}\right)$

6. $z = e^{-x}\tan y$

7. $w = x^2yz^2 + \sin yz$

8. $w = (x + y)/(z - 3y)$

 **Using a Differential as an Approximation** In Exercises 9–14, (a) find $f(2, 1)$ and $f(2.1, 1.05)$ and calculate Δz, and (b) use the total differential dz to approximate Δz.

9. $f(x, y) = 2x - 3y$

10. $f(x, y) = x^2 + y^2$

11. $f(x, y) = 16 - x^2 - y^2$

12. $f(x, y) = y/x$

13. $f(x, y) = ye^x$

14. $f(x, y) = x\cos y$

Approximating an Expression In Exercises 15–18, find $z = f(x, y)$ and use the total differential to approximate the quantity.

15. $(2.01)^2(9.02) - 2^2 \cdot 9$

16. $\dfrac{1 - (3.05)^2}{(5.95)^2} - \dfrac{1 - 3^2}{6^2}$

17. $\sin[(1.05)^2 + (0.95)^2] - \sin(1^2 + 1^2)$

18. $\sqrt{(4.03)^2 + (3.1)^2} - \sqrt{4^2 + 3^2}$

EXPLORING CONCEPTS

19. Continuity If f_x and f_y are each continuous in an open region R, is $f(x, y)$ continuous in R? Explain.

20. **HOW DO YOU SEE IT?** Which point has a greater differential, $(2, 2)$ or $\left(\frac{1}{2}, \frac{1}{2}\right)$? Explain. (Assume that dx and dy are the same for both points.)

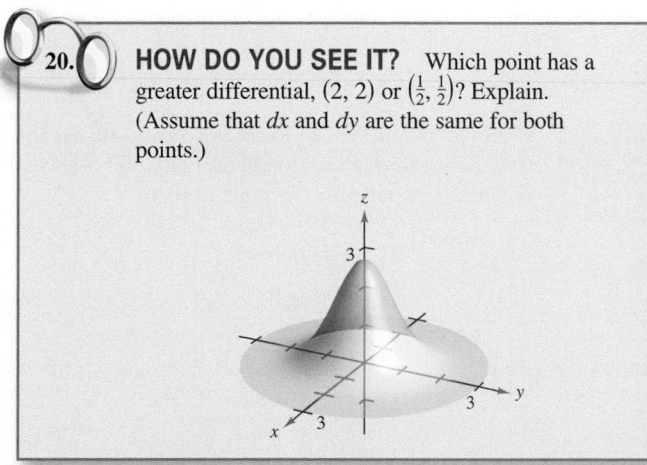

21. Area The area of the shaded rectangle in the figure is $A = lh$. The possible errors in the length and height are Δl and Δh, respectively. Find dA and identify the regions in the figure whose areas are given by the terms of dA. What region represents the difference between ΔA and dA?

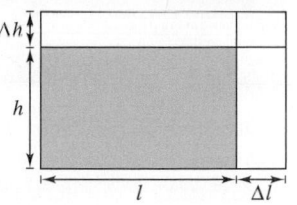

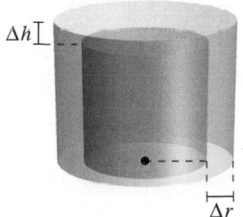

Figure for 21   Figure for 22

22. Volume The volume of the red right circular cylinder in the figure is $V = \pi r^2 h$. The possible errors in the radius and the height are Δr and Δh, respectively. Find dV and identify the solids in the figure whose volumes are given by the terms of dV. What solid represents the difference between ΔV and dV?

23. Volume The possible error involved in measuring each dimension of a rectangular box is ± 0.02 inch. The dimensions of the box are 8 inches by 5 inches by 12 inches. Approximate the propagated error and the relative error in the calculated volume of the box.

24. Volume The possible error involved in measuring each dimension of a right circular cylinder is ± 0.05 centimeter. The radius is 3 centimeters and the height is 10 centimeters. Approximate the propagated error and the relative error in the calculated volume of the cylinder.

25. Numerical Analysis A right circular cone of height $h = 8$ meters and radius $r = 4$ meters is constructed, and in the process, errors Δr and Δh are made in the radius and height, respectively. Let V be the volume of the cone. Complete the table to show the relationship between ΔV and dV for the indicated errors.

Δr	Δh	dV or dS	ΔV or ΔS	$\Delta V - dV$ or $\Delta S - dS$
0.1	0.1			
0.1	-0.1			
0.001	0.002			
-0.0001	0.0002			

Table for Exercises 25 and 26

26. Numerical Analysis A right circular cone of height $h = 16$ meters and radius $r = 6$ meters is constructed, and in the process, errors of Δr and Δh are made in the radius and height, respectively. Let S be the lateral surface area of the cone. Complete the table above to show the relationship between ΔS and dS for the indicated errors.

27. Wind Chill

The formula for wind chill C (in degrees Fahrenheit) is given by

$$C = 35.74 + 0.6215T - 35.75v^{0.16} + 0.4275Tv^{0.16}$$

where v is the wind speed in miles per hour and T is the temperature in degrees Fahrenheit. The wind speed is 23 ± 3 miles per hour and the temperature is $8° \pm 1°$. Use dC to estimate the maximum possible propagated error and relative error in calculating the wind chill. *(Source: National Oceanic and Atmospheric Administration)*

28. Resistance The total resistance R (in ohms) of two resistors connected in parallel is given by

$$\frac{1}{R} = \frac{1}{R_1} + \frac{1}{R_2}.$$

Approximate the change in R as R_1 is increased from 10 ohms to 10.5 ohms and R_2 is decreased from 15 ohms to 13 ohms.

29. Power Electrical power P is given by

$$P = \frac{E^2}{R}$$

where E is voltage and R is resistance. Approximate the maximum percent error in calculating power when 120 volts is applied to a 2000-ohm resistor and the possible percent errors in measuring E and R are 3% and 4%, respectively.

30. Acceleration The centripetal acceleration of a particle moving in a circle is $a = v^2/r$, where v is the velocity and r is the radius of the circle. Approximate the maximum percent error in measuring the acceleration due to errors of 3% in v and 2% in r.

31. Volume A trough is 16 feet long (see figure). Its cross sections are isosceles triangles with each of the two equal sides measuring 18 inches. The angle between the two equal sides is θ.

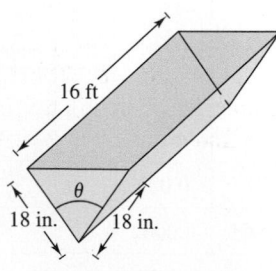

Not drawn to scale

(a) Write the volume of the trough as a function of θ and determine the value of θ such that the volume is a maximum.

(b) The maximum error in the linear measurements is one-half inch and the maximum error in the angle measure is 2°. Approximate the change in the maximum volume.

32. Sports A baseball player in center field is playing approximately 330 feet from a television camera that is behind home plate. A batter hits a fly ball that goes to the wall 420 feet from the camera (see figure).

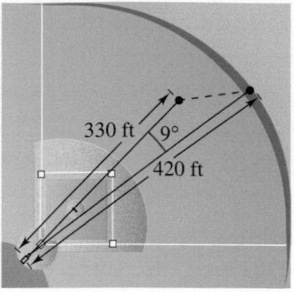

(a) The camera turns 9° to follow the play. Approximate the number of feet that the center fielder has to run to make the catch.

(b) The position of the center fielder could be in error by as much as 6 feet and the maximum error in measuring the rotation of the camera is 1°. Approximate the maximum possible error in the result of part (a).

33. Inductance The inductance L (in microhenrys) of a straight nonmagnetic wire in free space is

$$L = 0.00021\left(\ln\frac{2h}{r} - 0.75\right)$$

where h is the length of the wire in millimeters and r is the radius of a circular cross section. Approximate L when $r = 2 \pm \frac{1}{16}$ millimeters and $h = 100 \pm \frac{1}{100}$ millimeters.

34. Pendulum The period T of a pendulum of length L is $T = (2\pi\sqrt{L})/\sqrt{g}$, where g is the acceleration due to gravity. A pendulum is moved from the Canal Zone, where $g = 32.09$ feet per second per second, to Greenland, where $g = 32.23$ feet per second per second. Because of the change in temperature, the length of the pendulum changes from 2.5 feet to 2.48 feet. Approximate the change in the period of the pendulum.

Differentiability In Exercises 35–38, show that the function is differentiable by finding values of ε_1 and ε_2 as designated in the definition of differentiability, and verify that both ε_1 and ε_2 approach 0 as $(\Delta x, \Delta y) \to (0, 0)$.

35. $f(x, y) = x^2 - 2x + y$

36. $f(x, y) = x^2 + y^2$

37. $f(x, y) = x^2y$

38. $f(x, y) = 5x - 10y + y^3$

Differentiability In Exercises 39 and 40, use the function to show that $f_x(0, 0)$ and $f_y(0, 0)$ both exist but that f is not differentiable at $(0, 0)$.

39. $f(x, y) = \begin{cases} \dfrac{3x^2y}{x^4 + y^2}, & (x, y) \neq (0, 0) \\ 0, & (x, y) = (0, 0) \end{cases}$

40. $f(x, y) = \begin{cases} \dfrac{5x^2y}{x^3 + y^3}, & (x, y) \neq (0, 0) \\ 0, & (x, y) = (0, 0) \end{cases}$

13.5 Chain Rules for Functions of Several Variables

■ Use the Chain Rules for functions of several variables.
■ Find partial derivatives implicitly.

Chain Rules for Functions of Several Variables

Your work with differentials in the preceding section provides the basis for the extension of the Chain Rule to functions of two variables. There are two cases. The first case involves w as a function of x and y, where x and y are functions of a single independent variable t, as shown in Theorem 13.6.

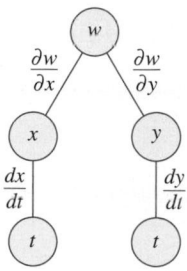

Chain Rule: one independent variable
w is a function of x and y, which are each functions of t. This diagram represents the derivative of w with respect to t.
Figure 13.39

THEOREM 13.6   Chain Rule: One Independent Variable

Let $w = f(x, y)$, where f is a differentiable function of x and y. If $x = g(t)$ and $y = h(t)$, where g and h are differentiable functions of t, then w is a differentiable function of t, and

$$\frac{dw}{dt} = \frac{\partial w}{\partial x}\frac{dx}{dt} + \frac{\partial w}{\partial y}\frac{dy}{dt}.$$

The Chain Rule is shown schematically in Figure 13.39. A proof of this theorem is given in Appendix A.

EXAMPLE 1   Chain Rule: One Independent Variable

Let $w = x^2y - y^2$, where $x = \sin t$ and $y = e^t$. Find dw/dt when $t = 0$.

Solution   By the Chain Rule for one independent variable, you have

$$\frac{dw}{dt} = \frac{\partial w}{\partial x}\frac{dx}{dt} + \frac{\partial w}{\partial y}\frac{dy}{dt}$$
$$= 2xy(\cos t) + (x^2 - 2y)e^t$$
$$= 2(\sin t)(e^t)(\cos t) + (\sin^2 t - 2e^t)e^t$$
$$= 2e^t \sin t \cos t + e^t \sin^2 t - 2e^{2t}.$$

When $t = 0$, it follows that

$$\frac{dw}{dt} = -2.$$

The Chain Rules presented in this section provide alternative techniques for solving many problems in single-variable calculus. For instance, in Example 1, you could have used single-variable techniques to find dw/dt by first writing w as a function of t,

$$w = x^2y - y^2$$
$$= (\sin t)^2(e^t) - (e^t)^2$$
$$= e^t \sin^2 t - e^{2t}$$

and then differentiating as usual.

$$\frac{dw}{dt} = 2e^t \sin t \cos t + e^t \sin^2 t - 2e^{2t}$$

The Chain Rule in Theorem 13.6 can be extended to any number of variables. For example, if each x_i is a differentiable function of a single variable t, then for

$$w = f(x_1, x_2, \ldots, x_n)$$

you have

$$\frac{dw}{dt} = \frac{\partial w}{\partial x_1}\frac{dx_1}{dt} + \frac{\partial w}{\partial x_2}\frac{dx_2}{dt} + \cdots + \frac{\partial w}{\partial x_n}\frac{dx_n}{dt}.$$

EXAMPLE 2 An Application of a Chain Rule to Related Rates

Two objects are traveling in elliptical paths given by the following parametric equations.

$$x_1 = 4\cos t \quad \text{and} \quad y_1 = 2\sin t \qquad \text{First object}$$
$$x_2 = 2\sin 2t \quad \text{and} \quad y_2 = 3\cos 2t \qquad \text{Second object}$$

At what rate is the distance between the two objects changing when $t = \pi$?

Solution From Figure 13.40, you can see that the distance s between the two objects is given by

$$s = \sqrt{(x_2 - x_1)^2 + (y_2 - y_1)^2}$$

and that when $t = \pi$, you have $x_1 = -4$, $y_1 = 0$, $x_2 = 0$, $y_2 = 3$, and

$$s = \sqrt{(0 + 4)^2 + (3 + 0)^2} = 5.$$

When $t = \pi$, the partial derivatives of s are as follows.

$$\frac{\partial s}{\partial x_1} = \frac{-(x_2 - x_1)}{\sqrt{(x_2 - x_1)^2 + (y_2 - y_1)^2}} = -\frac{1}{5}(0 + 4) = -\frac{4}{5}$$

$$\frac{\partial s}{\partial y_1} = \frac{-(y_2 - y_1)}{\sqrt{(x_2 - x_1)^2 + (y_2 - y_1)^2}} = -\frac{1}{5}(3 - 0) = -\frac{3}{5}$$

$$\frac{\partial s}{\partial x_2} = \frac{(x_2 - x_1)}{\sqrt{(x_2 - x_1)^2 + (y_2 - y_1)^2}} = \frac{1}{5}(0 + 4) = \frac{4}{5}$$

$$\frac{\partial s}{\partial y_2} = \frac{(y_2 - y_1)}{\sqrt{(x_2 - x_1)^2 + (y_2 - y_1)^2}} = \frac{1}{5}(3 - 0) = \frac{3}{5}$$

When $t = \pi$, the derivatives of x_1, y_1, x_2, and y_2 are

$$\frac{dx_1}{dt} = -4\sin t = 0$$

$$\frac{dy_1}{dt} = 2\cos t = -2$$

$$\frac{dx_2}{dt} = 4\cos 2t = 4$$

$$\frac{dy_2}{dt} = -6\sin 2t = 0.$$

So, using the appropriate Chain Rule, you know that the distance is changing at a rate of

$$\frac{ds}{dt} = \frac{\partial s}{\partial x_1}\frac{dx_1}{dt} + \frac{\partial s}{\partial y_1}\frac{dy_1}{dt} + \frac{\partial s}{\partial x_2}\frac{dx_2}{dt} + \frac{\partial s}{\partial y_2}\frac{dy_2}{dt}$$

$$= \left(-\frac{4}{5}\right)(0) + \left(-\frac{3}{5}\right)(-2) + \left(\frac{4}{5}\right)(4) + \left(\frac{3}{5}\right)(0)$$

$$= \frac{22}{5}.$$

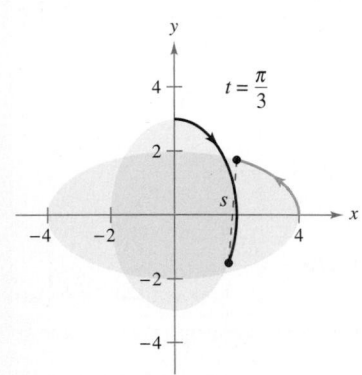

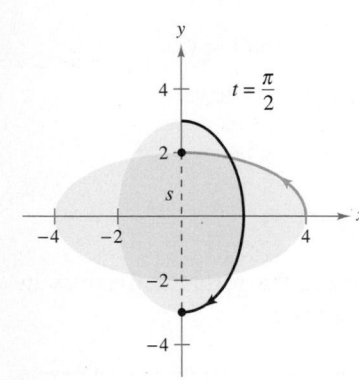

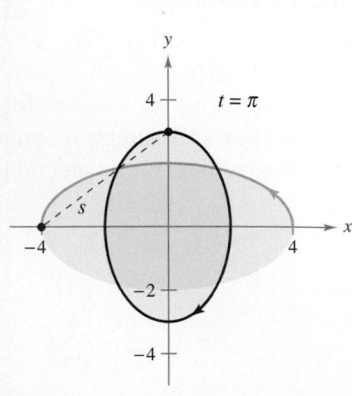

Paths of two objects traveling in elliptical orbits
Figure 13.40

In Example 2, note that s is the function of four *intermediate* variables, x_1, y_1, x_2, and y_2, each of which is a function of a single variable t. Another type of composite function is one in which the intermediate variables are themselves functions of more than one variable. For instance, for $w = f(x, y)$, where $x = g(s, t)$ and $y = h(s, t)$, it follows that w is a function of s and t, and you can consider the partial derivatives of w with respect to s and t. One way to find these partial derivatives is to write w as a function of s and t explicitly by substituting the equations $x = g(s, t)$ and $y = h(s, t)$ into the equation $w = f(x, y)$. Then you can find the partial derivatives in the usual way, as demonstrated in the next example.

EXAMPLE 3 Finding Partial Derivatives by Substitution

Find $\partial w/\partial s$ and $\partial w/\partial t$ for $w = 2xy$, where $x = s^2 + t^2$ and $y = s/t$.

Solution Begin by substituting $x = s^2 + t^2$ and $y = s/t$ into the equation $w = 2xy$ to obtain

$$w = 2xy = 2(s^2 + t^2)\left(\frac{s}{t}\right) = 2\left(\frac{s^3}{t} + st\right).$$

Then, to find $\partial w/\partial s$, hold t constant and differentiate with respect to s.

$$\frac{\partial w}{\partial s} = 2\left(\frac{3s^2}{t} + t\right)$$

$$= \frac{6s^2 + 2t^2}{t}$$

Similarly, to find $\partial w/\partial t$, hold s constant and differentiate with respect to t to obtain

$$\frac{\partial w}{\partial t} = 2\left(-\frac{s^3}{t^2} + s\right)$$

$$= 2\left(\frac{-s^3 + st^2}{t^2}\right)$$

$$= \frac{2st^2 - 2s^3}{t^2}.$$

Theorem 13.7 gives an alternative method for finding the partial derivatives in Example 3 without explicitly writing w as a function of s and t.

Chain Rule: two independent variables
Figure 13.41

> ### THEOREM 13.7 Chain Rule: Two Independent Variables
>
> Let $w = f(x, y)$, where f is a differentiable function of x and y. If $x = g(s, t)$ and $y = h(s, t)$ such that the first partials $\partial x/\partial s$, $\partial x/\partial t$, $\partial y/\partial s$, and $\partial y/\partial t$ all exist, then $\partial w/\partial s$ and $\partial w/\partial t$ exist and are given by
>
> $$\frac{\partial w}{\partial s} = \frac{\partial w}{\partial x}\frac{\partial x}{\partial s} + \frac{\partial w}{\partial y}\frac{\partial y}{\partial s}$$
>
> and
>
> $$\frac{\partial w}{\partial t} = \frac{\partial w}{\partial x}\frac{\partial x}{\partial t} + \frac{\partial w}{\partial y}\frac{\partial y}{\partial t}.$$
>
> The Chain Rule is shown schematically in Figure 13.41.

Proof To obtain $\partial w/\partial s$, hold t constant and apply Theorem 13.6 to obtain the desired result. Similarly, for $\partial w/\partial t$, hold s constant and apply Theorem 13.6. ∎

> **EXAMPLE 4** **The Chain Rule with Two Independent Variables**

: . . . ▷ *See LarsonCalculus.com for an interactive version of this type of example.*

Use the Chain Rule to find $\partial w/\partial s$ and $\partial w/\partial t$ for

$$w = 2xy$$

where $x = s^2 + t^2$ and $y = s/t$.

Solution Note that these same partials were found in Example 3. This time, using Theorem 13.7, you can hold t constant and differentiate with respect to s to obtain

$$\frac{\partial w}{\partial s} = \frac{\partial w}{\partial x}\frac{\partial x}{\partial s} + \frac{\partial w}{\partial y}\frac{\partial y}{\partial s}$$

$$= 2y(2s) + 2x\left(\frac{1}{t}\right)$$

$$= 2\left(\frac{s}{t}\right)(2s) + 2(s^2 + t^2)\left(\frac{1}{t}\right) \qquad \text{Substitute } \frac{s}{t} \text{ for } y \text{ and } s^2 + t^2 \text{ for } x.$$

$$= \frac{4s^2}{t} + \frac{2s^2 + 2t^2}{t}$$

$$= \frac{6s^2 + 2t^2}{t}.$$

Similarly, holding s constant gives

$$\frac{\partial w}{\partial t} = \frac{\partial w}{\partial x}\frac{\partial x}{\partial t} + \frac{\partial w}{\partial y}\frac{\partial y}{\partial t}$$

$$= 2y(2t) + 2x\left(\frac{-s}{t^2}\right)$$

$$= 2\left(\frac{s}{t}\right)(2t) + 2(s^2 + t^2)\left(\frac{-s}{t^2}\right) \qquad \text{Substitute } \frac{s}{t} \text{ for } y \text{ and } s^2 + t^2 \text{ for } x.$$

$$= 4s - \frac{2s^3 + 2st^2}{t^2}$$

$$= \frac{4st^2 - 2s^3 - 2st^2}{t^2}$$

$$= \frac{2st^2 - 2s^3}{t^2}.$$

The Chain Rule in Theorem 13.7 can also be extended to any number of variables. For example, if w is a differentiable function of the n variables

$$x_1, x_2, \ldots, x_n$$

where each x_i is a differentiable function of the m variables $t_1, t_2, \ldots, t_m$, then for

$$w = f(x_1, x_2, \ldots, x_n)$$

you obtain the following.

$$\frac{\partial w}{\partial t_1} = \frac{\partial w}{\partial x_1}\frac{\partial x_1}{\partial t_1} + \frac{\partial w}{\partial x_2}\frac{\partial x_2}{\partial t_1} + \cdots + \frac{\partial w}{\partial x_n}\frac{\partial x_n}{\partial t_1}$$

$$\frac{\partial w}{\partial t_2} = \frac{\partial w}{\partial x_1}\frac{\partial x_1}{\partial t_2} + \frac{\partial w}{\partial x_2}\frac{\partial x_2}{\partial t_2} + \cdots + \frac{\partial w}{\partial x_n}\frac{\partial x_n}{\partial t_2}$$

$$\vdots$$

$$\frac{\partial w}{\partial t_m} = \frac{\partial w}{\partial x_1}\frac{\partial x_1}{\partial t_m} + \frac{\partial w}{\partial x_2}\frac{\partial x_2}{\partial t_m} + \cdots + \frac{\partial w}{\partial x_n}\frac{\partial x_n}{\partial t_m}$$

EXAMPLE 5 The Chain Rule for a Function of Three Variables

Find $\partial w/\partial s$ and $\partial w/\partial t$ when $s = 1$ and $t = 2\pi$ for

$$w = xy + yz + xz$$

where $x = s \cos t$, $y = s \sin t$, and $z = t$.

Solution By extending the result of Theorem 13.7, you have

$$\frac{\partial w}{\partial s} = \frac{\partial w}{\partial x}\frac{\partial x}{\partial s} + \frac{\partial w}{\partial y}\frac{\partial y}{\partial s} + \frac{\partial w}{\partial z}\frac{\partial z}{\partial s}$$

$$= (y + z)(\cos t) + (x + z)(\sin t) + (y + x)(0)$$

$$= (y + z)(\cos t) + (x + z)(\sin t).$$

When $s = 1$ and $t = 2\pi$, you have $x = 1$, $y = 0$, and $z = 2\pi$. So,

$$\frac{\partial w}{\partial s} = (0 + 2\pi)(1) + (1 + 2\pi)(0) = 2\pi.$$

Furthermore,

$$\frac{\partial w}{\partial t} = \frac{\partial w}{\partial x}\frac{\partial x}{\partial t} + \frac{\partial w}{\partial y}\frac{\partial y}{\partial t} + \frac{\partial w}{\partial z}\frac{\partial z}{\partial t}$$

$$= (y + z)(-s \sin t) + (x + z)(s \cos t) + (y + x)(1)$$

and for $s = 1$ and $t = 2\pi$, it follows that

$$\frac{\partial w}{\partial t} = (0 + 2\pi)(0) + (1 + 2\pi)(1) + (0 + 1)(1)$$

$$= 2 + 2\pi.$$

Implicit Partial Differentiation

This section concludes with an application of the Chain Rule to determine the derivative of a function defined *implicitly*. Let x and y be related by the equation $F(x, y) = 0$, where $y = f(x)$ is a differentiable function of x. To find dy/dx, you could use the techniques discussed in Section 3.5. You will see, however, that the Chain Rule provides a convenient alternative. Consider the function

$$w = F(x, y) = F(x, f(x)).$$

You can apply Theorem 13.6 to obtain

$$\frac{dw}{dx} = F_x(x, y)\frac{dx}{dx} + F_y(x, y)\frac{dy}{dx}.$$

Because $w = F(x, y) = 0$ for all x in the domain of f, you know that

$$\frac{dw}{dx} = 0$$

and you have

$$F_x(x, y)\frac{dx}{dx} + F_y(x, y)\frac{dy}{dx} = 0.$$

Now, if $F_y(x, y) \neq 0$, you can use the fact that $dx/dx = 1$ to conclude that

$$\frac{dy}{dx} = -\frac{F_x(x, y)}{F_y(x, y)}.$$

A similar procedure can be used to find the partial derivatives of functions of several variables that are defined implicitly.

> **THEOREM 13.8** **Chain Rule: Implicit Differentiation**
>
> If the equation $F(x, y) = 0$ defines y implicitly as a differentiable function of x, then
>
> $$\frac{dy}{dx} = -\frac{F_x(x, y)}{F_y(x, y)}, \quad F_y(x, y) \neq 0.$$
>
> If the equation $F(x, y, z) = 0$ defines z implicitly as a differentiable function of x and y, then
>
> $$\frac{\partial z}{\partial x} = -\frac{F_x(x, y, z)}{F_z(x, y, z)} \quad \text{and} \quad \frac{\partial z}{\partial y} = -\frac{F_y(x, y, z)}{F_z(x, y, z)}, \quad F_z(x, y, z) \neq 0.$$

This theorem can be extended to differentiable functions defined implicitly with any number of variables.

EXAMPLE 6   Finding a Derivative Implicitly

Find dy/dx for

$$y^3 + y^2 - 5y - x^2 + 4 = 0.$$

Solution Begin by letting

$$F(x, y) = y^3 + y^2 - 5y - x^2 + 4.$$

Then

$$F_x(x, y) = -2x \quad \text{and} \quad F_y(x, y) = 3y^2 + 2y - 5.$$

Using Theorem 13.8, you have

$$\frac{dy}{dx} = -\frac{F_x(x, y)}{F_y(x, y)} = \frac{-(-2x)}{3y^2 + 2y - 5} = \frac{2x}{3y^2 + 2y - 5}.$$

• • **REMARK** Compare the
solution to Example 6 with
the solution to Example 2 in
Section 3.5.

EXAMPLE 7   Finding Partial Derivatives Implicitly

Find $\partial z/\partial x$ and $\partial z/\partial y$ for

$$3x^2z - x^2y^2 + 2z^3 + 3yz - 5 = 0.$$

Solution Begin by letting

$$F(x, y, z) = 3x^2z - x^2y^2 + 2z^3 + 3yz - 5.$$

Then

$$F_x(x, y, z) = 6xz - 2xy^2$$
$$F_y(x, y, z) = -2x^2y + 3z$$

and

$$F_z(x, y, z) = 3x^2 + 6z^2 + 3y.$$

Using Theorem 13.8, you have

$$\frac{\partial z}{\partial x} = -\frac{F_x(x, y, z)}{F_z(x, y, z)} = \frac{2xy^2 - 6xz}{3x^2 + 6z^2 + 3y}$$

and

$$\frac{\partial z}{\partial y} = -\frac{F_y(x, y, z)}{F_z(x, y, z)} = \frac{2x^2y - 3z}{3x^2 + 6z^2 + 3y}.$$

13.5 Exercises

CONCEPT CHECK

1. Chain Rule Consider $w = f(x, y)$, where $x = g(s, t)$ and $y = h(s, t)$. Describe two ways of finding the partial derivatives $\partial w/\partial s$ and $\partial w/\partial t$.

2. Implicit Differentiation Why is using the Chain Rule to determine the derivative of the equation $F(x, y) = 0$ implicitly easier than using the method you learned in Section 3.5?

 **Using the Chain Rule** In Exercises 3–6, find dw/dt using the appropriate Chain Rule. Evaluate dw/dt at the given value of t.

Function	Value
3. $w = x^2 + 5y$	$t = 2$
$x = 2t, \ y = t$	
4. $w = \sqrt{x^2 + y^2}$	$t = 0$
$x = \cos t, \ y = e^t$	
5. $w = x \sin y$	$t = 0$
$x = e^t, \ y = \pi - t$	
6. $w = \ln \dfrac{y}{x}$	$t = \dfrac{\pi}{4}$
$x = \cos t, \ y = \sin t$	

 **Using Different Methods** In Exercises 7–12, find dw/dt **(a)** by using the appropriate Chain Rule and **(b)** by converting w to a function of t before differentiating.

7. $w = x - \dfrac{1}{y}, \quad x = e^{2t}, \quad y = t^3$

8. $w = \cos(x - y), \quad x = t^2, \quad y = 1$

9. $w = x^2 + y^2 + z^2, \quad x = \cos t, \quad y = \sin t, \quad z = e^t$

10. $w = xy \cos z, \quad x = t, \quad y = t^2, \quad z = \arccos t$

11. $w = xy + xz + yz, \quad x = t - 1, \quad y = t^2 - 1, \quad z = t$

12. $w = xy^2 + x^2z + yz^2, \quad x = t^2, \quad y = 2t, \quad z = 2$

Projectile Motion In Exercises 13 and 14, the parametric equations for the paths of two objects are given. At what rate is the distance between the two objects changing at the given value of t?

13. $x_1 = 10 \cos 2t, \quad y_1 = 6 \sin 2t$ First object

$x_2 = 7 \cos t, \quad y_2 = 4 \sin t$ Second object

$t = \pi/2$

14. $x_1 = 48\sqrt{2}t, \quad y_1 = 48\sqrt{2}t - 16t^2$ First object

$x_2 = 48\sqrt{3}t, \quad y_2 = 48t - 16t^2$ Second object

$t = 1$

 **Finding Partial Derivatives** In Exercises 15–18, find $\partial w/\partial s$ and $\partial w/\partial t$ using the appropriate Chain Rule. Evaluate each partial derivative at the given values of s and t.

Function	Values
15. $w = x^2 + y^2$	$s = 1, \ t = 3$
$x = s + t, \ y = s - t$	
16. $w = y^3 - 3x^2y$	$s = -1, \ t = 2$
$x = e^s, \ y = e^t$	
17. $w = \sin(2x + 3y)$	$s = 0, \ t = \dfrac{\pi}{2}$
$x = s + t, \ y = s - t$	
18. $w = x^2 - y^2$	$s = 3, \ t = \dfrac{\pi}{4}$
$x = s \cos t, \ y = s \sin t$	

 **Using Different Methods** In Exercises 19–22, find $\partial w/\partial s$ and $\partial w/\partial t$ **(a)** by using the appropriate Chain Rule and **(b)** by converting w to a function of s and t before differentiating.

19. $w = xyz, \quad x = s + t, \quad y = s - t, \quad z = st^2$

20. $w = x^2 + y^2 + z^2, \quad x = t \sin s, \quad y = t \cos s, \quad z = st^2$

21. $w = ze^{xy}, \quad x = s - t, \quad y = s + t, \quad z = st$

22. $w = x \cos yz, \quad x = s^2, \quad y = t^2, \quad z = s - 2t$

 **Finding a Derivative Implicitly** In Exercises 23–26, differentiate implicitly to find dy/dx.

23. $x^2 - xy + y^2 - x + y = 0$

24. $\sec xy + \tan xy + 5 = 0$

25. $\ln\sqrt{x^2 + y^2} + x + y = 4$

26. $\dfrac{x}{x^2 + y^2} - y^2 = 6$

 **Finding Partial Derivatives Implicitly** In Exercises 27–34, differentiate implicitly to find the first partial derivatives of z.

27. $x^2 + y^2 + z^2 = 1$

28. $xz + yz + xy = 0$

29. $x^2 + 2yz + z^2 = 1$

30. $x + \sin(y + z) = 0$

31. $\tan(x + y) + \cos z = 2$

32. $z = e^x \sin(y + z)$

33. $e^{xz} + xy = 0$

34. $x \ln y + y^2z + z^2 = 8$

Finding Partial Derivatives Implicitly In Exercises 35–38, differentiate implicitly to find the first partial derivatives of w.

35. $7xy + yz^2 - 4wz + w^2z + w^2x - 6 = 0$

36. $x^2 + y^2 + z^2 - 5yw + 10w^2 = 2$

37. $\cos xy + \sin yz + wz = 20$

38. $w - \sqrt{x - y} - \sqrt{y - z} = 0$

Homogeneous Functions A function f is *homogeneous of degree n* when $f(tx, ty) = t^n f(x, y)$. In Exercises 39–42, (a) show that the function is homogeneous and determine n, and (b) show that $xf_x(x, y) + yf_y(x, y) = nf(x, y)$.

39. $f(x, y) = 2x^2 - 5xy$ **40.** $f(x, y) = x^3 - 3xy^2 + y^3$

41. $f(x, y) = e^{x/y}$ **42.** $f(x, y) = x \cos \dfrac{x + y}{y}$

43. Using a Table of Values Let $w = f(x, y)$, $x = g(t)$, and $y = h(t)$, where f, g, and h are differentiable. Use the appropriate Chain Rule and the table of values to find dw/dt when $t = 2$.

$g(2)$	$h(2)$	$g'(2)$	$h'(2)$	$f_x(4, 3)$	$f_y(4, 3)$
4	3	-1	6	-5	7

44. Using a Table of Values Let $w = f(x, y)$, $x = g(s, t)$, and $y = h(s, t)$, where f, g, and h are differentiable. Use the appropriate Chain Rule and the table of values to find $w_s(1, 2)$.

$g(1, 2)$	$h(1, 2)$	$g_s(1, 2)$	$h_s(1, 2)$	$f_x(4, 3)$	$f_y(4, 3)$
4	3	-3	5	-5	7

EXPLORING CONCEPTS

45. Using the Chain Rule Show that $\dfrac{\partial w}{\partial u} + \dfrac{\partial w}{\partial v} = 0$ for $w = f(x, y)$, $x = u - v$, and $y = v - u$.

46. Using the Chain Rule Demonstrate the result of Exercise 45 for $w = (x - y) \sin(y - x)$.

47. Using the Chain Rule Let $F(u, v)$ be a function of two variables. Find a formula for $f'(x)$ when (a) $f(x) = F(4x, 4)$ and (b) $f(x) = F(-2x, x^2)$.

48. **HOW DO YOU SEE IT?** The path of an object represented by $w = f(x, y)$ is shown, where x and y are functions of t. The point on the graph represents the position of the object.

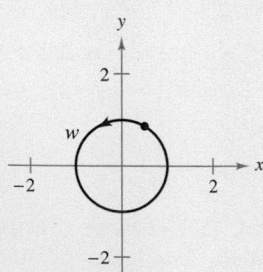

Determine whether each of the following is positive, negative, or zero.

(a) $\dfrac{dx}{dt}$ (b) $\dfrac{dy}{dt}$

49. Volume and Surface Area The radius of a right circular cylinder is increasing at a rate of 6 inches per minute, and the height is decreasing at a rate of 4 inches per minute. What are the rates of change of the volume and surface area when the radius is 12 inches and the height is 36 inches?

50. Ideal Gas Law The Ideal Gas Law is

$$PV = mRT$$

where P is the pressure, V is the volume, m is the constant mass, R is a constant, T is the temperature, and P and V are functions of time. Find dT/dt, the rate at which the temperature changes with respect to time.

51. Moment of Inertia An annular cylinder has an inside radius of r_1 and an outside radius of r_2 (see figure). Its moment of inertia is

$$I = \tfrac{1}{2}m(r_1^2 + r_2^2)$$

where m is the mass. The two radii are increasing at a rate of 2 centimeters per second. Find the rate at which I is changing at the instant the radii are 6 centimeters and 8 centimeters. (Assume mass is a constant.)

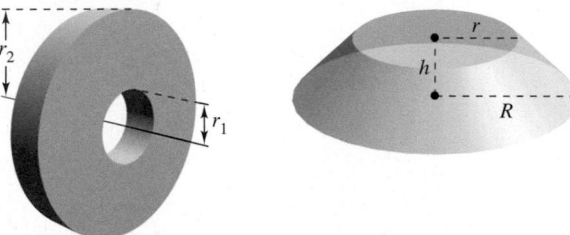

Figure for 51   Figure for 52

52. Volume and Surface Area The two radii of the frustum of a right circular cone are increasing at a rate of 4 centimeters per minute, and the height is increasing at a rate of 12 centimeters per minute (see figure). Find the rates at which the volume and surface area are changing when the two radii are 15 centimeters and 25 centimeters and the height is 10 centimeters.

53. Cauchy-Riemann Equations Given the functions $u(x, y)$ and $v(x, y)$, verify that the **Cauchy-Riemann equations**

$$\frac{\partial u}{\partial x} = \frac{\partial v}{\partial y} \quad \text{and} \quad \frac{\partial u}{\partial y} = -\frac{\partial v}{\partial x}$$

can be written in polar coordinate form as

$$\frac{\partial u}{\partial r} = \frac{1}{r} \cdot \frac{\partial v}{\partial \theta} \quad \text{and} \quad \frac{\partial v}{\partial r} = -\frac{1}{r} \cdot \frac{\partial u}{\partial \theta}.$$

54. Cauchy-Riemann Equations Demonstrate the result of Exercise 53 for the functions

$$u = \ln \sqrt{x^2 + y^2} \quad \text{and} \quad v = \arctan \frac{y}{x}.$$

55. Homogeneous Function Show that if $f(x, y)$ is homogeneous of degree n, then

$$xf_x(x, y) + yf_y(x, y) = nf(x, y).$$

[*Hint:* Let $g(t) = f(tx, ty) = t^n f(x, y)$. Find $g'(t)$ and then let $t = 1$.]

13.6 Directional Derivatives and Gradients

■ Find and use directional derivatives of a function of two variables.
■ Find the gradient of a function of two variables.
■ Use the gradient of a function of two variables in applications.
■ Find directional derivatives and gradients of functions of three variables.

Directional Derivative

You are standing on the hillside represented by $z = f(x, y)$ in Figure 13.42 and want to determine the hill's incline toward the z-axis. You already know how to determine the slopes in two different directions—the slope in the y-direction is given by the partial derivative $f_y(x, y)$, and the slope in the x-direction is given by the partial derivative $f_x(x, y)$. In this section, you will see that these two partial derivatives can be used to find the slope in *any* direction.

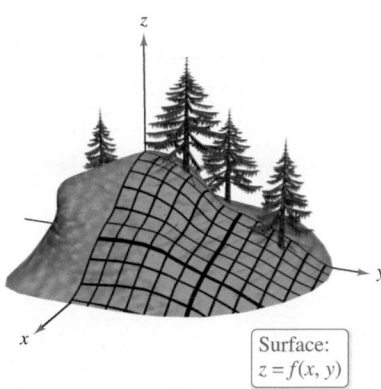

Surface:
$z = f(x, y)$

Figure 13.42

To determine the slope at a point on a surface, you will define a new type of derivative called a **directional derivative.** Begin by letting $z = f(x, y)$ be a *surface* and $P(x_0, y_0)$ be a *point* in the domain of f, as shown in Figure 13.43. The "direction" of the directional derivative is given by a unit vector

$$\mathbf{u} = \cos \theta \mathbf{i} + \sin \theta \mathbf{j}$$

where θ is the angle the vector makes with the positive x-axis. To find the desired slope, reduce the problem to two dimensions by intersecting the surface with a vertical plane passing through the point P and parallel to $\mathbf{u}$, as shown in Figure 13.44. This vertical plane intersects the surface to form a curve C. The slope of the surface at $(x_0, y_0, f(x_0, y_0))$ in the direction of $\mathbf{u}$ is defined as the slope of the curve C at that point.

Informally, you can write the slope of the curve C as a limit that looks much like those used in single-variable calculus. The vertical plane used to form C intersects the xy-plane in a line L, represented by the parametric equations

$$x = x_0 + t \cos \theta$$

and

$$y = y_0 + t \sin \theta$$

so that for any value of t, the point $Q(x, y)$ lies on the line L. For each of the points P and Q, there is a corresponding point on the surface.

$(x_0, y_0, f(x_0, y_0))$ Point above P
$(x, y, f(x, y))$ Point above Q

Moreover, because the distance between P and Q is

$$\sqrt{(x - x_0)^2 + (y - y_0)^2} = \sqrt{(t \cos \theta)^2 + (t \sin \theta)^2}$$
$$= |t|$$

you can write the slope of the secant line through $(x_0, y_0, f(x_0, y_0))$ and $(x, y, f(x, y))$ as

$$\frac{f(x, y) - f(x_0, y_0)}{t} = \frac{f(x_0 + t \cos \theta, y_0 + t \sin \theta) - f(x_0, y_0)}{t}.$$

Finally, by letting t approach 0, you arrive at the definition on the next page.

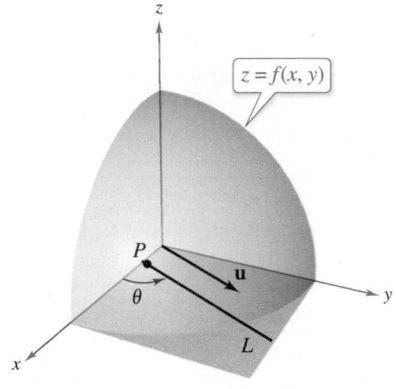

Figure 13.43

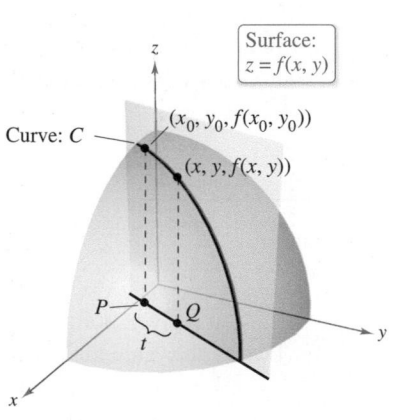

Surface:
$z = f(x, y)$

Curve: C

$(x_0, y_0, f(x_0, y_0))$

$(x, y, f(x, y))$

Figure 13.44

⋅ ⋅ ⋅ ⋅ ⋅ ⋅ ⋅ ⋅ ⋅ ⋅ ⋅ ⋅ ⋅ ⋅ ⋅ ⋅ ⋅ ⋅ ▷

⋅⋅REMARK Be sure you understand that the directional derivative represents the *rate of change of a function* in the direction of the unit vector $\mathbf{u} = \cos\theta\mathbf{i} + \sin\theta\mathbf{j}$. Geometrically, you can interpret the directional derivative as giving the *slope of a surface* in the direction of $\mathbf{u}$ at a point on the surface. (See Figure 13.46.)

Definition of Directional Derivative

Let f be a function of two variables x and y and let $\mathbf{u} = \cos\theta\mathbf{i} + \sin\theta\mathbf{j}$ be a unit vector. Then the **directional derivative of f in the direction of u,** denoted by $D_{\mathbf{u}}f$, is

$$D_{\mathbf{u}}f(x, y) = \lim_{t\to 0} \frac{f(x + t\cos\theta, y + t\sin\theta) - f(x, y)}{t}$$

provided this limit exists.

Calculating directional derivatives by this definition is similar to finding the derivative of a function of one variable by the limit process (see Section 3.1). A simpler formula for finding directional derivatives involves the partial derivatives f_x and f_y.

THEOREM 13.9 Directional Derivative

If f is a differentiable function of x and y, then the directional derivative of f in the direction of the unit vector $\mathbf{u} = \cos\theta\mathbf{i} + \sin\theta\mathbf{j}$ is

$$D_{\mathbf{u}}f(x, y) = f_x(x, y)\cos\theta + f_y(x, y)\sin\theta.$$

Proof For a fixed point (x_0, y_0), let

$$x = x_0 + t\cos\theta \quad \text{and} \quad y = y_0 + t\sin\theta.$$

Then, let $g(t) = f(x, y)$. Because f is differentiable, you can apply the Chain Rule given in Theorem 13.6 to obtain

$$g'(t) = f_x(x, y)x'(t) + f_y(x, y)y'(t) \qquad \text{Apply Chain Rule (Theorem 13.6).}$$
$$= f_x(x, y)\cos\theta + f_y(x, y)\sin\theta.$$

If $t = 0$, then $x = x_0$ and $y = y_0$, so

$$g'(0) = f_x(x_0, y_0)\cos\theta + f_y(x_0, y_0)\sin\theta.$$

By the definition of $g'(t)$, it is also true that

$$g'(0) = \lim_{t\to 0}\frac{g(t) - g(0)}{t}$$
$$= \lim_{t\to 0}\frac{f(x_0 + t\cos\theta, y_0 + t\sin\theta) - f(x_0, y_0)}{t}.$$

Consequently, $D_{\mathbf{u}}f(x_0, y_0) = f_x(x_0, y_0)\cos\theta + f_y(x_0, y_0)\sin\theta$. ∎

There are infinitely many directional derivatives of a surface at a given point—one for each direction specified by $\mathbf{u}$, as shown in Figure 13.45. Two of these are the partial derivatives f_x and f_y.

1. Direction of positive x-axis $(\theta = 0)$: $\mathbf{u} = \cos 0\mathbf{i} + \sin 0\mathbf{j} = \mathbf{i}$

$$D_{\mathbf{i}}f(x, y) = f_x(x, y)\cos 0 + f_y(x, y)\sin 0 = f_x(x, y)$$

2. Direction of positive y-axis $\left(\theta = \dfrac{\pi}{2}\right)$: $\mathbf{u} = \cos\dfrac{\pi}{2}\mathbf{i} + \sin\dfrac{\pi}{2}\mathbf{j} = \mathbf{j}$

$$D_{\mathbf{j}}f(x, y) = f_x(x, y)\cos\frac{\pi}{2} + f_y(x, y)\sin\frac{\pi}{2} = f_y(x, y)$$

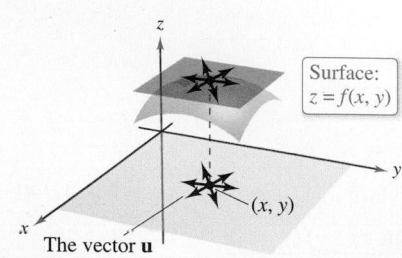

Surface:
$z = f(x, y)$

(x, y)

The vector $\mathbf{u}$

Figure 13.45

EXAMPLE 1 **Finding a Directional Derivative**

Find the directional derivative of

$$f(x, y) = 4 - x^2 - \frac{1}{4}y^2 \qquad \text{Surface}$$

at $(1, 2)$ in the direction of

$$\mathbf{u} = \left(\cos\frac{\pi}{3}\right)\mathbf{i} + \left(\sin\frac{\pi}{3}\right)\mathbf{j}. \qquad \text{Direction}$$

Solution Because $f_x(x, y) = -2x$ and $f_y(x, y) = -y/2$ are continuous, f is differentiable, and you can apply Theorem 13.9.

$$D_{\mathbf{u}}f(x, y) = f_x(x, y)\cos\theta + f_y(x, y)\sin\theta = (-2x)\cos\theta + \left(-\frac{y}{2}\right)\sin\theta$$

Evaluating at $\theta = \pi/3$, $x = 1$, and $y = 2$ produces

$$D_{\mathbf{u}}f(1, 2) = (-2)\left(\frac{1}{2}\right) + (-1)\left(\frac{\sqrt{3}}{2}\right)$$

$$= -1 - \frac{\sqrt{3}}{2}$$

$$\approx -1.866. \qquad \text{See Figure 13.46.}$$

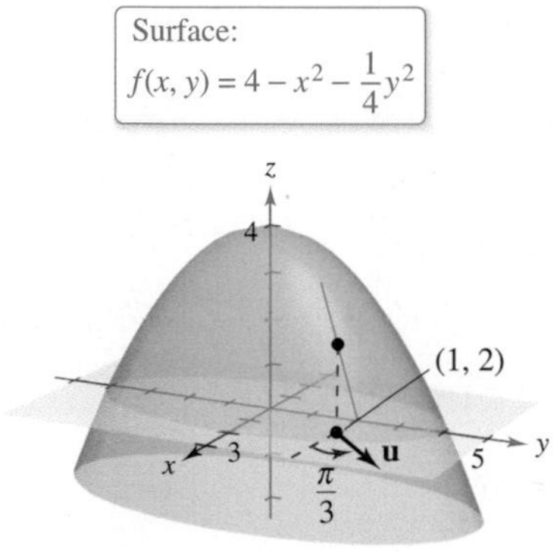

Note in Figure 13.46 that you can interpret the directional derivative as giving the slope of the surface at the point $(1, 2, 2)$ in the direction of the unit vector $\mathbf{u}$. ■

You have been specifying direction by a unit vector $\mathbf{u}$. When the direction is given by a vector whose length is not 1, you must normalize the vector before applying the formula in Theorem 13.9.

EXAMPLE 2 **Finding a Directional Derivative**

⋮ ····▷ *See LarsonCalculus.com for an interactive version of this type of example.*

Find the directional derivative of

$$f(x, y) = x^2\sin 2y \qquad \text{Surface}$$

at $(1, \pi/2)$ in the direction of

$$\mathbf{v} = 3\mathbf{i} - 4\mathbf{j}. \qquad \text{Direction}$$

Solution Because $f_x(x, y) = 2x\sin 2y$ and $f_y(x, y) = 2x^2\cos 2y$ are continuous, f is differentiable, and you can apply Theorem 13.9. Begin by finding a unit vector in the direction of $\mathbf{v}$.

$$\mathbf{u} = \frac{\mathbf{v}}{\|\mathbf{v}\|} = \frac{3}{5}\mathbf{i} - \frac{4}{5}\mathbf{j} = \cos\theta\mathbf{i} + \sin\theta\mathbf{j}$$

Using this unit vector, you have

$$D_{\mathbf{u}}f(x, y) = (2x\sin 2y)(\cos\theta) = (2x^2\cos 2y)(\sin\theta)$$

$$D_{\mathbf{u}}f\left(1, \frac{\pi}{2}\right) = (2\sin\pi)\left(\frac{3}{5}\right) + (2\cos\pi)\left(-\frac{4}{5}\right)$$

$$= (0)\left(\frac{3}{5}\right) + (-2)\left(-\frac{4}{5}\right)$$

$$= \frac{8}{5}. \qquad \text{See Figure 13.47.}$$ ■

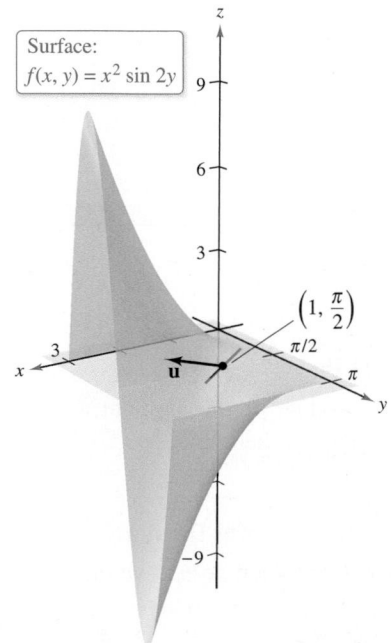

Surface:
$f(x, y) = 4 - x^2 - \frac{1}{4}y^2$

Figure 13.46

Surface:
$f(x, y) = x^2\sin 2y$

Figure 13.47

The Gradient of a Function of Two Variables

The **gradient** of a function of two variables is a vector-valued function of two variables. This function has many important uses, some of which are described later in this section.

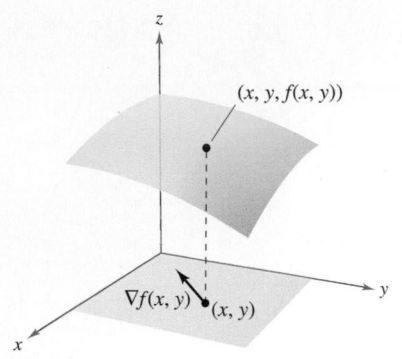

The gradient of f is a vector in the xy-plane.

Figure 13.48

> **Definition of Gradient of a Function of Two Variables**
>
> Let $z = f(x, y)$ be a function of x and y such that f_x and f_y exist. Then the **gradient of f,** denoted by $\nabla f(x, y)$, is the vector
>
> $$\nabla f(x, y) = f_x(x, y)\mathbf{i} + f_y(x, y)\mathbf{j}.$$
>
> (The symbol ∇f is read as "del f.") Another notation for the gradient is given by **grad** $f(x, y)$. In Figure 13.48, note that for each (x, y), the gradient $\nabla f(x, y)$ is a vector in the plane (not a vector in space).

Notice that no value is assigned to the symbol ∇ by itself. It is an operator in the same sense that d/dx is an operator. When ∇ operates on $f(x, y)$, it produces the vector $\nabla f(x, y)$.

EXAMPLE 3 **Finding the Gradient of a Function**

Find the gradient of

$$f(x, y) = y \ln x + xy^2$$

at the point $(1, 2)$.

Solution Using

$$f_x(x, y) = \frac{y}{x} + y^2 \quad \text{and} \quad f_y(x, y) = \ln x + 2xy$$

you have

$$\nabla f(x, y) = f_x(x, y)\mathbf{i} + f_y(x, y)\mathbf{j}$$
$$= \left(\frac{y}{x} + y^2\right)\mathbf{i} + (\ln x + 2xy)\mathbf{j}.$$

At the point $(1, 2)$, the gradient is

$$\nabla f(1, 2) = \left(\frac{2}{1} + 2^2\right)\mathbf{i} + [\ln 1 + 2(1)(2)]\mathbf{j}$$
$$= 6\mathbf{i} + 4\mathbf{j}.$$

Because the gradient of f is a vector, you can write the directional derivative of f in the direction of $\mathbf{u}$ as

$$D_{\mathbf{u}}f(x, y) = [f_x(x, y)\mathbf{i} + f_y(x, y)\mathbf{j}] \cdot (\cos\theta\mathbf{i} + \sin\theta\mathbf{j}).$$

In other words, the directional derivative is the dot product of the gradient and the direction vector. This useful result is summarized in the next theorem.

> **THEOREM 13.10** **Alternative Form of the Directional Derivative**
>
> If f is a differentiable function of x and y, then the directional derivative of f in the direction of the unit vector $\mathbf{u}$ is
>
> $$D_{\mathbf{u}}f(x, y) = \nabla f(x, y) \cdot \mathbf{u}.$$

EXAMPLE 4   **Using $\nabla f(x, y)$ to Find a Directional Derivative**

Find the directional derivative of $f(x, y) = 3x^2 - 2y^2$ at $\left(-\frac{3}{4}, 0\right)$ in the direction from $P\left(-\frac{3}{4}, 0\right)$ to $Q(0, 1)$.

Solution   Because the partials of f are continuous, f is differentiable and you can apply Theorem 13.10. A vector in the specified direction is

$$\overrightarrow{PQ} = \left(0 + \frac{3}{4}\right)\mathbf{i} + (1 - 0)\mathbf{j} = \frac{3}{4}\mathbf{i} + \mathbf{j}$$

and a unit vector in this direction is

$$\mathbf{u} = \frac{\overrightarrow{PQ}}{\|\overrightarrow{PQ}\|} = \frac{3}{5}\mathbf{i} + \frac{4}{5}\mathbf{j}. \qquad \text{Unit vector in direction of } \overrightarrow{PQ}$$

Because $\nabla f(x, y) = f_x(x, y)\mathbf{i} + f_y(x, y)\mathbf{j} = 6x\mathbf{i} - 4y\mathbf{j}$, the gradient at $\left(-\frac{3}{4}, 0\right)$ is

$$\nabla f\left(-\frac{3}{4}, 0\right) = -\frac{9}{2}\mathbf{i} + 0\mathbf{j}. \qquad \text{Gradient at } \left(-\frac{3}{4}, 0\right)$$

Consequently, at $\left(-\frac{3}{4}, 0\right)$, the directional derivative is

$$\begin{aligned} D_{\mathbf{u}}f\left(-\frac{3}{4}, 0\right) &= \nabla f\left(-\frac{3}{4}, 0\right) \cdot \mathbf{u} \\ &= \left(-\frac{9}{2}\mathbf{i} + 0\mathbf{j}\right) \cdot \left(\frac{3}{5}\mathbf{i} + \frac{4}{5}\mathbf{j}\right) \\ &= -\frac{27}{10}. \qquad \text{Directional derivative at } \left(-\frac{3}{4}, 0\right) \end{aligned}$$

See Figure 13.49.

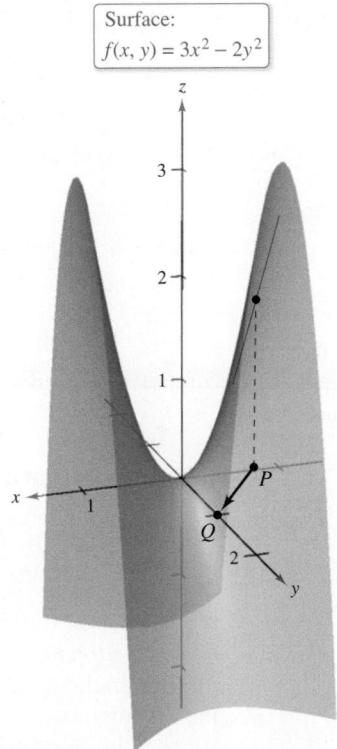

Figure 13.49

Applications of the Gradient

You have already seen that there are many directional derivatives at the point (x, y) on a surface. In many applications, you may want to know in which direction to move so that $f(x, y)$ increases most rapidly. This direction is called the direction of *steepest ascent*, and it is given by the gradient, as stated in the next theorem.

THEOREM 13.11   Properties of the Gradient

Let f be differentiable at the point (x, y).

1. If $\nabla f(x, y) = \mathbf{0}$, then $D_{\mathbf{u}} f(x, y) = 0$ for all $\mathbf{u}$.

2. The direction of *maximum* increase of f is given by $\nabla f(x, y)$. The maximum value of $D_{\mathbf{u}} f(x, y)$ is

$$\|\nabla f(x, y)\|. \qquad \text{Maximum value of } D_{\mathbf{u}} f(x, y)$$

3. The direction of *minimum* increase of f is given by $-\nabla f(x, y)$. The minimum value of $D_{\mathbf{u}} f(x, y)$ is

$$-\|\nabla f(x, y)\|. \qquad \text{Minimum value of } D_{\mathbf{u}} f(x, y)$$

> • • **REMARK** Property 2 of Theorem 13.11 says that at the point (x, y), f increases most rapidly in the direction of the gradient, $\nabla f(x, y)$.

Proof   If $\nabla f(x, y) = \mathbf{0}$, then for any direction (any $\mathbf{u}$), you have

$$
\begin{aligned}
D_{\mathbf{u}} f(x, y) &= \nabla f(x, y) \cdot \mathbf{u} \\
&= (0\mathbf{i} + 0\mathbf{j}) \cdot (\cos \theta \mathbf{i} + \sin \theta \mathbf{j}) \\
&= 0.
\end{aligned}
$$

If $\nabla f(x, y) \neq \mathbf{0}$, then let ϕ be the angle between $\nabla f(x, y)$ and a unit vector $\mathbf{u}$. Using the dot product, you can apply Theorem 11.5 to conclude that

$$
\begin{aligned}
D_{\mathbf{u}} f(x, y) &= \nabla f(x, y) \cdot \mathbf{u} \\
&= \|\nabla f(x, y)\| \, \|\mathbf{u}\| \cos \phi \\
&= \|\nabla f(x, y)\| \cos \phi
\end{aligned}
$$

and it follows that the maximum value of $D_{\mathbf{u}} f(x, y)$ will occur when

$$\cos \phi = 1.$$

So, $\phi = 0$, and the maximum value of the directional derivative occurs when $\mathbf{u}$ has the same direction as $\nabla f(x, y)$. Moreover, this largest value of $D_{\mathbf{u}} f(x, y)$ is precisely

$$\|\nabla f(x, y)\| \cos \phi = \|\nabla f(x, y)\|.$$

Similarly, the minimum value of $D_{\mathbf{u}} f(x, y)$ can be obtained by letting

$$\phi = \pi$$

so that $\mathbf{u}$ points in the direction opposite that of $\nabla f(x, y)$, as shown in Figure 13.50.

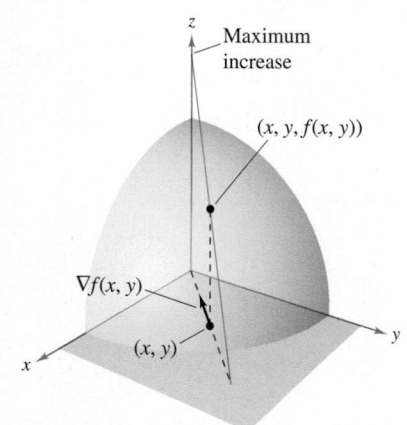

The gradient of f is a vector in the xy-plane that points in the direction of maximum increase on the surface given by $z = f(x, y)$.

Figure 13.50

To visualize one of the properties of the gradient, imagine a skier coming down a mountainside. If $f(x, y)$ denotes the altitude of the skier, then $-\nabla f(x, y)$ indicates the *compass direction* the skier should take to ski the path of steepest descent. (Remember that the gradient indicates direction in the xy-plane and does not itself point up or down the mountainside.)

As another illustration of the gradient, consider the temperature $T(x, y)$ at any point (x, y) on a flat metal plate. In this case, $\nabla T(x, y)$ gives the direction of greatest temperature increase at the point (x, y), as illustrated in the next example.

Level curves:
$T(x, y) = 20 - 4x^2 - y^2$

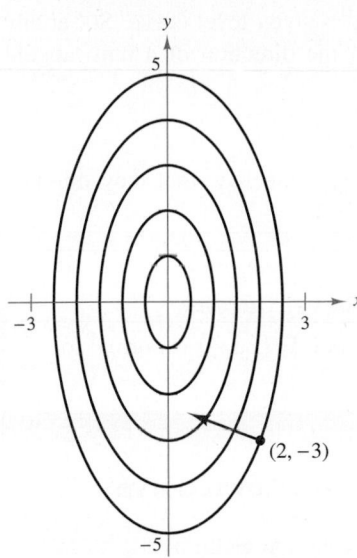

The direction of most rapid increase in temperature at $(2, -3)$ is given by $-16\mathbf{i} + 6\mathbf{j}$.

Figure 13.51

EXAMPLE 5 **Finding the Direction of Maximum Increase**

The temperature in degrees Celsius on the surface of a metal plate is

$$T(x, y) = 20 - 4x^2 - y^2$$

where x and y are measured in centimeters. In what direction from $(2, -3)$ does the temperature increase most rapidly? What is this rate of increase?

Solution The gradient is

$$\nabla T(x, y) = T_x(x, y)\mathbf{i} + T_y(x, y)\mathbf{j} = -8x\mathbf{i} - 2y\mathbf{j}.$$

It follows that the direction of maximum increase is given by

$$\nabla T(2, -3) = -16\mathbf{i} + 6\mathbf{j}$$

as shown in Figure 13.51, and the rate of increase is

$$\|\nabla T(2, -3)\| = \sqrt{256 + 36} = \sqrt{292} \approx 17.09° \text{ per centimeter.}$$

The solution presented in Example 5 can be misleading. Although the gradient points in the direction of maximum temperature increase, it does not necessarily point toward the hottest spot on the plate. In other words, the gradient provides a local solution to finding an increase relative to the temperature at the point $(2, -3)$. *Once you leave that position, the direction of maximum increase may change.*

EXAMPLE 6 **Finding the Path of a Heat-Seeking Particle**

A heat-seeking particle is located at the point $(2, -3)$ on a metal plate whose temperature at (x, y) is

$$T(x, y) = 20 - 4x^2 - y^2.$$

Find the path of the particle as it continuously moves in the direction of maximum temperature increase.

Solution Let the path be represented by the position vector

$$\mathbf{r}(t) = x(t)\mathbf{i} + y(t)\mathbf{j}.$$

A tangent vector at each point $(x(t), y(t))$ is given by

$$\mathbf{r}'(t) = \frac{dx}{dt}\mathbf{i} + \frac{dy}{dt}\mathbf{j}.$$

Because the particle seeks maximum temperature increase, the directions of $\mathbf{r}'(t)$ and $\nabla T(x, y) = -8x\mathbf{i} - 2y\mathbf{j}$ are the same at each point on the path. So,

$$-8x = k\frac{dx}{dt} \quad \text{and} \quad -2y = k\frac{dy}{dt}$$

where k depends on t. By solving each equation for dt/k and equating the results, you obtain

$$\frac{dx}{-8x} = \frac{dy}{-2y}.$$

The solution of this differential equation is $x = Cy^4$. Because the particle starts at the point $(2, -3)$, you can determine that $C = 2/81$. So, the path of the heat-seeking particle is

$$x = \frac{2}{81}y^4.$$

Level curves:
$T(x, y) = 20 - 4x^2 - y^2$

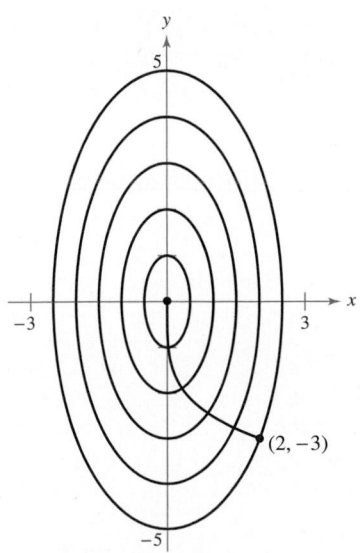

Path followed by a heat-seeking particle

Figure 13.52

The path is shown in Figure 13.52.

In Figure 13.52, the path of the particle (determined by the gradient at each point) appears to be orthogonal to each of the level curves. This becomes clear when you consider that the temperature $T(x, y)$ is constant along a given level curve. So, at any point (x, y) on the curve, the rate of change of T in the direction of a unit tangent vector $\mathbf{u}$ is 0, and you can write

$$\nabla f(x, y) \cdot \mathbf{u} = D_{\mathbf{u}}T(x, y) = 0. \qquad \text{\textbf{u} is a unit tangent vector.}$$

Because the dot product of $\nabla f(x, y)$ and $\mathbf{u}$ is 0, you can conclude that they must be orthogonal. This result is stated in the next theorem.

THEOREM 13.12   Gradient Is Normal to Level Curves

If f is differentiable at (x_0, y_0) and $\nabla f(x_0, y_0) \neq \mathbf{0}$, then $\nabla f(x_0, y_0)$ is normal to the level curve through (x_0, y_0).

EXAMPLE 7   Finding a Normal Vector to a Level Curve

Sketch the level curve corresponding to $c = 0$ for the function given by

$$f(x, y) = y - \sin x$$

and find a normal vector at several points on the curve.

Solution   The level curve for $c = 0$ is given by

$$0 = y - \sin x \quad \Longrightarrow \quad y = \sin x$$

as shown in Figure 13.53(a). Because the gradient of f at (x, y) is

$$\nabla f(x, y) = f_x(x, y)\mathbf{i} + f_y(x, y)\mathbf{j}$$
$$= -\cos x\,\mathbf{i} + \mathbf{j}$$

you can use Theorem 13.12 to conclude that $\nabla f(x, y)$ is normal to the level curve at the point (x, y). Some gradients are

$$\nabla f(-\pi, 0) = \mathbf{i} + \mathbf{j}$$

$$\nabla f\left(-\frac{2\pi}{3}, -\frac{\sqrt{3}}{2}\right) = \frac{1}{2}\mathbf{i} + \mathbf{j}$$

$$\nabla f\left(-\frac{\pi}{2}, -1\right) = \mathbf{j}$$

$$\nabla f\left(-\frac{\pi}{3}, -\frac{\sqrt{3}}{2}\right) = -\frac{1}{2}\mathbf{i} + \mathbf{j}$$

$$\nabla f(0, 0) = -\mathbf{i} + \mathbf{j}$$

$$\nabla f\left(\frac{\pi}{3}, \frac{\sqrt{3}}{2}\right) = -\frac{1}{2}\mathbf{i} + \mathbf{j}$$

$$\nabla f\left(\frac{\pi}{2}, 1\right) = \mathbf{j}$$

$$\nabla f\left(\frac{2\pi}{3}, \frac{\sqrt{3}}{2}\right) = \frac{1}{2}\mathbf{i} + \mathbf{j}$$

and

$$\nabla f(\pi, 0) = \mathbf{i} + \mathbf{j}.$$

These are shown in Figure 13.53(b).

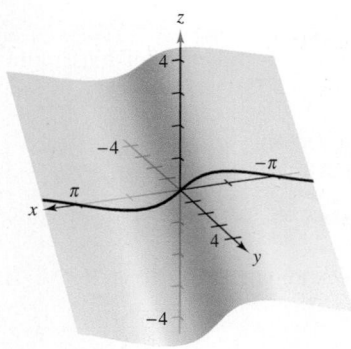

(a) The surface is given by $f(x, y) = y - \sin x$.

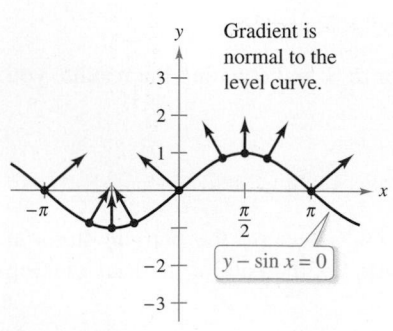

(b) The level curve is given by $f(x, y) = 0$.

Figure 13.53

Functions of Three Variables

The definitions of the directional derivative and the gradient can be extended naturally to functions of three or more variables. As often happens, some of the geometric interpretation is lost in the generalization from functions of two variables to those of three variables. For example, you cannot interpret the directional derivative of a function of three variables as representing slope.

The definitions and properties of the directional derivative and the gradient of a function of three variables are listed below.

Directional Derivative and Gradient for Three Variables

Let f be a function of x, y, and z with continuous first partial derivatives. The **directional derivative of f** in the direction of a unit vector

$$\mathbf{u} = a\mathbf{i} + b\mathbf{j} + c\mathbf{k}$$

is given by

$$D_{\mathbf{u}}f(x, y, z) = af_x(x, y, z) + bf_y(x, y, z) + cf_z(x, y, z).$$

The **gradient of f** is defined as

$$\nabla f(x, y, z) = f_x(x, y, z)\mathbf{i} + f_y(x, y, z)\mathbf{j} + f_z(x, y, z)\mathbf{k}.$$

Properties of the gradient are as follows.

1. $D_{\mathbf{u}}f(x, y, z) = \nabla f(x, y, z) \cdot \mathbf{u}$
2. If $\nabla f(x, y, z) = \mathbf{0}$, then $D_{\mathbf{u}}f(x, y, z) = 0$ for all $\mathbf{u}$.
3. The direction of *maximum* increase of f is given by $\nabla f(x, y, z)$. The maximum value of $D_{\mathbf{u}}f(x, y, z)$ is

    $$\|\nabla f(x, y, z)\|.$$ Maximum value of $D_{\mathbf{u}}f(x, y, z)$

4. The direction of *minimum* increase of f is given by $-\nabla f(x, y, z)$. The minimum value of $D_{\mathbf{u}}f(x, y, z)$ is

    $$-\|\nabla f(x, y, z)\|.$$ Minimum value of $D_{\mathbf{u}}f(x, y, z)$

You can generalize Theorem 13.12 to functions of three variables. Under suitable hypotheses,

$$\nabla f(x_0, y_0, z_0)$$

is normal to the level surface through (x_0, y_0, z_0).

EXAMPLE 8   **Finding the Gradient of a Function**

Find $\nabla f(x, y, z)$ for the function

$$f(x, y, z) = x^2 + y^2 - 4z$$

and find the direction of maximum increase of f at the point $(2, -1, 1)$.

Solution   The gradient is

$$\nabla f(x, y, z) = f_x(x, y, z)\mathbf{i} + f_y(x, y, z)\mathbf{j} + f_z(x, y, z)\mathbf{k}$$
$$= 2x\mathbf{i} + 2y\mathbf{j} - 4\mathbf{k}.$$

So, it follows that the direction of maximum increase at $(2, -1, 1)$ is

$$\nabla f(2, -1, 1) = 4\mathbf{i} - 2\mathbf{j} - 4\mathbf{k}. \qquad \text{See Figure 13.54.}$$

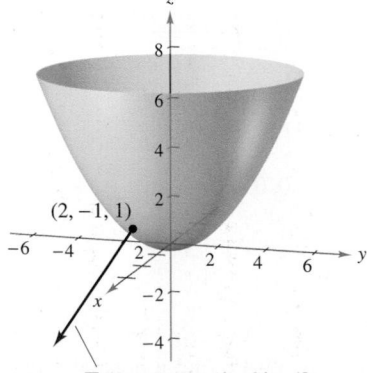

$\nabla f(2, -1, 1) = 4\mathbf{i} - 2\mathbf{j} - 4\mathbf{k}$

Level surface and gradient at $(2, -1, 1)$ for $f(x, y, z) = x^2 + y^2 - 4z$

Figure 13.54

13.6 Exercises

See CalcChat.com for tutorial help and worked-out solutions to odd-numbered exercises.

CONCEPT CHECK

1. Directional Derivative For a function $f(x, y)$, when does the directional derivative at the point (x_0, y_0) equal the partial derivative with respect to x at the point (x_0, y_0)? What does this mean graphically?

2. Gradient What is the meaning of the gradient of a function f at a point (x, y)?

 **Finding a Directional Derivative** In Exercises 3–6, use Theorem 13.9 to find the directional derivative of the function at P in the direction of the unit vector $\mathbf{u} = \cos\theta\mathbf{i} + \sin\theta\mathbf{j}$.

3. $f(x, y) = x^2 + y^2$, $P(1, -2)$, $\theta = \dfrac{\pi}{4}$

4. $f(x, y) = \dfrac{y}{x + y}$, $P(3, 0)$, $\theta = -\dfrac{\pi}{6}$

5. $f(x, y) = \sin(2x + y)$, $P(0, \pi)$, $\theta = -\dfrac{5\pi}{6}$

6. $g(x, y) = xe^y$, $P(0, 2)$, $\theta = \dfrac{2\pi}{3}$

 **Finding a Directional Derivative** In Exercises 7–10, use Theorem 13.9 to find the directional derivative of the function at P in the direction of $\mathbf{v}$.

7. $f(x, y) = 3x - 4xy + 9y$, $P(1, 2)$, $\mathbf{v} = \frac{3}{5}\mathbf{i} + \frac{4}{5}\mathbf{j}$

8. $f(x, y) = x^3 - y^3$, $P(4, 3)$, $\mathbf{v} = \dfrac{\sqrt{2}}{2}(\mathbf{i} + \mathbf{j})$

9. $g(x, y) = \sqrt{x^2 + y^2}$, $P(3, 4)$, $\mathbf{v} = 3\mathbf{i} - 4\mathbf{j}$

10. $h(x, y) = e^{-(x^2 + y^2)}$, $P(0, 0)$, $\mathbf{v} = \mathbf{i} + \mathbf{j}$

Finding a Directional Derivative In Exercises 11–14, use Theorem 13.9 to find the directional derivative of the function at P in the direction of $\overrightarrow{PQ}$.

11. $f(x, y) = x^2 + 3y^2$, $P(1, 1)$, $Q(4, 5)$

12. $f(x, y) = \cos(x + y)$, $P(0, \pi)$, $Q\left(\dfrac{\pi}{2}, 0\right)$

13. $f(x, y) = e^y \sin x$, $P(0, 0)$, $Q(2, 1)$

14. $f(x, y) = \sin 2x \cos y$, $P(\pi, 0)$, $Q\left(\dfrac{\pi}{2}, \pi\right)$

 **Finding the Gradient of a Function** In Exercises 15–20, find the gradient of the function at the given point.

15. $f(x, y) = 3x + 5y^2 + 1$, $(2, 1)$

16. $g(x, y) = 2xe^{y/x}$, $(2, 0)$

17. $z = \dfrac{\ln(x^2 - y)}{x} - 4$, $(2, 3)$

18. $z = \cos(x^2 + y^2)$, $(3, -4)$

19. $w = 6xy - y^2 + 2xyz^3$, $(-1, 5, -1)$

20. $w = x\tan(y + z)$, $(4, 3, -1)$

Finding a Directional Derivative In Exercises 21–24, use the gradient to find the directional derivative of the function at P in the direction of $\mathbf{v}$.

21. $f(x, y) = xy$, $P(0, -2)$, $\mathbf{v} = \frac{1}{2}(\mathbf{i} + \sqrt{3}\mathbf{j})$

22. $h(x, y) = e^{-3x}\sin y$, $P\left(1, \dfrac{\pi}{2}\right)$, $\mathbf{v} = -\mathbf{i}$

23. $f(x, y, z) = x^2 + y^2 + z^2$, $P(1, 1, 1)$, $\mathbf{v} = \dfrac{\sqrt{3}}{3}(\mathbf{i} - \mathbf{j} + \mathbf{k})$

24. $f(x, y, z) = xy + yz + xz$, $P(1, 2, -1)$, $\mathbf{v} = 2\mathbf{i} + \mathbf{j} - \mathbf{k}$

 **Finding a Directional Derivative** In Exercises 25–28, use the gradient to find the directional derivative of the function at P in the direction of $\overrightarrow{PQ}$.

25. $g(x, y) = x^2 + y^2 + 1$, $P(1, 2)$, $Q(2, 3)$

26. $f(x, y) = 3x^2 - y^2 + 4$, $P(-1, 4)$, $Q(3, 6)$

27. $g(x, y, z) = xye^z$, $P(2, 4, 0)$, $Q(0, 0, 0)$

28. $h(x, y, z) = \ln(x + y + z)$, $P(1, 0, 0)$, $Q(4, 3, 1)$

 **Using Properties of the Gradient** In Exercises 29–38, find the gradient of the function and the maximum value of the directional derivative at the given point.

29. $f(x, y) = y^2 - x\sqrt{y}$, $(0, 3)$

30. $f(x, y) = \dfrac{x + y}{y + 1}$, $(0, 1)$

31. $h(x, y) = x\tan y$, $\left(2, \dfrac{\pi}{4}\right)$

32. $h(x, y) = y\cos(x - y)$, $\left(0, \dfrac{\pi}{3}\right)$

33. $f(x, y) = \sin x^2 y^3$, $\left(\dfrac{1}{\pi}, \pi\right)$

34. $g(x, y) = \ln\sqrt[3]{x^2 + y^2}$, $(1, 2)$

35. $f(x, y, z) = \sqrt{x^2 + y^2 + z^2}$, $(1, 4, 2)$

36. $w = \dfrac{1}{\sqrt{1 - x^2 - y^2 - z^2}}$, $(0, 0, 0)$

37. $w = xy^2z^2$, $(2, 1, 1)$

38. $f(x, y, z) = xe^{yz}$, $(2, 0, -4)$

 **Finding a Normal Vector to a Level Curve** In Exercises 39–42, find a normal vector to the level curve $f(x, y) = c$ at P.

39. $f(x, y) = 6 - 2x - 3y$
$c = 6$, $P(0, 0)$

40. $f(x, y) = x^2 + y^2$
$c = 25$, $P(3, 4)$

41. $f(x, y) = xy$

$c = -3, \quad P(-1, 3)$

42. $f(x, y) = \dfrac{x}{x^2 + y^2}$

$c = \frac{1}{2}, \quad P(1, 1)$

Using a Function In Exercises 43–46, (a) find the gradient of the function at P, (b) find a unit normal vector to the level curve $f(x, y) = c$ at P, (c) find the tangent line to the level curve $f(x, y) = c$ at P, and (d) sketch the level curve, the unit normal vector, and the tangent line in the xy-plane.

43. $f(x, y) = 4x^2 - y$

$c = 6, \quad P(2, 10)$

44. $f(x, y) = x - y^2$

$c = 3, \quad P(4, -1)$

45. $f(x, y) = 3x^2 - 2y^2$

$c = 1, \quad P(1, 1)$

46. $f(x, y) = 9x^2 + 4y^2$

$c = 40, \quad P(2, -1)$

47. Using a Function Consider the function

$$f(x, y) = 3 - \frac{x}{3} - \frac{y}{2}.$$

(a) Sketch the graph of f in the first octant and plot the point $(3, 2, 1)$ on the surface.

(b) Find $D_{\mathbf{u}} f(3, 2)$, where $\mathbf{u} = \cos \theta \mathbf{i} + \sin \theta \mathbf{j}$, using each given value of θ.

   (i) $\theta = \dfrac{\pi}{4}$ (ii) $\theta = \dfrac{2\pi}{3}$ (iii) $\theta = \dfrac{4\pi}{3}$ (iv) $\theta = -\dfrac{\pi}{6}$

(c) Find $D_{\mathbf{u}} f(3, 2)$, where $\mathbf{u} = \dfrac{\mathbf{v}}{\|\mathbf{v}\|}$, using each given vector $\mathbf{v}$.

   (i) $\mathbf{v} = \mathbf{i} + \mathbf{j}$ (ii) $\mathbf{v} = -3\mathbf{i} - 4\mathbf{j}$

   (iii) $\mathbf{v}$ is the vector from $(1, 2)$ to $(-2, 6)$.

   (iv) $\mathbf{v}$ is the vector from $(3, 2)$ to $(4, 5)$.

(d) Find $\nabla f(x, y)$.

(e) Find the maximum value of the directional derivative at $(3, 2)$.

(f) Find a unit vector $\mathbf{u}$ orthogonal to $\nabla f(3, 2)$ and calculate $D_{\mathbf{u}} f(3, 2)$. Discuss the geometric meaning of the result.

48. Using a Function Consider the function

$$f(x, y) = 9 - x^2 - y^2.$$

(a) Sketch the graph of f in the first octant and plot the point $(1, 2, 4)$ on the surface.

(b) Find $D_{\mathbf{u}} f(1, 2)$, where $\mathbf{u} = \cos \theta \mathbf{i} + \sin \theta \mathbf{j}$, using each given value of θ.

   (i) $\theta = -\dfrac{\pi}{4}$ (ii) $\theta = \dfrac{\pi}{3}$ (iii) $\theta = \dfrac{3\pi}{4}$ (iv) $\theta = -\dfrac{\pi}{2}$

(c) Find $D_{\mathbf{u}} f(1, 2)$, where $\mathbf{u} = \dfrac{\mathbf{v}}{\|\mathbf{v}\|}$, using each given vector $\mathbf{v}$.

   (i) $\mathbf{v} = 3\mathbf{i} + \mathbf{j}$ (ii) $\mathbf{v} = -8\mathbf{i} - 6\mathbf{j}$

   (iii) $\mathbf{v}$ is the vector from $(-1, -1)$ to $(3, 5)$.

   (iv) $\mathbf{v}$ is the vector from $(-2, 0)$ to $(1, 3)$.

(d) Find $\nabla f(1, 2)$.

(e) Find the maximum value of the directional derivative at $(1, 2)$.

(f) Find a unit vector $\mathbf{u}$ orthogonal to $\nabla f(1, 2)$ and calculate $D_{\mathbf{u}} f(1, 2)$. Discuss the geometric meaning of the result.

49. Investigation Consider the function

$$f(x, y) = x^2 - y^2$$

at the point $(4, -3, 7)$.

(a) Use a computer algebra system to graph the surface represented by the function.

(b) Determine the directional derivative $D_{\mathbf{u}} f(4, -3)$ as a function of θ, where $\mathbf{u} = \cos \theta \mathbf{i} + \sin \theta \mathbf{j}$. Use a computer algebra system to graph the function on the interval $[0, 2\pi)$.

(c) Approximate the zeros of the function in part (b) and interpret each in the context of the problem.

(d) Approximate the critical numbers of the function in part (b) and interpret each in the context of the problem.

(e) Find $\|\nabla f(4, -3)\|$ and explain its relationship to your answers in part (d).

(f) Use a computer algebra system to graph the level curve of the function f at the level $c = 7$. On this curve, graph the vector in the direction of $\nabla f(4, -3)$ and state its relationship to the level curve.

50. Investigation Consider the function

$$f(x, y) = \frac{8y}{1 + x^2 + y^2}.$$

(a) Analytically verify that the level curve of $f(x, y)$ at the level $c = 2$ is a circle.

(b) At the point $\left(\sqrt{3}, 2\right)$ on the level curve for which $c = 2$, sketch the vector showing the direction of the greatest rate of increase of the function. To print a graph of the level curve, go to *MathGraphs.com*.

(c) At the point $\left(\sqrt{3}, 2\right)$ on the level curve for which $c = 2$, sketch a vector such that the directional derivative is 0.

(d) Use a computer algebra system to graph the surface to verify your answers in parts (a)–(c).

EXPLORING CONCEPTS

51. Think About It Consider $\mathbf{v} = 3\mathbf{u}$. Is the directional derivative of a differentiable function $f(x, y)$ in the direction of $\mathbf{v}$ at the point (x_0, y_0) three times the directional derivative of f in the direction of $\mathbf{u}$ at the point (x_0, y_0)? Explain.

52. Sketching a Graph and a Vector Sketch the graph of a surface and select a point P on the surface. Sketch a vector in the xy-plane giving the direction of steepest ascent on the surface at P.

---

53. Topography The surface of a mountain is modeled by the equation

$$h(x, y) = 5000 - 0.001x^2 - 0.004y^2.$$

A mountain climber is at the point $(500, 300, 4390)$. In what direction should the climber move in order to ascend at the greatest rate?

54. **HOW DO YOU SEE IT?** The figure shows a topographic map carried by a group of hikers. Sketch the paths of steepest descent when the hikers start at point A and when they start at point B. (To print an enlarged copy of the graph, go to *MathGraphs.com*.)

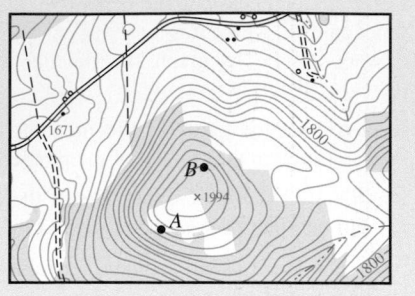

55. Temperature The temperature at the point (x, y) on a metal plate is $T(x, y) = x/(x^2 + y^2)$. Find the direction of greatest increase in heat from the point $(3, 4)$.

56. Temperature The temperature at the point (x, y) on a metal plate is $T(x, y) = 400e^{-(x^2+y)/2}$, $x \geq 0$, $y \geq 0$.

 (a) Use a computer algebra system to graph the temperature distribution function.

(b) Find the directions of no change in heat on the plate from the point $(3, 5)$.

(c) Find the direction of greatest increase in heat from the point $(3, 5)$.

 **Finding the Direction of Maximum Increase** In Exercises 57 and 58, the temperature in degrees Celsius on the surface of a metal plate is given by $T(x, y)$, where x and y are measured in centimeters. Find the direction from point P where the temperature increases most rapidly and this rate of increase.

57. $T(x, y) = 80 - 3x^2 - y^2$, $P(-1, 5)$

58. $T(x, y) = 50 - x^2 - 4y^2$, $P(2, -1)$

 **Finding the Path of a Heat-Seeking Particle** In Exercises 59 and 60, find the path of a heat-seeking particle placed at point P on a metal plate whose temperature at (x, y) is $T(x, y)$.

59. $T(x, y) = 400 - 2x^2 - y^2$, $P(10, 10)$

60. $T(x, y) = 100 - x^2 - 2y^2$, $P(4, 3)$

True or False? In Exercises 61–64, determine whether the statement is true or false. If it is false, explain why or give an example that shows it is false.

61. If $f(x, y) = \sqrt{1 - x^2 - y^2}$, then $D_{\mathbf{u}}f(0, 0) = 0$ for any unit vector $\mathbf{u}$.

62. If $f(x, y) = x + y$, then $-1 \leq D_{\mathbf{u}}f(x, y) \leq 1$.

63. If $D_{\mathbf{u}}f(x, y)$ exists, then $D_{\mathbf{u}}f(x, y) = -D_{\mathbf{u}}f(x, y)$.

64. If $D_{\mathbf{u}}f(x_0, y_0) = c$ for any unit vector $\mathbf{u}$, then $c = 0$.

65. Finding a Function Find a function f such that

$$\nabla f = e^x \cos y\mathbf{i} - e^x \sin y\mathbf{j} + z\mathbf{k}.$$

66. Ocean Floor

A team of oceanographers is mapping the ocean floor to assist in the recovery of a sunken ship. Using sonar, they develop the model

$$D = 250 + 30x^2 + 50 \sin \frac{\pi y}{2}, \quad 0 \leq x \leq 2, \quad 0 \leq y \leq 2$$

where D is the depth in meters, and x and y are the distances in kilometers.

(a) Use a computer algebra system to graph D.

(b) Because the graph in part (a) is showing depth, it is not a map of the ocean floor. How could the model be changed so that the graph of the ocean floor could be obtained?

(c) What is the depth of the ship if it is located at the coordinates $x = 1$ and $y = 0.5$?

(d) Determine the steepness of the ocean floor in the positive x-direction from the position of the ship.

(e) Determine the steepness of the ocean floor in the positive y-direction from the position of the ship.

(f) Determine the direction of the greatest rate of change of depth from the position of the ship.

67. Using a Function Consider the function

$$f(x, y) = \sqrt[3]{xy}.$$

(a) Show that f is continuous at the origin.

(b) Show that f_x and f_y exist at the origin but that the directional derivatives at the origin in all other directions do not exist.

 (c) Use a computer algebra system to graph f near the origin to verify your answers in parts (a) and (b). Explain.

68. Directional Derivative Consider the function

$$f(x, y) = \begin{cases} \dfrac{4xy}{x^2 + y^2}, & (x, y) \neq (0, 0) \\ 0, & (x, y) \neq (0, 0) \end{cases}$$

and the unit vector

$$\mathbf{u} = \frac{1}{\sqrt{2}}(\mathbf{i} + \mathbf{j}).$$

Does the directional derivative of f at $P(0, 0)$ in the direction of $\mathbf{u}$ exist? If $f(0, 0)$ were defined as 2 instead of 0, would the directional derivative exist? Explain.

13.7 Tangent Planes and Normal Lines

■ Find equations of tangent planes and normal lines to surfaces.
■ Find the angle of inclination of a plane in space.
■ Compare the gradients $\nabla f(x, y)$ and $\nabla F(x, y, z)$.

Tangent Plane and Normal Line to a Surface

So far, you have represented surfaces in space primarily by equations of the form

$$z = f(x, y). \qquad \text{Equation of a surface } S$$

In the development to follow, however, it is convenient to use the more general representation $F(x, y, z) = 0$. For a surface S given by $z = f(x, y)$, you can convert to the general form by defining F as

$$F(x, y, z) = f(x, y) - z.$$

Because $f(x, y) - z = 0$, you can consider S to be the level surface of F given by

$$F(x, y, z) = 0. \qquad \text{Alternative equation of surface } S$$

EXAMPLE 1 Writing an Equation of a Surface

For the function

$$F(x, y\, z) = x^2 + y^2 + z^2 - 4$$

describe the level surface given by

$$F(x, y, z) = 0.$$

Solution The level surface given by $F(x, y, z) = 0$ can be written as

$$x^2 + y^2 + z^2 = 4$$

which is a sphere of radius 2 whose center is at the origin.

You have seen many examples of the usefulness of normal lines in applications involving curves. Normal lines are equally important in analyzing surfaces and solids. For example, consider the collision of two billiard balls. When a stationary ball is struck at a point P on its surface, it moves along the **line of impact** determined by P and the center of the ball. The impact can occur in *two* ways. When the cue ball is moving along the line of impact, it stops dead and imparts all of its momentum to the stationary ball, as shown in Figure 13.55.

| Figure 13.55 | Figure 13.56 |

When the cue ball is not moving along the line of impact, it is deflected to one side or the other and retains part of its momentum. The part of the momentum that is transferred to the stationary ball occurs along the line of impact, *regardless* of the direction of the cue ball, as shown in Figure 13.56. This line of impact is called the **normal line** to the surface of the ball at the point P.

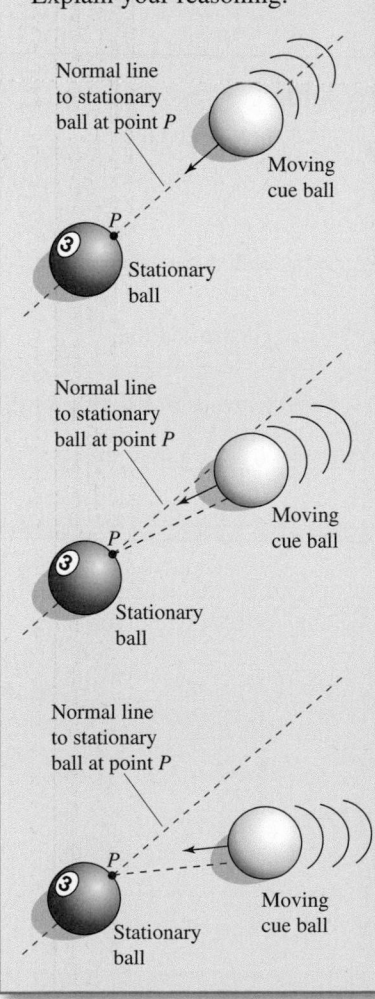

In the process of finding a normal line to a surface, you are also able to solve the problem of finding a **tangent plane** to the surface. Let S be a surface given by

$$F(x, y, z) = 0$$

and let $P(x_0, y_0, z_0)$ be a point on S. Let C be a curve on S through P that is defined by the vector-valued function

$$\mathbf{r}(t) = x(t)\mathbf{i} + y(t)\mathbf{j} + z(t)\mathbf{k}.$$

Then, for all t,

$$F(x(t), y(t), z(t)) = 0.$$

If F is differentiable and $x'(t)$, $y'(t)$, and $z'(t)$ all exist, then it follows from the Chain Rule that

$$0 = F'(t)$$
$$= F_x(x, y, z)x'(t) + F_y(x, y, z)y'(t) + F_z(x, y, z)z'(t).$$

At (x_0, y_0, z_0), the equivalent vector form is

$$0 = \underbrace{\nabla F(x_0, y_0, z_0)}_{\text{Gradient}} \cdot \underbrace{\mathbf{r}'(t_0)}_{\text{Tangent vector}}.$$

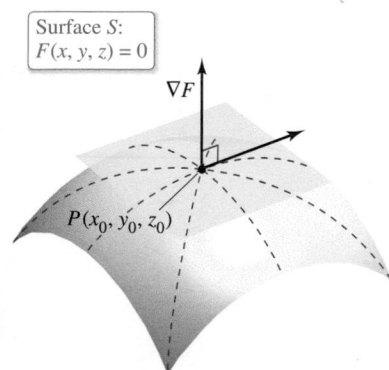

Surface S:
$F(x, y, z) = 0$

∇F

$P(x_0, y_0, z_0)$

Tangent plane to surface S at P
Figure 13.57

This result means that the gradient at P is orthogonal to the tangent vector of every curve on S through P. So, all tangent lines on S lie in a plane that is normal to $\nabla F(x_0, y_0, z_0)$ and contains P, as shown in Figure 13.57.

Definitions of Tangent Plane and Normal Line

Let F be differentiable at the point $P(x_0, y_0, z_0)$ on the surface S given by $F(x, y, z) = 0$ such that

$$\nabla F(x_0, y_0, z_0) \neq \mathbf{0}.$$

1. The plane through P that is normal to $\nabla F(x_0, y_0, z_0)$ is called the **tangent plane to S at P.**
2. The line through P having the direction of $\nabla F(x_0, y_0, z_0)$ is called the **normal line to S at P.**

▷

•• REMARK In the remainder of this section, assume $\nabla F(x_0, y_0, z_0)$ to be nonzero unless stated otherwise.

To find an equation for the tangent plane to S at (x_0, y_0, z_0), let (x, y, z) be an arbitrary point in the tangent plane. Then the vector

$$\mathbf{v} = (x - x_0)\mathbf{i} + (y - y_0)\mathbf{j} + (z - z_0)\mathbf{k}$$

lies in the tangent plane. Because $\nabla F(x_0, y_0, z_0)$, is normal to the tangent plane at (x_0, y_0, z_0), it must be orthogonal to every vector in the tangent plane, and you have

$$\nabla F(x_0, y_0, z_0) \cdot \mathbf{v} = 0$$

which leads to the next theorem.

THEOREM 13.13 Equation of Tangent Plane

If F is differentiable at (x_0, y_0, z_0), then an equation of the tangent plane to the surface given by $F(x, y, z) = 0$ at (x_0, y_0, z_0) is

$$F_x(x_0, y_0, z_0)(x - x_0) + F_y(x_0, y_0, z_0)(y - y_0) + F_z(x_0, y_0, z_0)(z - z_0) = 0.$$

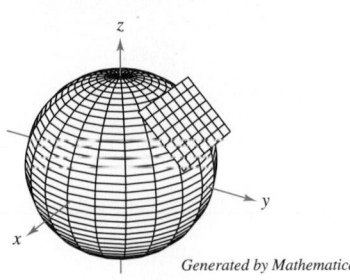
EXAMPLE 2 Finding an Equation of a Tangent Plane

Find an equation of the tangent plane to the hyperboloid

$$z^2 - 2x^2 - 2y^2 = 12$$

at the point $(1, -1, 4)$.

Solution Begin by writing the equation of the surface as

$$z^2 - 2x^2 - 2y^2 - 12 = 0.$$

Then, considering

$$F(x, y, z) = z^2 - 2x^2 - 2y^2 - 12$$

you have

$$F_x(x, y, z) = -4x, \quad F_y(x, y, z) = -4y, \quad \text{and} \quad F_z(x, y, z) = 2z.$$

At the point $(1, -1, 4)$, the partial derivatives are

$$F_x(1, -1, 4) = -4, \quad F_y(1, -1, 4) = 4, \quad \text{and} \quad F_z(1, -1, 4) = 8.$$

So, an equation of the tangent plane at $(1, -1, 4)$ is

$$-4(x - 1) + 4(y + 1) + 8(z - 4) = 0$$
$$-4x + 4 + 4y + 4 + 8z - 32 = 0$$
$$-4x + 4y + 8z - 24 = 0$$
$$x - y - 2z + 6 = 0.$$

Figure 13.58 shows a portion of the hyperboloid and the tangent plane.

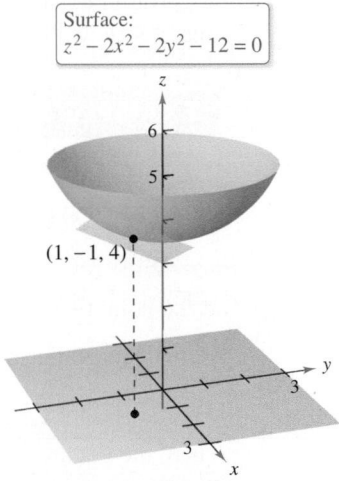

Surface:
$z^2 - 2x^2 - 2y^2 - 12 = 0$

$(1, -1, 4)$

Tangent plane to surface
Figure 13.58

To find an equation of the tangent plane at a point on a surface given by $z = f(x, y)$, you can define the function F by

$$F(x, y, z) = f(x, y) - z.$$

Then S is given by the level surface $F(x, y, z) = 0$, and by Theorem 13.13, an equation of the tangent plane to S at the point (x_0, y_0, z_0) is

$$f_x(x_0, y_0)(x - x_0) + f_y(x_0, y_0)(y - y_0) - (z - z_0) = 0.$$

EXAMPLE 3 **Finding an Equation of the Tangent Plane**

Find an equation of the tangent plane to the paraboloid

$$z = 1 - \frac{1}{10}(x^2 + 4y^2)$$

at the point $\left(1, 1, \frac{1}{2}\right)$.

Solution From $z = f(x, y) = 1 - \frac{1}{10}(x^2 + 4y^2)$, you obtain

$$f_x(x, y) = -\frac{x}{5} \implies f_x(1, 1) = -\frac{1}{5}$$

and

$$f_y(x, y) = -\frac{4y}{5} \implies f_y(1, 1) = -\frac{4}{5}.$$

So, an equation of the tangent plane at $\left(1, 1, \frac{1}{2}\right)$ is

$$f_x(1, 1)(x - 1) + f_y(1, 1)(y - 1) - \left(z - \frac{1}{2}\right) = 0$$

$$-\frac{1}{5}(x - 1) - \frac{4}{5}(y - 1) - \left(z - \frac{1}{2}\right) = 0$$

$$-\frac{1}{5}x - \frac{4}{5}y - z + \frac{3}{2} = 0.$$

This tangent plane is shown in Figure 13.59.

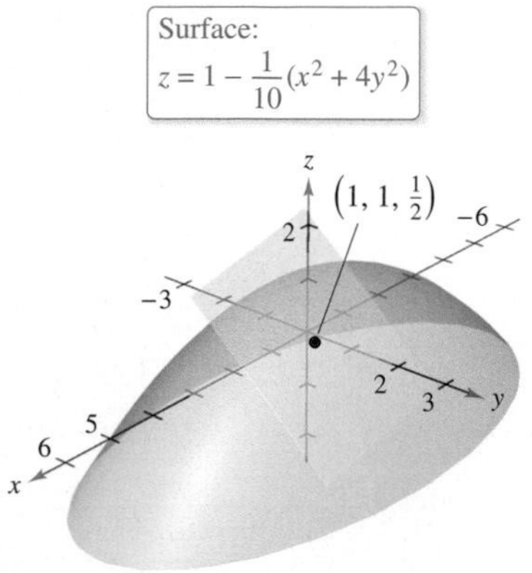

Surface:
$$z = 1 - \frac{1}{10}(x^2 + 4y^2)$$

Figure 13.59

The gradient $\nabla F(x, y, z)$ provides a convenient way to find equations of normal lines, as shown in Example 4.

EXAMPLE 4 **Finding an Equation of a Normal Line to a Surface**

$\cdots \triangleright$ *See LarsonCalculus.com for an interactive version of this type of example.*

Find a set of symmetric equations for the normal line to the surface

$$xyz = 12$$

at the point $(2, -2, -3)$.

Solution Begin by letting

$$F(x, y, z) = xyz - 12.$$

Then, the gradient is given by

$$\nabla F(x, y, z) = F_x(x, y, z)\mathbf{i} + F_y(x, y, z)\mathbf{j} + F_z(x, y, z)\mathbf{k}$$
$$= yz\mathbf{i} + xz\mathbf{j} + xy\mathbf{k}$$

and at the point $(2, -2, -3)$, you have

$$\nabla F(2, -2, -3) = (-2)(-3)\mathbf{i} + (2)(-3)\mathbf{j} + (2)(-2)\mathbf{k}$$
$$= 6\mathbf{i} - 6\mathbf{j} - 4\mathbf{k}.$$

The normal line at $(2, -2, -3)$ has direction numbers 6, -6, and -4, and the corresponding set of symmetric equations is

$$\frac{x - 2}{6} = \frac{y + 2}{-6} = \frac{z + 3}{-4}.$$

See Figure 13.60.

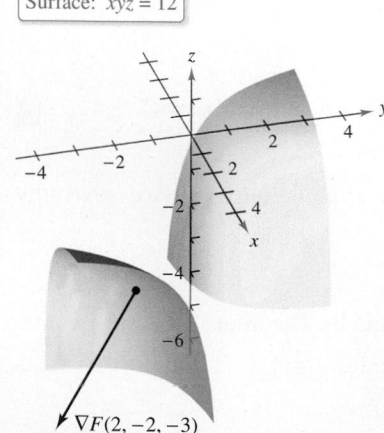

Surface: $xyz = 12$

$\nabla F(2, -2, -3)$

Figure 13.60

Knowing that the gradient $\nabla F(x, y, z)$ is normal to the surface given by $F(x, y, z) = 0$ allows you to solve a variety of problems dealing with surfaces and curves in space.

EXAMPLE 5   Finding the Equation of a Tangent Line to a Curve

Find a set of parametric equations for the tangent line to the curve of intersection of the ellipsoid

$$x^2 + 2y^2 + 2z^2 = 20 \qquad\qquad \text{Ellipsoid}$$

and the paraboloid

$$x^2 + y^2 + z = 4 \qquad\qquad \text{Paraboloid}$$

at the point $(0, 1, 3)$, as shown in Figure 13.61.

Solution   Begin by finding the gradients to both surfaces at the point $(0, 1, 3)$.

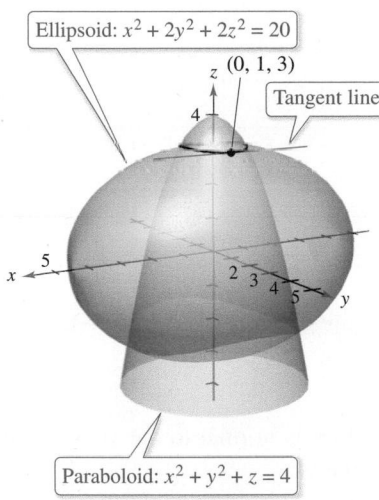

Ellipsoid: $x^2 + 2y^2 + 2z^2 = 20$

z $(0, 1, 3)$

Tangent line

Paraboloid: $x^2 + y^2 + z = 4$

Figure 13.61

Ellipsoid	**Paraboloid**
$F(x, y, z) = x^2 + 2y^2 + 2z^2 - 20$	$G(x, y, z) = x^2 + y^2 + z - 4$
$\nabla F(x, y, z) = 2x\mathbf{i} + 4y\mathbf{j} + 4z\mathbf{k}$	$\nabla G(x, y, z) = 2x\mathbf{i} + 2y\mathbf{j} + \mathbf{k}$
$\nabla F(0, 1, 3) = 4\mathbf{j} + 12\mathbf{k}$	$\nabla G(0, 1, 3) = 2\mathbf{j} + \mathbf{k}$

The cross product of these two gradients is a vector that is tangent to both surfaces at the point $(0, 1, 3)$.

$$\nabla F(0, 1, 3) \times \nabla G(0, 1, 3) = \begin{vmatrix} \mathbf{i} & \mathbf{j} & \mathbf{k} \\ 0 & 4 & 12 \\ 0 & 2 & 1 \end{vmatrix} = -20\mathbf{i}$$

So, the tangent line to the curve of intersection of the two surfaces at the point $(0, 1, 3)$ is a line that is parallel to the x-axis and passes through the point $(0, 1, 3)$. Because $-20\mathbf{i} = -20(\mathbf{i} + 0\mathbf{j} + 0\mathbf{k})$, the direction numbers are 1, 0, and 0. So a set of parametric equations for the tangent line passing through the point $(0, 1, 3)$ is $x = t$, $y = 1$, and $z = 3$.

The Angle of Inclination of a Plane

Another use of the gradient $\nabla F(x, y, z)$ is to determine the angle of inclination of the tangent plane to a surface. The **angle of inclination** of a plane is defined as the angle θ $(0 \le \theta \le \pi/2)$ between the given plane and the xy-plane, as shown in Figure 13.62. (The angle of inclination of a horizontal plane is defined as zero.) Because the vector $\mathbf{k}$ is normal to the xy-plane, you can use the formula for the cosine of the angle between two planes (given in Section 11.5) to conclude that the angle of inclination of a plane with normal vector $\mathbf{n}$ is

$$\cos \theta = \frac{|\mathbf{n} \cdot \mathbf{k}|}{\|\mathbf{n}\| \|\mathbf{k}\|} = \frac{|\mathbf{n} \cdot \mathbf{k}|}{\|\mathbf{n}\|}. \qquad\qquad \text{Angle of inclination of a plane}$$

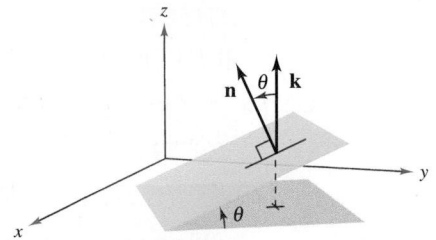

The angle of inclination
Figure 13.62

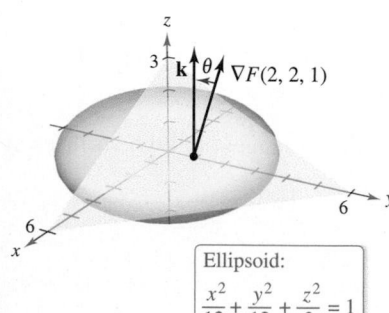

Figure 13.63

EXAMPLE 6 **Finding the Angle of Inclination of a Tangent Plane**

Find the angle of inclination of the tangent plane to the ellipsoid

$$\frac{x^2}{12} + \frac{y^2}{12} + \frac{z^2}{3} = 1$$

at the point $(2, 2, 1)$.

Solution Begin by letting

$$F(x, y, z) = \frac{x^2}{12} + \frac{y^2}{12} + \frac{z^2}{3} - 1.$$

Then, the gradient of F at the point $(2, 2, 1)$ is

$$\nabla F(x, y, z) = \frac{x}{6}\mathbf{i} + \frac{y}{6}\mathbf{j} + \frac{2z}{3}\mathbf{k}$$

$$\nabla F(2, 2, 1) = \frac{1}{3}\mathbf{i} + \frac{1}{3}\mathbf{j} + \frac{2}{3}\mathbf{k}.$$

Because $\nabla F(2, 2, 1)$ is normal to the tangent plane and $\mathbf{k}$ is normal to the xy-plane, it follows that the angle of inclination of the tangent plane is

$$\cos\theta = \frac{|\nabla F(2, 2, 1) \cdot \mathbf{k}|}{\|\nabla F(2, 2, 1)\|} = \frac{2/3}{\sqrt{(1/3)^2 + (1/3)^2 + (2/3)^2}} = \sqrt{\frac{2}{3}}$$

which implies that

$$\theta = \arccos\sqrt{\frac{2}{3}} \approx 35.3°$$

as shown in Figure 13.63. ∎

A special case of the procedure shown in Example 6 is worth noting. The angle of inclination θ of the tangent plane to the surface $z = f(x, y)$ at (x_0, y_0, z_0) is

$$\cos\theta = \frac{1}{\sqrt{[f_x(x_0, y_0)]^2 + [f_y(x_0, y_0)]^2 + 1}}.$$

Alternative formula for angle of inclination (See Exercise 63.)

A Comparison of the Gradients $\nabla f(x, y)$ and $\nabla F(x, y, z)$

This section concludes with a comparison of the gradients $\nabla f(x, y)$ and $\nabla F(x, y, z)$. In the preceding section, you saw that the gradient of a function f of two variables is normal to the level curves of f. Specifically, Theorem 13.12 states that if f is differentiable at (x_0, y_0) and $\nabla f(x_0, y_0) \neq \mathbf{0}$, then $\nabla f(x_0, y_0)$ is normal to the level curve through (x_0, y_0). Having developed normal lines to surfaces, you can now extend this result to a function of three variables. The proof of Theorem 13.14 is left as an exercise (see Exercise 64).

THEOREM 13.14 **Gradient Is Normal to Level Surfaces**

If F is differentiable at (x_0, y_0, z_0) and

$$\nabla F(x_0, y_0, z_0) \neq \mathbf{0}$$

then $\nabla F(x_0, y_0, z_0)$ is normal to the level surface through (x_0, y_0, z_0).

When working with the gradients $\nabla f(x, y)$ and $\nabla F(x, y, z)$, be sure you remember that $\nabla f(x, y)$ is a vector in the xy-plane and $\nabla F(x, y, z)$ is a vector in space.

CONCEPT CHECK

1. Tangent Vector Consider a point (x_0, y_0, z_0) on a surface given by $F(x, y, z) = 0$. What is the relationship between $\nabla F(x_0, y_0, z_0)$ and any tangent vector $\mathbf{v}$ at (x_0, y_0, z_0)? How do you represent this relationship mathematically?

2. Normal Line Consider a point (x_0, y_0, z_0) on a surface given by $F(x, y, z) = 0$. What is the relationship between $\nabla F(x_0, y_0, z_0)$ and the normal line through (x_0, y_0, z_0)?

 **Describing a Surface In Exercises 3–6,** describe the level surface $F(x, y, z) = 0$.

3. $F(x, y, z) = 3x - 5y + 3z - 15$

4. $F(x, y, z) = 36 - x^2 - y^2 - z^2$

5. $F(x, y, z) = 4x^2 + 9y^2 - 4z^2$

6. $F(x, y, z) = 16x^2 - 9y^2 + 36z$

 **Finding an Equation of a Tangent Plane In Exercises 7–16, find an equation of the tangent plane to the surface at the given point.**

7. $z = x^2 + y^2 + 3$

$(2, 1, 8)$

8. $f(x, y) = \dfrac{y}{x}$

$(1, 2, 2)$

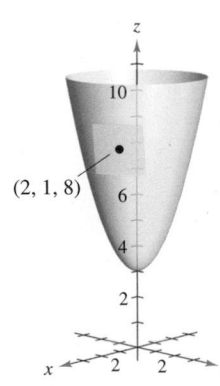

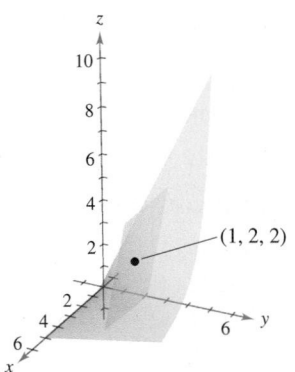

9. $z = \sqrt{x^2 + y^2}$, $(3, 4, 5)$

10. $g(x, y) = \arctan \dfrac{y}{x}$, $(1, 0, 0)$

11. $g(x, y) = x^2 + y^2$, $(1, -1, 2)$

12. $f(x, y) = x^2 - 2xy + y^2$, $(1, 2, 1)$

13. $h(x, y) = \ln \sqrt{x^2 + y^2}$, $(3, 4, \ln 5)$

14. $f(x, y) = \sin x \cos y$, $\left(\dfrac{\pi}{3}, \dfrac{\pi}{6}, \dfrac{3}{4}\right)$

15. $x^2 + y^2 - 5z^2 = 15$, $(-4, -2, 1)$

16. $x^2 + 2z^2 = y^2$, $(1, 3, -2)$

 Finding an Equation of a Tangent Plane and a Normal Line In Exercises 17–26, (a) find an equation of the tangent plane to the surface at the given point and (b) find a set of symmetric equations for the normal line to the surface at the given point.

17. $x + y + z = 9$, $(3, 3, 3)$

18. $x^2 + y^2 + z^2 = 9$, $(1, 2, 2)$

19. $x^2 + 2y^2 + z^2 = 7$, $(1, -1, 2)$

20. $z = 16 - x^2 - y^2$, $(2, 2, 8)$

21. $z = x^2 - y^2$, $(3, 2, 5)$

22. $xy - z = 0$, $(-2, -3, 6)$

23. $xyz = 10$, $(1, 2, 5)$

24. $6xy = z$, $(-1, 1, -6)$

25. $z = ye^{2xy}$, $(0, 2, 2)$

26. $y \ln xz^2 = 2$, $(e, 2, 1)$

 **Finding the Equation of a Tangent Line to a Curve In Exercises 27–32, find a set of parametric equations for the tangent line to the curve of intersection of the surfaces at the given point.**

27. $x^2 + y^2 = 2$, $z = x$, $(1, 1, 1)$

28. $z = x^2 + y^2$, $z = 4 - y$, $(2, -1, 5)$

29. $x^2 + z^2 = 25$, $y^2 + z^2 = 25$, $(3, 3, 4)$

30. $z = \sqrt{x^2 + y^2}$, $5x - 2y + 3z = 22$, $(3, 4, 5)$

31. $x^2 + y^2 + z^2 = 14$, $x - y - z = 0$, $(3, 1, 2)$

32. $z = x^2 + y^2$, $x + y + 6z = 33$, $(1, 2, 5)$

 **Finding the Angle of Inclination of a Tangent Plane In Exercises 33–36, find the angle of inclination of the tangent plane to the surface at the given point.**

33. $3x^2 + 2y^2 - z = 15$, $(2, 2, 5)$

34. $2xy - z^3 = 0$, $(2, 2, 2)$

35. $x^2 - y^2 + z = 0$, $(1, 2, 3)$

36. $x^2 + y^2 = 5$, $(2, 1, 3)$

Horizontal Tangent Plane In Exercises 37–42, find the point(s) on the surface at which the tangent plane is horizontal.

37. $z = 3 - x^2 - y^2 + 6y$

38. $z = 3x^2 + 2y^2 - 3x + 4y - 5$

39. $z = x^2 - xy + y^2 - 2x - 2y$

40. $z = 4x^2 + 4xy - 2y^2 + 8x - 5y - 4$

41. $z = 5xy$

42. $z = xy + \dfrac{1}{x} + \dfrac{1}{y}$

Tangent Surfaces In Exercises 43 and 44, show that the surfaces are tangent to each other at the given point by showing that the surfaces have the same tangent plane at this point.

43. $x^2 + 2y^2 + 3z^2 = 3$, $x^2 + y^2 + z^2 + 6x - 10y + 14 = 0$,
$(-1, 1, 0)$

44. $x^2 + y^2 + z^2 - 8x - 12y + 4z + 42 = 0$,
$x^2 + y^2 + 2z = 7$, $(2, 3, -3)$

Perpendicular Tangent Planes In Exercises 45 and 46, (a) show that the surfaces intersect at the given point and (b) show that the surfaces have perpendicular tangent planes at this point.

45. $z = 2xy^2$, $8x^2 - 5y^2 - 8z = -13$, $(1, 1, 2)$

46. $x^2 + y^2 + z^2 + 2x - 4y - 4z - 12 = 0$,
$4x^2 + y^2 + 16z^2 = 24$, $(1, -2, 1)$

EXPLORING CONCEPTS

47. Tangent Plane The tangent plane to the surface represented by $F(x, y, z) = 0$ at a point P is also tangent to the surface represented by $G(x, y, z) = 0$ at P. Is $\nabla F(x, y, z) = \nabla G(x, y, z)$ at P? Explain.

48. Normal Lines For some surfaces, the normal lines at any point pass through the same geometric object. What is the common geometric object for a sphere? What is the common geometric object for a right circular cylinder? Explain.

---

49. Using an Ellipsoid Find a point on the ellipsoid $3x^2 + y^2 + 3z^2 = 1$ where the tangent line is parallel to the plane $-12x + 2y + 6z = 0$.

50. Using a Hyperboloid Find a point on the hyperboloid $x^2 + 4y^2 - z^2 = 1$ where the tangent plane is parallel to the plane $x + 4y - z = 0$.

51. Using an Ellipsoid Find a point on the ellipsoid $x^2 + 4y^2 + z^2 = 9$ where the tangent plane is perpendicular to the line with parametric equations
$x = 2 - 4t$, $y = 1 + 8t$, and $z = 3 - 2t$.

52. HOW DO YOU SEE IT? The graph shows the ellipsoid $x^2 + 4y^2 + z^2 = 16$. Use the graph to determine the equation of the tangent plane at each of the given points.

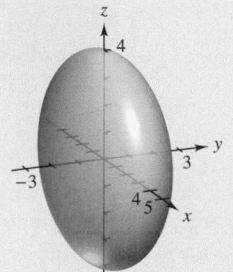

   (a) $(4, 0, 0)$ (b) $(0, -2, 0)$ (c) $(0, 0, -4)$

53. Investigation Consider the function
$$f(x, y) = \frac{4xy}{(x^2 + 1)(y^2 + 1)}$$
on the intervals $-2 \le x \le 2$ and $0 \le y \le 3$.

(a) Find a set of parametric equations of the normal line and an equation of the tangent plane to the surface at the point $(1, 1, 1)$.

(b) Repeat part (a) for the point $\left(-1, 2, -\frac{4}{5}\right)$.

(c) Use a computer algebra system to graph the surface, the normal lines, and the tangent planes found in parts (a) and (b).

54. Investigation Consider the function
$$f(x, y) = \frac{\sin y}{x}$$
on the intervals $-3 \le x \le 3$ and $0 \le y \le 2\pi$.

(a) Find a set of parametric equations of the normal line and an equation of the tangent plane to the surface at the point $\left(2, \frac{\pi}{2}, \frac{1}{2}\right)$.

(b) Repeat part (a) for the point $\left(-\frac{2}{3}, \frac{3\pi}{2}, \frac{3}{2}\right)$.

(c) Use a computer algebra system to graph the surface, the normal lines, and the tangent planes found in parts (a) and (b).

55. Using Functions Consider the functions
$$f(x, y) = 6 - x^2 - \frac{y^2}{4} \quad \text{and} \quad g(x, y) = 2x + y.$$

(a) Find a set of parametric equations of the tangent line to the curve of intersection of the surfaces at the point $(1, 2, 4)$ and find the angle between the gradients of f and g.

(b) Use a computer algebra system to graph the surfaces and the tangent line found in part (a).

56. Using Functions Consider the functions
$$f(x, y) = \sqrt{16 - x^2 - y^2 + 2x - 4y}$$
and
$$g(x, y) = \frac{\sqrt{2}}{2}\sqrt{1 - 3x^2 + y^2 + 6x + 4y}.$$

(a) Use a computer algebra system to graph the first-octant portion of the surfaces represented by f and g.

(b) Find one first-octant point on the curve of intersection and show that the surfaces are orthogonal at this point.

(c) These surfaces are orthogonal along the curve of intersection. Does part (b) prove this fact? Explain.

Writing a Tangent Plane In Exercises 57 and 58, show that the tangent plane to the quadric surface at the point (x_0, y_0, z_0) can be written in the given form.

57. Ellipsoid: $\dfrac{x^2}{a^2} + \dfrac{y^2}{b^2} + \dfrac{z^2}{c^2} = 1$

Tangent plane: $\dfrac{x_0 x}{a^2} + \dfrac{y_0 y}{b^2} + \dfrac{z_0 z}{c^2} = 1$

58. Hyperboloid: $\dfrac{x^2}{a^2} + \dfrac{y^2}{b^2} - \dfrac{z^2}{c^2} = 1$

Tangent plane: $\dfrac{x_0 x}{a^2} + \dfrac{y_0 y}{b^2} - \dfrac{z_0 z}{c^2} = 1$

59. Tangent Planes of a Cone Show that any tangent plane to the cone

$$z^2 = a^2 x^2 + b^2 y^2$$

passes through the origin.

60. Tangent Planes Let f be a differentiable function and consider the surface

$$z = xf\left(\frac{y}{x}\right).$$

Show that the tangent plane at any point $P(x_0, y_0, z_0)$ on the surface passes through the origin.

61. Approximation Consider the following approximations for a function $f(x, y)$ centered at $(0, 0)$.

Linear Approximation:

$$P_1(x, y) = f(0, 0) + f_x(0, 0)x + f_y(0, 0)y$$

Quadratic Approximation:

$$P_2(x, y) = f(0, 0) + f_x(0, 0)x + f_y(0, 0)y +$$
$$\tfrac{1}{2}f_{xx}(0, 0)x^2 + f_{xy}(0, 0)xy + \tfrac{1}{2}f_{yy}(0, 0)y^2$$

[Note that the linear approximation is the tangent plane to the surface at $(0, 0, f(0, 0))$.]

(a) Find the linear approximation of $f(x, y) = e^{x-y}$ centered at $(0, 0)$.

(b) Find the quadratic approximation of $f(x, y) = e^{x-y}$ centered at $(0, 0)$.

(c) When $x = 0$ in the quadratic approximation, you obtain the second-degree Taylor polynomial for what function? Answer the same question for $y = 0$.

(d) Complete the table.

x	y	$f(x, y)$	$P_1(x, y)$	$P_2(x, y)$
0	0			
0	0.1			
0.2	0.1			
0.2	0.5			
1	0.5			

(e) Use a computer algebra system to graph the surfaces $z = f(x, y)$, $z = P_1(x, y)$, and $z = P_2(x, y)$.

62. Approximation Repeat Exercise 61 for the function $f(x, y) = \cos(x + y)$.

63. Proof Prove that the angle of inclination θ of the tangent plane to the surface $z = f(x, y)$ at the point (x_0, y_0, z_0) is given by

$$\cos \theta = \frac{1}{\sqrt{[f_x(x_0, y_0)]^2 + [f_y(x_0, y_0)]^2 + 1}}.$$

64. Proof Prove Theorem 13.14.

Wildflowers

The diversity of wildflowers in a meadow can be measured by counting the numbers of daisies, buttercups, shooting stars, and so on. When there are n types of wildflowers, each with a proportion p_i of the total population, it follows that

$$p_1 + p_2 + \cdots + p_n = 1.$$

The measure of diversity of the population is defined as

$$H = -\sum_{i=1}^{n} p_i \log_2 p_i.$$

In this definition, it is understood that $p_i \log_2 p_i = 0$ when $p_i = 0$. The tables show proportions of wildflowers in a meadow in May, June, August, and September.

May

Flower type	1	2	3	4
Proportion	$\frac{5}{16}$	$\frac{5}{16}$	$\frac{5}{16}$	$\frac{1}{16}$

June

Flower type	1	2	3	4
Proportion	$\frac{1}{4}$	$\frac{1}{4}$	$\frac{1}{4}$	$\frac{1}{4}$

August

Flower type	1	2	3	4
Proportion	$\frac{1}{4}$	0	$\frac{1}{4}$	$\frac{1}{2}$

September

Flower type	1	2	3	4
Proportion	0	0	0	1

(a) Determine the wildflower diversity for each month. How would you interpret September's diversity? Which month had the greatest diversity?

(b) When the meadow contains 10 types of wildflowers in roughly equal proportions, is the diversity of the population greater than or less than the diversity of a similar distribution of 4 types of flowers? What type of distribution (of 10 types of wildflowers) would produce maximum diversity?

(c) Let H_n represent the maximum diversity of n types of wildflowers. Does H_n approach a limit as n approaches ∞?

■ **FOR FURTHER INFORMATION** Biologists use the concept of diversity to measure the proportions of different types of organisms within an environment. For more information on this technique, see the article "Information Theory and Biological Diversity" by Steven Kolmes and Kevin Mitchell in the *UMAP Modules*.

13.8 Extrema of Functions of Two Variables

- Find absolute and relative extrema of a function of two variables.
- Use the Second Partials Test to find relative extrema of a function of two variables.

Absolute Extrema and Relative Extrema

In Chapter 3, you studied techniques for finding the extreme values of a function of a single variable. In this section, you will extend these techniques to functions of two variables. For example, in Theorem 13.15 below, the Extreme Value Theorem for a function of a single variable is extended to a function of two variables.

Consider the continuous function f of two variables, defined on a closed bounded region R in the xy-plane. The values $f(a, b)$ and $f(c, d)$ such that

$$f(a, b) \le f(x, y) \le f(c, d) \qquad (a, b) \text{ and } (c, d) \text{ are in } R.$$

for all (x, y) in R are called the **minimum** and **maximum** of f in the region R, as shown in Figure 13.64. Recall from Section 13.2 that a region in the plane is *closed* when it contains all of its boundary points. The Extreme Value Theorem deals with a region in the plane that is both closed and *bounded*. A region in the plane is **bounded** when it is a subregion of a closed disk in the plane.

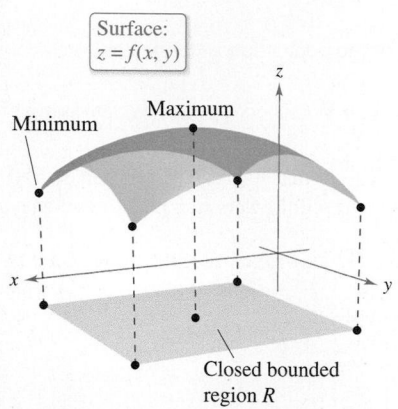

Surface:
$z = f(x, y)$

Minimum Maximum

Closed bounded region R

R contains point(s) at which $f(x, y)$ is a minimum and point(s) at which $f(x, y)$ is a maximum.

Figure 13.64

THEOREM 13.15 Extreme Value Theorem

Let f be a continuous function of two variables x and y defined on a closed bounded region R in the xy-plane.

1. There is at least one point in R at which f takes on a minimum value.
2. There is at least one point in R at which f takes on a maximum value.

A minimum is also called an **absolute minimum** and a maximum is also called an **absolute maximum**. As in single-variable calculus, there is a distinction made between absolute extrema and **relative extrema.**

Definition of Relative Extrema

Let f be a function defined on a region R containing (x_0, y_0).

1. The function f has a **relative minimum** at (x_0, y_0) if
$$f(x, y) \ge f(x_0, y_0)$$
for all (x, y) in an *open* disk containing (x_0, y_0).
2. The function f has a **relative maximum** at (x_0, y_0) if
$$f(x, y) \le f(x_0, y_0)$$
for all (x, y) in an *open* disk containing (x_0, y_0).

To say that f has a relative maximum at (x_0, y_0) means that the point (x_0, y_0, z_0) is at least as high as all nearby points on the graph of

$$z = f(x, y).$$

Similarly, f has a relative minimum at (x_0, y_0) when (x_0, y_0, z_0) is at least as low as all nearby points on the graph. (See Figure 13.65.)

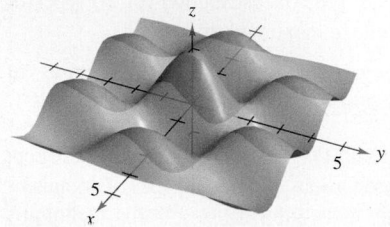

Relative extrema
Figure 13.65

To locate relative extrema of f, you can investigate the points at which the gradient of f is $\mathbf{0}$ or the points at which one of the partial derivatives does not exist. Such points are called **critical points** of f.

Definition of Critical Point

Let f be defined on an open region R containing (x_0, y_0). The point (x_0, y_0) is a **critical point** of f if one of the following is true.

1. $f_x(x_0, y_0) = 0$ and $f_y(x_0, y_0) = 0$
2. $f_x(x_0, y_0)$ or $f_y(x_0, y_0)$ does not exist.

KARL WEIERSTRASS
(1815–1897)

Although the Extreme Value Theorem had been used by earlier mathematicians, the first to provide a rigorous proof was the German mathematician Karl Weierstrass. Weierstrass also provided rigorous justifications for many other mathematical results already in common use. We are indebted to him for much of the logical foundation on which modern calculus is built.

See LarsonCalculus.com to read more of this biography.

Recall from Theorem 13.11 that if f is differentiable and

$$\nabla f(x_0, y_0) = f_x(x_0, y_0)\mathbf{i} + f_y(x_0, y_0)\mathbf{j} = 0\mathbf{i} + 0\mathbf{j}$$

then every directional derivative at (x_0, y_0) must be 0. This implies that the function has a horizontal tangent plane at the point (x_0, y_0), as shown in Figure 13.66. It appears that such a point is a likely location of a relative extremum. This is confirmed by Theorem 13.16.

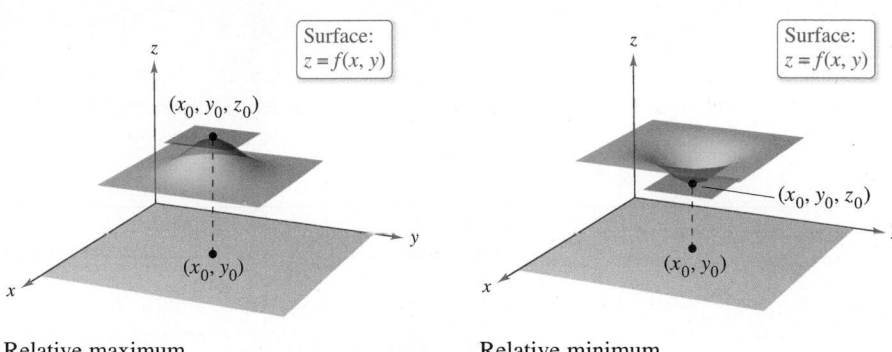

Relative maximum

Relative minimum
Figure 13.66

THEOREM 13.16 Relative Extrema Occur Only at Critical Points

If f has a relative extremum at (x_0, y_0) on an open region R, then (x_0, y_0) is a critical point of f.

Exploration

Use a graphing utility to graph $z = x^3 - 3xy + y^3$ using the bounds $0 \le x \le 3$, $0 \le y \le 3$, and $-3 \le z \le 3$. This view makes it appear as though the surface has an absolute minimum. Does the surface have an absolute minimum? Why or why not?

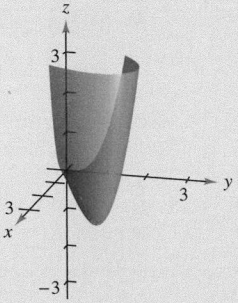

EXAMPLE 1 Finding a Relative Extremum

•••▷ *See LarsonCalculus.com for an interactive version of this type of example.*

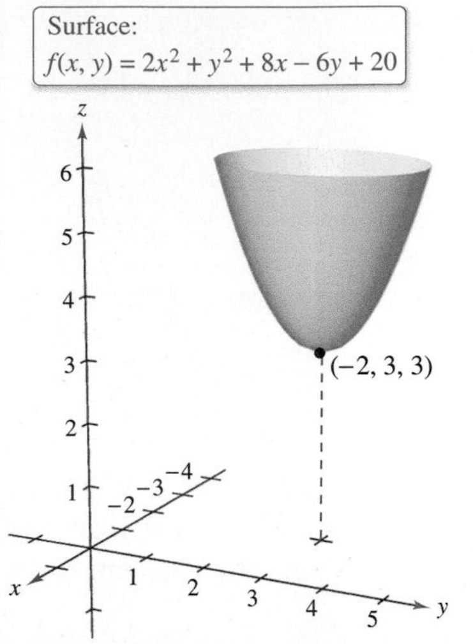

Determine the relative extrema of

$$f(x, y) = 2x^2 + y^2 + 8x - 6y + 20.$$

Solution Begin by finding the critical points of f. Because

$$f_x(x, y) = 4x + 8 \qquad \text{Partial with respect to } x$$

and

$$f_y(x, y) = 2y - 6 \qquad \text{Partial with respect to } y$$

are defined for all x and y, the only critical points are those for which both first partial derivatives are 0. To locate these points, set $f_x(x, y)$ and $f_y(x, y)$ equal to 0, and solve the equations

$$4x + 8 = 0 \quad \text{and} \quad 2y - 6 = 0$$

to obtain the critical point $(-2, 3)$. By completing the square for f, you can see that for all $(x, y) \neq (-2, 3)$

$$f(x, y) = 2(x + 2)^2 + (y - 3)^2 + 3 > 3.$$

So, a relative *minimum* of f occurs at $(-2, 3)$. The value of the relative minimum is $f(-2, 3) = 3$, as shown in Figure 13.67. ∎

Example 1 shows a relative minimum occurring at one type of critical point—the type for which both $f_x(x, y)$ and $f_y(x, y)$ are 0. The next example concerns a relative maximum that occurs at the other type of critical point—the type for which either $f_x(x, y)$ or $f_y(x, y)$ does not exist.

EXAMPLE 2 Finding a Relative Extremum

Determine the relative extrema of

$$f(x, y) = 1 - (x^2 + y^2)^{1/3}.$$

Solution Because

$$f_x(x, y) = -\frac{2x}{3(x^2 + y^2)^{2/3}} \qquad \text{Partial with respect to } x$$

and

$$f_y(x, y) = -\frac{2y}{3(x^2 + y^2)^{2/3}} \qquad \text{Partial with respect to } y$$

it follows that both partial derivatives exist for all points in the xy-plane except for $(0, 0)$. Moreover, because the partial derivatives cannot both be 0 unless both x and y are 0, you can conclude that $(0, 0)$ is the only critical point. In Figure 13.68, note that $f(0, 0)$ is 1. For all other (x, y), it is clear that

$$f(x, y) = 1 - (x^2 + y^2)^{1/3} < 1.$$

So, f has a relative *maximum* at $(0, 0)$. ∎

In Example 2, $f_x(x, y) = 0$ for every point on the y-axis other than $(0, 0)$. However, because $f_y(x, y)$ is nonzero, these are not critical points. Remember that *one* of the partials must not exist or *both* must be 0 in order to yield a critical point.

Surface:
$f(x, y) = 2x^2 + y^2 + 8x - 6y + 20$

The function $z = f(x, y)$ has a relative minimum at $(-2, 3)$.
Figure 13.67

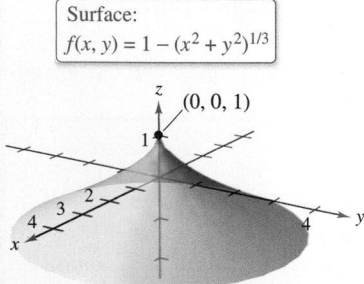

Surface:
$f(x, y) = 1 - (x^2 + y^2)^{1/3}$

$f_x(x, y)$ and $f_y(x, y)$ are undefined at $(0, 0)$.
Figure 13.68

The Second Partials Test

Theorem 13.16 tells you that to find relative extrema, you need only examine values of $f(x, y)$ at critical points. However, as is true for a function of one variable, the critical points of a function of two variables do not always yield relative maxima or minima. Some critical points yield **saddle points,** which are neither relative maxima nor relative minima.

As an example of a critical point that does not yield a relative extremum, consider the hyperbolic paraboloid

$$f(x, y) = y^2 - x^2$$

as shown in Figure 13.69. At the point $(0, 0)$, both partial derivatives

$$f_x(x, y) = -2x \quad \text{and} \quad f_y(x, y) = 2y$$

are 0. The function f does not, however, have a relative extremum at this point because in any open disk centered at $(0, 0)$, the function takes on both negative values (along the x-axis) *and* positive values (along the y-axis), So, the point $(0, 0, 0)$ is a saddle point of the surface. (The term "saddle point" comes from the fact that surfaces such as the one shown in Figure 13.69 resemble saddles.)

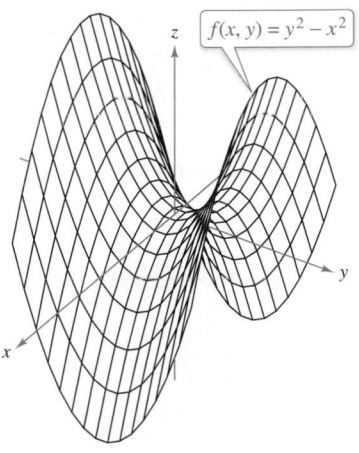

Saddle point at $(0, 0, 0)$:
$f_x(0, 0) = f_y(0, 0) = 0$
Figure 13.69

For the functions in Examples 1 and 2, it was relatively easy to determine the relative extrema, because each function was either given, or able to be written, in completed square form. For more complicated functions, algebraic arguments are less convenient and it is better to rely on the analytic means presented in the following Second Partials Test. This is the two-variable counterpart of the Second Derivative Test for functions of one variable. The proof of this theorem is best left to a course in advanced calculus.

THEOREM 13.17   Second Partials Test

Let f have continuous second partial derivatives on an open region containing a point (a, b) for which

$$f_x(a, b) = 0 \quad \text{and} \quad f_y(a, b) = 0.$$

To test for relative extrema of f, consider the quantity

$$d = f_{xx}(a, b)f_{yy}(a, b) - [f_{xy}(a, b)]^2.$$

1. If $d > 0$ and $f_{xx}(a, b) > 0$, then f has a **relative minimum** at (a, b).
2. If $d > 0$ and $f_{xx}(a, b) < 0$, then f has a **relative maximum** at (a, b).
3. If $d < 0$, then $(a, b, f(a, b))$ is a **saddle point.**
4. The test is inconclusive if $d = 0$.

• • REMARK   If $d > 0$, then $f_{xx}(a, b)$ and $f_{yy}(a, b)$ must have the same sign. This means that $f_{xx}(a, b)$ can be replaced by $f_{yy}(a, b)$ in the first two parts of the test.

A convenient device for remembering the formula for d in the Second Partials Test is given by the 2×2 determinant

$$d = \begin{vmatrix} f_{xx}(a, b) & f_{xy}(a, b) \\ f_{yx}(a, b) & f_{yy}(a, b) \end{vmatrix}$$

where $f_{xy}(a, b) = f_{yx}(a, b)$ by Theorem 13.3.

EXAMPLE 3 **Using the Second Partials Test**

Find the relative extrema of $f(x, y) = -x^3 + 4xy - 2y^2 + 1$.

Solution Begin by finding the critical points of f. Because

$$f_x(x, y) = -3x^2 + 4y \quad \text{and} \quad f_y(x, y) = 4x - 4y$$

exist for all x and y, the only critical points are those for which both first partial derivatives are 0. To locate these points, set $f_x(x, y)$ and $f_y(x, y)$ equal to 0 to obtain

$$-3x^2 + 4y = 0 \quad \text{and} \quad 4x - 4y = 0.$$

From the second equation, you know that $x = y$, and, by substitution into the first equation, you obtain two solutions: $y = x = 0$ and $y = x = \frac{4}{3}$. Because

$$f_{xx}(x, y) = -6x, \quad f_{yy}(x, y) = -4, \quad \text{and} \quad f_{xy}(x, y) = 4$$

it follows that, for the critical point $(0, 0)$,

$$d = f_{xx}(0, 0)f_{yy}(0, 0) - [f_{xy}(0, 0)]^2 = 0 - 16 < 0$$

and, by the Second Partials Test, you can conclude that $(0, 0, 1)$ is a saddle point of f. Furthermore, for the critical point $\left(\frac{4}{3}, \frac{4}{3}\right)$,

$$d = f_{xx}\left(\frac{4}{3}, \frac{4}{3}\right)f_{yy}\left(\frac{4}{3}, \frac{4}{3}\right) - \left[f_{xy}\left(\frac{4}{3}, \frac{4}{3}\right)\right]^2$$

$$= -8(-4) - 16$$

$$= 16$$

$$> 0$$

and because $f_{xx}\left(\frac{4}{3}, \frac{4}{3}\right) = -8 < 0$, you can conclude that f has a relative maximum at $\left(\frac{4}{3}, \frac{4}{3}\right)$, as shown in Figure 13.70. ■

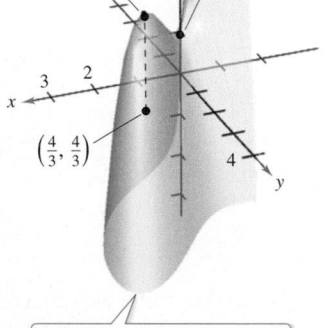

Relative maximum

Saddle point $(0, 0, 1)$

$\left(\frac{4}{3}, \frac{4}{3}\right)$

$f(x, y) = -x^3 + 4xy - 2y^2 + 1$

Figure 13.70

The Second Partials Test can fail to find relative extrema in two ways. If either of the first partial derivatives does not exist, you cannot use the test. Also, if

$$d = f_{xx}(a, b)f_{yy}(a, b) - [f_{xy}(a, b)]^2 = 0$$

the test fails. In such cases, you can try a sketch or some other approach, as demonstrated in the next example.

EXAMPLE 4 **Failure of the Second Partials Test**

Find the relative extrema of $f(x, y) = x^2y^2$.

Solution Because $f_x(x, y) = 2xy^2$ and $f_y(x, y) = 2x^2y$, you know that both partial derivatives are 0 when $x = 0$ or $y = 0$. That is, every point along the x- or y-axis is a critical point. Moreover, because

$$f_{xx}(x, y) = 2y^2, \quad f_{yy}(x, y) = 2x^2, \quad \text{and} \quad f_{xy}(x, y) = 4xy$$

you know that

$$d = f_{xx}(x, y)f_{yy}(x, y) - [f_{xy}(x, y)]^2$$

$$= 4x^2y^2 - 16x^2y^2$$

$$= -12x^2y^2$$

which is 0 when either $x = 0$ or $y = 0$. So, the Second Partials Test fails. However, because $f(x, y) = 0$ for every point along the x- or y-axis and $f(x, y) = x^2y^2 > 0$ for all other points, you can conclude that each of these critical points yields an absolute minimum, as shown in Figure 13.71. ■

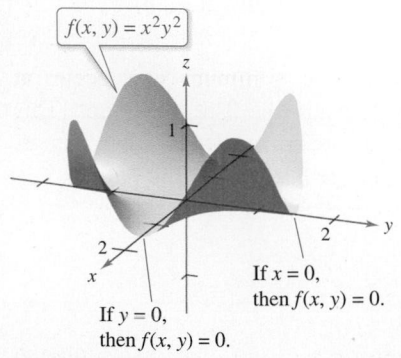

$f(x, y) = x^2y^2$

If $x = 0$, then $f(x, y) = 0$.

If $y = 0$, then $f(x, y) = 0$.

Figure 13.71

Absolute extrema of a function can occur in two ways. First, some relative extrema also happen to be absolute extrema. For instance, in Example 1, $f(-2, 3)$ is an absolute minimum of the function. (On the other hand, the relative maximum found in Example 3 is not an absolute maximum of the function.) Second, absolute extrema can occur at a boundary point of the domain. This is illustrated in Example 5.

EXAMPLE 5   **Finding Absolute Extrema**

Find the absolute extrema of the function

$$f(x, y) = \sin xy$$

on the closed region given by

$$0 \leq x \leq \pi \quad \text{and} \quad 0 \leq y \leq 1.$$

Solution   From the partial derivatives

$$f_x(x, y) = y \cos xy \quad \text{and} \quad f_y(x, y) = x \cos xy$$

you can see that each point lying on the hyperbola $xy = \pi/2$ is a critical point. These points each yield the value

$$f(x, y) = \sin \frac{\pi}{2} = 1$$

which you know is the absolute maximum, as shown in Figure 13.72. The only other critical point of f *lying in the given region* is $(0, 0)$. It yields an absolute minimum of 0, because

$$0 \leq xy \leq \pi$$

implies that

$$0 \leq \sin xy \leq 1.$$

To locate other absolute extrema, you should consider the four boundaries of the region formed by taking traces with the vertical planes $x = 0$, $x = \pi$, $y = 0$, and $y = 1$. In doing this, you will find that $\sin xy = 0$ at all points on the x-axis, at all points on the y-axis, and at the point $(\pi, 1)$. Each of these points yields an absolute minimum for the surface, as shown in Figure 13.72.

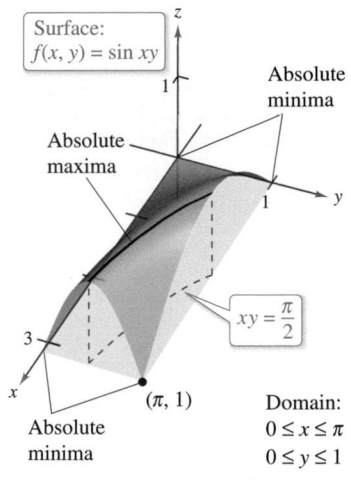

Surface:
$f(x, y) = \sin xy$

Absolute minima

Absolute maxima

$xy = \dfrac{\pi}{2}$

Absolute minima

$(\pi, 1)$

Domain:
$0 \leq x \leq \pi$
$0 \leq y \leq 1$

Figure 13.72

The concepts of relative extrema and critical points can be extended to functions of three or more variables. When all first partial derivatives of

$$w = f(x_1, x_2, x_3, \ldots, x_n)$$

exist, it can be shown that a relative maximum or minimum can occur at $(x_1, x_2, x_3, \ldots, x_n)$ only when every first partial derivative is 0 at that point. This means that the critical points are obtained by solving the following system of equations.

$$f_{x_1}(x_1, x_2, x_3, \ldots, x_n) = 0$$
$$f_{x_2}(x_1, x_2, x_3, \ldots, x_n) = 0$$
$$\vdots$$
$$f_{x_n}(x_1, x_2, x_3, \ldots, x_n) = 0$$

The extension of Theorem 13.17 to three or more variables is also possible, although you will not study such an extension in this text.

13.8 Exercises

See **CalcChat.com** for tutorial help and worked-out solutions to odd-numbered exercises.

CONCEPT CHECK

1. Function of Two Variables For a function of two variables, describe (a) relative minimum, (b) relative maximum, (c) critical point, and (d) saddle point.

2. Second Partials Test Under what condition does the Second Partials Test fail?

 **Finding Relative Extrema** In Exercises 3–8, identify any extrema of the function by recognizing its given form or its form after completing the square. Verify your results by using the partial derivatives to locate any critical points and test for relative extrema.

3. $g(x, y) = (x - 1)^2 + (y - 3)^2$

4. $g(x, y) = 5 - (x - 6)^2 - (y + 2)^2$

5. $f(x, y) = \sqrt{x^2 + y^2 + 1}$

6. $f(x, y) = \sqrt{49 - (x - 2)^2 - y^2}$

7. $f(x, y) = x^2 + y^2 + 2x - 6y + 6$

8. $f(x, y) = -x^2 - y^2 + 10x + 12y - 64$

 **Using the Second Partials Test** In Exercises 9–24, find all relative extrema and saddle points of the function. Use the Second Partials Test where applicable.

9. $f(x, y) = x^2 + y^2 + 8x - 12y - 3$

10. $g(x, y) = x^2 - y^2 - x - y$

11. $f(x, y) = -2x^4y^4$ **12.** $f(x, y) = \frac{1}{2}xy$

13. $f(x, y) = -3x^2 - 2y^2 + 3x - 4y + 5$

14. $h(x, y) = x^2 - 3xy - y^2$

15. $f(x, y) = 7x^2 + 2y^2 - 7x + 16y - 13$

16. $f(x, y) = x^5 + y^5$

17. $z = x^2 + xy + \frac{1}{2}y^2 - 2x + y$

18. $z = -5x^2 + 4xy - y^2 + 16x + 10$

19. $f(x, y) = -4(x^2 + y^2 + 81)^{1/4}$

20. $h(x, y) = (x^2 + y^2)^{1/3} + 2$

21. $f(x, y) = x^2 - xy - y^2 - 3x - y$

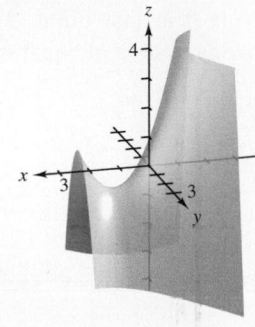

22. $f(x, y) = 2xy - \frac{1}{2}(x^4 + y^4) + 1$

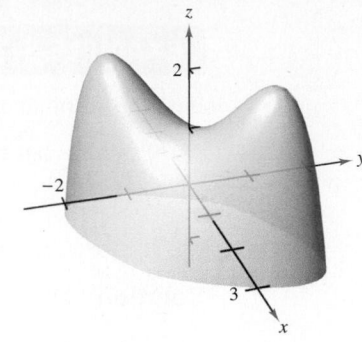

23. $z = e^{-x} \sin y$

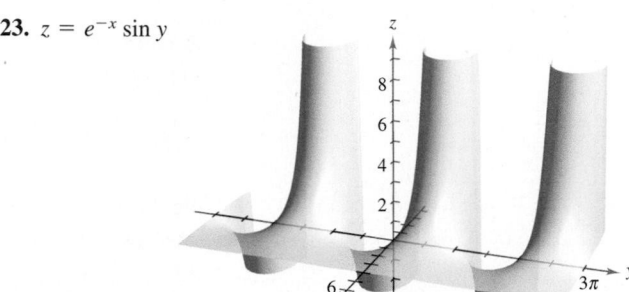

24. $z = \left(\frac{1}{2} - x^2 + y^2\right)e^{1 - x^2 - y^2}$

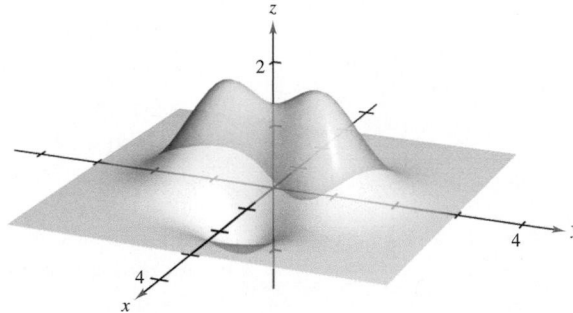

Finding Relative Extrema and Saddle Points Using Technology In Exercises 25–28, use a computer algebra system to graph the surface and locate any relative extrema and saddle points.

25. $z = \dfrac{-4x}{x^2 + y^2 + 1}$

26. $z = \cos x + \sin y, \ -\pi/2 < x < \pi/2, \ -\pi < y < \pi$

27. $z = (x^2 + 4y^2)e^{1 - x^2 - y^2}$ **28.** $z = e^{xy}$

Finding Relative Extrema In Exercises 29 and 30, examine the function for extrema without using the derivative tests and use a computer algebra system to graph the surface and verify your answers. (*Hint:* By observation, determine whether it is possible for z to be negative. When is z equal to 0?)

29. $z = \dfrac{(x - y)^4}{x^2 + y^2}$ **30.** $z = \dfrac{(x^2 - y^2)^2}{x^2 + y^2}$

Think About It In Exercises 31–34, determine whether there is a relative maximum, a relative minimum, a saddle point, or insufficient information to determine the nature of the function $f(x, y)$ at the critical point (x_0, y_0).

31. $f_{xx}(x_0, y_0) = 9, \quad f_{yy}(x_0, y_0) = 4, \quad f_{xy}(x_0, y_0) = 6$

32. $f_{xx}(x_0, y_0) = -3, \quad f_{yy}(x_0, y_0) = -8, \quad f_{xy}(x_0, y_0) = 2$

33. $f_{xx}(x_0, y_0) = -9, \quad f_{yy}(x_0, y_0) = 6, \quad f_{xy}(x_0, y_0) = 10$

34. $f_{xx}(x_0, y_0) = 25, \quad f_{yy}(x_0, y_0) = 8, \quad f_{xy}(x_0, y_0) = 10$

 **Finding Relative Extrema and Saddle Points** In Exercises 35–38, (a) find the critical points, (b) test for relative extrema, (c) list the critical points for which the Second Partials Test fails, and (d) use a computer algebra system to graph the function, labeling any extrema and saddle points.

35. $f(x, y) = x^3 + y^3$

36. $f(x, y) = x^3 + y^3 - 6x^2 + 9x^2 + 12x + 27y + 19$

37. $f(x, y) = (x - 1)^2(y + 4)^2$

38. $f(x, y) = x^{2/3} + y^{2/3}$

Finding Absolute Extrema In Exercises 39–46, find the absolute extrema of the function over the region R. (In each case, R contains the boundaries.) Use a computer algebra system to confirm your results.

39. $f(x, y) = x^2 - 4xy + 5$

$R = \{(x, y): 1 \le x \le 4, 0 \le y \le 2\}$

40. $f(x, y) = x^2 + xy, \quad R = \{(x, y): |x| \le 2, |y| \le 1\}$

41. $f(x, y) = 12 - 3x - 2y$

R: The triangular region in the xy-plane with vertices $(2, 0)$, $(0, 1)$, and $(1, 2)$

42. $f(x, y) = (2x - y)^2$

R: The triangular region in the xy-plane with vertices $(2, 0)$, $(0, 1)$, and $(1, 2)$

43. $f(x, y) = 3x^2 + 2y^2 - 4y$

R: The region in the xy-plane bounded by the graphs of $y = x^2$ and $y = 4$

44. $f(x, y) = 2x - 2xy + y^2$

R: The region in the xy-plane bounded by the graphs of $y = x^2$ and $y = 1$

45. $f(x, y) = x^2 + 2xy + y^2, \quad R = \{(x, y): |x| \le 2, |y| \le 1\}$

46. $f(x, y) = \dfrac{4xy}{(x^2 + 1)(y^2 + 1)}$

$R = \{(x, y): 0 \le x \le 1, 0 \le y \le 1\}$

Examining a Function In Exercises 47 and 48, find the critical points of the function and, from the form of the function, determine whether a relative maximum or a relative minimum occurs at each point.

47. $f(x, y, z) = x^2 + (y - 3)^2 + (z + 1)^2$

48. $f(x, y, z) = 9 - [x(y - 1)(z + 2)]^2$

EXPLORING CONCEPTS

49. Using the Second Partials Test A function f has continuous second partial derivatives on an open region containing the critical point $(3, 7)$. The function has a minimum at $(3, 7)$, and $d > 0$ for the Second Partials Test. Determine the interval for $f_{xy}(3, 7)$ when $f_{xx}(3, 7) = 2$ and $f_{yy}(3, 7) = 8$.

50. Using the Second Partials Test A function f has continuous second partial derivatives on an open region containing the critical point (a, b). If $f_{xx}(a, b)$ and $f_{yy}(a, b)$ have opposite signs, what is implied? Explain.

Sketching a Graph In Exercises 51 and 52, sketch the graph of an arbitrary function f satisfying the given conditions. State whether the function has any extrema or saddle points. (There are many correct answers.)

51. All of the first and second partial derivatives of f are 0.

52. $f_x(x, y) > 0$ and $f_y(x, y) < 0$ for all (x, y).

53. Comparing Functions Consider the functions

$$f(x, y) = x^2 - y^2 \quad \text{and} \quad g(x, y) = x^2 + y^2.$$

(a) Show that both functions have a critical point at $(0, 0)$.

(b) Explain how f and g behave differently at this critical point.

54. **HOW DO YOU SEE IT?** Determine whether each labeled point is an absolute maximum, an absolute minimum, or neither.

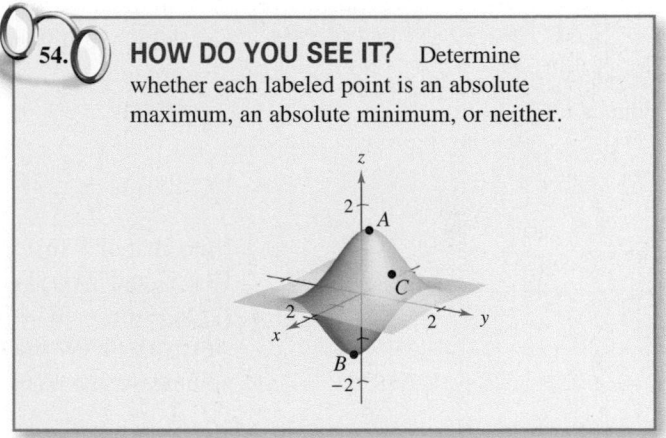

True or False? In Exercises 55–58, determine whether the statement is true or false. If it is false, explain why or give an example that shows it is false.

55. If f has a relative maximum at (x_0, y_0, z_0), then

$$f_x(x_0, y_0) = f_y(x_0, y_0) = 0.$$

56. If $f_x(x_0, y_0) = f_y(x_0, y_0) = 0$, then f has a relative extremum at (x_0, y_0, z_0).

57. Between any two relative minima of f, there must be at least one relative maximum of f.

58. If f is continuous for all x and y and has two relative minima, then f must have at least one relative maximum.

13.9 Applications of Extrema

■ Solve optimization problems involving functions of several variables.
■ Use the method of least squares.

Applied Optimization Problems

In this section, you will study a few of the many applications of extrema of functions of two (or more) variables.

EXAMPLE 1   Finding Maximum Volume

⋅⋅⋅▷ *See LarsonCalculus.com for an interactive version of this type of example.*

A rectangular box is resting on the xy-plane with one vertex at the origin. The opposite vertex lies in the plane

$$6x + 4y + 3z = 24$$

as shown in Figure 13.73. Find the maximum volume of the box.

Solution Let x, y, and z represent the length, width, and height of the box. Because one vertex of the box lies in the plane $6x + 4y + 3z = 24$, you know that $z = \frac{1}{3}(24 - 6x - 4y)$. So, you can write the volume xyz of the box as a function of two variables.

$$V(x, y) = (x)(y)\left[\tfrac{1}{3}(24 - 6x - 4y)\right]$$
$$= \tfrac{1}{3}(24xy - 6x^2y - 4xy^2)$$

Next, find the first partial derivatives of V.

$$V_x(x, y) = \frac{1}{3}(24y - 12xy - 4y^2) = \frac{y}{3}(24 - 12x - 4y)$$

$$V_y(x, y) = \frac{1}{3}(24x - 6x^2 - 8xy) = \frac{x}{3}(24 - 6x - 8y)$$

Note that the first partial derivatives are defined for all x and y. So, by setting $V_x(x, y)$ and $V_y(x, y)$ equal to 0 and solving the equations $\frac{1}{3}y(24 - 12x - 4y) = 0$ and $\frac{1}{3}x(24 - 6x - 8y) = 0$, you obtain the critical points $(0, 0)$, $(4, 0)$, $(0, 6)$, and $\left(\frac{4}{3}, 2\right)$. At $(0, 0)$, $(4, 0)$, and $(0, 6)$, the volume is 0, so these points do not yield a maximum volume. At the point $\left(\frac{4}{3}, 2\right)$, you can apply the Second Partials Test.

$$V_{xx}(x, y) = -4y$$

$$V_{yy}(x, y) = \frac{-8x}{3}$$

$$V_{xy}(x, y) = \frac{1}{3}(24 - 12x - 8y)$$

Because

$$V_{xx}\left(\tfrac{4}{3}, 2\right)V_{yy}\left(\tfrac{4}{3}, 2\right) - \left[V_{xy}\left(\tfrac{4}{3}, 2\right)\right]^2 = (-8)\left(-\tfrac{32}{9}\right) - \left(-\tfrac{8}{3}\right)^2 = \tfrac{64}{3} > 0$$

and

$$V_{xx}\left(\tfrac{4}{3}, 2\right) = -8 < 0$$

you can conclude from the Second Partials Test that the maximum volume is

$$V\left(\tfrac{4}{3}, 2\right) = \tfrac{1}{3}\left[24\left(\tfrac{4}{3}\right)(2) - 6\left(\tfrac{4}{3}\right)^2(2) - 4\left(\tfrac{4}{3}\right)(2^2)\right] = \tfrac{64}{9} \text{ cubic units.}$$

Note that the volume is 0 at the boundary points of the triangular domain of V.

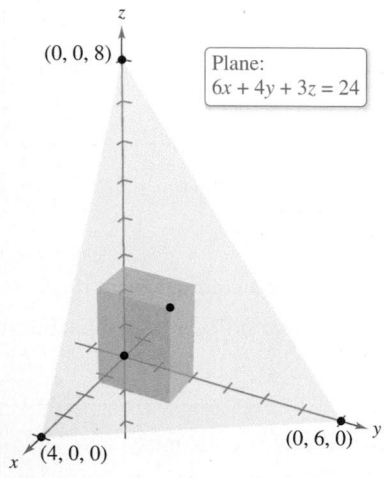

Plane:
$6x + 4y + 3z = 24$

$(0, 0, 8)$

$(4, 0, 0)$

$(0, 6, 0)$

Figure 13.73

⋅⋅**REMARK** In many applied problems, the domain of the function to be optimized is a closed bounded region. To find minimum or maximum points, you must not only test critical points, but also consider the values of the function at points on the boundary.

Applications of extrema in economics and business often involve more than one independent variable. For instance, a company may produce several models of one type of product. The price per unit and profit per unit are usually different for each model. Moreover, the demand for each model is often a function of the prices of the other models (as well as its own price). The next example illustrates an application involving two products.

EXAMPLE 2   Finding the Maximum Profit

A manufacturer determines that the profit P (in dollars) obtained by producing and selling x units of Product 1 and y units of Product 2 is approximated by the model

$$P(x, y) = 8x + 10y - (0.001)(x^2 + xy + y^2) - 10{,}000.$$

Find the production level that produces a maximum profit. What is the maximum profit?

Solution   The partial derivatives of the profit function are

$$P_x(x, y) = 8 - (0.001)(2x + y)$$

and

$$P_y(x, y) = 10 - (0.001)(x + 2y).$$

By setting these partial derivatives equal to 0, you obtain the following system of equations.

$$8 - (0.001)(2x + y) = 0$$
$$10 - (0.001)(x + 2y) = 0$$

After simplifying, this system of linear equations can be written as

$$2x + \ y = \ \ \ 8000$$
$$x + 2y = 10{,}000.$$

Solving this system produces $x = 2000$ and $y = 4000$. The second partial derivatives of P are

$$P_{xx}(2000, 4000) = -0.002$$
$$P_{yy}(2000, 4000) = -0.002$$
$$P_{xy}(2000, 4000) = -0.001.$$

Because $P_{xx} < 0$ and

$$P_{xx}(2000, 4000)P_{yy}(2000, 4000) - [P_{xy}(2000, 4000)]^2 = (-0.002)^2 - (-0.001)^2$$

is greater than 0, you can conclude that the production level of $x = 2000$ units and $y = 4000$ units yields a *maximum* profit. The maximum profit is

$$P(2000, 4000)$$
$$= 8(2000) + 10(4000) - (0.001)[2000^2 + 2000(4000) + 4000^2] - 10{,}000$$
$$= \$18{,}000. \qquad \blacksquare$$

In Example 2, it was assumed that the manufacturing plant is able to produce the required number of units to yield a maximum profit. In actual practice, the production would be bounded by physical constraints. You will study such constrained optimization problems in the next section.

■ **FOR FURTHER INFORMATION**   For more information on the use of mathematics in economics, see the article "Mathematical Methods of Economics" by Joel Franklin in *The American Mathematical Monthly*. To view this article, go to *MathArticles.com*.

The Method of Least Squares

Many of the examples in this text have involved **mathematical models.** For instance, Example 2 involves a quadratic model for profit. There are several ways to develop such models; one is called the **method of least squares.**

In constructing a model to represent a particular phenomenon, the goals are simplicity and accuracy. Of course, these goals often conflict. For instance, a simple linear model for the points in Figure 13.74 is

$$y = 1.9x - 5.$$

However, Figure 13.75 shows that by choosing the slightly more complicated quadratic model

$$y = 0.20x^2 - 0.7x + 1$$

you can achieve greater accuracy.

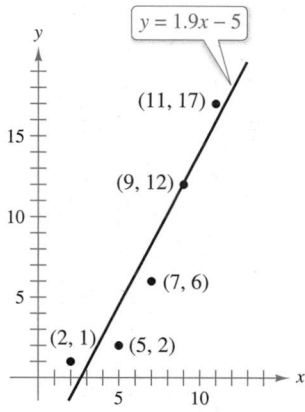

Figure 13.74

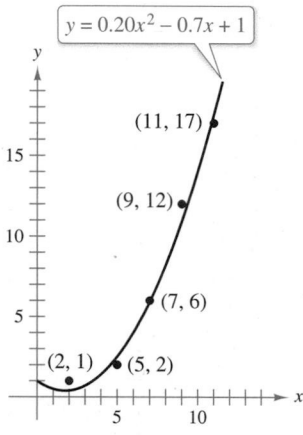

Figure 13.75

As a measure of how well the model $y = f(x)$ fits the collection of points

$$\{(x_1, y_1), (x_2, y_2), (x_3, y_3), \ldots, (x_n, y_n)\}$$

you can add the squares of the differences between the actual y-values and the values given by the model to obtain the **sum of the squared errors**

$$S = \sum_{i=1}^{n} [f(x_i) - y_i]^2. \qquad \text{Sum of the squared errors}$$

Graphically, S can be interpreted as the sum of the squares of the vertical distances between the graph of f and the given points in the plane, as shown in Figure 13.76. If the model is perfect, then $S = 0$. However, when perfection is not feasible, you can settle for a model that minimizes S. For instance, the sum of the squared errors for the linear model in Figure 13.74 is

$$S = 17.6.$$

Statisticians call the *linear model* that minimizes S the **least squares regression line.** The proof that this line actually minimizes S involves the minimizing of a function of two variables.

• • **REMARK**   A method
   for finding the least squares
   regression quadratic for
   a collection of data is
   described in Exercise 31.

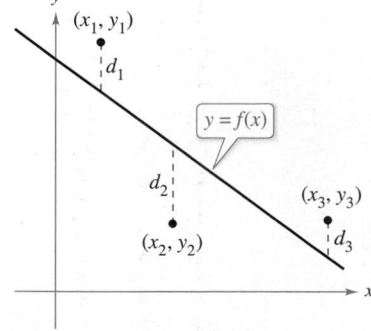

Sum of the squared errors:
$S = d_1^2 + d_2^2 + d_3^2$
Figure 13.76

THEOREM 13.18   Least Squares Regression Line

The **least squares regression line** for $\{(x_1, y_1), (x_2, y_2), \ldots, (x_n, y_n)\}$ is given by $f(x) = ax + b$, where

$$a = \frac{n\sum_{i=1}^{n} x_i y_i - \sum_{i=1}^{n} x_i \sum_{i=1}^{n} y_i}{n\sum_{i=1}^{n} x_i^2 - \left(\sum_{i=1}^{n} x_i\right)^2} \quad \text{and} \quad b = \frac{1}{n}\left(\sum_{i=1}^{n} y_i - a\sum_{i=1}^{n} x_i\right).$$

Proof   Let $S(a, b)$ represent the sum of the squared errors for the model

$$f(x) = ax + b$$

and the given set of points. That is,

$$S(a, b) = \sum_{i=1}^{n} [f(x_i) - y_i]^2$$

$$= \sum_{i=1}^{n} (ax_i + b - y_i)^2$$

where the points (x_i, y_i) represent constants. Because S is a function of a and b, you can use the methods discussed in the preceding section to find the minimum value of S. Specifically, the first partial derivatives of S are

$$S_a(a, b) = \sum_{i=1}^{n} 2x_i(ax_i + b - y_i)$$

$$= 2a\sum_{i=1}^{n} x_i^2 + 2b\sum_{i=1}^{n} x_i - 2\sum_{i=1}^{n} x_i y_i$$

and

$$S_b(a, b) = \sum_{i=1}^{n} 2(ax_i + b - y_i)$$

$$= 2a\sum_{i=1}^{n} x_i + 2nb - 2\sum_{i=1}^{n} y_i.$$

By setting these two partial derivatives equal to 0, you obtain the values of a and b that are listed in the theorem. It is left to you to apply the Second Partials Test (see Exercise 41) to verify that these values of a and b yield a minimum. ∎

If the x-values are symmetrically spaced about the y-axis, then $\sum x_i = 0$ and the formulas for a and b simplify to

$$a = \frac{\sum_{i=1}^{n} x_i y_i}{\sum_{i=1}^{n} x_i^2}$$

and

$$b = \frac{1}{n}\sum_{i=1}^{n} y_i.$$

This simplification is often possible with a translation of the x-values. For instance, given that the x-values in a data collection consist of the values 9, 10, 11, 12, and 13, you could let 11 be represented by 0.

EXAMPLE 3 **Finding the Least Squares Regression Line**

Find the least squares regression line for the points

$$(-3, 0), \quad (-1, 1), \quad (0, 2), \quad \text{and} \quad (2, 3).$$

Solution The table shows the calculations involved in finding the least squares regression line using $n = 4$.

x	y	xy	x^2
-3	0	0	9
-1	1	-1	1
0	2	0	0
2	3	6	4
$\displaystyle\sum_{i=1}^{n} x_i = -2$	$\displaystyle\sum_{i=1}^{n} y_i = 6$	$\displaystyle\sum_{i=1}^{n} x_i y_i = 5$	$\displaystyle\sum_{i=1}^{n} x_i^2 = 14$

Applying Theorem 13.18 produces

$$a = \frac{n\displaystyle\sum_{i=1}^{n} x_i y_i - \sum_{i=1}^{n} x_i \sum_{i=1}^{n} y_i}{n\displaystyle\sum_{i=1}^{n} x_i^2 - \left(\sum_{i=1}^{n} x_i\right)^2}$$

$$= \frac{4(5) - (-2)(6)}{4(14) - (-2)^2}$$

$$= \frac{8}{13}$$

and

$$b = \frac{1}{n}\left(\sum_{i=1}^{n} y_i - a\sum_{i=1}^{n} x_i\right)$$

$$= \frac{1}{4}\left[6 - \frac{8}{13}(-2)\right]$$

$$= \frac{47}{26}.$$

▷ **TECHNOLOGY** Many calculators have "built-in" least squares regression programs. If your calculator has such a program, use it to duplicate the results of Example 3.

The least squares regression line is

$$f(x) = \frac{8}{13}x + \frac{47}{26}$$

as shown in Figure 13.77.

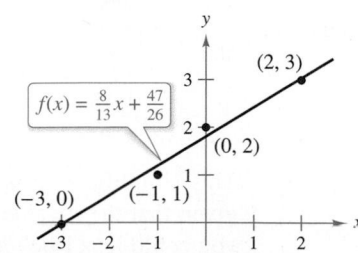

Least squares regression line
Figure 13.77

13.9 Exercises

See **CalcChat.com** for tutorial help and worked-out solutions to odd-numbered exercises.

CONCEPT CHECK

1. **Applied Optimization Problems** In your own words, state the problem-solving strategy for applied minimum and maximum problems.

2. **Method of Least Squares** In your own words, describe the method of least squares for finding mathematical models.

 **Finding Minimum Distance** In Exercises 3 and 4, find the minimum distance from the point to the plane $x - y + z = 3$. (*Hint:* To simplify the computations, minimize the square of the distance.)

3. $(1, -3, 2)$

4. $(4, 0, 6)$

Finding Minimum Distance In Exercises 5 and 6, find the minimum distance from the point to the surface $z = \sqrt{1 - 2x - 2y}$. (*Hint:* To simplify the computations, minimize the square of the distance.)

5. $(-2, -2, 0)$

6. $(-4, 1, 0)$

 **Finding Positive Numbers** In Exercises 7–10, find three positive integers x, y, and z that satisfy the given conditions.

7. The product is 27, and the sum is a minimum.

8. The sum is 32, and $P = xy^2z$ is a maximum.

9. The sum is 30, and the sum of the squares is a minimum.

10. The product is 1, and the sum of the squares is a minimum.

11. **Cost** A home improvement contractor is painting the walls and ceiling of a rectangular room. The volume of the room is 668.25 cubic feet. The cost of wall paint is $0.06 per square foot and the cost of ceiling paint is $0.11 per square foot. Find the room dimensions that result in a minimum cost for the paint. What is the minimum cost for the paint?

12. **Maximum Volume** The material for constructing the base of an open box costs 1.5 times as much per unit area as the material for constructing the sides. For a fixed amount of money C, find the dimensions of the box of largest volume that can be made.

13. **Volume and Surface Area** Show that a rectangular box of given volume and minimum surface area is a cube.

14. **Maximum Volume** Show that the rectangular box of maximum volume inscribed in a sphere of radius r is a cube.

15. **Maximum Revenue** A company manufactures running shoes and basketball shoes. The total revenue (in thousands of dollars) from x_1 units of running shoes and x_2 units of basketball shoes is

$$R = -5x_1^2 - 8x_2^2 - 2x_1x_2 + 42x_1 + 102x_2$$

where x_1 and x_2 are in thousands of units. Find x_1 and x_2 so as to maximize the revenue.

16. **Maximum Profit** A corporation manufactures candles at two locations. The cost of producing x_1 units at location 1 is $C_1 = 0.02x_1^2 + 4x_1 + 500$ and the cost of producing x_2 units at location 2 is $C_2 = 0.05x_2^2 + 4x_2 + 275$. The candles sell for $15 per unit. Find the quantity that should be produced at each location to maximize the profit $P = 15(x_1 + x_2) - C_1 - C_2$.

17. **Hardy-Weinberg Law** Common blood types are determined genetically by three alleles A, B, and O. (An allele is any of a group of possible mutational forms of a gene.) A person whose blood type is AA, BB, or OO is homozygous. A person whose blood type is AB, AO, or BO is heterozygous. The Hardy-Weinberg Law states that the proportion P of heterozygous individuals in any given population is

$$P(p, q, r) = 2pq + 2pr + 2qr$$

where p represents the percent of allele A in the population, q represents the percent of allele B in the population, and r represents the percent of allele O in the population. Use the fact that

$$p + q + r = 1$$

to show that the maximum proportion of heterozygous individuals in any population is $\frac{2}{3}$.

18. **Shannon Diversity Index** One way to measure species diversity is to use the Shannon diversity index H. If a habitat consists of three species, A, B, and C, then its Shannon diversity index is

$$H = -x \ln x - y \ln y - z \ln z$$

where x is the percent of species A in the habitat, y is the percent of species B in the habitat, and z is the percent of species C in the habitat. Use the fact that

$$x + y + z = 1$$

to show that the maximum value of H occurs when $x = y = z = \frac{1}{3}$. What is the maximum value of H?

19. **Minimum Cost** A water line is to be built from point P to point S and must pass through regions where construction costs differ (see figure). The cost per kilometer (in dollars) is $3k$ from P to Q, $2k$ from Q to R, and k from R to S. Find x and y such that the total cost C will be minimized.

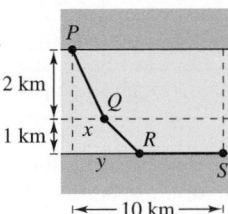

Figure for 19   Figure for 20

20. **Area** A trough with trapezoidal cross sections is formed by turning up the edges of a 30-inch-wide sheet of aluminum (see figure). Find the cross section of maximum area.

Finding the Least Squares Regression Line In Exercises 21–24, (a) find the least squares regression line and (b) calculate S, the sum of the squared errors. Use the regression capabilities of a graphing utility to verify your results.

21.

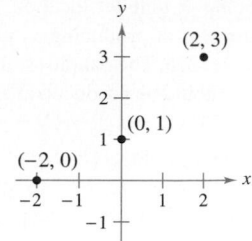

22.

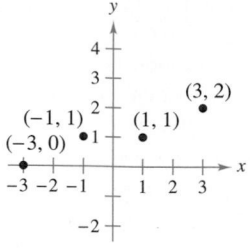

23.

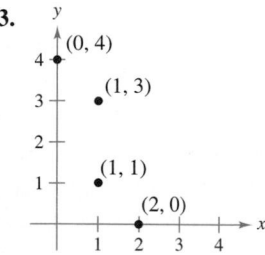

24.

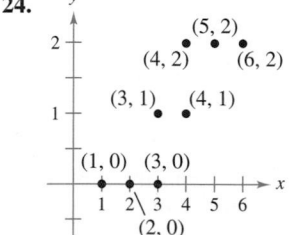

 Finding the Least Squares Regression Line In Exercises 25–28, find the least squares regression line for the points. Use the regression capabilities of a graphing utility to verify your results. Use the graphing utility to plot the points and graph the regression line.

25. $(0, 0)$, $(1, 1)$, $(3, 6)$, $(4, 8)$, $(5, 9)$

26. $(0, 4)$, $(4, 1)$, $(7, -3)$

27. $(0, 6)$, $(4, 3)$, $(5, 0)$, $(8, -4)$, $(10, -5)$

28. $(6, 4)$, $(1, 2)$, $(3, 3)$, $(8, 6)$, $(11, 8)$, $(13, 8)$

29. Modeling Data The table shows the gross income tax collections (in billions of dollars) by the Internal Revenue Service for individuals x and businesses y for selected years. *(Source: U.S. Internal Revenue Service)*

Year	1980	1985	1990	1995
Individual, x	288	397	540	676
Business, y	72	77	110	174

Year	2000	2005	2010	2015
Individual, x	1137	1108	1164	1760
Business, y	236	307	278	390

(a) Use the regression capabilities of a graphing utility to find the least squares regression line for the data.

(b) Use the model to estimate the business income taxes collected when the individual income taxes collected is $1300 billion.

(c) In 1975, the individual income taxes collected was $156 billion and the business income taxes collected was $46 billion. Describe how including this information would affect the model.

 30. Modeling Data The ages x (in years) and systolic blood pressures y (in mmHg) of seven men are shown in the table.

Age, x	16	25	39	45	49	64	70
Systolic Blood Pressure, y	109	122	150	165	159	183	199

(a) Use the regression capabilities of a graphing utility to find the least squares regression line for the data.

(b) Use a graphing utility to plot the data and graph the model.

(c) Use the model to approximate the change in systolic blood pressure for each one-year increase in age.

(d) A 30-year-old man has a systolic blood pressure of 180 mmHg. Describe how including this information would affect the model.

EXPLORING CONCEPTS

31. Method of Least Squares Find a system of equations whose solution yields the coefficients a, b, and c for the least squares regression quadratic

$$y = ax^2 + bx + c$$

for the points $(x_1, y_1), (x_2, y_2), \ldots, (x_n, y_n)$ by minimizing the sum

$$S(a, b, c) = \sum_{i=1}^{n} (y_i - ax_i^2 - bx_i - c)^2.$$

---

32. **HOW DO YOU SEE IT?** Match the regression equation with the appropriate graph. Explain your reasoning. (Note that the x- and y-axes are broken.)

(a) $y = 0.22x - 7.5$

(b) $y = -0.35x + 11.5$

(c) $y = 0.09x + 19.8$

(d) $y = -1.29x + 89.8$

(i)

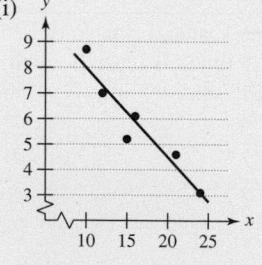

(ii)

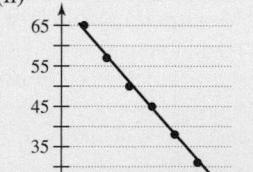

(iii)

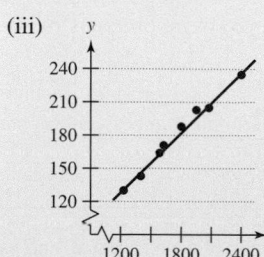

(iv)

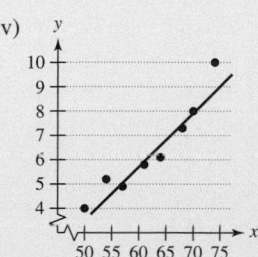

Finding the Least Squares Regression Quadratic In Exercises 33–36, use the result of Exercise 31 to find the least squares regression quadratic for the points. Use the regression capabilities of a graphing utility to verify your results. Use the graphing utility to plot the points and graph the least squares regression quadratic.

33. $(-2, 0)$, $(-1, 0)$, $(0, 1)$, $(1, 2)$, $(2, 5)$

34. $(-4, 5)$, $(-2, 6)$, $(2, 6)$, $(4, 2)$

35. $(0, 0)$, $(2, 2)$, $(3, 6)$, $(4, 12)$

36. $(0, 10)$, $(1, 9)$, $(2, 6)$, $(3, 0)$

37. Modeling Data After a new turbocharger for an automobile engine was developed, the following experimental data were obtained for speed y in miles per hour at two-second time intervals x.

Time, x	0	2	4	6	8	10
Speed, y	0	15	30	50	65	70

(a) Use the result of Exercise 31 to find the least squares regression quadratic for the data.

 (b) Use a graphing utility to plot the points and graph the model.

 38. Modeling Data The table shows the total numbers of enrollees y (in millions) for the Veterans Health Administration for 2010 through 2014. Let $x = 0$ represent the year 2010. *(Source: U.S. Department of Veterans Affairs)*

Year, x	2010	2011	2012	2013	2014
Total Enrollees, y	8.3	8.6	8.8	8.9	9.1

(a) Use the regression capabilities of a graphing utility to find the least squares regression line for the data.

(b) Use the regression capabilities of a graphing utility to find the least squares regression quadratic for the data.

(c) Use a graphing utility to plot the data and graph the models.

(d) Use both models to forecast the total number of enrollees for the year 2025. How do the two models differ as you extrapolate into the future?

 39. Modeling Data A meteorologist measures the atmospheric pressure P (in kilograms per square meter) at altitude h (in kilometers). The data are shown below.

Altitude, h	0	5	10	15	20
Pressure, P	10,332	5583	2376	1240	517

(a) Use the regression capabilities of a graphing utility to find the least squares regression line for the points $(h, \ln P)$.

(b) The result in part (a) is an equation of the form $\ln P = ah + b$. Write this logarithmic form in exponential form.

(c) Use a graphing utility to plot the original data and graph the exponential model in part (b).

 40. Modeling Data The endpoints of the interval over which distinct vision is possible are called the near point and far point of the eye. With increasing age, these points normally change. The table shows the approximate near points y (in inches) for various ages x (in years). *(Source: Ophthalmology & Physiological Optics)*

Age, x	16	32	44	50	60
Near Point, y	3.0	4.7	9.8	19.7	39.4

(a) Find a rational model for the data by taking the reciprocals of the near points to generate the points $(x, 1/y)$. Use the regression capabilities of a graphing utility to find the least squares regression line for the revised data. The resulting line has the form $1/y = ax + b$. Solve for y.

(b) Use a graphing utility to plot the data and graph the model.

(c) Do you think the model can be used to predict the near point for a person who is 70 years old? Explain.

41. Using the Second Partials Test Use the Second Partials Test to verify that the formulas for a and b given in Theorem 13.18 yield a minimum.

$$\left[\text{Hint: Use the fact that } n \sum_{i=1}^{n} x_i^2 \geq \left(\sum_{i=1}^{n} x_i \right)^2. \right]$$

SECTION PROJECT

Building a Pipeline

An oil company wishes to construct a pipeline from its offshore facility A to its refinery B. The offshore facility is 2 miles from shore, and the refinery is 1 mile inland. Furthermore, A and B are 5 miles apart, as shown in the figure.

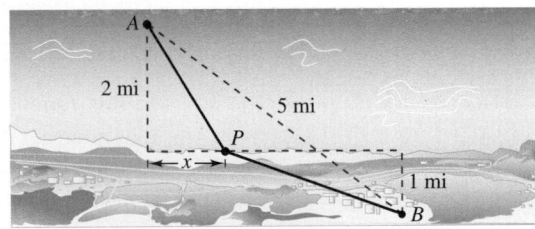

The cost of building the pipeline is $3 million per mile in the water and $4 million per mile on land. So, the cost of the pipeline depends on the location of point P, where it meets the shore. What would be the most economical route of the pipeline?

Imagine that you are to write a report to the oil company about this problem. Let x be the distance shown in the figure. Determine the cost of building the pipeline from A to P and the cost of building it from P to B. Analyze some sample pipeline routes and their corresponding costs. For instance, what is the cost of the most direct route? Then use calculus to determine the route of the pipeline that minimizes the cost. Explain all steps of your development and include any relevant graphs.

13.10 Lagrange Multipliers

- Understand the Method of Lagrange Multipliers.
- Use Lagrange multipliers to solve constrained optimization problems.
- Use the Method of Lagrange Multipliers with two constraints.

Lagrange Multipliers

LAGRANGE MULTIPLIERS

The Method of Lagrange Multipliers is named after the French mathematician Joseph-Louis Lagrange. Lagrange first introduced the method in his famous paper on mechanics, written when he was just 19 years old.

Many optimization problems have restrictions, or **constraints,** on the values that can be used to produce the optimal solution. Such constraints tend to complicate optimization problems because the optimal solution can occur at a boundary point of the domain. In this section, you will study an ingenious technique for solving such problems. It is called the **Method of Lagrange Multipliers.**

To see how this technique works, consider the problem of finding the rectangle of maximum area that can be inscribed in the ellipse

$$\frac{x^2}{3^2} + \frac{y^2}{4^2} = 1.$$

Let (x, y) be the vertex of the rectangle in the first quadrant, as shown in Figure 13.78. Because the rectangle has sides of lengths $2x$ and $2y$, its area is given by

$$f(x, y) = 4xy. \qquad \text{Objective function}$$

You want to find x and y such that $f(x, y)$ is a maximum. Your choice of (x, y) is restricted to first-quadrant points that lie on the ellipse

$$\frac{x^2}{3^2} + \frac{y^2}{4^2} = 1. \qquad \text{Constraint}$$

Now, consider the constraint equation to be a fixed level curve of

$$g(x, y) = \frac{x^2}{3^2} + \frac{y^2}{4^2}.$$

The level curves of f represent a family of hyperbolas

$$f(x, y) = 4xy = k.$$

In this family, the level curves that meet the constraint correspond to the hyperbolas that intersect the ellipse. Moreover, to maximize $f(x, y)$, you want to find the hyperbola that just barely satisfies the constraint. The level curve that does this is the one that is *tangent* to the ellipse, as shown in Figure 13.79.

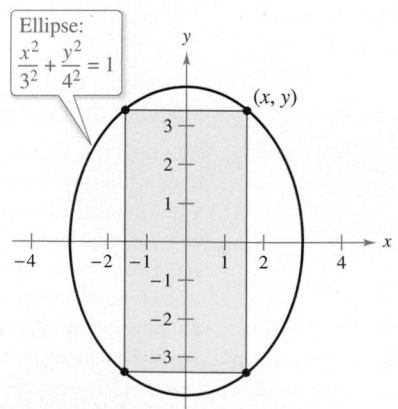

Objective function: $f(x, y) = 4xy$

Figure 13.78

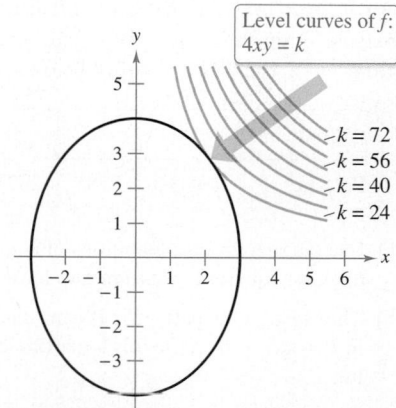

Constraint: $g(x, y) = \frac{x^2}{3^2} + \frac{y^2}{4^2} = 1$

Figure 13.79

To find the appropriate hyperbola, use the fact that two curves are tangent at a point if and only if their gradients are parallel. This means that $\nabla f(x, y)$ must be a scalar multiple of $\nabla g(x, y)$ at the point of tangency. In the context of constrained optimization problems, this scalar is denoted by λ (the lowercase Greek letter lambda).

$$\nabla f(x, y) = \lambda \nabla g(x, y)$$

The scalar λ is called a **Lagrange multiplier.** Theorem 13.19 gives the necessary conditions for the existence of such multipliers.

· · · · · · · · · · · · · · · · · ▷

REMARK Lagrange's Theorem can be shown to be true for functions of three variables, using a similar argument with level surfaces and Theorem 13.14.

THEOREM 13.19 Lagrange's Theorem

Let f and g have continuous first partial derivatives such that f has an extremum at a point (x_0, y_0) on the smooth constraint curve $g(x, y) = c$. If $\nabla g(x_0, y_0) \neq \mathbf{0}$, then there is a real number λ such that

$$\nabla f(x_0, y_0) = \lambda \nabla g(x_0, y_0).$$

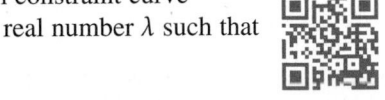

Proof To begin, represent the smooth curve given by $g(x, y) = c$ by the vector-valued function

$$\mathbf{r}(t) = x(t)\mathbf{i} + y(t)\mathbf{j}, \quad \mathbf{r}'(t) \neq \mathbf{0}$$

where x' and y' are continuous on an open interval I. Define the function h as $h(t) = f(x(t), y(t))$. Then, because $f(x_0, y_0)$ is an extreme value of f, you know that

$$h(t_0) = f(x(t_0), y(t_0)) = f(x_0, y_0)$$

is an extreme value of h. This implies that $h'(t_0) = 0$, and, by the Chain Rule,

$$h'(t_0) = f_x(x_0, y_0)x'(t_0) + f_y(x_0, y_0)y'(t_0) = \nabla f(x_0, y_0) \cdot \mathbf{r}'(t_0) = 0.$$

So, $\nabla f(x_0, y_0)$ is orthogonal to $\mathbf{r}'(t_0)$. Moreover, by Theorem 13.12, $\nabla g(x_0, y_0)$ is also orthogonal to $\mathbf{r}'(t_0)$. Consequently, the gradients $\nabla f(x_0, y_0)$ and $\nabla g(x_0, y_0)$ are parallel, and there must exist a scalar λ such that

$$\nabla f(x_0, y_0) = \lambda \nabla g(x_0, y_0). \qquad \blacksquare$$

The Method of Lagrange Multipliers uses Theorem 13.19 to find the extreme values of a function f subject to a constraint.

· · · · · · · · · · · · · · · · ▷

REMARK As you will see in Examples 1 and 2, the Method of Lagrange Multipliers requires solving systems of nonlinear equations. This often can require some tricky algebraic manipulation.

Method of Lagrange Multipliers

Let f and g satisfy the hypothesis of Lagrange's Theorem, and let f have a minimum or maximum subject to the constraint $g(x, y) = c$. To find the minimum or maximum of f, use these steps.

1. Simultaneously solve the equations $\nabla f(x, y) = \lambda \nabla g(x, y)$ and $g(x, y) = c$ by solving the following system of equations.

$$f_x(x, y) = \lambda g_x(x, y)$$
$$f_y(x, y) = \lambda g_y(x, y)$$
$$g(x, y) = c$$

2. Evaluate f at each solution point obtained in the first step. The greatest value yields the maximum of f subject to the constraint $g(x, y) = c$, and the least value yields the minimum of f subject to the constraint $g(x, y) = c$.

Constrained Optimization Problems

In the problem at the beginning of this section, you wanted to maximize the area of a rectangle that is inscribed in an ellipse. Example 1 shows how to use Lagrange multipliers to solve this problem.

EXAMPLE 1 **Using a Lagrange Multiplier with One Constraint**

Find the maximum value of $f(x, y) = 4xy$, where $x > 0$ and $y > 0$, subject to the constraint $(x^2/3^2) + (y^2/4^2) = 1$.

Solution To begin, let

$$g(x, y) = \frac{x^2}{3^2} + \frac{y^2}{4^2} = 1.$$

By equating $\nabla f(x, y) = 4y\mathbf{i} + 4x\mathbf{j}$ and $\lambda \nabla g(x, y) = (2\lambda x/9)\mathbf{i} + (\lambda y/8)\mathbf{j}$, you obtain the following system of equations.

$$4y = \frac{2}{9}\lambda x \qquad f_x(x, y) = \lambda g_x(x, y)$$

$$4x = \frac{1}{8}\lambda y \qquad f_y(x, y) = \lambda g_y(x, y)$$

$$\frac{x^2}{3^2} + \frac{y^2}{4^2} = 1 \qquad \text{Constraint}$$

From the first equation, you obtain $\lambda = 18y/x$, and substitution into the second equation produces

$$4x = \frac{1}{8}\left(\frac{18y}{x}\right)y \quad \Longrightarrow \quad x^2 = \frac{9}{16}y^2.$$

Substituting this value for x^2 into the third equation produces

$$\frac{1}{9}\left(\frac{9}{16}y^2\right) + \frac{1}{16}y^2 = 1 \quad \Longrightarrow \quad y^2 = 8 \quad \Longrightarrow \quad y = \pm 2\sqrt{2}.$$

Because $y > 0$, choose the positive value and find that

$$x^2 = \frac{9}{16}y^2$$

$$= \frac{9}{16}(8)$$

$$= \frac{9}{2}$$

$$x = \pm \frac{3}{\sqrt{2}}.$$

Because $x > 0$, choose the positive value. So, the maximum value of f is

$$f\left(\frac{3}{\sqrt{2}}, 2\sqrt{2}\right) = 4\left(\frac{3}{\sqrt{2}}\right)\left(2\sqrt{2}\right) = 24. \qquad \blacksquare$$

•••••••••••••••••▷
•
••**REMARK** Note in Example 1 that writing the constraint as

$$\frac{x^2}{3^2} + \frac{y^2}{4^2} = 1$$

or

$$\frac{x^2}{3^2} + \frac{y^2}{4^2} - 1 = 0$$

does not affect the solution—the constant is eliminated when you form ∇g.

Example 1 can also be solved using the techniques you learned in Chapter 3. To see how, try to find the maximum value of $A = 4xy$ given that $(x^2/3^2) + (y^2/4^2) = 1$. To begin, solve the second equation for y to obtain $y = \frac{4}{3}\sqrt{9 - x^2}$. Then substitute into the first equation to obtain $A = 4x\left(\frac{4}{3}\sqrt{9 - x^2}\right)$. Finally, use the techniques of Chapter 3 to maximize A.

For some industrial applications, a robot can cost more than the annual wages and benefits for one employee. So, manufacturers must carefully balance the amount of money spent on labor and capital.

EXAMPLE 2 **A Business Application**

The Cobb-Douglas production function (see Section 13.1) for a manufacturer is given by

$$f(x, y) = 100x^{3/4}y^{1/4} \qquad \text{Objective function}$$

where x represents the units of labor (at \$150 per unit) and y represents the units of capital (at \$250 per unit). The total cost of labor and capital is limited to \$50,000. Find the maximum production level for this manufacturer.

Solution The gradient of f is

$$\nabla f(x, y) = 75x^{-1/4}y^{1/4}\mathbf{i} + 25x^{3/4}y^{-3/4}\mathbf{j}.$$

The limit on the cost of labor and capital produces the constraint

$$g(x, y) = 150x + 250y = 50{,}000. \qquad \text{Constraint}$$

So, $\lambda\nabla g(x, y) = 150\lambda\mathbf{i} + 250\lambda\mathbf{j}$. This gives rise to the following system of equations.

$$75x^{-1/4}y^{1/4} = 150\lambda \qquad f_x(x, y) = \lambda g_x(x, y)$$
$$25x^{3/4}y^{-3/4} = 250\lambda \qquad f_y(x, y) = \lambda g_y(x, y)$$
$$150x + 250y = 50{,}000 \qquad \text{Constraint}$$

By solving for λ in the first equation

$$\lambda = \frac{75x^{-1/4}y^{1/4}}{150} = \frac{x^{-1/4}y^{1/4}}{2}$$

and substituting into the second equation, you obtain

$$25x^{3/4}y^{-3/4} = 250\left(\frac{x^{-1/4}y^{1/4}}{2}\right)$$
$$25x = 125y \qquad \text{Multiply by } x^{1/4}y^{3/4}.$$
$$x = 5y.$$

By substituting this value for x in the third equation, you have

$$150(5y) + 250y = 50{,}000$$
$$1000y = 50{,}000$$
$$y = 50 \text{ units of capital.}$$

This means that the value of x is

$$x = 5(50)$$
$$= 250 \text{ units of labor.}$$

So, the maximum production level is

$$f(250, 50) = 100(250)^{3/4}(50)^{1/4}$$
$$\approx 16{,}719 \text{ units of product.}$$

FOR FURTHER INFORMATION
For more information on the use of Lagrange multipliers in economics, see the article "Lagrange Multiplier Problems in Economics" by John V. Baxley and John C. Moorhouse in *The American Mathematical Monthly*. To view this article, go to *MathArticles.com*.

Economists call the Lagrange multiplier obtained in a production function the **marginal productivity of money.** For instance, in Example 2, the marginal productivity of money at $x = 250$ and $y = 50$ is

$$\lambda = \frac{x^{-1/4}y^{1/4}}{2} = \frac{(250)^{-1/4}(50)^{1/4}}{2} \approx 0.334$$

which means that for each additional dollar spent on production, an additional 0.334 unit of the product can be produced.

EXAMPLE 3 **Lagrange Multipliers and Three Variables**

⋮ ⋯▷ *See LarsonCalculus.com for an interactive version of this type of example.*

Find the minimum value of

$$f(x, y, z) = 2x^2 + y^2 + 3z^2 \qquad \text{Objective function}$$

subject to the constraint $2x - 3y - 4z = 49$.

Solution Let $g(x, y, z) = 2x - 3y - 4z = 49$. Then, because

$$\nabla f(x, y, z) = 4x\mathbf{i} + 2y\mathbf{j} + 6z\mathbf{k}$$

and

$$\lambda \nabla g(x, y, z) = 2\lambda\mathbf{i} - 3\lambda\mathbf{j} - 4\lambda\mathbf{k}$$

you obtain the following system of equations.

$$
\begin{aligned}
4x &= 2\lambda & f_x(x, y, z) &= \lambda g_x(x, y, z) \\
2y &= -3\lambda & f_y(x, y, z) &= \lambda g_y(x, y, z) \\
6z &= -4\lambda & f_z(x, y, z) &= \lambda g_z(x, y, z) \\
2x - 3y - 4z &= 49 & &\text{Constraint}
\end{aligned}
$$

The solution of this system is $x = 3$, $y = -9$, and $z = -4$. So, the optimum value of f is

$$
\begin{aligned}
f(3, -9, -4) &= 2(3)^2 + (-9)^2 + 3(-4)^2 \\
&= 147.
\end{aligned}
$$

From the original function and constraint, it is clear that $f(x, y, z)$ has no maximum. So, the optimum value of f determined above is a minimum. ■

A graphical interpretation of constrained optimization problems in two variables was given at the beginning of this section. In three variables, the interpretation is similar, except that level surfaces are used instead of level curves. For instance, in Example 3, the level surfaces of f are ellipsoids centered at the origin, and the constraint

$$2x - 3y - 4z = 49$$

is a plane. The minimum value of f is represented by the ellipsoid that is tangent to the constraint plane, as shown in Figure 13.80.

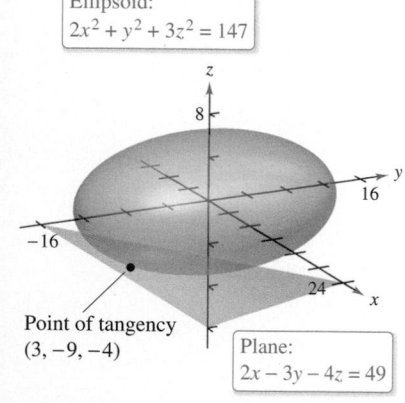

Ellipsoid:
$2x^2 + y^2 + 3z^2 = 147$

Point of tangency
$(3, -9, -4)$

Plane:
$2x - 3y - 4z = 49$

Figure 13.80

EXAMPLE 4 **Optimization Inside a Region**

Find the extreme values of

$$f(x, y) = x^2 + 2y^2 - 2x + 3 \qquad \text{Objective function}$$

subject to the constraint $x^2 + y^2 \le 10$.

Solution To solve this problem, you can break the constraint into two cases.

a. For points *on the circle* $x^2 + y^2 = 10$, you can use Lagrange multipliers to find that the maximum value of $f(x, y)$ is 24—this value occurs at $(-1, 3)$ and at $(-1, -3)$. In a similar way, you can determine that the minimum value of $f(x, y)$ is approximately 6.675—this value occurs at $(\sqrt{10}, 0)$.

b. For points *inside the circle,* you can use the techniques discussed in Section 13.8 to conclude that the function has a relative minimum of 2 at the point $(1, 0)$.

By combining these two results, you can conclude that f has a maximum of 24 at $(-1, \pm 3)$ and a minimum of 2 at $(1, 0)$, as shown in Figure 13.81.■

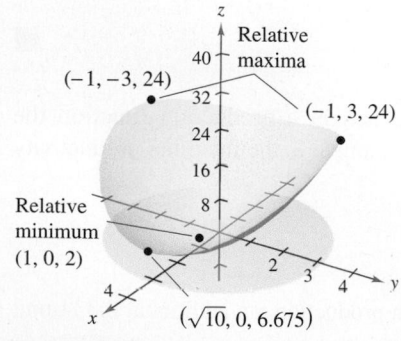

Relative maxima
$(-1, -3, 24)$
$(-1, 3, 24)$

Relative minimum
$(1, 0, 2)$

$(\sqrt{10}, 0, 6.675)$

Figure 13.81

The Method of Lagrange Multipliers with Two Constraints

For optimization problems involving *two* constraint functions g and h, you can introduce a second Lagrange multiplier, μ (the lowercase Greek letter mu), and then solve the equation

$$\nabla f = \lambda \nabla g + \mu \nabla h$$

where the gradients are not parallel, as illustrated in Example 5.

EXAMPLE 5   Optimization with Two Constraints

Let $T(x, y, z) = 20 + 2x + 2y + z^2$ represent the temperature at each point on the sphere

$$x^2 + y^2 + z^2 = 11.$$

Find the extreme temperatures on the curve formed by the intersection of the plane $x + y + z = 3$ and the sphere.

Solution   The two constraints are

$$g(x, y, z) = x^2 + y^2 + z^2 = 11 \quad \text{and} \quad h(x, y, z) = x + y + z = 3.$$

Using

$$\nabla T(x, y, z) = 2\mathbf{i} + 2\mathbf{j} + 2z\mathbf{k}$$
$$\lambda \nabla g(x, y, z) = 2\lambda x\mathbf{i} + 2\lambda y\mathbf{j} + 2\lambda z\mathbf{k}$$

and

$$\mu \nabla h(x, y, z) = \mu\mathbf{i} + \mu\mathbf{j} + \mu\mathbf{k}$$

you can write the following system of equations.

$2 = 2\lambda x + \mu$	$T_x(x, y, z) = \lambda g_x(x, y, z) + \mu h_x(x, y, z)$
$2 = 2\lambda y + \mu$	$T_y(x, y, z) = \lambda g_y(x, y, z) + \mu h_y(x, y, z)$
$2z = 2\lambda z + \mu$	$T_z(x, y, z) = \lambda g_z(x, y, z) + \mu h_z(x, y, z)$
$x^2 + y^2 + z^2 = 11$	Constraint 1
$x + y + z = 3$	Constraint 2

By subtracting the second equation from the first, you obtain the following system.

$$\lambda(x - y) = 0$$
$$2z(1 - \lambda) - \mu = 0$$
$$x^2 + y^2 + z^2 = 11$$
$$x + y + z = 3$$

> **REMARK**   The systems of equations that arise when the Method of Lagrange Multipliers is used are not, in general, linear systems, and finding the solutions often requires ingenuity.

From the first equation, you can conclude that $\lambda = 0$ or $x = y$. For $\lambda = 0$, you can show that the critical points are $(3, -1, 1)$ and $(-1, 3, 1)$. (Try doing this—it takes a little work.) For $\lambda \neq 0$, then $x = y$ and you can show that the critical points occur when $x = y = (3 \pm 2\sqrt{3})/3$ and $z = (3 \mp 4\sqrt{3})/3$. Finally, to find the optimal solutions, compare the temperatures at the four critical points.

$$T(3, -1, 1) = T(-1, 3, 1) = 25$$
$$T\left(\frac{3 - 2\sqrt{3}}{3}, \frac{3 - 2\sqrt{3}}{3}, \frac{3 + 4\sqrt{3}}{3}\right) = \frac{91}{3} \approx 30.33$$
$$T\left(\frac{3 + 2\sqrt{3}}{3}, \frac{3 + 2\sqrt{3}}{3}, \frac{3 - 4\sqrt{3}}{3}\right) = \frac{91}{3} \approx 30.33$$

So, $T = 25$ is the minimum temperature and $T = \frac{91}{3}$ is the maximum temperature on the curve.

13.10 Exercises

See **CalcChat.com** for tutorial help and worked-out solutions to odd-numbered exercises.

CONCEPT CHECK

1. **Constrained Optimization Problems** Explain what is meant by constrained optimization problems.

2. **Method of Lagrange Multipliers** In your own words, describe the Method of Lagrange Multipliers for solving constrained optimization problems.

 **Using Lagrange Multipliers** In Exercises 3–10, use Lagrange multipliers to find the indicated extrema, assuming that x and y are positive.

3. Maximize $f(x, y) = xy$

   Constraint: $x + y = 10$

4. Minimize $f(x, y) = 2x + y$

   Constraint: $xy = 32$

5. Minimize $f(x, y) = x^2 + y^2$

   Constraint: $x + 2y - 5 = 0$

6. Maximize $f(x, y) = x^2 - y^2$

   Constraint: $2y - x^2 = 0$

7. Maximize $f(x, y) = 2x + 2xy + y$

   Constraint: $2x + y = 100$

8. Minimize $f(x, y) = 3x + y + 10$

   Constraint: $x^2 y = 6$

9. Maximize $f(x, y) = \sqrt{6 - x^2 - y^2}$

   Constraint: $x + y - 2 = 0$

10. Minimize $f(x, y) = \sqrt{x^2 + y^2}$

    Constraint: $2x + 4y - 15 = 0$

 **Using Lagrange Multipliers** In Exercises 11–14, use Lagrange multipliers to find the indicated extrema, assuming that x, y, and z are positive.

11. Minimize $f(x, y, z) = x^2 + y^2 + z^2$

    Constraint: $x + y + z - 9 = 0$

12. Maximize $f(x, y, z) = xyz$

    Constraint: $x + y + z - 3 = 0$

13. Minimize $f(x, y, z) = x^2 + y^2 + z^2$

    Constraint: $x + y + z = 1$

14. Maximize $f(x, y, z) = x + y + z$

    Constraint: $x^2 + y^2 + z^2 = 1$

 **Using Lagrange Multipliers** In Exercises 15 and 16, use Lagrange multipliers to find any extrema of the function subject to the constraint $x^2 + y^2 \le 1$.

15. $f(x, y) = x^2 + 3xy + y^2$ 16. $f(x, y) = e^{-xy/4}$

 **Using Lagrange Multipliers** In Exercises 17 and 18, use Lagrange multipliers to find the indicated extrema of f subject to two constraints, assuming that x, y, and z are nonnegative.

17. Maximize $f(x, y, z) = xyz$

    Constraints: $x + y + z = 32$, $x - y + z = 0$

18. Minimize $f(x, y, z) = x^2 + y^2 + z^2$

    Constraints: $x + 2z = 6$, $x + y = 12$

 **Finding Minimum Distance** In Exercises 19–28, use Lagrange multipliers to find the minimum distance from the curve or surface to the indicated point. (*Hint:* To simplify the computations, minimize the square of the distance.)

Curve	Point
19. Line: $x + y = 1$	$(0, 0)$
20. Line: $2x + 3y = -1$	$(0, 0)$
21. Line: $x - y = 4$	$(0, 2)$
22. Line: $x + 4y = 3$	$(1, 0)$
23. Parabola: $y = x^2$	$(0, 3)$
24. Parabola: $y = x^2$	$(-3, 0)$
25. Circle: $x^2 + (y - 1)^2 = 9$	$(4, 4)$
26. Circle: $(x - 4)^2 + y^2 = 4$	$(0, 10)$
Surface	**Point**
27. Plane: $x + y + z = 1$	$(2, 1, 1)$
28. Cone: $z = \sqrt{x^2 + y^2}$	$(4, 0, 0)$

Intersection of Surfaces In Exercises 29 and 30, use Lagrange multipliers to find the highest point on the curve of intersection of the surfaces.

29. Cone: $x^2 + y^2 - z^2 = 0$ 30. Sphere: $x^2 + y^2 + z^2 = 36$

    Plane: $x + 2z = 4$ Plane: $2x + y - z = 2$

Using Lagrange Multipliers In Exercises 31–38, use Lagrange multipliers to solve the indicated exercise in Section 13.9.

31. Exercise 3   32. Exercise 4

33. Exercise 7   34. Exercise 8

35. Exercise 11   36. Exercise 12

37. Exercise 17   38. Exercise 18

39. **Maximum Volume** Use Lagrange multipliers to find the dimensions of a rectangular box of maximum volume that can be inscribed (with edges parallel to the coordinate axes) in the ellipsoid

$$\frac{x^2}{a^2} + \frac{y^2}{b^2} + \frac{z^2}{c^2} = 1.$$

40. HOW DO YOU SEE IT? The graphs show the constraint and several level curves of the objective function. Use the graph to approximate the indicated extrema.

(a) Maximize $z = xy$ (b) Minimize $z = x^2 + y^2$

Constraint: $2x + y = 4$ Constraint: $x + y - 4 = 0$

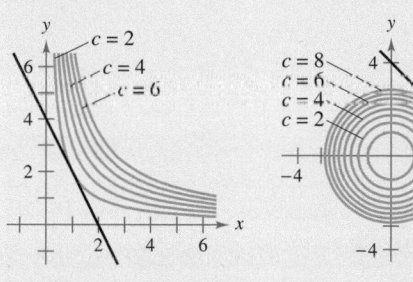

EXPLORING CONCEPTS

41. Method of Lagrange Multipliers Explain why you cannot use Lagrange multipliers to find the minimum of the function $f(x, y) = x$ subject to the constraint $y^2 + x^4 - x^3 = 0$.

42. Method of Lagrange Multipliers Draw the level curves for $f(x, y) = x^2 + y^2 = c$ for $c = 1, 2, 3,$ and 4, and sketch the constraint $x + y = 2$. Explain analytically how you know that the extremum of $f(x, y) = x^2 + y^2$ at $(1, 1)$ is a minimum instead of a maximum.

43. Minimum Cost A cargo container (in the shape of a rectangular solid) must have a volume of 480 cubic feet. The bottom will cost \$5 per square foot to construct, and the sides and the top will cost \$3 per square foot to construct. Use Lagrange multipliers to find the dimensions of the container of this size that has minimum cost.

44. Geometric and Arithmetic Means

(a) Use Lagrange multipliers to prove that the product of three positive numbers x, y, and z, whose sum has the constant value S, is a maximum when the three numbers are equal. Use this result to prove that

$$\sqrt[3]{xyz} \le \frac{x + y + z}{3}.$$

(b) Generalize the result of part (a) to prove that the product $x_1 x_2 x_3 \cdots x_n$ is a maximum when

$$x_1 = x_2 = x_3 = \cdots = x_n, \quad \sum_{i=1}^{n} x_i = S, \text{ and all } x_i \ge 0.$$

Then prove that

$$\sqrt[n]{x_1 x_2 x_3 \cdots x_n} \le \frac{x_1 + x_2 + x_3 + \cdots + x_n}{n}.$$

This shows that the geometric mean is never greater than the arithmetic mean.

45. Minimum Surface Area Use Lagrange multipliers to find the dimensions of a right circular cylinder with volume V_0 and minimum surface area.

46. Temperature Let $T(x, y, z) = 100 + x^2 + y^2$ represent the temperature at each point on the sphere

$$x^2 + y^2 + z^2 = 50.$$

Use Lagrange multipliers to find the maximum temperature on the curve formed by the intersection of the sphere and the plane $x - z = 0$.

47. Refraction of Light When light waves traveling in a transparent medium strike the surface of a second transparent medium, they tend to "bend" in order to follow the path of minimum time. This tendency is called *refraction* and is described by **Snell's Law of Refraction,**

$$\frac{\sin \theta_1}{v_1} = \frac{\sin \theta_2}{v_2}$$

where θ_1 and θ_2 are the magnitudes of the angles shown in the figure, and v_1 and v_2 are the velocities of light in the two media. Use Lagrange multipliers to derive this law using $x + y = a$.

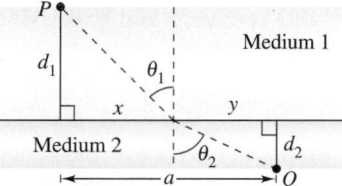 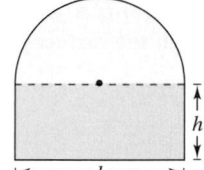

Figure for 47   Figure for 48

48. Area and Perimeter A semicircle is on top of a rectangle (see figure). When the area is fixed and the perimeter is a minimum, or when the perimeter is fixed and the area is a maximum, use Lagrange multipliers to verify that the length of the rectangle is twice its height.

 **Production Level** In Exercises 49 and 50, use Lagrange multipliers to find the maximum production level when the total cost of labor (at \$112 per unit) and capital (at \$60 per unit) is limited to \$250,000, where P is the production function, x is the number of units of labor, and y is the number of units of capital.

49. $P(x, y) = 100x^{0.25}y^{0.75}$ **50.** $P(x, y) = 100x^{0.4}y^{0.6}$

Cost In Exercises 51 and 52, use Lagrange multipliers to find the minimum cost of producing 50,000 units of a product, where P is the production function, x is the number of units of labor (at \$72 per unit), and y is the number of units of capital (at \$80 per unit).

51. $P(x, y) = 100x^{0.25}y^{0.75}$ **52.** $P(x, y) = 100x^{0.6}y^{0.4}$

PUTNAM EXAM CHALLENGE

53. A can buoy is to be made of three pieces, namely, a cylinder and two equal cones, the altitude of each cone being equal to the altitude of the cylinder. For a given area of surface, what shape will have the greatest volume?

Review Exercises See CalcChat.com for tutorial help and worked-out solutions to odd-numbered exercises.

Evaluating a Function In Exercises 1 and 2, evaluate the function at the given values of the independent variables. Simplify the results.

1. $f(x, y) = x^2y - 3$

   (a) $f(0, 4)$ (b) $f(2, -1)$ (c) $f(-3, 2)$ (d) $f(x, 7)$

2. $f(x, y) = 6 - 4x - 2y^2$

   (a) $f(0, 2)$ (b) $f(5, 0)$ (c) $f(-1, -2)$ (d) $f(-3, y)$

Finding the Domain and Range of a Function In Exercises 3 and 4, find the domain and range of the function.

3. $f(x, y) = \dfrac{\sqrt{x}}{y}$

4. $f(x, y) = \sqrt{36 - x^2 - y^2}$

Sketching a Surface In Exercises 5 and 6, describe and sketch the surface given by the function.

5. $f(x, y) = -2$ **6.** $g(x, y) = x$

Sketching a Contour Map In Exercises 7 and 8, describe the level curves of the function. Sketch a contour map of the surface using level curves for the given c-values.

7. $z = 3 - 2x + y, \quad c = 0, 2, 4, 6, 8$

8. $z = 2x^2 + y^2, \quad c = 1, 2, 3, 4, 5$

9. **Conjecture** Consider the function $f(x, y) = x^2 + y^2$.

   (a) Sketch the graph of the surface given by f.

   (b) Make a conjecture about the relationship between the graphs of f and $g(x, y) = f(x, y) + 2$. Explain your reasoning.

   (c) Make a conjecture about the relationship between the graphs of f and $g(x, y) = f(x, y - 2)$. Explain your reasoning.

   (d) On the surface in part (a), sketch the graphs of $z = f(1, y)$ and $z = f(x, 1)$.

10. **Cobb-Douglas Production Function** A manufacturer estimates that its production can be modeled by

$$f(x, y) = 100x^{0.8}y^{0.2}$$

where x is the number of units of labor and y is the number of units of capital.

   (a) Find the production level when $x = 100$ and $y = 200$.

   (b) Find the production level when $x = 500$ and $y = 1500$.

Sketching a Level Surface In Exercises 11 and 12, describe and sketch the graph of the level surface $f(x, y, z) = c$ at the given value of c.

11. $f(x, y, z) = x^2 - y + z^2, \quad c = 2$

12. $f(x, y, z) = 4x^2 - y^2 + 4z^2, \quad c = 0$

Limit and Continuity In Exercises 13–18, find the limit (if it exists) and discuss the continuity of the function.

13. $\displaystyle\lim_{(x, y)\to(1, 1)} \dfrac{xy}{x^2 + y^2}$ **14.** $\displaystyle\lim_{(x, y)\to(1, 1)} \dfrac{xy}{x^2 - y^2}$

15. $\displaystyle\lim_{(x, y)\to(0, 0)} \dfrac{y + xe^{-y^2}}{1 + x^2}$ **16.** $\displaystyle\lim_{(x, y)\to(0, 0)} \dfrac{x^2y}{x^4 + y^2}$

17. $\displaystyle\lim_{(x, y, z)\to(-3, 1, 2)} \dfrac{\ln z}{xy - z}$ **18.** $\displaystyle\lim_{(x, y, z)\to(1, 3, \pi)} \sin \dfrac{xz}{2y}$

Finding Partial Derivatives In Exercises 19–26, find all first partial derivatives.

19. $f(x, y) = 5x^3 + 7y - 3$ **20.** $f(x, y) = 4x^2 - 2xy + y^2$

21. $f(x, y) = e^x \cos y$ **22.** $f(x, y) = \dfrac{xy}{x + y}$

23. $f(x, y) = y^3e^{y/x}$ **24.** $z = \ln(x^2 + y^2 + 1)$

25. $f(x, y, z) = 2xz^2 + 6xyz$ **26.** $w = \sqrt{x^2 - y^2 - z^2}$

Finding and Evaluating Partial Derivatives In Exercises 27–30, find all first partial derivatives, and evaluate each at the given point.

27. $f(x, y) = x^2 - y, \quad (0, 2)$ **28.** $f(x, y) = xe^{2y}, \quad (-1, 1)$

29. $f(x, y, z) = xy \cos xz, \quad (2, 3, -\pi/3)$

30. $f(x, y, z) = \sqrt{x^2 + y - z^2}, \quad (-3, -3, 1)$

Finding Second Partial Derivatives In Exercises 31–34, find the four second partial derivatives. Observe that the second mixed partials are equal.

31. $f(x, y) = 3x^2 - xy + 2y^3$ **32.** $h(x, y) = \dfrac{x}{x + y}$

33. $h(x, y) = x \sin y + y \cos x$ **34.** $g(x, y) = \cos(x - 2y)$

35. **Finding the Slopes of a Surface** Find the slopes of the surface $z = x^2 \ln(y + 1)$ in the x- and y-directions at the point $(2, 0, 0)$.

36. **Marginal Revenue** A company has two plants that produce the same lawn mower. If x_1 and x_2 are the numbers of units produced at plant 1 and plant 2, respectively, then the total revenue for the product is given by

$$R = 300x_1 + 300x_2 - 5x_1^2 - 10x_1x_2 - 5x_2^2.$$

When $x_1 = 5$ and $x_2 = 8$, find (a) the marginal revenue for plant 1, $\partial R/\partial x_1$, and (b) the marginal revenue for plant 2, $\partial R/\partial x_2$.

Finding a Total Differential In Exercises 37–40, find the total differential.

37. $z = x \sin xy$ **38.** $z = 5x^4y^3$

39. $w = 3xy^2 - 2x^3yz^2$ **40.** $w = \dfrac{3x + 4y}{y + 3z}$

Using a Differential as an Approximation In Exercises 41 and 42, (a) find $f(2, 1)$ and $f(2.1, 1.05)$ and calculate Δz, and (b) use the total differential dz to approximate Δz.

41. $f(x, y) = 4x + 2y$ **42.** $f(x, y) = 36 - x^2 - y^2$

43. Volume The possible error involved in measuring each dimension of a right circular cone is $\pm\frac{1}{8}$ inch. The radius is 2 inches and the height is 5 inches. Approximate the propagated error and the relative error in the calculated volume of the cone.

44. Lateral Surface Area Approximate the propagated error and the relative error in the calculated lateral surface area of the cone in Exercise 43. $\left(\text{The lateral surface area is given by } A = \pi r \sqrt{r^2 + h^2}.\right)$

Differentiability In Exercises 45 and 46, show that the function is differentiable by finding values of ε_1 and ε_2 as designated in the definition of differentiability, and verify that both ε_1 and ε_2 approach 0 as $(\Delta x, \Delta y) \to (0, 0)$.

45. $f(x, y) = 6x - y^2$

46. $f(x, y) = xy^2$

Using Different Methods In Exercises 47–50, find dw/dt (a) by using the appropriate Chain Rule and (b) by converting w to a function of t before differentiating.

47. $w = \ln(x^2 + y)$, $x = 2t$, $y = 4 - t$

48. $w = y^2 - x$, $x = \cos t$, $y = \sin t$

49. $w = x^2z + y + z$, $x = e^t$, $y = t$, $z = t^2$

50. $w = \sin x + y^2z + 2z$, $x = \arcsin(t - 1)$, $y = t^3$, $z = 3$

Using Different Methods In Exercises 51 and 52, find $\partial w/\partial r$ and $\partial w/\partial t$ (a) by using the appropriate Chain Rule and (b) by converting w to a function of r and t before differentiating.

51. $w = \dfrac{xy}{z}$, $x = 2r + t$, $y = rt$, $z = 2r - t$

52. $w = x^2 + y^2 + z^2$, $x = r \cos t$, $y = r \sin t$, $z = t$

Finding a Derivative Implicitly In Exercises 53 and 54, differentiate implicitly to find dy/dx.

53. $x^3 - xy + 5y = 0$

54. $\dfrac{xy^2}{x + y} = 3$

Finding Partial Derivatives Implicitly In Exercises 55 and 56, differentiate implicitly to find the first partial derivatives of z.

55. $x^2 + xy + y^2 + yz + z^2 = 0$

56. $xz^2 - y \sin z = 0$

Finding a Directional Derivative In Exercises 57 and 58, use Theorem 13.9 to find the directional derivative of the function at P in the direction of $\mathbf{v}$.

57. $f(x, y) = x^2y$, $P(-5, 5)$, $\mathbf{v} = 3\mathbf{i} - 4\mathbf{j}$

58. $f(x, y) = \frac{1}{4}y^2 - x^2$, $P(1, 4)$, $\mathbf{v} = 2\mathbf{i} + \mathbf{j}$

Finding a Directional Derivative In Exercises 59 and 60, use the gradient to find the directional derivative of the function at P in the direction of $\mathbf{v}$.

59. $w = y^2 + xz$, $P(1, 2, 2)$, $\mathbf{v} = 2\mathbf{i} - \mathbf{j} + 2\mathbf{k}$

60. $w = 5x^2 + 2xy - 3y^2z$, $P(1, 0, 1)$, $\mathbf{v} = \mathbf{i} + \mathbf{j} - \mathbf{k}$

Using Properties of the Gradient In Exercises 61–66, find the gradient of the function and the maximum value of the directional derivative at the given point.

61. $z = x^2y$, $(2, 1)$ **62.** $z - e^{-x} \cos y$, $\left(0, \dfrac{\pi}{4}\right)$

63. $z = \dfrac{y}{x^2 + y^2}$, $(1, 1)$ **64.** $z = \dfrac{x^2}{x - y}$, $(2, 1)$

65. $w = x^4y - y^2z^2$, $\left(-1, \frac{1}{2}, 2\right)$

66. $w = e^{\sqrt{x+y+z^2}}$, $(5, 0, 2)$

Using a Function In Exercises 67 and 68, (a) find the gradient of the function at P, (b) find a unit normal vector to the level curve $f(x, y) = c$ at P, (c) find the tangent line to the level curve $f(x, y) = c$ at P, and (d) sketch the level curve, the unit normal vector, and the tangent line in the xy-plane.

67. $f(x, y) = 9x^2 - 4y^2$ **68.** $f(x, y) = 4y \sin x - y$

  $c = 65$, $P(3, 2)$ $c = 3$, $P\left(\dfrac{\pi}{2}, 1\right)$

Finding an Equation of a Tangent Plane In Exercises 69–72, find an equation of the tangent plane to the surface at the given point.

69. $z = x^2 + y^2 + 2$, $(1, 3, 12)$

70. $9x^2 + y^2 + 4z^2 = 25$, $(0, -3, 2)$

71. $z = -9 + 4x - 6y - x^2 - y^2$, $(2, -3, 4)$

72. $f(x, y) = \sqrt{25 - y^2}$, $(2, 3, 4)$

Finding an Equation of a Tangent Plane and a Normal Line In Exercises 73 and 74, (a) find an equation of the tangent plane to the surface at the given point and (b) find a set of symmetric equations for the normal line to the surface at the given point.

73. $f(x, y) = x^2y$, $(2, 1, 4)$

74. $z = \sqrt{9 - x^2 - y^2}$, $(1, 2, 2)$

Finding the Angle of Inclination of a Tangent Plane In Exercises 75 and 76, find the angle of inclination of the tangent plane to the surface at the given point.

75. $x^2 + y^2 + z^2 = 14$, $(2, 1, 3)$

76. $xy + yz^2 = 32$, $(-4, 1, 6)$

Horizontal Tangent Plane In Exercises 77 and 78, find the point(s) on the surface at which the tangent plane is horizontal.

77. $z = 9 - 2x^2 + y^3$

78. $z = 2xy + 3x + 5y$

Using the Second Partials Test In Exercises 79–84, find all relative extrema and saddle points of the function. Use the Second Partials Test where applicable.

79. $f(x, y) = -x^2 - 4y^2 + 8x - 8y - 11$

80. $f(x, y) = x^2 - y^2 - 16x - 16y$

81. $f(x, y) = 2x^2 + 6xy + 9y^2 + 8x + 14$

82. $f(x, y) = x^6 y^6$

83. $f(x, y) = xy + \dfrac{1}{x} + \dfrac{1}{y}$

84. $f(x, y) = -8x^2 + 4xy - y^2 + 12x + 7$

85. **Finding Minimum Distance** Find the minimum distance from the point $(2, 1, 4)$ to the surface $x + y + z = 4$. (*Hint:* To simplify the computations, minimize the square of the distance.)

86. **Finding Positive Numbers** Find three positive integers, x, y, and z, such that the product is 64 and the sum is a minimum.

87. **Maximum Revenue** A company manufactures two types of bicycles, a racing bicycle and a mountain bicycle. The total revenue (in thousands of dollars) from x_1 units of racing bicycles and x_2 units of mountain bicycles is

$$R = -6x_1^2 - 10x_2^2 - 2x_1 x_2 + 32x_1 + 84x_2$$

where x_1 and x_2 are in thousands of units. Find x_1 and x_2 so as to maximize the revenue.

88. **Maximum Profit** A corporation manufactures digital cameras at two locations. The cost of producing x_1 units at location 1 is $C_1 = 0.05x_1^2 + 15x_1 + 5400$ and the cost of producing x_2 units at location 2 is $C_2 = 0.03x_2^2 + 15x_2 + 6100$. The digital cameras sell for \$180 per unit. Find the quantity that should be produced at each location to maximize the profit $P = 180(x_1 + x_2) - C_1 - C_2$.

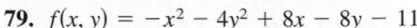 **Finding the Least Squares Regression Line** In Exercises 89 and 90, find the least squares regression line for the points. Use the regression capabilities of a graphing utility to verify your results. Use the graphing utility to plot the points and graph the regression line.

89. $(0, 4)$, $(1, 5)$, $(3, 6)$, $(6, 8)$, $(8, 10)$

90. $(0, 10)$, $(2, 8)$, $(4, 7)$, $(7, 5)$, $(9, 3)$, $(12, 0)$

91. **Modeling Data** An agronomist used four test plots to determine the relationship between the wheat yield y (in bushels per acre) and the amount of fertilizer x (in pounds per acre). The results are shown in the table.

Fertilizer, x	100	150	200	250
Yield, y	35	44	50	56

(a) Use the regression capabilities of a graphing utility to find the least squares regression line for the data.

(b) Use the model to approximate the wheat yield for a fertilizer application of 175 pounds per acre.

92. **Modeling Data** The table shows the yield y (in milligrams) of a chemical reaction after t minutes.

Minutes, t	1	2	3	4
Yield, y	1.2	7.1	9.9	13.1

Minutes, t	5	6	7	8
Yield, y	15.5	16.0	17.9	18.0

(a) Use the regression capabilities of a graphing utility to find the least squares regression line for the data. Then use the graphing utility to plot the data and graph the model.

(b) Use a graphing utility to plot the points $(\ln t, y)$. Do these points appear to follow a linear pattern more closely than the plot of the given data in part (a)?

(c) Use the regression capabilities of a graphing utility to find the least squares regression line for the points $(\ln t, y)$ and obtain the logarithmic model $y = a + b \ln t$.

(d) Use a graphing utility to plot the original data and graph the linear and logarithmic models. Which is a better model? Explain.

Using Lagrange Multipliers In Exercises 93–98, use Lagrange multipliers to find the indicated extrema, assuming that x and y are positive.

93. Minimize $f(x, y) = x^2 + y^2$

Constraint: $x + y - 8 = 0$

94. Maximize $f(x, y) = xy$

Constraint: $x + 3y - 6 = 0$

95. Maximize $f(x, y) = 2x + 3xy + y$

Constraint: $x + 2y = 29$

96. Minimize $f(x, y) = x^2 - y^2$

Constraint: $x - 2y + 6 = 0$

97. Maximize $f(x, y) = 2xy$

Constraint: $2x + y = 12$

98. Minimize $f(x, y) = 3x^2 - y^2$

Constraint: $2x - 2y + 5 = 0$

99. **Minimum Cost** A water line is to be built from point P to point S and must pass through regions where construction costs differ (see figure). The cost per kilometer (in dollars) is $3k$ from P to Q, $2k$ from Q to R, and k from R to S. For simplicity, let $k = 1$. Use Lagrange multipliers to find x, y, and z such that the total cost C will be minimized.

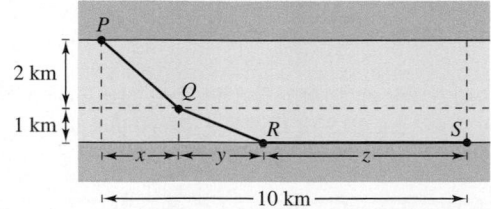

P.S. Problem Solving

See **CalcChat.com** for tutorial help and worked-out solutions to odd-numbered exercises.

1. **Area**   **Heron's Formula** states that the area of a triangle with sides of lengths a, b, and c is given by

$$A = \sqrt{s(s-a)(s-b)(s-c)}$$

where $s = \dfrac{a+b+c}{2}$, as shown in the figure.

(a) Use Heron's Formula to find the area of the triangle with vertices $(0, 0)$, $(3, 4)$, and $(6, 0)$.

(b) Show that among all triangles having a fixed perimeter, the triangle with the largest area is an equilateral triangle.

(c) Show that among all triangles having a fixed area, the triangle with the smallest perimeter is an equilateral triangle.

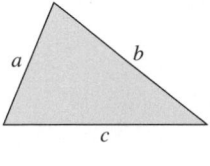
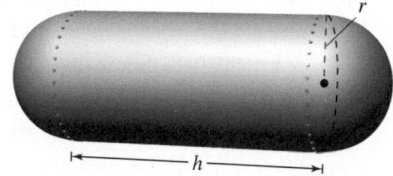

Figure for 1   Figure for 2

2. **Minimizing Material**   An industrial container is in the shape of a cylinder with hemispherical ends, as shown in the figure. The container must hold 1000 liters of fluid. Determine the radius r and length h that minimize the amount of material used in the construction of the tank.

3. **Tangent Plane**   Let $P(x_0, y_0, z_0)$ be a point in the first octant on the surface $xyz = 1$, as shown in the figure.

(a) Find the equation of the tangent plane to the surface at the point P.

(b) Show that the volume of the tetrahedron formed by the three coordinate planes and the tangent plane is constant, independent of the point of tangency (see figure).

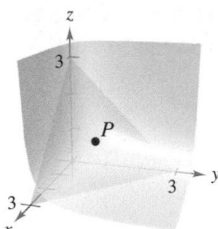

4. **Using Functions**   Use a graphing utility to graph the functions

$$f(x) = \sqrt[3]{x^3 - 1} \quad \text{and} \quad g(x) = x$$

in the same viewing window.

(a) Show that

$$\lim_{x \to \infty} [f(x) - g(x)] = 0 \quad \text{and} \quad \lim_{x \to -\infty} [f(x) - g(x)] = 0.$$

(b) Find the point on the graph of f that is farthest from the graph of g.

5. **Finding Maximum and Minimum Values**

(a) Let $f(x, y) = x - y$ and $g(x, y) = x^2 + y^2 = 4$. Graph various level curves of f and the constraint g in the xy-plane. Use the graph to determine the maximum value of f subject to the constraint $g = 4$. Then verify your answer using Lagrange multipliers.

(b) Let $f(x, y) = x - y$ and $g(x, y) = x^2 + y^2 = 0$. Find the maximum and minimum values of f subject to the constraint $g = 0$. Does the Method of Lagrange Multipliers work in this case? Explain.

6. **Minimizing Costs**   A heated storage room has the shape of a rectangular prism and has a volume of 1000 cubic feet, as shown in the figure. Because warm air rises, the heat loss per unit of area through the ceiling is five times as great as the heat loss through the floor. The heat loss through the four walls is three times as great as the heat loss through the floor. Determine the room dimensions that will minimize heat loss and therefore minimize heating costs.

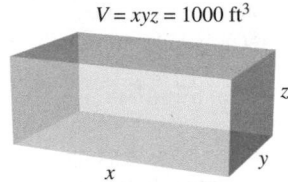

$$V = xyz = 1000 \text{ ft}^3$$

7. **Minimizing Costs**   Repeat Exercise 6 assuming that the heat loss through the walls and ceiling remain the same, but the floor is insulated so that there is no heat loss through the floor.

8. **Temperature**   Consider a circular plate of radius 1 given by $x^2 + y^2 \le 1$, as shown in the figure. The temperature at any point $P(x, y)$ on the plate is $T(x, y) = 2x^2 + y^2 - y + 10$.

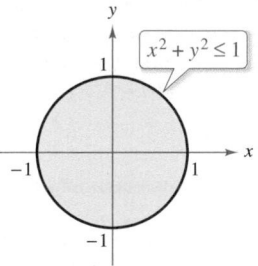

$$x^2 + y^2 \le 1$$

(a) Sketch the isotherm $T(x, y) = 10$. To print an enlarged copy of the graph, go to *MathGraphs.com*.

(b) Find the hottest and coldest points on the plate.

9. **Cobb-Douglas Production Function**   Consider the Cobb-Douglas production function

$$f(x, y) = Cx^a y^{1-a}, \quad 0 < a < 1.$$

(a) Show that f satisfies the equation $x\dfrac{\partial f}{\partial x} + y\dfrac{\partial f}{\partial y} = f.$

(b) Show that $f(tx, ty) = tf(x, y)$.

10. Minimizing Area   Consider the ellipse

$$\frac{x^2}{a^2} + \frac{y^2}{b^2} = 1$$

that encloses the circle $x^2 + y^2 = 2x$. Find values of a and b that minimize the area of the ellipse.

11. Projectile Motion   A projectile is launched at an angle of 45° with the horizontal and with an initial velocity of 64 feet per second. A television camera is located in the plane of the path of the projectile 50 feet behind the launch site (see figure).

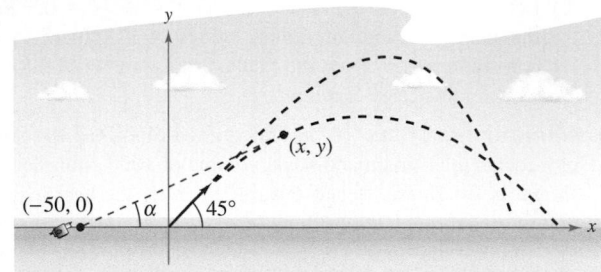

(a) Find parametric equations for the path of the projectile in terms of the parameter t representing time.

(b) Write the angle α that the camera makes with the horizontal in terms of x and y and in terms of t.

(c) Use the results of part (b) to find $\dfrac{d\alpha}{dt}$.

(d) Use a graphing utility to graph α in terms of t. Is the graph symmetric to the axis of the parabolic arch of the projectile? At what time is the rate of change of α greatest?

(e) At what time is the angle α maximum? Does this occur when the projectile is at its greatest height?

12. Distance   Consider the distance d between the launch site and the projectile in Exercise 11.

(a) Write the distance d in terms of x and y and in terms of the parameter t.

(b) Use the results of part (a) to find the rate of change of d.

(c) Find the rate of change of the distance when $t = 2$.

(d) When is the rate of change of d minimum during the flight of the projectile? Does this occur at the time when the projectile reaches its maximum height?

13. Finding Extrema and Saddle Points Using Technology   Consider the function

$$f(x, y) = (\alpha x^2 + \beta y^2)e^{-(x^2+y^2)}, \quad 0 < |\alpha| < \beta.$$

(a) Use a computer algebra system to graph the function for $\alpha = 1$ and $\beta = 2$, and identify any extrema or saddle points.

(b) Use a computer algebra system to graph the function for $\alpha = -1$ and $\beta = 2$, and identify any extrema or saddle points.

(c) Generalize the results in parts (a) and (b) for the function f.

14. Proof   Prove that if f is a differentiable function such that $\nabla f(x_0, y_0) = \mathbf{0}$, then the tangent plane at (x_0, y_0) is horizontal.

15. Area   The figure shows a rectangle that is approximately $l = 6$ centimeters long and $h = 1$ centimeter high.

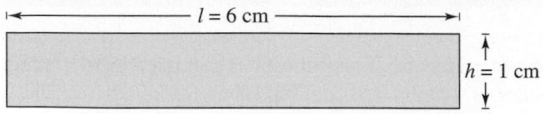

(a) Draw a rectangular strip along the rectangular region showing a small increase in length.

(b) Draw a rectangular strip along the rectangular region showing a small increase in height.

(c) Use the results in parts (a) and (b) to identify the measurement that has more effect on the area A of the rectangle.

(d) Verify your answer in part (c) analytically by comparing the value of dA when $dl = 0.01$ and when $dh = 0.01$.

16. Tangent Planes   Let f be a differentiable function of one variable. Show that all tangent planes to the surface $z = yf(x/y)$ intersect in a common point.

17. Wave Equation   Show that

$$u(x, t) = \frac{1}{2}[\sin(x - t) + \sin(x + t)]$$

is a solution to the one-dimensional wave equation

$$\frac{\partial^2 u}{\partial t^2} = \frac{\partial^2 u}{\partial x^2}.$$

18. Wave Equation   Show that

$$u(x, t) = \frac{1}{2}[f(x - ct) + f(x + ct)]$$

is a solution to the one-dimensional wave equation

$$\frac{\partial^2 u}{\partial t^2} = c^2\frac{\partial^2 u}{\partial x^2}.$$

(This equation describes the small transverse vibration of an elastic string such as those on certain musical instruments.)

19. Verifying Equations   Consider the function $w = f(x, y)$, where $x = r\cos\theta$ and $y = r\sin\theta$. Verify each of the following.

(a) $\dfrac{\partial w}{\partial x} = \dfrac{\partial w}{\partial r}\cos\theta - \dfrac{\partial w}{\partial \theta}\dfrac{\sin\theta}{r}$

   $\dfrac{\partial w}{\partial y} = \dfrac{\partial w}{\partial r}\sin\theta + \dfrac{\partial w}{\partial \theta}\dfrac{\cos\theta}{r}$

(b) $\left(\dfrac{\partial w}{\partial x}\right)^2 + \left(\dfrac{\partial w}{\partial y}\right)^2 = \left(\dfrac{\partial w}{\partial r}\right)^2 + \left(\dfrac{1}{r^2}\right)\left(\dfrac{\partial w}{\partial \theta}\right)^2$

20. Using a Function   Demonstrate the result of Exercise 19(b) for

$$w = \arctan\frac{y}{x}.$$

21. Laplace's Equation   Rewrite Laplace's equation

$$\frac{\partial^2 u}{\partial x^2} + \frac{\partial^2 u}{\partial y^2} + \frac{\partial^2 u}{\partial z^2} = 0$$

in cylindrical coordinates.

14 Multiple Integration

Modeling Data *(Exercise 36, p. 1012)*

Center of Pressure on a Sail
(Section Project, p. 1005)

Glacier *(Exercise 58, p. 997)*

Population
(Exercise 55, p. 996)

Average Production *(Exercise 57, p. 988)*

14.1 Iterated Integrals and Area in the Plane

■ Evaluate an iterated integral.
■ Use an iterated integral to find the area of a plane region.

Iterated Integrals

In Chapters 14 and 15, you will study several applications of integration involving functions of several variables. Chapter 14 is like Chapter 7 in that it surveys the use of integration to find plane areas, volumes, surface areas, moments, and centers of mass.

In Chapter 13, you saw that it is meaningful to differentiate functions of several variables with respect to one variable while holding the other variables constant. You can *integrate* functions of several variables by a similar procedure. For example, consider the partial derivative $f_x(x, y) = 2xy$. By considering y constant, you can integrate with respect to x to obtain

$$f(x, y) = \int f_x(x, y)\, dx \qquad \text{Integrate with respect to } x.$$

$$= \int 2xy\, dx \qquad \text{Hold } y \text{ constant.}$$

$$= y \int 2x\, dx \qquad \text{Factor out constant } y.$$

$$= y(x^2) + C(y) \qquad \text{Antiderivative of } 2x \text{ is } x^2.$$

$$= x^2 y + C(y). \qquad C(y) \text{ is a function of } y.$$

The "constant" of integration, $C(y)$, is a function of y. In other words, by integrating with respect to x, you are able to recover $f(x, y)$ only partially. The total recovery of a function of x and y from its partial derivatives is a topic you will study in Chapter 15. For now, you will focus on extending definite integrals to functions of several variables. For instance, by considering y constant, you can apply the Fundamental Theorem of Calculus to evaluate

$$\int_1^{2y} 2xy\, dx = x^2 y \Big]_1^{2y} = (2y)^2 y - (1)^2 y = 4y^3 - y.$$

x is the variable of integration and y is fixed.   Replace x by the limits of integration.   The result is a function of y.

Similarly, you can integrate with respect to y by holding x fixed. Both procedures are summarized as follows.

$$\int_{h_1(y)}^{h_2(y)} f_x(x, y)\, dx = f(x, y) \Big]_{h_1(y)}^{h_2(y)} = f(h_2(y), y) - f(h_1(y), y) \qquad \text{With respect to } x$$

$$\int_{g_1(x)}^{g_2(x)} f_y(x, y)\, dy = f(x, y) \Big]_{g_1(x)}^{g_2(x)} = f(x, g_2(x)) - f(x, g_1(x)) \qquad \text{With respect to } y$$

Note that the variable of integration cannot appear in either limit of integration. For instance, it makes no sense to write

$$\int_0^x y\, dx.$$

EXAMPLE 1   **Integrating with Respect to _y_**

Evaluate $\displaystyle\int_{1}^{x} (2xy + 3y^2)\, dy$.

Solution   Considering x to be constant and integrating with respect to y, you have

$$\int_{1}^{x} (2xy + 3y^2)\, dy = \left[\, xy^2 + y^3 \,\right]_{1}^{x} \qquad \text{Integrate with respect to } y.$$
$$= (2x^3) - (x + 1)$$
$$= 2x^3 - x - 1.$$

▷

·· REMARK   Remember that you can check an antiderivative using differentiation. For instance, in Example 1, you can verify that

$$xy^2 + y^3$$

is the correct anitderivative by finding

$$\frac{\partial}{\partial y}[xy^2 + y^3].$$

Notice in Example 1 that the integral defines a function of x and can *itself* be integrated, as shown in the next example.

EXAMPLE 2   **The Integral of an Integral**

Evaluate $\displaystyle\int_{1}^{2} \left[\int_{1}^{x} (2xy + 3y^2)\, dy \right] dx$.

Solution   Using the result of Example 1, you have

$$\int_{1}^{2} \left[\int_{1}^{x} (2xy + 3y^2)\, dy \right] dx = \int_{1}^{2} (2x^3 - x - 1)\, dx$$
$$= \left[\frac{x^4}{2} - \frac{x^2}{2} - x \right]_{1}^{2} \qquad \text{Integrate with respect to } x.$$
$$= 4 - (-1)$$
$$= 5.$$

The integral in Example 2 is an **iterated integral.** The brackets used in Example 2 are normally not written. Instead, iterated integrals are usually written simply as

$$\int_{a}^{b} \int_{g_1(x)}^{g_2(x)} f(x, y)\, dy\, dx \quad \text{and} \quad \int_{c}^{d} \int_{h_1(y)}^{h_2(y)} f(x, y)\, dx\, dy.$$

The **inside limits of integration** can be variable with respect to the outer variable of integration. However, the **outside limits of integration** *must be* constant with respect to both variables of integration. After performing the inside integration, you obtain a "standard" definite integral, and the second integration produces a real number. The limits of integration for an iterated integral identify two sets of boundary intervals for the variables. For instance, in Example 2, the outside limits indicate that x lies in the interval $1 \le x \le 2$ and the inside limits indicate that y lies in the interval $1 \le y \le x$. Together, these two intervals determine the **region of integration R** of the iterated integral, as shown in Figure 14.1.

Because an iterated integral is just a special type of definite integral—one in which the integrand is also an integral—you can use the properties of definite integrals to evaluate iterated integrals.

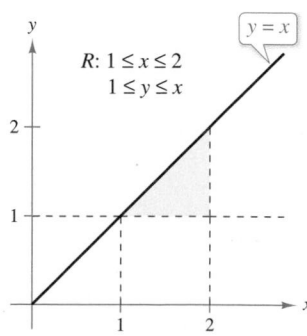

The region of integration for

$$\int_{1}^{2} \int_{1}^{x} f(x, y)\, dy\, dx$$

Figure 14.1

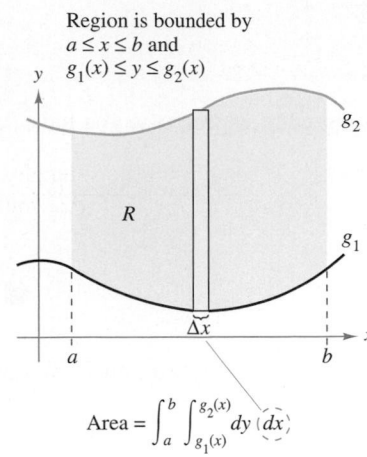

Region is bounded by
$a \le x \le b$ and
$g_1(x) \le y \le g_2(x)$

$$\text{Area} = \int_a^b \int_{g_1(x)}^{g_2(x)} dy \, (dx)$$

Vertically simple region
Figure 14.2

Area of a Plane Region

In the remainder of this section, you will take another look at the problem of finding the area of a plane region. Consider the plane region R bounded by $a \le x \le b$ and $g_1(x) \le y \le g_2(x)$, as shown in Figure 14.2. The area of R is

$$\int_a^b [g_2(x) - g_1(x)] \, dx. \qquad \text{Area of } R$$

Using the Fundamental Theorem of Calculus, you can rewrite the integrand $g_2(x) - g_1(x)$ as a definite integral. Specifically, consider x to be fixed and let y vary from $g_1(x)$ to $g_2(x)$, and you can write

$$\int_{g_1(x)}^{g_2(x)} dy = y \Big]_{g_1(x)}^{g_2(x)} = g_2(x) - g_1(x).$$

Combining these two integrals, you can write the area of the region R as an iterated integral

$$\int_a^b \int_{g_1(x)}^{g_2(x)} dy \, dx = \int_a^b y \Big]_{g_1(x)}^{g_2(x)} dx$$

$$= \int_a^b [g_2(x) - g_1(x)] \, dx. \qquad \text{Area of } R$$

Placing a representative rectangle in the region R helps determine both the order and the limits of integration. A vertical rectangle implies the order $dy \, dx$, with the inside limits of integration corresponding to the upper and lower bounds of the rectangle, as shown in Figure 14.2. This type of region is **vertically simple,** because the outside limits of integration represent the vertical lines

$$x = a \quad \text{and} \quad x = b.$$

Similarly, a horizontal rectangle implies the order $dx \, dy$, with the inside limits of integration determined by the left and right bounds of the rectangle, as shown in Figure 14.3. This type of region is **horizontally simple,** because the outside limits of integration represent the horizontal lines

$$y = c \quad \text{and} \quad y = d.$$

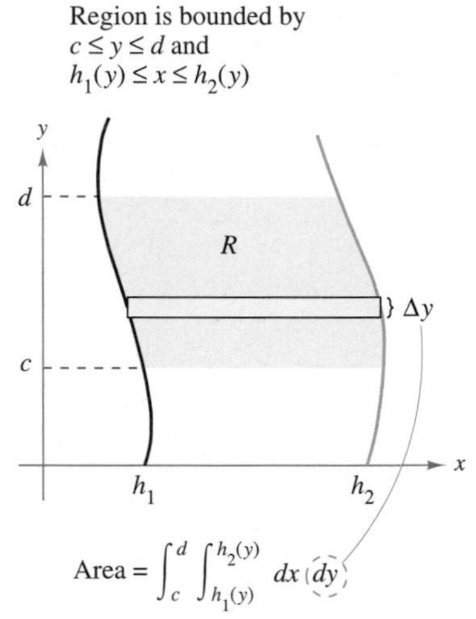

The iterated integrals used for these two types of simple regions are summarized as follows.

Region is bounded by
$c \le y \le d$ and
$h_1(y) \le x \le h_2(y)$

$$\text{Area} = \int_c^d \int_{h_1(y)}^{h_2(y)} dx \, (dy)$$

Horizontally simple region
Figure 14.3

⊳

REMARK Be sure you see that the orders of integration of these two integrals are different—the order $dy \, dx$ corresponds to a vertically simple region, and the order $dx \, dy$ corresponds to a horizontally simple region.

Area of a Region in the Plane

1. If R is defined by $a \le x \le b$ and $g_1(x) \le y \le g_2(x)$, where g_1 and g_2 are continuous on $[a, b]$, then the area of R is

$$A = \int_a^b \int_{g_1(x)}^{g_2(x)} dy \, dx. \qquad \text{Figure 14.2 (vertically simple)}$$

2. If R is defined by $c \le y \le d$ and $h_1(y) \le x \le h_2(y)$, where h_1 and h_2 are continuous on $[c, d]$, then the area of R is

$$A = \int_c^d \int_{h_1(y)}^{h_2(y)} dx \, dy. \qquad \text{Figure 14.3 (horizontally simple)}$$

If all four limits of integration happen to be constants, then the region of integration is rectangular, as shown in Example 3.

EXAMPLE 3 **The Area of a Rectangular Region**

Use an iterated integral to represent the area of the rectangle shown in Figure 14.4.

Solution The region shown in Figure 14.4 is both vertically simple and horizontally simple, so you can use either order of integration. By choosing the order $dy\,dx$, you obtain the following.

$$\int_a^b \int_c^d dy\,dx = \int_a^b \Big[y \Big]_c^d dx \qquad\qquad \text{Integrate with respect to } y.$$

$$= \int_a^b (d - c)\,dx$$

$$= \Big[(d - c)x \Big]_a^b \qquad\qquad \text{Integrate with respect to } x.$$

$$= (d - c)(b - a)$$

Notice that this answer is consistent with what you know from geometry.

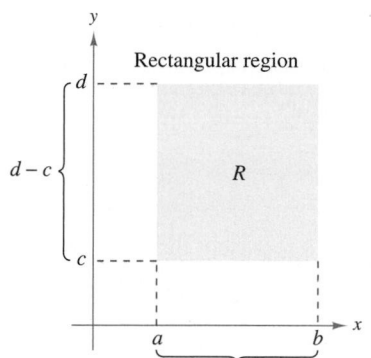

Figure 14.4

EXAMPLE 4 **Finding Area by an Iterated Integral**

Use an iterated integral to find the area of the region bounded by the graphs of

$$f(x) = \sin x \qquad\qquad\qquad \text{Sine curve forms upper boundary.}$$

and

$$g(x) = \cos x \qquad\qquad\qquad \text{Cosine curve forms lower boundary.}$$

between $x = \pi/4$ and $x = 5\pi/4$.

Solution Because f and g are given as functions of x, a vertical representative rectangle is convenient, and you can choose $dy\,dx$ as the order of integration, as shown in Figure 14.5. The outside limits of integration are

$$\frac{\pi}{4} \le x \le \frac{5\pi}{4}.$$

Moreover, because the rectangle is bounded above by $f(x) = \sin x$ and below by $g(x) = \cos x$, you have

$$\text{Area of } R = \int_{\pi/4}^{5\pi/4} \int_{\cos x}^{\sin x} dy\,dx$$

$$= \int_{\pi/4}^{5\pi/4} \Big[y \Big]_{\cos x}^{\sin x} dx \qquad\qquad \text{Integrate with respect to } y.$$

$$= \int_{\pi/4}^{5\pi/4} (\sin x - \cos x)\,dx$$

$$= \Big[-\cos x - \sin x \Big]_{\pi/4}^{5\pi/4} \qquad\qquad \text{Integrate with respect to } x.$$

$$= 2\sqrt{2}.$$

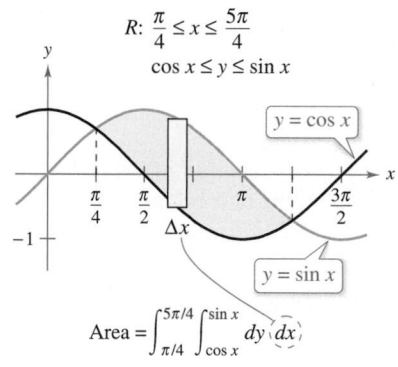

Figure 14.5

The region of integration of an iterated integral need not have any straight lines as boundaries. For instance, the region of integration shown in Figure 14.5 is *vertically simple* even though it has no vertical lines as left and right boundaries. The quality that makes the region vertically simple is that it is bounded above and below by the graphs of *functions of x.*

One order of integration will often produce a simpler integration problem than the other order. For instance, try reworking Example 4 with the order $dx\ dy$. You may be surprised to see that the task is formidable. However, if you succeed, you will see that the answer is the same. In other words, the order of integration affects the ease of integration but not the value of the integral.

EXAMPLE 5   Comparing Different Orders of Integration

⋮•••▷ *See LarsonCalculus.com for an interactive version of this type of example.*

Sketch the region whose area is represented by the integral

$$\int_0^2 \int_{y^2}^4 dx\ dy.$$

Then find another iterated integral using the order $dy\ dx$ to represent the same area and show that both integrals yield the same value.

Solution   From the given limits of integration, you know that

$$y^2 \le x \le 4 \qquad \text{Inner limits of integration}$$

which means that the region R is bounded on the left by the parabola $x = y^2$ and on the right by the line $x = 4$. Furthermore, because

$$0 \le y \le 2 \qquad \text{Outer limits of integration}$$

you know that R is bounded below by the x-axis, as shown in Figure 14.6(a). The value of this integral is

$$\int_0^2 \int_{y^2}^4 dx\ dy = \int_0^2 x\Big]_{y^2}^4 dy \qquad \text{Integrate with respect to } x.$$

$$= \int_0^2 (4 - y^2)\ dy$$

$$= \left[4y - \frac{y^3}{3}\right]_0^2 \qquad \text{Integrate with respect to } y.$$

$$= \frac{16}{3}.$$

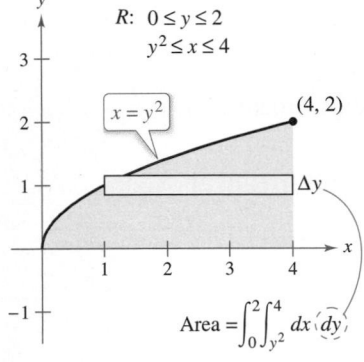

R: $0 \le y \le 2$
$y^2 \le x \le 4$

$x = y^2$

$(4, 2)$

Δy

Area $= \int_0^2 \int_{y^2}^4 dx\ dy$

(a)

To change the order of integration to $dy\ dx$, place a vertical rectangle in the region, as shown in Figure 14.6(b). From this, you can see that the constant bounds $0 \le x \le 4$ serve as the outer limits of integration. By solving for y in the equation $x = y^2$, you can conclude that the inner bounds are $0 \le y \le \sqrt{x}$. So, the area of the region can also be represented by

$$\int_0^4 \int_0^{\sqrt{x}} dy\ dx.$$

By evaluating this integral, you can see that it has the same value as the original integral.

$$\int_0^4 \int_0^{\sqrt{x}} dy\ dx = \int_0^4 y\Big]_0^{\sqrt{x}} dx \qquad \text{Integrate with respect to } y.$$

$$= \int_0^4 \sqrt{x}\ dx$$

$$= \frac{2}{3} x^{3/2}\Big]_0^4 \qquad \text{Integrate with respect to } x.$$

$$= \frac{16}{3}$$

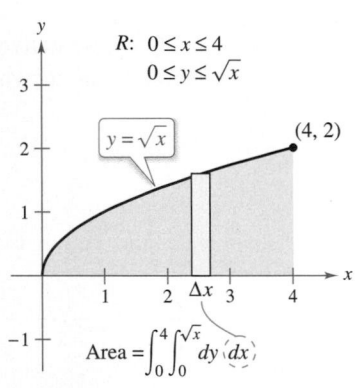

R: $0 \le x \le 4$
$0 \le y \le \sqrt{x}$

$y = \sqrt{x}$

$(4, 2)$

Δx

Area $= \int_0^4 \int_0^{\sqrt{x}} dy\ dx$

(b)

Figure 14.6

Sometimes it is not possible to calculate the area of a region with a single iterated integral. In these cases, you can divide the region into subregions such that the area of each subregion can be calculated by an iterated integral. The total area is then the sum of the iterated integrals.

EXAMPLE 6 An Area Represented by Two Iterated Integrals

Find the area of the region R that lies below the parabola

$$y = 4x - x^2 \qquad \text{Parabola forms upper boundary.}$$

above the x-axis, and above the line

$$y = -3x + 6. \qquad \text{Line and } x\text{-axis form lower boundary.}$$

Solution Begin by dividing R into the two subregions R_1 and R_2 shown in Figure 14.7.

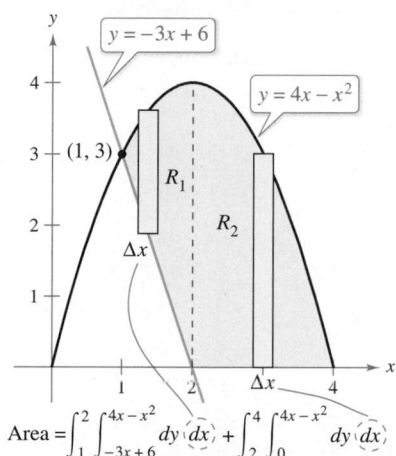

$$\text{Area} = \int_1^2 \int_{-3x+6}^{4x-x^2} dy \, dx + \int_2^4 \int_0^{4x-x^2} dy \, dx$$

Figure 14.7

▷ •• REMARK In Examples 3 through 6, be sure you see the benefit of sketching the region of integration. You should develop the habit of making sketches to help you determine the limits of integration for all iterated integrals in this chapter.

In both regions, it is convenient to use vertical rectangles, and you have

$$\begin{aligned}
\text{Area} &= \int_1^2 \int_{-3x+6}^{4x-x^2} dy \, dx + \int_2^4 \int_0^{4x-x^2} dy \, dx \\
&= \int_1^2 (4x - x^2 + 3x - 6) \, dx + \int_2^4 (4x - x^2) \, dx \\
&= \left[\frac{7x^2}{2} - \frac{x^3}{3} - 6x \right]_1^2 + \left[2x^2 - \frac{x^3}{3} \right]_2^4 \\
&= \left(14 - \frac{8}{3} - 12 - \frac{7}{2} + \frac{1}{3} + 6 \right) + \left(32 - \frac{64}{3} - 8 + \frac{8}{3} \right) \\
&= \frac{15}{2}.
\end{aligned}$$

The area of the region is $15/2$ square units. Try checking this using the procedure for finding the area between two curves, as presented in Section 7.1. ■

At this point, you may be wondering why you would need iterated integrals. After all, you already know how to use conventional integration to find the area of a region in the plane. (For instance, compare the solution to Example 4 in this section with that given in Example 3 in Section 7.1.) The need for iterated integrals will become clear in the next section. In this section, primary attention is given to procedures for finding the limits of integration of the region of an iterated integral, and the following exercise set is designed to develop skill in this important procedure.

14.1 Exercises

CONCEPT CHECK

1. Iterated Integral Explain what is meant by an iterated integral. How is it evaluated?

2. Region of Integration Sketch the region of integration for the iterated integral.

$$\int_1^2 \int_0^{2-x} f(x, y)\, dy\, dx.$$

 **Evaluating an Integral** In Exercises 3–10, evaluate the integral.

3. $\displaystyle\int_0^x (2x - y)\, dy$

4. $\displaystyle\int_x^{x^2} \frac{y}{x}\, dy$

5. $\displaystyle\int_0^{\sqrt{4-x^2}} x^2 y\, dy$

6. $\displaystyle\int_{x^3}^{\sqrt{x}} (x^2 + 3y^2)\, dy$

7. $\displaystyle\int_{e^y}^y \frac{y \ln x}{x}\, dx, \quad y > 0$

8. $\displaystyle\int_{-\sqrt{1-y^2}}^{\sqrt{1-y^2}} (x^2 + y^2)\, dx$

9. $\displaystyle\int_0^{x^3} y e^{-y/x}\, dy$

10. $\displaystyle\int_y^{\pi/2} \sin^3 x \cos y\, dx$

 **Evaluating an Iterated Integral** In Exercises 11–28, evaluate the iterated integral.

11. $\displaystyle\int_0^1 \int_0^2 (x + y)\, dy\, dx$

12. $\displaystyle\int_{-1}^1 \int_{-2}^2 (x^2 - y^2)\, dy\, dx$

13. $\displaystyle\int_0^{\pi/4} \int_0^1 y \cos x\, dy\, dx$

14. $\displaystyle\int_0^{\ln 4} \int_0^{\ln 3} e^{x+y}\, dy\, dx$

15. $\displaystyle\int_0^2 \int_0^{6x} x^3\, dy\, dx$

16. $\displaystyle\int_0^1 \int_0^y (6x + 5y^3)\, dx\, dy$

17. $\displaystyle\int_0^{\pi/2} \int_0^{\cos x} (1 + \sin x)\, dy\, dx$

18. $\displaystyle\int_1^4 \int_1^{\sqrt{x}} 2y e^{-x}\, dy\, dx$

19. $\displaystyle\int_0^1 \int_0^x \sqrt{1 - x^2}\, dy\, dx$

20. $\displaystyle\int_{-4}^4 \int_0^{x^2} \sqrt{64 - x^3}\, dy\, dx$

21. $\displaystyle\int_0^1 \int_0^{\sqrt{1-y^2}} (x + y)\, dx\, dy$

22. $\displaystyle\int_0^2 \int_{3y^2-6y}^{2y-y^2} 3y\, dx\, dy$

23. $\displaystyle\int_0^2 \int_0^{\sqrt{4-y^2}} \frac{2}{\sqrt{4 - y^2}}\, dx\, dy$

24. $\displaystyle\int_1^3 \int_0^y \frac{4}{x^2 + y^2}\, dx\, dy$

25. $\displaystyle\int_0^{\pi/2} \int_0^{2\cos\theta} r\, dr\, d\theta$

26. $\displaystyle\int_0^{\pi/4} \int_{\sqrt{3}}^{\sqrt{3}\cos\theta} r\, dr\, d\theta$

27. $\displaystyle\int_0^{\pi/2} \int_0^{\sin\theta} \theta r\, dr\, d\theta$

28. $\displaystyle\int_0^{\pi/4} \int_0^{\cos\theta} 3r^2 \sin\theta\, dr\, d\theta$

Evaluating an Improper Iterated Integral In Exercises 29–32, evaluate the improper iterated integral.

29. $\displaystyle\int_1^\infty \int_0^{1/x} y\, dy\, dx$

30. $\displaystyle\int_0^3 \int_0^\infty \frac{x^2}{1 + y^2}\, dy\, dx$

31. $\displaystyle\int_1^\infty \int_1^\infty \frac{1}{xy}\, dx\, dy$

32. $\displaystyle\int_0^\infty \int_0^\infty xy e^{-(x^2+y^2)}\, dx\, dy$

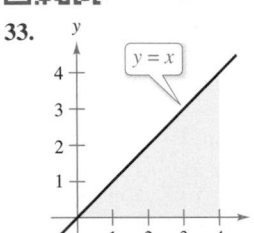 **Finding the Area of a Region** In Exercises 33–36, use an iterated integral to find the area of the region.

33.
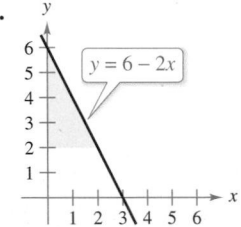

34.

35.
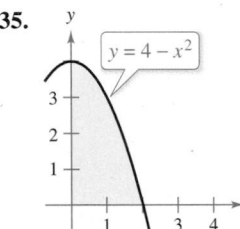

36.
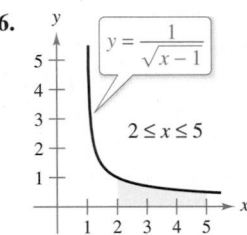

Finding the Area of a Region In Exercises 37–42, use an iterated integral to find the area of the region bounded by the graphs of the equations.

37. $y = 9 - x^2, \quad y = 0$

38. $2x - 3y = 0, \quad x + y = 5, \quad y = 0$

39. $\sqrt{x} + \sqrt{y} = 2, \quad x = 0, \quad y = 0$

40. $y = x^{3/2}, \quad y = 2x$

41. $y = 4 - x^2, \quad y = x + 2$

42. $y = x, \quad y = 2x, \quad x = 2$

Changing the Order of Integration In Exercises 43–50, sketch the region R of integration and change the order of integration.

43. $\displaystyle\int_0^4 \int_0^y f(x, y)\, dx\, dy$

44. $\displaystyle\int_0^4 \int_{\sqrt{y}}^2 f(x, y)\, dx\, dy$

45. $\displaystyle\int_{-2}^2 \int_0^{\sqrt{4-x^2}} f(x, y)\, dy\, dx$

46. $\displaystyle\int_0^2 \int_0^{4-x^2} f(x, y)\, dy\, dx$

47. $\displaystyle\int_1^{10} \int_0^{\ln y} f(x, y)\, dx\, dy$

48. $\displaystyle\int_{-1}^2 \int_0^{e^{-x}} f(x, y)\, dy\, dx$

49. $\displaystyle\int_{-1}^1 \int_{x^2}^1 f(x, y)\, dy\, dx$

50. $\displaystyle\int_{-\pi/2}^{\pi/2} \int_0^{\cos x} f(x, y)\, dy\, dx$

Changing the Order of Integration In Exercises 51–60, sketch the region R whose area is given by the iterated integral. Then change the order of integration and show that both orders yield the same area.

51. $\displaystyle\int_0^1 \int_0^2 dy\, dx$

52. $\displaystyle\int_1^2 \int_2^4 dx\, dy$

53. $\displaystyle\int_0^1 \int_{2y}^2 dx\, dy$

54. $\displaystyle\int_0^9 \int_{\sqrt{x}}^3 dy\, dx$

55. $\displaystyle\int_0^1 \int_{-\sqrt{1-y^2}}^{\sqrt{1-y^2}} dx\, dy$

56. $\displaystyle\int_{-2}^2 \int_{-\sqrt{4-x^2}}^{\sqrt{4-x^2}} dy\, dx$

57. $\displaystyle\int_0^2 \int_0^x dy\, dx + \int_2^4 \int_0^{4-x} dy\, dx$

58. $\displaystyle\int_0^4 \int_0^{x/2} dy\, dx + \int_4^6 \int_0^{6-x} dy\, dx$

59. $\displaystyle\int_0^1 \int_{y^2}^{\sqrt[3]{y}} dx\, dy$

60. $\displaystyle\int_{-2}^2 \int_0^{4-y^2} dx\, dy$

Changing the Order of Integration In Exercises 61–66, sketch the region of integration. Then evaluate the iterated integral. (*Hint:* Note that it is necessary to change the order of integration.)

61. $\displaystyle\int_0^2 \int_x^2 x\sqrt{1+y^3}\, dy\, dx$

62. $\displaystyle\int_0^4 \int_{\sqrt{x}}^2 \frac{3}{2+y^3}\, dy\, dx$

63. $\displaystyle\int_0^1 \int_{2x}^2 4e^{y^2}\, dy\, dx$

64. $\displaystyle\int_0^2 \int_x^2 e^{-y^2}\, dy\, dx$

65. $\displaystyle\int_0^1 \int_y^1 \sin x^2\, dx\, dy$

66. $\displaystyle\int_0^2 \int_{y^2}^4 \sqrt{x} \sin x\, dx\, dy$

EXPLORING CONCEPTS

67. Area of a Circle Write an iterated integral that represents the area of a circle of radius 5 centered at the origin. Verify that your integral produces the correct area.

68. Using Different Methods Express the area of the region bounded by $x = \sqrt{4-4y^2}$, $y = 1$, and $x = 2$ in at least two different ways, one of which is an iterated integral. Do not find the area of the region.

69. Think About It Determine whether each expression represents the area of the shaded region (see figure).

(a) $\displaystyle\int_0^5 \int_y^{\sqrt{50-y^2}} dy\, dx$

(b) $\displaystyle\int_0^5 \int_x^{\sqrt{50-x^2}} dy\, dx$

(c) $\displaystyle\int_0^5 \int_0^y dx\, dy + \int_5^{5\sqrt{2}} \int_0^{\sqrt{50-y^2}} dx\, dy$

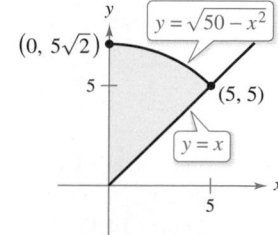

70. **HOW DO YOU SEE IT?** Use each order of integration to write an iterated integral that represents the area of the region R (see figure).

(a) Area $= \displaystyle\iint dx\, dy$

(b) Area $= \displaystyle\iint dy\, dx$

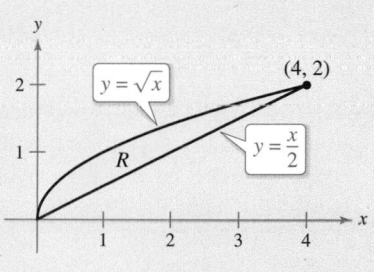

Evaluating an Iterated Integral Using Technology In Exercises 71–76, use a computer algebra system to evaluate the iterated integral.

71. $\displaystyle\int_0^1 \int_y^{2y} \sin(x+y)\, dx\, dy$

72. $\displaystyle\int_0^2 \int_0^{4-x^2} e^{xy}\, dy\, dx$

73. $\displaystyle\int_0^4 \int_0^y \frac{2}{(x+1)(y+1)}\, dx\, dy$

74. $\displaystyle\int_0^2 \int_x^2 \sqrt{16 - x^3 - y^3}\, dy\, dx$

75. $\displaystyle\int_0^{2\pi} \int_0^{1+\cos\theta} 6r^2 \cos\theta\, dr\, d\theta$

76. $\displaystyle\int_0^{\pi/2} \int_0^{1+\sin\theta} 15\theta r\, dr\, d\theta$

Comparing Different Orders of Integration Using Technology In Exercises 77 and 78, (a) sketch the region of integration, (b) change the order of integration, and (c) use a computer algebra system to show that both orders yield the same value.

77. $\displaystyle\int_0^2 \int_{y^3}^{4\sqrt{2y}} (x^2 y - xy^2)\, dx\, dy$

78. $\displaystyle\int_0^2 \int_{\sqrt{4-x^2}}^{4-(x^2/4)} \frac{xy}{x^2+y^2+1}\, dy\, dx$

True or False? In Exercises 79 and 80, determine whether the statement is true or false. If it is false, explain why or give an example that shows it is false.

79. $\displaystyle\int_a^b \int_c^d f(x, y)\, dy\, dx = \int_c^d \int_a^b f(x, y)\, dx\, dy$

80. $\displaystyle\int_0^1 \int_0^x f(x, y)\, dy\, dx = \int_0^1 \int_0^y f(x, y)\, dx\, dy$

14.2 Double Integrals and Volume

■ Use a double integral to represent the volume of a solid region and use properties of double integrals.
■ Evaluate a double integral as an iterated integral.
■ Find the average value of a function over a region.

Double Integrals and Volume of a Solid Region

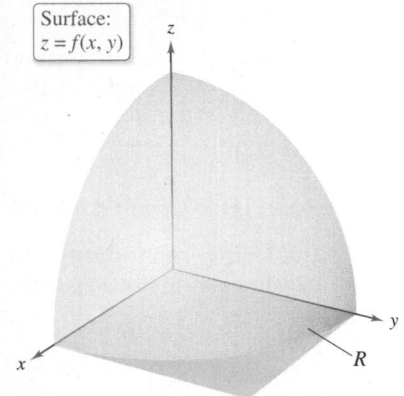

Surface:
$z = f(x, y)$

Figure 14.8

You already know that a definite integral over an *interval* uses a limit process to assign measures to quantities such as area, volume, arc length, and mass. In this section, you will use a similar process to define the **double integral** of a function of two variables over a *region in the plane*.

Consider a continuous function f such that $f(x, y) \geq 0$ for all (x, y) in a region R in the xy-plane. The goal is to find the volume of the solid region lying between the surface given by

$$z = f(x, y) \qquad \text{Surface lying above the } xy\text{-plane}$$

and the xy-plane, as shown in Figure 14.8. You can begin by superimposing a rectangular grid over the region, as shown in Figure 14.9. The rectangles lying entirely within R form an **inner partition** Δ, whose **norm** $\|\Delta\|$ is defined as the length of the longest diagonal of the n rectangles. Next, choose a point (x_i, y_i) in each rectangle and form the rectangular prism whose height is

$$f(x_i, y_i) \qquad \text{Height of } i\text{th prism}$$

as shown in Figure 14.10. Because the area of the ith rectangle is

$$\Delta A_i \qquad \text{Area of } i\text{th rectangle}$$

it follows that the volume of the ith prism is

$$f(x_i, y_i) \, \Delta A_i \qquad \text{Volume of } i\text{th prism}$$

and you can approximate the volume of the solid region by the Riemann sum of the volumes of all n prisms,

$$\sum_{i=1}^{n} f(x_i, y_i) \, \Delta A_i \qquad \text{Riemann sum}$$

as shown in Figure 14.11. This approximation can be improved by tightening the mesh of the grid to form smaller and smaller rectangles, as shown in Example 1.

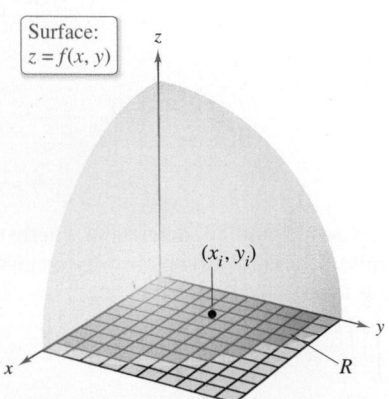

Surface:
$z = f(x, y)$

(x_i, y_i)

The rectangles lying within R form an inner partition of R.
Figure 14.9

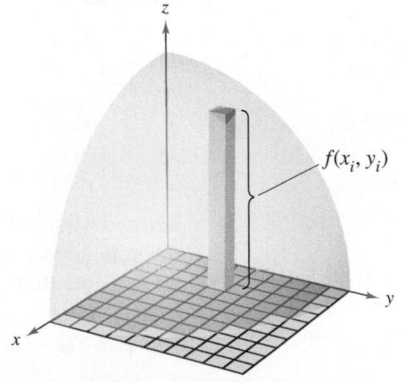

Rectangular prism whose base has an area of ΔA_i and whose height is $f(x_i, y_i)$
Figure 14.10

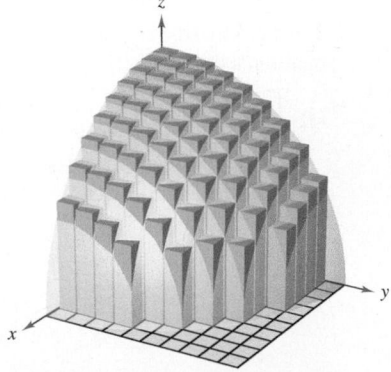

$f(x_i, y_i)$

Volume approximated by rectangular prisms
Figure 14.11

EXAMPLE 1   **Approximating the Volume of a Solid**

Approximate the volume of the solid lying between the paraboloid

$$f(x, y) = 1 - \frac{1}{2}x^2 - \frac{1}{2}y^2$$

and the square region R given by $0 \leq x \leq 1, 0 \leq y \leq 1$. Use a partition made up of squares whose sides have a length of $\frac{1}{4}$.

Solution   Begin by forming the specified partition of R. For this partition, it is convenient to choose the centers of the subregions as the points at which to evaluate $f(x, y)$.

$$\left(\tfrac{1}{8}, \tfrac{1}{8}\right) \quad \left(\tfrac{1}{8}, \tfrac{3}{8}\right) \quad \left(\tfrac{1}{8}, \tfrac{5}{8}\right) \quad \left(\tfrac{1}{8}, \tfrac{7}{8}\right)$$
$$\left(\tfrac{3}{8}, \tfrac{1}{8}\right) \quad \left(\tfrac{3}{8}, \tfrac{3}{8}\right) \quad \left(\tfrac{3}{8}, \tfrac{5}{8}\right) \quad \left(\tfrac{3}{8}, \tfrac{7}{8}\right)$$
$$\left(\tfrac{5}{8}, \tfrac{1}{8}\right) \quad \left(\tfrac{5}{8}, \tfrac{3}{8}\right) \quad \left(\tfrac{5}{8}, \tfrac{5}{8}\right) \quad \left(\tfrac{5}{8}, \tfrac{7}{8}\right)$$
$$\left(\tfrac{7}{8}, \tfrac{1}{8}\right) \quad \left(\tfrac{7}{8}, \tfrac{3}{8}\right) \quad \left(\tfrac{7}{8}, \tfrac{5}{8}\right) \quad \left(\tfrac{7}{8}, \tfrac{7}{8}\right)$$

Because the area of each square is $\Delta A_i = \frac{1}{16}$, you can approximate the volume by the sum

$$\sum_{i=1}^{16} f(x_i, y_i)\, \Delta A_i = \sum_{i=1}^{16} \left(1 - \frac{1}{2}x_i^2 - \frac{1}{2}y_i^2\right)\left(\frac{1}{16}\right) \approx 0.672.$$

This approximation is shown graphically in Figure 14.12. The exact volume of the solid is $\frac{2}{3}$ (see Example 2). You can obtain a better approximation by using a finer partition. For example, with a partition of squares with sides of length $\frac{1}{10}$, the approximation is 0.668.

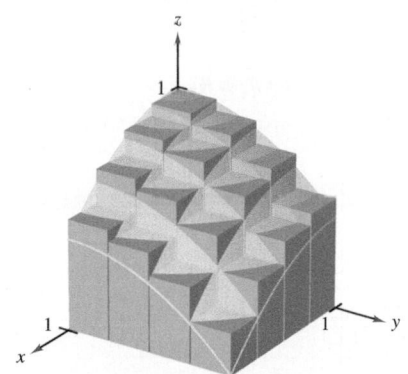

Surface:
$f(x, y) = 1 - \frac{1}{2}x^2 - \frac{1}{2}y^2$

Figure 14.12

▷ **TECHNOLOGY**   Some three-dimensional graphing utilities are capable of graphing figures such as that shown in Figure 14.12. For instance, the graph shown at the right was drawn with a computer program. In this graph, note that each of the rectangular prisms lies within the solid region.

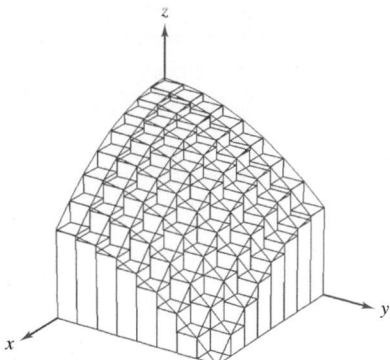

In Example 1, note that by using finer partitions, you obtain better approximations of the volume. This observation suggests that you could obtain the exact volume by taking a limit. That is,

$$\text{Volume} = \lim_{\|\Delta\| \to 0} \sum_{i=1}^{n} f(x_i, y_i)\, \Delta A_i.$$

The precise meaning of this limit is that the limit is equal to L if for every $\varepsilon > 0$, there exists a $\delta > 0$ such that

$$\left| L - \sum_{i=1}^{n} f(x_i, y_i)\, \Delta A_i \right| < \varepsilon$$

for all partitions Δ of the plane region R (that satisfy $\|\Delta\| < \delta$) and for all possible choices of x_i and y_i in the ith region.

Using the limit of a Riemann sum to define volume is a special case of using the limit to define a **double integral.** The general case, however, does not require that the function be positive or continuous.

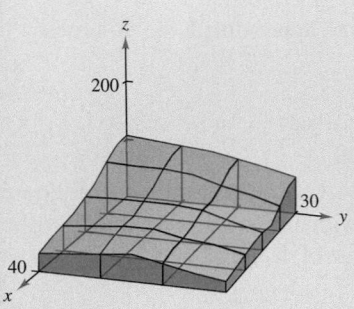
Definition of Double Integral

If f is defined on a closed, bounded region R in the xy-plane, then the **double integral of f over R** is

$$\int_R \int f(x, y)\, dA = \lim_{\|\Delta\| \to 0} \sum_{i=1}^{n} f(x_i, y_i)\, \Delta A_i$$

provided the limit exists. If the limit exists, then f is **integrable** over R.

Having defined a double integral, you will see that a definite integral is occasionally referred to as a **single integral**.

Sufficient conditions for the double integral of f on the region R to exist are that R can be written as a union of a finite number of nonoverlapping subregions (see Figure 14.13) that are vertically or horizontally simple *and* that f is continuous on the region R. This means that the intersection of two nonoverlapping regions is a set that has an area of 0. In Figure 14.13, the area of the line segment common to R_1 and R_2 is 0.

A double integral can be used to find the volume of a solid region that lies between the xy-plane and the surface given by $z = f(x, y)$.

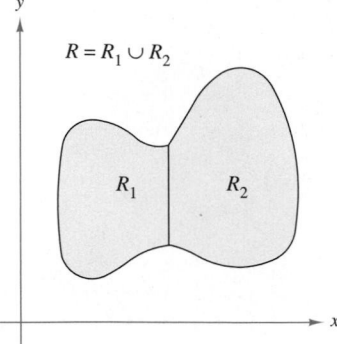

The two regions R_1 and R_2 are nonoverlapping.

Figure 14.13

Volume of a Solid Region

If f is integrable over a plane region R and $f(x, y) \geq 0$ for all (x, y) in R, then the volume of the solid region that lies above R and below the graph of f is

$$V = \int_R \int f(x, y)\, dA.$$

Double integrals share many properties of single integrals.

THEOREM 14.1 Properties of Double Integrals

Let f and g be continuous over a closed, bounded plane region R, and let c be a constant.

1. $\displaystyle \int_R \int cf(x, y)\, dA = c \int_R \int f(x, y)\, dA$

2. $\displaystyle \int_R \int [f(x, y) \pm g(x, y)]\, dA = \int_R \int f(x, y)\, dA \pm \int_R \int g(x, y)\, dA$

3. $\displaystyle \int_R \int f(x, y)\, dA \geq 0, \quad \text{if } f(x, y) \geq 0$

4. $\displaystyle \int_R \int f(x, y)\, dA \geq \int_R \int g(x, y)\, dA, \quad \text{if } f(x, y) \geq g(x, y)$

5. $\displaystyle \int_R \int f(x, y)\, dA = \int_{R_1} \int f(x, y)\, dA + \int_{R_2} \int f(x, y)\, dA$, where R is the union of two nonoverlapping subregions R_1 and R_2.

Evaluation of Double Integrals

Normally, the first step in evaluating a double integral is to rewrite it as an iterated integral. To show how this is done, a geometric model of a double integral is used as the volume of a solid.

Consider the solid region bounded by the plane $z = f(x, y) = 2 - x - 2y$ and the three coordinate planes, as shown in Figure 14.14. Each vertical cross section taken parallel to the yz-plane is a triangular region whose base has a length of $y = (2 - x)/2$ and whose height is $z = 2 - x$. This implies that for a fixed value of x, the area of the triangular cross section is

$$A(x) = \frac{1}{2}(\text{base})(\text{height}) = \frac{1}{2}\left(\frac{2 - x}{2}\right)(2 - x) = \frac{(2 - x)^2}{4}.$$

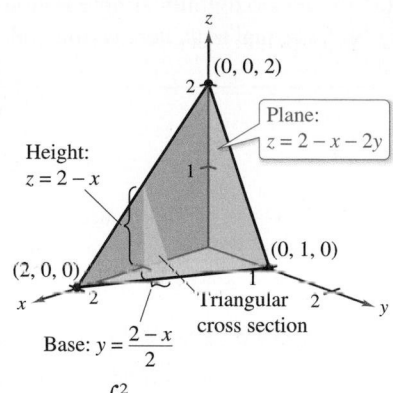

Height:
$z = 2 - x$

Plane:
$z = 2 - x - 2y$

(0, 0, 2)

(2, 0, 0)

(0, 1, 0)

Triangular cross section

Base: $y = \dfrac{2 - x}{2}$

Volume: $\displaystyle\int_0^2 A(x)\,dx$

Figure 14.14

By the formula for the volume of a solid with known cross sections (see Section 7.2), the volume of the solid is

$$
\begin{aligned}
\text{Volume} &= \int_a^b A(x)\,dx && \text{Formula for volume}\\[2mm]
&= \int_0^2 \frac{(2 - x)^2}{4}\,dx && \text{Substitute.}\\[2mm]
&= -\frac{(2 - x)^3}{12}\bigg]_0^2 && \text{Integrate with respect to } x.\\[2mm]
&= \frac{2}{3}. && \text{Volume of solid region (See Figure 14.14.)}
\end{aligned}
$$

This procedure works no matter how $A(x)$ is obtained. In particular, you can find $A(x)$ by integration, as shown in Figure 14.15. That is, you consider x to be constant and integrate $z = 2 - x - 2y$ from 0 to $(2 - x)/2$ to obtain

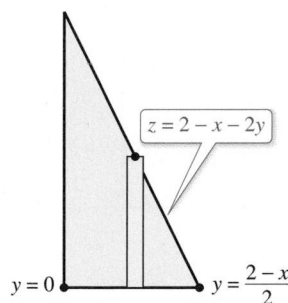

$z = 2 - x - 2y$

$y = 0$

$y = \dfrac{2 - x}{2}$

Triangular cross section
Figure 14.15

$$
\begin{aligned}
A(x) &= \int_0^{(2-x)/2} (2 - x - 2y)\,dy && \text{Apply formula for area.}\\[2mm]
&= \left[(2 - x)y - y^2\right]_0^{(2-x)/2} && \text{Integrate with respect to } y.\\[2mm]
&= \frac{(2 - x)^2}{4}. && \text{Area of triangular cross section (See Figure 14.15.)}
\end{aligned}
$$

Combining these results, you have the *iterated integral*

$$\text{Volume} = \int_R \int f(x, y)\,dA = \int_0^2 \int_0^{(2-x)/2} (2 - x - 2y)\,dy\,dx.$$

To understand this procedure better, it helps to imagine the integration as two sweeping motions. For the inner integration, a vertical line sweeps out the area of a cross section. For the outer integration, the triangular cross section sweeps out the volume, as shown in Figure 14.16.

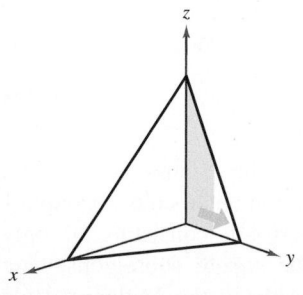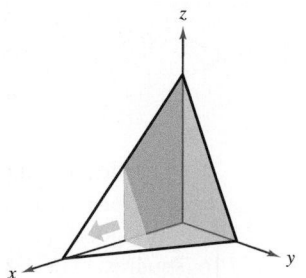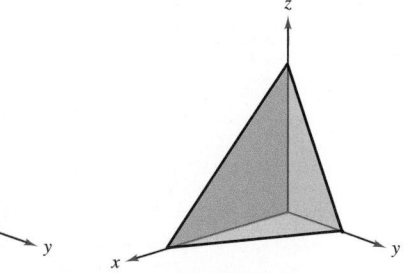

Integrate with respect to y to obtain the area of the cross section.   Integrate with respect to x to obtain the volume of the solid.

Figure 14.16

The next theorem was proved by the Italian mathematician Guido Fubini (1879–1943). The theorem states that if R is a vertically or horizontally simple region and f is continuous on R, then the double integral of f on R is equal to an iterated integral.

THEOREM 14.2 Fubini's Theorem

Let f be continuous on a plane region R.

1. If R is defined by $a \leq x \leq b$ and $g_1(x) \leq y \leq g_2(x)$, where g_1 and g_2 are continuous on $[a, b]$, then

$$\int_R \int f(x, y)\, dA = \int_a^b \int_{g_1(x)}^{g_2(x)} f(x, y)\, dy\, dx.$$

2. If R is defined by $c \leq y \leq d$ and $h_1(y) \leq x \leq h_2(y)$, where h_1 and h_2 are continuous on $[c, d]$, then

$$\int_R \int f(x, y)\, dA = \int_c^d \int_{h_1(y)}^{h_2(y)} f(x, y)\, dx\, dy.$$

EXAMPLE 2 **Evaluating a Double Integral as an Iterated Integral**

Evaluate

$$\int_R \int \left(1 - \frac{1}{2}x^2 - \frac{1}{2}y^2\right) dA$$

where R is the region given by

$$0 \leq x \leq 1, \quad 0 \leq y \leq 1.$$

Solution Because the region R is a square, it is both vertically and horizontally simple, and you can use either order of integration. Choose $dy\, dx$ by placing a vertical representative rectangle in the region (see the figure at the right). This produces the following.

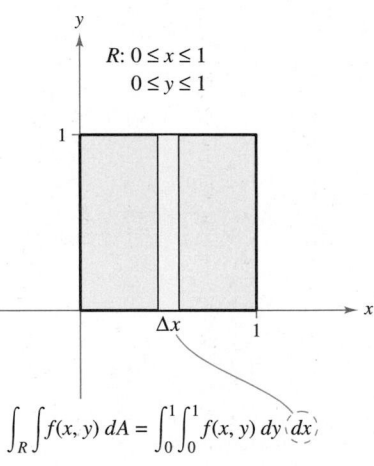

$$\int_R \int \left(1 - \frac{1}{2}x^2 - \frac{1}{2}y^2\right) dA = \int_0^1 \int_0^1 \left(1 - \frac{1}{2}x^2 - \frac{1}{2}y^2\right) dy\, dx$$

$$= \int_0^1 \left[\left(1 - \frac{1}{2}x^2\right)y - \frac{y^3}{6}\right]_0^1 dx$$

$$= \int_0^1 \left(\frac{5}{6} - \frac{1}{2}x^2\right) dx$$

$$= \left[\frac{5}{6}x - \frac{x^3}{6}\right]_0^1$$

$$= \frac{2}{3}$$

The double integral evaluated in Example 2 represents the volume of the solid region approximated in Example 1. Note that the approximation obtained in Example 1 is quite good $\left(0.672 \text{ vs. } \frac{2}{3}\right)$, even though you used a partition consisting of only 16 squares. The error resulted because the centers of the square subregions were used as the points in the approximation. This is comparable to the Midpoint Rule approximation of a single integral.

Exploration

Volume of a Paraboloid Sector   The solid in Example 3 has an elliptical (not a circular) base. Consider the region bounded by the circular paraboloid

$$z = a^2 - x^2 - y^2, \quad a > 0$$

and the *xy*-plane. How many ways of finding the volume of this solid do you now know? For instance, you could use the disk method to find the volume as a solid of revolution. Does each method involve integration?

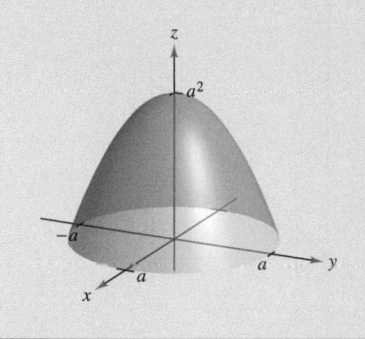

• • • • • • • • • • • • • • ▷
:
• • REMARK   In Example 3, note the usefulness of Wallis's Formula to evaluate $\int_0^{\pi/2} \cos^n \theta \, d\theta$. You may want to review this formula in Section 8.3.

The difficulty of evaluating a single integral $\int_a^b f(x) \, dx$ usually depends on the function f, not on the interval $[a, b]$. This is a major difference between single and double integrals. In the next example, you will integrate a function similar to the one in Examples 1 and 2. Notice that a change in the region R produces a much more difficult integration problem.

EXAMPLE 3   Finding Volume by a Double Integral

Find the volume of the solid region bounded by the paraboloid $z = 4 - x^2 - 2y^2$ and the *xy*-plane, as shown in Figure 14.17(a).

Solution   By letting $z = 0$, you can see that the base of the region in the *xy*-plane is the ellipse $x^2 + 2y^2 = 4$, as shown in Figure 14.17(b). This plane region is both vertically and horizontally simple, so the order $dy \, dx$ is appropriate.

Variable bounds for y: $-\sqrt{\dfrac{4 - x^2}{2}} \le y \le \sqrt{\dfrac{4 - x^2}{2}}$

Constant bounds for x: $-2 \le x \le 2$

The volume is

$$V = \int_{-2}^{2} \int_{-\sqrt{(4-x^2)/2}}^{\sqrt{(4-x^2)/2}} (4 - x^2 - 2y^2) \, dy \, dx \qquad \text{See Figure 14.17(b).}$$

$$= \int_{-2}^{2} \left[(4 - x^2)y - \frac{2y^3}{3} \right]_{-\sqrt{(4-x^2)/2}}^{\sqrt{(4-x^2)/2}} dx$$

$$= \frac{4}{3\sqrt{2}} \int_{-2}^{2} (4 - x^2)^{3/2} \, dx$$

$$= \frac{4}{3\sqrt{2}} \int_{-\pi/2}^{\pi/2} 16 \cos^4 \theta \, d\theta \qquad x = 2 \sin \theta$$

$$= \frac{64}{3\sqrt{2}} (2) \int_{0}^{\pi/2} \cos^4 \theta \, d\theta$$

$$= \frac{128}{3\sqrt{2}} \left(\frac{3\pi}{16} \right) \qquad \text{Wallis's Formula}$$

$$= 4\sqrt{2}\pi.$$

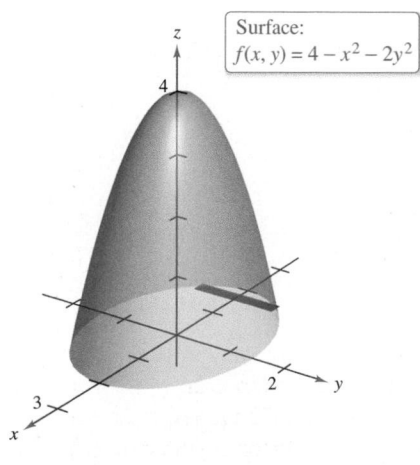

(a)

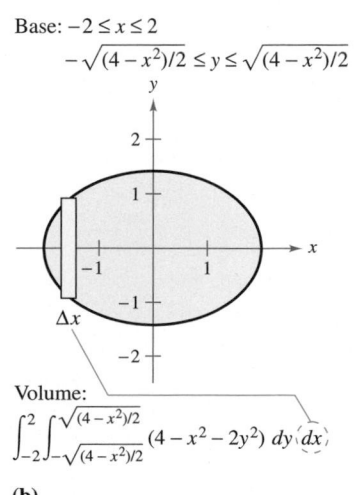

Surface:
$f(x, y) = 4 - x^2 - 2y^2$

Base: $-2 \le x \le 2$
$-\sqrt{(4-x^2)/2} \le y \le \sqrt{(4-x^2)/2}$

Volume:
$\int_{-2}^{2} \int_{-\sqrt{(4-x^2)/2}}^{\sqrt{(4-x^2)/2}} (4 - x^2 - 2y^2) \, dy \, dx$

(b)

Figure 14.17

In Examples 2 and 3, the problems could be solved with either order of integration because the regions were both vertically and horizontally simple. Moreover, had you used the order $dx\,dy$, you would have obtained integrals of comparable difficulty. There are, however, some occasions when one order of integration is much more convenient than the other. Example 4 shows such a case.

EXAMPLE 4 Comparing Different Orders of Integration

⋮ ⋯▷ *See LarsonCalculus.com for an interactive version of this type of example.*

Find the volume of the solid region bounded by the surface

$$f(x, y) = e^{-x^2} \qquad \text{Surface}$$

and the planes $z = 0$, $y = 0$, $y = x$, and $x = 1$, as shown in Figure 14.18.

Solution The base of the solid region in the xy-plane is bounded by the lines $y = 0$, $x = 1$, and $y = x$. The two possible orders of integration are shown in Figure 14.19.

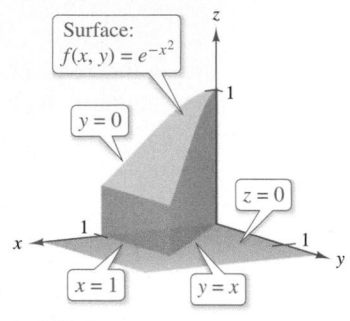

Base is bounded by $y = 0$, $y = x$, and $x = 1$.

Figure 14.18

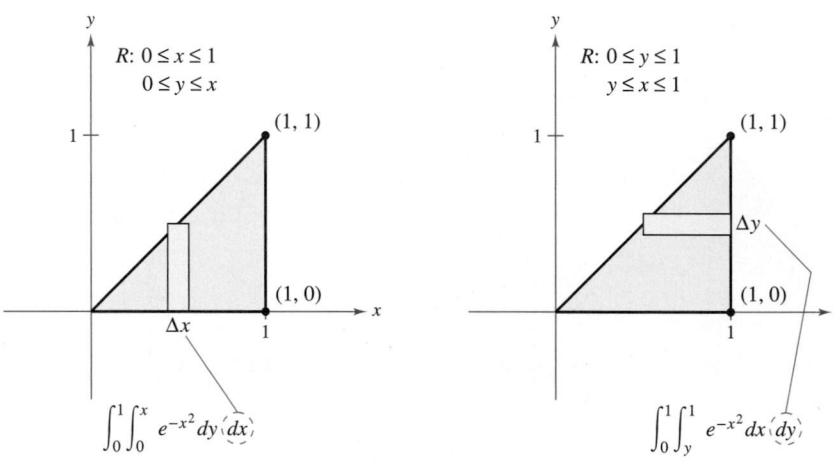

Figure 14.19

By setting up the corresponding iterated integrals, you can see that the order $dx\,dy$ requires the antiderivative

$$\int e^{-x^2}\,dx$$

which is not an elementary function. On the other hand, the order $dy\,dx$ produces

$$
\begin{aligned}
\int_0^1 \int_0^x e^{-x^2}\,dy\,dx &= \int_0^1 e^{-x^2}\,y\,\Big]_0^x\,dx && \text{Integrate with respect to } y. \\
&= \int_0^1 xe^{-x^2}\,dx \\
&= -\frac{1}{2}e^{-x^2}\,\Big]_0^1 && \text{Integrate with respect to } x. \\
&= -\frac{1}{2}\left(\frac{1}{e} - 1\right) \\
&= \frac{e - 1}{2e} && \text{Volume of solid region (See Figure 14.18.)} \\
&\approx 0.316.
\end{aligned}
$$

▷ **TECHNOLOGY** Try using a symbolic integration utility to evaluate the iterated integral in Example 4.

EXAMPLE 5   **Volume of a Region Bounded by Two Surfaces**

Find the volume of the solid region bounded above by the paraboloid

$$z = 1 - x^2 - y^2 \qquad \text{Paraboloid}$$

and below by the plane

$$z = 1 - y \qquad \text{Plane}$$

as shown in Figure 14.20.

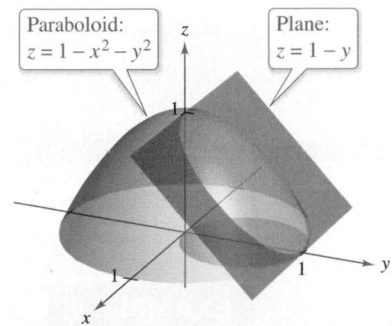

Figure 14.20

Solution   Equating z-values, you can determine that the intersection of the two surfaces occurs on the right circular cylinder given by

$$1 - y = 1 - x^2 - y^2 \quad \Longrightarrow \quad x^2 = y - y^2.$$

So, the region R in the xy-plane is a circle, as shown in Figure 14.21. Because the volume of the solid region is the difference between the volume under the paraboloid and the volume under the plane, you have

$$\text{Volume} = (\text{volume under paraboloid}) - (\text{volume under plane})$$

$$= \int_0^1 \int_{-\sqrt{y-y^2}}^{\sqrt{y-y^2}} (1 - x^2 - y^2)\, dx\, dy - \int_0^1 \int_{-\sqrt{y-y^2}}^{\sqrt{y-y^2}} (1 - y)\, dx\, dy$$

$$= \int_0^1 \int_{-\sqrt{y-y^2}}^{\sqrt{y-y^2}} (y - y^2 - x^2)\, dx\, dy$$

$$= \int_0^1 \left[(y - y^2)x - \frac{x^3}{3} \right]_{-\sqrt{y-y^2}}^{\sqrt{y-y^2}} dy$$

$$= \frac{4}{3} \int_0^1 (y - y^2)^{3/2}\, dy$$

$$= \left(\frac{4}{3}\right)\left(\frac{1}{8}\right) \int_0^1 [1 - (2y - 1)^2]^{3/2}\, dy$$

$$= \frac{1}{6} \int_{-\pi/2}^{\pi/2} \frac{\cos^4 \theta}{2}\, d\theta \qquad 2y - 1 = \sin\theta$$

$$= \frac{1}{6} \int_0^{\pi/2} \cos^4 \theta\, d\theta$$

$$= \left(\frac{1}{6}\right)\left(\frac{3\pi}{16}\right) \qquad \text{Wallis's Formula}$$

$$= \frac{\pi}{32}.$$

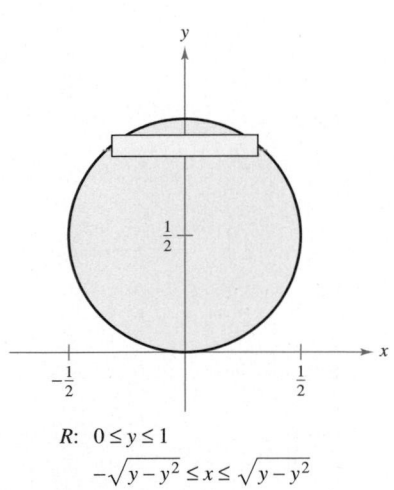

$R: \ 0 \le y \le 1$

$-\sqrt{y - y^2} \le x \le \sqrt{y - y^2}$

Figure 14.21

Average Value of a Function

Recall from Section 5.4 that for a function f in one variable, the average value of f on the interval $[a, b]$ is

$$\frac{1}{b-a}\int_a^b f(x)\,dx.$$

Given a function f in two variables, you can find the average value of f over the plane region R as shown in the following definition.

Definition of the Average Value of a Function Over a Region

If f is integrable over the plane region R, then the **average value** of f over R is

$$\text{Average value} = \frac{1}{A}\int_R\int f(x, y)\,dA$$

where A is the area of R.

EXAMPLE 6 **Finding the Average Value of a Function**

Find the average value of

$$f(x, y) = \frac{1}{2}xy$$

over the plane region R, where R is a rectangle with vertices

$(0, 0), (4, 0), (4, 3),$ and $(0, 3)$.

Solution The area of the rectangular region R is

$$A = (4)(3) = 12$$

as shown in Figure 14.22. The bounds for x are

$$0 \le x \le 4$$

and the bounds for y are

$$0 \le y \le 3.$$

So, the average value is

$$\text{Average value} = \frac{1}{A}\int_R\int f(x, y)\,dA$$

$$= \frac{1}{12}\int_0^4\int_0^3 \frac{1}{2}xy\,dy\,dx$$

$$= \frac{1}{12}\int_0^4 \frac{1}{4}xy^2\Big]_0^3\,dx$$

$$= \left(\frac{1}{12}\right)\left(\frac{9}{4}\right)\int_0^4 x\,dx$$

$$= \frac{3}{16}\left[\frac{1}{2}x^2\right]_0^4$$

$$= \left(\frac{3}{16}\right)(8)$$

$$= \frac{3}{2}.$$

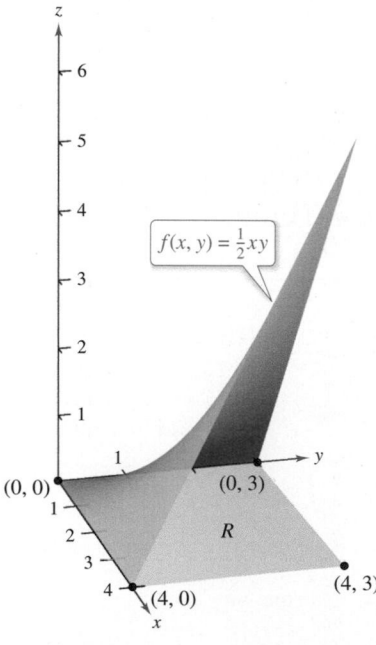

Figure 14.22

14.2 Exercises

See **CalcChat.com** for tutorial help and worked-out solutions to odd-numbered exercises.

CONCEPT CHECK

1. Approximating the Volume of a Solid In your own words, describe the process of using an inner partition to approximate the volume of a solid region lying above the xy-plane. How can the approximation be improved?

2. Fubini's Theorem What is the benefit of Fubini's Theorem when evaluating a double integral?

Approximation In Exercises 3–6, approximate the integral $\int_R \int f(x, y) \, dA$ by dividing the rectangle R with vertices $(0, 0)$, $(4, 0)$, $(4, 2)$, and $(0, 2)$ into eight equal squares and finding the sum

$$\sum_{i=1}^{8} f(x_i, y_i) \, \Delta A_i,$$

where (x_i, y_i) is the center of the ith square. Evaluate the iterated integral and compare it with the approximation.

3. $\displaystyle\int_0^4 \int_0^2 (x + y) \, dy \, dx$

4. $\displaystyle\frac{1}{2}\int_0^4 \int_0^2 x^2 y \, dy \, dx$

5. $\displaystyle\int_0^4 \int_0^2 (x^2 + y^2) \, dy \, dx$

6. $\displaystyle\int_0^4 \int_0^2 \frac{1}{(x + 1)(y + 1)} \, dy \, dx$

Evaluating a Double Integral In Exercises 7–12, sketch the region R and evaluate the iterated integral $\int_R \int f(x, y) \, dA$.

7. $\displaystyle\int_0^2 \int_0^1 (1 - 4x + 8y) \, dy \, dx$

8. $\displaystyle\int_0^\pi \int_0^{\pi/2} \sin^2 x \cos^2 y \, dy \, dx$

9. $\displaystyle\int_0^6 \int_{y/2}^3 (x + y) \, dx \, dy$

10. $\displaystyle\int_0^4 \int_{y/2}^{\sqrt{y}} x^2 y^2 \, dx \, dy$

11. $\displaystyle\int_{-3}^3 \int_{-\sqrt{9-x^2}}^{\sqrt{9-x^2}} (x + y) \, dy \, dx$

12. $\displaystyle\int_0^1 \int_{y-1}^0 e^{x+y} \, dx \, dy + \int_0^1 \int_0^{1-y} e^{x+y} \, dx \, dy$

Evaluating a Double Integral In Exercises 13–20, set up integrals for both orders of integration. Use the more convenient order to evaluate the integral over the plane region R.

13. $\displaystyle\int_R \int xy \, dA$

R: rectangle with vertices $(0, 0)$, $(0, 5)$, $(3, 5)$, $(3, 0)$

14. $\displaystyle\int_R \int \sin x \sin y \, dA$

R: rectangle with vertices $(-\pi, 0)$, $(\pi, 0)$, $(\pi, \pi/2)$, $(-\pi, \pi/2)$

15. $\displaystyle\int_R \int \frac{y}{x^2 + y^2} \, dA$

R: trapezoid bounded by $y = x$, $y = 2x$, $x = 1$, $x = 2$

16. $\displaystyle\int_R \int xe^y \, dA$

R: triangle bounded by $y = 4 - x$, $y = 0$, $x = 0$

17. $\displaystyle\int_R \int -2y \, dA$

R: region bounded by $y = 4 - x^2$, $y = 4 - x$

18. $\displaystyle\int_R \int \frac{y}{1 + x^2} \, dA$

R: region bounded by $y = 0$, $y = \sqrt{x}$, $x = 4$

19. $\displaystyle\int_R \int x \, dA$

R: sector of a circle in the first quadrant bounded by $y = \sqrt{25 - x^2}$, $3x - 4y = 0$, $y = 0$

20. $\displaystyle\int_R \int (x^2 + y^2) \, dA$

R: semicircle bounded by $y = \sqrt{4 - x^2}$, $y = 0$

Finding Volume In Exercises 21–26, use a double integral to find the volume of the indicated solid.

21.

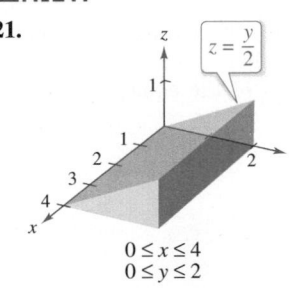

$0 \le x \le 4$
$0 \le y \le 2$

22.

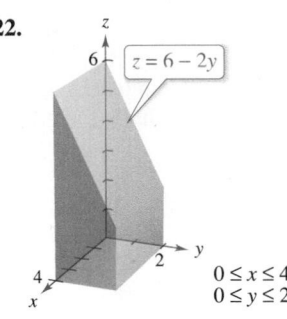

$0 \le x \le 4$
$0 \le y \le 2$

23. **24.**

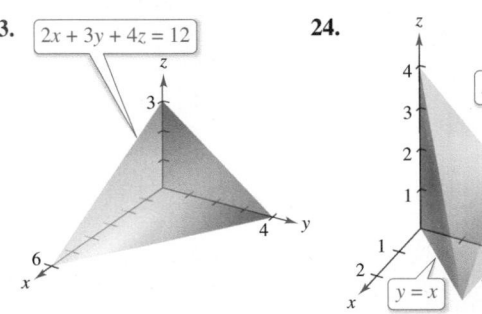

25. **26.**

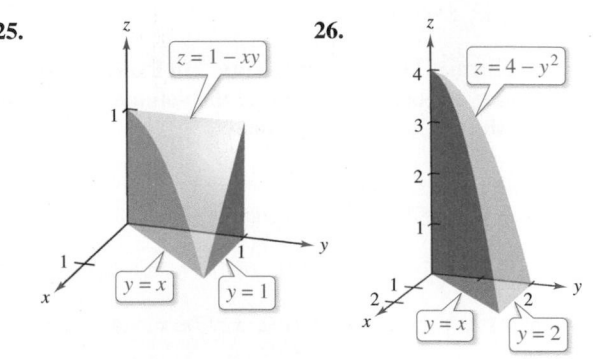

Finding Volume In Exercises 27 and 28, use an improper double integral to find the volume of the indicated solid.

27.
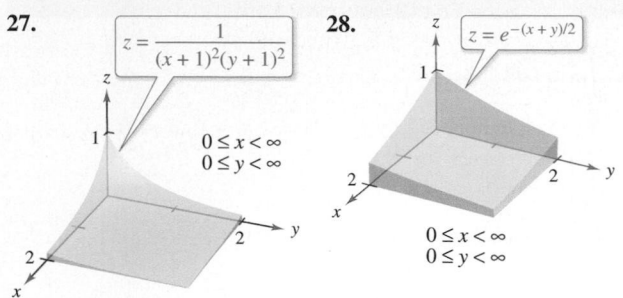
$z = \dfrac{1}{(x+1)^2(y+1)^2}$

$0 \le x < \infty$
$0 \le y < \infty$

28.
$z = e^{-(x+y)/2}$

$0 \le x < \infty$
$0 \le y < \infty$

Finding Volume In Exercises 29–34, set up and evaluate a double integral to find the volume of the solid bounded by the graphs of the equations.

29. $z = xy$, $z = 0$, $y = x^3$, $x = 1$, first octant

30. $z = 0$, $z = x^2$, $x = 0$, $x = 2$, $y = 0$, $y = 4$

31. $z = x + y$, $x^2 + y^2 = 4$, first octant

32. $z = \dfrac{1}{1 + y^2}$, $x = 0$, $x = 2$, $y \ge 0$

33. $y = 4 - x^2$, $z = 4 - x^2$, first octant

34. $x^2 + z^2 = 1$, $y^2 + z^2 = 1$, first octant

Volume of a Region Bounded by Two Surfaces In Exercises 35–40, set up a double integral to find the volume of the solid region bounded by the graphs of the equations. Do not evaluate the integral.

35.
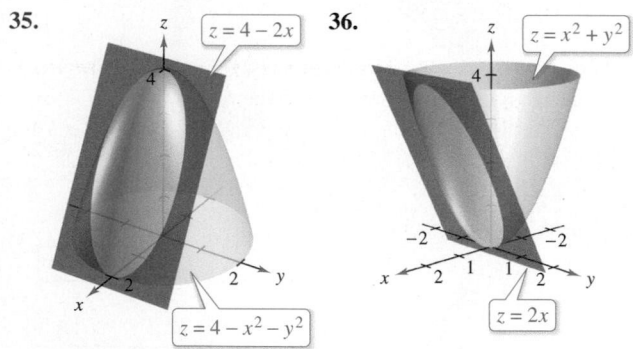
$z = 4 - 2x$

$z = 4 - x^2 - y^2$

36.
$z = x^2 + y^2$

$z = 2x$

37. $z = x^2 + y^2$, $x^2 + y^2 = 4$, $z = 0$

38. $z = \sin^2 x$, $z = 0$, $0 \le x \le \pi$, $0 \le y \le 5$

39. $z = x^2 + 2y^2$, $z = 4y$

40. $z = x^2 + y^2$, $z = 18 - x^2 - y^2$

Finding Volume Using Technology In Exercises 41–44, use a computer algebra system to find the volume of the solid bounded by the graphs of the equations.

41. $z = 9 - x^2 - y^2$, $z = 0$

42. $x^2 = 9 - y$, $z^2 = 9 - y$, first octant

43. $z = \dfrac{2}{1 + x^2 + y^2}$, $z = 0$, $y = 0$, $x = 0$, $y = -0.5x + 1$

44. $z = \ln(1 + x + y)$, $z = 0$, $y = 0$, $x = 0$, $x = 4 - \sqrt{y}$

Evaluating an Iterated Integral In Exercises 45–50, sketch the region of integration. Then evaluate the iterated integral, changing the order of integration if necessary.

45. $\displaystyle\int_0^1 \int_{y/2}^{1/2} e^{-x^2}\, dx\, dy$

46. $\displaystyle\int_0^{\ln 10} \int_{e^x}^{10} \dfrac{1}{\ln y}\, dy\, dx$

47. $\displaystyle\int_{-2}^2 \int_{-\sqrt{4-x^2}}^{\sqrt{4-x^2}} \sqrt{4 - y^2}\, dy\, dx$

48. $\displaystyle\int_0^3 \int_{y/3}^1 \dfrac{1}{1 + x^4}\, dx\, dy$

49. $\displaystyle\int_0^2 \int_{2x}^4 \sin y^2\, dy\, dx$

50. $\displaystyle\int_0^2 \int_{x^2/2}^2 \sqrt{y}\cos y\, dy\, dx$

Average Value In Exercises 51–56, find the average value of $f(x, y)$ over the plane region R.

51. $f(x, y) = x$

R: rectangle with vertices $(0, 0)$, $(4, 0)$, $(4, 2)$, $(0, 2)$

52. $f(x, y) = 2xy$

R: rectangle with vertices $(0, 1)$, $(1, 1)$, $(1, 6)$, $(0, 6)$

53. $f(x, y) = x^2 + y^2$

R: square with vertices $(0, 0)$, $(2, 0)$, $(2, 2)$, $(0, 2)$

54. $f(x, y) = \dfrac{1}{x + y}$, R: triangle with vertices $(0, 0)$, $(1, 0)$, $(1, 1)$

55. $f(x, y) = e^{x+y}$, R: triangle with vertices $(0, 0)$, $(0, 1)$, $(1, 1)$

56. $f(x, y) = \sin(x + y)$

R: rectangle with vertices $(0, 0)$, $(\pi, 0)$, (π, π), $(0, \pi)$

57. Average Production

The Cobb-Douglas production function for an automobile manufacturer is $f(x, y) = 100x^{0.6}y^{0.4}$, where x is the number of units of labor and y is the number of units of capital. Estimate the average production level when the number of units of labor x varies between 200 and 250 and the number of units of capital y varies between 300 and 325.

58. Average Temperature The temperature in degrees Celsius on the surface of a metal plate is $T(x, y) = 20 - 4x^2 - y^2$, where x and y are measured in centimeters. Estimate the average temperature when x varies between 0 and 2 centimeters and y varies between 0 and 4 centimeters.

EXPLORING CONCEPTS

59. Volume Let R be a region in the xy-plane whose area is B. When $f(x, y) = k$ for every point (x, y) in R, what is the value of $\int_R \int f(x, y)\, dA$? Explain.

60. Volume Let the plane region R be a unit circle and let the maximum value of f on R be 6. Is the greatest possible value of $\int_R \int f(x, y)\, dy\, dx$ equal to 6? Why or why not? If not, what is the greatest possible value?

Probability A *joint density function* of the continuous random variables x and y is a function $f(x, y)$ satisfying the following properties.

(a) $f(x, y) \geq 0$ for all (x, y)

(b) $\displaystyle\int_{-\infty}^{\infty} \int_{-\infty}^{\infty} f(x, y)\, dA = 1$

(c) $P[(x, y) \in R] = \displaystyle\int_{R} \int f(x, y)\, dA$

In Exercises 61–64, show that the function is a joint density function and find the required probability.

61. $f(x, y) = \begin{cases} \frac{1}{3}, & 0 \leq x \leq 1, 1 \leq y \leq 4 \\ 0, & \text{elsewhere} \end{cases}$

$P(0 \leq x \leq 1, 1 \leq y \leq 3)$

62. $f(x, y) = \begin{cases} \frac{1}{5}xy, & 0 \leq x \leq 2, 0 \leq y \leq \sqrt{5} \\ 0, & \text{elsewhere} \end{cases}$

$P(0 \leq x \leq 1, 0 \leq y \leq 2)$

63. $f(x, y) = \begin{cases} \frac{1}{27}(9 - x - y), & 0 \leq x \leq 3, 3 \leq y \leq 6 \\ 0, & \text{elsewhere} \end{cases}$

$P(0 \leq x \leq 1, 3 \leq y \leq 6)$

64. $f(x, y) = \begin{cases} e^{-x-y}, & x \geq 0, y \geq 0 \\ 0, & \text{elsewhere} \end{cases}$

$P(0 \leq x \leq 1, x \leq y \leq 1)$

65. Proof Let f be a continuous function such that $0 \leq f(x, y) \leq 1$ over a region R of area 1. Prove that $0 \leq \int_{R} \int f(x, y)\, dA \leq 1$.

66. Finding Volume Find the volume of the solid in the first octant bounded by the coordinate planes and the plane

$$\frac{x}{a} + \frac{y}{b} + \frac{z}{c} = 1$$

where $a > 0$, $b > 0$, and $c > 0$.

67. Approximation The table shows values of a function f over a square region R. Divide the region into 16 equal squares and select (x_i, y_i) to be the point in the ith square closest to the origin. Approximate the value of the integral below. Compare this approximation with that obtained by using the point in the ith square farthest from the origin.

$$\int_{0}^{4} \int_{0}^{4} f(x, y)\, dy\, dx$$

x \ y	0	1	2	3	4
0	32	31	28	23	16
1	31	30	27	22	15
2	28	27	24	19	12
3	23	22	19	14	7
4	16	15	12	7	0

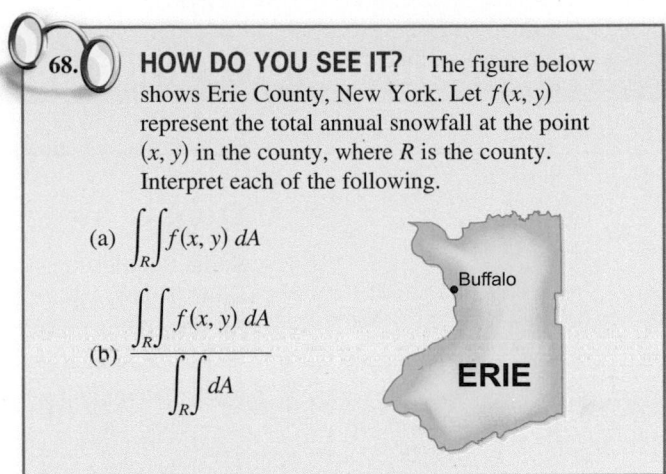

68. HOW DO YOU SEE IT? The figure below shows Erie County, New York. Let $f(x, y)$ represent the total annual snowfall at the point (x, y) in the county, where R is the county. Interpret each of the following.

(a) $\displaystyle\int_{R} \int f(x, y)\, dA$

(b) $\dfrac{\displaystyle\int_{R} \int f(x, y)\, dA}{\displaystyle\int_{R} \int dA}$

Buffalo

ERIE

True or False? In Exercises 69 and 70, determine whether the statement is true or false. If it is false, explain why or give an example that shows it is false.

69. The volume of the sphere $x^2 + y^2 + z^2 = 1$ is given by the integral

$$V = 8 \int_{0}^{1} \int_{0}^{1} \sqrt{1 - x^2 - y^2}\, dx\, dy.$$

70. If $f(x, y) \leq g(x, y)$ for all (x, y) in R, and both f and g are continuous over R, then $\int_{R} \int f(x, y)\, dA \leq \int_{R} \int g(x, y)\, dA$.

71. Maximizing a Double Integral Determine the region R in the xy-plane that maximizes the value of

$$\int_{R} \int (9 - x^2 - y^2)\, dA.$$

72. Minimizing a Double Integral Determine the region R in the xy-plane that minimizes the value of

$$\int_{R} \int (x^2 + y^2 - 4)\, dA.$$

73. Average Value Let

$$f(x) = \int_{1}^{x} e^{t^2}\, dt.$$

Find the average value of f on the interval $[0, 1]$.

74. Using Geometry Use a geometric argument to show that

$$\int_{0}^{3} \int_{0}^{\sqrt{9-y^2}} \sqrt{9 - x^2 - y^2}\, dx\, dy = \frac{9\pi}{2}.$$

PUTNAM EXAM CHALLENGE

75. Evaluate $\int_{0}^{a} \int_{0}^{b} e^{\max\{b^2 x^2,\, a^2 y^2\}}\, dy\, dx$, where a and b are positive.

76. Show that if $\lambda > \frac{1}{2}$ there does not exist a real-valued function u such that for all x in the closed interval $0 \leq x \leq 1$, $u(x) = 1 + \lambda \int_{x}^{1} u(y) u(y - x)\, dy$.

These problems were composed by the Committee on the Putnam Prize Competition. © The Mathematical Association of America. All rights reserved.

14.3 Change of Variables: Polar Coordinates

■ Write and evaluate double integrals in polar coordinates.

Double Integrals in Polar Coordinates

Some double integrals are *much* easier to evaluate in polar form than in rectangular form. This is especially true for regions such as circles, cardioids, and rose curves, and for integrands that involve $x^2 + y^2$.

In Section 10.4, you learned that the polar coordinates (r, θ) of a point are related to the rectangular coordinates (x, y) of the point as follows.

$$x = r \cos \theta \quad \text{and} \quad y = r \sin \theta$$

$$r^2 = x^2 + y^2 \quad \text{and} \quad \tan \theta = \frac{y}{x}$$

EXAMPLE 1 Using Polar Coordinates to Describe a Region

Use polar coordinates to describe each region shown in Figure 14.23.

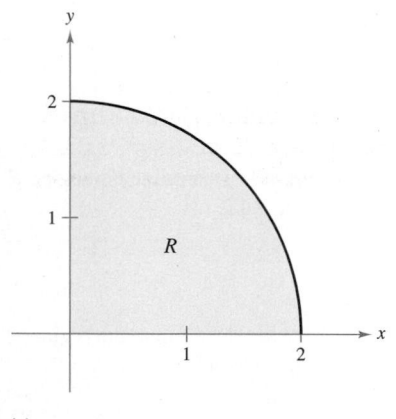

(a)

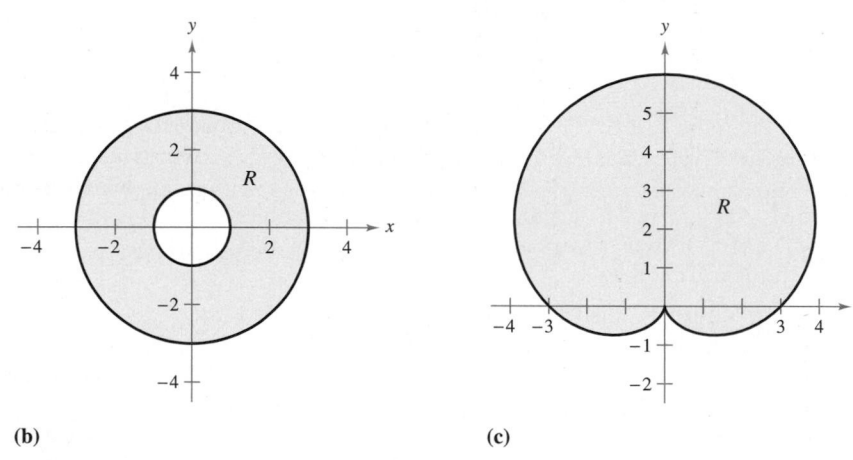

(b)   (c)

Figure 14.23

Solution

a. The region R is a quarter circle of radius 2. It can be described in polar coordinates as
$$R = \{(r, \theta): \ 0 \le r \le 2, \quad 0 \le \theta \le \pi/2\}.$$

b. The region R consists of all points between concentric circles of radii 1 and 3. It can be described in polar coordinates as
$$R = \{(r, \theta): \ 1 \le r \le 3, \quad 0 \le \theta \le 2\pi\}.$$

c. The region R is a cardioid with $a = b = 3$. It can be described in polar coordinates as
$$R = \{(r, \theta): \ 0 \le r \le 3 + 3 \sin \theta, \quad 0 \le \theta \le 2\pi\}. \quad ■$$

The regions in Example 1 are special cases of **polar sectors**

$$R = \{(r, \theta): r_1 \le r \le r_2, \quad \theta_1 \le \theta \le \theta_2\} \qquad \text{Polar sector}$$

as shown in Figure 14.24.

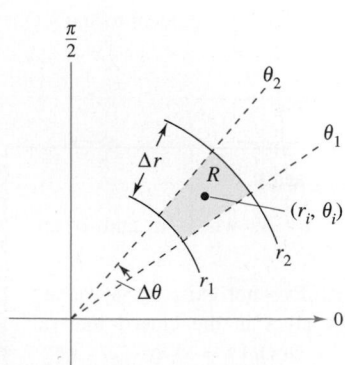

Polar sector
Figure 14.24

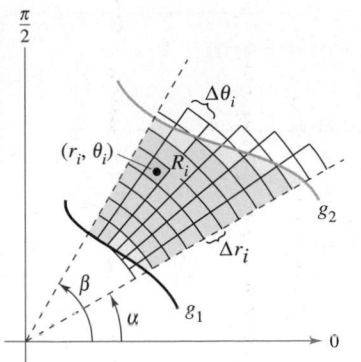

Polar grid superimposed over region R
Figure 14.25

To define a double integral of a continuous function $z = f(x, y)$ in polar coordinates, consider a region R bounded by the graphs of

$$r = g_1(\theta) \quad \text{and} \quad r = g_2(\theta)$$

and the lines $\theta = \alpha$ and $\theta = \beta$. Instead of partitioning R into small rectangles, use a partition of small polar sectors. On R, superimpose a polar grid made of rays and circular arcs, as shown in Figure 14.25. The polar sectors R_i lying entirely within R form an **inner polar partition** Δ, whose **norm** $\|\Delta\|$ is the length of the longest diagonal of the n polar sectors.

Consider a specific polar sector R_i, as shown in Figure 14.26. It can be shown (see Exercise 68) that the area of R_i is

$$\Delta A_i = r_i \Delta r_i \, \Delta \theta_i \qquad \text{Area of } R_i$$

where $\Delta r_i = r_2 - r_1$ and $\Delta \theta_i = \theta_2 - \theta_1$. This implies that the volume of the solid of height $f(r_i \cos \theta_i, r_i \sin \theta_i)$ above R_i is approximately

$$f(r_i \cos \theta_i, r_i \sin \theta_i) r_i \, \Delta r_i \, \Delta \theta_i$$

and you have

$$\int_R \int f(x, y) \, dA \approx \sum_{i=1}^{n} f(r_i \cos \theta_i, r_i \sin \theta_i) r_i \, \Delta r_i \, \Delta \theta_i.$$

The sum on the right can be interpreted as a Riemann sum for

$$f(r \cos \theta, r \sin \theta) r.$$

The region R corresponds to a *horizontally simple* region S in the $r\theta$-plane, as shown in Figure 14.27. The polar sectors R_i correspond to rectangles S_i, and the area ΔA_i of S_i is $\Delta r_i \, \Delta \theta_i$. So, the right-hand side of the equation corresponds to the double integral

$$\int_S \int f(r \cos \theta, r \sin \theta) r \, dA.$$

From this, you can apply Theorem 14.2 to write

$$\int_R \int f(x, y) \, dA = \int_S \int f(r \cos \theta, r \sin \theta) r \, dA$$

$$= \int_\alpha^\beta \int_{g_1(\theta)}^{g_2(\theta)} f(r \cos \theta, r \sin \theta) r \, dr \, d\theta.$$

This suggests the theorem on the next page, the proof of which is discussed in Section 14.8.

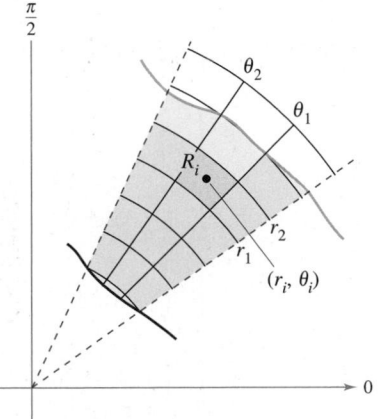

The polar sector R_i is the set of all points (r, θ) such that $r_1 \le r \le r_2$ and $\theta_1 \le \theta \le \theta_2$.
Figure 14.26

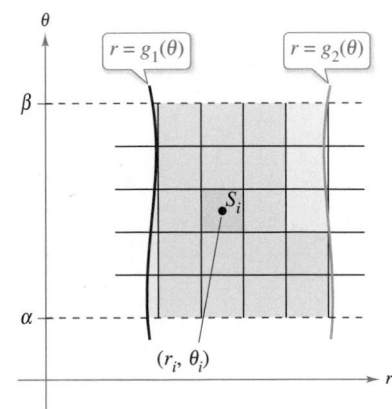

Horizontally simple region S
Figure 14.27

> **THEOREM 14.3 Change of Variables to Polar Form**
>
> Let R be a plane region consisting of all points $(x, y) = (r \cos \theta, r \sin \theta)$ satisfying the conditions $0 \le g_1(\theta) \le r \le g_2(\theta)$, $\alpha \le \theta \le \beta$, where $0 \le (\beta - \alpha) \le 2\pi$. If g_1 and g_2 are continuous on $[\alpha, \beta]$ and f is continuous on R, then
>
> $$\int_R \int f(x, y) \, dA = \int_\alpha^\beta \int_{g_1(\theta)}^{g_2(\theta)} f(r \cos \theta, r \sin \theta) r \, dr \, d\theta.$$

Exploration

Volume of a Paraboloid Sector In the Exploration on page 983, you were asked to summarize the different ways you know of finding the volume of the solid bounded by the paraboloid

$$z = a^2 - x^2 - y^2, a > 0$$

and the xy-plane. You now know another way. Use it to find the volume of the solid.

If $z = f(x, y)$ is nonnegative on R, then the integral in Theorem 14.3 can be interpreted as the volume of the solid region between the graph of f and the region R. When using the integral in Theorem 14.3, be certain not to omit the extra factor of r in the integrand.

The region R is restricted to two basic types, **r-simple** regions and **θ-simple** regions, as shown in Figure 14.28.

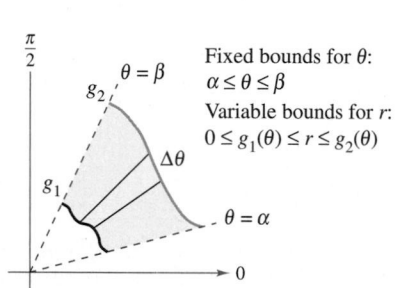

Fixed bounds for θ:
$\alpha \le \theta \le \beta$
Variable bounds for r:
$0 \le g_1(\theta) \le r \le g_2(\theta)$

r-Simple region

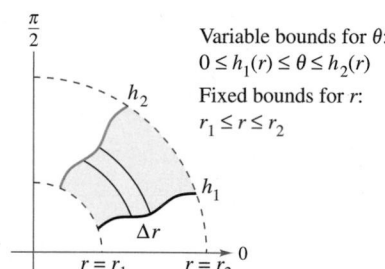

Variable bounds for θ:
$0 \le h_1(r) \le \theta \le h_2(r)$
Fixed bounds for r:
$r_1 \le r \le r_2$

θ-Simple region

Figure 14.28

EXAMPLE 2 **Evaluating a Double Polar Integral**

Let R be the annular region lying between the two circles $x^2 + y^2 = 1$ and $x^2 + y^2 = 5$. Evaluate the integral

$$\int_R \int (x^2 + y) \, dA.$$

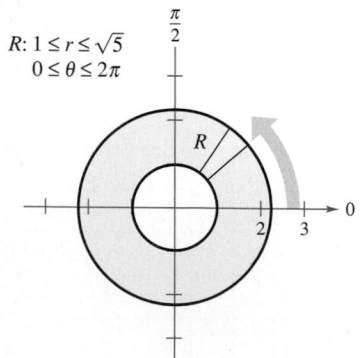

$R: 1 \le r \le \sqrt{5}$
$0 \le \theta \le 2\pi$

Figure 14.29

Solution The polar boundaries are $1 \le r \le \sqrt{5}$ and $0 \le \theta \le 2\pi$, as shown in Figure 14.29. Furthermore, $x^2 = (r \cos \theta)^2$ and $y = r \sin \theta$. So, you have

$$\int_R \int (x^2 + y) \, dA = \int_0^{2\pi} \int_1^{\sqrt{5}} (r^2 \cos^2 \theta + r \sin \theta) r \, dr \, d\theta$$

$$= \int_0^{2\pi} \int_1^{\sqrt{5}} (r^3 \cos^2 \theta + r^2 \sin \theta) \, dr \, d\theta$$

$$= \int_0^{2\pi} \left(\frac{r^4}{4} \cos^2 \theta + \frac{r^3}{3} \sin \theta \right) \Big]_1^{\sqrt{5}} \, d\theta$$

$$= \int_0^{2\pi} \left(6 \cos^2 \theta + \frac{5\sqrt{5} - 1}{3} \sin \theta \right) d\theta$$

$$= \int_0^{2\pi} \left(3 + 3 \cos 2\theta + \frac{5\sqrt{5} - 1}{3} \sin \theta \right) d\theta$$

$$= \left(3\theta + \frac{3 \sin 2\theta}{2} - \frac{5\sqrt{5} - 1}{3} \cos \theta \right) \Big]_0^{2\pi}$$

$$= 6\pi.$$

In Example 2, be sure to notice the factor of r with $dr\ d\theta$ in the integrand. This comes from the formula for the area of a polar sector. In differential notation, you can write

$$dA = r\ dr\ d\theta$$

which indicates that the area of a polar sector increases as you move away from the origin.

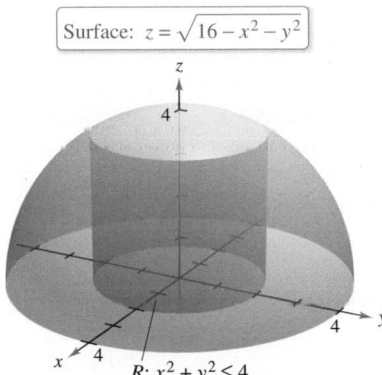

Surface: $z = \sqrt{16 - x^2 - y^2}$

$R: x^2 + y^2 \le 4$

Figure 14.30

EXAMPLE 3 Change of Variables to Polar Coordinates

Use polar coordinates to find the volume of the solid region bounded above by the hemisphere

$$z = \sqrt{16 - x^2 - y^2} \qquad \text{Hemisphere forms upper surface.}$$

and below by the circular region R given by

$$x^2 + y^2 \le 4 \qquad \text{Circular region forms lower surface.}$$

as shown in Figure 14.30.

Solution In Figure 14.30, you can see that R has the bounds

$$-\sqrt{4 - y^2} \le x \le \sqrt{4 - y^2}, \quad -2 \le y \le 2$$

and that $0 \le z \le \sqrt{16 - x^2 - y^2}$. In polar coordinates, the bounds are

$$0 \le r \le 2 \quad \text{and} \quad 0 \le \theta \le 2\pi$$

with height $z = \sqrt{16 - x^2 - y^2} = \sqrt{16 - r^2}$. Consequently, the volume V is

$$
\begin{aligned}
V &= \int_R\!\!\int f(x, y)\ dA & \text{Formula for volume} \\
&= \int_0^{2\pi}\!\!\int_0^2 \sqrt{16 - r^2}\ r\ dr\ d\theta & \text{Polar coordinates} \\
&= -\frac{1}{3}\int_0^{2\pi} (16 - r^2)^{3/2}\Big]_0^2\ d\theta & \text{Integrate with respect to } r. \\
&= -\frac{1}{3}\int_0^{2\pi} \left(24\sqrt{3} - 64\right)\ d\theta \\
&= -\frac{8}{3}\left(3\sqrt{3} - 8\right)\theta\Big]_0^{2\pi} & \text{Integrate with respect to } \theta. \\
&= \frac{16\pi}{3}\left(8 - 3\sqrt{3}\right) \\
&\approx 46.979.
\end{aligned}
$$

· · REMARK To see the benefit of polar coordinates in Example 3, you should try to evaluate the corresponding rectangular iterated integral

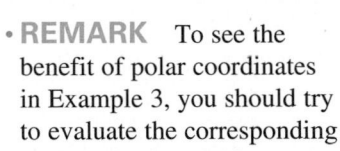

$$\int_{-2}^2\!\!\int_{-\sqrt{4 - y^2}}^{\sqrt{4 - y^2}} \sqrt{16 - x^2 - y^2}\ dx\ dy.$$

▷ **TECHNOLOGY** Any computer algebra system that can evaluate double integrals in rectangular coordinates can also evaluate double integrals in polar coordinates. The reason this is true is that once you have formed the iterated integral, its value is not changed by using different variables. In other words, if you use a computer algebra system to evaluate

$$\int_0^{2\pi}\!\!\int_0^2 \sqrt{16 - x^2}\ x\ dx\ dy$$

you should obtain the same value as that obtained in Example 3.

Just as with rectangular coordinates, the double integral

$$\int_R\!\!\int dA$$

can be used to find the area of a region in the plane.

EXAMPLE 4 Finding Areas of Polar Regions

See LarsonCalculus.com for an interactive version of this type of example.

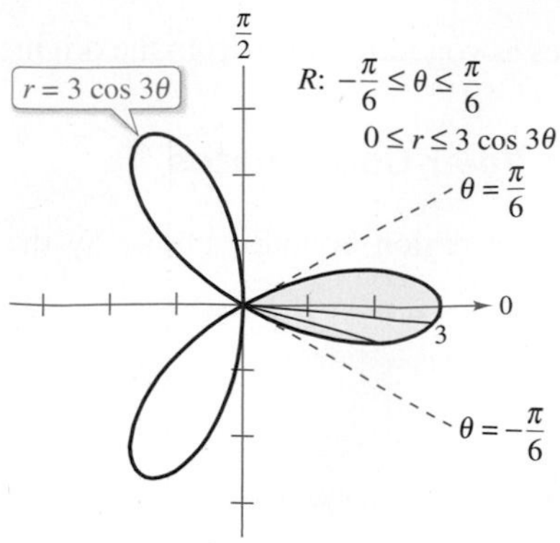

Figure 14.31

To use a double integral to find the area enclosed by the graph of $r = 3\cos 3\theta$, let R be one petal of the curve shown in Figure 14.31. This region is r-simple, and the boundaries are $-\pi/6 \le \theta \le \pi/6$ and $0 \le r \le 3\cos 3\theta$. So, the area of one petal is

$$\frac{1}{3}A = \iint_R dA = \int_{-\pi/6}^{\pi/6}\int_0^{3\cos 3\theta} r\,dr\,d\theta$$

$$= \int_{-\pi/6}^{\pi/6} \left[\frac{r^2}{2}\right]_0^{3\cos 3\theta} d\theta \qquad \text{Integrate with respect to } r.$$

$$= \frac{9}{2}\int_{-\pi/6}^{\pi/6} \cos^2 3\theta\,d\theta$$

$$= \frac{9}{4}\int_{-\pi/6}^{\pi/6} (1 + \cos 6\theta)\,d\theta$$

$$= \frac{9}{4}\left[\theta + \frac{1}{6}\sin 6\theta\right]_{-\pi/6}^{\pi/6} \qquad \text{Integrate with respect to } \theta.$$

$$= \frac{3\pi}{4}.$$

So, the total area is $A = 9\pi/4$. ■

As illustrated in Example 4, the area of a region in the plane can be represented by

$$A = \int_\alpha^\beta \int_{g_1(\theta)}^{g_2(\theta)} r\,dr\,d\theta.$$

For $g_1(\theta) = 0$, you obtain

$$A = \int_\alpha^\beta \int_0^{g_2(\theta)} r\,dr\,d\theta = \int_\alpha^\beta \left[\frac{r^2}{2}\right]_0^{g_2(\theta)} d\theta = \frac{1}{2}\int_\alpha^\beta [g_2(\theta)]^2\,d\theta$$

which agrees with Theorem 10.13.

So far in this section, all of the examples of iterated integrals in polar form have been of the form

$$\int_\alpha^\beta \int_{g_1(\theta)}^{g_2(\theta)} f(r\cos\theta, r\sin\theta)r\,dr\,d\theta$$

in which the order of integration is with respect to r first. Sometimes you can obtain a simpler integration problem by integrating with respect to θ first.

EXAMPLE 5 Integrating with Respect to θ First

Find the area of the region bounded above by the spiral $r = \pi/(3\theta)$ and below by the polar axis, between $r = 1$ and $r = 2$.

Solution The region is shown in Figure 14.32. The polar boundaries for the region are

$$1 \le r \le 2 \quad \text{and} \quad 0 \le \theta \le \frac{\pi}{3r}.$$

So, the area of the region can be evaluated as follows.

$$A = \int_1^2 \int_0^{\pi/(3r)} r\,d\theta\,dr = \int_1^2 \left[r\theta\right]_0^{\pi/(3r)} dr = \int_1^2 \frac{\pi}{3}\,dr = \left[\frac{\pi r}{3}\right]_1^2 = \frac{\pi}{3}$$ ■

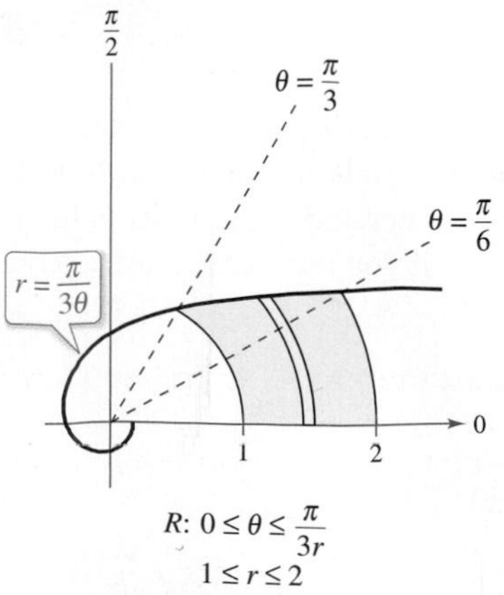

θ-Simple region
Figure 14.32

14.3   Exercises

CONCEPT CHECK

Choosing a Coordinate System In Exercises 1 and 2, the region R for the integral $\int_R \int f(x, y) \, dA$ is shown. State whether you would use rectangular or polar coordinates to evaluate the integral.

1.

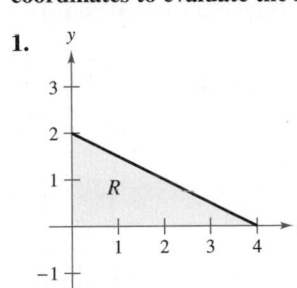

2.

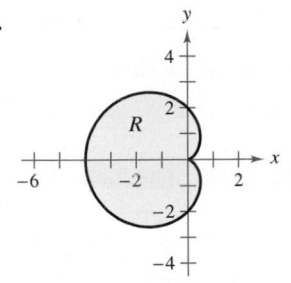

3. Describing Regions In your own words, describe r-simple regions and θ-simple regions.

4. Using Polar Coordinates Sketch the region of integration represented by the double integral

$$\int_0^{2\pi} \int_3^6 f(r, \theta) r \, dr \, d\theta.$$

Describing a Region In Exercises 5–8, use polar coordinates to describe the region shown.

5.

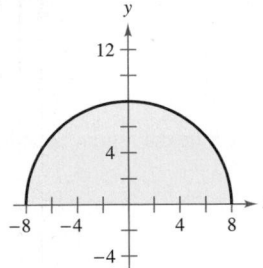

6.

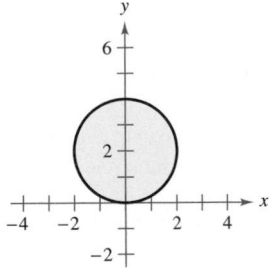

7.

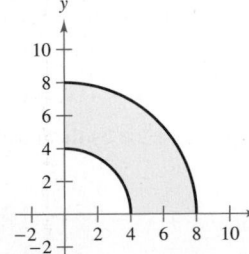

8.

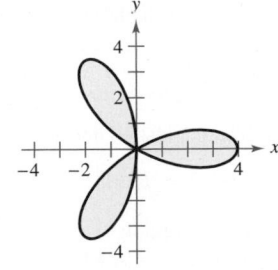

Evaluating a Double Integral In Exercises 9–16, evaluate the double integral $\int_R \int f(r, \theta) \, dA$ and sketch the region R.

9. $\displaystyle\int_0^\pi \int_0^{2\cos\theta} r \, dr \, d\theta$

10. $\displaystyle\int_0^{\pi/2} \int_0^{\sin\theta} r^2 \, dr \, d\theta$

11. $\displaystyle\int_0^{2\pi} \int_0^1 6r^2 \sin\theta \, dr \, d\theta$

12. $\displaystyle\int_0^{\pi/4} \int_0^4 r^2 \sin\theta \cos\theta \, dr \, d\theta$

13. $\displaystyle\int_0^{\pi/2} \int_1^3 \sqrt{9 - r^2} \, r \, dr \, d\theta$

14. $\displaystyle\int_0^{\pi/2} \int_0^3 re^{-r^2} \, dr \, d\theta$

15. $\displaystyle\int_0^{\pi/2} \int_0^{1+\sin\theta} \theta r \, dr \, d\theta$

16. $\displaystyle\int_0^{\pi/2} \int_0^{1-\cos\theta} (\sin\theta) r \, dr \, d\theta$

Converting to Polar Coordinates In Exercises 17–26, evaluate the iterated integral by converting to polar coordinates.

17. $\displaystyle\int_0^3 \int_0^{\sqrt{9-y^2}} y \, dx \, dy$

18. $\displaystyle\int_0^2 \int_0^{\sqrt{4-x^2}} x \, dy \, dx$

19. $\displaystyle\int_{-2}^2 \int_0^{\sqrt{4-x^2}} (x^2 + y^2) \, dy \, dx$

20. $\displaystyle\int_0^1 \int_{-\sqrt{x-x^2}}^{\sqrt{x-x^2}} (x^2 + y^2) \, dy \, dx$

21. $\displaystyle\int_0^1 \int_0^{\sqrt{1-x^2}} (x^2 + y^2)^{3/2} \, dy \, dx$

22. $\displaystyle\int_0^2 \int_y^{\sqrt{8-y^2}} \sqrt{x^2 + y^2} \, dx \, dy$

23. $\displaystyle\int_0^2 \int_0^{\sqrt{2x-x^2}} xy \, dy \, dx$

24. $\displaystyle\int_0^4 \int_0^{\sqrt{4y-y^2}} x^2 \, dx \, dy$

25. $\displaystyle\int_{-1}^1 \int_0^{\sqrt{1-x^2}} \cos(x^2 + y^2) \, dy \, dx$

26. $\displaystyle\int_0^{\sqrt{6}} \int_0^{\sqrt{6-x^2}} \sin\sqrt{x^2 + y^2} \, dy \, dx$

Converting to Polar Coordinates In Exercises 27 and 28, write the sum of the two iterated integrals as a single iterated integral by converting to polar coordinates. Evaluate the resulting iterated integral.

27. $\displaystyle\int_0^2 \int_0^x \sqrt{x^2 + y^2} \, dy \, dx + \int_2^{2\sqrt{2}} \int_0^{\sqrt{8-x^2}} \sqrt{x^2 + y^2} \, dy \, dx$

28. $\displaystyle\int_0^{(5\sqrt{2})/2} \int_0^x xy \, dy \, dx + \int_{(5\sqrt{2})/2}^5 \int_0^{\sqrt{25-x^2}} xy \, dy \, dx$

Converting to Polar Coordinates In Exercises 29–32, use polar coordinates to set up and evaluate the double integral $\int_R \int f(x, y) \, dA$.

29. $f(x, y) = x + y$

R: $x^2 + y^2 \leq 36$, $x \geq 0$, $y \geq 0$

30. $f(x, y) = e^{-(x^2+y^2)/2}$

R: $x^2 + y^2 \leq 25$, $x \geq 0$

31. $f(x, y) = \arctan \dfrac{y}{x}$

R: $x^2 + y^2 \geq 1$, $x^2 + y^2 \leq 4$, $0 \leq y \leq x$

32. $f(x, y) = 9 - x^2 - y^2$

R: $x^2 + y^2 \leq 9$, $x \geq 0$, $y \geq 0$

 **Volume** In Exercises 33–38, use a double integral in polar coordinates to find the volume of the solid bounded by the graphs of the equations.

33. $z = xy$, $x^2 + y^2 = 1$, first octant

34. $z = x^2 + y^2 + 3$, $z = 0$, $x^2 + y^2 = 1$

35. $z = \sqrt{x^2 + y^2}$, $z = 0$, $x^2 + y^2 = 25$

36. $z = \ln(x^2 + y^2)$, $z = 0$, $x^2 + y^2 \geq 1$, $x^2 + y^2 \leq 4$

37. Inside the hemisphere $z = \sqrt{16 - x^2 - y^2}$ and inside the cylinder $x^2 + y^2 - 4x = 0$

38. Inside the hemisphere $z = \sqrt{16 - x^2 - y^2}$ and outside the cylinder $x^2 + y^2 = 1$

39. Volume Use a double integral in polar coordinates to find a such that the volume inside the hemisphere $z = \sqrt{16 - x^2 - y^2}$ and outside the cylinder $x^2 + y^2 = a^2$ is one-half the volume of the hemisphere.

40. Volume Use a double integral in polar coordinates to find the volume of a sphere of radius a.

 **Area** In Exercises 41–46, use a double integral to find the area of the shaded region.

41.

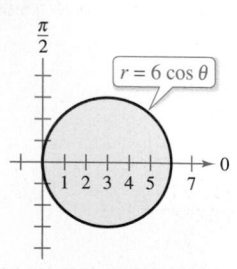

42.

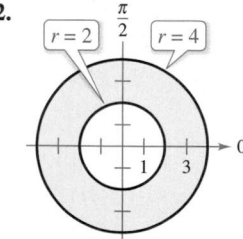

43.

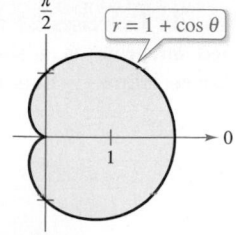

44.

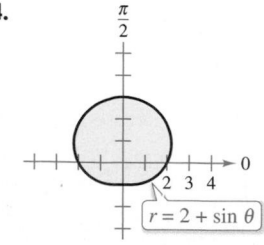

45.

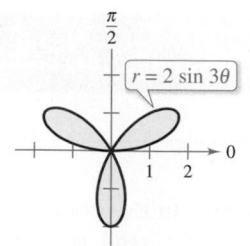

46.

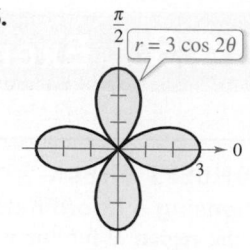

Area In Exercises 47–52, sketch a graph of the region bounded by the graphs of the equations. Then use a double integral to find the area of the region.

47. Inside the circle $r = 2 \cos \theta$ and outside the circle $r = 1$

48. Inside the cardioid $r = 2 + 2 \cos \theta$ and outside the circle $r = 1$

49. Inside the circle $r = 3 \cos \theta$ and outside the cardioid $r = 1 + \cos \theta$

50. Inside the cardioid $r = 1 + \cos \theta$ and outside the circle $r = 3 \cos \theta$

51. Inside the rose curve $r = 4 \sin 3\theta$ and outside the circle $r = 2$

52. Inside the circle $r = 2$ and outside the cardioid $r = 2 - 2 \cos \theta$

EXPLORING CONCEPTS

53. Area Express the area of the region in the figure using the sum of two double polar integrals. Then find the area of the region without using integrals.

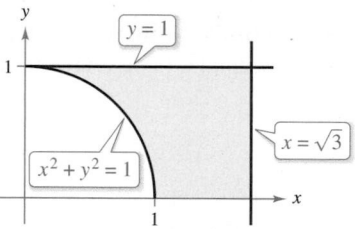

54. Comparing Integrals Let R be the region bounded by the circle $x^2 + y^2 = 9$.

(a) Set up the integral $\int_R \int f(x, y) \, dA$.

(b) Convert the integral in part (a) to polar coordinates.

(c) Which integral would you choose to evaluate? Why?

• • 55. Population • • • • • • • • •

The population density of a city is approximated by the model

$f(x, y) = 4000e^{-0.01(x^2+y^2)}$

for the region $x^2 + y^2 \leq 49$, where x and y are measured in miles. Integrate the density function over the indicated circular region to approximate the population of the city.

56. **HOW DO YOU SEE IT?** Each figure shows a region of integration for the double integral $\int_R\int f(x, y)\, dA$. For each region, state whether horizontal representative elements, vertical representative elements, or polar sectors would yield the easiest method for obtaining the limits of integration. Explain your reasoning.

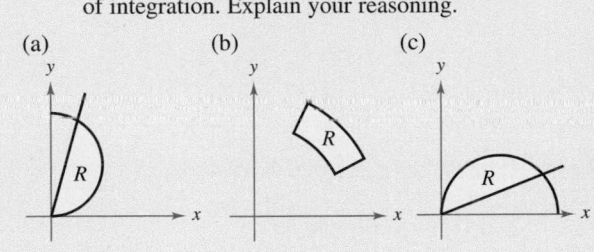

(a)   (b)   (c)

57. Volume Determine the diameter of a hole that is drilled vertically through the center of the solid bounded by the graphs of the equations $z = 25e^{-(x^2+y^2)/4}$, $z = 0$, and $x^2 + y^2 = 16$ when one-tenth of the volume of the solid is removed.

58. Glacier

Horizontal cross sections of a piece of ice that broke from a glacier are in the shape of a quarter of a circle with a radius of approximately 50 feet. The base is divided into 20 subregions, as shown in the figure. At the center of each subregion, the height of the ice is measured, yielding the following points in cylindrical coordinates.

$\left(5, \frac{\pi}{16}, 7\right)$, $\left(15, \frac{\pi}{16}, 8\right)$, $\left(25, \frac{\pi}{16}, 10\right)$, $\left(35, \frac{\pi}{16}, 12\right)$, $\left(45, \frac{\pi}{16}, 9\right)$,

$\left(5, \frac{3\pi}{16}, 9\right)$, $\left(15, \frac{3\pi}{16}, 10\right)$, $\left(25, \frac{3\pi}{16}, 14\right)$, $\left(35, \frac{3\pi}{16}, 15\right)$, $\left(45, \frac{3\pi}{16}, 10\right)$,

$\left(5, \frac{5\pi}{16}, 9\right)$, $\left(15, \frac{5\pi}{16}, 11\right)$, $\left(25, \frac{5\pi}{16}, 15\right)$, $\left(35, \frac{5\pi}{16}, 18\right)$, $\left(45, \frac{5\pi}{16}, 14\right)$,

$\left(5, \frac{7\pi}{16}, 5\right)$, $\left(15, \frac{7\pi}{16}, 8\right)$, $\left(25, \frac{7\pi}{16}, 11\right)$, $\left(35, \frac{7\pi}{16}, 16\right)$, $\left(45, \frac{7\pi}{16}, 12\right)$

(a) Approximate the volume of the piece of ice.

(b) Ice weighs approximately 57 pounds per cubic foot. Approximate the weight of the piece of ice.

(c) There are 7.48 gallons of water per cubic foot. Approximate the number of gallons of water in the piece of ice.

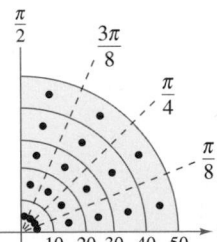

Approximation In Exercises 59 and 60, use a computer algebra system to approximate the iterated integral.

59. $\displaystyle\int_{\pi/4}^{\pi/2}\int_0^5 r\sqrt{1 + r^3}\sin\sqrt{\theta}\, dr\, d\theta$

60. $\displaystyle\int_0^{\pi/4}\int_0^4 5re^{\sqrt{r\theta}}\, dr\, d\theta$

True or False? In Exercises 61 and 62, determine whether the statement is true or false. If it is false, explain why or give an example that shows it is false.

61. If $\int_R\int f(r, \theta)\, dA > 0$, then $f(r, \theta) > 0$ for all (r, θ) in R.

62. If $f(r, \theta)$ is a constant function and the area of the region S is twice that of the region R, then

$$2\int_R\int f(r, \theta)\, dA = \int_S\int f(r, \theta)\, dA.$$

63. Probability The value of the integral

$$I = \int_{-\infty}^{\infty} e^{-x^2/2}\, dx$$

is required in the development of the normal probability density function.

(a) Use polar coordinates to evaluate the improper integral.

$$I^2 = \left(\int_{-\infty}^{\infty} e^{-x^2/2}\, dx\right)\left(\int_{-\infty}^{\infty} e^{-y^2/2}\, dy\right)$$

$$= \int_{-\infty}^{\infty}\int_{-\infty}^{\infty} e^{-(x^2+y^2)/2}\, dA$$

(b) Use the result of part (a) to determine I.

FOR FURTHER INFORMATION For more information on this problem, see the article "Integrating e^{-x^2} Without Polar Coordinates" by William Dunham in *Mathematics Teacher*. To view this article, go to *MathArticles.com*.

64. Evaluating Integrals Use the result of Exercise 63 and a change of variables to evaluate each integral. No integration is required.

(a) $\displaystyle\int_{-\infty}^{\infty} e^{-x^2}\, dx$ (b) $\displaystyle\int_{-\infty}^{\infty} e^{-4x^2}\, dx$

65. Think About It Consider the region R bounded by the graphs of $y = 2$, $y = 4$, $y = x$, and $y = \sqrt{3}x$ and the double integral $\int_R\int f(x, y)\, dA$. Determine the limits of integration when the region R is divided into (a) horizontal representative elements, (b) vertical representative elements, and (c) polar sectors.

66. Think About It Repeat Exercise 65 for a region R bounded by the graph of the equation $(x - 2)^2 + y^2 = 4$.

67. Probability Find k such that the function

$$f(x, y) = \begin{cases} ke^{-(x^2+y^2)}, & x \ge 0, y \ge 0 \\ 0, & \text{elsewhere} \end{cases}$$

is a probability density function. (*Hint:* Show that $\int_R\int f(x, y)\, dA = 1$.)

68. Area Show that the area A of the polar sector R (see figure) is $A = r\Delta r\Delta\theta$, where $r = (r_1 + r_2)/2$ is the average radius of R.

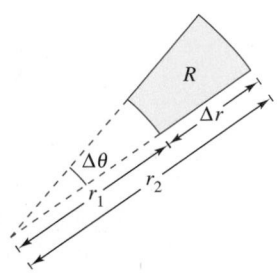

14.4 Center of Mass and Moments of Inertia

- Find the mass of a planar lamina using a double integral.
- Find the center of mass of a planar lamina using double integrals.
- Find moments of inertia using double integrals.

Mass

Section 7.6 discussed several applications of integration involving a lamina of *constant* density ρ. For example, if the lamina corresponding to the region R, as shown in Figure 14.33, has a constant density ρ, then the mass of the lamina is given by

$$\text{Mass} = \rho A = \rho \int_R \int dA = \int_R \int \rho \, dA. \qquad \text{Constant density}$$

If not otherwise stated, a lamina is assumed to have a constant density. In this section, however, you will extend the definition of the term *lamina* to include thin plates of *variable* density. Double integrals can be used to find the mass of a lamina of variable density, where the density at (x, y) is given by the **density function ρ.**

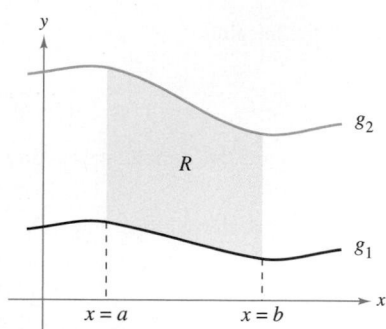

Lamina of constant density ρ
Figure 14.33

Definition of Mass of a Planar Lamina of Variable Density

If ρ is a continuous density function on the lamina corresponding to a plane region R, then the mass m of the lamina is given by

$$m = \int_R \int \rho(x, y) \, dA. \qquad \text{Variable density}$$

Density is normally expressed as mass per unit volume. For a planar lamina, however, density is mass per unit surface area.

EXAMPLE 1 Finding the Mass of a Planar Lamina

Find the mass of the triangular lamina with vertices $(0, 0)$, $(0, 3)$, and $(2, 3)$, given that the density at (x, y) is $\rho(x, y) = 2x + y$.

Solution As shown in Figure 14.34, region R has the boundaries $x = 0$, $y = 3$, and $y = 3x/2$ (or $x = 2y/3$). Therefore, the mass of the lamina is

$$
\begin{aligned}
m &= \int_R \int (2x + y) \, dA \\
&= \int_0^3 \int_0^{2y/3} (2x + y) \, dx \, dy \\
&= \int_0^3 \left[x^2 + xy \right]_0^{2y/3} dy \qquad \text{Integrate with respect to } x. \\
&= \frac{10}{9} \int_0^3 y^2 \, dy \\
&= \frac{10}{9} \left[\frac{y^3}{3} \right]_0^3 \qquad \text{Integrate with respect to } y. \\
&= 10.
\end{aligned}
$$

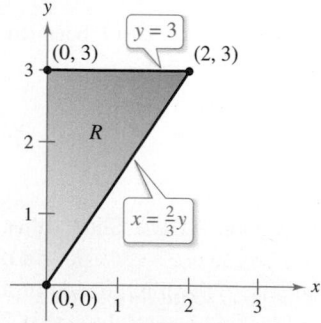

Lamina of variable density
$\rho(x, y) = 2x + y$
Figure 14.34

In Figure 14.34, note that the planar lamina is shaded so that the darkest shading corresponds to the densest part.

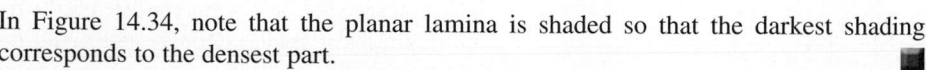

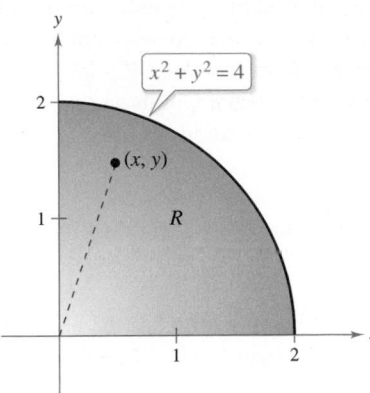

Density at (x, y): $\rho(x, y) = k\sqrt{x^2 + y^2}$
Figure 14.35

$x^2 + y^2 = 4$

| EXAMPLE 2 | **Finding Mass by Polar Coordinates** |

Find the mass of the lamina corresponding to the first-quadrant portion of the circle

$$x^2 + y^2 = 4$$

where the density at the point (x, y) is proportional to the distance between the point and the origin, as shown in Figure 14.35.

Solution At any point (x, y), the density of the lamina is

$$\rho(x, y) = k\sqrt{(x - 0)^2 + (y - 0)^2}$$
$$= k\sqrt{x^2 + y^2}$$

where k is the constant of proportionality. Because $0 \leq x \leq 2$ and $0 \leq y \leq \sqrt{4 - x^2}$, the mass is given by

$$m = \int_R \int k\sqrt{x^2 + y^2}\, dA$$
$$= \int_0^2 \int_0^{\sqrt{4 - x^2}} k\sqrt{x^2 + y^2}\, dy\, dx.$$

To simplify the integration, you can convert to polar coordinates, using the bounds

$$0 \leq \theta \leq \pi/2 \quad \text{and} \quad 0 \leq r \leq 2.$$

So, the mass is

$$m = \int_R \int k\sqrt{x^2 + y^2}\, dA$$

$$= \int_0^{\pi/2} \int_0^2 k\sqrt{r^2}\, r\, dr\, d\theta \qquad \text{Polar coordinates}$$

$$= \int_0^{\pi/2} \int_0^2 kr^2\, dr\, d\theta \qquad \text{Simplify integrand.}$$

$$= \int_0^{\pi/2} \frac{kr^3}{3}\Big]_0^2 d\theta \qquad \text{Integrate with respect to } r.$$

$$= \frac{8k}{3} \int_0^{\pi/2} d\theta$$

$$= \frac{8k}{3} \Big[\theta\Big]_0^{\pi/2} \qquad \text{Integrate with respect } \theta.$$

$$= \frac{4\pi k}{3}. \qquad \blacksquare$$

▷ **TECHNOLOGY** On many occasions, this text has mentioned the benefits of computer programs that perform symbolic integration. Even if you use such a program regularly, you should remember that its greatest benefit comes only in the hands of a knowledgeable user. For instance, notice how much simpler the integral in Example 2 becomes when it is converted to polar form.

Rectangular Form   **Polar Form**

$$\int_0^2 \int_0^{\sqrt{4 - x^2}} k\sqrt{x^2 + y^2}\, dy\, dx \qquad \int_0^{\pi/2} \int_0^2 kr^2\, dr\, d\theta$$

If you have access to software that performs symbolic integration, use it to evaluate both integrals. Some software programs cannot handle the first integral, but any program that can handle double integrals can evaluate the second integral.

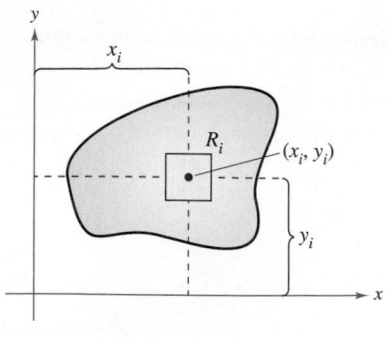

$M_x = (\text{mass})(y_i)$
$M_y = (\text{mass})(x_i)$
Figure 14.36

Moments and Center of Mass

For a lamina of variable density, moments of mass are defined in a manner similar to that used for the uniform density case. For a partition Δ of a lamina corresponding to a plane region R, consider the ith rectangle R_i of one area ΔA_i, as shown in Figure 14.36. Assume that the mass of R_i is concentrated at one of its interior points (x_i, y_i). The moment of mass of R_i with respect to the x-axis can be approximated by

$$(\text{Mass})(y_i) \approx [\rho(x_i, y_i) \, \Delta A_i](y_i).$$

Similarly, the moment of mass with respect to the y-axis can be approximated by

$$(\text{Mass})(x_i) \approx [\rho(x_i, y_i)\Delta A_i](x_i).$$

By forming the Riemann sum of all such products and taking the limits as the norm of Δ approaches 0, you obtain the following definitions of moments of mass with respect to the x- and y-axes.

Moments and Center of Mass of a Variable Density Planar Lamina

Let ρ be a continuous density function on the planar lamina R. The **moments of mass** with respect to the x- and y-axes are

$$M_x = \iint_R (y)\rho(x, y) \, dA$$

and

$$M_y = \iint_R (x)\rho(x, y) \, dA.$$

If m is the mass of the lamina, then the **center of mass** is

$$(\bar{x}, \bar{y}) = \left(\frac{M_y}{m}, \frac{M_x}{m}\right).$$

If R represents a simple plane region rather than a lamina, then the point $(\bar{x}, \bar{y})$ is called the **centroid** of the region.

For some planar laminas with a constant density ρ, you can determine the center of mass (or one of its coordinates) using symmetry rather than using integration. For instance, consider the laminas of constant density shown in Figure 14.37. Using symmetry, you can see that $\bar{y} = 0$ for the first lamina and $\bar{x} = 0$ for the second lamina.

$R: 0 \leq x \leq 1$
$\quad -\sqrt{1 - x^2} \leq y \leq \sqrt{1 - x^2}$

$R: -\sqrt{1 - y^2} \leq x \leq \sqrt{1 - y^2}$
$\quad 0 \leq y \leq 1$

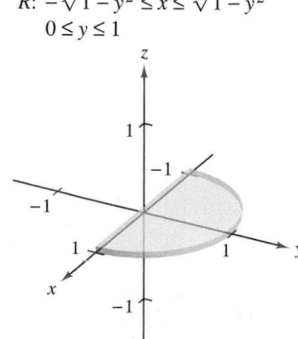

Lamina of constant density that is symmetric with respect to the x-axis
Figure 14.37

Lamina of constant density that is symmetric with respect to the y-axis

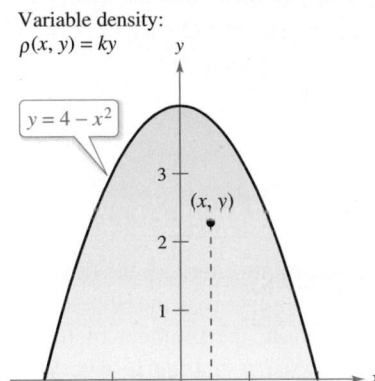

Variable density:
$\rho(x, y) = ky$

$y = 4 - x^2$

(x, y)

Parabolic region of variable density
Figure 14.38

EXAMPLE 3 Finding the Center of Mass

. . . .▷ See LarsonCalculus.com for an interactive version of this type of example.

Find the center of mass of the lamina corresponding to the parabolic region

$$0 \le y \le 4 - x^2 \qquad \text{Parabolic region}$$

where the density at the point (x, y) is proportional to the distance between (x, y) and the x-axis, as shown in Figure 14.38.

Solution The lamina is symmetric with respect to the y-axis and $\rho(x, y) = ky$, where k is the constant of proportionality. So, the center of mass lies on the y-axis and $\bar{x} = 0$. To find $\bar{y}$, first find the mass of the lamina.

$$m = \int_{-2}^{2} \int_{0}^{4-x^2} ky \, dy \, dx$$

$$= \frac{k}{2} \int_{-2}^{2} y^2 \Big]_{0}^{4-x^2} dx \qquad \text{Integrate with respect to } y.$$

$$= \frac{k}{2} \int_{-2}^{2} (16 - 8x^2 + x^4) \, dx$$

$$= \frac{k}{2} \left[16x - \frac{8x^3}{3} + \frac{x^5}{5} \right]_{-2}^{2} \qquad \text{Integrate with respect to } x.$$

$$= k\left(32 - \frac{64}{3} + \frac{32}{5} \right)$$

$$= \frac{256k}{15} \qquad \text{Mass of the limina}$$

Next, find the moment of mass about the x-axis.

$$M_x = \int_{-2}^{2} \int_{0}^{4-x^2} (y)(ky) \, dy \, dx$$

$$= \frac{k}{3} \int_{-2}^{2} y^3 \Big]_{0}^{4-x^2} dx \qquad \text{Integrate with respect to } y.$$

$$= \frac{k}{3} \int_{-2}^{2} (64 - 48x^2 + 12x^4 - x^6) \, dx$$

$$= \frac{k}{3} \left[64x - 16x^3 + \frac{12x^5}{5} - \frac{x^7}{7} \right]_{-2}^{2} \qquad \text{Integrate with respect to } x.$$

$$= \frac{4096k}{105} \qquad \text{Moment of mass about } x\text{-axis}$$

So,

$$\bar{y} = \frac{M_x}{m} = \frac{4096k/105}{256k/15} = \frac{16}{7}$$

and the center of mass is $\left(0, \frac{16}{7}\right)$.

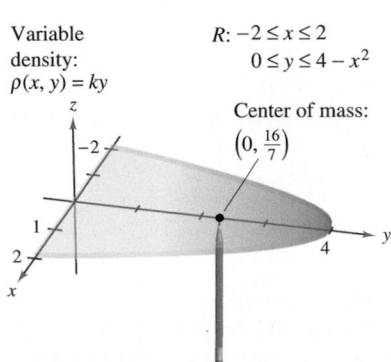

Variable
density:
$\rho(x, y) = ky$

$R: -2 \le x \le 2$
$0 \le y \le 4 - x^2$

Center of mass:
$\left(0, \frac{16}{7}\right)$

Figure 14.39

Although you can think of the moments M_x and M_y as measuring the tendency to rotate about the x- or y-axis, the calculation of moments is usually an intermediate step toward a more tangible goal. The use of the moments M_x and M_y is typical—to find the center of mass. Determination of the center of mass is useful in a variety of applications that allow you to treat a lamina as if its mass were concentrated at just one point. Intuitively, you can think of the center of mass as the balancing point of the lamina. For instance, the lamina in Example 3 should balance on the point of a pencil placed at $\left(0, \frac{16}{7}\right)$, as shown in Figure 14.39.

Moments of Inertia

The moments of M_x and M_y used in determining the center of mass of a lamina are sometimes called the **first moments** about the x- and y-axes. In each case, the moment is the product of a mass times a distance.

$$M_x = \int_R\!\!\int (y)\rho(x, y)\, dA \qquad\qquad M_y = \int_R\!\!\int (x)\rho(x, y)\, dA$$

| Distance to x-axis | Mass | Distance to y-axis | Mass |

You will now look at another type of moment—the **second moment,** or the **moment of inertia** of a lamina about a line. In the same way that mass is a measure of the tendency of matter to resist a change in straight-line motion, the moment of inertia about a line is a *measure of the tendency of matter to resist a change in rotational motion.* For example, when a particle of mass m is a distance d from a fixed line, its moment of inertia about the line is defined as

$$I = md^2 = (\text{mass})(\text{distance})^2.$$

As with moments of mass, you can generalize this concept to obtain the moments of inertia about the x- and y-axes of a lamina of variable density. These second moments are denoted by I_x and I_y, and in each case the moment is the product of a mass times the square of a distance.

$$I_x = \int_R\!\!\int (y^2)\rho(x, y)\, dA \qquad\qquad I_y = \int_R\!\!\int (x^2)\rho(x, y)\, dA$$

| Square of distance to x-axis | Mass | Square of distance to y-axis | Mass |

The sum of the moments I_x and I_y is called the **polar moment of inertia** and is denoted by I_0. For a lamina in the xy-plane, I_0 represents the moment of inertia of the lamina about the z-axis. The term "polar moment of inertia" stems from the fact that the square of the polar distance r is used in the calculation.

$$I_0 = \int_R\!\!\int (x^2 + y^2)\rho(x, y)\, dA = \int_R\!\!\int (r^2)\rho(x, y)\, dA$$

EXAMPLE 4 Finding the Moment of Inertia

Find the moment of inertia about the x-axis of the lamina in Example 3.

Solution From the definition of moment of inertia, you have

$$\begin{aligned}
I_x &= \int_{-2}^{2}\int_{0}^{4-x^2} (y^2)(ky)\, dy\, dx \\[2mm]
&= \frac{k}{4}\int_{-2}^{2} y^4 \Big]_0^{4-x^2} dx && \text{Integrate with respect to } y. \\[2mm]
&= \frac{k}{4}\int_{-2}^{2} (256 - 256x^2 + 96x^4 - 16x^6 + x^8)\, dx \\[2mm]
&= \frac{k}{4}\left[256x - \frac{256x^3}{3} + \frac{96x^5}{5} - \frac{16x^7}{7} + \frac{x^9}{9} \right]_{-2}^{2} && \text{Integrate with respect to } x. \\[2mm]
&= \frac{32{,}768k}{315}. && \text{Moment of inertia about } x\text{-axis}
\end{aligned}$$

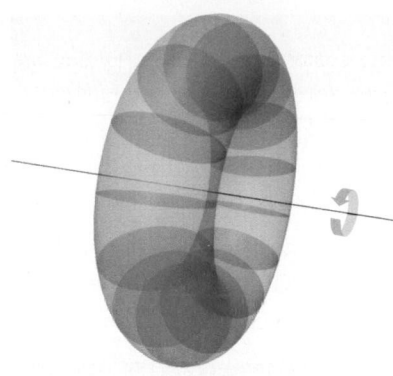

Planar lamina revolving at ω radians per second

Figure 14.40

The moment of inertia I of a revolving lamina can be used to measure its kinetic energy. For example, suppose a planar lamina is revolving about a line with an **angular speed** of ω radians per second, as shown in Figure 14.40. The kinetic energy E of the revolving lamina is

$$E = \frac{1}{2}I\omega^2. \qquad \text{Kinetic energy for rotational motion}$$

On the other hand, the kinetic energy E of a mass m moving in a straight line at a velocity v is

$$E = \frac{1}{2}mv^2. \qquad \text{Kinetic energy for linear motion}$$

So, the kinetic energy of a mass moving in a straight line is proportional to its mass, but the kinetic energy of a mass revolving about an axis is proportional to its moment of inertia.

The **radius of gyration** $\bar{\bar{r}}$ of a revolving mass m with moment of inertia I is defined as

$$\bar{\bar{r}} = \sqrt{\frac{I}{m}}. \qquad \text{Radius of gyration}$$

If the entire mass were located at a distance $\bar{\bar{r}}$ from its axis of revolution, it would have the same moment of inertia and, consequently, the same kinetic energy. For instance, the radius of gyration of the lamina in Example 4 about the x-axis is

$$\bar{\bar{y}} = \sqrt{\frac{I_x}{m}} = \sqrt{\frac{32{,}768k/315}{256k/15}} = \sqrt{\frac{128}{21}} \approx 2.469.$$

EXAMPLE 5   **Finding the Radius of Gyration**

Find the radius of gyration about the y-axis for the lamina corresponding to the region $R: 0 \le y \le \sin x, 0 \le x \le \pi$, where the density at (x, y) is given by $\rho(x, y) = x$.

Solution   The region R is shown in Figure 14.41. By integrating $\rho(x, y) = x$ over the region R, you can determine that the mass of the region is π. The moment of inertia about the y-axis is

$$I_y = \int_0^\pi \int_0^{\sin x} x^3 \, dy \, dx$$

$$= \int_0^\pi x^3 y \Big]_0^{\sin x} dx \qquad \text{Integrate with respect to } y.$$

$$= \int_0^\pi x^3 \sin x \, dx$$

$$= \left[(3x^2 - 6)(\sin x) - (x^3 - 6x)(\cos x) \right]_0^\pi \qquad \text{Integrate with respect to } x.$$

$$= \pi^3 - 6\pi. \qquad \text{Moment of inertia about } y\text{-axis}$$

So, the radius of gyration about the y-axis is

$$\bar{\bar{x}} = \sqrt{\frac{I_y}{m}}$$

$$= \sqrt{\frac{\pi^3 - 6\pi}{\pi}}$$

$$= \sqrt{\pi^2 - 6}$$

$$\approx 1.967. \qquad \text{Radius of gyration about } y\text{-axis}$$

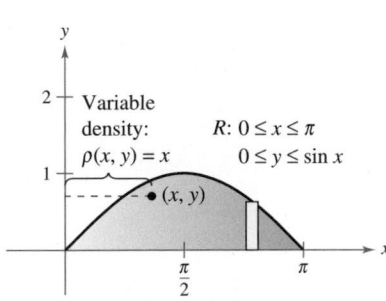

Figure 14.41

14.4   Exercises

CONCEPT CHECK

1. **Mass of a Planar Lamina**   Explain when you should use a double integral to find the mass of a planar lamina.

2. **Moment of Inertia**   Describe what the moment of inertia measures.

 **Finding the Mass of a Lamina**   In Exercises 3–6, find the mass of the lamina described by the inequalities, given that its density is $\rho(x, y) = xy$.

3. $0 \le x \le 2,\ 0 \le y \le 2$

4. $0 \le x \le 2,\ 0 \le y \le 4 - x^2$

5. $0 \le x \le 1,\ 0 \le y \le \sqrt{1 - x^2}$

6. $x \ge 0,\ 3 \le y \le 3 + \sqrt{9 - x^2}$

Finding the Center of Mass   In Exercises 7–10, find the mass and center of mass of the lamina corresponding to the region R for each density.

7. R: square with vertices $(0, 0)$, $(a, 0)$, $(0, a)$, (a, a)

   (a) $\rho = k$ (b) $\rho = ky$ (c) $\rho = kx$

8. R: rectangle with vertices $(0, 0)$, $(a, 0)$, $(0, b)$, (a, b)

   (a) $\rho = kxy$ (b) $\rho = k(x^2 + y^2)$

9. R: triangle with vertices $(0, 0)$, $(0, a)$, (a, a)

   (a) $\rho = k$ (b) $\rho = ky$ (c) $\rho = kx$

10. R: triangle with vertices $(0, 0)$, $(a/2, a)$, $(a, 0)$

    (a) $\rho = k$ (b) $\rho = kxy$

11. **Translations in the Plane**   Translate the lamina in Exercise 7 to the right five units and determine the resulting center of mass.

12. **Conjecture**   Use the result of Exercise 11 to make a conjecture about the change in the center of mass when a lamina of constant density is translated c units horizontally or d units vertically. Is the conjecture true when the density is not constant? Explain.

 **Finding the Center of Mass**   In Exercises 13–24, find the mass and center of mass of the lamina bounded by the graphs of the equations for the given density.

13. $y = \sqrt{x},\ y = 0,\ x = 1,\ \rho = ky$

14. $y = x^2,\ y = 0,\ x = 2,\ \rho = kxy$

15. $y = 4/x,\ y = 0,\ x = 1,\ x = 4,\ \rho = kx^2$

16. $y = \dfrac{1}{1 + x^2},\ y = 0,\ x = -1,\ x = 1,\ \rho = k$

17. $y = e^x,\ y = 0,\ x = 0,\ x = 1,\ \rho = k$

18. $y = e^{-x},\ y = 0,\ x = 0,\ x = 1,\ \rho = ky^2$

19. $y = 4 - x^2,\ y = 0,\ \rho = ky$

20. $x = 9 - y^2,\ x = 0,\ \rho = kx$

21. $y = \sin \dfrac{\pi x}{3},\ y = 0,\ x = 0,\ x = 3,\ \rho = k$

22. $y = \cos \dfrac{\pi x}{8},\ y = 0,\ x = 0,\ x = 4,\ \rho = ky$

23. $y = \sqrt{36 - x^2},\ 0 \le y \le x,\ \rho = k$

24. $x^2 + y^2 = 16,\ x \ge 0,\ y \ge 0,\ \rho = k(x^2 + y^2)$

Finding the Center of Mass Using Technology   In Exercises 25–28, use a computer algebra system to find the mass and center of mass of the lamina bounded by the graphs of the equations for the given density.

25. $y = e^{-x},\ y = 0,\ x = 0,\ x = 2,\ \rho = kxy$

26. $y = \ln x,\ y = 0,\ x = 1,\ x = e,\ \rho = k/x$

27. $r = 2 \cos 3\theta,\ -\dfrac{\pi}{6} \le \theta \le \dfrac{\pi}{6},\ \rho = k$

28. $r = 1 + \cos \theta,\ \rho = k$

Finding the Radius of Gyration About Each Axis   In Exercises 29–34, verify the given moment(s) of inertia and find $\bar{\bar{x}}$ and $\bar{\bar{y}}$. Assume that each lamina has a density of $\rho = 1$ gram per square centimeter. (These regions are common shapes used in engineering.)

29. Rectangle

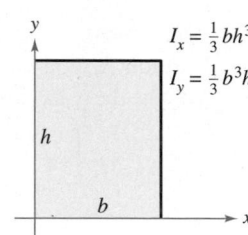

$I_x = \frac{1}{3} bh^3$

$I_y = \frac{1}{3} b^3 h$

30. Right triangle

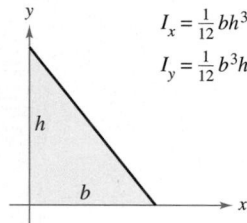

$I_x = \frac{1}{12} bh^3$

$I_y = \frac{1}{12} b^3 h$

31. Circle

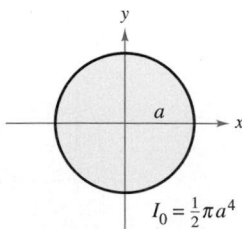

$I_0 = \frac{1}{2} \pi a^4$

32. Semicircle

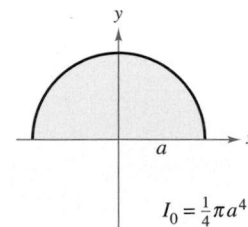

$I_0 = \frac{1}{4} \pi a^4$

33. Quarter circle

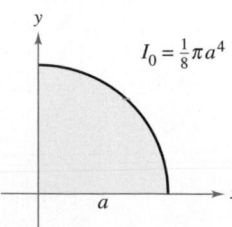

$I_0 = \frac{1}{8} \pi a^4$

34. Ellipse

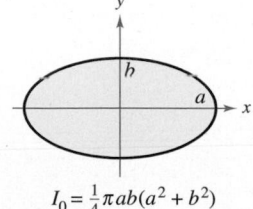

$I_0 = \frac{1}{4} \pi ab(a^2 + b^2)$

Finding Moments of Inertia and Radii of Gyration In Exercises 35–38, find I_x, I_y, I_0, $\bar{\bar{x}}$, and $\bar{\bar{y}}$ for the lamina bounded by the graphs of the equations.

35. $y = 4 - x^2$, $y = 0$, $x > 0$, $\rho = kx$

36. $y = x$, $y = x^2$, $\rho = kxy$

37. $y = \sqrt{x}$, $y = 0$, $x = 4$, $\rho = kxy$

38. $y = x^2$, $y^2 = x$, $\rho = kx$

Finding a Moment of Inertia Using Technology In Exercises 39–42, set up the double integral required to find the moment of inertia about the given line of the lamina bounded by the graphs of the equations for the given density. Use a computer algebra system to evaluate the double integral.

39. $x^2 + y^2 = b^2$, $\rho = k$, line: $x = a$ $(a > b)$

40. $y = \sqrt{x}$, $y = 0$, $x = 4$, $\rho = kx$, line: $x = 6$

41. $y = \sqrt{a^2 - x^2}$, $y = 0$, $\rho = ky$, line: $y = a$

42. $y = 4 - x^2$, $y = 0$, $\rho = k$, line: $y = 2$

Hydraulics In Exercises 43–46, determine the location of the horizontal axis y_a at which a vertical gate in a dam is to be hinged so that there is no moment causing rotation under the indicated loading (see figure). The model for y_a is

$$y_a = \bar{y} - \frac{I_{\bar{y}}}{hA}$$

where $\bar{y}$ is the y-coordinate of the centroid of the gate, $I_{\bar{y}}$ is the moment of inertia of the gate about the line $y = \bar{y}$, h is the depth of the centroid below the surface, and A is the area of the gate.

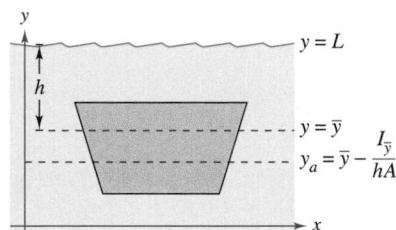

43.

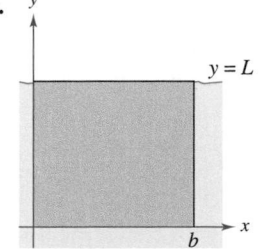

44.

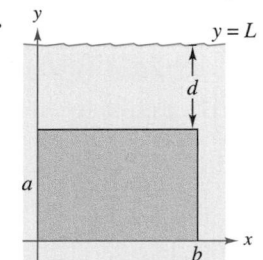

45.

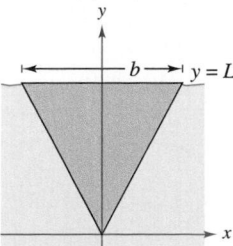

46.

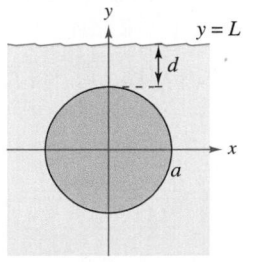

EXPLORING CONCEPTS

47. Polar Moment of Inertia What does it mean for an object to have a greater polar moment of inertia than another object?

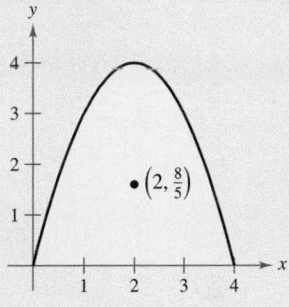

48. **HOW DO YOU SEE IT?** The center of mass of the lamina of constant density shown in the figure is $\left(2, \frac{8}{5}\right)$. Make a conjecture about how the center of mass $(\bar{x}, \bar{y})$ changes for each given nonconstant density $\rho(x, y)$. Explain. (Make your conjecture *without* performing any calculations.)

(a) $\rho(x, y) = ky$ (b) $\rho(x, y) = k|2 - x|$

(c) $\rho(x, y) = kxy$ (d) $\rho(x, y) = k(4 - x)(4 - y)$

49. Proof Prove the following Theorem of Pappus: Let R be a region in a plane and let L be a line in the same plane such that L does not intersect the interior of R. If r is the distance between the centroid of R and the line, then the volume V of the solid of revolution formed by revolving R about the line is $V = 2\pi r A$, where A is the area of R.

SECTION PROJECT

Center of Pressure on a Sail

The center of pressure on a sail is the point (x_p, y_p) at which the total aerodynamic force may be assumed to act. If the sail is represented by a plane region R, then the center of pressure is

$$x_p = \frac{\int_R \int xy \, dA}{\int_R \int y \, dA} \quad \text{and} \quad y_p = \frac{\int_R \int y^2 \, dA}{\int_R \int y \, dA}.$$

Consider a triangular sail with vertices at $(0, 0)$, $(2, 1)$, and $(0, 5)$. Verify the value of each integral.

(a) $\displaystyle\int_R \int y \, dA = 10$

(b) $\displaystyle\int_R \int xy \, dA = \frac{35}{6}$

(c) $\displaystyle\int_R \int y^2 \, dA = \frac{155}{6}$

Calculate the coordinates (x_p, y_p) of the center of pressure. Sketch a graph of the sail and indicate the location of the center of pressure.

14.5 Surface Area

■ Use a double integral to find the area of a surface.

Surface Area

At this point, you know a great deal about the solid region lying between a surface and a closed and bounded region R in the xy-plane, as shown in Figure 14.42. For example, you know how to find the extrema of f on R (Section 13.8), the area of the base R of the solid (Section 14.1), the volume of the solid (Section 14.2), and the centroid of the base R (Section 14.4).

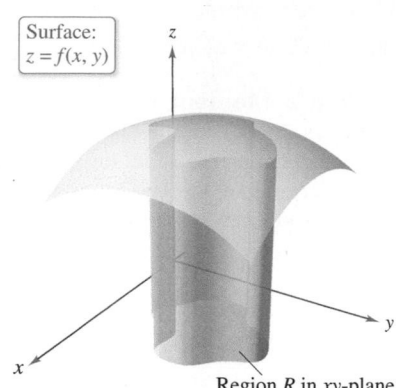

Surface:
$z = f(x, y)$

Region R in xy-plane

Figure 14.42

In this section, you will learn how to find the upper **surface area** of the solid. Later, you will learn how to find the centroid of the solid (Section 14.6) and the lateral surface area (Section 15.2).

To begin, consider a surface S given by

$$z = f(x, y) \qquad \text{Surface defined over a region } R$$

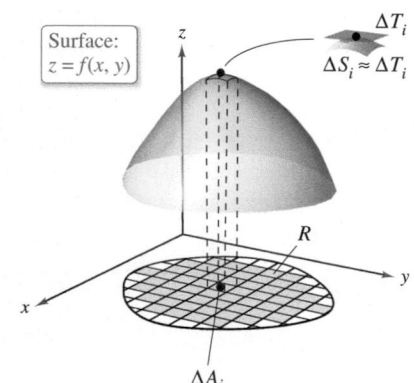

Surface:
$z = f(x, y)$

ΔT_i

$\Delta S_i \approx \Delta T_i$

R

ΔA_i

Figure 14.43

defined over a region R. Assume that R is closed and bounded and that f has continuous first partial derivatives. To find the surface area, construct an inner partition of R consisting of n rectangles, where the area of the ith rectangle R_i is $\Delta A_i = \Delta x_i \Delta y_i$, as shown in Figure 14.43. In each R_i, let (x_i, y_i) be the point that is closest to the origin. At the point $(x_i, y_i, z_i) = (x_i, y_i, f(x_i, y_i))$ on the surface S, construct a tangent plane T_i. The area of the portion of the tangent plane that lies directly above R_i is approximately equal to the area of the surface lying directly above R_i. That is, $\Delta T_i \approx \Delta S_i$. So, the surface area of S is approximated by

$$\sum_{i=1}^{n} \Delta S_i \approx \sum_{i=1}^{n} \Delta T_i.$$

To find the area of the parallelogram ΔT_i, note that its sides are given by the vectors

$$\mathbf{u} = \Delta x_i \mathbf{i} + f_x(x_i, y_i)\, \Delta x_i \mathbf{k}$$

and

$$\mathbf{v} = \Delta y_i \mathbf{j} + f_y(x_i, y_i)\, \Delta y_i \mathbf{k}.$$

From Theorem 11.8, the area of ΔT_i is given by $\|\mathbf{u} \times \mathbf{v}\|$, where

$$
\mathbf{u} \times \mathbf{v} =
\begin{vmatrix}
\mathbf{i} & \mathbf{j} & \mathbf{k} \\
\Delta x_i & 0 & f_x(x_i, y_i)\, \Delta x_i \\
0 & \Delta y_i & f_y(x_i, y_i)\, \Delta y_i
\end{vmatrix}
$$

$$= -f_x(x_i, y_i)\, \Delta x_i \Delta y_i \mathbf{i} - f_y(x_i, y_i)\, \Delta x_i \Delta y_i \mathbf{j} + \Delta x_i \Delta y_i \mathbf{k}$$

$$= [-f_x(x_i, y_i)\mathbf{i} - f_y(x_i, y_i)\mathbf{j} + \mathbf{k}]\, \Delta A_i.$$

So, the area of ΔT_i is $\|\mathbf{u} \times \mathbf{v}\| = \sqrt{[f_x(x_i, y_i)]^2 + [f_y(x_i, y_i)]^2 + 1}\, \Delta A_i$, and

$$\text{Surface area of } S \approx \sum_{i=1}^{n} \Delta S_i$$

$$\approx \sum_{i=1}^{n} \sqrt{1 + [f_x(x_i, y_i)]^2 + [f_y(x_i, y_i)]^2}\, \Delta A_i.$$

This suggests the definition of surface area on the next page.

Definition of Surface Area

If f and its first partial derivatives are continuous on the closed region R in the xy-plane, then the **area of the surface S** given by $z = f(x, y)$ over R is defined as

$$\text{Surface area} = \iint_R dS$$

$$= \iint_R \sqrt{1 + [f_x(x, y)]^2 + [f_y(x, y)]^2}\, dA.$$

As an aid to remembering the double integral for surface area, it is helpful to note its similarity to the integral for arc length.

• • **REMARK** Note that the differential ds of arc length in the xy-plane is

$$\sqrt{1 + [f'(x)]^2}\, dx$$

and the differential dS of surface area in space is

$$\sqrt{1 + [f_x(x, y)]^2 + [f_y(x, y)]^2}\, dA.$$

▷

Length on x-axis: $\quad \displaystyle\int_a^b dx$

Arc length in xy-plane: $\quad \displaystyle\int_a^b ds = \int_a^b \sqrt{1 + [f'(x)]^2}\, dx$

Area in xy-plane: $\quad \displaystyle\iint_R dA$

Surface area in space: $\quad \displaystyle\iint_R dS = \iint_R \sqrt{1 + [f_x(x, y)]^2 + [f_y(x, y)]^2}\, dA$

Like integrals for arc length, integrals for surface area are often very difficult to evaluate. However, one type that is easily evaluated is demonstrated in the next example.

EXAMPLE 1 **The Surface Area of a Plane Region**

Find the surface area of the portion of the plane

$$z = 2 - x - y$$

that lies above the circle $x^2 + y^2 \le 1$ in the first quadrant, as shown in Figure 14.44.

Solution Note that $f(x, y) = 2 - x - y$, $f_x(x, y) = -1$, and $f_y(x, y) = -1$ are continuous on the region R. So, the surface area is given by

$$S = \iint_R \sqrt{1 + [f_x(x, y)]^2 + [f_y(x, y)]^2}\, dA \qquad \text{Formula for surface area}$$

$$= \iint_R \sqrt{1 + (-1)^2 + (-1)^2}\, dA \qquad \text{Substitute.}$$

$$= \iint_R \sqrt{3}\, dA$$

$$= \sqrt{3} \iint_R dA.$$

Note that the last integral is $\sqrt{3}$ times the area of the region R. Because R is a quarter circle of radius 1, the area of R is $\frac{1}{4}\pi(1^2)$ or $\pi/4$. So, the area of S is

$$S = \sqrt{3}\,(\text{area of } R)$$

$$= \sqrt{3}\left(\frac{\pi}{4}\right)$$

$$= \frac{\sqrt{3}\,\pi}{4}.$$

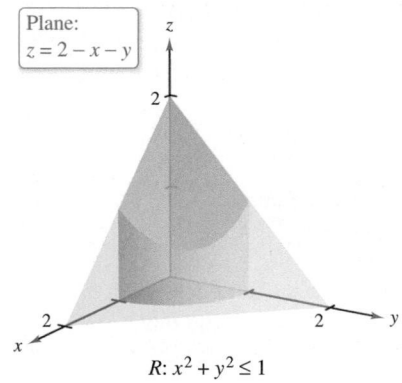

Plane:
$z = 2 - x - y$

$R: x^2 + y^2 \le 1$

Figure 14.44

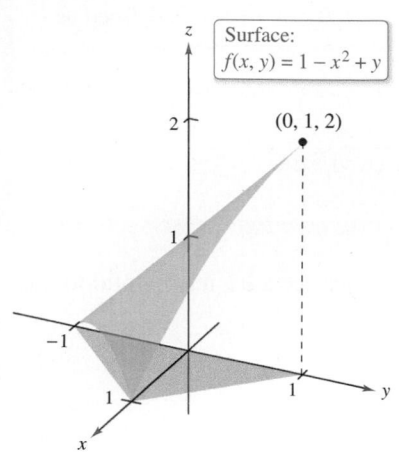

Figure 14.45

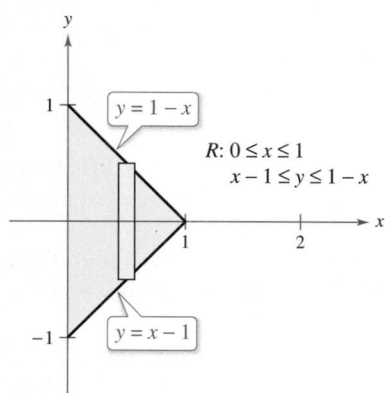

Figure 14.46

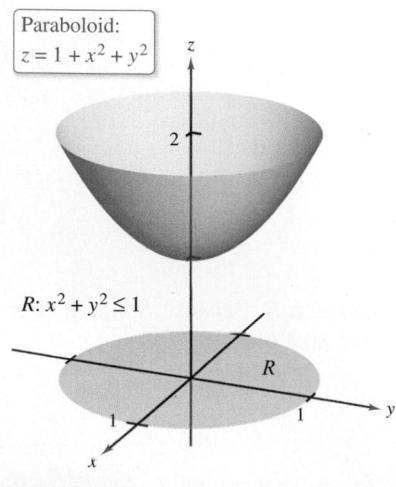

Figure 14.47

EXAMPLE 2 **Finding Surface Area**

∙∙∙∙▷ *See LarsonCalculus.com for an interactive version of this type of example.*

Find the area of the portion of the surface $f(x, y) = 1 - x^2 + y$ that lies above the triangular region with vertices $(1, 0, 0)$, $(0, -1, 0)$, and $(0, 1, 0)$, as shown in Figure 14.45.

Solution Because $f_x(x, y) = -2x$ and $f_y(x, y) = 1$, you have

$$S = \int_R\int \sqrt{1 + [f_x(x, y)]^2 + [f_y(x, y)]^2}\, dA = \int_R\int \sqrt{1 + 4x^2 + 1}\, dA.$$

In Figure 14.46, you can see that the bounds for R are $0 \le x \le 1$ and $x - 1 \le y \le 1 - x$. So, the integral becomes

$$S = \int_0^1\int_{x-1}^{1-x} \sqrt{2 + 4x^2}\, dy\, dx \qquad \text{Apply formula for surface area.}$$

$$= \int_0^1 y\sqrt{2 + 4x^2}\, \Big]_{x-1}^{1-x} dx$$

$$= \int_0^1 \left[(1 - x)\sqrt{2 + 4x^2} - (x - 1)\sqrt{2 + 4x^2}\right] dx$$

$$= \int_0^1 \left(2\sqrt{2 + 4x^2} - 2x\sqrt{2 + 4x^2}\right) dx \qquad \begin{array}{l}\text{Integration tables (Appendix B),}\\ \text{Formula 26 and Power Rule}\end{array}$$

$$= \left[x\sqrt{2 + 4x^2} + \ln\left(2x + \sqrt{2 + 4x^2}\right) - \frac{(2 + 4x^2)^{3/2}}{6}\right]_0^1$$

$$= \sqrt{6} + \ln\left(2 + \sqrt{6}\right) - \sqrt{6} - \ln\sqrt{2} + \frac{1}{3}\sqrt{2}$$

$$\approx 1.618.$$

EXAMPLE 3 **Change of Variables to Polar Coordinates**

Find the surface area of the paraboloid $z = 1 + x^2 + y^2$ that lies above the unit circle, as shown in Figure 14.47.

Solution Because $f_x(x, y) = 2x$ and $f_y(x, y) = 2y$, you have

$$S = \int_R\int \sqrt{1 + [f_x(x, y)]^2 + [f_y(x, y)]^2}\, dA = \int_R\int \sqrt{1 + 4x^2 + 4y^2}\, dA.$$

You can convert to polar coordinates by letting $x = r\cos\theta$ and $y = r\sin\theta$. Then, because the region R is bounded by $0 \le r \le 1$ and $0 \le \theta \le 2\pi$, you have

$$S = \int_0^{2\pi}\int_0^1 \sqrt{1 + 4r^2}\, r\, dr\, d\theta \qquad \text{Polar coordinates}$$

$$= \int_0^{2\pi} \frac{1}{12}(1 + 4r^2)^{3/2}\Big]_0^1 d\theta \qquad \text{Integrate with respect to } r.$$

$$= \int_0^{2\pi} \frac{5\sqrt{5} - 1}{12}\, d\theta$$

$$= \frac{5\sqrt{5} - 1}{12}\, \theta\, \Big]_0^{2\pi} \qquad \text{Integrate with respect to } \theta.$$

$$= \frac{\pi(5\sqrt{5} - 1)}{6}$$

$$\approx 5.33.$$

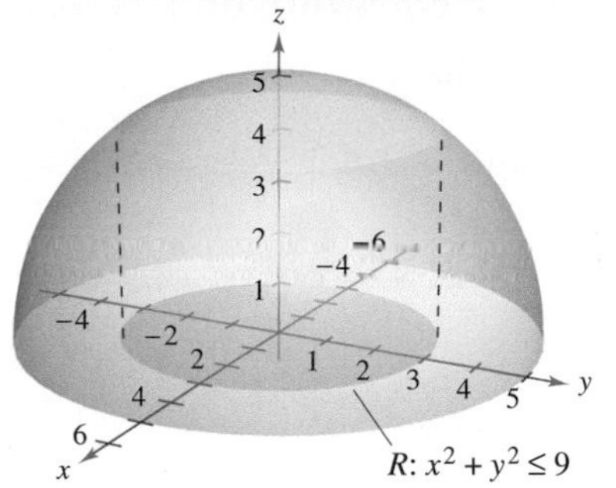

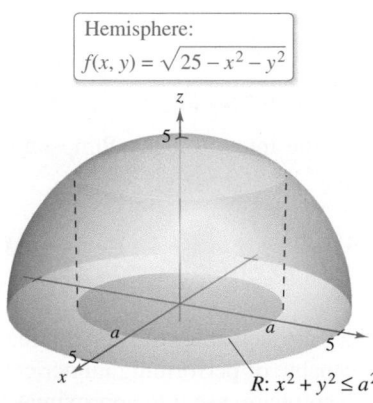

Hemisphere:
$f(x, y) = \sqrt{25 - x^2 - y^2}$

Figure 14.48

EXAMPLE 4   Finding Surface Area

Find the surface area S of the portion of the hemisphere

$$f(x, y) = \sqrt{25 - x^2 - y^2} \qquad \text{Hemisphere}$$

that lies above the region R bounded by the circle $x^2 + y^2 \le 9$, as shown in Figure 14.48.

Solution   The first partial derivatives of f are

$$f_x(x, y) = \frac{-x}{\sqrt{25 - x^2 - y^2}}$$

and

$$f_y(x, y) = \frac{-y}{\sqrt{25 - x^2 - y^2}}$$

and, from the formula for surface area, you have

$$dS = \sqrt{1 + [f_x(x, y)]^2 + [f_y(x, y)]^2} \, dA$$

$$= \sqrt{1 + \left(\frac{-x}{\sqrt{25 - x^2 - y^2}}\right)^2 + \left(\frac{-y}{\sqrt{25 - x^2 - y^2}}\right)^2} \, dA$$

$$= \frac{5}{\sqrt{25 - x^2 - y^2}} \, dA.$$

So, the surface area is

$$S = \int\int_R \frac{5}{\sqrt{25 - x^2 - y^2}} \, dA.$$

You can convert to polar coordinates by letting $x = r \cos \theta$ and $y = r \sin \theta$. Then, because the region R is bounded by $0 \le r \le 3$ and $0 \le \theta \le 2\pi$, you obtain

$$S = \int_0^{2\pi} \int_0^3 \frac{5}{\sqrt{25 - r^2}} \, r \, dr \, d\theta \qquad \text{Polar coordinates}$$

$$= 5 \int_0^{2\pi} -\sqrt{25 - r^2} \,\Big]_0^3 \, d\theta \qquad \text{Integrate with respect to } r.$$

$$= 5 \int_0^{2\pi} d\theta$$

$$= 10\pi. \qquad \text{Integrate with respect to } \theta.$$

The procedure used in Example 4 can be extended to find the surface area of a sphere by using the region R bounded by the circle $x^2 + y^2 \le a^2$, where $0 < a < 5$, as shown in Figure 14.49. The surface area of the portion of the hemisphere

$$f(x, y) = \sqrt{25 - x^2 - y^2}$$

lying above the circular region can be shown to be

$$S = \int\int_R \frac{5}{\sqrt{25 - x^2 - y^2}} \, dA$$

$$= \int_0^{2\pi} \int_0^a \frac{5}{\sqrt{25 - r^2}} \, r \, dr \, d\theta$$

$$= 10\pi\left(5 - \sqrt{25 - a^2}\right).$$

By taking the limit as a approaches 5 and doubling the result, you obtain a total area of 100π. (The surface area of a sphere of radius r is $S = 4\pi r^2$.)

Hemisphere:
$f(x, y) = \sqrt{25 - x^2 - y^2}$

Figure 14.49

You can use Simpson's Rule or the Trapezoidal Rule to approximate the value of a double integral, *provided* you can get through the first integration. This is demonstrated in the next example.

EXAMPLE 5 Approximating Surface Area by Simpson's Rule

Find the area of the surface of the paraboloid

$$f(x, y) = 2 - x^2 - y^2 \qquad \text{Paraboloid}$$

that lies above the square region bounded by

$$-1 \leq x \leq 1 \quad \text{and} \quad -1 \leq y \leq 1$$

as shown in Figure 14.50.

Solution Using the partial derivatives

$$f_x(x, y) = -2x \quad \text{and} \quad f_y(x, y) = -2y$$

you have a surface area of

$$S = \int_R \int \sqrt{1 + [f_x(x, y)]^2 + [f_y(x, y)]^2} \, dA \qquad \text{Formula for surface area}$$

$$= \int_R \int \sqrt{1 + (-2x)^2 + (-2y)^2} \, dA \qquad \text{Substitute.}$$

$$= \int_R \int \sqrt{1 + 4x^2 + 4y^2} \, dA. \qquad \text{Simplify.}$$

In polar coordinates, the line $x = 1$ is given by

$$r \cos \theta = 1 \quad \text{or} \quad r = \sec \theta$$

and you can determine from Figure 14.51 that one-fourth of the region R is bounded by

$$0 \leq r \leq \sec \theta \quad \text{and} \quad -\frac{\pi}{4} \leq \theta \leq \frac{\pi}{4}.$$

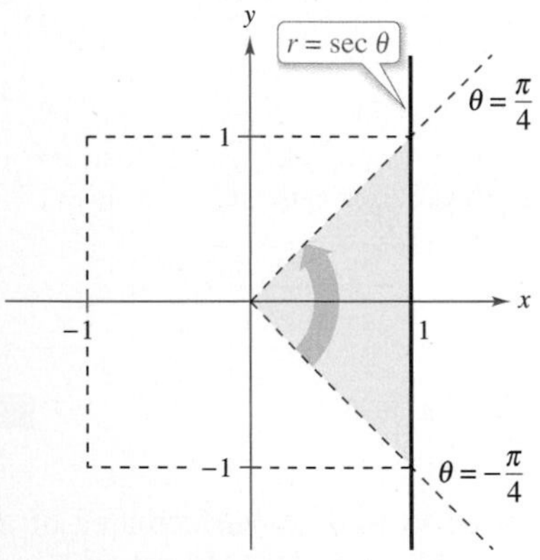

Letting $x = r \cos \theta$ and $y = r \sin \theta$ produces

$$\frac{1}{4} S = \frac{1}{4} \int_R \int \sqrt{1 + 4x^2 + 4y^2} \, dA \qquad \text{One-fourth of surface area}$$

$$= \int_{-\pi/4}^{\pi/4} \int_0^{\sec \theta} \sqrt{1 + 4r^2} \, r \, dr \, d\theta \qquad \text{Polar coordinates}$$

$$= \int_{-\pi/4}^{\pi/4} \frac{1}{12} (1 + 4r^2)^{3/2} \Big]_0^{\sec \theta} d\theta \qquad \text{Integrate with respect to } r.$$

$$= \frac{1}{12} \int_{-\pi/4}^{\pi/4} [(1 + 4 \sec^2 \theta)^{3/2} - 1] \, d\theta.$$

After multiplying each side by 4, you can approximate the integral using Simpson's Rule with $n = 10$ to find that the area of the surface is

$$S = 4 \left(\frac{1}{12} \right) \int_{-\pi/4}^{\pi/4} [(1 + 4 \sec^2 \theta)^{3/2} - 1] \, d\theta \approx 7.450. \qquad \blacksquare$$

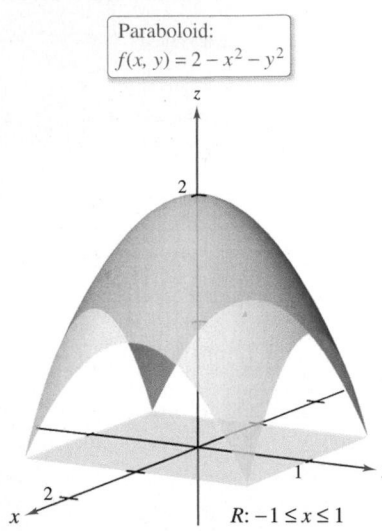

Paraboloid:

$$f(x, y) = 2 - x^2 - y^2$$

$R: -1 \leq x \leq 1$
$-1 \leq y \leq 1$

Figure 14.50

One-fourth of the region R is bounded by $0 \leq r \leq \sec \theta$ and $-\dfrac{\pi}{4} \leq \theta \leq \dfrac{\pi}{4}$.

Figure 14.51

▷ **TECHNOLOGY** Most computer programs that are capable of performing symbolic integration for multiple integrals are also capable of performing numerical approximation techniques. If you have access to such software, use it to approximate the value of the integral in Example 5.

14.5 Exercises

See **CalcChat.com** for tutorial help and worked-out solutions to odd-numbered exercises.

CONCEPT CHECK

1. **Surface Area** What is the differential of surface area, dS, in space?

2. **Numerical Integration** Write a double integral that represents the surface area of the portion of the plane $z = 3$ that lies above the rectangular region with vertices $(0, 0)$, $(4, 0)$, $(0, 5)$, and $(4, 5)$. Then find the surface area without integrating.

 **Finding Surface Area** In Exercises 3–16, find the area of the surface given by $z = f(x, y)$ that lies above the region R.

3. $f(x, y) = 2x + 2y$

   R: triangle with vertices $(0, 0)$, $(4, 0)$, $(0, 4)$

4. $f(x, y) = 15 + 2x - 3y$

   R: square with vertices $(0, 0)$, $(3, 0)$, $(0, 3)$, $(3, 3)$

5. $f(x, y) = 4 + 5x + 6y$, $R = \{(x, y): x^2 + y^2 \le 4\}$

6. $f(x, y) = 12 + 2x - 3y$, $R = \{(x, y): x^2 + y^2 \le 9\}$

7. $f(x, y) = 9 - x^2$

   R: square with vertices $(0, 0)$, $(2, 0)$, $(0, 2)$, $(2, 2)$

8. $f(x, y) = y^2$

   R: square with vertices $(0, 0)$, $(3, 0)$, $(0, 3)$, $(3, 3)$

9. $f(x, y) = 3 + 2x^{3/2}$

   R: rectangle with vertices $(0, 0)$, $(0, 4)$, $(1, 4)$, $(1, 0)$

10. $f(x, y) = 2 + \frac{2}{3}y^{3/2}$

    $R = \{(x, y): 0 \le x \le 2, 0 \le y \le 2 - x\}$

11. $f(x, y) = \ln|\sec x|$

    $R = \left\{(x, y): 0 \le x \le \frac{\pi}{4}, 0 \le y \le \tan x\right\}$

12. $f(x, y) = 13 + x^2 - y^2$, $R = \{(x, y): x^2 + y^2 \le 4\}$

13. $f(x, y) = \sqrt{x^2 + y^2}$, $R = \{(x, y): 0 \le f(x, y) \le 1\}$

14. $f(x, y) = xy$, $R = \{(x, y): x^2 + y^2 \le 16\}$

15. $f(x, y) = \sqrt{a^2 - x^2 - y^2}$

    $R = \{(x, y): x^2 + y^2 \le b^2, \ 0 < b < a\}$

16. $f(x, y) = \sqrt{a^2 - x^2 - y^2}$

    $R = \{(x, y): x^2 + y^2 \le a^2\}$

 **Finding Surface Area** In Exercises 17–20, find the area of the surface.

17. The portion of the plane $z = 12 - 3x - 2y$ in the first octant

18. The portion of the paraboloid $z = 16 - x^2 - y^2$ in the first octant

19. The portion of the sphere $x^2 + y^2 + z^2 = 25$ inside the cylinder $x^2 + y^2 = 9$

20. The portion of the cone $z = 2\sqrt{x^2 + y^2}$ inside the cylinder $x^2 + y^2 = 4$

 Finding Surface Area Using Technology In Exercises 21–26, write a double integral that represents the surface area of $z = f(x, y)$ that lies above the region R. Use a computer algebra system to evaluate the double integral.

21. $f(x, y) = 2y + x^2$, R: triangle with vertices $(0, 0)$, $(1, 0)$, $(1, 1)$

22. $f(x, y) = 2x + y^2$, R: triangle with vertices $(0, 0)$, $(2, 0)$, $(2, 2)$

23. $f(x, y) = 9 - x^2 - y^2$, $R = \{(x, y): 0 \le f(x, y)\}$

24. $f(x, y) = x^2 + y^2$, $R = \{(x, y): 0 \le f(x, y) \le 16\}$

25. $f(x, y) = 4 - x^2 - y^2$

    $R = \{(x, y): 0 \le x \le 1, 0 \le y \le 1\}$

26. $f(x, y) = \frac{2}{3}x^{3/2} + \cos x$

    $R = \{(x, y): 0 \le x \le 1, 0 \le y \le 1\}$

Setting Up a Double Integral In Exercises 27–30, set up a double integral that represents the area of the surface given by $z = f(x, y)$ that lies above the region R.

27. $f(x, y) = e^{xy}$, $R = \{(x, y): 0 \le x \le 4, 0 \le y \le 10\}$

28. $f(x, y) = x^2 - 3xy - y^2$

    $R = \{(x, y): 0 \le x \le 4, 0 \le y \le x\}$

29. $f(x, y) = e^{-x} \sin y$, $R = \{(x, y): x^2 + y^2 \le 4\}$

30. $f(x, y) = \cos(x^2 + y^2)$, $R = \left\{(x, y): x^2 + y^2 \le \frac{\pi}{2}\right\}$

EXPLORING CONCEPTS

31. **Surface Area** Will the surface area of the graph of a function $z = f(x, y)$ that lies above a region R increase when the graph is shifted k units vertically? Explain using the partial derivatives of z.

32. **HOW DO YOU SEE IT?** Consider the surface $f(x, y) = x^2 + y^2$ (see figure) and the surface area of f that lies above each region R. Without integrating, order the surface areas from least to greatest. Explain.

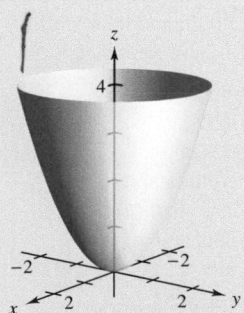

(a) R: rectangle with vertices $(0, 0)$, $(2, 0)$, $(2, 2)$, $(0, 2)$

(b) R: triangle with vertices $(0, 0)$, $(2, 0)$, $(0, 2)$

(c) $R = \{(x, y): x^2 + y^2 \le 4, \text{ first quadrant only}\}$

33. Surface Area Answer each question about the surface area S on a surface given by a positive function $z = f(x, y)$ that lies above a region R in the xy-plane. Explain each answer.

(a) Is it possible for S to equal the area of R?

(b) Can S be greater than the area of R?

(c) Can S be less than the area of R?

34. Surface Area Consider the surface $f(x, y) = x + y$. What is the relationship between the area of the surface that lies above the region

$$R_1 = \{(x, y): x^2 + y^2 \le 1\}$$

and the area of the surface that lies above the region

$$R_2 = \{(x, y): x^2 + y^2 \le 4\}?$$

35. Product Design A company produces a spherical object of radius 25 centimeters. A hole of radius 4 centimeters is drilled through the center of the object.

(a) Find the volume of the object.

(b) Find the outer surface area of the object.

• • **36. Modeling Data** • • • • • • • • • • • • • • • • •

A company builds a warehouse with dimensions 30 feet by 50 feet. The symmetrical shape and selected heights of the roof are shown in the figure.

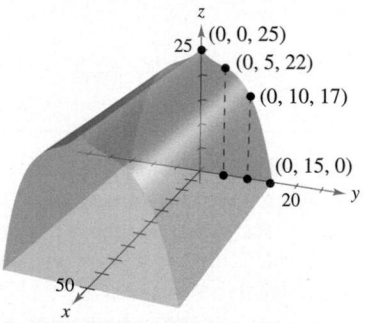

(a) Use the regression capabilities of a graphing utility to find a model of the form

$$z = ay^3 + by^2 + cy + d$$

for the roof line.

(b) Use the numerical integration capabilities of a graphing utility and the model in part (a) to approximate the volume of storage space in the warehouse.

(c) Use the numerical integration capabilities of a graphing utility and the model in part (a) to approximate the surface area of the roof.

(d) Approximate the arc length of the roof line and find the surface area of the roof by multiplying the arc length by the length of the warehouse. Compare the results and the integrations with those found in part (c).

37. Surface Area Find the surface area of the solid of intersection of the cylinders $x^2 + z^2 = 1$ and $y^2 + z^2 = 1$ (see figure).

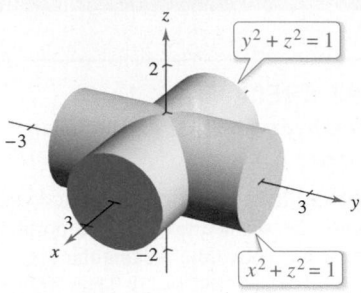

38. Surface Area Show that the surface area of the cone $z = k\sqrt{x^2 + y^2}$, $k > 0$, that lies above the circular region $x^2 + y^2 \le r^2$ in the xy-plane is $\pi r^2 \sqrt{k^2 + 1}$ (see figure).

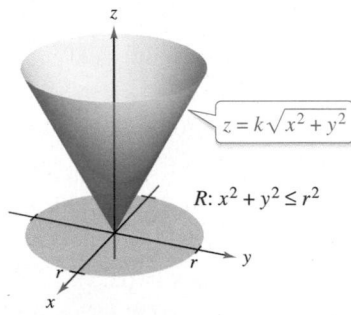

SECTION PROJECT

Surface Area in Polar Coordinates

(a) Use the formula for surface area in rectangular coordinates to derive the following formula for surface area in polar coordinates, where $z = f(x, y) = f(r \cos \theta, r \sin \theta)$. (*Hint:* You will need to use the Chain Rule for functions of two variables.)

$$S = \int\!\!\int_R \sqrt{1 + f_r^2 + \frac{1}{r^2} f_\theta^2} \; r \, dr \, d\theta$$

(b) Use the formula from part (a) to find the surface area of the paraboloid $z = x^2 + y^2$ that lies above the circular region $x^2 + y^2 \le 4$ in the xy-plane (see figure).

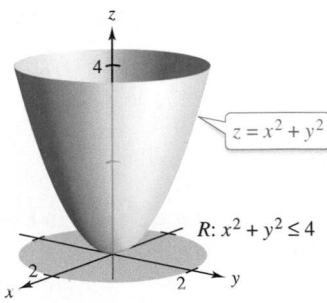

(c) Use the formula from part (a) to find the surface area of $z = xy$ that lies above the circular region $x^2 + y^2 \le 16$ in the xy-plane. Compare your answer with your answer to Exercise 14.

14.6 Triple Integrals and Applications

- Use a triple integral to find the volume of a solid region.
- Find the center of mass and moments of inertia of a solid region.

Triple Integrals

The procedure used to define a **triple integral** follows that used for double integrals. Consider a function f of three variables that is continuous over a bounded solid region Q. Then encompass Q with a network of boxes and form the **inner partition** consisting of all boxes lying entirely within Q, as shown in Figure 14.52. The volume of the ith box is

$$\Delta V_i = \Delta x_i \Delta y_i \Delta z_i. \qquad \text{Volume of } i\text{th box}$$

The **norm** $\|\Delta\|$ of the partition is the length of the longest diagonal of the n boxes in the partition. Choose a point (x_i, y_i, z_i) in each box and form the Riemann sum

$$\sum_{i=1}^{n} f(x_i, y_i, z_i) \, \Delta V_i.$$

Taking the limit as $\|\Delta\| \to 0$ leads to the following definition.

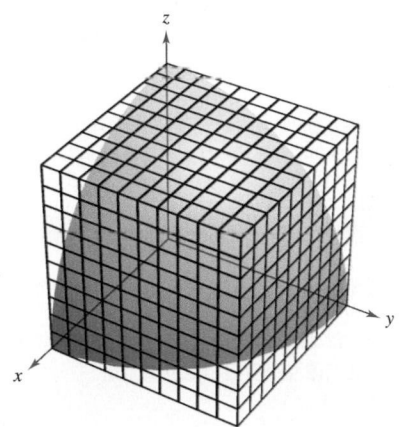

Solid region Q

Definition of Triple Integral

If f is continuous over a bounded solid region Q, then the **triple integral of f over Q** is defined as

$$\iiint_Q f(x, y, z) \, dV = \lim_{\|\Delta\| \to 0} \sum_{i=1}^{n} f(x_i, y_i, z_i) \, \Delta V_i$$

provided the limit exists. The **volume** of the solid region Q is given by

$$\text{Volume of } Q = \iiint_Q dV.$$

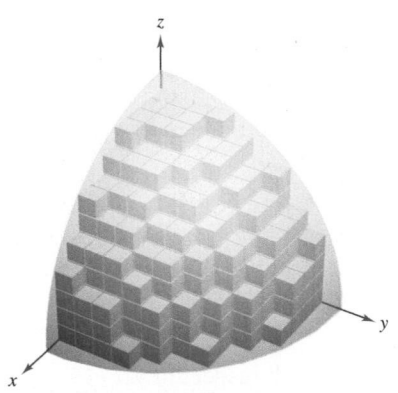

Volume of $Q \approx \displaystyle\sum_{i=1}^{n} \Delta V_i$

Figure 14.52

Some of the properties of double integrals in Theorem 14.1 can be restated in terms of triple integrals.

1. $\displaystyle\iiint_Q cf(x, y, z) \, dV = c \iiint_Q f(x, y, z) \, dV$

2. $\displaystyle\iiint_Q [f(x, y, z) \pm g(x, y, z)] \, dV = \iiint_Q f(x, y, z) \, dV \pm \iiint_Q g(x, y, z) \, dV$

3. $\displaystyle\iiint_Q f(x, y, z) \, dV = \iiint_{Q_1} f(x, y, z) \, dV + \iiint_{Q_2} f(x, y, z) \, dV$

In the properties above, Q is the union of two nonoverlapping solid subregions Q_1 and Q_2. If the solid region Q is simple, then the triple integral $\iiint f(x, y, z) \, dV$ can be evaluated with an iterated integral using one of the six possible orders of integration listed below.

$$dx \, dy \, dz \quad dy \, dx \, dz \quad dz \, dx \, dy$$
$$dx \, dz \, dy \quad dy \, dz \, dx \quad dz \, dy \, dx$$

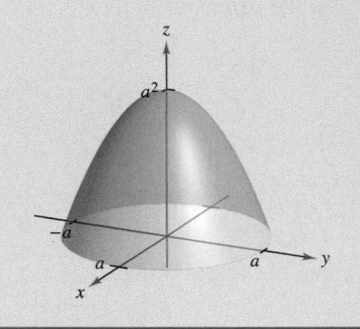
The following version of Fubini's Theorem describes a region that is considered simple with respect to the order *dz dy dx*. Similar versions of this theorem can be given for the other five orders.

THEOREM 14.4 Evaluation by Iterated Integrals

Let *f* be continuous on a solid region *Q* defined by

$$a \le x \le b,$$
$$h_1(x) \le y \le h_2(x),$$
$$g_1(x, y) \le z \le g_2(x, y)$$

where h_1, h_2, g_1, and g_2 are continuous functions. Then,

$$\iiint\limits_{Q} f(x, y, z)\, dV = \int_a^b \int_{h_1(x)}^{h_2(x)} \int_{g_1(x, y)}^{g_2(x, y)} f(x, y, z)\, dz\, dy\, dx.$$

To evaluate a triple iterated integral in the order *dz dy dx*, hold *both x* and *y* constant for the innermost integration. Then hold *x* constant for the second integration.

EXAMPLE 1 Evaluating a Triple Iterated Integral

Evaluate the triple iterated integral

$$\int_0^2 \int_0^x \int_0^{x+y} e^x(y + 2z)\, dz\, dy\, dx.$$

Solution For the first integration, hold *x* and *y* constant and integrate with respect to *z*.

$$\int_0^2 \int_0^x \int_0^{x+y} e^x(y + 2z)\, dz\, dy\, dx = \int_0^2 \int_0^x e^x(yz + z^2) \Big]_0^{x+y}\, dy\, dx$$

$$= \int_0^2 \int_0^x e^x(x^2 + 3xy + 2y^2)\, dy\, dx$$

For the second integration, hold *x* constant and integrate with respect to *y*.

$$\int_0^2 \int_0^x e^x(x^2 + 3xy + 2y^2)\, dy\, dx = \int_0^2 \left[e^x\left(x^2 y + \frac{3xy^2}{2} + \frac{2y^3}{3}\right) \right]_0^x\, dx$$

$$= \frac{19}{6} \int_0^2 x^3 e^x\, dx$$

Finally, integrate with respect to *x*.

$$\frac{19}{6} \int_0^2 x^3 e^x\, dx = \frac{19}{6} \left[e^x(x^3 - 3x^2 + 6x - 6) \right]_0^2$$

$$= 19\left(\frac{e^2}{3} + 1\right)$$

$$\approx 65.797$$

• • REMARK To do the last integration in Example 1, use integration by parts three times.

Example 1 demonstrates the integration order *dz dy dx*. For other orders, you can follow a similar procedure. For instance, to evaluate a triple iterated integral in the order *dx dy dz*, hold *both y* and *z* constant for the innermost integration and integrate with respect to *x*. Then, for the second integration, hold *z* constant and integrate with respect to *y*. Finally, for the third integration, integrate with respect to *z*.

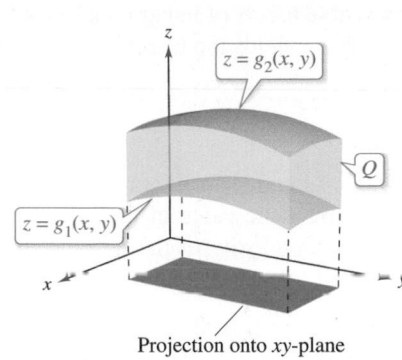

$z = g_2(x, y)$

$z = g_1(x, y)$

Projection onto xy-plane

Solid region Q lies between two surfaces.
Figure 14.53

To find the limits for a particular order of integration, it is generally advisable first to determine the innermost limits, which may be functions of the outer two variables. Then, by projecting the solid Q onto the coordinate plane of the outer two variables, you can determine their limits of integration by the methods used for double integrals. For instance, to evaluate

$$\iiint_Q f(x, y, z) \, dz \, dy \, dx$$

first determine the limits for z; the integral then has the form

$$\iint \left[\int_{g_1(x, y)}^{g_2(x, y)} f(x, y, z) \, dz \right] dy \, dx.$$

By projecting the solid Q onto the xy-plane, you can determine the limits for x and y as you did for double integrals, as shown in Figure 14.53.

EXAMPLE 2   **Using a Triple Integral to Find Volume**

Find the volume of the ellipsoid given by $4x^2 + 4y^2 + z^2 = 16$.

Solution   Because x, y, and z play similar roles in the equation, the order of integration is probably immaterial, and you can arbitrarily choose $dz \, dy \, dx$. Moreover, you can simplify the calculation by considering only the portion of the ellipsoid lying in the first octant, as shown in Figure 14.54. From the order $dz \, dy \, dx$, you first determine the bounds for z.

$$0 \leq z \leq 2\sqrt{4 - x^2 - y^2} \qquad \text{Bounds for } z$$

In Figure 14.55, you can see that the bounds for x and y are

$$0 \leq x \leq 2 \quad \text{and} \quad 0 \leq y \leq \sqrt{4 - x^2}. \qquad \text{Bounds for } x \text{ and } y$$

So, the volume of the ellipsoid is

$$V = \iiint_Q dV \qquad \text{Formula for volume}$$

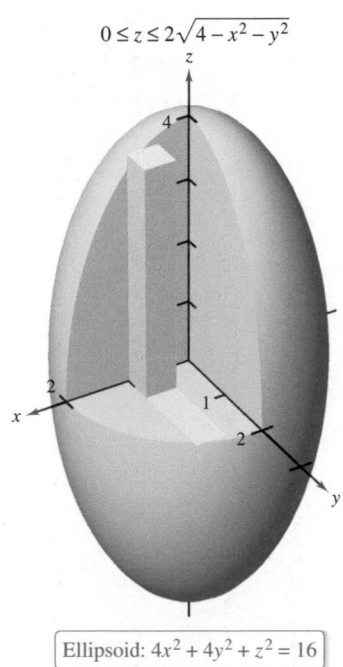

$0 \leq z \leq 2\sqrt{4 - x^2 - y^2}$

Ellipsoid: $4x^2 + 4y^2 + z^2 = 16$

Figure 14.54

$$= 8 \int_0^2 \int_0^{\sqrt{4-x^2}} \int_0^{2\sqrt{4-x^2-y^2}} dz \, dy \, dx \qquad \text{Convert to iterated integral.}$$

$$= 8 \int_0^2 \int_0^{\sqrt{4-x^2}} z \Big]_0^{2\sqrt{4-x^2-y^2}} dy \, dx$$

$$= 16 \int_0^2 \int_0^{\sqrt{4-x^2}} \sqrt{(4 - x^2) - y^2} \, dy \, dx \qquad \begin{array}{l}\text{Integration tables (Appendix B),}\\\text{Formula 37}\end{array}$$

$$= 8 \int_0^2 \left[y\sqrt{4 - x^2 - y^2} + (4 - x^2) \arcsin\left(\frac{y}{\sqrt{4 - x^2}}\right) \right]_0^{\sqrt{4-x^2}} dx$$

$$= 8 \int_0^2 [0 + (4 - x^2) \arcsin(1) - 0 - 0] \, dx$$

$$= 8 \int_0^2 (4 - x^2)\left(\frac{\pi}{2}\right) dx$$

$$= 4\pi \left[4x - \frac{x^3}{3} \right]_0^2$$

$$= \frac{64\pi}{3}.$$

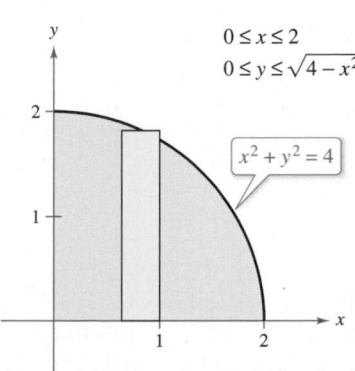

$0 \leq x \leq 2$
$0 \leq y \leq \sqrt{4 - x^2}$

$x^2 + y^2 = 4$

Figure 14.55

Example 2 is unusual in that all six possible orders of integration produce integrals of comparable difficulty. Try setting up some other possible orders of integration to find the volume of the elipsoid. For instance, the order $dx\,dy\,dz$ yields the integral

$$V = 8 \int_0^4 \int_0^{\sqrt{16-z^2}/2} \int_0^{\sqrt{16-4y^2-z^2}/2} dx\,dy\,dz.$$

The evaluation of this integral yields the same volume obtained in Example 2. This is always the case—the order of integration does not affect the value of the integral. However, the order of integration often does affect the complexity of the integral. In Example 3, the given order of integration is not convenient, so you can change the order to simplify the problem.

EXAMPLE 3   Changing the Order of Integration

Evaluate $\displaystyle\int_0^{\sqrt{\pi/2}} \int_x^{\sqrt{\pi/2}} \int_1^3 \sin(y^2)\,dz\,dy\,dx.$

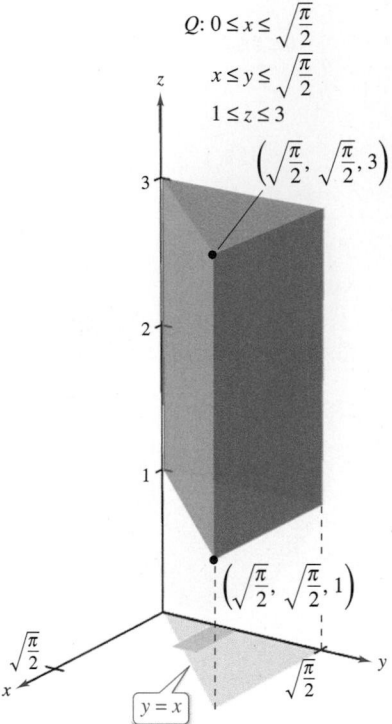

$Q: 0 \le x \le \sqrt{\dfrac{\pi}{2}}$

$x \le y \le \sqrt{\dfrac{\pi}{2}}$

$1 \le z \le 3$

$\left(\sqrt{\dfrac{\pi}{2}}, \sqrt{\dfrac{\pi}{2}}, 3\right)$

$\left(\sqrt{\dfrac{\pi}{2}}, \sqrt{\dfrac{\pi}{2}}, 1\right)$

$y = x$

Figure 14.56

Solution   Note that after one integration in the given order, you would encounter the integral $2\int \sin(y^2)\,dy$, which is not an elementary function. To avoid this problem, change the order of integration to $dz\,dx\,dy$ so that y is the outer variable. From Figure 14.56, you can see that the solid region Q is

$$0 \le x \le \sqrt{\frac{\pi}{2}}$$

$$x \le y \le \sqrt{\frac{\pi}{2}}$$

$$1 \le z \le 3$$

and the projection of Q in the xy-plane yields the bounds

$$0 \le y \le \sqrt{\frac{\pi}{2}}$$

and

$$0 \le x \le y.$$

So, evaluating the triple integral using the order $dz\,dx\,dy$ produces

$$\int_0^{\sqrt{\pi/2}} \int_0^y \int_1^3 \sin(y^2)\,dz\,dx\,dy = \int_0^{\sqrt{\pi/2}} \int_0^y z \sin(y^2)\Big]_1^3 dx\,dy$$

$$= 2 \int_0^{\sqrt{\pi/2}} \int_0^y \sin(y^2)\,dx\,dy$$

$$= 2 \int_0^{\sqrt{\pi/2}} x \sin(y^2)\Big]_0^y dy$$

$$= 2 \int_0^{\sqrt{\pi/2}} y \sin(y^2)\,dy$$

$$= -\cos(y^2)\Big]_0^{\sqrt{\pi/2}}$$

$$= 1.$$

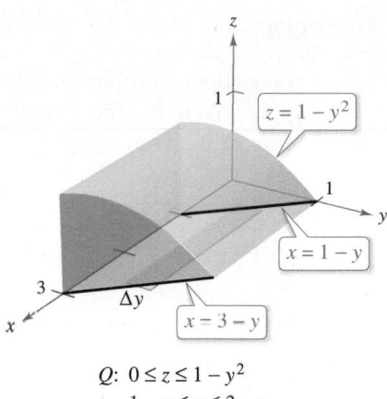

$Q: 0 \le z \le 1 - y^2$
$\quad 1 - y \le x \le 3 - y$
$\quad 0 \le y \le 1$

Figure 14.57

EXAMPLE 4 **Determining the Limits of Integration**

Set up a triple integral for the volume of each solid region.

a. The region in the first octant bounded above by the cylinder $z = 1 - y^2$ and lying between the vertical planes $x + y = 1$ and $x + y = 3$

b. The upper hemisphere $z = \sqrt{1 - x^2 - y^2}$

c. The region bounded below by the paraboloid $z = x^2 + y^2$ and above by the sphere $x^2 + y^2 + z^2 = 6$

Solution

a. In Figure 14.57, note that the solid is bounded below by the xy-plane ($z = 0$) and above by the cylinder $z = 1 - y^2$. So,

$$0 \le z \le 1 - y^2. \qquad \text{Bounds for } z$$

Projecting the region onto the xy-plane produces a parallelogram. Because two sides of the parallelogram are parallel to the x-axis, you have the following bounds:

$$1 - y \le x \le 3 - y \quad \text{and} \quad 0 \le y \le 1. \qquad \text{Bounds for } x \text{ and } y$$

So, the volume of the region is given by

$$V = \iiint_Q dV = \int_0^1 \int_{1-y}^{3-y} \int_0^{1-y^2} dz\, dx\, dy.$$

b. For the upper hemisphere $z = \sqrt{1 - x^2 - y^2}$, you have

$$0 \le z \le \sqrt{1 - x^2 - y^2}. \qquad \text{Bounds for } z$$

In Figure 14.58, note that the projection of the hemisphere onto the xy-plane is the circle

$$x^2 + y^2 = 1$$

and you can use either order $dx\, dy$ or $dy\, dx$. Choosing the first produces

$$-\sqrt{1 - y^2} \le x \le \sqrt{1 - y^2} \quad \text{and} \quad -1 \le y \le 1 \qquad \text{Bounds for } x \text{ and } y$$

which implies that the volume of the region is given by

$$V = \iiint_Q dV = \int_{-1}^1 \int_{-\sqrt{1-y^2}}^{\sqrt{1-y^2}} \int_0^{\sqrt{1-x^2-y^2}} dz\, dx\, dy.$$

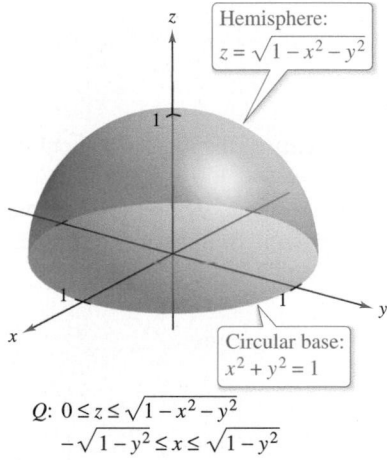

$Q: 0 \le z \le \sqrt{1 - x^2 - y^2}$
$\quad -\sqrt{1 - y^2} \le x \le \sqrt{1 - y^2}$
$\quad -1 \le y \le 1$

Figure 14.58

c. For the region bounded below by the paraboloid $z = x^2 + y^2$ and above by the sphere $x^2 + y^2 + z^2 = 6$, you have

$$x^2 + y^2 \le z \le \sqrt{6 - x^2 - y^2}. \qquad \text{Bounds for } z$$

The sphere and the paraboloid intersect at $z = 2$. Moreover, you can see in Figure 14.59 that the projection of the solid region onto the xy-plane is the circle

$$x^2 + y^2 = 2.$$

Using the order $dy\, dx$ produces

$$-\sqrt{2 - x^2} \le y \le \sqrt{2 - x^2} \quad \text{and} \quad -\sqrt{2} \le x \le \sqrt{2} \qquad \text{Bounds for } x \text{ and } y$$

which implies that the volume of the region is given by

$$V = \iiint_Q dV = \int_{-\sqrt{2}}^{\sqrt{2}} \int_{-\sqrt{2-x^2}}^{\sqrt{2-x^2}} \int_{x^2+y^2}^{\sqrt{6-x^2-y^2}} dz\, dy\, dx. \qquad \blacksquare$$

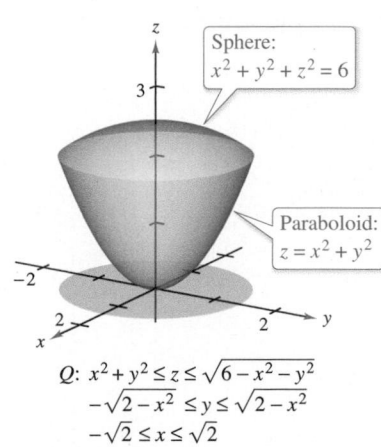

$Q: x^2 + y^2 \le z \le \sqrt{6 - x^2 - y^2}$
$\quad -\sqrt{2 - x^2} \le y \le \sqrt{2 - x^2}$
$\quad -\sqrt{2} \le x \le \sqrt{2}$

Figure 14.59

Center of Mass and Moments of Inertia

In the remainder of this section, two important engineering applications of triple integrals are discussed. Consider a solid region Q whose density is given by the **density function** ρ. The **center of mass** of a solid region Q of mass m is given by $(\bar{x}, \bar{y}, \bar{z})$, where

$$m = \iiint\limits_{Q} \rho(x, y, z) \, dV \qquad \text{Mass of the solid}$$

$$M_{yz} = \iiint\limits_{Q} x\rho(x, y, z) \, dV \qquad \text{First moment about } yz\text{-plane}$$

$$M_{xz} = \iiint\limits_{Q} y\rho(x, y, z) \, dV \qquad \text{First moment about } xz\text{-plane}$$

$$M_{xy} = \iiint\limits_{Q} z\rho(x, y, z) \, dV \qquad \text{First moment about } xy\text{-plane}$$

and

$$\bar{x} = \frac{M_{yz}}{m}, \quad \bar{y} = \frac{M_{xz}}{m}, \quad \bar{z} = \frac{M_{xy}}{m}.$$

The quantities M_{yz}, M_{xz}, and M_{xy} are called the **first moments** of the region Q about the yz-, xz-, and xy-planes, respectively.

The first moments for solid regions are taken about a plane, whereas the second moments for solids are taken about a line. The **second moments** (or **moments of inertia**) about the x-, y-, and z-axes are

$$I_x = \iiint\limits_{Q} (y^2 + z^2)\rho(x, y, z) \, dV \qquad \text{Moment of inertia about } x\text{-axis}$$

$$I_y = \iiint\limits_{Q} (x^2 + z^2)\rho(x, y, z) \, dV \qquad \text{Moment of inertia about } y\text{-axis}$$

and

$$I_z = \iiint\limits_{Q} (x^2 + y^2)\rho(x, y, z) \, dV. \qquad \text{Moment of inertia about } z\text{-axis}$$

For problems requiring the calculation of all three moments, considerable effort can be saved by applying the additive property of triple integrals and writing

$$I_x = I_{xz} + I_{xy}, \quad I_y = I_{yz} + I_{xy}, \quad \text{and} \quad I_z = I_{yz} + I_{xz}$$

where I_{xy}, I_{xz}, and I_{yz} are

$$I_{xy} = \iiint\limits_{Q} z^2\rho(x, y, z) \, dV$$

$$I_{xz} = \iiint\limits_{Q} y^2\rho(x, y, z) \, dV$$

and

$$I_{yz} = \iiint\limits_{Q} x^2\rho(x, y, z) \, dV.$$

• • REMARK In engineering and physics, the moment of inertia of a mass is used to find the time required for the mass to reach a given speed of rotation about an axis, as shown in Figure 14.60. The greater the moment of inertia, the longer a force must be applied for the mass to reach the given speed.

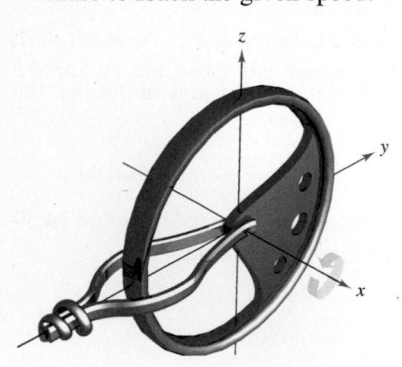

Figure 14.60

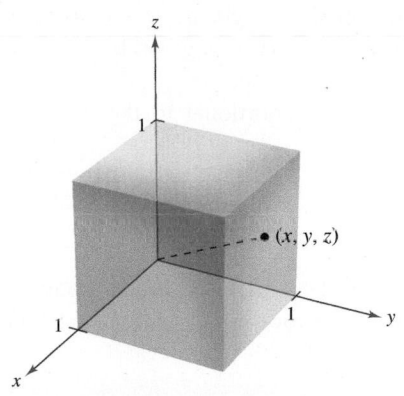

Variable density:
$\rho(x, y, z) = k(x^2 + y^2 + z^2)$
Figure 14.61

EXAMPLE 5 **Finding the Center of Mass of a Solid Region**

⋮ ⋅ ⋅ ⋅ ▷ *See LarsonCalculus.com for an interactive version of this type of example.*

Find the center of mass of the unit cube shown in Figure 14.61, given that the density at the point (x, y, z) is proportional to the square of its distance from the origin.

Solution Because the density at (x, y, z) is proportional to the square of the distance between $(0, 0, 0)$ and (x, y, z), you have

$$\rho(x, y, z) = k(x^2 + y^2 + z^2)$$

where k is the constant of proportionality. You can use this density function to find the mass of the cube. Because of the symmetry of the region, any order of integration will produce an integral of comparable difficulty.

$$m = \int_0^1 \int_0^1 \int_0^1 k(x^2 + y^2 + z^2) \, dz \, dy \, dx \qquad \text{Apply formula for mass of a solid.}$$

$$= k \int_0^1 \int_0^1 \left[(x^2 + y^2)z + \frac{z^3}{3} \right]_0^1 dy \, dx \qquad \text{Integrate with respect to } z.$$

$$= k \int_0^1 \int_0^1 \left(x^2 + y^2 + \frac{1}{3} \right) dy \, dx$$

$$= k \int_0^1 \left[\left(x^2 + \frac{1}{3} \right)y + \frac{y^3}{3} \right]_0^1 dx \qquad \text{Integrate with respect to } y.$$

$$= k \int_0^1 \left(x^2 + \frac{2}{3} \right) dx$$

$$= k \left[\frac{x^3}{3} + \frac{2x}{3} \right]_0^1 \qquad \text{Integrate with respect to } x.$$

$$= k$$

The first moment about the yz-plane is

$$M_{yz} = k \int_0^1 \int_0^1 \int_0^1 x(x^2 + y^2 + z^2) \, dz \, dy \, dx \qquad \begin{array}{l}\text{Apply formula for first moment}\\\text{about } yz\text{-plane.}\end{array}$$

$$= k \int_0^1 x \left[\int_0^1 \int_0^1 (x^2 + y^2 + z^2) \, dz \, dy \right] dx. \qquad \text{Factor.}$$

Note that x can be factored out of the two inner integrals, because it is constant with respect to y and z. After factoring, the two inner integrals are the same as for the mass m. Therefore, you have

$$M_{yz} = k \int_0^1 x \left(x^2 + \frac{2}{3} \right) dx$$

$$= k \left[\frac{x^4}{4} + \frac{x^2}{3} \right]_0^1 \qquad \text{Integrate with respect to } x.$$

$$= \frac{7k}{12}. \qquad \text{First moment about } yz\text{-plane}$$

So,

$$\bar{x} = \frac{M_{yz}}{m} = \frac{7k/12}{k} = \frac{7}{12}.$$

Finally, from the nature of ρ and the symmetry of x, y, and z in this solid region, you have $\bar{x} = \bar{y} = \bar{z}$, and the center of mass is $\left(\frac{7}{12}, \frac{7}{12}, \frac{7}{12} \right)$. ∎

EXAMPLE 6   **Moments of Inertia for a Solid Region**

Find the moments of inertia about the x- and y-axes for the solid region lying between the hemisphere

$$z = \sqrt{4 - x^2 - y^2}$$

and the xy-plane, given that the density at (x, y, z) is proportional to the distance between (x, y, z) and the xy-plane.

Solution   The density of the region is given by

$$\rho(x, y, z) = kz$$

where k is the constant of proportionality. Considering the symmetry of this problem, you know that $I_x = I_y$, and you need to find only one moment, such as I_x. From Figure 14.62, choose the order $dz\, dy\, dx$ and write

$$I_x = \iiint_Q (y^2 + z^2)\rho(x, y, z)\, dV \qquad \text{Moment of inertia about } x\text{-axis}$$

$$= \int_{-2}^{2} \int_{-\sqrt{4-x^2}}^{\sqrt{4-x^2}} \int_{0}^{\sqrt{4-x^2-y^2}} (y^2 + z^2)(kz)\, dz\, dy\, dx$$

$$= k \int_{-2}^{2} \int_{-\sqrt{4-x^2}}^{\sqrt{4-x^2}} \left[\frac{y^2 z^2}{2} + \frac{z^4}{4} \right]_{0}^{\sqrt{4-x^2-y^2}} dy\, dx$$

$$= k \int_{-2}^{2} \int_{-\sqrt{4-x^2}}^{\sqrt{4-x^2}} \left[\frac{y^2(4 - x^2 - y^2)}{2} + \frac{(4 - x^2 - y^2)^2}{4} \right] dy\, dx$$

$$= \frac{k}{4} \int_{-2}^{2} \int_{-\sqrt{4-x^2}}^{\sqrt{4-x^2}} [(4 - x^2)^2 - y^4]\, dy\, dx$$

$$= \frac{k}{4} \int_{-2}^{2} \left[(4 - x^2)^2 y - \frac{y^5}{5} \right]_{-\sqrt{4-x^2}}^{\sqrt{4-x^2}} dx$$

$$= \frac{k}{4} \int_{-2}^{2} \frac{8}{5}(4 - x^2)^{5/2}\, dx$$

$$= \frac{4k}{5} \int_{0}^{2} (4 - x^2)^{5/2}\, dx$$

$$= \frac{4k}{5} \int_{0}^{\pi/2} 64 \cos^6 \theta\, d\theta \qquad \text{Trigonometric substitution: } x = 2 \sin \theta$$

$$= \left(\frac{256k}{5} \right)\left(\frac{5\pi}{32} \right) \qquad \text{Wallis's Formula}$$

$$= 8k\pi.$$

So, $I_x = 8k\pi = I_y$.

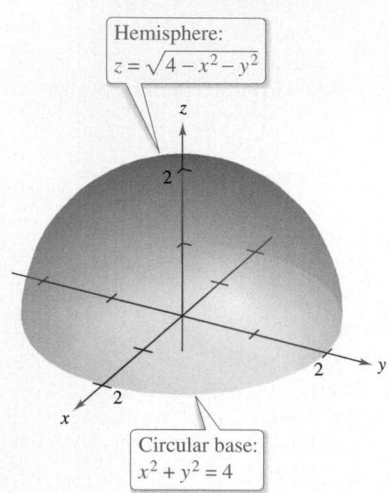

$$0 \leq z \leq \sqrt{4 - x^2 - y^2}$$
$$-\sqrt{4 - x^2} \leq y \leq \sqrt{4 - x^2}$$
$$-2 \leq x \leq 2$$

Hemisphere:
$$z = \sqrt{4 - x^2 - y^2}$$

Circular base:
$$x^2 + y^2 = 4$$

Variable density: $\rho(x, y, z) = kz$
Figure 14.62

In Example 6, notice that the moments of inertia about the x- and y-axes are equal to each other. The moment about the z-axis, however, is different. Does it seem that the moment of inertia about the z-axis should be less than or greater than the moments calculated in Example 6? By performing the calculations, you can determine that

$$I_z = \frac{16}{3}k\pi.$$

This tells you that the solid shown in Figure 14.62 has a greater resistance to rotation about the x- or y-axis than about the z-axis.

14.6 Exercises

See **CalcChat.com** for tutorial help and worked-out solutions to odd-numbered exercises.

CONCEPT CHECK

1. Triple Integrals What does $Q = \iiint_Q dV$ represent?

2. Changing the Order of Integration Why is it beneficial to be able to change the order of integration for a triple integral? Explain.

Evaluating a Triple Iterated Integral In Exercises 3–10, evaluate the triple iterated integral.

3. $\int_0^3 \int_0^2 \int_0^1 (x + y + z)\, dx\, dz\, dy$

4. $\int_0^2 \int_0^1 \int_{-1}^2 xyz^3\, dx\, dy\, dz$

5. $\int_0^1 \int_0^x \int_0^{\sqrt{xy}} x\, dz\, dy\, dx$

6. $\int_0^9 \int_0^{y/3} \int_0^{\sqrt{y^2 - 9x^2}} z\, dz\, dx\, dy$

7. $\int_1^4 \int_0^1 \int_0^x 2ze^{-x^2}\, dy\, dx\, dz$

8. $\int_1^4 \int_1^{e^2} \int_0^{1/xz} \ln z\, dy\, dz\, dx$

9. $\int_{-3}^4 \int_0^{\pi/2} \int_0^{1+3x} x \cos y\, dz\, dy\, dx$

10. $\int_0^{\pi/2} \int_0^{y/2} \int_0^{1/y} \sin y\, dz\, dx\, dy$

Evaluating a Triple Iterated Integral Using Technology In Exercises 11 and 12, use a computer algebra system to evaluate the triple iterated integral.

11. $\int_0^3 \int_{-\sqrt{9-y^2}}^{\sqrt{9-y^2}} \int_0^{y^2} y\, dz\, dx\, dy$

12. $\int_0^3 \int_0^{2-(2y/3)} \int_0^{6-2y-3z} ze^{-x^2 y^2}\, dx\, dz\, dy$

Setting Up a Triple Integral In Exercises 13–18, set up a triple integral for the volume of the solid. Do not evaluate the integral.

13. The solid in the first octant bounded by the coordinate planes and the plane $z = 7 - x - 2y$

14. The solid bounded by $z = 9 - x^2$, $z = 0$, $y = 0$, and $y = 2x$

15. The solid bounded by $z = 6 - x^2 - y^2$ and $z = 0$

16. The solid bounded by $z = \sqrt{1 - x^2 - y^2}$ and $z = 0$

17. The solid that is the common interior below the sphere $x^2 + y^2 + z^2 = 80$ and above the paraboloid $z = \frac{1}{2}(x^2 + y^2)$

18. The solid bounded above by the cylinder $z = 4 - x^2$ and below by the paraboloid $z = x^2 + 3y^2$

Volume In Exercises 19–24, use a triple integral to find the volume of the solid bounded by the graphs of the equations.

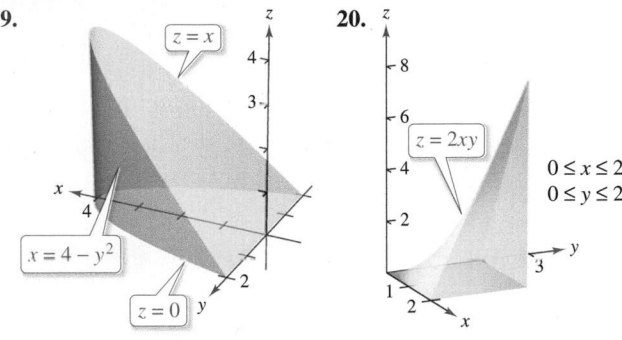

21. $z = 6x^2$, $y = 3 - 3x$, first octant

22. $z = 9 - x^3$, $y = -x^2 + 2$, $y = 0$, $z = 0$, $x \geq 0$

23. $z = 2 - y$, $z = 4 - y^2$, $x = 0$, $x = 3$, $y = 0$

24. $z = \sqrt{x}$, $y = x + 2$, $y = x^2$, first octant

Changing the Order of Integration In Exercises 25–30, sketch the solid whose volume is given by the iterated integral. Then rewrite the integral using the indicated order of integration.

25. $\int_0^1 \int_{-1}^0 \int_0^{y^2} dz\, dy\, dx$

Rewrite using $dy\, dz\, dx$.

26. $\int_{-1}^1 \int_{y^2}^1 \int_0^{1-x} dz\, dx\, dy$

Rewrite using $dx\, dz\, dy$.

27. $\int_0^4 \int_0^{(4-x)/2} \int_0^{(12-3x-6y)/4} dz\, dy\, dx$

Rewrite using $dy\, dx\, dz$.

28. $\int_0^3 \int_0^{\sqrt{9-x^2}} \int_0^{6-x-y} dz\, dy\, dx$

Rewrite using $dz\, dx\, dy$.

29. $\int_0^1 \int_y^1 \int_0^{\sqrt{1-y^2}} dz\, dx\, dy$

Rewrite using $dz\, dy\, dx$.

30. $\int_0^2 \int_{2x}^4 \int_0^{\sqrt{y^2-4x^2}} dz\, dy\, dx$

Rewrite using $dx\, dy\, dz$.

Orders of Integration In Exercises 31–34, write a triple integral for $f(x, y, z) = xyz$ over the solid region Q for each of the six possible orders of integration. Then evaluate one of the triple integrals.

31. $Q = \{(x, y, z): 0 \leq x \leq 1, 0 \leq y \leq 5x, 0 \leq z \leq 3\}$

32. $Q = \{(x, y, z): 0 \leq x \leq 2, x^2 \leq y \leq 4, 0 \leq z \leq 2 - x\}$

33. $Q = \{(x, y, z): x^2 + y^2 \leq 9, 0 \leq z \leq 4\}$

34. $Q = \{(x, y, z): 0 \leq x \leq 1, 0 \leq y \leq 1 - x^2, 0 \leq z \leq 6\}$

Orders of Integration In Exercises 35 and 36, the figure shows the region of integration for the given integral. Rewrite the integral as an equivalent iterated integral in the five other orders.

35. $\int_0^1 \int_0^{1-y^2} \int_0^{1-y} dz\, dx\, dy$ **36.** $\int_0^3 \int_0^x \int_0^{9-x^2} dz\, dy\, dx$

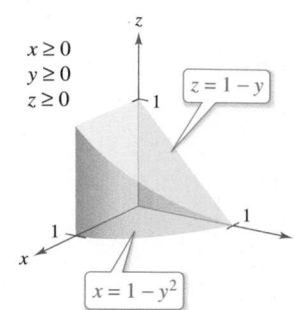

$x \geq 0$
$y \geq 0$
$z \geq 0$

$z = 1 - y$

$x = 1 - y^2$

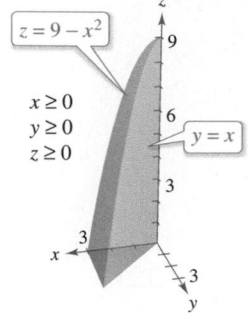

$z = 9 - x^2$

$x \geq 0$
$y \geq 0$
$z \geq 0$

$y = x$

Center of Mass In Exercises 37–40, find the mass and the indicated coordinate of the center of mass of the solid region Q of density ρ bounded by the graphs of the equations.

37. Find $\bar{x}$ using $\rho(x, y, z) = k$.

Q: $2x + 3y + 6z = 12$, $x = 0$, $y = 0$, $z = 0$

38. Find $\bar{y}$ using $\rho(x, y, z) = ky$.

Q: $3x + 3y + 5z = 15$, $x = 0$, $y = 0$, $z = 0$

39. Find $\bar{z}$ using $\rho(x, y, z) = kx$.

Q: $z = 4 - x$, $z = 0$, $y = 0$, $y = 4$, $x = 0$

40. Find $\bar{y}$ using $\rho(x, y, z) = k$.

Q: $\dfrac{x}{a} + \dfrac{y}{b} + \dfrac{z}{c} = 1$ $(a, b, c > 0)$, $x = 0$, $y = 0$, $z = 0$

Center of Mass In Exercises 41 and 42, set up the triple integrals for finding the mass and the center of mass of the solid of density ρ bounded by the graphs of the equations. Do not evaluate the integrals.

41. $x = 0$, $x = b$, $y = 0$, $y = b$, $z = 0$, $z = b$, $\rho(x, y, z) = kxy$
42. $x = 0$, $x = a$, $y = 0$, $y = b$, $z = 0$, $z = c$, $\rho(x, y, z) = kz$

Think About It The center of mass of a solid of constant density is shown in the figure. In Exercises 43–46, make a conjecture about how the center of mass $(\bar{x}, \bar{y}, \bar{z})$ will change for the nonconstant density $\rho(x, y, z)$. Explain. (Make your conjecture *without* performing any calculations.)

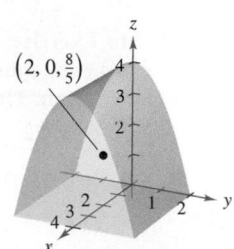

$\left(2, 0, \frac{8}{5}\right)$

43. $\rho(x, y, z) = kx$ **44.** $\rho(x, y, z) = kz$
45. $\rho(x, y, z) = k(y + 2)$ **46.** $\rho(x, y, z) = kxz^2(y + 2)^2$

Centroid In Exercises 47–52, find the centroid of the solid region bounded by the graphs of the equations or described by the figure. Use a computer algebra system to evaluate the triple integrals. (Assume uniform density and find the center of mass.)

47. $z = \dfrac{h}{r}\sqrt{x^2 + y^2}$, $z = h$

48. $y = \sqrt{9 - x^2}$, $z = y$, $z = 0$

49. $z = \sqrt{16 - x^2 - y^2}$, $z = 0$

50. $z = \dfrac{1}{y^2 + 1}$, $z = 0$, $x = -2$, $x = 2$, $y = 0$, $y = 1$

51.

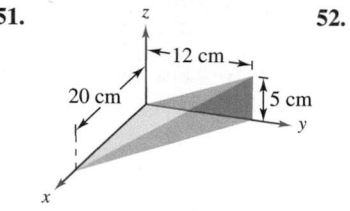

12 cm
20 cm
5 cm

52.

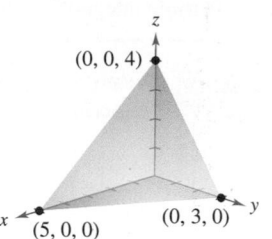

$(0, 0, 4)$
$(5, 0, 0)$
$(0, 3, 0)$

Moments of Inertia In Exercises 53–56, find I_x, I_y, and I_z for the solid of given density. Use a computer algebra system to evaluate the triple integrals.

53. (a) $\rho = k$ **54.** (a) $\rho(x, y, z) = k$
    (b) $\rho = kxyz$ (b) $\rho(x, y, z) = k(x^2 + y^2)$

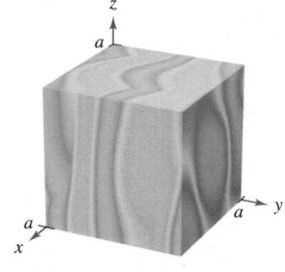
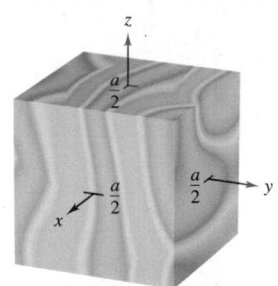

55. (a) $\rho(x, y, z) = k$ **56.** (a) $\rho = kz$
    (b) $\rho = ky$ (b) $\rho = k(4 - z)$

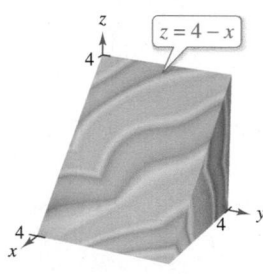

$z = 4 - x$

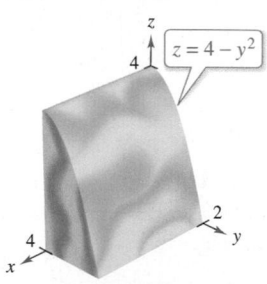

$z = 4 - y^2$

Moments of Inertia In Exercises 57 and 58, verify the moments of inertia for the solid of uniform density. Use a computer algebra system to evaluate the triple integrals.

57. $I_x = \frac{1}{12}m(3a^2 + L^2)$
    $I_y = \frac{1}{2}ma^2$
    $I_z = \frac{1}{12}m(3a^2 + L^2)$

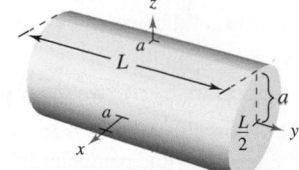

L
a
$\frac{L}{2}$

58. $I_x = \frac{1}{12}m(a^2 + b^2)$

$I_y = \frac{1}{12}m(b^2 + c^2)$

$I_z = \frac{1}{12}m(a^2 + c^2)$

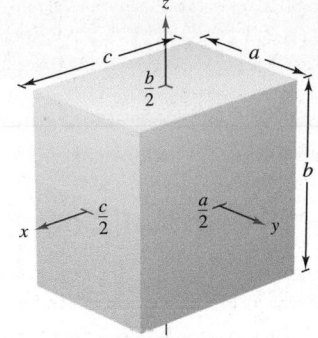

Moments of Inertia In Exercises 59 and 60, set up a triple integral for the moment of inertia about the z-axis of the solid region Q of density ρ. Do not evaluate the integral.

59. $Q = \{(x, y, z): -1 \le x \le 1, -1 \le y \le 1, 0 \le z \le 1 - x\}$

$\rho = \sqrt{x^2 + y^2 + z^2}$

60. $Q = \{(x, y, z): x^2 + y^2 \le 1, 0 \le z \le 4 - x^2 - y^2\}$

$\rho = kx^2$

Setting Up Triple Integrals In Exercises 61 and 62, use the description of the solid region to set up the triple integral for (a) the mass, (b) the center of mass, and (c) the moment of inertia about the z-axis. Do not evaluate the integrals.

61. The solid bounded by $z = 4 - x^2 - y^2$ and $z = 0$ with density $\rho(x, y, z) = kz$

62. The solid in the first octant bounded by the coordinate planes and $x^2 + y^2 + z^2 = 25$ with density $\rho(x, y, z) = kxy$

Average Value In Exercises 63–66, find the average value of the function over the given solid region. The average value of a continuous function $f(x, y, z)$ over a solid region Q is

$$\text{Average value} = \frac{1}{V}\iiint_Q f(x, y, z)\, dV$$

where V is the volume of the solid region Q.

63. $f(x, y, z) = z^2 + 4$ over the cube in the first octant bounded by the coordinate planes and the planes $x = 1$, $y = 1$, and $z = 1$

64. $f(x, y, z) = xyz$ over the cube in the first octant bounded by the coordinate planes and the planes $x = 4$, $y = 4$, and $z = 4$

65. $f(x, y, z) = x + y + z$ over the tetrahedron in the first octant with vertices $(0, 0, 0)$, $(2, 0, 0)$, $(0, 2, 0)$, and $(0, 0, 2)$

66. $f(x, y, z) = x + y$ over the solid bounded by the sphere $x^2 + y^2 + z^2 = 3$

EXPLORING CONCEPTS

67. Moment of Inertia Determine whether the moment of inertia about the y-axis of the cylinder in Exercise 57 will increase or decrease for the nonconstant density $\rho(x, y, z) = \sqrt{x^2 + z^2}$.

68. Using Different Methods Find the volume of the sphere $x^2 + y^2 + z^2 = 9$ using the shell method and using a triple integral. Compare your answers.

EXPLORING CONCEPTS (continued)

69. Think About It Which of the integrals below is equal to $\int_1^3 \int_0^2 \int_{-1}^1 f(x, y, z)\, dz\, dy\, dx$? Explain.

(a) $\int_1^3 \int_0^2 \int_{-1}^1 f(x, y, z)\, dz\, dx\, dy$

(b) $\int_{-1}^1 \int_0^2 \int_1^3 f(x, y, z)\, dx\, dy\, dz$

(c) $\int_0^2 \int_1^3 \int_{-1}^1 f(x, y, z)\, dy\, dx\, dz$

70. HOW DO YOU SEE IT? Consider two solids of equal weight, as shown below.

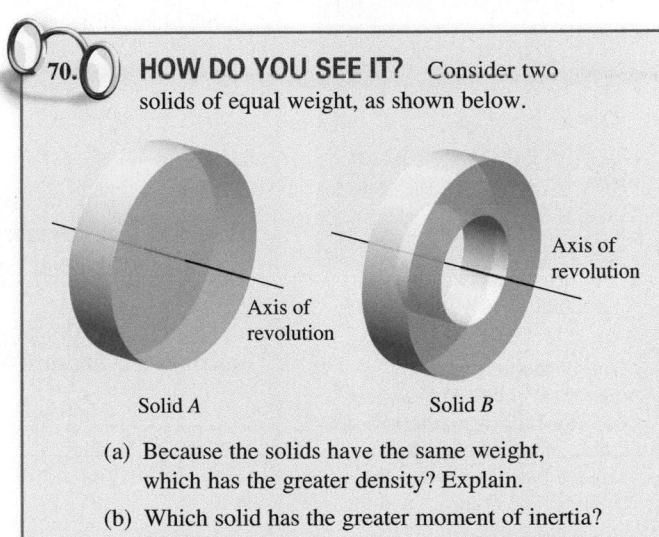

Solid A   Solid B

(a) Because the solids have the same weight, which has the greater density? Explain.

(b) Which solid has the greater moment of inertia? Explain.

(c) The solids are rolled down an inclined plane. They are started at the same time and at the same height. Which will reach the bottom first? Explain.

71. Maximizing a Triple Integral Find the solid region Q where the triple integral

$$\iiint_Q (1 - 2x^2 - y^2 - 3z^2)\, dV$$

is a maximum. Use a computer algebra system to approximate the maximum value. What is the exact maximum value?

72. Finding a Value Solve for a in the triple integral.

$$\int_0^1 \int_0^{3-a-y^2} \int_a^{4-x-y^2} dz\, dx\, dy = \frac{14}{15}$$

PUTNAM EXAM CHALLENGE

73. Evaluate

$$\lim_{n\to\infty} \int_0^1 \int_0^1 \cdots \int_0^1 \cos^2\left\{\frac{\pi}{2n}(x_1 + x_2 + \cdots + x_n)\right\} dx_1\, dx_2 \cdots dx_n.$$

14.7 Triple Integrals in Other Coordinates

■ Write and evaluate a triple integral in cylindrical coordinates.
■ Write and evaluate a triple integral in spherical coordinates.

Triple Integrals in Cylindrical Coordinates

Many common solid regions, such as spheres, ellipsoids, cones, and paraboloids, can yield difficult triple integrals in rectangular coordinates. In fact, it is precisely this difficulty that led to the introduction of nonrectangular coordinate systems. In this section, you will learn how to use *cylindrical* and *spherical* coordinates to evaluate triple integrals.

Recall from Section 11.7 that the rectangular conversion equations for cylindrical coordinates are

$$x = r \cos \theta$$
$$y = r \sin \theta$$
$$z = z.$$

An easy way to remember these conversions is to note that the equations for x and y are the same as in polar coordinates and z is unchanged.

In this coordinate system, the simplest solid region is a cylindrical block determined by

$$r_1 \le r \le r_2$$
$$\theta_1 \le \theta \le \theta_2$$

and

$$z_1 \le z \le z_2$$

as shown in Figure 14.63.

**PIERRE SIMON DE LAPLACE
(1749–1827)**

One of the first to use a cylindrical coordinate system was the French mathematician Pierre Simon de Laplace. Laplace has been called the "Newton of France," and he published many important works in mechanics, differential equations, and probability.
See LarsonCalculus.com to read more of this biography.

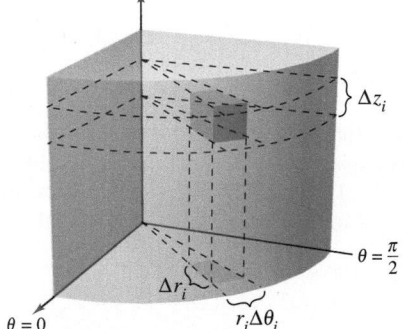

Volume of cylindrical block:
$\Delta V_i = r_i \Delta r_i \Delta \theta_i \Delta z_i$
Figure 14.63

To obtain the cylindrical coordinate form of a triple integral, consider a solid region Q whose projection R onto the xy-plane can be described in polar coordinates. That is,

$$Q = \{(x, y, z): (x, y) \text{ is in } R, \quad h_1(x, y) \le z \le h_2(x, y)\}$$

and

$$R = \{(r, \theta): \theta_1 \le \theta \le \theta_2, \quad g_1(\theta) \le r \le g_2(\theta)\}.$$

If f is a continuous function on the solid Q, then you can write the triple integral of f over Q as

$$\iiint_Q f(x, y, z) \, dV = \iint_R \left[\int_{h_1(x, y)}^{h_2(x, y)} f(x, y, z) \, dz \right] dA$$

where the double integral over R is evaluated in polar coordinates. That is, R is a plane region that is either r-simple or θ-simple (see Section 14.3). If R is r-simple, then the iterated form of the triple integral in cylindrical form is

$$\iiint_Q f(x, y, z) \, dV = \int_{\theta_1}^{\theta_2} \int_{g_1(\theta)}^{g_2(\theta)} \int_{h_1(r \cos \theta, r \sin \theta)}^{h_2(r \cos \theta, r \sin \theta)} f(r \cos \theta, r \sin \theta, z) r \, dz \, dr \, d\theta.$$

This is only one of six possible orders of integration. The other five are $dz \, d\theta \, dr$, $dr \, dz \, d\theta$, $dr \, d\theta \, dz$, $d\theta \, dz \, dr$, and $d\theta \, dr \, dz$.

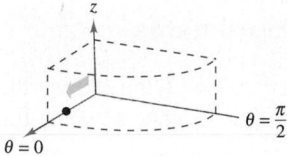

Integrate with respect to r.

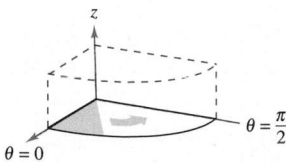

Integrate with respect to θ.

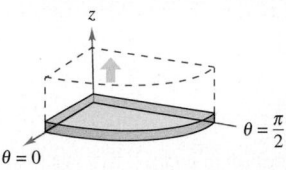

Integrate with respect to z.
Figure 14.64

To visualize a particular order of integration, it helps to view the iterated integral in terms of three sweeping motions, each adding another dimension to the solid. For instance, in the order $dr\, d\theta\, dz$, the first integration occurs in the r-direction as a point sweeps out a ray. Then, as θ increases, the line sweeps out a sector. Finally, as z increases, the sector sweeps out a solid wedge, as shown in Figure 14.64.

Exploration

Volume of a Paraboloid Sector In the Explorations on pages 983, 992, and 1014, you were asked to summarize the different ways you know of finding the volume of the solid bounded by the paraboloid

$$z = a^2 - x^2 - y^2, \quad a > 0$$

and the xy-plane. You now know one more way. Use it to find the volume of the solid. Compare the different methods. What are the advantages and disadvantages of each?

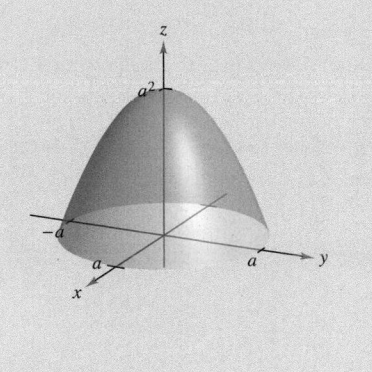

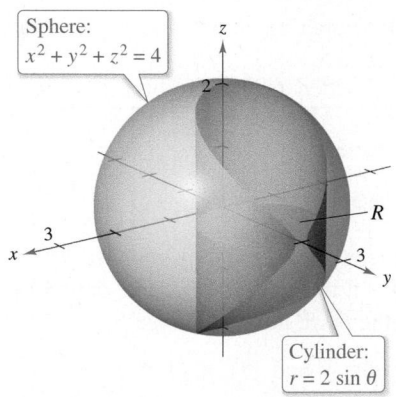

Figure 14.65

EXAMPLE 1 Finding Volume in Cylindrical Coordinates

Find the volume of the solid region Q cut from the sphere $x^2 + y^2 + z^2 = 4$ by the cylinder $r = 2 \sin \theta$, as shown in Figure 14.65.

Solution Because $x^2 + y^2 + z^2 = r^2 + z^2 = 4$, the bounds on z are

$$-\sqrt{4 - r^2} \le z \le \sqrt{4 - r^2}. \qquad \text{Bounds for } z$$

Let R be the circular projection of the solid onto the $r\theta$-plane. Then the bounds on R are

$$0 \le r \le 2 \sin \theta \quad \text{and} \quad 0 \le \theta \le \pi. \qquad \text{Bounds for } R$$

So, the volume of Q is

$$V = \int_0^\pi \int_0^{2 \sin \theta} \int_{-\sqrt{4-r^2}}^{\sqrt{4-r^2}} r\, dz\, dr\, d\theta \qquad \text{Apply formula for volume.}$$

$$= 2 \int_0^{\pi/2} \int_0^{2 \sin \theta} \int_{-\sqrt{4-r^2}}^{\sqrt{4-r^2}} r\, dz\, dr\, d\theta \qquad \text{Use symmetry to rewrite bounds for } \theta.$$

$$= 2 \int_0^{\pi/2} \int_0^{2 \sin \theta} 2r\sqrt{4 - r^2}\, dr\, d\theta \qquad \text{Integrate with respect to } z.$$

$$= 2 \int_0^{\pi/2} \left[-\frac{2}{3}(4 - r^2)^{3/2} \right]_0^{2 \sin \theta} d\theta \qquad \text{Integrate with respect to } r.$$

$$= \frac{4}{3} \int_0^{\pi/2} (8 - 8 \cos^3 \theta)\, d\theta$$

$$= \frac{32}{3} \int_0^{\pi/2} \left[1 - (\cos \theta)(1 - \sin^2 \theta) \right] d\theta \qquad \substack{\text{Factor and use trigonometric} \\ \text{identity } \cos^2 \theta = 1 - \sin^2 \theta.}$$

$$= \frac{32}{3} \left[\theta - \sin \theta + \frac{\sin^3 \theta}{3} \right]_0^{\pi/2} \qquad \text{Integrate with respect to } \theta.$$

$$= \frac{16}{9}(3\pi - 4)$$

$$\approx 9.644.$$

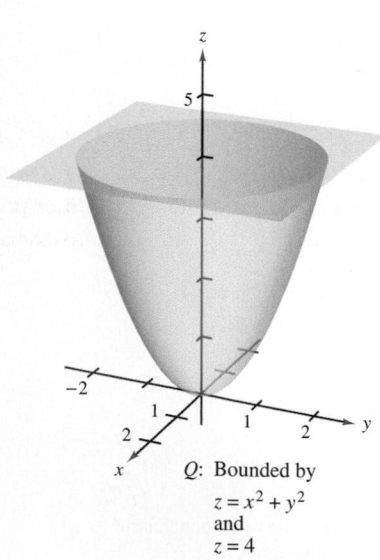

$0 \leq z \leq \sqrt{16 - 4r^2}$

Ellipsoid: $4x^2 + 4y^2 + z^2 = 16$

Figure 14.66

EXAMPLE 2 **Finding Mass in Cylindrical Coordinates**

Find the mass of the ellipsoid Q given by $4x^2 + 4y^2 + z^2 = 16$, lying above the xy-plane. The density at a point in the solid is proportional to the distance between the point and the xy-plane.

Solution The density function is $\rho(r, \theta, z) = kz$, where k is the constant of proportionality. The bounds on z are

$$0 \leq z \leq \sqrt{16 - 4x^2 - 4y^2} = \sqrt{16 - 4r^2}$$

where $0 \leq r \leq 2$ and $0 \leq \theta \leq 2\pi$, as shown in Figure 14.66. The mass of the solid is

$$
\begin{aligned}
m &= \int_0^{2\pi} \int_0^2 \int_0^{\sqrt{16-4r^2}} kzr \, dz \, dr \, d\theta && \text{Apply formula for mass of a solid.} \\
&= \frac{k}{2} \int_0^{2\pi} \int_0^2 z^2 r \Big]_0^{\sqrt{16-4r^2}} dr \, d\theta && \text{Integrate with respect to } z. \\
&= \frac{k}{2} \int_0^{2\pi} \int_0^2 (16r - 4r^3) \, dr \, d\theta && \\
&= \frac{k}{2} \int_0^{2\pi} \Big[8r^2 - r^4 \Big]_0^2 d\theta && \text{Integrate with respect to } r. \\
&= 8k \int_0^{2\pi} d\theta && \\
&= 16\pi k. && \text{Integrate with respect to } \theta.
\end{aligned}
$$

Integration in cylindrical coordinates is useful when factors involving $x^2 + y^2$ appear in the integrand, as illustrated in Example 3.

EXAMPLE 3 **Finding a Moment of Inertia**

Find the moment of inertia about the axis of symmetry of the solid Q bounded by the paraboloid $z = x^2 + y^2$ and the plane $z = 4$, as shown in Figure 14.67. The density at each point is proportional to the distance between the point and the z-axis.

Solution Because the z-axis is the axis of symmetry and $\rho(x, y, z) = k\sqrt{x^2 + y^2}$, where k is the constant of proportionality, it follows that

$$I_z = \iiint_Q k(x^2 + y^2)\sqrt{x^2 + y^2} \, dV. \qquad \text{Moment of inertia about } z\text{-axis}$$

In cylindrical coordinates, $0 \leq r \leq \sqrt{x^2 + y^2} = \sqrt{z}$ and $0 \leq \theta \leq 2\pi$. So, you have

$$
\begin{aligned}
I_z &= k \int_0^4 \int_0^{2\pi} \int_0^{\sqrt{z}} r^2(r)r \, dr \, d\theta \, dz && \text{Cylindrical coordinates} \\
&= k \int_0^4 \int_0^{2\pi} \frac{r^5}{5} \Big]_0^{\sqrt{z}} d\theta \, dz && \text{Integrate with respect to } r. \\
&= k \int_0^4 \int_0^{2\pi} \frac{z^{5/2}}{5} \, d\theta \, dz && \\
&= \frac{k}{5} \int_0^4 z^{5/2} (2\pi) \, dz && \text{Integrate with respect to } \theta. \\
&= \frac{2\pi k}{5} \Big[\frac{2}{7} z^{7/2} \Big]_0^4 && \text{Integrate with respect to } z. \\
&= \frac{512k\pi}{35}.
\end{aligned}
$$

Q: Bounded by
$z = x^2 + y^2$
and
$z = 4$

Figure 14.67

Triple Integrals in Spherical Coordinates

Triple integrals involving spheres or cones are often easier to evaluate by converting to spherical coordinates. Recall from Section 11.7 that the rectangular conversion equations for spherical coordinates are

$$x = \rho \sin \phi \cos \theta$$
$$y = \rho \sin \phi \sin \theta$$
$$z = \rho \cos \phi.$$

> **REMARK** The Greek letter ρ used in spherical coordinates is not related to density. Rather, it is the three-dimensional analog of the r used in polar coordinates. For problems involving spherical coordinates and a density function, this text uses a different symbol to denote density.

In this coordinate system, the simplest region is a spherical block determined by

$$\{(\rho, \theta, \phi): \rho_1 \leq \rho \leq \rho_2, \quad \theta_1 \leq \theta \leq \theta_2, \quad \phi_1 \leq \phi \leq \phi_2\}$$

where $\rho_1 \geq 0$, $\theta_2 - \theta_1 \leq 2\pi$, and $0 \leq \phi_1 \leq \phi_2 \leq \pi$, as shown in Figure 14.68. If (ρ, θ, ϕ) is a point in the interior of such a block, then the volume of the block can be approximated by $\Delta V \approx \rho^2 \sin \phi \Delta\rho \Delta\phi \Delta\theta$. (See Exercise 8 in the Problem Solving exercises at the end of this chapter.)

Using the usual process involving an inner partition, summation, and a limit, you can develop a triple integral in spherical coordinates for a continuous function f defined on the solid region Q. This formula, shown below, can be modified for different orders of integration and generalized to include regions with variable boundaries.

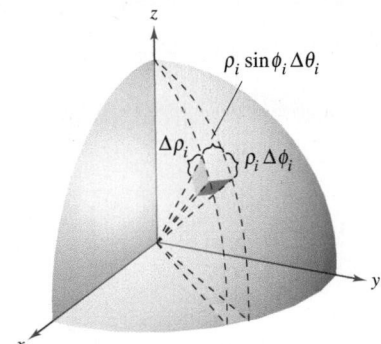

Spherical block: $\Delta V_i \approx \rho_i^2 \sin \phi_i \Delta\rho_i \Delta\phi_i \Delta\theta_i$
Figure 14.68

$$\iiint\limits_{Q} f(x, y, z) \, dV = \int_{\theta_1}^{\theta_2} \int_{\phi_1}^{\phi_2} \int_{\rho_1}^{\rho_2} f(\rho \sin \phi \cos \theta, \rho \sin \phi \sin \theta, \rho \cos \phi) \rho^2 \sin \phi \, d\rho \, d\phi \, d\theta$$

Like triple integrals in cylindrical coordinates, triple integrals in spherical coordinates are evaluated with iterated integrals. As with cylindrical coordinates, you can visualize a particular order of integration by viewing the iterated integral in terms of three sweeping motions, each adding another dimension to the solid. For instance, the iterated integral

$$\int_0^{2\pi} \int_0^{\pi/4} \int_0^3 \rho^2 \sin \phi \, d\rho \, d\phi \, d\theta$$

(which is used in Example 4) is illustrated in Figure 14.69.

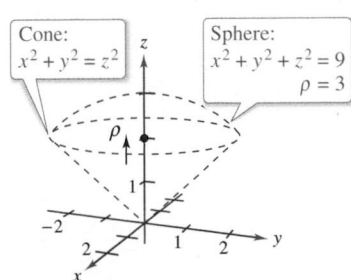

ρ varies from 0 to 3 with ϕ and θ held constant.

Figure 14.69

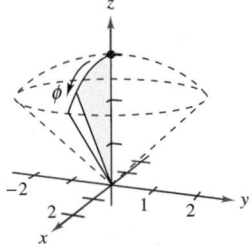

ϕ varies from 0 to $\pi/4$ with θ held constant.

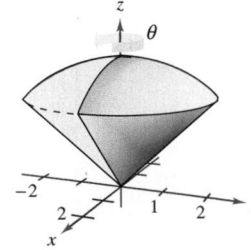

θ varies from 0 to 2π.

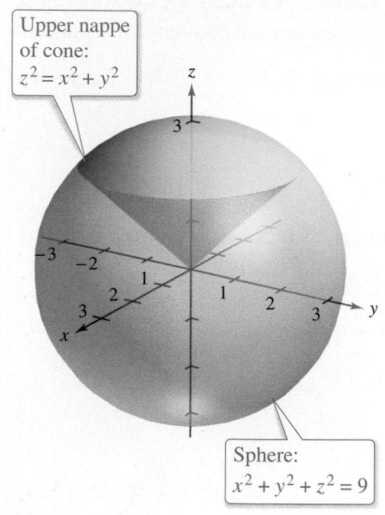

Upper nappe of cone:
$z^2 = x^2 + y^2$

Sphere:
$x^2 + y^2 + z^2 = 9$

Figure 14.70

EXAMPLE 4   **Finding Volume in Spherical Coordinates**

Find the volume of the solid region Q bounded below by the upper nappe of the cone $z^2 = x^2 + y^2$ and above by the sphere $x^2 + y^2 + z^2 = 9$, as shown in Figure 14.70.

Solution   In spherical coordinates, the equation of the sphere is

$$\rho^2 = x^2 + y^2 + z^2 = 9 \implies \rho = 3.$$

Furthermore, the sphere and cone intersect when

$$(x^2 + y^2) + z^2 = (z^2) + z^2 = 9 \implies z = \frac{3}{\sqrt{2}}$$

and, because $z = \rho \cos \phi$, it follows that

$$\left(\frac{3}{\sqrt{2}}\right)\left(\frac{1}{3}\right) = \cos \phi \implies \phi = \frac{\pi}{4}.$$

Consequently, you can use the integration order $d\rho \, d\phi \, d\theta$, where $0 \le \rho \le 3$, $0 \le \phi \le \pi/4$, and $0 \le \theta \le 2\pi$. The volume is

$$
\begin{aligned}
V &= \int_0^{2\pi} \int_0^{\pi/4} \int_0^3 \rho^2 \sin \phi \, d\rho \, d\phi \, d\theta && \text{Apply formula for volume.} \\
&= \int_0^{2\pi} \int_0^{\pi/4} 9 \sin \phi \, d\phi \, d\theta && \text{Integrate with respect to } \rho. \\
&= 9 \int_0^{2\pi} -\cos \phi \Big]_0^{\pi/4} d\theta && \text{Integrate with respect to } \phi. \\
&= 9 \int_0^{2\pi} \left(1 - \frac{\sqrt{2}}{2}\right) d\theta \\
&= 9\pi\left(2 - \sqrt{2}\right) && \text{Integrate with respect to } \theta. \\
&\approx 16.563.
\end{aligned}
$$

EXAMPLE 5   **Finding the Center of Mass of a Solid Region**

$\vdots \cdots \triangleright$ *See LarsonCalculus.com for an interactive version of this type of example.*

Find the center of mass of the solid region Q of uniform density from Example 4.

Solution   Because the density is uniform, you can consider the density at the point (x, y, z) to be k. By symmetry, the center of mass lies on the z-axis, and you need only calculate $\bar{z} = M_{xy}/m$, where $m = kV = 9k\pi(2 - \sqrt{2})$. Because $z = \rho \cos \phi$, it follows that

$$
\begin{aligned}
M_{xy} &= \iiint_Q kz \, dV = k \int_0^3 \int_0^{2\pi} \int_0^{\pi/4} (\rho \cos \phi)\rho^2 \sin \phi \, d\phi \, d\theta \, d\rho \\
&= k \int_0^3 \int_0^{2\pi} \rho^3 \frac{\sin^2 \phi}{2} \Big]_0^{\pi/4} d\theta \, d\rho \\
&= \frac{k}{4} \int_0^3 \int_0^{2\pi} \rho^3 \, d\theta \, d\rho = \frac{k\pi}{2} \int_0^3 \rho^3 \, d\rho = \frac{81k\pi}{8}.
\end{aligned}
$$

So,

$$\bar{z} = \frac{M_{xy}}{m} = \frac{81k\pi/8}{9k\pi(2 - \sqrt{2})} = \frac{9(2 + \sqrt{2})}{16} \approx 1.920$$

and the center of mass is approximately $(0, 0, 1.920)$.

14.7 Exercises

See CalcChat.com for tutorial help and worked-out solutions to odd-numbered exercises.

CONCEPT CHECK

1. Volume Explain why triple integrals that represent the volumes of solids are sometimes easier to evaluate in cylindrical or spherical coordinates instead of rectangular coordinates.

2. Differential of Volume What is the differential of volume, dV, for (a) cylindrical coordinates and (b) spherical coordinates? Choose one order of integration for each system.

Evaluating a Triple Iterated Integral In Exercises 3–8, evaluate the triple iterated integral.

3. $\displaystyle\int_{-1}^{5}\int_{0}^{\pi/2}\int_{0}^{3} r\cos\theta \, dr \, d\theta \, dz$
4. $\displaystyle\int_{0}^{\pi/4}\int_{0}^{6}\int_{0}^{6-r} rz \, dz \, dr \, d\theta$

5. $\displaystyle\int_{0}^{\pi/2}\int_{0}^{\cos\theta}\int_{0}^{3+r^2} 2r\sin\theta \, dz \, dr \, d\theta$

6. $\displaystyle\int_{0}^{\pi/2}\int_{0}^{\pi}\int_{0}^{2} e^{-\rho^3}\rho^2 \, d\rho \, d\theta \, d\phi$

7. $\displaystyle\int_{0}^{2\pi}\int_{0}^{\pi/2}\int_{0}^{\sin\phi} \rho\cos\phi \, d\rho \, d\phi \, d\theta$

8. $\displaystyle\int_{0}^{\pi/4}\int_{0}^{\pi/4}\int_{0}^{\cos\theta} \rho^2\sin\phi\cos\phi \, d\rho \, d\theta \, d\phi$

 **Evaluating a Triple Iterated Integral Using Technology** In Exercises 9 and 10, use a computer algebra system to evaluate the triple iterated integral.

9. $\displaystyle\int_{0}^{4}\int_{0}^{z}\int_{0}^{\pi/2} re^r \, d\theta \, dr \, dz$

10. $\displaystyle\int_{0}^{\pi/2}\int_{0}^{\pi}\int_{0}^{\sin\theta} 2\rho^2\cos\phi \, d\rho \, d\theta \, d\phi$

Volume In Exercises 11–14, sketch the solid region whose volume is given by the iterated integral and evaluate the iterated integral.

11. $\displaystyle\int_{0}^{\pi/2}\int_{0}^{3}\int_{0}^{e^{-r^2}} r \, dz \, dr \, d\theta$
12. $\displaystyle\int_{0}^{2\pi}\int_{0}^{2\sqrt{2}}\int_{r^2-2}^{6} r \, dz \, dr \, d\theta$

13. $\displaystyle\int_{0}^{2\pi}\int_{\pi/6}^{\pi/2}\int_{0}^{4} \rho^2\sin\phi \, d\rho \, d\phi \, d\theta$

14. $\displaystyle\int_{0}^{2\pi}\int_{0}^{\pi}\int_{2}^{5} \rho^2\sin\phi \, d\rho \, d\phi \, d\theta$

 Volume In Exercises 15–20, use cylindrical coordinates to find the volume of the solid.

15. Solid inside both $x^2 + y^2 + z^2 = 36$ and $(x - 3)^2 + y^2 = 9$

16. Solid inside $x^2 + y^2 + z^2 = 16$ and outside $z = \sqrt{x^2 + y^2}$

17. Solid bounded above by $z = 2x$ and below by $z = 2x^2 + 2y^2$

18. Solid bounded above by $z = 2 - x^2 - y^2$ and below by $z = x^2 + y^2$

19. Solid bounded by the graphs of the sphere $r^2 + z^2 = 25$ and the cylinder $r = 5\cos\theta$

20. Solid inside the sphere $x^2 + y^2 + z^2 = 4$ and above the upper nappe of the cone $z^2 = x^2 + y^2$

 **Mass** In Exercises 21 and 22, use cylindrical coordinates to find the mass of the solid Q of density ρ.

21. $Q = \{(x, y, z): 0 \le z \le 9 - x - 2y, x^2 + y^2 \le 4\}$
$\rho(x, y, z) = k\sqrt{x^2 + y^2}$

22. $Q = \{(x, y, z): 0 \le z \le 12e^{-(x^2+y^2)}, x^2 + y^2 \le 4, x \ge 0, y \ge 0\}$
$\rho(x, y, z) = k$

Using Cylindrical Coordinates In Exercises 23–28, use cylindrical coordinates to find the indicated characteristic of the cone shown in the figure.

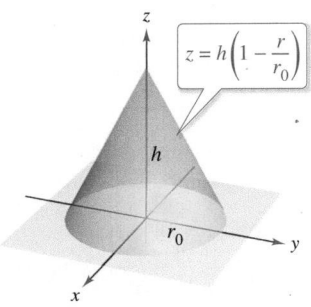

23. Find the volume of the cone.

24. Find the centroid of the cone.

25. Find the center of mass of the cone, assuming that its density at any point is proportional to the distance between the point and the axis of the cone. Use a computer algebra system to evaluate the triple integral.

26. Find the center of mass of the cone, assuming that its density at any point is proportional to the distance between the point and the base. Use a computer algebra system to evaluate the triple integral.

27. Assume that the cone has uniform density and show that the moment of inertia about the z-axis is

$$I_z = \tfrac{3}{10}mr_0^2.$$

28. Assume that the density of the cone is $\rho(x, y, z) = k\sqrt{x^2 + y^2}$ and find the moment of inertia about the z-axis.

Moment of Inertia In Exercises 29 and 30, use cylindrical coordinates to verify the given moment of inertia of the solid of uniform density.

29. Cylindrical shell: $I_z = \tfrac{1}{2}m(a^2 + b^2)$
$0 < a \le r \le b, \quad 0 \le z \le h$

30. Right circular cylinder: $I_z = \frac{3}{2}ma^2$

$r = 2a\sin\theta, \quad 0 \le z \le h$

(Use a computer algebra system to evaluate the triple integral.)

 Volume In Exercises 31–34, use spherical coordinates to find the volume of the solid.

31. Solid inside $x^2 + y^2 + z^2 = 9$, outside $z = \sqrt{x^2 + y^2}$, and above the xy-plane

32. Solid bounded above by $x^2 + y^2 + z^2 = z$ and below by $z = \sqrt{x^2 + y^2}$

 33. The torus given by $\rho = 4\sin\phi$ (Use a computer algebra system to evaluate the triple integral.)

34. The solid between the spheres

$x^2 + y^2 + z^2 = a^2$ and $x^2 + y^2 + z^2 = b^2, b > a,$

and inside the cone $z^2 = x^2 + y^2$

Mass In Exercises 35 and 36, use spherical coordinates to find the mass of the sphere $x^2 + y^2 + z^2 = a^2$ with the given density.

35. The density at any point is proportional to the distance between the point and the origin.

36. The density at any point is proportional to the distance between the point and the z-axis.

 **Center of Mass** In Exercises 37 and 38, use spherical coordinates to find the center of mass of the solid of uniform density.

37. Hemispherical solid of radius r

38. Solid lying between two concentric hemispheres of radii r and R, where $r < R$.

Moment of Inertia In Exercises 39 and 40, use spherical coordinates to find the moment of inertia about the z-axis of the solid of uniform density.

39. Solid bounded by the hemisphere $\rho = \cos\phi, \dfrac{\pi}{4} \le \phi \le \dfrac{\pi}{2}$, and the cone $\phi = \dfrac{\pi}{4}$

40. Solid lying between two concentric hemispheres of radii r and R, where $r < R$.

Converting Coordinates In Exercises 41–44, convert the integral from rectangular coordinates to both cylindrical and spherical coordinates, and evaluate the simplest iterated integral.

41. $\displaystyle\int_{-2}^{2}\int_{-\sqrt{4-x^2}}^{\sqrt{4-x^2}}\int_{x^2+y^2}^{4} x\, dz\, dy\, dx$

42. $\displaystyle\int_{0}^{2}\int_{0}^{\sqrt{4-x^2}}\int_{0}^{\sqrt{16-x^2-y^2}} \sqrt{x^2+y^2}\, dz\, dy\, dx$

43. $\displaystyle\int_{-1}^{1}\int_{-\sqrt{1-x^2}}^{\sqrt{1-x^2}}\int_{1}^{1+\sqrt{1-x^2-y^2}} x\, dz\, dy\, dx$

44. $\displaystyle\int_{0}^{3}\int_{0}^{\sqrt{9-x^2}}\int_{0}^{\sqrt{9-x^2-y^2}} \sqrt{x^2+y^2+z^2}\, dz\, dy\, dx$

EXPLORING CONCEPTS

45. Using Coordinates Describe the surface whose equation is a coordinate equal to a constant for each of the coordinates in (a) the cylindrical coordinate system and (b) the spherical coordinate system.

46. **HOW DO YOU SEE IT?** The solid is bounded below by the upper nappe of a cone and above by a sphere (see figure). Would it be easier to use cylindrical coordinates or spherical coordinates to find the volume of the solid? Explain.

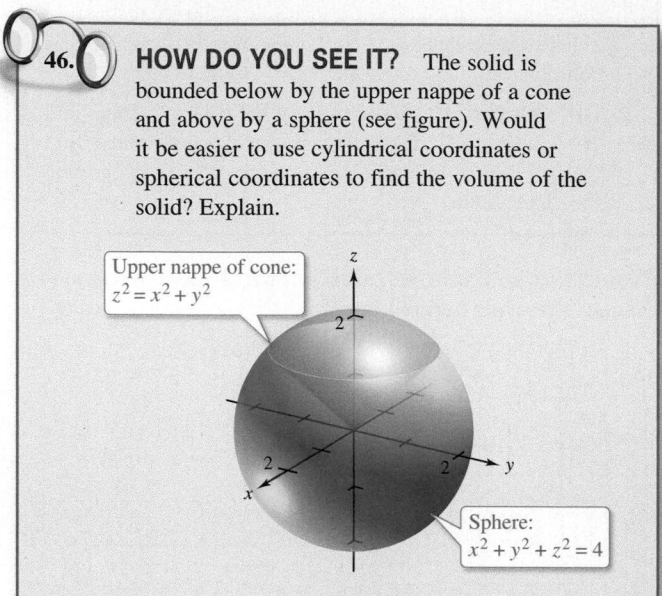

Upper nappe of cone: $z^2 = x^2 + y^2$

Sphere: $x^2 + y^2 + z^2 = 4$

PUTNAM EXAM CHALLENGE

47. Find the volume of the region of points (x, y, z) such that
$(x^2 + y^2 + z^2 + 8)^2 \le 36(x^2 + y^2).$

SECTION PROJECT

Wrinkled and Bumpy Spheres

In parts (a) and (b), find the volume of the wrinkled sphere or bumpy sphere. These solids are used as models for tumors.

(a) Wrinkled sphere

$\rho = 1 + 0.2\sin 8\theta\sin\phi$

$0 \le \theta \le 2\pi, 0 \le \phi \le \pi$

(b) Bumpy sphere

$\rho = 1 + 0.2\sin 8\theta\sin 4\phi$

$0 \le \theta \le 2\pi, 0 \le \phi \le \pi$

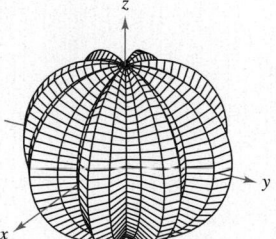

Generated by Maple   *Generated by Maple*

■ **FOR FURTHER INFORMATION** For more information on these types of spheres, see the article "Heat Therapy for Tumors" by Leah Edelstein-Keshet in *The UMAP Journal.*

14.8 Change of Variables: Jacobians

■ Understand the concept of a Jacobian.
■ Use a Jacobian to change variables in a double integral.

Jacobians

**CARL GUSTAV JACOBI
(1804–1851)**

The Jacobian is named after the German mathematician Carl Gustav Jacobi. Jacobi is known for his work in many areas of mathematics, but his interest in integration stemmed from the problem of finding the circumference of an ellipse.
See LarsonCalculus.com to read more of this biography.

For the single integral

$$\int_a^b f(x)\, dx$$

you can change variables by letting $x = g(u)$, so that $dx = g'(u)\, du$, and obtain

$$\int_a^b f(x)\, dx = \int_c^d f(g(u))g'(u)\, du$$

where $a = g(c)$ and $b = g(d)$. Note that the change of variables process introduces an additional factor $g'(u)$ into the integrand. This also occurs in the case of double integrals

$$\iint_R f(x, y)\, dA = \iint_S f(g(u, v), h(u, v)) \underbrace{\left| \frac{\partial x}{\partial u} \frac{\partial y}{\partial v} - \frac{\partial y}{\partial u} \frac{\partial x}{\partial v} \right|}_{\text{Jacobian}} du\, dv$$

where the change of variables $x = g(u, v)$ and $y = h(u, v)$ introduces a factor called the Jacobian of x and y with respect to u and v. In defining the Jacobian, it is convenient to use the determinant notation shown below.

Definition of the Jacobian

If $x = g(u, v)$ and $y = h(u, v)$, then the **Jacobian** of x and y with respect to u and v, denoted by $\partial(x, y)/\partial(u, v)$, is

$$\frac{\partial(x, y)}{\partial(u, v)} = \begin{vmatrix} \dfrac{\partial x}{\partial u} & \dfrac{\partial x}{\partial v} \\ \dfrac{\partial y}{\partial u} & \dfrac{\partial y}{\partial v} \end{vmatrix} = \frac{\partial x}{\partial u} \frac{\partial y}{\partial v} - \frac{\partial y}{\partial u} \frac{\partial x}{\partial v}.$$

EXAMPLE 1 **The Jacobian for Rectangular-to-Polar Conversion**

Find the Jacobian for the change of variables defined by

$$x = r \cos \theta \quad \text{and} \quad y = r \sin \theta.$$

Solution From the definition of the Jacobian, you obtain

$$\frac{\partial(x, y)}{\partial(r, \theta)} = \begin{vmatrix} \dfrac{\partial x}{\partial r} & \dfrac{\partial x}{\partial \theta} \\ \dfrac{\partial y}{\partial r} & \dfrac{\partial y}{\partial \theta} \end{vmatrix} \qquad \text{Definition of Jacobian}$$

$$= \begin{vmatrix} \cos \theta & -r \sin \theta \\ \sin \theta & r \cos \theta \end{vmatrix} \qquad \text{Substitute.}$$

$$= r \cos^2 \theta + r \sin^2 \theta \qquad \text{Find determinant.}$$

$$= r(\cos^2 \theta + \sin^2 \theta) \qquad \text{Factor.}$$

$$= r \qquad \text{Trigonometric identity}$$

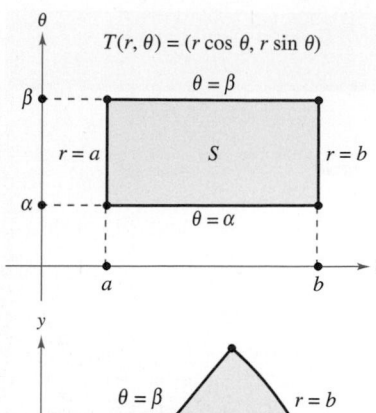

S in the region in the $r\theta$-plane that corresponds to R in the xy-plane.
Figure 14.71

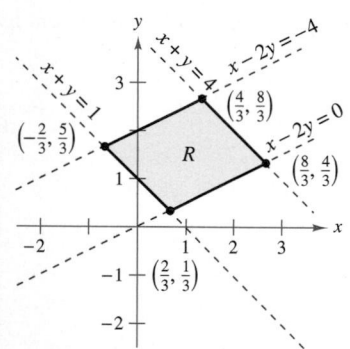

Region R in the xy-plane
Figure 14.72

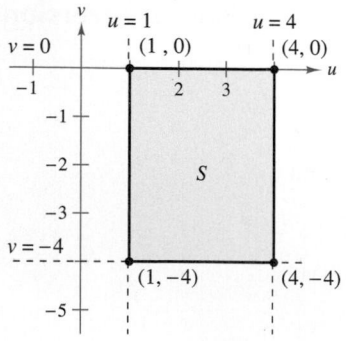

Region S in the uv-plane
Figure 14.73

Example 1 points out that the change of variables from rectangular to polar coordinates for a double integral can be written as

$$\iint_R f(x, y)\, dA = \iint_S f(r\cos\theta, r\sin\theta)r\, dr\, d\theta,\ r > 0$$

$$= \iint_S f(r\cos\theta, r\sin\theta)\left|\frac{\partial(x, y)}{\partial(r, \theta)}\right| dr\, d\theta$$

where S is the region in the $r\theta$-plane that corresponds to the region R in the xy-plane, as shown in Figure 14.71. This formula is similar to that found in Theorem 14.3 on page 992.

In general, a change of variables using a one-to-one **transformation** T from a region S in the uv-plane to a region R in the xy-plane is given by

$$T(u, v) = (x, y) = (g(u, v), h(u, v))$$

where g and h have continuous first partial derivatives in the region S. Note that the point (u, v) lies in S and the point (x, y) lies in R. In most cases, you are hunting for a transformation in which the region S is simpler than the region R.

EXAMPLE 2 **Finding a Change of Variables to Simplify a Region**

Let R be the region bounded by the lines

$$x - 2y = 0,\quad x - 2y = -4,\quad x + y = 4,\quad \text{and}\quad x + y = 1$$

as shown in Figure 14.72. Find a transformation T from a region S to R such that S is a rectangular region (with sides parallel to the u- or v-axis).

Solution To begin, let $u = x + y$ and $v = x - 2y$. Solving this system of equations for x and y produces $T(u, v) = (x, y)$, where

$$x = \frac{1}{3}(2u + v)\quad \text{and}\quad y = \frac{1}{3}(u - v).$$

The four boundaries for R in the xy-plane give rise to the following bounds for S in the uv-plane.

Bounds in the xy-Plane		Bounds in the uv-Plane
$x + y = 1$	⇨	$u = 1$
$x + y = 4$	⇨	$u = 4$
$x - 2y = 0$	⇨	$v = 0$
$x - 2y = -4$	⇨	$v = -4$

The region S is shown in Figure 14.73. Note that the transformation

$$T(u, v) = (x, y) = \left(\frac{1}{3}[2u + v], \frac{1}{3}[u - v]\right)$$

maps the vertices of the region S onto the vertices of the region R, as shown below.

$$T(1, 0) = \left(\frac{1}{3}[2(1) + 0], \frac{1}{3}[1 - 0]\right) = \left(\frac{2}{3}, \frac{1}{3}\right)$$

$$T(4, 0) = \left(\frac{1}{3}[2(4) + 0], \frac{1}{3}[4 - 0]\right) = \left(\frac{8}{3}, \frac{4}{3}\right)$$

$$T(4, -4) = \left(\frac{1}{3}[2(4) - 4], \frac{1}{3}[4 - (-4)]\right) = \left(\frac{4}{3}, \frac{8}{3}\right)$$

$$T(1, -4) = \left(\frac{1}{3}[2(1) - 4], \frac{1}{3}[1 - (-4)]\right) = \left(-\frac{2}{3}, \frac{5}{3}\right)$$

Change of Variables for Double Integrals

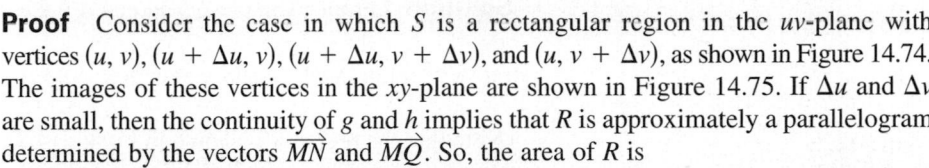

THEOREM 14.5   Change of Variables for Double Integrals

Let R be a vertically or horizontally simple region in the xy-plane, and let S be a vertically or horizontally simple region in the uv-plane. Let T from S to R be given by $T(u, v) = (x, y) = (g(u, v), h(u, v))$, where g and h have continuous first partial derivatives. Assume that T is one-to-one except possibly on the boundary of S. If f is continuous on R, and $\partial(x, y)/\partial(u, v)$ is nonzero on S, then

$$\iint_R f(x, y)\, dx\, dy = \iint_S f(g(u, v), h(u, v)) \left| \frac{\partial(x, y)}{\partial(u, v)} \right| du\, dv.$$

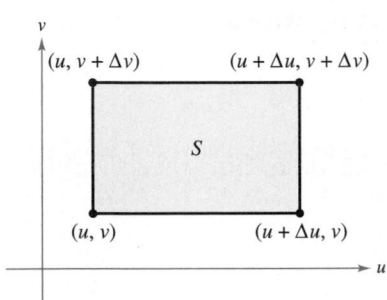

Area of $S = \Delta u\, \Delta v$
$\Delta u > 0,\ \Delta v > 0$
Figure 14.74

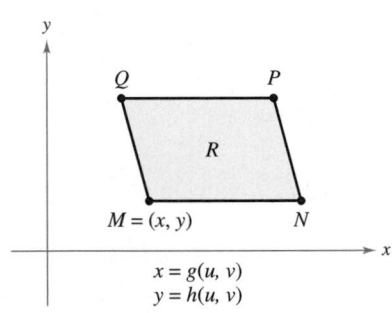

$x = g(u, v)$
$y = h(u, v)$

The vertices in the xy-plane are
$M(g(u, v), h(u, v))$,
$N(g(u + \Delta u, v), h(u + \Delta u, v))$,
$P(g(u + \Delta u, v + \Delta v)$,
$h(u + \Delta u, v + \Delta v))$, and
$Q(g(u, v + \Delta v), h(u, v + \Delta v))$.
Figure 14.75

Proof   Consider the case in which S is a rectangular region in the uv-plane with vertices $(u, v), (u + \Delta u, v), (u + \Delta u, v + \Delta v)$, and $(u, v + \Delta v)$, as shown in Figure 14.74. The images of these vertices in the xy-plane are shown in Figure 14.75. If Δu and Δv are small, then the continuity of g and h implies that R is approximately a parallelogram determined by the vectors $\overrightarrow{MN}$ and $\overrightarrow{MQ}$. So, the area of R is

$$\Delta A \approx \| \overrightarrow{MN} \times \overrightarrow{MQ} \|.$$

Moreover, for small Δu and Δv, the partial derivatives of g and h with respect to u can be approximated by

$$g_u(u, v) \approx \frac{g(u + \Delta u, v) - g(u, v)}{\Delta u} \quad \text{and} \quad h_u(u, v) \approx \frac{h(u + \Delta u, v) - h(u, v)}{\Delta u}.$$

Consequently,

$$\begin{aligned}
\overrightarrow{MN} &= [g(u + \Delta u, v) - g(u, v)]\mathbf{i} + [h(u + \Delta u, v) - h(u, v)]\mathbf{j} \\
&\approx [g_u(u, v)\, \Delta u]\mathbf{i} + [h_u(u, v)\, \Delta u]\mathbf{j} \\
&= \frac{\partial x}{\partial u} \Delta u \mathbf{i} + \frac{\partial y}{\partial u} \Delta u \mathbf{j}.
\end{aligned}$$

Similarly, you can approximate $\overrightarrow{MQ}$ by $\dfrac{\partial x}{\partial v} \Delta v \mathbf{i} + \dfrac{\partial y}{\partial v} \Delta v \mathbf{j}$, which implies that

$$\overrightarrow{MN} \times \overrightarrow{MQ} \approx
\begin{vmatrix}
\mathbf{i} & \mathbf{j} & \mathbf{k} \\
\dfrac{\partial x}{\partial u} \Delta u & \dfrac{\partial y}{\partial u} \Delta u & 0 \\
\dfrac{\partial x}{\partial v} \Delta v & \dfrac{\partial y}{\partial v} \Delta v & 0
\end{vmatrix}
=
\begin{vmatrix}
\dfrac{\partial x}{\partial u} & \dfrac{\partial y}{\partial u} \\
\dfrac{\partial x}{\partial v} & \dfrac{\partial y}{\partial v}
\end{vmatrix}
\Delta u\, \Delta v \mathbf{k}.$$

It follows that, in Jacobian notation,

$$\Delta A \approx \| \overrightarrow{MN} \times \overrightarrow{MQ} \| \approx \left| \frac{\partial(x, y)}{\partial(u, v)} \right| \Delta u\, \Delta v.$$

Because this approximation improves as Δu and Δv approach 0, the limiting case can be written as

$$dA \approx \| \overrightarrow{MN} \times \overrightarrow{MQ} \| \approx \left| \frac{\partial(x, y)}{\partial(u, v)} \right| du\, dv.$$

So,

$$\iint_R f(x, y)\, dx\, dy = \iint_S f(g(u, v), h(u, v)) \left| \frac{\partial(x, y)}{\partial(u, v)} \right| du\, dv. \qquad \blacksquare$$

The next two examples show how a change of variables can simplify the integration process. The simplification can occur in various ways. You can make a change of variables to simplify either the *region R* or the *integrand* $f(x, y)$, or both.

EXAMPLE 3 Using a Change of Variables to Simplify a Region

$\cdots\cdots\triangleright$ *See LarsonCalculus.com for an interactive version of this type of example.*

Let R be the region bounded by the lines

$$x - 2y = 0, \quad x - 2y = -4, \quad x + y = 4, \quad \text{and} \quad x + y = 1$$

as shown in Figure 14.76. Evaluate the double integral

$$\int_R \int 3xy \, dA.$$

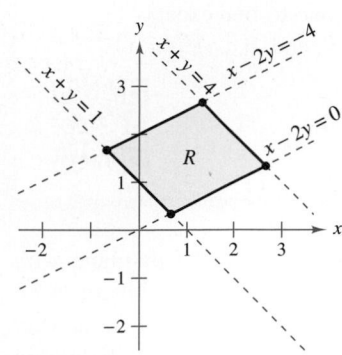

Figure 14.76

Solution From Example 2, you can use the change of variables

$$x = \frac{1}{3}(2u + v) \quad \text{and} \quad y = \frac{1}{3}(u - v).$$

(Note that the region S is shown in Figure 14.77.) The partial derivatives of x and y are

$$\frac{\partial x}{\partial u} = \frac{2}{3}, \quad \frac{\partial x}{\partial v} = \frac{1}{3}, \quad \frac{\partial y}{\partial u} = \frac{1}{3}, \quad \text{and} \quad \frac{\partial y}{\partial v} = -\frac{1}{3}$$

which implies that the Jacobian is

$$\begin{aligned}
\frac{\partial(x, y)}{\partial(u, v)} &= \begin{vmatrix} \dfrac{\partial x}{\partial u} & \dfrac{\partial x}{\partial v} \\[2mm] \dfrac{\partial y}{\partial u} & \dfrac{\partial y}{\partial v} \end{vmatrix} \\[4mm]
&= \begin{vmatrix} \dfrac{2}{3} & \dfrac{1}{3} \\[2mm] \dfrac{1}{3} & -\dfrac{1}{3} \end{vmatrix} \\[4mm]
&= -\frac{2}{9} - \frac{1}{9} \\[2mm]
&= -\frac{1}{3}.
\end{aligned}$$

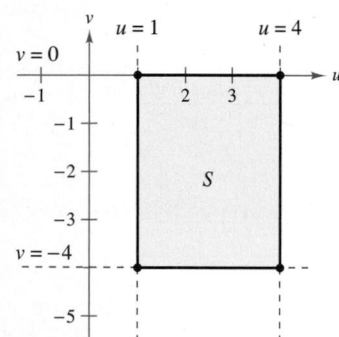

Figure 14.77

So, by Theorem 14.5, you obtain

$$\begin{aligned}
\int_R \int 3xy \, dA &= \int_S \int 3\left[\frac{1}{3}(2u + v)\frac{1}{3}(u - v)\right]\left|\frac{\partial(x, y)}{\partial(u, v)}\right| dv \, du \\[2mm]
&= \int_1^4 \int_{-4}^0 \frac{1}{9}(2u^2 - uv - v^2) \, dv \, du \\[2mm]
&= \frac{1}{9}\int_1^4 \left[2u^2v - \frac{uv^2}{2} - \frac{v^3}{3}\right]_{-4}^0 du \\[2mm]
&= \frac{1}{9}\int_1^4 \left(8u^2 + 8u - \frac{64}{3}\right) du \\[2mm]
&= \frac{1}{9}\left[\frac{8u^3}{3} + 4u^2 - \frac{64}{3}u\right]_1^4 \\[2mm]
&= \frac{164}{9}.
\end{aligned}$$

EXAMPLE 4 **Change of Variables: Simplifying an Integrand**

Let R be the region bounded by the square with vertices $(0, 1)$, $(1, 2)$, $(2, 1)$, and $(1, 0)$. Evaluate the integral

$$\int_R\int (x + y)^2 \sin^2(x - y)\, dA.$$

Solution Note that the sides of R lie on the lines $x + y = 1$, $x - y = 1$, $x + y = 3$, and $x - y = -1$, as shown in Figure 14.78. Letting $u = x + y$ and $v = x - y$, you can determine the bounds for region S in the uv plane to be

$$1 \le u \le 3 \quad \text{and} \quad -1 \le v \le 1$$

as shown in Figure 14.79. Solving for x and y in terms of u and v produces

$$x = \frac{1}{2}(u + v) \quad \text{and} \quad y = \frac{1}{2}(u - v).$$

The partial derivatives of x and y are

$$\frac{\partial x}{\partial u} = \frac{1}{2}, \quad \frac{\partial x}{\partial v} = \frac{1}{2}, \quad \frac{\partial y}{\partial u} = \frac{1}{2}, \quad \text{and} \quad \frac{\partial y}{\partial v} = -\frac{1}{2}$$

which implies that the Jacobian is

$$\frac{\partial(x, y)}{\partial(u, v)} = \begin{vmatrix} \dfrac{\partial x}{\partial u} & \dfrac{\partial x}{\partial v} \\[2mm] \dfrac{\partial y}{\partial u} & \dfrac{\partial y}{\partial v} \end{vmatrix} = \begin{vmatrix} \dfrac{1}{2} & \dfrac{1}{2} \\[2mm] \dfrac{1}{2} & -\dfrac{1}{2} \end{vmatrix} = -\frac{1}{4} - \frac{1}{4} = -\frac{1}{2}.$$

By Theorem 14.5, it follows that

$$\int_R\int (x + y)^2 \sin^2(x - y)\, dA = \int_{-1}^{1}\int_{1}^{3} u^2 (\sin^2 v)\left(\frac{1}{2}\right) du\, dv$$

$$= \frac{1}{2}\int_{-1}^{1} (\sin^2 v)\, \frac{u^3}{3}\bigg]_{1}^{3} dv$$

$$= \frac{13}{3}\int_{-1}^{1} \sin^2 v\, dv$$

$$= \frac{13}{6}\int_{-1}^{1} (1 - \cos 2v)\, dv$$

$$= \frac{13}{6}\left[v - \frac{1}{2}\sin 2v \right]_{-1}^{1}$$

$$= \frac{13}{6}\left[2 - \frac{1}{2}\sin 2 + \frac{1}{2}\sin(-2) \right]$$

$$= \frac{13}{6}(2 - \sin 2)$$

$$\approx 2.363.$$

In each of the change of variables examples in this section, the region S has been a rectangle with sides parallel to the u- or v-axis. Occasionally, a change of variables can be used for other types of regions. For instance, letting $T(u, v) = \left(x, \frac{1}{2}y\right)$ changes the circular region $u^2 + v^2 = 1$ to the elliptical region

$$x^2 + \frac{y^2}{4} = 1.$$

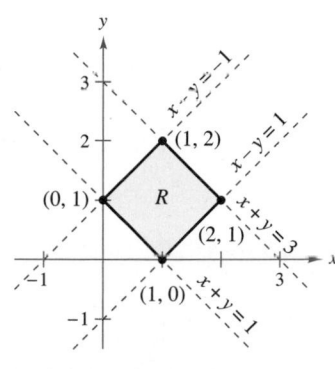

Region R in the xy-plane
Figure 14.78

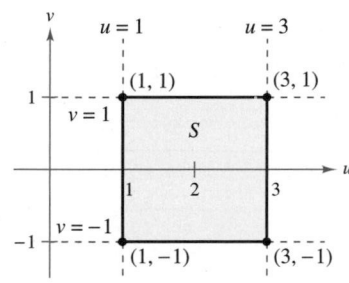

Region S in the uv-plane
Figure 14.79

14.8 Exercises

See **CalcChat.com** for tutorial help and worked-out solutions to odd-numbered exercises.

CONCEPT CHECK

1. Jacobian Describe how to find the Jacobian of x and y with respect to u and v for $x = g(u, v)$ and $y = h(u, v)$.

2. Change of Variable When is it beneficial to use the Jacobian to change variables in a double integral?

 **Finding a Jacobian** In Exercises 3–10, find the Jacobian $\partial(x, y)/\partial(u, v)$ for the indicated change of variables.

3. $x = -\frac{1}{2}(u - v)$, $y = \frac{1}{2}(u + v)$

4. $x = 5u - v$, $y = 3u + 4v$

5. $x = u - v^2$, $y = u + v$

6. $x = uv - 2u$, $y = uv$

7. $x = u \cos \theta - v \sin \theta$, $y = u \sin \theta + v \cos \theta$

8. $x = u + 1$, $y = 9v$

9. $x = e^u \sin v$, $y = e^u \cos v$

10. $x = u/v$, $y = u + v$

 **Using a Transformation** In Exercises 11–14, sketch the image S in the uv-plane of the region R in the xy-plane using the given transformations.

11. $x = 3u + 2v$
$y = 3v$

12. $x = \frac{1}{3}(4u - v)$
$y = \frac{1}{3}(u - v)$

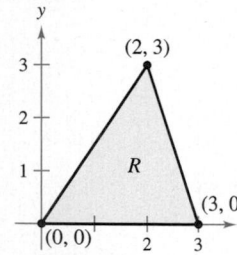

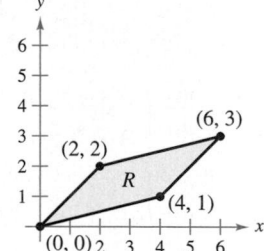

13. $x = \frac{1}{2}(u + v)$
$y = \frac{1}{2}(u - v)$

14. $x = \frac{1}{3}(v - u)$
$y = \frac{1}{3}(2v + u)$

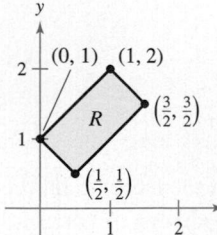

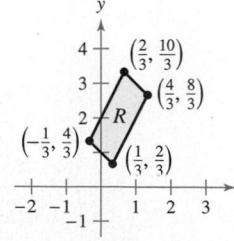

Verifying a Change of Variables In Exercises 15 and 16, verify the result of the indicated example by setting up the integral using $dy\,dx$ or $dx\,dy$ for dA. Then use a computer algebra system to evaluate the integral.

15. Example 3

16. Example 4

Evaluating a Double Integral Using a Change of Variables In Exercises 17–22, use the indicated change of variables to evaluate the double integral.

17. $\displaystyle\int_R\int 4(x^2 + y^2)\, dA$
$x = \frac{1}{2}(u + v)$
$y = \frac{1}{2}(u - v)$

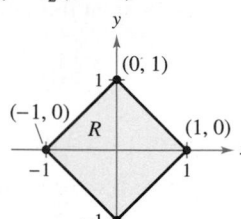

18. $\displaystyle\int_R\int (2y - x)\, dA$
$x = \frac{1}{2}(v - u)$
$y = \frac{1}{2}(3u - v)$

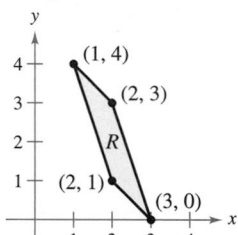

19. $\displaystyle\int_R\int y(x - y)\, dA$
$x = u + v$
$y = u$

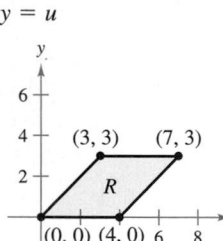

20. $\displaystyle\int_R\int 4(x + y)e^{x-y}\, dA$
$x = \frac{1}{2}(u + v)$
$y = \frac{1}{2}(u - v)$

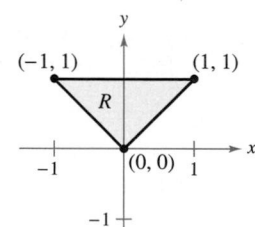

21. $\displaystyle\int_R\int e^{-xy/2}\, dA$

$x = \sqrt{\dfrac{v}{u}}, \quad y = \sqrt{uv}$

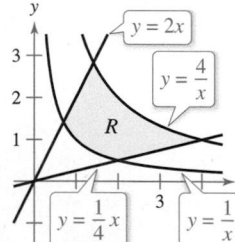

22. $\displaystyle\int_R\int y \sin xy\, dA$

$x = \dfrac{u}{v}, \quad y = v$

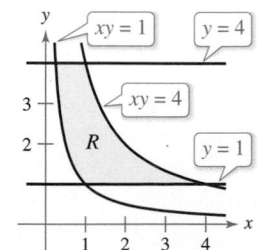

Finding Volume Using a Change of Variables In Exercises 23–30, use a change of variables to find the volume of the solid region lying below the surface $z = f(x, y)$ and above the plane region R.

23. $f(x, y) = 9xy$

R: region bounded by the square with vertices $(1, 0)$, $(0, 1)$, $(1, 2)$, $(2, 1)$

24. $f(x, y) = (3x + 2y)^2 \sqrt{2y - x}$

R: region bounded by the parallelogram with vertices $(0, 0)$, $(-2, 3)$, $(2, 5)$, $(4, 2)$

25. $f(x, y) = (x + y)e^{x-y}$

R: region bounded by the square with vertices $(4, 0)$, $(6, 2)$, $(4, 4)$, $(2, 2)$

26. $f(x, y) = (x + y)^2 \sin^2(x - y)$

R: region bounded by the square with vertices $(\pi, 0)$, $(3\pi/2, \pi/2)$, (π, π), $(\pi/2, \pi/2)$

27. $f(x, y) = \sqrt{(x - y)(x + 4y)}$

R: region bounded by the parallelogram with vertices $(0, 0)$, $(1, 1)$, $(5, 0)$, $(4, -1)$

28. $f(x, y) = (3x + 2y)(2y - x)^{3/2}$

R: region bounded by the parallelogram with vertices $(0, 0)$, $(-2, 3)$, $(2, 5)$, $(4, 2)$

29. $f(x, y) = \sqrt{x + y}$

R: region bounded by the triangle with vertices $(0, 0)$, $(a, 0)$, $(0, a)$, where $a > 0$

30. $f(x, y) = \dfrac{xy}{1 + x^2y^2}$

R: region bounded by the graphs of $xy = 1$, $xy = 4$, $x = 1$, $x = 4$ (*Hint:* Let $x = u$, $y = v/u$.)

EXPLORING CONCEPTS

31. Using a Transformation The substitutions $u = 2x - y$ and $v = x + y$ make the region R (see figure) into a simpler region S in the uv-plane. Determine the total number of sides of S that are parallel to either the u-axis or the v-axis.

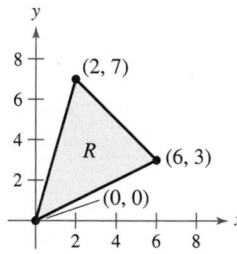

32. **HOW DO YOU SEE IT?** The region R is transformed into a simpler region S (see figure). Which substitution can be used to make the transformation?

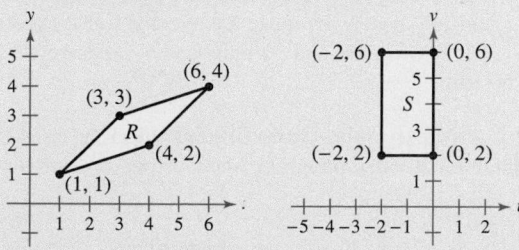

(a) $u = 3y - x$, $v = y - x$ (b) $u = y - x$, $v = 3y - x$

33. Using an Ellipse Consider the region R in the xy-plane bounded by the ellipse $(x^2/a^2) + (y^2/b^2) = 1$ and the transformations $x = au$ and $y = bv$.

(a) Sketch the graph of the region R and its image S under the given transformation.

(b) Find $\dfrac{\partial(x, y)}{\partial(u, v)}$.

(c) Find the area of the ellipse using the indicated change of variables.

34. Volume Use the result of Exercise 33 to find the volume of each dome-shaped solid lying below the surface $z = f(x, y)$ and above the elliptical region R. (*Hint:* After making the change of variables given by the results in Exercise 33, make a second change of variables to polar coordinates.)

(a) $f(x, y) = 16 - x^2 - y^2$; $R: \dfrac{x^2}{16} + \dfrac{y^2}{9} \leq 1$

(b) $f(x, y) = A \cos\left(\dfrac{\pi}{2}\sqrt{\dfrac{x^2}{a^2} + \dfrac{y^2}{b^2}}\right)$; $R: \dfrac{x^2}{a^2} + \dfrac{y^2}{b^2} \leq 1$

Finding a Jacobian In Exercises 35–40, find the Jacobian

$$\frac{\partial(x, y, z)}{\partial(u, v, w)}$$

for the indicated change of variables. If

$$x = f(u, v, w), \quad y = g(u, v, w), \quad \text{and} \quad z = h(u, v, w),$$

then the Jacobian of x, y, and z with respect to u, v, and w is

$$\frac{\partial(x, y, z)}{\partial(u, v, w)} = \begin{vmatrix} \dfrac{\partial x}{\partial u} & \dfrac{\partial x}{\partial v} & \dfrac{\partial x}{\partial w} \\ \dfrac{\partial y}{\partial u} & \dfrac{\partial y}{\partial v} & \dfrac{\partial y}{\partial w} \\ \dfrac{\partial z}{\partial u} & \dfrac{\partial z}{\partial v} & \dfrac{\partial z}{\partial w} \end{vmatrix}.$$

35. $x = u(1 - v)$, $y = uv(1 - w)$, $z = uvw$

36. $x = 4u - v$, $y = 4v - w$, $z = u + w$

37. $x = \frac{1}{2}(u + v)$, $y = \frac{1}{2}(u - v)$, $z = 2uvw$

38. $x = u - v + w$, $y = 2uv$, $z = u + v + w$

39. Spherical Coordinates

$x = \rho \sin \phi \cos \theta$, $y = \rho \sin \phi \sin \theta$, $z = \rho \cos \phi$

40. Cylindrical Coordinates

$x = r \cos \theta$, $y = r \sin \theta$, $z = z$

PUTNAM EXAM CHALLENGE

41. Let A be the area of the region in the first quadrant bounded by the line $y = \frac{1}{2}x$, the x-axis, and the ellipse $\frac{1}{9}x^2 + y^2 = 1$. Find the positive number m such that A is equal to the area of the region in the first quadrant bounded by the line $y = mx$, the y-axis, and the ellipse $\frac{1}{9}x^2 + y^2 = 1$.

This problem was composed by the Committee on the Putnam Prize Competition.
© The Mathematical Association of America. All rights reserved.

Evaluating an Integral In Exercises 1 and 2, evaluate the integral.

1. $\displaystyle\int_0^{3x} \sin(xy)\, dy$

2. $\displaystyle\int_y^{y^2} \frac{x}{y+1}\, dx$

Evaluating an Iterated Integral In Exercises 3–6, evaluate the iterated integral.

3. $\displaystyle\int_0^1 \int_0^{1+x} (3x + 2y)\, dy\, dx$

4. $\displaystyle\int_0^2 \int_{x^2}^{2x} (x^2 + 2y)\, dy\, dx$

5. $\displaystyle\int_0^1 \int_0^{\sqrt{1-x^4}} x^3\, dy\, dx$

6. $\displaystyle\int_0^1 \int_0^{2y} (9 + 3x^2 + 3y^2)\, dx\, dy$

Finding the Area of a Region In Exercises 7–10, use an iterated integral to find the area of the region bounded by the graphs of the equations.

7. $x + 3y = 3,\ x = 0,\ y = 0$

8. $y = 6x - x^2,\ y = x^2 - 2x$

9. $y = x,\ y = 2x + 2,\ x = 0,\ x = 4$

10. $x = y^2 + 1,\ x = 0,\ y = 0,\ y = 2$

Changing the Order of Integration In Exercises 11–14, sketch the region R whose area is given by the iterated integral. Then change the order of integration and show that both orders yield the same area.

11. $\displaystyle\int_1^5 \int_0^4 dy\, dx$

12. $\displaystyle\int_{-3}^3 \int_0^{9-y^2} dx\, dy$

13. $\displaystyle\int_0^2 \int_{y/2}^{3-y} dx\, dy$

14. $\displaystyle\int_0^3 \int_0^x dy\, dx + \int_3^6 \int_0^{6-x} dy\, dx$

Evaluating a Double Integral In Exercises 15 and 16, set up integrals for both orders of integration. Use the more convenient order to evaluate the integral over the plane region R.

15. $\displaystyle\iint_R 4xy\, dA$

R: rectangle with vertices $(0, 0),\ (0, 4),\ (2, 4),\ (2, 0)$

16. $\displaystyle\iint_R 6x^2\, dA$

R: region bounded by $y = 0,\ y = \sqrt{x},\ x = 1$

Finding Volume In Exercises 17–20, use a double integral to find the volume of the indicated solid.

17.

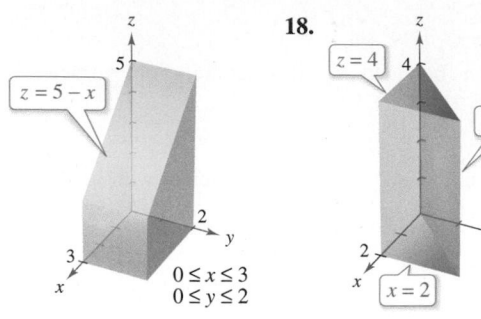

18.

19.

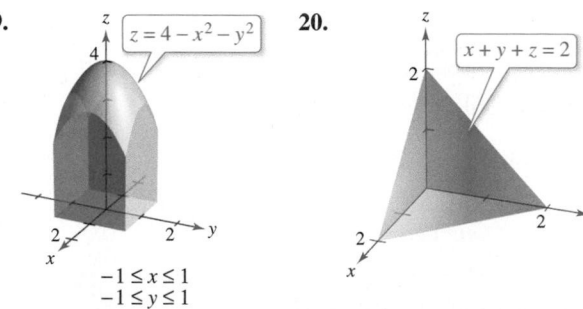

20.

Average Value In Exercises 21 and 22, find the average value of $f(x, y)$ over the plane region R.

21. $f(x) = 16 - x^2 - y^2$

R: rectangle with vertices $(2, 2),\ (-2, 2),\ (-2, -2),\ (2, -2)$

22. $f(x) = 2x^2 + y^2$

R: square with vertices $(0, 0),\ (3, 0),\ (3, 3),\ (0, 3)$

23. Average Temperature The temperature in degrees Celsius on the surface of a metal plate is

$$T(x, y) = 40 - 6x^2 - y^2$$

where x and y are measured in centimeters. Estimate the average temperature when x varies between 0 and 3 centimeters and y varies between 0 and 5 centimeters.

24. Average Profit A firm's profit P (in dollars) from marketing two television models is

$$P = 192x + 576y - x^2 - 5y^2 - 2xy - 5000$$

where x and y represent the numbers of units of the two television models. Estimate the average weekly profit when x varies between 40 and 50 units and y varies between 45 and 60 units.

Converting to Polar Coordinates In Exercises 25 and 26, evaluate the iterated integral by converting to polar coordinates.

25. $\displaystyle\int_0^{\sqrt{5}} \int_0^{\sqrt{5-x^2}} \sqrt{x^2 + y^2}\, dy\, dx$

26. $\displaystyle\int_0^4 \int_0^{\sqrt{16-y^2}} (x^2 + y^2)\, dx\, dy$

Volume In Exercises 27 and 28, use a double integral in polar coordinates to find the volume of the solid bounded by the graphs of the equations.

27. $z = xy^2$, $x^2 + y^2 = 9$, first octant

28. $z = \sqrt{25 - x^2 - y^2}$, $z = 0$, $x^2 + y^2 = 16$

Area In Exercises 29 and 30, use a double integral to find the area of the shaded region.

29.

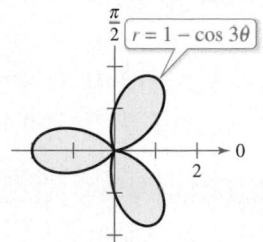

$\frac{\pi}{2}$ $\boxed{r = 1 - \cos 3\theta}$

30.

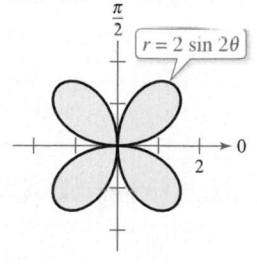

$\frac{\pi}{2}$ $\boxed{r = 2 \sin 2\theta}$

Area In Exercises 31 and 32, sketch a graph of the region bounded by the graphs of the equations. Then use a double integral to find the area of the region.

31. Inside the limaçon $r = 3 + 2 \cos \theta$ and outside the circle $r = 4$

32. Inside the circle $r = 3 \sin \theta$ and outside the cardioid $r = 1 + \sin \theta$

33. Area and Volume Consider the region R in the xy-plane bounded by $(x^2 + y^2)^2 = 9(x^2 - y^2)$.

(a) Convert the equation to polar coordinates. Use a graphing utility to graph the equation.

(b) Use a double integral to find the area of the region R.

(c) Use a computer algebra system to find the volume of the solid region bounded above by the hemisphere $z = \sqrt{9 - x^2 - y^2}$ and below by the region R.

34. Converting to Polar Coordinates Write the sum of the two iterated integrals as a single iterated integral by converting to polar coordinates. Evaluate the resulting iterated integral.

$$\int_0^{8/\sqrt{13}} \int_0^{3x/2} xy \, dy \, dx + \int_{8/\sqrt{13}}^4 \int_0^{\sqrt{16-x^2}} xy \, dy \, dx$$

Finding the Mass of a Lamina In Exercises 35 and 36, find the mass of the lamina described by the inequalities, given that its density is $\rho(x, y) = x + 3y$.

35. $0 \le x \le 1$, $0 \le y \le 2$

36. $x \ge 0$, $0 \le y \le \sqrt{4 - x^2}$

Finding the Center of Mass In Exercises 37–40, find the mass and center of mass of the lamina bounded by the graphs of the equations for the given density.

37. $y = x^3$, $y = 0$, $x = 2$, $\rho = kx$

38. $y = \dfrac{2}{x}$, $y = 0$, $x = 1$, $x = 2$, $\rho = ky$

39. $y = 2x$, $y = 2x^3$, $x \ge 0$, $y \ge 0$, $\rho = kxy$

40. $y = 6 - x$, $y = 0$, $x = 0$, $\rho = kx^2$

Finding Moments of Inertia and Radii of Gyration In Exercises 41 and 42, find $I_x, I_y, I_0, \bar{x}$, and $\bar{y}$ for the lamina bounded by the graphs of the equations.

41. $y = 0$, $y = 2$, $x = 0$, $x = 3$, $\rho = kx$

42. $y = 4 - x^2$, $y = 0$, $x > 0$, $\rho = ky$

Finding Surface Area In Exercises 43–46, find the area of the surface given by $z = f(x, y)$ that lies above the region R.

43. $f(x, y) = 25 - x^2 - y^2$

   $R = \{(x, y)\colon x^2 + y^2 \le 25\}$

44. $f(x, y) = 8 + 4x - 5y$

   $R = \{(x, y)\colon x^2 + y^2 \le 1\}$

45. $f(x, y) = 9 - y^2$

   R: triangle with vertices $(-3, 3)$, $(0, 0)$, $(3, 3)$

46. $f(x, y) = 4 - x^2$

   R: triangle with vertices $(-2, 2)$, $(0, 0)$, $(2, 2)$

47. Building Design A new auditorium is built with a foundation in the shape of one-fourth of a circle of radius 50 feet. So, it forms a region R bounded by the graph of $x^2 + y^2 = 50^2$ with $x \ge 0$ and $y \ge 0$. The following equations are models for the floor and ceiling.

   Floor: $z = \dfrac{x + y}{5}$

   Ceiling: $z = 20 + \dfrac{xy}{100}$

(a) Calculate the volume of the room, which is needed to determine the heating and cooling requirements.

(b) Find the surface area of the ceiling.

48. Surface Area The roof over the stage of an open air theater at a theme park is modeled by

$$f(x, y) = 25\left[1 + e^{-(x^2+y^2)/1000} \cos^2\left(\frac{x^2 + y^2}{1000}\right)\right]$$

where the stage is a semicircle bounded by the graphs of $y = \sqrt{50^2 - x^2}$ and $y = 0$.

(a) Use a computer algebra system to graph the surface.

(b) Use a computer algebra system to approximate the number of square feet of roofing required to cover the surface.

Evaluating a Triple Iterated Integral In Exercises 49–52, evaluate the triple iterated integral.

49. $\displaystyle\int_0^4 \int_0^1 \int_0^2 (2x + y + 4z) \, dy \, dz \, dx$

50. $\displaystyle\int_0^1 \int_0^{1+\sqrt{y}} \int_0^{xy} y \, dz \, dx \, dy$

51. $\displaystyle\int_0^2 \int_1^2 \int_0^1 (e^x + y^2 + z^2) \, dx \, dy \, dz$

52. $\displaystyle\int_0^3 \int_{\pi/2}^\pi \int_2^5 z \sin x \, dy \, dx \, dz$

Evaluating a Triple Iterated Integral Using Technology In Exercises 53 and 54, use a computer algebra system to evaluate the triple iterated integral.

53. $\int_{-1}^{1} \int_{-\sqrt{1-x^2}}^{\sqrt{1-x^2}} \int_{-\sqrt{1-x^2-y^2}}^{\sqrt{1-x^2-y^2}} (x^2 + y^2) \, dz \, dy \, dx$

54. $\int_{0}^{2} \int_{0}^{\sqrt{4-x^2}} \int_{0}^{\sqrt{4-x^2-y^2}} xyz \, dz \, dy \, dx$

Volume In Exercises 55 and 56, use a triple integral to find the volume of the solid bounded by the graphs of the equations.

55. $z = xy$, $z = 0$, $0 \le x \le 3$, $0 \le y \le 4$

56. $z = 8 - x - y$, $z = 0$, $y = x$, $y = 3$, $x = 0$

Changing the Order of Integration In Exercises 57 and 58, sketch the solid whose volume is given by the iterated integral. Then rewrite the integral using the indicated order of integration.

57. $\int_{0}^{1} \int_{0}^{y} \int_{0}^{\sqrt{1-x^2}} dz \, dx \, dy$

Rewrite using the order $dz \, dy \, dx$.

58. $\int_{0}^{6} \int_{0}^{6-x} \int_{0}^{6-x-y} dz \, dy \, dx$

Rewrite using the order $dy \, dx \, dz$.

Center of Mass In Exercises 59 and 60, find the mass and the indicated coordinate of the center of mass of the solid region Q of density ρ bounded by the graphs of the equations.

59. Find $\bar{x}$ using $\rho(x, y, z) = k$.

Q: $x + y + z = 10$, $x = 0$, $y = 0$, $z = 0$

60. Find $\bar{y}$ using $\rho(x, y, z) = kx$.

Q: $z = 5 - y$, $z = 0$, $y = 0$, $x = 0$, $x = 5$

Evaluating a Triple Iterated Integral In Exercises 61–64, evaluate the triple iterated integral.

61. $\int_{0}^{3} \int_{\pi/6}^{\pi/3} \int_{0}^{4} r \cos \theta \, dr \, d\theta \, dz$

62. $\int_{0}^{\pi/2} \int_{0}^{3} \int_{0}^{4-z} z \, dr \, dz \, d\theta$

63. $\int_{0}^{\pi} \int_{0}^{\pi/2} \int_{0}^{\sin \theta} \rho^2 \sin \theta \cos \theta \, d\rho \, d\theta \, d\phi$

64. $\int_{0}^{\pi/4} \int_{0}^{\pi/4} \int_{0}^{\cos \phi} \cos \theta \, d\rho \, d\phi \, d\theta$

Evaluating a Triple Iterated Integral Using Technology In Exercises 65 and 66, use a computer algebra system to evaluate the triple iterated integral.

65. $\int_{0}^{\pi} \int_{0}^{2} \int_{0}^{3} \sqrt{z^2 + 4} \, dz \, dr \, d\theta$

66. $\int_{0}^{\pi/2} \int_{0}^{\pi/2} \int_{0}^{\cos \phi} \rho^2 \cos \theta \, d\rho \, d\theta \, d\phi$

Volume In Exercises 67 and 68, use cylindrical coordinates to find the volume of the solid.

67. Solid bounded above by $z = 8 - x^2 - y^2$ and below by $z = x^2 + y^2$

68. Solid bounded above by $3x^2 + 3y^2 + z^2 = 45$ and below by the xy-plane

Volume In Exercises 69 and 70, use spherical coordinates to find the volume of the solid.

69. Solid bounded above by $x^2 + y^2 + z^2 = 4$ and below by $z^2 = 3x^2 + 3y^2$

70. Solid bounded above by $x^2 + y^2 + z^2 = 36$ and below by $z = \sqrt{x^2 + y^2}$

Finding a Jacobian In Exercises 71–74, find the Jacobian $\partial(x, y)/\partial(u, v)$ for the indicated change of variables.

71. $x = 3uv$, $y = 2(u - v)$

72. $x = u^2 + v^2$, $y = u^2 - v^2$

73. $x = u \sin \theta + v \cos \theta$, $y = u \cos \theta + v \sin \theta$

74. $x = uv$, $y = \dfrac{v}{u}$

Evaluating a Double Integral Using a Change of Variables In Exercises 75–78, use the indicated change of variables to evaluate the double integral.

75. $\displaystyle\int_{R}\int \ln(x + y) \, dA$

$x = \dfrac{1}{2}(u + v)$

$y = \dfrac{1}{2}(u - v)$

76. $\displaystyle\int_{R}\int 16xy \, dA$

$x = \dfrac{1}{4}(u + v)$

$y = \dfrac{1}{2}(v - u)$

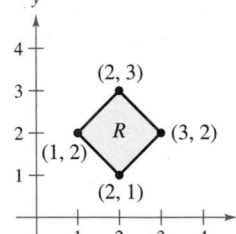

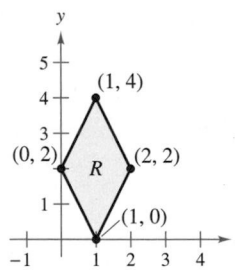

77. $\displaystyle\int_{R}\int (xy + x^2) \, dA$

$x = u$

$y = \dfrac{1}{3}(u - v)$

78. $\displaystyle\int_{R}\int \dfrac{x}{1 + x^2 y^2} \, dA$

$x = u$

$y = \dfrac{v}{u}$

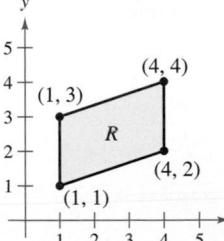

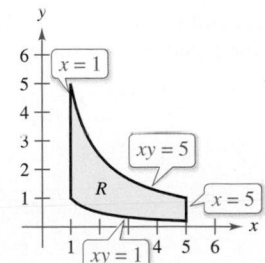

P.S. Problem Solving

See **CalcChat.com** for tutorial help and worked-out solutions to odd-numbered exercises.

1. Volume Find the volume of the solid of intersection of the three cylinders $x^2 + z^2 = 1$, $y^2 + z^2 = 1$, and $x^2 + y^2 = 1$ (see figure).

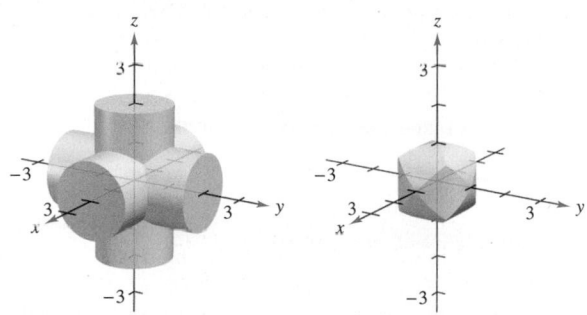

2. Surface Area Let a, b, c, and d be positive real numbers. The portion of the plane $ax + by + cz = d$ in the first octant is shown in the figure. Show that the surface area of this portion of the plane is equal to

$$\frac{A(R)}{c}\sqrt{a^2 + b^2 + c^2}$$

where $A(R)$ is the area of the triangular region R in the xy-plane, as shown in the figure.

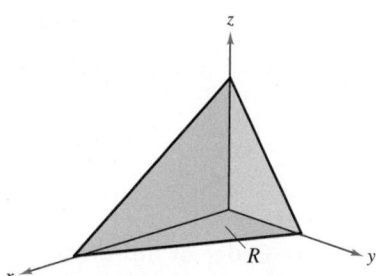

3. Using a Change of Variables The figure shows the region R bounded by the curves

$$y = \sqrt{x}, \ y = \sqrt{2x}, \ y = \frac{x^2}{3}, \text{ and } y = \frac{x^2}{4}.$$

Use the change of variables $x = u^{1/3}v^{2/3}$ and $y = u^{2/3}v^{1/3}$ to find the area of the region R.

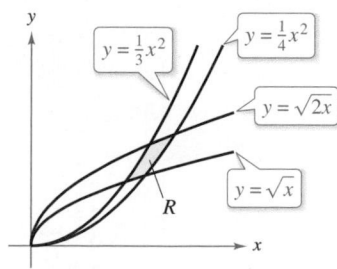

4. Proof Prove that $\displaystyle\lim_{n\to\infty} \int_0^1 \int_0^1 x^n y^n \, dx \, dy = 0$.

5. Deriving a Sum Derive Euler's famous result that was mentioned in Section 9.3,

$$\sum_{n=1}^{\infty} \frac{1}{n^2} = \frac{\pi^2}{6}$$

by completing each step.

(a) Prove that

$$\int \frac{dv}{2 - u^2 + v^2} = \frac{1}{\sqrt{2 - u^2}} \arctan \frac{v}{\sqrt{2 - u^2}} + C.$$

(b) Prove that $\displaystyle I_1 = \int_0^{\sqrt{2}/2} \int_{-u}^{u} \frac{2}{2 - u^2 + v^2} \, dv \, du = \frac{\pi^2}{18}$ by using the substitution $u = \sqrt{2} \sin \theta$.

(c) Prove that

$$I_2 = \int_{\sqrt{2}/2}^{\sqrt{2}} \int_{u-\sqrt{2}}^{-u+\sqrt{2}} \frac{2}{2 - u^2 + v^2} \, dv \, du$$

$$= 4 \int_{\pi/6}^{\pi/2} \arctan \frac{1 - \sin \theta}{\cos \theta} \, d\theta$$

by using the substitution $u = \sqrt{2} \sin \theta$.

(d) Prove the trigonometric identity

$$\frac{1 - \sin \theta}{\cos \theta} = \tan\left[\frac{(\pi/2) - \theta}{2}\right].$$

(e) Prove that $\displaystyle I_2 = \int_{\sqrt{2}/2}^{\sqrt{2}} \int_{u-\sqrt{2}}^{-u+\sqrt{2}} \frac{2}{2 - u^2 + v^2} \, dv \, du = \frac{\pi^2}{9}$.

(f) Use the formula for the sum of an infinite geometric series to verify that

$$\sum_{n=1}^{\infty} \frac{1}{n^2} = \int_0^1 \int_0^1 \frac{1}{1 - xy} \, dx \, dy.$$

(g) Use the change of variables

$$u = \frac{x + y}{\sqrt{2}} \quad \text{and} \quad v = \frac{y - x}{\sqrt{2}}$$

to prove that

$$\sum_{n=1}^{\infty} \frac{1}{n^2} = \int_0^1 \int_0^1 \frac{1}{1 - xy} \, dx \, dy = I_1 + I_2 = \frac{\pi^2}{6}.$$

6. Evaluating a Double Integral Evaluate the integral

$$\int_0^{\infty} \int_0^{\infty} \frac{1}{(1 + x^2 + y^2)^2} \, dx \, dy.$$

7. Evaluating Double Integrals Evaluate the integrals

$$\int_0^1 \int_0^1 \frac{x - y}{(x + y)^3} \, dx \, dy \quad \text{and} \quad \int_0^1 \int_0^1 \frac{x - y}{(x + y)^3} \, dy \, dx.$$

Are the results the same? Why or why not?

8. Volume Show that the volume of a spherical block can be approximated by $\Delta V \approx \rho^2 \sin \phi \, \Delta\rho \, \Delta\phi \, \Delta\theta$. (*Hint:* See Section 14.7, page 1027.)

Evaluating an Integral In Exercises 9 and 10, evaluate the integral. (*Hint:* See Exercise 63 in Section 14.3.)

9. $\displaystyle\int_0^\infty x^2 e^{-x^2}\,dx$

10. $\displaystyle\int_0^1 \sqrt{\ln\frac{1}{x}}\,dx$

11. Joint Density Function Consider the function

$$f(x, y) = \begin{cases} ke^{-(x+y)/a}, & x \ge 0, y \ge 0 \\ 0, & \text{elsewhere.} \end{cases}$$

Find the relationship between the positive constants a and k such that f is a joint density function of the continuous random variables x and y. (*Hint:* See Exercises 61–64 in Section 14.2)

12. Volume Find the volume of the solid generated by revolving the region in the first quadrant bounded by $y = e^{-x^2}$ about the y-axis. Use this result to find

$$\int_{-\infty}^\infty e^{-x^2}\,dx.$$

13. Volume and Surface Area From 1963 to 1986, the volume of the Great Salt Lake approximately tripled while its top surface area approximately doubled. Read the article "Relations between Surface Area and Volume in Lakes" by Daniel Cass and Gerald Wildenberg in *The College Mathematics Journal*. Then give examples of solids that have "water levels" a and b such that $V(b) = 3V(a)$ and $A(b) = 2A(a)$, where V is volume and A is area (see figure).

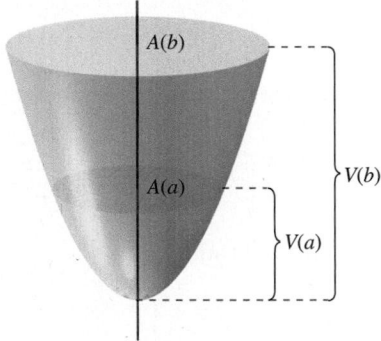

14. Proof The angle between a plane P and the xy-plane is θ, where $0 \le \theta < \pi/2$. The projection of a rectangular region in P onto the xy-plane is a rectangle whose sides have lengths Δx and Δy, as shown in the figure. Prove that the area of the rectangular region in P is $\sec \theta\, \Delta x\, \Delta y$.

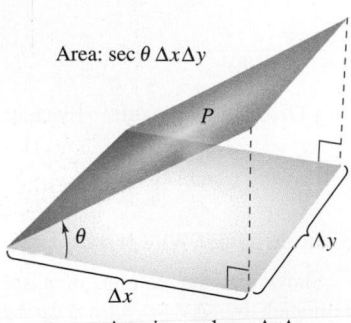

Area: $\sec \theta\, \Delta x \Delta y$

Area in xy-plane: $\Delta x \Delta y$

15. Surface Area Use the result of Exercise 14 to order the planes in ascending order of their surface areas for a fixed region R in the xy-plane. Explain your ordering without doing any calculations.

(a) $z_1 = 2 + x$

(b) $z_2 = 5$

(c) $z_3 = 10 - 5x + 9y$

(d) $z_4 = 3 + x - 2y$

16. Sprinkler Consider a circular lawn with a radius of 10 feet, as shown in the figure. Assume that a sprinkler distributes water in a radial fashion according to the formula

$$f(r) = \frac{r}{16} - \frac{r^2}{160}$$

(measured in cubic feet of water per hour per square foot of lawn), where r is the distance in feet from the sprinkler. Find the amount of water that is distributed in 1 hour in the following two annular regions.

$$A = \{(r, \theta)\colon 4 \le r \le 5, 0 \le \theta \le 2\pi\}$$

$$B = \{(r, \theta)\colon 9 \le r \le 10, 0 \le \theta \le 2\pi\}$$

Is the distribution of water uniform? Determine the amount of water the entire lawn receives in 1 hour.

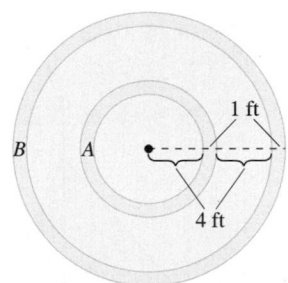

17. Changing the Order of Integration Sketch the solid whose volume is given by the sum of the iterated integrals

$$\int_0^6 \int_{z/2}^3 \int_{z/2}^y dx\,dy\,dz + \int_0^6 \int_3^{(12-z)/2} \int_{z/2}^{6-y} dx\,dy\,dz.$$

Then write the volume as a single iterated integral in the order $dy\,dz\,dx$ and find the volume of the solid.

18. Volume The figure shows a solid bounded below by the plane $z = 2$ and above by the sphere $x^2 + y^2 + z^2 = 8$.

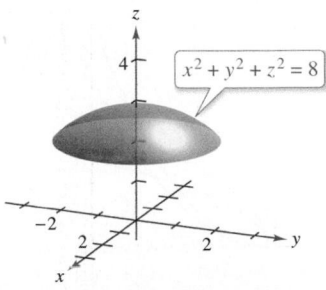

$x^2 + y^2 + z^2 = 8$

(a) Find the volume of the solid using cylindrical coordinates.

(b) Find the volume of the solid using spherical coordinates.

15 Vector Analysis

Finding Surface Area
(Example 6, p. 1093)

Work *(Exercise 35, p. 1077)*

Mass of a Spring *(Example 5, p. 1059)*

Building Design
(Exercise 74, p. 1068)

Earth's Magnetic Field *(Exercise 78, p. 1054)*

15.1   Vector Fields

■ Understand the concept of a vector field.
■ Determine whether a vector field is conservative.
■ Find the curl of a vector field.
■ Find the divergence of a vector field.

Vector Fields

In Chapter 12, you studied vector-valued functions—functions that assign a vector to a *real number*. There you saw that vector-valued functions of real numbers are useful in representing curves and motion along a curve. In this chapter, you will study two other types of vector-valued functions—functions that assign a vector to a *point in the plane* or a *point in space*. Such functions are called **vector fields,** and they are useful in representing various types of **force fields** and **velocity fields.**

> ### Definition of Vector Field
>
> A **vector field over a plane region R** is a function $\mathbf{F}$ that assigns a vector $\mathbf{F}(x, y)$ to each point in R.
>
> A **vector field over a solid region Q in space** is a function $\mathbf{F}$ that assigns a vector $\mathbf{F}(x, y, z)$ to each point in Q.

Although a vector field consists of infinitely many vectors, you can get a good idea of what the vector field looks like by sketching several representative vectors $\mathbf{F}(x, y)$ whose initial points are (x, y).

The *gradient* is one example of a vector field. For instance, if

$$f(x, y) = x^2y + 3xy^3$$

then the gradient of f

$$\nabla f(x, y) = f_x(x, y)\mathbf{i} + f_y(x, y)\mathbf{j}$$
$$= (2xy + 3y^3)\mathbf{i} + (x^2 + 9xy^2)\mathbf{j} \qquad \text{Vector field in the plane}$$

is a vector field in the plane. From Chapter 13, the graphical interpretation of this field is a family of vectors, each of which points in the direction of maximum increase along the surface given by $z = f(x, y)$.

Similarly, if

$$f(x, y, z) = x^2 + y^2 + z^2$$

then the gradient of f

$$\nabla f(x, y, z) = f_x(x, y, z)\mathbf{i} + f_y(x, y, z)\mathbf{j} + f_z(x, y, z)\mathbf{k}$$
$$= 2x\mathbf{i} + 2y\mathbf{j} + 2z\mathbf{k} \qquad \text{Vector field in space}$$

is a vector field in space. Note that the component functions for this particular vector field are $2x$, $2y$, and $2z$.

A vector field

$$\mathbf{F}(x, y, z) = M(x, y, z)\mathbf{i} + N(x, y, z)\mathbf{j} + P(x, y, z)\mathbf{k}$$

is **continuous** at a point if and only if each of its component functions M, N, and P is continuous at that point.

Some common *physical* examples of vector fields are **velocity fields, gravitational fields,** and **electric force fields.**

Velocity field

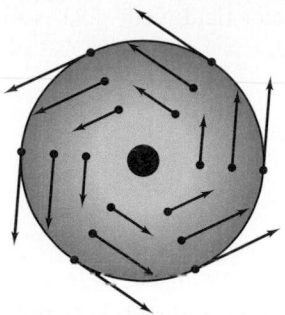

Rotating wheel
Figure 15.1

Air flow vector field
Figure 15.2

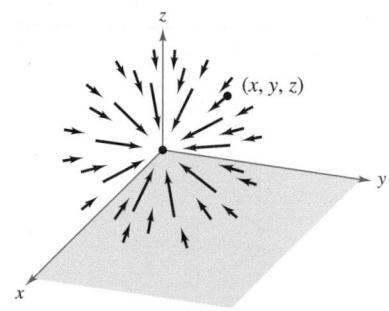

m_1 is located at (x, y, z).
m_2 is located at $(0, 0, 0)$.

Gravitational force field
Figure 15.3

1. *Velocity fields* describe the motions of systems of particles in the plane or in space. For instance, Figure 15.1 shows the vector field determined by a wheel rotating on an axle. Notice that the velocity vectors are determined by the locations of their initial points—the farther a point is from the axle, the greater its velocity. Velocity fields are also determined by the flow of liquids through a container or by the flow of air currents around a moving object, as shown in Figure 15.2.

2. *Gravitational fields* are defined by **Newton's Law of Gravitation,** which states that the force of attraction exerted on a particle of mass m_1 located at (x, y, z) by a particle of mass m_2 located at $(0, 0, 0)$ is

$$\mathbf{F}(x, y, z) = \frac{-Gm_1m_2}{x^2 + y^2 + z^2}\mathbf{u}$$

where G is the gravitational constant and $\mathbf{u}$ is the unit vector in the direction from the origin to (x, y, z). In Figure 15.3, you can see that the gravitational field $\mathbf{F}$ has the properties that $\mathbf{F}(x, y, z)$ always points toward the origin, and that the magnitude of $\mathbf{F}(x, y, z)$ is the same at all points equidistant from the origin. A vector field with these two properties is called a **central force field.** Using the position vector

$$\mathbf{r} = x\mathbf{i} + y\mathbf{j} + z\mathbf{k}$$

for the point (x, y, z), you can write the gravitational field $\mathbf{F}$ as

$$\mathbf{F}(x, y, z) = \frac{-Gm_1m_2}{\|\mathbf{r}\|^2}\left(\frac{\mathbf{r}}{\|\mathbf{r}\|}\right) = \frac{-Gm_1m_2}{\|\mathbf{r}\|^2}\mathbf{u}.$$

3. *Electric force fields* are defined by **Coulomb's Law,** which states that the force exerted on a particle with electric charge q_1 located at (x, y, z) by a particle with electric charge q_2 located at $(0, 0, 0)$ is

$$\mathbf{F}(x, y, z) = \frac{cq_1q_2}{\|\mathbf{r}\|^2}\mathbf{u}$$

where $\mathbf{r} = x\mathbf{i} + y\mathbf{j} + z\mathbf{k}$, $\mathbf{u} = \mathbf{r}/\|\mathbf{r}\|$, and c is a constant that depends on the choice of units for $\|\mathbf{r}\|$, q_1, and q_2.

Note that an electric force field has the same form as a gravitational field. That is,

$$\mathbf{F}(x, y, z) = \frac{k}{\|\mathbf{r}\|^2}\mathbf{u}.$$

Such a force field is called an **inverse square field.**

Definition of Inverse Square Field

Let $\mathbf{r}(t) = x(t)\mathbf{i} + y(t)\mathbf{j} + z(t)\mathbf{k}$ be a position vector. The vector field $\mathbf{F}$ is an **inverse square field** if

$$\mathbf{F}(x, y, z) = \frac{k}{\|\mathbf{r}\|^2}\mathbf{u}$$

where k is a real number and

$$\mathbf{u} = \frac{\mathbf{r}}{\|\mathbf{r}\|}$$

is a unit vector in the direction of $\mathbf{r}$.

Because vector fields consist of infinitely many vectors, it is not possible to create a sketch of the entire field. Instead, when you sketch a vector field, your goal is to sketch representative vectors that help you visualize the field.

EXAMPLE 1 Sketching a Vector Field

Sketch some vectors in the vector field

$$\mathbf{F}(x, y) = -y\mathbf{i} + x\mathbf{j}.$$

Solution You could plot vectors at several random points in the plane. It is more enlightening, however, to plot vectors of equal magnitude. This corresponds to finding level curves in scalar fields. In this case, vectors of equal magnitude lie on circles.

$$\|\mathbf{F}\| = c \qquad \text{Vectors of length } c$$
$$\sqrt{x^2 + y^2} = c$$
$$x^2 + y^2 = c^2 \qquad \text{Equation of circle}$$

To begin making the sketch, choose a value for c and plot several vectors on the resulting circle. For instance, the following vectors occur on the unit circle.

Point	Vector
$(1, 0)$	$\mathbf{F}(1, 0) = \mathbf{j}$
$(0, 1)$	$\mathbf{F}(0, 1) = -\mathbf{i}$
$(-1, 0)$	$\mathbf{F}(-1, 0) = -\mathbf{j}$
$(0, -1)$	$\mathbf{F}(0, -1) = \mathbf{i}$

These and several other vectors in the vector field are shown in Figure 15.4. Note in the figure that this vector field is similar to that given by the rotating wheel shown in Figure 15.1.

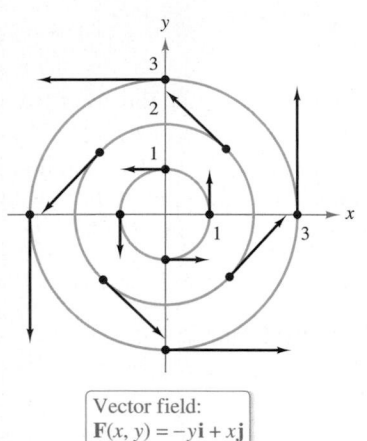

Vector field:
$\mathbf{F}(x, y) = -y\mathbf{i} + x\mathbf{j}$

Figure 15.4

EXAMPLE 2 Sketching a Vector Field

Sketch some vectors in the vector field

$$\mathbf{F}(x, y) = 2x\mathbf{i} + y\mathbf{j}.$$

Solution For this vector field, vectors of equal magnitude lie on ellipses given by

$$\|\mathbf{F}\| = c$$
$$\sqrt{(2x)^2 + (y)^2} = c$$

which implies that

$$4x^2 + y^2 = c^2. \qquad \text{Equation of ellipse}$$

For $c = 1$, sketch several vectors $2x\mathbf{i} + y\mathbf{j}$ of magnitude 1 at points on the ellipse given by

$$4x^2 + y^2 = 1.$$

For $c = 2$, sketch several vectors $2x\mathbf{i} + y\mathbf{j}$ of magnitude 2 at points on the ellipse given by

$$4x^2 + y^2 = 4.$$

These vectors are shown in Figure 15.5.

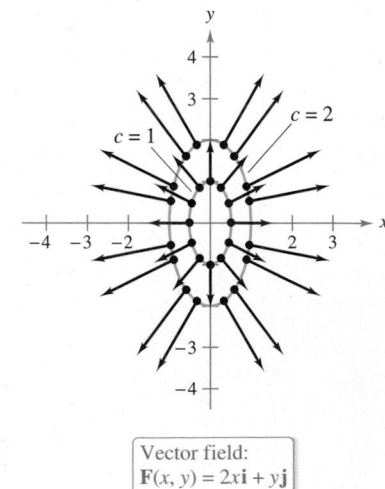

Vector field:
$\mathbf{F}(x, y) = 2x\mathbf{i} + y\mathbf{j}$

Figure 15.5

▷ **TECHNOLOGY** A computer algebra system can be used to graph vectors in a vector field. If you have access to a computer algebra system, use it to graph several representative vectors for the vector field in Example 2.

EXAMPLE 3   Sketching a Velocity Field

Sketch some vectors in the velocity field

$$\mathbf{v}(x, y, z) = (16 - x^2 - y^2)\mathbf{k}$$

where $x^2 + y^2 \le 16$.

Solution You can imagine that $\mathbf{v}$ describes the velocity of a liquid flowing through a tube of radius 4. Vectors near the z-axis are longer than those near the edge of the tube. For instance, at the point $(0, 0, 0)$, the velocity vector is $\mathbf{v}(0, 0, 0) = 16\mathbf{k}$, whereas at the point $(0, 3, 0)$, the velocity vector is $\mathbf{v}(0, 3, 0) = 7\mathbf{k}$. Figure 15.6 shows these and several other vectors for the velocity field. From the figure, you can see that the speed of the liquid is greater near the center of the tube than near the edges of the tube.

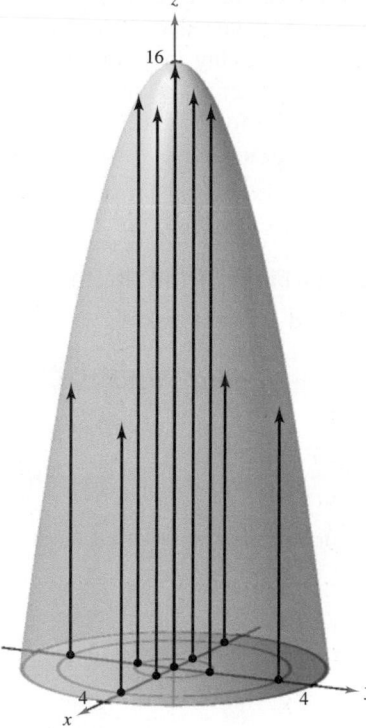

Velocity field:
$\mathbf{v}(x, y, z) = (16 - x^2 - y^2)\mathbf{k}$

Figure 15.6

Conservative Vector Fields

Notice in Figure 15.5 that all the vectors appear to be normal to the level curve from which they emanate. Because this is a property of gradients, it is natural to ask whether the vector field

$$\mathbf{F}(x, y) = 2x\mathbf{i} + y\mathbf{j}$$

is the *gradient* of some differentiable function f. The answer is that some vector fields can be represented as the gradients of differentiable functions and some cannot—those that can are called **conservative** vector fields.

Definition of Conservative Vector Field

A vector field $\mathbf{F}$ is called **conservative** when there exists a differentiable function f such that $\mathbf{F} = \nabla f$. The function f is called the **potential function** for $\mathbf{F}$.

EXAMPLE 4   Conservative Vector Fields

a. The vector field given by $\mathbf{F}(x, y) = 2x\mathbf{i} + y\mathbf{j}$ is conservative. To see this, consider the potential function $f(x, y) = x^2 + \frac{1}{2}y^2$. Because

$$\nabla f = 2x\mathbf{i} + y\mathbf{j} = \mathbf{F}$$

it follows that $\mathbf{F}$ is conservative.

b. Every inverse square field is conservative. To see this, let

$$\mathbf{F}(x, y, z) = \frac{k}{\|\mathbf{r}\|^2}\mathbf{u} \quad \text{and} \quad f(x, y, z) = \frac{-k}{\sqrt{x^2 + y^2 + z^2}}$$

where $\mathbf{u} = \mathbf{r}/\|\mathbf{r}\|$. Because

$$\nabla f = \frac{kx}{(x^2 + y^2 + z^2)^{3/2}}\mathbf{i} + \frac{ky}{(x^2 + y^2 + z^2)^{3/2}}\mathbf{j} + \frac{kz}{(x^2 + y^2 + z^2)^{3/2}}\mathbf{k}$$

$$= \frac{k}{x^2 + y^2 + z^2}\left(\frac{x\mathbf{i} + y\mathbf{j} + z\mathbf{k}}{\sqrt{x^2 + y^2 + z^2}}\right)$$

$$= \frac{k}{\|\mathbf{r}\|^2}\left(\frac{\mathbf{r}}{\|\mathbf{r}\|}\right)$$

$$= \frac{k}{\|\mathbf{r}\|^2}\mathbf{u}$$

it follows that $\mathbf{F}$ is conservative.

As can be seen in Example 4(b), many important vector fields, including gravitational fields and electric force fields, are conservative. Most of the terminology in this chapter comes from physics. For example, the term "conservative" is derived from the classic physical law regarding the conservation of energy. This law states that the sum of the kinetic energy and the potential energy of a particle moving in a conservative force field is constant. (The kinetic energy of a particle is the energy due to its motion, and the potential energy is the energy due to its position in the force field.)

The next theorem gives a necessary and sufficient condition for a vector field *in the plane* to be conservative.

REMARK Theorem 15.1 is valid on *simply connected* domains. A plane region R is simply connected when every simple closed curve in R encloses only points that are in R. (See Figure 15.26 in Section 15.4.)

THEOREM 15.1 Test for Conservative Vector Field in the Plane

Let M and N have continuous first partial derivatives on an open disk R. The vector field $\mathbf{F}(x, y) = M\mathbf{i} + N\mathbf{j}$ is conservative if and only if

$$\frac{\partial N}{\partial x} = \frac{\partial M}{\partial y}.$$

Proof   To prove that the given condition is necessary for $\mathbf{F}$ to be conservative, suppose there exists a potential function f such that

$$\mathbf{F}(x, y) = \nabla f(x, y) = M\mathbf{i} + N\mathbf{j}.$$

Then you have

$$f_x(x, y) = M \quad \Longrightarrow \quad f_{xy}(x, y) = \frac{\partial M}{\partial y}$$

$$f_y(x, y) = N \quad \Longrightarrow \quad f_{yx}(x, y) = \frac{\partial N}{\partial x}$$

and, by the equivalence of the mixed partials f_{xy} and f_{yx}, you can conclude that $\partial N/\partial x = \partial M/\partial y$ for all (x, y) in R. The sufficiency of this condition is proved in Section 15.4. ∎

EXAMPLE 5   **Testing for Conservative Vector Fields in the Plane**

Determine whether the vector field given by $\mathbf{F}$ is conservative.

a. $\mathbf{F}(x, y) = x^2 y\mathbf{i} + xy\mathbf{j}$

b. $\mathbf{F}(x, y) = 2x\mathbf{i} + y\mathbf{j}$

Solution

a. The vector field

$$\mathbf{F}(x, y) = x^2 y\mathbf{i} + xy\mathbf{j}$$

is not conservative because

$$\frac{\partial M}{\partial y} = \frac{\partial}{\partial y}[x^2 y] = x^2 \quad \text{and} \quad \frac{\partial N}{\partial x} = \frac{\partial}{\partial x}[xy] = y.$$

b. The vector field

$$\mathbf{F}(x, y) = 2x\mathbf{i} + y\mathbf{j}$$

is conservative because

$$\frac{\partial M}{\partial y} = \frac{\partial}{\partial y}[2x] = 0 \quad \text{and} \quad \frac{\partial N}{\partial x} = \frac{\partial}{\partial x}[y] = 0.$$

Theorem 15.1 tells you whether a vector field **F** is conservative. It does not tell you how to find a potential function of **F**. The problem is comparable to antidifferentiation. Sometimes you will be able to find a potential function by simple inspection. For instance, in Example 4, you observed that

$$f(x, y) = x^2 + \frac{1}{2}y^2$$

has the property that

$$\nabla f(x, y) = 2x\mathbf{i} + y\mathbf{j}.$$

EXAMPLE 6 **Finding a Potential Function for F(x, y)**

Find a potential function for

$$\mathbf{F}(x, y) = 2xy\mathbf{i} + (x^2 - y)\mathbf{j}.$$

Solution From Theorem 15.1, it follows that **F** is conservative because

$$\frac{\partial}{\partial y}[2xy] = 2x \quad \text{and} \quad \frac{\partial}{\partial x}[x^2 - y] = 2x.$$

If f is a function whose gradient is equal to $\mathbf{F}(x, y)$, then

$$\nabla f(x, y) = 2xy\mathbf{i} + (x^2 - y)\mathbf{j}$$

which implies that

$$f_x(x, y) = 2xy$$

and

$$f_y(x, y) = x^2 - y.$$

To reconstruct the function f from these two partial derivatives, integrate $f_x(x, y)$ with respect to x

$$f(x, y) = \int f_x(x, y)\, dx = \int 2xy\, dx = x^2y + g(y)$$

and integrate $f_y(x, y)$ with respect to y

$$f(x, y) = \int f_y(x, y)\, dy = \int (x^2 - y)\, dy = x^2y - \frac{y^2}{2} + h(x).$$

Notice that $g(y)$ is constant with respect to x and $h(x)$ is constant with respect to y. To find a single expression that represents $f(x, y)$, let

$$g(y) = -\frac{y^2}{2} + K_1 \quad \text{and} \quad h(x) = K_2.$$

Then you can write

$$f(x, y) = x^2y - \frac{y^2}{2} + K. \qquad K = K_1 + K_2$$

You can check this result by forming the gradient of f. You will see that it is equal to the original function **F**. ∎

Notice that the solution to Example 6 is comparable to that given by an indefinite integral. That is, the solution represents a family of potential functions, any two of which differ by a constant. To find a unique solution, you would have to be given an initial condition that is satisfied by the potential function.

Curl of a Vector Field

Theorem 15.1 has a counterpart for vector fields in space. Before stating that result, the definition of the **curl of a vector field** in space is given.

Definition of Curl of a Vector Field

The curl of $\mathbf{F}(x, y, z) = M\mathbf{i} + N\mathbf{j} + P\mathbf{k}$ is

$$\text{curl } \mathbf{F}(x, y, z) = \nabla \times \mathbf{F}(x, y, z)$$

$$= \left(\frac{\partial P}{\partial y} - \frac{\partial N}{\partial z}\right)\mathbf{i} - \left(\frac{\partial P}{\partial x} - \frac{\partial M}{\partial z}\right)\mathbf{j} + \left(\frac{\partial N}{\partial x} - \frac{\partial M}{\partial y}\right)\mathbf{k}.$$

If curl $\mathbf{F} = \mathbf{0}$, then $\mathbf{F}$ is said to be **irrotational.**

The cross product notation used for curl comes from viewing the gradient ∇f as the result of the **differential operator** ∇ acting on the function f. In this context, you can use the following determinant form as an aid in remembering the formula for curl.

$$\text{curl } \mathbf{F}(x, y, z) = \nabla \times \mathbf{F}(x, y, z)$$

$$= \begin{vmatrix} \mathbf{i} & \mathbf{j} & \mathbf{k} \\ \dfrac{\partial}{\partial x} & \dfrac{\partial}{\partial y} & \dfrac{\partial}{\partial z} \\ M & N & P \end{vmatrix}$$

$$= \left(\frac{\partial P}{\partial y} - \frac{\partial N}{\partial z}\right)\mathbf{i} - \left(\frac{\partial P}{\partial x} - \frac{\partial M}{\partial z}\right)\mathbf{j} + \left(\frac{\partial N}{\partial x} - \frac{\partial M}{\partial y}\right)\mathbf{k}$$

EXAMPLE 7 **Finding the Curl of a Vector Field**

····▷ *See LarsonCalculus.com for an interactive version of this type of example.*

Find curl $\mathbf{F}$ of the vector field

$$\mathbf{F}(x, y, z) = 2xy\mathbf{i} + (x^2 + z^2)\mathbf{j} + 2yz\mathbf{k}.$$

Is $\mathbf{F}$ irrotational?

Solution The curl of $\mathbf{F}$ is

$$\text{curl } \mathbf{F}(x, y, z) = \nabla \times \mathbf{F}(x, y, z)$$

$$= \begin{vmatrix} \mathbf{i} & \mathbf{j} & \mathbf{k} \\ \dfrac{\partial}{\partial x} & \dfrac{\partial}{\partial y} & \dfrac{\partial}{\partial z} \\ 2xy & x^2 + z^2 & 2yz \end{vmatrix}$$

$$= \begin{vmatrix} \dfrac{\partial}{\partial y} & \dfrac{\partial}{\partial z} \\ x^2 + z^2 & 2yz \end{vmatrix}\mathbf{i} - \begin{vmatrix} \dfrac{\partial}{\partial x} & \dfrac{\partial}{\partial z} \\ 2xy & 2yz \end{vmatrix}\mathbf{j} + \begin{vmatrix} \dfrac{\partial}{\partial x} & \dfrac{\partial}{\partial y} \\ 2xy & x^2 + z^2 \end{vmatrix}\mathbf{k}$$

$$= (2z - 2z)\mathbf{i} - (0 - 0)\mathbf{j} + (2x - 2x)\mathbf{k}$$

$$= \mathbf{0}.$$

Because curl $\mathbf{F} = \mathbf{0}$, $\mathbf{F}$ is irrotational.

▷ **TECHNOLOGY** Some computer algebra systems have a command that can be used to find the curl of a vector field. If you have access to a computer algebra system that has such a command, use it to find the curl of the vector field in Example 7.

Later in this chapter, you will assign a physical interpretation to the curl of a vector field. But for now, the primary use of curl is shown in the following test for conservative vector fields in space. The test states that for a vector field in space, the curl is **0** at every point in its domain if and only if **F** is conservative. The proof is similar to that given for Theorem 15.1.

·· REMARK   Theorem 15.2 is valid for *simply connected* domains in space. A simply connected domain in space is a domain D for which every simple closed curve in D can be shrunk to a point in D without leaving D.

> **THEOREM 15.2   Test for Conservative Vector Field in Space**
>
> Suppose that M, N, and P have continuous first partial derivatives in an open sphere Q in space. The vector field
>
> $$\mathbf{F}(x, y, z) = M\mathbf{i} + N\mathbf{j} + P\mathbf{k}$$
>
> is conservative if and only if
>
> $$\operatorname{curl} \mathbf{F}(x, y, z) = \mathbf{0}.$$
>
> That is, **F** is conservative if and only if
>
> $$\frac{\partial P}{\partial y} = \frac{\partial N}{\partial z}, \quad \frac{\partial P}{\partial x} = \frac{\partial M}{\partial z}, \quad \text{and} \quad \frac{\partial N}{\partial x} = \frac{\partial M}{\partial y}.$$

From Theorem 15.2, you can see that the vector field given in Example 7 is conservative because curl $\mathbf{F}(x, y, z) = \mathbf{0}$. Try showing that the vector field

$$\mathbf{F}(x, y, z) = x^3 y^2 z\mathbf{i} + x^2 z\mathbf{j} + x^2 y\mathbf{k}$$

is not conservative—you can do this by showing that its curl is

$$\operatorname{curl} \mathbf{F}(x, y, z) = (x^3 y^2 - 2xy)\mathbf{j} + (2xz - 2x^3 yz)\mathbf{k} \neq \mathbf{0}.$$

For vector fields in space that pass the test for being conservative, you can find a potential function by following the same pattern used in the plane (as demonstrated in Example 6).

·· REMARK   Examples 6 and 8 are illustrations of a type of problem called *recovering a function from its gradient.* If you go on to take a course in differential equations, you will study other methods for solving this type of problem. One popular method gives an interplay between successive "partial integrations" and partial differentiations.

EXAMPLE 8   **Finding a Potential Function for F(x, y, z)**

Find a potential function for

$$\mathbf{F}(x, y, z) = 2xy\mathbf{i} + (x^2 + z^2)\mathbf{j} + 2yz\mathbf{k}.$$

Solution   From Example 7, you know that the vector field given by **F** is conservative. If f is a function such that $\mathbf{F}(x, y, z) = \nabla f(x, y, z)$, then

$$f_x(x, y, z) = 2xy, \quad f_y(x, y, z) = x^2 + z^2, \quad \text{and} \quad f_z(x, y, z) = 2yz$$

and integrating with respect to x, y, and z separately produces

$$f(x, y, z) = \int M\, dx = \int 2xy\, dx = x^2 y + g(y, z)$$

$$f(x, y, z) = \int N\, dy = \int (x^2 + z^2)\, dy = x^2 y + yz^2 + h(x, z)$$

$$f(x, y, z) = \int P\, dz = \int 2yz\, dz = yz^2 + k(x, y).$$

Comparing these three versions of $f(x, y, z)$, you can conclude that

$$g(y, z) = yz^2 + K_1, \quad h(x, z) = K_2, \quad \text{and} \quad k(x, y) = x^2 y + K_3.$$

So, $f(x, y, z)$ is given by

$$f(x, y, z) = x^2 y + yz^2 + K. \qquad K = K_1 + K_2 + K_3$$

Divergence of a Vector Field

You have seen that the curl of a vector field **F** is itself a vector field. Another important function defined on a vector field is **divergence,** which is a scalar function.

Definition of Divergence of a Vector Field

The **divergence** of $\mathbf{F}(x, y) = M\mathbf{i} + N\mathbf{j}$ is

$$\text{div } \mathbf{F}(x, y) = \nabla \cdot \mathbf{F}(x, y) = \frac{\partial M}{\partial x} + \frac{\partial N}{\partial y}. \qquad \text{Plane}$$

The **divergence** of $\mathbf{F}(x, y, z) = M\mathbf{i} + N\mathbf{j} + P\mathbf{k}$ is

$$\text{div } \mathbf{F}(x, y, z) = \nabla \cdot \mathbf{F}(x, y, z) = \frac{\partial M}{\partial x} + \frac{\partial N}{\partial y} + \frac{\partial P}{\partial z}. \qquad \text{Space}$$

If div $\mathbf{F} = 0$, then $\mathbf{F}$ is said to be **divergence free.**

The dot product notation used for divergence comes from considering ∇ as a **differential operator,** as follows.

$$\nabla \cdot \mathbf{F}(x, y, z) = \left[\left(\frac{\partial}{\partial x} \right)\mathbf{i} + \left(\frac{\partial}{\partial y} \right)\mathbf{j} + \left(\frac{\partial}{\partial z} \right)\mathbf{k} \right] \cdot (M\mathbf{i} + N\mathbf{j} + P\mathbf{k})$$

$$= \frac{\partial M}{\partial x} + \frac{\partial N}{\partial y} + \frac{\partial P}{\partial z}$$

▷ **TECHNOLOGY** Some computer algebra systems have a command that can be used to find the divergence of a vector field. If you have access to a computer algebra system that has such a command, use it to find the divergence of the vector field in Example 9.

EXAMPLE 9 **Finding the Divergence of a Vector Field**

Find the divergence at $(2, 1, -1)$ for the vector field

$$\mathbf{F}(x, y, z) = x^3y^2z\mathbf{i} + x^2z\mathbf{j} + x^2y\mathbf{k}.$$

Solution The divergence of **F** is

$$\text{div } \mathbf{F}(x, y, z) = \frac{\partial}{\partial x}[x^3y^2z] + \frac{\partial}{\partial y}[x^2z] + \frac{\partial}{\partial z}[x^2y] = 3x^2y^2z.$$

At the point $(2, 1, -1)$, the divergence is

$$\text{div } \mathbf{F}(2, 1, -1) = 3(2^2)(1^2)(-1) = -12. \qquad ■$$

Divergence can be viewed as a type of derivative of **F** in that, for vector fields representing velocities of moving particles, the divergence measures the rate of particle flow per unit volume at a point. In hydrodynamics (the study of fluid motion), a velocity field that is divergence free is called **incompressible.** In the study of electricity and magnetism, a vector field that is divergence free is called **solenoidal.**

There are many important properties of the divergence and curl of a vector field **F** [see Exercise 77(a)–(g)]. One that is used often is described in Theorem 15.3. You are asked to prove this theorem in Exercise 77(h).

THEOREM 15.3 **Divergence and Curl**

If $\mathbf{F}(x, y, z) = M\mathbf{i} + N\mathbf{j} + P\mathbf{k}$ is a vector field and M, N, and P have continuous second partial derivatives, then

$$\text{div}(\text{curl } \mathbf{F}) = 0.$$

15.1 Exercises

CONCEPT CHECK

1. **Vector Field** Define a vector field in the plane and in space. Give some physical examples of vector fields.

2. **Conservative Vector Field** What is a conservative vector field? How do you test whether a vector field is conservative in the plane and in space?

3. **Potential Function** Describe how to find a potential function for a vector field that is conservative.

4. **Vector Field** A vector field in space is conservative. Is the vector field irrotational? Explain.

Matching In Exercises 5–8, match the vector field with its graph. [The graphs are labeled (a), (b), (c), and (d).]

(a)
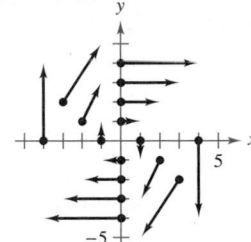

(b)
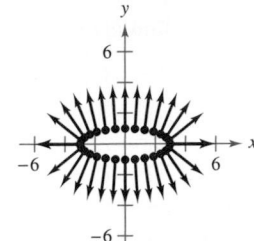

(c)
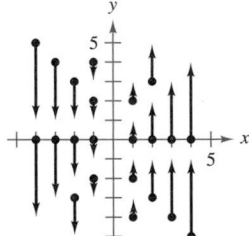

(d)
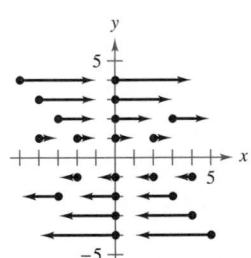

5. $\mathbf{F}(x, y) = y\mathbf{i}$

6. $\mathbf{F}(x, y) = x\mathbf{j}$

7. $\mathbf{F}(x, y) = y\mathbf{i} - x\mathbf{j}$

8. $\mathbf{F}(x, y) = x\mathbf{i} + 3y\mathbf{j}$

 **Sketching a Vector Field** In Exercises 9–14, find $\|\mathbf{F}\|$ and sketch several representative vectors in the vector field.

9. $\mathbf{F}(x, y) = \mathbf{i} + \mathbf{j}$ 10. $\mathbf{F}(x, y) = y\mathbf{i} - 2x\mathbf{j}$

11. $\mathbf{F}(x, y) = -\mathbf{i} + 3y\mathbf{j}$ 12. $\mathbf{F}(x, y) = y\mathbf{i} + x\mathbf{j}$

13. $\mathbf{F}(x, y, z) = \mathbf{i} + \mathbf{j} + \mathbf{k}$ 14. $\mathbf{F}(x, y, z) = x\mathbf{i} + y\mathbf{j} + z\mathbf{k}$

Graphing a Vector Field Using Technology In Exercises 15–18, use a computer algebra system to graph several representative vectors in the vector field.

15. $\mathbf{F}(x, y) = \frac{1}{8}(2xy\mathbf{i} + y^2\mathbf{j})$

16. $\mathbf{F}(x, y) = \langle 2y - x, 2y + x \rangle$

17. $\mathbf{F}(x, y, z) = \dfrac{x\mathbf{i} + y\mathbf{j} + z\mathbf{k}}{\sqrt{x^2 + y^2 + z^2}}$

18. $\mathbf{F}(x, y, z) = \langle x, -y, z \rangle$

 Finding a Conservative Vector Field In Exercises 19–28, find the conservative vector field for the potential function by finding its gradient.

19. $f(x, y) = x^2 + 2y^2$ 20. $f(x, y) = x^3 - 2xy$

21. $g(x, y) = 5x^2 + 3xy + y^2$ 22. $g(x, y) = \sin 3x \cos 4y$

23. $f(x, y, z) = 6xyz$ 24. $f(x, y, z) = \sqrt{x^2 y + z^2}$

25. $g(x, y, z) = z + ye^{x^2}$ 26. $g(x, y, z) = \dfrac{y}{z} + \dfrac{z}{x} - \dfrac{xz}{y}$

27. $h(x, y, z) = xy \ln(x + y)$ 28. $h(x, y, z) = x \arcsin yz$

 **Testing for a Conservative Vector Field** In Exercises 29–36, determine whether the vector field is conservative.

29. $\mathbf{F}(x, y) = xy^2\mathbf{i} + x^2y\mathbf{j}$ 30. $\mathbf{F}(x, y) = \dfrac{1}{x^2}(y\mathbf{i} - x\mathbf{j})$

31. $\mathbf{F}(x, y) = \sin y\mathbf{i} + x \sin y\mathbf{j}$ 32. $\mathbf{F}(x, y) = 5y^2(y\mathbf{i} + 2x\mathbf{j})$

33. $\mathbf{F}(x, y) = \dfrac{1}{xy}(y\mathbf{i} - x\mathbf{j})$ 34. $\mathbf{F}(x, y) = \dfrac{2}{y^2}e^{2x/y}(y\mathbf{i} - x\mathbf{j})$

35. $\mathbf{F}(x, y) = \dfrac{\mathbf{i} + \mathbf{j}}{\sqrt{x^2 + y^2}}$ 36. $\mathbf{F}(x, y) = \dfrac{y\mathbf{i} + x\mathbf{j}}{\sqrt{1 + xy}}$

 **Finding a Potential Function** In Exercises 37–44, determine whether the vector field is conservative. If it is, find a potential function for the vector field.

37. $\mathbf{F}(x, y) = (3y - x^2)\mathbf{i} + (3x + y)\mathbf{j}$

38. $\mathbf{F}(x, y) = (x^3 + e^y)\mathbf{i} + (xe^y - 6)\mathbf{j}$

39. $\mathbf{F}(x, y) = xe^{x^2y}(2y\mathbf{i} + x\mathbf{j})$ 40. $\mathbf{F}(x, y) = \dfrac{1}{y^2}(y\mathbf{i} - 2x\mathbf{j})$

41. $\mathbf{F}(x, y) = \dfrac{2y}{x}\mathbf{i} - \dfrac{x^2}{y^2}\mathbf{j}$ 42. $\mathbf{F}(x, y) = \dfrac{x\mathbf{i} + y\mathbf{j}}{x^2 + y^2}$

43. $\mathbf{F}(x, y) = \sin y\mathbf{i} + x \cos y\mathbf{j}$ 44. $\mathbf{F}(x, y) = (\ln y + 2)\mathbf{i} + \dfrac{x}{y}\mathbf{j}$

 Finding the Curl of a Vector Field In Exercises 45–48, find the curl of the vector field at the given point.

45. $\mathbf{F}(x, y, z) = xyz\mathbf{i} + xyz\mathbf{j} + xyz\mathbf{k}$; $(2, 1, 3)$

46. $\mathbf{F}(x, y, z) = x^2z\mathbf{i} - 2xz\mathbf{j} + yz\mathbf{k}$; $(2, -1, 3)$

47. $\mathbf{F}(x, y, z) = e^x \sin y\mathbf{i} - e^x \cos y\mathbf{j}$; $(0, 0, 1)$

48. $\mathbf{F}(x, y, z) = e^{-xyz}(\mathbf{i} + \mathbf{j} + \mathbf{k})$; $(3, 2, 0)$

Finding the Curl of a Vector Field Using Technology In Exercises 49 and 50, use a computer algebra system to find the curl of the vector field.

49. $\mathbf{F}(x, y, z) = \arctan\left(\dfrac{x}{y}\right)\mathbf{i} + \ln\sqrt{x^2 + y^2}\,\mathbf{j} + \mathbf{k}$

50. $\mathbf{F}(x, y, z) = \dfrac{yz}{y - z}\mathbf{i} + \dfrac{xz}{x - z}\mathbf{j} + \dfrac{xy}{x - y}\mathbf{k}$

Finding a Potential Function In Exercises 51–56, determine whether the vector field is conservative. If it is, find a potential function for the vector field.

51. $F(x, y, z) = (3x^2 + yz)\mathbf{i} + (3y^2 + xz)\mathbf{j} + (3z^2 + xy)\mathbf{k}$

52. $F(x, y, z) = y^2z^3\mathbf{i} + 2xyz^3\mathbf{j} + 3xy^2z^2\mathbf{k}$

53. $F(x, y, z) = \sin z\mathbf{i} + \sin x\mathbf{j} + \sin y\mathbf{k}$

54. $F(x, y, z) = ye^z\mathbf{i} + ze^x\mathbf{j} + xe^y\mathbf{k}$

55. $F(x, y, z) = \dfrac{z}{y}\mathbf{i} - \dfrac{xz}{y^2}\mathbf{j} + \left(\dfrac{x}{y} - 1\right)\mathbf{k}$

56. $F(x, y, z) = \dfrac{x}{x^2 + y^2}\mathbf{i} + \dfrac{y}{x^2 + y^2}\mathbf{j} + \mathbf{k}$

Finding the Divergence of a Vector Field In Exercises 57–60, find the divergence of the vector field.

57. $F(x, y) = x^2\mathbf{i} + 2y^2\mathbf{j}$ **58.** $F(x, y) = xe^x\mathbf{i} - x^2y^2\mathbf{j}$

59. $F(x, y, z) = \sin^2 x\mathbf{i} + z\cos z\mathbf{j} + z^3\mathbf{k}$

60. $F(x, y, z) = \ln(x^2 + y^2)\mathbf{i} + xy\mathbf{j} + \ln(y^2 + z^2)\mathbf{k}$

Finding the Divergence of a Vector Field In Exercises 61–64, find the divergence of the vector field at the given point.

61. $F(x, y, z) = xyz\mathbf{i} + xz^2\mathbf{j} + 3yz^2\mathbf{k}$; $(2, 4, 1)$

62. $F(x, y, z) = x^2z\mathbf{i} - 2xz\mathbf{j} + yz\mathbf{k}$; $(2, -1, 3)$

63. $F(x, y, z) = e^x \sin y\mathbf{i} - e^x \cos y\mathbf{j} + z^2\mathbf{k}$; $(3, 0, 0)$

64. $F(x, y, z) = \ln(xyz)(\mathbf{i} + \mathbf{j} + \mathbf{k})$; $(3, 2, 1)$

EXPLORING CONCEPTS

Think About It In Exercises 65–67, consider a scalar function f and a vector field F in space. Determine whether the expression is a vector field, a scalar function, or neither. Explain.

65. $\text{curl}(\nabla f)$ **66.** $\text{div}[\text{curl}(\nabla f)]$

67. $\text{curl}(\text{div } F)$

68. **HOW DO YOU SEE IT?** Several representative vectors in the vector fields

$$F(x, y) = \frac{x\mathbf{i} + y\mathbf{j}}{\sqrt{x^2 + y^2}} \quad \text{and} \quad G(x, y) = \frac{x\mathbf{i} - y\mathbf{j}}{\sqrt{x^2 + y^2}}$$

are shown below. Match each vector field with its graph. Explain your reasoning.

(a)   (b)

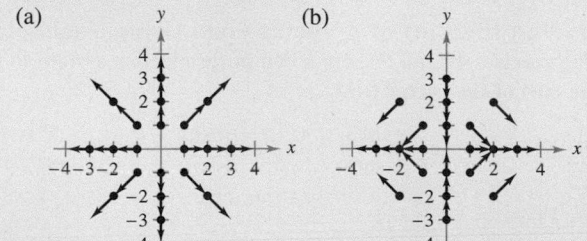

Curl of a Cross Product In Exercises 69 and 70, find $\text{curl}(F \times G) = \nabla \times (F \times G)$.

69. $F(x, y, z) = \mathbf{i} + 3x\mathbf{j} + 2y\mathbf{k}$ **70.** $F(x, y, z) = x\mathbf{i} - z\mathbf{k}$
     $G(x, y, z) = x\mathbf{i} - y\mathbf{j} + z\mathbf{k}$ $G(x, y, z) = x^2\mathbf{i} + y\mathbf{j} + z^2\mathbf{k}$

Curl of the Curl of a Vector Field In Exercises 71 and 72, find $\text{curl}(\text{curl } F) = \nabla \times (\nabla \times F)$.

71. $F(x, y, z) = xyz\mathbf{i} + y\mathbf{j} + z\mathbf{k}$

72. $F(x, y, z) = x^2z\mathbf{i} - 2xz\mathbf{j} + yz\mathbf{k}$

Divergence of a Cross Product In Exercises 73 and 74, find $\text{div}(F \times G) = \nabla \cdot (F \times G)$.

73. $F(x, y, z) = \mathbf{i} + 3x\mathbf{j} + 2y\mathbf{k}$
     $G(x, y, z) = x\mathbf{i} - y\mathbf{j} + z\mathbf{k}$

74. $F(x, y, z) = x\mathbf{i} - z\mathbf{k}$
     $G(x, y, z) = x^2\mathbf{i} + y\mathbf{j} + z^2\mathbf{k}$

Divergence of the Curl of a Vector Field In Exercises 75 and 76, find $\text{div}(\text{curl } F) = \nabla \cdot (\nabla \times F)$.

75. $F(x, y, z) = xyz\mathbf{i} + y\mathbf{j} + z\mathbf{k}$

76. $F(x, y, z) = x^2z\mathbf{i} - 2xz\mathbf{j} + yz\mathbf{k}$

77. **Proof** In parts (a)–(h), prove the property for vector fields F and G and scalar function f. (Assume that the required partial derivatives are continuous.)

(a) $\text{curl}(F + G) = \text{curl } F + \text{curl } G$

(b) $\text{curl}(\nabla f) = \nabla \times (\nabla f) = \mathbf{0}$

(c) $\text{div}(F + G) = \text{div } F + \text{div } G$

(d) $\text{div}(F \times G) = (\text{curl } F) \cdot G - F \cdot (\text{curl } G)$

(e) $\nabla \times [\nabla f + (\nabla \times F)] = \nabla \times (\nabla \times F)$

(f) $\nabla \times (f F) = f(\nabla \times F) + (\nabla f) \times F$

(g) $\text{div}(f F) = f\text{div } F + \nabla f \cdot F$

(h) $\text{div}(\text{curl } F) = 0$ (Theorem 15.3)

78. Earth's Magnetic Field

A cross section of Earth's magnetic field can be represented as a vector field in which the center of Earth is located at the origin and the positive y-axis points in the direction of the magnetic north pole. The equation for this field is

$$F(x, y) = M(x, y)\mathbf{i} + N(x, y)\mathbf{j}$$

$$= \frac{m}{(x^2 + y^2)^{5/2}}[3xy\mathbf{i} + (2y^2 - x^2)\mathbf{j}]$$

where m is the magnetic moment of Earth. Show that this vector field is conservative.

15.2 Line Integrals

■ Understand and use the concept of a piecewise smooth curve.
■ Write and evaluate a line integral.
■ Write and evaluate a line integral of a vector field.
■ Write and evaluate a line integral in differential form.

Piecewise Smooth Curves

A classic property of gravitational fields is that, subject to certain physical constraints, the work done by gravity on an object moving between two points in the field is independent of the path taken by the object. One of the constraints is that the **path** must be a piecewise smooth curve. Recall that a plane curve C given by

$$\mathbf{r}(t) = x(t)\mathbf{i} + y(t)\mathbf{j}, \quad a \le t \le b$$

is **smooth** when

$$\frac{dx}{dt} \quad \text{and} \quad \frac{dy}{dt}$$

are continuous on $[a, b]$ and not simultaneously 0 on (a, b). Similarly, a space curve C given by

$$\mathbf{r}(t) = x(t)\mathbf{i} + y(t)\mathbf{j} + z(t)\mathbf{k}, \quad a \le t \le b$$

is **smooth** when

$$\frac{dx}{dt}, \quad \frac{dy}{dt}, \quad \text{and} \quad \frac{dz}{dt}$$

are continuous on $[a, b]$ and not simultaneously 0 on (a, b). A curve C is **piecewise smooth** when the interval $[a, b]$ can be partitioned into a finite number of subintervals, on each of which C is smooth.

EXAMPLE 1 **Finding a Piecewise Smooth Parametrization**

Find a piecewise smooth parametrization of the graph of C shown in Figure 15.7.

Solution Because C consists of three line segments C_1, C_2, and C_3, you can construct a smooth parametrization for each segment and piece them together by making the last t-value in C_i correspond to the first t-value in C_{i+1}.

C_1: $x(t) = 0$,	$y(t) = 2t$,	$z(t) = 0$,	$0 \le t \le 1$
C_2: $x(t) = t - 1$,	$y(t) = 2$,	$z(t) = 0$,	$1 \le t \le 2$
C_3: $x(t) = 1$,	$y(t) = 2$,	$z(t) = t - 2$,	$2 \le t \le 3$

So, C is given by

$$\mathbf{r}(t) = \begin{cases} 2t\mathbf{j}, & 0 \le t \le 1 \\ (t - 1)\mathbf{i} + 2\mathbf{j}, & 1 \le t \le 2. \\ \mathbf{i} + 2\mathbf{j} + (t - 2)\mathbf{k}, & 2 \le t \le 3 \end{cases}$$

Because C_1, C_2, and C_3 are smooth, it follows that C is piecewise smooth. ■

Recall that parametrization of a curve induces an **orientation** to the curve. For instance, in Example 1, the curve is oriented such that the positive direction is from $(0, 0, 0)$, following the curve to $(1, 2, 1)$. Try finding a parametrization that induces the opposite orientation.

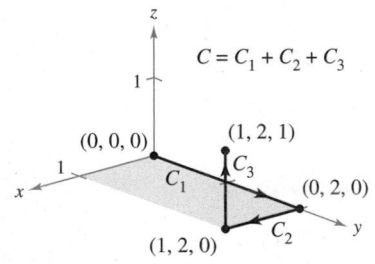

$C = C_1 + C_2 + C_3$

Figure 15.7

Line Integrals

Up to this point in the text, you have studied various types of integrals. For a single integral

$$\int_a^b f(x)\, dx \qquad \text{Integrate over interval } [a, b].$$

you integrated over the interval $[a, b]$. Similarly, for a double integral

$$\iint_R f(x, y)\, dA \qquad \text{Integrate over region } R.$$

you integrated over the region R in the plane. In this section, you will study a new type of integral called a **line integral**

$$\int_C f(x, y)\, ds \qquad \text{Integrate over curve } C.$$

for which you integrate over a piecewise smooth curve C. (The terminology is somewhat unfortunate—this type of integral might be better described as a "curve integral.")

To introduce the concept of a line integral, consider the mass of a wire of finite length, given by a curve C in space. The density (mass per unit length) of the wire at the point (x, y, z) is given by $f(x, y, z)$. Partition the curve C by the points

$$P_0, P_1, \ldots, P_n$$

producing n subarcs, as shown in Figure 15.8. The length of the ith subarc is given by Δs_i. Next, choose a point (x_i, y_i, z_i) in each subarc. If the length of each subarc is small, then the total mass of the wire can be approximated by the sum

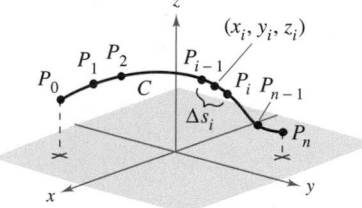

$$\text{Mass of wire} \approx \sum_{i=1}^{n} f(x_i, y_i, z_i)\, \Delta s_i.$$

Partitioning of curve C
Figure 15.8

By letting $\|\Delta\|$ denote the length of the longest subarc and letting $\|\Delta\|$ approach 0, it seems reasonable that the limit of this sum approaches the mass of the wire. This leads to the next definition.

Definition of Line Integral

If f is defined in a region containing a smooth curve C of finite length, then the **line integral of f along C** is given by

$$\int_C f(x, y)\, ds = \lim_{\|\Delta\| \to 0} \sum_{i=1}^{n} f(x_i, y_i)\, \Delta s_i \qquad \text{Plane}$$

or

$$\int_C f(x, y, z)\, ds = \lim_{\|\Delta\| \to 0} \sum_{i=1}^{n} f(x_i, y_i, z_i)\, \Delta s_i \qquad \text{Space}$$

provided this limit exists.

As with the integrals discussed in Chapter 14, evaluation of a line integral is best accomplished by converting it to a definite integral. It can be shown that if f is *continuous*, then the limit given above exists and is the same for all smooth parametrizations of C.

JOSIAH WILLARD GIBBS
(1839–1903)

Many physicists and mathematicians have contributed to the theory and applications described in this chapter—Newton, Gauss, Laplace, Hamilton, and Maxwell, among others. However, the use of vector analysis to describe these results is attributed primarily to the American mathematical physicist Josiah Willard Gibbs. *See LarsonCalculus.com to read more of this biography.*

To evaluate a line integral over a plane curve C given by $\mathbf{r}(t) = x(t)\mathbf{i} + y(t)\mathbf{j}$, use the fact that

$$ds = \|\mathbf{r}'(t)\| \, dt = \sqrt{[x'(t)]^2 + [y'(t)]^2} \, dt.$$

A similar formula holds for a space curve, as indicated in Theorem 15.4.

THEOREM 15.4 Evaluation of a Line Integral as a Definite Integral

Let f be continuous in a region containing a smooth curve C. If C is given by $\mathbf{r}(t) = x(t)\mathbf{i} + y(t)\mathbf{j}$, where $a \leq t \leq b$, then

$$\int_C f(x, y) \, ds = \int_a^b f(x(t), y(t)) \sqrt{[x'(t)]^2 + [y'(t)]^2} \, dt.$$

If C is given by $\mathbf{r}(t) = x(t)\mathbf{i} + y(t)\mathbf{j} + z(t)\mathbf{k}$, where $a \leq t \leq b$, then

$$\int_C f(x, y, z) \, ds = \int_a^b f(x(t), y(t), z(t)) \sqrt{[x'(t)]^2 + [y'(t)]^2 + [z'(t)]^2} \, dt.$$

Note that if $f(x, y, z) = 1$, then the line integral gives the arc length of the curve C, as defined in Section 12.5. That is,

$$\int_C 1 \, ds = \int_a^b \|\mathbf{r}'(t)\| \, dt = \text{length of curve } C.$$

EXAMPLE 2 Evaluating a Line Integral

Evaluate

$$\int_C (x^2 - y + 3z) \, ds$$

where C is the line segment shown in Figure 15.9.

Solution Begin by writing a parametric form of the equation of the line segment:

$$x = t, \quad y = 2t, \quad \text{and} \quad z = t, \quad 0 \leq t \leq 1.$$

Therefore, $x'(t) = 1$, $y'(t) = 2$, and $z'(t) = 1$, which implies that

$$\sqrt{[x'(t)]^2 + [y'(t)]^2 + [z'(t)]^2} = \sqrt{1^2 + 2^2 + 1^2} = \sqrt{6}.$$

So, the line integral takes the following form.

$$\int_C (x^2 - y + 3z) \, ds = \int_0^1 (t^2 - 2t + 3t)\sqrt{6} \, dt$$
$$= \sqrt{6} \int_0^1 (t^2 + t) \, dt$$
$$= \sqrt{6} \left[\frac{t^3}{3} + \frac{t^2}{2} \right]_0^1$$
$$= \frac{5\sqrt{6}}{6}$$

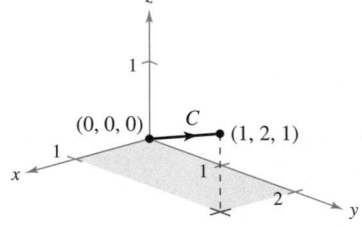

Figure 15.9

The value of the line integral in Example 2 does not depend on the parametrization of the line segment C; any smooth parametrization will produce the same value. To convince yourself of this, try some other parametrizations, such as $x = 1 + 2t$, $y = 2 + 4t$, and $z = 1 + 2t$, $-\frac{1}{2} \leq t \leq 0$, or $x = -t$, $y = -2t$, and $z = -t$, $-1 \leq t \leq 0$.

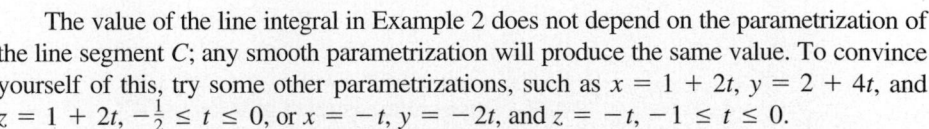

Let C be a path composed of smooth curves $C_1, C_2, \ldots, C_n$. If f is continuous on C, then it can be shown that

$$\int_C f(x, y)\,ds = \int_{C_1} f(x, y)\,ds + \int_{C_2} f(x, y)\,ds + \cdots + \int_{C_n} f(x, y)\,ds.$$

This property is used in Example 3.

EXAMPLE 3 Evaluating a Line Integral Over a Path

Evaluate

$$\int_C x\,ds$$

where C is the piecewise smooth curve shown in Figure 15.10.

Solution Begin by integrating up the line $y = x$, using the following parametrization.

$$C_1\colon\ x = t, \quad y = t, \quad 0 \le t \le 1$$

For this curve, $\mathbf{r}(t) = t\mathbf{i} + t\mathbf{j}$, which implies that $x'(t) = 1$ and $y'(t) = 1$. So,

$$\sqrt{[x'(t)]^2 + [y'(t)]^2} = \sqrt{2}$$

and you have

$$\int_{C_1} x\,ds = \int_0^1 t\sqrt{2}\,dt = \frac{\sqrt{2}}{2}t^2\bigg]_0^1 = \frac{\sqrt{2}}{2}.$$

Next, integrate down the parabola $y = x^2$, using the parametrization

$$C_2\colon x = 1 - t, \quad y = (1 - t)^2, \quad 0 \le t \le 1.$$

For this curve,

$$\mathbf{r}(t) = (1 - t)\mathbf{i} + (1 - t)^2\mathbf{j}$$

which implies that $x'(t) = -1$ and $y'(t) = -2(1 - t)$. So,

$$\sqrt{[x'(t)]^2 + [y'(t)]^2} = \sqrt{1 + 4(1 - t)^2}$$

and you have

$$\begin{aligned}
\int_{C_2} x\,ds &= \int_0^1 (1 - t)\sqrt{1 + 4(1 - t)^2}\,dt \\
&= -\frac{1}{8}\left[\frac{2}{3}[1 + 4(1 - t)^2]^{3/2}\right]_0^1 \\
&= \frac{1}{12}(5^{3/2} - 1).
\end{aligned}$$

Consequently,

$$\int_C x\,ds = \int_{C_1} x\,ds + \int_{C_2} x\,ds = \frac{\sqrt{2}}{2} + \frac{1}{12}(5^{3/2} - 1) \approx 1.56. \qquad \blacksquare$$

For parametrizations given by $\mathbf{r}(t) = x(t)\mathbf{i} + y(t)\mathbf{j} + z(t)\mathbf{k}$, it is helpful to remember the form of ds as

$$ds = \|\mathbf{r}'(t)\|\,dt = \sqrt{[x'(t)]^2 + [y'(t)]^2 + [z'(t)]^2}\,dt.$$

This is demonstrated in Example 4.

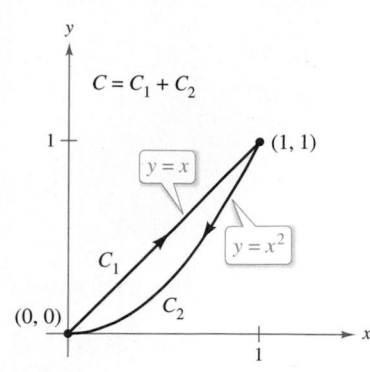

Figure 15.10

EXAMPLE 4   **Evaluating a Line Integral**

Evaluate $\int_C (x + 2)\, ds$, where C is the curve represented by

$$\mathbf{r}(t) = t\mathbf{i} + \frac{4}{3}t^{3/2}\mathbf{j} + \frac{1}{2}t^2\mathbf{k}, \quad 0 \le t \le 2.$$

Solution   Because $\mathbf{r}'(t) = \mathbf{i} + 2t^{1/2}\mathbf{j} + t\mathbf{k}$ and

$$\|\mathbf{r}'(t)\| = \sqrt{[x'(t)^2] + [y'(t)]^2 + [z'(t)]^2} = \sqrt{1 + 4t + t^2}$$

it follows that

$$
\begin{aligned}
\int_C (x + 2)\, ds &= \int_0^2 (t + 2)\sqrt{1 + 4t + t^2}\, dt \\
&= \frac{1}{2}\int_0^2 2(t + 2)(1 + 4t + t^2)^{1/2}\, dt \\
&= \frac{1}{3}\Big[(1 + 4t + t^2)^{3/2}\Big]_0^2 \\
&= \frac{1}{3}\big(13\sqrt{13} - 1\big) \\
&\approx 15.29.
\end{aligned}
$$

The next example shows how a line integral can be used to find the mass of a spring whose density varies. In Figure 15.11, note that the density of this spring increases as the spring spirals up the z-axis.

EXAMPLE 5   **Finding the Mass of a Spring**

Find the mass of a spring in the shape of the circular helix

$$\mathbf{r}(t) = \frac{1}{\sqrt{2}}(\cos t\mathbf{i} + \sin t\mathbf{j} + t\mathbf{k})$$

where $0 \le t \le 6\pi$ and the density of the spring is

$$\rho(x, y, z) = 1 + z$$

as shown in Figure 15.11.

Solution   Because

$$\|\mathbf{r}'(t)\| = \frac{1}{\sqrt{2}}\sqrt{(-\sin t)^2 + (\cos t)^2 + (1)^2} = 1$$

it follows that the mass of the spring is

$$
\begin{aligned}
\text{Mass} &= \int_C (1 + z)\, ds \\
&= \int_0^{6\pi}\left(1 + \frac{t}{\sqrt{2}}\right) dt \\
&= \left[t + \frac{t^2}{2\sqrt{2}}\right]_0^{6\pi} \\
&= 6\pi\left(1 + \frac{3\pi}{\sqrt{2}}\right) \\
&\approx 144.47.
\end{aligned}
$$

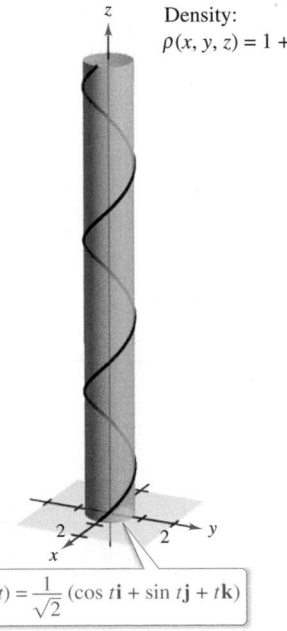

Density:
$\rho(x, y, z) = 1 + z$

$\mathbf{r}(t) = \dfrac{1}{\sqrt{2}}(\cos t\mathbf{i} + \sin t\mathbf{j} + t\mathbf{k})$

Figure 15.11

Line Integrals of Vector Fields

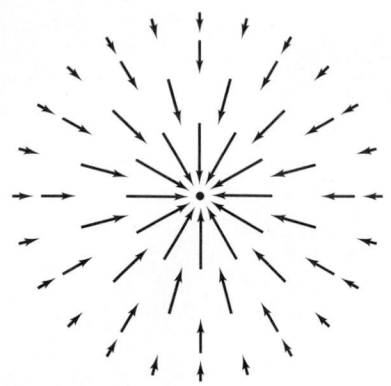

Inverse square force field **F**

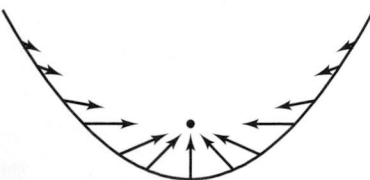

Vectors along a parabolic path in the force field **F**

Figure 15.12

One of the most important physical applications of line integrals is that of finding the **work** done on an object moving in a force field. For example, Figure 15.12 shows an inverse square force field similar to the gravitational field of the sun. Note that the magnitude of the force along a circular path about the center is constant, whereas the magnitude of the force along a parabolic path varies from point to point.

To see how a line integral can be used to find work done in a force field **F**, consider an object moving along a path C in the field, as shown in Figure 15.13. To determine the work done by the force, you need consider only that part of the force that is acting in the same direction as that in which the object is moving (or the opposite direction). This means that at each point on C, you can consider the projection **F · T** of the force vector **F** onto the unit tangent vector **T**. On a small subarc of length Δs_i, the increment of work is

$$\Delta W_i = (\text{force})(\text{distance})$$
$$\approx \left[\mathbf{F}(x_i, y_i, z_i) \cdot \mathbf{T}(x_i, y_i, z_i)\right] \Delta s_i$$

where (x_i, y_i, z_i) is a point in the ith subarc. Consequently, the total work done is given by the integral

$$W = \int_C \mathbf{F}(x, y, z) \cdot \mathbf{T}(x, y, z) \, ds.$$

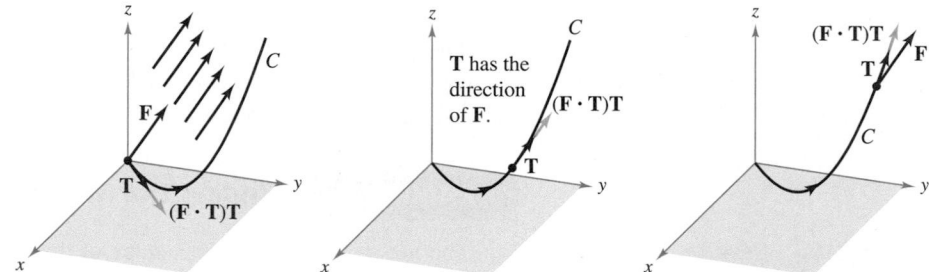

At each point on C, the force in the direction of motion is $(\mathbf{F} \cdot \mathbf{T})\mathbf{T}$.
Figure 15.13

This line integral appears in other contexts and is the basis of the definition of the **line integral of a vector field** shown below. Note in the definition that

$$\mathbf{F} \cdot \mathbf{T} \, ds = \mathbf{F} \cdot \frac{\mathbf{r}'(t)}{\|\mathbf{r}'(t)\|} \|\mathbf{r}'(t)\| \, dt$$

$$= \mathbf{F} \cdot \mathbf{r}'(t) \, dt$$

$$= \mathbf{F} \cdot d\mathbf{r}.$$

Definition of the Line Integral of a Vector Field

Let **F** be a continuous vector field defined on a smooth curve C given by

$$\mathbf{r}(t), \quad a \le t \le b.$$

The **line integral** of **F** on C is given by

$$\int_C \mathbf{F} \cdot d\mathbf{r} = \int_C \mathbf{F} \cdot \mathbf{T} \, ds$$

$$= \int_a^b \mathbf{F}(x(t), y(t), z(t)) \cdot \mathbf{r}'(t) \, dt.$$

EXAMPLE 6   **Work Done by a Force**

⋯⋯▷ *See LarsonCalculus.com for an interactive version of this type of example.*

Find the work done by the force field

$$\mathbf{F}(x, y, z) = -\frac{1}{2}x\mathbf{i} - \frac{1}{2}y\mathbf{j} + \frac{1}{4}\mathbf{k} \qquad \text{Force field } \mathbf{F}$$

on a particle as it moves along the helix given by

$$\mathbf{r}(t) = \cos t\mathbf{i} + \sin t\mathbf{j} + t\mathbf{k} \qquad \text{Space curve } C$$

from the point $(1, 0, 0)$ to the point $(-1, 0, 3\pi)$, as shown in Figure 15.14.

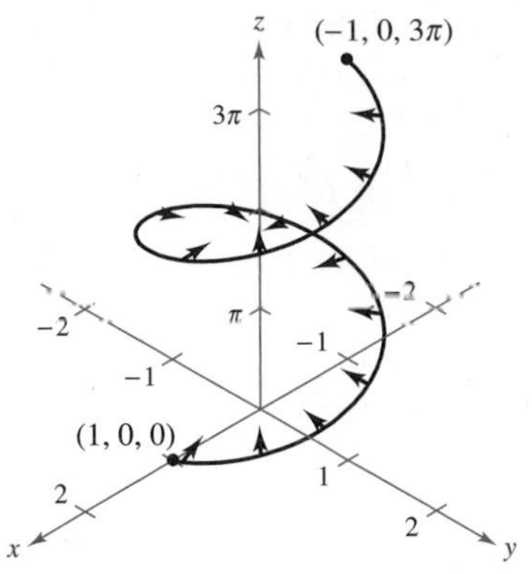

Solution   Because

$$\mathbf{r}(t) = x(t)\mathbf{i} + y(t)\mathbf{j} + z(t)\mathbf{k}$$
$$= \cos t\mathbf{i} + \sin t\mathbf{j} + t\mathbf{k}$$

it follows that

$$x(t) = \cos t, \quad y(t) = \sin t, \quad \text{and} \quad z(t) = t.$$

So, the force field can be written as

$$\mathbf{F}(x(t), y(t), z(t)) = -\frac{1}{2}\cos t\mathbf{i} - \frac{1}{2}\sin t\mathbf{j} + \frac{1}{4}\mathbf{k}.$$

To find the work done by the force field in moving a particle along the curve C, use the fact that

$$\mathbf{r}'(t) = -\sin t\mathbf{i} + \cos t\mathbf{j} + \mathbf{k}$$

and write the following.

$$
\begin{aligned}
W &= \int_C \mathbf{F} \cdot d\mathbf{r} \\
&= \int_a^b \mathbf{F}(x(t), y(t), z(t)) \cdot \mathbf{r}'(t)\, dt \\
&= \int_0^{3\pi} \left(-\frac{1}{2}\cos t\mathbf{i} - \frac{1}{2}\sin t\mathbf{j} + \frac{1}{4}\mathbf{k}\right) \cdot (-\sin t\mathbf{i} + \cos t\mathbf{j} + \mathbf{k})\, dt \\
&= \int_0^{3\pi} \left(\frac{1}{2}\sin t \cos t - \frac{1}{2}\sin t \cos t + \frac{1}{4}\right) dt \\
&= \int_0^{3\pi} \frac{1}{4}\, dt \\
&= \frac{1}{4}t \Big]_0^{3\pi} \\
&= \frac{3\pi}{4}
\end{aligned}
$$

In Example 6, note that the x- and y-components of the force field end up contributing nothing to the total work. This occurs because *in this particular example*, the z-component of the force field is the only portion of the force that is acting in the same (or opposite) direction in which the particle is moving (see Figure 15.15).

▷ **TECHNOLOGY**   Figure 15.15 shows a computer-generated view of the force field in Example 6. The figure indicates that each vector in the force field points toward the z-axis.

Figure 15.14

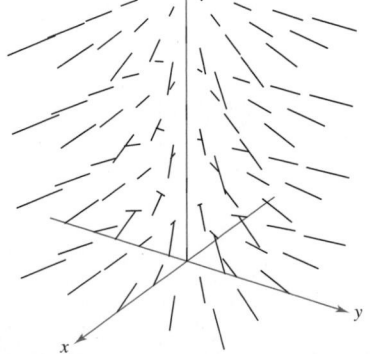

Generated by Mathematica

Figure 15.15

For line integrals of vector functions, the orientation of the curve C is important. When the orientation of the curve is reversed, the unit tangent vector $\mathbf{T}(t)$ is changed to $-\mathbf{T}(t)$, and you obtain

$$\int_{-C} \mathbf{F} \cdot d\mathbf{r} = -\int_C \mathbf{F} \cdot d\mathbf{r}.$$

EXAMPLE 7   Orientation and Parametrization of a Curve

Let $\mathbf{F}(x, y) = y\mathbf{i} + x^2\mathbf{j}$ and evaluate the line integral

$$\int_C \mathbf{F} \cdot d\mathbf{r}$$

for each parabolic curve shown in Figure 15.16.

a. C_1: $\mathbf{r}_1(t) = (4 - t)\mathbf{i} + (4t - t^2)\mathbf{j}, \quad 0 \le t \le 3$

b. C_2: $\mathbf{r}_2(t) = t\mathbf{i} + (4t - t^2)\mathbf{j}, \quad 1 \le t \le 4$

Solution

a. Because $\mathbf{r}_1'(t) = -\mathbf{i} + (4 - 2t)\mathbf{j}$ and

$$\mathbf{F}(x(t), y(t)) = (4t - t^2)\mathbf{i} + (4 - t)^2\mathbf{j}$$

the line integral is

$$\int_{C_1} \mathbf{F} \cdot d\mathbf{r} = \int_0^3 [(4t - t^2)\mathbf{i} + (4 - t)^2\mathbf{j}] \cdot [-\mathbf{i} + (4 - 2t)\mathbf{j}] \, dt$$

$$= \int_0^3 (-4t + t^2 + 64 - 64t + 20t^2 - 2t^3) \, dt$$

$$= \int_0^3 (-2t^3 + 21t^2 - 68t + 64) \, dt$$

$$= \left[-\frac{t^4}{2} + 7t^3 - 34t^2 + 64t \right]_0^3$$

$$= \frac{69}{2}.$$

b. Because $\mathbf{r}_2'(t) = \mathbf{i} + (4 - 2t)\mathbf{j}$ and

$$\mathbf{F}(x(t), y(t)) = (4t - t^2)\mathbf{i} + t^2\mathbf{j}$$

the line integral is

$$\int_{C_2} \mathbf{F} \cdot d\mathbf{r} = \int_1^4 [(4t - t^2)\mathbf{i} + t^2\mathbf{j}] \cdot [\mathbf{i} + (4 - 2t)\mathbf{j}] \, dt$$

$$= \int_1^4 (4t - t^2 + 4t^2 - 2t^3) \, dt$$

$$= \int_1^4 (-2t^3 + 3t^2 + 4t) \, dt$$

$$= \left[-\frac{t^4}{2} + t^3 + 2t^2 \right]_1^4$$

$$= -\frac{69}{2}.$$

The answer in part (b) is the negative of that in part (a) because C_1 and C_2 represent opposite orientations of the same parabolic segment.

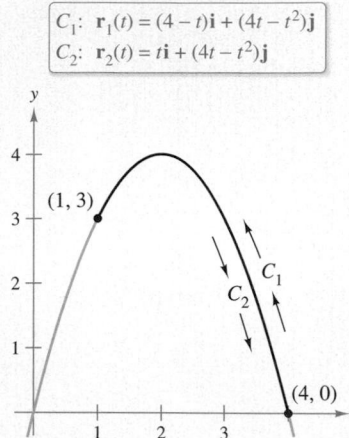

C_1: $\mathbf{r}_1(t) = (4 - t)\mathbf{i} + (4t - t^2)\mathbf{j}$
C_2: $\mathbf{r}_2(t) = t\mathbf{i} + (4t - t^2)\mathbf{j}$

Figure 15.16

Line Integrals in Differential Form

A second commonly used form of line integrals is derived from the vector field notation used in Section 15.1. If $\mathbf{F}$ is a vector field of the form $\mathbf{F}(x, y) = M\mathbf{i} + N\mathbf{j}$ and C is given by $\mathbf{r}(t) = x(t)\mathbf{i} + y(t)\mathbf{j}$, then $\mathbf{F} \cdot d\mathbf{r}$ is often written as $M\, dx + N\, dy$.

$$\int_C \mathbf{F} \cdot d\mathbf{r} = \int_C \mathbf{F} \cdot \frac{d\mathbf{r}}{dt}\, dt$$

$$= \int_a^b (M\mathbf{i} + N\mathbf{j}) \cdot (x'(t)\mathbf{i} + y'(t)\mathbf{j})\, dt$$

$$= \int_a^b \left(M\frac{dx}{dt} + N\frac{dy}{dt} \right) dt$$

$$= \int_C (M\, dx + N\, dy)$$

This **differential form** can be extended to three variables.

• • REMARK The parentheses are often omitted from this differential form, as shown below.

$$\int_C M\, dx + N\, dy$$

In three variables, the differential form is

$$\int_C M\, dx + N\, dy + P\, dz.$$

EXAMPLE 8 **Evaluating a Line Integral in Differential Form**

Let C be the circle of radius 3 given by

$$\mathbf{r}(t) = 3 \cos t\mathbf{i} + 3 \sin t\mathbf{j}, \quad 0 \le t \le 2\pi$$

as shown in Figure 15.17. Evaluate the line integral

$$\int_C y^3\, dx + (x^3 + 3xy^2)\, dy.$$

Solution Because $x = 3 \cos t$ and $y = 3 \sin t$, you have $dx = -3 \sin t\, dt$ and $dy = 3 \cos t\, dt$. So, the line integral is

$$\int_C M\, dx + N\, dy$$

$$= \int_C y^3\, dx + (x^3 + 3xy^2)\, dy$$

$$= \int_0^{2\pi} [(27 \sin^3 t)(-3 \sin t) + (27 \cos^3 t + 81 \cos t \sin^2 t)(3 \cos t)]\, dt$$

$$= 81 \int_0^{2\pi} (\cos^4 t - \sin^4 t + 3 \cos^2 t \sin^2 t)\, dt$$

$$= 81 \int_0^{2\pi} \left(\cos^2 t - \sin^2 t + \frac{3}{4} \sin^2 2t \right) dt-$$

$$= 81 \int_0^{2\pi} \left[\cos 2t + \frac{3}{4}\left(\frac{1 - \cos 4t}{2} \right) \right] dt$$

$$= 81 \left[\frac{\sin 2t}{2} + \frac{3}{8}t - \frac{3 \sin 4t}{32} \right]_0^{2\pi}$$

$$= \frac{243\pi}{4}.$$

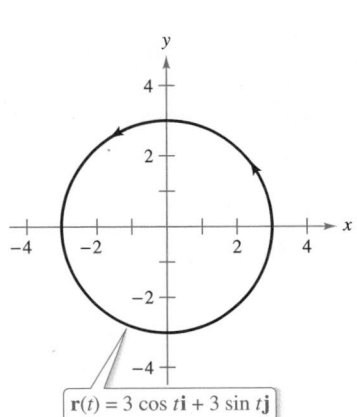

$\mathbf{r}(t) = 3 \cos t\mathbf{i} + 3 \sin t\mathbf{j}$

Figure 15.17

The orientation of C affects the value of the differential form of a line integral. Specifically, if $-C$ has the orientation opposite to that of C, then

$$\int_{-C} M\, dx + N\, dy = -\int_C M\, dx + N\, dy.$$

So, of the three line integral forms presented in this section, the orientation of C does not affect the form $\int_C f(x, y)\, ds$, but it does affect the vector form and the differential form.

For curves represented by $y = g(x)$, $a \le x \le b$, you can let $x = t$ and obtain the parametric form

$$x = t \quad \text{and} \quad y = g(t), \quad a \le t \le b.$$

Because $dx = dt$ for this form, you have the option of evaluating the line integral in the variable x or the variable t. This is demonstrated in Example 9.

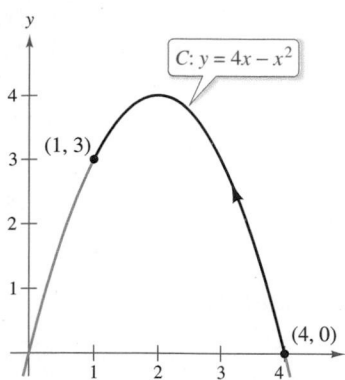

EXAMPLE 9 **Evaluating a Line Integral in Differential Form**

Evaluate

$$\int_C y \, dx + x^2 \, dy$$

where C is the parabolic arc given by $y = 4x - x^2$ from $(4, 0)$ to $(1, 3)$, as shown in Figure 15.18.

Solution Rather than converting to the parameter t, you can simply retain the variable x and write

$$y = 4x - x^2 \quad \implies \quad dy = (4 - 2x) \, dx.$$

Then, in the direction from $(4, 0)$ to $(1, 3)$, the line integral is

$$\int_C y \, dx + x^2 \, dy = \int_4^1 \left[(4x - x^2) \, dx + x^2(4 - 2x) \, dx \right]$$

$$= \int_4^1 (4x + 3x^2 - 2x^3) \, dx$$

$$= \left[2x^2 + x^3 - \frac{x^4}{2} \right]_4^1$$

$$= \frac{69}{2}. \qquad \text{See Example 7.}$$

C: $y = 4x - x^2$

$(1, 3)$

$(4, 0)$

Figure 15.18

Exploration

Finding Lateral Surface Area The figure below shows a piece of tin that has been cut from a circular cylinder. The base of the circular cylinder is modeled by $x^2 + y^2 = 9$. At any point (x, y) on the base, the height of the object is

$$f(x, y) = 1 + \cos \frac{\pi x}{4}.$$

Explain how to use a line integral to find the surface area of the piece of tin.

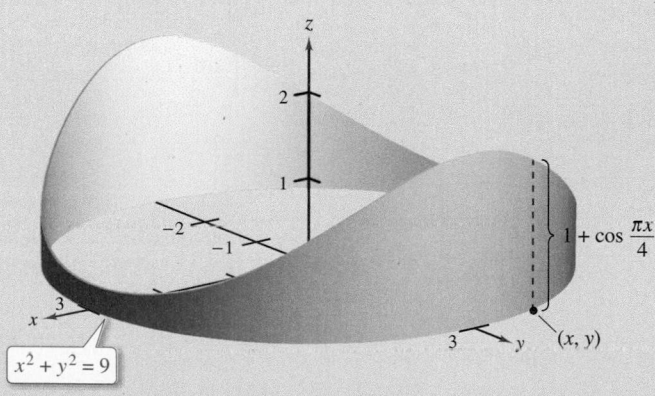

15.2 Exercises

CONCEPT CHECK

1. Line Integral What is the physical interpretation of each line integral?

(a) $\int_C 1 \, ds$

(b) $\int_C f(x, y, z) \, ds$, where $f(x, y, z)$ is the density of a string of finite length

2. Orientation of a Curve Describe how reversing the orientation of a curve C affects $\int_C \mathbf{F} \cdot d\mathbf{r}$.

Finding a Piecewise Smooth Parametrization In Exercises 3–8, find a piecewise smooth parametrization of the path C. (There is more than one correct answer.)

3.
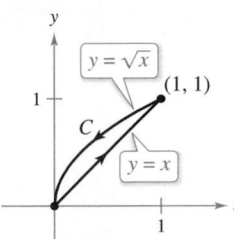

4.
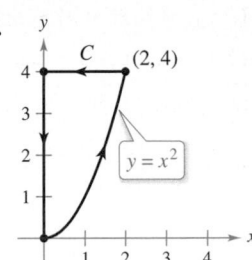

5.
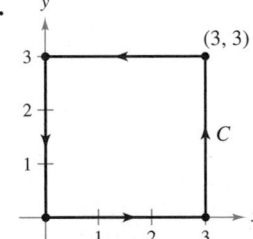

6.
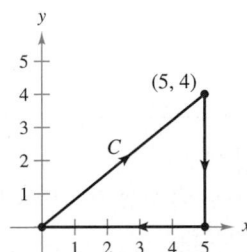

7.
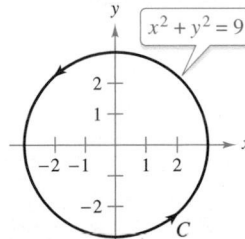

8.
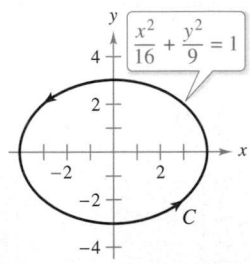

Evaluating a Line Integral In Exercises 9–12, (a) find a parametrization of the path C, and (b) evaluate $\int_C (x^2 + y^2) \, ds$.

9. C: line segment from $(0, 0)$ to $(1, 1)$

10. C: line segment from $(0, 0)$ to $(2, 4)$

11. C: counterclockwise around the circle $x^2 + y^2 = 1$ from $(1, 0)$ to $(0, 1)$

12. C: counterclockwise around the circle $x^2 + y^2 = 4$ from $(2, 0)$ to $(-2, 0)$

Evaluating a Line Integral In Exercises 13–16, (a) find a piecewise smooth parametrization of the path C, and (b) evaluate $\int_C (2x + 3\sqrt{y}) \, ds$.

13. C: line segments from $(0, 0)$ to $(1, 0)$ and $(1, 0)$ to $(2, 4)$

14. C: line segments from $(0, 1)$ to $(0, 4)$ and $(0, 4)$ to $(3, 3)$

15. C: counterclockwise around the triangle with vertices $(0, 0)$, $(1, 0)$, and $(0, 1)$

16. C: counterclockwise around the square with vertices $(0, 0)$, $(2, 0)$, $(2, 2)$, and $(0, 2)$

Evaluating a Line Integral In Exercises 17 and 18, (a) find a piecewise smooth parametrization of the path C shown in the figure and (b) evaluate $\int_C (2x + y^2 - z) \, ds$.

17. **18.**
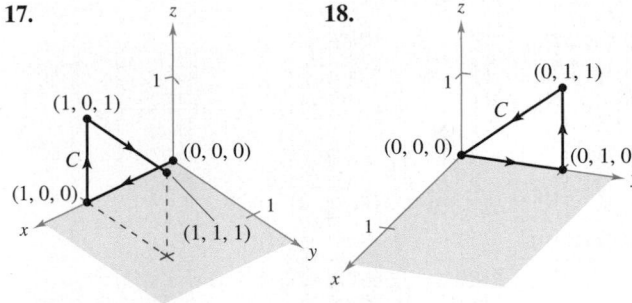

Evaluating a Line Integral In Exercises 19–22, evaluate the line integral along the given path.

19. $\int_C xy \, ds$

C: $\mathbf{r}(t) = 4t\mathbf{i} + 3t\mathbf{j}$

$0 \leq t \leq 1$

20. $\int_C 3(x - y) \, ds$

C: $\mathbf{r}(t) = t\mathbf{i} + (2 - t)\mathbf{j}$

$0 \leq t \leq 2$

21. $\int_C (x^2 + y^2 + z^2) \, ds$

C: $\mathbf{r}(t) = \sin t\mathbf{i} + \cos t\mathbf{j} + 2\mathbf{k}$

$0 \leq t \leq \dfrac{\pi}{2}$

22. $\int_C 2xyz \, ds$

C: $\mathbf{r}(t) = 12t\mathbf{i} + 5t\mathbf{j} + 84t\mathbf{k}$

$0 \leq t \leq 1$

Mass In Exercises 23 and 24, find the total mass of a spring with density ρ in the shape of the circular helix

$$\mathbf{r}(t) = 2\cos t\mathbf{i} + 2\sin t\mathbf{j} + t\mathbf{k}, \quad 0 \leq t \leq 4\pi.$$

23. $\rho(x, y, z) = \frac{1}{2}(x^2 + y^2 + z^2)$

24. $\rho(x, y, z) = z$

Mass   In Exercises 25–28, find the total mass of the wire with density ρ whose shape is modeled by **r.**

25. $\mathbf{r}(t) = \cos t\mathbf{i} + \sin t\mathbf{j}, \ 0 \le t \le \pi, \ \rho(x, y) = x + y + 2$

26. $\mathbf{r}(t) = t^2\mathbf{i} + 2t\mathbf{j}, \ 0 \le t \le 1, \ \rho(x, y) = \frac{3}{4}y$

27. $\mathbf{r}(t) = t^2\mathbf{i} + 2t\mathbf{j} + t\mathbf{k}, \ 1 \le t \le 3, \ \rho(x, y, z) = kz \ (k > 0)$

28. $\mathbf{r}(t) = 2\cos t\mathbf{i} + 2\sin t\mathbf{j} + 3t\mathbf{k}, \ 0 \le t \le 2\pi,$
$\rho(x, y, z) = k + z \ (k > 0)$

Evaluating a Line Integral of a Vector Field
In Exercises 29–34, evaluate $\int_C \mathbf{F} \cdot d\mathbf{r}.$

29. $\mathbf{F}(x, y) = x\mathbf{i} + y\mathbf{j}$

   $C:$ $\mathbf{r}(t) = (3t + 1)\mathbf{i} + t\mathbf{j}, \ 0 \le t \le 1$

30. $\mathbf{F}(x, y) = xy\mathbf{i} + y\mathbf{j}$

   $C:$ $\mathbf{r}(t) = 4\cos t\mathbf{i} + 4\sin t\mathbf{j}, \ 0 \le t \le \dfrac{\pi}{2}$

31. $\mathbf{F}(x, y) = x^2\mathbf{i} + 4y\mathbf{j}$

   $C:$ $\mathbf{r}(t) = e^t\mathbf{i} + t^2\mathbf{j}, \ 0 \le t \le 2$

32. $\mathbf{F}(x, y) = 3x\mathbf{i} + 4y\mathbf{j}$

   $C:$ $\mathbf{r}(t) = t\mathbf{i} + \sqrt{4 - t^2}\mathbf{j}, \ -2 \le t \le 2$

33. $\mathbf{F}(x, y, z) = xy\mathbf{i} + xz\mathbf{j} + yz\mathbf{k}$

   $C:$ $\mathbf{r}(t) = t\mathbf{i} + t^2\mathbf{j} + 2t\mathbf{k}, \ 0 \le t \le 1$

34. $\mathbf{F}(x, y, z) = x^2\mathbf{i} + y^2\mathbf{j} + z^2\mathbf{k}$

   $C:$ $\mathbf{r}(t) = 2\sin t\mathbf{i} + 2\cos t\mathbf{j} + \frac{1}{2}t^2\mathbf{k}, \ 0 \le t \le \pi$

Evaluating a Line Integral of a Vector Field Using Technology   In Exercises 35 and 36, use a computer algebra system to evaluate $\int_C \mathbf{F} \cdot d\mathbf{r}.$

35. $\mathbf{F}(x, y, z) = x^2 z\mathbf{i} + 6y\mathbf{j} + yz^2\mathbf{k}$

   $C:$ $\mathbf{r}(t) = t\mathbf{i} + t^2\mathbf{j} + \ln t\mathbf{k}, \ 1 \le t \le 3$

36. $\mathbf{F}(x, y, z) = \dfrac{x\mathbf{i} + y\mathbf{j} + z\mathbf{k}}{\sqrt{x^2 + y^2 + z^2}}$

   $C:$ $\mathbf{r}(t) = t\mathbf{i} + t\mathbf{j} + e^t\mathbf{k}, \ 0 \le t \le 2$

Work   In Exercises 37–42, find the work done by the force field **F** on a particle moving along the given path.

37. $\mathbf{F}(x, y) = x\mathbf{i} + 2y\mathbf{j}$

   $C:$ $x = t, y = t^3$ from $(0, 0)$ to $(2, 8)$

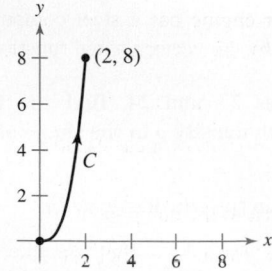

Figure for 37   Figure for 38

38. $\mathbf{F}(x, y) = x^2\mathbf{i} - xy\mathbf{j}$

   $C:$ $x = \cos^3 t, y = \sin^3 t$ from $(1, 0)$ to $(0, 1)$

39. $\mathbf{F}(x, y) = x\mathbf{i} + y\mathbf{j}$

   $C:$ counterclockwise around the triangle with vertices $(0, 0)$, $(1, 0)$, and $(0, 1)$

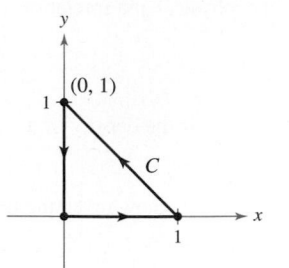
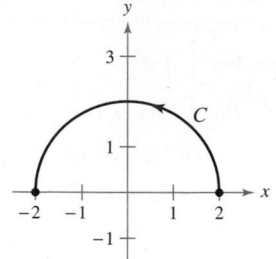

Figure for 39   Figure for 40

40. $\mathbf{F}(x, y) = -y\mathbf{i} - x\mathbf{j}$

   $C:$ counterclockwise around the semicircle $y = \sqrt{4 - x^2}$ from $(2, 0)$ to $(-2, 0)$

41. $\mathbf{F}(x, y, z) = x\mathbf{i} + y\mathbf{j} - 5z\mathbf{k}$

   $C:$ $\mathbf{r}(t) = 2\cos t\mathbf{i} + 2\sin t\mathbf{j} + t\mathbf{k}, \ 0 \le t \le 2\pi$

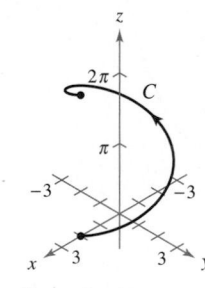
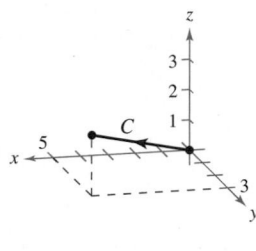

Figure for 41   Figure for 42

42. $\mathbf{F}(x, y, z) = yz\mathbf{i} + xz\mathbf{j} + xy\mathbf{k}$

   $C:$ line from $(0, 0, 0)$ to $(5, 3, 2)$

Work   In Exercises 43–46, determine whether the work done along the path C is positive, negative, or zero. Explain.

43.

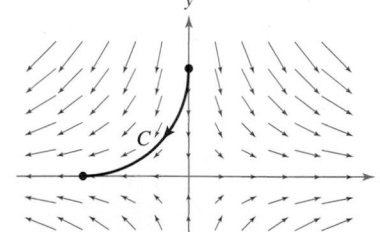

44.

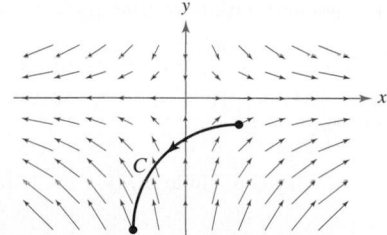

45.

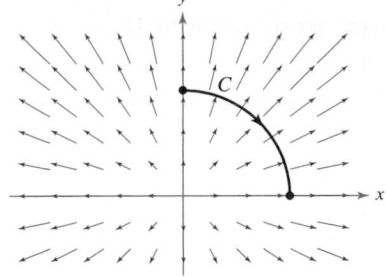

46.

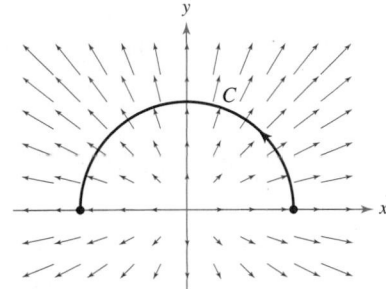

 **Evaluating a Line Integral of a Vector Field**   In Exercises 47 and 48, evaluate $\int_C \mathbf{F} \cdot d\mathbf{r}$ for each curve. Discuss the orientation of the curve and its effect on the value of the integral.

47. $\mathbf{F}(x, y) = x^2\mathbf{i} + xy\mathbf{j}$

  (a) C_1: $\mathbf{r}_1(t) = 2t\mathbf{i} + (t - 1)\mathbf{j}$, $1 \le t \le 3$

  (b) C_2: $\mathbf{r}_2(t) = 2(3 - t)\mathbf{i} + (2 - t)\mathbf{j}$, $0 \le t \le 2$

48. $\mathbf{F}(x, y) = x^2 y\mathbf{i} + xy^{3/2}\mathbf{j}$

  (a) C_1: $\mathbf{r}_1(t) = (t + 1)\mathbf{i} + t^2\mathbf{j}$, $0 \le t \le 2$

  (b) C_2: $\mathbf{r}_2(t) = (1 + 2 \cos t)\mathbf{i} + (4 \cos^2 t)\mathbf{j}$, $0 \le t \le \pi/2$

Demonstrating a Property   In Exercises 49–52, demonstrate the property that $\int_C \mathbf{F} \cdot d\mathbf{r} = 0$ regardless of the initial and terminal points of C, where the tangent vector $\mathbf{r}'(t)$ is orthogonal to the force field $\mathbf{F}$.

49. $\mathbf{F}(x, y) = y\mathbf{i} - x\mathbf{j}$

  C: $\mathbf{r}(t) = t\mathbf{i} - 2t\mathbf{j}$

50. $\mathbf{F}(x, y) = -3y\mathbf{i} + x\mathbf{j}$

  C: $\mathbf{r}(t) = t\mathbf{i} - t^3\mathbf{j}$

51. $\mathbf{F}(x, y) = (x^3 - 2x^2)\mathbf{i} + \left(x - \dfrac{y}{2}\right)\mathbf{j}$

  C: $\mathbf{r}(t) = t\mathbf{i} + t^2\mathbf{j}$

52. $\mathbf{F}(x, y) = x\mathbf{i} + y\mathbf{j}$

  C: $\mathbf{r}(t) = 3 \sin t\mathbf{i} + 3 \cos t\mathbf{j}$

Evaluating a Line Integral in Differential Form   In Exercises 53–56, evaluate the line integral along the path C given by $x = 2t$, $y = 4t$, where $0 \le t \le 1$.

53. $\displaystyle\int_C (x + 3y^2)\, dy$

54. $\displaystyle\int_C (x^3 + 2y)\, dx$

55. $\displaystyle\int_C xy\, dx + y\, dy$

56. $\displaystyle\int_C (y - x)\, dx + 5x^2y^2\, dy$

 **Evaluating a Line Integral in Differential Form**   In Exercises 57–64, evaluate

$$\int_C (2x - y)\, dx + (x + 3y)\, dy.$$

57. C: x-axis from $x = 0$ to $x = 5$

58. C: y-axis from $y = 0$ to $y = 2$

59. C: line segments from $(0, 0)$ to $(3, 0)$ and $(3, 0)$ to $(3, 3)$

60. C: line segments from $(0, 0)$ to $(0, -3)$ and $(0, -3)$ to $(2, -3)$

61. C: arc on $y = 1 - x^2$ from $(0, 1)$ to $(1, 0)$

62. C: arc on $y = x^{3/2}$ from $(0, 0)$ to $(4, 8)$

63. C: parabolic path $x = t$, $y = 2t^2$ from $(0, 0)$ to $(2, 8)$

64. C: elliptic path $x = 4 \sin t$, $y = 3 \cos t$ from $(0, 3)$ to $(4, 0)$

Lateral Surface Area   In Exercises 65–72, find the area of the lateral surface (see figure) over the curve C in the xy-plane and under the surface $z = f(x, y)$, where

$$\text{Lateral surface area} = \int_C f(x, y)\, ds.$$

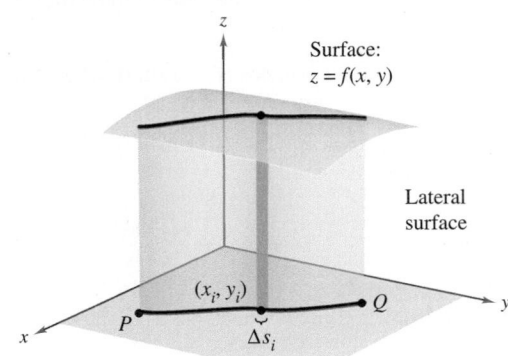

C: Curve in xy-plane

65. $f(x, y) = h$, C: line from $(0, 0)$ to $(3, 4)$

66. $f(x, y) = y$, C: line from $(0, 0)$ to $(4, 4)$

67. $f(x, y) = xy$, C: $x^2 + y^2 = 1$ from $(1, 0)$ to $(0, 1)$

68. $f(x, y) = x + y$, C: $x^2 + y^2 = 1$ from $(1, 0)$ to $(0, 1)$

69. $f(x, y) = h$, C: $y = 1 - x^2$ from $(1, 0)$ to $(0, 1)$

70. $f(x, y) = y + 1$, C: $y = 1 - x^2$ from $(1, 0)$ to $(0, 1)$

71. $f(x, y) = xy$, C: $y = 1 - x^2$ from $(1, 0)$ to $(0, 1)$

72. $f(x, y) = x^2 - y^2 + 4$, C: $x^2 + y^2 = 4$

73. Engine Design   A tractor engine has a steel component with a circular base modeled by the vector-valued function

$$\mathbf{r}(t) = 2 \cos t\mathbf{i} + 2 \sin t\mathbf{j}.$$

Its height is given by $z = 1 + y^2$. (All measurements of the component are in centimeters.)

  (a) Find the lateral surface area of the component.

  (b) The component is in the form of a shell of thickness 0.2 centimeter. Use the result of part (a) to approximate the amount of steel used to manufacture the component.

  (c) Draw a sketch of the component.

74. Building Design

The ceiling of a building has a height above the floor given by $z = 20 + \frac{1}{4}x$. One of the walls follows a path modeled by $y = x^{3/2}$. Find the surface area of the wall for $0 \le x \le 40$. (All measurements are in feet.)

Moments of Inertia Consider a wire of density $\rho(x, y)$ given by the space curve

$$C: \mathbf{r}(t) = x(t)\mathbf{i} + y(t)\mathbf{j}, \quad 0 \le t \le b.$$

The moments of inertia about the x- and y-axes are given by

$$I_x = \int_C y^2 \rho(x, y)\, ds \quad \text{and} \quad I_y = \int_C x^2 \rho(x, y)\, ds.$$

In Exercises 75 and 76, find the moments of inertia for the wire of density ρ.

75. A wire lies along $\mathbf{r}(t) = a \cos t\mathbf{i} + a \sin t\mathbf{j}$, where $0 \le t \le 2\pi$ and $a > 0$, with density $\rho(x, y) = 1$.

76. A wire lies along $\mathbf{r}(t) = a \cos t\mathbf{i} + a \sin t\mathbf{j}$, where $0 \le t \le 2\pi$ and $a > 0$, with density $\rho(x, y) = y$.

77. Investigation The top outer edge of a solid with vertical sides that is resting on the xy-plane is modeled by $\mathbf{r}(t) = 3 \cos t\mathbf{i} + 3 \sin t\mathbf{j} + (1 + \sin^2 2t)\mathbf{k}$, where all measurements are in centimeters. The intersection of the plane $y = b$, where $-3 < b < 3$, with the top of the solid is a horizontal line.

(a) Use a computer algebra system to graph the solid.

(b) Use a computer algebra system to approximate the lateral surface area of the solid.

(c) Find (if possible) the volume of the solid.

78. Work A particle moves along the path $y = x^2$ from the point $(0, 0)$ to the point $(1, 1)$. The force field $\mathbf{F}$ is measured at five points along the path, and the results are shown in the table. Use Simpson's Rule or a graphing utility to approximate the work done by the force field.

(x, y)	$(0, 0)$	$\left(\frac{1}{4}, \frac{1}{16}\right)$	$\left(\frac{1}{2}, \frac{1}{4}\right)$	$\left(\frac{3}{4}, \frac{9}{16}\right)$	$(1, 1)$
$\mathbf{F}(x, y)$	$\langle 5, 0 \rangle$	$\langle 3.5, 1 \rangle$	$\langle 2, 2 \rangle$	$\langle 1.5, 3 \rangle$	$\langle 1, 5 \rangle$

79. Work Find the work done by a person weighing 175 pounds walking exactly one revolution up a circular helical staircase of radius 3 feet when the person rises 10 feet.

80. Investigation Determine the value of c such that the work done by the force field $\mathbf{F}(x, y) = 15[(4 - x^2 y)\mathbf{i} - xy\mathbf{j}]$ on an object moving along the parabolic path $y = c(1 - x^2)$ between the points $(-1, 0)$ and $(1, 0)$ is a minimum. Compare the result with the work required to move the object along the straight-line path connecting the points.

84. 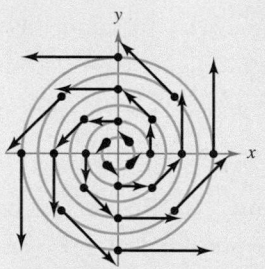 **HOW DO YOU SEE IT?** For each of the following, determine whether the work done in moving an object from the first to the second point through the force field shown in the figure is positive, negative, or zero. Explain your answer. (In the figure, the circles have radii 1, 2, 3, 4, 5, and 6.)

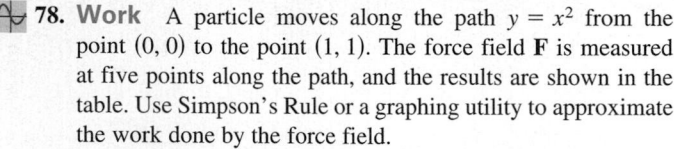

(a) From $(-3, -3)$ to $(3, 3)$

(b) From $(-3, 0)$ to $(0, 3)$

(c) From $(5, 0)$ to $(0, 3)$

True or False? In Exercises 85 and 86, determine whether the statement is true or false. If it is false, explain why or give an example that shows it is false.

85. If C is given by $x = t$, $y = t$, where $0 \le t \le 1$, then

$$\int_C xy\, ds = \int_0^1 t^2\, dt.$$

86. If $C_2 = -C_1$, then $\displaystyle\int_{C_1} f(x, y)\, ds + \int_{C_2} f(x, y)\, ds = 0$.

87. Work Consider a particle that moves through the force field

$$\mathbf{F}(x, y) = (y - x)\mathbf{i} + xy\mathbf{j}$$

from the point $(0, 0)$ to the point $(0, 1)$ along the curve $x = kt(1 - t)$, $y = t$. Find the value of k such that the work done by the force field is 1.

15.3 Conservative Vector Fields and Independence of Path

- ■ Understand and use the Fundamental Theorem of Line Integrals.
- ■ Understand the concept of independence of path.
- ■ Understand the concept of conservation of energy.

Fundamental Theorem of Line Integrals

The discussion at the beginning of Section 15.2 pointed out that in a gravitational field, the work done by gravity on an object moving between two points in the field is independent of the path taken by the object. In this section, you will study an important generalization of this result—it is called the **Fundamental Theorem of Line Integrals.** To begin, an example is presented in which the line integral of a *conservative vector field* is evaluated over three different paths.

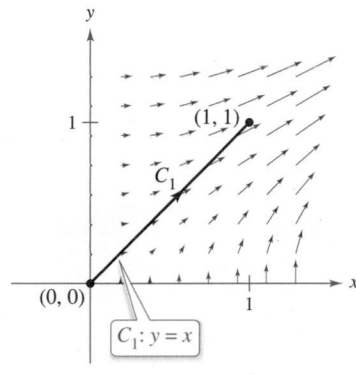

$C_1: y = x$

(a)

EXAMPLE 1   Line Integral of a Conservative Vector Field

Find the work done by the force field

$$\mathbf{F}(x, y) = \frac{1}{2}xy\mathbf{i} + \frac{1}{4}x^2\mathbf{j}$$

on a particle that moves from $(0, 0)$ to $(1, 1)$ along each path, as shown in Figure 15.19.

a. $C_1: y = x$ **b.** $C_2: x = y^2$ **c.** $C_3: y = x^3$

Solution   Note that $\mathbf{F}$ is conservative because the first partial derivatives are equal.

$$\frac{\partial}{\partial y}\left[\frac{1}{2}xy\right] = \frac{1}{2}x \quad \text{and} \quad \frac{\partial}{\partial x}\left[\frac{1}{4}x^2\right] = \frac{1}{2}x$$

a. Let $\mathbf{r}(t) = t\mathbf{i} + t\mathbf{j}$ for $0 \le t \le 1$, so that

$$d\mathbf{r} = (\mathbf{i} + \mathbf{j})\, dt \quad \text{and} \quad \mathbf{F}(x, y) = \frac{1}{2}t^2\mathbf{i} + \frac{1}{4}t^2\mathbf{j}.$$

Then the work done is

$$W = \int_{C_1} \mathbf{F} \cdot d\mathbf{r} = \int_0^1 \frac{3}{4}t^2 \, dt = \frac{1}{4}t^3 \Big]_0^1 = \frac{1}{4}.$$

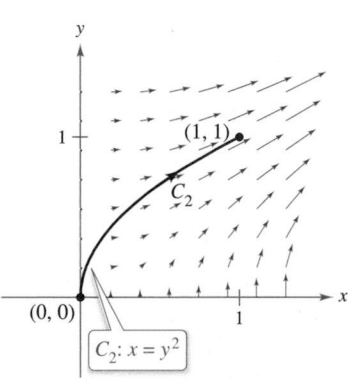

$C_2: x = y^2$

(b)

b. Let $\mathbf{r}(t) = t\mathbf{i} + \sqrt{t}\mathbf{j}$ for $0 \le t \le 1$, so that

$$d\mathbf{r} = \left(\mathbf{i} + \frac{1}{2\sqrt{t}}\mathbf{j}\right) dt \quad \text{and} \quad \mathbf{F}(x, y) = \frac{1}{2}t^{3/2}\mathbf{i} + \frac{1}{4}t^2\mathbf{j}.$$

Then the work done is

$$W = \int_{C_2} \mathbf{F} \cdot d\mathbf{r} = \int_0^1 \frac{5}{8}t^{3/2} \, dt = \frac{1}{4}t^{5/2} \Big]_0^1 = \frac{1}{4}.$$

c. Let $\mathbf{r}(t) = \frac{1}{2}t\mathbf{i} + \frac{1}{8}t^3\mathbf{j}$ for $0 \le t \le 2$, so that

$$d\mathbf{r} = \left(\frac{1}{2}\mathbf{i} + \frac{3}{8}t^2\mathbf{j}\right) dt \quad \text{and} \quad \mathbf{F}(x, y) = \frac{1}{32}t^4\mathbf{i} + \frac{1}{16}t^2\mathbf{j}.$$

Then the work done is

$$W = \int_{C_3} \mathbf{F} \cdot d\mathbf{r} = \int_0^2 \frac{5}{128}t^4 \, dt = \frac{1}{128}t^5 \Big]_0^2 = \frac{1}{4}.$$

So, the work done by the conservative vector field $\mathbf{F}$ is the same for each path.

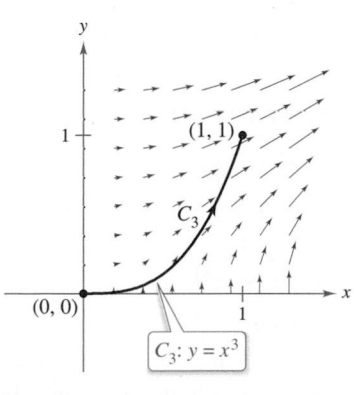

$C_3: y = x^3$

(c)

Figure 15.19

In Example 1, note that the vector field $\mathbf{F}(x, y) = \frac{1}{2}xy\mathbf{i} + \frac{1}{4}x^2\mathbf{j}$ is conservative because $\mathbf{F}(x, y) = \nabla f(x, y)$, where $f(x, y) = \frac{1}{4}x^2y$. In such cases, the next theorem states that the value of $\int_C \mathbf{F} \cdot d\mathbf{r}$ is given by

$$\int_C \mathbf{F} \cdot d\mathbf{r} = f(x(1), y(1)) - f(x(0), y(0))$$

$$= \frac{1}{4} - 0$$

$$= \frac{1}{4}.$$

• • • • • • • • • • • • • • ▷

• • REMARK Notice how the Fundamental Theorem of Line Integrals is similar to the Fundamental Theorem of Calculus (see Section 5.4), which states that

$$\int_a^b f(x) \, dx = F(b) - F(a)$$

where $F'(x) = f(x)$.

THEOREM 15.5 Fundamental Theorem of Line Integrals

Let C be a piecewise smooth curve lying in an open region R and given by

$$\mathbf{r}(t) = x(t)\mathbf{i} + y(t)\mathbf{j}, \quad a \le t \le b.$$

If $\mathbf{F}(x, y) = M\mathbf{i} + N\mathbf{j}$ is conservative in R, and M and N are continuous in R, then

$$\int_C \mathbf{F} \cdot d\mathbf{r} = \int_C \nabla f \cdot d\mathbf{r} = f(x(b), y(b)) - f(x(a), y(a))$$

where f is a potential function of $\mathbf{F}$. That is, $\mathbf{F}(x, y) = \nabla f(x, y)$.

Proof A proof is provided only for a smooth curve. For piecewise smooth curves, the procedure is carried out separately on each smooth portion. Because

$$\mathbf{F}(x, y) = \nabla f(x, y) = f_x(x, y)\mathbf{i} + f_y(x, y)\mathbf{j}$$

it follows that

$$\int_C \mathbf{F} \cdot d\mathbf{r} = \int_a^b \mathbf{F} \cdot \frac{d\mathbf{r}}{dt} \, dt$$

$$= \int_a^b \left[f_x(x, y) \frac{dx}{dt} + f_y(x, y) \frac{dy}{dt} \right] dt$$

and, by the Chain Rule (see Theorem 13.6 in Section 13.5), you have

$$\int_C \mathbf{F} \cdot d\mathbf{r} = \int_a^b \frac{d}{dt}[f(x(t), y(t))] \, dt$$

$$= f(x(b), y(b)) - f(x(a), y(a)).$$

The last step is an application of the Fundamental Theorem of Calculus.

In space, the Fundamental Theorem of Line Integrals takes the following form. Let C be a piecewise smooth curve lying in an open region Q and given by

$$\mathbf{r}(t) = x(t)\mathbf{i} + y(t)\mathbf{j} + z(t)\mathbf{k}, \quad a \le t \le b.$$

If $\mathbf{F}(x, y, z) = M\mathbf{i} + N\mathbf{j} + P\mathbf{k}$ is conservative and M, N, and P are continuous, then

$$\int_C \mathbf{F} \cdot d\mathbf{r} = \int_C \nabla f \cdot d\mathbf{r} = f(x(b), y(b), z(b)) - f(x(a), y(a), z(a))$$

where $\mathbf{F}(x, y, z) = \nabla f(x, y, z)$.

The Fundamental Theorem of Line Integrals states that if the vector field $\mathbf{F}$ is conservative, then the line integral between any two points is simply the difference in the values of the *potential* function f at these points.

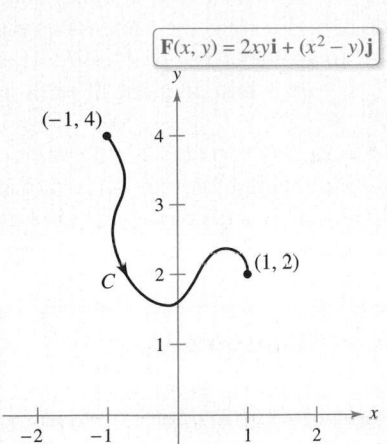

Figure 15.20

| EXAMPLE 2 | **Using the Fundamental Theorem of Line Integrals** |

Evaluate $\displaystyle\int_C \mathbf{F} \cdot d\mathbf{r}$, where C is a piecewise smooth curve from $(-1, 4)$ to $(1, 2)$ and

$$\mathbf{F}(x, y) = 2xy\mathbf{i} + (x^2 - y)\mathbf{j}$$

as shown in Figure 15.20.

Solution From Example 6 in Section 15.1, you know that $\mathbf{F}$ is the gradient of f, where

$$f(x, y) = x^2 y - \frac{y^2}{2} + K.$$

Consequently, $\mathbf{F}$ is conservative, and by the Fundamental Theorem of Line Integrals, it follows that

$$\int_C \mathbf{F} \cdot d\mathbf{r} = f(1, 2) - f(-1, 4)$$

$$= \left[1^2(2) - \frac{2^2}{2} \right] - \left[(-1)^2(4) - \frac{4^2}{2} \right]$$

$$= 4.$$

Note that it is unnecessary to include a constant K as part of f, because it is canceled by subtraction.

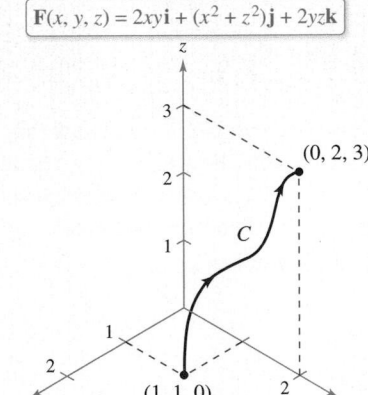

Figure 15.21

| EXAMPLE 3 | **Using the Fundamental Theorem of Line Integrals** |

Evaluate $\displaystyle\int_C \mathbf{F} \cdot d\mathbf{r}$, where C is a piecewise smooth curve from $(1, 1, 0)$ to $(0, 2, 3)$ and

$$\mathbf{F}(x, y, z) = 2xy\mathbf{i} + (x^2 + z^2)\mathbf{j} + 2yz\mathbf{k}$$

as shown in Figure 15.21.

Solution From Example 8 in Section 15.1, you know that $\mathbf{F}$ is the gradient of f, where

$$f(x, y, z) = x^2 y + yz^2 + K.$$

Consequently, $\mathbf{F}$ is conservative, and by the Fundamental Theorem of Line Integrals, it follows that

$$\int_C \mathbf{F} \cdot d\mathbf{r} = f(0, 2, 3) - f(1, 1, 0)$$

$$= [(0)^2(2) + (2)(3)^2] - [(1)^2(1) + (1)(0)^2]$$

$$= 17.$$

In Examples 2 and 3, be sure you see that the value of the line integral is the same for any smooth curve C that has the given initial and terminal points. For instance, in Example 3, try evaluating the line integral for the curve given by

$$\mathbf{r}(t) = (1 - t)\mathbf{i} + (1 + t)\mathbf{j} + 3t\mathbf{k}.$$

You should obtain

$$\int_C \mathbf{F} \cdot d\mathbf{r} = \int_0^1 (30t^2 + 16t - 1) \, dt$$

$$= 17.$$

Independence of Path

From the Fundamental Theorem of Line Integrals, it is clear that if $\mathbf{F}$ is continuous and conservative in an open region R, then the value of $\int_C \mathbf{F} \cdot d\mathbf{r}$ is the same for every piecewise smooth curve C from one fixed point in R to another fixed point in R. This result is described by saying that the line integral $\int_C \mathbf{F} \cdot d\mathbf{r}$ is **independent of path** in the region R.

A region in the plane (or in space) is **connected** when any two points in the region can be joined by a piecewise smooth curve lying entirely within the region, as shown in Figure 15.22. In open regions that are *connected,* the path independence of $\int_C \mathbf{F} \cdot d\mathbf{r}$ is equivalent to the condition that $\mathbf{F}$ is conservative.

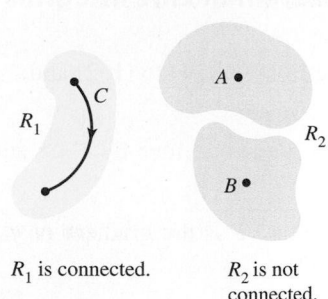

R_1 is connected. R_2 is not connected.

Figure 15.22

THEOREM 15.6   Independence of Path and Conservative Vector Fields

If $\mathbf{F}$ is continuous on an open connected region, then the line integral

$$\int_C \mathbf{F} \cdot d\mathbf{r}$$

is independent of path if and only if $\mathbf{F}$ is conservative.

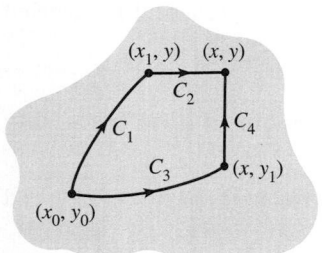

Figure 15.23

Proof   If $\mathbf{F}$ is conservative, then, by the Fundamental Theorem of Line Integrals, the line integral is independent of path. Now establish the converse for a plane region R. Let $\mathbf{F}(x, y) = M\mathbf{i} + N\mathbf{j}$, and let (x_0, y_0) be a fixed point in R. For any point (x, y) in R, choose a piecewise smooth curve C running from (x_0, y_0) to (x, y), and define f by

$$f(x, y) = \int_C \mathbf{F} \cdot d\mathbf{r}$$
$$= \int_C M \, dx + N \, dy.$$

The existence of C in R is guaranteed by the fact that R is connected. You can show that f is a potential function of $\mathbf{F}$ by considering two different paths between (x_0, y_0) and (x, y). For the *first* path, choose (x_1, y) in R such that $x \neq x_1$. This is possible because R is open. Then choose C_1 and C_2, as shown in Figure 15.23. Using the independence of path, it follows that

$$f(x, y) = \int_C M \, dx + N \, dy$$
$$= \int_{C_1} M \, dx + N \, dy + \int_{C_2} M \, dx + N \, dy.$$

Because the first integral does not depend on x and because $dy = 0$ in the second integral, you have

$$f(x, y) = g(y) + \int_{C_2} M \, dx$$

and it follows that the partial derivative of f with respect to x is $f_x(x, y) = M$. For the *second* path, choose a point (x, y_1). Using reasoning similar to that used for the first path, you can conclude that $f_y(x, y) = N$. Therefore,

$$\nabla f(x, y) = f_x(x, y)\mathbf{i} + f_y(x, y)\mathbf{j}$$
$$= M\mathbf{i} + N\mathbf{j}$$
$$= \mathbf{F}(x, y)$$

and it follows that $\mathbf{F}$ is conservative.

EXAMPLE 4   **Finding Work in a Conservative Force Field**

For the force field given by

$$\mathbf{F}(x, y, z) = e^x \cos y\mathbf{i} - e^x \sin y\mathbf{j} + 2\mathbf{k}$$

show that $\int_C \mathbf{F} \cdot d\mathbf{r}$ is independent of path, and calculate the work done by $\mathbf{F}$ on an object moving along a curve C from $(0, \pi/2, 1)$ to $(1, \pi, 3)$.

Solution   Writing the force field in the form $\mathbf{F}(x, y, z) = M\mathbf{i} + N\mathbf{j} + P\mathbf{k}$, you have $M = e^x \cos y$, $N = -e^x \sin y$, and $P = 2$, and it follows that

$$\frac{\partial P}{\partial y} = 0 = \frac{\partial N}{\partial z}$$

$$\frac{\partial P}{\partial x} = 0 = \frac{\partial M}{\partial z}$$

and

$$\frac{\partial N}{\partial x} = -e^x \sin y = \frac{\partial M}{\partial y}.$$

So, $\mathbf{F}$ is conservative. If f is a potential function of $\mathbf{F}$, then

$$f_x(x, y, z) = e^x \cos y$$
$$f_y(x, y, z) = -e^x \sin y$$

and

$$f_z(x, y, z) = 2.$$

By integrating with respect to x, y, and z separately, you obtain

$$f(x, y, z) = \int f_x(x, y, z) \, dx = \int e^x \cos y \, dx = e^x \cos y + g(y, z)$$

$$f(x, y, z) = \int f_y(x, y, z) \, dy = \int -e^x \sin y \, dy = e^x \cos y + h(x, z)$$

and

$$f(x, y, z) = \int f_z(x, y, z) \, dz = \int 2 \, dz = 2z + k(x, y).$$

By comparing these three versions of $f(x, y, z)$, you can conclude that

$$f(x, y, z) = e^x \cos y + 2z + K.$$

Therefore, the work done by $\mathbf{F}$ along *any* curve C from $(0, \pi/2, 1)$ to $(1, \pi, 3)$ is

$$W = \int_C \mathbf{F} \cdot d\mathbf{r}$$

$$= f(1, \pi, 3) - f\left(0, \frac{\pi}{2}, 1\right)$$

$$= (-e + 6) - (0 + 2)$$

$$= 4 - e. \qquad \blacksquare$$

For the object in Example 4, how much work is done when the object moves on a curve from $(0, \pi/2, 1)$ to $(1, \pi, 3)$ and then back to the starting point $(0, \pi/2, 1)$? The Fundamental Theorem of Line Integrals states that there is zero work done. Remember that, by definition, work can be negative. So, by the time the object gets back to its starting point, the amount of work that registers positively is canceled out by the amount of work that registers negatively.

A curve C given by $\mathbf{r}(t)$ for $a \le t \le b$ is **closed** when $\mathbf{r}(a) = \mathbf{r}(b)$. By the Fundamental Theorem of Line Integrals, you can conclude that if $\mathbf{F}$ is continuous and conservative on an open region R, then the line integral over every closed curve C is 0.

• • • • • • • • • • • • • • • ▷

• • REMARK Theorem 15.7 gives you options for evaluating a line integral involving a conservative vector field. You can use a potential function, or it might be more convenient to choose a particularly simple path, such as a straight line.

> **THEOREM 15.7 Equivalent Conditions**
>
> Let $\mathbf{F}(x, y, z) = M\mathbf{i} + N\mathbf{j} + P\mathbf{k}$ have continuous first partial derivatives in an open connected region R, and let C be a piecewise smooth curve in R. The conditions listed below are equivalent.
>
> 1. $\mathbf{F}$ is conservative. That is, $\mathbf{F} = \nabla f$ for some function f.
>
> 2. $\displaystyle\int_C \mathbf{F} \cdot d\mathbf{r}$ is independent of path.
>
> 3. $\displaystyle\int_C \mathbf{F} \cdot d\mathbf{r} = 0$ for every *closed* curve C in R.

EXAMPLE 5   Evaluating a Line Integral

• • • • ▷ *See LarsonCalculus.com for an interactive version of this type of example.*

Evaluate $\displaystyle\int_{C_1} \mathbf{F} \cdot d\mathbf{r}$, where

$$\mathbf{F}(x, y) = (y^3 + 1)\mathbf{i} + (3xy^2 + 1)\mathbf{j}$$

and C_1 is the semicircular path from $(0, 0)$ to $(2, 0)$, as shown in Figure 15.24.

Solution   You have the following three options.

a. You can use the method presented in Section 15.2 to evaluate the line integral along the *given curve*. To do this, you can use the parametrization $\mathbf{r}(t) = (1 - \cos t)\mathbf{i} + \sin t\mathbf{j}$, where $0 \le t \le \pi$. For this parametrization, it follows that

$$d\mathbf{r} = \mathbf{r}'(t)\, dt = (\sin t\, \mathbf{i} + \cos t\, \mathbf{j})\, dt$$

and

$$\int_{C_1} \mathbf{F} \cdot d\mathbf{r} = \int_0^\pi (\sin t + \sin^4 t + \cos t + 3\sin^2 t \cos t - 3\sin^2 t \cos^2 t)\, dt.$$

This integral should dampen your enthusiasm for this option.

b. You can try to find a *potential function* and evaluate the line integral by the Fundamental Theorem of Line Integrals. Using the technique demonstrated in Example 4, you can find the potential function to be $f(x, y) = xy^3 + x + y + K$, and, by the Fundamental Theorem,

$$W = \int_{C_1} \mathbf{F} \cdot d\mathbf{r} = f(2, 0) - f(0, 0) = 2.$$

c. Knowing that $\mathbf{F}$ is conservative, you have a third option. Because the value of the line integral is independent of path, you can replace the semicircular path with a *simpler path*. Choose the straight-line path C_2 from $(0, 0)$ to $(2, 0)$. Let $\mathbf{r}(t) = t\mathbf{i}$ for $0 \le t \le 2$, so that

$$d\mathbf{r} = \mathbf{i}\, dt \quad \text{and} \quad \mathbf{F}(x, y) = \mathbf{i} + \mathbf{j}.$$

Then the integral is

$$\int_{C_1} \mathbf{F} \cdot d\mathbf{r} = \int_{C_2} \mathbf{F} \cdot d\mathbf{r} = \int_0^2 1\, dt = t\Big]_0^2 = 2.$$

Of the three options, the third one is obviously the easiest.

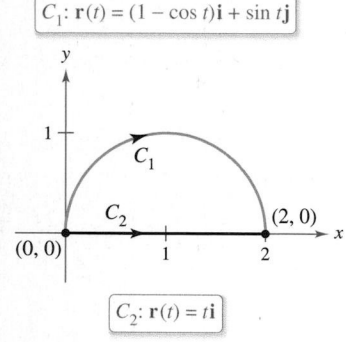

$C_1: \mathbf{r}(t) = (1 - \cos t)\mathbf{i} + \sin t\mathbf{j}$

$C_2: \mathbf{r}(t) = t\mathbf{i}$

Figure 15.24

Conservation of Energy

In 1840, the English physicist Michael Faraday wrote, "Nowhere is there a pure creation or production of power without a corresponding exhaustion of something to supply it." This statement represents the first formulation of one of the most important laws of physics—the **Law of Conservation of Energy.** In modern terminology, the law is stated as follows: *In a conservative force field, the sum of the potential and kinetic energies of an object remains constant from point to point.*

You can use the Fundamental Theorem of Line Integrals to derive this law. From physics, the **kinetic energy** of a particle of mass m and speed v is

$$k = \frac{1}{2}mv^2. \qquad \text{Kinetic energy}$$

The **potential energy** p of a particle at point (x, y, z) in a conservative vector field $\mathbf{F}$ is defined as $p(x, y, z) = -f(x, y, z)$, where f is the potential function for $\mathbf{F}$. Consequently, the work done by $\mathbf{F}$ along a smooth curve C from A to B is

$$W = \int_C \mathbf{F} \cdot d\mathbf{r} = f(x, y, z)\Big]_A^B = -p(x, y, z)\Big]_A^B = p(A) - p(B)$$

as shown in Figure 15.25. In other words, work W is equal to the difference in the potential energies of A and B. Now, suppose that $\mathbf{r}(t)$ is the position vector for a particle moving along C from $A = \mathbf{r}(a)$ to $B = \mathbf{r}(b)$. At any time t, the particle's velocity, acceleration, and speed are $\mathbf{v}(t) = \mathbf{r}'(t)$, $\mathbf{a}(t) = \mathbf{r}''(t)$, and $v(t) = \|\mathbf{v}(t)\|$, respectively. So, by Newton's Second Law of Motion, $\mathbf{F} = m\mathbf{a}(t) = m(\mathbf{v}'(t))$, and the work done by $\mathbf{F}$ is

$$
\begin{aligned}
W &= \int_C \mathbf{F} \cdot d\mathbf{r} \\
&= \int_a^b \mathbf{F} \cdot \mathbf{r}'(t)\, dt \\
&= \int_a^b \mathbf{F} \cdot \mathbf{v}(t)\, dt \\
&= \int_a^b [m\mathbf{v}'(t)] \cdot \mathbf{v}(t)\, dt \\
&= \int_a^b m[\mathbf{v}'(t) \cdot \mathbf{v}(t)]\, dt \\
&= \frac{m}{2} \int_a^b \frac{d}{dt}[\mathbf{v}(t) \cdot \mathbf{v}(t)]\, dt \\
&= \frac{m}{2} \int_a^b \frac{d}{dt}[\|\mathbf{v}(t)\|^2]\, dt \\
&= \frac{m}{2}\Big[\|\mathbf{v}(t)\|^2\Big]_a^b \\
&= \frac{m}{2}\Big[[v(t)]^2\Big]_a^b \\
&= \frac{1}{2}m[v(b)]^2 - \frac{1}{2}m[v(a)]^2 \\
&= k(B) - k(A).
\end{aligned}
$$

Equating these two results for W produces

$$p(A) - p(B) = k(B) - k(A)$$
$$p(A) + k(A) = p(B) + k(B)$$

which implies that the sum of the potential and kinetic energies remains constant from point to point.

MICHAEL FARADAY (1791–1867)

Several philosophers of science have considered Faraday's Law of Conservation of Energy to be the greatest generalization ever conceived by humankind. Many physicists have contributed to our knowledge of this law. Two early and influential ones were James Prescott Joule (1818–1889) and Hermann Ludwig Helmholtz (1821–1894).

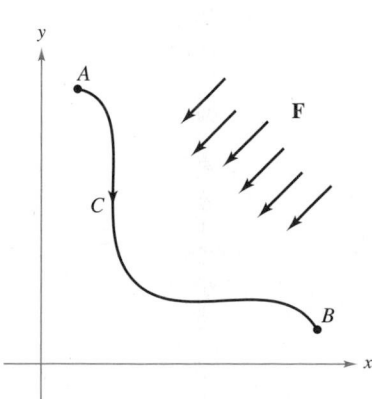

The work done by $\mathbf{F}$ along C is
$$W = \int_C \mathbf{F} \cdot d\mathbf{r} = p(A) - p(B).$$
Figure 15.25

CONCEPT CHECK

1. **Fundamental Theorem of Line Integrals** Explain how to evaluate a line integral using the Fundamental Theorem of Line Integrals.

2. **Independence of Path** What does it mean for a line integral to be independent of path? State the method for determining whether a line integral is independent of path.

 **Line Integral of a Conservative Vector Field** In Exercises 3–8, (a) show that F is conservative and (b) verify that the value of

$$\int_C \mathbf{F} \cdot d\mathbf{r}$$

is the same for each parametric representation of C.

3. $\mathbf{F}(x, y) = x^2\mathbf{i} + y\mathbf{j}$

  (i) C_1: $\mathbf{r}_1(t) = t\mathbf{i} + t^2\mathbf{j}, \quad 0 \le t \le 1$

  (ii) C_2: $\mathbf{r}_2(\theta) = \sin\theta\mathbf{i} + \sin^2\theta\mathbf{j}, \quad 0 \le \theta \le \pi/2$

4. $\mathbf{F}(x, y) = (x^2 - y^2)\mathbf{i} - 2xy\mathbf{j}$

  (i) C_1: $\mathbf{r}_1(t) = t\mathbf{i} + \sqrt{t}\mathbf{j}, \quad 0 \le t \le 4$

  (ii) C_2: $\mathbf{r}_2(w) = w^2\mathbf{i} + w\mathbf{j}, \quad 0 \le w \le 2$

5. $\mathbf{F}(x, y) = 3y\mathbf{i} + 3x\mathbf{j}$

  (i) C_1: $\mathbf{r}_1(\theta) = \sec\theta\mathbf{i} + \tan\theta\mathbf{j}, \quad 0 \le \theta \le \pi/3$

  (ii) C_2: $\mathbf{r}_2(t) = \sqrt{t+1}\mathbf{i} + \sqrt{t}\mathbf{j}, \quad 0 \le t \le 3$

6. $\mathbf{F}(x, y) = y\mathbf{i} + x\mathbf{j}$

  (i) C_1: $\mathbf{r}_1(t) = (2 + t)\mathbf{i} + (3 - t)\mathbf{j}, \quad 0 \le t \le 1$

  (ii) C_2: $\mathbf{r}_2(w) = (2 + \ln w)\mathbf{i} + (3 - \ln w)\mathbf{j}, \quad 1 \le w \le e$

7. $\mathbf{F}(x, y, z) = y^2 z\mathbf{i} + 2xyz\mathbf{j} + xy^2\mathbf{k}$

  (i) C_1: $\mathbf{r}_1(t) = t\mathbf{i} + 2t\mathbf{j} + 4t\mathbf{k}, \quad 0 \le t \le 1$

  (ii) C_2: $\mathbf{r}_2(\theta) = \sin\theta\mathbf{i} + 2\sin\theta\mathbf{j} + 4\sin\theta\mathbf{k}, \quad 0 \le \theta \le \pi/2$

8. $\mathbf{F}(x, y, z) = 2yz\mathbf{i} + 2xz\mathbf{j} + 2xy\mathbf{k}$

  (i) C_1: $\mathbf{r}_1(t) = t\mathbf{i} - 4t\mathbf{j} + t^2\mathbf{k}, \quad 0 \le t \le 3$

  (ii) C_2: $\mathbf{r}_2(s) = s^2\mathbf{i} - \frac{4}{3}s^4\mathbf{j} + s^4\mathbf{k}, \quad 0 \le s \le \sqrt{3}$

 **Using the Fundamental Theorem of Line Integrals** In Exercises 9–18, evaluate

$$\int_C \mathbf{F} \cdot d\mathbf{r}$$

using the Fundamental Theorem of Line Integrals. Use a computer algebra system to verify your results.

9. $\mathbf{F}(x, y) = 3y\mathbf{i} + 3x\mathbf{j}$

  C: smooth curve from $(0, 0)$ to $(3, 8)$

10. $\mathbf{F}(x, y) = 2(x + y)\mathbf{i} + 2(x + y)\mathbf{j}$

  C: smooth curve from $(-1, 1)$ to $(3, 2)$

11. $\mathbf{F}(x, y) = \cos x \sin y\mathbf{i} + \sin x \cos y\mathbf{j}$

  C: line segment from $(0, -\pi)$ to $\left(\dfrac{3\pi}{2}, \dfrac{\pi}{2}\right)$

12. $\mathbf{F}(x, y) = \dfrac{y}{x^2 + y^2}\mathbf{i} - \dfrac{x}{x^2 + y^2}\mathbf{j}$

  C: line segment from $(1, 1)$ to $(2\sqrt{3}, 2)$

13. $\mathbf{F}(x, y) = e^x \sin y\mathbf{i} + e^x \cos y\mathbf{j}$

  C: cycloid $x = \theta - \sin\theta$, $y = 1 - \cos\theta$ from $(0, 0)$ to $(2\pi, 0)$

14. $\mathbf{F}(x, y) = \dfrac{2x}{(x^2 + y^2)^2}\mathbf{i} + \dfrac{2y}{(x^2 + y^2)^2}\mathbf{j}$

  C: clockwise around the circle $(x - 4)^2 + (y - 5)^2 = 9$ from $(7, 5)$ to $(1, 5)$

15. $\mathbf{F}(x, y, z) = (z + 2y)\mathbf{i} + (2x - z)\mathbf{j} + (x - y)\mathbf{k}$

  (a) C_1: line segment from $(0, 0, 0)$ to $(1, 1, 1)$

  (b) C_2: line segments from $(0, 0, 0)$ to $(0, 0, 1)$ and $(0, 0, 1)$ to $(1, 1, 1)$

  (c) C_3: line segments from $(0, 0, 0)$ to $(1, 0, 0)$, from $(1, 0, 0)$ to $(1, 1, 0)$, and from $(1, 1, 0)$ to $(1, 1, 1)$

16. Repeat Exercise 15 using

  $\mathbf{F}(x, y, z) = zy\mathbf{i} + xz\mathbf{j} + xy\mathbf{k}$.

17. $\mathbf{F}(x, y, z) = -\sin x\mathbf{i} + z\mathbf{j} + y\mathbf{k}$

  C: smooth curve from $(0, 0, 0)$ to $\left(\dfrac{\pi}{2}, 3, 4\right)$

18. $\mathbf{F}(x, y, z) = 6x\mathbf{i} - 4z\mathbf{j} - (4y - 20z)\mathbf{k}$

  C: smooth curve from $(0, 0, 0)$ to $(3, 4, 0)$

 **Finding Work in a Conservative Force Field** In Exercises 19–22, (a) show that $\int_C \mathbf{F} \cdot d\mathbf{r}$ is independent of path and (b) calculate the work done by the force field F on an object moving along a curve from P to Q.

19. $\mathbf{F}(x, y) = 9x^2y^2\mathbf{i} + (6x^3y - 1)\mathbf{j}$

  $P(0, 0)$, $Q(5, 9)$

20. $\mathbf{F}(x, y) = \dfrac{2x}{y}\mathbf{i} - \dfrac{x^2}{y^2}\mathbf{j}$

  $P(-1, 1)$, $Q(3, 2)$

21. $\mathbf{F}(x, y, z) = 3\mathbf{i} + 4y\mathbf{j} - \sin z\mathbf{k}$

  $P\left(0, 1, \dfrac{\pi}{2}\right)$, $Q(1, 4, \pi)$

22. $\mathbf{F}(x, y, z) = 8x^3\mathbf{i} + z^2 \cos 2y\mathbf{j} + z \sin 2y\mathbf{k}$

  $P\left(0, \dfrac{\pi}{4}, 1\right)$, $Q(-2, 0, -1)$

Evaluating a Line Integral In Exercises 23–32, evaluate

$$\int_C \mathbf{F} \cdot d\mathbf{r}$$

along each path. (*Hint:* If **F** is conservative, the integration may be easier on an alternative path.)

23. $\mathbf{F}(x, y) = 2xy\mathbf{i} + x^2\mathbf{j}$

(a) C_1: $\mathbf{r}_1(t) = t\mathbf{i} + t^2\mathbf{j}$, $0 \le t \le 1$

(b) C_2: $\mathbf{r}_2(t) = t\mathbf{i} + t^3\mathbf{j}$, $0 \le t \le 1$

24. $\mathbf{F}(x, y) = ye^{xy}\mathbf{i} + xe^{xy}\mathbf{j}$

(a) C_1: $\mathbf{r}_1(t) = t\mathbf{i} - (t - 3)\mathbf{j}$, $0 \le t \le 3$

(b) C_2: The closed path consisting of line segments from $(0, 3)$ to $(0, 0)$, from $(0, 0)$ to $(3, 0)$, and then from $(3, 0)$ to $(0, 3)$

25. $\int_C y^2 \, dx + 2xy \, dy$

(a)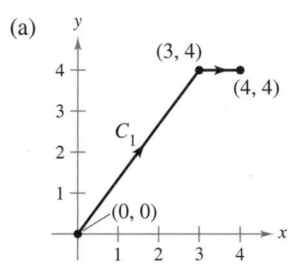

(b)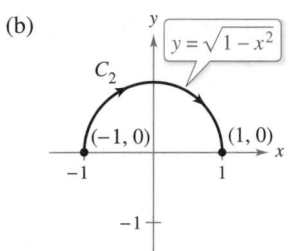

(c)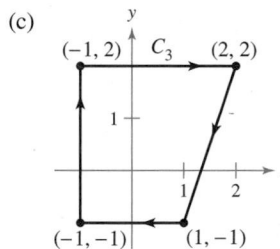

(d)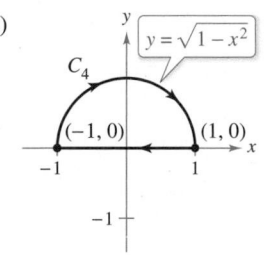

26. $\int_C (2x - 3y + 1) \, dx - (3x + y - 5) \, dy$

(a)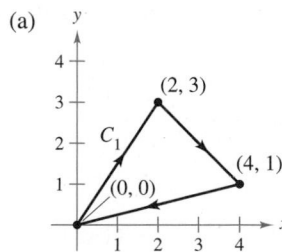

(b)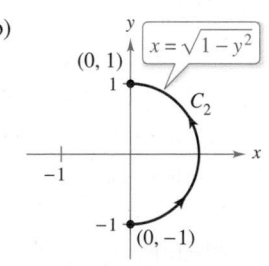

(c)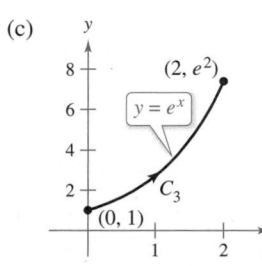

(d)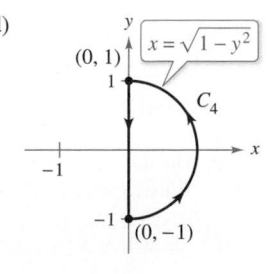

27. $\mathbf{F}(x, y, z) = yz\mathbf{i} + xz\mathbf{j} + xy\mathbf{k}$

(a) C_1: $\mathbf{r}_1(t) = t\mathbf{i} + 2\mathbf{j} + t\mathbf{k}$, $0 \le t \le 4$

(b) C_2: $\mathbf{r}_2(t) = t^2\mathbf{i} + t\mathbf{j} + t^2\mathbf{k}$, $0 \le t \le 2$

28. $\mathbf{F}(x, y, z) = \mathbf{i} + z\mathbf{j} + y\mathbf{k}$

(a) C_1: $\mathbf{r}_1(t) = \cos t\mathbf{i} + \sin t\mathbf{j} + t^2\mathbf{k}$, $0 \le t \le \pi$

(b) C_2: $\mathbf{r}_2(t) = (1 - 2t)\mathbf{i} + \pi^2 t\mathbf{k}$, $0 \le t \le 1$

29. $\mathbf{F}(x, y, z) = (2y + x)\mathbf{i} + (x^2 - z)\mathbf{j} + (2y - 4z)\mathbf{k}$

(a) C_1: $\mathbf{r}_1(t) = t\mathbf{i} + t^2\mathbf{j} + \mathbf{k}$, $0 \le t \le 1$

(b) C_2: $\mathbf{r}_2(t) = t\mathbf{i} + t\mathbf{j} + (2t - 1)^2\mathbf{k}$, $0 \le t \le 1$

30. $\mathbf{F}(x, y, z) = -y\mathbf{i} + x\mathbf{j} + 3xz^2\mathbf{k}$

(a) C_1: $\mathbf{r}_1(t) = \cos t\mathbf{i} + \sin t\mathbf{j} + t\mathbf{k}$, $0 \le t \le \pi$

(b) C_2: $\mathbf{r}_2(t) = (1 - 2t)\mathbf{i} + \pi t\mathbf{k}$, $0 \le t \le 1$

31. $\mathbf{F}(x, y, z) = e^z(y\mathbf{i} + x\mathbf{j} + xy\mathbf{k})$

(a) C_1: $\mathbf{r}_1(t) = 4\cos t\mathbf{i} + 4\sin t\mathbf{j} + 3\mathbf{k}$, $0 \le t \le \pi$

(b) C_2: $\mathbf{r}_2(t) = (4 - 8t)\mathbf{i} + 3\mathbf{k}$, $0 \le t \le 1$

32. $\mathbf{F}(x, y, z) = y\sin z\mathbf{i} + x\sin z\mathbf{j} + xy\cos x\mathbf{k}$

(a) C_1: $\mathbf{r}_1(t) = t^2\mathbf{i} + t^2\mathbf{j}$, $0 \le t \le 2$

(b) C_2: $\mathbf{r}_2(t) = 4t\mathbf{i} + 4t\mathbf{j}$, $0 \le t \le 1$

33. Work A stone weighing 1 pound is attached to the end of a two-foot string and is whirled horizontally with one end held fixed. It makes 1 revolution per second. Find the work done by the force **F** that keeps the stone moving in a circular path. [*Hint:* Use Force = (mass)(centripetal acceleration).]

34. Work A grappling hook weighing 1 kilogram is attached to the end of a five-meter rope and is whirled horizontally with one end held fixed. It makes 0.5 revolution per second. Find the work done by the force **F** that keeps the grappling hook moving in a circular path. [*Hint:* Use Force = (mass)(centripetal acceleration).]

35. Work

A zip line is installed 50 meters above ground level. It runs to a point on the ground 50 meters away from the base of the installation. Show that the work done by the gravitational force field for a 175-pound person moving the length of the zip line is the same for each path.

(a) C_1: $\mathbf{r}_1(t) = t\mathbf{i} + (50 - t)\mathbf{j}$

(b) C_2: $\mathbf{r}_2(t) = t\mathbf{i} + \frac{1}{50}(50 - t)^2\mathbf{j}$

36. Work Can you find a path for the zip line in Exercise 35 such that the work done by the gravitational force field would differ from the amounts of work done for the two paths given? Explain why or why not.

EXPLORING CONCEPTS

37. Think About It Consider

$$\mathbf{F}(x, y) = \frac{y}{x^2 + y^2}\mathbf{i} - \frac{x}{x^2 + y^2}\mathbf{j}.$$

Sketch an open connected region around the smooth curve C shown in the figure such that you can use Theorem 15.7 to evaluate $\int_C \mathbf{F} \cdot d\mathbf{r}$. Explain how you created your sketch.

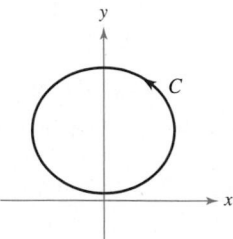

38. Work Let $\mathbf{F}(x, y, z) = a_1\mathbf{i} + a_2\mathbf{j} + a_3\mathbf{k}$ be a constant force vector field. Show that the work done in moving a particle along any path from P to Q is $W = \mathbf{F} \cdot \overrightarrow{PQ}$.

39. Using Different Methods Use two different methods to evaluate $\int_C \mathbf{F} \cdot d\mathbf{r}$ along the path

$$\mathbf{r}(t) = \frac{1}{t}\mathbf{i} + 3t\mathbf{j}, \quad 0.5 \le t \le 2$$

where $\mathbf{F}(x, y) = (x^2y^2 - 3x)\mathbf{i} + \frac{2}{3}x^3y\mathbf{j}$.

---

40. HOW DO YOU SEE IT? Consider the force field shown in the figure. To print an enlarged copy of the graph, go to *MathGraphs.com*.

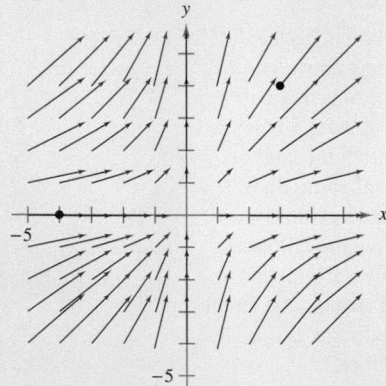

(a) Give a verbal argument that the force field is not conservative because you can identify two paths that require different amounts of work to move an object from $(-4, 0)$ to $(3, 4)$. Of the two paths, which requires the greater amount of work?

(b) Give a verbal argument that the force field is not conservative because you can find a closed curve C such that $\int_C \mathbf{F} \cdot d\mathbf{r} \ne 0$.

Graphical Reasoning In Exercises 41 and 42, consider the force field shown in the figure. Is the force field conservative? Explain why or why not.

41. 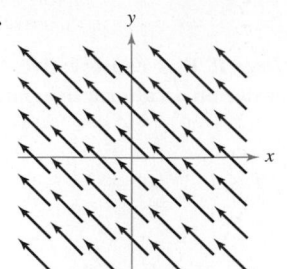 42.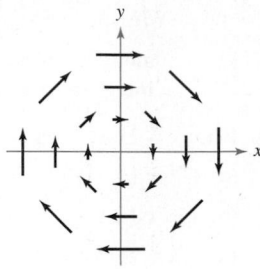

True or False? In Exercises 43–46, determine whether the statement is true or false. If it is false, explain why or give an example that shows it is false.

43. If C_1, C_2, and C_3 have the same initial and terminal points and $\int_{C_1} \mathbf{F} \cdot d\mathbf{r}_1 = \int_{C_2} \mathbf{F} \cdot d\mathbf{r}_2$, then $\int_{C_1} \mathbf{F} \cdot d\mathbf{r}_1 = \int_{C_3} \mathbf{F} \cdot d\mathbf{r}_3$.

44. If $\mathbf{F} = y\mathbf{i} + x\mathbf{j}$ and C is given by $\mathbf{r}(t) = 4\sin t\mathbf{i} + 3\cos t\mathbf{j}$ for $0 \le t \le \pi$, then

$$\int_C \mathbf{F} \cdot d\mathbf{r} = 0.$$

45. If $\mathbf{F}$ is conservative in a region R bounded by a simple closed path and C lies within R, then $\int_C \mathbf{F} \cdot d\mathbf{r}$ is independent of path.

46. If $\mathbf{F} = M\mathbf{i} + N\mathbf{j}$ and $\frac{\partial M}{\partial x} = \frac{\partial N}{\partial y}$, then $\mathbf{F}$ is conservative.

47. Harmonic Function A function f is called *harmonic* when

$$\frac{\partial^2 f}{\partial x^2} + \frac{\partial^2 f}{\partial y^2} = 0.$$

Prove that if f is harmonic, then

$$\int_C \left(\frac{\partial f}{\partial y}dx - \frac{\partial f}{\partial x}dy\right) = 0$$

where C is a smooth closed curve in the plane.

48. Kinetic and Potential Energy The kinetic energy of an object moving through a conservative force field is decreasing at a rate of 15 units per minute. At what rate is the potential energy changing? Explain.

49. Investigation Let $\mathbf{F}(x, y) = \frac{y}{x^2 + y^2}\mathbf{i} - \frac{x}{x^2 + y^2}\mathbf{j}$.

(a) Show that $\frac{\partial N}{\partial x} = \frac{\partial M}{\partial y}$.

(b) Let $\mathbf{r}(t) = \cos t\mathbf{i} + \sin t\mathbf{j}$ for $0 \le t \le \pi$. Find $\int_C \mathbf{F} \cdot d\mathbf{r}$.

(c) Let $\mathbf{r}(t) = \cos t\mathbf{i} - \sin t\mathbf{j}$ for $0 \le t \le \pi$. Find $\int_C \mathbf{F} \cdot d\mathbf{r}$.

(d) Let $\mathbf{r}(t) = \cos t\mathbf{i} + \sin t\mathbf{j}$ for $0 \le t \le 2\pi$. Find $\int_C \mathbf{F} \cdot d\mathbf{r}$.

(e) Do the results of parts (b)–(d) contradict Theorem 15.7? Why or why not?

(f) Show that $\nabla\left(\arctan\frac{x}{y}\right) = \mathbf{F}$.

15.4 Green's Theorem

■ Use Green's Theorem to evaluate a line integral.
■ Use alternative forms of Green's Theorem.

Green's Theorem

In this section, you will study **Green's Theorem,** named after the English mathematician George Green (1793–1841). This theorem states that the value of a double integral over a *simply connected* plane region R is determined by the value of a line integral around the boundary of R.

A curve C given by $\mathbf{r}(t) = x(t)\mathbf{i} + y(t)\mathbf{j}$, where $a \le t \le b$, is **simple** when it does not cross itself—that is, $\mathbf{r}(c) \ne \mathbf{r}(d)$ for all c and d in the open interval (a, b). A connected plane region R is **simply connected** when every simple closed curve in R encloses only points that are in R (see Figure 15.26). Informally, a simply connected region cannot consist of separate parts or holes.

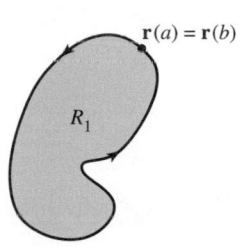

Simply connected

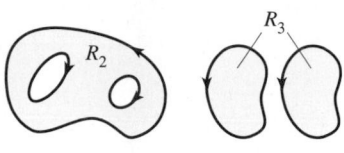

Not simply connected

Figure 15.26

THEOREM 15.8 Green's Theorem

Let R be a simply connected region with a piecewise smooth boundary C, oriented counterclockwise (that is, C is traversed *once* so that the region R always lies to the *left*). If M and N have continuous first partial derivatives in an open region containing R, then

$$\int_C M\, dx + N\, dy = \int\!\!\int_R \left(\frac{\partial N}{\partial x} - \frac{\partial M}{\partial y} \right) dA.$$

Proof A proof is given only for a region that is both vertically simple and horizontally simple, as shown in Figure 15.27.

$$\int_C M\, dx = \int_{C_1} M\, dx + \int_{C_2} M\, dx$$

$$= \int_a^b M(x, f_1(x))\, dx + \int_b^a M(x, f_2(x))\, dx$$

$$= \int_a^b [M(x, f_1(x)) - M(x, f_2(x))]\, dx$$

On the other hand,

$$\int\!\!\int_R \frac{\partial M}{\partial y}\, dA = \int_a^b \int_{f_1(x)}^{f_2(x)} \frac{\partial M}{\partial y}\, dy\, dx$$

$$= \int_a^b M(x, y)\Big]_{f_1(x)}^{f_2(x)}\, dx$$

$$= \int_a^b [M(x, f_2(x)) - M(x, f_1(x))]\, dx.$$

Consequently,

$$\int_C M\, dx = -\int\!\!\int_R \frac{\partial M}{\partial y}\, dA.$$

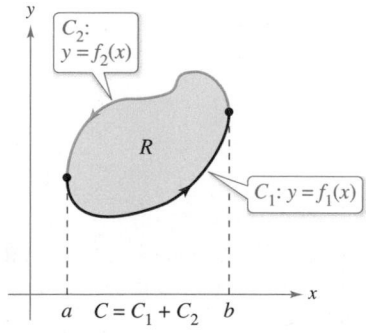

R is vertically simple.

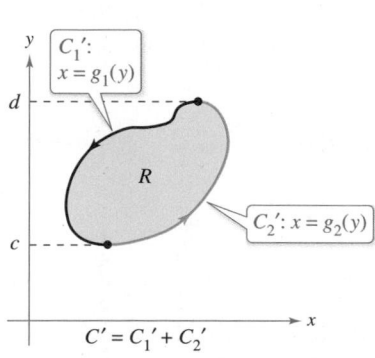

R is horizontally simple.
Figure 15.27

Similarly, you can use $g_1(y)$ and $g_2(y)$ to show that $\int_C N\, dy = \int_R\!\!\int (\partial N/\partial x)\, dA$. By adding the integrals $\int_C M\, dx$ and $\int_C N\, dy$, you obtain the conclusion stated in the theorem.

An integral sign with a circle is sometimes used to indicate a line integral around a simple closed curve, as shown below. To indicate the orientation of the boundary, an arrow can be used. For instance, in the second integral, the arrow indicates that the boundary C is oriented counterclockwise.

1. $\oint_C M\,dx + N\,dy$ **2.** $\oint_C M\,dx + N\,dy$

EXAMPLE 1 Using Green's Theorem

Use Green's Theorem to evaluate the line integral

$$\int_C y^3\,dx + (x^3 + 3xy^2)\,dy$$

where C is the path from $(0, 0)$ to $(1, 1)$ along the graph of $y = x^3$ and from $(1, 1)$ to $(0, 0)$ along the graph of $y = x$, as shown in Figure 15.28.

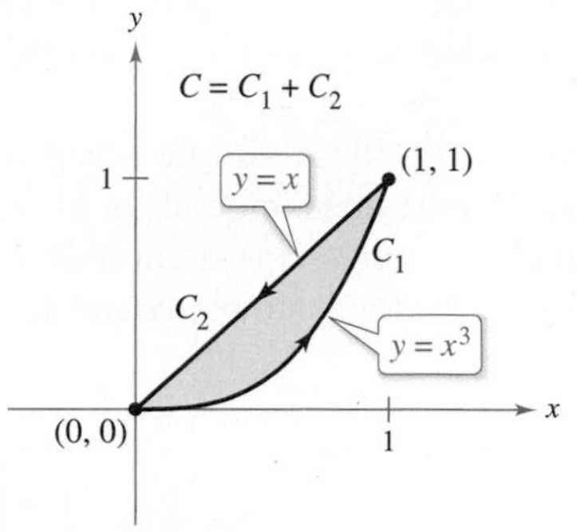

C is simple and closed, and the region R always lies to the left of C.
Figure 15.28

Solution Because $M = y^3$ and $N = x^3 + 3xy^2$, it follows that

$$\frac{\partial N}{\partial x} = 3x^2 + 3y^2 \quad \text{and} \quad \frac{\partial M}{\partial y} = 3y^2.$$

Applying Green's Theorem, you then have

$$\begin{aligned}\int_C y^3\,dx + (x^3 + 3xy^2)\,dy &= \int_R\!\int \left(\frac{\partial N}{\partial x} - \frac{\partial M}{\partial y}\right) dA\\ &= \int_0^1 \int_{x^3}^{x} [(3x^2 + 3y^2) - 3y^2]\,dy\,dx\\ &= \int_0^1 \int_{x^3}^{x} 3x^2\,dy\,dx\\ &= \int_0^1 3x^2y\Big]_{x^3}^{x} dx\\ &= \int_0^1 (3x^3 - 3x^5)\,dx\\ &= \left[\frac{3x^4}{4} - \frac{x^6}{2}\right]_0^1\\ &= \frac{1}{4}.\end{aligned}$$

GEORGE GREEN (1793–1841)

Green, a self-educated miller's son, first published the theorem that bears his name in 1828 in an essay on electricity and magnetism. At that time, there was almost no mathematical theory to explain electrical phenomena. "Considering how desirable it was that a power of universal agency, like electricity, should, as far as possible, be submitted to calculation, . . . I was induced to try whether it would be possible to discover any general relations existing between this function and the quantities of electricity in the bodies producing it."

Green's Theorem cannot be applied to every line integral. Among other restrictions stated in Theorem 15.8, the curve C must be simple and closed. When Green's Theorem does apply, however, it can save time. To see this, try using the techniques described in Section 15.2 to evaluate the line integral in Example 1. To do this, you would need to write the line integral as

$$\begin{aligned}&\int_C y^3\,dx + (x^3 + 3xy^2)\,dy\\ &\quad = \int_{C_1} y^3\,dx + (x^3 + 3xy^2)\,dy + \int_{C_2} y^3\,dx + (x^3 + 3xy^2)\,dy\end{aligned}$$

where C_1 is the cubic path given by

$$\mathbf{r}(t) = t\mathbf{i} + t^3\mathbf{j}$$

from $t = 0$ to $t = 1$, and C_2 is the line segment given by

$$\mathbf{r}(t) = (1 - t)\mathbf{i} + (1 - t)\mathbf{j}$$

from $t = 0$ to $t = 1$.

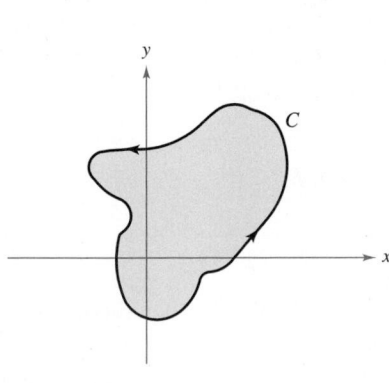

$$\mathbf{F}(x, y) = y^3\mathbf{i} + (x^3 + 3xy^2)\mathbf{j}$$

Figure 15.29

EXAMPLE 2   Using Green's Theorem to Calculate Work

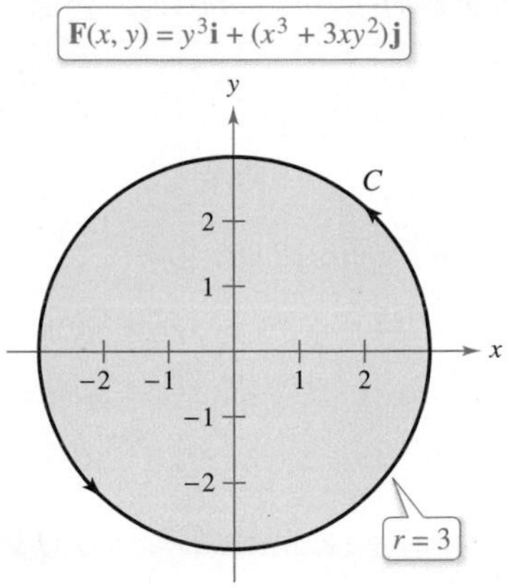

While subject to the force

$$\mathbf{F}(x, y) = y^3\mathbf{i} + (x^3 + 3xy^2)\mathbf{j}$$

a particle travels once around the circle of radius 3 shown in Figure 15.29. Use Green's Theorem to find the work done by $\mathbf{F}$.

Solution   From Example 1, you know by Green's Theorem that

$$\int_C y^3\, dx + (x^3 + 3xy^2)\, dy = \int_R\int 3x^2\, dA.$$

In polar coordinates, using $x = r \cos \theta$ and $dA = r\, dr\, d\theta$, the work done is

$$
\begin{aligned}
W &= \int_R\int 3x^2\, dA \\
&= \int_0^{2\pi}\int_0^3 3(r \cos \theta)^2 r\, dr\, d\theta \\
&= 3\int_0^{2\pi}\int_0^3 r^3 \cos^2 \theta\, dr\, d\theta \\
&= 3\int_0^{2\pi} \frac{r^4}{4} \cos^2 \theta \Big]_0^3 d\theta \\
&= 3\int_0^{2\pi} \frac{81}{4} \cos^2 \theta\, d\theta \\
&= \frac{243}{8} \int_0^{2\pi} (1 + \cos 2\theta)\, d\theta \\
&= \frac{243}{8} \left[\theta + \frac{\sin 2\theta}{2}\right]_0^{2\pi} \\
&= \frac{243\pi}{4}.
\end{aligned}
$$

When evaluating line integrals over closed curves, remember that for conservative vector fields (those for which $\partial N/\partial x = \partial M/\partial y$), the value of the line integral is 0. This is easily seen from the statement of Green's Theorem:

$$\int_C M\, dx + N\, dy = \int_R\int \left(\frac{\partial N}{\partial x} - \frac{\partial M}{\partial y}\right) dA = 0.$$

EXAMPLE 3   Green's Theorem and Conservative Vector Fields

Evaluate the line integral

$$\int_C y^3\, dx + 3xy^2\, dy$$

where C is the path shown in Figure 15.30.

Solution   From this line integral, $M = y^3$ and $N = 3xy^2$. So, $\partial N/\partial x = 3y^2$ and $\partial M/\partial y = 3y^2$. This implies that the vector field $\mathbf{F} = M\mathbf{i} + N\mathbf{j}$ is conservative, and because C is closed, you can conclude that

$$\int_C y^3\, dx + 3xy^2\, dy = 0.$$

C is closed.

Figure 15.30

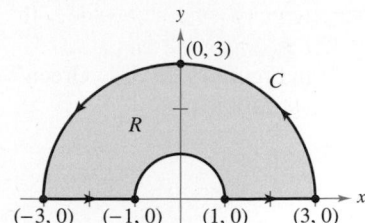

C is piecewise smooth.
Figure 15.31

EXAMPLE 4   Using Green's Theorem

$\vdots$ $\cdots\triangleright$ *See LarsonCalculus.com for an interactive version of this type of example.*

Evaluate

$$\int_C (\arctan x + y^2)\, dx + (e^y - x^2)\, dy$$

where C is the path enclosing the annular region shown in Figure 15.31.

Solution   In polar coordinates, R is given by $1 \le r \le 3$ for $0 \le \theta \le \pi$. Moreover,

$$\frac{\partial N}{\partial x} - \frac{\partial M}{\partial y} = -2x - 2y = -2(r\cos\theta + r\sin\theta).$$

So, by Green's Theorem,

$$
\begin{aligned}
\int_C (\arctan x + y^2)\, dx + (e^y - x^2)\, dy &= \iint_R -2(x+y)\, dA \\
&= \int_0^\pi \int_1^3 -2r(\cos\theta + \sin\theta) r\, dr\, d\theta \\
&= \int_0^\pi -2(\cos\theta + \sin\theta)\frac{r^3}{3}\bigg]_1^3 d\theta \\
&= \int_0^\pi -\frac{52}{3}(\cos\theta + \sin\theta)\, d\theta \\
&= -\frac{52}{3}\Big[\sin\theta - \cos\theta\Big]_0^\pi \\
&= -\frac{104}{3}.
\end{aligned}
$$

In Examples 1, 2, and 4, Green's Theorem was used to evaluate line integrals as double integrals. You can also use the theorem to evaluate double integrals as line integrals. One useful application occurs when $\partial N/\partial x - \partial M/\partial y = 1$.

$$
\begin{aligned}
\int_C M\, dx + N\, dy &= \iint_R \left(\frac{\partial N}{\partial x} - \frac{\partial M}{\partial y}\right) dA \\
&= \iint_R 1\, dA \qquad\qquad \frac{\partial N}{\partial x} - \frac{\partial M}{\partial y} = 1 \\
&= \text{area of region } R
\end{aligned}
$$

Among the many choices for M and N satisfying the stated condition, the choice of

$$M = -\frac{y}{2} \quad \text{and} \quad N = \frac{x}{2}$$

produces the following line integral for the area of region R.

THEOREM 15.9   Line Integral for Area

If R is a plane region bounded by a piecewise smooth simple closed curve C, oriented counterclockwise, then the area of R is given by

$$A = \frac{1}{2}\int_C x\, dy - y\, dx.$$

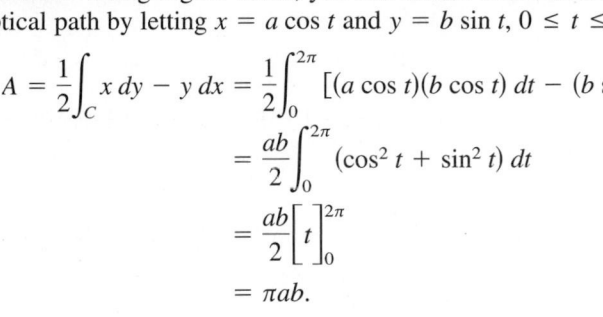

EXAMPLE 5 **Finding Area by a Line Integral**

Use a line integral to find the area of the ellipse $(x^2/a^2) + (y^2/b^2) = 1$.

Solution Using Figure 15.32, you can induce a counterclockwise orientation to the elliptical path by letting $x = a \cos t$ and $y = b \sin t$, $0 \le t \le 2\pi$. So, the area is

$$
\begin{aligned}
A &= \frac{1}{2}\int_C x\, dy - y\, dx = \frac{1}{2}\int_0^{2\pi} [(a\cos t)(b\cos t)\, dt - (b\sin t)(-a\sin t)\, dt] \\
&= \frac{ab}{2}\int_0^{2\pi} (\cos^2 t + \sin^2 t)\, dt \\
&= \frac{ab}{2}\Big[\, t \,\Big]_0^{2\pi} \\
&= \pi ab.
\end{aligned}
$$

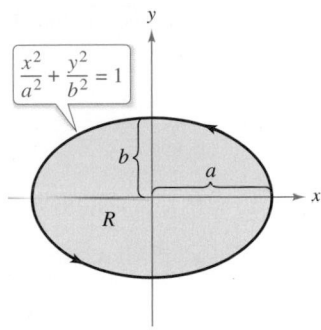

Figure 15.32

Green's Theorem can be extended to cover some regions that are not simply connected. This is demonstrated in the next example.

EXAMPLE 6 **Green's Theorem Extended to a Region with a Hole**

Let R be the region inside the ellipse $(x^2/9) + (y^2/4) = 1$ and outside the circle $x^2 + y^2 = 1$. Evaluate the line integral

$$\int_C 2xy\, dx + (x^2 + 2x)\, dy$$

where $C = C_1 + C_2$ is the boundary of R, as shown in Figure 15.33.

Solution To begin, introduce the line segments C_3 and C_4, as shown in Figure 15.33. Note that because the curves C_3 and C_4 have opposite orientations, the line integrals over them cancel. Furthermore, apply Green's Theorem to the region R using the boundary $C_1 + C_4 + C_2 + C_3$ to obtain

$$
\begin{aligned}
\int_C 2xy\, dx + (x^2 + 2x)\, dy &= \iint_R \left(\frac{\partial N}{\partial x} - \frac{\partial M}{\partial y}\right) dA \\
&= \iint_R (2x + 2 - 2x)\, dA \\
&= 2\iint_R dA \\
&= 2(\text{area of } R) \\
&= 2(\pi ab - \pi r^2) \\
&= 2[\pi(3)(2) - \pi(1^2)] \\
&= 10\pi.
\end{aligned}
$$

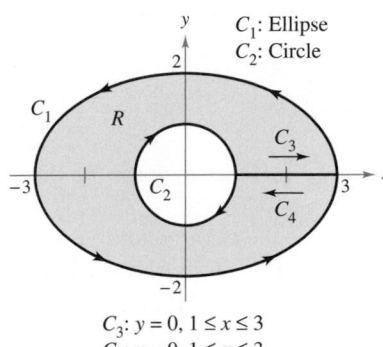

$C_3: y = 0, 1 \le x \le 3$
$C_4: y = 0, 1 \le x \le 3$

Figure 15.33

In Section 15.1, a necessary and sufficient condition for conservative vector fields was listed. There, only one direction of the proof was shown. You can now outline the other direction, using Green's Theorem. Let $\mathbf{F}(x, y) = M\mathbf{i} + N\mathbf{j}$ be defined on an open disk R. You want to show that if M and N have continuous first partial derivatives and $\partial M/\partial y = \partial N/\partial x$, then $\mathbf{F}$ is conservative. Let C be a closed path forming the boundary of a connected region lying in R. Then, using the fact that $\partial M/\partial y = \partial N/\partial x$, apply Green's Theorem to conclude that

$$\int_C \mathbf{F} \cdot d\mathbf{r} = \int_C M\, dx + N\, dy = \iint_R \left(\frac{\partial N}{\partial x} - \frac{\partial M}{\partial y}\right) dA = 0.$$

This, in turn, is equivalent to showing that $\mathbf{F}$ is conservative (see Theorem 15.7).

Alternative Forms of Green's Theorem

This section concludes with the derivation of two vector forms of Green's Theorem for regions in the plane. The extension of these vector forms to three dimensions is the basis for the discussion in the remaining sections of this chapter. For a vector field $\mathbf{F}$ in the plane, you can write

$$\mathbf{F}(x, y, z) = M\mathbf{i} + N\mathbf{j} + 0\mathbf{k}$$

so that the curl of $\mathbf{F}$, as described in Section 15.1, is given by

$$\operatorname{curl} \mathbf{F} = \nabla \times \mathbf{F} = \begin{vmatrix} \mathbf{i} & \mathbf{j} & \mathbf{k} \\ \dfrac{\partial}{\partial x} & \dfrac{\partial}{\partial y} & \dfrac{\partial}{\partial z} \\ M & N & 0 \end{vmatrix} = -\frac{\partial N}{\partial z}\mathbf{i} + \frac{\partial M}{\partial z}\mathbf{j} + \left(\frac{\partial N}{\partial x} - \frac{\partial M}{\partial y}\right)\mathbf{k}.$$

Consequently,

$$(\operatorname{curl} \mathbf{F}) \cdot \mathbf{k} = \left[-\frac{\partial N}{\partial z}\mathbf{i} + \frac{\partial M}{\partial z}\mathbf{j} + \left(\frac{\partial N}{\partial x} - \frac{\partial M}{\partial y}\right)\mathbf{k}\right] \cdot \mathbf{k} = \frac{\partial N}{\partial x} - \frac{\partial M}{\partial y}.$$

With appropriate conditions on $\mathbf{F}$, C, and R, you can write Green's Theorem in the vector form

$$\int_C \mathbf{F} \cdot d\mathbf{r} = \iint_R \left(\frac{\partial N}{\partial x} - \frac{\partial M}{\partial y}\right) dA$$

$$= \iint_R (\operatorname{curl} \mathbf{F}) \cdot \mathbf{k} \, dA. \qquad \text{First alternative form}$$

The extension of this vector form of Green's Theorem to surfaces in space produces **Stokes's Theorem,** discussed in Section 15.8.

For the second vector form of Green's Theorem, assume the same conditions for $\mathbf{F}$, C, and R. Using the arc length parameter s for C, you have $\mathbf{r}(s) = x(s)\mathbf{i} + y(s)\mathbf{j}$. So, a unit tangent vector $\mathbf{T}$ to curve C is given by $\mathbf{r}'(s) = \mathbf{T} = x'(s)\mathbf{i} + y'(s)\mathbf{j}$. From Figure 15.34, you can see that the *outward* unit normal vector $\mathbf{N}$ can then be written as

$$\mathbf{N} = y'(s)\mathbf{i} - x'(s)\mathbf{j}.$$

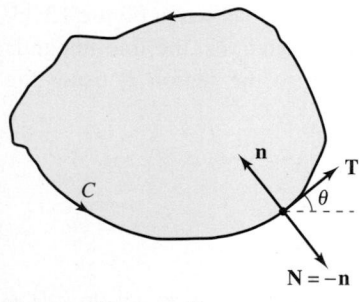

$\mathbf{T} = \cos\theta\mathbf{i} + \sin\theta\mathbf{j}$

$\mathbf{n} = \cos\left(\theta + \dfrac{\pi}{2}\right)\mathbf{i} + \sin\left(\theta + \dfrac{\pi}{2}\right)\mathbf{j}$

$\quad = -\sin\theta\mathbf{i} + \cos\theta\mathbf{j}$

$\mathbf{N} = \sin\theta\mathbf{i} - \cos\theta\mathbf{j}$

Figure 15.34

Consequently, for $\mathbf{F}(x, y) = M\mathbf{i} + N\mathbf{j}$, you can apply Green's Theorem to obtain

$$\int_C \mathbf{F} \cdot \mathbf{N} \, ds = \int_a^b (M\mathbf{i} + N\mathbf{j}) \cdot (y'(s)\mathbf{i} - x'(s)\mathbf{j}) \, ds$$

$$= \int_a^b \left(M\frac{dy}{ds} - N\frac{dx}{ds}\right) ds$$

$$= \int_C M \, dy - N \, dx$$

$$= \int_C -N \, dx + M \, dy$$

$$= \iint_R \left(\frac{\partial M}{\partial x} + \frac{\partial N}{\partial y}\right) dA \qquad \text{Green's Theorem}$$

$$= \iint_R \operatorname{div} \mathbf{F} \, dA.$$

Therefore,

$$\int_C \mathbf{F} \cdot \mathbf{N} \, ds = \iint_R \operatorname{div} \mathbf{F} \, dA. \qquad \text{Second alternative form}$$

The extension of this form to three dimensions is called the **Divergence Theorem** and will be discussed in Section 15.7. The physical interpretations of divergence and curl will be discussed in Sections 15.7 and 15.8.

15.4 Exercises

CONCEPT CHECK

1. **Writing** What does it mean for a curve to be simple? What does it mean for a plane region to be simply connected?

2. **Green's Theorem** Explain the usefulness of Green's Theorem.

3. **Integral Sign** What information do you learn from the integral sign $\oint_C$?

4. **Area** Describe how to find the area of a plane region bounded by a piecewise smooth simple closed curve that is oriented counterclockwise.

 **Verifying Green's Theorem** In Exercises **5–8, verify Green's Theorem by evaluating both integrals**

$$\int_C y^2\,dx + x^2\,dy = \int_R\int \left(\frac{\partial N}{\partial x} - \frac{\partial M}{\partial y}\right) dA$$

for the given path.

5. C: boundary of the region lying between the graphs of $y = x$ and $y = x^2$

6. C: boundary of the region lying between the graphs of $y = x$ and $y = \sqrt{x}$

7. C: square with vertices $(0, 0)$, $(1, 0)$, $(1, 1)$, and $(0, 1)$

8. C: rectangle with vertices $(0, 0)$, $(3, 0)$, $(3, 4)$, and $(0, 4)$

Verifying Green's Theorem In Exercises **9 and 10, verify Green's Theorem by using a computer algebra system to evaluate both integrals** $\int_C xe^y\,dx + e^x\,dy = \int_R\int \left(\frac{\partial N}{\partial x} - \frac{\partial M}{\partial y}\right) dA$ **for the given path.**

9. C: circle given by $x^2 + y^2 = 4$

10. C: boundary of the region lying between the graphs of $y = x$ and $y = x^3$ in the first quadrant

 **Evaluating a Line Integral Using Green's Theorem** In Exercises **11–14, use Green's Theorem to evaluate the line integral** $\int_C (y - x)\,dx + (2x - y)\,dy$ **for the given path.**

11. C: boundary of the region lying between the graphs of $y = x$ and $y = x^2 - 2x$

12. C: $x = 2\cos\theta$, $y = \sin\theta$

13. C: boundary of the region lying inside the rectangle with vertices $(5, 3)$, $(-5, 3)$, $(-5, -3)$, and $(5, -3)$, and outside the square with vertices $(1, 1)$, $(-1, 1)$, $(-1, -1)$, and $(1, -1)$

14. C: boundary of the region lying inside the semicircle $y = \sqrt{25 - x^2}$ and outside the semicircle $y = \sqrt{9 - x^2}$

 **Evaluating a Line Integral Using Green's Theorem** In Exercises **15–24, use Green's Theorem to evaluate the line integral.**

15. $\int_C 2xy\,dx + (x + y)\,dy$

    C: boundary of the region lying between the graphs of $y = 0$ and $y = 1 - x^2$

16. $\int_C y^2\,dx + xy\,dy$

    C: boundary of the region lying between the graphs of $y = 0$, $y = \sqrt{x}$, and $x = 9$

17. $\int_C (x^2 - y^2)\,dx + 2xy\,dy$

    C: $x^2 + y^2 = 16$

18. $\int_C (x^2 - y^2)\,dx + 2xy\,dy$

    C: $r = 1 + \cos\theta$

19. $\int_C e^x \cos 2y\,dx - 2e^x \sin 2y\,dy$

    C: $x^2 + y^2 = a^2$

20. $\int_C 2\arctan\frac{y}{x}\,dx + \ln(x^2 + y^2)\,dy$

    C: $x = 4 + 2\cos\theta$, $y = 4 + \sin\theta$

21. $\int_C \cos y\,dx + (xy - x\sin y)\,dy$

    C: boundary of the region lying between the graphs of $y = x$ and $y = \sqrt{x}$

22. $\int_C (e^{-x^2/2} - y)\,dx + (e^{-y^2/2} + x)\,dy$

    C: boundary of the region lying between the graphs of the circle $x = 6\cos\theta$, $y = 6\sin\theta$ and the ellipse $x = 3\cos\theta$, $y = 2\sin\theta$

23. $\int_C (x - 3y)\,dx + (x + y)\,dy$

    C: boundary of the region lying between the graphs of $x^2 + y^2 = 1$ and $x^2 + y^2 = 9$

24. $\int_C 3x^2 e^y\,dx + e^y\,dy$

    C: boundary of the region lying between the squares with vertices $(1, 1)$, $(-1, 1)$, $(-1, -1)$, and $(1, -1)$, and $(2, 2)$, $(-2, 2)$, $(-2, -2)$, and $(2, -2)$

 **Work** In Exercises **25–28, use Green's Theorem to calculate the work done by the force F on a particle that is moving counterclockwise around the closed path C.**

25. $\mathbf{F}(x, y) = xy\mathbf{i} + (x + y)\mathbf{j}$

    C: $x^2 + y^2 = 1$

26. $\mathbf{F}(x, y) = (e^x - 3y)\mathbf{i} + (e^y + 6x)\mathbf{j}$

   $C: r = 2 \cos \theta$

27. $\mathbf{F}(x, y) = (x^{3/2} - 3y)\mathbf{i} + (6x + 5\sqrt{y})\mathbf{j}$

   $C:$ triangle with vertices $(0, 0)$, $(5, 0)$, and $(0, 5)$

28. $\mathbf{F}(x, y) = (3x^2 + y)\mathbf{i} + 4xy^2\mathbf{j}$

   $C:$ boundary of the region lying between the graphs of $y = \sqrt{x}$, $y = 0$, and $x = 9$

 **Area**   In Exercises 29–32, use a line integral to find the area of the region R.

29. $R:$ region bounded by the graph of $x^2 + y^2 = 4$

30. $R:$ triangle bounded by the graphs of $x = 0$, $3x - 2y = 0$, and $x + 2y = 8$

31. $R:$ region bounded by the graphs of $y = 5x - 3$ and $y = x^2 + 1$

32. $R:$ region inside the loop of the folium of Descartes bounded by the graph of

$$x = \frac{3t}{t^3 + 1}, \quad y = \frac{3t^2}{t^3 + 1}$$

Using Green's Theorem to Verify a Formula   In Exercises 33 and 34, use Green's Theorem to verify the line integral formula(s).

33. The centroid of the region having area A bounded by the simple closed path C has coordinates

$$\bar{x} = \frac{1}{2A} \int_C x^2 \, dy \quad \text{and} \quad \bar{y} = -\frac{1}{2A} \int_C y^2 \, dx.$$

34. The area of a plane region bounded by the simple closed path C given in polar coordinates is

$$A = \frac{1}{2} \int_C r^2 \, d\theta.$$

Centroid   In Exercises 35–38, use the results of Exercise 33 to find the centroid of the region.

35. $R:$ region bounded by the graphs of $y = 0$ and $4 - x^2$

36. $R:$ region bounded by the graphs of $y = \sqrt{1 - x^2}$ and $y = 0$

37. $R:$ region bounded by the graphs of $y = x^3$ and $y = x$, $0 \le x \le 1$

38. $R:$ triangle with vertices $(-a, 0)$, $(a, 0)$, and (b, c), where $-a \le b \le a$

Area   In Exercises 39–42, use the result of Exercise 34 to find the area of the region bounded by the graph of the polar equation.

39. $r = 6(1 - \cos \theta)$

40. $r = a \cos 3\theta$

41. $r = 1 + 2 \cos \theta$ (inner loop)

42. $r = \dfrac{3}{2 - \cos \theta}$

43. Maximum Value

(a) Evaluate $\displaystyle\int_{C_1} y^3 \, dx + (27x - x^3) \, dy$, where C_1 is the unit circle given by $\mathbf{r}(t) = \cos t\mathbf{i} + \sin t\mathbf{j}$, for $0 \le t \le 2\pi$.

(b) Find the maximum value of $\displaystyle\int_C y^3 \, dx + (27x - x^3) \, dy$, where C is any circle centered at the origin in the xy-plane, oriented counterclockwise.

44.  **HOW DO YOU SEE IT?**   The figure shows a region R bounded by a piecewise smooth simple closed path C.

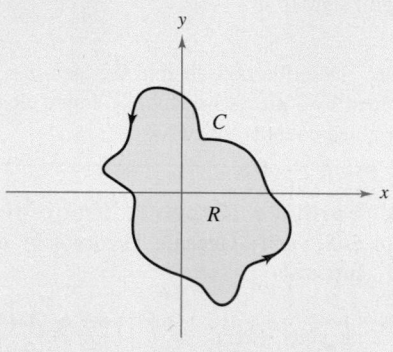

(a) Is R simply connected? Explain.

(b) Explain why $\displaystyle\int_C f(x) \, dx + g(y) \, dy = 0$, where f and g are differentiable functions.

45. Green's Theorem: Region with a Hole   Let R be the region inside the circle $x = 5 \cos \theta$, $y = 5 \sin \theta$ and outside the ellipse $x = 2 \cos \theta$, $y = \sin \theta$. Evaluate the line integral

$$\int_C (e^{-x^2/2} - y) \, dx + (e^{-y^2/2} + x) \, dy$$

where $C = C_1 + C_2$ is the boundary of R, as shown in the figure.

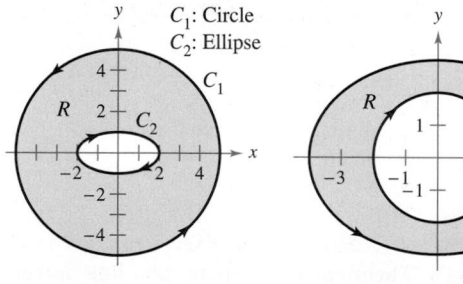

Figure for 45   Figure for 46

46. Green's Theorem: Region with a Hole   Let R be the region inside the ellipse $x = 4 \cos \theta$, $y = 3 \sin \theta$ and outside the circle $x = 2 \cos \theta$, $y = 2 \sin \theta$. Evaluate the line integral

$$\int_C (3x^2y + 1) \, dx + (x^3 + 4x) \, dy$$

where $C = C_1 + C_2$ is the boundary of R, as shown in the figure.

EXPLORING CONCEPTS

47. Think About It Let

$$I = \int_C \frac{y \, dx - x \, dy}{x^2 + y^2}$$

where C is a circle oriented counterclockwise.

(a) Show that $I = 0$ when C does not contain the origin.

(b) What is I when C does contain the origin?

48. Think About It For each given path, verify Green's Theorem by showing that

$$\int_C y^2 \, dx + x^2 \, dy = \int_R \int \left(\frac{\partial N}{\partial x} - \frac{\partial M}{\partial y} \right) dA.$$

For each path, which integral is easier to evaluate? Explain.

(a) C: triangle with vertices $(0, 0)$, $(4, 0)$, and $(4, 4)$

(b) C: circle given by $x^2 + y^2 = 1$

49. Proof

(a) Let C be the line segment joining (x_1, y_1) and (x_2, y_2). Show that $\int_C -y \, dx + x \, dy = x_1 y_2 - x_2 y_1$.

(b) Let (x_1, y_1), (x_2, y_2), . . . , (x_n, y_n) be the vertices of a polygon. Prove that the area enclosed is

$$\frac{1}{2}[(x_1 y_2 - x_2 y_1) + (x_2 y_3 - x_3 y_2) + \cdots +$$
$$(x_{n-1} y_n - x_n y_{n-1}) + (x_n y_1 - x_1 y_n)].$$

50. Area Use the result of Exercise 49(b) to find the area enclosed by the polygon with the given vertices.

(a) Pentagon: $(0, 0)$, $(2, 0)$, $(3, 2)$, $(1, 4)$, and $(-1, 1)$

(b) Hexagon: $(0, 0)$, $(2, 0)$, $(3, 2)$, $(2, 4)$, $(0, 3)$, and $(-1, 1)$

Proof In Exercises 51 and 52, prove the identity, where R is a simply connected region with piecewise smooth boundary C. Assume that the required partial derivatives of the scalar functions f and g are continuous. The expressions $D_N f$ and $D_N g$ are the derivatives in the direction of the outward normal vector N of C and are defined by $D_N f = \nabla f \cdot N$ and $D_N g = \nabla g \cdot N$.

51. Green's first identity:

$$\int_R \int (f \nabla^2 g + \nabla f \cdot \nabla g) \, dA = \int_C f D_N g \, ds$$

[*Hint:* Use the second alternative form of Green's Theorem and the property $\text{div}(f\mathbf{G}) = f \, \text{div} \, \mathbf{G} + \nabla f \cdot \mathbf{G}$.]

52. Green's second identity:

$$\int_R \int (f \nabla^2 g - g \nabla^2 f) \, dA = \int_C (f D_N g - g D_N f) \, ds$$

(*Hint:* Use Green's first identity from Exercise 51 twice.)

53. Proof Let $\mathbf{F} = M\mathbf{i} + N\mathbf{j}$, where M and N have continuous first partial derivatives in a simply connected region R. Prove that if C is simple, smooth, and closed, and $N_x = M_y$, then $\int_C \mathbf{F} \cdot d\mathbf{r} = 0$.

PUTNAM EXAM CHALLENGE

54. Find the least possible area of a convex set in the plane that intersects both branches of the hyperbola $xy = 1$ and both branches of the hyperbola $xy = -1$. (A set S in the plane is called *convex* if for any two points in S the line segment connecting them is contained in S.)

SECTION PROJECT

Hyperbolic and Trigonometric Functions

(a) Sketch the plane curve represented by the vector-valued function $\mathbf{r}(t) = \cosh t \mathbf{i} + \sinh t \mathbf{j}$ on the interval $0 \le t \le 5$. Show that the rectangular equation corresponding to $\mathbf{r}(t)$ is the hyperbola $x^2 - y^2 = 1$. Verify your sketch by using a graphing utility to graph the hyperbola.

(b) Let $P = (\cosh \phi, \sinh \phi)$ be the point on the hyperbola corresponding to $\mathbf{r}(\phi)$ for $\phi > 0$. Use the formula for area

$$A = \frac{1}{2} \int_C x \, dy - y \, dx$$

to verify that the area of the region shown in the figure is $\frac{1}{2}\phi$.

(c) Show that the area of the region shown in the figure is also given by the integral

$$A = \int_0^{\sinh \phi} \left[\sqrt{1 + y^2} - (\coth \phi) y \right] dy.$$

Confirm your answer in part (b) by evaluating this integral for $\phi = 1, 2, 4$, and 10.

(d) Consider the unit circle given by $x^2 + y^2 = 1$. Let θ be the angle formed by the x-axis and the radius to (x, y). The area of the corresponding sector is $\frac{1}{2}\theta$. That is, the trigonometric functions $f(\theta) = \cos \theta$ and $g(\theta) = \sin \theta$ could have been defined as the coordinates of the point $(\cos \theta, \sin \theta)$ on the unit circle that determines a sector of area $\frac{1}{2}\theta$. Write a short paragraph explaining how you could define the hyperbolic functions in a similar manner, using the "unit hyperbola" $x^2 - y^2 = 1$.

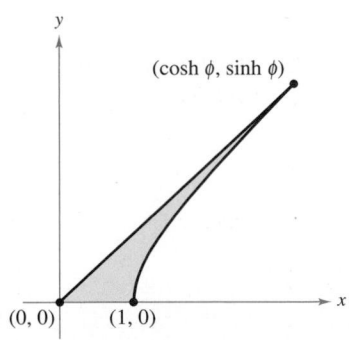

15.5 Parametric Surfaces

■ Understand the definition of a parametric surface, and sketch the surface.
■ Find a set of parametric equations to represent a surface.
■ Find a normal vector and a tangent plane to a parametric surface.
■ Find the area of a parametric surface.

Parametric Surfaces

You already know how to represent a curve in the plane or in space by a set of parametric equations—or, equivalently, by a vector-valued function.

$$\mathbf{r}(t) = x(t)\mathbf{i} + y(t)\mathbf{j} \qquad \text{Plane curve}$$

$$\mathbf{r}(t) = x(t)\mathbf{i} + y(t)\mathbf{j} + z(t)\mathbf{k} \qquad \text{Space curve}$$

In this section, you will learn how to represent a surface in space by a set of parametric equations—or by a vector-valued function. For curves, note that the vector-valued function $\mathbf{r}$ is a function of a *single* parameter t. For surfaces, the vector-valued function is a function of *two* parameters u and v.

Definition of Parametric Surface

Let x, y, and z be functions of u and v that are continuous on a domain D in the uv-plane. The set of points (x, y, z) given by

$$\mathbf{r}(u, v) = x(u, v)\mathbf{i} + y(u, v)\mathbf{j} + z(u, v)\mathbf{k} \qquad \text{Parametric surface}$$

is called a **parametric surface.** The equations

$$x = x(u, v), \quad y = y(u, v), \quad \text{and} \quad z = z(u, v) \qquad \text{Parametric equations}$$

are the **parametric equations** for the surface.

If S is a parametric surface given by the vector-valued function $\mathbf{r}$, then S is traced out by the position vector $\mathbf{r}(u, v)$ as the point (u, v) moves throughout the domain D, as shown in Figure 15.35.

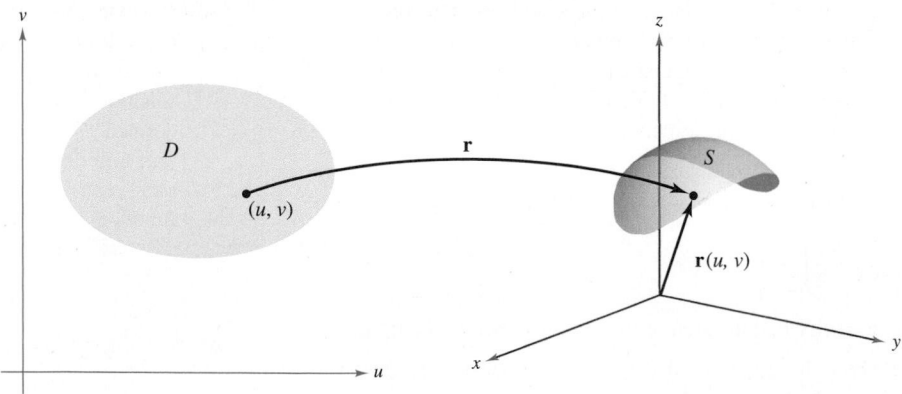

The parametric surface S given by the vector-valued function $\mathbf{r}$, where $\mathbf{r}$ is a function of two variables u and v defined on a domain D
Figure 15.35

▷ **TECHNOLOGY** Some computer algebra systems are capable of graphing surfaces that are represented parametrically. If you have access to such software, use it to graph some of the surfaces in the examples and exercises in this section.

EXAMPLE 1 **Sketching a Parametric Surface**

Identify and sketch the parametric surface S given by

$$\mathbf{r}(u, v) = 3 \cos u\mathbf{i} + 3 \sin u\mathbf{j} + v\mathbf{k}$$

where $0 \le u \le 2\pi$ and $0 \le v \le 4$.

Solution Because $x = 3 \cos u$ and $y = 3 \sin u$, you know that for each point (x, y, z) on the surface, x and y are related by the equation

$$x^2 + y^2 = 3^2.$$

In other words, each cross section of S taken parallel to the xy-plane is a circle of radius 3, centered on the z-axis. Because $z = v$, where

$$0 \le v \le 4$$

you can see that the surface is a right circular cylinder of height 4. The radius of the cylinder is 3, and the z-axis forms the axis of the cylinder, as shown in Figure 15.36.

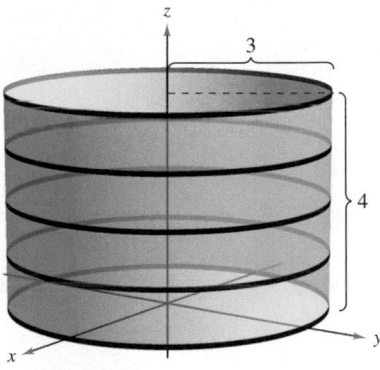

Figure 15.36

As with parametric representations of curves, parametric representations of surfaces are not unique. That is, there are many other sets of parametric equations that could be used to represent the surface shown in Figure 15.36.

EXAMPLE 2 **Sketching a Parametric Surface**

Identify and sketch the parametric surface S given by

$$\mathbf{r}(u, v) = \sin u \cos v\mathbf{i} + \sin u \sin v\mathbf{j} + \cos u\mathbf{k}$$

where $0 \le u \le \pi$ and $0 \le v \le 2\pi$.

Solution To identify the surface, you can try to use trigonometric identities to eliminate the parameters. After some experimentation, you can discover that

$$\begin{aligned}
x^2 + y^2 + z^2 &= (\sin u \cos v)^2 + (\sin u \sin v)^2 + (\cos u)^2 \\
&= \sin^2 u \cos^2 v + \sin^2 u \sin^2 v + \cos^2 u \\
&= (\sin^2 u)(\cos^2 v + \sin^2 v) + \cos^2 u \\
&= \sin^2 u + \cos^2 u \\
&= 1.
\end{aligned}$$

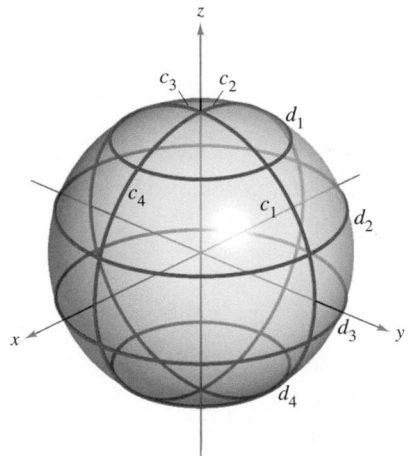

Figure 15.37

So, each point on S lies on the unit sphere, centered at the origin, as shown in Figure 15.37. For fixed $u = d_i$, $\mathbf{r}(u, v)$ traces out latitude circles

$$x^2 + y^2 = \sin^2 d_i, \quad 0 \le d_i \le \pi$$

that are parallel to the xy-plane, and for fixed $v = c_i$, $\mathbf{r}(u, v)$ traces out longitude (or meridian) half-circles.

To convince yourself further that $\mathbf{r}(u, v)$ traces out the entire unit sphere, recall that the parametric equations

$$x = \rho \sin \phi \cos \theta, \quad y = \rho \sin \phi \sin \theta, \quad \text{and} \quad z = \rho \cos \phi$$

where $0 \le \theta \le 2\pi$ and $0 \le \phi \le \pi$, describe the conversion from spherical to rectangular coordinates, as discussed in Section 11.7.

Finding Parametric Equations for Surfaces

In Examples 1 and 2, you were asked to identify the surface described by a given set of parametric equations. The reverse problem—that of writing a set of parametric equations for a given surface—is generally more difficult. One type of surface for which this problem is straightforward, however, is a surface that is given by $z = f(x, y)$. You can parametrize such a surface as

$$\mathbf{r}(x, y) = x\mathbf{i} + y\mathbf{j} + f(x, y)\mathbf{k}.$$

EXAMPLE 3 **Representing a Surface Parametrically**

Write a set of parametric equations for the cone given by

$$z = \sqrt{x^2 + y^2}$$

as shown in Figure 15.38.

Solution Because this surface is given in the form $z = f(x, y)$, you can let x and y be the parameters. Then the cone is represented by the vector-valued function

$$\mathbf{r}(x, y) = x\mathbf{i} + y\mathbf{j} + \sqrt{x^2 + y^2}\mathbf{k}$$

where (x, y) varies over the entire xy-plane.

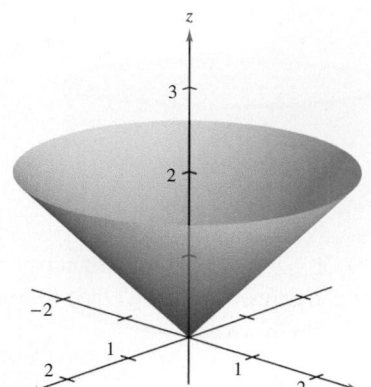

Figure 15.38

A second type of surface that is easily represented parametrically is a surface of revolution. For instance, to represent the surface formed by revolving the graph of

$$y = f(x), \quad a \le x \le b$$

about the x-axis, use

$$x = u, \quad y = f(u) \cos v, \quad \text{and} \quad z = f(u) \sin v$$

where $a \le u \le b$ and $0 \le v \le 2\pi$.

EXAMPLE 4 **Representing a Surface of Revolution Parametrically**

⋮ ⋯▷ *See LarsonCalculus.com for an interactive version of this type of example.*

Write a set of parametric equations for the surface of revolution obtained by revolving

$$f(x) = \frac{1}{x}, \quad 1 \le x \le 10$$

about the x-axis.

Solution Use the parameters u and v as described above to write

$$x = u, \quad y = f(u) \cos v = \frac{1}{u} \cos v, \quad \text{and} \quad z = f(u) \sin v = \frac{1}{u} \sin v$$

where

$$1 \le u \le 10 \quad \text{and} \quad 0 \le v \le 2\pi.$$

The resulting surface is a portion of *Gabriel's Horn,* as shown in Figure 15.39.

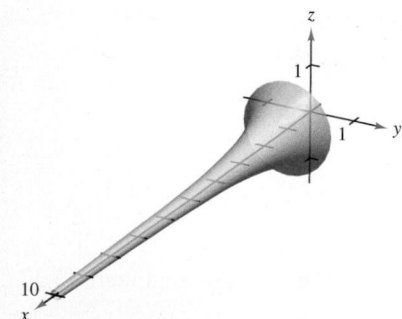

Figure 15.39

The surface of revolution in Example 4 is formed by revolving the graph of $y = f(x)$ about the x-axis. For other types of surfaces of revolution, a similar parametrization can be used. For instance, to parametrize the surface formed by revolving the graph of $x = f(z)$ about the z-axis, you can use

$$z = u, \quad x = f(u) \cos v, \quad \text{and} \quad y = f(u) \sin v.$$

Normal Vectors and Tangent Planes

Let S be a parametric surface given by

$$\mathbf{r}(u, v) = x(u, v)\mathbf{i} + y(u, v)\mathbf{j} + z(u, v)\mathbf{k}$$

over an open region D such that x, y, and z have continuous partial derivatives on D. The **partial derivatives of r** with respect to u and v are defined as

$$\mathbf{r}_u = \frac{\partial x}{\partial u}(u, v)\mathbf{i} + \frac{\partial y}{\partial u}(u, v)\mathbf{j} + \frac{\partial z}{\partial u}(u, v)\mathbf{k}$$

and

$$\mathbf{r}_v = \frac{\partial x}{\partial v}(u, v)\mathbf{i} + \frac{\partial y}{\partial v}(u, v)\mathbf{j} + \frac{\partial z}{\partial v}(u, v)\mathbf{k}.$$

Each of these partial derivatives is a vector-valued function that can be interpreted geometrically in terms of tangent vectors. For instance, if $v = v_0$ is held constant, then $\mathbf{r}(u, v_0)$ is a vector-valued function of a single parameter and defines a curve C_1 that lies on the surface S. The tangent vector to C_1 at the point

$$(x(u_0, v_0), y(u_0, v_0), z(u_0, v_0))$$

is given by

$$\mathbf{r}_u(u_0, v_0) = \frac{\partial x}{\partial u}(u_0, v_0)\mathbf{i} + \frac{\partial y}{\partial u}(u_0, v_0)\mathbf{j} + \frac{\partial z}{\partial u}(u_0, v_0)\mathbf{k}$$

as shown in Figure 15.40. In a similar way, if $u = u_0$ is held constant, then $\mathbf{r}(u_0, v)$ is a vector-valued function of a single parameter and defines a curve C_2 that lies on the surface S. The tangent vector to C_2 at the point $(x(u_0, v_0), y(u_0, v_0), z(u_0, v_0))$ is given by

$$\mathbf{r}_v(u_0, v_0) = \frac{\partial x}{\partial v}(u_0, v_0)\mathbf{i} + \frac{\partial y}{\partial v}(u_0, v_0)\mathbf{j} + \frac{\partial z}{\partial v}(u_0, v_0)\mathbf{k}.$$

If the normal vector $\mathbf{r}_u \times \mathbf{r}_v$ is not $\mathbf{0}$ for any (u, v) in D, then the surface S is called **smooth** and will have a tangent plane. Informally, a smooth surface is one that has no sharp points or cusps. For instance, spheres, ellipsoids, and paraboloids are smooth, whereas the cone given in Example 3 is not smooth.

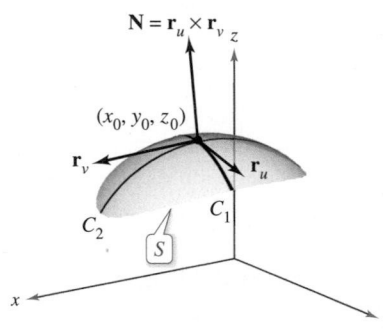

Figure 15.40

Normal Vector to a Smooth Parametric Surface

Let S be a smooth parametric surface

$$\mathbf{r}(u, v) = x(u, v)\mathbf{i} + y(u, v)\mathbf{j} + z(u, v)\mathbf{k}$$

defined over an open region D in the uv-plane. Let (u_0, v_0) be a point in D. A normal vector at the point

$$(x_0, y_0, z_0) = (x(u_0, v_0), y(u_0, v_0), z(u_0, v_0))$$

is given by

$$\mathbf{N} = \mathbf{r}_u(u_0, v_0) \times \mathbf{r}_v(u_0, v_0) = \begin{vmatrix} \mathbf{i} & \mathbf{j} & \mathbf{k} \\ \dfrac{\partial x}{\partial u} & \dfrac{\partial y}{\partial u} & \dfrac{\partial z}{\partial u} \\ \dfrac{\partial x}{\partial v} & \dfrac{\partial y}{\partial v} & \dfrac{\partial z}{\partial v} \end{vmatrix}.$$

Figure 15.40 shows the normal vector $\mathbf{r}_u \times \mathbf{r}_v$. The vector $\mathbf{r}_v \times \mathbf{r}_u$ is also normal to S and points in the opposite direction.

> **EXAMPLE 5** **Finding a Tangent Plane to a Parametric Surface**
>
> Find an equation of the tangent plane to the paraboloid
>
> $$\mathbf{r}(u, v) = u\mathbf{i} + v\mathbf{j} + (u^2 + v^2)\mathbf{k}$$
>
> at the point $(1, 2, 5)$.
>
> **Solution** The point in the uv-plane that is mapped to the point $(x, y, z) = (1, 2, 5)$ is $(u, v) = (1, 2)$. The partial derivatives of $\mathbf{r}$ are
>
> $$\mathbf{r}_u = \mathbf{i} + 2u\mathbf{k} \quad \text{and} \quad \mathbf{r}_v = \mathbf{j} + 2v\mathbf{k}.$$
>
> The normal vector is given by
>
> $$\mathbf{r}_u \times \mathbf{r}_v = \begin{vmatrix} \mathbf{i} & \mathbf{j} & \mathbf{k} \\ 1 & 0 & 2u \\ 0 & 1 & 2v \end{vmatrix} = -2u\mathbf{i} - 2v\mathbf{j} + \mathbf{k}$$
>
> which implies that the normal vector at $(1, 2, 5)$ is
>
> $$\mathbf{r}_u \times \mathbf{r}_v = -2\mathbf{i} - 4\mathbf{j} + \mathbf{k}.$$
>
> So, an equation of the tangent plane at $(1, 2, 5)$ is
>
> $$-2(x - 1) - 4(y - 2) + (z - 5) = 0$$
> $$-2x - 4y + z = -5.$$
>
> The tangent plane is shown in Figure 15.41.

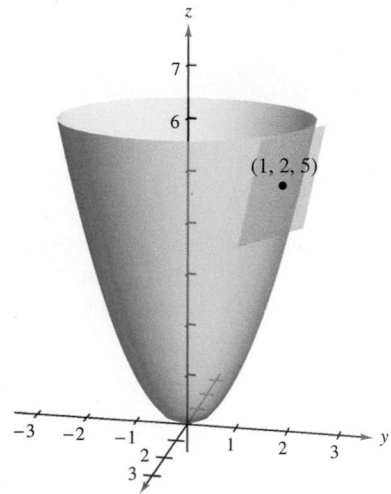

Figure 15.41

Area of a Parametric Surface

To define the area of a parametric surface, you can use a development that is similar to that given in Section 14.5. Begin by constructing an inner partition of D consisting of n rectangles, where the area of the ith rectangle D_i is $\Delta A_i = \Delta u_i \Delta v_i$, as shown in Figure 15.42. In each D_i, let (u_i, v_i) be the point that is closest to the origin. At the point $(x_i, y_i, z_i) = (x(u_i, v_i), y(u_i, v_i), z(u_i, v_i))$ on the surface S, construct a tangent plane T_i. The area of the portion of S that corresponds to D_i, ΔS_i, can be approximated by a parallelogram ΔT_i in the tangent plane. That is, $\Delta T_i \approx \Delta S_i$. So, the surface area of S is given by $\sum \Delta S_i \approx \sum \Delta T_i$. The area of the parallelogram in the tangent plane is

$$\text{Area of } \Delta T_i = \|\Delta u_i \mathbf{r}_u \times \Delta v_i \mathbf{r}_v\| = \|\mathbf{r}_u \times \mathbf{r}_v\| \Delta u_i \Delta v_i$$

which leads to the next definition.

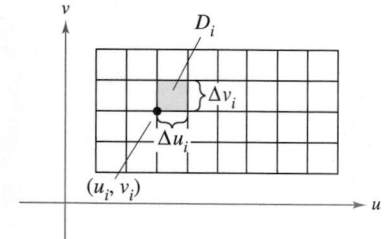

Figure 15.42

> ### Area of a Parametric Surface
>
> Let S be a smooth parametric surface
>
> $$\mathbf{r}(u, v) = x(u, v)\mathbf{i} + y(u, v)\mathbf{j} + z(u, v)\mathbf{k}$$
>
> defined over an open region D in the uv-plane. If each point on the surface S corresponds to exactly one point in the domain D, then the **surface area** of S is given by
>
> $$\text{Surface area} = \iint_S dS = \iint_D \|\mathbf{r}_u \times \mathbf{r}_v\| \, dA$$
>
> where
>
> $$\mathbf{r}_u = \frac{\partial x}{\partial u}\mathbf{i} + \frac{\partial y}{\partial u}\mathbf{j} + \frac{\partial z}{\partial u}\mathbf{k} \quad \text{and} \quad \mathbf{r}_v = \frac{\partial x}{\partial v}\mathbf{i} + \frac{\partial y}{\partial v}\mathbf{j} + \frac{\partial z}{\partial v}\mathbf{k}.$$

For a surface S given by $z = f(x, y)$, this formula for surface area corresponds to that given in Section 14.5. To see this, you can parametrize the surface using the vector-valued function $\mathbf{r}(x, y) = x\mathbf{i} + y\mathbf{j} + f(x, y)\mathbf{k}$ defined over the region R in the xy-plane. Using $\mathbf{r}_x = \mathbf{i} + f_x(x, y)\mathbf{k}$ and $\mathbf{r}_y = \mathbf{j} + f_y(x, y)\mathbf{k}$, you have

$$\mathbf{r}_x \times \mathbf{r}_y = \begin{vmatrix} \mathbf{i} & \mathbf{j} & \mathbf{k} \\ 1 & 0 & f_x(x, y) \\ 0 & 1 & f_y(x, y) \end{vmatrix} = -f_x(x, y)\mathbf{i} - f_y(x, y)\mathbf{j} + \mathbf{k}$$

and

$$\|\mathbf{r}_x \times \mathbf{r}_y\| = \sqrt{[f_x(x, y)]^2 + [f_y(x, y)]^2 + 1}.$$

This implies that the surface area of S is

$$\text{Surface area} = \int\int_R \|\mathbf{r}_x \times \mathbf{r}_y\| \, dA$$
$$= \int\int_R \sqrt{1 + [f_x(x, y)]^2 + [f_y(x, y)]^2} \, dA.$$

EXAMPLE 6 Finding Surface Area

Find the surface area of the unit sphere

$$\mathbf{r}(u, v) = \sin u \cos v\mathbf{i} + \sin u \sin v\mathbf{j} + \cos u\mathbf{k}$$

where the domain D is $0 \le u \le \pi$ and $0 \le v \le 2\pi$.

Solution Begin by calculating $\mathbf{r}_u$ and $\mathbf{r}_v$.

$$\mathbf{r}_u = \cos u \cos v\mathbf{i} + \cos u \sin v\mathbf{j} - \sin u\mathbf{k}$$
$$\mathbf{r}_v = -\sin u \sin v\mathbf{i} + \sin u \cos v\mathbf{j}$$

The cross product of these two vectors is

$$\mathbf{r}_u \times \mathbf{r}_v = \begin{vmatrix} \mathbf{i} & \mathbf{j} & \mathbf{k} \\ \cos u \cos v & \cos u \sin v & -\sin u \\ -\sin u \sin v & \sin u \cos v & 0 \end{vmatrix}$$
$$= \sin^2 u \cos v\mathbf{i} + \sin^2 u \sin v\mathbf{j} + \sin u \cos u\mathbf{k}$$

which implies that

$$\|\mathbf{r}_u \times \mathbf{r}_v\| = \sqrt{(\sin^2 u \cos v)^2 + (\sin^2 u \sin v)^2 + (\sin u \cos u)^2}$$
$$= \sqrt{\sin^4 u + \sin^2 u \cos^2 u}$$
$$= \sqrt{\sin^2 u}$$
$$= \sin u. \qquad \sin u > 0 \text{ for } 0 \le u \le \pi$$

Because of high surface gravity, the shape of a neutron star is almost a perfect sphere. Using the surface area along with other data, scientists can estimate the mass and radius of the star.

Finally, the surface area of the sphere is

$$A = \int\int_D \|\mathbf{r}_u \times \mathbf{r}_v\| \, dA$$
$$= \int_0^{2\pi} \int_0^\pi \sin u \, du \, dv$$
$$= \int_0^{2\pi} 2 \, dv$$
$$= 4\pi.$$

The surface in Example 6 does not quite fulfill the hypothesis that each point on the surface corresponds to exactly one point in D. For this surface, $\mathbf{r}(u, 0) = \mathbf{r}(u, 2\pi)$ for any fixed value of u. However, because the overlap consists of only a semicircle (which has no area), you can still apply the formula for the area of a parametric surface.

EXAMPLE 7 Finding Surface Area

Find the surface area of the torus given by

$$\mathbf{r}(u, v) = (2 + \cos u) \cos v\mathbf{i} + (2 + \cos u) \sin v\mathbf{j} + \sin u\mathbf{k}$$

where the domain D is given by $0 \leq u \leq 2\pi$ and $0 \leq v \leq 2\pi$. (See Figure 15.43.)

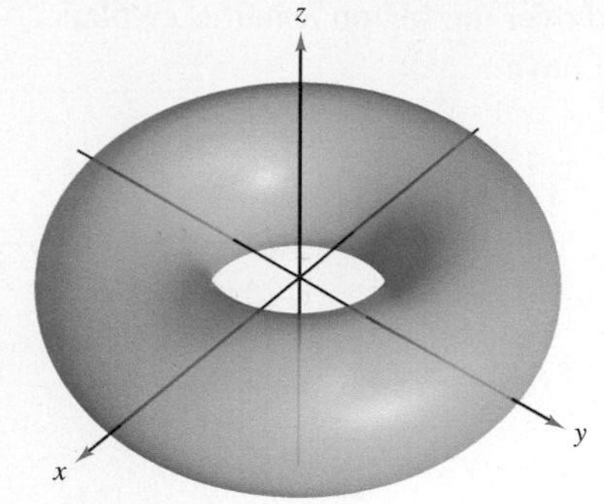

Figure 15.43

Solution Begin by calculating $\mathbf{r}_u$ and $\mathbf{r}_v$.

$$\mathbf{r}_u = -\sin u \cos v\mathbf{i} - \sin u \sin v\mathbf{j} + \cos u\mathbf{k}$$
$$\mathbf{r}_v = -(2 + \cos u) \sin v\mathbf{i} + (2 + \cos u) \cos v\mathbf{j}$$

The cross product of these two vectors is

$$\begin{aligned}\mathbf{r}_u \times \mathbf{r}_v &= \begin{vmatrix} \mathbf{i} & \mathbf{j} & \mathbf{k} \\ -\sin u \cos v & -\sin u \sin v & \cos u \\ -(2 + \cos u) \sin v & (2 + \cos u) \cos v & 0 \end{vmatrix} \\ &= -(2 + \cos u)(\cos v \cos u\mathbf{i} + \sin v \cos u\mathbf{j} + \sin u\mathbf{k})\end{aligned}$$

which implies that

$$\begin{aligned}\|\mathbf{r}_u \times \mathbf{r}_v\| &= (2 + \cos u)\sqrt{(\cos v \cos u)^2 + (\sin v \cos u)^2 + \sin^2 u} \\ &= (2 + \cos u)\sqrt{\cos^2 u(\cos^2 v + \sin^2 v) + \sin^2 u} \\ &= (2 + \cos u)\sqrt{\cos^2 u + \sin^2 u} \\ &= 2 + \cos u.\end{aligned}$$

Finally, the surface area of the torus is

$$\begin{aligned}A &= \iint_D \|\mathbf{r}_u \times \mathbf{r}_v\|\, dA \\ &= \int_0^{2\pi}\int_0^{2\pi} (2 + \cos u)\, du\, dv \\ &= \int_0^{2\pi} 4\pi\, dv \\ &= 8\pi^2.\end{aligned}$$

> **Exploration**
>
> For the torus in Example 7, describe the function $\mathbf{r}(u, v)$ for fixed u. Then describe the function $\mathbf{r}(u, v)$ for fixed v.

For a surface of revolution, you can show that the formula for surface area given in Section 7.4 is equivalent to the formula given in this section. For instance, suppose f is a nonnegative function such that f' is continuous over the interval $[a, b]$. Let S be the surface of revolution formed by revolving the graph of f, where $a \leq x \leq b$, about the x-axis. From Section 7.4, you know that the surface area is given by

$$\text{Surface area} = 2\pi \int_a^b f(x)\sqrt{1 + [f'(x)]^2}\, dx.$$

To represent S parametrically, let

$$x = u, \quad y = f(u) \cos v, \quad \text{and} \quad z = f(u) \sin v$$

where $a \leq u \leq b$ and $0 \leq v \leq 2\pi$. Then

$$\mathbf{r}(u, v) - u\mathbf{i} + f(u) \cos v\mathbf{j} + f(u) \sin v\mathbf{k}.$$

Try showing that the formula

$$\text{Surface area} = \iint_D \|\mathbf{r}_u \times \mathbf{r}_v\|\, dA$$

is equivalent to the formula given above (see Exercise 56).

15.5 Exercises

See **CalcChat.com** for tutorial help and worked-out solutions to odd-numbered exercises.

CONCEPT CHECK

1. **Parametric Surface** Explain how a parametric surface is represented by a vector-valued function and how the vector-valued function is used to sketch the parametric surface.

2. **Surface Area** A surface S is represented by $z = f(x, y)$. What are the parametric equations for S?

Matching In Exercises 3–8, match the vector-valued function with its graph. [The graphs are labeled (a), (b), (c), (d), (e), and (f).]

(a)
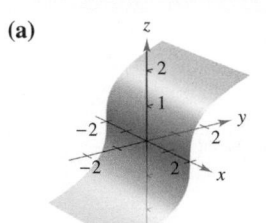

(b)
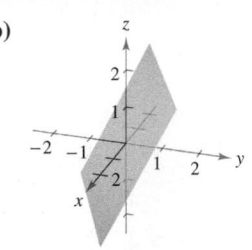

(c)
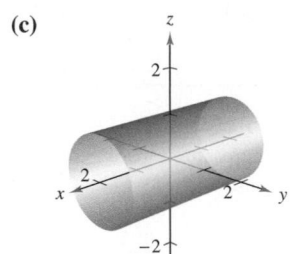

(d)
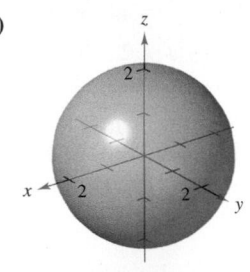

(e)
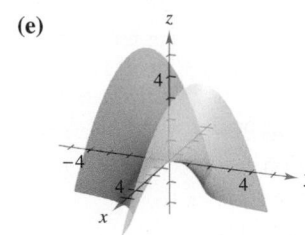

(f)
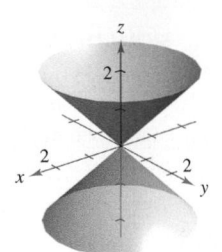

3. $\mathbf{r}(u, v) = u\mathbf{i} + v\mathbf{j} + uv\mathbf{k}$

4. $\mathbf{r}(u, v) = u \cos v\mathbf{i} + u \sin v\mathbf{j} + u\mathbf{k}$

5. $\mathbf{r}(u, v) = u\mathbf{i} + \frac{1}{2}(u + v)\mathbf{j} + v\mathbf{k}$

6. $\mathbf{r}(u, v) = v\mathbf{i} + \cos u\mathbf{j} + \sin u\mathbf{k}$

7. $\mathbf{r}(u, v) = 2 \cos v \cos u\mathbf{i} + 2 \cos v \sin u\mathbf{j} + 2 \sin v\mathbf{k}$

8. $\mathbf{r}(u, v) = u\mathbf{i} + \frac{1}{4}v^3\mathbf{j} + v\mathbf{k}$

 **Sketching a Parametric Surface** In Exercises 9–12, find the rectangular equation for the surface by eliminating the parameters from the vector-valued function. Identify the surface and sketch its graph.

9. $\mathbf{r}(u, v) = u\mathbf{i} + v\mathbf{j} + \frac{v}{2}\mathbf{k}$

10. $\mathbf{r}(u, v) = 2u \cos v\mathbf{i} + 2u \sin v\mathbf{j} + \frac{1}{2}u^2\mathbf{k}$

11. $\mathbf{r}(u, v) = 2 \cos u\mathbf{i} + v\mathbf{j} + 2 \sin u\mathbf{k}$

12. $\mathbf{r}(u, v) = 3 \cos v \cos u\mathbf{i} + 3 \cos v \sin u\mathbf{j} + 5 \sin v\mathbf{k}$

 **Graphing a Parametric Surface** In Exercises 13–16, use a computer algebra system to graph the surface represented by the vector-valued function.

13. $\mathbf{r}(u, v) = 2u \cos v\mathbf{i} + 2u \sin v\mathbf{j} + u^4\mathbf{k}$

$0 \le u \le 1, \quad 0 \le v \le 2\pi$

14. $\mathbf{r}(u, v) = 2u \cos v\mathbf{i} + 2u \sin v\mathbf{j} + v\mathbf{k}$

$0 \le u \le 1, \quad 0 \le v \le 3\pi$

15. $\mathbf{r}(u, v) = (u - \sin u) \cos v\mathbf{i} + (1 - \cos u) \sin v\mathbf{j} + u\mathbf{k}$

$0 \le u \le \pi, \quad 0 \le v \le 2\pi$

16. $\mathbf{r}(u, v) = \cos^3 u \cos v\mathbf{i} + \sin^3 u \sin v\mathbf{j} + u\mathbf{k}$

$0 \le u \le \frac{\pi}{2}, \quad 0 \le v \le 2\pi$

 **Representing a Surface Parametrically** In Exercises 17–26, find a vector-valued function whose graph is the indicated surface.

17. The plane $z = 3y$

18. The plane $x + y + z = 6$

19. The cone $y = \sqrt{4x^2 + 9z^2}$

20. The cone $x = \sqrt{16y^2 + z^2}$

21. The cylinder $x^2 + y^2 = 25$

22. The cylinder $4x^2 + y^2 = 16$

23. The paraboloid $x = y^2 + z^2 + 7$

24. The ellipsoid $\frac{x^2}{9} + \frac{y^2}{4} + \frac{z^2}{1} = 1$

25. The part of the plane $z = 4$ that lies inside the cylinder $x^2 + y^2 = 9$

26. The part of the paraboloid $z = x^2 + y^2$ that lies inside the cylinder $x^2 + y^2 = 9$

Representing a Surface of Revolution Parametrically In Exercises 27–32, write a set of parametric equations for the surface of revolution obtained by revolving the graph of the function about the given axis.

Function	Axis of Revolution
27. $y = \dfrac{x}{2}, \quad 0 \le x \le 6$	x-axis
28. $y = \sqrt{x}, \quad 0 \le x \le 4$	x-axis
29. $x = \sin z, \quad 0 \le z \le \pi$	z-axis
30. $x = z - 2, \quad 2 \le z \le 5$	z-axis
31. $z = \cos^2 y, \quad \dfrac{\pi}{2} \le y \le \pi$	y-axis
32. $z = y^2 + 1, \quad 0 \le y \le 2$	y-axis

Finding a Tangent Plane In Exercises 33–36, find an equation of the tangent plane to the surface represented by the vector-valued function at the given point.

33. $\mathbf{r}(u, v) = 3 \cos v \cos u \mathbf{i} + 2 \cos v \sin u \mathbf{j} + 4 \sin v \mathbf{k}$, $\left(0, \sqrt{3}, 2\right)$

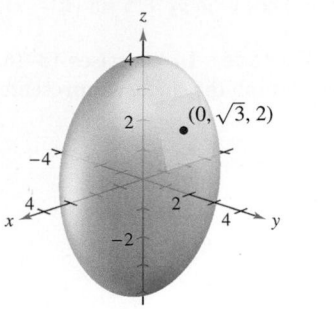

Figure for 33   Figure for 34

34. $\mathbf{r}(u, v) = u \mathbf{i} + v \mathbf{j} + \sqrt{uv} \mathbf{k}$, $(1, 1, 1)$

35. $\mathbf{r}(u, v) = 2u \cos v \mathbf{i} + 3u \sin v \mathbf{j} + u^2 \mathbf{k}$, $(0, 6, 4)$

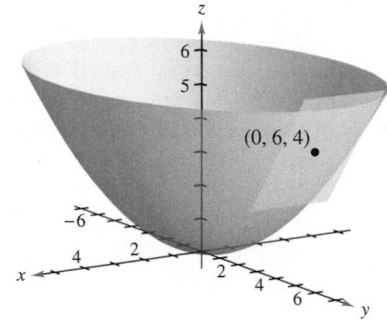

36. $\mathbf{r}(u, v) = 2u \cosh v \mathbf{i} + 2u \sinh v \mathbf{j} + \frac{1}{2} u^2 \mathbf{k}$, $(-4, 0, 2)$

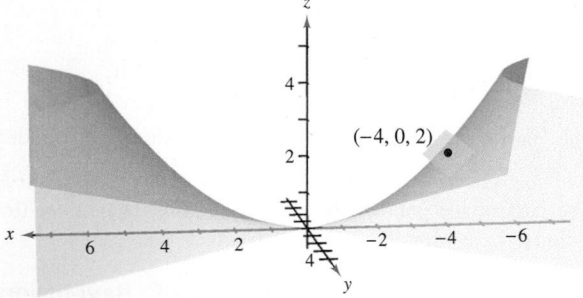

Finding Surface Area In Exercises 37–42, find the area of the surface over the given region. Use a computer algebra system to verify your results.

37. $\mathbf{r}(u, v) = 4u \mathbf{i} - v \mathbf{j} + v \mathbf{k}$, $0 \le u \le 2$, $0 \le v \le 1$

38. $\mathbf{r}(u, v) = 2u \cos v \mathbf{i} + 2u \sin v \mathbf{j} + u^2 \mathbf{k}$, $0 \le u \le 2$, $0 \le v \le 2\pi$

39. $\mathbf{r}(u, v) = au \cos v \mathbf{i} + au \sin v \mathbf{j} + u \mathbf{k}$, $0 \le u \le b$, $0 \le v \le 2\pi$

40. $\mathbf{r}(u, v) = (a + b \cos v) \cos u \mathbf{i} + (a + b \cos v) \sin u \mathbf{j} + b \sin v \mathbf{k}$, $a > b$, $0 \le u \le 2\pi$, $0 \le v \le 2\pi$

41. $\mathbf{r}(u, v) = \sqrt{u} \cos v \mathbf{i} + \sqrt{u} \sin v \mathbf{j} + u \mathbf{k}$, $0 \le u \le 4$, $0 \le v \le 2\pi$

42. $\mathbf{r}(u, v) = \sin u \cos v \mathbf{i} + u \mathbf{j} + \sin u \sin v \mathbf{k}$, $0 \le u \le \pi$, $0 \le v \le 2\pi$

EXPLORING CONCEPTS

Think About It In Exercises 43–46, determine how the graph of the surface $s(u, v)$ differs from the graph of $\mathbf{r}(u, v) = u \cos v \mathbf{i} + u \sin v \mathbf{j} + u^2 \mathbf{k}$, where $0 \le u \le 2$ and $0 \le v \le 2\pi$, as shown in the figure. (It is not necessary to graph s.)

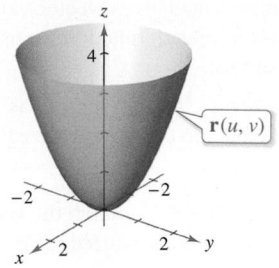

43. $s(u, v) = u \cos v \mathbf{i} + u \sin v \mathbf{j} - u^2 \mathbf{k}$
$0 \le u \le 2$, $0 \le v \le 2\pi$

44. $s(u, v) = u \cos v \mathbf{i} + u^2 \mathbf{j} + u \sin v \mathbf{k}$
$0 \le u \le 2$, $0 \le v \le 2\pi$

45. $s(u, v) = u \cos v \mathbf{i} + u \sin v \mathbf{j} + u^2 \mathbf{k}$
$0 \le u \le 3$, $0 \le v \le 2\pi$

46. $s(u, v) = 4u \cos v \mathbf{i} + 4u \sin v \mathbf{j} + u^2 \mathbf{k}$
$0 \le u \le 2$, $0 \le v \le 2\pi$

47. Representing a Cone Parametrically Show that the cone in Example 3 can be represented parametrically by $\mathbf{r}(u, v) = u \cos v \mathbf{i} + u \sin v \mathbf{j} + u \mathbf{k}$, where $u \ge 0$ and $0 \le v \le 2\pi$.

48. **HOW DO YOU SEE IT?** The figures below are graphs of $\mathbf{r}(u, v) = u \mathbf{i} + \sin u \cos v \mathbf{j} + \sin u \sin v \mathbf{k}$, where $0 \le u \le \pi/2$ and $0 \le v \le 2\pi$. Match each of the four graphs with the point in space from which the surface is viewed.

(a)   (b)

(c)   (d)

(i) $(10, 0, 0)$ (ii) $(-10, 10, 0)$

(iii) $(0, 10, 0)$ (iv) $(10, 10, 10)$

49. Astroidal Sphere An equation of an **astroidal sphere** in x, y, and z is

$$x^{2/3} + y^{2/3} + z^{2/3} = a^{2/3}.$$

A graph of an astroidal sphere is shown below. Show that this surface can be represented parametrically by

$$\mathbf{r}(u, v) = a \sin^3 u \cos^3 v\mathbf{i} + a \sin^3 u \sin^3 v\mathbf{j} + a \cos^3 u\mathbf{k}$$

where $0 \le u \le \pi$ and $0 \le v \le 2\pi$.

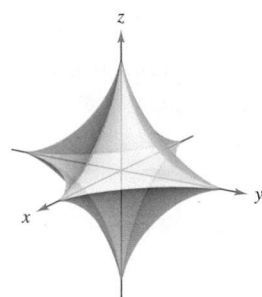

50. Different Views of a Surface Use a computer algebra system to graph the vector-valued function

$$\mathbf{r}(u, v) = u \cos v\mathbf{i} + u \sin v\mathbf{j} + v\mathbf{k}, \ 0 \le u \le \pi, \ 0 \le v \le \pi$$

from each of the points $(10, 0, 0)$, $(0, 0, 10)$, and $(10, 10, 10)$.

51. Investigation Use a computer algebra system to graph the torus

$$\mathbf{r}(u, v) = (a + b \cos v) \cos u\mathbf{i} + (a + b \cos v) \sin u\mathbf{j} + b \sin v\mathbf{k}$$

for each set of values of a and b, where $0 \le u \le 2\pi$ and $0 \le v \le 2\pi$. Use the results to describe the effects of a and b on the shape of the torus.

(a) $a = 4$, $b = 1$

(b) $a = 4$, $b = 2$

(c) $a = 8$, $b = 1$

(d) $a = 8$, $b = 3$

52. Investigation Consider the function in Exercise 14.

(a) Sketch a graph of the function where u is held constant at $u = 1$. Identify the graph.

(b) Sketch a graph of the function where v is held constant at $v = 2\pi/3$. Identify the graph.

(c) Assume that a surface is represented by the vector-valued function $\mathbf{r} = \mathbf{r}(u, v)$. What generalization can you make about the graph of the function when one of the parameters is held constant?

53. Surface Area The surface of the dome on a new museum is given by

$$\mathbf{r}(u, v) = 20 \sin u \cos v\mathbf{i} + 20 \sin u \sin v\mathbf{j} + 20 \cos u\mathbf{k}$$

where $0 \le u \le \pi/3$, $0 \le v \le 2\pi$, and $\mathbf{r}$ is in meters. Find the surface area of the dome.

54. Hyperboloid Find a vector-valued function for the hyperboloid

$$x^2 + y^2 - z^2 = 1$$

and determine the tangent plane at $(1, 0, 0)$.

55. Area Use a computer algebra system to graph one turn of the spiral ramp $\mathbf{r}(u, v) = u \cos v\mathbf{i} + u \sin v\mathbf{j} + 2v\mathbf{k}$, where $0 \le u \le 3$ and $0 \le v \le 2\pi$. Then analytically find the area of one turn of the spiral ramp.

56. Surface Area Let f be a nonnegative function such that f' is continuous over the interval $[a, b]$. Let S be the surface of revolution formed by revolving the graph of f, where $a \le x \le b$, about the x-axis. Let $x = u$, $y = f(u) \cos v$, and $z = f(u) \sin v$, where $a \le u \le b$ and $0 \le v \le 2\pi$. Then S is represented parametrically by

$$\mathbf{r}(u, v) = u\mathbf{i} + f(u) \cos v\mathbf{j} + f(u) \sin v\mathbf{k}.$$

Show that the following formulas are equivalent.

$$\text{Surface area} = 2\pi \int_a^b f(x) \sqrt{1 + [f'(x)]^2} \, dx$$

$$\text{Surface area} = \int\int_D \|\mathbf{r}_u \times \mathbf{r}_v\| \, dA$$

57. Open-Ended Project The parametric equations

$$x = 3 + [7 - \cos(3u - 2v) - 2\cos(3u + v)]\sin u$$

$$y = 3 + [7 - \cos(3u - 2v) - 2\cos(3u + v)]\cos u$$

$$z = \sin(3u - 2v) + 2\sin(3u + v)$$

where $-\pi \le u \le \pi$ and $-\pi \le v \le \pi$, represent the surface shown below. Try to create your own parametric surface using a computer algebra system.

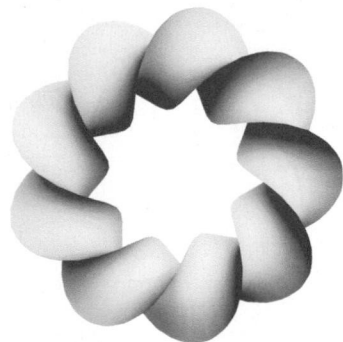

58. Möbius Strip The surface shown in the figure is called a **Möbius strip** and can be represented by the parametric equations

$$x = \left(a + u \cos\frac{v}{2}\right) \cos v, \ y = \left(a + u \cos\frac{v}{2}\right) \sin v, \ z = u \sin\frac{v}{2}$$

where $-1 \le u \le 1$, $0 \le v \le 2\pi$, and $a = 3$. Try to graph other Möbius strips for different values of a using a computer algebra system.

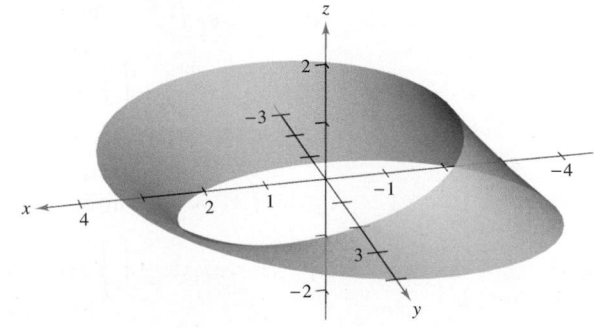

15.6 Surface Integrals

- Evaluate a surface integral as a double integral.
- Evaluate a surface integral for a parametric surface.
- Determine the orientation of a surface.
- Understand the concept of a flux integral.

Surface Integrals

The remainder of this chapter deals primarily with **surface integrals.** You will first consider surfaces given by $z = g(x, y)$. Later in this section, you will consider more general surfaces given in parametric form.

Let S be a surface given by $z = g(x, y)$ and let R be its projection onto the xy-plane, as shown in Figure 15.44. Let g, g_x, and g_y be continuous at all points in R and let f be a scalar function defined on S. Employing the procedure used to find surface area in Section 14.5, evaluate f at (x_i, y_i, z_i) and form the sum

$$\sum_{i=1}^{n} f(x_i, y_i, z_i) \, \Delta S_i$$

where

$$\Delta S_i \approx \sqrt{1 + [g_x(x_i, y_i)]^2 + [g_y(x_i, y_i)]^2} \, \Delta A_i.$$

Provided the limit of this sum as $\|\Delta\|$ approaches 0 exists, the **surface integral of f over S** is defined as

$$\iint_S f(x, y, z) \, dS = \lim_{\|\Delta\| \to 0} \sum_{i=1}^{n} f(x_i, y_i, z_i) \, \Delta S_i.$$

This integral can be evaluated by a double integral.

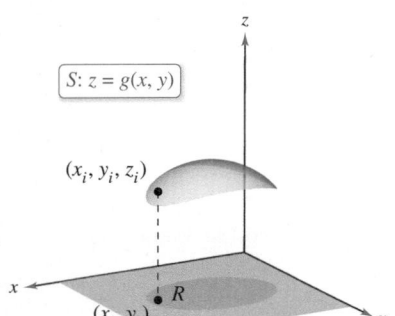

Scalar function f assigns a number to each point of S.
Figure 15.44

THEOREM 15.10 Evaluating a Surface Integral

Let S be a surface given by $z = g(x, y)$ and let R be its projection onto the xy-plane. If g, g_x, and g_y are continuous on R and f is continuous on S, then the surface integral of f over S is

$$\iint_S f(x, y, z) \, dS = \iint_R f(x, y, g(x, y)) \sqrt{1 + [g_x(x, y)]^2 + [g_y(x, y)]^2} \, dA.$$

For surfaces described by functions of x and z (or y and z), you can make the following adjustments to Theorem 15.10. If S is the graph of $y = g(x, z)$ and R is its projection onto the xz-plane, then

$$\iint_S f(x, y, z) \, dS = \iint_R f(x, g(x, z), z) \sqrt{1 + [g_x(x, z)]^2 + [g_z(x, z)]^2} \, dA.$$

If S is the graph of $x = g(y, z)$ and R is its projection onto the yz-plane, then

$$\iint_S f(x, y, z) \, dS = \iint_R f(g(y, z), y, z) \sqrt{1 + [g_y(y, z)]^2 + [g_z(y, z)]^2} \, dA.$$

If $f(x, y, z) = 1$, the surface integral over S yields the surface area of S. For instance, suppose the surface S is the plane given by $z = x$, where $0 \le x \le 1$ and $0 \le y \le 1$. The surface area of S is $\sqrt{2}$ square units. Try verifying that

$$\iint_S f(x, y, z) \, dS = \sqrt{2}.$$

EXAMPLE 1 **Evaluating a Surface Integral**

Evaluate the surface integral

$$\iint_S (y^2 + 2yz)\, dS$$

where S is the first-octant portion of the plane

$$2x + y + 2z = 6.$$

Solution Begin by writing S as

$$z = g(x, y) = \frac{1}{2}(6 - 2x - y).$$

Using the partial derivatives $g_x(x, y) = -1$ and $g_y(x, y) = -\frac{1}{2}$, you can write

$$\sqrt{1 + [g_x(x, y)]^2 + [g_y(x, y)]^2} = \sqrt{1 + 1 + \frac{1}{4}} = \frac{3}{2}.$$

Using Figure 15.45 and Theorem 15.10, you obtain

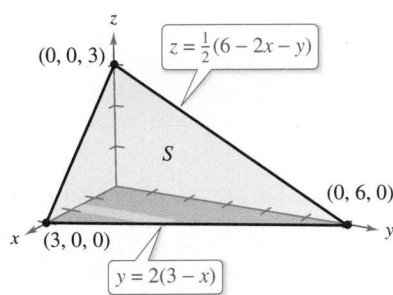

(0, 0, 3) $z = \frac{1}{2}(6 - 2x - y)$

S

(0, 6, 0)

x (3, 0, 0) y

$y = 2(3 - x)$

Figure 15.45

$$\iint_S (y^2 + 2yz)\, dS = \iint_R f(x, y, g(x, y))\sqrt{1 + [g_x(x, y)]^2 + [g_y(x, y)]^2}\, dA$$

$$= \iint_R \left[y^2 + 2y\left(\frac{1}{2}\right)(6 - 2x - y) \right]\left(\frac{3}{2}\right) dA$$

$$= 3 \int_0^3 \int_0^{2(3-x)} y(3 - x)\, dy\, dx \qquad \text{Convert to iterated integral.}$$

$$= 3 \int_0^3 \frac{y^2}{2}(3 - x) \Big]_0^{2(3-x)} dx \qquad \text{Integrate with respect to } y.$$

$$= 6 \int_0^3 (3 - x)^3\, dx$$

$$= -\frac{3}{2}(3 - x)^4 \Big]_0^3 \qquad \text{Integrate with respect to } x.$$

$$= \frac{243}{2}.$$

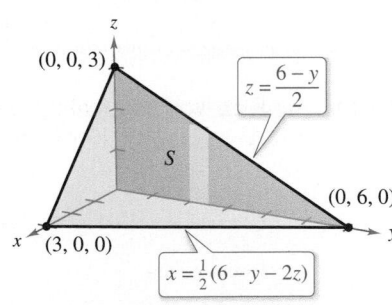

(0, 0, 3) $z = \frac{6 - y}{2}$

S

(0, 6, 0)

x (3, 0, 0) y

$x = \frac{1}{2}(6 - y - 2z)$

Figure 15.46

An alternative solution to Example 1 would be to project S onto the yz-plane, as shown in Figure 15.46. Then $x = \frac{1}{2}(6 - y - 2z)$, and

$$\sqrt{1 + [g_y(y, z)]^2 + [g_z(y, z)]^2} = \sqrt{1 + \frac{1}{4} + 1} = \frac{3}{2}.$$

So, the surface integral is

$$\iint_S (y^2 + 2yz)\, dS = \iint_R f(g(y, z), y, z)\sqrt{1 + [g_y(y, z)]^2 + [g_z(y, z)]^2}\, dA$$

$$= \int_0^6 \int_0^{(6-y)/2} (y^2 + 2yz)\left(\frac{3}{2}\right) dz\, dy$$

$$= \frac{3}{8} \int_0^6 (36y - y^3)\, dy$$

$$= \frac{243}{2}.$$

Try reworking Example 1 by projecting S onto the xz-plane.

In Example 1, you could have projected the surface S onto any one of the three coordinate planes. In Example 2, S is a portion of a cylinder centered about the x-axis, and you can project it onto either the xz-plane or the xy-plane.

EXAMPLE 2 Evaluating a Surface Integral

•••▷ *See LarsonCalculus.com for an interactive version of this type of example.*

Evaluate the surface integral

$$\iint_S (x + z)\, dS$$

where S is the first-octant portion of the cylinder

$$y^2 + z^2 = 9$$

between $x = 0$ and $x = 4$, as shown in Figure 15.47.

Solution Project S onto the xy-plane so that

$$z = g(x, y) = \sqrt{9 - y^2}$$

and obtain

$$\sqrt{1 + [g_x(x, y)]^2 + [g_y(x, y)]^2} = \sqrt{1 + \left(\frac{-y}{\sqrt{9 - y^2}}\right)^2}$$

$$= \frac{3}{\sqrt{9 - y^2}}.$$

Theorem 15.10 does not apply directly, because g_y is not continuous when $y = 3$. However, you can apply Theorem 15.10 for $0 \le b < 3$ and then take the limit as b approaches 3, as follows.

$$\iint_S (x + z)\, dS = \lim_{b \to 3^-} \int_0^b \int_0^4 (x + \sqrt{9 - y^2}) \frac{3}{\sqrt{9 - y^2}}\, dx\, dy$$

$$= \lim_{b \to 3^-} 3 \int_0^b \int_0^4 \left(\frac{x}{\sqrt{9 - y^2}} + 1\right) dx\, dy$$

$$= \lim_{b \to 3^-} 3 \int_0^b \left[\frac{x^2}{2\sqrt{9 - y^2}} + x\right]_0^4 dy \qquad \text{Integrate with respect to } x.$$

$$= \lim_{b \to 3^-} 3 \int_0^b \left(\frac{8}{\sqrt{9 - y^2}} + 4\right) dy$$

$$= \lim_{b \to 3^-} 3 \left[4y + 8 \arcsin \frac{y}{3}\right]_0^b \qquad \text{Integrate with respect to } y.$$

$$= \lim_{b \to 3^-} 3 \left(4b + 8 \arcsin \frac{b}{3}\right)$$

$$= 36 + 24\left(\frac{\pi}{2}\right) \qquad \text{Evaluate limit.}$$

$$= 36 + 12\pi$$

▷ **TECHNOLOGY** Some computer algebra systems are capable of evaluating improper integrals. If you have access to such software, use it to evaluate the improper integral

$$\int_0^3 \int_0^4 (x + \sqrt{9 - y^2}) \frac{3}{\sqrt{9 - y^2}}\, dx\, dy.$$

Do you obtain the same result as in Example 2?

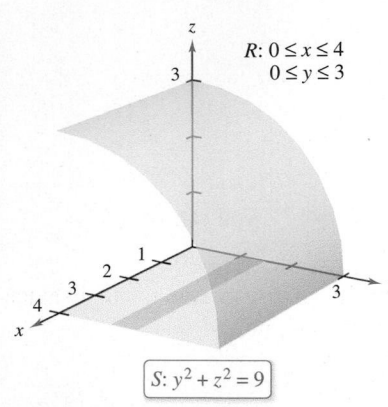

$R: 0 \le x \le 4$
$\quad 0 \le y \le 3$

$S: y^2 + z^2 = 9$

Figure 15.47

You have already seen that when the function f defined on the surface S is simply $f(x, y, z) = 1$, the surface integral yields the *surface area* of S.

$$\text{Area of surface} = \iint_S 1 \, dS$$

On the other hand, when S is a lamina of variable density and $\rho(x, y, z)$ is the density at the point (x, y, z), then the *mass* of the lamina is given by

$$\text{Mass of lamina} = \iint_S \rho(x, y, z) \, dS.$$

EXAMPLE 3 Finding the Mass of a Surface Lamina

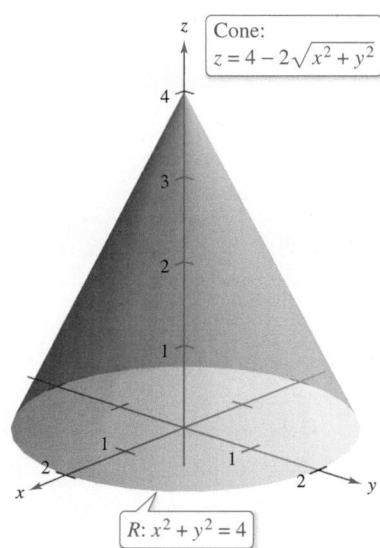

z Cone:
$z = 4 - 2\sqrt{x^2 + y^2}$

$R: x^2 + y^2 = 4$

Figure 15.48

A cone-shaped surface lamina S is given by

$$z = 4 - 2\sqrt{x^2 + y^2}, \quad 0 \le z \le 4$$

as shown in Figure 15.48. At each point on S, the density is proportional to the distance between the point and the z-axis. Find the mass m of the lamina.

Solution Projecting S onto the xy-plane produces

$$S: z = 4 - 2\sqrt{x^2 + y^2} = g(x, y), \quad 0 \le z \le 4$$
$$R: x^2 + y^2 \le 4$$

with a density of $\rho(x, y, z) = k\sqrt{x^2 + y^2}$, where k is the constant of proportionality. Using a surface integral, you can find the mass to be

$$
\begin{aligned}
m &= \iint_S \rho(x, y, z) \, dS \\[4pt]
&= \iint_R k\sqrt{x^2 + y^2} \sqrt{1 + [g_x(x, y)]^2 + [g_y(x, y)]^2} \, dA \\[4pt]
&= k \iint_R \sqrt{x^2 + y^2} \sqrt{1 + \frac{4x^2}{x^2 + y^2} + \frac{4y^2}{x^2 + y^2}} \, dA \\[4pt]
&= k \iint_R \sqrt{5}\sqrt{x^2 + y^2} \, dA \\[4pt]
&= k \int_0^{2\pi} \int_0^2 (\sqrt{5}\,r) r \, dr \, d\theta && \text{Polar coordinates} \\[4pt]
&= \frac{\sqrt{5}k}{3} \int_0^{2\pi} r^3 \Big]_0^2 \, d\theta && \text{Integrate with respect to } r. \\[4pt]
&= \frac{8\sqrt{5}k}{3} \int_0^{2\pi} d\theta \\[4pt]
&= \frac{8\sqrt{5}k}{3} \Big[\theta\Big]_0^{2\pi} && \text{Integrate with respect to } \theta. \\[4pt]
&= \frac{16\sqrt{5}k\pi}{3}.
\end{aligned}
$$

▷ **TECHNOLOGY** Use a computer algebra system to confirm the result shown in Example 3. The computer algebra system *Mathematica* evaluated the integral as follows.

$$k \int_{-2}^{2} \int_{-\sqrt{4-y^2}}^{\sqrt{4-y^2}} \sqrt{5}\sqrt{x^2 + y^2} \, dx \, dy = k \int_0^{2\pi} \int_0^2 (\sqrt{5}\,r) r \, dr \, d\theta = \frac{16\sqrt{5}k\pi}{3}$$

Parametric Surfaces and Surface Integrals

For a surface S given by the vector-valued function

$$\mathbf{r}(u, v) = x(u, v)\mathbf{i} + y(u, v)\mathbf{j} + z(u, v)\mathbf{k} \qquad \text{Parametric surface}$$

defined over a region D in the uv-plane, you can show that the surface integral of $f(x, y, z)$ over S is given by

$$\iint_S f(x, y, z)\, dS = \iint_D f(x(u, v), y(u, v), z(u, v))\|\mathbf{r}_u(u, v) \times \mathbf{r}_v(u, v)\|\, dA.$$

Note the similarity to a line integral over a space curve C.

$$\int_C f(x, y, z)\, ds = \int_a^b f(x(t), y(t), z(t))\|\mathbf{r}'(t)\|\, dt \qquad \text{Line integral}$$

Also, notice that ds and dS can be written as

$$ds = \|\mathbf{r}'(t)\|\, dt \quad \text{and} \quad dS = \|\mathbf{r}_u(u, v) \times \mathbf{r}_v(u, v)\|\, dA.$$

EXAMPLE 4 Evaluating a Surface Integral

Example 2 demonstrated an evaluation of the surface integral

$$\iint_S (x + z)\, dS$$

where S is the first-octant portion of the cylinder

$$y^2 + z^2 = 9$$

between $x = 0$ and $x = 4$, as shown in Figure 15.49. Reevaluate this integral in parametric form.

Solution In parametric form, the surface is given by

$$\mathbf{r}(x, \theta) = x\mathbf{i} + 3 \cos \theta \mathbf{j} + 3 \sin \theta \mathbf{k}$$

where $0 \le x \le 4$ and $0 \le \theta \le \pi/2$. To evaluate the surface integral in parametric form, begin by calculating the following.

$$\mathbf{r}_x = \mathbf{i}$$
$$\mathbf{r}_\theta = -3 \sin \theta \mathbf{j} + 3 \cos \theta \mathbf{k}$$

$$\mathbf{r}_x \times \mathbf{r}_\theta = \begin{vmatrix} \mathbf{i} & \mathbf{j} & \mathbf{k} \\ 1 & 0 & 0 \\ 0 & -3 \sin \theta & 3 \cos \theta \end{vmatrix} = -3 \cos \theta \mathbf{j} - 3 \sin \theta \mathbf{k}$$

$$\|\mathbf{r}_x \times \mathbf{r}_\theta\| = \sqrt{9 \cos^2 \theta + 9 \sin^2 \theta} = 3$$

So, the surface integral can be evaluated as follows.

$$\iint_D (x + 3 \sin \theta)3\, dA = \int_0^4 \int_0^{\pi/2} (3x + 9 \sin \theta)\, d\theta\, dx$$

$$= \int_0^4 \left[3x\theta - 9 \cos \theta \right]_0^{\pi/2} dx$$

$$= \int_0^4 \left(\frac{3\pi}{2}x + 9 \right) dx$$

$$= \left[\frac{3\pi}{4}x^2 + 9x \right]_0^4$$

$$= 12\pi + 36$$

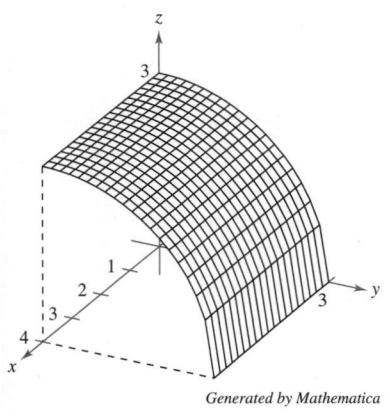

Generated by Mathematica

Figure 15.49

Orientation of a Surface

Unit normal vectors are used to induce an orientation to a surface S in space. A surface is **orientable** when a unit normal vector $\mathbf{N}$ can be defined at every nonboundary point of S in such a way that the normal vectors vary continuously over the surface S. The surface S is called an **oriented surface.**

An orientable surface S has two distinct sides. So, when you orient a surface, you are selecting one of the two possible unit normal vectors. For a closed surface such as a sphere, it is customary to choose the unit normal vector $\mathbf{N}$ to be the one that points outward from the sphere.

Most common surfaces, such as spheres, paraboloids, ellipses, and planes, are orientable. (See Exercise 43 for an example of a surface that is *not* orientable.) Moreover, for an orientable surface, the gradient provides a convenient way to find a unit normal vector. That is, for an orientable surface S given by

$$z = g(x, y) \qquad \text{Orientable surface}$$

let

$$G(x, y, z) = z - g(x, y).$$

Then, S can be oriented by either the unit normal vector

$$\mathbf{N} = \frac{\nabla G(x, y, z)}{\|\nabla G(x, y, z)\|}$$

$$= \frac{-g_x(x, y)\mathbf{i} - g_y(x, y)\mathbf{j} + \mathbf{k}}{\sqrt{1 + [g_x(x, y)]^2 + [g_y(x, y)]^2}} \qquad \text{Upward unit normal vector}$$

or the unit normal vector

$$\mathbf{N} = \frac{-\nabla G(x, y, z)}{\|\nabla G(x, y, z)\|}$$

$$= \frac{g_x(x, y)\mathbf{i} + g_y(x, y)\mathbf{j} - \mathbf{k}}{\sqrt{1 + [g_x(x, y)]^2 + [g_y(x, y)]^2}} \qquad \text{Downward unit normal vector}$$

as shown in Figure 15.50. If the smooth orientable surface S is given in parametric form by

$$\mathbf{r}(u, v) = x(u, v)\mathbf{i} + y(u, v)\mathbf{j} + z(u, v)\mathbf{k} \qquad \text{Parametric surface}$$

then the unit normal vectors are given by

$$\mathbf{N} = \frac{\mathbf{r}_u \times \mathbf{r}_v}{\|\mathbf{r}_u \times \mathbf{r}_v\|} \qquad \text{Upward unit normal vector}$$

and

$$\mathbf{N} = \frac{\mathbf{r}_v \times \mathbf{r}_u}{\|\mathbf{r}_v \times \mathbf{r}_u\|}. \qquad \text{Downward unit normal vector}$$

For an orientable surface given by

$$y = g(x, z) \quad \text{or} \quad x = g(y, z)$$

you can use the gradient

$$\nabla G(x, y, z) = -g_x(x, z)\mathbf{i} + \mathbf{j} - g_z(x, z)\mathbf{k} \qquad G(x, y, z) = y - g(x, z)$$

or

$$\nabla G(x, y, z) = \mathbf{i} - g_y(y, z)\mathbf{j} - g_z(y, z)\mathbf{k} \qquad G(x, y, z) = x - g(y, z)$$

to orient the surface.

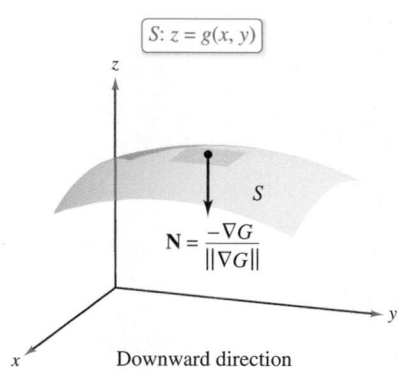

$S: z = g(x, y)$

$\mathbf{N} = \dfrac{\nabla G}{\|\nabla G\|}$

Upward direction

S is oriented in an upward direction.

$S: z = g(x, y)$

$\mathbf{N} = \dfrac{-\nabla G}{\|\nabla G\|}$

Downward direction

S is oriented in a downward direction.
Figure 15.50

Flux Integrals

One of the principal applications involving the vector form of a surface integral relates to the flow of a fluid through a surface. Consider an oriented surface S submerged in a fluid having a continuous velocity field $\mathbf{F}$. Let ΔS be the area of a small patch of the surface S over which $\mathbf{F}$ is nearly constant. Then the amount of fluid crossing this region per unit of time is approximated by the volume of the column of height $\mathbf{F} \cdot \mathbf{N}$, as shown in Figure 15.51. That is,

$$\Delta V = (\text{height})(\text{area of base})$$
$$= (\mathbf{F} \cdot \mathbf{N})\,\Delta S.$$

Consequently, the volume of fluid crossing the surface S per unit of time (called the **flux of F across S**) is given by the surface integral in the next definition.

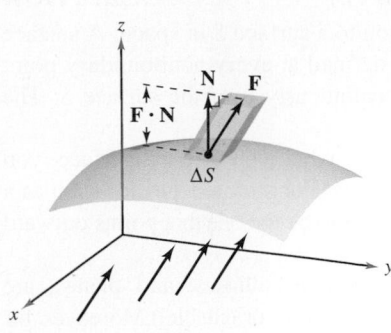

The velocity field $\mathbf{F}$ indicates the direction of the fluid flow.

Figure 15.51

Definition of Flux Integral

Let $\mathbf{F}(x, y, z) = M\mathbf{i} + N\mathbf{j} + P\mathbf{k}$, where M, N, and P have continuous first partial derivatives on the surface S oriented by a unit normal vector $\mathbf{N}$. The **flux integral of F across S** is given by

$$\iint_S \mathbf{F} \cdot \mathbf{N}\, dS.$$

Geometrically, a flux integral is the surface integral over S of the *normal component* of $\mathbf{F}$. If $\rho(x, y, z)$ is the density of the fluid at (x, y, z), then the flux integral

$$\iint_S \rho \mathbf{F} \cdot \mathbf{N}\, dS$$

represents the *mass* of the fluid flowing across S per unit of time.

To evaluate a flux integral for a surface given by $z = g(x, y)$, let

$$G(x, y, z) = z - g(x, y).$$

Then $\mathbf{N}\, dS$ can be written as follows.

$$\mathbf{N}\, dS = \frac{\nabla G(x, y, z)}{\|\nabla G(x, y, z)\|}\, dS$$

$$= \frac{\nabla G(x, y, z)}{\sqrt{(g_x)^2 + (g_y)^2 + 1}}\sqrt{(g_x)^2 + (g_y)^2 + 1}\, dA$$

$$= \nabla G(x, y, z)\, dA$$

THEOREM 15.11   Evaluating a Flux Integral

Let S be an oriented surface given by $z = g(x, y)$ and let R be its projection onto the xy-plane.

$$\iint_S \mathbf{F} \cdot \mathbf{N}\, dS = \iint_R \mathbf{F} \cdot [-g_x(x, y)\mathbf{i} - g_y(x, y)\mathbf{j} + \mathbf{k}]\, dA \qquad \text{Oriented upward}$$

$$\iint_S \mathbf{F} \cdot \mathbf{N}\, dS = \iint_R \mathbf{F} \cdot [g_x(x, y)\mathbf{i} + g_y(x, y)\mathbf{j} - \mathbf{k}]\, dA \qquad \text{Oriented downward}$$

For the first integral, the surface is oriented upward, and for the second integral, the surface is oriented downward.

EXAMPLE 5 **Using a Flux Integral to Find the Rate of Mass Flow**

Let S be the portion of the paraboloid

$$z = g(x, y) = 4 - x^2 - y^2$$

lying above the xy-plane, oriented by an upward unit normal vector, as shown in Figure 15.52. A fluid of constant density ρ is flowing through the surface S according to the vector field

$$\mathbf{F}(x, y, z) = x\mathbf{i} + y\mathbf{j} + z\mathbf{k}.$$

Find the rate of mass flow through S.

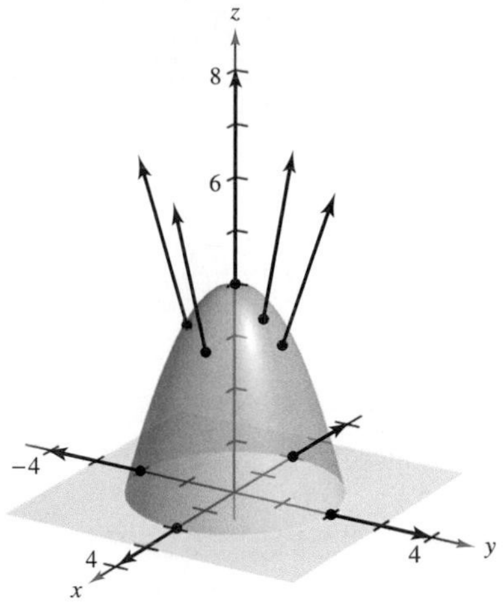

Solution Note that S is oriented upward and the partial derivatives of g are

$$g_x(x, y) = -2x$$

and

$$g_y(x, y) = -2y.$$

So, the rate of mass flow through the surface S is

$$\iint_S \rho\mathbf{F} \cdot \mathbf{N}\, dS = \rho \iint_R \mathbf{F} \cdot [-g_x(x, y)\mathbf{i} - g_y(x, y)\mathbf{j} + \mathbf{k}]\, dA$$

$$= \rho \iint_R [x\mathbf{i} + y\mathbf{j} + (4 - x^2 - y^2)\mathbf{k}] \cdot (2x\mathbf{i} + 2y\mathbf{j} + \mathbf{k})\, dA$$

$$= \rho \iint_R [2x^2 + 2y^2 + (4 - x^2 - y^2)]\, dA$$

$$= \rho \iint_R (4 + x^2 + y^2)\, dA$$

$$= \rho \int_0^{2\pi} \int_0^2 (4 + r^2)r\, dr\, d\theta \qquad \text{Polar coordinates}$$

$$= \rho \int_0^{2\pi} \left[2r^2 + \frac{r^4}{4} \right]_0^2 d\theta$$

$$= \rho \int_0^{2\pi} 12\, d\theta$$

$$= 24\pi\rho.$$

Figure 15.52

For an oriented upward surface S given by the vector-valued function

$$\mathbf{r}(u, v) = x(u, v)\mathbf{i} + y(u, v)\mathbf{j} + z(u, v)\mathbf{k} \qquad \text{Parametric surface}$$

defined over a region D in the uv-plane, you can define the flux integral of $\mathbf{F}$ across S as

$$\iint_S \mathbf{F} \cdot \mathbf{N}\, dS = \iint_D \mathbf{F} \cdot \left(\frac{\mathbf{r}_u \times \mathbf{r}_v}{\|\mathbf{r}_u \times \mathbf{r}_v\|} \right) \|\mathbf{r}_u \times \mathbf{r}_v\|\, dA = \iint_D \mathbf{F} \cdot (\mathbf{r}_u \times \mathbf{r}_v)\, dA.$$

Note the similarity of this integral to the line integral

$$\int_C \mathbf{F} \cdot d\mathbf{r} = \int_C \mathbf{F} \cdot \mathbf{T}\, ds.$$

A summary of formulas for line and surface integrals is presented on page 1107.

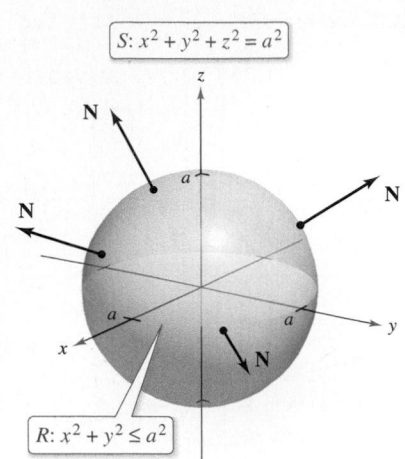

$S: x^2 + y^2 + z^2 = a^2$

$R: x^2 + y^2 \le a^2$

Figure 15.53

EXAMPLE 6   Finding the Flux of an Inverse Square Field

Find the flux over the sphere S given by

$$x^2 + y^2 + z^2 = a^2 \qquad \text{Sphere } S$$

where $\mathbf{F}$ is an inverse square field given by

$$\mathbf{F}(x, y, z) = \frac{kq}{\|\mathbf{r}\|^2} \frac{\mathbf{r}}{\|\mathbf{r}\|} = \frac{kq\mathbf{r}}{\|\mathbf{r}\|^3} \qquad \text{Inverse square field } \mathbf{F}$$

and

$$\mathbf{r} = x\mathbf{i} + y\mathbf{j} + z\mathbf{k}.$$

Assume S is oriented outward, as shown in Figure 15.53.

Solution   The sphere is given by

$$\begin{aligned}
\mathbf{r}(u, v) &= x(u, v)\mathbf{i} + y(u, v)\mathbf{j} + z(u, v)\mathbf{k} \\
&= a \sin u \cos v\mathbf{i} + a \sin u \sin v\mathbf{j} + a \cos u\mathbf{k}
\end{aligned}$$

where $0 \le u \le \pi$ and $0 \le v \le 2\pi$. The partial derivatives of $\mathbf{r}$ are

$$\mathbf{r}_u(u, v) = a \cos u \cos v\mathbf{i} + a \cos u \sin v\mathbf{j} - a \sin u\mathbf{k}$$

and

$$\mathbf{r}_v(u, v) = -a \sin u \sin v\mathbf{i} + a \sin u \cos v\mathbf{j}$$

which implies that the normal vector $\mathbf{r}_u \times \mathbf{r}_v$ is

$$\begin{aligned}
\mathbf{r}_u \times \mathbf{r}_v &= \begin{vmatrix} \mathbf{i} & \mathbf{j} & \mathbf{k} \\ a \cos u \cos v & a \cos u \sin v & -a \sin u \\ -a \sin u \sin v & a \sin u \cos v & 0 \end{vmatrix} \\
&= a^2(\sin^2 u \cos v\mathbf{i} + \sin^2 u \sin v\mathbf{j} + \sin u \cos u\mathbf{k}).
\end{aligned}$$

Now, using

$$\begin{aligned}
\mathbf{F}(x, y, z) &= \frac{kq\mathbf{r}}{\|\mathbf{r}\|^3} \\
&= kq \frac{x\mathbf{i} + y\mathbf{j} + z\mathbf{k}}{\|x\mathbf{i} + y\mathbf{j} + z\mathbf{k}\|^3} \\
&= \frac{kq}{a^3}(a \sin u \cos v\mathbf{i} + a \sin u \sin v\mathbf{j} + a \cos u\mathbf{k})
\end{aligned}$$

it follows that

$$\begin{aligned}
\mathbf{F} \cdot (\mathbf{r}_u \times \mathbf{r}_v) &= \frac{kq}{a^3}[(a \sin u \cos v\mathbf{i} + a \sin u \sin v\mathbf{j} + a \cos u\mathbf{k}) \cdot \\
&\qquad a^2(\sin^2 u \cos v\mathbf{i} + \sin^2 u \sin v\mathbf{j} + \sin u \cos u\mathbf{k})] \\
&= kq(\sin^3 u \cos^2 v + \sin^3 u \sin^2 v + \sin u \cos^2 u) \\
&= kq \sin u.
\end{aligned}$$

Finally, the flux over the sphere S is given by

$$\begin{aligned}
\iint_S \mathbf{F} \cdot \mathbf{N} \, dS &= \iint_D kq \sin u \, dA \\
&= kq \int_0^{2\pi} \int_0^{\pi} \sin u \, du \, dv \\
&= kq \int_0^{2\pi} 2 \, dv \\
&= 4\pi kq.
\end{aligned}$$

The result in Example 6 shows that the flux across a sphere S in an inverse square field is independent of the radius of S. In particular, if $\mathbf{E}$ is an electric field, then the result in Example 6, along with Coulomb's Law (see Section 15.1), yields one of the basic laws of electrostatics, known as **Gauss's Law:**

$$\iint_S \mathbf{E} \cdot \mathbf{N} \, dS = 4\pi k q \qquad \text{Gauss's Law}$$

where q is a point charge located at the center of the sphere and k is the Coulomb constant. Gauss's Law is valid for more general closed surfaces that enclose the origin, and relates the flux out of the surface to the total charge inside the surface.

Surface integrals are also used in the study of **heat flow.** Heat flows from areas of higher temperature to areas of lower temperature in the direction of greatest change. As a result, measuring **heat flux** involves the gradient of the temperature. The flux depends on the area of the surface. It is the normal direction to the surface that is important, because heat that flows in directions tangential to the surface will produce no heat loss. So, assume that the heat flux across a portion of the surface of area ΔS is given by $\Delta H \approx -k\nabla T \cdot \mathbf{N} \, dS$, where T is the temperature, $\mathbf{N}$ is the unit normal vector to the surface in the direction of the heat flow, and k is the thermal diffusivity of the material. The heat flux across the surface is given by

$$H = \iint_S -k\nabla T \cdot \mathbf{N} \, dS. \qquad \text{Heat flux across } S$$

This section concludes with a summary of different forms of line integrals and surface integrals.

SUMMARY OF LINE AND SURFACE INTEGRALS

Line Integrals

$$ds = \|\mathbf{r}'(t)\| \, dt$$
$$= \sqrt{[x'(t)]^2 + [y'(t)]^2 + [z'(t)]^2} \, dt$$

$$\int_C f(x, y, z) \, ds = \int_a^b f(x(t), y(t), z(t)) \sqrt{[x'(t)]^2 + [y'(t)]^2 + [z'(t)]^2} \, dt \qquad \text{Scalar form}$$

$$\int_C \mathbf{F} \cdot d\mathbf{r} = \int_C \mathbf{F} \cdot \mathbf{T} \, ds$$

$$= \int_a^b \mathbf{F}(x(t), y(t), z(t)) \cdot \mathbf{r}'(t) \, dt \qquad \text{Vector form}$$

Surface Integrals $[z = g(x, y)]$

$$dS = \sqrt{1 + [g_x(x, y)]^2 + [g_y(x, y)]^2} \, dA$$

$$\iint_S f(x, y, z) \, dS = \iint_R f(x, y, g(x, y)) \sqrt{1 + [g_x(x, y)]^2 + [g_y(x, y)]^2} \, dA \qquad \text{Scalar form}$$

$$\iint_S \mathbf{F} \cdot \mathbf{N} \, dS = \iint_R \mathbf{F} \cdot [-g_x(x, y)\mathbf{i} - g_y(x, y)\mathbf{j} + \mathbf{k}] \, dA \qquad \text{Vector form (upward normal)}$$

Surface Integrals (parametric form)

$$dS = \|\mathbf{r}_u(u, v) \times \mathbf{r}_v(u, v)\| \, dA$$

$$\iint_S f(x, y, z) \, dS = \iint_D f(x(u, v), y(u, v), z(u, v)) \|\mathbf{r}_u(u, v) \times \mathbf{r}_v(u, v)\| \, dA \qquad \text{Scalar form}$$

$$\iint_S \mathbf{F} \cdot \mathbf{N} \, dS = \iint_D \mathbf{F} \cdot (\mathbf{r}_u \times \mathbf{r}_v) \, dA \qquad \text{Vector form (upward normal)}$$

15.6 Exercises

See CalcChat.com for tutorial help and worked-out solutions to odd-numbered exercises.

CONCEPT CHECK

1. **Surface Integral** Explain how to set up a surface integral given that you will project the surface onto the xz-plane.

2. **Surface Integral** For what condition does the surface integral over S yield the surface area of S?

3. **Orientation of a Surface** Describe a physical characteristic of an orientable surface.

4. **Flux** What is the physical interpretation of the flux of $\mathbf{F}$ across S? How do you calculate it?

 **Evaluating a Surface Integral** In Exercises 5–8, evaluate $\displaystyle\iint_S (x - 2y + z)\, dS$.

5. S: $z = 4 - x$, $0 \le x \le 4$, $0 \le y \le 3$
6. S: $z = 15 - 2x + 3y$, $0 \le x \le 2$, $0 \le y \le 4$
7. S: $z = 2$, $x^2 + y^2 \le 1$
8. S: $z = 3y$, $0 \le x \le 2$, $0 \le y \le x$

Evaluating a Surface Integral In Exercises 9 and 10, evaluate $\displaystyle\iint_S xy\, dS$.

9. S: $z = 3 - x - y$, first octant
10. S: $z = \frac{1}{4}x^4$, $0 \le x \le 1$, $0 \le y \le x^2$

 **Evaluating a Surface Integral** In Exercises 11 and 12, use a computer algebra system to evaluate $\displaystyle\iint_S (x^2 - 2xy)\, dS$.

11. S: $z = 10 - x^2 - y^2$, $0 \le x \le 2$, $0 \le y \le 2$
12. S: $z = \cos x$, $0 \le x \le \dfrac{\pi}{2}$, $0 \le y \le \dfrac{1}{2}x$

 **Mass** In Exercises 13 and 14, find the mass of the surface lamina S of density ρ.

13. S: $2x + 3y + 6z = 12$, first octant, $\rho(x, y, z) = x^2 + y^2$
14. S: $z = \sqrt{a^2 - x^2 - y^2}$, $\rho(x, y, z) = kz$

 **Evaluating a Surface Integral** In Exercises 15–18, evaluate $\displaystyle\iint_S f(x, y)\, dS$.

15. $f(x, y) = y + 5$
   S: $\mathbf{r}(u, v) = u\mathbf{i} + v\mathbf{j} + 2v\mathbf{k}$
   $0 \le u \le 1$, $0 \le v \le 2$

16. $f(x, y) = xy$
   S: $\mathbf{r}(u, v) = 2\cos u\mathbf{i} + 2\sin u\mathbf{j} + v\mathbf{k}$
   $0 \le u \le \dfrac{\pi}{2}$, $0 \le v \le 1$

17. $f(x, y) = 3y - x$
   S: $\mathbf{r}(u, v) = \cos u\mathbf{i} + \sin u\mathbf{j} + v\mathbf{k}$
   $0 \le u \le \dfrac{\pi}{3}$, $0 \le v \le 1$

18. $f(x, y) = x + y$
   S: $\mathbf{r}(u, v) = 4u\cos v\mathbf{i} + 4u\sin v\mathbf{j} + 3u\mathbf{k}$
   $0 \le u \le 4$, $0 \le v \le \pi$

Evaluating a Surface Integral In Exercises 19–24, evaluate $\displaystyle\iint_S f(x, y, z)\, dS$.

19. $f(x, y, z) = x^2 + y^2 + z^2$
   S: $z = x + y$, $x^2 + y^2 \le 1$

20. $f(x, y, z) = \dfrac{xy}{z}$
   S: $z = x^2 + y^2$, $4 \le x^2 + y^2 \le 16$

21. $f(x, y, z) = \sqrt{x^2 + y^2 + z^2}$
   S: $z = \sqrt{x^2 + y^2}$, $x^2 + y^2 \le 4$

22. $f(x, y, z) = \sqrt{x^2 + y^2 + z^2}$
   S: $z = \sqrt{x^2 + y^2}$, $(x - 1)^2 + y^2 \le 1$

23. $f(x, y, z) = x^2 + y^2 + z^2$
   S: $x^2 + y^2 = 9$, $0 \le x \le 3$, $0 \le y \le 3$, $0 \le z \le 9$

24. $f(x, y, z) = x^2 + y^2 + z^2$
   S: $x^2 + y^2 = 9$, $0 \le x \le 3$, $0 \le z \le x$

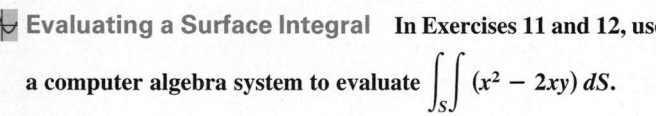 **Evaluating a Flux Integral** In Exercises 25–30, find the flux of $\mathbf{F}$ across S,
$$\iint_S \mathbf{F} \cdot \mathbf{N}\, dS$$
where $\mathbf{N}$ is the upward unit normal vector to S.

25. $\mathbf{F}(x, y, z) = 3z\mathbf{i} - 4\mathbf{j} + y\mathbf{k}$; S: $z = 1 - x - y$, first octant
26. $\mathbf{F}(x, y, z) = x\mathbf{i} + 2y\mathbf{j}$; S: $z = 6 - 3x - 2y$, first octant
27. $\mathbf{F}(x, y, z) = x\mathbf{i} + y\mathbf{j} + z\mathbf{k}$; S: $z = 1 - x^2 - y^2$, $z \ge 0$
28. $\mathbf{F}(x, y, z) = x\mathbf{i} + y\mathbf{j} + z\mathbf{k}$
   S: $x^2 + y^2 + z^2 = 36$, first octant
29. $\mathbf{F}(x, y, z) = 4\mathbf{i} - 3\mathbf{j} + 5\mathbf{k}$
   S: $z = x^2 + y^2$, $x^2 + y^2 \le 4$
30. $\mathbf{F}(x, y, z) = x\mathbf{i} + y\mathbf{j} - 2z\mathbf{k}$
   S: $z = \sqrt{a^2 - x^2 - y^2}$

Evaluating a Flux Integral In Exercises 31 and 32, find the flux of F over the closed surface. (Let N be the outward unit normal vector of the surface.)

31. $\mathbf{F}(x, y, z) = (x + y)\mathbf{i} + y\mathbf{j} + z\mathbf{k}$

   S: $z = 16 - x^2 - y^2$, $z = 0$

32. $\mathbf{F}(x, y, z) = 4xy\mathbf{i} + z^2\mathbf{j} + yz\mathbf{k}$

   S: unit cube bounded by the planes $x = 0$, $x = 1$, $y = 0$, $y = 1$, $z = 0$, $z = 1$.

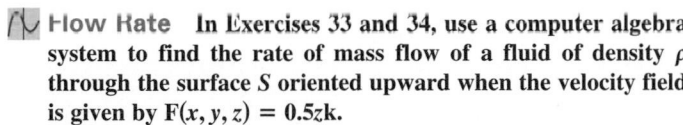 **Flow Rate** In Exercises 33 and 34, use a computer algebra system to find the rate of mass flow of a fluid of density ρ through the surface S oriented upward when the velocity field is given by $\mathbf{F}(x, y, z) = 0.5z\mathbf{k}$.

33. S: $z = 16 - x^2 - y^2$, $z \geq 0$

34. S: $z = \sqrt{16 - x^2 - y^2}$

Gauss's Law In Exercises 35 and 36, evaluate $\displaystyle\int_S\!\!\int \mathbf{E} \cdot \mathbf{N}\, dS$ to find the total charge of the electrostatic field E enclosed by the closed surface consisting of the hemisphere $z = \sqrt{1 - x^2 - y^2}$ and its circular base in the xy-plane.

35. $\mathbf{E} = yz\mathbf{i} + xz\mathbf{j} + xy\mathbf{k}$ **36.** $\mathbf{E} = x\mathbf{i} + y\mathbf{j} + 2z\mathbf{k}$

Moments of Inertia In Exercises 37–40, use the following formulas for the moments of inertia about the coordinate axes of a surface lamina of density ρ.

$$I_x = \int_S\!\!\int (y^2 + z^2)\rho(x, y, z)\, dS \qquad I_y = \int_S\!\!\int (x^2 + z^2)\rho(x, y, z)\, dS$$

$$I_z = \int_S\!\!\int (x^2 + y^2)\rho(x, y, z)\, dS$$

37. Verify that the moment of inertia of a conical shell of uniform density about its axis is $\frac{1}{2}ma^2$, where m is the mass and a is the radius and height.

38. Verify that the moment of inertia of a spherical shell of uniform density about its diameter is $\frac{2}{3}ma^2$, where m is the mass and a is the radius.

39. Find the moment of inertia about the z-axis for the surface lamina $x^2 + y^2 = a^2$, where $0 \leq z \leq h$, with a uniform density of 1.

40. Find the moment of inertia about the z-axis for the surface lamina $z = x^2 + y^2$, where $0 \leq z \leq h$, with a uniform density of 1.

EXPLORING CONCEPTS

41. Using Different Methods Evaluate

$$\int_S\!\!\int (x + 2y)\, dS$$

where S is the first-octant portion of the plane

$$2x + 2y + z = 4$$

by projecting S onto (a) the xy-plane, (b) the xz-plane, and (c) the yz-plane. Verify that all answers are the same.

42. **HOW DO YOU SEE IT?** Is the surface shown in the figure orientable? Explain why or why not.

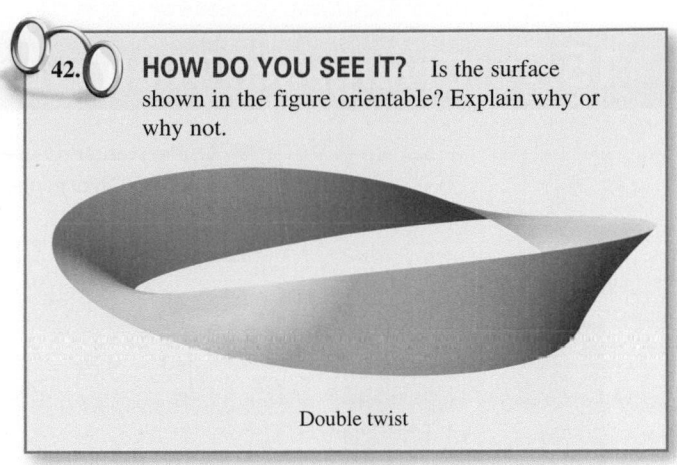

Double twist

43. Investigation

(a) Use a computer algebra system to graph the vector-valued function

$$\mathbf{r}(u, v) = (4 - v \sin u)\cos(2u)\mathbf{i} + (4 - v \sin u)\sin(2u)\mathbf{j} + v \cos u\mathbf{k}$$

where $0 \leq u \leq \pi$ and $-1 \leq v \leq 1$. This surface is called a Möbius strip.

(b) Is the surface orientable? Explain why or why not.

(c) Use a computer algebra system to graph the space curve represented by $\mathbf{r}(u, 0)$. Identify the curve.

(d) Cut a strip of paper and draw a line lengthwise through the center. Construct a Möbius strip by making a single twist and pasting the ends of the strip of paper together.

(e) Cut the Möbius strip along the line you drew in part (c), and describe the result.

SECTION PROJECT

Hyperboloid of One Sheet

Consider the parametric surface given by the function

$$\mathbf{r}(u, v) = a \cosh u \cos v\mathbf{i} + a \cosh u \sin v\mathbf{j} + b \sinh u\mathbf{k}.$$

(a) Use a graphing utility to graph $\mathbf{r}$ for various values of the constants a and b. Describe the effect of the constants on the shape of the surface.

(b) Show that the surface is a hyperboloid of one sheet given by

$$\frac{x^2}{a^2} + \frac{y^2}{a^2} - \frac{z^2}{b^2} = 1.$$

(c) For fixed values $u = u_0$, describe the curves given by

$$\mathbf{r}(u_0, v) = a \cosh u_0 \cos v\mathbf{i} + a \cosh u_0 \sin v\mathbf{j} + b \sinh u_0\mathbf{k}.$$

(d) For fixed values $v = v_0$, describe the curves given by

$$\mathbf{r}(u, v_0) = a \cosh u \cos v_0\mathbf{i} + a \cosh u \sin v_0\mathbf{j} + b \sinh u\mathbf{k}.$$

(e) Find a normal vector to the surface at $(u, v) = (0, 0)$.

15.7 Divergence Theorem

■ Understand and use the Divergence Theorem.
■ Use the Divergence Theorem to calculate flux.

Divergence Theorem

Recall from Section 15.4 that an alternative form of Green's Theorem is

$$\int_C \mathbf{F} \cdot \mathbf{N}\, ds = \int\int_R \left(\frac{\partial M}{\partial x} + \frac{\partial N}{\partial y} \right) dA$$

$$= \int\int_R \text{div } \mathbf{F}\, dA.$$

In an analogous way, the **Divergence Theorem** gives the relationship between a triple integral over a solid region Q and a surface integral over the surface of Q. In the statement of the theorem, the surface S is **closed** in the sense that it forms the complete boundary of the solid Q. Regions bounded by spheres, ellipsoids, cubes, tetrahedrons, or combinations of these surfaces are typical examples of closed surfaces. Let Q be a solid region on which a triple integral can be evaluated, and let S be a closed surface that is oriented by *outward* unit normal vectors, as shown in Figure 15.54. With these restrictions on S and Q, the Divergence Theorem can be stated as shown below the figure.

**CARL FRIEDRICH GAUSS
(1777–1855)**

The *Divergence Theorem* is also called *Gauss's Theorem*, after the famous German mathematician Carl Friedrich Gauss. Gauss is recognized, with Newton and Archimedes, as one of the three greatest mathematicians in history. One of his many contributions to mathematics was made at the age of 22, when, as part of his doctoral dissertation, he proved the *Fundamental Theorem of Algebra*. *See LarsonCalculus.com to read more of this biography.*

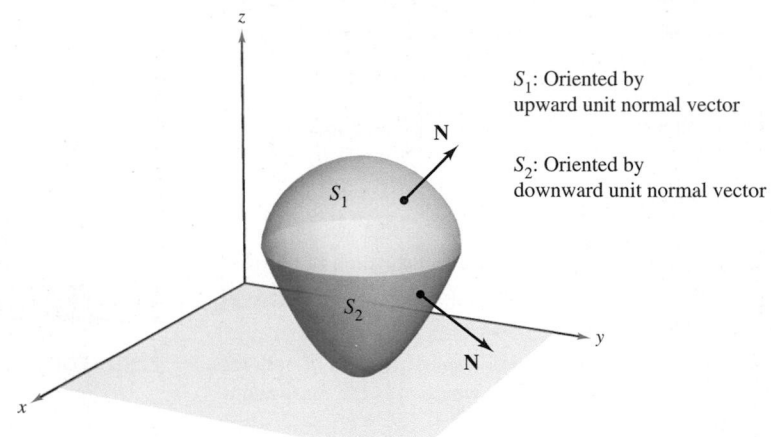

S_1: Oriented by
upward unit normal vector

S_2: Oriented by
downward unit normal vector

Figure 15.54

> **THEOREM 15.12 The Divergence Theorem**
>
> Let Q be a solid region bounded by a closed surface S oriented by a unit normal vector directed outward from Q. If $\mathbf{F}$ is a vector field whose component functions have continuous first partial derivatives in Q, then
>
> $$\int\int_S \mathbf{F} \cdot \mathbf{N}\, dS = \int\int\int_Q \text{div } \mathbf{F}\, dV.$$

REMARK As noted at the left above, the Divergence Theorem is sometimes called Gauss's Theorem. It is also sometimes called Ostrogradsky's Theorem, after the Russian mathematician Michel Ostrogradsky (1801–1861).

•••••••••••••••••••▷

•• **REMARK** This proof is restricted to a *simple* solid region. The general proof is best left to a course in advanced calculus.

Proof For $\mathbf{F}(x, y, z) = M\mathbf{i} + N\mathbf{j} + P\mathbf{k}$, the theorem takes the form

$$\iint_S \mathbf{F} \cdot \mathbf{N} \, dS = \iint_S (M\mathbf{i} \cdot \mathbf{N} + N\mathbf{j} \cdot \mathbf{N} + P\mathbf{k} \cdot \mathbf{N}) \, dS$$

$$= \iiint_Q \left(\frac{\partial M}{\partial x} + \frac{\partial N}{\partial y} + \frac{\partial P}{\partial z} \right) dV.$$

You can prove this by verifying that the following three equations are valid.

$$\iint_S M\mathbf{i} \cdot \mathbf{N} \, dS = \iiint_Q \frac{\partial M}{\partial x} \, dV$$

$$\iint_S N\mathbf{j} \cdot \mathbf{N} \, dS = \iiint_Q \frac{\partial N}{\partial y} \, dV$$

$$\iint_S P\mathbf{k} \cdot \mathbf{N} \, dS = \iiint_Q \frac{\partial P}{\partial z} \, dV$$

Because the verifications of the three equations are similar, only the third is discussed. Restrict the proof to a **simple solid** region with upper surface

$$z = g_2(x, y) \qquad \text{Upper surface}$$

and lower surface

$$z = g_1(x, y) \qquad \text{Lower surface}$$

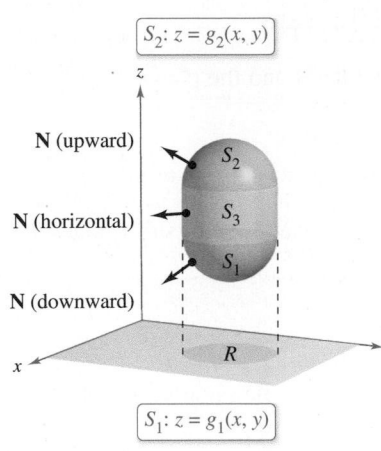

$S_2: z = g_2(x, y)$

N (upward)

S_2

N (horizontal)

S_3

S_1

N (downward)

R

$S_1: z = g_1(x, y)$

Figure 15.55

whose projections onto the xy-plane coincide and form region R. If Q has a lateral surface like S_3 in Figure 15.55, then a normal vector is horizontal, which implies that $P\mathbf{k} \cdot \mathbf{N} = 0$. Consequently, you have

$$\iint_S P\mathbf{k} \cdot \mathbf{N} \, dS = \iint_{S_1} P\mathbf{k} \cdot \mathbf{N} \, dS + \iint_{S_2} P\mathbf{k} \cdot \mathbf{N} \, dS + 0.$$

On the upper surface S_2, the outward normal vector is upward, whereas on the lower surface S_1, the outward normal vector is downward. So, by Theorem 15.11, you have

$$\iint_{S_1} P\mathbf{k} \cdot \mathbf{N} \, dS = \iint_R P(x, y, g_1(x, y))\mathbf{k} \cdot \left(\frac{\partial g_1}{\partial x}\mathbf{i} + \frac{\partial g_1}{\partial y}\mathbf{j} - \mathbf{k} \right) dA$$

$$= -\iint_R P(x, y, g_1(x, y)) \, dA$$

and

$$\iint_{S_2} P\mathbf{k} \cdot \mathbf{N} \, dS = \iint_R P(x, y, g_2(x, y))\mathbf{k} \cdot \left(-\frac{\partial g_2}{\partial x}\mathbf{i} - \frac{\partial g_2}{\partial y}\mathbf{j} + \mathbf{k} \right) dA$$

$$= \iint_R P(x, y, g_2(x, y)) \, dA.$$

Adding these results, you obtain

$$\iint_S P\mathbf{k} \cdot \mathbf{N} \, dS = \iint_R [P(x, y, g_2(x, y)) - P(x, y, g_1(x, y))] \, dA$$

$$= \iint_R \left[\int_{g_1(x, y)}^{g_2(x, y)} \frac{\partial P}{\partial z} \, dz \right] dA$$

$$= \iiint_Q \frac{\partial P}{\partial z} \, dV.$$

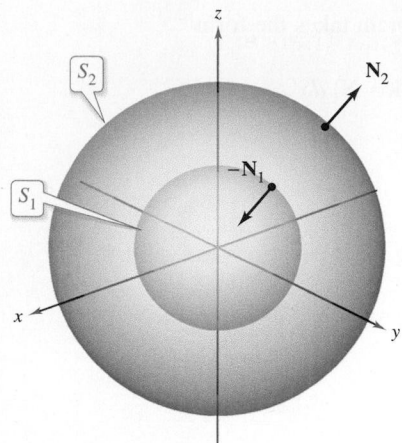

Figure 15.56

Even though the Divergence Theorem was stated for a simple solid region Q bounded by a closed surface, the theorem is also valid for regions that are the finite unions of simple solid regions. For example, let Q be the solid bounded by the closed surfaces S_1 and S_2, as shown in Figure 15.56. To apply the Divergence Theorem to this solid, let $S = S_1 \cup S_2$. The normal vector $\mathbf{N}$ to S is given by $-\mathbf{N}_1$ on S_1 and by $\mathbf{N}_2$ on S_2. So, you can write

$$\iiint_Q \operatorname{div} \mathbf{F} \, dV = \iint_S \mathbf{F} \cdot \mathbf{N} \, dS$$

$$= \iint_{S_1} \mathbf{F} \cdot (-\mathbf{N}_1) \, dS + \iint_{S_2} \mathbf{F} \cdot \mathbf{N}_2 \, dS$$

$$= -\iint_{S_1} \mathbf{F} \cdot \mathbf{N}_1 \, dS + \iint_{S_2} \mathbf{F} \cdot \mathbf{N}_2 \, dS.$$

For the remainder of this section, you will apply the Divergence Theorem to simple solid regions bounded by closed surfaces.

EXAMPLE 1   **Using the Divergence Theorem**

Let Q be the solid region bounded by the coordinate planes and the plane

$$2x + 2y + z = 6$$

and let $\mathbf{F} = x\mathbf{i} + y^2\mathbf{j} + z\mathbf{k}$. Find $\iint_S \mathbf{F} \cdot \mathbf{N} \, dS$, where S is the surface of Q.

Solution   From Figure 15.57, you can see that Q is bounded by four subsurfaces. So, you would need four *surface integrals* to evaluate

$$\iint_S \mathbf{F} \cdot \mathbf{N} \, dS.$$

However, by the Divergence Theorem, you need only one triple integral. Because

$$\operatorname{div} \mathbf{F} = \frac{\partial M}{\partial x} + \frac{\partial N}{\partial y} + \frac{\partial P}{\partial z} = 1 + 2y + 1 = 2 + 2y$$

you have

$$\iint_S \mathbf{F} \cdot \mathbf{N} \, dS = \iiint_Q \operatorname{div} \mathbf{F} \, dV$$

$$= \int_0^3 \int_0^{3-y} \int_0^{6-2x-2y} (2 + 2y) \, dz \, dx \, dy$$

$$= \int_0^3 \int_0^{3-y} (2z + 2yz) \Big]_0^{6-2x-2y} \, dx \, dy$$

$$= \int_0^3 \int_0^{3-y} (12 - 4x + 8y - 4xy - 4y^2) \, dx \, dy$$

$$= \int_0^3 \left[12x - 2x^2 + 8xy - 2x^2y - 4xy^2 \right]_0^{3-y} \, dy$$

$$= \int_0^3 (18 + 6y - 10y^2 + 2y^3) \, dy$$

$$= \left[18y + 3y^2 - \frac{10y^3}{3} + \frac{y^4}{2} \right]_0^3$$

$$= \frac{63}{2}.$$

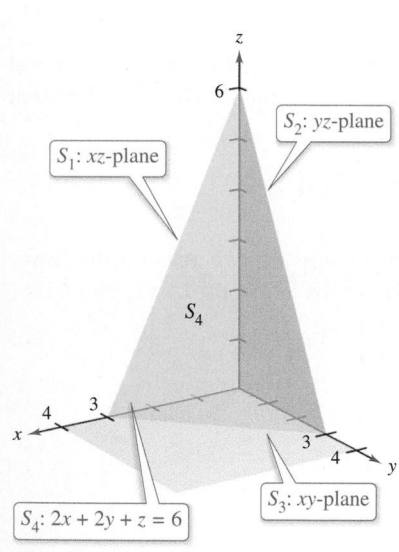

Figure 15.57

S_1: xz-plane
S_2: yz-plane
S_4
S_3: xy-plane
S_4: $2x + 2y + z = 6$

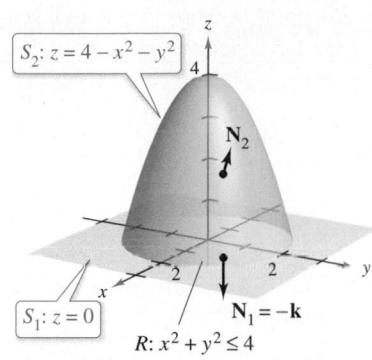

$S_2: z = 4 - x^2 - y^2$

N_2

$N_1 = -k$

$S_1: z = 0$

$R: x^2 + y^2 \leq 4$

Figure 15.58

EXAMPLE 2   Verifying the Divergence Theorem

Let Q be the solid region between the paraboloid

$$z = 4 - x^2 - y^2$$

and the xy-plane. Verify the Divergence Theorem for

$$\mathbf{F}(x, y, z) = 2z\mathbf{i} + x\mathbf{j} + y^2\mathbf{k}.$$

Solution   From Figure 15.58, you can see that the outward normal vector for the surface S_1 is $\mathbf{N}_1 = -\mathbf{k}$, whereas the outward normal vector for the surface S_2 is

$$\mathbf{N}_2 = \frac{2x\mathbf{i} + 2y\mathbf{j} + \mathbf{k}}{\sqrt{4x^2 + 4y^2 + 1}}.$$

So, by Theorem 15.11, you have

$$\iint_S \mathbf{F} \cdot \mathbf{N} \, dS = \iint_{S_1} \mathbf{F} \cdot \mathbf{N}_1 \, dS + \iint_{S_2} \mathbf{F} \cdot \mathbf{N}_2 \, dS$$

$$= \iint_{S_1} \mathbf{F} \cdot (-\mathbf{k}) \, dS + \iint_{S_2} \mathbf{F} \cdot \frac{(2x\mathbf{i} + 2y\mathbf{j} + \mathbf{k})}{\sqrt{4x^2 + 4y^2 + 1}} \, dS$$

$$= \iint_R -y^2 \, dA + \iint_R (4xz + 2xy + y^2) \, dA$$

$$= -\int_{-2}^{2} \int_{-\sqrt{4-y^2}}^{\sqrt{4-y^2}} y^2 \, dx \, dy + \int_{-2}^{2} \int_{-\sqrt{4-y^2}}^{\sqrt{4-y^2}} (4xz + 2xy + y^2) \, dx \, dy$$

$$= \int_{-2}^{2} \int_{-\sqrt{4-y^2}}^{\sqrt{4-y^2}} (4xz + 2xy) \, dx \, dy$$

$$= \int_{-2}^{2} \int_{-\sqrt{4-y^2}}^{\sqrt{4-y^2}} [4x(4 - x^2 - y^2) + 2xy] \, dx \, dy$$

$$= \int_{-2}^{2} \int_{-\sqrt{4-y^2}}^{\sqrt{4-y^2}} (16x - 4x^3 - 4xy^2 + 2xy) \, dx \, dy$$

$$= \int_{-2}^{2} \left[8x^2 - x^4 - 2x^2y^2 + x^2y \right]_{-\sqrt{4-y^2}}^{\sqrt{4-y^2}} \, dy$$

$$= \int_{-2}^{2} 0 \, dy$$

$$= 0.$$

On the other hand, because

$$\text{div } \mathbf{F} = \frac{\partial}{\partial x}[2z] + \frac{\partial}{\partial y}[x] + \frac{\partial}{\partial z}[y^2]$$

$$= 0 + 0 + 0$$

$$= 0$$

you can apply the Divergence Theorem to obtain the equivalent result

$$\iint_S \mathbf{F} \cdot \mathbf{N} \, dS = \iiint_Q \text{div } \mathbf{F} \, dV$$

$$= \iiint_Q 0 \, dV$$

$$= 0.$$

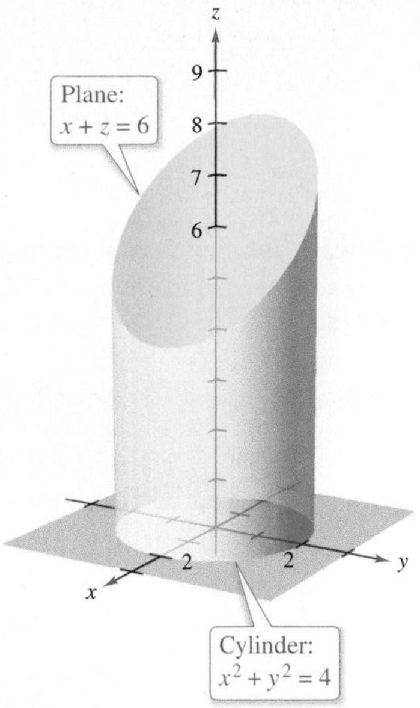

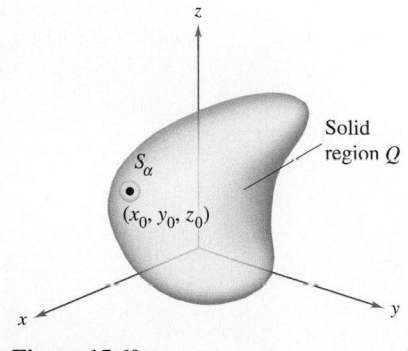

Plane:
x + z = 6

Cylinder:
$x^2 + y^2 = 4$

Figure 15.59

EXAMPLE 3 **Using the Divergence Theorem**

Let Q be the solid bounded by the cylinder $x^2 + y^2 = 4$, the plane $x + z = 6$, and the xy-plane, as shown in Figure 15.59. Find

$$\iint_S \mathbf{F} \cdot \mathbf{N} \, dS$$

where S is the surface of Q and

$$\mathbf{F}(x, y, z) = (x^2 + \sin z)\mathbf{i} + (xy + \cos z)\mathbf{j} + e^y\mathbf{k}.$$

Solution Direct evaluation of this surface integral would be difficult. However, by the Divergence Theorem, you can evaluate the integral as follows.

$$\iint_S \mathbf{F} \cdot \mathbf{N} \, dS = \iiint_Q \text{div } \mathbf{F} \, dV = \iiint_Q (2x + x + 0) \, dV = \iiint_Q 3x \, dV$$

Next, use cylindrical coordinates with $x = r \cos \theta$ and $dV = r \, dz \, dr \, d\theta$.

$$\iiint_Q 3x \, dV = \int_0^{2\pi} \int_0^2 \int_0^{6 - r \cos \theta} (3r \cos \theta) r \, dz \, dr \, d\theta \qquad \text{Cylindrical coordinates}$$

$$= \int_0^{2\pi} \int_0^2 (18r^2 \cos \theta - 3r^3 \cos^2 \theta) \, dr \, d\theta$$

$$= \int_0^{2\pi} (48 \cos \theta - 12 \cos^2 \theta) \, d\theta$$

$$= \left[48 \sin \theta - 6\left(\theta + \frac{1}{2} \sin 2\theta \right) \right]_0^{2\pi}$$

$$= -12\pi$$

Flux and the Divergence Theorem

To help understand the Divergence Theorem, consider the two sides of the equation

$$\iint_S \mathbf{F} \cdot \mathbf{N} \, dS = \iiint_Q \text{div } \mathbf{F} \, dV.$$

You know from Section 15.6 that the flux integral on the left determines the total fluid flow across the surface S per unit of time. This can be approximated by summing the fluid flow across small patches of the surface. The triple integral on the right measures this same fluid flow across S but from a very different perspective—namely, by calculating the flow of fluid into (or out of) small *cubes* of volume ΔV_i. The flux of the ith cube is approximately div $\mathbf{F}(x_i, y_i, z_i) \, \Delta V_i$ for some point (x_i, y_i, z_i) in the ith cube. Note that for a cube in the interior of Q, the gain (or loss) of fluid through any one of its six sides is offset by a corresponding loss (or gain) through one of the sides of an adjacent cube. After summing over all the cubes in Q, the only fluid flow that is not canceled by adjoining cubes is that on the outside edges of the cubes on the boundary. So, the sum

$$\sum_{i=1}^n \text{div } \mathbf{F}(x_i, y_i, z_i) \, \Delta V_i$$

approximates the total flux into (or out of) Q and therefore through the surface S.

To see what is meant by the divergence of $\mathbf{F}$ at a point, consider ΔV_α to be the volume of a small sphere S_α of radius α and center (x_0, y_0, z_0) contained in region Q, as shown in Figure 15.60.

Solid region Q

S_α

(x_0, y_0, z_0)

Figure 15.60

Applying the Divergence Theorem to S_α produces

$$\text{Flux of } \mathbf{F} \text{ across } S_\alpha = \iiint_{Q_\alpha} \text{div } \mathbf{F} \, dV \approx \text{div } \mathbf{F}(x_0, y_0, z_0) \, \Delta V_\alpha$$

where Q_α is the interior of S_α. Consequently, you have

$$\text{div } \mathbf{F}(x_0, y_0, z_0) \approx \frac{\text{flux of } \mathbf{F} \text{ across } S_\alpha}{\Delta V_\alpha}.$$

By taking the limit as $\alpha \to 0$, you obtain the divergence of $\mathbf{F}$ at the point (x_0, y_0, z_0).

$$\text{div } \mathbf{F}(x_0, y_0, z_0) = \lim_{\alpha \to 0} \frac{\text{flux of } \mathbf{F} \text{ across } S_\alpha}{\Delta V_\alpha} = \text{flux per unit volume at } (x_0, y_0, z_0)$$

The point (x_0, y_0, z_0) in a vector field is classified as a source, a sink, or incompressible, as shown in the list below.

1. **Source,** for div $\mathbf{F} > 0$ See Figure 15.61(a).
2. **Sink,** for div $\mathbf{F} < 0$ See Figure 15.61(b).
3. **Incompressible,** for div $\mathbf{F} = 0$ See Figure 15.61(c).

> **REMARK** In hydrodynamics, a *source* is a point at which additional fluid is considered as being introduced to the region occupied by the fluid. A *sink* is a point at which fluid is considered as being removed.

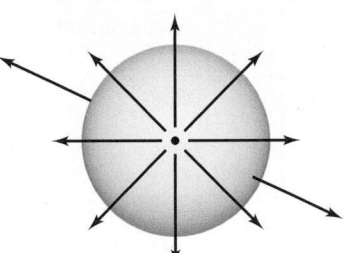
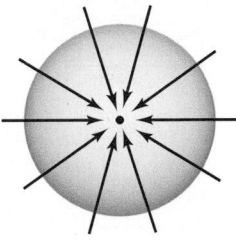
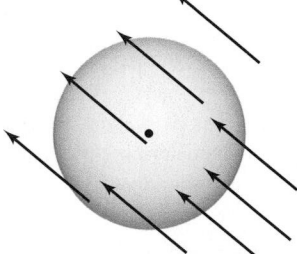

(a) Source: div $\mathbf{F} > 0$ **(b)** Sink: div $\mathbf{F} < 0$ **(c)** Incompressible: div $\mathbf{F} = 0$

Figure 15.61

EXAMPLE 4   **Calculating Flux by the Divergence Theorem**

$\vdots \cdots \triangleright$ *See LarsonCalculus.com for an interactive version of this type of example.*

Let Q be the region bounded by the sphere $x^2 + y^2 + z^2 = 4$. Find the outward flux of the vector field $\mathbf{F}(x, y, z) = 2x^3\mathbf{i} + 2y^3\mathbf{j} + 2z^3\mathbf{k}$ through the sphere.

Solution   By the Divergence Theorem, you have

$$\text{Flux across } S = \iint_S \mathbf{F} \cdot \mathbf{N} \, dS = \iiint_Q \text{div } \mathbf{F} \, dV = \iiint_Q 6(x^2 + y^2 + z^2) \, dV.$$

Next, use spherical coordinates with $\rho^2 = x^2 + y^2 + z^2$ and $dV = \rho^2 \sin \phi \, d\theta \, d\phi \, d\rho$.

$$\iiint_Q 6(x^2 + y^2 + z^2) \, dV = 6 \int_0^2 \int_0^\pi \int_0^{2\pi} \rho^4 \sin \phi \, d\theta \, d\phi \, d\rho \qquad \text{Spherical coordinates}$$

$$= 6 \int_0^2 \int_0^\pi 2\pi\rho^4 \sin \phi \, d\phi \, d\rho$$

$$= 12\pi \int_0^2 2\rho^4 \, d\rho$$

$$= 24\pi \left(\frac{32}{5} \right)$$

$$= \frac{768\pi}{5}.$$

15.7 Exercises

CONCEPT CHECK

1. Using Different Methods Suppose that a solid region Q is bounded by $z = x^2 + y^2$ and $z = 2$, as shown in the figure. What methods can you use to evaluate $\int_S \int \mathbf{F} \cdot \mathbf{N} \, dS$, where $\mathbf{F} = 2x\mathbf{i} + 3y\mathbf{j} - z^2\mathbf{k}$? Which method do you prefer?

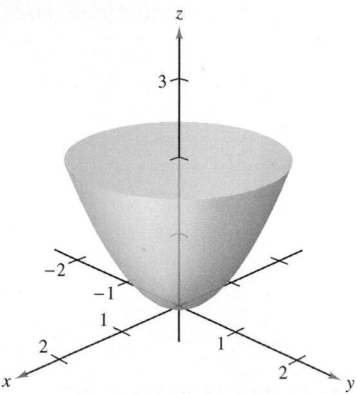

2. Classifying a Point in a Vector Field How do you determine whether a point (x_0, y_0, z_0) in a vector field is a source, a sink, or incompressible?

 **Verifying the Divergence Theorem In Exercises 3–8, verify the Divergence Theorem by evaluating**

$$\int_S \int \mathbf{F} \cdot \mathbf{N} \, dS$$

as a surface integral and as a triple integral.

3. $\mathbf{F}(x, y, z) = 2x\mathbf{i} - 2y\mathbf{j} + z^2\mathbf{k}$

S: cube bounded by the planes $x = 0$, $x = 1$, $y = 0$, $y = 1$, $z = 0$, $z = 1$

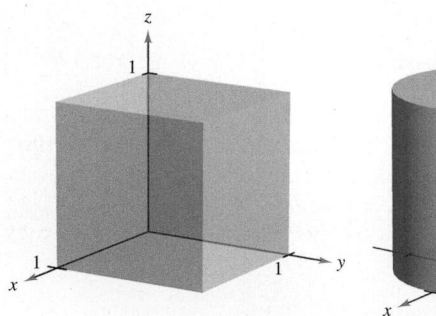

Figure for 3   Figure for 4

4. $\mathbf{F}(x, y, z) = 2x\mathbf{i} - 2y\mathbf{j} + z^2\mathbf{k}$

S: cylinder $x^2 + y^2 = 4$, $0 \le z \le 3$

5. $\mathbf{F}(x, y, z) = (2x - y)\mathbf{i} - (2y - z)\mathbf{j} + z\mathbf{k}$

S: surface bounded by the plane $2x + 4y + 2z = 12$ and the coordinate planes

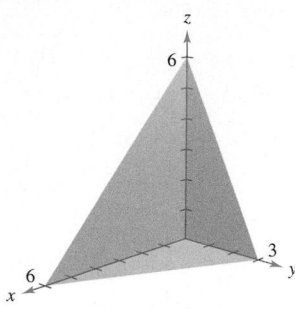

6. $\mathbf{F}(x, y, z) = xy\mathbf{i} + z\mathbf{j} + (x + y)\mathbf{k}$

S: surface bounded by the planes $y = 4$ and $z = 4 - x$ and the coordinate planes

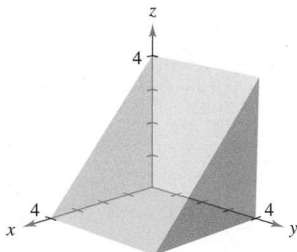

7. $\mathbf{F}(x, y, z) = xz\mathbf{i} + zy\mathbf{j} + 2z^2\mathbf{k}$

S: surface bounded by $z = 1 - x^2 - y^2$ and $z = 0$

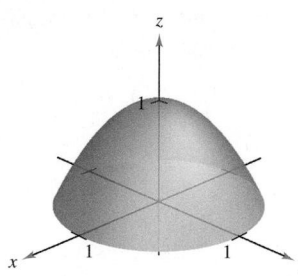

8. $\mathbf{F}(x, y, z) = xy^2\mathbf{i} + yx^2\mathbf{j} + e\mathbf{k}$

S: surface bounded by $z = \sqrt{x^2 + y^2}$ and $z = 4$

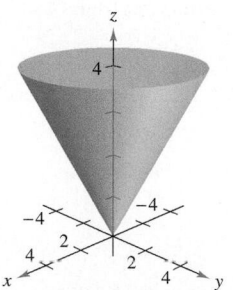

Using the Divergence Theorem In Exercises 9–18, use the Divergence Theorem to evaluate

$$\iint_S \mathbf{F} \cdot \mathbf{N} \, dS$$

and find the outward flux of **F** through the surface of the solid S bounded by the graphs of the equations. Use a computer algebra system to verify your results.

9. $\mathbf{F}(x, y, z) = x^2\mathbf{i} + y^2\mathbf{j} + z^2\mathbf{k}$

   S: $x = 0, x = a, y = 0, y = a, z = 0, z = a$

10. $\mathbf{F}(x, y, z) = x^2z^2\mathbf{i} - 2y\mathbf{j} + 3xyz\mathbf{k}$

    S: $x = 0, x = a, y = 0, y = a, z = 0, z = a$

11. $\mathbf{F}(x, y, z) = x^2\mathbf{i} - 2xy\mathbf{j} + xyz^2\mathbf{k}$

    S: $z = \sqrt{a^2 - x^2 - y^2}, z = 0$

12. $\mathbf{F}(x, y, z) = xy\mathbf{i} + yz\mathbf{j} - yz\mathbf{k}$

    S: $z = \sqrt{a^2 - x^2 - y^2}, z = 0$

13. $\mathbf{F}(x, y, z) = x\mathbf{i} + y\mathbf{j} + z\mathbf{k}$ 14. $\mathbf{F}(x, y, z) = xyz\mathbf{j}$

    S: $x^2 + y^2 + z^2 = 9$ S: $x^2 + y^2 = 4, z = 0, z = 5$

15. $\mathbf{F}(x, y, z) = x\mathbf{i} + y^2\mathbf{j} - z\mathbf{k}$

    S: $x^2 + y^2 = 25, z = 0, z = 7$

16. $\mathbf{F}(x, y, z) = (xy^2 + \cos z)\mathbf{i} + (x^2y + \sin z)\mathbf{j} + e^z\mathbf{k}$

    S: $z = \frac{1}{2}\sqrt{x^2 + y^2}, z = 8$

17. $\mathbf{F}(x, y, z) = xe^z\mathbf{i} + ye^z\mathbf{j} + e^z\mathbf{k}$

    S: $z = 4 - y, z = 0, x = 0, x = 6, y = 0$

18. $\mathbf{F}(x, y, z) = xy\mathbf{i} + 4y\mathbf{j} + xz\mathbf{k}$

    S: $x^2 + y^2 + z^2 = 16$

Classifying a Point In Exercises 19–22, a vector field and a point in the vector field are given. Determine whether the point is a source, a sink, or incompressible.

19. $\mathbf{F}(x, y, z) = 2\mathbf{i} + y\mathbf{j} + \mathbf{k}$, $(2, 2, 1)$

20. $\mathbf{F}(x, y, z) = e^{-x}\mathbf{i} - xy^2\mathbf{j} + \ln z\mathbf{k}$, $(0, -3, 1)$

21. $\mathbf{F}(x, y, z) = \sin x\mathbf{i} + \cos y\mathbf{j} + z^3 \sin y\mathbf{k}$, $\left(\dfrac{\pi}{2}, \pi, 4\right)$

22. $\mathbf{F}(x, y, z) = (4xy + z^2)\mathbf{i} + (2x^2 + 6yz)\mathbf{j} + 2xz\mathbf{k}$, $(1, -4, 2)$

23. **Source** Find a point that is a source in the vector field

    $\mathbf{F}(x, y, z) = x^2yz\mathbf{i} + x\mathbf{j} - z\mathbf{k}$.

24. **Sink** Find a point that is a sink in the vector field

    $\mathbf{F}(x, y, z) = e^{-x}\mathbf{i} + 4y\mathbf{j} + xyz^2\mathbf{k}$.

EXPLORING CONCEPTS

25. **Closed Surface** What is the value of

    $$\iint_S \text{curl } \mathbf{F} \cdot \mathbf{N} \, dS$$

    for any closed surface S? Explain.

---

26. **HOW DO YOU SEE IT?** The graph of a vector field **F** is shown. Does the graph suggest that the divergence of **F** at P is positive, negative, or zero?

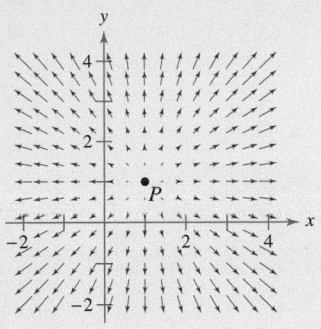

27. **Volume**

    (a) Use the Divergence Theorem to verify that the volume of the solid bounded by a surface S is

    $$\iint_S x \, dy \, dz = \iint_S y \, dz \, dx = \iint_S z \, dx \, dy.$$

    (b) Verify the result of part (a) for the cube bounded by $x = 0$, $x = a, y = 0, y = a, z = 0$, and $z = a$.

28. **Constant Vector Field** For the constant vector field $\mathbf{F}(x, y, z) = a_1\mathbf{i} + a_2\mathbf{j} + a_3\mathbf{k}$, verify the following integral for any closed surface S.

    $$\iint_S \mathbf{F} \cdot \mathbf{N} \, dS = 0$$

29. **Volume** For the vector field $\mathbf{F}(x, y, z) = x\mathbf{i} + y\mathbf{j} + z\mathbf{k}$, verify the following integral, where V is the volume of the solid bounded by the closed surface S.

    $$\iint_S \mathbf{F} \cdot \mathbf{N} \, dS = 3V$$

30. **Verifying an Identity** For the vector field $\mathbf{F}(x, y, z) = x\mathbf{i} + y\mathbf{j} + z\mathbf{k}$, verify that

    $$\frac{1}{\|\mathbf{F}\|}\iint_S \mathbf{F} \cdot \mathbf{N} \, dS = \frac{3}{\|\mathbf{F}\|}\iiint_Q dV.$$

Proof In Exercises 31 and 32, prove the identity, assuming that Q, S, and $\mathbf{N}$ meet the conditions of the Divergence Theorem and that the required partial derivatives of the scalar functions f and g are continuous. The expressions $D_\mathbf{N}f$ and $D_\mathbf{N}g$ are the derivatives in the direction of the vector $\mathbf{N}$ and are defined by $D_\mathbf{N}f = \nabla f \cdot \mathbf{N}$ and $D_\mathbf{N}g = \nabla g \cdot \mathbf{N}$.

31. $\displaystyle\iiint_Q (f\nabla^2 g + \nabla f \cdot \nabla g) \, dV = \iint_S fD_\mathbf{N}g \, dS$

    [*Hint:* Use div$(f\mathbf{G}) = f$ div $\mathbf{G} + \nabla f \cdot \mathbf{G}$.]

32. $\displaystyle\iiint_Q (f\nabla^2 g - g\nabla^2 f) \, dV = \iint_S (fD_\mathbf{N}g - gD_\mathbf{N}f) \, dS$

    (*Hint:* Use Exercise 31 twice.)

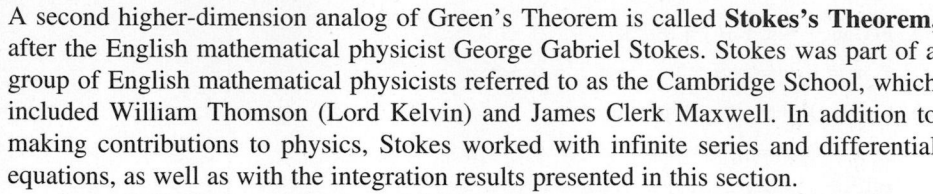

15.8 Stokes's Theorem

■ Understand and use Stokes's Theorem.
■ Use curl to analyze the motion of a rotating liquid.

Stokes's Theorem

GEORGE GABRIEL STOKES
(1819–1903)

Stokes became a Lucasian professor of mathematics at Cambridge in 1849. Five years later, he published the theorem that bears his name as a prize examination question there.
See LarsonCalculus.com to read more of this biography.

A second higher-dimension analog of Green's Theorem is called **Stokes's Theorem,** after the English mathematical physicist George Gabriel Stokes. Stokes was part of a group of English mathematical physicists referred to as the Cambridge School, which included William Thomson (Lord Kelvin) and James Clerk Maxwell. In addition to making contributions to physics, Stokes worked with infinite series and differential equations, as well as with the integration results presented in this section.

Stokes's Theorem gives the relationship between a surface integral over an oriented surface S and a line integral along a closed space curve C forming the boundary of S, as shown in Figure 15.62. The positive direction along C is counterclockwise relative to the normal vector $\mathbf{N}$. That is, if you imagine grasping the normal vector $\mathbf{N}$ with your right hand, with your thumb pointing in the direction of $\mathbf{N}$, then your fingers will point in the positive direction C, as shown in Figure 15.63.

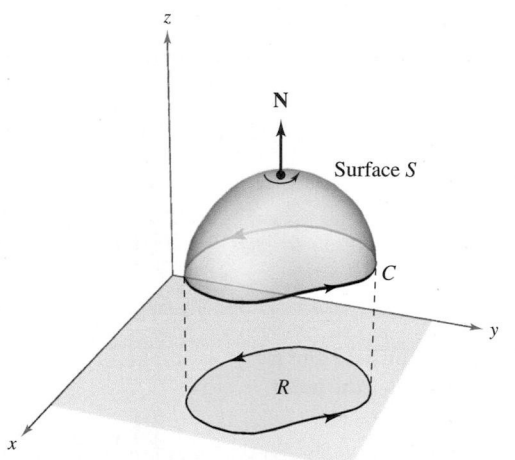

An oriented surface S bounded by a closed space curve C
Figure 15.62

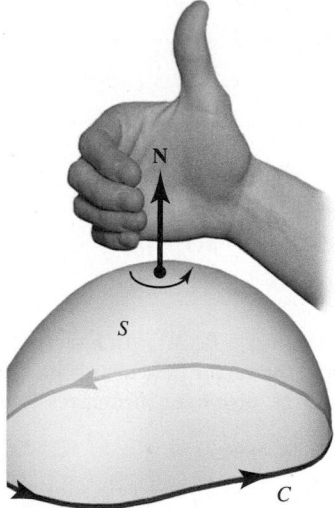

The positive direction along C is counterclockwise relative to $\mathbf{N}$.
Figure 15.63

THEOREM 15.13 Stokes's Theorem

Let S be an oriented surface with unit normal vector $\mathbf{N}$, bounded by a piecewise smooth simple closed curve C with a positive orientation. If $\mathbf{F}$ is a vector field whose component functions have continuous first partial derivatives on an open region containing S and C, then

$$\int_C \mathbf{F} \cdot d\mathbf{r} = \iint_S (\text{curl } \mathbf{F}) \cdot \mathbf{N} \, dS.$$

In Theorem 15.13, note that the line integral may be written in the differential form $\int_C M \, dx + N \, dy + P \, dz$ or in the vector form $\int_C \mathbf{F} \cdot \mathbf{T} \, ds$.

$S: 2x + 2y + z = 6$

C_3

C_2

N (upward)

R

C_1

$x + y = 3$

Figure 15.64

EXAMPLE 1 **Using Stokes's Theorem**

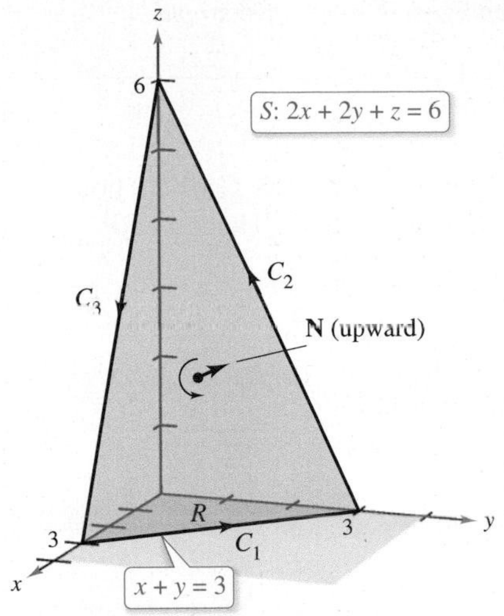

Let C be the oriented triangle lying in the plane

$$2x + 2y + z = 6$$

as shown in Figure 15.64. Evaluate

$$\int_C \mathbf{F} \cdot d\mathbf{r}$$

where $\mathbf{F}(x, y, z) = -y^2\mathbf{i} + z\mathbf{j} + x\mathbf{k}$.

Solution Using Stokes's Theorem, begin by finding the curl of $\mathbf{F}$.

$$\text{curl } \mathbf{F} = \begin{vmatrix} \mathbf{i} & \mathbf{j} & \mathbf{k} \\ \dfrac{\partial}{\partial x} & \dfrac{\partial}{\partial y} & \dfrac{\partial}{\partial z} \\ -y^2 & z & x \end{vmatrix} = -\mathbf{i} - \mathbf{j} + 2y\mathbf{k}$$

Considering

$$z = g(x, y) = 6 - 2x - 2y$$

you can use Theorem 15.11 for an upward normal vector to obtain

$$\int_C \mathbf{F} \cdot d\mathbf{r} = \iint_S (\text{curl } \mathbf{F}) \cdot \mathbf{N} \, dS$$

$$= \iint_R (-\mathbf{i} - \mathbf{j} + 2y\mathbf{k}) \cdot [-g_x(x, y)\mathbf{i} - g_y(x, y)\mathbf{j} + \mathbf{k}] \, dA$$

$$= \iint_R (-\mathbf{i} - \mathbf{j} + 2y\mathbf{k}) \cdot (2\mathbf{i} + 2\mathbf{j} + \mathbf{k}) \, dA$$

$$= \int_0^3 \int_0^{3-y} (2y - 4) \, dx \, dy$$

$$= \int_0^3 (-2y^2 + 10y - 12) \, dy$$

$$= \left[-\frac{2y^3}{3} + 5y^2 - 12y \right]_0^3$$

$$= -9.$$

Try evaluating the line integral in Example 1 directly, *without* using Stokes's Theorem. One way to do this would be to consider C as the union of C_1, C_2, and C_3, as follows.

C_1: $\mathbf{r}_1(t) = (3 - t)\mathbf{i} + t\mathbf{j}, \quad 0 \le t \le 3$

C_2: $\mathbf{r}_2(t) = (6 - t)\mathbf{j} + (2t - 6)\mathbf{k}, \quad 3 \le t \le 6$

C_3: $\mathbf{r}_3(t) = (t - 6)\mathbf{i} + (18 - 2t)\mathbf{k}, \quad 6 \le t \le 9$

The value of the line integral is

$$\int_C \mathbf{F} \cdot d\mathbf{r} = \int_{C_1} \mathbf{F} \cdot \mathbf{r}_1'(t) \, dt + \int_{C_2} \mathbf{F} \cdot \mathbf{r}_2'(t) \, dt + \int_{C_3} \mathbf{F} \cdot \mathbf{r}_3'(t) \, dt$$

$$= \int_0^3 t^2 \, dt + \int_3^6 (-2t + 6) \, dt + \int_6^9 (-2t + 12) \, dt$$

$$= 9 - 9 - 9$$

$$= -9.$$

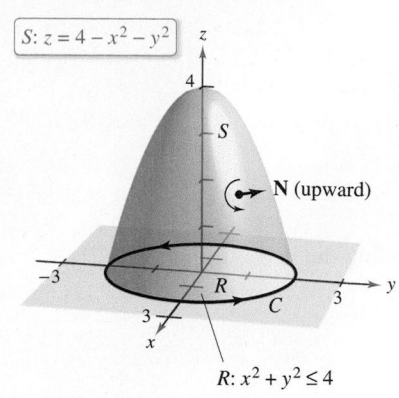

$S: z = 4 - x^2 - y^2$

$R: x^2 + y^2 \le 4$

Figure 15.65

EXAMPLE 2 **Verifying Stokes's Theorem**

⚬ ⚬ ⚬ ⚬ ▷ *See LarsonCalculus.com for an interactive version of this type of example.*

Let S be the portion of the paraboloid

$$z = 4 - x^2 - y^2$$

lying above the xy-plane, oriented upward (see Figure 15.65). Let C be its boundary curve in the xy-plane, oriented counterclockwise. Verify Stokes's Theorem for

$$\mathbf{F}(x, y, z) = 2z\mathbf{i} + x\mathbf{j} + y^2\mathbf{k}$$

by evaluating the surface integral and the equivalent line integral.

Solution As a *surface integral*, you have $z = g(x, y) = 4 - x^2 - y^2$, $g_x = -2x$, $g_y = -2y$, and

$$\text{curl } \mathbf{F} = \begin{vmatrix} \mathbf{i} & \mathbf{j} & \mathbf{k} \\ \dfrac{\partial}{\partial x} & \dfrac{\partial}{\partial y} & \dfrac{\partial}{\partial z} \\ 2z & x & y^2 \end{vmatrix} = 2y\mathbf{i} + 2\mathbf{j} + \mathbf{k}.$$

By Theorem 15.11 (for an upward normal vector), you obtain

$$\iint_S (\text{curl } \mathbf{F}) \cdot \mathbf{N} \, dS = \iint_R (2y\mathbf{i} + 2\mathbf{j} + \mathbf{k}) \cdot (2x\mathbf{i} + 2y\mathbf{j} + \mathbf{k}) \, dA$$

$$= \int_{-2}^{2} \int_{-\sqrt{4-x^2}}^{\sqrt{4-x^2}} (4xy + 4y + 1) \, dy \, dx$$

$$= \int_{-2}^{2} \left[2xy^2 + 2y^2 + y \right]_{-\sqrt{4-x^2}}^{\sqrt{4-x^2}} dx$$

$$= \int_{-2}^{2} 2\sqrt{4 - x^2} \, dx$$

$$= \text{Area of circle of radius 2}$$

$$= 4\pi.$$

As a *line integral*, you can parametrize C as

$$\mathbf{r}(t) = 2 \cos t\mathbf{i} + 2 \sin t\mathbf{j} + 0\mathbf{k}, \quad 0 \le t \le 2\pi.$$

For $\mathbf{F}(x, y, z) = 2z\mathbf{i} + x\mathbf{j} + y^2\mathbf{k}$, you obtain

$$\int_C \mathbf{F} \cdot d\mathbf{r} = \int_C M \, dx + N \, dy + P \, dz$$

$$= \int_C 2z \, dx + x \, dy + y^2 \, dz$$

$$= \int_0^{2\pi} [0 + (2 \cos t)(2 \cos t) + 0] \, dt$$

$$= \int_0^{2\pi} 4 \cos^2 t \, dt$$

$$= 2\int_0^{2\pi} (1 + \cos 2t) \, dt$$

$$= 2\left[t + \frac{1}{2} \sin 2t \right]_0^{2\pi}$$

$$= 4\pi.$$

Physical Interpretation of Curl

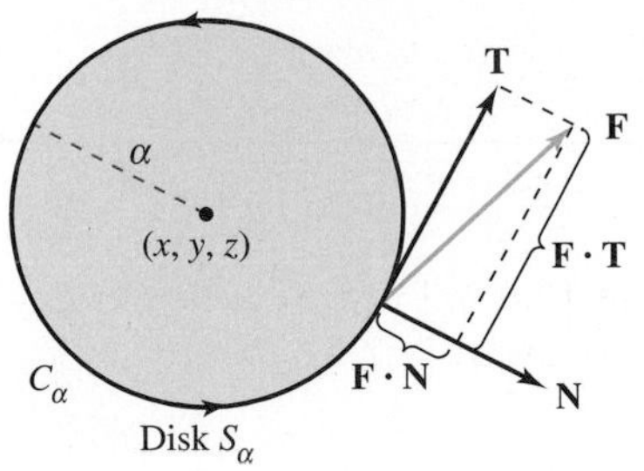

Stokes's Theorem provides insight into a physical interpretation of curl. In a vector field **F**, let S_α be a *small* circular disk of radius α, centered at (x, y, z) and with boundary C_α, as shown in Figure 15.66. At each point on the circle C_α, **F** has a normal component **F** · **N** and a tangential component **F** · **T**. The more closely **F** and **T** are aligned, the greater the value of **F** · **T**. So, a fluid tends to move along the circle rather than across it. Consequently, you say that the line integral around C_α measures the **circulation of F around C_α**. That is,

$$\int_{C_\alpha} \mathbf{F} \cdot \mathbf{T}\, ds = \text{circulation of } \mathbf{F} \text{ around } C_\alpha.$$

Figure 15.66

Now consider a small disk S_α to be centered at some point (x, y, z) on the surface S, as shown in Figure 15.67. On such a small disk, curl **F** is nearly constant, because it varies little from its value at (x, y, z). Moreover, (curl **F**) · **N** is also nearly constant on S_α because all unit normals to S_α are about the same. Consequently, Stokes's Theorem yields

$$\int_{C_\alpha} \mathbf{F} \cdot \mathbf{T}\, ds = \iint_{S_\alpha} (\text{curl } \mathbf{F}) \cdot \mathbf{N}\, dS$$

$$\approx (\text{curl } \mathbf{F}) \cdot \mathbf{N} \iint_{S_\alpha} dS$$

$$\approx (\text{curl } \mathbf{F}) \cdot \mathbf{N}(\pi\alpha^2).$$

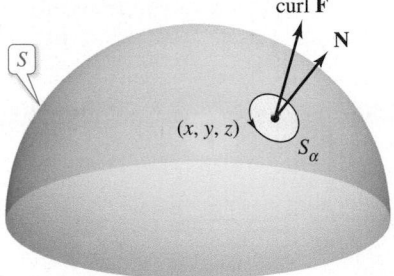

Figure 15.67

So,

$$(\text{curl } \mathbf{F}) \cdot \mathbf{N} \approx \frac{\displaystyle\int_{C_\alpha} \mathbf{F} \cdot \mathbf{T}\, ds}{\pi\alpha^2}$$

$$= \frac{\text{circulation of } \mathbf{F} \text{ around } C_\alpha}{\text{area of disk } S_\alpha}$$

$$= \text{rate of circulation.}$$

Assuming conditions are such that the approximation improves for smaller and smaller disks $(\alpha \to 0)$, it follows that

$$(\text{curl } \mathbf{F}) \cdot \mathbf{N} = \lim_{\alpha \to 0} \frac{1}{\pi\alpha^2} \int_{C_\alpha} \mathbf{F} \cdot \mathbf{T}\, ds$$

which is referred to as the **rotation of F about N**. That is,

$$\text{curl } \mathbf{F}(x, y, z) \cdot \mathbf{N} = \text{rotation of } \mathbf{F} \text{ about } \mathbf{N} \text{ at } (x, y, z).$$

In this case, the rotation of **F** is maximum when curl **F** and **N** have the same direction. Normally, this tendency to rotate will vary from point to point on the surface S, and Stokes's Theorem

$$\underbrace{\iint_S (\text{curl } \mathbf{F}) \cdot \mathbf{N}\, dS}_{\text{Surface integral}} = \underbrace{\int_C \mathbf{F} \cdot d\mathbf{r}}_{\text{Line integral}}$$

says that the collective measure of this *rotational* tendency taken over the entire surface S (surface integral) is equal to the tendency of a fluid to *circulate* around the boundary C (line integral).

EXAMPLE 3 **An Application of Curl**

A liquid is swirling around in a cylindrical container of radius 2, so that its motion is described by the velocity field

$$\mathbf{F}(x, y, z) = -y\sqrt{x^2 + y^2}\,\mathbf{i} + x\sqrt{x^2 + y^2}\,\mathbf{j}$$

as shown in the figure. Find

$$\iint_S (\text{curl } \mathbf{F}) \cdot \mathbf{N}\, dS$$

where S is the upper surface of the cylindrical container.

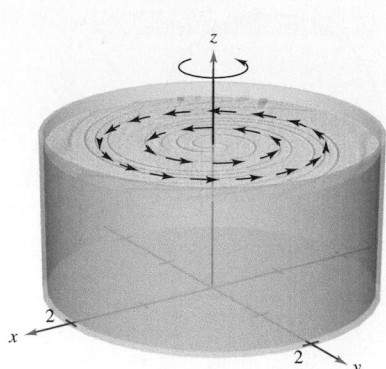

Solution The curl of $\mathbf{F}$ is given by

$$\text{curl } \mathbf{F} = \begin{vmatrix} \mathbf{i} & \mathbf{j} & \mathbf{k} \\ \dfrac{\partial}{\partial x} & \dfrac{\partial}{\partial y} & \dfrac{\partial}{\partial z} \\ -y\sqrt{x^2 + y^2} & x\sqrt{x^2 + y^2} & 0 \end{vmatrix} = 3\sqrt{x^2 + y^2}\,\mathbf{k}.$$

Letting $\mathbf{N} = \mathbf{k}$, you have

$$\iint_S (\text{curl } \mathbf{F}) \cdot \mathbf{N}\, dS = \iint_R 3\sqrt{x^2 + y^2}\, dA$$

$$= \int_0^{2\pi} \int_0^2 (3r)r\, dr\, d\theta$$

$$= \int_0^{2\pi} r^3 \Big]_0^2 \, d\theta$$

$$= \int_0^{2\pi} 8\, d\theta$$

$$= 16\pi.$$

If curl $\mathbf{F} = \mathbf{0}$ throughout region Q, then the rotation of $\mathbf{F}$ about each unit normal $\mathbf{N}$ is 0. That is, $\mathbf{F}$ is irrotational. From Section 15.1, you know that this is a characteristic of conservative vector fields.

SUMMARY OF INTEGRATION FORMULAS

Fundamental Theorem of Calculus

$$\int_a^b F'(x)\, dx = F(b) - F(a)$$

Green's Theorem

$$\int_C M\, dx + N\, dy = \iint_R \left(\frac{\partial N}{\partial x} - \frac{\partial M}{\partial y}\right) dA = \int_C \mathbf{F} \cdot \mathbf{T}\, ds = \int_C \mathbf{F} \cdot d\mathbf{r} = \iint_R (\text{curl } \mathbf{F}) \cdot \mathbf{k}\, dA$$

$$\int_C \mathbf{F} \cdot \mathbf{N}\, ds = \iint_R \text{div } \mathbf{F}\, dA$$

Divergence Theorem

$$\iint_S \mathbf{F} \cdot \mathbf{N}\, dS = \iiint_Q \text{div } \mathbf{F}\, dV$$

Fundamental Theorem of Line Integrals

$$\int_C \mathbf{F} \cdot d\mathbf{r} = \int_C \nabla f \cdot d\mathbf{r} = f(x(b), y(b)) - f(x(a), y(a))$$

Stokes's Theorem

$$\int_C \mathbf{F} \cdot d\mathbf{r} = \iint_S (\text{curl } \mathbf{F}) \cdot \mathbf{N}\, dS$$

15.8 Exercises

CONCEPT CHECK

1. Stokes's Theorem Explain the benefit of Stokes's Theorem when the boundary of the surface is a piecewise curve.

2. Curl What is the physical interpretation of curl?

Verifying Stokes's Theorem In Exercises 3–6, verify Stokes's Theorem by evaluating $\int_C \mathbf{F} \cdot d\mathbf{r}$ as a line integral and as a double integral.

3. $\mathbf{F}(x, y, z) = (-y + z)\mathbf{i} + (x - z)\mathbf{j} + (x - y)\mathbf{k}$

   S: $z = 9 - x^2 - y^2$, $z \geq 0$

4. $\mathbf{F}(x, y, z) = (-y + z)\mathbf{i} + (x - z)\mathbf{j} + (x - y)\mathbf{k}$

   S: $z = \sqrt{1 - x^2 - y^2}$

5. $\mathbf{F}(x, y, z) = xyz\mathbf{i} + y\mathbf{j} + z\mathbf{k}$

   S: $6x + 6y + z = 12$, first octant

6. $\mathbf{F}(x, y, z) = z^2\mathbf{i} + x^2\mathbf{j} + y^2\mathbf{k}$

   S: $z = y^2$, $0 \leq x \leq a$, $0 \leq y \leq a$

Using Stokes's Theorem In Exercises 7–16, use Stokes's Theorem to evaluate $\int_C \mathbf{F} \cdot d\mathbf{r}$. In each case, C is oriented counterclockwise as viewed from above.

7. $\mathbf{F}(x, y, z) = 2y\mathbf{i} + 3z\mathbf{j} + x\mathbf{k}$

   C: triangle with vertices $(2, 0, 0)$, $(0, 2, 0)$, and $(0, 0, 2)$

8. $\mathbf{F}(x, y, z) = 4z\mathbf{i} + x^2\mathbf{j} + e^y\mathbf{k}$

   C: triangle with vertices $(4, 0, 0)$, $(0, 2, 0)$, and $(0, 0, 8)$

9. $\mathbf{F}(x, y, z) = z^2\mathbf{i} + 2x\mathbf{j} + y^2\mathbf{k}$

   S: $z = 1 - x^2 - y^2$, $z \geq 0$

10. $\mathbf{F}(x, y, z) = 4xz\mathbf{i} + y\mathbf{j} + 4xy\mathbf{k}$

   S: $z = 9 - x^2 - y^2$, $z \geq 0$

11. $\mathbf{F}(x, y, z) = z^2\mathbf{i} + y\mathbf{j} + z\mathbf{k}$

   S: $z = \sqrt{4 - x^2 - y^2}$

12. $\mathbf{F}(x, y, z) = x^2\mathbf{i} + z^2\mathbf{j} - xyz\mathbf{k}$

   S: $z = \sqrt{4 - x^2 - y^2}$

13. $\mathbf{F}(x, y, z) = -\ln\sqrt{x^2 + y^2}\,\mathbf{i} + \arctan\dfrac{x}{y}\mathbf{j} + \mathbf{k}$

   S: $z = 9 - 2x - 3y$ over $r = 2\sin 2\theta$ in the first octant

14. $\mathbf{F}(x, y, z) = yz\mathbf{i} + (2 - 3y)\mathbf{j} + (x^2 + y^2)\mathbf{k}$, $x^2 + y^2 \leq 16$

   S: the first-octant portion of $x^2 + z^2 = 16$ over $x^2 + y^2 = 16$

15. $\mathbf{F}(x, y, z) = xyz\mathbf{i} + y\mathbf{j} + z\mathbf{k}$

   S: $z = x^2$, $0 \leq x \leq a$, $0 \leq y \leq a$

16. $\mathbf{F}(x, y, z) = xyz\mathbf{i} + y\mathbf{j} + z\mathbf{k}$, $x^2 + y^2 \leq a^2$

   S: the first-octant portion of $z = x^2$ over $x^2 + y^2 = a^2$

Motion of a Liquid In Exercises 17 and 18, the motion of a liquid in a cylindrical container of radius 3 is described by the velocity field $\mathbf{F}(x, y, z)$. Find $\int_S\!\int (\text{curl }\mathbf{F}) \cdot \mathbf{N}\, dS$, where S is the upper surface of the cylindrical container.

17. $\mathbf{F}(x, y, z) = -\frac{1}{6}y^3\mathbf{i} + \frac{1}{6}x^3\mathbf{j} + 5\mathbf{k}$

18. $\mathbf{F}(x, y, z) = -z\mathbf{i} + y^2\mathbf{k}$

EXPLORING CONCEPTS

19. Think About It Let $\mathbf{K}$ be a constant vector. Let S be an oriented surface with a unit normal vector $\mathbf{N}$, bounded by a smooth curve C. Determine whether

$$\iint_S \mathbf{K} \cdot \mathbf{N}\, dS = \frac{1}{2}\int_C (\mathbf{K} \times \mathbf{r}) \cdot d\mathbf{r}.$$

Explain. (*Hint:* Use $\mathbf{r} = x\mathbf{i} + y\mathbf{j} + z\mathbf{k}$.)

20. HOW DO YOU SEE IT? Let S_1 be the portion of the paraboloid lying above the xy-plane, and let S_2 be the hemisphere, as shown in the figures. Both surfaces are oriented upward. For a vector field $\mathbf{F}(x, y, z)$ with continuous partial derivatives, does

$$\iint_{S_1} (\text{curl }\mathbf{F}) \cdot \mathbf{N}\, dS_1 = \iint_{S_2} (\text{curl }\mathbf{F}) \cdot \mathbf{N}\, dS_2?$$

Explain your reasoning.

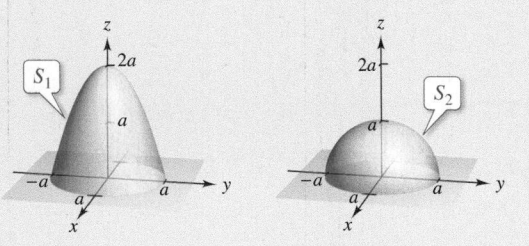

PUTNAM EXAM CHALLENGE

21. Let $\mathbf{G}(x, y) = \left(\dfrac{-y}{x^2 + 4y^2}, \dfrac{x}{x^2 + 4y^2}, 0\right)$.

Prove or disprove that there is a vector-valued function $\mathbf{F}(x, y, z) = (M(x, y, z), N(x, y, z), P(x, y, z))$ with the following properties:

(i) M, N, P have continuous partial derivatives for all $(x, y, z) \neq (0, 0, 0)$;

(ii) Curl $\mathbf{F} = \mathbf{0}$ for all $(x, y, z) \neq (0, 0, 0)$;

(iii) $\mathbf{F}(x, y, 0) = \mathbf{G}(x, y)$.

Sketching a Vector Field In Exercises 1 and 2, find $\|\mathbf{F}\|$ and sketch several representative vectors in the vector field. Use a computer algebra system to verify your results.

1. $\mathbf{F}(x, y, z) = x\mathbf{i} + \mathbf{j} + 2\mathbf{k}$ **2.** $\mathbf{F}(x, y) = \mathbf{i} - 2y\mathbf{j}$

Finding a Conservative Vector Field In Exercises 3–6, find the conservative vector field for the potential function by finding its gradient.

3. $f(x, y) = \sin xy - y^2$

4. $f(x, y) = \sqrt{xy}$

5. $f(x, y, z) = 2x^2 + xy + z^2$

6. $f(x, y, z) = x^2 e^{yz}$

Testing for a Conservative Vector Field In Exercises 7–10, determine whether the vector field is conservative.

7. $\mathbf{F}(x, y) = \cosh y\mathbf{i} + x \sinh x\mathbf{j}$

8. $\mathbf{F}(x, y) = \dfrac{y \ln x}{x}\mathbf{i} + (\ln x)^2\mathbf{j}$

9. $\mathbf{F}(x, y, z) = y^2\mathbf{i} + 2xy\mathbf{j} + \cos z\mathbf{k}$

10. $\mathbf{F}(x, y, z) = 3e^{xy}\mathbf{i} + 3e^{x+y}\mathbf{j} + e^{3yz}\mathbf{k}$

Finding a Potential Function In Exercises 11–18, determine whether the vector field is conservative. If it is, find a potential function for the vector field.

11. $\mathbf{F}(x, y) = -\dfrac{y}{x^2}\mathbf{i} + \dfrac{1}{x}\mathbf{j}$ **12.** $\mathbf{F}(x, y) = \dfrac{1}{y}\mathbf{i} - \dfrac{y}{x^2}\mathbf{j}$

13. $\mathbf{F}(x, y) = (xy^2 - x^2)\mathbf{i} + (x^2y + y^2)\mathbf{j}$

14. $\mathbf{F}(x, y) = (-2y^3 \sin 2x)\mathbf{i} + 3y^2(1 + \cos 2x)\mathbf{j}$

15. $\mathbf{F}(x, y, z) = 4xy^2\mathbf{i} + 2x^2\mathbf{j} + 2z\mathbf{k}$

16. $\mathbf{F}(x, y, z) = (4xy + z^2)\mathbf{i} + (2x^2 + 6yz)\mathbf{j} + 2xz\mathbf{k}$

17. $\mathbf{F}(x, y, z) = \dfrac{yz\mathbf{i} - xz\mathbf{j} - xy\mathbf{k}}{y^2z^2}$

18. $\mathbf{F}(x, y, z) = (\sin z)(y\mathbf{i} + x\mathbf{j} + \mathbf{k})$

Divergence and Curl In Exercises 19–26, find (a) the divergence of the vector field and (b) the curl of the vector field.

19. $\mathbf{F}(x, y, z) = x^2\mathbf{i} + xy^2\mathbf{j} + x^2z\mathbf{k}$

20. $\mathbf{F}(x, y, z) = y^2\mathbf{j} - z^2\mathbf{k}$

21. $\mathbf{F}(x, y, z) = (\cos y + y \cos x)\mathbf{i} + (\sin x - x \sin y)\mathbf{j} + xyz\mathbf{k}$

22. $\mathbf{F}(x, y, z) = (3x - y)\mathbf{i} + (y - 2z)\mathbf{j} + (z - 3x)\mathbf{k}$

23. $\mathbf{F}(x, y, z) = \arcsin x\mathbf{i} + xy^2\mathbf{j} + yz^2\mathbf{k}$

24. $\mathbf{F}(x, y, z) = (x^2 - y)\mathbf{i} - (x + \sin^2 y)\mathbf{j}$

25. $\mathbf{F}(x, y, z) = \ln(x^2 + y^2)\mathbf{i} + \ln(x^2 + y^2)\mathbf{j} + z\mathbf{k}$

26. $\mathbf{F}(x, y, z) = \dfrac{z}{x}\mathbf{i} + \dfrac{z}{y}\mathbf{j} + z^2\mathbf{k}$

Evaluating a Line Integral In Exercises 27–30, evaluate the line integral along the given path(s).

27. $\displaystyle\int_C (x^2 + y^2)\, ds$

(a) C: line segment from $(0, 0)$ to $(3, 4)$

(b) C: one revolution counterclockwise around the circle $x^2 + y^2 = 1$, starting at $(1, 0)$

28. $\displaystyle\int_C xy\, ds$

(a) C: line segment from $(0, 0)$ to $(5, 4)$

(b) C: counterclockwise around the triangle with vertices $(0, 0)$, $(4, 0)$, and $(0, 2)$

29. $\displaystyle\int_C (x^2 + y^2)\, ds$

C: $\mathbf{r}(t) = (1 - \sin t)\mathbf{i} + (1 - \cos t)\mathbf{j}, \quad 0 \le t \le 2\pi$

30. $\displaystyle\int_C (x^2 + y^2)\, ds$

C: $\mathbf{r}(t) = (\cos t + t \sin t)\mathbf{i} + (\sin t - t \cos t)\mathbf{j}, \quad 0 \le t \le 2\pi$

Evaluating a Line Integral Using Technology In Exercises 31 and 32, use a computer algebra system to evaluate the line integral along the given path.

31. $\displaystyle\int_C (2x + y)\, ds$

C: $\mathbf{r}(t) = a\cos^3 t\mathbf{i} + a\sin^3 t\mathbf{j}, \quad 0 \le t \le \dfrac{\pi}{2}$

32. $\displaystyle\int_C (x^2 + y^2 + z^2)\, ds$

C: $\mathbf{r}(t) = t\mathbf{i} + t^2\mathbf{j} + t^{3/2}\mathbf{k}, \quad 0 \le t \le 4$

Mass In Exercises 33 and 34, find the total mass of the wire with density ρ whose shape is modeled by r.

33. $\mathbf{r}(t) = 3\cos t\mathbf{i} + 3\sin t\mathbf{j}, \quad 0 \le t \le \pi, \quad \rho(x, y) = 1 + x$

34. $\mathbf{r}(t) = 3\mathbf{i} + t^2\mathbf{j} + 2t\mathbf{k}, \quad 2 \le t \le 4, \quad \rho(x, y, z) = xz$

Evaluating a Line Integral of a Vector Field In Exercises 35–38, evaluate $\displaystyle\int_C \mathbf{F} \cdot d\mathbf{r}$.

35. $\mathbf{F}(x, y) = xy\mathbf{i} + 2xy\mathbf{j}$

C: $\mathbf{r}(t) = t^2\mathbf{i} + t^2\mathbf{j}, \quad 0 \le t \le 1$

36. $\mathbf{F}(x, y) = (x - y)\mathbf{i} + (x + y)\mathbf{j}$

C: $\mathbf{r}(t) = 4\cos t\mathbf{i} + 3\sin t\mathbf{j}, \quad 0 \le t \le 2\pi$

37. $\mathbf{F}(x, y, z) = x\mathbf{i} + y\mathbf{j} + z\mathbf{k}$

C: $\mathbf{r}(t) = 2\cos t\mathbf{i} + 2\sin t\mathbf{j} + t\mathbf{k}, \quad 0 \le t \le 2\pi$

38. $\mathbf{F}(x, y, z) = (2y - z)\mathbf{i} + (z - x)\mathbf{j} + (x - y)\mathbf{k}$

C: $\mathbf{r}(t) = -3t\mathbf{i} + (2t + 1)\mathbf{j} + 4\mathbf{k}, \quad 0 \le t \le 2$

Work In Exercises 39 and 40, find the work done by the force field **F** on a particle moving along the given path.

39. $\mathbf{F}(x, y) = x\mathbf{i} - \sqrt{y}\mathbf{j}$

  C: $x = t$, $y = t^{3/2}$ from $(0, 0)$ to $(4, 8)$

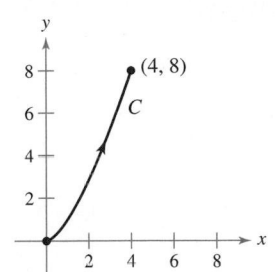

Figure for 39

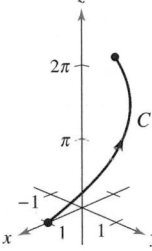

Figure for 40

40. $\mathbf{F}(x, y, z) = 2\mathbf{i} + y\mathbf{j} + z\mathbf{k}$

  C: $\mathbf{r}(t) = \cos t\mathbf{i} + \sin t\mathbf{j} + 2t\mathbf{k}$, $\quad 0 \le t \le \pi$

Evaluating a Line Integral in Differential Form In Exercises 41 and 42, evaluate $\displaystyle\int_C (y - x)\, dx + (2x + 5y)\, dy$.

41. *C*: line segments from $(0, 0)$ to $(2, -4)$ and $(2, -4)$ to $(4, -4)$

42. *C*: arc on $y = \sqrt{x}$ from $(0, 0)$ to $(9, 3)$

Lateral Surface Area In Exercises 43 and 44, find the area of the lateral surface over the curve *C* in the *xy*-plane and under the surface $z = f(x, y)$, where

Lateral surface area $= \displaystyle\int_C f(x, y)\, ds$.

43. $f(x, y) = 3 + \sin(x + y)$; *C*: $y = 2x$ from $(0, 0)$ to $(2, 4)$

44. $f(x, y) = 12 - x - y$; *C*: $y = x^2$ from $(0, 0)$ to $(2, 4)$

Line Integral of a Conservative Vector Field In Exercises 45 and 46, (a) show that **F** is conservative and (b) verify that the value of $\displaystyle\int_C \mathbf{F} \cdot d\mathbf{r}$ is the same for each parametric representation of *C*.

45. $\mathbf{F}(x, y) = (3x + 4)\mathbf{i} + y^3\mathbf{j}$

  (i) $\mathbf{r}(t) = t\mathbf{i} + t\mathbf{j}$, $0 \le t \le 4$

  (ii) $\mathbf{r}(w) = w^2\mathbf{i} + w^2\mathbf{j}$, $0 \le w \le 2$

46. $\mathbf{F}(x, y) = xy\mathbf{i} + \frac{1}{2}x^2\mathbf{j}$

  (i) $\mathbf{r}(\theta) = \sin \theta\mathbf{i} + \cos \theta\mathbf{j}$, $0 \le \theta \le \dfrac{\pi}{2}$

  (ii) $\mathbf{r}(t) = t\mathbf{i} + (1 - t)\mathbf{j}$, $0 \le t \le 1$

Using the Fundamental Theorem of Line Integrals In Exercises 47–50, evaluate $\displaystyle\int_C \mathbf{F} \cdot d\mathbf{r}$ using the Fundamental Theorem of Line Integrals.

47. $\mathbf{F}(x, y) = e^{2x}\mathbf{i} + e^{2y}\mathbf{j}$

  C: line segment from $(-1, -1)$ to $(0, 0)$

48. $\mathbf{F}(x, y) = -\sin y\mathbf{i} - x \cos y\mathbf{j}$

  C: clockwise around the circle $(x + 1)^2 + y^2 = 16$ from $(-1, 4)$ to $(3, 0)$

49. $\mathbf{F}(x, y, z) = 2xyz\mathbf{i} + x^2z\mathbf{j} + x^2y\mathbf{k}$

  C: smooth curve from $(0, 0, 0)$ to $(1, 3, 2)$

50. $\mathbf{F}(x, y, z) = y\mathbf{i} + x\mathbf{j} + \dfrac{1}{z}\mathbf{k}$

  C: smooth curve from $(0, 0, 1)$ to $(4, 4, 4)$

Finding Work in a Conservative Force Field In Exercises 51 and 52, (a) show that $\displaystyle\int_C \mathbf{F} \cdot d\mathbf{r}$ is independent of path and (b) calculate the work done by the force field **F** on an object moving along a curve from *P* to *Q*.

51. $\mathbf{F}(x, y) = (1 - 3xy^2)\mathbf{i} - 3x^2y\mathbf{j}$; $\quad P(4, 2)$, $Q(0, 1)$

52. $\mathbf{F}(x, y) = e^{2y}\mathbf{i} + 2xe^{2y}\mathbf{j}$; $\quad P(-1, 3)$, $Q(4, 5)$

Evaluating a Line Integral Using Green's Theorem In Exercises 53–58, use Green's Theorem to evaluate the line integral.

53. $\displaystyle\int_C y\, dx + 2x\, dy$

  C: square with vertices $(0, 0)$, $(0, 1)$, $(1, 0)$, and $(1, 1)$

54. $\displaystyle\int_C xy\, dx + (x^2 + y^2)\, dy$

  C: square with vertices $(0, 0)$, $(0, 2)$, $(2, 0)$, and $(2, 2)$

55. $\displaystyle\int_C xy^2\, dx + x^2y\, dy$

  C: $x = 4 \cos t$, $y = 4 \sin t$

56. $\displaystyle\int_C (x^2 - y^2)\, dx + 3y^2\, dy$

  C: $x^2 + y^2 = 9$

57. $\displaystyle\int_C xy\, dx + x^2\, dy$

  C: boundary of the region between the graphs of $y = x^2$ and $y = 1$

58. $\displaystyle\int_C y^2\, dx + x^{4/3}\, dy$

  C: $x^{2/3} + y^{2/3} = 1$

Work In Exercises 59 and 60, use Green's Theorem to calculate the work done by the force **F** on a particle that is moving counterclockwise around the closed path *C*.

59. $\mathbf{F}(x, y) = y^2\mathbf{i} + 2xy\mathbf{j}$

  C: $x^2 + y^2 = 36$

60. $\mathbf{F}(x, y) = 3\mathbf{i} + (x^3 + 1)\mathbf{j}$

  C: boundary of the region lying between the graphs of $y = x^2$ and $y = 4$

Area In Exercises 61 and 62, use a line integral to find the area of the region *R*.

61. *R*: triangle bounded by the graphs of $y = \frac{1}{2}x$, $y = 6 - x$, and $y = x$

62. *R*: region bounded by the graphs of $y = 3x$ and $y = 4 - x^2$

Sketching a Parametric Surface In Exercises 63 and 64, find the rectangular equation for the surface by eliminating the parameters from the vector-valued function. Identify the surface and sketch its graph.

63. $\mathbf{r}(u, v) = 3u \cos v\mathbf{i} + 3u \sin v\mathbf{j} + 18u^2\mathbf{k}$

64. $\mathbf{r}(u, v) = 3(u + v)\mathbf{i} + u\mathbf{j} - 6v\mathbf{k}$

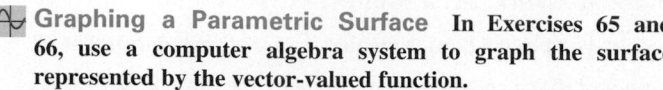 **Graphing a Parametric Surface** In Exercises 65 and 66, use a computer algebra system to graph the surface represented by the vector-valued function.

65. $\mathbf{r}(u, v) = \sec u \cos v\mathbf{i} + (1 + 2 \tan u) \sin v\mathbf{j} + 2u\mathbf{k}$

$$0 \le u \le \frac{\pi}{3}, \quad 0 \le v \le 2\pi$$

66. $\mathbf{r}(u, v) = e^{-u/4} \cos v\mathbf{i} + e^{-u/4} \sin v\mathbf{j} + \frac{u}{6}\mathbf{k}$

$$0 \le u \le 4, \quad 0 \le v \le 2\pi$$

Representing a Surface Parametrically In Exercises 67 and 68, find a vector-valued function whose graph is the indicated surface.

67. The ellipsoid $\dfrac{x^2}{1} + \dfrac{y^2}{8} + \dfrac{z^2}{9} = 1$

68. The part of the plane $z = 2$ that lies inside the cylinder $x^2 + y^2 = 25$

Representing a Surface of Revolution Parametrically In Exercises 69 and 70, write a set of parametric equations for the surface of revolution obtained by revolving the graph of the function about the given axis.

Function	Axis of Revolution
69. $y = 2x^3, 0 \le x \le 2$	x-axis
70. $z = \sqrt{y + 1}, 0 \le y \le 3$	y-axis

Finding Surface Area In Exercises 71 and 72, find the area of the surface over the given region. Use a computer algebra system to verify your results.

71. $\mathbf{r}(u, v) = 4u\mathbf{i} + (3u - v)\mathbf{j} + v\mathbf{k}$

$$0 \le u \le 3, \quad 0 \le v \le 1$$

72. $\mathbf{r}(u, v) = 3u \cos v\mathbf{i} + 3u \sin v\mathbf{j} + u\mathbf{k}$

$$0 \le u \le 2, \quad 0 \le v \le 2\pi$$

Evaluating a Surface Integral In Exercises 73 and 74, evaluate

$$\iint_S (5x + y - 2z) \, dS.$$

73. S: $z = x + \dfrac{y}{2}, \quad 0 \le x \le 2, \quad 0 \le y \le 5$

74. S: $z = e^2 - x, \quad 0 \le x \le 4, \quad 0 \le y \le \sqrt{x}$

Mass In Exercises 75 and 76, find the mass of the surface lamina S of density ρ.

75. S: $2y + 6x + z = 18$, first octant, $\rho(x, y, z) = 2x$

76. S: $z = 20 - 4x - 5y$, first octant, $\rho(x, y, z) = ky$

Evaluating a Surface Integral In Exercises 77 and 78, evaluate $\displaystyle\iint_S f(x, y) \, dS.$

77. $f(x, y) = x + y$

S: $\mathbf{r}(u, v) = u\mathbf{i} + v\mathbf{j} + 5v\mathbf{k}, \quad 0 \le u \le 1, \quad 0 \le v \le 3$

78. $f(x, y) = x^2y$

S: $\mathbf{r}(u, v) = 5 \cos u\mathbf{i} + 5 \sin u\mathbf{j} + v\mathbf{k}$

$$0 \le u \le \frac{\pi}{2}, \quad 0 \le v \le 1$$

Evaluating a Flux Integral In Exercises 79 and 80, find the flux of F across S,

$$\iint_S \mathbf{F} \cdot \mathbf{N} \, dS$$

where N is the upward unit normal vector to S.

79. $\mathbf{F}(x, y, z) = -2\mathbf{i} - 2\mathbf{j} + \mathbf{k}$

S: $z = 25 - x^2 - y^2, \quad z \ge 0$

80. $\mathbf{F}(x, y, z) = x\mathbf{i} + 2y\mathbf{j} + 2z\mathbf{k}$

S: $x + y + 3z = 3$, first octant

Using the Divergence Theorem In Exercises 81 and 82, use the Divergence Theorem to evaluate

$$\iint_S \mathbf{F} \cdot \mathbf{N} \, dS$$

and find the outward flux of F through the surface of the solid bounded by the graphs of the equations.

81. $\mathbf{F}(x, y, z) = x^2\mathbf{i} + xy\mathbf{j} + z\mathbf{k}$

Q: solid region bounded by the coordinate planes and the plane $2x + 3y + 4z = 12$

82. $\mathbf{F}(x, y, z) = x\mathbf{i} + y\mathbf{j} + z\mathbf{k}$

Q: solid region bounded by the coordinate planes and the plane $2x + 3y + 4z = 12$

Using Stokes's Theorem In Exercises 83 and 84, use Stokes's Theorem to evaluate

$$\int_C \mathbf{F} \cdot d\mathbf{r}.$$

In each case, C is oriented counterclockwise as viewed from above.

83. $\mathbf{F}(x, y, z) = (\cos y + y \cos x)\mathbf{i} + (\sin x - x \sin y)\mathbf{j} + xyz\mathbf{k}$

S: portion of $z = y^2$ over the square in the xy-plane with vertices $(0, 0), (a, 0), (a, a)$, and $(0, a)$

84. $\mathbf{F}(x, y, z) = (x - z)\mathbf{i} + (y - z)\mathbf{j} + x^2\mathbf{k}$

S: first-octant portion of the plane $3x + y + 2z = 12$

Motion of a Liquid In Exercises 85 and 86, the motion of a liquid in a cylindrical container of radius 4 is described by the velocity field $\mathbf{F}(x, y, z)$. Find $\displaystyle\iint_S (\operatorname{curl} \mathbf{F}) \cdot \mathbf{N} \, dS$, where S is the upper surface of the cylindrical container.

85. $\mathbf{F}(x, y, z) = \mathbf{i} + x\mathbf{j} - \mathbf{k}$ **86.** $\mathbf{F}(x, y, z) = y^2\mathbf{i} + 3z\mathbf{j} + \mathbf{k}$

P.S. Problem Solving

See **CalcChat.com** for tutorial help and worked-out solutions to odd-numbered exercises.

1. Heat Flux Consider a single heat source located at the origin with temperature

$$T(x, y, z) = \frac{25}{\sqrt{x^2 + y^2 + z^2}}.$$

(a) Calculate the heat flux across the surface

$$S = \left\{ (x, y, z): z = \sqrt{1 - x^2}, \ -\frac{1}{2} \le x \le \frac{1}{2}, 0 \le y \le 1 \right\}$$

as shown in the figure.

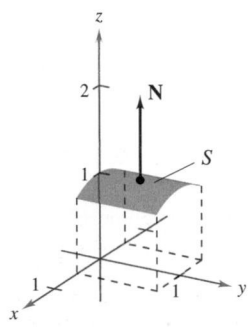

(b) Repeat the calculation in part (a) using the parametrization

$$x = \cos u, \quad y = v, \quad z = \sin u$$

where

$$\frac{\pi}{3} \le u \le \frac{2\pi}{3} \quad \text{and} \quad 0 \le v \le 1.$$

2. Heat Flux Consider a single heat source located at the origin with temperature

$$T(x, y, z) = \frac{25}{\sqrt{x^2 + y^2 + z^2}}.$$

(a) Calculate the heat flux across the surface

$$S = \left\{ (x, y, z): z = \sqrt{1 - x^2 - y^2}, x^2 + y^2 \le 1 \right\}$$

as shown in the figure.

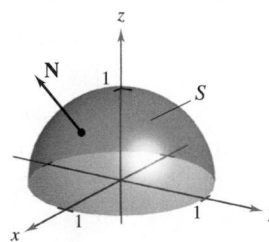

(b) Repeat the calculation in part (a) using the parametrization

$$x = \sin u \cos v, \quad y = \sin u \sin v, \quad z = \cos u$$

where

$$0 \le u \le \frac{\pi}{2} \quad \text{and} \quad 0 \le v \le 2\pi.$$

3. Moments of Inertia Consider a wire of density $\rho(x, y, z)$ given by the space curve

$$C: \mathbf{r}(t) = x(t)\mathbf{i} + y(t)\mathbf{j} + z(t)\mathbf{k}, \quad a \le t \le b.$$

The **moments of inertia** about the x-, y-, and z-axes are given by

$$I_x = \int_C (y^2 + z^2)\rho(x, y, z) \, ds$$

$$I_y = \int_C (x^2 + z^2)\rho(x, y, z) \, ds$$

$$I_z = \int_C (x^2 + y^2)\rho(x, y, z) \, ds.$$

Find the moments of inertia for a wire of uniform density $\rho = 1$ in the shape of the helix

$$\mathbf{r}(t) = 3 \cos t\mathbf{i} + 3 \sin t\mathbf{j} + 2t\mathbf{k}, \quad 0 \le t \le 2\pi \text{ (see figure)}.$$

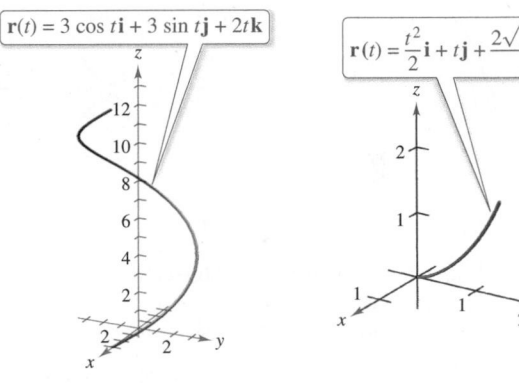

Figure for 3   Figure for 4

4. Moments of Inertia Using the formulas from Exercise 3, find the moments of inertia for a wire of density $\rho = \dfrac{1}{1 + t}$ given by the curve

$$C: \mathbf{r}(t) = \frac{t^2}{2}\mathbf{i} + t\mathbf{j} + \frac{2\sqrt{2}t^{3/2}}{3}\mathbf{k}, \quad 0 \le t \le 1 \text{ (see figure)}.$$

5. Laplace's Equation Let $\mathbf{F}(x, y, z) = x\mathbf{i} + y\mathbf{j} + z\mathbf{k}$, and let $f(x, y, z) = \|\mathbf{F}(x, y, z)\|$.

(a) Show that $\nabla(\ln f) = \dfrac{\mathbf{F}}{f^2}$.

(b) Show that $\nabla\left(\dfrac{1}{f}\right) = -\dfrac{\mathbf{F}}{f^3}$.

(c) Show that $\nabla f^n = nf^{n-2}\mathbf{F}$.

(d) The **Laplacian** is the differential operator

$$\nabla^2 = \nabla \cdot \nabla = \frac{\partial^2}{\partial x^2} + \frac{\partial^2}{\partial y^2} + \frac{\partial^2}{\partial z^2}$$

and **Laplace's equation** is

$$\nabla^2 w = \frac{\partial^2 w}{\partial x^2} + \frac{\partial^2 w}{\partial y^2} + \frac{\partial^2 w}{\partial z^2} = 0.$$

Any function that satisfies this equation is called **harmonic.** Show that the function $w = 1/f$ is harmonic.

6. Green's Theorem Consider the line integral

$$\int_C y^n \, dx + x^n \, dy$$

where C is the boundary of the region lying between the graphs of $y = \sqrt{a^2 - x^2}$, $a > 0$, and $y = 0$.

(a) Use a computer algebra system to verify Green's Theorem for n, an odd integer from 1 through 7.

(b) Use a computer algebra system to verify Green's Theorem for n, an even integer from 2 through 8.

(c) For n an odd integer, make a conjecture about the value of the integral.

7. Area Use a line integral to find the area bounded by one arch of the cycloid $x(\theta) = a(\theta - \sin \theta)$, $y(\theta) = a(1 - \cos \theta)$, $0 \le \theta \le 2\pi$, as shown in the figure.

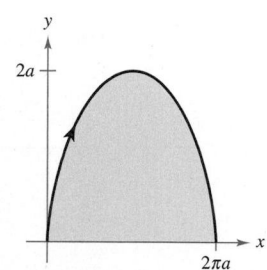

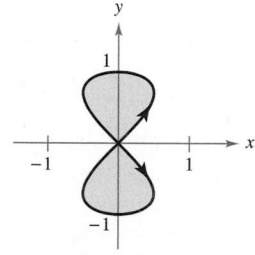

Figure for 7   Figure for 8

8. Area Use a line integral to find the area bounded by the two loops of the eight curve

$$x(t) = \frac{1}{2} \sin 2t, \quad y(t) = \sin t, \quad 0 \le t \le 2\pi$$

as shown in the figure.

9. Work The force field $\mathbf{F}(x, y) = (x + y)\mathbf{i} + (x^2 + 1)\mathbf{j}$ acts on an object moving from the point $(0, 0)$ to the point $(0, 1)$, as shown in the figure.

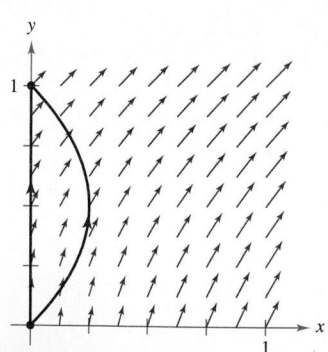

(a) Find the work done when the object moves along the path $x = 0, 0 \le y \le 1$.

(b) Find the work done when the object moves along the path $x = y - y^2, 0 \le y \le 1$.

(c) The object moves along the path $x = c(y - y^2), 0 \le y \le 1$, $c > 0$. Find the value of the constant c that minimizes the work.

10. Work The force field $\mathbf{F}(x, y) = 3x^2y^2\mathbf{i} + 2x^3y\mathbf{j}$ is shown in the figure below. Three particles move from the point $(1, 1)$ to the point $(2, 4)$ along different paths. Explain why the work done is the same for each particle and find the value of the work.

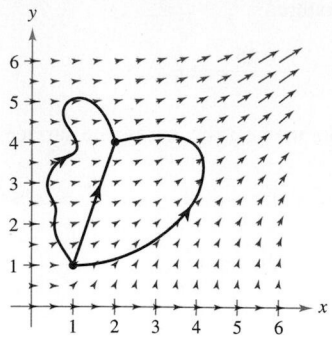

11. Area and Work How does the area of the ellipse

$$\frac{x^2}{a^2} + \frac{y^2}{b^2} = 1$$

compare with the magnitude of the work done by the force field

$$\mathbf{F}(x, y) = -\frac{1}{2}y\mathbf{i} + \frac{1}{2}x\mathbf{j}$$

on a particle that moves once around the ellipse (see figure)?

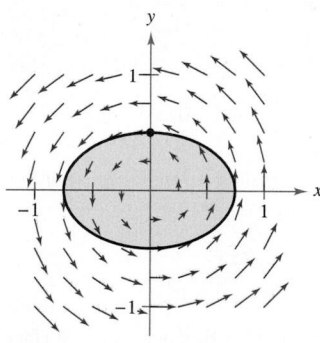

12. Verifying Identities

(a) Let f and g be scalar functions with continuous partial derivatives, and let C and S satisfy the conditions of Stokes's Theorem. Verify each identity.

(i) $\int_C (f\nabla g) \cdot d\mathbf{r} = \iint_S (\nabla f \times \nabla g) \cdot \mathbf{N} \, dS$

(ii) $\int_C (f\nabla f) \cdot d\mathbf{r} = 0$

(iii) $\int_C (f\nabla g + g\nabla f) \cdot d\mathbf{r} = 0$

(b) Demonstrate the results of part (a) for the functions

$$f(x, y, z) = xyz \quad \text{and} \quad g(x, y, z) = z.$$

Let S be the hemisphere $z = \sqrt{4 - x^2 - y^2}$.

Appendices

A   Proofs of Selected Theorems

The text version of Appendix A, Proofs of Selected Theorems, is available at *CengageBrain.com*. Also, to enhance your study of calculus, each proof is available in video format at *LarsonCalculus.com*. At this website, you can watch videos of Bruce Edwards explaining each proof in the text and in Appendix A. To access a video, visit the website at *LarsonCalculus.com* or scan the code near the proof or the proof's reference.

Sample Video: Bruce Edwards's Proof of the Power Rule at *LarsonCalculus.com*

The Power Rule

Before proving the next rule, it is important to review the procedure for expanding a binomial.

$$(x + \Delta x)^2 = x^2 + 2x\Delta x + (\Delta x)^2$$
$$(x + \Delta x)^3 = x^3 + 3x^2\Delta x + 3x(\Delta x)^2 + (\Delta x)^3$$
$$(x + \Delta x)^4 = x^4 + 4x^3\Delta x + 6x^2(\Delta x)^2 + 4x(\Delta x)^3 + (\Delta x)^4$$
$$(x + \Delta x)^5 = x^5 + 5x^4\Delta x + 10x^3(\Delta x)^2 + 10x^2(\Delta x)^3 + 5x(\Delta x)^4 + (\Delta x)^5$$

The general binomial expansion for a positive integer n is

$$(x + \Delta x)^n = x^n + nx^{n-1}(\Delta x) + \underbrace{\frac{n(n-1)x^{n-2}}{2}(\Delta x)^2 + \cdots + (\Delta x)^n}_{(\Delta x)^2 \text{ is a factor of these terms.}}$$

This binomial expansion is used in proving a special case of the Power Rule.

> **THEOREM 3.3   The Power Rule**
>
> If n is a rational number, then the function $f(x) = x^n$ is differentiable and
>
> $$\frac{d}{dx}[x^n] = nx^{n-1}.$$
>
> For f to be differentiable at $x = 0$, n must be a number such that x^{n-1} is defined on an interval containing 0.

• • **REMARK** From Example 7 in Section 3.1, you know that the function $f(x) = x^{1/3}$ is defined at $x = 0$ but is not differentiable at $x = 0$. This is because $x^{-2/3}$ is not defined on an interval containing 0.

Proof   If n is a positive integer greater than 1, then the binomial expansion produces

$$\frac{d}{dx}[x^n] = \lim_{\Delta x \to 0} \frac{(x + \Delta x)^n - x^n}{\Delta x}$$

$$= \lim_{\Delta x \to 0} \frac{x^n + nx^{n-1}(\Delta x) + \frac{n(n-}{}}{}$$

$$= \lim_{\Delta x \to 0} \left[nx^{n-1} + \frac{n(n-1)x^{n-2}}{2} \right(}$$

$$= nx^{n-1} + 0 + \cdots + 0$$

$$= nx^{n-1}.$$

This proves the case for which n is a positive inte the case for $n = 1$. Example 7 in Section 3.3 p integer. The cases for which n is rational and Section 3.5, Exercise 92).

When using the Power Rule, the case fo separate differentiation rule. That is,

$$\frac{d}{dx}[x] = 1.$$

This rule is consistent with the fact that the s Figure 3.15.

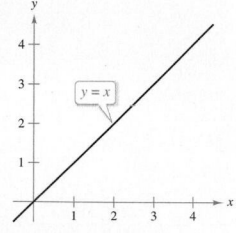

The slope of the line $y = x$ is 1.
Figure 3.15

Proof: The Power Rule

B   Integration Tables

Forms Involving u^n

1. $\displaystyle\int u^n \, du = \frac{u^{n+1}}{n+1} + C, \quad n \neq -1$

2. $\displaystyle\int \frac{1}{u} \, du = \ln|u| + C$

Forms Involving $a + bu$

3. $\displaystyle\int \frac{u}{a+bu} \, du = \frac{1}{b^2}(bu - a\ln|a+bu|) + C$

4. $\displaystyle\int \frac{u}{(a+bu)^2} \, du = \frac{1}{b^2}\left(\frac{a}{a+bu} + \ln|a+bu|\right) + C$

5. $\displaystyle\int \frac{u}{(a+bu)^n} \, du = \frac{1}{b^2}\left[\frac{-1}{(n-2)(a+bu)^{n-2}} + \frac{a}{(n-1)(a+bu)^{n-1}}\right] + C, \quad n \neq 1, 2$

6. $\displaystyle\int \frac{u^2}{a+bu} \, du = \frac{1}{b^3}\left[-\frac{bu}{2}(2a-bu) + a^2\ln|a+bu|\right] + C$

7. $\displaystyle\int \frac{u^2}{(a+bu)^2} \, du = \frac{1}{b^3}\left(bu - \frac{a^2}{a+bu} - 2a\ln|a+bu|\right) + C$

8. $\displaystyle\int \frac{u^2}{(a+bu)^3} \, du = \frac{1}{b^3}\left[\frac{2a}{a+bu} - \frac{a^2}{2(a+bu)^2} + \ln|a+bu|\right] + C$

9. $\displaystyle\int \frac{u^2}{(a+bu)^n} \, du = \frac{1}{b^3}\left[\frac{-1}{(n-3)(a+bu)^{n-3}} + \frac{2a}{(n-2)(a+bu)^{n-2}} - \frac{a^2}{(n-1)(a+bu)^{n-1}}\right] + C, \quad n \neq 1, 2, 3$

10. $\displaystyle\int \frac{1}{u(a+bu)} \, du = \frac{1}{a}\ln\left|\frac{u}{a+bu}\right| + C$

11. $\displaystyle\int \frac{1}{u(a+bu)^2} \, du = \frac{1}{a}\left(\frac{1}{a+bu} + \frac{1}{a}\ln\left|\frac{u}{a+bu}\right|\right) + C$

12. $\displaystyle\int \frac{1}{u^2(a+bu)} \, du = -\frac{1}{a}\left(\frac{1}{u} + \frac{b}{a}\ln\left|\frac{u}{a+bu}\right|\right) + C$

13. $\displaystyle\int \frac{1}{u^2(a+bu)^2} \, du = -\frac{1}{a^2}\left[\frac{a+2bu}{u(a+bu)} + \frac{2b}{a}\ln\left|\frac{u}{a+bu}\right|\right] + C$

Forms Involving $a + bu + cu^2, b^2 \neq 4ac$

14. $\displaystyle\int \frac{1}{a+bu+cu^2} \, du = \begin{cases} \dfrac{2}{\sqrt{4ac-b^2}} \arctan \dfrac{2cu+b}{\sqrt{4ac-b^2}} + C, & b^2 < 4ac \\[4mm] \dfrac{1}{\sqrt{b^2-4ac}} \ln\left|\dfrac{2cu+b-\sqrt{b^2-4ac}}{2cu+b+\sqrt{b^2-4ac}}\right| + C, & b^2 > 4ac \end{cases}$

15. $\displaystyle\int \frac{u}{a+bu+cu^2} \, du = \frac{1}{2c}\left(\ln|a+bu+cu^2| - b\int \frac{1}{a+bu+cu^2} \, du\right)$

Forms Involving $\sqrt{a+bu}$

16. $\displaystyle\int u^n\sqrt{a+bu} \, du = \frac{2}{b(2n+3)}\left[u^n(a+bu)^{3/2} - na\int u^{n-1}\sqrt{a+bu} \, du\right]$

17. $\displaystyle\int \frac{1}{u\sqrt{a+bu}} \, du = \begin{cases} \dfrac{1}{\sqrt{a}}\ln\left|\dfrac{\sqrt{a+bu}-\sqrt{a}}{\sqrt{a+bu}+\sqrt{a}}\right| + C, & a > 0 \\[4mm] \dfrac{2}{\sqrt{-a}}\arctan\sqrt{\dfrac{a+bu}{-a}} + C, & a < 0 \end{cases}$

18. $\displaystyle\int \frac{1}{u^n\sqrt{a+bu}} \, du = \frac{-1}{a(n-1)}\left[\frac{\sqrt{a+bu}}{u^{n-1}} + \frac{(2n-3)b}{2}\int \frac{1}{u^{n-1}\sqrt{a+bu}} \, du\right], \quad n \neq 1$

19. $\displaystyle\int \frac{\sqrt{a + bu}}{u}\,du = 2\sqrt{a + bu} + a\int \frac{1}{u\sqrt{a + bu}}\,du$

20. $\displaystyle\int \frac{\sqrt{a + bu}}{u^n}\,du = \frac{-1}{a(n-1)}\left[\frac{(a + bu)^{3/2}}{u^{n-1}} + \frac{(2n-5)b}{2}\int \frac{\sqrt{a + bu}}{u^{n-1}}\,du\right],\ n \neq 1$

21. $\displaystyle\int \frac{u}{\sqrt{a + bu}}\,du = \frac{-2(2a - bu)}{3b^2}\sqrt{a + bu} + C$

22. $\displaystyle\int \frac{u^n}{\sqrt{a + bu}}\,du = \frac{2}{(2n+1)b}\left(u^n\sqrt{a + bu} - na\int \frac{u^{n-1}}{\sqrt{a + bu}}\,du\right)$

Forms Involving $a^2 \pm u^2$, $a > 0$

23. $\displaystyle\int \frac{1}{a^2 + u^2}\,du = \frac{1}{a}\arctan\frac{u}{a} + C$

24. $\displaystyle\int \frac{1}{u^2 - a^2}\,du = -\int \frac{1}{a^2 - u^2}\,du = \frac{1}{2a}\ln\left|\frac{u - a}{u + a}\right| + C$

25. $\displaystyle\int \frac{1}{(a^2 \pm u^2)^n}\,du = \frac{1}{2a^2(n-1)}\left[\frac{u}{(a^2 \pm u^2)^{n-1}} + (2n-3)\int \frac{1}{(a^2 \pm u^2)^{n-1}}\,du\right],\ n \neq 1$

Forms Involving $\sqrt{u^2 \pm a^2}$, $a > 0$

26. $\displaystyle\int \sqrt{u^2 \pm a^2}\,du = \frac{1}{2}\left(u\sqrt{u^2 \pm a^2} \pm a^2\ln\left|u + \sqrt{u^2 \pm a^2}\right|\right) + C$

27. $\displaystyle\int u^2\sqrt{u^2 \pm a^2}\,du = \frac{1}{8}\left[u(2u^2 \pm a^2)\sqrt{u^2 \pm a^2} - a^4\ln\left|u + \sqrt{u^2 \pm a^2}\right|\right] + C$

28. $\displaystyle\int \frac{\sqrt{u^2 + a^2}}{u}\,du = \sqrt{u^2 + a^2} - a\ln\left|\frac{a + \sqrt{u^2 + a^2}}{u}\right| + C$

29. $\displaystyle\int \frac{\sqrt{u^2 - a^2}}{u}\,du = \sqrt{u^2 - a^2} - a\,\mathrm{arcsec}\frac{|u|}{a} + C$

30. $\displaystyle\int \frac{\sqrt{u^2 \pm a^2}}{u^2}\,du = \frac{-\sqrt{u^2 \pm a^2}}{u} + \ln\left|u + \sqrt{u^2 \pm a^2}\right| + C$

31. $\displaystyle\int \frac{1}{\sqrt{u^2 \pm a^2}}\,du = \ln\left|u + \sqrt{u^2 \pm a^2}\right| + C$

32. $\displaystyle\int \frac{1}{u\sqrt{u^2 + a^2}}\,du = \frac{-1}{a}\ln\left|\frac{a + \sqrt{u^2 + a^2}}{u}\right| + C$ **33.** $\displaystyle\int \frac{1}{u\sqrt{u^2 - a^2}}\,du = \frac{1}{a}\,\mathrm{arcsec}\frac{|u|}{a} + C$

34. $\displaystyle\int \frac{u^2}{\sqrt{u^2 \pm a^2}}\,du = \frac{1}{2}\left(u\sqrt{u^2 \pm a^2} \mp a^2\ln\left|u + \sqrt{u^2 \pm a^2}\right|\right) + C$

35. $\displaystyle\int \frac{1}{u^2\sqrt{u^2 \pm a^2}}\,du = \mp\frac{\sqrt{u^2 \pm a^2}}{a^2 u} + C$ **36.** $\displaystyle\int \frac{1}{(u^2 \pm a^2)^{3/2}}\,du = \frac{\pm u}{a^2\sqrt{u^2 \pm a^2}} + C$

Forms Involving $\sqrt{a^2 - u^2}$, $a > 0$

37. $\displaystyle\int \sqrt{a^2 - u^2}\,du = \frac{1}{2}\left(u\sqrt{a^2 - u^2} + a^2\arcsin\frac{u}{a}\right) + C$

38. $\displaystyle\int u^2\sqrt{a^2 - u^2}\,du = \frac{1}{8}\left[u(2u^2 - a^2)\sqrt{a^2 - u^2} + a^4\arcsin\frac{u}{a}\right] + C$

39. $\displaystyle\int \frac{\sqrt{a^2 - u^2}}{u}\, du = \sqrt{a^2 - u^2} - a \ln\left|\frac{a + \sqrt{a^2 - u^2}}{u}\right| + C$

40. $\displaystyle\int \frac{\sqrt{a^2 - u^2}}{u^2}\, du = \frac{-\sqrt{a^2 - u^2}}{u} - \arcsin\frac{u}{a} + C$

41. $\displaystyle\int \frac{1}{\sqrt{a^2 - u^2}}\, du = \arcsin\frac{u}{a} + C$

42. $\displaystyle\int \frac{1}{u\sqrt{a^2 - u^2}}\, du = \frac{-1}{a} \ln\left|\frac{a + \sqrt{a^2 - u^2}}{u}\right| + C$

43. $\displaystyle\int \frac{u^2}{\sqrt{a^2 - u^2}}\, du = \frac{1}{2}\left(-u\sqrt{a^2 - u^2} + a^2 \arcsin\frac{u}{a}\right) + C$

44. $\displaystyle\int \frac{1}{u^2\sqrt{a^2 - u^2}}\, du = \frac{-\sqrt{a^2 - u^2}}{a^2 u} + C$

45. $\displaystyle\int \frac{1}{(a^2 - u^2)^{3/2}}\, du = \frac{u}{a^2\sqrt{a^2 - u^2}} + C$

Forms Involving sin u or cos u

46. $\displaystyle\int \sin u\, du = -\cos u + C$

47. $\displaystyle\int \cos u\, du = \sin u + C$

48. $\displaystyle\int \sin^2 u\, du = \frac{1}{2}(u - \sin u \cos u) + C$

49. $\displaystyle\int \cos^2 u\, du = \frac{1}{2}(u + \sin u \cos u) + C$

50. $\displaystyle\int \sin^n u\, du = -\frac{\sin^{n-1} u \cos u}{n} + \frac{n-1}{n}\int \sin^{n-2} u\, du$

51. $\displaystyle\int \cos^n u\, du = \frac{\cos^{n-1} u \sin u}{n} + \frac{n-1}{n}\int \cos^{n-2} u\, du$

52. $\displaystyle\int u \sin u\, du = \sin u - u \cos u + C$

53. $\displaystyle\int u \cos u\, du = \cos u + u \sin u + C$

54. $\displaystyle\int u^n \sin u\, du = -u^n \cos u + n\int u^{n-1} \cos u\, du$

55. $\displaystyle\int u^n \cos u\, du = u^n \sin u - n\int u^{n-1} \sin u\, du$

56. $\displaystyle\int \frac{1}{1 \pm \sin u}\, du = \tan u \mp \sec u + C$

57. $\displaystyle\int \frac{1}{1 \pm \cos u}\, du = -\cot u \pm \csc u + C$

58. $\displaystyle\int \frac{1}{\sin u \cos u}\, du = \ln|\tan u| + C$

Forms Involving tan u, cot u, sec u, or csc u

59. $\displaystyle\int \tan u\, du = -\ln|\cos u| + C$

60. $\displaystyle\int \cot u\, du = \ln|\sin u| + C$

61. $\displaystyle\int \sec u\, du = \ln|\sec u + \tan u| + C$

62. $\displaystyle\int \csc u\, du = \ln|\csc u - \cot u| + C$ or $\displaystyle\int \csc u\, du = -\ln|\csc u + \cot u| + C$

63. $\displaystyle\int \tan^2 u\, du = -u + \tan u + C$

64. $\displaystyle\int \cot^2 u\, du = -u - \cot u + C$

65. $\displaystyle\int \sec^2 u\, du = \tan u + C$

66. $\displaystyle\int \csc^2 u\, du = -\cot u + C$

67. $\displaystyle\int \tan^n u\, du = \frac{\tan^{n-1} u}{n-1} - \int \tan^{n-2} u\, du,\ n \neq 1$

68. $\displaystyle\int \cot^n u\, du = -\frac{\cot^{n-1} u}{n-1} - \int \cot^{n-2} u\, du,\ n \neq 1$

69. $\displaystyle\int \sec^n u\, du = \frac{\sec^{n-2} u \tan u}{n-1} + \frac{n-2}{n-1}\int \sec^{n-2} u\, du,\ n \neq 1$

70. $\displaystyle\int \csc^n u\, du = -\frac{\csc^{n-2} u \cot u}{n-1} + \frac{n-2}{n-1}\int \csc^{n-2} u\, du,\ n \neq 1$

71. $\int \dfrac{1}{1 \pm \tan u} \, du = \dfrac{1}{2}(u \pm \ln|\cos u \pm \sin u|) + C$

72. $\int \dfrac{1}{1 \pm \cot u} \, du = \dfrac{1}{2}(u \mp \ln|\sin u \pm \cos u|) + C$

73. $\int \dfrac{1}{1 \pm \sec u} \, du = u + \cot u \mp \csc u + C$

74. $\int \dfrac{1}{1 \pm \csc u} \, du = u - \tan u \pm \sec u + C$

Forms Involving Inverse Trigonometric Functions

75. $\int \arcsin u \, du = u \arcsin u + \sqrt{1 - u^2} + C$

76. $\int \arccos u \, du = u \arccos u - \sqrt{1 - u^2} + C$

77. $\int \arctan u \, du = u \arctan u - \ln\sqrt{1 + u^2} + C$

78. $\int \text{arccot } u \, du = u \text{ arccot } u + \ln\sqrt{1 + u^2} + C$

79. $\int \text{arcsec } u \, du = u \text{ arcsec } u - \ln\left|u + \sqrt{u^2 - 1}\right| + C$

80. $\int \text{arccsc } u \, du = u \text{ arccsc } u + \ln\left|u + \sqrt{u^2 - 1}\right| + C$

Forms Involving e^u

81. $\int e^u \, du = e^u + C$

82. $\int u e^u \, du = (u - 1)e^u + C$

83. $\int u^n e^u \, du = u^n e^u - n \int u^{n-1} e^u \, du$

84. $\int \dfrac{1}{1 + e^u} \, du = u - \ln(1 + e^u) + C$

85. $\int e^{au} \sin bu \, du = \dfrac{e^{au}}{a^2 + b^2}(a \sin bu - b \cos bu) + C$

86. $\int e^{au} \cos bu \, du = \dfrac{e^{au}}{a^2 + b^2}(a \cos bu + b \sin bu) + C$

Forms Involving $\ln u$

87. $\int \ln u \, du = u(-1 + \ln u) + C$

88. $\int u \ln u \, du = \dfrac{u^2}{4}(-1 + 2 \ln u) + C$

89. $\int u^n \ln u \, du = \dfrac{u^{n+1}}{(n+1)^2}[-1 + (n+1)\ln u] + C, \; n \neq -1$

90. $\int (\ln u)^2 \, du = u[2 - 2 \ln u + (\ln u)^2] + C$

91. $\int (\ln u)^n \, du = u(\ln u)^n - n \int (\ln u)^{n-1} \, du$

Forms Involving Hyperbolic Functions

92. $\int \cosh u \, du = \sinh u + C$

93. $\int \sinh u \, du = \cosh u + C$

94. $\int \text{sech}^2 u \, du = \tanh u + C$

95. $\int \text{csch}^2 u \, du = -\coth u + C$

96. $\int \text{sech } u \tanh u \, du = -\text{sech } u + C$

97. $\int \text{csch } u \coth u \, du = -\text{csch } u + C$

Forms Involving Inverse Hyperbolic Functions (in logarithmic form)

98. $\int \dfrac{du}{\sqrt{u^2 \pm a^2}} = \ln\left(u + \sqrt{u^2 \pm a^2}\right) + C$

99. $\int \dfrac{du}{a^2 - u^2} = \dfrac{1}{2a} \ln\left|\dfrac{a + u}{a - u}\right| + C$

100. $\int \dfrac{du}{u\sqrt{a^2 \pm u^2}} = -\dfrac{1}{a} \ln \dfrac{a + \sqrt{a^2 \pm u^2}}{|u|} + C$

C Precalculus Review

C.1   Real Numbers and the Real Number Line

■ Represent and classify real numbers.
■ Order real numbers and use inequalities.
■ Find the absolute values of real numbers and find the distance between two real numbers.

Real Numbers and the Real Number Line

Real numbers can be represented by a coordinate system called the **real number line** or *x*-axis (see Figure C.1). The real number corresponding to a point on the real number line is the **coordinate** of the point. As Figure C.1 shows, it is customary to identify those points whose coordinates are integers.

$$\overset{x}{\underset{-4\ -3\ -2\ -1\ \ 0\ \ 1\ \ 2\ \ 3\ \ 4}{\longmapsto}}$$

The real number line
Figure C.1

The point on the real number line corresponding to zero is the **origin** and is denoted by 0. The **positive direction** (to the right) is denoted by an arrowhead and is the direction of increasing values of *x*. Numbers to the right of the origin are **positive.** Numbers to the left of the origin are **negative.** The term **nonnegative** describes a number that is either positive or zero. The term **nonpositive** describes a number that is either negative or zero.

Each point on the real number line corresponds to one and only one real number, and each real number corresponds to one and only one point on the real number line. This type of relationship is called a **one-to-one correspondence.**

Each of the four points in Figure C.2 corresponds to a **rational number**—one that can be written as the ratio of two integers. $\left(\text{Note that } 4.5 = \frac{9}{2} \text{ and } -2.6 = -\frac{13}{5}.\right)$ Rational numbers can be represented either by *terminating decimals* such as $\frac{2}{5} = 0.4$ or by *repeating decimals* such as $\frac{1}{3} = 0.333\ldots = 0.\overline{3}$.

Real numbers that are not rational are **irrational.** Irrational numbers cannot be represented as terminating or repeating decimals. In computations, irrational numbers are represented by decimal approximations. Here are three familiar examples.

$$\sqrt{2} \approx 1.414213562$$
$$\pi \approx 3.141592654$$
$$e \approx 2.718281828$$

(See Figure C.3.)

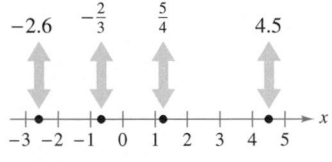

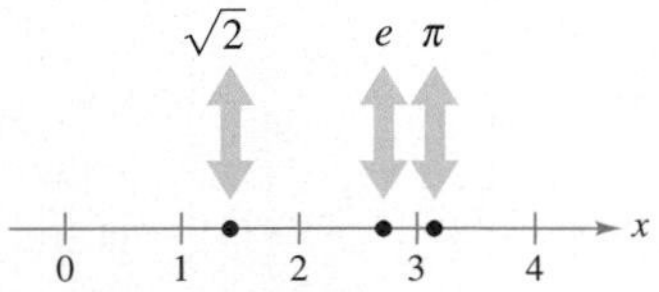

Rational numbers
Figure C.2

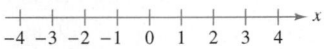

Irrational numbers
Figure C.3

Order and Inequalities

One important property of real numbers is that they are **ordered.** For two real numbers a and b, a is **less than** b when $b - a$ is positive. This order is denoted by the **inequality**

$$a < b.$$

This relationship can also be described by saying that b is **greater than** a and writing $b > a$. If three real numbers a, b, and c are ordered such that $a < b$ and $b < c$, then b is **between** a and c and $a < b < c$.

Geometrically, $a < b$ if and only if a lies to the *left* of b on the real number line (see Figure C.4). For example, $1 < 2$ because 1 lies to the left of 2 on the real number line.

Several properties used in working with inequalities are listed below. Similar properties are obtained when $<$ is replaced by $\leq$ and $>$ is replaced by $\geq$. (The symbols $\leq$ and $\geq$ mean **less than or equal to** and **greater than or equal to,** respectively.)

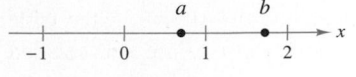

$a < b$ if and only if a lies to the left of b.

Figure C.4

Properties of Inequalities

Let a, b, c, d, and k be real numbers.

1. If $a < b$ and $b < c$, then $a < c$.   Transitive Property

2. If $a < b$ and $c < d$, then $a + c < b + d$.   Add inequalities.

3. If $a < b$, then $a + k < b + k$.   Add a constant.

4. If $a < b$ and $k > 0$, then $ak < bk$.   Multiply by a positive constant.

5. If $a < b$ and $k < 0$, then $ak > bk$.   Multiply by a negative constant.

Note that you *reverse the inequality* when you multiply the inequality by a negative number. For example, if $x < 3$, then $-4x > -12$. This also applies to division by a negative number. So, if $-2x > 4$, then $x < -2$.

A **set** is a collection of elements. Two common sets are the set of real numbers and the set of points on the real number line. Many problems in calculus involve **subsets** of one of these two sets. In such cases, it is convenient to use **set notation** of the form $\{x\colon \text{condition on } x\}$, which is read as follows.

The set of all x such that a certain condition is true.

$$\{\quad x \quad : \quad \text{condition on } x\}$$

For example, you can describe the set of positive real numbers as

$$\{x\colon x > 0\}. \qquad \text{Set of positive real numbers}$$

Similarly, you can describe the set of nonnegative real numbers as

$$\{x\colon x \geq 0\}. \qquad \text{Set of nonnegative real numbers}$$

The **union** of two sets A and B, denoted by $A \cup B$, is the set of elements that are members of A *or* B *or both*. The **intersection** of two sets A and B, denoted by $A \cap B$, is the set of elements that are members of A *and* B. Two sets are **disjoint** when they have no elements in common.

The most commonly used subsets are **intervals** on the real number line. For example, the **open** interval

$$(a, b) = \{x: a < x < b\} \qquad \text{Open interval}$$

is the set of all real numbers greater than a and less than b, where a and b are the **endpoints** of the interval. Note that the endpoints are not included in an open interval. Intervals that include their endpoints are **closed** and are denoted by

$$[a, b] = \{x: a \le x \le b\}. \qquad \text{Closed interval}$$

The nine basic types of intervals on the real number line are shown in the table below. The first four are **bounded intervals** and the remaining five are **unbounded intervals.** Unbounded intervals are also classified as open or closed. The intervals $(-\infty, b)$ and (a, ∞) are open, the intervals $(-\infty, b]$ and $[a, \infty)$ are closed, and the interval $(-\infty, \infty)$ is considered to be both open *and* closed.

Intervals on the Real Number Line

	Interval Notation	Set Notation	Graph
Bounded open interval	(a, b)	$\{x: a < x < b\}$	
Bounded closed interval	$[a, b]$	$\{x: a \le x \le b\}$	
Bounded intervals (neither open nor closed)	$[a, b)$	$\{x: a \le x < b\}$	
	$(a, b]$	$\{x: a < x \le b\}$	
Unbounded open intervals	$(-\infty, b)$	$\{x: x < b\}$	
	(a, ∞)	$\{x: x > a\}$	
Unbounded closed intervals	$(-\infty, b]$	$\{x: x \le b\}$	
	$[a, \infty)$	$\{x: x \ge a\}$	
Entire real line	$(-\infty, \infty)$	$\{x: x \text{ is a real number}\}$	

Note that the symbols ∞ and $-\infty$ refer to positive and negative infinity, respectively. These symbols do not denote real numbers. They simply enable you to describe unbounded conditions more concisely. For instance, the interval $[a, \infty)$ is unbounded to the right because it includes *all* real numbers that are greater than or equal to a.

EXAMPLE 1 **Liquid and Gaseous States of Water**

Describe the intervals on the real number line that correspond to the temperatures x (in degrees Celsius) of water in

a. a liquid state.   **b.** a gaseous state.

Solution

a. Water is in a liquid state at temperatures greater than 0°C and less than 100°C, as shown in Figure C.5(a).

$$(0, 100) = \{x: 0 < x < 100\}$$

b. Water is in a gaseous state (steam) at temperatures greater than or equal to 100°C, as shown in Figure C.5(b).

$$[100, \infty) = \{x: x \geq 100\}$$

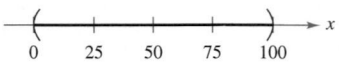

(a) Temperature range of water
     (in degrees Celsius)

(b) Temperature range of steam
     (in degrees Celsius)

Figure C.5

If a real number a is a **solution** of an inequality, then the inequality is **satisfied** (is true) when a is substituted for x. The set of all solutions is the **solution set** of the inequality.

EXAMPLE 2 **Solving an Inequality**

Solve $2x - 5 < 7$.

Solution

$$2x - 5 < 7 \qquad \text{Write original inequality.}$$
$$2x - 5 + 5 < 7 + 5 \qquad \text{Add 5 to each side.}$$
$$2x < 12 \qquad \text{Simplify.}$$
$$\frac{2x}{2} < \frac{12}{2} \qquad \text{Divide each side by 2.}$$
$$x < 6 \qquad \text{Simplify.}$$

The solution set is $(-\infty, 6)$.

In Example 2, all five inequalities listed as steps in the solution are called **equivalent** because they have the same solution set.

Once you have solved an inequality, check some x-values in your solution set to verify that they satisfy the original inequality. You should also check some values outside your solution set to verify that they *do not* satisfy the inequality. For example, Figure C.6 shows that when $x = 0$ or $x = 5$ the inequality $2x - 5 < 7$ is satisfied, but when $x = 7$ the inequality $2x - 5 < 7$ is not satisfied.

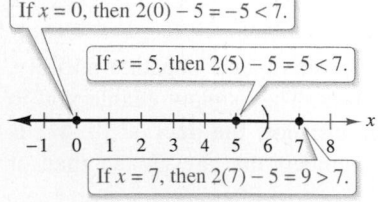

If $x = 0$, then $2(0) - 5 = -5 < 7$.

If $x = 5$, then $2(5) - 5 = 5 < 7$.

If $x = 7$, then $2(7) - 5 = 9 > 7$.

Checking solutions of $2x - 5 < 7$

Figure C.6

EXAMPLE 3 Solving a Double Inequality

Solve $-3 \le 2 - 5x \le 12$.

Solution

$$
\begin{array}{llll}
-3 \le & 2 - 5x & \le 12 & \text{Write original inequality.} \\
-3 - 2 \le & 2 - 5x - 2 & \le 12 - 2 & \text{Subtract 2 from each part.} \\
-5 \le & -5x & \le 10 & \text{Simplify.} \\
\dfrac{-5}{-5} \ge & \dfrac{-5x}{-5} & \ge \dfrac{10}{-5} & \text{Divide each part by } -5 \text{ and} \\
& & & \text{reverse both inequalities.} \\
1 \ge & x & \ge -2 & \text{Simplify.}
\end{array}
$$

The solution set is $[-2, 1]$, as shown in Figure C.7.

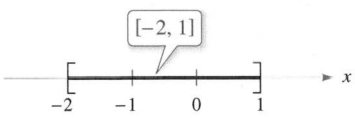

Solution set of $-3 \le 2 - 5x \le 12$

Figure C.7

The inequalities in Examples 2 and 3 are **linear inequalities**—that is, they involve first-degree polynomials. To solve inequalities involving polynomials of higher degree, use the fact that a polynomial can change signs *only* at its real **zeros** (the x-values that make the polynomial equal to zero). Between two consecutive real zeros, a polynomial must be either entirely positive or entirely negative. This means that when the real zeros of a polynomial are put in order, they divide the real number line into **test intervals** in which the polynomial has no sign changes. So, if a polynomial has the factored form

$$(x - r_1)(x - r_2) \cdots (x - r_n), \qquad r_1 < r_2 < r_3 < \cdots < r_n$$

then the test intervals are

$$(-\infty, r_1), \quad (r_1, r_2), \ldots, \quad (r_{n-1}, r_n), \quad \text{and} \quad (r_n, \infty).$$

To determine the sign of the polynomial in each test interval, you need to test only *one value* from the interval.

EXAMPLE 4 Solving a Quadratic Inequality

Solve $x^2 < x + 6$.

Solution

$$
\begin{array}{ll}
x^2 < x + 6 & \text{Write original inequality.} \\
x^2 - x - 6 < 0 & \text{Write in general form.} \\
(x - 3)(x + 2) < 0 & \text{Factor.}
\end{array}
$$

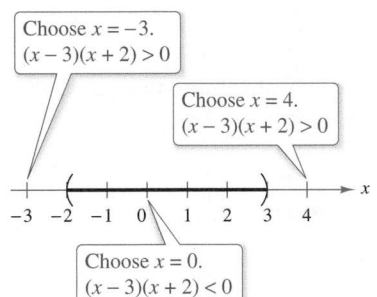

Testing an interval

Figure C.8

The polynomial $x^2 - x - 6$ has $x = -2$ and $x = 3$ as its zeros. So, you can solve the inequality by testing the sign of $x^2 - x - 6$ in each of the test intervals $(-\infty, -2)$, $(-2, 3)$, and $(3, \infty)$. To test an interval, choose any number in the interval and determine the sign of $x^2 - x - 6$. After doing this, you will find that the polynomial is positive for all real numbers in the first and third intervals and negative for all real numbers in the second interval. The solution of the original inequality is therefore $(-2, 3)$, as shown in Figure C.8.

Absolute Value and Distance

If a is a real number, then the **absolute value** of a is

$$|a| = \begin{cases} a, & a \geq 0 \\ -a, & a < 0 \end{cases}.$$

The absolute value of a number cannot be negative. For example, let $a = -4$. Then, because $-4 < 0$, you have

$$|a| = |-4| = -(-4) = 4.$$

Remember that the symbol $-a$ does not necessarily mean that $-a$ is negative.

$\triangleright$

• • • • • • • • • • • • • • • •
•
•• REMARK   You are asked
to prove these properties in
Exercises 73, 75, 76, and 77.

Operations with Absolute Value

Let a and b be real numbers and let n be a positive integer.

1. $|ab| = |a|\,|b|$ 2. $\left|\dfrac{a}{b}\right| = \dfrac{|a|}{|b|}, \quad b \neq 0$

3. $|a| = \sqrt{a^2}$ 4. $|a^n| = |a|^n$

Properties of Inequalities and Absolute Value

Let a and b be real numbers and let k be a positive real number.

1. $-|a| \leq a \leq |a|$
2. $|a| \leq k$ if and only if $-k \leq a \leq k$.
3. $|a| \geq k$ if and only if $a \leq -k$ or $a \geq k$.
4. *Triangle Inequality:* $|a + b| \leq |a| + |b|$

Properties 2 and 3 are also true when $\leq$ is replaced by $<$ and $\geq$ is replaced by $>$.

EXAMPLE 5   **Solving an Absolute Value Inequality**

Solve $|x - 3| \leq 2$.

Solution   Using the second property of inequalities and absolute value, you can rewrite the original inequality as a double inequality.

$$\begin{array}{ll} -2 \leq \quad x - 3 \quad \leq 2 & \text{Write as double inequality.} \\ -2 + 3 \leq x - 3 + 3 \leq 2 + 3 & \text{Add 3 to each part.} \\ 1 \leq \quad\quad x \quad\quad \leq 5 & \text{Simplify.} \end{array}$$

The solution set is $[1, 5]$, as shown in Figure C.9.

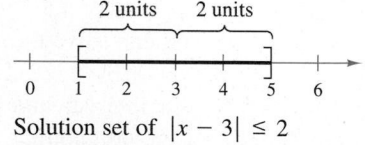

2 units   2 units

Solution set of $|x - 3| \leq 2$
Figure C.9

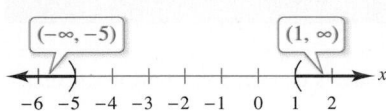

Solution set of $|x + 2| > 3$
Figure C.10

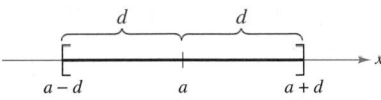

Solution set of $|x - a| \leq d$

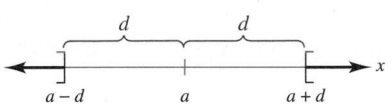

Solution set of $|x - a| \geq d$
Figure C.11

EXAMPLE 6   **A Two-Interval Solution Set**

Solve $|x + 2| > 3$.

Solution   Using the third property of inequalities and absolute value, you can rewrite the original inequality as two linear inequalities.

$$x + 2 < -3 \quad \text{or} \quad x + 2 > 3$$
$$x < -5 \qquad\qquad\quad x > 1$$

The solution set is the union of the disjoint intervals $(-\infty, -5)$ and $(1, \infty)$, as shown in Figure C.10.   ◼

Examples 5 and 6 illustrate the general results shown in Figure C.11. Note that for $d > 0$, the solution set for the inequality $|x - a| \leq d$ is a *single* interval, whereas the solution set for the inequality $|x - a| \geq d$ is the union of *two* disjoint intervals.

The **distance between two points** a and b on the real number line is given by

$$d = |a - b| = |b - a|.$$

The **directed distance from a to b** is $b - a$ and the **directed distance from b to a** is $a - b$, as shown in Figure C.12.

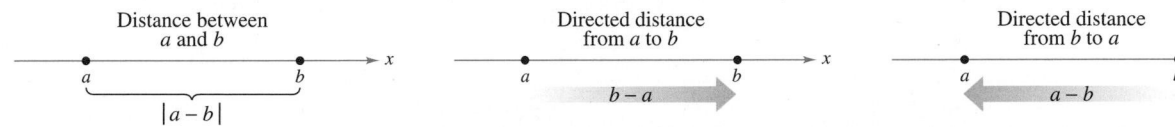

Figure C.12

EXAMPLE 7   **Distance on the Real Number Line**

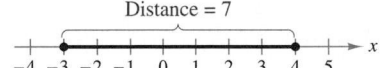

Figure C.13

a. The distance between -3 and 4 is

$$|4 - (-3)| = |7| = 7 \quad \text{or} \quad |-3 - 4| = |-7| = 7.$$

(See Figure C.13.)

b. The directed distance from -3 to 4 is

$$4 - (-3) = 7.$$

c. The directed distance from 4 to -3 is

$$-3 - 4 = -7.$$ ◼

The **midpoint** of an interval with endpoints a and b is the average value of a and b. That is,

$$\text{Midpoint of interval } (a, b) = \frac{a + b}{2}.$$

To show that this is the midpoint, you need only show that $(a + b)/2$ is equidistant from a and b.

C.1 Exercises

Rational or Irrational? In Exercises 1–10, determine whether the real number is rational or irrational.

1. 0.7

2. -3678

3. $\dfrac{3\pi}{2}$

4. $3\sqrt{2} - 1$

5. $4.3\overline{451}$

6. $\dfrac{22}{7}$

7. $\sqrt[3]{64}$

8. $0.8\overline{177}$

9. $4\tfrac{5}{8}$

10. $\left(\sqrt{2}\right)^3$

Repeating Decimal In Exercises 11–14, write the repeating decimal as a ratio of two integers using the following procedure. If $x = 0.6363\ldots$, then $100x = 63.6363\ldots$. Subtracting the first equation from the second produces $99x = 63$ or $x = \dfrac{63}{99} = \dfrac{7}{11}$.

11. $0.\overline{36}$

12. $0.3\overline{18}$

13. $0.\overline{297}$

14. $0.\overline{9900}$

15. Using Properties of Inequalities Given $a < b$, determine which of the following are true.

(a) $a + 2 < b + 2$

(b) $5b < 5a$

(c) $5 - a > 5 - b$

(d) $\dfrac{1}{a} < \dfrac{1}{b}$

(e) $(a - b)(b - a) > 0$

(f) $a^2 < b^2$

16. Intervals and Graphs on the Real Number Line Complete the table with the appropriate interval notation, set notation, and graph on the real number line.

Interval Notation	Set Notation	Graph
$(-\infty, -4]$		
	$\left\{x: 3 \le x \le \tfrac{11}{2}\right\}$	
$(-1, 7)$		

Analyzing an Inequality In Exercises 17–20, verbally describe the subset of real numbers represented by the inequality. Sketch the subset on the real number line, and state whether the interval is bounded or unbounded.

17. $-3 < x < 3$

18. $x \ge 4$

19. $x \le 5$

20. $0 \le x < 8$

Using Inequality and Interval Notation In Exercises 21–24, use inequality and interval notation to describe the set.

21. y is at least 4.

22. q is nonnegative.

23. The interest rate r on loans is expected to be greater than 3% and no more than 7%.

24. The temperature T is forecast to be above 90°F today.

Solving an Inequality In Exercises 25–44, solve the inequality and graph the solution on the real number line.

25. $2x - 1 \ge 0$

26. $3x + 1 \ge 2x + 2$

27. $-4 < 2x - 3 < 4$

28. $0 \le x + 3 < 5$

29. $\dfrac{x}{2} + \dfrac{x}{3} > 5$

30. $x > \dfrac{1}{x}$

31. $|x| < 1$

32. $\dfrac{x}{2} - \dfrac{x}{3} > 5$

33. $\left|\dfrac{x-3}{2}\right| \ge 5$

34. $\left|\dfrac{x}{2}\right| > 3$

35. $|x - a| < b,\, b > 0$

36. $|x + 2| < 5$

37. $|2x + 1| < 5$

38. $|3x + 1| \ge 4$

39. $\left|1 - \dfrac{2}{3}x\right| < 1$

40. $|9 - 2x| < 1$

41. $x^2 \le 3 - 2x$

42. $x^4 - x \le 0$

43. $x^2 + x - 1 \le 5$

44. $2x^2 + 1 < 9x - 3$

Distance on the Real Number Line In Exercises 45–48, find the directed distance from a to b, the directed distance from b to a, and the distance between a and b.

45.

46.

47. (a) $a = 126, b = 75$ (b) $a = -126, b = -75$

48. (a) $a = 9.34, b = -5.65$ (b) $a = \tfrac{16}{5}, b = \tfrac{112}{75}$

Using Absolute Value Notation In Exercises 49–54, use absolute value notation to define the interval or pair of intervals on the real number line.

49.

50.

51.

52.

53. (a) All numbers that are at most 10 units from 12

   (b) All numbers that are at least 10 units from 12

54. (a) y is at most two units from a.

   (b) y is less than δ units from c.

Finding the Midpoint In Exercises 55–58, find the midpoint of the interval.

55.

$a = -1$ $b = 3$

56.

$a = -5$ $b = -\frac{3}{2}$

57. (a) $[7, 21]$

   (b) $[8.6, 11.4]$

58. (a) $[-6.85, 9.35]$

   (b) $[-4.6, -1.3]$

59. **Profit** The revenue R from selling x units of a product is

   $R = 115.95x$

   and the cost C of producing x units is

   $C = 95x + 750.$

   To make a (positive) profit, R must be greater than C. For what values of x will the product return a profit?

60. **Fleet Costs** A utility company has a fleet of vans. The annual operating cost C (in dollars) of each van is estimated to be

   $C = 0.32m + 2300$

   where m is measured in miles. The company wants the annual operating cost of each van to be less than \$10,000. To do this, m must be less than what value?

61. **Fair Coin** To determine whether a coin is fair (has an equal probability of landing tails up or heads up), you toss the coin 100 times and record the number of heads x. The coin is declared unfair when

   $\left| \dfrac{x - 50}{5} \right| \geq 1.645.$

   For what values of x will the coin be declared unfair?

62. **Daily Production** The estimated daily oil production p at a refinery is

   $|p - 2{,}250{,}000| < 125{,}000$

   where p is measured in barrels. Determine the high and low production levels.

Which Number Is Greater? In Exercises 63 and 64, determine which of the two real numbers is greater.

63. (a) π or $\frac{355}{113}$

   (b) π or $\frac{22}{7}$

64. (a) $\frac{224}{151}$ or $\frac{144}{97}$

   (b) $\frac{73}{81}$ or $\frac{6427}{7132}$

65. **Approximation—Powers of 10** Light travels at the speed of 2.998×10^8 meters per second. Which best estimates the distance in meters that light travels in a year?

   (a) 9.5×10^5

   (b) 9.5×10^{15}

   (c) 9.5×10^{12}

   (d) 9.6×10^{16}

66. **Writing** The accuracy of an approximation of a number is related to how many significant digits there are in the approximation. Write a definition of significant digits and illustrate the concept with examples.

True or False? In Exercises 67–72, determine whether the statement is true or false. If it is false, explain why or give an example that shows it is false.

67. The reciprocal of a nonzero integer is an integer.

68. The reciprocal of a nonzero rational number is a rational number.

69. Each real number is either rational or irrational.

70. The absolute value of each real number is positive.

71. If $x < 0$, then $\sqrt{x^2} = -x$.

72. If a and b are any two distinct real numbers, then $a < b$ or $a > b$.

Proof In Exercises 73–80, prove the property.

73. $|ab| = |a||b|$

74. $|a - b| = |b - a|$

   $[Hint: (a - b) = (-1)(b - a)]$

75. $\left| \dfrac{a}{b} \right| = \dfrac{|a|}{|b|}, \quad b \neq 0$

76. $|a| = \sqrt{a^2}$

77. $|a^n| = |a|^n, \quad n = 1, 2, 3, \ldots$

78. $-|a| \leq a \leq |a|$

79. $|a| \leq k$ if and only if $-k \leq a \leq k, \quad k > 0.$

80. $|a| \geq k$ if and only if $a \leq -k$ or $a \geq k, \quad k > 0.$

81. **Proof** Find an example for which $|a - b| > |a| - |b|$, and an example for which $|a - b| = |a| - |b|$. Then prove that $|a - b| \geq |a| - |b|$ for all a, b.

82. **Maximum and Minimum** Show that the maximum of two numbers a and b is given by the formula

   $\max(a, b) = \frac{1}{2}(a + b + |a - b|).$

   Derive a similar formula for $\min(a, b)$.

C.2 The Cartesian Plane

- Understand the Cartesian plane.
- Use the Distance Formula to find the distance between two points and use the Midpoint Formula to find the midpoint of a line segment.
- Find equations of circles and sketch the graphs of circles.

The Cartesian Plane

Just as you can represent real numbers by points on a real number line, you can represent ordered pairs of real numbers by points in a plane called the **rectangular coordinate system,** or the **Cartesian plane,** after the French mathematician René Descartes.

The Cartesian plane is formed by using two real number lines intersecting at right angles, as shown in Figure C.14. The horizontal real number line is usually called the **x-axis,** and the vertical real number line is usually called the **y-axis.** The point of intersection of these two axes is the **origin.** The two axes divide the plane into four parts called **quadrants.**

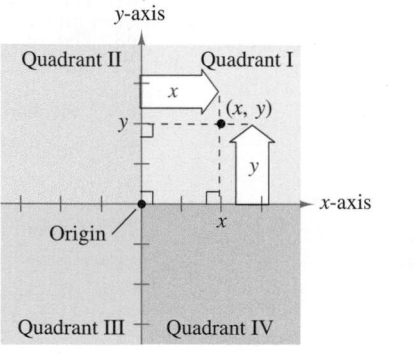

The Cartesian plane
Figure C.14

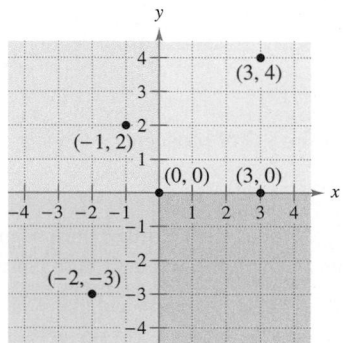

Points represented by ordered pairs
Figure C.15

Each point in the plane is identified by an **ordered pair** (x, y) of real numbers x and y, called the **coordinates** of the point. The number x represents the directed distance from the y-axis to the point, and the number y represents the directed distance from the x-axis to the point (see Figure C.14). For the point (x, y), the first coordinate is the **x-coordinate** or **abscissa,** and the second coordinate is the **y-coordinate** or **ordinate.** For example, Figure C.15 shows the locations of the points $(-1, 2)$, $(3, 4)$, $(0, 0)$, $(3, 0)$, and $(-2, -3)$ in the Cartesian plane. The signs of the coordinates of a point determine the quadrant in which the point lies. For instance, if $x > 0$ and $y < 0$, then the point (x, y) lies in Quadrant IV.

Note that an ordered pair (a, b) is used to denote either a point in the plane *or* an open interval on the real number line. This, however, should not be confusing—the nature of the problem should clarify whether a point in the plane or an open interval is being discussed.

The Distance and Midpoint Formulas

Recall from the Pythagorean Theorem that, in a right triangle, the hypotenuse c and sides a and b are related by $a^2 + b^2 = c^2$. Conversely, if $a^2 + b^2 = c^2$, then the triangle is a right triangle (see Figure C.16).

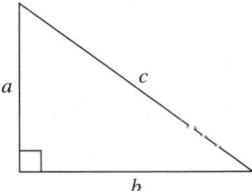

The Pythagorean Theorem:
$a^2 + b^2 = c^2$

Figure C.16

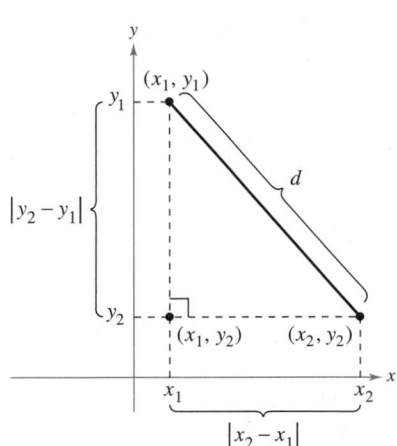

The distance between two points

Figure C.17

Now, consider the problem of determining the distance d between the two points (x_1, y_1) and (x_2, y_2) in the plane. If the points lie on a horizontal line, then $y_1 = y_2$ and the distance between the points is $|x_2 - x_1|$. If the points lie on a vertical line, then $x_1 = x_2$ and the distance between the points is $|y_2 - y_1|$. When the two points do not lie on a horizontal or vertical line, they can be used to form a right triangle, as shown in Figure C.17. The length of the vertical side of the triangle is $|y_2 - y_1|$, and the length of the horizontal side is $|x_2 - x_1|$. By the Pythagorean Theorem, it follows that

$$d^2 = |x_2 - x_1|^2 + |y_2 - y_1|^2$$
$$d = \sqrt{|x_2 - x_1|^2 + |y_2 - y_1|^2}.$$

Replacing $|x_2 - x_1|^2$ and $|y_2 - y_1|^2$ by the equivalent expressions $(x_2 - x_1)^2$ and $(y_2 - y_1)^2$ produces the **Distance Formula.**

Distance Formula

The distance d between the points (x_1, y_1) and (x_2, y_2) in the plane is given by

$$d = \sqrt{(x_2 - x_1)^2 + (y_2 - y_1)^2}.$$

EXAMPLE 1 **Finding the Distance Between Two Points**

Find the distance between the points $(-2, 1)$ and $(3, 4)$.

Solution

$$
\begin{aligned}
d &= \sqrt{(x_2 - x_1)^2 + (y_2 - y_1)^2} && \text{Distance Formula} \\
&= \sqrt{[3 - (-2)]^2 + (4 - 1)^2} && \text{Substitute for } x_1, y_1, x_2, \text{ and } y_2. \\
&= \sqrt{5^2 + 3^2} \\
&= \sqrt{25 + 9} \\
&= \sqrt{34} \\
&\approx 5.83
\end{aligned}
$$

EXAMPLE 2   Verifying a Right Triangle

Verify that the points $(2, 1)$, $(4, 0)$, and $(5, 7)$ form the vertices of a right triangle.

Solution   Figure C.18 shows the triangle formed by the three points. The lengths of the three sides are as follows.

$$d_1 = \sqrt{(5-2)^2 + (7-1)^2} = \sqrt{9+36} = \sqrt{45}$$
$$d_2 = \sqrt{(4-2)^2 + (0-1)^2} = \sqrt{4+1} = \sqrt{5}$$
$$d_3 = \sqrt{(5-4)^2 + (7-0)^2} = \sqrt{1+49} = \sqrt{50}$$

Because

$$d_1^2 + d_2^2 = 45 + 5 = 50 \qquad \text{Sum of squares of sides}$$

and

$$d_3^2 = 50 \qquad \text{Square of hypotenuse}$$

you can apply the Pythagorean Theorem to conclude that the triangle is a right triangle.

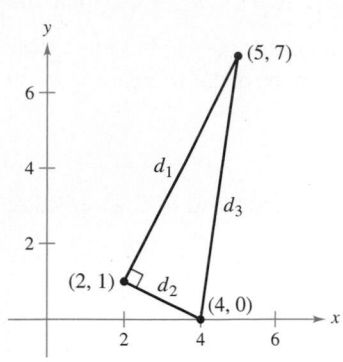

Verifying a right triangle
Figure C.18

EXAMPLE 3   Using the Distance Formula

Find x such that the distance between $(x, 3)$ and $(2, -1)$ is 5.

Solution   Using the Distance Formula, you can write the following.

$$5 = \sqrt{(x-2)^2 + [3-(-1)]^2} \qquad \text{Distance Formula}$$
$$25 = (x^2 - 4x + 4) + 16 \qquad \text{Square each side.}$$
$$0 = x^2 - 4x - 5 \qquad \text{Write in general form.}$$
$$0 = (x - 5)(x + 1) \qquad \text{Factor.}$$

So, $x = 5$ or $x = -1$, and you can conclude that there are two solutions. That is, each of the points $(5, 3)$ and $(-1, 3)$ lies five units from the point $(2, -1)$, as shown in Figure C.19.

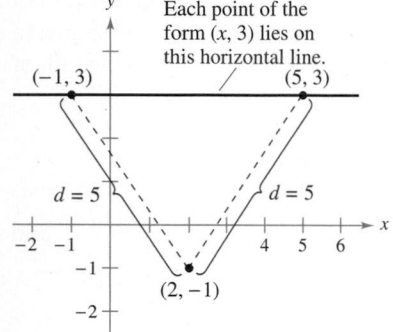

Each point of the form $(x, 3)$ lies on this horizontal line.

Figure C.19

The coordinates of the **midpoint** of the line segment joining two points can be found by "averaging" the x-coordinates of the two points and "averaging" the y-coordinates of the two points. That is, the midpoint of the line segment joining the points (x_1, y_1) and (x_2, y_2) in the plane is

$$\left(\frac{x_1 + x_2}{2}, \frac{y_1 + y_2}{2} \right). \qquad \text{Midpoint Formula}$$

For instance, the midpoint of the line segment joining the points $(-5, -3)$ and $(9, 3)$ is

$$\left(\frac{-5 + 9}{2}, \frac{-3 + 3}{2} \right) = (2, 0)$$

as shown in Figure C.20.

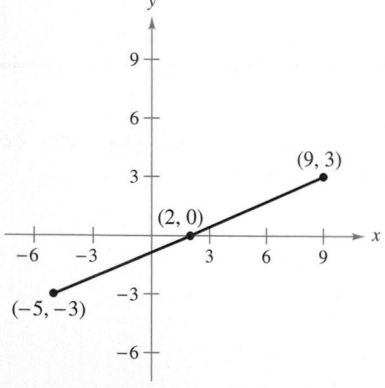

Midpoint of a line segment
Figure C.20

Equations of Circles

A **circle** can be defined as the set of all points in a plane that are equidistant from a fixed point. The fixed point is the **center** of the circle, and the distance between the center and a point on the circle is the **radius** (see Figure C.21).

You can use the Distance Formula to write an equation for the circle with center (h, k) and radius r. Let (x, y) be any point on the circle. Then the distance between (x, y) and the center (h, k) is given by

$$\sqrt{(x - h)^2 + (y - k)^2} = r.$$

By squaring each side of this equation, you obtain the **standard form of the equation of a circle.**

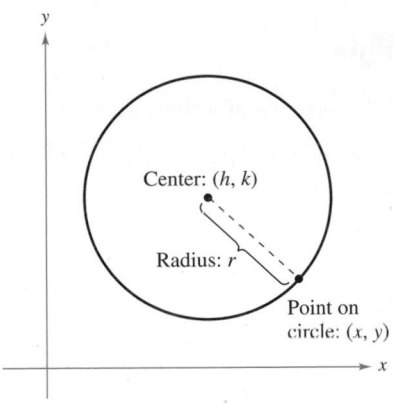

Definition of a circle
Figure C.21

Standard Form of the Equation of a Circle

The point (x, y) lies on the circle of radius r and center (h, k) if and only if

$$(x - h)^2 + (y - k)^2 = r^2.$$

The standard form of the equation of a circle with center at the origin, $(h, k) = (0, 0)$, is

$$x^2 + y^2 = r^2.$$

If $r = 1$, then the circle is called the **unit circle.**

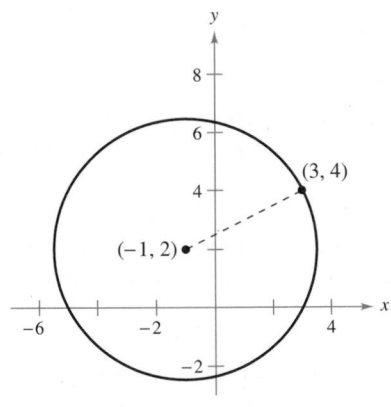

Figure C.22

EXAMPLE 4   Writing the Equation of a Circle

The point $(3, 4)$ lies on a circle whose center is at $(-1, 2)$, as shown in Figure C.22. Write the standard form of the equation of this circle.

Solution   The radius of the circle is the distance between $(-1, 2)$ and $(3, 4)$.

$$r = \sqrt{[3 - (-1)]^2 + (4 - 2)^2} = \sqrt{16 + 4} = \sqrt{20}$$

You can write the standard form of the equation of this circle as

$$[x - (-1)]^2 + (y - 2)^2 = \left(\sqrt{20}\right)^2$$
$$(x + 1)^2 + (y - 2)^2 = 20. \qquad \text{Write in standard form.} \qquad ■$$

By squaring and simplifying, the equation $(x - h)^2 + (y - k)^2 = r^2$ can be written in the following **general form of the equation of a circle.**

$$Ax^2 + Ay^2 + Dx + Ey + F = 0, \quad A \neq 0$$

To convert such an equation to the standard form

$$(x - h)^2 + (y - k)^2 = p$$

you can use a process called **completing the square.** If $p > 0$, then the graph of the equation is a circle. If $p = 0$, then the graph is the single point (h, k). If $p < 0$, then the equation has no graph.

EXAMPLE 5   **Completing the Square**

Sketch the graph of the circle whose general equation is

$$4x^2 + 4y^2 + 20x - 16y + 37 = 0.$$

Solution   To complete the square, first divide by 4 so that the coefficients of x^2 and y^2 are both 1.

$4x^2 + 4y^2 + 20x - 16y + 37 = 0$	Write original equation.
$x^2 + y^2 + 5x - 4y + \dfrac{37}{4} = 0$	Divide by 4.
$(x^2 + 5x + \quad) + (y^2 - 4y + \quad) = -\dfrac{37}{4}$	Group terms.
$\left(x^2 + 5x + \dfrac{25}{4}\right) + (y^2 - 4y + 4) = -\dfrac{37}{4} + \dfrac{25}{4} + 4$	Complete the square by adding $\left(\frac{5}{2}\right)^2 = \frac{25}{4}$ and $\left(\frac{4}{2}\right)^2 = 4$ to each side.

$$\underbrace{\quad}_{(\text{half})^2} \qquad \underbrace{\quad}_{(\text{half})^2}$$

$\left(x + \dfrac{5}{2}\right)^2 + (y - 2)^2 = 1$	Write in standard form.

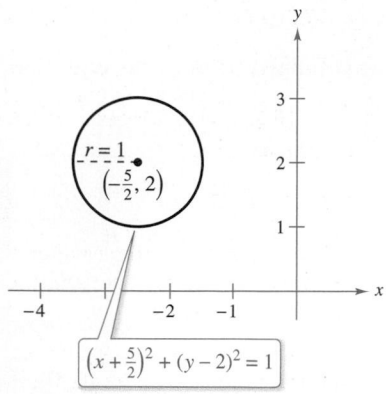

A circle with a radius of 1 and center at $\left(-\frac{5}{2}, 2\right)$

Figure C.23

Note that you complete the square by adding the square of half the coefficient of x *and* the square of half the coefficient of y to each side of the equation. The circle is centered at $\left(-\frac{5}{2}, 2\right)$ and its radius is 1, as shown in Figure C.23.

You have now reviewed some fundamental concepts of *analytic geometry*. Because these concepts are in common use today, it is easy to overlook their revolutionary nature. At the time analytic geometry was being developed by Pierre de Fermat and René Descartes, the two major branches of mathematics—geometry and algebra—were largely independent of each other. Circles belonged to geometry, and equations belonged to algebra. The coordination of the points on a circle and the solutions of an equation belongs to what is now called analytic geometry.

It is important to become skilled in analytic geometry so that you can move easily between geometry and algebra. For instance, in Example 4, you were given a geometric description of a circle and were asked to find an algebraic equation for the circle. So, you were moving from geometry to algebra. Similarly, in Example 5, you were given an algebraic equation and asked to sketch a geometric picture. In this case, you were moving from algebra to geometry. These two examples illustrate the two most common problems in analytic geometry.

1. Given a graph, find its equation.

> Geometry ⇒ Algebra

2. Given an equation, find its graph.

> Algebra ⇒ Geometry

C.2 Exercises

Using the Distance and Midpoint Formulas In Exercises 1–6, (a) plot the points, (b) find the distance between the points, and (c) find the midpoint of the line segment joining the points.

1. $(2, 1), (4, 5)$
2. $(-3, 2), (3, -2)$
3. $\left(\frac{1}{2}, 1\right), \left(-\frac{3}{2}, -5\right)$
4. $\left(\frac{2}{3}, -\frac{1}{3}\right), \left(\frac{5}{6}, 1\right)$
5. $\left(1, \sqrt{3}\right), (-1, 1)$
6. $(-2, 0), \left(0, \sqrt{2}\right)$

Locating a Point In Exercises 7–10, determine the quadrant(s) in which (x, y) is located so that the condition(s) is (are) satisfied.

7. $x = -2$ and $y > 0$
8. $y < -2$
9. $xy > 0$
10. $(x, -y)$ is in Quadrant II.

Vertices of a Polygon In Exercises 11–14, show that the points are the vertices of the polygon. (A rhombus is a quadrilateral whose sides are all the same length.)

Vertices	Polygon
11. $(4, 0), (2, 1), (-1, -5)$	Right triangle
12. $(1, -3), (3, 2), (-2, 4)$	Isosceles triangle
13. $(0, 0), (1, 2), (2, 1), (3, 3)$	Rhombus
14. $(0, 1), (3, 7), (4, 4), (1, -2)$	Parallelogram

15. **Number of Stores** The table shows the number y of Target stores for each year x from 2006 through 2015. Select reasonable scales on the coordinate axes and plot the points (x, y). (*Source: Target Corp.*)

Year, x	2006	2007	2008	2009	2010
Number, y	1488	1591	1682	1740	1750

Year, x	2011	2012	2013	2014	2015
Number, y	1763	1778	1917	1790	1792

16. **Conjecture** Plot the points $(2, 1), (-3, 5)$, and $(7, -3)$ in a rectangular coordinate system. Then change the sign of the x-coordinate of each point and plot the three new points in the same rectangular coordinate system. What conjecture can you make about the location of a point when the sign of the x-coordinate is changed? Repeat the exercise for the case in which the signs of the y-coordinates are changed.

Collinear Points? In Exercises 17–20, use the Distance Formula to determine whether the points lie on the same line.

17. $(0, -4), (2, 0), (3, 2)$
18. $(0, 4), (7, -6), (-5, 11)$
19. $(-2, 1), (-1, 0), (2, -2)$
20. $(-1, 1), (3, 3), (5, 5)$

Using the Distance Formula In Exercises 21 and 22, find x such that the distance between the points is 5.

21. $(0, 0), (x, -4)$
22. $(2, -1), (x, 2)$

Using the Distance Formula In Exercises 23 and 24, find y such that the distance between the points is 8.

23. $(0, 0), (3, y)$
24. $(5, 1), (5, y)$

25. **Using the Midpoint Formula** Use the Midpoint Formula to find the three points that divide the line segment joining (x_1, y_1) and (x_2, y_2) into four equal parts.

26. **Using the Midpoint Formula** Use the result of Exercise 25 to find the points that divide the line segment joining the given points into four equal parts.

   (a) $(1, -2), (4, -1)$ (b) $(-2, -3), (0, 0)$

Matching In Exercises 27–30, match the equation with its graph. [The graphs are labeled (a), (b), (c), and (d).]

(a)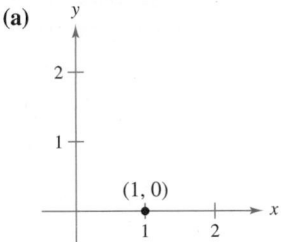

(b)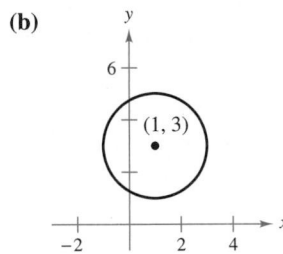

(c)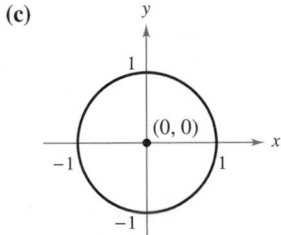

(d)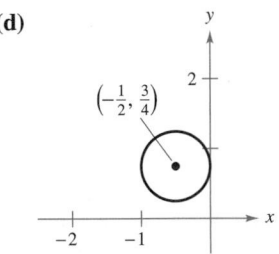

27. $x^2 + y^2 = 1$
28. $(x - 1)^2 + (y - 3)^2 = 4$
29. $(x - 1)^2 + y^2 = 0$
30. $\left(x + \frac{1}{2}\right)^2 + \left(y - \frac{3}{4}\right)^2 = \frac{1}{4}$

Writing the Equation of a Circle In Exercises 31–38, write the standard form of the equation of the circle.

31. Center: $(0, 0)$
    Radius: 3
32. Center: $(0, 0)$
    Radius: 5
33. Center: $(2, -1)$
    Radius: 4
34. Center: $(-4, 3)$
    Radius: $\frac{5}{8}$

35. Center: $(-1, 2)$

Point on circle: $(0, 0)$

36. Center: $(3, -2)$

Point on circle: $(-1, 1)$

37. Endpoints of a diameter: $(2, 5), (4, -1)$

38. Endpoints of a diameter: $(1, 1), (-1, -1)$

39. Satellite Communication Write the standard form of the equation for the path of a communications satellite in a circular orbit 22,000 miles above Earth. (Assume that the radius of Earth is 4000 miles.)

40. Building Design A circular air duct of diameter D is fit firmly into the right-angle corner where a basement wall meets the floor (see figure). Find the diameter of the largest water pipe that can be run in the right-angle corner behind the air duct.

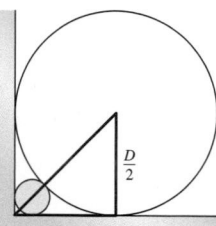

Writing the Equation of a Circle In Exercises 41–48, write the standard form of the equation of the circle and sketch its graph.

41. $x^2 + y^2 - 2x + 6y + 6 = 0$

42. $x^2 + y^2 - 2x + 6y - 15 = 0$

43. $x^2 + y^2 - 2x + 6y + 10 = 0$

44. $3x^2 + 3y^2 - 6y - 1 = 0$

45. $2x^2 + 2y^2 - 2x - 2y - 3 = 0$

46. $4x^2 + 4y^2 - 4x + 2y - 1 = 0$

47. $16x^2 + 16y^2 + 16x + 40y - 7 = 0$

48. $x^2 + y^2 - 4x + 2y + 3 = 0$

Graphing a Circle In Exercises 49 and 50, use a graphing utility to graph the equation. Use a *square setting*. (*Hint:* It may be necessary to solve the equation for y and graph the resulting two equations.)

49. $4x^2 + 4y^2 - 4x + 24y - 63 = 0$

50. $x^2 + y^2 - 8x - 6y - 11 = 0$

Sketching a Graph of an Inequality In Exercises 51 and 52, sketch the set of all points satisfying the inequality. Use a graphing utility to verify your result.

51. $x^2 + y^2 - 4x + 2y + 1 \le 0$

52. $(x - 1)^2 + \left(y - \frac{1}{2}\right)^2 > 1$

53. Proof Prove that

$$\left(\frac{2x_1 + x_2}{3}, \frac{2y_1 + y_2}{3}\right)$$

is one of the points of trisection of the line segment joining (x_1, y_1) and (x_2, y_2). Find the midpoint of the line segment joining

$$\left(\frac{2x_1 + x_2}{3}, \frac{2y_1 + y_2}{3}\right)$$

and (x_2, y_2) to find the second point of trisection.

54. Finding Points of Trisection Use the results of Exercise 53 to find the points of trisection of the line segment joining each pair of points.

(a) $(1, -2), (4, 1)$

(b) $(-2, -3), (0, 0)$

True or False? In Exercises 55–58, determine whether the statement is true or false. If it is false, explain why or give an example that shows it is false.

55. If $ab < 0$, then the point (a, b) lies in either Quadrant II or Quadrant IV.

56. The distance between the points $(a + b, a)$ and $(a - b, a)$ is $2b$.

57. If the distance between two points is zero, then the two points must coincide.

58. If $ab = 0$, then the point (a, b) lies on the x-axis or on the y-axis.

Proof In Exercises 59–62, prove the statement.

59. The line segments joining the midpoints of the opposite sides of a quadrilateral bisect each other.

60. The perpendicular bisector of a chord of a circle passes through the center of the circle.

61. An angle inscribed in a semicircle is a right angle.

62. The midpoint of the line segment joining the points (x_1, y_1) and (x_2, y_2) is

$$\left(\frac{x_1 + x_2}{2}, \frac{y_1 + y_2}{2}\right).$$

Chapter 1

Section 1.1   *(page 8)*

1. To find the x-intercepts of the graph of an equation, let y be zero and solve the equation for x. To find the y-intercepts of the graph of an equation, let x be zero and solve the equation for y.

3. b   4. d   5. a   6. c

7.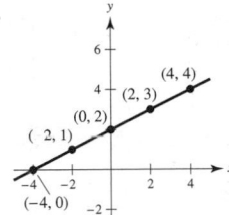

9.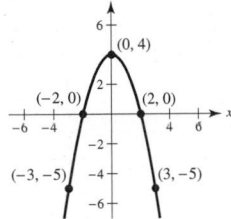

11.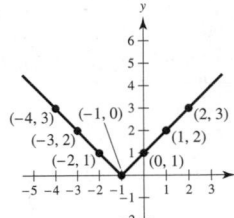

13.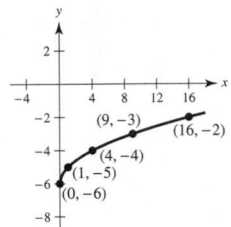

15.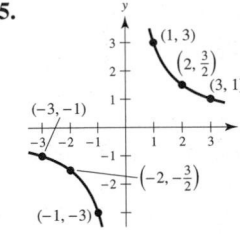

17.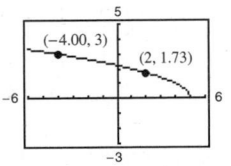

(a) $y \approx 1.73$ (b) $x = -4$

19. $(0, -5), \left(\frac{5}{2}, 0\right)$ 21. $(0, -2), (-2, 0), (1, 0)$

23. $(0, 0), (4, 0), (-4, 0)$ 25. $(0, 2), (4, 0)$ 27. $(0, 0)$

29. Symmetric with respect to the y-axis

31. Symmetric with respect to the x-axis

33. Symmetric with respect to the origin   35. No symmetry

37. Symmetric with respect to the origin

39. Symmetric with respect to the y-axis

41.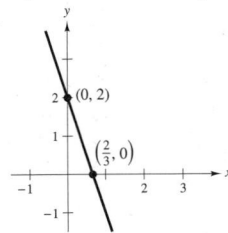

Symmetry: none

43.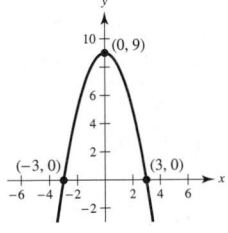

Symmetry: y-axis

45.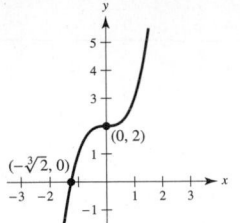

Symmetry: none

47.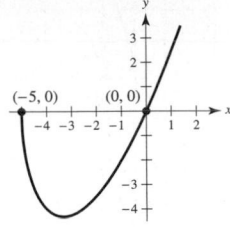

Symmetry: none

49.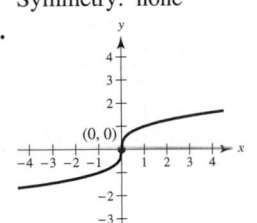

Symmetry: origin

51.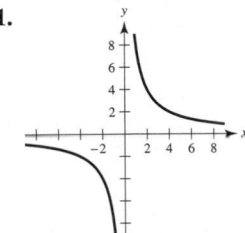

Symmetry: origin

53.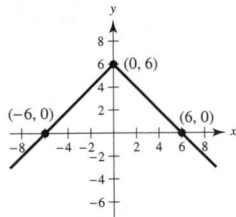

Symmetry: y-axis

55.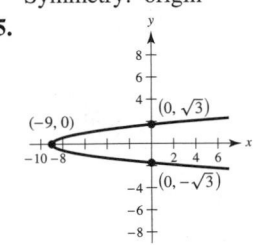

Symmetry: x-axis

57. $(3, 5)$ 59. $(-4, -1), (1, 14)$ 61. $(-1, -2), (2, 1)$

63. $(-1, -5), (0, -1), (2, 1)$ 65. $(-2, 2), \left(-3, \sqrt{3}\right)$

67. (a) $y = 0.58t + 9.2$

(b)

The model is a good fit for the data.

(c) $23.1 trillion

69. 4480 units

71. Answers will vary. Sample answer:
$y = (2x + 3)(x - 4)(2x - 5)$

73. Yes. Assume that the graph has x-axis and origin symmetry. If (x, y) is on the graph, so is $(x, -y)$ by x-axis symmetry. Because $(x, -y)$ is on the graph, then so is $(-x, -(-y)) = (-x, y)$ by origin symmetry. Therefore, the graph is symmetric with respect to the y-axis. The argument is similar for y-axis and origin symmetry.

75. False. $(4, -5)$ is not a point on the graph of $x = y^2 - 29$.

77. True

Section 1.2   *(page 16)*

1. Slope; y-intercept   3. $m = 2$ 5. $m = -1$

7.
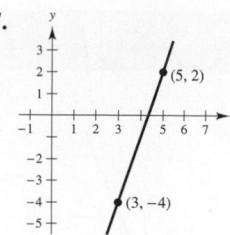
$m = 3$

9.
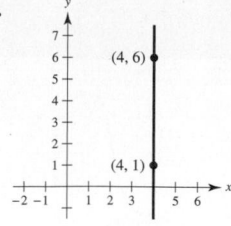
m is undefined.

11.
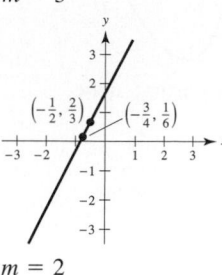
$m = 2$

13.
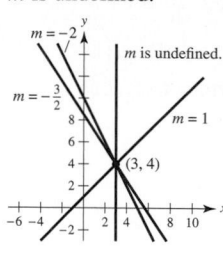

15. Answers will vary. Sample answers: $(0, 2), (1, 2), (5, 2)$

17. Answers will vary. Sample answers: $(0, 10), (2, 4), (3, 1)$

19. $3x - 4y + 12 = 0$ **21.** $x = 1$

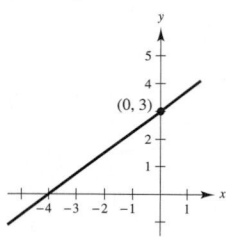
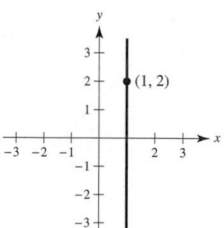

23. $3x - y - 11 = 0$

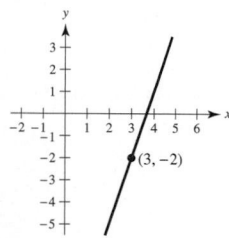

25. 12 ft

27. (a)
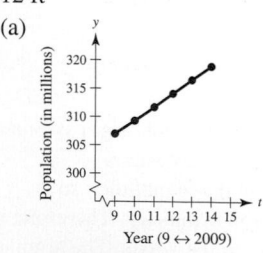
From 2009 to 2010

(b) 2.38 million people per year   (c) 345.1 million people

29. $m = 4, (0, -3)$ **31.** $m = -5, (0, 20)$

33. m is undefined, no y-intercept

35.

37.
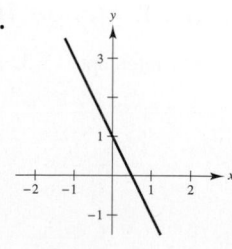

39.
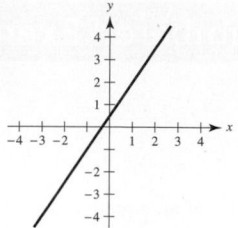

41.
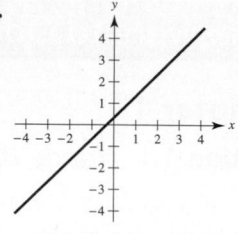

43. $2x - y - 5 = 0$ **45.** $8x + 3y - 40 = 0$

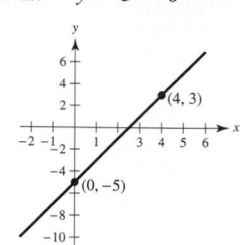
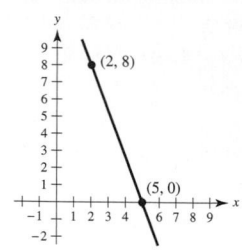

47. $x - 6 = 0$ **49.** $y - 1 = 0$

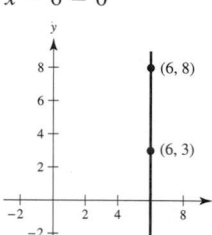
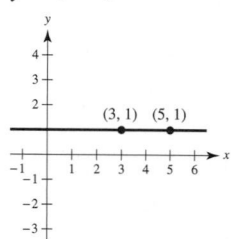

51. $y = \left(\dfrac{1 - b}{3}\right)x + b$ **53.** $3x + 2y - 6 = 0$

55. $x + 2y - 5 = 0$ **57.** (a) $x + 7 = 0$ (b) $y + 2 = 0$

59. (a) $x + y + 1 = 0$ (b) $x - y + 5 = 0$

61. (a) $40x - 24y - 9 = 0$ (b) $24x + 40y - 53 = 0$

63. $V = 250t + 350$ **65.** Not collinear, because $m_1 \neq m_2$

67. The adjacent line segments are perpendicular and each line segment has a length of $\sqrt{8} = 2\sqrt{2}$ units.

69. $12y + 5x - 169 = 0$

71. (a) $\left(0, \dfrac{-a^2 + b^2 + c^2}{2c}\right)$ (b) $\left(\dfrac{b}{3}, \dfrac{c}{3}\right)$

73. $5F - 9C - 160 = 0; 72°F \approx 22.2°C$

75. (a) $x = (1530 - p)/15$

(b)
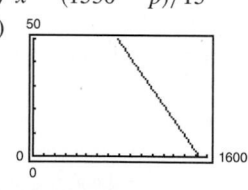
45 units

(c) 49 units

77. Proof   **79.** $\dfrac{5\sqrt{2}}{2}$ **81–83.** Proofs   **85.** True

Section 1.3 *(page 27)*

1. A relation between two sets X and Y is a set of ordered pairs of the form (x, y), where x is a member of X and y is a member of Y.

A function from X to Y is a relation between X and Y that has the property that any two ordered pairs with the same x-value also have the same y-value.

3. Vertical shifts, horizontal shifts, reflections

5. (a) -2 (b) 13   (c) $3b - 2$ (d) $3x - 5$

7. (a) $2\sqrt{2}$ (b) $\sqrt{13}$ (c) $2\sqrt{2}$
   (d) $\sqrt{x^2 + 2bx^2 + b^2x^2 + 4}$

9. (a) 5   (b) 0   (c) 1   (d) $4 + 2t - t^2$

11. $3x^2 + 3x\Delta x + (\Delta x)^2,\ \Delta x \neq 0$

13. Domain: $(-\infty, \infty)$; Range: $[0, \infty)$

15. Domain: $(-\infty, \infty)$; Range: $(-\infty, \infty)$

17. Domain: $[0, \infty)$; Range: $[0, \infty)$

19. Domain: $[-4, 4]$; Range: $[0, 4]$

21. Domain: $(-\infty, 0) \cup (0, \infty)$; Range: $(-\infty, 0) \cup (0, \infty)$

23. Domain: $[0, 1]$

25. Domain: $(-\infty, -3) \cup (-3, \infty)$

27. (a) -1 (b) 2   (c) 6   (d) $2t^2 + 4$
    Domain: $(-\infty, \infty)$; Range: $(-\infty, 1) \cup [2, \infty)$

29. (a) 4   (b) 0   (c) -2 (d) $-b^2$
    Domain: $(-\infty, \infty)$; Range: $(-\infty, 0] \cup [1, \infty)$

31.
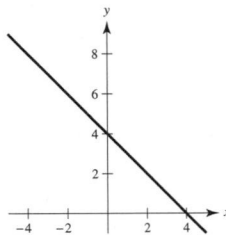
Domain: $(-\infty, \infty)$
Range: $(-\infty, \infty)$

33.
Domain: $(-\infty, \infty)$
Range: $\left(0, \frac{1}{2}\right]$

35.
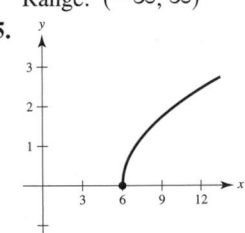
Domain: $[6, \infty)$
Range: $[0, \infty)$

37.
Domain: $[-3, 3]$
Range: $[0, 3]$

39. y is not a function of x.   41. y is a function of x.

43. y is not a function of x.   45. y is not a function of x.

47. Horizontal shift to the right two units
    $y = \sqrt{x - 2}$

49. Horizontal shift to the right two units and vertical shift down one unit
    $y = (x - 2)^2 - 1$

51. d 52. b 53. c 54. a 55. e 56. g

57. (a)
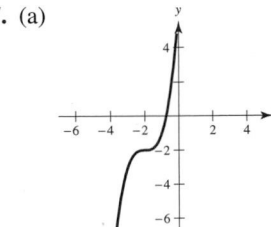
(b)
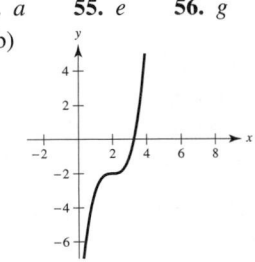

(c)
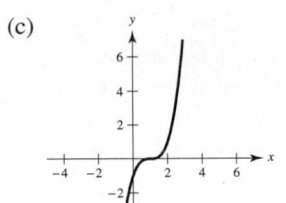

(d)
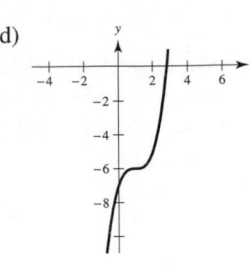

(e)
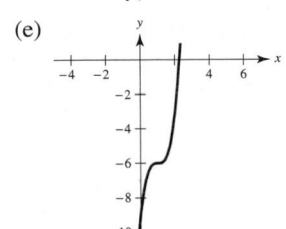

(f)
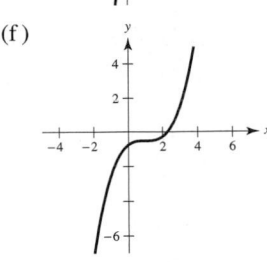

(g)
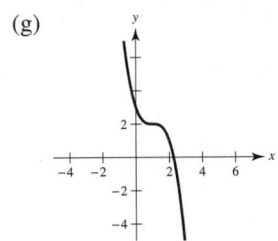

(h)
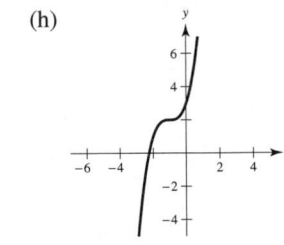

59. (a) $-x - 1$ (b) $5x - 9$
    (c) $-6x^2 + 23x - 20$ (d) $\dfrac{2x - 5}{4 - 3x}$

61. (a) 0   (b) 0   (c) -1 (d) $\sqrt{15}$
    (e) $\sqrt{x^2 - 1}$ (f) $x - 1\ (x \geq 0)$

63. $(f \circ g)(x) = x$; Domain: $[0, \infty)$
    $(g \circ f)(x) = |x|$; Domain: $(-\infty, \infty)$
    No, their domains are different.

65. $(f \circ g)(x) = \dfrac{3}{x^2 - 1}$; Domain: $(-\infty, -1) \cup (-1, 1) \cup (1, \infty)$
    $(g \circ f)(x) = \dfrac{9}{x^2} - 1$; Domain: $(-\infty, 0) \cup (0, \infty)$
    No

67. (a) 4   (b) -2
    (c) Undefined. The graph of g does not exist at $x = -5$.
    (d) 3   (e) 2
    (f) Undefined. The graph of f does not exist at $x = -4$.

69. Answers will vary.
    Sample answer: $f(x) = \sqrt{x},\ g(x) = x - 2,\ h(x) = 2x$

71. (a) $\left(\frac{3}{2}, 4\right)$ (b) $\left(\frac{3}{2}, -4\right)$

73. f is even. g is neither even nor odd. h is odd.

75. Even; zeros: $x = -2, 0, 2$ 77. Neither; zeros: $x = 0$

79. $f(x) = -5x - 6,\ -2 \leq x \leq 0$ 81. $y = -\sqrt{-x}$

83. Answers will vary.   85. Answers will vary.
    Sample answer: Sample answer:

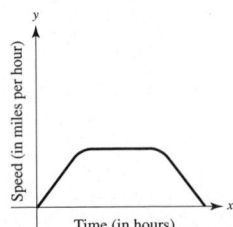

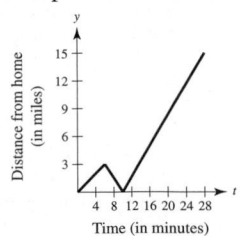

87. $c = 25$

89. No. A graph of a function that is intersected by a horizontal line more than once would mean that there is more than one x-value corresponding to the same y-value.

91. No. Consider $y = x^3 + x + 2$.

$f(-x) \neq -f(x)$

This is an odd-degree function that is not odd.

93. (a) $T(4) = 16°C$, $T(15) \approx 23°C$

(b) The changes in temperature occur 1 hour later.

(c) The temperatures are 1° lower.

95. (a)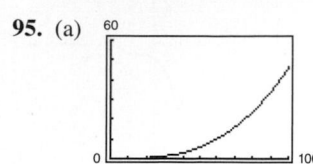

(b) $H\left(\dfrac{x}{1.6}\right) = 0.00001132x^3$

97–99. Proofs   **101.** $L = \sqrt{x^2 + \left(\dfrac{2x}{x-3}\right)^2}$

103. False. For example, if $f(x) = x^2$, then $f(-1) = f(1)$.

105. True

107. False. $f(x) = 0$ is symmetric with respect to the x-axis.

109. Putnam Problem A1, 1988

Section 1.4   *(page 38)*

1. In general, if θ is any angle, then the angle $\theta + n(360°)$, n is a nonzero integer, is coterminal with θ.

3. $\sin\theta = \dfrac{7}{25}$

$\cos\theta = \dfrac{24}{25}$

$\tan\theta = \dfrac{7}{24}$

5. (a) $396°, -324°$ (b) $240°, -480°$

7. (a) $\dfrac{19\pi}{9}, -\dfrac{17\pi}{9}$ (b) $\dfrac{10\pi}{3}, -\dfrac{2\pi}{3}$

9. (a) $\dfrac{\pi}{6}; 0.524$ (b) $\dfrac{5\pi}{6}; 2.618$

(c) $\dfrac{7\pi}{4}; 5.498$ (d) $\dfrac{2\pi}{3}; 2.094$

11. (a) $270°$ (b) $210°$ (c) $-105°$ (d) $-135.62°$

13.

r	8 ft	15 in.	85 cm	24 in.	$\dfrac{12{,}963}{\pi}$ mi
s	12 ft	24 in.	63.75π cm	96 in.	8642 mi
θ	1.5	1.6	$\dfrac{3\pi}{4}$	4	$\dfrac{2\pi}{3}$

15. (a) $\sin\theta = \frac{4}{5}$ $\csc\theta = \frac{5}{4}$ (b) $\sin\theta = -\frac{5}{13}$ $\csc\theta = -\frac{13}{5}$

$\cos\theta = \frac{3}{5}$ $\sec\theta = \frac{5}{3}$ $\cos\theta = -\frac{12}{13}$ $\sec\theta = -\frac{13}{12}$

$\tan\theta = \frac{4}{3}$ $\cot\theta = \frac{3}{4}$ $\tan\theta = \frac{5}{12}$ $\cot\theta = \frac{12}{5}$

17.

$\cos\theta = \dfrac{\sqrt{3}}{2}$

$\tan\theta = \dfrac{\sqrt{3}}{3}$

$\csc\theta = 2$

$\sec\theta = \dfrac{2\sqrt{3}}{3}$

$\cot\theta = \sqrt{3}$

19.

$\sin\theta = \dfrac{3}{5}$

$\tan\theta = \dfrac{3}{4}$

$\csc\theta = \dfrac{5}{3}$

$\sec\theta = \dfrac{5}{4}$

$\cot\theta = \dfrac{4}{3}$

21. (a) $\sin 60° = \dfrac{\sqrt{3}}{2}$ (b) $\sin 120° = \dfrac{\sqrt{3}}{2}$

$\cos 60° = \dfrac{1}{2}$ $\cos 120° = -\dfrac{1}{2}$

$\tan 60° = \sqrt{3}$ $\tan 120° = -\sqrt{3}$

(c) $\sin\dfrac{\pi}{4} = \dfrac{\sqrt{2}}{2}$ (d) $\sin\dfrac{5\pi}{4} = -\dfrac{\sqrt{2}}{2}$

$\cos\dfrac{\pi}{4} = \dfrac{\sqrt{2}}{2}$ $\cos\dfrac{5\pi}{4} = -\dfrac{\sqrt{2}}{2}$

$\tan\dfrac{\pi}{4} = 1$ $\tan\dfrac{5\pi}{4} = 1$

23. (a) $\sin 225° = -\dfrac{\sqrt{2}}{2}$ (b) $\sin(-225°) = \dfrac{\sqrt{2}}{2}$

$\cos 225° = -\dfrac{\sqrt{2}}{2}$ $\cos(-225°) = -\dfrac{\sqrt{2}}{2}$

$\tan 225° = 1$ $\tan(-225°) = -1$

(c) $\sin\dfrac{5\pi}{3} = -\dfrac{\sqrt{3}}{2}$ (d) $\sin\dfrac{11\pi}{6} = -\dfrac{1}{2}$

$\cos\dfrac{5\pi}{3} = \dfrac{1}{2}$ $\cos\dfrac{11\pi}{6} = \dfrac{\sqrt{3}}{2}$

$\tan\dfrac{5\pi}{3} = -\sqrt{3}$ $\tan\dfrac{11\pi}{6} = -\dfrac{\sqrt{3}}{3}$

25. (a) 0.1736 (b) 5.7588 **27.** (a) 0.3640 (b) 0.3640

29. (a) Quadrant III   (b) Quadrant IV

31. (a) $\theta = \dfrac{\pi}{4}, \dfrac{7\pi}{4}$ (b) $\theta = \dfrac{3\pi}{4}, \dfrac{5\pi}{4}$

33. (a) $\theta = \dfrac{\pi}{4}, \dfrac{5\pi}{4}$ (b) $\theta = \dfrac{5\pi}{6}, \dfrac{11\pi}{6}$

35. $\theta = \dfrac{\pi}{4}, \dfrac{3\pi}{4}, \dfrac{5\pi}{4}, \dfrac{7\pi}{4}$ **37.** $\theta = 0, \dfrac{\pi}{4}, \pi, \dfrac{5\pi}{4}, 2\pi$

39. $\theta = \dfrac{\pi}{3}, \dfrac{5\pi}{3}$ **41.** $\theta = 0, \dfrac{\pi}{2}, \pi, 2\pi$ **43.** 5099 ft

45. Period: π **47.** Period: $\frac{1}{2}$

Amplitude: 2   Amplitude: 3

49. Period: $\dfrac{\pi}{2}$ **51.** Period: $\dfrac{2\pi}{5}$

53. (a)

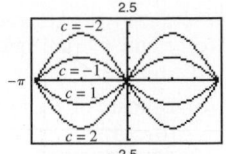

Change in amplitude

(b)

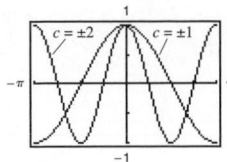

Change in period

(c)

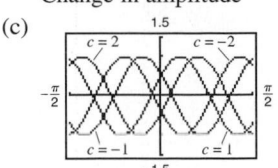

Horizontal translation

55.

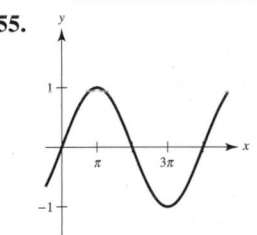

57.

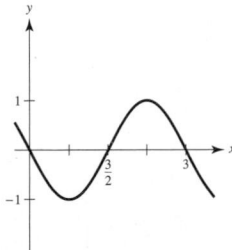

59.

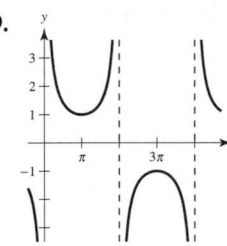

61.

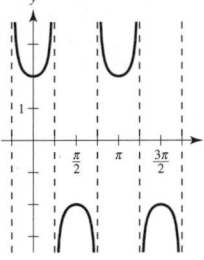

63.

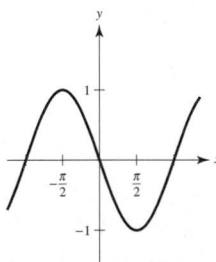

65.

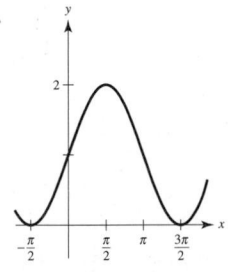

67. $a = 3$, $b = \dfrac{1}{2}$, $c = \dfrac{\pi}{2}$

69. No. You can use $1 + \tan^2 \theta = \sec^2 \theta$, but you are unable to determine the sign.

71. The range of the cosine function is $-1 \leq y \leq 1$. The range of the secant function is $y \leq -1$ or $y \geq 1$.

73.

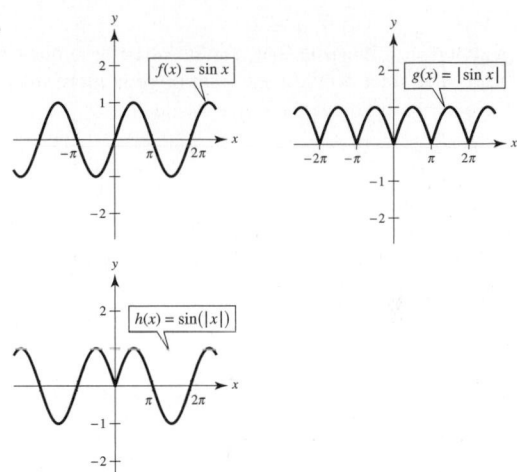

The graph of $|f(x)|$ will reflect any parts of the graph of $f(x)$ below the x-axis about the x-axis. The graph of $f(|x|)$ will reflect the part of the graph of $f(x)$ right of the y-axis about the y-axis.

75.

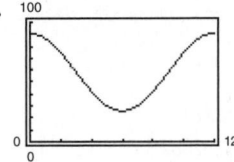

January, November, December

77. False. 4π radians (not 4 radians) corresponds to two complete revolutions from the initial side to the terminal side of an angle.

79. False. The amplitude of the function $y = \frac{1}{2} \sin 2x$ is one-half the amplitude of the function $y = \sin x$.

Section 1.5   *(page 48)*

1. The domain of f is equal to the range of f^{-1}. The domain of f^{-1} is equal to the range of f. The graph of f^{-1} is a reflection of the graph of f in the line $y = x$.

3. arccos x is the angle, $0 \leq \theta \leq \pi$, whose cosine is x.

5. c   **6.** b   **7.** a   **8.** d

9. (a) $f(g(x)) = 5[(x - 1)/5] + 1 = x$;
   $g(f(x)) = [(5x + 1) - 1]/5 = x$
   (b)

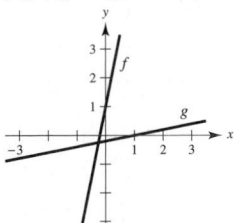

11. (a) $f(g(x)) = \left(\sqrt[3]{x}\right)^3 = x$; $g(f(x)) = \sqrt[3]{x^3} = x$
   (b)

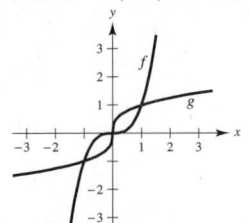

13. (a) $f(g(x)) = \sqrt{x^2 + 4 - 4} = x;$
$g(f(x)) = \left(\sqrt{x - 4}\right)^2 + 4 = x$

(b)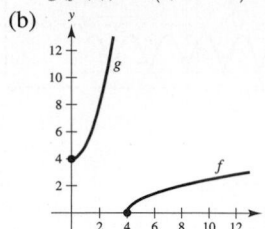

15. (a) $f(g(x)) = \dfrac{1}{1/x} = x;$ $g(f(x)) = \dfrac{1}{1/x} = x$

(b)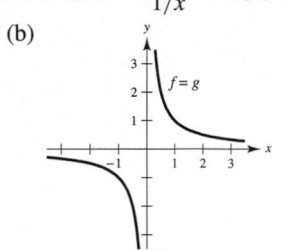

17. Not one-to-one, inverse does not exist.
19. One-to-one, inverse exists.
21. One-to-one, inverse exists.
23. Not one-to-one, inverse does not exist.

25.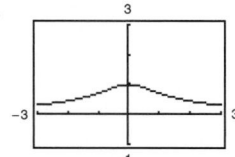

One-to-one, inverse exists.

27.

Not one-to-one, inverse does
not exist.

29.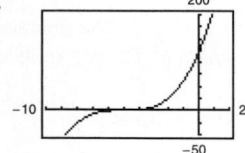

One-to-one, inverse exists.

31. (a) $f^{-1}(x) = (x + 3)/2$

(b)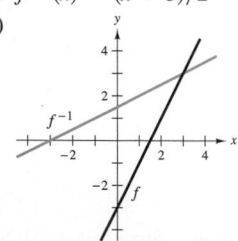

(c) f and f^{-1} are symmetric
about $y = x$.
(d) Domain of f and f^{-1}:
$(-\infty, \infty)$
Range of f and f^{-1}:
$(-\infty, \infty)$

33. (a) $f^{-1}(x) = x^{1/5}$

(b)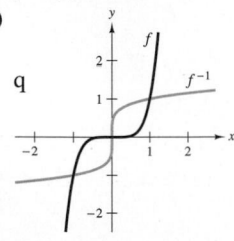

(c) f and f^{-1} are symmetric
about $y = x$.
(d) Domain of f and f^{-1}:
$(-\infty, \infty)$
Range of f and f^{-1}:
$(-\infty, \infty)$

35. (a) $f^{-1}(x) = x^2$, $x \geq 0$

(b)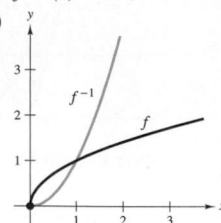

(c) f and f^{-1} are symmetric
about $y = x$.
(d) Domain of f and f^{-1}:
$[0, \infty)$
Range of f and f^{-1}:
$[0, \infty)$

37. (a) $f^{-1}(x) = \sqrt{4 - x^2}$, $0 \leq x \leq 2$

(b)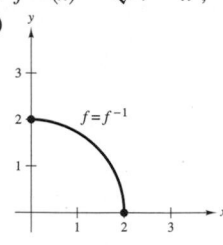

(c) f and f^{-1} are symmetric
about $y = x$.
(d) Domain of f and f^{-1}:
$[0, 2]$
Range of f and f^{-1}:
$[0, 2]$

39. (a) $f^{-1}(x) = x^3 + 1$

(b)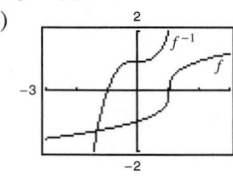

(c) f and f^{-1} are symmetric
about $y = x$.
(d) Domain of f and f^{-1}:
$(-\infty, \infty)$
Range of f and f^{-1}:
$(-\infty, \infty)$

41. (a) $f^{-1}(x) = x^{3/2}$, $x \geq 0$

(b)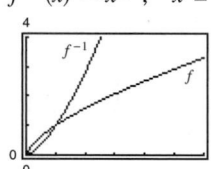

(c) f and f^{-1} are symmetric
about $y = x$.
(d) Domain of f and f^{-1}:
$[0, \infty)$
Range of f and f^{-1}:
$[0, \infty)$

43. (a) $f^{-1}(x) = \sqrt{7}x/\sqrt{1 - x^2}$, $-1 < x < 1$

(b)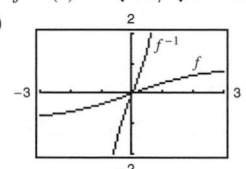

(c) f and f^{-1} are symmetric
about $y = x$.
(d) Domain of f: $(-\infty, \infty)$
Range of f: $(-1, 1)$
Domain of f^{-1}: $(-1, 1)$
Range of f^{-1}:
$(-\infty, \infty)$

45.

x	0	1	2	4
$f(x)$	1	2	3	4

x	1	2	3	4
$f^{-1}(x)$	0	1	2	4

47. (a) Proof
(b) $y = \frac{2}{3}(137.5 - x)$
x: total cost
y: number of pounds of the less expensive commodity
(c) $[62.5, 137.5]$; $50(1.25) = 62.5$ gives the total cost when
purchasing 50 pounds of the less expensive commodity,
and $50(2.75) = 137.5$ gives the total cost when purchasing
50 pounds of the more expensive commodity.
(d) 43 lb

49. One-to-one; $f^{-1}(x) = x^2 + 2$, $x \geq 0$ **51.** Not one-to-one

53. One-to-one; $f^{-1}(x) = \dfrac{x - b}{a}$, $a \neq 0$

55. The function f passes the Horizontal Line Test on $[4, \infty)$, so it is one-to-one on $[4, \infty)$.

57. The function f passes the Horizontal Line Test on $(0, \infty)$, so it is one-to-one on $(0, \infty)$.

59. The function f passes the Horizontal Line Test on $[0, \pi]$, so it is one-to-one on $[0, \pi]$.

61. Answers will vary. Sample answer: $f^{-1}(x) = \sqrt{x} + 3$, $x \geq 0$

63. (a)

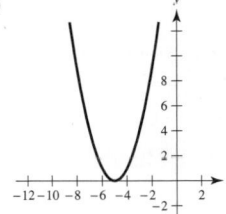

65. (a)

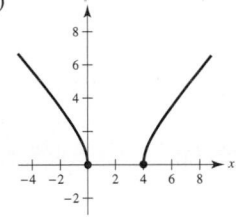

(b) Answers will vary.
Sample answer: $[-5, \infty)$
(c) $f^{-1}(x) = \sqrt{x} - 5$
(d) Domain of f^{-1}: $[0, \infty)$

(b) Answers will vary.
Sample answer: $[4, \infty)$
(c) $f^{-1}(x) = 2 + \sqrt{x^2 + 4}$
(d) Domain of f^{-1}: $[0, \infty)$

67. (a)

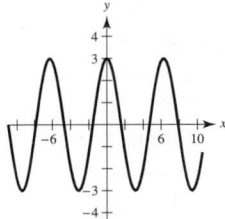

(b) Answers will vary. Sample answer: $[0, \pi]$

(c) $f^{-1}(x) = \arccos\left(\dfrac{x}{3}\right)$

(d) Domain of f^{-1}: $[-3, 3]$

69. 1   **71.** -0.5236 **73.** 2   **75.** 32   **77.** 88

79. $(g^{-1} \circ f^{-1})(x) = \sqrt[3]{6 - x}$ **81.** $(f \circ g)^{-1}(x) = \sqrt[3]{6 - x}$

83. (a) f is one-to-one because it passes the Horizontal Line Test.
(b) $[-2, 2]$
(c) -4

85.

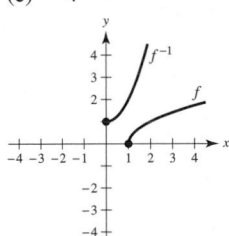

87. (a)

x	-1	-0.8	-0.6	-0.4	-0.2
y	-1.57	-0.93	-0.64	-0.41	-0.20

x	0	0.2	0.4	0.6	0.8	1
y	0	0.20	0.41	0.64	0.93	1.57

(b)

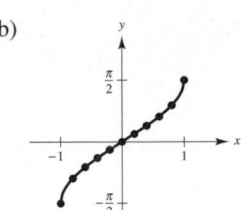

(c)

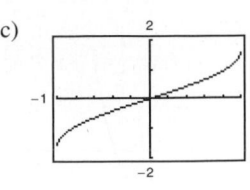

(d) Intercept: $(0, 0)$; Symmetry: origin

89. $\left(-\dfrac{\sqrt{2}}{2}, \dfrac{3\pi}{4}\right), \left(\dfrac{1}{2}, \dfrac{\pi}{3}\right), \left(\dfrac{\sqrt{3}}{2}, \dfrac{\pi}{6}\right)$

91.

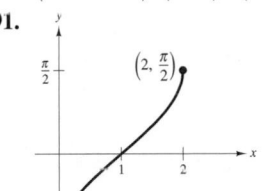

93.

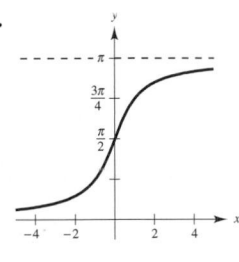

95. $\dfrac{\pi}{6}$ **97.** $\dfrac{\pi}{3}$ **99.** $\dfrac{\pi}{6}$ **101.** $-\dfrac{\pi}{4}$ **103.** 1.52

105. 0.66   **107.** -0.1

109. No. Graphically, adding a constant shifts the graph vertically.

111. The trigonometric functions are not one-to-one, so their domains must be restricted to define the inverse trigonometric functions.

113. $x = \frac{1}{3}\left[\sin\left(\frac{1}{2}\right) + \pi\right] \approx 1.207$ **115.** $x = \frac{1}{3}$

117. x **119.** $\dfrac{\sqrt{1 - x^2}}{x}$ **121.** $\dfrac{1}{x}$

123. (a) $\dfrac{3}{5}$ (b) $\dfrac{5}{3}$ **125.** (a) $-\sqrt{3}$ (b) $-\dfrac{13}{5}$

127. $\sqrt{1 - 4x^2}$ **129.** $\dfrac{\sqrt{x^2 - 1}}{|x|}$ **131.** $\dfrac{\sqrt{x^2 - 9}}{3}$

133. $\arcsin\left(\dfrac{9}{\sqrt{x^2 + 81}}\right)$ **135.** Answers will vary.

137. Answers will vary.   **139.** False. Let $f(x) = x^2$.

141. False. $\arcsin^2 0 + \arccos^2 0 = \left(\dfrac{\pi}{2}\right)^2 \neq 1$

143. True   **145–149.** Proofs   **151.** $k \geq 1$ and $k \leq -1$

153. $ad - bc \neq 0$; $f^{-1}(x) = \dfrac{b - dx}{cx - a}$

Section 1.6   (page 57)

1. Answers will vary.

3. $f(x) = e^x$ and $g(x) = \ln x$ are inverse functions of each other. So, $\ln e^x = g(f(x)) = x$.

5. (a) 125   (b) 9   (c) $\frac{1}{9}$ (d) $\frac{1}{3}$

7. (a) 5^5 (b) $\frac{1}{5}$ (c) $\frac{1}{5}$ (d) 2^2

9. (a) e^6 (b) e^{12} (c) $\dfrac{1}{e^6}$ (d) $\dfrac{1}{e^8}$

11. $x = 4$ **13.** $x = 4$ **15.** $x = -5$ **17.** $x = -2$

19. $x = 2$ **21.** $x = 16$ **23.** $x = -1$ **25.** $x = -\frac{5}{2}$

27.

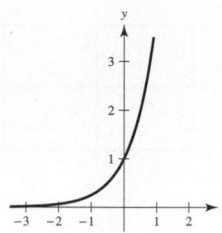

29.

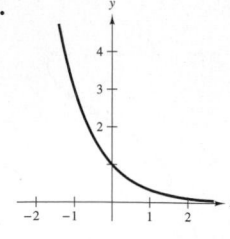

31.

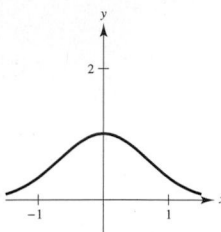

33.

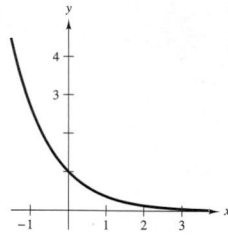

35.

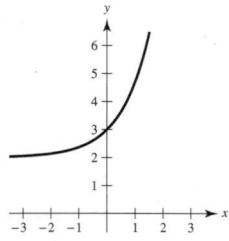

37.

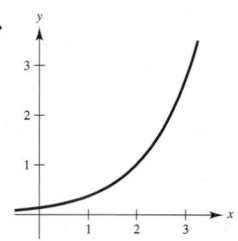

39.

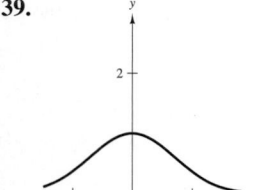

41. Domain: $(-\infty, \infty)$
43. Domain: $(-\infty, 0]$
45. Domain: $(-\infty, \infty)$

47. c   **48.** d   **49.** a   **50.** b   **51.** $y = 2(3^x)$
53. $\ln 1 = 0$ **55.** $e^{1.4231\ldots} = 4.15$
57. b   **58.** d   **59.** a   **60.** c

61.

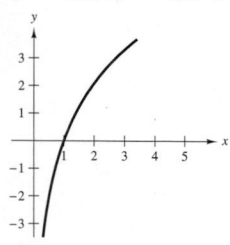

Domain: $x > 0$

63.

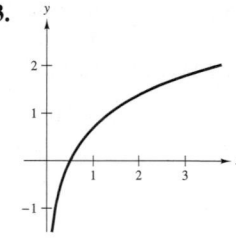

Domain: $x > 0$

65.

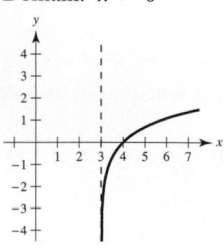

Domain: $x > 3$

67.

Domain: $x > -2$

69. $g(x) = -e^x - 8$ **71.** $g(x) = \ln(x - 5) - 1$
73.

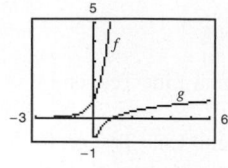

75.

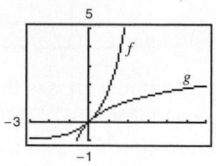

77. (a) $f^{-1}(x) = \dfrac{\ln x + 1}{4}$

(b)

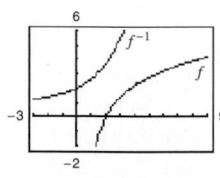

(c) Answers will vary.
79. (a) $f^{-1}(x) = e^{x/2} + 1$

(b)

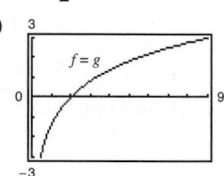

(c) Answers will vary.
81. x^2 **83.** $5x + 2$ **85.** $-1 + 2x$
87. (a) 1.7917   (b) -0.4055 (c) 4.3944   (d) 0.5493
89. $\ln x - \ln 4$ **91.** $\ln x + \ln y - \ln z$
93. $\ln x + \frac{1}{2}\ln(x^2 + 5)$ **95.** $\frac{1}{2}[\ln(x - 1) - \ln x]$
97. $2 + \ln 3$ **99.** $\ln(7x)$
101. $\ln \dfrac{x - 2}{x + 2}$ **103.** $\ln \sqrt[3]{\dfrac{x(x + 3)^2}{x^2 - 1}}$ **105.** $\ln \dfrac{9}{\sqrt{x^2 + 1}}$
107. $x \approx 2.485$ **109.** $x = 0$ **111.** $x \approx 0.511$
113. $x \approx 7.389$ **115.** $x \approx 10.389$ **117.** $x \approx 5.389$
119. $x > \dfrac{\ln 3 - 1}{2}$ **121.** $e^{-2} < x < 1$
123. (a)

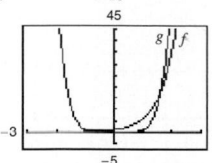

(b) Answers will vary.
125. No; $\ln(a^b) = b \ln a$ is only true when $a > 0$ because this follows the properties of logarithms.
127. (a) False   (b) True. $2^y = x$
    (c) True. $y = \log_2 x$ (d) False
129. $\beta = 10 \log_{10} I + 160$
131.

$(-0.7899, 0.2429)$, $(1.6242, 18.3615)$, and $(6, 46{,}656)$;
As x increases, $f(x) = 6^x$ grows more rapidly.

133. (a) 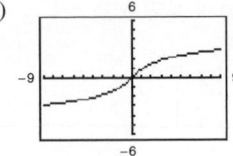   Domain: $(-\infty, \infty)$

(b) Proof

(c) $f^{-1}(x) = \dfrac{e^{2x} - 1}{2e^x}$

135. $12! = 479{,}001{,}600$
Stirling's Formula: $12! \approx 475{,}687{,}487$

137. Proof

Review Exercises for Chapter 1 *(page 60)*

1. $\left(\frac{8}{5}, 0\right), (0, -8)$ **3.** $(3, 0), \left(0, \frac{3}{4}\right)$ **5.** Not symmetric

7. Symmetric with respect to the x-axis, the y-axis, and the origin

9. 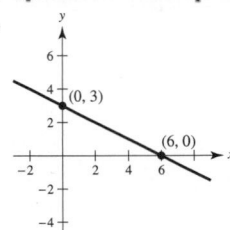   **11.**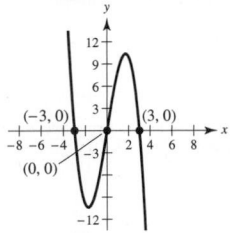

Symmetry: none   Symmetry: origin

13.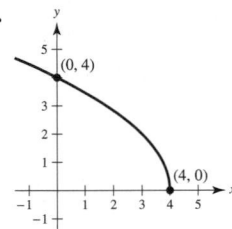

Symmetry: none

15. $(-2, 3)$ **17.** $(-2, 3), (3, 8)$

19. 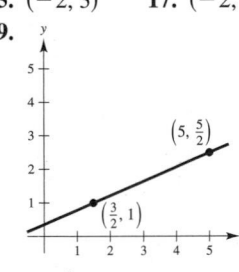   **21.** $7x - 4y - 41 = 0$

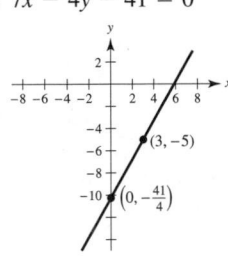

$m = \frac{3}{7}$

23. $2x + 3y + 6 = 0$

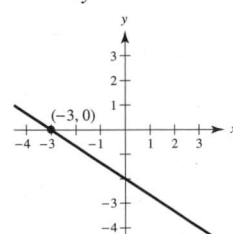

25. Slope: 3
y-intercept: $(0, 5)$

27. 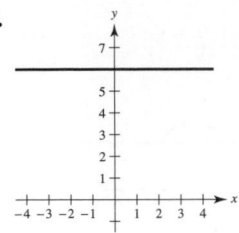   **29.**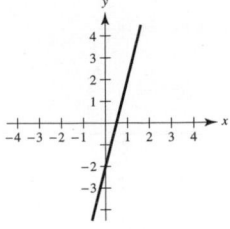

31. $x - 4y = 0$ **33.** (a) $7x - 16y + 101 = 0$
(b) $5x - 3y + 30 = 0$
(c) $4x - 3y + 27 = 0$
(d) $x + 3 = 0$

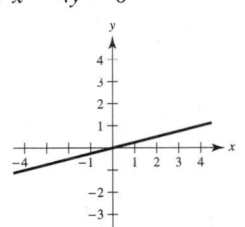

35. $V = 12{,}500 - 850t$; \$9950
37. (a) 4   (b) 29   (c) -11 (d) $5t + 9$
39. $8x + 4\Delta x, \Delta x \neq 0$
41. Domain: $(-\infty, \infty)$; Range: $[3, \infty)$
43.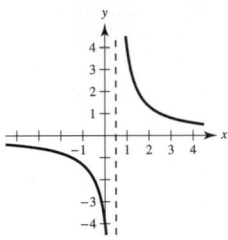

Domain: $\left(-\infty, \frac{1}{2}\right) \cup \left(\frac{1}{2}, \infty\right)$
Range: $(-\infty, 0) \cup (0, \infty)$
45. Not a function   **47.** Function
49. $f(x) = x^3 - 3x^2$

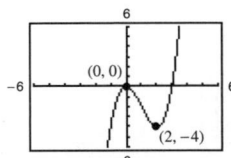

(a) $g(x) = -x^3 + 3x^2 + 1$
(b) $g(x) = (x - 2)^3 - 3(x - 2)^2 + 1$
51. $f(g(x)) = -3x + 1$; Domain: $(-\infty, \infty)$
$g(f(x)) = -3x - 1$; Domain: $(-\infty, \infty)$
No
53. Even; zeros: $x = -1, 0, 1$
55. $\dfrac{17\pi}{9} \approx 5.934$ **57.** $-\dfrac{8\pi}{3} \approx -8.378$
59. 30°   **61.** $-120°$
63. $\sin(-45°) = -\dfrac{\sqrt{2}}{2}$ **65.** $\sin\dfrac{13\pi}{6} = \dfrac{1}{2}$
$\cos(-45°) = \dfrac{\sqrt{2}}{2}$ $\cos\dfrac{13\pi}{6} = \dfrac{\sqrt{3}}{2}$
$\tan(-45°) = -1$ $\tan\dfrac{13\pi}{6} = \dfrac{\sqrt{3}}{3}$

67. $\sin 405° = \dfrac{\sqrt{2}}{2}$

$\cos 405° = \dfrac{\sqrt{2}}{2}$

$\tan 405° = 1$

69. 0.6494   **71.** 3.2361   **73.** -0.3420 **75.** $\dfrac{2\pi}{3}, \dfrac{4\pi}{3}$

77. $\dfrac{7\pi}{6}, \dfrac{3\pi}{2}, \dfrac{11\pi}{6}$ **79.** $\dfrac{\pi}{3}, \pi, \dfrac{5\pi}{3}$

81.
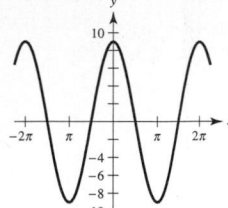

83.
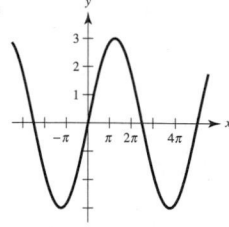

85.
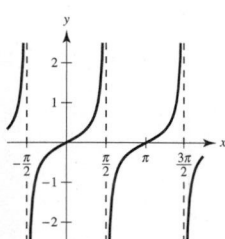

87.
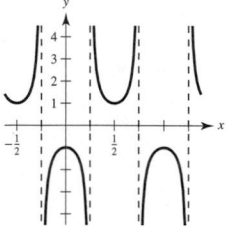

89. (a) $f(g(x)) = 4[(x + 1)/4] - 1 = x;$
$g(f(x)) = [(4x - 1) + 1]/4 = x$
(b)
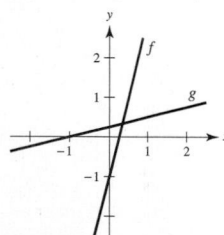

91. Not one-to-one, inverse does not exist
93. Not one-to-one, inverse does not exist
95. (a) $f^{-1}(x) = 2x + 6$
(b)   (c) Proof
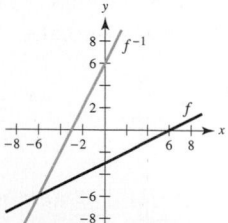

(d) Domain of f and f^{-1}: all real numbers
Range of f and f^{-1}: all real numbers
97. (a) $f^{-1}(x) = x^2 - 1, \quad x \geq 0$
(b)   (c) Proof
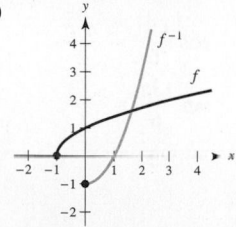

(d) Domain of f: $x \geq -1$, Domain of f^{-1}: $x \geq 0$

Range of f: $y \geq 0$, Range of f^{-1}: $y \geq -1$
99. (a) $f^{-1}(x) = x^3 - 1$
(b)   (c) Proof
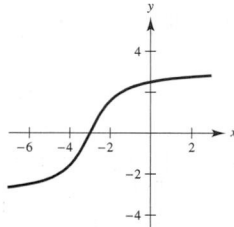

(d) Domain of f and f^{-1}: all real numbers
Range of f and f^{-1}: all real numbers
101. Not one-to-one, inverse does not exist
103. The function f passes the Horizontal Line Test on $[0, \infty)$,
so it is one-to-one on $[0, \infty)$.
105.
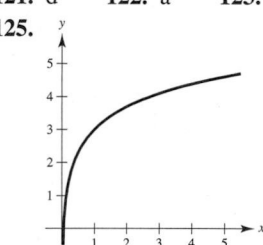

107. $\dfrac{\pi}{4} + n\pi$ **109.** $x = \dfrac{\cos 2 - 1}{2} \approx -0.708$

111. (a) $\dfrac{1}{2}$ (b) $\dfrac{\sqrt{3}}{2}$ **113.** $\dfrac{2x}{\sqrt{4x^2 + 1}}$

115. (a) 3^2 (b) 3^8 (c) 3^6 (d) 3^2 **117.** $x = 8$
119.

121. d   **122.** a   **123.** c   **124.** b
125.

Domain: $x > 0$
127. $\frac{1}{5}[\ln(2x + 1) + \ln(2x - 1) - \ln(4x^2 + 1)]$
129. $\ln\left(\dfrac{3\sqrt[3]{4 - x^2}}{x}\right)$ **131.** $x \approx -0.602$

P.S. Problem Solving   *(page 63)*

1. (a) Center: $(3, 4)$; Radius: 5
(b) $y = -\frac{3}{4}x$ (c) $y = \frac{3}{4}x - \frac{9}{2}$ (d) $\left(3, -\frac{9}{4}\right)$

3.

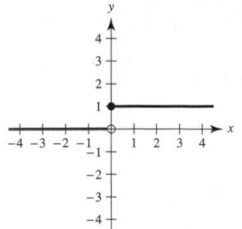

(a)

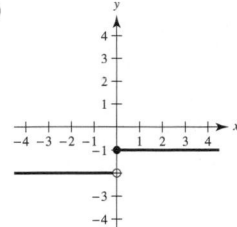

(b)

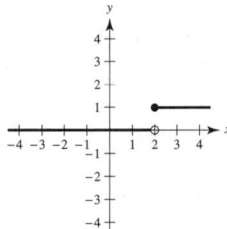

(c)

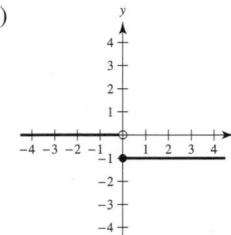

(d)

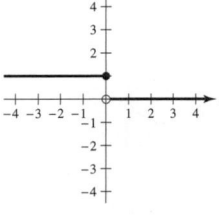

(e)

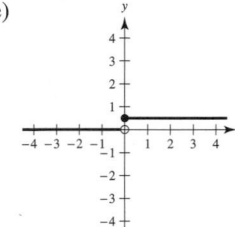

(f)

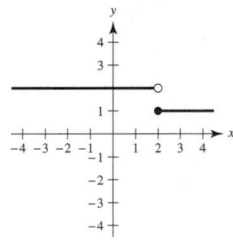

5. (a) $A(x) = x\left(\dfrac{100 - x}{2}\right)$; Domain: $(0, 100)$

(b)

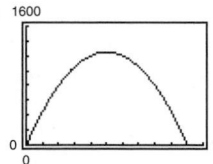

Dimensions 50 m $\times$ 25 m yield maximum area of 1250 m².

(c) 50 m $\times$ 25 m; Area $= 1250$ m²

7. $T(x) = \dfrac{2\sqrt{4 + x^2} + \sqrt{(3 - x)^2 + 1}}{4}$

9. (a) 5; less   (b) 3; greater   (c) 4.1; less
(d) $4 + h$ (e) 4; Answers will vary.

11. (a) Domain: $(-\infty, 1) \cup (1, \infty)$; Range: $(-\infty, 0) \cup (0, \infty)$

(b) $f(f(x)) = \dfrac{x - 1}{x}$

Domain: $(-\infty, 0) \cup (0, 1) \cup (1, \infty)$

(c) $f(f(f(x))) = x$

Domain: $(-\infty, 0) \cup (0, 1) \cup (1, \infty)$

(d)

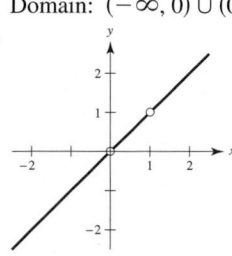

The graph is not a line because there are holes at $x = 0$ and $x = 1$.

13. (a) $x \approx 1.2426, -7.2426$ **15.** Proof
(b) $(x + 3)^2 + y^2 = 18$

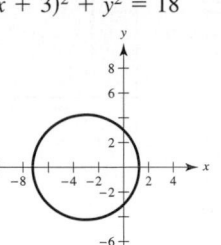

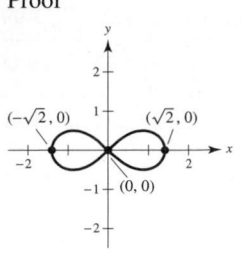

Chapter 2

Section 2.1   *(page 71)*

1. Calculus is the mathematics of change. Precalculus mathematics is more static.

Answers will vary. Sample answer:

Precalculus	**Calculus**
Area of a rectangle	Area under a curve
Work done by a constant force	Work done by a variable force
Center of a rectangle	Centroid of a region

3. Precalculus: 300 ft

5. Calculus: Slope of the tangent line at $x = 2$ is 0.16.

7. (a)

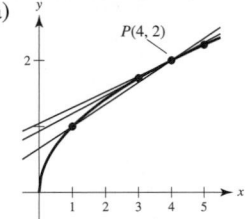

(b) $x = 1$: $m = \frac{1}{3}$

$x = 3$: $m = \dfrac{1}{\sqrt{3} + 2} \approx 0.2679$

$x = 5$: $m = \dfrac{1}{\sqrt{5} + 2} \approx 0.2361$

(c) $\frac{1}{4}$; You can improve your approximation of the slope at $x = 4$ by considering x-values very close to 4.

9. Area ≈ 10.417; Area ≈ 9.145; Use more rectangles.

11. (a) About 5.66   (b) About 6.11
(c) Increase the number of line segments.

Section 2.2   *(page 79)*

1. As the graph of the function approaches 8 on the horizontal axis, the graph approaches 25 on the vertical axis.

3.

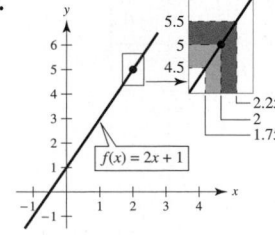

5.

x	3.9	3.99	3.999	4
$f(x)$	0.3448	0.3344	0.3334	?

x	4.001	4.01	4.1
$f(x)$	0.3332	0.3322	0.3226

$$\lim_{x\to 4}\frac{x-4}{x^2-5x+4}\approx 0.3333 \left(\text{Actual limit is }\frac{1}{3}.\right)$$

7.

x	-0.1	-0.01	-0.001	0
$f(x)$	0.9983	0.99998	1.0000	?

x	0.001	0.01	0.1
$f(x)$	1.0000	0.99998	0.9983

$$\lim_{x\to 0}\frac{\sin x}{x}\approx 1.0000 \text{ (Actual limit is 1.)}$$

9.

x	-0.1	-0.01	-0.001	0
$f(x)$	0.9516	0.9950	0.9995	?

x	0.001	0.01	0.1
$f(x)$	1.0005	1.0050	1.0517

$$\lim_{x\to 0}\frac{e^x-1}{x}\approx 1.0000 \text{ (Actual limit is 1.)}$$

11.

x	0.9	0.99	0.999	1
$f(x)$	0.2564	0.2506	0.2501	?

x	1.001	1.01	1.1
$f(x)$	0.2499	0.2494	0.2439

$$\lim_{x\to 1}\frac{x-2}{x^2+x-6}\approx 0.2500 \left(\text{Actual limit is }\frac{1}{4}.\right)$$

13.

x	0.9	0.99	0.999	1
$f(x)$	0.7340	0.6733	0.6673	?

x	1.001	1.01	1.1
$f(x)$	0.6660	0.6600	0.6015

$$\lim_{x\to 1}\frac{x^4-1}{x^6-1}\approx 0.6666 \left(\text{Actual limit is }\frac{2}{3}.\right)$$

15.

x	-6.1	-6.01	-6.001	-6
$f(x)$	-0.1248	-0.1250	-0.1250	?

x	-5.999	-5.99	-5.9
$f(x)$	-0.1250	-0.1250	-0.1252

$$\lim_{x\to -6}\frac{\sqrt{10-x}-4}{x+6}\approx -0.1250 \left(\text{Actual limit is }-\frac{1}{8}.\right)$$

17.

x	-0.1	-0.01	-0.001	0
$f(x)$	1.9867	1.9999	2.0000	?

x	0.001	0.01	0.1
$f(x)$	2.0000	1.9999	1.9867

$$\lim_{x\to 0}\frac{\sin 2x}{x}\approx 2.0000 \text{ (Actual limit is 2.)}$$

19.

x	1.9	1.99	1.999	2
$f(x)$	0.5129	0.5013	0.5001	?

x	2.001	2.01	2.1
$f(x)$	0.4999	0.4988	0.4879

$$\lim_{x\to 2}\frac{\ln x-\ln 2}{x-2}\approx 0.5000 \left(\text{Actual limit is }\frac{1}{2}.\right)$$

21.

x	-0.1	-0.01	-0.001	0
$f(x)$	3.9998	4	4	?

x	0.001	0.01	0.1
$f(x)$	0	0	0.0002

The function approaches 0 from the right side of 0, but it approaches 4 from the left side of 0.

23. 1   **25.** 2

27. Limit does not exist. The function approaches 1 from the right side of 2, but it approaches -1 from the left side of 2.

29. Limit does not exist. The function oscillates between 1 and -1 as x approaches 0.

31. (a) 2

(b) Limit does not exist. The function approaches 1 from the right side of 1, but it approaches 3.5 from the left side of 1.

(c) Value does not exist. The function is undefined at $x=4$.

(d) 2

33. 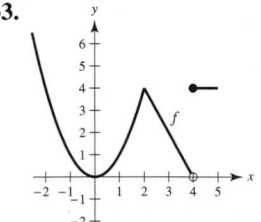   **35.**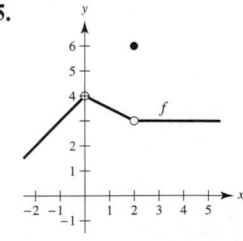

$\lim_{x\to c} f(x)$ exists for all points on the graph except where $c=4$.

37. $\delta=0.4$ **39.** $\delta=\frac{1}{11}\approx 0.091$

41. $L=8$

Answers will vary. Sample answers:

(a) $\delta\approx 0.0033$ (b) $\delta\approx 0.00167$

43. $L = 1$
Answers will vary.
Sample answers:
(a) $\delta = 0.002$
(b) $\delta = 0.001$

45. $L = 12$
Answers will vary.
Sample answers:
(a) $\delta = 0.00125$
(b) $\delta = 0.000625$

47. 6   **49.** -3 **51.** 3   **53.** 0   **55.** 10
57. 2   **59.** 4

61.
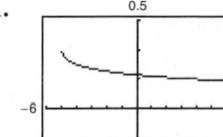

$\lim_{x \to 4} f(x) = \frac{1}{6}$
Domain: $[-5, 4) \cup (4, \infty)$
The graph has a hole at $x = 4$.

63. (a) $17.89; the cost of a 10-minute, 45-second phone call
(b)
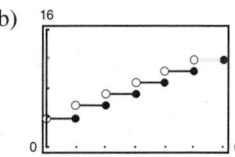
The limit does not exist because the limits from the right and left are not equal.

65. Choosing a smaller positive value of δ will still satisfy the inequality $|f(x) - L| < \varepsilon$.

67. No. The fact that $f(2) = 4$ has no bearing on the existence of the limit of $f(x)$ as x approaches 2.

69. (a) $r = \frac{3}{\pi} \approx 0.9549$ cm
(b) $\frac{5.5}{2\pi} \le r \le \frac{6.5}{2\pi}$, or approximately $0.8754 < r < 1.0345$
(c) $\lim_{r \to 3/\pi} 2\pi r = 6$; $\varepsilon = 0.5$; $\delta \approx 0.0796$

71.

x	-0.001	-0.0001	-0.00001
$f(x)$	2.7196	2.7184	2.7183

x	0.00001	0.0001	0.001
$f(x)$	2.7183	2.7181	2.7169

$\lim_{x \to 0} f(x) \approx 2.7183$

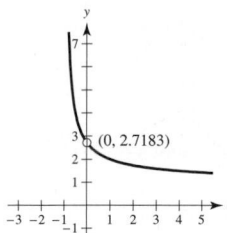
$(0, 2.7183)$

73.
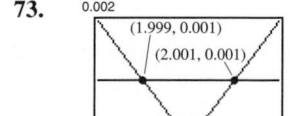

$\delta = 0.001$, $(1.999, 2.001)$

75. False. The existence or nonexistence of $f(x)$ at $x = c$ has no bearing on the existence of the limit of $f(x)$ as $x \to c$.

77. False. See Exercise 25.

79. Yes. As x approaches 0.25 from either side, $\sqrt{x}$ becomes arbitrarily close to 0.5.

81. $\lim_{x \to 0} \dfrac{\sin nx}{x} = n$ **83–85.** Proofs

87. Putnam Problem B1, 1986

Section 2.3   (page 91)

1. Substitute c for x and simplify.

3. If a function f is squeezed between two functions h and g, $h(x) \le f(x) \le g(x)$, and h and g have the same limit L as $x \to c$, then $\lim_{x \to c} f(x)$ exists and equals L.

5. -1 **7.** 0   **9.** $\sqrt{11}$ **11.** 125   **13.** $\frac{3}{5}$
15. $\frac{1}{5}$ **17.** 7   **19.** (a) 4   (b) 64   (c) 64
21. (a) 3   (b) 2   (c) 2   **23.** 1   **25.** $\frac{1}{2}$ **27.** 1
29. $\frac{1}{2}$ **31.** -1 **33.** 1   **35.** $\ln 3 + e$
37. (a) 10   (b) $\frac{12}{5}$ (c) $\frac{4}{5}$ (d) $\frac{1}{5}$
39. (a) 256   (b) 4   (c) 48   (d) 64

41. $f(x) = \dfrac{x^2 - 1}{x + 1}$ and $g(x) = x - 1$ agree except at $x = -1$.
$\lim_{x \to -1} f(x) = \lim_{x \to -1} g(x) = -2$

43. $f(x) = \dfrac{x^3 - 8}{x - 2}$ and $g(x) = x^2 + 2x + 4$ agree except at $x = 2$.
$\lim_{x \to 2} f(x) = \lim_{x \to 2} g(x) = 12$

45. $-\dfrac{\ln 2}{8} \approx -0.0866$

$f(x) = \dfrac{(x + 4)\ln(x + 6)}{x^2 - 16}$ and $g(x) = \dfrac{\ln(x + 6)}{x - 4}$ agree except at $x = -4$.

47. -1 **49.** $\dfrac{1}{8}$ **51.** $\dfrac{5}{6}$ **53.** $\dfrac{1}{6}$ **55.** $\dfrac{\sqrt{5}}{10}$
57. $-\frac{1}{9}$ **59.** 2   **61.** $2x - 2$ **63.** $\frac{1}{5}$ **65.** 0
67. 0   **69.** 0   **71.** 0   **73.** 1   **75.** $\frac{3}{2}$

77.
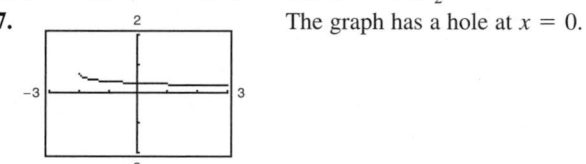
The graph has a hole at $x = 0$.

Answers will vary. Sample answer:

x	-0.1	-0.01	-0.001	0.001	0.01	0.1
$f(x)$	0.358	0.354	0.354	0.354	0.353	0.349

$\lim_{x \to 0} \dfrac{\sqrt{x + 2} - \sqrt{2}}{x} \approx 0.354$; Actual limit is $\dfrac{1}{2\sqrt{2}} = \dfrac{\sqrt{2}}{4}$.

79.
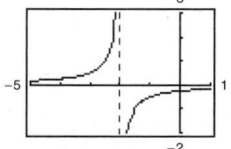
The graph has a hole at $x = 0$.

Answers will vary. Sample answer:

x	-0.1	-0.01	-0.001
$f(x)$	-0.263	-0.251	-0.250

x	0.001	0.01	0.1
$f(x)$	-0.250	-0.249	-0.238

$\lim_{x \to 0} \dfrac{[1/(2 + x)] - (1/2)}{x} \approx -0.250$; Actual limit is $-\dfrac{1}{4}$.

81.

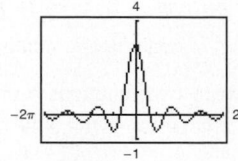

The graph has a hole at $t = 0$.

Answers will vary. Sample answer:

t	-0.1	-0.01	0	0.01	0.1
$f(t)$	2.96	2.9996	?	2.9996	2.96

$\lim\limits_{t \to 0} \dfrac{\sin 3t}{t} \approx 3.0000$; Actual limit is 3.

83.

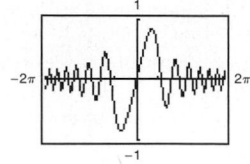

The graph has a hole at $x = 0$.

Answers will vary. Sample answer:

x	-0.1	-0.01	-0.001	0	0.001	0.01	0.1
$f(x)$	-0.1	-0.01	-0.001	?	0.001	0.01	0.1

$\lim\limits_{x \to 0} \dfrac{\sin x^2}{x} = 0$; Actual limit is 0.

85.

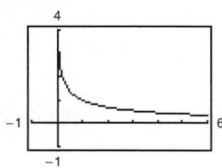

The graph has a hole at $x = 1$.

Answers will vary. Sample answer:

x	0.5	0.9	0.99	1
$f(x)$	1.3863	1.0536	1.0050	?

x	1.01	1.1	1.5
$f(x)$	0.9950	0.9531	0.8109

$\lim\limits_{x \to 1} \dfrac{\ln x}{x - 1} \approx 1$; Actual limit is 1.

87. 3   **89.** $2x - 4$ **91.** $x^{-1/2}$
93. $-1/(x + 3)^2$ **95.** 4
97.

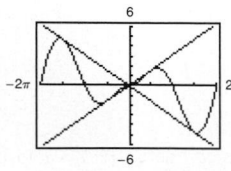

0

99.

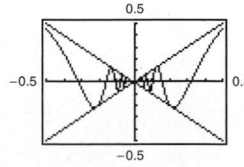

0

The graph has a hole at $x = 0$.

101. (a) f and g agree at all but one point if c is a real number such that $f(x) = g(x)$ for all $x \neq c$.

(b) Sample answer: $f(x) = \dfrac{x^2 - 1}{x - 1}$ and $g(x) = x + 1$ agree at all points except $x = 1$.

103.

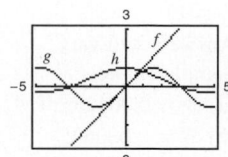

The magnitudes of $f(x)$ and $g(x)$ are approximately equal when x is close to 0. Therefore, their ratio is approximately 1.

105. -64 ft/sec (speed $= 64$ ft/sec)   **107.** -29.4 m/sec
109. Let $f(x) = 1/x$ and $g(x) = -1/x$.

$\lim\limits_{x \to 0} f(x)$ and $\lim\limits_{x \to 0} g(x)$ do not exist. However,

$\lim\limits_{x \to 0} [f(x) + g(x)] = \lim\limits_{x \to 0} \left[\dfrac{1}{x} + \left(-\dfrac{1}{x} \right) \right] = \lim\limits_{x \to 0} 0 = 0$

and therefore does exist.
111–115. Proofs
117. Let $f(x) = \begin{cases} 4, & x \geq 0 \\ -4, & x < 0 \end{cases}$.

$\lim\limits_{x \to 0} |f(x)| = \lim\limits_{x \to 0} 4 = 4$

$\lim\limits_{x \to 0} f(x)$ does not exist because for $x < 0$, $f(x) = -4$ and for $x \geq 0$, $f(x) = 4$.
119. False. The limit does not exist because the function approaches 1 from the right side of 0 and approaches -1 from the left side of 0.
121. True.
123. False. The limit does not exist because $f(x)$ approaches 3 from the left side of 2 and approaches 0 from the right side of 2.
125. Proof
127. (a) All $x \neq 0$, $\dfrac{\pi}{2} + n\pi$

(b)

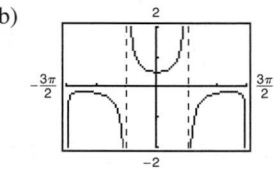

The domain is not obvious. The hole at $x = 0$ is not apparent from the graph.

(c) $\dfrac{1}{2}$ (d) $\dfrac{1}{2}$

Section 2.4   *(page 103)*

1. A function is continuous at a point c if there is no interruption of the graph at c.
3. The limit exists because the limit from the left and the limit from the right are equivalent.
5. (a) 3   (b) 3   (c) 3; $f(x)$ is continuous on $(-\infty, \infty)$.
7. (a) 0   (b) 0   (c) 0; Discontinuity at $x = 3$
9. (a) -3 (b) 3   (c) Limit does not exist.
Discontinuity at $x = 2$
11. $\dfrac{1}{16}$ **13.** $\dfrac{1}{10}$
15. Limit does not exist. The function decreases without bound as x approaches -3 from the left.
17. -1 **19.** $-\dfrac{1}{x^2}$ **21.** $\dfrac{5}{2}$
23. Limit does not exist. The function decreases without bound as x approaches π from the left and increases without bound as x approaches π from the right.
25. 8   **27.** 2
29. Limit does not exist. The function decreases without bound as x approaches 3 from the right.

31. $\ln 4$

33. Discontinuities at $x = -2$ and $x = 2$

35. Discontinuities at every integer

37. Continuous on $[-7, 7]$ **39.** Continuous on $[-1, 4]$

41. Nonremovable discontinuity at $x = 6$

43. Continuous for all real x

45. Nonremovable discontinuity at $x = 1$
Removable discontinuity at $x = 0$

47. Removable discontinuity at $x = -2$
Nonremovable discontinuity at $x = 5$

49. Nonremovable discontinuity at $x = -7$

51. Nonremovable discontinuity at $x = 2$

53. Continuous for all real x

55. Nonremovable discontinuity at $x = 0$

57. Nonremovable discontinuities at integer multiples of $\dfrac{\pi}{2}$

59. Nonremovable discontinuities at each integer

61. $a = 2$ **63.** $a = 4$ **65.** $a = -1$

67. Nonremovable discontinuities at $x = 1$ and $x = -1$

69. Continuous on the open intervals
$. . . , (-3\pi, -\pi), (-\pi, \pi), (\pi, 3\pi), . . .$

71. 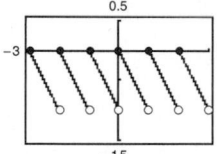   **73.**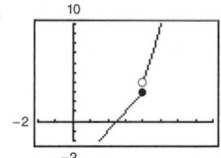

Nonremovable discontinuity   Nonremovable discontinuity
at each integer   at $x = 4$

75. Continuous on $(-\infty, \infty)$ **77.** Continuous on $[0, \infty)$

79. Continuous on the open intervals $. . . , (-6, -2), (-2, 2), (2, 6), . . .$

81. Continuous on $(-\infty, \infty)$

83. Because $f(x)$ is continuous on the interval $[1, 2]$ and $f(1) = \frac{37}{12}$ and $f(2) = -\frac{8}{3}$, by the Intermediate Value Theorem there exists a real number c in $[1, 2]$ such that $f(c) = 0$.

85. Because $f(x)$ is continuous on the interval $[0, \pi]$ and $f(0) = -3$ and $f(\pi) \approx 8.87$, by the Intermediate Value Theorem there exists a real number c in $[0, \pi]$ such that $f(c) = 0$.

87. Because $h(x)$ is continuous on the interval $[0, \pi/2]$ and $h(0) = -2$ and $h(\pi/2) \approx 0.9119$, by the Intermediate Value Theorem there exists a real number c in $[0, \pi/2]$ such that $f(c) = 0$.

89. Consider the intervals $[1, 3]$ and $[3, 5]$.
$f(1) = 2 > 0$ and $f(3) = -2 < 0$. So, there is at least one zero in the interval $[1, 3]$.
$f(3) = -2 < 0$ and $f(5) = 2 > 0$. So, there is at least one zero in the interval $[3, 5]$.

91. $0.68, 0.6823$ **93.** $0.95, 0.9472$ **95.** $0.56, 0.5636$

97. $0.79, 0.7921$ **99.** $f(3) = 11; c = 3$

101. $f(0) \approx 0.6458, f(5) \approx 1.464; c = 2$

103. $f(1) = 0, f(3) = 24; c = 2$

105. Answers will vary. Sample answer:
$$f(x) = \frac{1}{(x - a)(x - b)}$$

107. If f and g are continuous for all real x, then so is $f + g$ (Theorem 2.11, part 2). However, $\dfrac{f}{g}$ might not be continuous when $g(x) = 0$. For example, let $f(x) = x$ and $g(x) = x^2 - 1$. Then f and g are continuous for all real x, but $\dfrac{f}{g}$ is not continuous at $x = \pm 1$.

109. True

111. False. $f(x) = \cos x$ has two zeros in $[0, 2\pi]$. However, $f(0)$ and $f(2\pi)$ have the same sign.

113. False. A rational function can be written as $\dfrac{P(x)}{Q(x)}$, where P and Q are polynomials of degree m and n, respectively. It can have, at most, n discontinuities.

115. The functions differ by 1 for non-integer values of x.

117.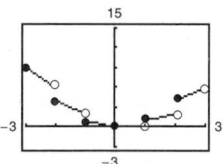
There is a jump discontinuity every gigabyte.

119–121. Proofs   **123.** Answers will vary.

125. (a)   (b) No. The frequency is oscillating.

127. $c = \dfrac{-1 \pm \sqrt{5}}{2}$

129. Domain: $[-c^2, 0) \cup (0, \infty)$; Let $f(0) = \dfrac{1}{2c}$.

131. $h(x)$ has a nonremovable discontinuity at every integer except 0.

133. Putnam Problem B2, 1988

Section 2.5   *(page 112)*

1. A limit in which $f(x)$ increases or decreases without bound as x approaches c is called an infinite limit. ∞ is not a number. Rather, the symbol $\lim\limits_{x \to c} f(x) = \infty$ says how the limit fails to exist.

3. $\lim\limits_{x \to -2^+} 2\left| \dfrac{x}{x^2 - 4} \right| = \infty$, $\lim\limits_{x \to -2^-} 2\left| \dfrac{x}{x^2 - 4} \right| = \infty$

5. $\lim\limits_{x \to -2^+} \tan \dfrac{\pi x}{4} = -\infty$, $\lim\limits_{x \to -2^-} \tan \dfrac{\pi x}{4} = \infty$

7. $\lim\limits_{x \to 4^+} \dfrac{1}{x - 4} = \infty$, $\lim\limits_{x \to 4^-} \dfrac{1}{x - 4} = -\infty$

9. $\lim\limits_{x\to4^+}\dfrac{1}{(x-4)^2}=\infty,\quad \lim\limits_{x\to4^-}\dfrac{1}{(x-4)^2}=\infty$

11.

x	-3.5	-3.1	-3.01	-3.001	-3
$f(x)$	0.31	1.64	16.6	167	?

x	-2.999	-2.99	-2.9	-2.5
$f(x)$	-167	-16.7	-1.69	-0.36

$\lim\limits_{x\to-3^+}f(x)=-\infty;\quad \lim\limits_{x\to-3^-}f(x)=\infty$

13.

x	-3.5	-3.1	-3.01	-3.001	-3
$f(x)$	3.8	16	151	1501	?

x	-2.999	-2.99	-2.9	-2.5
$f(x)$	-1499	-149	-14	-2.3

$\lim\limits_{x\to-3^+}f(x)=-\infty;\quad \lim\limits_{x\to-3^-}f(x)=\infty$

15.

x	-3.5	-3.1	-3.01	-3.001	-3
$f(x)$	-1.7321	-9.514	-95.49	-954.9	?

x	-2.999	-2.99	-2.9	-2.5
$f(x)$	954.9	95.49	9.514	1.7321

$\lim\limits_{x\to-3^-}f(x)=-\infty;\quad \lim\limits_{x\to-3^+}f(x)=\infty$

17. $x=\pm2$ **19.** $x=-2, x=1$ **21.** $x=0, x=3$

23. $x=1$ **25.** $t=-2$ **27.** $x=0$

29. $x=n$, n is an integer **31.** $t=n\pi$, n is a nonzero integer

33. Removable discontinuity at $x=-1$

35. Vertical asymptote at $x=-1$ **37.** ∞

39. $-\dfrac{1}{5}$ **41.** $-\infty$ **43.** $-\infty$ **45.** ∞

47. $-\infty$ **49.** $-\infty$ **51.** ∞

53.

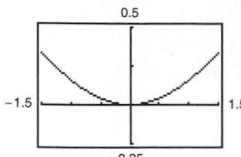

$\lim\limits_{x\to1^+}f(x)=\infty$

55. (a) ∞ (b) $-\infty$ (c) 0

57. Answers will vary. Sample answer: $f(x)=\dfrac{x-3}{x^2-4x-12}$

59.

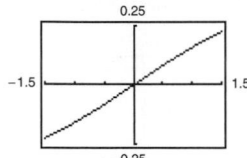

61. (a)

x	1	0.5	0.2	0.1
$f(x)$	0.1585	0.0411	0.0067	0.0017

x	0.01	0.001	0.0001
$f(x)$	≈ 0	≈ 0	≈ 0

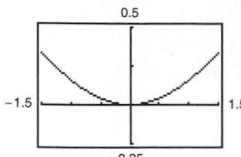

$\lim\limits_{x\to0^+}\dfrac{x-\sin x}{x}=0$

(b)

x	1	0.5	0.2	0.1
$f(x)$	0.1585	0.0823	0.0333	0.0167

x	0.01	0.001	0.0001
$f(x)$	0.0017	≈ 0	≈ 0

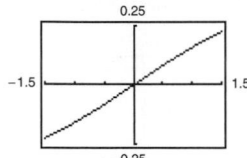

$\lim\limits_{x\to0^+}\dfrac{x-\sin x}{x^2}=0$

(c)

x	1	0.5	0.2	0.1
$f(x)$	0.1585	0.1646	0.1663	0.1666

x	0.01	0.001	0.0001
$f(x)$	0.1667	0.1667	0.1667

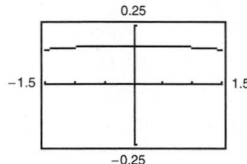

$\lim\limits_{x\to0^+}\dfrac{x-\sin x}{x^3}=0.1\overline{6}$ or $\dfrac{1}{6}$

(d)

x	1	0.5	0.2	0.1
$f(x)$	0.1585	0.3292	0.8317	1.6658

x	0.01	0.001	0.0001
$f(x)$	16.67	166.7	1667.0

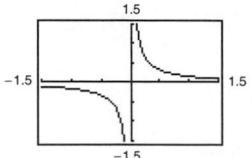

$\lim\limits_{x\to0^+}\dfrac{x-\sin x}{x^4}=\infty$

For $n>3$, $\lim\limits_{x\to0^+}\dfrac{x-\sin x}{x^n}=\infty$.

63. (a) $\frac{7}{12}$ ft/sec   (b) $\frac{3}{2}$ ft/sec

    (c) $\lim\limits_{x\to 25^-} \dfrac{2x}{\sqrt{625 - x^2}} = \infty$

65. (a) $A = 50\tan\theta - 50\theta$; Domain: $\left(0, \dfrac{\pi}{2}\right)$

    (b)

θ	0.3	0.6	0.9	1.2	1.5
$f(\theta)$	0.47	4.21	18.0	68.6	630.1

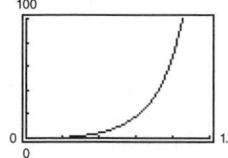

    (c) $\lim\limits_{\theta \to \pi/2} A = \infty$

67. True   **69.** False. Let $f(x) = \tan x$.

71. Let $f(x) = \dfrac{1}{x^2}$ and $g(x) = \dfrac{1}{x^4}$, and let $c = 0$. $\lim\limits_{x\to 0}\dfrac{1}{x^2} = \infty$ and

   $\lim\limits_{x\to 0}\dfrac{1}{x^4} = \infty$, but $\lim\limits_{x\to 0}\left(\dfrac{1}{x^2} - \dfrac{1}{x^4}\right) = \lim\limits_{x\to 0}\left(\dfrac{x^2 - 1}{x^4}\right) = -\infty \neq 0$.

73. Given $\lim\limits_{x\to c} f(x) = \infty$, let $g(x) = 1$. Then $\lim\limits_{x\to c}\dfrac{g(x)}{f(x)} = 0$ by

   Theorem 2.15.

75–77. Proofs

Review Exercises for Chapter 2   *(page 115)*

1. Calculus

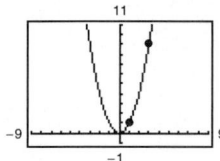

   Estimate: 8.3

3.

x	2.9	2.99	2.999	3
$f(x)$	-0.9091	-0.9901	-0.9990	?

x	3.001	3.01	3.1
$f(x)$	-1.0010	-1.0101	-1.1111

$\lim\limits_{x\to 0}\dfrac{x - 3}{x^2 - 7x + 12} \approx -1.0000$

5. (a) Limit does not exist. The function approaches 3 from the
    left side of 2, but it approaches 2 from the right side of 2.
    (b) 0

7. 5   **9.** -3 **11.** 36   **13.** 16   **15.** $\frac{4}{3}$ **17.** -1

19. $\frac{1}{2}$ **21.** -1 **23.** 0   **25.** 1

27. $\sqrt{3}/2$ **29.** -3 **31.** -5

33.

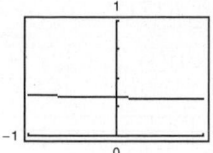

The graph has a hole at $x = 0$.

x	-0.1	-0.01	-0.001	0
$f(x)$	0.3352	0.3335	0.3334	?

x	0.001	0.01	0.1
$f(x)$	0.3333	0.3331	0.3315

$\lim\limits_{x\to 0}\dfrac{\sqrt{2x + 9} - 3}{x} \approx 0.3333$;   Actual limit is $\dfrac{1}{3}$.

35.

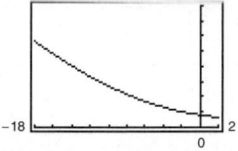

The graph has a hole at $x = -9$.

x	-9.1	-9.01	-9.001	-9
$f(x)$	245.7100	243.2701	243.0270	?

x	-8.999	-8.99	-8.9
$f(x)$	242.9730	242.7301	240.3100

$\lim\limits_{x\to -9}\dfrac{x^3 + 729}{x + 9} \approx 243.00$;   Actual limit is 243.

37. -39.2 m/sec   **39.** $\frac{1}{6}$ **41.** $\frac{1}{10}$ **43.** 0

45. Limit does not exist. The function approaches 2 from the left
    side of 1, but it approaches 1 from the right side of 1.

47. 3   **49.** -4 **51.** Continuous on $[-2, 2]$

53. No discontinuities

55. Nonremovable discontinuity at $x = 5$

57. Nonremovable discontinuities at $x = -1$ and $x = 1$
    Removable discontinuity at $x = 0$

59. $c = -\frac{1}{2}$ **61.** Continuous for all real x

63. Continuous on $[0, \infty)$

65. Continuous on $(k, k + 1)$ for all integers k

67. Removable discontinuity at $x = 1$
    Continuous on $(-\infty, 1) \cup (1, \infty)$

69. Proof

71. $f(-1) = -8$, $f(2) = 10$; $c = 1$

73. $\lim\limits_{x\to 6^+}\dfrac{1}{x - 6} = \infty$, $\lim\limits_{x\to 6^-}\dfrac{1}{x - 6} = -\infty$

75. $x = 0$ **77.** $x = \pm 3$

79. $x = 2n + 1$, where n is an integer

81. $x = \pm 5$ **83.** $-\infty$ **85.** $\frac{1}{3}$ **87.** $-\infty$ **89.** $\frac{4}{5}$

91. ∞ **93.** $-\infty$

95. (a) $\$80{,}000.00$ (b) $\$720{,}000.00$
    (c) ∞; No matter how much the company spends, the company
    will never be able to remove 100% of the pollutants.

P.S. Problem Solving   *(page 117)*

1. (a) Perimeter $\triangle PAO = 1 + \sqrt{(x^2 - 1)^2 + x^2} + \sqrt{x^4 + x^2}$
Perimeter $\triangle PBO = 1 + \sqrt{x^4 + (x - 1)^2} + \sqrt{x^4 + x^2}$

(b)

x	4	2	1
Perimeter $\triangle PAO$	33.0166	9.0777	3.4142
Perimeter $\triangle PBO$	33.7712	9.5952	3.4142
$r(x)$	0.9777	0.9461	1.0000

x	0.1	0.01
Perimeter $\triangle PAO$	2.0955	2.0100
Perimeter $\triangle PBO$	2.0006	2.0000
$r(x)$	1.0475	1.0050

1

3. (a) Area (hexagon) $= (3\sqrt{3})/2 \approx 2.5981$
Area (circle) $= \pi \approx 3.1416$
Area (circle) $-$ Area (hexagon) ≈ 0.5435

(b) $A_n = (n/2) \sin(2\pi/n)$

(c)

n	6	12	24	48	96
A_n	2.5981	3.0000	3.1058	3.1326	3.1394

3.1416 or π

5. (a) $m = -\frac{12}{5}$ (b) $y = \frac{5}{12}x - \frac{169}{12}$

(c) $m_x = \dfrac{-\sqrt{169 - x^2} + 12}{x - 5}$

(d) $\frac{5}{12}$; It is the same as the slope of the tangent line found in part (b).

7. (a) Domain: $[-27, 1) \cup (1, \infty)$

(b)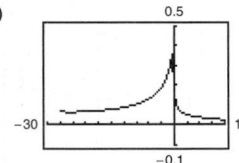

(c) $\frac{1}{14}$ (d) $\frac{1}{12}$

The graph has a hole at $x = 1$.

9. (a) g_1, g_4 (b) g_1 (c) g_1, g_3, g_4

11. 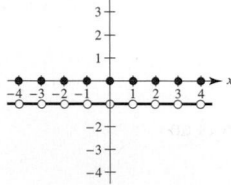   The graph jumps at every integer.

(a) $f(1) = 0$, $f(0) = 0$, $f(\frac{1}{2}) = -1$, $f(-2.7) = -1$

(b) $\lim_{x \to 1^-} f(x) = -1$, $\lim_{x \to 1^+} f(x) = -1$, $\lim_{x \to 1/2} f(x) = -1$

(c) There is a discontinuity at each integer.

13. (a)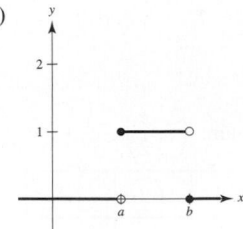

(b)   (i) $\lim_{x \to a^+} P_{a,\,b}(x) = 1$

(ii) $\lim_{x \to a^-} P_{a,\,b}(x) = 0$

(iii) $\lim_{x \to b^+} P_{a,\,b}(x) = 0$

(iv) $\lim_{x \to b^-} P_{a,\,b}(x) = 1$

(c) Continuous for all positive real numbers except a and b

(d) The area under the graph of U and above the x-axis is 1.

Chapter 3

Section 3.1   *(page 127)*

1. Let $(c, f(c))$ represent an arbitrary point on the graph of f. Then the slope of the tangent line at $(c, f(c))$ is

$$m = \lim_{\Delta x \to 0} \frac{f(c + \Delta x) - f(c)}{\Delta x}.$$

3. The limit used to define the slope of a tangent line is also used to define differentiation. The key is to rewrite the difference quotient so that Δx does not occur as a factor of the denominator.

5. $m_1 = 0, m_2 = 5/2$

7. (a)–(d)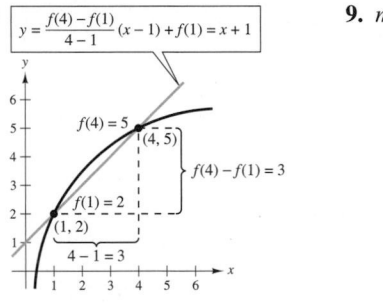

9. $m = -5$

11. $m = 8$ **13.** $m = 3$ **15.** $f'(x) = 0$

17. $f'(x) = -5$ **19.** $h'(s) = \frac{2}{3}$ **21.** $f'(x) = 2x + 1$

23. $f'(x) = 3x^2 - 12$ **25.** $f'(x) = -\dfrac{1}{(x - 1)^2}$

27. $f'(x) = \dfrac{1}{2\sqrt{x + 4}}$

29. (a) Tangent line:
$y = -2x + 2$

(b)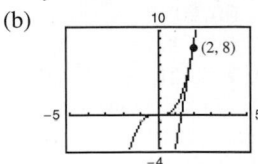

31. (a) Tangent line:
$y = 12x - 16$

(b)

33. (a) Tangent line:
$y = \frac{1}{2}x + \frac{1}{2}$
(b)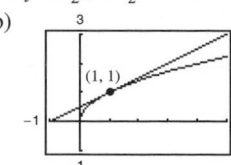

35. (a) Tangent line:
$y = \frac{3}{4}x - 2$
(b)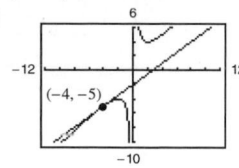

37. $y = -x + 1$ **39.** $y = 3x - 2, \ y = 3x + 2$

41. $y = -\frac{1}{2}x + \frac{3}{2}$

43.

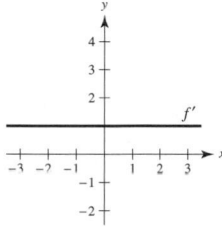

The slope of the graph of f is 1 for all x-values.

45.

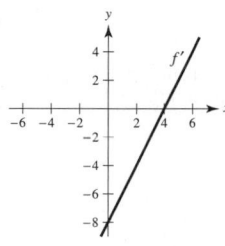

The slope of the graph of f is negative for $x < 4$, positive for $x > 4$, and 0 at $x = 4$.

47.

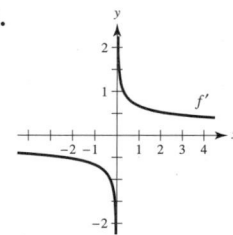

The slope of the graph of f is negative for $x < 0$ and positive for $x > 0$. The slope is undefined at $x = 0$.

49. Answers will vary.
Sample answer: $y = -x$

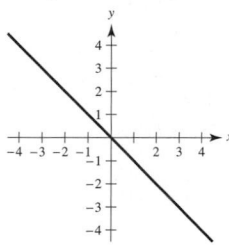

51. No. Consider $f(x) = \sqrt{x}$ and its derivative.

53. $g(4) = 5, \ g'(4) = -\frac{5}{3}$

55. $f(x) = 5 - 3x$
$c = 1$

57. $f(x) = -x^2$
$c = 6$

59. $f(x) = -3x + 2$

61. $y = 2x + 1, \ y = -2x + 9$

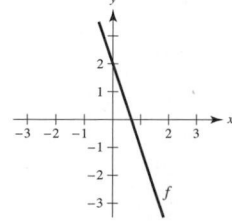

63. (a)

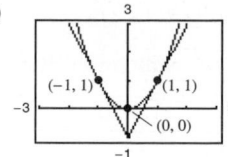

For this function, the slopes of the tangent lines are always distinct for different values of x.

(b)

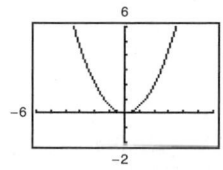

For this function, the slopes of the tangent lines are sometimes the same.

65. (a)

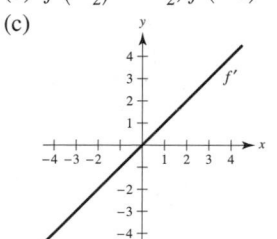

$f'(0) = 0, \ f'\left(\frac{1}{2}\right) = \frac{1}{2}, \ f'(1) = 1, \ f'(2) = 2$

(b) $f'\left(-\frac{1}{2}\right) = -\frac{1}{2}, \ f'(-1) = -1, \ f'(-2) = -2$

(c)

The slope of the graph of f'.

(d) $f'(x) = x$

67. $f(2) = 4, \ f(2.1) = 3.99, \ f'(2) \approx -0.1$ **69.** 4

71. $g(x)$ is not differentiable at $x = 0$.

73. $f(x)$ is not differentiable at $x = 6$.

75. $h(x)$ is not differentiable at $x = -7$.

77. $(-\infty, -4) \cup (-4, \infty)$ **79.** $(-1, \infty)$

81.

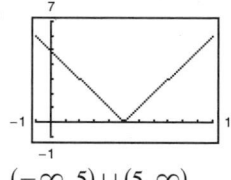

$(-\infty, 5) \cup (5, \infty)$

83.

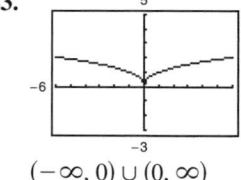

$(-\infty, 0) \cup (0, \infty)$

85. The derivative from the left is -1 and the derivative from the right is 1, so f is not differentiable at $x = 1$.

87. The derivatives from both the right and the left are 0, so $f'(1) = 0$.

89. f is differentiable at $x = 2$.

91.

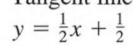

Yes, f is differentiable for all $x \neq n$, n is an integer.

93. False. The slope is $\lim\limits_{\Delta x \to 0} \dfrac{f(2 + \Delta x) - f(2)}{\Delta x}$.

95. False. For example, $f(x) = |x|$. The derivative from the left and the derivative from the right both exist but are not equal.

97. Proof

Section 3.2 *(page 139)*

1. Use the Constant Multiple Rule and Power Rule to get $f'(x) = cnx^{n-1}$.

3. $f(x) = ce^x$

5. (a) $\frac{1}{2}$ (b) 3   **7.** 0   **9.** $7x^6$ **11.** $-5/x^6$

13. $1/(9x^{8/9})$ **15.** 1   **17.** $-6t + 2$ **19.** $3t^2 + 10t - 3$

21. $\frac{\pi}{2}\cos\theta$ **23.** $2x + \frac{1}{2}\sin x$ **25.** $\frac{1}{2}e^x - 3\cos x$

	Function	Rewrite	Differentiate	Simplify
27.	$y = \dfrac{2}{7x^4}$	$y = \dfrac{2}{7}x^{-4}$	$y' = -\dfrac{8}{7}x^{-5}$	$y' = -\dfrac{8}{7x^5}$
29.	$y = \dfrac{6}{(5x)^3}$	$y = \dfrac{6}{125}x^{-3}$	$y' = -\dfrac{18}{125}x^{-4}$	$y' = -\dfrac{18}{125x^4}$

31. -2 **33.** 0   **35.** 8   **37.** 3   **39.** $\frac{3}{4}$

41. $2x + \dfrac{6}{x^3}$ **43.** $2t + \dfrac{12}{t^4}$ **45.** $\dfrac{x^3 - 8}{x^3}$

47. $\dfrac{3t^2 - 4t + 24}{2t^{5/2}}$ **49.** $3x^2 + 1$ **51.** $\dfrac{1}{2\sqrt{x}} - \dfrac{2}{x^{2/3}}$

53. $\dfrac{3}{\sqrt{x}} - 5\sin x$ **55.** $18x + 5\sin x$ **57.** $\dfrac{-2}{x^3} - 2e^x$

59. (a) $y = 2x - 2$
(b)
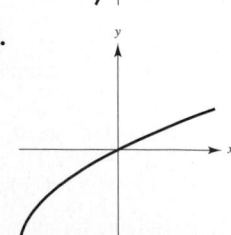

61. (a) $2x - y + 1 = 0$
(b)

63. $(-1, 2), (0, 3), (1, 2)$ **65.** No horizontal tangents

67. $(\ln 4, 4 - 4\ln 4)$ **69.** (π, π) **71.** $k = -8$

73. $k = 3$ **75.** $g'(x) = f'(x)$

77.

The rate of change of f is constant, and therefore f' is a constant function.

79.

81.

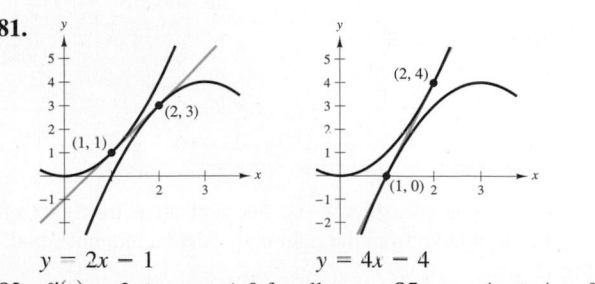

$y = 2x - 1$ $y = 4x - 4$

83. $f'(x) = 3 + \cos x \neq 0$ for all x.   **85.** $x - 4y + 4 = 0$

87. (a)

(3.9, 7.7019),
$S(x) = 2.981x - 3.924$

(b) $T(x) = 3(x - 4) + 8 = 3x - 4$
The slope (and equation) of the secant line approaches that of the tangent line at $(4, 8)$ as you choose points closer and closer to $(4, 8)$.

(c)

The approximation becomes less accurate.

(d)

Δx	-3	-2	-1	-0.5	-0.1	0
$f(4 + \Delta x)$	1	2.828	5.196	6.548	7.702	8
$T(4 + \Delta x)$	-1	2	5	6.5	7.7	8

Δx	0.1	0.5	1	2	3
$f(4 + \Delta x)$	8.302	9.546	11.180	14.697	18.520
$T(4 + \Delta x)$	8.3	9.5	11	14	17

89. False. Let $f(x) = x$ and $g(x) = x + 1$.

91. False. $\dfrac{dy}{dx} = 0$ **93.** False. $f'(x) = 0$

95. Average rate: 3   **97.** Average rate: $\frac{1}{2}$
Instantaneous rates: Instantaneous rates:
$f'(1) = 3, f'(2) = 3$ $f'(1) = 1, f'(2) = \frac{1}{4}$

99. Average rate: $e \approx 2.718$
Instantaneous rates: $g'(0) = 1, g'(1) = 2 + e \approx 4.718$

101. (a) $s(t) = -16t^2 + 1362, v(t) = -32t$ (b) -48 ft/sec
(c) $s'(1) = -32$ ft/sec, $s'(2) = -64$ ft/sec
(d) $t = \dfrac{\sqrt{1362}}{4} \approx 9.226$ sec   (e) -295.242 ft/sec

103. $v(5) = 71$ m/sec; $v(10) = 22$ m/sec

105.

107. $V'(6) = 108$ cm^3/cm

109. (a) $R(v) = 0.417v - 0.02$
(b) $B(v) = 0.0056v^2 + 0.001v + 0.04$
(c) $T(v) = 0.0056v^2 + 0.418v + 0.02$
(d)

(e) $T'(v) = 0.0112v + 0.418$
$T'(40) = 0.866$
$T'(80) = 1.314$
$T'(100) = 1.538$

(f) Stopping distance increases at an increasing rate.

111. Proof   **113.** $y = 2x^2 - 3x + 1$

115. $9x + y = 0,\ 9x + 4y + 27 = 0$ **117.** $a = \frac{1}{3},\ b = -\frac{4}{3}$

119. $f_1(x) = |\sin x|$ is differentiable for all $x \neq n\pi$, n an integer.
$f_2(x) = \sin|x|$ is differentiable for all $x \neq 0$.

121. Putnam Problem A2, 2010

Section 3.3   *(page 150)*

1. To find the derivative of the product of two differentiable functions f and g, multiply the first function f by the derivative of the second function g, and then add the second function g times the derivative of the first function f.

3. $\dfrac{d}{dx}\tan x = \sec^2 x$

$\dfrac{d}{dx}\cot x = -\csc^2 x$

$\dfrac{d}{dx}\sec x = \sec x \tan x$

$\dfrac{d}{dx}\csc x = -\csc x \cot x$

5. $-20x + 17$ **7.** $\dfrac{1 - 5t^2}{2\sqrt{t}}$ **9.** $e^x(\cos x - \sin x)$

11. $-\dfrac{5}{(x-5)^2}$ **13.** $\dfrac{1 - 5x^3}{2\sqrt{x}(x^3 + 1)^2}$ **15.** $\dfrac{\cos x - \sin x}{e^x}$

17. $f'(x) = (x^3 + 4x)(6x + 2) + (3x^2 + 2x - 5)(3x^2 + 4)$
$= 15x^4 + 8x^3 + 21x^2 + 16x - 20$
$f'(0) = -20$

19. $f'(x) = \dfrac{x^2 - 6x + 4}{(x - 3)^2}$ **21.** $f'(x) = \cos x - x \sin x$

$f'(1) = -\dfrac{1}{4}$ $f'\left(\dfrac{\pi}{4}\right) = \dfrac{\sqrt{2}}{8}(4 - \pi)$

23. $f'(x) = e^x(\cos x + \sin x)$
$f'(0) = 1$

	Function	Rewrite	Differentiate	Simplify
25.	$y = \dfrac{x^3 + 6x}{3}$	$y = \dfrac{1}{3}x^3 + 2x$	$y' = \dfrac{1}{3}(3x^2) + 2$	$y' = x^2 + 2$
27.	$y = \dfrac{6}{7x^2}$	$y = \dfrac{6}{7}x^{-2}$	$y' = -\dfrac{12}{7}x^{-3}$	$y' = -\dfrac{12}{7x^3}$
29.	$y = \dfrac{4x^{3/2}}{x}$	$y = 4x^{1/2},$ $x > 0$	$y' = 2x^{-1/2}$	$y' = \dfrac{2}{\sqrt{x}},$ $x > 0$

31. $\dfrac{3}{(x + 1)^2},\ x \neq 1$ **33.** $\dfrac{x^2 + 6x - 3}{(x + 3)^2}$ **35.** $\dfrac{3x + 1}{2x^{3/2}}$

37. $-\dfrac{2x^2 - 2x + 3}{x^2(x - 3)^2}$ **39.** $\dfrac{4s^2(3s^2 + 13s + 15)}{(s + 2)^2}$

41. $10x^4 - 8x^3 - 21x^2 - 10x - 30$ **43.** $t(t \cos t + 2 \sin t)$

45. $\dfrac{-(t \sin t + \cos t)}{t^2}$ **47.** $-e^x + \sec^2 x$

49. $\dfrac{1}{4t^{3/4}} - 6 \csc t \cot t$ **51.** $\dfrac{3}{2}\sec x(\tan x - \sec x)$

53. $\cos x \cot^2 x$ **55.** $x(x \sec^2 x + 2 \tan x)$

57. $2x \cos x + 2 \sin x + x^2 e^x + 2xe^x$ **59.** $\dfrac{e^x}{(8x^{3/2})(2x - 1)}$

61. $\dfrac{2x^2 + 8x - 1}{(x + 2)^2}$ **63.** $-4\sqrt{3}$ **65.** $\dfrac{1}{\pi^2}$

67. (a) $y = -3x - 1$
(b)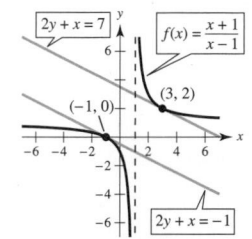

69. (a) $y = 4x + 25$
(b)

71. (a) $4x - 2y - \pi + 2 = 0$ **73.** (a) $y = e(x - 1)$
(b) 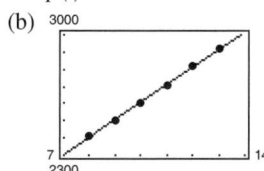   (b)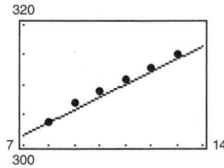

75. $2y + x - 4 = 0$ **77.** $25y - 12x + 16 = 0$

79. $(0, 0),\ (2, 4)$ **81.** $(3, 8e^{-3})$

83. $2y + x = 7,\ 2y + x = -1$

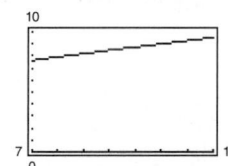

85. $f(x) + 2 = g(x)$ **87.** (a) $p'(1) = 1$ (b) $q'(4) = -\dfrac{1}{3}$

89. $\dfrac{18t + 5}{2\sqrt{t}}$ cm²/sec

91. (a) $-\$38.13$ thousand/100 components
(b) $-\$10.37$ thousand/100 components
(c) $-\$3.80$ thousand/100 components
The cost decreases with increasing order size.

93. (a)–(c) Proofs

95. (a) $h(t) = 101.7t + 1593$
$p(t) = 2.1t + 287$
(b)

(c) $A = \dfrac{101.7t + 1593}{2.1t + 287}$

A represents the average health care expenditures per person (in thousands of dollars).

(d) $A'(t) = \dfrac{25,842.6}{4.41t^2 + 1205.4t + 82,369}$

$A'(t)$ represents the rate of change of the average health care expenditures per person for the given year t.

97. 2   **99.** $\dfrac{3}{\sqrt{x}}$ **101.** $\dfrac{2}{(x - 1)^3}$ **103.** $2 \cos x - x \sin x$

105. $\csc^3 x + \csc x \cot^2 x$ **107.** $(e^x/x^3)(x^2 - 2x + 2)$

109. $6x + \dfrac{6}{25x^{8/5}}$ **111.** $\sin x$ **113.** 0 **115.** -10

117. $n - 1$ or lower; Answers will vary. Sample answer:
$f(x) = x^3,\ f'(x) = 3x^2,\ f''(x) = 6x,\ f'''(x) = 6,\ f^{(4)}(x) = 0$

119.

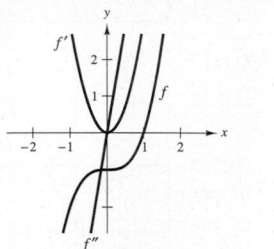

It appears that f is cubic, so f' would be quadratic and f'' would be linear.

121.

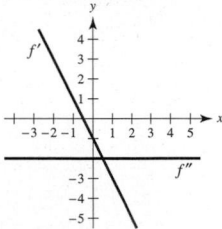

123. Answers will vary.
Sample answer: $f(x) = (x - 2)^2$

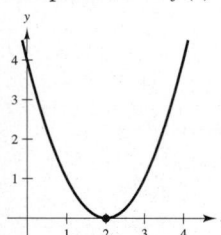

125. $v(3) = 27$ m/sec
$a(3) = -6$ m/sec^2
The speed of the object is decreasing.

127.

t	0	1	2	3	4
$s(t)$	0	57.75	99	123.75	132
$v(t)$	66	49.5	33	16.5	0
$a(t)$	-16.5	-16.5	-16.5	-16.5	-16.5

The average velocity on $[0, 1]$ is 57.75, on $[1, 2]$ is 41.25, on $[2, 3]$ is 24.75, and on $[3, 4]$ is 8.25.

129. $f^{(n)}(x) = n(n - 1)(n - 2) \cdots (2)(1) = n!$

131. (a) $f''(x) = g(x)h''(x) + 2g'(x)h'(x) + g''(x)h(x)$
$f'''(x) = g(x)h'''(x) + 3g'(x)h''(x)$
$\qquad\qquad + 3g''(x)h'(x) + g'''(x)h(x)$
$f^{(4)}(x) = g(x)h^{(4)}(x) + 4g'(x)h'''(x) + 6g''(x)h''(x)$
$\qquad\qquad + 4g'''(x)h'(x) + g^{(4)}(x)h(x)$

(b) $f^{(n)}(x) = g(x)h^{(n)}(x) + \dfrac{n!}{1!(n - 1)!}g'(x)h^{(n-1)}(x)$

$\qquad + \dfrac{n!}{2!(n - 2)!}g''(x)h^{(n-2)}(x) + \cdots$

$\qquad + \dfrac{n!}{(n - 2)!1!}g^{(n-1)}(x)h'(x) + g^{(n)}(x)h(x)$

133. $n = 1$: $f'(x) = x \cos x + \sin x$
$n = 2$: $f'(x) = x^2 \cos x + 2x \sin x$
$n = 3$: $f'(x) = x^3 \cos x + 3x^2 \sin x$
$n = 4$: $f'(x) = x^4 \cos x + 4x^3 \sin x$
General rule: $f'(x) = x^n \cos x + nx^{(n-1)} \sin x$

135. $y' = -\dfrac{1}{x^2},\ y'' = \dfrac{2}{x^3},$

$x^3 y'' + 2x^2 y' = x^3 \left(\dfrac{2}{x^3}\right) + 2x^2 \left(\dfrac{-1}{x^2}\right)$

$\qquad\qquad\quad = 2 - 2$

$\qquad\qquad\quad = 0$

137. $y' = 2 \cos x,\ y'' = -2 \sin x,$
$y'' + y = -2 \sin x + 2 \sin x + 3 = 3$

139. False. $\dfrac{dy}{dx} = f(x)g'(x) + g(x)f'(x)$ **141.** True

143. True   **145.** Proof

Section 3.4   *(page 164)*

1. To find the derivative of the composition of two differentiable functions, take the derivative of the outer function and keep the inner function the same. Then multiply by the derivative of the inner function.

3. Because $d/dx = u'/u$ for $\ln u$, you get $d/dx = 2/2x$ for $\ln 2x$ and $d/dx = 3/3x$ for $\ln 3x$. So, both derivatives simplify to $1/x$.

$y = f(g(x))$	$u = g(x)$	$y = f(u)$
5. $y = (6x - 5)^4$	$u = 6x - 5$	$y = u^4$
7. $y = \dfrac{1}{3x + 5}$	$u = 3x + 5$	$y = \dfrac{1}{u}$
9. $y = \csc^3 x$	$u = \csc x$	$y = u^3$
11. $y = e^{-2x}$	$u = -2x$	$y = e^u$

13. $6(2x - 7)^2$ **15.** $-\dfrac{45}{2(4 - 9x)^{1/6}}$ **17.** $-\dfrac{10s}{\sqrt{5s^2 + 3}}$

19. $-\dfrac{1}{(x - 2)^2}$ **21.** $-\dfrac{54s^2}{(s^3 - 2)^4}$ **23.** $-\dfrac{3}{2\sqrt{(3x + 5)^3}}$

25. $(2x - 5)^2(8x - 5)$ **27.** $\dfrac{1}{\sqrt{(x^2 + 1)^3}}$

29. $\dfrac{-2(x + 5)(x^2 + 10x - 2)}{(x^2 + 2)^3}$

31. $20x(x^2 + 3)^9 + 2(x^2 + 3)^5 + 20x^2(x^2 + 3)^4 + 2x$

33. $-4 \sin 4x$ **35.** $15 \sec^2 3x$ **37.** $2\pi^2 x \cos(\pi x)^2$

39. $2 \cos 4x$ **41.** $\dfrac{-1 - \cos^2 x}{\sin^3 x}$

43. $\sin 2\theta \cos 2\theta$, or $\frac{1}{2} \sin 4\theta$

45. $6\pi(\pi t - 1) \sec(\pi t - 1)^2 \tan(\pi t - 1)^2$

47. $(6x - \sin x) \cos(3x^2 + \cos x)$

49. $-\dfrac{3\pi \cos \sqrt{\cot 3\pi x}\ \csc^2(3\pi x)}{2\sqrt{\cot 3\pi x}}$

51. $\dfrac{1 - 3x^2 - 4x^{3/2}}{2\sqrt{x}(x^2 + 1)^2}$

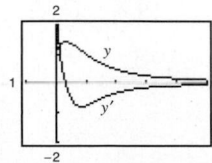

The zero of y' corresponds to the point on the graph of the function where the tangent line is horizontal.

53. $-\dfrac{\sqrt{\dfrac{x+1}{x}}}{2x(x+1)}$

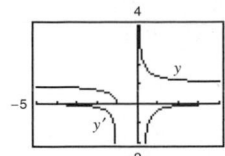

y' has no zeros.

55. $-\dfrac{\pi x \sin(\pi x) + \cos(\pi x) + 1}{x^2}$

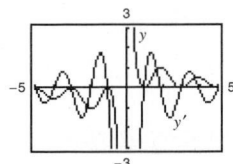

The zeros of y' correspond to the points on the graph of the function where the tangent lines are horizontal.

57. $5e^{5x}$ **59.** $e^{\sqrt{x}}/(2\sqrt{x})$ **61.** $3(e^{-t} + e^t)^2(e^t - e^{-t})$

63. $x^2 e^x$ **65.** $\dfrac{-2(e^x - e^{-x})}{(e^x + e^{-x})^2}$ **67.** $-\dfrac{2e^x}{(e^x - 1)^2}$

69. $2e^x \cos x$ **71.** $-e^{\csc x} \csc x \cot x$

73. $\dfrac{2x}{x^2 + 3}$ **75.** $\dfrac{4(\ln x)^3}{x}$ **77.** $\dfrac{2}{t+1}$ **79.** $\dfrac{2x^2 - 1}{x(x^2 - 1)}$

81. $\dfrac{1 - x^2}{x(x^2 + 1)}$ **83.** $\dfrac{1 - 2\ln t}{t^3}$ **85.** $\dfrac{2}{x \ln x^2} = \dfrac{1}{x \ln x}$

87. $\dfrac{1}{1 - x^2}$ **89.** $\cot x$ **91.** $-\tan x + \dfrac{\sin x}{\cos x - 1}$

93. $-\dfrac{3}{e^{3x} + 1}$ **95.** 3; 3 cycles in $[0, 2\pi]$ **97.** 3

99. 3 **101.** $\frac{5}{3}$ **103.** $-\frac{3}{5}$ **105.** -1 **107.** 0

109. (a) $24x + y + 23 = 0$ **111.** (a) $y = 8x - 8\pi$

(b)
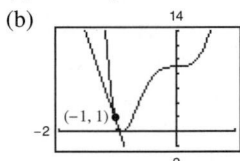

(b)
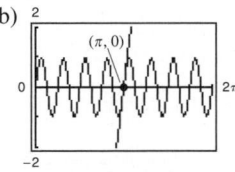

113. (a) $4x - y + (1 - \pi) = 0$ **115.** (a) $x + 2y - 8 = 0$

(b)
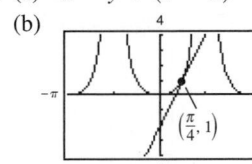

(b)
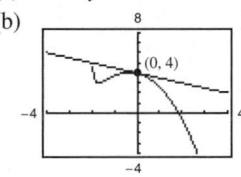

117. $3x + 4y - 25 = 0$

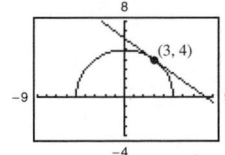

119. $\left(\dfrac{\pi}{6}, \dfrac{3\sqrt{3}}{2}\right), \left(\dfrac{5\pi}{6}, -\dfrac{3\sqrt{3}}{2}\right), \left(\dfrac{3\pi}{2}, 0\right)$ **121.** $2940(2 - 7x)^2$

123. $\dfrac{242}{(11x - 6)^3}$ **125.** $2(\cos x^2 - 2x^2 \sin x^2)$

127. $3(6x + 5)e^{-3x}$ **129.** $h''(x) = 18x + 6, 24$

131. $f''(x) = -4x^2 \cos x^2 - 2 \sin x^2, 0$

133. $(\ln 4)4^x$ **135.** $t2^t(t \ln 2 + 2)$

137. $\dfrac{2t^2 \ln 8 - 4t}{8^t}$ **139.** $-2^{-\theta}[(\ln 2) \cos \pi\theta + \pi \sin \pi\theta]$

141. $\dfrac{1}{x(\ln 3)}$ **143.** $\dfrac{x}{(\ln 5)(x^2 - 1)}$ **145.** $\dfrac{x - 2}{(\ln 2)x(x - 1)}$

147. $\dfrac{5}{(\ln 2)t^2}(1 - \ln t)$

149.
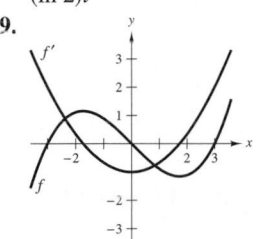

The zeros of f' correspond to the points where the graph of f has horizontal tangents.

151. (a) The rate of change of g is three times as fast as the rate of change of f.

(b) The rate of change of g is $2x$ times as fast as the rate of change of f.

153. (a) $g'(x) = f'(x)$ (b) $h'(x) = 2f'(x)$

(c) $r'(x) = -3f'(-3x)$ (d) $s'(x) = f'(x + 2)$

x	-2	-1	0	1	2	3
$f'(x)$	4	$\frac{2}{3}$	$-\frac{1}{3}$	-1	-2	-4
$g'(x)$	4	$\frac{2}{3}$	$-\frac{1}{3}$	-1	-2	-4
$h'(x)$	8	$\frac{4}{3}$	$-\frac{2}{3}$	-2	-4	-8
$r'(x)$		12	1			
$s'(x)$	$-\frac{1}{3}$	-1	-2	-4		

155. (a) $\frac{1}{2}$

(b) $s'(5)$ does not exist because g is not differentiable at 6.

157. 0.2 rad, 1.45 rad/sec   **159.** (a) 1.461   (b) -1.016

161. (a) $48.79

(b) When $t = 1$, $dC/dt \approx 0.051P$.

When $t = 8$, $dC/dt \approx 0.072P$.

(c) $dC/dt = \ln(1.05)C(t)$; $\ln 1.05$

163. (a)
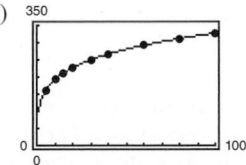
    (c)
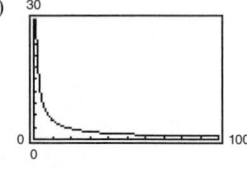

(b) $T'(10) \approx 4.75°/\text{lb/in.}^2$ $\lim\limits_{p \to \infty} T'(p) = 0$

$T'(70) \approx 0.97°/\text{lb/in.}^2$ Answers will vary.

165. (a) 0 bacteria per day   (b) 177.8 bacteria per day

(c) 44.4 bacteria per day   (d) 10.8 bacteria per day

(e) 3.3 bacteria per day

(f) The rate of change of the population is decreasing as time passes.

167. (a) $f'(x) = \beta \cos \beta x$

$f''(x) = -\beta^2 \sin \beta x$

$f'''(x) = -\beta^3 \cos \beta x$

$f^{(4)}(x) = \beta^4 \sin \beta x$

(b) $f''(x) + \beta^2 f(x) = -\beta^2 \sin \beta x + \beta^2 (\sin \beta x) = 0$

(c) $f^{(2k)}(x) = (-1)^k \beta^{2k} \sin \beta x$

$f^{(2k-1)}(x) = (-1)^{k+1} \beta^{2k-1} \cos \beta x$

169. (a) $r'(1) = 0$ (b) $s'(4) = \frac{5}{8}$

171. (a) and (b) Proofs

173. $g'(x) = 3\left(\dfrac{3x - 5}{|3x - 5|}\right), \quad x \neq \dfrac{5}{3}$

175. $h'(x) = -|x|\sin x + \dfrac{x}{|x|}\cos x, \quad x \neq 0$

177. (a) $P_1(x) = 2\left(x - \dfrac{\pi}{4}\right) + 1$

$P_2(x) = 2\left(x - \dfrac{\pi}{4}\right)^2 + 2\left(x - \dfrac{\pi}{4}\right) + 1$

(b) 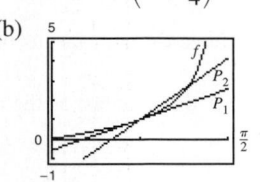   (c) P_2

(d) The accuracy worsens as you move away from $x = \dfrac{\pi}{4}$.

179. (a) $P_1(x) = x + 1$

$P_2(x) = \frac{1}{2}x^2 + x + 1$

(b) 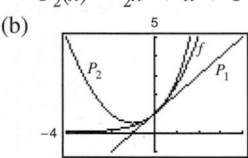   (c) P_2

(d) The accuracy worsens as you move away from $x = 0$.

181. True.   **183.** True   **185.** Putnam Problem A1, 1967

Section 3.5 *(page 175)*

1. Answers will vary. Sample answer: In the explicit form of a function, the dependent variable y is explicitly written as a function of the independent variable x $[y = f(x)]$. In an implicit equation, the dependent variable y is not necessarily written in the form $y = f(x)$. An example of an implicit function is $x^2 + xy = 5$. In explicit form, it would be

$y = \dfrac{5 - x^2}{x}$.

3. If y is an implicit function of x, then to compute y', you differentiate the equation with respect to x. For example, if $xy^2 = 1$, then $y^2 + 2xyy' = 0$. Here, the derivative of y^2 is $2yy'$.

5. $-\dfrac{x}{y}$ **7.** $-\dfrac{x^4}{y^4}$ **9.** $\dfrac{y - 3x^2}{2y - x}$

11. $\dfrac{1 - 3x^2y^3}{3x^3y^2 - 1}$ **13.** $\dfrac{6xy - 3x^2 - 2y^2}{4xy - 3x^2}$ **15.** $\dfrac{10 - e^y}{xe^y + 3}$

17. $\dfrac{\cos x}{4 \sin 2y}$ **19.** $-\dfrac{\cot x \csc x + \tan y + 1}{x \sec^2 y}$

21. $\dfrac{y \cos xy}{1 - x \cos xy}$ **23.** $\dfrac{2xy}{3 - 2y^2}$ **25.** $\dfrac{y(1 - 6x^2)}{1 + y}$

27. (a) $y_1 = \sqrt{64 - x^2}, \ y_2 = -\sqrt{64 - x^2}$

(b)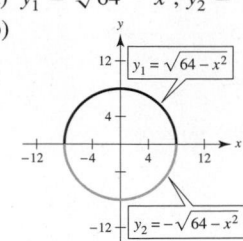

(c) $y' = \mp \dfrac{x}{\sqrt{64 - x^2}} = -\dfrac{x}{y}$ (d) $y' = -\dfrac{x}{y}$

29. (a) $y_1 = \dfrac{\sqrt{x^2 + 16}}{4}, \ y_2 = \dfrac{-\sqrt{x^2 + 16}}{4}$

(b)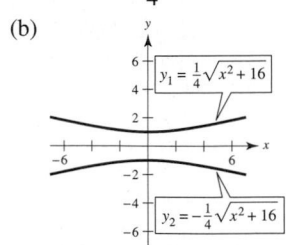

(c) $y' = \dfrac{\pm x}{4\sqrt{x^2 + 16}} = \dfrac{x}{16y}$ (d) $y' = \dfrac{x}{16y}$

31. $-\dfrac{y}{x}; -\dfrac{1}{6}$ **33.** $\dfrac{98x}{y(x^2 + 49)^2}$; Undefined

35. $-\sin^2(x + y)$ or $-\dfrac{x^2}{x^2 + 1}; 0$ **37.** $\dfrac{1 - 3ye^{xy}}{3xe^{xy}}, \dfrac{1}{9}$

39. $-\dfrac{1}{2}$ **41.** 0   **43.** $y = -x + 7$

45. $y = \dfrac{\sqrt{3}x}{6} + \dfrac{8\sqrt{3}}{3}$ **47.** $y = -\dfrac{2}{11}x + \dfrac{30}{11}$

49. $y = -\dfrac{9}{4}x + \dfrac{9}{2}$ **51.** $y = x - 1$

53. Answers will vary. Sample answers:

$xy = 2, yx^2 + x = 2; x^2 + y^2 + y = 4, xy + y^2 = 2$

55. (a) $y = -2x + 4$ (b) Answers will vary.

57. $\cos^2 y, -\dfrac{\pi}{2} < y < \dfrac{\pi}{2}, \dfrac{1}{1 + x^2}$ **59.** $\dfrac{6xy - 16}{x^3}$

61. $\dfrac{x \sin x + 2 \cos x + 14y}{7x^2}$

63. At $(4, 3)$:

Tangent line: $4x + 3y - 25 = 0$

Normal line: $3x - 4y = 0$

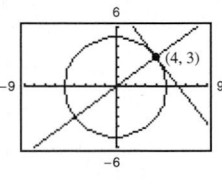

At $(-3, 4)$:

Tangent line: $3x - 4y + 25 = 0$

Normal line: $4x + 3y = 0$

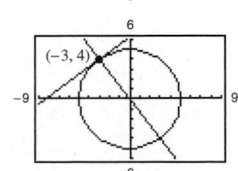

65. $x^2 + y^2 = r^2 \Rightarrow y' = -\dfrac{x}{y} \Rightarrow \dfrac{y}{x} = $ slope of normal line.

Then for (x_0, y_0) on the circle, $x_0 \neq 0$, an equation of the

normal line is $y = \left(\dfrac{y_0}{x_0}\right)x$, which passes through the origin. If

$x_0 = 0$, the normal line is vertical and passes through the origin.

67. Horizontal tangents: $(-4, 0)$, $(-4, 10)$
Vertical tangents: $(0, 5)$, $(-8, 5)$

69. $\dfrac{2x^2 + 1}{\sqrt{x^2 + 1}}$ **71.** $\dfrac{3x^3 + 15x^2 - 8x}{2(x + 1)^3 \sqrt{3x - 2}}$

73. $\dfrac{(2x^2 + 2x - 1)\sqrt{x - 1}}{(x + 1)^{3/2}}$ **75.** $2(1 - \ln x)x^{(2/x) - 2}$

77. $(x - 2)^{x+1}\left[\dfrac{x + 1}{x - 2} + \ln(x - 2)\right]$ **79.** $\dfrac{2x^{\ln x}\ln x}{x}$

81.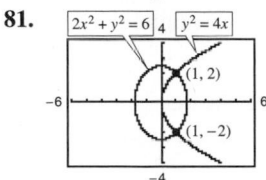

At $(1, 2)$:
Slope of ellipse: -1
Slope of parabola: 1
At $(1, -2)$:
Slope of ellipse: 1
Slope of parabola: -1

85.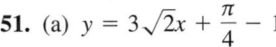

At $(0, 0)$:
Slope of line: -1
Slope of sine curve: 1

85. Derivatives: $\dfrac{dy}{dx} = -\dfrac{y}{x}$, $\dfrac{dy}{dx} = \dfrac{x}{y}$

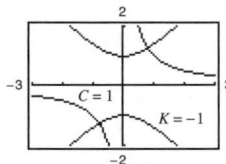   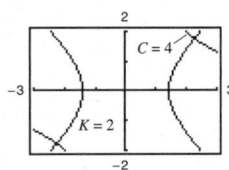

87. (a) True

(b) False. $\dfrac{d}{dy}\cos(y^2) = -2y\sin(y^2)$

(c) False. $\dfrac{d}{dx}\cos(y^2) = -2yy'\sin(y^2)$

89. (a) 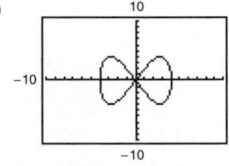   (b)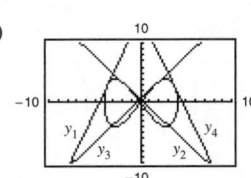

$y_1 = \tfrac{1}{3}\left[(\sqrt{7} + 7)x + (8\sqrt{7} + 23)\right]$
$y_2 = -\tfrac{1}{3}\left[(-\sqrt{7} + 7)x - (23 - 8\sqrt{7})\right]$
$y_3 = -\tfrac{1}{3}\left[(\sqrt{7} - 7)x - (23 - 8\sqrt{7})\right]$
$y_4 = -\tfrac{1}{3}\left[(\sqrt{7} + 7)x - (8\sqrt{7} + 23)\right]$

(c) $\left(\dfrac{8\sqrt{7}}{7}, 5\right)$

91. $(6, -8)$, $(-6, 8)$

93. $y = -\dfrac{\sqrt{3}}{2}x + 2\sqrt{3}$, $y = \dfrac{\sqrt{3}}{2}x - 2\sqrt{3}$

95. (a) $y = 2x - 6$
(b)   (c) $\left(\dfrac{28}{17}, -\dfrac{46}{17}\right)$

Section 3.6   *(page 182)*

1. Because you know that f^{-1} exists and $y_1 = f(x_1)$, by Theorem 3.17 you know that $(f^{-1})'(y_1) = \dfrac{1}{f'(x_1)}$, $f'(x_1) \neq 0$.

3. $-\dfrac{1}{6}$ **5.** $\dfrac{1}{27}$ **7.** $\dfrac{2\sqrt{3}}{3}$ **9.** -2 **11.** $\dfrac{1}{13}$

13. $f'\left(\tfrac{1}{2}\right) = \tfrac{3}{4}$, $(f^{-1})'\left(\tfrac{1}{8}\right) = \tfrac{4}{3}$ **15.** $f'(5) = \tfrac{1}{2}$, $(f^{-1})'(1) = 2$

17. $\dfrac{1}{\sqrt{2x - x^2}}$ **19.** $-\dfrac{3}{\sqrt{4 - x^2}}$ **21.** $\dfrac{e^x}{1 + e^{2x}}$

23. $\dfrac{3x - \sqrt{1 - 9x^2}\arcsin 3x}{x^2\sqrt{1 - 9x^2}}$

25. $e^{2x}\left[2\arcsin x + \dfrac{1}{\sqrt{1 - x^2}}\right]$ **27.** $-\dfrac{6}{1 + 36x^2}$

29. $-\dfrac{t}{\sqrt{1 - t^2}}$ **31.** $\arccos x - \dfrac{x}{\sqrt{1 - x^2}} - \dfrac{1}{\sqrt{x}}$

33. $\dfrac{x^2 + 3}{1 - x^4}$ **35.** $\dfrac{1}{(1 - t^2)^{3/2}}$ **37.** $\arcsin x$

39. $\dfrac{x^2}{\sqrt{16 - x^2}}$ **41.** $\dfrac{2}{(1 + x^2)^2}$

43. $y = \tfrac{1}{3}\left(4\sqrt{3}x - 2\sqrt{3} + \pi\right)$

45. $y = \tfrac{1}{4}x + (\pi - 2)/4$ **47.** $y = (2\pi - 4)x + 4$

49. (a) $y = \dfrac{\pi}{2}$ **51.** (a) $y = 3\sqrt{2}x + \dfrac{\pi}{4} - 1$

(b) 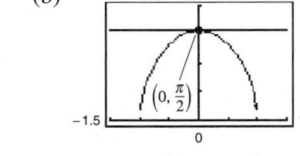   (b)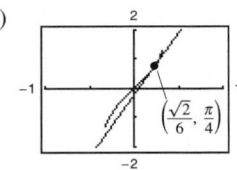

53. $y = -2x + \left(\dfrac{\pi}{6} + \sqrt{3}\right)$

$y = -2x + \left(\dfrac{5\pi}{6} - \sqrt{3}\right)$

55. $P_1(x) = x$, $P_2(x) = x$

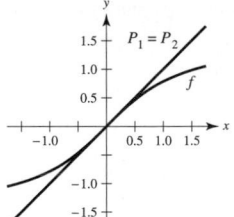

57. $P_1(x) = \dfrac{\pi}{6} + \dfrac{2\sqrt{3}}{3}\left(x - \dfrac{1}{2}\right)$

$P_2(x) = \dfrac{\pi}{6} + \dfrac{2\sqrt{3}}{3}\left(x - \dfrac{1}{2}\right) + \dfrac{2\sqrt{3}}{9}\left(x - \dfrac{1}{2}\right)^2$

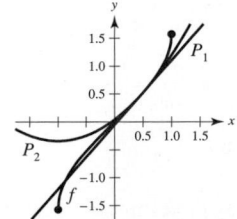

59. $\dfrac{dy}{dx} = \dfrac{1}{3y^2 - 14y}$; $m = -\dfrac{1}{11}$

61. $\dfrac{dy}{dx} = \dfrac{1}{e^y}\left(\dfrac{x}{x^2+1} + \arctan x\right)$; $m = \dfrac{\pi+2}{\pi}$

63. $y = \dfrac{-2\pi x}{\pi+8} + 1 - \dfrac{\pi^2}{2\pi+16}$ **65.** $y = -x + \sqrt{2}$

67. (a)   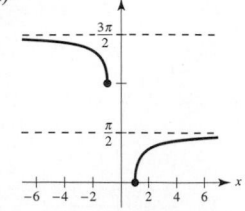   (b) Proof

69. $y = x$; The tangent lines have reciprocal slopes.

71. Because the slope of f at $(1, 3)$ is $m = 2$, the slope of f^{-1} at $(3, 1)$ is $\frac{1}{2}$.

73. (a) $\theta = \operatorname{arccot}\left(\dfrac{x}{5}\right)$

(b) $x = 10$: 16 rad/h; $x = 3$: 58.824 rad/h

75. (a) $h(t) = -16t^2 + 256$; $t = 4$ sec

(b) $t = 1$: -0.0520 rad/sec; $t = 2$: -0.1116 rad/sec

77. 0.015 rad/sec   **79.** True   **81.** True   **83–85.** Proofs

Section 3.7 (page 190)

1. A related-rate equation is an equation that relates the rates of change of various quantities.

3. (a) $\frac{3}{4}$ (b) 20 **5.** (a) $-\frac{5}{8}$ (b) $\frac{3}{2}$

7. (a) -8 cm/sec   (b) 0 cm/sec   (c) 8 cm/sec

9. (a) 12 ft/sec   (b) 6 ft/sec   (c) 3 ft/sec

11. 296π cm²/min

13. (a) 972π in.³/min, $15{,}552\pi$ in.³/min

(b) If $\dfrac{dr}{dt}$ is constant, $\dfrac{dV}{dt}$ is proportional to r^2.

15. (a) 72 cm³/sec   (b) 1800 cm³/sec

17. $\dfrac{8}{405\pi}$ ft/min   **19.** (a) 12.5% (b) $\dfrac{1}{144}$ m/min

21. (a) $-\frac{7}{12}$ ft/sec, $-\frac{3}{2}$ ft/sec, $-\frac{48}{7}$ ft/sec

(b) $\frac{527}{24}$ ft²/sec   (c) $\frac{1}{12}$ rad/sec

23. Rate of vertical change: $\dfrac{1}{5}$ m/sec

Rate of horizontal change: $-\dfrac{\sqrt{3}}{15}$ m/sec

25. (a) -750 mi/h   (b) 30 min

27. $-\dfrac{50}{\sqrt{85}} \approx -5.42$ ft/sec

29. (a) $\frac{25}{3}$ ft/sec   (b) $\frac{10}{3}$ ft/sec

31. (a) 12 sec   (b) $\dfrac{1}{2}\sqrt{3}$ m   (c) $\dfrac{\sqrt{5}\pi}{120}$ m/sec

33. Evaporation rate proportional to $S \Rightarrow \dfrac{dV}{dt} = k(4\pi r^2)$

$V = \left(\dfrac{4}{3}\right)\pi r^3 \Rightarrow \dfrac{dV}{dt} = 4\pi r^2 \dfrac{dr}{dt}$. So $k = \dfrac{dr}{dt}$.

35. (a) $\dfrac{dy}{dt} = 3\dfrac{dx}{dt}$ means that y changes three times as fast as x changes.

(b) y changes slowly when $x \approx 0$ or $x \approx L$. y changes more rapidly when x is near the middle of the interval.

37. 0.6 ohm/sec   **39.** About 84.9797 mi/h

41. $\dfrac{2\sqrt{21}}{525} \approx 0.017$ rad/sec

43. (a) $\dfrac{200\pi}{3}$ ft/sec   (b) 200π ft/sec

(c) About 427.43π ft/sec

45. (a) $A = 2xe^{-x^2/2}$ (b) $\dfrac{dA}{dt} = -3.25$ cm²/min

47. (a) $t = 65°$: $H \approx 99.8\%$ (b) -4.7%/h

$t = 80°$: $H \approx 60.2\%$

49. -0.1808 ft/sec²   **51.** -97.96 m/sec

Section 3.8 (page 198)

1. Answers will vary. Sample answer:
If f is a function continuous on $[a, b]$ and differentiable on (a, b), where $c \in [a, b]$ and $f(c) = 0$, then Newton's Method uses tangent lines to approximate c. First, estimate an initial x_1 close to c. (See graph.) Then determine

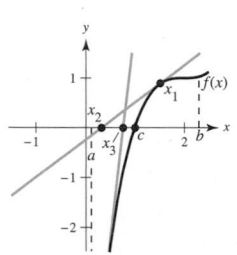

x_2 using $x_2 = x_1 - \dfrac{f(x_1)}{f'(x_1)}$. Calculate a third estimate x_3 using

$x_3 = x_2 - \dfrac{f(x_2)}{f'(x_2)}$. Continue this process until $|x_n - x_{n+1}|$ is

within the desired accuracy, and let x_{n+1} be the final approximation of c.

In the answers for Exercises 3 and 5, the values in the tables have been rounded for convenience. Because a calculator and a computer program calculate internally using more digits than they display, you may produce slightly different values from those shown in the tables.

3.

n	x_n	$f(x_n)$	$f'(x_n)$	$\dfrac{f(x_n)}{f'(x_n)}$	$x_n - \dfrac{f(x_n)}{f'(x_n)}$
1	2	-1	4	-0.25	2.25
2	2.25	0.0625	4.5	0.0139	2.2361

5.

n	x_n	$f(x_n)$	$f'(x_n)$	$\dfrac{f(x_n)}{f'(x_n)}$	$x_n - \dfrac{f(x_n)}{f'(x_n)}$
1	1.6	-0.0292	-0.9996	0.0292	1.5708
2	1.5708	0	-1	0	1.5708

7. -1.587 **9.** 0.682 **11.** $1.250, 5.000$

13. 0.567 **15.** $0.900, 1.100, 1.900$ **17.** 1.935

19. 0.569 **21.** 4.493

23. (a)   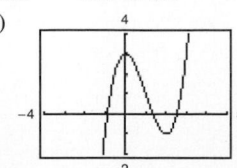   (b) 1.347 (c) 2.532

(d)

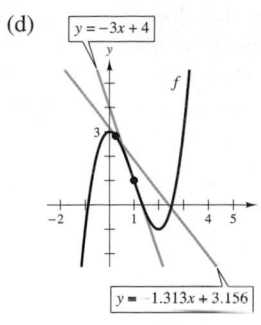

$y = -3x + 4$

$y = -1.313x + 3.156$

If the initial estimate $x = x_1$ is not sufficiently close to the desired zero of a function, then the x-intercept of the corresponding tangent line to the function may approximate a second zero of the function.

25. $f'(x_1) = 0$ **27.** 0.74   **29.** 1.12
31. \$384,356   **33.** The values would be identical.
35. Not always; Let $p(x) = x^2 - 1$ and $q(x) = x - 1$.
37. (a) Proof   (b) $\sqrt{5} \approx 2.236$, $\sqrt{7} \approx 2.646$ **39.** Proof
41. True   **43.** True   **45.** 0.217

Review Exercises for Chapter 3 *(page 200)*

1. 0   **3.** $3x^2 - 2$ **5.** 5
7. f is differentiable at all $x \neq 3$.   **9.** 0   **11.** $3x^2 - 22x$
13. $\dfrac{3}{\sqrt{x}} + \dfrac{1}{\sqrt[3]{x^2}}$ **15.** $-\dfrac{4}{3t^3}$ **17.** $4 - 5\cos\theta$
19. $-3\sin t - 4e^t$ **21.** -1 **23.** 2
25. (a) 50 vibrations/sec/lb   (b) 33.33 vibrations/sec/lb
27. (a) $s(t) = -16t^2 - 30t + 600$
    $v(t) = -32t - 30$
    (b) -94 ft/sec
    (c) $v'(1) = -62$ ft/sec, $v'(3) = -126$ ft/sec
    (d) About 5.258 sec   (e) About -198.256 ft/sec
29. $4(5x^3 - 15x^2 - 11x - 8)$
31. $9x\cos x - \cos x + 9\sin x$
33. $\dfrac{-(x^2 + 1)}{(x^2 - 1)^2}$ **35.** $\dfrac{4x^3\cos x + x^4\sin x}{\cos^2 x}$
37. $3x^2\sec x\tan x + 6x\sec x$ **39.** $-x\sin x$
41. $4xe^x + 4e^x + \csc^2 x$
43. $y = 4x + 10$ **45.** $y = -8x + 1$ **47.** $-48t$
49. $\dfrac{225}{4}\sqrt{x}$ **51.** $6\sec^2\theta\tan\theta$ **53.** $8\cot x\csc^2 x$
55. $v(3) = 11$ m/sec, $a(3) = -6$ m/sec^2 **57.** $28(7x + 3)^3$
59. $-\dfrac{6x}{(x^2 + 5)^4}$ **61.** $-45\sin(9x + 1)$
63. $\frac{1}{2}(1 - \cos 2x)$, or $\sin^2 x$ **65.** $(36x + 1)(6x + 1)^4$
67. $\dfrac{3x^2(x + 10)}{2(x + 5)^{5/2}}$ **69.** $-ze^{-z^2/2}$ **71.** $\frac{1}{4}te^{t/4}(t + 8)$
73. $\dfrac{e^{2x} - e^{-2x}}{\sqrt{e^{2x} + e^{-2x}}}$ **75.** $\dfrac{1}{2x}$ **77.** $\dfrac{1 + 2\ln x}{2\sqrt{\ln x}}$
79. $\dfrac{14x + 3}{4x^2 + x}$ **81.** $-\dfrac{x^2 + 8x + 20}{x^3 + 9x^2 + 20x}$
83. -2 **85.** -11 **87.** 0
89. $384(8x + 5)$ **91.** $2\csc^2 x\cot x$
93. (a) $-18.667°$/h   (b) $-7.284°$/h
    (c) $-3.240°$/h   (d) $-0.747°$/h
95. (a) $h = 0$ is not in the domain of the function.
    (b) $h = 0.8627 - 6.4474\ln p$

(c)

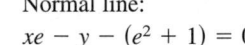

(d) 2.72 km
(e) 0.15 atm

(f) $h = 5$: $\dfrac{dp}{dh} = -0.0816$ atm/km

$h = 20$: $\dfrac{dp}{dh} = -0.008$ atm/km

As the altitude increases, the rate of change of pressure decreases.

97. $-\dfrac{x}{y}$ **99.** $\dfrac{y(y^2 - 3x^2)}{x(x^2 - 3y^2)}$ **101.** $\dfrac{y\sin x + \sin y}{\cos x - x\cos y}$

103. Tangent line: **105.** Tangent line:
    $3x + y - 10 = 0$ $xe^{-1} + y = 0$
    Normal line: Normal line:
    $x - 3y = 0$ $xe - y - (e^2 + 1) = 0$

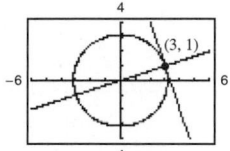   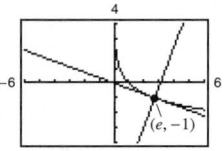

107. $\dfrac{x^3 + 8x^2 + 4}{(x + 4)^2\sqrt{x^2 + 1}}$ **109.** $\dfrac{1}{3(\sqrt[3]{-3})^2} \approx 0.160$ **111.** $\dfrac{3}{4}$

113. $\dfrac{2}{(4x^2 + 1)^{3/2}}$ **115.** $\dfrac{x}{|x|\sqrt{x^2 - 1}} + \operatorname{arcsec} x$

117. $(\arcsin x)^2$
119. (a) $2\sqrt{2}$ units/sec   (b) 4 units/sec   (c) 8 units/sec
121. 450π km/h   **123.** $-0.347, -1.532, 1.879$
125. 1.202   **127.** $-2.182, -0.795$ **129.** 0.567

P.S. Problem Solving *(page 203)*

1. (a) $r = \frac{1}{2}$; $x^2 + \left(y - \frac{1}{2}\right)^2 = \frac{1}{4}$
    (b) Center: $\left(0, \frac{5}{4}\right)$; $x^2 + \left(y - \frac{5}{4}\right)^2 = 1$
3. $p(x) = 2x^3 + 4x^2 - 5$
5. (a) $y = 4x - 4$ (b) $y = -\frac{1}{4}x + \frac{9}{2}$; $\left(-\frac{9}{4}, \frac{81}{16}\right)$
    (c) Tangent line: $y = 0$ (d) Proof
    Normal line: $x = 0$
7. (a) Graph $\begin{cases} y_1 = \dfrac{1}{a}\sqrt{x^2(a^2 - x^2)} \\ y_2 = -\dfrac{1}{a}\sqrt{x^2(a^2 - x^2)} \end{cases}$ as separate equations.

    (b) Answers will vary. Sample answer:

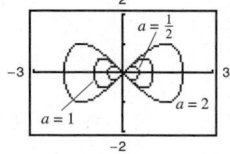

    The intercepts will always be $(0, 0)$, $(a, 0)$, and $(-a, 0)$, and the maximum and minimum y-values appear to be $\pm\frac{1}{2}a$.

    (c) $\left(\dfrac{a\sqrt{2}}{2}, \dfrac{a}{2}\right)$, $\left(\dfrac{a\sqrt{2}}{2}, -\dfrac{a}{2}\right)$, $\left(-\dfrac{a\sqrt{2}}{2}, \dfrac{a}{2}\right)$, $\left(-\dfrac{a\sqrt{2}}{2}, -\dfrac{a}{2}\right)$

9. (a) When the man is 90 ft from the light, the tip of his shadow is $112\frac{1}{2}$ ft from the light. The tip of the child's shadow is $111\frac{1}{9}$ ft from the light, so the man's shadow extends $1\frac{7}{18}$ ft beyond the child's shadow.

(b) When the man is 60 ft from the light, the tip of his shadow is 75 ft from the light. The tip of the child's shadow is $77\frac{7}{9}$ ft from the light, so the child's shadow extends $2\frac{7}{9}$ ft beyond the man's shadow.

(c) $d = 80$ ft

(d) Let x be the distance of the man from the light, and let s be the distance from the light to the tip of the shadow.

If $0 < x < 80$, then $\dfrac{ds}{dt} = -\dfrac{50}{9}$.

If $x > 80$, then $\dfrac{ds}{dt} = -\dfrac{25}{4}$.

There is a discontinuity at $x = 80$.

11. (a) $v(t) = -\frac{27}{5}t + 27$ ft/sec (b) 5 sec; 73.5 ft

$a(t) = -\frac{27}{5}$ ft/sec^2

(c) The acceleration due to gravity on Earth is greater in magnitude than that on the moon.

13. $a = 1$, $b = \frac{1}{2}$, $c = -\frac{1}{2}$

$f(x) = \dfrac{1 + \frac{1}{2}x}{1 - \frac{1}{2}x}$

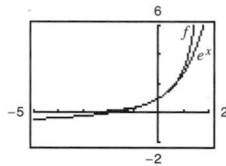

15. (a) j would be the rate of change of acceleration.

(b) $j = 0$. Acceleration is constant, so there is no change in acceleration.

(c) a: position function, d: velocity function, b: acceleration function, c: jerk function

Chapter 4

Section 4.1 *(page 211)*

1. The Extreme Value Theorem states that if f is continuous on a closed interval $[a, b]$, then f has both a minimum and a maximum on the interval.

3.

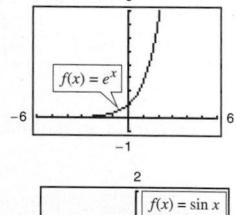

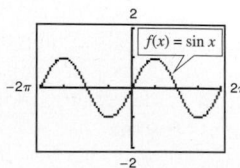

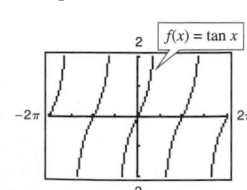

$f(x) = \sin x$; From the graph, you can see that f is defined and $f'(x) = 0$ for $f(x) = \sin x$ when $x = \pi/2 + n\pi$ (n is an integer).

5. $f'(0) = 0$ **7.** $f'(2) = 0$ **9.** $f'(-2)$ is undefined.

11. 2, absolute maximum (and relative maximum)

13. 1, absolute maximum (and relative maximum);
2, absolute minimum (and relative minimum);
3, absolute maximum (and relative maximum)

15. $x = \dfrac{3}{4}$ **17.** $t = \dfrac{8}{3}$ **19.** $x = \dfrac{\pi}{3}, \pi, \dfrac{5\pi}{3}$

21. $t = \frac{1}{2}$ **23.** $x = 0$

25. Minimum: $(2, 1)$ **27.** Minimum: $(-3, -13)$
Maximum: $(-1, 4)$ Maximum: $(0, 5)$

29. Minimum: $\left(-1, -\frac{5}{2}\right)$ **31.** Minimum: $(0, 0)$
Maximum: $(2, 2)$ Maximum: $(-1, 5)$

33. Minima: $(1, -6)$ and $(-2, -6)$
Maximum: $(0, 0)$

35. Minimum: $(-1, -1)$
Maximum: $(3, 3)$

37. Minimum value is -2 for $-2 \le x < -1$.
Maximum: $(2, 2)$

39. Minimum: $(\pi, -3)$
Maxima: $(0, 3)$ and $(2\pi, 3)$

41. Minimum: $(0, 0)$ **43.** Minimum: $(2, 5e^2 - e^4)$
Maximum: $(-2, \arctan 4)$ Maximum: $\left(\ln \frac{5}{2}, \frac{25}{4}\right)$

45. Minima: $(0, 0)$ and $(\pi, 0)$

Maximum: $\left(\dfrac{3\pi}{4}, \dfrac{\sqrt{2}}{2}e^{3\pi/4}\right)$

47. (a) Minimum: $(0, -3)$ **49.** (a) Minimum: $(1, -1)$
Maximum: $(2, 1)$ Maximum: $(-1, 3)$

(b) Minimum: $(0, -3)$ (b) Maximum: $(3, 3)$

(c) Maximum: $(2, 1)$ (c) Minimum: $(1, -1)$

(d) No extrema (d) Minimum: $(1, -1)$

51.

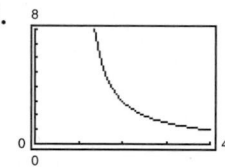

Minimum: $(4, 1)$

53.

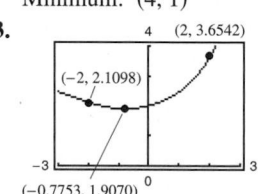

Minimum: $(-0.7753, 1.9070)$

55. (a)

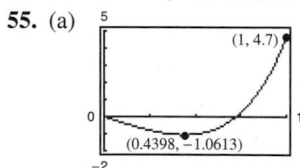

(b) Minimum: $(0.4398, -1.0613)$

57. (a)

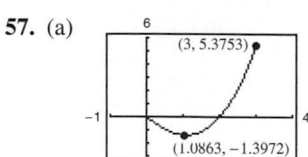

(b) Minimum: $(1.0863, -1.3972)$

59. Maximum: $\left|f''\left(\sqrt[3]{-10 + \sqrt{108}}\right)\right| = f''(\sqrt{3} - 1) \approx 1.47$

61. Maximum: $|f''(0)| = 1$ **63.** Maximum: $\left|f^{(4)}(0)\right| - \frac{56}{81}$

65. f is continuous on $\left[0, \dfrac{\pi}{4}\right]$ but not on $[0, \pi]$.

67. (a) Yes. The value is defined.

(b) No. The value is undefined.

69. No. The function is not defined at $x = -2$.

71. Maximum: $P(12) = 72$; No. P is decreasing for $I > 12$.

73. $\theta = \text{arcsec } \sqrt{3} \approx 0.9553$ rad

75. False. The maximum would be 9 if the interval was closed.

77. True   **79.** Proof   **81.** Putnam Problem B3, 2004

Section 4.2   *(page 218)*

1. Rolle's Theorem gives conditions that guarantee the existence of an extreme value in the interior of a closed interval.

3. $f(-1) = f(1) = 1$; f is not continuous on $[-1, 1]$.

5. $f(0) = f(2) = 0$; f is not differentiable on $(0, 2)$.

7. $(2, 0), (-1, 0)$; $f'\left(\frac{1}{2}\right) = 0$ **9.** $(0, 0), (-4, 0)$; $f'\left(-\frac{8}{3}\right) = 0$

11. $f'\left(\frac{3}{2}\right) = 0$ **13.** $f'\left(\frac{6 - \sqrt{3}}{3}\right) = 0, f'\left(\frac{6 + \sqrt{3}}{3}\right) = 0$

15. Not differentiable at $x = 0$ **17.** $f'\left(-2 + \sqrt{5}\right) = 0$

19. $f'\left(\frac{\pi}{2}\right) = 0, f'\left(\frac{3\pi}{2}\right) = 0$ **21.** $f'(1) = 0$

23. Not continuous on $[0, \pi]$ **25.** $f'\left(\sqrt{2}\right) = 0$

27.

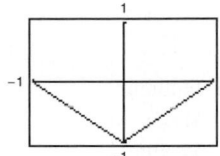

Rolle's Theorem does not apply.

29.

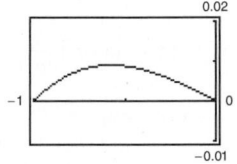

$f'\left(-\frac{6}{\pi} \text{ arccos } \frac{3}{\pi}\right) = 0$

31.

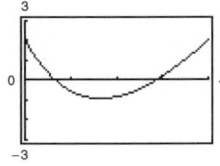

$f'(1.6633) = 0$

33. (a) $f(1) = f(2) = 38$

(b) Velocity $= 0$ for some t in $(1, 2)$; $t = \frac{3}{2}$ sec

35.

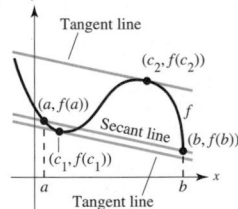

37. The function is not continuous on $[0, 6]$.

39. The function is not continuous on $[0, 6]$.

41. (a) $x + y - 3 = 0$ (b) $c = \frac{1}{2}$

(c) $4x + 4y - 21 = 0$

(d)

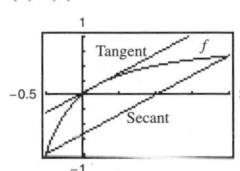

43. $f'\left(\frac{\sqrt{21}}{3}\right) = 42$ **45.** $f'\left(\frac{1}{\sqrt{3}}\right) = 3, f'\left(-\frac{1}{\sqrt{3}}\right) = 3$

47. f is not continuous at $x = 1$.

49. f is not differentiable at $x = -\frac{1}{2}$.   **51.** $f'\left(\frac{\pi}{2}\right) = 0$

53. f is not continuous at $x = \frac{\pi}{2}$.

55. $f'(4e^{-1}) = 2$

57. (a)–(c)

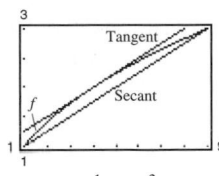

(b) $y = \frac{2}{3}(x - 1)$

(c) $y = \frac{1}{3}\left(2x + 5 - 2\sqrt{6}\right)$

59. (a)–(c)

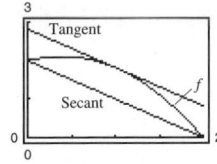

(b) $y = \frac{1}{4}x + \frac{3}{4}$

(c) $y = \frac{1}{4}x + 1$

61. (a)–(c)

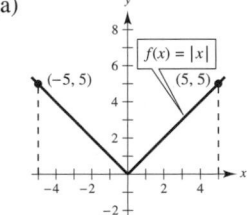

(b) $y = -x + 2$ (c) $y = -x + 2.8161$

63. (a) -14.7 m/sec   (b) 1.5 sec

65. No. Let $f(x) = x^2$ on $[-1, 2]$.

67. No. $f(x)$ is not continuous on $[0, 1]$, so it does not satisfy the hypothesis of Rolle's Theorem.

69. By the Mean Value Theorem, there is a time when the speed of the plane must equal the average speed of 454.5 miles/hour. The speed was 400 miles/hour when the plane was accelerating to 454.5 miles/hour and decelerating from 454.5 miles/hour.

71. Proof

73. (a)   (b)

75–77. Proofs   **79.** $f(x) = 5$; $f(x) = c$ and $f(2) = 5$.

81. $f(x) = x^2 - 1$; $f(x) = x^2 + c$ and $f(1) = 0$, so $c = -1$.

83. False. f is not continuous on $[-1, 1]$.   **85.** True

87–95. Proofs

Section 4.3   *(page 227)*

1. A positive derivative of a function on an open interval implies that the function is increasing on the interval. A negative derivative implies that the function is decreasing. A zero derivative implies that the function is constant.

3. (a) $(0, 6)$ (b) $(6, 8)$

5. Increasing on $(-\infty, -1)$; Decreasing on $(-1, \infty)$

7. Increasing on $(-\infty, -2)$ and $(2, \infty)$; Decreasing on $(-2, 2)$

9. Increasing on $(-\infty, -1)$; Decreasing on $(-1, \infty)$

11. Increasing on $(1, \infty)$; Decreasing on $(-\infty, 1)$

13. Increasing on $\left(-2\sqrt{2}, 2\sqrt{2}\right)$;
Decreasing on $\left(-4, -2\sqrt{2}\right)$ and $\left(2\sqrt{2}, 4\right)$

15. Increasing on $\left(0, \dfrac{\pi}{2}\right)$ and $\left(\dfrac{3\pi}{2}, 2\pi\right)$;

Decreasing on $\left(\dfrac{\pi}{2}, \dfrac{3\pi}{2}\right)$

17. Increasing on $\left(0, \dfrac{7\pi}{6}\right)$ and $\left(\dfrac{11\pi}{6}, 2\pi\right)$;

Decreasing on $\left(\dfrac{7\pi}{6}, \dfrac{11\pi}{6}\right)$

19. Increasing on $\left(-\frac{1}{4}\ln 3, \infty\right)$; Decreasing on $\left(-\infty, -\frac{1}{4}\ln 3\right)$

21. Increasing on $\left(\dfrac{2}{\sqrt{e}}, \infty\right)$; Decreasing on $\left(0, \dfrac{2}{\sqrt{e}}\right)$

23. (a) Critical number: $x = 4$
   (b) Increasing on $(4, \infty)$; Decreasing on $(-\infty, 4)$
   (c) Relative minimum: $(4, -16)$

25. (a) Critical number: $x = 1$
   (b) Increasing on $(-\infty, 1)$; Decreasing on $(1, \infty)$
   (c) Relative maximum: $(1, 5)$

27. (a) Critical numbers: $x = -1, 1$
   (b) Increasing on $(-1, 1)$;
   Decreasing on $(-\infty, -1)$ and $(1, \infty)$
   (c) Relative maximum: $(1, 17)$;
   Relative minimum: $(-1, -11)$

29. (a) Critical numbers: $x = -\frac{5}{3}, 1$
   (b) Increasing on $\left(-\infty, -\frac{5}{3}\right)$, $(1, \infty)$;
   Decreasing on $\left(-\frac{5}{3}, 1\right)$
   (c) Relative maximum: $\left(-\frac{5}{3}, \frac{256}{27}\right)$;
   Relative minimum: $(1, 0)$

31. (a) Critical numbers: $x = \pm 1$
   (b) Increasing on $(-\infty, -1)$ and $(1, \infty)$;
   Decreasing on $(-1, 1)$
   (c) Relative maximum: $\left(-1, \frac{4}{5}\right)$; Relative minimum: $\left(1, -\frac{4}{5}\right)$

33. (a) Critical number: $x = 0$
   (b) Increasing on $(-\infty, \infty)$
   (c) No relative extrema

35. (a) Critical number: $x = -2$
   (b) Increasing on $(-2, \infty)$; Decreasing on $(-\infty, -2)$
   (c) Relative minimum: $(-2, 0)$

37. (a) Critical number: $x = 5$
   (b) Increasing on $(-\infty, 5)$; Decreasing on $(5, \infty)$
   (c) Relative maximum: $(5, 5)$

39. (a) Critical numbers: $x = \pm \dfrac{\sqrt{2}}{2}$; Discontinuity: $x = 0$
   (b) Increasing on $\left(-\infty, -\dfrac{\sqrt{2}}{2}\right)$ and $\left(\dfrac{\sqrt{2}}{2}, \infty\right)$;
   Decreasing on $\left(-\dfrac{\sqrt{2}}{2}, 0\right)$ and $\left(0, \dfrac{\sqrt{2}}{2}\right)$
   (c) Relative maximum: $\left(-\dfrac{\sqrt{2}}{2}, -2\sqrt{2}\right)$;
   Relative minimum: $\left(\dfrac{\sqrt{2}}{2}, 2\sqrt{2}\right)$

41. (a) Critical number: $x = 0$; Discontinuities: $x = \pm 3$
   (b) Increasing on $(-\infty, -3)$ and $(-3, 0)$;
   Decreasing on $(0, 3)$ and $(3, \infty)$
   (c) Relative maximum: $(0, 0)$

43. (a) Critical number: $x = 0$
   (b) Increasing on $(-\infty, 0)$; Decreasing on $(0, \infty)$
   (c) Relative maximum: $(0, 4)$

45. (a) Critical number: $x = 2$
   (b) Increasing on $(-\infty, 2)$;
   Decreasing on $(2, \infty)$
   (c) Relative maximum: $(2, e^{-1})$

47. (a) Critical number: $x = 0$
   (b) Decreasing on $[-1, 1]$
   (c) No relative extrema

49. (a) Critical number: $x = \dfrac{1}{\ln 3}$
   (b) Increasing on $\left(-\infty, \dfrac{1}{\ln 3}\right)$;
   Decreasing on $\left(\dfrac{1}{\ln 3}, \infty\right)$
   (c) Relative maximum: $\left(\dfrac{1}{\ln 3}, \dfrac{3^{-1/\ln 3}}{\ln 3}\right)$ or
   $\left(\dfrac{1}{\ln 3}, \dfrac{1}{e \ln 3}\right)$

51. (a) Critical number: $x = \dfrac{1}{\ln 4}$
   (b) Increasing on $\left(\dfrac{1}{\ln 4}, \infty\right)$;
   Decreasing on $\left(0, \dfrac{1}{\ln 4}\right)$
   (c) Relative minimum: $\left(\dfrac{1}{\ln 4}, \dfrac{\ln(\ln 4) + 1}{\ln 4}\right)$

53. (a) No critical numbers
   (b) Increasing on $(-\infty, \infty)$
   (c) No relative extrema

55. (a) No critical numbers
   (b) Increasing on $(-\infty, 2)$ and $(2, \infty)$
   (c) No relative extrema

57. (a) Critical numbers: $x = \dfrac{\pi}{3}, \dfrac{5\pi}{3}$; Increasing on $\left(\dfrac{\pi}{3}, \dfrac{5\pi}{3}\right)$;
   Decreasing on $\left(0, \dfrac{\pi}{3}\right)$ and $\left(\dfrac{5\pi}{3}, 2\pi\right)$
   (b) Relative maximum: $\left(\dfrac{5\pi}{3}, \dfrac{5\pi}{3} + \sqrt{3}\right)$;
   Relative minimum: $\left(\dfrac{\pi}{3}, \dfrac{\pi}{3} - \sqrt{3}\right)$

59. (a) Critical numbers: $x = \dfrac{\pi}{4}, \dfrac{5\pi}{4}$;
   Increasing on $\left(0, \dfrac{\pi}{4}\right)$, $\left(\dfrac{5\pi}{4}, 2\pi\right)$;
   Decreasing on $\left(\dfrac{\pi}{4}, \dfrac{5\pi}{4}\right)$
   (b) Relative maximum: $\left(\dfrac{\pi}{4}, \sqrt{2}\right)$;
   Relative minimum: $\left(\dfrac{5\pi}{4}, -\sqrt{2}\right)$

61. (a) Critical numbers:
   $x = \dfrac{\pi}{4}, \dfrac{\pi}{2}, \dfrac{3\pi}{4}, \pi, \dfrac{5\pi}{4}, \dfrac{3\pi}{2}, \dfrac{7\pi}{4}$;
   Increasing on $\left(\dfrac{\pi}{4}, \dfrac{\pi}{2}\right)$, $\left(\dfrac{3\pi}{4}, \pi\right)$, $\left(\dfrac{5\pi}{4}, \dfrac{3\pi}{2}\right)$, $\left(\dfrac{7\pi}{4}, 2\pi\right)$;
   Decreasing on $\left(0, \dfrac{\pi}{4}\right)$, $\left(\dfrac{\pi}{2}, \dfrac{3\pi}{4}\right)$, $\left(\pi, \dfrac{5\pi}{4}\right)$, $\left(\dfrac{3\pi}{2}, \dfrac{7\pi}{4}\right)$

(b) Relative maxima: $\left(\dfrac{\pi}{2}, 1\right), (\pi, 1), \left(\dfrac{3\pi}{2}, 1\right)$;

     Relative minima: $\left(\dfrac{\pi}{4}, 0\right), \left(\dfrac{3\pi}{4}, 0\right), \left(\dfrac{5\pi}{4}, 0\right), \left(\dfrac{7\pi}{4}, 0\right)$

63. (a) $f'(x) = \dfrac{2(9 - 2x^2)}{\sqrt{9 - x^2}}$

(b)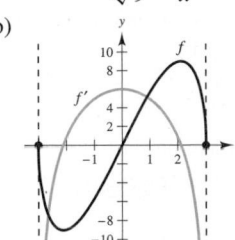
     (c) Critical numbers:
$$x = \pm\frac{3\sqrt{2}}{2}$$

(d) $f' > 0$ on $\left(-\dfrac{3\sqrt{2}}{2}, \dfrac{3\sqrt{2}}{2}\right)$;

     $f' < 0$ on $\left(-3, -\dfrac{3\sqrt{2}}{2}\right), \left(\dfrac{3\sqrt{2}}{2}, 3\right)$;

     f is increasing when f' is positive and decreasing when f' is negative.

65. (a) $f'(t) = t(t \cos t + 2 \sin t)$

(b)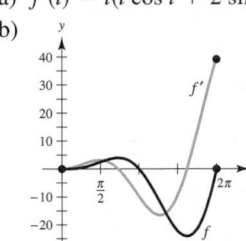
     (c) Critical numbers:
     $t = 2.2889, 5.0870$

(d) $f' > 0$ on $(0, 2.2889)$, $(5.0870, 2\pi)$;

     $f' < 0$ on $(2.2889, 5.0870)$;

     f is increasing when f' is positive and decreasing when f' is negative.

67. (a) $f'(x) = -\cos \dfrac{x}{3}$

(b)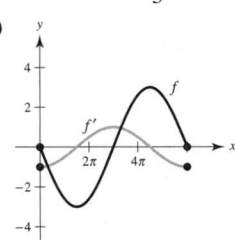

(c) Critical numbers: $x = \dfrac{3\pi}{2}, \dfrac{9\pi}{2}$

(d) $f' > 0$ on $\left(\dfrac{3\pi}{2}, \dfrac{9\pi}{2}\right)$; $f' < 0$ on $\left(0, \dfrac{3\pi}{2}\right), \left(\dfrac{9\pi}{2}, 6\pi\right)$;

     f is increasing when f' is positive and decreasing when f' is negative.

69. (a) $f'(x) = \dfrac{2x^2 - 1}{2x}$

(b)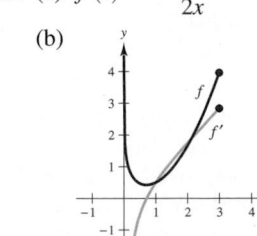
     (c) Critical number: $x = \dfrac{\sqrt{2}}{2}$

(d) $f' > 0$ on $\left(\dfrac{\sqrt{2}}{2}, 3\right)$;

     $f' < 0$ on $\left(0, \dfrac{\sqrt{2}}{2}\right)$;

     f is increasing when f' is positive and decreasing when f' is negative.

71. $f(x)$ is symmetric with respect to the origin.
Zeros: $(0, 0), (\pm\sqrt{3}, 0)$

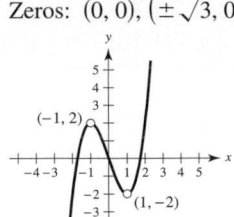

$g(x)$ is continuous on $(-\infty, \infty)$, and $f(x)$ has holes at $x = 1$ and $x = -1$.

73.
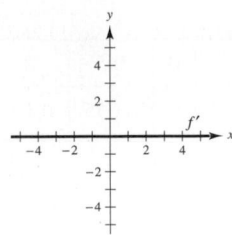

75.
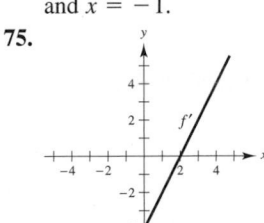

77.
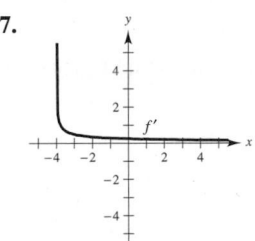

79. $g'(0) < 0$ **81.** $g'(-6) < 0$

83. Answers will vary. Sample answer:
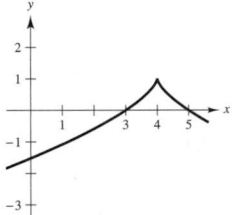

85. (a) Yes. If $h(x) = f(x) + g(x)$ where f and g are increasing, then $h'(x) = f'(x) + g'(x) > 0$. So, h is increasing.

(b) No. For example, the product of $f(x) = x$ and $g(x) = x$ is $f(x) \cdot g(x) = x^2$, which is decreasing on $(-\infty, 0)$ and increasing on $(0, \infty)$.

87. $(5, f(5))$ is a relative minimum.

89. (a)
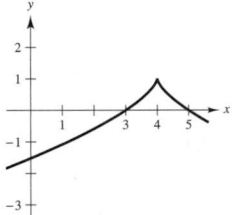

(b) Critical numbers: $x \approx -0.40$ and $x \approx 0.48$

(c) Relative maximum: $(0.48, 1.25)$;
     Relative minimum: $(-0.40, 0.75)$

91. (a) $s'(t) = 9.8(\sin \theta)t$; speed $= |9.8(\sin \theta)t|$

(b)

θ	0	$\dfrac{\pi}{4}$	$\dfrac{\pi}{3}$	$\dfrac{\pi}{2}$	$\dfrac{2\pi}{3}$	$\dfrac{3\pi}{4}$	π
$s'(t)$	0	$4.9\sqrt{2}\,t$	$4.9\sqrt{3}\,t$	$9.8t$	$4.9\sqrt{3}\,t$	$4.9\sqrt{2}\,t$	0

The speed is maximum at $\theta = \dfrac{\pi}{2}$.

93. (a)

t	0	0.5	1	1.5	2	2.5	3
$C(t)$	0	0.055	0.107	0.148	0.171	0.176	0.167

$t = 2.5$ h

(b) 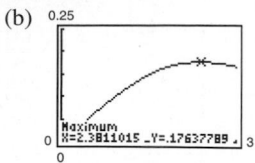 $t \approx 2.38$ h

(c) $t \approx 2.38$ h

95. $r = \dfrac{2R}{3}$

97. (a) $v(t) = 6 - 2t$ (b) $[0, 3)$ (c) $(3, \infty)$ (d) $t = 3$

99. (a) $v(t) = 3t^2 - 10t + 4$

(b) $\left[0, \dfrac{5 - \sqrt{13}}{3}\right)$ and $\left(\dfrac{5 + \sqrt{13}}{3}, \infty\right)$

(c) $\left(\dfrac{5 - \sqrt{13}}{3}, \dfrac{5 + \sqrt{13}}{3}\right)$ (d) $t = \dfrac{5 \pm \sqrt{13}}{3}$

101. Answers will vary.

103. (a) Minimum degree: 3

(b) $a_3(0)^3 + a_2(0)^2 + a_1(0) + a_0 = 0$
$a_3(2)^3 + a_2(2)^2 + a_1(2) + a_0 = 2$
$3a_3(0)^2 + 2a_2(0) + a_1 = 0$
$3a_3(2)^2 + 2a_2(2) + a_1 = 0$

(c) $f(x) = -\frac{1}{2}x^3 + \frac{3}{2}x^2$

105. (a) Minimum degree: 4

(b) $a_4(0)^4 + a_3(0)^3 + a_2(0)^2 + a_1(0) + a_0 = 0$
$a_4(2)^4 + a_3(2)^3 + a_2(2)^2 + a_1(2) + a_0 = 4$
$a_4(4)^4 + a_3(4)^3 + a_2(4)^2 + a_1(4) + a_0 = 0$
$4a_4(0)^3 + 3a_3(0)^2 + 2a_2(0) + a_1 = 0$
$4a_4(2)^3 + 3a_3(2)^2 + 2a_2(2) + a_1 = 0$
$4a_4(4)^3 + 3a_3(4)^2 + 2a_2(4) + a_1 = 0$

(c) $f(x) = \frac{1}{4}x^4 - 2x^3 + 4x^2$

107. False. Let $f(x) = \sin x$.   **109.** False. Let $f(x) = x^3$.

111. False. Let $f(x) = x^3$. There is a critical number at $x = 0$, but not a relative extremum.

113–117. Proofs   **119.** Putnam Problem A3, 2003

Section 4.4   *(page 236)*

1. Find the second derivative of a function and form test intervals by using the values for which the second derivative is zero or does not exist and the values at which the function is not continuous. Determine the sign of the second derivative on these test intervals. If the second derivative is positive, then the graph is concave upward. If the second derivative is negative, then the graph is concave downward.

3. Concave upward: $(-\infty, \infty)$

5. Concave upward: $(-\infty, 0)$, $\left(\frac{3}{2}, \infty\right)$;
Concave downward: $\left(0, \frac{3}{2}\right)$

7. Concave upward: $(-\infty, -2)$, $(2, \infty)$;
Concave downward: $(-2, 2)$

9. Concave upward: $\left(-\infty, -\frac{1}{6}\right)$;
Concave downward: $\left(-\frac{1}{6}, \infty\right)$

11. Concave upward: $(-\infty, -1)$, $(1, \infty)$;
Concave downward: $(-1, 1)$

13. Concave upward: $\left(-\dfrac{\pi}{2}, 0\right)$; Concave downward: $\left(0, \dfrac{\pi}{2}\right)$

15. Point of inflection: $(3, 0)$; Concave downward: $(-\infty, 3)$;
Concave upward: $(3, \infty)$

17. Points of inflection: None; Concave downward: $(-\infty, \infty)$

19. Points of inflection: $(2, -16)$, $(4, 0)$;
Concave upward: $(-\infty, 2)$, $(4, \infty)$;
Concave downward: $(2, 4)$

21. Points of inflection: None; Concave upward: $(-3, \infty)$

23. Points of inflection: None; Concave upward: $(0, \infty)$

25. Point of inflection: $(2\pi, 0)$;
Concave upward: $(2\pi, 4\pi)$; Concave downward: $(0, 2\pi)$

27. Concave upward: $(0, \pi)$, $(2\pi, 3\pi)$;
Concave downward: $(\pi, 2\pi)$, $(3\pi, 4\pi)$

29. Points of inflection: $(\pi, 0)$, $(1.823, 1.452)$, $(4.46, -1.452)$;
Concave upward: $(1.823, \pi)$, $(4.46, 2\pi)$;
Concave downward: $(0, 1.823)$, $(\pi, 4.46)$

31. Point of inflection: $\left(\frac{3}{2}, e^{-2}\right)$;
Concave upward: $(-\infty, 0)$, $\left(0, \frac{3}{2}\right)$;
Concave downward: $\left(\frac{3}{2}, \infty\right)$

33. Concave upward: $(0, \infty)$

35. Points of inflection:
$\left(-\left(\dfrac{1}{5}\right)^{5/8}, \arcsin\dfrac{\sqrt{5}}{5}\right), \left(\left(\dfrac{1}{5}\right)^{5/8}, \arcsin\dfrac{\sqrt{5}}{5}\right)$;
Concave upward: $\left(-1, -\left(\frac{1}{5}\right)^{5/8}\right), \left(\left(\frac{1}{5}\right)^{5/8}, 1\right)$;
Concave downward: $\left(-\left(\frac{1}{5}\right)^{5/8}, 0\right), \left(0, \left(\frac{1}{5}\right)^{5/8}\right)$

37. Relative maximum: $(3, 9)$

39. Relative minimum: $(3, -25)$

41. Relative minimum: $(0, -3)$

43. Relative maximum: $(-2, -4)$; Relative minimum: $(2, 4)$

45. No relative extrema, because f is nonincreasing.

47. Relative minimum: $\left(\frac{1}{4}, \frac{1}{2} + \ln 4\right)$

49. Relative minimum: (e, e) **51.** Relative minimum: $(0, 1)$

53. Relative minimum: $(0, 0)$;
Relative maximum: $(2, 4e^{-2})$

55. Relative maximum: $\left(\dfrac{1}{\ln 4}, \dfrac{4e^{-1}}{\ln 2}\right)$

57. Relative minimum: $(-1.272, 3.747)$;
Relative maximum: $(1.272, -0.606)$

59. (a) $f'(x) = 0.2x(x - 3)^2(5x - 6)$;
$f''(x) = 0.4(x - 3)(10x^2 - 24x + 9)$

(b) Relative maximum: $(0, 0)$;
Relative minimum: $(1.2, -1.6796)$;
Points of inflection: $(0.4652, -0.7048)$,
$(1.9348, -0.9048)$, $(3, 0)$

(c)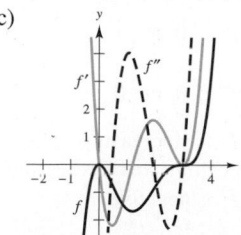

f is increasing when f' is positive and decreasing when f' is negative. f is concave upward when f'' is positive and concave downward when f'' is negative.

61. (a) $f'(x) = \cos x - \cos 3x + \cos 5x$;
$f''(x) = -\sin x + 3\sin 3x - 5\sin 5x$

(b) Relative maximum: $\left(\dfrac{\pi}{2}, 1.53333\right)$;

Points of inflection: $\left(\dfrac{\pi}{6}, 0.2667\right)$, $(1.1731, 0.9637)$,

$(1.9685, 0.9637)$, $\left(\dfrac{5\pi}{6}, 0.2667\right)$

(c)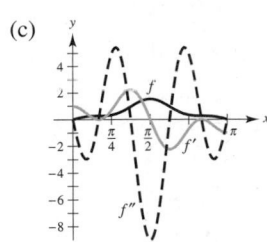

f is increasing when f' is positive and decreasing when f' is negative. f is concave upward when f'' is positive and concave downward when f'' is negative.

63. (a) 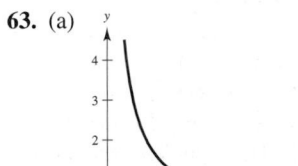   (b)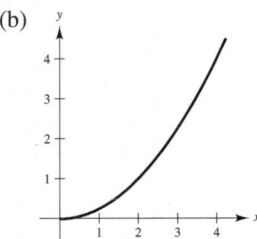

65. (a) 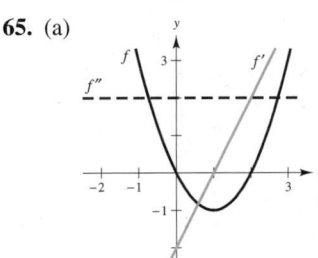   (b)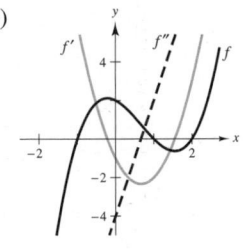

67. 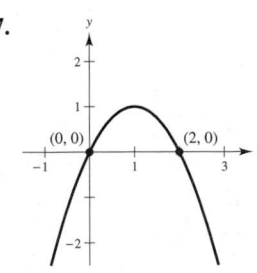   **69.**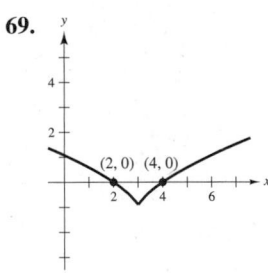

71. Sample answer:
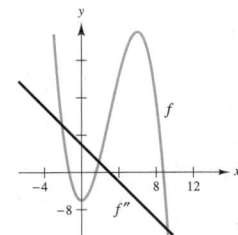

73. (a)
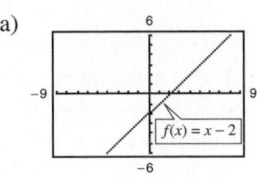 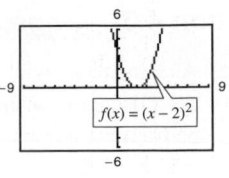
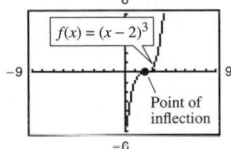 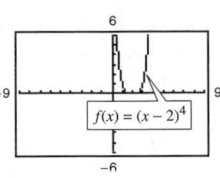

$f(x) = (x-2)^n$ has a point of inflection at $(2, 0)$ if n is odd and $n \geq 3$.

(b) Proof

75. $f(x) = \frac{1}{2}x^3 - 6x^2 + \frac{45}{2}x - 24$

77. (a) $f(x) = \frac{1}{32}x^3 + \frac{3}{16}x^2$ (b) Two miles from touchdown

79. $x = 100$ units

81. (a)

t	0.5	1	1.5	2	2.5	3
S	151.5	555.6	1097.6	1666.7	2193.0	2647.1

$1.5 < t < 2$

(b) 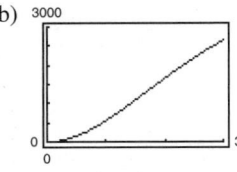   (c) About 1.633 yr

$t \approx 1.5$

83. $P_1(x) = 2\sqrt{2}$

$P_2(x) = 2\sqrt{2} - \sqrt{2}\left(x - \dfrac{\pi}{4}\right)^2$

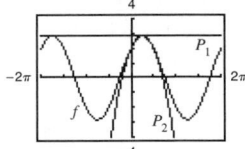

The values of f, P_1, and P_2 and their first derivatives are equal when $x = \dfrac{\pi}{4}$. The approximations worsen as you move away from $x = \dfrac{\pi}{4}$.

85. $P_1(x) = -\dfrac{\pi}{4} + \dfrac{1}{2}(x + 1)$

$P_2(x) = -\dfrac{\pi}{4} + \dfrac{1}{2}(x + 1) + \dfrac{1}{4}(x + 1)^2$

The values of f, P_1, and P_2 and their first derivatives are equal when $x = -1$. The approximations worsen as you move away from $x = -1$.

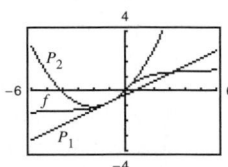

87. 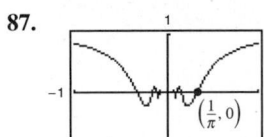   **89.** True

91. False. f is concave upward at $x = c$ if $f''(c) > 0$.

93. Proof

Section 4.5   *(page 246)*

1. (a) As x increases without bound, $f(x)$ approaches -5.
   (b) As x decreases without bound, $f(x)$ approaches 3.
3. 2; one from the left and one from the right
5. f   6. c   7. d   8. a   9. b   10. e
11. (a) ∞ (b) 5   (c) 0   13. (a) 0   (b) 1   (c) ∞
15. (a) 0   (b) $-\frac{2}{3}$ (c) $-\infty$ 17. 4   19. $\frac{7}{9}$ 21. 0
23. 0   25. -1 27. -2 29. $\frac{1}{2}$ 31. ∞
33. 0   35. 0   37. 2   39. 0   41. $-\dfrac{\pi}{2}$

43. 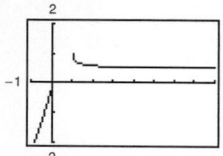   45.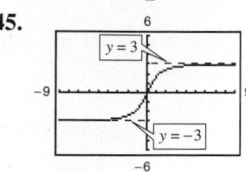

47. 1   49. 0   51. $\frac{1}{6}$
53.

x	10^0	10^1	10^2	10^3	10^4	10^5	10^6
$f(x)$	1.000	0.513	0.501	0.500	0.500	0.500	0.500

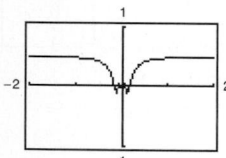

$$\lim_{x \to \infty} \left[x - \sqrt{x(x-1)} \right] = \frac{1}{2}$$

55.

x	10^0	10^1	10^2	10^3	10^4	10^5	10^6
$f(x)$	0.479	0.500	0.500	0.500	0.500	0.500	0.500

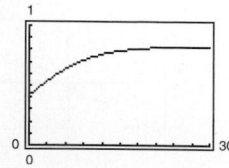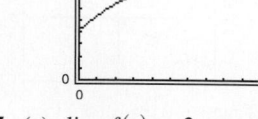

The graph has a hole at $x = 0$.

$$\lim_{x \to \infty} x \sin \frac{1}{2x} = \frac{1}{2}$$

57. 100%
59. An infinite limit is a description of how a limit fails to exist. A limit at infinity deals with the end behavior of a function.
61. (a) 5   (b) -5
63. (a)   (b) 83%

65. (a) $\lim_{x \to \infty} f(x) = 2$

(b) $x_1 = \sqrt{\dfrac{4 - 2\varepsilon}{\varepsilon}}, x_2 = -\sqrt{\dfrac{4 - 2\varepsilon}{\varepsilon}}$

(c) $M = \sqrt{\dfrac{4 - 2\varepsilon}{\varepsilon}}$ (d) $N = -\sqrt{\dfrac{4 - 2\varepsilon}{\varepsilon}}$

67. (a) Answers will vary. $M = \dfrac{5\sqrt{33}}{11}$
   (b) Answers will vary. $M = \dfrac{29\sqrt{177}}{59}$
69–71. Proofs
73. (a) $d(m) = \dfrac{|3m + 3|}{\sqrt{m^2 + 1}}$

(b)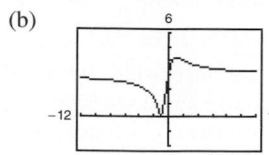

(c) $\lim_{m \to \infty} d(m) = 3$;
   $\lim_{m \to -\infty} d(m) = 3$;
   As m approaches $\pm\infty$, the distance approaches 3.

75. Proof

Section 4.6   *(page 256)*

1. Domain, range, intercepts, asymptotes, symmetry, end behavior, differentiability, relative extrema, points of inflection, concavity, increasing and decreasing, infinite limits at infinity
3. Rational function; Use long division to rewrite the rational function as the sum of a first-degree polynomial and another rational function.

5. 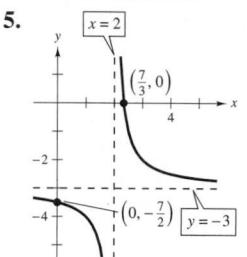   7.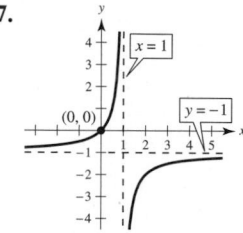

9. 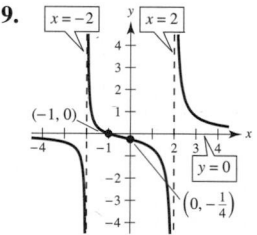   11.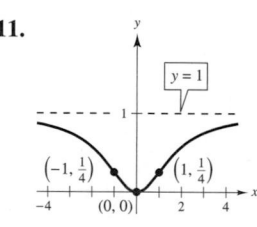

13. 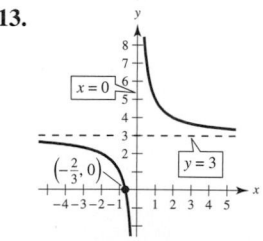   15.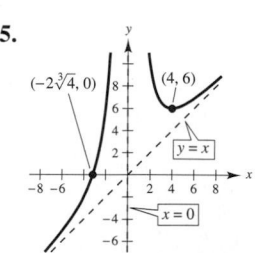

17. 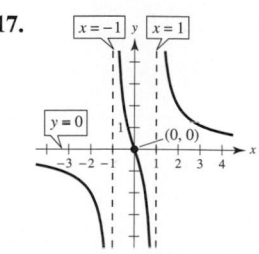   19.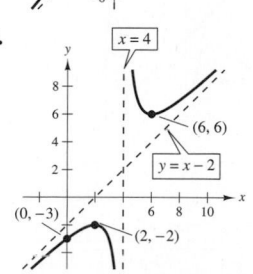

21.
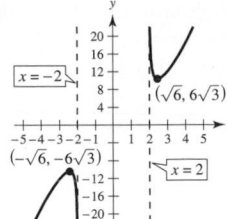
$x = -2$ $(\sqrt{6}, 6\sqrt{3})$
$(-\sqrt{6}, -6\sqrt{3})$ $x = 2$

23.
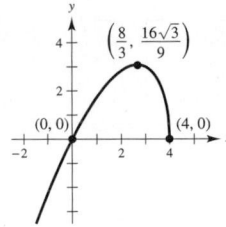
$\left(\dfrac{8}{3}, \dfrac{16\sqrt{3}}{9}\right)$
$(0, 0)$ $(4, 0)$

25.
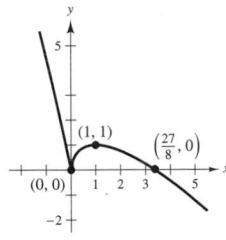
$(1, 1)$ $\left(\dfrac{27}{8}, 0\right)$
$(0, 0)$

27.
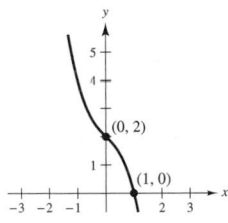
$(0, 2)$
$(1, 0)$

29.
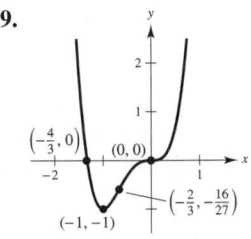
$\left(-\dfrac{4}{3}, 0\right)$ $(0, 0)$
$(-1, -1)$ $\left(-\dfrac{2}{3}, -\dfrac{16}{27}\right)$

31.
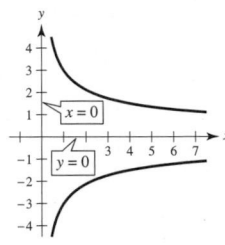
$x = 0$
$y = 0$

33.
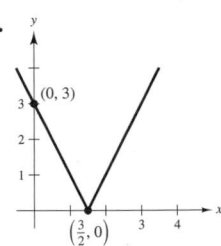
$(0, 3)$
$\left(\dfrac{3}{2}, 0\right)$

35.
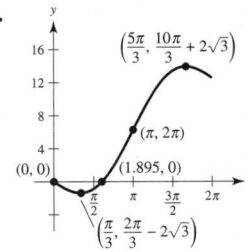
$\left(\dfrac{5\pi}{3}, \dfrac{10\pi}{3} + 2\sqrt{3}\right)$
$(\pi, 2\pi)$
$(0, 0)$ $(1.895, 0)$
$\left(\dfrac{\pi}{3}, \dfrac{2\pi}{3} - 2\sqrt{3}\right)$

37.
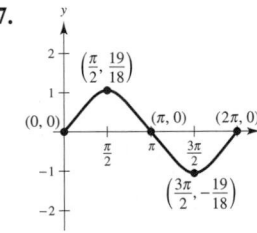
$\left(\dfrac{\pi}{2}, \dfrac{19}{18}\right)$
$(0, 0)$ $(\pi, 0)$ $(2\pi, 0)$
$\left(\dfrac{3\pi}{2}, -\dfrac{19}{18}\right)$

39.
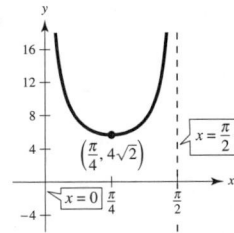
$\left(\dfrac{\pi}{4}, 4\sqrt{2}\right)$ $x = \dfrac{\pi}{2}$
$x = 0$

41.
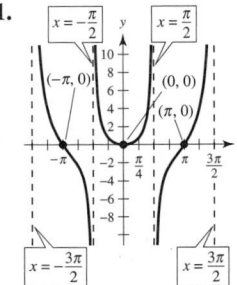
$x = -\dfrac{\pi}{2}$ $x = \dfrac{\pi}{2}$
$(-\pi, 0)$ $(0, 0)$ $(\pi, 0)$
$x = -\dfrac{3\pi}{2}$ $x = \dfrac{3\pi}{2}$

43.
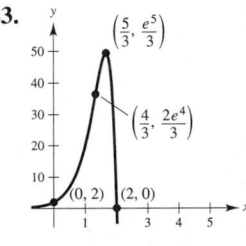
$\left(\dfrac{5}{3}, \dfrac{e^5}{3}\right)$
$\left(\dfrac{4}{3}, \dfrac{2e^4}{3}\right)$
$(0, 2)$ $(2, 0)$

45.
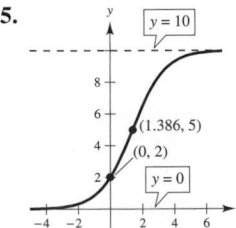
$y = 10$
$(1.386, 5)$
$(0, 2)$
$y = 0$

47.
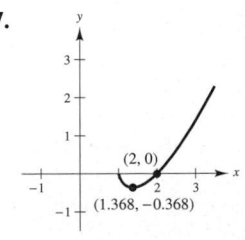
$(2, 0)$
$(1.368, -0.368)$

49.
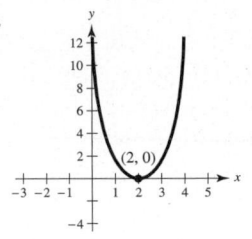
$(2, 0)$

51.
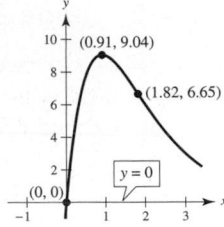
$(0.91, 9.04)$
$(1.82, 6.65)$
$y = 0$
$(0, 0)$

53.
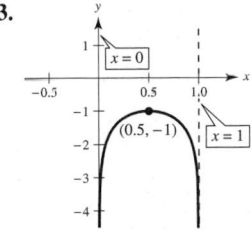
$x = 0$
$(0.5, -1)$ $x = 1$

55.
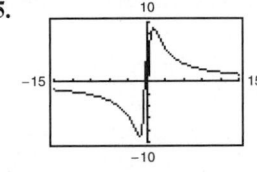
Minimum: $(-1.10, -9.05)$;
Maximum: $(1.10, 9.05)$;
Points of inflection:
$(-1.84, -7.86)$, $(1.84, 7.86)$;
Vertical asymptote: $x = 0$;
Horizontal asymptote: $y = 0$

57.
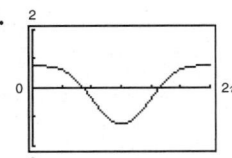
Relative minimum: $\left(\pi, -\dfrac{5}{4}\right)$;
Points of inflection:
$\left(\dfrac{2\pi}{3}, -\dfrac{3}{8}\right)$, $\left(\dfrac{4\pi}{3}, -\dfrac{3}{8}\right)$

59.
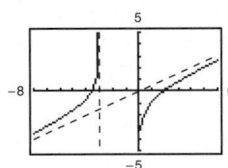
Vertical asymptotes:
$x = -3$, $x = 0$
Slant asymptote: $y = \dfrac{x}{2}$

61.
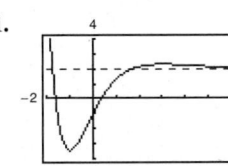
Relative minimum: $(-1, 2 - 2e)$
Relative maximum: $(3, 2 + 6e^{-3})$
Horizontal asymptote: $y = 2$

63.
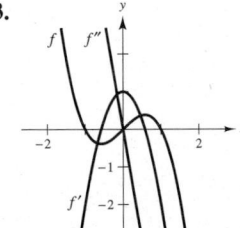
The zeros of f' correspond to the points where the graph of f has horizontal tangents. The zero of f'' corresponds to the point where the graph of f' has a horizontal tangent.

65.
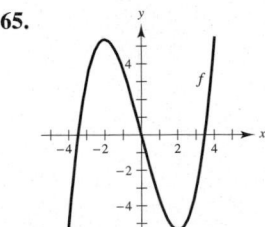
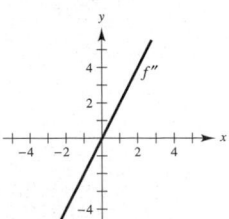

67.

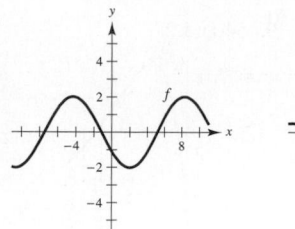

69. (a)

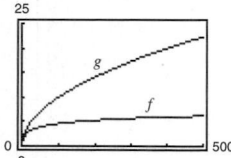

The graph has holes at $x = 0$ and at $x = 4$.

Visually approximated critical numbers: $\frac{1}{2}$, 1, $\frac{3}{2}$, 2, $\frac{5}{2}$, 3, $\frac{7}{2}$

(b) $f'(x) = \dfrac{-x \cos^2 \pi x}{(x^2 + 1)^{3/2}} - \dfrac{2\pi \sin \pi x \cos \pi x}{\sqrt{x^2 + 1}}$;

Approximate critical numbers: $\frac{1}{2}$, 0.97, $\frac{3}{2}$, 1.98, $\frac{5}{2}$, 2.98, $\frac{7}{2}$; The critical numbers where maxima occur appear to be integers in part (a), but by approximating them using f', you can see that they are not integers.

71. (a)

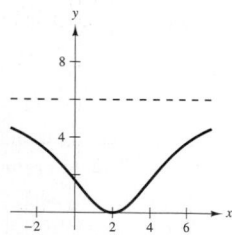

g; $f(x) = \ln x$ increases very slowly for "large" values of x.

(b)

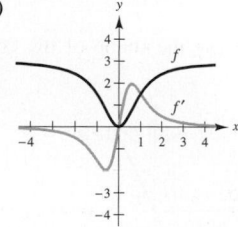

g; $f(x) = \ln x$ increases very slowly for "large" values of x.

73. Answers will vary. Sample answer: Let

$$f(x) = \dfrac{-6}{0.1(x - 2)^2 + 1} + 6.$$

75. f is decreasing on $(2, 8)$, and therefore $f(3) > f(5)$.

77. (a)

(b) $\displaystyle\lim_{x \to \infty} f(x) = 3$, $\displaystyle\lim_{x \to \infty} f'(x) = 0$

(c) Because $\displaystyle\lim_{x \to \infty} f(x) = 3$, the graph approaches that of a horizontal line, $\displaystyle\lim_{x \to \infty} f'(x) = 0$.

79.

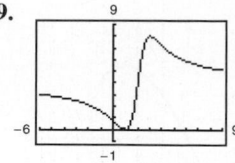

The graph crosses the horizontal asymptote $y = 4$.

The graph of a function f does not cross its vertical asymptote $x = c$ because $f(c)$ does not exist.

81.

The graph has a hole at $x = 0$. The graph crosses the horizontal asymptote $y = 0$.

The graph of a function f does not cross its vertical asymptote $x = c$ because $f(c)$ does not exist.

83.

The graph has a hole at $x = 3$. The rational function is not reduced to lowest terms.

85.

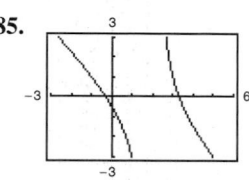

The graph appears to approach the line $y = -x + 1$, which is the slant asymptote.

87.

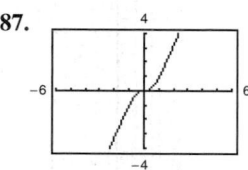

The graph appears to approach the line $y = 2x$, which is the slant asymptote.

89.

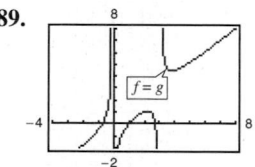

The graph appears to approach the line $y = x$, which is the slant asymptote.

91. (a)–(h) Proofs

93. Answers will vary. Sample answer: $y = \dfrac{1}{x - 3}$

95. Answers will vary.

Sample answer: $y = \dfrac{3x^2 - 7x - 5}{x - 3}$

97. False. Let $f(x) = \dfrac{2x}{\sqrt{x^2 + 2}}$, $f'(x) > 0$ for all real numbers.

99. False. For example,

$$y = \dfrac{x^3 - 1}{x}$$

does not have a slant asymptote.

101. (a) $(-3, 1)$ **(b)** $(-7, -1)$
(c) Relative maximum at $x = -3$, relative minimum at $x = 1$
(d) $x = -1$

103. Answers will vary. Sample answer: The graph has a vertical asymptote at $x = b$. If a and b are both positive or both negative, then the graph of f approaches ∞ as x approaches b, and the graph has a minimum at $x = -b$. If a and b have opposite signs, then the graph of f approaches $-\infty$ as x approaches b, and the graph has a maximum at $x = -b$.

105. $y = 4x, \ y = -4x$

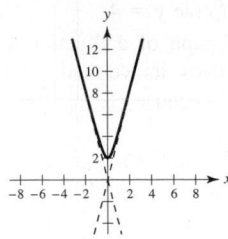

107. (a) When n is even, f is symmetric about the y-axis. When n is odd, f is symmetric about the origin.

(b) $n = 0, 1, 2, 3$ (c) $n = 4$ (d) $y = 2x$

(e)

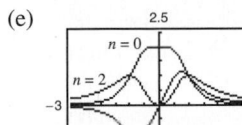

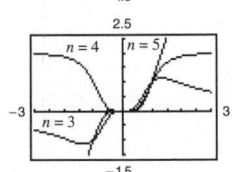

n	0	1	2	3	4	5
M	1	2	3	2	1	0
N	2	3	4	5	2	3

Section 4.7 *(page 266)*

1. A primary equation is a formula for the quantity to be optimized. A secondary equation can be solved for a variable and then substituted into the primary equation to obtain a function of just one variable. A feasible domain is the set of input values that makes sense in an optimization problem.

3. (a)

First Number, x	Second Number	Product, P
10	$110 - 10$	$10(110 - 10) = 1000$
20	$110 - 20$	$20(110 - 20) = 1800$
30	$110 - 30$	$30(110 - 30) = 2400$
40	$110 - 40$	$40(110 - 40) = 2800$
50	$110 - 50$	$50(110 - 50) = 3000$
60	$110 - 60$	$60(110 - 60) = 3000$
70	$110 - 70$	$70(110 - 70) = 2800$
80	$110 - 80$	$80(110 - 80) = 2400$
90	$110 - 90$	$90(110 - 90) = 1800$
100	$110 - 100$	$100(110 - 100) = 1000$

The maximum is attained near $x = 50$ and 60.

(b) $P = x(110 - x)$ (c) 55 and 55

(d)

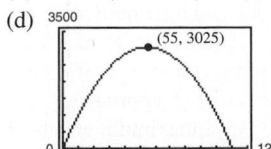

5. $\dfrac{S}{2}$ and $\dfrac{S}{2}$ **7.** 21 and 7   **9.** 54 and 27

11. $\ell = w = 20$ m   **13.** $\ell = w = 7$ ft

15. $\left(-\sqrt{\dfrac{5}{2}}, \dfrac{5}{2}\right), \left(\sqrt{\dfrac{5}{2}}, \dfrac{5}{2}\right)$ **17.** 40 in. $\times$ 20 in.

19. 900 m $\times$ 450 m

21. Rectangular portion: $\dfrac{16}{\pi + 4} \times \dfrac{32}{\pi + 4}$ ft

23. (a) $L = \sqrt{x^2 + 4 + \dfrac{8}{x - 1} + \dfrac{4}{(x - 1)^2}}, \quad x > 1$

(b)

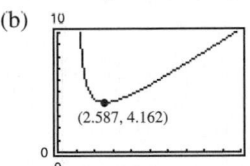

Minimum when $x \approx 2.587$

(c) $(0, 0), (2, 0), (0, 4)$

25. Width: $\dfrac{5\sqrt{2}}{2}$; Length: $5\sqrt{2}$

27. (a)

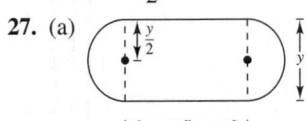

(b)

Length, x	Width, y	Area, xy
10	$(2/\pi)(100 - 10)$	$(10)(2/\pi)(100 - 10) \approx 573$
20	$(2/\pi)(100 - 20)$	$(20)(2/\pi)(100 - 20) \approx 1019$
30	$(2/\pi)(100 - 30)$	$(30)(2/\pi)(100 - 30) \approx 1337$
40	$(2/\pi)(100 - 40)$	$(40)(2/\pi)(100 - 40) \approx 1528$
50	$(2/\pi)(100 - 50)$	$(50)(2/\pi)(100 - 50) \approx 1592$
60	$(2/\pi)(100 - 60)$	$(60)(2/\pi)(100 - 60) \approx 1528$

The maximum area of the rectangle is approximately 1592 m².

(c) $A = \dfrac{2}{\pi}(100x - x^2), \ 0 < x < 100$

(d) $\dfrac{dA}{dx} = \dfrac{2}{\pi}(100 - 2x)$

$= 0$ when $x = 50$;

The maximum value is approximately 1592 when $x = 50$.

(e)

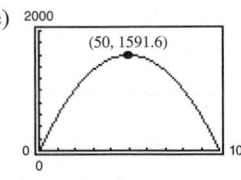

29. 18 in. $\times$ 18 in. $\times$ 36 in.

31. No. The volume changes because the shape of the container changes when it is squeezed.

33. $r = \sqrt[3]{\dfrac{21}{2\pi}} \approx 1.50$ ($h = 0$, so the solid is a sphere.)

35. Side of square: $\dfrac{10\sqrt{3}}{9 + 4\sqrt{3}}$; Side of triangle: $\dfrac{30}{9 + 4\sqrt{3}}$

37. $w = \dfrac{20\sqrt{3}}{3}$ in., $h = \dfrac{20\sqrt{6}}{3}$ in.

39.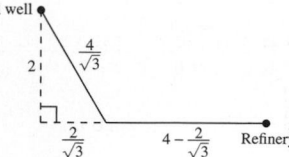
Oil well

The path of the pipe should go underwater from the oil well to the coast following the hypotenuse of a right triangle with leg lengths of 2 miles and $\dfrac{2}{\sqrt{3}}$ miles for a distance of $\dfrac{4}{\sqrt{3}}$ miles. Then the pipe should go down the coast to the refinery for a distance of $\left(4 - \dfrac{2}{\sqrt{3}}\right)$ miles.

41. (a) One mile from the nearest point on the coast   (b) Proof

43.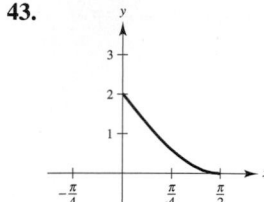

(a) Origin to y-intercept: 2;
Origin to x-intercept: $\dfrac{\pi}{2}$
(b) $d = \sqrt{x^2 + (2 - 2\sin x)^2}$
(c) Minimum distance is 0.9795 when $x \approx 0.7967$.

45. About 1.153 radians or 66°   **47.** 5.$\overline{3}$%   **49.** Proof

51. $y = \dfrac{64}{141}x$, $S \approx 6.1$ mi   **53.** $y = \dfrac{3}{10}x$, $S_3 \approx 4.50$ mi

55. $(0, 0)$ **57.** Putnam Problem A1, 1986

Section 4.8   *(page 276)*

1. $y = f(c) + f'(c)(x - c)$

3. Propagated error $= f(x + \Delta x) - f(x)$, relative error $= \left|\dfrac{dy}{y}\right|$,

percent error $= \left|\dfrac{dy}{y}\right| \cdot 100$

5. $T(x) = 4x - 4$

x	1.9	1.99	2	2.01	2.1
$f(x)$	3.610	3.960	4	4.040	4.410
$T(x)$	3.600	3.960	4	4.040	4.400

7. $T(x) = 80x - 128$

x	1.9	1.99	2	2.01	2.1
$f(x)$	24.761	31.208	32	32.808	40.841
$T(x)$	24.000	31.200	32	32.800	40.000

9. $T(x) = (\cos 2)(x - 2) + \sin 2$

x	1.9	1.99	2	2.01	2.1
$f(x)$	0.946	0.913	0.909	0.905	0.863
$T(x)$	0.951	0.913	0.909	0.905	0.868

11. $T(x) = 9(\ln 3)x - 18 \ln 3 + 9$

x	1.9	1.99	2	2.01	2.1
$f(x)$	8.064	8.902	9	9.099	10.045
$T(x)$	8.011	8.901	9	9.099	9.989

13. $y - f(0) = f'(0)(x - 0)$
$y - 2 = \frac{1}{4}x$
$y = 2 + \dfrac{x}{4}$

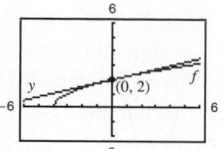

15. $\Delta y = 0.1655$, $dy = 0.15$
17. $\Delta y = -0.039$, $dy = -0.040$
19. $\Delta y \approx -0.053018$, $dy = -0.053$
21. $6x\,dx$ **23.** $(x \sec^2 x + \tan x)\,dx$
25. $-\dfrac{3}{(2x - 1)^2}\,dx$ **27.** $-\dfrac{x}{\sqrt{9 - x^2}}\,dx$
29. $\dfrac{x}{x^2 - 4}\,dx$ **31.** $\left(\arcsin x + \dfrac{x}{\sqrt{1 - x^2}}\right)dx$
33. (a) 0.9   (b) 1.04   **35.** (a) 8.035   (b) 7.95
37. (a) $\pm\frac{5}{8}$ in.^{2} (b) 0.625%
39. (a) ± 20.25 in.^{3} (b) ± 5.4 in.^{2} (c) 0.6%, 0.4%
41. 27.5 mi, about 7.3%   **43.** (a) $\frac{1}{4}$%   (b) 216 sec = 3.6 min
45. About -2.65%

47. $f(x) = \sqrt{x}$, $dy = \dfrac{1}{2\sqrt{x}}\,dx$

$f(99.4) \approx \sqrt{100} + \dfrac{1}{2\sqrt{100}}(-0.6) = 9.97$

Calculator: 9.97

49. $f(x) = \sqrt[4]{x}$, $dy = \dfrac{1}{4x^{3/4}}\,dx$

$f(624) \approx \sqrt[4]{625} + \dfrac{1}{4(625)^{3/4}}(-1) = 4.998$

Calculator: 4.998

51. The value of dy becomes closer to the value of Δy as Δx approaches 0. Graphs will vary.

53. $f(x) = \sqrt{x}$, $dy = \dfrac{1}{2\sqrt{x}}\,dx$

$f(4 + 0.02) \approx \sqrt{4} + \dfrac{1}{2\sqrt{4}}(0.02) = 2 + \dfrac{1}{4}(0.02)$

55. True   **57.** True   **59.** True

Review Exercises for Chapter 4   *(page 278)*

1. Maximum: $(0, 0)$
Minimum: $\left(-\frac{5}{2}, -\frac{25}{4}\right)$
3. Maximum: $(4, 0)$
Minimum: $(0, -2)$
5. Maximum: $\left(3, \frac{2}{3}\right)$
Minimum: $\left(-3, -\frac{2}{3}\right)$
7. Maximum: $(2\pi, 17.57)$
Minimum: $(2.73, 0.88)$
9. $f'(1) = 0$ **11.** Not continuous on $[-2, 2]$
13. $f'\left(\frac{2744}{729}\right) = \frac{3}{7}$ **15.** f is not differentiable at $x = 5$.
17. $f'(0) = 1$
19. No; The function has a discontinuity at $x = 0$, which is in the interval $[-2, 1]$.
21. Increasing on $\left(-\frac{3}{2}, \infty\right)$; Decreasing on $\left(-\infty, -\frac{3}{2}\right)$
23. Increasing on $(-\infty, 1)$, $(2, \infty)$; Decreasing on $(1, 2)$
25. Increasing on $(1, \infty)$; Decreasing on $(0, 1)$
27. Increasing on $\left(-\infty, 2 - \dfrac{1}{\ln 2}\right)$;

Decreasing on $\left(2 - \dfrac{1}{\ln 2}, \infty\right)$

29. (a) Critical number: $x = 3$
   (b) Increasing on $(3, \infty)$; Decreasing on $(-\infty, 3)$
   (c) Relative minimum: $(3, -4)$

31. (a) Critical number: $t = 2$
   (b) Increasing on $(2, \infty)$; Decreasing on $(-\infty, 2)$
   (c) Relative minimum: $(2, -12)$

33. (a) Critical number: $x = -8$; Discontinuity: $x = 0$
   (b) Increasing on $(-8, 0)$;
    Decreasing on $(-\infty, -8)$ and $(0, \infty)$
   (c) Relative minimum: $\left(-8, -\frac{1}{16}\right)$

35. (a) Critical numbers: $x = \dfrac{3\pi}{4}, \dfrac{7\pi}{4}$
   (b) Increasing on $\left(\dfrac{3\pi}{4}, \dfrac{7\pi}{4}\right)$;
    Decreasing on $\left(0, \dfrac{3\pi}{4}\right)$ and $\left(\dfrac{7\pi}{4}, 2\pi\right)$
   (c) Relative minimum: $\left(\dfrac{3\pi}{4}, -\sqrt{2}\right)$;
    Relative maximum: $\left(\dfrac{7\pi}{4}, \sqrt{2}\right)$

37. (a) Critical number: $x = \dfrac{\sqrt{2}}{2}$
   (b) Increasing on $\left(0, \dfrac{\sqrt{2}}{2}\right)$; Decreasing on $\left(\dfrac{\sqrt{2}}{2}, \infty\right)$
   (c) Relative maximum: $\left(\dfrac{\sqrt{2}}{2}, \ln\dfrac{\sqrt{2}}{2} - \dfrac{1}{2}\right)$

39. (a) $v(t) = 3 - 4t$ (b) $\left[0, \frac{3}{4}\right)$ (c) $\left(\frac{3}{4}, \infty\right)$ (d) $t = \frac{3}{4}$

41. Point of inflection: $(3, -54)$; Concave upward: $(3, \infty)$;
   Concave downward: $(-\infty, 3)$

43. Points of inflection: None; Concave upward: $(-5, \infty)$

45. Points of inflection: $\left(\dfrac{\pi}{2}, \dfrac{\pi}{2}\right), \left(\dfrac{3\pi}{2}, \dfrac{3\pi}{2}\right)$;
   Concave upward: $\left(\dfrac{\pi}{2}, \dfrac{3\pi}{2}\right)$;
   Concave downward: $\left(0, \dfrac{\pi}{2}\right), \left(\dfrac{3\pi}{2}, 2\pi\right)$

47. Point of inflection: $(-2, e^{-2})$;
   Concave upward: $(-2, 0), (0, \infty)$;
   Concave downward: $(-\infty, -2)$

49. Relative minimum: $(-9, 0)$

51. Relative maxima: $\left(\dfrac{\sqrt{2}}{2}, \dfrac{1}{2}\right), \left(-\dfrac{\sqrt{2}}{2}, \dfrac{1}{2}\right)$;
   Relative minimum: $(0, 0)$

53. Relative maximum: $(-3, -12)$; Relative minimum: $(3, 12)$

55. Relative maximum: $\left(\dfrac{2}{e}, \dfrac{2}{e \ln 5}\right)$

57.
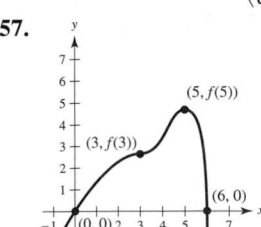

59. The first derivative is positive and the second derivative is negative. The graph is increasing and concave downward.

61. (a)
$$D = 0.41489t^4 - 17.1307t^3 + 249.888t^2 - 1499.45t + 3684.8$$
   (b)
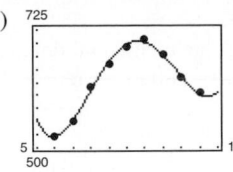
   (c) 2011; 2006 (d) 2008

63. 8   **65.** $-\dfrac{1}{8}$ **67.** $-\infty$ **69.** 0   **71.** 6   **73.** 0

75.
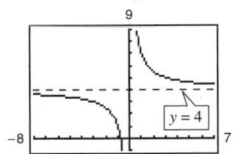
77.
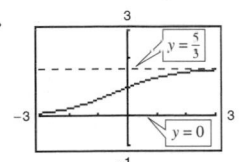

79. (a)
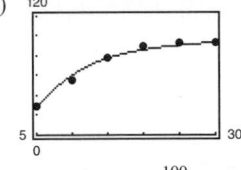
   (b) Yes. $\lim\limits_{t \to \infty} S = \dfrac{100}{1} = 100$

81.
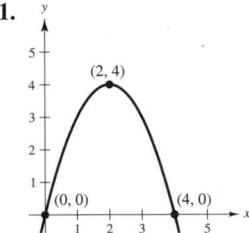
83.
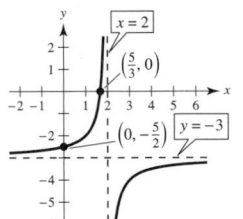

85.
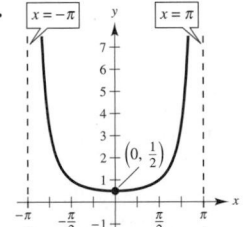
87.

89.   **91.**

93. 54 and 36   **95.** $50 \text{ ft} \times \dfrac{200}{3} \text{ ft}$

97. $(0, 0), (5, 0), (0, 10)$ **99.** 14.05 ft   **101.** $\dfrac{32\pi r^3}{81}$

103. $\Delta y = 5.044$, $dy = 4.8$ **105.** $(1 - \cos x + x \sin x)\, dx$

107. (a) $\pm 8.1\pi \text{ cm}^3$ (b) $\pm 1.8\pi \text{ cm}^2$
   (c) About 0.83%, about 0.56%

109. 267.24, 3.1%

P.S. Problem Solving   (page 281)

1. Choices of a may vary.

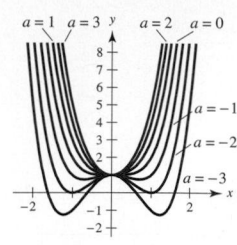

(a) One relative minimum at $(0, 1)$ for $a \geq 0$

(b) One relative maximum at $(0, 1)$ for $a < 0$

(c) Two relative minima for
$a < 0$ when $x = \pm \sqrt{-\dfrac{a}{2}}$

(d) If $a < 0$, then there are three critical points. If $a \geq 0$, then there is only one critical point.

3. All c, where c is a real number   **5.** Proof

7. The bug should head toward the midpoint of the opposite side. Without calculus, imagine opening up the cube. The shortest distance is the line PQ, passing through the midpoint as shown.

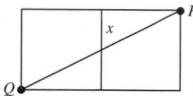

9. $a = 6$, $b = 1$, $c = 2$ **11.** Proof

13. Greatest slope: $\left(-\dfrac{\sqrt{3}}{3}, \dfrac{3}{4}\right)$, Least slope: $\left(\dfrac{\sqrt{3}}{3}, \dfrac{3}{4}\right)$

15. Proof   **17.** Proof; Point of inflection: $(1, 0)$

19. (a) $(0, \infty)$

(b) Answers will vary. Sample answer: $x = e^{\pi/2}$, $x = e^{(\pi/2)+2\pi}$

(c) Answers will vary. Sample answer: $x = e^{-\pi/2}$, $x = e^{3\pi/2}$

(d) $[-1, 1]$ (e) $f'(x) = \dfrac{\cos(\ln x)}{x}$; Maximum $= e^{\pi/2}$

(f)

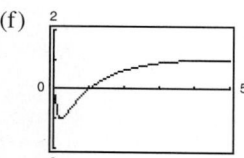

$\lim\limits_{x \to 0^+} f(x)$ seems to be $-\dfrac{1}{2}$. (This is incorrect.)

(g) The limit does not exist.

Chapter 5

Section 5.1   (page 291)

1. A function F is an antiderivative of f on an interval I when $F'(x) = f(x)$ for all x in I.

3. The particular solution results from knowing the value of $y = F(x)$ for one value of x. Using the initial condition in the general solution, you can solve for C to obtain the particular solution.

5. Proof   **7.** $y = 3t^3 + C$ **9.** $y = \frac{2}{5}x^{5/2} + C$

Original Integral	Rewrite	Integrate	Simplify
11. $\int \sqrt[3]{x}\,dx$	$\int x^{1/3}\,dx$	$\dfrac{x^{4/3}}{4/3} + C$	$\dfrac{3}{4}x^{4/3} + C$
13. $\int \dfrac{1}{x\sqrt{x}}\,dx$	$\int x^{-3/2}\,dx$	$\dfrac{x^{-1/2}}{-1/2} + C$	$-\dfrac{2}{\sqrt{x}} + C$

15. $\frac{1}{2}x^2 + 7x + C$ **17.** $\frac{2}{5}x^{5/2} + x^2 + x + C$

19. $x^3 + \frac{1}{2}x^2 - 2x + C$

21. $-\dfrac{1}{4x^4} + C$ **23.** $\frac{2}{3}x^{3/2} + 12x^{1/2} + C$

25. $5\sin x - 4\cos x + C$ **27.** $\tan \theta + \cos \theta + C$

29. $\tan y + C$ **31.** $-2\cos x - 5e^x + C$

33. $x^2 - \dfrac{4^x}{\ln 4} + C$ **35.** $\frac{1}{2}x^2 - 5\ln|x| + C$

37. $f(x) = 3x^2 + 8$ **39.** $f(x) = x^2 + x + 4$

41. $f(x) = -4\sqrt{x} + 3x$ **43.** $f(x) = e^x + x + 4$

45. (a) Answers will vary. Sample answer: (b) $y = \dfrac{x^3}{3} - x + \dfrac{7}{3}$

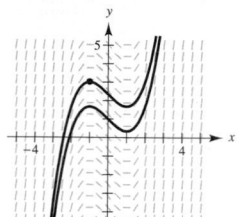   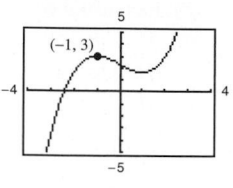

47. (a)   (b) $y = x^2 - 6$ (c)

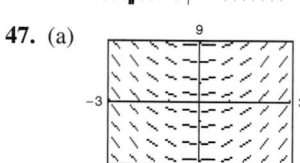   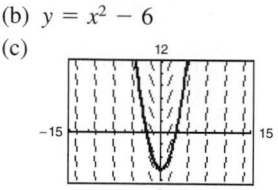

49. Answers will vary. Sample answer:

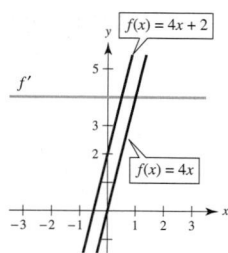

51. $f(x) = \tan^2 x \implies f'(x) = 2\tan x \cdot \sec^2 x$
$g(x) = \sec^2 x \implies g'(x) = 2\sec x \cdot \sec x \tan x = f'(x)$
The derivatives are the same, so f and g differ by a constant.

53. $f(x) = \dfrac{x^3}{3} - 4x + \dfrac{16}{3}$

55. (a) $h(t) = \frac{3}{4}t^2 + 5t + 12$ (b) 69 cm   **57.** 62.25 ft

59. (a) $t \approx 2.562$ sec   (b) $v(t) \approx -65.970$ ft/sec

61. $v_0 \approx 62.3$ m/sec   **63.** 320 m; -32 m/sec

65. (a) $v(t) = 3t^2 - 12t + 9$, $a(t) = 6t - 12$
(b) $(0, 1), (3, 5)$ (c) -3

67. $a(t) = -\dfrac{1}{2t^{3/2}}$, $x(t) = 2\sqrt{t} + 2$

69. (a) 1.18 m/sec^2 (b) 190 m

71. (a) 300 ft   (b) 60 ft/sec ≈ 41 mi/h

73. False. f has an infinite number of antiderivatives, each differing by a constant.

75–77. Proofs

79. Putnam Problem B2, 1991

Section 5.2 *(page 303)*

1. The index of summation is i, the upper bound of summation is 8, and the lower bound of summation is 3.

3. You can use the line $y = x$ bounded by $x = a$ and $x = b$. The sum of the areas of the inscribed rectangles in the figure below is the lower sum.

The sum of the areas of the circumscribed rectangles in the figure below is the upper sum.

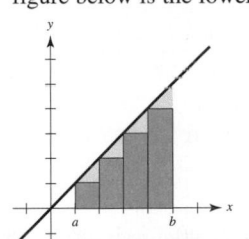 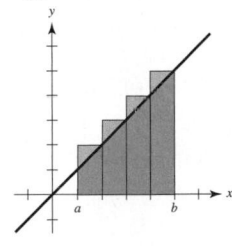

The rectangles in the first graph do not contain all of the area of the region, and the rectangles in the second graph cover more than the area of the region. The exact value of the area lies between these two sums.

5. 75   **7.** $\dfrac{158}{85}$ **9.** $8c$ **11.** $\displaystyle\sum_{i=1}^{11} \dfrac{1}{5i}$

13. $\displaystyle\sum_{j=1}^{6}\left[7\left(\dfrac{j}{6}\right) + 5\right]$ **15.** $\dfrac{2}{n}\displaystyle\sum_{i=1}^{n}\left[\left(\dfrac{2i}{n}\right)^3 - \left(\dfrac{2i}{n}\right)\right]$ **17.** 84

19. 1200   **21.** 2470   **23.** 1876

25. $\dfrac{n+2}{n}$ **27.** $\dfrac{2(n+1)(n-1)}{n^2}$

$n = 10:\ S = 1.2$ $n = 10:\ S = 1.98$
$n = 100:\ S = 1.02$ $n = 100:\ S = 1.9998$
$n = 1000:\ S = 1.002$ $n = 1000:\ S = 1.999998$
$n = 10{,}000:\ S = 1.0002$ $n = 10{,}000:\ S = 1.99999998$

29. $13 < (\text{Area of region}) < 15$

31. $55 < (\text{Area of region}) < 74.5$

33. $0.7908 < (\text{Area of region}) < 1.1835$

35. The area of the shaded region falls between 12.5 square units and 16.5 square units.

37. $A \approx S \approx 0.768$ **39.** $A \approx S \approx 0.746$
$\ \ \ \ A \approx s \approx 0.518$ $\ \ \ \ A \approx s \approx 0.646$

41. $s(n) = 24 - \dfrac{24}{n},\ S(n) = 24 + \dfrac{24}{n}$

43. $s(n) = \dfrac{5(2n^2 - 3n + 1)}{6n^2},\ S(n) = \dfrac{5(2n^2 + 3n + 1)}{6n^2}$

45. (a)

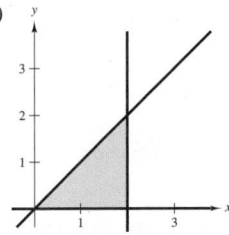

(b) $\Delta x = \dfrac{2 - 0}{n} = \dfrac{2}{n}$

(c) $s(n) = \displaystyle\sum_{i=1}^{n} f(x_{i-1})\,\Delta x = \sum_{i=1}^{n}\left[(i - 1)\left(\dfrac{2}{n}\right)\right]\left(\dfrac{2}{n}\right)$

(d) $S(n) = \displaystyle\sum_{i=1}^{n} f(x_i)\,\Delta x = \sum_{i=1}^{n}\left[i\left(\dfrac{2}{n}\right)\right]\left(\dfrac{2}{n}\right)$

(e)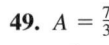

n	5	10	50	100
$s(n)$	1.6	1.8	1.96	1.98
$S(n)$	2.4	2.2	2.04	2.02

(f) $\displaystyle\lim_{n\to\infty}\sum_{i=1}^{n}\left[(i - 1)\left(\dfrac{2}{n}\right)\right]\left(\dfrac{2}{n}\right) = 2$

$\displaystyle\lim_{n\to\infty}\sum_{i=1}^{n}\left[i\left(\dfrac{2}{n}\right)\right]\left(\dfrac{2}{n}\right) = 2$

47. $A = 3$ **49.** $A = \dfrac{7}{3}$

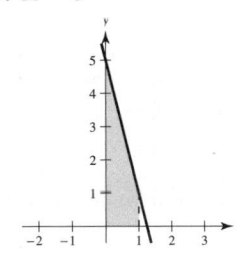 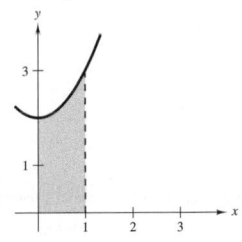

51. $A = 54$ **53.** $A = 34$

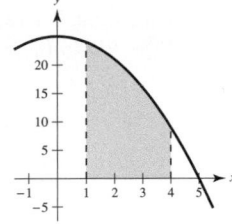 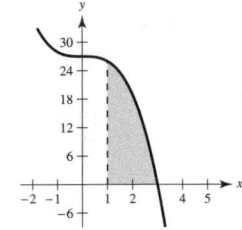

55. $A = \dfrac{2}{3}$ **57.** $A = 8$

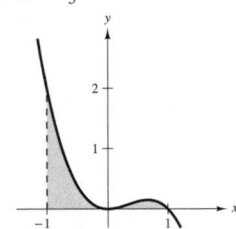 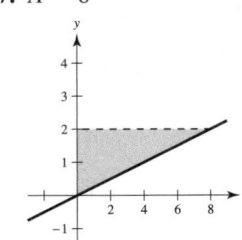

59. $A = \dfrac{125}{3}$ **61.** $A = \dfrac{44}{3}$

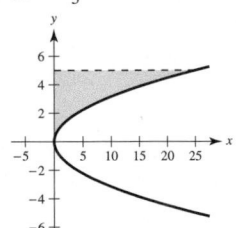 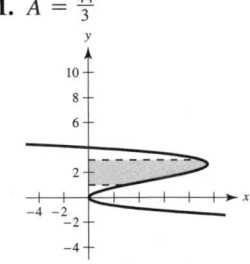

63. $\dfrac{69}{8}$ **65.** 0.345   **67.** 4.0786   **69.** b

71. An overestimate on one side of the midpoint compensates for an underestimate on the other side of the midpoint.

73. (a)   (b)

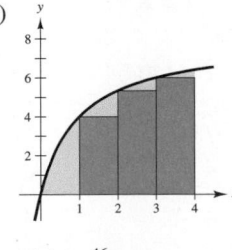 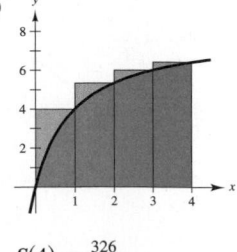

$s(4) = \dfrac{46}{3}$ $S(4) = \dfrac{326}{15}$

(c)

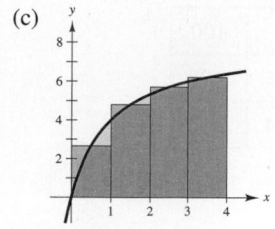

(d) Proof

$M(4) = \frac{6112}{315}$

(e)

n	4	8	20	100	200
$s(n)$	15.333	17.368	18.459	18.995	19.060
$S(n)$	21.733	20.568	19.739	19.251	19.188
$M(n)$	19.403	19.201	19.137	19.125	19.125

(f) Because f is an increasing function, $s(n)$ is always increasing and $S(n)$ is always decreasing.

75. True

77. Suppose there are n rows and $n + 1$ columns. The stars on the left total $1 + 2 + \cdots + n$, as do the stars on the right. There are $n(n + 1)$ stars in total. So, $2[1 + 2 + \cdots + n] = n(n + 1)$

and $1 + 2 + \cdots + n = \dfrac{n(n + 1)}{2}$.

79. When n is odd, there are $\left(\dfrac{n + 1}{2}\right)^2$ seats. When n is even,

there are $\dfrac{n^2 + 2n}{4}$ seats.

81. Putnam Problem B1, 1989

Section 5.3   (page 313)

1. A Riemann sum represents the addition of all of the subregions for a function f on an interval $[a, b]$.

3. $2\sqrt{3} \approx 3.464$ **5.** 32   **7.** 0   **9.** $\frac{10}{3}$

11. $\displaystyle\int_{-1}^{5} (3x + 10)\, dx$ **13.** $\displaystyle\int_{1}^{5} \left(1 + \frac{3}{x}\right) dx$ **15.** $\displaystyle\int_{0}^{4} 5\, dx$

17. $\displaystyle\int_{-4}^{4} \left(4 - |x|\right) dx$ **19.** $\displaystyle\int_{-5}^{5} (25 - x^2)\, dx$

21. $\displaystyle\int_{0}^{\pi/2} \cos x\, dx$ **23.** $\displaystyle\int_{0}^{2} y^3\, dy$ **25.** $\displaystyle\int_{1}^{4} \frac{2}{x}\, dx$

27.

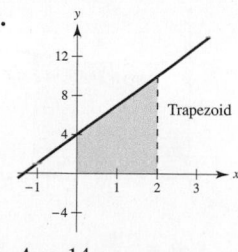

$A = 12$

29.

Triangle

$A = 8$

31.

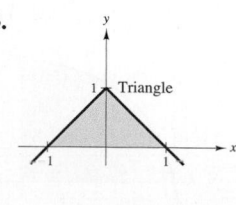

Trapezoid

$A = 14$

33.

Triangle

$A = 1$

35.

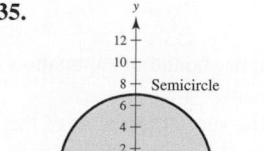

Semicircle

$A = \dfrac{49\pi}{2}$

37. -320 **39.** 80   **41.** -40 **43.** 508

45. (a) 13   (b) -10 (c) 0   (d) 30

47. (a) 8   (b) -12 (c) -4 (d) 30   **49.** $-48, 88$

51. (a) $-\pi$ (b) 4   (c) $-(1 + 2\pi)$ (d) $3 - 2\pi$

(e) $5 + 2\pi$ (f) $23 - 2\pi$

53. (a) 14   (b) 4   (c) 8   (d) 0   **55.** 40   **57.** a   **59.** c

61. Answers will vary. Sample answer:

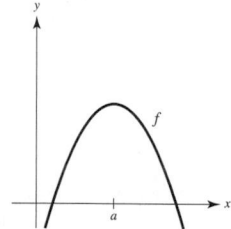

There is no region.

63. Geometric method:

$\displaystyle\int_{-1}^{3} (x + 2)\, dx =$ Area of large triangle $-$ Area of small triangle

$= \dfrac{25}{2} - \dfrac{1}{2} = 12$

Limit definition:

$\displaystyle\int_{-1}^{3} (x + 2)\, dx = \lim_{n\to\infty} \sum_{i=1}^{n} \left[\left(-1 + \frac{4i}{n} + 2\right)\left(\frac{4}{n}\right)\right] = 12$

65. $a = -2, b = 5$

67. Answers will vary. Sample answer: $a = \pi, b = 2\pi$

$\displaystyle\int_{\pi}^{2\pi} \sin x\, dx < 0$

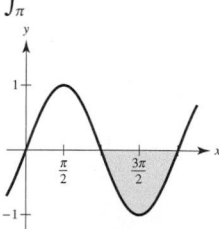

69. True   **71.** True   **73.** False. $\displaystyle\int_{0}^{2} (-x)\, dx = -2$

75. 272   **77.** Proof

79. No. No matter how small the subintervals, the number of both rational and irrational numbers within each subinterval is infinite, and $f(c_i) = 0$ or $f(c_i) = 1$.

81. $a = -1$ and $b = 1$; The function is nonnegative between $x = -1$ and $x = 1$.

83. Answers will vary. Sample answer:

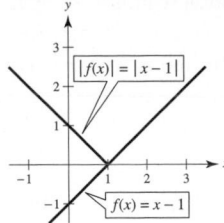

The integrals are equal when f is always greater than or equal to 0 on $[a, b]$.

85. $\frac{1}{3}$

Section 5.4 *(page 328)*

1. Find an antiderivative of the function and evaluate the difference of the antiderivative at the upper limit of integration and the lower limit of integration.

3. The average value of a function on an interval is the integral of the function on $[a, b]$ times $\dfrac{1}{b - a}$.

5. 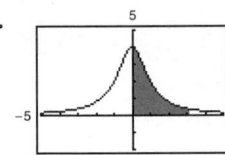 **7.**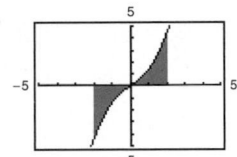

Positive   Zero

9. -2 **11.** $\frac{1}{3}$ **13.** $\frac{1}{2}$ **15.** $\frac{2}{3}$ **17.** -4

19. $-\frac{27}{20}$ **21.** $\frac{25}{2}$ **23.** $2 - 7\pi$

25. $\dfrac{\pi}{4}$ **27.** $\dfrac{2\sqrt{3}}{3}$ **29.** 0 **31.** $\dfrac{3}{\ln 2} + 12$

33. $e - e^{-1}$ **35.** $\frac{1}{6}$ **37.** 1 **39.** $\frac{52}{3}$ **41.** 20

43. 4 **45.** $\dfrac{3\sqrt[3]{2}}{2} \approx 1.8899$ **47.** $\dfrac{3}{\ln 4} \approx 2.1640$

49. $\pm\arccos \dfrac{\sqrt{\pi}}{2} \approx \pm 0.4817$

51. Average value $= \frac{8}{3}$

$x = \pm \dfrac{2\sqrt{3}}{3}$

53. Average value $= e - e^{-1} \approx 2.3504$

$x = \ln \dfrac{e - e^{-1}}{2} \approx 0.1614$

55. Average value $= \dfrac{2}{\pi}$

$x \approx 0.690,\ x \approx 2.451$

57. (a) $F(x) = 500 \sec^2 x$

(b) $\dfrac{1500\sqrt{3}}{\pi} \approx 827$ N

59. $\dfrac{2}{\pi} \approx 63.7\%$

61. $F(x) = -\dfrac{20}{x} + 20$

$F(2) = 10$

$F(5) = 16$

$F(8) = \dfrac{35}{2}$

63. $F(x) = \sin x$

$F(0) = 0$

$F\left(\dfrac{\pi}{4}\right) = \dfrac{\sqrt{2}}{2}$

$F\left(\dfrac{\pi}{2}\right) = 1$

65. (a) $g(0) = 0$, $g(2) \approx 7$, $g(4) \approx 9$, $g(6) \approx 8$, $g(8) \approx 5$

(b) Increasing: $(0, 4)$; Decreasing: $(4, 8)$

(c) A maximum occurs at $x = 4$.

(d)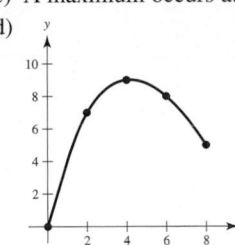

67. $\frac{1}{2}x^2 + 2x$ **69.** $\frac{3}{4}x^{4/3} - 12$ **71.** $\tan x - 1$

73. $e^x - e^{-1}$ **75.** $x^2 - 2x$ **77.** $\sqrt{x^4 + 1}$

79. $\sqrt{x} \csc x$ **81.** 8 **83.** $\cos x \sqrt{\sin x}$ **85.** $3x^2 \sin x^6$

87. 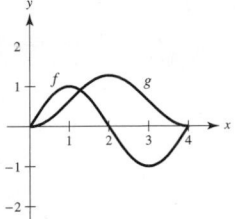   **89.** 8190 L

An extremum of g occurs at $x = 2$.

91. About 540 ft   **93.** (a) $\frac{3}{2}$ ft to the right   (b) $\frac{113}{10}$ ft

95. (a) 0 ft   (b) $\frac{63}{2}$ ft   **97.** (a) 2 ft to the right   (b) 2 ft

99. The displacement and total distance traveled are equal when the particle is always moving in the same direction on an interval.

101. The Fundamental Theorem of Calculus requires that f be continuous on $[a, b]$ and that F be an antiderivative for f on the entire interval. On an interval containing c, the function

$f(x) = \dfrac{1}{x - c}$ is not continuous at c.

103. 28 units

105. $f(x) = x^{-2}$ has a nonremovable discontinuity at $x = 0$.

107. $f(x) = \sec^2 x$ has a nonremovable discontinuity at $x = \dfrac{\pi}{2}$.

109. True

111. $f'(x) = \dfrac{1}{(1/x)^2 + 1}\left(-\dfrac{1}{x^2}\right) + \dfrac{1}{x^2 + 1} = 0$

Because $f'(x) = 0$, $f(x)$ is constant.

113. (a) 0 (b) 0 (c) $xf(x) + \int_0^x f(t)\, dt$ (d) 0

115. Putnam Problem B5, 2006

Section 5.5 *(page 341)*

1. You can move constant multiples outside the integral sign.

$\displaystyle \int kf(x)\, dx = k \int f(x)\, dx$

3. The integral of $[g(x)]^n g'(x)$ is $\dfrac{[g(x)]^{n+1}}{n + 1} + C,\ n \neq -1$.

Recall the power rule for polynomials.

$\displaystyle \int f(g(x))g'(x)\, dx$	$u = g(x)$	$du = g'(x)\, dx$
5. $\displaystyle \int (5x^2 + 1)^2 (10x)\, dx$	$5x^2 + 1$	$10x\, dx$
7. $\displaystyle \int \tan^2 x \sec^2 x\, dx$	$\tan x$	$\sec^2 x\, dx$

9. $\frac{1}{5}(1 + 6x)^5 + C$ **11.** $\frac{2}{3}(25 - x^2)^{3/2} + C$

13. $\frac{1}{12}(x^4 + 3)^3 + C$ **15.** $\frac{1}{30}(2x^3 - 1)^5 + C$

17. $\frac{1}{3}(t^2 + 2)^{3/2} + C$ **19.** $\dfrac{7}{4(1 - x^2)^2} + C$

21. $-\dfrac{1}{3(1 + x^3)} + C$ **23.** $-\sqrt{1 - x^2} + C$

25. $-\dfrac{1}{4}\left(1 + \dfrac{1}{t}\right)^4 + C$ **27.** $\sqrt{2x} + C$

29. $2x^2 - 4\sqrt{16 - x^2} + C$ **31.** $-\dfrac{1}{2(x^2 + 2x - 3)} + C$

33. (a) Answers will vary.   (b) $y = -\frac{1}{3}(4 - x^2)^{3/2} + 2$

Sample answer:

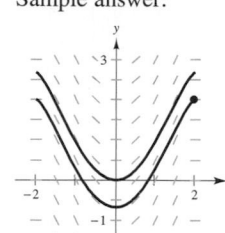   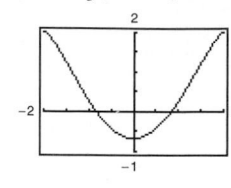

35. $-\cos \pi x + C$ **37.** $\frac{1}{6} \sin 6x + C$ **39.** $-\sin\dfrac{1}{\theta} + C$

41. $\frac{1}{4} \sin^2 2x + C$ or $-\frac{1}{4} \cos^2 2x + C_1$ or $-\frac{1}{8} \cos 4x + C_2$

43. $\frac{1}{2} \tan^2 x + C$ or $\frac{1}{2} \sec^2 x + C_1$ **45.** $e^{7x} + C$

47. $\frac{1}{3}(e^x + 1)^3 + C$ **49.** $-\frac{5}{2}e^{-2x} + e^{-x} + C$

51. $\dfrac{1}{\pi e^{\sin \pi x}} + C$ **53.** $-\tan(e^{-x}) + C$ **55.** $\dfrac{2}{\ln 3 (3^{x/2})} + C$

57. $f(x) = \dfrac{1}{12}(4x^2 - 10)^3 - 8$ **59.** $f(x) = 2 \cos\dfrac{x}{2} + 4$

61. $f(x) = -8e^{-x/4} + 9$

63. $f(x) = \dfrac{3}{\ln 0.4}0.4^{x/3} + \dfrac{1}{2} - \dfrac{3}{\ln 0.4}$

65. $\frac{2}{5}(x + 6)^{5/2} - 4(x + 6)^{3/2} + C = \frac{2}{5}(x + 6)^{3/2}(x - 4) + C$

67. $-\left[\frac{2}{3}(1 - x)^{3/2} - \frac{4}{5}(1 - x)^{5/2} + \frac{2}{7}(1 - x)^{7/2}\right] + C =$

$-\frac{2}{105}(1 - x)^{3/2}(15x^2 + 12x + 8) + C$

69. $\frac{1}{8}\left[\frac{2}{5}(2x - 1)^{5/2} + \frac{4}{3}(2x - 1)^{3/2} - 6(2x - 1)^{1/2}\right] + C =$

$\dfrac{\sqrt{2x - 1}}{15}(3x^2 + 2x - 13) + C$

71. $-\frac{1}{8} \cos^4 2x + C$ **73.** 0   **75.** $12 - \frac{8}{9}\sqrt{2}$

77. 2   **79.** $\dfrac{1}{2}$ **81.** $2e^9(e^7 - 1)$ **83.** $\dfrac{e}{3(e^2 - 1)}$

85. $\frac{1209}{28}$ **87.** $2(\sqrt{3} - 1)$

89. $e^5 - 1 \approx 147.413$ **91.** $2(1 - e^{-3/2}) \approx 1.554$

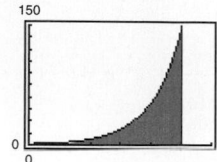   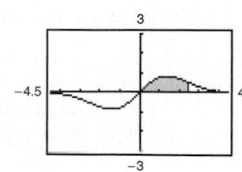

93. $\frac{272}{15}$ **95.** 0

97. (a) 144   (b) 72   (c) -144 (d) 432

99. $2\displaystyle\int_0^3 (4x^2 - 6)\, dx = 36$

101. (a) $\displaystyle\int x^2\sqrt{x^3 + 1}\, dx$; Use substitution with $u = x^3 + 1$.

(b) $\displaystyle\int \cot^3(2x)\, \csc^2(2x)\, dx$; Use substitution with $u = \cot 2x$.

(c) $\displaystyle\int e^{4x-3}\, dx$; Use substitution with $u = 4x - 3$.

103. \$340,000

105. (a) 102.532 thousand units   (b) 102.352 thousand units

(c) 74.5 thousand units

107. (a)

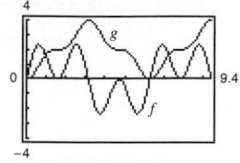

(b) g is nonnegative, because the graph of f is positive at the beginning and generally has more positive sections than negative ones.

(c) The points on g that correspond to the extrema of f are points of inflection of g.

(d) No, some zeros of f, such as $x = \dfrac{\pi}{2}$, do not correspond to extrema of g. The graph of g continues to increase after $x = \dfrac{\pi}{2}$, because f remains above the x-axis.

(e)

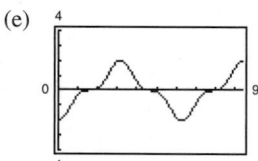

The graph of h is that of g shifted 2 units downward.

109. (a) and (b) Proofs

111. (a) $P_{0.50,\, 0.75} \approx 35.3\%$ (b) $b \approx 58.6\%$

113. True   **115.** True   **117.** True   **119–121.** Proofs

123. Putnam Problem A1, 1958

Section 5.6   *(page 352)*

1. L'Hôpital's Rule allows you to address limits of the form $0/0$ and ∞/∞.

3.

x	-0.1	-0.01	-0.001	0
$f(x)$	1.3177	1.3332	1.3333	?

x	0.001	0.01	0.1
$f(x)$	1.3333	1.3332	1.3177

$\frac{4}{3}$

5.

x	1	10	10^2
$f(x)$	0.9900	90,483.7	3.7×10^9

x	10^3	10^4	10^5
$f(x)$	4.5×10^{10}	0	0

0

7. $\frac{3}{8}$ **9.** $\frac{1}{8}$ **11.** 0   **13.** $\frac{5}{3}$ **15.** 4   **17.** 0

19. ∞ **21.** $\frac{11}{4}$ **23.** $\frac{3}{5}$ **25.** $\frac{7}{6}$ **27.** ∞ **29.** 0

31. 1   **33.** 0   **35.** 0   **37.** ∞ **39.** $\frac{5}{9}$ **41.** sin 3

43. (a) Not indeterminate

(b) ∞

(c)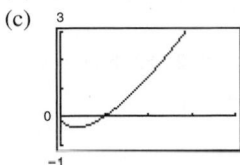

45. (a) $0 \cdot \infty$

(b) 1

(c)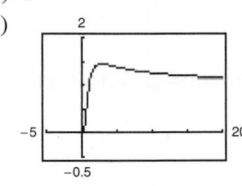

47. (a) 1^{∞}

(b) e^4

(c)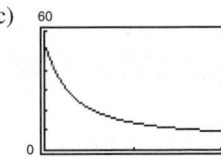

49. (a) ∞^0

(b) 1

(c)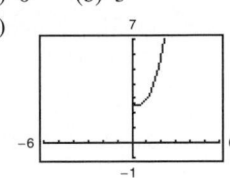

51. (a) 1^{∞} (b) e

(c)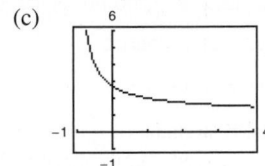

53. (a) 0^0 (b) 3

(c)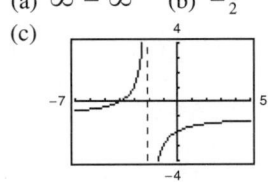

55. (a) 0^0 (b) 1

(c)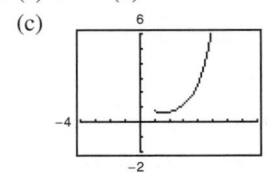

57. (a) $\infty - \infty$ (b) $-\frac{3}{2}$

(c)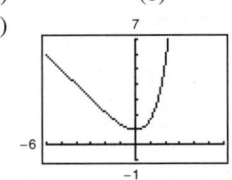

59. (a) $\infty - \infty$ (b) ∞

(c)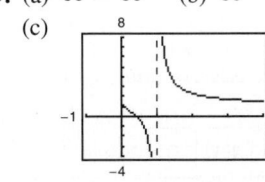

61. (a) $\infty - \infty$ (b) ∞

(c)

63. Answers will vary. Sample answers:

(a) $f(x) = x^2 - 25$, $g(x) = x - 5$

(b) $f(x) = (x - 5)^2$, $g(x) = x^2 - 25$

(c) $f(x) = x^2 - 25$, $g(x) = (x - 5)^3$

65. (a) Yes; $\frac{0}{0}$ (b) No; $\frac{0}{-1}$ (c) Yes; $\frac{\infty}{\infty}$ (d) Yes; $\frac{0}{0}$

(e) No; $\frac{-1}{0}$ (f) Yes; $\frac{0}{0}$

67.

x	10	10^2	10^4	10^6	10^8	10^{10}
$\dfrac{(\ln x)^4}{x}$	2.811	4.498	0.720	0.036	0.001	0.000

69. 0   **71.** 0   **73.** 0

75. Horizontal asymptote:
   $y = 1$

   Relative maximum: $(e, e^{1/e})$

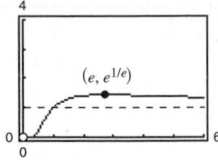

77. Horizontal asymptote:
   $y = 0$

   Relative maximum: $\left(1, \dfrac{2}{e}\right)$

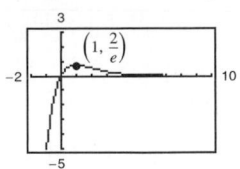

79. Limit is not of the form $\dfrac{0}{0}$ or $\dfrac{\infty}{\infty}$.

81. Limit is not of the form $\dfrac{0}{0}$ or $\dfrac{\infty}{\infty}$.

83. (a) $\displaystyle\lim_{x\to\infty} \frac{x}{\sqrt{x^2 + 1}} = \lim_{x\to\infty} \frac{\sqrt{x^2 + 1}}{x} = \lim_{x\to\infty} \frac{x}{\sqrt{x^2 + 1}}$

Applying L'Hôpital's Rule twice results in the original limit, so L'Hôpital's Rule fails.

(b) 1

(c)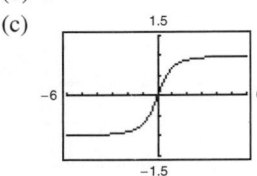

85.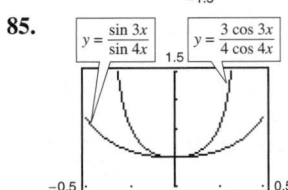

As $x \to 0$, the graphs get closer together (they both approach 0.75). By L'Hôpital's Rule,

$$\lim_{x\to 0} \frac{\sin 3x}{\sin 4x} = \lim_{x\to 0} \frac{3\cos 3x}{4\cos 4x} = \frac{3}{4}.$$

87. $\dfrac{Vt}{L}$ **89.** Proof   **91.** $c = \dfrac{2}{3}$ **93.** $c = \dfrac{\pi}{4}$

95. False. $\dfrac{\infty}{0} = \pm\infty$ **97.** True   **99.** True   **101.** $\dfrac{3}{4}$

103. $c = \dfrac{4}{3}$ **105.** $a = 1, b = \pm 2$ **107.** Proof

109. (a) $0 \cdot \infty$ (b) 0   **111.** Proof   **113.** (a)–(c) 2

115. (a)

(b) $\displaystyle\lim_{x\to\infty} h(x) = 1$

(c) No. $h(x)$ is not an indeterminate form.

117. Putnam Problem A1, 1956

Section 5.7 *(page 362)*

1. No. To use the Log Rule, look for quotients in which the numerator is the derivative of the denominator, with rewriting in mind.

3. Ways to alter an integrand are to rewrite using a trigonometric identity, multiply and divide by the same quantity, add and subtract the same quantity, or use long division.

5. $5 \ln|x| + C$ **7.** $\frac{1}{2} \ln|2x + 5| + C$

9. $\frac{1}{2} \ln|x^2 - 3| + C$ **11.** $\ln|x^4 + 3x| + C$

13. $\frac{x^2}{14} - \ln|x| + C$ **15.** $\frac{1}{3} \ln|x^3 + 3x^2 + 9x| + C$

17. $\frac{1}{2}x^2 - 4x + 6 \ln|x + 1| + C$

19. $\frac{1}{3}x^3 + 5 \ln|x - 3| + C$

21. $\frac{1}{3}x^3 - 2x + \ln\sqrt{x^2 + 2} + C$ **23.** $\frac{1}{3}(\ln x)^3 + C$

25. $-\frac{2}{3} \ln|1 - 3\sqrt{x}| + C$ **27.** $6 \ln|x - 5| - \frac{30}{x - 5} + C$

29. $\sqrt{2x} - \ln|1 + \sqrt{2x}| + C$

31. $x + 6\sqrt{x} + 18 \ln|\sqrt{x} - 3| + C$ **33.** $3 \ln\left|\sin\frac{\theta}{3}\right| + C$

35. $-\frac{1}{2} \ln|\csc 2x + \cot 2x| + C$ **37.** $\ln|1 + \sin t| + C$

39. $\ln|\sec x - 1| + C$ **41.** $\ln|\cos(e^{-x})| + C$

43. $y = -3 \ln|2 - x| + C$ **45.** $y = \ln|x^2 - 9| + C$

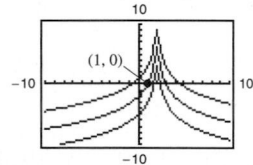   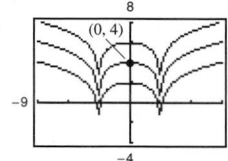

47. $f(x) = -2 \ln x + 3x - 2$

49. (a)   (b) $y = \ln\left(\frac{x + 2}{2}\right) + 1$

51. $\frac{5}{3} \ln 13 \approx 4.275$ **53.** $\frac{7}{3}$ **55.** $-\ln 3 \approx 1.099$

57. $\ln\left|\frac{2 - \sin 2}{1 - \sin 1}\right| \approx 1.929$

59. $4\sqrt{x} - x - 4 \ln(1 + \sqrt{x}) + C$ **61.** $\frac{1}{x}$

63. $4 \cot 4x$ **65.** $6 \ln 3 \approx 6.592$

67. $\ln|\csc 1 + \cot 1| - \ln|\csc 2 + \cot 2| \approx 1.048$

69. $\frac{15}{2} + 8 \ln 2 \approx 13.045$ **71.** $\frac{12}{\pi} \ln(2 + \sqrt{3}) \approx 5.03$

73. 1   **75.** $\frac{1}{e - 1} \approx 0.582$ **77.** About 13.077

79. d   **81.** Proof   **83.** $x = 2$ **85.** Proof

87. $-\ln|\cos x| + C = \ln\left|\frac{1}{\cos x}\right| + C = \ln|\sec x| + C$

89. $\ln|\sec x + \tan x| + C = \ln\left|\frac{\sec^2 x - \tan^2 x}{\sec x - \tan x}\right| + C$
$= -\ln|\sec x - \tan x| + C$

91. (a) $P(t) = 1000(12 \ln|1 + 0.25t| + 1)$
(b) $P(3) \approx 7715$

93. About 4.15 min

95.

(a) $A = \frac{1}{2} \ln 2 - \frac{1}{4}$
(b) $0 < m < 1$
(c) $A = \frac{1}{2}(m - \ln m - 1)$

97. True   **99.** True   **101.** Proof

Section 5.8   *(page 370)*

1. (a) No
(b) Yes. Use the rule involving the arcsecant function.

3. $\arcsin\frac{x}{3} + C$ **5.** $\operatorname{arcsec}|2x| + C$

7. $\arcsin(x + 1) + C$ **9.** $\frac{1}{2} \arcsin t^2 + C$

11. $\frac{1}{10} \arctan\frac{t^2}{5} + C$ **13.** $\frac{1}{4} \arctan\frac{e^{2x}}{2} + C$

15. $\arcsin\frac{\csc x}{5} + C$ **17.** $2 \arcsin\sqrt{x} + C$

19. $\frac{1}{2} \ln(x^2 + 1) - 3 \arctan x + C$

21. $8 \arcsin\frac{x - 3}{3} - \sqrt{6x - x^2} + C$ **23.** $\frac{\pi}{6}$ **25.** $\frac{\pi}{6}$

27. $\frac{1}{3}\left(\arctan 3 - \frac{\pi}{4}\right) \approx 0.155$ **29.** $\arctan 5 - \frac{\pi}{4} \approx 0.588$

31. $\frac{\pi}{4}$ **33.** $\frac{1}{32}\pi^2 \approx 0.308$ **35.** $\frac{\pi}{2}$

37. $\frac{\sqrt{2}}{2} \arcsin\left[\frac{\sqrt{6}}{6}(x - 2)\right] + C$ **39.** $\arcsin\frac{x + 2}{2} + C$

41. $4 - 2\sqrt{3} + \frac{1}{6}\pi \approx 1.059$

43. $2\sqrt{e^t - 3} - 2\sqrt{3} \arctan\frac{\sqrt{e^t - 3}}{\sqrt{3}} + C$ **45.** $\frac{\pi}{6}$

47. (a) $\arcsin x + C$ (b) $-\sqrt{1 - x^2} + C$ (c) Not possible

49. (a) $\frac{2}{3}(x - 1)^{3/2} + C$ (b) $\frac{2}{15}(x - 1)^{3/2}(3x + 2) + C$
(c) $\frac{2}{3}\sqrt{x - 1}(x + 2) + C$

51. Proof

53. No. Graphing $f(x) = \arcsin x$ and $g(x) = -\arccos x$, you can see that the graph of f is the graph of g shifted vertically.

55. (a)   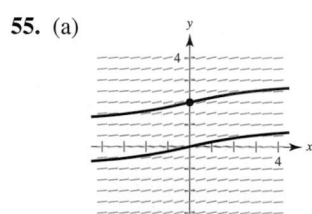   (b) $y = \frac{2}{3} \arctan\frac{x}{3} + 2$

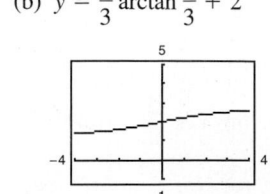

57.   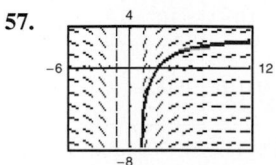   **59.**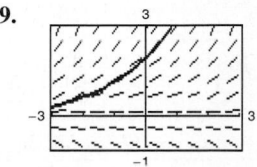

61. $y = \arcsin\frac{x}{2} + \pi$ **63.** $\frac{\pi}{3}$ **65.** $\frac{3\pi}{2}$

67. (a)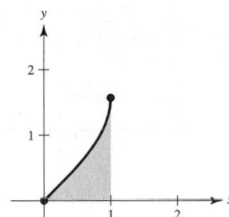

(b) 0.5708

(c) $\dfrac{\pi - 2}{2}$

69. (a) $F(x)$ represents the average value of $f(x)$ over the interval $[x, x + 2]$; Maximum at $x = -1$

(b) Maximum at $x = -1$

71. False. $\displaystyle\int \dfrac{dx}{3x\sqrt{9x^2 - 16}} = \dfrac{1}{12} \operatorname{arcsec} \dfrac{|3x|}{4} + C$

73–75. Proofs

77. (a) $\displaystyle\int_0^1 \dfrac{1}{1 + x^2}\, dx$ (b) About 0.7857

(c) Because $\displaystyle\int_0^1 \dfrac{1}{1 + x^2}\, dx = \dfrac{\pi}{4}$, you can use the Midpoint Rule to approximate $\dfrac{\pi}{4}$. Multiplying the result by 4 gives an estimation of π.

Section 5.9 *(page 380)*

1. Hyperbolic function came from the comparison of the area of a semicircular region with the area of a region under a hyperbola.

3. $\sinh^2 x = \dfrac{-1 + \cosh 2x}{2}$ **5.** (a) 10.018   (b) -0.964

7. (a) $\frac{4}{3}$ (b) $\frac{13}{12}$ **9.** (a) 1.317   (b) 0.962

11–17. Proofs

19. $\cosh x = \dfrac{\sqrt{13}}{2}$, $\tanh x = \dfrac{3\sqrt{13}}{13}$, $\operatorname{csch} x = \dfrac{2}{3}$,

$\operatorname{sech} x = \dfrac{2\sqrt{13}}{13}$, $\coth x = \dfrac{\sqrt{13}}{3}$

21. ∞ **23.** 1   **25.** $9 \cosh 9x$

27. $-10x(\operatorname{sech} 5x^2 \tanh 5x^2)$ **29.** $\coth x$

31. $-\dfrac{t}{2}\cosh(-3t) + \dfrac{\sinh(-3t)}{6}$ **33.** $\operatorname{sech} t$

35. $y = -2x + 2$ **37.** $y = 1 - 2x$

39. Relative maximum: $(1.20, 0.66)$

Relative minimum: $(-1.20, -0.66)$

41. Relative maxima: $(\pm\pi, \cosh \pi)$, Relative minimum: $(0, -1)$

43. (a)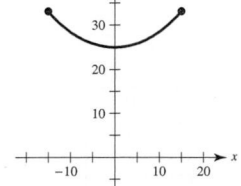

(b) 33.146 units, 25 units

(c) $m = \sinh 1 \approx 1.175$

45. $\frac{1}{4}\sinh 4x + C$ **47.** $-\frac{1}{2}\cosh(1 - 2x) + C$

49. $\frac{1}{3}\cosh^3(x - 1) + C$ **51.** $\ln|\sinh x| + C$

53. $-\coth\dfrac{x^2}{2} + C$ **55.** $\ln\dfrac{5}{4}$ **57.** $\coth 1 - \coth 2$

59. $-\frac{1}{3}(\operatorname{csch} 2 - \operatorname{csch} 1)$

61.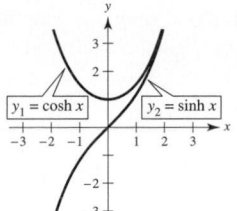

The graphs do not intersect.

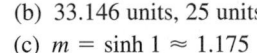

63. Proof   **65.** $\dfrac{3}{\sqrt{9x^2 - 1}}$ **67.** $\dfrac{1}{2\sqrt{x}(1 - x)}$

69. $|\sec x|$ **71.** $-\csc x$ **73.** $2\sinh^{-1}(2x)$

75. $\dfrac{\sqrt{3}}{18}\ln\left|\dfrac{1 + \sqrt{3}x}{1 - \sqrt{3}x}\right| + C$ **77.** $\ln\left(\sqrt{e^{2x} + 1} - 1\right) - x + C$

79. $2\sinh^{-1}\sqrt{x} + C = 2\ln\left(\sqrt{x} + \sqrt{1 + x}\right) + C$

81. $\dfrac{1}{4}\ln\left|\dfrac{x - 4}{4}\right| + C$ **83.** $\ln\left(\dfrac{3 + \sqrt{5}}{2}\right)$ **85.** $\dfrac{\ln 7}{12}$

87. $-\dfrac{x^2}{2} - 4x - \dfrac{10}{3}\ln\left|\dfrac{x - 5}{x + 1}\right| + C$

89. $8\arctan e^2 - 2\pi \approx 5.207$ **91.** $\frac{5}{2}\ln\left(\sqrt{17} + 4\right) \approx 5.237$

93. (a) $-\dfrac{\sqrt{a^2 - x^2}}{x}$ (b) Proof

95–103. Proofs   **105.** Putnam Problem 8, 1939

Review Exercises for Chapter 5 *(page 383)*

1. $\frac{4}{3}x^3 + \frac{1}{2}x^2 + 3x + C$ **3.** $\dfrac{x^2}{2} - \dfrac{4}{x^2} + C$

5. $5x - e^x + C$ **7.** $f(x) = 1 - 3x^2$

9. $f(x) = 4x^3 - 5x - 3$

11. (a) 3 sec; 144 ft   (b) $\frac{3}{2}$ sec   (c) 108 ft

13. 60   **15.** $\displaystyle\sum_{i=1}^{10}\dfrac{i}{5(i + 2)}$ **17.** 420   **19.** 3310

21. $s(n) = 11 - \dfrac{2}{n}$, $S(n) = 11 + \dfrac{2}{n}$

23. $A = 15$ **25.** $A = 12$

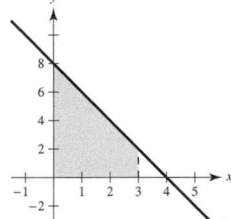   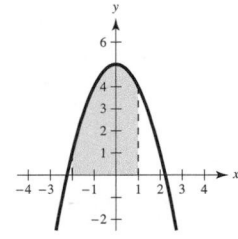

27. 43   **29.** 48

31.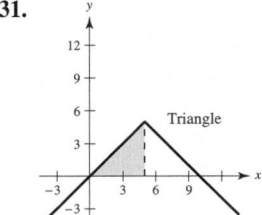

Triangle

$A = \dfrac{25}{2}$

33. (a) 7   (b) 9

35. 56   **37.** $\dfrac{422}{5}$ **39.** $e^2 + 1$ **41.** 30

43. $2\ln 3 \approx 2.1972$ **45.** $\sqrt{\dfrac{13}{3}}$

47. Average value $= \frac{2}{5}$; $x = \dfrac{25}{4}$ **49.** $x^2\sqrt{1 + x^3}$

51. $x^2 + 3x + 2$ **53.** $\frac{2}{3}\sqrt{x^3 + 3} + C$

55. $-\frac{1}{30}(1 - 3x^2)^5 + C = \frac{1}{30}(3x^2 - 1)^5 + C$

57. $-2\sqrt{1 - \sin\theta} + C$ **59.** $-\frac{1}{6}e^{-3x^2} + C$

61. $\frac{1}{2\ln 5}(5^{(x+1)^2}) + C$ **63.** $\frac{455}{2}$ **65.** 2   **67.** $\frac{28\pi}{15}$

69. 2   **71.** 0   **73.** ∞ **75.** 1

77. $1000e^{0.09} \approx 1094.17$ **79.** $\frac{1}{7}\ln|7x - 2| + C$

81. $-\ln|1 + \cos x| + C$ **83.** $\frac{1}{2}\ln(e^{2x} + e^{-2x}) + C$

85. $3 + \ln 2$ **87.** $\ln(2 + \sqrt{3})$ **89.** $\frac{1}{2}\arctan(e^{2x}) + C$

91. $\frac{1}{2}\arcsin x^2 + C$ **93.** $\frac{1}{4}\left(\arctan\frac{x}{2}\right)^2 + C$

95. $-4\,\mathrm{sech}(4x - 1)\tanh(4x - 1)$ **97.** $\dfrac{4}{\sqrt{16x^2 + 1}}$

99. $\frac{1}{3}\tanh x^3 + C$ **101.** $\frac{1}{12}\ln\left|\dfrac{3 + 2x}{3 - 2x}\right| + C$

P.S. Problem Solving   *(page 385)*

1. (a) $L(1) = 0$ (b) $L'(x) = \dfrac{1}{x}$, $L'(1) = 1$

   (c) $x \approx 2.718$ (d) Proof

3. (a) Proof   (b) $\frac{1}{2}$ (c) $\frac{3}{2}$

5. (a) 1.6758; Error of approximation ≈ 0.0071

   (b) $\frac{3}{2}$ (c) Proof

7–9. Proofs   **11.** $\displaystyle\lim_{n\to\infty}\sum_{t=1}^{n}\left(\frac{t}{n}\right)^5\left(\frac{1}{n}\right) = \frac{1}{6}$

13. (a)

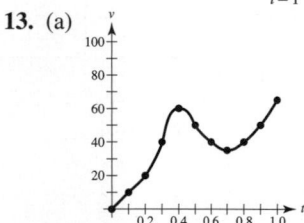

   (b) $(0, 0.4)$ and $(0.7, 1.0)$ (c) 150 mi/h²

   (d) Total distance traveled in miles; 37 mi

   (e) Sample answer: 100 mi/h²

15. (a)–(c) Proofs   **17.** $2\ln\frac{3}{2} \approx 0.8109$

19. (a) (i)

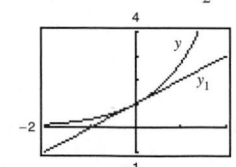

   (ii)

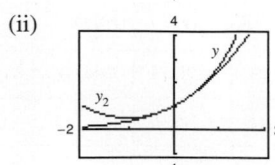

   (iii)

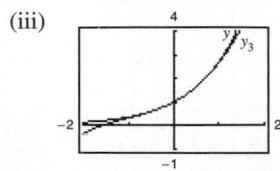

(b) Pattern: $y_n = 1 + \dfrac{x}{1!} + \dfrac{x^2}{2!} + \cdots + \dfrac{x^n}{n!} + \cdots$

$$y_4 = 1 + \frac{x}{1!} + \frac{x^2}{2!} + \frac{x^3}{3!} + \frac{x^4}{4!}$$

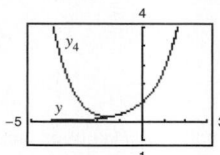

(c) The pattern implies that $e^x = 1 + \dfrac{x}{1!} + \dfrac{x^2}{2!} + \dfrac{x^3}{3!} + \cdots$.

Chapter 6

Section 6.1   *(page 393)*

1. Substitute $f(x)$ and its derivatives into the differential equation. If the equation is satisfied, then $f(x)$ is a solution.

3. The line segments show the general shape of all the solutions of a differential equation and give a visual perspective of the directions of the solutions of the differential equation.

5–13. Proofs   **15.** Solution   **17.** Not a solution

19. Solution   **21.** Not a solution   **23.** Not a solution

25. Solution   **27.** Not a solution

29. Not a solution   **31.** $y = 3e^{-x/2}$ **33.** $4y^2 = x^3$

35.

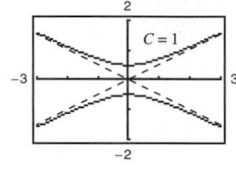

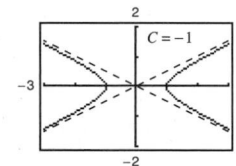

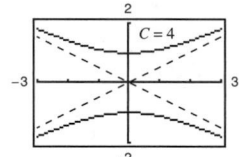

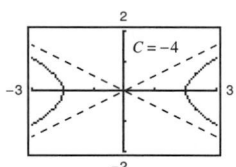

37. $y = 3e^{-6x}$ **39.** $y = 2\sin 3x - \frac{1}{3}\cos 3x$

41. $y = -2x + \frac{1}{2}x^3$ **43.** $4x^3 + C$

45. $y = \frac{1}{2}\ln(1 + x^2) + C$ **47.** $y = -\frac{1}{2}\cos 2x + C$

49. $y = \frac{2}{5}(x - 6)^{5/2} + 4(x - 6)^{3/2} + C$ **51.** $y = \frac{1}{2}e^{x^2} + C$

53.

x	-4	-2	0	2	4	8
y	2	0	4	4	6	8
dy/dx	-4	Undef.	0	1	$\frac{4}{3}$	2

55.

x	-4	-2	0	2	4	8
y	2	0	4	4	6	8
dy/dx	$-2\sqrt{2}$	-2	0	0	$-2\sqrt{2}$	-8

57. b   **58.** c   **59.** d   **60.** a

61. (a) and (b)

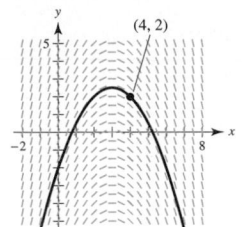

(c) As $x \to \infty$, $y \to -\infty$

   As $x \to -\infty$, $y \to -\infty$

63. (a) and (b)

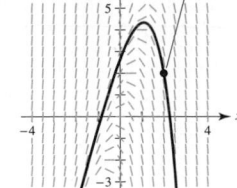

(c) As $x \to \infty$, $y \to -\infty$

   As $x \to -\infty$, $y \to -\infty$

65. (a)

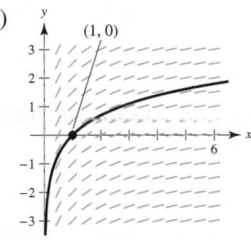

As $x \to \infty$, $y \to \infty$

(b)

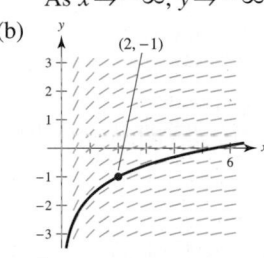

As $x \to \infty$, $y \to \infty$

67. (a) and (b)

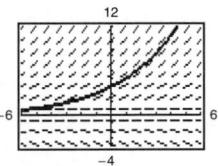

69. (a) and (b)

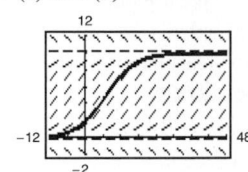

71. (a) and (b)

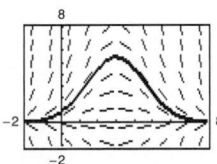

73.

n	0	1	2	3	4	5	6
x_n	0	0.1	0.2	0.3	0.4	0.5	0.6
y_n	2	2.2	2.43	2.693	2.992	3.332	3.715

n	7	8	9	10
x_n	0.7	0.8	0.9	1.0
y_n	4.146	4.631	5.174	5.781

75.

n	0	1	2	3	4	5	6
x_n	0	0.05	0.1	0.15	0.2	0.25	0.3
y_n	3	2.7	2.438	2.209	2.010	1.839	1.693

n	7	8	9	10
x_n	0.35	0.4	0.45	0.5
y_n	1.569	1.464	1.378	1.308

77.

n	0	1	2	3	4	5	6
x_n	0	0.1	0.2	0.3	0.4	0.5	0.6
y_n	1	1.1	1.212	1.339	1.488	1.670	1.900

n	7	8	9	10
x_n	0.7	0.8	0.9	1.0
y_n	2.213	2.684	3.540	5.958

79.

x	0	0.2	0.4	0.6	0.8	1
$y(x)$ (exact)	3.0000	3.6642	4.4755	5.4664	6.6766	8.1548
$y(x)$ ($h = 0.2$)	3.0000	3.6000	4.3200	5.1840	6.2208	7.4650
$y(x)$ ($h = 0.1$)	3.0000	3.6300	4.3923	5.3147	6.4308	7.7812

81.

x	0	0.2	0.4	0.6	0.8	1
$y(x)$ (exact)	0.0000	0.2200	0.4801	0.7807	1.1231	1.5097
$y(x)$ ($h = 0.2$)	0.0000	0.2000	0.4360	0.7074	1.0140	1.3561
$y(x)$ ($h = 0.1$)	0.0000	0.2095	0.4568	0.7418	1.0649	1.4273

83. (a) $y(1) = 112.7141°$, $y(2) = 96.3770°$, $y(3) = 86.5954°$

   (b) $y(1) = 113.2441°$, $y(2) = 97.0158°$, $y(3) = 87.1729°$

   (c) Euler's Method: $y(1) = 112.9828°$, $y(2) = 96.6998°$,

   $\quad\quad\quad\quad\quad\quad\quad y(3) = 86.8863°$

   Exact solution: $y(1) = 113.2441°$, $y(2) = 97.0158°$,

   $\quad\quad\quad\quad\quad\quad\quad y(3) = 87.1729°$

   The approximations are better using $h = 0.05$.

85. Euler's Method produces an exact solution to an initial value problem when the exact solution is a line.

87. False. $y = x^3$ is a solution of $xy' - 3y = 0$, but $y = x^3 + 1$ is not a solution.

89. (a)

x	0	0.2	0.4	0.6	0.8	1
y	4	2.6813	1.7973	1.2048	0.8076	0.5413
y_1	4	2.56	1.6384	1.0486	0.6711	0.4295
y_2	4	2.4	1.44	0.864	0.5184	0.3110
e_1	0	0.1213	0.1589	0.1562	0.1365	0.1118
e_2	0	0.2813	0.3573	0.3408	0.2892	0.2303
r		0.4312	0.4447	0.4583	0.4720	0.4855

(b) If h is halved, then the error is approximately halved because r is approximately 0.5.

(c) The error will again be halved.

91. (a)

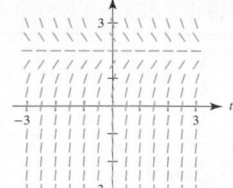

(b) $\lim\limits_{t\to\infty} I(t) = 2$

93. $\omega = \pm 4$ **95.** Putnam Problem B2, 1997

Section 6.2 *(page 402)*

1. C is the initial value of y, and k is the proportionality constant.

3. $y = \frac{1}{2}x^2 + 3x + C$ **5.** $y = Ce^x - 3$

7. $y^2 - 5x^2 = C$ **9.** $y = Ce^{(2x^{3/2})/3}$ **11.** $y = C(1 + x^2)$

13. $\dfrac{dQ}{dt} = \dfrac{k}{t^2}$

$Q = -\dfrac{k}{t} + C$

15. (a)

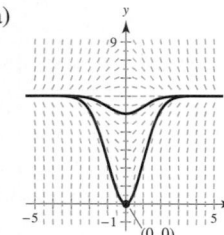

(b) $y = 6 - 6e^{-x^2/2}$

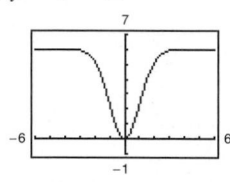

17. $y = \frac{1}{4}t^2 + 10$

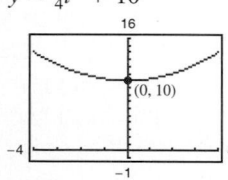

19. $y = 10e^{-t/2}$

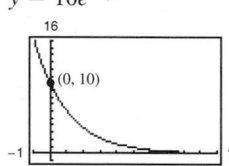

21. $N = 250e^{\ln(8/5)t};\ N = \frac{8192}{5}$

23. $y = 2e^{[(1/4)\ln(3/2)]t} \approx 2e^{0.1014t}$

25. $y = 5\left(\frac{5}{2}\right)^{1/4}e^{[\ln(2/5)/4]t} \approx 6.2872e^{-0.2291t}$

27. Quadrants I and III; $\dfrac{dy}{dx}$ is positive when both x and y are positive (Quadrant I) or when both x and y are negative (Quadrant III).

29. Amount after 1000 yr: 12.96 g
Amount after 10,000 yr: 0.26 g

31. Initial quantity: 7.63 g
Amount after 1000 yr: 4.95 g

33. Amount after 1000 yr: 4.43 g
Amount after 10,000 yr: 1.49 g

35. Initial quantity: 2.16 g
Amount after 10,000 yr: 1.62 g

37. 95.76%

39. Time to double: 5.78 yr
Amount after 10 yr: $3320.12

41. Annual rate: 4.62%
Amount after 10 yr: $238.09

43. Annual rate: 7.18%
Time to double: 9.65 yr

45. $224,174.18   **47.** $61,377.75

49. (a) 10.24 yr   (b) 9.93 yr   (c) 9.90 yr   (d) 9.90 yr

51. (a) $P = 2.113e^{-0.011t}$ (b) 1.70 million people
(c) Because $k < 0$, the population is decreasing.

53. (a) $P = 6.404e^{0.012t}$ (b) 8.14 million people
(c) Because $k > 0$, the population is increasing.

55. (a) $N = 100.1596(1.2455)^t$ (b) 6.3 h

57. (a) $N \approx 30(1 - e^{-0.0502t})$ (b) 36 days

59. (a) Because the population increases by a constant each month, the rate of change from month to month will always be the same. So, the slope is constant, and the model is linear.

(b) Although the percentage increase is constant each month, the rate of growth is not constant. The rate of change of y is $\dfrac{dy}{dt} = ry$, which is an exponential model.

61. (a) $M_1 = 2335.3e^{0.0407t}$ (b) $M_2 = 206.9t + 1685$
(c) The exponential model fits the data better because the graph is closer to the data values than is the graph of the linear model.

(d) 2026 ($t \approx 46$); Yes. The exponential model indicates a reasonably slow growth rate.

63. (a) 20 dB   (b) 70 dB   (c) 120 dB

65. (a) $y = 1420e^{[\ln(52/71)]t} + 80 \approx 1420e^{-0.3114t} + 80$
(b) 299.2°F

67. False. It takes 1599 years.

Section 6.3 *(page 413)*

1. (a) Separable   (b) Not separable

3. $y^2 - x^2 = C$ **5.** $y^4 - 2x^2 + 4x = C$

7. $r = Ce^{(4/9)s}$ **9.** $y = C(x + 2)^3$

11. $y^3 = C - \frac{1}{3}\cos 9x$ **13.** $y = -\frac{1}{4}\sqrt{1 - 4x^2} + C$

15. $y = Ce^{(\ln x)^2/2}$ **17.** $y^2 = 4e^x + 5$

19. $y = e^{-(x^2 + 2x)/2}$ **21.** $y^2 = 4x^2 + 3$

23. $u = e^{(5 - \cos v^2)/2}$ **25.** $P = P_0e^{kt}$

27. $4y^2 - x^2 = 16$ **29.** $y = \frac{1}{3}\sqrt{x}$ **31.** $f(x) = Ce^{-x/2}$

33.

[figure: slope field with parabola curves]

$y = \frac{1}{2}x^2 + C$

35.

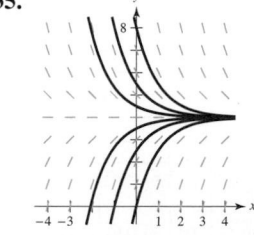

$y = 4 + Ce^{-x}$

37. (a) $y = 0.1602$ (b) $y = 5e^{-3x^2}$ (c) $y = 0.2489$

39. 97.9% of the original amount

41. (a) $\dfrac{dy}{dx} = k(y - 4)$ (b) i   (c) Proof

42. (a) $\dfrac{dy}{dx} = k(x - 4)$ (b) ii   (c) Proof

43. (a) $\dfrac{dy}{dx} = ky(y - 4)$ (b) iii   (c) Proof

44. (a) $\dfrac{dy}{dx} = ky^2$ (b) iv   (c) Proof

45. (a) $w = 1200 - 1140e^{-kt}$

(b) $w = 1200 - 1140e^{-0.8t}$ $w = 1200 - 1140e^{-0.9t}$

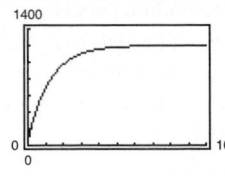   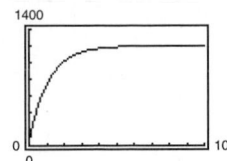

$w = 1200 - 1140e^{-t}$

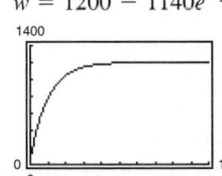

(c) 1.31 yr; 1.16 yr; 1.05 yr   (d) 1200 lb

47. Hyperbolas: $3x^2 - y^2 = C$

Orthogonal trajectory: $y = \dfrac{K}{\sqrt[3]{x}}$

Graphs will vary.

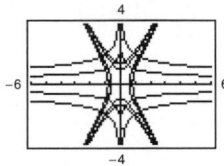

49. Parabolas: $x^2 = Cy$ **51.** Curves: $y^2 = Cx^3$
Ellipses: $x^2 + 2y^2 = K$ Ellipses: $2x^2 + 3y^2 = K$
Graphs will vary.   Graphs will vary.

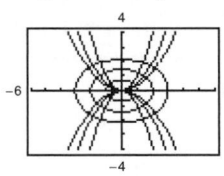   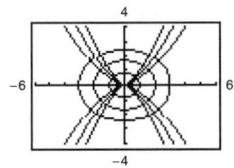

53. $N = \dfrac{500}{1 + 4e^{-0.2452t}}$ **55.** $y = 1 - e^{-1.386t}$

57. $y = \dfrac{360}{8 + 41t}$ **59.** $y = 500e^{-1.6094e^{-0.1451t}}$

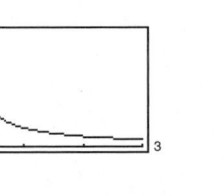

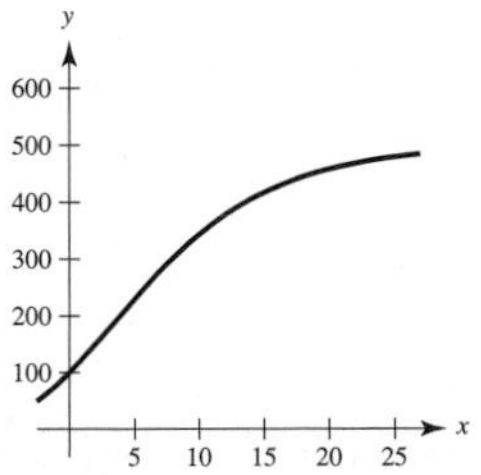

61. 34 beavers   **63.** 92%
65. (a) $Q = 25e^{-(1/20)t}$ (b) About 10.2 min
67. (a) $y = Ce^{kt}$ (b) About 6.2 h   **69.** About 3.15 h

71. $P = Ce^{kt} - \dfrac{N}{k}$ **73.** $A = \dfrac{P}{r}(e^{rt} - 1)$

75. \$23,981,015.77

77. (a)

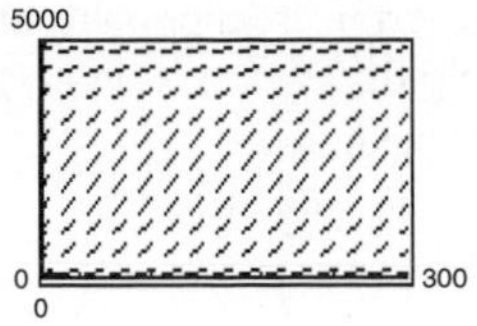

The graph at top right shows a slope field.

(b) As $t \to \infty$, $y \to L$.
(c) $y = 5000e^{-2.303e^{-0.02t}}$

(d)

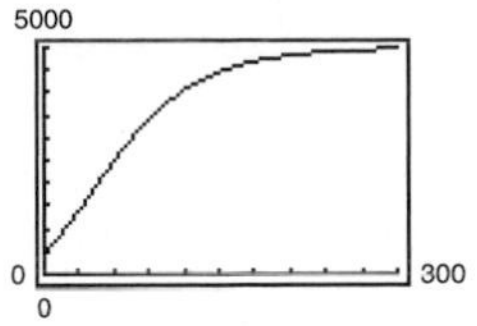

The graph is concave upward on $(0, 41.7)$ and concave downward on $(41.7, \infty)$.

79. Yes. Rewrite the equation as $\dfrac{1}{g(y) - h(y)}\, dy = f(x)\, dx$.

81. Separable; $\dfrac{1}{y}\, dy = -\dfrac{(1 + x)}{x}\, dx$

83. Not separable
85. (a) $v = 20(1 - e^{-1.386t})$
(b) $s \approx 20t + 14.43(e^{-1.386t} - 1)$
87. Homogeneous of degree 3
89. Homogeneous of degree 0   **91.** Not homogeneous
93. Homogeneous of degree 0   **95.** $|x| = C(x - y)^2$
97. $|y^2 + 2xy - x^2| = C$ **99.** $y = Ce^{-x^2/(2y^2)}$
101. False. $y' = x/y$ is separable, but $y = 0$ is not a solution.
103. True

Section 6.4   *(page 422)*

1. The carrying capacity is the maximum population that can be sustained over time.
3. d   **4.** a   **5.** b   **6.** c
7. $y(0) = 4$ **9.** $y(0) = \frac{12}{7}$
11. (a) 0.75   (b) 2100   (c) 70   (d) 4.49 yr

(e) $\dfrac{dP}{dt} = 0.75P\left(1 - \dfrac{P}{2100}\right)$

13. (a) 0.8   (b) 6000   (c) 1.2   (d) 10.65 yr

(e) $\dfrac{dP}{dt} = 0.8P\left(1 - \dfrac{P}{6000}\right)$

15. (a) 3   (b) 100   **17.** (a) 0.1   (b) 250
(c)   (c)

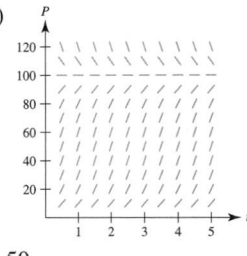   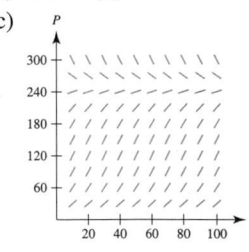

(d) 50   (d) 125

19. $y = \dfrac{36}{1 + 8e^{-t}}$; 34.16; 36

21. $y = \dfrac{120}{1 + 14e^{-0.8t}}$; 95.51; 120

23. c   **24.** d   **25.** b   **26.** a

27. (a) 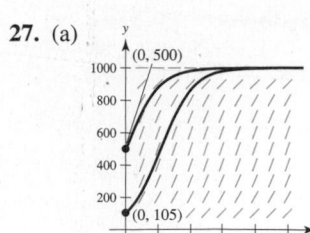   (b) $y = \dfrac{1000}{1 + (179/21)e^{-0.2t}}$

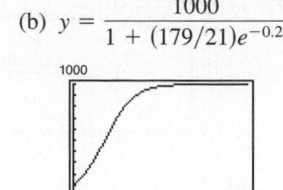

29. No, it is not possible to determine b. However, $L = 2500$ and $k = 0.75$. You need an initial condition to determine b.

31. Proof

33. (a) $P = \dfrac{200}{1 + 7e^{-0.2640t}}$ (b) 70 panthers   (c) 7.37 yr

(d) $\dfrac{dP}{dt} = 0.2640P\left(1 - \dfrac{P}{200}\right)$; 69.25 panthers

(e) 100 yr

35. False. $\dfrac{dy}{dt} < 0$ and the population decreases to approach L.

37. $y = 0$, $y = L$; A population of 0 or the carrying capacity L will not change.

Section 6.5   *(page 428)*

1. The derivative in the equation is first order.

3. Linear; can be written in the form $\dfrac{dy}{dx} + P(x)y = Q(x)$

5. Not linear; cannot be written in the form $\dfrac{dy}{dx} + P(x)y = Q(x)$

7. $y = 2x^2 + x + \dfrac{C}{x}$ **9.** $y = 5 + Ce^{-x^2}$

11. $y = -1 + Ce^{\sin x}$ **13.** $y = \frac{1}{6}e^{3x} + Ce^{-3x}$

15. (a) Answers will vary.   (b) $y = \frac{1}{2}(e^x + e^{-x})$
Sample answer:

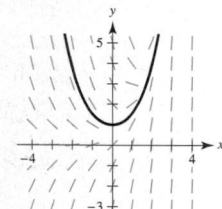

(c)

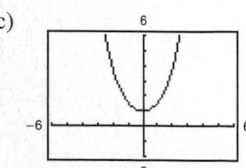

17. $y = 3e^x$ **19.** $y = \sin x + (x + 1)\cos x$ **21.** $xy = 4$

23. $y = -2 + x \ln|x| + 12x$ **25.** $P = -\dfrac{N}{k} + \left(\dfrac{N}{k} + P_0\right)e^{kt}$

27. (a) $\$4,212,796.94$ (b) $\$31,424,909.75$

29. (a) $\dfrac{dN}{dt} = k(75 - N)$ (b) $N = 75 + Ce^{-kt}$

(c) $N = 75 - 55.9296e^{-0.0168t}$

31. $v(t) = -49.1(1 - e^{-0.1996t})$; -49.1 m/sec

33. $I = \dfrac{E_0}{R} + Ce^{-Rt/L}$ **35.** Proof

37. (a) $Q = 25e^{-t/20}$ (b) $-20 \ln\left(\frac{3}{5}\right) \approx 10.2$ min   (c) 0

39. a   **41.** Use separation of variables or an integrating factor.

43. c   **44.** d   **45.** a   **46.** b

47. (a) 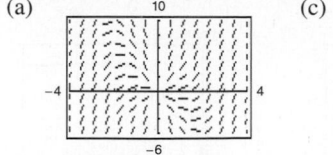   (c)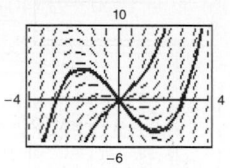

(b) $(-2, 4)$: $y = \frac{1}{2}x(x^2 - 8)$
$(2, 8)$: $y = \frac{1}{2}x(x^2 + 4)$

49. $2e^x + e^{-2y} = C$ **51.** $y = Ce^{-\sin x} + 1$

53. $y = \dfrac{e^x(x - 1) + C}{x^2}$ **55.** $y = \dfrac{12}{5}x^2 + \dfrac{C}{x^3}$

57. $\dfrac{1}{y^2} = Ce^{2x^3} + \dfrac{1}{3}$ **59.** $y = \dfrac{1}{Cx - x^2}$

61. $\dfrac{1}{y^2} = 2x + Cx^2$ **63.** $y^{2/3} = 2e^x + Ce^{2x/3}$

65. False. $y' + xy = x^2$ is linear.

Section 6.6   *(page 436)*

1. An autonomous differential equation does not depend explicitly on time t.

3. $\dfrac{dx}{dt} = 0.9x + 0.05xy$,

$\dfrac{dy}{dt} = -0.6y + 0.008xy$; $(0, 0)$ and $(75, 18)$

5. (a) 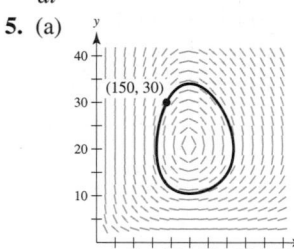   (b)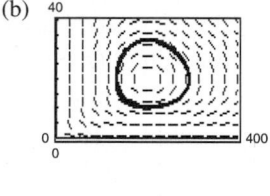

7. (a) $(40, 20)$ **9.** $(0, 0)$, $(50, 20)$

(b)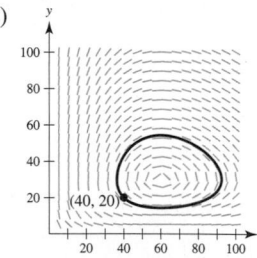

11. 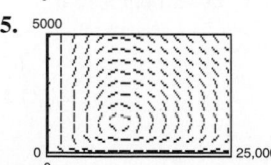   **13.** $(0, 0)$, $(10{,}000, 1250)$

15.

17. 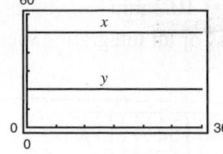   As t increases, both x and y are constant.

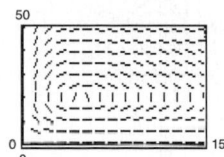

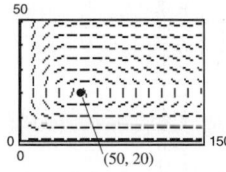

The solution curve reduces to a single point at $(50, 20)$.

19. $\dfrac{dx}{dt} = 2x - 3x^2 - 2xy$, $\dfrac{dy}{dt} = 2y - 3y^2 - 2xy$;

$(0, 0)$, $\left(\dfrac{2}{5}, \dfrac{2}{5}\right)$, $\left(0, \dfrac{2}{3}\right)$, and $\left(\dfrac{2}{3}, 0\right)$

21. $\dfrac{dx}{dt} = 0.15x - 0.6x^2 - 0.75xy$,

$\dfrac{dy}{dt} = 0.15y - 1.2y^2 - 0.45xy$;

$(0, 0)$, $\left(0, \dfrac{1}{8}\right)$, $\left(\dfrac{1}{4}, 0\right)$, and $\left(\dfrac{3}{17}, \dfrac{1}{17}\right)$

23. $(0, 0)$, $(0, 0.5)$, $(2, 0)$, and $\left(\dfrac{45}{23}, \dfrac{4}{23}\right)$

25. $(0, 0)$, $(0, 0.5)$, $(2, 0)$, and $\left(-\dfrac{9}{38}, \dfrac{17}{19}\right)$

27.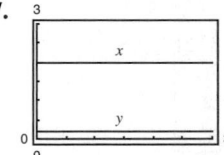

As t increases, both x and y are constant.

29. Use the coordinates of a critical point as the initial values.

31. (a) $\dfrac{dx}{dt} = ax\left(1 - \dfrac{x}{L}\right)$. The equation is logistic.

(b) $\dfrac{dx}{dt} = 0.4x\left(1 - \dfrac{x}{100}\right) - 0.01xy$,

$\dfrac{dy}{dt} = -0.3y + 0.005xy$

Critical points: $(0, 0)$, $(60, 16)$, $(100, 0)$

(c) 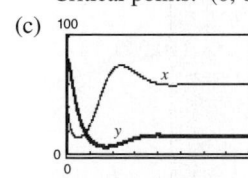   (d)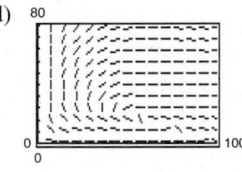

Answers will vary.

(e)

Answers will vary.

1. Solution   **3.** $y = \frac{4}{3}x^3 + 7x + C$ **5.** $y = \frac{1}{2}\sin 2x + C$

7. $y = -e^{2-x} + C$

9.

x	-4	-2	0	2	4	8
y	2	0	4	4	6	8
dy/dx	-10	-4	-4	0	2	8

11. (a) and (b)

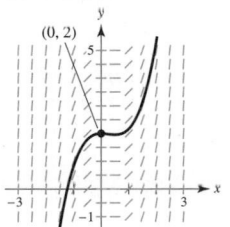

13.

n	0	1	2	3	4	5
x_n	0	0.05	0.1	0.15	0.2	0.25
y_n	4	3.8	3.6125	3.4369	3.2726	3.1190

n	6	7	8	9	10
x_n	0.3	0.35	0.4	0.45	0.5
y_n	2.9756	2.8418	2.7172	2.6038	2.4986

15. $y = 3x^2 - \dfrac{1}{4}x^4 + C$ **17.** $y = 1 - \dfrac{1}{x + C}$

19. $y = \dfrac{Ce^x}{(2 + x)^2}$ **21.** $y = Ce^{2\sqrt{x+1}}$

23. $\dfrac{dy}{dt} = \dfrac{k}{t^3}$, $y = -\dfrac{k}{2t^2} + C$ **25.** $y \approx \dfrac{3}{4}e^{0.379t}$

27. $y = \frac{9}{20}e^{(1/2)\ln(10/3)t}$ **29.** About 7.79 in.

31. About 37.5 yr

33. (a) $S \approx 30e^{-1.7918/t}$ (b) 20,965 units

35. $y^2 = 5x^2 + C$ **37.** $y = -\ln\left(C - \dfrac{e^{4x}}{4}\right)$

39. $y^4 = 6x^2 - 8$ **41.** $y^4 = 2x^4 + 1$

43. $4x^2 + y^2 = C$

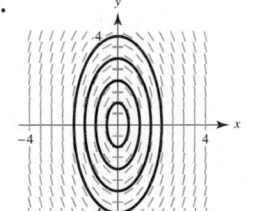

45. $y^2 = 2x^2 + 7$

47. Hyperbolas: $5x^2 - 4y^2 = C$

Orthogonal trajectory: $y = Kx^{-4/5}$

Graphs will vary.

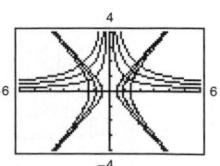

49. (a) $k = 0.55$ (b) 5250   (c) 150   (d) 6.41 yr

(e) $\dfrac{dP}{dt} = 0.55P\left(1 - \dfrac{P}{5250}\right)$

51. $y = \dfrac{80}{1 + 9e^{-t}}$ **53.** 184 racoons

55. $\dfrac{dS}{dt} = k(L - S)$; $S = L(1 - e^{-kt})$

57. $\dfrac{dP}{dn} = kP(L - P)$; $P = \dfrac{CL}{e^{-Lkn} + C}$

59. $y = -10 + Ce^x$ **61.** $y = e^{x/4}\left(\frac{1}{4}x + C\right)$

63. $y = \dfrac{x + C}{x - 2}$ **65.** $y = \dfrac{1}{10}e^{5x} + Ce^{-5x}$

67. (a) Answers will vary.   (b) $y = \frac{1}{3}(2e^{x/2} - 5e^{-x})$
Sample answer: (c)

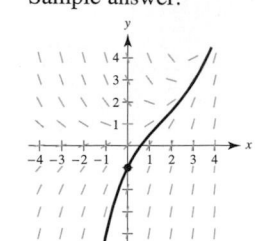   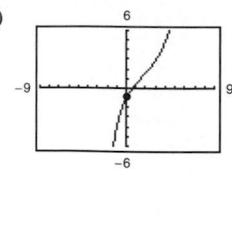

69. (a) Answers will vary.   (b) $y = -\cos x + 1.8305 \sin x$
Sample answer: (c)

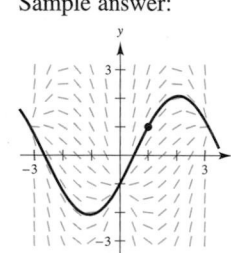   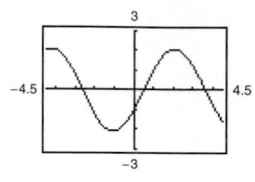

71. $y = \frac{1}{10}e^{5x} + \frac{29}{10}e^{-5x}$

73. $y = -\dfrac{5}{3} + \dfrac{5}{3}e^{3\sin x}$ **75.** $A = \dfrac{P}{r} + \left(A_0 - \dfrac{P}{r}\right)e^{rt}$

77. $v(t) = -166.67(1 - e^{-0.192t})$; -166.67 ft/sec

79. (a) Prey: $\dfrac{dx}{dt} = 0.3x - 0.02xy$

Predator: $\dfrac{dy}{dt} = -0.4y + 0.01xy$

(b) $(0, 0)$ and $(40, 14)$
(c)

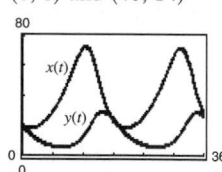

As t increases, both x and y oscillate.

81. (a) Species 1: $\dfrac{dx}{dt} = 3x - x^2 - xy$

Species 2: $\dfrac{dy}{dt} = 2y - y^2 - 0.5xy$

(b) $(0, 0)$, $(0, 2)$, $(3, 0)$, and $(2, 1)$
(c)

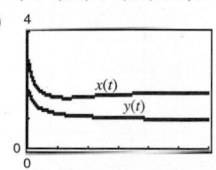

As t increases, x remains constant at approximately 2 and y remains constant at approximately 1.

P.S. Problem Solving *(page 441)*

1. (a) $y = \dfrac{1}{(1 - 0.01t)^{100}}$; $T = 100$

(b) $y = 1 \Big/ \left[\left(\dfrac{1}{y_0}\right)^{\varepsilon} - k\varepsilon t\right]^{1/\varepsilon}$; Explanations will vary.

3. (a) $(0, 1)$, $(0.1, 0.91)$, $(0.2, 0.83805)$, $(0.3, 0.78244)$, $(0.4, 0.74160)$, $(0.5, 0.71415)$, $(0.6, 0.69881)$, $(0.7, 0.69442)$, $(0.8, 0.69995)$, $(0.9, 0.71446)$, $(1, 0.73708)$

(b)

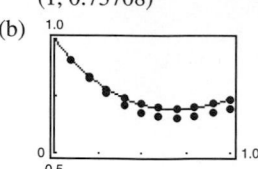

The modified Euler Method is more accurate.

5. 1481.45 sec $\approx$ 24 min, 41 sec
7. 2575.95 sec $\approx$ 42 min, 56 sec
9. (a) $s = 184.21 - Ce^{-0.019t}$

(b)

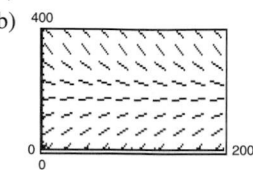

(c) As $t \to \infty$, $Ce^{-0.019t} \to 0$, and $s \to 184.21$.

11. (a) $C = C_0 e^{-Rt/V}$ (b) 0
13. (a) $C = \dfrac{Q}{R}(1 - e^{-Rt/V})$ (b) $\dfrac{Q}{R}$

Chapter 7

Section 7.1   *(page 450)*

1. In variable x, the area of the region between two graphs is the area under the graph of the top function minus the area under the graph of the bottom function.

3. The points of intersection are used to determine the vertical lines that bound the region.

5. $-\displaystyle\int_0^6 (x^2 - 6x)\, dx$ **7.** $\displaystyle\int_0^3 (-2x^2 + 6x)\, dx$

9. $-6\displaystyle\int_0^1 (x^3 - x)\, dx$

11.   **13.**

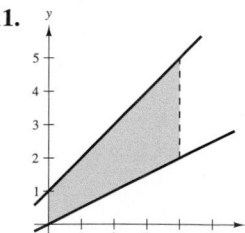   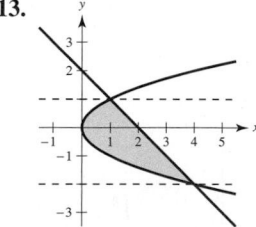

15.   **17.**

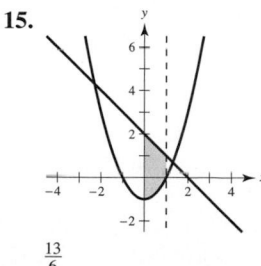   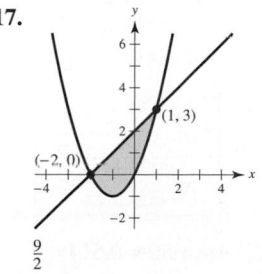

$\dfrac{13}{6}$ $\dfrac{9}{2}$

19.

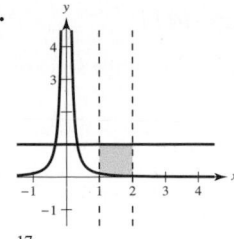

$\frac{17}{18}$

21.

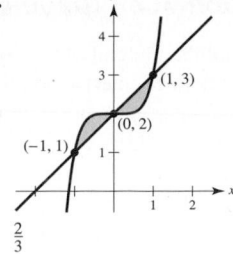

$\frac{2}{3}$

23.

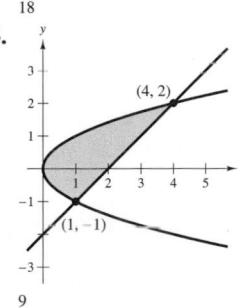

$\frac{9}{2}$

25.

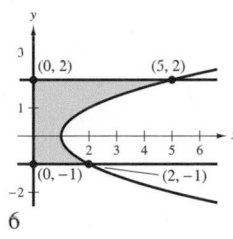

6

27.

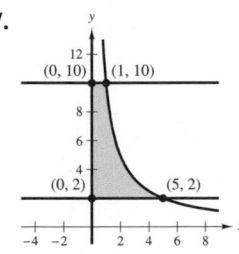

$10 \ln 5 \approx 16.094$

29. (a) $\frac{125}{6}$ (b) $\frac{125}{6}$
(c) Integrating with respect to y; Answers will vary.

31. (a)

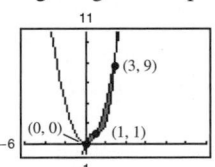

(b) $\frac{37}{12}$

33. (a)

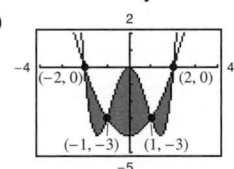

(b) 8

35. (a)

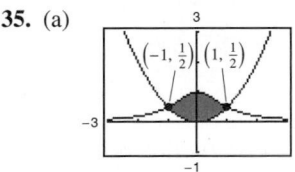

(b) $\frac{\pi}{2} - \frac{1}{3} \approx 1.237$

37.

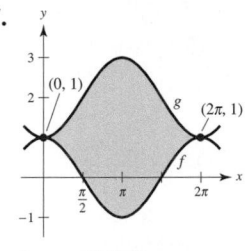

$4\pi \approx 12.566$

39.

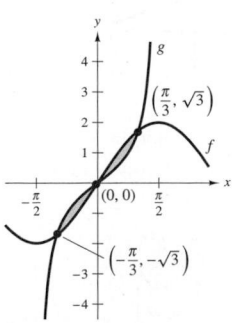

$2(1 - \ln 2) \approx 0.614$

41.

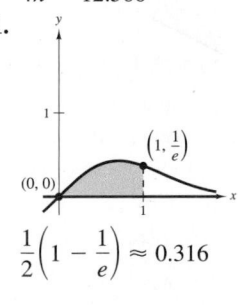

$\frac{1}{2}\left(1 - \frac{1}{e}\right) \approx 0.316$

43. (a)

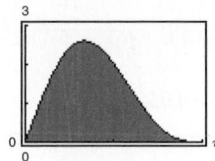

(b) 4

45. (a)

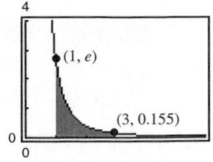

(b) About 1.323

47. (a)

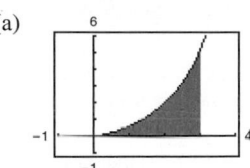

(b) The function is difficult
to integrate.
(c) About 4.7721

49. (a)

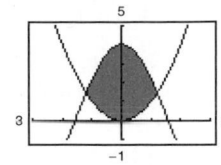

(b) The intersections are
difficult to find.
(c) About 6.3043

51. 2

53. $F(x) = \frac{1}{4}x^2 + x$
(a) $F(0) = 0$

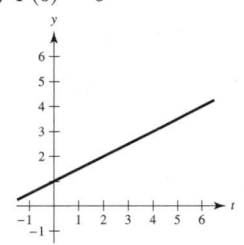

(b) $F(2) = 3$

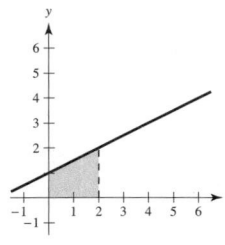

(c) $F(6) = 15$

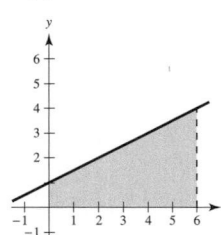

55. $F(\alpha) = \frac{2}{\pi}\left(\sin \frac{\pi\alpha}{2} + 1\right)$

(a) $F(-1) = 0$

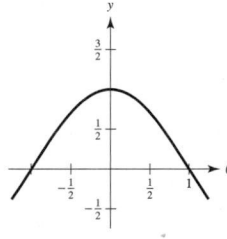

(b) $F(0) = \frac{2}{\pi} \approx 0.6366$

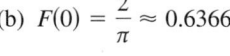

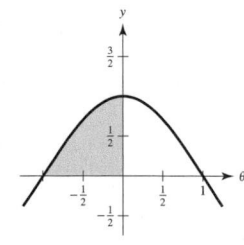

(c) $F\left(\frac{1}{2}\right) = \frac{\sqrt{2} + 2}{\pi} \approx 1.0868$

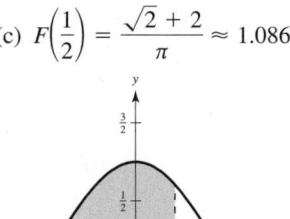

57. 3   **59.** 16   **61.** $\int_{-2}^{1} [(2x^3 - 1) - (6x - 5)]\, dx = \frac{27}{2}$

63. $\int_{0}^{1} \left[\frac{1}{x^2 + 1} - \left(-\frac{1}{2}x + 1\right) \right] dx \approx 0.0354$

65. Answers will vary.

Example: $x^4 - 2x^2 + 1 \le 1 - x^2$ on $[-1, 1]$

$\int_{-1}^{1} [(1 - x^2) - (x^4 - 2x^2 + 1)]\, dx = \frac{4}{15}$

67. (a) The integral $\int_{0}^{5} [v_1(t) - v_2(t)]\, dt = 10$ means that the first car traveled 10 more meters than the second car between 0 and 5 seconds.

The integral $\int_{0}^{10} [v_1(t) - v_2(t)]\, dt = 30$ means that the first car traveled 30 more meters than the second car between 0 and 10 seconds.

The integral $\int_{20}^{30} [v_1(t) - v_2(t)]\, dt = -5$ means that the second car traveled 5 more meters than the first car between 20 and 30 seconds.

(b) No. You do not know when both cars started or the initial distance between the cars.

(c) The car with velocity v_1 is ahead by 30 meters.

(d) Car 1 is ahead by 8 meters.

69. $b = 9\left(1 - \frac{1}{\sqrt[3]{4}}\right) \approx 3.330$ **71.** $a = 4 - 2\sqrt{2} \approx 1.172$

73. Answers will vary. Sample answer: $\frac{1}{6}$

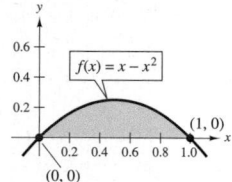

75. R_1; \$1.625 million

77. (a) $y = 0.0124x^2 - 0.385x + 7.85$

(b)

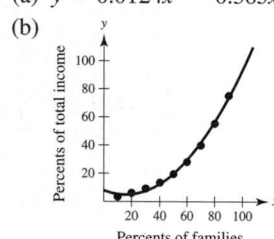

(c)

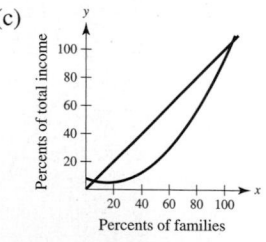

For $6 \le x \le 100$, the values of y are larger for the model $y = x$.

(d) About 2006.7

79. (a) About 6.031 m²   (b) About 12.062 m³   (c) 60,310 lb

81. $\frac{\sqrt{3}}{2} + \frac{7\pi}{24} + 1 \approx 2.7823$ **83.** True

85. False. Let $f(x) = x$ and $g(x) = 2x - x^2$. Then f and g intersect at $(1, 1)$, the midpoint of $[0, 2]$, but

$\int_{a}^{b} [f(x) - g(x)]\, dx = \int_{0}^{2} [x - (2x - x^2)]\, dx = \frac{2}{3} \neq 0.$

87. Putnam Problem A1, 1993

Section 7.2   *(page 461)*

1. Find the integral of the square of the radius of the solid over the defined interval and then multiply by π.

3. When the solid of revolution is formed by two or more distinct solids.

5. $\pi \int_{1}^{4} (\sqrt{x})^2\, dx = \frac{15\pi}{2}$ **7.** $\pi \int_{0}^{1} [(x^2)^2 - (x^5)^2]\, dx = \frac{6\pi}{55}$

9. $\pi \int_{0}^{4} (\sqrt{y})^2\, dy = 8\pi$ **11.** $\pi \int_{0}^{1} (y^{3/2})^2\, dy = \frac{\pi}{4}$

13. (a) $\frac{9\pi}{2}$ (b) $\frac{36\pi\sqrt{3}}{5}$ (c) $\frac{24\pi\sqrt{3}}{5}$ (d) $\frac{84\pi\sqrt{3}}{5}$

15. (a) $\frac{32\pi}{3}$ (b) $\frac{64\pi}{3}$ **17.** 18π **19.** $\pi\left(16 \ln 5 - \frac{16}{5}\right)$

21. $\frac{124\pi}{3}$ **23.** $\frac{832\pi}{15}$ **25.** $\frac{\pi}{3} \ln \frac{11}{5}$ **27.** 24π

29. $\pi\left(\frac{1 - e^{-12}}{6}\right)$ **31.** $\frac{277\pi}{3}$ **33.** 8π **35.** $\frac{25\pi}{2}$

37. $\frac{\pi^2}{2} \approx 4.935$ **39.** $\frac{\pi}{2}(e^2 - 1) \approx 10.036$ **41.** $\frac{\pi}{3}$

43. $\frac{\pi}{3}$ **45.** $\frac{2\pi}{15}$ **47.** $\frac{\pi}{2}$ **49.** 1.969   **51.** 15.4115

53. (a) A sine curve on $\left[0, \frac{\pi}{2}\right]$ revolved about the x-axis

(b) A polynomial function on $[2, 4]$ revolved about the y-axis

55. $b < c < a$ **57.** $\sqrt{5}$ **59.** $V = \frac{4}{3}\pi(R^2 - r^2)^{3/2}$

61. Proof   **63.** $\pi r^2 h\left(1 - \frac{h}{H} + \frac{h^2}{3H^2}\right)$

65.

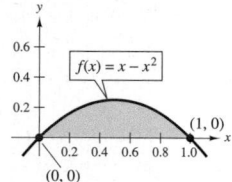

$\frac{\pi}{30}$

67. (a) 60π (b) 50π

69. (a) $V = \pi\left(4b^2 - \frac{64}{3}b + \frac{512}{15}\right)$

(b)

(c) $b = \frac{8}{3} \approx 2.67$

$b \approx 2.67$

71. (a) ii; right circular cylinder of radius r and height h

(b) iv; ellipsoid whose underlying ellipse has the equation

$\left(\frac{x}{b}\right)^2 + \left(\frac{y}{a}\right)^2 = 1$

(c) iii; sphere of radius r

(d) i; right circular cone of radius r and height h

(e) v; torus of cross-sectional radius r and other radius R

73. (a) $\frac{81}{10}$ (b) $\frac{9}{2}$ **75.** $\frac{16}{3}r^3$

77. (a) $\frac{2}{3}r^3$ (b) $\frac{2}{3}r^3 \tan\theta$; As $\theta \to 90°$, $V \to \infty$.

Section 7.3 (page 470)

1. Determine the distance from the center of a representative rectangle to the axis of revolution, and find the height of the rectangle. Then use the formula $V = 2\pi \int_c^d p(y)h(y)\,dy$ for a horizontal axis of revolution or $V = 2\pi \int_a^b p(x)h(x)\,dx$ for a vertical axis of revolution.

3. $2\pi \displaystyle\int_0^2 x^2\,dx = \dfrac{16\pi}{3}$ **5.** $2\pi \displaystyle\int_0^4 x\sqrt{x}\,dx = \dfrac{128\pi}{5}$

7. $2\pi \displaystyle\int_0^4 \dfrac{1}{4}x^3\,dx = 32\pi$ **9.** $2\pi \displaystyle\int_0^2 x(4x - 2x^2)\,dx = \dfrac{16\pi}{3}$

11. $2\pi \displaystyle\int_{5/2}^4 x\sqrt{2x - 5}\,dx = \dfrac{34\pi\sqrt{3}}{5}$

13. $2\pi \displaystyle\int_0^2 y(2 - y)\,dy = \dfrac{8\pi}{3}$

15. $2\pi \left[\displaystyle\int_0^{1/2} y\,dy + \int_{1/2}^1 y\left(\dfrac{1}{y} - 1\right) dy \right] = \dfrac{\pi}{2}$

17. $2\pi \displaystyle\int_0^8 y^{4/3}\,dy = \dfrac{768\pi}{7}$ **19.** $2\pi \displaystyle\int_0^2 y(4 - 2y)\,dy = \dfrac{16\pi}{3}$

21. $\displaystyle\int_0^1 y(y^2 - 3y + 2)\,dy = \dfrac{\pi}{2}$

23. 8π **25.** $\dfrac{45\pi}{16}$

27. Shell method; It is much easier to put x in terms of y rather than vice versa.

29. (a) $\dfrac{128\pi}{7}$ (b) $\dfrac{64\pi}{5}$ (c) $\dfrac{96\pi}{5}$

31. (a) $\dfrac{\pi a^3}{15}$ (b) $\dfrac{\pi a^3}{15}$ (c) $\dfrac{4\pi a^3}{15}$

33. (a) 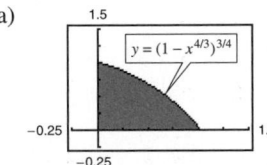   (b) 1.506

35. (a) 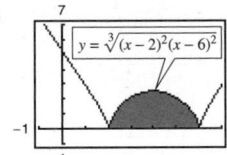   (b) 187.25

37. (a) Height: b, radius: k (b) Height: k, radius: b

39. Both integrals yield the volume of the solid generated by revolving the region bounded by the graphs of $y = \sqrt{x - 1}$, $y = 0$, and $x = 5$ about the x-axis.

41. a, c, b

43. (a) Region bounded by $y = x^2$, $y = 0$, $x = 0$, $x = 2$
    (b) Revolved about the y-axis

45. (a) Region bounded by $x = \sqrt{6 - y}$, $y = 0$, $x = 0$
    (b) Revolved about $y = -2$

47. Diameter $= 2\sqrt{4 - 2\sqrt{3}} \approx 1.464$ **49.** $4\pi^2$

51. (a) Proof   (b) (i) $V = 2\pi$ (ii) $V = 6\pi^2$ **53.** Proof

55. (a) $R_1(n) = \dfrac{n}{n + 1}$ (b) $\displaystyle\lim_{n\to\infty} R_1(n) = 1$

(c) $V = \pi ab^{n+2}\left(\dfrac{n}{n + 2}\right); \ R_2(n) = \dfrac{n}{n + 2}$

(d) $\displaystyle\lim_{n\to\infty} R_2(n) = 1$

(e) As $n \to \infty$, the graph approaches the line $x = b$.

57. About 121,475 ft³   **59.** $c = 2$

61. (a) $\dfrac{64\pi}{3}$ (b) $\dfrac{2048\pi}{35}$ (c) $\dfrac{8192\pi}{105}$

Section 7.4 (page 481)

1. The graph of a function f is rectifiable between $(a, f(a))$ and $(b, f(b))$ if f' is continuous on $[a, b]$.

3. Answers will vary by a constant. Sample answer: $f(x) = 2x^2$

5. (a) and (b) $\sqrt{13}$ **7.** $\dfrac{5}{3}$ **9.** $\dfrac{2}{3}(2\sqrt{2} - 1) \approx 1.219$

11. $5\sqrt{5} - 2\sqrt{2} \approx 8.352$ **13.** 309.3195

15. $\ln \dfrac{\sqrt{2} + 1}{\sqrt{2} - 1} \approx 1.763$ **17.** $\dfrac{1}{2}\left(e^2 - \dfrac{1}{e^2}\right) \approx 3.627$

19. $\dfrac{76}{3}$

21. (a)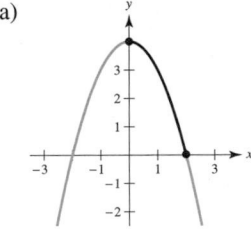

(b) $\displaystyle\int_0^2 \sqrt{1 + 4x^2}\,dx$

(c) About 4.647

23. (a)

(b) $\displaystyle\int_1^3 \sqrt{1 + \dfrac{1}{x^4}}\,dx$

(c) About 2.147

25. (a)

(b) $\displaystyle\int_0^\pi \sqrt{1 + \cos^2 x}\,dx$

(c) About 3.820

27. (a)

(b) $\displaystyle\int_0^1 \sqrt{1 + \left(\dfrac{2}{1 + x^2}\right)^2}\,dx$

(c) About 1.871

29. (a)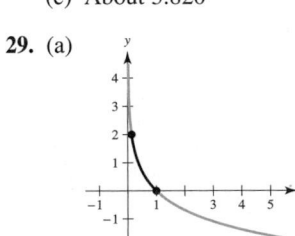

(b) $\displaystyle\int_0^2 \sqrt{1 + e^{-2y}}\,dy$

$= \displaystyle\int_{e^{-2}}^1 \sqrt{1 + \dfrac{1}{x^2}}\,dx$

(c) About 2.221

31. (a) 64.125   (b) 64.525   (c) 64.672

33. $\dfrac{20(e^2 - 1)}{e} \approx 47.0$ m   **35.** About 1480

37. $3 \arcsin \dfrac{2}{3} \approx 2.1892$

39. $2\pi \displaystyle\int_0^3 \dfrac{1}{3}x^3 \sqrt{1 + x^4}\,dx = \dfrac{\pi}{9}(82\sqrt{82} - 1) \approx 258.85$

41. $2\pi \int_{1}^{2} \left(\dfrac{x^3}{6} + \dfrac{1}{2x} \right) \left(\dfrac{x^2}{2} + \dfrac{1}{2x^2} \right) dx = \dfrac{47\pi}{16} \approx 9.23$

43. $2\pi \int_{-1}^{1} 2 \, dx = 8\pi \approx 25.13$

45. $2\pi \int_{1}^{8} x \sqrt{1 + \dfrac{1}{9x^{4/3}}} \, dx = \dfrac{\pi}{27}\left(145\sqrt{145} - 10\sqrt{10} \right) \approx 199.48$

47. $2\pi \int_{0}^{2} x \sqrt{1 + \dfrac{x^2}{4}} \, dx = \dfrac{\pi}{3}\left(16\sqrt{2} - 8 \right) \approx 15.318$

49. 14.424   **51.** b   **53.** They have the same value.

55. (a)

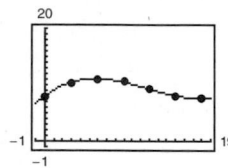

(b) y_1, y_2, y_3, y_4
(c) $s_1 \approx 5.657$, $s_2 \approx 5.759$, $s_3 \approx 5.916$, $s_4 \approx 6.063$

57. 20π **59.** $6\pi(3 - \sqrt{5}) \approx 14.40$

61. (a) Answers will vary. Sample answer: 5207.62 in.³
(b) Answers will vary. Sample answer: 1168.64 in.²
(c) $r = 0.0040y^3 - 0.142y^2 + 1.23y + 7.9$

(d) $V = 5279.64$ in.³, $S = 1179.5$ in.²

63. (a) $\pi\left(1 - \dfrac{1}{b} \right)$ (b) $2\pi \int_{1}^{b} \dfrac{\sqrt{x^4 + 1}}{x^3} \, dx$

(c) $\displaystyle\lim_{b \to \infty} V = \lim_{b \to \infty} \pi\left(1 - \dfrac{1}{b} \right) = \pi$

(d) Because $\dfrac{\sqrt{x^4 + 1}}{x^3} > \dfrac{\sqrt{x^4}}{x^3} = \dfrac{1}{x} > 0$ on $[1, b]$,

you have $\displaystyle\int_{1}^{b} \dfrac{\sqrt{x^4 + 1}}{x^3} \, dx > \int_{1}^{b} \dfrac{1}{x} \, dx = \left[\ln x \right]_{1}^{b} = \ln b$

and $\displaystyle\lim_{b \to \infty} \ln b \to \infty$. So, $\displaystyle\lim_{b \to \infty} 2\pi \int_{1}^{b} \dfrac{\sqrt{x^4 + 1}}{x^3} \, dx = \infty$.

65. Fleeing object: $\dfrac{2}{3}$ unit

Pursuer: $\dfrac{1}{2} \displaystyle\int_{0}^{1} \dfrac{x + 1}{\sqrt{x}} \, dx = \dfrac{4}{3} = 2\left(\dfrac{2}{3} \right)$

67. $\dfrac{384\pi}{5}$ **69.** Proof   **71.** Proof; $g(x) = 1$

Section 7.5   *(page 491)*

1. Work is done by a force when it moves an object.
3. The force needed to extend or compress a spring by some distance is proportional to that distance.
5. 48,000 ft-lb   **7.** 896 N-m   **9.** 40.833 in.-lb ≈ 3.403 ft-lb
11. 160 in.-lb ≈ 13.3 ft-lb   **13.** 37.125 ft-lb
15. (a) 487.8 mile-tons $\approx 5.679 \times 10^9$ ft-lb
(b) 1395.3 mile-tons $\approx 1.624 \times 10^{10}$ ft-lb
17. (a) 29,333.3 mile-tons $\approx 3.415 \times 10^{11}$ ft-lb
(b) 33,846.2 mile-tons $\approx 3.941 \times 10^{11}$ ft-lb
19. (a) 2496 ft-lb   (b) 9984 ft-lb   **21.** 470,400π N-m

23. 2995.2π ft-lb   **25.** 20,217.6π ft-lb   **27.** 2457π ft-lb
29. 600 ft-lb   **31.** 450 ft-lb   **33.** 168.75 ft-lb
35. No. Something can require a lot of physical effort but take no work. There is no work because there is no change in distance.

37. $Gm_1m_2\left(\dfrac{1}{a} - \dfrac{1}{b} \right)$ **39.** $\dfrac{3k}{4}$

41. (a) 54 ft-lb   (b) 160 ft-lb   (c) 9 ft-lb   (d) 18 ft-lb
43. 2000 ln $\dfrac{3}{2} \approx 810.93$ ft-lb   **45.** 3249.4 ft-lb
47. 10,330.3 ft-lb
49. (a) 16,000π ft-lb

(b) $F(x) = -16,261.36x^4 + 82,295.45x^3 - 157,738.64x^2 + 104,386.36x - 32.4675$

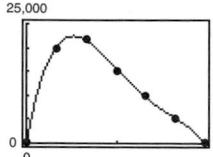

(c) 0.524 ft   (d) 25,180.5 ft-lb

Section 7.6   *(page 502)*

1. Weight is a force that is dependent on gravity. Mass is a measure of a body's resistance to changes in motion and is independent of the gravitational system in which the body is located. The weight (or force) of an object is its mass times the acceleration due to gravity.
3. A planar lamina is a flat plate of material of constant density. The center of mass of a lamina is its balancing point.

5. $\bar{x} = -\dfrac{4}{3}$ **7.** $\bar{x} = 4$ **9.** $x = 6$ ft

11. $(\bar{x}, \bar{y}) = \left(\dfrac{10}{9}, -\dfrac{1}{9} \right)$ **13.** $(\bar{x}, \bar{y}) = \left(2, \dfrac{48}{25} \right)$

15. $M_x = \dfrac{\rho}{3}, M_y = \dfrac{4\rho}{3}, \quad (\bar{x}, \bar{y}) = \left(\dfrac{4}{3}, \dfrac{1}{3} \right)$

17. $M_x = 4\rho, M_y = \dfrac{64\rho}{5}, \quad (\bar{x}, \bar{y}) = \left(\dfrac{12}{5}, \dfrac{3}{4} \right)$

19. $M_x = \dfrac{\rho}{35}, M_y = \dfrac{\rho}{20}, \quad (\bar{x}, \bar{y}) = \left(\dfrac{3}{5}, \dfrac{12}{35} \right)$

21. $M_x = \dfrac{99\rho}{5}, M_y = \dfrac{27\rho}{4}, \quad (\bar{x}, \bar{y}) = \left(\dfrac{3}{2}, \dfrac{22}{5} \right)$

23. $M_x = \dfrac{192\rho}{7}, M_y = 96\rho, \quad (\bar{x}, \bar{y}) = \left(5, \dfrac{10}{7} \right)$

25. $M_x = 0, M_y = \dfrac{256\rho}{15}, \quad (\bar{x}, \bar{y}) = \left(\dfrac{8}{5}, 0 \right)$

27. $M_x = \dfrac{27\rho}{4}, M_y = -\dfrac{27\rho}{10}, \quad (\bar{x}, \bar{y}) = \left(-\dfrac{3}{5}, \dfrac{3}{2} \right)$

29.

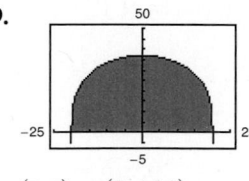

$(\bar{x}, \bar{y}) = (0, 16.2)$

31.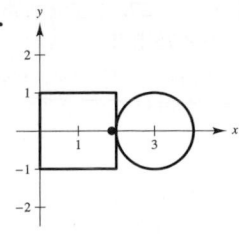

33.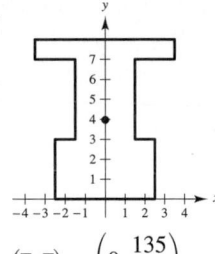

$$(\bar{x}, \bar{y}) = \left(\frac{4 + 3\pi}{4 + \pi}, 0\right)$$

$$(\bar{x}, \bar{y}) = \left(0, \frac{135}{34}\right)$$

35. $(\bar{x}, \bar{y}) = \left(\dfrac{2 + 3\pi}{2 + \pi}, 0\right)$ **37.** $160\pi^2 \approx 1579.14$

39. $\dfrac{128\pi}{3} \approx 134.04$

41. The center of mass is translated k units as well.

43. Answers will vary. Sample answer: Use three rectangles with width 1 and length 4 and place them as follows.

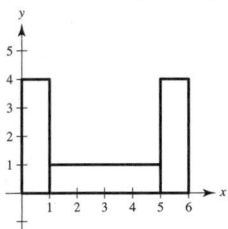

$$(\bar{x}, \bar{y}) = (3, 1.5)$$

45. $(\bar{x}, \bar{y}) = \left(\dfrac{b}{3}, \dfrac{c}{3}\right)$ **47.** $(\bar{x}, \bar{y}) = \left(\dfrac{(a + 2b)c}{3(a + b)}, \dfrac{a^2 + ab + b^2}{3(a + b)}\right)$

49. $(\bar{x}, \bar{y}) = \left(0, \dfrac{4b}{3\pi}\right)$

51. (a)

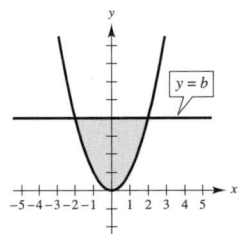

(b) $M_y = \displaystyle\int_{-\sqrt{b}}^{\sqrt{b}} x(b - x^2)\, dx = 0$ because $x(b - x^2)$ is an odd function; $\bar{x} = 0$ by symmetry.

(c) $\bar{y} > \dfrac{b}{2}$ because the area is greater for $y > \dfrac{b}{2}$.

(d) $\bar{y} = \dfrac{3}{5}b$

53. (a) $y = (-1.02 \times 10^{-5})x^4 - 0.0019x^2 + 29.28$

(b) $(\bar{x}, \bar{y}) = (0, 12.85)$

55. $9\pi\sqrt{2}$

57. $(\bar{x}, \bar{y}) = \left(\dfrac{n + 1}{n + 2}, \dfrac{n + 1}{4n + 2}\right)$; As $n \to \infty$, the region shrinks toward the line segments $y = 0$ for $0 \le x \le 1$ and $x = 1$ for $0 \le y \le 1$; $(\bar{x}, \bar{y}) \to \left(1, \dfrac{1}{4}\right)$.

59. Putnam Problem A1, 1982

Section 7.7 *(page 509)*

1. Fluid pressure is the force per unit area over the surface of a body submerged in a fluid.

3. 1497.6 lb   **5.** 4992 lb   **7.** 2223 lb   **9.** 1123.2 lb

11. 748.8 lb   **13.** 1064.96 lb   **15.** 117,600 N

17. 2,381,400 N   **19.** 2814 lb   **21.** 6753.6 lb

23. 94.5 lb

25. Because you are measuring total force against a region between two depths

27. $\dfrac{3\sqrt{2}}{2} \approx 2.12$ ft; The pressure increases with increasing depth.

29–31. Proofs   **33.** 960 lb   **35.** 2936 lb

Review Exercises for Chapter 7 *(page 511)*

1.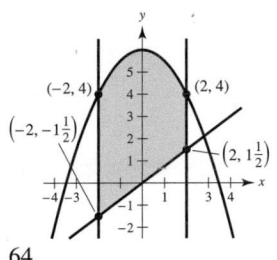

$$\dfrac{64}{3}$$

3.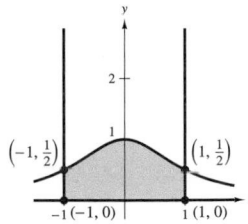

$$\dfrac{\pi}{2}$$

5.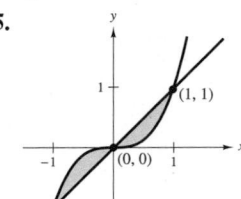

$$\dfrac{1}{2}$$

7.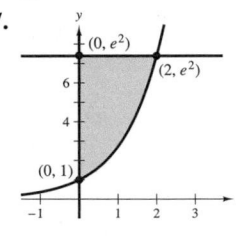

$$e^2 + 1$$

9.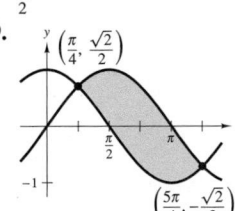

$$2\sqrt{2}$$

11. (a)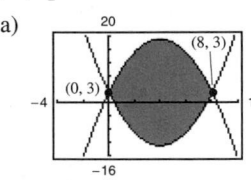

(b) 170.6667

13. (a)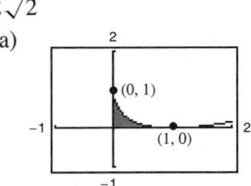

(b) 0.1667

15. $F(x) = \dfrac{3}{2}x^2 + x$

(a) $F(0) = 0$ (b) $F(2) = 8$

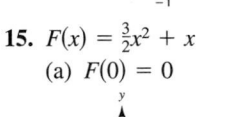
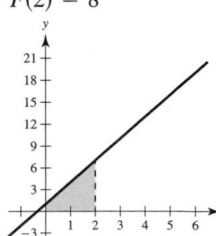

(c) $F(6) = 60$

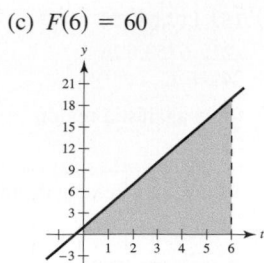

17. R_1; \$1.125 million   **19.** $\dfrac{\pi^2}{2}$ **21.** $\dfrac{\pi^2}{4}$

23. (a) 9π (b) 18π (c) 9π (d) 36π

25. 3 ft³   **27.** $\dfrac{8}{15}\left(1 + 6\sqrt{3}\right) \approx 6.076$

29. $2\pi \displaystyle\int_3^6 \dfrac{x^3}{18}\sqrt{1 + \dfrac{x^4}{36}}\,dx \approx 459.098$

31. $2\pi \displaystyle\int_0^2 x\sqrt{1 + x^2}\,dx \approx 21.322$ **33.** 5.208 ft-lb

35. 952.4 mile-tons $\approx 1.109 \times 10^{10}$ ft-lb   **37.** 200 ft-lb

39. 693.15 ft-lb

41. 3.6   **43.** $M_x = \dfrac{544\rho}{15}$, $M_y = \dfrac{32\rho}{3}$, $(\bar{x}, \bar{y}) = \left(1, \dfrac{17}{5}\right)$

45. Answers will vary. Sample answer:

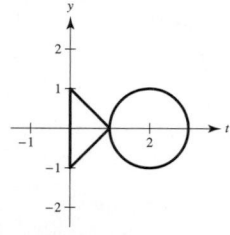

$(\bar{x}, \bar{y}) = (1.596, 0)$

47. 374.4 lb   **49.** 3072 lb   **51.** 723,822.95 lb

P.S. Problem Solving   *(page 513)*

1. 3   **3.** $y = 0.2063x$ **5.** $(\bar{x}, \bar{y}) = \left(\dfrac{2(9\pi + 49)}{3(\pi + 9)}, 0\right)$

7. $V = 2\pi\left(d + \frac{1}{2}\sqrt{w^2 + l^2}\right)lw$

9. $f(x) = 2e^{x/2} - 2$ **11.** 89.3%

13.

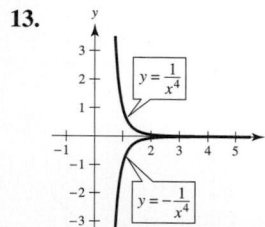

(a) $(\bar{x}, \bar{y}) = \left(\dfrac{63}{43}, 0\right)$

(b) $(\bar{x}, \bar{y}) = \left(\dfrac{3b(b + 1)}{2(b^2 + b + 1)}, 0\right)$

(c) $\left(\dfrac{3}{2}, 0\right)$

15. Consumer surplus: 1600, Producer surplus: 400

17. Wall at shallow end: 9984 lb

Wall at deep end: 39,936 lb

Side wall: $19{,}968 + 26{,}624 = 46{,}592$ lb

Chapter 8

Section 8.1   *(page 520)*

1. Use long division to rewrite the function as the sum of a polynomial and a proper rational function.

3. b

5. $\displaystyle\int u^n\,du$

$u = 5x - 3, n = 4$

7. $\displaystyle\int \dfrac{du}{u}$

$u = 1 - 2\sqrt{x}$

9. $\displaystyle\int \dfrac{du}{\sqrt{a^2 - u^2}}$

$u = t, a = 1$

11. $\displaystyle\int \sin u\,du$

$u = t^2$

13. $\displaystyle\int e^u\,du$

$u = \sin x$

15. $2(x - 5)^7 + C$

17. $-\dfrac{7}{6(z - 10)^6} + C$ **19.** $\dfrac{z^3}{3} - \dfrac{1}{5(z - 1)^5} + C$

21. $-\frac{1}{3}\ln\left|-t^3 + 9t + 1\right| + C$ **23.** $\frac{1}{2}x^2 + x + \ln|x - 1| + C$

25. $x + \ln|x + 1| + C$ **27.** $\dfrac{x}{15}(48x^4 + 200x^2 + 375) + C$

29. $\dfrac{\sin 2\pi x^2}{4\pi} + C$ **31.** $-2\sqrt{\cos x} + C$

33. $2\ln(1 + e^x) + C$ **35.** $(\ln x)^2 + C$

37. $-\ln|\csc \alpha + \cot \alpha| + \ln|\sin \alpha| + C$

39. $-\dfrac{1}{4}\arcsin(4t + 1) + C$ **41.** $\dfrac{1}{2}\ln\left|\cos\dfrac{2}{t}\right| + C$

43. $\dfrac{6}{5}\operatorname{arcsec}\dfrac{|3z|}{5} + C$ **45.** $\dfrac{1}{4}\arctan\dfrac{2x + 1}{8} + C$

47. (a)

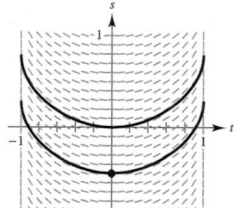

(b) $\frac{1}{2}\arcsin t^2 - \frac{1}{2}$

49. $y = 4e^{0.8x}$

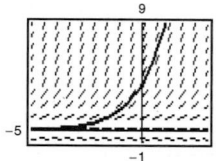

51. $y = \frac{1}{2}e^{2x} + 10e^x + 25x + C$ **53.** $r = 10\arcsin e^t + C$

55. $y = \dfrac{1}{2}\arctan\dfrac{\tan x}{2} + C$ **57.** $\dfrac{1}{15}$ **59.** $\dfrac{1}{2}$

61. $\dfrac{1}{2}(1 - e^{-1}) \approx 0.316$ **63.** $\dfrac{3}{2}[(\ln 4)^2 - (\ln 3)^2] \approx 1.072$

65. 8   **67.** $\ln 9 + \dfrac{8}{3} \approx 4.864$ **69.** $\dfrac{\pi}{18}$

71. $\dfrac{240}{\ln 3} \approx 218.457$ **73.** $\dfrac{18\sqrt{6}}{5} \approx 8.82$ **75.** $\dfrac{4}{3} \approx 1.333$

77. $\frac{1}{3}\arctan\left[\frac{1}{3}(x + 2)\right] + C$ **79.** $\tan \theta - \sec \theta + C$

Graphs will vary.   Graphs will vary.

Example: Example:

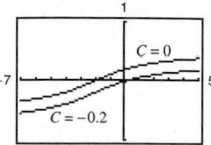

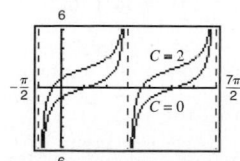

One graph is a vertical   One graph is a vertical

translation of the other.   translation of the other.

81. No. When $u = x^2$, it does not follow that $x = \sqrt{u}$ because x is negative on $[-1, 0)$.

83. $a = \sqrt{2}, b = \dfrac{\pi}{4}; \ -\dfrac{1}{\sqrt{2}} \ln\left|\csc\left(x + \dfrac{\pi}{4}\right) + \cot\left(x + \dfrac{\pi}{4}\right)\right| + C$

85. (a) They are equivalent because
$$e^{x+C_1} = e^x \cdot e^{C_1} = Ce^x, \ C = e^{C_1}.$$
(b) They differ by a constant.
$$\sec^2 x + C_1 = (\tan^2 x + 1) + C_1 = \tan^2 x + C$$

87. a

89. (a)

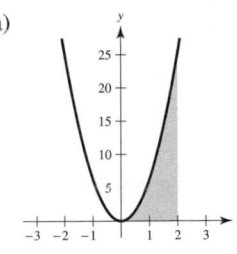

(b)

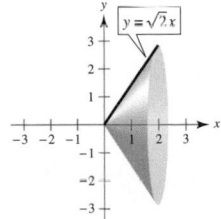

(c)

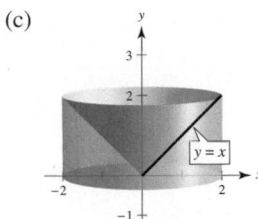

91. (a) $\pi(1 - e^{-1}) \approx 1.986$

(b) $b = \sqrt{\ln \dfrac{3\pi}{3\pi - 4}} \approx 0.743$

93. $\ln(\sqrt{2} + 1) \approx 0.8814$

95. $\dfrac{8\pi}{3}(10\sqrt{10} - 1) \approx 256.545$ **97.** $\dfrac{1}{3}\arctan 3 \approx 0.416$

99. About 1.0320

101. (a) $\frac{1}{3}\sin x(\cos^2 x + 2)$

(b) $\frac{1}{15}\sin x(3\cos^4 x + 4\cos^2 x + 8)$

(c) $\frac{1}{35}\sin x(5\cos^6 x + 6\cos^4 x + 8\cos^2 x + 16)$

(d) $\displaystyle\int \cos^{15} x \, dx = \int (1 - \sin^2 x)^7 \cos x \, dx$
You would expand $(1 - \sin^2 x)^7$.

103. Proof

Section 8.2   *(page 529)*

1. The formula for the derivative of a product

3. Let $dv = dx$.   **5.** $u = x, dv = e^{9x} \, dx$

7. $u = (\ln x)^2, dv = dx$ **9.** $u = x, dv = \sec^2 x \, dx$

11. $\frac{1}{16}x^4(4 \ln x - 1) + C$

13. $-\frac{1}{4}(2x + 1) \cos 4x + \frac{1}{8}\sin 4x + C$

15. $\dfrac{e^{4x}}{16}(4x - 1) + C$ **17.** $e^x(x^3 - 3x^2 + 6x - 6) + C$

19. $\frac{1}{4}[2(t^2 - 1) \ln|t + 1| - t^2 + 2t] + C$ **21.** $\frac{1}{3}(\ln x)^3 + C$

23. $\dfrac{e^{2x}}{4(2x + 1)} + C$ **25.** $\dfrac{2}{15}(x - 5)^{3/2}(3x + 10) + C$

27. $-x \cot x + \ln|\sin x| + C$

29. $(6x - x^3)\cos x + (3x^2 - 6)\sin x + C$

31. $x \arctan x - \frac{1}{2}\ln(1 + x^2) + C$

33. $-\frac{3}{34}e^{-3x}\sin 5x - \frac{5}{34}e^{-3x}\cos 5x + C$

35. $x \ln x - x + C$

37. $y = \dfrac{2}{5}t^2\sqrt{3 + 5t} - \dfrac{8t}{75}(3 + 5t)^{3/2} + \dfrac{16}{1875}(3 + 5t)^{5/2} + C$

$= \dfrac{2}{625}\sqrt{3 + 5t}(25t^2 - 20t + 24) + C$

39. (a)

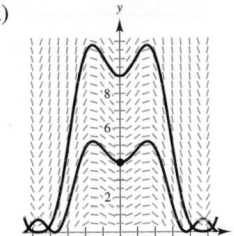

(b) $2\sqrt{y} - \cos x - x \sin x = 3$

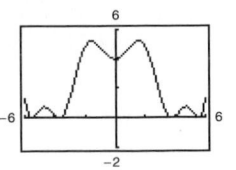

41.

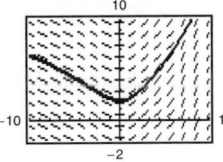

43. $2e^{3/2} + 4 \approx 12.963$ **45.** $\dfrac{\pi}{8} - \dfrac{1}{4} \approx 0.143$

47. $\dfrac{\pi - 3\sqrt{3} + 6}{6} \approx 0.658$

49. $\frac{1}{2}[e(\sin 1 - \cos 1) + 1] \approx 0.909$

51. $8 \arcsec 4 + \dfrac{\sqrt{3}}{2} - \dfrac{\sqrt{15}}{2} - \dfrac{2\pi}{3} \approx 7.380$

53. $\dfrac{e^{2x}}{4}(2x^2 - 2x + 1) + C$

55. $-\cos x(x + 2)^2 + 2 \sin x(x + 2) + 2 \cos x + C$

57. $\frac{1}{20}(4x + 9)^{3/2}(2x + 17) + C$

59. Answers will vary. Sample answer: $\int x^3 \sin x \, dx$
It takes three applications until the algebraic factor becomes a constant.

61. (a) No, substitution   (b) Yes, $u = \ln x, dv = x \, dx$
(c) Yes, $u = x^2, dv = e^{-3x} \, dx$ (d) No, substitution
(e) Yes, $u = x$ and $dv = \dfrac{1}{\sqrt{x + 1}} \, dx$ (f) No, substitution

63. $2(\sin \sqrt{x} - \sqrt{x} \cos \sqrt{x}) + C$

65. $\frac{1}{2}(x^4 e^{x^2} - 2x^2 e^{x^2} + 2e^{x^2}) + C$

67. (a) and (b) $\frac{1}{3}\sqrt{4 + x^2}(x^2 - 8) + C$

69. $n = 0$: $x(\ln x - 1) + C$
$n = 1$: $\frac{1}{4}x^2(2 \ln x - 1) + C$
$n = 2$: $\frac{1}{9}x^3(3 \ln x - 1) + C$
$n = 3$: $\frac{1}{16}x^4(4 \ln x - 1) + C$
$n = 4$: $\frac{1}{25}x^5(5 \ln x - 1) + C$
$\displaystyle\int x^n \ln x \, dx = \dfrac{x^{n+1}}{(n + 1)^2}[(n + 1) \ln x - 1] + C$

71–75. Proofs   **77.** $-x^2 \cos x + 2x \sin x + 2 \cos x + C$

79. $\frac{1}{36}x^6(6 \ln x - 1) + C$

81. $\dfrac{e^{-3x}(-3 \sin 4x - 4 \cos 4x)}{25} + C$

83.

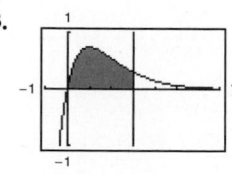

$$2 - \frac{8}{e^3} \approx 1.602$$

85.

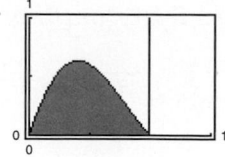

$$\frac{\pi}{1 + \pi^2}\left(\frac{1}{e} + 1\right) \approx 0.395$$

87. (a) 1   (b) $\pi(e - 2) \approx 2.257$ (c) $\frac{1}{2}\pi(e^2 + 1) \approx 13.177$

(d) $\left(\dfrac{e^2 + 1}{4}, \dfrac{e - 2}{2}\right) \approx (2.097, 0.359)$

89. In Example 6, we showed that the centroid of an equivalent region was $\left(1, \dfrac{\pi}{8}\right)$. By symmetry, the centroid of this region is $\left(\dfrac{\pi}{8}, 1\right)$.

91. $\dfrac{7}{10\pi}(1 - e^{-4\pi}) \approx 0.223$ **93.** \$931,265

95. Proof   **97.** $b_n = \dfrac{8h}{(n\pi)^2} \sin \dfrac{n\pi}{2}$

99. For any integrable function, $\int f(x)\,dx = C + \int f(x)\,dx$, but this cannot be used to imply that $C = 0$.

Section 8.3   (page 538)

1. $\int \sin^8 x\,dx$; The other integral can be found using u-substitution.
3. $-\frac{1}{6}\cos^6 x + C$ **5.** $\frac{1}{5}\sin^5 x - \frac{1}{7}\sin^7 x + C$
7. $-\frac{1}{3}\cos^3 x + \frac{1}{5}\cos^5 x + C$
9. $-\frac{1}{3}(\cos 2\theta)^{3/2} + \frac{1}{7}(\cos 2\theta)^{7/2} + C$
11. $\frac{1}{12}(6x + \sin 6x) + C$ **13.** $2x^2 + 2x \sin 2x + \cos 2x + C$

15. $\dfrac{2}{3}$ **17.** $\dfrac{\pi}{4}$ **19.** $\dfrac{63\pi}{512}$ **21.** $\frac{1}{4}\ln|\sec 4x + \tan 4x| + C$

23. $\dfrac{\sec \pi x \tan \pi x + \ln|\sec \pi x + \tan \pi x|}{2\pi} + C$

25. $\dfrac{1}{2}\tan^4 \dfrac{x}{2} - \tan^2 \dfrac{x}{2} - 2 \ln\left|\cos \dfrac{x}{2}\right| + C$

27. $\dfrac{1}{2}\left[\dfrac{\sec^5 2t}{5} - \dfrac{\sec^3 2t}{3}\right] + C$ **29.** $\dfrac{1}{24}\sec^6 4x + C$

31. $\frac{1}{7}\sec^7 x - \frac{1}{5}\sec^5 x + C$
33. $\ln|\sec x + \tan x| - \sin x + C$
35. $\dfrac{12\pi\theta - 8 \sin 2\pi\theta + \sin 4\pi\theta}{32\pi} + C$

37. $y = \frac{1}{9}\sec^3 3x - \frac{1}{3}\sec 3x + C$

39. (a)

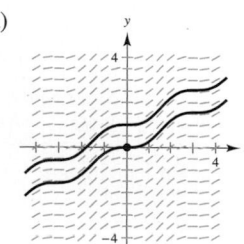

(b) $y = \frac{1}{2}x - \frac{1}{4}\sin 2x$

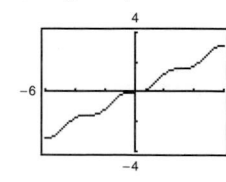

41.

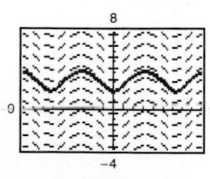

43. $\frac{1}{16}(2 \sin 4x + \sin 8x) + C$

45. $\frac{1}{14}\cos 7t - \frac{1}{22}\cos 11t + C$
47. $\frac{1}{8}(2 \sin 2\theta - \sin 4\theta) + C$
49. $\frac{1}{4}(\ln|\csc^2 2x| - \cot^2 2x) + C$
51. $-\frac{1}{3}\cot 3x - \frac{1}{9}\cot^3 3x + C$
53. $\ln|\csc t - \cot t| + \cos t + C$
55. $\ln|\csc x - \cot x| + \cos x + C$ **57.** $t - 2 \tan t + C$
59. π **61.** $3(1 - \ln 2)$ **63.** $\ln 2$ **65.** 4
67. (a) $\frac{1}{18}\tan^6 3x + \frac{1}{12}\tan^4 3x + C_1, \frac{1}{18}\sec^6 3x - \frac{1}{12}\sec^4 3x + C_2$
(b)

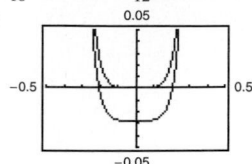

(c) Proof

69. (a) $\frac{1}{2}\sin^2 x + C$ (b) $-\frac{1}{2}\cos^2 x + C$
(c) $\frac{1}{2}\sin^2 x + C$ (d) $-\frac{1}{4}\cos 2x + C$
The answers are all the same, but they are written in different forms. Using trigonometric identities, you can rewrite each answer in the same form.

71. $\dfrac{1}{3}$ **73.** 1   **75.** $2\pi\left(1 - \dfrac{\pi}{4}\right) \approx 1.348$

77. (a) $\dfrac{\pi^2}{2}$ (b) $(\bar{x}, \bar{y}) = \left(\dfrac{\pi}{2}, \dfrac{\pi}{8}\right)$ **79–81.** Proofs

83. $-\frac{1}{15}\cos x(3 \sin^4 x + 4 \sin^2 x + 8) + C$
85. $-\frac{1}{48}(8 \cos^3 x \sin^3 x + 6 \cos^3 x \sin x - 3 \cos x \sin x - 3x) + C$
87. (a) and (b) Proofs
89. (a) Proof   (b) $a_1 = 2, a_2 = -1, a_3 = \frac{2}{3}$

Section 8.4   (page 547)

1. (a) $x = 3 \tan \theta$ (b) $x = 2 \sin \theta$
(c) $x = 5 \sin \theta$ (d) $x = 5 \sec \theta$

3. $\dfrac{x}{16\sqrt{16 - x^2}} + C$

5. $4 \ln\left|\dfrac{4 - \sqrt{16 - x^2}}{x}\right| + \sqrt{16 - x^2} + C$

7. $\ln\left|x + \sqrt{x^2 - 25}\right| + C$
9. $\frac{1}{15}(x^2 - 25)^{3/2}(3x^2 + 50) + C$

11. $\dfrac{(4 + x^2)^{3/2}}{6} + C$ **13.** $\dfrac{1}{4}\left(\arctan \dfrac{x}{2} + \dfrac{2x}{4 + x^2}\right) + C$

15. $\dfrac{1}{2}x\sqrt{49 - 16x^2} + \dfrac{49}{8}\arcsin \dfrac{4x}{7} + C$

17. $\dfrac{1}{2\sqrt{5}}\left(\sqrt{5}x\sqrt{36 - 5x^2} + 36 \arcsin \dfrac{\sqrt{5}x}{6}\right) + C$

19. $4 \arcsin \dfrac{x}{2} + x\sqrt{4 - x^2} + C$ **21.** $-\dfrac{(1 - x^2)^{3/2}}{3x^3} + C$

23. $-\dfrac{1}{3}\ln\left|\dfrac{\sqrt{4x^2 + 9} + 3}{2x}\right| + C$ **25.** $-\dfrac{x}{\sqrt{x^2 + 3}} + C$

27. $\dfrac{1}{2}\left(\arcsin e^x + e^x\sqrt{1 - e^{2x}}\right) + C$

29. $\dfrac{1}{4}\left(\dfrac{x}{x^2 + 2} + \dfrac{1}{\sqrt{2}}\arctan \dfrac{x}{\sqrt{2}}\right) + C$

31. $x \,\text{arcsec}\, 2x - \frac{1}{2}\ln\left|2x + \sqrt{4x^2 - 1}\right| + C$

33. $2 \arcsin \dfrac{x - 2}{2} - \sqrt{4x - x^2} + C$

35. $\sqrt{x^2 + 6x + 12} - 3 \ln\left|\sqrt{x^2 + 6x + 12} + (x + 3)\right| + C$

37. (a) and (b) $\sqrt{3} - \dfrac{\pi}{3} \approx 0.685$

39. (a) and (b) $9(2 - \sqrt{2}) \approx 5.272$

41. (a) and (b) $-\dfrac{9}{2} \ln\left(\dfrac{2\sqrt{7}}{3} - \dfrac{4\sqrt{3}}{3} - \dfrac{\sqrt{21}}{3} + \dfrac{8}{3}\right)$
$+ 9\sqrt{3} - 2\sqrt{7} \approx 12.644$

43. Substitution: $u = x^2 + 1$, $du = 2x\,dx$

45. (a) $-\sqrt{1 - x^2} + C$; The answers are equivalent.

(b) $x - 3 \arctan\dfrac{x}{3} + C$; The answers are equivalent.

47. True   **49.** False. $\displaystyle\int_0^{\sqrt{3}} \dfrac{dx}{(1 + x^2)^{3/2}} = \int_0^{\pi/3} \cos\theta\,d\theta$

51. πab

53. $\ln \dfrac{5(\sqrt{2} + 1)}{\sqrt{26} + 1} + \sqrt{26} - \sqrt{2} \approx 4.367$ **55.** $6\pi^2$

57. $(0, 0.422)$

59. (a) $V = \dfrac{3\pi}{2} + 3\arcsin(d - 1) + 3(d - 1)\sqrt{2d - d^2}$

(b)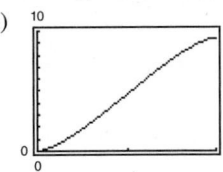

(c) The full tank holds $3\pi \approx 9.4248$ cubic meters. The horizontal lines
$$y = \dfrac{3\pi}{4}, \ y = \dfrac{3\pi}{2}, \ y = \dfrac{9\pi}{4}$$
intersect the curve at $d = 0.596$, 1.0, 1.404. The dipstick would have these markings on it.

(d) $d'(t) = \dfrac{1}{24\sqrt{1 - (d - 1)^2}}$

(e)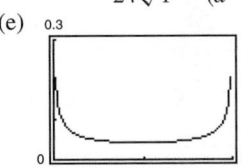

The minimum occurs at $d = 1$, which is the widest part of the tank.

61. (a) Proof

(b) $y = -12 \ln \dfrac{12 - \sqrt{144 - x^2}}{x} - \sqrt{144 - x^2}$

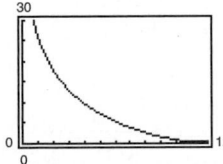

(c) Vertical asymptote: $x = 0$ (d) About 5.2 m

63. (a) 187.2π lb   (b) $62.4\pi d$ lb   **65.** Proof

67. $12 + \dfrac{9\pi}{2} - 25 \arcsin\dfrac{3}{5} \approx 10.050$

69. Putnam Problem A5, 2005

Section 8.5   (page 557)

1. (a) $\dfrac{A}{x} + \dfrac{B}{x - 8}$ (b) $\dfrac{A}{x - 3} + \dfrac{B}{(x - 3)^2} + \dfrac{C}{(x - 3)^3}$

(c) $\dfrac{A}{x} + \dfrac{Bx + C}{x^2 + 10}$ (d) $\dfrac{A}{x} + \dfrac{Bx + C}{x^2 + 1} + \dfrac{Dx + E}{(x^2 + 1)^2}$

3. $\dfrac{1}{6} \ln\left|\dfrac{x - 3}{x + 3}\right| + C$ **5.** $\ln\left|\dfrac{x - 1}{x + 4}\right| + C$

7. $5 \ln|x - 2| - \ln|x + 2| - 3 \ln|x| + C$

9. $x^2 + \dfrac{3}{2} \ln|x - 4| - \dfrac{1}{2} \ln|x + 2| + C$

11. $\dfrac{1}{x} + \ln|x^4 + x^3| + C$

13. $\dfrac{9}{x + 1} + 2 \ln|x| - \ln|x + 1| + C$

15. $9 \ln|x| - \dfrac{32}{7} \ln(7x^2 + 1) + C$

17. $\dfrac{1}{6}\left(\ln\left|\dfrac{x - 2}{x + 2}\right| + \sqrt{2} \arctan\dfrac{x}{\sqrt{2}}\right) + C$

19. $\ln|x + 1| + \sqrt{2} \arctan\dfrac{x - 1}{\sqrt{2}} + C$ **21.** $\ln 3$

23. $\dfrac{1}{2} \ln\dfrac{8}{5} - \dfrac{\pi}{4} + \arctan 2 \approx 0.557$ **25.** $\ln|1 + \sec x| + C$

27. $\ln\left|\dfrac{\tan x + 2}{\tan x + 3}\right| + C$ **29.** $\dfrac{1}{5} \ln\left|\dfrac{e^x - 1}{e^x + 4}\right| + C$

31. $2\sqrt{x} + 2 \ln\left|\dfrac{\sqrt{x} - 2}{\sqrt{x} + 2}\right| + C$ **33–35.** Proofs

37. Substitution: $u = x^2 + 2x - 8$

39. Trigonometric substitution (tan) or inverse tangent rule

41. $12 \ln\dfrac{9}{7}$ **43.** $\dfrac{5}{2} \ln 5$ **45.** 4.90, or $\$490{,}000$

47. (a) $V = 2\pi\left(\arctan 3 - \dfrac{3}{10}\right) \approx 5.963$

(b) $(\bar{x}, \bar{y}) \approx (1.521, 0.412)$

49. $x = \dfrac{n[e^{(n + 1)kt} - 1]}{n + e^{(n + 1)kt}}$ **51.** $\dfrac{\pi}{8}$

53. Putnam Problem B4, 1992

Section 8.6   (page 564)

1. No. The integral can be easily evaluated using basic integration rules.

	Trapezoidal	Simpson's	Exact
3.	2.7500	2.6667	2.6667
5.	0.6970	0.6933	0.6931
7.	20.2222	20.0000	20.0000
9.	12.6640	12.6667	12.6667
11.	0.3352	0.3334	0.3333
13.	0.5706	0.5930	0.5940

	Trapezoidal	Simpson's	Graphing Utility
15.	3.2833	3.2396	3.2413
17.	0.7828	0.7854	0.7854
19.	102.5553	93.3752	92.7437
21.	0.5495	0.5483	0.5493
23.	0.1940	0.1860	0.1858

25. (a) $\dfrac{1}{12}$ (b) 0 **27.** (a) $\dfrac{1}{4}$ (b) $\dfrac{1}{12}$

29. (a) $n = 366$ (b) $n = 26$

31. (a) $n = 77$ (b) $n = 8$

33. (a) $n = 643$ (b) $n = 48$

35. Trapezoidal Rule: 24.5

Simpson's Rule: 25.67

37. 0.701   **39.** $T_n = \frac{1}{2}(L_n + R_n)$ **41.** 89,250 m^2

43. 10,233.58 ft-lb   **45.** 2.477   **47.** Proof

Section 8.7 *(page 570)*

1. Formula 40   **3.** $-\frac{1}{2}x(10 - x) + 25 \ln|5 + x| + C$

5. $-\dfrac{\sqrt{1 - x^2}}{x} + C$

7. $\frac{1}{24}(3x + \sin 3x \cos 3x + 2 \cos^3 3x \sin 3x) + C$

9. $-2(\cot \sqrt{x} + \csc \sqrt{x}) + C$ **11.** $x - \frac{1}{2}\ln(1 + e^{2x}) + C$

13. $\dfrac{x^7}{49}(7 \ln x - 1) + C$ **15.** (a) and (b) $x\left(\ln \dfrac{x}{3} - 1\right) + C$

17. (a) and (b) $\ln\left|\dfrac{x - 1}{x}\right| + \dfrac{1}{x} + C$

19. $\frac{1}{2}\left[(x^2 + 1) \operatorname{arccsc}(x^2 + 1) + \ln\left(x^2 + 1 + \sqrt{x^4 + 2x^2}\right)\right] + C$

21. $\dfrac{\sqrt{x^4 - 1}}{x^2} + C$ **23.** $\dfrac{1}{36}\left(\dfrac{7}{7 - 6x} + \ln|7 - 6x|\right) + C$

25. $e^x \arccos(e^x) - \sqrt{1 - e^{2x}} + C$

27. $\frac{1}{2}(x^2 + \cot x^2 + \csc x^2) + C$

29. $\dfrac{\sqrt{2}}{2} \arctan \dfrac{1 + \sin \theta}{\sqrt{2}} + C$ **31.** $-\dfrac{\sqrt{2 + 9x^2}}{2x} + C$

33. $\frac{1}{4}\left(2 \ln|x| - 3 \ln|3 + 2 \ln|x||\right) + C$

35. $\dfrac{3x - 10}{2(x^2 - 6x + 10)} + \dfrac{3}{2} \arctan(x - 3) + C$

37. $\frac{1}{2}\ln\left|x^2 - 3 + \sqrt{x^4 - 6x^2 + 5}\right| + C$

39. $\dfrac{2}{1 + e^x} - \dfrac{1}{2(1 + e^x)^2} + \ln(1 + e^x) + C$

41. $\frac{2}{3}(2 - \sqrt{2}) \approx 0.3905$ **43.** $\frac{32}{5} \ln 2 - \frac{31}{25} \approx 3.1961$

45. $\dfrac{\pi}{2}$ **47.** $\dfrac{\pi^3}{8} - 3\pi + 6 \approx 0.4510$ **49–53.** Proofs

55. $\dfrac{1}{\sqrt{5}} \ln\left|\dfrac{2 \tan(\theta/2) - 3 - \sqrt{5}}{2 \tan(\theta/2) - 3 + \sqrt{5}}\right| + C$ **57.** $\ln 2$

59. $\frac{1}{2}\ln(3 - 2 \cos \theta) + C$ **61.** $-2 \cos \sqrt{\theta} + C$

63. $4\sqrt{3}$

65. (a) $\displaystyle\int x \ln x \, dx = \frac{1}{2}x^2 \ln x - \frac{1}{4}x^2 + C$

$\displaystyle\int x^2 \ln x \, dx = \frac{1}{3}x^3 \ln x - \frac{1}{9}x^3 + C$

$\displaystyle\int x^3 \ln x \, dx = \frac{1}{4}x^4 \ln x - \frac{1}{16}x^4 + C$

(b) $\displaystyle\int x^n \ln x \, dx = \frac{x^{n+1}}{n + 1} \ln x - \frac{x^{n+1}}{(n + 1)^2} + C$

(c) Proof

67. 1919.145 ft-lb   **69.** About 401.4   **71.** $32\pi^2$

73. Putnam Problem A3, 1980

Section 8.8 *(page 579)*

1. One or both of the limits of integration are infinite, or the function has a finite number of infinite discontinuities on the interval you are considering.

3. To evaluate the improper integral $\int_a^\infty f(x) \, dx$, find the limit as $b \to \infty$ when f is continuous on $[a, \infty)$ or find the limit as $a \to -\infty$ when f is continuous on $(-\infty, b]$.

5. Improper; $0 \le \frac{2}{5} \le 1$

7. Not improper; continuous on $[0, 1]$

9. Not improper; continuous on $[0, 2]$

11. Improper; infinite limits of integration

13. Infinite discontinuity at $x = 0$; 4

15. Infinite discontinuity at $x = 1$; diverges

17. $\frac{1}{8}$ **19.** Diverges   **21.** Diverges   **23.** 2

25. $\dfrac{1}{2(\ln 4)^2}$ **27.** π **29.** $\dfrac{\pi}{4}$ **31.** Diverges

33. Diverges   **35.** 0   **37.** $-\dfrac{1}{4}$ **39.** Diverges

41. $\dfrac{\pi}{3}$ **43.** $\ln 3$ **45.** $\dfrac{\pi}{6}$ **47.** $\dfrac{2\pi \sqrt{6}}{3}$ **49.** $p > 1$

51. Proof   **53.** Converges   **55.** Converges

57. Converges   **59.** Converges

61. The improper integral diverges.   **63.** $\frac{7}{8}$ **65.** π

67. (a) 1   (b) $\dfrac{\pi}{2}$ (c) 2π **69.** 2π

71. (a) $W = 20,000$ mile-tons   (b) 4000 mi

73. (a) Proof   (b) 48.66%

75. (a)

(b) About 0.1587   (c) 0.1587; same by symmetry

77. (a) \$807,992.41   (b) \$887,995.15   (c) \$1,116,666.67

79. $P = \dfrac{2\pi NI\left(\sqrt{r^2 + c^2} - c\right)}{kr\sqrt{r^2 + c^2}}$

81. False. Let $f(x) = \dfrac{1}{x + 1}$.   **83.** True   **85.** True

87. (a) and (b) Proofs

(c) The definition of the improper integral $\displaystyle\int_{-\infty}^\infty$ is not $\displaystyle\lim_{a \to \infty} \int_{-a}^a$ but rather that if you rewrite the integral that diverges, you can find that the integral converges.

89. Proof

91. $\dfrac{1}{s}, s > 0$ **93.** $\dfrac{2}{s^3}, s > 0$ **95.** $\dfrac{s}{s^2 + a^2}, s > 0$

97. $\dfrac{s}{s^2 - a^2}, s > |a|$

99. (a) $\Gamma(1) = 1, \Gamma(2) = 1, \Gamma(3) = 2$ (b) Proof

(c) $\Gamma(n) = (n - 1)!$

101. $c = 1$; $\ln 2$ **103.** $8\pi\left[\dfrac{(\ln 2)^2}{3} - \dfrac{\ln 4}{9} + \dfrac{2}{27}\right] \approx 2.01545$

105. $\displaystyle\int_0^1 2 \sin u^2 \, du$; 0.6278   **107.** Proof

Review Exercises for Chapter 8 *(page 583)*

1. $\dfrac{2}{9}(x^3 - 27)^{3/2} + C$ **3.** $-4 \cot \dfrac{x + 8}{4} + C$

5. $\dfrac{1}{2} + \ln 2 \approx 1.1931$ **7.** $100 \arcsin \dfrac{x}{10} + C$

9. $-xe^{1-x} - e^{1-x} + C$

11. $\frac{1}{13}e^{2x}(2 \sin 3x - 3 \cos 3x) + C$

13. $x \tan x + \ln|\cos x| + C$

15. $\frac{1}{16}\left[(8x^2 - 1) \arcsin 2x + 2x\sqrt{1 - 4x^2}\right] + C$

17. $-\dfrac{\cos^5 x}{5} + C$　　**19.** $\dfrac{\sin(\pi x - 1)[\cos^2(\pi x - 1) + 2]}{3\pi} + C$

21. $\dfrac{2}{3}\left(\tan^3\dfrac{x}{2} + 3\tan\dfrac{x}{2}\right) + C$

23. $\dfrac{\tan^3 x^2}{6} - \dfrac{\tan x^2}{2} + \dfrac{x^2}{2} + C$　　**25.** $\tan\theta + \sec\theta + C$

27. $\dfrac{3\pi}{16} + \dfrac{1}{2} \approx 1.0890$　　**29.** $\dfrac{3\sqrt{4 - x^2}}{x} + C$

31. $\frac{1}{3}(x^2 + 4)^{1/2}(x^2 - 8) + C$　　**33.** $256 - 62\sqrt{17} \approx 0.3675$

35. (a), (b), and (c) $\frac{1}{3}\sqrt{4 - x^2}(x^2 - 8) + C$

37. $2\ln|x + 2| - \ln|x - 3| + C$

39. $\frac{1}{4}[6\ln|x - 1| - \ln(x^2 + 1) + 6\arctan x] + C$

41. $x + \dfrac{1}{1 - x} + 2\ln|x - 1| + C$

43. $-\ln|e^x + 1| + \frac{1}{2}\ln|e^x + 3| + \frac{1}{2}\ln|e^x - 1| + C$

	Trapezoidal	Simpson's	Graphing Utility
45.	0.2848	0.2838	0.2838
47.	0.6366	0.6847	0.7041

49. $\dfrac{1}{25}\left(\dfrac{4}{4 + 5x} + \ln|4 + 5x|\right) + C$　　**51.** $1 - \dfrac{\sqrt{2}}{2}$

53. $\dfrac{1}{2}\ln|x^2 + 4x + 8| - \arctan\dfrac{x + 2}{2} + C$

55. $\dfrac{\ln|\tan \pi x|}{\pi} + C$　　**57.** $\frac{1}{8}(\sin 2\theta - 2\theta\cos 2\theta) + C$

59. $\frac{4}{3}(x^{3/4} - 3x^{1/4} + 3\arctan x^{1/4}) + C$

61. $2\sqrt{1 - \cos x} + C$　　**63.** $\sin x\ln(\sin x) - \sin x + C$

65. $\dfrac{5}{2}\ln\left|\dfrac{x - 5}{x + 5}\right| + C$

67. $y = x\ln|x^2 + x| - 2x + \ln|x + 1| + C$　　**69.** $\frac{1}{5}$

71. $\frac{1}{2}(\ln 4)^2 \approx 0.961$

73. $\pi^2 - 4\sin 2 - 2\cos 2 - 6 \approx 1.0647$　　**75.** $\dfrac{\sqrt{27}}{5} \approx 1.0392$

77. $(\bar{x}, \bar{y}) = \left(0, \dfrac{4}{3\pi}\right)$　　**79.** $\dfrac{32}{3}$　　**81.** Diverges　　**83.** 1

85. $\dfrac{\pi}{4}$　　**87.** (a) \$6,321,205.59　　(b) \$10,000,000

89. (a) 0.4581　　(b) 0.0135

P.S. Problem Solving　*(page 585)*

1. (a) $\frac{4}{3}, \frac{16}{15}$　　(b) Proof

3. (a) $R(n), I, T(n), L(n)$
(b) $S(4) = \frac{1}{3}[f(0) + 4f(1) + 2f(2) + 4f(3) + f(4)] \approx 5.42$

5. $\dfrac{\pi\sqrt{3}}{9} \approx 0.6046$　　**7.** $(\bar{x}, \bar{y}) = \left(0, \dfrac{\sqrt{2}}{4}\right)$

9. (a) Proof　　(b) $x\arcsin x + \sqrt{1 - x^2} + C$　　(c) 1

11. Proof　　**13.** (a) $\dfrac{\pi}{4}$　　(b) $\dfrac{\pi}{4}$

15. $s(t) = -16t^2 + 12,000t\left(1 + \ln\dfrac{50,000}{50,000 - 400t}\right)$
$+ 1,500,000\ln\dfrac{50,000 - 400t}{50,000}$; 557,168.626 ft

17. Proof　　**19.** (a) $\dfrac{2}{\pi}$　　(b) 0

Chapter 9

Section 9.1　*(page 596)*

1. You need to be given one or more of the first few terms of a sequence, and then all other terms are defined using previous terms.

3. g; Factorial functions grow faster than exponential functions.

5. 3, 9, 27, 81, 243　　**7.** 1, 0, −1, 0, 1　　**9.** $2, -1, \frac{2}{3}, -\frac{1}{2}, \frac{2}{5}$

11. 3, 4, 6, 10, 18　　**13.** c　　**14.** a　　**15.** d　　**16.** b

17. $n^2 + n$　　**19.** $n(n - 1)(n - 2)$　　**21.** 1　　**23.** 2

25. 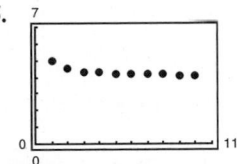　　**27.**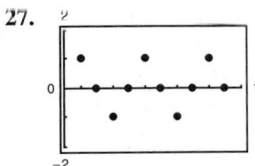

Converges to 4　　　　　Diverges

29. Converges to 0　　**31.** Diverges　　**33.** Converges to $\frac{3}{4}$

35. Converges to 0　　**37.** Diverges　　**39.** Converges to 0

41. Converges to 1　　**43.** Converges to 0

45. $6n - 4$; diverges　　**47.** $n^2 - 3$; diverges

49. $\dfrac{n + 1}{n + 2}$; converges　　**51.** $\dfrac{n + 1}{n}$; converges

53. Monotonic, bounded　　**55.** Not monotonic, bounded

57. Monotonic, bounded　　**59.** Not monotonic, bounded

61. (a) $\left|7 + \dfrac{1}{n}\right| \geq 7 \Rightarrow$ bounded

$a_n > a_{n+1} \Rightarrow$ monotonic
So, $\{a_n\}$ converges.

(b) 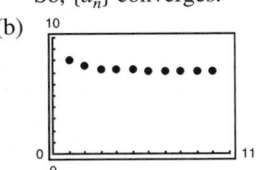　　Limit $= 7$

63. (a) $\left|\dfrac{1}{3}\left(1 - \dfrac{1}{3^n}\right)\right| < \dfrac{1}{3} \Rightarrow$ bounded

$a_n < a_{n+1} \Rightarrow$ monotonic
So, $\{a_n\}$ converges.

(b) 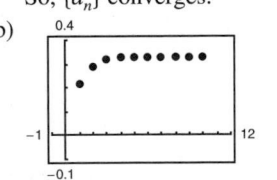　　Limit $= \frac{1}{3}$

65. (a) No. $\lim\limits_{n\to\infty} A_n$ does not exist.
(b)

n	1	2	3	4
A_n	\$10,045.83	\$10,091.88	\$10,138.13	\$10,184.60

n	5	6	7
A_n	\$10,231.28	\$10,278.17	\$10,325.28

n	8	9	10
A_n	\$10,372.60	\$10,420.14	\$10,467.90

67. $26,125.00, $27,300.63, $28,529.15, $29,812.97, $31,154.55

69. Answers will vary. Sample answers:

(a) $a_n = 10 - \dfrac{1}{n}$ (b) $a_n = \dfrac{3n}{4n + 1}$

71. The sequence $\{a_n\}$ could converge or diverge. If $\{a_n\}$ is increasing, then it converges to a limit less than or equal to 1. If $\{a_n\}$ is decreasing, then it could converge (example: $a_n = 1/n$) or diverge (example: $a_n = -n$).

73. 1, 1.4142, 1.4422, 1.4142, 1.3797, 1.3480; Converges to 1

75. Proof

77. False. The sequence could also alternate between two values.

79. True

81. (a) 1, 1, 2, 3, 5, 8, 13, 21, 34, 55, 89, 144

(b) 1, 2, 1.5, 1.6667, 1.6, 1.6250, 1.6154, 1.6190, 1.6176, 1.6182   (c) Proof

(d) $\rho = \dfrac{1 + \sqrt{5}}{2} \approx 1.6180$

83. (a) 1.4142, 1.8478, 1.9616, 1.9904, 1.9976

(b) $a_n = \sqrt{2 + a_{n-1}}$ (c) $\lim\limits_{n \to \infty} a_n = 2$

85. (a) Proof

(b)

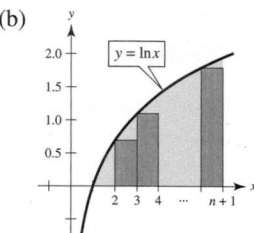

(c) and (d) Proofs

(e) $\dfrac{\sqrt[20]{20!}}{20} \approx 0.4152$

$\dfrac{\sqrt[50]{50!}}{50} \approx 0.3897$

$\dfrac{\sqrt[100]{100!}}{100} \approx 0.3799$

87–89. Proofs   **91.** Putnam Problem A1, 1990

Section 9.2   (page 605)

1. $\lim\limits_{n \to \infty} a_n = 5$ means that the limit of the sequence $\{a_n\}$ is 5. $\sum\limits_{n=1}^{\infty} a_n = a_1 + a_2 + \cdots = 5$ means that the limit of the partial sums is 5.

3. You cannot make a conclusion. The series may either converge or diverge.

5. 1, 1.25, 1.361, 1.424, 1.464

7. 3, -1.5, 5.25, -4.875, 10.3125

9. 3, 4.5, 5.25, 5.625, 5.8125

11. Geometric series: $r = \frac{5}{2} > 1$ **13.** $\lim\limits_{n \to \infty} a_n = 1 \neq 0$

15. $\lim\limits_{n \to \infty} a_n = 1 \neq 0$ **17.** $\lim\limits_{n \to \infty} a_n = \frac{1}{4} \neq 0$

19. Geometric series: $r = \frac{5}{6} < 1$

21. Geometric series: $r = 0.9 < 1$

23. Telescoping series: $a_n = 1/n - 1/(n + 1)$; Converges to 1.

25. (a) $\frac{11}{3}$

(b)

n	5	10	20	50	100
S_n	2.7976	3.1643	3.3936	3.5513	3.6078

(c)

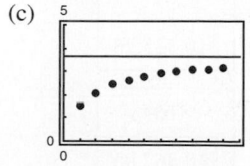

(d) The terms of the series decrease in magnitude relatively slowly, and the sequence of partial sums approaches the sum of the series relatively slowly.

27. (a) 20

(b)

n	5	10	20	50	100
S_n	8.1902	13.0264	17.5685	19.8969	19.9995

(c)

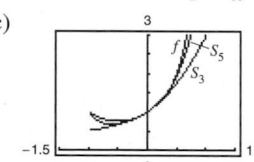

(d) The terms of the series decrease in magnitude relatively slowly, and the sequence of partial sums approaches the sum of the series relatively slowly.

29. 15   **31.** 3   **33.** 32   **35.** $\dfrac{1}{2}$ **37.** $\dfrac{\sin 1}{1 - \sin 1}$

39. (a) $\sum\limits_{n=0}^{\infty} \dfrac{4}{10}(0.1)^n$ **41.** (a) $\sum\limits_{n=0}^{\infty} \dfrac{12}{100}\left(\dfrac{1}{100}\right)^n$

(b) $\dfrac{4}{9}$ (b) $\dfrac{4}{33}$

43. (a) $\sum\limits_{n=0}^{\infty} \dfrac{3}{40}(0.01)^n$ (b) $\dfrac{5}{66}$

45. Diverges   **47.** Diverges   **49.** Converges

51. Diverges   **53.** Diverges   **55.** Diverges

57. Diverges

59. Yes. If you remove a finite number of terms, the sum of the sequence of partial sums still diverges.

61. $|x| < \dfrac{1}{3}; \dfrac{3x}{1 - 3x}$ **63.** $0 < x < 2; \dfrac{x - 1}{2 - x}$

65. $-1 < x < 1; \dfrac{1}{1 + x}$

67. (a) x (b) $f(x) = \dfrac{1}{1 - x}, \quad |x| < 1$

(c)

Answers will vary.

69. The required terms for the two series are $n = 100$ and $n = 5$, respectively. The second series converges at a higher rate.

71. $160,000(1 - 0.95^n)$ units

73. $\sum\limits_{i=0}^{\infty} 200(0.75)^i$; Sum $= \$800$ million   **75.** 152.42 ft

77. $\dfrac{1}{8}; \sum\limits_{n=0}^{\infty} \dfrac{1}{2}\left(\dfrac{1}{2}\right)^n = \dfrac{1/2}{1 - 1/2} = 1$

79. (a) $-1 + \sum\limits_{n=0}^{\infty} \left(\dfrac{1}{2}\right)^n = -1 + \dfrac{a}{1 - r} = -1 + \dfrac{1}{1 - 1/2} = 1$

(b) No   (c) 2

81. (a) 126 in.^{2} (b) 128 in.^{2}

83. The $2,000,000 sweepstakes has a present value of $1,146,992.12. After accruing interest over the 20-year period, it attains its full value.

85. (a) $5,368,709.11   (b) $10,737,418.23   (c) $21,474,836.47

87. (a) $14,739.84   (b) $14,742.45

89. (a) $518,136.56   (b) $518,168.67

91. False. $\lim\limits_{n \to \infty} \dfrac{1}{n} = 0$, but $\sum\limits_{n=1}^{\infty} \dfrac{1}{n}$ diverges.

93. False. $\displaystyle\sum_{n=1}^{\infty} ar^n = \frac{a}{1-r} - a$; The formula requires that the geometric series begins with $n = 0$.

95. True

97. Answers will vary. Sample answer: $\displaystyle\sum_{n=0}^{\infty} 1$, $\displaystyle\sum_{n=0}^{\infty} (-1)$

99–101. Proofs   **103.** Putnam Problem A2, 1984

Section 9.3   (page 613)

1. f must be positive, continuous, and decreasing for $x \geq 1$ and $a_n = f(n)$.

3. Diverges   **5.** Converges   **7.** Converges

9. Diverges   **11.** Diverges   **13.** Converges

15. Converges   **17.** Converges   **19.** Diverges

21. Converges   **23.** Diverges

25. $f(x)$ is not positive for $x \geq 1$.

27. $f(x)$ is not always decreasing.   **29.** Converges

31. Diverges   **33.** Diverges   **35.** Converges

37. Converges

39. (a)

n	5	10	20	50	100
S_n	3.7488	3.75	3.75	3.75	3.75

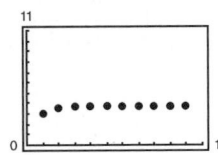

The partial sums approach the sum 3.75 very quickly.

(b)

n	5	10	20	50	100
S_n	1.4636	1.5498	1.5962	1.6251	1.635

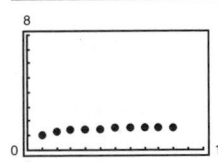

The partial sums approach the sum $\dfrac{\pi^2}{6} \approx 1.6449$ more slowly than the series in part (a).

41. No. Because $\displaystyle\sum_{n=1}^{\infty} \frac{1}{n}$ diverges, $\displaystyle\sum_{n=10,000}^{\infty} \frac{1}{n}$ also diverges. The convergence or divergence of a series is not determined by the first finite number of terms of the series.

43. (a)

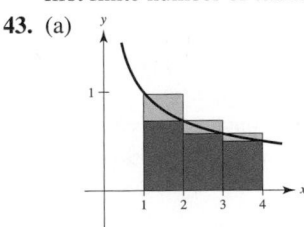

The area under the rectangles is greater than the area under the curve. Because $\displaystyle\int_1^{\infty} \frac{1}{\sqrt{x}}\,dx = \left[2\sqrt{x}\right]_1^{\infty} = \infty$ diverges, $\displaystyle\sum_{n=1}^{\infty} \frac{1}{\sqrt{n}}$ diverges.

(b)

The area under the rectangles is less than the area under the curve. Because $\displaystyle\int_1^{\infty} \frac{1}{x^2}\,dx = \left[-\frac{1}{x}\right]_1^{\infty} = 1$ converges, $\displaystyle\sum_{n=2}^{\infty} \frac{1}{n^2}$ converges $\left(\text{and so does } \displaystyle\sum_{n=1}^{\infty} \frac{1}{n^2}\right)$.

45. $p > 1$ **47.** $p > 1$ **49.** $p > 3$ **51.** Proof

53. $S_3 \approx 1.0748$ **55.** $S_8 \approx 0.9597$
    $R_3 \approx 0.0123$ $R_8 \approx 0.1244$

57. $S_4 \approx 0.4049$
    $R_4 \approx 5.6 \times 10^{-8}$

59. $N \geq 7$ **61.** $N \geq 16$

63. (a) $\displaystyle\sum_{n=2}^{\infty} \frac{1}{n^{1.1}}$ converges by the p-Series Test because $1.1 > 1$.

$\displaystyle\sum_{n=2}^{\infty} \frac{1}{n \ln n}$ diverges by the Integral Test because

$\displaystyle\int_2^{\infty} \frac{1}{x \ln x}\,dx$ diverges.

(b) $\displaystyle\sum_{n=2}^{\infty} \frac{1}{n^{1.1}} = 0.4665 + 0.2987 + 0.2176 + 0.1703$
$+ 0.1393 + \cdots$

$\displaystyle\sum_{n=2}^{\infty} \frac{1}{n \ln n} = 0.7213 + 0.3034 + 0.1803 + 0.1243$
$+ 0.0930 + \cdots$

(c) $n \geq 3.431 \times 10^{15}$

65. (a) Let $f(x) = \dfrac{1}{x}$. f is positive, continuous, and decreasing on $[1, \infty)$.

$S_n - 1 \leq \displaystyle\int_1^n \frac{1}{x}\,dx = \ln n$

$S_n \geq \displaystyle\int_1^{n+1} \frac{1}{x}\,dx \ln(n + 1)$

So, $\ln(n + 1) \leq S_n \leq 1 + \ln n$.

(b) $\ln(n + 1) - \ln n \leq S_n - \ln n \leq 1$
Also, $\ln(n + 1) - \ln n > 0$ for $n \geq 1$. So,
$0 \leq S_n - \ln n \leq 1$, and the sequence $\{a_n\}$ is bounded.

(c) $a_n - a_{n+1} = [S_n - \ln n] - [S_{n+1} - \ln(n + 1)]$
$= \displaystyle\int_n^{n+1} \frac{1}{x}\,dx - \frac{1}{n + 1} \geq 0$
So, $a_n \geq a_{n+1}$.

(d) Because the sequence is bounded and monotonic, it converges to a limit, γ.

(e) 0.5822

67. (a) Diverges   (b) Diverges

(c) $\displaystyle\sum_{n=2}^{\infty} x^{\ln n}$ converges for $x < \dfrac{1}{e}$.

69. Diverges   **71.** Converges   **73.** Converges

75. Diverges   **77.** Diverges   **79.** Converges

Section 9.4 (page 620)

1. Yes. The test requires that $0 \le a_n \le b_n$ for all n greater than some integer N. The beginning terms do not affect the convergence or divergence of a series.

3. (a)

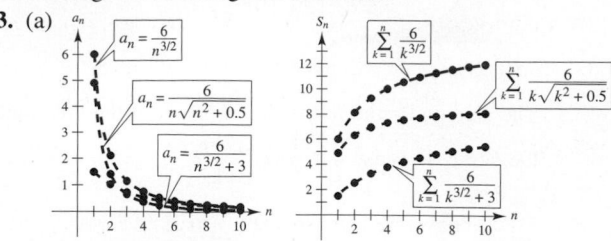

(b) $\displaystyle\sum_{n=1}^{\infty} \frac{6}{n^{3/2}}$; Converges

(c) The magnitudes of the terms are less than the magnitudes of the terms of the p-series. Therefore, the series converges.

(d) The smaller the magnitudes of the terms, the smaller the magnitudes of the terms of the sequence of partial sums.

5. Diverges   **7.** Diverges   **9.** Diverges

11. Converges   **13.** Converges   **15.** Converges

17. Diverges   **19.** Diverges   **21.** Converges

23. Converges   **25.** Diverges

27. Diverges; p-Series Test

29. Converges; Direct Comparison Test with $\displaystyle\sum_{n=1}^{\infty} \left(\frac{1}{5}\right)^n$

31. Diverges; nth-Term Test   **33.** Converges; Integral Test

35. $\displaystyle\lim_{n\to\infty} \frac{a_n}{1/n} = \lim_{n\to\infty} na_n$; $\displaystyle\lim_{n\to\infty} na_n \ne 0$ but is finite.

The series diverges by the Limit Comparison Test.

37. Diverges   **39.** Converges

41. $\displaystyle\lim_{n\to\infty} n\left(\frac{n^3}{5n^4 + 3}\right) = \frac{1}{5} \ne 0$

43. Diverges   **45.** Converges

47. Convergence or divergence is dependent on the form of the general term for the series and not necessarily on the magnitudes of the terms.

49. (a) Proof

(b)

n	5	10	20	50	100
S_n	1.1839	1.2087	1.2212	1.2287	1.2312

(c) 0.1226   (d) 0.0277

51. Proof   **53.** False. Let $a_n = \dfrac{1}{n^3}$ and $b_n = \dfrac{1}{n^2}$.

55. True   **57.** True   **59.** Proof   **61.** $\displaystyle\sum_{n=1}^{\infty} \frac{1}{n^2}, \sum_{n=1}^{\infty} \frac{1}{n^3}$

63–69. Proofs

71. Putnam Problem 1, afternoon session, 1953

Section 9.5 (page 629)

1. The series diverges because of the nth-Term Test for Divergence.

3. Σa_n is absolutely convergent if $\Sigma |a_n|$ converges. Σa_n is conditionally convergent if $\Sigma |a_n|$ diverges, but Σa_n converges.

5. (a)

n	1	2	3	4	5
S_n	1.0000	0.6667	0.8667	0.7238	0.8349

n	6	7	8	9	10
S_n	0.7440	0.8209	0.7543	0.8131	0.7605

(b)

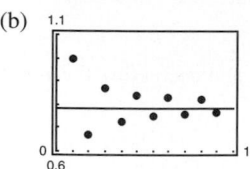

(c) The points alternate sides of the horizontal line $y = \dfrac{\pi}{4}$ that represents the sum of the series. The distances between the successive points and the line decrease.

(d) The distance in part (c) is always less than the magnitude of the next term of the series.

7. (a)

n	1	2	3	4	5
S_n	1.0000	0.7500	0.8611	0.7986	0.8386

n	6	7	8	9	10
S_n	0.8108	0.8312	0.8156	0.8280	0.8180

(b)

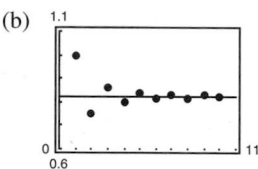

(c) The points alternate sides of the horizontal line $y = \dfrac{\pi^2}{12}$ that represents the sum of the series. The distances between the successive points and the line decrease.

(d) The distance in part (c) is always less than the magnitude of the next term of the series.

9. Converges   **11.** Converges   **13.** Diverges

15. Diverges   **17.** Converges   **19.** Diverges

21. Diverges   **23.** Converges   **25.** Converges

27. Converges   **29.** Converges   **31.** $1.8264 \le S \le 1.8403$

33. $1.7938 \le S \le 1.8054$ **35.** 10   **37.** 7

39. 7 terms (Note that the sum begins with $n = 0$.)

41. Converges absolutely   **43.** Converges absolutely

45. Converges conditionally   **47.** Diverges

49. Converges conditionally   **51.** Converges absolutely

53. Converges absolutely   **55.** Converges conditionally

57. Converges absolutely

59. Overestimate; The next term is negative.

61. (a) False. For example, let $a_n = \dfrac{(-1)^n}{n}$.

Then $\sum a_n = \sum \dfrac{(-1)^n}{n}$ converges

and $\sum (-a_n) = \sum \dfrac{(-1)^{n+1}}{n}$ converges.

But, $\sum |a_n| = \sum \dfrac{1}{n}$ diverges.

(b) True. For if $\sum |a_n|$ converged, then so would $\sum a_n$ by Theorem 9.16.

63. $p > 0$

65. Proof; The converse is false. For example: Let $a_n = \dfrac{1}{n}$.

67. $\displaystyle\sum_{n=1}^{\infty} \dfrac{1}{n^2}$ converges, and so does $\displaystyle\sum_{n=1}^{\infty} \dfrac{1}{n^4}$.

69. (a) No; $a_{n+1} \le a_n$ is not satisfied for all n. For example, $\frac{1}{9} < \frac{1}{8}$.
(b) Yes; 0.5

71. Diverges; p-Series Test   **73.** Diverges; nth-Term Test

75. Diverges; Geometric Series Test

77. Converges; Integral Test

79. Converges; Alternating Series Test

81. You cannot arbitrarily change 0 to $1 - 1$.

Section 9.6   (page 637)

1. Conveges   **3.** Diverges   **5.** Inconclusive   **7.** Proof

9. d   **10.** c   **11.** f   **12.** b   **13.** a   **14.** e

15. (a) Proof

(b)

n	5	10	15	20	25
S_n	13.7813	24.2363	25.8468	25.9897	25.9994

(c) 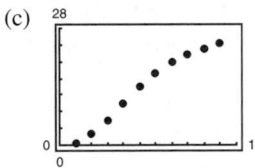   (d) 26

(e) The more rapidly the terms of the series approach 0, the more rapidly the sequence of partial sums approaches the sum of the series.

17. Converges   **19.** Diverges   **21.** Diverges

23. Diverges   **25.** Converges   **27.** Converges

29. Diverges   **31.** Converges   **33.** Converges

35. Diverges   **37.** Converges   **39.** Converges

41. Diverges   **43.** Converges   **45.** Diverges

47. Converges   **49.** Converges   **51.** Converges

53. Converges; Alternating Series Test

55. Converges; p-Series Test   **57.** Diverges; nth-Term Test

59. Diverges; Geometric Series Test

61. Converges; Limit Comparison Test with $b_n = \dfrac{1}{2^n}$

63. Converges; Direct Comparison Test with $b_n = \dfrac{1}{3^n}$

65. Diverges; Ratio Test   **67.** Converges; Ratio Test

69. Converges; Ratio Test   **71.** a and c   **73.** a and b

75. $\displaystyle\sum_{n=0}^{\infty} \dfrac{n+1}{7^{n+1}}$ **77.** Diverges; $\displaystyle\lim_{n\to\infty} \left|\dfrac{a_{n+1}}{a_n}\right| > 1$

79. Converges; $\displaystyle\lim_{n\to\infty} \left|\dfrac{a_{n+1}}{a_n}\right| < 1$ **81.** Diverges; $\lim a_n \ne 0$

83. Converges   **85.** Converges   **87.** $(-3, 3)$

89. $(-2, 0]$ **91.** $x = 0$ **93.** The test is inconclusive.

95. No; The series $\displaystyle\sum_{n=1}^{\infty} \dfrac{1}{n + 10,000}$ diverges.

97–103. Proofs

105. (a) Diverges   (b) Converges   (c) Converges
(d) Converges for all integers $x \ge 2$

107. Putnam Problem 7, morning session, 1951

Section 9.7   (page 648)

1. The graphs of the approximating polynomial P and the elementary function f both pass through the point $(c, f(c))$, and the slope of the graph of P is the same as the slope of the graph of f at the point $(c, f(c))$. If P is of degree n, then the first n derivatives of f and P agree at c. This allows the graph of P to resemble the graph of f near the point $(c, f(c))$.

3. The accuracy is represented by the remainder of the Taylor polynomial. The remainder is $R_n(x) = \dfrac{f^{(n+1)}(z)(x - c)^{n+1}}{(n + 1)!}$.

5. d   **6.** c   **7.** a   **8.** b

9. $P_1 = \dfrac{1}{16}x + \dfrac{1}{4}$ **11.** $P_1 = \dfrac{2\sqrt{3}}{3} + \dfrac{2}{3}\left(x - \dfrac{\pi}{6}\right)$

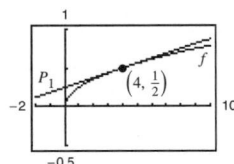   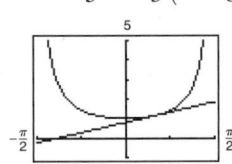

13.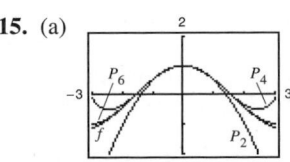

x	0	0.8	0.9	1	1.1
$f(x)$	Error	4.4721	4.2164	4.0000	3.8139
$P_2(x)$	7.5000	4.4600	4.2150	4.0000	3.8150

x	1.2	2
$f(x)$	3.6515	2.8284
$P_2(x)$	3.6600	3.5000

15. (a)

(b) $f^{(2)}(0) = -1$ $P_2^{(2)}(0) = -1$
$f^{(4)}(0) = 1$ $P_4^{(4)}(0) = 1$
$f^{(6)}(0) = -1$ $P_6^{(6)}(0) = -1$

(c) $f^{(n)}(0) = P_n^{(n)}(0)$

17. $1 + 4x + 8x^2 + \frac{32}{3}x^3 + \frac{32}{3}x^4$ **19.** $x - \frac{1}{6}x^3 + \frac{1}{120}x^5$

21. $x + x^2 + \frac{1}{2}x^3 + \frac{1}{6}x^4$ **23.** $1 + x + x^2 + x^3 + x^4 + x^5$

25. $1 + \frac{1}{2}x^2$ **27.** $2 - 2(x - 1) + 2(x - 1)^2 - 2(x - 1)^3$

29. $2 + \frac{1}{4}(x - 4) - \frac{1}{64}(x - 4)^2$

31. $\ln 2 + \frac{1}{2}(x - 2) - \frac{1}{8}(x - 2)^2 + \frac{1}{24}(x - 2)^3 - \frac{1}{64}(x - 2)^4$

33. (a) $P_3(x) = \pi x + \frac{\pi^3}{3}x^3$

(b) $Q_3(x) = 1 + 2\pi\left(x - \frac{1}{4}\right) + 2\pi^2\left(x - \frac{1}{4}\right)^2 + \frac{8\pi^3}{3}\left(x - \frac{1}{4}\right)^3$

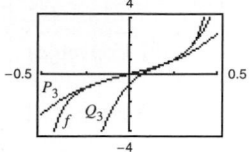

35. (a)

x	0	0.25	0.50	0.75	1.00
$\sin x$	0	0.2474	0.4794	0.6816	0.8415
$P_1(x)$	0	0.25	0.50	0.75	1.00
$P_3(x)$	0	0.2474	0.4792	0.6797	0.8333
$P_5(x)$	0	0.2474	0.4794	0.6817	0.8417

(b)

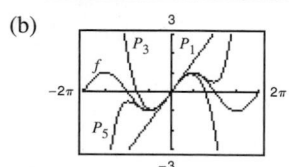

(c) As the distance increases, the polynomial approximation becomes less accurate.

37. (a) $P_3(x) = x + \frac{1}{6}x^3$

(b)

x	-0.75	-0.50	-0.25	0	0.25
$f(x)$	-0.848	-0.524	-0.253	0	0.253
$P_3(x)$	-0.820	-0.521	-0.253	0	0.253

x	0.50	0.75
$f(x)$	0.524	0.848
$P_3(x)$	0.521	0.820

(c)

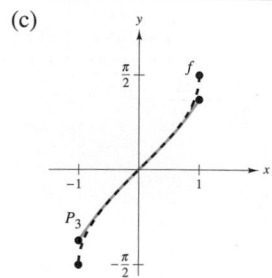

39. 2.7083   **41.** 0.227   **43.** 0.7419

45. $R_4 \le 2.03 \times 10^{-5}$; 0.000001

47. $R_5 \le 1.8 \times 10^{-8}$; 2.5×10^{-9}

49. $R_3 \le 7.82 \times 10^{-3}$; 0.00085   **51.** 3   **53.** 5   **55.** 2

57. $n = 9$; $\ln(1.5) \approx 0.4055$ **59.** $-0.3936 < x < 0$

61. $-0.9467 < x < 0.9467$

63. The tangent line to a function at a point is the first Taylor polynomial for the function at the point.

65. Substitute $2x$ into the polynomial for $f(x) = e^x$ to obtain the polynomial for $g(x) = e^{2x}$.

67. (a) $f(x) \approx P_4(x) = 1 + x + \frac{1}{2}x^2 + \frac{1}{6}x^3 + \frac{1}{24}x^4$

$g(x) \approx Q_5(x) = x + x^2 + \frac{1}{2}x^3 + \frac{1}{6}x^4 + \frac{1}{24}x^5$

$Q_5(x) = xP_4(x)$

(b) $g(x) \approx P_6(x) = x^2 - \frac{x^4}{3!} + \frac{x^6}{5!}$

(c) $g(x) \approx P_4(x) = 1 - \frac{x^2}{3!} + \frac{x^4}{5!}$

69. (a) $Q_2(x) = -1 + \left(\frac{\pi^2}{32}\right)(x + 2)^2$

(b) $R_2(x) = -1 + \left(\frac{\pi^2}{32}\right)(x - 6)^2$

(c) No. Horizontal translations of the result in part (a) are possible only at $x = -2 + 8n$ (where n is an integer) because the period of f is 8.

71–73. Proofs

Section 9.8   *(page 658)*

1. A Maclaurin polynomial approximates a function, whereas a power series exactly represents a function. The Maclaurin polynomial has a finite number of terms and the power series has an infinite number of terms.

3. $R = 5$ **5.** 0   **7.** 2   **9.** $R = 1$ **11.** $R = \frac{1}{4}$

13. $R = \infty$ **15.** $(-4, 4)$ **17.** $(-1, 1]$ **19.** $(-\infty, \infty)$

21. $x = 0$ **23.** $(-6, 6)$ **25.** $(-5, 13]$ **27.** $(0, 2]$

29. $(0, 6)$ **31.** $\left(-\frac{1}{2}, \frac{1}{2}\right)$ **33.** $(-\infty, \infty)$ **35.** $(-1, 1)$

37. $x = 3$ **39.** $R = c$ **41.** $(-k, k)$ **43.** $(-1, 1)$

45. $\displaystyle\sum_{n=1}^{\infty} \frac{x^{n-1}}{(n - 1)!}$ **47.** $\displaystyle\sum_{n=1}^{\infty} \frac{x^n}{(7n + 6)!}$

49. (a) $(-3, 3)$ (b) $(-3, 3)$ (c) $(-3, 3)$ (d) $[-3, 3]$

51. (a) $(0, 2]$ (b) $(0, 2)$ (c) $(0, 2)$ (d) $[0, 2]$

53. $\displaystyle\sum_{n=1}^{\infty} \left(\frac{x}{3}\right)^n$; Answers will vary.

55. Answers will vary. Sample answer:

$\displaystyle\sum_{n=1}^{\infty} \frac{x^n}{n}$ converges for $-1 \le x < 1$. At $x = -1$, the convergence is conditional because $\sum \frac{1}{n}$ diverges.

$\displaystyle\sum_{n=1}^{\infty} \frac{x^n}{n^2}$ converges for $-1 \le x \le 1$. At $x = \pm 1$, the convergence is absolute.

57. (a) For $f(x)$: $(-\infty, \infty)$; For $g(x)$: $(-\infty, \infty)$

(b) Proofs   (c) $f(x) = \sin x$, $g(x) = \cos x$

59–63. Proofs

65. (a) and (b) Proofs

(c)   (d) 0.92

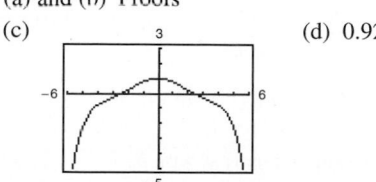

67. (a) $\frac{8}{3}$ (b) $\frac{8}{13}$

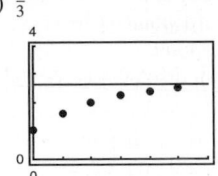
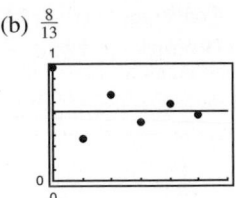

(c) The alternating series converges more rapidly. The partial sums of the series of positive terms approach the sum from below. The partial sums of the alternating series alternate sides of the horizontal line representing the sum.

(d)

M	10	100	1000	10,000
N	5	14	24	35

69.
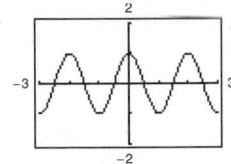

71.
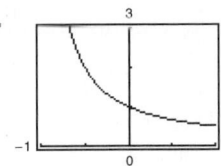

$f(x) = \cos \pi x$ $f(x) = \dfrac{1}{1+x}$

73. False. Let $a_n = \dfrac{(-1)^n}{n2^n}$.   **75.** True   **77.** Proof

79. (a) $(-1, 1)$ (b) $f(x) = \dfrac{c_0 + c_1 x + c_2 x^2}{1 - x^3}$

81. Proof

Section 9.9   *(page 666)*

1. You need to algebraically manipulate $\dfrac{b}{c-x}$ so that it resembles the form $\dfrac{a}{1-r}$.

3. $\displaystyle\sum_{n=0}^{\infty} \dfrac{x^n}{4^{n+1}}$ **5.** $\displaystyle\sum_{n=0}^{\infty} \dfrac{4}{3}\left(\dfrac{-x}{3}\right)^n$

7. $\displaystyle\sum_{n=0}^{\infty} \left(\dfrac{1}{5}\right)\left(\dfrac{x-1}{5}\right)^n$ **9.** $\displaystyle\sum_{n=0}^{\infty} (3x)^n$

$(-4, 6)$ $\left(-\dfrac{1}{3}, \dfrac{1}{3}\right)$

11. $-\dfrac{5}{9}\displaystyle\sum_{n=0}^{\infty} \left[\dfrac{2}{9}(x+3)\right]^n$ **13.** $-2\displaystyle\sum_{n=0}^{\infty} [5(x+1)]^n$

$\left(-\dfrac{15}{2}, \dfrac{3}{2}\right)$ $\left(-\dfrac{6}{5}, -\dfrac{4}{5}\right)$

15. $\displaystyle\sum_{n=0}^{\infty} \left[\dfrac{1}{(-3)^n} - 1\right]x^n$ **17.** $\displaystyle\sum_{n=0}^{\infty} x^n[1 + (-1)^n] = 2\displaystyle\sum_{n=0}^{\infty} x^{2n}$

$(-1, 1)$ $(-1, 1)$

19. $2\displaystyle\sum_{n=0}^{\infty} x^{2n}$ **21.** $\displaystyle\sum_{n=1}^{\infty} n(-1)^n x^{n-1}$ **23.** $\displaystyle\sum_{n=0}^{\infty} \dfrac{(-1)^n x^{n+1}}{n+1}$

$(-1, 1)$ $(-1, 1)$ $(-1, 1]$

25. $\displaystyle\sum_{n=0}^{\infty} (-1)^n x^{2n}$ **27.** $\displaystyle\sum_{n=0}^{\infty} (-1)^n (2x)^{2n}$

$(-1, 1)$ $\left(-\dfrac{1}{2}, \dfrac{1}{2}\right)$

29.
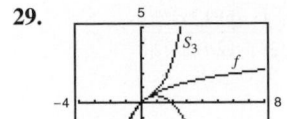

x	0.0	0.2	0.4	0.6	0.8	1.0
S_2	0.000	0.180	0.320	0.420	0.480	0.500
$\ln(x+1)$	0.000	0.182	0.336	0.470	0.588	0.693
S_3	0.000	0.183	0.341	0.492	0.651	0.833

31. (a) 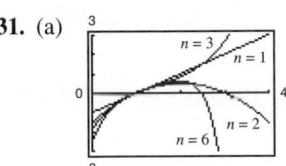   (b) $\ln x,\ 0 < x \le 2,\ R = 1$
(c) -0.6931
(d) $\ln(0.5)$; The error is approximately 0.

33. 0.245   **35.** 0.125   **37.** $\displaystyle\sum_{n=1}^{\infty} nx^{n-1},\ -1 < x < 1$

39. $\displaystyle\sum_{n=0}^{\infty} (2n+1)x^n,\ -1 < x < 1$

41. $E(n) = 2$; Yes. Because the probability of obtaining a head on a single toss is $\frac{1}{2}$, it is expected that, on average, a head will be obtained in two tosses.

43. Proof   **45.** (a) Proof   (b) 3.14

47. $\ln\frac{3}{2} \approx 0.4055$; See Exercise 23.

49. $\ln\frac{7}{5} \approx 0.3365$; See Exercise 47.

51. $\arctan\frac{1}{2} \approx 0.4636$; See Exercise 50.

53. The series in Exercise 50 converges to its sum at a lower rate because its terms approach 0 at a much lower rate.

55. The series converges on the interval $(-5, 3)$ and perhaps also at one or both endpoints.

57. $S_1 = 0.3183098862,\ \dfrac{1}{\pi} \approx 0.3183098862$

Section 9.10   *(page 677)*

1. The Taylor series converges to $f(x)$ if and only if $R_n(x) \to 0$ as $n \to \infty$.

3. Multiply and divide as you would polynomials.

5. $\displaystyle\sum_{n=0}^{\infty} \dfrac{(2x)^n}{n!}$ **7.** $\dfrac{\sqrt{2}}{2}\displaystyle\sum_{n=0}^{\infty} \dfrac{(-1)^{n(n+1)/2}}{n!}\left(x - \dfrac{\pi}{4}\right)^n$

9. $\displaystyle\sum_{n=0}^{\infty} (-1)^n (x-1)^n$ **11.** $\displaystyle\sum_{n=0}^{\infty} \dfrac{(-1)^n (x-1)^{n+1}}{n+1}$

13. $\displaystyle\sum_{n=0}^{\infty} \dfrac{(-1)^n (3x)^{2n+1}}{(2n+1)!}$ **15.** $1 + \dfrac{x^2}{2!} + \dfrac{5x^4}{4!} + \cdots$

17–19. Proofs   **21.** $1 + \displaystyle\sum_{n=1}^{\infty} \dfrac{1 \cdot 3 \cdot 5 \cdots (2n-1)x^n}{2^n n!}$

23. $1 + \displaystyle\sum_{n=1}^{\infty} \dfrac{1 \cdot 3 \cdot 5 \cdots (2n-1)x^{2n}}{2^n n!}$

25. $1 + \dfrac{1}{4}x + \displaystyle\sum_{n=2}^{\infty} \dfrac{(-1)^{n+1} 3 \cdot 7 \cdot 11 \cdots (4n-5)x^n}{4^n n!}$

27. $\displaystyle\sum_{n=0}^{\infty} \dfrac{x^{2n}}{2^n n!}$ **29.** $\displaystyle\sum_{n=1}^{\infty} \dfrac{(-1)^{n-1} x^n}{n}$ **31.** $\displaystyle\sum_{n=0}^{\infty} \dfrac{(-1)^n 4^{2n} x^{2n}}{(2n)!}$

33. $\displaystyle\sum_{n=0}^{\infty} \dfrac{(-1)^n (5x)^{2n+1}}{2n+1}$ **35.** $\displaystyle\sum_{n=0}^{\infty} \dfrac{(-1)^n x^{3n}}{(2n)!}$

37. $\displaystyle\sum_{n=0}^{\infty} \frac{x^{2n+1}}{(2n+1)!}$ **39.** $\displaystyle\frac{1}{2}\left[1 + \sum_{n=0}^{\infty} \frac{(-1)^n(2x)^{2n}}{(2n)!}\right]$

41. Proof   **43.** $\displaystyle\sum_{n=0}^{\infty} \frac{(-1)^n x^{2n+2}}{(2n+1)!}$

45. $\begin{cases} \displaystyle\sum_{n=0}^{\infty} \frac{(-1)^n x^{2n}}{(2n+1)!}, & x \neq 0 \\ 1, & x = 0 \end{cases}$

47. $P_5(x) = x + x^2 + \frac{1}{3}x^3 - \frac{1}{30}x^5$

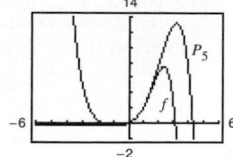

49. $P_5(x) = x - \frac{1}{2}x^2 - \frac{1}{6}x^3 + \frac{3}{40}x^5$

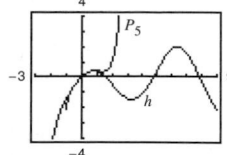

51. $P_4(x) = x - x^2 + \frac{5}{6}x^3 - \frac{5}{6}x^4$

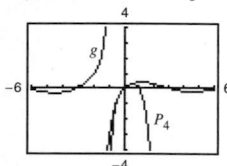

53. $\displaystyle\sum_{n=0}^{\infty} \frac{(-1)^{(n+1)}x^{2n+3}}{(2n+3)(n+1)!}$ **55.** 0.6931   **57.** 7.3891

59. 0   **61.** 1   **63.** 0.8075   **65.** 0.9461   **67.** 0.4872

69. 0.2010   **71.** 0.7040   **73.** 0.3412

75. Square the series for $\cos x$, use a half-angle identity, or compute the coefficients using the definition.

First three terms: $1, \dfrac{x^2}{2}, \dfrac{x^4}{3}$

77. $f(x) = \dfrac{\sin(x+3)}{4}$; Answers will vary.

79. Proof

81. (a)

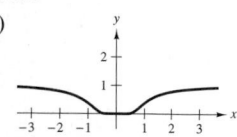

(b) Proof

(c) $\displaystyle\sum_{n=0}^{\infty} 0x^n = 0 \neq f(x)$; The series converges to f at $x = 0$ only.

83. Proof   **85.** 20   **87.** -0.612864 **89.** $\displaystyle\sum_{n=0}^{\infty} \binom{k}{n}x^n$

91. Proof   **93.** Putnam Problem 4, morning session, 1962

Review Exercises for Chapter 9 *(page 680)*

1. 4, 34, 214, 1294, 7774   **3.** $-\frac{1}{4}, \frac{1}{16}, -\frac{1}{64}, \frac{1}{256}, -\frac{1}{1024}$

5. a   **6.** c   **7.** d   **8.** b

9.

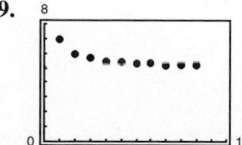

Converges to 5

11. Converges to 0   **13.** Converges to 5   **15.** Diverges

17. Diverges   **19.** $a_n = 5n - 2$; diverges

21. $a_n = \dfrac{1}{(n!+1)}$; converges   **23.** Monotonic, bounded

25. (a)

n	1	2	3	4
A_n	\$8100.00	\$8201.25	\$8303.77	\$8407.56

n	5	6	7	8
A_n	\$8512.66	\$8619.07	\$8726.80	\$8835.89

(b) \$13,148.96

27. 3, 4.5, 5.5, 6.25, 6.85

29. (a)

n	5	10	15	20	25
S_n	13.2	113.3	873.8	6648.5	50,500.3

(b)

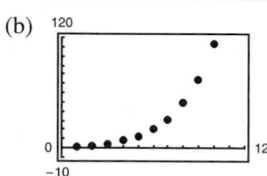

31. $\dfrac{5}{3}$ **33.** $\dfrac{35}{3}$ **35.** (a) $\displaystyle\sum_{n=0}^{\infty}(0.09)(0.01)^n$ (b) $\dfrac{1}{11}$

37. Diverges   **39.** Diverges

41. $120{,}000[1 - 0.92^n], n > 0$ **43.** Diverges

45. Converges   **47.** Diverges   **49.** Diverges

51. Converges   **53.** Diverges   **55.** Converges

57. Converges   **59.** Diverges   **61.** 10   **63.** Diverges

65. Diverges   **67.** Converges

69. (a) Proof

(b)

n	5	10	15	20	25
S_n	2.8752	3.6366	3.7377	3.7488	3.7499

(c)

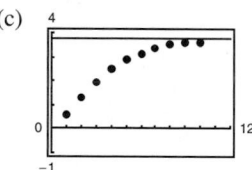

(d) 3.75

71. Converges; p-Series Test   **73.** Diverges; nth-Term Test

75. Diverges; Limit Comparison Test

77. $P_3(x) = 1 - 2x + 2x^2 - \frac{4}{3}x^3$

79. $P_3(x) = 1 - 3(x-1) + 6(x-1)^2 - 10(x-1)^3$

81. 3   **83.** $(-10, 10)$ **85.** $[1, 3]$

87. Converges only at $x = 2$

89. (a) $(-5, 5)$ (b) $(-5, 5)$ (c) $(-5, 5)$ (d) $[-5, 5)$

91. Proof   **93.** $\displaystyle\sum_{n=0}^{\infty} \frac{2}{3}\left(\frac{x}{3}\right)^n$ **95.** $\displaystyle\sum_{n=0}^{\infty} 2\left(\frac{x-1}{3}\right)^n$; $(-2, 4)$

97. $\ln\frac{5}{4} \approx 0.2231$ **99.** $e^{1/2} \approx 1.6487$

101. $\cos\frac{2}{3} \approx 0.7859$ **103.** $\dfrac{\sqrt{2}}{2}\displaystyle\sum_{n=0}^{\infty} \frac{(-1)^{n(n+1)/2}}{n!}\left(x - \frac{3\pi}{4}\right)^n$

105. $\displaystyle\sum_{n=0}^{\infty} \frac{(x\ln 3)^n}{n!}$ **107.** $-\displaystyle\sum_{n=0}^{\infty} (x+1)^n$

109. $1 + \dfrac{x}{5} - \dfrac{2x^2}{25} + \dfrac{6x^3}{125} - \dfrac{21x^4}{625} + \cdots$

111. (a)–(c) $1 + 2x + 2x^2 + \dfrac{4}{3}x^3$ **113.** $\displaystyle\sum_{n=0}^{\infty} \dfrac{(6x)^n}{n!}$

115. $\displaystyle\sum_{n=0}^{\infty} \dfrac{(-1)^n (5x)^{2n+1}}{(2n+1)!}$ **117.** 0.5

P.S. Problem Solving   (page 683)

1. (a) 1 (b) Answers will vary. Sample answer: $0, \frac{1}{3}, \frac{2}{3}$
(c) 0

3. Proof   **5.** (a) Proof   (b) Yes   (c) Any distance

7. (a) $\displaystyle\sum_{n=0}^{\infty} \dfrac{x^{n+2}}{(n+2)n!}; \dfrac{1}{2}$ (b) $\displaystyle\sum_{n=0}^{\infty} \dfrac{(n+1)x^n}{n!}; 5.4366$

9. For $a = b$, the series converges conditionally. For no values of a and b does the series converge absolutely.

11. Proof   **13.** (a) and (b) Proofs

15. (a) The height is infinite.   (b) The surface area is infinite.
(c) Proof

Chapter 10

Section 10.1   (page 696)

1. A parabola is the set of all points that are equidistant from a fixed line, the directrix, and a fixed point, the focus, not on the line. An ellipse is the set of all points the sum of whose distances from two distinct fixed points called foci is constant. A hyperbola is the set of all points whose absolute value of the difference between the distances from two distinct fixed points called foci is constant.

3. (a) $0 < e < 1$
(b) As e gets closer to 1, the graph of the ellipse flattens.

5. a   **6.** e   **7.** c   **8.** b   **9.** f   **10.** d

11. Vertex: $(-5, 3)$
Focus: $\left(-\frac{21}{4}, 3\right)$
Directrix: $x = -\frac{19}{4}$

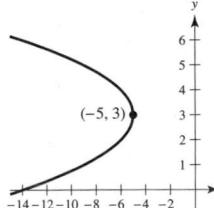

13. Vertex: $(-1, 2)$
Focus: $(0, 2)$
Directrix: $x = -2$

15. Vertex: $(-2, 2)$
Focus: $(-2, 1)$
Directrix: $y = 3$

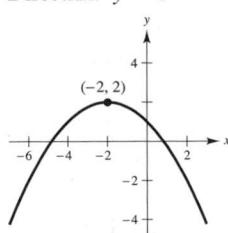

17. $(y - 4)^2 = 4(-2)(x - 5)$ **19.** $(x - 0)^2 = 4(8)(y - 5)$

21. $(x - 1)^2 = 4\left(-\frac{1}{3}\right)(y + 1)$ **23.** $\left(x - \frac{7}{5}\right)^2 = 4\left(\frac{3}{20}\right)\left(y + \frac{4}{15}\right)$

25. Center: $(0, 0)$
Foci: $\left(0, \pm\sqrt{15}\right)$
Vertices: $(0, \pm 4)$
$e = \dfrac{\sqrt{15}}{4}$

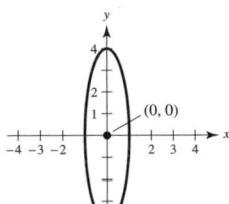

27. Center: $(3, 1)$
Foci: $(3, 4), (3, -2)$
Vertices: $(3, 6), (3, -4)$
$e = \dfrac{3}{5}$

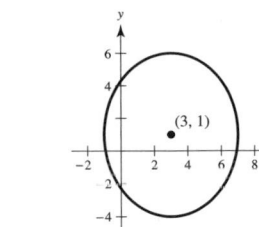

29. Center: $(-2, 3)$
Foci: $\left(-2, 3 \pm \sqrt{5}\right)$
Vertices: $(-2, 6), (-2, 0)$
$e = \dfrac{\sqrt{5}}{3}$

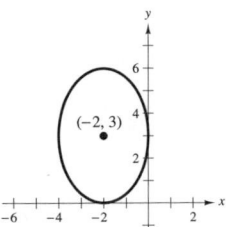

31. $\dfrac{x^2}{36} + \dfrac{y^2}{11} = 1$ **33.** $\dfrac{(x-3)^2}{9} + \dfrac{(y-5)^2}{16} = 1$

35. $\dfrac{x^2}{16} + \dfrac{7y^2}{16} = 1$

37. Center: $(0, 0)$
Foci: $\left(\pm\sqrt{41}, 0\right)$
Vertices: $(\pm 5, 0)$
$e = \dfrac{\sqrt{41}}{5}$
Asymptotes: $y = \pm\dfrac{4}{5}x$

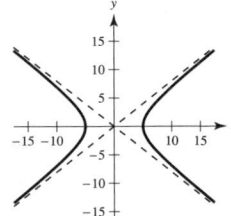

39. Center: $(2, -3)$
Foci: $\left(2 \pm \sqrt{10}, -3\right)$
Vertices: $(1, -3), (3, -3)$
$e = \sqrt{10}$
Asymptotes: $y = -3 \pm 3(x - 2)$

41. $\dfrac{x^2}{1} - \dfrac{y^2}{25} = 1$ **43.** $\dfrac{y^2}{9} - \dfrac{(x-2)^2}{9/4} = 1$

45. $\dfrac{y^2}{4} - \dfrac{x^2}{12} = 1$ **47.** $\dfrac{(x-3)^2}{9} - \dfrac{(y-2)^2}{4} = 1$

49. (a) $\left(6, \sqrt{3}\right)$: $2x - 3\sqrt{3}y - 3 = 0$
$\left(6, -\sqrt{3}\right)$: $2x + 3\sqrt{3}y - 3 = 0$
(b) $\left(6, \sqrt{3}\right)$: $9x + 2\sqrt{3}y - 60 = 0$
$\left(6, -\sqrt{3}\right)$: $9x - 2\sqrt{3}y - 60 = 0$

51. Parabola   **53.** Hyperbola   **55.** Circle

57. (a) Ellipse   (b) Hyperbola   (c) Circle
(d) Answers will vary. Sample answer: Eliminate the y^2-term.

59. Recall that $0 \leq \sin^2 \theta \leq 1$. The circumference is given by

$$C = 4 \int_0^{\pi/2} \sqrt{a^2 - (a^2 - b^2) \sin^2 \theta} \, d\theta$$

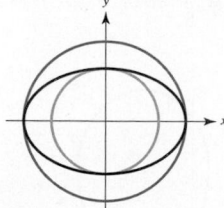

61. $\frac{9}{4}$ m   **63.** (a) Proof   (b) $(3, -3)$ **65.** $y = \frac{1}{180}x^2$

67. $\dfrac{16(4 + 3\sqrt{3} - 2\pi)}{3} \approx 15.536 \text{ ft}^2$

69. Minimum distance: 147,099,713.4 km
        Maximum distance: 152,096,286.6 km

71. $e \approx 0.1373$ **73.** $e \approx 0.9671$

75. (a) Area $= 2\pi$

   (b) Volume $= \dfrac{8\pi}{3}$

   Surface area $= \dfrac{2\pi(9 + 4\sqrt{3}\pi)}{9} \approx 21.48$

   (c) Volume $= \dfrac{16\pi}{3}$

   Surface area $= \dfrac{4\pi[6 + \sqrt{3} \ln(2 + \sqrt{3})]}{3} \approx 34.69$

77. 37.96   **79.** 40   **81.** $\dfrac{(x - 6)^2}{9} - \dfrac{(y - 2)^2}{7} = 1$

83. 110.3 mi   **85.** Proof

87. False. See the definition of a parabola.   **89.** True

91. True   **93.** Putnam Problem B4, 1976

Section 10.2 *(page 707)*

1. The position, direction, and speed at a given time

3. Different parametric representations can be used to represent various speeds at which objects travel along a given path.

5.

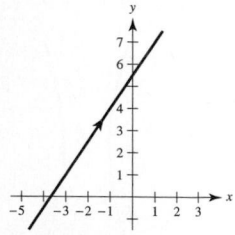

$3x - 2y + 11 = 0$

7.

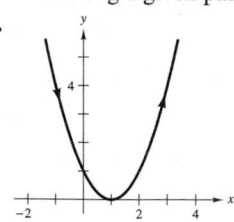

$y = (x - 1)^2$

9.

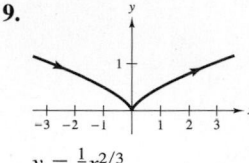

$y = \frac{1}{2}x^{2/3}$

11.

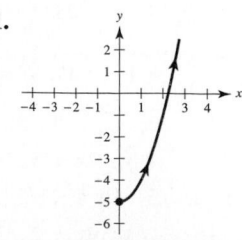

$y = x^2 - 5, \quad x \geq 0$

13.

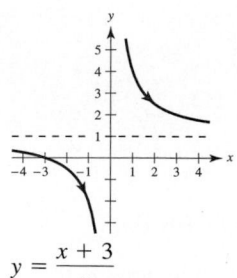

$y = \dfrac{x + 3}{x}$

15.

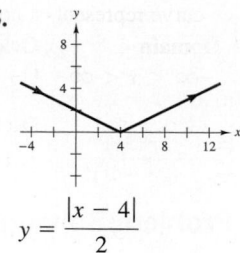

$y = \dfrac{|x - 4|}{2}$

17.

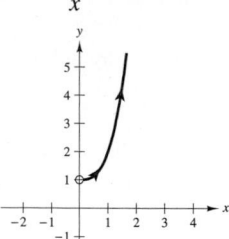

$y = x^3 + 1, \quad x > 0$

19.

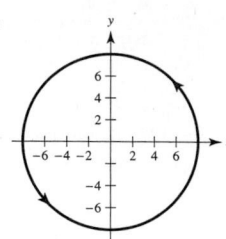

$x^2 + y^2 = 64$

21.

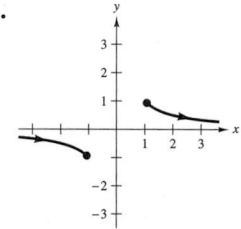

$y = \dfrac{1}{x}, \quad |x| \geq 1$

23.

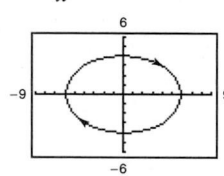

$\dfrac{x^2}{36} + \dfrac{y^2}{16} = 1$

25.

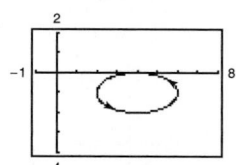

$\dfrac{(x - 4)^2}{4} + \dfrac{(y + 1)^2}{1} = 1$

27.

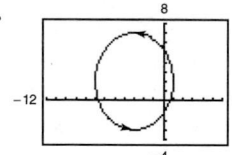

$\dfrac{(x + 3)^2}{16} + \dfrac{(y - 2)^2}{25} = 1$

29.

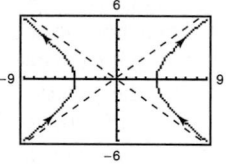

$\dfrac{x^2}{16} - \dfrac{y^2}{9} = 1$

31.

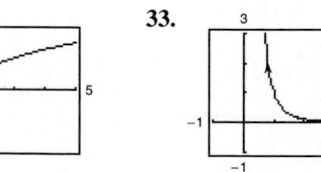

$y = \ln x$

33.

$y = \dfrac{1}{x^3}, \quad x > 0$

35. Both curves represent the parabola $y = x^2$.

	Domain	Orientation	Smooth
(a)	$-\infty < x < \infty$	Left to right	Yes
(b)	$-\infty < x < \infty$	Right to left	Yes

37. Each curve represents a portion of the line $y = 2x + 1$.

	Domain	Orientation	Smooth
(a)	$-\infty < x < \infty$	Up	Yes
(b)	$-1 \le x \le 1$	Oscillates	No, $\dfrac{dx}{d\theta} = \dfrac{dy}{d\theta} = 0$ when $\theta = 0, \pm\pi, \pm2\pi, \ldots$
(c)	$0 < x < \infty$	Down	Yes
(d)	$0 < x < \infty$	Up	Yes

39. $y - y_1 = \dfrac{y_2 - y_1}{x_2 - x_1}(x - x_1)$ **41.** $\dfrac{(x - h)^2}{a^2} + \dfrac{(y - k)^2}{b^2} = 1$

43. $x = 4t$
$y = -7t$
(Solution is not unique.)

45. $x = 1 + 2\cos\theta$
$y = 1 + 2\sin\theta$
(Solution is not unique.)

47. $x = 2 + 5\cos\theta$
$y = 4\sin\theta$
(Solution is not unique.)

49. $x = 2\tan\theta$
$y = \sec\theta$
(Solution is not unique.)

51. $x = t$
$y = 6t - 5$;
$x = t + 1$
$y = 6t + 1$
(Solution is not unique.)

53. $x = t$
$y = t^3$;
$x = \tan t$
$y = \tan^3 t$
(Solution is not unique.)

55. $x = t + 3, \; y = 2t + 1$ **57.** $x = t, \; y = t^2$

59.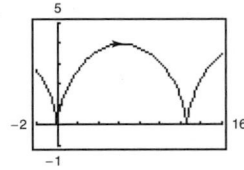
Not smooth at $\theta = 2n\pi$

61.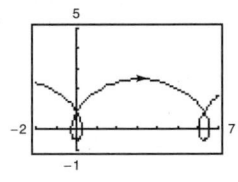
Smooth everywhere

63.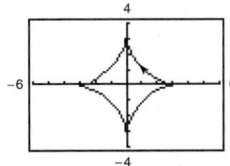
Not smooth at $\theta = \frac{1}{2}n\pi$

65.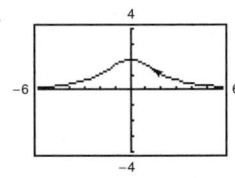
Smooth everywhere

67. The orientation moves right to left on $[-1, 0]$ and left to right on $[0, 1]$, failing to determine a definite direction.

69. No. In the interval $0 < \theta < \pi$, $\cos\theta = \cos(-\theta)$ and $\sin^2\theta = \sin^2(-\theta)$. So, the parameter was not changed.

71. d; $(4, 0)$ is on the graph.   **73.** b; $(1, 0)$ is on the graph.

75. $x = a\theta - b\sin\theta, \; y = a - b\cos\theta$

77. False. The graph of the parametric equations is the portion of the line $y = x$ when $x \ge 0$.

79. True

81. (a) $x = \left(\frac{440}{3}\cos\theta\right)t, \; y = 3 + \left(\frac{440}{3}\sin\theta\right)t - 16t^2$

(b)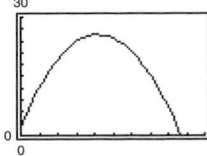
Not a home run

(c)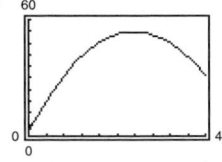
Home run

(d) $19.4°$

Section 10.3   (page 715)

1. The slope of the curve at (x, y)

3. Horizontal tangent lines when $dy/dt = 0$ and $dx/dt \ne 0$ for some value of t; vertical tangent lines when $dx/dt = 0$ and $dy/dt \ne 0$ for some value of t

5. $-\dfrac{3}{t}$ **7.** -1

9. $\dfrac{dy}{dx} = \dfrac{3}{4}, \; \dfrac{d^2y}{dx^2} = 0$; Neither concave upward nor concave downward

11. $\dfrac{dy}{dx} = 2t + 3, \; \dfrac{d^2y}{dx^2} = 2$

At $t = -2, \; \dfrac{dy}{dx} = -1, \; \dfrac{d^2y}{dx^2} = 2$; Concave upward

13. $\dfrac{dy}{dx} = -\cot\theta, \; \dfrac{d^2y}{dx^2} = -\dfrac{(\csc\theta)^3}{4}$

At $\theta = \dfrac{\pi}{4}, \; \dfrac{dy}{dx} = -1, \; \dfrac{d^2y}{dx^2} = -\dfrac{\sqrt{2}}{2}$; Concave downward

15. $\dfrac{dy}{dx} = 2\csc\theta, \; \dfrac{d^2y}{dx^2} = -2\cot^3\theta$

At $\theta = -\dfrac{\pi}{3}, \; \dfrac{dy}{dx} = -\dfrac{4\sqrt{3}}{3}, \; \dfrac{d^2y}{dx^2} = \dfrac{2\sqrt{3}}{9}$;
Concave upward

17. $\dfrac{dy}{dx} = -\tan\theta, \; \dfrac{d^2y}{dx^2} = \sec^4\theta \csc\dfrac{\theta}{3}$

At $\theta = \dfrac{\pi}{4}, \; \dfrac{dy}{dx} = -1, \; \dfrac{d^2y}{dx^2} = \dfrac{4\sqrt{2}}{3}$; Concave upward

19. $\left(-\dfrac{2}{\sqrt{3}}, \dfrac{3}{2}\right)$: $3\sqrt{3}x - 8y + 18 = 0$
$(0, 2)$: $y - 2 = 0$
$\left(2\sqrt{3}, \dfrac{1}{2}\right)$: $\sqrt{3}x + 8y - 10 = 0$

21. $(0, 0)$: $2y - x = 0$
$(-3, -1)$: $y + 1 = 0$
$(-3, 3)$: $2x - y + 9 = 0$

23. (a) and (d)

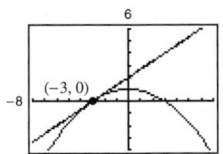

(b) At $t = -\dfrac{1}{2}, \; \dfrac{dx}{dt} = 6, \; \dfrac{dy}{dt} = 4$,
and $\dfrac{dy}{dx} = \dfrac{2}{3}$.

(c) $y = \dfrac{2}{3}x + 2$

25. (a) and (d)

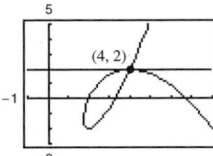

(b) At $t = -1, \; \dfrac{dx}{dt} = -3$,
$\dfrac{dy}{dt} = 0$, and $\dfrac{dy}{dx} = 0$.

(c) $y = 2$

27. $y = \pm\frac{3}{4}x$ **29.** $y = 3x - 5$ and $y = 1$

31. Horizontal: $(-1, -\pi), (-1, \pi), (1, 2\pi), (1, -2\pi)$

Vertical: $\left(\dfrac{\pi}{2}, 1\right), \left(\dfrac{\pi}{2}, -1\right), \left(-\dfrac{3\pi}{2}, 1\right), \left(-\dfrac{3\pi}{2}, -1\right)$

33. Horizontal: $(9, 0)$ **35.** Horizontal: $(2, 22), (6, -10)$
Vertical: None   Vertical: None

37. Horizontal: $(0, 7), (0, -7)$
Vertical: $(7, 0), (-7, 0)$

39. Horizontal: $(5, -1), (5, -3)$ **41.** Horizontal: None
Vertical: $(8, -2), (2, -2)$ Vertical: $(4, 0)$

43. Concave downward: $-\infty < t < 0$
Concave upward: $0 < t < \infty$

45. Concave upward: $t > 0$

47. Concave downward: $0 < t < \dfrac{\pi}{2}$

Concave upward: $\dfrac{\pi}{2} < t < \pi$

49. $4\sqrt{13} \approx 14.422$ **51.** $\sqrt{2}(1 - e^{-\pi/2}) \approx 1.12$

53. $\dfrac{1}{12}\left[\ln\left(\sqrt{37} + 6\right) + 6\sqrt{37}\right] \approx 3.249$ **55.** $6a$ **57.** $8a$

59. (a) 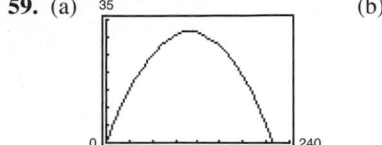   (b) 219.2 ft   (c) 230.8 ft

61. (a) 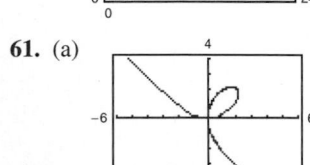   (b) $(0, 0), \left(\dfrac{4\sqrt[3]{2}}{3}, \dfrac{4\sqrt[3]{4}}{3}\right)$
(c) About 6.557

63. (a) $27\pi\sqrt{13}$ (b) $18\pi\sqrt{13}$ **65.** 50π **67.** $\dfrac{12\pi a^2}{5}$

69. $S = 2\pi \displaystyle\int_0^2 (t + 2)\sqrt{9t^4 + 1}\, dt \approx 185.78$

71. $S = 2\pi \displaystyle\int_0^{\pi/2} \left(\sin\theta \cos\theta \sqrt{4\cos^2\theta + 1}\right) d\theta$

$= \dfrac{(5\sqrt{5} - 1)\pi}{6}$

≈ 5.330

73. (a)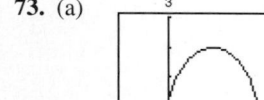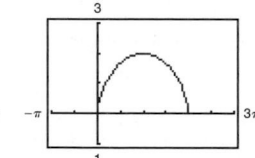

(b) The average speed of the particle on the second path is twice the average speed of the particle on the first path.
(c) 4π

75. Answers will vary. Sample answer: Let $x = -3t$, $y = -4t$.

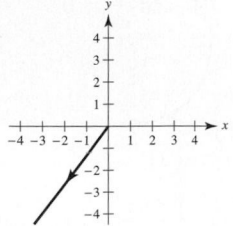

77. Proof   **79.** $\dfrac{3\pi}{2}$ **81.** d   **82.** b   **83.** f   **84.** c

85. a   **86.** e   **87.** $\left(\dfrac{3}{4}, \dfrac{8}{5}\right)$ **89.** 288π

91. (a) $\dfrac{dy}{dx} = \dfrac{\sin\theta}{1 - \cos\theta}$, $\dfrac{d^2y}{dx^2} = -\dfrac{1}{a(\cos\theta - 1)^2}$

(b) $y = (2 + \sqrt{3})\left[x - a\left(\dfrac{\pi}{6} - \dfrac{1}{2}\right)\right] + a\left(1 - \dfrac{\sqrt{3}}{2}\right)$

(c) $(a(2n + 1)\pi, 2a)$

(d) Concave downward on $(0, 2\pi), (2\pi, 4\pi)$, etc.

(e) $s = 8a$

93. Proof

95. (a)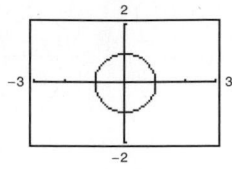

(b) Circle of radius 1 and center at $(0, 0)$ except the point $(-1, 0)$

(c) As t increases from -20 to 0, the speed increases, and as t increases from 0 to 20, the speed decreases.

97. False. $\dfrac{d^2y}{dx^2} = \dfrac{\dfrac{d}{dt}\left[\dfrac{g'(t)}{f'(t)}\right]}{f'(t)} = \dfrac{f'(t)g''(t) - g'(t)f''(t)}{[f'(t)]^3}.$

99. False. The resulting rectangular equation is a line.

Section 10.4   (page 726)

1. r is the directed distance from the origin to the point in the plane. θ is the directed angle, counterclockwise from the polar axis to the segment from the origin to the point in the plane.

3. The rectangular coordinate system is a collection of points of the form (x, y), where x is the directed distance from the y-axis to the point and y is the directed distance from the x-axis to the point. Every point has a unique representation.

The polar coordinate system is a collection of points of the form (r, θ), where r is the directed distance from the origin O to a point P and θ is the directed angle, measured counterclockwise, from the polar axis to the segment $\overline{OP}$. Polar coordinates do not have unique representations.

5. 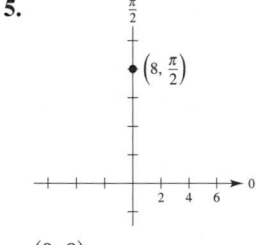   **7.**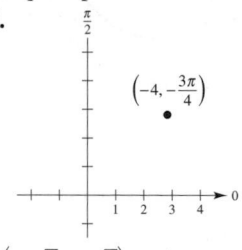

$(0, 8)$ $(2\sqrt{2}, 2\sqrt{2}) \approx (2.828, 2.828)$

9. 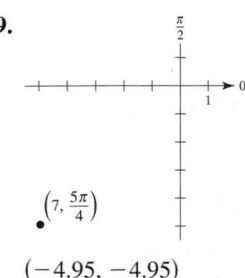   **11.**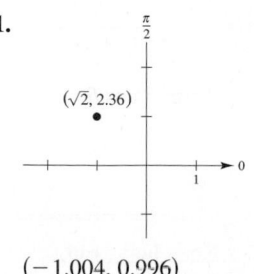

$(-4.95, -4.95)$ $(-1.004, 0.996)$

13.

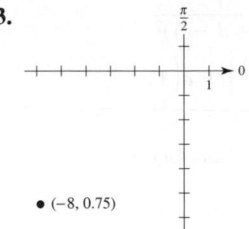

$(-5.854, -5.453)$

15.

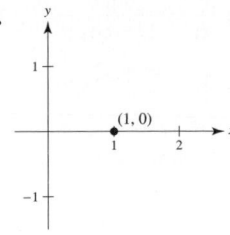

$(1, 0), (-1, \pi)$

17.

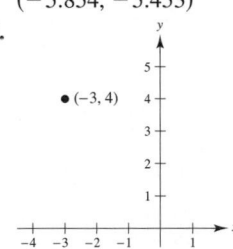

$(5, 2.214), (-5, 5.356)$

19.

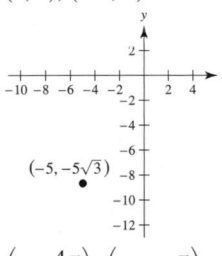

$\left(10, \dfrac{4\pi}{3}\right), \left(-10, \dfrac{\pi}{3}\right)$

21.

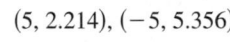

$\left(\sqrt{14}, \dfrac{7\pi}{4}\right), \left(-\sqrt{14}, \dfrac{3\pi}{4}\right)$

23.

$\left(\sqrt{41}, 0.8961\right),$
$\left(-\sqrt{41}, 4.0376\right)$

25. $r = 3$

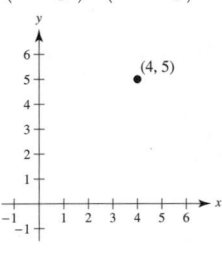

27. $r = a$

29. $r = 8 \csc \theta$

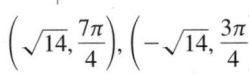

31. $r = \dfrac{-2}{3 \cos \theta - \sin \theta}$

33. $r = 9 \csc^2 \theta \cos \theta$

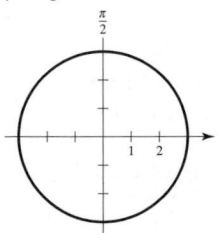

35. $x^2 + y^2 = 16$

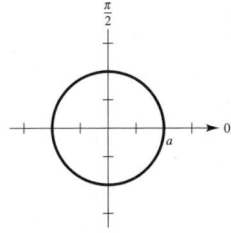

37. $x^2 + y^2 - 3y = 0$

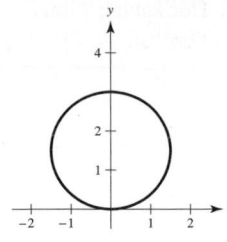

39. $\sqrt{x^2 + y^2} = \arctan \dfrac{y}{x}$

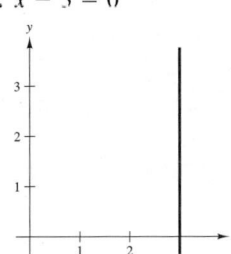

41. $x - 3 = 0$

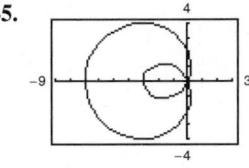

43. $x^2 - y = 0$

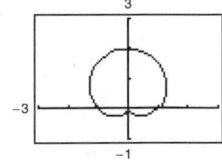

45.

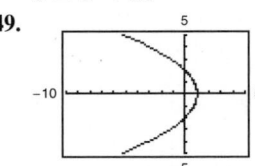

$0 \le \theta < 2\pi$

47.

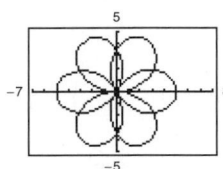

$0 \le \theta < 2\pi$

49.

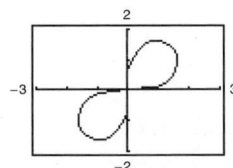

$-\pi < \theta < \pi$

51.

$0 \le \theta < 4\pi$

53.

$0 \le \theta < \dfrac{\pi}{2}$

55. $(x - h)^2 + (y - k)^2 = h^2 + k^2$
Radius: $\sqrt{h^2 + k^2}$
Center: (h, k)

57. (a)

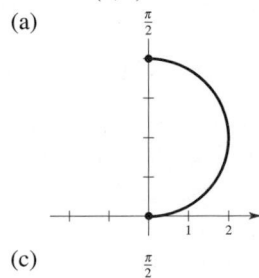

(b)

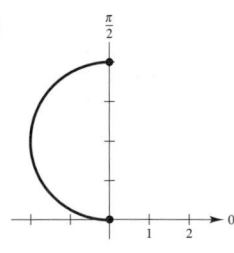

(c)

59. $\sqrt{17}$ **61.** About 5.6

63. $\dfrac{dy}{dx} = \dfrac{-2\cos\theta\sin\theta + 2\cos\theta(1-\sin\theta)}{-2\cos\theta\cos\theta - 2\sin\theta(1-\sin\theta)}$

$(2, 0)$: $\dfrac{dy}{dx} = -1$

$\left(3, \dfrac{7\pi}{6}\right)$: $\dfrac{dy}{dx}$ is undefined.

$\left(4, \dfrac{3\pi}{2}\right)$: $\dfrac{dy}{dx} = 0$

65. (a) and (b)

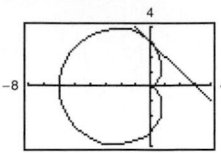

(c) $\dfrac{dy}{dx} = -1$

67. (a) and (b)

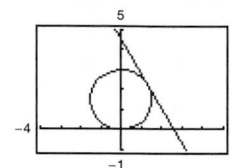

(c) $\dfrac{dy}{dx} = -\sqrt{3}$

69. Horizontal: $\left(2, \dfrac{3\pi}{2}\right), \left(\dfrac{1}{2}, \dfrac{\pi}{6}\right), \left(\dfrac{1}{2}, \dfrac{5\pi}{6}\right)$

Vertical: $\left(\dfrac{3}{2}, \dfrac{7\pi}{6}\right), \left(\dfrac{3}{2}, \dfrac{11\pi}{6}\right)$

71. $\left(5, \dfrac{\pi}{2}\right), \left(1, \dfrac{3\pi}{2}\right)$

73.

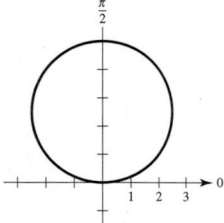

$\theta = 0$

75.

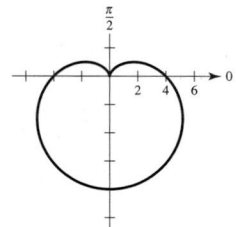

77.

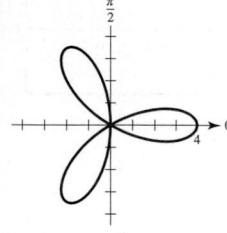

$\theta = \dfrac{\pi}{6}, \dfrac{\pi}{2}, \dfrac{5\pi}{6}$

79.

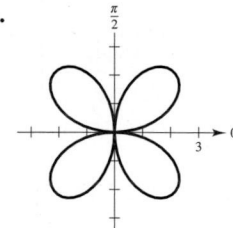

$\theta = 0, \dfrac{\pi}{2}$

81.

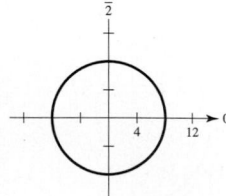

83.

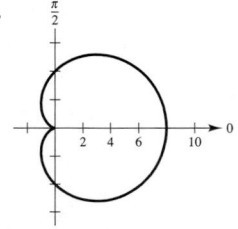

85.

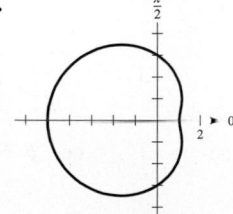

87.

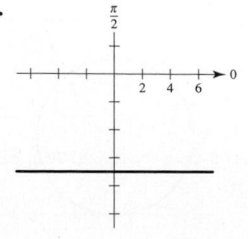

89.

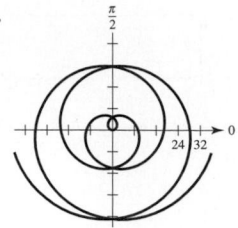

91.

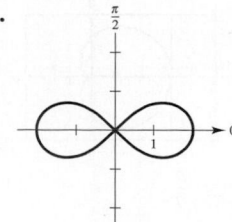

93.

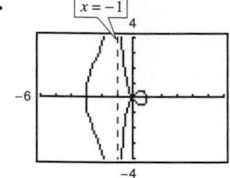

95.

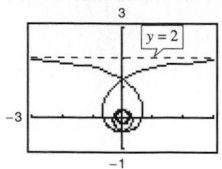

97.

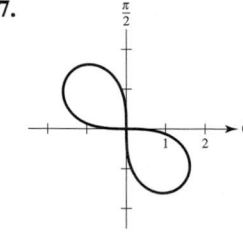

99. (a) To test for symmetry about the x-axis, replace (r, θ) by $(r, -\theta)$ or $(-r, \pi - \theta)$. If the substitution yields an equivalent equation, then the graph is symmetric about the x-axis.

(b) To test for symmetry about the y-axis, replace (r, θ) by $(r, \pi - \theta)$ or $(-r, -\theta)$. If the substitution yields an equivalent equation, then the graph is symmetric about the y-axis.

101. Proof

103. (a) $r = 2 - \sin\left(\theta - \dfrac{\pi}{4}\right)$

$= 2 - \dfrac{\sqrt{2}(\sin\theta - \cos\theta)}{2}$

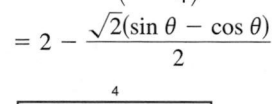
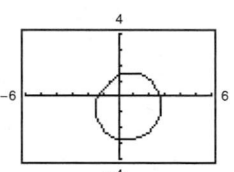

(b) $r = 2 + \cos\theta$

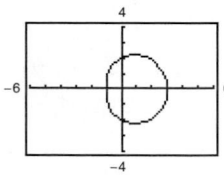

(c) $r = 2 + \sin\theta$

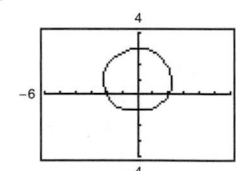

(d) $r = 2 - \cos\theta$

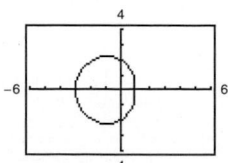

105. (a)

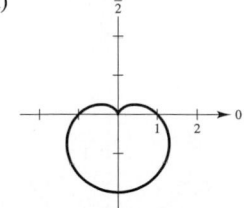

(b)

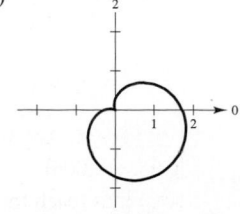

107.

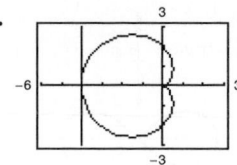

$$\psi = \frac{\pi}{2}$$

109.

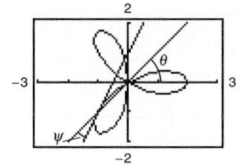

$$\psi = \arctan \frac{1}{3} \approx 18.4°$$

111.

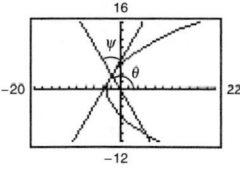

$$\psi = \frac{\pi}{3}, 60°$$

113. True   **115.** True

Section 10.5   *(page 735)*

1. Check that f is continuous and either nonnegative or nonpositive on the interval of consideration.

3. $8\displaystyle\int_0^{\pi/2} \sin^2 \theta \, d\theta$ **5.** $\dfrac{1}{2}\displaystyle\int_{\pi/2}^{3\pi/2} (3 - 2\sin\theta)^2 \, d\theta$ **7.** 9π

9. $\dfrac{\pi}{3}$ **11.** $\dfrac{\pi}{16}$ **13.** $\dfrac{97\pi}{4} - 60 \approx 16.184$ **15.** $\dfrac{33\pi}{2}$

17. 4

19.

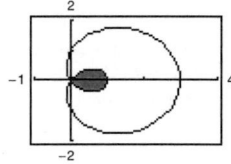

$$\frac{2\pi - 3\sqrt{3}}{2}$$

21.

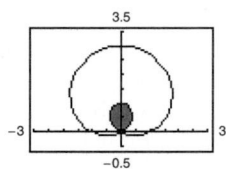

$$\frac{2\pi - 3\sqrt{3}}{2}$$

23.

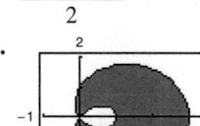

$$\pi + 3\sqrt{3}$$

25.

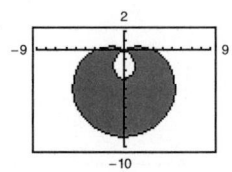

$$9\pi + 27\sqrt{3}$$

27. $\left(1, \dfrac{\pi}{2}\right), \left(1, \dfrac{3\pi}{2}\right), (0, 0)$

29. $\left(\dfrac{2 - \sqrt{2}}{2}, \dfrac{3\pi}{4}\right), \left(\dfrac{2 + \sqrt{2}}{2}, \dfrac{7\pi}{4}\right), (0, 0)$

31. $\left(\dfrac{3}{2}, \dfrac{\pi}{6}\right), \left(\dfrac{3}{2}, \dfrac{5\pi}{6}\right), (0, 0)$ **33.** $(2, 4), (-2, -4)$

35.

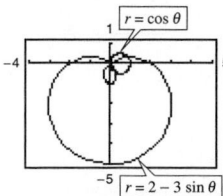

$(0, 0), (0.935, 0.363),$
$(0.535, -1.006)$
The graphs reach the pole
at different times (θ-values).

37.

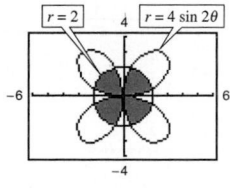

$$\frac{4}{3}\left(4\pi - 3\sqrt{3}\right)$$

39.

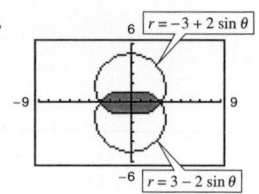

$$11\pi - 24$$

41.

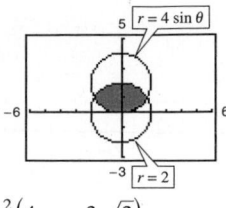

$$\frac{2}{3}\left(4\pi - 3\sqrt{3}\right)$$

43.

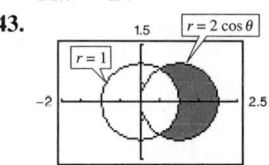

$$\frac{\pi}{3} + \frac{\sqrt{3}}{2}$$

45. $\dfrac{5\pi a^2}{4}$ **47.** $\dfrac{a^2}{2}(\pi - 2)$

49. (a) $(x^2 + y^2)^{3/2} = ax^2$

(b)

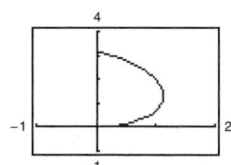

(c) $\dfrac{15\pi}{2}$

51. The area enclosed by the function is $\dfrac{\pi a^2}{4}$ if n is odd and is $\dfrac{\pi a^2}{2}$ if n is even.

53. $\dfrac{4\pi}{3}$ **55.** 4π **57.** 8

59.

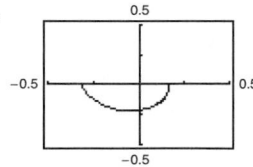

About 4.16

61.

About 0.71

63.

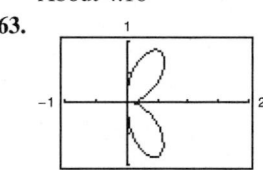

About 4.39

65. 36π **67.** $\dfrac{2\pi\sqrt{1 + a^2}}{1 + 4a^2}(e^{\pi a} - 2a)$ **69.** 21.87

71. (a)

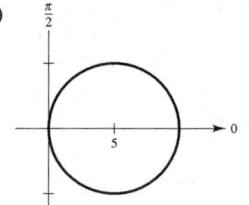

(b) $0 \le \theta < \pi$
(c) and (d) 25π

73. Answers will vary. Sample answer: $f(\theta) = \cos^2 \theta + 1$, $g(\theta) = -\frac{3}{2}$

75. $40\pi^2$

77. (a) 16π

(b)

θ	0.2	0.4	0.6	0.8	1.0	1.2	1.4
A	6.32	12.14	17.06	20.80	23.27	24.60	25.08

(c) and (d) For $\frac{1}{4}$ of area $(4\pi \approx 12.57)$: 0.42

For $\frac{1}{2}$ of area $(8\pi \approx 25.13)$: $1.57\left(\dfrac{\pi}{2}\right)$

For $\frac{3}{4}$ of area $(12\pi \approx 37.70)$: 2.73

(e) No; Answers will vary.

79. (a)
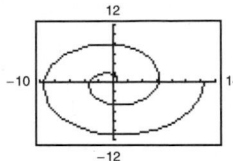

The graph becomes larger and more spread out. The graph is reflected over the y-axis.

(b) $(an\pi, n\pi)$, where $n = 1, 2, 3, \ldots$

(c) About 21.26   (d) $\dfrac{4}{3\pi^3}$

81. $r = \sqrt{2}\cos\theta$

83. (a) $r = \dfrac{3\cos\theta\sin\theta}{\cos^3\theta + \sin^3\theta}$

(b)
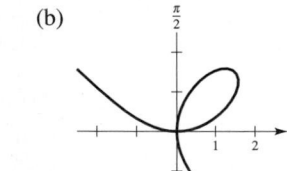

(c) $\dfrac{3}{2}$

Section 10.6 *(page 743)*

1. (a) Hyperbola   (b) Parabola
(c) Ellipse   (d) Hyperbola

3.
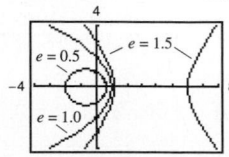

(a) Parabola
(b) Ellipse
(c) Hyperbola

5.
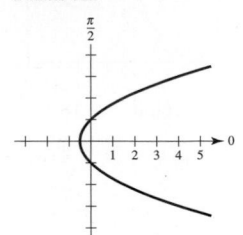

Ellipse
As $e \to 1^-$, the ellipse becomes more elliptical, and as $e \to 0^+$, it becomes more circular.

7. c   **8.** f   **9.** a   **10.** e   **11.** b   **12.** d

13. $e = 1$
Distance = 1
Parabola
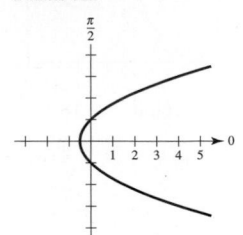

15. $e = 2$
Distance = $\frac{7}{8}$
Hyperbola
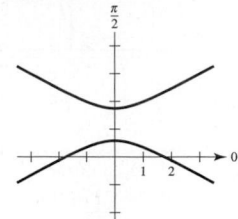

17. $e = \frac{3}{2}$
Distance = 2
Hyperbola
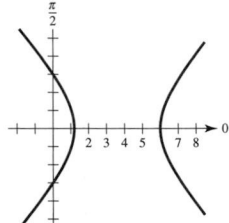

19. $e = \frac{1}{2}$
Distance = 6
Ellipse
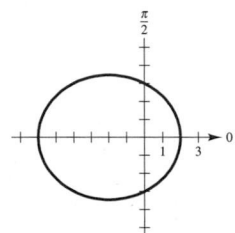

21. $e = \frac{1}{2}$
Distance = 50
Ellipse
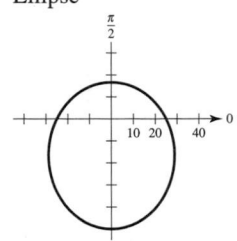

23.
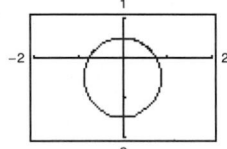

Ellipse
$e = \frac{1}{2}$

25.
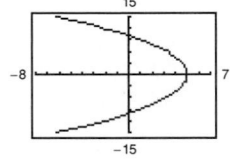

Parabola
$e = 1$

27.
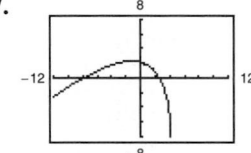

Rotated $\dfrac{\pi}{3}$ radian counterclockwise.

29.
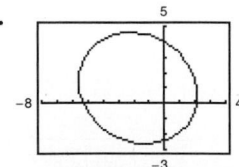

Rotated $\dfrac{\pi}{6}$ radian clockwise.

31. $r = \dfrac{8}{8 + 5\cos\left(\theta + \dfrac{\pi}{6}\right)}$

33. $r = \dfrac{3}{1 - \cos\theta}$

35. $r = \dfrac{1}{4 + \sin\theta}$

37. $r = \dfrac{8}{3 + 4\cos\theta}$

39. $r = \dfrac{2}{1 - \sin\theta}$

41. $r = \dfrac{16}{5 + 3\cos\theta}$

43. $r = \dfrac{9}{4 - 5\sin\theta}$

45. No. The flatness of the ellipse does not depend on the distance between foci.

47. $r = \dfrac{4}{2 + \cos\theta}$ **49.** Proof

51. $r^2 = \dfrac{9}{1 - (16/25)\cos^2\theta}$ **53.** $r^2 = \dfrac{-16}{1 - (25/9)\cos^2\theta}$

55. 10.88   **57.** 1.88

59. $r = \dfrac{7979.21}{1 - 0.9372\cos\theta}$, 11,015 mi

61. $r = \dfrac{149{,}558{,}278.0560}{1 - 0.0167\cos\theta}$

Perihelion: 147,101,680 km

Aphelion: 152,098,320 km

63. $r = \dfrac{4{,}494{,}426{,}033}{1 - 0.0113\cos\theta}$

Perihelion: 4,444,206,500 km

Aphelion: 4,545,793,500 km

65. Answers will vary. Sample answers:

(a) 3.591×10^{18} km²; 9.322 yr

(b) $\alpha \approx 0.361 + \pi$; Larger angle with the smaller ray to generate an equal area

(c) Part (a): 1.583×10^9 km; 1.698×10^8 km/yr

Part (b): 1.610×10^9 km; 1.727×10^8 km/yr

67. Proof

Review Exercises for Chapter 10   *(page 746)*

1. e   **2.** c   **3.** b   **4.** d   **5.** a   **6.** f

7. Circle

Center: $(1, 4)$

Radius: 5

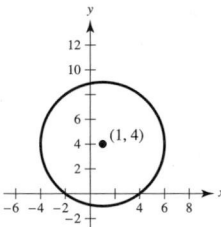

9. Hyperbola

Center: $(-4, 3)$

Vertices: $\left(-4 \pm \sqrt{2}, 3\right)$

Foci: $\left(-4 \pm \sqrt{5}, 3\right)$

$e = \sqrt{\dfrac{5}{2}}$

Asymptotes:

$y = 3 \pm \dfrac{\sqrt{3}}{\sqrt{2}}(x + 4)$

11. Circle

Center: $\left(\tfrac{1}{2}, -\tfrac{3}{4}\right)$

Radius: 1

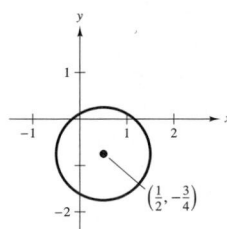

13. Parabola

Vertex: $(-5, -1)$

Directrix: $y = -4$

Focus: $(-5, 2)$

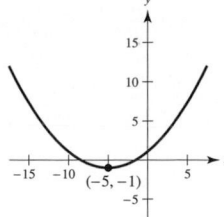

15. $(y - 0)^2 = 4(2)(x - 7)$ **17.** $\dfrac{x^2}{36} + \dfrac{(y - 1)^2}{20} = 1$

19. $\dfrac{(x - 3)^2}{5} + \dfrac{(y - 4)^2}{9} - 1$ **21.** $\dfrac{y^2}{64} - \dfrac{x^2}{16} = 1$

23. $\dfrac{x^2}{49} - \dfrac{(y + 1)^2}{32} = 1$

25. (a) $(0, 50)$ (b) About 38,294.49

27.

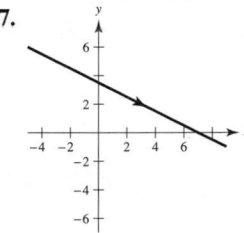

$x + 2y - 7 = 0$

29.

$y = (x - 1)^2 - 3, \; x \geq 1$

31.

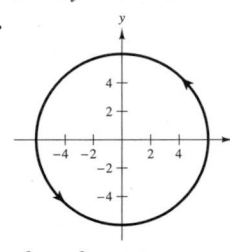

$x^2 + y^2 = 36$

33.

$(x - 2)^2 - (y - 3)^2 = 1$

35. $x = t, \; y = 4t + 3; \; x = t + 1, \; y = 4t + 7$

(Solution is not unique.)

37.

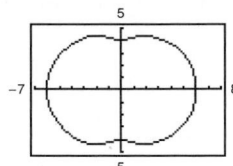

39. $\dfrac{dy}{dx} = -\dfrac{5}{6}, \; \dfrac{d^2y}{dx^2} = 0$

At $t = 3, \; \dfrac{dy}{dx} = -\dfrac{5}{6}, \; \dfrac{d^2y}{dx^2} = 0$; Neither concave upward nor downward.

41. $\dfrac{dy}{dx} = -2t^3, \; \dfrac{d^2y}{dx^2} = 6t^4$

At $t = -2, \; \dfrac{dy}{dx} = 16, \; \dfrac{d^2y}{dx^2} = 96$; Concave upward

43. $\dfrac{dy}{dx} = -e^{-2t}, \; \dfrac{d^2y}{dx^2} = \dfrac{2}{e^{3t}}$

At $t = 1, \; \dfrac{dy}{dx} = -\dfrac{1}{e^2}, \; \dfrac{d^2y}{dx^2} = \dfrac{2}{e^3}$; Concave upward

45. $\dfrac{dy}{dx} = -\cot\theta$, $\dfrac{d^2y}{dx^2} = -\dfrac{1}{10}\csc^3\theta$

At $\theta = \dfrac{\pi}{4}$, $\dfrac{dy}{dx} = -1$, $\dfrac{d^2y}{dx^2} = -\dfrac{\sqrt{2}}{5}$; Concave downward

47. (a) and (d)

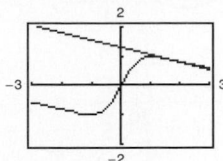

(b) $\dfrac{dx}{d\theta} = -4$, $\dfrac{dy}{d\theta} = 1$, $\dfrac{dy}{dx} = -\dfrac{1}{4}$ (c) $y = -\dfrac{1}{4}x + \dfrac{3\sqrt{3}}{4}$

49. Horizontal: $(5, 0)$ **51.** Horizontal: $(2, 2)$, $(2, 0)$
Vertical: None   Vertical: $(4, 1)$, $(0, 1)$

53. $\dfrac{1}{54}(145^{3/2} - 1) \approx 32.315$

55. (a) 25π (b) 20π **57.** $A = 3\pi$

59.

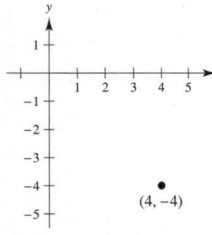

$(0, -5)$

61.

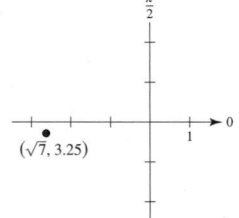

$(-2.6302, -0.2863)$

63.

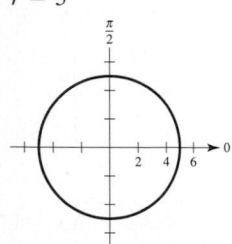

$\left(4\sqrt{2}, \dfrac{7\pi}{4}\right)$, $\left(-4\sqrt{2}, \dfrac{3\pi}{4}\right)$

65.

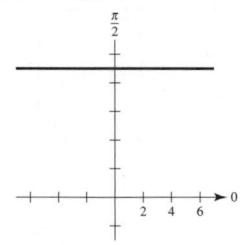

$\left(\sqrt{10}, 1.89\right)$, $\left(-\sqrt{10}, 5.03\right)$

67. $r = 5$

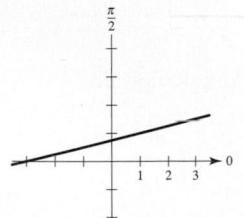

69. $r = 9\csc\theta$

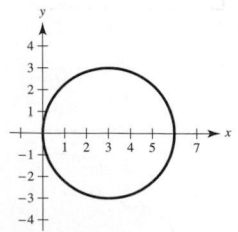

71. $r = \dfrac{3}{4\sin\theta - \cos\theta}$

73. $x^2 + y^2 - 6x = 0$

75. $x = -4$

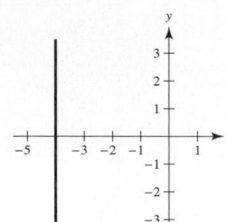

77. $y = -x$

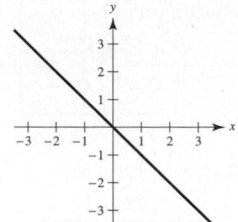

79.

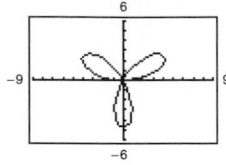

$0 \le \theta \le \pi$

81.

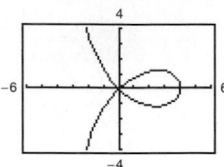

$0 \le \theta \le \pi$

83. Horizontal: $\left(\dfrac{3}{2}, \dfrac{2\pi}{3}\right)$, $\left(\dfrac{3}{2}, \dfrac{4\pi}{3}\right)$

Vertical: $\left(\dfrac{1}{2}, \dfrac{\pi}{3}\right)$, $(2, \pi)$, $\left(\dfrac{1}{2}, \dfrac{5\pi}{3}\right)$

85.

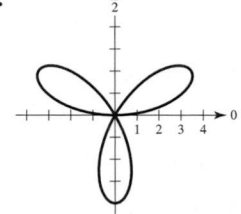

$\theta = 0, \dfrac{\pi}{3}, \dfrac{2\pi}{3}$

87. Circle

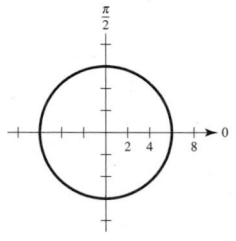

89. Line

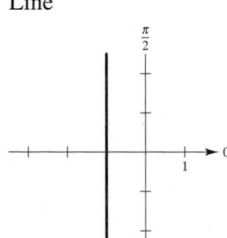

91. Limaçon

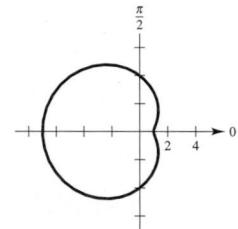

93. Spiral

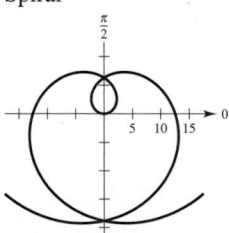

95. Lemniscate

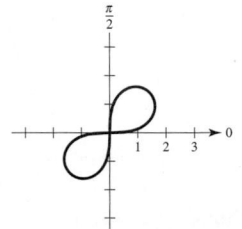

97. $\dfrac{9\pi}{20}$ **99.** $\dfrac{9\pi}{2}$

101. $\left(1 + \dfrac{\sqrt{2}}{2}, \dfrac{3\pi}{4}\right)$, $\left(1 - \dfrac{\sqrt{2}}{2}, \dfrac{7\pi}{4}\right)$, $(0, 0)$

103.

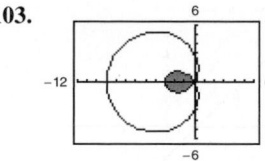

$$9\pi - \frac{27\sqrt{3}}{2}$$

105.

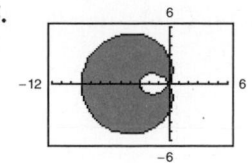

$$9\pi + 27\sqrt{3}$$

107.

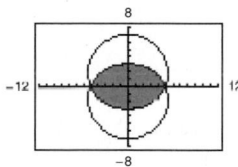

$$27\pi - 40$$

109. $\dfrac{5\pi}{2}$ **111.** $4\pi^2$

113. $e = 1$
Distance $= 6$
Parabola

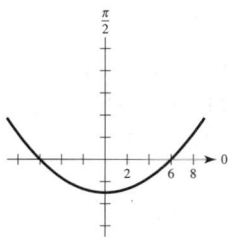

115. $e = \frac{2}{3}$
Distance $= 3$
Ellipse

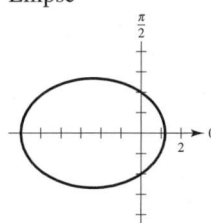

117. $e = \frac{3}{2}$
Distance $= \frac{4}{3}$
Hyperbola

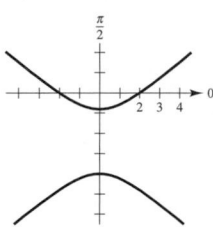

119. $r = \dfrac{5}{1 + \cos\theta}$ **121.** $r = \dfrac{9}{1 + 3\sin\theta}$

123. $r = \dfrac{4}{1 + \sin\theta}$ **125.** $r = \dfrac{5}{3 - 2\cos\theta}$

P.S. Problem Solving *(page 749)*

1. (a)

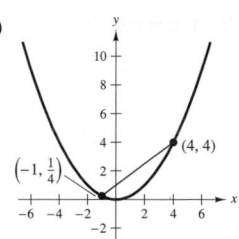

3. Proof

(b) and (c) Proofs

5. (a) $y^2 = x^2\left(\dfrac{1 - x}{1 + x}\right)$ (b) $r = \cos 2\theta \cdot \sec\theta$

(c)

(d) $y = x,\ y = -x$

(e) $\left(\dfrac{\sqrt{5} - 1}{2},\ \pm\dfrac{\sqrt{5} - 1}{2}\sqrt{-2 + \sqrt{5}}\right)$

7. (a)

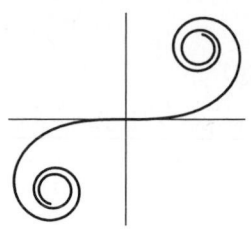

Generated by Mathematica

(b) Proof   (c) $a;\ 2\pi$

9. $A = \frac{1}{2}ab$ **11.** $r^2 = 2\cos 2\theta$

13. $r = \dfrac{d}{\sqrt{2}}e^{((\pi/4) - \theta)},\ \theta \geq \dfrac{\pi}{4}$

15. (a) $r = 2a\tan\theta\sin\theta$

(b) $x = \dfrac{2at^2}{1 + t^2}$

$y = \dfrac{2at^3}{1 + t^2}$

(c) $y^2 = \dfrac{x^3}{2a - x}$

17.

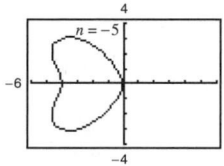

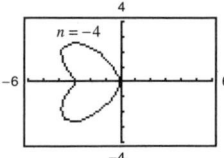

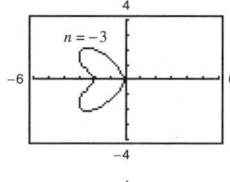

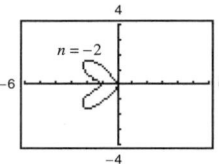

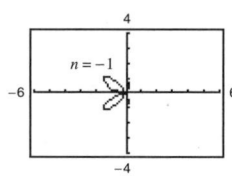

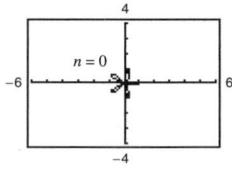

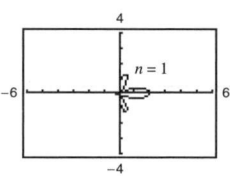

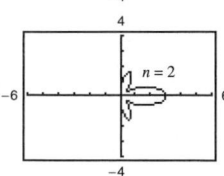

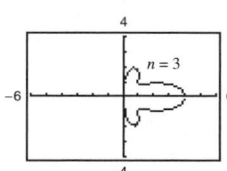

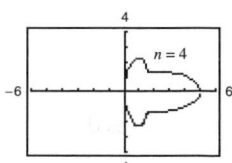

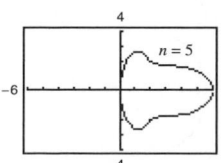

$n = -1, -2, -3, -4, -5$ produce "hearts"; $n = 1, 2, 3, 4, 5$ produce "bells"

Chapter 11

Section 11.1   *(page 759)*

1. Answers will vary. Sample answer: A scalar is a single real number, such as 2. A vector is a line segment having both direction and magnitude. The vector $\langle \sqrt{3}, 1 \rangle$, given in component form, has a direction of $\pi/6$ and a magnitude of 2.

3. (a) $\langle 4, 2 \rangle$

(b)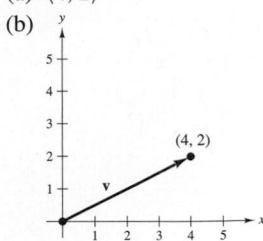

5. $\mathbf{u} = \mathbf{v} = \langle 2, 4 \rangle$ **7.** $\mathbf{u} = \mathbf{v} = \langle 6, -5 \rangle$

9. (a) and (d)   **11.** (a) and (d)

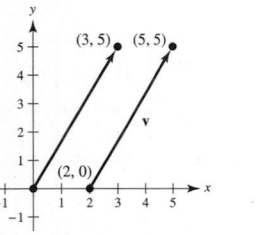

(b) $\langle 3, 5 \rangle$ (b) $\langle -2, -4 \rangle$
(c) $\mathbf{v} = 3\mathbf{i} + 5\mathbf{j}$ (c) $\mathbf{v} = -2\mathbf{i} - 4\mathbf{j}$

13. (a) and (d)   **15.** (a) and (d)

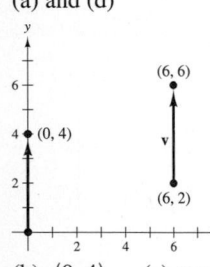

(b) $\langle 0, 4 \rangle$ (c) $\mathbf{v} = 4\mathbf{j}$ (b) $\langle -1, \frac{5}{3} \rangle$
                      (c) $\mathbf{v} = -\mathbf{i} + \frac{5}{3}\mathbf{j}$

17. $(3, 5)$ **19.** 4   **21.** 17   **23.** $\sqrt{26}$

25. (a) $\langle 6, 10 \rangle$ (b) $\langle -9, -15 \rangle$

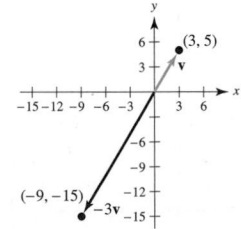

(c) $\langle \frac{21}{2}, \frac{35}{2} \rangle$ (d) $\langle 2, \frac{10}{3} \rangle$

27. (a) $\langle \frac{8}{3}, 6 \rangle$ (b) $\langle 6, -15 \rangle$
(c) $\langle -2, -14 \rangle$ (d) $\langle 18, -7 \rangle$

29.   **31.**

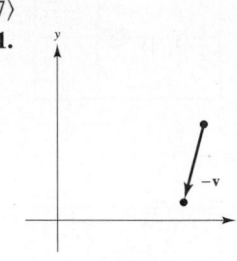

33.

35. $\left\langle \dfrac{\sqrt{17}}{17}, \dfrac{4\sqrt{17}}{17} \right\rangle$ **37.** $\left\langle \dfrac{3\sqrt{34}}{34}, \dfrac{5\sqrt{34}}{34} \right\rangle$

39. (a) $\sqrt{2}$ (b) $\sqrt{5}$ (c) 1   (d) 1   (e) 1   (f) 1

41. (a) $\dfrac{\sqrt{5}}{2}$ (b) $\sqrt{13}$ (c) $\dfrac{\sqrt{85}}{2}$ (d) 1   (e) 1   (f) 1

43.

$\|\mathbf{u}\| + \|\mathbf{v}\| = \sqrt{5} + \sqrt{41}$ and $\|\mathbf{u} + \mathbf{v}\| = \sqrt{74}$
$\sqrt{74} \le \sqrt{5} + \sqrt{41}$

45. $\langle 0, 6 \rangle$ **47.** $\langle -\sqrt{5}, 2\sqrt{5} \rangle$ **49.** $\langle 3, 0 \rangle$

51. $\langle -\sqrt{3}, 1 \rangle$ **53.** $\left\langle \dfrac{2 + 3\sqrt{2}}{2}, \dfrac{3\sqrt{2}}{2} \right\rangle$

55. $\langle 2\cos 4 + \cos 2, 2\sin 4 + \sin 2 \rangle$

57. $\theta = 0°$

59. 0; Vectors that start and end at the same point have a magnitude of 0.

61. $a = 3, b = 1$ **63.** $a = -2, b = -4$

65. $a = -\frac{2}{3}, b = \frac{5}{3}$

67. (a) $\pm \dfrac{1}{\sqrt{37}} \langle 1, 6 \rangle$ **69.** (a) $\pm \dfrac{1}{\sqrt{10}} \langle 1, 3 \rangle$

(b) $\pm \dfrac{1}{\sqrt{37}} \langle 6, -1 \rangle$ (b) $\pm \dfrac{1}{\sqrt{10}} \langle 3, -1 \rangle$

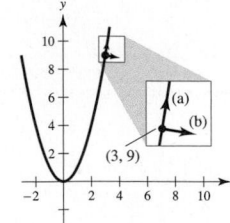

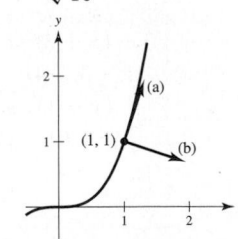

71. (a) $\pm\frac{1}{5}\langle -4, 3\rangle$

(b) $\pm\frac{1}{5}\langle 3, 4\rangle$

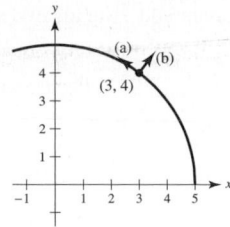

73. $\left\langle -\dfrac{\sqrt{2}}{2}, \dfrac{\sqrt{2}}{2}\right\rangle$ **75.** 10.7°, 584.6 lb   **77.** 71.3°, 228.5 lb

79. Tension in cable CB: 1958.1 lb
Tension in cable CA: 2638.2 lb

81. Horizontal: 1193.43 ft/sec
Vertical: 125.43 ft/sec

83. 38.3° north of west, 882.9 km/h

85. False. Weight has direction.   **87.** True

89. True   **91.** True   **93.** False. $\|a\mathbf{i} + b\mathbf{j}\| = \sqrt{2}|a|$

95–97. Proofs   **99.** $x^2 + y^2 = 25$

Section 11.2   *(page 767)*

1. x_0 is directed distance to yz-plane.
y_0 is directed distance to xz-plane.
z_0 is directed distance to xy-plane.

3. (a) Point   (b) Vertical line   (c) Plane

5. 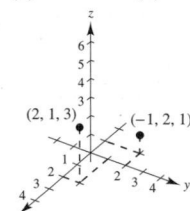   **7.**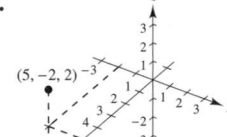

9. $(-3, 4, 5)$ **11.** $(12, 0, 0)$

13. One unit above the xy-plane

15. Three units behind the yz-plane

17. To the left of the xz-plane

19. Within three units of the xz-plane

21. Three units below the xy-plane and below either Quadrant I or Quadrant III

23. Above the xy-plane and above Quadrants II or IV *or* below the xy-plane and below Quadrants I or III

25. $3\sqrt{2}$ **27.** 5   **29.** 7, $7\sqrt{5}$, 14; Right triangle

31. $\sqrt{41}$, $\sqrt{41}$, $\sqrt{14}$; Isosceles triangle   **33.** $(6, 4, 7)$

35. $(2, 6, 3)$

37. $(x - 7)^2 + (y - 1)^2 + (z + 2)^2 = 1$

39. $\left(x - \frac{3}{2}\right)^2 + (y - 2)^2 + (z - 1)^2 = \frac{21}{4}$

41. $(x + 7)^2 + (y - 7)^2 + (z - 6)^2 = 36$

43. $(x - 1)^2 + (y + 3)^2 + (z + 4)^2 = 25$
Center: $(1, -3, -4)$ Radius: 5

45. $\left(x - \frac{1}{3}\right)^2 + (y + 1)^2 + z^2 = 1$
Center: $\left(\frac{1}{3}, -1, 0\right)$ Radius: 1

47. (a) $\langle -2, 2, 2\rangle$ **49.** (a) and (d)
(b) $\mathbf{v} = -2\mathbf{i} + 2\mathbf{j} + 2\mathbf{k}$
(c)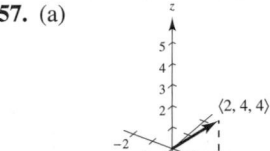

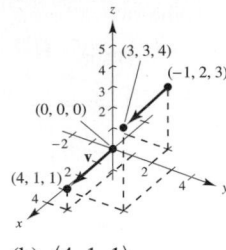

(b) $\langle 4, 1, 1\rangle$
(c) $\mathbf{v} = 4\mathbf{i} + \mathbf{j} + \mathbf{k}$

51. $\mathbf{v} = \langle 1, -1, 6\rangle$ **53.** $\mathbf{v} = \langle -4, 3, 2\rangle$
$\|\mathbf{v}\| = \sqrt{38}$ $\|\mathbf{v}\| = \sqrt{29}$
$\mathbf{u} = \dfrac{1}{\sqrt{38}}\langle 1, -1, 6\rangle$ $\mathbf{u} = \dfrac{1}{\sqrt{29}}\langle -4, 3, 2\rangle$

55. $(3, 1, 8)$

57. (a) 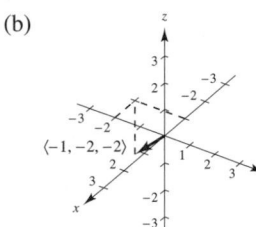   (b)

(c) 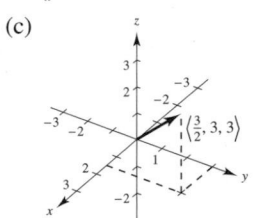   (d)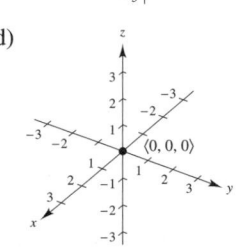

59. $\langle 3, 0, 0\rangle$ **61.** $\langle 21, 18, 15\rangle$ **63.** a and b

65. a   **67.** Collinear   **69.** Not collinear

71. $\overrightarrow{AB} = \langle 1, 2, 3\rangle$, $\overrightarrow{CD} = \langle 1, 2, 3\rangle$, $\overrightarrow{BD} = \langle -2, 1, 1\rangle$,
$\overrightarrow{AC} = \langle -2, 1, 1\rangle$; Because $\overrightarrow{AB} = \overrightarrow{CD}$ and $\overrightarrow{BD} = \overrightarrow{AC}$,
the given points form the vertices of a parallelogram.

73. $\sqrt{2}$ **75.** $\sqrt{34}$ **77.** $\sqrt{14}$

79. (a) $\frac{1}{3}\langle 2, -1, 2\rangle$ (b) $-\frac{1}{3}\langle 2, -1, 2\rangle$

81. (a) $\dfrac{2\sqrt{2}}{5}\mathbf{i} - \dfrac{\sqrt{2}}{2}\mathbf{j} + \dfrac{3\sqrt{2}}{10}\mathbf{k}$
(b) $-\dfrac{2\sqrt{2}}{5}\mathbf{i} + \dfrac{\sqrt{2}}{2}\mathbf{j} - \dfrac{3\sqrt{2}}{10}\mathbf{k}$

83. $\left\langle 0, \dfrac{10}{\sqrt{2}}, \dfrac{10}{\sqrt{2}}\right\rangle$ **85.** $\left\langle 1, -1, \dfrac{1}{2}\right\rangle$

87. 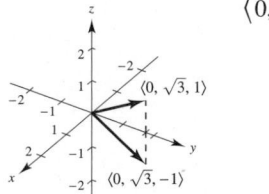 $\left\langle 0, \sqrt{3}, \pm 1\right\rangle$ **89.** $(2, -1, 2)$

91. A sphere of radius 4 centered at (x_1, y_1, z_1):
$(x - x_1)^2 + (y - y_1)^2 + (z - z_1)^2 = 16$

93. The set of points outside a sphere of radius 1 centered at the origin

95. The terminal points of the vectors $t\mathbf{u}$, $\mathbf{u} + t\mathbf{v}$, and $s\mathbf{u} + t\mathbf{v}$ are collinear.

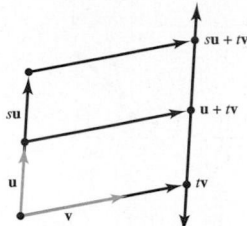

97. $\dfrac{\sqrt{3}}{3}\langle 1, 1, 1 \rangle$

99. (a) $T = \dfrac{8L}{\sqrt{L^2 - 18^2}}, \; L > 18$

(b)

L	20	25	30	35	40	45	50
T	18.4	11.5	10	9.3	9.0	8.7	8.6

(c) 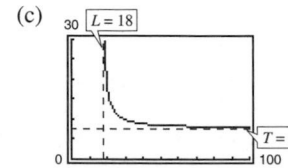   (d) Proof   (e) 30 in.

101. Tension in cable AB: 202.919 N
Tension in cable AC: 157.909 N
Tension in cable AD: 226.521 N

103. $\left(x - \frac{4}{3}\right)^2 + (y - 3)^2 + \left(z + \frac{1}{3}\right)^2 = \frac{44}{9}$

Sphere; center: $\left(\frac{4}{3}, 3, -\frac{1}{3}\right)$, radius: $\dfrac{2\sqrt{11}}{3}$

Section 11.3 (page 777)

1. The vectors are orthogonal.
3. (a) 17   (b) 25   (c) 26   (d) $\langle -17, 85 \rangle$ (e) 51
5. (a) -26 (b) 52   (c) 13   (d) $\langle 78, -52 \rangle$ (e) -78
7. (a) 2   (b) 29   (c) 61   (d) $\langle 0, 12, 10 \rangle$ (e) 6
9. (a) 1   (b) 6   (c) 2   (d) $\mathbf{i} - \mathbf{k}$ (e) 3

11. (a) $\dfrac{\pi}{2}$ (b) $90°$ **13.** (a) 1.7127   (b) $98.1°$

15. (a) 1.0799   (b) $61.9°$ **17.** (a) 2.0306   (b) $116.3°$

19. 20   **21.** Orthogonal   **23.** Neither   **25.** Orthogonal
27. Right triangle; Answers will vary.
29. Acute triangle; Answers will vary.

31. $\cos \alpha = \dfrac{1}{3}, \; \alpha \approx 70.5°$ **33.** $\cos \alpha = \dfrac{7}{\sqrt{51}}, \; \alpha \approx 11.4°$

$\cos \beta = \dfrac{2}{3}, \; \beta \approx 48.2°$ $\cos \beta = \dfrac{1}{\sqrt{51}}, \; \beta \approx 82.0°$

$\cos \gamma = \dfrac{2}{3}, \; \gamma \approx 48.2°$ $\cos \gamma = -\dfrac{1}{\sqrt{51}}, \; \gamma \approx 98.0°$

35. $\cos \alpha = 0, \; \alpha \approx 90°$

$\cos \beta = \dfrac{3}{\sqrt{13}}, \; \beta \approx 33.7°$

$\cos \gamma = -\dfrac{2}{\sqrt{13}}, \; \gamma \approx 123.7°$

37. (a) $\langle 2, 8 \rangle$ (b) $\langle 4, -1 \rangle$ **39.** (a) $\langle \frac{5}{2}, \frac{1}{2} \rangle$ (b) $\langle -\frac{1}{2}, \frac{5}{2} \rangle$
41. (a) $\langle -2, 2, 2 \rangle$ (b) $\langle 2, 1, 1 \rangle$
43. (a) $\langle 0, -3, -3 \rangle$ (b) $\langle -9, 1, -1 \rangle$

45. You cannot add a vector to a scalar.
47. Yes.   **49.** \$17,490.25; Total revenue

$$\left\| \frac{\mathbf{u} \cdot \mathbf{v}}{\|\mathbf{v}\|^2} \mathbf{v} \right\| = \left\| \frac{\mathbf{v} \cdot \mathbf{u}}{\|\mathbf{u}\|^2} \mathbf{u} \right\|$$

$$|\mathbf{u} \cdot \mathbf{v}| \frac{\|\mathbf{v}\|}{\|\mathbf{v}\|^2} = |\mathbf{v} \cdot \mathbf{u}| \frac{\|\mathbf{u}\|}{\|\mathbf{u}\|^2}$$

$$\frac{1}{\|\mathbf{v}\|} = \frac{1}{\|\mathbf{u}\|}$$

$$\|\mathbf{u}\| = \|\mathbf{v}\|$$

51. Answers will vary. Sample answer: $\langle 12, 2 \rangle$ and $\langle -12, -2 \rangle$
53. Answers will vary. Sample answer: $\langle 2, 0, 3 \rangle$ and $\langle -2, 0, -3 \rangle$

55. $\arccos \dfrac{1}{\sqrt{3}} \approx 54.7°$

57. (a) 8335.1 lb   (b) 47,270.8 lb
59. 425 ft-lb   **61.** 2900.2 km-N
63. False. For example, $\langle 1, 1 \rangle \cdot \langle 2, 3 \rangle = 5$ and $\langle 1, 1 \rangle \cdot \langle 1, 4 \rangle = 5$, but $\langle 2, 3 \rangle \neq \langle 1, 4 \rangle$.
65. (a) $(0, 0), (1, 1)$

(b) To $y = x^2$ at $(1, 1)$: $\left\langle \pm\dfrac{\sqrt{5}}{5}, \pm\dfrac{2\sqrt{5}}{5} \right\rangle$

To $y = x^{1/3}$ at $(1, 1)$: $\left\langle \pm\dfrac{3\sqrt{10}}{10}, \pm\dfrac{\sqrt{10}}{10} \right\rangle$

To $y = x^2$ at $(0, 0)$: $\langle \pm 1, 0 \rangle$
To $y = x^{1/3}$ at $(0, 0)$: $\langle 0, \pm 1 \rangle$

(c) At $(1, 1)$: $\theta = 45°$
At $(0, 0)$: $\theta = 90°$

67. (a) $(-1, 0), (1, 0)$

(b) To $y = 1 - x^2$ at $(1, 0)$: $\left\langle \pm\dfrac{\sqrt{5}}{5}, \mp\dfrac{2\sqrt{5}}{5} \right\rangle$

To $y = x^2 - 1$ at $(1, 0)$: $\left\langle \pm\dfrac{\sqrt{5}}{5}, \pm\dfrac{2\sqrt{5}}{5} \right\rangle$

To $y = 1 - x^2$ at $(-1, 0)$: $\left\langle \pm\dfrac{\sqrt{5}}{5}, \pm\dfrac{2\sqrt{5}}{5} \right\rangle$

To $y = x^2 - 1$ at $(-1, 0)$: $\left\langle \pm\dfrac{\sqrt{5}}{5}, \mp\dfrac{2\sqrt{5}}{5} \right\rangle$

(c) At $(1, 0)$: $\theta = 53.13°$
At $(-1, 0)$: $\theta = 53.13°$

69. Proof
71. (a)

(b) $k\sqrt{2}$ (c) $60°$ (d) $109.5°$
73–75. Proofs

Section 11.4 (page 785)

1. $\mathbf{u} \times \mathbf{v}$ is a vector that is perpendicular to both $\mathbf{u}$ and $\mathbf{v}$.

3. $-\mathbf{k}$ **5.** $-\mathbf{j}$

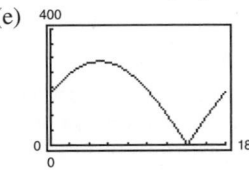

7. (a) $20\mathbf{i} + 10\mathbf{j} - 16\mathbf{k}$ (b) $-20\mathbf{i} - 10\mathbf{j} + 16\mathbf{k}$ (c) $\mathbf{0}$

9. (a) $17\mathbf{i} - 33\mathbf{j} - 10\mathbf{k}$ (b) $-17\mathbf{i} + 33\mathbf{j} + 10\mathbf{k}$ (c) $\mathbf{0}$

11. $\langle 0, 0, 6 \rangle$ **13.** $\langle -2, 3, -1 \rangle$

15. $\left\langle -\dfrac{7}{9\sqrt{3}}, -\dfrac{5}{9\sqrt{3}}, \dfrac{13}{9\sqrt{3}} \right\rangle$ or $\left\langle \dfrac{7}{9\sqrt{3}}, \dfrac{5}{9\sqrt{3}}, -\dfrac{13}{9\sqrt{3}} \right\rangle$

17. $\left\langle \dfrac{3}{\sqrt{59}}, \dfrac{7}{\sqrt{59}}, \dfrac{1}{\sqrt{59}} \right\rangle$ or $\left\langle -\dfrac{3}{\sqrt{59}}, -\dfrac{7}{\sqrt{59}}, -\dfrac{1}{\sqrt{59}} \right\rangle$

19. 1 **21.** $6\sqrt{5}$ **23.** $9\sqrt{5}$ **25.** $\dfrac{11}{2}$

27. $10 \cos 40° \approx 7.66$ ft-lb

29. (a) $\mathbf{F} = -180(\cos \theta \mathbf{j} + \sin \theta \mathbf{k})$

(b) $\|\overrightarrow{AB} \times \mathbf{F}\| = |225 \sin \theta + 180 \cos \theta|$

(c) $\|\overrightarrow{AB} \times \mathbf{F}\| = 225\left(\dfrac{1}{2}\right) + 180\left(\dfrac{\sqrt{3}}{2}\right) \approx 268.38$

(d) $\theta = 141.34°$

$\overrightarrow{AB}$ and $\mathbf{F}$ are perpendicular.

(e)
From part (d), the zero is $\theta \approx 141.34°$ when the vectors are parallel.

31. 1 **33.** 6 **35.** 2 **37.** 75

39. $a = b = c = h$ and $e = f = g$

41. On the x-axis; The cross product has the form $\langle k, 0, 0 \rangle$.

43. False. The cross product of two vectors is not defined in a two-dimensional coordinate system.

45. False. Let $\mathbf{u} = \langle 1, 0, 0 \rangle$, $\mathbf{v} = \langle 1, 0, 0 \rangle$, and $\mathbf{w} = \langle -1, 0, 0 \rangle$. Then $\mathbf{u} \times \mathbf{v} = \mathbf{u} \times \mathbf{w} = \mathbf{0}$, but $\mathbf{v} \neq \mathbf{w}$.

47–55. Proofs

Section 11.5   (page 794)

1. Parametric equations: $x = x_1 + at$, $y = y_1 + bt$, $z = z_1 + ct$

Symmetric equations: $\dfrac{x - x_1}{a} = \dfrac{y - y_1}{b} = \dfrac{z - z_1}{c}$

You need a vector $\mathbf{v} = \langle a, b, c \rangle$ parallel to the line and a point $P(x_1, y_1, z_1)$ on the line.

3. Answers will vary. Sample answer: $3y - z = 5$

5. (a) Yes   (b) No   (c) Yes

Parametric Equations (a)	Symmetric Equations (b)	Direction Numbers
7. $x = 3t$ $y = t$ $z = 5t$	$\dfrac{x}{3} = y = \dfrac{x}{5}$	$3, 1, 5$
9. $x = -2 + 2t$ $y = 4t$ $z = 3 - 2t$	$\dfrac{x + 2}{2} = \dfrac{y}{4} = \dfrac{z - 3}{-2}$	$2, 4, -2$

Parametric Equations (a)	Symmetric Equations (b)	Direction Numbers
11. $x = 1 + 3t$ $y = -2t$ $z = 1 + t$	$\dfrac{x - 1}{3} = \dfrac{y}{-2} = \dfrac{z - 1}{1}$	$3, -2, 1$
13. $x = 5 + 17t$ $y = -3 - 11t$ $z = -2 - 9t$	$\dfrac{x - 5}{17} = \dfrac{y + 3}{-11} = \dfrac{z + 2}{-9}$	$17, -11, -9$
15. $x = 7 - 10t$ $y = -2 + 2t$ $z = 6$	Not possible	$-10, 2, 0$

17. $x = 2$ $y = 3$ $z = 4 + t$ **19.** $x = 2 + 3t$ $y = 3 + 2t$ $z = 4 - t$ **21.** $x = 5 + 2t$ $y = -3 - t$ $z = -4 + 3t$

23. $x = 2 - t$ $y = 1 + t$ $z = 2 + t$ **25.** $P(3, -1, -2)$ $\mathbf{v} = \langle -1, 2, 0 \rangle$ **27.** $P(7, -6, -2)$ $\mathbf{v} = \langle 4, 2, 1 \rangle$

29. Identical   **31.** Identical   **33.** $(2, 3, 1)$, $55.5°$

35. Not intersecting   **37.** (a) Yes   (b) Yes   (c) No

39. $y - 3 = 0$ **41.** $2x + 3y - z = 10$

43. $2x - y - 2z + 6 = 0$ **45.** $3x - 19y - 2z = 0$

47. $4x - 3y + 4z = 10$ **49.** $z = 3$ **51.** $x + y + z = 5$

53. $7x + y - 11z = 5$ **55.** $y - z = -1$ **57.** $x - z = 0$

59. $9x - 3y + 2z - 21 = 0$ **61.** Parallel   **63.** Identical

65. (a) $\theta \approx 65.91°$ (b) $x = 2$ $y = 1 + t$ $z = 1 + 2t$ **67.** (a) $\theta \approx 69.67°$ (b) $x = 2 - 9t$ $y = -1 - 5t$ $z = 22t$

69. Orthogonal   **71.** Neither; $83.5°$ **73.** Parallel

75. 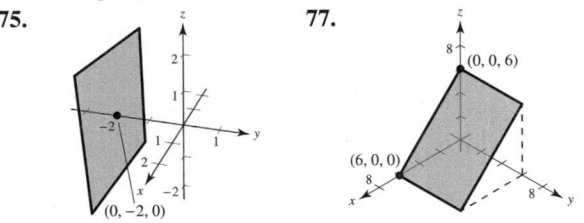   **77.**

79. 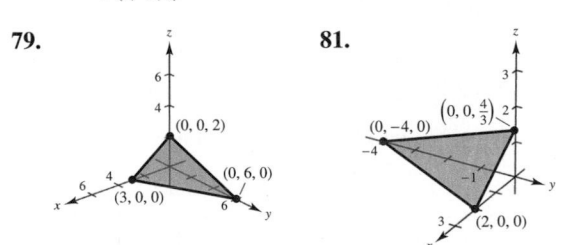   **81.**

83. The line lies in the plane.   **85.** Not intersecting

87. $\dfrac{6\sqrt{14}}{7}$ **89.** $\dfrac{11\sqrt{6}}{6}$ **91.** $\dfrac{2\sqrt{26}}{13}$ **93.** $\dfrac{27\sqrt{94}}{188}$

95. $\dfrac{\sqrt{2533}}{17}$ **97.** $\dfrac{7\sqrt{3}}{3}$ **99.** $\dfrac{\sqrt{66}}{3}$ **101.** Exactly 1

103. Yes. Consider three points, two on one line and one on the second line. A unique plane contains all three points.

105. (a)

Year	2009	2010	2011	2012	2013	2014
z (approx.)	18.93	19.46	20.31	21.10	21.58	22.62

The approximations are close to the actual values.

(b) An increase

107. (a) $\sqrt{70}$ in.

(b)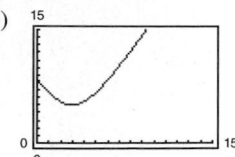

(c) The distance is never zero.

(d) 5 in.

109. $\left(\frac{77}{13}, \frac{48}{13}, -\frac{23}{13}\right)$ **111.** $x = 21t,\ y = 1 + 11t,\ z = 4 + 13t$

113. True **115.** True

117. False. Plane $7x + y - 11z = 5$ and plane $5x + 2y - 4z = 1$ are both perpendicular to plane $2x - 3y + z = 3$ but are not parallel.

Section 11.6 *(page 806)*

1. Quadric surfaces are the three-dimensional analogs of conic sections.

3. The trace of a surface is the intersection of the surface with a plane. You find a trace by setting one variable equal to a constant, such as $x = 0$ or $z = 2$.

5. c **6.** e **7.** f **8.** b **9.** d **10.** a

11. Right circular cylinder **13.** Elliptic cylinder

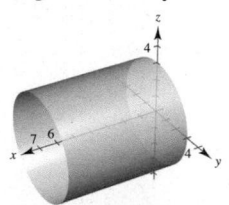 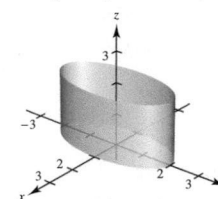

15. Hyperboloid of two sheets **17.** Hyperboloid of one sheet

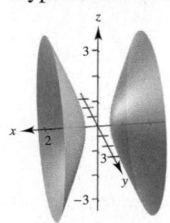 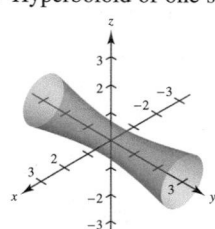

19. Ellipsoid **21.** Elliptic cone

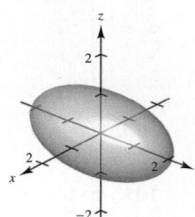 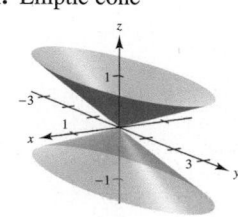

23. Hyperbolic paraboloid **25.** Elliptic paraboloid

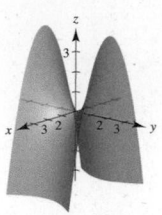 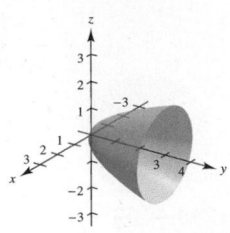

27. There have to be two minus signs to have a hyperboloid of two sheets. The number of sheets is the same as the number of minus signs.

29. No. See table on pages 800 and 801.

31. $x^2 + z^2 = 25y^2$ **33.** $x^2 + y^2 = 4z^{2/3}$

35. $y^2 + z^2 = \dfrac{4}{x^2}$ **37.** $y = \sqrt{2z}\ \left(\text{or } x = \sqrt{2z}\right)$

39. $y = \sqrt{5 - 8x^2}\ \left(\text{or } z = \sqrt{5 - 8x^2}\right)$ **41.** $\dfrac{128\pi}{3}$

43. (a) Major axis: $4\sqrt{2}$ (b) Major axis: $8\sqrt{2}$

Minor axis: 4   Minor axis: 8

Foci: $(0, \pm 2, 2)$ Foci: $(0, \pm 4, 8)$

45. $x^2 + z^2 = 8y$, elliptic paraboloid

47. $\dfrac{x^2}{3963^2} + \dfrac{y^2}{3963^2} + \dfrac{z^2}{3950^2} = 1$

49. $x = at,\ y = -bt,\ z = 0$;

$x = at,\ y = bt + ab^2,\ z = 2abt + a^2b^2$

51. The Klein bottle does not have both an "inside" and an "outside." It is formed by inserting the small open end through the side of the bottle and making it contiguous with the top of the bottle.

Section 11.7 *(page 813)*

1. The cylindrical coordinate system is an extension of the polar coordinate system. In this system, a point P in space is represented by an ordered triple (r, θ, z). (r, θ) is a polar representation of the projection of P in the xy-plane, and z is the directed distance from (r, θ) to P.

3. $(-7, 0, 5)$ **5.** $\left(\dfrac{3\sqrt{2}}{2}, \dfrac{3\sqrt{2}}{2}, 1\right)$ **7.** $\left(-2\sqrt{3}, -2, -3\right)$

9. $\left(5, \dfrac{\pi}{2}, 1\right)$ **11.** $\left(2\sqrt{2}, -\dfrac{\pi}{4}, -4\right)$ **13.** $\left(2, \dfrac{\pi}{3}, 4\right)$

15. $z = 4$ **17.** $r^2 - 2z^2 = 5$ **19.** $r = \sec\theta\tan\theta$

21. $r^2\sin^2\theta = 10 - z^2$

23. $x^2 + y^2 = 9$ **25.** $x - \sqrt{3}y = 0$

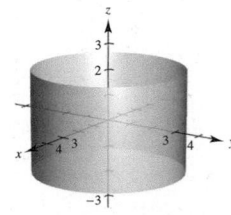 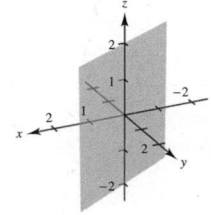

27. $x^2 + y^2 + z^2 = 5$

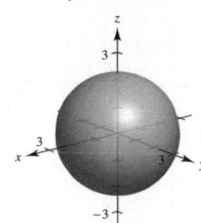

29. $x^2 + (y - 2)^2 = 4$

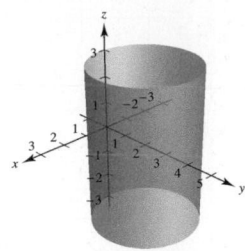

31. $\left(4, 0, \dfrac{\pi}{2}\right)$ **33.** $\left(4\sqrt{2}, \dfrac{2\pi}{3}, \dfrac{\pi}{4}\right)$ **35.** $\left(4, \dfrac{\pi}{6}, \dfrac{\pi}{6}\right)$

37. $\left(\sqrt{6}, \sqrt{2}, 2\sqrt{2}\right)$ **39.** $(0, 0, 12)$

41. $(0.915, 0.915, 4.830)$ **43.** $\rho = 2 \csc \phi \csc \theta$

45. $\rho = 7$ **47.** $\rho = 4 \csc \phi$ **49.** $\tan^2 \phi = 2$

51. $x^2 + y^2 + z^2 = 1$ **53.** $3x^2 + 3y^2 - z^2 = 0$

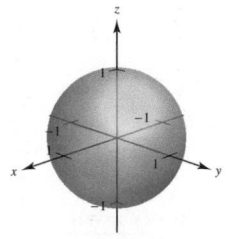   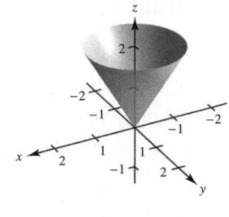

55. $x^2 + y^2 + (z - 2)^2 = 4$ **57.** $x^2 + y^2 = 1$

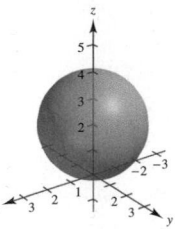   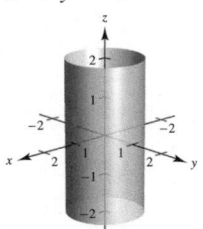

59. $\left(4, \dfrac{\pi}{4}, \dfrac{\pi}{2}\right)$ **61.** $\left(6\sqrt{2}, \dfrac{\pi}{2}, \dfrac{3\pi}{4}\right)$

63. $\left(13, \pi, \arccos \dfrac{5}{13}\right)$ **65.** $\left(10, \dfrac{\pi}{6}, 0\right)$

67. $\left(3\sqrt{3}, -\dfrac{\pi}{6}, 3\right)$ **69.** $\left(4, \dfrac{7\pi}{6}, 4\sqrt{3}\right)$

71. d   **72.** e   **73.** c   **74.** a   **75.** f   **76.** b

77. Because of the restriction $r \geq 0$

79. (a) $r^2 + z^2 = 27$ (b) $\rho = 3\sqrt{3}$

81. (a) $r^2 + (z - 1)^2 = 1$ (b) $\rho = 2 \cos \phi$

83. (a) $r = 4 \sin \theta$ (b) $\rho = \dfrac{4 \sin \theta}{\sin \phi} = 4 \sin \theta \csc \phi$

85. (a) $r^2 = \dfrac{9}{\cos^2 \theta - \sin^2 \theta}$

  (b) $\rho^2 = \dfrac{9 \csc^2 \phi}{\cos^2 \theta - \sin^2 \theta}$

87.

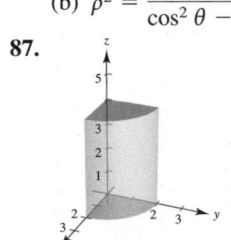

89.

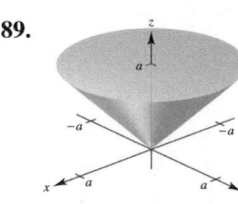

91.

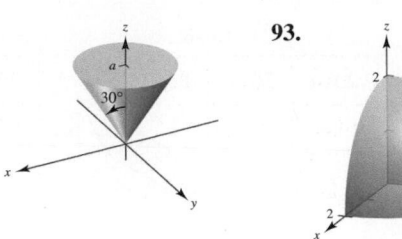

93.

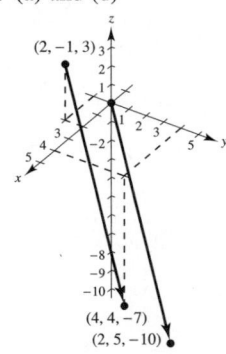

95. Rectangular: $0 \leq x \leq 10, 0 \leq y \leq 10, 0 \leq z \leq 10$

97. Spherical: $4 \leq \rho \leq 6$

99. Cylindrical: $r^2 + z^2 < 9, r \leq 3 \cos \theta, 0 \leq \theta \leq \pi$

101. False. See page 809.   **103.** Ellipse

Review Exercises for Chapter 11   *(page 815)*

1. (a) $\mathbf{u} = \langle 3, -1 \rangle$, $\mathbf{v} = \langle 4, 2 \rangle$ (b) $\mathbf{u} = 3\mathbf{i} - \mathbf{j}$, $\mathbf{v} = 4\mathbf{i} + 2\mathbf{j}$

  (c) $\|\mathbf{u}\| = \sqrt{10}$, $\|\mathbf{v}\| = 2\sqrt{5}$ (d) $\langle -5, 5 \rangle$

3. $\mathbf{v} = \left\langle 4, 4\sqrt{3} \right\rangle$ **5.** $(-5, 4, 0)$ **7.** $\sqrt{22}$

9. $(x - 3)^2 + (y + 2)^2 + (z - 6)^2 = 16$

11. $(x - 2)^2 + (y - 3)^2 + z^2 = 9$

  Center: $(2, 3, 0)$ Radius: 3

13. (a) and (d)

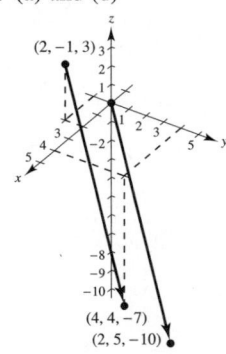

  (b) $\mathbf{u} = \langle 2, 5, -10 \rangle$ (c) $\mathbf{u} = 2\mathbf{i} + 5\mathbf{j} - 10\mathbf{k}$

15. $\langle -8, 5, 1 \rangle$ **17.** Collinear   **19.** $\dfrac{1}{\sqrt{38}} \langle 2, 3, 5 \rangle$

21. (a) $\mathbf{u} = \langle -1, 4, 0 \rangle$

    $\mathbf{v} = \langle -3, 0, 6 \rangle$

  (b) 3   (c) 45

23. (a) $\dfrac{\pi}{12}$ (b) $15°$ **25.** Orthogonal

27. (a) $\left\langle \dfrac{12}{5}, \dfrac{16}{5} \right\rangle$ (b) $\left\langle \dfrac{8}{5}, -\dfrac{6}{5} \right\rangle$

29. Answers will vary. Sample answer: $\langle -6, 5, 0 \rangle, \langle 6, -5, 0 \rangle$

31. (a) $-9\mathbf{i} + 26\mathbf{j} - 7\mathbf{k}$ (b) $9\mathbf{i} - 26\mathbf{j} + 7\mathbf{k}$ (c) $\mathbf{0}$

33. $\left\langle \dfrac{8}{\sqrt{377}}, \dfrac{12}{\sqrt{377}}, \dfrac{13}{\sqrt{377}} \right\rangle$ or $\left\langle -\dfrac{8}{\sqrt{377}}, -\dfrac{12}{\sqrt{377}}, -\dfrac{13}{\sqrt{377}} \right\rangle$

35. 15 ft-lb

37. (a) $x = 3 + 6t, y = 11t, z = 2 + 4t$

  (b) $\dfrac{x - 3}{6} = \dfrac{y}{11} = \dfrac{z - 2}{4}$

39. $x = -6, y = -8 + t, z = 2$

41. $27x + 4y + 32z + 33 = 0$ **43.** $x + 2y = 1$ **45.** $\dfrac{8}{7}$

47. $\dfrac{\sqrt{35}}{7}$

49. Plane

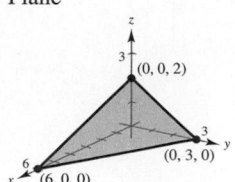

51. Plane

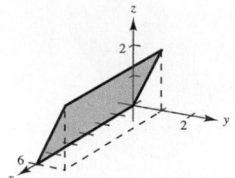

53. Ellipsoid

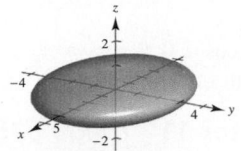

55. Hyperboloid of two sheets

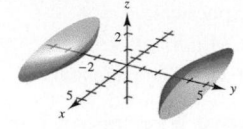

57. Cylinder

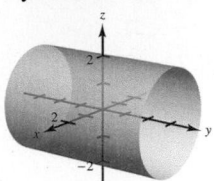

59. $x^2 + z^2 = 2y$

61. (a) $\left(2\sqrt{3}, -\dfrac{\pi}{3}, -5\right)$ (b) $\left(\sqrt{37}, -\dfrac{\pi}{3}, \arccos\left(-\dfrac{5\sqrt{37}}{37}\right)\right)$

63. $(-5, 0, 1)$ **65.** $\left(-2\sqrt{2}, 0, 2\sqrt{2}\right)$

67. (a) $r^2 \cos 2\theta = 2z$ (b) $\rho = 2 \sec 2\theta \cos \phi \csc^2 \phi$

69. $z = y^2 + 3x$ **71.** $x^2 + y^2 - z^2 = 0$

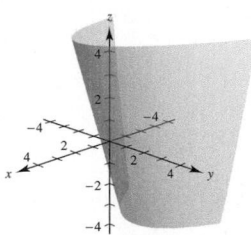
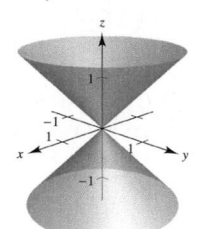

P.S. Problem Solving   *(page 817)*

1–3. Proofs   **5.** (a) $\dfrac{3\sqrt{2}}{2} \approx 2.12$ (b) $\sqrt{5} \approx 2.24$

7. (a) $\dfrac{\pi}{2}$ (b) $\dfrac{1}{2}(\pi abk)k$

   (c) $V = \dfrac{1}{2}(\pi ab)k^2$

   $V = \dfrac{1}{2}(\text{area of base})\text{height}$

9. Proof

11. (a)   (b)

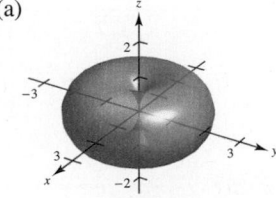
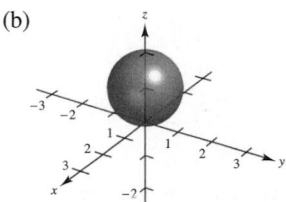

13. (a) Tension: $\dfrac{2\sqrt{3}}{3} \approx 1.1547$ lb

   Magnitude of **u**: $\dfrac{\sqrt{3}}{3} \approx 0.5774$ lb

   (b) $T = \sec \theta$, $\|\mathbf{u}\| = \tan \theta$; Domain: $0° \le \theta \le 90°$

(c)

θ	0°	10°	20°	30°
T	1	1.0154	1.0642	1.1547
$\|\mathbf{u}\|$	0	0.1763	0.3640	0.5774

θ	40°	50°	60°
T	1.3054	1.5557	2
$\|\mathbf{u}\|$	0.8391	1.1918	1.7321

(d)

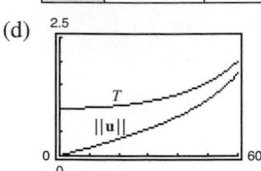

(e) Both are increasing functions.

(f) $\lim\limits_{\theta \to \pi/2^-} T = \infty$ and $\lim\limits_{\theta \to \pi/2^-} \|\mathbf{u}\| = \infty$

   Yes. As θ increases, both T and $\|\mathbf{u}\|$ increase.

15. $\langle 0, 0, \cos \alpha \sin \beta - \cos \beta \sin \alpha \rangle$; Proof

17. $D = \dfrac{|\overrightarrow{PQ} \cdot \mathbf{n}|}{\|\mathbf{n}\|}$

   $= \dfrac{|\mathbf{w} \cdot (\mathbf{u} \times \mathbf{v})|}{\|\mathbf{u} \times \mathbf{v}\|} = \dfrac{|(\mathbf{u} \times \mathbf{v}) \cdot \mathbf{w}|}{\|\mathbf{u} \times \mathbf{v}\|} = \dfrac{|\mathbf{u} \cdot (\mathbf{v} \times \mathbf{w})|}{\|\mathbf{u} \times \mathbf{v}\|}$

19. Proof

Chapter 12

Section 12.1   *(page 825)*

1. You can use a vector-valued function to trace the graph of a curve. Recall that the terminal point of the position vector $\mathbf{r}(t)$ coincides with a point on the curve.

3. $(-\infty, -1) \cup (-1, \infty)$ **5.** $(0, \infty)$

7. $[0, \infty)$ **9.** $(-\infty, \infty)$

11. (a) $\frac{1}{2}\mathbf{i}$ (b) $\mathbf{j}$ (c) $\frac{1}{2}(s + 1)^2\mathbf{i} - s\mathbf{j}$

   (d) $\frac{1}{2}\Delta t(\Delta t + 4)\mathbf{i} - \Delta t\mathbf{j}$

13. $\mathbf{r}(t) = 5t\mathbf{i} + 2t\mathbf{j} + 2t\mathbf{k}, 0 \le t \le 1$

   $x = 5t, y = 2t, z = 2t, 0 \le t \le 1$

15. $\mathbf{r}(t) = (-3 + 2t)\mathbf{i} + (-6 - 3t)\mathbf{j} + (-1 - 7t)\mathbf{k}, 0 \le t \le 1$

   $x = -3 + 2t, y = -6 - 3t, z = -1 - 7t, 0 \le t \le 1$

17. $t^2(5t - 1)$; No, the dot product is a scalar.

19. b   **20.** c   **21.** d   **22.** a

23.   **25.**

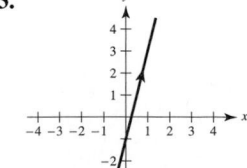
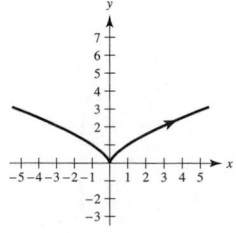

27.   **29.**

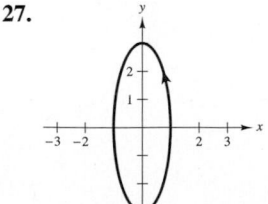
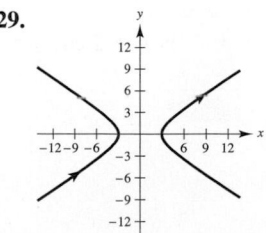

31.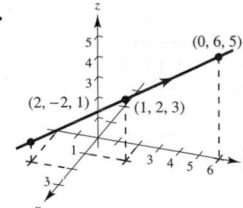

33.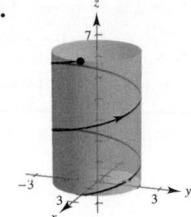

35.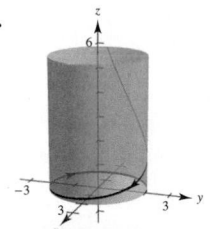

37.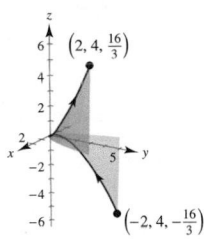

39.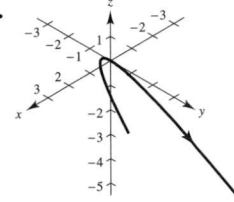

Parabola

41.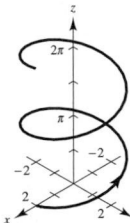

(a) The helix is translated two units back on the x-axis.

(b) The height of the helix increases at a greater rate.

(c) The orientation of the graph is reversed.

(d) The radius of the helix is increased from 2 to 6.

43. $\mathbf{u}(t) = 3t^2\mathbf{i} + (t-1)\mathbf{j} + (t+2)\mathbf{k}$

45. $\mathbf{u}(t) = 3t^2\mathbf{i} + 2(t-1)\mathbf{j} + t\mathbf{k}$

47–53. Answers will vary.

55.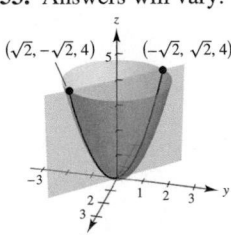

$\mathbf{r}(t) = t\mathbf{i} - t\mathbf{j} + 2t^2\mathbf{k}$

57.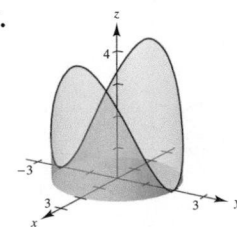

$\mathbf{r}(t) = 2\sin t\mathbf{i} + 2\cos t\mathbf{j} + 4\sin^2 t\mathbf{k}$

59.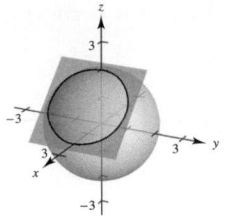

$\mathbf{r}(t) = (1+\sin t)\mathbf{i} + \sqrt{2}\cos t\mathbf{j} + (1-\sin t)\mathbf{k}$ and
$\mathbf{r}(t) = (1+\sin t)\mathbf{i} - \sqrt{2}\cos t\mathbf{j} + (1-\sin t)\mathbf{k}$

61.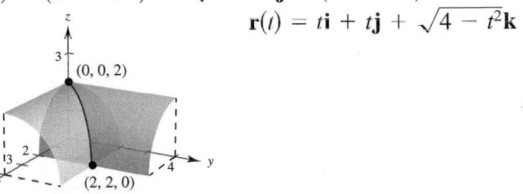

$\mathbf{r}(t) = t\mathbf{i} + t\mathbf{j} + \sqrt{4-t^2}\mathbf{k}$

63. Let $x = t$, $y = 2t\cos t$, and $z = 2t\sin t$. Then
$$y^2 + z^2 = (2t\cos t)^2 + (2t\sin t)^2$$
$$= 4t^2\cos^2 t + 4t^2\sin^2 t$$
$$= 4t^2(\cos^2 t + \sin^2 t)$$
$$= 4t^2.$$
Because $x = t$, $y^2 + z^2 = 4x^2$.

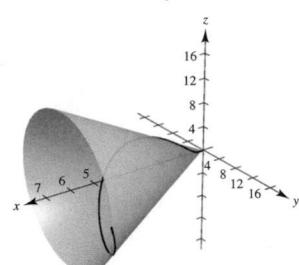

65. $\pi\mathbf{i} - \mathbf{j}$ **67.** $\mathbf{0}$ **69.** $\mathbf{i} + \mathbf{j} + \mathbf{k}$

71. $\left(-\infty, -\frac{1}{2}\right), \left(-\frac{1}{2}, 0\right), (0, \infty)$ **73.** $[-1, 1]$

75. $\left(-\frac{\pi}{2} + n\pi, \frac{\pi}{2} + n\pi\right)$, n is an integer.

77. It is a line; Answers will vary.

79. $\mathbf{r}(t) = \begin{cases} \mathbf{i} + \mathbf{j}, & t \geq 3 \\ -\mathbf{i} + \mathbf{j}, & t < 3 \end{cases}$

81. $\mathbf{r}(t) = \cos t\mathbf{i} + \sin t\mathbf{j} + \frac{1}{\pi}t\mathbf{k}$, $0 \leq t \leq 4\pi$

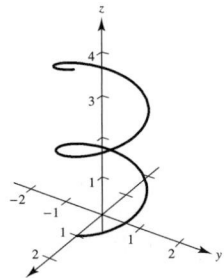

83–85. Proofs   **87.** Not necessarily   **89.** Yes; yes

Section 12.2   *(page 834)*

1. $\mathbf{r}'(t_0)$ represents the vector that is tangent to the curve represented by $\mathbf{r}(t)$ at the point t_0.

3. $\mathbf{r}'(t) = -2t\mathbf{i} + \mathbf{j}$
$\mathbf{r}(3) = -8\mathbf{i} + 3\mathbf{j}$
$\mathbf{r}'(3) = -6\mathbf{i} + \mathbf{j}$

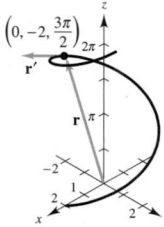

5. $\mathbf{r}'(t) = -\sin t\mathbf{i} + \cos t\mathbf{j}$
$\mathbf{r}\left(\dfrac{\pi}{2}\right) = \mathbf{j}$
$\mathbf{r}'\left(\dfrac{\pi}{2}\right) = -\mathbf{i}$

7. $\mathbf{r}'(t) = \langle e^t, 2e^{2t}\rangle$
$\mathbf{r}(0) = \mathbf{i} + \mathbf{j}$
$\mathbf{r}'(0) = \mathbf{i} + 2\mathbf{j}$

9. $\mathbf{r}'(t) = -2\sin t\mathbf{i} + 2\cos t\mathbf{j} + \mathbf{k}$
$\mathbf{r}\left(\dfrac{3\pi}{2}\right) = -2\mathbf{j} + \left(\dfrac{3\pi}{2}\right)\mathbf{k}$
$\mathbf{r}'\left(\dfrac{3\pi}{2}\right) = 2\mathbf{i} + \mathbf{k}$

11. $4t^3\mathbf{i} - 5\mathbf{j}$ **13.** $-9\sin t\cos^2 t\mathbf{i} + 6\sin^2 t\cos t\mathbf{j}$

15. $-e^{-t}\mathbf{i} + (5te^t + 5e^t)\mathbf{k}$

17. $\langle\sin t + t\cos t, \cos t - t\sin t, 1\rangle$

19. (a) $3t^2\mathbf{i} + t\mathbf{j}$ (b) $6t\mathbf{i} + \mathbf{j}$ (c) $18t^3 + t$

21. (a) $-4\sin t\mathbf{i} + 4\cos t\mathbf{j}$ (b) $-4\cos t\mathbf{i} - 4\sin t\mathbf{j}$ (c) 0

23. (a) $t\mathbf{i} - \mathbf{j} + \frac{1}{2}t^2\mathbf{k}$ (b) $\mathbf{i} + t\mathbf{k}$ (c) $\dfrac{t^3}{2} + t$

     (d) $-t\mathbf{i} - \frac{1}{2}t^2\mathbf{j} + \mathbf{k}$

25. (a) $\langle t\cos t, t\sin t, 1\rangle$
     (b) $\langle\cos t - t\sin t, \sin t + t\cos t, 0\rangle$ (c) t
     (d) $\langle-\sin t - t\cos t, \cos t - t\sin t, t^2\rangle$

27. $(-\infty, 0), (0, \infty)$ **29.** $\left(\dfrac{\pi}{2}, 2\pi\right)$

31. $(-\infty, -2), (-2, \infty)$

33. $\left(-\dfrac{\pi}{2} + n\pi, \dfrac{\pi}{2} + n\pi\right)$, n is an integer

35. (a) $\mathbf{i} + 3\mathbf{j} + 2t\mathbf{k}$ (b) $-\mathbf{i} + (9 - 2t)\mathbf{j} + (6t - 3t^2)\mathbf{k}$
     (c) $40t\mathbf{i} + 15t^2\mathbf{j} + 20t^3\mathbf{k}$ (d) $8t + 9t^2 + 5t^4$
     (e) $8t^3\mathbf{i} + (12t^2 - 4t^3)\mathbf{j} + (3t^2 - 24t)\mathbf{k}$
     (f) $2\mathbf{i} + 6\mathbf{j} + 8t\mathbf{k}$

37. (a) $7t^6$ (b) $12t^5\mathbf{i} - 5t^4\mathbf{j}$ **39.** $t^2\mathbf{i} + t\mathbf{j} + 9t\mathbf{k} + \mathbf{C}$

41. $\ln|t|\mathbf{i} + t\mathbf{j} - \frac{2}{5}t^{5/2}\mathbf{k} + \mathbf{C}$ **43.** $t\mathbf{i} + t^4\mathbf{j} + \dfrac{5^t}{\ln 5}\mathbf{k} + \mathbf{C}$

45. $e^t\mathbf{i} + t\mathbf{j} + (t\sin t + \cos t)\mathbf{k} + \mathbf{C}$

47. $4\mathbf{i} + \dfrac{1}{2}\mathbf{j} - \mathbf{k}$ **49.** $5\mathbf{i} + 6\mathbf{j} + \dfrac{\pi}{2}\mathbf{k}$

51. $2\mathbf{i} + (e^2 - 1)\mathbf{j} - (e^2 + 1)\mathbf{k}$

53. $2e^{2t}\mathbf{i} + 3(e^t - 1)\mathbf{j}$ **55.** $600\sqrt{3}t\mathbf{i} + (-16t^2 + 600t)\mathbf{j}$

57. $\dfrac{2 - e^{-t^2}}{2}\mathbf{i} + (e^{-t} - 2)\mathbf{j} + (t + 1)\mathbf{k}$

59. The three components of $\mathbf{u}$ are increasing functions of t at
     $t = t_0$.

61–67. Proofs

69. (a)

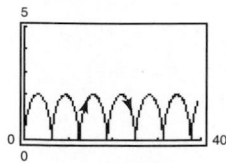

     The curve is a cycloid.

     (b) The maximum of $\|\mathbf{r}'\|$ is 2 and the minimum of $\|\mathbf{r}'\|$ is 0.
     The maximum and the minimum of $\|\mathbf{r}''\|$ are 1.

71. Proof   **73.** True

75. False. Let $\mathbf{r}(t) = \cos t\mathbf{i} + \sin t\mathbf{j} + \mathbf{k}$, then $\dfrac{d}{dt}[\|\mathbf{r}(t)\|] = 0$,
     but $\|\mathbf{r}'(t)\| = 1$.

Section 12.3   *(page 842)*

1. The direction of the velocity vector provides the direction
of motion at time t and the magnitude of the velocity vector
provides the speed of the object.

3. (a) $\mathbf{v}(t) = 3\mathbf{i} + \mathbf{j}$
     $\|\mathbf{v}(t)\| = \sqrt{10}$
     $\mathbf{a}(t) = \mathbf{0}$
     (b) $\mathbf{v}(1) = 3\mathbf{i} + \mathbf{j}$
     $\mathbf{a}(1) = \mathbf{0}$
     (c)

5. (a) $\mathbf{v}(t) = 2t\mathbf{i} + \mathbf{j}$
     $\|\mathbf{v}(t)\| = \sqrt{4t^2 + 1}$
     $\mathbf{a}(t) = 2\mathbf{i}$
     (b) $\mathbf{v}(2) = 4\mathbf{i} + \mathbf{j}$
     $\mathbf{a}(2) = 2\mathbf{i}$
     (c)

7. (a) $\mathbf{v}(t) = -2\sin t\mathbf{i} + 2\cos t\mathbf{j}$ (c)
     $\|\mathbf{v}(t)\| = 2$
     $\mathbf{a}(t) = -2\cos t\mathbf{i} - 2\sin t\mathbf{j}$
     (b) $\mathbf{v}\left(\dfrac{\pi}{4}\right) = -\sqrt{2}\mathbf{i} + \sqrt{2}\mathbf{j}$
     $\mathbf{a}\left(\dfrac{\pi}{4}\right) = -\sqrt{2}\mathbf{i} - \sqrt{2}\mathbf{j}$

9. (a) $\mathbf{v}(t) = \langle 1 - \cos t, \sin t\rangle$ (c)
     $\|\mathbf{v}(t)\| = \sqrt{2 - 2\cos t}$
     $\mathbf{a}(t) = \langle\sin t, \cos t\rangle$
     (b) $\mathbf{v}(\pi) = \langle 2, 0\rangle$
     $\mathbf{a}(\pi) = \langle 0, -1\rangle$

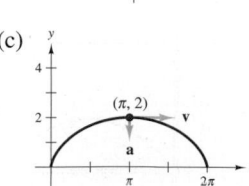

11. (a) $\mathbf{v}(t) = \mathbf{i} + 5\mathbf{j} + 3\mathbf{k}$
     $\|\mathbf{v}(t)\| = \sqrt{35}$
     $\mathbf{a}(t) = \mathbf{0}$
     (b) $\mathbf{v}(1) = \mathbf{i} + 5\mathbf{j} + 3\mathbf{k}$
     $\mathbf{a}(1) = \mathbf{0}$

13. (a) $\mathbf{v}(t) = \mathbf{i} + 2t\mathbf{j} + t\mathbf{k}$
     $\|\mathbf{v}(t)\| = \sqrt{1 + 5t^2}$
     $\mathbf{a}(t) = 2\mathbf{j} + \mathbf{k}$
     (b) $\mathbf{v}(4) = \mathbf{i} + 8\mathbf{j} + 4\mathbf{k}$
     $\mathbf{a}(4) = 2\mathbf{j} + \mathbf{k}$

15. (a) $\mathbf{v}(t) = \mathbf{i} - \mathbf{j} - \dfrac{t}{\sqrt{9 - t^2}}\mathbf{k}$

$\|\mathbf{v}(t)\| = \sqrt{\dfrac{18 - t^2}{9 - t^2}}$

$\mathbf{a}(t) = -\dfrac{9}{(9 - t^2)^{3/2}}\mathbf{k}$

(b) $\mathbf{v}(0) = \mathbf{i} - \mathbf{j}$

$\mathbf{a}(0) = -\tfrac{1}{3}\mathbf{k}$

17. (a) $\mathbf{v}(t) = 4\mathbf{i} - 3\sin t\mathbf{j} + 3\cos t\mathbf{k}$

$\|\mathbf{v}(t)\| = 5$

$\mathbf{a}(t) = -3\cos t\mathbf{j} - 3\sin t\mathbf{k}$

(b) $\mathbf{v}(\pi) = \langle 4, 0, -3 \rangle$

$\mathbf{a}(\pi) = \langle 0, 3, 0 \rangle$

19. (a) $\mathbf{v}(t) = (e^t\cos t - e^t\sin t)\mathbf{i} + (e^t\sin t + e^t\cos t)\mathbf{j} + e^t\mathbf{k}$

$\|\mathbf{v}(t)\| = e^t\sqrt{3}$

$\mathbf{a}(t) = -2e^t\sin t\mathbf{i} + 2e^t\cos t\mathbf{j} + e^t\mathbf{k}$

(b) $\mathbf{v}(0) = \langle 1, 1, 1 \rangle$

$\mathbf{a}(0) = \langle 0, 2, 1 \rangle$

21. $\mathbf{v}(t) = t(\mathbf{i} + \mathbf{j} + \mathbf{k})$

$\mathbf{r}(t) = \dfrac{t^2}{2}(\mathbf{i} + \mathbf{j} + \mathbf{k})$

$\mathbf{r}(2) = 2(\mathbf{i} + \mathbf{j} + \mathbf{k})$

23. $\mathbf{v}(t) = \left(\dfrac{t^2}{2} + \dfrac{9}{2}\right)\mathbf{j} + \left(\dfrac{t^2}{2} - \dfrac{1}{2}\right)\mathbf{k}$

$\mathbf{r}(t) = \left(\dfrac{t^3}{6} + \dfrac{9}{2}t - \dfrac{14}{3}\right)\mathbf{j} + \left(\dfrac{t^3}{6} - \dfrac{1}{2}t + \dfrac{1}{3}\right)\mathbf{k}$

$\mathbf{r}(2) = \dfrac{17}{3}\mathbf{j} + \dfrac{2}{3}\mathbf{k}$

25. $\mathbf{v}(t) = -\sin t\mathbf{i} + \cos t\mathbf{j} + \mathbf{k}$

$\mathbf{r}(t) = \cos t\mathbf{i} + \sin t\mathbf{j} + t\mathbf{k}$

$\mathbf{r}(2) = (\cos 2)\mathbf{i} + (\sin 2)\mathbf{j} + 2\mathbf{k}$

27. 45.5 ft; The ball will clear the fence.

29. $v_0 = 40\sqrt{6}$ ft/sec; 78 ft　　**31.** Proof

33. (a) $\mathbf{r}(t) = \left(\tfrac{440}{3}\cos\theta_0\right)t\mathbf{i} + \left[3 + \left(\tfrac{440}{3}\sin\theta_0\right)t - 16t^2\right]\mathbf{j}$

(b)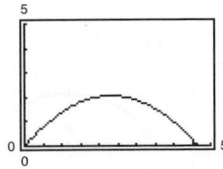

The minimum angle appears to be $\theta_0 = 20°$.

(c) $\theta_0 \approx 19.38°$

35. (a) $v_0 = 28.78$ ft/sec, $\theta = 58.28°$　(b) $v_0 \approx 32$ ft/sec

37. 1.91°

39. (a) 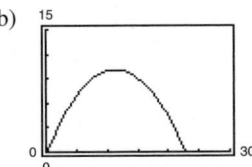　(b)

Maximum height: 2.1 ft　　　Maximum height: 10.0 ft
Range: 46.6 ft　　　　　　　Range: 227.8 ft

(c) 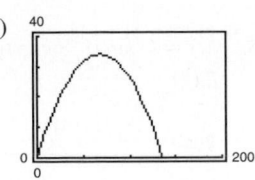　(d)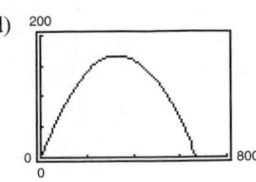

Maximum height: 34.0 ft　　Maximum height: 166.5 ft
Range: 136.1 ft　　　　　　Range: 666.1 ft

(e) 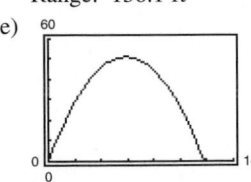　(f)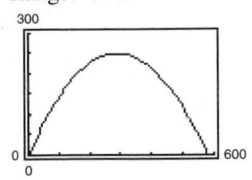

Maximum height: 51.0 ft　　Maximum height: 249.8 ft
Range: 117.9 ft　　　　　　Range: 576.9 ft

41. Maximum height: 129.1 m; Range: 886.3 m　　**43.** Proof

45. $\mathbf{v}(t) = b\omega[(1 - \cos\omega t)\mathbf{i} + \sin\omega t\mathbf{j}]$

$\mathbf{a}(t) = b\omega^2(\sin\omega t\mathbf{i} + \cos\omega t\mathbf{j})$

(a) $\|\mathbf{v}(t)\| = 0$ when $\omega t = 0, 2\pi, 4\pi, \ldots$

(b) $\|\mathbf{v}(t)\|$ is maximum when $\omega t = \pi, 3\pi, \ldots$

47. $\mathbf{v}(t) = -b\omega\sin\omega t\mathbf{i} + b\omega\cos\omega t\mathbf{j}$

$\mathbf{v}(t) \cdot \mathbf{r}(t) = 0$

49. $\mathbf{a}(t) = -b\omega^2(\cos\omega t\mathbf{i} + \sin\omega t\mathbf{j}) = -\omega^2\mathbf{r}(t)$; $\mathbf{a}(t)$ is a negative multiple of a unit vector from $(0, 0)$ to $(\cos\omega t, \sin\omega t)$, so $\mathbf{a}(t)$ is directed toward the origin.

51. $8\sqrt{2}$ ft/sec

53. The particle could be changing direction.

55. This is true for uniform circular motion but not true for non-uniform circular motion.

57–59. Proofs　　**61.** True

63. False. Consider $\mathbf{r}(t) = \langle t^2, -t^2 \rangle$. Then $\mathbf{v}(t) = \langle 2t, -2t \rangle$ and $\|\mathbf{v}(t)\| = \sqrt{8t^2}$.

Section 12.4　*(page 852)*

1. The unit tangent vector points in the direction of motion.

3. $\mathbf{T}(1) = \dfrac{\sqrt{2}}{2}(\mathbf{i} + \mathbf{j})$　**5.** $\mathbf{T}\left(\dfrac{\pi}{3}\right) = -\dfrac{\sqrt{3}}{2}\mathbf{i} + \dfrac{1}{2}\mathbf{j}$

7. $\mathbf{T}(e) = \dfrac{3e\mathbf{i} - \mathbf{j}}{\sqrt{9e^2 + 1}} \approx 0.9926\mathbf{i} - 0.1217\mathbf{j}$

9. $\mathbf{T}(0) = \dfrac{\sqrt{2}}{2}(\mathbf{i} + \mathbf{k})$　　**11.** $\mathbf{T}(0) = \dfrac{\sqrt{2}}{2}(\mathbf{j} + \mathbf{k})$

$x = t$　　　　　　　　　　　　　$x = 1$

$y = 0$　　　　　　　　　　　　　$y = 3t$

$z = t$　　　　　　　　　　　　　$z = -4 + 3t$

13. $\mathbf{T}\left(\dfrac{\pi}{4}\right) = \dfrac{1}{2}\langle -\sqrt{2}, \sqrt{2}, 0 \rangle$

$x = \sqrt{2} - \sqrt{2}\,t$

$y = \sqrt{2} + \sqrt{2}\,t$

$z = 4$

15. $\mathbf{N}(2) = \dfrac{\sqrt{5}}{5}(-2\mathbf{i} + \mathbf{j})$

17. $\mathbf{N}(1) = -\dfrac{\sqrt{14}}{14}(\mathbf{i} - 2\mathbf{j} + 3\mathbf{k})$

19. $\mathbf{N}\left(\dfrac{3\pi}{4}\right) = \dfrac{\sqrt{2}}{2}(\mathbf{i} - \mathbf{j})$

21. $\mathbf{r}(2) = 2\mathbf{i} + \frac{1}{2}\mathbf{j}$

$\mathbf{T}(2) = \frac{\sqrt{17}}{17}(4\mathbf{i} - \mathbf{j})$

$\mathbf{N}(2) = \frac{\sqrt{17}}{17}(\mathbf{i} + 4\mathbf{j})$

23. $\mathbf{r}(2) = 5\mathbf{i} - 4\mathbf{j}$

$\mathbf{T}(2) = \frac{\mathbf{i} - 2\mathbf{j}}{\sqrt{5}}$

$\mathbf{N}(2) = \frac{-2\mathbf{i} - \mathbf{j}}{\sqrt{5}}$,

perpendicular to $\mathbf{T}(2)$

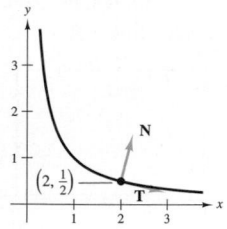

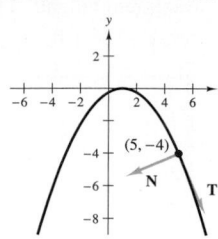

25. $a_{\mathbf{T}} = -\sqrt{2}$
$a_{\mathbf{N}} = \sqrt{2}$

27. $a_{\mathbf{T}} = -\frac{7\sqrt{5}}{5}$
$a_{\mathbf{N}} = \frac{6\sqrt{5}}{5}$

29. $a_{\mathbf{T}} = \sqrt{2}e^{\pi/2}$
$a_{\mathbf{N}} = \sqrt{2}e^{\pi/2}$

31. $\mathbf{T}(t) = -\sin \omega t\mathbf{i} + \cos \omega t\mathbf{j}$
$\mathbf{N}(t) = -\cos \omega t\mathbf{i} - \sin \omega t\mathbf{j}$
$a_{\mathbf{T}} = 0$
$a_{\mathbf{N}} = a\omega^2$

33. $\|\mathbf{v}(t)\| = a\omega$; The speed is constant because $a_{\mathbf{T}} = 0$.

35. $a_{\mathbf{T}}$ is undefined.

$a_{\mathbf{N}}$ is undefined.

37. $a_{\mathbf{T}} = \frac{5\sqrt{6}}{6}$

$a_{\mathbf{N}} = \frac{\sqrt{30}}{6}$

39. $a_{\mathbf{T}} = \sqrt{3}$
$a_{\mathbf{N}} = \sqrt{2}$

41. The particle's motion is in a straight line.

43. $\mathbf{v}(t) = \mathbf{r}'(t) = 3\mathbf{i} + 4\mathbf{j}$

$\|\mathbf{v}(t)\| = \sqrt{9 + 16} = 5$

$\mathbf{a}(t) = \mathbf{v}'(t) = \mathbf{0}$

$\mathbf{T}(t) = \frac{\mathbf{v}(t)}{\|\mathbf{v}(t)\|} = \frac{3}{5}\mathbf{i} + \frac{4}{5}\mathbf{j}$

$\mathbf{T}'(t) = \mathbf{0} \implies \mathbf{N}(t)$ does not exist.

The path is a line. The speed is constant (5).

45. (a) $t = \frac{1}{2}$: $a_{\mathbf{T}} = \frac{\sqrt{2}\pi^2}{2}$, $a_{\mathbf{N}} = \frac{\sqrt{2}\pi^2}{2}$

$t = 1$: $a_{\mathbf{T}} = 0$, $a_{\mathbf{N}} = \pi^2$

$t = \frac{3}{2}$: $a_{\mathbf{T}} = -\frac{\sqrt{2}\pi^2}{2}$, $a_{\mathbf{N}} = \frac{\sqrt{2}\pi^2}{2}$

(b) $t = \frac{1}{2}$: Increasing because $a_{\mathbf{T}} > 0$.

$t = 1$: Maximum because $a_{\mathbf{T}} = 0$.

$t = \frac{3}{2}$: Decreasing because $a_{\mathbf{T}} < 0$.

47. $\mathbf{T}\left(\frac{\pi}{2}\right) = \frac{\sqrt{17}}{17}(-4\mathbf{i} + \mathbf{k})$

$\mathbf{N}\left(\frac{\pi}{2}\right) = -\mathbf{j}$

$\mathbf{B}\left(\frac{\pi}{2}\right) = \frac{\sqrt{17}}{17}(\mathbf{i} + 4\mathbf{k})$

49. $\mathbf{T}\left(\frac{\pi}{4}\right) = \frac{\sqrt{2}}{2}(\mathbf{j} - \mathbf{k})$

$\mathbf{N}\left(\frac{\pi}{4}\right) = -\frac{\sqrt{2}}{2}(\mathbf{j} + \mathbf{k})$

$\mathbf{B}\left(\frac{\pi}{4}\right) = -\mathbf{i}$

51. $\mathbf{T}\left(\frac{\pi}{3}\right) = \frac{\sqrt{5}}{5}(\mathbf{i} - \sqrt{3}\mathbf{j} + \mathbf{k})$

$\mathbf{N}\left(\frac{\pi}{3}\right) = -\frac{1}{2}(\sqrt{3}\mathbf{i} + \mathbf{j})$

$\mathbf{B}\left(\frac{\pi}{3}\right) = \frac{\sqrt{5}}{10}(\mathbf{i} - \sqrt{3}\mathbf{j} - 4\mathbf{k})$

53. $\mathbf{N}(t) = \frac{1}{\sqrt{16t^2 + 9}}(-4t\mathbf{i} + 3\mathbf{j})$

55. $\mathbf{N}(t) = \frac{1}{\sqrt{5t^2 + 25}}(-t\mathbf{i} - 2t\mathbf{j} + 5\mathbf{k})$

57. $a_{\mathbf{T}} = \frac{-32(v_0 \sin \theta - 32t)}{\sqrt{v_0^2 \cos^2 \theta + (v_0 \sin \theta - 32t)^2}}$

$a_{\mathbf{N}} = \frac{32v_0 \cos \theta}{\sqrt{v_0^2 \cos^2 \theta + (v_0 \sin \theta - 32t)^2}}$

At maximum height, $a_{\mathbf{T}} = 0$ and $a_{\mathbf{N}} = 32$.

59. (a) $\mathbf{r}(t) = 60\sqrt{3}t\mathbf{i} + (5 + 60t - 16t^2)\mathbf{j}$

(b)

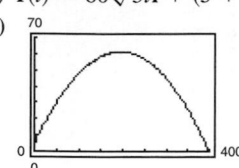

Maximum height ≈ 61.245 ft
Range ≈ 398.186 ft

(c) $\mathbf{v}(t) = 60\sqrt{3}\mathbf{i} + (60 - 32t)\mathbf{j}$
$\|\mathbf{v}(t)\| = 8\sqrt{16t^2 - 60t + 225}$
$\mathbf{a}(t) = -32\mathbf{j}$

(d)

t	0.5	1.0	1.5
Speed	112.85	107.63	104.61

t	2.0	2.5	3.0
Speed	104	105.83	109.98

(e)

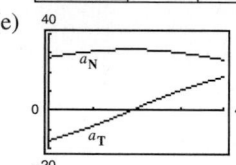

The speed is decreasing when $a_{\mathbf{T}}$ and $a_{\mathbf{N}}$ have opposite signs.

61. (a) $4\sqrt{625\pi^2 + 1} \approx 314$ mi/h

(b) $a_{\mathbf{T}} = 0$, $a_{\mathbf{N}} = 1000\pi^2$

$a_{\mathbf{T}} = 0$ because the speed is constant.

63. (a) The centripetal component is quadrupled.

(b) The centripetal component is halved.

65. 4.74 mi/sec   **67.** 4.67 mi/sec

69. False. These vectors are perpendicular for an object traveling at a constant speed but not for an object traveling at a variable speed.

71. (a) and (b) Proofs   **73–75.** Proofs

Section 12.5   (page 864)

1. The curve bends more sharply at Q than at P.

3.

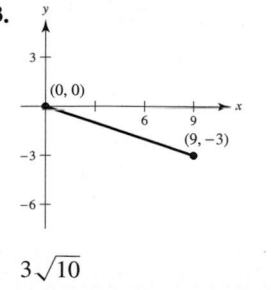

$3\sqrt{10}$

5.

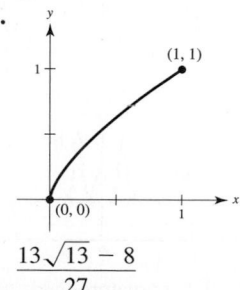

$\frac{13\sqrt{13} - 8}{27}$

7.

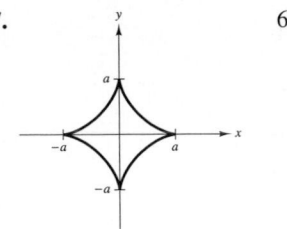

6a

9. 362.9 ft

11.

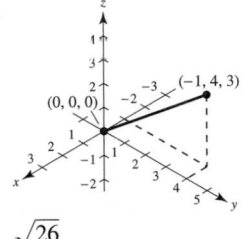

$\sqrt{26}$

13.

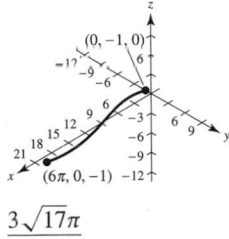

$\dfrac{3\sqrt{17}\pi}{2}$

$2\pi\sqrt{a^2 + b^2}$

15.

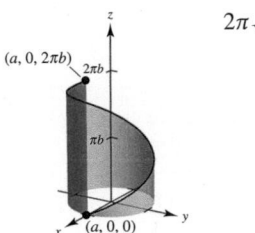

17. (a) $2\sqrt{21} \approx 9.165$ (b) 9.529
(c) Increase the number of line segments. (d) 9.571

19. 0 **21.** $\dfrac{1}{4}$ **23.** 0 **25.** $\dfrac{\sqrt{2}}{2}$ **27.** 1 **29.** $\dfrac{1}{4}$

31. $\dfrac{1}{a}$ **33.** $\dfrac{\sqrt{5}}{(1 + 5t^2)^{3/2}}$ **35.** $\dfrac{3}{25}$ **37.** $\dfrac{12}{125}$

39. $\dfrac{7\sqrt{26}}{676}$ **41.** $K = 0$, $\dfrac{1}{K}$ is undefined.

43. $K = \dfrac{10}{101^{3/2}}, \dfrac{1}{K} = \dfrac{101^{3/2}}{10}$ **45.** $K = 4, \dfrac{1}{K} = \dfrac{1}{4}$

47. $K = \dfrac{12}{145^{3/2}}, \dfrac{1}{K} = \dfrac{145^{3/2}}{12}$ **49.** (a) $(1, 3)$ (b) 0

51. (a) $K \to \infty$ as $x \to 0$ (No maximum) (b) 0

53. (a) $\left(\dfrac{1}{\sqrt{2}}, -\dfrac{\ln 2}{2}\right)$ (b) 0 **55.** $(0, 1)$

57. $(\pi + 2n\pi, 0)$ **59.** $c = \pm\sqrt{2}$

61. (a) $K = \dfrac{2|6x^2 - 1|}{(16x^6 - 16x^4 + 4x^2 + 1)^{3/2}}$

(b) $x = 0$: $x^2 + \left(y + \dfrac{1}{2}\right)^2 = \dfrac{1}{4}$

$x = 1$: $x^2 + \left(y - \dfrac{1}{2}\right)^2 = \dfrac{5}{4}$

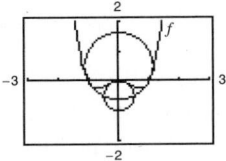

(c)

The curvature tends to be greatest near the extrema of the function and decreases as $x \to \pm\infty$. However, f and K do not have the same critical numbers.

Critical numbers of f: $x = 0, \pm\dfrac{\sqrt{2}}{2} \approx \pm 0.7071$

Critical numbers of K: $x = 0, \pm 0.7647, \pm 0.4082$

63. $a = \dfrac{1}{4}, b = 2$

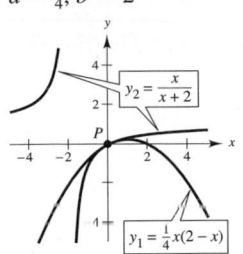

65. (a) 12.25 units (b) $\dfrac{1}{2}$ **67–69.** Proofs
71. (a) 0 (b) 0 **73.** $\dfrac{1}{4}$ **75.** Proof

77. $K = \dfrac{1}{4a}\left|\csc\dfrac{\theta}{2}\right|$

Minimum: $K = \dfrac{1}{4a}$

There is no maximum.

79. 3327.5 lb **81.** Proof
83. False. See Exploration on page 855.
85. True **87–93.** Proofs

Review Exercises for Chapter 12 (page 867)

1. (a) All reals except $\dfrac{\pi}{2} + n\pi$, n is an integer.

(b) Continuous except at $t = \dfrac{\pi}{2} + n\pi$, n is an integer.

3. (a) $[3, \infty)$ (b) Continuous for all $t \geq 3$
5. (a) $\mathbf{i} - \sqrt{2}\mathbf{k}$ (b) $-3\mathbf{i} + 4\mathbf{j}$
(c) $(2c - 1)\mathbf{i} + (c - 1)^2\mathbf{j} - \sqrt{c + 1}\mathbf{k}$
(d) $2\Delta t\mathbf{i} + \Delta t(\Delta t + 2)\mathbf{j} - \left(\sqrt{\Delta t + 3} - \sqrt{3}\right)\mathbf{k}$
7. $\mathbf{r}(t) = (3 - t)\mathbf{i} - 2t\mathbf{j} + (5 - 2t)\mathbf{k}$, $0 \leq t \leq 1$
$x = 3 - t, y = -2t, z = 5 - 2t$, $0 \leq t \leq 1$

9.

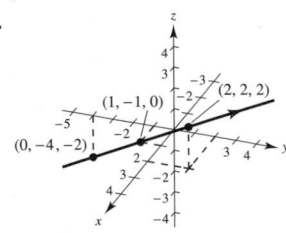

11.

13. $\mathbf{r}(t) = t\mathbf{i} + \left(-\dfrac{3}{4}t + 3\right)\mathbf{j}$ **15.**

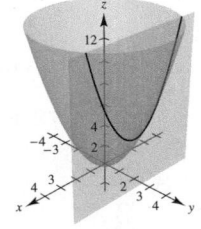

$x = t, y = 2, z = t^2 + 4$

17. $\ln 3\mathbf{j} - \dfrac{1}{3}\mathbf{k}$
19. (a) $(2t + 4)\mathbf{i} - 6t\mathbf{j}$
(b) $2\mathbf{i} - 6\mathbf{j}$
(c) $40t + 8$
21. (a) $6t^2\mathbf{i} + 4\mathbf{j} - 2t\mathbf{k}$
(b) $12t\mathbf{i} - 2\mathbf{k}$
(c) $72t^3 + 4t$
(d) $-8\mathbf{i} - 12t^2\mathbf{j} - 48t\mathbf{k}$
23. $(-\infty, 1), (1, \infty)$

25. (a) $3\mathbf{i} + \mathbf{j}$ (b) $-5\mathbf{i} + (2t - 2)\mathbf{j} + 2t^2\mathbf{k}$
(c) $18t\mathbf{i} + (6t - 3)\mathbf{j}$ (d) $4t + 3t^2$
(e) $\left(\frac{8}{3}t^3 - 2t^2\right)\mathbf{i} - 8t^3\mathbf{j} + (9t^2 - 2t + 1)\mathbf{k}$
(f) $2\mathbf{i} + 8t\mathbf{j} + 16t^2\mathbf{k}$

27. $\frac{1}{3}t^3\mathbf{i} + \frac{5}{2}t^2\mathbf{j} + 2t^4\mathbf{k} + \mathbf{C}$ **29.** $2t^{3/2}\mathbf{i} + 2\ln|t|\mathbf{j} + t\mathbf{k} + \mathbf{C}$

31. $\frac{32}{3}\mathbf{j}$ **33.** $2(e - 1)\mathbf{i} - 8\mathbf{j} - 2\mathbf{k}$

35. $\mathbf{r}(t) = (t^2 + 1)\mathbf{i} + (e^t + 2)\mathbf{j} - (e^{-t} + 4)\mathbf{k}$

37. (a) $\mathbf{v}(t) = 4\mathbf{i} + 3t^2\mathbf{j} - \mathbf{k}$
$\|\mathbf{v}(t)\| = \sqrt{17 + 9t^4}$
$\mathbf{a}(t) = 6t\mathbf{j}$
(b) $\mathbf{v}(1) = 4\mathbf{i} + 3\mathbf{j} - \mathbf{k}$
$\mathbf{a}(1) = 6\mathbf{j}$

39. (a) $\mathbf{v}(t) = \langle -3\cos^2 t \sin t, 3 \sin^2 t \cos t, 3 \rangle$
$\|\mathbf{v}(t)\| = 3\sqrt{\sin^2 t \cos^2 t + 1}$
$\mathbf{a}(t) = \langle 3\cos t(2\sin^2 t - \cos^2 t), 3\sin t(2\cos^2 t - \sin^2 t), 0 \rangle$
(b) $\mathbf{v}(\pi) = \langle 0, 0.3 \rangle$
$\mathbf{a}(\pi) = \langle 3, 0, 0 \rangle$

41. 11.67 ft; The ball will clear the fence.

43. $\mathbf{T}(2) = \dfrac{3\mathbf{i} - 2\mathbf{j}}{\sqrt{13}}$

45. $\mathbf{T}(0) = \dfrac{2\mathbf{i} - 3\mathbf{k}}{\sqrt{13}}$; $x = 1 + 2t$, $y = 1$, $z = -3t$

47. $\mathbf{N}(1) = -\dfrac{3\sqrt{10}}{10}\mathbf{i} + \dfrac{\sqrt{10}}{10}\mathbf{j}$ **49.** $\mathbf{N}\left(\dfrac{\pi}{4}\right) = -\mathbf{j}$

51. $a_{\mathbf{T}} = -\dfrac{2\sqrt{13}}{585}$ **53.** $a_{\mathbf{T}} = 0$
$a_{\mathbf{N}} = \dfrac{4\sqrt{13}}{65}$ $a_{\mathbf{N}} = 1$

55.

$5\sqrt{13}$

57.

2

59.

$3\sqrt{29}$

61.

$\dfrac{\sqrt{65}\pi}{2}$

63. 0   **65.** $\dfrac{2\sqrt{5}}{(4 + 5t^2)^{3/2}}$ **67.** $\dfrac{\sqrt{2}}{3}$

69. $K = \dfrac{1}{26^{3/2}}$, $\dfrac{1}{K} = 26\sqrt{26}$ **71.** $K = \dfrac{\sqrt{2}}{4}$, $r = 2\sqrt{2}$

73. 2016.7 lb

P.S. Problem Solving   *(page 869)*

1. (a) a (b) πa (c) $K = \pi a$
3. Initial speed: 447.21 ft/sec; $\theta \approx 63.43°$ **5–7.** Proofs

9. Unit tangent: $\left\langle -\frac{4}{5}, 0, \frac{3}{5} \right\rangle$
Principal unit normal: $\langle 0, -1, 0 \rangle$
Binormal: $\left\langle \frac{3}{5}, 0, \frac{4}{5} \right\rangle$

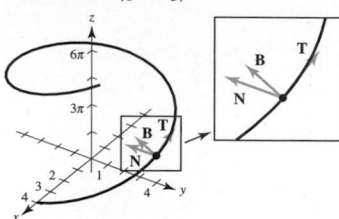

11. (a) and (b) Proofs
13. (a)

(b) 6.766

(c) $K = \dfrac{\pi(\pi^2 t^2 + 2)}{(\pi^2 t^2 + 1)^{3/2}}$
$K(0) = 2\pi$
$K(1) = \dfrac{\pi(\pi^2 + 2)}{(\pi^2 + 1)^{3/2}} \approx 1.04$
$K(2) \approx 0.51$

(d)

(e) $\displaystyle\lim_{t\to\infty} K = 0$

(f) As $t \to \infty$, the graph spirals outward and the curvature decreases.

Chapter 13
Section 13.1   *(page 880)*

1. There is not a unique value of z for each ordered pair.
3. z is a function of x and y.   **5.** z is a function of x and y.
7. z is not a function of x and y.
9. (a) 1   (b) 1   (c) -17
(d) $9 - y$ (e) $2x - 1$ (f) $13 - t$
11. (a) -1 (b) 0   (c) xe^3 (d) te^{-y}
13. (a) 3   (b) 2   (c) $\dfrac{16}{t}$ (d) $-\dfrac{6}{5}$
15. (a) $\sqrt{2}$ (b) $3\sin 1$ (c) 0   (d) 4
17. (a) -4 (b) -6 (c) $-\frac{25}{4}$ (d) $\frac{9}{4}$
19. (a) $2, \Delta x \neq 0$ (b) $2y + \Delta y, \Delta y \neq 0$
21. Domain: $\{(x, y): x \text{ is any real number}, y \text{ is any real number}\}$
Range: all real numbers
23. Domain: $\{(x, y): y \geq 0\}$
Range: all real numbers
25. Domain: $\{(x, y): x \neq 0, y \neq 0\}$
Range: all real numbers
27. Domain: $\{(x, y): x^2 + y^2 \leq 4\}$
Range: $0 \leq z \leq 2$
29. Domain: $\{(x, y): -1 \leq x + y \leq 1\}$
Range: $0 \leq z \leq \pi$

31. Domain: $\{(x, y): y < -x + 5\}$
Range: all real numbers

33. (a) $(20, 0, 0)$ (b) $(-15, 10, 20)$
(c) $(20, 15, 25)$ (d) $(20, 20, 0)$

35. Plane

37. Cylinder with rulings parallel to the x-axis

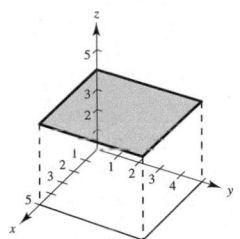

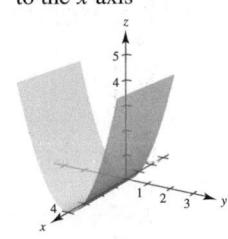

39. Paraboloid

41. Cylinder with rulings parallel to the y-axis

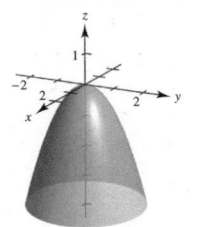

43.

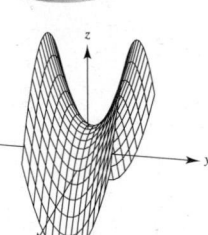

45.

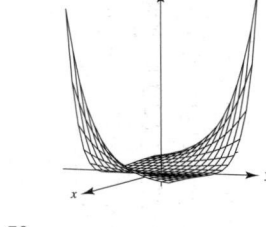

47. c **48.** d **49.** b **50.** a

51. Lines: $x + y = c$

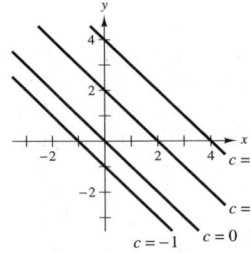

53. Ellipses: $x^2 + 4y^2 = c$
[except $x^2 + 4y^2 = 0$ is the point $(0, 0)$]

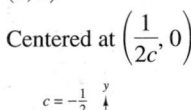

55. Hyperbolas: $xy = c$

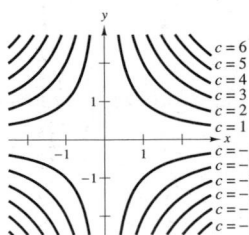

57. Circles passing through $(0, 0)$

Centered at $\left(\dfrac{1}{2c}, 0\right)$

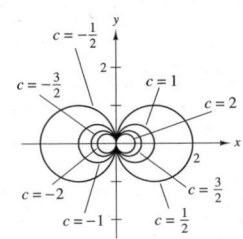

59.

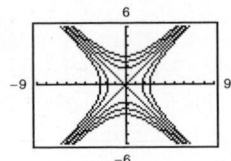

61.

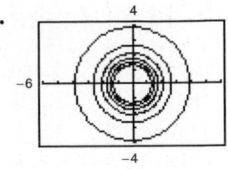

63. Yes; The definition of a function of two variables requires that z be unique for each ordered pair (x, y) in the domain.

65. $f(x, y) = \dfrac{x}{y}$ $\left(\text{The level curves are the lines } y = \dfrac{x}{c}.\right)$

67. The surface may be shaped like a saddle. For example, let $f(x, y) = xy$. The graph is not unique because any vertical translation will produce the same level curves.

69.

Tax Rate	Inflation Rate		
	0	0.03	0.05
0	\$1790.85	\$1332.56	\$1099.43
0.28	\$1526.43	\$1135.80	\$937.09
0.35	\$1466.07	\$1090.90	\$900.04

71. Plane

73. Sphere

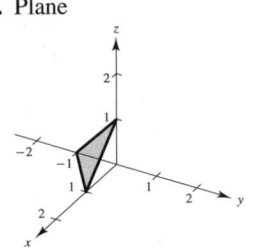

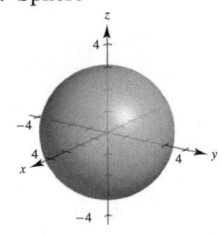

75. Elliptic cone

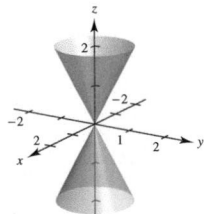

77. (a) 243 board-ft (b) 507 board-ft

79.

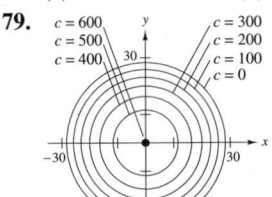

81. 36,661 units

83. Proof

85. (a) $k = \dfrac{520}{3}$

(b) $P = \dfrac{520T}{3V}$

The level curves are lines.

87. (a) C (b) A (c) B

89. $C = 4.50xy + 5.00(xz + yz)$ **91.** False. Let $f(x, y) = 4$.

93. False. The equation of a sphere is not a function.

95. Putnam Problem A1, 2008

Section 13.2 (page 891)

1. As x approaches -1 and y approaches 3, z approaches 1.
3–5. Proofs   7. 9   9. -20 11. 7, continuous
13. e^2, continuous   15. 0, continuous for $y \neq 0$
17. $\frac{1}{2}$, continuous except at $(0, 0)$ 19. -1, continuous
21. 0, continuous for $xy \neq 1$, $|xy| \leq 1$
23. $2\sqrt{2}$, continuous for $x + y + z \geq 0$ 25. 0
27. Limit does not exist.   29. Limit does not exist.
31. Limit does not exist.   33. 0
35. Limit does not exist.
37. No. The existence of $f(2, 3)$ has no bearing on the existence of the limit as $(x, y) \to (2, 3)$.
39. $\lim_{x \to 0} f(x, 0) = 0$ if $f(x, 0)$ exists.   41. Continuous, 1

43.

(x, y)	$(1, 0)$	$(0.5, 0)$	$(0.1, 0)$	$(0.01, 0)$	$(0.001, 0)$
$f(x, y)$	0	0	0	0	0

$y = 0$: 0

(x, y)	$(1, 1)$	$(0.5, 0.5)$	$(0.1, 0.1)$
$f(x, y)$	$\frac{1}{2}$	$\frac{1}{2}$	$\frac{1}{2}$

(x, y)	$(0.01, 0.01)$	$(0.001, 0.001)$
$f(x, y)$	$\frac{1}{2}$	$\frac{1}{2}$

$y = x$: $\frac{1}{2}$
Limit does not exist.
Continuous except at $(0, 0)$

45.

(x, y)	$(1, 0)$	$(0.5, 0)$	$(0.1, 0)$	$(0.01, 0)$	$(0.001, 0)$
$f(x, y)$	0	0	0	0	0

$y = 0$: 0

(x, y)	$(1, 1)$	$(0.5, 0.5)$	$(0.1, 0.1)$
$f(x, y)$	$\frac{1}{2}$	1	5

(x, y)	$(0.01, 0.01)$	$(0.001, 0.001)$
$f(x, y)$	50	500

$y = x$: ∞
The limit does not exist.
Continuous except at $(0, 0)$

47. (a) $\dfrac{1 + a^2}{a}$, $a \neq 0$ (b) Limit does not exist.
    (c) No; Different paths result in different limits.
49. f is continuous. g is continuous except at $(0, 0)$. g has a removable discontinuity at $(0, 0)$.
51. 0   53. 0   55. 1   57. 1   59. 0
61. Continuous except at $(0, 0, 0)$ 63. Continuous
65. Continuous   67. Continuous
69. Continuous for $y \neq \dfrac{2x}{3}$ 71. (a) $2x$ (b) -4
73. (a) $\dfrac{1}{y}$ (b) $-\dfrac{x}{y^2}$ 75. (a) $3 + y$ (b) $x - 2$
77. 0

79. True
81. False. Let $f(x, y) = \begin{cases} \ln(x^2 + y^2), & x \neq 0, y \neq 0 \\ 0, & x = 0, y = 0 \end{cases}$
83. $\dfrac{\pi}{2}$ 85. Proof

Section 13.3 (page 900)

1. z_x, $f_x(x, y)$, $\dfrac{\partial z}{\partial x}$
3. (a) Differentiate first with respect to y, then with respect to x, and last with respect to z.
   (b) Differentiate first with respect to z and then with respect to x.
5. No. Because you are finding the partial derivative with respect to x, you consider y to be constant. So, the denominator is considered a constant and does not contain any variables.
7. No. Because you are finding the partial derivative with respect to y, you consider x to be constant. So, the denominator is considered a constant and does not contain any variables.
9. Yes. Because you are finding the partial derivative with respect to x, you consider y to be constant. So, both the numerator and denominator contain variables.
11. $f_x(x, y) = 2$ 13. $\dfrac{\partial z}{\partial x} = 6 - 2xy$
    $f_y(x, y) = -5$ $\dfrac{\partial z}{\partial y} = -x^2 + 16y$
15. $\dfrac{\partial z}{\partial x} = \sqrt{y}$ 17. $\dfrac{\partial z}{\partial x} = ye^{xy}$
    $\dfrac{\partial z}{\partial y} = \dfrac{x}{2\sqrt{y}}$ $\dfrac{\partial z}{\partial y} = xe^{xy}$
19. $\dfrac{\partial z}{\partial x} = 2xe^{2y}$ 21. $\dfrac{\partial z}{\partial x} = \dfrac{1}{x}$
    $\dfrac{\partial z}{\partial y} = 2x^2e^{2y}$ $\dfrac{\partial z}{\partial y} = -\dfrac{1}{y}$
23. $\dfrac{\partial z}{\partial x} = \dfrac{2x}{x^2 + y^2}$ 25. $\dfrac{\partial z}{\partial x} = \dfrac{x^3 - 3y^3}{x^2 y}$
    $\dfrac{\partial z}{\partial y} = \dfrac{2y}{x^2 + y^2}$ $\dfrac{\partial z}{\partial y} = \dfrac{-x^3 + 12y^3}{2xy^2}$
27. $h_x(x, y) = -2xe^{-(x^2+y^2)}$ 29. $f_x(x, y) = \dfrac{x}{\sqrt{x^2 + y^2}}$
    $h_y(x, y) = -2ye^{-(x^2+y^2)}$ $f_y(x, y) = \dfrac{y}{\sqrt{x^2 + y^2}}$
31. $\dfrac{\partial z}{\partial x} = -y \sin xy$ 33. $\dfrac{\partial z}{\partial x} = 2 \sec^2(2x - y)$
    $\dfrac{\partial z}{\partial y} = -x \sin xy$ $\dfrac{\partial z}{\partial y} = -\sec^2(2x - y)$
35. $\dfrac{\partial z}{\partial x} = 8ye^y \cos 8xy$
    $\dfrac{\partial z}{\partial y} = e^y(8x \cos 8xy + \sin 8xy)$
37. $\dfrac{\partial z}{\partial x} = 2 \cosh(2x + 3y)$
    $\dfrac{\partial z}{\partial y} = 3 \cosh(2x + 3y)$
39. $f_x(x, y) = 1 - x^2$ 41. $f_x(x, y) = 3$
    $f_y(x, y) = y^2 - 1$ $f_y(x, y) = 2$

43. $f_x(x, y) = \dfrac{1}{2\sqrt{x + y}}$

$f_y(x, y) = \dfrac{1}{2\sqrt{x + y}}$

45. $f_x = 12$

$f_y = 12$

47. $f_x = -1$

$f_y = \frac{1}{2}$

49. $f_x = \frac{1}{4}$

$f_y = \frac{1}{4}$

51. $f_x = -\dfrac{1}{4}$

$f_y = \dfrac{1}{4}$

53. $\dfrac{\partial z}{\partial x}(1, 2) = 2$

$\dfrac{\partial z}{\partial y}(1, 2) = 1$

55. $g_x(1, 1) = -2$

$g_y(1, 1) = -2$

57. $H_x(x, y, z) = \cos(x + 2y + 3z)$

$H_y(x, y, z) = 2\cos(x + 2y + 3z)$

$H_z(x, y, z) = 3\cos(x + 2y + 3z)$

59. $\dfrac{\partial w}{\partial x} = \dfrac{x}{\sqrt{x^2 + y^2 + z^2}}$

$\dfrac{\partial w}{\partial y} = \dfrac{y}{\sqrt{x^2 + y^2 + z^2}}$

$\dfrac{\partial w}{\partial z} = \dfrac{z}{\sqrt{x^2 + y^2 + z^2}}$

61. $F_x(x, y, z) = \dfrac{x}{x^2 + y^2 + z^2}$

$F_y(x, y, z) = \dfrac{y}{x^2 + y^2 + z^2}$

$F_z(x, y, z) = \dfrac{z}{x^2 + y^2 + z^2}$

63. $f_x = 3, f_y = 1, f_z = 2$

65. $f_x = 1, f_y = 0, f_z = 0$

67. $f_x = 4, f_y = 24, f_z = 0$

69. $x = 2, y = -2$

71. $x = -6, y = 4$ **73.** $x = 1, y = 1$ **75.** $x = 0, y = 0$

77. $\dfrac{\partial^2 z}{\partial x^2} = 0$

$\dfrac{\partial^2 z}{\partial y^2} = 6x$

$\dfrac{\partial^2 z}{\partial y \partial x} = \dfrac{\partial^2 z}{\partial x \partial y} = 6y$

79. $\dfrac{\partial^2 z}{\partial x^2} = 12x^2$

$\dfrac{\partial^2 z}{\partial y^2} = 18y$

$\dfrac{\partial^2 z}{\partial y \partial x} = \dfrac{\partial^2 z}{\partial x \partial y} = -2$

81. $\dfrac{\partial^2 z}{\partial x^2} = \dfrac{y^2}{(x^2 + y^2)^{3/2}}$

$\dfrac{\partial^2 z}{\partial y^2} = \dfrac{x^2}{(x^2 + y^2)^{3/2}}$

$\dfrac{\partial^2 z}{\partial y \partial x} = \dfrac{\partial^2 z}{\partial x \partial y} = \dfrac{-xy}{(x^2 + y^2)^{3/2}}$

83. $\dfrac{\partial^2 z}{\partial x^2} = e^x \tan y$

$\dfrac{\partial^2 z}{\partial y^2} = 2e^x \sec^2 y \tan y$

$\dfrac{\partial^2 z}{\partial y \partial x} = \dfrac{\partial^2 z}{\partial x \partial y} = e^x \sec^2 y$

85. $\dfrac{\partial^2 z}{\partial x^2} = -y^2 \cos xy$

$\dfrac{\partial^2 z}{\partial y^2} = -x^2 \cos xy$

$\dfrac{\partial^2 z}{\partial y \partial x} = \dfrac{\partial^2 z}{\partial x \partial y} = -xy \cos xy - \sin xy$

87. $\dfrac{\partial z}{\partial x} = \sec y$

$\dfrac{\partial z}{\partial y} = x \sec y \tan y$

$\dfrac{\partial^2 z}{\partial x^2} = 0$

$\dfrac{\partial^2 z}{\partial y^2} = x \sec y (\sec^2 y + \tan^2 y)$

$\dfrac{\partial^2 z}{\partial y \partial x} = \dfrac{\partial^2 z}{\partial x \partial y} = \sec y \tan y$

No values of x and y exist such that $f_x(x, y) = f_y(x, y) = 0$.

89. $\dfrac{\partial z}{\partial x} = \dfrac{y^2 - x^2}{x(x^2 + y^2)}$

$\dfrac{\partial z}{\partial y} = \dfrac{-2y}{x^2 + y^2}$

$\dfrac{\partial^2 z}{\partial x^2} = \dfrac{x^4 - 4x^2y^2 - y^4}{x^2(x^2 + y^2)^2}$

$\dfrac{\partial^2 z}{\partial y^2} = \dfrac{2(y^2 - x^2)}{(x^2 + y^2)^2}$

$\dfrac{\partial^2 z}{\partial y \partial x} = \dfrac{\partial^2 z}{\partial x \partial y} = \dfrac{4xy}{(x^2 + y^2)^2}$

No values of x and y exist such that $f_x(x, y) = f_y(x, y) = 0$.

91. $f_{xyy}(x, y, z) = f_{yxy}(x, y, z) = f_{yyx}(x, y, z) = 0$

93. $f_{xyy}(x, y, z) = f_{yxy}(x, y, z) = f_{yyx}(x, y, z) = z^2 e^{-x} \sin yz$

95. $\dfrac{\partial^2 z}{\partial x^2} + \dfrac{\partial^2 z}{\partial y^2} = 0 + 0 = 0$

97. $\dfrac{\partial^2 z}{\partial x^2} + \dfrac{\partial^2 z}{\partial y^2} = e^x \sin y - e^x \sin y = 0$

99. $\dfrac{\partial^2 z}{\partial t^2} = -c^2 \sin(x - ct) = c^2\left(\dfrac{\partial^2 z}{\partial x^2}\right)$

101. $\dfrac{\partial^2 z}{\partial t^2} = \dfrac{-c^2}{(x + ct)^2} = c^2\left(\dfrac{\partial^2 z}{\partial x^2}\right)$

103. $\dfrac{\partial z}{\partial t} = \dfrac{-e^{-t} \cos x}{c} = c^2\left(\dfrac{\partial^2 z}{\partial x^2}\right)$ **105.** Proof

107. Yes; $f(x, y) = \cos(3x - 2y)$

109. No. Let $z = x + y + 1$.

111.

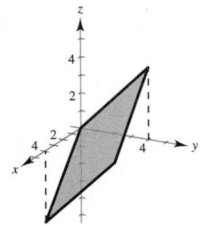

113. Dollars/yr; negative; You expect the influence that age has on the cost of the car to be negative.

115. (a) $\sqrt{2}$ (b) $\frac{5}{2}$ **117.** (a) 72   (b) 72

119. $IQ_M = \dfrac{100}{C}, IQ_M(12, 10) = 10$

IQ increases at a rate of 10 points per year of mental age when the mental age is 12 and the chronological age is 10.

$IQ_C = -\dfrac{100M}{C^2}, IQ_C(12, 10) = -12$

IQ decreases at a rate of 12 points per year of chronological age when the mental age is 12 and the chronological age is 10.

121. An increase in either the charge for food and housing or the tuition will cause a decrease in the number of applicants.

123. $\dfrac{\partial T}{\partial x} = -2.4°/\text{m}, \dfrac{\partial T}{\partial y} = -9°/\text{m}$

125. $T = \dfrac{PV}{nR} \Rightarrow \dfrac{\partial T}{\partial P} = \dfrac{v}{nR}$

$P = \dfrac{nRT}{V} \Rightarrow \dfrac{\partial P}{\partial V} = \dfrac{-nRT}{V^2}$

$V = \dfrac{nRT}{P} \Rightarrow \dfrac{\partial V}{\partial T} = \dfrac{nR}{P}$

$\dfrac{\partial T}{\partial P} \cdot \dfrac{\partial P}{\partial V} \cdot \dfrac{\partial V}{\partial T} = -\dfrac{nRT}{VP} = -\dfrac{nRT}{nRT} = -1$

127. (a) $\dfrac{\partial z}{\partial x} = 0.23, \dfrac{\partial z}{\partial y} = 0.14$

(b) As the expenditures on amusement parks and campgrounds (x) increase, the expenditures on spectator sports (z) increase. As the expenditures on live entertainment (y) increase, the expenditures on spectator sports (z) also increase.

129. (a) $f_x(x, y) = \dfrac{y(x^4 + 4x^2y^2 - y^4)}{(x^2 + y^2)^2}$

$f_y(x, y) = \dfrac{x(x^4 - 4x^2y^2 - y^4)}{(x^2 + y^2)^2}$

(b) $f_x(0, 0) = 0, f_y(0, 0) = 0$

(c) $f_{xy}(0, 0) = -1, f_{yx}(0, 0) = 1$

(d) f_{xy} or f_{yx} or both are not continuous at $(0, 0)$.

131. Proof

Section 13.4   *(page 909)*

1. In general, the accuracy worsens as Δx and Δy increase.

3. $dz = 15x^2y^2\, dx + 10x^3y\, dy$

5. $dz = (e^{x^2+y^2} + e^{-x^2-y^2})(x\, dx + y\, dy)$

7. $dw = 2xyz^2\, dx + (x^2z^2 + z\cos yz)\, dy + (2x^2yz + y\cos yz)\, dz$

9. (a) $f(2, 1) = 1, f(2.1, 1.05) = 1.05, \Delta z = 0.05$

(b) $dz = 0.05$

11. (a) $f(2, 1) = 11, f(2.1, 1.05) = 10.4875, \Delta z = -0.5125$

(b) $dz = -0.5$

13. (a) $f(2, 1) = e^2 \approx 7.3891, f(2.1, 1.05) = 1.05e^{2.1} \approx 8.5745,$
$\Delta z \approx 1.1854$

(b) $dz \approx 1.1084$

15. 0.44   **17.** 0

19. Yes. Because f_x and f_y are continuous on R, you know that f is differentiable on R. Because f is differentiable on R, you know that f is continuous on R.

21. $dA = h\, dl + l\, dh$

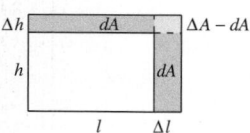

$\Delta A - dA = dl\, dh$

23. $dV = \pm 3.92$ in.^{3}, $\dfrac{dV}{V} = 0.82\%$

25.

Δr	Δh	dV	ΔV	$\Delta V - dV$
0.1	0.1	8.3776	8.5462	0.1686
0.1	-0.1	5.0265	5.0255	-0.0010
0.001	0.002	0.1005	0.1006	0.0001
-0.0001	0.0002	-0.0034	-0.0034	0.0000

27. $dC = \pm 2.4418, \dfrac{dC}{C} = 19\%$ **29.** 10%

31. (a) $V = 18 \sin\theta$ ft^3, $\theta = \dfrac{\pi}{2}$ (b) 1.047 ft^3

33. $L \approx 8.096 \times 10^{-4} \pm 6.6 \times 10^{-6}$ microhenrys

35. Answers will vary.
Sample answer:
$\varepsilon_1 = \Delta x$
$\varepsilon_2 = 0$

37. Answers will vary.
Sample answer:
$\varepsilon_1 = y\,\Delta x$
$\varepsilon_2 = 2x\,\Delta x + (\Delta x)^2$

39. Proof

Section 13.5   *(page 917)*

1. You can convert w into a function of s and t, or you can use the Chain Rule given in Theorem 13.7.

3. $8t + 5$; 21   **5.** $e^t(\sin t + \cos t)$; 1

7. (a) and (b) $2e^{2t} + \dfrac{3}{t^4}$ **9.** (a) and (b) $2e^{2t}$

11. (a) and (b) $3(2t^2 - 1)$ **13.** $\dfrac{-11\sqrt{29}}{29} \approx -2.04$

15. $\dfrac{\partial w}{\partial s} = 4s, 4$

$\dfrac{\partial w}{\partial t} = 4t, 12$

17. $\dfrac{\partial w}{\partial s} = 5\cos(5s - t), 0$

$\dfrac{\partial w}{\partial t} = -\cos(5s - t), 0$

19. (a) and (b)

$\dfrac{\partial w}{\partial s} = t^2(3s^2 - t^2)$

$\dfrac{\partial w}{\partial t} = 2st(s^2 - 2t^2)$

21. (a) and (b)

$\dfrac{\partial w}{\partial s} = te^{s^2-t^2}(2s^2 + 1)$

$\dfrac{\partial w}{\partial t} = se^{s^2-t^2}(1 - 2t^2)$

23. $\dfrac{y - 2x + 1}{2y - x + 1}$ **25.** $\dfrac{x^2 + y^2 + x}{x^2 + y^2 + y}$

27. $\dfrac{\partial z}{\partial x} = -\dfrac{x}{z}$

$\dfrac{\partial z}{\partial y} = -\dfrac{y}{z}$

29. $\dfrac{\partial z}{\partial x} = -\dfrac{x}{y + z}$

$\dfrac{\partial z}{\partial y} = -\dfrac{z}{y + z}$

31. $\dfrac{\partial z}{\partial x} = \dfrac{\partial z}{\partial y} = \dfrac{\sec^2(x + y)}{\sin z}$

33. $\dfrac{\partial z}{\partial x} = -\dfrac{(ze^{xz} + y)}{xe^{xz}}$

$\dfrac{\partial z}{\partial y} = -e^{-xz}$

35. $\dfrac{\partial w}{\partial x} = \dfrac{7y + w^2}{4z - 2wz - 2wx}$

$\dfrac{\partial w}{\partial y} = \dfrac{7x + z^2}{4z - 2wz - 2wx}$

$\dfrac{\partial w}{\partial z} = \dfrac{2yz - 4w + w^2}{4z - 2wz - 2wx}$

37. $\dfrac{\partial w}{\partial x} = \dfrac{y\sin xy}{z}$

$\dfrac{\partial w}{\partial y} = \dfrac{x\sin xy - z\cos yz}{z}$

$\dfrac{\partial w}{\partial z} = -\dfrac{y\cos yz + w}{z}$

39. (a) $f(tx, ty) = 2(tx)^2 - 5(tx)(ty)$
$= t^2(2x^2 - 5xy) = t^2 f(x, y); n = 2$

(b) $xf_x(x, y) + yf_y(x, y) = 4x^2 - 10xy = 2f(x, y)$

41. (a) $f(tx, ty) = e^{tx/ty} = e^{x/y} = f(x, y); n = 0$

(b) $xf_x(x, y) + yf_y(x, y) = \dfrac{xe^{x/y}}{y} - \dfrac{xe^{x/y}}{y} = 0$

43. 47   **45.** Proof

47. (a) $\dfrac{\partial F}{\partial u}\dfrac{\partial u}{\partial x} + \dfrac{\partial F}{\partial v}\dfrac{\partial v}{\partial x} = 4\dfrac{\partial F}{\partial u}$

(b) $\dfrac{\partial F}{\partial u}\dfrac{\partial u}{\partial x} + \dfrac{\partial F}{\partial v}\dfrac{\partial v}{\partial x} = -2\dfrac{\partial F}{\partial u} + 2x\dfrac{\partial F}{\partial v}$

49. 4608π in.^{3}/min, 624π in.^{2}/min   **51.** $28m$ cm^2/sec

53–55. Proofs

Section 13.6 *(page 928)*

1. The partial derivative with respect to x is the directional derivative in the direction of the positive x-axis. That is, the directional derivative for $\theta = 0$.

3. $-\sqrt{2}$ **5.** $\frac{1}{2} + \sqrt{3}$ **7.** 1   **9.** $-\frac{7}{25}$ **11.** 6

13. $\dfrac{2\sqrt{5}}{5}$ **15.** $3\mathbf{i} + 10\mathbf{j}$ **17.** $2\mathbf{i} - \dfrac{1}{2}\mathbf{j}$

19. $20\mathbf{i} - 14\mathbf{j} - 30\mathbf{k}$ **21.** -1 **23.** $\dfrac{2\sqrt{3}}{3}$ **25.** $3\sqrt{2}$

27. $-\dfrac{8}{\sqrt{5}}$ **29.** $-\sqrt{y}\,\mathbf{i} + \left(2y - \dfrac{x}{2\sqrt{y}}\right)\mathbf{j}$; $\sqrt{39}$

31. $\tan y\,\mathbf{i} + x\sec^2 y\,\mathbf{j}$; $\sqrt{17}$

33. $\cos x^2 y^3 (2x\mathbf{i} + 3y^2\mathbf{j})$; $\dfrac{1}{\pi}\sqrt{4 + 9\pi^6}$

35. $\dfrac{x\mathbf{i} + y\mathbf{j} + z\mathbf{k}}{\sqrt{x^2 + y^2 + z^2}}$; 1   **37.** $yz(yz\mathbf{i} + 2xz\mathbf{j} + 2xy\mathbf{k})$; $\sqrt{33}$

39. $-2\mathbf{i} - 3\mathbf{j}$ **41.** $3\mathbf{i} - \mathbf{j}$

43. (a) $16\mathbf{i} - \mathbf{j}$ (b) $\dfrac{\sqrt{257}}{257}(16\mathbf{i} - \mathbf{j})$ (c) $y = 16x - 22$

(d)

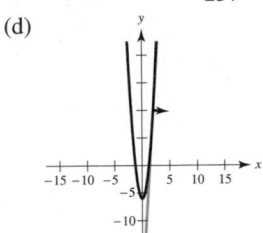

45. (a) $6\mathbf{i} - 4\mathbf{j}$ (b) $\dfrac{\sqrt{13}}{13}(3\mathbf{i} - 2\mathbf{j})$ (c) $y = \dfrac{3}{2}x - \dfrac{1}{2}$

(d)

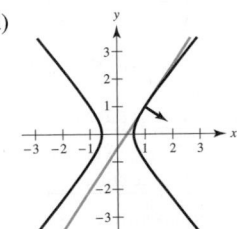

47. (a)

(b) (i) $-\dfrac{5\sqrt{2}}{12}$ (ii) $\dfrac{2 - 3\sqrt{3}}{12}$

(iii) $\dfrac{2 + 3\sqrt{3}}{12}$ (iv) $\dfrac{3 - 2\sqrt{3}}{12}$

(c) (i) $-\dfrac{5\sqrt{2}}{12}$ (ii) $\dfrac{3}{5}$ (iii) $-\dfrac{1}{5}$ (iv) $-\dfrac{11\sqrt{10}}{60}$

(d) $-\dfrac{1}{3}\mathbf{i} - \dfrac{1}{2}\mathbf{j}$ (e) $\dfrac{\sqrt{13}}{6}$

(f) $\mathbf{u} = \dfrac{1}{\sqrt{13}}(3\mathbf{i} - 2\mathbf{j})$

$D_{\mathbf{u}}f(3, 2) = \nabla f \cdot \mathbf{u} = 0$

∇f is the direction of the greatest rate of change of f. So, in a direction orthogonal to ∇f, the rate of change of f is 0.

49. (a)

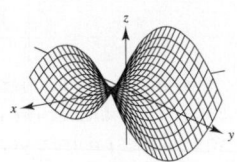

(b) $D_{\mathbf{u}}f(4, -3) = 8\cos\theta + 6\sin\theta$

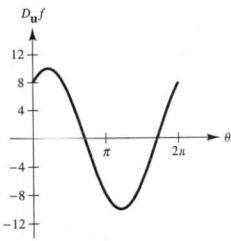

Generated by Mathematica

(c) $\theta \approx 2.21$, $\theta \approx 5.36$
Directions in which there is no change in f

(d) $\theta \approx 0.64$, $\theta \approx 3.79$
Directions of greatest rate of change in f

(e) 10; Magnitude of the greatest rate of change

(f)

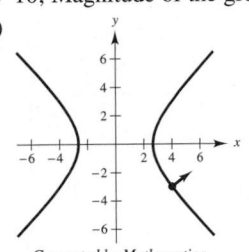

Generated by Mathematica

Orthogonal to the level curve

51. No; Answers will vary.   **53.** $5\nabla h = -(5\mathbf{i} + 12\mathbf{j})$

55. $\dfrac{1}{625}(7\mathbf{i} - 24\mathbf{j})$ **57.** $6\mathbf{i} - 10\mathbf{j}$; $11.66°/\text{cm}$ **59.** $y^2 = 10x$

61. True   **63.** True   **65.** $f(x, y, z) = e^x \cos y + \dfrac{1}{2}z^2 + C$

67. (a) and (b) Proofs

(c)

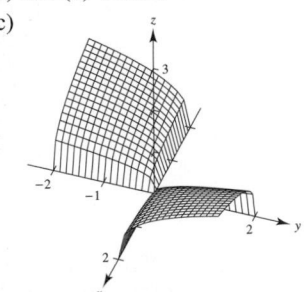

Section 13.7 *(page 937)*

1. $\nabla F(x_0, y_0, z_0)$ and any tangent vector $\mathbf{v}$ at (x_0, y_0, z_0) are orthogonal. So, $\nabla F(x_0, y_0, z_0) \cdot \mathbf{v} = 0$.

3. The level surface can be written as $3x - 5y + 3z = 15$, which is an equation of a plane in space.

5. The level surface can be written as $4x^2 + 9y^2 - 4z^2 = 0$, which is an elliptic cone that lies on the z-axis.

7. $4x + 2y - z = 2$ **9.** $3x + 4y - 5z = 0$

11. $2x - 2y - z = 2$ **13.** $3x + 4y - 25z = 25(1 - \ln 5)$

15. $4x + 2y + 5z = -15$

17. (a) $x + y + z = 9$ (b) $x - 3 = y - 3 = z - 3$

19. (a) $x - 2y + 2z = 7$ (b) $x - 1 = \dfrac{y + 1}{-2} = \dfrac{z - 2}{2}$

21. (a) $6x - 4y - z = 5$ (b) $\dfrac{x - 3}{6} = \dfrac{y - 2}{-4} = \dfrac{z - 5}{-1}$

23. (a) $10x + 5y + 2z = 30$ (b) $\dfrac{x - 1}{10} = \dfrac{y - 2}{5} = \dfrac{z - 5}{2}$

25. (a) $8x + y - z = 0$ (b) $\dfrac{x}{8} = \dfrac{y - 2}{1} = \dfrac{z - 2}{-1}$

27. $x = t + 1, y = 1 - t, z = t + 1$

29. $x = 4t + 3, y = 4t + 3, z = 4 - 3t$

31. $x = t + 3, y = 5t + 1, z = 2 - 4t$

33. $86.0°$ **35.** $77.4°$ **37.** $(0, 3, 12)$ **39.** $(2, 2, -4)$

41. $(0, 0, 0)$ **43.** Proof   **45.** (a) and (b) Proofs

47. Not necessarily; They only need to be parallel.

49. $\left(-\frac{1}{2}, \frac{1}{4}, \frac{1}{4}\right)$ or $\left(\frac{1}{2}, -\frac{1}{4}, -\frac{1}{4}\right)$ **51.** $(-2, 1, -1)$ or $(2, -1, 1)$

53. (a) Line: $x = 1, y = 1, z = 1 - t$

   Plane: $z = 1$

   (b) Line: $x = -1, y = 2 + \frac{6}{25}t, z = -\frac{4}{5} - t$

   Plane: $6y - 25z - 32 = 0$

   (c)

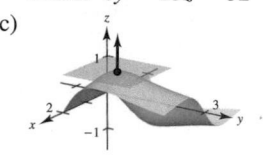
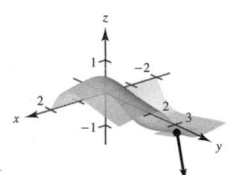

55. (a) $x = 1 + t$ (b)

   $y = 2 - 2t$

   $z = 4$

   $\theta \approx 48.2°$

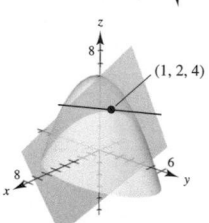

57. $F(x, y, z) = \dfrac{x^2}{a^2} + \dfrac{y^2}{b^2} + \dfrac{z^2}{c^2} - 1$

   $F_x(x, y, z) = \dfrac{2x}{a^2}$

   $F_y(x, y, z) = \dfrac{2y}{b^2}$

   $F_z(x, y, z) = \dfrac{2z}{c^2}$

   Plane: $\dfrac{2x_0}{a^2}(x - x_0) + \dfrac{2y_0}{b^2}(y - y_0) + \dfrac{2z_0}{c^2}(z - z_0) = 0$

   $\dfrac{x_0 x}{a^2} + \dfrac{y_0 y}{b^2} + \dfrac{z_0 z}{c^2} = 1$

59. $F(x, y, z) = a^2x^2 + b^2y^2 - z^2$

   $F_x(x, y, z) = 2a^2x$

   $F_y(x, y, z) = 2b^2y$

   $F_z(x, y, z) = -2z$

   Plane: $2a^2x_0(x - x_0) + 2b^2y_0(y - y_0) - 2z_0(z - z_0) = 0$

   $a^2x_0x + b^2y_0y - z_0z = 0$

   Therefore, the plane passes through the origin.

61. (a) $P_1(x, y) = 1 + x - y$

   (b) $P_2(x, y) = 1 + x - y + \frac{1}{2}x^2 - xy + \frac{1}{2}y^2$

   (c) If $x = 0$, $P_2(0, y) = 1 - y + \frac{1}{2}y^2$.

   This is the second-degree Taylor polynomial for e^{-y}.

   If $y = 0$, $P_2(x, 0) = 1 + x + \frac{1}{2}x^2$.

   This is the second-degree Taylor polynomial for e^x.

(d)

x	y	$f(x, y)$	$P_1(x, y)$	$P_2(x, y)$
0	0	1	1	1
0	0.1	0.9048	0.9000	0.9050
0.2	0.1	1.1052	1.1000	1.1050
0.2	0.5	0.7408	0.7000	0.7450
1	0.5	1.6487	1.5000	1.6250

(e)

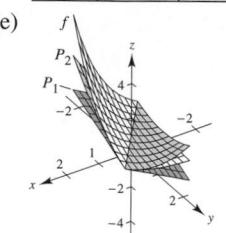

63. Proof

Section 13.8   *(page 946)*

1. (a) To say that f has a relative minimum at (x_0, y_0) means that the point (x_0, y_0, z_0) is at least as low as all nearby points on the graph of $z = f(x, y)$.

   (b) To say that f has a relative maximum at (x_0, y_0) means that the point (x_0, y_0, z_0) is at least as high as all nearby points in the graph of $z = f(x, y)$.

   (c) Critical points of f are the points at which the gradient of f is 0 or the points at which one of the partial derivatives does not exist.

   (d) A critical point is a saddle point if it is neither a relative minimum nor a relative maximum.

3. Relative minimum: **5.** Relative minimum:

   $(1, 3, 0)$ $(0, 0, 1)$

7. Relative minimum: **9.** Relative minimum:

   $(-1, 3, -4)$ $(-4, 6, -55)$

11. Every point along the x- or y-axis is a critical point. Each of the critical points yields an absolute maximum.

13. Relative maximum: **15.** Relative minimum:

   $\left(\frac{1}{2}, -1, \frac{31}{4}\right)$ $\left(\frac{1}{2}, -4, -\frac{187}{4}\right)$

17. Relative minimum: **19.** Relative maximum:

   $(3, -4, -5)$ $(0, 0, -12)$

21. Saddle point: **23.** No critical numbers

   $(1, -1, -1)$

25.   **27.**

Relative maximum: $(-1, 0, 2)$ Relative minimum: $(0, 0, 0)$

Relative minimum: $(1, 0, -2)$ Relative maxima: $(0, \pm 1, 4)$

   Saddle points: $(\pm 1, 0, 1)$

29. z is never negative. Minimum: $z = 0$ when $x = y \ne 0$.

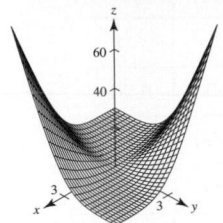

31. Insufficient information   **33.** Saddle point
35. (a) $(0, 0)$ (b) Saddle point: $(0, 0, 0)$ (c) $(0, 0)$
(d)

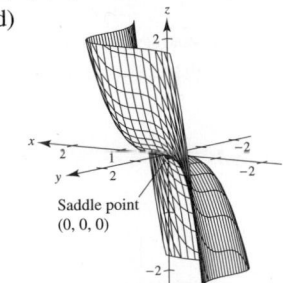

Saddle point
$(0, 0, 0)$

37. (a) $(1, a)$, $(b, -4)$
(b) Absolute minima: $(1, a, 0)$, $(b, -4, 0)$
(c) $(1, a)$, $(b, -4)$
(d)

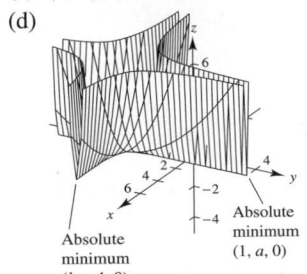

Absolute minimum $(b, -4, 0)$
Absolute minimum $(1, a, 0)$

39. Absolute maximum:
$(4, 0, 21)$
Absolute minimum:
$(4, 2, -11)$

41. Absolute maximum:
$(0, 1, 10)$
Absolute minimum:
$(1, 2, 5)$

43. Absolute maxima:
$(\pm 2, 4, 28)$
Absolute minimum:
$(0, 1, -2)$

45. Absolute maxima:
$(-2, -1, 9)$, $(2, 1, 9)$
Absolute minima:
$(x, -x, 0)$, $|x| \le 1$

47. Relative minimum: $(0, 3, -1)$ **49.** $-4 < f_{xy}(3, 7) < 4$
51.

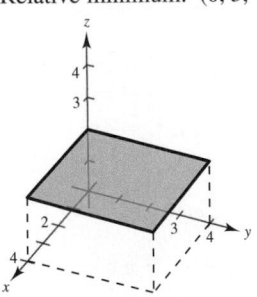

Extrema at all (x, y)

53. (a) $f_x = 2x = 0$, $f_y = -2y = 0 \implies (0, 0)$ is a critical point.
$g_x = 2x = 0$, $g_y = 2y = 0 \implies (0, 0)$ is a critical point.
(b) $d = 2(-2) - 0 < 0 \implies (0, 0)$ is a saddle point.
$d = 2(2) - 0 > 0 \implies (0, 0)$ is a relative minimum.

55. False. Let $f(x, y) = 1 - |x| - |y|$ at the point $(0, 0, 1)$.

57. False. Let $f(x, y) = x^2 y^2$ (see Example 4 on page 944).

Section 13.9   *(page 953)*

1. Write the equation to be maximized or minimized as a function of two variables. Take the partial derivatives and set them equal to zero or undefined to obtain the critical points. Use the Second Partials Test to test for relative extrema using the critical points. Check the boundary points.
3. $\sqrt{3}$ **5.** $\sqrt{7}$ **7.** $x = y = z = 3$
9. $x = y = z = 10$ **11.** 9 ft $\times$ 9 ft $\times$ 8.25 ft; \$26.73
13. Let x, y, and z be the length, width, and height, respectively, and let V_0 be the given volume. Then $V_0 = xyz$ and $z = \dfrac{V_0}{xy}$.

The surface area is
$$S = 2xy + 2yz + 2xz = 2\left(xy + \frac{V_0}{x} + \frac{V_0}{y}\right).$$
$$\left. \begin{array}{l} S_x = 2\left(y - \dfrac{V_0}{x^2}\right) = 0 \\[2mm] S_y = 2\left(x - \dfrac{V_0}{y^2}\right) = 0 \end{array} \right\} \begin{array}{l} x^2 y - V_0 = 0 \\[2mm] xy^2 - V_0 = 0 \end{array}$$

So, $x = \sqrt[3]{V_0}$, $y = \sqrt[3]{V_0}$, and $z = \sqrt[3]{V_0}$.
15. $x_1 = 3$, $x_2 = 6$ **17.** Proof
19. $x = \dfrac{\sqrt{2}}{2} \approx 0.707$ km
$y = \dfrac{3\sqrt{2} + 2\sqrt{3}}{6} \approx 1.284$ km
21. (a) $y = \frac{3}{4}x + \frac{4}{3}$ (b) $\frac{1}{6}$ **23.** (a) $y = -2x + 4$ (b) 2
25. $y = \frac{84}{43}x - \frac{12}{43}$ **27.** $y = -\frac{175}{148}x + \frac{945}{148}$

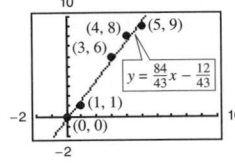

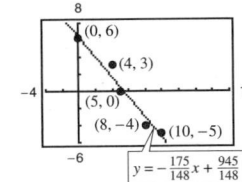

29. (a) $y = 0.23x + 2.38$ (b) \$301.4 billion
(c) The new model is $y = 0.23x + 5.09$, so the constant increases.

31. $a\sum\limits_{i=1}^{n} x_i^4 + b\sum\limits_{i=1}^{n} x_i^3 + c\sum\limits_{i=1}^{n} x_i^2 = \sum\limits_{i=1}^{n} x_i^2 y_i$
$a\sum\limits_{i=1}^{n} x_i^3 + b\sum\limits_{i=1}^{n} x_i^2 + c\sum\limits_{i=1}^{n} x_i = \sum\limits_{i=1}^{n} x_i y_i$
$a\sum\limits_{i=1}^{n} x_i^2 + b\sum\limits_{i=1}^{n} x_i + cn = \sum\limits_{i=1}^{n} y_i$
33. $y = \frac{3}{7}x^2 + \frac{6}{5}x + \frac{26}{35}$ **35.** $y = x^2 - x$

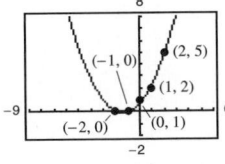

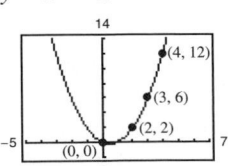

37. (a) $y = -0.22x^2 + 9.66x - 1.79$
(b)

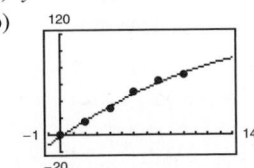

39. (a) $\ln P = -0.1499h + 9.3018$ (b) $P = 10{,}957.7e^{-0.1499h}$

(c)

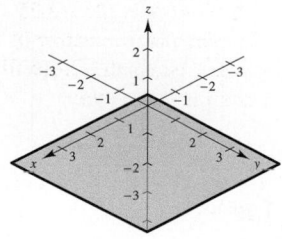

41. Proof

Section 13.10   *(page 962)*

1. Optimization problems that have restrictions or constraints on the values that can be used to produce the optimal solutions are called constrained optimization problems.

3. $f(5, 5) = 25$ **5.** $f(1, 2) = 5$ **7.** $f(25, 50) = 2600$

9. $f(1, 1) = 2$ **11.** $f(3, 3, 3) = 27$ **13.** $f\left(\frac{1}{3}, \frac{1}{3}, \frac{1}{3}\right) = \frac{1}{3}$

15. Maxima: $f\left(\frac{\sqrt{2}}{2}, \frac{\sqrt{2}}{2}\right) = \frac{5}{2}$

$f\left(-\frac{\sqrt{2}}{2}, -\frac{\sqrt{2}}{2}\right) = \frac{5}{2}$

Minima: $f\left(-\frac{\sqrt{2}}{2}, \frac{\sqrt{2}}{2}\right) = -\frac{1}{2}$

$f\left(\frac{\sqrt{2}}{2}, -\frac{\sqrt{2}}{2}\right) = -\frac{1}{2}$

17. $f(8, 16, 8) = 1024$ **19.** $\frac{\sqrt{2}}{2}$ **21.** $3\sqrt{2}$ **23.** $\frac{\sqrt{11}}{2}$

25. 2   **27.** $\sqrt{3}$ **29.** $(-4, 0, 4)$ **31.** $\sqrt{3}$

33. $x = y = z = 3$ **35.** 9 ft $\times$ 9 ft $\times$ 8.25 ft; $26.73

37. Proof   **39.** $\frac{2\sqrt{3}a}{3} \times \frac{2\sqrt{3}b}{3} \times \frac{2\sqrt{3}c}{3}$

41. At $(0, 0)$, the Lagrange equations are inconsistent.

43. $\sqrt[3]{360} \times \sqrt[3]{360} \times \frac{4}{3}\sqrt[3]{360}$ ft

45. $r = \sqrt[3]{\frac{v_0}{2\pi}}$ and $h = 2\sqrt[3]{\frac{v_0}{2\pi}}$ **47.** Proof

49. $P\left(\frac{15{,}625}{28}, 3125\right) \approx 203{,}144$

51. $x \approx 237.4$

$y \approx 640.9$

Cost $\approx$ $68,364.80

53. Putnam Problem 2, morning session, 1938

Review Exercises for Chapter 13   *(page 964)*

1. (a) -3 (b) -7 (c) 15   (d) $7x^2 - 3$

3. Domain: $\{(x, y): x \geq 0 \text{ and } y \neq 0\}$

Range: all real numbers

5.

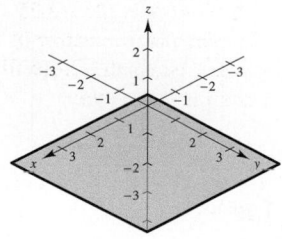

Plane

7. Lines: $y = 2x - 3 + c$

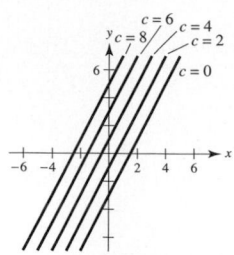

9. (a)

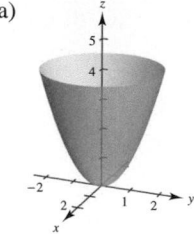

(b) g is a vertical translation of f two units upward.

(c) g is a horizontal translation of f two units to the right.

(d)

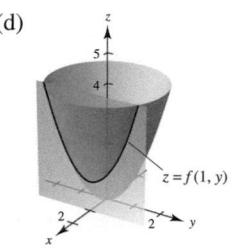   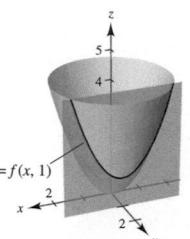

$z = f(1, y)$ $z = f(x, 1)$

11. Elliptic paraboloid

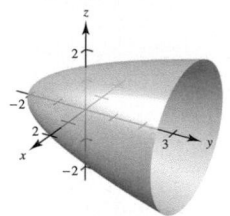

13. Limit: $\frac{1}{2}$

Continuous except at $(0, 0)$

15. Limit: 0

Continuous

17. Limit: $-\frac{\ln 2}{5}$

Continuous for $x \neq \frac{z}{y}$

19. $f_x(x, y) = 15x^2$

$f_y(x, y) = 7$

21. $f_x(x, y) = e^x \cos y$

$f_y(x, y) = -e^x \sin y$

23. $f_x(x, y) = -\frac{y^4}{x^2}e^{y/x}$

$f_y(x, y) = \frac{y^3}{x}e^{y/x} + 3y^2e^{y/x}$

25. $f_x(x, y, z) = 2z^2 + 6yz$

$f_y(x, y, z) = 6xz$

$f_z(x, y, z) = 4xz + 6xy$

27. $f_x(0, 2) = 0$

$f_y(0, 2) = -1$

29. $f_x\left(2, 3, -\frac{\pi}{3}\right) = -\sqrt{3}\pi - \frac{3}{2}$

$f_y\left(2, 3, -\frac{\pi}{3}\right) = -1$

$f_z\left(2, 3, -\frac{\pi}{3}\right) = 6\sqrt{3}$

31. $f_{xx}(x, y) = 6$

$f_{yy}(x, y) = 12y$

$f_{xy}(x, y) = f_{yx}(x, y) = -1$

33. $h_{xx}(x, y) = -y \cos x$
$h_{yy}(x, y) = -x \sin y$
$h_{xy}(x, y) = h_{yx}(x, y) = \cos y - \sin x$

35. Slope in x-direction: 0
Slope in y-direction: 4

37. $(xy \cos xy + \sin xy) \, dx + (x^2 \cos xy) \, dy$

39. $dw = (3y^2 - 6x^2yz^2) \, dx + (6xy - 2x^3z^2) \, dy + (-4x^3yz) \, dz$

41. (a) $f(2, 1) = 10$ (b) $dz = 0.5$
$f(2.1, 1.05) = 10.5$
$\Delta z = 0.5$

43. $dV = \pm\pi$ in.^{3}, $\dfrac{dV}{V} = 15\%$ **45.** Proof

47. (a) and (b) $\dfrac{dw}{dt} = \dfrac{8t - 1}{4t^2 - t + 4}$

49. (a) and (b) $\dfrac{dw}{dt} = 2t^2e^{2t} + 2te^{2t} + 2t + 1$

51. (a) and (b) $\dfrac{\partial w}{\partial r} = \dfrac{4r^2t - 4rt^2 - t^3}{(2r - t)^2}$
$\dfrac{\partial w}{\partial t} = \dfrac{4r^2t - rt^2 - 4r^3}{(2r - t)^2}$

53. $\dfrac{-3x^2 + y}{-x + 5}$

55. $\dfrac{\partial z}{\partial x} = \dfrac{-2x - y}{y + 2z}$
$\dfrac{\partial z}{\partial y} = \dfrac{-x - 2y - z}{y + 2z}$

57. -50 **59.** $\frac{2}{3}$ **61.** $\langle 4, 4 \rangle$, $4\sqrt{2}$ **63.** $\langle -\frac{1}{2}, 0 \rangle$, $\frac{1}{2}$

65. $\langle -2, -3, -1 \rangle$, $\sqrt{14}$

67. (a) $54\mathbf{i} - 16\mathbf{j}$ (b) $\dfrac{27}{\sqrt{793}}\mathbf{i} - \dfrac{8}{\sqrt{793}}\mathbf{j}$ (c) $y = \dfrac{27}{8}x - \dfrac{65}{8}$

(d)

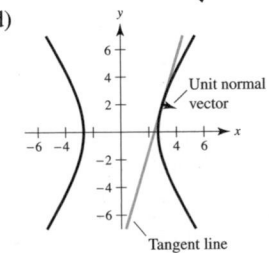

69. $2x + 6y - z = 8$ **71.** $z = 4$

73. (a) $4x + 4y - z = 8$
(b) $x = 2 + 4t, y = 1 + 4t, z = 4 - t$

75. $36.7°$ **77.** $(0, 0, 9)$

79. Relative maximum: $(4, -1, 9)$

81. Relative minimum: $\left(-4, \frac{4}{3}, -2\right)$

83. Relative minimum: $(1, 1, 3)$ **85.** $\sqrt{3}$

87. $x_1 = 2, x_2 = 4$

89. $y = \frac{161}{226}x + \frac{456}{113}$

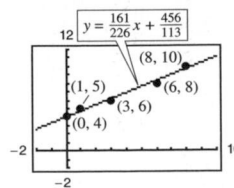

91. (a) $y = 0.138x + 22.1$ (b) 46.25 bushels/acre

93. $f(4, 4) = 32$ **95.** $f(15, 7) = 352$ **97.** $f(3, 6) = 36$

99. $x = \dfrac{\sqrt{2}}{2} \approx 0.707$ km, $y = \dfrac{\sqrt{3}}{3} \approx 0.577$ km,
$z = (60 - 3\sqrt{2} - 2\sqrt{3})6 \approx 8.716$ km

P.S. Problem Solving   *(page 967)*

1. (a) 12 square units   (b) and (c) Proofs

3. (a) $y_0z_0(x - x_0) + x_0z_0(y - y_0) + x_0y_0(z - z_0) = 0$

(b) $x_0y_0z_0 = 1 \implies z_0 = \dfrac{1}{x_0y_0}$
Then the tangent plane is
$y_0\left(\dfrac{1}{x_0y_0}\right)(x - x_0) + x_0\left(\dfrac{1}{x_0y_0}\right)(y - y_0) + x_0y_0\left(z - \dfrac{1}{x_0y_0}\right) = 0.$
Intercepts: $(3x_0, 0, 0), (0, 3y_0, 0), \left(0, 0, \dfrac{3}{x_0y_0}\right)$

5. (a)

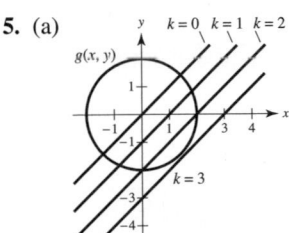

(b)

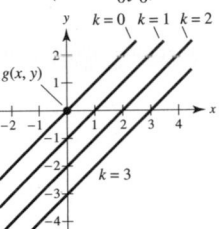

Maximum value: $2\sqrt{2}$ Maximum and minimum value: 0

The method of Lagrange multipliers does not work because $\nabla g(x_0, y_0) = \mathbf{0}$.

7. $2\sqrt[3]{150}$ ft $\times 2\sqrt[3]{150}$ ft $\times \dfrac{5\sqrt[3]{150}}{3}$ ft

9. (a) $x\dfrac{\partial f}{\partial x} + y\dfrac{\partial f}{\partial y} = xCy^{1-a}ax^{a-1} + yCx^a(1 - a)y^{1-a-1}$
$= ax^aCy^{1-a} + (1 - a)x^aC(y^{1-a})$
$= Cx^ay^{1-a}[a + (1 - a)]$
$= Cx^ay^{1-a}$
$= f(x, y)$

(b) $f(tx, ty) = C(tx)^a(ty)^{1-a}$
$= Ctx^ay^{1-a}$
$= tCx^ay^{1-a}$
$= tf(x, y)$

11. (a) $x = 32\sqrt{2}t$
$y = 32\sqrt{2}t - 16t^2$

(b) $\alpha = \arctan\left(\dfrac{y}{x + 50}\right) = \arctan\left(\dfrac{32\sqrt{2}t - 16t^2}{32\sqrt{2}t + 50}\right)$

(c) $\dfrac{d\alpha}{dt} = \dfrac{-16(8\sqrt{2}t^2 + 25t - 25\sqrt{2})}{64t^4 - 256\sqrt{2}t^3 + 1024t^2 + 800\sqrt{2}t + 625}$

(d)

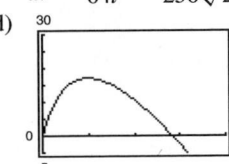

No; The rate of change of α is greatest when the projectile is closest to the camera.

(e) α is maximum when $t = 0.98$ second.
No, the projectile is at its maximum height when $t = \sqrt{2} \approx 1.41$ seconds.

13. (a) 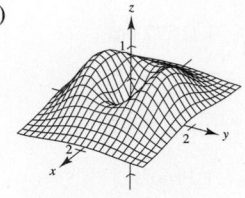   (b)

Minimum: $(0, 0, 0)$ Minima: $(\pm 1, 0, -e^{-1})$
Maxima: $(0, \pm 1, 2e^{-1})$ Maxima: $(0, \pm 1, 2e^{-1})$
Saddle points: $(\pm 1, 0, e^{-1})$ Saddle point: $(0, 0, 0)$

(c) $\alpha > 0$ $\alpha < 0$
Minimum: $(0, 0, 0)$ Minima: $(\pm 1, 0, \alpha e^{-1})$
Maxima: $(0, \pm 1, \beta e^{-1})$ Maxima: $(0, \pm 1, \beta e^{-1})$
Saddle points: Saddle point: $(0, 0, 0)$
$(\pm 1, 0, \alpha e^{-1})$

15. (a)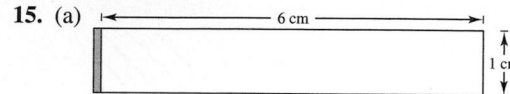

(b)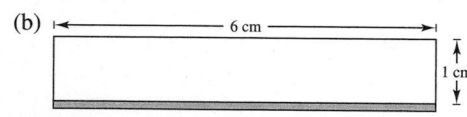

(c) Height
(d) $dl = 0.01,\ dh = 0:\ dA = 0.01$
    $dl = 0,\ dh = 0.01:\ dA = 0.06$

17–21. Proofs

Chapter 14

Section 14.1   *(page 976)*

1. An iterated integral is an integral of a function of several variables. Integrate with respect to one variable while holding the other variables constant.

3. $\dfrac{3x^2}{2}$ **5.** $\dfrac{4x^2 - x^4}{2}$ **7.** $\dfrac{y}{2}\left[(\ln y)^2 - y^2\right]$

9. $x^2(1 - e^{-x^2} - x^2 e^{-x^2})$ **11.** 3   **13.** $\dfrac{\sqrt{2}}{4}$ **15.** 64

17. $\dfrac{3}{2}$ **19.** $\dfrac{1}{3}$ **21.** $\dfrac{2}{3}$ **23.** 4   **25.** $\dfrac{\pi}{2}$ **27.** $\dfrac{\pi^2}{32} + \dfrac{1}{8}$

29. $\dfrac{1}{2}$ **31.** Diverges   **33.** 8   **35.** $\dfrac{16}{3}$ **37.** 36

39. $\dfrac{8}{3}$ **41.** $\dfrac{9}{2}$

43.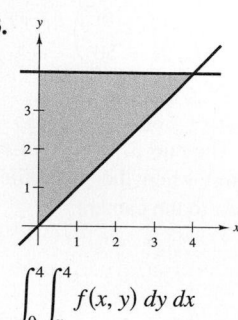

$$\int_0^4 \int_x^4 f(x, y)\, dy\, dx$$

45.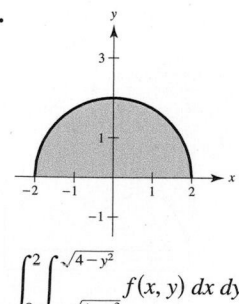

$$\int_0^2 \int_{-\sqrt{4-y^2}}^{\sqrt{4-y^2}} f(x, y)\, dx\, dy$$

47.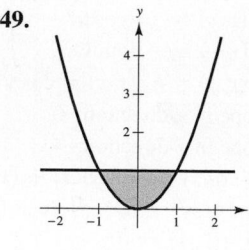

$$\int_0^{\ln 10} \int_{e^x}^{10} f(x, y)\, dy\, dx$$

49.

$$\int_0^1 \int_{-\sqrt{y}}^{\sqrt{y}} f(x, y)\, dx\, dy$$

51.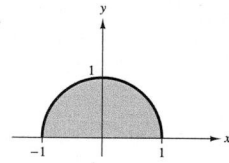

$$\int_0^1 \int_0^2 dy\, dx = \int_0^2 \int_0^1 dx\, dy = 2$$

53.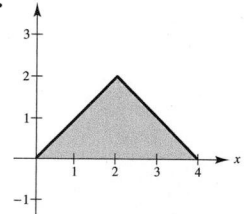

$(2, 1)$

$$\int_0^1 \int_{2y}^2 dx\, dy = \int_0^2 \int_0^{x/2} dy\, dx = 1$$

55.

$$\int_0^1 \int_{-\sqrt{1-y^2}}^{\sqrt{1-y^2}} dx\, dy = \int_{-1}^1 \int_0^{\sqrt{1-x^2}} dy\, dx = \frac{\pi}{2}$$

57.

$$\int_0^2 \int_0^x dy\, dx + \int_2^4 \int_0^{4-x} dy\, dx = \int_0^2 \int_y^{4-y} dx\, dy = 4$$

59.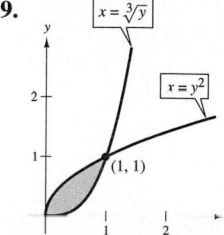

$x = \sqrt[3]{y}$
$x = y^2$
$(1, 1)$

$$\int_0^1 \int_{y^2}^{\sqrt[3]{y}} dx\, dy = \int_0^1 \int_{x^3}^{\sqrt{x}} dy\, dx = \frac{5}{12}$$

61.

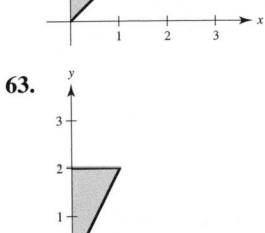

$$\int_0^2 \int_x^2 x\sqrt{1+y^3}\, dy\, dx = \frac{26}{9}$$

63.

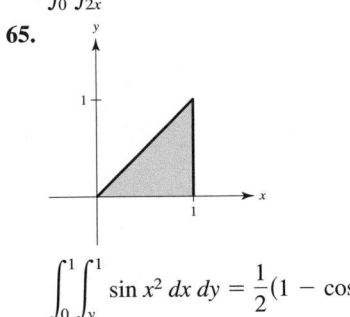

$$\int_0^1 \int_{2x}^2 4e^{y^2}\, dy\, dx = e^4 - 1 \approx 53.598$$

65.

$$\int_0^1 \int_y^1 \sin x^2\, dx\, dy = \frac{1}{2}(1 - \cos 1) \approx 0.230$$

67. $4\int_0^5 \int_0^{\sqrt{25-x^2}} dy\, dx = 25\pi$ square units

69. (a) No   (b) Yes   (c) Yes   **71.** $\dfrac{\sin 2}{2} - \dfrac{\sin 3}{3}$

73. $(\ln 5)^2$ **75.** $\dfrac{15\pi}{2}$

77. (a)

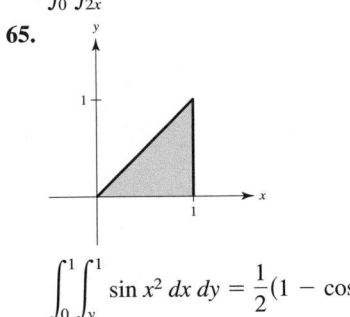

(b) $\int_0^8 \int_{x^2/32}^{\sqrt[3]{x}} (x^2 y - xy^2)\, dy\, dx$ (c) $\dfrac{67,520}{693}$

79. True

Section 14.2   *(page 987)*

1. Use rectangular prisms to approximate the volume, where $f(x_i, y_i)$ is the height of prism i and ΔA_i is the area of the rectangular base of the prism. You can improve the approximation by using more rectangular prisms of smaller rectangular bases.

3. 24 (approximation is exact)

5. Approximation: 52; Exact: $\frac{160}{3}$

7.

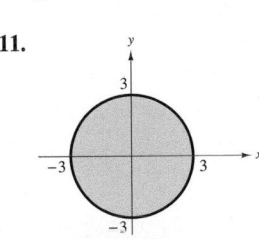

2

9.

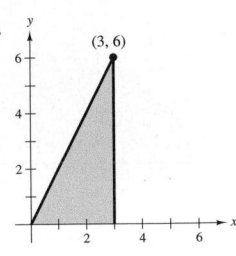

(3, 6)

36

11.

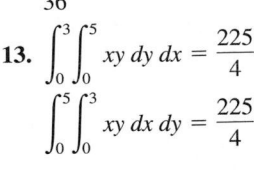

0

13. $\int_0^3 \int_0^5 xy\, dy\, dx = \dfrac{225}{4}$

$\int_0^5 \int_0^3 xy\, dx\, dy = \dfrac{225}{4}$

15. $\int_1^2 \int_x^{2x} \dfrac{y}{x^2+y^2}\, dy\, dx = \dfrac{1}{2}\ln\dfrac{5}{2}$

$\int_1^2 \int_1^y \dfrac{y}{x^2+y^2}\, dx\, dy + \int_2^4 \int_{y/2}^2 \dfrac{y}{x^2+y^2}\, dx\, dy = \dfrac{1}{2}\ln\dfrac{5}{2}$

17. $\int_0^1 \int_{4-x}^{4-x^2} -2y\, dy\, dx = -\dfrac{6}{5}$

$\int_3^4 \int_{4-y}^{\sqrt{4-y}} -2y\, dx\, dy = -\dfrac{6}{5}$

19. $\int_0^3 \int_{4y/3}^{\sqrt{25-y^2}} x\, dx\, dy = 25$

$\int_0^4 \int_0^{3x/4} x\, dy\, dx + \int_4^5 \int_0^{\sqrt{25-x^2}} x\, dy\, dx = 25$

21. 4   **23.** 12   **25.** $\frac{3}{8}$ **27.** 1

29. $\int_0^1 \int_0^{x^3} xy\, dy\, dx = \dfrac{1}{16}$ **31.** $\int_0^2 \int_0^{\sqrt{4-x^2}} (x+y)\, dy\, dx = \dfrac{16}{3}$

33. $\int_0^2 \int_0^{4-x^2} (4-x^2)\, dy\, dx = \dfrac{256}{15}$

35. $2\int_0^2 \int_0^{\sqrt{1-(x-1)^2}} (2x - x^2 + y^2)\, dy\, dx$

37. $4\int_0^2 \int_0^{\sqrt{4-x^2}} (x^2 + y^2)\, dy\, dx$

39. $\int_0^2 \int_{-\sqrt{2-2(y-1)^2}}^{\sqrt{2-2(y-1)^2}} (4y - x^2 - 2y^2)\, dx\, dy$

41. $\dfrac{81\pi}{2}$ **43.** 1.2315

45.

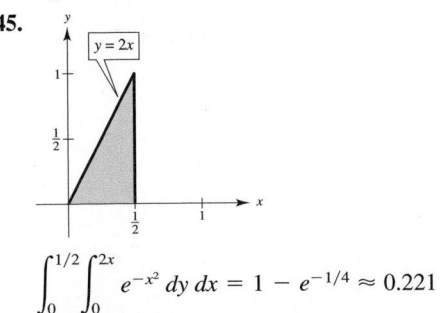

$$\int_0^{1/2} \int_0^{2x} e^{-x^2}\, dy\, dx = 1 - e^{-1/4} \approx 0.221$$

47.

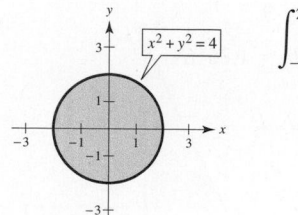

$$\int_{-2}^{2}\int_{-\sqrt{4-y^2}}^{\sqrt{4-y^2}} \sqrt{4-y^2}\, dx\, dy = \frac{64}{3}$$

49.

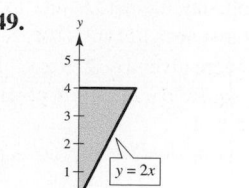

$$\int_{0}^{4}\int_{0}^{y/2} \sin y^2 \, dx\, dy = \frac{1 - \cos 16}{4} \approx 0.489$$

51. 2   **53.** $\frac{8}{3}$ **55.** $(e-1)^2$ **57.** 25,645.24

59. kB; Answers will vary.   **61.** Proof; $\frac{2}{3}$ **63.** Proof; $\frac{4}{9}$

65. Proof   **67.** 400; 272

69. False. $V = 8\int_{0}^{1}\int_{0}^{\sqrt{1-y^2}} \sqrt{1-x^2-y^2}\, dx\, dy$

71. R: $x^2 + y^2 \leq 9$ **73.** $\frac{1}{2}(1-e)$

75. Putnam Problem A2, 1989

Section 14.3 (page 995)

1. Rectangular

3. r-simple regions have fixed bounds for θ and variable bounds for r. θ-simple regions have variable bounds for θ and fixed bounds for r.

5. $R = \{(r, \theta): 0 \leq r \leq 8, 0 \leq \theta \leq \pi\}$

7. $R = \left\{(r, \theta): 4 \leq r \leq 8, 0 \leq \theta \leq \frac{\pi}{2}\right\}$

9. π **11.** 0

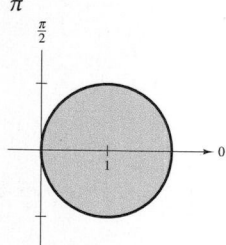   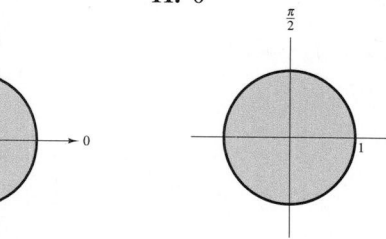

13. $\dfrac{8\sqrt{2}\pi}{3}$ **15.** $\dfrac{9}{8} + \dfrac{3\pi^2}{32}$

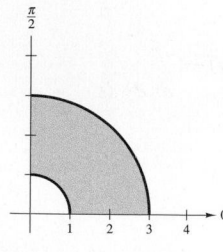   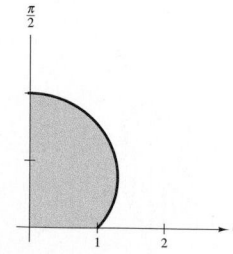

17. 9   **19.** 4π **21.** $\dfrac{\pi}{10}$ **23.** $\dfrac{2}{3}$

25. $\dfrac{\pi}{2}\sin 1$ **27.** $\int_{0}^{\pi/4}\int_{0}^{2\sqrt{2}} r^2 \, dr\, d\theta = \dfrac{4\sqrt{2}\pi}{3}$

29. $\int_{0}^{\pi/2}\int_{0}^{6} (\cos\theta + \sin\theta) r^2 \, dr\, d\theta = 144$

31. $\int_{0}^{\pi/4}\int_{1}^{2} r\theta \, dr\, d\theta = \dfrac{3\pi^2}{64}$ **33.** $\dfrac{1}{8}$ **35.** $\dfrac{250\pi}{3}$

37. $\dfrac{64}{9}(3\pi - 4)$ **39.** $2\sqrt{4 - 2\sqrt[3]{2}}$ **41.** 9π

43. $\dfrac{3\pi}{2}$ **45.** π

47.   **49.**

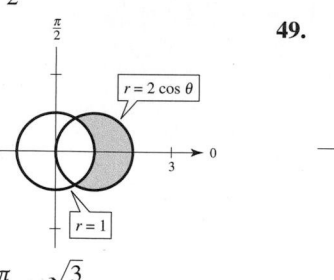   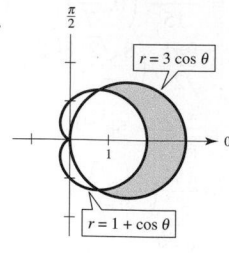

$\dfrac{\pi}{3} + \dfrac{\sqrt{3}}{2}$ π

51.

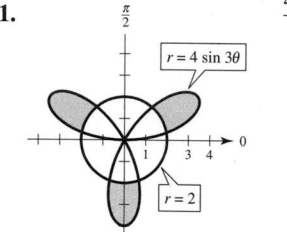

$\dfrac{4\pi}{3} + 2\sqrt{3}$

53. $\int_{0}^{\pi/6}\int_{1}^{\sqrt{3}\sec\theta} r \, dr\, d\theta + \int_{\pi/6}^{\pi/2}\int_{1}^{\csc\theta} r \, dr\, d\theta = \sqrt{3} - \dfrac{\pi}{4}$

55. 486,788   **57.** 1.2858   **59.** 56.051

61. False. Let $f(r, \theta) = r - 1$ and let R be a sector where $0 \leq r \leq 6$ and $0 \leq \theta \leq \pi$.

63. (a) 2π (b) $\sqrt{2\pi}$

65. (a) $\int_{2}^{4}\int_{y/\sqrt{3}}^{y} f \, dx\, dy$

(b) $\int_{2/\sqrt{3}}^{2}\int_{2}^{\sqrt{3}x} f \, dy\, dx + \int_{2}^{4/\sqrt{3}}\int_{x}^{\sqrt{3}x} f \, dy\, dx + \int_{4/\sqrt{3}}^{4}\int_{x}^{4} f \, dy\, dx$

(c) $\int_{\pi/4}^{\pi/3}\int_{2\csc\theta}^{4\csc\theta} fr \, dr\, d\theta$

67. $\dfrac{4}{\pi}$

Section 14.4 (page 1004)

1. Use a double integral when the density of the lamina is not constant.

3. $m = 4$ **5.** $m = \dfrac{1}{8}$

7. (a) $m = ka^2, \left(\dfrac{a}{2}, \dfrac{a}{2}\right)$ (b) $m = \dfrac{ka^3}{2}, \left(\dfrac{a}{2}, \dfrac{2a}{3}\right)$

(c) $m = \dfrac{ka^3}{2}, \left(\dfrac{2a}{3}, \dfrac{a}{2}\right)$

9. (a) $m = \dfrac{ka^2}{2}, \left(\dfrac{a}{3}, \dfrac{2a}{3}\right)$ (b) $m = \dfrac{ka^3}{3}, \left(\dfrac{3a}{8}, \dfrac{3a}{4}\right)$

(c) $m = \dfrac{ka^3}{6}, \left(\dfrac{a}{2}, \dfrac{3a}{4}\right)$

11. (a) $\left(\dfrac{a}{2} + 5, \dfrac{a}{2}\right)$ (b) $\left(\dfrac{a}{2} + 5, \dfrac{2a}{3}\right)$

(c) $\left(\dfrac{2(a^2 + 15a + 75)}{3(a + 10)}, \dfrac{a}{2}\right)$

13. $m = \dfrac{k}{4}, \left(\dfrac{2}{3}, \dfrac{8}{15}\right)$ **15.** $m = 30k, \left(\dfrac{14}{5}, \dfrac{4}{5}\right)$

17. $m = k(e - 1), \left(\dfrac{1}{e - 1}, \dfrac{e + 1}{4}\right)$

19. $m = \dfrac{256k}{15}, \left(0, \dfrac{16}{7}\right)$ **21.** $m = \dfrac{6k}{\pi}, \left(\dfrac{3}{2}, \dfrac{\pi}{8}\right)$

23. $m = \dfrac{9\pi k}{2}, \left(\dfrac{8\sqrt{2}}{\pi}, \dfrac{8(2 - \sqrt{2})}{\pi}\right)$

25. $m = \dfrac{k}{8}(1 - 5e^{-4}), \left(\dfrac{e^4 - 13}{e^4 - 5}, \dfrac{8}{27}\left[\dfrac{e^6 - 7}{e^6 - 5e^2}\right]\right)$

27. $m - \dfrac{k\pi}{3}, \left(\dfrac{81\sqrt{3}}{40\pi}, 0\right)$

29. $\bar{\bar{x}} = \dfrac{\sqrt{3}b}{3}$ **31.** $\bar{\bar{x}} = \dfrac{a}{2}$ **33.** $\bar{\bar{x}} = \dfrac{a}{2}$

$\bar{\bar{y}} = \dfrac{\sqrt{3}h}{3}$ $\bar{\bar{y}} = \dfrac{a}{2}$ $\bar{\bar{y}} = \dfrac{a}{2}$

35. $I_x = \dfrac{32k}{3}$ **37.** $I_x = 16k$

$I_y = \dfrac{16k}{3}$ $I_y = \dfrac{512k}{5}$

$I_0 = 16k$ $I_0 = \dfrac{592k}{5}$

$\bar{\bar{x}} = \dfrac{2\sqrt{3}}{3}$ $\bar{\bar{x}} = \dfrac{4\sqrt{15}}{5}$

$\bar{\bar{y}} = \dfrac{2\sqrt{6}}{3}$ $\bar{\bar{y}} = \dfrac{\sqrt{6}}{2}$

39. $2k\displaystyle\int_{-b}^{b}\int_{0}^{\sqrt{b^2 - x^2}} (x - a)^2 \, dy \, dx = \dfrac{k\pi b^2}{4}(b^2 + 4a^2)$

41. $\displaystyle\int_{-a}^{a}\int_{0}^{\sqrt{a^2 - x^2}} ky(y - a)^2 \, dy \, dx = ka^5\left(\dfrac{56 - 15\pi}{60}\right)$

43. $\dfrac{L}{3}$ **45.** $\dfrac{L}{2}$

47. The object with a greater polar moment of inertia has more resistance, so more torque is required to twist the object.

49. Proof

Section 14.5   *(page 1011)*

1. If f and its first partial derivatives are continuous on the closed region R in the xy-plane, then the differential of the surface area given by $z = f(x, y)$ over R is $dS = \sqrt{1 + [f_x(x, y)]^2 + [f_y(x, y)]^2} \, dA$.

3. 24   **5.** $4\pi\sqrt{62}$ **7.** $\dfrac{1}{2}\left[4\sqrt{17} + \ln\left(4 + \sqrt{17}\right)\right]$

9. $\dfrac{8}{27}\left(10\sqrt{10} - 1\right)$ **11.** $\sqrt{2} - 1$ **13.** $\sqrt{2}\pi$

15. $2\pi a\left(a - \sqrt{a^2 - b^2}\right)$ **17.** $12\sqrt{14}$ **19.** 20π

21. $\displaystyle\int_{0}^{1}\int_{0}^{x} \sqrt{5 + 4x^2} \, dy \, dx = \dfrac{27 - 5\sqrt{5}}{12} \approx 1.3183$

23. $\displaystyle\int_{-3}^{3}\int_{-\sqrt{9-x^2}}^{\sqrt{9-x^2}} \sqrt{1 + 4x^2 + 4y^2} \, dy \, dx$

$= \dfrac{\pi}{6}\left(37\sqrt{37} - 1\right) \approx 117.3187$

25. $\displaystyle\int_{0}^{1}\int_{0}^{1} \sqrt{1 + 4x^2 + 4y^2} \, dy \, dx \approx 1.8616$

27. $\displaystyle\int_{0}^{4}\int_{0}^{10} \sqrt{1 + e^{2xy}(x^2 + y^2)} \, dy \, dx$

29. $\displaystyle\int_{-2}^{2}\int_{-\sqrt{4-x^2}}^{\sqrt{4-x^2}} \sqrt{1 + e^{-2x}} \, dy \, dx$

31. No. The size and shape of the graph stay the same, just the position is changed. So, the surface area does not increase.

33. (a) Yes. For example, let R be the square given by $0 \le x \le 1$ and $0 \le y \le 1$, and let S be the square parallel to R given by $0 \le x \le 1, 0 \le y \le 1$, and $z = 1$.

(b) Yes. Let R be the region in part (a) and let S be the surface given by $f(x, y) = xy$.

(c) No

35. (a) $812\pi\sqrt{609}$ cm³   (b) $100\pi\sqrt{609}$ cm²   **37.** 16

Section 14.6   *(page 1021)*

1. The volume of the solid region Q **3.** 18   **5.** $\dfrac{1}{9}$

7. $\dfrac{15}{2}\left(1 - \dfrac{1}{e}\right)$ **9.** $\dfrac{189}{2}$ **11.** $\dfrac{324}{5}$

13. $V = \displaystyle\int_{0}^{7}\int_{0}^{(7-x)/2}\int_{0}^{7-x-2y} dz \, dy \, dx$

15. $V = \displaystyle\int_{-\sqrt{6}}^{\sqrt{6}}\int_{-\sqrt{6-y^2}}^{\sqrt{6-y^2}}\int_{0}^{6-x^2-y^2} dz \, dx \, dy$

17. $V = \displaystyle\int_{-4}^{4}\int_{-\sqrt{16-x^2}}^{\sqrt{16-x^2}}\int_{(x^2+y^2)/2}^{\sqrt{80-x^2-y^2}} dz \, dy \, dx$ **19.** $\dfrac{256}{15}$

21. $\dfrac{3}{2}$ **23.** 10

25.

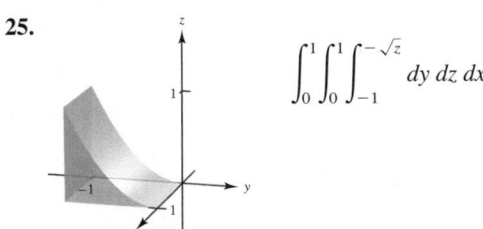

$\displaystyle\int_{0}^{1}\int_{0}^{1}\int_{-1}^{-\sqrt{z}} dy \, dz \, dx$

27.

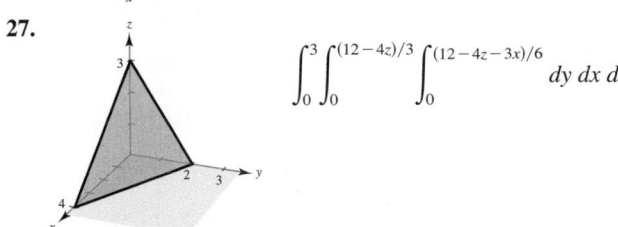

$\displaystyle\int_{0}^{3}\int_{0}^{(12-4z)/3}\int_{0}^{(12-4z-3x)/6} dy \, dx \, dz$

29.

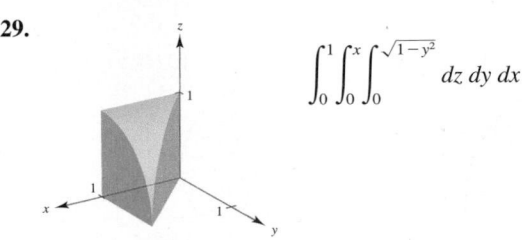

$\displaystyle\int_{0}^{1}\int_{0}^{x}\int_{0}^{\sqrt{1-y^2}} dz \, dy \, dx$

31. $\int_0^3 \int_0^5 \int_{y/5}^1 xyz \, dx \, dy \, dz$, $\int_0^3 \int_0^1 \int_0^{5x} xyz \, dy \, dx \, dz$,

$\int_0^5 \int_0^3 \int_{y/5}^1 xyz \, dx \, dz \, dy$, $\int_0^1 \int_0^3 \int_0^{5x} xyz \, dy \, dz \, dx$,

$\int_0^5 \int_{y/5}^1 \int_0^3 xyz \, dz \, dx \, dy$, $\int_0^1 \int_0^{5x} \int_0^3 xyz \, dz \, dy \, dx$; $\dfrac{225}{16}$

33. $\int_{-3}^3 \int_{-\sqrt{9-x^2}}^{\sqrt{9-x^2}} \int_0^4 xyz \, dz \, dy \, dx$, $\int_{-3}^3 \int_{-\sqrt{9-y^2}}^{\sqrt{9-y^2}} \int_0^4 xyz \, dz \, dx \, dy$,

$\int_{-3}^3 \int_0^4 \int_{-\sqrt{9-x^2}}^{\sqrt{9-x^2}} xyz \, dy \, dz \, dx$, $\int_0^4 \int_{-3}^3 \int_{-\sqrt{9-x^2}}^{\sqrt{9-x^2}} xyz \, dy \, dx \, dz$,

$\int_0^4 \int_{-3}^3 \int_{-\sqrt{9-y^2}}^{\sqrt{9-y^2}} xyz \, dx \, dy \, dz$, $\int_{-3}^3 \int_0^4 \int_{-\sqrt{9-y^2}}^{\sqrt{9-y^2}} xyz \, dx \, dz \, dy$; 0

35. $\int_0^1 \int_0^{1-z} \int_0^{1-y^2} dx \, dy \, dz$, $\int_0^1 \int_0^{1-y} \int_0^{1-y^2} dx \, dz \, dy$,

$\int_0^1 \int_0^{2z-z^2} \int_0^{1-z} 1 \, dy \, dx \, dz + \int_0^1 \int_{2z-z^2}^1 \int_0^{\sqrt{1-x}} 1 \, dy \, dx \, dz$,

$\int_0^1 \int_{1-\sqrt{1-x}}^1 \int_0^{1-z} 1 \, dy \, dz \, dx + \int_0^1 \int_0^{1-\sqrt{1-x}} \int_0^{\sqrt{1-x}} 1 \, dy \, dz \, dx$,

$\int_0^1 \int_0^{\sqrt{1-x}} \int_0^{1-y} dz \, dy \, dx$

37. $m = 8k$, $\bar{x} = \dfrac{3}{2}$ **39.** $m = \dfrac{128k}{3}$, $\bar{z} = 1$

41. $m = k \int_0^b \int_0^b \int_0^b xy \, dz \, dy \, dx$

$M_{yz} = k \int_0^b \int_0^b \int_0^b x^2y \, dz \, dy \, dx$

$M_{xz} = k \int_0^b \int_0^b \int_0^b xy^2 \, dz \, dy \, dx$

$M_{xy} = k \int_0^b \int_0^b \int_0^b xyz \, dz \, dy \, dx$

43. $\bar{x}$ will be greater than 2, and $\bar{y}$ and $\bar{z}$ will be unchanged.

45. $\bar{x}$ and $\bar{z}$ will be unchanged, and $\bar{y}$ will be greater than 0.

47. $\left(0, 0, \dfrac{3h}{4}\right)$ **49.** $\left(0, 0, \dfrac{3}{2}\right)$ **51.** $\left(5, 6, \dfrac{5}{4}\right)$

53. (a) $I_x = \dfrac{2ka^5}{3}$ **55.** (a) $I_x = 256k$

$I_y = \dfrac{2ka^5}{3}$ $I_y = \dfrac{512k}{3}$

$I_z = \dfrac{2ka^5}{3}$ $I_z = 256k$

(b) $I_x = \dfrac{ka^8}{8}$ (b) $I_x = \dfrac{2048k}{3}$

$I_y = \dfrac{ka^8}{8}$ $I_y = \dfrac{1024k}{3}$

$I_z = \dfrac{ka^8}{8}$ $I_z = \dfrac{2048k}{3}$

57. Proof

59. $\int_{-1}^1 \int_{-1}^1 \int_0^{1-x} (x^2 + y^2)\sqrt{x^2 + y^2 + z^2} \, dz \, dy \, dx$

61. (a) $m = \int_{-2}^2 \int_{-\sqrt{4-x^2}}^{\sqrt{4-x^2}} \int_0^{4-x^2-y^2} kz \, dz \, dy \, dx$

(b) $\bar{x} = \bar{y} = 0$ by symmetry.

$\bar{z} = \dfrac{1}{m} \int_{-2}^2 \int_{-\sqrt{4-x^2}}^{\sqrt{4-x^2}} \int_0^{4-x^2-y^2} kz^2 \, dz \, dy \, dx$

(c) $I_z = \int_{-2}^2 \int_{-\sqrt{4-x^2}}^{\sqrt{4-x^2}} \int_0^{4-x^2-y^2} kz(x^2 + y^2) \, dz \, dy \, dx$

63. $\dfrac{13}{3}$ **65.** $\dfrac{3}{2}$ **67.** Increase

69. b **71.** Q: $2x^2 + y^2 + 3z^2 \le 1$; 0.684; $\dfrac{4\sqrt{6}\pi}{45}$

73. Putnam Problem B1, 1965

Section 14.7 *(page 1029)*

1. Some solids are represented by equations involving x^2 and y^2. Often, converting these equations to cylindrical or spherical coordinates yields equations you can work with more easily.

3. 27 **5.** $\dfrac{11}{10}$ **7.** $\dfrac{\pi}{3}$ **9.** $\pi(e^4 + 3)$

11.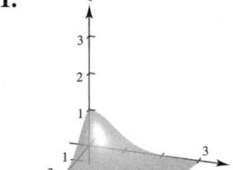

13.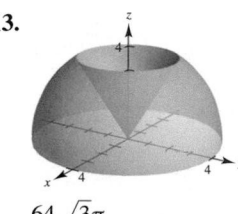

$\dfrac{64\sqrt{3}\pi}{3}$

$\dfrac{\pi}{4}(1 - e^{-9})$

15. $48(3\pi - 4)$ **17.** $\dfrac{\pi}{6}$ **19.** $\dfrac{250}{9}(3\pi - 4)$ **21.** $48k\pi$

23. $\dfrac{\pi r_0^2 h}{3}$ **25.** $\left(0, 0, \dfrac{h}{5}\right)$

27. $I_z = 4k \int_0^{\pi/2} \int_0^{r_0} \int_0^{h(r_0-r)/r_0} r^3 \, dz \, dr \, d\theta = \dfrac{3mr_0^2}{10}$

29. Proof **31.** $9\pi\sqrt{2}$ **33.** $16\pi^2$ **35.** $k\pi a^4$

37. $\left(0, 0, \dfrac{3r}{8}\right)$ **39.** $\dfrac{k\pi}{192}$

41. Cylindrical: $\int_0^{2\pi} \int_0^2 \int_{r^2}^4 r^2 \cos\theta \, dz \, dr \, d\theta = 0$

Spherical: $\int_0^{2\pi} \int_0^{\arctan(1/2)} \int_0^{4\sec\phi} \rho^3 \sin^2\phi \cos\theta \, d\rho \, d\phi \, d\theta$

$+ \int_0^{2\pi} \int_{\arctan(1/2)}^{\pi/2} \int_0^{\cot\phi \csc\phi} \rho^3 \sin^2\phi \cos\phi \, d\rho \, d\phi \, d\theta = 0$

43. Cylindrical: $\int_0^{2\pi} \int_0^1 \int_1^{1+\sqrt{1-r^2}} r^2 \cos\theta \, dz \, dr \, d\theta = 0$

Spherical: $\int_0^{\pi/4} \int_0^{2\pi} \int_{\sec\phi}^{2\cos\phi} \rho^3 \sin^2\phi \cos\theta \, d\rho \, d\theta \, d\phi = 0$

45. (a) r constant: right circular cylinder about z-axis

θ constant: plane parallel to z-axis

z constant: plane parallel to xy-plane

(b) ρ constant: sphere

θ constant: plane parallel to z-axis

ϕ constant: cone

47. Putnam Problem A1, 2006

Section 14.8   *(page 1036)*

1. $\dfrac{\partial x}{\partial u}\dfrac{\partial y}{\partial v} - \dfrac{\partial y}{\partial u}\dfrac{\partial x}{\partial v}$ **3.** $-\dfrac{1}{2}$ **5.** $1 + 2v$

7. 1 **9.** $-e^{2u}$

11.

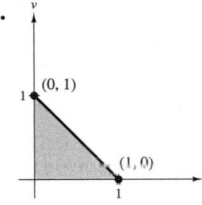

13.

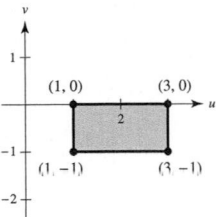

15. $\displaystyle\int_{R}\int 3xy\,dA = \int_{-2/3}^{2/3}\int_{1-x}^{(1/2)x+2} 3xy\,dy\,dx$

$\displaystyle + \int_{2/3}^{4/3}\int_{(1/2)x}^{(1/2)x+2} 3xy\,dy\,dx + \int_{4/3}^{8/3}\int_{(1/2)x}^{4-x} 3xy\,dy\,dx = \dfrac{164}{9}$

17. $\dfrac{8}{3}$ **19.** 36 **21.** $(e^{-1/2} - e^{-2})\ln 8 \approx 0.9798$ **23.** 18

25. $12(e^4 - 1)$ **27.** $\dfrac{100}{9}$ **29.** $\dfrac{2}{5}a^{5/2}$ **31.** One

33. (a)

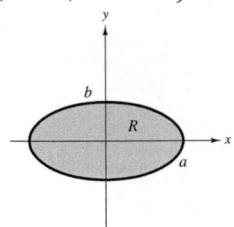   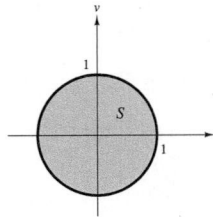

(b) ab (c) πab

35. $u^2 v$ **37.** $-uv$ **39.** $-\rho^2 \sin\phi$

41. Putnam Problem A2, 1994

Review Exercises for Chapter 14   *(page 1038)*

1. $\dfrac{1 - \cos 3x^2}{x}$ **3.** $\dfrac{29}{6}$ **5.** $\dfrac{1}{6}$ **7.** $\dfrac{3}{2}$ **9.** 16

11.

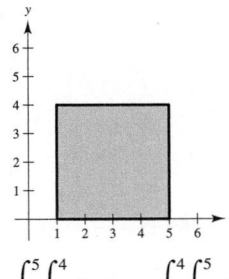

$\displaystyle\int_{1}^{5}\int_{0}^{4} dy\,dx = \int_{0}^{4}\int_{1}^{5} dx\,dy = 16$

13.

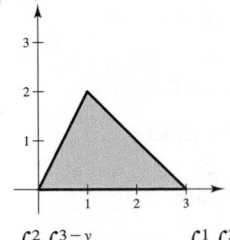

$\displaystyle\int_{0}^{2}\int_{y/2}^{3-y} dx\,dy = \int_{0}^{1}\int_{0}^{2x} dy\,dx + \int_{1}^{3}\int_{0}^{3-x} dy\,dx = 3$

15. $\displaystyle\int_{0}^{2}\int_{0}^{4} 4xy\,dy\,dx = \int_{0}^{4}\int_{0}^{2} 4xy\,dx\,dy = 64$ **17.** 21

19. $\dfrac{40}{3}$ **21.** $\dfrac{40}{3}$ **23.** $13.67°C$ **25.** $\dfrac{5\sqrt{5}\pi}{6}$

27. $\dfrac{81}{5}$ **29.** $\dfrac{3\pi}{2}$

31.

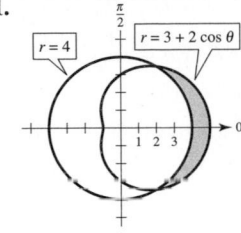

$\dfrac{13\sqrt{3}}{2} - \dfrac{5\pi}{3}$

33. (a) $r = 3\sqrt{\cos 2\theta}$

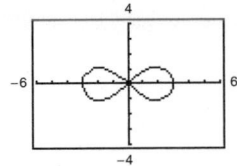

(b) 9 (c) $3(3\pi - 16\sqrt{2} + 20) \approx 20.392$

35. 7 **37.** $m = \dfrac{32k}{5}, \left(\dfrac{5}{3}, \dfrac{5}{2}\right)$ **39.** $m = \dfrac{k}{4}, \left(\dfrac{32}{45}, \dfrac{64}{55}\right)$

41. $I_x = 12k$

$I_y = \dfrac{81k}{2}$

$I_0 = \dfrac{105k}{2}$

$\bar{\bar{x}} = \dfrac{3\sqrt{2}}{2}$

$\bar{\bar{y}} = \dfrac{2\sqrt{3}}{3}$

43. $\dfrac{\pi}{6}(101\sqrt{101} - 1)$ **45.** $\dfrac{1}{6}(37\sqrt{37} - 1)$

47. (a) $30{,}415.74$ ft^3 (b) 2081.53 ft^2 **49.** 56

51. $\dfrac{16}{3} + 2e$ **53.** $\dfrac{8\pi}{5}$ **55.** 36

57.

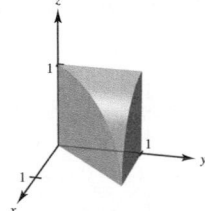

$\displaystyle\int_{0}^{1}\int_{x}^{1}\int_{0}^{\sqrt{1-x^2}} dz\,dy\,dx$

59. $m = \dfrac{500k}{3}, \bar{x} = \dfrac{5}{2}$ **61.** $12(\sqrt{3} - 1)$ **63.** $\dfrac{\pi}{15}$

65. $\pi\left(3\sqrt{13} + 4\ln\dfrac{3 + \sqrt{13}}{2}\right) \approx 48.995$ **67.** 16π

69. $\dfrac{8\pi}{3}(2 - \sqrt{3})$ **71.** $-6(v + u)$ **73.** $\sin^2\theta - \cos^2\theta$

75. $5\ln 5 - 3\ln 3 - 2 \approx 2.751$ **77.** 81

P.S. Problem Solving *(page 1041)*

1. $8\left(2 - \sqrt{2}\right)$ **3.** $\frac{1}{3}$ **5.** (a)–(g) Proofs

7. $-\frac{1}{2}$; $\frac{1}{2}$; No; Fubini's Theorem is not valid because f is not continuous on the region $0 \le x \le 1, 0 \le y \le 1$.

9. $\frac{\sqrt{\pi}}{4}$ **11.** If $a, k > 0$, then $1 = ka^2$ or $a = \frac{1}{\sqrt{k}}$.

13. Answers will vary.

15. The greater the angle between the given plane and the xy-plane, the greater the surface area. So, $z_2 < z_1 < z_4 < z_3$.

17.

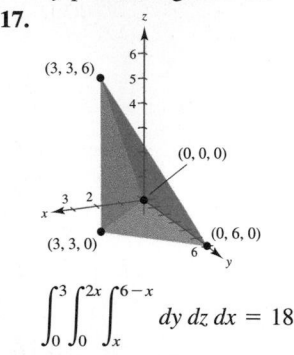

$$\int_0^3 \int_0^{2x} \int_x^{6-x} dy \, dz \, dx = 18$$

Chapter 15

Section 15.1 *(page 1053)*

1. See "Definition of Vector Field" on page 1044. Some physical examples of vector fields include velocity fields, gravitational fields, and electric force fields.

3. Reconstruct a function from its partial derivatives by integrating and comparing versions of the function to determine constants.

5. d **6.** c **7.** a **8.** b

9. $\sqrt{2}$ **11.** $\sqrt{1 + 9y^2}$

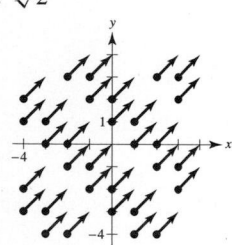

13. $\sqrt{3}$ **15.**

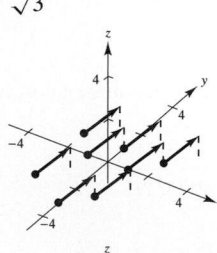

17.

19. $2x\mathbf{i} + 4y\mathbf{j}$ **21.** $(10x + 3y)\mathbf{i} + (3x + 2y)\mathbf{j}$

23. $6yz\mathbf{i} + 6xz\mathbf{j} + 6xy\mathbf{k}$ **25.** $2xye^{x^2}\mathbf{i} + e^{x^2}\mathbf{j} + \mathbf{k}$

27. $\left[\dfrac{xy}{x + y} + y \ln(x + y)\right]\mathbf{i} + \left[\dfrac{xy}{x + y} + x \ln(x + y)\right]\mathbf{j}$

29. Conservative **31.** Not conservative **33.** Conservative

35. Not conservative

37. Conservative; $f(x, y) = 3xy - \dfrac{x^3}{3} + \dfrac{y^2}{2} + K$

39. Conservative; $f(x, y) = e^{x^2y} + K$ **41.** Not conservative

43. Conservative; $f(x, y) = x \sin y + K$ **45.** $4\mathbf{i} - \mathbf{j} - 3\mathbf{k}$

47. $-2\mathbf{k}$ **49.** $\dfrac{2x}{x^2 + y^2}\mathbf{k}$

51. Conservative; $f(x, y, z) = x^3 + y^3 + z^3 + xyz + K$

53. Not conservative

55. Conservative; $f(x, y, z) = \dfrac{xz}{y} - z + K$ **57.** $2x + 4y$

59. $2 \sin x \cos x + 3z^2$ **61.** 28 **63.** 0

65. Vector field; The curl of a vector field is a vector field.

67. Neither; The expression is meaningless because you can only take the curl of a vector field.

69. $9x\mathbf{j} - 2y\mathbf{k}$ **71.** $z\mathbf{j} + y\mathbf{k}$ **73.** $3z + 2x$ **75.** 0

77. (a)–(h) Proofs

Section 15.2 *(page 1065)*

1. (a) The arc length of C (b) The mass of the string

3. $\mathbf{r}(t) = \begin{cases} t\mathbf{i} + t\mathbf{j}, & 0 \le t \le 1 \\ (2 - t)\mathbf{i} + \sqrt{2 - t}\,\mathbf{j}, & 1 \le t \le 2 \end{cases}$

5. $\mathbf{r}(t) = \begin{cases} t\mathbf{i}, & 0 \le t \le 3 \\ 3\mathbf{i} + (t - 3)\mathbf{j}, & 3 \le t \le 6 \\ (9 - t)\mathbf{i} + 3\mathbf{j}, & 6 \le t \le 9 \\ (12 - t)\mathbf{j}, & 9 \le t \le 12 \end{cases}$

7. $\mathbf{r}(t) = 3 \cos t\mathbf{i} + 3 \sin t\mathbf{j}, 0 \le t \le 2\pi$

9. (a) C: $\mathbf{r}(t) = t\mathbf{i} + t\mathbf{j}, 0 \le t \le 1$ (b) $\dfrac{2\sqrt{2}}{3}$

11. (a) C: $\mathbf{r}(t) = \cos t\mathbf{i} + \sin t\mathbf{j}, 0 \le t \le \dfrac{\pi}{2}$ (b) $\dfrac{\pi}{2}$

13. (a) C: $\mathbf{r}(t) = \begin{cases} t\mathbf{i}, & 0 \le t \le 1 \\ t\mathbf{i} + (4t - 4)\mathbf{j}, & 1 \le t \le 2 \end{cases}$

(b) $1 + 7\sqrt{17}$

15. (a) C: $\mathbf{r}(t) = \begin{cases} t\mathbf{i}, & 0 \le t \le 1 \\ (2 - t)\mathbf{i} + (t - 1)\mathbf{j}, & 1 \le t \le 2 \\ (3 - t)\mathbf{j}, & 2 \le t \le 3 \end{cases}$

(b) $3 + 3\sqrt{2}$

17. (a) C: $\mathbf{r}(t) = \begin{cases} t\mathbf{i}, & 0 \le t \le 1 \\ \mathbf{i} + (t - 1)\mathbf{k}, & 1 \le t \le 2 \\ \mathbf{i} + (t - 2)\mathbf{j} + \mathbf{k}, & 2 \le t \le 3 \end{cases}$ (b) $\dfrac{23}{6}$

19. 20 **21.** $\dfrac{5\pi}{2}$ **23.** $8\sqrt{5}\pi\left(1 + \dfrac{4\pi^2}{3}\right) \approx 795.7$

25. $2\pi + 2$ **27.** $\dfrac{k}{12}\left(41\sqrt{41} - 27\right)$ **29.** 8

31. $\frac{1}{3}e^6 + \frac{95}{3}$ **33.** $\frac{9}{4}$ **35.** About 249.49 **37.** 66

39. 0 **41.** $-10\pi^2$

43. Positive; The vector field determined by $\mathbf{F}$ points in the general direction of the path C, so $\mathbf{F} \cdot \mathbf{T} > 0$.

45. Zero; The vector field determined by **F** is perpendicular to the path C.

47. (a) $\frac{236}{3}$; Orientation is from left to right, so the value is positive.

(b) $-\frac{236}{3}$; Orientation is from right to left, so the value is negative.

49. $\mathbf{F}(t) = -2t\mathbf{i} - t\mathbf{j}$

$\mathbf{r}'(t) = \mathbf{i} - 2\mathbf{j}$

$\mathbf{F}(t) \cdot \mathbf{r}'(t) = -2t + 2t = 0$

$\int_C \mathbf{F} \cdot d\mathbf{r} = 0$

51. $\mathbf{F}(t) = (t^3 - 2t^2)\mathbf{i} + \left(t - \frac{t^2}{2}\right)\mathbf{j}$

$\mathbf{r}'(t) = \mathbf{i} + 2t\mathbf{j}$

$\mathbf{F}(t) \cdot \mathbf{r}'(t) = t^3 - 2t^2 + 2t^2 - t^3 = 0$

$\int_C \mathbf{F} \cdot d\mathbf{r} = 0$

53. 68   **55.** $\frac{40}{3}$ **57.** 25   **59.** $\frac{63}{2}$ **61.** $-\frac{11}{6}$

63. $\frac{316}{3}$ **65.** $5h$ **67.** $\frac{1}{2}$ **69.** $\frac{h}{4}\left[2\sqrt{5} + \ln\left(2 + \sqrt{5}\right)\right]$

71. $\frac{1}{120}\left(25\sqrt{5} - 11\right)$

73. (a) $12\pi \approx 37.70 \text{ cm}^2$ (b) $\frac{12\pi}{5} \approx 7.54 \text{ cm}^3$

(c)

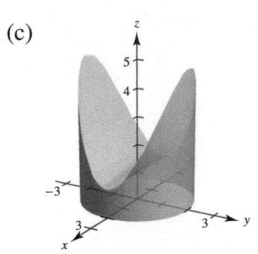

75. $I_x = I_y = a^3\pi$

77. (a)

(b) $9\pi \text{ cm}^2 \approx 28.274 \text{ cm}^2$ (c) $\frac{27\pi}{2} \text{ cm}^2 \approx 42.412 \text{ cm}^3$

79. 1750 ft-lb   **81.** No. $y = 2x$, so $dy = 2 \, dx$.

83. z_3, z_1, z_2, z_4; The greater the height of the surface over the curve $y = \sqrt{x}$, the greater the lateral surface area.

85. False. $\int_C xy \, ds = \sqrt{2}\int_0^1 t^2 \, dt$ **87.** -12

Section 15.3 *(page 1076)*

1. Verify that the vector field is conservative. Find a potential function. Calculate the difference of the values of the function evaluated at the endpoints.

3. (a) Proof

(b) $\int_{C_1} \mathbf{F} \cdot d\mathbf{r} = \int_0^1 (t^2 + 2t^3) \, dt = \frac{5}{6}$

$\int_{C_2} \mathbf{F} \cdot d\mathbf{r} = \int_0^{\pi/2} (\sin^2\theta \cos\theta + 2\sin^3\theta \cos\theta) \, d\theta = \frac{5}{6}$

5. (a) Proof

(b) $\int_{C_1} \mathbf{F} \cdot d\mathbf{r} = \int_0^{\pi/3} (3\tan^2\theta \sec\theta + 3\sec^3\theta) \, d\theta$

≈ 10.392

$\int_{C_2} \mathbf{F} \cdot d\mathbf{r} = \int_0^3 \left(\frac{3\sqrt{t}}{2\sqrt{t+1}} + \frac{3\sqrt{t+1}}{2\sqrt{t}}\right) dt \approx 10.392$

7. (a) Proof

(b) $\int_{C_1} \mathbf{F} \cdot d\mathbf{r} = \int_0^1 64t^3 \, dt = 16$

$\int_{C_2} \mathbf{F} \cdot d\mathbf{r} = \int_0^{\pi/2} 64\sin^3\theta \cos\theta \, d\theta = 16$

9. 72   **11.** -1 **13.** 0   **15.** (a) 2   (b) 2   (c) 2

17. 11   **19.** (a) Proof   (b) 30,366

21. (a) Proof   (b) 32   **23.** (a) 1   (b) 1

25. (a) 64   (b) 0   (c) 0   (d) 0

27. (a) 32   (b) 32   **29.** (a) $\frac{2}{3}$ (b) $\frac{17}{6}$ **31.** (a) 0   (b) 0

33. 0

35. (a) $d\mathbf{r} = (\mathbf{i} - \mathbf{j}) \, dt \Rightarrow \int_0^{50} 175 \, dt = 8750$ ft-lb

(b) $d\mathbf{r} = \left(\mathbf{i} - \frac{1}{25}(50 - t)\mathbf{j}\right) dt$

$7\int_0^{50} (50 - t) \, dt = 8750$ ft-lb

37. The partial derivatives of **F** are not continuous at $(0, 0)$. Draw an open connected region that excludes that point.

39. 1.125

41. Yes, because the work required to get from point to point is independent of the path taken.

43. False. It would be true if **F** were conservative.

45. True   **47.** Proof

49. (a) Proof   (b) $-\pi$ (c) π (d) -2π

(e) No, because **F** is not continuous at $(0, 0)$ in R enclosed by C.

(f) $\nabla\left(\arctan\frac{x}{y}\right) = \frac{1/y}{1 + (x/y)^2}\mathbf{i} + \frac{-x/y^2}{1 + (x/y)^2}\mathbf{j}$

Section 15.4 *(page 1085)*

1. A curve is simple when it does not cross itself. A connected plane region is simply connected when every simple closed curve in the region encloses only points that are in the region. For example, a region with a hole is not simply connected.

3. You are working with a simple closed curve with a boundary whose orientation is counterclockwise.

5. $\frac{1}{30}$ **7.** 0   **9.** About 19.99   **11.** $\frac{9}{2}$ **13.** 56

15. $\frac{4}{3}$ **17.** 0   **19.** 0   **21.** $\frac{1}{12}$ **23.** 32π

25. π **27.** $\frac{225}{2}$ **29.** 4π **31.** $\frac{9}{2}$ **33.** Proof

35. $\left(0, \frac{8}{5}\right)$ **37.** $\left(\frac{8}{15}, \frac{8}{21}\right)$ **39.** 54π **41.** $\pi - \frac{3\sqrt{3}}{2}$

43. (a) $\dfrac{51\pi}{2}$ (b) $\dfrac{243\pi}{2}$ **45.** 46π

47. (a) $\displaystyle\int_C \mathbf{F} \cdot d\mathbf{r} = \int_C M\,dx + N\,dy = \int\!\!\int_R \left(\frac{\partial N}{\partial x} - \frac{\partial M}{\partial y}\right) dA = 0$

(b) $I = -2\pi$ when C is a circle that contains the origin.

49–53. Proofs

Section 15.5 *(page 1095)*

1. S is traced out by the position vector $\mathbf{r}(u, v)$ as the point (u, v) moves throughout the domain. To sketch the surface, it is helpful to relate x, y, and z, where x, y, and z are functions of u and v.

3. e   **4.** f   **5.** b   **6.** c   **7.** d   **8.** a

9. $y - 2z = 0$
Plane

11. $x^2 + z^2 = 4$
Cylinder

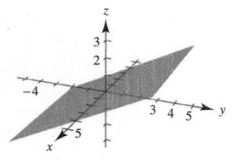

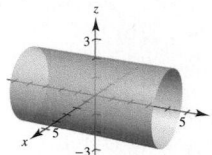

13.

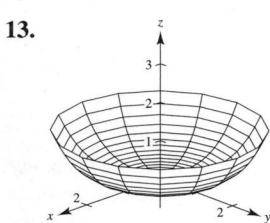

15.

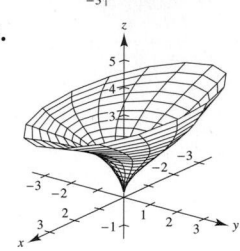

17. $\mathbf{r}(u, v) = u\mathbf{i} + v\mathbf{j} + 3v\mathbf{k}$

19. $\mathbf{r}(u, v) = \frac{1}{3}u \cos v\mathbf{i} + u\mathbf{j} + \frac{1}{3}u \sin v\mathbf{k},\ u \geq 0,\ 0 \leq v \leq 2\pi$ or
$\mathbf{r}(x, y) = x\mathbf{i} + \sqrt{4x^2 + 9y^2}\mathbf{j} + z\mathbf{k}$

21. $\mathbf{r}(u, v) = 5 \cos u\mathbf{i} + 5 \sin u\mathbf{j} + v\mathbf{k}$

23. $\mathbf{r}(u, v) = u\mathbf{i} + \sqrt{u - 7} \cos v\mathbf{j} + \sqrt{u - 7} \sin v\mathbf{k}$ or
$\mathbf{r}(y, z) = (y^2 + z^2 + 7)\mathbf{i} + y\mathbf{j} + z\mathbf{k}$

25. $\mathbf{r}(u, v) = v \cos u\mathbf{i} + v \sin u\mathbf{j} + 4\mathbf{k},\quad 0 \leq v \leq 3$

27. $x = u,\ y = \frac{u}{2}\cos v,\ z = \frac{u}{2}\sin v,\ 0 \leq u \leq 6,\ 0 \leq v \leq 2\pi$

29. $x = \sin u \cos v,\ y = \sin u \sin v,\ z = u$
$0 \leq u \leq \pi,\ 0 \leq v \leq 2\pi$

31. $x = \cos^2 u \cos v,\ y = u,\ z = \cos^2 u \sin v$
$\frac{\pi}{2} \leq u \leq \pi,\ 0 \leq v \leq 2\pi$

33. $9y + \dfrac{3\sqrt{3}}{2}z = 12\sqrt{3}$ **35.** $4y - 3z = 12$ **37.** $8\sqrt{2}$

39. $\pi ab^2 \sqrt{a^2 + 1}$ **41.** $\dfrac{\pi}{6}\left(17\sqrt{17} - 1\right) \approx 36.177$

43. The paraboloid is reflected (inverted) through the xy-plane.

45. The height of the paraboloid is increased from 4 to 9.

47–49. Proofs

51. (a)   (b)
(c)   (d)

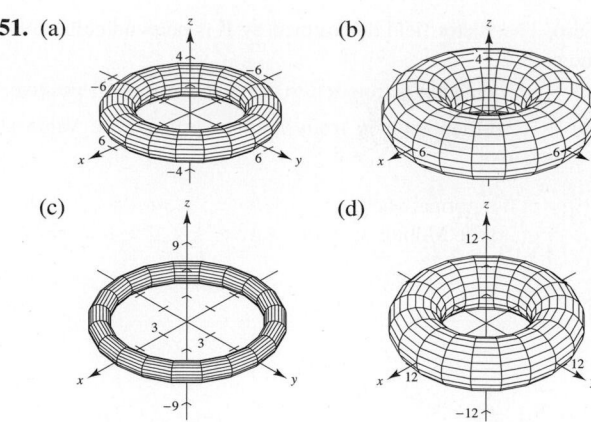

The radius of the generating circle that is revolved about the z-axis is b, and its center is a units from the axis of revolution.

53. 400π m²

55.

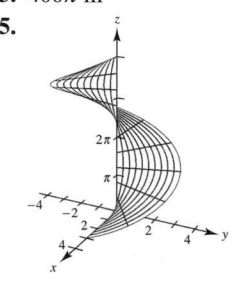

$2\pi\left[\frac{3}{2}\sqrt{13} + 2\ln\left(3 + \sqrt{13}\right) - 2\ln 2\right]$

57. Answers will vary. Sample answer: Let
$x = (2 - u)(5 + \cos v) \cos 3\pi u$
$y = (2 - u)(5 + \cos v) \sin 3\pi u$
$z = 5u + (2 - u) \sin v$
where $-\pi \leq u \leq \pi$ and $-\pi \leq v \leq \pi$.

Section 15.6 *(page 1108)*

1. Solve for y in the equation of the surface. Then use the integral
$$\int\!\!\int_S f(x, g(x, z), z)\sqrt{1 + [g_x(x, z)]^2 + [g_z(x, z)]^2}\,dA.$$

3. An orientable surface has two distinct sides.

5. $12\sqrt{2}$ **7.** 2π **9.** $\dfrac{27\sqrt{3}}{8}$ **11.** About -11.47

13. $\frac{364}{3}$ **15.** $12\sqrt{5}$ **17.** $\dfrac{3 - \sqrt{3}}{2}$ **19.** $\sqrt{3}\pi$

21. $\dfrac{32\pi}{3}$ **23.** 486π **25.** $-\dfrac{4}{3}$ **27.** $\dfrac{3\pi}{2}$ **29.** 20π

31. 384π **33.** $64\pi\rho$ **35.** 0   **37.** Proof   **39.** $2\pi a^3 h$

41. (a) 12   (b) 12   (c) 12

43. (a)

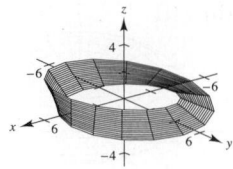

(b) No. If a normal vector at a point P on the surface is moved around the Möbius strip once, it will point in the opposite direction.

(c)

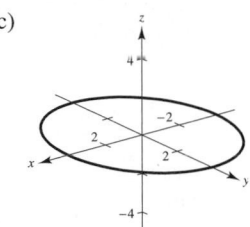

Circle

(d) Construction

(e) You obtain a strip with a double twist that is twice as long as the Möbius strip.

Section 15.7   *(page 1116)*

1. Divergence Theorem or two surface integrals; In this case, it is easier to use the Divergence Theorem.

3. 1   **5.** 18   **7.** π **9.** $3a^4$ **11.** 0   **13.** 108π

15. 0   **17.** $18(e^4 - 5)$ **19.** Source

21. Incompressible   **23.** Any point that satisfies $xyz > \frac{1}{2}$

25. 0; Proof   **27–31.** Proofs

Section 15.8   *(page 1123)*

1. Stokes's Theorem allows you to evaluate a line integral using a single double integral.

3. 18π **5.** 0   **7.** -12 **9.** 2π **11.** 0   **13.** $\frac{8}{3}$

15. $-\dfrac{a^5}{4}$ **17.** $\dfrac{81\pi}{4}$ **19.** Yes; Proof

21. Putnam Problem A5, 1987

Review Exercises for Chapter 15   *(page 1124)*

1. $\sqrt{x^2 + 5}$

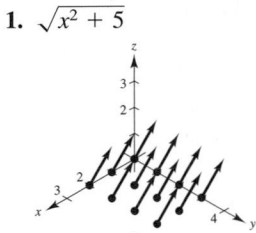

3. $y \cos xy\mathbf{i} + (x \cos xy - 2y)\mathbf{j}$ **5.** $(4x + y)\mathbf{i} + x\mathbf{j} + 2z\mathbf{k}$

7. Not conservative   **9.** Conservative

11. Conservative; $f(x, y) = \dfrac{y}{x} + K$

13. Conservative; $f(x, y) = \frac{1}{2}x^2y^2 - \frac{1}{3}x^3 + \frac{1}{3}y^3 + K$

15. Not conservative   **17.** Conservative; $f(x, y, z) = \dfrac{x}{yz} + K$

19. (a) div $\mathbf{F} = 2x + 2xy + x^2$ (b) curl $\mathbf{F} = -2xz\mathbf{j} + y^2\mathbf{k}$

21. (a) div $\mathbf{F} = -y \sin x - x \cos y + xy$

(b) curl $\mathbf{F} = xz\mathbf{i} - yz\mathbf{j}$

23. (a) div $\mathbf{F} = \dfrac{1}{\sqrt{1 - x^2}} + 2xy + 2yz$

(b) curl $\mathbf{F} = z^2\mathbf{i} + y^2\mathbf{k}$

25. (a) div $\mathbf{F} = \dfrac{2x + 2y}{x^2 + y^2} + 1$ (b) curl $\mathbf{F} = \dfrac{2x - 2y}{x^2 + y^2}\mathbf{k}$

27. (a) $\frac{125}{3}$ (b) 2π **29.** 6π **31.** $\dfrac{9a^2}{5}$ **33.** 3π

35. 1   **37.** $2\pi^2$ **39.** $\frac{8}{3}(3 - 4\sqrt{2}) \approx -7.085$ **41.** 12

43. $\dfrac{\sqrt{5}}{3}(19 - \cos 6) \approx 13.446$

45. (a) Proof

(b) (i) $\displaystyle\int_C \mathbf{F} \cdot d\mathbf{r} = \int_0^4 (3t + 4 + t^3)\, dt = 104$

(ii) $\displaystyle\int_C \mathbf{F} \cdot d\mathbf{r} = \int_0^2 \left[(3w^2 + 4)(2w) + w^6(2w)\right] dw = 104$

47. $1 - \dfrac{1}{e^2}$ **49.** 6   **51.** (a) Proof   (b) 92   **53.** 1

55. 0   **57.** 0   **59.** 0   **61.** 3

63. $z = 2(x^2 + y^2)$ **65.**

Paraboloid

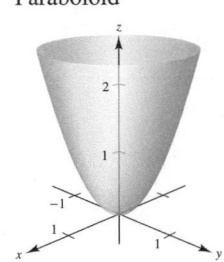

67. $\mathbf{r}(u, v) = \cos v \cos u\mathbf{i} + 2\sqrt{2} \cos v \sin u\mathbf{j} + 3 \sin v\mathbf{k}$

69. $x = u, y = 2u^3 \cos v, z = 2u^3 \sin v, 0 \le u \le 2, 0 \le v \le 2\pi$

71. $3\sqrt{41}$ **73.** 45   **75.** $27\sqrt{41}$ **77.** $6\sqrt{26}$

79. 25π **81.** 66   **83.** $\dfrac{2a^6}{5}$ **85.** 16π

P.S. Problem Solving   *(page 1127)*

1. (a) and (b) $\dfrac{25\sqrt{2}k\pi}{6}$

3. $I_x = \dfrac{\sqrt{13}\pi}{3}(27 + 32\pi^2)$

$I_y = \dfrac{\sqrt{13}\pi}{3}(27 + 32\pi^2)$

$I_z = 18\sqrt{13}\pi$

5. (a)–(d) Proofs   **7.** $3a^2\pi$ **9.** (a) 1   (b) $\frac{13}{15}$ (c) $\frac{5}{2}$

11. The area is the same as the magnitude.

Appendix C

Appendix C.1 *(page A14)*

1. Rational   **3.** Irrational   **5.** Rational   **7.** Rational
9. Rational   **11.** $\frac{4}{11}$ **13.** $\frac{11}{37}$
15. (a) True   (b) False   (c) True   (d) False
(e) False   (f) False
17. x is greater than -3 and less than 3.

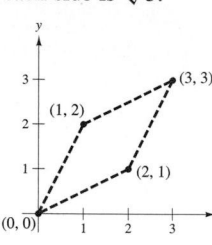

The interval is bounded.
19. x is no more than 5.

The interval is unbounded.
21. $y \geq 4$; $[4, \infty)$ **23.** $0.03 < r \leq 0.07$; $(0.03, 0.07]$
25. $x \geq \frac{1}{2}$ **27.** $-\frac{1}{2} < x < \frac{7}{2}$

29. $x > 6$ **31.** $-1 < x < 1$

33. $x \geq 13$, $x \leq -7$ **35.** $a - b < x < a + b$

37. $-3 < x < 2$ **39.** $0 < x < 3$

41. $-3 \leq x \leq 1$ **43.** $-3 \leq x \leq 2$

45. 4; -4; 4 **47.** (a) -51; 51; 51 (b) 51; -51; 51
49. $|x| \leq 2$ **51.** $|x - 2| > 2$
53. (a) $|x - 12| \leq 10$ (b) $|x - 12| \geq 10$
55. 1   **57.** (a) 14   (b) 10   **59.** $x \geq 36$ units
61. $x \leq 41$ or $x \geq 59$ **63.** (a) $\frac{355}{113} > \pi$ (b) $\frac{22}{7} > \pi$ **65.** b
67. False. The reciprocal of 2 is $\frac{1}{2}$. which is not an integer.
69. True   **71.** True   **73–79.** Proofs
81. $|-3 - 1| > |-3| - |1|$; $|3 - 1| = |3| - |1|$; Proof

Appendix C.2 *(page A21)*

1. (a)

(b) $2\sqrt{5}$
(c) $(3, 3)$
3. (a)

(b) $2\sqrt{10}$
(c) $\left(-\frac{1}{2}, -2\right)$
5. (a)

(b) $\sqrt{8 - 2\sqrt{3}}$
(c) $\left(0, \frac{1 + \sqrt{3}}{2}\right)$

7. Quadrant II   **9.** Quadrants I and III
11. Right triangle:
$d_1 = \sqrt{45}$
$d_2 = \sqrt{5}$
$d_3 = \sqrt{50}$
$(d_1)^2 + (d_2)^2 = (d_3)^2$

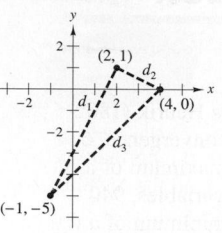

13. Rhombus: the length of each side is $\sqrt{5}$.   **15.**

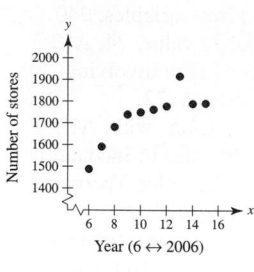

17. $d_1 = 2\sqrt{5}$, $d_2 = \sqrt{5}$, $d_3 = 3\sqrt{5}$
Collinear, because $d_1 + d_2 = d_3$.
19. $d_1 = \sqrt{2}$, $d_2 = \sqrt{13}$, $d_3 = 5$
Not collinear, because $d_1 + d_2 > d_3$.
21. $x = \pm 3$ **23.** $y = \pm\sqrt{55}$
25. $\left(\dfrac{3x_1 + x_2}{4}, \dfrac{3y_1 + y_2}{4}\right)$, $\left(\dfrac{x_1 + x_2}{2}, \dfrac{y_1 + y_2}{2}\right)$, $\left(\dfrac{x_1 + 3x_2}{4}, \dfrac{y_1 + 3y_2}{4}\right)$
27. c   **28.** b   **29.** a   **30.** d   **31.** $x^2 + y^2 - 9 = 0$
33. $x^2 + y^2 - 4x + 2y - 11 = 0$
35. $x^2 + y^2 + 2x - 4y = 0$
37. $x^2 + y^2 - 6x - 4y + 3 = 0$ **39.** $x^2 + y^2 = 26{,}000^2$
41. $(x - 1)^2 + (y + 3)^2 = 4$ **43.** $(x - 1)^2 + (y + 3)^2 = 0$

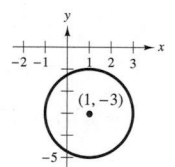   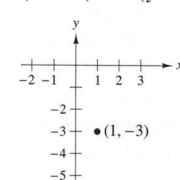

45. $\left(x - \frac{1}{2}\right)^2 + \left(y - \frac{1}{2}\right)^2 = 2$ **47.** $\left(x + \frac{1}{2}\right)^2 + \left(y + \frac{5}{4}\right)^2 = \frac{9}{4}$

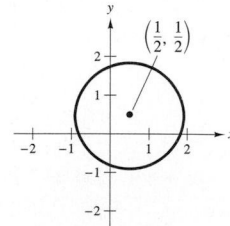   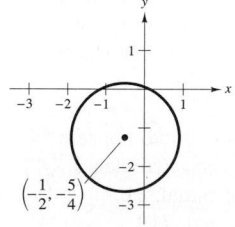

49.   **51.**

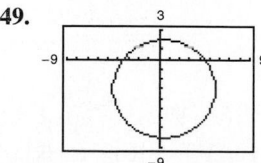   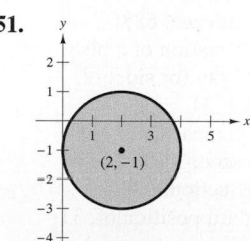

53. Proof   **55.** True   **57.** True   **59–61.** Proofs

Index